W9-AEA-619

Becker's

10th Edition

WORLD OF THE CELL

JEFF HARDIN
University of Wisconsin–Madison

JAMES P. LODOLCE
Loyola University Chicago

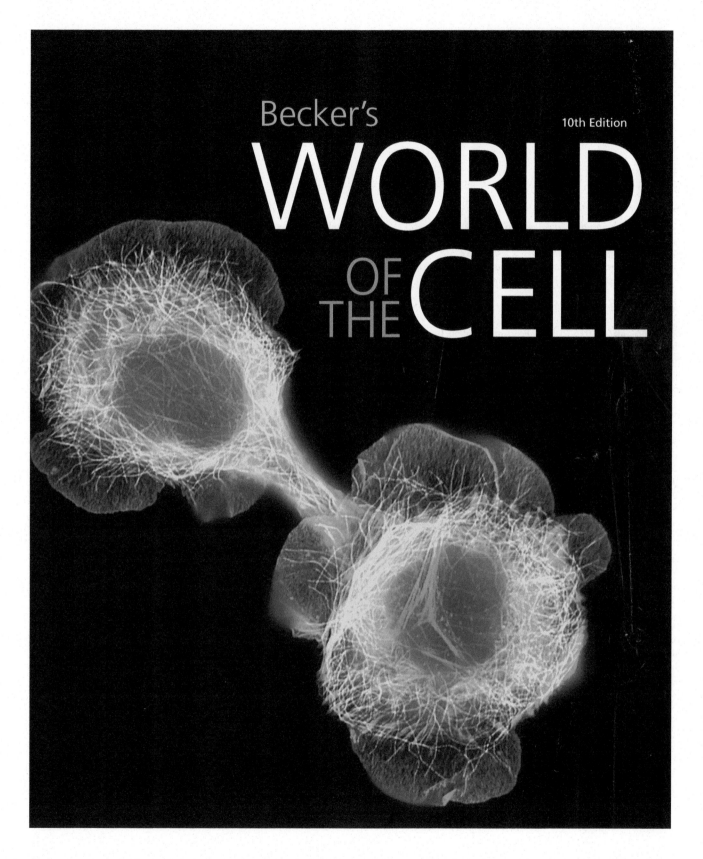

Content Development: *Evelyn Dahlgren, Ginnie Simione Jutson, Imagineering Art*
Content Management: *Jeanne Zalesky, Joshua Frost*
Content Production: *Suddha Sen, Titas Basu, Mike Early, Lucinda Bingham, Tod Regan, Marianne Peters-Riordan, Integra Software Services*
Product Management: *Michael Gillespie, Rebecca Berardy-Schwartz*
Product Marketing: *Kate Gittins, Kelly Galli*
Rights and Permissions: *Ben Ferrini, Matthew Perry, SPi Global, Kristin Piljay*

Cover Image: *Dr. Torsten Wittmann/Science Source*

2 2021

Library of Congress Cataloging-in-Publication Data

Names: Hardin, Jeff, author. | Lodolce, James P., author.
Title: Becker's World of the Cell/Jeff Hardin, University of Wisconsin-Madison; James P. Lodolce, Loyola University Chicago.
Other titles: World of the Cell
Description: Tenth Edition. | Hoboken: PEARSON, 2022.
Identifiers: LCCN 2020037805 | ISBN 9780135259498 (hardcover)
Subjects: LCSH: Cytology. | Molecular biology.
Classification: LCC QH581.2 .B43 2021 | DDC 571.9/36–dc23
LC record available at https://lccn.loc.gov/2020037805

ISBN 10: 0-13-525949-5; ISBN 13: 978-0-13-525949-8 (Student edition)
ISBN 10: 0-13-583216-0; ISBN 13: 978-0-13-583216-5 (Print Offer)

www.pearsonhighered.com

ABOUT THE AUTHORS

JEFF HARDIN received his Ph.D. in Biophysics from the University of California–Berkeley. He is the Raymond E. Keller Professor and Chair of the Department of Integrative Biology at the University of Wisconsin–Madison, where he has been since 1991. For 18 years he was Faculty Director of the Biology Core Curriculum, a four-semester honors biology sequence for undergraduates at Wisconsin known for its teaching innovations. Jeff's research focuses on how cells migrate and adhere to one another during early embryonic development. Jeff's teaching is enhanced by his extensive use of digital microscopy and his web-based teaching materials, which are used on many campuses in the United States and in other countries. Jeff was a founding member of the UW Teaching Academy, and has received several teaching awards, including a Lily Teaching Fellowship, a National Science Foundation Young Investigator Award, and a Chancellor's Distinguished Teaching Award.

JAMES P. LODOLCE earned his Ph.D. in Immunology from the University of Chicago in 2002. His thesis examined the signals that promote the survival of memory lymphocytes. As a postdoctoral fellow in the laboratory of Dr. David Boone, he studied the genetics and regulation of inflammation in autoimmunity. Cell biology was the first class that James taught when he arrived at Loyola University Chicago in 2010. He currently holds the title of Senior Lecturer and teaches a variety of courses ranging from molecular biology to virology. James is an active member of the Department of Biology and was appointed Co-Chairperson of Loyola's 2021 Pre-Health Professions Advisory Committee. In his career at Loyola, James has received several teaching honors, including a nomination for the 2014 Ignatius Loyola Award for Excellence in Teaching, the 2016 Master Teacher Award in the College of Arts and Sciences, and the 2020 Edwin T. and Vivijeanne F. Sujack Award for Teaching Excellence.

WAYNE M. BECKER taught cell biology at the University of Wisconsin–Madison for 30 years until his retirement. His interest in textbook writing grew out of notes, outlines, and problem sets that he assembled for his students, culminating in *Energy and the Living Cell*, a paperback text on bioenergetics published in 1977, and *The World of the Cell*, the first edition of which appeared in 1986. All his degrees are in biochemistry from the University of Wisconsin–Madison, an orientation that is readily discernible in his writing. His research interests were in plant molecular biology, focused on the expression of genes that encode enzymes of the photorespiratory pathway. Later in his career he focused on teaching, especially students from underrepresented groups. His honors include a Chancellor's Award for Distinguished Teaching, Guggenheim and Fulbright Fellowships, and a Visiting Scholar Award from the Royal Society of London. This text builds on his foundation and is inspired by his legacy.

DETAILED CONTENTS

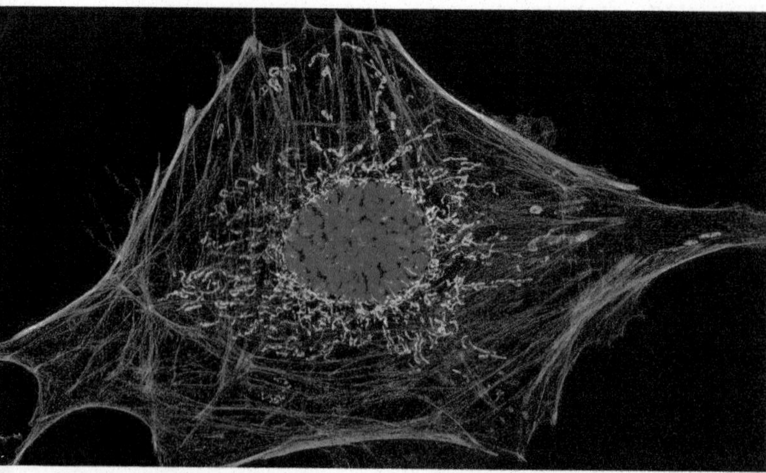

7 Membranes: Their Structure, Function, and Chemistry *154*

8 Transport Across Membranes: Overcoming the Permeability Barrier *185*

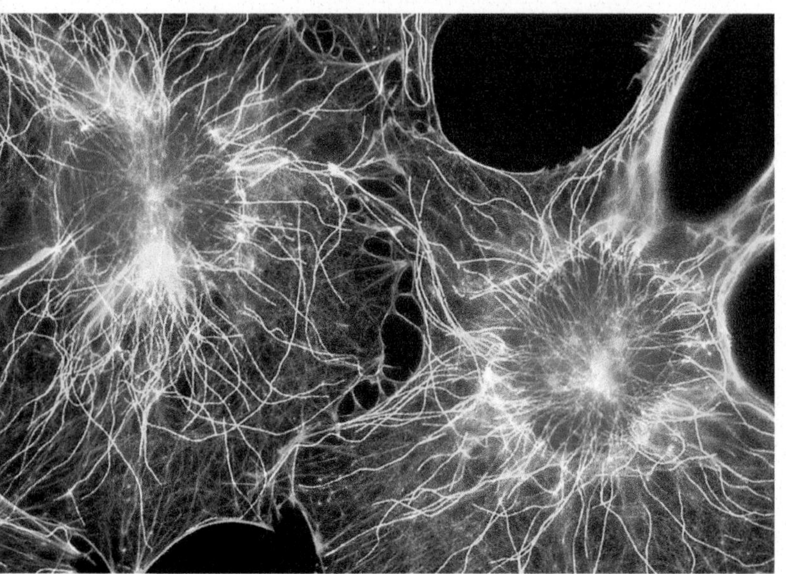

13 Cytoskeletal Systems 358

14 Cellular Movement: Motility and Contractility 384

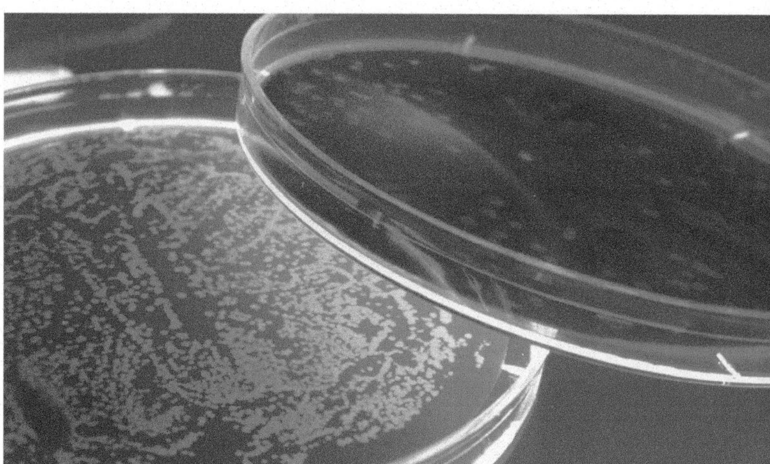

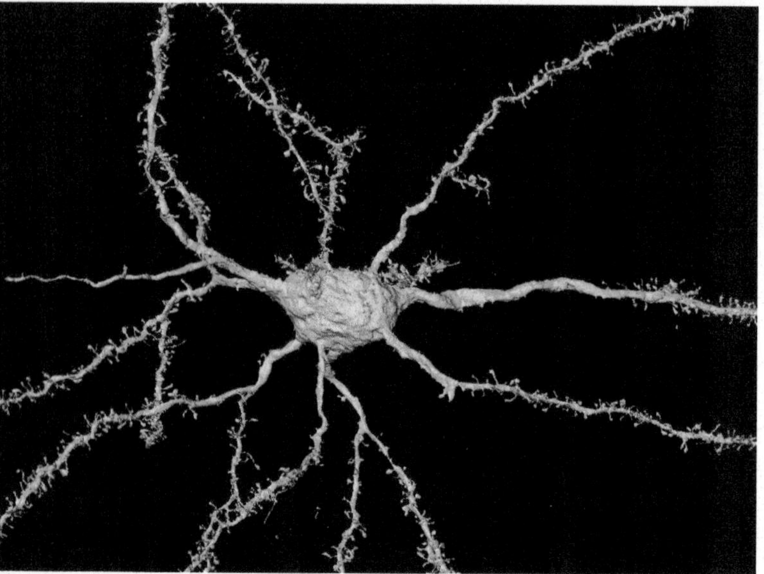

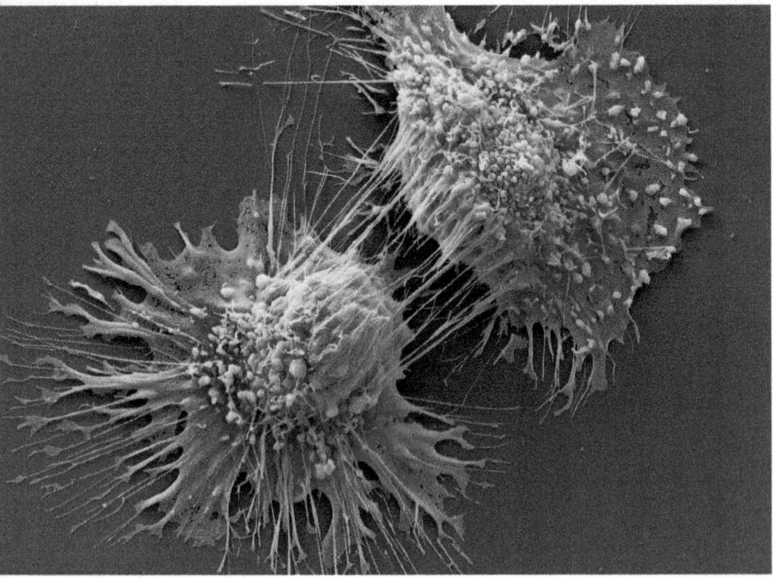

PREFACE

Cells are the fundamental building blocks of life on this planet. Despite their tiny size, they are wonders of intricacy. Moment by moment, the cells of our bodies are engaged in a dazzling repertoire of biochemical events, including signaling processes, transmission of genetic information, and delicately choreographed movements. Understanding the basic functions of cells also gives us insight when something goes wrong, as in the case of a disease, or when the cell is highjacked, as in the case of a viral infection. Helping our students to appreciate the complexities of this amazing cellular world lies at the heart of our goals as authors of *Becker's World of the Cell*. The motivations that drove our colleague, Wayne Becker, to write the first edition of this book continue to drive us today. We believe that our students should have biology textbooks that are clearly written, make the subject matter relevant, and help them to appreciate not only how much we already know about cell biology but also the exciting journey of continued discovery that lies ahead. We, as authors, have an extensive history of teaching undergraduate courses in cell biology and related areas, and we treasure our contact with students as one of the most rewarding aspects of being faculty members.

The amazing success of modern cell biology creates both exciting opportunities and central challenges in our teaching. How can we capture the core elements of modern cell biology in a way that draws our students in without overwhelming them? The enormous profusion of information challenges us to keep *Becker's World of the Cell* up to date while ensuring that it remains both manageable in length and readily comprehensible to students studying cell and molecular biology for the first time.

This tenth edition engages students with new innovative features in each chapter and an exciting, fresh look. In addition, a major goal of this edition has been to reorganize the presentation of several key topics. We hope that the often-requested consolidation of translation of secreted and plasma membrane-associated proteins with the larger discussion of the endomembrane system has led to an even more compelling presentation of these important topics. We also hope students and instructors will find that the continued emphasis on molecular biology throughout the tenth edition reinforces how indispensable these techniques are in the everyday work of modern cell biologists.

As with the previous editions, we remain committed to three central goals. First, our primary goal is to introduce students to the fundamental principles that guide cellular organization and function. Second, we want students to understand some of the key scientific evidence that has helped us formulate these central concepts. And third, we have sought to accomplish these goals in a book of manageable length that is easily read and understood by beginning cell biology students—and that still fits in their backpacks! We have therefore been necessarily selective both in the examples chosen to illustrate key concepts and in the quantity of scientific evidence included. The result is an update that we hope students and instructors will be as excited about as we are.

What's New in This Edition

- **Make Connection questions:** Two new questions in every chapter ask students to make connections across concepts and chapters in the text. By reinforcing fundamental conceptual connections throughout cell biology, these features help overcome students' tendencies to compartmentalize information. These questions are also assignable and automatically graded in **Mastering Biology.**
- **Data Analysis questions:** Every chapter of *World of the Cell* now has a Data Analysis question for students to practice their ability to interpret data. Students must be able to analyze data in order to make informed decisions, generate well-formed, testable hypotheses, design follow-up experiments, and provide compelling evidence for results. These questions are also assignable and automatically graded in **Mastering Biology.**
- **Figure Walkthroughs:** In the *World of the Cell* e-text, Figure Walkthroughs guide students through key figures with narrated explanations and figure mark-ups that reinforce important points. All walkthroughs are also assignable in **Mastering Biology** and paired with several auto-gradable questions for student assessment.

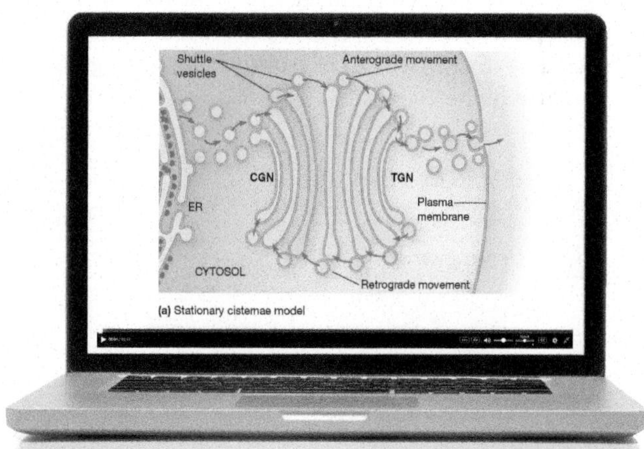

- **Reorganization of material on translation and intracellular trafficking:** Because the molecular genetics material comes earlier in the book, topics that relate to translation of secreted and plasma membrane-associated proteins are now more naturally integrated into the discussion of intracellular trafficking. These topics are now combined in Chapter 12, which focuses on the endomembrane system, including cotranslational import into the endoplasmic reticulum of proteins destined for secretion or insertion into the plasma membrane.

Hallmark Features

- **Key Technique boxes in every chapter:** Twenty-six Key Technique boxes are integrated throughout the text, demonstrating how cutting-edge technologies can be used to answer key questions in cell biology.

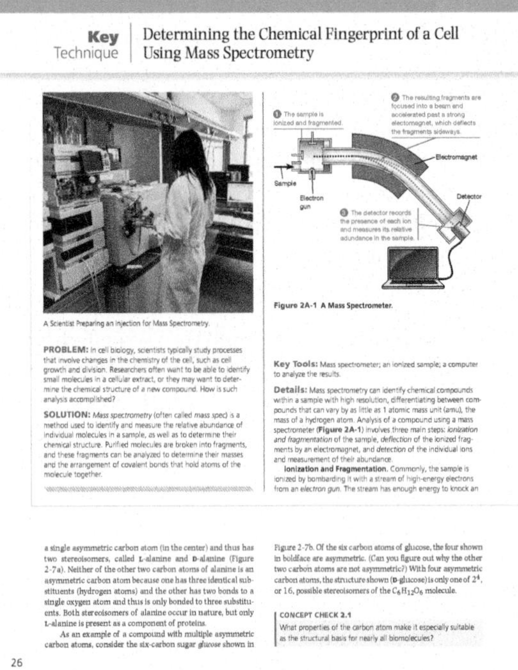
Key Technique | Determining the Chemical Fingerprint of a Cell Using Mass Spectrometry

Figure 2A-1 A Mass Spectrometer.

- **Human Connections boxes in every chapter:** Twenty-six Human Connections boxes emphasize the relevance of cell biology to human health and society, from the story of Henrietta Lacks and the HeLa cell line to the relevance of biochemical pathways to our diet, to the many cases in which cell biology helps us diagnose and treat human disease.

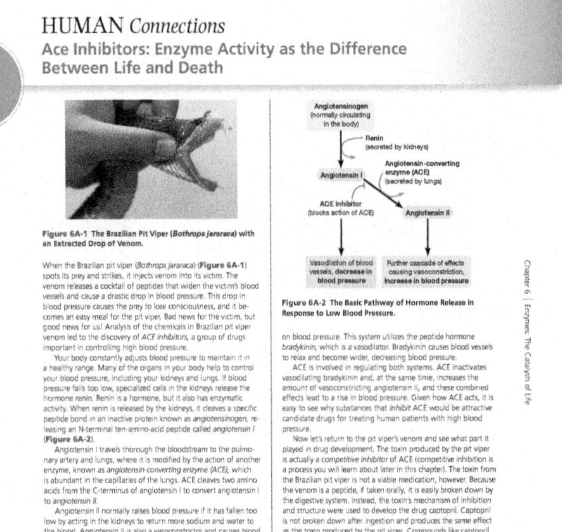
HUMAN *Connections*
Ace Inhibitors: Enzyme Activity as the Difference Between Life and Death

- **Concept Check questions:** Each main section of a chapter ends with a Concept Check question. These questions provide students with numerous opportunities to assess their understanding as they read. Answers to these questions are available at the back of the book.
- **Quantitative questions in every end-of-chapter Problem Set:** New and existing quantitative questions are flagged at the end of each chapter to encourage students to work on developing their ability to perform calculations or to interpret quantitative information. Most of these questions are assignable through **Mastering Biology.**
- **Content updates:** Updated information highlighting the most recent advances in cell and molecular biology has been added throughout the book (see Content Highlights of the Tenth Edition below).

Mastering Biology is an innovative online homework, tutorial, and assessment system that delivers self-paced tutorials with individualized coaching, hints, and feedback. The Mastering system helps instructors and students with customizable, easy-to-assign, and automatically graded assignments.

Integrated links in every chapter of the textbook point students to a variety of interactive online materials, including the following:

- 52 assignable Make Connection questions help students make connections across chapters and concepts
- 10 figure walkthrough tutorials walk students through key figures and then assess their understanding
- More than 100 tutorials and activities that teach complex cell processes
- More than 100 molecular and microscopy videos, which provide vivid images of cellular processes
- 240 Reading Quiz questions, which encourage students to read before class
- Many end-of-chapter questions and problems that are assignable and automatically gradable
- Test Bank questions for every chapter
- The e-text, also available through **Mastering Biology**, which provides both access to the complete textbook and powerful interactive and customizable functions
- A suite of Instructor Resources, including PowerPoint lecture outlines containing all the figures and photos and five to ten personal response system (PRS) clicker questions per chapter
- Learning Catalytics is a "bring your own device" assessment and active classroom system that expands the possibilities for student engagement beyond standard clickers where instructors can deliver a wide range of auto-gradable or open-ended questions that test content knowledge and build critical thinking skills

Content Highlights of the Tenth Edition

Updated material and new information have been added throughout the book in both the text and art. Major topics that have been altered, updated, or added include the following:

Chapter 1: Created new Figure 1-1 (Hooke's microscope and drawing of cork). Added CRISPR genome editing to Figure 1-3 and added a new subsection on CRISPR to Section 1.2. Condensed the three microscopy subsections (The Light Microscope, Specialized Light Microscopes, and The Electron Microscope) into one large subsection subtitled "Microscopy." Modified Figure 1-8 to better illustrate the central dogma in a cell. Added a new Data Analysis question.

Chapter 2: Added reference to organic carbon discoveries made by Mars rover to Section 2.1 and the importance of water transport to Section 2.2. Added new subsection on prion self-assembly to Section 2.5. Added a new Data Analysis question.

Chapter 3: Added reference to gecko pad and van der Waals interactions. Added information about the Folding @ Home initiative. Added subsection on chaperones in protein folding to Section 3.1. Added a new figure to the Human Connections box on Tau tangle formation. Added a new Data Analysis question.

Chapter 4: Significantly updated the discussion of the endosymbiont theory, including discussion of "inside-out" and "outside-in" proposals in a largely revised figure. Moved three domains of life discussion and figure from 9e Ch. 21 to Ch. 4.

Chapter 5: Added a new Data Analysis question; updated Figure 5-1 to add improved concentration work diagram.

Chapter 6: Majorly revised Figure 6-11 and relevant text to conform to the majority of advanced biochemistry texts regarding inhibitors. Removed sucrase discussion to comport with deletion of the relevant figure in the previous edition and generated a new figure showing the catalytic site of lysozyme accordingly. Shortened the discussion of ACE inhibitors in the Human Connections box. Replaced one Problem Set question on biological relevance with another graphical analysis problem on competitive inhibitors.

Chapter 7: Moved SDS-PAGE material to Ch. 21. Reduced treatment of lipid rafts to reflect ongoing controversy in the field. Added a new Key Technique box on fluorescence recovery after photobleaching (FRAP). Added a Human Connections box, adapted from 9e Ch. 12. Reinstated a more detailed structure diagram in Figure 7-6.

Chapter 8: Improved clarity of Figure 8-7. Added panel to Figure 8-10 to show frog oocytes. Added a new Data Analysis question.

Chapter 9: Shortened discussion of other uses of glycolytic enzymes. Improved several biochemical pathway diagrams for clarity.

Chapter 10: Revised the discussion of ATP yield in aerobic respiration while retaining the theoretical yield discussion as a *via media*. Substantially revised electron transport details in several figures. Substantially revised and improved Q cycle discussion and the relevant figure. Trimmed discussion and figure coverage of cristae and added a light micrograph showing mitochondria. Integrated TIM/TOM discussion into this chapter, moving it out of 9e Ch. 19 to join the discussion of the structure of mitochondria. Added figure on location of ATP synthesis in bacteria to compare to mitochondria. Replaced problem on thermogenin with Data Analysis question.

Chapter 11: Added information and figure about carboxysomes in cyanobacteria. Improved the molecular model presentation of light-harvesting complexes. Improved the treatment of electron flow in the chloroplast, including improving and shortening the discussion of the Q cycle. Updated information on protons per ATP. Improved depiction of the glycolate pathway and C_3/C_4 plant leaf anatomy. Added Quantitative and Data Analysis questions.

Chapter 12: Added an update on the types of models used to explain movement through the Golgi. Provided some rationale for grouping peroxisomes into the endomembrane system. Moved protein trafficking/sorting sections from 9E Ch. 19 to here. Added paragraph on how viruses can co-opt endosomes for infection. Combined 9e Sections 12.7 and 12.8 into one section (since the plant vacuole is a digestive compartment). Authored new Human Connections box on the role of autophagy in human disease.

Chapter 13: Updated *MreB* discussion to match current understanding of MreB function. Changed microtubule figures to show curved protofilaments at plus ends as per recent TEM work. Updated discussion of MT minus-end binding proteins; added information on augmin and branched MTs. Added info on CRWN proteins in higher plants to the IF section.

Chapter 14: Made minor changes to Figure 14A-2. Added a new Data Analysis question.

Chapter 15: Added brief mention of mechanotransduction via α-catenin. Added a new Data Analysis question.

Chapter 16: Added a purines/pyrimidine column in Table 16-1 on Chargaff's rules. Added detail on new studies on how histone H1 interacts with the nucleosome. Included an introduction to epigenetics in the section on chromatin remodeling. Mentioned how mRNA modifications are important in nuclear export of mRNA. Added possibility of NMCPs functioning as lamins in plant cells. Mentioned telomere dysfunction as a potential cause of premature aging in HGPS. Added detail about how charges in the histone tails affect DNA packaging. Moved section and figure on retroviruses from 9e Ch. 18 into this chapter.

Chapter 17: Added oxidation damage to Section 17.2. Authored a new Key Technique box on CRISPR genome modification. Updated the mutagenic mechanism of BrdU. Added a description of heteroduplex DNA to the homologous recombination section. Added a note on most likely mechanism of strand discrimination in eukaryotic mismatch repair. Updated nucleotide excision repair figure with more recent mechanism (Figure 17-27). Added quote from Francis Crick about the importance of DNA repair. Moved section and figure on retrotransposons from 9e Ch. 18 into this chapter. Added a new Data Analysis question.

Chapter 18: Improved the flow and organization of the chapter by moving discussion/figure about retroviruses to Ch. 16, moving retrotransposon discussion/figure to Ch. 17, and moving genetic code discussion/figures to Ch. 19. Authored a new Human Connections box on death cap mushrooms. Modified figure on the central dogma to include advances since Francis Crick's first proposal. Authored a new Concept Check question for Section 18.1. Added a note on the discovery of ribozymes, a subsection on mature mRNA nuclear export to Section 18.3, and a new Data Analysis question.

Chapter 19: Significantly reorganized the chapter flow by moving genetic code section from 9e Ch. 18 into new Section 19.1 and moving 9e Section 19.5 on protein targeting and sorting into Ch. 12. Added a new subsection on codon usage bias to Section 19.1.

Chapter 20: Added reference to temperature-sensitive riboswitches in Section 20.1, a paragraph on histone modifications to epigenetics in Section 20.2, and a new Data Analysis question.

Chapter 21: Moved 9e Figure 21-13 (tree of life) to Ch. 4 (new Figure 4-3) and the Key Technique box from 9e Ch. 17 (PCR) into this chapter. Worked the 9e Key Technique box from this chapter (DNA cloning) into the text in Section 21.1. Updated Southern blotting and Western blotting techniques for modern approach of not using film. Reorganized the techniques in Section 21.1 in a more logical way and moved all sequencing techniques into Section 21.2. Updated the description of next-generation and third-generation sequencing techniques to include state of the art in the field. Added a new subsection on quantitative PCR (qPCR). Expanded RNA-seq with details on single-cell RNAseq. Added description of conditional knockout mice engineering.

Chapter 22: Added a Make Connections question on *shibire* mutants in *Drosophila* that was needed in the synaptic transmission section.

Chapter 23: Changed title of Section 23.3 from Protein Kinase-Associated Receptors to Enzyme-Coupled Receptors. Added a subsection to the end of this section on other enzyme-coupled receptors (phosphatase receptor and guanylyl receptor families).

Chapter 24: Updated some sections with more modern treatment at the molecular level, including kinetochore (including revised Figure 24-4), chromosomal congression, FtsZ/divisome in bacteria, and spindle assembly checkpoint. Altered Figure 24-25 (9e 24-23) to improve clarity and moved to a later position.

Chapter 25: Added a paragraph on the potential role of the double-strand break repair model of homologous recombination in meiotic recombination in Section 25.6.

Chapter 26: Updated smoking statistics in Figure 26-7a through 2015 and added gender-specific data. Improved HPV figure (Figure 26-17). Changed emphasis to reflect replication/repair errors as a major cause of cancer, including discussion of the recent work by the Vogelstein group. Updated hallmarks of cancer discussion to correspond to the revised Weinberg paper from 2011. Added more detail in the immunotherapy section on Nobel Prize-winning work and CAR T cells, including a new small figure on CAR T cells. Added detail on Cdk4/6 therapy in the Key Technique box.

Appendix: Added explicit mention of GCaMP proteins in the calcium imaging section. Added a discussion of serial block-face SEM. Updated the cryoEM example image.

Building on the Strengths of Previous Editions

We have retained and built upon the strengths of prior editions in four key areas:

1. **The chapter organization focuses on main concepts.**
   - Each chapter is divided into sections that begin with a numbered *concept statement heading*, which summarizes the material and helps students focus on the main points to study and review.
   - Chapters are written and organized to allow instructors to assign chapters and chapter sections in different sequences, making the book adaptable to a wide variety of course plans.
   - Each chapter ends with a bulleted *Summary of Key Points* that briefly describes the main points covered in each section of the chapter.
2. **The illustrations teach concepts at an appropriate level of detail.**
   - Many of the more complex figures incorporate *mini-captions* to help students grasp concepts more quickly by drawing their focus to the body of an illustration rather than depending solely on a separate figure legend to describe what is taking place.
   - *Overview figures* outline complicated structures or processes in broad strokes and are followed by text and figures that present supporting details.
   - Carefully selected micrographs showing key cellular structures are accompanied by scale bars to indicate magnification.
3. **Important terminology is highlighted and defined in several ways.**
   - **Boldface type** is used to highlight the most important terms in each chapter, all of which are defined in the Glossary.

- *Italic type* is used to identify additional technical terms that are less important than boldfaced terms but significant in their own right. Occasionally, italics is also used to highlight important phrases or sentences.
- The Glossary includes definitions and page references for all boldfaced key terms and acronyms in every chapter—more than 1500 terms in all, a veritable dictionary of cell biology in its own right.

4. **Each chapter helps students learn the process of science, not just facts.**
   - Text discussions emphasize the experimental evidence that underlies our understanding of cell structure and function, to remind readers that advances in cell biology, as in all branches of science, come not from lecturers in their classrooms or textbook authors at their computers but from researchers in their laboratories.
   - The inclusion of a *Problem Set* at the end of each chapter reflects our conviction that we learn science not just by reading or hearing about it but by working with it. The problems are designed to emphasize understanding and application, rather than rote recall. These include highlighted questions that involve quantitative analysis and data analysis. Many are class-tested, having been selected from problem sets and exams we have used in our own courses. Detailed answers for all problems are available for students in the *Solutions Manual*, available on **Mastering Biology**.

Supplementary Learning Aids

Instructor Resources Area for *Becker's World of the Cell* (See Instructor Resource Area of Mastering Biology)

- PowerPoint lecture tools, including lecture outlines containing all of the figures, photos, and embedded animations, with five to ten personal response system clicker questions per chapter.
- JPEG images of all textbook figures and photos.
- Videos and animations of key concepts, organized by chapter for ease of use in the classroom.

Test Bank for *Becker's World of the Cell* (See Instructor Resource Area of Mastering Biology)

The test bank provides more than 1000 multiple-choice, short-answer, and inquiry/activity questions.

Solutions Manual for *Becker's World of the Cell* (See Instructor Resource Area of Mastering Biology)

Written by the authors, this manual includes complete, detailed answers for all of the Make Connections questions and end-of-chapter problems.

We Welcome Your Comments and Suggestions

The ultimate test of any textbook is how effectively it helps instructors teach and students learn. We welcome feedback and suggestions from readers and will try to acknowledge all correspondence. Please send your comments, criticisms, and suggestions to the appropriate authors listed here.

Chapters 1–3, 12, 16–21, 23, 25: James P. Lodolce
Department of Biology
Loyola University Chicago
1032 W. Sheridan Rd.
Chicago, IL 60660
e-mail: jlodolce@luc.edu

Chapters 4–11, 13–15, 22, 24, 26, Appendix, and Glossary: Jeff Hardin
Department of Integrative Biology
University of Wisconsin–Madison
Madison, WI 53706
e-mail: jdhardin@wisc.edu

ACKNOWLEDGMENTS

We want to acknowledge the contributions of the numerous people who have made this book possible. We are indebted especially to the many students whose words of encouragement catalyzed the writing of these chapters and whose thoughtful comments and criticisms have contributed much to the reader-friendliness of the text. Extending a long tradition started by Wayne Becker, many of these chapters are used every year in teaching Biocore students at the University of Wisconsin–Madison. These students continue to inspire us to do our very best.

We owe a special debt of gratitude to our colleagues, from whose insights and suggestions we have benefited greatly and borrowed freely. We also acknowledge those who have contributed to previous editions of the book, including David Deamer, Martin Poenie, Jane Reece, John Raasch, and Valerie Kish, as well as Peter Armstrong, John Carson, Ed Clark, Joel Goodman, David Gunn, Jeanette Natzle, Mary Jane Niles, Timothy Ryan, Beth Schaefer, Lisa Smit, David Spiegel, Akif Uzman, Karen Valentine, Deb Pires, and Ann Sturtevant. Most important, we are grateful to Wayne Becker for his incisive writing and vision, which led to the creation of this book and which featured so prominently in previous editions, to Lewis Kleinsmith, who played a key role in the 4th–8th editions, and to Greg Bertoni, who made important contributions to the 6th–9th editions. We have tried to carry on their tradition of excellence. In addition, we want to express our appreciation to the many colleagues who graciously consented to contribute micrographs to this endeavor, as well as the authors and publishers who have kindly granted permission to reproduce copyrighted material.

Many reviewers have graciously provided helpful criticisms and suggestions at various stages of manuscript development and revision. A special thanks goes to Catherine Putonti and Michael Burns for their help updating the molecular techniques in Chapter 21. The words of appraisal and counsel of all our reviewers were gratefully received and greatly appreciated. Indeed, the extensive review process for each new edition is a significant feature of the book. Nonetheless, the final responsibility for what you read here remains ours, and you may confidently attribute to us any errors of omission or commission encountered in these pages.

We are also deeply indebted to the many publishing professionals whose consistent encouragement, hard work, and careful attention to detail contributed much to the clarity of both the text and the art. This edition in particular has required the unflagging efforts of a remarkable publishing team, including Josh Frost, Content Strategy Manager; Rebecca Berardy Schwartz, Product Manager; Evelyn Dahlgren and Sonia DiVittorio, Developmental Editors; Chelsea Noack, Senior Associate Content Analyst; Suddha Satwa Sen and Margaret Young, Content Producers; Chloe Veylit, Lucinda Bingham, and Sarah Shefveland, Rich Media Producers; Ben Ferrini, Rights and Permissions Manager; and Kristin Piljay, Photo Researcher. We would also like to thank the Product Management, Content Strategy, and Digital Studio directors and managers for their support: Mike Early, Michael Gillespie, Ginnie Simione Jutson, Tod Regan, and Jeanne Zalesky.

James would like to dedicate this book to the memory of Amy E. Lodolce, a lifetime educator and scholar.

Finally, we are grateful beyond measure to our families and students, without whose patience, understanding, and forbearance this book could not have been written.

1 A Preview of Cell Biology

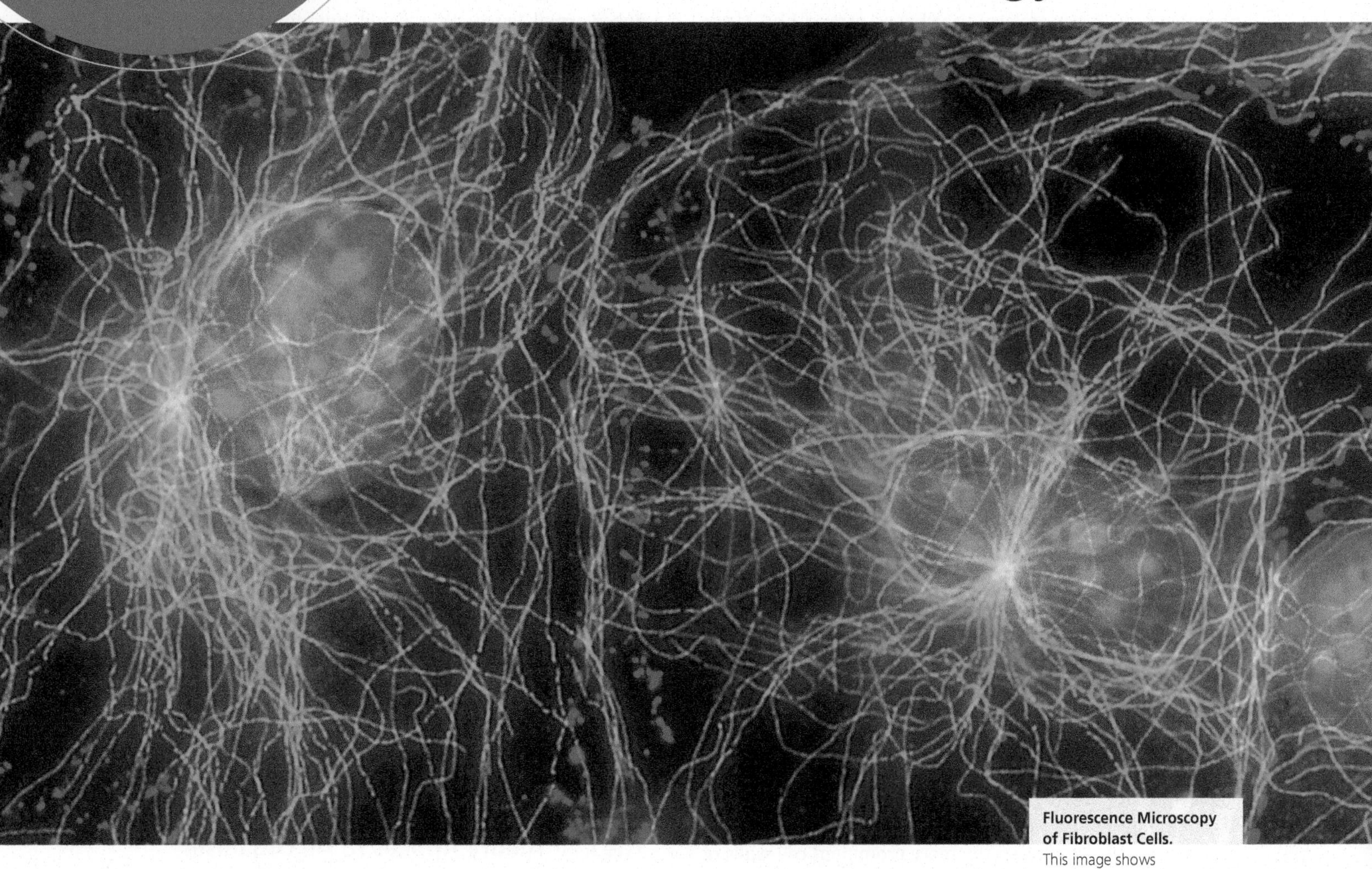

Fluorescence Microscopy of Fibroblast Cells. This image shows fluorescently labeled cell nuclei (red), microtubules (green), and cell-cell contacts (blue).

The **cell** is the basic unit of biology. Every organism either consists of cells or is itself a single cell. Therefore, it is only by understanding the structure and function of cells that we can appreciate both the capabilities and the limitations of living organisms, whether they are animals, plants, fungi, or microorganisms.

The field of cell biology is rapidly changing as scientists from a variety of related disciplines work together to gain a better understanding of how cells are constructed and how they carry out all the intricate functions necessary for life. Particularly significant is the dynamic nature of the cell. Cells are constantly changing; they have the capacity to grow, reproduce, and become specialized. In addition, once specialized, they have the ability to respond to stimuli and adapt to changes in the environment. The convergence of cytology, genetics, and biochemistry has made modern cell biology one of the most exciting and dynamic disciplines in all of biology. Nowhere is this excitement more evident than in the recent advances being made in our ability to modify genomes. If this text helps you appreciate the marvels and diversity of cellular

Mastering™ Biology www.masteringbiology.com

functions and allows you to experience the excitement of discovery, then one of our main goals in writing this book for you will have been met.

In this introductory chapter, we will look briefly at the origin of cell biology as a discipline. Then we will consider the three main historical strands of cytology, genetics, and biochemistry that have formed our current understanding of what cells are and how they work. The chapter concludes with a brief discussion of the nature of scientific knowledge itself by considering biological facts, the scientific method, experimental design, and the use of some common model organisms to answer important questions in modern cell biology.

1.1 The Cell Theory: A Brief History

The story of cell biology started to unfold more than 300 years ago, as European scientists began to focus their crude microscopes on a variety of biological material ranging from tree bark to bacteria to human sperm. One such scientist was Robert Hooke, Curator of Instruments for the Royal Society of London. In 1665, Hooke built a microscope and examined thin slices of cork (**Figure 1-1**). He observed and sketched a network of tiny boxlike compartments that reminded him of a honeycomb and called these little compartments *cells*, from the Latin word *cellula*, meaning "little room."

What Hooke observed were not cells at all. Those empty boxlike compartments were formed by the cell walls of dead plant tissue, which is what cork is. However, Hooke would not have thought of these cells as dead because he did not understand that they could be alive. Although he noticed that cells in other plant tissues were filled with what he called "juices," he concentrated instead on the more prominent cell walls of the dead cork cells that he had first encountered.

(a) Hooke's microscope

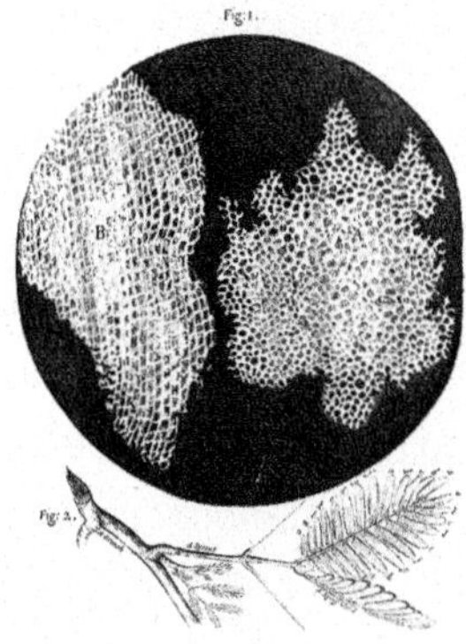

(b) Hooke's drawing of cork

Figure 1-1 The Birth of Microscopy. (a) Pictured is a reconstruction of Robert Hooke's original microscope, which he used to observe cork. **(b)** Hooke then sketched his observations.

Advances in Microscopy Allowed Detailed Studies of Cells

Hooke's observations were limited by the *magnification power* of his microscope, which enlarged objects to only 30 times (30×) their normal size. This made it difficult to learn much about the internal organization of cells. A few years later, Antonie van Leeuwenhoek, a Dutch textile merchant, produced small lenses that could magnify objects to almost 300 times (300×) their size. Using these superior lenses, van Leeuwenhoek became the first to observe living cells, including blood cells, sperm cells, bacteria, and single-celled organisms (algae and protozoa) found in pond water. He reported his observations to the Royal Society of London in a series of letters during the late 1600s. His detailed reports attest to both the high quality of his lenses and his keen powers of observation.

Two factors restricted further understanding of the nature of cells. First, the microscopes of the day had limited *resolution (resolving power)*—the ability to see fine details of structure. Even van Leeuwenhoek's superior instruments could push this limit only so far. The second factor was the descriptive nature of seventeenth-century biology. It was an age of observation, with little thought given to explaining the intriguing architectural details being discovered in biological materials.

More than a century passed before the combination of improved microscopes and more experimentally minded microscopists resulted in a series of developments that led to an understanding of the importance of cells in biological organization. By the 1830s, important optical improvements were made in lens quality and in the *compound microscope*, an instrument in which one lens (the eyepiece) magnifies the image created by a second lens (the objective). This allowed both higher magnification and better resolution. At that point, structures only 1 micrometer (μm) in size could be seen clearly.

The Cell Theory Applies to All Organisms

Aided by such improved lenses, the Scottish botanist Robert Brown found that every plant cell he looked at contained a rounded structure, which he called a *nucleus*, a term derived from the Latin word for "kernel." In 1838, his German colleague Matthias Schleiden came to the important conclusion that all plant tissues are composed of cells and that an embryonic plant always arises from a single cell. A year later, German cytologist Theodor Schwann reported similar conclusions concerning animal tissue, thereby discrediting earlier speculations that plants and animals do not resemble each other structurally. These speculations arose because plant cell walls form conspicuous boundaries between cells that are readily visible even with a crude microscope, whereas individual animal cells, which lack cell walls, are much harder to distinguish in a tissue sample. However, when Schwann examined animal cartilage cells, he saw that they were unlike most other animal cells because they have boundaries that are well defined by thick deposits of collagen fibers. Thus, he became convinced of the fundamental similarity between plant and animal tissue. Based on his astute observations, Schwann developed a single unified theory of cellular organization. This theory has stood the test of time and continues to be the basis for our own understanding of the importance of cells and cell biology. (The recent discovery of certain giant viruses has led some to speculate that this definition may someday be expanded.)

As originally postulated by Schwann in 1839, the **cell theory** had two basic principles:

1. All organisms consist of one or more cells.
2. The cell is the basic unit of structure for all organisms.

Less than 20 years later, a third principle was added. This grew out of Brown's original description of nuclei, which Swiss botanist Karl Nägeli extended to include observations on the nature of cell division. By 1855 Rudolf Virchow, a German physiologist, concluded that cells arose only by the division of other, preexisting cells. Virchow encapsulated this conclusion in the now-famous Latin phrase *omnis cellula e cellula*, which in translation becomes the third principle of the modern cell theory:

3. All cells arise only from preexisting cells.

Thus, the cell is not only the basic unit of structure for all organisms but also the basic unit of reproduction. No wonder, then, that we must understand cells and their properties to appreciate all other aspects of biology. Because many of you have seen examples of "typical" cells in textbooks that may give the false impression that there are relatively few different types of cells, let's take a look at a few examples of the diversity of cells that exist in our world (**Figure 1-2**).

Cells exist in a wide variety of shapes and sizes, ranging from filamentous fungal cells to spiral-shaped *Treponema* bacteria to the differently shaped cells of the human blood system (Figure 1-2a–c). Other cells have much more exotic shapes,

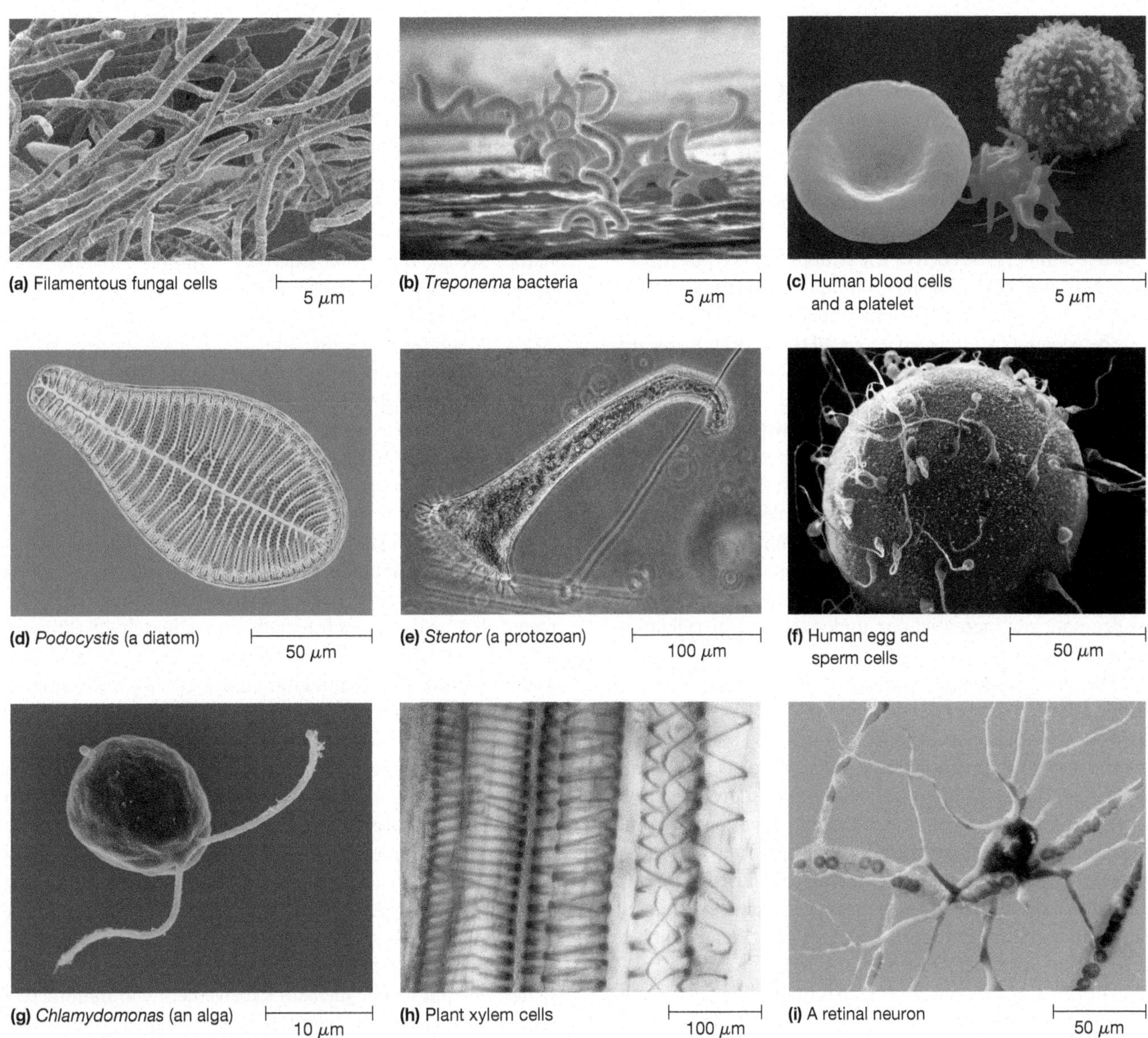

Figure 1-2 The Cells of the World. The diversity of cell types existing all around us includes the examples shown in this figure and thousands upon thousands more.

such as the diatom and the protozoan shown in Figure 1-2d and 1-2e. Note how the two human single-celled gametes, the egg and the sperm, differ greatly in size and shape (Figure 1-2f). As in leaves, the green chlorophyll in a *Chlamydomonas* cell shows that these algae carry out photosynthesis (Figure 1-2g). Often, a cell's shape and structure give clues about its function. For example, the spiral thickenings in the cell walls of plant xylem tissue give strength to these water-conducting vessels in wood (Figure 1-2h), and the highly branched cells of a human neuron allow it to interact with numerous other neurons (Figure 1-2i). In our studies throughout this textbook, we will see many other interesting examples of diversity in cell structure and function. First, though, let's examine the historical roots leading to the development of contemporary cell biology.

CONCEPT CHECK 1.1

What evidence led scientists to develop the basic principles of the cell theory? Note how technology played a role in its development.

1.2 The Emergence of Modern Cell Biology

Modern cell biology results from the weaving together of three different strands of biological inquiry—cytology, biochemistry, and genetics—into a single cord. As the timeline in **Figure 1-3** illustrates, each of the strands had its own historical origins, and each one makes unique and significant contributions to modern cell biology. Contemporary cell biologists must be adequately informed about all three strands, regardless of their own immediate interests.

Historically, the first of the strands to emerge was **cytology**, which is concerned primarily with cellular structure. In biological studies, you will often encounter words containing the Greek prefix *cyto–* or the suffix *–cyte*, both of which mean "hollow vessel" and refer to cells. Cytology had its origins more than three centuries ago and depended heavily on the light microscope for its initial impetus. The advent of electron microscopy and other advanced optical techniques has dramatically increased our understanding of cell structure and function.

The second strand represents the contributions of **biochemistry** to our understanding of cellular structure and function. Most of the developments in this field have occurred over the past 95 years, though the roots go back at least a century earlier. Especially important has been the development of laboratory techniques such as ultracentrifugation, chromatography, radioactive labeling, electrophoresis, and mass spectrometry for separating and identifying cellular components. You will encounter these and other techniques later in your studies as you learn how specific details of cellular structure and function were discovered using these techniques.

The third strand contributing to the development of modern cell biology is **genetics**. Although the timeline for genetics stretches back more than 150 years, most of our present understanding has been gained within the past 75 years. An especially important discovery was the demonstration that, in all organisms, DNA (deoxyribonucleic acid) is the bearer of genetic information. It encodes the tremendous variety of proteins and RNA (ribonucleic acid) molecules responsible for most of the functional and structural features of cells. Recent accomplishments on the genetic strand include the sequencing of the entire **genome** (all of the DNA) of humans and other species, the *cloning* (production of genetically identical organisms) of mammals, including livestock, pets, and primates, and the editing of genomes.

Therefore, an understanding of present-day cell biology requires an appreciation of its diverse roots and the important contributions made by each of its component strands to our current understanding of what a cell is and what it can do. Each of the three historical strands of cell biology is discussed briefly here; a deeper appreciation of these historical strands will come in later chapters as we explore cells in detail. Keep in mind also that in addition to developments in cytology, biochemistry, and genetics, the field of cell biology has benefited greatly from advancements in other fields of study such as chemistry, physics, computer science, and engineering.

The Cytological Strand Deals with Cellular Structure

Strictly speaking, cytology is the study of cells. Historically, however, cytology has dealt primarily with cellular structure, mainly through the use of optical techniques. Here we will describe briefly some of the microscopy that is important in cell biology. (For more detailed discussion of microscopic techniques, see the Appendix.) Microscopy has been invaluable in helping cell biologists overcome a fundamental problem—the problem of small size.

Cellular Dimensions. One challenge involved in understanding cellular structure and organization is the fact that most cells and their organelles are too small to be seen by the unaided eye. The cellular structures that microscopists routinely deal with are measured using units that may not be familiar to you.

The **micrometer** (μm) is the most useful unit for expressing the size of cells and organelles (**Figure 1-4**, on page 6). A micrometer (historically called a *micron*) is one-millionth of a meter (10^{-6} m). One inch equals approximately 25,000 μm. In general, bacterial cells are a few micrometers in diameter, and the cells of plants and animals are 10 to 20 times larger. Organelles such as mitochondria and chloroplasts tend to be a few micrometers in size and are thus comparable in size to whole bacterial cells. In general, if you can see it with a light microscope, you can express its dimensions conveniently in micrometers (Figure 1-4a).

The **nanometer** (nm) is the unit of choice for molecules and subcellular structures that are too small to be seen using the light microscope. A nanometer is one-billionth of a meter (10^{-9} m), so it takes 1000 nanometers to equal 1 micrometer. A ribosome has a diameter of about 25 to 30 nm. Other structures that can be measured conveniently in nanometers are cell membranes, microtubules, microfilaments, and DNA molecules (Figure 1-4b). A slightly smaller unit, the angstrom (Å), is used in cell biology when measuring dimensions within proteins and DNA molecules. An angstrom equals 0.1 nm, which is about the size of a hydrogen atom.

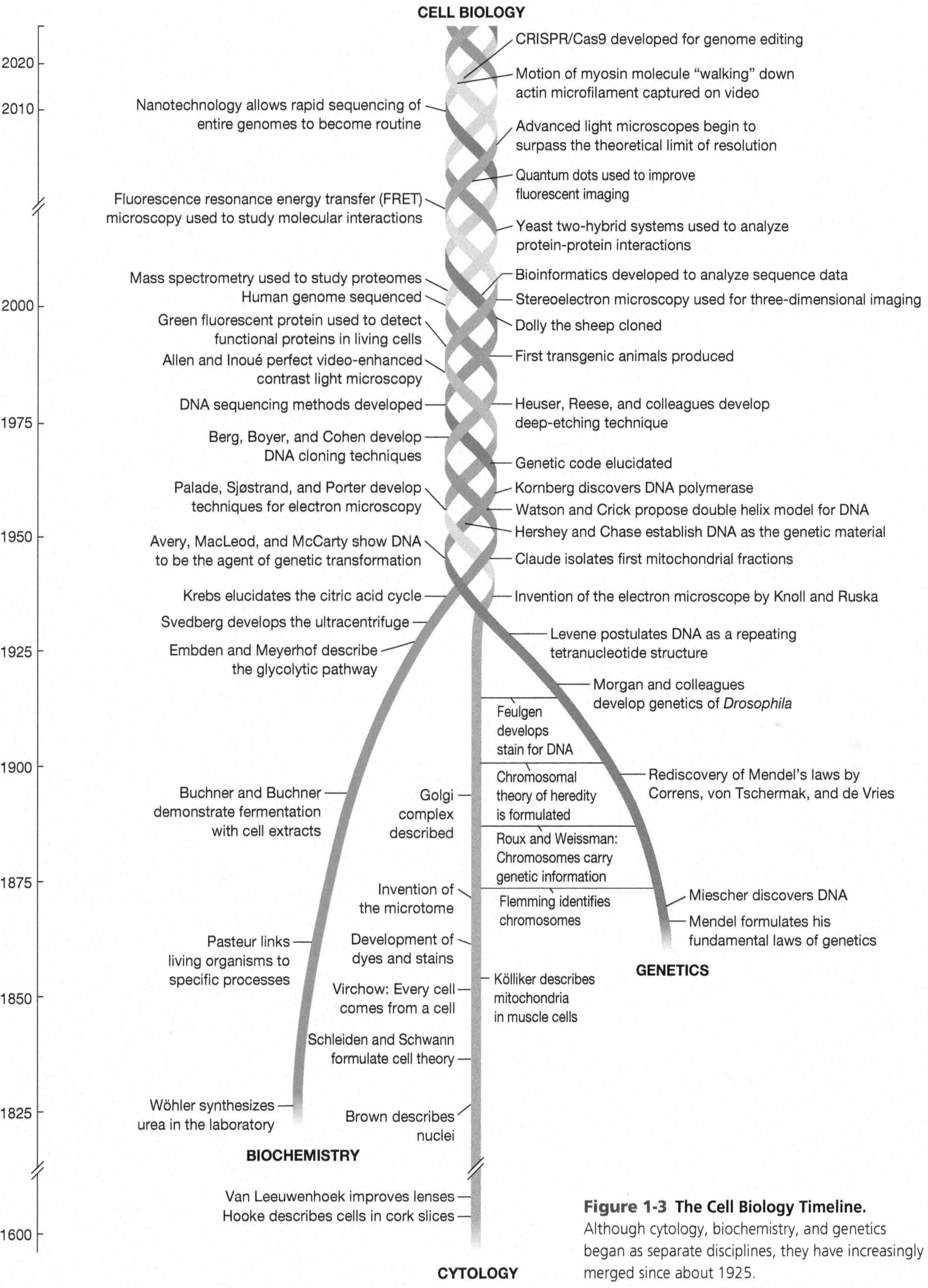
CELL BIOLOGY
2020
2010
2000
1975
1950
1925
1900
1875
1850
1825
1600
CRISPR/Cas9 developed for genome editing
Motion of myosin molecule "walking" down actin microfilament captured on video
Nanotechnology allows rapid sequencing of entire genomes to become routine
Advanced light microscopes begin to surpass the theoretical limit of resolution
Quantum dots used to improve fluorescent imaging
Fluorescence resonance energy transfer (FRET) microscopy used to study molecular interactions
Yeast two-hybrid systems used to analyze protein-protein interactions
Mass spectrometry used to study proteomes
Bioinformatics developed to analyze sequence data
Human genome sequenced
Stereoelectron microscopy used for three-dimensional imaging
Green fluorescent protein used to detect functional proteins in living cells
Dolly the sheep cloned
Allen and Inoué perfect video-enhanced contrast light microscopy
First transgenic animals produced
DNA sequencing methods developed
Heuser, Reese, and colleagues develop deep-etching technique
Berg, Boyer, and Cohen develop DNA cloning techniques
Genetic code elucidated
Palade, Sjøstrand, and Porter develop techniques for electron microscopy
Kornberg discovers DNA polymerase
Watson and Crick propose double helix model for DNA
Hershey and Chase establish DNA as the genetic material
Avery, MacLeod, and McCarty show DNA to be the agent of genetic transformation
Claude isolates first mitochondrial fractions
Krebs elucidates the citric acid cycle
Invention of the electron microscope by Knoll and Ruska
Svedberg develops the ultracentrifuge
Levene postulates DNA as a repeating tetranucleotide structure
Embden and Meyerhof describe the glycolytic pathway
Morgan and colleagues develop genetics of *Drosophila*
Feulgen develops stain for DNA
Chromosomal theory of heredity is formulated
Rediscovery of Mendel's laws by Correns, von Tschermak, and de Vries
Buchner and Buchner demonstrate fermentation with cell extracts
Golgi complex described
Roux and Weissman: Chromosomes carry genetic information
Invention of the microtome
Flemming identifies chromosomes
Miescher discovers DNA
Mendel formulates his fundamental laws of genetics
Pasteur links living organisms to specific processes
Development of dyes and stains
GENETICS
Virchow: Every cell comes from a cell
Kölliker describes mitochondria in muscle cells
Schleiden and Schwann formulate cell theory
Wöhler synthesizes urea in the laboratory
Brown describes nuclei
BIOCHEMISTRY
Van Leeuwenhoek improves lenses
Hooke describes cells in cork slices
CYTOLOGY

Figure 1-3 The Cell Biology Timeline. Although cytology, biochemistry, and genetics began as separate disciplines, they have increasingly merged since about 1925.

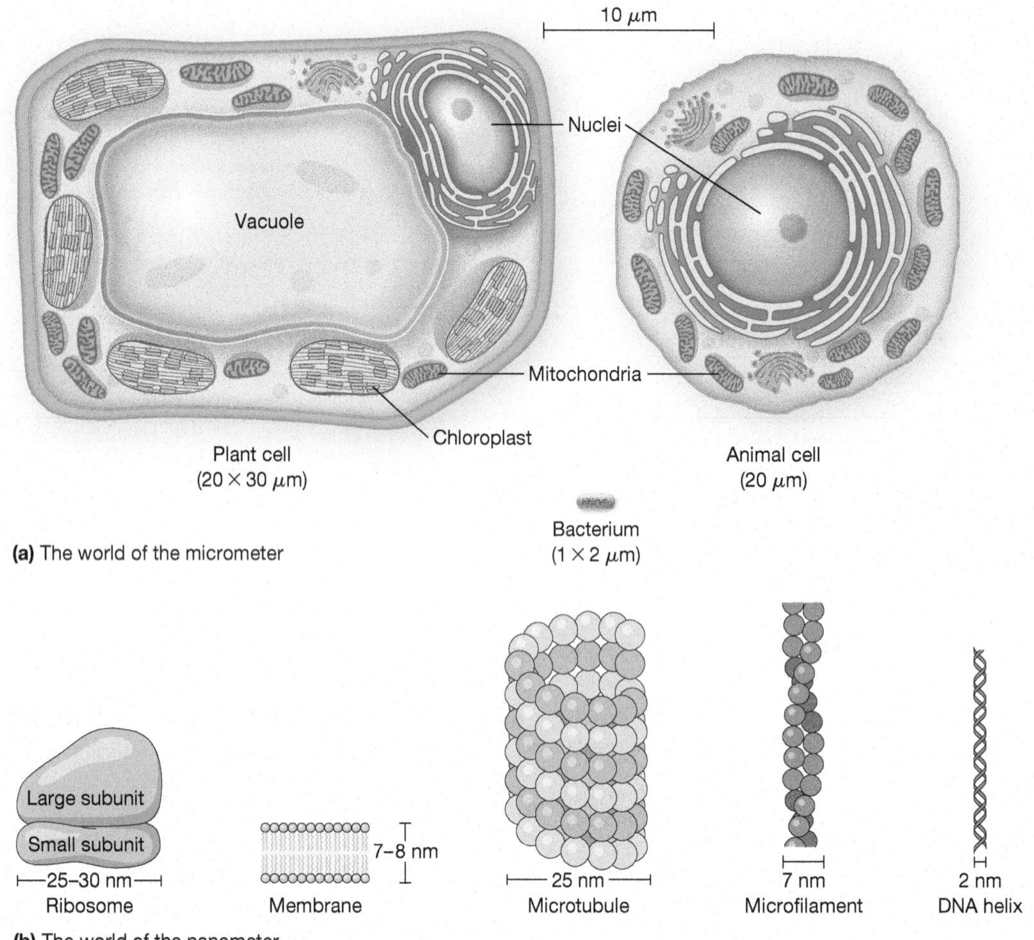

Figure 1-4 The Worlds of the Micrometer and Nanometer. Illustrations show **(a)** typical cells and **(b)** common cellular structures.

Microscopy. The most important technique within the cytological strand is microscopy. This technique allows scientists to visualize cells and cellular components at the previously mentioned cellular dimensions. Depending on the level of resolution required, the two major forms of microscopy used are light microscopy and electron microscopy.

The **light microscope** was the earliest tool of the cytologists and continues to play an important role in our elucidation of cellular structure. Light microscopy allowed cytologists to identify membrane-bounded structures such as *nuclei, mitochondria,* and *chloroplasts* within a variety of cell types. Such structures are called *organelles* ("little organs") and are prominent features of most plant and animal (but not bacterial) cells. (Chapter 4 presents an overview of organelle types, and later chapters investigate their structure and function in more detail.)

The basic type of light microscopy is called *brightfield microscopy* because white light is passed directly through a specimen that is either stained or unstained and the background (the field) is illuminated. A significant limitation of this approach is that specimens often must be chemically fixed (preserved), dehydrated, embedded in paraffin or plastic for slicing into thin sections, and stained to highlight otherwise transparent features. Fixed and stained specimens are no longer alive; therefore, features observed using this method could be distortions caused by slide preparation processes and might not be typical of living cells.

To overcome the limitations of a brightfield microscope, a variety of specialized light microscopes have been developed for observing living cells directly. These techniques include phase-contrast microscopy, differential interference contrast microscopy, fluorescence microscopy, and confocal microscopy. Each of these forms of microscopy is introduced below. (More detail on these techniques, including sample images using them, can be found in the Appendix.)

Phase-contrast and *differential interference contrast* microscopy make it possible to see living cells clearly. Like water waves, light waves have crests and troughs, and the precise positions of these maxima and minima as light travels are known as the *phase* of the light. Both techniques enhance and amplify slight changes in the phase of transmitted light as it passes through a structure having a different density than the surrounding medium.

Fluorescence microscopy is a powerful method that enables researchers to detect specific proteins, DNA sequences, or other molecules that are made fluorescent by coupling them to a fluorescent dye or a fluorescent protein or by binding them to a fluorescently labeled antibody. An **antibody** is a protein molecule produced by the immune system that binds one particular target molecule, known as its antigen.

By simultaneously using two or more such dyes or antibodies, each emitting light of a different color, researchers can follow the distributions of different kinds of molecules in the same cell. Antibody labeling is a powerful method to both visualize and identify specific molecules within cells and is described in more detail (see Key Technique, pages 8–9). In recent years, green fluorescent protein (GFP) from the bioluminescent jellyfish *Aequorea victoria* has become an invaluable tool for studying the temporal and spatial distribution of particular proteins in a cell. When a protein of interest is fused with GFP, its synthesis and movement can be followed in living cells using a fluorescence microscope.

An inherent limitation of fluorescence microscopy is that the viewer can focus on only a single plane of the specimen at a time, yet fluorescent light is emitted throughout the specimen, blurring the image. This problem is largely overcome by *confocal microscopy*, which uses a laser beam to illuminate just one plane of the specimen at a time. When used with thick specimens such as whole cells, this approach gives much better resolution.

Another recent development in light microscopy is *digital video microscopy*, which allows researchers to observe cells for extended periods of time using very low levels of light. This image intensification is particularly useful to visualize fluorescent molecules present at low levels in living cells and even to see and identify individual *macromolecules* such as DNA and protein molecules. In fact, extremely powerful *superresolution* light microscopy methods have been developed that use imaging and computational methods so advanced that they can visualize structures 50–100 nm in size, which, until the past few years, were believed impossible to see with any light microscope. However, despite recent significant advances, light microscopy is inevitably subject to the limit of resolution imposed by the wavelength of the light used to view the sample.

As used in microscopy, the **limit of resolution** refers to how far apart adjacent objects must be to appear as separate entities. For example, if the limit of resolution of a microscope is 400 nm, objects must be at least 400 nm apart to be recognizable as separate entities. The smaller the limit of resolution, the greater the **resolving power**, or ability to see fine details of structure, of the microscope. Therefore, a better microscope might have a resolution of 200 nm, meaning that objects only 200 nm apart can be distinguished from each other.

Because of the physical nature of light itself, the theoretical limit of resolution for the light microscope is approximately half the size of the wavelength of light used for illumination, allowing maximum magnifications of about 1000–1400×. For *visible light* (wavelengths of 400–700 nm), the limit of resolution is about 200–350 nm. **Figure 1-5** illustrates the useful range of the light microscope and compares its resolving power with that of the human eye and the electron microscope.

A major breakthrough in resolving power came with the development of the **electron microscope**, which was invented in Germany in 1931 by Max Knoll and Ernst Ruska. In place of visible light and optical lenses, the electron microscope uses a beam of electrons that is deflected and focused by an electromagnetic field. Because the wavelength of electrons is so much shorter than the wavelengths of visible light, the practical limit of resolution for the electron microscope is much better—generally about 100 times better than a light microscope, or 2 nm (see Figure 1-5). As a result, the useful magnification of the electron microscope is also much higher—up to 100,000×.

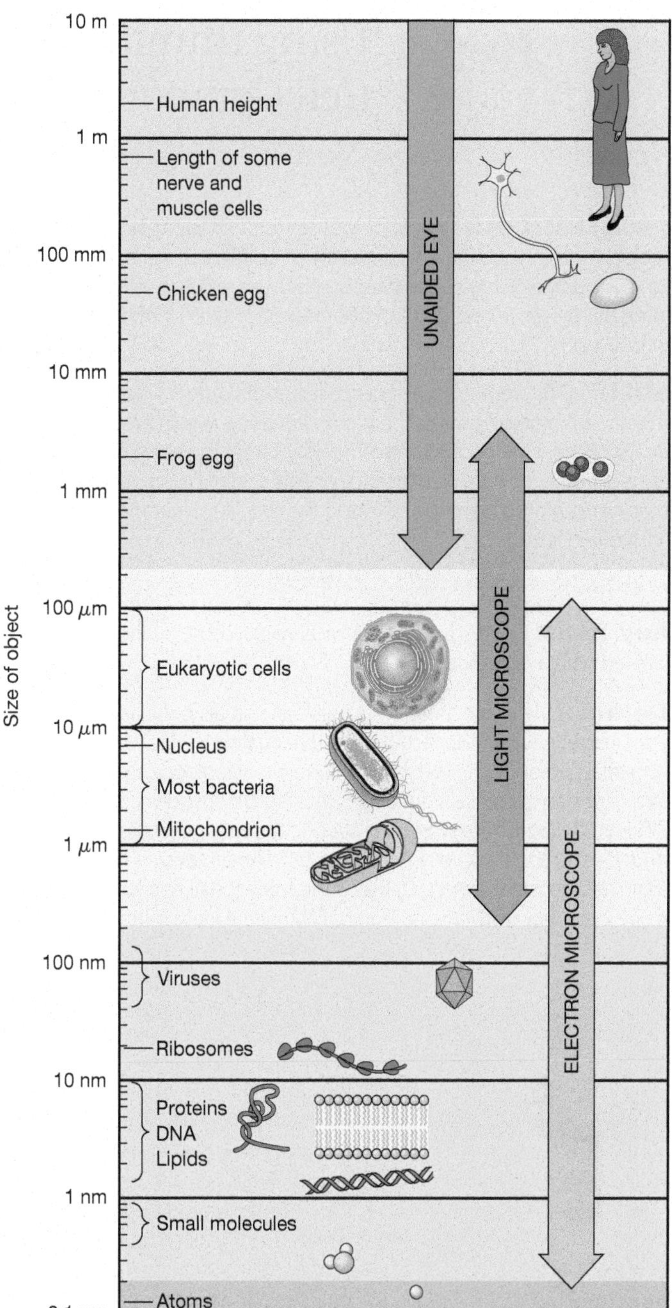

Figure 1-5 Relative Resolving Power of the Human Eye, the Light Microscope, and the Electron Microscope. Notice that the vertical axis is on a logarithmic scale to accommodate the wide range of sizes shown (based on powers of 10).

Electron microscopy continues to revolutionize our understanding of cellular architecture by making detailed ultrastructural investigations possible. Whereas organelles such as nuclei or mitochondria are large enough to be seen with a light microscope, they can be studied in much greater detail with an electron microscope. In addition, electron microscopy has revealed cellular structures that are too small to be seen with a light microscope. These include ribosomes, cell membranes,

Key Technique | Using Immunofluorescence to Identify Specific Cell Components

PROBLEM: Cells are made of thousands of different types of molecules that make up a wide variety of cellular structures. With so many different molecules present, how can researchers determine the presence and location of one specific type of molecule within a cell?

SOLUTION: *Immunofluorescence* is a technique in which a fluorescent molecule is attached to an *antibody,* which recognizes and binds to one specific complementary target molecule, known as its *antigen*. Using a fluorescence or confocal microscope, a researcher can then identify and locate the specific target molecule within the cell.

Key Tools: Fluorescence or confocal microscope; antibodies labeled with a fluorescent dye.

Details: One of the amazing features of animals is the ability of their immune systems to recognize and neutralize a wide variety of potential pathogens. In vertebrates, certain white blood cells, known as *B lymphocytes,* secrete antibodies into the bloodstream, and each different antibody recognizes one specific type of antigen, targeting it for destruction by other white blood cells. An antibody is a protein that has a constant region (C) that is the same for all antibodies of a particular type and variable regions (V) that are identical to each other but unique for each antibody (**Figure 1A-1**). The unique V regions at the tips of the Y contain a binding pocket into which only one specific antigen will fit.

Variable regions
Antigen-binding site
Constant region

Figure 1A-1 Antibody Structure.

Immunofluorescence exploits the specificity of antibodies for their antigen targets. Rather than targeting antigens for destruction, however, immunofluorescence is used to detect where the antigen is located within a cell. Antibodies can be generated in the laboratory by injecting a foreign protein or other macromolecule into an animal host, such as a rabbit or mouse, producing antibodies that will bind selectively to virtually any protein that a scientist wishes to study. Using *primary* (or *direct*) *immunofluorescence,* antibody molecules are labeled with a fluorescent dye, known as a *fluorophore,* that is covalently linked to the C region of each antibody molecule (**Figure 1A-2**). The antibody recognizes and binds to the target molecule, which can then be detected using fluorescence or confocal microscopy.

More commonly, researchers use *secondary* (or *indirect*) *immunofluorescence.* In this case, a tissue or cell is treated with an antibody that is not labeled with dye (**Figure 1A-3**). This antibody, called the *primary antibody,* attaches to specific antigenic sites within the tissue or cell. A second type of antibody, called the *secondary antibody,* is labeled with a fluorescent dye and then added to the sample, where it attaches to the primary antibody. Because more than one primary antibody molecule can attach to an antigen and more than one secondary antibody molecule can

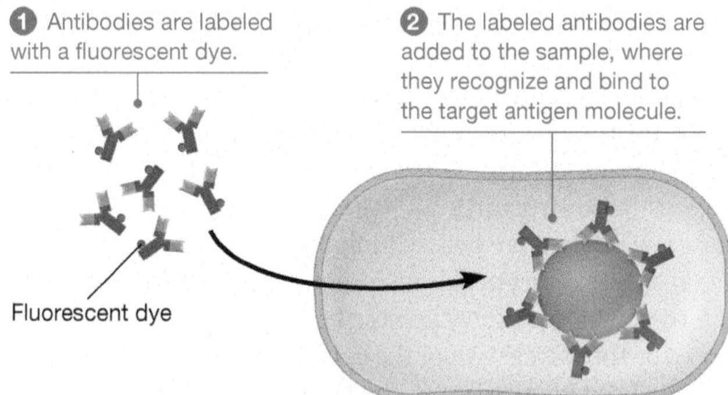

Figure 1A-2 Primary Immunofluorescence. In primary immunofluorescence, an antibody that binds to a specific antigen in a tissue or cell is labeled with a fluorescent dye. The labeled antibody is then added to the sample, where it binds to its target molecule. The pattern of fluorescence that results is visualized using fluorescence or confocal microscopy.

microtubules, and microfilaments (see Figure 1-4b), as well as some *macromolecules* such as DNA and protein molecules.

Most electron microscopes have one of two basic designs: the **transmission electron microscope (TEM)** and the **scanning electron microscope (SEM)**. Images from each are shown in **Figure 1-6** on page 10. Transmission and scanning electron microscopes are similar in that each employs a beam of electrons, but they use quite different mechanisms to form the image. As the name implies, a TEM forms an image from electrons that are transmitted through the specimen. An SEM, on the other hand, scans the surface of the specimen and forms an image by detecting electrons that are deflected from its outer surface. Scanning electron microscopy is an especially spectacular technique because of the sense of depth it gives to biological structures.

Electron microscopy is constantly evolving. Several specialized techniques for electron microscopy allow visualization of specimens in three dimensions and can determine structures of some macromolecules such as proteins. Still other techniques combine some of the principles of TEM and SEM and even allow visualization of cells in liquid without the need for a vacuum. (All of these microscopy techniques are described in detail in the Appendix.)

The Biochemical Strand Concerns the Chemistry of Biological Structure and Function

At about the same time that cytologists started exploring cellular structure with their microscopes, other scientists were making observations that began to explain and clarify cellular function. Using techniques derived from classical chemistry,

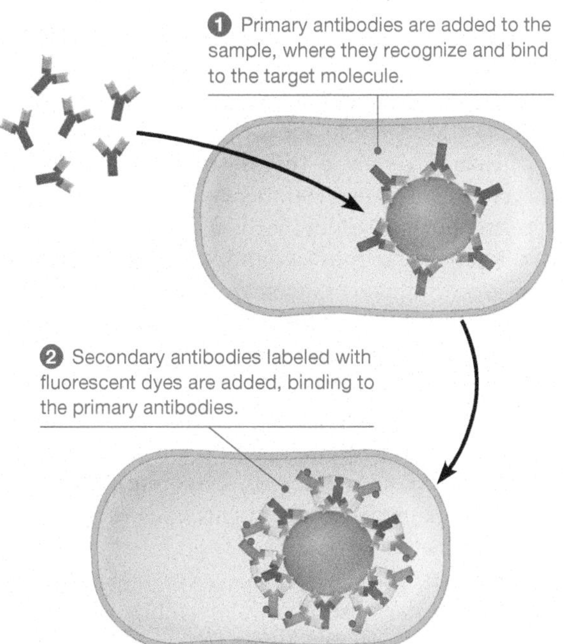

Figure 1A-3 Secondary Immunofluorescence. In secondary immunofluorescence, a primary antibody is added to a tissue or cell. A secondary antibody that is labeled with a fluorescent dye is then added to the sample, where it binds to the primary antibody, amplifying the fluorescence signal.

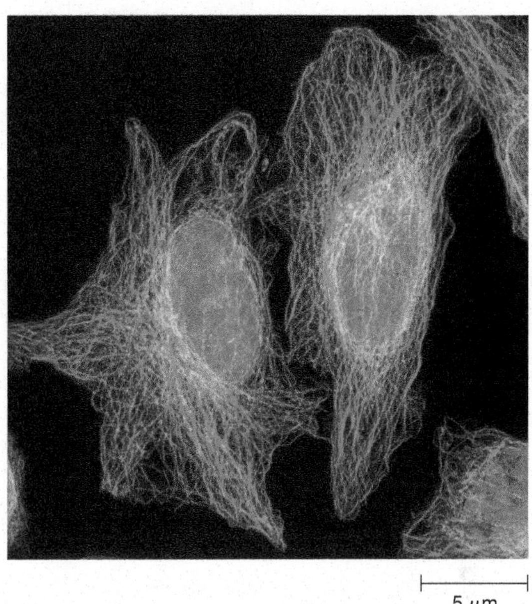

Figure 1A-4 Labeling Cells Using Two Different Colors. HeLa cells were stained using indirect immunofluorescence for microtubules (green) and a dye to label DNA (purple).

attach to each primary antibody, more fluorescent molecules are concentrated near the target molecule. As a result, indirect immunofluorescence amplifies the fluorescence signal and is much more sensitive than the use of a primary antibody alone.

Antibodies that recognize any one of thousands of specific antigens are commercially available. By using different combinations of antibodies and dyes, more than one molecule in a cell can be labeled at the same time. Different dyes can be imaged using different combinations of fluorescent filters, and the different images can be combined to generate striking pictures of cellular structures (**Figure 1A-4**). In some cases, instead of a fluorescent dye, antibodies can be linked to an enzyme performing a chemical reaction, resulting in a colored precipitation product that can be seen using a standard microscope.

QUESTION: Why is a high degree of antibody-antigen specificity so important for identification of cellular components?

these scientists began to understand the structure and function of biological molecules, launching a field that came to be known as biochemistry.

Biochemical Reactions and Pathways. Much of what is now called biochemistry dates from a discovery reported in 1828 by German chemist Friedrich Wöhler, a contemporary and fellow countryman of Schleiden and Schwann. Wöhler revolutionized our thinking about biology and chemistry by demonstrating that urea, an organic compound of biological origin, could be synthesized in the laboratory from an inorganic starting material, ammonium cyanate.

Until Wöhler reported his results, it had been widely held that living organisms were unique, not governed by the laws of chemistry and physics that apply to the nonliving world. By showing that a compound made by living organisms—a "biochemical"—could be synthesized in a laboratory just like any other chemical, Wöhler helped dispel the notion that biochemical processes were somehow exempt from the laws of chemistry and physics.

Another major advance came about 30 years later, when French chemist and biologist Louis Pasteur showed that living yeast cells were responsible for the fermentation of sugar into alcohol. In 1897, German bacteriologists Eduard and Hans Buchner found that fermentation could take place with isolated extracts from yeast cells—that is, the intact cells themselves were not required. Gradually, it became clear that the active agents in the extracts were specific biological catalysts that have since come to be called **enzymes**—from *zyme*, a Greek word meaning "yeast."

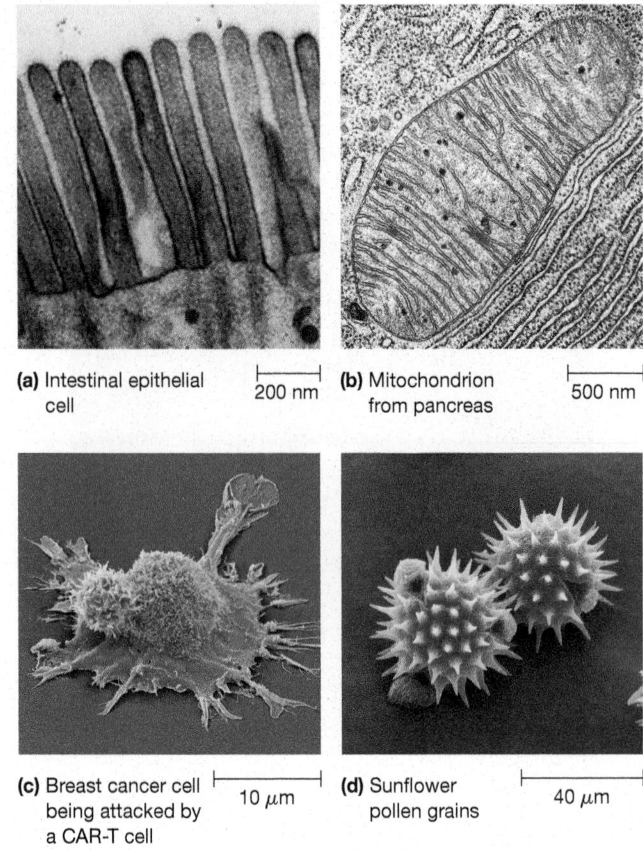

Figure 1-6 Electron Microscopy. A transmission electron microscope was used to produce the images in parts **(a)** and **(b)** (TEMs). A scanning electron microscope was used to make the images in parts **(c)** and **(d)** (colorized SEMs).

In the 1920s and 1930s, individual steps in the complex, multistep biochemical pathways for fermentation and related cellular processes were elucidated. German biochemists, such as Gustav Embden, Otto Meyerhof, Otto Warburg, and Hans Krebs, described the enzymatic steps in the Embden–Meyerhof pathway for *glycolysis* for glucose breakdown and the *Krebs cycle* for energy production. Both of these pathways are important because of their role in the process by which cells extract energy from glucose and other foodstuffs. (For a detailed examination of these biochemical pathways, see Chapters 9 and 10.) At about the same time, Fritz Lipmann, an American biochemist, showed that the high-energy compound *adenosine triphosphate (ATP)* is the principal energy storage compound in most cells.

An important advance in the study of biochemical reactions and pathways came as radioactive isotopes such as ^{3}H, ^{14}C, and ^{32}P were first used to trace the metabolic fate of specific atoms and molecules. American chemist Melvin Calvin and his colleagues at the University of California, Berkeley, were pioneers in this field as they traced the fate of ^{14}C-labeled carbon dioxide ($^{14}CO_2$) in illuminated algal cells that were actively photosynthesizing. Their work, carried out in the late 1940s and early 1950s, led to the elucidation of the *Calvin cycle*—the most common pathway for photosynthetic carbon metabolism. The Calvin cycle was the first metabolic pathway to be elucidated using a radioisotope.

Biochemistry Methods. Biochemistry took another major step forward with the development of techniques for isolating, purifying, and analyzing subcomponents of cells.

Centrifugation is a means of separating and isolating subcellular structures and macromolecules based on their size, shape, and/or density—a process called **subcellular fractionation**. This process helps us study specific parts of the cell, such as the nucleus or specific proteins.

Especially useful for resolving small organelles and macromolecules is the **ultracentrifuge**, developed by Swedish chemist Theodor Svedberg in the late 1920s. An ultracentrifuge is capable of very high speeds—more than 100,000 revolutions per minute—and can thereby subject samples to forces exceeding 500,000 times the force of gravity. In many ways, the ultracentrifuge is as significant to biochemistry as the electron microscope is to cytology. In fact, both instruments were developed at about the same time, so the ability to see organelles and other subcellular structures coincided with the capability to isolate and purify them.

Other biochemical techniques that have proven useful for separating and purifying subcellular components include chromatography and electrophoresis. **Chromatography** is a general term describing a variety of techniques by which a mixture of molecules in solution is separated into individual components. Chromatographic techniques separate molecules based on their size, charge, or affinity for specific molecules or functional groups (**Figure 1-7**). In fact, this technique gets its name from its early use in separating differently colored plant pigments, as shown in part a.

Electrophoresis refers to several related techniques that use an electrical field to separate macromolecules based on their mobility through a semisolid gel. Different molecules move at different rates depending on their size and charge.

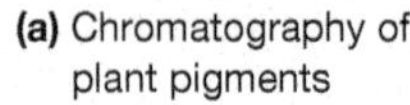

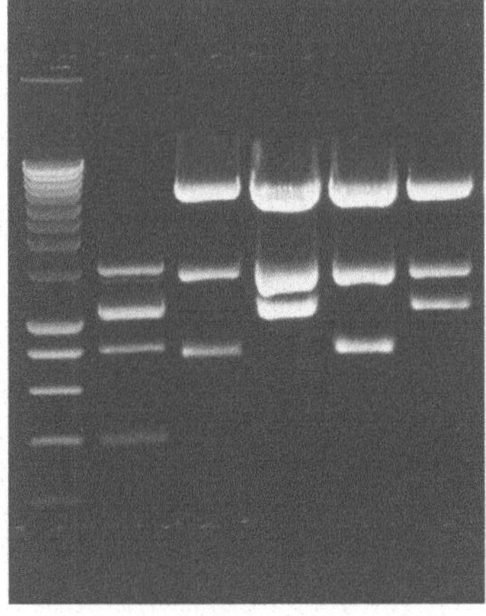

(b) Electrophoresis of DNA samples

Figure 1-7 Separation of Molecules by Chromatography and Electrophoresis. (a) Filter paper is used as the stationary medium in this chromatogram of plant pigments. As the solvent containing the pigments moves across the filter, different pigments move at different rates and can be separated and then purified. **(b)** An agarose gel containing samples of mixed DNA molecules is separated by electrophoresis, stained with ethidium bromide, and illuminated with ultraviolet light to show each DNA fragment as a visible band.

Electrophoresis is used extensively to isolate and characterize DNA, RNA, and protein molecules. An example of separation of different DNA fragments using gel electrophoresis is shown in Figure 1-7b. After proteins have been separated by electrophoresis, **mass spectrometry** is commonly used to determine the size and composition of individual proteins. This technique allows researchers to determine the identity and characteristics of individual proteins.

To summarize, with the enhanced ability to see subcellular structures, to fractionate them, and to isolate them, cytologists and biochemists began to realize how well their respective observations on cellular structure and function could complement each other. These scientists were laying the foundations for modern cell biology.

The Genetic Strand Focuses on Information Flow

The third strand in the historical cord of cell biology is genetics, the study of the inheritance of characteristics from generation to generation. Although more than 2000 years ago the Greek philosopher Aristotle referred to a physical entity he called the "germ" and stated that it "springs forth from a definite parent and gives rise to a predictable progeny," it was not until the nineteenth century that scientists discovered the nature of these inherited physical entities, now known as genes.

Classical Genetics. The genetic strand begins with Gregor Mendel, whose studies with the pea plants he grew in a monastery garden must surely rank among the most famous experiments in all of biology. His findings were published in 1866, laying out the principles of segregation and independent assortment of the "hereditary factors" known today as **genes**. But Mendel was clearly a man ahead of his time. His work went almost unnoticed when it was first published and was not fully appreciated until its rediscovery nearly 35 years later.

In the decade following publication of Mendel's work, the role of the nucleus in the genetic continuity of cells came to be appreciated. In 1880, German biologist Walther Flemming identified **chromosomes**, threadlike bodies seen in dividing cells. Flemming called the division process *mitosis*, from the Greek word for "thread." Chromosome number soon came to be recognized as a distinctive characteristic of a species and was shown to remain constant from generation to generation. That the chromosomes themselves might be the actual bearers of genetic information was suggested by German anatomist Wilhelm Roux as early as 1883 and expressed more formally by his countryman, biologist August Weissman, shortly thereafter.

With the roles of the nucleus and chromosomes established and appreciated, the stage was set for the rediscovery of Mendel's initial observations. This came in 1900, when three plant geneticists working independently cited his studies almost simultaneously: Carl Correns in Germany, Ernst von Tschermak in Austria, and Hugo de Vries in Holland. Within three years, the **chromosome theory of heredity** was formulated, following work by American physician Walter Sutton and German biologist Theodor Boveri. The chromosome theory of heredity proposed that the hereditary factors responsible for Mendelian inheritance are located on the chromosomes within the nucleus. This hypothesis received its strongest confirmation from the work of American biologist Thomas Hunt Morgan and his students, Calvin Bridges and Alfred Sturtevant, at Columbia University during the first two decades of the twentieth century. Using *Drosophila melanogaster*, the common fruit fly, as their experimental model organism, they identified a variety of morphological mutants of *Drosophila* and were able to link specific traits to specific chromosomes.

Meanwhile, the foundation for our understanding of the chemical basis of inheritance was slowly being laid. An important milestone was the discovery of DNA by Swiss biologist Johann Friedrich Miescher in 1869. Using such unlikely sources as salmon sperm and human pus from surgical bandages, Miescher isolated and described what he called "nuclein." But, like Mendel, Miescher was ahead of his time. It would be about 75 years before the role of his nuclein as the genetic information of the cell came to be fully appreciated.

As early as 1914, DNA was implicated as an important component of chromosomes by the German chemist Robert Feulgen's staining technique, a method to identify DNA that is still in use today. But it was considered quite unlikely that DNA could be the bearer of genetic information due to its apparently monotonous structure. By 1930, DNA was known to be composed of only four different nucleotides—and this did not seem to be enough variety to account for all the diversity seen in living organisms. Proteins, on the other hand, were much more diverse, being composed of 20 different amino acids. In fact, until the middle of the twentieth century, it was widely thought that proteins were the carriers of genetic information from generation to generation because they seemed to be the only nuclear components with enough variety to account for the obvious diversity of genes.

A landmark experiment clearly pointing to DNA as the genetic material was reported in 1944 by Canadian scientists Oswald Avery and Colin MacLeod and American scientist Maclyn McCarty, working together at Rockefeller University. Their work (discussed in more detail in Chapter 16) showed that DNA could "transform" a nonpathogenic strain of bacteria into a pathogenic strain, causing a heritable genetic change. Eight years later, American biochemists Alfred Hershey and Martha Chase showed that DNA, and not protein, enters a bacterial cell when it is infected and genetically altered by a bacterial virus. Meanwhile, American biologists George Beadle and Edward Tatum, working in the 1940s with the bread mold *Neurospora crassa*, formulated the "one gene–one enzyme" concept, asserting that each gene controls the production of a single, specific protein.

Molecular Genetics. Shortly thereafter, in 1953, the unlikely team of former ornithology student James Watson and physicist Francis Crick, using images provided by X-ray crystallographer Rosalind Franklin, proposed their now-famous *double helix model* for the structure of DNA, which immediately suggested how replication during cell division could occur by precise base pairing between complementary strands. The 1960s brought more significant developments, including the discovery of the polymerase enzymes that synthesize DNA and RNA and the "cracking" of the genetic code, which

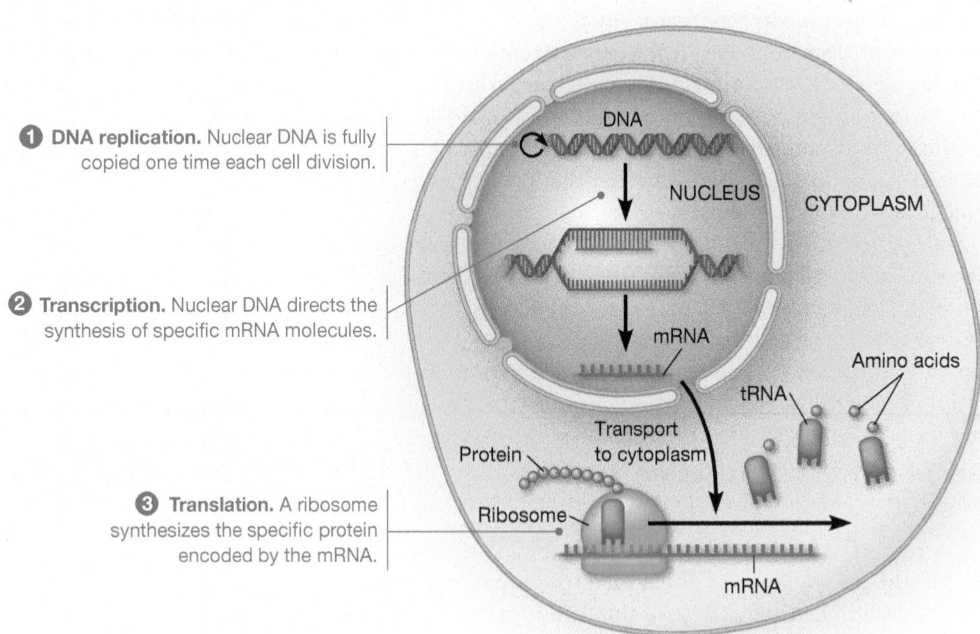

1 DNA replication. Nuclear DNA is fully copied one time each cell division.
2 Transcription. Nuclear DNA directs the synthesis of specific mRNA molecules.
3 Translation. A ribosome synthesizes the specific protein encoded by the mRNA.
DNA
NUCLEUS
CYTOPLASM
mRNA
Amino acids
tRNA
Transport to cytoplasm
Protein
Ribosome
mRNA

Figure 1-8 Central Dogma: The Flow of Genetic Information in the Cell. In eukaryotes, most of the DNA in a cell is located in the nucleus. ❶ This DNA is copied each time the cell replicates. ❷ DNA also contains instructions for the synthesis of a complementary messenger RNA (mRNA) in the process of transcription. ❸ The mRNA then travels to the cytoplasm, where it is used by ribosomes to synthesize protein in the process of translation.

specifies the relationship between the order of nucleotides in a DNA or RNA molecule and the order of amino acids in a protein. At about the same time, biochemist Jacques Monod and geneticist François Jacob of France deduced the mechanism responsible for regulating bacterial gene expression.

Soon after the double-helical model of DNA was proposed in 1953, Francis Crick articulated a molecularly based model of genetic information flow, which he christened the *central dogma of molecular biology*. The steps shown in **Figure 1-8** summarize this model. Notice how the flow of genetic information involves *replication* of DNA to produce two identical copies, *transcription* of information carried by DNA into the form of RNA, and *translation* of this information from RNA into protein. The term *transcription* refers to RNA synthesis using DNA as a template to emphasize that this phase of gene expression is simply a transfer of information from one nucleic acid to another, so the basic "language" remains the same. In contrast, protein synthesis is called *translation* because it involves a language change—from the nucleotide sequence of an RNA molecule to the amino acid sequence of a polypeptide chain.

The discovery that three different kinds of RNA molecules serve as intermediates in protein synthesis completed our basic understanding of how the central dogma operates in cells (Figure 1-8). RNA that is translated into protein is called *messenger RNA (mRNA)* because it carries a genetic message from DNA to macromolecular complexes known as ribosomes, where protein synthesis actually takes place. *Ribosomal RNA (rRNA)* molecules are integral components of the ribosome itself. *Transfer RNA (tRNA)* molecules serve as intermediaries that recognize the coded base sequence of an mRNA and bring the appropriate amino acids to the ribosome for protein synthesis.

In the years since it was first formulated by Crick, the central dogma has been refined in various ways. For example, many viruses with RNA genomes have been found to synthesize mRNA molecules using RNA as a template. Other RNA viruses, such as HIV, carry out reverse transcription, whereby the viral RNA is used as a template for DNA synthesis—a "backward" flow of genetic information. But despite these variations on the original model, the principle that information flows from DNA to RNA to protein remains the main operating principle by which all cells express their genetic information. (The roles of DNA and RNA in the storage, transmission, and expression of genetic information will be considered in full detail in Chapters 16–20.)

Our current understanding of gene expression has relied heavily on the development of **recombinant DNA technology** since the 1970s. This technology was made possible by the discovery of *restriction enzymes*, which have the ability to cleave DNA molecules at specific sequences so that scientists can create *recombinant DNA molecules* containing DNA sequences from two different sources. This capability led quickly to the development of *DNA cloning*, a process used to generate many copies of specific DNA sequences for detailed study and further manipulation, and *DNA transformation*, the process of introducing DNA into cells. (These important techniques are explained and explored in detail in Chapter 21.)

At about the same time, **DNA sequencing** methods were devised for rapidly determining the base sequences of DNA molecules. This technology is now routinely applied not just to individual genes but also to entire genomes. Initially, genome sequencing was applied mainly to bacterial genomes because they are relatively small—a few million bases, typically. But DNA sequencing has long since been successfully applied to much larger genomes, including those from species of yeast, roundworm, plants, and animals that are of special interest to researchers. A major triumph was the sequencing of the entire human genome, which contains about 3.2 billion bases. This feat was accomplished by the *Human Genome Project*, a cooperative international effort that began in 1990, involved hundreds of scientists, cost billions of dollars, and established the complete sequence of the human genome by 2003.

Bioinformatics and "–Omics." The challenge of analyzing the vast amount of data generated by DNA sequencing has led to a new discipline, called **bioinformatics**, which merges computer science and biology as a means of making sense of sequence data. This approach has led to the recognition that the human genome contains approximately 20,000 protein-coding genes, about half of which were not characterized before genome sequencing. *Genomics*, the study of all the genes of an organism, is providing remarkable insights into cell biology and human health. Similarly, using modern techniques and bioinformatics, scientists can also study the *proteome*, the total protein content of a cell. In the emerging field of *proteomics*, researchers are attempting to understand the functions and interactions of all of the proteins present in a particular cell. Proteomic studies aim to understand the structure and properties of every protein produced by a genome and to learn how these proteins interact with each other in biological networks to regulate cellular functions.

Numerous bioinformatic tools are publicly available through the National Center for Biotechnology Information (NCBI), which is operated by the U.S. National Institutes of Health (NIH). In addition to housing PubMed, a searchable archive of more than 30 million citations from life science journals, NCBI maintains GenBank, a comprehensive database of all publicly available DNA sequences (over 217 million as of mid-2020). Similarly, the Protein Knowledgebase (UniProtKB) maintains a database of more than 560,000 protein sequences. Also available are numerous tools to compare gene and protein sequences from all organisms and to analyze their structure and function. For example, using a program known as BLAST, or Basic Local Alignment Search Tool, a researcher can compare the sequence of a newly discovered gene to all known gene sequences in a matter of minutes! In addition, NCBI provides a wealth of biological information, such as the OMIM (Online Mendelian Inheritance in Man) database, which is an encyclopedic collection of information regarding human genetic disorders and mutations involving almost 16,000 genes.

The past decade has seen major advances in technology that have miniaturized and automated molecular analyses, allowing them to be performed much faster. These *high-throughput* methods have resulted in dramatic increases in speed. The first human genome took 13 years to sequence, but now a genome can be sequenced in just a few hours and for a fraction of the cost. Likewise, the expression levels of hundreds or even thousands of genes can be monitored simultaneously, making it possible to study all the genes in a genome at the same time.

The ability to simultaneously analyze thousands of molecules on a global basis throughout the cell has led to a proliferation of "–omics" studies in addition to genomics and proteomics. For example, recent advanced methods of RNA sequencing allow us to determine the complete set of genes transcribed in a cell. This type of study is called *transcriptomics*. Scientists can also conduct *metabolomics*, the analysis of all metabolic reactions happening at a given time in a cell, *lipidomics*, the study of all the lipids in a cell, and even *ionomics*, the global study of all the ions in a cell. Keep your eyes open in the next few years, as this explosion of biological information will likely lead to a host of new fields of "–omics" studies.

CRISPR Genome Editing. Since the discovery of restriction enzymes, scientists have been trying to manipulate genomes. The key to these manipulations is having the right tools to modify DNA at specific sequences in specific ways. For example, precise sequence manipulation at the genome level is now being achieved through a new technique known as CRISPR/Cas9 genome editing (see Key Technique in Chapter 17 pages 500–501).

CRISPR (pronounced "crisper") is the abbreviation of **clustered regularly interspaced short palindromic repeats**. Although the CRISPR/Cas9 system is being used extensively for genome editing, it was first discovered as a prokaryotic defense mechanism against viral infection. Think of it as a bacterial immune system. (You will learn more details about this bacterial defense system in Chapter 20 [Figure 20-10].)

So how is a system designed to protect bacteria from invading viruses useful for editing genomes? The key is the generation of a double-stranded break at a precise location in the genome, which is targeted by a short nucleotide called a **guide RNA (gRNA)**. Double-stranded breaks in DNA are notoriously difficult for the cell to repair. Often, errors occur in this repair process, leading to the inactivation of the target gene. It is also possible to provide a piece of DNA with an alternative sequence that the cell can use to repair the break (called a *repair template*) using a process known as homology-directed repair. Both genome-editing strategies are depicted in **Figure 1-9**.

These and other techniques helped launch an era of molecular analysis that continues to revolutionize biology. In the process, the historical strand of genetics that dates back to Mendel became intimately entwined with the strands of cytology and biochemistry, and the discipline of cell biology came into being.

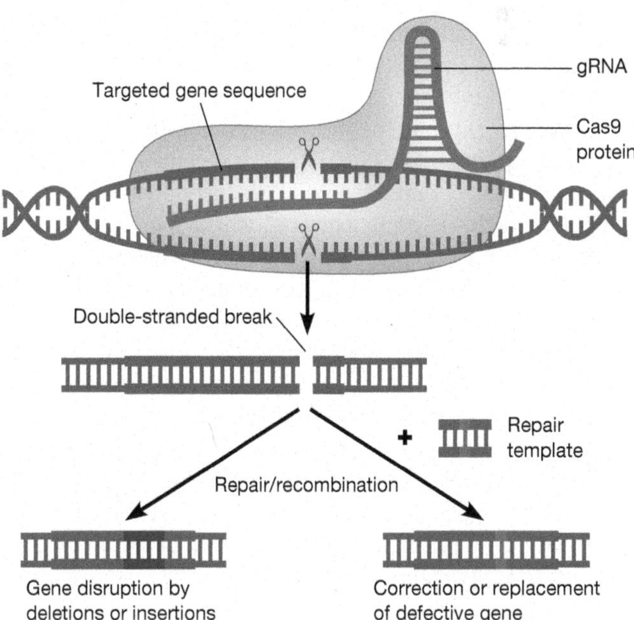

Figure 1-9 CRISPR Genome Editing. Guide RNA (gRNA) helps Cas9 target a specific place within the genome, where Cas9 causes a double-stranded break in the DNA. Repairs, which are often error prone, can lead to a gene disruption. If a repair template is added, homologous recombination will edit the genome using the DNA information from this alternative sequence.

CONCEPT CHECK 1.2

You have discovered a defective gene that allows cancer cells to use glucose more efficiently and therefore outgrow normal cells to form a tumor. Explain how your discovery likely involves contributions from all three historical strands of cell biology.

1.3 How Do We Know What We Know?

If asked what you expect to get out of this textbook, you might reply that you intend to learn the facts about cell biology. If pressed to explain what a "fact" is, most people would probably reply that a fact is "something that is known to be true." Saying, for example, that "DNA is the bearer of genetic information in cells" suggests that this statement is a fact of cell biology. But it must also be recognized that this statement actually replaced an earlier misconception that proteins carried the genetic information in cells.

Biological "Facts" May Turn Out to Be Incorrect

Cell biology is rich with examples of "facts" that were once widely held but have since been altered or even been discarded as cell biologists gained a better understanding of the phenomena those facts attempted to explain. As noted earlier, the early-nineteenth-century "fact" that living matter consisted of substances quite different from those in nonliving matter was discredited following work by Wöhler, who synthesized the biological compound urea from an inorganic compound, and by the Buchners, who showed that nonliving extracts from yeast cells could ferment sugar into alcohol. Thus, views held as fact by generations of scientists were eventually replaced by the new fact that living matter follows the same laws of chemistry and physics that inorganic materials do.

For a more contemporary example, until recently it was regarded as a fact that the sun is the ultimate source of all energy in the biosphere. Then came the discovery of *deep-sea thermal vents* and the thriving communities of organisms that live around them, none of which depend on solar energy. Instead, these organisms depend on energy derived from hydrogen sulfide (H_2S) by bacteria, which use this energy to synthesize organic compounds from carbon dioxide.

Thus, as you can see, sometimes biological "facts" are really much more provisional pieces of information than our everyday sense of the word might imply. Just like cells themselves, these "facts" are dynamic and subject to change, sometimes abruptly. Some results of biological research are, of course, not provisional in this sense. That most organisms are composed of cells is established beyond dispute. To a scientist, a "fact" is simply an attempt to state our best current understanding of the natural world around us, based on observations and experiments.

Experiments Test Specific Hypotheses

How does a new and deeper understanding of a cell biological process become available? First, a researcher will typically conduct a search of the scientific literature to determine what is known in the specific area of interest. Such information usually comes from a *peer-reviewed* scientific or medical journal, rather than from a nonreviewed site on the Internet. "Peer-reviewed" means that after scientists conducted their research and submitted a written article to a journal, the article was examined in detail by several experts in the particular field and found to be sound in methodology, experimental design, and analysis of results.

Once a cell biologist has assessed the current state of knowledge in her field, then she can formulate a **hypothesis**, a tentative explanation that can be tested experimentally or via further observation. Often, a hypothesis takes the form of a *model* that appears to provide a reasonable explanation of the phenomenon in question. Next, the investigator designs a controlled experiment to test the hypothesis by varying specific conditions while keeping other variables constant. The scientist then collects the data, interprets the results, and accepts or rejects the hypothesis, which must be consistent not only with the results of this particular experiment but also with prior knowledge.

Rather than try to prove a hypothesis, scientists typically try to rephrase the hypothesis as its opposite, known as the **null hypothesis**, and seek to prove the latter. Failure to confirm this null hypothesis in a sufficiently large number of attempts is indirect evidence that the hypothesis is correct. The certainty of its correctness increases with the number of experimental samples and the number of times the results are replicated.

For example, suppose you proposed the hypothesis that every person on Earth was under 20 feet tall. To prove this proposal rigorously, you would have to measure the height of everyone on Earth and determine that all were under 20 feet tall. The null hypothesis would be that there is indeed a person more than 20 feet tall. Failure to find such a person after years, perhaps centuries, of trying would provide reasonably solid evidence that the hypothesis is true—there is indeed no one on Earth more than 20 feet tall.

Experiments are often conducted in the laboratory using purified chemicals and cellular components. This type of experiment is described as in vitro, which literally means "in glass." But to fully understand how cells work, hypotheses may need to be tested in vivo ("in life"), meaning in live cells and organisms. Often, this is done using one of a variety of popular *model organisms*, which will be discussed in the next section. More recently, experiments using computers to test new hypotheses involving vast amounts of data have been described as in silico, referring to the silicon used to make computer chips. All of these approaches use the same basic steps for hypothesis formulation and testing as outlined above.

Model Organisms Play a Key Role in Modern Cell Biology Research

Although many studies of basic cellular functions can be carried out in the laboratory using isolated cellular components such as membranes, enzymes, and DNA molecules, the results obtained from such experiments may not directly reflect the corresponding process in an intact, living system. Scientists have developed a number of *model systems* to study cellular processes directly in living cells and organisms.

Cell and Tissue Cultures. Scientists make extensive use of *cell cultures* as model systems. Many types of cells can be grown in the laboratory outside their tissue of origin—such as skin cells, muscle cells, and cancer cells. Some of the first human cells ever grown in defined culture conditions in the laboratory were HeLa cells taken from cervical cancer tissue obtained from a woman named Henrietta Lacks in 1951. Descendants of her cells are still being grown today and are commonly used in cancer and virus research (see Human Connections, page 16).

A variety of other cells are commonly used as model systems: egg cells from the frog *Xenopus* to study channel proteins; Chinese hamster ovary cells for cell signaling studies and commercial production of proteins; mouse 3T3 fibroblast cells to assess the carcinogenicity of new compounds; and undifferentiated embryonic stem cells to study cellular differentiation. In addition, cell cultures are indispensable for growing and studying viruses, which are tiny infectious noncellular particles that cannot multiply outside their host cells. But because what we learn by studying isolated cells in culture may not always reflect what happens in the intact organism, it is important to perform studies in the living organisms themselves.

Model Organisms. A **model organism** is a species that is widely studied, well characterized, and easy to manipulate and has particular advantages, making it useful for experimental studies. A few examples are the bacterium *Escherichia coli*, the yeast *Saccharomyces cerevisiae*, and the fruit fly *Drosophila melanogaster* (**Figure 1-10**).

Much of what is known about basic cellular processes, such as DNA replication, membrane function, and protein synthesis, was learned using cells of the bacterium *E. coli* as a model system. *E. coli* is easy to grow in the lab, divides rapidly (it has a generation time of 20 minutes), and is easily mutagenized for studies of gene function. In 1997, it became the first bacterium to have its complete genome sequenced. Because it readily takes up DNA by transformation from virtually any organism, it has become the workhorse of cellular and molecular biology for isolating and cloning genes. It is routinely used for the analysis and production of genes and proteins for research, industrial, and medical uses.

To study processes unique to eukaryotic cells, such as chromosome pairing during cell division, organelle development, or cell signaling, cells of *S. cerevisiae* (bakers' and brewers' yeast) offer several similar experimental advantages. They are unicellular, as well as easy to grow and mutagenize, and well-characterized mutant lines are available. Using yeast cells as a model organism to isolate and characterize genetic mutants following *mutagenesis* by chemicals or radiation, scientists have learned much about how eukaryotic cells function. By studying mutant yeast strains that are deficient in particular genes and their protein products, researchers can determine the normal cellular function of those genes and proteins. For example, mutant yeast cells that display abnormal cell division because they are deficient in a particular gene necessary for proper cell division have been invaluable in helping us understand how normal cells divide. For example, much of what we know about human cell division originally came from the

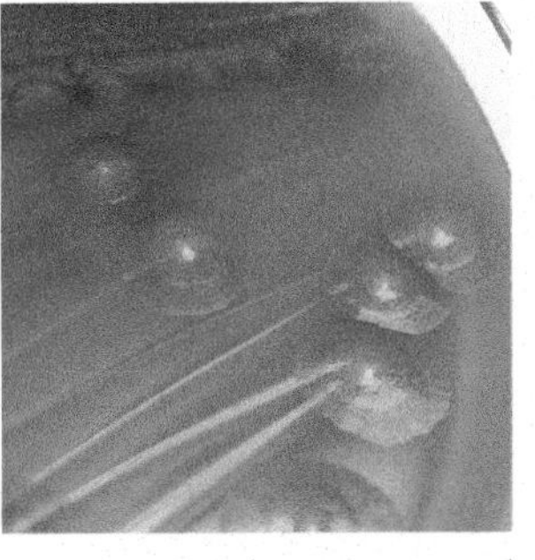

(a) *E. coli* colonies on blood agar
- Unicellular bacterium with short generation time
- Easily mutagenized and transformed to study gene function
- Widely used for gene cloning and protein function

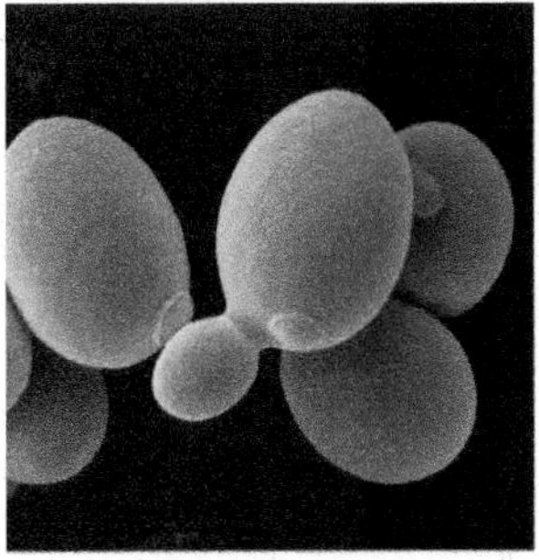

(b) *S. cerevisiae* cells (yeast; SEM)
- Unicellular eukaryotic cells that are easy to grow in the lab
- Thousands of mutant lines available to study gene function
- Widely used to study the cell cycle and protein-protein interactions

(c) Wild-type *D. melanogaster* (fruit fly)
- Multicellular eukaryotic organism with a short generation time
- Many mutant strains available with well characterized genetic defects
- Widely used for studies of embryogenesis and development

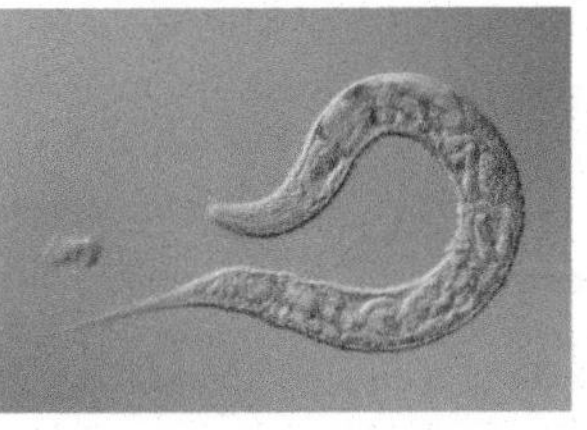

(d) *C. elegans* (roundworm)
- Multicellular eukaryotic organism with a short life cycle
- Nearly transparent with cell fates mapped out
- Widely used for cell differentiation and development studies

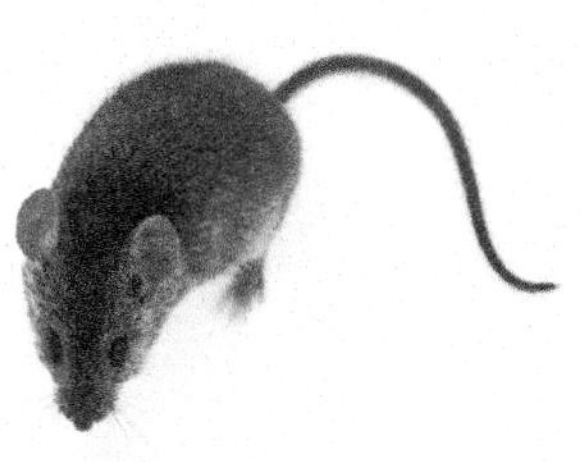

(e) *M. musculus* (house mouse)
- Mammalian organism with many similarities to humans
- Numerous strains with genetic "knockouts"
- Widely used as models for human disease

(f) *A. thaliana* (mustard plant)
- Rapid life cycle and one of the smallest plant genomes
- Easily mutagenized with thousands of mutant lines available
- Widely used for studies of plant-specific processes

Figure 1-10 Common Model Organisms. Examples of some model organisms frequently used in cell biology research are shown with descriptions of the advantages of each.

HUMAN *Connections*

The Immortal Cells of Henrietta Lacks

George Otto Gey was an anatomist at heart, who was fascinated by cancer and how cancer cells divided. He and his wife, Margaret, coordinated the cell culture lab at Johns Hopkins University and for decades attempted to produce a line of human cells that could live indefinitely if given the necessary environment and supplemental nutrition. Although the Geys were given cells from multiple human tissues, they had been unable to culture a line that would not age and eventually die. This would all change one day in 1951, when they were given a tissue sample biopsied from a cervical tumor growing in a woman reporting abnormal bleeding after the birth of her fifth child. After a few weeks the cells from this biopsy were obviously different. They grew well in culture and didn't seem to age and die. These cells, named *HeLa cells,* would become the first immortal human cell line. Almost 70 years later, they remain one of the most widely used human cell lines.

Creation of the HeLa line was a milestone in biomedical research. Using animals to study human diseases is time consuming and expensive, and animals do not always respond to disease in the same way that human cells do. HeLa cells have made an enormous contribution to science; they have been cited in more than 70,000 scientific papers, and they have been instrumental in research on viruses, cancer, and AIDS.

The name "HeLa" comes from the name of the woman from whom the cells were taken, Henrietta Lacks (**Figure 1B-1**). But who was the woman behind the cell line? Henrietta Lacks was born in 1920 to a poor African American family in rural Virginia. Henrietta moved to Baltimore in 1941 with her husband and started a family. In 1951, after the birth of her fifth child, Henrietta was diagnosed with cervical cancer. Tragically, Henrietta had a very aggressive form of cancer that did not respond to the treatment available at the time, and she died at the age of 31. At the time of her diagnosis and biopsy, there were no laws requiring informed consent before a doctor took tissue or blood samples, nor were there laws governing what the tissue could be used for after it was removed from the patient's body. After the HeLa cell culture line (shown in Figure 1B-1b) was created, it was freely disseminated to other scientists for use in their research. The Lacks family did not learn that Henrietta's cells were grown in laboratories around the world until two decades later.

(a) Henrietta Lacks

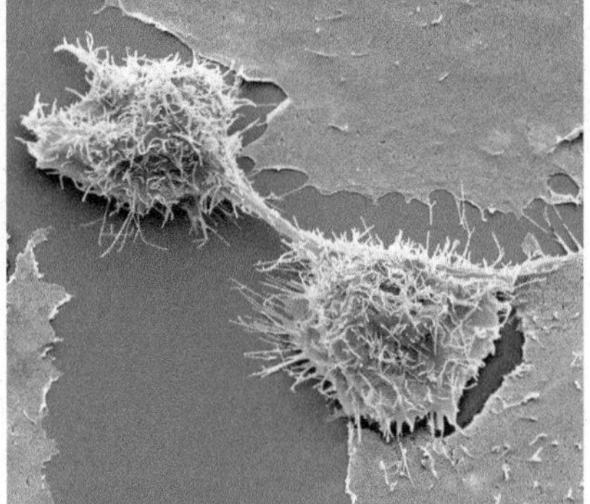

(b) HeLa cells in culture

10 μm

Figure 1B-1 Henrietta Lacks and Her Immortal Cells.
(a) Henrietta Lacks before her diagnosis. **(b)** HeLa cells in the process of dividing (colorized SEM).

From the beginning, Henrietta's cells were unique. Not only did they divide endlessly in culture, but they grew so vigorously that they would double in number every 24 hours! What made Henrietta's cells so different? Normal cells can divide only about 50 times before undergoing cell death because the ends of their chromosomes, known as *telomeres,* shorten (see Chapter 17). HeLa cells, however, are derived from cancer cells, which abnormally express the gene encoding *telomerase,* an enzyme that allows telomeres to be synthesized anew with each round of cell division. As a result, HeLa cells can divide indefinitely.

One cause of cervical cancer is the *human papilloma virus (HPV).* HeLa cells contain a copy of HPV inserted into the chromosome near the *myc* gene, causing it to be overexpressed. The Myc protein normally controls cell growth, so overexpression of *myc* is associated with uncontrolled cell division and cancer (see Chapter 24). The presence of HPV near the *myc* locus may explain why Henrietta's cancer was so aggressive and why HeLa cells grow so vigorously in culture. In fact, HeLa cells grow so rapidly in culture that they frequently contaminate other cell lines. It is estimated that 10–20% of existing cell lines are actually HeLa contaminants that have overgrown and replaced the original cell culture.

HeLa cells played an important role in the development of a vaccine for polio. During the early 1900s, polio was the most dreaded disease in America. In 1952 the worst polio epidemic ever recorded swept across the United States. Of the 58,000 people who contracted polio, 21,000 were left with some form of paralysis and 3000 died. During the 1952 epidemic, HeLa cells were shipped to Jonas Salk and other polio researchers when it was discovered that HeLa cells could be infected with viruses and cellular responses to treatments could be easily monitored. Beginning with the introduction of the Salk vaccine in the mid-1950s and a later oral vaccine developed by Albert Sabin, polio, once a dreaded disease, was quickly eradicated in the United States and, eventually across most of the globe.

Once they were used to study human disease, the demand for HeLa cells exploded, and companies were established to grow HeLa cells and provide them to researchers. Ironically, the Lacks family never received any money from the sale of HeLa cells.

The Lacks family's plight raises several important ethical considerations. First, once a blood or tissue sample is removed from your body, is it still yours? Many people believe that donated tissue should be used in medical research for the common good. However, after the tissue has been donated, should the donor have control over the kinds of research his or her cells are used for? Second, if the donated tissue is used to develop a commercial product, should the donor be granted a share of the profits?

A third issue was raised in 2013, when a group of scientists published the DNA sequence of HeLa cells without the family's knowledge or permission. Requests for access to HeLa DNA sequence information are now reviewed by a committee consisting of scientists, physicians, and members of the Lacks family. Researchers who use HeLa cells or the genomic data are asked to acknowledge the contribution of Mrs. Lacks and her family in their publications. The use of HeLa cells, and HeLa sequence information, in biomedical research continues to benefit millions of people. But Henrietta's family finally has a say in how her genetic information will be used.

study of cell cycle mutants in yeast. Recently, use of the yeast *two-hybrid system* (see Chapter 21), which allows researchers to determine whether and how specific proteins interact within a living cell, has contributed greatly to our understanding of the complex molecular interactions involved in cellular function.

However, if you want to study processes such as communication between cells, differentiation of cells, or embryonic development, you may need to use a *multicellular* organism as a model system. You may have heard biologists talk about experiments using "flies and worms." They are referring to the tiny fruit fly *Drosophila melanogaster* and the roundworm *Caenorhabditis elegans*, both of which are extensively used for studies of the cell biology of multicellular, eukaryotic organisms.

Much of our basic understanding of genetics and gene function comes from using mutants of *Drosophila*, which has many experimental advantages: The flies are easy to grow and manipulate in the lab, have a short (2-week) generation time, produce numerous progeny, and have easily observable physical characteristics, such as eye color and wing shape. Thousands of mutant strains are available, each defective in a particular gene; this makes *Drosophila* quite valuable for studies of embryogenesis, developmental biology, and cell signaling.

Similarly, *C. elegans* is a widely used model organism for studies of cell differentiation and development in multicellular organisms. Its advantages include its ease of manipulation, relatively short life cycle, and small genome, the first of any multicellular organism to be sequenced. It is also one of the simplest animals to possess a nervous system. Its development from a fertilized egg is remarkably predictable, and the origin and fate of each of its approximately 1000 cells have been mapped out, as have the hundreds of connections among the roughly 200 nerve cells. In addition, the tiny worms are transparent, making it is easy to see individual cells under the microscope and to view fluorescently labeled molecules in the living organism.

For studies of cellular and physiological processes specific to mammals (including humans), the common laboratory mouse *(Mus musculus)* has become the primary model organism. It shares many cellular, anatomical, and physiological similarities with humans and is widely used for research in medicine, immunology, and aging. It is subject to, and therefore useful for the study of, a variety of diseases that also affect humans, such as cancer, diabetes, and osteoporosis. Numerous mouse strains have been bred or engineered in which particular genes have either been "knocked out" or introduced, making them extremely valuable in biomedical research.

For studies of processes in plants, such as photosynthesis and light perception, and some processes common to all organisms, *Chlamydomonas reinhardtii*, a unicellular green algae, is often used (see Figure 1-2g). Like *E. coli* and yeast, "*Chlamy*" is easily grown in the lab on Petri plates and has been used to study photosynthesis, light perception, mating type, cellular motility, and DNA methylation. For flowering plant studies, *Arabidopsis thaliana* is a powerful model organism. It has one of the smallest genomes of any plant and a rapid (6-week) life cycle, facilitating genetic studies. Its complete genome has been sequenced, and thousands of mutant strains have been created, enabling detailed studies of plant gene function.

Many other model organisms are currently being used in biology to study a wide variety of cellular, genetic, and biochemical processes. As you proceed through your studies in cell biology, keep in mind that much of our knowledge is based on research using relatively few of the millions of living organisms on Earth, and remember that it is always important to understand how this knowledge was obtained.

MAKE CONNECTIONS 1.1

What type of microscopy is *best* suited for taking advantage of the transparency of *C. elegans* in order to visualize a specific internal structure? (Ch. 1.2)

Well-Designed Experiments Alter Only One Variable at a Time

Modern cell biologists have an extensive array of tools to use as they consider how to perform experiments. How can they use model organisms, powerful microscopes, and genetic and biochemical techniques to meaningfully answer questions in cell biology? In a typical experiment in cell biology, many individual conditions can be varied, such as length of treatment or temperature, but it is best to vary, or perturb, only one condition, called the *independent variable*, and hold all others constant. The outcome of the change that is measured (which depends on the independent variable) is called the *dependent variable*. For example, if you wanted to test the rate of growth of cells at different temperatures, you would hold all culture conditions constant except temperature, which you would vary. Temperature is the independent variable that you set, and the growth you measure is the dependent variable whose value depends on the particular temperature.

The fact that there should be only one independent variable in an experiment is a key reason why genetic mutants are so valuable for studying gene function. For in vivo studies of gene function, for example, the classical genetic approach is to isolate a naturally occurring mutant form of an organism. It is now possible to artificially alter the DNA of an organism as well. In either case, an organism (*E. coli*, yeast, a mouse, or other model organism) with unaltered DNA is called the *wild type*. The mutant strain is identical to the wild type except that it lacks one particular gene's function.

Scientists can also perform in vitro experiments in which purified cellular components are added to a test tube to simulate a process that occurs in intact, living cells. Then the system can be perturbed systematically—adding, deleting, or modifying one ingredient of the mixture at a time—to test the hypothesis that a particular component is necessary. A powerful method (one you will see used throughout this text) is to introduce inhibitors or antibodies to block the function of a particular component in a reaction. Thus, by changing only one component or other variable at a time, scientists can determine the specific function of that component and the effect of that particular variable.

As you have now seen, the "facts" presented in biology textbooks such as this one are simply our best current attempts to describe and explain the biological world we live in. As you proceed through this text, notice how the process of science has been used to advance our knowledge of cell biology and pay attention to how this new knowledge was gained by experimentation. You will likely find that, regardless of the approach, the conclusions from an experiment add to our knowledge of how cells work but usually lead to more questions as well, continuing the cycle of scientific inquiry. We hope, as you gain greater understanding of cells and how they are studied throughout this book, that you will learn to appreciate what experiments tell us and that you will learn how to design your own experiments to answer questions and test hypotheses.

MAKE CONNECTIONS 1.2

When generating knockout mice, scientists use inbred mouse stains. Based on what you know about well-designed experiments, why is this important? (Ch. 1.3)

CONCEPT CHECK 1.3

Suppose you often have heartburn on nights after you have eaten your favorite pepperoni, anchovy, and onion pizza. You wonder whether the pizza or one of its toppings might be causing the heartburn. Describe how you might determine whether the heartburn is due to the pizza and, if so, to which of the toppings.

Summary of Key Points

Mastering™ Biology For activities, animations, and review quizzes, go to the study area at www.masteringbiology.com.

1.1 The Cell Theory: A Brief History

- The biological world is a world of cells. The cell theory states that all organisms are made of cells, the basic units of biological structure, and that cells arise only from preexisting cells.
- The cell theory was developed through the work of many different scientists, including Hooke, van Leeuwenhoek, Brown, Schleiden, Schwann, Nägeli, and Virchow.
- Although the importance of cells in biological organization has been appreciated for about 150 years, the discipline of cell biology as we know it today is of much more recent origin.

1.2 The Emergence of Modern Cell Biology

- Modern cell biology has come about by the interweaving of three historically distinct strands—cytology, biochemistry, and genetics—which in their early development probably did not seem at all related.
- The contemporary cell biologist must understand all three strands because they complement one another in the quest to learn what cells are made of and how they function.
- The cytological strand deals with cellular structure.
- The cytological strand is best studied using microscopes, which include both light and electron microscopy. The light microscope has allowed us to visualize individual cells. Several types of light microscopes allow us to view preserved or living specimens, including brightfield, phase-contrast, differential interference contrast, fluorescence, confocal, and digital video microscopes. Historically, the limited resolving power of the light microscope did not allow us to see the finer details of cellular structure, but electron microscopes and modern light microscopes have solved this limitation. The electron microscope uses a beam of electrons, rather than visible light, for imaging specimens. It can magnify objects with a resolving power of less than 1 nm, enabling us to view subcellular structures such as membranes, ribosomes, organelles, and even individual DNA and protein molecules.
- The biochemical strand concerns the chemistry of biological structure and function.
- Discoveries in biochemistry have revealed how many of the chemical processes in cells are carried out, greatly expanding our knowledge of how cells function.
- Major discoveries in biochemistry were the identification of enzymes as biological catalysts, the discovery of adenosine triphosphate (ATP) as the main carrier of energy in living organisms, and the description of the major metabolic pathways cells use to harness energy and synthesize cellular components.
- Several important biochemical techniques that have allowed us to understand cell structure and function are subcellular fractionation, ultracentrifugation, chromatography, electrophoresis, and mass spectrometry.
- The genetic strand focuses on information flow.
- The chromosome theory of heredity states that the characteristics of organisms passed down from generation to generation result from the inheritance of chromosomes carrying discrete physical units known as genes.
- Each gene is a specific sequence of DNA that contains the information to direct the synthesis of one cellular protein.
- DNA itself is a double helix of complementary strands held together by precise base pairing. This structure allows the DNA to be accurately duplicated as it is passed down to successive generations.
- The flow of genetic information in cells is typically from DNA to RNA to protein, although exceptions such as reverse transcription exist. Expression of this genetic information to produce a protein requires several important types of RNA: mRNA, tRNA, and rRNA.
- Bioinformatics allows us to compare and analyze thousands of genes or other molecules simultaneously, causing a revolution in genomic, proteomic, and numerous other fields of "–omics" research.
- CRISPR genome editing is an exciting new technique that allows precise changes to genomic sequences.

1.3 How Do We Know What We Know?

- Science is not a collection of facts but a process of discovering answers to questions about our natural world. Scientists gain knowledge by using the scientific method, which involves creating a hypothesis that can be tested for validity by collecting data through well-designed, controlled experiments.
- A well-designed experiment will vary and test only one condition at a time to test a hypothesis. This can involve the use of mutants, experiments in which one component at a time is changed, and the use of inhibitors of specific cellular processes.
- Progress in science is based on the consistency and reproducibility of experimental results. These results are often presented in the form of peer-reviewed journal articles.
- Scientists use a variety of well-studied cell cultures and model organisms to test new hypotheses, develop new theories, and advance our knowledge of cell biology.

Problem Set

Mastering™ Biology For activities, animations, and review quizzes, go to the study area at www.masteringbiology.com.

1-1 The Historical Strands of Cell Biology. For each of the following events, deduce whether it belongs mainly to the cytological (C), biochemical (B), or genetic (G) strand in the historical development of cell biology.

(a) Köllicker describes "sarcosomes" (now called mitochondria) in muscle cells (1857).

(b) Hoppe-Seyler isolates the protein hemoglobin in crystalline form (1864).

(c) Haeckel postulates that the nucleus is responsible for heredity (1868).

(d) Ostwald proves that enzymes are catalysts (1893).

(e) Muller discovers that X-rays induce mutations (1927).

(f) Davson and Danielli postulate a model for the structure of cell membranes (1935).

(g) Beadle and Tatum formulate the one gene–one enzyme hypothesis (1940).

(h) Claude isolates the first mitochondrial fractions from rat liver (1940).

(i) Lipmann postulates the central importance of ATP in cellular energy transactions (1940).

(j) Avery, MacLeod, and McCarty demonstrate that bacterial transformation is attributable to DNA, not protein (1944).

(k) Palade, Porter, and Sjøstrand each develop techniques for fixing and sectioning biological tissue for electron microscopy (1952–1953).

(l) Lehninger demonstrates that oxidative phosphorylation depends for its immediate energy source on the transport of electrons in the mitochondrion (1957).

1-2 QUANTITATIVE Cell Sizes. To appreciate the differences in cell size illustrated in Figure 1-4a, consider these specific examples. *Escherichia coli*, a typical bacterial cell, is cylindrical in shape, with a diameter of about 1 μm and a length of about 2 μm. As a typical animal cell, consider a human liver cell, which is roughly spherical and has a diameter of about 20 μm. For a typical plant cell, consider the columnar palisade cells located just beneath the upper surface of many plant leaves. These cells are cylindrical, with a diameter of about 20 μm and a length of about 35 μm.

(a) Calculate the approximate volume of each of these three cell types in cubic micrometers. (Recall that $V = \pi r^2 h$ for a cylinder and that $V = 4\pi r^3/3$ for a sphere.)

(b) Approximately how many bacterial cells would fit in the internal volume of a human liver cell?

(c) Approximately how many liver cells would fit inside a palisade cell?

1-3 QUANTITATIVE Sizing Things Up. To appreciate the sizes of the subcellular structures shown in Figure 1-4b, consider the following calculations.

(a) All cells and many subcellular structures are surrounded by a membrane. Assuming a typical membrane to be about 8 nm wide, how many such membranes would have to be aligned side by side before the structure could be seen with the light microscope? How many with the electron microscope?

(b) Ribosomes are the cell structures in which the process of protein synthesis takes place. A human ribosome is a roughly spherical structure with a diameter of about 30 nm. How many ribosomes would fit in the internal volume of the human liver cell described in Problem 1-2 if the entire volume of the cell were filled with ribosomes?

(c) The genetic material of the *Escherichia coli* cell described in Problem 1-2 consists of a circular DNA molecule with a strand diameter of 2 nm and a total length of 1.36 mm. To be accommodated in a cell that is only a few micrometers long, this large DNA molecule is tightly coiled and folded into a *nucleoid* that occupies a small proportion of the cell's internal volume. Approximating the DNA molecule as a very thin cylinder, calculate the smallest possible volume the DNA molecule could fit into, and express it as a percentage of the internal volume of the bacterial cell that you calculated in Problem 1-2a.

1-4 QUANTITATIVE Limits of Resolution Then and Now. Based on what you learned in this chapter about the limit of resolution of a light microscope, answer each of the following questions. Assume that the unaided human eye has a limit of resolution of about 0.25 mm and that a modern light microscope has a useful magnification of about 1000×.

(a) Define *limit of resolution* in your own words. What was the limit of resolution of Hooke's microscope? What about van Leeuwenhoek's microscope?

(b) What are the approximate dimensions of the smallest structure that Hooke would have been able to observe with his microscope? Would he have been able to see any of the structures shown in Figure 1-4a? If so, which ones? And if not, why not?

(c) What are the approximate dimensions of the smallest structure that van Leeuwenhoek would have been able to observe with his microscope? Would he have been able to see any of the structures shown in Figure 1-4a? If so, which ones? And if not, why not?

(d) What are the approximate dimensions of the smallest structure that a contemporary cell biologist should be able to observe with a modern light microscope?

(e) Consider the eight structures shown in Figure 1-4a and 1-4b. Which of these structures would both Hooke and van Leeuwenhoek have been able to see with their respective microscopes? Which, if any, would van Leeuwenhoek have been able to see that Hooke could not? Explain your reasoning. Which, if any, that neither Hooke nor van Leeuwenhoek could see would a contemporary cell biologist be able to see using a modern light microscope?

1-5 The Contemporary Strands of Cell Biology. For each pair of techniques listed, indicate whether its members belong to the cytological (C), biochemical (B), or genetic (G) strand of cell biology (see Figure 1-3). Suggest one advantage that the second technique has over the first technique.

(a) Light microscopy/electron microscopy

(b) Centrifugation/ultracentrifugation

(c) Cell cultures/model organisms

(d) Sequencing of a genome/bioinformatics

(e) Transmission electron microscopy/scanning electron microscopy

(f) Chromatography/electrophoresis

1-6 The "Facts" of Life. Each of the following statements was once regarded as a biological fact but is now understood to be untrue. In each case, indicate why the statement was once thought to be true and why it is no longer considered a fact.

(a) Plant and animal tissues are constructed quite differently because animal tissues do not have conspicuous boundaries that divide them into cells.

(b) Living organisms are not governed by the laws of chemistry and physics, as is nonliving matter, but are subject to different laws that are responsible for the formation of organic compounds.

(c) Genes most likely consist of proteins because the only other likely candidate, DNA, is a relatively uninteresting molecule consisting of only four kinds of monomers (nucleotides) arranged in a relatively repetitive sequence.

(d) The fermentation of sugar to alcohol can take place only if living yeast cells are present.

1-7 Wrong Again. Explain why each of the following statements is false.

(a) Because of the wavelength of light, resolution of cellular structures smaller than 200 nm can never be achieved.

(b) Fluorescence microscopy can allow us to visualize cells but cannot help identify them.

(c) Because all DNA molecules have similar chemical composition, it is not possible to separate and characterize individual DNA molecules.

(d) The best way to carry out a scientific experiment is to try to prove a hypothesis by varying all the relevant conditions.

(e) The flow of genetic information is always from DNA to RNA to protein.

1-8 A New Biofuel. As a recent cell biology graduate, you have just been hired by a biotechnology company to develop a biofuel using algal cells that produce an oil very similar to diesel fuel. What model system(s) might you use for each of the following aspects of this project? What tools and techniques would you use?

(a) Determining which genes and enzymes are required to produce the oil

(b) Producing large amounts of a certain enzyme for further research

(c) Studying whether any of the cellular enzymes interact with each other

(d) Examining the involvement of any cellular organelles in storage or secretion of the oil

1-9 DATA ANALYSIS Worm Microscopy. With the use of microscopy, the localization of two proteins (A and B) using different fluorescent dyes can be examined in the developing *C. elegans* embryo images shown in **Figure 1-11**. The images of A and B are overlaid in the third image. Which type of microscopy is being used—light microscopy or electron microscopy? Are the two proteins likely to be interacting with each other? Provide evidence for your answers.

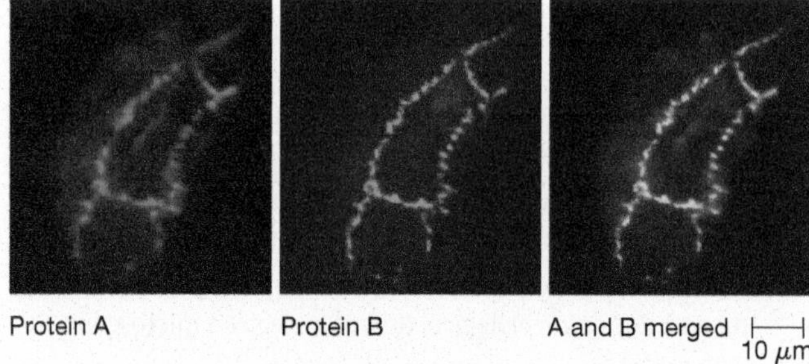

Figure 1-11 Localization of *C. elegans* Proteins Using Microscopy. See Problem 1-9.

2 The Chemistry of the Cell

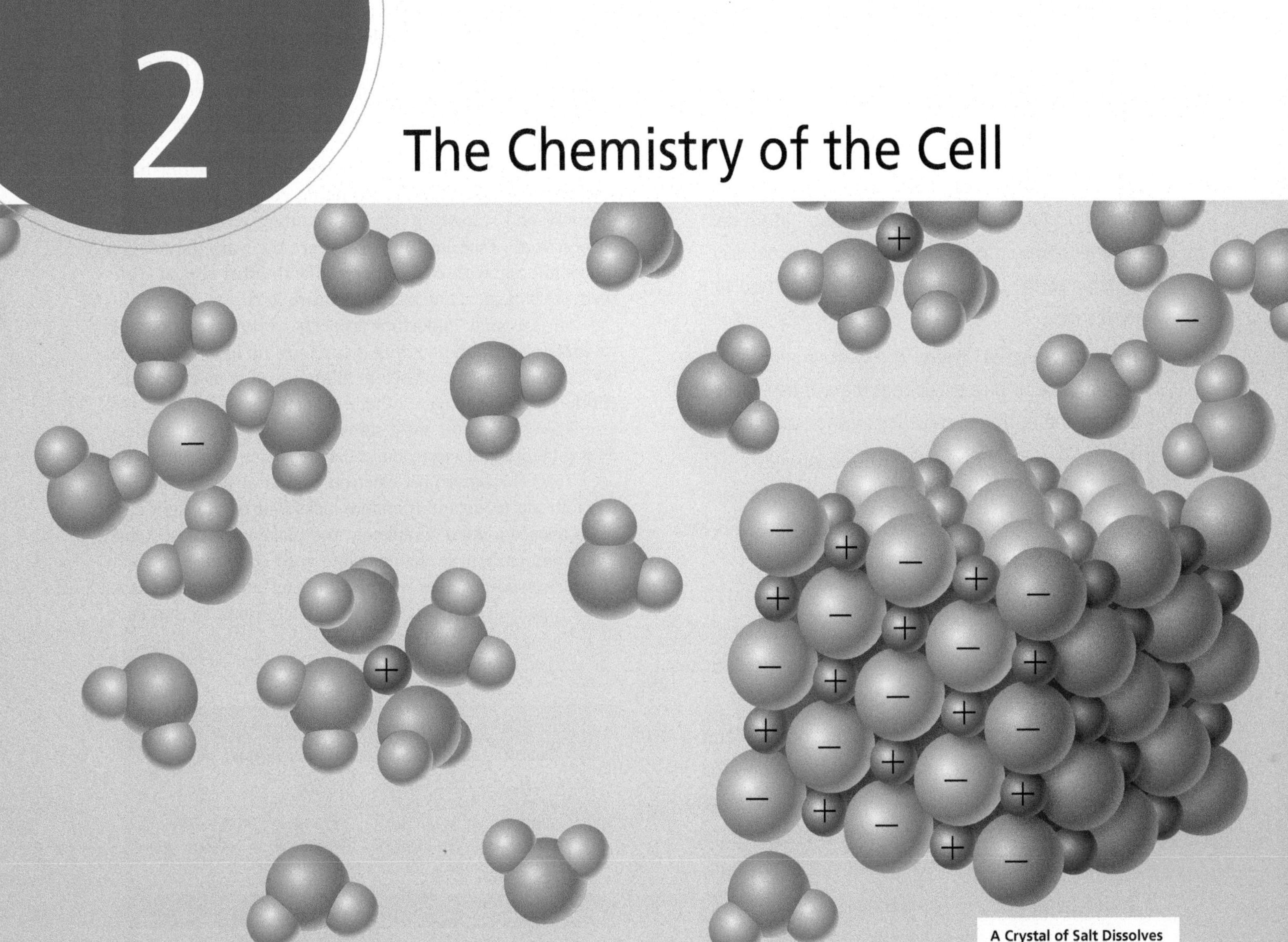

A Crystal of Salt Dissolves in Water.
As a crystal of salt (green and purple structure) dissolves in water, the oxygen atoms of water (red) surround the positive ions, and its hydrogen atoms (blue) surround the negative ions.

Students just beginning in cell biology are sometimes surprised—and perhaps even dismayed—to find that courses and textbooks dealing with cell biology involve a substantial amount of chemistry. Yet biology in general and cell biology in particular depend heavily on both chemistry and physics. After all, cells and organisms follow all the laws of the physical universe, so biology is really just the study of chemistry and physics in systems that are alive. In fact, everything cells are and do has a molecular and chemical basis. Therefore, we can truly understand and appreciate cellular structure and function only when we can describe cellular structure in molecular terms and express cellular function in terms of chemical reactions and events.

Trying to appreciate cellular biology without a knowledge of chemistry would be like trying to appreciate a translation of Chekhov without a knowledge of Russian. Most of the meaning would probably get through, but much of the beauty and depth of appreciation would be lost in the translation. For this reason, we will consider the chemical background necessary for the cell biologist. Specifically, this chapter will provide an overview of several chemical principles critical for understanding cellular biology and will introduce the major classes of macromolecules in cells: proteins, nucleic acids, carbohydrates, and lipids.

The main points of this chapter can conveniently be structured around five principles:

1. *The importance of carbon.* The carbon atom has several unique properties that make it especially suitable as the backbone of biologically important molecules.
2. *The importance of water.* The water molecule has several unique properties that make it especially suitable as the universal solvent of living systems.
3. *The importance of selectively permeable membranes.* Membranes define cellular compartments and control the movements of molecules and ions into and out of cells and organelles.
4. *The importance of synthesis by polymerization of small molecules.* Biological macromolecules are polymers formed by linking many similar or identical small molecules known as monomers.
5. *The importance of self-assembly.* Biological macromolecules are often capable of self-assembly into higher levels of structural organization because the information needed to specify the spatial configuration of the molecule is contained in the polymer.

Understanding these five principles will help you to appreciate the cellular chemistry necessary before venturing further into our exploration of what it means to be a cell.

2.1 The Importance of Carbon

To study cellular molecules really means to study compounds containing carbon. Almost without exception, molecules of importance to the cell biologist have a backbone, or skeleton, of carbon atoms linked together covalently in chains or rings. Actually, the study of carbon-containing compounds is the domain of **organic chemistry.** In its early days, organic chemistry was synonymous with *biological chemistry* because the carbon-containing compounds that chemists first investigated were obtained from biological sources (hence the word *organic*, acknowledging the organismal origins of the compounds).

The terms *organic chemistry* and *biological chemistry* have long since gone their separate ways, however, because organic chemists have now synthesized an incredible variety of carbon-containing compounds that do not occur naturally in the biological world. Organic chemistry therefore is the study of all classes of naturally occurring and synthetic carbon-containing compounds. **Biological chemistry** (*biochemistry* for short) deals specifically with the chemistry of living systems and is, as we have already seen, one of the several historical strands that form an integral part of modern cell biology (see Figure 1-3).

The **carbon atom (C)** is the most important atom in biological molecules. Carbon-containing compounds owe their diversity and stability to specific bonding properties of the carbon atom. Especially important are the ways that carbon atoms interact with each other and with other chemical elements of biological importance (**Figure 2-1**).

An extremely important property of the carbon atom is its **valence** of four, meaning it can form up to four chemical bonds with other atoms before filling its outer electron shell. Atoms can bond to each other via their outer electrons, and atoms are usually the most stable when they are surrounded by a total of eight electrons, satisfying what is known as the *octet rule.* The outermost electron orbital of a carbon atom has four electrons and therefore lacks four of the eight electrons needed to fill it completely and make it the most stable. Therefore, carbon atoms associate with each other or with other electron-deficient atoms, allowing adjacent atoms to share a pair of electrons, one from each atom, so that each

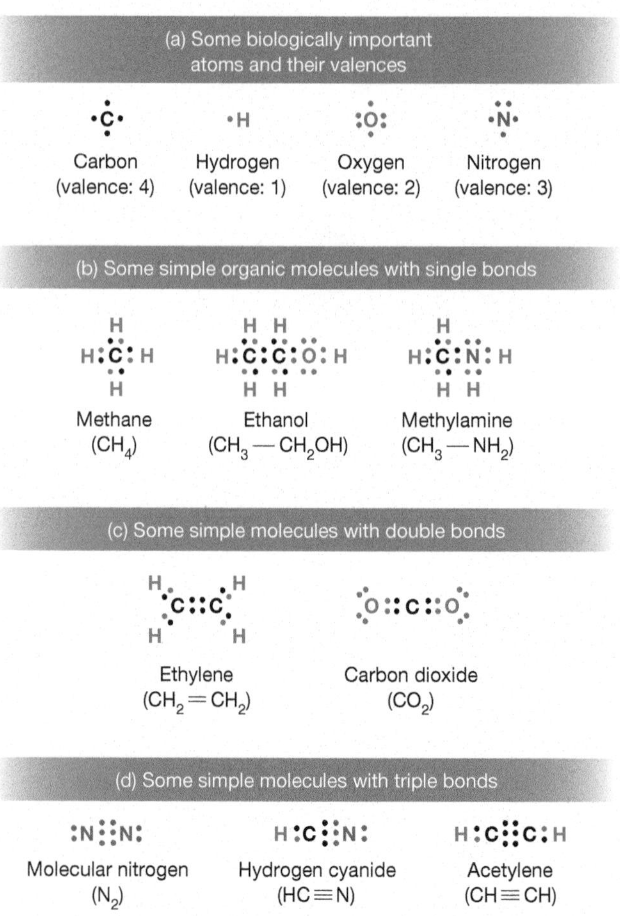

Figure 2-1 Electron Configurations of Some Biologically Important Atoms and Molecules. Electronic configurations are shown for **(a)** individual atoms and **(b–d)** some simple molecules. Only electrons in the outermost electron orbital are shown.

atom's outer orbital has a full set of eight electrons including shared electrons. (For hydrogen alone, a full set is only two electrons.) Atoms that share electrons in this way are held together and are said to be joined by a **covalent bond.** *Because four additional electrons are required to fill the outer orbital of carbon, stable organic compounds have four covalent bonds for every carbon atom.* This gives carbon-containing molecules great diversity in molecular structure and function.

Carbon atoms are most likely to form covalent bonds with other carbon atoms and with atoms of oxygen (O), hydrogen (H), nitrogen (N), and sulfur (S). The electronic configurations of several of these atoms are shown in Figure 2-1a. Sulfur, like oxygen, has six outer electrons and a valence of two. Notice that, in each case, one or more electrons are required to complete the outer orbital to a total of eight electrons. The number of "missing" electrons corresponds in each case to the valence of the atom, which is the number of covalent bonds the atom can form. Because hydrogen's outermost electron orbital can hold only two electrons, it has a valence of one and forms only one covalent bond.

The sharing of one pair of electrons between atoms results in a **single bond.** Methane, ethanol, and methylamine are simple examples of carbon-containing compounds containing only single bonds between atoms (Figure 2-1b), as represented by the pair of electrons between any two chemical symbols. Sometimes two or even three pairs of electrons can be shared by two atoms, giving rise to **double bonds** or **triple bonds.** Ethylene and carbon dioxide are examples of double-bonded compounds (Figure 2-1c). Notice that, in these compounds, each carbon atom still forms a total of four covalent bonds, either one double bond and two single bonds or two double bonds. Triple bonds are rare but can be found in molecular nitrogen, hydrogen cyanide, and acetylene (Figure 2-1d). Thus, both the valence and the low atomic weight of carbon give it unique properties that account for the diversity and stability of carbon-containing compounds, giving carbon a preeminent role in biological molecules.

Carbon-Containing Molecules Are Stable

The stability of organic molecules is a property of the favorable electronic configuration of each carbon atom in the molecule. This stability is expressed as **bond energy**—the amount of energy required to break 1 *mole* (about 6×10^{23}) of such bonds. (The term *bond energy* is a frequent source of confusion. Be careful not to think of it as energy that is somehow "stored" in the bond but rather as the amount of energy needed to *break* the bond.) Bond energies are usually expressed in *calories per mole (cal/mol)*, where a **calorie** is the amount of energy needed to raise the temperature of 1 gram of water by 1°C and a *kilocalorie (kcal)* is equal to 1000 calories.

It takes a large amount of energy to break a covalent bond. For example, the carbon-carbon (C – C) bond has a bond energy of 83 kilocalories per mole (kcal/mol). The bond energies for carbon-nitrogen (C – N), carbon-oxygen (C – O), and carbon-hydrogen (C – H) bonds are all in the same range: 70, 84, and 99 kcal/mol, respectively. Even more energy is required to break a carbon-carbon double bond (C = C; 146 kcal/mol) or a carbon-carbon triple bond (C ≡ C; 212 kcal/mol), so these compounds are even more stable.

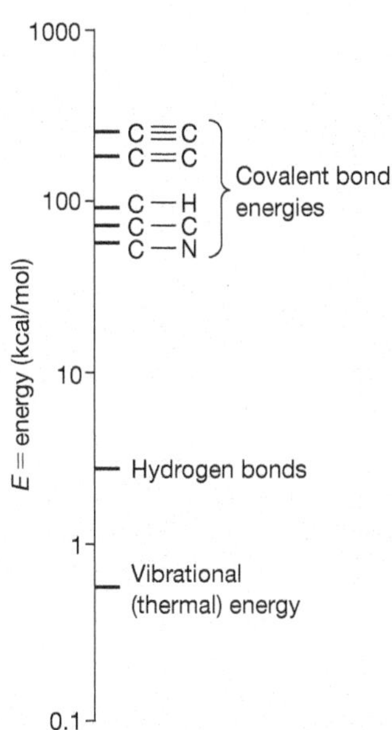

Figure 2-2 Energies of Biologically Important Bonds. Notice that energy is plotted on a logarithmic scale to accommodate the wide range of values shown.

We can appreciate the significance of these bond energies by comparing them with other relevant energy values, as shown in **Figure 2-2**. Most noncovalent bonds in biologically important molecules, such as the hydrogen bonds we will see later in this chapter, have energies of only a few kilocalories per mole. The energy of thermal vibration is even lower—only about 0.6 kcal/mol. Covalent bonds are much higher in energy than noncovalent bonds and are therefore much more stable.

The fitness of the carbon-carbon bond for biological chemistry on Earth is especially clear when we compare its energy with that of solar radiation. As shown in **Figure 2-3** on page 24, an inverse relationship exists between the wavelength of electromagnetic radiation and its energy content. This figure shows that the visible portion of sunlight (wavelengths of 380–750 nm) is lower in energy than the carbon-carbon bond is. If this were not the case, visible light would break covalent bonds spontaneously, and life as we know it would not exist.

Figure 2-3 illustrates another important point: the hazard that ultraviolet radiation poses to biological molecules due to its high energy. At a wavelength of 300 nm, for example, ultraviolet light has an energy content of about 95 kcal/mol. This is enough to break carbon-carbon bonds spontaneously. This threat underlies the current concern about pollutants that destroy the ozone layer in the upper atmosphere because the ozone layer filters out much of the ultraviolet radiation that would otherwise reach Earth's surface and disrupt the covalent bonds that hold biological molecules together.

The stability of carbon is evident not only through our study of organic chemistry here on Earth but also through exciting discoveries on other planets in our solar system. Recent work made possible by the Mars rover Curiosity has demonstrated the presence of carbon-containing compounds near the planet surface. The organic matter was detected despite the high levels of ionizing radiation bombarding the Martian planet's surface, highlighting carbon's stable nature and suggesting

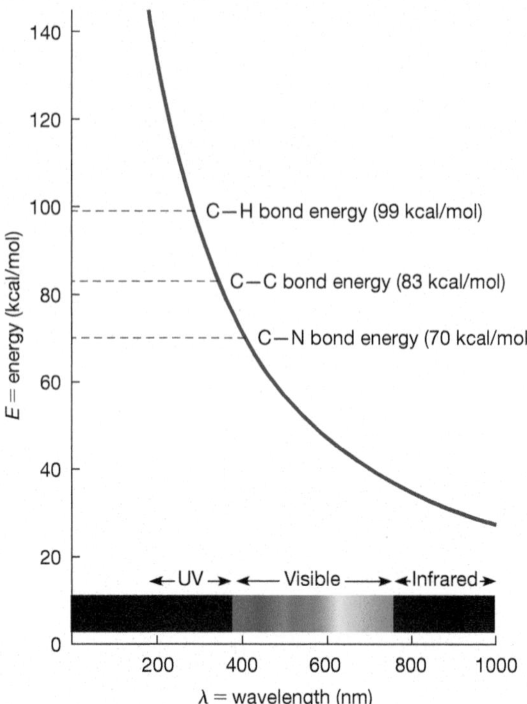

Figure 2-3 The Relationship Between Energy (*E*) and Wavelength (λ) for Electromagnetic Radiation. The dashed lines mark the bond energies of the C—H, the C—C, and the C—N single bonds. The bottom of the graph shows the approximate range of wavelengths for ultraviolet (UV), visible, and infrared radiation.

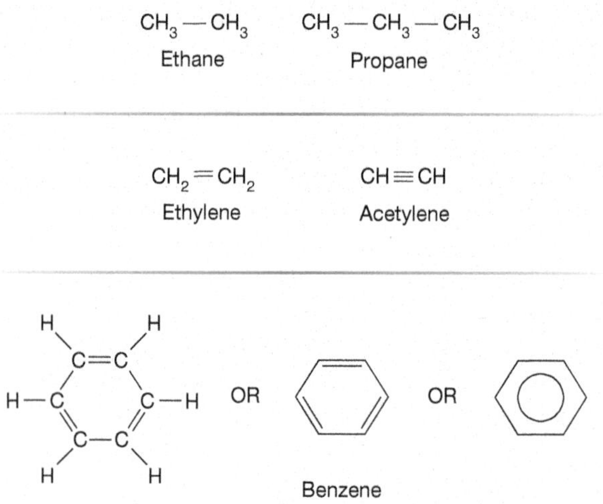

Figure 2-4 Some Simple Hydrocarbon Compounds. Compounds in the top row have single bonds only, whereas those in the second and third rows have double or triple bonds.

that more exciting organic compounds could be discovered farther below the surface of Mars where less radiation exists.

Carbon-Containing Molecules Are Diverse

In addition to their inherent stability, carbon-containing compounds are characterized by the great diversity of molecules that can be generated from relatively few different kinds of atoms. Again, this diversity is due to the tetravalent nature of the carbon atom and the resulting ability of each carbon atom to form covalent bonds to four other atoms. Because one or more of these bonds can be bonds to other carbon atoms, molecules consisting of long chains of carbon atoms can be built up. Ring compounds are also common. Further variety is possible by the introduction of branching and the presence of double bonds in the carbon-carbon chains.

When only hydrogen atoms are bonded to carbon atoms in linear or branched chains or in rings, the resulting compounds are called **hydrocarbons** (**Figure 2-4**). Economically important hydrocarbons such as hexane (C_6H_{14}), octane (C_8H_{18}), and decane ($C_{10}H_{22}$) are found in gasoline and other petroleum products. The natural gas that many of us use for fuel is a mixture of methane, ethane, propane, and butane, which are hydrocarbons with one to four carbon atoms, respectively. Benzene (C_6H_6) is a common industrial solvent.

In biology, on the other hand, hydrocarbons play only a limited role because they are essentially insoluble in water, the universal solvent in biological systems. One exception is ethylene (C_2H_4), a gas that acts as a plant hormone and is used commercially to promote fruit ripening. However, hydrocarbons do play an important role in the structure of biological membranes. The interior of every biological membrane is a nonaqueous environment consisting of the long hydrocarbon "tails" of phospholipid molecules that project into the interior of the membrane from either surface. This feature of membranes has important implications for their role as permeability barriers, as we will see shortly.

Most biological compounds contain, in addition to carbon and hydrogen, one or more atoms of oxygen and often nitrogen, phosphorus, or sulfur as well. These atoms are usually part of various **functional groups,** which are specific arrangements of atoms that confer characteristic chemical properties on the molecules to which they are attached. Some of the more common functional groups present in biological molecules are shown in **Figure 2-5**. At the near-neutral pH of most cells, several of these groups form **ions,** which are atoms or molecules that are charged because they have gained or lost an electron or a *proton* (a hydrogen atom without its electron).

For example, the *carboxyl* and *phosphate* groups, which are considered acidic because they have given up a proton, are negatively charged. By contrast, the *amino* group, which is considered basic because it has gained a proton, is positively charged. Other groups, such as the *hydroxyl, sulfhydryl, carbonyl,* and *aldehyde* groups, are uncharged at pH values near neutrality.

However, the presence of any oxygen or sulfur atoms bound to carbon or hydrogen results in a *polar bond* due to unequal sharing of electrons. This is because oxygen and sulfur have higher *electronegativity*, or affinity for electrons, than carbon and hydrogen. Therefore, when "sharing" electrons with carbon or hydrogen, an oxygen (or nitrogen) atom will have the electron more than half the time, giving it a slightly negative charge and giving the hydrogen (or carbon) a slightly positive charge. The resulting polar bonds have higher water solubility and chemical reactivity than nonpolar C – C or C – H bonds, in which electrons are equally shared.

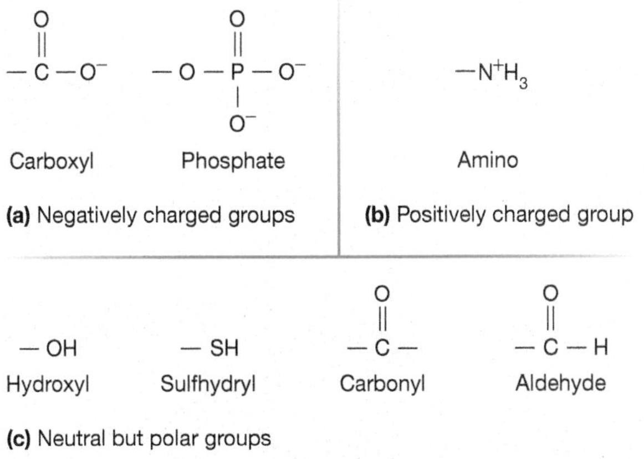

$-C(=O)-O^-$ Carboxyl

$-O-P(=O)(O^-)-O^-$ Phosphate

(a) Negatively charged groups

$-N^+H_3$ Amino

(b) Positively charged group

$-OH$ Hydroxyl

$-SH$ Sulfhydryl

$-C(=O)-$ Carbonyl

$-C(=O)-H$ Aldehyde

(c) Neutral but polar groups

Figure 2-5 Some Common Functional Groups Found in Biological Molecules. Each functional group is shown in the form that predominates at the near-neutral pH of most cells. They are separated into **(a)** negatively charged, **(b)** positively charged, and **(c)** neutral but polar groups.

Often, carbon-containing compounds will lose electrons to other molecules such as molecular oxygen. This process is called *oxidation* and typically involves degradation and releases energy, as in the oxidation of glucose to carbon dioxide and water. The reverse process, by which carbon-containing compounds gain electrons, is known as *reduction* and typically is biosynthetic and requires energy, as in the photosynthetic reduction of carbon dioxide to glucose. (Oxidation and reduction will be discussed in more detail in Chapter 9; see Equations 9-7 to 9-10.)

Given the incredible diversity of chemical compounds in cells, you may wonder how it is possible to study individual compounds to determine their structure. Key Technique, pages 26–27, describes the use of mass spectrometry to determine the chemical structure and identity of individual chemical compounds in cells.

MAKE CONNECTIONS 2.1

Which strand of cell biology does mass spectrometry align best with? (Fig. 1-3)

Carbon-Containing Molecules Can Form Stereoisomers

Carbon-containing molecules are capable of still greater diversity because the carbon atom is a **tetrahedral** structure. When four different atoms or groups of atoms are bonded to the four corners of such a tetrahedral structure, two different spatial configurations are possible. Although both forms have the same structural formula, they are not superimposable but are, in fact, mirror images of each other as shown by the plane of symmetry, which represents the mirror. Such mirror-image forms of the same compound are called **stereoisomers** (**Figure 2-6**).

A carbon atom that has four different substituents (atoms or groups attached) is called an **asymmetric carbon atom (Figure 2-7).** Because two stereoisomers are possible for each asymmetric carbon atom, a compound with n asymmetric carbon atoms will have 2^n possible stereoisomers. As shown in Figure 2-7, the three-carbon amino acid *alanine* has

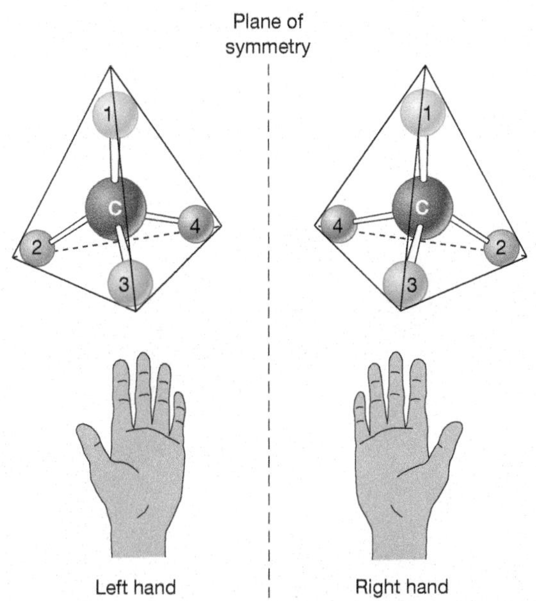

Figure 2-6 Stereoisomers. Stereoisomers of organic compounds occur when four different groups are attached to a tetrahedral carbon atom. Stereoisomers, like left and right hands, are mirror images of each other and cannot be superimposed on one another.

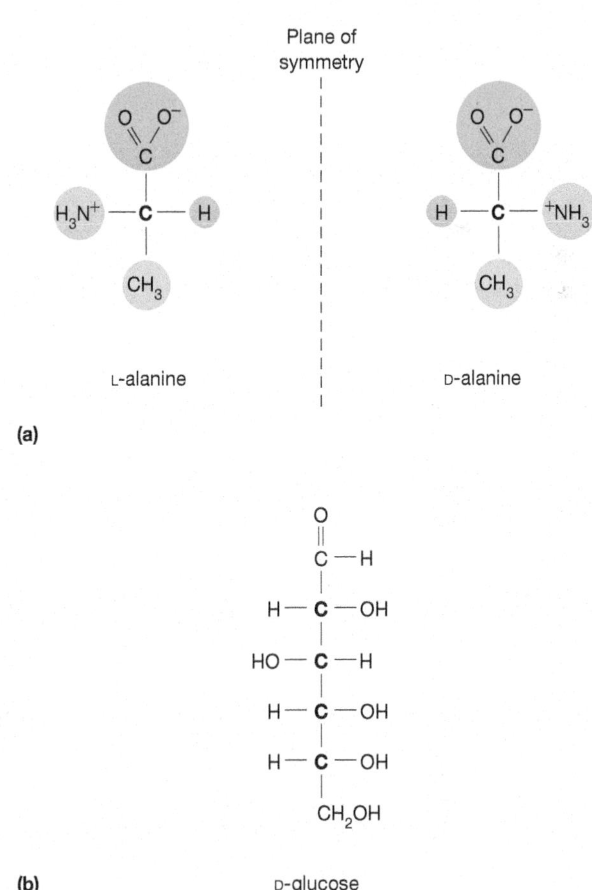

Figure 2-7 Stereoisomers of Biological Molecules. (a) The amino acid alanine has a single asymmetric carbon atom (in boldface) and can therefore exist in two spatially different forms, designated as L- and D-alanine. **(b)** The six-carbon sugar glucose has four asymmetric carbon atoms (in boldface).

Key Technique | Determining the Chemical Fingerprint of a Cell Using Mass Spectrometry

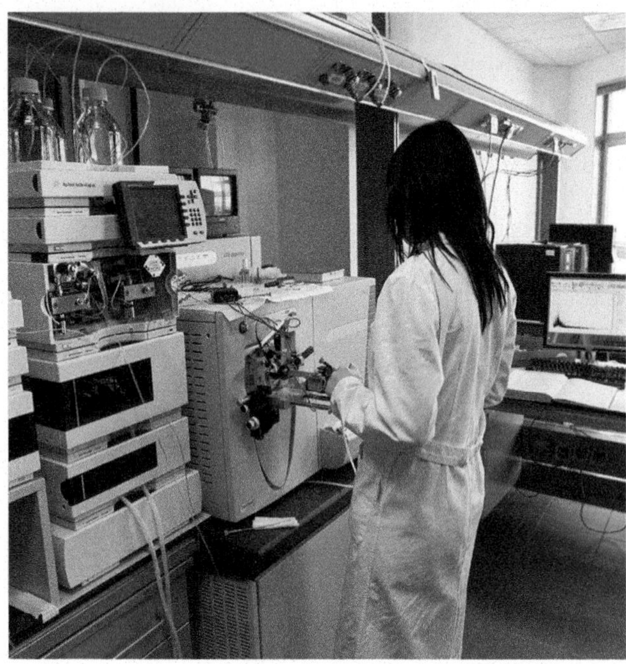

A Scientist Preparing an Injection for Mass Spectrometry.

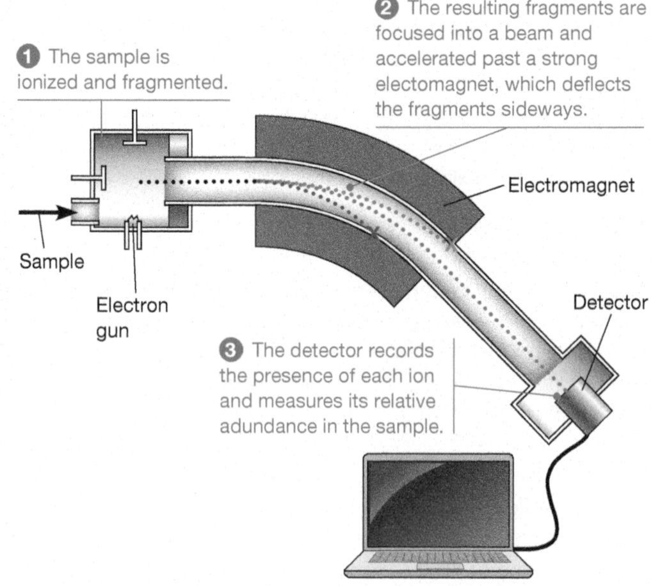

1 The sample is ionized and fragmented.
2 The resulting fragments are focused into a beam and accelerated past a strong electomagnet, which deflects the fragments sideways.
Electromagnet
Sample
Electron gun
Detector
3 The detector records the presence of each ion and measures its relative adundance in the sample.

Figure 2A-1 A Mass Spectrometer.

PROBLEM: In cell biology, scientists typically study processes that involve changes in the chemistry of the cell, such as cell growth and division. Researchers often want to be able to identify small molecules in a cellular extract, or they may want to determine the chemical structure of a new compound. How is such analysis accomplished?

SOLUTION: *Mass spectrometry* (often called *mass spec*) is a method used to identify and measure the relative abundance of individual molecules in a sample, as well as to determine their chemical structure. Purified molecules are broken into fragments, and these fragments can be analyzed to determine their masses and the arrangement of covalent bonds that hold atoms of the molecule together.

Key Tools: Mass spectrometer; an ionized sample; a computer to analyze the results.

Details: Mass spectrometry can identify chemical compounds within a sample with high resolution, differentiating between compounds that can vary by as little as 1 atomic mass unit (amu), the mass of a hydrogen atom. Analysis of a compound using a mass spectrometer (**Figure 2A-1**) involves three main steps: *ionization and fragmentation* of the sample, *deflection* of the ionized fragments by an electromagnet, and *detection* of the individual ions and measurement of their abundance.

Ionization and Fragmentation. Commonly, the sample is ionized by bombarding it with a stream of high-energy electrons from an *electron gun*. The stream has enough energy to knock an

a single asymmetric carbon atom (in the center) and thus has two stereoisomers, called L-alanine and D-alanine (Figure 2-7a). Neither of the other two carbon atoms of alanine is an asymmetric carbon atom because one has three identical substituents (hydrogen atoms) and the other has two bonds to a single oxygen atom and thus is only bonded to three substituents. Both stereoisomers of alanine occur in nature, but only L-alanine is present as a component of proteins.

As an example of a compound with multiple asymmetric carbon atoms, consider the six-carbon sugar *glucose* shown in Figure 2-7b. Of the six carbon atoms of glucose, the four shown in boldface are asymmetric. (Can you figure out why the other two carbon atoms are not asymmetric?) With four asymmetric carbon atoms, the structure shown (D-glucose) is only one of 2^4, or 16, possible stereoisomers of the $C_6H_{12}O_6$ molecule.

CONCEPT CHECK 2.1

What properties of the carbon atom make it especially suitable as the structural basis for nearly all biomolecules?

electron off the sample to form a positively charged *molecular ion* (M^+). Molecular ions are generally unstable and will break up into smaller fragments, some of which will be positively charged and others neutral.

Deflection. The resulting fragments are focused into a fine beam and accelerated past a powerful electromagnet that separates them by mass. The mass of each fragment will depend on which particular covalent bond in the parent ion (M^+) was broken. Because each molecule fragments in a predictable pattern, the specific pattern of fragment masses can be used to identify a compound.

As the beam passes by the electromagnet, individual ions are pulled sideways, deflected from a straight path. The amount of deflection depends on the mass of the ion, with lighter ions being deflected more. The effect is like dropping a large cannonball and a small steel ball (such as can be found in a pinball machine) near a large magnet. The heavy cannonball will be pulled sideways during its flight much less than the smaller ball. The strength of the electromagnet can be increased (to deflect heavier ions) or decreased (to deflect lighter ions) so that ions of different masses are focused on a detector at the end of the spectrometer. (Only positively charged particles reach the detector.)

Detection and Analysis. The detector records the presence of each ion and, based on the strength of the magnetic field, can determine the ion's mass. It also records the number of ions having each different mass and calculates their abundance in the sample. A computer then converts this information into a graph showing vertical lines representing a *spectrum* of mass-to-charge ratios (m/z) across the *x*-axis, with the heights of the lines (also called peaks) showing each ion's relative abundance (*y*-axis). **Figure 2A-2** shows mass spectral results for a simple amino acid, glycine.

Data Interpretation. The heaviest ion in a mass spectrum (the one with the highest m/z value) is often the molecular ion. In Figure 2A-2, this is the peak at m/z = 75. (To confirm that a molecular weight of 75 corresponds to glycine, start with its elemental formula, $C_2H_5NO_2$, and sum the atomic weights of its carbons, 12 amu × 2; nitrogen, 14 amu × 1; oxygens, 16 amu × 2; and hydrogens, 1 amu × 5.)

The tallest peak in a mass spectrum, called the base peak, is assigned a *y*-axis value of 100%. The heights of all other lines are plotted relative to the base peak. The base peak typically corresponds to the most stable, hence most abundant, fragment ion. In this example, the base peak is at m/z = 30, which, for glycine, is a fragment with a single carbon, a nitrogen, and four hydrogens (CH_2NH_2). By comparing the pattern of the lines to the patterns of known compounds, a compound can often be identified. For novel compounds, the pattern of lines helps identify the fragments. By determining how the fragments would fit together to form an intact molecule, a researcher can predict the types and arrangement of the covalent bonds and determine the overall chemical structure of the novel compound.

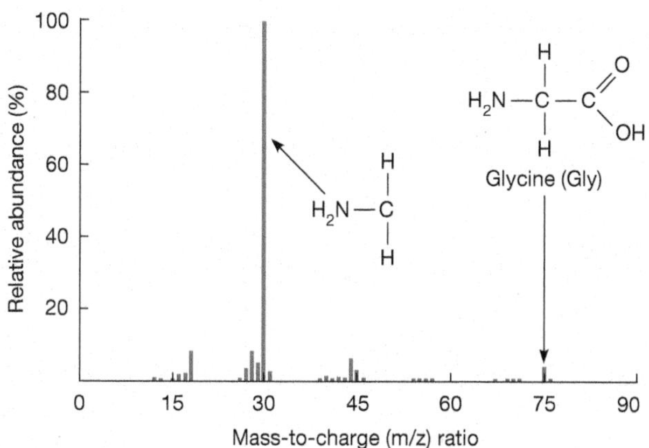

Figure 2A-2 The Mass Spectrum of Glycine. The non-ionized form of glycine is shown.

QUESTION: Compare the flights of the ionized molecular fragments whose paths are shown in blue and red in Figure 2A-1. Which fragment has the higher m/z ratio? What factors might explain why the deflection of that fragment differs from the other fragment?

2.2 The Importance of Water

Just as the carbon atom is uniquely significant because of its role as the universal backbone of biologically important molecules, the water molecule commands special attention because of its indispensable role as the universal solvent in biological systems. Water is, in fact, the single most abundant component of cells and organisms. Typically, about 75–85% of a cell by weight is water, and most cellular processes (like protein folding) take place in this aqueous environment. In addition, many cells depend on an extracellular environment that is essentially aqueous as well. In some cases, this is a body of water—whether an ocean, a lake, or a river—where the cell or organism lives, and in other cases, it may be the body fluids with which the cell is bathed. Therefore, we must always take into account the presence of water when considering the ways that a cell functions.

Water is indispensable for life. True, there are life forms that can become dormant and survive periods of severe water scarcity. Seeds of plants and spores of bacteria and fungi are clearly in this category. Some plants and animals—notably certain mosses, lichens, nematodes, and rotifers—can also undergo physiological adaptations that allow them to dry out and survive in a highly dehydrated form, sometimes for surprisingly long periods of time. Such adaptations are clearly an advantage in environments characterized by periods of drought. Yet all of these are, at best, temporary survival mechanisms. Resumption of normal activity always requires rehydration.

The successful transport of water into and out of cells, as well between cells is also critical. Water can move across cellular membranes based on the concentration of solutes present in a process called **osmosis.** Whereas osmosis can be a slow process, water is able to move much more quickly through a specialized channel protein known as an **aquaporin (AQP).** For example, aquaporins allow water to flow rapidly between cells within certain organs like the kidneys. (You will learn more detail about the movement of water using both these mechanisms in Chapter 8.)

To understand why water is so uniquely suitable for its role, we need to look at its chemical properties. The most critical attribute of water is its *polarity* because this property accounts for its *cohesiveness*, its *temperature-stabilizing capacity*, and its *solvent properties*, all of which have important consequences for biological chemistry.

Water Molecules Are Polar

An unequal distribution of electrons gives the water molecule its **polarity,** which we can define as an uneven distribution of charge within a molecule. To understand the polar nature of water, we need to consider the shape of the molecule (**Figure 2-8**). As shown in Figure 2-8a, the water molecule is

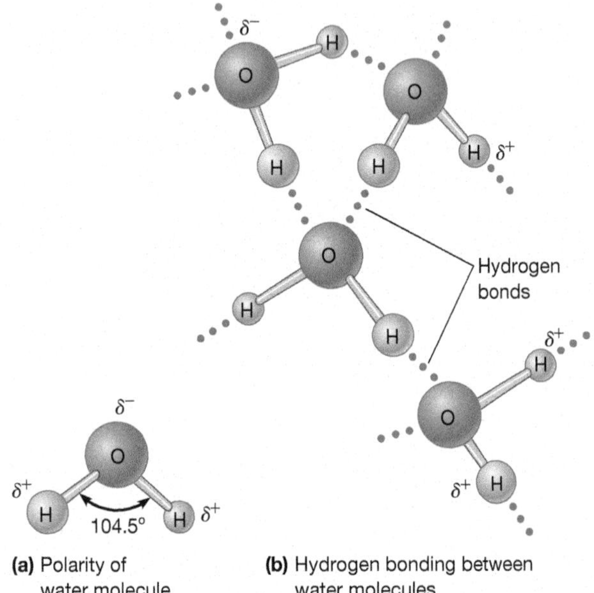

Figure 2-8 Hydrogen Bonding Among Water Molecules. (a) The water molecule is polar because it has an asymmetric charge distribution, partly due to the high electronegativity of the oxygen atom. It has a partial negative charge (δ^-), and each of the two hydrogen atoms has a partial positive charge (δ^+). **(b)** The extensive association of water molecules with one another in either the liquid or the solid state is due to hydrogen bonds (blue dots).

bent rather than linear in shape, with the two hydrogen atoms bonded to the oxygen at an angle of 104.5° rather than 180°. It is no overstatement to say that life depends critically on this angle because of the distinctive properties that the resulting asymmetry produces in the water molecule.

Although the water molecule as a whole is uncharged, its electrons are unevenly distributed. The oxygen atom is highly **electronegative**—it tends to draw electrons toward it. Therefore, the oxygen atom has a partial negative charge (denoted as δ^-, with the Greek letter delta standing for "partial"), and each of the two hydrogen atoms has a partial positive charge (δ^+). This unequal distribution of charge makes any O – H bond polar and, along with the presence of two lone electron pairs, helps explain why water is such a highly polar molecule.

Water Molecules Are Cohesive

Because of their polarity, water molecules are attracted to each other so that the electronegative oxygen atom of one molecule is associated with the electropositive hydrogen atoms of adjacent molecules. This forms a **hydrogen bond** (dotted lines in Figure 2-8b), which is a type of noncovalent interaction that is about one-tenth as strong as a covalent bond.

Each oxygen atom can bond to two hydrogens, and both of the hydrogen atoms can associate in this way with the oxygen atoms of adjacent molecules. As a result, water is characterized by an extensive three-dimensional network of hydrogen-bonded molecules. Although individual hydrogen bonds are weak, the combined effect of large numbers of them can be quite significant. In liquid water, the hydrogen bonds between adjacent molecules are constantly being broken and re-formed, with a typical bond having a half-life of a few microseconds. On average, however, each molecule of water in the liquid state is hydrogen-bonded to at least three neighbor molecules at any given time. In ice, the hydrogen bonding is still more extensive, giving rise to a rigid, hexagonal crystalline lattice with every oxygen hydrogen-bonded to hydrogens of two adjacent molecules and every water molecule therefore hydrogen-bonded to four neighboring molecules.

It is this tendency to form hydrogen bonds between adjacent molecules that makes water so highly *cohesive*. This cohesiveness accounts for the high *surface tension* of water, as well as for its high *boiling point*, high *specific heat*, and high *heat of vaporization*. The high surface tension of water allows some insects to move across the surface of a pond without breaking the surface (**Figure 2-9**). High surface tension is also important in allowing water to move upward through the conducting tissues of plants.

Water Has a High Temperature-Stabilizing Capacity

An important property of water that stems directly from the hydrogen bonding between adjacent molecules is the high specific heat that gives water its *temperature-stabilizing capacity*. *Specific heat* is the amount of heat a substance must absorb per gram to increase its temperature 1°C. The specific heat of water is 1.0 calorie per gram.

Because of its extensive hydrogen bonding, the specific heat of water is much higher than that of most other liquids.

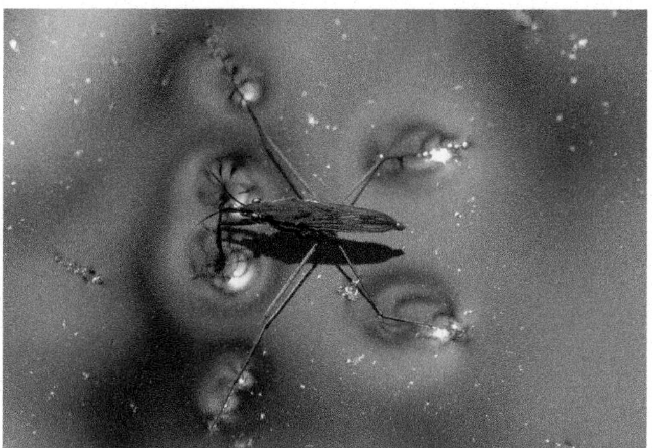

Figure 2-9 Walking on Water. The high surface tension of water results from the collective strength of vast numbers of hydrogen bonds. It enables insects such as this water strider to walk on the surface of a pond without breaking the surface.

In other liquids, much of the energy would cause an increase in the motion of solvent molecules and therefore an increase in temperature. In water, the energy is used instead to break hydrogen bonds between neighboring water molecules, buffering aqueous solutions against large changes in temperature. This capability is an important consideration for the cell biologist because cells release large amounts of energy as heat during metabolic reactions. If not for the extensive hydrogen bonding and the resulting high specific heat of water molecules, this release of energy would pose a serious overheating problem for cells, and life would not be possible.

Water also has a high *heat of vaporization*, which is defined as the amount of energy required to convert 1 gram of a liquid into vapor. This value is high for water because of the hydrogen bonds that must be disrupted in the process. This property makes water an excellent coolant and explains why people perspire, why dogs pant, and why plants lose water through transpiration. In each case, the heat required to evaporate water is drawn from the organism, which is therefore cooled in the process.

Water Is an Excellent Solvent

From a biological perspective, one of the most important properties of water is its excellence as a general solvent. A **solvent** is a fluid in which another substance, called the **solute**, can be dissolved. Water is an especially good solvent for biological purposes because of its remarkable capacity to dissolve a great variety of solutes.

It is the polarity of water that makes it so useful as a solvent. Many of the molecules in cells are also polar and therefore form hydrogen bonds with water molecules. Solutes that have an affinity for water and therefore dissolve readily in water are called **hydrophilic** ("water-loving"). Most small organic molecules found in cells are hydrophilic. Examples are sugars, organic acids, and some of the amino acids. Molecules that are not very soluble in water are termed **hydrophobic** ("water-fearing"). Among the more important hydrophobic compounds found in cells are the lipids and proteins found in biological membranes. In general, polar molecules and ions are hydrophilic, and nonpolar molecules are hydrophobic. Some biological macromolecules, notably proteins, have both hydrophobic and hydrophilic regions, so some parts of the molecule have an affinity for water whereas other parts of the molecule do not.

To understand why polar substances and ions dissolve so readily in water, let's consider a salt such as sodium chloride (NaCl) (**Figure 2-10**). Because it is a salt, NaCl exists in crystalline form as a lattice of positively charged sodium *cations* (Na^+) and negatively charged chloride *anions* (Cl^-). For NaCl to dissolve in a liquid, solvent molecules must overcome the attraction of the Na^+ cations and Cl^- anions for each other. When NaCl is placed in water, both the sodium and chloride ions become involved in *electrostatic interactions* with the water molecules instead of with each other, and the Na^+ and Cl^- ions separate and become dissolved. Because of their polarity, water molecules can form *spheres of hydration* around both Na^+ and Cl^-, thus neutralizing their attraction for each other and decreasing their likelihood of reassociation.

As Figure 2-10a shows, the sphere of hydration around a cation such as Na^+ involves water molecules clustered around the ion with their negative (oxygen) ends pointing toward it. For an anion such as Cl^-, the orientation of the

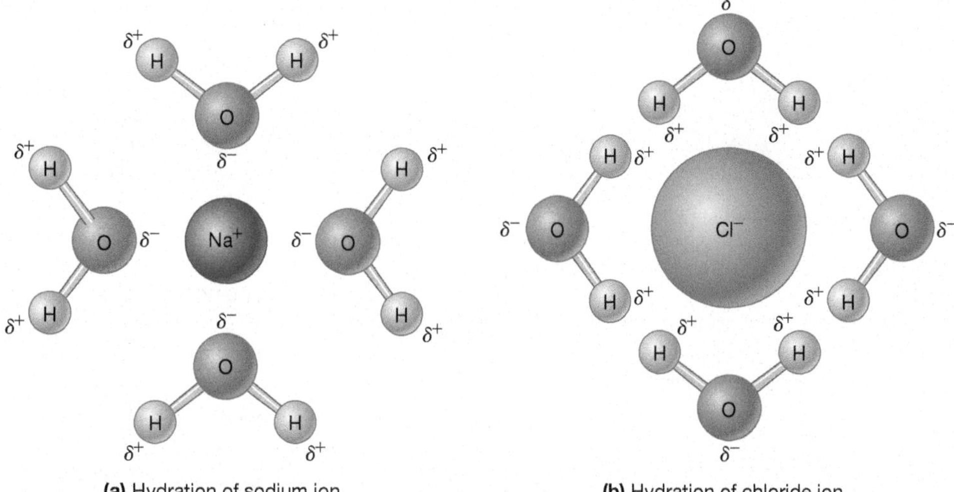

Figure 2-10 The Solubilization of Sodium Chloride. Sodium chloride (NaCl) dissolves in water because spheres of hydration are formed around both **(a)** the sodium ions and **(b)** the chloride ions. The oxygen atom and the sodium and chloride ions are drawn to scale.

HUMAN *Connections*

Taking a Deeper Look: Magnetic Resonance Imaging (MRI)

We often take the water content of our bodies for granted. On average, the human body is 55–65% water, depending on age and weight. But the chemistry of water also provides an opportunity: each water molecule possesses two hydrogen atoms, which are constantly breaking and re-forming hydrogen bonds with surrounding molecules. In addition, most biological molecules contain hydrogen, and the products of many biological reactions release hydrogen ions into surrounding tissues. It is these hydrogen ions, so plentiful in water and tissue, especially the circulatory system, that make one of the safest imaging technologies possible: *magnetic resonance imaging,* or *MRI.*

It is very likely that you or someone you know has had an MRI to diagnose a medical condition, but how does an MRI device exploit water in tissues to produce images of the body? MRI is a noninvasive method that uses basic principles of chemistry and physics to visualize internal structures of the body. An MRI machine capitalizes on the ability of protons to align with an externally applied magnetic field by placing the patient inside the field of a powerful electromagnet. In the absence of an applied magnetic field, the magnetic fields of individual hydrogens are randomly oriented. When a magnetic field is applied, however, most of the hydrogen atoms in the patient's body will align in one of two directions: either "up" or "down" with respect to the applied field (**Figure 2B-1**). Then a second, oscillating magnetic field is applied, which causes the aligned hydrogen atoms to absorb energy, which they subsequently release in the form of radio waves as the magnetic field changes. By rapidly varying the main magnetic field using a special set of magnetic coils, the release of these radio waves is detected by a receiver and provides relevant spatial information. The rapid switching of these coils on and off gives rise to the familiar repetitive clicking sounds made by an MRI machine. Once the radio waves are detected, a computer creates a spatial map of energy differences to produce an image of the tissue.

All tissues in the body vary in density and water content. Because an MRI requires a change in orientation of hydrogen ions in a magnetic field, these differences throughout the body actually provide the perfect contrast for visualization. By changing magnetic field strength or the frequency of the radio waves used during an MRI, a specific organ or tissue can be highlighted. For example, water and fluid-containing tissues are bright, whereas fatty tissues with little water are darker. These differences produce striking images of individual organs that can be examined for tumors or other abnormalities (**Figure 2B-2**). For many purposes, the images produced in an MRI scan are far better than those produced by taking multiple X-ray images and reconstructing them (a technique known as *computer-aided tomography,* or a *CAT scan*). With the addition of contrast agents injected into a patient's bloodstream or other tissues, specific structures such as blood vessels can be imaged with extraordinary detail using MRI.

The cost of an MRI procedure in the United States can range from $1000 to $2000. While more expensive than an X-ray, this procedure may be a better choice for many patients due to its potential for more accurate diagnoses. Unlike traditional X-rays, which can image only dense structures like bone and can cause damage to DNA, an MRI can give a detailed image of soft tissue such as internal organs, blood vessels, and nerves without posing a significant health risk.

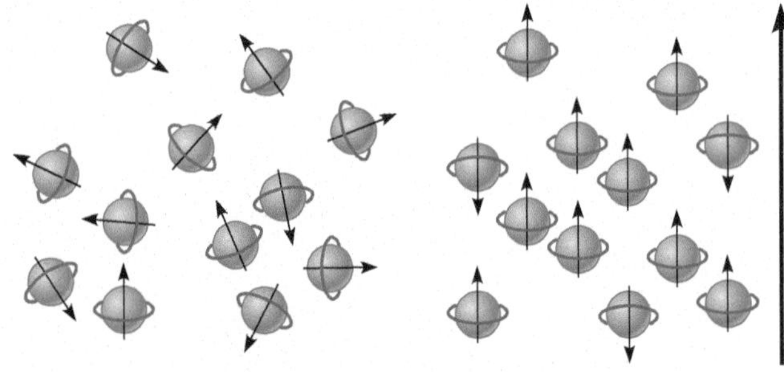

(a) Before magnetic field applied: random orientation of proton spins
(b) After magnetic field (large arrow) applied: proton spins oriented

Figure 2B-1 Hydrogen Atom Spin Patterns. (a) The magnetic field orientation of hydrogen atoms in the absence of a magnetic field is random. **(b)** The magnetic fields of hydrogen atoms align in the presence of an applied magnetic field.

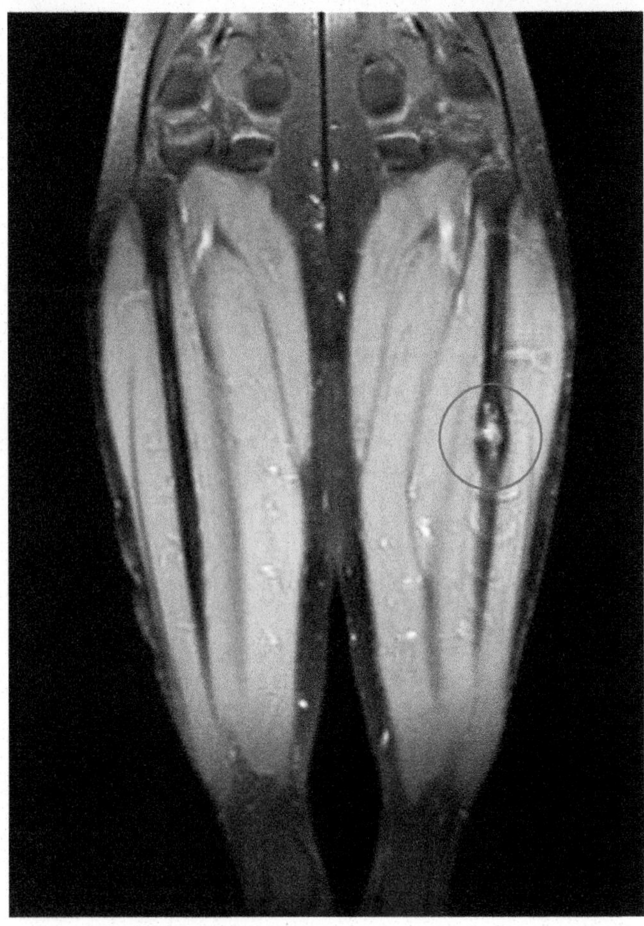

Figure 2B-2 MRI of Tumor on the Femur. The small tumor is easily visible (red circle).

water molecules is reversed, with the positive (hydrogen) ends of the solvent molecules pointing in toward the ion (Figure 2-10b). Similar spheres of hydration develop around charged functional groups (see Figure 2-5a, b), increasing their solubility. Even uncharged polar functional groups, such as aldehyde or sulfhydryl groups (see Figure 2-5c), will have a sphere of hydration, as the polar oxygen or sulfur atoms attract the positively charged ends of the polar water molecule and increase solubility.

Some biological compounds are soluble in water because they exist as ions at the near neutral pH of the cell and are therefore solubilized and hydrated like the ions of Figure 2-10. Compounds containing carboxyl, phosphate, or amino groups are in this category (see Figure 2-5). Most organic acids, for example, are almost completely ionized by deprotonation at a pH near 7 and therefore exist as anions that are kept in solution by spheres of hydration, as we have just seen with the chloride ion in Figure 2-10b. Amines, on the other hand, are usually protonated at cellular pH and thus exist as hydrated cations, and behave like the sodium ion in Figure 2-10a.

Often, organic molecules have no net charge but are nonetheless hydrophilic because they have some regions that are positively charged and other regions that are negatively charged and thus are soluble in water. Also, compounds containing the polar hydroxyl, sulfhydryl, carbonyl, or aldehyde groups shown in Figure 2-5 are also water soluble due to hydrogen bonding of these polar groups with water molecules.

Hydrophobic molecules such as hydrocarbons, on the other hand, have no such polar regions and therefore show no tendency to interact electrostatically with water molecules. In fact, because they disrupt the hydrogen-bonded structure of water, they tend to be excluded by the water molecules. Therefore, hydrophobic molecules tend to coalesce as they associate with one another rather than with the water. As we will see later in the chapter, such associations of hydrophobic molecules (or parts of molecules) are a major driving force in the folding of proteins, the assembly of cellular structures, and the organization of membranes.

Because typical cells are primarily water and most other biological molecules in cells contain hydrogen, the chemical and physical properties of the hydrogen atom can be exploited for medical imaging purposes. The common procedure known as magnetic resonance imaging (MRI) takes advantage of this abundance of water to image internal body tissues in a noninvasive manner (see Human Connections, page 30).

CONCEPT CHECK 2.2

How would the properties of water change if the water molecule were linear rather than bent? Why would it be less satisfactory as the basis for living cells?

2.3 The Importance of Selectively Permeable Membranes

Every cell and organelle needs some sort of physical barrier to keep its contents in and external materials out. A cell also needs some means of controlling exchange between its internal environment and the extracellular environment. Ideally, such a barrier should be impermeable to most of the molecules and ions found in cells and their surroundings. Otherwise, substances could diffuse freely in and out, and the cell would not really have a defined content at all. On the other hand, the barrier cannot be completely impermeable. If it were, necessary exchanges of material between the cell and its environment could not take place. The barrier must be insoluble in water so that it will not be dissolved by the aqueous medium of the cell. At the same time, it must be readily permeable to water because water is the basic solvent system of the cell and must be able to flow into and out of the cell as needed.

As you might expect, the membranes that surround cells and organelles satisfy these criteria admirably. A cellular **membrane** is essentially a hydrophobic permeability barrier that consists of *phospholipids, glycolipids,* and *membrane proteins.* In most organisms other than bacteria, the membranes also contain *sterols—cholesterol* in the case of animal cells, *ergosterols* in fungi, and *phytosterols* in the membranes of plant cells. (Don't be concerned if you haven't encountered these kinds of molecules before; we'll meet them all again in Chapter 3.)

Most membrane lipids and proteins are not simply hydrophobic or hydrophilic. They typically have both hydrophilic and hydrophobic regions and are therefore referred to as **amphipathic molecules** (the Greek prefix *amphi–* means "of both kinds" and pathic means "to feel"). The amphipathic nature of membrane phospholipids is illustrated in **Figure 2-11** on page 32, which shows the structure of *phosphatidylethanolamine,* a prominent phospholipid in many kinds of membranes. The distinguishing feature of amphipathic phospholipids is that each molecule consists of a polar *head* and two nonpolar hydrocarbon *tails.* The polarity of the hydrophilic head is due to the presence of a negatively charged phosphate group that is often linked to a positively charged group—an amino group, in the case of phosphatidylethanolamine and most other phosphoglycerides. (We will learn more about phospholipids in Chapter 3.)

Soap is a familiar amphipathic molecule that you all have likely used to dissolve grease, oil, and other nonpolar substances. The nonpolar hydrocarbon tails of the soap molecules interact with and surround the oil or grease, and the polar heads interact with water, enabling the oil or grease to be washed away. In the lab, we often use the amphipathic detergent sodium dodecyl sulfate (SDS) to isolate insoluble proteins and lipids. SDS has a negatively charged sulfate group attached to a single hydrocarbon chain of 12 carbons and acts much like the soap described above to solubilize nonpolar and amphipathic molecules.

A Membrane Is a Lipid Bilayer with Proteins Embedded in It

When exposed to an aqueous environment, amphipathic molecules undergo hydrophobic interactions. In a membrane, for example, phospholipids are organized into two layers: their polar heads face outward toward the aqueous environment on both sides, and their hydrophobic tails are hidden from the water by interacting with the tails of other molecules oriented in the opposite direction. The resulting structure is the

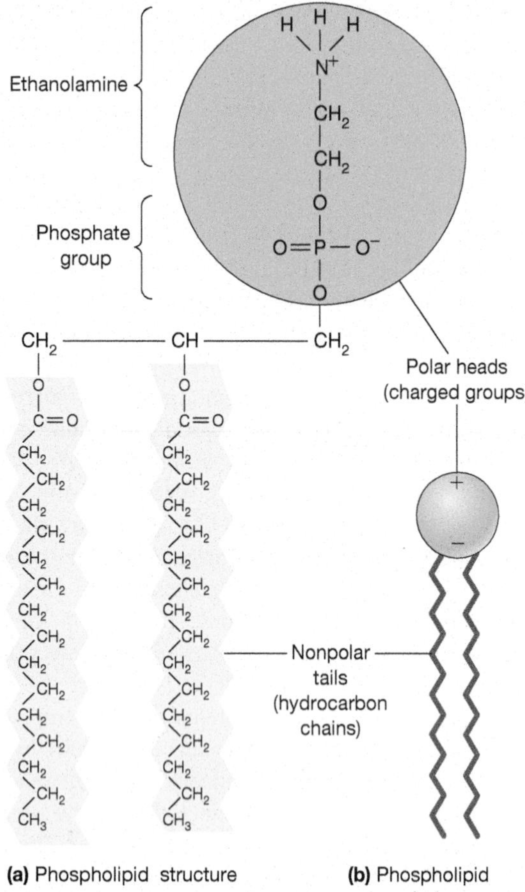

Figure 2-11 The Amphipathic Nature of Membrane Phospholipids. (a) A phospholipid molecule consists of two long nonpolar tails (yellow) and a polar head (orange). The polarity of the head of a phospholipid molecule results from a negatively charged phosphate group linked to a positively charged group—an amino group, in the case of the phosphatidylethanolamine shown here. **(b)** A phospholipid molecule is often represented schematically by a circle for the charged polar head and two zigzag lines for the nonpolar hydrocarbon chains.

lipid bilayer, shown in **Figure 2-12**. The heads of both layers face outward, and the hydrocarbon tails extend inward, forming the continuous hydrophobic interior of the membrane.

Every known biological membrane has such a lipid bilayer as its basic structure. Each of the lipid layers is typically 3–4 nm thick, so the bilayer has a width of ~7–8 nm.

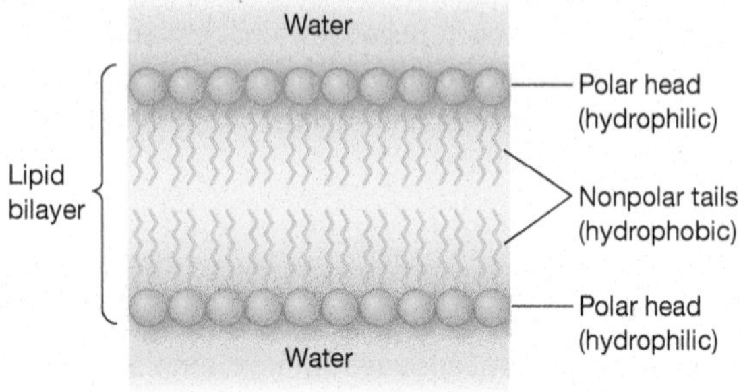

Figure 2-12 The Lipid Bilayer as the Basis of Membrane Structure. Phospholipids in an aqueous environment orient themselves in a double layer, with the hydrophobic tails buried on the inside and the hydrophilic heads interacting with the aqueous environment on either side of the membrane.

Lipid Bilayers Are Selectively Permeable

Because of its hydrophobic interior, a lipid bilayer is readily permeable to nonpolar molecules. However, it is quite impermeable to most polar molecules and is highly impermeable to all ions (**Figure 2-13**). Because most cellular constituents are either polar or charged, they have little or no affinity for the membrane interior and are effectively prevented from entering or escaping from the cell. Small, uncharged molecules are an exception, however. Compounds with molecular weights below about 100 readily diffuse across membranes, meaning they can freely and spontaneously pass through membranes, regardless of whether they are nonpolar (O_2 and CO_2) or polar (water and ethanol). Water is an especially important example of a very small molecule that, though polar, diffuses rapidly across membranes and can readily enter or leave cells.

Large, uncharged polar molecules, such as glucose and sucrose, can diffuse across the membrane, but to a lesser extent than small molecules. In contrast, even the smallest ions are effectively excluded from the hydrophobic interior of the membrane. For example, a lipid bilayer is at least 10^8 times less permeable to small cations such as Na^+ or K^+ than to water. This striking difference is due to both the charge on an ion and the sphere of hydration surrounding the ion.

Of course, it is essential that cells have ways of transferring not only ions such as $Na^+$and K^+ but also a wide variety of polar molecules across membranes that are not otherwise permeable to these substances. To transport these substances into and out of the cell, biological membranes are equipped with a wide variety of transport proteins (which we will discuss in detail in Chapter 8). A transport protein is a specialized transmembrane protein that serves either as a *hydrophilic channel* through an otherwise hydrophobic membrane or as a *carrier* that binds a specific solute on one side of the membrane and then undergoes a change to its **conformation**, or three-dimensional shape, in order to move the solute across the membrane.

Whether a channel or a carrier, each transport protein is specific for a particular molecule or ion (or, in some cases, for a class of closely related molecules or ions). Moreover, the

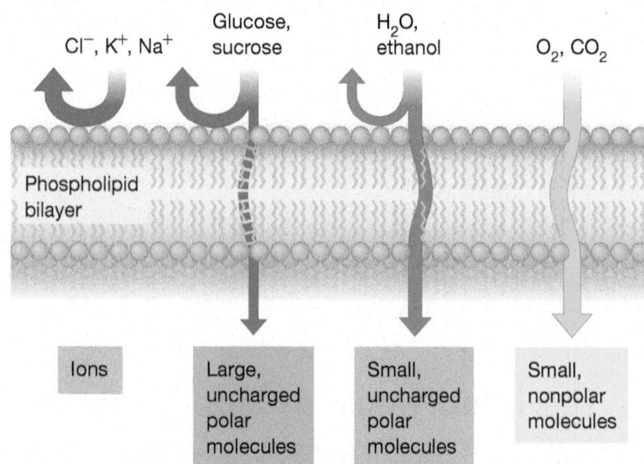

Figure 2-13 Permeability of Membranes to Various Classes of Solutes. The relative thicknesses of the arrows show the proportion of each substance that can freely cross the membrane (wavy arrows) or is repelled by the membrane (curved arrows.)

activities of these proteins can be carefully regulated to meet cellular needs. As a result, biological membranes are best described as *selectively permeable*. Except for very small molecules and nonpolar molecules, molecules and ions can move across a particular membrane only if the membrane contains an appropriate transport protein.

CONCEPT CHECK 2.3

Why is the amphipathic nature of membrane phospholipids necessary for a membrane to function as a barrier to polar molecules and ions? How do polar molecules and ions pass through membranes?

2.4 The Importance of Synthesis by Polymerization

For the most part, cellular structures such as ribosomes, chromosomes, flagella, and cell walls are made up of ordered arrays of linear *polymers* that are called **macromolecules.** Important macromolecules in cells include *proteins*, *nucleic acids* (both DNA and RNA), and *polysaccharides* such as starch, glycogen, and cellulose. *Lipids* are often regarded as macromolecules as well; however, they differ somewhat from the other classes of macromolecules in the way they are synthesized. Macromolecules are important in both the structure and the function of cells. To understand the biochemical basis of cell biology, therefore, really means to understand macromolecules—how they are made, how they are assembled, and how they function.

Macromolecules Are Critical for Cellular Form and Function

The importance of macromolecules in cell biology is emphasized by the cellular hierarchy shown in **Figure 2-14** on page 34. We use the term *hierarchy* here to denote how biological molecules and structures can be organized into a series of levels, each level building on the preceding one. In each level, note that the chemical properties of the constituent atoms and functional groups determine the types of bonds formed, the characteristics of the component molecules, and the interactions of different molecules and structures.

Most cellular structures are composed of small, water-soluble *organic molecules* (level 1) that cells either obtain from other cells or synthesize from simple nonbiological molecules such as carbon dioxide, ammonia, or phosphate ions. Small organic molecules known as *monomers* polymerize to form *biological macromolecules* (level 2) such as polysaccharides, proteins, or nucleic acids. These macromolecules may function on their own, or they can then be assembled into a variety of *supramolecular structures* (level 3). These supramolecular structures are themselves components of *organelles* (level 4) that make up the *cell* itself (level 5).

From such examples, a general principle emerges: *the macromolecules that are responsible for most of the form and order characteristic of living systems are generated by the polymerization of small organic molecules into long chains.* This strategy of forming large molecules by joining smaller units in a repetitive manner is illustrated in **Figure 2-15** on page 35. The importance of this strategy can hardly be overemphasized because it is a fundamental principle of cellular chemistry. The many enzymes responsible for catalyzing cellular reactions, the nucleic acids involved in storing and expressing genetic information, the glycogen stored by your liver, and the cellulose that gives rigidity to a plant cell wall are all variations on the same design theme. Each is a macromolecule made by linking together small repeating units, forming a polymer.

Examples of such repeating units, or **monomers,** are *monosaccharides* such as the glucose present in cellulose or starch, the 20 different *amino acids* needed to make proteins, and *nucleotides* (represented as A, C, G, T, and U) that make up DNA and RNA. In general, these are small, water-soluble organic molecules with molecular weights less than about 350. They can be transported across most biological membranes, provided the appropriate transport proteins are present in the membrane. By contrast, most macromolecules that are synthesized from these monomers are too large to traverse membranes and therefore must be made in the cell or compartment where they are needed.

MAKE CONNECTIONS 2.2

What form of microscopy is best suited for observing the fourth level of cellular hierarchical structure, the organelles? (Ch. 1.2)

Cells Contain Three Different Kinds of Macromolecular Polymers

The three major kinds of macromolecular **polymers** found in the cell are proteins, nucleic acids, and polysaccharides. (Lipids, though considered macromolecules, are not polymers consisting of a long chain of monomers.) For nucleic acids and proteins, the exact order of the different monomers is critical for function. In both nucleic acids and proteins, there are nearly limitless ways in which the different monomers can be arranged in a linear order. Thus, there are millions of different DNA, RNA, and protein sequences. In contrast, polysaccharides are typically composed of a single monomer or of two different monomers. These monomers occur in a repetitious pattern and, thus, there are relatively few different types of polysaccharides.

Nucleic acids (both DNA and RNA) are often called *informational macromolecules* because the order of the four kinds of nucleotide monomers in each is nonrandom and carries important information. A critical role of many DNA and RNA molecules is to serve a *coding* function, meaning that they contain the information necessary to specify the precise amino acid sequences of particular proteins. In this manner, the specific order of the monomeric units in nucleic acids stores and transmits the genetic information of the cell. As we will learn later, other DNA and RNA sequences have regulatory or enzymatic functions.

Proteins are a second type of macromolecule composed of a nonrandom series of monomers—in this case, amino acids. In contrast to nucleic acids, the monomer sequence does not transmit information but rather determines the three-dimensional structure of the protein, which—as we shall soon see—dictates its biological activity. Because there are 20 different amino acids found in proteins, the variety of possible protein sequences is nearly infinite. Thus, proteins have a wide range of functions in the cell, including roles in structure, defense, transport, catalysis, and signaling. Because the precise sequence of amino acids in a protein is so important,

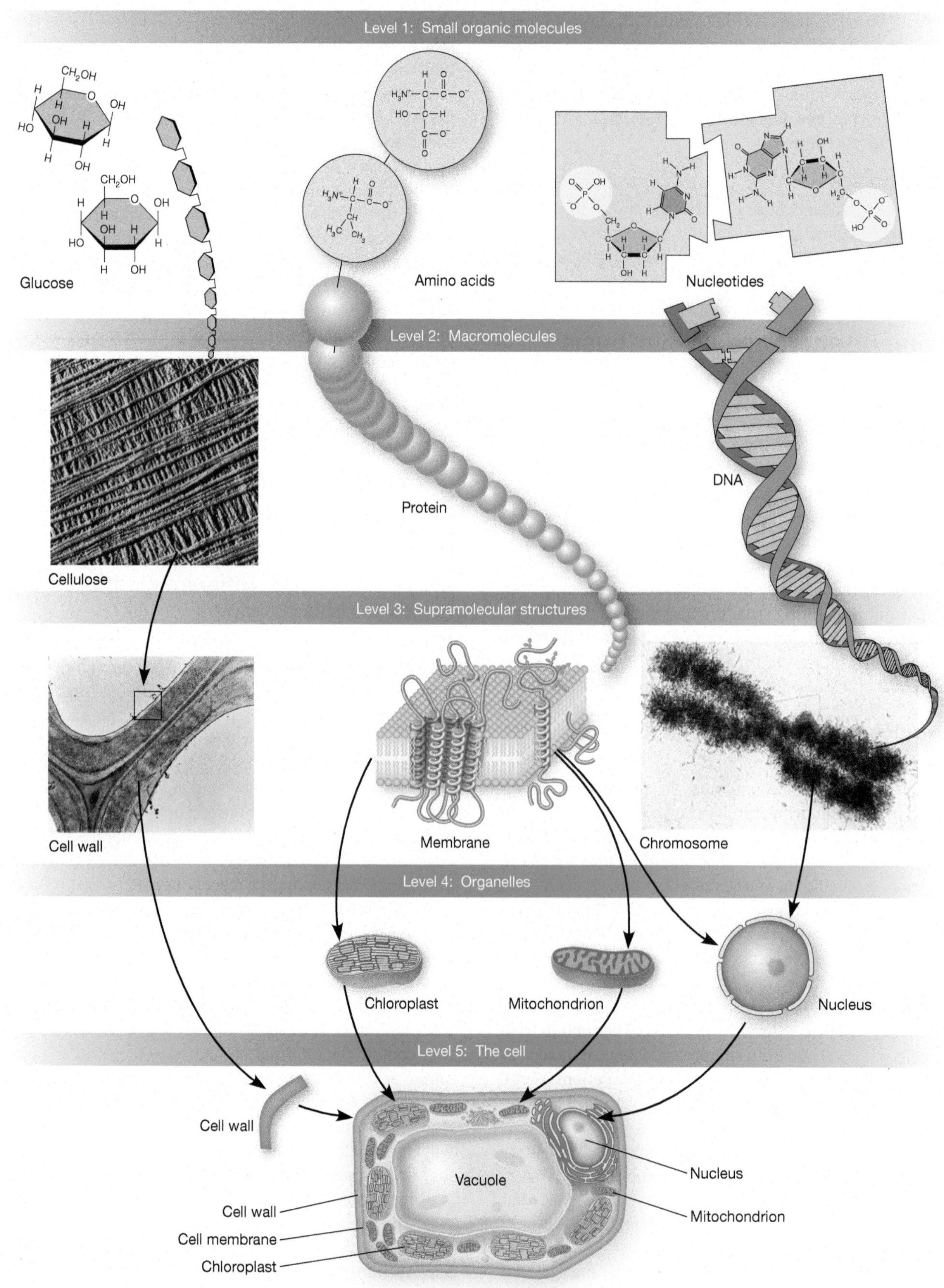

Figure 2-14 The Hierarchical Nature of Cellular Structures and Their Assembly. Small organic molecules known as monomers (level 1) are synthesized from simple inorganic substances and polymerize to form macromolecules (level 2). The macromolecules then assemble into the supramolecular structures (level 3) that make up organelles and other subcellular structures (level 4) and, ultimately, the cell itself (level 5).

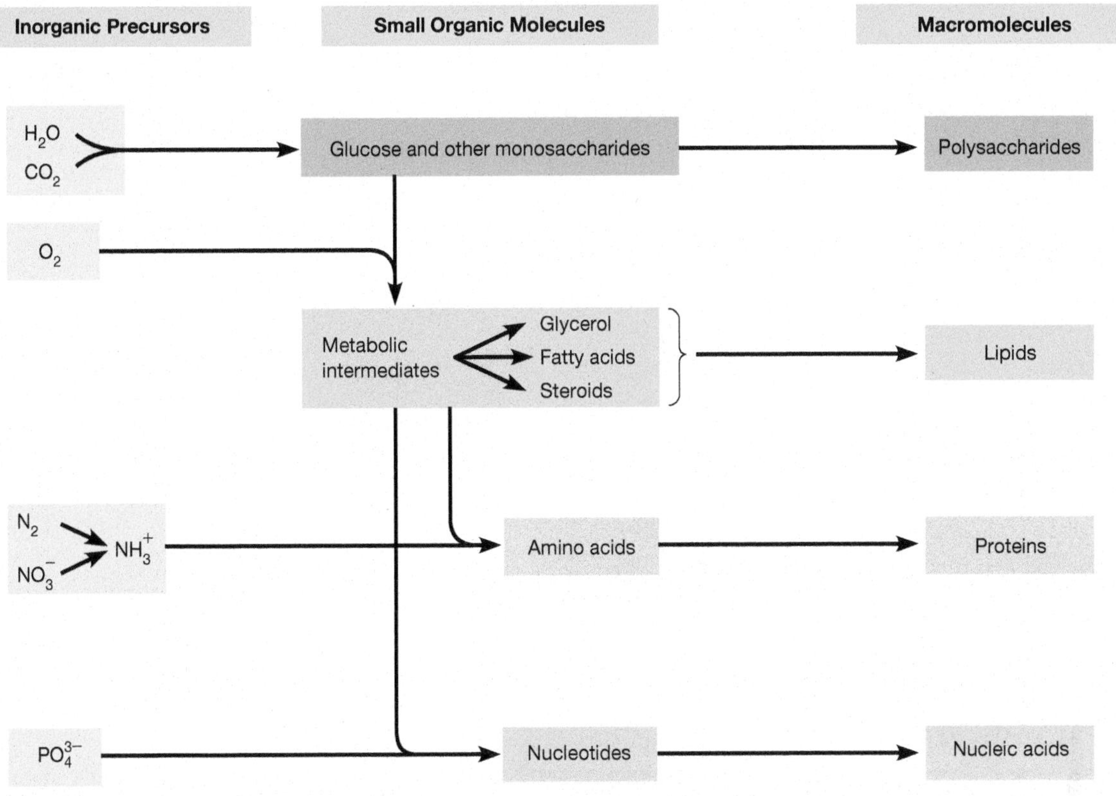

Figure 2-15 The Synthesis of Biological Macromolecules. Simple inorganic precursors (left) react to form small organic molecules (center), which are the monomers used in synthesizing the macromolecules (right) that make up most cellular structures.

variations in that sequence often negatively affect the ability of the protein to perform its function.

Polysaccharides, on the other hand, typically consist of either a single repeating subunit or two subunits occurring in strict alternation and usually do not carry specific information. However, in some cases, such as heparin's blood-thinning activity, the specific order of sugars is critical. Most polysaccharides are either *storage macromolecules*, which act as an energy source, or *structural macromolecules*, which provide physical support to cells. The most familiar storage polysaccharides are the *starch* of plant cells and the *glycogen* found in bacteria and animal cells. Both starch and glycogen consist of a single repeating glucose monomer, and both function in energy storage. Two examples of structural polysaccharides are the *cellulose* present in plant cell walls, cotton fibers, and woody tissues and the *chitin* in fungal cell walls, insect exoskeletons, and crab shells.

In future chapters, we will see how addition of short sugar chains called *oligosaccharides* (*oligo* meaning "a few") to proteins and lipids is common, forming the hybrid macromolecules known as *glycoproteins* and *glycolipids*, which have very important roles in the cell. The emerging field of *glycobiology* studies these important roles in cell-cell recognition, lipid signaling, protein folding, immune cell function, and blood type incompatibility interactions.

Macromolecules Are Synthesized by Stepwise Polymerization of Monomers

Before we look at each of the major kinds of biologically important macromolecules in detail (in Chapter 3), it will be useful to consider several important principles that underlie the polymerization processes by which all of these macromolecules arise. Although the chemistry of the monomeric units, and hence of the resulting polymers, differs markedly among macromolecules, the following basic principles apply in each case:

1. Macromolecules are always synthesized by the stepwise polymerization of similar or identical small molecules called *monomers.*
2. The addition of each monomer occurs with the removal of a water molecule and is therefore termed a **condensation reaction.**
3. The monomeric units that are to be joined together must be present as **activated monomers** before condensation can occur.
4. Activation usually involves coupling of the monomer to a **carrier molecule,** forming an activated monomer.
5. The energy needed to couple the monomer to the carrier molecule is provided by a molecule called *adenosine triphosphate (ATP)* or a related high-energy compound.
6. Because of the way they are synthesized, macromolecules have an inherent **directionality.** This means that the two ends of the polymer chain are chemically different from each other.

Because the elimination of a water molecule is essential in all biological polymerization reactions, each monomer must have both an available hydrogen (H) and an available hydroxyl group (–OH) on the molecule. This structural

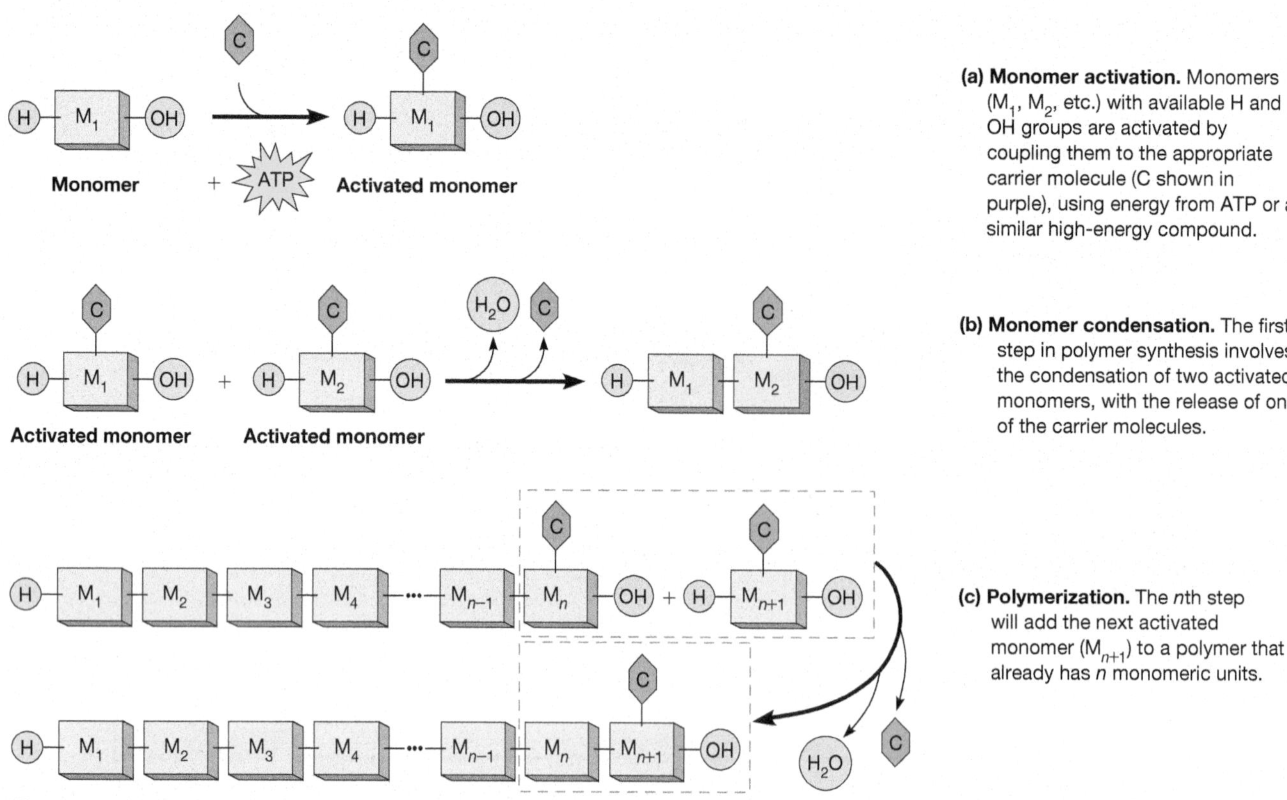

C
H
M1
OH
Monomer
+
ATP
Activated monomer
(a) Monomer activation. Monomers (M1, M2, etc.) with available H and OH groups are activated by coupling them to the appropriate carrier molecule (C shown in purple), using energy from ATP or a similar high-energy compound.
H2O
M2
Activated monomer
Activated monomer
(b) Monomer condensation. The first step in polymer synthesis involves the condensation of two activated monomers, with the release of one of the carrier molecules.
M3
M4
Mn−1
Mn
Mn+1
(c) Polymerization. The nth step will add the next activated monomer (Mn+1) to a polymer that already has n monomeric units.
Elongating polymer

Figure 2-16 Macromolecule Biosynthesis. Biological macromolecules are synthesized in a process involving **(a)** monomer activation, **(b)** condensation, and **(c)** polymerization. Depending on the polymer, the monomers may be identical (some polysaccharides) or may differ from one another (proteins and nucleic acids).

feature is depicted schematically in **Figure 2-16**; in which the monomeric units (M) are represented as boxes labeled with the relevant hydrogen and hydroxyl group. For a given kind of polymer, the monomers may differ from one another in other aspects of their structure; but each monomer possesses an available hydrogen and hydroxyl group.

Figure 2-16a shows monomer activation, an energy-requiring process. Regardless of the polymer type, the addition of each monomer to a growing chain is always energetically unfavorable unless the incoming monomer is in an activated, energized form. Activation usually, but not always, involves coupling of the monomer to a carrier molecule. The energy to drive this activation process is provided by ATP or a closely related *high-energy compound*—one that releases a large amount of energy upon hydrolysis (see Figure 9-1).

A different kind of carrier molecule is used for each kind of polymer. For protein synthesis, amino acids are activated by linking them to carriers called *transfer RNA* (or *tRNA*) molecules. Polysaccharides are synthesized from sugar molecules that are activated by linking them to derivatives of nucleotides. Starch synthesis uses ADP-glucose, and sucrose (a disaccharide) synthesis uses UDP-glucose. However, it is important to note that, for nucleic acid synthesis there is no need for specific carrier molecules because the nucleotides themselves (for example, ATP or GTP) are high-energy molecules.

Once activated, monomers are capable of reacting with each other in a condensation reaction that is followed or accompanied by the release of the carrier molecule from one of the two monomers (see Figure 2-16b). This is also known as dehydration synthesis because, as each monomer is added during polymer synthesis, one molecule of water is removed. Subsequent elongation of the polymer is a sequential, stepwise process. One activated monomeric unit is added at a time, thus lengthening the elongating polymer by one unit. Figure 2-16c illustrates the *n*th step in such a process, during which the next monomer unit is added to an elongating polymer that already contains *n* monomeric units. The chemical nature of the energized monomers, the carrier, and the actual activation process differ for each biological polymer, but the general principle is always the same: *polymer synthesis always involves activated monomers joined via condensation reactions, and ATP or a similar high-energy compound is required to activate the monomers by linking them to an appropriate carrier molecule.*

The *degradation*, or breakdown, of macromolecules occurs in a manner opposite to the condensation reactions used for synthesis. During degradation, monomers are removed from the polymer by **hydrolysis**—the addition of a water molecule that breaks the bond between adjacent monomers. In the process, one monomer receives a hydroxyl group from the water molecule and the other receives a hydrogen atom, re-forming the original monomers.

CONCEPT CHECK 2.4

What characteristics of proteins, nucleic acids, and carbohydrates are similar? Which are different?

2.5 The Importance of Self-Assembly

So far, we have seen that the macromolecules that characterize biological organization and function are polymers of small, hydrophilic, organic molecules. The only requirements for polymerization are an adequate supply of the monomeric subunits and a source of energy—and, in the case of proteins and nucleic acids, sufficient information to specify the order of the monomers. But how are these macromolecules organized into the supramolecular assemblies and organelles that are readily recognizable as cellular structures like those shown in Figure 2-14?

Crucial to our understanding of these higher-level structures is the principle of molecular **self-assembly**, which asserts that the information required to specify the spontaneous folding of macromolecules and the interactions of these macromolecules to form more complex structures is inherent in the polymers themselves. Once macromolecules are synthesized in the cell, their assembly into more complex structures will occur spontaneously without further input of energy or information. As we will see shortly, proteins called *molecular chaperones* are sometimes needed during protein folding to prevent incorrect molecular interactions that would give rise to inactive structures. But the chaperone molecules do not provide additional information; they simply assist the assembly process.

Noncovalent Bonds and Interactions Are Important in the Folding of Macromolecules

Polypeptides can fold and self-assemble without the input of further information, and the three-dimensional structure of a protein or protein complex is remarkably stable once it has been attained. To understand the folding and self-assembly of proteins (and, as it turns out, of other biological molecules and structures as well), we need to consider the covalent as well as noncovalent bonds that hold polypeptides and other macromolecules together.

Covalent bonds are easy to understand. Every protein or other macromolecule in the cell is held together by strong covalent bonds such as those discussed earlier in the chapter. But, as important as covalent bonds are in cellular chemistry, the complexity of molecular structure cannot be described by patterns of covalent bonding alone. Many cellular structures are held together by **noncovalent bonds and interactions**—much weaker forces that occur within and among macromolecules.

The most important noncovalent bonds and interactions in biological macromolecules are *hydrogen bonds, ionic bonds, van der Waals interactions,* and *hydrophobic interactions.* Each of these interactions is introduced here and discussed later in more detail (see Figure 3-5). As we've already noted, *hydrogen bonds* involve weak, attractive interactions between an electronegative atom such as oxygen (or nitrogen) and a hydrogen atom that is covalently bonded to a second electronegative atom. We will soon see that hydrogen bonds are extremely important in maintaining the three-dimensional structure of proteins and in holding together the two strands of the DNA double helix.

Ionic bonds are noncovalent electrostatic interactions between two oppositely charged ions. Typically, in the case of macromolecules, ionic bonds form between positively charged and negatively charged functional groups such as amino groups, carboxyl groups, and phosphate groups. Ionic bonds between functional groups of amino acids in the same protein molecule play a major role in determining and maintaining the structure of the protein. They are also important in the binding of different macromolecules to each other as, for example, in the binding of positively charged proteins to negatively charged DNA molecules.

Van der Waals interactions (or *forces*) are weak attractive interactions between two atoms that occur only if the atoms are very close to one another and are oriented appropriately. However, when atoms or groups of atoms get too close together, they begin to repel each other because of their overlapping outer electron orbitals. The *van der Waals radius* of a specific atom specifies the "personal space" around it that limits how close other atoms can come. This radius is typically about 0.12 to 0.19 nm. Van der Waals radii provide the basis for *space-filling models* of biological macromolecules (such as that for the DNA molecule shown in Figure 3-19b).

The term *hydrophobic interactions* describes the tendency of nonpolar groups within a macromolecule to associate with each other as they minimize their contact with surrounding water molecules and with any hydrophilic groups in the same or another macromolecule. Hydrophobic interactions among nonpolar groups are common in proteins and will cause nonpolar groups to be found in the interior of a protein molecule or embedded in the nonpolar interior of a membrane.

Many Proteins Spontaneously Fold into Their Biologically Functional State

To understand self-assembly processes, let's consider the coiling and folding necessary to form a functional three-dimensional protein from one or more linear chains of amino acids. Although the distinction is not always properly made, the immediate product of amino acid polymerization is actually not a protein but a **polypeptide.** To become a functional protein, one or more such linear polypeptide chains must coil and fold in a precise, predetermined manner to assume the unique three-dimensional structure, or **conformation,** necessary for biological activity.

The evidence for self-assembly of proteins comes largely from studies in which the *native,* or natural, conformation of a protein is disrupted by changing the environmental conditions. Such disruption, or unfolding, can be achieved by raising the temperature, by making the pH of the solution highly acidic or highly alkaline, or by adding certain chemical agents such as urea or any of several alcohols. The unfolding of a polypeptide under such conditions is called **denaturation** because it results in the loss of the natural three-dimensional structure of the protein—and the loss of its function as well. In the case of an enzyme, denaturation results in the loss of catalytic activity.

When the denatured polypeptide is returned to conditions in which the native conformation is stable, the polypeptide may undergo **renaturation,** the return to its correct three-dimensional conformation. In at least some cases, the renatured protein also regains its biological function—catalytic activity, in the case of an enzyme.

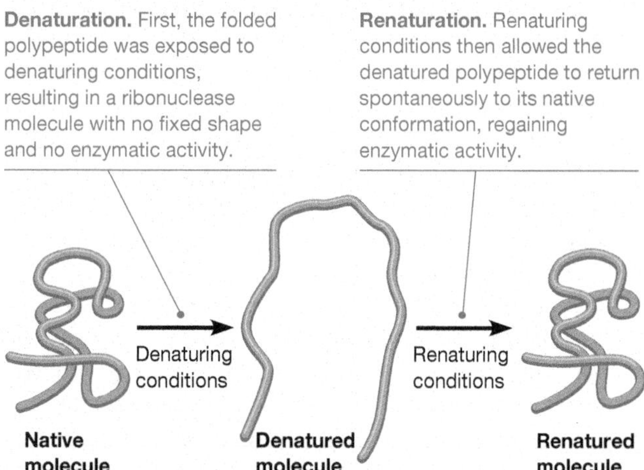

Denaturation. First, the folded polypeptide was exposed to denaturing conditions, resulting in a ribonuclease molecule with no fixed shape and no enzymatic activity.
Renaturation. Renaturing conditions then allowed the denatured polypeptide to return spontaneously to its native conformation, regaining enzymatic activity.
Denaturing conditions
Renaturing conditions
Native molecule
Denatured molecule
Renatured molecule

Figure 2-17 The Spontaneity of Polypeptide Folding. The polypeptide depicted here is ribonuclease, the enzyme that was used in Anfinsen's in vitro protein-folding experiments. These experiments showed that all the information needed for the proper three-dimensional folding of this polypeptide is present in its amino acid sequence.

Figure 2-17 depicts the denaturation and subsequent renaturation of the protein *ribonuclease,* an enzyme used by Christian Anfinsen and his colleagues in their classic studies of protein folding. Denaturation of a solution of ribonuclease results in an unfolded, randomly coiled polypeptide with no fixed shape and no catalytic activity. If the solution of the denatured molecules is then allowed to cool slowly, the ribonuclease molecules regain their original conformation and again have catalytic activity. Thus, all of the information necessary to specify the three-dimensional structure of the ribonuclease molecule is inherent in its amino acid sequence.

Molecular Chaperones Assist the Assembly of Some Proteins

Based on the ability of some denatured proteins to return to their original configuration and to regain biological function, biologists initially assumed that proteins and protein-containing structures fold and self-assemble in cells as well. However, this model for self-assembly in vivo (in the cell) is based entirely on studies with isolated proteins, and, even under laboratory conditions, not all isolated proteins regain their native structure. John Ellis and his colleagues at Warwick University concluded that the self-assembly model may not be adequate for many proteins. They proposed that the interactions that drive protein folding may need to be assisted and controlled by proteins known as **molecular chaperones** to reduce the probability of the formation of incorrect structures having no biological activity.

Molecular chaperones are proteins that facilitate the correct folding of proteins and assembly of protein-containing structures but are not themselves components of the assembled structures. (Their role in protein folding is discussed in much more detail in Chapter 19; see Figure 19-20.) The molecular chaperones that have been identified to date do not convey information either for polypeptide folding or for the assembly of multiple polypeptides into a single protein. Instead, they function by binding to specific regions that are exposed only in the early stages of assembly, thereby inhibiting unproductive assembly pathways that would lead to incorrect structures. Commenting on the term *molecular chaperone,* Ellis and colleagues observed that "the term chaperone is appropriate for this family of proteins because the role of the human chaperone is to prevent incorrect interactions between people, not to provide steric information for those interactions."*

Self-Assembly Also Occurs in Other Cellular Structures

The same principle of self-assembly that accounts for the folding and interactions of polypeptides also applies to more complex cellular structures. Many of the characteristic structures of the cell are complexes of two or more different kinds of polymers and involve interactions that are chemically distinct from those of polypeptide folding and association. However, the principle of self-assembly may nonetheless apply. For example, *ribosomes* contain both RNA and proteins and can be reconstituted from the individual molecular components by self-assembly. The phospholipids that make up *membranes* can self-assemble into a bilayer when mixed with water because the hydrophobic hydrocarbon tails do not interact with water but instead associate with each other. Despite the chemical differences among such polymers as proteins, nucleic acids, and polysaccharides, the noncovalent interactions that drive these supramolecular assembly processes are similar to those that dictate the folding of individual protein molecules.

A rare example of self-assembly in proteins occurs in diseases caused by **prions,** infectious protein molecules (which you will read more about in Chapter 4). The term *prion* was coined from the words *proteinaceous infectious particles.* When a prion protein (PrP) folds in an abnormal way, it serves as the "seed" for long fibrils of PrP that form through self-assembly. The exact mechanism by which these fibrils form is not well understood, but it must involve the conversion of normally folded PrP into the abnormally folded form. It is also likely that long fibrils fragment into pieces, providing seeds for the formation of more prion fibrils. This process of nucleation and fragmentation has been compared to the process whereby crystals form. A similar kind of self-assembly and fibril formation can be observed in Alzheimer's disease (see **Human Connections in Chapter 3, pages 48–49**).

The Tobacco Mosaic Virus Is a Case Study in Self-Assembly

Some of the most definitive findings concerning the self-assembly of complex biological structures have come from studies of viruses. A *virus* is a complex of proteins and nucleic acid, either DNA or RNA. A virus is not itself alive, but it can invade and infect a specific living host cell, take over the synthetic machinery of this cell, and use the host cell to produce more virus components. When the components—viral nucleic acid and viral proteins—are synthesized, these macromolecules spontaneously assemble into the mature virus particles.

An especially good example is tobacco mosaic virus (TMV), a plant RNA virus that molecular biologists have studied for many years. TMV is a rodlike particle about 18 nm in diameter

* Ellis, R. J., and S. M. Van der Vies. Molecular chaperones. *Annu. Rev. Biochem.* 60 (1991): 321.

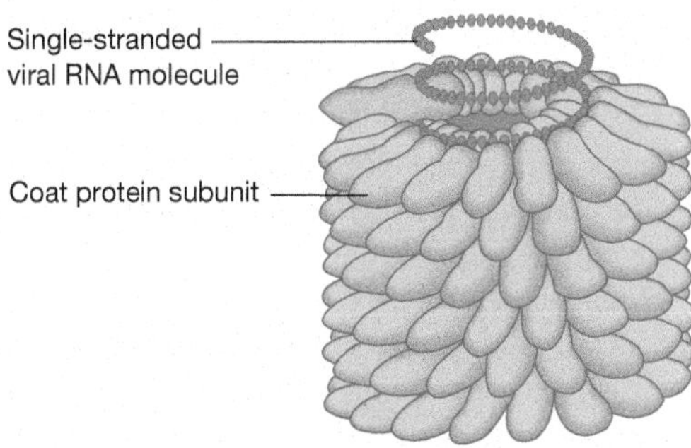

Figure 2-18 A Structural Model of a Portion of the Tobacco Mosaic Virus (TMV). A single-stranded RNA molecule is coiled into a helix, surrounded by a coat consisting of 2130 identical protein subunits.

and 300 nm in length. It consists of a single strand of RNA with 6395 nucleotides and about 2130 copies of a single kind of polypeptide—the *coat protein*, each containing 158 amino acids. The RNA molecule forms a helical core, with a cylinder of protein subunits clustered around it (**Figure 2-18**). Note that only a portion of the entire TMV virion is shown here, and several layers of protein have been omitted from the upper end of the structure to reveal the helical RNA molecule inside.

Heinz Fraenkel-Conrat and his colleagues carried out several important experiments that contributed significantly to our understanding of self-assembly. They separated TMV into its RNA and protein components and then allowed them to reassemble in vitro. Functional, infectious viral particles formed spontaneously, providing one of the first and most convincing demonstrations that the components of a complex biological structure can reassemble spontaneously without external information. Especially interesting was the finding that mixing the RNA from one strain of virus with the protein from another formed a hybrid virus that was also infectious.

The assembly process has since been studied in detail and is known to be surprisingly complex (**Figure 2-19**). The basic unit of assembly is a two-layered disk of 34 identical coat protein molecules that undergoes a conformational change to a helical shape as the protein molecules interact with a short segment (102 nucleotides) of the RNA molecule. This transition allows another disk to bind, and each successive disk undergoes the same conformational change from a cylinder to a helix as it binds another segment of RNA. The process continues until the end of the RNA molecule is reached, producing the mature virus particle, its RNA completely covered with coat protein.

Recently, scientists have begun exploiting the self-assembly ability of viral particles to produce custom-made synthetic molecular complexes for use in nanotechnology and biomedicine. Individual coat protein molecules can be modified or engineered to have specific properties, such as electrical conductivity, then allowed to self-assemble into predictable structures. These self-assembling structures are being tested for use in a variety of applications, including construction of nanometer-sized biosensors, drug delivery, and quantum computing.

Self-Assembly Has Limits

In many cases, the information required to specify the exact conformation of a cellular structure seems to lie entirely within the polymers that contribute to the structure. Such self-assembling systems achieve stable three-dimensional conformations without additional information input because the information content of the component polymers is adequate to specify the complete assembly process. Even in assisted self-assembly, the molecular chaperones that are involved provide no additional information.

However, some assembly systems appear to depend, additionally, on information supplied by a preexisting structure.

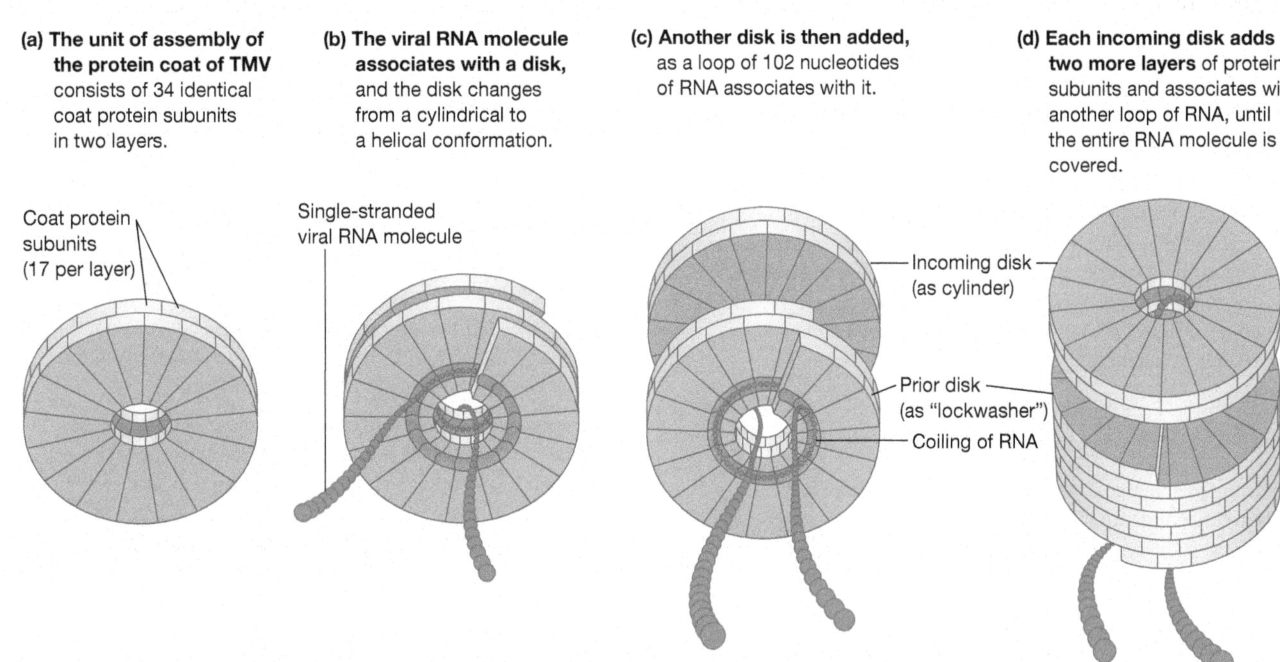

(a) The unit of assembly of the protein coat of TMV consists of 34 identical coat protein subunits in two layers.
(b) The viral RNA molecule associates with a disk, and the disk changes from a cylindrical to a helical conformation.
(c) Another disk is then added, as a loop of 102 nucleotides of RNA associates with it.
(d) Each incoming disk adds two more layers of protein subunits and associates with another loop of RNA, until the entire RNA molecule is covered.
Coat protein subunits (17 per layer)
Single-stranded viral RNA molecule
Incoming disk (as cylinder)
Prior disk (as "lockwasher")
Coiling of RNA

Figure 2-19 Spontaneous Self-Assembly of the Tobacco Mosaic Virus (TMV). Parts **(a)** through **(d)** show how the TMV virion will assemble spontaneously from a mixture of its coat protein and RNA constituents. All the information needed for proper assembly is contained in the proteins and RNA themselves, and no energy input is required.

In such cases, the ultimate structure arises not by assembling the components into a new structure but rather by ordering the components into the matrix of an existing structure. Examples of cellular structures that are routinely built up by adding new material to existing structures include cell membranes and cell walls.

Hierarchical Assembly Provides Advantages for the Cell

Each of the assembly processes we have examined exemplifies the basic cellular strategy of **hierarchical assembly** illustrated in Figure 2-14. Biological structures are almost always constructed hierarchically, with subassemblies acting as important intermediates en route from simple starting molecules to the end products of organelles, cells, and organisms. Consider how cellular structures are made. First, large numbers of similar, or even identical, monomeric subunits are assembled by condensation into polymers. These polymers then aggregate spontaneously but specifically into characteristic multimeric units. The multimeric units can, in turn, give rise to still more complex structures and eventually to assemblies that are recognizable as distinctive subcellular structures.

This hierarchical process has the double advantage of chemical simplicity and efficiency of assembly. To appreciate the chemical simplicity, we need only recognize that almost all structures found in cells and organisms are synthesized from about 30 small precursor molecules. These include the 20 typical amino acids that are found in proteins, the 5 aromatic bases present in nucleic acids, and a few sugar and lipid molecules. Given these building blocks and the polymers that can be derived from them through just a few different kinds of condensation reactions, most of the structural complexity of life can be readily elaborated by hierarchical assembly into successively more complex structures.

The second advantage of hierarchical assembly lies in the "quality control" that can be exerted at each level of assembly. This allows defective components to be discarded at an early stage rather than being built into a more complex structure that would be more costly to reject and replace. Thus, if the wrong subunit becomes inserted into a polymer at some critical point in the chain, that particular molecule may have to be discarded; but the cell will be spared the cost of synthesizing a more complicated supramolecular assembly or even an entire organelle before the defect is discovered. As we proceed through this course, we will often see how cells have exquisite means for ensuring the proper structure and function of cellular components. We will also see how cells can recognize and replace defective components.

CONCEPT CHECK 2.5

TMV is an RNA-containing virus. Mixing RNA from TMV strain A with coat protein from strain B results in infectious virus particles. How were these particles formed without outside information? Would these hybrid viruses infect the normal host of strain A or strain B?

Summary of Key Points

Mastering™ Biology For activities, animations, and review quizzes, go to the study area at www.masteringbiology.com.

2.1 The Importance of Carbon

- The carbon atom has several properties that make it uniquely suited as the basis for life. Each atom forms four stable covalent bonds and can participate in single, double, or triple bonds with one or more other carbon atoms. This creates a wide variety of linear, branched, and ring-containing compounds.
- Carbon also readily bonds to several other types of atoms commonly found in cellular compounds, including hydrogen, oxygen, nitrogen, phosphorus, and sulfur. These elements are frequently found in functional groups attached to carbon skeletons—groups such as the hydroxyl, sulfhydryl, carboxyl, amino, phosphate, carbonyl, and aldehyde groups.

2.2 The Importance of Water

- The unique chemical properties of water help make it the most abundant compound in cells. These properties include its cohesiveness, its high specific heat, its temperature-stabilizing capacity, and its ability to act as a solvent for most biological molecules.
- These chemical properties are a result of the polarity of the water molecule. Its polarity arises from the unequal sharing of electrons in the oxygen-hydrogen covalent bonds and its bent shape, causing the oxygen end of the atom to be slightly negative and the hydrogen atoms to be slightly positive.
- This polarity leads to extensive hydrogen bonding between water molecules, as well as between water and other polar molecules, rendering them soluble in aqueous solutions.

2.3 The Importance of Selectively Permeable Membranes

- All cells are surrounded by a defining cell membrane that is selectively permeable, controlling the flow of materials into and out of the cell.
- Biological membranes have a phospholipid bilayer structure with a hydrophobic interior that blocks the direct passage of large polar molecules, charged molecules, and ions.
- However, large polar and charged molecules and ions can cross membranes via membrane-spanning transport proteins that form hydrophilic channels to allow passage of specific molecules.

2.4 The Importance of Synthesis by Polymerization

- The molecular building blocks of the cell are small, organic molecules put together by stepwise polymerization to form the macromolecules so important to cellular structure and function.

Relatively few kinds of monomeric units comprise most of the polysaccharide, protein, and nucleic acid polymers in cells.

- Polymerization of monomers requires that they first be activated, usually at the expense of ATP. Monomers are then linked together by removal of water in condensation reactions. Polymers are degraded by the reverse reaction—hydrolysis.
- Because they are added to only one specific end of these macromolecules, all biological polymers have directionality.

2.5 The Importance of Self-Assembly

- Although energy input is required for polymerization of monomers into macromolecules, most macromolecules fold into their final three-dimensional conformations spontaneously. The information needed for this folding is inherent in the chemical nature of the monomeric units and the order in which the monomers are put together.
- For example, the unique three-dimensional conformation of a protein forms by spontaneous folding of the linear polypeptide chain, and this conformation depends only on the specific order of amino acids in the protein. The final structure is the result of several covalent and noncovalent interactions, including disulfide bonding, hydrogen bonding, ionic bonding, van der Waals interactions, and hydrophobic interactions.
- Individual polymers can interact with each other in a unique and predictable manner to generate successively more complex structures. This hierarchical assembly process has the dual advantages of chemical simplicity and efficiency of assembly. It also allows quality control at multiple steps to ensure proper production of cell components. Self-assembly of proteins into fibrils can also be pathogenic, such as in the case of prions.

Problem Set

Mastering™ Biology For activities, animations, and review quizzes, go to the study area at www.masteringbiology.com.

2-1 The Fitness of Carbon. Each of the properties that follow is characteristic of the carbon atom. In each case, indicate how the property contributes to the role of the carbon atom as the most important atom in biological molecules.

(a) The carbon atom has a valence of four.

(b) The carbon-carbon bond has a bond energy that is above the energy of photons of light in the visible range (380–750 nm).

(c) A carbon atom can bond simultaneously to two other carbon atoms.

(d) Carbon atoms can bond readily to hydrogen, nitrogen, and sulfur atoms.

(e) Carbon-containing compounds can contain asymmetric carbon atoms.

2-2 The Fitness of Water. For each of these statements about water, decide whether the statement is true and describes a property that makes water a desirable component of cells (T); is true but describes a property that has no bearing on water as a cellular constituent (X); or is false (F). For each true statement, indicate a possible benefit to living organisms.

(a) Water is a polar molecule and hence an excellent solvent for polar compounds.

(b) Water can be formed by the reduction of molecular oxygen (O_2).

(c) The density of water is less than the density of ice.

(d) The molecules of liquid water are extensively hydrogen-bonded to one another.

(e) Water does not absorb visible light.

(f) Water is odorless and tasteless.

(g) Water has a high specific heat.

(h) Water has a high heat of vaporization.

2-3 Wrong Again. For each of the following false statements, change the statement to make it true.

(a) Hydrogen, nitrogen, and carbon are the elements that most readily form strong multiple bonds.

(b) Water has a higher specific heat than most other liquids because of its low molecular weight.

(c) Oil droplets in water coalesce to form a separate phase because of the strong attraction of hydrophobic molecules for each other.

(d) In a hydrogen bond, electrons are shared between a hydrogen atom and two neighboring electronegative atoms.

(e) Biological membranes are called selectively permeable because large polar molecules and ions can never pass through them.

2-4 Solubility in Benzene. Consider the chemical structure of benzene that is shown in Figure 2-4. Which cellular molecules would you expect to be soluble in a container of benzene?

2-5 Drug Targeting. Your company has developed a new anticancer drug, but it will not cross through the membrane of target cells because it is large and contains many polar functional groups. Can you design a strategy to get this new drug into the cancer cells?

2-6 It's All About Membranes. Answer each of the following questions about biological membranes in 50 or fewer words.

(a) What is an amphipathic molecule? Why are amphipathic molecules important constituents of membranes?

(b) Why do membranes consist of lipid bilayers rather than lipid monolayers?

(c) What does it mean to say that a membrane is selectively permeable?

(d) If a lipid bilayer is at least 10^8 times less permeable to K^+ ions than it is to water, how do the K^+ ions that are needed for many cellular functions get into the cell?

(e) It is often possible to predict the number of times a membrane protein crosses the membrane by determining how many short sequences of hydrophobic amino acids are present in the amino acid sequence of the protein. Explain.

2-7 The Polarity of Water. Defend the assertion that all of life as we know it depends critically on the fact that the angle between the two hydrogen atoms in the water molecule is 104.5° and not 180°.

2-8 QUANTITATIVE Bond Energies. A single covalent bond has a bond energy of approximately 90 kcal/mol, and a typical hydrogen bond has a bond energy of about 5 kcal/mol. Although weak, hydrogen bonds can be a major structural force when present in large numbers as in DNA. In double-stranded DNA, each AT base pair is held together by two hydrogen bonds, and each GC base pair is held together by three hydrogen bonds.

(a) What is the total bond energy in a propane molecule (see Figure 2-4)? The C—C bond energy is 83 kcal/mol, and the C—C bond energy is 99 kcal/mol.

(b) In a short 15 base-pair molecule of DNA having 60% GC pairs and 40% AT pairs, what is the total bond energy of all the hydrogen bonds? How does this compare to the bond energy of a carbon-carbon bond?

(c) In a typical gene consisting of 1000 base pairs with the same relative GC versus AT content, what is the total bond energy of all the hydrogen bonds? How does this compare to the bond energy of a carbon-carbon bond?

2-9 TMV Assembly. Each of these statements is an experimental observation concerning the reassembly of tobacco mosaic virus (TMV) virions from TMV RNA and coat protein subunits. In each case, carefully state a reasonable conclusion that can be drawn from the experimental finding.

(a) When RNA from a specific strain of TMV is mixed with coat protein from the same strain, infectious virions are formed.

(b) When RNA from strain A of TMV is mixed with coat protein from strain B, the reassembled virions are infectious, giving rise to strain A virus particles in the infected tobacco cells.

(c) Isolated coat protein monomers can polymerize into a viruslike helix in the absence of RNA.

(d) In infected plant cells, the TMV virions that form contain only TMV RNA and never any of the various kinds of cellular RNAs present in the host cell.

(e) Regardless of the ratio of RNA to coat protein in the starting mixture, the reassembled virions always contain RNA and coat protein in the ratio of three nucleotides of RNA per coat protein monomer.

2-10 Mars Is Alive? Imagine this futuristic scenario: life has been discovered on Mars and shown to contain a new type of macromolecule. You have been hired to study this new compound and want to determine whether it is a structural or an informational macromolecule. What features would you look for?

2-11 DATA ANALYSIS Infectious Self-Assembly. Table 2.1 is adapted from some of the early work done by Heinz Fraenkel-Conrat and colleagues demonstrating the spontaneous self-assembly of the tobacco mosaic virus (TMV). They mixed together TMV protein and RNA to test whether infectious TMV particles could spontaneously form. The test was quantified by applying the reaction mixtures on healthy leaves and counting the number of lesions formed per half-leaf. A higher number of lesions represents a larger amount of infectious virus. Use the data in the table to answer the following questions.

(a) Did infectious TMV particles likely form in this experiment? Why or why not?

(b) Is RNA required for the production of infectious TMV particles? Why or why not?

(c) Does the source of the RNA in the reaction mixture affect the production of infectious TMV particles? Why or why not?

Table 2.1 Effect of Reaction Conditions on Regeneration of Virus Activity

Reaction Conditions	Lesions/Half-Leaf
Protein + TMV-RNA (1 minute)	0.6
Protein + TMV-RNA (24 hours)	10.2
Protein + TMV-RNA (96 hours)	13.0
Protein + RNase-digested TMV-RNA	0.0
Protein + TYMV*-RNA	0.3
Protein alone	0.0
TMV-RNA alone	0.0

* Turnip yellow mosaic virus.

3 The Macromolecules of the Cell

Crystals Grown in Outer Space. These colorful protein crystals can be used in X-ray crystallography to determine the three-dimensional structure of individual protein molecules.

We have already looked at some of the basic chemical principles of cellular organization (Chapter 2). We saw that each of the major kinds of biological macromolecules—proteins, nucleic acids, and polysaccharides—consists of a relatively small number (from 1 to 20) of repeating monomeric units. These polymers are synthesized by condensation reactions in which activated monomers are linked together by the removal of water. Once synthesized, the individual polymer molecules fold and coil spontaneously into stable, three-dimensional shapes. These folded molecules then associate with one another in a hierarchical manner to generate higher levels of structural complexity, usually without further input of energy or information. We are now ready to examine the major kinds of biological macromolecules in greater detail. In each case, we will focus first on the chemical nature of the monomeric components and then on the synthesis and properties of the polymer itself. As we will see shortly, most biological macromolecules in cells are synthesized from about 30 common small molecules. Our survey begins with proteins because they play such important and widespread roles in cellular structure and function. We then move on to nucleic acids and polysaccharides. The tour concludes with lipids, which do not quite fit the definition of a polymer but are important cellular components whose synthesis resembles that of true polymers.

Mastering™ Biology www.masteringbiology.com

3.1 Proteins

Proteins are a class of extremely important and ubiquitous macromolecules in all organisms, occurring nearly everywhere in the cell. In fact, their importance is implied by their name, which comes from the Greek word *proteios*, meaning "first place." Whether we are talking about conversion of carbon dioxide to sugar in photosynthesis, oxygen transport in the blood, the regulation of gene expression by transcription factors, cell-to-cell communication, or the motility of a flagellated bacterium, we are dealing with the processes that depend crucially on particular proteins with specific properties and functions.

Proteins can be classified in different ways, including either functionally or structurally. Based on function, proteins fall into nine major classes. Many proteins are *enzymes*, serving as catalysts that greatly increase the rates of the thousands of chemical reactions on which life depends. *Structural proteins*, on the other hand, provide physical support and shape to cells and organelles, giving them their characteristic appearances. *Motility proteins* play key roles in the contraction and movement of cells and intracellular materials. *Regulatory proteins* are responsible for control and coordination of cellular functions, ensuring that cellular activities are regulated to meet cellular needs. *Transport proteins* are involved in the movement of other substances into, out of, and within the cell. *Signaling proteins* mediate communication between cells in an organism, and *receptor proteins* enable cells to respond to chemical stimuli from their environment. Finally, *defensive proteins* provide protection against disease, and *storage proteins* serve as reservoirs of amino acids.

Proteins can also be classified based on their structure. When we examine the various levels of protein structure later in this chapter, you will see that the two main structural classes of proteins are *globular* and *fibrous*. Generally, globular proteins are either involved in cellular structure or are enzymatic, whereas fibrous proteins tend to be purely structural in nature within the cell.

Because virtually everything that a cell is or does depends on the proteins it contains, it is clearly important that we understand what proteins are and why they have the properties they do. We begin our discussion by looking at the amino acids present in the proteins, and then we will consider some properties of the proteins themselves.

The Monomers Are Amino Acids

Proteins are linear polymers of **amino acids.** Although more than 60 different kinds of amino acids are typically present in a cell, only 20 kinds are used in protein synthesis, as indicated in **Table 3-1**. Some proteins contain more than 20 different kinds of amino acids, but the additional ones usually are a result of modifications that occur after the protein has been synthesized. Although most proteins contain all or most of the 20 amino acids, the proportions vary greatly, and no two different proteins have the same amino acid sequence.

Every amino acid has the basic structure illustrated in **Figure 3-1**, with a carboxyl group, an amino group, a hydrogen atom, and a side chain known as an **R group,** all attached to a central carbon atom known as the *α carbon.* The R group is different for each amino acid and gives each amino acid its distinctive properties. Except for glycine, for which the R group is just a hydrogen atom, all amino acids have four different groups attached to the α carbon. This means that all amino acids except glycine exist in two stereoisomeric forms that are mirror images of each other and cannot be superimposed. These two mirror-image forms are called D- and L-amino acids. Both forms exist in nature, but only L-amino acids occur in proteins. The reason for this bias is not known but remains an interesting evolutionary question. What is clear is that cellular enzymes have evolved to use L-amino acids. Whether this was due to the relative abundance of L- compared with D-amino acids in the prebiotic environment or some other reason is not well understood. Note that, as the figure shows, both the carboxyl and the amino groups are ionized at typical cellular pH.

Because the carboxyl and amino groups shown in Figure 3-1 are common features of all amino acids, the

Table 3-1 Common Small Molecules in Cells

Kind of Molecules	Number Present	Names of Molecules	Role in Cell	Figure Number for Structures
Amino acids	20	See list in Table 3-2	Monomeric units of all proteins	3-2
Aromatic bases	5	Adenine	Components of nucleic acids	3-15
		Cytosine		3-15
		Guanine		3-15
		Thymine		3-15
		Uracil		3-15
Sugars	Varies	Ribose	Component of RNA	3-15
		Deoxyribose	Component of DNA	3-15
		Glucose	Energy metabolism; component of starch and glycogen	3-21
Lipids	Varies	Fatty acids	Components of phospholipids and membranes	3-27a
		Cholesterol		3-27e

Source: Adapted from Wald, G. The origins of life. *Proc. Natl. Acad. Sci. USA* 52 (1994): 595.

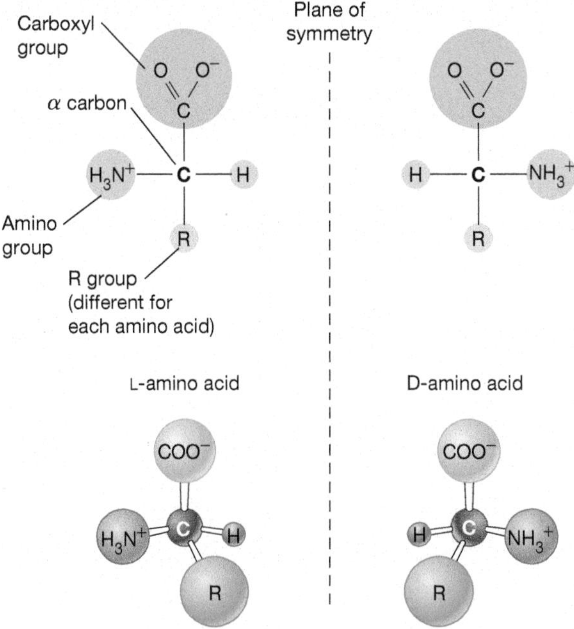

Figure 3-1 The Structure and Stereochemistry of an Amino Acid. Most amino acids can exist in two isomeric forms, designated L and D and shown here as (top) conventional structural formulas and (bottom) ball-and-stick models.

specific properties of the various amino acids differ depending on the chemical nature of their R groups, which range from a single hydrogen atom in glycine to relatively complex aromatic (ring-containing) groups as found in tryptophan and tyrosine. **Figure 3-2** on page 46 shows the structures of the 20 L-amino acids found in proteins, and **Table 3-2** lists the three-letter and one-letter abbreviations for the amino acids that have been standardized for use by the scientific community.

Nine of these amino acids have nonpolar, *hydrophobic* R groups (Group A). As you look at their structures, you will notice the hydrocarbon nature of the R groups, with few or no oxygen and nitrogen atoms. These hydrophobic amino acids are usually found in the interior of proteins when the proteins are in an aqueous environment. If a protein is destined for a cellular membrane, it will have a preponderance of hydrophobic amino acids in its membrane-spanning domains.

The remaining 11 amino acids have *hydrophilic* R groups that are either distinctly polar (Group B) or actually charged at the neutral pH values characteristic of cells (Group C). Notice that the two acidic amino acids are negatively charged and that the three basic amino acids are positively charged. Hydrophilic amino acids tend to occur on the surface of proteins in solution, thereby maximizing their interactions with water molecules and other polar or charged substances in this environment.

MAKE CONNECTIONS 3.1

What type of bond would be most likely to form between the polar, uncharged amino acids and water molecules? (Ch. 2.2)

Table 3-2 Abbreviations for Amino Acids

Amino Acid	Three-Letter Abbreviation	One-Letter Abbreviation
Alanine	Ala	A
Arginine	Arg	R
Asparagine	Asn	N
Aspartate	Asp	D
Cysteine	Cys	C
Glutamate	Glu	E
Glutamine	Gln	Q
Glycine	Gly	G
Histidine	His	H
Isoleucine	Ile	I
Leucine	Leu	L
Lysine	Lys	K
Methionine	Met	M
Phenylalanine	Phe	F
Proline	Pro	P
Serine	Ser	S
Threonine	Thr	T
Tryptophan	Trp	W
Tyrosine	Tyr	Y
Valine	Val	V

The Polymers Are Polypeptides and Proteins

The process of stringing individual amino acids together into a linear polymer involves the stepwise addition of each new amino acid to the growing chain by a *condensation* (or *dehydration*) *reaction* (see Figure 2-16). As the three atoms comprising H_2O are removed, the carboxyl carbon of one amino acid and the amino nitrogen of a second are linked directly by a covalent bond. This C—N bond linking two amino acids is known as a **peptide bond,** shown below in bold:

$$H_3N^+\text{—}\underset{R_1}{CH}\text{—}\overset{O}{\overset{\|}{C}}\text{—}O^- \; + \; H_3N^+\text{—}\underset{R_2}{CH}\text{—}\overset{O}{\overset{\|}{C}}\text{—}O^- \longrightarrow$$

Amino acid 1   Amino acid 2

$$H_3N^+\text{—}\underset{R_1}{CH}\text{—}\overset{O}{\overset{\|}{\mathbf{C}}}\mathbf{—}\underset{H}{\mathbf{N}}\text{—}\overset{R_2}{CH}\text{—}\underset{O}{\underset{\|}{C}}\text{—}O^- \; + \; H_2O$$

Polypeptide

As each new peptide bond is formed by condensation, the growing chain of amino acids is lengthened by one amino acid. Peptide bond formation is illustrated schematically in **Figure 3-3** on page 47 using ball-and-stick models of the amino acids glycine and alanine. Peptide bonds have a partial double-bond character, and thus the six nearest atoms are nearly planar (see the shaded rectangle in Figure 3-3).

Notice that the chain of amino acids formed in this way has an intrinsic *directionality* because it always has an amino group at one end and a carboxyl group at the other end. The end of the chain with the amino group is known as the **N- (or amino) terminus,** and the end with the carboxyl group is known as the **C- (or carboxyl) terminus.**

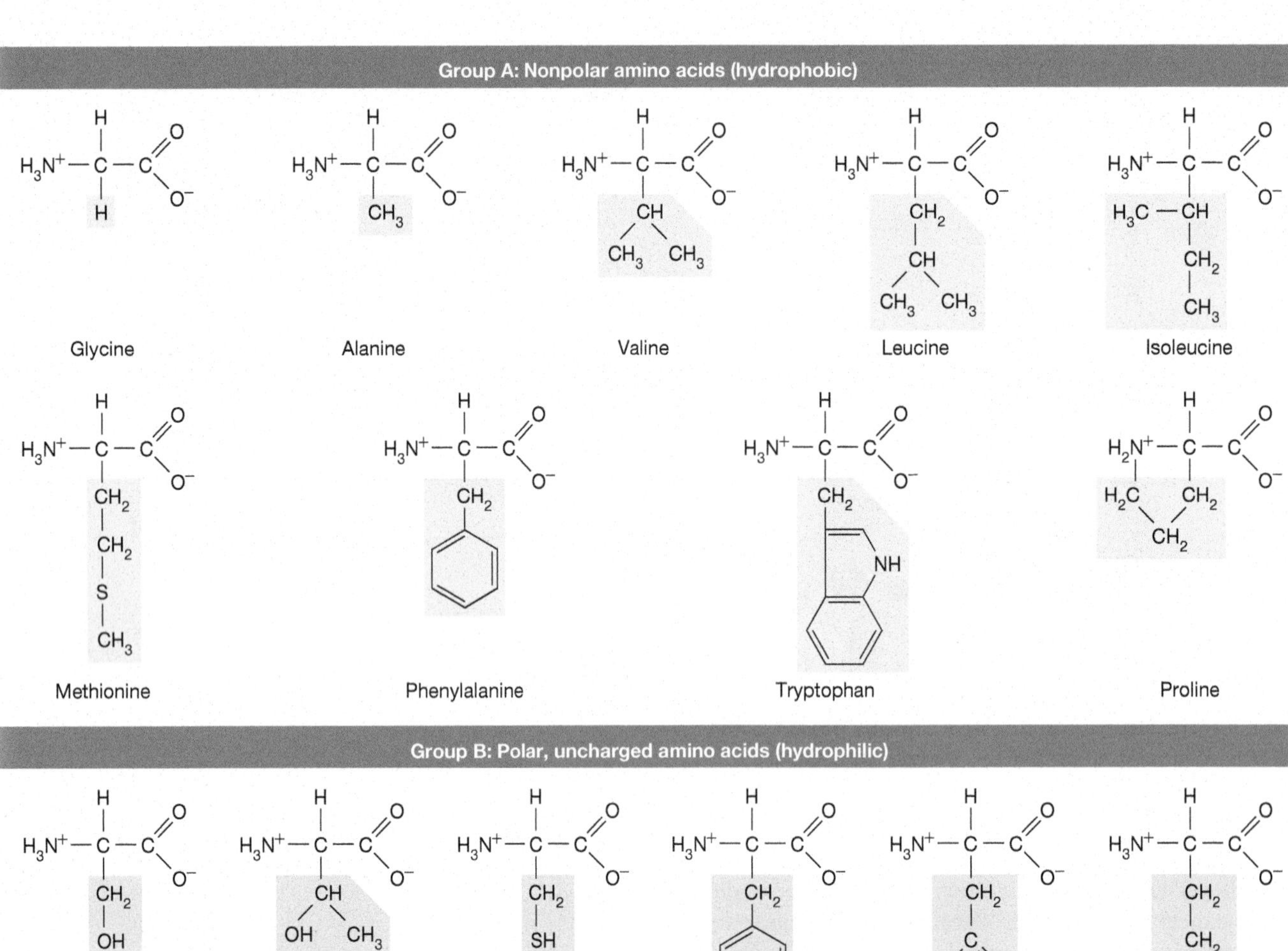

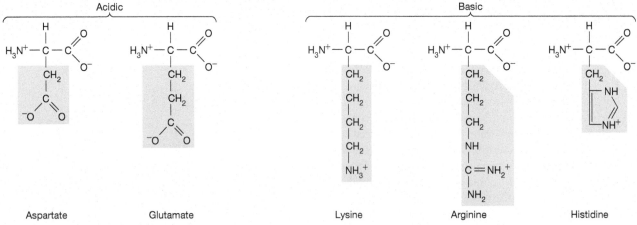

Figure 3-2 The Structures of the 20 Amino Acids Found in Proteins. All amino acids have a carboxyl group and an amino group attached to the central (α) carbon, but each has its own distinctive R group (shaded boxes). Those in Group A have nonpolar, hydrocarbon R groups and are therefore hydrophobic. The others are hydrophilic, either because the R group is polar (Group B) or because the R group is acidic or basic and thus is charged at cellular pH (Group C).

This process of elongating a chain of amino acids is often called *protein synthesis*, but the term is not entirely accurate because the immediate product of amino acid polymerization is not a protein but a **polypeptide.** A protein is a polypeptide chain (or a complex of several polypeptides) that has attained a unique, stable, three-dimensional shape and is biologically active as a result. Some proteins consist of a single polypeptide, and their final shape is due to the folding

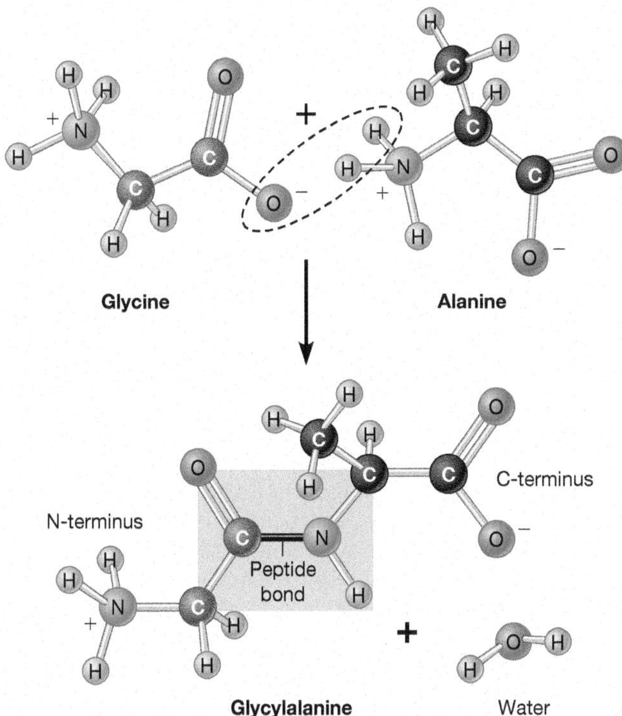

Figure 3-3 Peptide Bond Formation. A peptide bond is formed between the carboxyl group of glycine and the amino group of alanine as water is removed (dotted oval). The six atoms adjacent to the peptide bond in the shaded rectangle are nearly planar.

and coiling that occur spontaneously as the chain is being formed. Such proteins are called **monomeric proteins** (the word *monomer* literally means "one part"). Many other proteins are **multimeric proteins,** consisting of two or more polypeptides that are often called polypeptide subunits.

Although a polypeptide is itself a *polymer,* the entire polypeptide is sometimes a monomeric unit of a multimeric protein. If a multimeric protein is composed of two polypeptides, it is referred to as a *dimer,* and if it has three polypeptides, it is known as a *trimer.* The hemoglobin that carries oxygen in your bloodstream is a multimeric protein known as a *tetramer* because it contains four polypeptides, two each of two different types known as the *α and β subunits* (**Figure 3-4**). In the case of multimeric proteins, protein synthesis involves not only elongation and folding of the individual polypeptide subunits but also their subsequent interaction and assembly into the multimeric protein.

Several Kinds of Bonds and Interactions Are Important in Protein Folding and Stability

The initial folding of a polypeptide into its proper shape, or conformation, depends on several different kinds of bonds and interactions (see Chapter 2), including the covalent disulfide bond and several **noncovalent bonds and interactions.** Although typically much weaker than covalent bonds, noncovalent forces are diverse, are numerous, and collectively exert a powerful influence on protein structure and stability. These include *hydrogen bonds, ionic bonds, van der Waals interactions,* and *hydrophobic interactions.* In addition, the association of individual polypeptides to form a multimeric protein such as hemoglobin relies on these same bonds and interactions, which are depicted in **Figure 3-5**. These interactions primarily involve the R groups

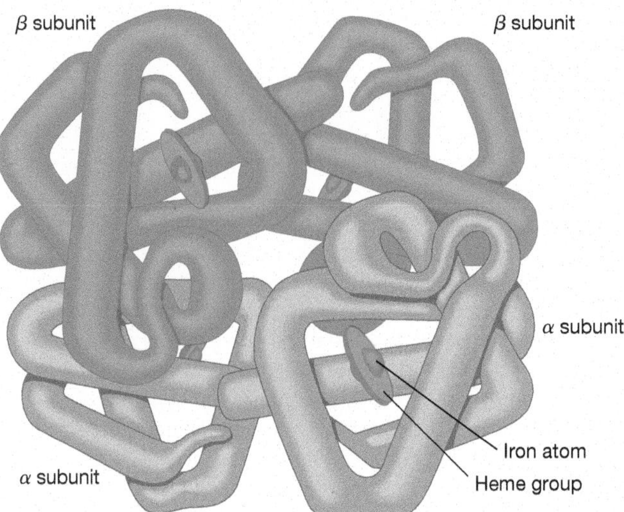

Figure 3-4 The Structure of Hemoglobin. Hemoglobin is a multimeric protein composed of four polypeptide subunits (two α subunits and two β subunits). Each subunit contains a heme group with an iron atom that can bind a single oxygen molecule.

of the individual *amino acid residues,* the name given to the amino acids once they are incorporated into the polypeptide.

Disruption of these interactions by heat, high salt, or chemical treatment can result in *denaturation,* or unfolding, of the polypeptide (see Figure 2-17). Usually, a protein is inactive in this denatured, unfolded state. Similarly, incorrect folding of polypeptides due to incorrect interactions can have serious biological effects. In fact, the presence of misfolded proteins in cells can cause human diseases such as Alzheimer disease (see Human Connections, pages 48–49).

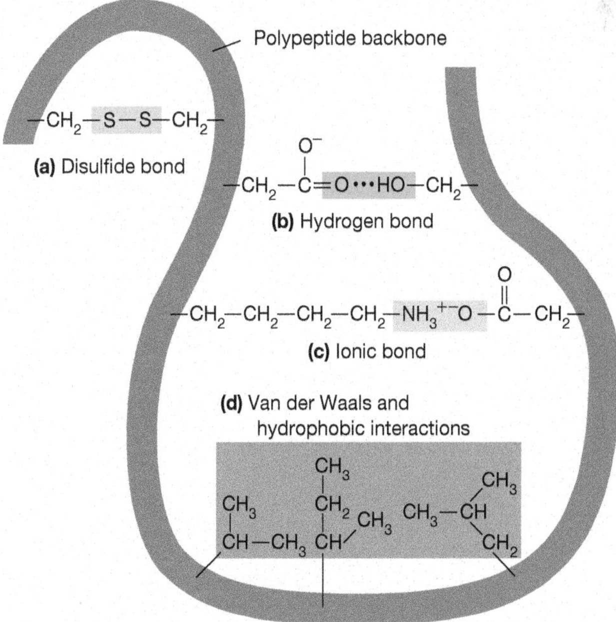

Figure 3-5 Bonds and Interactions Involved in Protein Folding and Stability. The initial folding and subsequent stability of a polypeptide depend on **(a)** covalent disulfide bonds as well as on several kinds of noncovalent bonds and interactions, including **(b)** hydrogen bonds, **(c)** ionic bonds, and **(d)** van der Waals interactions and hydrophobic interactions.

HUMAN *Connections*

Aggregated Proteins and Alzheimer's

Mrs. Petersen got a new bracelet today. The front contains important medical information, and her daughter's phone number is inscribed on the back should someone need to call her. Mrs. Petersen is not sure if she likes the bracelet or why she even has it. She sometimes gets lost on her way to the store, forgets why she walked into a room, and can't recognize people she has known for years. For the next 5 to 10 years her memory will continue to decline. She is in the early stages of Alzheimer disease.

Those familiar with its devastating effects know all too well the suffering and emotional strain of Alzheimer's on both Alzheimer's patients and their caregivers. The impact of Alzheimer's is far-reaching: one in ten Americans over the age of 65 has the disease. It does not discriminate: past presidents, actors, and athletes have all succumbed to the disease. Within the next 40 years, Alzheimer's is projected to result in ~$1.2 trillion in annual payouts to health-care and health-related services in the United States alone.

The symptoms of Alzheimer's are caused by the degeneration of brain cells due to excessive association of proteins both outside and within brain cells. Alzheimer's patients exhibit two kinds of structural abnormalities. The first, known as *amyloid plaques,* are found outside of brain cells (**Figure 3A-1**). These structures contain fibrils made of a protein fragment 40 to 42 amino acids long called *amyloid-beta* peptide (Aβ), produced via the action of enzymes on a protein normally embedded in the plasma membrane called amyloid precursor protein (APP). Aβ fibrils are not soluble in the extracellular environment and lead to the amyloid plaques that accumulate at synapses between brain cells.

Inherited mutations in the *APP* gene or in the genes that encode the enzymes responsible for cleaving APP into Aβ can produce hereditary forms of Alzheimer's. Most people with Alzheimer's, however, do not carry these mutations. Some people who have a high likelihood of developing Alzheimer's produce different forms of a protein known as apolipoprotein E (apoE). ApoE functions primarily in cholesterol transport, but some forms of apoE stimulate amyloid plaque formation. Other noninherited forms of Alzheimer's have been linked to repetitive mild and major brain injuries, such as those suffered by professional football players and boxers.

Amyloid accumulation leads to a second major type of alteration in the brain tissue of Alzheimer's patients. Known as *neurofibrillary tangles,* these abnormal structures are largely composed of a polymerized form of a protein called Tau (**Figure 3A-2**). Tau is abundant in the central nervous system, and its normal role is to stabilize microtubules, which are key structural elements inside of many cells, including nerve cells (see Chapters 4 and 13). The Tau protein in tangles is excessively phosphorylated. Due to the accumulation of amyloid plaques and neurofibrillary tangles, Alzheimer's patients experience progressive cell death in the brain and consequent memory loss (**Figure 3A-3**).

Is it possible that one day we could prevent or cure Alzheimer's? The relationship between Aβ and Alzheimer's suggests that the disease might eventually be cured using treatments that either inhibit the formation of Aβ at a number of different steps in the process

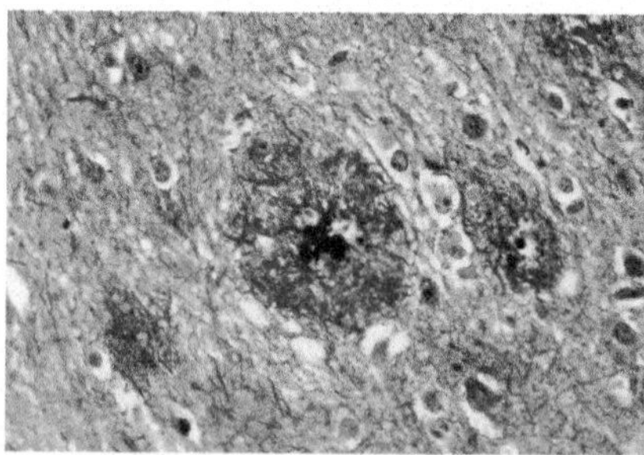

Figure 3A-1 Protein Aggregates in Alzheimer's Patients. Diseased brain tissue from an Alzheimer's patient showing several amyloid plaques.

Disulfide Bonds. A special type of covalent bond that helps stabilize protein conformation is the **disulfide bond,** which forms between the sulfur atoms of two cysteine amino acid residues. These become covalently linked following an oxidation reaction that removes the two hydrogen atoms from the sulfhydryl groups of the two cysteines, forming a disulfide bond, shown below in bold below and in Figure 3-5a:

$$\text{HC}-\text{CH}_2-\text{SH} \;+\; \text{HS}-\text{CH}_2-\text{CH} \xrightarrow{-2\text{H}} \text{HC}-\text{CH}_2-\mathbf{S-S}-\text{CH}_2-\text{CH}$$

Once formed, a disulfide bond confers considerable stability to the structure of the protein because of its covalent nature. It can be broken only by reducing it again—by adding two hydrogen atoms and regenerating the two sulfhydryl groups in the reverse of the reaction above. In many cases, the cysteine residues involved in a particular disulfide bond are part of the same polypeptide. They may be distant from each other along the polypeptide, but they are brought close together by the folding process. Such *intramolecular disulfide bonds* stabilize the conformation of the polypeptide. In the case of multimeric proteins, a disulfide bond may form between cysteine residues located in two different polypeptides. Such *intermolecular disulfide bonds* link the two polypeptides to one another covalently. The hormone insulin is a dimeric protein that has its two subunits linked in this manner.

Hydrogen Bonds. Hydrogen bonds are familiar from our discussion of the properties of water in Chapter 2. In water, a **hydrogen bond** forms between a covalently bonded hydrogen atom on one water molecule and an oxygen atom on another molecule (see Figure 2-8b). In addition, the R groups of many amino acids have functional groups that are able to participate in hydrogen bonding. This allows hydrogen bonds to form between amino acid residues that may be distant from one another along the amino acid sequence but are brought into

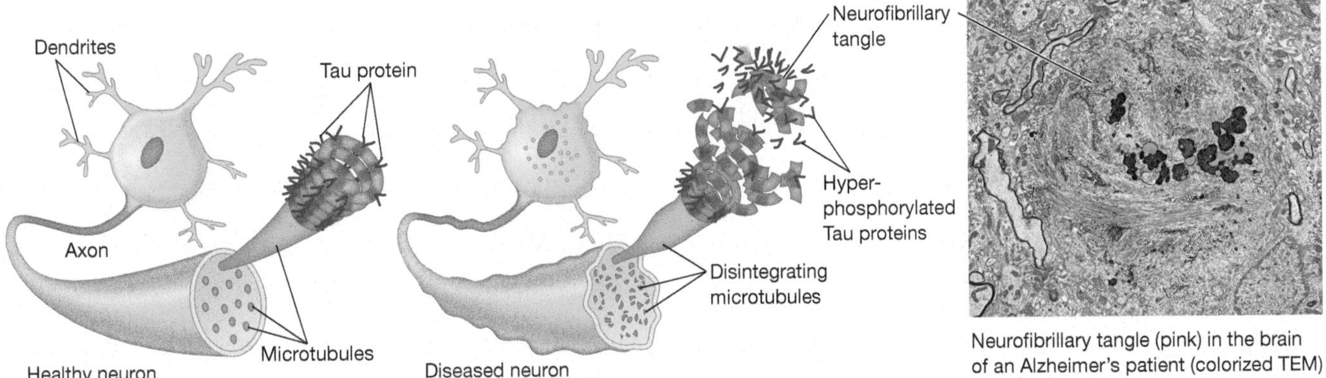

Figure 3A-2 Neurofibrillary Tangle Formation. Breakdown of microtubules within neurons in Alzheimer's patients occurs when hyperphosphorylated Tau protein begins to form neurofibrillary tangles.

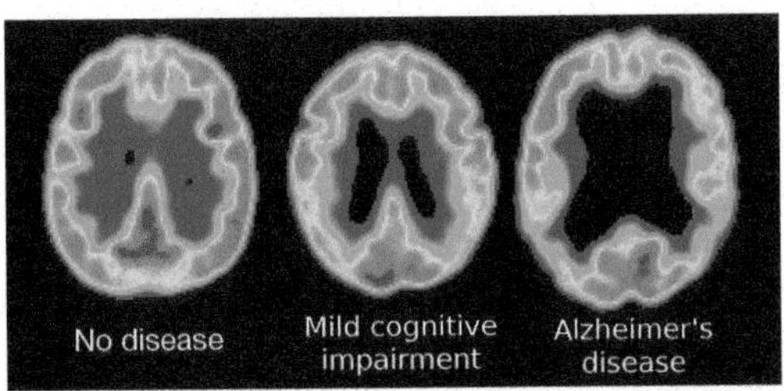

Figure 3A-3 Brain Scans. Positron emission tomography (PET) was used to image changes in the brain as Alzheimer's progresses.

or promote its elimination from the brain. These treatments include (1) enzyme inhibitors that block the cleavage of Aβ from its precursor APP, (2) small molecules that disrupt amyloid plaques or prevent their formation, (3) knocking down levels of RNA using a technique called RNAi (see Chapter 20) that in turn reduces the translation of the proteins that lead to plaques, and (4) a Aβ vaccine that stimulates the immune system to clean up amyloid plaques and/or prevent them from forming. Such vaccines are already capable of protecting mice with Alzheimer's symptoms from suffering further memory loss, providing hope that this devastating illness will be conquered in the not too distant future.

close proximity by the folding of the polypeptide (Figure 3-5b). In polypeptides, hydrogen bonding is particularly important in stabilizing helical and sheet structures that are prominent parts of many proteins, as we will soon see.

Hydrogen bond *donors* have a hydrogen atom that is covalently linked to a more electronegative atom (one with a higher affinity for electrons and thus a partial negative charge), such as oxygen or nitrogen, so the hydrogen atom carries a partial positive charge. Hydrogen bond *acceptors* have an electronegative atom that attracts this hydrogen atom. Examples of good donors include the hydroxyl groups of several amino acids and the amino groups of others (see Figure 3-2). The carbonyl and sulfhydryl groups of several other amino acids are examples of good acceptors. An individual hydrogen bond is quite weak, with an energy of about 5 kcal/mol, compared to 70–100 kcal/mol for single covalent bonds and 150–200 kcal/mol for double and triple covalent bonds. But because hydrogen bonds are abundant in biological macromolecules such as proteins and DNA, they become a formidable force when present in large numbers (see Problem 2-8).

Ionic Bonds. The role of **ionic bonds** (or *electrostatic interactions*) in protein structure is easy to understand. Because the R groups of some amino acids are positively charged and the R groups of others are negatively charged, polypeptide folding is dictated in part by the tendency of charged groups to repel groups with the same charge and to attract groups with the opposite charge (Figure 3-5c). Several features of ionic bonds are particularly significant. The strength of such interactions—about 3 kcal/mol—allows them to exert an attractive force over greater distances than some of the other noncovalent interactions. Moreover, the attractive force is nondirectional, so ionic bonds are not limited to discrete angles, as is the case with covalent bonds. Because ionic bonds depend on both groups remaining charged, they will be disrupted if the pH value becomes so high or so low that either of the groups loses its charge. This loss of ionic bonds accounts in part for the denaturation that most proteins undergo at high or low pH.

Van Der Waals Interactions. Interactions based on charge are not limited to ions that carry a discrete charge. Even molecules with nonpolar covalent bonds may have regions that, transiently, are partially charged positively and negatively. Momentary asymmetries in the distribution of electrons, and hence in the separation of charge within a molecule, are called *dipoles*. When two molecules that have such transient dipoles are very close to each other and are oriented appropriately, they are attracted to each other. However, the attraction lasts only as long as the asymmetric electron distribution persists in both molecules. This transient attraction of two nonpolar molecules is called a **van der Waals interaction,** or *van der Waals force.* A single such interaction is transient and very weak—typically 0.1–0.2 kcal/mol—and is effective only when the two molecules are quite close together—within 0.2 nm of each other. Despite this interaction being one of the weakest noncovalent forces, there is tremendous strength in numbers. For instance, a 2002 study demonstrated that the millions of hairs on the feet of the gecko likely use van der Waals interactions for their adhesive properties. Van der Waals interactions may be weak but are nonetheless important in the structure of proteins (Figure 3-5d) and other biological macromolecules, as well as in the binding together of two molecules with complementary surfaces that fit closely together.

Hydrophobic Interactions. The fourth type of noncovalent interaction that plays a role in maintaining protein conformation is usually called a **hydrophobic interaction,** but it is not really a bond or interaction at all. Rather, it is the tendency of hydrophobic molecules or parts of molecules to be excluded from interactions with water (Figure 3-5d). As already noted, the side chains of the 20 different amino acids vary greatly in their affinity for water. Amino acids with hydrophilic R groups tend to be located near the surface of a folded polypeptide, where they can interact maximally with the surrounding water molecules. In contrast, amino acids with hydrophobic R groups are essentially nonpolar and are usually located on the inside of the polypeptide, where they interact with one another because they are excluded by water (see Figure 3-2, Group A).

Thus, polypeptide folding to form the final protein structure is, in part, a balance between the tendency of hydrophilic groups to seek an aqueous environment near the surface of the molecule and the tendency of hydrophobic groups to minimize contact with water by associating with each other in the interior of the molecule. If most of the amino acids in a protein were hydrophobic, the protein would be virtually insoluble in water and would be found instead in a nonpolar environment. Membrane proteins, which have many hydrophobic residues, are localized in membranes for this very reason. Similarly, if all or most of the amino acids were hydrophilic, the polypeptide would most likely remain in a fairly distended, random shape, allowing maximum access of each amino acid to an aqueous environment. But precisely because most polypeptide chains contain both hydrophobic and hydrophilic amino acids, hydrophilic regions are drawn toward the surface, whereas hydrophobic regions are driven toward the interior.

Overall, then, the stability of the folded structure of a polypeptide depends on an interplay of covalent disulfide bonds and four noncovalent factors: hydrogen bonds between R groups that are good hydrogen bond donors and acceptors, ionic bonds between charged amino acid R groups, transient van der Waals interactions between nonpolar molecules in very close proximity, and hydrophobic interactions that drive nonpolar groups to the interior of the molecule.

Chaperone Interactions. Whereas many proteins adopt their final conformation spontaneously using the forces mentioned above, others require the aid of other proteins for proper folding. These **molecular chaperones** (briefly mentioned in Chapter 2) are responsible for ensuring accurate protein folding. Although chaperones most often act during protein synthesis, they also facilitate the refolding of incorrectly folded mature proteins. In general, the mechanism used by most chaperones is to shield parts of a newly synthesized protein from participating in the interactions mentioned above until more of the protein can be made. (We will review this mechanism in more detail in Chapter 19, Figure 19-20.)

The final conformation of the fully folded polypeptide is the net result of the forces, tendencies, and interactions described above. Individually, each of these noncovalent interactions is quite low in energy. However, the cumulative effect of many of them—involving the side groups of the hundreds of amino acids that make up a typical polypeptide—greatly stabilizes the conformation of the folded polypeptide.

Protein Structure Depends on Amino Acid Sequence and Interactions

The overall shape and structure of a protein are usually described in terms of four hierarchical levels of organization, each building on the previous one: the *primary*, *secondary*, *tertiary*, and *quaternary* structures (**Table 3-3**). Primary structure refers to the amino acid sequence, whereas the higher levels of organization concern the interactions between the amino acid residues. These interactions give the protein its characteristic *conformation*, or three-dimensional arrangement of atoms in space (**Figure 3-6**). Secondary structure involves local interactions between amino acid residues that are close together along the chain. Tertiary structure results from long-distance interactions between stretches of amino

Table 3-3 Levels of Organization of Protein Structure

Level of Structure	Basis of Structure	Kinds of Bonds and Interactions Involved
Primary	Amino acid sequence	Covalent peptide bonds
Secondary	Folding into α helix, β sheet, or random coil	Hydrogen bonds between NH and CO groups of peptide bonds in the backbone
Tertiary	Three-dimensional folding of a single polypeptide chain	Disulfide bonds, hydrogen bonds, ionic bonds, van der Waals interactions, hydrophobic interactions
Quaternary	Association of multiple polypeptides to form a multimeric protein	Same as for tertiary structure

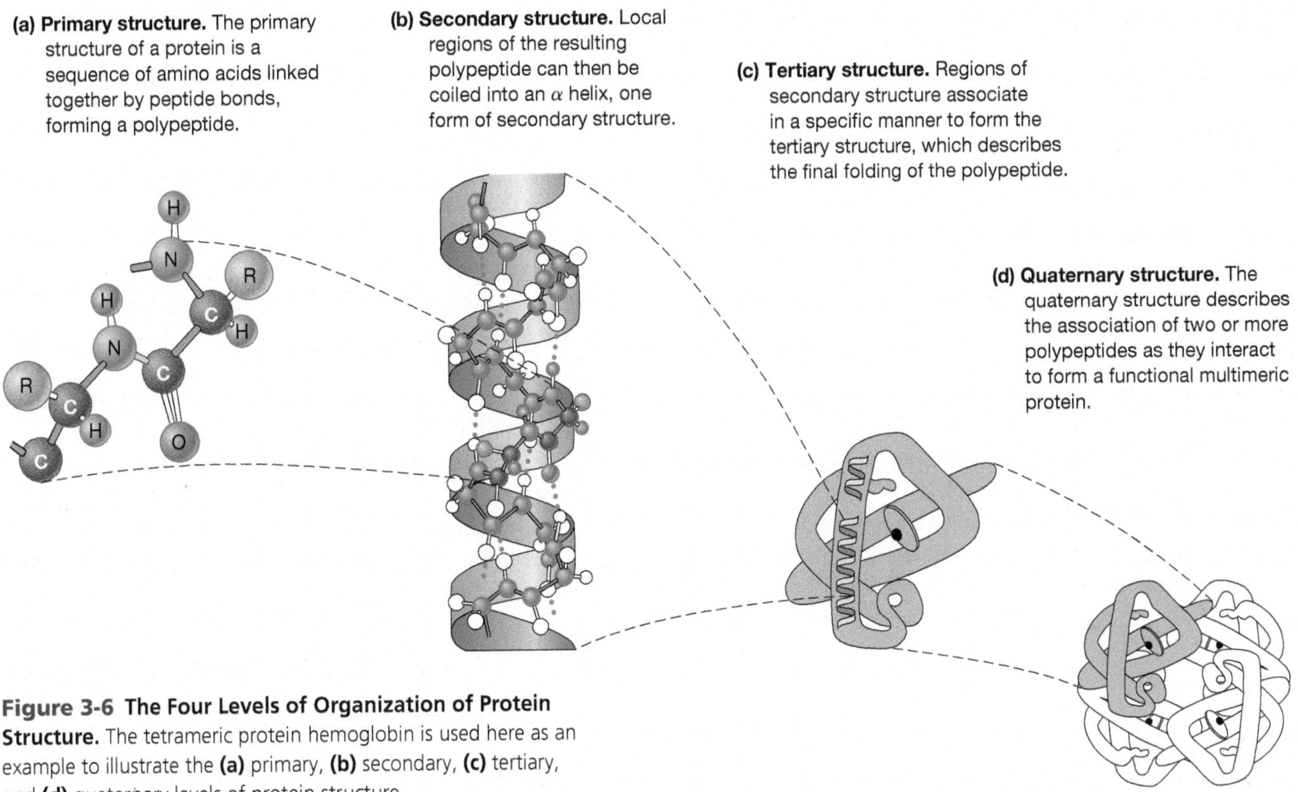

(a) Primary structure. The primary structure of a protein is a sequence of amino acids linked together by peptide bonds, forming a polypeptide.
(b) Secondary structure. Local regions of the resulting polypeptide can then be coiled into an α helix, one form of secondary structure.
(c) Tertiary structure. Regions of secondary structure associate in a specific manner to form the tertiary structure, which describes the final folding of the polypeptide.
(d) Quaternary structure. The quaternary structure describes the association of two or more polypeptides as they interact to form a functional multimeric protein.
H
N
R
C
H
H
N
C
R
C
H
O
C

Figure 3-6 The Four Levels of Organization of Protein Structure. The tetrameric protein hemoglobin is used here as an example to illustrate the **(a)** primary, **(b)** secondary, **(c)** tertiary, and **(d)** quaternary levels of protein structure.

acid residues from different parts of a polypeptide molecule. Quaternary structure describes the interaction of two or more individual folded polypeptides to form a single multimeric protein. All three of these higher-level structures are dictated by the primary structure.

Primary Structure. As noted, the **primary structure** of a protein is a formal designation for the amino acid sequence (Figure 3-6a). When we describe the primary structure, we are simply specifying the order in which its amino acids appear from one end of the molecule to the other. By convention, amino acid sequences are always written from the N-terminus to the C-terminus of the polypeptide, which is also the direction in which the polypeptide is synthesized.

The first protein to have its complete amino acid sequence determined was the hormone *insulin*. This important technical advance was reported in 1953 by Frederick Sanger, who received a Nobel Prize in 1958 for the work. To determine the sequence of the insulin molecule, Sanger cleaved it into smaller fragments and analyzed the amino acid order within individual, overlapping fragments. Insulin consists of two polypeptides, called the *A subunit* and the *B subunit*, with 21 and 30 amino acid residues, respectively. **Figure 3-7** shows the structure of insulin, illustrating the primary sequence of each subunit in sequence from its N-terminus (left) to its C-terminus (right). Notice also the covalent disulfide (—S—S—) bond between two cysteine residues within the A chain and the two disulfide bonds linking the A and B chains. Disulfide bonds play an important role in stabilizing the tertiary structure of many proteins.

Sanger's techniques paved the way for the sequencing of hundreds of other proteins and led ultimately to the design of machines that can determine an amino acid sequence automatically. However, with current technology, it is much easier to purify a DNA molecule and determine its nucleotide sequence than it is to purify a protein and analyze its amino acid sequence. Once a DNA nucleotide sequence has been determined, the amino acid sequence of the polypeptide encoded by that DNA segment can be easily inferred

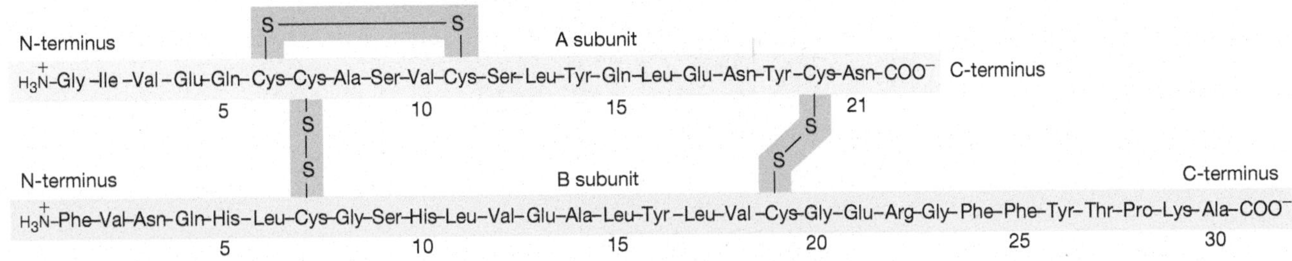

Figure 3-7 The Structure of Insulin. Insulin consists of two polypeptides, called the A and B subunits, which are covalently linked by two disulfide bonds. (For abbreviations of amino acids, see Table 3-2.)

using the genetic code. Computerized data banks are now available that contain thousands of polypeptide sequences, making it easy to compare sequences and look for regions of similarity between polypeptides.

The primary structure of a protein is important both genetically and structurally. Genetically, it is significant because the amino acid sequence of the polypeptide is determined by the order of nucleotides in the corresponding messenger RNA. The messenger RNA in turn reflects DNA sequences in the gene that encodes the protein. Therefore, the primary structure of a protein is the result of the order of nucleotides in the DNA of the gene.

Of more immediate significance are the implications of the primary structure for higher levels of protein structure. In essence, all three higher levels of protein organization are direct consequences of the primary structure. Although protein denaturation by heating unfolds a polypeptide and eliminates all but the primary structure, the information in the primary sequence specifies these higher levels of structure, and often the protein can refold into its native conformation, as we saw for ribonuclease (see Figure 2-17). Similarly, if synthetic polypeptides are made that correspond in sequence to the α and β subunits of hemoglobin, they will assume the native three-dimensional conformations of these subunits and will then interact spontaneously to form the native $\alpha_2\beta_2$ tetramer that we recognize as hemoglobin (see Figure 3-4).

Secondary Structure. The **secondary structure** of a protein describes local regions of structure that result from hydrogen bonding between NH and CO groups along the polypeptide backbone. These local interactions result in two major structural patterns, referred to as the **α helix** and **β-sheet** conformations (**Figure 3-8**).

The α helix structure was proposed in 1951 by Linus Pauling and Robert Corey. As shown in Figure 3-8a, an α helix is spiral in shape, consisting of a backbone of amino acids linked by peptide bonds with the specific R groups of the individual amino acid residues sticking out from it. In the

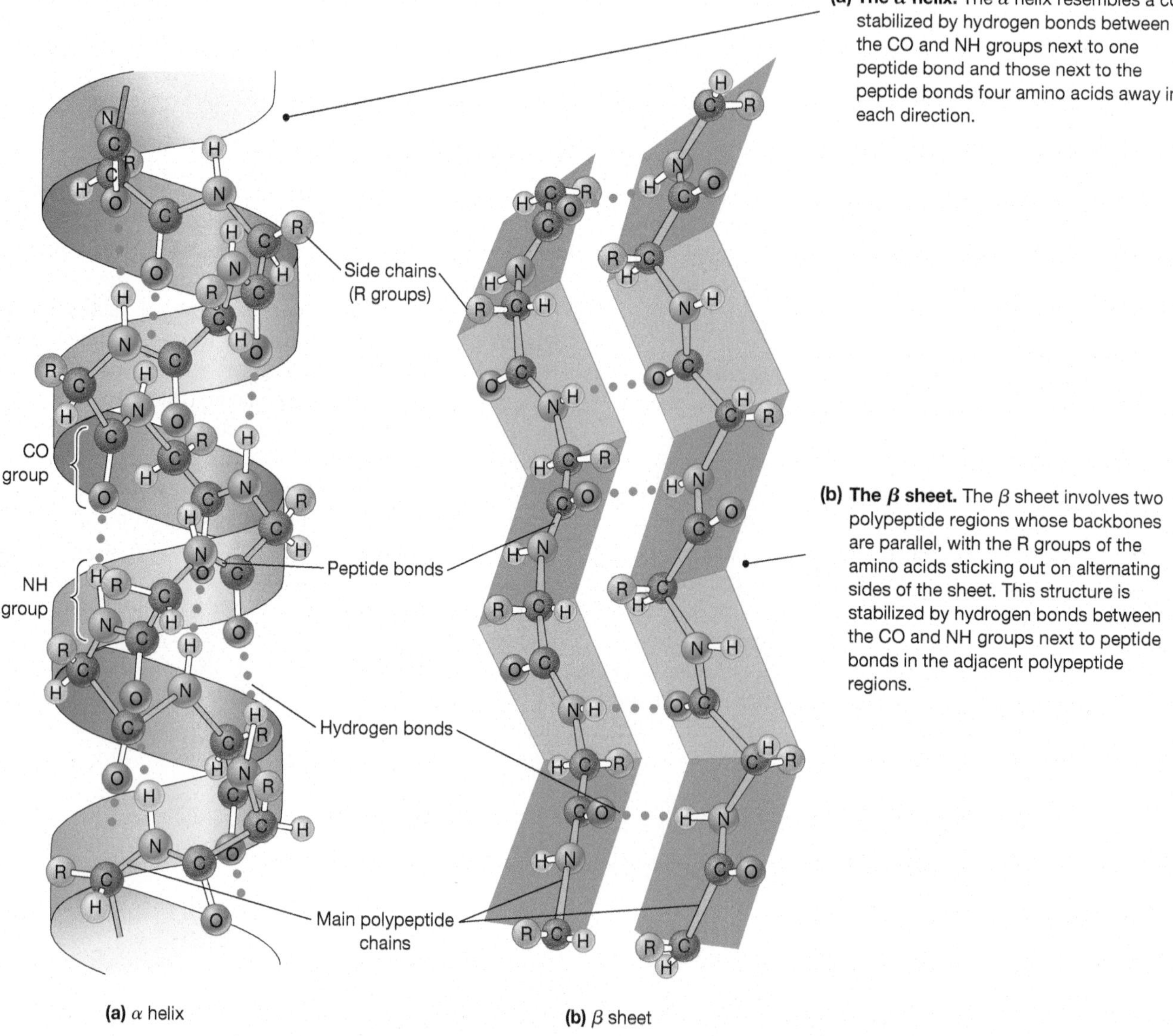

(a) The α helix. The α helix resembles a coil stabilized by hydrogen bonds between the CO and NH groups next to one peptide bond and those next to the peptide bonds four amino acids away in each direction.
Side chains (R groups)
CO group
NH group
Peptide bonds
Hydrogen bonds
Main polypeptide chains
(b) The β sheet. The β sheet involves two polypeptide regions whose backbones are parallel, with the R groups of the amino acids sticking out on alternating sides of the sheet. This structure is stabilized by hydrogen bonds between the CO and NH groups next to peptide bonds in the adjacent polypeptide regions.
(a) α helix
(b) β sheet

Figure 3-8 The α Helix and β Sheet. The α helix shown in **(a)** and the β sheet shown in **(b)** are both stabilized by hydrogen bonds (blue dots), either within a local region of primary sequence (α helix) or between two separate regions (β sheet).

α helix there are 3.6 amino acids per turn, bringing the peptide bonds of every fourth amino acid in close proximity. The distance between these peptide bonds is, in fact, just right for the formation of a hydrogen bond between the NH group adjacent to one peptide bond and the CO group adjacent to the other, as shown in Figure 3-8a.

As a result, every peptide bond in the helix is hydrogen-bonded through its CO group to the peptide bond immediately "below" it in the spiral and through its NH group to the peptide bond just "above" it, even though the amino acid residues involved are not directly adjacent. These hydrogen bonds are all nearly parallel to the main axis of the helix and therefore tend to stabilize the spiral structure by holding successive turns of the helix together. In addition, two or more α helices can coil together in a ropelike fashion to form a bundle of α helices called a *coiled coil*, as we will see shortly in the keratin protein that makes up hair.

Another form of common secondary structure in proteins is the β sheet, also initially proposed by Pauling and Corey. As shown in Figure 3-8b, this structure is an extended sheetlike conformation with successive atoms in the polypeptide chain located at the "peaks" and "troughs" of the pleats. The R groups of successive amino acids stick out on alternating sides of the sheet. Because the carbon atoms that make up the backbone of the polypeptide chain are successively located a little above and a little below the plane of the β sheet, such structures are sometimes called *β-pleated sheets.*

Like the α helix, the β sheet is characterized by a maximum of hydrogen bonding. In both cases, all of the CO groups and NH groups adjacent to the peptide bonds are involved. However, hydrogen bonding in an α helix is invariably intramolecular (within the same polypeptide), whereas hydrogen bonding in the β sheet can be either intramolecular (between two segments of the same polypeptide) or intermolecular (linking two different polypeptides). The protein regions that form β sheets can interact with each other in two different ways. If the two interacting regions run in the same N-terminus-to-C-terminus direction, the structure is called a *parallel β sheet;* if the two strands run in opposite N-terminus-to-C-terminus directions, the structure is called an *antiparallel β sheet.* In some proteins, several antiparallel β sheets associate symmetrically around a central axis in a structure known as a β propeller.

Whether a specific segment of a polypeptide will form an α helix, a β sheet, or neither depends on the amino acids present in that segment. For example, leucine, methionine, and glutamate are strong "α helix formers" and are commonly found in α-helical regions. Isoleucine, valine, and phenylalanine are strong "β-sheet formers" and are often found in β-sheet regions. Proline is considered a "helix breaker" because its R group is covalently bonded to its amino nitrogen, which therefore lacks the hydrogen atom needed for hydrogen bonding. Proline is rarely found in an α helix and, when present, introduces a bend in the helix.

To depict localized regions of structure within a protein, biochemists have adopted the conventions shown in **Figure 3-9**. An α-helical region is represented as either a spiral or a cylinder, whereas a β-sheet region is drawn as a flat ribbon or arrow with the arrowhead pointing in the direction of the C-terminus. Depending on their relative

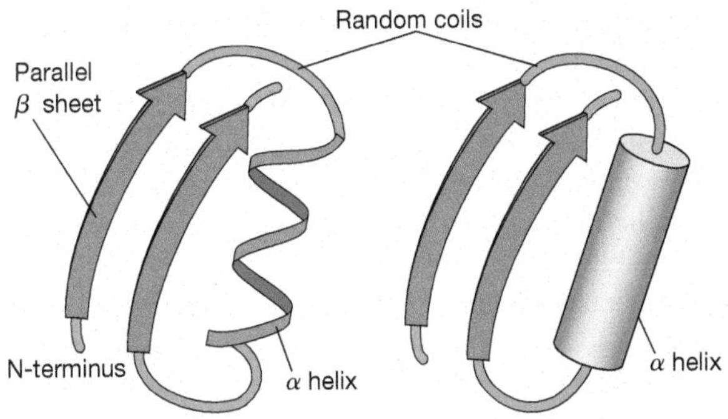

(a) β-α-β motif with α helix represented as a spiral (left) or a cylinder (right)

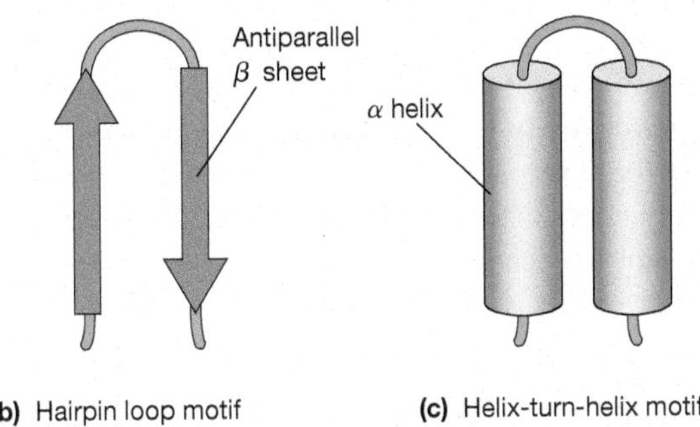

(b) Hairpin loop motif **(c)** Helix-turn-helix motif

Figure 3-9 Common Structural Motifs. These short sections of polypeptides show three common units of secondary structure: the **(a)** β-α-β, **(b)** hairpin loop, and **(c)** helix-turn-helix motifs. Note that the sheets can be either parallel (as in part a) or antiparallel (as in part b).

orientation, β sheets can be parallel (Figure 3-9a) or antiparallel (Figure 3-9b). A looped segment that connects α-helical and/or β-sheet regions is called a *random coil* that is intrinsically disordered. This segment has no defined secondary structure and is depicted as a narrow cord.

Certain combinations of α helices and β sheets have been identified in many proteins. These combinations of secondary structure, called **motifs**, consist of small segments of an α helix and/or a β sheet connected to each other by looped regions of varying length. Among the most commonly encountered motifs are the β-α-β motif shown in Figure 3-9a and the hairpin loop and helix-turn-helix motifs depicted in Figure 3-9b and c, respectively. When the same motif is present in different proteins, it usually serves the same purpose in each. (For example, the helix-turn-helix motif is one of several secondary structure motifs that are characteristic of the DNA-binding proteins we will encounter in Chapter 20 when we consider the regulation of gene expression.)

Tertiary Structure. The **tertiary structure** of a protein can probably be best understood by contrasting it with the secondary structure (Figure 3-6b, c). Secondary structure is a predictable, repeating conformational pattern that derives from the repetitive nature of the polypeptide because it involves hydrogen bonding between NH and CO groups adjacent to peptide bonds—the common structural elements along every polypeptide chain. If proteins contained only one or a few kinds of similar amino acids, virtually all aspects of protein conformation could probably be understood in terms of secondary structure, with only modest variations among proteins.

Tertiary structure comes about precisely because of the variety of amino acids present in proteins and the very different chemical properties of their R groups. In fact, tertiary structure depends almost entirely on the previously mentioned interactions (including hydrogen bonds, van der Waals interactions, and hydrophobic interactions) between the various R groups, regardless of where along the primary sequence they happen to be. Tertiary structure therefore reflects the nonrepetitive and unique aspect of each polypeptide because it depends not on the CO and NH groups common to all of the amino acids in the chain but instead on the very feature that makes each amino acid distinctive—its R group.

Tertiary structure is neither repetitive nor readily predictable; it involves competing interactions between side groups with different properties. Hydrophobic R groups, for example, are spontaneously drawn into the nonaqueous environment in the interior of the molecule, whereas polar amino acids are drawn to the surface. Oppositely charged R groups can form ionic bonds, whereas similarly charged groups will repel each other. As a result, the polypeptide chain will be folded, coiled, and twisted into its **native conformation**—the most stable three-dimensional structure for that particular sequence of amino acids.

The relative contributions of secondary and tertiary structures to the overall shape of a polypeptide vary from protein to protein and depend critically on the relative proportions and sequence of amino acids in the chain. Broadly speaking, proteins can be divided into two categories: *fibrous proteins* and *globular proteins.*

Fibrous proteins have extensive secondary structure (either α helix or β sheet) throughout the molecule, giving them a highly ordered, repetitive structure. In general, secondary structure is much more important than tertiary interactions in determining the shape of fibrous proteins, which often have an extended, filamentous structure. Especially prominent examples of fibrous proteins include the *fibroin* protein of silk and the *keratins* of hair and wool, as well as *collagen* (found in tendons and skin) and *elastin* (present in ligaments and blood vessels).

The amino acid sequence of each of these proteins favors a particular kind of secondary structure, which in turn confers a specific set of desirable mechanical properties on the protein. Fibroin, for example, consists mainly of long stretches of antiparallel β sheets, with the polypeptide chains running parallel to the axis of the silk fiber but in opposite directions (**Figure 3-10**). The most prevalent amino acids in fibroin are glycine, alanine, and serine. These amino acids have small R groups (see Figure 3-2) that pack together well. The result is a silk fiber that is strong and relatively inextensible because the polypeptide chains in a β-sheet conformation are already stretched to nearly their maximum possible length.

Hair and wool fibers, on the other hand, consist of the protein α-keratin, which is almost entirely α helical. The individual keratin molecules are very long and lie with the axes of their helices nearly parallel to the fiber axis. As a result, hair is quite extensible because stretching of the fiber is opposed not by the covalent bonds of the polypeptide chain, as in β sheets, but by the hydrogen bonds that stabilize the α-helical structure. The individual α helices in a hair are wound together in a coiled coil to form a strong, ropelike structure, as shown in **Figure 3-11.** First, two keratin α helices are coiled around each other, and two of these coiled pairs associate to form a protofilament containing four α helices. Groups of eight protofilaments then interact to form intermediate filaments, which bundle together to form the actual hair fiber. Not surprisingly, the α-keratin polypeptides in hair are rich in hydrophobic

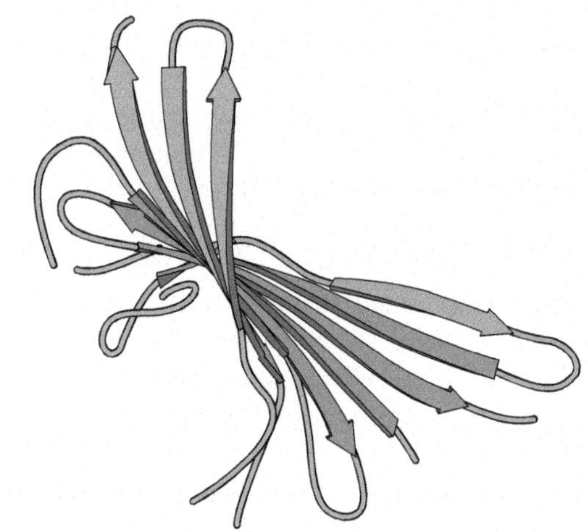

Figure 3-10 Fibroin Structure. Silk is composed primarily of fibroin, a fibrous protein containing mainly of regions of antiparallel β sheets.

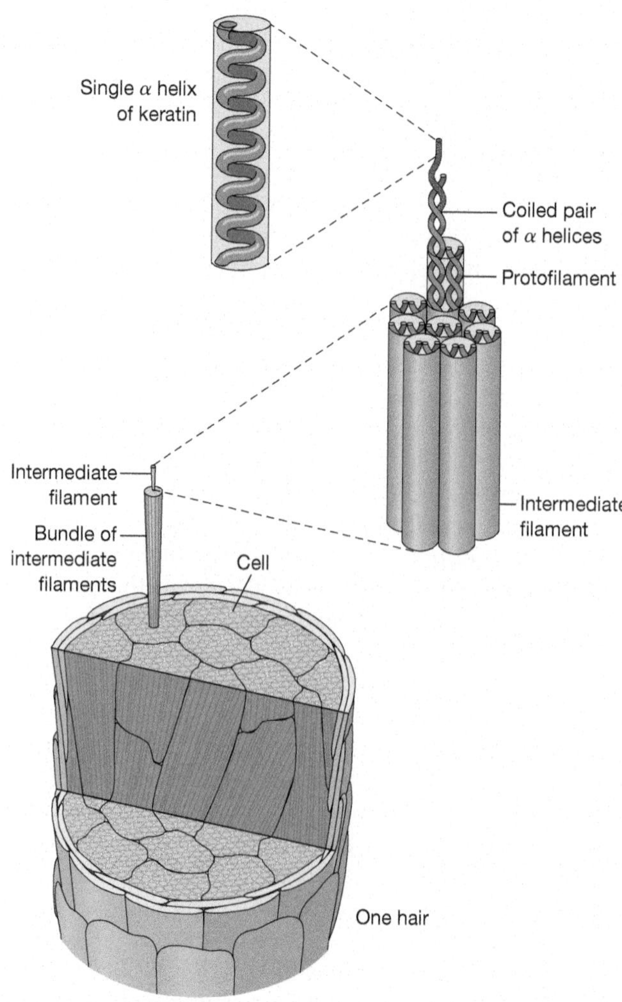

Figure 3-11 The Structure of Hair. The main structural protein of hair is α-keratin, a fibrous protein with an α-helical shape.

residues that interact with each other where the helices touch, allowing the tight packing of the filaments in hair.

As important as fibrous proteins may be, they represent only a small fraction of the kinds of proteins present in most cells. Most of the proteins involved in cellular structure are **globular proteins,** so named because their polypeptide chains are folded into compact structures rather than extended filaments (see Figure 3-4). The polypeptide chain of a globular protein is often folded locally into regions with α-helical or α-sheet structures, and these regions of secondary structure are themselves folded on one another to give the protein its compact, globular shape. This folding is possible because regions of the β helix or β sheet are interspersed with random coils, irregularly structured regions that allow the polypeptide chain to loop and fold (see Figure 3-9). Thus, every globular protein has its own unique tertiary structure, made up of secondary structural elements (helices and sheets) folded in a specific way that is especially suited to the particular functional role of that protein.

Most enzymes are globular proteins, and their enzymatic function depends critically on their proper structure. **Figure 3-12** shows the native tertiary structure of

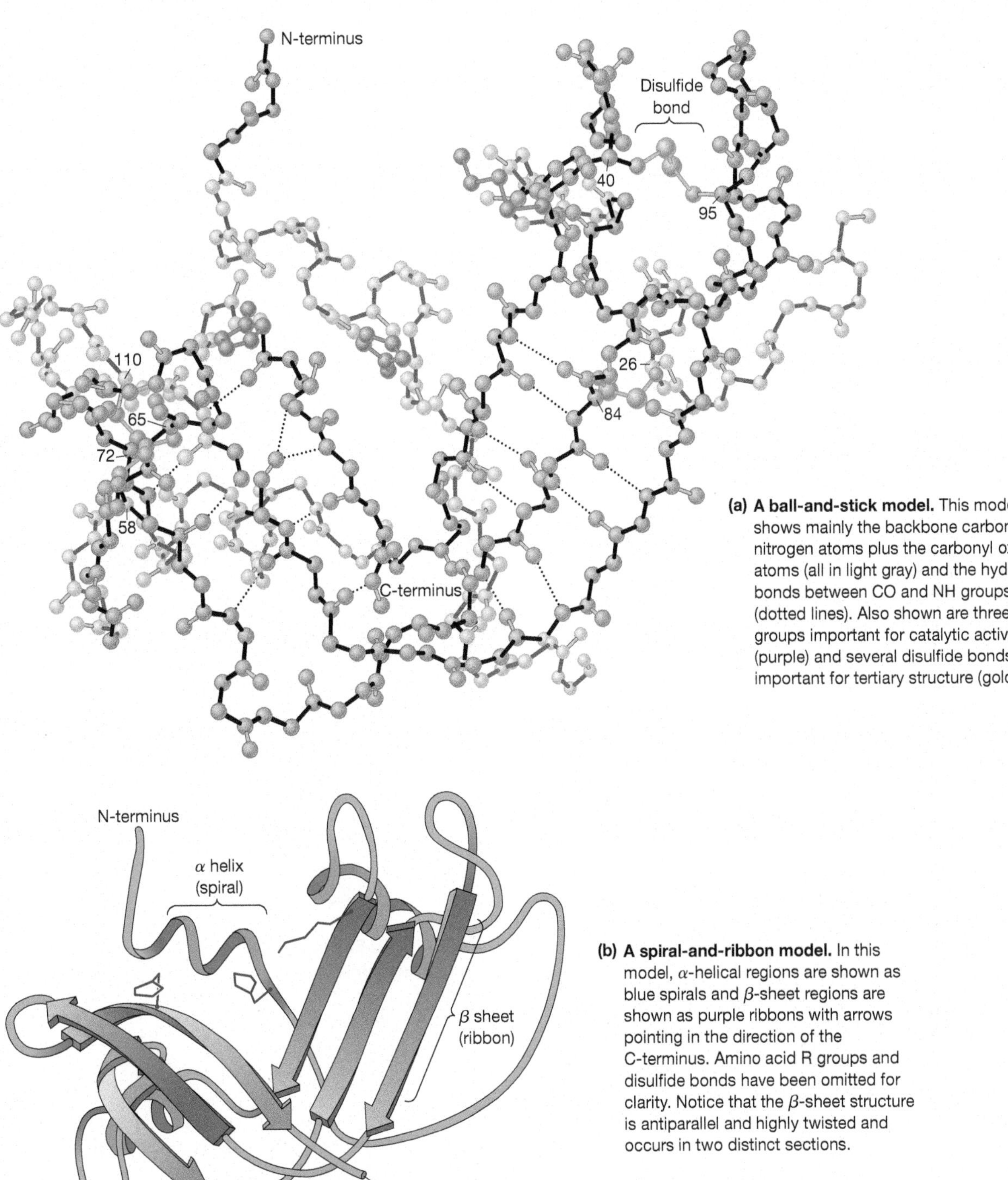

(a) A ball-and-stick model. This model shows mainly the backbone carbon and nitrogen atoms plus the carbonyl oxygen atoms (all in light gray) and the hydrogen bonds between CO and NH groups (dotted lines). Also shown are three R groups important for catalytic activity (purple) and several disulfide bonds important for tertiary structure (gold).

(b) A spiral-and-ribbon model. In this model, α-helical regions are shown as blue spirals and β-sheet regions are shown as purple ribbons with arrows pointing in the direction of the C-terminus. Amino acid R groups and disulfide bonds have been omitted for clarity. Notice that the β-sheet structure is antiparallel and highly twisted and occurs in two distinct sections.

Figure 3-12 The Three-Dimensional Structure of Ribonuclease. Ribonuclease is a monomeric globular protein with significant α-helical and β-sheet regions connected by random coils. Its tertiary structure can be represented by either **(a)** a ball-and-stick model or **(b)** a spiral-and-ribbon model.

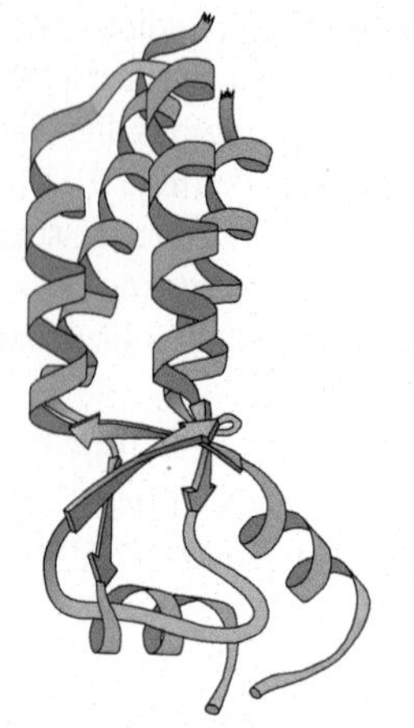

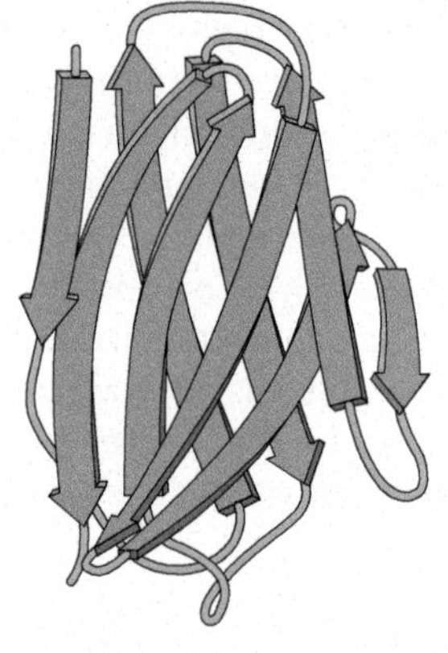

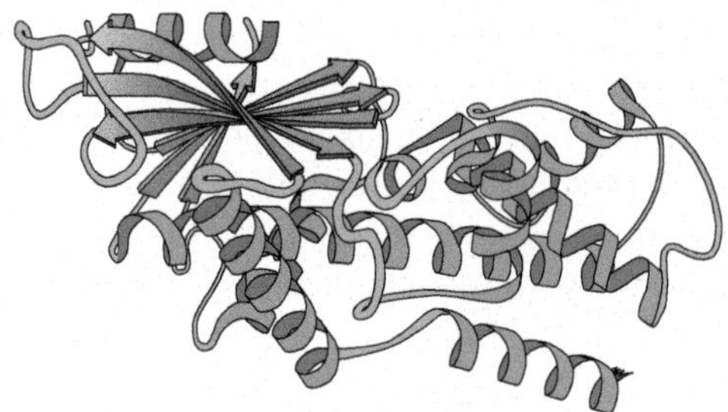

Figure 3-13 Structures of Several Globular Proteins. Shown here are proteins with different tertiary structures: **(a)** the predominantly α-helical structure of the coat protein of tobacco mosaic virus (TMV), **(b)** the mainly β-sheet structure of the V_2 domain of immunoglobulin, and **(c)** the mixture of α helices and β sheets seen in domain 2 of hexokinase.

ribonuclease, a typical globular protein. (We encountered ribonuclease in Figure 2-17, as an example of the denaturation and renaturation of a polypeptide and the spontaneity of its folding.) Figure 3-12 uses two different conventions to represent the structure of ribonuclease: the *ball-and-stick* model used in Figure 3-8 and the *spiral-and-ribbon* model used in Figure 3-9. For clarity, most of the side chains of ribonuclease have been omitted in both models. The groups shown in gold in Figure 3-12a are the four disulfide bonds that help stabilize the tertiary structure of ribonuclease.

Globular proteins can be mainly α helical, mainly β sheet, or a mixture of both structures. These categories are illustrated in **Figure 3-13** by the coat protein of tobacco mosaic virus (TMV), a portion of an immunoglobulin (antibody) molecule, and a portion of the enzyme hexokinase, respectively. Helical segments of globular proteins often consist of bundles of helices, as seen for the coat protein of TMV in Figure 3-13a. Segments with mainly β-sheet structure are usually characterized by a barrel-like configuration (Figure 3-13b) or by a twisted sheet (Figure 3-13c).

Many globular proteins consist of a number of segments called domains. A **domain** is a discrete, locally folded unit of tertiary structure that usually has a specific function. Each domain typically includes 50–350 amino acids, with regions of α helices and β sheets packed together compactly. Small globular proteins are usually folded into a single domain (for example, ribonuclease; see Figure 3-12b). Large globular proteins usually have multiple domains. The portions of the immunoglobulin and hexokinase molecules shown in Figure 3-13b and c are, in fact, specific domains of these proteins. **Figure 3-14** shows an example of a protein that consists of a single polypeptide folded into two functional domains.

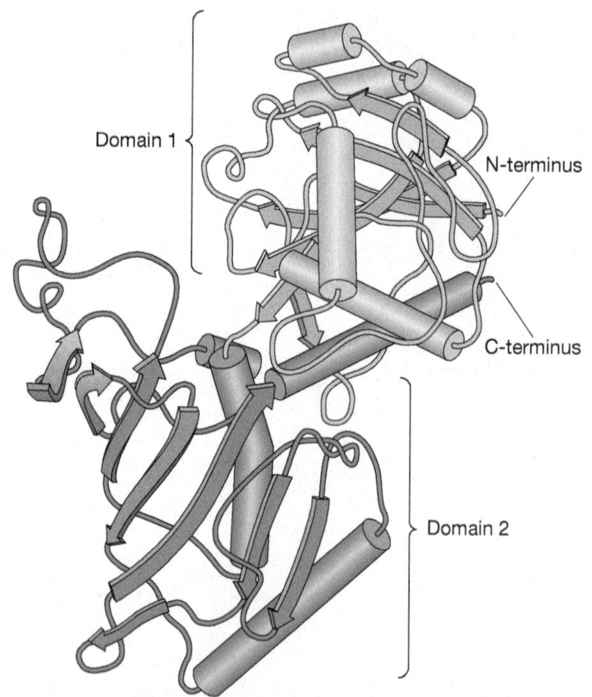

Figure 3-14 An Example of a Protein Containing Two Functional Domains. The enzyme glyceraldehyde phosphate dehydrogenase is a single polypeptide chain folded into two domains, which are indicated by different shadings.

Proteins that have similar functions (such as binding a specific ion or recognizing a specific molecule) usually have a common domain containing a sequence of identical or very similar amino acid residues. Moreover, proteins with multiple functions usually have a separate domain for each function. Thus, domains can be thought of as the modular units of function from which globular proteins are constructed. Many different types of domains have been described in proteins and given names such as the immunoglobulin domain, the kringle domain, and the death domain. Each type is composed of a particular combination of α-helix and β-sheet regions and random coil loops that give the domain a specific function. Often, for proteins that are enzymes, one domain carries out the catalytic activity and another regulates enzymatic activity.

Before leaving the topic of tertiary structure, we should emphasize again the dependence of these higher levels of organization on the primary structure of the polypeptide. The significance of the primary structure is exemplified especially well by the inherited condition *sickle-cell anemia.* People with this trait have red blood cells that are distorted from their normal disk shape into a "sickle" shape, which causes the abnormal cells to clog blood vessels and impede blood flow, limiting oxygen availability in the tissues.

This condition is caused by a slight change in the hemoglobin molecule within the red blood cells. In people with sickle-cell anemia, the hemoglobin molecules have normal α-polypeptide chains, but their β chains have a single amino acid that is different. At one specific position in the chain (the sixth amino acid residue from the N-terminus), the glutamate (E) normally present is replaced by valine (V). This single substitution (written as E6V) causes enough of a difference in the tertiary structure of the β chain that the hemoglobin molecules tend to crystallize, deforming the cell into a sickle shape. Not all amino acid substitutions cause such dramatic changes in structure and function, but this example underscores the crucial relationship between the amino acid sequence of a polypeptide and the final shape and biological activity of the molecule.

Although we know that the primary sequence of a protein determines its final folded shape, we still are not able to predict exactly how a given protein will fold, especially for large proteins (more than 100 amino acids). In fact, one of the most challenging unsolved problems in structural biochemistry is to predict the final folded tertiary structure of a protein from its known primary structure. This task is further complicated by the need to factor in the numerous interactions that the folding protein has with the surrounding water molecules in the aqueous environment of the cell. Even with all our knowledge of the factors and forces involved in folding, and the availability of supercomputers to do billions of calculations per second, we often cannot predict the most stable conformation for a given protein. In 2000, Vijay Pande, at Stanford University, started the Folding@home project. This effort uses crowd-sourced computing power to help perform these calculations. Anyone can download a program that will help researchers improve our protein-folding predictive power.

In addition, in every other year since 1994, protein modelers worldwide test their predictive methods in a major modeling experiment known as CASP—the critical assessment of techniques for protein structure prediction. Their predictions are compared to subsequently released three-dimensional protein structures, and the results are published in a special issue of the journal *Proteins: Structure, Function and Bioinformatics.* One goal of these efforts in modeling research is drug discovery—the ability to design therapeutic agents able to bind to specific regions of a protein involved in human disease.

Although we can deduce the primary structure of a polypeptide from the nucleotide sequence of the DNA encoding it, determining its overall three-dimensional conformation is much more complicated. **Key Technique, pages 58–59,** describes how researchers use a technique known as X-ray crystallography to determine accurate three-dimensional structures of polypeptides. In some cases, this technique can determine the precise position of nearly every atom in the polypeptide.

Quaternary Structure. The **quaternary structure** of a protein is the level of organization concerned with subunit interactions and assembly (see Figure 3-6d). Quaternary structure therefore applies only to multimeric proteins. Many proteins are included in this category, particularly those with molecular weights above 50,000. Hemoglobin, for example, is a multimeric protein with two α subunits and two β subunits (see Figure 3-4). Some multimeric proteins contain identical polypeptide subunits; others, such as hemoglobin, contain two or more different kinds of polypeptides.

The bonds and forces that maintain quaternary structure are the same as those responsible for tertiary structure: hydrogen bonds, electrostatic interactions, van der Waals interactions, hydrophobic interactions, and covalent disulfide bonds. As noted earlier, disulfide bonds may be either within a polypeptide chain or between chains. When they occur within a polypeptide, they stabilize tertiary structure. When they occur between polypeptides, they help maintain quaternary structure, holding the individual polypeptides together (see Figure 3-7). As in the case of polypeptide folding, the process of subunit assembly is often, though not always, spontaneous. Most, if not all, of the requisite information is provided by the amino acid sequence of the individual polypeptides, but often molecular chaperones are required to ensure proper assembly.

In some cases, a still higher level of assembly is possible in the sense that two or more proteins (often enzymes) are organized into a **multiprotein complex,** with each protein involved sequentially in a common multistep process. An example of such a complex is an enzyme system called the *pyruvate dehydrogenase complex.* This complex catalyzes the oxidative removal of a carbon atom (as CO_2) from the three-carbon compound pyruvate, the product of glycolysis (Chapter 10). Three individual enzymes and five kinds of molecules called coenzymes constitute a highly organized *multienzyme complex.* The pyruvate dehydrogenase complex is one of the best understood examples of how cells can achieve economy of function by ordering the enzymes that catalyze sequential reactions into a single multienzyme complex. Other multiprotein complexes we will encounter in our studies include ribosomes, proteasomes, the photosystems, and the DNA replication complex.

CONCEPT CHECK 3.1

Suppose you have a patient suffering from anemia and find that her hemoglobin is missing three amino acids from the primary sequence. How might this affect each of the three higher levels of protein structure, thus causing this condition?

Key Technique | Using X-Ray Crystallography to Determine Protein Structure

PROBLEM: The correct three-dimensional conformation of a protein is critical for its proper function. Knowing the precise positions of the protein backbone and amino acid side chains can give important clues to function. But how can scientists determine the conformation of a single protein molecule with sufficient resolution to locate the individual atoms?

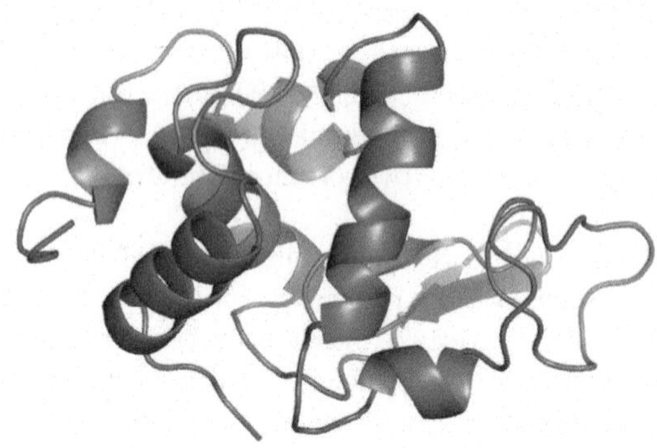

This protein's three-dimensional structure was determined using X-Ray Crystallography.

SOLUTION: *X-ray crystallography* allows researchers to determine the positions of most of the thousands of atoms in a typical protein molecule. By forming a protein molecule into a crystal, in which the atoms align in a well-ordered and repetitive manner, researchers can measure the diffraction of X-rays by these atoms to determine their positions within the protein molecule.

Key Tools: An X-ray instrument; a crystallized sample; a rotating mount to hold the sample; a detector to record the diffraction pattern; and a computer to analyze the results.

Details: Even with the most powerful microscopes, it is not possible to see atoms using visible light. Therefore, in order to visualize proteins in atomic detail, researchers use electromagnetic radiation in the form of X-rays. In X-ray crystallography, a beam of X-rays is directed at a crystal, where it diffracts at particular angles producing a *diffraction pattern* of spots. The pattern shows the density of electrons within the crystal and can then be converted into a three-dimensional model of the protein. X-ray crystallography has been key to advances in many fields. Notably, X-ray diffraction data produced from DNA crystals by Rosalind Franklin and R. G. Gosling (**Figure 3B-1**) were instrumental in allowing James Watson and Francis Crick to determine the helical structure of double-stranded DNA.

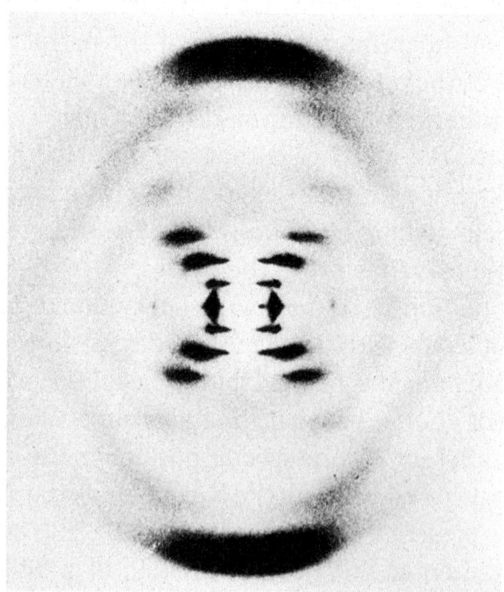

Figure 3B-1 X-ray Diffraction Pattern of DNA. This image was obtained by Rosalind Franklin in 1953, the year in which Watson and Crick determined the structure of DNA.

Determining the structure of a purified protein by X-ray crystallography typically involves three phases: (1) production of a suitable pure protein crystal, (2) irradiation of the crystal with X-rays at various angles to produce a diffraction pattern, and (3) analysis of this diffraction pattern using a computer to produce an electron density map and a three-dimensional model of the protein.

Protein Crystallization. The first and often most difficult step is the creation of a suitable crystal. The crystal must be sufficiently large (20–100 μm on each side) and free of imperfections, such as fractures that would scatter the X-rays and interfere with the analysis. Crystals are grown by gradually decreasing the solubility of the protein by adding precipitants. The precipitants bind with water molecules, thus reducing the amount of free water in the solution in which the proteins are dissolved. As the solution becomes supersaturated, crystals form. Typically, a number of different solvents and concentrations are tested—often using robots to automate the process—to produce the largest and most defect-free crystals (**Figure 3B-2**).

3.2 Nucleic Acids

Next, we come to the **nucleic acids**, macromolecules of paramount importance to the cell because of their role in storing, transmitting, and expressing genetic information. Nucleic acids are linear polymers of nucleotides strung together in a genetically determined order that is critical to their role as informational macromolecules. The two major types of nucleic acids are **DNA (deoxyribonucleic acid)** and **RNA (ribonucleic acid).** DNA and RNA differ

Irradiation with X-Rays. Once a suitable crystal is obtained, it is irradiated with X-rays (**Figure 3B-3**). As the X-rays hit the electron cloud of an atom, they are diffracted at a specific angle. Because the atoms in a crystal are arranged in a regular pattern at particular angles, the X-rays will reinforce or cancel each other, resulting in a pattern of spots that is recorded by a detector. This pattern of spots depends on the precise arrangement of the atoms in the crystal. The crystal is rotated during irradiation, and diffraction data are collected at all possible angles. (For more information on diffraction patterns, see the "X-ray Crystallography" section of the Appendix.)

Model Construction. The diffraction data are converted into an *electron density map* for the protein using specialized mathematical procedures. The electron density map is then used to produce a three-dimensional model of the protein, atom by atom, and can show the positions of nearly all of the atoms in a protein. The model is then refined by repeatedly using the structure to predict a diffraction pattern, comparing it to the actual pattern, and adjusting the model to achieve a better fit.

QUESTION: What structural features of protein molecules allow them to form the crystals required for imaging by X-ray crystallography?

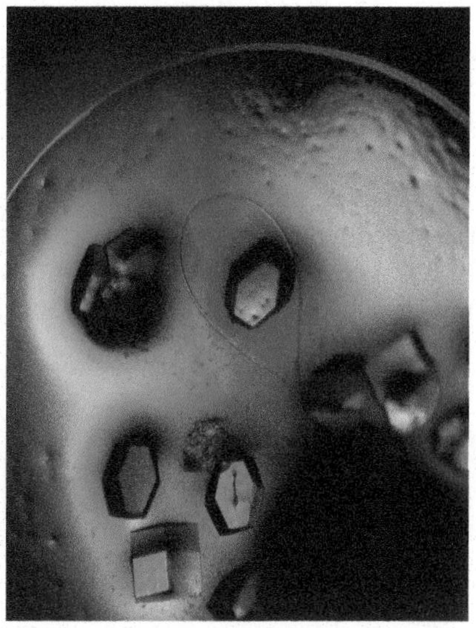

Figure 3B-2 Protein Crystals for X-ray Crystallography. These crystals of pure lysozyme are ready for analysis. Note the small wire loop used to remove one of the crystals.

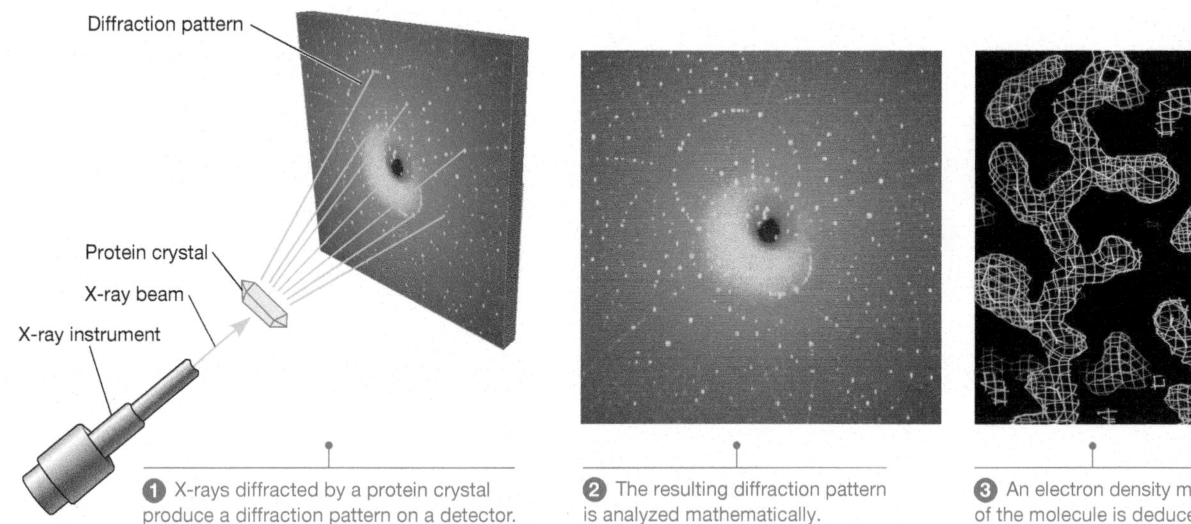
Diffraction pattern
Protein crystal
X-ray beam
X-ray instrument
1 X-rays diffracted by a protein crystal produce a diffraction pattern on a detector.
2 The resulting diffraction pattern is analyzed mathematically.
3 An electron density map of the molecule is deduced.

Figure 3B-3 X-ray Crystallography. ❶ X-rays directed through a lysozyme crystal produce ❷ a diffraction pattern that is used to create ❸ an electron density map. The electron density map is then used to produce a three-dimensional model of the protein.

in their chemistry and their role in the cell. As the names suggest, RNA contains the five-carbon sugar **ribose** in each of its nucleotides, whereas DNA contains the closely related sugar **deoxyribose.** DNA serves primarily as the repository of genetic information (as we discussed in Chapter 1), whereas RNA molecules play several different roles in expressing that information—that is, in gene regulation and protein synthesis.

The Monomers Are Nucleotides

Nucleic acids are informational macromolecules and contain nonidentical monomeric units in a specified sequence. The monomeric units of nucleic acids are called **nucleotides.** Nucleotides exhibit less variety than amino acids do; DNA and RNA each contain only four different kinds of nucleotides. (Actually, there is more variety than this suggests, especially in some RNA molecules in which some nucleotides have been chemically modified after insertion into the chain.)

As shown in **Figure 3-15**, each nucleotide consists of a five-carbon sugar to which is attached a phosphate group and a nitrogen-containing aromatic base. The sugar is either D-ribose (in RNA) or D-deoxyribose (in DNA). The phosphate is joined by a phosphoester bond to the 5′ carbon of the sugar, and the base is attached to the 1′ carbon. The base may be either a **purine** or a **pyrimidine.** DNA contains the purines **adenine (A)** and **guanine (G)** and the pyrimidines **cytosine (C)** and **thymine (T).** RNA also has adenine, guanine, and cytosine, but it contains the pyrimidine **uracil (U)** in place of thymine. Like the 20 amino acids present in proteins, these five aromatic bases are among the most common small molecules in cells (see Table 3-1).

Without the phosphate, the remaining base-sugar unit is called a **nucleoside.** Each pyrimidine and purine may therefore occur as the free base, the nucleoside, or the nucleotide. The appropriate names for these compounds are given in **Table 3-4**. Notice that nucleotides and nucleosides containing deoxyribose are specified by a lowercase "d" preceding the letters identifying the base.

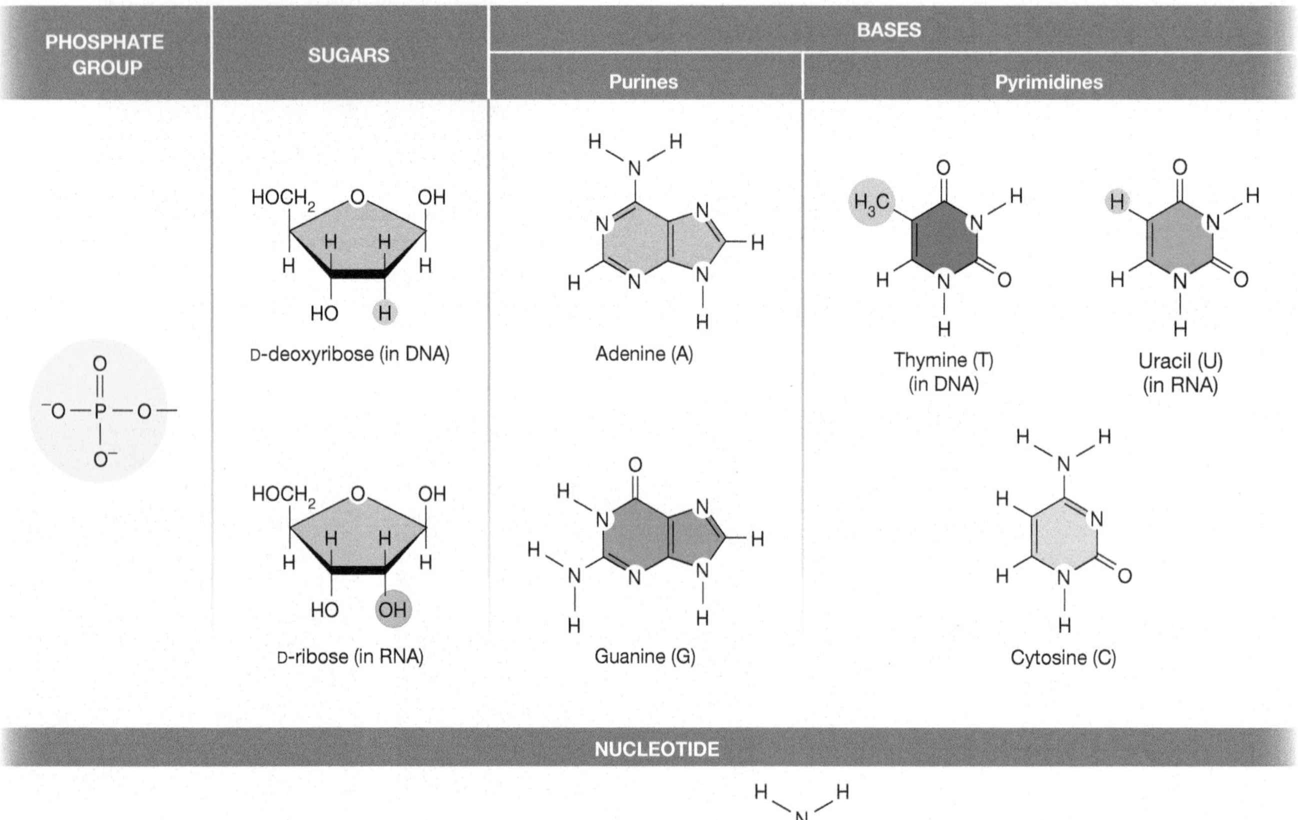

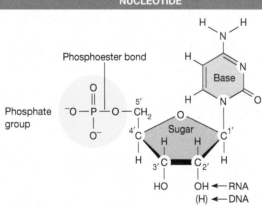

Figure 3-15 The Structure of a Nucleotide. In RNA, a nucleotide consists of the five-carbon sugar D-ribose with an aromatic nitrogen-containing base attached to the 1′ carbon and a phosphate group linked to the 5′ carbon by a phosphoester bond. Carbon atoms in the sugar of a nucleotide are numbered from 1′ to 5′ to distinguish them from those in the base, which are numbered without the prime. In DNA, the hydroxyl group on the 2′ carbon is replaced by a hydrogen atom, so the sugar is D-deoxyribose.

Table 3-4 The Bases, Nucleosides, and Nucleotides of RNA and DNA

	RNA		DNA	
Bases	**Nucleoside**	**Nucleotide**	**Deoxynucleoside**	**Deoxynucleotide**
Purines				
Adenine (A)	Adenosine	Adenosine monophosphate (AMP)	Deoxyadenosine	Deoxyadenosine monophosphate (dAMP)
Guanine (G)	Guanosine	Guanosine monophosphate (GMP)	Deoxyguanosine	Deoxyguanosine monophosphate (dGMP)
Pyrimidines				
Cytosine (C)	Cytidine	Cytidine monophosphate (CMP)	Deoxycytidine	Deoxycytidine monophosphate (dCMP)
Uracil (U)	Uridine	Uridine monophosphate (UMP)	—	—
Thymine (T)	—	—	Deoxythymidine	Deoxythymidine monophosphate (dTMP)

As the nomenclature indicates, a nucleotide can be thought of as a **nucleoside monophosphate** because it is a nucleoside with a single phosphate group attached to it. This terminology can be readily extended to molecules with two or three phosphate groups attached to the 5′ carbon. For example, the nucleoside adenosine (adenine plus ribose) can have one, two, or three phosphates attached and is designated accordingly as **adenosine monophosphate (AMP)**, **adenosine diphosphate (ADP)**, or **adenosine triphosphate (ATP)**. The relationships among these compounds are shown in **Figure 3-16.**

You probably recognize ATP as the energy-rich compound used to drive various reactions in the cell, including the activation of monomers for polymer formation that we encountered in the previous chapter (see Figure 2-16). As this example suggests, nucleotides play two roles in cells: they are the monomeric units of nucleic acids, and several of them—ATP most notably—serve as intermediates in various energy-transferring reactions.

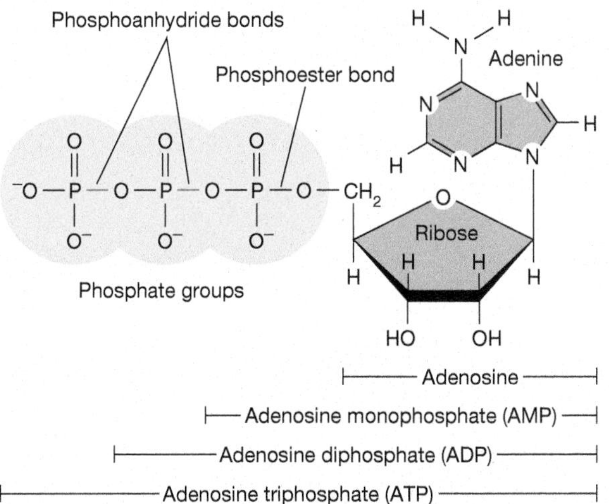

Figure 3-16 The Phosphorylated Forms of Adenosine. Adenosine occurs as the free nucleoside (a sugar linked to a base) and can also form part of AMP, ADP, and ATP. The bond linking the first phosphate to the ribose of adenosine is a phosphoester bond, whereas the bonds linking the second and third phosphate groups to the molecule are phosphoanhydride bonds, which liberate two to three times as much energy as a phosphoester bond.

The Polymers Are DNA and RNA

Nucleic acids are linear polymers formed by linking each nucleotide to the next through a phosphate group, as shown in **Figure 3-17** on page 62. Specifically, the phosphate group already attached by a phosphoester bond to the 5′ carbon of one nucleotide becomes linked by a second phosphoester bond to the 3′ carbon of the next nucleotide. The resulting linkage is known as a 3′, 5′ **phosphodiester bridge,** which consists of a phosphate group linked to two adjacent nucleotides via two phosphoester bonds (one bond to each nucleotide). The **polynucleotide** formed by this process has an intrinsic directionality, with a 5′ phosphate group at one end and a 3′ hydroxyl group at the other end. By convention, nucleotide sequences are always written from the 5′ end to the 3′ end of the polynucleotide because this is the direction of nucleic acid synthesis in cells (see Chapter 17).

Nucleic acid synthesis requires both energy and information. To provide the energy needed to form each new phosphodiester bridge, each successive nucleotide enters as a high-energy nucleoside triphosphate. The precursors for DNA synthesis are therefore dATP, dCTP, dGTP, and dTTP. For RNA synthesis, rATP, rCTP, rGTP, and rUTP are needed. Information is required for nucleic acid synthesis because successive incoming nucleotides must be added in a specific, genetically determined sequence. For this purpose, a preexisting molecule is used as a **template** to specify nucleotide order. For both DNA and RNA synthesis, the template is usually DNA. Template-directed nucleic acid synthesis relies on precise and predictable *base pairing* between a template nucleotide and the specific incoming nucleotide that can pair with the template nucleotide.

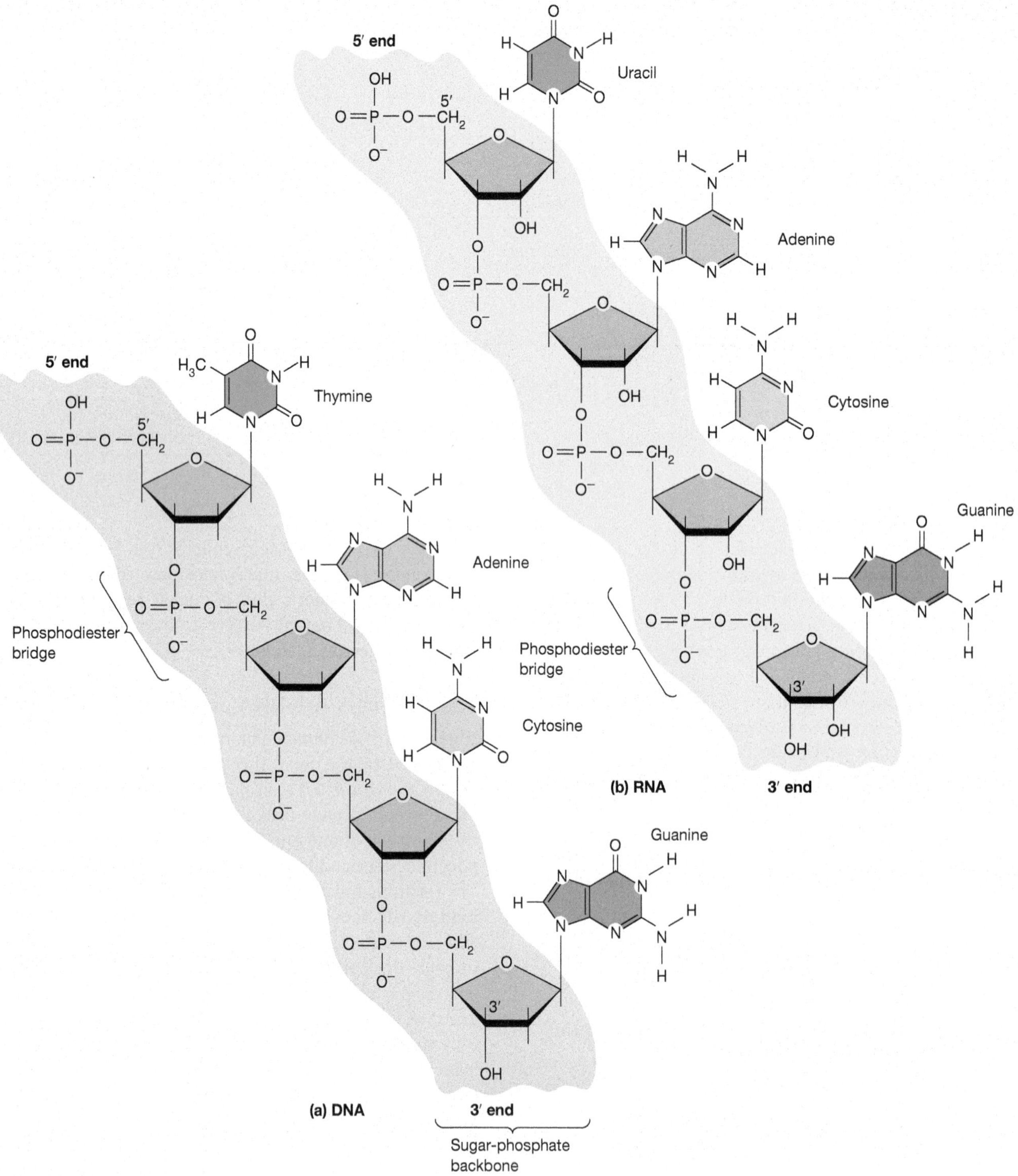

Figure 3-17 The Structure of Nucleic Acids. Nucleic acids are linear chains of nucleotides joined together by 3′, 5′ phosphodiester bridges. The resulting polynucleotide has an intrinsic directionality, with a 5′ end and a 3′ end. For both DNA and RNA, the backbone of the chain is an alternating sugar-phosphate sequence, from which the bases stick out.

This recognition process depends on an important chemical feature of the purine and pyrimidine bases shown in **Figure 3-18**. These bases have carbonyl groups and nitrogen atoms capable of hydrogen bond formation under appropriate conditions. Complementary relationships between purines and pyrimidines allow A to form two hydrogen bonds with T (or U) and G to form three hydrogen bonds with C, as shown in Figure 3-18. This pairing of A with T (or U) and G with C is a fundamental property of nucleic acids. Genetically, this **base pairing** provides a mechanism for nucleic acids to recognize one another (as we will see in Chapter 18). For now, however, let's concentrate on the structural implications.

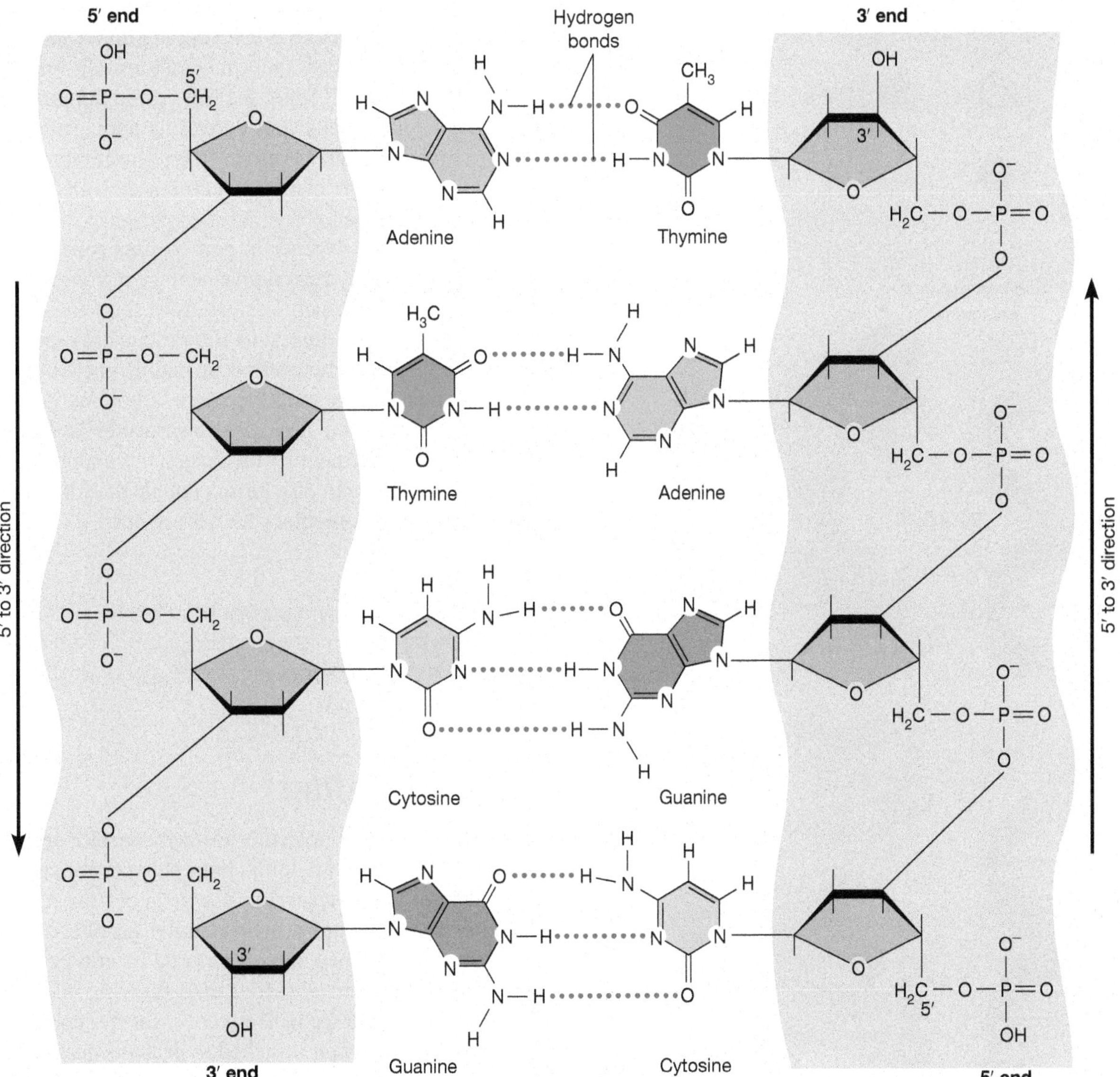

Figure 3-18 Hydrogen Bonding in DNA Nucleic Acid Structure. Two hydrogen bonds (blue dots) between the complementary bases adenine and thymine and three hydrogen bonds between the complementary bases cytosine and guanine account for the AT and CG base pairing of DNA. Because the strands run in the opposite orientation to each other, they are said to be antiparallel.

A DNA Molecule Is a Double-Stranded Helix

One of the most significant biological advances of the twentieth century came in 1953 in a two-page article in the scientific journal *Nature*. In the article, Francis Crick and James Watson postulated a double-stranded helical structure for DNA—the now-famous **double helix**—that not only accounted for the known physical and chemical properties of DNA but also suggested a mechanism for replication of the DNA.

The double helix consists of two complementary chains of DNA twisted together around a common axis to form a right-handed helical structure that resembles a spiral staircase (**Figure 3-19**, on page 64). The two chains are said to be antiparallel, which means they are oriented in opposite directions along the helix, with one running in the 5′ → 3′ direction and the other in the 3′ → 5′ direction. The backbone of each chain consists of sugar molecules alternating with phosphate groups (see Figure 3-18). The phosphate groups are charged, and the sugar molecules contain polar hydroxyl groups. Therefore, it is not surprising that the hydrophilic sugar-phosphate backbones of the two strands are on the outside of the DNA helix, where their interaction with the surrounding aqueous environment can be maximized. The pyrimidine and purine bases, on the other hand, are aromatic compounds with less affinity for water (more hydrophobic). Accordingly, they are oriented inward, away from water, forming the base pairs that hold the two chains together. Hydrophobic interactions among the aromatic rings result in *base stacking*, which helps stabilize the structure of the DNA molecule.

To form a stable DNA double helix, the two component strands must be both antiparallel and *complementary*, that is, each base in one strand will pair with one specific base

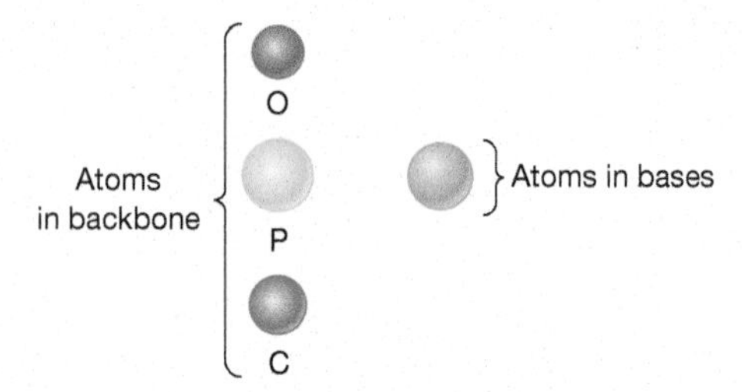

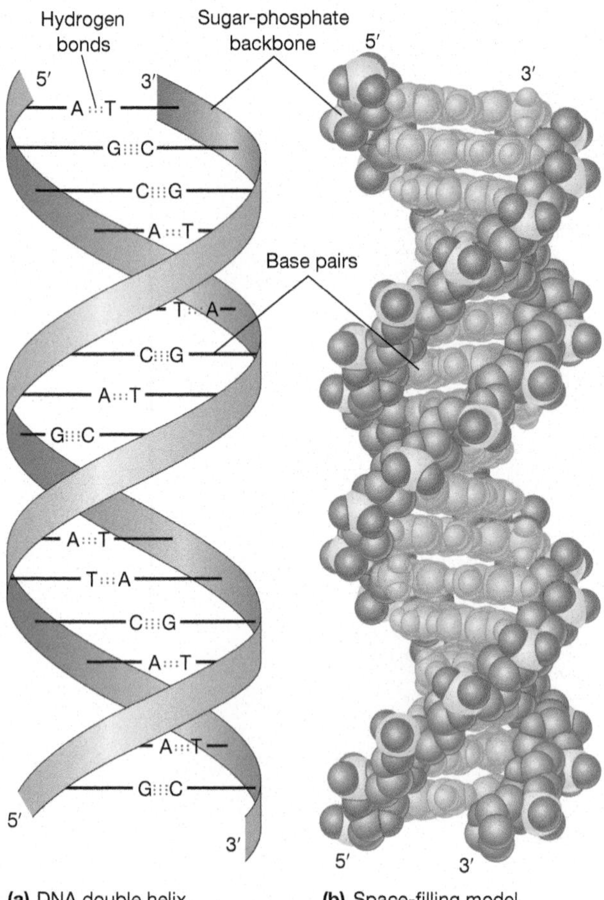

Figure 3-19 The Structure of Double-Stranded DNA. (a) A schematic representation of the double-helical structure of DNA. The continuously turning strips represent the sugar-phosphate backbones of the molecule, and the horizontal bars represent paired bases of the two strands. **(b)** A space-filling model of the DNA double helix, with color-coded atoms as shown at the top of the figure.

directly across from it in the other strand. From the pairing possibilities shown in Figure 3-18, this means that each A must be paired with a T, and each G with a C. In both cases, one member of the pair is a pyrimidine (T or C) and the other is a purine (A or G). The distance between the two sugar-phosphate backbones in the double helix is just sufficient to accommodate one of each kind of base. If we envision the sugar-phosphate backbones of the two strands as the sides of a circular staircase, then each step or rung of the stairway corresponds to a pair of bases held in place by hydrogen bonding (Figure 3-19).

The right-handed Watson–Crick helix shown in Figure 3-19 is actually an idealized version of what is called *B-DNA*. B-DNA is the main form of DNA in cells, but two other rare forms may also exist, perhaps in short segments interspersed within molecules consisting mainly of B-DNA. *A-DNA* has a right-handed, helical configuration that is shorter and thicker than B-DNA. *Z-DNA*, on the other hand, is a left-handed double helix that derives its name from the zigzag pattern of its longer, thinner sugar-phosphate backbone. (For a comparison of the structures of B-DNA and Z-DNA, see Figure 16-9.)

RNA structure also depends in part on base pairing, but this pairing is usually between complementary regions within the same strand and is much less extensive than the interstrand pairing of the DNA duplex. Of the various RNA species, secondary and tertiary structures occur mainly in rRNA and tRNA (see Chapter 19). In addition, some infectious viruses consist of double-stranded RNA held together by hydrogen bonding between complementary base pairs. It is important to keep in mind that RNA can also form a double helix if a long enough stretch of complementary RNA is present.

CONCEPT CHECK 3.2

Like proteins, nucleotides are important informational macromolecules. How are they similar to proteins, and how do they differ in terms of monomer types and assembly, polymer structure, and cellular functions?

3.3 Polysaccharides

The next group of macromolecules we will consider are the **polysaccharides,** which are long-chain polymers of sugars and sugar derivatives. Polysaccharides usually consist of a single kind of repeating unit, or sometimes an alternating pattern of two kinds. They serve primarily in energy storage and as cellular structures rather than carrying information. However (as we will see in Chapter 7), shorter polymers called *oligosaccharides,* when attached to proteins on the cell surface, play important roles in cellular recognition of extracellular signal molecules and of other cells. As noted earlier, polysaccharides include the storage polysaccharides starch and glycogen and the structural polysaccharide cellulose. Each of these polymers contains the six-carbon sugar glucose as its single repeating unit, but they differ in the nature of the bond between successive glucose units as well as in the presence and extent of side branches on the chains.

The Monomers Are Monosaccharides

The repeating units of polysaccharides are simple sugars called **monosaccharides** (from the Greek *mono,* meaning "single," and *sakkharon,* meaning "sugar"). A sugar can be defined as an aldehyde or ketone that has two or more hydroxyl groups (**Figure 3-20**). Thus, there are two categories of sugars: the *aldosugars,* with a terminal carbonyl group (Figure 3-20a), and the *ketosugars,* with an internal carbonyl group at carbon 2 (Figure 3-20b). Within these categories, sugars are named generically according to the number of carbon atoms they contain. Most sugars have between three and seven carbon atoms and thus are classified as *trioses* (three carbons), *tetroses* (four carbons), *pentoses* (five carbons), *hexoses* (six carbons), or *heptoses* (seven carbons). We have already encountered two pentoses—the ribose of RNA and the deoxyribose of DNA.

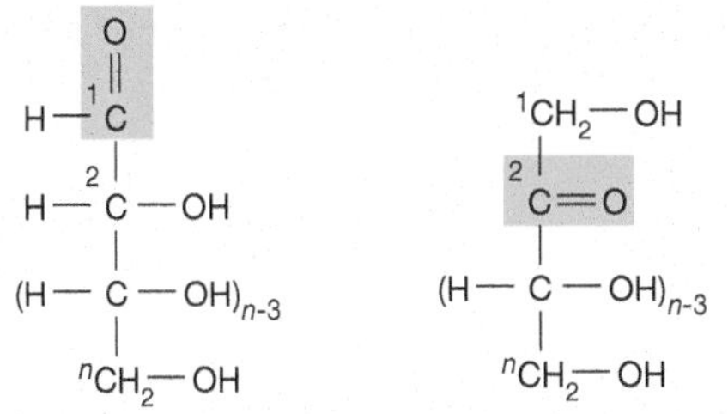

(a) Aldosugar

(b) Ketosugar

Figure 3-20 Structures of Monosaccharides. (a) Aldosugars have a carbonyl group on carbon atom 1. **(b)** Ketosugars have a carbonyl group on carbon atom 2. The number of carbon atoms in a monosaccharide (n) varies from three to seven.

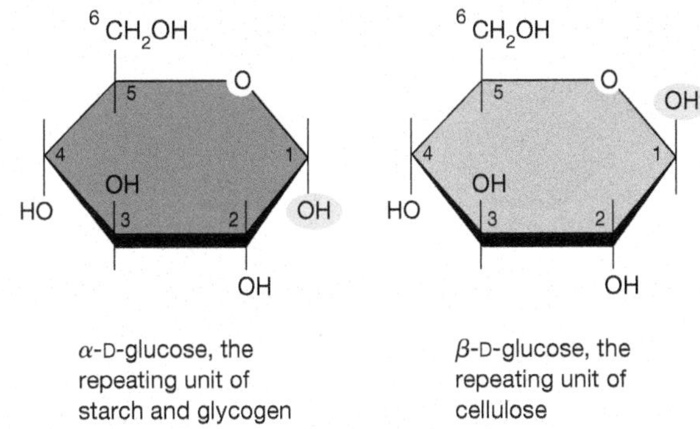

Figure 3-22 The Ring Forms of D-Glucose. The hydroxyl group on carbon atom 1 (blue oval) points downward in the α form and upward in the β form.

The single most common monosaccharide in the biological world is the aldohexose D-glucose, represented by the formula $C_6H_{12}O_6$ and by the structure shown in **Figure 3-21**. The formula $C_nH_{2n}O_n$ is characteristic of sugars and gave rise to the general term **carbohydrate** because compounds of this sort were originally thought of as "hydrates of carbon"—$C_n(H_2O)_n$. Although carbohydrates are not simply hydrated carbons, for every CO_2 molecule incorporated into sugar during photosynthesis, one water molecule is also added (see Reaction 11-2, page 286).

In keeping with the general rule for numbering carbon atoms in organic molecules, *the carbons of glucose are numbered beginning with the more oxidized end of the molecule,* the carbonyl group. Because glucose has four asymmetric carbon atoms (carbon atoms 2, 3, 4, and 5), there are $2^4 = 16$ different possible stereoisomers of the aldosugar $C_6H_{12}O_6$. Here, we will concern ourselves only with D-glucose, which is the most stable of the 16 isomers.

Figure 3-21a illustrates D-glucose as it appears in what chemists call a **Fischer projection,** with the —H and —OH groups intended to project slightly out of the plane of the paper. This structure depicts glucose as a linear molecule, and it is often a useful representation of glucose for teaching purposes. Note that the carbon atoms are numbered from the more oxidized end of the molecule.

In reality, however, glucose exists in the cell in a dynamic equilibrium between the linear (or straight-chain) configuration of Figure 3-21a and the ring form shown in Figure 3-21b. This ring forms when the oxygen atom of the hydroxyl group on carbon atom 5 forms a bond with carbon atom 1. This six-membered ring, formed by five carbon atoms and one oxygen atom, is called a *pyranose* ring. This pyranose ring form is the predominant structure because it is more stable energetically than the linear form.

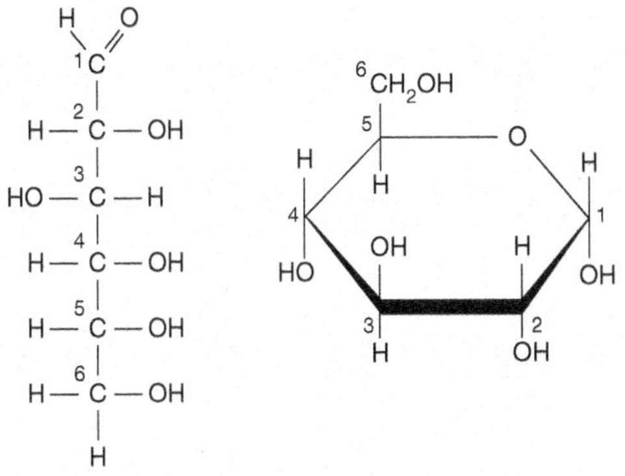

(a) Fischer projection

(b) Haworth projection

Figure 3-21 The Structure of D-Glucose. The D-glucose molecule can be represented by **(a)** the Fischer projection of the straight-chain form or **(b)** the Haworth projection of the ring form.

Therefore, the more satisfactory representation of glucose is the **Haworth projection** shown in Figure 3-21b. This view shows the spatial relationship of different parts of the molecule and makes the spontaneous formation of a bond between an oxygen atom and carbon atoms 1 and 5 appear more likely. In the Haworth projection, carbon atoms 2 and 3 are intended to stick out of the plane of the paper, and carbon atoms 5 and 6 are behind the plane of the paper. The —H and —OH groups then project upward or downward, as indicated. Either of the representations of glucose shown in Figure 3-21 is valid, but the Haworth projection is generally preferred because it indicates both the ring form and the spatial relationship of the carbon atoms.

Notice that formation of the pyranose ring structure results in the generation of one of two alternative forms of the molecule, depending on the spatial orientation of the hydroxyl group on carbon atom 1. These alternative forms of glucose are designated α and β. As shown in **Figure 3-22,** α-D-glucose has the hydroxyl group on carbon atom 1 pointing downward in the Haworth projection, and β-D-glucose has the hydroxyl group on carbon atom 1 pointing upward. Starch and glycogen both have α-D-glucose as their repeating unit, whereas cellulose consists of strings of β-D-glucose.

In addition to the free monosaccharide and the long-chain polysaccharides, glucose also occurs in **disaccharides,** which consist of two monosaccharide units linked covalently. Three common disaccharides are shown in **Figure 3-23** on page 66. *Maltose* (malt sugar) consists of two glucose units linked together, whereas *lactose* (milk sugar) contains a glucose linked to a galactose and *sucrose* (table sugar) has a glucose linked to a fructose. Note that fructose contains a five-membered ring known as a *furanose* ring, a type also found in ribose and deoxyribose.

Each of these disaccharides is formed by a condensation reaction in which two monosaccharides are linked together via an oxygen atom following the elimination of water. The

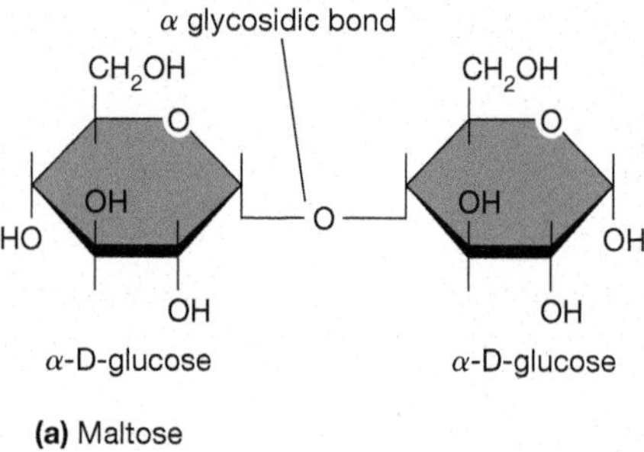

(a) Maltose

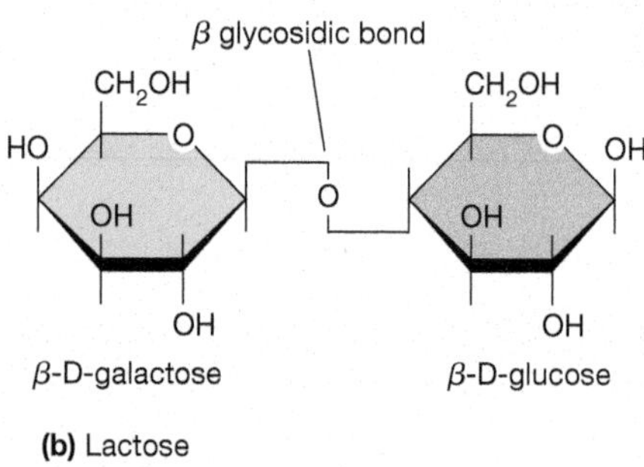

(b) Lactose

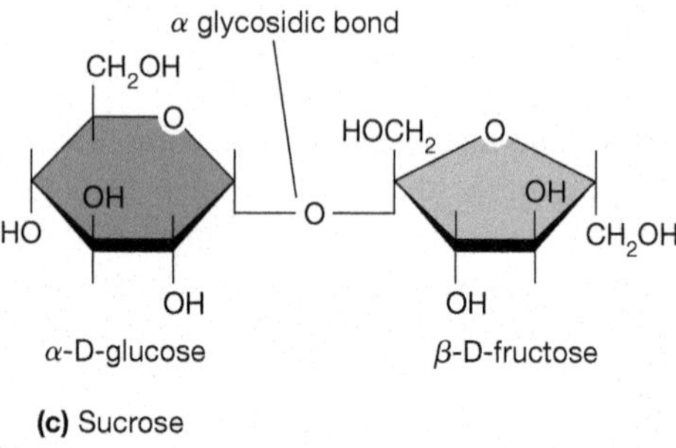

(c) Sucrose

Figure 3-23 Some Common Disaccharides. (a) Maltose consists of two molecules of α-D-glucose, **(b)** lactose consists of a molecule of β-D-galactose linked to β-D-glucose, and **(c)** sucrose consists of a molecule of α-D-glucose linked to β-D-fructose. Note that monomers in maltose and lactose are joined by α glycosidic bonds, but by a β glycosidic bond in sucrose.

resulting **glycosidic bond** is characteristic of linkages between sugars. Note that the "corners" on the glycosidic bonds do not imply the presence of additional atoms—there is only a single oxygen atom connecting the two monomers. In maltose, both of the constituent glucose molecules are in the α form, and the glycosidic bond forms between carbon atom 1 of one glucose and carbon atom 4 of the other (Figure 3-23a). This is called an *α glycosidic bond* because it involves a carbon atom 1 with its hydroxyl group in the α configuration. Lactose, on the other hand, is characterized by a *β glycosidic bond* because the hydroxyl group on carbon atom 1 of the galactose is in the β configuration (Figure 3-23b). Some people lack the enzyme needed to hydrolyze this β glycosidic bond and are considered *lactose intolerant* due to their difficulty in metabolizing this disaccharide. Sucrose consists of a molecule of α-D-glucose linked to a molecule of β-D-fructose by an α glycosidic bond (Figure 3-23c). The distinction between α and β again becomes critical when we get to the polysaccharides because both the three-dimensional configuration and the biological role of the polymer depend critically on the nature of the bond between the repeating monosaccharide units.

The Polymers Are Storage and Structural Polysaccharides

Polysaccharides typically perform either storage or structural roles in cells (**Figure 3-24**). The most familiar **storage polysaccharides** are **starch,** found in plant cells (Figure 3-24a), and **glycogen,** found in animal cells and bacteria (Figure 3-24b). Both of these polymers consist of α-D-glucose units linked together by α glycosidic bonds. In addition to $\alpha(1 \rightarrow 4)$ bonds that link carbon atoms 1 and 4 of adjacent glucose units, these polysaccharides may contain occasional $\alpha(1 \rightarrow 6)$ linkages along the backbone, giving rise to side chains (Figure 3-24c). Storage polysaccharides can therefore be branched or unbranched polymers, depending on the presence or absence of $\alpha(1 \rightarrow 6)$ linkages.

Glycogen is highly branched, with $\alpha(1 \rightarrow 6)$ linkages occurring every 8 to 10 glucose units along the backbone and giving rise to short side chains of about 8 to 12 glucose units (Figure 3-24b). In our bodies, glycogen is stored mainly in the liver and in muscle tissue. In the liver, it is used as a source of glucose to maintain blood sugar levels. In muscle, it serves as a fuel source to generate ATP for muscle contraction. Bacteria also commonly store glycogen as a glucose reserve.

Starch, the glucose reserve commonly found in plant tissue, occurs both as unbranched **amylose** and as branched **amylopectin.** Like glycogen, amylopectin has $\alpha(1 \rightarrow 6)$ branches, but these occur less frequently along the backbone (once every 12 to 25 glucose units) and give rise to longer side chains (lengths of 20 to 25 glucose units are common; Figure 3-24a). Starch deposits are usually 10–30% amylose and 70–90% amylopectin. Starch is stored in plant cells as *starch grains* within the plastids—either within the *chloroplasts* that are the sites of carbon fixation and sugar synthesis in photosynthetic tissue or within the *amyloplasts,* which are specialized plastids for starch storage. The potato tuber, for example, is filled with starch-laden amyloplasts.

A common example of a **structural polysaccharide** is the **cellulose** found in plant cell walls (**Figure 3-25**, on page 68). Cellulose is an important polymer quantitatively—more than half of the carbon in many plants is typically present in cellulose. Like starch and glycogen, cellulose is a polymer of glucose; however, the repeating monomer is β-D-glucose, and the linkage is therefore $\beta(1 \rightarrow 4)$. This bond has structural consequences that we will get to shortly, but it also has nutritional implications. Mammals do not possess an enzyme that can hydrolyze this $\beta(1 \rightarrow 4)$ bond and therefore cannot use cellulose as food. As a result, you can digest potatoes (starch) but not grass and wood (cellulose).

Animals such as cows and sheep might seem to be exceptions because they do eat grass and similar plant products. But they cannot cleave β glycosidic bonds either; they rely on microorganisms (bacteria and protozoa) in their digestive systems to do this for them. The microorganisms digest the cellulose, and the host animal then obtains the end-products of microbial digestion, which are in a form the animal can use (glucose). Even termites do not actually digest wood, but simply chew it into small pieces that are then hydrolyzed to glucose monomers by microorganisms in the termite's digestive tract.

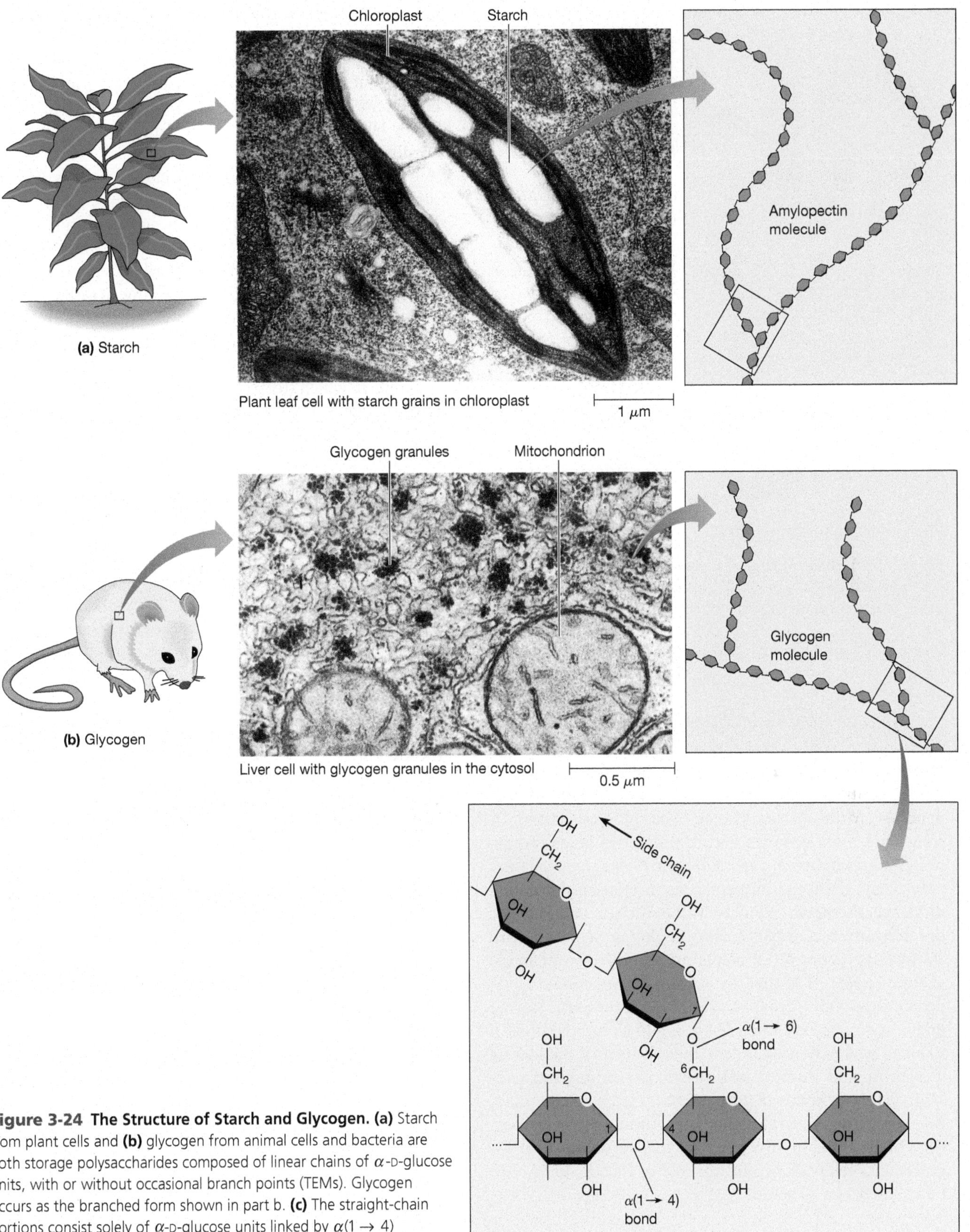

Figure 3-24 The Structure of Starch and Glycogen. (a) Starch from plant cells and **(b)** glycogen from animal cells and bacteria are both storage polysaccharides composed of linear chains of α-D-glucose units, with or without occasional branch points (TEMs). Glycogen occurs as the branched form shown in part b. **(c)** The straight-chain portions consist solely of α-D-glucose units linked by α(1 → 4) glycosidic bonds, whereas branch chains originate at α(1 → 6) glycosidic bonds.

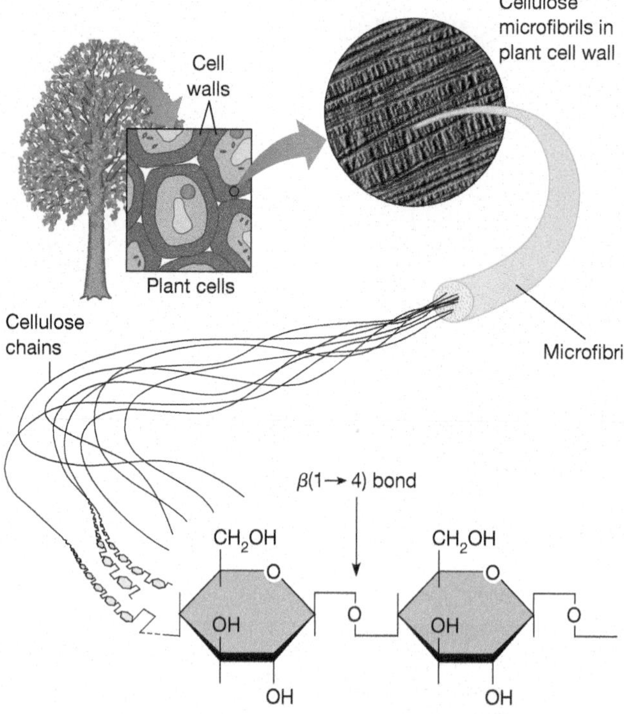

Figure 3-25 The Structure of Cellulose. Cellulose consists of long, unbranched chains of β-D-glucose units linked together by $\beta(1 \rightarrow 4)$ glycosidic bonds. Many such chains associate laterally and are held together by hydrogen bonds to form microfibrils, which can be seen in the micrograph of a primary plant cell wall shown here (TEM).

Although $\beta(1 \rightarrow 4)$-linked cellulose is the most abundant structural polysaccharide, others are also known (**Figure 3-26**). The celluloses of fungal cell walls, for example, contain either $\beta(1 \rightarrow 4)$ or $\beta(1 \rightarrow 3)$ linkages, depending on the species. The cell wall of most bacteria is somewhat more complex and contains two kinds of sugars, *N-acetylglucosamine (GlcNAc)* and *N-acetylmuramic acid (MurNAc)*. These two sugars occur in a strict alternating sequence. As shown in Figure 3-26a, GlcNAc and MurNAc are derivatives of *β-glucosamine,* a glucose molecule with the hydroxyl group on carbon atom 2 replaced by an amino group. GlcNAc is formed by adding a two-carbon acetyl group to the amino group, and MurNAc requires the further addition of a three-carbon lactyl group to carbon atom 3. The cell wall polysaccharide is then formed by the linking of GlcNAc and MurNAc in a strictly alternating sequence with $\beta(1 \rightarrow 4)$ bonds (Figure 3-26b). Figure 3-26c shows the structure of yet another structural polysaccharide, the **chitin** found in insect exoskeletons, crustacean shells, and fungal cell walls. Chitin consists of GlcNAc units only, joined by $\beta(1 \rightarrow 4)$ bonds.

Polysaccharide Structure Depends on the Kinds of Glycosidic Bonds Involved

The distinction between the α and β glycosidic bonds of storage and structural polysaccharides has more than just nutritional significance. Because of the difference in linkages and therefore in the spatial relationship between successive glucose units, the two classes of polysaccharides differ markedly in secondary structure. The helical shape already established as a characteristic of both proteins and nucleic acids is also found in polysaccharides. Both starch and glycogen coil spontaneously into loose helices, but often the structure is not highly ordered due to the numerous side chains of amylopectin and glycogen.

Cellulose, by contrast, forms rigid, linear rods. These in turn aggregate laterally into *microfibrils* (see Figure 3-25). Microfibrils are about 5–20 nm in diameter and are composed of about 36 cellulose chains. Plant and fungal cell walls consist of these rigid microfibrils of cellulose embedded in a *noncellulosic matrix* containing a rather variable mixture of several other polymers (*hemicellulose* and *pectin,* mainly) and a protein called *extensin* that occurs only in the cell wall. Cell walls have been aptly compared to reinforced concrete, in which steel rods are embedded in the cement before it hardens to add strength. In cell walls, the cellulose microfibrils are the "rods" and the noncellulosic matrix is the "cement."

CONCEPT CHECK 3.3

Polysaccharides are also important macromolecules in cell structure and function. How are they similar to proteins and nucleic acids, and how do they differ?

3.4 Lipids

Strictly speaking, **lipids** differ from the macromolecules discussed so far in this chapter because they are not formed by the kind of linear polymerization that gives rise to proteins, nucleic acids, and polysaccharides. However, they are commonly regarded as macromolecules because of their high molecular weights and their presence in important cellular structures, particularly membranes. Also, the final steps in the synthesis of triglycerides, phospholipids, and other large lipid molecules involve condensation reactions similar to those used in polymer synthesis.

Lipids constitute a rather heterogeneous category of cellular components that resemble one another more in their solubility properties than in their chemical structures. *The distinguishing feature of lipids is their hydrophobic nature.* Although they have little, if any, affinity for water, they are readily soluble in nonpolar solvents such as chloroform or ether. Accordingly, we can expect to find that they are rich in nonpolar hydrocarbon regions and have relatively few polar groups. Some lipids, however, are *amphipathic,* having both a polar and a nonpolar region. This characteristic has important implications for membrane structure (as we saw in Figures 2-11 and 2-12).

Because they are defined in terms of solubility characteristics rather than chemical structure, it should not be surprising that lipids as a group include molecules that are diverse in terms of structure, chemistry, and function. Functionally, lipids play at least three main roles in cells. Some serve as forms of *energy storage,* others are involved in *membrane structure,* and still others have *specific biological*

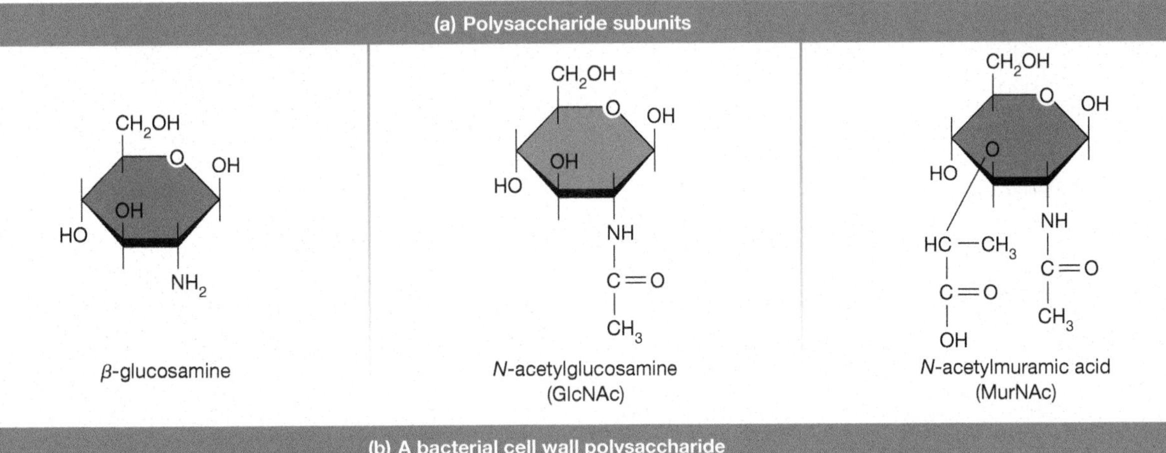

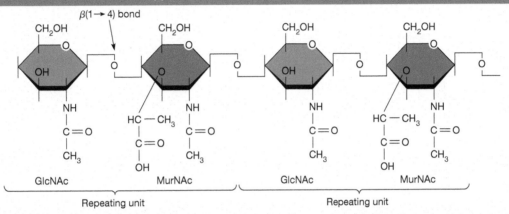

(b) A bacterial cell wall polysaccharide

β(1→4) bond

GlcNAc MurNAc GlcNAc MurNAc

Repeating unit Repeating unit

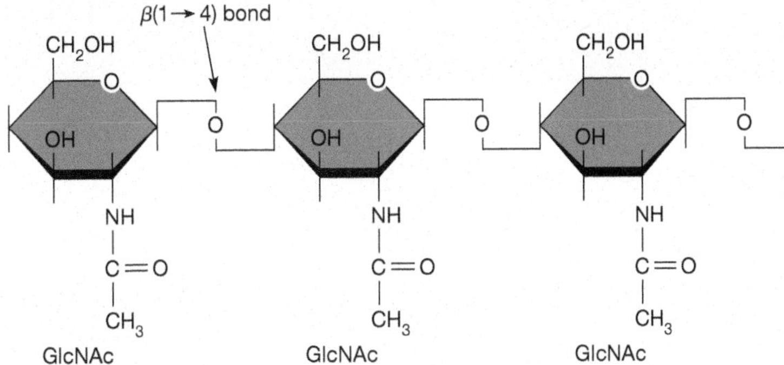

Figure 3-26 Polysaccharides of Bacterial Cell Walls and Insect Exoskeletons. (a) Chemical structures of the monosaccharide subunits glucosamine, *N*-acetylglucosamine (GlcNAc), and *N*-acetylmuramic acid (MurNAc). **(b)** A bacterial cell wall polysaccharide, consisting of alternating GlcNAc and MurNAc units linked by β(1 → 4) bonds. **(c)** Chitin has GlcNAc as its single repeating unit, with successive GlcNAc units linked by β(1 → 4) bonds.

functions, such as the transmission of chemical signals into and within the cell. We will discuss lipids in terms of six main classes, based on their chemical structure: *fatty acids, triacylglycerols, phospholipids, glycolipids, steroids,* and *terpenes.* Note that because of the wide variety of lipids and the fact that members of different classes sometimes share structural and chemical similarities, this is only one of several different ways to classify lipids. The six main classes of lipids discussed here are illustrated in **Figure 3-27** on page 70, which includes representative examples of each class. We will look briefly at each of these six kinds of lipids, pointing out their functional roles in the process.

Fatty Acids Are the Building Blocks of Several Classes of Lipids

Our discussion begins with **fatty acids** because they are components of several other kinds of lipids. A fatty acid is a long, unbranched hydrocarbon chain with a carboxyl group at one end (Figure 3-27a). The fatty acid molecule is therefore amphipathic; the carboxyl group renders one end (often called the "head") polar, whereas the hydrocarbon "tail" is nonpolar. Fatty acids contain a variable, but usually an even, number of carbon atoms. The usual range is from 12 to 20 carbon atoms per chain, with 16- and 18-carbon fatty acids especially common.

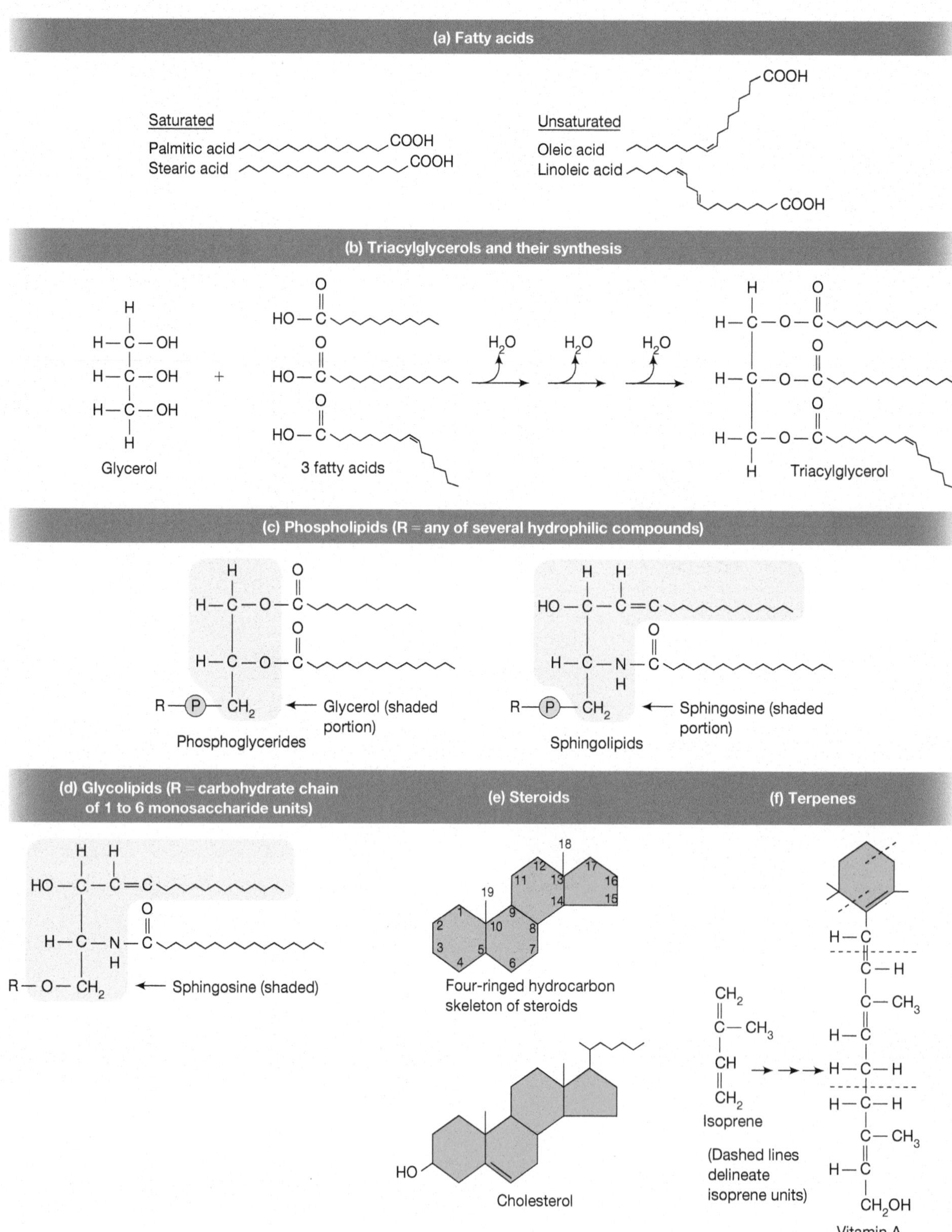

Figure 3-27 The Main Classes of Lipids. The zigzag lines in parts a–d represent the long hydrocarbon chains of fatty acids. Each corner of the zigzag lines represents a methylene (—CH_2—) group.

Table 3-5 lists some common fatty acids. Fatty acids with even numbers of carbon atoms are greatly favored because fatty acid synthesis involves the stepwise addition of two-carbon units to a growing fatty acid chain. Because they are highly reduced, having many hydrogen atoms but few oxygen atoms, fatty acids yield a great deal of energy upon oxidation and are therefore efficient forms of energy storage—a gram of fat contains more than twice as much usable energy as a gram of sugar or polysaccharide.

There is also variability in fatty acids due to the presence of double bonds between carbons (**Figure 3-28**). Fatty acids without double bonds are referred to as **saturated fatty acids** because every carbon atom in the chain has the maximum number of hydrogen atoms attached to it (Figure 3-28a). The

Table 3-5 Some Common Fatty Acids in Cells		
Number of Carbons	Number of Double Bonds	Common Name*
12	0	Laurate
14	0	Myristate
16	0	Palmitate
18	0	Stearate
20	0	Arachidate
16	1	Palmitoleate
18	1	Oleate
18	2	Linoleate
18	3	Linolenate
20	4	Arachidonate

*Shown are the names for the ionized forms of the fatty acids as they exist at the near neutral pH of most cells. For the names of the free fatty acids, simply replace the *-ate* ending with *-ic acid*.

general formula for a saturated fatty acid with *n* carbon atoms is $C_nH_{2n}O_2$. Saturated fatty acids have long, straight chains that pack together well. By contrast, **unsaturated fatty acids** contain one or more double bonds, resulting in a bend or kink in the chain that prevents tight packing (Figure 3-28b; also see Figure 7-13). (Structures and models of several of these fatty acids are shown in Figure 7-8.)

Much concern has arisen recently over a particular type of unsaturated fatty acid known as a *trans fat*. *Trans* fats contain unsaturated fatty acids with a particular type of double bond that causes less of a bend in the fatty acid chain (see Figure 7-13). This causes them to resemble saturated fatty acids in both their shape and their ability to pack together more tightly than typical unsaturated fatty acids. Although they are naturally present in small amounts in meat and dairy products, *trans* fats are produced artificially during the commercial production of shortening and margarine. *Trans* fats have been linked to changes in blood cholesterol that are associated with increased risk of heart disease.

Triacylglycerols Are Storage Lipids

The **triacylglycerols**, also called *triglycerides*, consist of a glycerol molecule with three fatty acids linked to it. As shown in Figure 3-27b, **glycerol** is a three-carbon alcohol with a hydroxyl group on each carbon. Fatty acids are linked to glycerol by *ester bonds*, which are formed between a carboxyl and a hydroxyl group by the removal of water. Triacylglycerols are synthesized stepwise, with one fatty acid added at a time. *Monoacylglycerols* contain a single fatty acid, *diacylglycerols* have two, and *triacylglycerols* have three. The three fatty acids of a given triacylglycerol generally vary in chain length, degree of unsaturation, or both. Each fatty acid in a triacylglycerol is linked to a carbon atom of glycerol by means of a condensation reaction.

The main function of triacylglycerols is to store energy. In some animals, triacylglycerols also provide insulation against low temperatures. Animals that live in very cold climates, such as walruses, seals, and penguins, store triacylglycerols under their skin and depend on the insulating properties of this fat for survival.

Triacylglycerols containing mostly saturated fatty acids are usually solid or semisolid at room temperature and are called *fats*. Fats are prominent in the bodies of animals, as evidenced by the fat that comes with most cuts of meat, by the large quantity of lard that is obtained as a by-product of the meat-packing industry, and by the widespread concern people have that they are "getting fat." In plants, most triacylglycerols are liquid at room temperature, as the term *vegetable oil* suggests. Because the fatty acids of oils are predominantly unsaturated, their hydrocarbon chains have kinks that prevent an orderly packing of the molecules. As a result, vegetable oils have lower melting temperatures than most animal fats do. Soybean oil and corn oil are two familiar vegetable oils. Vegetable oils can be converted into solid products such as margarine and shortening by hydrogenation (saturation) of the double bonds, a process explored further in Problem 3-15 at the end of the chapter.

Phospholipids Are Important in Membrane Structure

Phospholipids make up a third class of lipids (see Figure 3-27c). They are similar to triacylglycerols in some chemical details but differ strikingly in their properties and their role in the cell. First and foremost, phospholipids are important in membrane structure due to their amphipathic nature and are a key component of the bilayer structure found in all membranes (see Figure 2-12). Based on their chemistry, phospholipids are classified as *phosphoglycerides* or *sphingolipids* (see Figure 3-27c).

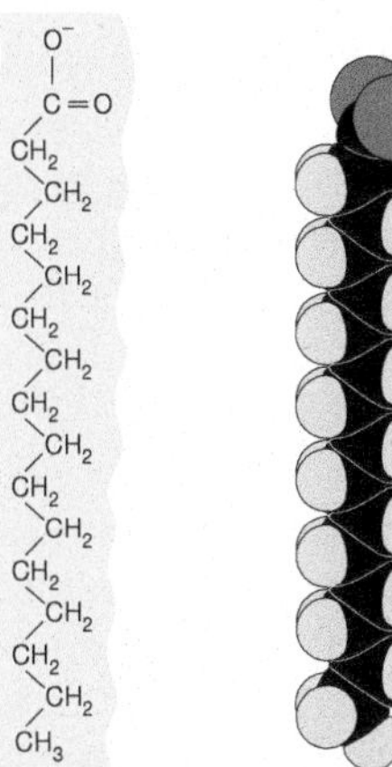

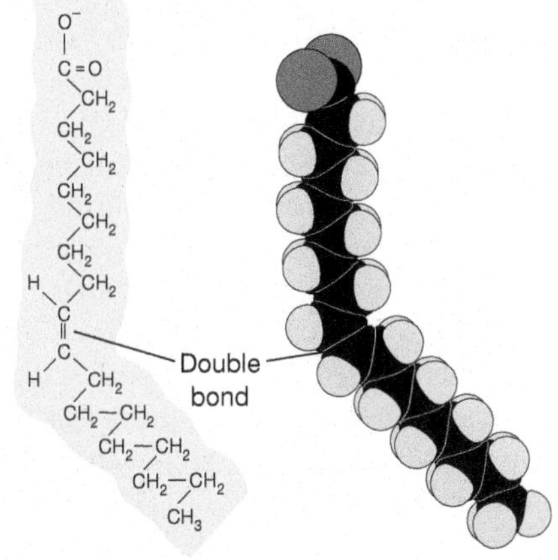

Figure 3-28 Structures of Saturated and Unsaturated Fatty Acids. (a) The saturated 16-carbon fatty acid palmitate. **(b)** The unsaturated 18-carbon fatty acid oleate. The space-filling models are intended to emphasize the overall shape of the molecules. Notice the kink that the double bond creates in the oleate molecule.

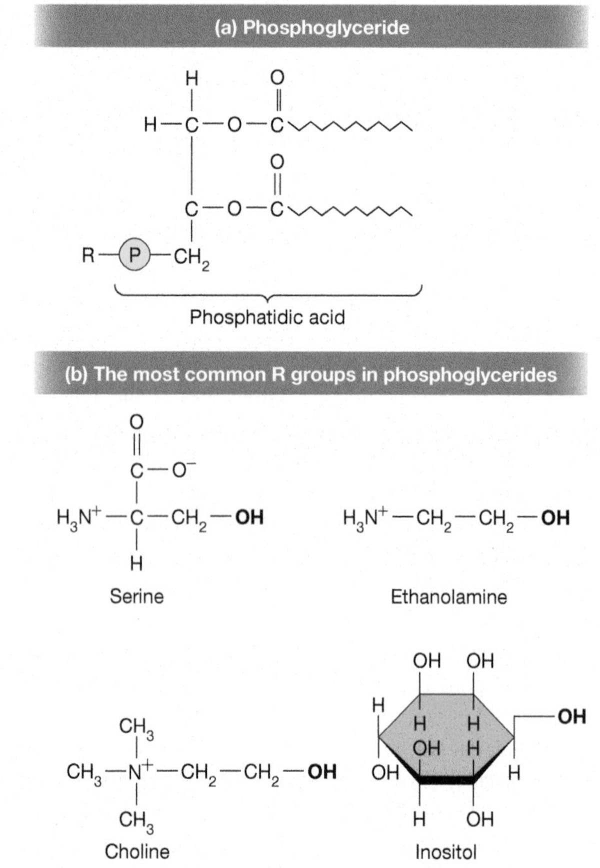

(a) Phosphoglyceride

Phosphatidic acid

(b) The most common R groups in phosphoglycerides

Serine

Ethanolamine

Choline

Inositol

Figure 3-29 Structures of Common Phosphoglycerides.
(a) A phosphoglyceride consists of a molecule of phosphatidic acid with a small polar alcohol (R) attached to the phosphate group.
(b) The four most common R groups found in phosphoglycerides are serine, ethanolamine, choline, and inositol. (The bold hydroxyl group highlights where each R group bonds with phosphatidic acid.)

Phosphoglycerides are the predominant phospholipids present in most membranes. Like triacylglycerols, a phosphoglyceride consists of fatty acids that are esterified to a glycerol molecule (**Figure 3-29**). However, the basic component of a phosphoglyceride is **phosphatidic acid,** which has just two fatty acids and a phosphate group attached to a glycerol backbone (Figure 3-29a). Phosphatidic acid is a key intermediate in the synthesis of other phosphoglycerides but is itself not at all prominent in membranes. Instead, membrane phosphoglycerides invariably have, in addition, a small hydrophilic alcohol linked to the phosphate by an ester bond. The alcohol is usually *serine, ethanolamine, choline,* or *inositol,* groups that contribute to the polar nature of the phospholipid head group (Figure 3-29b). The first three of these groups contain a positively charged amino group or nitrogen atom, and inositol is a sugar derivative containing a number of polar oxygen atoms.

The combination of a highly polar head and two long nonpolar chains gives phosphoglycerides the characteristic amphipathic nature that is so critical to their role in membrane structure. As shown earlier, the fatty acids can vary considerably in both length and the presence and position of sites of unsaturation. In membranes, 16- and 18-carbon fatty acids are most common, and a typical phosphoglyceride molecule is likely to have one saturated and one unsaturated fatty acid. The length and the degree of unsaturation of fatty acid chains in membrane phospholipids profoundly affect membrane fluidity and can, in fact, be regulated by the cells of some organisms.

In addition to the phosphoglycerides, some membranes contain another class of phospholipid called **sphingolipids,** which are important in membrane structure and cell signaling. In the nineteenth century, when Johann Thudicum first discovered sphingolipids, their biological role seemed as enigmatic as the Sphinx, after which he named them. As the name suggests, these lipids are based not on glycerol but on the amine alcohol **sphingosine.** As shown in Figure 3-27c, sphingosine has a long hydrocarbon chain with a single site of unsaturation near the polar end. Through its amino group, sphingosine can form a bond to a long-chain fatty acid (up to 34 carbons). The resulting molecule is called a *ceramide* and consists of a polar region flanked by two long, nonpolar tails, giving it a shape approximating that of the phospholipids.

The hydroxyl group on carbon atom 1 of sphingosine juts out from what is effectively the head of this hairpin molecule. A sphingolipid is formed when any of several polar groups becomes linked to this hydroxyl group. A whole family of sphingolipids exists, differing only in the chemical nature of the polar group attached to the hydroxyl group of the ceramide (the R group of Figure 3-27c). Sphingolipids are present predominantly in the outer monolayer of the plasma membrane bilayer, where they often are found in *lipid rafts,* which are localized microdomains within a membrane that facilitate communication with the external environment of the cell (see Chapter 7).

MAKE CONNECTIONS 3.2

The ends of amphipathic phospholipids are critical to forming lipid bilayers. How would membrane formation differ if you added a polar group to the end of the nonpolar tail of all phospholipids? (Ch. 2.3)

Glycolipids Are Specialized Membrane Components

Glycolipids are lipids containing a carbohydrate group instead of a phosphate group and are typically derivatives of sphingosine or glycerol (see Figure 3-27d). Those containing sphingosine are called *glycosphingolipids.* The carbohydrate group attached to a glycolipid may contain one to six sugar units, which can be D-glucose, D-galactose, or N-acetyl-D-galactosamine. These carbohydrate groups, like phosphate groups, are hydrophilic, giving the glycolipid an amphipathic nature. Glycolipids are specialized constituents of some membranes, especially those found in certain plant cells and in the cells of the nervous system. Glycolipids occur largely in the outer monolayer of the plasma membrane, and the glycosphingolipids are often sites of biological recognition on the surface of the plasma membrane.

Steroids Are Lipids with a Variety of Functions

The **steroids** constitute yet another distinctive class of lipids. Steroids are derivatives of a four-ringed hydrocarbon skeleton (see Figure 3-27e), which makes them structurally distinct from other lipids. In fact, the only property linking them to the other classes of lipids is that they are relatively nonpolar and therefore hydrophobic. As (**Figure 3-30**) illustrates, steroids differ from one another in the number and positions of double bonds and functional groups.

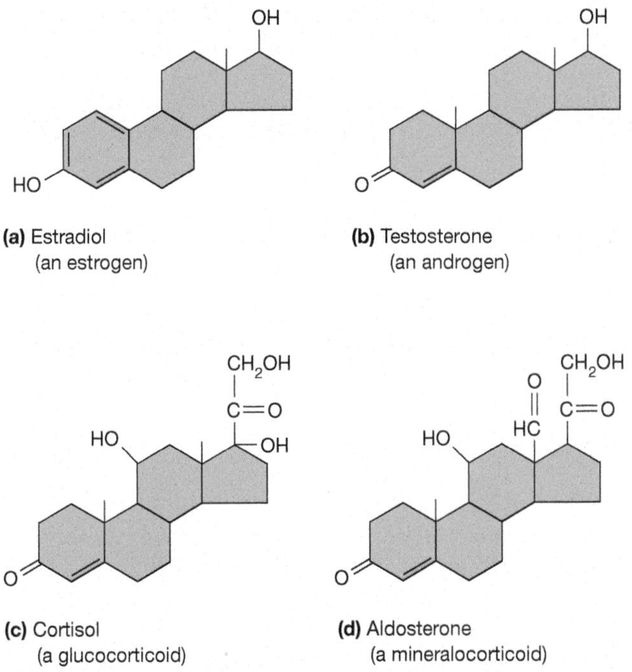

(a) Estradiol (an estrogen)

(b) Testosterone (an androgen)

(c) Cortisol (a glucocorticoid)

(d) Aldosterone (a mineralocorticoid)

Figure 3-30 Structures of Several Common Steroid Hormones. Among the many steroids that are synthesized from cholesterol are the hormones **(a)** estradiol, an estrogen; **(b)** testosterone, an androgen; **(c)** cortisol, a glucocorticoid; and **(d)** aldosterone, a mineralocorticoid.

Steroids are found almost exclusively in eukaryotic cells. The most common steroid in animal cells is **cholesterol,** the structure of which is shown in Figure 3-27e. Cholesterol is an amphipathic molecule, with a polar head group (the hydroxyl group at position 3) and a nonpolar hydrocarbon body and tail (the four-ringed skeleton and the hydrocarbon side chain at position 17). Because most of the molecule is hydrophobic, cholesterol is insoluble in water and is found primarily in membranes. It occurs in the plasma membrane of animal cells and in most of the membranes of organelles, except the inner membranes of mitochondria and chloroplasts. Similar membrane steroids also occur in other cells, including *stigmasterol* and *sitosterol* in plant cells, *ergosterol* in fungal cells, and *hopanoids* in *Mycoplasma* bacteria.

Cholesterol is the starting point for the synthesis of all the **steroid hormones** (Figure 3-30), which include the male and female *sex hormones*, the *glucocorticoids*, and the *mineralocorticoids*. The sex hormones include the *estrogens* produced by the ovaries of females (*estradiol*, for example) and the *androgens* produced by the testes of males (*testosterone*, for example). The glucocorticoids (*cortisol*, for example) are a family of hormones that promote gluconeogenesis (synthesis of glucose) and suppress inflammation reactions. They also play an important role in stress responses. Mineralocorticoids such as *aldosterone* regulate ion balance by promoting the reabsorption of sodium, chloride, and bicarbonate ions in the kidney. Plants also use a sterol backbone to create hormones called *brassinosteroids*. In this case, the sterol is *campesterol*, a phytosterol similar in structure to cholesterol.

Terpenes Are Formed from Isoprene

The final class of lipids shown in Figure 3-27 consists of the **terpenes.** Terpenes are synthesized from the five-carbon compound *isoprene* and are therefore also called *isoprenoids*. Isoprene and its derivatives are joined together in various combinations to produce such substances as *vitamin A* (see Figure 3-27f), a required nutrient in our bodies, and *carotenoid pigments*, which are involved in light harvesting in plants during photosynthesis. Other isoprene-based compounds are *dolichols*, which are involved in activating sugar derivatives, and electron carriers such as *coenzyme Q* and *plastoquinone* (which we will encounter when we study respiration and photosynthesis in detail in Chapters 10 and 11). Finally, polymers of isoprene units known as *polyisoprenoids* are found in the cell membranes of Archaea, a unique domain of organisms distinct from eukaryotes and bacteria (see Chapter 4).

CONCEPT CHECK 3.4

Lipids, our final class of macromolecules, are quite different from the other three classes, yet have some similarities. Explain.

Summary of Key Points

Mastering™ Biology For activities, animations, and review quizzes, go to the study area at www.masteringbiology.com.

3.1 Proteins

- Four major classes of macromolecules are prominent in cells: proteins, nucleic acids, polysaccharides, and lipids. Proteins, nucleic acids, and polysaccharides are macromolecular polymers composed of a long series of monomers linked in a specific order. Lipids are not long, polymeric macromolecules, but they are discussed here because of their general importance as constituents of cells, especially membranes.
- All of the thousands of different proteins in a cell are linear chains of amino acid monomers. Each of the 20 different amino acids found in proteins has a different R group, which can be hydrophobic, hydrophilic uncharged, or hydrophilic charged. These amino acids can be linked together in any sequence via peptide bonds to form the wide variety of polypeptides that make up monomeric and multimeric proteins.
- The precise order of the monomers carries information. The amino acid sequence of a polypeptide (its primary structure) usually contains all of the information necessary to specify local folding into α helices and β sheets (secondary structure), overall three-dimensional shape (tertiary structure), and, for multimeric proteins, association with other polypeptides (quaternary structure).
- Major forces influencing polypeptide folding and stability are covalent disulfide bond formation and several types of noncovalent interactions: hydrogen bonds, ionic bonds, van der Waals interactions, and hydrophobic interactions.
- Despite extensive knowledge of the forces involved in protein folding, we are still not able to predict the final folded tertiary structure of a protein from its primary amino acid sequence, except in the case of peptides and relatively small proteins.

3.2 Nucleic Acids

- The nucleic acids DNA and RNA are informational macromolecules composed of nucleotide monomers linked together by phosphodiester bridges in a specific order. Each nucleotide is composed of a deoxyribose or ribose sugar, a phosphate, and a purine or pyrimidine base.
- Whereas RNA is mainly single stranded, DNA forms a double-stranded helix based on complementary base pairing (A with T and C with G) that is stabilized by hydrogen bonding. Elucidation of the double-helical structure of DNA and its significance for DNA as the carrier of genetic information was a defining biological breakthrough of the twentieth century.
- As in proteins, the precise order of monomers in nucleic acids carries information. The base sequence of the nucleotides in a particular segment of DNA known as a gene determines the sequence of the amino acids in the protein encoded by that gene.

3.3 Polysaccharides

- In contrast to nucleic acids and proteins, polysaccharides show little variation in sequence and do not carry information, but instead serve storage or structural roles. They typically consist of either a single type of monosaccharide or two alternating monosaccharides linked together by either α or β glycosidic bonds. The type of glycosidic bond determines whether the polysaccharide serves as energy storage or as a structural polysaccharide.
- The α linkages are readily digestible by animals and are found in storage polysaccharides such as starch and glycogen, which consist solely of glucose monomers. In contrast, the β glycosidic bonds of cellulose and chitin are generally not digestible by animals and give these molecules a rigid shape suitable to their functions as structural molecules.

3.4 Lipids

- Lipids are not true polymers but often considered macromolecules due to their high molecular weight and frequent association with proteins. Lipids vary substantially in structure but are grouped together because they share the property of being hydrophobic and nearly insoluble in water.
- Fatty acids are lipids consisting of a long hydrocarbon chain of 12–20 carbon atoms with a carboxylic acid group at one end. They are energy-rich molecules found in the triacylglycerols that make up animal fats and vegetable oils, as well as in the phospholipids found in all cellular membranes.
- Phosphoglycerides and sphingolipids are types of phospholipids that make up the lipid bilayer of biological membranes. They are amphipathic molecules with two hydrophobic fatty acid chains and a polar phosphate-containing head group.
- Glycolipids are similar to phospholipids but have a polar carbohydrate group instead of phosphate. They are often found on the outer surface of membranes, and play a role in cell recognition.
- Other important cellular lipids are steroids (cholesterol and steroid hormones) and terpenes (vitamin A and some coenzymes).

Problem Set

3-1 Polymers and Their Properties. For each of the six biological polymers listed, indicate which of the properties apply. Each polymer has multiple properties, and a given property may be used more than once.

Polymers

(a) Cellulose

(b) Messenger RNA

(c) Globular protein

(d) Amylopectin

(e) DNA

(f) Fibrous protein

Properties

1. Branched-chain polymer
2. Extracellular location
3. Glycosidic bonds
4. Informational macromolecule
5. Peptide bond
6. β linkage
7. Phosphodiester bridge
8. Nucleoside triphosphates
9. Helical structure possible
10. Synthesis requires a template

3-2 Stability of Protein Structure. Several different kinds of bonds or interactions are involved in generating and maintaining the structure of proteins. List five such bonds or interactions, give an example of an amino acid that might be involved in each, and indicate which level(s) of protein structure might be generated or stabilized by that particular kind of bond or interaction.

3-3 Amino Acid Localization in Proteins. Amino acids tend to be localized either in the interior or on the exterior of a globular protein molecule, depending on their relative affinities for water.

(a) For each of the following pairs of amino acids, choose the one that is more likely to be found in the interior of a protein molecule, and explain why.

Alanine; glycine — Glutamate; aspartate

Tyrosine; phenylalanine — Methionine; cysteine

(b) Explain why cysteine residues with free sulfhydryl groups tend to be localized on the exterior of a protein molecule, whereas those involved in disulfide bonds are more likely to be buried in the interior of the molecule.

3-4 Sickle-Cell Anemia. Sickle-cell anemia is a striking example of the drastic effect a single amino acid substitution can have on the structure and function of a protein.

(a) Given the chemical nature of glutamate and valine, can you suggest why substitution of valine for glutamate at position 6 of the β chain would be especially deleterious?

But assuming that RNA existed before DNA still leaves us with the question of how RNA was generated and replicated in the absence of enzymes. Scientists have shown that ribonucleotides, the building blocks of RNA polymers, can form abiotically and that in some cases RNA polymers can form, although this process is not nearly as efficient as in cells. But what about replication of RNA? A major step toward answering this question came in the early 1980s, when Sidney Altman and Thomas Cech discovered that some RNAs can fold into shapes that resemble protein enzymes, allowing them to catalyze chemical reactions. These catalytic RNAs became known as *ribozymes*.

Others have since shown that ribozymes can catalyze an array of different reactions, including synthesis of short pieces of RNA by complementary base pairing when supplied with nucleotide building blocks. These findings support the hypothesis that there was an "RNA world" before the appearance of DNA. Indeed, the nature of modern RNAs supports the idea of a prebiotic RNA world. For example, in the ribosome, the site of protein synthesis in all known organisms, it is the ribosomal RNA, not ribosomal proteins, that catalyzes peptide bond formation between amino acids to produce proteins.

There are limitations to the conclusions that can be drawn from current experiments, however. First, for RNAs to have been the first macromolecules, they would need to catalyze the formation of new RNA polymers as long as themselves. So far, RNAs have not been shown to do so, although some researchers are now close to achieving this in laboratory settings. Second, it remains unknown how the genetic code—which is universal to all living organisms and requires intricate interplay between triplet codons, tRNA anticodons, and amino acid synthesis (see Chapter 18)—could have arisen. Many scientists believe that part of the answer may lie in the selective stabilization of specific associations between proteins and RNAs from among their many possible interactions.

Liposomes May Have Defined the First Primitive Protocells

Even if catalytic RNAs capable of replicating an RNA molecule existed on early Earth, they alone would not constitute a cell. How could the biochemical contents of the precursors to the first cells have become separated from the environment? One clue is that all cells today are surrounded by a lipid bilayer membrane, which encloses the aqueous space inside the cell.

Using lipids similar to those found in cellular membranes, scientists have produced artificial structures called **liposomes**—hollow membrane-bound vesicles of varying sizes that form spontaneously when lipids are mixed with water (**Figure 4-2**). Abiotically produced vesicles can "reproduce" by splitting on their own (Figure 4-2a), and they can increase in size under appropriate conditions. By creating liposomes in a solution containing enzymes and their substrates, scientists have created liposomes that can even carry out simple metabolic reactions. Primordial lipids may have similarly formed in aqueous environments in the early Earth, trapping nearby RNAs and in the process forming the first "protocells." In 2013, Jack Szostak and colleagues were able to create vesicles in the laboratory capable of copying a template strand of RNA. If such vesicles existed in the early Earth, they might have been able to grow, split, and pass along RNA to "daughter" protocells.

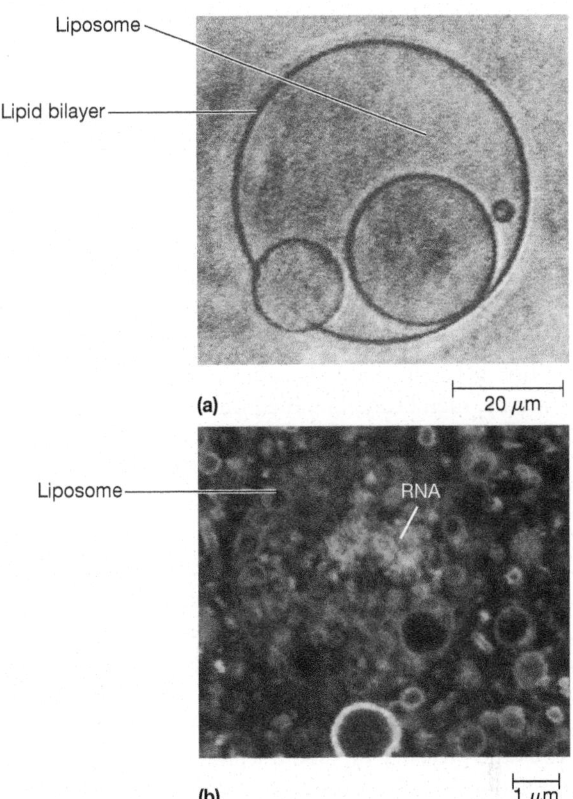

Figure 4-2 Artificial Liposomes. (a) Light micrograph of liposomes formed when membrane lipids are added to water. **(b)** Encapsulation of RNA inside a liposome could have formed precursors to the first primitive cells.

One process that may have fostered the trapping of RNA into vesicles involves the soft mineral clay *montmorillonite*, derived from volcanic ash that many scientists believe was abundant on the early Earth, and may have greatly increased the rate of vesicle self-assembly (see Problem 4-1, on page 105). Montmorillonite can concentrate organic molecules, increasing the likelihood of chemical reactions. In addition, vesicles can absorb montmorillonite particles, including those to which RNA has become attached (Figure 4-2b). Others have suggested that heating, drying, and rewetting events could have produced protocell structures that fostered complex chemical reactions.

Although we may never know with certainty how the first cells *did* form, we are learning about how chemical processes essential to all life *can* arise, at least in the laboratory. This is an active area of research that is likely to reveal future surprises and insights into the origin of life itself.

CONCEPT CHECK 4.1

Why do many scientists currently believe that RNA, rather than DNA, was the first informational molecule?

4.2 Basic Properties of Cells

As we begin to consider what cells are and how they function, several general characteristics of cells quickly emerge. These include the organizational complexity and molecular components of cells, their sizes and shapes, and the specializations they exhibit.

The Three Domains of Life Are Bacteria, Archaea, and Eukaryotes

With improvements in microscopy, biologists came to recognize two fundamentally different types of cellular organization: a simpler one characteristic of bacteria and a more complex one found in all other kinds of cells. Based on the structural differences of their cells, organisms have been traditionally divided into two broad groups, the **prokaryotes** (bacteria and archaea) and the **eukaryotes** (plants, animals, fungi, algae, and protozoa). The most fundamental distinction between the two groups is that eukaryotic cells have a true, membrane-bounded nucleus (*eu–* is Greek for "true" or "genuine"; *–karyon* means "nucleus"), whereas prokaryotic cells do not (*pro–* means "before," suggesting an evolutionarily earlier form of life).

Recently, however, the term *prokaryote* is becoming less satisfactory to describe these non-nucleated cells, partly because this is a negative classification based on what cells do not have and partly because it wrongly implies a fundamental similarity among all organisms whose cells lack a nucleus. The fact that two organisms lack a particular gross structural feature does not necessarily imply evolutionary relatedness, nor does the presence of such a feature necessarily mean a close relationship. For example, although plants and most bacteria have cell walls, they are only distantly related.

Molecular and biochemical criteria are proving to be more reliable than structural criteria in describing evolutionary relationships among organisms. The more closely related two different organisms are, the more similar are the sequences of their particular DNA, RNA, and protein molecules. Especially useful for comparative studies are molecules common to all living organisms that are necessary for universal, basic processes, in which even slight changes in component molecules are not well tolerated. This includes molecules such as the ribosomal RNAs used in protein synthesis and the cytochrome proteins used in energy metabolism.

Based on the pioneering ribosomal RNA sequencing work of Carl Woese, Ralph Wolfe, and co-workers starting in the 1970s, it is now clear that the group traditionally called the prokaryotes actually includes two widely divergent groups—the **bacteria** and the **archaea**, which in many ways are as different from each other as humans are from bacteria. Rather than the prokaryote-eukaryote dichotomy, **Figure 4-3** illustrates a more biologically accurate way to describe all organisms as belonging to one of three *domains*—the Bacteria, the Archaea, or the **Eukarya** (eukaryotes). As shown in **Table 4-1**, cells of each of these domains share some structural, biochemical, and genetic characteristics with cells of the other domains, but all three domains possess unique characteristics as well.

The bacteria include most of the commonly encountered single-celled, non-nucleated organisms that were traditionally referred to as prokaryotes—for example, *Escherichia coli, Pseudomonas aeruginosa,* and *Streptococcus lactis*. The archaea (which were called *archaebacteria* before investigators realized how different they are from bacteria) include many species that live in extreme habitats on Earth and have very diverse metabolic strategies. Members of the archaea include the *methanogens,* which obtain energy from hydrogen while converting carbon dioxide into methane; the *halophiles,* which can grow in extremely salty environments; and the *thermacidophiles,* which thrive in acidic hot springs where the pH can be as low as 2 and the temperature can exceed 100°C. Rather than being considered an evolutionarily ancient form of prokaryote (*archae–* is a Greek prefix meaning "ancient"), archaea are now considered

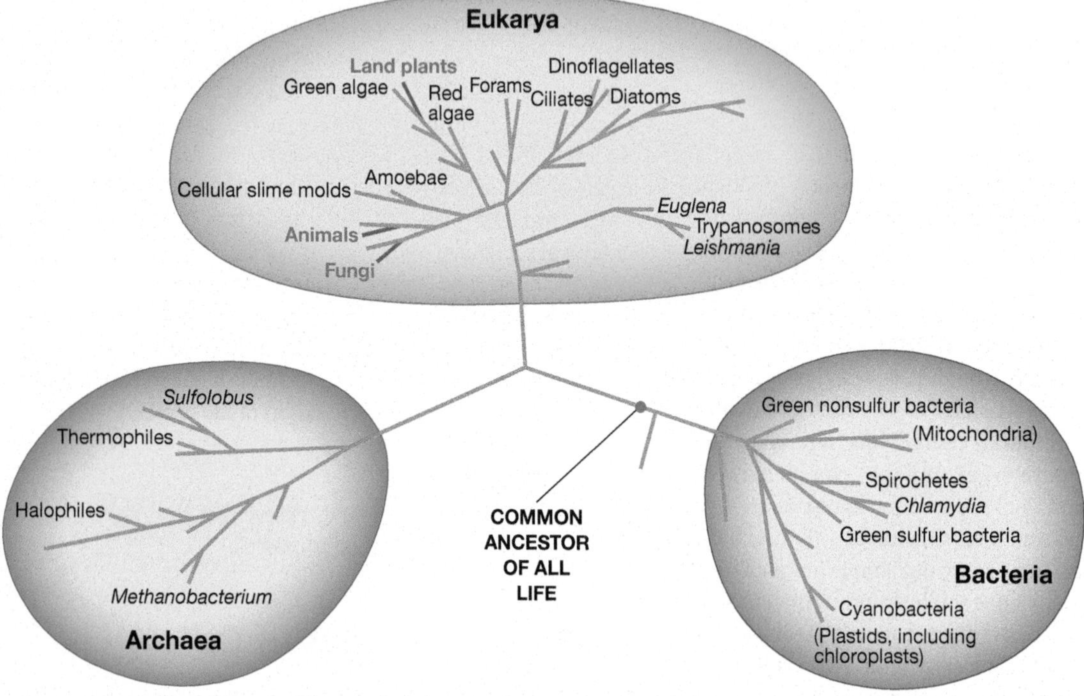

Figure 4-3 The Three Domains of Life. The three domains of life have been determined via numerous DNA sequence comparisons, including ribosomal RNA genes. Representative organisms within the three domains are shown.

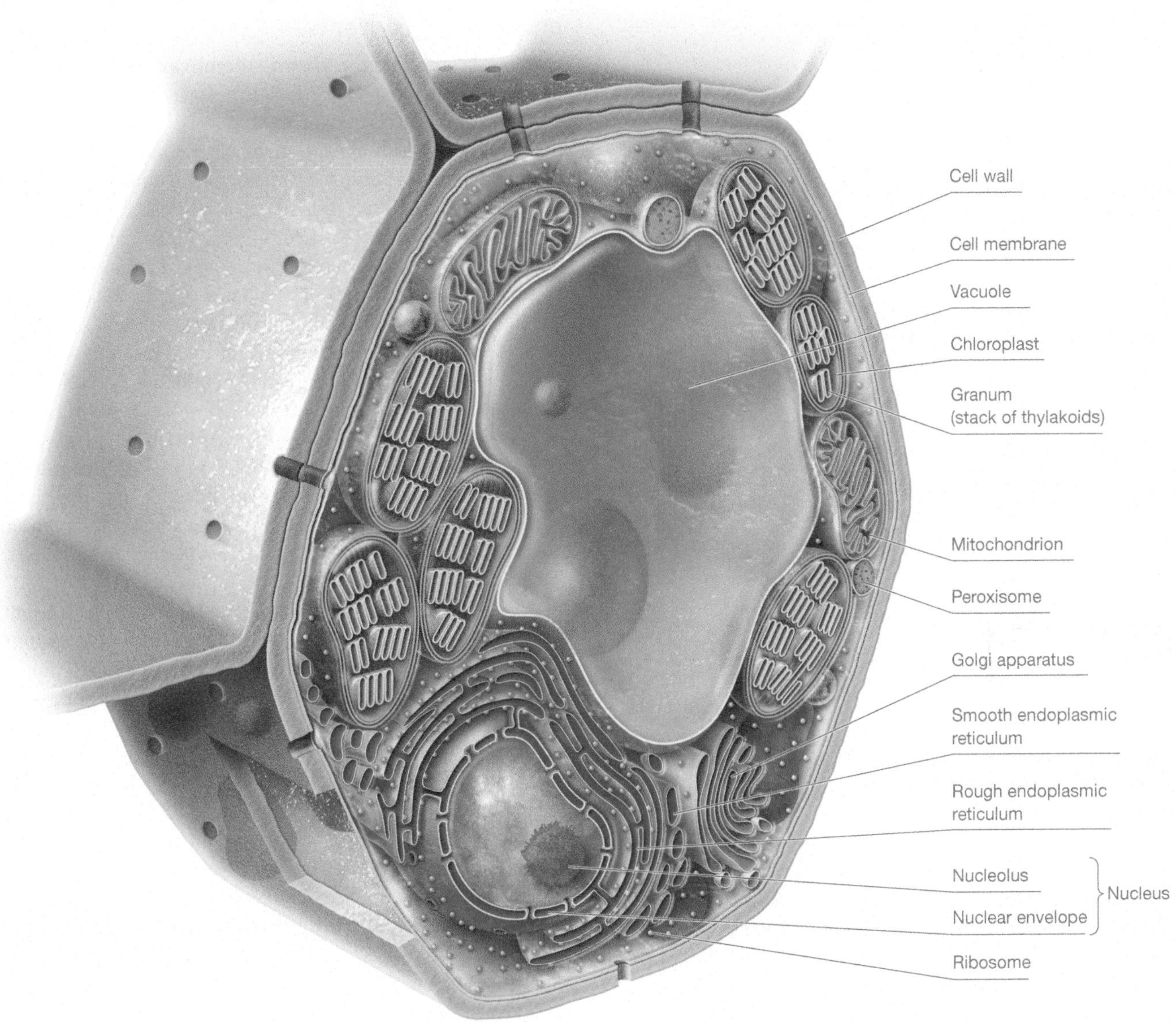

Figure 4-8 A Plant Cell. A schematic drawing of a plant cell.

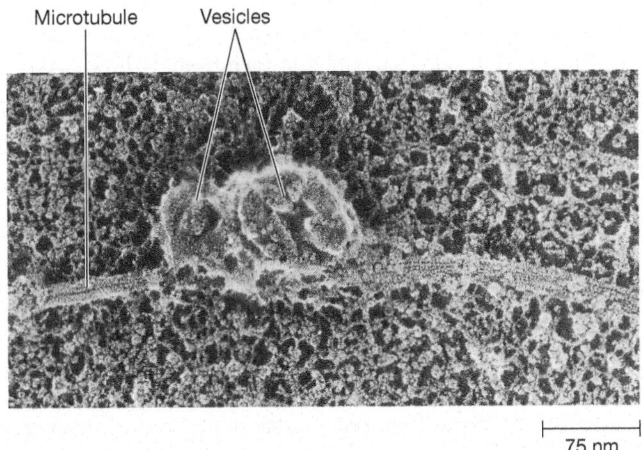

Figure 4-9 Vesicle Transport. This SEM of a squid giant axon shows two neurotransmitter-containing vesicles attached to a microtubule in the cytoplasm. The microtubule provides a "track" to move the molecules in these vesicles through the cell to the axon tips, where they will aid in nerve cell signaling.

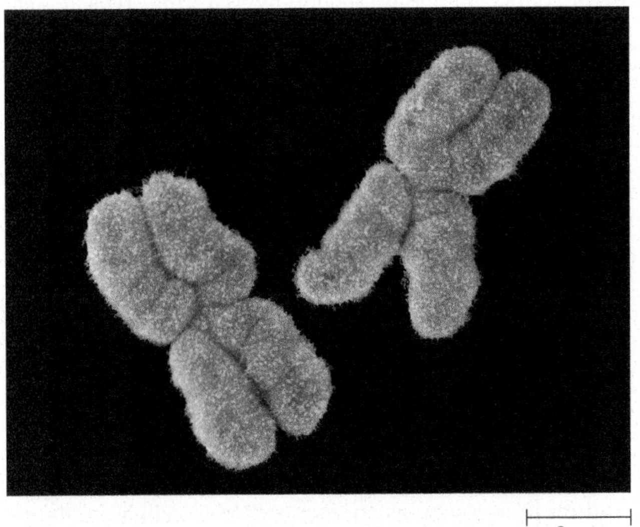

Figure 4-10 A Pair of Eukaryotic Chromosomes. These human chromosomes were obtained from a cell during the process of cell division and thus are highly coiled and condensed (SEM).

mature messenger RNA (mRNA) molecules to the cytoplasm for protein synthesis. Each mature mRNA molecule typically encodes one polypeptide.

By contrast, bacteria transcribe very specific segments of genetic information into RNA messages, and often a single mRNA molecule contains the information to produce several polypeptides. In bacteria, little or no processing of RNA occurs; a moderate amount is seen in archaea, though less than in eukaryotes. In general, mechanisms of expression of DNA in archaea are more similar to those in eukaryotes than to those in bacteria. The absence of a nuclear membrane in bacteria and archaea makes it possible for mRNA molecules to become involved in the process of protein synthesis even before they are completely synthesized. Bacteria, archaea, and eukaryotes also differ in the size and composition of the ribosomes and ribosomal RNAs used to synthesize proteins (see Table 4-1). We will return to this distinction in more detail later in the chapter.

CONCEPT CHECK 4.2

You have discovered a new organism living in marine sediments. What tools or techniques would you use to determine if it is bacterial, archaeal, or eukaryotic? Perhaps consider the three historical strands of cell biology—cytology, biochemistry, and genetics.

4.3 The Eukaryotic Cell in Overview: Structure and Function

From the preceding discussion, it should be clear that all cells must carry out many of the same basic functions and have some of the same basic structural features. However, the cells of eukaryotic organisms are far more complicated structurally than bacterial or archaeal cells, primarily because of the organelles and other intracellular structures that eukaryotes use to compartmentalize various functions. The structural complexity of eukaryotic cells is illustrated by the typical animal and plant cells shown in Figures 4-7 and 4-8. In reality, of course, there is no such thing as a truly "typical" cell. Nearly all eukaryotic cells have features that distinguish them from the generalized cells shown in Figures 4-7 and 4-8. Nonetheless, most eukaryotic cells are sufficiently similar to warrant a general overview of their structural features.

Most eukaryotic cells share at least four major structural features: an external *plasma membrane* to define the boundary of the cell and to retain its contents, a *nucleus* to house the DNA that directs cellular activities, *membrane-bounded organelles* in which various cellular functions are localized, and a semifluid *cytosol* interlaced by *cytoskeletal fibers*. In addition, plant and fungal cells have a rigid *cell wall* external to the plasma membrane. Animal cells do not have a cell wall; they are usually surrounded by an *extracellular matrix* consisting primarily of proteins that provide structural support.

Our intention here is to look at each of these structural features in overview, as an introduction to cellular architecture and function. Each structure will be considered in detail in later chapters when we consider the dynamic cellular processes in which these organelles and other structures are involved.

The Plasma Membrane Defines Cell Boundaries and Retains Contents

Our tour begins with the **plasma membrane** that surrounds every cell (**Figure 4-11**). The plasma membrane defines the boundaries of the cell, ensuring that its contents are retained (Figure 4-11a). The plasma membrane consists of *phospholipids*, other lipids, and *membrane proteins* and is organized into two layers (Figure 4-11b). Typically, each phospholipid molecule consists of two hydrophobic "tails" and a hydrophilic "head" and is therefore an *amphipathic molecule* (see Figures 2-11 and 2-12). The phospholipid molecules orient themselves in the two layers of the membrane such that the hydrophobic hydrocarbon tails of each molecule face inward and the hydrophilic phosphate-containing heads of the molecules face outward (Figure 4-11c). The resulting **lipid bilayer** is the basic structural unit of virtually all membranes and serves as a permeability barrier to most water-soluble substances. Some archaea, however, have an unusual phospholipid membrane that has long hydrophobic tails (twice the normal length) linked to a polar head group on both ends, forming a monolayer.

Membrane proteins are also amphipathic. They orient themselves in the membrane such that hydrophobic regions of the protein are located within the hydrophobic interior of the membrane, whereas hydrophilic regions protrude into the aqueous environment at the surfaces of the membrane. Many of the proteins with hydrophilic regions exposed on the external side of the plasma membrane have carbohydrate side chains known as oligosaccharides attached to them and are therefore called *glycoproteins* (Figure 4-11b and c).

The proteins present in the plasma membrane play a variety of roles. Some are *enzymes*, which catalyze reactions known to be associated with the membrane—reactions such as cell wall synthesis. Others serve as *anchors* for structural elements of the cytoskeleton. Still others are *transport proteins*, responsible for moving specific substances (ions and hydrophilic solutes, usually) across the membrane. Membrane proteins are also important as *receptors* for external chemical signals that trigger specific processes within the cell. Transport proteins, receptor proteins, and most other membrane proteins are *transmembrane proteins* that have hydrophilic regions protruding from both sides of the membrane. These hydrophilic regions are connected by one or more hydrophobic, membrane-spanning domains.

The Nucleus Is the Information Center of the Eukaryotic Cell

Perhaps the most prominent organelle in a eukaryotic cell is the **nucleus** (**Figure 4-12** on page 88). The nucleus serves as the cell's information center. Inside are the DNA-bearing chromosomes of the cell, separated from the rest of the cell by two membranes, called the *inner* and *outer nuclear membranes*. Taken together, the two membranes make up the **nuclear envelope** (Figure 4-12a).

The number of chromosomes within the nucleus is characteristic of the species. It can be as low as two (in the sperm and egg cells of some grasshoppers, for example), or it can run into the hundreds. Chromosomes are most readily visualized during mitosis, when they are highly condensed and can

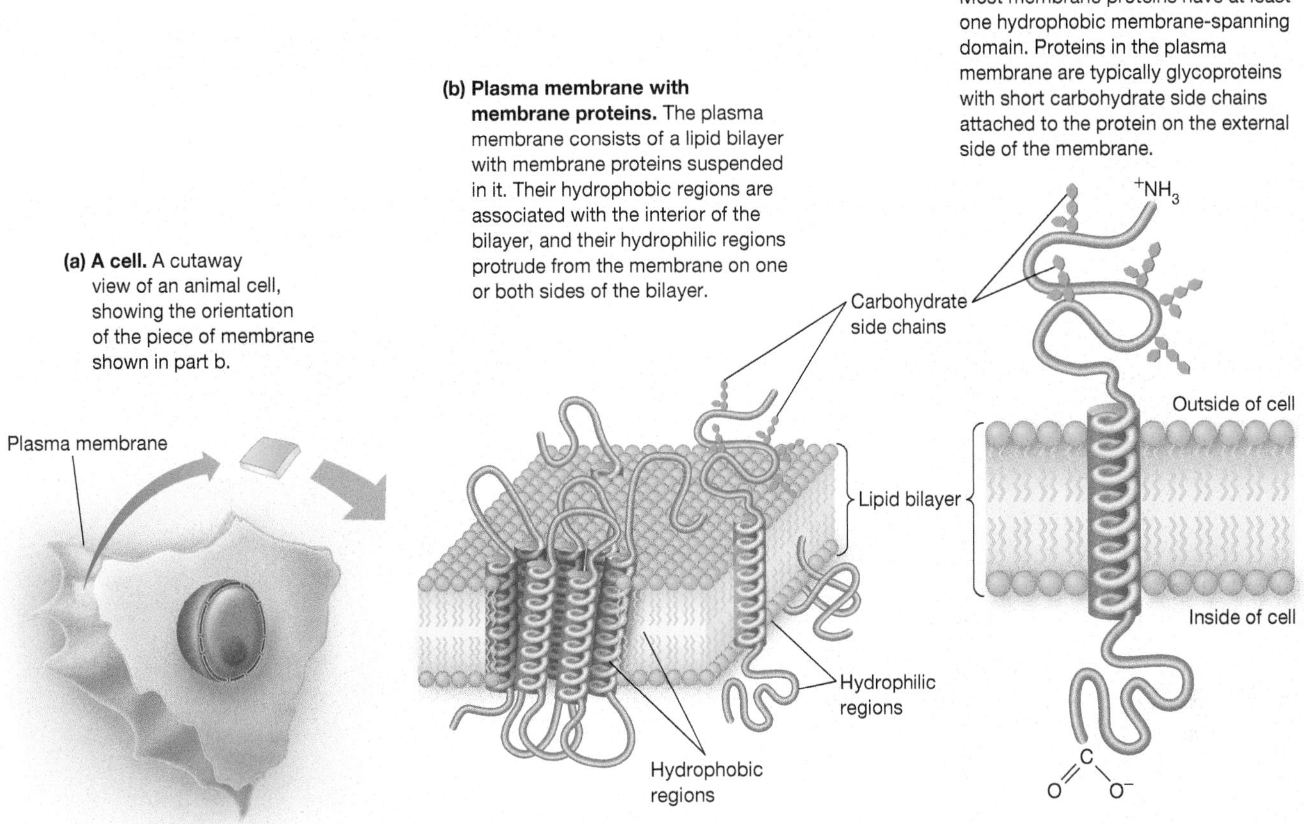

(c) Lipid bilayer with a glycoprotein. Most membrane proteins have at least one hydrophobic membrane-spanning domain. Proteins in the plasma membrane are typically glycoproteins with short carbohydrate side chains attached to the protein on the external side of the membrane.
(b) Plasma membrane with membrane proteins. The plasma membrane consists of a lipid bilayer with membrane proteins suspended in it. Their hydrophobic regions are associated with the interior of the bilayer, and their hydrophilic regions protrude from the membrane on one or both sides of the bilayer.
(a) A cell. A cutaway view of an animal cell, showing the orientation of the piece of membrane shown in part b.
Plasma membrane
Carbohydrate side chains
$^{+}NH_3$
Outside of cell
Lipid bilayer
Inside of cell
Hydrophilic regions
Hydrophobic regions
C
O
O^-

Figure 4-11 Organization of the Plasma Membrane. (a) A cutaway view of a cell and **(b)** a close-up view of a section of the plasma membrane with imbedded proteins and **(c)** a membrane section containing an individual glycoprotein.

easily be seen (see Figure 4-10). During the *interphase* between cell divisions, on the other hand, chromosomes are dispersed as fibers of DNA complexed with histone proteins, called **chromatin**, and are not easy to visualize (Figure 4-12b).

Also present in the nucleus is the **nucleolus** (plural: **nucleoli**), the structure responsible for synthesizing ribosomal RNA and beginning the assembly of the protein components needed to form ribosomes. Nucleoli are associated with specific regions of particular chromosomes that contain the genes encoding ribosomal RNAs.

Unique to the membranes of the nuclear envelope are numerous small openings called *pores* (Figure 4-12c). Each pore is a channel through which water-soluble molecules can move between the nucleus and cytoplasm. This channel is lined with transport machinery known as a *pore complex* that regulates the movement of macromolecules through the nuclear envelope. Ribosomal proteins, messenger RNA molecules, chromosomal proteins, and enzymes needed for nuclear activities are transported across the nuclear envelope through these nuclear pores.

Mitochondria and Chloroplasts Provide Energy for the Cell

Our tour of the eukaryotic organelles continues in this section with two important organelles—the *mitochondrion* and the *chloroplast*. Both are involved in energy production in cells. The mitochondrion provides energy to the cell by assisting in the degradation of sugars (as we will explore in detail in Chapter 10). The chloroplast also provides energy to some cells by harvesting solar energy and converting it to chemical energy, which is then used to convert CO_2 to sugars in the process of photosynthesis (Chapter 11).

The Mitochondrion. The **mitochondrion** (plural: **mitochondria**; **Figure 4-13** on page 88) is found in all eukaryotic cells and is the site of aerobic respiration (the topic of Chapter 10).

Mitochondria are large by cellular standards—up to a micrometer across and usually a few micrometers long. A mitochondrion is therefore comparable in size to a whole bacterial cell. Most eukaryotic cells contain hundreds of mitochondria, and each is surrounded by two membranes, designated the *inner* and *outer mitochondrial membranes*. The inner membrane encloses the mitochondrial **matrix**, a semifluid material that fills the mitochondria. In addition to numerous enzymes, the matrix contains small, circular molecules of DNA that encode some of the RNAs and proteins needed in mitochondria, along with ribosomes involved in protein synthesis. In humans and most animals, mitochondria are inherited only through the mother. Therefore, analysis of mitochondrial DNA sequences has been quite useful in establishing genetic lineages, such as retracing the geographic region(s) of origin and subsequent dispersal of modern humans.

Nucleus

Nuclear envelope

(a) Cell

Nucleus

Chromatin

Nuclear envelope

Nuclear pores

Nucleolus

Nucleoplasm

(b) Nucleus

1 μm

(c) Nuclear envelope

2.5 μm

Figure 4-12 The Nucleus. (a) A cutaway view of an animal cell, highlighting the nucleus. **(b)** This electron micrograph shows the nucleus of a rat liver cell in interphase, with the chromosomes dispersed as chromatin (TEM). **(c)** This freeze-fracture electron micrograph of a cell from a *Drosophila* (fruit fly) embryo provides a surface view of the nuclear envelope with its prominent pores.

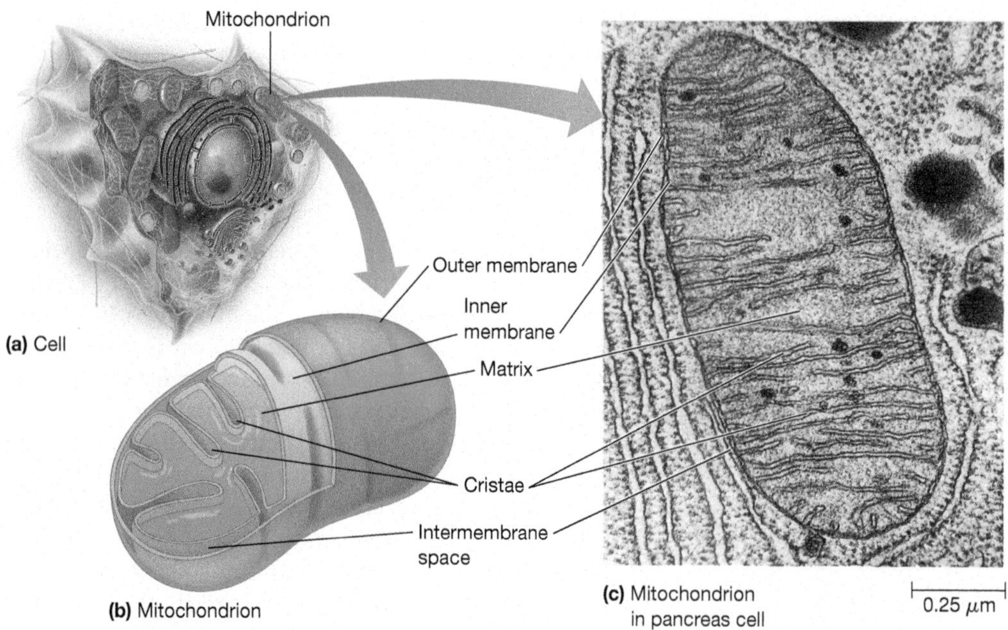

Figure 4-13 The Mitochondrion. (a) A cutaway view showing the relative size of the mitochondrion within a typical animal cell. Remember that plant cells and all other eukaryotic cells also have mitochondria. **(b)** A schematic illustration of mitochondrial structure. **(c)** A mitochondrion in a rat pancreas cell (TEM).

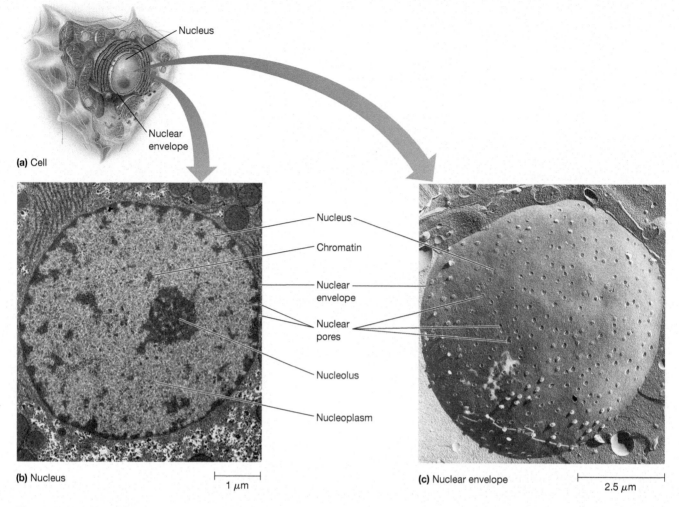

It is within the mitochondrion that the enzymes and intermediates involved in such important metabolic processes as the citric acid cycle and ATP generation are found (topics that will be covered in detail in Chapter 10). Oxidation of sugars and other cellular "fuel" molecules to carbon dioxide in mitochondria extracts energy from food molecules and conserves it as *adenosine triphosphate (ATP)*. (Remember from Chapter 2 that oxidation involves the removal of electrons, often with hydrogen ions, a process that releases energy.) Many of the intermediates involved in this process are located in or on **cristae** (singular: **crista**), infoldings of the inner mitochondrial membrane. Other reaction sequences, particularly those of the citric acid cycle and those involved in oxidation of fatty acids, occur in the matrix.

The number and location of mitochondria within a cell can often be related directly to their role in that cell. Tissues with an especially heavy demand for ATP as an energy source have cells that are well endowed with mitochondria, and they are usually located within the cell just where the energy need is greatest. For example, mitochondria cluster at the base of the flagellum in the tail of sperm cells and therefore are just where the ATP is actually needed to propel the sperm cell. Muscle cells also have numerous mitochondria located strategically near the contracting muscle fibrils to meet the special energy needs of such cells (see Figure 14-21).

The Chloroplast. The next organelle on our tour is the **chloroplast**, the site of photosynthesis in plants and algae (**Figure 4-14**). Chloroplasts are large organelles—typically a few micrometers in diameter and 5–10 μm long—and can be quite numerous in the leaves of green plants. Like mitochondria, chloroplasts are surrounded by both an inner and an outer membrane. Inside the chloroplast there is a third membrane system consisting of flattened sacs called **thylakoids** and the tubular membranes (**stroma thylakoids**) that interconnect them. Thylakoids are stacked together to form the **grana** (singular: **granum**) that characterize most chloroplasts (Figure 4-14b).

Chloroplasts are the site of *photosynthesis*, the light-driven process that uses solar energy and carbon dioxide to manufacture the sugars and other organic compounds from which all life is ultimately fabricated. Chloroplasts are found in leaves and other photosynthetic tissues of higher plants, as well as in the photosynthetic algae. Located within this organelle are most of the enzymes, intermediates, and light-absorbing pigments needed for photosynthesis—the enzymatic reduction of carbon dioxide to sugar, an energy-requiring process. (Remember from Chapter 2 that *reduction* involves the addition of electrons, often with hydrogen ions, a process that requires energy.) Reactions that depend directly on solar energy are localized in or on the thylakoid membrane system.

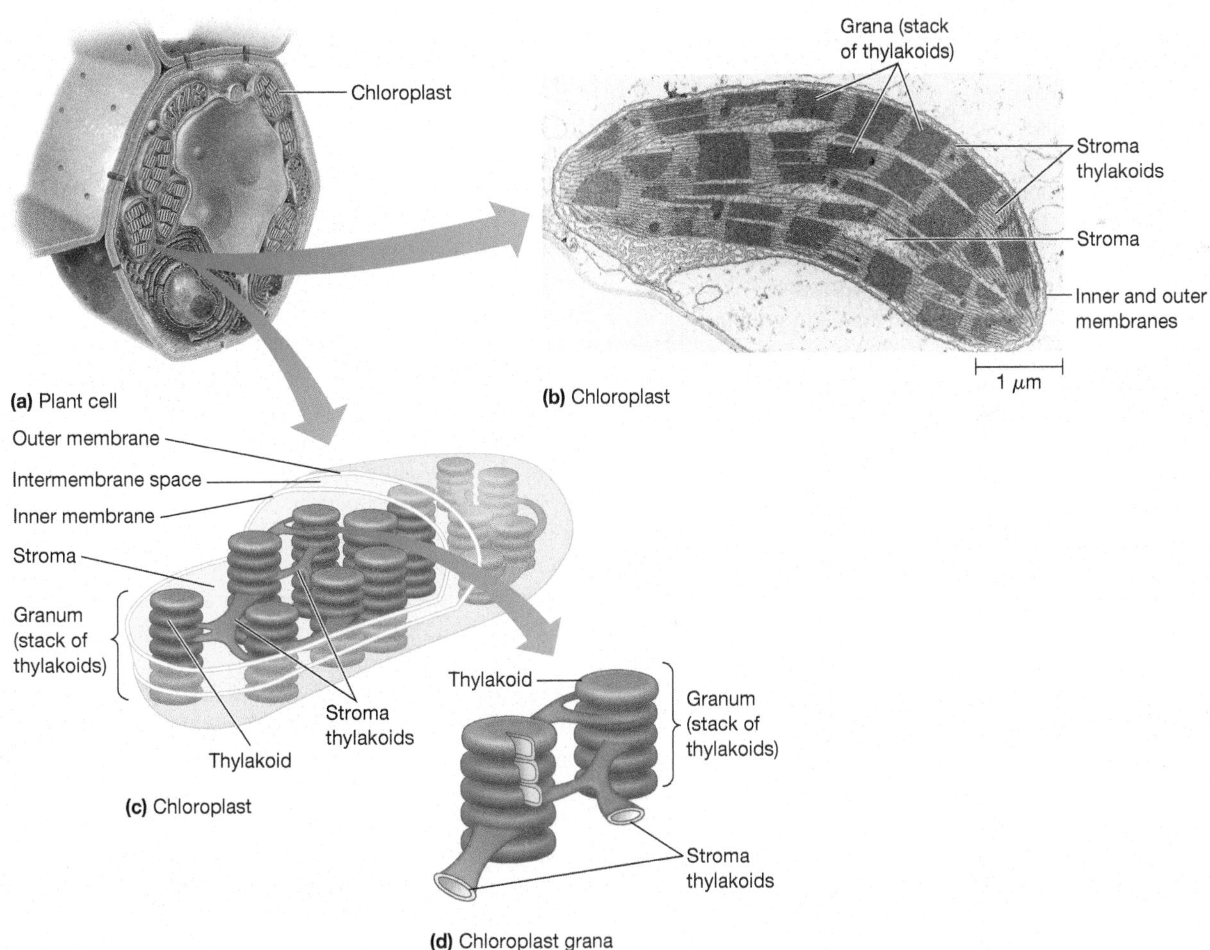

Figure 4-14 The Chloroplast. (a) A cutaway view of a plant cell showing the relative size and orientation of the chloroplasts. **(b)** A chloroplast as seen by electron microscopy (TEM). **(c)** A schematic illustration of chloroplast structure. **(d)** A cutaway view of two grana.

Reactions involved in the conversion of carbon dioxide to sugar molecules occur within the semifluid **stroma** that fills the interior of the chloroplast. Also found in the stroma are chloroplast ribosomes along with small, circular molecules of DNA that encode some of the RNAs and proteins needed in the chloroplast.

The chloroplast is the most prominent example of a class of plant organelles known as **plastids**, organelles that are found in nearly all plant cells. Plastids other than chloroplasts perform a variety of functions in plant cells. *Chromoplasts*, for example, are pigment-containing plastids responsible for the characteristic coloration of flowers, fruits, and other plant parts. *Amyloplasts* are plastids specialized for the storage of starch (amylose and amylopectin).

The Endosymbiont Theory Proposes That Mitochondria and Chloroplasts Were Derived from Bacteria

Having just met the mitochondrion and chloroplast as eukaryotic organelles, we pause here for a brief digression concerning their possible evolutionary origins. The evolutionary origins of mitochondria and chloroplasts have a long history of debate. As early as 1883, Andreas F. W. Schimper suggested that chloroplasts arose from a symbiotic relationship between photosynthetic bacteria and nonphotosynthetic cells. By the mid-1920s, other investigators had extended Schimper's idea by proposing a symbiotic origin for mitochondria. Such ideas were neglected for decades, however, until the 1960s, when it was discovered that mitochondria and chloroplasts contain their own DNA molecules and that these molecules are circular like the DNA in bacteria. Further research revealed that mitochondria and chloroplasts are **semiautonomous organelles**, organelles that can divide on their own and contain not only their own DNA but also their own mRNA, tRNAs, and ribosomes (all involved in protein synthesis).

As molecular biologists studied nucleic acid and protein synthesis in these organelles, they were further struck by the many similarities between these processes in mitochondria and chloroplasts and the comparable processes in bacterial cells. All show similarities in ribosomal RNA sequences, ribosome size, and sensitivity to inhibitors of RNA and protein synthesis. In addition to these molecular features, mitochondria and chloroplasts resemble bacterial cells in size and shape, and they have a double membrane in which the inner membrane has bacterial-type lipids.

The realization that the DNA, RNA, and mechanisms of protein synthesis in mitochondria and chloroplasts are most similar to those found in bacterial cells led biologists to formulate the **endosymbiont theory**. This theory, put forth by Lynn Margulis in 1967, proposes that mitochondria and chloroplasts evolved from ancient bacteria that established a **symbiotic relationship** (a mutually beneficial association) with primitive nucleated cells 1 to 2 billion years ago. Additional support for the endosymbiont theory comes from two observations. First, like gram-negative bacteria, mitochondria and chloroplasts have a double membrane. Second, the membranes of mitochondria and chloroplasts have bacterial-type lipids and transmembrane proteins. Mitochondria, for example, have a specialized lipid called *cardiolipin*, and they have proteins in their outer membranes called *porins* that are strikingly similar to those in the outer membranes of gram-negative bacterial plasma membranes (see Figure 10-6b).

Outside-In Versus Inside-Out Theories of Eukaryotic Origins. As we have already seen, two striking features of eukaryotic cells are the nuclear envelope and mitochondria. Two basic classes of theories have been suggested to account for these key features of the first eukaryotic cells. *Outside-in theories* suggest that in the ancestor of eukaryotic cells (called a **protoeukaryote**), the nuclear envelope formed by an invagination (infolding) of the plasma membrane, which eventually came to surround the cell's DNA (**Figure 4-15**). These cells also developed the ability to engulf and ingest nutrients and particles—including smaller cells of bacteria and cyanobacteria—from the environment by *phagocytosis* ("cell eating"). Following engulfment, the smaller cells were not digested. Instead, surrounded by host cell membrane, they took up residence in the cytoplasm of the protoeukaryote and eventually evolved into mitochondria and chloroplasts, which are each surrounded by a double membrane.

As an alternative to "outside-in" theories, others have suggested *inside-out theories*. Some existing archaeal cells can produce fingerlike protrusions that poke through their cell walls. In inside-out scenarios, the protoeukaryotic cell is thought to have had similar capabilities (Figure 4-15). These protrusions are proposed to have led ultimately to two innovations. First, they allowed the protoeukaryote to interact with and ultimately envelop nearby gram-negative bacteria. These captured bacteria could then evolve into mitochondria. Second, as the protrusions expanded outward, they may have folded over and fused, forming the primitive endoplasmic reticulum and other structures of the endomembrane system. In such inside-out theories, the nuclear membrane originated from what was the outer plasma membrane of the host cell.

The idea that an archaean-like protoeukaryote might have been the predecessor to modern eukaryotes received a boost beginning in 2015 when DNA isolated from environmental samples off the coast of Norway identified the first of a group of archaea called *Lokiarchaeota* (after the Norse god Loki), with many similarities to eukaryotes. In 2020, a member of this "Asgard" superphylum isolated off the coast of Japan was grown in the laboratory. It bears striking similarities to the intermediate form imagined in the inside-out model. These cells lack an internal membrane-bounded compartment, but have a "nuclear-like" region, as well as external bulges and long finger-like protrusions.

In either the outside-in or inside-out scenarios, once the bacterium was engulfed, it should have had *three* membranes surrounding it (its own double membrane plus one from the host), but mitochondria today only have two membranes. Where did the outer membrane go? One idea is that the engulfed bacteria, now in the cytosol, broke free from the surrounding host membrane and then divided within the host.

Mitochondrial Evolution from Ancient Aerobic Bacteria. What sort of bacteria did mitochondria come from? The first step toward the evolution of mitochondria may have occurred when an anaerobic protoeukaryote that depended on glucose

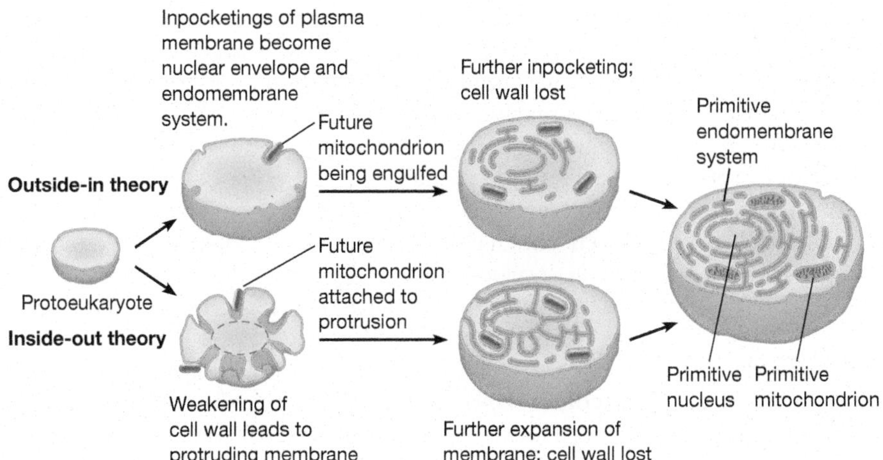

Figure 4-15 Theories of Endosymbiosis. Most biologists agree that mitochondria developed from primitive aerobic purple bacteria, which have double outer membranes. This is based on similarities in size, membrane lipid and protein composition, comparisons of rRNA base sequences, the presence of circular DNA molecules and bacterial-type ribosomes, and the ability to reproduce autonomously. In "outside-in" theories, primitive protoeukaryotic cells developed inpocketings of the plasma membrane that eventually engulfed the cell's DNA to form the nuclear envelope. Aerobic bacteria were ingested but not digested by protoeukaryotes and developed into mitochondria, possibly by escaping from the surrounding phagosome. In "inside-out" theories, extended projections from the cell membrane of an ancient protoeukaryotic cell captured surface bacteria. These bacteria then evolved into mitochondria. The expansion and fusion of the projections eventually led to the formation of internal membranes that became the endoplasmic reticulum. The membrane from the original protoeukaryotic cell's plasma membrane became the nuclear envelope.

for energy developed a symbiotic relationship with smaller aerobic bacteria. The aerobic bacteria were able to metabolize glucose more efficiently in the presence of oxygen (as we will see in Chapter 10). The ingested aerobic bacteria provided additional energy to the anaerobic host cell in the form of ATP. In turn, the host cell provided protection and nutrients to the bacteria residing in its cytoplasm. Gradually, over hundreds of millions of years, the bacteria lost functions that were not essential in their new cytoplasmic environment and developed into mitochondria.

When the base sequences of mitochondrial ribosomal RNAs (rRNAs) are compared with those of various bacterial rRNAs, the closest matches occur in modern *aerobic purple bacteria*, suggesting that the ingested ancestor of mitochondria was an ancient member of this group. Purple bacteria are gram-negative bacteria and so have a double outer membrane, as do mitochondria. Mitochondria today are dependent on their hosts in many ways; one way is that the information needed to produce many proteins in mitochondria, including some in electron transport (see Chapter 10), is encoded in the nuclear DNA of the "host" cell.

Chloroplast Evolution from Ancient Photosynthetic Bacteria. The first step toward the evolution of chloroplasts likely occurred when members of a subgroup of early eukaryotes, perhaps already equipped with primitive mitochondria, internalized primitive photosynthetic cells. As just described for the evolution of mitochondria, the internalized organisms probably provided energy for the host cell, in the form of carbohydrates, in exchange for shelter and nutrients. The photosynthetic cells gradually lost functions that were not essential in their new environment and evolved into an integral component of the eukaryotic host.

When the base sequences of chloroplast rRNAs are compared with those of various bacterial rRNAs, the closest matches occur among photosynthetic, gram-negative bacteria called *cyanobacteria* (these are sometimes colloquially called "blue-green algae," but they are bacteria), suggesting that the ancestor of chloroplasts was an ancient member of this group. Cyanobacteria have double outer membranes, as is true of chloroplasts. Cyanobacteria also have thylakoids, consistent with the chloroplasts arising from a free-living cyanobacterial ancestor.

MAKE CONNECTIONS 4.2

Suppose researchers discovered a new species of photosynthetic unicellular eukaryote that has several large chloroplasts. Consistent with the endosymbiosis hypothesis, they found ribosomes in the chloroplasts, mitochondria, and cytosol of these cells. What technique(s) besides subcellular fractionation could they use to confirm the locations in which the ribosomes were found? (Ch. 1)

Endosymbiosis Today. Support for the endosymbiont theory also comes from contemporary symbiotic relationships that resemble what might have occurred in the distant past. Algae, dinoflagellates, diatoms, and photosynthetic bacteria live as endosymbionts in the cytoplasm of cells occurring in more than 150 kinds of existing protists and invertebrates.

In the final analysis, our ideas about how eukaryotic cells acquired their complex array of organelles over billions of years of evolution must remain speculative. We can never test them directly because the proposed events are inaccessible to direct laboratory experimentation. However, that endosymbiosis seems to occur in cells today strengthens the case for the endosymbiotic origin of mitochondria and chloroplasts.

In addition to mitochondria and chloroplasts, which are involved primarily in energy production for the cell, eukaryotic cells have a variety of other organelles, each with one or more specific functions. We will next preview these organelles, which will be discussed in detail in future chapters, and then tour the cytoplasm of the cell and its contents.

The Endomembrane System Synthesizes Proteins for a Variety of Cellular Destinations

The next set of organelles we will preview constitutes a dynamic system known as the endomembrane system, which, among other functions, synthesizes proteins that are destined for various organelles, cellular membranes, or secretion. These proteins are synthesized in the endoplasmic reticulum, processed and packaged in the Golgi apparatus, and then directed to various destinations in and out of the cell in small membrane-bound vesicles. The lysosome, which is also derived from the endomembrane system, is involved in degradation of food ingested by phagocytosis and in recycling cellular components.

The Endoplasmic Reticulum. Extending throughout the cytoplasm of almost every eukaryotic cell is a prominent network of membranes called the **endoplasmic reticulum**, or **ER** (**Figure 4-16**). The name sounds complicated, but endoplasmic just means "within the plasm" (of the cell), and reticulum is simply a fancy word for "network." The endoplasmic reticulum consists of tubular membranes and flattened sacs, or **cisternae** (singular: **cisterna**), that are interconnected. The internal space enclosed by the ER membranes is called the **lumen**. The ER is continuous with the outer membrane of the nuclear envelope (Figure 4-16a). The space between the two nuclear membranes is therefore a part of the same compartment as the lumen of the ER.

The ER can be either *rough* or *smooth*. **Rough endoplasmic reticulum (rough ER)** appears "rough" in the electron microscope because it is studded with ribosomes on the side of the membrane that faces the cytosol (Figure 4-16b, d). These ribosomes are actively synthesizing polypeptides that either accumulate within the membrane or are transported across the ER membrane to accumulate in the space inside the ER (its lumen). Many membrane proteins and secreted proteins are synthesized in this way. These proteins then travel to the appropriate membrane or to the cell surface via the Golgi apparatus and secretory vesicles, as described in the following subsection (and in more detail in Chapter 12).

Not all proteins are synthesized by ribosomes associated with membranes of the rough ER, however. Many proteins are synthesized on ribosomes that are not attached to the ER

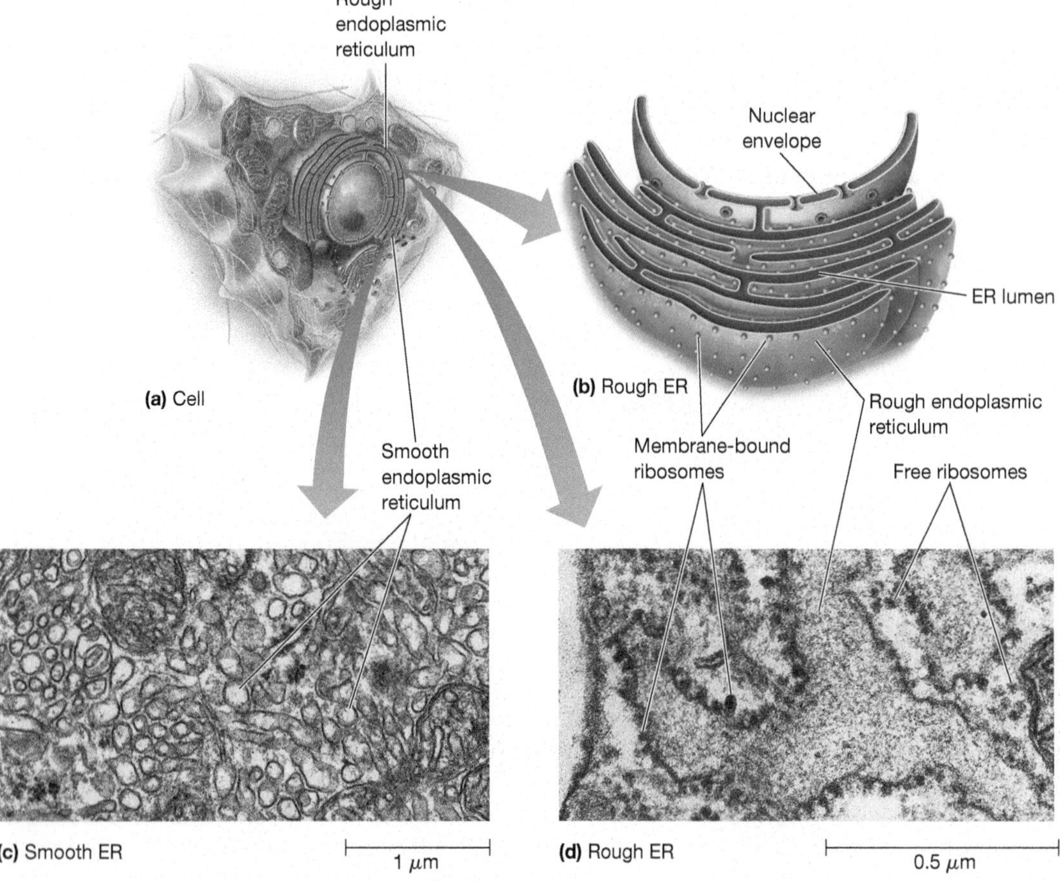

Figure 4-16 The Endoplasmic Reticulum. (a) A cutaway view of a typical animal cell showing the location and relative size of the rough and the smooth endoplasmic reticulum (ER). **(b)** A schematic illustration depicting the organization of the rough ER as layers of flattened membranes. **(c)** An electron micrograph of smooth ER in a cell from a guinea pig testis (TEM). **(d)** An electron micrograph of rough ER in a rat pancreas cell (TEM).

but are instead found free in the cytosol, unassociated with a membrane (Figure 4-16d). In general, secretory proteins and membrane proteins are made by ribosomes on the rough ER, whereas proteins intended for use within the cytosol or for import into organelles are made on free ribosomes. We will preview the structure and function of ribosomes later in this chapter.

The **smooth endoplasmic reticulum (smooth ER)** has no role in protein synthesis and hence no ribosomes. It therefore has a characteristically smooth appearance when viewed by electron microscopy (Figure 4-16c). Smooth ER is involved in the synthesis of lipids and steroids, such as cholesterol and the steroid hormones derived from it. In addition, smooth ER is responsible for inactivating and detoxifying drugs such as barbiturates and other compounds that might otherwise be toxic or harmful to the cell. When we discuss muscle contraction in detail (in Chapter 16), we will see how a specialized type of smooth ER known as the *sarcoplasmic reticulum* is critical for storage and release of the calcium ions that trigger contraction.

The Golgi Apparatus. Closely related to the ER in both proximity and function is the **Golgi apparatus** (or *Golgi complex*), named after its Italian discoverer, Camillo Golgi (**Figure 4-17**). The Golgi apparatus, shown in Figure 4-17a, consists of a stack of flattened vesicles (also known as *cisternae*; Figure 4-17b). The Golgi apparatus plays an important role in processing and packaging secretory proteins and in synthesizing complex polysaccharides. *Transition vesicles* that arise by budding off the ER are accepted by the Golgi apparatus. Here, the contents of the vesicles (proteins, for the most part) and sometimes the vesicle membranes are further modified and processed. The processed contents are then passed on to other components of the cell in vesicles that arise by budding off the Golgi apparatus (Figure 4-17c).

Most membrane and secretory proteins are *glycoproteins*, meaning they bear one or more covalently linked carbohydrates. The initial steps in *glycosylation* (addition of short-chain carbohydrates) take place within the lumen of the rough ER, but the process is usually completed within the Golgi apparatus. The Golgi apparatus should therefore be understood as a processing station, with vesicles both fusing with it and arising from it. Almost everything that goes into it comes back out—but in a modified, packaged form, often ready for export from the cell.

Secretory Vesicles. Once processed by the Golgi apparatus, secretory proteins and other substances intended for export from the cell are packaged into **secretory vesicles**. The cells of your pancreas, for example, contain many such vesicles because the pancreas is responsible for the synthesis of several important digestive enzymes. These enzymes are synthesized on the rough ER, packaged by the Golgi apparatus, and then released from the cell via secretory vesicles, as shown

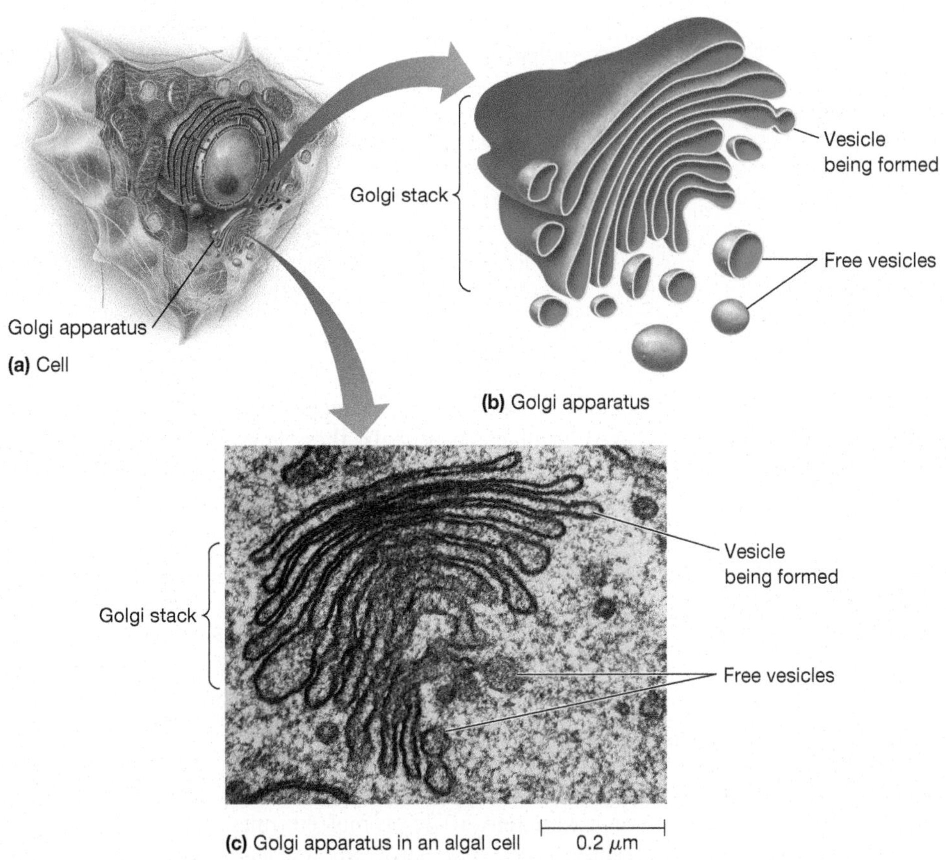

Figure 4-17 The Golgi Apparatus. (a) A cutaway view showing the relative orientation and size of a Golgi apparatus within a cell. **(b)** A schematic drawing of a Golgi apparatus, showing vesicle formation by budding. **(c)** An electron micrograph of a Golgi apparatus in an algal cell (TEM).

HUMAN *Connections*

When Cellular "Breakdown" Breaks Down

Coffee, tomato juice, and beer: this is a list of items you may not think are related. What do they have in common? They all have a pH of about 5.0. Interestingly, so does the interior of lysosomes. Lysosomes play a crucial role in cellular digestion, functioning as part of the cell's recycling system. Molecules marked for degradation can be broken down and released back into the cytosol, and the monomers can be reused to build new polymers.

In any given lysosome, approximately 40 different lysosomal enzymes are found, and there are more than 40 different lysosomal storage diseases. Each disease is characterized by a mutation in a specific *hydrolase* in the lysosome. Each hydrolase is responsible for breaking down specific covalent bonds, allowing the dismantling of many different molecules into their constituent parts. When even one of these hydrolases fails to function, its target molecule can accumulate, producing swollen lysosomes and cellular dysfunction. **Figure 4A-1** shows a normal lymphocyte and a lymphocyte from a patient with *α-mannosidosis,* which arises due to an inactivating mutation in the gene that encodes the enzyme α-mannosidase, which, as the name implies, is needed to breakdown the monosaccharide mannose.

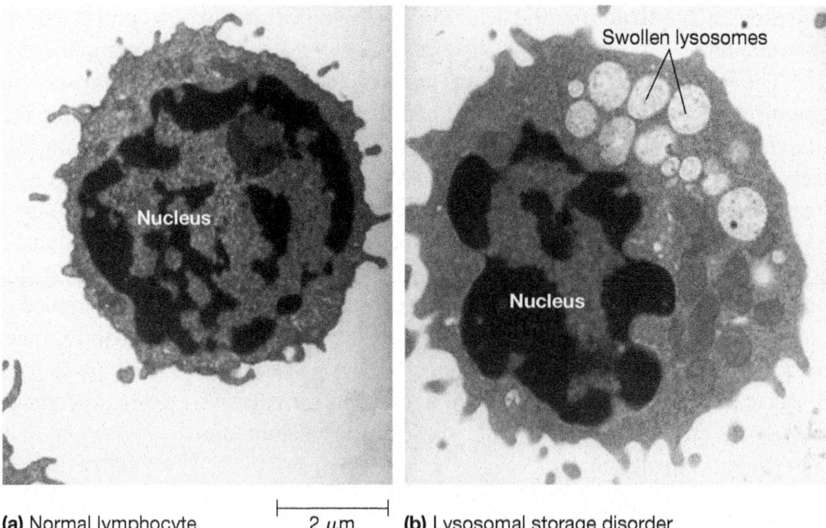

Figure 4A-1 Normal and Lysosomal Storage Disorder Lymphocytes. (a) A normal human lymphocyte. **(b)** A lymphocyte from a patient with a mutation in the enzyme α-mannosidase, which catalyzes breakdown of mannose in lysosomes (TEMs).

α-mannosidosis involves a single lysosomal hydrolase. Imagine what would happen if most hydrolases failed to find their way into lysosomes. The results would be devastating. In fact, human patients can be afflicted by this very problem. One of the most pernicious lysosomal disorders is *I-cell disease*. To understand I-cell disease, we need to consider the path that lysosomal proteins take before arriving at the lysosome. (You will learn much more about this pathway and other mechanisms of protein sorting in Chapter 12, so we need only discuss the basics of this specific pathway here.)

Like many other proteins, lysosomal enzymes are processed in the rough endoplasmic reticulum (RER), where the sugar *mannose* is attached to specific amino acids within these enzymes (**Figure 4A-2**). Subsequent modifications occur in the Golgi apparatus, where phosphoryl groups are added on to the mannose residues. This phosphorylation is catalyzed by an enzyme, known as a *phosphotransferase,* which produces proteins that are "tagged" with mannose-6-phosphate. The tag binds to a specific receptor in the Golgi, which allows the tagged proteins to be packaged into vesicles. In normal individuals, these vesicles eventually fuse with another type of vesicle known as an *endosome*; after fusion, the resulting vesicles ultimately contribute to functional lysosomes (Figure 4A-2).

in **Figure 4-18** on page 96. These vesicles from the Golgi region move to and fuse with the plasma membrane and discharge their contents to the exterior of the cell by the process of *exocytosis*. Because of their close functional relationships, the ER, Golgi, secretory vesicles, and lysosomes (which we will discuss next) are collectively said to constitute the cell's *endomembrane system*. The endomembrane system is responsible for trafficking of molecules through the cell (a process we will consider in more detail in Chapter 12).

The Lysosome. The next stop on our tour is the **lysosome**, an organelle about 0.5–1.0 μm in diameter and surrounded by a single membrane (**Figure 4-19** on page 96). Lysosomes are used by the cell as storage containers for *hydrolases*, enzymes capable of digesting specific biological molecules such as proteins, carbohydrates, and fats. Although cells need such enzymes, both to digest food molecules and to break down damaged cellular constituents, these enzymes must be carefully sequestered until actually needed lest they digest the normal components that were not scheduled for destruction in the cell.

Like secretory proteins, lysosomal enzymes are synthesized on the rough ER, transported to the Golgi apparatus, and then packaged into vesicles that can become lysosomes. The inner face of the lysosomal membrane is highly glycosylated, forming a special carbohydrate covering that protects this membrane (and the cell) from the hydrolytic activities of these enzymes until they are needed. Equipped with its repertoire of hydrolases, the lysosome is able to break down virtually any kind of biological molecule. However, a defect in any of the approximately 40 lysosomal enzymes can disrupt normal cellular degradation processes and lead to serious human disease, as discussed above in **Human Connections**.

If a mutation renders this Golgi-specific phosphotransferase nonfunctional, then packaging of hydrolytic enzymes into lysosomes is incomplete. Without the phosphoryl tag, the enzymes normally delivered to the lysosome are instead eventually routed to different vesicles, which fuse with the plasma membrane in a process known as *exocytosis* (see Chapter 12 for more details). In this case, the hydrolytic enzymes are released into the extracellular environment. The released enzymes are not functional because they normally function at the highly acidic pH of the lysosome (pH 5), whereas the extracellular environment and circulatory system have a pH of 7.2–7.4. Nevertheless, the presence of these enzymes in the blood is one accurate indicator of I-cell disease.

Failure to transport hydrolytic enzymes into lysosomes is devastating. Macromolecules destined for degradation accumulate in lysosomes and cause them to become engorged (Figure 4A-2b), forming *inclusions* (hence the letter "I" in the name of the disease). Defective lysosomes also lead to a host of defects in affected patients. Afflicted individuals typically fail to grow normally as infants, have stiff joints, and in some cases show enlarged livers and spleens. They rarely live beyond 7 years of age.

There is currently no effective treatment for I-cell disease. It is hoped that one day prenatal gene therapy may be used to provide a functioning copy of the gene to a fetus lacking it, thereby preventing the devastating effects of this disease.

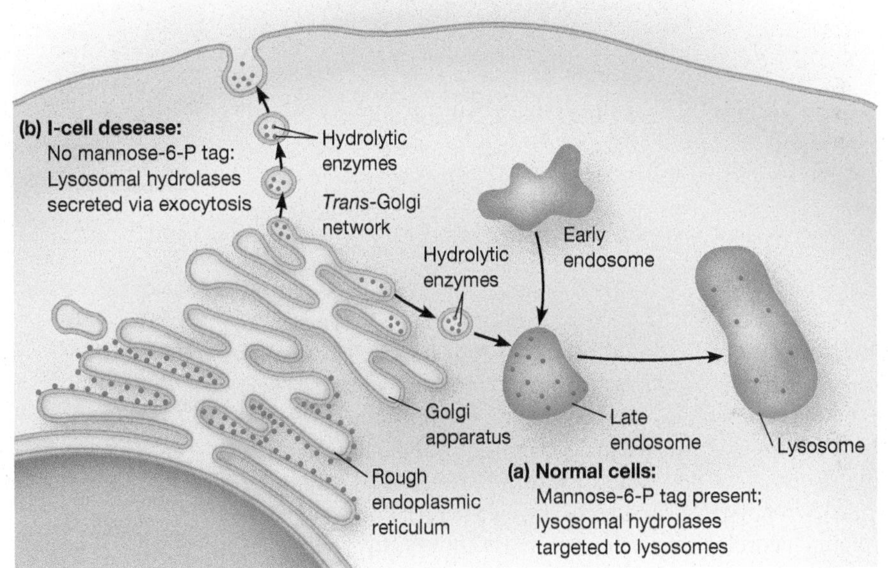

Figure 4A-2 Normal and I-Cell Disease Transport Pathways for Hydrolytic Enzymes. (a) Normal trafficking of lysosomal enzymes depends on addition of mannose to hydrolases in the rough ER. Mannose is then phosphorylated to mannose-6-phosphate (mannose-6-P) in the Golgi. Mannose-6-P-tagged enzymes move through the Golgi, eventually fuse with endosomes, and are ultimately incorporated into lysosomes. Note that many other aspects of endosomal trafficking are omitted from this diagram for clarity. **(b)** In cells of I-cell patients, the enzyme in the Golgi that adds a phosphoryl group to mannose is absent, so the enzymes are misrouted to the plasma membrane. Ultimately, lysosomes become swollen in such cells.

Other Organelles Also Have Specific Functions

The last two organelles on our tour are the peroxisome and the vacuole. Depending on the organism and cell type, these two organelles can have different functions.

The Peroxisome. The **peroxisome** resembles the lysosome in size and general lack of obvious internal structure. Like lysosomes, peroxisomes are surrounded by a single rather than a double membrane. Unlike lysosomes, however, peroxisomes are not part of the endomembrane system.

Peroxisomes are found in plant and animal cells, as well as in fungi, protozoa, and algae. They perform several distinctive functions that differ with cell type, but they have the common property of both generating and degrading hydrogen peroxide (H_2O_2). Hydrogen peroxide, a by-product of several normal metabolic reactions, is highly toxic to cells but can be decomposed into water and oxygen by the enzyme catalase. Eukaryotic cells protect themselves from the detrimental effects of hydrogen peroxide by packaging peroxide-generating reactions together with catalase in a single compartment, the peroxisome.

In animals, peroxisomes are found in most cell types but are especially prominent in liver and kidney cells (**Figure 4-20** on page 96). In addition to detoxifying hydrogen peroxide, animal peroxisomes detoxify other harmful compounds (such as methanol, ethanol, formate, and formaldehyde). Animal peroxisomes also play a role in the oxidative breakdown of fatty acids, which are components of triacylglycerols, phospholipids, and glycolipids (see Figure 3-27). Fatty acid breakdown occurs primarily in the mitochondrion. However, fatty acids with more than 12 carbon atoms or branched fatty acids are oxidized relatively slowly by mitochondria. Peroxisomes break

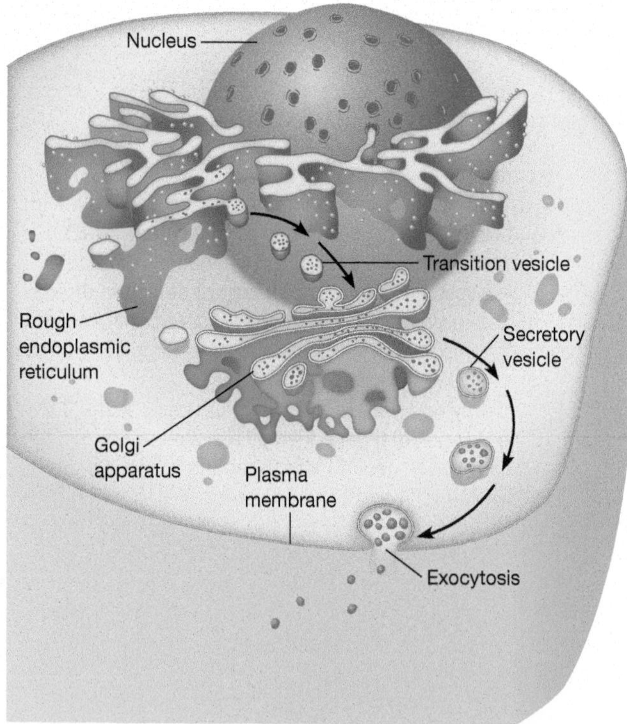

Figure 4-18 The Process of Secretion in Eukaryotic Cells. Proteins for export are synthesized on the rough ER, passed to the Golgi apparatus for processing, and packaged into secretory vesicles. These vesicles then move to and fuse with the plasma membrane, where they release the proteins to the exterior of the cell.

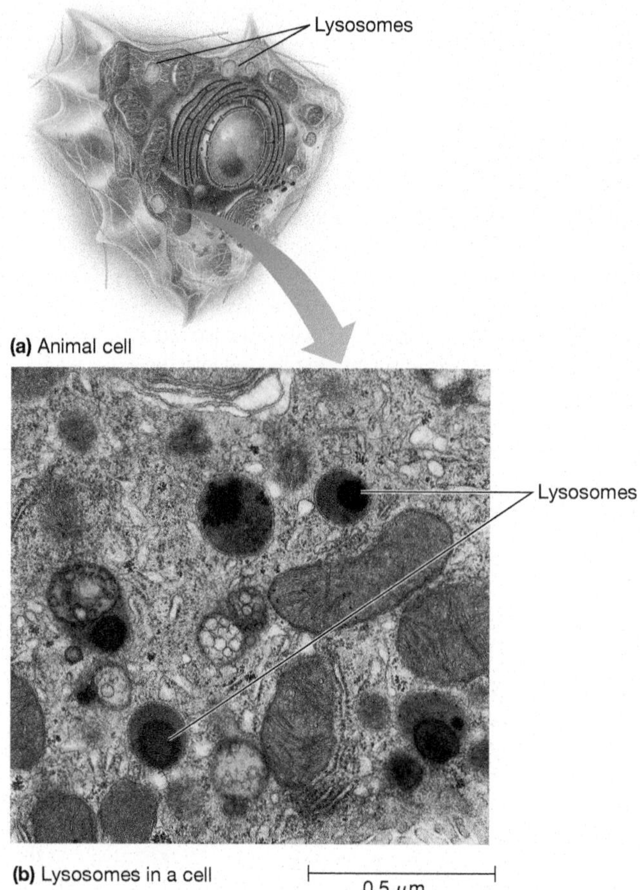

Figure 4-19 Lysosomes. (a) A cutaway view showing lysosomes within an animal cell. **(b)** Lysosomes in an animal cell stained cytochemically for acid phosphatase, a lysosomal enzyme. The cytochemical staining technique results in dense deposits of lead phosphate at the site of acid phosphatase activity (TEM).

such fatty acids down until they get to a length (10–12 carbon atoms) that the mitochondria can efficiently handle. The vital role of peroxisomes in fatty acid breakdown is underscored by the serious human diseases that result when one or more peroxisomal enzymes involved in degrading long-chain fatty acids are defective or absent. One such disease is neonatal adrenoleukodystrophy (NALD), a peroxisomal disorder that leads to profound neurological debilitation and eventually to death.

Peroxisomes have key metabolic roles in plant cells as well. During the germination of fat-storing seeds, specialized peroxisomes called **glyoxysomes** play a key role in converting stored fat into carbohydrates. In photosynthetic tissue, **leaf peroxisomes** are prominent due to their role in *photorespiration*, the light-dependent uptake of oxygen and release of carbon dioxide (to be discussed further in Chapter 11), which requires an intimate association of peroxisomes with mitochondria and chloroplasts in many leaf cells, as shown in **Figure 4-21**.

Vacuoles. Some cells contain another type of membrane-bounded organelle called a **vacuole**. In animal and yeast cells, vacuoles are used for temporary storage or transport. Some protozoa take up food particles or other materials from their environment by phagocytosis. Phagocytosis is a form of endocytosis that involves an infolding of the plasma membrane around the desired substance. This infolding is followed by a pinching-off process that internalizes the membrane-bounded particle as a type of vacuole, known as a *phagosome*. Following fusion with a lysosome, the contents of a phagosome are hydrolyzed by lysosomal enzymes to provide nutrients for the cell.

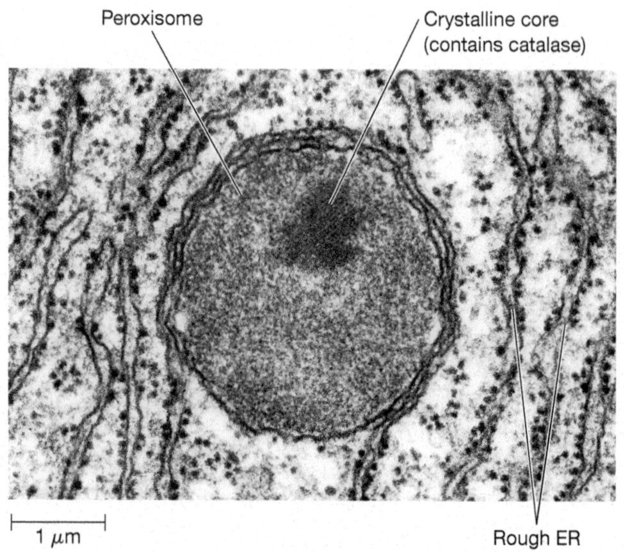

Figure 4-20 A Peroxisome in a Liver Cell. A peroxisome can be seen in this cross section of a liver cell (TEM). Peroxisomes are found in most animal cells but are especially prominent features of liver and kidney cells.

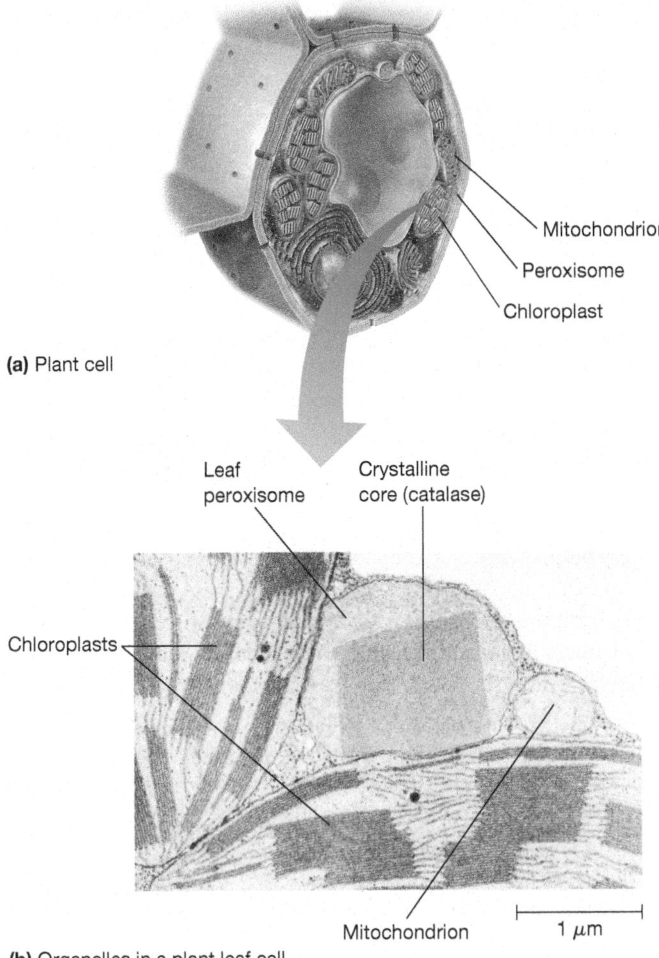

Figure 4-21 A Leaf Peroxisome and Its Relationship to Other Organelles in a Plant Cell. (a) A cutaway view showing a peroxisome, a mitochondrion, and a chloroplast within a plant cell. **(b)** A peroxisome in close proximity to chloroplasts and to a mitochondrion within a tobacco leaf cell (TEM). The crystalline core frequently observed in leaf peroxisomes is the enzyme catalase.

Plant cells also contain vacuoles. In fact, most mature plant cells contain a single large vacuole, which occupies much of the internal volume of the cell, leaving the cytoplasm sandwiched into a thin region between the vacuole and the plasma membrane (**Figure 4-22**). This vacuole, sometimes called the **central vacuole**, may play a limited role in storage and in intracellular digestion. However, its main importance is its role in maintaining the *turgor pressure* that keeps plant tissue from wilting. The vacuole has a high concentration of solutes; thus, water tends to move into the vacuole, causing it to swell. As a result, the vacuole presses the rest of the cell constituents against the cell wall, thereby maintaining the turgor pressure. The limp appearance of wilted tissue results when the central vacuole does not provide adequate pressure.

Ribosomes Synthesize Proteins in the Cytoplasm

In our tour of the cell we now move to components within the cytosol, beginning with the ribosome. Recall that **ribosomes** are the site of protein synthesis. Ribosomes are far more numerous than most other intracellular structures. Prokaryotic cells usually contain thousands of ribosomes in their cytoplasm, and eukaryotic cells may have hundreds of thousands or even millions of them. Ribosomes are also found in both mitochondria and chloroplasts, where they function in organelle-specific protein synthesis. The ribosomes of these two organelles differ in size and composition from the ribosomes found in the cytoplasm of the same cell, but, as noted in our discussion of the endosymbiont theory, they are strikingly similar to those found in bacteria and cyanobacteria.

Ribosomes are found in all cells, but bacteria, archaea, and eukaryotes differ from each other in ribosome size and in the number and kinds of ribosomal protein and rRNA molecules (see Table 4-1). Bacteria and archaea contain smaller ribosomes than those found in eukaryotes, and all three have different numbers of ribosomal proteins, although there is some similarity between the ribosomal proteins of archaeal and eukaryotic cells. A defining feature is that each of these three cell types has its own unique type of ribosomal RNA, the very molecule whose characteristics were used to define the archaea as a domain distinct from bacteria.

Compared with even the smallest organelles, ribosomes are tiny structures. The ribosomes of eukaryotic and prokaryotic cells have diameters of about 30 and 25 nm, respectively. An electron microscope is therefore required to visualize ribosomes (see Figure 4-16d). To appreciate how small ribosomes

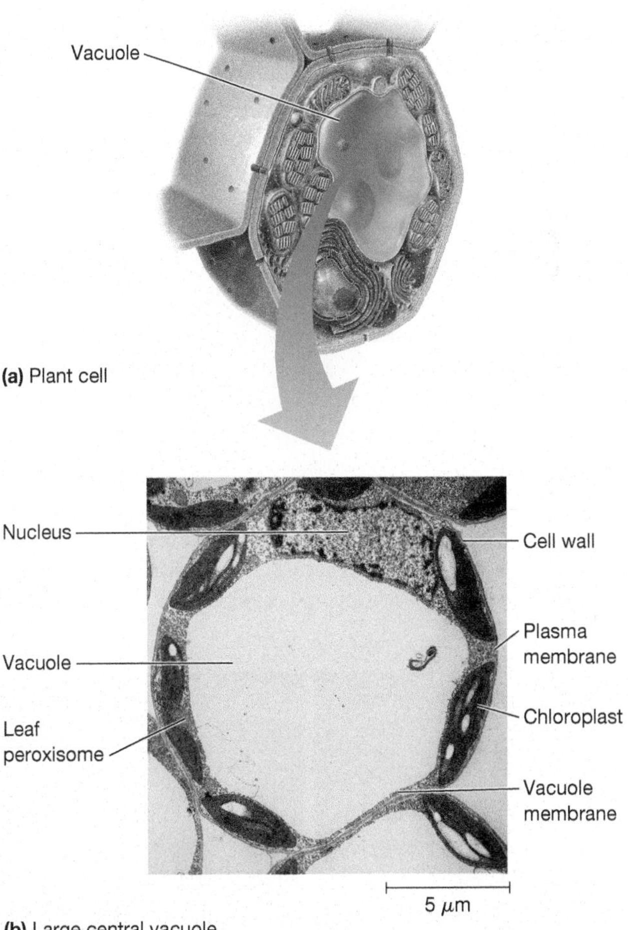

Figure 4-22 The Vacuole in a Plant Cell. (a) A cutaway view showing the vacuole in a plant cell. **(b)** An electron micrograph of a bean leaf cell with a large central vacuole (TEM).

are, consider that more than 350,000 ribosomes could fit inside a typical bacterial cell, with room to spare!

Another way to express the size of such a small particle is to refer to its **sedimentation coefficient**. The sedimentation coefficient of a particle or macromolecule is a measure of how rapidly the particle sediments when subjected to centrifugation in a standard piece of lab equipment known as a **centrifuge** (see Key Technique, pages 100–101). The rate of sedimentation is expressed in **Svedberg units (S)**, named in honor of Theodor Svedberg, the Swedish chemist who developed the **ultracentrifuge** between 1920 and 1940. Sedimentation coefficients are widely used to indicate relative size, especially for large macromolecules such as proteins and nucleic acids and for small particles such as ribosomes. Ribosomes from eukaryotic cells have sedimentation coefficients of about 80S, and those from bacteria and archaea are about 70S. Differences in sedimentation coefficients allow ribosomes to be separated and purified from other components in cell extracts, such as membranes and organelles, for detailed study.

A ribosome consists of two subunits differing in size, shape, and composition (**Figure 4-23**). In eukaryotic cells, the **large** and **small ribosomal subunits** have sedimentation coefficients of about 60S and 40S, respectively. For bacterial and archaeal ribosomes, the corresponding values are about 50S and 30S. (Note that the sedimentation coefficients of the subunits do not add up to that of the intact ribosome. This is because sedimentation coefficients depend on both size and shape and are therefore not linearly related to molecular weight.) In early 2009, scientists successfully extracted all the ribosomal proteins and rRNAs from an *E. coli* cell and then were able to reconstruct a functional artificial ribosome capable of synthesizing protein. This is a significant step toward the creation of artificial cells in vitro that have custom-made ribosomes tailored for industrial uses to produce particular desired proteins.

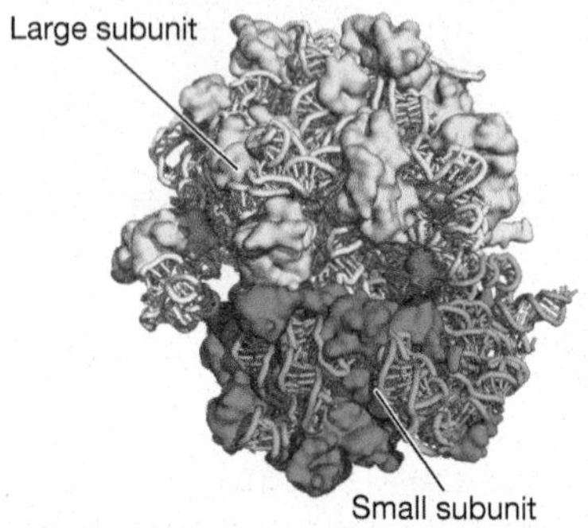

Figure 4-23 Structure of a Ribosome. Each ribosome is made up of a large subunit and a small subunit that join together when they attach to a messenger RNA and begin to make a protein. The fully assembled ribosome of a eukaryotic cell has a diameter of about 30 nm.

The Cytoskeleton Provides Structure to the Cytoplasm

In the early days of cell biology, the cytosol was regarded as a structureless fluid. Its proteins were thought to be soluble and freely diffusible. However, several modern techniques have shown that the cytoplasm of eukaryotic cells contains an intricate, organized three-dimensional array of interconnected proteinaceous structures collectively called the **cytoskeleton**. As the name suggests, the cytoskeleton is an internal framework that gives a cell its distinctive shape and high level of internal organization. Its elaborate array of fibers forms a highly structured yet dynamic matrix that helps establish and maintain cell shape. In addition, the cytoskeleton plays an important role in cell movement and cell division. The cytoskeleton was initially thought to exist only in eukaryotes; however, proteins related to the eukaryotic cytoskeletal proteins have recently been discovered in bacteria and archaea and appear to play a role in maintaining cell shape.

In eukaryotic cells, the cytoskeleton serves as a framework for positioning and actively moving organelles and macromolecules within the cytosol. Some researchers estimate that up to 80% of the proteins and enzymes of the cytosol are not freely diffusible but are instead associated with the cytoskeleton. The three major structural elements of the eukaryotic cytoskeleton—*microtubules*, composed of tubulin subunits, *microfilaments*, composed of actin subunits, and *intermediate filaments*—are shown in **Figure 4-24**. Each is a network of protein polymers that can be visualized by various types of specialized light microscopy, including immunostaining and subsequent fluorescence microscopy, as well as by electron microscopy. Some of the structures in which these elements are found can be seen by ordinary light microscopy—structures such as cilia, flagella (both formed mainly by microtubules), and muscle fibrils (a key component of which is microfilaments), which are involved in cell movement and contractility. Less dynamic intermediate

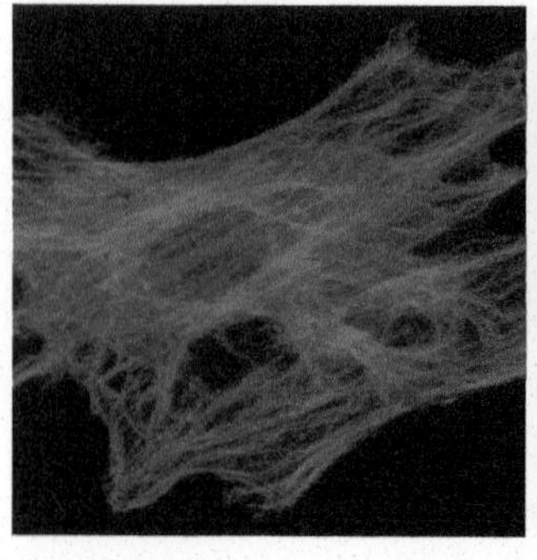

(a) Microtubules

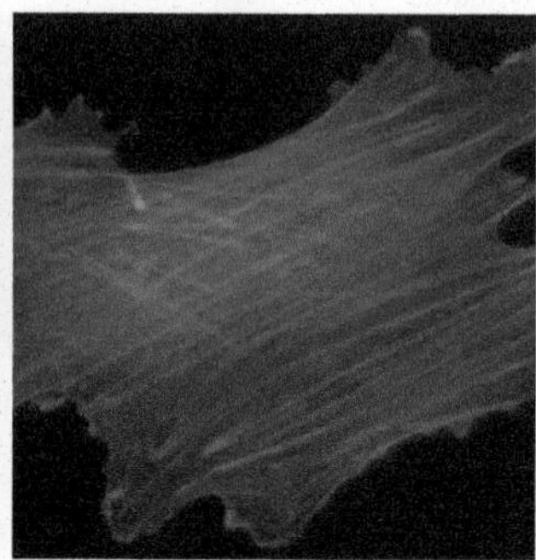

(b) Microfilaments

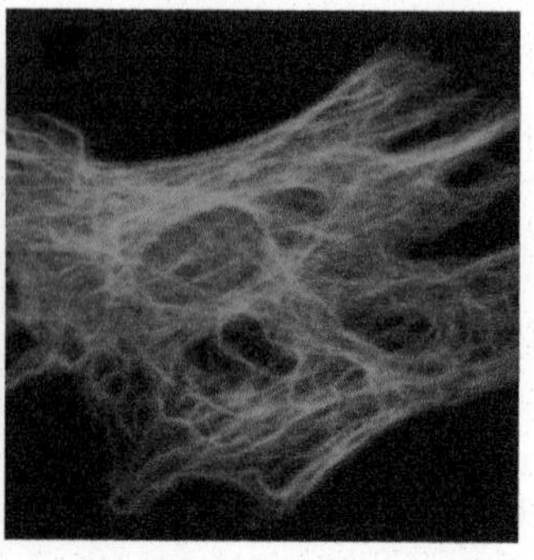

(c) Intermediate filaments

Figure 4-24 The Cytoskeleton. Immunofluorescence microscopy reveals the cytoskeleton of osteoblast (bone-producing) cells (Chapter 15). In these images of the same cell, one fluorescent antibody is directed against **(a)** microtubules (red), another against **(b)** microfilaments (blue), and a third against **(c)** intermediate filaments (green).

filaments play primarily a structural role within cells. (We will explore all three cytoskeletal elements in detail in Chapter 13.)

The Extracellular Matrix and Cell Walls Are Outside the Plasma Membrane

So far, we have considered the plasma membrane that surrounds every cell, the nucleus, and the variety of organelles, membrane systems, ribosomes, and cytoskeletal fibers found in the cytoplasm of most eukaryotic cells. Although it may seem that our tour of the cell is complete, most cells, both eukaryotic and prokaryotic, are also characterized by extracellular structures formed from materials that the cells transport outward across the plasma membrane. These structures often give physical support to cells. For many animal cells, these structures are called the **extracellular matrix (ECM)**, consisting primarily of *collagen fibrils* and *proteoglycans*. The primary function of the ECM is support, but the kinds of extracellular materials and the patterns in which they are deposited regulate such diverse processes as cell motility and migration, cell division, cell recognition and adhesion, and cell differentiation during embryonic development. The ECM of animal cells can vary in composition depending on the cell type. In vertebrates, collagen is such a prominent part of tendons, cartilage, and bone that it is the single most abundant protein of the animal body. (We will learn more about the ECM in Chapter 15.)

For plant and fungal cells, the extracellular structure is the rigid **cell wall**, which consists mainly of *cellulose microfibrils* embedded in a matrix of other polysaccharides and small amounts of protein. **Figure 4-25** illustrates the prominence of the cell wall as a structural feature of a typical plant cell. Plant cell walls actually consist of two layers. The wall that is laid down during cell division, called the *primary cell wall*, consists mainly of cellulose fibrils embedded in a gel-like polysaccharide matrix. Primary walls are quite flexible and extensible, which allows them to expand somewhat in response to cell enlargement and elongation. As a cell reaches its final size and shape, a much thicker and more rigid *secondary cell wall* may form by deposition of additional cell wall material on the inner surface of the primary wall. The secondary wall usually contains more cellulose than the primary wall and may have a high content of lignin, a major component of wood. Deposition of a secondary cell wall renders the cell inextensible and therefore defines the final size and shape of the cell. (We will return to primary and secondary cell walls in more detail in Chapter 15.)

Figure 4-25 also shows that neighboring plant cells, though separated by the cell walls between them, are actually connected by numerous openings, called **plasmodesmata** (singular: **plasmodesma**), which pass through the fused cell walls. (Plasmodesmata will be discussed in more detail in Chapter 15.)

Most bacteria and archaea are surrounded by an extracellular structure, also called a cell wall. Bacteria and archaea usually live in hypotonic environments (lower concentrations of external solutes than found in the cells), and their cell walls provide rigid support and protection from bursting due to osmotic pressure as water enters the cell.

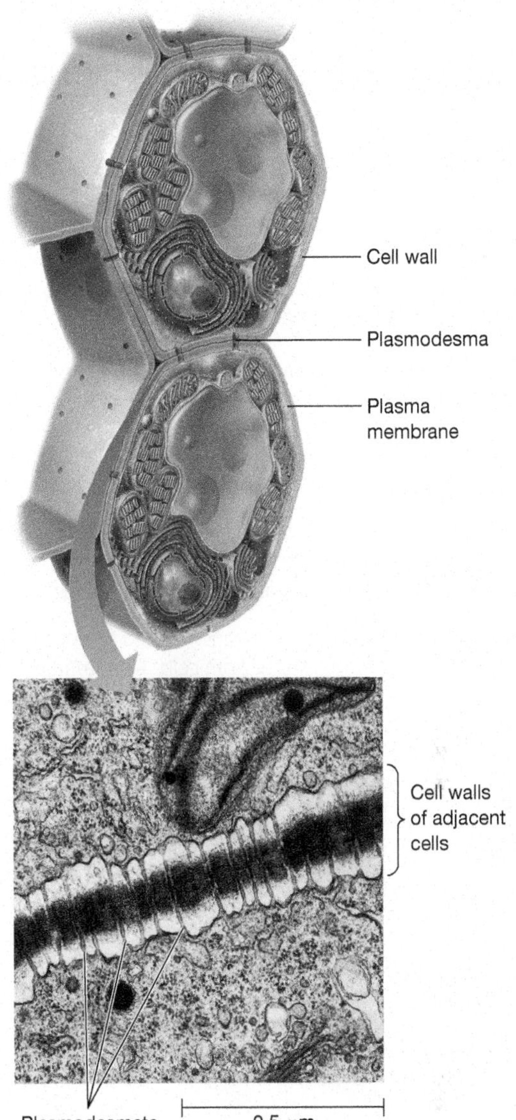

Figure 4-25 The Plant Cell Wall. Notice how the cell walls of neighboring cells are connected by plasmodesmata (TEM).

Unlike plant cell walls, bacterial cell walls consist not of cellulose but mainly of *peptidoglycans*. Peptidoglycans contain long chains of repeating units of *N*-acetylglucosamine (GlcNAc) and *N*-acetylmuramic acid (MurNAc), which are amino sugars (see Figure 3-26). These chains are held together to form a netlike structure by crosslinks composed of about a dozen amino acids linked by peptide bonds—thus the name peptidoglycan. In addition, bacterial cell walls contain a variety of other constituents, some of which are unique to each of the major groups of bacteria. Archaeal cell walls vary considerably from species to species—some are mainly proteinaceous, whereas others have peptidoglycan-like components.

CONCEPT CHECK 4.3

Imagine that you are building an artificial eukaryotic cell starting with an artificial bacterial cell. What organelles would you need to add?

Key Technique | Using Centrifugation to Isolate Organelles

PROBLEM: Eukaryotic cells contain a wide variety of organelles. Many studies require purified preparations of a particular organelle. But how can one type of organelle be isolated from all other organelles and cellular components?

SOLUTION: Although some organelles may appear to be similar, each has a characteristic density based on its unique composition. Researchers can take advantage of these density differences to isolate purified preparations using subcellular fractionation and centrifugation.

Key Tools: Centrifuge; centrifuge tubes; a cellular extract.

Details: *Centrifugation* is a technique used to isolate and purify organelles and macromolecules from cells. A *centrifuge* consists of a rotor—often housed in a refrigerated chamber—spun extremely rapidly by an electric motor. The rotor holds centrifuge tubes containing solutions or suspensions of particles. Centrifugation separates different organelles and macromolecules based on differences in their physical properties in a solution.

A centrifuge.

When a particle is subjected to centrifugal force, the rate of movement of the particle through a solution (its *sedimentation rate*) will depend on its size and density, as well as on the solution's density and viscosity. The larger or denser a particle is, the higher its sedimentation rate. Centrifugation is used routinely in laboratories throughout the world to separate dissolved molecules from cell debris in suspensions of broken cells, to collect precipitated macromolecules such as DNA and proteins from cell suspensions, and to separate different cellular components.

Because most organelles and macromolecules differ significantly from one another in size and density, centrifuging a mixture of cellular components will separate the faster-moving components from the slower-moving ones. This procedure, called *subcellular fractionation,* enables researchers to isolate and purify specific organelles and macromolecules for further manipulation and study in vitro.

Homogenization. Tissues must first undergo *homogenization,* or *disruption,* before the cellular components can be separated by centrifugation. The resulting *homogenate* is a suspension of organelles, smaller cellular components, membranes, and molecules. Homogenization is usually done in a cold isotonic solution, such as 0.25 *M* sucrose, to preserve organelle integrity. Disruption can be achieved by grinding tissue in a mortar and pestle, by forcing cells through a narrow orifice, or by subjecting tissue to ultrasonic vibration, osmotic shock, or enzymatic digestion. If tissue is homogenized gently enough, most organelles and other structures remain intact and retain their original biochemical properties.

Differential Centrifugation. Following homogenization, *differential centrifugation* is then used to separate organelles and other cellular components based on their size and density differences. As illustrated in **Figure 4B-1**, particles that are large or dense (large spheres) sediment rapidly, those that are intermediate in size or density (medium-sized spheres) sediment less rapidly, and the smallest or least dense particles (small spheres) sediment very slowly.

The relative size and density of an organelle or macromolecule are expressed in Svedberg units (S), which describe its *sedimentation coefficient.* Larger and/or denser particles have higher sedimentation coefficients. For example, human ribosomes are slightly larger than bacterial ribosomes and have larger sedimentation coefficients. The steps in a typical differential centrifugation experiment are illustrated in **Figure 4B-2**. A homogenate of the tissue is first suspended in solution (step ❶). Subcellular fractions are then isolated by subjecting the initial suspension and subsequent supernatant fractions to successively higher centrifugal forces and longer centrifugation times (steps ❷–❺). The *supernatant* is the clarified suspension of homogenate that remains after particles of a given size and density are separated from the sample, in the form of a *pellet,* following each step of the centrifugation process. The supernatant from one step is poured off into a new centrifuge tube and then returned to the centrifuge and subjected to greater

● Particles with large sedimentation coefficients
• Particles with intermediate sedimentation coefficients
· Particles with small sedimentation coefficients

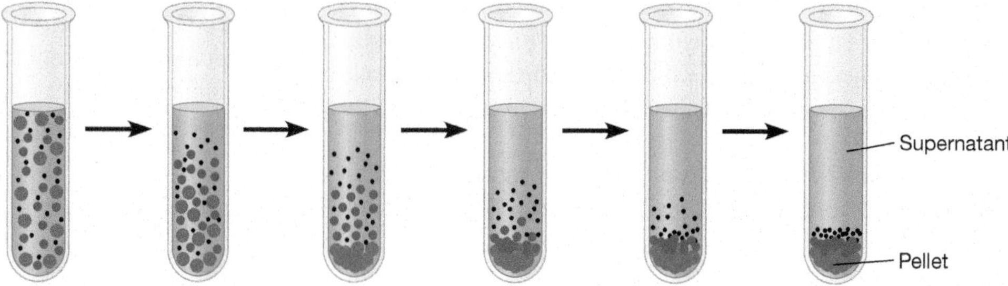

Figure 4B-1 Separation of Particles by Differential Centrifugation.

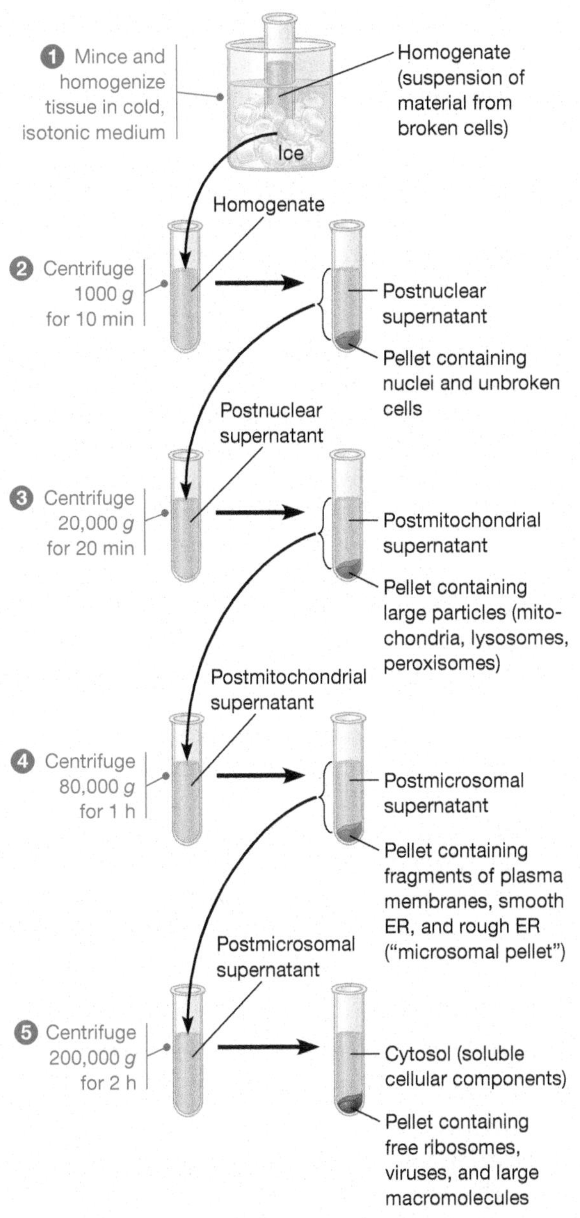

Figure 4B-2 Differential Centrifugation and the Isolation of Cellular Components.

centrifugal force to obtain the next pellet. The material in each pellet can be resuspended and used for electron microscopy or biochemical studies. The final supernatant consists mainly of soluble cellular components, and the final pellet contains free ribosomes and large macromolecules.

Density Gradient Centrifugation. Although each fraction obtained via differential centrifugation is enriched for a particular organelle, it is also likely contaminated with other organelles and cellular components. These contaminants can be removed by a technique known as *density gradient centrifugation* (sometimes called *rate-zonal centrifugation*).

Whereas in the differential centrifugation example, the particles about to be separated are uniformly distributed throughout the solution prior to centrifugation, the homogenated sample in density gradient centrifugation is placed as a thin layer on top of a gradient of solute, typically sucrose. This gradient has an increasing concentration of solute—and therefore buoyant density—from the top of the tube to the bottom. When centrifuged, particles differing in size and density move downward as discrete *zones*, or *bands*, that migrate at different rates. Each zone remains very compact, maximizing the ability to separate different particles.

Using density gradient centrifugation, the components of a single pellet can be separated—for example, the lysosomes, mitochondria, and peroxisomes pelleted in step 3 of Figure 4B-2—because each has a slightly different density (**Figure 4B-3**). The particles that are largest or densest (peroxisomes; purple) move into the gradient as a rapidly sedimenting band. The particles that are intermediate in size or density (mitochondria; blue) sediment less rapidly. The smallest particles (lysosomes; black) move quite slowly. The centrifuge is stopped after the bands of interest have moved far enough into the gradient to be resolved from each other but before any of the bands reach the bottom of the tube. These can be collected in separate fractions, and by assaying each fraction for marker enzymes unique to each organelle, the fractions containing these organelles can be identified.

QUESTION: If a scientist were studying a disease in which membrane composition was altered, how would she isolate these membranes?

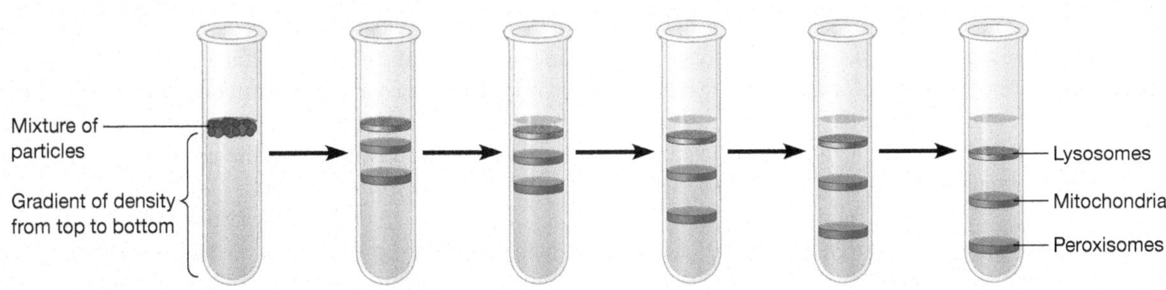

Figure 4B-3 Density Gradient Centrifugation.

4.4 Viruses, Viroids, and Prions: Agents That Invade Cells

Before concluding this preview of cellular biology, we will look at several kinds of agents that invade cells, subverting normal cellular functions and often killing their unwilling hosts. These include the viruses, which have been studied for more than 100 years, and two other agents we know much less about—the *viroids* and *prions*.

A Virus Consists of a DNA or RNA Core Surrounded by a Protein Coat

Viruses are acellular, parasitic particles that are incapable of a free-living existence and must therefore depend on the cells they invade for their replication. A virus particle has no cytoplasm, no ribosomes, no organelles, and few or no enzymes and typically consists of only a few different molecules of nucleic acid and protein. However, viruses can invade and infect cells and redirect the synthetic machinery of the infected host—host cell enzymes and ribosomes—toward the production of more virus particles. After replication, the viral progeny typically erupt from the host cell and, in the process, destroy it.

Viruses are responsible for many diseases in plants and animals (including humans) and for certain cancers (Chapter 26); they are thus important in their own right. They are also significant research tools for cell and molecular biologists. Viruses have been used to study self-assembly of complex biological structures (Chapter 2) and genetic recombination (Chapter 17). Engineered viruses are currently used as tools in gene therapy (see Human Connections in Chapter 8, pages 200–201).

Some viruses are named for the diseases they cause. Polio virus, influenza virus, herpes simplex virus, and tobacco mosaic virus (TMV) are several examples. Other viruses are named based on their structure. Coronaviruses are good examples ("corona" is Latin for "crown"; their coat structure is reminiscent of a crown). Other viruses have more cryptic laboratory names (such as T4, $Q\beta$, and λ). Viruses that infect bacterial cells are called *bacteriophages*, or often just *phages* for short. (Bacteriophages and other viruses will figure prominently in our discussion of molecular genetics in Chapters 16–21.)

Viruses are smaller than all but the tiniest cells, with most ranging in size from about 25 to 300 nm. They are small enough to pass through a filter that traps bacteria, and the collected filtrate will still be infectious. Because of this, their existence was postulated by Louis Pasteur 50 years before they were able to be seen in the electron microscope! The smallest viruses are about the size of a ribosome, whereas the largest ones are about one-quarter the diameter of a typical bacterial cell. Each virus has a characteristic shape, as defined by its protein capsid. Some examples are shown in **Figure 4-26**.

Despite their morphological diversity, viruses are chemically quite simple. Most viruses consist of little more than a *coat* (or *capsid*) of protein surrounding a *core* that contains one or more molecules of either RNA or DNA, depending on the type of virus. This is another feature distinguishing viruses from cells, which always have both DNA and RNA. The simplest viruses, such as TMV, have a single nucleic acid molecule surrounded by a capsid consisting of proteins of a single type (see Figure 2-19 and Figure 4-26a). More complex viruses have cores that contain several nucleic acid molecules and capsids consisting of several (or even many) different kinds of proteins. Some viruses are surrounded by a membrane that is derived from the plasma membrane of the host cell in which the viral particles were previously synthesized and assembled. Such viruses are called *enveloped viruses*. Human immunodeficiency virus (HIV), the virus that causes AIDS (acquired immune deficiency syndrome), is an example of an enveloped virus; it is covered by a membrane that it received from a previously infected white blood cell (see Figure 18-4).

Students sometimes ask whether viruses are living. The answer depends crucially on what is meant by "living," and it is probably worth pondering only to the extent that it helps us more fully understand what viruses are—and what they are not. The most fundamental properties of living things are *metabolism* (cellular reactions organized into coherent pathways), *irritability* (perception of, and response to, environmental stimuli), and the *ability to reproduce*. Viruses do not satisfy the first two criteria. Outside their host cells, viruses are inert and inactive. They can, in fact, be isolated and crystallized almost like a chemical compound. It is only in an appropriate host cell that a virus becomes functional, undergoing a cycle of synthesis and assembly that gives rise to more viruses.

Even the ability of viruses to reproduce has to be qualified carefully. A basic tenet of the cell theory is that cells arise only from preexisting cells, but this is not true of viruses. No virus can give rise to another virus by any sort of self-duplication process. Rather, the virus must take over the metabolic and genetic machinery of a host cell, reprogramming it for synthesis of the proteins necessary to package the DNA or RNA molecules that arise by copying the genetic information of the parent virus.

Viroids Are Small, Circular RNA Molecules That Can Cause Plant Diseases

As simple as viruses are, there are even simpler acellular agents that can infect eukaryotic cells (though apparently not prokaryotic cells, as far as we currently know). The **viroids** found in some plant cells represent one class of such agents. Viroids are small, circular RNA molecules, and they are the smallest known infectious agents. These RNA molecules are only about 250–400 nucleotides long, do not encode any protein, and are replicated in the host cell. Some have been shown to have enzymatic properties that aid in their own replication. Viroids do not occur in free form but can be transmitted from one plant cell to another when the surfaces of adjacent cells are damaged and there is no membrane barrier to prevent RNA molecules from crossing.

Viroids are responsible for diseases of several crop plants, including potatoes and tobacco. A viroid disease that has severe economic consequences is *cadang-cadang disease* of the coconut palm. It is not yet clear how viroids cause disease. They may enter the nucleus of an infected plant cell and interfere

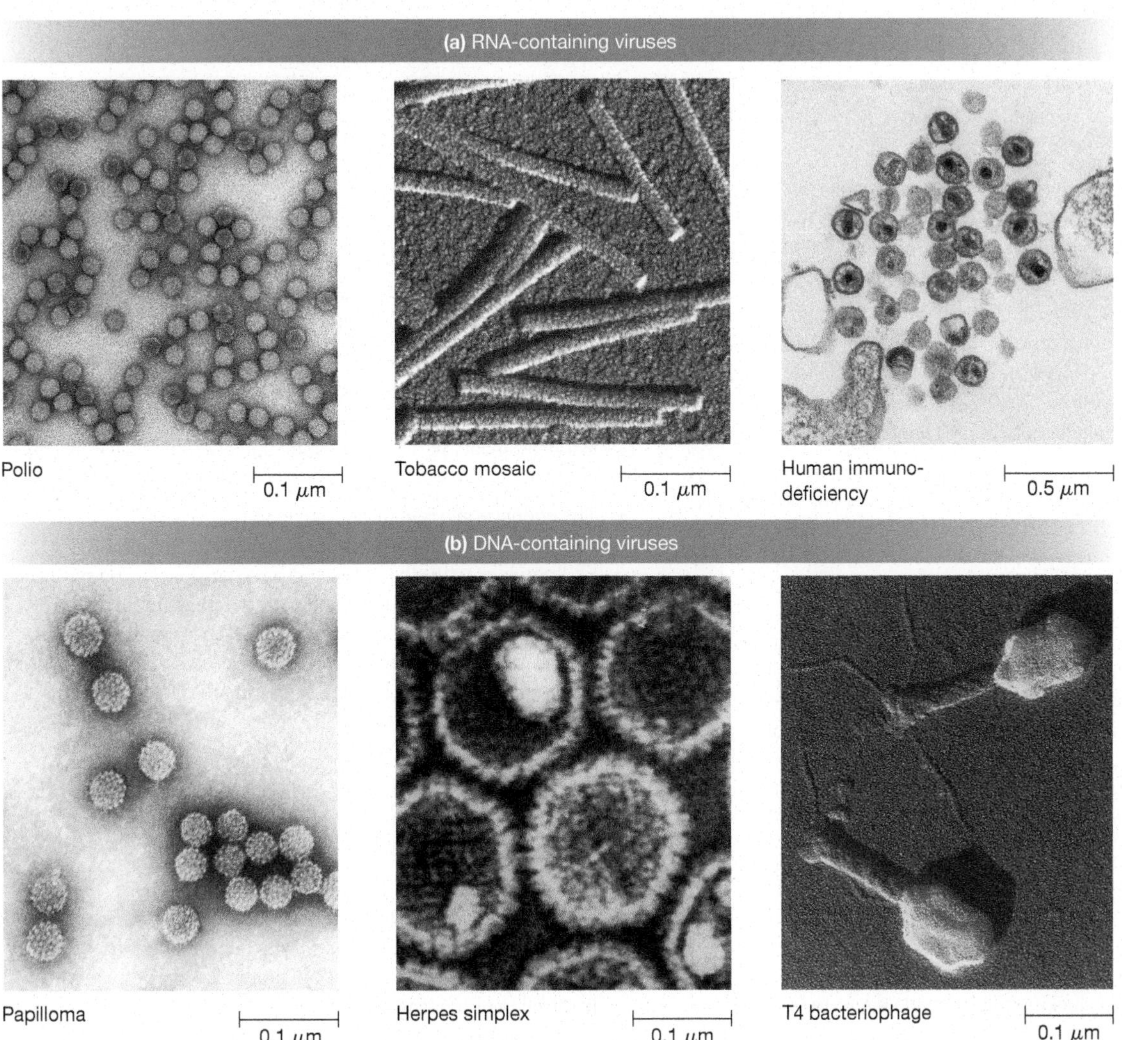

Figure 4-26 Sizes and Shapes of Viruses. These electron micrographs illustrate the morphological diversity of viruses.

with the transcription of DNA into RNA in a process known as *gene silencing*. Alternatively, they may interfere with the subsequent processing required of most eukaryotic mRNAs, disrupting subsequent protein synthesis.

Prions Are Infectious Protein Molecules

Prions represent another class of acellular infectious agents. The term was coined from the words "*pro*tein" and "*in*fectious." Prions are responsible for neurological diseases such as scrapie in sheep and goats, bovine spongiform encephalopathy (BSE, or "mad cow" disease) in cattle, and kuru in humans. *Scrapie* is so named because infected animals rub incessantly against trees or other objects, scraping off most of their wool in the process. *BSE* leads to degenerative anatomical defects in the brains of affected cattle, leading to behavioral and motor deficits. *Kuru* is a degenerative disease of the central nervous system originally reported among native peoples in New Guinea who consumed infected brain tissue. Patients with this or other prion-based diseases suffer initially from mild physical weakness and dementia, but the effects slowly become more severe, and the diseases are eventually fatal.

Prion proteins are abnormally folded versions of normal cellular proteins. In the diseases mentioned above, both the normal and variant forms of the prion protein are found within neurons. Some biologists speculate that a similar mechanism is responsible for the plaques of misfolded proteins that are found in the brains of deceased Alzheimer disease patients (see Human Connections in Chapter 3, pages 48–49).

Prions are not destroyed by cooking or boiling, so special precautions are recommended in areas where prion-caused diseases are known to occur. For example, a prion disease known as *chronic wasting disease* is found in some deer and elk, especially in the Rocky Mountain region of the western United States. Hunters in areas where this prion is known to occur are cautioned to have meat tested for this prion and to avoid eating meat from any animal that tests positive or appears sick.

CONCEPT CHECK 4.4

If viruses are nonliving particles, how can they reproduce and cause serious diseases?

Summary of Key Points

Mastering™ Biology For activities, animations, and review quizzes, go to the study area at www.masteringbiology.com.

4.1 The Origins of the First Cells

- It is theorized that organic building block molecules such as amino acids, simple sugars, and nucleotide bases formed abiotically in the young Earth. How these smaller subunits first polymerized into the prebiotic precursors of today's proteins and nucleic acids is not clear.
- Although genetic information currently flows from DNA to RNA to protein, it is believed that RNA existed before DNA as an informational molecule because RNA is capable of the catalysis required for its own replication, in the absence of protein enzymes.
- Catalytic self-replicating RNA likely became trapped in liposomes made of simple fatty acids, forming the precursors of today's cells.

4.2 Basic Properties of Cells

- Cells have traditionally been described as either eukaryotic (animals, plants, fungi, protozoa, and algae) or prokaryotic (bacteria and archaea) based on a gross morphological feature—the presence or absence of a membrane-bounded nucleus. More recent analysis of ribosomal RNA sequences and other molecular data suggests that all organisms fall into one of three domains of life: eukaryotes, bacteria, or archaea.
- A plasma membrane is a structural feature common to cells of all three groups. Organelles, however, are typically found only in eukaryotic cells, where they play indispensable roles in the compartmentalization of function. Bacterial and archaeal cells are relatively small and structurally less complex than eukaryotic cells, most lacking the internal membrane systems and organelles of eukaryotes.
- Cell size is limited by the need for an adequate amount of surface area for exchange of materials with the environment and the need for high enough concentrations of the compounds necessary to sustain life. Eukaryotic cells are much larger than prokaryotic cells and compensate for their lower surface area/volume ratio by the compartmentalization of materials within membrane-bounded organelles.

4.3 The Eukaryotic Cell in Overview: Structure and Function

- Besides the plasma membrane and ribosomes common to all cells, eukaryotic cells have a nucleus that houses most of the cell's DNA, a variety of organelles, and the cytoplasm with its cytoskeletal filaments. Exterior to the plasma membrane, plant cells have a rigid cell wall, and animal cells are usually surrounded by a strong but flexible extracellular matrix of collagen and proteoglycans.
- The nucleus contains chromosomes, which are the eukaryotic cell's DNA complexed with protein in the form of chromatin; chromosomes condense during cell division to form visible structures. The nucleus is surrounded by a double membrane called the nuclear envelope, which has pores that allow the regulated exchange of macromolecules with the cytoplasm.
- Mitochondria, which are surrounded by a double membrane, degrade food molecules to provide the energy used to make ATP. Mitochondria also contain ribosomes and their own circular DNA molecules.
- Chloroplasts trap solar energy and use it to convert carbon dioxide and water into sugar and oxygen. Chloroplasts are surrounded by a double membrane and have an extensive system of internal membranes called the thylakoids. Chloroplasts also contain ribosomes and circular DNA molecules.
- Based on biochemical and genetic evidence, such as the presence of circular DNA molecules, ribosome size, rRNA sequence, and double membranes with particular lipid and protein composition, the endosymbiont theory proposes that mitochondria originated from aerobic bacteria and that chloroplasts originated from cyanobacteria.
- The endoplasmic reticulum is an extensive network of membranes that are known as either rough ER or smooth ER. Rough ER is studded with ribosomes and is responsible for the synthesis of secretory and membrane proteins, whereas smooth ER is involved in lipid synthesis and drug detoxification. Proteins synthesized on the rough ER are further processed and packaged in the Golgi apparatus and are then transported either to membranes or to the surface of the cell via secretory vesicles.
- Lysosomes contain hydrolytic enzymes and are involved in cellular digestion. Because of their close functional relationships, the ER, Golgi apparatus, secretory vesicles, and lysosomes are collectively called the endomembrane system.
- Peroxisomes are about the same size as lysosomes, and both generate and degrade hydrogen peroxide. Animal peroxisomes play an important role in catabolizing long-chain fatty acids. In plants, specialized peroxisomes known as glyoxysomes are involved in photorespiration and in converting stored fat into carbohydrate during seed germination.
- Ribosomes are sites of protein synthesis in all cells. The striking similarities between mitochondrial and chloroplast ribosomes and those of bacteria and cyanobacteria, respectively, lend strong support to the endosymbiont theory that these organelles are of bacterial origin.
- The cytoskeleton provides structural support to cells that allows them to adopt distinctive shapes. In eukaryotes it is an extensive network of microtubules, microfilaments, and intermediate filaments. The cytoskeleton is also important in cellular motility and the intracellular movement of cellular structures and materials.
- Plant, fungal, bacterial, and archaeal cells are surrounded by a rigid cell wall, which gives the cell shape and physical support. Animal cells, which lack a cell wall, often are surrounded by an extracellular matrix, which similarly provides support between adjacent cells within tissues.

4.4 Viruses, Viroids, and Prions: Agents That Invade Cells

- Viruses do not satisfy the basic criteria of living things. They are important both as infectious agents that cause diseases in plants and animals and as laboratory tools, particularly for geneticists.
- Viroids and prions are infectious agents that are even smaller (and less well understood) than viruses. Viroids are small RNA molecules that are replicated in the host cytoplasm and can cause plant diseases. Prions are misfolded proteins that can cause other protein molecules to misfold.

Problem Set

4-1 DATA ANALYSIS **Lipids in the mix.** Scientists have proposed that the soft clay montmorillonite could have had significant effects on chemical processes on the prebiotic Earth. The graph in **Figure 4-27** shows the effects of adding montmorillonite to solutions of two different fatty acids. An increase in absorbance at 400 nm is an indirect measure of the concentration of liposomes (lipid vesicles) that form in the mixture.

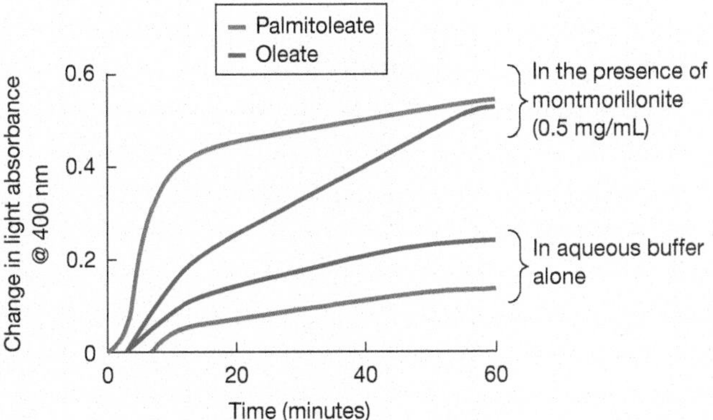

Figure 4-27 Effects of Adding Clay to Fatty Acids in Aqueous Buffer. Liposome formation over time was indirectly assessed by measuring absorbance of 400-nm light by a solution of 10 m*M* fatty acid (palmitoleate, red; oleate, blue) in the presence of montmorillonite (0.5 mg/mL) or in aqueous buffer alone.

(a) What do you conclude about the effect of montmorillonite on liposome formation?

(b) Which fatty acid is more sensitive to the additional of montmorillonite, and why?

4-2 Wrong Again. For each of the following false statements, change the statement to make it true.

(a) Archaea are ancient bacteria that are the ancestors of modern bacteria.

(b) Bacteria differ from eukaryotes in having no nucleus, mitochondria, chloroplasts, or ribosomes.

(c) Instead of a cell wall, eukaryotic cells have an extracellular matrix for structural support.

(d) All the ribosomes found in a typical human muscle cell are identical.

(e) DNA is found only in the nucleus of a cell.

(f) Because bacterial cells have no organelles, they cannot carry out either ATP synthesis or photosynthesis.

(g) Even the simplest types of infectious agents must have DNA, RNA, and protein.

4-3 QUANTITATIVE **As Cells Get Larger ...** Consider three spherical cells with diameters of 2 μm, 10 μm, and 20 μm.

(a) For each cell, calculate the total volume using the equation $V = 4\pi r^3/3$.

(b) Now calculate the surface area of each cell using the equation $A = 4\pi r^2$.

(c) Finally, determine the surface area/volume ratio for each cell.

(d) How does the surface area/volume ratio for a cell change as the cell increases in size?

(e) Why does it change in this manner?

4-4 Sentence Completion. Complete each of the following statements about cellular structure in ten or fewer words.

(a) If you were shown an electron micrograph of a section of a cell and were asked to identify the cell as plant or animal, one thing you might do is ...

(b) A slice of raw apple placed in a concentrated sugar solution will ...

(c) A cellular structure that is visible with an electron microscope but not with a light microscope is ...

(d) Several environments in which you are more likely to find archaea than bacteria are ...

(e) One reason it might be difficult to separate lysosomes from peroxisomes by centrifugation techniques is that ...

(f) The nucleic acid of a virus is composed of ...

4-5 Telling Them Apart. Suggest a way to distinguish between the two elements in each of the following pairs.

(a) Bacterial cells; archaeal cells

(b) Rough ER; smooth ER

(c) Animal peroxisomes; leaf peroxisomes

(d) Peroxisomes; lysosomes

(e) Viruses; viroids

(f) Eukaryotic ribosomes; bacterial ribosomes

4-6 Structural Relationships. For each pair of structural elements, indicate with an A if the first element is a constituent part of the second, with a B if the second element is a constituent part of the first, and with an N if they are separate structures with no particular relationship to each other.

(a) Mitochondrion; crista

(b) Golgi apparatus; nucleus

(c) Cytoplasm; cytoskeleton

(d) Cell wall; extracellular matrix

(e) Nucleolus; nucleus

(f) Smooth ER; ribosome

(g) Lipid bilayer; plasma membrane

(h) Peroxisome; thylakoid

(i) Chloroplast; granum

4-7 Malfunctions at the Organelle Level. Each of the following medical problems involves a malfunction of an organelle or other cell structure. In each case, identify the organelle or structure involved, and indicate whether it is likely to be underactive or overactive.

(a) A girl inadvertently consumes cyanide and dies almost immediately because ATP production ceases.

(b) A boy is diagnosed with neonatal adrenoleukodystrophy (NALD), which is characterized by an inability of his body to break down very-long-chain fatty acids.

(c) A young man learns that he is infertile because his sperm are nonmotile.

(d) A young child dies because her cells lack an enzyme that normally breaks down a membrane glycolipid that therefore accumulated in the membranes of her nervous system.

(e) A young child is placed on a milk-free diet because the mucosal cells that line his small intestine do not secrete the enzyme necessary to hydrolyze lactose, the disaccharide present in milk.

4-8 Are They Alive? Biologists sometimes debate whether viruses should be considered alive. Let's join in the debate.

(a) What are some ways in which viruses resemble cells?

(b) What are some ways in which viruses differ from cells?

(c) Choose either of the two following positions and defend it. (1) Viruses are alive. (2) Viruses are not alive.

(d) Why do you suppose that viral illnesses are more difficult to treat than bacterial illnesses?

(e) Design a strategy to cure a viral disease without harming the patient.

4-9 The Endosymbiont Theory. The endosymbiont theory suggests that mitochondria and chloroplasts evolved from ancient bacteria that were ingested by primitive nucleated cells. Over hundreds of millions of years, the ingested bacteria lost features not essential for survival inside the host cell.

(a) Describe one structure or metabolic process that purple bacteria might have dispensed with once they became endosymbionts in a eukaryotic cell. How would loss of this feature prevent the bacteria from living outside the host?

(b) Describe one structure or metabolic process that cyanobacteria might have dispensed with once they became endosymbionts in a eukaryotic cell. How would loss of this feature prevent the bacteria from living outside the host?

(c) Peroxisomes, unlike mitochondria or chloroplasts, scarcely resemble free-living organisms, yet biologists have suggested that peroxisomes may also have arisen via endosymbiosis. Assuming peroxisomes evolved from ingested bacteria, describe three features mitochondria retain but peroxisomes apparently lost over hundreds of millions of years. Describe one advantage peroxisomes might have conferred on ancient nucleated cells.

5 Bioenergetics: The Flow of Energy in the Cell

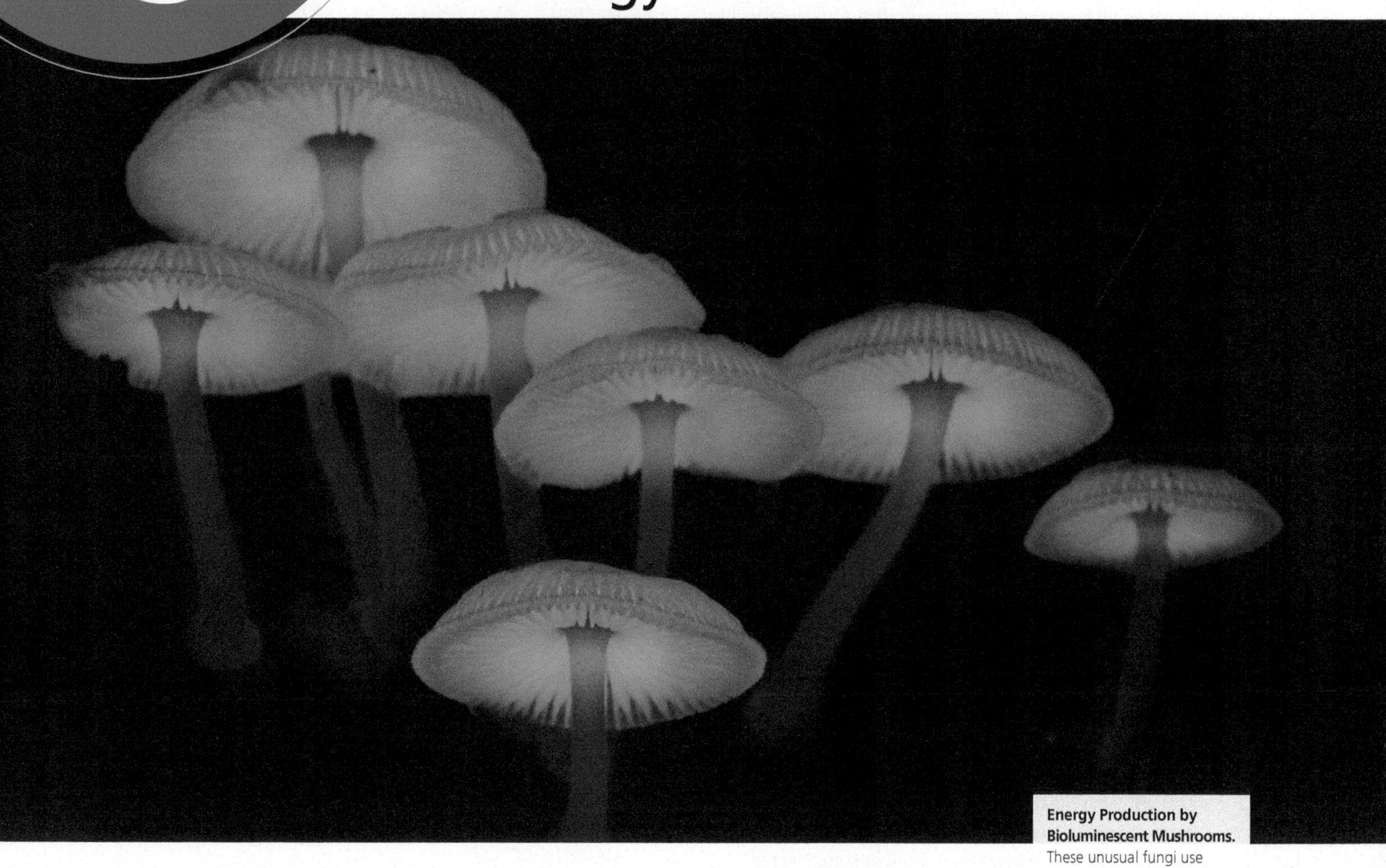

Energy Production by Bioluminescent Mushrooms. These unusual fungi use energy to produce light, possibly to attract nocturnal animals that spread their reproductive spores.

Broadly speaking, every cell has four essential needs: *molecular building blocks* for biosynthesis, *chemical catalysts* called enzymes (the topic of Chapter 6), *information* to guide all its activities, and *energy* to drive the various reactions and processes that are essential to life and biological function. All cells need amino acids, nucleotides, sugars, and lipids to synthesize the macromolecules used to build cellular structures. All cells need *enzymes* to speed up chemical reactions that would otherwise occur much too slowly to maintain life as we know it. Cells also need information encoded within the nucleotide sequences of DNA and RNA and expressed in the ordered synthesis of specific proteins.

In addition, all cells require energy to drive biosynthetic chemical reactions and to power cellular activities such as movement, nutrient uptake, and production of heat or light. The capacity to obtain, store, and use energy is, in fact, one of the obvious features of most living organisms and helps define life itself. Like the flow of information, the flow of energy is a major theme of this text and will be considered in more detail in Chapters 9–11. In this chapter, we will learn the basics of *bioenergetics* and see how the thermodynamic concept of free energy can allow us to predict whether specific chemical reactions can occur spontaneously in the cell.

5.1 The Importance of Energy

All living systems require an ongoing supply of **energy,** which is often defined as the capacity to do work. A more useful definition is that *energy is the capacity to cause specific physical or chemical changes.* Because life is characterized first and foremost by change, this definition emphasizes the total dependence of all forms of life on the continuous availability of energy.

Cells Need Energy to Perform Six Different Kinds of Work

If energy is the capacity to cause specific physical or chemical changes, then what kinds of cellular activities give rise to these changes? Six categories of change define six kinds of work that require an input of energy: synthetic work, mechanical work, concentration work, and electrical work, as well as the generation of heat and light (**Figure 5-1**).

1. *Synthetic Work: Changes in Chemical Bonds.* An important activity of virtually every cell at all times is the work of **biosynthesis,** which results in the formation of new chemical bonds and the synthesis of new molecules. This activity is especially obvious in a population of growing cells, where additional molecules must be synthesized for cells to increase in size or number or both. Also, synthetic work is required to maintain existing cellular structures. Because most structural components of the cell are in a state of constant turnover, the molecules that make up these structures are continuously being degraded and replaced. Almost all of the energy that cells require for biosynthetic work is used to make energy-rich organic molecules from simpler starting materials and to incorporate them into macromolecules.
2. *Mechanical Work: Changes in the Location or Orientation of a Cell or a Subcellular Structure.* Cells often need energy for **mechanical work,** which involves a physical change in the position or orientation of a cell or some part of the cell. An especially good example is the movement of a cell with respect to its environment. This movement often requires one or more appendages such as cilia or flagella (**Figure 5-2**), which require energy to propel the cell forward. In other cases, the environment is moved past the cell, as when the ciliated cells that line your trachea beat upward to sweep inhaled particles upward and away from the lungs. This motion also requires energy. Muscle contraction is another good example of mechanical work, involving not just a single cell but also a large number of muscle cells working together (Chapter 14). Other examples of mechanical work that occur inside cells include the movement of chromosomes along the spindle fibers during mitosis, the streaming of cytoplasm, the movement of organelles and vesicles along microtubules, and the translocation of a ribosome along a strand of messenger RNA.
3. *Concentration Work: Moving Molecules Across a Membrane Against a Concentration Gradient.* Less conspicuous than either of the previous two categories but every bit as important to the cell is the work of moving molecules or ions against a concentration gradient. The purpose

(a) Synthetic work: the process of photosynthesis

(b) Mechanical work: the contraction of a weight lifter's muscles

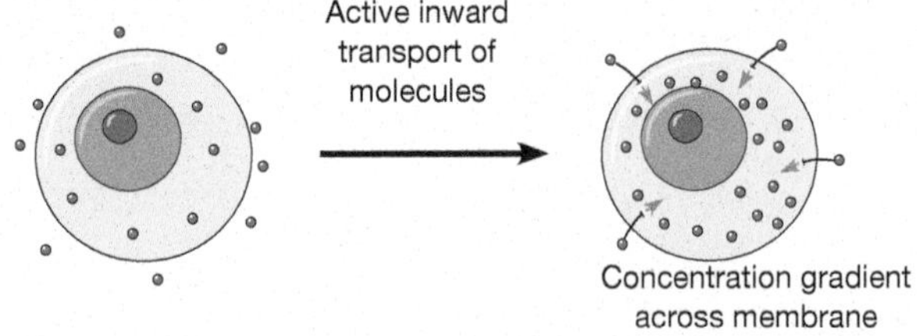

(c) Concentration work: the accumulation of molecules in a cell

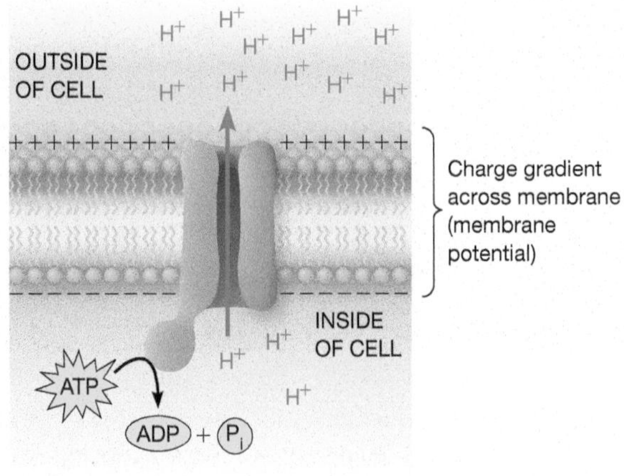

(d) Electrical work: the membrane potential of a plant cell

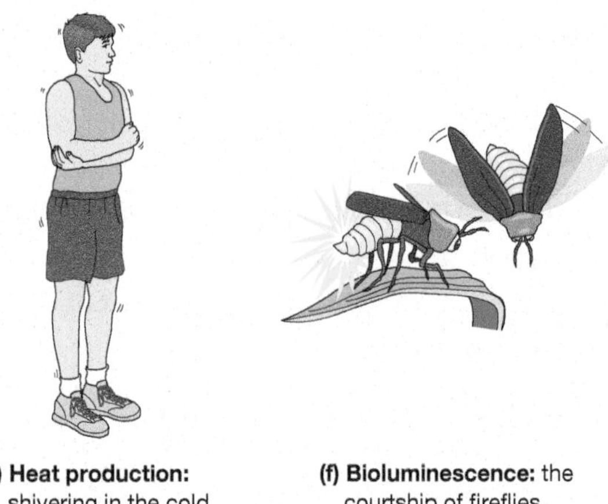

(e) Heat production: shivering in the cold

(f) Bioluminescence: the courtship of fireflies

Figure 5-1 Several Kinds of Biological Work. The six major categories of biological work are shown here **(a–f).** All require an input of energy to cause a physical or chemical change.

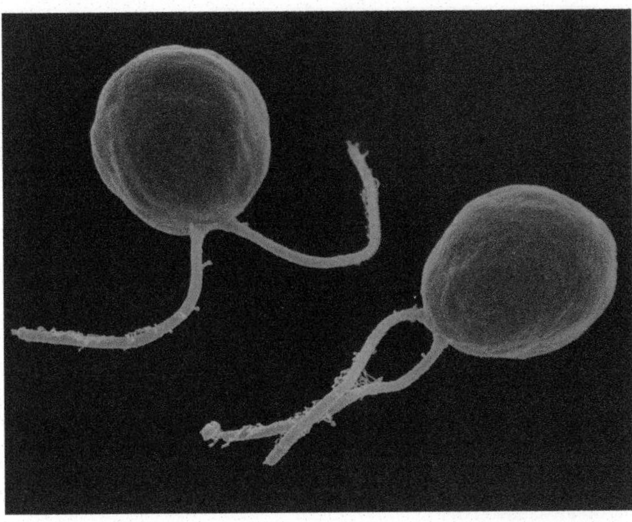

Figure 5-2 Cell Motility. The movement of flagella by these *Chlamydomonas* algal cells is an example of mechanical work (colorized SEM).

of **concentration work** is either to accumulate substances within a cell or organelle or to remove potentially toxic by-products of cellular activity. Diffusion of molecules, which always proceeds from areas of high concentration to areas of low concentration, is a spontaneous process and requires no added energy (Chapter 2). Therefore, the opposite process, concentration, requires an input of energy. Examples of concentration work include the import of certain sugar and amino acid molecules from low to high concentration across the plasma membrane, the concentration of specific molecules and enzymes within organelles, and the accumulation of digestive enzymes into secretory vesicles to be released as you digest your food.

4. *Electrical Work: Moving Ions Across a Membrane Against an Electrochemical Gradient.* Often, **electrical work** is considered a specialized case of concentration work because it also involves movement across membranes. In this case, however, ions are transported, and the result is not just a change in concentration. A charge difference, known as an *electrical potential* or a *membrane potential*, is also established across the membrane. Every cellular membrane has some characteristic electrical potential that is generated in this way, such as the plant cell membrane in Figure 5-1d. A difference in the concentration of protons on either side of the mitochondrial or chloroplast membrane forms an electrical potential that is essential for the production of ATP in both cellular respiration (Chapter 10) and photosynthesis (Chapter 11). Electrical work is also important in the transmission of impulses in nerve cells. This involves the establishment of a membrane potential, resulting from pumping Na^+ and K^+ ions into and out of the cell (Chapter 22). An especially dramatic example of electrical work is found in *Electrophorus electricus*, the electric eel. Individual cells in its electric organ use energy to generate a membrane potential of about 150 millivolts (mV) in one cell, and thousands of such cells arranged in series allow the electric eel to develop electric potentials of several hundred volts (ouch!).

5. *Heat: An Increase in Temperature That Is Useful to Homeothermic Animals.* Although living organisms do not use **heat** as a form of energy in the same way that a steam engine does, heat production is a major use of energy in *homeotherms* (animals that regulate and maintain their body temperature independent of the environment). Heat is released as a by-product of many chemical reactions, and we, as homeotherms, take advantage of this by-product every day. In fact, as you read these lines, about two-thirds of your metabolic energy is being used just to maintain your body at 37°C. At this temperature, your body functions most efficiently because it is the optimum temperature for enzyme activity and cellular metabolism in humans. When you are cold, your muscles use energy to shiver, generating heat to keep you warm.

6. *Bioluminescence and Fluorescence: The Production of Light.* To complete this list, we will include **bioluminescence,** the production of light using ATP or chemical oxidation as an energy source, and **fluorescence,** the production of light following absorption of light of a shorter wavelength. Light is produced by a number of bioluminescent organisms, such as fireflies, certain jellyfish, and luminous mushrooms (see the chapter-opening photo). Studies of the bioluminescent jellyfish *Aequorea victoria* in the 1960s by Osamu Shimomura led to the isolation of *aequorin,* a protein that, upon binding calcium ions, is able to carry out an oxidation reaction that releases energy in the form of pale blue light (**Figure 5-3a** on page 110). Further study of this organism led to the discovery of green fluorescent protein (GFP), which absorbs this pale blue light and, in turn, produces a pale green fluorescence. GFP has become an extremely powerful tool for studying the locations and functions of specific proteins inside cells. When GFP is fused to a particular protein of interest, the fusion protein fluoresces when irradiated with blue light (Figure 5-3b), and thus the distribution and movement of the protein of interest can be followed in live cells using fluorescence microscopy.

Organisms Obtain Energy Either from Sunlight or from the Oxidation of Chemical Compounds

Nearly all life on Earth is sustained, directly or indirectly, by the sunlight that continuously floods our planet with energy. Not all organisms can obtain energy from sunlight directly, of course. In fact, based on their energy sources, organisms (and cells) can be classified as either *phototrophs* (literally, "light-feeders") or *chemotrophs* ("chemical-feeders"). Organisms can also be classified as *autotrophs* ("self-feeders") or *heterotrophs* ("other-feeders"), depending on whether their source of carbon is CO_2 or organic molecules, respectively. Most organisms are either photoautotrophs (plants, algae, some bacteria) or chemoheterotrophs (all animals, fungi, protozoa, and most bacteria).

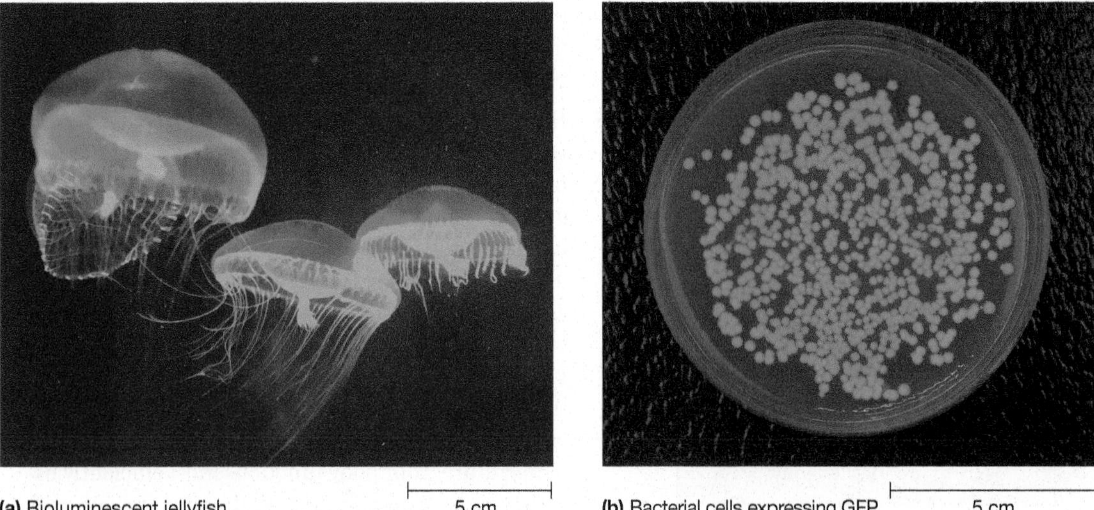

Figure 5-3 Bioluminescence and Fluorescence. (a) These *Aequorea* jellyfish contain the protein aequorin, which uses the energy of a chemical oxidation reaction to produce a pale blue light by bioluminescence. **(b)** These colonies of bacterial cells have been engineered to produce a GFP fusion protein, resulting in a pale green fluorescence when irradiated with pale blue light.

Phototrophs capture light energy from the sun using light-absorbing pigments and then transform this light energy into chemical energy, storing the energy in the form of ATP. *Photoautotrophs* use solar energy to produce all their necessary carbon compounds from CO_2 during photosynthesis. Examples of photoautotrophs include plants, algae, cyanobacteria, and photosynthetic bacteria. *Photoheterotrophs* (some bacteria) harvest solar energy to power cellular activities, but they must rely on the intake of organic molecules for their carbon needs.

In contrast, **chemotrophs** get energy by oxidizing chemical bonds in organic or inorganic molecules. *Chemoautotrophs* (a few bacteria) oxidize inorganic compounds such as H_2S, H_2 gas, or inorganic ions for energy and synthesize all their organic compounds from CO_2. *Chemoheterotrophs*, on the other hand, ingest and use chemical compounds such as carbohydrates, fats, and proteins to provide both energy and carbon for cellular needs. All animals, protozoa, fungi, and many bacteria are chemoheterotrophs.

Keep in mind that, although phototrophs can use solar energy when it is available, they must function as chemotrophs whenever they are not illuminated. Most plants are a mixture of phototrophic and chemotrophic cells. For example, neither a plant root cell underground nor a plant leaf cell in the dark can carry out photosynthesis. Although these cells are clearly part of a phototrophic organism, they themselves are chemotrophic like animal cells.

Energy Flows Through the Biosphere Continuously

Before we continue our discussion of energy flow, first let's review the chemical concepts of oxidation and reduction because they are critical to understanding energy flow in cells and organisms. *Oxidation* is the removal of electrons from a substance. In biology, oxidation usually involves the removal of hydrogen atoms (a hydrogen ion plus an electron) and the addition of oxygen atoms. Oxidation reactions release energy, as when either glucose or methane is oxidized to carbon dioxide, shown below in Reactions 5-1 and 5-2. Note that each carbon atom in glucose or methane has lost hydrogen atoms and gained oxygen atoms as carbon dioxide is formed and energy is released.

Glucose oxidation to carbon dioxide:

$$C_6H_{12}O_6 + 6O_2 \rightarrow 6CO_2 + 6H_2O + \text{energy} \qquad \textbf{(5-1)}$$

Methane oxidation to carbon dioxide:

$$CH_4 + 2O_2 \rightarrow CO_2 + 2H_2O + \text{energy} \qquad \textbf{(5-2)}$$

Reduction is the reverse reaction—the addition of electrons to a substance—and it usually involves the addition of hydrogen atoms (and a loss of oxygen atoms). Reduction reactions require an input of energy, as when carbon dioxide is reduced to glucose during photosynthesis, shown below in Reaction 5-3. The carbon atoms in six molecules of carbon dioxide have all gained hydrogen atoms and lost oxygen atoms during their energy-requiring reduction to glucose.

Carbon dioxide reduction to glucose:

$$\text{energy} + 6CO_2 + 6H_2O \rightarrow C_6H_{12}O_6 + 6O_2 \qquad \textbf{(5-3)}$$

The flow of energy and matter through the biosphere is depicted in **Figure 5-4**. Phototrophs are producers that use solar energy to convert carbon dioxide and water into more reduced cellular compounds such as glucose during photosynthesis. These reduced compounds are converted into other carbohydrates, proteins, lipids, nucleic acids, and all the materials a cell needs to survive. In a sense, we can consider the entire phototrophic organism to be the product of photosynthesis because nearly every carbon atom in every molecule of that organism is derived from carbon dioxide that is fixed into organic form by the photosynthetic process.

Chemotrophs, on the other hand, are consumers that are unable to use solar energy directly and therefore must depend completely on energy that has been packaged into oxidizable

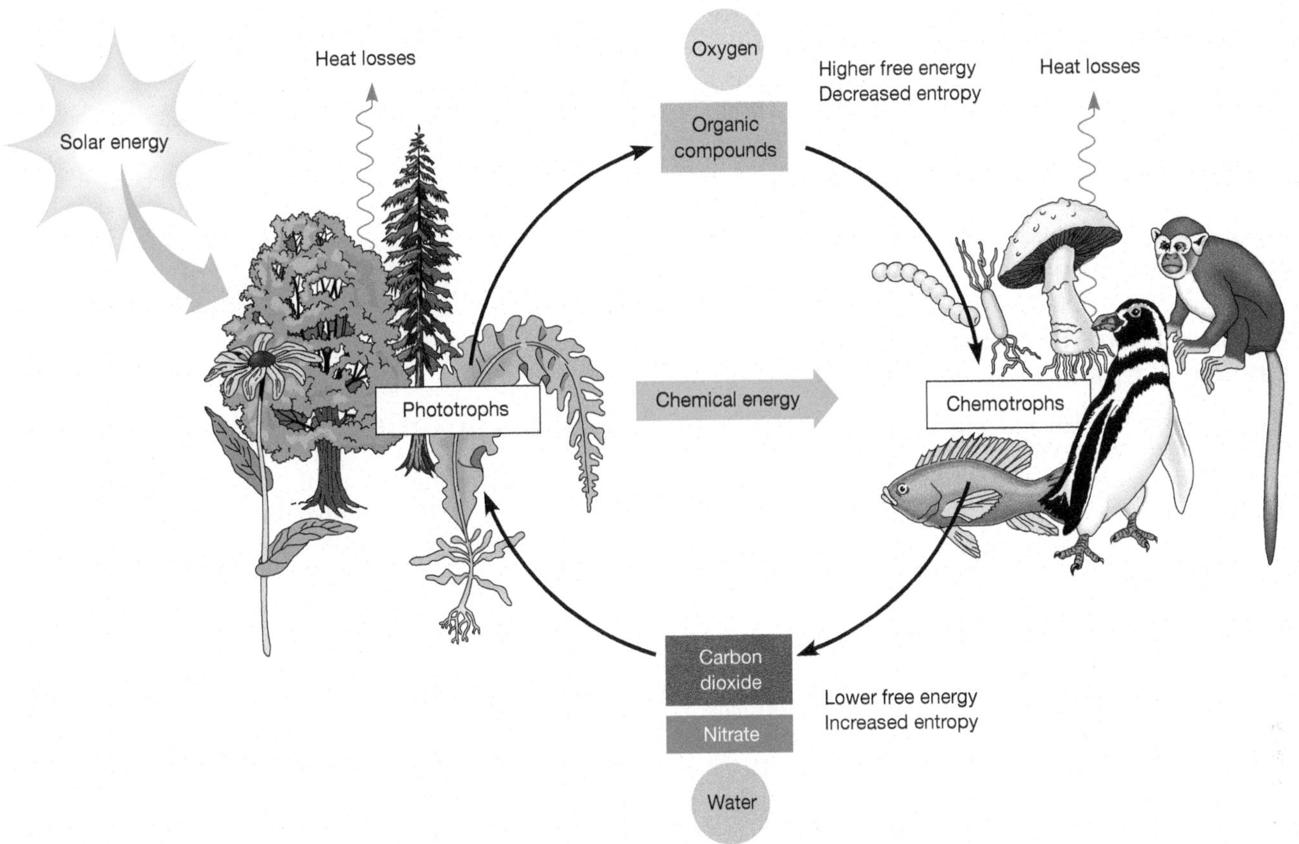

Figure 5-4 The Flow of Energy and Matter Through the Biosphere. Most of the energy in the biosphere originates in the sun and is eventually released to the environment as heat. Accompanying the unidirectional flow of energy from phototrophs to chemotrophs is a cyclic flow of matter between the two groups of organisms. Solar energy is used to reduce low-energy inorganic compounds to high-energy organic compounds, which are used by both phototrophs and chemotrophs.

food molecules by phototrophs. A world composed solely of chemoheterotrophs would last only so long as food supplies lasted, for even though we live on a planet that is flooded with solar energy each day, this energy is in a form that chemotrophs such as humans cannot use to meet our energy needs.

Both phototrophs and chemotrophs use energy to carry out the six kinds of work we have already discussed (see Figure 5-1). An important principle of energy conversion is that no chemical or physical process occurs with 100% efficiency—some energy is released as heat. Although many biological processes are remarkably efficient in energy conversion, loss of energy as heat is inevitable in biological processes. Sometimes the heat that is liberated during cellular processes is utilized, as when homeothermic animals use heat to maintain a constant body temperature. Some plants use metabolically generated heat to attract pollinators or to melt overlying snow (**Figure 5-5**). In general, however, the heat is simply dissipated into the environment, representing a loss of energy from the organism.

Viewed on a cosmic scale, there is a continuous, massive, and unidirectional flow of energy on Earth from its source in the nuclear fusion reactions of the sun to its eventual sink, the environment. We here in the biosphere are the transient custodians of an infinitesimally small portion of that energy. It is

Figure 5-5 Skunk Cabbage, a Plant That Depends on Metabolically Generated Heat. The skunk cabbage plant (*Symplocarpus foetidus*) is one of the earliest-flowering plants in the eastern United States. The heat that it generates enables it to melt through overlying snow in late winter and begin growing when most other plants are still dormant.

precisely that small but critical fraction of energy and its flow through the living systems of the biosphere that is of concern to us. The flow of energy in the biosphere begins with phototrophs, which use light energy to drive electrons energetically "uphill" to create high-energy reduced compounds. This energy is then released by both phototrophs and chemotrophs in "downhill" oxidation reactions. This flux of energy through living matter—from the sun to phototrophs to chemotrophs to heat—drives all life processes.

The Flow of Energy Through the Biosphere Is Accompanied by a Flow of Matter

Energy enters the biosphere as photons of light unaccompanied by matter and leaves the biosphere as heat similarly unaccompanied by matter. However, while it is passing through the biosphere, energy exists primarily in the form of chemical bond energies of oxidizable organic molecules in cells and organisms. As a result, the flow of energy in the biosphere is coupled to a correspondingly immense flow of matter.

Whereas energy flows unidirectionally from the sun through phototrophs to chemotrophs and to the environment, matter flows in a cyclic fashion between phototrophs and chemotrophs (Figure 5-4). Phototrophs use solar energy to create organic nutrients from inorganic starting materials such as carbon dioxide and water, releasing oxygen in the process. Phototrophs use some of these high-energy, reduced nutrients themselves, and some of them become available to chemotrophs that consume these phototrophs. Chemotrophs typically take in organic nutrients from their surroundings and use oxygen to oxidize them back to carbon dioxide and water, providing energy. Low-energy, oxidized molecules are returned to the environment and then become the raw materials that phototrophic organisms use to make new organic molecules, returning oxygen to the environment in the process and completing the cycle.

In addition, there is an accompanying cycle of nitrogen. Phototrophs obtain nitrogen from the environment in an oxidized, inorganic form (as nitrate from the soil or, in some cases, as N_2 from the atmosphere). They reduce it to ammonia (NH_3), a high-energy form of nitrogen used in the synthesis of amino acids, proteins, nucleotides, and nucleic acids. Eventually, these cellular molecules, like other components of phototrophic cells, are consumed by chemotrophs. The nitrogen in these molecules is then converted back into ammonia and eventually oxidized to nitrate, mostly by soil microorganisms.

Carbon, oxygen, nitrogen, and water thus cycle continuously between the phototrophic and chemotrophic worlds. They enter the chemotrophic sphere as reduced, energy-rich compounds and leave in an oxidized, energy-poor form. The two great groups of organisms can therefore be thought of as living in a symbiotic relationship with each other, with a cyclic flow of matter and a unidirectional flow of energy as components of that symbiosis.

When we deal with the overall macroscopic flux of energy and matter through living organisms, we find cellular biology interfacing with ecology. Ecologists study cycles of energy and nutrients, the roles of various species in these cycles, and environmental factors affecting the flow. In contrast, our ultimate concern in cell biology is how the flux of energy and matter functions on a microscopic and molecular scale. We will be able to understand these energy transactions and the chemical processes that occur within cells as soon as we acquaint ourselves with the physical principles underlying energy transfer in cells. For that we turn to the topic of bioenergetics.

CONCEPT CHECK 5.1

How do phototrophs and chemotrophs depend intimately on each other? How does the flow of energy differ from the flow of matter?

5.2 Bioenergetics

The principles governing energy flow are the subject of an area of science known as **thermodynamics.** Although the prefix *thermo–* suggests that the term is limited to heat (and that is indeed its historical origin), thermodynamics also takes into account other forms of energy and processes that converted energy from one form to another. Specifically, thermodynamics describes the energy transactions that accompany most physical processes and all chemical reactions. **Bioenergetics,** in turn, can be thought of as *applied thermodynamics*—the application of thermodynamic principles to reactions and processes in the biological world.

Understanding Energy Flow Requires Knowledge of Systems, Heat, and Work

It is useful to define energy not simply as the capacity to do work, but specifically as the ability to cause change. Without energy, all processes would be at a standstill, including those associated with living cells.

Energy exists in various forms, many of them of interest to biologists. Think, for example, of the energy represented by a ray of sunlight, a teaspoon of sugar, a moving flagellum, an excited electron, or the concentration of ions or small molecules within a cell or an organelle. These phenomena are diverse, but they are all governed by certain basic principles of energetics.

Energy is distributed throughout the universe—and for some purposes it is necessary to consider the total energy of the universe, at least in a theoretical way. Whereas the total energy of the universe remains constant, we are usually interested in the energy content of only a small portion of it. We might, for example, be concerned with a reaction or process occurring in a beaker of chemicals, in a cell, or in a particular organism. By convention, the restricted portion of the universe that is being considered at any given moment is called the **system,** and all the rest of the universe is referred to as the **surroundings.** Sometimes the system has a physical boundary, such as a glass beaker or a cell membrane. In other cases, the boundary between the system and its surroundings is just a hypothetical one used only for convenience of discussion, such as the imaginary boundary around 1 mole of glucose molecules in a solution.

Systems can be either open or closed, depending on whether they can exchange energy with their surroundings (**Figure 5-6**). A **closed system** is sealed from its environment and can neither take in nor release energy in any form. An

fact proceed depends not only on its favorable (negative) ΔG but also on the availability of a mechanism or pathway to get from the initial state to the final state. Usually, an initial input of activation energy is required as well, such as the heat energy from the match that was used to ignite the piece of paper.

Thermodynamic spontaneity is therefore a necessary but insufficient criterion for determining whether a reaction will actually occur. (In Chapter 6, we will explore the subject of reaction rates in the context of enzyme-catalyzed reactions.) For the moment, we need only note that when we designate a reaction as thermodynamically spontaneous, we simply mean that it is an energetically feasible event that will liberate free energy if and when it actually takes place.

CONCEPT CHECK 5.2

How does the spontaneity of a chemical reaction depend on changes in entropy and enthalpy? Why do these changes have opposite effects?

5.3 Understanding ΔG and K_{eq}

Our final task in this chapter will be to understand how ΔG is calculated and how it can then be used to assess the thermodynamic feasibility of reactions under specified conditions. For that, we come back to the reaction that converts glucose-6-phosphate into fructose-6-phosphate (see Reaction 5-9) and ask what we can learn about the spontaneity of the conversion in the direction written (from left to right). Experience and familiarity provide no clues here, nor is it obvious how the entropy of the universe would be affected if the reaction were to proceed. Clearly, we need to be able to calculate ΔG and to determine whether it is positive or negative under the particular conditions we specify for the reaction.

The Equilibrium Constant K_{eq} Is a Measure of Directionality

To assess whether a reaction can proceed in a given direction under specified conditions, we must first understand the **equilibrium constant** K_{eq}, which is the ratio of product concentrations to reactant concentrations at *equilibrium.* When a reversible reaction (see Reaction 5-14) is at equilibrium, there is no net change in the concentrations of either products or reactants with time. A is being converted into B and B is being converted into A, but these processes are happening at the same rate. For the general reaction in which A is converted reversibly into B, the equilibrium constant is simply the ratio of the equilibrium concentrations of A and B,

$$A \rightleftharpoons B \tag{5-14}$$

$$K_{eq} = \frac{[B]_{eq}}{[A]_{eq}} \tag{5-15}$$

where $[A]_{eq}$ and $[B]_{eq}$ are the concentrations of A and B, in moles per liter, when Reaction 5-14 is at equilibrium at 25°C. Given the equilibrium constant for a reaction, you can easily tell whether a specific mixture of products and reactants is at equilibrium. If it is not, it is easy to tell how far away the reaction is from equilibrium and the direction it must proceed to reach equilibrium. (But how can we determine the K_{eq} for a given chemical reaction or a biological process? **Key Technique, pages 122–123**, shows how isothermal titration calorimetry can be used to determine thermodynamic parameters such as the K_{eq} and changes in enthalpy, entropy, and free energy during a biological reaction or process.)

For example, the equilibrium constant for Reaction 5-9 at 25°C is known to be 0.5. This means that, at equilibrium, there will be one-half as much fructose-6-phosphate as glucose-6-phosphate, regardless of the actual magnitudes of the concentrations:

$$K_{eq} = \frac{[\text{fructose-6-phosphate}]_{eq}}{[\text{glucose-6-phosphate}]_{eq}} = 0.5 \tag{5-16}$$

If the two compounds are present in any other concentration ratio, the reaction will not be at equilibrium and will move toward equilibrium. Thus, a concentration ratio *less* than K_{eq} means that there is too little fructose-6-phosphate present, and the reaction will tend to proceed to the right to generate more fructose-6-phosphate at the expense of glucose-6-phosphate. Conversely, a concentration ratio *greater* than K_{eq} indicates that the relative concentration of fructose-6-phosphate is too high, and the reaction will tend to proceed to the left.

Figure 5-10 illustrates this concept for the interconversion of A and B (Reaction 5-14), showing the relationship between the free energy of the reaction and how far the concentrations of A and B are from equilibrium. (Notice that K_{eq} is assumed to be 1.0 in this illustration; for other values of K_{eq}, the curve would be the same shape but still centered over K_{eq}.) The point of Figure 5-10 is clear: the free energy is lowest at equilibrium and increases as the system is displaced from equilibrium in either direction. Moreover, if we know how the ratio of prevailing concentrations compares with the equilibrium concentration ratio, we can predict in which direction a reaction will tend to proceed *and* how much free energy will be released as it does so. Thus, *the tendency toward equilibrium*

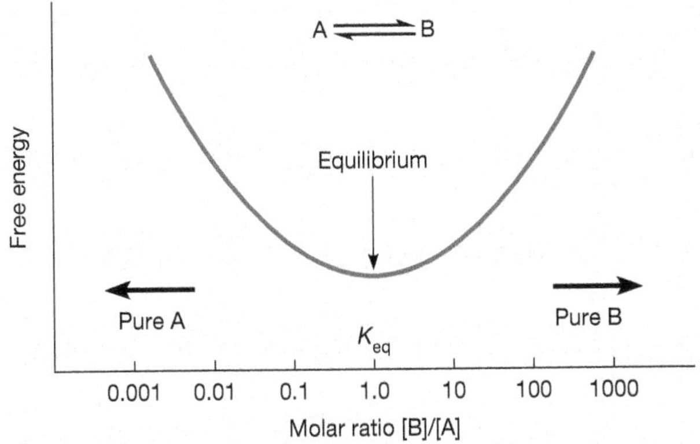

Figure 5-10 Free Energy and Chemical Equilibrium. The amount of free energy available from a chemical reaction depends on how far the components are from equilibrium. This principle is illustrated here for a reaction that interconverts A and B and has an equilibrium constant, K_{eq}, of 1.0. The free energy of the system increases as the [B]/[A] ratio changes on either side of the equilibrium point. For a reaction with a K_{eq} value other than 1.0, the graph would have the same shape but would be centered over the K_{eq} value for that reaction.

provides the driving force for every chemical reaction, and a comparison of prevailing and equilibrium concentration ratios provides one measure of that tendency.

ΔG Can Be Calculated Readily

It should come as no surprise that ΔG is really just a means of calculating how far from equilibrium a reaction lies under specified conditions and how much energy will be released as the reaction proceeds toward equilibrium. Nor should it be surprising that both the equilibrium constant and the prevailing concentrations of reactants (A) and products (B) are needed to calculate ΔG. For Reaction 5-14,

$$\begin{aligned}\Delta G &= RT \ln\frac{[\mathrm{B}]_{\mathrm{pr}}}{[\mathrm{A}]_{\mathrm{pr}}} - RT\ln\frac{[\mathrm{B}]_{\mathrm{eq}}}{[\mathrm{A}]_{\mathrm{eq}}}\\ &= RT\ln\frac{[\mathrm{B}]_{\mathrm{pr}}}{[\mathrm{A}]_{\mathrm{pr}}} - RT\ln \mathrm{K}_{\mathrm{eq}}\\ &= -RT\ln \mathrm{K}_{\mathrm{eq}} - RT\ln\frac{[\mathrm{B}]_{\mathrm{pr}}}{[\mathrm{A}]_{\mathrm{pr}}} \qquad \textbf{(5-17)}\end{aligned}$$

where ΔG is the free energy change in cal/mol under the specified conditions; R is the gas constant (1.987 cal/mol-K); T is the temperature in kelvins (use 25°C = 298 K unless otherwise specified); $[\mathrm{A}]_{\mathrm{pr}}$ and $[\mathrm{B}]_{\mathrm{pr}}$ are the prevailing concentrations of A and B in moles per liter; $[\mathrm{A}]_{\mathrm{eq}}$ and $[\mathrm{B}]_{\mathrm{eq}}$ are the equilibrium concentrations of A and B in moles per liter; K_{eq} is the equilibrium constant at the standard temperature of 298 K (25°C); and ln stands for "the natural logarithm of" (that is, the logarithm of a quantity to the base of the natural logarithm system, e, which equals approximately 2.718). Natural logarithms are used because they describe processes in which the rate of change of the process is directly related to the quantity of material undergoing the change (for example, as in describing radioactive decay).

More generally, for a reaction in which a molecules of reactant A combine with b molecules of reactant B to form c molecules of product C plus d molecules of product D,

$$a\mathrm{A} + b\mathrm{B} \rightleftharpoons c\mathrm{C} + d\mathrm{D} \qquad \textbf{(5-18)}$$

ΔG is calculated as

$$\Delta G = -RT\ln K_{\mathrm{eq}} + RT\ln\frac{[\mathrm{C}]^{c}_{\mathrm{pr}}\,[\mathrm{D}]^{d}_{\mathrm{pr}}}{[\mathrm{A}]^{a}_{\mathrm{pr}}\,[\mathrm{B}]^{b}_{\mathrm{pr}}} \qquad \textbf{(5-19)}$$

where all the constants and variables are as previously defined, and K_{eq} is the equilibrium constant for Reaction 5-18.

Returning to Reaction 5-9, assume that the prevailing concentrations of glucose-6-phosphate and fructose-6-phosphate in a cell are 10 μM (10×10^{-6} M) and 1 μM (1×10^{-6} M), respectively, at 25°C. Because the ratio of prevailing product concentrations to reactant concentrations is 0.1 and the equilibrium constant is 0.5, there is clearly too little fructose-6-phosphate present relative to glucose-6-phosphate for the reaction to be at equilibrium. The reaction should therefore tend toward the right in the direction of fructose-6-phosphate generation. In other words, the reaction is thermodynamically favorable in the direction written. This, in turn, means that ΔG must be negative under these conditions.

The actual value for ΔG is calculated as follows:

$$\begin{aligned}\Delta G &= -(1.987\text{ cal/mol-}K)(298\ K)\ln(0.5)\\ &\quad +(1.978\text{ cal/mol-}K)(298\ K)\ln\frac{1\times10^{-6}\ M}{10\times10^{-6}\ M}\\ &= -(592\text{ cal/mol})\ln(0.5) + (592\text{ cal/mol})\ln(0.1)\\ &= -(592\text{ cal/mol})(-0.693) + (592\text{ cal/mol})(-2.303)\\ &= +410\text{ cal/mol} - 1364\text{ cal/mol}\\ &= -954\text{ cal/mol} \qquad \textbf{(5-20)}\end{aligned}$$

Notice that our expectation of a negative ΔG is confirmed, and we now know exactly how much free energy is liberated upon the spontaneous conversion of 1 mole of glucose-6-phosphate into 1 mole of fructose-6-phosphate under the specified conditions. The free energy liberated in this or some other exergonic reaction can be "harnessed" to do work, stored in the chemical bonds of ATP, or released as heat.

It is important to understand exactly what this calculated value for ΔG means and under what conditions it is valid. Because it is a thermodynamic parameter, ΔG can tell us whether a reaction is thermodynamically possible as written, but it says nothing about the rate or the mechanism of the reaction. It simply says that if the reaction does occur, it will proceed to the right and will liberate 954 calories (0.954 kcal) of free energy for every mole of glucose-6-phosphate that is converted to fructose-6-phosphate, *provided* that the concentrations of both the reactant and the product are *maintained at the initial values* (10 and 1μM, respectively) throughout the course of the reaction.

In upcoming chapters, we will see examples of how some of the energy provided by exergonic reactions can be harnessed by the use of *coupled reactions.* In coupled reactions, an endergonic reaction happens simultaneously with an exergonic reaction, and the free energy of the exergonic reaction is used to permit the endergonic reaction to occur. For example, although the synthesis of ATP is endergonic, often this reaction is coupled to an exergonic oxidative reaction such that the overall sum of the two reactions has a negative ΔG and both reactions can proceed spontaneously. Rather than being released as heat, some of the energy of the oxidation has been conserved in the chemical bonds of the high-energy ATP molecule.

More generally, *ΔG is a measure of thermodynamic spontaneity for a reaction in the direction in which it is written (from left to right), at the specified concentrations of reactants and products.* In a beaker or test tube, this requirement for constant reactant and product concentrations means that reactants must be added continuously and products must be removed continuously. In the cell, each reaction is part of a metabolic pathway, and its reactants and products are maintained at fairly constant, nonequilibrium concentrations by the reactions that precede and follow it in the sequence.

The Standard Free Energy Change Is ΔG Measured Under Standard Conditions

Because it is a thermodynamic parameter, ΔG is independent of the actual mechanism or pathway of a reaction, but it depends crucially on the conditions under which the reaction occurs. A reaction characterized by a large decrease in free energy under one set of conditions may have a much smaller (but still negative) ΔG or may even have a positive ΔG under a different set of conditions. The melting of ice, for example, depends on temperature; it proceeds spontaneously above 0°C but goes in the opposite direction (freezing) below that temperature. It is therefore important to identify the conditions under which a given measurement of ΔG is made.

By convention, biochemists have agreed on certain arbitrary conditions to define the **standard state** of a system for convenience in reporting, comparing, and tabulating free energy changes in chemical reactions. For systems consisting of dilute aqueous solutions, these are usually a standard temperature of 25°C (298 K), a pressure of 1 atmosphere, and all products and reactants present at a concentration of 1 *M*.

The only common exception to this standard concentration rule is water. The concentration of water in a dilute aqueous solution is approximately 55.5 *M* and does not change significantly during the course of reactions, even when water is itself a reactant or product. By convention, biochemists do not include the concentration of water in calculations of free energy changes, even though the reaction may indicate a net consumption or production of water.

In addition to standard conditions of temperature, pressure, and concentration, biochemists also frequently specify a standard pH of 7.0 because most biological reactions occur at or near neutrality. The concentration of hydrogen ions (and of hydroxyl ions) is therefore 10^{-7} *M*, so the standard concentration of 1.0 *M* does not apply to H^+ or OH^- ions when a pH of 7.0 is specified. Values of K_{eq}, ΔG, or other thermodynamic parameters determined or calculated at pH 7.0 are always written with a prime (as K'_{eq}, $\Delta G'$, and so on) to indicate this exception to standard conditions.

Energy changes for reactions are usually reported in standardized form as the change that would occur if the reactions were run under the standard conditions. More precisely, the standard change in any thermodynamic parameter refers to the conversion of a mole of a specified reactant to products or the formation of a mole of a specified product from the reactants under conditions where the temperature, pressure, pH, and concentrations of all relevant species are maintained at their standard values.

The free energy change calculated under these conditions is called the **standard free energy change**, designated $\Delta G^{\circ\prime}$, where the superscript (°) refers to standard conditions of temperature, pressure, and concentration, and the prime (′) emphasizes that the standard hydrogen ion concentration is 10^{-7} *M*, not 1.0 *M*.

It turns out that $\Delta G^{\circ\prime}$ bears a simple linear relationship to the natural logarithm of the equilibrium constant K'_{eq} (**Figure 5-11**). This relationship can readily be seen by rewriting Equation 5-19 with primes and then assuming standard concentrations for all reactants and products. All concentration terms are now 1.0 and the natural logarithm of 1.0 is zero, so the second term in the general expression for $\Delta G^{\circ\prime}$ is eliminated, and what remains is an equation for $\Delta G^{\circ\prime}$, the free energy change under standard conditions:

$$\begin{aligned}\Delta G^{\circ\prime} &= -RT \ln K'_{eq} + RT \ln 1 \\ &= -RT \ln K'_{eq}\end{aligned} \quad \textbf{(5-21)}$$

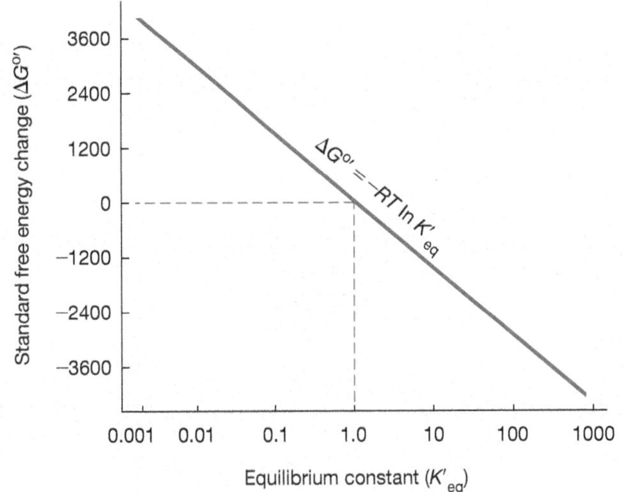

Figure 5-11 The Relationship Between $\Delta G^{\circ\prime}$ and K'_{eq}. The standard free energy change and the equilibrium constant are related by the equation $\Delta G^{\circ\prime} = -RT \ln K'_{eq}$. Note that if the equilibrium constant is 1.0, the standard free energy change is zero.

In other words, $\Delta G^{\circ\prime}$ can be calculated directly from the equilibrium constant, provided that the latter has also been determined under the same standard conditions of temperature, pressure, and pH. This, in turn, allows Equation 5-19 to be simplified as

$$\Delta G' = \Delta G^{\circ\prime} + RT \ln \frac{[C]^c_{pr}[D]^d_{pr}}{[A]^a_{pr}[B]^b_{pr}} \quad \textbf{(5-22)}$$

At the standard temperature of 25°C (298 K), the term *RT* becomes (1.987)(298) = 592 cal/mol, so Equations 5-21 and 5-22 can be rewritten as follows, in what are the most useful formulas for our purposes:

$$\Delta G^{\circ\prime} = -592 \ln K'_{eq} \quad \textbf{(5-23)}$$

$$\Delta G' = \Delta G^{\circ\prime} + 592 \ln \frac{[C]^c_{pr}[D]^d_{pr}}{[A]^a_{pr}[B]^b_{pr}} \quad \textbf{(5-24)}$$

Summing Up: The Meaning of $\Delta G'$ and $\Delta G^{\circ\prime}$

Equations 5-23 and 5-24 represent the most important contribution of thermodynamics to biochemistry and cell biology—a means of assessing the feasibility of a chemical reaction based on the prevailing concentrations of products and reactants and a knowledge of the equilibrium constant. Equation 5-23 expresses the relationship between the standard free energy change $\Delta G^{\circ\prime}$ and the equilibrium constant K'_{eq} and enables us to calculate the free energy change that would be associated with any reaction of interest if all reactants and products were maintained at a standard concentration of 1.0 *M*.

Key Technique | Measuring How Molecules Bind to One Another Using Isothermal Titration Calorimetry

PROBLEM: Cell biologists may often know, in a qualitative way, that one molecule, a *ligand,* binds to another molecule, its *target.* But how can researchers measure—*quantitatively*—how much affinity the ligand and target molecules have for one another? Such information is often crucial in drug design and for understanding how strongly proteins bind one another during normal cellular events.

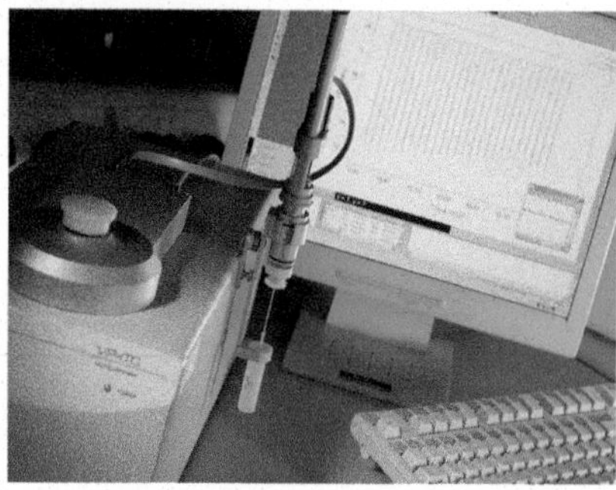

An isothermal calorimeter.

SOLUTION: *Isothermal titration calorimetry (ITC)* allows scientists to measure thermodynamic parameters of specific biological processes. An *isothermal calorimeter* is an instrument consisting of two identical containers called "cells"—a sample cell that holds the target molecule and a reference cell containing buffer only, acting as a control. The two cells are kept at the same temperature ("isothermal"), and as the ligand molecule that binds to the target molecule is gradually added in small, discrete amounts to the sample cell ("titration"), the amount of heat input needed to keep the sample cell and reference cell at the same temperature is measured ("calorimetry").

Key Tools: Isothermal calorimeter; a target molecule; a ligand that binds to the target molecule; a buffer; and a computer to analyze the results.

Details: The binding between a ligand and its target, like other biological processes, involves changes in enthalpy (ΔH), entropy (ΔS), and free energy (ΔG). These changes in energy can be measured using ITC (Figure 5B-1). The sample and reference cells of the isothermal calorimeter are connected by a sensitive electrical circuit that can detect and correct for minute temperature differences between the two cells (Figure 5B-1a). The cells are surrounded by a special insulating jacket that will not allow heat transfer between the two cells or between the cells and the environment. In this way, the calorimeter functions as a closed system.

In ITC, the reaction or process—such as a drug binding to a target molecule—takes place in the sample cell, where small amounts of the drug are progressively added using an attached syringe (Figure 5B-1b). Constant power is applied to the reference cell to keep it at a constant temperature, and variable power is applied to the sample cell to keep it at the same temperature as the reference cell. If the reaction has a negative ΔH, as in the exothermic binding reaction shown in Figure 5B-1b, heat will be released and the temperature in the sample cell will increase. Therefore, less power will be needed to keep it isothermic with the reference cell. Conversely, an endothermic reaction will absorb heat in the sample cell, requiring more power to keep it isothermal with the reference cell.

A computer-controlled injector operates the syringe to repeatedly inject small amounts of the drug into the sample cell with the target molecule, and it continually monitors the amount of power needed to keep the cells isothermal (Figure 5B-2). The output is a series of spikes, one per ligand injection, showing the changes in power (Figure 5B-2a). The area enclosed by each of the spikes represents the amount of heat released or absorbed per injection.

This information, along with the molar ratio of ligand to target molecule at each injection, is used to determine the change in enthalpy as the process proceeds and is displayed as a graph of heat change versus the molar ratio of ligand to target molecule. The beginning and end points of this curve allow the calculation of the overall change in enthalpy (ΔH) for the reaction (Figure 5B-2b).

Additionally, the maximum slope of this line can be used to determine K_{eq} for the binding reaction, giving important information about the strength of ligand-target binding. Using K_{eq} and ΔH, we can calculate ΔG and ΔS for a binding reaction. For a drug binding to its target, this information can identify the number of binding sites on the target molecule.

Isothermal titration calorimetry thus provides a powerful analytical method for investigating many molecular binding interactions that involve changes in enthalpy. It is increasingly being used to analyze enzymatic reactions, protein-protein interactions, and drug-target binding and is a key technique in the screening of novel chemotherapeutic agents.

QUESTION: Why does the height of the spike get shorter after successive injections of ligand in Figure 5B-2a?

It will be helpful to refer to **Table 5-1** periodically during the following discussion. If K'_{eq} is greater than 1.0, then ln K'_{eq} will be positive and $\Delta G^{\circ\prime}$ will be negative, and the reaction can proceed to the right under standard conditions. This makes sense because if K'_{eq} is greater than 1.0, products will predominate over reactants at equilibrium. A predominance of products can be achieved from the standard state only by the conversion of reactants to products, so the reaction will tend to proceed spontaneously to the right. Conversely, if K'_{eq} is less than 1.0, then $\Delta G^{\circ\prime}$ will be positive and the reaction cannot proceed to the right. Instead, it will tend toward the left because $\Delta G^{\circ\prime}$ for the reverse reaction will have the same absolute value but will be opposite in sign. This is in keeping with the small value for K'_{eq}, which specifies that reactants are favored over products (that is, the equilibrium lies to the left).

The $\Delta G^{\circ\prime}$ values are convenient both because they can easily be determined from the equilibrium constant and because they provide a uniform convention for reporting free energy changes. But keep in mind that a $\Delta G^{\circ\prime}$ value is an arbitrary standard in that it refers to an arbitrary state specifying conditions of concentration that cannot be achieved with

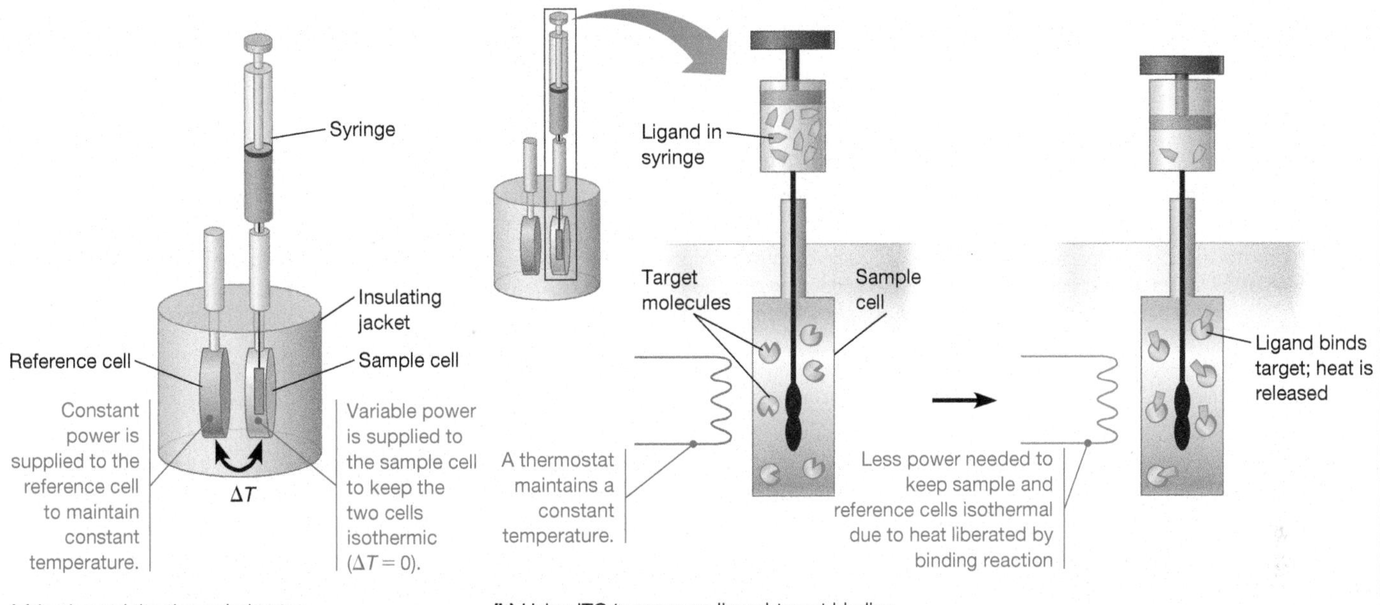

Figure 5B-1 Measuring Ligand-Target Binding Using ITC. Diagrams show **(a)** the insulated jacket containing the sample and reference cells and **(b)** heat release during an exothermic binding reaction.

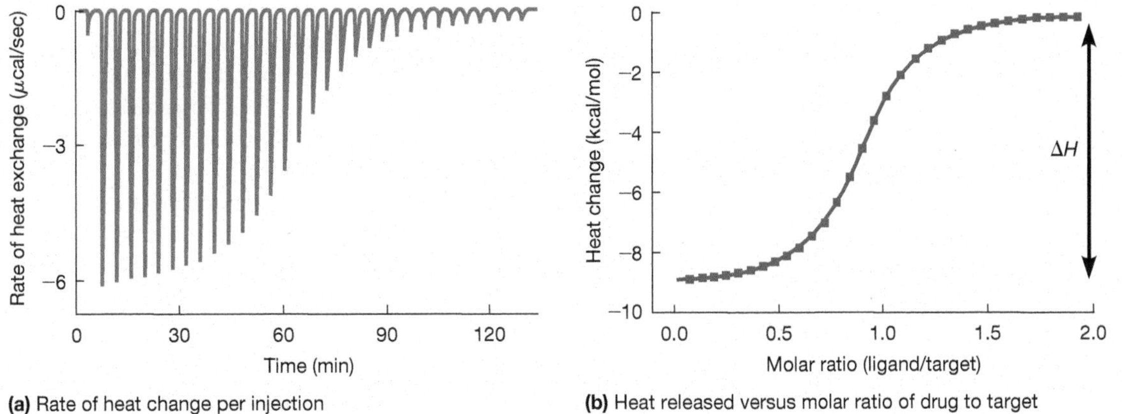

Figure 5B-2 Data Obtained from an ITC Experiment. During a series of injections of ligand into a sample cell containing the target molecule, **(a)** the rate of heat change is recorded for each injection (red spikes) and **(b)** this information is used to determine ΔH.

most biologically important compounds. Because of this, $\Delta G^{\circ\prime}$ is useful for standardized reporting, but *it is not a valid measure of the thermodynamic spontaneity of reactions as they occur under real conditions.*

For real-life situations in cell biology, we will use $\Delta G'$, which provides a direct measure of how far from equilibrium a reaction is at the concentrations of reactants and products that actually prevail in the cell (Equation 5-24). Therefore, *$\Delta G'$ is the most useful measure of thermodynamic spontaneity.* If it is negative, the reaction in question is thermodynamically spontaneous and can proceed as written under the conditions for which the calculations were made. Its magnitude serves as a measure of how much free energy will be liberated as the reaction occurs under the specified conditions. This, in turn, determines the maximum amount of work that can be performed on the surroundings, provided a mechanism is available to conserve and use the energy as it is liberated.

A positive $\Delta G'$, on the other hand, indicates that the reaction cannot occur in the direction written under the conditions for which the calculations were made. Such reactions can sometimes be rendered spontaneous, however, by

Table 5-1 The Meaning of $\Delta G^{\circ\prime}$ and ΔG°

The Meaning of $\Delta G^{\circ\prime}$		
$\Delta G^{\circ\prime}$ Negative ($K'_{eq} > 1.0$)	**$\Delta G^{\circ\prime}$ Positive ($K'_{eq} < 1.0$)**	**$\Delta G^{\circ\prime} = 0$ ($K'_{eq} = 1.0$)**
Products predominate over reactants at equilibrium at standard temperature, pressure, and pH.	Reactants predominate over products at equilibrium at standard temperature, pressure, and pH.	Products and reactants are present equally at equilibrium at standard temperature, pressure, and pH.
Reaction goes spontaneously to the right under standard conditions.	Reaction goes spontaneously to the left under standard conditions.	Reaction is at equilibrium under standard conditions.
The Meaning of $\Delta G'$		
$\Delta G'$ Negative	**$\Delta G'$ Positive**	**$\Delta G' = 0$**
Reaction is thermodynamically feasible as written under the conditions for which $\Delta G'$ was calculated.	Reaction is not feasible as written under the conditions for which $\Delta G'$ was calculated.	Reaction is at equilibrium under the conditions for which $\Delta G'$ was calculated.
Work can be done by the reaction under the conditions for which $\Delta G'$ was calculated.	Energy must be supplied to drive the reaction under the conditions for which $\Delta G'$ was calculated.	No work can be done nor is energy required by the reaction under the conditions for which $\Delta G'$ was calculated.

increasing the concentrations of reactants or by decreasing the concentrations of products. Throughout the text, you will often see reactions that are unfavorable under standard conditions but are able to proceed because the products are rapidly removed by a subsequent reaction.

For the special case where $\Delta G' = 0$, the reaction is at equilibrium, and no net energy change accompanies the conversion of reactant molecules into product molecules or vice versa. As we will see shortly, reactions in living cells are rarely at equilibrium. Table 5-1 summarizes the basic features of K'_{eq}, $\Delta G^{\circ\prime}$, and $\Delta G'$.

Free Energy Change: Sample Calculations

To illustrate the calculation and utility of $\Delta G'$ and $\Delta G^{\circ\prime}$, we return once more to the interconversion of glucose-6-phosphate and fructose-6-phosphate (see Reaction 5-9). We already know that the equilibrium constant for this reaction under standard conditions of temperature, pH, and pressure is 0.5 (see Equation 5-16). This means that if the enzyme that catalyzes this reaction in cells is added to a solution of glucose-6-phosphate at 25°C, 1 atmosphere, and pH 7.0, and the solution is incubated until no further reaction occurs, fructose-6-phosphate and glucose-6-phosphate will be present in an equilibrium ratio of 0.5. (Notice that this ratio is independent of the actual starting concentration of glucose-6-phosphate and could have been achieved equally well by starting with any concentration of fructose-6-phosphate or any mixture of both, in any starting concentrations.)

The standard free energy change $\Delta G^{\circ\prime}$ can be calculated from K'_{eq} as follows:

$$\begin{aligned}\Delta G^{\circ\prime} &= -RT \ln K'_{eq} = -592 \ln K'_{eq} \\ &= -592 \ln 0.5 = -592(-0.693) \\ &= +410 \text{ cal/mol}\end{aligned} \quad \textbf{(5-25)}$$

The positive value for $\Delta G^{\circ\prime}$ is therefore another way of expressing that the reactant (glucose-6-phosphate) is the predominant species at equilibrium. A positive $\Delta G^{\circ\prime}$ value also means that under standard conditions of concentration, the reaction is nonspontaneous (thermodynamically impossible) in the direction written. In other words, if we begin with both glucose-6-phosphate and fructose-6-phosphate present at concentrations of 1.0 *M*, no net conversion of glucose-6-phosphate to fructose-6-phosphate can occur.

As a matter of fact, the reaction will proceed to the *left* under standard conditions. Fructose-6-phosphate will be converted into glucose-6-phosphate until the equilibrium ratio of 0.5 is reached. Alternatively, if both species were added or removed continuously as necessary to maintain the concentrations of both at 1.0 *M*, the reaction would proceed continuously and spontaneously to the left, with the liberation of 410 cal of free energy per mole of fructose-6-phosphate converted to glucose-6-phosphate. In the absence of any provision for conserving this energy, it would be dissipated as heat.

In a real cell, neither of these phosphorylated sugars would ever be present at a concentration even approaching 1.0 *M*. In fact, experimental values for the actual concentrations of these substances in human red blood cells are as follows:

$$\text{[glucose-6-phosphate]}: 83\mu M (83 \times 10^{-6} M)$$
$$\text{[fructose-6-phosphate]}: 14\mu M (14 \times 10^{-6} M)$$

Using these values, we can calculate the actual $\Delta G'$ for the interconversion of these sugars in red blood cells as follows:

$$\begin{aligned}\Delta G' &= \Delta G^{\circ\prime} + 592 \ln \frac{\text{[fructose-6-phosphate]}_{pr}}{\text{[glucose-6-phosphate]}_{pr}} \\ &= +410 + 592 \ln \frac{14 \times 10^{-6}}{83 \times 10^{-6}} \\ &= +410 + 592 \ln 0.169 \\ &= +410 + 592(-1.78) = +410 - 1054 \\ &= -644 \text{ cal/mol}\end{aligned} \quad \textbf{(5-26)}$$

The negative value for $\Delta G'$ means that the conversion of glucose-6-phosphate into fructose-6-phosphate is thermodynamically possible under the actual conditions of concentration prevailing in red blood cells and that the reaction will yield 644 cal of free energy per mole of reactant converted to product. Thus, the conversion of reactant to product is thermodynamically impossible under standard conditions, but the red blood cell maintains these two phosphorylated sugars at concentrations adequate to offset the positive $\Delta G^{\circ\prime}$, thereby rendering the reaction possible. This adaptation, of course, is essential if the red blood cell is to be successful in carrying out the glucose-degrading process of glycolysis of which this reaction is a part.

MAKE CONNECTIONS 5.2

A polypeptide in solution usually folds spontaneously into its proper three-dimensional shape. (a) What is the sign of ΔG for the folding process? How do you know? (b) What is the sign for ΔS for the folding process? How do you know? (Fig. 2-18)

Jumping Beans Provide a Useful Analogy for Bioenergetics

If you are finding the concepts of free energy, entropy, and enthalpy difficult to grasp, perhaps a simple analogy might help. For this we will need an imaginary supply of jumping beans, which are really seeds of certain Mexican shrubs with larvae of the moth *Laspeyresia saltitans* inside. Whenever the larvae inside the seed move, the seeds move, too.

The Jumping Reaction. For purposes of illustration, imagine that we have some high-powered jumping beans in two chambers at the same level that are separated by a low partition, as shown below. As soon as we place a handful of jumping beans in chamber 1, they begin jumping about randomly. Although most of the beans jump to only a modest height most of the time, occasionally one of them gives a more energetic leap, surmounting the barrier and falling into chamber 2. We can write this as the *jumping reaction*:

$$\text{Beans in chamber 1} \rightleftharpoons \text{Beans in chamber 2}$$

Occasionally, one of the beans that has reached chamber 2 will happen to jump back into chamber 1, which is the *back reaction*. At first, of course, there will be more beans jumping from chamber 1 to chamber 2 because there are more beans in chamber 1, but things will eventually even out so that, on average, there will be the same number of beans in both compartments. The system will then be at *equilibrium*. Beans will still continue to jump between the two chambers, but the numbers jumping in both directions will be equal.

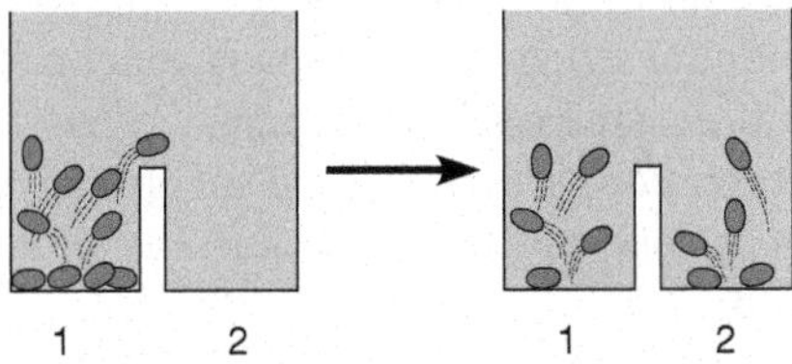

The Equilibrium Constant. Once our system is at equilibrium, we can count up the number of beans in each chamber and express the results as the ratio of the number of beans in chamber 2 to the number in chamber 1. This is simply the *equilibrium constant* (K_{eq}) for the jumping reaction:

$$K_{eq} = \frac{\text{number of beans in chamber 2 at equilibrium}}{\text{number of beans in chamber 1 at equilibrium}}$$

For the specific case shown above, the numbers of beans in the two chambers are equal at equilibrium, so the equilibrium constant for the jumping reaction under these conditions is 1.0.

Enthalpy Change (ΔH). Now suppose that the level of chamber 1 is somewhat higher than that of chamber 2, as shown in the next diagram. Jumping beans placed in chamber 1 will again tend to distribute themselves between chambers 1 and 2, but this time a higher jump is required to get from 2 to 1, so this will occur less frequently. At equilibrium, there will be more beans in chamber 2 than in chamber 1, and the equilibrium constant will therefore be greater than 1.

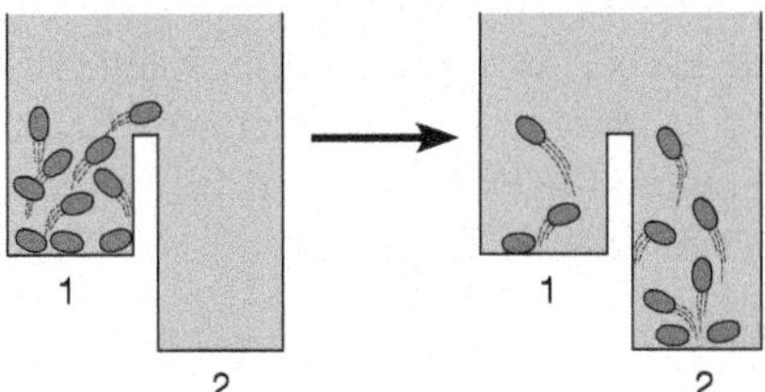

The relative heights of the two chambers can be thought of as measures of the *enthalpy*, or *heat content* (H), of the chambers, such that chamber 1 has a higher H value than chamber 2, and the difference between them is represented by ΔH. Because it is a "downhill" jump from chamber 1 to chamber 2, it makes sense that ΔH has a negative value for the jumping reaction from chamber 1 to chamber 2. Similarly, it seems reasonable that ΔH for the reverse reaction should have a positive value because that jump is "uphill."

Entropy Change (ΔS). Now imagine instead the situation shown below, where the two chambers are again at the same height, but chamber 2 now has a greater floor area than chamber 1. The probability of a bean finding itself in chamber 2 is therefore correspondingly greater, so there will be more beans in chamber 2 than in chamber 1 at equilibrium, and the equilibrium constant will be greater than 1 in this case also. This means that the equilibrium position of the jumping reaction has been shifted to the right, even though there is no change in enthalpy.

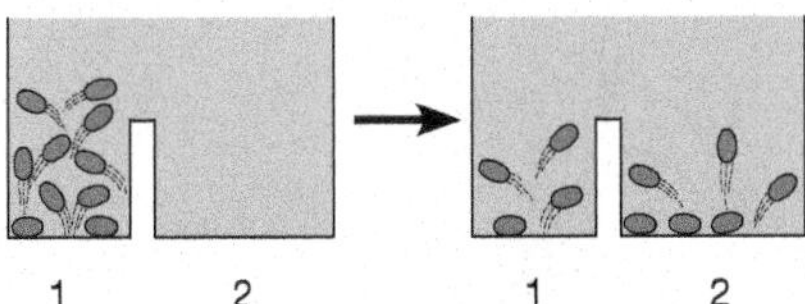

The floor area of the chambers can be thought of as a measure of the *entropy*, or randomness, of the system, S, and the *difference* between the two chambers can be represented by ΔS. Because chamber 2 has a greater floor area than chamber 1, the entropy change is positive for the jumping reaction as it proceeds from left to right under these conditions. Note that for ΔH, negative values are associated with favorable reactions, whereas for ΔS, favorable reactions are indicated by positive values.

Free Energy Change (ΔG). So far, we have encountered two different factors that affect the distribution of beans: the difference in levels of the two chambers (ΔH) and the difference in floor area (ΔS). Moreover, it should be clear that neither of these factors by itself is an adequate indicator of how the beans will be distributed at equilibrium because a favorable (negative) ΔH could be more than offset by an unfavorable (negative) ΔS, and vice versa.

Clearly, what we need is a way of summing these two effects algebraically to see what the net tendency will be. The new measure we come up with is called the *free energy change*, ΔG, which turns out to be the most important thermodynamic parameter for our purposes. ΔG is defined so that *negative* values correspond to favorable (that is, thermodynamically spontaneous) reactions, and *positive* values represent unfavorable reactions. Thus, ΔG will have the *same* sign as ΔH (because a negative ΔH is also favorable) but the *opposite* sign of ΔS (because for ΔS, a positive value is favorable). Based on real-life thermodynamics, the expression for ΔG in terms of ΔH and ΔS and the temperature T is

$$\Delta G = \Delta H - T\Delta S$$

ΔG and the Capacity to Do Work. As long as ΔG is negative, beans will continue to jump from chamber 1 to chamber 2, whether driven primarily by changes in entropy, changes in enthalpy, or both. This means that if some sort of bean-powered "bean wheel" is placed in the second chamber as shown below, then the movement of beans from one chamber to the other can be harnessed to do work until equilibrium is reached, at which point no further work is possible.

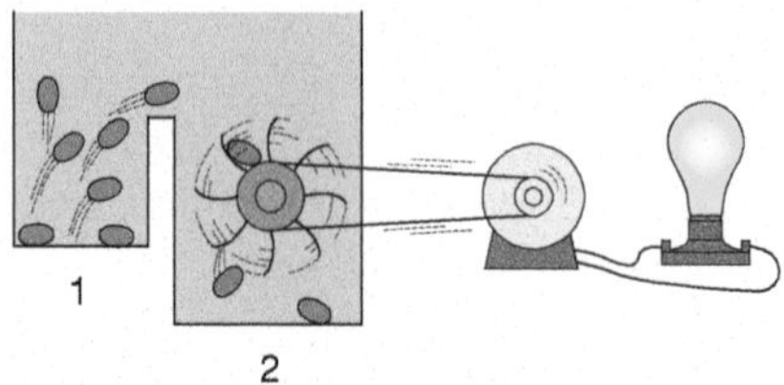

Furthermore, the greater the difference in free energy between the two chambers (that is, the more highly negative ΔG is), the more work the system can do.

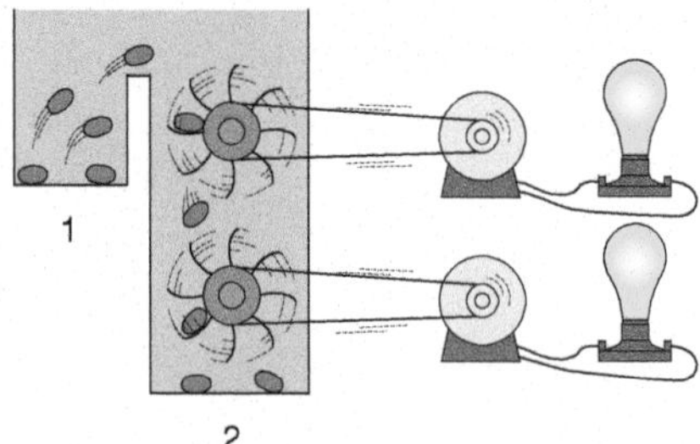

You might, then, want to think of ΔG as free energy in the sense of *energy that is free or available to do useful work.* Moreover, if we contrive to keep ΔG negative by continuously adding beans to chamber 1 and removing them from chamber 2, we have a dynamic *steady state*, a condition that effectively harnesses the inexorable drive to achieve equilibrium. Work can then be performed continuously by beans that are forever jumping toward equilibrium but that never actually reach it.

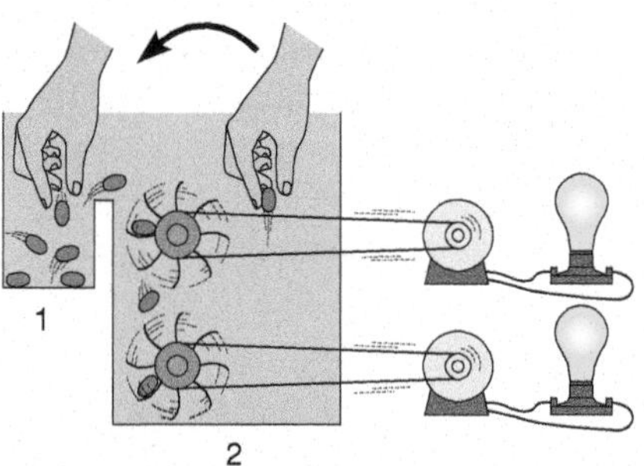

Looking Ahead. To anticipate the next chapter, begin thinking about the *rate* at which beans actually proceed from chamber 1 to chamber 2. ΔG measures how much energy will be released if beans do jump, but it says nothing at all about the rate. That would appear to depend critically on how high the barrier between the two chambers is. Label this the *activation energy barrier* and then contemplate how you might get the beans to move over the barrier more rapidly. One approach might be to heat the chambers; this would be effective because the larvae inside the seeds wiggle more vigorously if they are warmed. Cells have a far more effective and specific means of speeding up reactions: they lower the activation barrier by using catalysts called enzymes.*

Life Requires Steady-State Reactions That Move Toward Equilibrium Without Ever Getting There

As this chapter has emphasized, the driving force in all reactions is their tendency to move toward equilibrium. But to understand how cells really function, we must appreciate the importance of reactions that move toward equilibrium without ever achieving it. At equilibrium, the forward and backward rates are the same for a reaction, and there is therefore no net flow of matter in either direction. Most importantly, no further energy can be extracted from the reaction because $\Delta G'$ is zero for a reaction at equilibrium. Perhaps you can use Equations 5-21 and 5-22 to prove mathematically that every reaction at equilibrium has a $\Delta G' = 0$.

For all practical purposes, then, a reaction that is at equilibrium is a reaction that has stopped. But a living cell is characterized by reactions that are continuous, not stopped. A cell at equilibrium would be a dead cell. We might, in fact, define life as a continual struggle to maintain many cellular reactions in positions far from equilibrium—because at equilibrium, no net reactions are possible, no energy can be released, no work can be done, and the order of the living state cannot be maintained.

Thus, life is possible only because living cells maintain themselves in a **steady state**, with most of their reactions far from thermodynamic equilibrium. The levels of glucose-6-phosphate and fructose-6-phosphate found in red blood

*We are indebted to Princeton University Press for permission to use this analogy, which was first developed by Harold F. Blum in the book Time's Arrow and Evolution, 3rd ed., 1968, pp. 17–26.

cells illustrate this point. As we have seen, these compounds are maintained in the cell at steady-state concentrations far from the equilibrium condition predicted by the K'_{eq} value of 0.5. In fact, the levels are so far from equilibrium concentrations that the conversion of glucose-6-phosphate to fructose-6-phosphate can occur continuously in the cell, even though the equilibrium state has a positive $\Delta G^{\circ\prime}$ and actually favors glucose-6-phosphate. The same is true of most reactions and pathways in the cell. They can proceed and can be harnessed to perform various kinds of cellular work because reactants, products, and intermediates are maintained at steady-state concentrations far from the thermodynamic equilibrium.

This state, in turn, is possible only because a cell is an open system and receives large amounts of energy from its environment. If the cell were a closed system, all its reactions would gradually run to equilibrium. The cell would come inevitably to a state of minimum free energy, after which no further changes could occur, no work could be accomplished, and life would cease. The steady state so vital to life is possible only because the cell is able to take up energy continuously from its environment, whether in the form of light or preformed organic food molecules. This continuous uptake of energy and the accompanying flow of matter make possible the maintenance of a steady state in which all the reactants and products of cellular chemistry are kept far enough from equilibrium to ensure that the thermodynamic drive toward equilibrium can be harnessed by the cell to perform useful work, thereby maintaining and extending its activities and structural complexity.

We will focus on how this is accomplished in later chapters. In Chapter 6, we will look at principles of enzyme catalysis that determine the rates of cellular reactions—that is, we will translate the *can go* of thermodynamics into the *will go* of kinetics. Then we will be ready to move on to subsequent chapters, where we will encounter functional metabolic pathways that result from a series of such reactions acting in concert.

CONCEPT CHECK 5.3

For a chemical reaction happening in a cell, what is the difference between ΔG, ΔG°, and $\Delta G^{\circ\prime}$? Which is the most relevant value to predict the spontaneity of a process in a cell?

Summary of Key Points

Mastering™ Biology For activities, animations, and review quizzes, go to the study area at www.masteringbiology.com.

5.1 The Importance of Energy

- The complexity of cells is possible only due to the availability of energy from the environment. All cells require energy, the capacity to cause physical or chemical changes in cells. These energy-requiring changes include biosynthesis, movement, concentration or electrical work, and the generation of heat or bioluminescence.
- Phototrophs obtain energy directly from the sun and use it to reduce low-energy inorganic molecules such as carbon dioxide and nitrate to high-energy molecules such as carbohydrates, proteins, and lipids. These molecules are used both to build cellular structures and to provide energy in the absence of sunlight.
- Chemotrophs cannot harvest solar energy directly but must obtain their energy by oxidizing the high-energy molecules synthesized by phototrophs.
- There is a unidirectional flow of energy in the biosphere as energy moves from the sun to phototrophs to chemotrophs and is ultimately released into the environment as heat. Matter flows in a cyclic manner between phototrophs and chemotrophs as carbon and nitrogen atoms are alternately reduced and oxidized.

5.2 Bioenergetics

- All living cells and organisms are open systems that exchange energy with their environment. The flow of energy through these living systems is governed by the laws of thermodynamics.
- The first law specifies that energy can change form but must always be conserved. The second law provides a measure of thermodynamic spontaneity, although this means only that a reaction can occur and says nothing about whether it will actually occur or at what rate.
- Spontaneous processes are always accompanied by an *increase* in the entropy of the universe and a *decrease* in the free energy of the system. Free energy is a far more practical indicator of spontaneity because it can be calculated readily from the equilibrium constant, the prevailing concentrations of reactants and products, and the temperature.

5.3 Understanding ΔG and K_{eq}

- The equilibrium constant K_{eq} is a measure of the directionality of a particular chemical reaction. It can be used to calculate $\Delta G^{\circ\prime}$, the free energy change under standard conditions.
- $\Delta G'$, which describes the free energy change under specified conditions, is a measure of how far a reaction is from equilibrium. It represents how much energy will be released as the reaction moves toward equilibrium.
- An exergonic reaction has a negative $\Delta G'$ and proceeds spontaneously in the direction written, whereas an endergonic reaction has a positive $\Delta G'$ and requires the input of energy to proceed as written.
- A negative $\Delta G'$ is a necessary prerequisite for a reaction to proceed spontaneously, but it does not guarantee that the reaction will actually occur at a reasonable rate. Even a reaction with a positive standard free energy change, $\Delta G^{\circ\prime}$, can proceed if the actual concentration of product is kept low enough to provide a negative $\Delta G'$.
- Cells obtain the energy they need to carry out their activities by maintaining the many reactants and products of the various reactions at steady-state concentrations far from equilibrium. This allows the reactions to move exergonically toward equilibrium without ever actually reaching it.
- A reaction at equilibrium has a $\Delta G' = 0$, and no useful work can be done by this reaction. Therefore, a cell with all reactions at equilibrium is a dead cell.

Problem Set

Mastering™ Biology For activities, animations, and review quizzes, go to the study area at www.masteringbiology.com.

5-1 QUANTITATIVE Solar Energy. Although we sometimes hear concerns about a global energy crisis, we live on a planet that is flooded continuously with an extravagant amount of energy in the form of solar radiation. Every day, year in and year out, solar energy arrives at the upper surface of the Earth's atmosphere at the rate of 1.94 cal / min-cm^2 of cross-sectional area.

(a) Assuming the cross-sectional area of the Earth to be 1.28×10^{18} cm^2, what is the total annual amount of incoming energy?

(b) A substantial portion of that energy, particularly in wavelengths below 300 nm and above 800 nm, never reaches the Earth's surface. What happens to it?

(c) Of the radiation that reaches the Earth's surface, only a small proportion is actually trapped photosynthetically by phototrophs. (You can calculate the actual value in Problem 5-2.) Why is the efficiency of utilization so low?

5-2 QUANTITATIVE Photosynthetic Energy Transduction. The amount of energy trapped and the volume of carbon converted to organic form by photosynthetic energy transducers are mind-boggling: about 5×10^{16} g of carbon per year over the entire Earth's surface.

(a) Assuming that the average organic molecule in a cell has about the same proportion of carbon as glucose does, how many grams of organic matter are produced annually by phototrophs?

(b) Assuming that all the organic matter in part a is glucose (or any molecule with an energy content equivalent to that of glucose), how much energy is represented by that quantity of organic matter? Assume that glucose has a free energy content (free energy of combustion) of 3.8 kcal/g.

(c) Referring to the answer for Problem 5-1a, what is the average efficiency with which the radiant energy incident on the upper atmosphere is trapped photosynthetically on the Earth's surface?

(d) What proportion of the net annual phototrophic production of organic matter calculated in part a is used by chemotrophs?

5-3 Energy Conversion. Most cellular activities involve converting energy from one form to another. For each of the following cases, give a biological example and explain the significance of the conversion.

(a) Chemical energy into mechanical energy

(b) Chemical energy into light energy

(c) Solar (light) energy into chemical energy

(d) Chemical energy into electrical energy

(e) Chemical energy into the potential energy of a concentration gradient

5-4 Enthalpy, Entropy, and Free Energy. The oxidation of glucose to carbon dioxide and water is represented by the following reaction, whether the oxidation occurs by combustion in the laboratory or by biological oxidation in living cells:

$$C_6H_{12}O_6 + 6O_2 \rightarrow 6CO_2 + 6H_2O \qquad \textbf{(5-27)}$$

This reaction is highly exothermic, with an enthalpy change (ΔH) of −673 kcal/mol. As you know from Figure 5-9, ΔG for this reaction at 25°C is −686 kcal/mol, so the reaction is also highly exergonic.

(a) Explain in your own words what the ΔH and ΔG values mean. What do the negative signs mean in each case?

(b) What does it mean to say that the difference between the ΔG and ΔH values is due to entropy?

(c) Without doing any calculations, would ΔS for this reaction be positive or negative? Explain your answer.

(d) Now calculate ΔS for this reaction at 25°C. Does the calculated value agree in sign with your prediction in part c?

(e) What are the values of ΔG, ΔH, and ΔS for the reverse of the above reaction as carried out by a photosynthetic algal cell that is using CO_2 and H_2O to make $C_6H_{12}O_6$?

5-5 Violating the Second Law? The second law of thermodynamics states that all chemical and physical changes result in increased randomness. It is sometimes argued that the creation of complex, nonrandom arrangements of molecules and organisms violates this law. Why is this argument incorrect?

5-6 QUANTITATIVE The Equilibrium Constant. The following reaction is one of the steps in the glycolytic pathway (which we will encounter again in Chapter 9). You should recognize it already, however, because we used it as an example earlier (see Reaction 5-9):

$$\text{glucose-6-phosphate} \rightleftharpoons \text{fructose-6-phosphate} \qquad \textbf{(5-28)}$$

The equilibrium constant K_{eq} for this reaction at 25°C is 0.5.

(a) Assume that you incubate a solution containing 0.15 *M* glucose-6-phosphate (G6P) overnight at 25°C with the enzyme phosphoglucoisomerase that catalyzes Reaction 5-28. How many millimoles of fructose-6-phosphate (F6P) will you recover from 10 mL of the incubation mixture the next morning?

(b) What answer would you get for part a if you had started with a solution containing 0.15 *M* F6P instead?

(c) What answer would you expect for part a if you had started with a solution containing 0.15 *M* G6P but forgot to add phosphoglucoisomerase to the incubation mixture?

(d) Would you be able to answer the question in part a if you had incubated at 15°C instead of 25°C? Why or why not?

5-7 QUANTITATIVE Calculating $\Delta G^{\circ\prime}$ and $\Delta G'$. Like Reaction 5-28, the conversion of 3-phosphoglycerate (3PG) to 2-phosphoglycerate (2PG) is an important cellular reaction because it is one of the steps in the glycolytic pathway (see Chapter 9):

$$\text{3-phosphoglycerate} \rightleftharpoons \text{2-phosphoglycerate} \qquad \textbf{(5-29)}$$

If the enzyme that catalyzes this reaction is added to a solution of 3PG at 25°C and pH 7.0, the equilibrium ratio between the two species will be 0.165:

$$K'_{eq} = \frac{[\text{2-phosphoglycerate}]_{eq}}{[\text{3-phosphoglycerate}]_{eq}} = 0.165 \qquad \textbf{(5-30)}$$

Experimental values for the actual steady-state concentrations of these compounds in human red blood cells are 61 μM for 3PG and 4.3 μM for 2PG.

(a) Calculate $\Delta G^{\circ\prime}$. Explain in your own words what this value means.

(b) Calculate $\Delta G'$. Explain in your own words what this value means. Why are $\Delta G'$ and $\Delta G^{\circ\prime}$ different?

(c) If conditions in the cell change such that the concentration of 3PG remains at 61 μM but the concentration of 2PG rises, how high can the 2PG concentration get before Reaction 5-29 stops because it is no longer thermodynamically feasible?

5-8 Backward or Forward? The interconversion of dihydroxyacetone phosphate (DHAP) and glyceraldehyde-3-phosphate (G3P) is a part of both the glycolytic pathway (see Chapter 9) and the Calvin cycle for photosynthetic carbon fixation (see Chapter 11):

$$\underset{\text{DHAP}}{\text{dihydroxyacetone phosphate}} \rightleftharpoons \underset{\text{G3P}}{\text{glyceraldehyde-3-phosphate}} \quad \textbf{(5-31)}$$

The value of $G^{\circ\prime}$ for this reaction is +1.8 kcal/mol at 25°C. In the glycolytic pathway, this reaction goes to the right, converting DHAP to G3P. In the Calvin cycle, this reaction proceeds to the left, converting G3P to DHAP.

(a) In which direction does the equilibrium lie? What is the equilibrium constant at 25°C?

(b) In which direction does this reaction tend to proceed under standard conditions? What is $\Delta G'$ for the reaction *in that direction?*

(c) In the glycolytic pathway, this reaction is driven to the right because G3P is consumed by the next reaction in the sequence, thereby maintaining a low G3P concentration. What will $\Delta G'$ be (at 25°C) if the concentration of G3P is maintained at 1% of the DHAP concentration (that is, if [G3P]/[DHAP]=0.01)?

(d) In the Calvin cycle, this reaction proceeds to the left. How high must the [G3P]/[DHAP] ratio be to ensure that the reaction is exergonic by at least −3.0 kcal/mol (at 25°C)?

5-9 QUANTITATIVE Succinate Oxidation. The oxidation of succinate to fumarate is an important cellular reaction because it is one of the steps in the citric acid cycle (see Figure 10-9). The two hydrogen atoms that are removed from succinate are accepted by a coenzyme molecule called flavin adenine dinucleotide (FAD), which is thereby reduced to $FADH_2$

$$\text{Succinate} + \text{FAD} \rightleftharpoons \text{Fumarate} + \text{FADH}_2 \quad \textbf{(5-32)}$$

$\Delta G^{\circ\prime}$ for Reaction 5-32 is 0 cal/mol.

(a) If you start with a solution containing 0.01 *M* each of succinate and FAD and add an appropriate amount of the enzyme that catalyzes this reaction, will any fumarate be formed? If so, calculate the resulting equilibrium concentrations of all four species. If not, explain why not.

(b) Answer part a assuming that 0.01 *M* $FADH_2$ is also present initially.

(c) If the steady-state conditions in a cell are such that the $FADH_2$/FAD ratio is 5 and the fumarate concentration is 2.5 μM, what steady-state concentration of succinate is needed to maintain $\Delta G'$ for succinate oxidation at −1.5 kcal/mol?

5-10 DATA ANALYSIS The "Line" on Energy. Analyze the graph in **Figure 5-12**. Match each of the following possible energy combinations to the correct line in the figure. For each, explain your answer.

(a) $\Delta H < 0$, $\Delta S > 0$

(b) $\Delta H > 0$, $\Delta S < 0$

(c) $\Delta H < 0$, $\Delta S < 0$

(d) $\Delta H > 0$, $\Delta S > 0$

5-11 Protein Folding Revisited. A folded polypeptide can be induced to unfold (that is, will undergo denaturation) if the solution is heated or made acidic or alkaline. The process of unfolding of a polypeptide can be represented as an equation:

$$\text{Folded protein} \rightleftharpoons \text{Unfolded protein}$$

Values for the unfolding of one part of the enzyme lysozyme from horses under a particular set of conditions were measured to be ΔH = 250 kcal/mol; ΔS = 0.54 kcal/mol-K.

(a) Calculate the ΔG for the unfolding of horse lysozyme at 25°C.

(b) Do you predict horse lysozyme to be stable at 25°C? Explain your reasoning.

(c) What are the main kinds of bonds and interactions that must be broken or disrupted if a folded polypeptide is to be unfolded? Why do heat and extremes of pH cause unfolding?

5-12 Proof of Additivity. A useful property of thermodynamic parameters such as $\Delta G'$ and $\Delta G^{\circ\prime}$ is that they are additive for sequential reactions. Assume that K'_{AB}, K'_{BC}, and K'_{CD} are the respective equilibrium constants for Reactions 1, 2, and 3 of the following sequence:

$$\text{A} \underset{\text{Reaction 1}}{\rightleftharpoons} \text{B} \underset{\text{Reaction 2}}{\rightleftharpoons} \text{C} \underset{\text{Reaction 3}}{\rightleftharpoons} \text{D} \quad \textbf{(5-33)}$$

(a) Prove that the equilibrium constant K'_{AD} for the overall conversion of A to D is the *product* of the three component equilibrium constants:

$$K'_{AD} = K'_{AB}K'_{BC}K'_{CD} \quad \textbf{(5-34)}$$

(b) Prove that the $\Delta G^{\circ\prime}$ for the overall conversion of A to D is the *sum* of the three component $\Delta G^{\circ\prime}$ values:

$$\Delta G^{\circ\prime}_{AD} = \Delta G^{\circ\prime}_{AB} + \Delta G^{\circ\prime}_{BC} + \Delta G^{\circ\prime}_{CD} \quad \textbf{(5-35)}$$

(c) Prove that the $\Delta G'$ values are similarly additive.

5-13 Utilizing Additivity. The additivity of thermodynamic parameters discussed in Problem 5-12 applies not just to sequential reactions in a pathway, but to *any* reactions or processes. Moreover, it applies to subtraction of reactions. Use this information to answer the following questions.

(a) The phosphorylation of glucose using inorganic phosphate (abbreviated P_i) is endergonic ($\Delta G^{\circ\prime}$ = +3.3 kcal/mol), whereas the dephosphorylation (hydrolysis) of ATP is exergonic ($\Delta G^{\circ\prime}$ = −7.3 kcal/mol):

$$\text{glucose} + \text{P}_i \rightleftharpoons \text{glucose-6-phosphate} + \text{H}_2\text{O} \quad \textbf{(5-36)}$$

$$\text{ATP} + \text{H}_2\text{O} \rightleftharpoons \text{ADP} + \text{P}_i \quad \textbf{(5-37)}$$

Write a reaction for the phosphorylation of glucose by the transfer of a phosphate group from ATP, and calculate $\Delta G^{\circ\prime}$ for the reaction.

(b) Phosphocreatine is used by your muscle cells to store energy. The dephosphorylation of phosphocreatine (Reaction 5-38), like that of ATP, is a highly exergonic reaction with $\Delta G^{\circ\prime}$ = −10.3 kcal/mol.

$$\text{phosphocreatine} + \text{H}_2\text{O} \rightleftharpoons \text{creatine} + \text{P}_i \quad \textbf{(5-38)}$$

Write a reaction for the transfer of phosphate from phosphocreatine to ADP to generate creatine and ATP, and calculate $\Delta G^{\circ\prime}$ for the reaction.

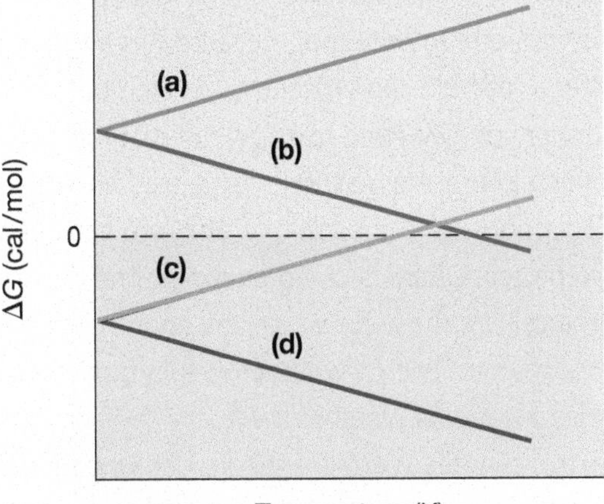

Figure 5-12 ΔG as a function of increasing temperature for four different reactions. See Problem 5-10.

6 Enzymes: The Catalysts of Life

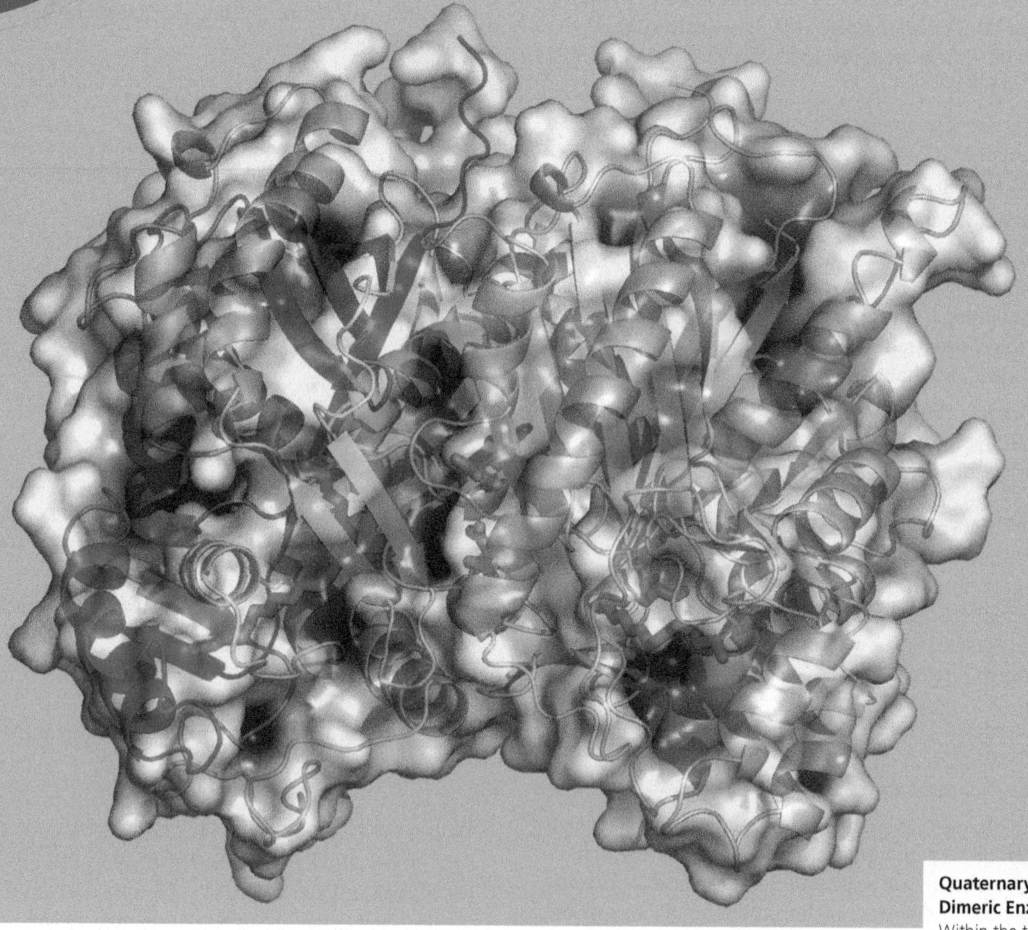

Quaternary Structure of a Dimeric Enzyme. Within the transparent surface model of the enzyme (white) are ribbon models of the two monomers that make up the dimer. One monomer is shown in dark gray and the other in rainbow colors from the N-terminus (blue) to the C-terminus (red).

Earlier (in Chapter 5) we encountered $\Delta G'$, the change in free energy under biological conditions. We saw how the sign of $\Delta G'$ tells us whether a reaction is possible in the indicated direction, and how the magnitude of $\Delta G'$ indicates how much energy will be released (or must be provided) as the reaction proceeds in that direction. At the same time, we were careful to note that, because it is a thermodynamic parameter, $\Delta G'$ tells us only whether a reaction *can* proceed as written but not whether it actually *will* proceed. For that distinction, we need to know not just the direction and energetics of the reaction, but something about the reaction mechanism and its rate.

This brings us to the topic of **enzyme catalysis**, because nearly all cellular reactions involve protein catalysts called *enzymes*. The only reactions that occur at any appreciable rate in a cell are those for which the appropriate enzymes are present and active. Whereas thermodynamics tells us whether a reaction "can proceed," the presence and activity of enzymes determine whether it actually "will proceed."

In this chapter, we will first consider why thermodynamically spontaneous reactions rarely occur at appreciable rates without a catalyst and then look at the role of enzymes as specific biological catalysts. We will also see how the rate of an enzyme-catalyzed reaction is affected by the concentration of the molecule(s) upon which it acts, called its **substrate**, the affinity of the enzyme for its substrate, covalent modification of the enzyme itself, and also how reaction rates are regulated to meet the needs of the cell.

6.1 Activation Energy and the Metastable State

We saw (in Chapter 5) that reactions having a negative change in free energy ($\Delta G' < 0$) are thermodynamically spontaneous in the direction written. However, just because a reaction is thermodynamically spontaneous does not necessarily mean that it actually will proceed under specific cellular conditions. If you stop to think about it, you are already familiar with many reactions that are thermodynamically feasible yet do not proceed to any appreciable extent. An obvious example is the oxidation of glucose (see Chapter 5, Reaction 5-1). This reaction (or series of reactions, really) is highly exergonic ($\Delta G^{\circ\prime} = -686$kcal/mol) and yet does not take place on its own. In fact, glucose, either as crystals or in solution, can be exposed to the oxygen in the air indefinitely, and little or no oxidation will occur. Likewise, the cellulose in the pages these words are printed on in the paper edition of this book does not spontaneously combust—and neither, for that matter, do *you*, consisting of a complex collection of thermodynamically unstable molecules.

Equally important to cellular chemistry are the many thermodynamically feasible reactions that could proceed but do not occur at an appreciable rate on their own. As an example, consider the high-energy molecule adenosine triphosphate (ATP), which has a favorable $\Delta G^{\circ\prime}$ (−7.3kcal/mol) for the hydrolysis of its terminal phosphoryl group to form the corresponding diphosphate (ADP) and inorganic phosphate (P_i):

$$ATP + H_2O \rightleftharpoons ADP + P_i \qquad \textbf{(6-1)}$$

This reaction is exergonic under standard conditions and is even more so under the conditions that prevail in cells. Yet despite the favorable free energy change, this reaction occurs only slowly on its own, so that ATP remains stable for several days when dissolved in pure water. This property turns out to be shared by many biologically important molecules and reactions, and it is important to understand why.

Before a Chemical Reaction Can Occur, the Activation Energy Barrier Must Be Overcome

Molecules that could react with one another often do not because they lack sufficient energy. For every reaction, there is a specific **activation energy (E_A)**, which is the minimum amount of energy that reactants must contain before collisions between them will be successful in giving rise to products. More specifically, reactants need to reach an intermediate chemical stage called the **transition state**, which has a free energy higher than that of the initial reactants.

Figure 6-1 on page 132 shows how temperature and enzymes affect activation energy for an important reaction, the hydrolysis of ATP. Figure 6-1a shows the activation energy required for molecules of ATP and H_2O to reach their transition state before being converted into ADP and P_i. $\Delta G^{\circ\prime}$ measures the difference in free energy between reactants and products (−7.3 kcal/mol for this particular reaction), whereas E_A indicates the minimum energy required for the reactants to reach the transition state and be converted to products.

The actual rate of a reaction is always proportional to the fraction of molecules that have an energy content equal to or greater than E_A. When in solution at room temperature, molecules of ATP and water move about readily, each possessing a certain amount of *kinetic energy* (energy of motion) at any given moment. As Figure 6-1b shows, the energy distribution among molecules at a given temperature (T_1 or T_2, for example) will be symmetrically distributed around a mean value. Some molecules will have very little energy, some will have a lot, and most will be somewhere near the average. The important point is that the only molecules that are capable of reacting are those with enough energy to exceed the *activation energy barrier*, E_A. In Figure 6-1b, these fractions fall in the shaded area to the right of the dashed line.

The Metastable State Is a Result of the Activation Barrier

For most biologically important reactions at normal cellular temperatures, the activation energy is sufficiently high that the proportion of molecules possessing enough energy to react is extremely small. Accordingly, the rates of uncatalyzed reactions in cells are very low, and most molecules appear to be stable even though they are potential reactants in thermodynamically favorable reactions. Such seemingly stable molecules are said to be in a **metastable state**, a state in which they are thermodynamically unstable but do not have enough energy to exceed the activation energy barrier.

For cells, high activation energies and the resulting metastable state of cellular constituents are crucial because life by its very nature is a system maintained in a steady state far from equilibrium. Were it not for the metastable state, all reactions would proceed quickly to equilibrium, and life as we know it would be impossible. Life, then, depends critically on the high activation energies that prevent most cellular reactions from occurring at appreciable rates in the absence of a suitable catalyst.

An analogy might help you to understand and appreciate the metastable state. Imagine an egg in a bowl near the edge of a table—its static position represents the metastable state. Although energy would be released if the egg hit the floor, it cannot do so because the edge of the bowl acts as a barrier. A small amount of energy must be applied to lift it up out of the bowl and over the table edge. Then a much greater amount of energy is released as the egg spontaneously drops to the floor and breaks.

Catalysts Overcome the Activation Energy Barrier

The activation energy requirement is a barrier that must be overcome if desirable reactions are to proceed at reasonable rates. Because the energy content of a given molecule must exceed E_A before that molecule is capable of undergoing a reaction, the only way a reaction involving metastable reactants would proceed at an appreciable rate is to increase the

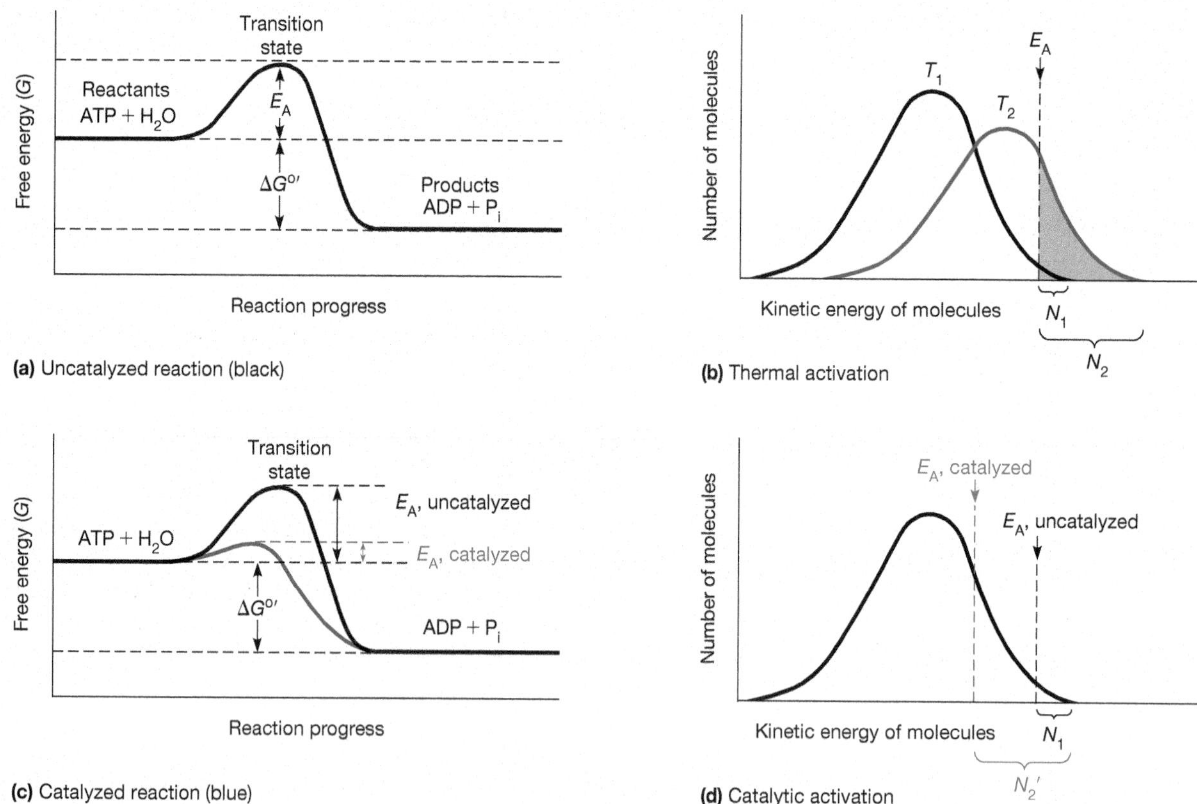

Figure 6-1 The Effect of Catalysis on Activation Energy and Number of Molecules Capable of Reaction. (a) The reactants (in this example, ATP and H_2O) must possess sufficient energy to exceed the activation energy barrier (E_A) and reach the transition state leading to formation of the products (ADP and P_i). **(b)** The number of molecules N_1 that have sufficient energy to exceed E_A can be increased to N_2 by raising the temperature from T_1 to T_2. **(c)** Alternatively, E_A can be lowered by a catalyst (blue curve), thereby **(d)** increasing the number of molecules able to exceed the activation energy barrier from N_1 to N_2' with no change in temperature.

proportion of molecules capable of interacting with sufficient energy. This can be achieved either by increasing the average energy content of all molecules or by lowering the activation energy requirement.

One way to increase the energy content of the system is by the input of heat. As Figure 6-1b illustrated, simply increasing the temperature of the system from T_1 to T_2 will increase the kinetic energy of the average molecule, thereby ensuring a greater number of reactive molecules (N_2 instead of N_1). Thus, the hydrolysis of ATP could be facilitated by heating the solution, giving each ATP and each water molecule more kinetic energy. The problem with using an elevated temperature is that such an approach is not compatible with life because, at a given moment, most biological systems are *isothermal*, meaning they are at a relatively constant temperature.

The alternative to an increase in temperature is to lower the activation energy requirement. This ensures that a greater proportion of molecules will have sufficient energy to collide successfully and undergo reaction. This would be like changing the bowl holding the egg described earlier to a shallow dish. Now, less energy is needed to lift the egg over the edge of the dish. If the reactants can be bound on some sort of surface in an arrangement that brings potentially reactive portions of adjacent molecules into close juxtaposition, their interaction will be greatly favored and the activation energy effectively reduced.

Providing such a reactive surface is the task of a **catalyst**—an agent that enhances the rate of a reaction by lowering the activation energy requirement (Figure 6-1c), thereby ensuring that a higher proportion of the molecules possess sufficient energy to undergo reaction without the input of heat (Figure 6-1d). A corollary feature of a catalyst is that *it is not permanently changed or consumed as the reaction proceeds.* It simply provides a suitable surface and environment to facilitate the reaction.

Recent work suggests an additional mechanism to overcome the activation energy barrier. This mechanism is known as "quantum tunneling" and sounds like something from a science fiction novel. It is based in part on the realization that matter has both particle-like and wave-like properties. In dehydrogenation reactions, the catalyst may allow a hydrogen atom to tunnel *through* the barrier, effectively ending up on the other side without actually going over the top. Unlike most catalyzed reactions, these quantum tunneling reactions are temperature independent because an input of thermal energy is not required to surmount the activation energy barrier.

For a specific example of catalysis, let's consider the decomposition of hydrogen peroxide (H_2O_2) into water and oxygen:

$$2H_2O_2 \rightleftharpoons 2H_2O + O_2 \qquad \textbf{(6-2)}$$

This is a thermodynamically favorable reaction, yet hydrogen peroxide exists in a metastable state because of the high activation energy of the reaction. However, if we add a small number of ferric ions (Fe^{3+}) to a hydrogen peroxide solution, the rate of conversion to products per second increases dramatically and the decomposition reaction proceeds about 30,000 times faster than without these ions. The ferric ions are not consumed in the process and thus are needed in very small amounts relative to the amount of substrate to be converted. Clearly, Fe^{3+} is a catalyst for this reaction, lowering the activation energy (as shown in Figure 6-1c) and thereby ensuring that a significantly greater proportion of the hydrogen peroxide molecules possesses adequate energy to decompose at the existing temperature without the input of added energy.

In cells, the solution to hydrogen peroxide breakdown is not the addition of ferric ions but the enzyme *catalase*, an iron-containing protein. In the presence of catalase, the reaction proceeds about 100,000,000 times faster than the uncatalyzed reaction. Catalase contains iron ions bound to the enzyme, thus taking advantage of inorganic catalysis within the context of a protein molecule. This combination is obviously a much more effective catalyst for hydrogen peroxide decomposition than ferric ions by themselves. The rate enhancement (catalyzed rate ÷ uncatalyzed rate) of about 10^8 for catalase is not at all an atypical value. The rate enhancements of enzyme-catalyzed reactions range from 10^7 to as high as 10^{17} compared with the uncatalyzed reaction. These values underscore the extraordinary importance of enzymes as catalysts and bring us to the main theme of this chapter.

CONCEPT CHECK 6.1

Gasoline is highly combustible yet doesn't burst into flame spontaneously in the presence of oxygen. Why not? How is the action of a match to promote this combustion reaction different from the action of a catalyst?

6.2 Enzymes as Biological Catalysts

Regardless of their chemical nature, all catalysts share the following three basic properties:

1. A catalyst increases the rate of a reaction by lowering the activation energy requirement, thereby allowing a thermodynamically feasible reaction to occur at a reasonable rate in the absence of thermal activation.
2. A catalyst acts by forming transient, reversible complexes with substrate molecules, binding them in a manner that facilitates their interaction and stabilizes the intermediate transition state.
3. A catalyst changes only the *rate* at which equilibrium is achieved; it has no effect on the *position* of the equilibrium. This means that a catalyst can enhance the rate of exergonic reactions but cannot somehow change the $\Delta G'$ to allow an endergonic reaction to become spontaneous.

These properties are common to all catalysts, organic and inorganic alike. In terms of our example, they apply equally to ferric ions and to catalase molecules. However, biological systems rarely use inorganic catalysts. Instead, essentially all catalysis in cells is carried out by organic molecules (proteins, in most cases) called **enzymes**. Because enzymes are organic molecules, they are much more specific than inorganic catalysts, and their activities can be regulated much more carefully.

Most Enzymes Are Proteins

The capacity of cellular extracts to catalyze chemical reactions has been known since the fermentation studies of Eduard and Hans Buchner in 1897. In fact, one of the first terms for what we now call enzymes was *ferments*. However, it was not until 1926 that a specific enzyme, *urease*, was crystallized (from jack beans, by James B. Sumner) and shown to be a protein. Since the early 1980s, biologists have recognized that in addition to proteins, certain RNA molecules, known as *ribozymes*, also have catalytic activity. Ribozymes will be discussed in a later section. Here, we will consider enzymes as proteins because most enzymes are proteins.

The Active Site. One of the most important concepts to emerge from our understanding of enzymes as proteins is the **active site**. Every enzyme contains a set of amino acids that form the active site where the substrates bind and the catalytic event occurs. Usually, the active site is an actual groove or pocket with chemical and structural properties that accommodate the intended substrate or substrates with high specificity. The active site is formed by a small number of amino acids that are not necessarily adjacent to one another along the primary sequence of the protein. Instead, they are brought together in just the right arrangement by the specific three-dimensional folding of the polypeptide chain as it assumes its characteristic tertiary structure.

Figure 6-2 shows the unfolded and folded structures of the enzyme *lysozyme*. Lysozyme breaks down components of bacterial cell walls by catalyzing cleavage of peptidoglycans. The active site of lysozyme is a small groove in the enzyme surface into which a peptidoglycan molecule fits. Lysozyme is a single polypeptide consisting of 129 amino acid residues, but relatively few of these are directly involved in substrate binding and catalysis. Substrate binding in the active site depends on amino acid residues from various positions along the polypeptide, including residues at positions 33–36, 46, 52, 60–64, and 102–110, shown in red. Subsequent catalysis involves two specific residues: a glutamate at position 35 (Glu-35) and

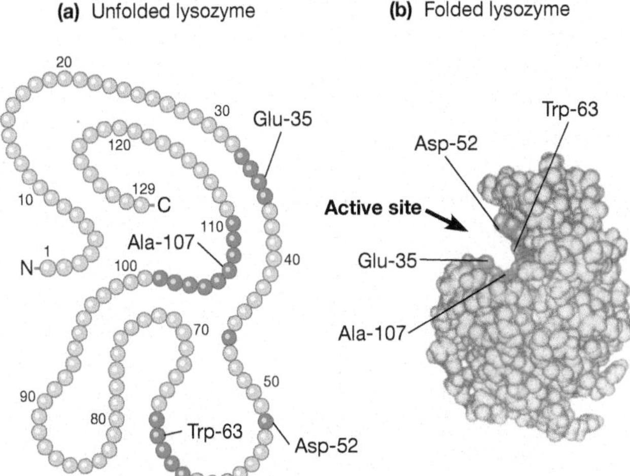

Figure 6-2 The Active Site of Lysozyme. (a) Amino acids involved in substrate binding are shown in red. Four amino acid residues that are especially important for catalysis are labeled. They are far apart in the primary structure of unfolded lysozyme. **(b)** These residues are brought together to form part of the active site as lysozyme folds into its active tertiary structure.

an aspartic acid at position 52 (Asp-52). Only as the lysozyme molecule folds to attain its stable three-dimensional conformation are these specific amino acids brought together to form the active site (Figure 6-2b).

Of the 20 different amino acids that make up proteins, only a few are actually involved in the active sites of the many proteins that have been studied. Often, these are cysteine, histidine, serine, aspartate, glutamate, and lysine. All of these residues can participate in binding the substrate to the active site during catalysis, and several also serve as donors or acceptors of protons.

Some enzymes contain specific nonprotein cofactors that are tightly bound at the active site and are indispensable for catalytic activity. These cofactors, also called **prosthetic groups**, are usually either metal ions or small organic molecules known as *coenzymes* that are derivatives of vitamins. Frequently, prosthetic groups (especially positively charged metal ions) function as electron acceptors because none of the amino acid side chains are good electron acceptors. Where present, prosthetic groups often are located at the active site and are indispensable for the catalytic activity of the enzyme. For example, each catalase enzyme molecule contains a multi-ring structure known as a *porphyrin* ring, to which an iron ion necessary for catalysis is bound (see Figure 10-16).

The requirement for various prosthetic groups on some enzymes explains our nutritional requirements for trace amounts of vitamins and certain metals. Oxidation of glucose for energy requires two specific coenzymes that are derivatives of the vitamins niacin and riboflavin (as we will see in Chapters 9 and 10). Both niacin and riboflavin are essential nutrients in the human diet because our cells cannot synthesize them. These coenzymes, which are bound to the active site of certain enzymes, accept electrons and hydrogen ions from glucose as it is oxidized. Likewise, carboxypeptidase A, a digestive enzyme that degrades proteins, requires a single zinc ion bound to the active site, as we will see later in the chapter. Other enzymes may require iron, copper, molybdenum, or even lithium. Like enzymes, prosthetic groups are not consumed during chemical reactions, so cells require only minute amounts of them.

Enzyme Specificity. Due to the structure of the active site, enzymes display a very high degree of **substrate specificity**, which is the ability to discriminate between very similar molecules. Specificity is one of the most characteristic properties of living systems, and enzymes are excellent examples of biological specificity.

We can illustrate their specificity by comparing enzymes with inorganic catalysts. Most inorganic catalysts are quite nonspecific in that they will act on a variety of compounds that share some general chemical feature. Consider, for example, the addition of hydrogen to a C=C bond, a process known as *hydrogenation*.

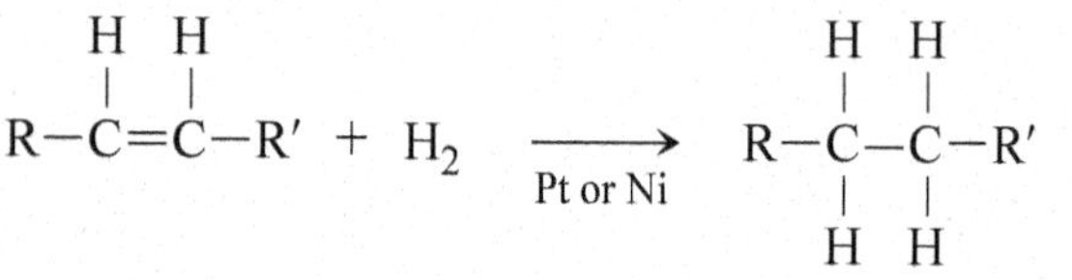

$$\mathrm{R{-}\overset{H}{\overset{|}{C}}{=}\overset{H}{\overset{|}{C}}{-}R' + H_2 \xrightarrow[\text{Pt or Ni}]{} R{-}\underset{H}{\underset{|}{\overset{H}{\overset{|}{C}}}}{-}\underset{H}{\underset{|}{\overset{H}{\overset{|}{C}}}}{-}R'} \qquad \textbf{(6-3)}$$

The C=C bond is said to be *unsaturated*, and the hydrogenation of this bond is also called a *saturation* reaction because, following hydrogenation, no additional hydrogen atoms can be added to the carbon atoms. This reaction can be carried out in the laboratory using a platinum (Pt) or nickel (Ni) catalyst, as indicated. These inorganic catalysts are very nonspecific, however; they can catalyze the hydrogenation of a wide variety of unsaturated compounds.

In practice, nickel and platinum are used commercially to hydrogenate liquid polyunsaturated vegetable oils in the manufacture of solid cooking fats or shortenings. Any C=C bond in an unsaturated compound will be hydrogenated in the presence of nickel or platinum. This lack of specificity of inorganic catalysts during hydrogenation is responsible for the formation of certain *trans* fats (see Chapter 7) in artificially hydrogenated oils that are rare in nature.

By contrast, consider the biological example of the hydrogenation of fumarate as it is converted to succinate (a reaction we will encounter again in the citric acid cycle in Chapter 10):

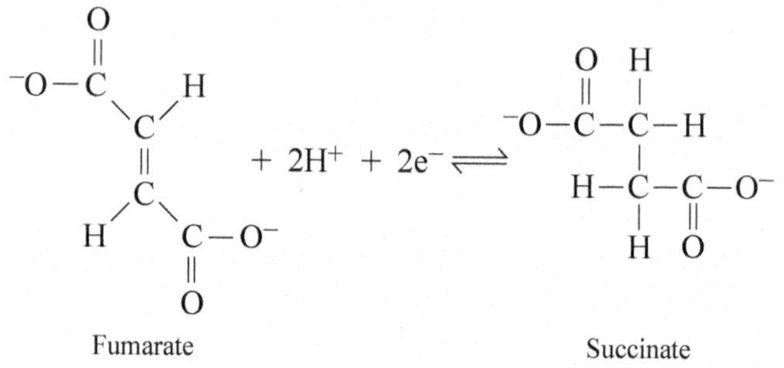

$$\underset{\text{Fumarate}}{\mathrm{{}^{-}O{-}\overset{O}{\overset{\|}{C}}{-}CH{=}CH{-}\underset{O}{\underset{\|}{C}}{-}O^{-}}} + 2H^{+} + 2e^{-} \rightleftharpoons \underset{\text{Succinate}}{\mathrm{{}^{-}O{-}\overset{O}{\overset{\|}{C}}{-}CH_2{-}CH_2{-}\underset{O}{\underset{\|}{C}}{-}O^{-}}} \qquad \textbf{(6-4)}$$

This particular reaction is catalyzed in cells by the enzyme *succinate dehydrogenase* (so named because it normally functions in the opposite direction during energy metabolism). This dehydrogenase, like most enzymes, is highly specific. It will not add or remove hydrogen atoms from any compounds except those shown in Reaction 6-4. In fact, this particular enzyme is so specific that it will not even recognize maleate, which is an isomer of fumarate (**Figure 6-3**).

Not all enzymes are quite that specific. Some accept a number of closely related substrates, and others accept any of a whole group of substrates as long as they possess some common structural feature that is recognized and bound by the

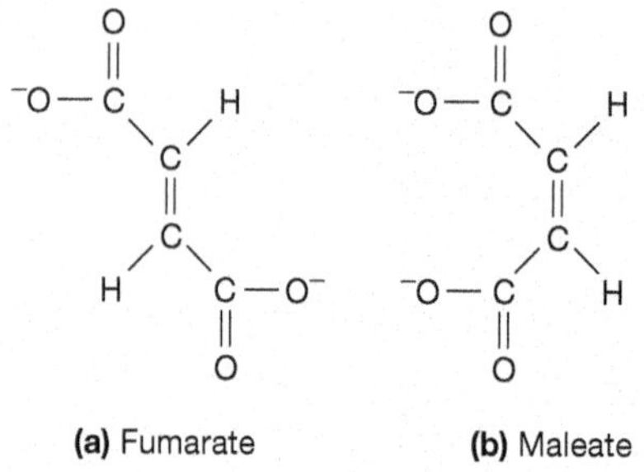

(a) Fumarate: $\mathrm{{}^{-}O{-}\overset{O}{\overset{\|}{C}}{-}CH{=}CH{-}\underset{O}{\underset{\|}{C}}{-}O^{-}}$ (trans)

(b) Maleate: $\mathrm{{}^{-}O{-}\overset{O}{\overset{\|}{C}}{-}CH{=}CH{-}\underset{O}{\underset{\|}{C}}{-}O^{-}}$ (cis)

Figure 6-3 Substrate Specificity in Enzyme-Catalyzed Reactions. Unlike most inorganic catalysts, enzymes can distinguish between closely related isomers. For example, the enzyme succinate dehydrogenase uses **(a)** fumarate as a substrate but not **(b)** its isomer, maleate.

enzyme active site. Such **group specificity** is seen most often with enzymes involved in the synthesis or degradation of polymers. Because the purpose of carboxypeptidase A is to degrade dietary polypeptide chains by removing the C-terminal amino acid, it makes sense for the enzyme to accept any of a wide variety of polypeptides as substrates. It would be needlessly extravagant of the cell to require a separate enzyme for every different amino acid residue that has to be removed during polypeptide degradation.

In general, however, enzymes are highly specific with respect to substrates, such that a cell must possess almost as many different kinds of enzymes as it has reactions to catalyze. For a typical cell, this means that thousands of different enzymes are necessary to carry out its full metabolic program.

Enzyme Diversity and Nomenclature. Given the specificity of enzymes and the large number of reactions occurring within a cell, it is not surprising that thousands of different enzymes have been identified. This enormous diversity of enzymes led to a variety of schemes for naming enzymes as they were discovered and characterized. Some were given names based on the substrate; *ribonuclease, protease,* and *amylase* are examples. Others, such as *succinate dehydrogenase,* were named to describe their function. Still other enzymes have names like *trypsin* and *catalase* that tell us little about either their substrates or their functions.

The resulting confusion prompted the International Union of Biochemistry to appoint an Enzyme Commission (EC) to devise a rational system for naming enzymes. Using the EC system, enzymes are divided into the following six major classes based on their general functions: *oxidoreductases, transferases, hydrolases, lyases, isomerases,* and *ligases.* The EC system assigns every known enzyme a unique four-part number based on its function—for example, EC 3.2.1.17 is the number for lysozyme. **Table 6-1** provides one representative example within each class of enzymes and the reaction catalyzed by that enzyme.

Sensitivity to Temperature. Besides their specificity and diversity, as shown in **Figure 6-4** on page 136, enzymes are characterized by an optimal temperature and pH at which they operate. Temperature dependence is not usually a practical concern for enzymes in the cells of mammals or birds because these organisms are *homeotherms,* "warm-blooded" animals that are capable of regulating body temperature independently from the environment. However, many organisms (for example, insects, reptiles, worms, plants, protozoa, algae, and bacteria) function at the temperature of their environment, which can vary greatly. For such organisms, the dependence of enzyme activity on temperature is significant.

In general, the rate of an enzyme-catalyzed reaction increases with temperature. This occurs because the greater kinetic energy of both enzyme and substrate molecules ensures more frequent and more energetic collisions, thereby increasing the likelihood of correct substrate binding and sufficient energy to undergo reaction. At some point above a critical temperature, however, further increases in temperature result in *denaturation* of the enzyme molecule. It loses its defined tertiary shape as hydrogen and ionic bonds are broken and the native polypeptide assumes a random, extended conformation. During denaturation, the active site is disrupted, abolishing enzyme activity.

The temperature range over which an enzyme denatures varies from enzyme to enzyme and especially from organism to organism. Figure 6-4a contrasts the temperature dependence of a typical enzyme from the human body with that of a typical enzyme from a thermophilic bacterium. Not surprisingly, the reaction rate of the human enzyme is maximum at about 37°C (the *optimal temperature* for the enzyme), which is normal body temperature. The sharp decrease in activity at higher temperatures reflects the denaturation of the enzyme molecules. Most enzymes of homeotherms are inactivated by temperatures above about 50–55°C. However, some enzymes are remarkably sensitive to heat and are denatured and inactivated at temperatures lower than this—in some cases, even by body temperatures encountered in people with high fevers (40°C). This is thought to be part of the beneficial effect of fever when you are ill from an infection—the denaturation of heat-sensitive enzymes in the pathogen.

Some enzymes, however, retain activity at unusually high temperatures. The green curve in Figure 6-4a depicts the temperature dependence of an enzyme from a thermophilic bacterium. Some of these bacteria thrive in hot springs at temperatures as

Table 6-1 The Major Classes of Enzymes with an Example of Each

Class	Reaction Type	Example	Reaction Catalyzed
1. Oxidoreductases	Oxidation-reduction reactions (electron transfer)	Alcohol dehydrogenase (EC 1.1.1.1)	Oxidation of ethanol to acetaldehyde as NAD^+ is reduced to NADH (Figure 9-8)
2. Transferases	Transfer of functional groups from one molecule to another	Hexokinase (EC 2.7.1.1)	Phosphorylation of glucose to glucose-6-phosphate using the terminal phosphoryl group from ATP (Figure 9-7)
3. Hydrolases	Hydrolytic cleavage of one molecule into two molecules	Glucose-6-phosphatase (EC 3.1.3.9)	Cleavage of glucose-6-phosphate into glucose and inorganic phosphate using a water molecule (Figure 9-12)
4. Lyases	Removal of a group from, or addition of a group to, a molecule	Pyruvate decarboxylase (EC 4.1.1.1)	Removal of a carboxyl group from pyruvate to produce acetaldehyde and CO_2 (Figure 9-8)
5. Isomerases	Movement of a functional group within a molecule	Maleate isomerase (EC 5.2.1.1)	A cis-trans isomerization of maleate to fumarate (Figure 6-3)
6. Ligases	Joining of two molecules to form a single molecule	Pyruvate carboxylase (EC 6.4.1.1)	Addition of CO_2 to pyruvate to produce oxaloacetate (Figure 9-12)

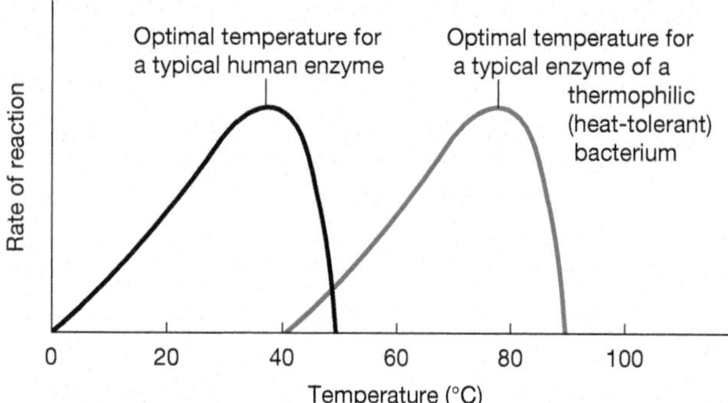

(a) Temperature dependence. The reaction rate for both a typical human enzyme (black) and a typical enzyme from a thermophilic bacterium (green) varies with temperature.

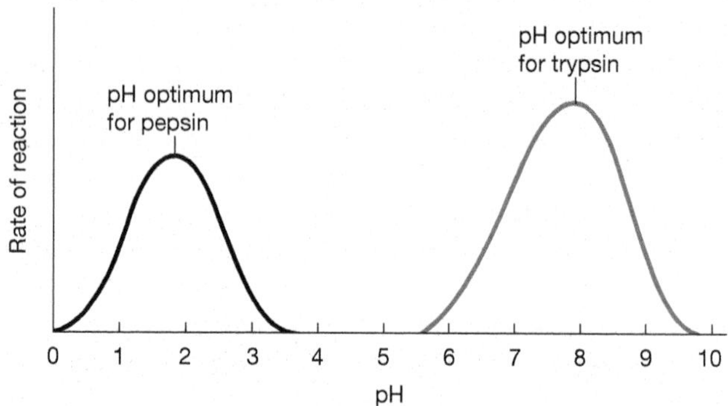

(b) pH dependence. The reaction rate of an enzyme is highest at its optimal pH, which is about 2 for pepsin (stomach pH) and near 8 for trypsin (intestinal pH).

Figure 6-4 The Effect of Temperature and pH on the Reaction Rate of Enzyme-Catalyzed Reactions. Every enzyme has an optimum temperature and pH that usually reflect the environment where that enzyme is found in nature.

high as 80°C. Other enzymes, such as those of psychrophilic ("cold-loving") *Listeria* bacteria and certain yeasts and molds, can function at low temperatures, allowing these organisms to grow slowly even at refrigerator temperatures (4–6°C).

> **MAKE CONNECTIONS 6.1**
>
> If you compared an enzyme in a typical homeotherm and the equivalent in a thermophilic bacterium, in which organism's enzyme might you expect to find more disulfide linkages? Why? (Fig. 3-5)

Sensitivity to pH. Enzymes are also sensitive to pH. In fact, most of them are active ***only within a pH range of about 3–4 pH units. This pH dependence is usually due to the presence*** of one or more charged amino acids at the active site and/or on the substrate itself. Activity is usually dependent on having one or more such groups present in a specific, either charged or uncharged form. For example, the active site of carboxypeptidase A includes the carboxyl groups from two glutamate residues. These carboxyl groups must be present in the charged (ionized) form, so the enzyme becomes inactive if the pH is decreased to the point where the glutamate carboxyl groups on the enzyme molecules are protonated and therefore uncharged. Extreme changes in pH also disrupt ionic and hydrogen bonds, altering tertiary structure and function.

As you might expect, the pH dependence of an enzyme usually reflects the environment in which that enzyme is normally active. Figure 6-4b shows the pH dependence of two protein-degrading enzymes found in the human digestive tract. Pepsin (black line) is present in the stomach, where the pH is usually about 2, whereas trypsin (red line) is secreted into the small intestine, which has a pH close to 8. Both enzymes are active over a range of almost 4 pH units but differ greatly in their pH optima, consistent with the conditions in their respective locations within the body.

Sensitivity to Other Factors. In addition to temperature and pH, enzymes are sensitive to other factors, including molecules and ions that act as inhibitors or activators of the enzyme. For example, several enzymes involved in energy production via glucose degradation are inhibited by ATP, which inactivates these enzymes when energy is plentiful. Other enzymes in the glucose degradation pathway are activated by adenosine mono- and diphosphate (AMP and ADP), which act as signals that the ATP concentration is low and that more glucose should be degraded to provide more ATP.

Most enzymes are also sensitive to the ionic strength (concentration of dissolved ions) of the environment, which affects the noncovalent hydrogen bonding and ionic interactions that help to maintain the tertiary conformation of the enzyme. Because these same ionic interactions are often involved in the interaction between the substrate and the active site, the ionic environment may also affect binding of the substrate.

Substrate Binding, Activation, and Catalysis Occur at the Active Site

Because of the precise chemical fit between the active site of an enzyme and its substrate(s), enzymes are highly specific and much more effective than inorganic catalysts. As noted previously, enzyme-catalyzed reactions proceed 10^7 to 10^{17} times more quickly than uncatalyzed reactions, versus a rate increase of 10^3 to 10^4 times for inorganic catalysts. As you might guess, most of the interest in enzymes focuses on the active site, where binding, activation, and chemical transformation of the substrate occur.

The sequence of events at the active site of a typical enzyme begins with the initial random collision of a substrate molecule with the active site. This results in its binding to amino acid residues that are strategically positioned in the active site. Substrate binding induces a change in the enzyme conformation that tightens the fit between the substrate molecule and the active site and lowers the free energy of the transition state. These conformational effects facilitate the conversion of substrate into products. The products are then released from the active site, enabling the enzyme molecule to return to its original conformation, with the active site now available for another molecule of substrate. This entire sequence of events takes place in a sufficiently short time to allow hundreds or even thousands of such reactions to occur per second at the active site of a single enzyme molecule.

Substrate Binding. Initial contact between the active site of an enzyme and a potential substrate molecule depends on their collision. Once in the active site, the substrate molecule is bound to the enzyme surface in just the right orientation so

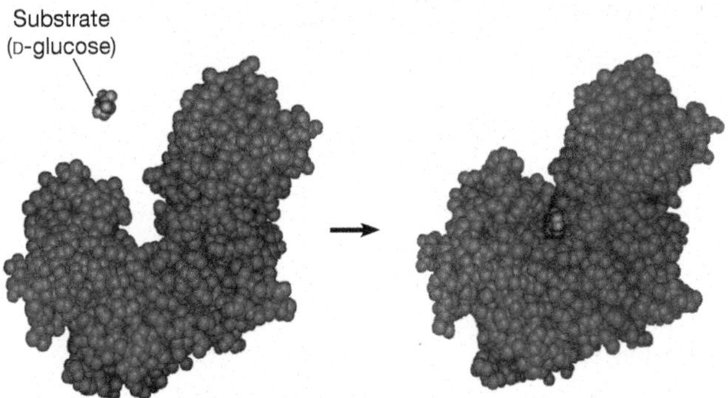

Figure 6-5 Induced Fit Following Substrate Binding. This figure shows a space-filling model of the enzyme hexokinase along with its substrate, a molecule of D-glucose. Substrate binding induces a conformational change in hexokinase, known as induced fit, that improves the catalytic activity of the enzyme.

that specific catalytic groups on the enzyme can facilitate a reaction. Substrate binding usually involves hydrogen bonds or ionic bonds (or both) to charged or polar amino acids. These are generally weak bonds, but several bonds may hold a single molecule in place. The strength of the bonds between an enzyme and a substrate molecule is often in the range of 3–12 kcal/mol. This may be less than one-tenth the strength of a single covalent bond (see Figure 2-2). Substrate binding is therefore readily reversible.

For many years, enzymologists regarded an enzyme as a rigid structure, with a specific substrate fitting into the active site as a key fits into a lock. This *lock-and-key model,* first suggested in 1894 by the German biochemist Emil Fischer, explained enzyme specificity but did little to enhance our understanding of the catalytic event. A more refined view of the enzyme-substrate interaction is provided by the **induced-fit model,** first proposed in 1958 by Daniel Koshland. According to the induced-fit model, substrate binding at the active site distorts both the enzyme and the substrate, thereby stabilizing the substrate molecules in their transition state and rendering certain substrate bonds more susceptible to catalytic attack.

Induced fit involves a conformational change in the shape of the enzyme molecule following substrate binding. This conformational change alters the configuration of the active site and optimally positions the proper reactive groups of the enzyme for the catalytic reaction. Evidence for such conformational changes upon binding of substrate has come from X-ray diffraction studies of crystallized proteins and nuclear magnetic resonance (NMR) studies of proteins in solution, which can determine the shape of an enzyme molecule with and without bound substrate. **Figure 6-5** illustrates the conformational change that occurs upon substrate binding to the enzyme hexokinase, which catalyzes the addition of a phosphoryl group to D-glucose. As D-glucose binds to the active site, it induces a conformational change that brings the upper and lower lobes of hexokinase toward each other, closing the active site around the substrate to facilitate catalysis.

Often, the induced conformational change brings critical amino acid side chains into the active site even if they are not nearby in the absence of substrate. In the active site of carboxypeptidase A (**Figure 6-6**), a zinc ion is tightly bound to three residues of the enzyme (Glu-72, His-69, and His-196). Substrate binding to the zinc ion induces a conformational change in the enzyme that brings other amino acid side chains into the active site, including Arg-145, Tyr-248, and Glu-270. These amino acid residues are then in position to participate in catalysis.

Once in the active site, the substrate is held in place by very specific noncovalent interactions such as hydrogen bonds that not only position the substrate optimally for catalysis but also discriminate between the genuine substrate and similar molecules. To see how specific, let's consider lysozyme again. The first X-ray crystal structure that was obtained of an enzyme was that of hen lysozyme, reported in 1965 by David Phillips and colleagues. The active site of lysozyme forms a deep pocket, clearly visible in **Figure 6-7** on page 138, that binds specifically with amino sugars of the substrate.

Substrate Activation. The role of the active site is not just to recognize and bind the appropriate substrate but also to *activate* it by subjecting it to the right chemical environment

Figure 6-6 The Change in Active Site Structure Induced by Substrate Binding. (a) The unoccupied active site of carboxypeptidase A contains a zinc ion (reddish brown) tightly bound to side chains of three amino acids (cyan). **(b)** Binding of the substrate (the dipeptide shown in orange) to this zinc ion induces a conformation change in the enzyme that brings other amino acid side chains (purple) into the active site to participate in catalysis.

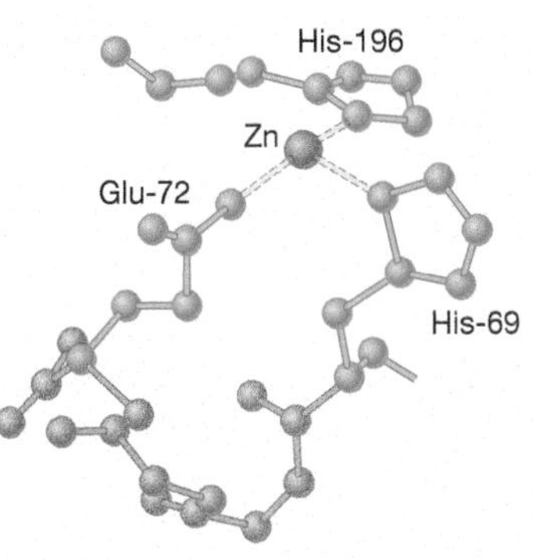

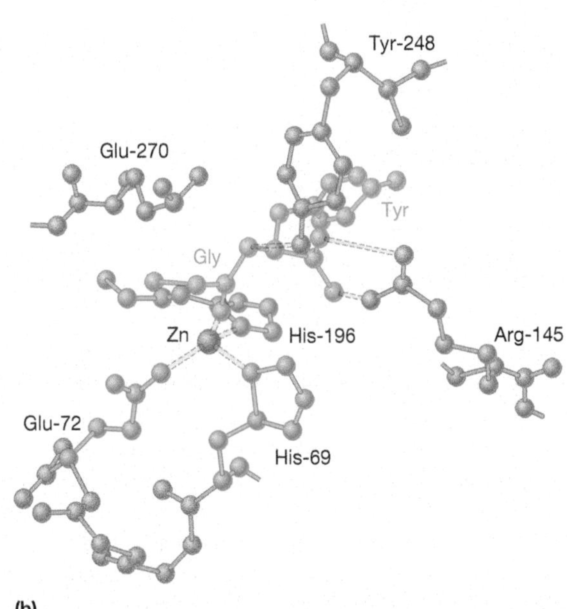

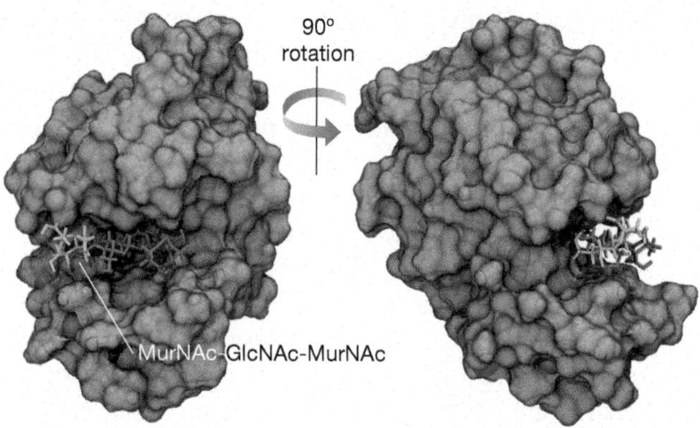

Figure 6-7 Hen Egg White Lysozyme with Peptidoglycan Bound at the Active Site. The surface of lysozyme is shown in blue. A trisaccharide substrate consisting of *N*-acetylmuramic acid (MurNAc) and *N*-acetylglucosamine (GlcNAc), is shown in stick representation bound to the active site.

for catalysis. A given enzyme-catalyzed reaction may involve one or more means of **substrate activation**.

Three of the most common methods of substrate activation are the following:

1. ***Bond distortion.*** The change in enzyme conformation induced by initial substrate binding to the active site causes a tighter enzyme-substrate fit and may also distort one or more of the substrate's bonds, weakening the bonds and making them more susceptible to catalytic attack.
2. ***Proton transfer.*** The enzyme may also accept protons from or donate protons to the substrate, thereby increasing the substrate's chemical reactivity. Proton transfer accounts for the importance of charged amino acids in active-site chemistry and explains why enzyme activity is often pH dependent.
3. ***Electron transfer.*** As a further means of substrate activation, the enzyme may also exchange (accept or donate) electrons with its substrate, thereby forming temporary covalent bonds between the enzyme and the substrate.

The Catalytic Event. Once bound at the active site, activated substrate can undergo conversion to product(s). Let's again consider lysozyme, an enzyme whose mechanism is fairly well understood. Specific amino acids of the active site interact with the bound substrate to destabilize a glycosidic bond in the peptidoglycan, leading to formation of product—in this case, a precisely cleaved peptidoglycan.

Failure of even a single enzyme to carry out its intended reaction can lead to disease. For example, the human disease phenylketonuria results from a defect in the liver enzyme phenylalanine hydroxylase, which normally converts the amino acid phenylalanine into the related amino acid tyrosine. This leads to abnormal accumulation of phenylalanine in the body and can result in mental retardation. In other cases, health can be profoundly affected when enzymes that normally catalyze the removal of amino acids from proteins cannot act. One example in humans is angiotensin-converting enzyme (ACE) and its inhibitors (see Human Connections, page 139).

Ribozymes Are Catalytic RNA Molecules

Until the early 1980s, it was thought that all enzymes were proteins because every enzyme isolated in the 55 years following Sumner's purification of urease in 1926 was a protein. But biology is full of surprises, and in the 1980s, catalytic RNA molecules were discovered. These RNA catalysts were called **ribozymes.** Many scientists now hypothesize that the earliest enzymes were molecules of catalytic, self-replicating RNA and that these molecules were present in primitive cells even before the existence of DNA (Chapter 4).

Tetrahymena RNA. In 1981, Thomas Cech and his colleagues at the University of Colorado discovered an apparent exception to the "all enzymes are proteins" rule. They were studying the removal of an internal segment of RNA known as an intron from a specific ribosomal RNA precursor (pre-rRNA) in *Tetrahymena thermophila*, a single-celled eukaryote. The researchers made the remarkable observation that the process proceeded in the absence of any proteins. The removal of the 413-nucleotide intron from the *Tetrahymena* pre-rRNA turned out to be catalyzed by the pre-rRNA molecule itself and is therefore an example of *autocatalysis.*

Ribonuclease P. Two years later, another RNA-based catalyst was discovered in the laboratory of Sidney Altman, at Yale University, which was studying *ribonuclease P*, an enzyme that cleaves transfer RNA (tRNA) precursors to yield functional tRNA molecules. Ribonuclease P consists of a protein component and an RNA component, and it was assumed that the active site was on the protein component. By isolating the components and studying them separately, however, Altman and his colleagues showed unequivocally that only the isolated RNA component was capable of catalyzing the specific cleavage of tRNA precursors on its own. Also, the RNA-catalyzed reaction followed typical enzyme kinetics, further evidence that the RNA component was acting like a true enzyme. The significance of these findings was recognized by the Nobel Prize that Cech and Altman shared in 1989 for their discoveries.

Ribosomal RNA. Since Cech and Altman's initial discoveries, additional examples of ribozymes have been reported. Of special significance is the active site for a crucial step in protein synthesis by ribosomes. The large ribosomal subunit (see Figure 4-23) is the site of the peptidyl transferase activity that catalyzes peptide bond formation (a subject covered in detail in Chapter 19). The active site for this peptidyl transferase activity was for a long time assumed to be located on one of the protein molecules of the large subunit, with ribosomal RNA (rRNA) merely providing a scaffold for structural support. However, in 1992, Harry Noller and his colleagues at the University of California, Santa Cruz, demonstrated that, despite the removal of 95% of the protein from the large ribosomal subunit, it retained 80% of the peptidyl transferase activity of the intact subunit. This finding strongly suggested that one of the rRNA molecules was the catalyst. Moreover, the catalytic activity of the ribosome was destroyed by treatment with ribonuclease, an enzyme that degrades RNA, but was not affected by proteinase K, an enzyme that degrades protein. Thus, the peptidyl transferase activity responsible for peptide bond formation during protein synthesis is due to

HUMAN *Connections*

Ace Inhibitors: Enzyme Activity as the Difference Between Life and Death

Figure 6A-1 The Brazilian Pit Viper (*Bothrops jararaca*) with an Extracted Drop of Venom.

When the Brazilian pit viper (*Bothrops jararaca*) (**Figure 6A-1**) spots its prey and strikes, it injects venom into its victim. The venom releases a cocktail of peptides that widen the victim's blood vessels and cause a drastic drop in blood pressure. This drop in blood pressure causes the prey to lose consciousness, and it becomes an easy meal for the pit viper. Bad news for the victim, but good news for us! Analysis of the chemicals in Brazilian pit viper venom led to the discovery of *ACE inhibitors,* a group of drugs important in controlling high blood pressure.

Your body constantly adjusts blood pressure to maintain it in a healthy range. Many of the organs in your body help to control your blood pressure, including your kidneys and lungs. If blood pressure falls too low, specialized cells in the kidneys release the hormone *renin.* Renin is a hormone, but it also has enzymatic activity. When renin is released by the kidneys, it cleaves a specific peptide bond in an inactive protein known as *angiotensinogen,* releasing an N-terminal ten-amino-acid peptide called *angiotensin I* (**Figure 6A-2**).

Angiotensin I travels thorough the bloodstream to the pulmonary artery and lungs, where it is modified by the action of another enzyme, known as *angiotensin-converting enzyme (ACE),* which is abundant in the capillaries of the lungs. ACE cleaves two amino acids from the C-terminus of angiotensin I to convert angiotensin I to *angiotensin II.*

Angiotensin II normally raises blood pressure if it has fallen too low by acting in the kidneys to return more sodium and water to the blood. Angiotensin II is also a vasoconstrictor and causes blood vessels to narrow, further increasing blood pressure.

Like many tightly regulated events in the body, there is a regulatory pathway that has the opposite effect of angiotensin II

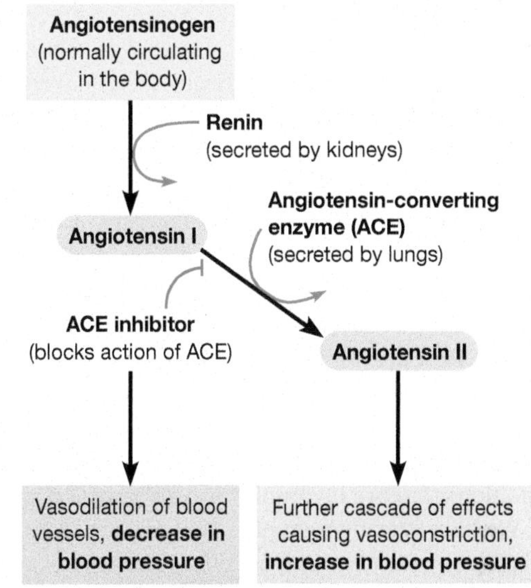

Figure 6A-2 The Basic Pathway of Hormone Release in Response to Low Blood Pressure.

on blood pressure. This system utilizes the peptide hormone *bradykinin,* which is a vasodilator. Bradykinin causes blood vessels to relax and become wider, decreasing blood pressure.

ACE is involved in regulating both systems. ACE inactivates vasodilating bradykinin and, at the same time, increases the amount of vasoconstricting angiotensin II, and these combined effects lead to a rise in blood pressure. Given how ACE acts, it is easy to see why substances that *inhibit* ACE would be attractive candidate drugs for treating human patients with high blood pressure.

Now let's return to the pit viper's venom and see what part it played in drug development. The toxin produced by the pit viper is actually a *competitive inhibitor* of ACE (competitive inhibition is a process you will learn about later in this chapter). The toxin from the Brazilian pit viper is not a viable medication, however. Because the venom is a peptide, if taken orally, it is easily broken down by the digestive system. Instead, the toxin's mechanism of inhibition and structure were used to develop the drug captopril. Captopril is not broken down after ingestion and produces the same effect as the toxin produced by the pit viper. Compounds like captopril allow for treatments that decrease high blood pressure and prevent secondary heart attacks, congestive heart failure, and complications from diabetes.

rRNA. That is, the rRNA is a ribozyme. The ribosomal proteins appear to support and stabilize the catalytic RNA, not the other way around.

The discovery of ribozymes has markedly changed the way we think about the origin of life on Earth. For many years, scientists had speculated that the first catalytic macromolecules must have been amino acid polymers resembling proteins. But this concept immediately ran into difficulty because there was no obvious way for a primitive protein to carry information or to replicate itself, which are two primary attributes of life. However, if the first catalysts were RNA rather than protein molecules, it becomes conceptually easier to imagine an "RNA world" in which RNA molecules acted both as catalysts and as replicating systems capable of transferring information from generation to generation.

CONCEPT CHECK 6.2

How can an enzyme recognize and bind one specific substrate in a cell containing thousands of different molecules? How will extremes of temperature and pH affect this specificity?

6.3 Enzyme Kinetics

So far, our discussion of enzymes has been basically qualitative. We have dealt with the activation energy requirement that prevents thermodynamically feasible reactions from occurring unassisted. We have also encountered enzymes as biological catalysts and have examined their structure and function in some detail. Moreover, we realize that the only reactions likely to occur in cells at reasonable rates are those for which specific enzymes are on hand.

Still lacking, however, is a means of assessing—in a *quantitative* fashion—the actual rates at which enzyme-catalyzed reactions proceed, as well as an appreciation for the factors that influence reaction rates. The mere presence of the appropriate enzyme in a cell does not ensure that a given reaction will occur at an adequate rate. We need to understand the cellular conditions that are favorable for activity of a particular enzyme. We have already seen how factors such as temperature and pH can affect enzyme activity. Now we are ready to appreciate how enzyme activity also depends critically on the concentrations of substrates, products, and inhibitors that prevail in the cell. In addition, we will see how at least some of these effects can be defined quantitatively.

Enzyme kinetics describes quantitative aspects of enzyme catalysis and the rate of substrate conversion into products. (The word *kinetics* is from the Greek word *kinetikos*, meaning "moving.") Specifically, enzyme kinetics concerns reaction rates and the manner in which reaction rates are influenced by factors such as the concentrations of substrates, products, and inhibitors. Much of our attention will focus on the effects of substrate concentration on the kinetics of enzyme-catalyzed reactions.

Understanding enzyme kinetics is not merely a theoretical exercise. In cells of all organisms, numerous processes crucially depend on enzyme kinetics. Having detailed knowledge of an enzyme's kinetic properties can help researchers understand the nature of enzyme activity in certain human diseases. This knowledge can also aid the design of drugs that act as enzyme inhibitors. Understanding enzyme kinetics also has important commercial applications; examples include optimizing industrial processes, maximizing production of proteins for research, and developing enzyme-based assays for diagnosis of human disease.

Monkeys and Peanuts Provide a Useful Analogy for Understanding Enzyme Kinetics

Enzyme kinetics can seem quite complex at first. If you found the Mexican jumping beans helpful in understanding free energy in Chapter 5, then you might like another analogy. In this case, we will consider a roomful of monkeys ("enzymes") shelling peanuts ("substrates"), with the peanuts, present in varying abundance, representing differences in substrate concentration. Try to understand each step first in terms of monkeys shelling peanuts and then in terms of an actual enzyme-catalyzed reaction.

The Peanut Gallery. Imagine ten monkeys, all equally adept at finding and shelling peanuts; also imagine a Peanut Gallery, a room with peanuts scattered equally about on the floor. The number of peanuts will be varied as we proceed, but in all of our "assays," there will be vastly more peanuts than monkeys in the room. And because we know the number of peanuts and the total floor space, we can use peanut density as the "concentration" of peanuts in the room. In each case, the monkeys start out in a room adjacent to the Peanut Gallery. To start an assay, we simply open the door and allow the eager monkeys to enter.

The Shelling Begins. For our first assay, we will start with an initial peanut concentration of 1 peanut per square meter, and we will assume that, at this concentration of peanuts, the average monkey spends 9 seconds looking for a peanut to shell and 1 second shelling it. This means that each monkey requires 10 seconds per peanut and can thus shell peanuts at the rate of 0.1 peanut per second. Then, because there are ten monkeys in the gallery, the velocity (v) of peanut shelling for all the monkeys is 1 peanut per second at this peanut concentration, which we will call [S]. All of this can be tabulated as follows:

[S] Concentration of peanuts (peanuts/m^2)	1
Time required per peanut:	
To find (sec/peanut)	9
To shell (sec/peanut)	1
Total (sec/peanut)	10
Rate of shelling (v):	
Per monkey (peanut/sec)	0.10
Total (peanut/sec)	1.0

The Peanuts Become More Abundant. For our second assay, we will triple the number of peanuts to a concentration of 3 peanuts per square meter. Now the average monkey will find a peanut three times more quickly than before, in only 3 seconds. But each peanut still takes 1 second to shell, so the total time per peanut is now 4 seconds, and the velocity of shelling is 0.25 peanut per second for each monkey, or 2.5 peanuts per second for the roomful of monkeys. This generates another column of entries for our data table:

[S] Concentration of peanuts (peanuts/m^2)	1	3
Time required per peanut:		
To find (sec/peanut)	9	3
To shell (sec/peanut)	1	1
Total (sec/peanut)	10	4
Rate of shelling (v):		
Per monkey (peanut/sec)	0.10	0.25
Total (peanut/sec)	1.0	2.5

What Happens to v as [S] Continues to Increase? So what eventually happens to the velocity of peanut shelling as the peanut concentration in the room gets higher and higher? All you need do is extend the data table by assuming ever-increasing values for [S] and calculating the corresponding values for v. Notice that a tripling of peanut concentration increased the rate by only 2.5-fold. A second tripling ([S] = 9) will result in only a further doubling of the rate. There seems, in other words, to be a diminishing return on additional peanuts.

As we increase [S], the time to find a peanut decreases proportionally, but the time to shell it doesn't change. So no matter how high [S] is, it will always take at least 1 second for the shelling reaction, and the overall rate for the ten monkeys can never exceed 10 peanuts per second. Problem 6-13 at the end of the chapter will allow you to further explore the kinetic parameters of this monkey-catalyzed shelling reaction.

Most Enzymes Display Michaelis–Menten Kinetics

Now that you have learned about how "reaction velocity" depends on substrate concentration in the fictional case of monkeys and peanuts, you should be ready for a more formal treatment of enzyme kinetics. Here we will consider how the **initial reaction velocity (v_0)** changes depending on the **substrate concentration ([S])**.

The initial reaction velocity is rigorously defined as the rate of change in product concentration measured over a brief initial period of time during which the substrate concentration has not yet decreased enough to affect the rate of the reaction and the accumulation of product is still too small for the reverse reaction (conversion of product back into substrate) to occur at a significant rate. Reaction velocities are often experimentally measured in a constant assay volume of 1 mL and are reported as μmol of product per minute. At low [S], a doubling of [S] will double v_0. But as [S] increases, each additional increment of substrate results in a smaller increase in initial reaction rate. As [S] becomes very large, increases in [S] increase initial velocity only slightly, and the value of v_0 reaches a maximum.

By determining v_0 in a series of experiments at varying substrate concentrations, the dependence of v_0 on [S] can be shown experimentally to be that of a hyperbola (**Figure 6-8**). An important property of this hyperbolic relationship is that as [S] tends toward infinity, v_0 approaches an upper limiting value known as the **maximum velocity (V_{max})**.

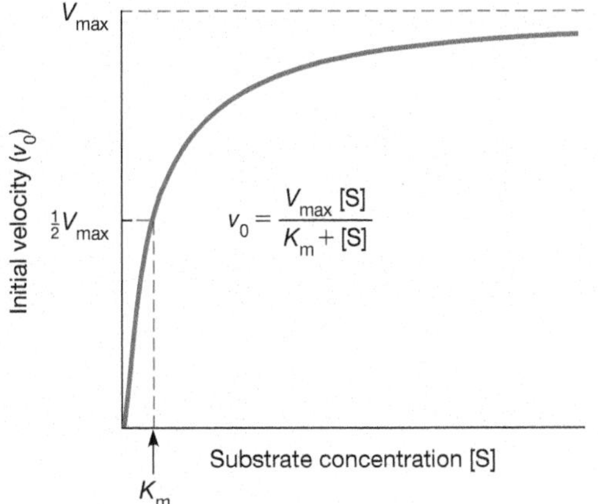

Figure 6-8 The Relationship Between Reaction Velocity and Substrate Concentration. The initial velocity of an enzyme-catalyzed reaction tends toward an upper limiting velocity V_{max} as the substrate concentration [S] tends toward infinity. K_m is the substrate concentration at which the reaction is proceeding at one-half of the maximum velocity.

This value depends on the number of enzyme molecules and can therefore be increased only by adding more enzyme. The inability of increasingly higher substrate concentrations to increase the reaction velocity beyond a finite upper value is called **saturation**. At saturation, all available enzyme molecules are operating at maximum capacity. Saturation is a fundamental, universal property of enzyme-catalyzed reactions. Catalyzed reactions always become saturated at high substrate concentrations, whereas uncatalyzed reactions do not.

Much of our understanding of the hyperbolic relationship between concentration and velocity is due to the pioneering work of two German enzymologists, Leonor Michaelis and Maud Menten. In 1913, they postulated a general theory of enzyme action that has helped us quantify almost every aspect of enzyme kinetics. To understand their approach, consider one of the simplest possible enzyme-catalyzed reactions, a reaction in which a single substrate S is converted into a single product P:

$$S \xrightarrow[\text{enzyme(E)}]{} P \qquad \textbf{(6-5)}$$

According to the Michaelis–Menten hypothesis, the enzyme E that catalyzes this reaction first reacts with the substrate S, forming the transient enzyme-substrate complex ES, which then undergoes the actual catalytic reaction to form free enzyme and product P, as shown in the sequence

$$E_f + S \underset{k_2}{\overset{k_1}{\rightleftharpoons}} ES \underset{k_4}{\overset{k_3}{\rightleftharpoons}} E_f + P \qquad \textbf{(6-6)}$$

where E_f is the free form of the enzyme, S is the substrate, ES is the enzyme-substrate complex, P is the product, and k_1, k_2, k_3, and k_4 are the **rate constants** for the indicated reactions. A rate constant is the proportionality constant relating the rate of a reaction to the concentrations of reactants.

Starting with this model and several simplifying assumptions, including steady-state conditions (described near the end of Chapter 5), Michaelis and Menten arrived at the relationship between the velocity of an enzyme-catalyzed reaction and the substrate concentration, as follows:

$$v_0 = \frac{V_{max}[S]}{K_m + [S]} \qquad \textbf{(6-7)}$$

Here, v_0 is the initial reaction velocity, [S] is the initial substrate concentration, V_{max} is the maximum velocity, and K_m (also known as the **Michaelis constant**) is the concentration of substrate that gives exactly half the maximum velocity. V_{max} and K_m are important kinetic parameters that we will consider in more detail in the next section. Equation 6-7 is known as the **Michaelis–Menten equation**, a central relationship of enzyme kinetics. (Problem 6-12 gives you an opportunity to derive the Michaelis–Menten equation yourself.)

What Is the Meaning of V_{max} and K_m?

To appreciate the implications of the relationship between v_0 and [S] and to examine the meaning of the parameters V_{max} and K_m, we can consider three special cases of substrate concentration: very low substrate concentration, very high substrate concentration, and the special case of [S] = K_m.

Case 1: Very Low Substrate Concentration ($[S] \ll K_m$). At very low substrate concentration, [S] becomes negligibly small compared with the constant K_m in the denominator of the Michaelis–Menten equation and can be ignored, so we can write

$$v_0 = \frac{V_{max}[S]}{K_m + [S]} \cong \frac{V_{max}[S]}{K_m} \tag{6-8}$$

Thus, at very low substrate concentration, the initial reaction velocity is roughly proportional to the substrate concentration. This can be seen at the extreme left side of the graph in Figure 6-8. As long as the substrate concentration is much lower than the K_m value, the velocity of an enzyme-catalyzed reaction increases linearly with substrate concentration.

Case 2: Very High Substrate Concentration ($[S] \gg K_m$). At very high substrate concentration, K_m becomes negligibly small compared with [S] in the denominator of the Michaelis–Menten equation, so we can write

$$v_0 = \frac{V_{max}[S]}{K_m + [S]} \cong \frac{V_{max}[S]}{[S]} = V_{max} \tag{6-9}$$

Therefore, at very high substrate concentrations, the velocity of an enzyme-catalyzed reaction is essentially independent of the variation in [S] and is approximately constant at a value close to V_{max} (see the right side of Figure 6-8).

Equation 6-9 provides us with a mathematical definition of V_{max}, which is one of the two kinetic parameters in the Michaelis–Menten equation. V_{max} is the upper limit of v_0 as the substrate concentration [S] approaches infinity. In other words, V_{max} is the velocity at saturating substrate concentrations. Under these conditions, the enzyme molecules are active almost all of the time because the substrate concentration is so high that as soon as a product molecule is released, another substrate molecule is bound at the active site.

V_{max} is therefore an upper limit determined by (1) the time required for the actual catalytic reaction and (2) how many such enzyme molecules are present. Because the actual reaction rate is fixed, the only way that V_{max} can be increased is to increase enzyme concentration. In fact, V_{max} is linearly proportional to the amount of enzyme present, as shown in **Figure 6-9**, where k_3 represents the rate constant.

Case 3: ($[S] = K_m$). To explore the meaning of K_m more precisely, consider the special case where [S] is equal to K_m. If $[S] = K_m$, then $K_m + [s] = 2[s]$, and the Michaelis–Menten equation can be written as

$$v_0 = \frac{V_{max}[S]}{K_m + [S]} \cong \frac{V_{max}[S]}{2[S]} = \frac{V_{max}}{2} \tag{6-10}$$

This equation demonstrates mathematically that K_m *is the specific substrate concentration at which the reaction proceeds at one-half of its maximum velocity*. K_m is a constant value for a given enzyme-substrate combination catalyzing a reaction under specified conditions. Figure 6-8 illustrates the meaning of both V_{max} and K_m.

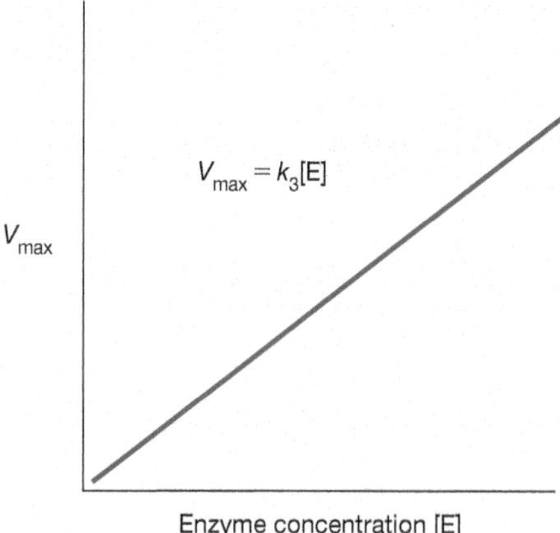

Figure 6-9 The Linear Relationship Between V_{max} and Enzyme Concentration. The linear increase in reaction velocity with enzyme concentration provides the basis for determining enzyme concentrations experimentally.

Why Are K_m and V_{max} Important to Cell Biologists?

Now that we understand what K_m and V_{max} mean, it is fair to ask why these kinetic parameters are important to cell biologists. The K_m value is useful because it allows us to estimate where along a Michaelis–Menten plot (see Figure 6-8) an enzyme is functioning in a cell—providing, of course, that the normal substrate concentration in the cell is known. We can then estimate at what fraction of the maximum velocity the enzyme-catalyzed reaction is likely to be proceeding in the cell. The lower the K_m value for a given enzyme and substrate, the lower the substrate concentration range in which the enzyme is effective. As we will soon see, enzyme activity in the cell can be modulated by regulatory molecules that bind to the enzyme and alter the K_m for a particular substrate. K_m values for several enzyme-substrate combinations are given in **Table 6-2** and, as you can see, can vary over several orders of magnitude.

The V_{max} for a particular reaction is important because it provides a measure of the potential maximum rate of the reaction. Few enzymes actually encounter saturating substrate concentrations in cells, so enzymes are not likely to be functioning at their maximum rate under cellular conditions. However, by knowing the V_{max} value, the K_m value, and the substrate concentration in vivo, we can at least estimate the likely rate of the reaction under cellular conditions.

V_{max} can also be used to determine another useful parameter called the **turnover number (k_{cat})**, which expresses the rate at which substrate molecules are converted to product by a single enzyme molecule when the enzyme is operating at its maximum velocity. The constant k_{cat} is the quotient of V_{max} divided by [E], the concentration of the enzyme:

$$k_{cat} = \frac{V_{max}}{[E]} \tag{6-11}$$

Because V_{max} is expressed as concentration/time and [E] is a concentration, k_{cat} has units of reciprocal time (sec^{-1} or min^{-1}). Turnover numbers vary greatly among enzymes, as is clear from the examples given in Table 6-2.

Table 6-2 K_m and k_{cat} Values for Some Enzymes

Enzyme Name	Substrate	K_m (*M*)	k_{cat} (sec^{-1})
Acetylcholinesterase	Acetylcholine	9×10^{-5}	1.4×10^{4}
Carbonic anhydrase	CO_2	1×10^{-2}	1×10^{6}
Fumarase	Fumarate	5×10^{-6}	8×10^{2}
Triose phosphate isomerase	Glyceraldehyde-3-phosphate	5×10^{-4}	4.3×10^{3}
β-lactamase	Benzylpenicillin	2×10^{-5}	2×10^{3}

The Double-Reciprocal Plot Is a Useful Means of Visualizing Kinetic Data

The classic Michaelis–Menten plot of v_0 versus [S] shown in Figure 6-8 illustrates the dependence of velocity on substrate concentration. However, it is not an especially useful tool for the quantitative determination of the key kinetic parameters K_m and V_{max}. Its hyperbolic shape makes it difficult to extrapolate accurately to infinite substrate concentration in order to determine the critical parameter V_{max}. Also, if V_{max} is not known accurately, K_m cannot be determined.

To find a way around this problem and provide a more useful graphic approach, Hans Lineweaver and Dean Burk in 1934 converted the hyperbolic relationship of the Michaelis–Menten equation into a linear function by inverting both sides of Equation 6-7 and simplifying the resulting expression into the form of an equation for a straight line:

$$\frac{1}{v_0} = \frac{K_m + [S]}{V_{max}[S]} = \frac{K_m}{V_{max}[S]} + \frac{[S]}{V_{max}[S]}$$

$$= \frac{K_m}{V_{max}}\left(\frac{1}{[S]}\right) + \frac{1}{V_{max}} \qquad \textbf{(6-12)}$$

Equation 6-12 is known as the **Lineweaver–Burk equation**. When it is plotted as $1/v_0$ versus $1/[S]$, as in **Figure 6-10**, the resulting **double-reciprocal plot** is linear in the general algebraic form $y = mx + b$, where m is the slope and b is the y-intercept. Therefore, it has a slope (m) of K_m/V_{max}, a y-intercept (b) of $1/V_{max}$, and an x-intercept ($y = 0$) of $-1/K_m$. (You should be able to convince yourself of these intercept values by setting first $1/[S]$ and then $1/v$ equal to zero in Equation 6-12 and solving for the other value.) Therefore, once the double-reciprocal plot has been constructed, V_{max} can be determined directly from the reciprocal of the y-intercept and K_m from the negative reciprocal of the x-intercept. Furthermore, the slope can be used to check both values.

The Lineweaver–Burk plot is useful experimentally because it allows us to determine the parameters K_m *and* V_{max} without the complication of a hyperbolic shape. **Key Technique, pages 144–145,** shows an experimental set-up for determining K_m and V_{max} for hexokinase, the enzyme that catalyzes the phosphorylation of glucose to begin the degradation of glucose in the glycolytic pathway.

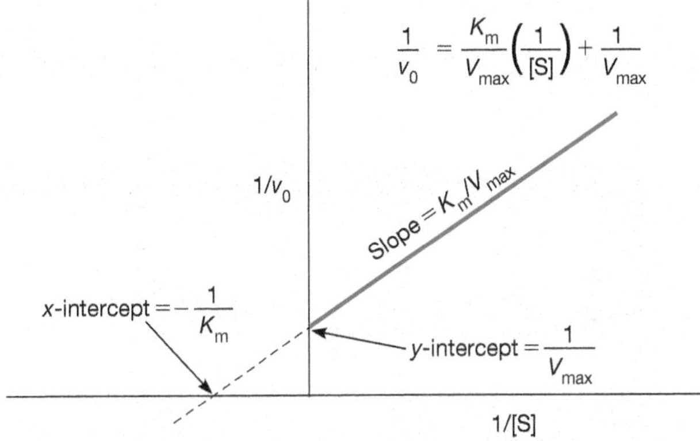

Figure 6-10 The Lineweaver–Burk Double-Reciprocal Plot. The reciprocal of the initial velocity, $1/v_0$, is plotted as a function of the reciprocal of the substrate concentration, 1/[S]. K_m can be calculated from the x-intercept and V_{max} from the y-intercept.

Enzyme Inhibitors Act Either Irreversibly or Reversibly

Thus far, we have assumed that substrates are the only substances in cells that affect the enzyme activity. However, enzymes are also influenced by products, alternative substrates, substrate analogues, drugs, toxins, and a very important class of regulators called *allosteric effectors*. Most of these substances have an inhibitory effect on enzyme activity, reducing the reaction rate with the desired substrate or sometimes blocking the reaction completely.

This **inhibition** of enzyme activity is important for several reasons. First and foremost, enzyme inhibition plays a vital role as a control mechanism in cells. Many enzymes are subject to regulation by specific small molecules other than their substrates. Often this is a means of sensing their immediate environment to respond to specific cellular conditions.

Enzyme inhibition is also important in the action of drugs and poisons, which frequently exert their effects by inhibiting specific enzymes. Inhibitors are also useful to enzymologists as tools in their studies of reaction mechanisms and to doctors for treatment of disease. Especially important inhibitors are **substrate analogues** and **transition state analogues**. These are compounds that resemble the usual substrate or the

Key Technique Determining K_m and V_{max} Using Enzyme Assays

PROBLEM: Hexokinase is an important enzyme in energy metabolism because it catalyzes the first step in the exothermic degradation of glucose. Understanding the kinetics of this reaction will help us to better understand its role in energy production in the cell. But how do scientists experimentally determine the kinetic parameters of K_m and V_{max} for an enzyme such as this?

SOLUTION: Researchers can determine K_m and V_{max} by performing an experiment in which the enzyme concentration and all other variables are held constant as the substrate concentration is varied in a series of enzyme assays (individual measurements). By measuring the initial velocity (v_0) at each substrate concentration [S] and graphing the results, we can find the numerical values of K_m and V_{max}.

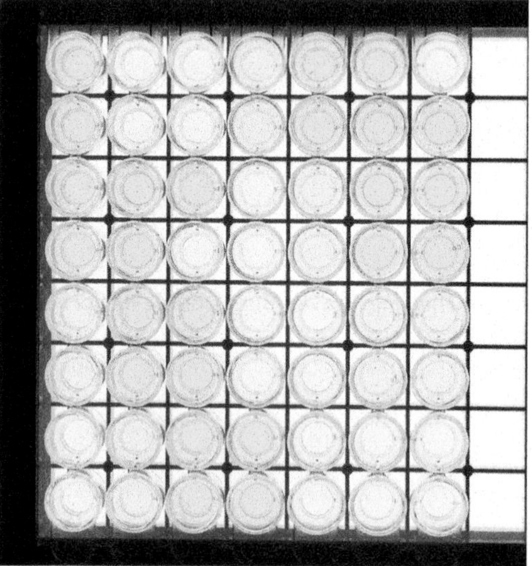

The enzyme in these assays produces a yellow product, so the intensity of the color is proportional to the total enzyme activity in each assay.

Key Tools: Purified enzyme and substrate; a color indicator to allow measurement of the amount of product; a spectrophotometer for measuring the accumulation of product.

Details: Hexokinase catalyzes the phosphorylation of glucose on carbon atom 6 as shown in the following equation:

$$\text{glucose} + \text{ATP} \xrightarrow[\text{hexokinase}]{} \text{glucose-6-phosphate} + \text{ADP}$$

To analyze the kinetics of this reaction, we must determine the initial velocity at each of several substrate concentrations. When an enzyme has two substrates, the usual approach is to vary the concentration of one substrate at a time while holding that of the other one constant at a level near saturation to ensure that it does not become rate limiting.

In the experimental approach shown in **Figure 6B-1**, glucose is the variable substrate, with ATP present at a saturating

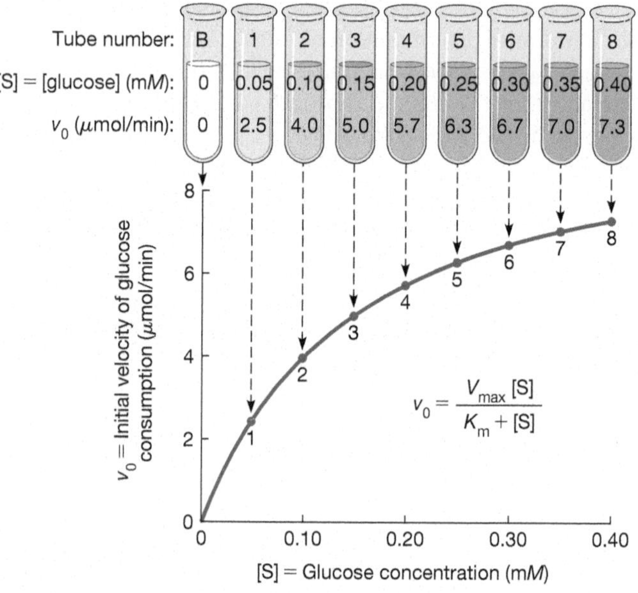

Figure 6B-1 Experimental Procedure for Studying the Kinetics of the Hexokinase Reaction. Test tubes containing varying concentrations of glucose and a saturating concentration of ATP were incubated with a standard amount of hexokinase. The initial velocity, v_0, was then plotted as a function of the substrate concentration [S]. The curve approaches V_{max} as the substrate concentration increases.

transition state closely enough to bind to the active site but cannot be converted to a functional product.

Substrate analogues are important tools in fighting infectious disease, and many have been developed to inhibit specific enzymes in pathogenic bacteria and viruses, usually targeting enzymes that we humans lack. For example, sulfa drugs resemble the folic acid precursor *para*-aminobenzoic acid. They can bind to and block the active site of the enzyme used by the bacterium to synthesize folic acid, which the bacterium requires for DNA synthesis. Likewise, azidothymidine (AZT), which is an antiviral medication, resembles the deoxythymidine molecule normally used by the human immunodeficiency virus (HIV) to synthesize DNA using viral reverse transcriptase. After binding to the active site of viral reverse transcriptase, AZT is incorporated into a growing strand of DNA but forms a "dead-end" molecule of DNA that cannot be elongated.

Inhibitors may be either *reversible* or *irreversible*. An **irreversible inhibitor** binds to the enzyme covalently, causing permanent loss of catalytic activity. Not surprisingly, irreversible inhibitors are usually toxic to cells. Ions of heavy metals are often irreversible inhibitors, as are nerve gas poisons and some insecticides. These substances can bind irreversibly to enzymes such as concentration in each tube. Of the nine reaction mixtures set up for this experiment, one tube is a negative control, designated the "reagent blank" (B), because it contains no glucose. The other tubes contain concentrations of glucose ranging from 0.05 to 0.40 m*M*. With all tubes prepared and maintained at some favorable temperature (25°C is often used), the reaction in each tube is initiated by adding a fixed amount of hexokinase.

The rate of product formation in each of the reaction mixtures can then be determined in several ways. In one common approach, a *spectrophotometer* is used to measure progress of the reaction. The spectrophotometer measures how much light of a particular wavelength is absorbed by a solution. If a reactant or product absorbs light at a characteristic wavelength, an increase in the products or a decrease in the reactants provides a measure of the progress of the reaction. Another common approach is to allow each reaction mixture to incubate for some short, fixed period of time, followed by a chemical assay for either substrate depletion or product accumulation. In this approach, the reaction's progress is often indirectly assessed using a colored indicator that can again be measured using a spectrophotometer. The measurement in Figure 6B-1 uses this second approach.

As Figure 6B-1 indicates, the initial velocity of the glucose consumption reaction for tubes 1–8 ranged from 2.5 to 7.3 μmol of glucose consumed per minute, with no detectable reaction in the blank (tube B). Plotting these initial reaction velocities as a function of glucose concentration generates the hyperbolic curve shown in Figure 6B-1. Although these data are idealized for illustrative purposes, most kinetic data generated by this approach do, in fact, fit a hyperbolic curve unless the enzyme has some special properties that cause departure from Michaelis–Menten kinetics.

Because neither V_{max} nor K_m can be easily determined from the values as plotted, reciprocals were calculated for each value of [S] and v_0 in Figure 6B-1 by dividing 1 by each value. Thus, the [S] values of 0.05–0.40 m*M* generate reciprocals of 20–2.5 mM^{-1}, and the v_0 values of 2.5–7.3 μmol/min give rise to reciprocals of 0.4–0.14 min/μmol. Because these are reciprocals, the data point representing the lowest concentration (tube 1) is farthest from the origin (0, 0), and each successive tube is represented by a point closer to the origin (**Figure 6B-2**).

When these data points are connected by a straight line, the *y*-intercept is found to be 0.1 min/μmol, and the *x*-intercept

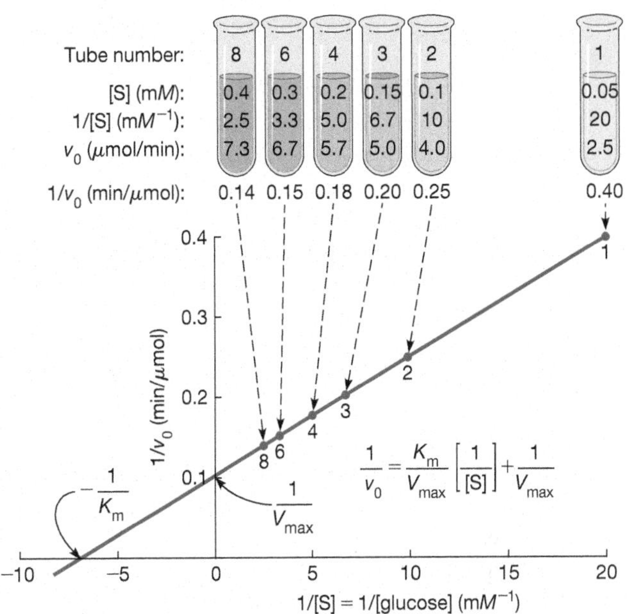

Figure 6B-2 Double-Reciprocal Plot for the Hexokinase Data. For each test tube in Figure 6B-1, $1/v_0$ was then plotted as a function of 1/[S].

is −6.7 mM^{-1}. From these intercepts, we can calculate that $V_{max} = 1/0.1 = 10\ \mu$mol/min and $K_m = -(1/-6.7) = 0.15$ m*M*. If we now go back to the Michaelis–Menten plot of Figure 6B-1, we will notice that the graph reaches one-half of the calculated value for V_{max} at a substrate concentration of 0.15 m*M*, so we can visually verify that 0.15 m*M* is the K_m of hexokinase for glucose.

QUESTION: The points on the graph representing [S] in Figure 6B-1 are equally spaced along the *x*-axis. Why may this not be the best set of concentrations for finding K_m and V_{max}? Can you suggest an improvement? Feel free to consider both the curve in Figure 6B-1 and the double-reciprocal plot in Figure 6B-2 in providing your answer.

acetylcholinesterase, an enzyme that is vital to the transmission of nerve impulses (see Chapter 22). Inhibition of acetylcholinesterase activity leads to rapid paralysis of vital functions and therefore to death. One such inhibitor is *diisopropyl fluorophosphate*, a nerve gas that binds covalently to the hydroxyl group of a critical serine at the active site of the enzyme, thereby rendering the enzyme molecule permanently inactive.

Some irreversible inhibitors can be used as therapeutic agents. For example, aspirin binds irreversibly to the enzyme cyclooxygenase-1 (COX-1), which produces prostaglandins and other signaling chemicals that cause inflammation, constriction of blood vessels, and platelet aggregation. Thus, aspirin is effective in relieving minor inflammation and headaches and has been recommended in low doses as a cardiovascular protectant. The antibiotic *penicillin* is an irreversible inhibitor of the enzyme needed for bacterial cell wall synthesis. Penicillin is therefore effective in treating bacterial infections because it prevents the bacterial cells from forming cell walls, thus blocking their growth and division. And because our cells lack a cell wall (and the enzyme that synthesizes it), penicillin is nontoxic to humans.

In contrast, a **reversible inhibitor** binds to an enzyme in a noncovalent, dissociable manner, such that the free and bound forms of the inhibitor exist in equilibrium with each other. We can represent such binding as

$$E + I \rightleftharpoons EI \qquad \textbf{(6-13)}$$

with E as the active free enzyme, I as the inhibitor, and EI as the inactive enzyme-inhibitor complex. Clearly, the fraction of the enzyme that is available to the cell in active form depends on the concentration of the inhibitor and the strength of the enzyme-inhibitor complex.

Reversible inhibitors vary in where and, in some cases, when they act on their target enzymes (**Figure 6-11**). **Competitive inhibitors** (see Figure 6-11a) bind to the active site of an enzyme and therefore compete directly with substrate molecules for the same site on the enzyme. In the language of Michaelis–Menten kinetics, this means that although V_{max} for the enzyme remains unchanged, the effective affinity of the enzyme for the normal substrate is reduced, leading to an apparent increase in K_m. This reduces enzyme efficiency because many of the active sites of the enzyme molecules are blocked by bound inhibitor molecules and thus cannot bind substrate molecules at the active site.

Other types of inhibitors bind the enzyme surface at a location *other* than the active site. They inhibit enzyme activity indirectly by causing a change in protein conformation that can either prevent substrate from binding to the active site or greatly reduce catalytic activity. The details of how such inhibitors act vary depending on when the inhibitor can bind the enzyme. **Uncompetitive inhibitors** only bind the enzyme after it is bound to substrate. They reduce the apparent V_{max} for the enzyme, while increasing the effective K_m. **Noncompetitive inhibitors** (Figure 6-11b), on the other hand, can bind free enzyme or enzyme-substrate complexes. Classical noncompetitive inhibitors decrease the apparent V_{max} for the reaction without affecting K_m. More commonly, however, noncompetitive inhibitors can not only decrease V_{max} but also increase K_m; this type of noncompetitive inhibitor is called a **mixed inhibitor**.

The ability to engineer chemicals that act as specific inhibitors of enzymes is an important component of modern drug design and has been aided greatly by the computational power of modern computers. In this approach, computers are used to analyze the three-dimensional structure of an enzyme active site in order to predict what types of molecules are likely to bind tightly to it and act as inhibitors. Scientists can then design a number of hypothetical inhibitors and test their likelihood of binding using computer models. In this way, it is not necessary to rely only upon inhibitors that are found in nature. Hundreds or even thousands of potential inhibitors can be designed and tested, and only the most promising are actually synthesized and evaluated experimentally.

CONCEPT CHECK 6.3

You work at a biotechnology company and are working with an enzyme that produces an anti-cancer compound from a proprietary substrate. To maximize production, how would you want to manipulate V_{max}, K_m, and k_{cat}?

6.4 Enzyme Regulation

To understand the role of enzymes in cellular function, we need to recognize that it is rarely in the cell's best interest to allow an enzyme to function at an indiscriminately high rate. Instead, the rates of enzyme-catalyzed reactions must be continuously adjusted to keep them finely tuned to the needs of the cell. An important aspect of that adjustment lies in the cell's ability to control enzyme activities with specificity and precision.

We have already encountered a variety of regulatory mechanisms, including alterations in temperature and pH, changes in substrate and product concentrations, and the

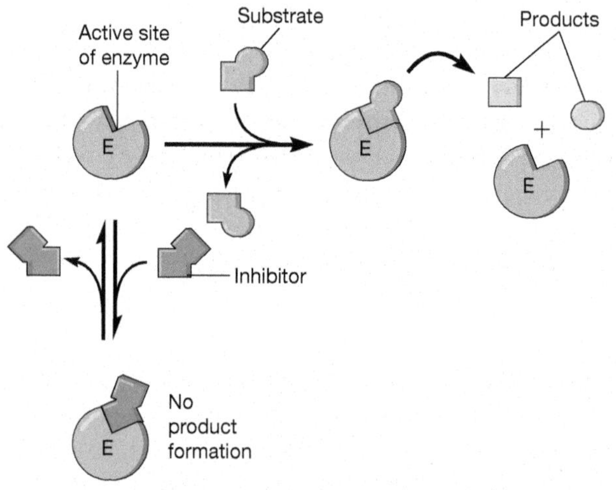

(a) Competitive inhibition. Inhibitor and substrate both bind to the active site of the enzyme. Binding of an inhibitor prevents substrate binding, thereby inhibiting enzyme activity.

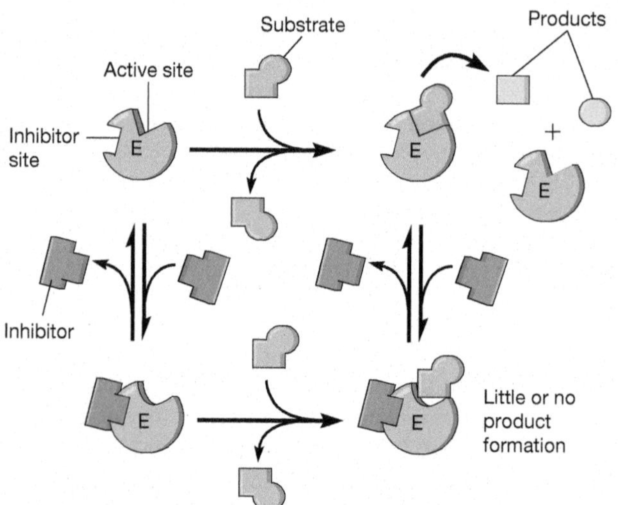

(b) Noncompetitive inhibition. Inhibitor and substrate bind to different sites on the enzyme. Binding of an inhibitor distorts the enzyme, inhibiting substrate binding or reducing catalytic activity.

Figure 6-11 Modes of Action of Competitive and Noncompetitive Inhibitors. Both **(a)** competitive inhibitors and **(b)** noncompetitive inhibitors bind reversibly to the enzyme (E). The two kinds of inhibitors differ in which site on the enzyme they bind to.

presence and concentration of inhibitors. Regulation that depends directly on the interactions of substrates and products with the enzyme is called **substrate-level regulation**. As the Michaelis–Menten equation makes clear, increases in substrate concentration result in higher reaction rates (see Figure 6-8). Conversely, increases in product concentration reduce the rate at which substrate is converted to product. (This inhibitory effect of product concentration is why v needs to be identified as the *initial* reaction velocity (v_0) in the Michaelis–Menten equation, as given by Equation 6-7.)

Substrate-level regulation is an important control mechanism in cells, but it is not sufficient for the regulation of multistep reaction sequences. For such pathways, enzymes are regulated by other mechanisms as well. Two of the most important of these are *allosteric regulation* and *covalent modification.* These mechanisms allow cells to turn enzymes on or off or to fine-tune their reaction rates.

Almost invariably, the enzyme that is most highly regulated catalyzes the first step of a multistep sequence and acts like a gatekeeper for an entire biochemical pathway. By increasing or reducing the rate at which the first step functions, the whole sequence is effectively controlled. Pathways that are regulated in this way include those required to break down large molecules (such as sugars, fats, or amino acids) and pathways that lead to the synthesis of substances needed by the cell (such as amino acids and nucleotides). For now, we will discuss allosteric regulation and covalent modification at an introductory level. We will return to these mechanisms as we encounter specific examples in later chapters.

Allosteric Enzymes Are Regulated by Molecules Other than Reactants and Products

The single most important control mechanism for adjusting the rates of enzyme-catalyzed reactions to meet cellular needs is *allosteric regulation.* To understand this mode of regulation, consider the pathway by which a cell converts some precursor A into some final product P via a series of intermediates B, C, and D in a sequence of reactions catalyzed by enzymes E_1, E_2, E_3, and E_4, respectively:

$$A \xrightarrow[E_1]{} B \xrightarrow[E_2]{} C \xrightarrow[E_3]{} D \xrightarrow[E_4]{} P \qquad \textbf{(6-14)}$$

Product P could, for example, be an amino acid needed by the cell for protein synthesis, and A could be some common cellular component that serves as the starting point for the specific reaction sequence leading to P.

Feedback Inhibition. If allowed to proceed at a constant, unrestrained rate, the pathway shown in reaction sequence 6-14 can convert large amounts of A to P, with possible adverse effects resulting from a depletion of A or an excessive accumulation of P. Clearly, the best interests of the cell are served when the pathway is functioning instead at a rate that is carefully tuned to the cellular need for P.

Somehow, the enzymes of this pathway must be responsive to the cellular level of the product P in somewhat the same way that a furnace needs to be responsive to the temperature of the rooms it is intended to heat. In the latter case, a thermostat provides the necessary regulatory link between the furnace and its "product," heat. If there is too much heat, the thermostat turns the furnace off, inhibiting heat production. If heat is needed, this inhibition is relieved due to the lack of heat. In our enzyme example, the desired regulation is possible because the product P is a specific inhibitor of E_1, the enzyme that catalyzes the first reaction in the sequence.

This phenomenon is called **feedback inhibition** and is represented by the dashed arrow that connects the product P to enzyme E_1 in the following reaction sequence:

$$A \xrightarrow[E_1]{} B \xrightarrow[E_2]{} C \xrightarrow[E_3]{} D \xrightarrow[E_4]{} P \qquad \textbf{(6-15)}$$

Feedback inhibition of E_1 by P

Feedback inhibition is one of the most common mechanisms used by cells to ensure that the activities of reaction sequences are adjusted to cellular needs.

Figure 6-12 on the next page provides a specific example of such a pathway—the five-step sequence whereby the amino acid *isoleucine* is synthesized from *threonine,* another amino acid. In this case, the first enzyme in the pathway, *threonine deaminase,* is regulated by the cellular concentration of isoleucine—the pathway end-product. If isoleucine is being used by the cell (in the synthesis of proteins, most likely), the isoleucine concentration will be low and the cell will need more. Under these conditions, threonine deaminase is active, and the pathway functions to produce more isoleucine. If the need for isoleucine decreases, isoleucine will begin to accumulate in the cell. This increase in its concentration will lead to inhibition of threonine deaminase and hence to a reduced rate of isoleucine synthesis.

Allosteric Regulation. How can the first enzyme in a pathway (for example, enzyme E_1 in reaction sequence 6-15) be sensitive to the concentration of a substance P that is neither its substrate nor its immediate product? The answer to this question was first proposed in 1963 by Jacques Monod, Jean-Pierre Changeux, and François Jacob, who studied the regulation of lactose uptake and degradation in bacteria. Their model went on to become the foundation for our understanding of **allosteric regulation**. The term *allosteric* derives from the Greek for "another shape (or form)," thereby indicating that all enzymes capable of allosteric regulation can exist in two different forms.

In one of the two forms, the enzyme has a high affinity for its substrate(s), leading to high activity. In the other form, it has little or no affinity for its substrate, and little or no catalytic activity. Enzymes with this property are called **allosteric enzymes**. Whether the active or inactive form of an allosteric enzyme is favored depends on the cellular concentration of the appropriate regulatory substance, called an **allosteric effector**. In the case of isoleucine synthesis, the allosteric effector is isoleucine, and the allosteric enzyme is threonine deaminase. More generally, an allosteric effector is a small organic molecule that regulates the activity of an enzyme for which it is neither the substrate nor the immediate product.

An allosteric effector influences enzyme activity by binding to the enzyme. The effector binds to the enzyme because

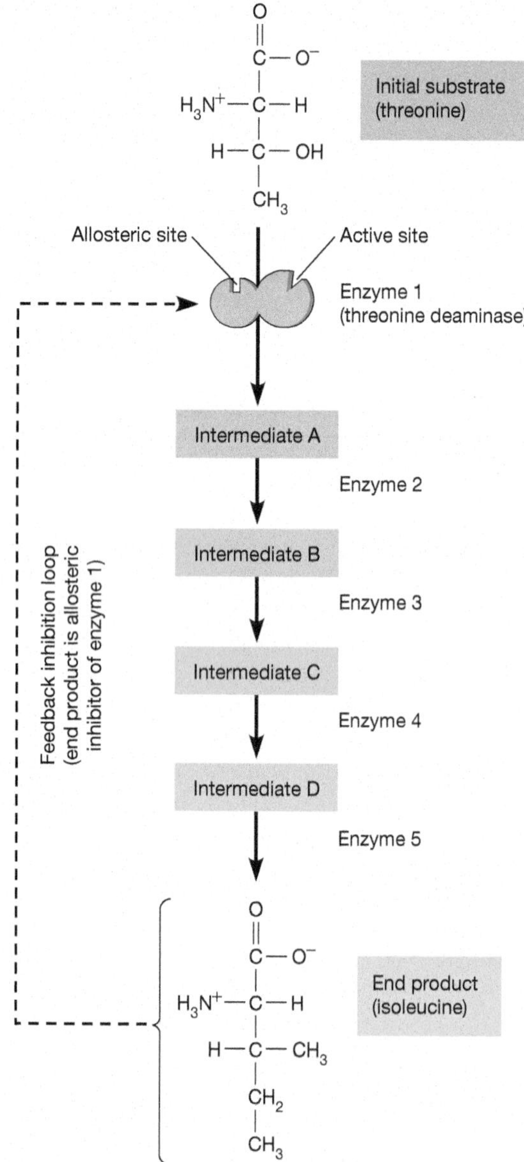

Figure 6-12 Allosteric Regulation of Enzyme Activity. Feedback inhibition is seen in the biosynthetic pathway by which the amino acid isoleucine is synthesized from the amino acid threonine. The first enzyme in the pathway, threonine deaminase, is allosterically inhibited by the end-product isoleucine, which binds to the enzyme at an allosteric site distinct from the active site.

of the presence on the enzyme surface of an **allosteric (or regulatory) site** that is distinct from the active site at which the catalytic event occurs. Thus, a distinguishing feature of all allosteric enzymes (and other allosteric proteins, as well) is the presence on the enzyme surface of an *active site* to which the substrate binds and an *allosteric site* to which the effector binds. In fact, some allosteric enzymes have multiple allosteric sites, each capable of recognizing a different effector.

An effector may be either an **allosteric inhibitor** or an **allosteric activator**, depending on the effect it has when bound to the allosteric site on the enzyme (**Figure 6-13**). The binding of an allosteric inhibitor shifts the enzyme into the low-affinity state (Figure 6-13a). The binding of an allosteric activator, on the other hand, shifts the enzyme to the high-affinity state (Figure 6-13b). In this way, binding of the effector to the allosteric site either decreases or increases the catalytic activity of the enzyme.

Most allosteric enzymes are large, multisubunit proteins with an active site or an allosteric site on each subunit. Thus, quaternary protein structure is important for these enzymes. Typically, the active sites and allosteric sites are on different subunits of the protein, which are referred to as **catalytic subunits** and **regulatory subunits**, respectively (notice the C and R subunits of the enzyme molecules shown in Figure 6-13). This means, in turn, that the binding of effector molecules to the allosteric sites affects not just the shape of the regulatory subunits but also that of the catalytic subunits.

Allosteric Enzymes Exhibit Cooperative Interactions Between Subunits

Many allosteric enzymes exhibit a property known as **cooperativity**. This means that, as multiple catalytic sites on an enzyme bind substrate molecules, the enzyme undergoes conformational changes that affect the affinity of the remaining catalytic sites for substrate. Some enzymes show *positive cooperativity*, in which the binding of a substrate molecule to one subunit increases the affinity of other subunits for substrate. For example, when one subunit of hemoglobin binds oxygen, this increases the binding of oxygen to the other subunits. Other enzymes show *negative cooperativity*, in which substrate binding to one subunit reduces the affinity of the other sites for substrate.

The cooperativity effect enables cells to produce enzymes that are more sensitive or less sensitive to changes in substrate concentration than would otherwise be predicted by Michaelis–Menten kinetics. Positive cooperativity causes an enzyme's catalytic activity to increase more rapidly than expected as the substrate concentration is increased, whereas negative cooperativity means that enzyme activity increases more slowly than expected.

Enzymes Can Also Be Regulated by the Addition or Removal of Chemical Groups

In addition to allosteric regulation, many enzymes are subject to control by **covalent modification**. In this form of regulation, an enzyme's activity is affected by the addition or removal of specific chemical groups via covalent bonding. Common modifications include the addition of phosphoryl groups, methyl groups, acetyl groups, or derivatives of nucleotides. Some of these modifications can be reversed, whereas others cannot. In each case, the effect of the modification is to activate or to inactivate the enzyme—or at least to adjust its activity upward or downward.

Phosphorylation/Dephosphorylation. One of the most frequently encountered and best understood covalent modifications involves the reversible addition of phosphoryl groups. The addition of phosphoryl groups is called **phosphorylation** and occurs most commonly by transfer of the phosphoryl group from ATP to the hydroxyl group of a serine, threonine, or tyrosine residue in the protein. Enzymes that catalyze the phosphorylation of other enzymes (or of other proteins) are called **protein kinases**. The reversal of this process is called **dephosphorylation**, which removes a phosphoryl group

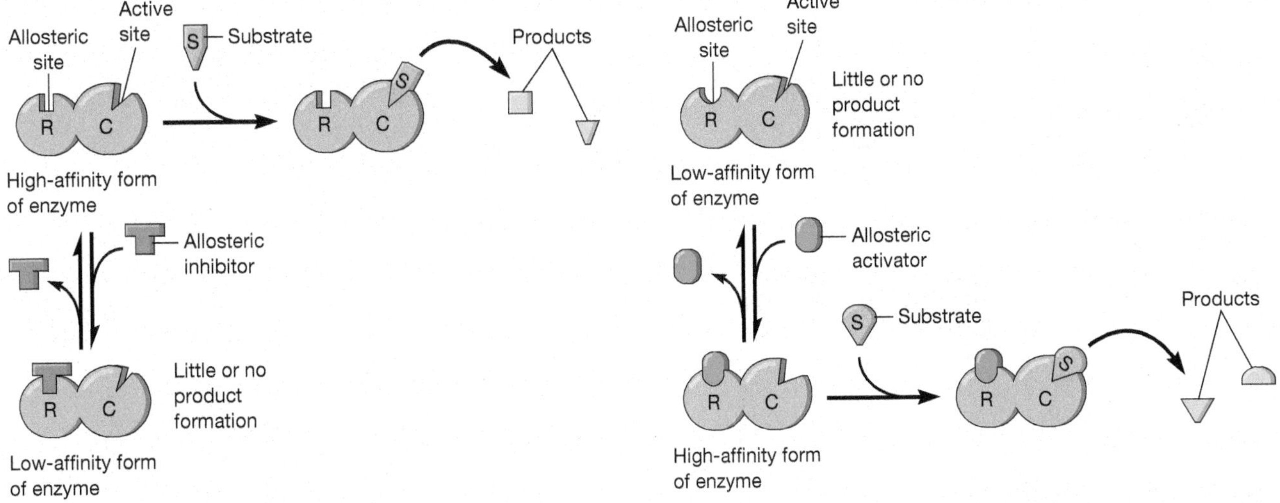

(a) Allosteric inhibition. An allosterically inhibited enzyme is active in the uncomplexed form, which has a high affinity for substrate (S). Binding of an allosteric inhibitor (red) stabilizes the enzyme in its low-affinity form, resulting in little or no activity.

(b) Allosteric activation. An allosterically activated enzyme is inactive in the uncomplexed form, which has a low affinity for substrate (S). Binding of an allosteric activator (green) stabilizes the enzyme in its high-affinity form, resulting in enzyme activity.

Figure 6-13 Mechanisms of Allosteric Inhibition and Activation. An allosteric enzyme has one or more catalytic subunits (C) with an active site and one or more regulatory subunits (R) with an allosteric site. One form of the enzyme has a high affinity for its substrate and therefore a high likelihood of product formation. The other site has a low affinity for substrate and a correspondingly low likelihood of product formation. The predominant form of the enzyme depends on the concentration of its allosteric effector(s).

from a phosphorylated protein and is catalyzed by enzymes called **protein phosphatases**. Depending on the enzyme, phosphorylation may activate or inhibit the enzyme.

Enzyme regulation by reversible phosphorylation/dephosphorylation was discovered by Edmond Fischer and Edwin Krebs (not Hans Krebs, namesake of the Krebs cycle) at the University of Washington in the 1950s. They were awarded the 1992 Nobel Prize in Physiology or Medicine for their groundbreaking work with *glycogen phosphorylase*, a glycogen-degrading enzyme found in liver and skeletal muscle cells described in **Figure 6-14** on page 150. In muscle cells, glycogen phosphorylase activity provides glucose as an energy source for muscle contraction, and in liver cells it provides glucose for secretion to help maintain a constant blood glucose level. Glycogen phosphorylase breaks down glycogen by successive removal of glucose units as glucose-1-phosphate (Figure 6-14a). Regulation of this dimeric enzyme is possible because glycogen phosphorylase can exist in one of two interconvertible forms—an active *a* form and an inactive *b* form (Figure 6-14b).

When glycogen breakdown is required in the cell, the inactive *b* form of the enzyme is converted into the active *a* form by the addition of a phosphoryl group to a particular serine on each of the two subunits of the phosphorylase molecule. The reaction is catalyzed by *phosphorylase kinase* and results in a conformational change of phosphorylase to the active form. When glycogen breakdown is no longer needed, the phosphoryl groups are removed from phosphorylase *a* by the enzyme *phosphorylase phosphatase.*

The muscle and liver forms of glycogen phosphorylase, known as *isozymes*, differ subtly in their manner of regulation. Liver glycogen phosphorylase is an allosteric enzyme. So besides being regulated by the phosphorylation/dephosphorylation (covalent modification) mechanism shown in Figure 6-14b, liver glycogen phosphorylase can be inhibited by glucose and ATP and activated by AMP. For example, glucose binding to active liver glycogen phosphorylase *a* will inactivate it, blocking glycogen breakdown when glucose accumulates faster than needed. The existence of two levels of regulation for glycogen phosphorylase illustrates an important aspect of enzyme regulation. Many enzymes are controlled by two or more regulatory mechanisms, thereby enabling the cell to make appropriate responses to a variety of situations.

Proteolytic Cleavage. A different kind of covalent regulation of enzymes involves the one-time, irreversible removal of a portion of the polypeptide chain by an appropriate proteolytic (protein-degrading) enzyme to activate the enzyme. This kind of modification, called **proteolytic cleavage**, is exemplified especially well by trypsin, chymotrypsin, and carboxypeptidase—proteolytic enzymes of the pancreas. After being synthesized in the pancreas, these enzymes are secreted into the duodenum of the small intestine in response to a hormonal signal. These proteases can digest almost all ingested proteins into free amino acids, which are then absorbed by the intestinal epithelial cells.

Pancreatic proteases are not synthesized in their active form. That would likely cause problems for the cells of the pancreas, which must protect themselves against their own proteolytic enzymes. Instead, each of these enzymes is synthesized and secreted as a slightly larger, catalytically inactive molecule called a *zymogen* (**Figure 6-15** on page 150). Zymogens must themselves be cleaved proteolytically to yield active enzymes. For example, trypsin is synthesized initially as a zymogen called *trypsinogen*. When trypsinogen reaches the duodenum, it is activated by removal of six amino acids from its N-terminus

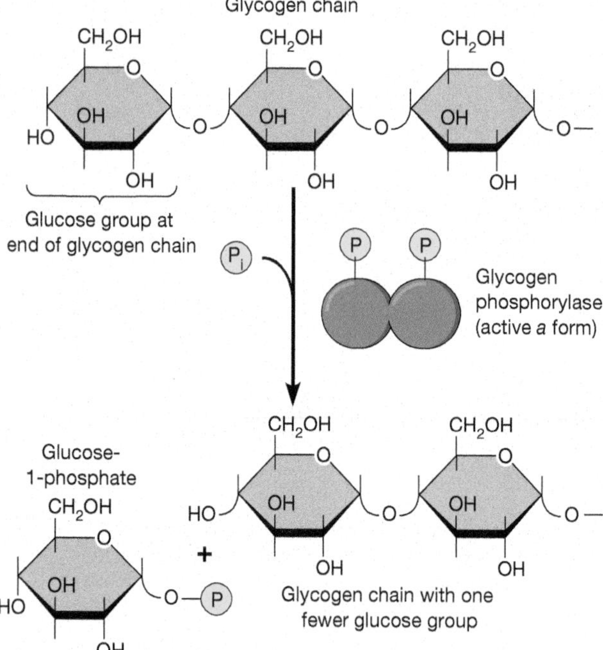

(a) Activated glycogen phosphorylase cleaves glycogen.

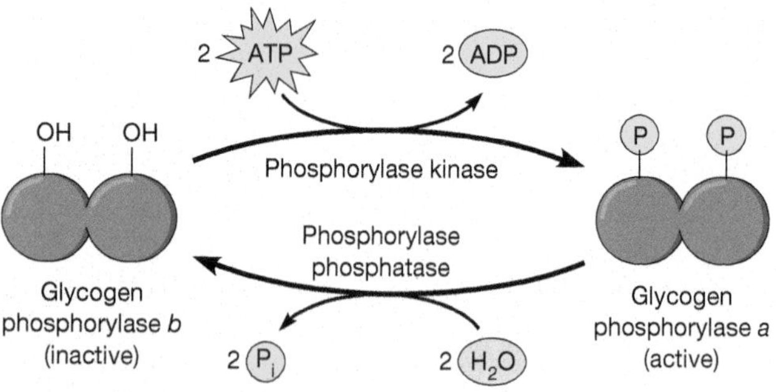

(b) Phosphorylation (top) or dephosphorylation (bottom) reversibly activates or inhibits glycogen phosphorylase.

Figure 6-14 The Action and Regulation of Glycogen Phosphorylase. (a) Activated glycogen phosphorylase catalyzes cleavage of glucose-1-phosphate from glycogen. **(b)** Phosphorylation by phosphorylase kinase converts glycogen phosphorylase to its active *a* form. Dephosphorylation by phosphorylase phosphatase converts glycogen phosphorylase to its inactive *b* form.

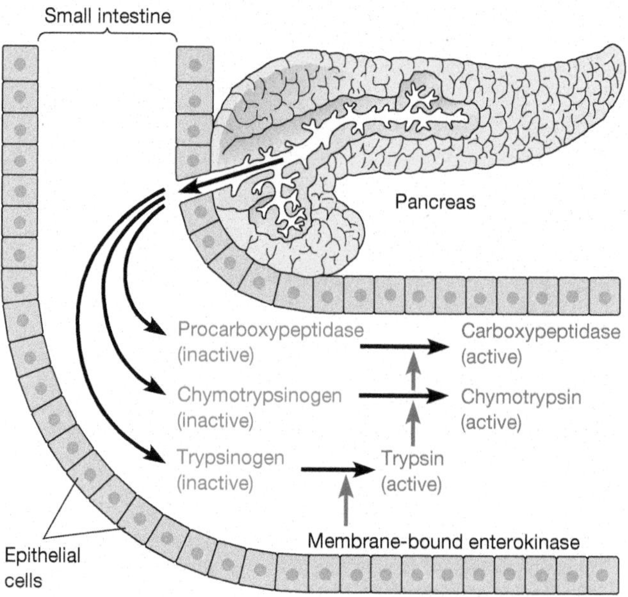

Figure 6-15 Activation of Pancreatic Zymogens by Proteolytic Cleavage. Pancreatic proteases are synthesized and secreted into the small intestine as inactive precursors known as zymogens. Procarboxypeptidase, trypsinogen, and chymotrypsinogen are zymogens. Activation of trypsinogen to trypsin requires removal of a hexapeptide segment by enterokinase, a membrane-bound duodenal enzyme. Trypsin then activates other zymogens by proteolytic cleavage. Procarboxypeptidase is activated by a single cleavage event, whereas the activation of chymotrypsinogen is a somewhat more complicated two-step process, the details of which are not shown here.

catalyzed by *enterokinase*, a membrane-bound protease produced by duodenal cells. The now active trypsin then activates other zymogens by specific proteolytic cleavages.

MAKE CONNECTIONS 6.2

What type of chemical reaction is involved in breakdown of glycogen and proteolysis? (Fig. 2-16)

CONCEPT CHECK 6.4

Why do enzymes need to be regulated? By what different mechanisms can this regulation be achieved?

Summary of Key Points

Mastering™ Biology For activities, animations, and review quizzes, go to the study area at www.masteringbiology.com.

6.1 Activation Energy and the Metastable State

- Although thermodynamics allows us to assess the feasibility of a reaction, it says nothing about the likelihood that the reaction will actually occur at a reasonable rate in the cell.
- For a given chemical reaction to occur in the cell, substrates must reach the transition state, which has a higher free energy than either the substrates or products. Reaching the transition state requires the input of enough energy to overcome the activation energy barrier.
- Because of the activation energy barrier, most biological compounds exist in an unreactive, metastable state. To ensure that the activation energy requirement is met and the transition state is achieved, a catalyst is required; in biological systems the catalyst is always an enzyme.

6.2 Enzymes as Biological Catalysts

- Catalysts, whether inorganic or organic, act by forming transient complexes with substrate molecules that lower the activation energy barrier and rapidly increase the rate of the particular reaction.
- Chemical reactions in cells are catalyzed by enzymes, which in some cases require organic or inorganic cofactors for activity. The vast majority of enzymes are proteins, but a few are composed of RNA and are known as ribozymes.
- Enzymes are exquisitely specific, either for a single specific substrate or for a class of closely related compounds. This is because the actual catalytic process takes place at the active site—a pocket or groove on the enzyme surface that only the correct substrates will fit into.

- The active site is composed of specific, noncontiguous amino acids that become positioned near each other as the protein folds into its tertiary structure. The amino acids that make up the active site are responsible for substrate binding, substrate activation, and catalysis.
- Binding of the appropriate substrate at the active site causes a change in the shape of the enzyme and substrate known as induced fit. These shape changes facilitate substrate activation, often by distorting one or more bonds in the substrate, by bringing necessary amino acid side chains into the active site, or by transferring protons and/or electrons between the enzyme and substrate.
- Although it was long thought that all enzymes were proteins, we now recognize that certain RNA molecules known as ribozymes can carry out enzymatic reactions. Examples include catalytic pre-rRNA molecules, RNA components of protein enzymes, and the rRNA component of the large ribosomal subunit.
- The discovery of ribozymes has changed the way biologists think about the origin of life on Earth because RNA molecules, unlike DNA or proteins, may be able to both carry information and replicate themselves.

6.3 Enzyme Kinetics

- Many enzyme-catalyzed reactions follow Michaelis–Menten kinetics, characterized by a hyperbolic relationship between the initial reaction velocity v_0 and the substrate concentration [S].
- The upper limit on velocity is called V_{max}, and the substrate concentration needed to reach one-half of this maximum velocity is termed the Michaelis constant, K_m. The hyperbolic relationship between v_0 and [S] can be linearized by a double-reciprocal equation and plot, from which V_{max} and K_m can be determined graphically.
- Enzyme activity is sensitive to temperature, pH, and the ionic environment. Enzyme activity is also influenced by substrate availability, products, alternative substrates, substrate analogues, drugs, and toxins, most of which have an inhibitory effect.
- Irreversible inhibition involves covalent bonding of the inhibitor to the enzyme surface, permanently disabling the enzyme. A reversible inhibitor, on the other hand, binds noncovalently to an enzyme in a reversible manner, either at the active site (competitive inhibition) or elsewhere on the enzyme surface (noncompetitive inhibition).

6.4 Enzyme Regulation

- Enzymes must be regulated to adjust their activity levels to cellular needs. Substrate-level regulation involves the effects of substrate and product concentrations on the reaction rate. Additional control mechanisms include allosteric regulation and covalent modification.
- Most allosterically regulated enzymes catalyze the first step in a reaction sequence; many are multisubunit proteins with both catalytic subunits and regulatory subunits. Each of the catalytic subunits has an active site that recognizes substrates, whereas each regulatory subunit has one or more allosteric sites that recognize specific effector molecules.
- Allosteric enzymes can exist in a low-activity or a high-activity form. Allosteric effectors bind to allosteric enzymes and will either inhibit or activate the enzyme.
- Enzyme inhibitors can act either irreversibly or reversibly. Irreversible inhibitors bind covalently to the enzyme, causing a permanent loss of activity. Reversible inhibitors bind noncovalently, either to the active site (competitive inhibitors such as substrate analogues) or to a separate site on the enzyme (noncompetitive inhibitors).
- Enzymes can also be regulated by covalent modification. The most common covalent modifications include phosphorylation, as seen with glycogen phosphorylase, and proteolytic cleavage, as occurs in the activation of proteolytic pancreatic zymogens.

Mastering™ Biology For activities, animations, and review quizzes, go to the study area at www.masteringbiology.com.

Problem Set

6-1 The Need for Enzymes. You should now be in a position to appreciate the difference between the thermodynamic feasibility of a reaction and the likelihood that it will actually proceed.

(a) Many reactions that are thermodynamically possible do not occur at an appreciable rate because of the activation energy required for the reactants to achieve the transition state. In molecular terms, what does this mean?

(b) One way to meet the activation energy requirement is by an input of heat, which in some cases need only be an initial, transient input. Give an example, and explain what this accomplishes in molecular terms.

(c) An alternative solution is to lower the activation energy barrier. What does it mean in molecular terms to say that a catalyst lowers the activation energy barrier of a reaction?

(d) Organic chemists often use inorganic catalysts such as nickel, platinum, or cations in their reactions, whereas cells use proteins called enzymes. What advantages can you see to the use of enzymes? Can you think of any disadvantages?

(e) Some enzymes can transfer a hydrogen atom without lowering the activation energy barrier. How is this possible?

6-2 Activation Energy. As shown in Reaction 6-2, hydrogen peroxide, H_2O_2, decomposes to H_2O and O_2. The activation energy, E_A, for the uncatalyzed reaction at 20°C is 18 kcal/mol. The reaction can be catalyzed either by an inorganic catalyst such as platinum (E_A = 13 kcal/mol) or by the enzyme *catalase* (E_A = 7 kcal/mol).

(a) Draw an activation energy diagram for this reaction under catalyzed and uncatalyzed conditions, and explain what it means for the activation energy to be lowered from 18 to 13 kcal/mol by platinum but from 18 to 7 kcal/mol by catalase.

(b) Suggest two properties of catalase that make it a more suitable intracellular catalyst than platinum.

(c) Suggest yet another way that the rate of hydrogen peroxide decomposition can be accelerated. Is this a suitable means of increasing reaction rates within cells? Why or why not?

6-3 QUANTITATIVE Rate Enhancement by Catalysts. The decomposition of H_2O_2 to H_2O and O_2 shown in Reaction 6-2 can be catalyzed either by an inorganic catalyst (such as ferric ions) or by the enzyme catalase. Compared with the uncatalyzed rate, this reaction proceeds about 30,000 times faster in the presence of ferric ions but about 100,000,000 times faster in the presence of catalase, an iron-containing enzyme. Assume that 1 μg of catalase can decompose a given quantity of H_2O_2 in 1 minute at 25°C and that all reactions are carried out under standard conditions.

(a) How long would it take for the same quantity of H_2O_2 to be decomposed in the presence of an amount of ferric ions equivalent to the iron content of 1 μg of catalase?

(b) How long would it take for the same quantity of H_2O_2 to decompose in the absence of a catalyst?

(c) Explain how these calculations illustrate the indispensability of catalysts and the superiority of enzymes over inorganic catalysts.

6-4 Temperature and pH Effects. Figure 6-4 illustrates the effects of temperature and pH on enzyme activity. In general, the activity of a specific enzyme is highest at the temperature and pH that are characteristic of the environment in which the enzyme normally functions.

(a) Explain the shapes of the curves in Figure 6-4 in terms of the major chemical or physical factors that affect enzyme activity.

(b) For each enzyme in Figure 6-4, suggest the adaptive advantage of having the enzyme activity profile shown in the figure.

(c) Some enzymes have a very flat pH profile—that is, they have essentially the same activity over a broad pH range. How might you explain this observation?

6-5 Meaning of Michaelis–Menten Plots. Figure 6-16 represents a Michaelis–Menten plot for a typical enzyme, with initial reaction velocity plotted as a function of substrate concentration. Three regions of the curve are identified by the letters A, B, and C. For each of the statements that follow, indicate with a single letter which of the three regions of the curve fits the statement best. A given letter can be used more than once.

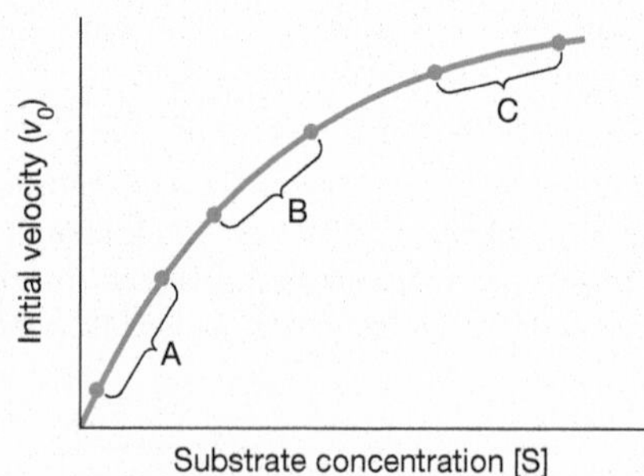

Figure 6-16 Analysis of the Michaelis–Menten Plot. See Problem 6-5.

(a) The active site of an enzyme molecule is occupied by substrate most of the time.

(b) The active site of an enzyme molecule is free most of the time.

(c) This is the range of substrate concentration in which most enzymes usually function in normal cells.

(d) This range includes the point (K_m, $V_{max}/2$).

(e) Reaction velocity is limited mainly by the number of enzyme molecules present.

(f) Reaction velocity is limited mainly by the number of substrate molecules present.

6-6 QUANTITATIVE Enzyme Kinetics. The enzyme β-galactosidase catalyzes the hydrolysis of the disaccharide lactose into its component monosaccharides:

$$\text{lactose} + H_2O \xrightarrow[\beta\text{-galactosidase}]{} \text{glucose} + \text{galactose} \qquad \textbf{(6-16)}$$

To determine V_{max} and K_m of β-galactosidase for lactose, the same amount of enzyme (1 μg per test tube) was incubated with a series of lactose concentrations under conditions where product concentrations remained negligible. At each lactose concentration, the initial reaction velocity was determined by assaying for the amount of lactose remaining at the end of the assay. The following data were obtained.

Lactose concentration (mM)	Rate of lactose consumption (μmol/min)
1	10.0
2	16.7
4	25.0
8	33.3
16	40.0
32	44.4

(a) Why is it necessary to specify that product concentrations remained negligible during the course of the reaction?

(b) Plot v_0 (initial rate of lactose consumption) versus [S] (lactose concentration). Why is it that when the lactose concentration is doubled, the increase in velocity is always less than twofold?

(c) Calculate $1/v_0$ and $1/[S]$ for each entry on the data table, and plot $1/v_0$ versus $1/[S]$.

(d) Determine K_m and V_{max} from your double-reciprocal plot.

(e) On the same graph as part b, plot the results you would expect if each test tube contained only 0.5 μg of enzyme. Explain your graph.

6-7 QUANTITATIVE More Enzyme Kinetics. The galactose formed in Reaction 6-16 can be phosphorylated by the transfer of a phosphoryl group from ATP, a reaction catalyzed by the enzyme galactokinase:

$$\text{galactose} + \text{ATP} \xrightarrow[\text{galactokinase}]{} \text{galactose-1-phosphate} + \text{ADP} \qquad \textbf{(6-17)}$$

Assume that you have isolated the galactokinase enzyme and have determined its kinetic parameters by varying the concentration of galactose in the presence of a constant, high (that is, saturating) concentration of ATP. The double-reciprocal (Lineweaver–Burk) plot of the data is shown as **Figure 6-17**.

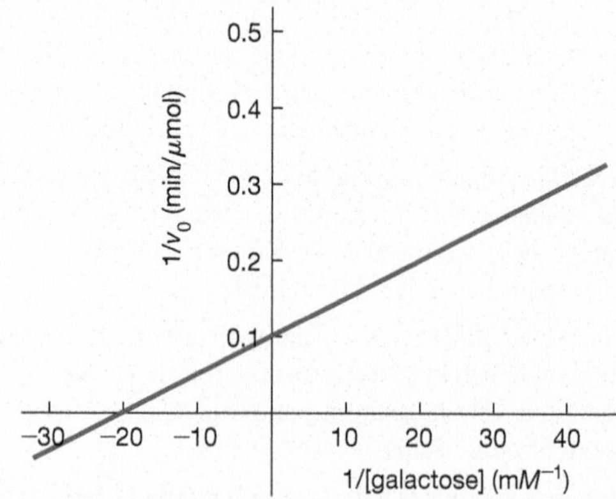

Figure 6-17 Double-Reciprocal Plot for the Enzyme Galactokinase. See Problem 6-7.

(a) What is the K_m of galactokinase for galactose under these assay conditions? What does K_m tell us about the enzyme?

(b) What is the V_{max} of the enzyme under these assay conditions? What does V_{max} tell us about the enzyme?

(c) Assume that you now repeat the experiment, but with the ATP concentration varied and galactose present at a constant, high concentration. Assuming that all other conditions are maintained as before, would you expect to get the same V_{max} value as in part b? Why or why not?

(d) In the experiment described in part c, the K_m value turned out to be very different from the value determined in part b. Can you explain why?

6-8 QUANTITATIVE **Turnover Number.** Carbonic anhydrase catalyzes the reversible hydration of carbon dioxide to form bicarbonate ion:

$$CO_2 + H_2O \rightleftharpoons HCO_3^- + H^+ \quad \textbf{(6-18)}$$

This reaction is important in the transport of carbon dioxide from body tissues to the lungs by red blood cells (see Chapter 8). Carbonic anhydrase has a molecular weight of 30,000 daltons and a turnover number (k_{cat} value) of 1×10^6 sec^{-1}. Assume that you are given 1 mL of a solution containing 2.0 μg of pure carbonic anhydrase.

(a) At what rate (in millimoles of CO_2 consumed per second) will you expect this reaction to proceed under optimal conditions?

(b) Assuming standard temperature and pressure, how much CO_2 is that in milliliters per second?

6-9 Inhibitors: Wrong Again. For each of the following false statements, change the statement to make it true.

(a) Diisopropyl fluorophosphate binds covalently to the hydroxyl group of a specific amino acid residue of its target enzyme and is therefore almost certainly an allosteric effector.

(b) The enzyme hexokinase is inhibited by its own product, glucose-6-phosphate, and is therefore an example of feedback inhibition.

(c) Glycogen synthase, like glycogen phosphorylase, is active in the phosphorylated form and inactive in the dephosphorylated form.

(d) An enzyme that is subject to allosteric activation is most likely to catalyze the final reaction in a biosynthetic pathway.

(e) If researchers claim that an enzyme is allosterically activated by compound A and allosterically inhibited by compound B, one of these claims must be wrong.

6-10 DATA ANALYSIS **What Type of Inhibition?** A Lineweaver–Burk plot was generated for an enzyme and its substrate in the presence and absence of an inhibitor (**Figure 6-18**). What kind of inhibitor is this likely to be? Explain your reasoning.

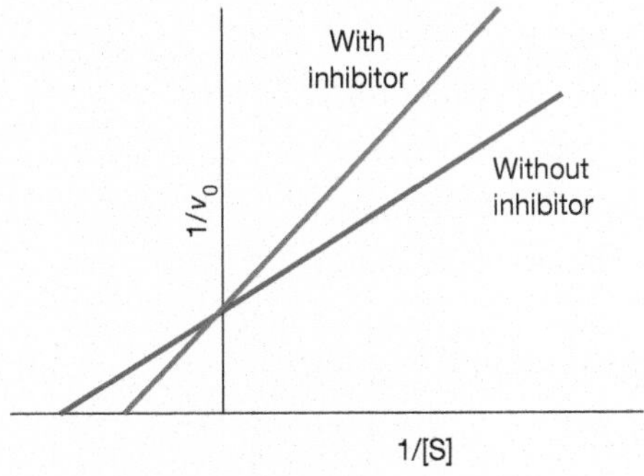

Figure 6-18 Lineweaver–Burk Plots for an Enzyme with and without Inhibitor. See Problem 6-10.

6-11 Mucking Up Mucinase. A mucinase enzyme was discovered that breaks down a glycoprotein in mucous membranes and contributes to bacterial vaginosis (*J. Clin. Microbiol.* 43:5504). You are a research pathologist testing a new inhibitor of this enzyme that you have discovered, and you want to design experiments to understand the nature of this inhibition. You have a supply of the normal glycoprotein substrate, the inhibitor, and an assay to measure product formation.

(a) How might you determine whether the inhibition is reversible or irreversible?

(b) If you find that the inhibition is reversible, how would you determine whether the inhibition is competitive or noncompetitive?

6-12 Derivation of the Michaelis–Menten Equation. For the enzyme-catalyzed reaction in which a substrate S is converted into a product P (see Reaction 6-5), velocity can be defined as the disappearance of substrate or the appearance of product per unit time:

$$v_0 = -\frac{d[S]}{dt} + \frac{d[P]}{dt} \quad \textbf{(6-19)}$$

Beginning with this definition and restricting your consideration to the initial stage of the reaction when [P] is essentially zero, derive the Michaelis–Menten equation (see Equation 6-7). The following points may help you in your derivation:

- Begin by expressing the rate equations for $d[S]/dt$, $d[P]/dt$, and $d[ES]/dt$ in terms of concentrations and rate constants.
- Assume a steady state at which the enzyme-substrate complex shown in Reaction 6-6 is being broken down at the same rate as it is being formed such that the net rate of change, $d[ES]/dt$, is zero.
- Note that the total amount of enzyme present, E_t, is the sum of the free form, E_f, plus the amount of complexed enzyme ES: $E_t = E_f + ES$.
- When you get that far, note that V_{max} and K_m can be defined as follows:

$$V_{max} = k_3[E_t] \text{ and } K_m = \frac{k_2 + k_3}{k_1} \quad \textbf{(6-20)}$$

6-13 QUANTITATIVE **Monkeys and Peanuts Revisited.** Recall the "enzyme-catalyzed reaction" of monkeys shelling peanuts earlier in the chapter. Here we will further increase the peanut concentration and determine the effect on the kinetic parameters of this animal "enzyme."

(a) Let's again triple the concentration of peanuts to 9 peanuts per square meter. Add another column to the following table and determine the overall rate of shelling for the ten monkeys.

[S] = Concentration of peanuts (peanuts/m^2)	1	3
Time required per peanut:		
To find (sec/peanut)	9	3
To shell (sec/peanut)	1	1
Total (sec/peanut)	10	4
Rate of shelling (v):		
Per monkey (peanut/sec)	0.10	0.25
Total (peanut/sec)	1.0	2.5

(b) How did this second tripling of [S] affect the overall rate? How does this compare to the effect of the first tripling?

(c) Now increase [S] by another tenfold to 90 peanuts per square meter. Add another column to the table and determine the overall rate of shelling for the ten monkeys.

(d) How did the tenfold increase in [S] affect the overall rate compared to the effect of the previous increases?

(e) Graph all the results as v on the y-axis (suggested scale: 0–10 peanuts/sec) versus [S] on the x-axis (suggested scale: 0–100 peanuts/m^2). What do you predict will be the V_{max}? Can you explain this logically? What is the K_m?

(f) What is the K_m for this shelling reaction?

7 Membranes: Their Structure, Function, and Chemistry

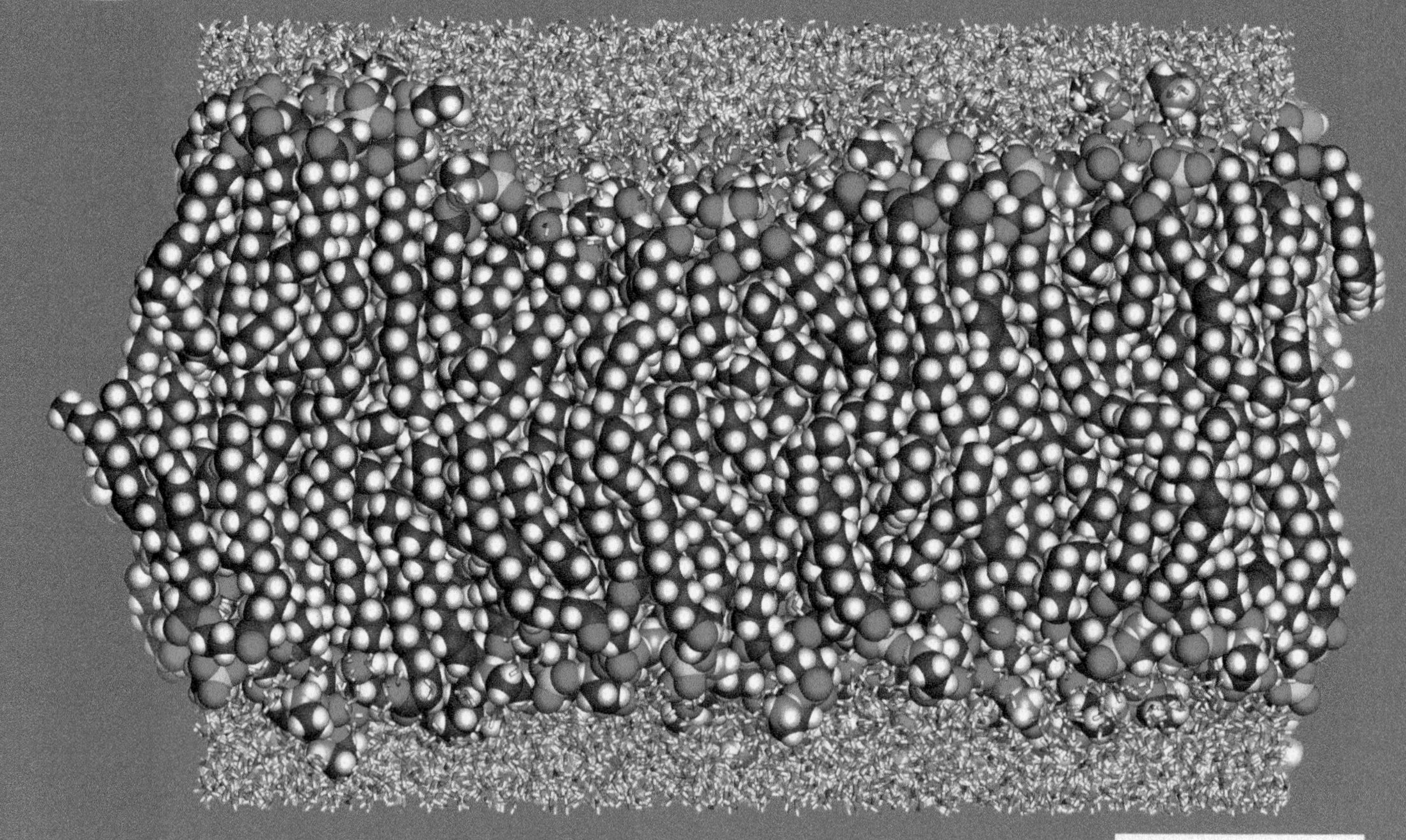

A Lipid Bilayer. This molecular model shows a section of a typical lipid bilayer that forms the basis of biological membranes. Nonpolar hydrocarbon chains form the interior of the membrane, with polar head groups and associated water molecules (shown using stick representation) at both surfaces.

What defines a cell and its organelles? An essential feature of every cell is the presence of **membranes,** structures that define the boundaries of the cell and any internal compartments such as the nucleus, mitochondria, and chloroplasts of eukaryotic cells (**Figure 7-1**). Even the casual observer of electron micrographs is likely to be struck by the prominence of membranes around all cells and within cells of eukaryotic organisms. Note how the membranes of the endoplasmic reticulum occupy a large area within the cells shown in Figure 7-1a and how all of the internal compartments shown are surrounded by membranes.

In previous chapters we encountered the structural molecules that allow membranes to be formed, and we began to study membranes and membrane-bounded organelles (Chapters 2 to 4). Now we are ready to look at membrane structure and function in greater detail. In this chapter, we will examine the molecular structure of membranes and explore the multiple functional roles that membranes play in the life of the cell.

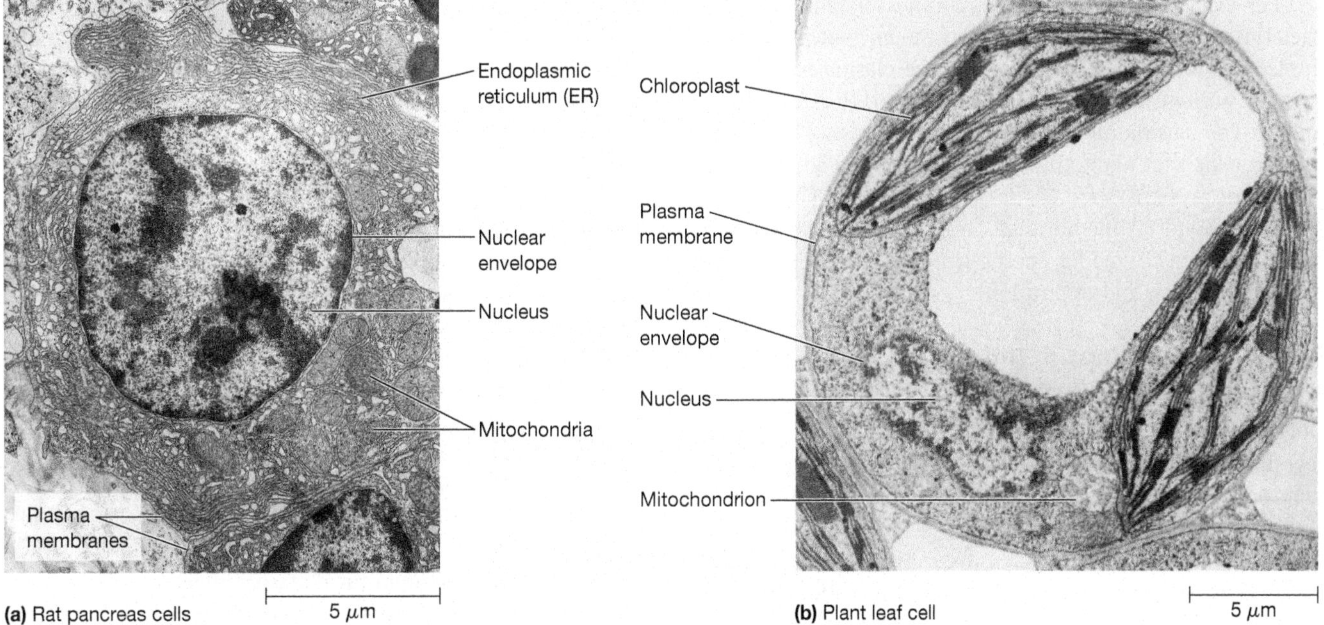

Figure 7-1 The Prominence of Membranes Around and Within Eukaryotic Cells. Among the structures of eukaryotic cells that involve membranes are the plasma membrane, nucleus, chloroplasts, mitochondria, and endoplasmic reticulum (ER). These structures are shown here in **(a)** portions of three cells from a rat pancreas and **(b)** a plant leaf cell (TEMs).

7.1 The Functions of Membranes

We begin by noting that biological membranes play five related yet distinct roles, as illustrated in **Figure 7-2.** ❶ They define the boundaries of the cell and its organelles and act as permeability barriers. ❷ They serve as sites for specific biochemical functions, such as electron transport during mitochondrial respiration or protein processing and folding in the ER. ❸ Membranes also possess transport proteins that regulate the movement of substances into and out of the cell and its organelles. ❹ In addition, membranes contain the protein molecules that act as receptors to detect extracellular signals. ❺ Finally, they provide mechanisms for cell-to-cell contact, adhesion, and communication. Each of these functions is described briefly in the following five sections.

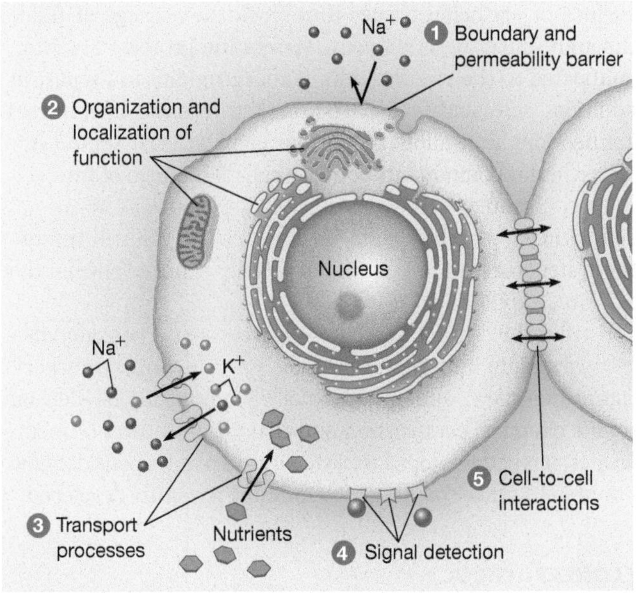

Figure 7-2 Functions of Membranes. Membranes not only define the cell and its organelles but also have a number of important functions, including transport, signaling, and adhesion.

Membranes Define Boundaries and Serve as Permeability Barriers

Two of the most obvious functions of membranes are to define the boundaries of the cell and its compartments and to serve as permeability barriers. The interior of the cell must be physically separated from the surrounding environment, not only to keep desirable substances in the cell but also to keep undesirable substances out. Membranes serve this function well because the hydrophobic interior of the membrane's phospholipid bilayer (see Figure 2-12) blocks the passage of polar molecules and ions and is thus an effective permeability barrier for these substances. The permeability barrier for the cell as a whole is the **plasma** (or **cell**) **membrane,** a membrane that surrounds the cell and regulates the passage of materials both into and out of cells. In addition to the plasma membrane, various **intracellular membranes** serve to compartmentalize functions within eukaryotic cells.

Membranes Contain Specific Proteins and Therefore Have Specific Functions

Membranes have specific functions associated with them because the molecules and structures responsible for those functions—proteins, in most cases—are either embedded in or localized on membranes. One of the most useful ways to characterize a specific membrane, in fact, is to describe the particular enzymes, transport proteins, receptors, and other molecules associated with it that can give the membrane a specific function.

For example, the plasma membranes of plant, fungal, bacterial, and archaeal cells contain the enzymes that synthesize cell walls. In vertebrate cells, the plasma membrane contains proteins that allow cells to attach to the extracellular matrix. The plasma membranes of sperm and egg cells contain proteins that allow them to recognize one other. Other membrane proteins, such as the ATP synthase in chloroplast and mitochondrial membranes or in the bacterial plasma membrane, are critical for energy-generating processes such as photosynthesis and respiration.

Membrane Proteins Regulate the Transport of Solutes

One key function of membrane proteins is to carry out and regulate the *transport* of substances into and out of cells and their organelles. Nutrients, ions, gases, water, and other substances are taken up into various compartments, and various products and wastes must be removed. Although substances such as gases and very small or lipophilic molecules can typically diffuse directly across cellular membranes, most substances needed by the cell are hydrophilic (polar or ionic) and require transport proteins that recognize and transport a specific molecule (for example, glucose) or a group of chemical species (for example, cations).

Let's consider several examples we will come back to in later chapters. Cells may have specific transporters to import glucose, amino acids, or other nutrients. Your nerve cells transmit electrical signals when sodium and potassium ions are transported across the plasma membrane by specific ion channel proteins. Transport proteins in muscle cells move calcium ions across membranes to facilitate muscle contraction. The chloroplast membrane has a transporter specific for the phosphate ions needed for ATP synthesis, and the mitochondrion has transporters for intermediates involved in aerobic respiration. There is even a specific transporter for water—known as an aquaporin—that can rapidly transport water molecules through membranes of kidney cells to facilitate urine production.

Molecules as large as proteins and RNA can be transferred across membranes, often by transporters that consist of many protein subunits—for example, the nuclear pore complexes in the nuclear envelope through which mRNA molecules and partially assembled ribosomes can move from the nucleus to the cytosol. In some cases, proteins synthesized on the endoplasmic reticulum or in the cytosol can be imported into lysosomes, peroxisomes, or mitochondria via transport proteins. In other cases, proteins in the membranes of intracellular vesicles facilitate the movement of molecules such as neurotransmitters either into or out of the cell. (The involvement of membrane proteins in transport of materials will be discussed in greater detail in Chapter 8.)

Membrane Proteins Detect and Transmit Electrical and Chemical Signals

Cells receive information from their environment, usually in the form of electrical or chemical signals that contact the outer surface of the cell. The nerve impulses being sent from your eyes to your brain as you read these words are examples of such signals, as are the various hormones present in your circulatory system. *Signal transduction* is the term used to describe the specific mechanisms used to transmit such signals from the outer surface of cells to the cell interior.

Many chemical signal molecules bind to specific membrane proteins known as *receptors* on the outer surface of the plasma membrane. Binding of these signal molecules to their receptors triggers specific chemical events on the inner surface of the membrane that lead to changes in cell function. For example, muscle and liver cell membranes contain insulin receptors and can therefore respond to this hormone, which helps cells take in glucose. White blood cells have specific receptors that recognize foreign molecules from infectious agents and initiate a cellular defense response.

Many plant cells have a transmembrane receptor protein that detects the gaseous hormone ethylene and transmits a signal to the cell that can affect a variety of processes including seed germination, fruit ripening, and defense against pathogens. Bacteria often have plasma membrane receptors that sense nutrients in the environment and can signal the cell to move toward these nutrients. Thus, membrane receptors allow cells to recognize, transmit, and respond to a variety of specific signals in nearly all types of cells. (Electrical and chemical signaling mechanisms will be the subject of Chapters 22 and 23.)

Membrane Proteins Mediate Cell Adhesion and Cell-to-Cell Communication

Membrane proteins also mediate adhesion and communication between adjacent cells. Although textbooks often depict cells as separate, isolated entities, most cells in multicellular organisms are in contact with other cells. Some membrane proteins in animal tissues form *adhesive junctions.* During embryonic development, for example, specific cell-to-cell contacts are critical and, in animals, are often mediated by membrane proteins known as *cadherins.* Other membrane proteins form *tight junctions,* which form seals along the surface of epithelial tissues that block the passage of fluids through spaces between cells. Membrane proteins are often connected to the cytoskeleton, lending rigidity to tissues. In addition, cells within a particular tissue often have direct connections that allow the exchange of at least some cytosolic cellular components between cells. This form of intercellular communication is provided by *gap junctions* in animal cells and by *plasmodesmata* in plant cells. (We will discuss these structures in greater detail when we move "beyond the cell" in Chapter 15.)

All the functions we have just considered—compartmentalization, localization of function, transport, signal detection, and intercellular communication—depend on the chemical composition and structural features of membranes. It is to these topics that we now turn as we consider how our present understanding of membrane structure developed.

CONCEPT CHECK 7.1

Which functions of a membrane would be compromised if the membrane consisted of a phospholipid bilayer without proteins?

7.2 Models of Membrane Structure: An Experimental Perspective

Researchers have been trying to understand the molecular organization of membranes for more than a century. Until electron microscopy was applied to the study of cell structure in the 1940s and 1950s, however, no one had ever seen a membrane. Nevertheless, indirect evidence led biologists to postulate the existence of membranes long before they could actually be seen. This intensive research led eventually to the *fluid mosaic model* of membrane structure, which was proposed in the 1970s. This model, which is now thought to be descriptive of almost all biological membranes, envisions a membrane as two fluid layers of phospholipids, with proteins localized within and on the bilayers. The model is described first as *fluid* because the lipids and proteins can easily move laterally in the membrane and also as a *mosaic* because of the presence of proteins within the membrane.

Before looking at the model in detail, we will consider some of the central experiments leading to this view of membrane structure and function. As we do so, you may gain some insight into how such developments come about as well as a greater respect for the diversity of approaches and techniques that are often important in advancing our understanding of biological phenomena. **Figure 7-3** presents a chronology of membrane studies that began over a century ago and led eventually to our current understanding of membranes as fluid mosaics.

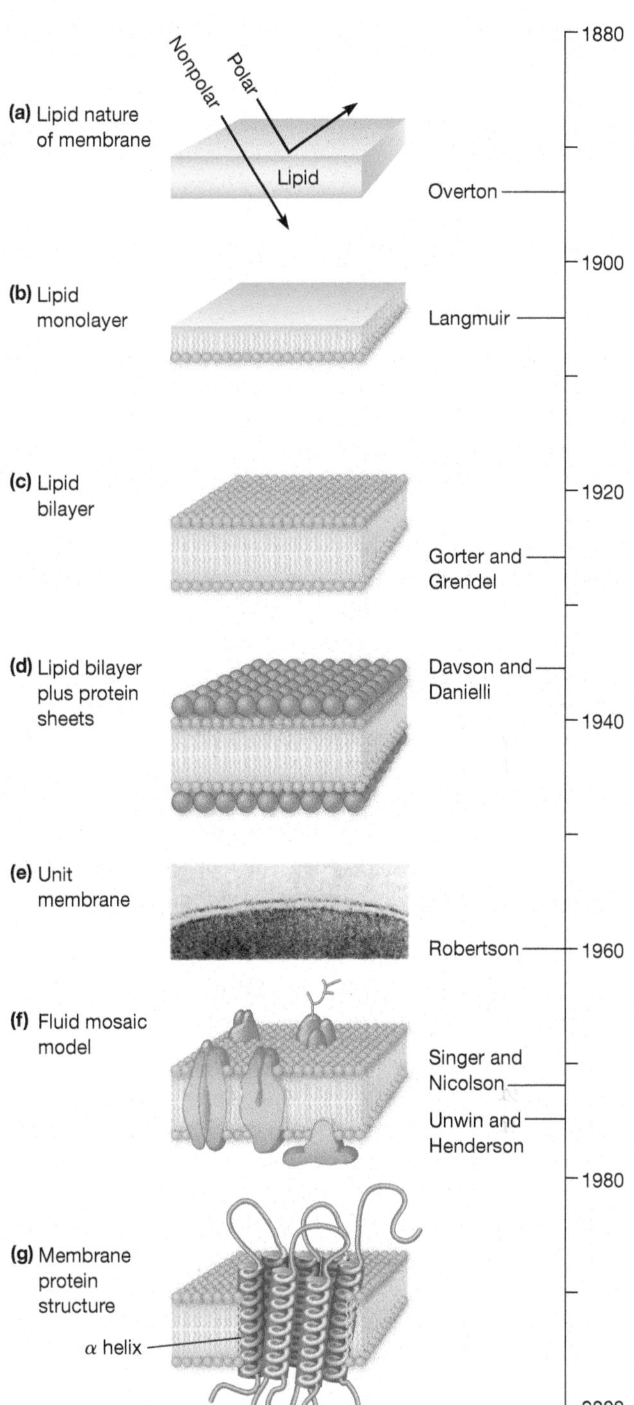

Figure 7-3 Timeline for Development of the Fluid Mosaic Model. The fluid mosaic model of membrane structure that Singer and Nicolson proposed in 1972 was the culmination of studies dating back to the 1890s **(a–e)**. This model **(f)** has been significantly refined by subsequent studies **(g)**.

Overton and Langmuir: Lipids Are Important Components of Membranes

A good starting point for our experimental overview is the pioneering work of British scientist Charles Ernest Overton in the 1890s. Working with cells of plant root hairs, he observed that nonpolar, lipid-soluble substances penetrate readily into cells, whereas polar, water-soluble substances do not. Overton concluded that lipids are present on the cell surface as some sort of "coat" (Figure 7-3a). He even suggested that the coats were probably mixtures of phospholipids and cholesterol, an insight that proved to be remarkably farsighted.

A second important advance came about a decade later, when Irving Langmuir dissolved purified phospholipids in benzene (a nonpolar organic solvent) and layered samples of them onto a water surface. The phospholipid molecules formed a lipid film one molecule thick—that is, a "monolayer." Because phospholipids are *amphipathic* molecules, which have one polar, hydrophilic end and one nonpolar, hydrophobic end (see Figure 2-11), Langmuir reasoned that the phospholipids orient themselves on water such that their hydrophilic heads face the water and their hydrophobic tails protrude away from the water (Figure 7-3b).

Gorter and Grendel: The Basis of Membrane Structure Is a Lipid Bilayer

The next major advance came in 1925 when two Dutch physiologists, Evert Gorter and François Grendel, extracted the lipids from a known number of erythrocytes (red blood cells) and used Langmuir's method to spread the lipids as a monolayer onto a water surface. They found that the area of the lipid film on the water was about twice the estimated total surface area of the erythrocyte. Therefore, they concluded that the erythrocyte plasma membrane consists of not one but *two* layers of lipids.

Hypothesizing a bilayer structure, Gorter and Grendel reasoned that it would be thermodynamically favorable for the nonpolar hydrocarbon chains of each layer to face inward, away from the aqueous milieu on either side of the membrane. The polar hydrophilic groups of each layer would then face outward, toward the aqueous environment on either side of the membrane (Figure 7-3c). The **lipid bilayer** first proposed by Gorter and Grendel became the basic underlying assumption for each successive refinement in our understanding of membrane structure.

Davson and Danielli: Membranes Also Contain Proteins

Shortly after Gorter and Grendel's work, it became clear that a simple lipid bilayer could not explain all the properties of membranes—particularly those related to *surface tension, solute permeability,* and *electrical resistance.* For example, the surface tension of a lipid film was significantly higher than that of cellular membranes but could be lowered by adding protein to the lipid film. Moreover, sugars, ions, and other hydrophilic solutes readily moved into and out of cells even though pure lipid bilayers are nearly impermeable to water-soluble substances.

To explain such differences, Hugh Davson and James Danielli suggested that proteins are present in membranes. They proposed in 1935 that biological membranes consist of lipid bilayers that are coated on both sides with thin sheets of protein (Figure 7-3d). Their model, a protein-lipid-protein "sandwich," was the first detailed representation of membrane organization and dominated the thinking of cell biologists for the next several decades. The real significance of the Davson–Danielli model was that it recognized the importance of proteins in membrane structure.

Robertson: All Membranes Share a Common Underlying Structure

With the increasing use of electron microscopy in the 1940s and 1950s, cell biologists could finally verify the presence of a plasma membrane around each cell. They could also observe that most subcellular organelles are bounded by similar membranes. Furthermore, when membranes were stained with osmium, a heavy metal that binds to membranes, they appeared as pairs of parallel dark lines separated by a lightly stained central zone, with an overall thickness of 6–8 nm. This "railroad track" pattern is seen in **Figure 7-4** for the plasma membranes of two adjacent cells that are separated from each other by a thin intercellular space. Because this same staining pattern was observed with many different kinds of membranes, J. David Robertson suggested that all cellular membranes share a common underlying structure, which he called the *unit membrane* (Figure 7-3e).

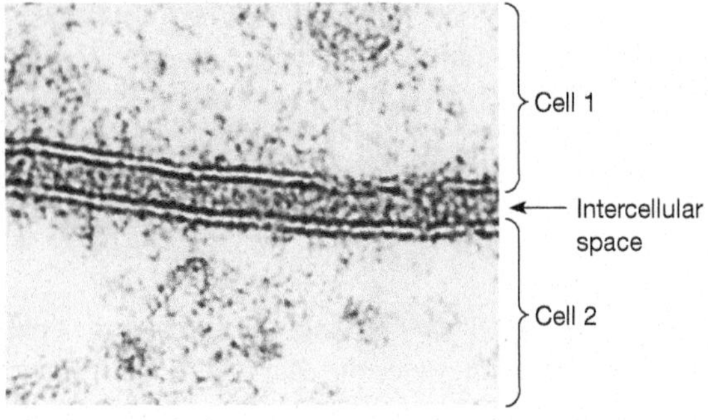

Figure 7-4 Trilaminar Appearance of Cellular Membranes. This electron micrograph of a thin section through two adjacent cells shows their plasma membranes separated by a small intercellular space. Each membrane appears as two dark lines separated by a lightly stained central zone (TEM).

When first proposed, the unit membrane structure seemed to agree remarkably well with the Davson–Danielli model. Robertson suggested that the space between the two dark lines of the trilaminar pattern, which resists staining, comprised the hydrophobic region of the lipid molecules. Conversely, the two dark lines were thought to be made up of phospholipid head groups and the thin sheets of protein bound to the membrane surfaces, both of which appear dark because of their affinity for heavy-metal stains. This interpretation appeared to provide strong support for the Davson–Danielli view that a membrane consists of a lipid bilayer coated on both surfaces with thin sheets of protein.

Further Research Revealed Major Shortcomings of the Davson–Danielli Model

Despite its apparent confirmation by electron microscopy and its extension to all membranes by Robertson, the Davson–Danielli model encountered difficulties in the 1960s as more and more data emerged that could not be reconciled with it. Based on electron microscopy, most membranes were reported to be about 6–8 nm thick—and, of this, the lipid bilayer accounted for about 4–5 nm. That left only about 1–2 nm of space on either surface of the bilayer, a space that could at best accommodate only a thin monolayer of protein. Yet after membrane proteins were isolated and studied, it became apparent that most of them were globular proteins with sizes and shapes inconsistent with the concept of thin sheets of protein on the two surfaces of the membrane.

As a further complication, the Davson–Danielli model did not readily account for the distinctiveness of different kinds of membranes. Depending on their source, membranes vary considerably in chemical composition, especially in the ratio of protein to lipid, which can vary from 3:1 or more in some bacterial cells to only 0.23:1 for the myelin sheath surrounding nerve axons (**Table 7-1**). Even the two membranes of the mitochondrion differ significantly: the protein/lipid ratio is about 1.2:1 for the outer membrane and about 3.5:1 for the inner membrane, which contains all the enzymes and proteins related to electron transport and ATP synthesis. Yet all of these membranes look essentially the same when visualized using electron microscopy.

The Davson–Danielli model was also called into question by studies in which membranes were exposed to *phospholipases,* enzymes that degrade phospholipids by removing their head groups. According to the model, the hydrophilic head groups of membrane lipids should be covered by a layer of protein and therefore protected from phospholipase digestion. However, up to 75% of the membrane phospholipid can be degraded when the membrane is exposed to phospholipases, suggesting that many of the phospholipid head groups are exposed at the membrane surface and not covered by a layer of protein.

Moreover, the surface localization of membrane proteins specified by the Davson–Danielli model was not supported by

Table 7-1 Protein and Lipid Content of Biological Membranes

Membrane	Approximate Percentage by Weight		
	Protein	Lipid	Protein/ Lipid Ratio
Plasma membrane			
Human erythrocyte	49	43	1.14:1
Mammalian liver cell	54	36	1.50:1
Amoeba	54	42	1.29:1
Myelin sheath of nerve axon	18	79	0.23:1
Nuclear envelope	66	32	2.06:1
Endoplasmic reticulum	63	27	2.33:1
Golgi apparatus	64	26	2.46:1
Chloroplast thylakoids	70	30	2.33:1
Mitochondrial outer membrane	55	45	1.22:1
Mitochondrial inner membrane	78	22	3.54:1
Gram-positive bacterium	75	25	3.00:1

the experience of scientists who tried to isolate such proteins. Most membrane proteins turned out to be quite insoluble in water and could be extracted only by using organic solvents or detergents. This observation indicated that many membrane proteins are hydrophobic (or at least amphipathic) and suggested that they are located, at least partially, within the hydrophobic interior of the membrane rather than on either of its surfaces.

Singer and Nicolson: A Membrane Consists of a Mosaic of Proteins in a Fluid Lipid Bilayer

The preceding problems with the Davson–Danielli model stimulated considerable interest in the development of new ideas about membrane organization, culminating in 1972 with the **fluid mosaic model** proposed by S. Jonathan Singer and Garth Nicolson. This model, which now dominates our view of membrane organization, has two key features, both implied by its name. Simply put, the model envisions a membrane as a *mosaic* of proteins embedded in or attached to a *fluid* lipid bilayer (Figure 7-3f). This model retained the basic lipid bilayer structure of earlier models but viewed membrane proteins in an entirely different way—not as thin sheets on the membrane surface but as discrete globular entities within the lipid bilayer (compare Figure 7-3d and 7-3f).

The fluid nature of the membrane is a critical feature of the Singer–Nicolson model. Rather than being rigidly locked in place, most of the lipid components of a membrane are in constant motion, capable of lateral mobility (that is, movement parallel to the membrane surface). Many membrane proteins are also able to move laterally within the membrane, although some proteins are anchored to structural elements such as the cytoskeleton and are therefore restricted in their mobility.

The major strength of the fluid mosaic model is that it readily explains most of the problems with the Davson–Danielli model. In particular, the concept of proteins partially embedded within the lipid bilayer accords well with the hydrophobic nature and globular structure of most membrane proteins and eliminates the need to accommodate membrane proteins in thin surface layers of unvarying thickness.

Unwin and Henderson: Most Membrane Proteins Contain Transmembrane Segments

The final illustration in the timeline (Figure 7-3g) depicts an important property of integral membrane proteins that cell biologists began to understand in the 1970s: most such proteins have in their primary structure one or more hydrophobic sequences that span the lipid bilayer. These *transmembrane segments* anchor the protein to the membrane and hold it in proper alignment within the lipid bilayer.

The example in Figure 7-3g is *bacteriorhodopsin*, the first membrane protein shown to possess this structural feature. Bacteriorhodopsin is a plasma membrane protein found in archaea of the genus *Halobacterium*, where its presence allows cells to obtain energy directly from sunlight (as you will see in Chapter 8). Nigel Unwin and Richard Henderson used electron microscopy to determine the three-dimensional structure of bacteriorhodopsin at low resolution and to reveal its orientation in the membrane.

Their remarkable finding, reported in 1975, was that a bacteriorhodopsin protein consists of a single polypeptide chain folded back and forth across the lipid bilayer a total of seven times. Each of the seven closely packed transmembrane segments of the protein is an α helix composed mainly of hydrophobic amino acids. Successive transmembrane segments are linked to each other by short loops of hydrophilic amino acids that protrude from the polar surfaces of the membrane (see Figure 8-16). Based on subsequent work in many laboratories, membrane biologists currently believe that all transmembrane proteins are integrated into the lipid bilayer via one or more transmembrane segments.

Almost from the moment Singer and Nicolson proposed the fluid mosaic model, it revolutionized the way scientists think about membrane structure. The model launched a new era in membrane research that not only confirmed the basic model but also refined and extended it. Recent developments emphasize the concept that membranes are not homogenous. In fact, most cellular processes that involve membranes depend critically on specific structural complexes of lipids and proteins within the membrane. Interaction between a membrane protein and a particular lipid can be highly specific and is often critical for proper membrane protein structure and function.

So, to understand membrane-associated processes, we need more than the original fluid mosaic model with lipids and proteins simply floating around randomly. Our understanding of membrane structure continues to expand as new research findings further improve the basic model. A summary of our current understanding of membranes is shown in **Figure 7-5** on page 160. Nevertheless, the fluid mosaic model is still basic to our understanding of membrane structure, so it is important for us to closely examine its essential features.

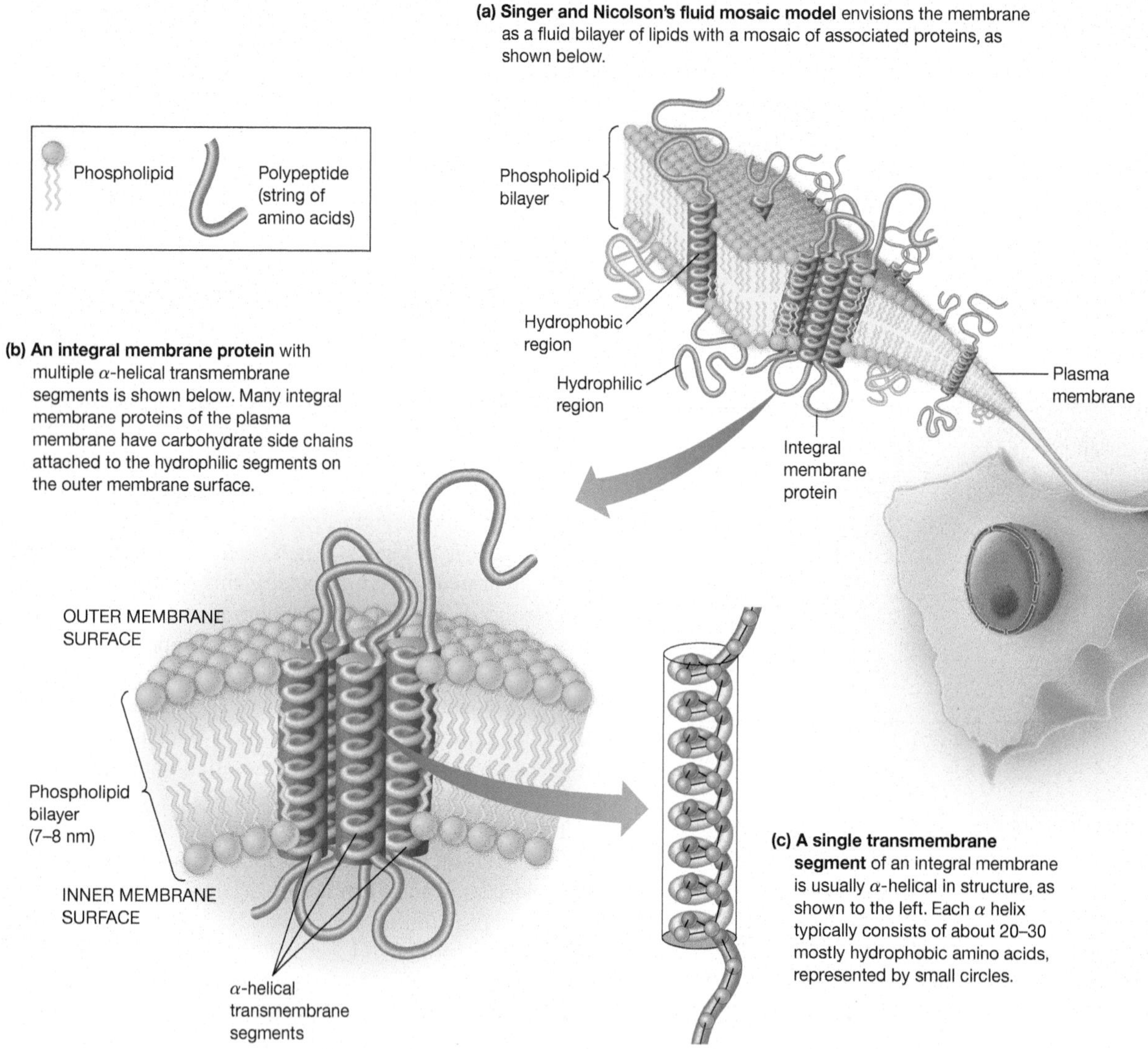

(a) Singer and Nicolson's fluid mosaic model envisions the membrane as a fluid bilayer of lipids with a mosaic of associated proteins, as shown below.

(b) An integral membrane protein with multiple α-helical transmembrane segments is shown below. Many integral membrane proteins of the plasma membrane have carbohydrate side chains attached to the hydrophilic segments on the outer membrane surface.

(c) A single transmembrane segment of an integral membrane is usually α-helical in structure, as shown to the left. Each α helix typically consists of about 20–30 mostly hydrophobic amino acids, represented by small circles.

Figure 7-5 The Fluid Mosaic Model of Membrane Structure. These drawings show **(a)** representative phospholipids (orange) and proteins (purple) in a typical plasma membrane. Enlargements show **(b)** an integral membrane protein and **(c)** one of its transmembrane segments.

CONCEPT CHECK 7.2

Which of the three historical strands forming modern cell biology (cytology, biochemistry, genetics; see Chapter 1) do you think has been the most important in helping us formulate our current concept of membrane structure?

7.3 Membrane Lipids: The "Fluid" Part of the Model

We will begin our detailed look at membranes by considering membrane lipids, which are important components of the "fluid" part of the fluid mosaic model.

Membranes Contain Several Major Classes of Lipids

One important feature of the fluid mosaic model is that it retains the lipid bilayer, though with a greater diversity and fluidity of lipid components than early investigators recognized. The main classes of membrane lipids are *phospholipids, glycolipids,* and *sterols.* **Figure 7-6** lists the main lipids in each of these categories and depicts some of their structures.

Phospholipids. The most abundant lipids found in cell membranes are the **phospholipids** (Figure 7-6a). As you may recall (Chapter 2), a phospholipid molecule consists of a backbone moiety to which are attached two fatty acids, a negatively charged phosphate group, and a charged or polar head group that is attached to the phosphate (see Figures 2-11 and 3-27c). The combination of a highly polar head and two nonpolar tails gives phospholipids their *amphipathic* character, which is so critical to their role in membrane structure. The fatty acid components form a hydrophobic barrier, whereas the remainder of the molecule has hydrophilic properties that enable it to interact with an aqueous environment.

The backbone of most membrane phospholipids is either glycerol, a three-carbon alcohol, or sphingosine, a derivative of the amino acid serine with a fatty acid attached (see Figure 3-27c). Membranes contain many different kinds of

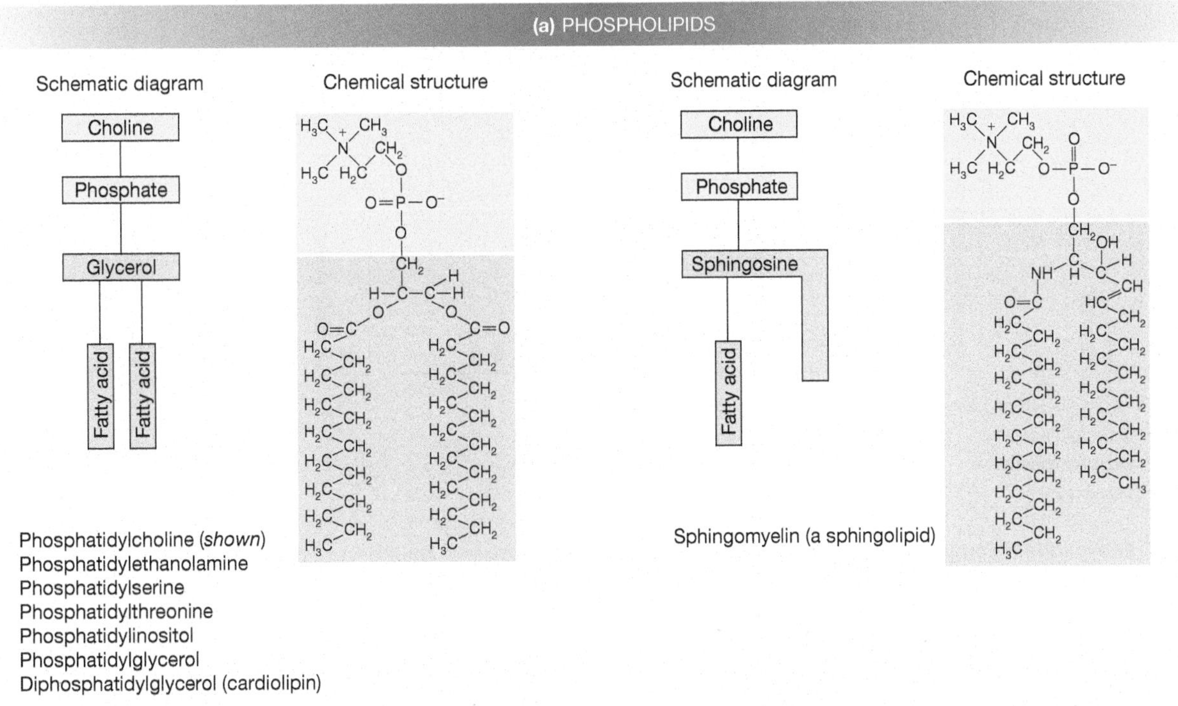

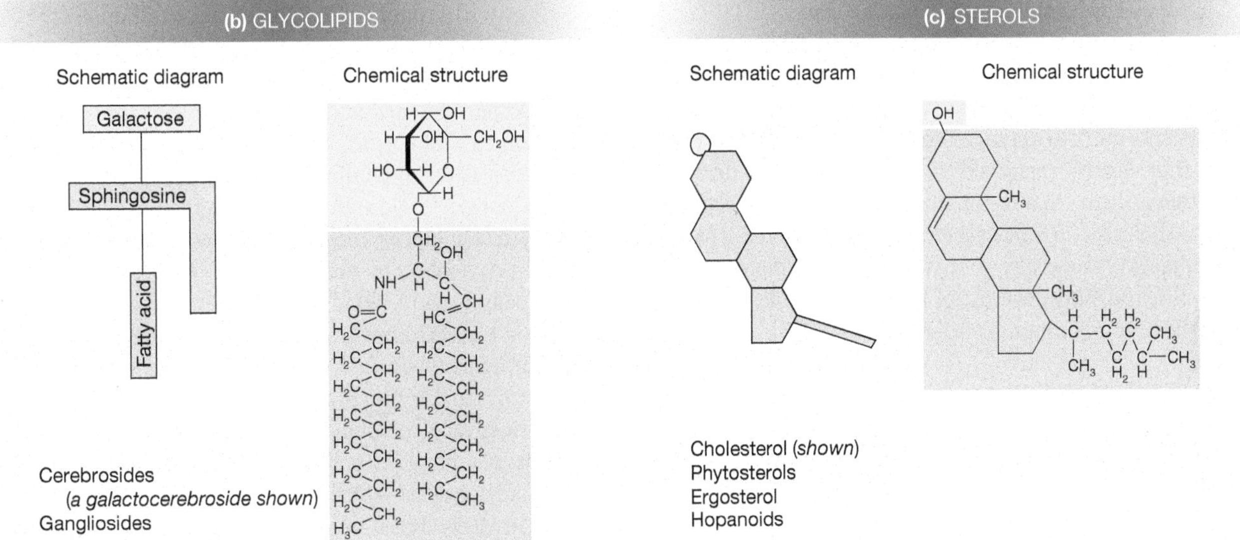

Figure 7-6 The Three Major Classes of Membrane Lipids. (a) Phospholipids have a small polar head group (such as choline) attached via a phosphate to a fatty acid–containing glycerol or sphingosine backbone, forming either phosphoglycerolipids (phosphoglycerides) or phosphosphingolipids. Carbons 1, 2, and 3 are labeled for glycerol and sphingosine. **(b)** Glycolipids have one or more monosaccharides attached to a fatty acid–containing backbone and also can be either glycerol- or sphingosine-based. **(c)** Sterols are multi-ring molecules that are related to cholesterol and steroid hormones.

phospholipids, including both the glycerol-based *phosphoglycerolipids*, more commonly called **phosphoglycerides**, and the sphingosine-based **phosphosphingolipids.** The most common phosphoglycerides have a head group such as choline, serine, ethanolamine, or inositol. These are therefore called *phosphatidylcholine, phosphatidylserine, phosphatidylethanolamine*, or *phosphatidylinositol*, respectively. A common phosphosphingolipid is *sphingomyelin*, which is one of the main phospholipids of animal plasma membranes but is absent from the plasma membranes of plants and most bacteria. The kinds and relative proportions of membrane lipids present vary significantly among membranes from different sources. **Figure 7-7** on page 162 illustrates this variability for phospholipids.

It is the amphipathic nature of phospholipids in the membrane bilayer that causes them to form a bilayer with a hydrophobic interior. This makes the plasma membrane an effective permeability barrier for the cell. However, amphipathic molecules such as detergents mimic and can replace membrane phospholipids and thus disrupt the integrity of the membrane bilayer and kill the cell. For example, the amphipathic detergent sodium dodecyl sulfate (SDS) is commonly used in the laboratory to disrupt and solubilize membranes and membrane proteins for biochemical analysis.

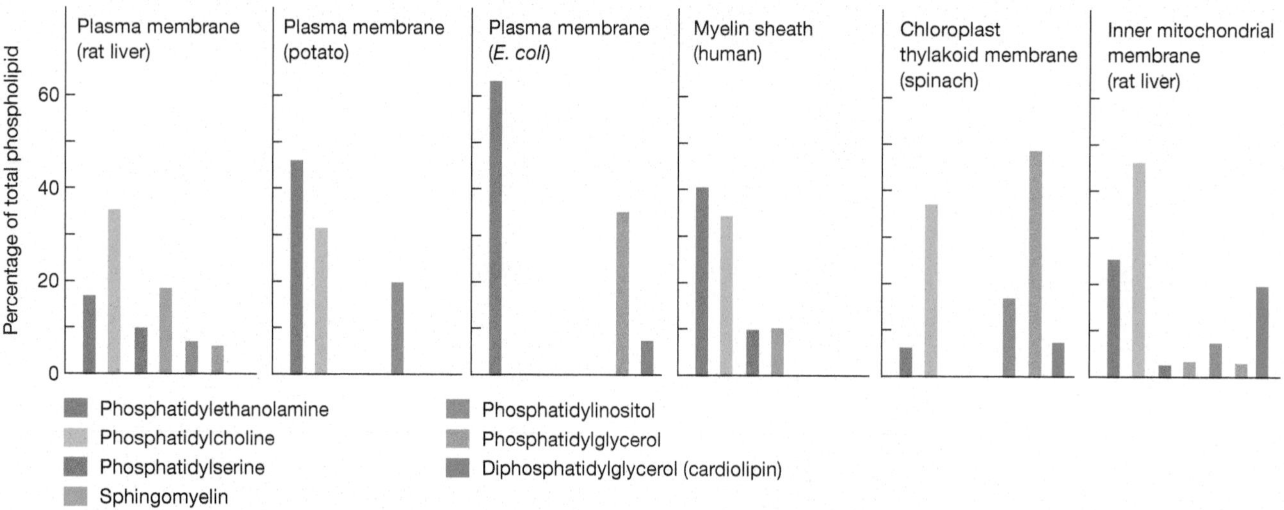

Figure 7-7 Phospholipid Composition of Several Kinds of Membranes. The relative abundance of different kinds of phospholipids in biological membranes varies greatly with the source of the membrane.

Recently, there has been much interest in a class of *antimicrobial peptides (AMPs)*, which are oligopeptides of 10–50 amino acids that can affect membrane permeability in bacteria. More than 1200 different AMPs are known, with more than 20 being produced by human skin alone. One class of AMPs consists of cationic, amphipathic molecules that disrupt bacterial membrane structure by interacting with the negatively charged phospholipids in their cell membranes. These AMPs thus act like detergents and disrupt membrane structure, causing holes to form that destroy bacterial cells' permeability barrier, ultimately killing them. As a result, host cells like human skin can use this mechanism to protect themselves from bacterial pathogens. Some AMPs have also shown promise as antiviral agents in disrupting the outer covering of membrane-enclosed viruses such as the human immunodeficiency virus (HIV).

MAKE CONNECTIONS 7.1

Fats also use glycerol and fatty acids as key components. How do fats differ from phospholipids chemically? How does this affect how amphipathic they are? (Figs. 3.27, 3.29)

Glycolipids. As their name indicates, **glycolipids** (Figure 7-6b) are formed by adding carbohydrate groups to lipids. Like phospholipids, some glycolipids are glycerol based and are called *glycoglycerolipids*. Others are derivatives of sphingosine and are therefore called *glycosphingolipids*. The most common examples of glycosphingolipids are **cerebrosides** and **gangliosides.** Cerebrosides are called *neutral glycolipids* because each molecule has a single uncharged sugar as its head group—galactose, in the case of the galactocerebroside shown in Figure 7-6b. A ganglioside, on the other hand, has an oligosaccharide head group that contains one or more negatively charged sialic acid residues, giving gangliosides a net negative charge.

Cerebroside and ganglioside glycosphingolipids are especially prominent in the membranes of brain and nerve cells. Several serious human diseases are known to result from impaired metabolism of glycosphingolipids. The best-known example is *Tay-Sachs disease*, a genetic disorder that is caused by the absence of a lysosomal enzyme, hexosaminidase A, that is responsible for one of the steps in ganglioside degradation. As a result, gangliosides accumulate in the brain and other nervous tissue, leading to impaired nerve and brain function and eventually to paralysis, severe mental deterioration, and death.

Recall that *antibodies* (introduced in Chapter 1) are immune system proteins that recognize and bind to specific molecular markers, known as *antigens*, on the cell surface. Gangliosides exposed on the surface of the plasma membrane function as antigens that are recognized by antibodies in immune reactions, including those responsible for blood group interactions. The human ABO blood groups, for example, involve surface glycosphingolipids known as A antigen and B antigen that serve as markers of the different groups of red blood cells. Cells of blood type A have the A antigen, and cells of blood type B have the B antigen. Type AB blood cells have both antigen types, and type O blood cells have neither. We will return to ABO blood groups later in this chapter.

Two glycoglycerolipids that are abundant in the thylakoid membranes of plant and algal chloroplasts are monogalactosyldiacylglycerol (MGDG) and digalactosyldiacylglycerol (DGDG). MGDG has one galactose molecule attached to a glycerol backbone bearing two fatty acid groups, and DGDG has two galactose molecules attached. These two glycolipids may constitute up to 75% of the total membrane lipids in leaves. Together, these two lipids are sometimes considered to be the most abundant glycolipids on Earth.

Sterols. Besides phospholipids and glycolipids, the membranes of most eukaryotic cells contain significant amounts of **sterols** (Figure 7-6c). The main sterol in animal cell membranes is **cholesterol,** a four-ringed molecule that is necessary for maintaining and stabilizing membranes in our bodies by acting as a fluidity buffer, as explained later in the chapter. The membranes of plant cells contain small amounts of cholesterol but larger amounts of **phytosterols.** Fungal membranes contain a sterol known as **ergosterol** that is similar in structure to cholesterol. Ergosterol is the target of antifungal medications such as nystatin, which selectively kill fungi but do not harm human cells because human cells lack ergosterol.

Sterols are not found in the membranes of most bacteria. Sterols are also absent from the inner membranes of both mitochondria and chloroplasts, which are believed to be derived evolutionarily from the plasma membranes of bacterial cells via endosymbiosis. Some bacteria contain sterol-like molecules called **hopanoids,** which are rigid, five-ringed compounds that appear to act like sterols in adding stability to membrane structure.

Fatty Acids Are Essential to Membrane Structure and Function

Fatty acids are components of all membrane lipids except the sterols. They are essential to membrane structure because their long hydrocarbon tails form an effective hydrophobic barrier to the diffusion of polar solutes. Most fatty acids in membranes are between 12 and 20 carbon atoms in length, with 16- and 18-carbon fatty acids especially common. This size range appears to be optimal for bilayer formation because chains with fewer than 12 or more than 20 carbons are less able to form a stable bilayer. Thus, the thickness of membranes (about 6–8 nm, depending on the source) is dictated primarily by the chain length of the fatty acids required for bilayer stability.

In addition to differences in length, the fatty acids found in membrane lipids vary considerably regarding the presence and number of double bonds. **Figure 7-8** shows the structures of several fatty acids that are especially common in membrane lipids. *Palmitate* and *stearate* are **saturated fatty acids,** with 16 and 18 carbon atoms, respectively. Recall (from Chapter 3) that saturated fatty acids contain no double bonds because every carbon atom in their tails is bound to the maximum number of hydrogen atoms. They are abbreviated as 16:0 and 18:0, respectively, to show both the number of carbons and the number of double bonds. *Oleate* and *linoleate* are 18-carbon **unsaturated fatty acids** with one and two double bonds, respectively, and are abbreviated as 18:1 and 18:2. Because some carbons are double-bonded to each other in unsaturated fatty acids, fewer hydrogens can be attached to carbons, hence the name "unsaturated."

Other *poly*unsaturated fatty acids (more than one double bond) commonly found in membranes are *linolenate,* with 18 carbons and three double bonds (18:3), and *arachidonate,* with 20 carbons and four double bonds (20:4). *Omega-3 fatty acids* are a type of polyunsaturated fatty acid with one of the double bonds on the third carbon. They are essential for normal human development, and recent studies suggest that consumption of omega-3 fatty acids may reduce the risk of heart disease.

Thin-Layer Chromatography Is an Important Technique for Lipid Analysis

How do we know so much about the lipid components of membranes? As you may suspect, they are difficult to isolate and study due to their largely hydrophobic nature. However, using nonpolar organic solvents such as acetone and chloroform, biologists and biochemists have been isolating, separating, and studying membrane lipids for more than a

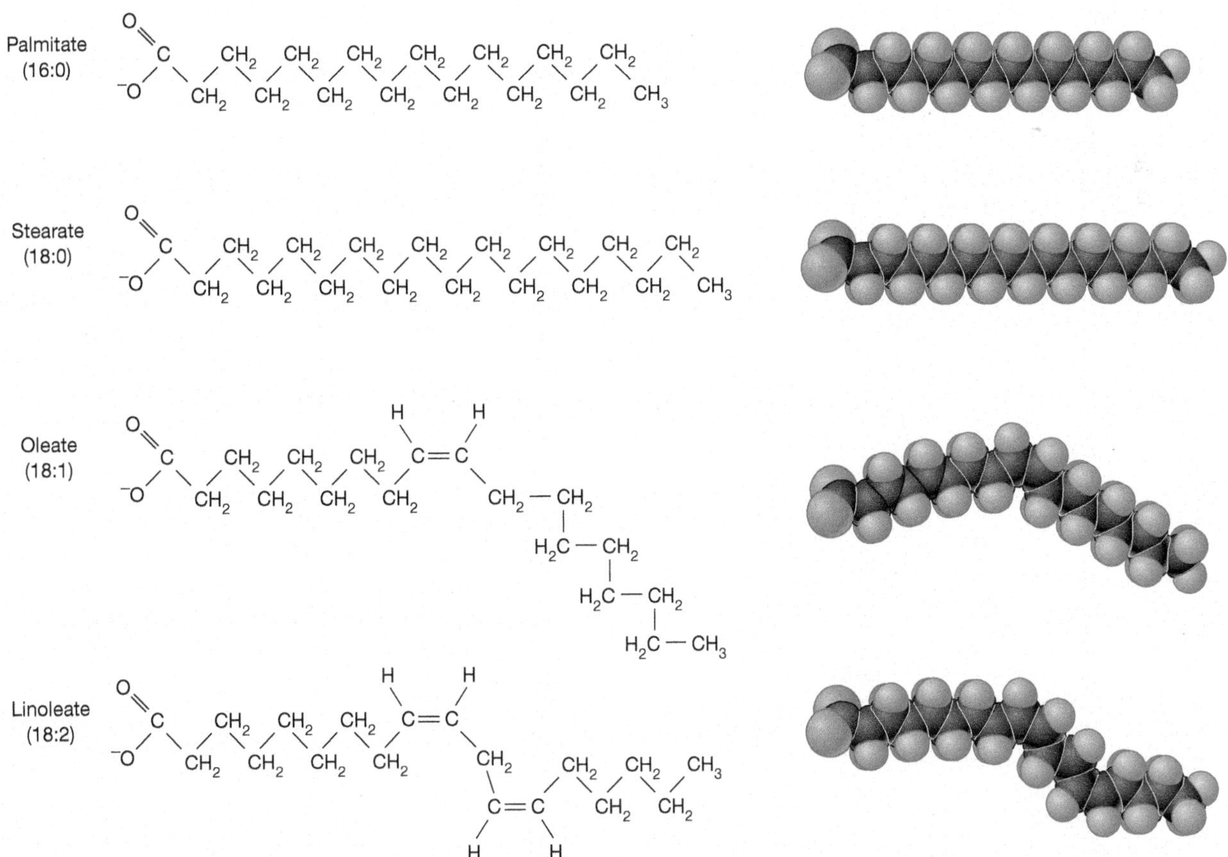

Figure 7-8 Structures of Some Common Fatty Acids Found in Membranes. Phospholipids in biological membranes vary greatly with the source of the membrane. Numbers in parentheses under the fatty acid names show number of carbons: number of double bonds.

century. One important technique for the analysis of lipids is **thin-layer chromatography (TLC),** depicted schematically in **Figure 7-9**. This technique is used to separate different kinds of lipids based on their relative polarities.

In this procedure, the lipids are solubilized from a membrane preparation using a mixture of nonpolar organic solvents and separated using a glass plate coated with silicic acid, a polar compound that dries to form a thin film on the glass plate. A sample of the membrane extract is applied to one end of the TLC plate by spotting the extract onto a small area called the *origin* (Figure 7-9a). After the solvent in the sample has evaporated, the edge of the plate is dipped into a weakly polar solvent system that typically consists of chloroform, methanol, and water. As the solvent moves past the origin and up the plate by capillary action, the lipids are separated based on their polarity—that is, by their relative affinities for the polar silicic acid plate and the less polar solvent.

Nonpolar lipids such as cholesterol have little affinity for the polar silicic acid (the *stationary phase*) on the plate and therefore move rapidly up the plate with the less polar solvent system (the *mobile phase*). Lipids that are more polar, such as phospholipids, interact more strongly with the silicic acid, which slows their movement. In this way, the various lipids are separated progressively as the leading edge (or *solvent front*) of the mobile phase continues to move up the plate. When the solvent front approaches the top, the plate is removed from the solvent system and dried. The separated lipids are then recovered from the plate by scraping each spot from the plate and dissolving each one in a nonpolar solvent such as chloroform for identification and further study.

Figure 7-9b shows the TLC pattern seen for the lipids of the erythrocyte plasma membrane. The main components of this membrane are cholesterol (25%) and phospholipids (55%), with phosphatidylethanolamine (PE), phosphatidylcholine (PC), and phosphatidylserine (PS) being the most prominent phospholipids. Other minor components, such as phosphatidylinositol and sphingolipids, are not shown. Control plates using small amounts of known lipids are run simultaneously for comparison and identification of lipid spots on the experimental plate.

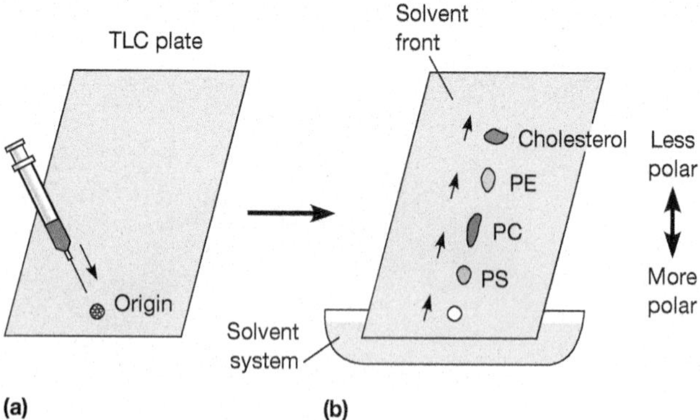

Figure 7-9 Using Thin-Layer Chromatography to Separate and Analyze Membrane Lipids. Thin-layer chromatography (TLC) is a useful technique for separating membrane lipids according to their degree of polarity. **(a)** A sample is spotted onto a small area of a glass TLC plate coated with a thin layer of silicic acid, which represents the stationary phase. **(b)** Components of the sample are then carried upward by the solvent (the mobile phase) into which the TLC plate is placed. The pattern shown is for lipids of the erythrocyte plasma membrane, whose main components are cholesterol, phosphatidylethanolamine (PE), phosphatidylcholine (PC), and phosphatidylserine (PS).

Membrane Asymmetry: Most Lipids Are Distributed Unequally Between the Two Monolayers

Chemical studies involving membranes derived from a variety of cell types have revealed that most lipids are unequally distributed between the two monolayers that constitute the lipid bilayer. This **membrane asymmetry** includes differences in both the kinds of lipids present and the degree of unsaturation of the fatty acids in the phospholipid molecules.

For example, most of the glycolipids present in the plasma membrane of an animal cell are restricted to the outer monolayer. As a result, their carbohydrate groups protrude from the outer membrane surface, where they are involved in various signaling and recognition events. Phosphatidylcholine is more common in the outer monolayer, whereas phosphatidylethanolamine, phosphatidylinositol, and phosphatidylserine are more prominent in the inner monolayer, where they are involved in transmitting various kinds of signals from the plasma membrane to the interior of the cell. (See Chapters 22 and 23.)

Membrane asymmetry is established during membrane biogenesis by the insertion of different proportions of the various lipids into each of the two monolayers. Once established, asymmetry tends to be maintained because the movement of lipids from one monolayer to the other requires the passage of hydrophilic head groups through the hydrophobic interior of the membrane—an event that is thermodynamically unfavorable. Although such "flip-flop," or **transverse diffusion,** of membrane lipids does occur occasionally, it is relatively rare. For instance, a typical phospholipid molecule flip-flops less than once a week in a pure phospholipid bilayer and only once every few hours in natural membranes. In striking contrast are the **rotation** of phospholipid molecules about their long axes and the **lateral diffusion** of phospholipids in the plane of the membrane, both of which occur freely, rapidly, and randomly. **Figure 7-10** illustrates all three types of lipid movements. In a pure phospholipid bilayer at 37°C, a typical lipid molecule exchanges places laterally with neighboring molecules about 10 million times per second and can move laterally at a rate of about several micrometers per second.

Although phospholipid flip-flop is relatively rare, it occurs more frequently in natural membranes than in artificial lipid bilayers. This is because some membranes—the smooth endoplasmic reticulum (ER), in particular—have proteins called **phospholipid translocators,** or **flippases,** that catalyze flip-flop of membrane lipids from one monolayer to the other. Such proteins act only on specific kinds of lipids. For example, one flippase in the smooth ER membrane catalyzes the translocation of phosphatidylcholine from one side of the membrane to the other but does not recognize other phospholipids. This ability to move lipid molecules selectively from one side of the bilayer to the other contributes further to the asymmetric distribution of phospholipids across the membrane. (The role of smooth ER in the synthesis and selective flip-flop of membrane phospholipids is a topic we will return to in Chapter 12.)

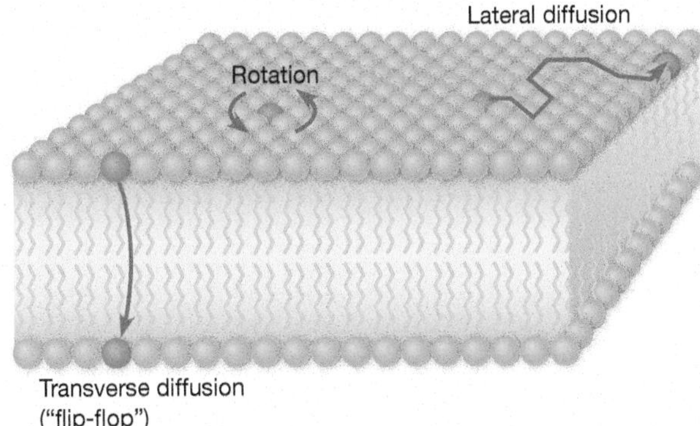

Figure 7-10 Movements of Phospholipid Molecules Within Membranes. A phospholipid molecule is capable of three kinds of movement in a membrane: rotation about its long axis, lateral diffusion by exchanging places with neighboring molecules in the same monolayer, and transverse diffusion, or "flip-flop," from one monolayer to the other.

The Lipid Bilayer Is Fluid

One of the most striking properties of membrane lipids is that rather than being fixed in place within the membrane, they form a fluid bilayer that permits lateral diffusion of membrane lipids as well as proteins. As we will see, cells can vary the composition of their lipid bilayers to ensure that their membranes are in the fluid state at physiological temperatures. Lipid molecules move especially fast because they are much smaller than proteins. Proteins move much more slowly than lipids, partly because they are much larger molecules and partly due to their interactions with cytoskeletal proteins on the inside of the cell. The lateral diffusion of membrane lipids can be demonstrated and quantified by a technique called **fluorescence recovery after photobleaching (FRAP)** (see Key Technique, pages 166–167). FRAP also allows measurements of the movements of proteins in the membranes. We will examine other classic evidence for protein mobility in membranes later in this chapter.

As you might guess, membrane fluidity changes with temperature, decreasing as the temperature drops and increasing as it rises. In fact, we know from studies with artificial lipid bilayers that every lipid bilayer has a characteristic **transition temperature (T_m)** at which it becomes fluid ("melts") when warmed from a solid gel-like state. This change in the state of the membrane is called a **phase transition,** and you have probably seen this yourself if you have ever accidentally left a stick of butter on the stove. To function properly, a membrane must be maintained in the fluid state—that is, at a temperature above its T_m value. At a temperature below the T_m value, all functions that depend on the mobility or conformational changes of membrane proteins will be impaired or disrupted. This includes such vital processes as transport of solutes across the membrane, detection and transmission of signals, and cell-to-cell communication.

Figure 7-11 shows one means of determining the transition temperature of a given membrane, the technique of **differential scanning calorimetry.** This procedure is similar to isothermal titration calorimetry (see Key Technique in Chapter 5, pages 122–123), which also measures heat uptake and release during biological processes. Differential scanning calorimetry monitors the uptake of heat that occurs during the transition from one physical state to another—the gel-to-fluid transition, in the case of membranes. The membrane of interest is placed in a sealed chamber, the *calorimeter,* and its uptake of heat is measured as the temperature is slowly increased. The point of maximum heat absorption corresponds to the T_m (Figure 7-11a).

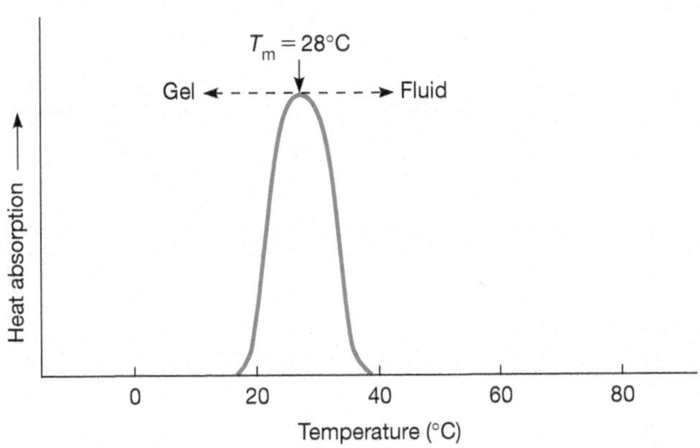

(a) Normal membrane. When the temperature of a typical membrane preparation is increased slowly in a calorimeter chamber, a peak of heat absorption marks the gel-to-fluid transition temperature, T_m.

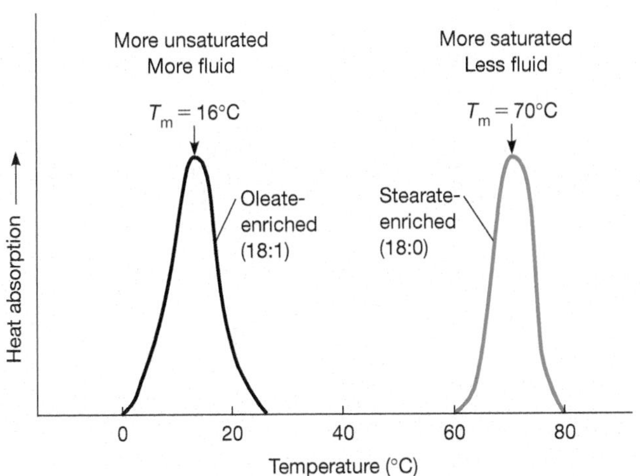

(b) Membranes enriched in unsaturated or saturated fatty acids. Membranes from cells grown in media enriched in the unsaturated fatty acid oleate (left) are more fluid than normal membranes (lower T_m) under similar conditions. Membranes from cells grown in media enriched in the saturated fatty acid stearate (right) are less fluid than normal membranes (higher T_m).

Figure 7-11 Determination of Membrane Transition Temperature by Differential Scanning Calorimetry. These graphs show T_m determinations for **(a)** a normal membrane from a homeotherm and **(b)** membranes enriched in specific unsaturated or saturated 18-carbon fatty acids.

Key Technique | Fluorescence Recovery After Photobleaching (FRAP)

PROBLEM: Although many experiments have verified the fluid mosaic nature of the plasma membrane, scientists often want to quantify how dynamic the movements of a lipid or protein are in the membrane of a living cell. They can then compare the mobility of a membrane component in various cell types or when a cell is subjected to experimental conditions. Such measurements are also valuable for inferring what fraction of a membrane's components is mobile. The challenge is that such measurements must be carried out on living cells.

SOLUTION: If the desired lipid or protein can be labeled with a fluorescent molecule, then a small patch of membrane can be subjected to *photobleaching* with a laser beam. The kinetics of the recovery of fluorescence in the bleached region, called *fluorescence recovery after photobleaching (FRAP),* can be used to infer aspects of the mobility of the fluorescent molecule in the membrane.

Key Tools: Fluorescently labeled lipids or proteins; a fluorescence microscope; a sensitive camera; a computer to analyze the data.

Details: FRAP provides the data cell biologists need to make quantitative measurements of membrane dynamics. In this technique, the investigator *tags,* or labels, molecules in the membrane of a living cell by covalently linking them to a fluorescent dye or, in the case of proteins, genetically tagging them using a fluorescent protein such as the green fluorescent protein (GFP; **see Key Technique in Chapter 19, pages 566–567**). A high-intensity laser beam is then used to bleach the dye in a tiny spot (a few square micrometers) on the cell surface. If the cell surface is examined immediately thereafter with a fluorescence microscope (see Appendix: Visualizing Cells and Molecules, pages A1–A25, for how such microscopes operate), a dark, nonfluorescent spot is seen on the membrane. Within seconds, however, the edges of the spot become fluorescent as bleached molecules diffuse out of the laser-treated area and fluorescent molecules from adjacent regions of the membrane diffuse in. Eventually, the spot is indistinguishable from the rest of the cell surface (**Figure 7A-1**).

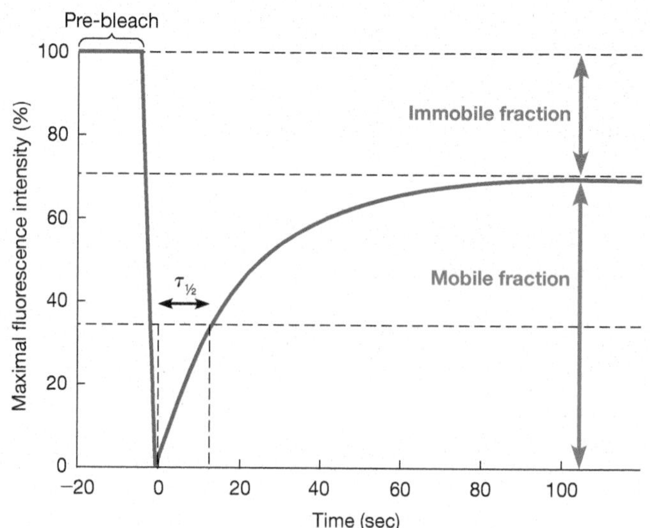

Figure 7A-2 Analyzing a FRAP Experiment. The fluorescence intensity in a region of a fluorescent specimen is measured using a computer before and after photobleaching with a laser or other intense light source. The data are then fit to an equation that allows determination of the half-life of recovery ($\tau_{1/2}$) and the percent of protein that remains immobile (the immobile fraction) versus that which is mobile (mobile fraction).

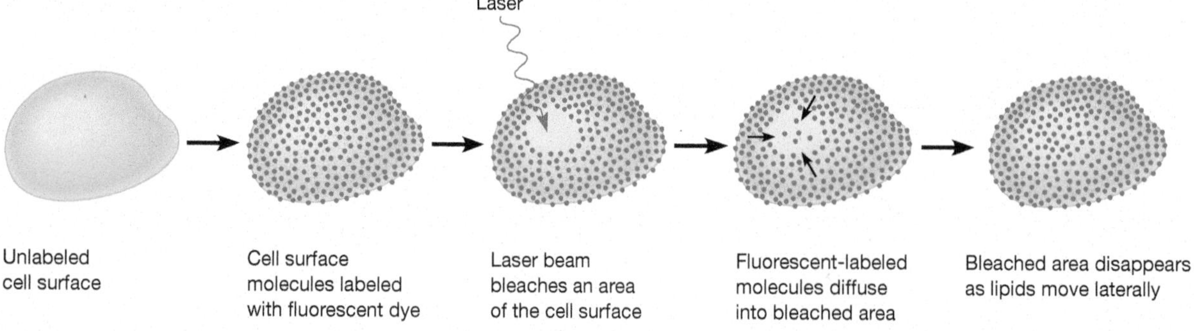

Figure 7A-1 Using Fluorescence Recovery After Photobleaching (FRAP) to Assess Lipid Mobility in a Membrane. Lateral movement of membrane lipids rapidly fills in a photobleached area.

Effects of Fatty Acid Composition on Membrane Fluidity. A membrane's fluidity depends primarily on the types of lipids it contains. Two properties of a membrane's lipid makeup are especially important in determining fluidity: the length of the fatty acid side chains and their degree of unsaturation. Long-chain fatty acids have higher transition temperatures (and thus are less fluid) than do short-chain fatty acids. Similarly, completely saturated fatty acids have higher transition temperatures than unsaturated fatty acids (see Figure 7-11b). This is why saturated fats such as butter or lard tend to be solid at room temperature, whereas unsaturated fats such as olive oil or canola oil are liquid.

FRAP demonstrates that membranes are in a fluid rather than a static state, and it provides a direct means of measuring the lateral movement of specific molecules.

To measure the recovery of fluorescence quantitatively, scientists measure the fluorescence intensity by capturing images of the bleached region over time. They then use computer software to measure the signal intensity at each time point and plot the measurements as a percentage of the pre-bleach intensity (which is normalized to 100%). The essential features of a recovery curve are shown in **Figure 7A-2**. After bleaching, the fluorescence intensity gradually recovers, and plateaus over time. Because image capture can cause some overall bleaching of a specimen, recovery data is usually normalized by measuring fluorescence intensity in another region of the same specimen that has not been subjected to the intense laser light of the FRAP bleach spot. After these corrections are made, typically the fluorescence in the bleach spot does not fully recover to pre-bleach levels. The difference in fluorescence intensity, known as the *immobile fraction,* is a measure, as the name implies, of the percentage of molecules that are not mobile. The rate of recovery of the remaining molecules in the *mobile fraction* can often be fitted using an exponential recovery equation. The time it takes for the fluorescence to recover to half of its maximal value, the *half-life* ($\tau_{1/2}$), is an indicator of how quickly the mobile fluorescent molecules can diffuse in the membrane. The longer the half-life, the slower the molecules are moving. Together, the half-life and immobile fraction are very useful quantitative measures of the mobility of specific membrane components. **Figure 7A-3** shows data from an actual FRAP experiment involving a protein within adherens junctions of cells in a *Caenorhabditis elegans* embryo.

QUESTION: Suppose you perform FRAP on fluorescently labeled lipids in cultured mammalian cells at 37°C and perform the same measurement on cells chilled to 15°C. In which cells would you expect the shorter half-life, and why?

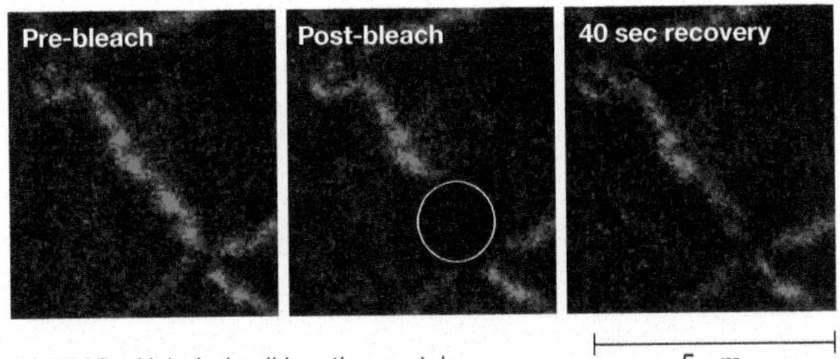

(a) FRAP of labeled cell junction protein

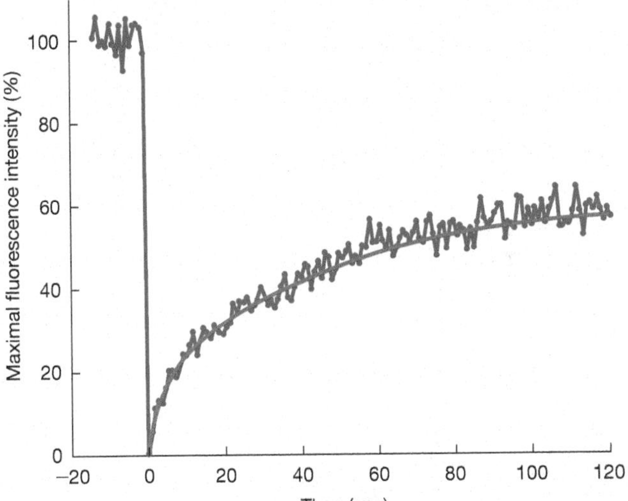

(b) Fluorescence signal over time as a percentage of pre-bleach intensity

Figure 7A-3 Results of an Actual FRAP Experiment on a Membrane Protein in a Nematode Worm Embryo. (a) A small region of a *C. elegans* embryo genetically engineered to express a green fluorescent protein–tagged version of a protein that allows cells to adhere to one another was subjected to photobleaching by a laser. Immediately after bleaching, a dark spot (circle) is visible. After 40 seconds, the fluorescence has recovered significantly, but not to pre-bleach levels. **(b)** Graphical analysis of the experiment shown in part a. The red curve is a single exponential curve fit of the data, from which half-life and immobile fraction can be calculated.

As **Figure 7-12** on page 168 shows, the chain length of fatty acids also affects the transition temperature. For example, as the chain length of saturated fatty acids increases from 10 to 20 carbon atoms, the T_m rises from 32°C to 76°C, and the membrane thus becomes progressively less fluid (Figure 7-12a). The presence of unsaturation affects the T_m even more markedly. For fatty acids with 18 carbon atoms, as shown in Figure 7-12b, the transition temperatures are 70°C, 16°C, 5°C, and −11°C for zero, one, two, and three double bonds, respectively. Thus, membranes containing many unsaturated fatty acids tend to have lower transition temperatures and thus under the same

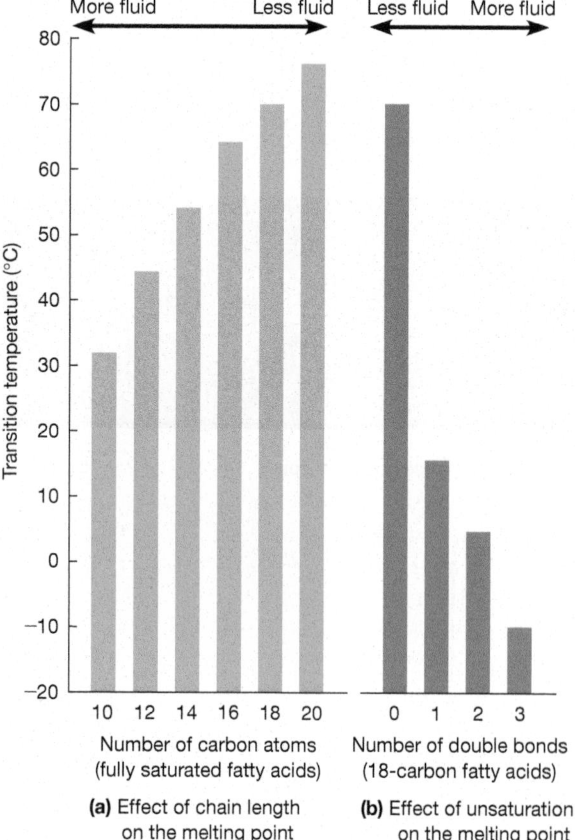

Figure 7-12 Chain Length and Degree of Unsaturation Affect the Melting Point of Fatty Acids. (a) The transition temperature, a measure of fluidity of fatty acids, increases with chain length for saturated fatty acids. **(b)** The transition temperature decreases dramatically with the number of double bonds for fatty acids with a fixed chain length (18 carbons as shown here).

conditions will be more fluid than membranes with many saturated fatty acids.

The effect of unsaturation on membrane fluidity is so dramatic because double bonds in fatty acids cause kinks that prevent their hydrocarbon chains from fitting together snugly. Membrane lipids with unkinked saturated fatty acids pack together tightly, whereas lipids with kinked unsaturated fatty acids do not (**Figure 7-13**). The lipids of most plasma membranes contain fatty acids that vary in both chain length and degree of unsaturation. In fact, the variability is often intramolecular because membrane lipids commonly contain one saturated and one unsaturated fatty acid. This property helps ensure that membranes are in the fluid state at physiological temperatures.

Most unsaturated fatty acids found in nature contain *cis* double bonds. In contrast, many commercially processed fats and hydrogenated oils contain significant numbers of *trans* double bonds. Compared to a *cis* double bond, the *trans* double bond does not introduce as much of a bend in the fatty acid chain, as shown below:

H H
C=C
R R

cis configuration

H R
C=C
R H

trans configuration

Thus, in their overall shape and ability to pack together closely, *trans* fats resemble saturated fats more than *cis* unsaturated fats do. The presence of *trans* fats in membranes increases the transition temperature and decreases the membrane fluidity. As with saturated fats, consuming *trans* fats has been correlated with high blood cholesterol levels and increased risk of heart disease (a topic we'll return to in Human Connections, page 170). In 2006, the U.S. Food and Drug Administration (FDA) required food manufacturers to declare the amount of *trans* fat on Nutrition Facts labels. Soon afterward, New York City and the state of California passed bans on the use of partially hydrogenated oils, the major source of *trans* fats in the diet, in restaurant foods. In 2013, the FDA issued a preliminary statement that *trans* fats may no longer be categorized as "generally recognized as safe" for use in food.

Effects of Sterols on Membrane Fluidity. For eukaryotic cells, membrane fluidity is also affected by the presence of sterols—mainly cholesterol in animal cell membranes, phytosterols in plant cell membranes, and ergosterol in fungi. A typical animal cell contains large amounts of cholesterol (**Figure 7-14**)—up to 50% of the total membrane lipid on a molar basis. Cholesterol molecules are usually found in both layers of the plasma membrane, but a given molecule is localized to one of the two layers (Figure 7-14a). The molecule orients itself in the layer with its single hydroxyl group—the only polar part of an otherwise hydrophobic molecule—close to the polar head group of a neighboring phospholipid molecule, where it can form a hydrogen bond (Figure 7-14b). The rigid hydrophobic steroid rings and the hydrocarbon side chain of the cholesterol molecule interact with the portions of adjacent hydrocarbon chains that are closest to the phospholipid head groups.

This intercalation of rigid cholesterol molecules into the membrane of an animal cell makes the membrane less

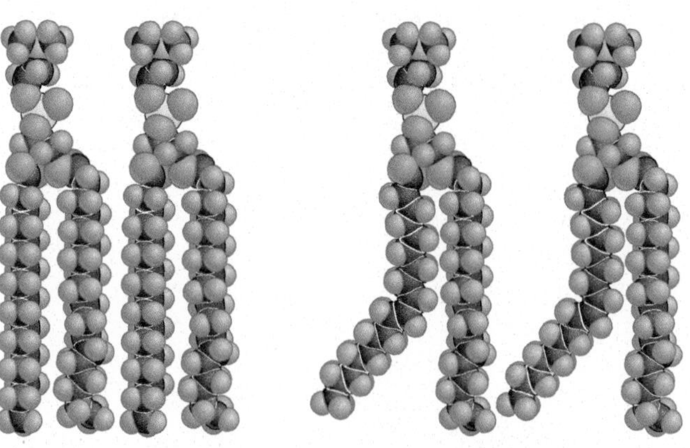

(a) Lipids with saturated fatty acids pack together well.

(b) Lipids with unsaturated fatty acids do not pack together well.

Figure 7-13 The Effect of Unsaturated Fatty Acids on the Packing of Membrane Lipids. (a) Membrane phospholipids with no unsaturated fatty acids fit together tightly because the fatty acid chains lie parallel to each other. **(b)** Membrane lipids with one or more unsaturated fatty acids do not fit together as tightly because their *cis* double bonds cause bends in the chains that interfere with packing.

separate charged molecules according to size. In one common version of this technique, *SDS-PAGE*, proteins are treated with *sodium dodecyl sulfate (SDS)* and then passed through a polymerized gel made of *polyacrylamide* (see Chapter 21, pages 645–646, for more detail).

Following electrophoresis, individual polypeptides can be detected and identified using a procedure known as *Western blotting* (described in detail in Chapter 21). In this procedure, the polypeptides in a standard protein gel are transferred to a nylon or nitrocellulose membrane, where they remain in the same relative positions that they occupied in the gel. By using antibodies that are known to bind to specific polypeptides, researchers can identify and quantify the polypeptides from the gel. Western blotting is very useful in determining which proteins are present in a particular cell and in comparing their relative abundances in different cell types.

Affinity Labeling. Biochemical approaches to studying membranes often take advantage of specific functions of different membrane proteins. One such approach, called *affinity labeling*, utilizes radioactive molecules that bind to specific proteins because of known functions of the proteins. For example, *cytochalasin B* is known to be a potent inhibitor of glucose transport. Membranes that have been exposed to radioactive cytochalasin B are therefore likely to contain radioactivity bound specifically to protein molecules involved in glucose transport.

Membrane Reconstitution. Another biochemical approach to studying membrane protein function involves the formation of artificial membrane vesicles from specific purified components. In this approach, called *membrane reconstitution*, proteins are extracted from membranes with detergent solutions. The purified proteins are then mixed together with phospholipids to form liquid-filled membrane vesicles called *liposomes* that can be "loaded" with particular molecules. These reconstituted vesicles can then be tested for their ability to carry out specific membrane protein functions, such as nutrient transport or cell-cell communication.

Determining the Three-Dimensional Structure of Membrane Proteins Is Becoming Easier

Determining the three-dimensional structure of integral membrane proteins has posed special challenges, primarily because these proteins are generally difficult to isolate and purify due to their hydrophobicity. However, they are proving increasingly amenable to study by *X-ray crystallography*, which determines the structure of proteins that can be isolated in crystalline form. For the many membrane proteins for which no crystal structure is available, an alternative approach called *hydropathy analysis* can be used, provided that the protein or its gene can at least be isolated and sequenced. We will look briefly at each of these techniques.

X-Ray Crystallography. **X-ray crystallography** is widely used to determine the three-dimensional structure of proteins. (A detailed description of this technique is included in Key Technique in Chapter 3, pages 58–59.) Due to their hydrophobicity, the difficulty of solubilizing and isolating integral membrane proteins in crystalline form virtually excluded these proteins from crystallographic analysis for many years. The first success was reported in 1985 by Hartmut Michel, Johann Deisenhofer, and Robert Huber, who crystallized the photosynthetic reaction center from the purple bacterium *Rhodopseudomonas viridis* and determined its molecular structure by X-ray crystallography. Based on their detailed three-dimensional structure of the protein, these investigators also provided the first detailed look at how pigment molecules are arranged to capture light energy (a topic we will return to in Chapter 11). In recognition of this work, Michel, Deisenhofer, and Huber shared the Nobel Prize for Chemistry in 1988.

Despite this breakthrough, the application of X-ray crystallography to the study of hydrophobic integral membrane proteins progressed very slowly until the late 1990s, particularly at the level of resolution required to identify transmembrane helices. More recently, however, there has been a veritable explosion of X-ray crystallographic data for integral membrane proteins. According to the database maintained by the Stephen White Laboratory at the University of California, Irvine, there were only 18 membrane proteins whose three-dimensional structures had been determined by 1997. Since then, almost 900 unique protein structures have been added to the list. Initially, most of these proteins were from bacterial sources, reflecting the relative ease with which membrane proteins are able to be isolated from microorganisms. Increasingly, however, membrane proteins from eukaryotic sources are proving amenable to X-ray crystallographic analysis as well.

Hydropathy Analysis. For the many integral membrane proteins that have not yet yielded to X-ray crystallography, the likely number and locations of transmembrane segments can often be inferred, provided that the protein or its gene can at least be isolated and sequenced. Once the amino acid sequence of a membrane protein is known, the number and positions of transmembrane segments can be inferred from a **hydropathy** (or **hydrophobicity**) **plot**, as shown in **Figure 7-19** on page 176. Such a plot is constructed by using a computer program to identify clusters of hydrophobic amino acids in a protein sequence. The amino acid sequence is scanned through a series of "windows," each representing a region of about 10 amino acids, with each successive window one amino acid farther along in the sequence.

Based on the known hydrophobicity values for the various amino acids, a **hydropathy index** is calculated for each successive window by averaging the hydrophobicity values of the amino acids in the window. (By convention, hydrophobic amino acids have positive hydropathy values and hydrophilic residues have negative values.) The hydropathy index is then plotted against the positions of the windows along the sequence of the protein. The resulting hydropathy plot predicts how many membrane-spanning regions might be present in the protein, based on the number of positive peaks. The hydropathy plot shown in Figure 7-19a is for a plasma membrane protein called *connexin*. The plot shows four positive peaks and therefore predicts that connexin has four stretches of hydrophobic amino acids and hence four transmembrane segments, as shown in Figure 7-19b.

(a) Hydropathy plot of connexin. The hydropathy index on the vertical axis is a numerical measure of the relative hydrophobicity of successive segments of the polypeptide chain based on its amino acid sequence.

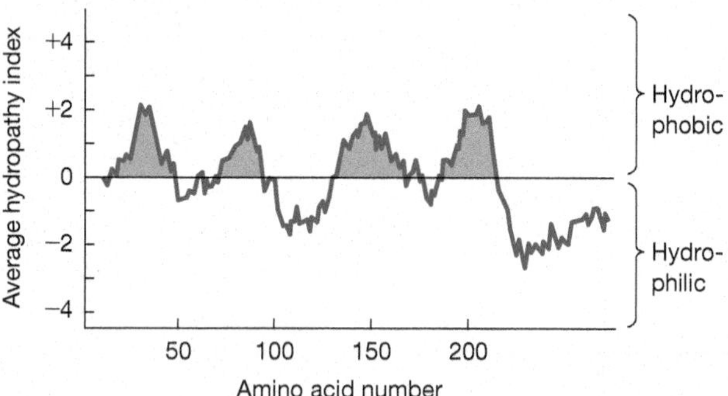

(b) Transmembrane structure of connexin. Connexin has four distinct hydrophobic regions, which correspond to the four α-helical segments that span the plasma membrane.

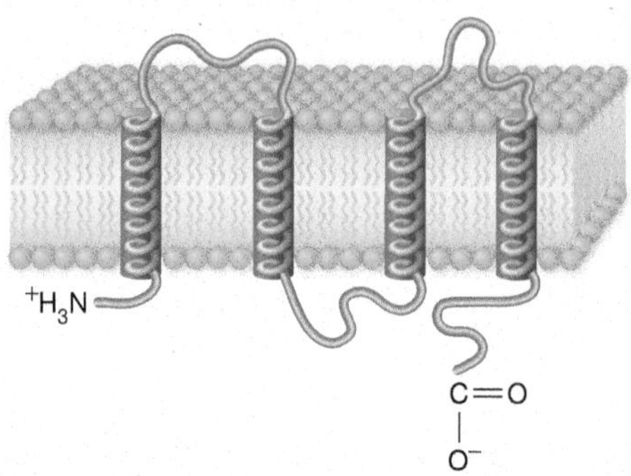

Figure 7-19 Hydropathy Analysis of an Integral Membrane Protein. A hydropathy plot is a means of representing hydrophobic regions (positive values) and hydrophilic regions (negative values) along the length of a protein. This example uses hydropathy data to analyze the plasma membrane protein *connexin*.

Molecular Biology Has Contributed Greatly to Our Understanding of Membrane Proteins

Membrane proteins have not yielded as well as other proteins to biochemical techniques, mainly because of the problems involved in isolating and purifying hydrophobic proteins in physiologically active form. Procedures such as SDS-PAGE and hydropathy analysis have certainly been useful, as have labeling techniques involving radioisotopes or fluorescent antibodies, which we will discuss later in the chapter. Within the past three decades, however, the study of membrane proteins has been revolutionized by the techniques of molecular biology, especially DNA sequencing and recombinant DNA technology.

DNA sequencing makes it possible to deduce the amino acid sequence of a protein without the need to isolate the protein in pure form for amino acid sequencing. In addition, sequence comparisons between proteins often reveal evolutionary and functional relationships that might not otherwise have been appreciated. For example, recombinant genes or gene segments can be used as probes to identify and isolate similar DNA sequences that encode related proteins. Moreover, the DNA sequence for a particular protein can be altered at specific nucleotide positions, a technique called *site-specific mutagenesis*, and then expressed in cultured cells to determine the effects of changing specific amino acids on the activity of the mutant protein it codes for. (Chapter 21 describes some of these exciting developments in more detail.)

Membrane Proteins Have a Variety of Functions

The functions of a membrane are really just those of its chemical components, especially its proteins. What types of proteins are found in membranes? Some of the proteins associated with membranes are *enzymes*, which accounts for the localization of specific functions to specific membranes. As we will see in coming chapters, each of the organelles in a eukaryotic cell is characterized by its own distinctive set of membrane-bound enzymes. In fact, such proteins are often useful as *markers* to identify particular membranes during the isolation of organelles and organelle membranes from suspensions of disrupted cells. For example, *glucose-6-phosphatase* is a membrane-bound enzyme found in the endoplasmic reticulum. Its presence in, say, a preparation of mitochondria would demonstrate contamination with ER membrane.

Also associated with membranes are *electron transport proteins* such as the cytochromes and iron-sulfur proteins that are involved in energy production in mitochondria, chloroplasts, and the plasma membranes of bacterial cells (see Chapters 10 and 11).

Other membrane proteins function in solute transport. These include *transport proteins*, which facilitate the movement of nutrients such as sugars and amino acids across membranes, and *channel proteins*, which provide hydrophilic passageways through otherwise hydrophobic membranes. Also in this category are *transport ATPases*, which use the energy of ATP to pump ions across membranes.

Numerous membrane proteins are *receptors* involved in recognizing and mediating the effects of specific chemical signals that impinge on the surface of the cell. Hormones, neurotransmitters, and growth-promoting substances are examples of chemical signals that interact with specific protein receptors on the plasma membrane of target cells. In most cases, the binding of a hormone or other signal molecule to the appropriate protein receptor on the membrane surface triggers an intracellular response. Membrane proteins are also involved in intercellular communication. Examples include the proteins that form structures called *connexons* at *gap junctions* between animal cells and those that make up the *plasmodesmata* between plant cells.

Other cellular functions in which membrane proteins play key roles include uptake and secretion of various substances by endocytosis and exocytosis; targeting, sorting, and modification of proteins within the endoplasmic reticulum and the Golgi apparatus; and the detection of light, whether by the human eye, a bacterial cell, or a plant leaf. Membrane proteins are also vital components of various structures, including the links between the plasma membrane and the extracellular matrix located outside of the cell, the pores found in the outer membranes of mitochondria and chloroplasts,

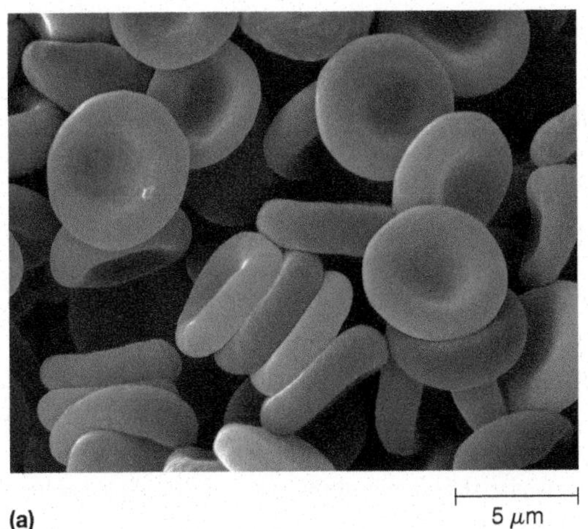

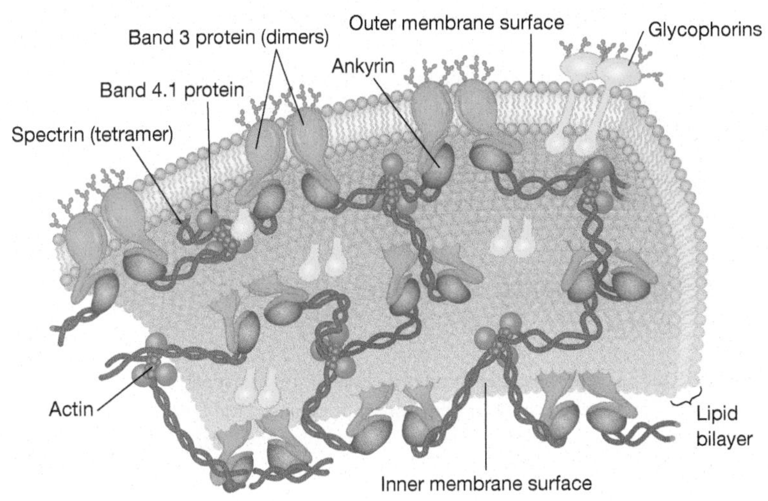

Figure 7-24 Structural Features of the Erythrocyte Plasma Membrane. (a) Human erythrocytes are a small, disk-shaped cells with a diameter of about 5 to 7 μm (colorized SEM). **(b)** The erythrocyte plasma membrane as seen from inside the cell, showing its protein composition.

An intricate meshwork of peripheral and integral membrane proteins is associated with the inner surface of the erythrocyte plasma membrane. This meshwork supports the plasma membrane, helping to maintain the distinctive biconcave shape of the human erythrocyte (Figure 7-24a) and enabling the cell to withstand stress on its membrane as it is forced through narrow capillaries in the circulatory system. The main peripheral proteins that lie immediately underneath the erythrocyte plasma membrane are *spectrin*, *ankyrin*, and a protein called *band 4.1 protein* (Figure 7-24b). Long, thin tetramers of α and β spectrin $(\alpha\beta)_2$, are linked to the glycosylated integral membrane protein *glycophorin* by band 4.1 protein and short actin filaments and to band 3 proteins by ankyrin and another protein, called band 4.2 protein. (Appropriately enough, *ankyrin* is derived from the Greek word for "anchor.") In this way, spectrin and its associated proteins provide mechanical support to the plasma membrane.

Mutations in genes for spectrin or ankyrin can cause a hereditary human condition known as spherocytosis, in which erythrocytes are spherical rather than concave. This results in anemia, a shortage of erythrocytes, and may be accompanied by jaundice and an enlarged spleen.

Not all peripheral membrane proteins at the erythrocyte membrane play structural roles. One example is the enzyme *glyceraldehyde-3-phosphate dehydrogenase (GAPDH)*, a peripheral membrane protein that associates with the cytoplasmic face of the erythrocyte membrane, where it is involved in the catabolism of blood glucose. The association is thought to depend on interaction of GAPDH with band 3.

By now, you should have an appreciation for the molecular basis of and the experimental evidence for our current understanding of membrane structure. In the next chapter, you can apply your knowledge of membrane structure and function as we investigate in detail the mechanisms by which cells can transport molecules across membranes.

CONCEPT CHECK 7.4

Proteins can be attached to the membrane in a variety of ways. How does the mode of attachment relate to their function?

Summary of Key Points

Mastering™ Biology For activities, animations, and review quizzes, go to the study area at www.masteringbiology.com.

7.1 The Functions of Membranes

- Cells have a variety of membranes that define the boundaries of the cell and its internal compartments. All biological membranes have the same general structure: a fluid phospholipid bilayer containing a mosaic of embedded proteins.
- Whereas the lipid component of membranes provides a permeability barrier, specific proteins in the membrane regulate transport of materials into and out of cells and organelles.
- Membrane proteins can detect and transduce external signals, mediate contact and adhesion between neighboring cells, or participate in cell-to-cell communication. They also produce or interact with external structures such as the cell wall or extracellular matrix.

7.2 Models of Membrane Structure: An Experimental Perspective

- Our current understanding of membrane structure represents the culmination of more than a century of studies, beginning with the recognition that lipids are an important membrane component.

- In place of earlier models, Singer and Nicolson's fluid mosaic model emerged and is now the universally accepted description of membrane structure. According to this model, proteins with varying affinities for the hydrophobic membrane interior float in and on a fluid lipid bilayer.
- Lipids may be found in microdomains known as lipid rafts that are potentially involved in cell signaling and other interactions.

7.3 Membrane Lipids: The "Fluid" Part of the Model

- Prominent lipids in most membranes include numerous types of phospholipids and glycolipids. The proportion of each lipid type can vary considerably depending on the particular membrane or monolayer.
- In eukaryotic cells, sterols–cholesterol in animal cells and phytosterols in plant cells – are important membrane components. Sterols are not found in the membranes of most bacteria, but some species contain similar compounds called hopanoids.
- Proper fluidity of a membrane is critical to its function. Cells often can vary the fluidity of membranes by changing the length and degree of unsaturation of the fatty acid chains of their membrane lipids or by the addition of cholesterol or other sterols.
- Long-chain fatty acids pack together well and decrease fluidity. Unsaturated fatty acids contain *cis* double bonds that interfere with packing and increase fluidity.
- Most membrane phospholipids are free to move within the plane of the membrane unless they are specifically anchored to structures on the inner or outer membrane surface. Transverse diffusion, or "flip-flop," of phospholipids between monolayers is not generally possible, except for when catalyzed by enzymes called phospholipid translocators, or flippases.
- Most membranes are characterized by an asymmetric distribution of lipids between the two monolayers so that the two sides of the membrane are structurally and functionally dissimilar.

7.4 Membrane Proteins: The "Mosaic" Part of the Model

- Proteins are major components of all cellular membranes. Membrane proteins are classified as integral, peripheral, or lipid anchored based on how they are associated with the lipid bilayer.
- Integral membrane proteins have one or more short segments of predominantly hydrophobic amino acids that pass through the membrane. Most transmembrane segments are α-helical sequences of 20–30 predominantly hydrophobic amino acids.
- Peripheral membrane proteins are hydrophilic and remain on the membrane surface. They are typically attached to the polar head groups of phospholipids by ionic and hydrogen bonding.
- Lipid-anchored proteins are also hydrophilic in nature but are covalently linked to the membrane by any of several lipid anchors that are embedded in the lipid bilayer.
- Membrane proteins function as enzymes, electron carriers, transport molecules, and receptor sites for chemical signals such as neurotransmitters and hormones. Membrane proteins also stabilize and shape the membrane and mediate intercellular communication and cell-cell adhesion.
- Many membrane proteins freely move within the membrane, whereas others are specifically anchored to structures on the inner or outer membrane surface.
- Many proteins in the plasma membrane are glycoproteins, with carbohydrate side chains in their extracellular regions that play important roles as recognition markers on the cell surface.
- Thanks to current advances in biochemistry, molecular biology, X-ray crystallography, affinity labeling, and the use of specific antibodies, we are learning much about the structure and function of membrane proteins that were difficult to study in the recent past.

Problem Set

7-1 Functions of Membranes. For each of the following statements, specify which one of the five general membrane functions (permeability barrier, localization of function, regulation of transport, detection of signals, or intercellular communication) the statement illustrates.

(a) When cells are disrupted and fractionated into subcellular components, the enzyme cytochrome P-450 is recovered with the endoplasmic reticulum fraction.

(b) On their outer surface, cells of multicellular organisms carry specific glycoproteins that are responsible for cell-cell adhesion.

(c) The interior of a membrane consists primarily of the hydrophobic portions of phospholipids and amphipathic proteins.

(d) Photosystems I and II are embedded in the thylakoid membrane of the chloroplast.

(e) All of the acid phosphatase in a mammalian cell is found within the lysosomes.

(f) The membrane of a plant root cell has an ion pump that exchanges phosphate inward for bicarbonate outward.

(g) Ions and large polar molecules cannot cross the membrane without the aid of a transport protein.

(h) Insulin does not enter a target cell but instead binds to a specific membrane receptor on the external surface of the membrane, thereby activating the enzyme adenylyl cyclase on the inner membrane surface.

(i) Adjacent plant cells frequently exchange cytoplasmic components through membrane-lined channels called plasmodesmata.

7-2 Elucidation of Membrane Structure. Each of the following observations played an important role in enhancing our understanding of membrane structure. Explain the significance of each, and indicate in what decade of the timeline shown in Figure 7-3 the observation was most likely made.

(a) When a membrane is observed in the electron microscope, both of the thin, electron-dense lines are about 2 nm thick, but the two lines are often distinctly different from each other in appearance.

(b) Ethylurea penetrates much more readily into a membrane than does urea, and diethylurea penetrates still more readily.

(c) The addition of phospholipase to living cells causes rapid digestion of the lipid bilayers of the membranes, which suggests that the enzyme has access to the membrane phospholipids.

(d) When artificial lipid bilayers are subjected to freeze-fracture analysis, no particles are seen on either face.

(e) The electrical resistivity of artificial lipid bilayers is several orders of magnitude greater than that of real membranes.

(f) Some membrane proteins can be readily extracted with 1 *M* NaCl, whereas others require the use of an organic solvent or a detergent.

(g) When halobacteria are grown in the absence of oxygen, they produce a purple pigment that is embedded in their plasma membranes and that has the ability to pump protons outward when illuminated. If the purple membranes are isolated and viewed by freeze-fracture electron microscopy, they are found to contain patches of crystalline particles.

7-3 Wrong Again. For each of the following false statements, change the statement to make it true and explain your reasoning.

(a) Because membranes have a hydrophobic interior, polar and charged molecules cannot pass through membranes.

(b) Proteins typically transmit signals from the outside of the cell to the cytoplasm by flip-flopping from the outer membrane monolayer to the inner monolayer.

(c) O-linked and N-linked glycoproteins are formed when sugar chains are attached to the oxygen and nitrogen atoms of the peptide bonds in proteins.

(d) The three-dimensional structure of a protein cannot be determined unless the protein can be isolated from cells in pure form.

(e) You would expect membrane lipids from tropical plants such as palm and coconut to have short-chain fatty acids with multiple C=C double bonds.

7-4 QUANTITATIVE **Gorter and Grendel Revisited.** Gorter and Grendel's classic conclusion that the plasma membrane of the human erythrocyte consists of a lipid bilayer was based on the following observations: (1) the lipids that they extracted with acetone from 4.74×10^9 erythrocytes formed a monolayer 0.89 m^2 in area when spread out on a water surface and (2) the surface area of one erythrocyte was about 100 μm^2, according to their measurements.

(a) Show from these data how Gorter and Grendel came to the conclusion that the erythrocyte membrane is a bilayer.

(b) We now know that the surface area of a human erythrocyte is about 145 μm^2. Explain how Gorter and Grendel could have come to the right conclusion when one of their measurements was only about two-thirds of the correct value.

7-5 Martian Membranes. Imagine that a new type of cell was discovered on Mars in an organism growing in benzene, a nonpolar liquid. The cell has a lipid bilayer made of phospholipids, but the bilayer's structure is very different from that of our cell membranes.

(a) Draw what might be a possible structure for this new type of membrane. What might be characteristic features of the phospholipid head groups?

(b) What properties would you expect to find in membrane proteins embedded in this membrane?

(c) How might you isolate and visualize these unusual membranes?

7-6 QUANTITATIVE **That's About the Size of It.** From chemistry, we know that each methylene ($-CH_2-$) group in a straight-chain hydrocarbon advances the chain length by about 0.13 nm. And from studies of protein structure, we know that one turn of an α helix includes 3.6 amino acid residues and extends the long axis of the helix by about 0.56 nm. Use this information to answer the following.

(a) How long is a single molecule of palmitate (16 carbon atoms) in its fully extended form? What about molecules of laurate (12 carbon atoms) and arachidate (20 carbon atoms)?

(b) How does the thickness of the hydrophobic interior of a typical membrane compare with the length of two palmitate molecules laid end to end? What about two molecules of laurate or arachidate?

(c) Approximately how many amino acids must a helical transmembrane segment of an integral membrane protein have if the segment is to span the lipid bilayer defined by two palmitate molecules laid end to end?

(d) The protein bacteriorhodopsin has 248 amino acids and seven transmembrane segments. Approximately what portion of the amino acids are part of the transmembrane segments? Assuming that most of the remaining amino acids are present in the hydrophilic loops linking the transmembrane segments together, approximately how many amino acids are present in each of these loops, on average?

7-7 Temperature and Membrane Composition. Which of the following responses are *not* likely to be seen when a bacterial culture growing at 37°C is transferred to a culture room maintained at 25°C? Explain your reasoning.

(a) Initial decrease in membrane fluidity

(b) Gradual replacement of shorter-chain fatty acids by longer-chain fatty acids in the membrane phospholipids

(c) Gradual replacement of stearate by oleate in the membrane phospholipids

(d) Enhanced rate of synthesis of unsaturated fatty acids

(e) Incorporation of more cholesterol into the membrane

7-8 Membrane Fluidity and Temperature. The effects of temperature and lipid composition on membrane fluidity are often studied by using artificial membranes containing only one or a few kinds of lipids and no proteins. Assume that you and your lab partner have made the following artificial membranes:

Membrane 1: Made entirely from phosphatidylcholine with saturated 16-carbon fatty acids.

Membrane 2: Same as membrane 1, except that each of the 16-carbon fatty acids has a single *cis* double bond.

Membrane 3: Same as membrane 1, except that each of the saturated fatty acids has only 14 carbon atoms.

After determining the transition temperatures of samples representing each of the membranes, you discover that your lab partner failed to record which membranes the samples correspond to. The three values you determined are −36°C, 23°C, and 41°C. Assign each of these transition temperatures to the correct artificial membrane, and explain your reasoning.

7-9 The Little Bacterium That Can't. *Acholeplasma laidlawii* is a small bacterium that cannot synthesize its own fatty acids and must therefore construct its plasma membrane from whatever fatty acids are available in the environment. As a result, the *Acholeplasma* membrane takes on the physical characteristics of the fatty acids available at the time.

(a) If you give *Acholeplasma* cells access to a mixture of saturated and unsaturated fatty acids, they will thrive at room temperature. Can you explain why?

(b) If you transfer the bacteria of part a to a medium containing only saturated fatty acids but make no other changes in culture conditions, they will stop growing shortly after the change in medium. Propose an explanation.

(c) What is one way you could get the bacteria of part b growing again without changing the medium? Explain your reasoning.

(d) What result would you predict if you were to transfer the bacteria of part a to a medium containing only unsaturated fatty acids without making any other changes in the culture conditions? Explain your reasoning.

7-10 DATA ANALYSIS Hydropathy: The Plot Thickens. A hydropathy plot can be used to predict the structure of a membrane protein based on its amino acid sequence and the hydrophobicity values of the amino acids. Hydrophobicity is measured as the standard free energy change, $\Delta G^{o\prime}$, for the transfer of a given amino acid residue from a hydrophobic solvent into water, in kilojoules per mole (kJ/mol). The hydropathy index is calculated by averaging the hydrophobicity values for a series of short segments of the polypeptide, with each segment displaced one amino acid farther from the N-terminus. The hydropathy index of each successive segment is then plotted as a function of the location of that segment in the amino acid sequence, and the plot is examined for regions of high hydropathy index.

(a) Why do scientists try to predict the structure of a membrane protein by this indirect means when the technique of X-ray crystallography would reveal the structure directly?

(b) Given the way it is defined, would you expect the hydrophobicity index of a hydrophobic residue such as valine or isoleucine to be positive or negative? What about a hydrophilic residue such as aspartic acid or arginine?

(c) Listed below are four amino acids and four hydrophobicity values. Match the hydrophobicity values with the correct amino acids, and explain your reasoning.

Amino acids: alanine, arginine, isoleucine, serine

Hydrophobicity (in kJ/mol): 3.1, 1.0,–1.1, –7.5

(d) Shown in **Figure 7-25** is a hydropathy plot for a specific integral membrane protein. Draw a horizontal bar over each transmembrane segment as identified by the plot. How long is the average transmembrane segment? How well does that value compare with the number you calculated in Problem 7-6c? How many transmembrane segments do you think the protein has? Can you guess which protein this might be?

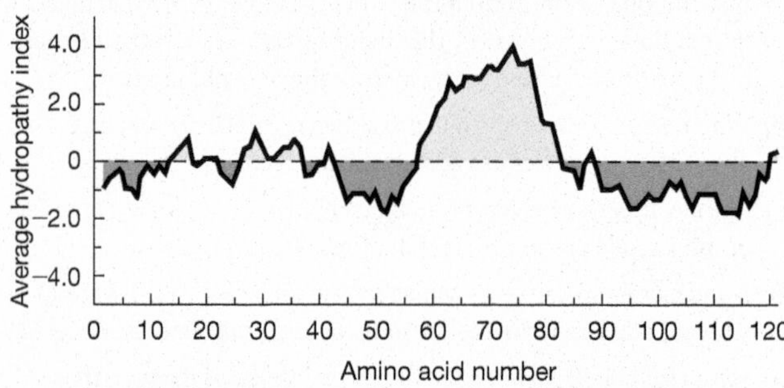

Figure 7-25 Hydropathy Plot for an Integral Membrane Protein. See Problem 7-10d.

7-11 Inside or Outside? From Figure 7-20, we know that exposed regions of membrane proteins can be labeled with ^{125}I by the lactoperoxidase (LP) reaction. Similarly, carbohydrate side chains of membrane glycoproteins can be labeled with ^{3}H by oxidation of galactose groups by galactose oxidase (GO) followed by reduction with tritiated borohydride ($^{3}H-BH_4$). Noting that both LP and GO are too large to penetrate into the interior of an intact cell, explain each of the following observations made with intact erythrocytes.

(a) When intact cells are incubated with LP in the presence of ^{125}I and the membrane proteins are then extracted and analyzed on SDS-polyacrylamide gels, several of the bands on the gel are found to be radioactive.

(b) When intact cells are incubated with GO and then reduced with $^{3}H-BH_4$, several of the bands on the gel are found to be radioactive.

(c) All of the proteins of the plasma membrane that are known to contain carbohydrates are labeled by the GO/$^{3}H-BH_4$ method.

(d) None of the proteins of the erythrocyte plasma membrane that are known to be devoid of carbohydrate is labeled by the LP/^{125}I method.

(e) If the erythrocytes are ruptured before the labeling procedure, the LP procedure labels virtually all of the major membrane proteins.

7-12 Inside-Out Membranes. It is technically possible to prepare sealed vesicles from erythrocyte membranes in which the original orientation of the membrane is inverted. Such vesicles have what was originally the cytoplasmic side of the membrane facing outward.

(a) What results would you expect if such inside-out vesicles were subjected to the GO/$^{3}H-BH_4$ procedure described in Problem 7-11?

(b) What results would you expect if such inside-out vesicles were subjected to the LP/^{125}I procedure of Problem 7-11?

(c) What conclusion would you draw if some of the proteins that become labeled by the LP/^{125}I method of part b were among those that had been labeled when intact cells were treated in the same way in Problem 7-11a?

(d) Knowing that it is possible to prepare inside-out vesicles from erythrocyte plasma membranes, can you think of a way to label a transmembrane protein with ^{3}H on one side of the membrane and with ^{125}I on the other side?

8 Transport Across Membranes: Overcoming the Permeability Barrier

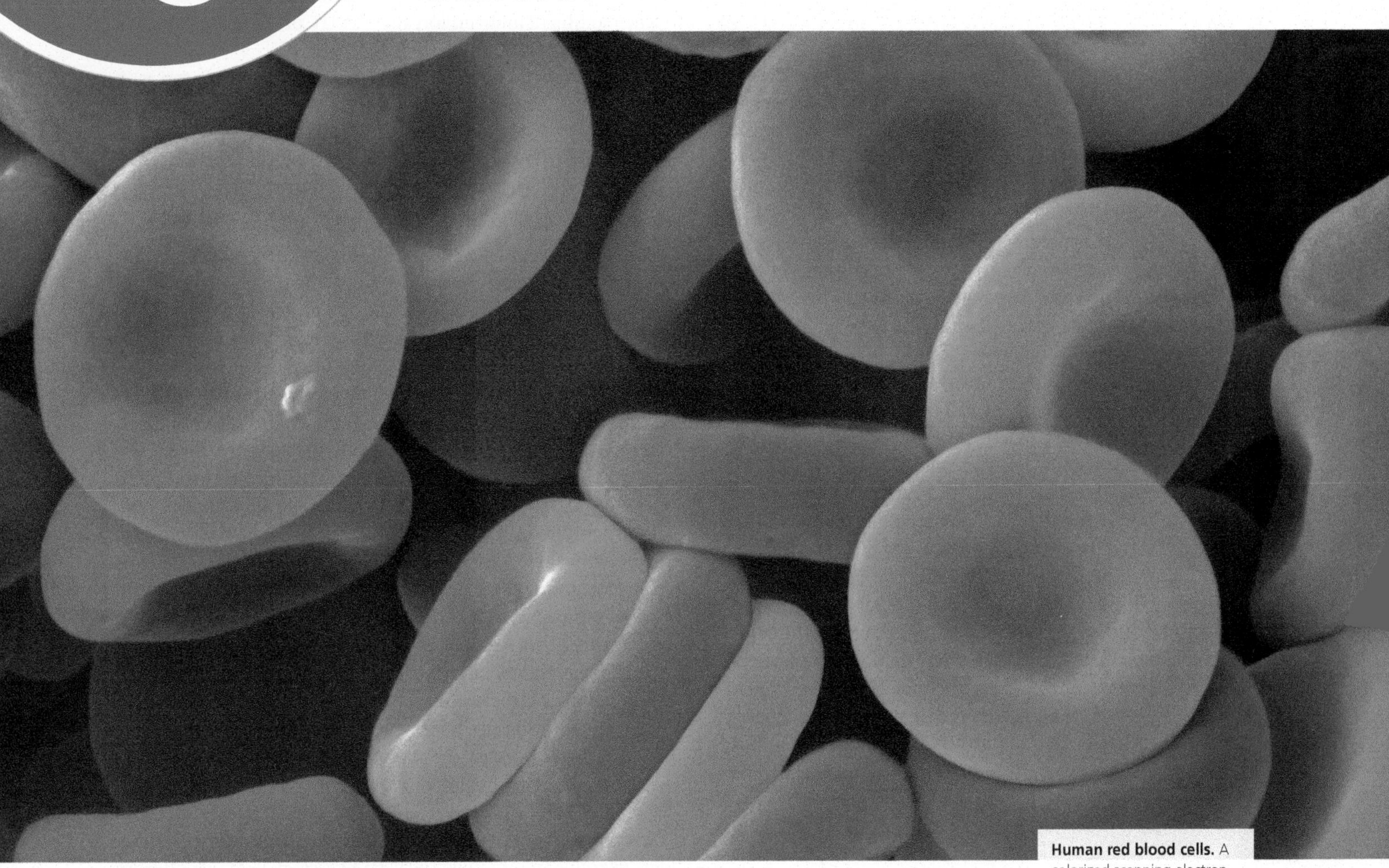

Human red blood cells. A colorized scanning electron micrograph (SEM) of red blood cells (erythrocytes).

We have previously (Chapter 7) focused on the structure and chemistry of membranes. We have noted that its hydrophobic interior makes a membrane an effective barrier to the passage of most molecules and ions, thereby keeping some substances inside the cell and others out. Within eukaryotic cells, membranes also delineate organelles by retaining the appropriate molecules and ions needed for specific functions.

However, it is not enough to think of membranes simply as permeability barriers. Crucial to the proper functioning of a cell or organelle is the ability to overcome the permeability barriers for specific molecules and ions so that they can be moved into and out of the cell or organelle selectively. In other words, membranes are not simply barriers to the indiscriminate movement of substances into and out of cells and organelles. They are also *selectively permeable*, or *semipermeable*, because they allow the controlled passage of specific molecules and ions from one side of the membrane to the other. In this chapter we will look at the ways substances are moved selectively across membranes, and we will consider the significance of such transport processes to the life of the cell.

8.1 Cells and Transport Processes

An essential feature of every cell is its ability to accumulate a variety of substances at concentrations that are often strikingly different from those in the surrounding environment. For cells to survive and function properly, they typically must maintain a constant internal environment very different from the extracellular environment, a property of living things known as *homeostasis.* Proper regulation of cellular transport is essential to ensure homeostasis and proper cell function.

Most of the substances that move across membranes are dissolved gases, ions, and small organic molecules—*solutes,* in other words. Gases include oxygen (O_2), carbon dioxide (CO_2), and nitrogen (N_2). Common ions transported across membranes are sodium (Na^+), potassium (K^+), calcium (Ca^{2+}), chloride (Cl^-), and hydrogen (H^+) ions. Most of the small organic molecules are *metabolites*—substrates, intermediates, and products in the various metabolic pathways that operate within cells or specific organelles. Sugars, amino acids, and nucleotides are some common examples.

These solutes are almost always present at higher concentrations on the inside of the cell or organelle than on the outside. Very few cellular reactions or processes could occur if they had to depend on the low concentrations at which essential substrates are present in the cell's surroundings. In some cases, such as electrical signaling in nerve and muscle tissue, the controlled movement of ions across the membrane is central to the function of the cell. In addition, many drugs, including prescribed medications, have intracellular targets and therefore must be able to cross membranes to enter the cell.

A central aspect of cell function, then, is **transport**—the ability to move ions and organic molecules across membranes selectively. The importance of membrane transport is evidenced by the fact that about 20% of the genes that have been identified in the bacterium *Escherichia coli* are involved in some aspect of transport. **Figure 8-1** summarizes a few of the many transport processes that occur within eukaryotic cells.

Solutes Cross Membranes by Simple Diffusion, Facilitated Diffusion, and Active Transport

Three fundamentally different mechanisms are involved in the movement of solutes across membranes, as shown in **Table 8-1** on page 188. A few types of solutes move across membranes by *simple diffusion*—direct, unaided movement into and through the lipid bilayer in the direction dictated by the difference in the concentrations of the solute on the two sides of the membrane.

For most solutes, however, movement across biological membranes at a significant rate is possible only because of the presence of *transport proteins*—integral membrane proteins that recognize substances with great specificity and speed their movement across the membrane. (**Key Technique, pages 196–197,** will explore how transport proteins can be studied in live cells.)

In some cases, transport proteins move solutes down their free energy gradient in the direction of thermodynamic equilibrium. The gradient represents the difference, on opposite sides of the membrane, in concentration, charge, or both concentration and charge. This mode of transport is known as *facilitated diffusion* (sometimes called *passive transport*) and requires no input of energy.

In other cases, transport proteins mediate the *active transport* of solutes, moving them against their respective free energy gradients in an energy-requiring process. Active transport must be driven by an energy-yielding process such as the hydrolysis of ATP or the simultaneous transport of another solute, usually an ion such as H^+ or Na^+, down its free energy gradient.

As we discuss each of these three transport processes in turn, it will be useful to refer back to Table 8-1.

The Movement of a Solute Across a Membrane Is Determined by Its Concentration Gradient or Its Electrochemical Potential

The movement of a solute that has no net charge is determined by the **concentration gradient** of that substance across the membrane. The concentration gradient is the magnitude of the difference in concentration of a substance on opposite sides of a membrane. A larger concentration difference represents a greater concentration gradient, which creates a larger driving force for that substance. Simple or facilitated diffusion involves exergonic and spontaneous movement "down" the concentration gradient from higher to lower concentration (a negative ΔG), whereas active transport involves endergonic (nonspontaneous) movement "up" the concentration gradient (a positive ΔG). Because it is not spontaneous, active transport requires some driving force.

The movement of a charged solute, on the other hand, is determined by its **electrochemical potential**, which is the sum of the chemical and the electrical driving forces. More specifically, electrochemical potential is the combined effect of the concentration gradient of a substance and the net difference in charge across the membrane. Movement may entail either facilitated diffusion or active transport.

Facilitated diffusion involves exergonic movement in the direction dictated by the electrochemical potential for a solute. For example, many cells have an excess of negative charges inside the cell, so that if a positive ion such as Na^+ were in equal concentration both inside and outside the cell, the charge gradient would drive the inward movement of Na^+, even though its concentration inside the cell becomes higher than outside. At thermodynamic equilibrium, there would be more Na^+ inside the cell than outside, and the concentration gradient favoring outward movement of Na^+ would exactly balance the charge gradient favoring inward movement.

In contrast, active transport involves endergonic movement of a solute *against* its electrochemical potential. Active transport creates the charge gradient, or **membrane potential (V_m)**, across the membrane that is present in most cells. Most cells have a negative V_m, which by convention means they have an excess of negatively charged solutes inside the cell. For example, a resting nerve cell has a V_m of approximately -60 mV. This charge difference favors the inward movement of cations such as Na^+ and the outward movement of anions such as Cl^-. It also opposes the outward movement of cations and the inward movement of anions. (In Chapter 22, you will learn about the role of cellular membrane potential in neuron function in detail.)

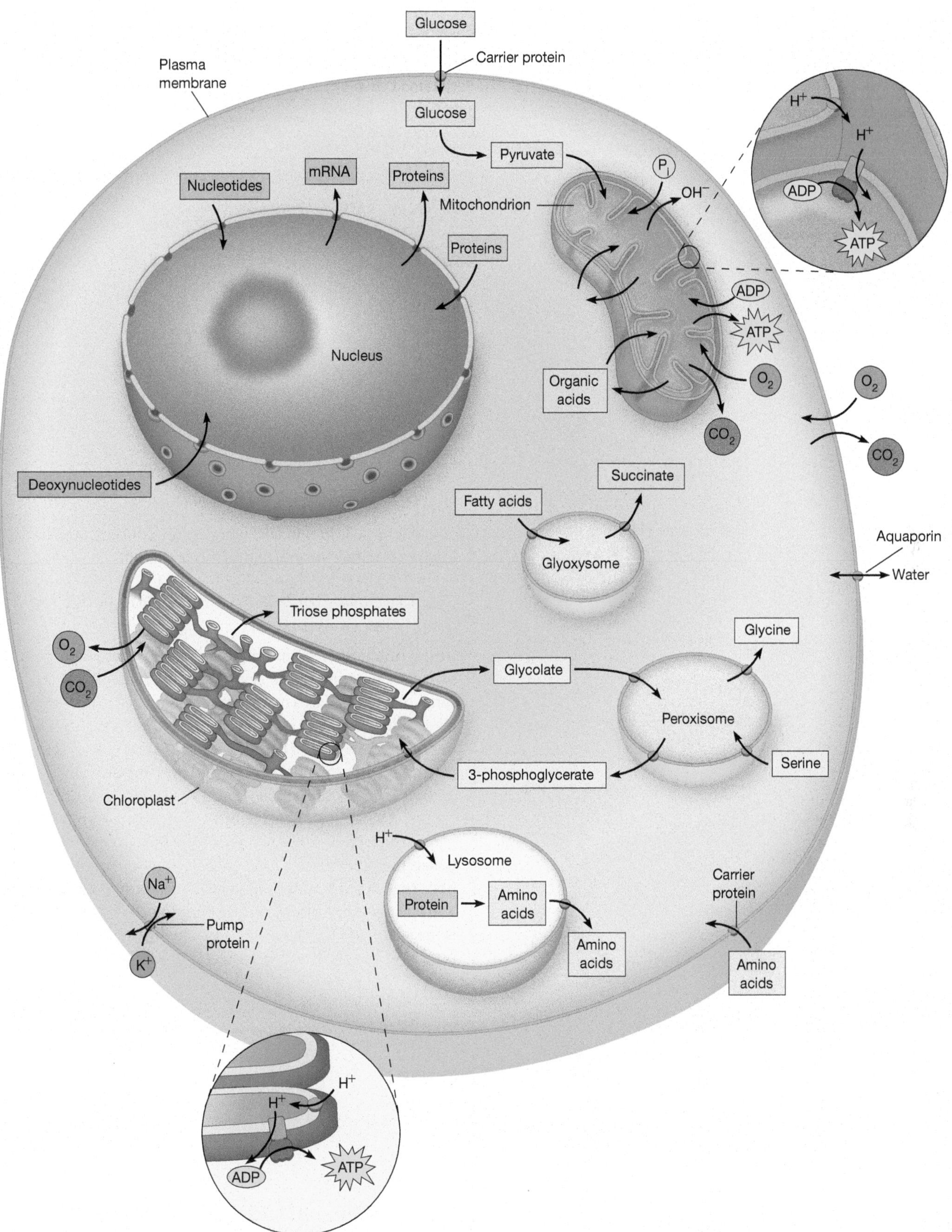

Figure 8-1 Transport Processes Within a Composite Eukaryotic Cell. The molecules and ions shown in this composite plant/animal cell are some of the many kinds of solutes that are transported across the membranes of eukaryotic cells. The enlargements depict a small portion of a mitochondrion (upper right) and a chloroplast (lower left), illustrating the movement of protons across membranes during electron transport and the use of the resulting electrochemical potential to drive ATP synthesis in these organelles.

Table 8-1 Comparison of Simple Diffusion, Facilitated Diffusion, and Active Transport

	Simple Diffusion	Facilitated Diffusion	Active Transport
Substances transported			
	Small polar molecules (H_2O, glycerol)	Small polar molecules (H_2O, glycerol)	
	Small nonpolar molecules (O_2, CO_2)	Large polar molecules (glucose)	Large polar molecules (glucose)
	Large nonpolar molecules (oils, steroids)	Ions (Na^+, K^+, Ca^{2+})	Ions (Na^+, K^+, Ca^{2+})
Thermodynamic properties			
Direction relative to electrochemical gradient	Down	Down	Up
Metabolic energy required	No	No	Yes
Intrinsic directionality	No	No	Yes
Kinetic properties			
Membrane protein required	No	Yes	Yes
Saturation kinetics	No	Yes	Yes
Competitive inhibition	No	Yes	Yes

(a) Simple diffusion. Oxygen, carbon dioxide, and water diffuse directly across the plasma membrane in response to their relative concentrations inside and outside the cell. No transport protein is required.

(b) Facilitated diffusion using carrier proteins. GLUT1 transports glucose into the erythrocyte, where the glucose concentration is lower. An anion exchange protein transports chloride (Cl^-) and bicarbonate (HCO_3^-) in opposite directions.

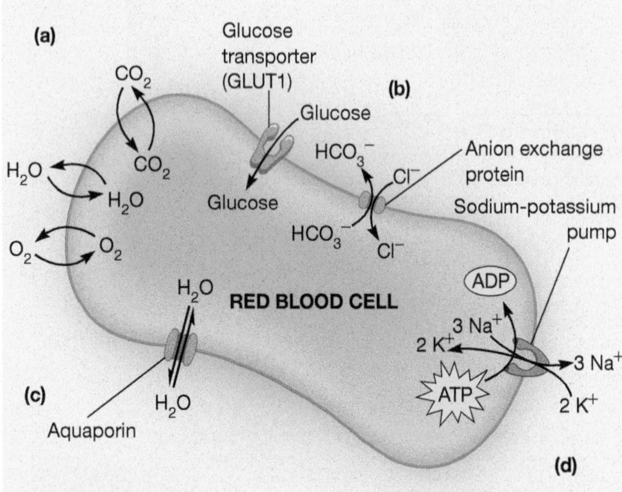

(c) Facilitated diffusion using channel proteins. Aquaporin channel proteins can facilitate the rapid inward or outward movement of water depending on the relative solute concentration on opposite sides of the membrane.

(d) Active transport using ATP-requiring pumps. Driven by the hydrolysis of ATP, the Na^+/K^+ pump moves sodium ions outward and potassium ions inward, establishing an electrochemical potential across the plasma membrane for both ions.

Figure 8-2 Important Transport Processes of the Erythrocyte. Depicted here are several types of transport processes that are vital to erythrocyte function.

An important result of active transport is distinctive asymmetric distributions of ions inside and outside of different cell types. For example, in human skeletal muscle, the concentrations of Na^+ and Cl^- are more than 10-fold higher outside the cell than inside, whereas the concentration of K^+ is approximately 40-fold higher inside the cell.

The Erythrocyte Plasma Membrane Provides Examples of Transport

In our discussion of membrane transport, we will use the transport proteins of the erythrocyte as examples. These are some of the most extensively studied and best understood of all cellular transport proteins. Vital to the erythrocyte's role in providing oxygen to body tissues is the movement across its plasma membrane of O_2, CO_2, and bicarbonate ion (HCO_3^-), as well as glucose, which serves as the cell's main energy source. Also important is the membrane potential maintained across the cell's plasma membrane by the active transport of potassium ions inward and sodium ions outward. In addition, special pores, or *channels*, allow water and ions to enter and leave the cell rapidly in response to cellular needs. These transport activities are summarized in **Figure 8-2** and will be used as examples throughout Sections 8.2 and 8.3.

CONCEPT CHECK 8.1

What is the difference between the concentration gradient and the electrochemical potential of a solute? For which of the three types of transport mechanisms is the magnitude of the concentration gradient relevant? For which is the electrochemical gradient important?

8.2 Simple Diffusion: Unassisted Movement down the Gradient

The most straightforward way for a substance to get from one side of a membrane to the other is **simple diffusion**, which is the unassisted net movement of a solute from a region where its concentration is higher to a region where its concentration is lower (Figure 8-2a). Because membranes have a hydrophobic interior, simple diffusion is typically a means of transport only for nonpolar molecules and very small polar molecules such as water, glycerol, and ethanol. Large polar molecules and all ions require an intrinsic membrane protein, known as a transporter, to move across the membrane, a subject we will return to later in the chapter.

Oxygen gas (O_2) is a small nonpolar molecule that traverses the hydrophobic lipid bilayer readily and therefore moves across membranes by simple diffusion. As **Figure 8-3** shows, this behavior enables erythrocytes in the circulatory system to take up oxygen in the lungs and release it in metabolically active body tissues. In the capillaries of body tissues, where the oxygen concentration is low, oxygen is released from hemoglobin (Chapter 3) and diffuses passively from the cytoplasm of the erythrocyte into the blood plasma and from there into the cells lining the capillaries (Figure 8-3a, left).

In the capillaries of the lungs, the opposite occurs: oxygen diffuses from the inhaled air in the lungs, where its concentration is higher, into the cytoplasm of the erythrocytes, where its concentration is lower (Figure 8-3b, left). Carbon dioxide is also able to cross membranes by simple diffusion. Not surprisingly, carbon dioxide and oxygen move across the erythrocyte membrane in opposite directions. Carbon dioxide diffuses inward in body tissues and outward in the lungs.

(a) In the capillaries of body tissues (low [O_2] and high [CO_2] relative to the erythrocytes), O_2 is released by hemoglobin within the erythrocytes and diffuses outward to meet tissue needs. CO_2 diffuses inward and is converted to bicarbonate by carbonic anhydrase in the cytosol. Bicarbonate ions are transported outward by the anion exchange protein, accompanied by the inward movement of chloride ions to to the lungs as bicarbonate ions.

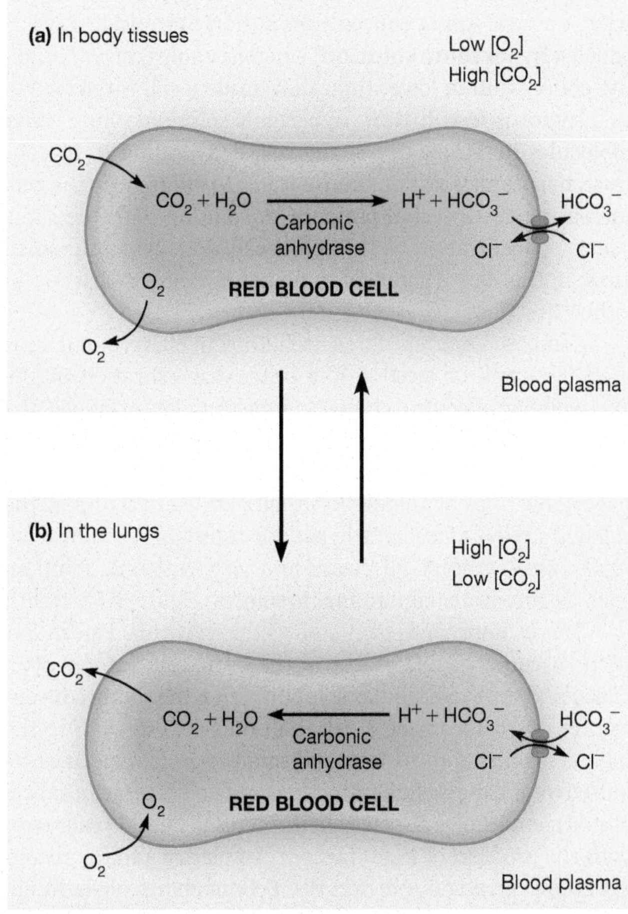

(b) In the capillaries of the lungs (high [O_2] and low [CO_2] relative to the erythrocytes), O_2 diffuses inward and binds to hemoglobin. Bicarbonate moves inward from the blood plasma, accompanied by an outward movement of chloride ions. Incoming bicarbonate is converted into CO_2, which diffuses out of the erythrocytes and into the cells lining the capillaries of the lungs. The CO_2 is now ready to be expelled from the body.

Figure 8-3 Directions of Oxygen, Carbon Dioxide, and Bicarbonate Transport in Erythrocytes. The directions in which O_2, CO_2, and HCO_3^- move across the plasma membrane of the erythrocyte depend on the location of the erythrocyte in the body.

Simple Diffusion Always Moves Solutes Toward Equilibrium

No matter how a population of solutes is distributed initially, simple diffusion always tends to create a uniform solution in which the concentration of all solutes is the same everywhere. To illustrate this point, consider the apparatus shown in **Figure 8-4** on page 190, consisting of two chambers separated by a membrane. The membrane in Figure 8-4a is freely permeable to molecules of an uncharged solute, represented by the black dots. Initially, the concentration of the solute is higher in chamber A than in chamber B. Other conditions being equal, random movements of solute molecules back and forth through the membrane will lead to a net movement of solute from chamber A to chamber B. When the concentration of solute is equal on both sides of the membrane, the system is at equilibrium. Random back-and-forth movement of individual molecules continues, but no further net change in concentration occurs. Thus, *simple diffusion is always movement toward equilibrium* and is therefore a spontaneous process. No additional energy input is required and no transport protein is needed.

Another way to express this is to say that simple diffusion always tends toward minimum free energy. Another kind of diffusion, facilitated diffusion, also tends toward minimum free energy, but, as we'll see, requires a transporter protein to move the relevant substance across the membrane. Chemical reactions and physical processes always proceed in the direction of decreasing free energy, in accordance with the second law of thermodynamics (see Chapter 5). Simple diffusion through membranes is no exception: free energy is minimized as molecules move down their concentration gradient. The driving force for simple diffusion is entropy, the randomization of molecules as concentrations equalize on both sides of the membrane. We will apply this principle later in the chapter when we calculate the free energy change (ΔG) that accompanies the transport of molecules and ions across membranes. At thermodynamic equilibrium, no further net movement occurs because the free energy of the system is at a minimum.

Osmosis Is the Simple Diffusion of Water Across a Selectively Permeable Membrane

Most of the discussion in this chapter focuses on the transport of *solutes*—ions and small molecules that are dissolved in the aqueous environment of cells, their organelles, and their

(a) Simple diffusion takes place when the membrane separating chambers A and B is permeable to molecules of dissolved solute, represented by the black dots. Net movement of solute molecules across the membrane is from chamber A to B (high to low solute concentration). Equilibrium is reached when the solute concentration is the same in both chambers.

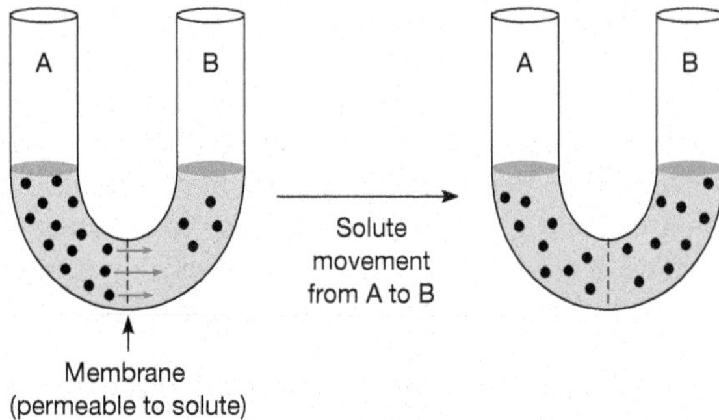

(b) Osmosis occurs when the membrane between the two chambers is not permeable to the dissoved solute, represented by the black triangles. Because solute cannot cross the membrane, water diffuses from chamber B, where the solute concentration is lower (more water), to chamber A, where the solute concentration is higher (less water). At equilibrium, the solute concentration will be equal on both sides of the membrane.

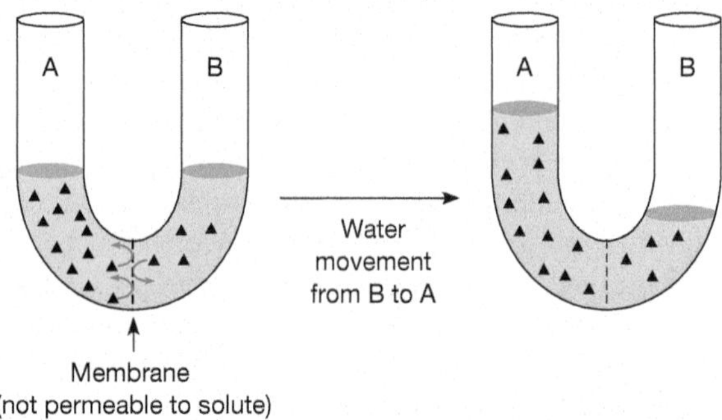

Figure 8-4 Comparison of Simple Diffusion and Osmosis. In both examples shown, there is initially more solute in chamber A than in chamber B. The membrane in example **(a)** is permeable to the solute, and the membrane in example **(b),** like a typical cell membrane, is not.

surroundings. That emphasis is quite appropriate because most of the traffic across membranes involves ions such as K^+, Na^+, and H^+ and hydrophilic molecules such as sugars, amino acids, and a variety of metabolic intermediates. But to understand solute transport fully, we also need to understand the forces that act on the water in which the solute is dissolved, thereby determining its movement into and out of cells.

Because most solutes cannot cross cell membranes by simple diffusion, water tends to move across membranes in response to differences in solute concentrations on the two sides of the membrane. Specifically, water will diffuse from the side of the membrane with the lower solute concentration (and slightly higher water concentration) to the side with the higher solute concentration (and slightly lower water concentration). This simple diffusion of water in response to differences in solute concentration is called **osmosis** and is readily observed when a selectively permeable membrane separates two compartments, one of them containing a solute that cannot cross the membrane. Water will move across the membrane to equalize the concentration of solutes on both sides of the membrane.

This principle is illustrated by Figure 8-4b. Solutions of differing solute concentrations are placed in chambers A and B, as in Figure 8-4a, but with the two chambers now separated by a *selectively permeable membrane* that is permeable to water but *not* to the dissolved solute. Under these conditions, water diffuses across the membrane from chamber B to chamber A. Although looking very different from the tube shown in Figure 8-4a, cells behave in much the same way. Because the concentration of solutes is almost always higher inside a cell than outside, for most cells this means that water will tend to move inward. If not controlled, that inward movement of water will cause cells that lack a cell wall to swell and perhaps burst.

Osmotic movement of water into and out of a cell is related to the relative *osmolarity*, or total solute concentration, of the cytoplasm versus the extracellular solution. A solution with a higher solute concentration than that inside a cell is called a **hypertonic solution,** whereas a solution with a solute concentration lower than that inside a cell is referred to as a **hypotonic solution.** Hypertonic solutions cause water molecules to diffuse out of the cell, dehydrating it. In contrast, hypotonic solutions cause water to diffuse into the cell, increasing the internal pressure. A solution with the same solute concentration as the cell is called an **isotonic solution;** in this case, there will be no net movement of water in either direction.

Osmosis accounts for a well-known observation: cells tend to shrink or swell as the solute concentration of the extracellular medium changes. Consider, for example, the scenario of **Figure 8-5**. An animal cell that starts out in an isotonic solution will shrink and shrivel if it is transferred to a hypertonic solution (Figure 8-5a, left). On the other hand, the cell will swell if placed in a hypotonic solution. And it will actually *lyse*—or burst—if placed in a very hypotonic solution, such as pure water containing no solutes (Figure 8-5a, right).

The osmotic movements of water shown in Figure 8-5 occur because of differences in the osmolarity of the cytoplasm and the extracellular solution. Often the solute concentration is greater inside a cell than outside, due both to the high concentrations of ions and small organic molecules required for normal cellular functions and to the large numbers of macromolecules dissolved in the cytosol. How do cells cope with the problem of high internal osmolarity and the resulting influx of water due to osmosis? Cells of plants, algae, fungi, and many bacteria have rigid cell walls that keep the cells from swelling and bursting in a hypotonic solution (Figure 8-5b). In the case of a non-woody plant, hypotonic conditions are in fact normal. Instead of bursting, the cells become very firm from the **turgor pressure** that builds up due to the inward movement of water. The resulting turgidity accounts for the firmness, or *turgor,* of fully hydrated plant tissue (Figure 8-5b, right). Without this turgidity, the tissue will wilt.

In a hypertonic solution, on the other hand, the outward movement of water causes the plasma membrane to pull away from the cell wall by a process called **plasmolysis** (Figure 8-5b, left). You can demonstrate plasmolysis readily by

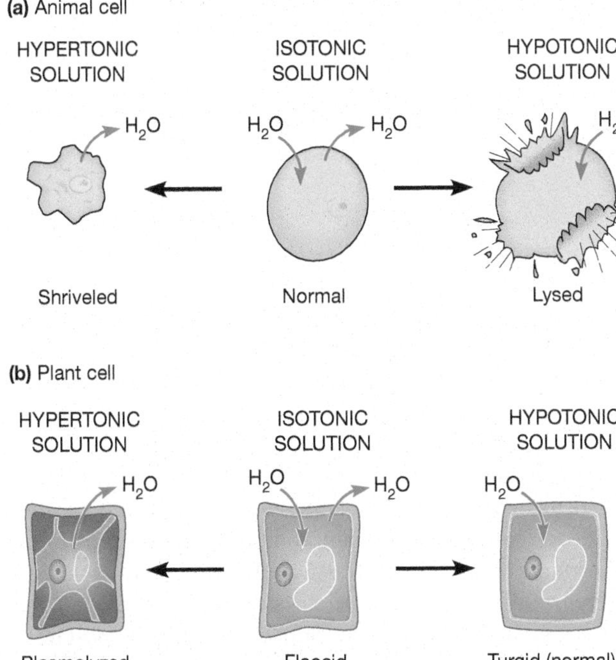

Figure 8-5 Responses of Animal and Plant Cells to Changes in Osmolarity. (a) If an animal cell (or other cell not surrounded by a cell wall) is transferred from an isotonic solution to a hypertonic solution, water leaves the cell, and the cell shrivels (left arrow). If the cell is transferred to a hypotonic solution, water enters the cell, and the cell swells, sometimes to the point that it bursts (right arrow). **(b)** Plant cells (or other cells with rigid cell walls) also shrivel (plasmolyze) in a hypertonic solution (left arrow), but they will only become turgid—their normal state—in a hypotonic solution (right arrow).

dropping a piece of celery into a solution with a high concentration of salt or sugar. Plasmolysis can be a practical problem when plants are grown under conditions of high salinity, as is sometimes the case in locations near an ocean.

Cells without cell walls solve the osmolarity problem by continuously and actively pumping out inorganic ions, thereby reducing the intracellular osmolarity and thus minimizing the difference in solute concentration between the cell and its surroundings. Animal cells continuously remove sodium ions. This is, in fact, one important purpose of the *Na^+/K^+ pump* (discussed in detail in Section 8.5). Animal cells swell and sometimes even lyse when treated with *ouabain*, an inhibitor of the Na^+/K^+ pump. When medications are given intravenously in a hospital, they are typically dissolved in *phosphate-buffered saline*, which has the same osmolarity as the blood, avoiding potential problems of cell lysis or dehydration.

Simple Diffusion Is Typically Limited to Small, Uncharged Molecules

To investigate the factors influencing the diffusion of solutes through membranes, scientists frequently use membrane models. An important advance in developing such models was provided in 1961 by Alec Bangham and his colleagues, who found that when lipids are extracted from cell membranes and dispersed in water, they form *liposomes*. Liposomes are small vesicles about 0.1 μm in diameter, each consisting of a closed, spherical lipid bilayer that is devoid of membrane proteins (see Figure 4-2). Bangham showed that it is possible to trap solutes such as potassium ions in the liposomes as they form and then measure the rate at which the solute escapes by diffusion across the liposome bilayer.

The results of Bangham's liposome experiment were remarkable. Ions such as potassium and sodium were trapped in the vesicles for days, whereas small uncharged molecules such as oxygen exchanged so rapidly that the rates could hardly be measured. The inescapable conclusion was that the lipid bilayer represents the primary permeability barrier of a membrane. Small, uncharged molecules can pass through the barrier by simple diffusion, whereas sodium and potassium ions can barely pass at all. Based on subsequent experiments by many investigators using a variety of lipid bilayer systems and thousands of different solutes, we can predict with considerable confidence how readily a solute will diffuse across a lipid bilayer. The three main factors affecting diffusion of solutes are *size*, *polarity*, and *charge*. We will consider each of these factors in turn.

Solute Size. Generally speaking, lipid bilayers are more permeable to smaller molecules than to larger molecules. The smallest molecules relevant to cell function are water, oxygen, and carbon dioxide. Membranes are quite permeable to these molecules, which require no specialized transport processes for moving into and out of cells, although we will see later that there is a specialized transporter in some cells allowing rapid transport of water molecules across certain membranes. But, in the absence of a transporter, even such small molecules do not move across membranes freely. Water molecules, for example, diffuse across a bilayer 10,000 times more slowly than they move when allowed to diffuse freely in the absence of a membrane.

Still, water diffuses across membranes at rates much higher than would be expected for such a polar molecule. The reason for this behavior is not well understood. One proposal is that membranes contain tiny pores that allow the passage of water molecules but are too small for any other polar substance. An alternative suggestion is that in the continual movements of membrane lipids, transient "holes" are created in the lipid monolayers that allow water molecules to move first through one monolayer and then through the other. There is little experimental evidence to support either of these hypotheses, however, and simple diffusion of water across membranes remains an enigma.

In addition to water, small polar molecules with a molecular weight (MW) of up to about 100—such as ethanol (CH_3CH_2OH; MW = 46) and glycerol ($C_3H_8O_3$; MW = 92)—are able to diffuse across membranes. However, larger polar molecules such as glucose ($C_6H_{12}O_6$; MW = 180) cannot. Cells therefore need specialized proteins in their plasma membranes to facilitate the entry of glucose and most other polar solutes.

Solute Polarity. There is a correlation between the lipid solubility of a molecule and its membrane permeability. In general, lipid bilayers are more permeable to nonpolar molecules and less permeable to polar molecules. This is because nonpolar molecules dissolve more readily in the hydrophobic phase of the lipid bilayer and can therefore cross the membrane much more rapidly than can polar molecules of similar size. For example, the steroid hormones estrogen and testosterone are largely nonpolar and therefore can diffuse across membranes despite having molecular weights of 370 and 288, respectively.

A simple measure of the polarity (or nonpolarity) of a solute is its *partition coefficient*, which is the ratio of its solubility in an organic solvent (such as vegetable oil or octanol) to its solubility in water. In general, the more nonpolar—or hydrophobic—a substance is, the higher its partition coefficient and the more readily and rapidly it can move across (or dissolve in) a membrane. For example, the partition coefficients of the various amino acids can be used to calculate the hydropathy index of a protein (see Figure 7-19).

When incorporated into proteins as amino acid residues, the amino acids no longer have two charged ends as do free amino acids, and the degree of polarity is determined solely by the side chain. Amino acids with nonpolar side chains (see Figure 3-2), such as tryptophan, leucine, and valine, have high partition coefficients and are likely to be found in transmembrane regions of a membrane protein, in contrast to those with polar side chains and low partition coefficients (serine and threonine).

Solute Charge. The relative impermeability of polar substances in general and of ions in particular is due to their strong association with water molecules, which form a *shell of hydration* around the ion. For such solutes to move into the membrane, the associated water molecules must be stripped off, and an input of energy is required to eliminate the bonding between the ions and the water molecules. Therefore, the association of ions with water molecules to form shells of hydration dramatically restricts ion transport across membranes.

The impermeability of membranes to ions is important to cell activity because for it to function, every cell must maintain an electrochemical potential across its plasma membrane. In most cases this potential is a gradient of either sodium ions (animal cells) or protons (most other cells). On the other hand, membranes must also allow ions to cross the barrier in a controlled manner. As we will see later in the chapter, the proteins that facilitate ion transport serve as hydrophilic channels that provide a low-energy pathway for movement of the ions across the membrane.

The Rate of Simple Diffusion Is Directly Proportional to the Concentration Gradient

So far, we have focused on qualitative aspects of simple diffusion. We can be more quantitative by considering the thermodynamic and kinetic properties of the process (see Table 8-1). Thermodynamically, simple diffusion is always an exergonic process, requiring no input of additional energy. Individual molecules simply diffuse randomly in both directions, but net flux will always be in the direction of minimum free energy—which in the case of uncharged molecules means down the concentration gradient.

Kinetically, a key feature of simple diffusion is that the net rate of transport for a specific substance is directly proportional to the concentration difference for that substance on opposite sides of the membrane. For the diffusion of solute S from the outside to the inside of a cell, the expression for the rate, or velocity, of inward diffusion through the membrane, v_{inward}, is

$$v_{inward} = P\,\Delta[S] \qquad \textbf{(8-1)}$$

where v_{inward} is the rate of inward diffusion (in moles/second-cm^2 of membrane surface), and $\Delta[S]$ is the concentration gradient of the solute across the membrane ($\Delta[S] = [S]_{outside} - [S]_{inside}$). *P* is the permeability coefficient, an experimentally determined parameter that depends on the thickness and viscosity of the membrane; the size, shape, and polarity of S; and the equilibrium distribution of S in the membrane and aqueous phases. As either the permeability or the concentration gradient increases, the rate of inward transport increases.

As Equation 8-1 indicates, simple diffusion is characterized by a linear relationship between the rate of inward flux of the solute across the membrane and the concentration gradient of the solute, with no evidence of saturation at high concentrations. This relationship is seen as the straight red line in **Figure 8-6**. Simple diffusion differs in this respect from facilitated diffusion, which is subject to saturation and generally follows hyperbolic Michaelis–Menten kinetics, as we will see shortly.

We can summarize simple diffusion by noting that it is relevant only to molecules such as ethanol and O_2 that are small enough and/or nonpolar enough to cross membranes without the aid of transport proteins. Simple diffusion proceeds exergonically in the direction dictated by the concentration gradient, with a linear, nonsaturating relationship between the diffusion rate and the concentration gradient.

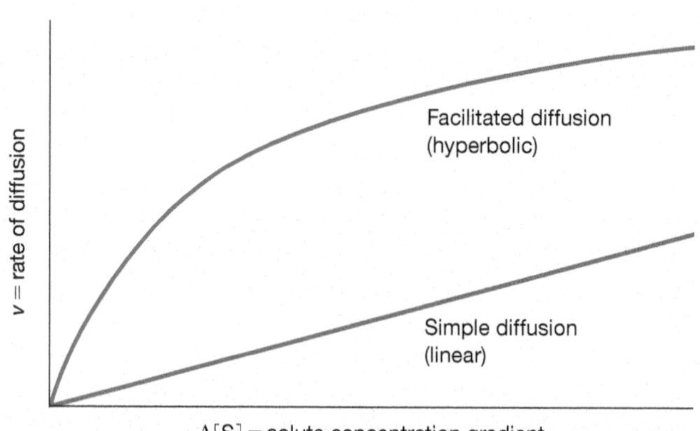

Figure 8-6 Comparison of the Kinetics of Simple Diffusion and Facilitated Diffusion. For simple diffusion across a membrane, the relationship between *v*, the rate of diffusion, and $\Delta[S]$, the solute concentration gradient, is linear over a broad concentration range (red line). For facilitated diffusion, the relationship is linear when the concentration gradient is small but exhibits saturation kinetics and is therefore hyperbolic (blue line), eventually reaching a maximum value at very high $\Delta[S]$.

CONCEPT CHECK 8.2

How is osmosis different from the simple diffusion of molecular oxygen (O_2) across a membrane? How are they similar?

8.3 Facilitated Diffusion: Protein-Mediated Movement down the Gradient

Most substances in cells are too large or too polar to cross membranes at reasonable rates by simple diffusion, even if the process is exergonic. Such solutes can move into and out of cells and organelles at appreciable rates only with the assistance of **transport proteins** that mediate the movement of solute molecules across the membrane. If such movement is exergonic, it is called **facilitated diffusion** because the solute still diffuses in the direction dictated by the concentration gradient (for uncharged molecules) or by the electrochemical gradient (for ions), with no input of additional energy needed. The role of the transport protein in facilitated diffusion is simply to provide a path through the hydrophobic lipid bilayer, facilitating the "downhill" diffusion of a large, polar, or charged solute across an otherwise impermeable barrier.

Unlike simple diffusion, whose rate increases proportionally with the concentration difference across the membrane, facilitated diffusion can become saturated at high solute concentrations because there are a limited number of transport proteins. Thus, whereas a graph of rate versus concentration difference will be linear for simple diffusion, it will be hyperbolic for facilitated diffusion (Figure 8-6, blue curve). Note that this resembles the hyperbolic shape of the curve for the initial velocity of an enzyme versus substrate concentration (see Figure 6-8).

As an example of facilitated diffusion, consider the movement of glucose across the plasma membrane of a cell in your body. The concentration of glucose is typically higher in the blood than in the cell, so the inward transport of glucose is exergonic—that is, it does not require the input of energy. However, glucose is too large and too polar to diffuse across the membrane unaided. A transport protein is required to facilitate its inward movement.

Carrier Proteins and Channel Proteins Facilitate Diffusion by Different Mechanisms

Transport proteins involved in facilitated diffusion of small molecules and ions are integral membrane proteins that contain several, or even many, transmembrane segments and therefore traverse the membrane multiple times. Functionally, these proteins fall into two main classes that transport solutes in quite different ways. **Carrier proteins** (also called *transporters* or *permeases*) bind one or more solute molecules on one side of the membrane and then undergo a conformational change that transfers the solute to the other side of the membrane. In so doing, a carrier protein binds the solute molecules in such a way as to shield the polar or charged groups of the solute from the nonpolar interior of the membrane.

Channel proteins, on the other hand, form hydrophilic *channels* through the membrane that allow the passage of solutes without a major change in the conformation of the protein. A key difference between carriers and channels is that when channels are not blocked, they are open to the inner and outer surface of a membrane at the same time. Some of these channels are relatively large and nonspecific, such as the *pores* found in the outer membranes of bacteria, mitochondria, and chloroplasts. Pores are formed by transmembrane proteins called *porins* and allow selected hydrophilic solutes with molecular weights up to about 600 to diffuse across the membrane. However, most channels are small and highly selective. Most of these smaller channels are involved in the transport of ions rather than molecules and are therefore referred to as *ion channels*. The movement of solutes through ion channels is much more rapid than transport by carrier proteins, presumably because complex conformational changes are not necessary.

MAKE CONNECTIONS 8.1

Recall what you learned about R groups and how they affect the hydrophobic or hydrophilic character of an amino acid. Where would you expect to find amino acids with hydrophobic side chains in a channel protein? Where would you expect to find amino acids with hydrophilic/polar side chains? Explain your answer. (Fig. 3-2)

Carrier Proteins Alternate Between Two Conformational States

The **alternating conformation model** states that a carrier protein is an allosteric membrane-spanning protein that alternates between two conformational states. In one state, the solute-binding site of the protein is open or accessible on one side of the membrane. Following solute binding, the protein changes to an alternate conformation in which the solute-binding site is on the other side of the membrane, triggering its release. We will encounter an example of this mechanism shortly when we discuss the facilitated diffusion of glucose into erythrocytes.

Carrier Proteins Are Analogous to Enzymes in Their Specificity and Kinetics

As we noted earlier, carrier proteins are sometimes called *permeases*. This term is apt because the suffix *–ase* suggests a similarity between carrier proteins and enzymes. Like an enzyme-catalyzed reaction, carrier-facilitated diffusion involves an initial binding of the "substrate" (the solute to be transported) to a specific site on a protein surface (the solute's binding site on the carrier protein) and the eventual release of the "product" (the transported solute), with an "enzyme-substrate" complex (solute bound to carrier protein) as an intermediate. Like enzymes, carrier proteins can be regulated by external factors that bind and modulate their activity.

Specificity of Carrier Proteins. Another property that carriers share with enzymes is *specificity*. Like enzymes, carrier proteins are highly specific, often for one compound or a small group of closely related compounds and sometimes even for a specific stereoisomer. A good example is the carrier protein that facilitates the diffusion of glucose into erythrocytes (see Figure 8-2b). This protein recognizes only glucose and a few closely related monosaccharides, such as galactose and mannose. Moreover, the protein is *stereospecific*: it accepts the D- but not the L-isomer of these sugars. This specificity is presumably a result of a precise stereochemical fit between the solute and its binding site on the carrier protein.

Thus, the properties of carrier proteins explain the characteristic features of carrier-facilitated diffusion: movement of polar molecules and ions down a concentration gradient, specificity for the particular substrate transported, the ability to be saturated at high levels of substrate, and sensitivity to specific inhibitors of transport.

Kinetics of Carrier Protein Function. As you might expect from the analogy with enzymes, carrier proteins become saturated as the concentration of the transportable solute is raised. This is because the number of transport proteins is limited, and each transport protein functions at some finite maximum velocity. As a result, carrier-facilitated transport, like enzyme catalysis, exhibits *saturation kinetics.* This type of transport has an upper limiting velocity V_{max} and a constant K_m corresponding to the concentration of transportable solute needed to achieve one-half of the maximum rate of transport. This means that the initial rate of solute transport, v_0, can be described mathematically by the same equation we used to describe enzyme kinetics (see Equation 6-7, on page 142):

$$v_0 = \frac{V_{max}[S]}{K_m + [S]} \quad \textbf{(8-2)}$$

where [S] is the initial concentration of solute on one side of the membrane (for example, on the outside of the membrane if the initial rate of inward transport is to be determined). A plot of transport rate versus initial solute concentration is therefore hyperbolic for facilitated diffusion instead of linear as for simple diffusion (see Figure 8-6, blue curve). This difference in saturation kinetics is an important means of distinguishing between simple and facilitated diffusion (see Table 8-1).

A further similarity between enzymes and carrier proteins is that carriers are often subject to *competitive inhibition* by molecules or ions that are structurally related to the intended substrate. For example, the transport of glucose by a glucose carrier protein is competitively inhibited by the other monosaccharides that the protein also accepts—that is, the rate of glucose transport is reduced in the presence of other transportable sugars.

Carrier Proteins Transport Either One or Two Solutes

Although carrier proteins are similar in their kinetics and their presumed mechanism of action involving alternate conformations, they may differ in significant ways. The most important differences concern the number of solutes transported and the direction they move (**Figure 8-7**). When a carrier protein transports a single solute across the membrane, the process is called **uniport** (Figure 8-7a). The glucose carrier protein we will discuss shortly is a *uniporter.* When two solutes are transported simultaneously and their transport is coupled such that transport of either stops if the other is absent, the process is called **coupled transport** (Figure 8-7b). Coupled transport is referred to as **symport** (or *cotransport*) if the two solutes are moved in the same direction or as **antiport** (or *countertransport*) if the two solutes are moved in opposite directions across the membrane. The transport proteins that mediate these processes are called *symporters* and *antiporters*, respectively.

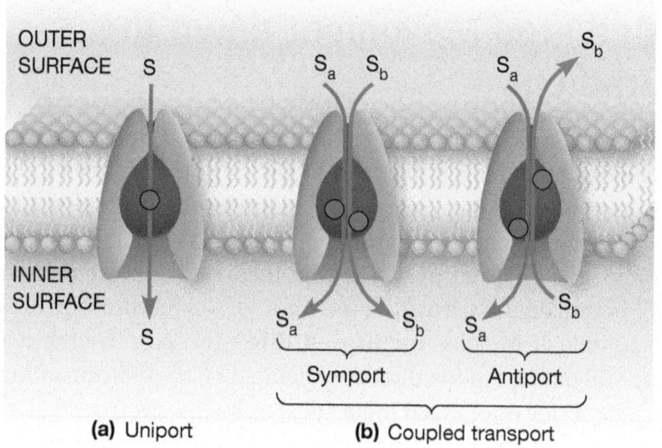

Figure 8-7 Comparison of Uniport, Symport, and Antiport Transport by Carrier Proteins. (a) In uniport transport, a membrane carrier protein moves a single solute (S) across a membrane. **(b)** In coupled transport, a membrane carrier protein simultaneously moves two solutes (S_a and S_b) either in the same direction (symport) or in opposite directions (antiport).

The Erythrocyte Glucose Transporter and Anion Exchange Protein Are Examples of Carrier Proteins

Now that we have described the general properties of carrier proteins, we will consider two specific examples: the uniport carrier for glucose and the antiport anion carrier for chloride (Cl^-) and bicarbonate (HCO_3^-). Both of these transporters are present in the plasma membranes of erythrocytes.

The Glucose Transporter: A Uniport Carrier. As we noted earlier, the movement of glucose into an erythrocyte is an example of facilitated diffusion mediated by a uniport carrier protein (see Figure 8-2b). The concentration of glucose in blood plasma is usually in the range of 65–90 mg/100 mL, or about 3.6–5.0 m*M*. The erythrocyte (or almost any other cell in contact with the blood, for that matter) is capable of glucose uptake by facilitated diffusion because of its low intracellular glucose concentration and the presence in its plasma membrane of a glucose carrier protein, or **glucose transporter (GLUT).** The GLUT of erythrocytes is called GLUT1 to distinguish it from related GLUTs in other mammalian tissues. GLUT1 allows glucose to enter the cell about 50,000 times faster than it would enter by free diffusion through a lipid bilayer.

GLUT1-mediated uptake of glucose displays all the classic features of facilitated diffusion: it is specific for glucose (and a few related sugars, such as galactose and mannose), it proceeds down a concentration gradient without energy input, it exhibits saturation kinetics, and it is susceptible to competitive inhibition by related monosaccharides. GLUT1 is an integral membrane protein with 12 hydrophobic transmembrane segments. These are presumably folded and assembled in the membrane to form a cavity lined with hydrophilic side chains that form hydrogen bonds with glucose molecules as they move through the membrane.

GLUT1 is thought to transport glucose by an alternating conformation mechanism, as illustrated in **Figure 8-8**. The alternate conformational states are called T_1, which has the

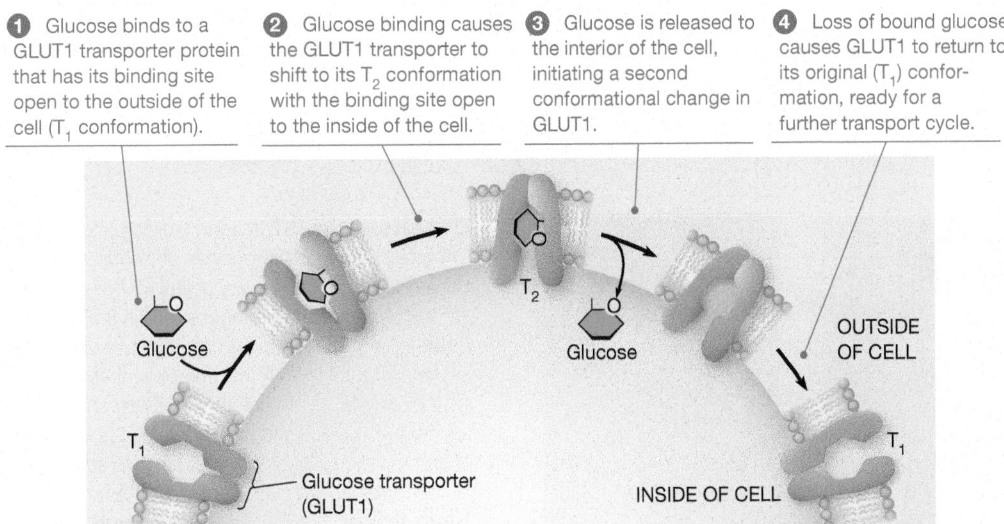

Figure 8-8 The Alternating Conformation Model for Facilitated Diffusion of Glucose by the Glucose Transporter GLUT1 in the Erythrocyte Membrane. The inward transport of glucose by GLUT1 is shown here in four steps, arranged at the periphery of a cell.

binding site for glucose open to the outside of the cell, and T_2, with the binding site open to the interior of the cell. The process begins when ❶ a molecule of D-glucose collides with and binds to a GLUT1 molecule that is in its T_1 conformation. ❷ With glucose bound, GLUT1 now shifts to its T_2 conformation. ❸ The conformational change allows the release of the glucose molecule to the interior of the cell, after which ❹ the GLUT1 molecule returns to its original conformation, with the glucose-binding site again facing outward.

The example shown in Figure 8-8 is for inward transport, but the process is readily reversible because carrier proteins function equally well in either direction. A carrier protein is really just a gate in an otherwise impenetrable wall, and, like most gates, it facilitates traffic in either direction. Individual solute molecules may be transported either inward or outward, depending on the relative concentrations of the solute on the two sides of the membrane. If the concentration is lower inside, net flow will be inward. If the lower concentration occurs outside, net flow will be outward.

The low intracellular glucose concentration that makes facilitated diffusion possible for most animal cells exists because incoming glucose is quickly phosphorylated to glucose-6-phosphate by the enzyme hexokinase, with ATP as the phosphate donor and energy source:

$$\text{glucose} \xrightarrow[\text{hexokinase}]{\text{ATP} \;\; \text{ADP}} \text{glucose-6-phosphate} \tag{8-3}$$

This hexokinase reaction is the first step in glucose metabolism (which we will discuss in Chapter 9). The low K_m of hexokinase for glucose (1.5 m*M*) and the highly exergonic nature of the reaction ($\Delta G^{\circ\prime} = -4.0$ kcal/mol) ensure that the concentration of free glucose within the cell is kept low, maintaining the concentration gradient across the cell membrane. For many mammalian cells, the intracellular glucose concentration ranges from 0.5 to 1.0 m*M*, which is about 15–20% of the glucose level in the blood plasma outside the cell.

The phosphorylation of glucose also has the effect of locking glucose in the cell, because the plasma membrane of the erythrocyte does not have a transport protein for glucose-6-phosphate. GLUT1, like most sugar transporters, does not recognize the phosphorylated form of the sugar. In addition, due to phosphorylation, the level of free glucose is kept low in the cell, ensuring that equilibrium is not reached and allowing the cell to continue to import glucose.

The GLUT1 of erythrocytes is just one of several glucose transporters in mammals. In humans there are 14 different GLUT proteins, each encoded by a separate gene. Each transporter has distinct physical and kinetic characteristics appropriate for its function in the particular cell type where it is found. For example, GLUT3 and GLUT4 are found in cerebral neurons and skeletal muscle, respectively, where they import glucose for energy. On the other hand, GLUT2 is the glucose transporter in liver cells, which break down glycogen to produce glucose for the blood. GLUT2 has properties that facilitate glucose transport *out* of liver cells to maintain blood glucose at a constant level.

Later in the chapter we will discuss another type of glucose carrier protein, the Na^+/glucose symporter. In contrast to GLUT1, which carries only glucose, the Na^+/glucose symporter simultaneously transports Na^+ and glucose molecules across a membrane in the same direction. This provides a means to transport glucose into a cell that already has a higher concentration of glucose than its environment. Typically, the facilitated diffusion of Na^+ down its electrochemical gradient provides the energy to transport glucose up its concentration gradient.

The Erythrocyte Anion Exchange Protein: An Antiport Carrier. Another well-studied example of facilitated diffusion is the **anion exchange protein** of the erythrocyte plasma membrane (see Figure 8-2b). This antiport protein, also called the *chloride-bicarbonate exchanger,* facilitates the reciprocal exchange of chloride and bicarbonate ions in opposite directions across the plasma membrane. The coupling of chloride and bicarbonate transport is obligatory, and transport will stop if

Key Technique

Expression of Heterologous Membrane Proteins in Frog Oocytes

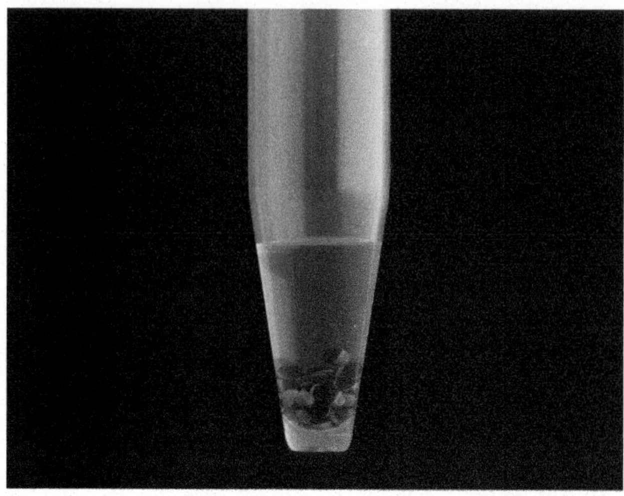

Xenopus oocytes in a microcentrifuge tube.

PROBLEM: Studying the function of transport proteins is difficult because, by nature, they are integral membrane proteins embedded in the membrane. As a consequence, they are difficult to purify and study in an intact, functional form. So how can the function of a specific membrane protein be studied? The answer is by expressing them in vivo. That is, in a living cell.

SOLUTION: Frog oocytes (egg cell precursors) are a common cell type that is used to produce functional integral membrane proteins. The mRNA encoding the membrane protein of interest can be microinjected into an individual oocyte, where the cell's ribosomes translate the mRNA into the desired protein. For multisubunit membrane proteins, the different mRNAs of the individual subunits can be co-injected.

Key Tools: *Xenopus laevis* oocytes; a micropipette; mRNA for the desired membrane protein; a fine, hollow needle to inject the mRNA into the oocyte, a voltage recorder and computer to measure the effects of ion channel proteins on the cell's electrical potential.

Details: In the 1970s, John Gurdon and colleagues at Oxford University showed that oocytes of *Xenopus laevis,* the African clawed toad, provide a very useful in vivo experimental system for the translation of heterologous mRNA (mRNA from other organisms). In their pioneering work, they showed that, following microinjection of rabbit mRNA into the cells, these oocytes could efficiently synthesize rabbit hemoglobin. This powerful system has been used for more than 40 years to express a wide variety of heterologous proteins, including the often difficult-to-study membrane proteins.

Xenopus oocytes are large cells, having a diameter of about 1 mm, making them visible to the naked eye. Using a dissecting microscope, the eggs can be injected with mRNA of a protein under study. For multisubunit proteins, several mRNAs—each representing one of the subunits—can be co-injected, often producing an intact, multisubunit protein in a functional form.

Xenopus oocytes have a number of practical advantages that make them extremely useful as an experimental system. Their large size makes them easily microinjected. The frogs themselves are relatively easy to raise in captivity, and harvesting the oocytes is a relatively routine procedure. Once isolated from female frogs, oocytes can be maintained without additional nutrients in little more than fresh water.

About a decade after Gurdon and colleagues showed that *Xenopus* oocytes could be used to express rabbit hemoglobin, they became a popular system for the expression of functional ion channels for studying electrophysiology in neurons and other cells.

To study the expression of heterologous proteins, an oocyte is typically injected under a dissecting microscope using a very fine, hollow needle (**Figure 8A-1**, step ❶). Once the tip of the needle is inside the cell, the mRNA is injected in a very small volume of buffer, typically about 50 nL.

The mRNA (or mRNAs) is then translated by the endogenous *Xenopus* ribosomes to synthesize an intact, often functional protein. In the case of membrane proteins, the protein is usually inserted into the membrane in the proper orientation dictated by the protein's structure (see step ❷). If the protein inserted is an ion channel, the voltage difference across the oocyte membrane can be measured to study the properties of the ion channel (see step ❸). The power of this approach was demonstrated by Peter Agre and colleagues, in collaboration with Bill Guggino, who used

either anion is absent. Moreover, the anion exchange protein is very selective. It exchanges bicarbonate for chloride in a strict 1:1 ratio, and it accepts no other anions.

The anion exchange protein is thought to function by alternating between two conformational states in what is termed a "ping-pong" mechanism. In one conformation, the anion exchange protein binds chloride on one side of the membrane. Chloride binding causes a change in conformation such that chloride is released on the other side of the membrane, where the protein binds bicarbonate. This causes a second conformational change, releasing bicarbonate on the other side of the membrane, where the protein again binds chloride. Repeated cycles of binding and release transport the two ions in opposite directions.

Either anion can bind to the protein on either side of the membrane, so the direction of transport depends upon the relative concentrations of the ions on opposite sides of the membrane. In cells with a high bicarbonate concentration, bicarbonate leaves the cell and chloride enters. In cells with a low bicarbonate concentration, the reciprocal process occurs—bicarbonate enters the cell and chloride leaves.

The anion exchange protein plays a key role in the process by which waste CO_2 produced in metabolically active tissues is delivered to the lungs to be expelled. In these tissues, CO_2 diffuses into erythrocytes, where the cytosolic enzyme *carbonic anhydrase* converts it to bicarbonate ions (see Figure 8-3a). As the concentration of bicarbonate in the erythrocyte rises, it moves out of the cell via the anion exchange protein (Figure 8-3a, right). To

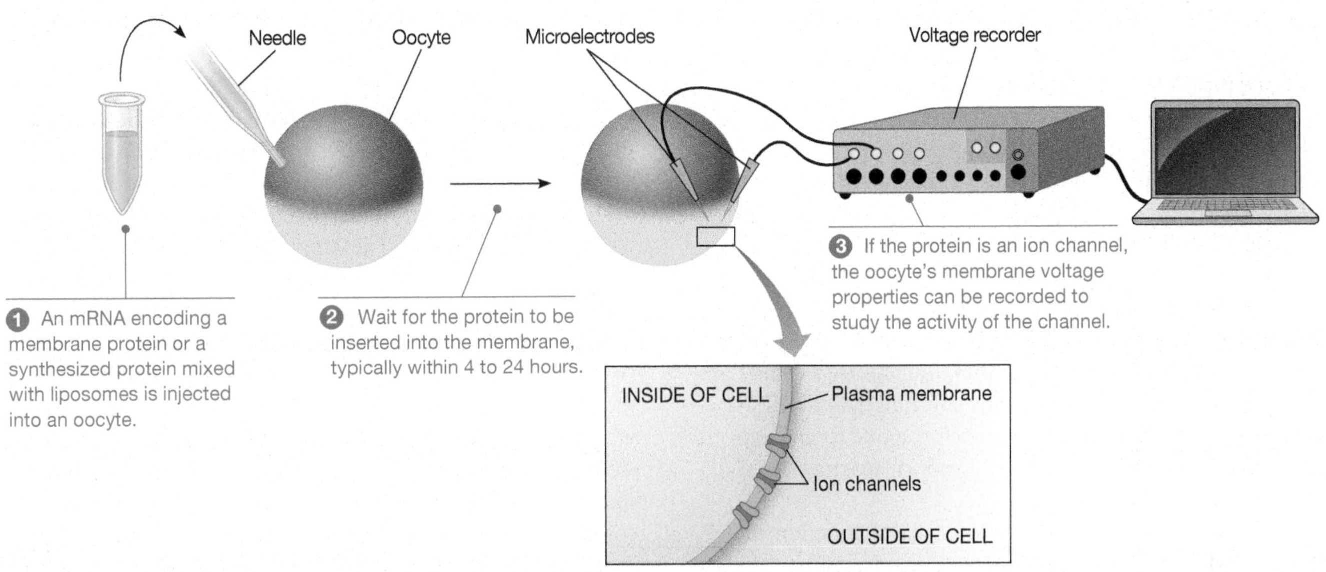

Needle
Oocyte
Microelectrodes
Voltage recorder
1 An mRNA encoding a membrane protein or a synthesized protein mixed with liposomes is injected into an oocyte.
2 Wait for the protein to be inserted into the membrane, typically within 4 to 24 hours.
3 If the protein is an ion channel, the oocyte's membrane voltage properties can be recorded to study the activity of the channel.
INSIDE OF CELL
Plasma membrane
Ion channels
OUTSIDE OF CELL

Figure 8A-1 Microinjection of an Oocyte to Produce a Membrane Protein. In this example, an ion channel is produced and inserted into the oocyte membrane, whose electrical properties can then be measured.

Xenopus oocytes to show that the aquaporin membrane protein was a water channel, as had been predicted by numerous physiological studies (see Figure 8-10a).

Although injecting mRNA into oocytes is common, in other cases, rather than allowing the oocyte to synthesize the corresponding protein, it is easier to synthesize a membrane protein directly in the laboratory. The protein is then mixed with liposomes, and the protein-carrying liposomes are then injected into the oocyte.

Although they are convenient, there are several limitations to use of *Xenopus* oocytes for expression of heterologous proteins. They are best kept and used at 20°C, which may not be the optimum temperature for mammalian proteins, which typically function optimally at 37°C. In addition, endogenous proteins and channels normally expressed by *Xenopus* oocytes may cause some interference with the production or function of the desired protein. In addition, *Xenopus* oocytes may lack specific signaling molecules that interact with the membrane protein in its native cell type that are necessary for its proper function. Despite a few inevitable shortcomings, *Xenopus* oocytes have proven to be an extremely valuable "living test tube" for the expression and functional study of integral membrane proteins.

QUESTION: A researcher is studying a fern that is shown to have a membrane protein that transports the arsenate ion ($HAsO_4^{2-}$), a common pollutant in ground water, across membranes and thus may be useful to clean polluted sites. However, after having injected the correct mRNA for the protein, the researcher failed to detect arsenate transport in the oocytes. What are the possible reasons for this?

prevent a net charge imbalance, the outward movement of each negatively charged bicarbonate ion is accompanied by the uptake of one negatively charged chloride ion.

In the lungs, this entire process is reversed: chloride ions are transported out of erythrocytes accompanied by the uptake of bicarbonate ions (Figure 8-3b, right), which are then converted back to CO_2 by carbonic anhydrase.

The net result is the movement of CO_2 (in the form of bicarbonate ions) from metabolically active tissues to the lungs, where the CO_2 is exhaled from the body. In addition, the import of bicarbonate into erythrocytes in the lungs increases the cellular pH, which enhances oxygen binding to hemoglobin in the lungs. When the erythrocytes reach the tissues and bicarbonate is exported, the cellular pH drops and oxygen binding decreases, allowing more rapid O_2 release to the tissues.

Channel Proteins Facilitate Diffusion by Forming Hydrophilic Transmembrane Channels

Whereas some transport proteins facilitate diffusion by functioning as carrier proteins that alternate between different conformational states, others do so by forming hydrophilic *transmembrane channels* that allow specific solutes to move across the membrane directly. We will consider three kinds of transmembrane protein channels: *ion channels, porins,* and *aquaporins.* Because these transport proteins are integral membrane proteins, they are often difficult to purify and analyze. **Key Technique, pages 196–197,** shows how the function of a transmembrane channel protein can be studied in vivo by expressing the protein in frog egg cells.

Ion Channels: Transmembrane Proteins That Allow Rapid Passage of Specific Ions. Despite their apparently simple design—a tiny pore lined with hydrophilic atoms—**ion channels** are strikingly selective. Most allow passage of only one kind of ion, so separate channels are needed for transporting such ions as Na^+, K^+, Ca^{2+}, and Cl^-. This selectivity is remarkable given the small differences in size and charge among some of these ions. Selectivity results both from ion-specific binding sites involving specific amino acid side chains and polypeptide backbone atoms inside the channel and from a constricted center of the channel that serves as a size filter. The rate of transport is equally impressive in some cases: a single channel can conduct almost a million ions of a specific type per second!

Most ion channels are *gated*, which means that the pore opens and closes in response to some stimulus. In animal cells, three kinds of stimuli control the opening and closing of gated channels: *voltage-gated channels* open and close in response to changes in the membrane potential, *ligand-gated channels* are triggered by the binding of specific substances to the channel protein, and *mechanosensitive channels* respond to mechanical forces that act on the membrane. (You will learn about gated channels in detail in Chapter 22.)

How do researchers study the cellular function of ion channels such as these? A technique known as *patch clamping* allows researchers to monitor the ion flow through individual ion channels in a small patch of membrane on the surface of a cell such as a neuron (see Key Technique in Chapter 22, pages 670–671). This technique is sensitive enough that it can record the opening and closing of a single ion channel. It has also allowed structural studies to determine which parts of the channel protein are essential for proper functioning in living cells.

Regulation of ion movement across membranes plays an important role in many types of cellular processes. Muscle contraction and many cellular responses require regulation of Ca^{2+} levels via calcium-specific ion channels. The transmission of electrical signals by nerve cells depends critically on rapid, controlled changes in the movement of Na^+ and K^+ through their respective channels. These changes are so rapid that they are measured in milliseconds. In addition to such short-term regulation, most ion channels are subject to longer-term regulation, usually in response to external stimuli such as hormones.

Ion channels are also necessary for maintaining the proper salt balance in the cells and airways lining our lungs. In lung epithelial cells, a specific chloride ion channel known as the *cystic fibrosis transmembrane conductance regulator (CFTR)* protein helps maintain the proper Cl^- concentration in these airways. We have long known that people with the life-threatening disease known as *cystic fibrosis* accumulate unusually thick mucus in their lungs—a condition that often leads to pneumonia and other lung disorders. Now we understand that the underlying problem is an inability to secrete chloride ions and that the genetic defect is in a protein that functions as a chloride ion channel (a topic we discuss in detail in Human Connections, page 200–201).

An especially important protein involved in ion transport down a concentration gradient is the ATP synthase enzyme (which we will encounter in more detail in Chapters 10 and 11; see Figure 10-23). ATP synthase acts as a proton channel that allows protons to travel down their concentration gradient by facilitated diffusion. This protein is found in mitochondria and chloroplasts, which use energy either from the oxidation of sugars (mitochondria) or from the sun (chloroplasts) to create a transmembrane proton gradient. The exergonic flow of protons down their concentration gradient through the ATP synthase allows the endergonic conversion of ADP and inorganic phosphate into the high-energy ATP molecule.

Porins: Transmembrane Proteins That Allow Rapid Passage of Various Solutes. Compared with ion channels, the pores found in the outer membranes of mitochondria, chloroplasts, and many bacteria are somewhat larger and much less specific. These pores are formed by multipass transmembrane proteins called **porins.** The transmembrane segments of porin molecules cross the membrane not as α helices but as a closed cylindrical *β sheet* called a *β barrel* (**Figure 8-9**). The β barrel forms a water-filled pore at its center. Polar side chains (not shown) line the inside of the pore, whereas the outside of the barrel consists mainly of nonpolar side chains that interact with the hydrophobic interior of the membrane. The pore allows passage of various hydrophilic solutes. The upper size limit for the solute molecules is determined by the pore size of the particular porin—only solutes smaller than about 600 Da can pass through the *Escherichia coli* porin shown in Figure 8-9. Mutations in certain bacterial porins can lead to antibiotic resistance in these bacteria by effectively blocking entry of the antibiotics that would normally be used to fight an infection.

MAKE CONNECTIONS 8.2

How are the places porins are found consistent with the proposed origin of eukaryotes? (Fig. 4-16)

Aquaporins: Transmembrane Channels That Allow Rapid Passage of Water. Whereas water can diffuse slowly across cell membranes in the absence of a protein transporter, movement of water across membranes in some tissues is much more rapid than can be accounted for by diffusion alone. In fact, the existence of water channels in cell membranes was

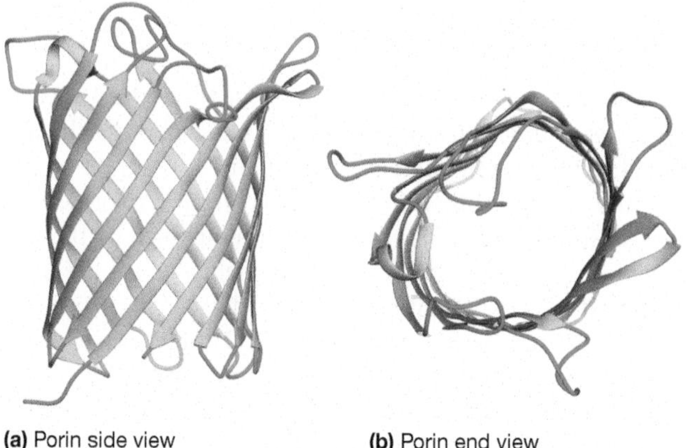

Figure 8-9 Porin Channels. **(a)** Side view and **(b)** end view of an *E. coli* porin protein showing the 14-stranded transmembrane β barrel and the N- and C-termini.

postulated as early as the mid- to late 1800s. Despite a century of experiments suggesting their existence, water channels remained elusive, and their very existence was sometimes in doubt.

It was not until 1992 that Peter Agre and colleagues at Johns Hopkins University finally isolated the long-sought-after water channel protein, which they named **aquaporin (AQP).** As **Figure 8-10** shows, they found that this particular erythrocyte protein, when introduced into membranes of relatively water-impermeable frog egg cells, would cause the cells to explode when placed in pure water due to the rapid influx of water (Figure 8-10a). Control eggs without the protein placed in the pure water were unchanged. In 2003, Agre shared a Nobel Prize in chemistry with Roderick MacKinnon, who determined the first three-dimensional structure of an ion channel.

Aquaporins facilitate the rapid movement of water molecules into or out of cells in specific tissues that require this capability. For example, your kidneys reabsorb water as part of urine formation, and specialized cells in this organ have a high density of AQPs in their plasma membrane, allowing the kidney to filter more than 100 L of blood plasma every day. Aquaporins are also abundant in erythrocytes, which must be able to expand or shrink rapidly in response to sudden changes in osmotic pressure as they move through the kidney or other arterial passages. Erythrocytes have approximately 200,000 aquaporin molecules per cell.

In plants, AQPs are a prominent feature of root cell plasma membranes and the vacuolar membrane, reflecting the rapid movement of water that is required to develop turgor pressure, as discussed previously. Bacteria have recently been shown to contain AQPs. One type of bacterial AQP has a slightly larger channel and can transport glycerol as well as water and is thus termed an *aquaglyceroporin.*

Aquaporins are tetrameric integral membrane proteins formed of four identical monomer subunits, with each monomer containing six helical transmembrane segments. The four monomers pack together in the membrane to form four identical water channels lined with amino acid residues having hydrophilic side chains (Figure 8-10b). The space in the center of the tetramer is blocked by a lipid molecule, preventing passage of gas or ions. The diameter of each of the four water channels narrows to about 0.3 nm (3 Å) (Figure 8-10c), which is just large enough for water molecules to pass through one at a time in single file. The specific amino acid residues lining the water channel allow only H_2O molecules through and discriminate against the similar-sized OH^- and H_3O^+ ions. In addition, the narrowest region of the water channel contains positively charged arginine residues, which repel and block the passage of any H^+ ions. Even with these constraints, water molecules flow through an aquaporin channel at the rate of several billion per second.

CONCEPT CHECK 8.3

How are carrier proteins and channel proteins similar? How do they differ? Consider both structure and mechanism.

8.4 Active Transport: Protein-Mediated Movement up the Gradient

Facilitated diffusion is an important mechanism for speeding up the movement of substances across cellular membranes. But it accounts for the transport of molecules only *toward* equilibrium, which means *down* a concentration or electrochemical gradient. What happens when a substance needs to

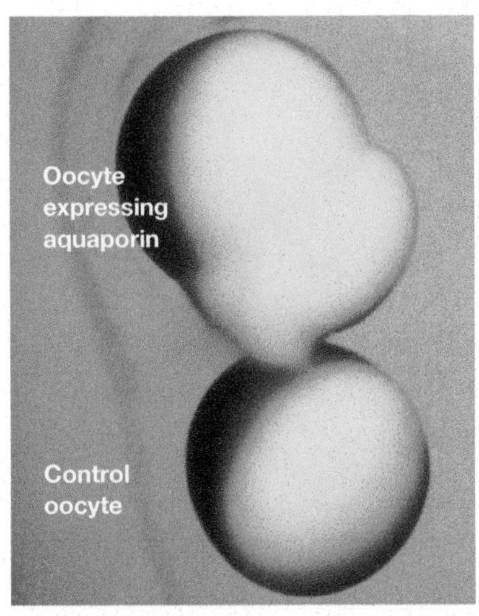

(a) Aquaporin expression in frog oocytes

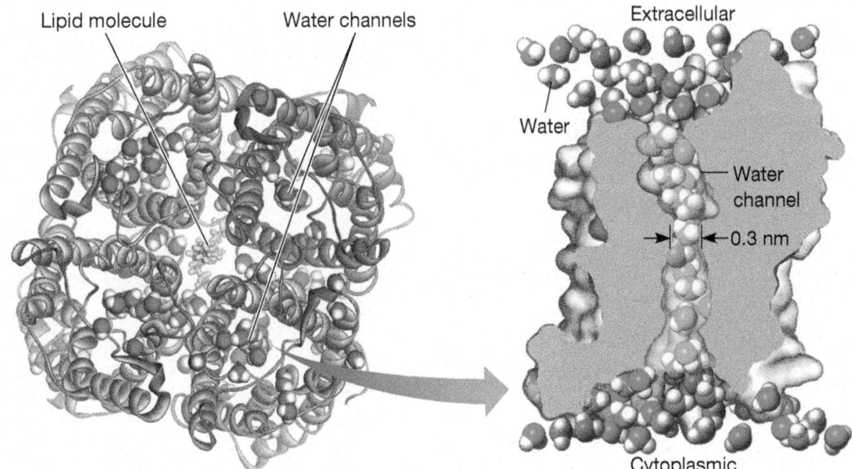

(b) Aquaporin tetramer (end view)

(c) Aquaporin monomer (side view)

Figure 8-10 Aquaporin Channels. (a) Expressing aquaporin-1 in *Xenopus* oocytes causes a dramatic change in water permeability. When transferred to a hypoosmotic solution for 2 minutes, a control *Xenopus* oocyte (bottom) exhibits little change in appearance. An oocyte previously injected with aquaporin-1 mRNA (top) rapidly swells and explodes due to movement of water into the cytosol. **(b)** End view of the tetrameric human aquaporin channel, showing the six α helices in each of the four identical monomers. Two of the four water channels in the tetramer are labeled. **(c)** Space-filling model of the side view of a single aquaporin channel, showing the entry and exit sites for water molecules (red spheres).

HUMAN *Connections*

Membrane Transport, Cystic Fibrosis, and the Prospects for Gene Therapy

A young mother sits in the waiting room of her doctor's office with her young daughter. Her daughter often has difficulty breathing and hasn't been gaining weight like most other children her age. Moreover, when the mother kisses the girl's forehead, she is left with a salty taste on her lips. As she will soon discover, the young girl has cystic fibrosis (CF). One hundred years ago the girl's outlook would have been bleak; she would have suffered from chronic weakness, difficulty breathing, and malnourishment and likely would have died by the age of 10. Today, with modern therapies and treatments, patients with CF have an average life span of about 40 years.

CF is a currently incurable disease that affects roughly one out of every 9000 individuals in the United States. It primarily affects the secretory cells in epithelial tissues of the lungs, pancreas, and skin (although the disease manifests itself in many different tissues of the body). The disease is named for its typical pancreatic cysts and fibrosis, or hardening, of lung tissue in affected individuals.

Common to all of the affected tissues is a transmembrane protein integral to the transport of chloride ions across the membrane: the cystic fibrosis transmembrane conductance regulator (CFTR), a cyclic AMP (cAMP)-regulated ion channel (you will learn more about cAMP in Chapter 23). In normal individuals, the channel pumps chloride ions into the lumen of lung airways and other tissues (**Figure 8B-1a**). Sodium ions then flow passively across the membrane into the lumen, and water follows via osmosis. This results in relatively thin mucus that keeps the tissue moist and can be expelled from the lungs and bronchial passages through the beating of cilia, microtubule-based structures that generate gentle fluid currents.

Patients with CF, however, have mutations in the gene encoding CFTR that render the ion channel nonfunctional. To date, more than 1000 different mutations have been identified, all with a similar effect: CF patients are unable to move chloride ions across the membrane. Ultimately, this leads to less water flow as well, so CF patients have unusually thick mucus in their lungs (Figure 8B-1b). Bacteria are able to colonize and grow within the thickened mucus layer, leading to chronic bacterial infections. In addition, gas exchange in the lungs is impaired by the thick mucus, so patients are unable to achieve normal levels of physical activity. In the pancreas, a different secretory problem results: digestive enzymes that should reach the small intestine fail to do so. This accounts for the inability of CF patients to gain weight.

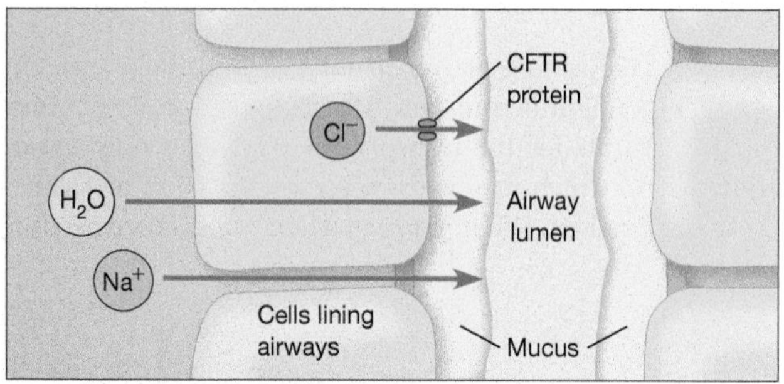

(a) Normal cells lining airways; hydrated mucus

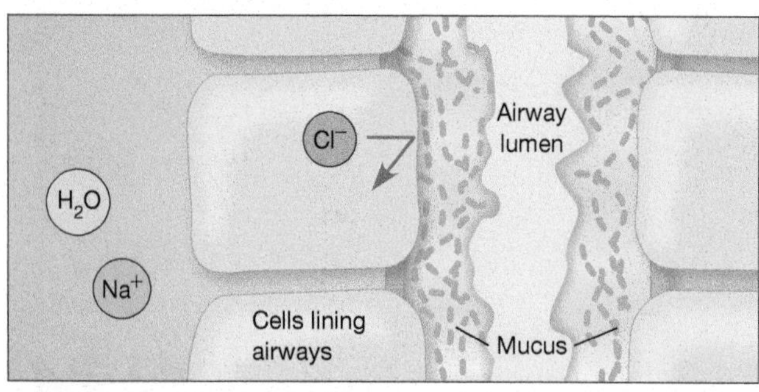

(b) Cells with cystic fibrosis infected with bacteria

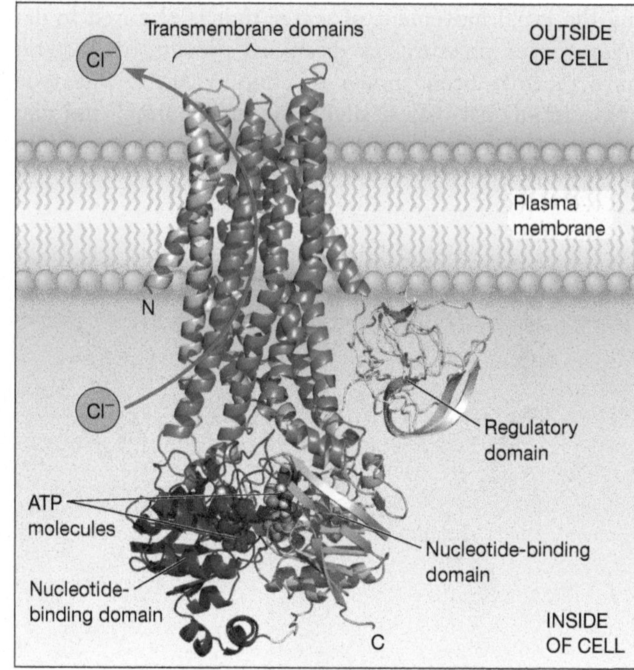

(c) Proposed CFTR structure

Figure 8B-1 Cystic Fibrosis and Chloride Ion Secretion. (a) Normal lungs have hydrated mucus lining the airways. **(b)** Cystic fibrosis is caused by a defect in the secretion of chloride ions in cells lining the lungs, leading to insufficient hydration, thick mucus, and the promotion of bacterial growth. **(c)** Proposed structure and orientation of CFTR in the lung cell membrane.

can transport a variety of antibiotics out of the cell, giving these cells resistance to multiple antibiotics.

Medical interest in this class of transport proteins was heightened when cystic fibrosis was shown to be caused by a genetic defect in a plasma membrane protein that is structurally related to the ABC transporters, the cystic fibrosis transmembrane conductance regulator (CFTR) that we encountered earlier in this chapter. However, CFTR is an ion channel, and, unlike most of the ABC-type ATPases, it does not use ATP to drive transport. Instead, ATP hydrolysis appears to be involved in opening the channel. (Recent developments in our understanding of cystic fibrosis at the molecular level and attempts at gene therapy to treat this disease are reported in Human Connections, page 200–201.)

Indirect Active Transport Is Driven by Ion Gradients

In contrast to direct active transport, which is powered by energy released from a chemical reaction such as ATP hydrolysis, indirect active transport (also called *secondary active transport*) is driven by the movement of an ion down its electrochemical gradient. This principle has emerged from studies of the active uptake of sugars, amino acids, and other organic molecules into cells where the inward transport of molecules *against* their electrochemical gradients is often coupled to, and driven by, the simultaneous inward movement of either sodium ions (for animal cells) or protons (for most plant, fungal, and bacterial cells) *down* their respective electrochemical gradients.

The widespread existence of such symport mechanisms explains why most cells continuously pump either sodium ions or protons out of the cell to maintain ion-specific gradients across the plasma membrane. In animals, for example, the relatively high extracellular concentration of sodium ions maintained by the Na^+/K^+ pump serves as the driving force for the uptake of a variety of sugars and amino acids. For example, to obtain the data shown in **Figure 8-12**, investigators varied the extracellular concentration of sodium ions and measured the transport rate of the amino acid glycine into erythrocytes or the sugar 7-deoxy-D-glucoheptose into cells lining the intestine. There is a clear correlation between increasing extracellular Na^+ concentration and the rate of transport of the relevant molecule into each tissue. Such experiments showed that transport of amino acids and sugars into cells can be stimulated by a high concentration of sodium ions in the extracellular medium.

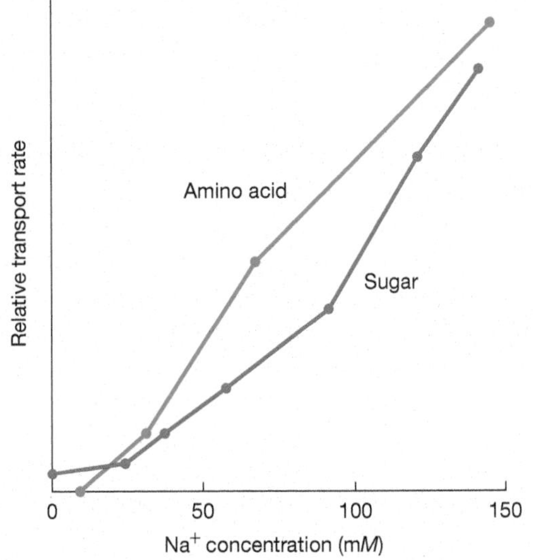

Figure 8-12 Effect of External Sodium Ion Concentration on Amino Acid and Sugar Transport. As the extracellular Na^+ concentration is increased, the rate of transport of certain nutrients into the cell increases proportionately due to indirect active transport.

The uptake of sugars and amino acids in this manner is regarded as indirect active transport because it is not directly driven by the hydrolysis of ATP or a related "high-energy" compound. Ultimately, however, uptake still depends on ATP because the Na^+/K^+ pump that maintains the sodium ion gradient is itself driven by ATP hydrolysis. The continuous outward pumping of Na^+ by the ATP-driven Na^+/K^+ pump and inward movement of Na^+ by symport (coupled to the uptake of another solute) establish a circulation of sodium ions across the plasma membrane of every animal cell, similar to the cycling of H^+ ions shown in Figure 8-11.

Whereas animal cells use sodium ions to drive indirect active transport, most other organisms rely on a proton gradient instead. For example, fungi and plants utilize proton symport for the uptake of organic solutes, with an ATP-driven proton pump responsible for the generation and maintenance of the proton electrochemical potential. Many kinds of bacterial cells also make extensive use of proton cotransport to drive the uptake of solute molecules, as do mitochondria. In all such cases, however, the proton gradient is maintained by an electron transfer process that accompanies cellular respiration (as we will see in Chapter 10).

In addition to the symport *uptake* of organic molecules such as sugars and amino acids, sodium ions or proton gradients can be used to drive the *export* of other ions, including Ca^{2+} and K^+. This type of indirect active transport is usually antiport; examples include the exchange of potassium ions for protons and the exchange of calcium ions for sodium ions.

CONCEPT CHECK 8.4

How would you determine whether a specific integral membrane transporter is operating by facilitated diffusion or active transport?

8.5 Examples of Active Transport

Having considered some general features of active transport, we are now ready to look at three specific examples: one example each of direct and indirect active transport in animal cells, and an unusual type of light-driven transport in a bacterium. In each case we will note what kinds of solutes are transported, what the driving force is, and how the energy source is coupled to the transport mechanism. We will focus first on the *Na^+/K^+ ATPase* (or *Na^+/K^+ pump*) present in all animal cells because it is a well-understood example of direct active transport by a P-type ATPase. Then we will consider a second example from animal cells: the indirect active transport of glucose by a *Na^+/glucose symporter*, using the energy of the sodium ion gradient established by the Na^+/K^+ ATPase. Finally, we will look briefly at light-driven *proton transport* in certain bacteria.

Direct Active Transport: The Na^+/K^+ Pump Maintains Electrochemical Ion Gradients

A characteristic feature of most animal cells is a high intracellular level of potassium ions and a low intracellular level of sodium ions. In a mammalian neuron, the $[K^+]_{inside}/[K^+]_{outside}$ ratio is about 30:1, and the $[Na^+]_{inside}/[Na^+]_{outside}$ ratio is about 0.08:1. The resulting electrochemical potentials for potassium and sodium ions are essential as the driving force for coupled transport as well as for the transmission of nerve impulses. The inward pumping of potassium ions and the outward pumping of sodium ions are therefore both energy-requiring processes, as both ions are being moved up their electrochemical gradients.

The protein responsible for this pumping was discovered in 1957 by Jens Skou, a Danish physiologist, who was awarded a Nobel Prize in chemistry in 1997. The discovery of this ion-transporting enzyme was in fact the first documented case of active transport. The **Na^+/K^+ ATPase**, or **Na^+/K^+ pump**, as this P-type ATPase is usually called, uses the exergonic hydrolysis of ATP to drive the endergonic inward transport of potassium ions and the outward transport of sodium ions against their concentration gradients. The Na^+/K^+ pump is primarily responsible for the asymmetric distribution of these ions across the plasma membrane in animal cells. Like most other active transport systems, this pump has inherent directionality: potassium ions are always pumped inward and sodium ions are always pumped outward. In fact, sodium and potassium ions activate the ATPase only on the side of the membrane from which they are transported—sodium ions from the inside, potassium ions from the outside. Three sodium ions are moved out and two potassium ions are moved in per molecule of ATP hydrolyzed.

Figure 8-13 is a schematic illustration of the Na^+/K^+ pump. The pump is a trimeric transmembrane protein with one α, one β, and one γ subunit. The α subunit, which forms the ion channel itself, contains binding sites for ATP and sodium ions on the inner side of the membrane and binding sites for potassium ions on the external side. The functions of the β subunit, which is glycosylated, and the γ subunit are not yet clear.

The Na^+/K^+ pump is an allosteric protein exhibiting two alternative conformational states, referred to as E_1 and E_2. The E_1 *conformation* is open to the inside of the cell and has a high affinity for sodium ions, whereas the E_2 *conformation* is open to the outside, with a high affinity for potassium ions. Phosphorylation of the pump by ATP, a sodium-triggered event, stabilizes it in the E_2 conformation. Dephosphorylation, on the other hand, is triggered by K^+ and stabilizes the enzyme in the E_1 conformation.

As illustrated schematically in **Figure 8-14**, the actual transport mechanism involves ❶ an initial binding of three sodium ions to E_1 on the inner side of the membrane. The binding of sodium ions ❷ triggers autophosphorylation of the α subunit of the pump using bound ATP as the phosphate donor, resulting in a conformational change from E_1 to E_2. As a result, ❸ the bound sodium ions are moved through the membrane to the external surface, where they are released to the outside. Then ❹ potassium ions from the outside bind to the α subunit, triggering ❺ dephosphorylation and a return to the original conformation. During this process, the potassium ions are moved to the inner surface, where ❻ ATP binding results in their exit from the pump, leaving the carrier ready to accept more sodium ions.

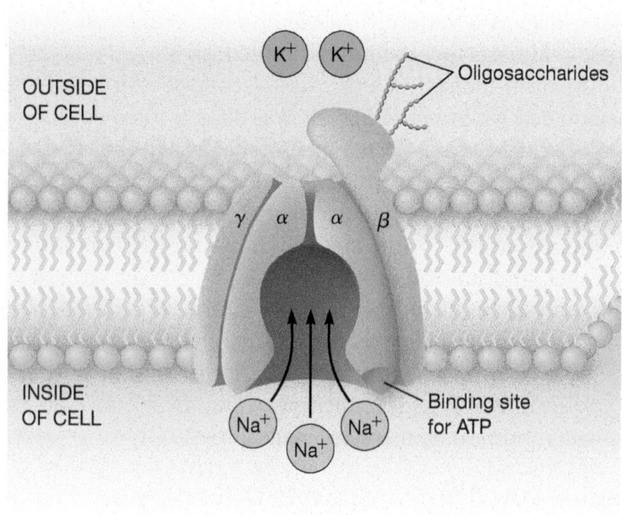

Figure 8-13 The Na^+/K^+ Pump. The trimeric Na^+/K^+ pump found in most animal cells contains α, β, and γ transmembrane subunits. The pump is shown in the E_1 conformation, which is open to the inside of the cell. Binding of sodium ions causes a conformational change to the E_2 form, which opens to the outside to accept potassium ions, which are transported inward. The transport mechanism in this pump is shown in detail in Figure 8-14.

The Na^+/K^+ pump is not only one of the best-understood transport systems but also one of the most important for animal cells. Besides maintaining the appropriate intracellular concentrations of both potassium and sodium ions, it is responsible for maintaining the membrane potential that exists across the plasma membrane. The Na^+/K^+ pump assumes still more significance when we take into account the vital role that sodium ions play in the inward transport of organic substrates, a topic we now come to as we consider sodium symport.

Indirect Active Transport: Sodium Symport Drives the Uptake of Glucose

As an example of indirect active transport, let's consider the uptake of glucose by the **Na^+/glucose symporter**. Although most glucose transport into or out of cells in your body occurs by facilitated diffusion, as shown in Figure 8-8, the epithelial cells lining your intestines have transport proteins that are able to take up glucose and certain amino acids from the intestines even when their concentrations there are much lower than in the epithelial cells. This energy-requiring process is driven by the simultaneous uptake of sodium ions, which is exergonic because of the steep electrochemical gradient for sodium that is maintained across the plasma membrane by the Na^+/K^+ pump (higher $[Na^+]$ outside the cell). These *sodium-dependent glucose transporters* are often referred to as *SGLT* proteins.

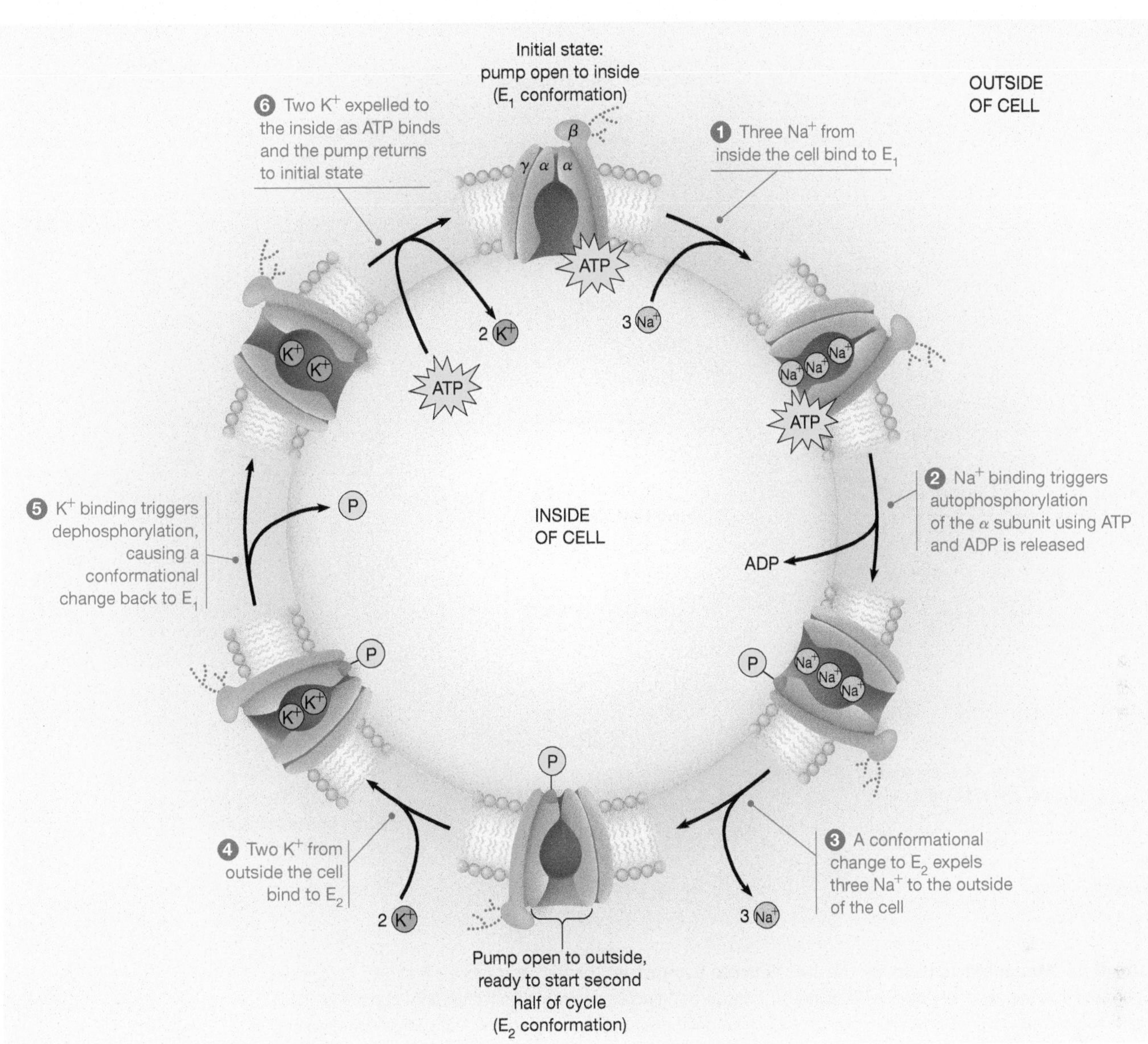
Initial state:
pump open to inside
(E_1 conformation)
OUTSIDE
OF CELL
6 Two K^+ expelled to the inside as ATP binds and the pump returns to initial state
1 Three Na^+ from inside the cell bind to E_1
β
γ α α
ATP
3 Na^+
2 K^+
2 Na^+ binding triggers autophosphorylation of the α subunit using ATP and ADP is released
ADP
INSIDE
OF CELL
5 K^+ binding triggers dephosphorylation, causing a conformational change back to E_1
P
4 Two K^+ from outside the cell bind to E_2
3 A conformational change to E_2 expels three Na^+ to the outside of the cell
Pump open to outside,
ready to start second
half of cycle
(E_2 conformation)

Figure 8-14 Model Mechanism for the Na^+/K^+ Pump. The transport process is shown here in six steps arranged at the periphery of a cell. The outward transport of sodium ions is coupled to the inward transport of potassium ions, both against their respective electrochemical potentials. The driving force is provided by ATP hydrolysis, which is required for phosphorylation of the α subunit of the pump at step 2.

Figure 8-15 on page 208 depicts the transport mechanism for the Na^+/glucose symporter, which requires the inward movement of two sodium ions to drive the simultaneous uptake of one glucose molecule. Transport is initiated by 1 the binding of two external sodium ions to their binding sites on the symporter, which is open to the outer surface of the membrane. This allows 2 a molecule of glucose to bind, followed by 3 a conformational change in the protein that exposes the sodium ions and the glucose molecule to the inner surface of the membrane. There 4 the two sodium ions dissociate in response to the low intracellular sodium ion concentration. This 5 locks the transporter in its inward-facing conformation until the glucose molecule dissociates. Then 6 the empty transporter is free to return to its outward-facing conformation.

The sodium ion gradient is in turn maintained by the continuous outward extrusion of sodium ions (dashed arrow) by the Na^+/K^+ pump shown in Figure 8-14. As a result, sodium ions circulate across the plasma membrane, pumped outward by the Na^+/K^+ pump and flowing back into the cell as the driving force for sodium symport of molecules such as glucose. Similar mechanisms are involved in the uptake of amino acids and other organic substrates by sodium symport in animal cells and by proton symport in plant, fungal, and bacterial cells.

This detailed understanding of the Na^+/glucose symport mechanism has helped in the treatment of the disease cholera. Cholera is caused by the bacterium *Vibrio cholera*, which produces a toxin that paralyzes intestinal cells and can lead to death via dehydration. Oral rehydration is the standard

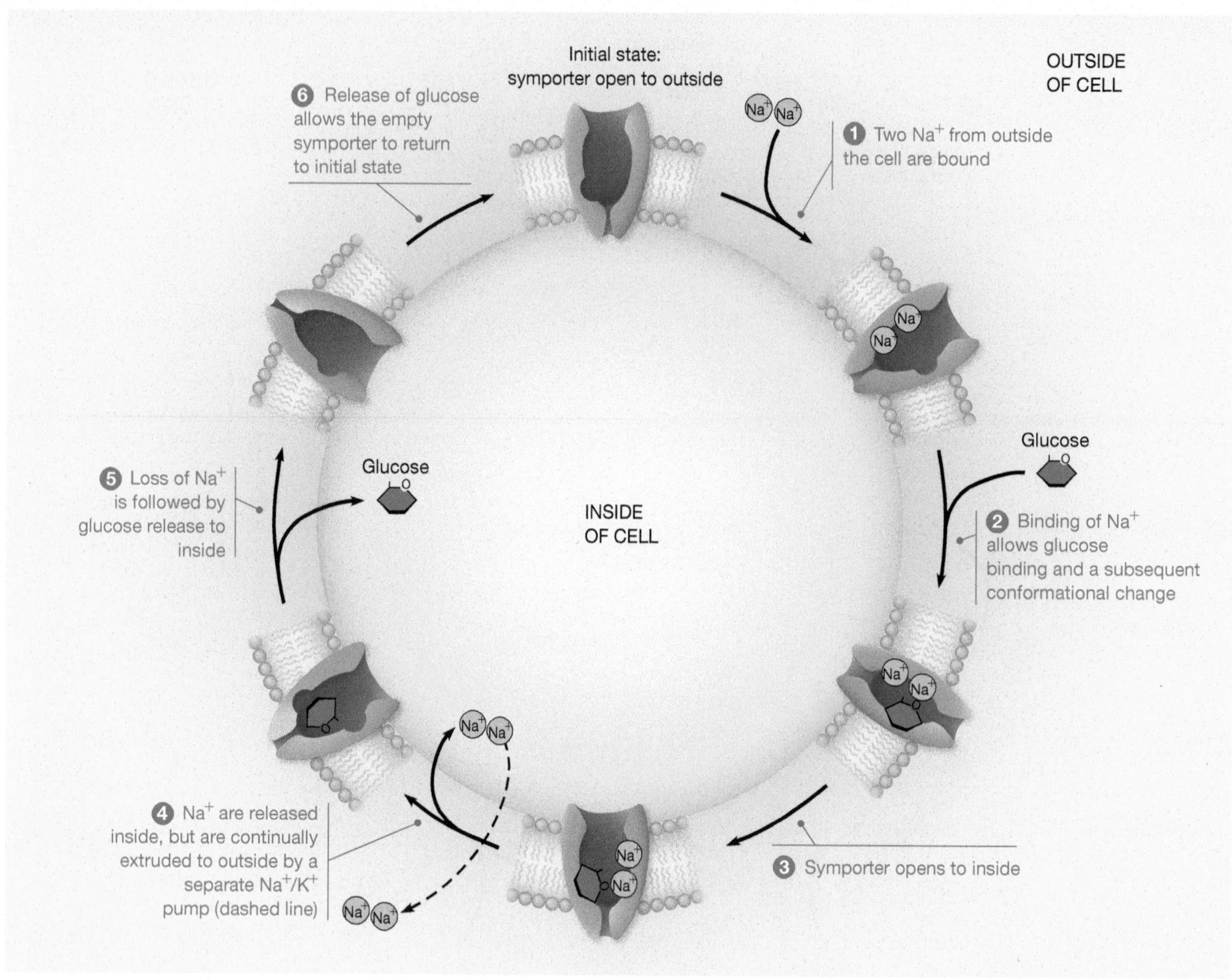

Figure 8-15 Model Mechanism for the Na^+/Glucose Symporter. The transport process is shown here in six steps arranged at the periphery of a cell. The inward transport of glucose against its concentration gradient is driven by the simultaneous inward transport of sodium ions down their electrochemical gradient. The sodium ion gradient is in turn maintained by the continuous outward transport of sodium ions (dashed arrow at step 4) by the Na^+/K^+ pump of Figure 8-14.

treatment for cholera, and it has been shown that rehydration with a salt and sugar solution is the most effective remedy. Administration of NaCl helps the body retain water as it strives to maintain salt balance in tissues; providing glucose along with the salt in turn allows for more efficient uptake of salt via the Na^+/glucose symporter.

Note that both types of glucose transporters that we have discussed, the GLUT uniporter and the Na^+/glucose symporter, can be found in the same cell. For example, cells lining the small intestine may have a Na^+/glucose symporter on the intestinal side to transport glucose by active transport from the intestinal lumen, where its concentration is high, into the cells, which have a lower glucose concentration. Then on the opposite side of the cell, a GLUT transporter can transport glucose into the bloodstream by facilitated diffusion, where its concentration is lower than in the cell.

The Bacteriorhodopsin Proton Pump Uses Light Energy to Transport Protons

The final active transport system we will consider is the simplest. Nothing more is involved than a small integral membrane protein called **bacteriorhodopsin** (**Figure 8-16**). This protein (briefly introduced in Chapter 7) is a proton pump found in the plasma membrane of halophilic ("salt-loving") archaea belonging to the genus *Halobacterium*. In contrast to transport proteins that utilize energy derived from ATP or ion gradients, bacteriorhodopsin uses energy derived from photons of light to drive active transport. Bacteriorhodopsin contains a pigment molecule that traps light energy and uses it to drive the active transport of protons outward across the plasma membrane. It thereby creates an electrochemical proton gradient that powers the synthesis of ATP by an ATP

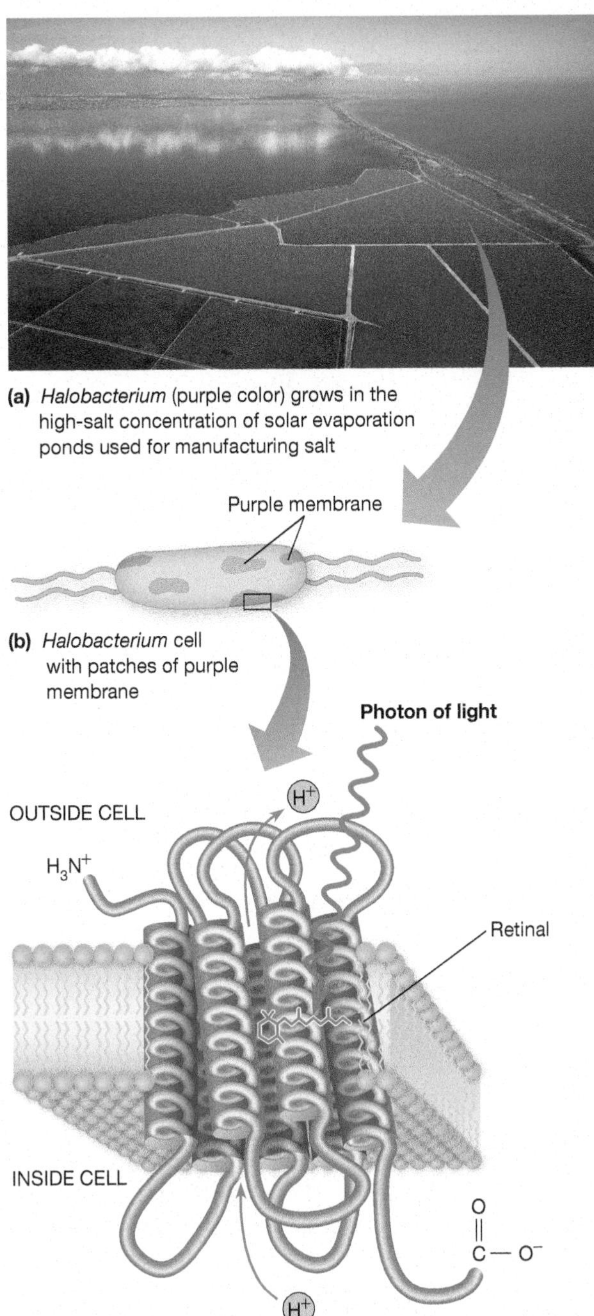

Figure 8-16 The Bacteriorhodopsin Proton Pump of Halobacteria. (a) Archaea belonging to the genus *Halobacterium* appear purple due to the protein bacteriorhodopsin. **(b)** Bacteriorhodopsin is a light-activated proton pump in the plasma membrane of *Halobacterium* cells that is found in bright purple patches known as purple membrane. **(c)** The seven α-helical transmembrane segments of bacteriorhodopsin form a cylinder that contains a molecule of the light-absorbing pigment retinal.

synthase analogous to the F-type ATP synthases found in chloroplasts and mitochondria.

The light-absorbing pigment of bacteriorhodopsin is *retinal,* a carotenoid derivative related to vitamin A that also serves as the visual pigment in the retina of your eyes. Retinal makes bacteriorhodopsin bright purple in color, which is why halobacteria have been called *purple photosynthetic bacteria* (Figure 8-16a). Bacteriorhodopsin appears in the plasma membrane of the *Halobacterium* cell as colored patches called *purple membrane* (Figure 8-16b).

Bacteriorhodopsin has seven α-helical membrane-spanning segments oriented in the membrane to form an overall cylindrical shape (Figure 8-16c). A molecule of retinal is attached to a lysine residue of the protein. When the retinal, which is a *chromophore* (a colored, light-absorbing pigment), absorbs a photon of light, it becomes photoactivated. The activated bacteriorhodopsin protein then becomes capable of transferring protons from the inside to the outside of the cell. The electrochemical potential produced by pumping protons out of the cell in this way is used by membrane-located halobacterial ATPases to synthesize ATP as the protons flow down their concentration gradient back into the cell.

Energy-dependent proton pumping is one of the most basic concepts in cellular energetics. Proton pumping, which occurs in all bacteria, mitochondria, and chloroplasts, represents the driving force for all of life on Earth because it is an absolute requirement for the efficient synthesis of ATP. (We will discuss these topics in more detail when we discuss respiration and photosynthesis in Chapters 10 and 11.)

CONCEPT CHECK 8.5

Both the Na^+/glucose symporter and the Na^+/K^+ pump move sodium ions across a membrane. How is the movement different for the two types of transporters?

8.6 The Energetics of Transport

Every transport event in the cell is an energy transaction. Energy either is released as transport occurs or is required to drive transport. To understand the energetics of transport, we must recognize that two different factors may be involved. For uncharged solutes, the only variable is the concentration gradient across the membrane, which determines whether transport is "down" a gradient (exergonic) or "up" a gradient (endergonic). For charged solutes, however, there is both a concentration gradient and an electrical potential across the membrane. The two may either reinforce each other or oppose each other, depending on the charge on the ion and the direction of transport.

For Uncharged Solutes, the ΔG of Transport Depends Only on the Concentration Gradient

For solutes with no net charge, we need only be concerned with the concentration gradient across the membrane. We can therefore treat the transport process as we would treat a simple chemical reaction, and we can calculate ΔG accordingly.

Calculating ΔG for the Transport of Molecules. The general "reaction" for the transport of molecules of solute S from the outside of a membrane-bounded compartment to the inside can be represented as

$$S_{outside} \rightarrow S_{inside} \quad \textbf{(8-4)}$$

The free energy change for this inward transport reaction (see Chapter 5) can be written as

$$\Delta G_{\text{inward}} = \Delta G^\circ + RT \ln \frac{[\text{S}]_{\text{inside}}}{[\text{S}]_{\text{outside}}} \tag{8-5}$$

where ΔG is the free energy change, ΔG° is the standard free energy change, R is the gas constant (1.987 cal/mol-K), T is the absolute temperature, and $[\text{S}]_{\text{inside}}$ and $[\text{S}]_{\text{outside}}$ are the prevailing concentrations of S on the inside and outside of the membrane, respectively. However, the equilibrium constant K_{eq} for the transport of an uncharged solute is always 1 because at equilibrium the solute concentrations on the two sides of the membrane will be the same:

$$K_{\text{eq}} = \frac{[\text{S}]_{\text{inside}}}{[\text{S}]_{\text{outside}}} = 1.0 \tag{8-6}$$

This means that ΔG° is always zero:

$$\Delta G^\circ = -RT \ln K_{\text{eq}} = -RT \ln 1 = 0 \tag{8-7}$$

So the expression for ΔG of inward transport of an uncharged solute shown in Equation 8-5 simplifies to

$$\Delta G_{\text{inward}} = RT \ln \frac{[\text{S}]_{\text{inside}}}{[\text{S}]_{\text{outside}}} \tag{8-8}$$

Notice that if $[\text{S}]_{\text{inside}}$ is less than $[\text{S}]_{\text{outside}}$, ΔG is negative, indicating that the inward transport of substance S is exergonic. It can therefore occur spontaneously, as would be expected for facilitated diffusion down a concentration gradient. But if $[\text{S}]_{\text{inside}}$ is greater than $[\text{S}]_{\text{outside}}$, inward transport of S is against the concentration gradient, and ΔG will be positive. In this case, inward transport of S is endergonic, and the amount of energy required to drive the active transport of S into the cell is indicated by the magnitude of the positive value of ΔG.

An Example: The Uptake of Lactose. Suppose that the concentration of lactose within a bacterial cell is to be maintained at 10 mM, whereas the external lactose concentration is only 0.20 mM. The energy requirement for the inward transport of lactose at 25°C can be calculated from Equation 8-8 as

$$\begin{aligned}\Delta G_{\text{inward}} &= RT \ln \frac{[\text{lactose}]_{\text{inside}}}{[\text{lactose}]_{\text{outside}}} \\ &= (1.987)(25+273) \ln \frac{0.010}{0.0002} \\ &= 592 \ln 50 = 2316 \text{ cal/mol} \\ &= 2.32 \text{kcal/mol}\end{aligned} \tag{8-9}$$

In many bacterial cells, the energy to drive lactose uptake is provided by an electrochemical proton gradient, so lactose uptake in such cells is an example of indirect active transport.

As written, Equation 8-8 applies to inward transport. For outward transport, the positions of S_{inside} and $\text{S}_{\text{outside}}$ are simply interchanged within the logarithm. As a result, the absolute value of ΔG remains the same, but the sign is changed. As for any other process, a transport reaction that is exergonic in one direction will be endergonic to the same degree in the opposite direction. The equations for calculating ΔG of inward and outward transport of uncharged solutes are summarized in **Table 8-3**.

Table 8-3 Calculation of ΔG for the Transport of Charged and Uncharged Solutes

Transport Process

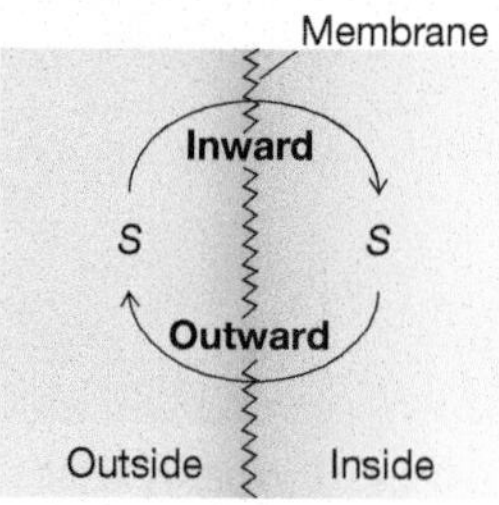

ΔG for Transport of Uncharged Solutes:

$\Delta G_{\text{inward}} = RT \ln \frac{[\text{S}]_{\text{inside}}}{[\text{S}]_{\text{outside}}}$	R = 1.987 cal/mol-K T = K = °C + 273
$\Delta G_{\text{outward}} = -\Delta G_{\text{inward}}$	

ΔG for Transport of Charged Solutes:

$\Delta G_{\text{inward}} = RT \ln \frac{[\text{S}]_{\text{inside}}}{[\text{S}]_{\text{outside}}} + zFV_{\text{m}}$	z = charge on ion F = 23,062 cal/mol–V V_{m} = membrane potential (in volts)
$\Delta G_{\text{outward}} = -\Delta G_{\text{inward}}$	

For Charged Solutes, the ΔG of Transport Depends on the Electrochemical Potential

For charged solutes—ions, in other words—we need to consider the electrochemical potential for that ion, which is influenced by both the concentration gradient for that ion and the membrane potential of the cell itself, V_{m}. For animal cells, V_{m} usually falls in the range of −60 to −90 mV. In bacterial and plant cells, it is significantly more negative, often about −150 mV in bacteria and between −200 and −300 mV in plants. By convention, *the minus sign indicates that the excess of negative charge is on the inside of the cell.* Thus, the V_{m} value indicates how negative (or positive, in the case of a plus sign) the *inside* of the cell is compared with the *outside*.

The membrane potential obviously has no effect on uncharged solutes, but it significantly affects the energetics of ion transport. Because it is usually negative, the membrane potential typically *favors* the inward movement of cations and *opposes* their outward movement. As we mentioned earlier, the net effect of both the concentration gradient and the potential gradient for an ion is called the *electrochemical potential* for that ion.

Calculating ΔG for the Transport of Ions. Both components of the electrochemical potential must be considered when determining the energetics of ion transport. To calculate ΔG for the transport of ions therefore requires an equation with two terms, one to express the effect of the concentration gradient across the membrane and the other to take into account the membrane potential.

If we let S^z represent a solute with a charge z, then we can calculate ΔG for the inward transport of S^z as

$$\Delta G_{\text{inward}} = RT \ln \frac{[S]_{\text{inside}}}{[S]_{\text{outside}}} + zFV_{\text{m}} \qquad \textbf{(8-10)}$$

where R, T, $[S]_{\text{inside}}$, and $[S]_{\text{outside}}$ are defined as before; z is the charge on S (such as +1, +2, −1, −2); F is the Faraday constant (23,062 cal/mol-V); and V_{m} is the membrane potential (in volts). Note that Equation 8-8 is just the simplified version of Equation 8-10 for solutes with a z value of zero—that is, for molecules with no net charge.

For a typical cell, the membrane potential V_{m} is negative (an excess of negative charge inside the cell). A positive ion ($z > 0$) will cause the term zFV_{m} to be negative, lowering the ΔG for inward transport. This indicates that uptake of the positive ion is energetically favorable, as you would expect if the interior of the cell has a net negative charge. The uptake of positive ions becomes more exergonic and more favorable as the membrane potential of the cell decreases. The opposite is true for a negative ion ($z < 0$), which gives a positive value for zFV_{m} and a positive change in free energy, indicating an unfavorable process that will not occur spontaneously.

For the outward transport of S, ΔG has the same value as for inward transport but is opposite in sign, so we can write

$$\Delta G_{\text{outward}} = -\Delta G_{\text{inward}} \qquad \textbf{(8-11)}$$

An Example: The Uptake of Chloride Ions. To illustrate the use of Equation 8-10 and to point out that intuition does not always serve us well in predicting the direction of ion transport, consider what happens when nerve cells with an intracellular chloride ion concentration of 50 mM are placed in a solution containing 100 mM Cl^-. Because the Cl^- concentration is twice as high outside the cell as inside, you might expect chloride ions to diffuse passively into the cell without the need for active transport. However, this expectation ignores the membrane potential of about −60 mV (−0.06V) that exists across the plasma membrane of nerve cells. The minus sign reminds us that the inside of the cell is negative with respect to the outside, which means that the inward movement of negatively charged anions such as Cl^- will be *against* the membrane potential. Thus, the inward movement of chloride ions is *down* the concentration gradient but *up* the charge gradient.

To quantify the relative magnitudes of these two opposing forces at 25°C, we can use Equation 8-10 with the relevant values inserted:

$$\begin{aligned}
\Delta G_{\text{inward}} &= \text{RT} \ln \frac{[S]_{\text{inside}}}{[S]_{\text{outside}}} + z\text{FV}_{\text{m}} \\
&= (1.987)(25+273) \ln\left(\frac{0.05}{0.10}\right) + (-1)(23{,}062)(-0.06) \\
&= 592 \ln(0.5) + (23{,}062)(0.06) \\
&= -410 + 1384 \\
&= 974 \text{ cal/mol} \\
&= +0.97 \text{ kcal/mol} \qquad \textbf{(8-12)}
\end{aligned}$$

The sign of ΔG turns out to be positive, meaning that even though the chloride ion concentration is twice as high outside the cell as inside, energy will still be required to drive the movement of chloride ions into the cell. This is due to the excess of negative charge inside the cell, represented by the negative value of V_{m}. The movement of a mole of chloride ions up the charge gradient represented by the membrane potential requires more energy (+1384 calories) than is released by the movement of the mole of chloride ions down their concentration gradient (−410 calories). To guide you in such calculations, the equations for inward and outward transport of charged solutes are included along with those for the transport of uncharged solutes in Table 8-3, which summarizes the thermodynamic properties of each of these processes.

At equilibrium, $\Delta G = 0$. By performing a few algebraic manipulations, Equation 8-10 then allows us to determine the magnitude of the membrane potential, V_{m}, if the ion concentrations on opposite sides of the membrane are known. This equation is very similar in basic form to the equation that allows us to calculate the free energy change of a chemical reaction by knowing the concentrations of reactants and products (see Equation 5-17 on page 120). Extensions of this equation are important for determining the contributions of various ions to the overall resting membrane potential of a cell (discussed in detail in Chapter 22).

In this chapter, we have focused on the movement of ions and small molecules into and out of cells and organelles and have seen that these solutes pass directly through the membrane, though often with the aid of specific transport proteins. In addition to such traffic, however, many eukaryotic cells are able to take up and release substances that are too large to pass through a membrane regardless of its permeability properties. *Exocytosis* is the process whereby cells release proteins that are synthesized within the cell and sequestered within membrane-bounded vesicles, whereas *endocytosis* involves the uptake of macromolecules and other substances by trapping and engulfing them within an invagination of the plasma membrane. (We will consider exocytosis and endocytosis in detail in Chapter 12.)

CONCEPT CHECK 8.6

You are studying the energetics of transport of the amino acid aspartic acid, whose side chain can exist in the neutral state (= COOH) or the ionized, negatively charged state ($= COO^-$). To calculate the ΔG of transport into a cell, why would you need to know the ionization state of the side chain?

Summary of Key Points

Mastering™ Biology For activities, animations, and review quizzes, go to the study area at www.masteringbiology.com.

8.1 Cells and Transport Processes

- The selective transport of molecules and ions across membrane barriers ensures that necessary substances are moved into and out of cells and cell compartments at the appropriate time and at useful rates.
- Nonpolar molecules and small, polar molecules cross the membrane by simple diffusion. Transport of all other solutes, including ions and many molecules of biological relevance, is mediated by specific transport proteins that provide passage through an otherwise impermeable membrane.
- Each such transport protein has at least one, and frequently several, hydrophobic membrane-spanning sequences that embed the protein within the membrane and often act as the channel itself. Typically, a separate regulatory domain controls channel opening and closing.

8.2 Simple Diffusion: Unassisted Movement down the Gradient

- Simple diffusion through biological membranes is limited to small or nonpolar molecules such as O_2, CO_2, and lipids. Water molecules, although polar, are small enough to diffuse across membranes in a manner that is not entirely understood.
- Membranes are permeable to lipids, which can pass through the nonpolar interior of the lipid bilayer. Membrane permeability of most compounds is directly proportional to their partition coefficient—their relative solubility in oil versus water.
- The direction of diffusion of a solute across a membrane is determined by its concentration gradient and always moves toward equilibrium. The solute will diffuse down the gradient from a region of higher concentration to a region of lower concentration.
- If the membrane is impermeable to the solute, water will move by osmosis from the area of lower solute concentration (higher $[H_2O]$) to the area of higher solute concentration (lower $[H_2O]$).

8.3 Facilitated Diffusion: Protein-Mediated Movement down the Gradient

- Transport can be either downhill or uphill in relation to an uncharged solute's concentration gradient. For an ion, we must consider its electrochemical potential—the combined effect of the ion's concentration gradient and the charge gradient across the membrane.
- Downhill transport of large, polar molecules and ions, called facilitated diffusion, must be mediated by carrier proteins or channel proteins because these molecules and ions cannot diffuse through the membrane directly.
- Carrier proteins function by alternating between two conformational states. Examples include the glucose transporter and the anion exchange protein found in the plasma membrane of the erythrocyte.
- Transport of a single kind of molecule or ion is called uniport. The coupled transport of two or more molecules or ions at a time may involve movement of both solutes in the same direction (symport) or in opposite directions (antiport).
- Channel proteins facilitate diffusion by forming transmembrane channels lined with hydrophilic amino acids. Three important categories of channel proteins are ion channels (used mainly for transport of H^+, Na^+, K^+, Ca^{2+}, Cl^-, and HCO_3^-), porins (for various high-molecular-weight solutes), and aquaporins (for water).

8.4 Active Transport: Protein-Mediated Movement up the Gradient

- Active transport—the uphill transport of large, polar molecules and ions—requires a protein transporter and an input of energy. It may be powered by ATP hydrolysis, the electrochemical potential of an ion gradient, or light energy.
- Active transport powered by ATP hydrolysis utilizes four major classes of transport proteins: P-type, V-type, F-type, and ABC-type ATPases. One widely encountered example is the ATP-powered Na^+/K^+ pump (a P-type ATPase), which maintains electrochemical potentials for sodium and potassium ions across the plasma membrane of animal cells.
- Active transport driven by an electrochemical potential usually depends on a gradient of either sodium ions (in animal cells) or protons (in plant, fungal, and many bacterial cells). For example, the inward transport of nutrients across the plasma membrane is often driven by the symport of sodium ions that were pumped outward by the Na^+/K^+ pump. As they flow back into the cell, they drive inward transport of sugars, amino acids, and other organic molecules.

8.5 Examples of Active Transport

- The Na^+/K^+ pump uses direct active transport to move sodium and potassium ions in opposite directions against their concentration gradients, with ATP hydrolysis providing the necessary input of energy.
- The Na^+/glucose symporter uses indirect active transport to move glucose into the cell, along with Na^+. The driving force is the movement of Na^+ down a steep concentration gradient maintained by the Na^+/K^+ pump.
- In *Halobacterium*, active transport is powered by light energy. As photons of light are absorbed by bacteriorhodopsin, protons are pumped across the cell membrane and out of the cell. As the protons flow back into the cell, ATP is synthesized.

8.6 The Energetics of Transport

- The ΔG for transport can be readily calculated. If $\Delta G < 0$, transport will be spontaneous. If $\Delta G > 0$, an input of energy will be required to drive transport. If $\Delta G = 0$, there will be no net movement of the solute.
- For uncharged solutes, ΔG depends only on the concentration gradient. For charged solutes, both the concentration gradient and the membrane potential must be taken into account.

Problem Set

Mastering™ Biology For activities, animations, and review quizzes, go to the study area at www.masteringbiology.com.

8-1 True or False? Indicate whether each of the following statements about membrane transport is true (T) or false (F). If false, reword the statement to make it true.

(a) Facilitated diffusion of a cation occurs only from a compartment of higher concentration to a compartment of lower concentration.

(b) Active transport is always driven by the hydrolysis of high-energy phosphate bonds.

(c) The K_{eq} value for the diffusion of polar molecules out of the cell is less than 1 because membranes are essentially impermeable to such molecules.

(d) A 0.25 *M* sucrose solution would not be isotonic for a mammalian cell if the cell had sucrose carrier proteins in its plasma membrane.

(e) The permeability coefficient for a particular solute is likely to be orders of magnitude lower if a transport protein for that solute is present in the membrane.

(f) Plasma membranes have few, if any, transport proteins that are specific for phosphorylated compounds.

(g) Carbon dioxide and bicarbonate anions usually move in the same direction across the plasma membrane of an erythrocyte.

(h) Treatment of an animal cell with an inhibitor that is specific for the Na^+/K^+ pump is not likely to affect the uptake of glucose by sodium cotransport.

8-2 Telling Them Apart. From the following list of properties, indicate which one(s) can be used to distinguish between each of the following pairs of transport mechanisms.

Transport Mechanisms

(a) Simple diffusion; facilitated diffusion

(b) Facilitated diffusion; active transport

(c) Simple diffusion; active transport

(d) Direct active transport; indirect active transport

(e) Symport; antiport

(f) Uniport; coupled transport

(g) P-type ATPase; V-type ATPase

Properties

1. Directions in which two transported solutes move
2. Direction the solute moves relative to its concentration gradient or its electrochemical potential
3. Kinetics of solute transport
4. Requirement for metabolic energy
5. Requirement for simultaneous transport of two solutes
6. Intrinsic directionality
7. Competitive inhibition
8. Sensitivity to the inhibitor vanadate

8-3 "Facilitating" Thinking About Transport. For each of the following statements, answer with a D if the statement is true of simple diffusion, with an F if it is true of facilitated diffusion, and with an A if it is true of active transport. Any, all, or none (N) of the choices may be appropriate for a given statement.

(a) Requires the presence of an integral membrane protein

(b) Depends primarily on solubility properties of the solute

(c) Doubling the concentration gradient of the molecule to be transported will double the rate of transport over a broad range of concentrations.

(d) Involves proteins called ATPases

(e) Work is done during the transport process.

(f) Applies only to small, nonpolar solutes

(g) Applies only to ions

(h) Transport can occur in either direction across the membrane, depending on the prevailing concentration gradient.

(i) $\Delta G° = 0$

(j) A Michaelis constant (K_m) can be calculated.

8-4 DATA ANALYSIS Glucose Uptake. *Helicobacter pylori* is a bacterium that is responsible for ulcers in the stomach and small intestine in humans. **Figure 8-17** shows the rate of uptake of radioactively labeled D-glucose by *H. pylori* in the presence of 150 m*M* NaCl (red) or 150 m*M* KCl (blue). Based on this information, what sort of transport into *H. pylori* cells does D-glucose exhibit, and why?

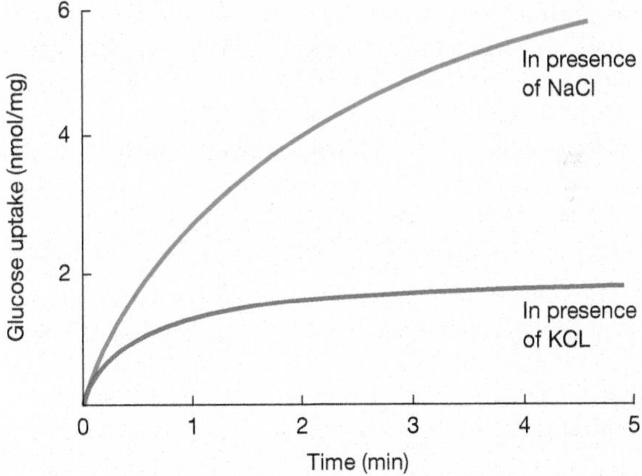

Figure 8-17 Glucose transport in *Helicobacter pylori*. Transport of radioactively labeled 0.8 m*M* D-glucose into *H. pylori* cells was measured in the presence of 150 m*M* NaCl (red) or 150 m*M* KCl (blue). See Problem 8-4.

8-5 QUANTITATIVE Potassium Ion Transport. Most of the cells of your body pump potassium ions inward to maintain an internal K^+ concentration that is 30 to 40 times the external concentration.

(a) What is ΔG (at 37°C) for the transport of potassium ions into a cell that maintains no membrane potential across its plasma membrane?

(b) For a nerve cell with a membrane potential of −60mV, what is ΔG for the inward transport of potassium ions at 37°C?

(c) For the nerve cell in part b, what is the maximum number of potassium ions that can be pumped inward by the hydrolysis of one ATP molecule if the ATP/ADP ratio in the cell is 5:1 and the inorganic phosphate concentration is 10 m*M*? (Assume $\Delta G°' = -7.3$ kcal/mol for ATP hydrolysis.)

8-6 QUANTITATIVE Ion Gradients and ATP Synthesis. The ion gradients maintained across the plasma membranes of most cells play a significant role in cellular energetics. Ion gradients are often either generated by the hydrolysis of ATP or used to make ATP by the phosphorylation of ADP.

(a) Cite an example in which ATP is used to generate and maintain an ion gradient. What is another way that an ion gradient can be generated and maintained?

(b) Cite an example in which an ion gradient is used to make ATP. What is another use for ion gradients?

(c) Assume that the sodium ion concentration is 12 m*M* inside a cell and 145 m*M* outside the cell and that the membrane potential is –90 mV. Can a cell use ATP hydrolysis to drive the outward transport of sodium ions on a 2:1 basis (two sodium ions transported per ATP hydrolyzed) if the ATP/ADP ratio is 5:1, the inorganic phosphate concentration is 50 m*M*, and the temperature is 37°C? What about on a 3:1 basis? Explain your answers.

(d) Assume that a bacterial cell maintains a proton gradient across its plasma membrane such that the pH inside the cell is 8.0 when the outside pH is 7.0. Can the cell use the proton gradient to drive ATP synthesis on a 1:1 basis (one ATP synthesized per proton transported) if the membrane potential is –180 mV, the temperature is 25°C, and the ATP, ADP, and inorganic phosphate concentrations are as in part c? What about on a 1:2 basis? Explain your answers.

8-7 QUANTITATIVE **Sodium Ion Transport.** A marine protozoan is known to pump sodium ions outward by a simple ATP-driven Na^+ pump that operates independently of potassium ions. The intracellular concentrations of ATP, ADP, and P_i are 20, 2, and 1 m*M*, respectively, and the membrane potential is –75 mV.

(a) Assuming that the pump transports three sodium ions outward per molecule of ATP hydrolyzed, what is the lowest internal sodium ion concentration that can be maintained at 25°C when the external sodium ion concentration is 150 m*M*?

(b) If you were dealing with an uncharged molecule rather than an ion, would your answer for part a be higher or lower, assuming all other conditions remained the same? Explain.

8-8 QUANTITATIVE **The Case of the Acid Stomach.** The gastric juice in your stomach has a pH of 2.0. This acidity is due to the secretion of protons into the stomach by the epithelial cells of the gastric mucosa. Epithelial cells have an internal pH of 7.0 and a membrane potential of –70 mV (inside is negative) and function at body temperature (37°C).

(a) What is the concentration gradient of protons across the epithelial membrane?

(b) Calculate the free energy change associated with the secretion of 1 mole of protons into gastric juice at 37°C.

(c) Do you think that proton transport can be driven by ATP hydrolysis at the ratio of one molecule of ATP per proton transported?

(d) If protons were free to move back into the cell, calculate the membrane potential that would be required to prevent them from doing so.

8-9 Charged or Not: Does It Make a Difference? Many solutes that must move into and out of cells exist in either a protonated or ionized form, or they have functional groups that can be either protonated or ionized. Simple molecules such as CO_2, H_3PO_4 (phosphoric acid), and NH_3 (ammonia) are in this category, as are organic molecules with carboxylic acid groups, phosphate groups, and/or amino groups.

(a) Consider ammonia as a simple example of such compounds. What is the charged form of ammonia called? What is its chemical formula?

(b) Which of these two forms will predominate in a solution with a highly acidic pH? Explain your answer.

(c) For which of these two forms will the uptake across the plasma membrane of a cell be affected by the concentration gradient of that form on the inside versus the outside of the cell? For which, if either, of these two forms will uptake be affected by the membrane potential of the plasma membrane? Explain your answers.

(d) For a cell that must take up ammonia from its environment, will uptake of the charged form require more or less energy than uptake of the uncharged form, assuming that the cell has a negative membrane potential? Explain your answer.

(e) Instead of ammonia, consider the uptake of acetic acid, CH_3COOH, an important intermediate in several biological pathways. What is the charged form in this case? For which of the two forms will the uptake across the plasma membrane of a cell be affected by the concentration gradient of that form on the inside versus the outside of the cell? For which, if either, of these two forms will uptake be affected by the membrane potential of the plasma membrane? Explain your answers.

(f) For a cell that must take up acetic acid from its environment, will uptake of the charged form require more or less energy than uptake of the uncharged form, assuming that the cell has a negative membrane potential? Explain your answer.

8-10 QUANTITATIVE **The Calcium Pump of the Sarcoplasmic Reticulum.** Muscle cells use calcium ions to regulate the contractile process. Calcium is both released and taken up by the sarcoplasmic reticulum (SR). Release of calcium from the SR activates muscle contraction, and ATP-driven calcium uptake causes the muscle cell to relax afterward. When muscle tissue is disrupted by homogenization, the SR forms small vesicles called *microsomes* that can still take up calcium. To obtain the data shown in **Figure 8-18**, a solution was prepared containing 5 m*M* ATP and 0.1 *M* KCl at pH 7.5. An aliquot of SR microsomes containing 1.0 mg protein was added to 1 mL of the solution, followed by 0.4 mmol of calcium. Two minutes later, a calcium ionophore was added. (An ionophore is a substance that facilitates the movement of an ion across a membrane.) ATPase activity was monitored during the additions, with the results shown in the figure.

(a) What is the ATPase activity, calculated as micromoles of ATP hydrolyzed per milligram of protein per minute?

(b) The ATPase is calcium activated, as shown by the increase in ATP hydrolysis when the calcium was added and the decrease in hydrolysis when all the added calcium was taken up into the vesicles 1 minute after it was added. How many calcium ions are taken up for each ATP hydrolyzed?

(c) The final addition is an ionophore that carries calcium ions across membranes. Why does ATP hydrolysis begin again?

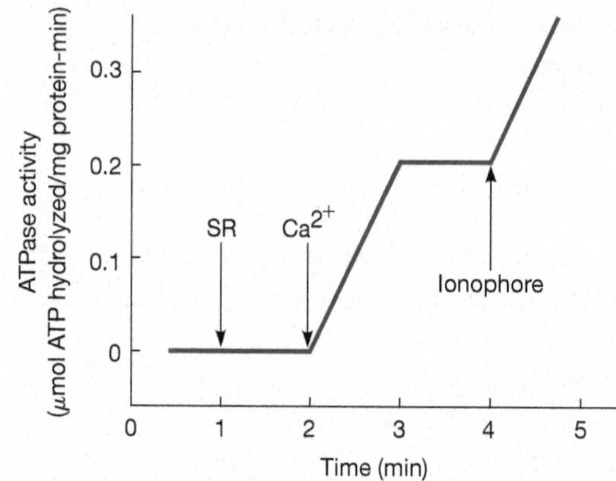

Figure 8-18 Calcium Uptake by the Sarcoplasmic Reticulum. See Problem 8-10.

8-11 Ouabain Inhibition. Ouabain is a specific inhibitor of the active transport of sodium ions out of the cell and is therefore a valuable tool in studies of membrane transport mechanisms. Which of the following processes in your own body would you expect to be sensitive to inhibition by ouabain? Explain your answer in each case.

(a) Facilitated diffusion of glucose into a muscle cell

(b) Active transport of dietary phenylalanine across the intestinal mucosa

(c) Uptake of potassium ions by red blood cells

(d) Active uptake of lactose by the bacteria in your intestine

9 Chemotrophic Energy Metabolism: Glycolysis and Fermentation

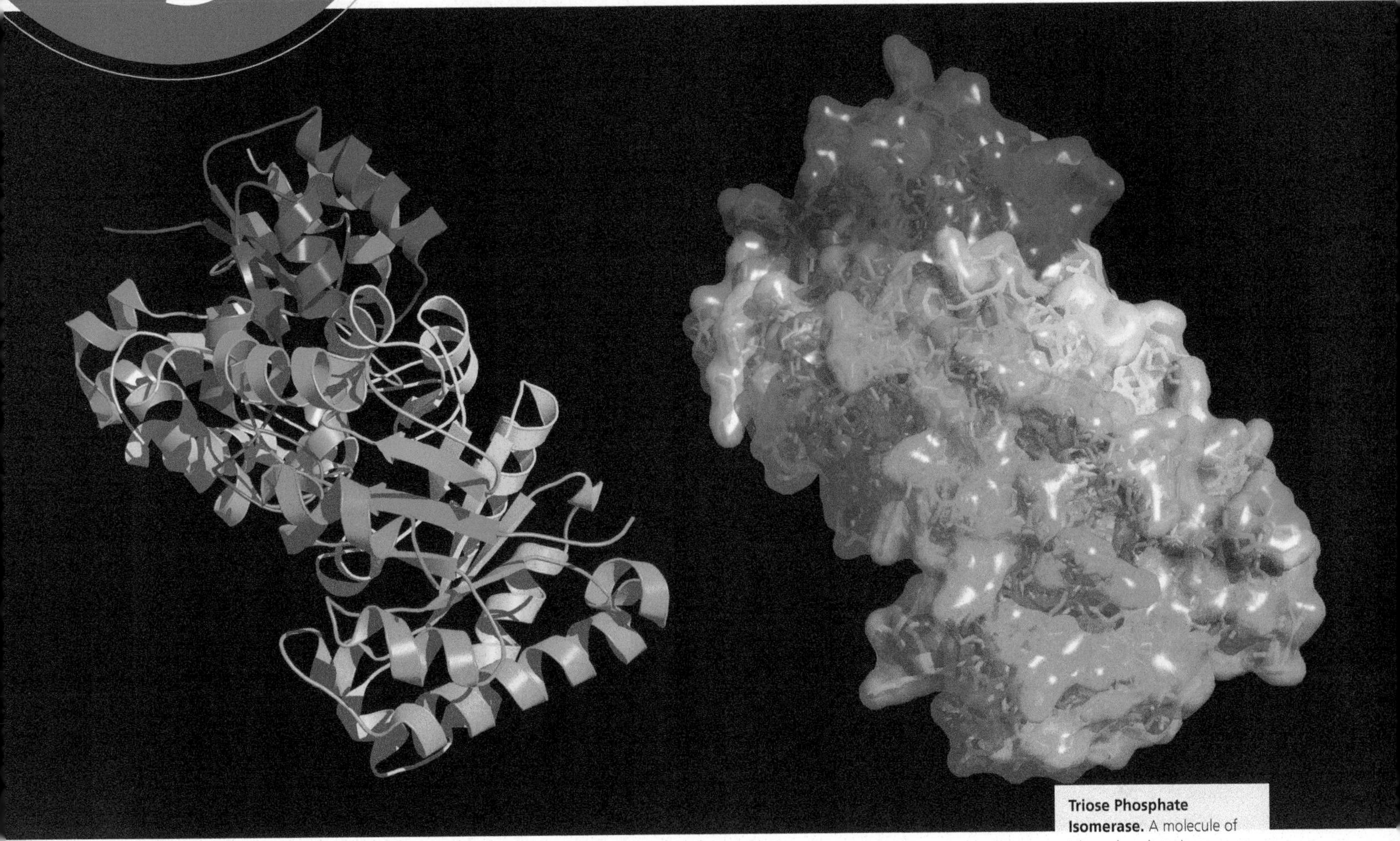

Triose Phosphate Isomerase. A molecule of triose phosphate isomerase, a key enzyme in glycolysis, is shown as a secondary structure model (left) and a space-filling model (right).

As you learned in earlier chapters, cells cannot survive without a source of energy and a source of chemical "building blocks"—the small molecules from which necessary cellular components such as proteins, nucleic acids, polysaccharides, and lipids are synthesized. In many organisms, including us, these two requirements are related. The desired energy and small molecules are both present in the food molecules that these organisms produce or ingest.

In this chapter and the next, we will consider how *chemotrophs,* such as animals and most microorganisms, obtain energy from the food they engulf or ingest, focusing especially on the oxidative breakdown of sugar molecules. Then (in Chapter 11) we will discuss the process by which *phototrophs,* such as green plants, algae, and some bacteria and archaea, tap the solar radiation that is the ultimate energy source for almost all living organisms. They will use this energy to reduce carbon dioxide (add electrons and hydrogens) in order to produce sugar molecules. Keep in mind that the reactions whereby cells obtain energy also can provide the various small molecules that cells need for synthesis of macromolecules and other cellular constituents.

9.1 Metabolic Pathways

When we encountered enzymes in Chapter 6, we considered individual chemical reactions catalyzed by individual enzymes functioning in isolation—but that is not the way cells really operate. To accomplish any major task, a cell requires a series of reactions occurring in an ordered sequence, and this, in turn, requires many different enzymes because most enzymes catalyze only a single reaction, and many such reactions are usually needed to accomplish a major biochemical operation.

When we consider all the chemical reactions that occur within a cell, we are talking about **metabolism** (from the Greek word *metaballein,* meaning "to change"). The overall metabolism of a cell consists, in turn, of numerous specific **metabolic pathways,** each of which accomplishes a particular task. From a biochemist's perspective, *life at the cellular level depends on a network of integrated and carefully regulated metabolic pathways, each contributing to the sum of activities that a cell must carry out.*

Metabolic pathways are of two general types. Synthetic pathways that produce cellular components are known as **anabolic pathways** (from the Greek prefix *ana–*, meaning "up"), whereas degradative pathways involved in the breakdown of cellular constituents are called **catabolic pathways** (from the Greek prefix *kata–*, meaning "down"). Anabolic pathways usually involve a substantial increase in molecular order (and therefore a local decrease in entropy) and are *endergonic* (energy-requiring). Polymer synthesis and the biological reduction of carbon dioxide to sugar are examples of anabolic pathways. Often, anabolic pathways synthesize polymers such as starch and glycogen from glucose units in order to store energy for future use. Certain steroid hormones, for example, are called anabolic steroids because they stimulate the synthesis of muscle proteins from amino acids.

Catabolic pathways, by contrast, are degradative pathways that typically involve a decrease in molecular order (increase in entropy) and are *exergonic* (energy-liberating). These reactions often involve hydrolysis of macromolecules or biological oxidations. Catabolic pathways play two roles in cells: they release the free energy needed to drive cellular functions, and they give rise to the small organic molecules, or *metabolites,* that are the building blocks for biosynthesis. However, a catabolic pathway is not simply the reverse of the corresponding anabolic pathway. For example, the catabolic pathway for glucose degradation and the anabolic pathway for glucose synthesis use slightly different enzymes and intermediates.

As we will see shortly, catabolism can be carried out in either the presence or the absence of oxygen (that is, under either *aerobic* or *anaerobic* conditions). The energy yield per glucose molecule is much greater in the presence of oxygen. However, anaerobic catabolism is also important, not only for organisms in environments that are always devoid of oxygen but also for organisms and cells that are temporarily deprived of oxygen.

CONCEPT CHECK 9.1

How are catabolic and anabolic pathways similar? How are they different?

9.2 ATP: The Primary Energy Molecule in Cells

The anabolic reactions of cells are responsible for biosynthesis during growth and repair processes and require an input of energy, whereas the catabolic reactions that degrade food molecules release the energy needed to drive the anabolic reactions. And, as you know, cells also require energy for a number of other critical cellular processes besides biosynthesis (see Figure 5-1).

In virtually all cells, the molecule most commonly used as an energy source to carry out cellular work is **adenosine triphosphate (ATP).** ATP is, in other words, the primary energy "currency" of the biological world. First, an external source of energy such as glucose (chemotrophs) or the sun (phototrophs) is required to produce ATP. Subsequently, ATP can be used as an energy source to power numerous necessary cellular activities, including movement of cells and materials within cells, active transport of molecules and ions across membranes, and a number of enzyme-catalyzed reactions.

Keep in mind, however, that ATP synthesis is not the only way that cells store chemical energy. Other high-energy molecules, such as GTP and creatine phosphate, store chemical energy that can be converted to ATP. In addition, as we will soon see, chemical energy can be stored as *reduced coenzymes* such as NADH that are a source of reducing power for anabolic reactions in cells. Because ATP is involved in most cellular energy transactions, it is essential that we first understand its structure and function and appreciate the properties that make this molecule so suitable for its role as the universal energy coupler.

ATP Contains Two Energy-Rich Phosphoanhydride Bonds

ATP is a complex molecule containing the aromatic base adenine, the five-carbon sugar ribose, and a chain of three phosphate groups (see Chapter 3). The phosphate groups are linked to each other by **phosphoanhydride bonds** and to the ribose by a **phosphoester bond,** as shown for the ATP molecule in **Figure 9-1**. The compound formed by linking adenine and ribose is called *adenosine.* Adenosine may occur in the cell in the unphosphorylated form or with one, two, or three phosphates attached to carbon atom 5 of the ribose, forming *adenosine monophosphate (AMP), diphosphate (ADP),* and *triphosphate (ATP),* respectively.

The ATP molecule serves well as an intermediate in cellular energy metabolism because energy is released when ATP undergoes *hydrolysis*—water is used to break the phosphoanhydride bond that links the third (outermost) phosphate to the second. Two products are formed: the terminal phosphate receives an –OH from the water molecule and is released as inorganic phosphate (HPO_4^{2-}, often written as P_i), and the resulting ADP molecule gets a hydrogen atom, which immediately loses a proton by ionization. Thus, the hydrolysis of ATP to form ADP and P_i is exergonic, with a standard free energy change ($\Delta G^{\circ\prime}$) of −7.3 kcal/mol (Figure 9-1, reaction 1). The reverse reaction, whereby ATP is synthesized from ADP and P_i with the loss of a water molecule by condensation, is correspondingly endergonic, with a $\Delta G^{\circ\prime}$ of +7.3 kcal/mol (Figure 9-1, reaction 2). As you can see, energy is required to drive ATP synthesis from ADP and P_i, and energy is released upon ATP hydrolysis.

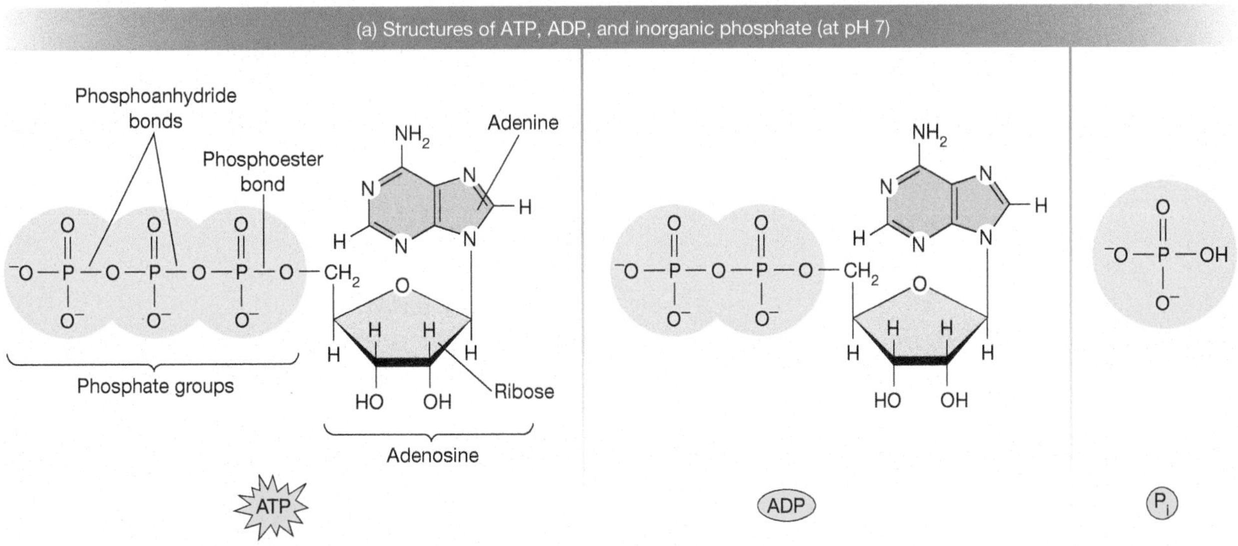

(a) Structures of ATP, ADP, and inorganic phosphate (at pH 7)

Phosphoanhydride bonds | Phosphoester bond | Adenine | Phosphate groups | Ribose | Adenosine | ATP | ADP | P_i

(b) Balanced chemical equation for ATP hydrolysis and synthesis

Reaction 1: Hydrolysis ($\Delta G^{\circ\prime} = -7.3$ kcal/mol)

$$ATP^{4-} + H_2O \rightleftharpoons ADP^{3-} + P_i^{2-} + H^+$$

Reaction 2: ATP synthesis ($\Delta G^{\circ\prime} = +7.3$ kcal/mol)

Figure 9-1 ATP Hydrolysis and Synthesis. (a) ATP consists of adenosine (adenine + ribose) plus three phosphate groups attached to ribose. **(b)** *Reaction 1*: ATP hydrolysis to ADP and inorganic phosphate (P_i) is highly exergonic, with a standard free energy change of −7.3 kcal/mol. *Reaction 2*: ATP synthesis by phosphorylation of ADP is equally endergonic, with a standard free energy change of +7.3 kcal/mol.

Biochemists sometimes refer to bonds such as the phosphoanhydride bonds of ATP as "high-energy" or "energy-rich" bonds, a very useful convention introduced in 1941 by Fritz Lipmann, a leading bioenergetics researcher of the time. However, these terms need to be understood correctly to avoid the erroneous impression that the bond somehow contains energy that can be released. All chemical bonds *require* energy to be broken and *release* energy when they form.

What we really mean by energy-rich bond is that free energy is released when the bond is hydrolyzed by addition of water. During the hydrolysis of ATP, more energy is released when the new bonds are formed in the products ADP and P_i than was required to break the bonds in ATP and H_2O. The energy is therefore a *characteristic of the reaction* the molecule is involved in and *not of a particular bond* within that molecule. Thus, calling ATP or any other molecule a high-energy compound or energy-rich compound should always be understood as a shorthand way of saying that the hydrolysis of one or more of its bonds is exergonic.

ATP Hydrolysis Is Exergonic Due to Several Factors

What is it about the ATP molecule that makes the hydrolysis of its phosphoanhydride bonds so exergonic? The answer to this question has three parts: hydrolysis of ATP to ADP and P_i is exergonic because of *charge repulsion* between the adjacent negatively charged phosphate groups, because of *resonance stabilization* of both products of hydrolysis, and because of their *increased entropy* and solubility.

Charge Repulsion. Opposite charges attract each other, as in an ionic bond, and similar charges repel each other. The strong **charge repulsion** between the negative charges in the phosphate groups of ATP is part of the reason that ATP hydrolysis is exergonic. By way of analogy, imagine forcing two magnets together with like poles touching (both north or both south). The like poles repel each other, and you need to make an effort and provide energy to force them together. If you let go, the magnets spring apart, releasing the energy.

Now consider the three phosphate groups of ATP (see Figure 9-1). Each group is negatively charged due to its ionization at the near neutral pH of the cell. These negative charges repel one another, straining the covalent bond linking the phosphate groups. Conversely, ATP synthesis requires the joining of two negatively charged molecules (ADP^{3-} and P_i^{2-}) that naturally repel each other, thus requiring an input of energy to overcome this repulsion.

Resonance Stabilization. As shown in **Figure 9-2** on page 218, a second important contribution to ATP bond energy is **resonance stabilization.** Consider the carboxylate group shown in Figure 9-2a. (A carboxylate group is the deprotonated form of an acid containing a carboxyl group.) Although the structure of the carboxylate group ($R{=}COO^-$) is formally written with one C=O double bond and one C—O single bond, (contributing structures) (resonance hybrid)

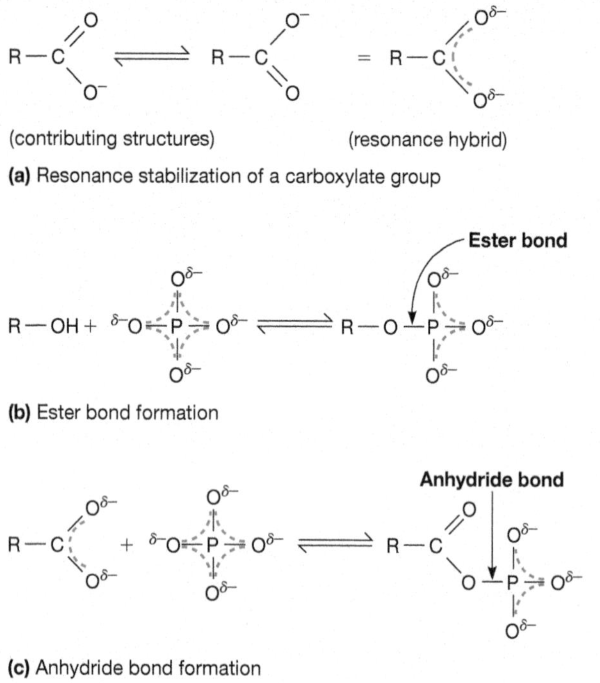

(a) Resonance stabilization of a carboxylate group

Ester bond

(b) Ester bond formation

Anhydride bond

(c) Anhydride bond formation

Figure 9-2 Decreased Resonance Stabilization of the Carboxylate and Phosphate Groups Following Bond Formation. Resonance stabilization due to electron delocalization (blue dashed lines) is an important feature of both the carboxylate group in **(a)** and the phosphate groups in **(b)** and **(c).** Creation of either **(b)** an ester bond or **(c)** an anhydride bond (by removal of water) decreases the opportunity for electron delocalization. As a result, the ester or anhydride product is a higher-energy compound than the reactants, and energy will be released when the bond is broken by hydrolysis.

it actually has one electron pair that is *delocalized* (equally distributed) over both of the C bonds to oxygen. The true structure of the carboxylate group is actually an average of the two contributing structures shown in Figure 9-2a and is called a *resonance hybrid.* Each C—O bond is the equivalent of one and a half bonds, and each O atom has only a partial negative charge (represented by the Greek letter δ). When electrons are delocalized in this way, a molecule is in its most stable (lowest-energy) configuration and is said to be *resonance-stabilized.*

Similarly, the phosphate ion (PO_4^{3-}) is resonance-stabilized because the extra electron pair formally shown as part of a P═O double bond is delocalized over all four O atoms adjacent to the central P. Consider what happens when a *phosphoester bond* forms between an alcohol and a phosphate ion: the extra electrons are delocalized over only three O atoms (Figure 9-2b). The product is less resonance-stabilized and thus has higher energy. Similarly, when a *phosphoanhydride* bond forms between a carboxylate group and a phosphate group (Figure 9-2c), the decrease in electron delocalization results in a higher-energy product.

Increased Entropy. A third important factor contributing to the exergonic nature of ATP hydrolysis is the overall increase in entropy as a phosphate group is removed from ATP and is no longer fixed in position. This spatial randomization of the ADP and phosphate decreases their free energy and makes the reaction more exergonic. Although a water molecule is added during hydrolysis (a decrease of entropy), it loses a proton, which also has increased entropy as it is randomized in solution. In addition, the ADP and phosphate become more soluble because they are more highly hydrated—the increased interactions with water molecules lead to a decrease in free energy, adding to the exergonic nature of ATP hydrolysis.

For esters, only a moderate amount of energy is liberated upon hydrolysis, whereas for anhydrides, the hydrolysis reaction is more exergonic. Unlike esters, both products of anhydride hydrolysis have increased resonance stabilization. In addition, both products of anhydride hydrolysis are charged and therefore repel each other, unlike the case with ester hydrolysis, in which only one product is charged. Hydrolysis of anhydride and phosphoanhydride bonds releases roughly twice the amount of free energy as does hydrolysis of ester and phosphoester bonds. ATP illustrates this difference. Hydrolysis of either of the phosphoanhydride bonds that link the second and third phosphate groups to the rest of the molecule has a standard free energy change of about −7.3 kcal/mol, whereas hydrolysis of the phosphoester bond that links the first (innermost) phosphate group to the ribose has a $\Delta G^{\circ\prime}$ of only about −3.6 kcal/mol:

$$\text{ATP} + H_2O \rightarrow \text{ADP} + P_i + H^+ \quad \Delta G^{\circ\prime} = -7.3 \text{ kcal/mol} \tag{9-1}$$

$$\text{ADP} + H_2O \rightarrow \text{AMP} + P_i + H^+ \quad \Delta G^{\circ\prime} = -7.3 \text{ kcal/mol} \tag{9-2}$$

$$\text{AMP} + H_2O \rightarrow \text{adenosine} + P_i \quad \Delta G^{\circ\prime} = -3.6 \text{ kcal/mol} \tag{9-3}$$

Thus, ATP and ADP are both "higher-energy compounds" than AMP is, to use the shorthand of biochemists.

In fact, because the standard $\Delta G^{\circ\prime}$ value of −7.3 kcal/mol is based on equal concentrations of ATP and ADP (1 *M*), it typically underestimates the actual free energy change associated with the hydrolysis of ATP to ADP under most biological conditions, in which the concentration of ATP is greater. The actual free energy change, $\Delta G'$, depends on the prevailing concentrations of reactants and products (see Chapter 5, Equation 5-22, on page 121). For the hydrolysis of ATP (Reaction 9-1), $\Delta G'$ is calculated as

$$\Delta G' = \Delta G^{\circ\prime} + RT \ln \frac{[\text{ADP}][P_i]}{[\text{ATP}]} \tag{9-4}$$

In most cells, the ATP/ADP ratio is significantly greater than 1:1, often in the range of about 5:1. As a result, the term $\ln([\text{ADP}][P_i]/[\text{ATP}])$ is negative, and $\Delta G'$ is therefore more negative than −7.3 kcal/mol, usually in the range of −10 to −14 kcal/mol.

ATP Is Extremely Important in Cellular Energy Metabolism

ATP occupies an intermediate position among energy-rich phosphorylated compounds in the cell. Some of the more common phosphorylated intermediates in cellular energy metabolism are ranked according to their $\Delta G^{\circ\prime}$ values for hydrolysis in **Table 9-1**. The values are negative, so the

Table 9-1 Standard Free Energies of Hydrolysis for Phosphorylated Compounds Involved in Energy Metabolism

Phosphorylated Compound and Its Hydrolysis Reaction	$\Delta G^{\circ\prime}$ (kcal/mol)
Phosphoenruvaolpyte (PEP) $+ H_2O \rightarrow$ pyruvate $+ P_i$	−14.8
1,3-Bisphosphoglycerate $+ H_2O \rightarrow$ 3-phosphoglycerate $+ P_i$	−11.8*
Phosphocreatine $+ H_2O \rightarrow$ creatine $+ P_i$	−10.3
Adenosine triphosphate (ATP) $+ H_2O \rightarrow$ adenosine diphosphate $+ P_i$	−7.3
Glucose-1-phosphate $+ H_2O \rightarrow$ glucose $+ P_i$	−5.0
Glucose-6-phosphate $+ H_2O \rightarrow$ glucose $+ P_i$	−3.3
Glycerol phosphate $+ H_2O \rightarrow$ glycerol $+ P_i$	−2.2

*The $\Delta G^{\circ\prime}$ value for 1,3-bisphosphoglycerate is for the hydrolysis of the phosphoanhydride bond on carbon atom 1.

compounds closest to the top of the table release the most energy upon hydrolysis of the phosphate group. This means that, under standard conditions, a phosphorylated compound is capable of phosphorylating any unphosphorylated compound below it but none of the unphosphorylated compounds above it.

For the hydrolysis of phosphate groups, phosphoenolpyruvate (PEP; −14.8 kcal/mol) is considered a high-energy compound, ATP (−7.3 kcal/mol) is an intermediate-energy compound, and glucose-6-phosphate (−3.3 kcal/mol) is a low-energy compound. Thus, ATP can be formed from ADP by the transfer of a phosphate group from PEP to ATP but not from glucose-6-phosphate to ATP, as **Figure 9-3** illustrates. Similarly, ATP can be used to phosphorylate glucose to glucose-6-phosphate but not pyruvate to PEP.

The following reactions (9-5 and 9-6) have negative $\Delta G^{\circ\prime}$ values, which we will designate as values of $\Delta G^{\circ\prime}_{transfer}$ to emphasize that they represent the standard free energy change that accompanies the transfer of a phosphate group from a donor to an acceptor molecule. $\Delta G^{\circ\prime}_{transfer}$ values can be predicted from Table 9-1 and calculated as shown below:

$$\begin{aligned} &\text{PEP} + \text{ADP} + \text{H}^+ \longrightarrow \text{pyruvate} + \text{ATP} \\ &\Delta G^{\circ\prime}_{\text{transfer}} = \Delta G^{\circ\prime}_{\text{donor}} - G^{\circ\prime}_{\text{acceptor}} \\ &\Delta G^{\circ\prime}_{\text{transfer}} = -14.8 - (-7.3) = -7.5 \text{ kcal/mol} \end{aligned} \quad \textbf{(9-5)}$$

$$\begin{aligned} &\text{glucose} + \text{ATP} \longrightarrow \text{glucose-6-phosphate} + \text{ADP} + \text{H}^+ \\ &\Delta G^{\circ\prime}_{\text{transfer}} = -7.3 - (-3.3) = -4.0 \text{ kcal/mol} \end{aligned} \quad \textbf{(9-6)}$$

PEP can therefore transfer its phosphate group exergonically onto ADP to form ATP ($\Delta G^{\circ\prime}_{transfer} = -7.5$ kcal/mol), and ATP can be used to phosphorylate glucose exergonically ($\Delta G^{\circ\prime}_{transfer} = -4.0$ kcal/mol), but the reverse reactions are not feasible under standard conditions. In fact,

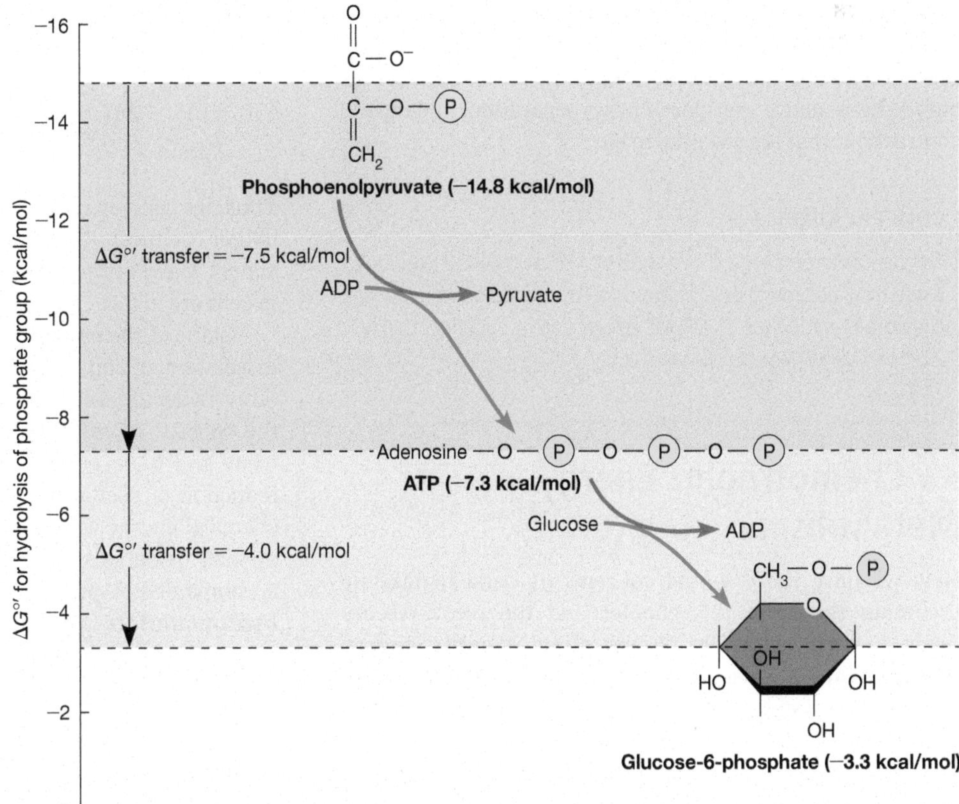

Figure 9-3 Examples of Exergonic Transfer of Phosphate Groups. These examples are based on the $\Delta G^{\circ\prime}$ values for the hydrolysis of the compounds listed in Table 9-1. Phosphoenolpyruvate can therefore transfer its phosphate group exergonically onto ADP to form ATP, and ATP can phosphorylate glucose exergonically, but the reverse reactions are not possible under standard conditions.

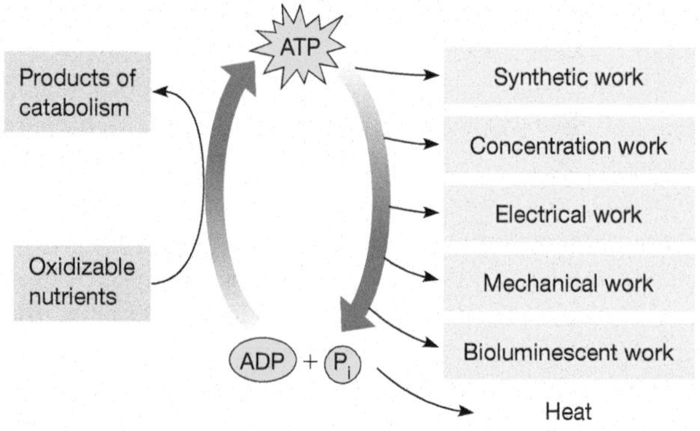

Figure 9-4 The Reversible ATP/ADP System for Conserving and Releasing Energy in Cells. The energy released when nutrients such as glucose are oxidized during catabolism (left side) is used to synthesize ATP. Subsequently, this ATP can be used to do cellular work (right side).

the $\Delta G^{\circ\prime}_{transfer}$ values for these reactions are so negative that both reactions are irreversible under typical cellular conditions.

The most important point to understand from Table 9-1 and Figure 9-3 is that the ATP/ADP pair occupies a crucial *intermediate* position in terms of bond energies. This means that ATP can serve as a phosphate *donor* in some biological reactions and its dephosphorylated form, ADP, can serve as a phosphate *acceptor* in other reactions because there are compounds both above and below the ATP/ADP pair in energy.

In summary, the ATP/ADP pair represents a reversible means of conserving, transferring, and releasing energy within the cell (**Figure 9-4**). As nutrients are oxidized by catabolic pathways in the cell, the energy-liberating catabolic reactions drive the formation of ATP from ADP. The free energy released upon hydrolysis of ATP then provides the driving force for the many energy-requiring processes (such as biosynthesis, active transport, charge separation, and muscle contraction) that are essential to life.

CONCEPT CHECK 9.2

The standard free energy of hydrolysis of ATP ($\Delta G^{\circ\prime} = -7.3$ kcal/mol) is intermediate among cellular phosphorylated compounds. Why is it important for the use of ATP as "cellular energy currency" that its $\Delta G^{\circ\prime}$ be neither very low nor very high?

9.3 Chemotrophic Energy Metabolism

Now we have the essential concepts in hand to take up the main theme of this chapter and the next. We are ready, in other words, to discuss **chemotrophic energy metabolism**—the reactions and pathways by which cells catabolize nutrients and conserve as ATP some of the free energy that is released in this breakdown. Or, to put it more personally, we will look at the specific metabolic processes by which the cells of your own body make use of the food you eat to meet your energy needs. We begin our discussion by considering *oxidation* because much of chemotrophic energy metabolism involves energy-yielding oxidative reactions.

Biological Oxidations Usually Involve the Removal of Both Electrons and Protons and Are Exergonic

To say that nutrients such as carbohydrates, fats, and proteins are sources of energy for cells means that these are oxidizable organic compounds and that their oxidation is exergonic. Recall from chemistry that **oxidation** is the removal of electrons. Thus, for example, a ferrous ion (Fe^{2+}) is oxidizable because it readily gives up an electron as it is converted to a ferric ion (Fe^{3+}):

$$Fe^{2+} \rightarrow Fe^{3+} + e^- \qquad \textbf{(9-7)}$$

The only difference in biological chemistry is that the oxidation of organic molecules frequently involves the removal not just of electrons but also of hydrogen ions (protons), so that the process is often also one of **dehydrogenation.** Consider, for example, the oxidation of ethanol to the corresponding aldehyde:

$$\underset{\text{Ethanol}}{CH_3{-}CH_2{-}OH} \xrightarrow{\text{oxidation}} \underset{\text{Acetaldehyde}}{CH_3{-}\overset{\overset{\displaystyle H}{|}}{C}{=}O} + 2e^- + 2H^+ \qquad \textbf{(9-8)}$$

Electrons are removed, so this is clearly an oxidation. But protons are liberated as well, and an electron plus a proton is the equivalent of a hydrogen atom. Therefore, what happens, in effect, is the removal of the equivalent of two hydrogen atoms:

$$\underset{\text{Ethanol}}{CH_3{-}CH_2{-}OH} \xrightarrow[\text{(dehydrogenation)}]{\text{oxidation}} \underset{\text{Acetaldehyde}}{CH_3{-}\overset{\overset{\displaystyle H}{|}}{C}{=}O} + [2H] \qquad \textbf{(9-9)}$$

Thus, for cellular reactions involving organic molecules, oxidation is almost always manifested as a dehydrogenation reaction. Many of the enzymes that catalyze oxidative reactions in cells are in fact called *dehydrogenases.*

None of the preceding oxidation reactions can take place in isolation, of course; the electrons must be transferred to another molecule, which is *reduced* in the process. **Reduction,** the opposite of oxidation, is defined as the addition of electrons and is an endergonic process. We will soon see that reduction of coenzymes is an important way that cells store chemical energy as reducing power. In biological reductions, as with oxidations, the electrons transferred are frequently accompanied by protons. The overall reaction is therefore a **hydrogenation:**

$$\underset{\text{Acetaldehyde}}{CH_3{-}\overset{\overset{\displaystyle H}{|}}{C}{=}O} + [2H] \xrightarrow[\text{(hydrogenation)}]{\text{reduction}} \underset{\text{Ethanol}}{CH_3{-}CH_2{-}OH} \qquad \textbf{(9-10)}$$

Reactions 9-9 and 9-10 also illustrate the general feature that biological oxidation-reduction reactions almost always involve *two-electron* (and therefore two-proton) transfers.

As written, Reactions 9-9 and 9-10 are only *half reactions*, representing an oxidation and a reduction event, respectively. In real reactions, however, *oxidation and reduction always take place simultaneously*. Any time an oxidation occurs, a reduction must occur as well because the electrons (and protons) removed from one molecule must be added to another molecule. The brackets around the 2H in Reactions 9-9 and 9-10 are meant to show that *hydrogen atoms are never actually released into solution but are instead transferred to another molecule.*

Coenzymes Such as NAD+ Serve as Electron Acceptors in Biological Oxidations

In most biological oxidations, electrons and hydrogens removed from the substrate being oxidized are transferred to one of several **coenzymes.** In general, coenzymes are small molecules that function along with enzymes, often by serving as carriers of electrons and hydrogens. As we will see shortly, coenzymes are not consumed but are recycled within the cell, so the relatively low intracellular concentration of a given coenzyme is adequate to meet the needs of the cell.

As glucose is being partially oxidized, the electrons removed from glucose are transferred to the coenzyme **nicotinamide adenine dinucleotide (NAD^+),** whose structure is shown in **Figure 9-5**. Despite its formidable name and structure, its function is straightforward: NAD^+ serves as an electron acceptor by adding two electrons and one proton to its aromatic ring, thereby generating the reduced form, NADH, plus a proton:

$$\underset{(\text{oxidized})}{NAD^+} + [2H] \rightarrow \underset{(\text{reduced})}{NADH} + H^+ \qquad \textbf{(9-11)}$$

As a nutritional note, the nicotinamide of NAD^+ is a derivative of *niacin*, which we recognize as a *B vitamin*—one of a family of water-soluble compounds essential in the diet of humans and other vertebrates unable to synthesize these compounds for themselves. FAD and CoA are two other coenzymes that also have B vitamin derivatives as part of their structures (see Chapter 10). It is precisely because these vitamins are components of indispensable coenzymes that they are essential in the diet of any organism that cannot manufacture them. Also, they are required only in small amounts because they are not consumed during the reactions but are recycled as they are alternatively reduced and oxidized.

Most Chemotrophs Meet Their Energy Needs by Oxidizing Organic Food Molecules

We are interested in oxidation because it is the means by which chemotrophs such as humans meet their energy needs. Many different kinds of substances serve as substrates for biological oxidation. For example, a wide variety of microorganisms can use inorganic compounds such as hydrogen gas or reduced forms of iron, sulfur, or nitrogen as their energy sources. These organisms, which utilize rather specialized oxidative pathways, play important roles in the geochemical

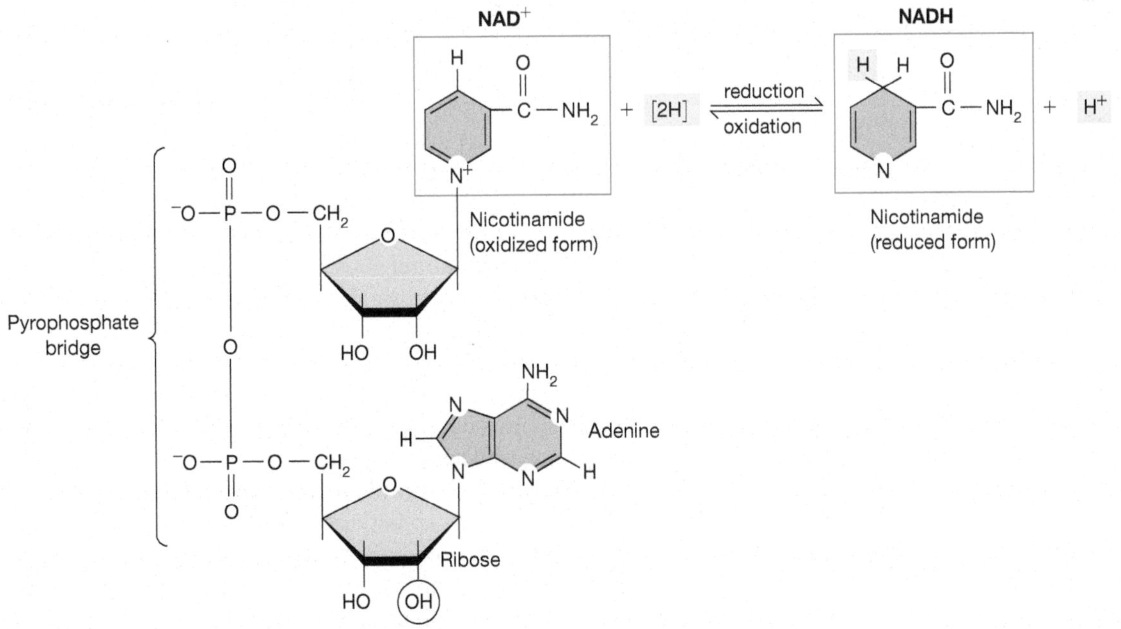

Figure 9-5 The Structure of NAD^+ and Its Oxidation and Reduction. The portion of the coenzyme enclosed in the red box is nicotinamide, derived from the B vitamin niacin. The hydrogen atoms removed from an oxidizable substrate are shown in light blue. When NAD^+ is the electron acceptor, two electrons and one proton from the oxidizable substrate are transferred to one of the carbon atoms of nicotinamide, and the other proton is released into solution. In $NADP^+$, a related coenzyme (which we will encounter in Chapter 11), the hydroxyl group circled here is replaced by a phosphate group.

cycling of nutrients in the biosphere. They also produce a significant amount of biomass—by some estimates, 50% as much as global plant production—and they are important food sources for numerous other organisms in the food chain. However, we humans and most other chemotrophs depend on organic food molecules as oxidizable substrates, namely, carbohydrates, fats, and proteins. Keep in mind that oxidation of these organic food molecules produces energy for the cell as both ATP and reduced coenzymes.

Glucose Is One of the Most Important Oxidizable Substrates in Energy Metabolism

To simplify our discussion initially and to provide a unifying metabolic theme, we will concentrate on the biological oxidation of the six-carbon sugar **glucose** ($C_6H_{12}O_6$). Glucose is a good choice for several reasons. In many vertebrates, including humans, glucose is the main sugar in the blood and hence the main energy source for most of the cells in the body. Blood glucose comes primarily from dietary carbohydrates such as sucrose or starch and from the breakdown of stored glycogen (see Figures 3-21 to 3-24). Current guidelines recommend a diet of approximately 50% carbohydrate, 30% lipid, and 20% protein. Glucose is therefore an especially important molecule for each of us personally.

Glucose is also important to plants because it is the monosaccharide released upon starch breakdown. In addition, glucose makes up one-half of the disaccharide sucrose (glucose + fructose), the major sugar in the vascular system of most plants. Moreover, the catabolism of most other energy-rich substances in plants, animals, and microorganisms alike begins with their conversion into one of the intermediates in the pathway for glucose catabolism. Rather than looking at the fate of a single compound, then, we are considering a metabolic pathway that is at the very heart of chemotrophic energy metabolism.

The Oxidation of Glucose Is Highly Exergonic

Glucose is a good potential source of energy because its oxidation is a highly exergonic process, with a $\Delta G^{\circ\prime}$ of −686 kcal/mol for the complete oxidation of glucose to carbon dioxide using oxygen as the *final electron acceptor.* Note that, as glucose is oxidized to CO_2, oxygen is reduced to water.

$$C_6H_{12}O_6 + 6O_2 \rightarrow 6CO_2 + 6H_2O \qquad \textbf{(9-12)}$$

As a thermodynamic parameter, $\Delta G^{\circ\prime}$ is unaffected by the route from substrates to products. Therefore, it will have the same value whether the oxidation is by direct combustion, with all of the energy released as heat, or by biological oxidation, with some of the energy conserved in the bonds of ATP. Thus, oxidation of the sugar molecules in a marshmallow releases the same amount of free energy whether you burn the marshmallow over a campfire or eat it and catabolize the sugar molecules in your body. Biologically, however, the distinction is critical: *uncontrolled combustion occurs at temperatures that are incompatible with life, and most of the free energy is lost as heat. Biological oxidations involve enzyme-catalyzed reactions that occur without significant temperature changes, and much of the free energy is conserved in chemical form as ATP.*

Glucose Catabolism Yields Much More Energy in the Presence of Oxygen Than in Its Absence

Access to the full 686 kcal/mol of free energy in glucose is possible only if glucose is completely oxidized to carbon dioxide and water. Even then, because no energy conversion process is 100% efficient, only part of the energy can be recovered. The complete oxidation of glucose to carbon dioxide and water in the presence of oxygen is called **aerobic respiration,** a complex, multistep process (which we will discuss in detail in Chapter 10). Many organisms, typically bacteria, can carry out **anaerobic respiration,** using inorganic electron acceptors other than oxygen. Examples of alternative acceptors include elemental sulfur (S), protons (H^+), and ferric ions (Fe^{3+}), which are reduced to H_2S, H_2, and Fe^{2+}, respectively.

Most organisms that rely on anaerobic respiration can still extract limited amounts of energy from the *partial* oxidation of glucose in the absence of oxygen, but with lower energy yields per glucose molecule. They do so by means of **glycolysis,** an ancient and ubiquitous pathway found in virtually all organisms. Glycolysis does not require oxygen as a final electron acceptor. In the absence of oxygen, electrons that are removed during glucose oxidation are instead typically accepted by an organic molecule formed later in the pathway, such as pyruvate. This process is called **fermentation** and is identified in terms of the principal end-product. In some animal cells and many bacteria, the end-product is lactate, so the process of anaerobic glucose catabolism is called *lactate fermentation.* In most plant cells and in microorganisms such as yeast, the process is termed *alcoholic fermentation* because the end-products are ethanol (an alcohol) plus carbon dioxide.

Based on Their Need for Oxygen, Organisms Are Aerobic, Anaerobic, or Facultative

Organisms can be classified in terms of their need for and use of oxygen as an electron acceptor in energy metabolism. Most organisms we see on a daily basis have an absolute requirement for oxygen and are called **obligate aerobes.** You look at such an organism every time you look in the mirror. On the other hand, some organisms, including many bacteria, cannot use oxygen as an electron acceptor and are called **obligate anaerobes.** In fact, oxygen is toxic to these organisms. Not surprisingly, such organisms occupy environments from which oxygen is excluded, such as deep puncture wounds, the sludge at the bottoms of ponds, or deep-sea hydrothermal vents. Most obligate anaerobes are bacteria or archaea, including organisms responsible for gangrene, food poisoning, and methane production.

Facultative organisms can function under either aerobic or anaerobic conditions. When oxygen is present, most facultative organisms carry out aerobic respiration. However, they can switch to anaerobic respiration or fermentation if oxygen is limiting or absent. Many bacteria and fungi are facultative organisms, as are most mollusks and annelid worms. Some cells or tissues of otherwise aerobic organisms can function in the temporary absence or scarcity of oxygen if required to do so. Your muscle cells are an example; they normally function aerobically but switch to lactate fermentation whenever the oxygen supply becomes limiting—during periods of prolonged or strenuous exercise, for example.

The rest of this chapter is devoted mainly to exploring the generation of ATP during glycolysis and fermentation. Aerobic energy metabolism will be the focus of the next chapter. The glycolytic pathway is common to both fermentation and aerobic respiration. As we discuss fermentation, we will be considering the ways that energy can be extracted from glucose without net oxidation, but we will also be laying the foundation for discussing aerobic processes in the next chapter.

CONCEPT CHECK 9.3

Why do oxidation and reduction reactions always happen simultaneously? When glucose is oxidized to CO_2, what is reduced?

9.4 Glycolysis: ATP Generation Without the Involvement of Oxygen

Glycolysis is common to both aerobic and anaerobic metabolism and is present in virtually all organisms. During glycolysis, the six-carbon glucose molecule is split into two three-carbon molecules, each of which is then partially oxidized by a reaction sequence that is sufficiently exergonic to generate two ATP molecules per molecule of glucose fermented. For most cells, this is the maximum possible energy yield that can be achieved without access to oxygen or to an alternative electron acceptor. However, there are exceptions to this limit. Certain microorganisms can get up to five ATP molecules per glucose by using specialized enzyme systems. Also, some plants and animals that are adapted to anoxia (oxygen deprivation) can produce more than two ATP molecules per glucose.

Glycolysis Generates ATP by Catabolizing Glucose to Pyruvate

Glycolysis, sometimes called the *glycolytic pathway*, is a complex, ten-step reaction sequence that converts one six-carbon molecule of glucose into two molecules of pyruvate, a three-carbon compound. During the partial oxidation of glucose to pyruvate in glycolysis, energy and reducing power are conserved in the form of ATP and NADH, respectively. In most cells, the enzymes responsible for glycolysis occur in the cytosol. In some parasitic protozoans called trypanosomes, however, the first seven glycolytic enzymes are compartmentalized in membrane-bounded organelles called *glycosomes*.

Historically, the glycolytic pathway was the first major metabolic sequence to be elucidated. Much of the definitive work was done in the 1930s by the German biochemists Gustav Embden, Otto Meyerhof, Otto Warburg, and colleagues, who, working with frog muscle tissue, identified key components of the pathway, such as ATP and other phosphorylated compounds. In fact, an alternative name for the glycolytic pathway is the *Embden–Meyerhof pathway*.

Glycolysis in Overview. In the presence of oxygen, glycolysis typically leads to aerobic respiration (the subject of Chapter 10). In the absence of oxygen, glycolysis leads to fermentation (discussed later in this chapter). The essence of glycolysis is suggested by its very name because the term *glycolysis* derives from two Greek roots: *glykos*, meaning "sweet," and *lysis*, meaning "loosening" or "splitting." As outlined in **Figure 9-6**, the overall glycolytic pathway can be thought of as consisting of three phases:

Phase 1: An initial input of two ATP and the glucose-splitting reaction for which glycolysis is named (Figure 9-6a)

Phase 2: an oxidation reaction that generates NADH and ATP (Figure 9-6b)

Phase 3: a final step that generates more ATP while producing pyruvate (Figure 9-6c)

Figure 9-7 on page 224 zooms in on the details of the specific reactions involved in each phase. Refer to this figure carefully as you read the next three sections.

Phase 1: Preparation and Cleavage. To begin our consideration of glycolysis, note that the net result of the first three reactions in Figure 9-7 is to convert an unphosphorylated molecule

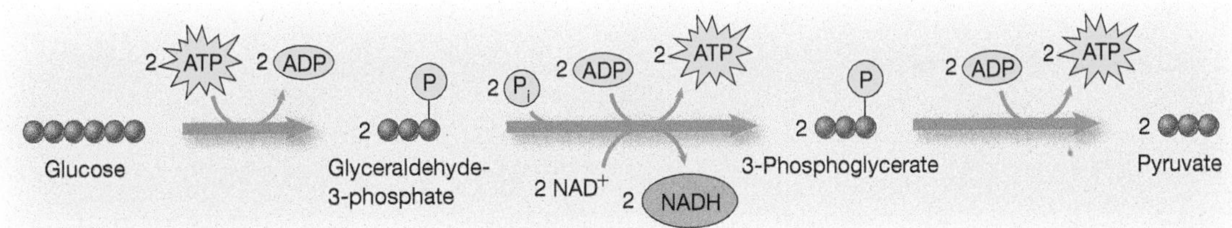

(a) Phase 1: Preparation and cleavage. The six-carbon glucose molecule is phosphorylated twice by ATP and split to form two molecules of glyceraldehyde-3-phosphate. This requires an input of two ATP per glucose.

(b) Phase 2: Oxidation and ATP generation. The two molecules of glyceraldehyde-3-phosphate are oxidized to two 3-phosphoglycerate molecules. Some of the energy from this oxidation is conserved as two ATP and two NADH molecules are produced.

(c) Phase 3: Pyruvate formation and ATP generation. The two 3-phosphoglycerate molecules are converted to pyruvate, with accompanying synthesis of two more ATP molecules, resulting in a net gain of two ATP per glucose.

Figure 9-6 An Overview of the Glycolytic Pathway. During glycolysis, one molecule of glucose is **(a)** split in half, **(b)** partially oxidized, and **(c)** converted to two molecules of pyruvate. Energy is conserved as a net gain of two molecules of ATP and two molecules of NADH. This ten-step process occurs in three main phases, as shown. Simplified structures show only carbon atoms (gray) and phosphate groups (yellow).

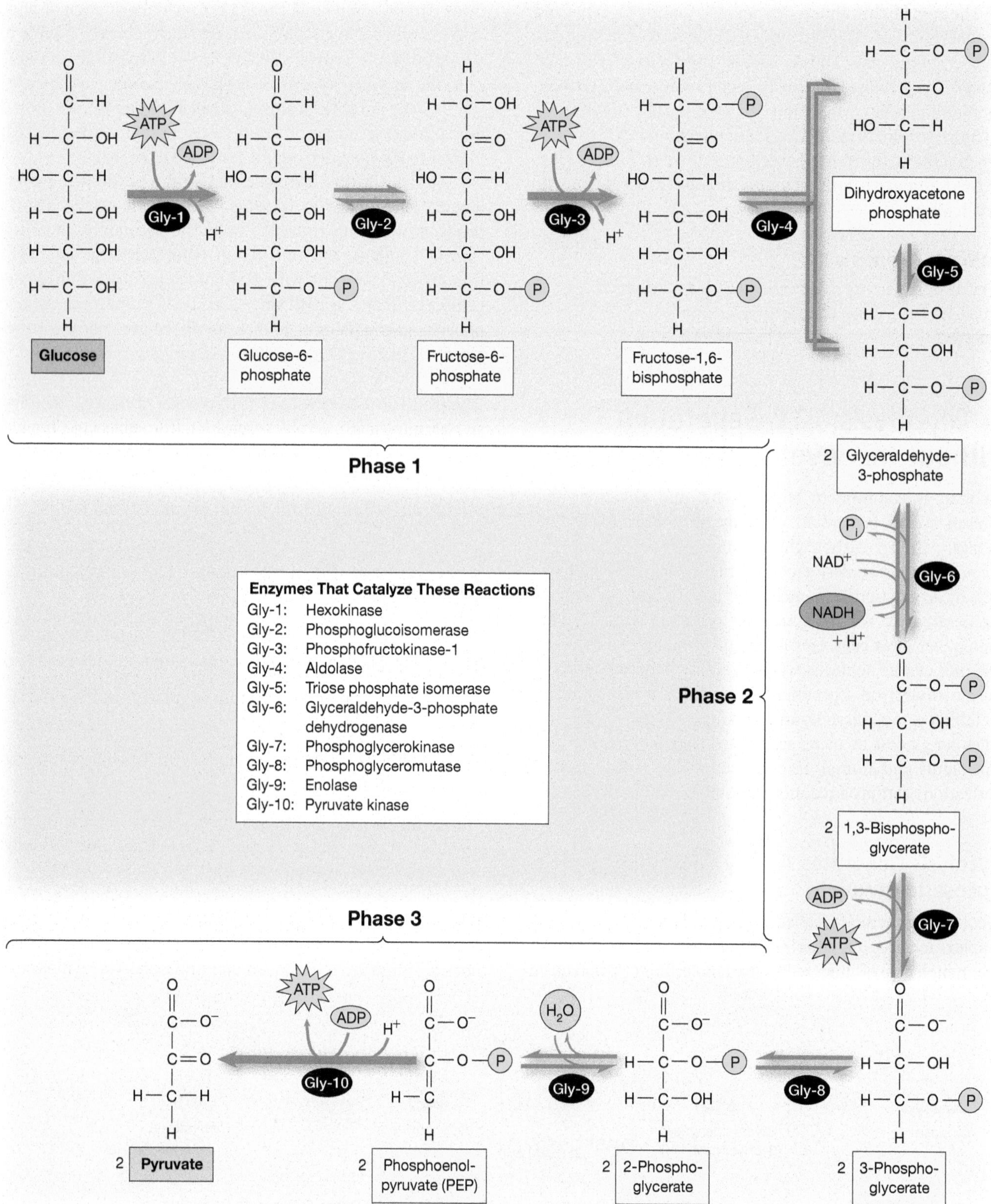

Figure 9-7 The Glycolytic Pathway from Glucose to Pyruvate. Glycolysis is a sequence of ten reactions (Gly-1 through Gly-10) in which glucose is catabolized to pyruvate, with a single oxidative reaction (Gly-6) and two ATP-generating steps (Gly-7 and Gly-10). The enzymes that catalyze each of these reactions are identified in the center box. The three reactions that are essentially irreversible (Gly-1, Gly-3, and Gly-10) are shown as single, thick forward arrows.

(glucose) into a doubly phosphorylated molecule (*fructose-1,6-bisphosphate*). This requires the transfer of a total of two phosphate groups from ATP to glucose, one on each terminal carbon. Looking at glucose, it is easy to see how phosphorylation can take place on carbon atom 6 in reaction Gly-1: the hydroxyl group there can be readily linked to a phosphate group to form glucose-6-phosphate. ATP hydrolysis provides not only the phosphate group but also the free energy that renders the phosphorylation reaction strongly exergonic ($\Delta G^{\circ\prime} = -4.0$ kcal/mol), making it essentially irreversible in the direction of glucose phosphorylation. Reaction Gly-1 and two other reactions of glycolysis (Gly-3 and Gly-10) that are essentially irreversible are shown in Figure 9-7 as single, thick forward arrows.

Notice, by the way, that the bond formed when glucose is phosphorylated is a *phosphoester bond,* a lower-energy bond than the phosphoanhydride bond linking the terminal phosphate to ATP. This difference is what makes the transfer of the phosphate group from ATP to glucose exergonic ($\Delta G^{\circ\prime} = -4.0$ kcal/mol; see bottom half of Figure 9-3 and Reaction 9-6). The enzyme that catalyzes this first reaction (Gly-1) is called *hexokinase*; as the name suggests, it is not specific for glucose but catalyzes the phosphorylation of other hexoses (six-carbon sugars) as well. (Liver cells contain an additional enzyme, *glucokinase,* that phosphorylates only glucose.)

The carbonyl group on carbon atom 1 of the glucose molecule is not as readily phosphorylated as the hydroxyl group on carbon atom 6. But in the next reaction (Gly-2), the aldosugar glucose-6-phosphate is converted to the corresponding ketosugar, fructose-6-phosphate, which has a hydroxyl group on carbon atom 1. That hydroxyl group can then be phosphorylated using another molecule of ATP, yielding the doubly phosphorylated sugar fructose-1,6-bisphosphate (reaction Gly-3).

Again, the energy difference between the anhydride bond of ATP and the phosphoester bond on the fructose molecule renders the reaction exergonic and therefore essentially irreversible in the glycolytic direction ($\Delta G^{\circ\prime} = -3.4$ kcal/mol). This reaction is catalyzed by *phosphofructokinase-1 (PFK-1),* an enzyme that is especially important in the regulation of glycolysis, as we will see later. The designation PFK-1 is to distinguish this enzyme from PFK-2, a similar enzyme also involved in regulating glycolysis.

Next comes the actual cleavage reaction from which glycolysis derives its name. Fructose-1,6-bisphosphate is split reversibly by the enzyme *aldolase* to yield two trioses (three-carbon sugars) called dihydroxyacetone phosphate and glyceraldehyde-3-phosphate (reaction Gly-4), which are readily interconvertible (reaction Gly-5). Because only the latter compound is directly oxidizable in the next phase of glycolysis, interconversion of the two trioses enables dihydroxyacetone phosphate to be catabolized simply by conversion of dihydroxyacetone phosphate to glyceraldehyde-3-phosphate.

We can summarize this first phase of the glycolytic pathway (Gly-1 to Gly-5) as follows:

$$\text{glucose} + 2\,\text{ATP} \rightarrow 2\ \text{glyceraldehyde-3-phosphate} + 2\,\text{ADP} \quad \textbf{(9-13)}$$

Note that in this and subsequent reactions, hydrogen ions and H_2O molecules are not necessarily included if not needed for the overall understanding of the reaction.

Phase 2: Oxidation and ATP Generation. So far, the original glucose molecule has been doubly phosphorylated and cleaved into two interconvertible triose phosphates. Notice, however, that energy has been expended, rather than generated, thus far: two molecules of ATP have been *consumed* per molecule of glucose in phase 1. But the ATP debt is about to be repaid with interest as we encounter the two energy-yielding phases of glycolysis. In phase 2, ATP production is linked directly to an oxidative event, and then in phase 3, an energy-rich phosphorylated form of the pyruvate molecule serves as the driving force behind ATP generation.

The oxidation of glyceraldehyde-3-phosphate to the corresponding three-carbon acid, 3-phosphoglycerate, is exergonic—sufficiently so, in fact, to drive both the reduction of the coenzyme NAD^+ to NADH (Gly-6) and the phosphorylation of ADP with inorganic phosphate, P_i (Gly-7). Historically, this was the first example of a reaction sequence in which the coupling of ATP generation to an oxidative event was understood.

One important feature of this exergonic sequence is the involvement of NAD^+ as the electron acceptor. In fact, *glycolysis cannot proceed without a steady supply of* NAD^+. Because there is a limited supply of this coenzyme, the NADH must be oxidized back to NAD^+ so glycolysis can continue. Under aerobic conditions, it is O_2 that oxidizes NADH by accepting two electrons and one proton, reducing the O_2 molecule to water. Under anaerobic conditions, an organic molecule oxidizes the NADH back to NAD^+ and itself is reduced.

Another important feature of glycolysis is the coupling of the oxidation of glyceraldehyde-3-phosphate to the formation of a high-energy, doubly phosphorylated intermediate, 1,3-bisphosphoglycerate. The phosphoanhydride bond on carbon atom 1 of this intermediate has such a sufficiently negative $\Delta G^{\circ\prime}$ of hydrolysis (−11.8 kcal/mol; see Table 9-1) that the transfer of the phosphate to ADP, catalyzed by the enzyme *phosphoglycerate kinase,* is an exergonic reaction. ATP generation by the direct transfer of a high-energy phosphate group to ADP from a phosphorylated substrate such as 1,3-bisphosphoglycerate is called **substrate-level phosphorylation.**

To summarize the substrate-level phosphorylation of reaction Gly-6 plus Gly-7, we can write an overall reaction that accounts for one of the two glyceraldehyde-3-phosphate molecules generated from each glucose molecule in the first phase of glycolysis:

$$\text{glyceraldehyde-3-phosphate} + NAD^+ + ADP + P_i \rightarrow 3\text{-phosphoglycerate} + NADH + H^+ + ATP \quad \textbf{(9-14)}$$

Keep in mind that each reaction in the glycolytic pathway beyond glyceraldehyde-3-phosphate occurs twice per starting molecule of glucose so that, on a per-glucose basis, *two* molecules of NADH need to be reoxidized to regenerate the NAD^+ that is needed for continual oxidation of glyceraldehyde-3-phosphate. It also means that the initial investment of two ATP molecules in phase 1 is recovered here in phase 2, so the net ATP yield to this point is now zero.

Next, in the final phase of glycolysis, we will see the generation of two more ATP molecules. Thus, for glycolysis overall, there is a net gain of two ATP molecules per glucose metabolized to pyruvate.

MAKE CONNECTIONS 9.1

The full reaction for Gly-6 is the following:

$$\text{glyceraldehyde-3-phosphate} + NAD^+ + P_i \rightarrow 1,3\text{-bisphosphoglycerate} + NADH + H^+$$

$\Delta G^{\circ\prime}$ for Gly-6 is +1.5 kcal/mol. Based on what you have learned, why does this reaction move in the direction written in Figure 9-7? (Ch. 5.3)

Phase 3: Pyruvate Formation and ATP Generation. Generating another molecule of ATP from 3-phosphoglycerate depends on the phosphate group on carbon atom 3. At this stage, the phosphate group is linked to the carbon atom by a phosphoester bond with a low free energy of hydrolysis ($\Delta G^{\circ\prime} = -3.3$ kcal/mol). In the final phase of the glycolytic pathway, this phosphoester bond is converted to a *phosphoenol bond,* the hydrolysis of which is exergonic ($\Delta G^{\circ\prime} = -14.8$ kcal/mol; see Table 9-1). This increase in the amount of stored free energy involves a rearrangement of internal energy within the molecule. To accomplish this, the phosphate group of 3-phosphoglycerate is moved to the adjacent carbon-2 atom, forming 2-phosphoglycerate (reaction Gly-8). Water is then removed from 2- phosphoglycerate by the enzyme *enolase* (reaction Gly-9), thereby generating the high-energy compound phosphoenolpyruvate (PEP).

If you look carefully at the structure of PEP, you will notice that, unlike the phosphoester bonds of either 3- or 2-phosphoglycerate, the phosphoenol bond of PEP has a phosphate group on a carbon atom that is linked by a double bond to another carbon atom. This characteristic makes the hydrolysis of the phosphoenol bond of PEP one of the most exergonic hydrolytic reactions known in biological systems.

PEP hydrolysis is exergonic enough to drive ATP synthesis in reaction Gly-10, which involves the transfer of a phosphate group from PEP to ADP, generating another molecule of ATP (see top half of Figure 9-3) in another substrate-level phosphorylation. This transfer, catalyzed by the enzyme *pyruvate kinase,* is exergonic ($\Delta G^{\circ\prime} = -7.5$ kcal/mol; see Reaction 9-5) and is therefore essentially irreversible in the direction of pyruvate and ATP formation.

To summarize the third phase of glycolysis, Gly-8 to Gly-10, we can write an overall reaction for pyruvate formation:

$$\text{3-phosphoglycerate} + \text{ADP} \rightarrow \text{pyruvate} + \text{ATP} \qquad \textbf{(9-15)}$$

Summary of Glycolysis. Two molecules of ATP were initially invested in reactions Gly-1 and Gly-3, and two were returned in the first phosphorylation event (Gly-7), so the two molecules of ATP formed per molecule of glucose by the second phosphorylation event (Gly-10) represent the net ATP yield of the glycolytic pathway. This becomes clear when we sum the three reactions that summarize the three phases of the pathway (Reactions 9-13, 9-14, and 9-15). The latter two reactions are multiplied by 2 to account for both triose molecules generated in Reaction 9-13. The result is an overall expression for the pathway from glucose to pyruvate:

$$\text{glucose} + 2\,\text{NAD}^+ + 2\,\text{ADP} + 2\,\text{P}_\text{i} \xrightarrow{\text{reactions Gly-1 through Gly-10}} 2\,\text{pyruvate} + 2\,\text{NADH} + 2\,\text{H}^+ + 2\,\text{ATP} \qquad \textbf{(9-16)}$$

This pathway is exergonic in the direction of pyruvate formation. Under typical intracellular conditions in your body, for example, $\Delta G'$ for the overall pathway from glucose to pyruvate with the generation of two molecules each of ATP and NADH is about –20 kcal/mol.

The glycolytic pathway is one of the most common and highly conserved metabolic pathways known. Virtually all cells possess the ability to extract energy from glucose by oxidizing it to pyruvate. Some of this energy is conserved in the form of two molecules of ATP per molecule of glucose. What happens next, however, usually depends on the availability of oxygen because catabolism beyond pyruvate is quite different under aerobic conditions than it is under anaerobic conditions.

CONCEPT CHECK 9.4

How can glucose be oxidized in the absence of oxygen? How is the energy of this oxidation conserved?

9.5 Fermentation

Pyruvate occupies a key position as a branching point in chemotrophic energy metabolism (**Figure 9-8**). Its fate depends on the kind of organism involved, the specific cell type, and whether oxygen is available. An important feature of glycolysis is that it does not require oxygen and therefore can also take place in its absence. Under anaerobic conditions, no further oxidation of pyruvate occurs because pyruvate must be used to oxidize NADH, and no additional ATP can be generated. Instead, the energy needs of the cell are met by the modest ATP yield of two ATP per glucose in the glycolytic pathway. Therefore, as cells switch from aerobic to anaerobic catabolism of glucose, they must consume glucose much more rapidly to maintain steady-state cellular ATP levels. This effect was noted by the famed microbiologist Louis Pasteur working with yeast in the 1880s and is called the *Pasteur effect.* Under anaerobic conditions, rather than being oxidized, pyruvate is *reduced* by accepting the electrons (and protons) that must be removed from NADH during the process of fermentation.

In the Absence of Oxygen, Pyruvate Undergoes Fermentation to Regenerate NAD^+

As usually defined, the glycolytic pathway ends with pyruvate. Fermentative processes cannot end there, however, *because of the need to regenerate NAD^+*, the oxidized form of the coenzyme. As Reaction 9-16 indicates, the conversion of glucose to pyruvate requires that one molecule of NAD^+ is reduced to NADH per molecule of pyruvate generated. Coenzymes are present in cells at only modest concentrations, however, so the conversion of NAD^+ to NADH during glycolysis would cause cells to run out of NAD^+ very quickly if there were not some mechanism for regenerating NAD^+. Cells also have mechanisms to continually monitor and stabilize the NAD^+/NADH ratio, which is an indicator of the cell's *redox* state, the general level of oxidation of cellular components. Excessive oxidation of cellular components can be damaging to the cell, as free radicals and other harmful compounds are produced. Large changes in this ratio are a signal that the cell is under oxidative stress, and cellular mechanisms will try to keep it relatively constant.

During aerobic respiration, NADH is reoxidized by the transfer of its electrons to oxygen (as we will see in Chapter 10). Under anaerobic conditions, however, the electrons are transferred to pyruvate, which has a carbonyl group that can be readily reduced to a hydroxyl group (see Figure 9-8b, c). The two most common pathways of fermentation use pyruvate as the electron acceptor, converting it either to lactate

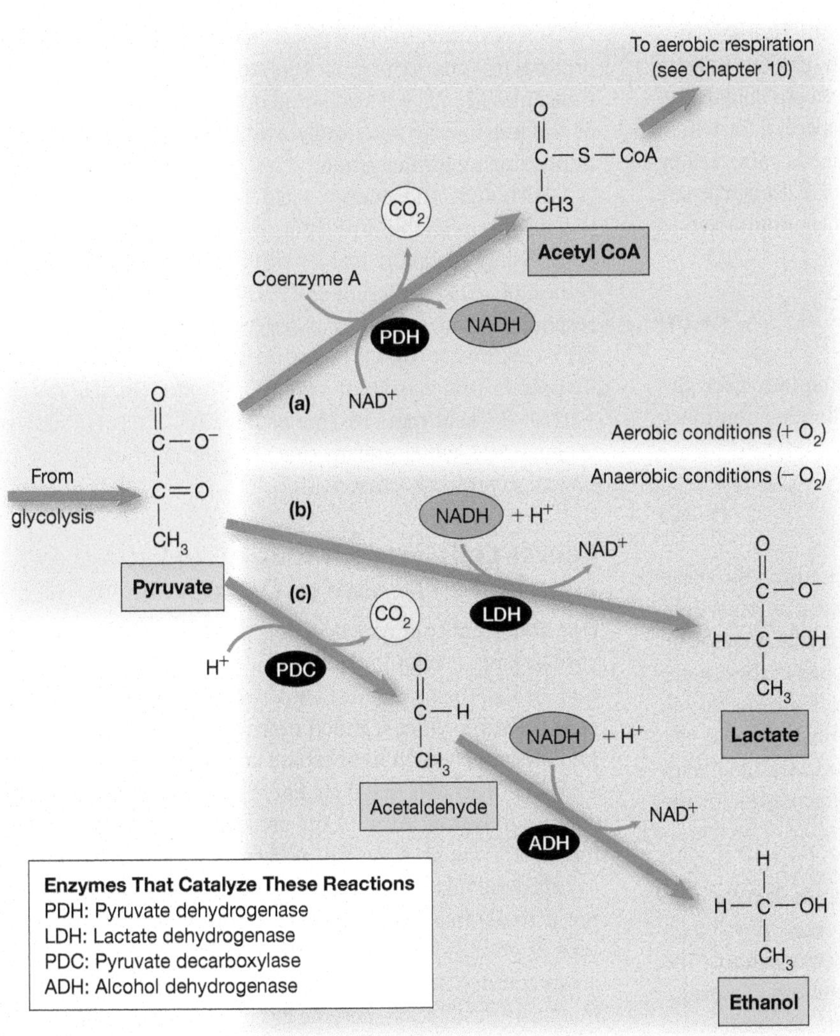

(a) Aerobic conditions. In the presence of oxygen, many organisms convert pyruvate to an activated form of acetate known as acetyl CoA. In this reaction, pyruvate is both oxidized (with NAD^+ being reduced to NADH) and decarboxylated (liberation of a carbon atom as CO_2). Acetyl CoA then becomes the substrate for aerobic respiration, where NADH is oxidized back to NAD^+ by molecular oxygen (see Chapter 10).

(b) and (c) Anaerobic conditions. When oxygen is absent, pyruvate is reduced so that NADH can be oxidized to NAD^+, the form of this coenzyme required in reaction Gly-6 of glycolysis. Common products of pyruvate reduction are **(b)** lactate (in most animal cells and many bacteria) or **(c)** ethanol and CO_2 (in many plant cells and in yeasts and other microorganisms).

Figure 9-8 The Fate of Pyruvate Under Aerobic and Anaerobic Conditions. The fate of pyruvate depends on the organism involved and on whether oxygen is available. The enzymes that catalyze these reactions are identified in the box at the bottom of the figure.

or to CO_2 and ethanol. Both of these fermentation pathways achieve the goal of generating modest amounts of ATP while ensuring an adequate supply of NAD^+.

Lactate Fermentation. The anaerobic process that terminates in lactate is called **lactate fermentation.** As Figure 9-8b indicates, lactate is generated by the direct transfer of electrons from NADH to the carbonyl group of pyruvate, reducing it to the hydroxyl group of lactate. On a per-glucose basis, this reaction can be represented as

$$2 \text{ pyruvate} + 2\text{ NADH} + 2\text{ H}^+ \rightleftharpoons 2 \text{ lactate} + 2\text{ NAD}^+ \quad \textbf{(9-17)}$$

This reaction is readily reversible; in fact, the enzyme that catalyzes it is called *lactate dehydrogenase* because of its ability to catalyze the oxidation, or dehydrogenation, of lactate to pyruvate.

By adding Reactions 9-16 and 9-17, we can write an overall reaction for the metabolism of glucose to lactate under anaerobic conditions:

$$\text{glucose} + 2\text{ ADP} + 2\text{ P}_i \rightarrow 2 \text{ lactate} + 2\text{ ATP} \quad \textbf{(9-18)}$$

Lactate fermentation is the major energy-yielding pathway in many anaerobic bacteria, as well as in animal cells operating under anaerobic or hypoxic (oxygen-deprived) conditions. Lactate fermentation is important to us commercially because the production of cheese, yogurt, and other dairy products depends on microbial fermentation of lactose, the main sugar found in milk.

A more personal example of lactate fermentation involves your own muscles during periods of strenuous exertion. Whenever muscle cells use oxygen faster than it can be supplied by the circulatory system, the cells become temporarily hypoxic. Pyruvate is then reduced to lactate instead of being further oxidized, as it is under aerobic conditions. The lactate produced in this way is transported by the circulatory system from the muscle to the liver. There, it is converted to glucose again by the process of *gluconeogenesis.* As we will see later in this chapter, gluconeogenesis is essentially the reverse of lactate fermentation but with several critical differences that enable it to proceed exergonically in the direction of glucose formation.

Alcoholic Fermentation. Under anaerobic conditions, plant cells can carry out **alcoholic fermentation** (in waterlogged roots, for example), as do yeasts and other microorganisms. In this process, pyruvate loses a carbon atom (as CO_2) to form the two-carbon compound acetaldehyde. Acetaldehyde reduction by NADH gives rise to ethanol, the alcohol for which the process is named. This reductive sequence is catalyzed by two enzymes, *pyruvate decarboxylase* and *alcohol dehydrogenase* (see Figure 9-8c). The overall reaction can be summarized as follows:

$$\begin{aligned} &2 \text{ pyruvate} + 2 \text{ NADH} + 4 \text{ H}^+ \rightarrow \\ &\qquad 2 \text{ ethanol} + 2 \text{ CO}_2 + 2 \text{ NAD}^+ \end{aligned} \quad \textbf{(9-19)}$$

By adding this reductive step to the overall equation for glycolysis (Reaction 9-16), we arrive at the following summary equation for alcoholic fermentation:

$$\begin{aligned} &\text{glucose} + 2 \text{ ADP} + 2 \text{ P}_i + 2 \text{ H}^+ \rightarrow \\ &\qquad 2 \text{ ethanol} + 2 \text{ CO}_2 + 2 \text{ ATP} \end{aligned} \quad \textbf{(9-20)}$$

Alcoholic fermentation by yeast cells is a key process in the baking, brewing, and winemaking industries. The yeast cells in bread dough break down glucose anaerobically, generating both CO_2 and ethanol. Carbon dioxide is trapped in the dough, causing it to rise, and the alcohol is driven off during baking and becomes part of the pleasant aroma of baking bread. For the brewer, both CO_2 and ethanol are essential; ethanol makes the product an alcoholic beverage, and CO_2 accounts for the carbonation.

Other Fermentation Pathways. Although lactate and ethanol are the fermentation products of greatest physiological or economic significance, they by no means exhaust the microbial fermentation repertoire. In *propionate fermentation*, for example, bacteria reduce pyruvate to propionate ($CH_3—CH_2—COO^-$), an important reaction in the production of Swiss cheese. Many bacteria that cause food spoilage do so by *butylene glycol fermentation*. Other fermentation processes yield acetone, isopropyl alcohol, or butyrate, the last of these being responsible for the rotten smell of rancid food and vomit. However, all these reactions are just metabolic variations on the common theme of reoxidizing NADH by the transfer of electrons to some organic acceptor.

Fermentation Taps Only a Fraction of the Substrate's Free Energy but Conserves That Energy Efficiently as ATP

An essential feature of every fermentative process is that *no external electron acceptor is involved and no net oxidation occurs.* In both lactate and alcoholic fermentation, for example, the NADH generated by the single oxidative step of glycolysis (reaction Gly-6) is reoxidized in the final reaction of the sequence (Reactions 9-17 and 9-19). Because no net oxidation occurs, fermentation gives a modest ATP yield—two molecules of ATP per molecule of glucose, in the case of either lactate (Reaction 9-18) or alcoholic fermentation (Reaction 9-20).

Most of the free energy of the glucose molecule is still present in the two lactate or ethanol molecules. In the case of lactate fermentation, for example, the two lactate molecules produced from every glucose molecule contain most of the 686 kcal of free energy present per mole of glucose because the complete aerobic oxidation of lactate has a $\Delta G°'$ of -319.5 kcal/mol. In other words, about 93% (639 kcal) of the original free energy of glucose is still present in the two lactate molecules (2×319.5 kcal), and only about 7% (47 kcal/mol) of the free energy potentially available from glucose was obtained during fermentation.

Although the energy yield from lactate fermentation is low, the available free energy is conserved efficiently as ATP. Using standard free energy changes, these two molecules of ATP represent $2 \times 7.3 = 14.6$ kcal/mol. This corresponds to an efficiency of energy conservation of about 30% ($14.6/47 \times 100\%$). Based on actual $\Delta G'$ values for ATP hydrolysis under cellular conditions (often in the range of -10 to -14 kcal/mol), two molecules of ATP represent at least 20 kcal/mol, which means that the efficiency of energy conservation probably exceeds 40%.

Cancer Cells Ferment Glucose to Lactate Even in the Presence of Oxygen

One abnormality of cancer cells is that they often ferment glucose to lactate even in the presence of oxygen. In the 1920s and 1930s, the biochemist Otto Warburg showed that cancer cells consume glucose much more rapidly than normal cells, and this observation has become known as the *Warburg effect*. Many cancer cells get their energy primarily by fermenting glucose to lactate, even in the presence of oxygen, producing only 2 ATP per glucose. This process is often called *aerobic glycolysis*, which is somewhat of a misnomer because oxygen is not actually involved. However, because a typically anaerobic type of glycolysis is still operating in the presence of oxygen, it has come to be known by this name. Despite its inefficiency in energy production, aerobic glycolysis allows cancer cells to outgrow normal cells, which are carrying out the more efficient process of aerobic respiration.

How can cancer cells do this? One way is by dramatically increasing the amount of glucose consumed, sometimes by a factor of 100 or more. Cancer cells often display a large increase in the activity of nutrient transporters, which tend to be down-regulated in nonproliferating cells. For example, the activity of the glucose transporter GLUT1 is often dramatically increased. A chemical inhibitor of GLUT1 decreases the rate of glycolysis in cancer cells and has been found to inhibit their growth, both in culture and in a mouse model system.

What are the advantages of aerobic glycolysis for a cancer cell? In the human body, glucose is rarely a limiting nutrient, so for cancer cells, energy efficiency is not a primary concern. Instead, it appears that cancer cells carry out aerobic glycolysis not primarily for energy production but for production of carbon skeletons for biosynthesis of the nucleic acids, phospholipids, and amino acids that are in high demand by rapidly growing and proliferating cells. Organic carbon that normally would be converted to pyruvate and then completely oxidized to carbon dioxide to maximize energy efficiency is instead diverted into anabolic pathways to fuel cancer cell growth. Consistent with this idea, many cancer cells accumulate mutations in genes encoding enzymes necessary for oxidation of pyruvate to carbon dioxide; whether this is a cause or a result of cancer is not clear.

The excessive glucose consumption by cancer cells described above is the basis for a common diagnostic procedure

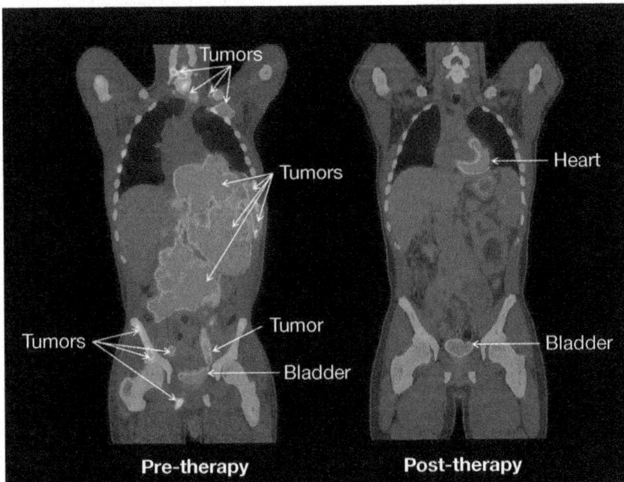

Figure 9-9 FDG-PET Scan Reveals Success of Anti-Cancer Chemotherapy. Because cancer cells consume excessive amounts of glucose, radioactively labeled fluorodeoxyglucose (FDG) is easily visualized on a PET scan prior to therapy (left side, bright red areas). After therapy (right), there is evidence of complete metabolic response at the prior tumor sites. Normal physiologic distribution of FDG is seen in the myocardium and in the bladder.

Courtesy of Annick D. Van den Abbeele, MD, FACR, FICIS, Dana-Farber Cancer Institute, Boston, MA, USA

known as *PET (positron emission tomography)*. Combined with a CT scan (PET/CT), the location, metabolic activity, and size of tumors in the human body can be imaged (**Figure 9-9**). In this procedure, fluorodeoxyglucose (FDG), a radioactive glucose analogue, is given to the patient intravenously and then is imaged as it accumulates in cancer cells. Thus, PET scans before and after cancer chemotherapy can be an important aid in determining the presence, location, and activity of tumors and the effectiveness of the anti-cancer chemotherapy.

The use of PET scans in this way highlights a general principle in biology and medicine: we can use radioactively labeled compounds known as *radiotracers* to follow the fate of individual types of molecules (such as glucose) both in the human body and during in vitro experiments. The use of radiotracers relies on the existence of naturally occurring **isotopes** of common chemical elements, that is, atoms of the same element differing slightly in atomic weight. (The use of **isotopic labeling** to follow the fate of biochemical intermediates in metabolic studies is explained in more detail in **Key Technique, page 236–237.**)

CONCEPT CHECK 9.5

In the absence of oxygen, why is it necessary to ferment pyruvate to lactate or ethanol plus CO_2?

9.6 Alternative Substrates for Glycolysis

So far, we have assumed glucose to be the starting point for glycolysis and thus, by implication, for all of cellular energy metabolism. Glucose is certainly a major substrate for both fermentation and respiration in a variety of organisms and tissues. It is not the only such substrate, however. For many organisms and some tissues within organisms, glucose is not significant at all. So it is important to ask two questions: what are some of the major alternatives to glucose, and how are they handled by cells?

One principle quickly emerges: *regardless of the chemical nature of the alternative substrate, it is often converted into an intermediate in the main pathway for glucose catabolism.* Most carbohydrates, for example, are converted to intermediates in the glycolytic pathway. To emphasize this point, we will briefly consider two classes of alternative carbohydrate substrates—other sugars and storage carbohydrates. (In Chapter 10, we will see how proteins and lipids can be converted into intermediates in the citric acid cycle, the next stage of aerobic respiration.)

Other Sugars and Glycerol Are Also Catabolized by the Glycolytic Pathway

Many sugars other than glucose are available to cells, depending on the food sources of the organism in question. Most of them are either monosaccharides (usually hexoses or pentoses) or disaccharides that can be readily hydrolyzed into their component monosaccharides. Ordinary table sugar (sucrose) is a disaccharide consisting of the hexoses glucose and fructose. Milk sugar (lactose) contains glucose and galactose, and maltose contains two glucose molecules (see Figure 3-23). Besides glucose, fructose, and galactose, mannose is another relatively common dietary hexose.

Figure 9-10 on page 230 illustrates the reactions that bring various carbohydrates into the glycolytic pathway. In general, disaccharides (for example, lactose, maltose, and sucrose) are hydrolyzed into their component monosaccharides, and each monosaccharide is converted to a glycolytic intermediate in one or a few steps. Glucose and fructose enter most directly after phosphorylation on carbon atom 6. Mannose is converted to mannose-6-phosphate and then to fructose-6-phosphate, a glycolytic intermediate. The entry of galactose requires a somewhat more complex reaction sequence involving five steps to convert it to glucose-6-phosphate.

Phosphorylated pentoses can also be channeled into the glycolytic pathway but only after being converted to hexose phosphates. That conversion is accomplished by a metabolic pathway called the *phosphogluconate pathway*, also known as the *pentose phosphate pathway*. Glycerol, a three-carbon molecule resulting from lipid breakdown, enters glycolysis after conversion to dihydroxyacetone phosphate. Thus, the typical cell has metabolic capabilities to convert most naturally occurring sugars (and a variety of other compounds as well) to one of the glycolytic intermediates for further catabolism under either anaerobic or aerobic conditions.

Polysaccharides Are Cleaved to Form Sugar Phosphates That Also Enter the Glycolytic Pathway

Although glucose is the immediate substrate for both fermentation and respiration in many cells and tissues, the concentration of the free monosaccharide in cells is low. Instead, it occurs primarily in the form of storage polysaccharides, most commonly starch in plants and glycogen in animals (green boxes in Figure 9-10). One advantage of storing glucose as

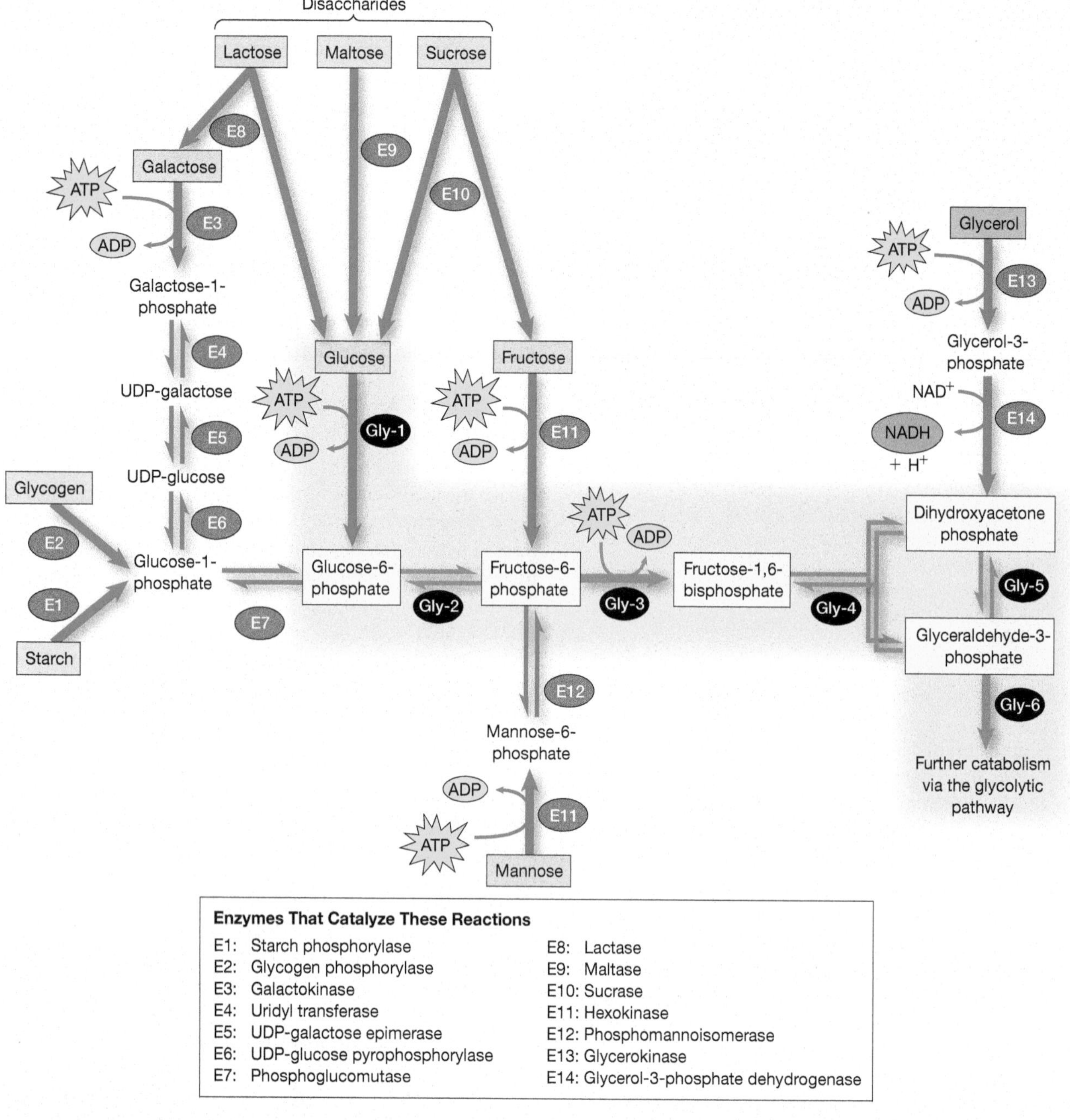

Figure 9-10 Carbohydrate Catabolism by the Glycolytic Pathway. Carbohydrate substrates that can be metabolized by conversion to an intermediate in the glycolytic pathway include hexoses (galactose, glucose, fructose, and mannose), disaccharides (lactose, maltose, and sucrose), polysaccharides (glycogen and starch), and glycerol. The enzymes that catalyze these reactions are identified in the box at the bottom. The first six reactions of the glycolytic pathway are highlighted in the central orange-tinted region (see Figure 9-7 for glycolytic enzyme names).

starch and glycogen is that these two polymerized forms of glucose are insoluble in water and thus do not overload the limited solute capacity of the cell. As indicated in **Figure 9-11**, these storage polysaccharides can be mobilized by a process called **phosphorolysis.** Phosphorolysis (or phosphorolytic cleavage) resembles hydrolysis but uses inorganic phosphate rather than water to break a chemical bond. Inorganic phosphate is used to break the α $(1 \rightarrow 4)$ glycosidic linkage between successive glucose units, liberating glucose monomers as glucose-1-phosphate. The glucose-1-phosphate that is formed in this way can be converted to glucose-6-phosphate, which is then catabolized by the glycolytic pathway.

Looking back at Figure 9-10, notice that glucose stored in polymerized form enters the glycolytic pathway as glucose-6-phosphate, without the input of the ATP that would be required for the initial phosphorylation of the free sugar. Consequently, the overall energy yield for glucose is greater by one molecule of ATP when it is catabolized from the polysaccharide level than when it is catabolized with the free sugar as the starting substrate. This is not a case of getting something for nothing, however, because energy was required to activate the glucose units that were added to the polysaccharide chain during starch or glycogen synthesis.

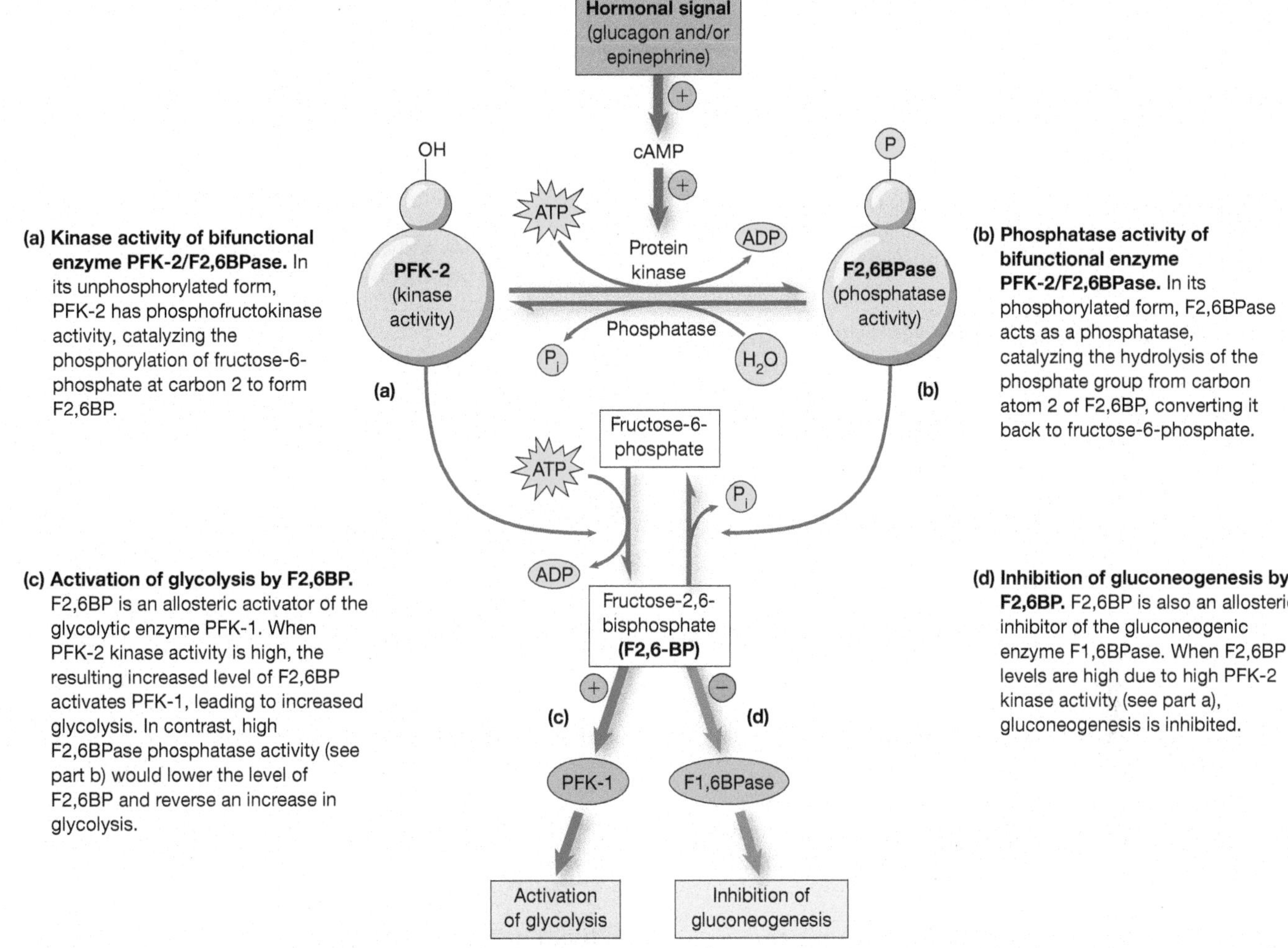
Hormonal signal (glucagon and/or epinephrine)
+
cAMP
+
OH
P
ATP
Protein kinase
ADP
PFK-2 (kinase activity)
F2,6BPase (phosphatase activity)
Phosphatase
Pi
H2O
(a)
(b)
(a) Kinase activity of bifunctional enzyme PFK-2/F2,6BPase. In its unphosphorylated form, PFK-2 has phosphofructokinase activity, catalyzing the phosphorylation of fructose-6-phosphate at carbon 2 to form F2,6BP.
(b) Phosphatase activity of bifunctional enzyme PFK-2/F2,6BPase. In its phosphorylated form, F2,6BPase acts as a phosphatase, catalyzing the hydrolysis of the phosphate group from carbon atom 2 of F2,6BP, converting it back to fructose-6-phosphate.
Fructose-6-phosphate
ATP
Pi
ADP
Fructose-2,6-bisphosphate (F2,6-BP)
(c) Activation of glycolysis by F2,6BP. F2,6BP is an allosteric activator of the glycolytic enzyme PFK-1. When PFK-2 kinase activity is high, the resulting increased level of F2,6BP activates PFK-1, leading to increased glycolysis. In contrast, high F2,6BPase phosphatase activity (see part b) would lower the level of F2,6BP and reverse an increase in glycolysis.
(d) Inhibition of gluconeogenesis by F2,6BP. F2,6BP is also an allosteric inhibitor of the gluconeogenic enzyme F1,6BPase. When F2,6BP levels are high due to high PFK-2 kinase activity (see part a), gluconeogenesis is inhibited.
+
−
(c)
(d)
PFK-1
F1,6BPase
Activation of glycolysis
Inhibition of gluconeogenesis

Figure 9-14 The Regulatory Roles of PFK-2 and F2,6BP. Phosphofructokinase-2 (PFK-2) is a bifunctional enzyme with two opposing catalytic activities that depend upon its own phosphorylation state **(a** or **b).** Which of the two activities it has depends on the presence of hormonal signals and therefore determines the level of fructose-2, 6-bisphosphate (F2,6BP), an activator of glycolysis **(c)** and inhibitor of gluconeogenesis **(d).**

an increase in liver cAMP concentration leads to a decrease in the rate of glycogen formation (see Human Connections, page 234). (For further details on hormonal regulation and the role of cAMP in mediating hormonal effects, see the discussion on hormonal signal transduction in Chapter 23.)

Glycolytic Enzymes May Have Functions Beyond Glycolysis

It is easy to get the impression that, for a fundamental and ubiquitous pathway like glycolysis, we have fully described its individual steps and learned "all there is to know" about the enzymes involved. But, as we often discover in biology, there may be surprises in store. One such surprise has been the discovery that several of the glycolytic enzymes may have regulatory functions that, at first glance, seem unrelated to their roles as catalysts in specific reactions in glycolysis. These functions unrelated to glycolysis are sometimes called "moonlighting," in much the same way that someone who has a job during the day can work at a second, unrelated job at night.

One example of moonlighting is in the enzyme hexokinase, which catalyzes the ATP-dependent phosphorylation of glucose to initiate the glycolytic sequence (Gly-1 in Figure 9-7). In yeast, an isoform known as hexokinase-2 becomes localized in the nucleus in response to high levels of glucose. In the nucleus, it acts to down-regulate the expression of genes required for catabolism of sugars other than glucose.

Other glycolytic enzymes may also moonlight. Phosphoglucoisomerase (PGI; Gly-2) has been shown to be secreted from tumor cells, and the secreted form of PGI stimulates high levels of cell migration and proliferation. Glyceraldehyde-3-phosphate dehydrogenase (GAPDH; Gly-6) is overexpressed in cells undergoing programmed cell death in response to cellular injury, where it accumulates in nuclei. GADPH has been proposed to be an intracellular sensor of oxidative stress, with a possible link to neurodegenerative diseases.

As you can see, we have much more to learn about genes once thought to serve merely a "housekeeping" function in supplying cellular energy.

CONCEPT CHECK 9.8

Why are the key regulatory enzymes in glycolysis and gluconeogenesis unique to each pathway? Why is there reciprocal regulation of these enzymes?

Summary of Key Points

Mastering™ Biology For activities, animations, and review quizzes, go to the study area at www.masteringbiology.com.

9.1 Metabolic Pathways

- Metabolic pathways in cells are usually either anabolic (synthetic) or catabolic (degradative). Catabolic reactions provide the energy necessary to drive the anabolic reactions. Catabolism often involves exergonic oxidation reactions, whereas anabolism often involves endergonic reduction reactions.

9.2 ATP: The Primary Energy Molecule in Cells

- ATP is useful for conserving chemical energy in cells because its terminal anhydride bond has an intermediate free energy of hydrolysis. This allows ATP to serve as a donor of phosphate groups to a number of biologically important molecules such as glucose. It also allows ADP to serve as an acceptor of phosphate groups from molecules such as PEP.

9.3 Chemotrophic Energy Metabolism

- Most chemotrophs derive the energy needed to generate ATP from the catabolism of organic nutrients such as carbohydrates, fats, and proteins. They do so either by fermentative processes in the absence of oxygen or by aerobic respiratory metabolism in the presence of oxygen.
- Glycolysis provides a mechanism by which glucose can be degraded in dilute solution at temperatures compatible with life, with a large portion of the free energy yield conserved as ATP.

9.4 Glycolysis: ATP Generation Without the Involvement of Oxygen

- Using glucose as a prototype substrate, catabolism under both anaerobic and aerobic conditions begins with glycolysis, a ten-step pathway that converts glucose into pyruvate. This leads to the production of two molecules of ATP per molecule of glucose.

9.5 Fermentation

- In the absence of oxygen, the reduced coenzyme NADH generated during glycolysis is oxidized by pyruvate or another organic molecule that is a product of glucose catabolism, producing fermentation end-products such as lactate or ethanol plus carbon dioxide. Thus, only the two ATP from glycolysis are produced during fermentation, and most of the energy of glucose is still found in the fermentation end-products (for example, lactate or ethanol).

9.6 Alternative Substrates for Glycolysis

- Although usually written with glucose as the starting substrate, the glycolytic sequence is also the mainstream pathway for catabolizing a variety of related sugars, such as fructose, galactose, and mannose.
- Glycolysis is also used to metabolize the glucose-1-phosphate derived by phosphorolytic cleavage of storage polysaccharides such as starch or glycogen.

9.7 Gluconeogenesis

- Gluconeogenesis is, in a sense, the opposite of glycolysis because it is the pathway used by some cells to synthesize glucose from three- and four-carbon starting materials such as pyruvate. However, the gluconeogenic pathway is not just glycolysis in reverse.
- The two pathways have seven enzyme-catalyzed reactions in common, but the three most exergonic reactions of glycolysis are bypassed in gluconeogenesis by reactions that render the pathway exergonic in the gluconeogenic direction by the input of energy from ATP and GTP.

9.8 The Regulation of Glycolysis and Gluconeogenesis

- Glycolysis and gluconeogenesis are regulated by modifying the activity of enzymes that are unique to each pathway. These enzymes are regulated by one or more key intermediates in aerobic respiration, including ATP, ADP, AMP, acetyl CoA, and citrate.
- An important allosteric regulator of both glycolysis and gluconeogenesis is fructose-2,6-bisphosphate (F2,6BP). Its concentration depends on the relative kinase and phosphatase activities of the bifunctional enzyme PFK-2/F2,6BPase.
- PFK-2 in turn is regulated by the hormones glucagon and epinephrine via their effects on the cAMP concentration in cells.
- In addition to their roles as catalysts, several well-known glycolytic enzymes have recently been shown to play regulatory roles in cells, such as regulated gene expression, programmed cell death, and cancer cell migration.

Problem Set

Mastering™ Biology For activities, animations, and review quizzes, go to the study area at www.masteringbiology.com.

9-1 High-Energy Bonds. When first introduced by Fritz Lipmann in 1941, the term *high-energy bond* was considered a useful concept for describing the energetics of biochemical molecules and reactions. However, the term can lead to confusion when relating ideas about cellular energy metabolism to those of physical chemistry. To check out your own understanding, indicate whether each of the following statements is true (T) or false (F). If false, reword the statement to make it true.

(a) Energy is stored in special high-energy bonds in molecules such as ATP and is released when these bonds are broken.

(b) Energy is always released whenever a covalent bond is formed and is always required to break a covalent bond.

(c) To a physical chemist, *high-energy bond* means a very stable bond that requires a lot of energy to break, whereas to a biochemist, the term is likely to mean a bond that releases a lot of energy upon hydrolysis.

(d) The terminal phosphate of ATP is a high-energy phosphate that takes its high energy with it when it is hydrolyzed.

(e) Phosphoester bonds are low-energy bonds because they require less energy to break than the high-energy bonds of phosphoanhydrides.

(f) The term *high-energy molecule* should be thought of as a characteristic of the reaction the molecule is involved in—not as an intrinsic property of a particular bond within that molecule.

9-2 The History of Glycolysis. Following are several historical observations that led to the elucidation of the glycolytic pathway. In each case, suggest a metabolic basis for the observed effect, and explain the significance of the observation for the elucidation of the pathway.

(a) Alcoholic fermentation in yeast extracts requires a heat-labile fraction originally called *zymase* and a heat-stable fraction *(cozymase)* that is necessary for the activity of zymase.

(b) Alcoholic fermentation does not take place in the absence of inorganic phosphate.

(c) In the presence of iodoacetate, a known inhibitor of glycolysis, fermenting yeast extracts accumulate a doubly phosphorylated hexose.

(d) In the presence of fluoride ion, another known glycolytic inhibitor, fermenting yeast extracts accumulate two phosphorylated three-carbon acids.

9-3 Glycolysis in 25 Words or Fewer. Complete each of the following statements about the glycolytic pathway in 25 words or fewer.

(a) Although the brain is an obligately aerobic organ, it still depends on glycolysis because . . .

(b) Although one of its reactions is an oxidation, glycolysis can proceed in the absence of oxygen because . . .

(c) What happens to the pyruvate generated by the glycolytic pathway depends on ...

(d) If you bake bread or brew beer, you depend on glycolysis for . . .

(e) Two organs in your body that can use lactate are . . .

(f) The synthesis of glucose from lactate in a liver cell requires more molecules of nucleoside triphosphates (ATP and GTP) than are formed during the catabolism of glucose to lactate in a muscle cell because . . .

9-4 Energetics of Carbohydrate Utilization. The anaerobic fermentation of free glucose has an ATP yield of two ATP molecules per molecule of glucose. For glucose units in a glycogen molecule, the yield is three ATP molecules per molecule of glucose. The corresponding value for the disaccharide sucrose is two ATP molecules per molecule of monosaccharide if the sucrose is eaten by an animal, but two and a half ATP molecules per molecule of monosaccharide if the sucrose is metabolized by a bacterium.

(a) Explain why the glucose units present in glycogen have a higher ATP yield than free glucose molecules.

(b) What is the likely mechanism for sucrose breakdown in the gut of an animal to explain the energy yield of two ATP molecules per molecule of monosaccharide?

(c) Based on what you know about the process of glycogen breakdown, suggest a mechanism for bacterial sucrose metabolism that is consistent with an energy yield of two and a half ATP molecules per molecule of monosaccharide.

(d) What energy yield (in molecules of ATP per molecule of monosaccharide) would you predict for the bacterial catabolism of raffinose, a trisaccharide?

9-5 QUANTITATIVE Glucose Phosphorylation. The direct phosphorylation of glucose by inorganic phosphate is a thermodynamically unfavorable reaction:

$$\text{glucose} + P_i \rightarrow \text{glucose-6-phosphate} + H_2O \qquad \textbf{(9-21)}$$
$$\Delta G^{\circ\prime} = 3.3 \text{ kcal/mol}$$

In the cell, glucose phosphorylation is accomplished by coupling the reaction to the hydrolysis of ATP, an exergonic reaction:

$$ATP + H_2O \rightarrow ADP + P_i \qquad \textbf{(9-22)}$$
$$\Delta G^{\circ\prime} = -7.3 \text{ kcal/mol}$$

Typical concentrations of these intermediates in yeast cells are

$$[\text{glucose-6-phosphate}] = 0.08\ mM$$
$$[ATP] = 1.8\ mM$$
$$[ADP] = 0.15\ mM$$
$$[P_i] = 1.0\ mM$$

Assume a temperature of 25°C for all calculations.

(a) What minimum concentration of glucose would have to be maintained in a yeast cell for direct phosphorylation (Reaction 9-21) to be thermodynamically spontaneous? Is this physiologically reasonable? Explain your reasoning.

(b) What is the overall equation for the coupled (ATP-driven) phosphorylation of glucose? What is its $\Delta G^{\circ\prime}$ value?

(c) What minimum concentration of glucose would have to be maintained in a yeast cell for the coupled reaction to be thermodynamically spontaneous? Is this physiologically reasonable?

(d) By about how many orders of magnitude is the minimum required glucose concentration reduced when the phosphorylation of glucose is coupled to the hydrolysis of ATP?

(e) Assuming a yeast cell to have a glucose concentration of 5.0 m*M*, what is $\Delta G'$ for the coupled phosphorylation reaction?

9-6 Ethanol Intoxication and Methanol Toxicity. The enzyme alcohol dehydrogenase was mentioned in this chapter because of its role in the final step of alcoholic fermentation. However, the enzyme also occurs commonly in aerobic organisms, including humans. The ability of the human body to catabolize the ethanol in alcoholic beverages depends on the presence of alcohol dehydrogenase in the liver. One effect of ethanol intoxication is a dramatic decrease in the NAD^+ concentration in liver cells, which decreases the aerobic utilization of glucose. Methanol, on the other hand, is not just an intoxicant; it is a deadly poison due to the toxic effect of the formaldehyde to which it is converted in the liver.

(a) Why does ethanol consumption lead to a reduction in NAD^+ concentration and to a decrease in aerobic respiration?

(b) Most of the unpleasant effects of hangovers result from an accumulation of acetaldehyde and its metabolites. Where does the acetaldehyde come from?

(c) The medical treatment for methanol poisoning usually involves administration of large doses of ethanol. Why is this treatment effective?

9-7 Propionate Fermentation. Although lactate and ethanol are the best-known products of fermentation, other pathways are also known, some with important commercial applications. Swiss cheese production, for example, depends on the bacterium *Propionibacterium freudenreichii*, which converts pyruvate to propionate ($CH_3—CH_2—COO^-$). Fermentation of glucose to propionate always generates at least one other product as well.

(a) Why is it not possible to devise a scheme for the fermentation of glucose with propionate as the sole end-product?

(b) Suggest an overall scheme for propionate production that generates only one additional product, and indicate what that product might be.

(c) If you know that Swiss cheese production actually requires both propionate and carbon dioxide and that both are produced by *Propionibacterium* fermentation, what else can you now say about the fermentation process that this bacterium carries out?

9-8 QUANTITATIVE Glycolysis and Gluconeogenesis. As Figure 9-12 indicates, gluconeogenesis is accomplished by what is essentially the reverse of the glycolytic pathway but with bypass reactions in place of the first, third, and tenth reactions in glycolysis.

(a) Explain why it is not possible to accomplish gluconeogenesis by a simple reversal of all the reactions in glycolysis.

(b) Write an overall reaction for gluconeogenesis that is comparable to Reaction 9-16 for glycolysis.

(c) Explain why gluconeogenesis requires the input of six molecules of nucleoside triphosphates (four ATPs and two GTPs) per molecule of glucose synthesized, whereas glycolysis yields only two molecules of ATP per molecule of glucose.

(d) Assuming concentrations of ATP, ADP, and P_i are such that $\Delta G'$ for the hydrolysis of ATP is about −10 kcal/mol, what is the approximate $\Delta G'$ value for the overall reaction for gluconeogenesis that you wrote in part b?

(e) With all of the enzymes for glycolysis and gluconeogenesis present in a liver cell, how does the cell "know" whether it should be synthesizing or catabolizing glucose at any given time?

9-9 Wrong Again. For each of the following false statements, change the statement to make it true, and explain why it was false.

(a) Although the reactions of gluconeogenesis are simply the reverse of the reactions of glycolysis, gluconeogenesis requires more energy than glycolysis releases.

(b) Because glycolysis involves the partial oxidation of glucose, it cannot proceed in the absence of oxygen.

(c) ATP is ideal for the energy currency of the cell because it has the most negative standard free energy change ($\Delta G^{\circ\prime}$) of hydrolysis of any phosphorylated compound in the cell.

(d) Because energy production is so important in the cell, glycolytic enzymes function solely to degrade glucose.

9-10 You've Got Some Explaining to Do. Explain each of the following observations.

(a) In his classic studies of glucose fermentation by yeast cells, Louis Pasteur observed that the rate of glucose consumption by yeast cells was much higher under anaerobic conditions than under aerobic conditions.

(b) In 1905, Arthur Harden and William Young found that addition of inorganic phosphate to a yeast extract stimulated and prolonged the fermentation of glucose.

(c) An alligator is normally very sluggish but, if provoked, is capable of rapidly moving its legs, jaws, and tail. However, such bursts of activity must be followed by long periods of recovery.

(d) Fermentation of glucose to lactate is an energy-yielding process, although it involves no net oxidation (i.e., even though the oxidation of glyceraldehyde-3-phosphate to glycerate is accompanied by the reduction of pyruvate to lactate and no net accumulation of NADH occurs).

9-11 Arsenate Poisoning. Arsenate ($HAsO_4^{2-}$) is a potent poison to almost all living systems. Among other effects, arsenate is known to uncouple the phosphorylation event from the oxidation of glyceraldehyde-3-phosphate. This uncoupling occurs because the enzyme involved, glyceraldehyde-3-phosphate dehydrogenase, can utilize arsenate instead of inorganic phosphate, forming glycerate-1-arseno-3-phosphate. This product is a highly unstable compound that immediately undergoes nonenzymatic hydrolysis into glycerate-3-phosphate and free arsenate.

(a) In what sense might arsenate be called an *uncoupler* of substrate-level phosphorylation?

(b) Why is arsenate such a toxic substance for an organism that depends critically on glycolysis to meet its energy needs?

(c) Can you think of other reactions that are likely to be uncoupled by arsenate in the same way as the glyceraldehyde-3-phosphate dehydrogenase reaction?

9-12 Life Without Phosphofructokinase. Many bacteria do not have phosphofructokinase-1 (Gly-3) and thus cannot convert glucose to fructose-1,6-bisphosphate. Instead, they use a pathway known as the Entner–Doudoroff pathway to partially oxidize glucose and convert it to two three-carbon molecules. See if you can draw the products of the first three steps of the pathway based on the following description.

(a) In the first step, the ring form of glucose-6-phosphate is oxidized at carbon 1 to form 6-phosphogluconolactone, as the coenzyme $NADP^+$ is reduced to $NADPH + H^+$.

(b) Next, the ring is broken by hydrolysis to form the carboxylic acid 6-phosphogluconate, which resembles glucose-6-phosphate but is more oxidized at carbon 1.

(c) After a molecule of water is removed, a molecule of 2-keto-3-deoxy-6-phosphogluconate is formed.

(d) Because an aldolase splits this six-carbon molecule into pyruvate plus glyceraldehyde-3-phosphate (which is then converted to another molecule of pyruvate), how will the final ATP yield per glucose compare with typical glycolysis?

(e) Given an alternate source of cellular ATP, how would you predict lactate production might be affected by the addition of arsenate (see Problem 9-11) in these bacteria compared to bacteria performing standard glycolysis?

9-13 DATA ANALYSIS Regulation of Phosphofructokinase-1. Shown in **Figure 9-15** are plots of initial reaction velocity (expressed as % of V_{max}) versus fructose-6-phosphate concentration for liver phosphofructokinase (PFK-1) in the presence and absence of fructose-2,6-bisphosphate (F2,6BP) (Figure 9-15a) and in the presence of a low or high concentration of ATP (Figure 9-15b).

(a) Explain the effect of F2,6BP on enzyme activity as shown in Figure 9-15a.

(b) Explain the effect of the ATP concentration on the data shown in Figure 9-15b.

(c) What assumptions do you have to make about the concentration of ATP in Figure 9-15a and about the concentration of F2,6BP in Figure 9-15b? Explain.

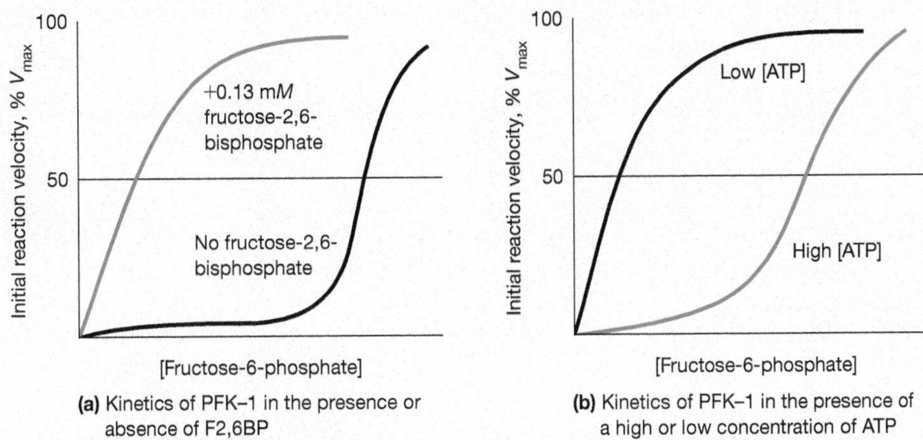

Figure 9-15 Allosteric Regulation of Phosphofructokinase-1. Shown here are Michaelis–Menten plots of liver phosphofructokinase (PFK-1) activity, depicting **(a)** the dependence of initial reaction velocity on concentration of the substrate fructose-6-phosphate in the presence (red line) or absence (black line) of fructose-2,6-bisphosphate and **(b)** the dependence of initial reaction velocity on fructose-6-phosphate concentration at high (red line) or low (black line) ATP concentrations. In both cases, initial reaction velocity is expressed as a percentage of V_{max}, the maximum velocity. See Problem 9-13.

10 Chemotrophic Energy Metabolism: Aerobic Respiration

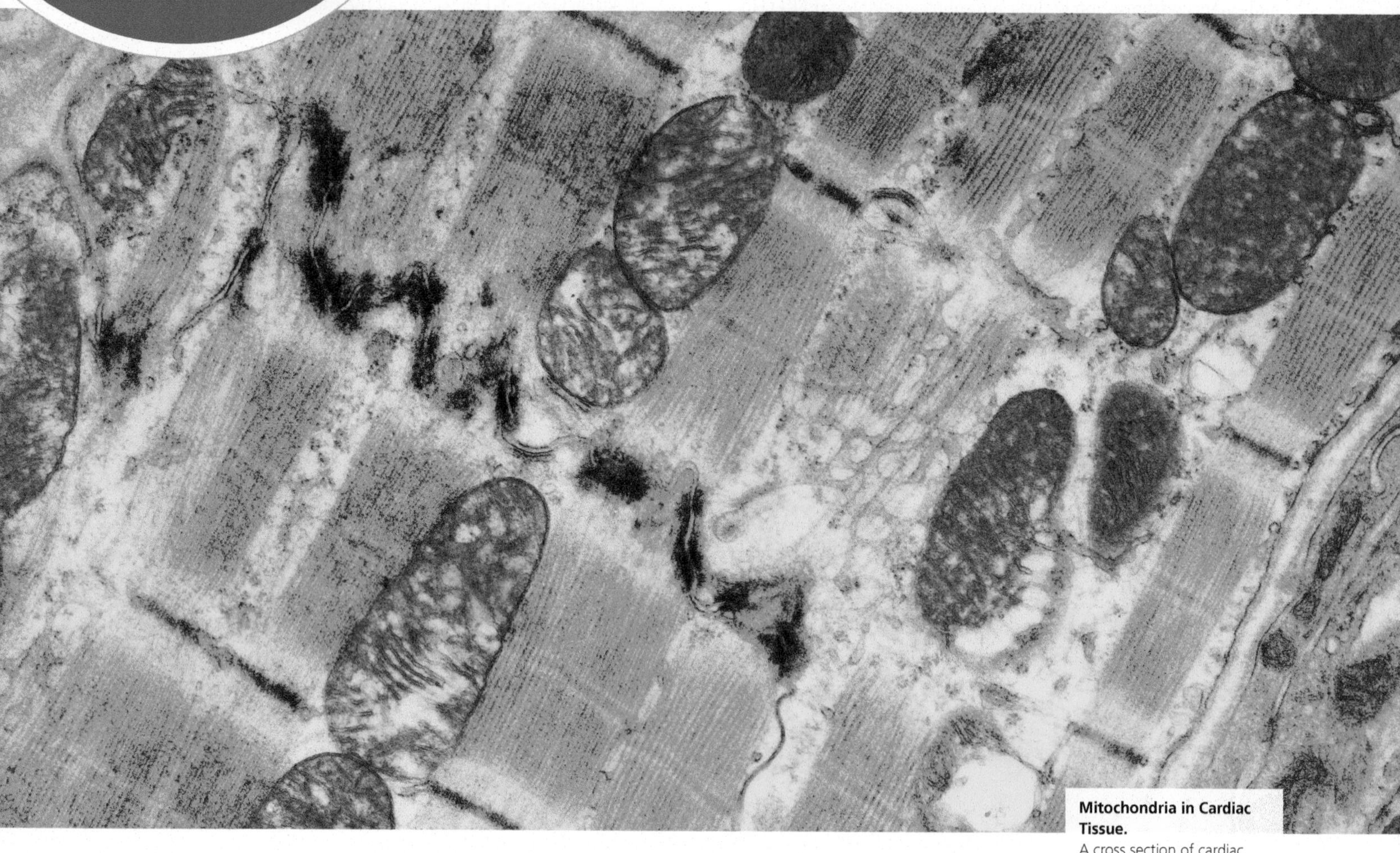

Mitochondria in Cardiac Tissue.
A cross section of cardiac muscle cells (red) showing the abundant mitochondria (blue) that are needed to provide a continual source of energy to keep the heart pumping.

In the previous chapter, we learned that some cells meet their energy needs by anaerobic fermentation either because they are strict anaerobes or because they are facultative cells functioning temporarily in the absence or scarcity of oxygen. However, we also noted that fermentation yields only modest amounts of energy due to the absence of an external electron acceptor. The electrons that are removed from glucose as it is partially oxidized during fermentation are transferred to pyruvate, and only two molecules of ATP can be generated per molecule of glucose.

In short, fermentation can meet energy needs, but the ATP yield is low because the cell has access to only a limited portion of the total free energy potentially available from the oxidizable molecules it uses as substrates. In addition, fermentation often results in the accumulation of waste products such as ethanol or lactate, which can be toxic to the cells if they accumulate.

10.1 Cellular Respiration: Maximizing ATP Yields

Cellular respiration, or **respiration** for short, dramatically improves the metabolic energy yield per molecule of glucose. With an *external* electron acceptor available, complete oxidation of substrates to CO_2 becomes possible, and ATP yields are much higher. By "external" we mean an electron acceptor that is not a by-product of glucose catabolism. When an "internal" electron acceptor such as pyruvate is used (see Figure 9-8a) (or acetaldehyde, as during alcoholic fermentation; Figure 9-8b) to accept electrons from NADH, the by-products are not completely oxidized to CO_2.

As a formal definition, *cellular respiration is the flow of electrons, through or within a membrane, from reduced coenzymes to an external electron acceptor, usually accompanied by the generation of ATP.* We have already encountered NADH as the reduced coenzyme generated by the glycolytic catabolism of sugars or related compounds. As we will see shortly, two other coenzymes, *FAD* (for *flavin adenine dinucleotide*) and *coenzyme Q* (or *ubiquinone*), also collect the electrons that are removed from oxidizable organic substrates and pass them to the terminal electron acceptor via a series of electron carriers, generating ATP in the process.

For many organisms, including us, the terminal electron acceptor is molecular *oxygen*, which is reduced to water. The overall oxygen-requiring process is known as **aerobic respiration.** (Note that in medical terminology at an organismal level, respiration refers to breathing, the uptake of oxygen.) We will concentrate on aerobic respiration because it is the basis of energy metabolism in the aerobic world of which we and most other familiar organisms are a part.

Aerobic Respiration Yields Much More Energy than Fermentation Does

With oxygen available as the terminal electron acceptor, pyruvate can be oxidized completely to CO_2 instead of being used to accept electrons from NADH. This yields much more ATP than glycolysis alone. Oxygen makes all of this possible by serving as the terminal electron acceptor, thereby providing a means for the continuous reoxidation of NADH and other reduced coenzymes. After these coenzyme molecules accept electrons from pyruvate and other oxidizable substrates, they can then transfer these electrons to oxygen. Aerobic respiration therefore involves oxidative pathways in which electrons are removed from organic substrates and transferred to coenzyme carriers, which then transfer these electrons to oxygen, accompanied by the generation of ATP.

Respiration Includes Glycolysis, Pyruvate Oxidation, the Citric Acid Cycle, Electron Transport, and ATP Synthesis

We will consider aerobic respiration in five stages, as shown in **Figure 10-1** on page 244. The first three stages involve substrate oxidation (loss of electrons) and the simultaneous reduction (gain of electrons) of coenzymes, and the second two involve coenzyme reoxidation and the generation of ATP. In cells, of course, all five stages occur continuously and simultaneously.

Stage ❶ is the *glycolytic pathway* (which we encountered in Chapter 9). The steps of glycolysis are the same under aerobic and anaerobic conditions and result in the oxidation of glucose to pyruvate. However, in the presence of oxygen, the fate of the pyruvate is different (see Figure 9-8a)—instead of serving as an electron acceptor, as occurs in anaerobic fermentation, pyruvate is further oxidized to generate *acetyl coenzyme A (acetyl CoA)* (stage ❷), which enters the *citric acid cycle* (stage ❸). The citric acid cycle completely oxidizes acetyl CoA to CO_2 and conserves most of the energy as high-energy reduced coenzyme molecules ($FADH_2$ and NADH).

Stage ❹ involves *electron transport*, or the transfer of electrons from reduced coenzymes to oxygen, coupled to the *pumping* of protons across a membrane. The transfer of electrons from coenzymes to oxygen is exergonic and provides the energy that drives the pumping of protons across the membrane containing the carrier proteins. This generates an *electrochemical proton gradient* across the membrane. In stage ❺, the energy of this proton gradient is used to drive ATP synthesis in a process known as *oxidative phosphorylation.*

Our goal in this chapter is to understand the processes following glycolysis under aerobic conditions—the complete oxidation of pyruvate in the citric acid cycle, electron transport from pyruvate via coenzymes to oxygen, and the formation of a proton gradient that drives ATP synthesis. We will begin our discussion of aerobic energy metabolism with a close look at the *mitochondrion* because of its prominent role in eukaryotic energy metabolism.

CONCEPT CHECK 10.1

Compared to anaerobic fermentation, aerobic respiration uses an external electron acceptor rather than an internal electron acceptor. Explain what this means. Why does using an external electron acceptor provide so much more energy to the cell?

10.2 The Mitochondrion: Where the Action Takes Place

Our discussion of aerobic respiration starts with the **mitochondrion** because most of aerobic energy metabolism in eukaryotic cells takes place within this organelle. Because of this, the mitochondrion is often called the "energy powerhouse" of the eukaryotic cell. In later chapters, we will discuss additional aspects of mitochondrial biology, such as destruction of damaged mitochondria by autophagy, their role in programed cell death, and the mitochondrial genome. Here, however, we will focus on the key role of mitochondria in cellular energy production.

As early as 1850, the German biologist Rudolph Kölliker described the presence of what he called "ordered arrays of particles" in muscle cells. Isolated particles were found to swell in water, leading Kölliker to conclude that each particle was surrounded by a semipermeable membrane. These particles are now called *mitochondria* (singular: *mitochondrion*) and are believed to have arisen when bacterial cells were engulfed by larger cells but survived, taking up permanent residence in the host cell cytoplasm via endosymbiosis (as discussed in Chapter 4).

Evidence suggesting a role for this organelle in oxidative events began to accumulate more than a century ago. In 1913,

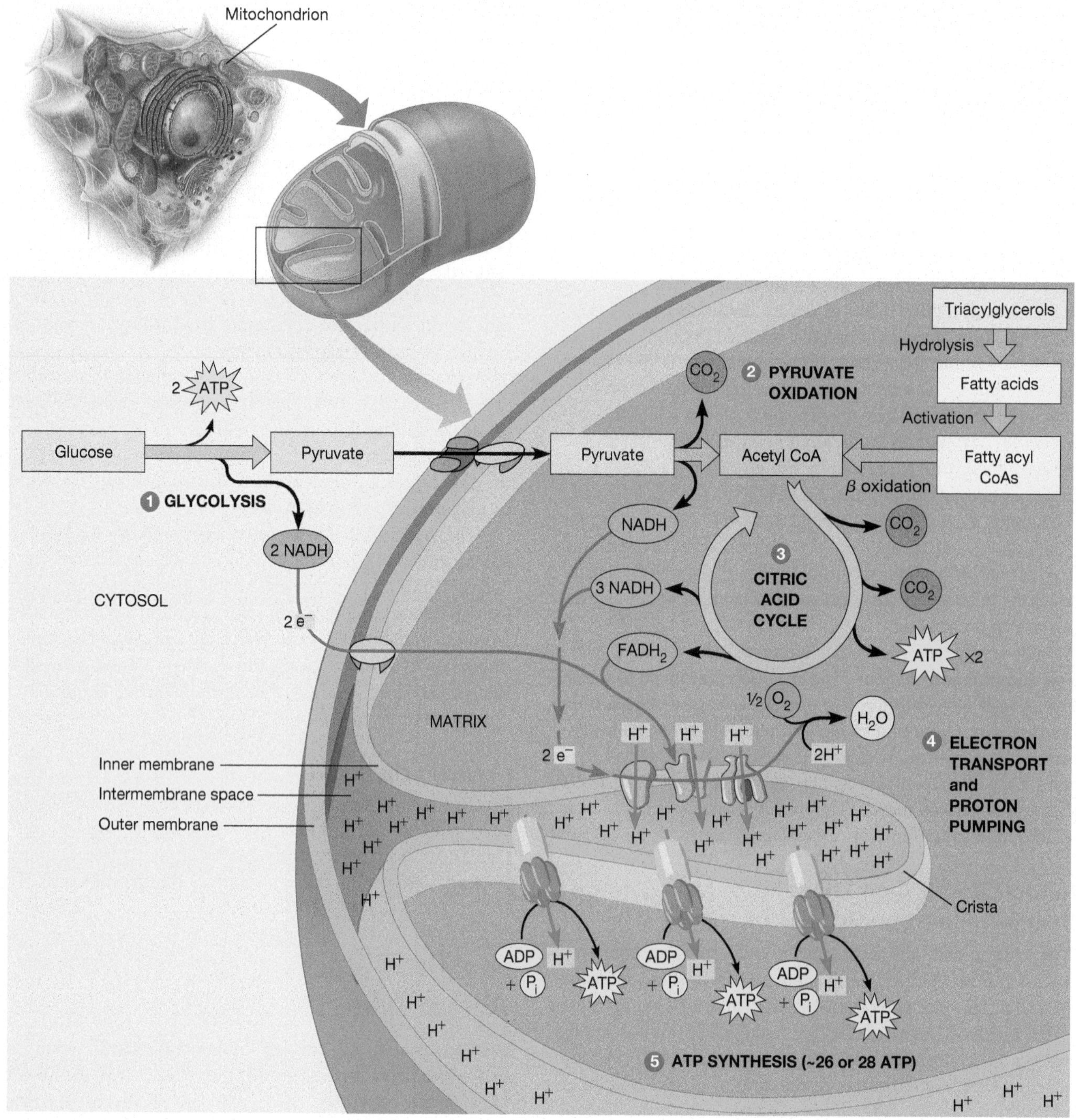

Figure 10-1 The Role of the Mitochondrion in Aerobic Respiration. The mitochondrion plays a central role in aerobic respiration. Most respiratory ATP production in eukaryotic cells occurs in this organelle. Oxidation of glucose and other sugars begins in the cytosol with glycolysis (stage ➊), producing pyruvate. Pyruvate is transported into the mitochondrion, where it is oxidized within the matrix to acetyl CoA (stage ➋), the primary substrate of the citric acid cycle (stage ➌). Acetyl CoA can also be formed by β oxidation of fatty acids. Electron transport is coupled to proton pumping (stage ➍) to produce an electrochemical proton gradient across the inner membrane of the mitochondrion. The energy of the proton gradient drives the synthesis of ATP from ADP and inorganic phosphate (stage ➎).

for example, Otto Warburg showed that these particles could consume oxygen. However, most of our understanding of the role of mitochondria in energy metabolism came since the development of differential centrifugation, pioneered by Albert Claude (see **Key Technique in Chapter 4, pages 100–101**). Intact, functionally active mitochondria were first isolated by this technique in 1948 and were subsequently shown by Eugene Kennedy, Albert Lehninger, and others to be capable of carrying out not only all the reactions of the citric acid cycle, but also electron transport and oxidative phosphorylation.

Mitochondria Are Often Present Where the ATP Needs Are Greatest

Mitochondria are found in virtually all aerobic cells of eukaryotes and are prominent features of many cell types when examined by electron microscopy. Mitochondria are present in both chemotrophic and phototrophic cells and are therefore found not only in animals but also in plants. Their occurrence in phototrophic cells reminds us that even photosynthetic organisms rely on respiration to meet energy needs.

The crucial role of the mitochondrion in meeting cellular ATP needs is often reflected in the localization of mitochondria within the cell. Frequently, mitochondria are clustered in regions of cells with the most intense metabolic activity and the greatest need for ATP. An especially good example is in muscle cells. The mitochondria in muscle cells are organized in rows along the fibrils responsible for contraction (see chapter-opening photo). The importance of mitochondria to muscle cells is underscored by human diseases known as *mitochondrial myopathies*, which arise due to defects in mitochondrial function. A similar strategic localization of mitochondria occurs in flagella and cilia, which need large quantities of ATP for microtubule sliding (see Chapter 14).

Mitochondria Can Adopt Complex Shapes and Vary in Number in Different Cell Types

Figure 10-2 shows a classic, bean-shaped mitochondrion. When seen in electron micrographs such as that in Figure 10-2a, mitochondria appear in outline as oval structures measuring one or more micrometers in length and 0.5–1.0 μm across; such images lead to diagrams like Figure 10-2b. Some mitochondria do in fact have tidy shapes like this; however, fluorescent staining of mitochondria in various cell types shows that the picture is often more complex. As shown in **Figure 10-3** on page 246, mitochondria in living cells can be large and highly elongated. They are also in a dynamic state of flux, with segments of one mitochondrion frequently pinching off and fusing with another (Figure 10-3a). Recent work using electron microscopy (EM) tomography, which allows three-dimensional reconstruction of a cellular structure from a parallel series of cross sections, shows just how complex the shapes of some mitochondria are (Figure 10-3b). (The technique of EM tomography is discussed in more detail in **Key Technique, pages 248–249.**)

The number of mitochondria per cell is highly variable, ranging from one or a few per cell in many protists, fungi, and algae (and in some mammalian cells as well) to a few thousand per cell in some tissues of higher plants and animals. Mammalian liver cells, for example, are thought to contain about 500–1000 mitochondria each.

The Outer and Inner Membranes Define Two Separate Mitochondrial Compartments and Three Regions

Figure 10-2 shows both an electron micrograph and an illustration of a mitochondrion. A distinctive feature is the presence of two membranes, called the outer and inner membranes. The **outer membrane** is not a significant permeability barrier for ions and small molecules because it contains transmembrane channel proteins called **porins** (see Figure 8-9) that permit the passage of solutes with molecular weights up to about 5000. Similar proteins are found in the outer membrane of gram-negative bacteria, supporting the endosymbiosis hypothesis (see Chapter 4). Because porins allow the free movement of small molecules and ions across the outer membrane, the **intermembrane space** between the inner and outer membranes of the

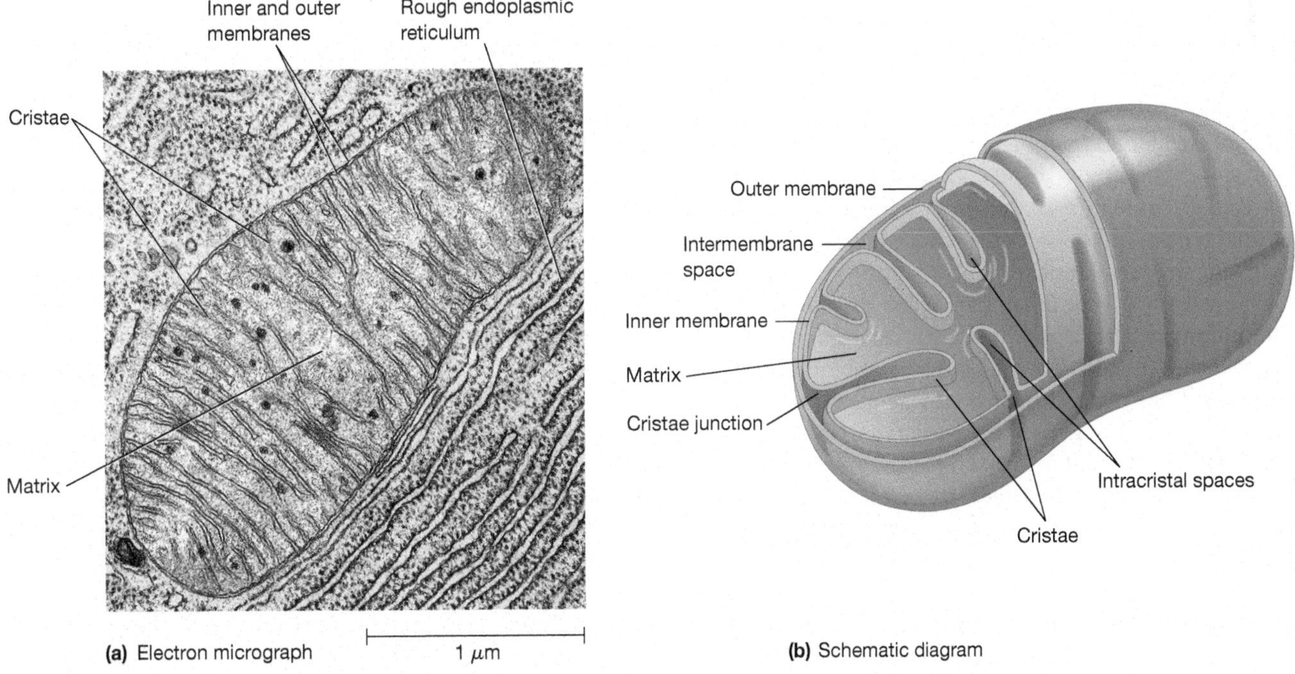

Figure 10-2 Mitochondrial Structure. (a) A mitochondrion of a bat pancreas cell as seen by electron microscopy (TEM). The cristae are formed by infoldings of the inner membrane. **(b)** A mitochondrion is illustrated schematically in this cutaway view that shows the traditional "baffle" model of cristae structure. As noted in the text, however, not all mitochondria have this shape.

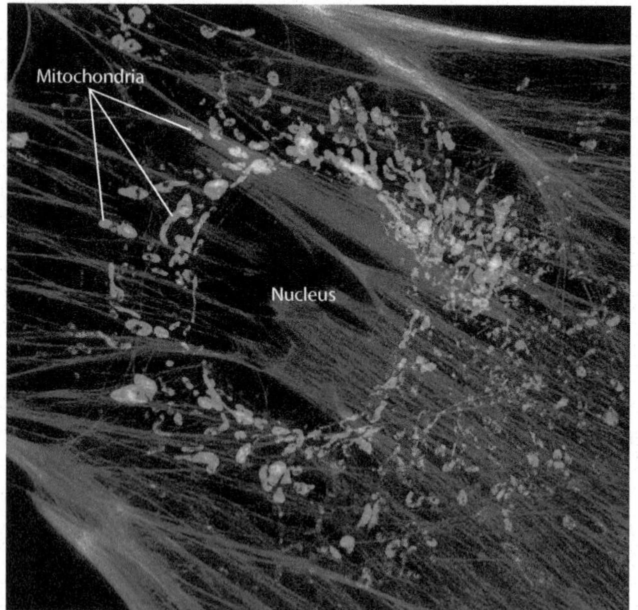

(a) Fluorescently stained mitochondria

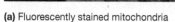

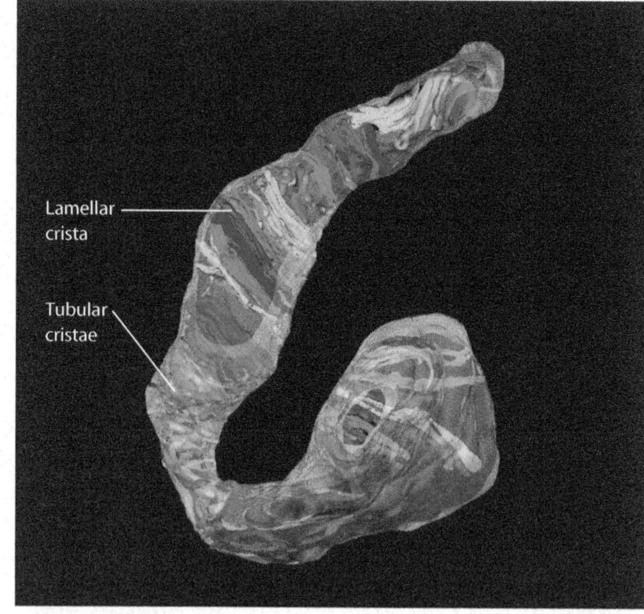

(b) TEM reconstruction of a single mitochondrion

Figure 10-3 Varied Shapes of Mitochondria. (a) Fluorescent labeling of mitochondria (green) in a bovine pulmonary artery endothelial cell. F-actin is stained in purple. Note the varied shapes and sizes of the mitochondria. **(b)** This tomogram of a mouse embryonic fibroblast shows the extended shape of the intact mitochondrion. Individual cristae, mostly lamellar, are colored separately. Note the tubular cristae (light blue) at the bottom left.

mitochondrion is continuous with the cytosol with respect to small solutes. However, enzymes targeted to the intermembrane space are effectively confined there because enzymes and other soluble proteins are too large to pass through the porin channels.

In contrast to the outer membrane, the **inner membrane** of the mitochondrion presents a permeability barrier to most solutes, thereby partitioning the mitochondrion into two separate compartments—the *intermembrane space* versus the interior of the organelle (the *mitochondrial matrix*). The inner and outer membranes are each about 7 nm thick and are typically separated by an intermembrane space of about 7 nm. However, in certain spots the membranes are in contact (visible if you look at Figure 10-2a closely), and it is in these regions that proteins destined for the mitochondrial matrix pass through the two membranes.

The portion of the inner membrane directly adjacent to the intermembrane space is known as the *inner boundary membrane.* In addition, the inner membrane of most mitochondria has many distinctive infoldings called **cristae** (singular: **crista**) that greatly increase its surface area. In a typical liver mitochondrion, for example, the area of the inner membrane (including the cristae) is about five times greater than that of the outer membrane. Because of its large surface area, the inner membrane can accommodate large numbers of the protein complexes needed for electron transport and ATP synthesis, thereby enhancing the mitochondrion's capacity for ATP generation. The inner membrane is about 75% protein by weight, which is a higher proportion of protein than in other cellular membranes. The cristae also provide numerous localized regions, the *intracristal spaces,* where protons can accumulate between the folded inner membranes during the electron transport process, which we will discuss later in the chapter.

In two-dimensional thin-section TEMs, cristae often appear to be arrays of flattened structures with broad connections to the inner boundary membrane, which gave rise to the "baffle" model of cristae structure shown in Figure 10-2b. Thanks, however, to EM tomography (see Key Technique, pages 248–249), we now see that cristae can adopt a variety of other shapes. In addition to lamellar (layered) cristae, in other mitochondria cristae may be tubular structures (Figure 10-3b). Cristae appear to have only limited connections to the inner boundary membrane through small tubular openings known as *crista junctions.* The small size of these openings (approximately 20 nm) is thought to limit diffusion of materials between the intracristal space and the intermembrane space, effectively creating a third, nearly enclosed region in the mitochondrion.

The relative prominence of cristae within the mitochondrion frequently reflects the relative metabolic activity of the cell or tissue in which the organelle is located. Heart, kidney, and muscle cells have high respiratory activities, and thus their mitochondria have correspondingly large numbers of prominent cristae. Plant cells, by contrast, have lower rates of respiratory activity than do most animal cells and have correspondingly fewer cristae within their mitochondria.

The interior of the mitochondrion is filled with a semifluid **matrix.** Within the matrix are many of the enzymes involved in mitochondrial function, as well as ribosomes and the short, circular DNA molecule of the mitochondrion. In most mammals, the mitochondrial genome is about 15,000–20,000 base pairs that encode ribosomal RNAs, transfer RNAs, and about a dozen polypeptide subunits of inner-membrane proteins. Mutations in mitochondrial DNA can lead to human disease and have been associated with aging and neurodegeneration.

Many Mitochondrial Proteins Originate in the Cytosol

Although mitochondria contain their own DNA and protein-synthesizing machinery, they synthesize only a few of the polypeptides they require. More than 95% of the proteins residing in mitochondria are encoded by nuclear genes and are synthesized in the cytosol. The small number of polypeptides synthesized in the mitochondrial matrix are targeted mainly to the inner mitochondrial membrane. Almost without exception, polypeptides encoded by mitochondrial genes are subunits of multimeric proteins, with one or more of the other subunits being imported from the cytosol.

How do cytosolic proteins destined for mitochondria arrive at their final destination? The targeting signal for such polypeptides is a special sequence called a *transit sequence*, located at the N-terminus of the polypeptide. Once inside the mitochondrion, the transit sequence is removed by a *transit peptidase*, an enzyme located within the organelle. Removal of the transit sequence often occurs before transport is complete. The uptake of polypeptide chains possessing transit sequences is mediated by specialized transport complexes located in the outer and inner mitochondrial membranes. As shown in **Figure 10-4**, the transport complexes are called **TOM** (translocase of the *o*uter *m*embrane) and **TIM** (translocase of the *i*nner *m*embrane). Polypeptides are initially selected for transport into mitochondria by a component of the TOM complex known as a *transit sequence receptor.* After a transit sequence has bound to its receptor, the polypeptide containing this sequence is translocated across the outer membrane through a pore component of TOM. If the polypeptide is destined for the interior of the organelle, movement through the TOM complex is quickly followed by passage through the TIM complex of the inner membrane, presumably at a contact site where the outer and inner membranes lie close together.

Polypeptides entering mitochondria must generally be in an unfolded state before they can pass across the membranes bounding these organelles. To maintain the necessary unfolded state, polypeptides targeted for mitochondria are usually bound to *chaperone proteins*. Figure 10-4 shows a current model for this chaperone-mediated import of polypeptides into the mitochondrial matrix. ❶ To start the process, cytosolic *Hsp70 chaperone proteins* (see page 555, especially Figure 19-15a) bind to a polypeptide, keeping it in a loosely folded state. ❷ Next, the transit sequence at the N-terminus of the polypeptide binds to the receptor component of TOM, which protrudes from the surface of the outer mitochondrial membrane. ❸ The chaperone proteins then are released, accompanied by ATP hydrolysis, as the polypeptide is translocated through the TOM and TIM pores and into the mitochondrial matrix. ❹ When the transit sequence emerges into the matrix, it is removed by transit peptidase. ❺ As the rest of the polypeptide subsequently enters the matrix, mitochondrial Hsp70 chaperones bind to it temporarily. The subsequent release of Hsp70 requires ATP hydrolysis, which is thought to drive the translocation process. ❻ Finally, in many cases mitochondrial *Hsp60 chaperones* (see Figure 19-15b) bind to the polypeptide and help it achieve its fully folded conformation.

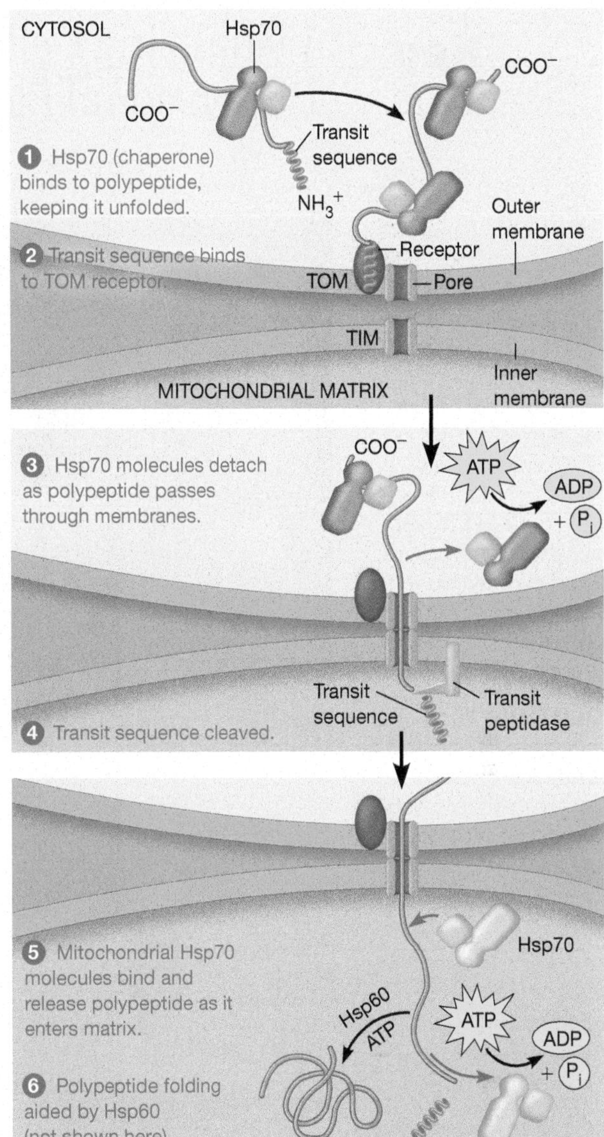

Figure 10-4 Polypeptide Transport Complexes of the Outer and Inner Mitochondrial Membranes. Mitochondrial polypeptides synthesized in the cytosol are transported into mitochondria by specialized transport complexes located in their outer and inner membranes. The outer and inner membrane complexes are called TOM and TIM, respectively. The TOM complex consists of two types of components: receptor proteins that recognize and bind to polypeptides targeted for uptake and pore proteins that form channels through which the polypeptides are translocated. Import of proteins into a mitochondrion involves a transit sequence, a membrane receptor, pore-forming membrane proteins, and a peptidase. Chaperone proteins play several crucial roles in the mitochondrial import process: they keep the polypeptide partially unfolded after synthesis in the cytosol so that binding of the transit sequence and translocation can occur (❶–❸); they drive the translocation itself by binding to and releasing from the polypeptide within the matrix, which is an ATP-requiring process (❺); and they help the polypeptide fold into its final conformation (❻). The chaperones included here are cytosolic (dark green) and mitochondrial (light blue) versions of Hsp70 and a mitochondrial Hsp60 (not illustrated).

Key Technique

Visualizing Cellular Structures with Three-Dimensional Electron Microscopy

PROBLEM: Transmission electron microscopy (TEM) is a powerful tool for examining the fine structure of cells and organelles in thin slices of tissue. While visualizing cellular structures in cross sections in two dimensions is very informative, to better understand the function of internal cellular structures, researchers often want to visualize these structures in three dimensions. How is this achieved?

SOLUTION: *Three-dimensional transmission electron microscopy (3-D TEM)* allows visualization of both surface contours and fine cellular details in three dimensions. 3-D TEM uses transmission electron microscopy (see page A-17) to collect images in a series of parallel or tilted slices through a cellular structure. The images are then combined using computer algorithms, creating a three-dimensional reconstruction of the entire structure.

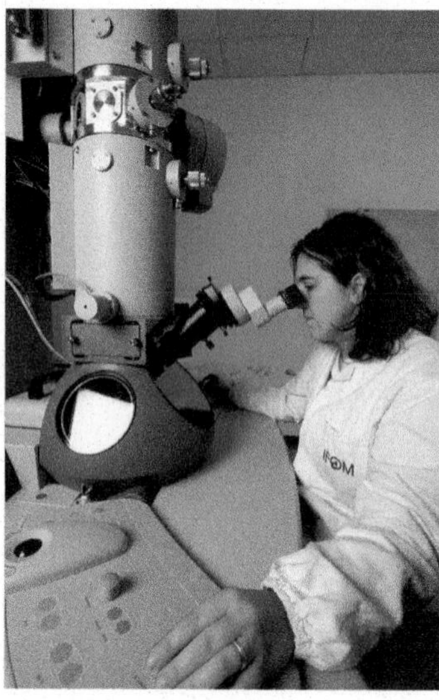

A technician using a transmission electron microscope.

Key Tools: Transmission electron microscope; a cellular sample; a computer for three-dimensional modeling

Details: There are two main ways to perform 3-D TEM, and both are useful. A classical approach, known as *serial-section TEM,* is often used for three-dimensional imaging of larger structures. In this case, a device known as an ultramicrotome cuts thin slices (50–100 nm thick) from a specimen embedded in a plastic resin (see page A-20). Each slice is then imaged using transmission electron microscopy, and structures of interest are traced from each slice using a computer. The tracings can then be stacked and reconstructed into a three-dimensional model (**Figure 10A-1**).

For somewhat smaller structures, *electron tomography (ET)* is often used. The word "tomography" is derived from the Greek words *tomos* ("slice, section") and *graphein* ("to write") and describes a

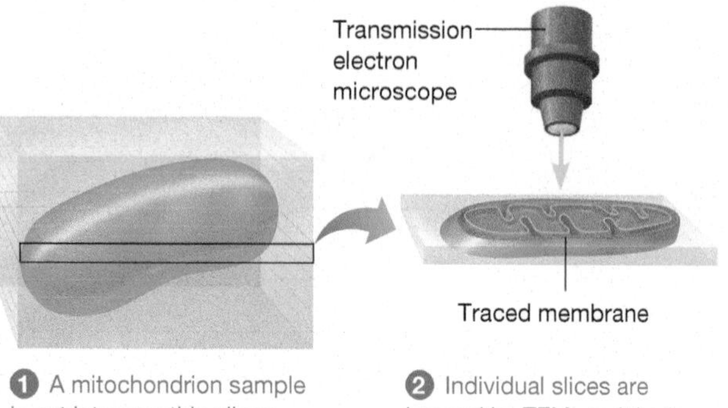

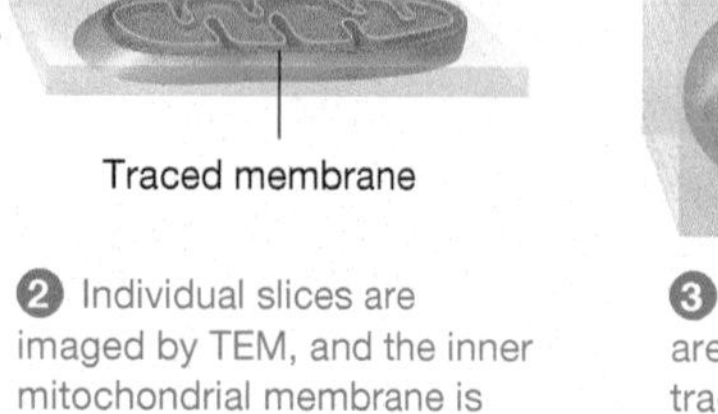

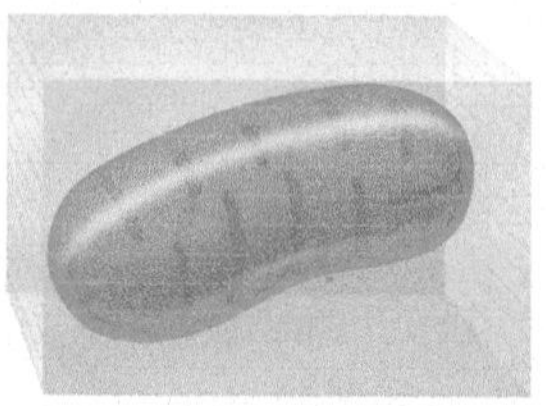

❶ A mitochondrion sample is cut into very thin slices that are 50–100 nm thick.

❷ Individual slices are imaged by TEM, and the inner mitochondrial membrane is traced (green).

❸ The images of the slices are stacked, and the tracings are reconstructed into a 3-D model.

Figure 10A-1 Serial-Section TEM. ❶ A series of very thin slices of a specimen, such as a region of a cell containing a mitochondrion, is made. ❷ Each slice is imaged using a transmission electron microscope, and structures of interest, such as the inner mitochondrial membrane, are traced using a computer. ❸ Tracings of each acquired image are used to generate a three-dimensional image of the structure of interest.

Mitochondrial Functions Occur in or on Specific Membranes and Compartments

Mitochondria are highly organized structures. Specific functions and pathways have been localized within the mitochondrion by disruption of the organelle followed by *subcellular fractionation* of the various components (see page 100). **Table 10-1** on page 250 lists some of the main functions localized to each compartment of the mitochondrion.

Most of the mitochondrial enzymes involved in pyruvate oxidation during the citric acid cycle, and in the catabolism of fatty acids and amino acids, are matrix enzymes. In fact, when mitochondria are disrupted very gently, six of the eight enzymes of the citric acid cycle are released as a single large multiprotein complex, suggesting that the product of one enzyme can pass directly to the next enzyme without having to diffuse through the matrix.

On the other hand, most of the intermediates in the electron transport chain are integral components of the inner membrane, where they are organized into large complexes. Protruding from the inner membrane into the matrix are knoblike spheres called **F_1 complexes** that are involved in ATP synthesis (**Figure 10-5** on page 250). Each complex is an assembly of several different polypeptides. Individual F_1 complexes can be seen in Figure 10-5a, an electron micrograph taken at high magnification using a technique called *negative staining,* which results in a light image against a dark background. F_1 complexes are about 9 nm in diameter and are especially abundant along the cristae (Figure 10-5b).

technique in which a series of two-dimensional images of a fairly thick section (~100–500 nm) of a specimen are acquired at different tilt angles, producing "digital slices" that can then be combined mathematically to produce a three-dimensional model (**Figure 10A-2**).

Whereas a typical TEM uses physical sections 50–100 nm thick, ET can mathematically "slice" thicker specimens—over 1 μm in thickness—much more finely, generating digital slices as closely spaced as 2–10 nm. Using a computer, structures under investigation can be highlighted in each individual section by tracing their edges in different colors, as shown for the membranes and cristae in Figure 10A-2b. Because ET can resolve details measuring as little as 5–10 nm across, the sizes and shapes of individual cristae in the mitochondrion can be determined (see Figure 10A-2b and c).

EM tomography is sometimes performed on sections of material that are frozen in a special way to avoid the formation of damaging ice crystals. This specialized type of TEM is called *cryoelectron microscopy*, or *cryoEM* (see Figure A-36). This technique bridges the gap between conventional electron microscopy and other techniques, such as X-ray crystallography (**see Key Technique in Chapter 3, pages 58–59**), and can reveal the structures of ribosomes, cilia, and flagella, as well as other large macromolecular complexes in great detail.

QUESTION: Working with 3-D TEM often involves the difficult activity of mentally reconstructing a three-dimensional object from a series of two-dimensional sections of that object. Using serial-section TEM as a guide, from top to bottom, what would you expect would be the shapes of a series of two-dimensional slices through a cylinder resting on (1) its side and (2) its end?

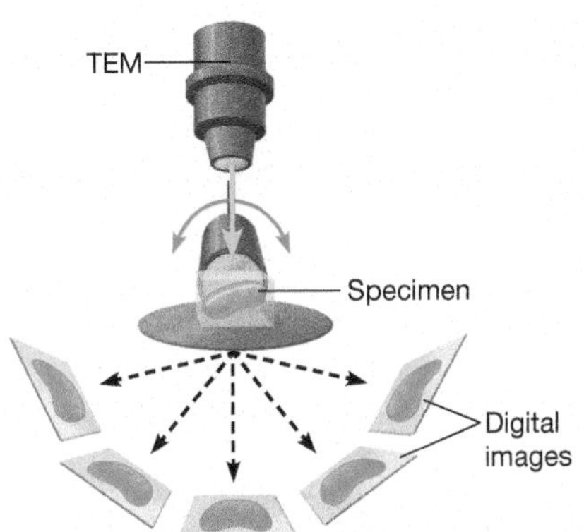

(a) A tilt series of a mitochondrion acquired using ET

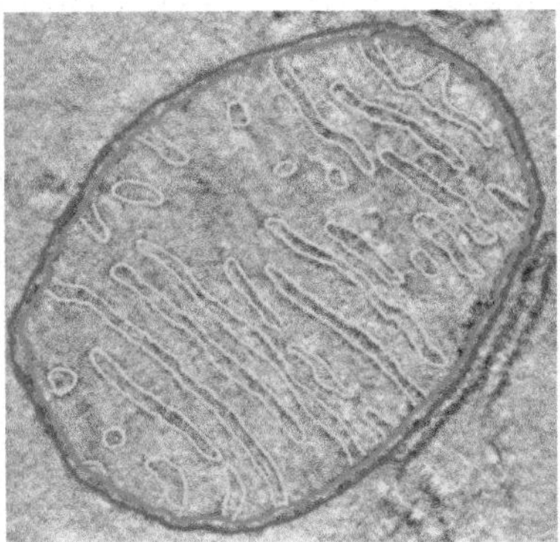

(b) A single ET section through the mitochondrion

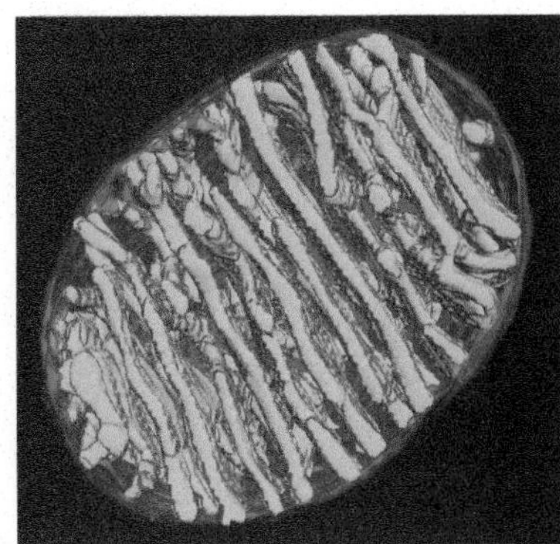

(c) A 3-D reconstruction of the mitochondrion

Figure 10A-2 Electron Tomography. (a) A transmission electron microscope with a tilting stage is used to capture a series of ultrathin digital images of a relatively thick specimen, which can then be reconstructed using special mathematical procedures. **(b)** One individual TEM section used to create a three-dimensional model of a mitochondrion. Structures are traced in color: outer membrane, dark blue; inner membrane, light blue; cristae membranes, yellow. **(c)** A three-dimensional reconstruction created from a series of cross sections, as shown in part b.

Each F_1 complex is attached by a short protein stalk to an **F_o complex**, which is embedded within the mitochondrial inner membrane (Figure 10-5c) or in the plasma membrane of bacteria. (Note that the subscript in F_o is the letter "o" rather than a zero. The F_o complex is sensitive to the antibiotic oligomycin.) The combination of an F_1 complex linked to an F_o complex is called an **F_oF_1 complex** and is more commonly referred as an **ATP synthase** because that is its normal role in energy metabolism. The ATP synthase complex is in fact responsible for most of the ATP generation that occurs in the mitochondrion and in bacterial cells—and (as we will see in Chapter 11) in chloroplasts as well. In each case, ATP generation by the ATP synthase is driven by an electrochemical gradient of protons across the membrane in which the ATP synthase complex is anchored, a topic we will explore in detail later in this chapter.

In Bacteria, Respiratory Functions Are Localized to the Plasma Membrane and the Cytoplasm

Bacteria do not have mitochondria, yet most bacterial cells are capable of aerobic respiration. Where in the bacterial cell are components of respiratory metabolism localized? Essentially, the cytoplasm and plasma membrane of a bacterial cell perform the same functions as the mitochondrial matrix and inner membrane, respectively. Therefore, in bacteria, most of the enzymes of the citric acid cycle are in the cytoplasm, whereas electron transport proteins are in the plasma membrane. ATP synthase is also localized to the plasma membrane in bacteria, with the F_o component embedded in the membrane and the F_1 component protruding from the membrane into the cytoplasm. The similarity to the mitochondrial inner membrane is consistent with the endosymbiosis hypothesis (see Chapter 4) (**Figure 10-6** on page 250).

Table 10-1 Localization of Metabolic Functions Within the Mitochondrion

Membrane or Compartment	Metabolic Functions
Outer membrane	Phospholipid synthesis Fatty acid desaturation Fatty acid elongation
Inner membrane	Electron transport Proton translocation for ATP synthesis Oxidative phosphorylation Pyruvate import Fatty acyl CoA import Metabolite transport
Matrix	Pyruvate oxidation Citric acid cycle ATP synthesis β oxidation of fats mtDNA replication mtRNA synthesis (transcription) Protein synthesis (translation)

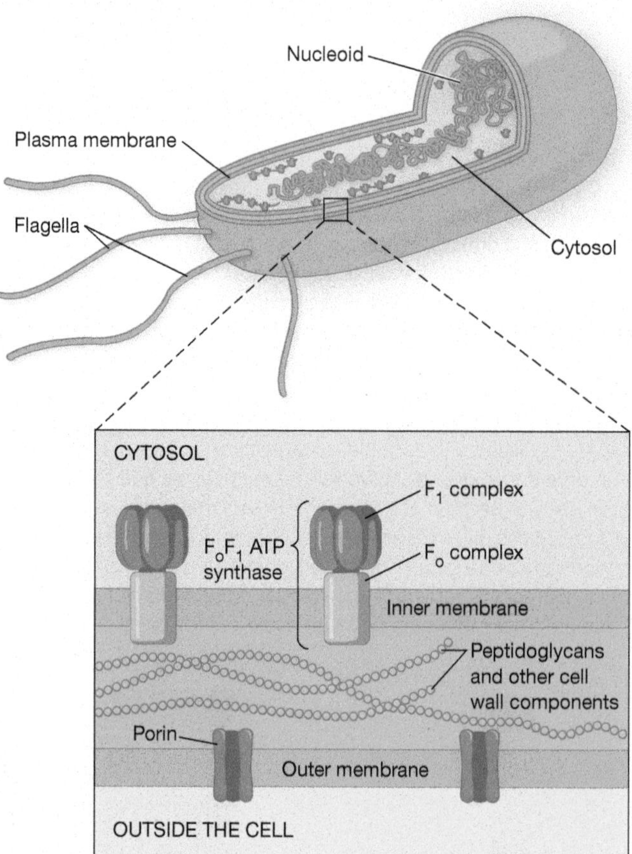

Figure 10-6 The F_1 and F_o Complexes of the Bacterial Membrane. The example shown is a gram-negative bacterium, which has a double outer membrane. As the enlarged region shows, F_1 and F_o complexes project from the inner membrane into the cytosol.

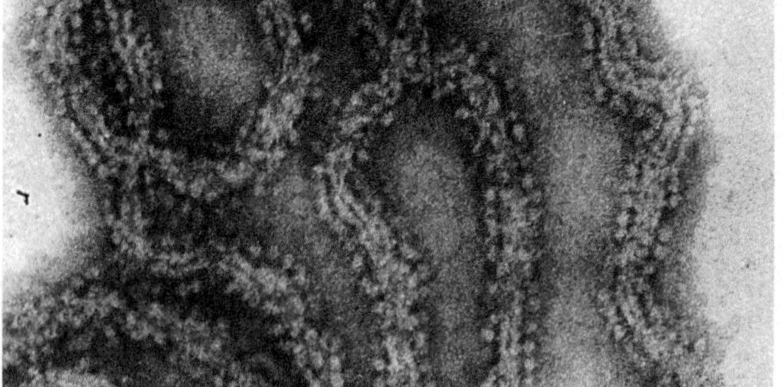

(a) Mitochondrial inner membrane

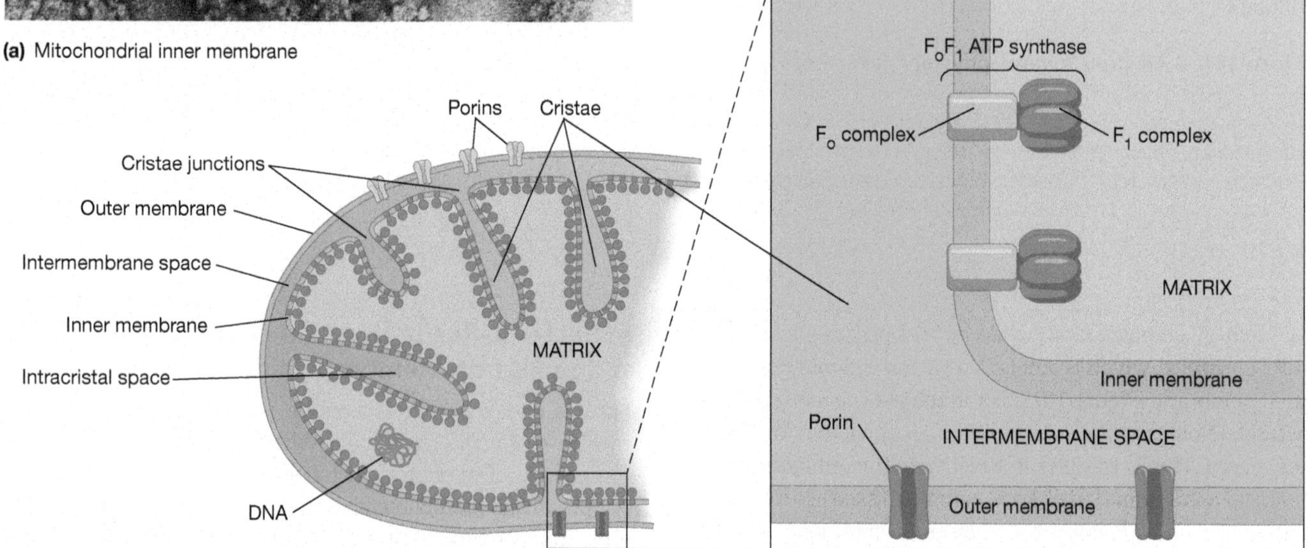

(b) Cross-sectional diagram of a mitochondrion

(c) Cross-sectional diagram of a portion of a crista showing F_oF_1 complexes

Figure 10-5 The F_1 and F_o Complexes of the Inner Mitochondrial Membrane. (a) This electron micrograph was prepared by negative staining to show the spherical F_1 complexes that line the matrix side of the inner membrane of a bovine heart mitochondrion (TEM). **(b)** Cross section of a mitochondrion, showing major structural features. **(c)** An enlargement of a small portion of a crista, showing the F_1 complexes that project from the inner membrane on the matrix side and the F_o complexes embedded in the inner membrane.

10.4 Electron Transport: Electron Flow from Coenzymes to Oxygen

Having considered the first three stages of aerobic respiration—glycolysis, pyruvate oxidation, and the citric acid cycle—let's pause briefly to ask what has been achieved thus far. As Reaction 10-3 indicates, chemotrophic energy metabolism through the citric acid cycle accounts for the synthesis of four ATP molecules per glucose. Two arise from glycolysis (Chapter 9, page 223), and two from the citric acid cycle (page 256). Complete oxidation of glucose to CO_2 could yield 686 kcal/mol (Chapter 9, page 222), but we have recovered less than 10% of that amount (only four ATP molecules × approximately 10 kcal/mol each, based on the $\Delta G'$ value for ATP synthesis in a typical cell). Where is the rest of the free energy? And when will we get to the substantially greater ATP yield that is characteristic of aerobic respiration?

The answer is straightforward: the free energy is right there in Reaction 10-3, represented by the reduced coenzyme molecules NADH and $FADH_2$. As we will see shortly, large amounts of free energy are released when these reduced coenzymes are reoxidized as their electrons are transferred to molecular oxygen. In fact, about 90% of the potential free energy present in one glucose molecule is conserved in the 12 molecules of NADH and $FADH_2$ that are formed when a molecule of glucose is oxidized to CO_2.

The Electron Transport Chain Conveys Electrons from Reduced Coenzymes to Oxygen

The process of coenzyme reoxidation by the transfer of electrons to oxygen is called **electron transport.** Electron transport is the fourth stage of respiratory metabolism (see Figure 10-1, stage ④). The accompanying process of ATP synthesis (Figure 10-1, stage ⑤) will be discussed later in this chapter. Keep in mind, however, that electron transport and ATP synthesis are not isolated processes. They are both integral parts of cellular respiration, functionally linked to each other by the electrochemical proton gradient that is the result of electron transport as well as the source of the energy that drives ATP synthesis.

Electron Transport and Coenzyme Oxidation. Electron transport involves the highly exergonic oxidation of NADH and $FADH_2$ with O_2 as the *terminal electron acceptor,* so we can write summary reactions as follows:

$$\mathrm{NADH} + \mathrm{H}^+ + \frac{1}{2}\mathrm{O}_2 \rightarrow \mathrm{NAD}^+ + \mathrm{H}_2\mathrm{O}$$
$$\Delta G^{\circ\prime} = -52.4 \text{ kcal/mol} \qquad \textbf{(10-4)}$$

$$\mathrm{FADH}_2 + \frac{1}{2}\mathrm{O}_2 \rightarrow \mathrm{FAD} + \mathrm{H}_2\mathrm{O}$$
$$\Delta G^{\circ\prime} = -45.9 \text{ kcal/mol} \qquad \textbf{(10-5)}$$

Electron transport therefore accounts not only for the reoxidation of coenzymes and the consumption of oxygen but also for the formation of water, which is the reduced form of oxygen and, along with CO_2, one of the two end-products of aerobic energy metabolism.

The Electron Transport Chain. The most important aspect of Reactions 10-4 and 10-5 is the large amount of free energy released upon oxidation of NADH and $FADH_2$ by the transfer of electrons to oxygen. The highly negative $\Delta G^{\circ\prime}$ values for these reactions make it clear that the oxidation of a coenzyme is an extraordinarily exergonic process, enough to synthesize several molecules of ATP. Electron transfer is accomplished as a multistep process that involves an ordered series of reversibly oxidizable electron carriers functioning together in what is called the **electron transport chain (ETC).** The ETC contains a number of integral membrane proteins that are found in the inner mitochondrial membrane of eukaryotes (or the plasma membrane of bacteria).

Our discussion of the electron transport chain will focus on three questions:

1. What are the major *electron carriers* in the ETC?
2. What is the *sequence* of these carriers in the ETC?
3. How are these carriers *organized* in the membrane to ensure that the flow of electrons from reduced coenzymes to oxygen is coupled to the pumping of protons across the membrane, thereby producing the electrochemical proton gradient on which ATP synthesis depends?

The Electron Transport Chain Consists of Five Kinds of Carriers

The carriers that make up the ETC include *flavoproteins, iron-sulfur proteins, cytochromes, copper-containing cytochromes,* and a quinone known as *coenzyme Q.* The flavoproteins and coenzyme Q transport protons along with electrons. Except for coenzyme Q, all the carriers are proteins with specific prosthetic groups capable of being reversibly oxidized and reduced. Almost all the events of electron transport occur within membranes, so it is not surprising that most of these carriers are hydrophobic molecules. In fact, most occur in the membrane as parts of large assemblies of proteins called *respiratory complexes.* We will first look briefly at the chemistry of these electron carriers and then see how they are organized into respiratory complexes and ordered into a sequence that transfers electrons from reduced coenzymes to oxygen.

Flavoproteins. Several membrane-bound **flavoproteins** participate in electron transport, using either *flavin adenine dinucleotide (FAD)* or *flavin mononucleotide (FMN)* as the prosthetic group. FMN is essentially the flavin-containing half of the FAD molecule shown in Figure 10-10. One FMN-containing flavoprotein is *NADH dehydrogenase,* which is part of the protein complex that accepts pairs of electrons from NADH. Another example, already familiar to us from the citric acid cycle, is the enzyme succinate dehydrogenase, which has $FADH_2$ as its prosthetic group and is part of the membrane-bound respiratory complex that accepts pairs of electrons from succinate via $FADH_2$. An important characteristic of the flavoproteins (and of the coenzyme NADH) is that they transfer both electrons and protons as they are reversibly oxidized and reduced.

Iron-Sulfur Proteins. Iron-sulfur proteins, also called *nonheme iron proteins,* are a family of proteins, each with an *iron-sulfur (Fe-S) center* that consists of iron and sulfur ions complexed with cysteine groups of the protein. At least a dozen different Fe-S centers are known to be involved in the mitochondrial transport system. The iron ions of these centers are the actual electron carriers. Each iron ion alternates between the more oxidized state (Fe^{3+}) and the more reduced state (Fe^{2+}) during electron transport, which, in this case, involves transfer of only one electron at a time and no protons.

Cytochromes. Like the iron-sulfur proteins, **cytochromes** also contain iron ions but as part of a porphyrin prosthetic group called *heme* (**Figure 10-16**), which you probably recognize as a component of hemoglobin (see Figure 3-4). There are at least five different kinds of cytochromes in the electron transport chain, designated as cytochromes *b*, *c*, c_1, *a*, and a_3. The iron ion of the heme prosthetic group, like that of the iron-sulfur center, serves as the electron carrier for the cytochromes. Thus, cytochromes are also one-electron carriers that do not transfer protons. Whereas cytochromes *b*, c_1, *a*, and a_3 are integral membrane proteins, cytochrome *c* is a peripheral membrane protein that is loosely associated with the outer surface of the membrane. Moreover, cytochrome *c* is not a part of a large complex and can therefore diffuse much more rapidly, a key property in its role in transferring electrons between protein complexes.

Copper-Containing Cytochromes. In addition to their iron ions, cytochromes *a* and a_3 also contain a single copper ion bound to the heme group of the cytochrome, where it associates with an iron ion to form a **bimetallic iron-copper (Fe-Cu) center.** Like iron ions, copper ions can be reversibly converted from the oxidized (Cu^{2+}) to the reduced (Cu^{+}) form by accepting or donating single electrons. The iron-copper center plays a critical role in keeping an O_2 molecule bound to the cytochrome oxidase complex until the O_2 molecule has picked up the requisite four electrons and four protons, at which point the oxygen atoms are released as two molecules of water. (A nutritional note: have you ever wondered why you require the elements iron and copper in your diet? Their roles in electron transport are a major part of the reason.)

Coenzyme Q. The only nonprotein component of the ETC is **coenzyme Q (CoQ),** a quinone with the structure shown in **Figure 10-17**. Because of its ubiquitous occurrence in nature, coenzyme Q is also known as *ubiquinone*. Figure 10-17 also illustrates the reversible reduction, in two successive one-electron (plus one-proton) steps, from the quinone form (CoQ) via the *semiquinone* form (CoQH•, which contains an oxygen radical indicated by a dot) to the *dihydroquinone* form ($CoQH_2$).

Unlike the proteins of the ETC, most coenzyme Q is freely mobile in the nonpolar interior of the inner mitochondrial membrane (or the plasma membrane, in the case of bacteria). CoQ molecules are the most abundant electron carriers in the membrane and occupy a central position in the ETC, serving as a collection point for electrons from the reduced prosthetic groups of FMN- and FAD-linked dehydrogenases in the membrane. Although most is mobile, recent work suggests that some CoQ is tightly bound to specific respiratory complexes and participates in the mechanism of proton pumping.

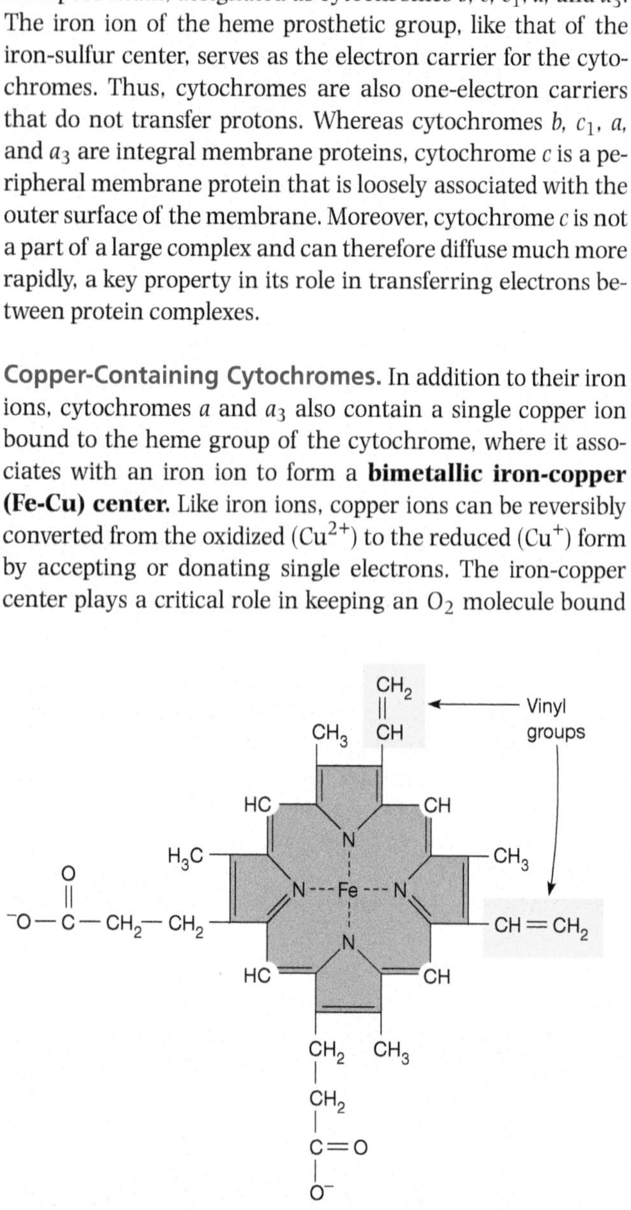

Figure 10-16 The Structure of Heme. Heme, also called iron-protoporphyrin IX, is the prosthetic group in cytochromes *b*, *c*, and c_1. A similar molecule, called heme A, is present in cytochromes a_1 and a_3. The heme of cytochromes *c* and c_1 is covalently attached to the protein by thioether bonds between the sulfhydryl groups of two cysteines in the protein and the vinyl (—CH═CH$_2$) groups of the heme (highlighted in yellow). In other cytochromes, the heme prosthetic group is linked noncovalently to the protein.

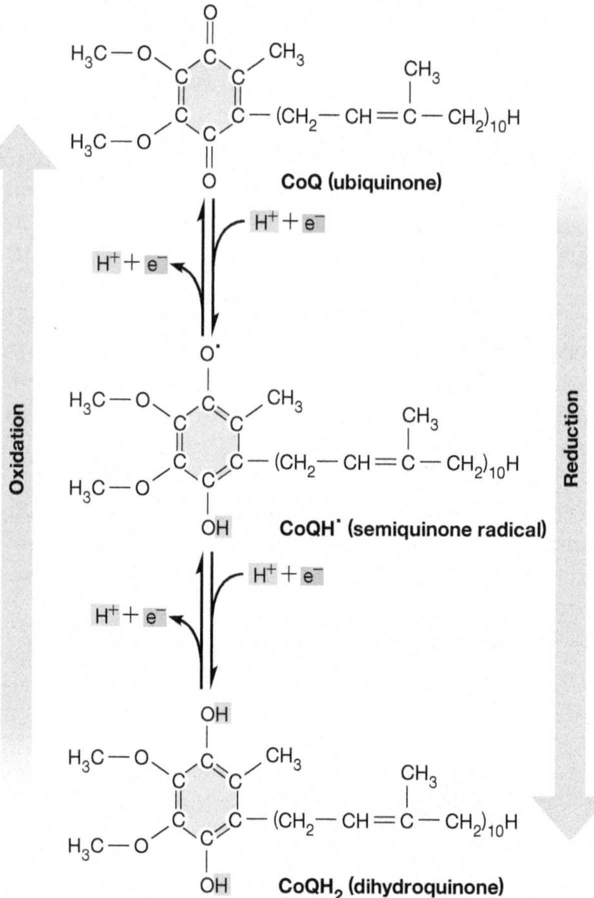

Figure 10-17 Oxidized and Reduced Forms of Coenzyme Q. Coenzyme Q (also called ubiquinone) accepts both electrons and protons as it is reversibly reduced in two successive one-electron steps to form first CoQH• (the semiquinone form) and then $CoQH_2$ (the dihydroquinone form).

Note that coenzyme Q accepts electrons as well as protons when it is reduced and that it releases both electrons and protons when it is oxidized. This property is vital to the role of coenzyme Q in the active transport, or "pumping," of protons across the inner mitochondrial membrane. When CoQ is reduced to $CoQH_2$, it accepts protons from one side of the membrane and then diffuses across the membrane to the outer surface, where it is oxidized to CoQ, with the protons ejected to the other side of the membrane. This motion provides a proton pump coupled to electron transport, which is thought to be one of the mechanisms whereby mitochondria, chloroplasts, and bacteria establish and maintain the electrochemical proton gradients that are used to store the energy of electron transport.

The Electron Carriers Function in a Sequence Determined by Their Reduction Potentials

Now that we are acquainted with the kinds of electron carriers that make up the ETC, our next question concerns the sequence in which these carriers operate. To answer that question, we need to understand the *reduction potential*, E_0, which is a measure, in volts (V), of the affinity a substance has for electrons. In other words, it describes how easily a substance will gain electrons and become reduced. Reduction potentials are determined experimentally for a **redox (reduction-oxidation) pair**, which consists of two molecules or ions that are interconvertible by the loss or gain of electrons. For example, NAD^+ and NADH constitute a redox pair, as do Fe^{3+} and Fe^{2+} as well as $\frac{1}{2}O_2$ and H_2O, as shown in Reactions 10-6 to 10-8:

$$NAD^+ + H^+ + 2e^- \rightarrow NADH \quad \textbf{(10-6)}$$

$$Fe^{3+} + e^- \rightarrow Fe^{2+} \quad \textbf{(10-7)}$$

$$\frac{1}{2}O_2 + 2H^+ + 2e^- \rightarrow H_2O \quad \textbf{(10-8)}$$

Relative E_0 values allow us to compare redox pairs and to predict the direction electrons will tend to flow when several redox pairs are present in the same system, as is clearly the case for the electron transport chain.

As described above, the reduction potential is a measure of the affinity that the oxidized form of a redox pair has for electrons. If a redox pair has a positive E_0, this means that the oxidized form has a high affinity for electrons and that it is therefore a good electron *acceptor*. For example, the E_0 value for the O_2/H_2O pair is highly positive (+0.8 V), meaning that O_2 is a very good electron acceptor. On the other hand, the NAD^+/NADH pair has a highly negative E'_0 value, meaning that NAD^+ is a poor electron acceptor. Alternatively, a negative E_0 value can be thought of as a measure of how good an electron *donor* the reduced form of a redox pair is. Thus, the highly negative E_0 value for the NAD^+/NADH pair means that NADH is a good electron donor.

Understanding Standard Reduction Potentials. To standardize calculations and comparisons of reduction potentials for various redox pairs, we clearly need values determined under specified conditions (much as we did for standard free energy in Chapter 5). For this purpose, we will use the **standard reduction potential** (E'_0), which is the reduction potential for a redox pair under standard conditions (25°C, 1 *M* concentration, 1 atmosphere pressure, and pH 7.0). The standard reduction potentials of redox pairs relevant to energy metabolism are given in **Table 10-2**. Note that in many cases, such as NADH, $FADH_2$, and oxygen, *two* electrons are transferred at a time.

By convention, the $2H^+/H_2$ redox pair is used as a reference and is assigned the value 0.00 V (Table 10-2, boldface line). A redox pair having a positive standard reduction

Table 10-2 Standard Reduction Potentials for Redox Pairs of Biological Relevance*

Redox Pair (oxidized form → reduced form)	Number of Electrons	E'_0(V)
acetate → pyrurate	2	−0.70
succinate → α-ketoglutarate	2	−0.67
acetate → acetaldehyde	2	−0.60
3-phosphoglycerate → glyceraldehyde-3-P	2	−0.55
α-ketoglutarate → isocitrate	2	−0.38
NAD^+ → NADH	2	−0.32
FMN → $FMNH_2$	2	−0.30
1,3-bisphosphoglycerate → glyceraldehyde-3-P	2	−0.29
acetaldehyde → ethanol	2	−0.20
pyruvate → lactate	2	−0.19
FAD → $FADH_2$	2	−0.18
oxaloacetate → malate	2	−0.17
fumarate → succinate	2	−0.03
$2H^+$ → H_2	**2**	**0.00****
CoQ → $CoQH_2$	2	+0.04
cytochrome *b* (Fe^{3+} → Fe^{2+})	1	+0.07
cytochrome *c* (Fe^{3+} → Fe^{2+})	1	+0.25
cytochrome *a* (Fe^{3+} → Fe^{2+})	1	+0.29
cytochrome a_3 (Fe^{3+} → Fe^{2+})	1	+0.55
Fe^{3+} → Fe^{2+} (inorganic iron)	1	+0.77
$\frac{1}{2}O_2$ → H_2O	2	+0.816

*Each E'_0 value is for the following half-reaction, where *n* is the number of electrons transferred:

oxidized form + nH^+ + ne^- → reduced form

**By definition, the $2H^+/H_2$ redox pair is the reference point for determining values of all other redox pairs. It requires that $[H^+] = 1.0M$ and therefore specifies pH 0.0. At pH 7.0, the value for the $2H^+/H_2$ pair is −0.42 V.

potential means that, under standard conditions, the oxidized form of the pair has a higher affinity for electrons than H^+ and will accept electrons from H_2. Conversely, a negative reduction potential means that the oxidized form of the pair has less affinity for electrons than H^+, and its reduced form will donate electrons to H^+ to form H_2.

The redox pairs of Table 10-2 are arranged in order, with the most negative E'_0 values (that is, the best electron donors and hence the strongest reducing agents) at the top. Any redox pair shown in Table 10-2 can undergo a redox reaction with any other pair. The direction of such a reaction under standard conditions can be predicted by inspection because, under standard conditions, *the reduced form of any redox pair will spontaneously reduce the oxidized form of any pair below it on the table.* Thus, NADH can reduce pyruvate to lactate but cannot reduce α-ketoglutarate to isocitrate.

The tendency of the reduced form of one pair to reduce the oxidized form of another pair can be quantified by determining $\Delta E'_0$ the difference in the E'_0 values between the two pairs:

$$\Delta E'_0 = E'_{0,\,\text{acceptor}} - E'_{0,\,\text{donor}} \quad \textbf{(10-9)}$$

For example, $\Delta E'_0$ for the transfer of electrons from NADH to O_2 (Reaction 10-4) is calculated as follows, with NADH as the donor and O_2 as the acceptor:

$$\begin{aligned}\Delta E'_0 &= E'_{0,\,\text{acceptor}} - E'_{0,\,\text{donor}} \\ &= +0.816 - (-0.32) = +1.136 \text{ V}\end{aligned} \quad \textbf{(10-10)}$$

The Relationship Between $\Delta G^{\circ\prime}$ and $\Delta E'_0$. As you may already have guessed, $\Delta E'_0$ is a measure of thermodynamic spontaneity for the redox reaction between any two redox pairs under standard conditions. The thermodynamic spontaneity of a redox reaction under standard conditions can therefore be expressed as either $\Delta G^{\circ\prime}$ or $\Delta E'_0$. The sign convention for $\Delta E'_0$ is the opposite of that for $\Delta G^{\circ\prime}$, so an exergonic reaction is one with a negative $\Delta G^{\circ\prime}$ and a positive $\Delta E'_0$. For any oxidation-reduction reaction, $\Delta G^{\circ\prime}$ is related to $\Delta E'_0$ by the equation

$$\Delta G^{\circ\prime} = -nF\Delta E'_0 \quad \textbf{(10-11)}$$

where n is the number of electrons transferred and F is the Faraday constant (23,062 cal/mol-V). For example, the reaction of NADH with oxygen (Reaction 10-4) involves the transfer of two electrons, so $\Delta G^{\circ\prime}$ for the reaction can be calculated as

$$\begin{aligned}\Delta G^{\circ\prime} &= -2F\Delta E'_0 = -2(23{,}062)(+1.136) \\ &= -52{,}400 \text{ cal/mol} = -52.4 \text{ kcal/mol}\end{aligned} \quad \textbf{(10-12)}$$

The $\Delta G^{\circ\prime}$ for this reaction is highly negative, so the transfer of electrons from NADH to O_2 is thermodynamically spontaneous under standard conditions. This difference in reduction potentials between the NAD^+/NADH and O_2/H_2O redox pairs drives the ETC and, as we will soon see, creates a proton gradient whose electrochemical potential will drive ATP synthesis.

Ordering of the Electron Carriers. We now have the information needed to put the pieces of the ETC together. As we already know, the respiratory ETC consists of several FMN- and FAD-linked dehydrogenases, iron-sulfur proteins with a total of 12 or more Fe-S centers, five cytochromes (including two with Fe-Cu centers), and a pool of coenzyme Q molecules. Shown in **Figure 10-18** are the major ETC components from free NADH ($E'_0 = -0.32$V) and the $FADH_2$ of succinate dehydrogenase ($E'_0 = -0.18$V) to oxygen ($E'_0 = +0.816$V), arranged according to their standard reduction potentials (E'_0 values as obtained from Table 10-2).

In terms of energy, the key point of Figure 10-18 is that *the position of each carrier is determined by its standard reduction potential.* The electron transport chain, in other words, consists of a series of chemically diverse electron carriers, with their order of participation in electron transfer determined by their relative reduction potentials. This means, in turn, that electron transfer from NADH or $FADH_2$ at the top of the figure to O_2 at the bottom is spontaneous and exergonic, with some of the transfers between successive carriers characterized by quite large differences in E'_0 values and hence by large changes in free energy.

Most of the Carriers Are Organized into Four Large Respiratory Complexes

Although there are many electron carriers involved in the ETC, most are not present in the membrane as separate entities. Instead, the behavior of these carriers in mitochondrial extraction experiments indicates that they are organized into *multiprotein complexes.* For example, when mitochondrial membranes are extracted under gentle conditions, only cytochrome *c* is readily removed. The other electron carriers remain bound to the inner membrane and are not released unless the membrane is exposed to detergents or concentrated salt solutions. Such treatments then extract the remaining electron carriers from the membrane not as individual molecules but as four large complexes.

Based on these and similar findings, most of the electron carriers in the ETC are thought to be organized within the inner mitochondrial membrane into four different kinds of **respiratory complexes.** These complexes are identified in Figure 10-18 by the large ovals and Roman numerals, which indicate the ordering of the various electron carriers based on their E'_0 values as well as their organization within the membrane.

Properties of the Respiratory Complexes. Each of the four respiratory complexes consists of a distinctive assembly of polypeptides and prosthetic groups, and each complex has a unique role to play in the electron transport process. **Table 10-3** on page 266 summarizes some properties of these complexes.

Figure 10-19 on page 266 places the three complexes needed for NADH oxidation (complexes I, III, and IV) in their proper eukaryotic context, the inner mitochondrial membrane. Also shown in the figure is the outward pumping of protons across the membrane that accompanies electron transport within the membrane. Each of these complexes represents one site of proton pumping. *For each pair of electrons transported through complexes I, III, and IV, ten protons are*

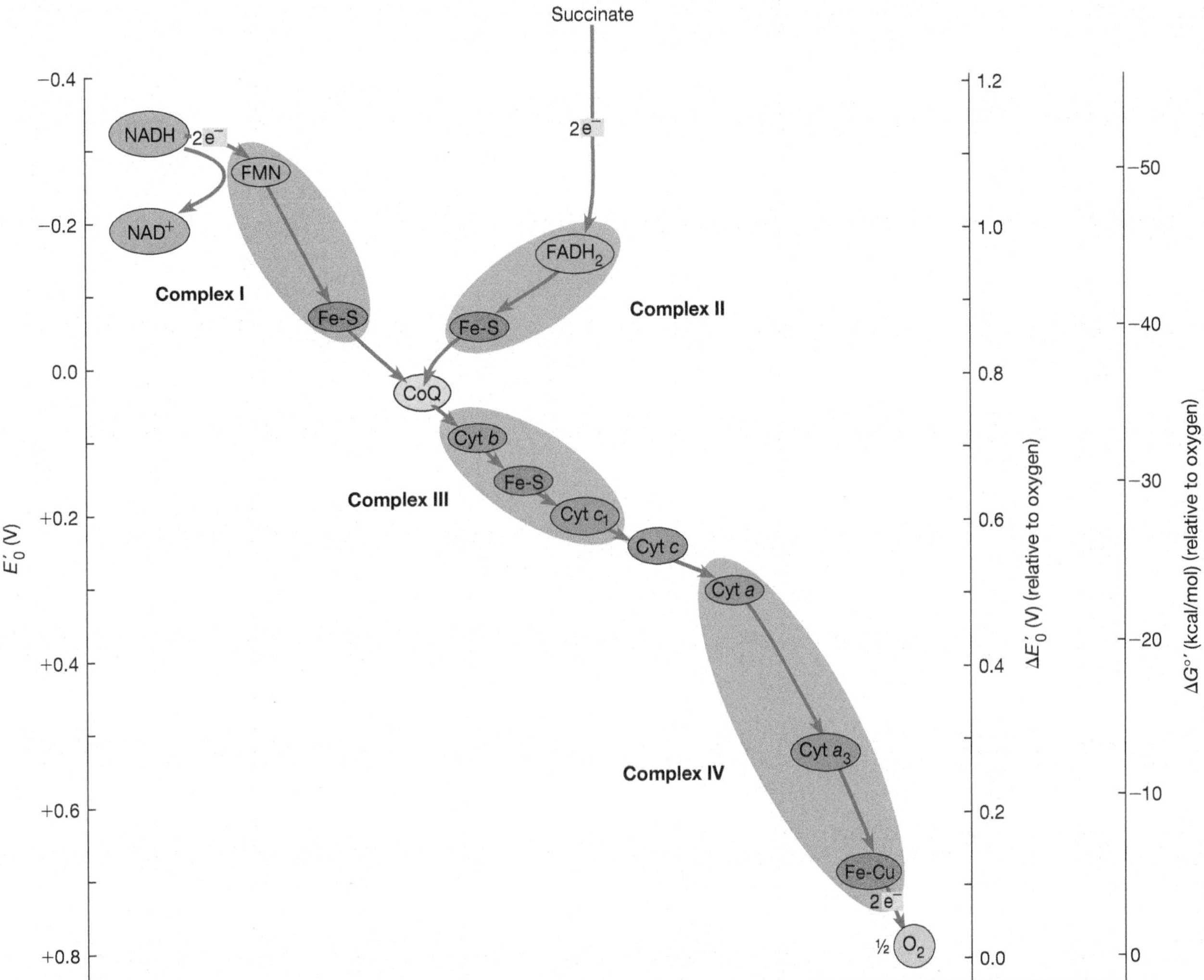

Figure 10-18 Major Components of the Electron Transport Chain and Their Energetics. Major intermediates in the transport of electrons from NADH (−0.32 V) and $FADH_2$ (−0.18 V) to oxygen (+0.816 V) are positioned vertically according to their energy levels, as measured by their standard reduction potentials (E'_0, left axis). The four respiratory complexes are shown as large brown ovals, with the major electron carriers in each complex enclosed within inset ovals. Coenzyme Q and cytochrome *c* are small, mobile intermediates that transfer electrons between the several complexes. The red arrows trace the exergonic flow of electrons through the system. On the right axes are the $\Delta E'_0$ and $\Delta G^{\circ\prime}$ values relative to oxygen (i.e., the changes in the standard reduction potential and the standard free energy for the transfer of two electrons to O_2).

pumped from the matrix into the intermembrane space. When these protons move back across the membrane, they travel through the ATP synthase complex where ATP is synthesized from ADP and inorganic phosphate in the matrix.

Complex I (Figure 10-19a) transfers electrons from NADH to coenzyme Q and is called *NADH dehydrogenase* (or the *NADH–coenzyme Q oxidoreductase complex*). It is a complex assembly of more than 40 polypeptides, depending on species. It receives electrons from NADH and transfers them to a bound FMN cofactor. FMN then transfers the electrons to an Fe-S center, which passes them to a mobile pool of CoQ in the interior of the mitochondrial inner membrane. During transfer of two electrons, four protons are pumped from the matrix across the inner membrane to the intermembrane space.

Complex II (Figure 10-19b) transfers electrons from succinate to FAD to generate $FADH_2$ (reaction CAC-6 in Figure 10-9); electrons from $FADH_2$ are then transferred through three Fe-S centers to CoQ. This complex is called *succinate dehydrogenase* (or the *succinate–coenzyme Q oxidoreductase complex*). No protons are pumped during succinate oxidation.

Complex III (Figure 10-19c) is called the *cytochrome complex* because those two cytochromes are its most prominent components. It is also called the *coenzyme Q–cytochrome c oxidoreductase complex* because it accepts electrons from CoQ and transfers them to cytochrome *c*. As a pair of electrons travels through this complex, four more protons are pumped into the intermembrane space.

Based on detailed studies of complex III, electron flow through the complex can be explained by the operation of a mitochondrial **Q cycle,** a variant of which was first proposed by Peter Mitchell in 1975. The net reaction is the following:

$$\begin{aligned}&CoQH_2 + 2\,\text{cyt}\,c_1(\text{oxidized}) + 2H^+(\text{matrix}) \rightarrow \\ &\quad CoQ + 2\,\text{cyt}\,c_1(\text{reduced}) + 4H^+(\text{intramembrane})\end{aligned} \quad \textbf{(10-13)}$$

Table 10-3 Properties of the Mitochondrial Respiratory Complexes

Respiratory Complex				Electron Flow		
Number	Name(s)	Number of Polypeptides*	Cofactors	Accepted from	Passed to	Protons Translocated (per electron pair)
I	NADH dehydrogenase (NADH–coenzyme Q oxidoreductase)	43	1 FMN	NADH	Coenzyme Q	4
		(7)	6–9 Fe-S centers			
II	Succinate dehydrogenase (Succinate–coenzyme Q oxidoreductase)	4	1 $FADH_2$	Succinate (via enzyme- bound $FADH_2$)	Coenzyme Q	0
		(0)	3 Fe-S centers			
III	Cytochrome b/c_1 complex (Coenzyme Q–cytochrome *c* oxidoreductase)	11	2 cytochrome *b*	Coenzyme Q	Cytochrome *c*	4**
		(1)	1 cytochrome c_1			
			1 Fe-S center			
IV	Cytochrome *c* oxidase	13	1 cytochrome *a*	Cytochrome *c*	Oxygen (O_2)	2
		(3)	1 cytochrome a_3			
			2 Cu centers (as Fe-Cu centers with cytochrome a_3)			

*The number of polypeptides encoded by the mitochondrial genome is indicated in parentheses for each complex.
**The value for complex III includes two protons translocated by coenzyme Q.

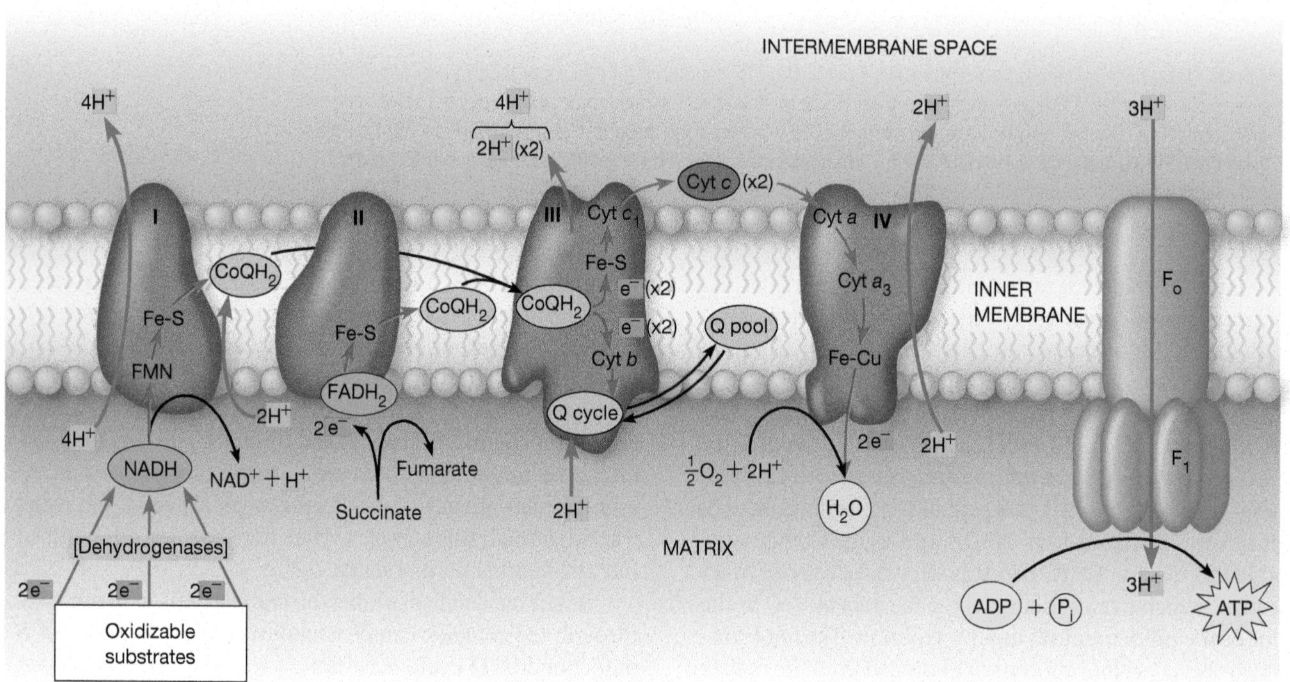

(a) **Complex I** receives 2 electrons from NADH and passes them to CoQ via FMN and an Fe-S protein. During this process, 4H+ are pumped out of the matrix by complex I.
(b) **Complex II** receives 2 electrons from succinate, reducing FAD to FADH2, which passes them to CoQ via Fe-S proteins.
(c) **Complex III** passes electrons from CoQH2 to cytochrome c1 and an Fe-S protein. As a result of the Q cycle, CoQH2 carries 4H+ across the inner membrane and 2 more H+ are pumped out of the matrix.
(d) **Complex IV** receives electrons from cytochrome c and, via cytochrome a and a3, passes them to molecular oxygen, which is reduced to water as 2 more H+ are pumped from the matrix by complex IV.
(e) **ATP synthase** uses the energy from the proton gradient generated during electron transport to synthesize ATP from ADP and Pi.
INTERMEMBRANE SPACE
4H+
4H+
2H+ (x2)
2H+
3H+
Cyt c (x2)
I
II
III
Cyt c1
Cyt a
IV
CoQH2
Fe-S
FMN
Fe-S
CoQH2
FADH2
Fe-S
CoQH2
e (x2)
e (x2)
Q pool
Cyt b
Q cycle
Cyt a3
Fe-Cu
INNER MEMBRANE
Fo
F1
4H+
2H+
2e
NADH
NAD+ + H+
Fumarate
Succinate
2H+
1/2 O2 + 2H+
2e
2H+
H2O
[Dehydrogenases]
MATRIX
2e
2e
2e
Oxidizable substrates
ADP
+
Pi
3H+
ATP

Figure 10-19 The Flow of Electrons Through Respiratory Complexes Causes Directional Proton Pumping. (a–d) Electrons derived from oxidizable substrates in the mitochondrial matrix flow exergonically from NADH (through complexes I, III, and IV) and $FADH_2$ (through complexes II, III, and IV) to oxygen. **(e)** During the transport of two electrons, 10 H^+ are transported across the inner membrane, and 3 ATP are synthesized by the F_oF_1 ATP synthase. Two extra H^+ are pumped due to the Q cycle.

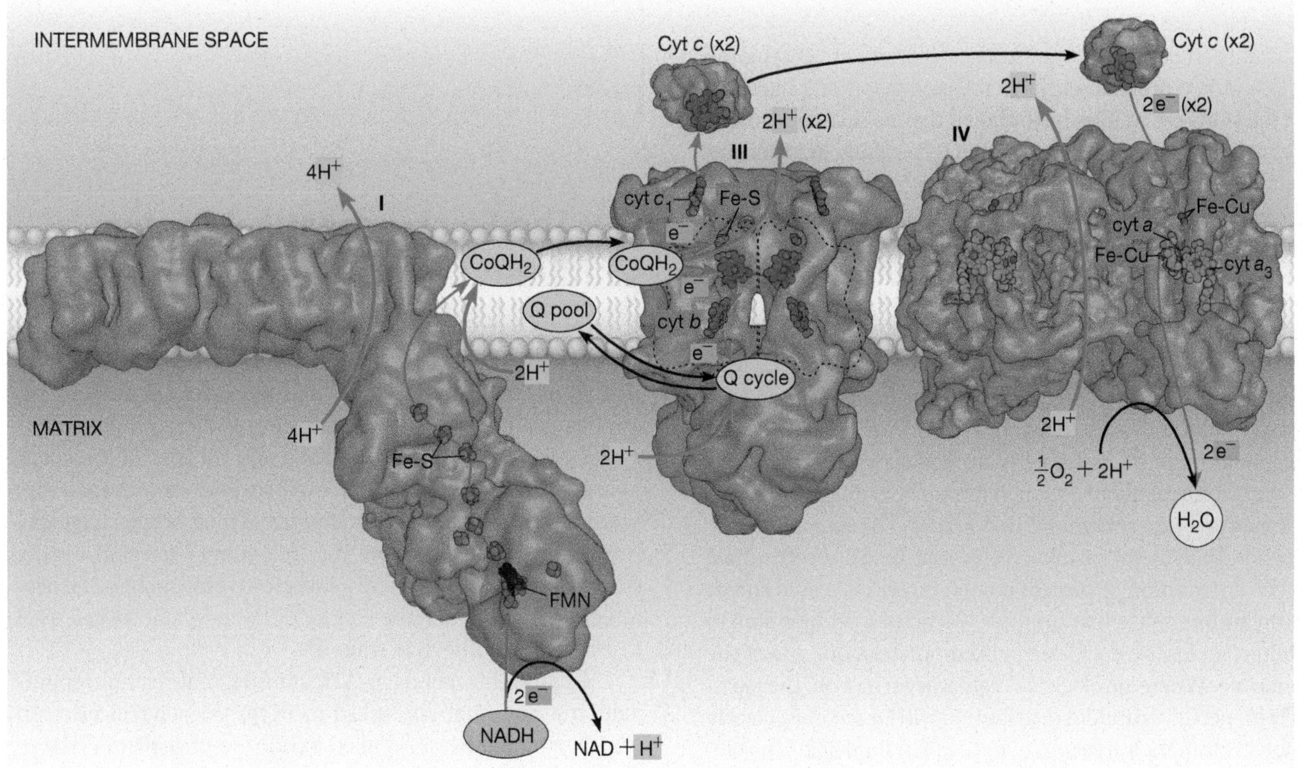

Figure 10-20 Respiratory Complexes Involved in Electron Transfer During NADH Oxidation. Accurate molecular renderings of respiratory complexes I, III, and IV showing the positions of the electron carriers in each complex and the flow of electrons from NADH to molecular oxygen. Also shown are sites of proton pumping across the inner mitochondrial membrane from the matrix to the intermembrane space.

Although the net reaction is straightforward, the details are somewhat complicated. The Q cycle helps explain how $CoQH_2$, which carries *two* electrons, can hand its electrons off to *single* electron carriers (cytochromes *b* and c_1, and ultimately cytochrome *c*). The current Q cycle model involves two steps. In the first step, a molecule of $CoQH_2$ transfers an electron to cytochrome c_1. Its other electron is passed via the cytochrome *b* complex to another CoQ molecule bound elsewhere within complex III, creating a semiquinone ($CoQ^{-\bullet}$):

$$CoQH_2 + \text{cytochrome } c_1(\text{oxidized}) + CoQ \rightarrow CoQ^{-\bullet} + \text{cytochrome } c_1(\text{reduced}) + 2H^+(\text{intermembrane}) \quad \textbf{(10-14)}$$

The CoQ generated in this first step is then released back into the CoQ pool in the mitochondrial inner membrane. In the second step, another $CoQH_2$ transfers an electron to cytochrome c_1. Its other electron is passed via the cytochrome *b* complex and transferred to the semiquinone to complete its reduction:

$$CoQH_2 + \text{cytochrome } c_1(\text{oxidized}) + CoQ^{-\bullet} + 2H^+(\text{matrix}) \rightarrow CoQ + \text{cytochrome } c_1(\text{reduced}) + 2H^+(\text{intermembrane}) + CoQH_2 \quad \textbf{(10-15)}$$

The CoQ and $CoQH_2$ in this second step are released into the CoQ pool in the mitochondrial inner membrane. For simplicity, these details are not shown in Figure 10-19, where the focus is on the overall flow of electrons and protons.

Complex IV (Figure 10-19d) transfers electrons from cytochrome *c* to oxygen and is called *cytochrome c oxidase*. Electrons from cytochrome *c* are transferred to an Fe atom in the heme A cofactor of cytochrome *a* and then to cytochrome a_3. This complex contains two pairs of copper atoms (see Figure 10-20, right), each of which can receive an electron, for a total of the four electrons needed to reduce an O_2 molecule to H_2O. Concurrently, two more protons are pumped into the intermembrane space for each pair of electrons transferred to oxygen, bringing the total number of protons pumped to ten for every pair of electrons transported through the ETC. As we will see in detail later, the asymmetric release of protons is used by ATP synthase to drive ATP synthesis (Figure 10-19e).

As **Figure 10-20** shows, in each of the complexes required for transfer of electrons from NADH to oxygen in Figure 10-19, the cofactors are held in specific positions within the complexes. This positioning facilitates the proper flow of electrons from carrier to carrier and complex to complex.

The Role of Cytochrome *c* Oxidase. Of the several respiratory complexes involved in aerobic respiration in humans, cytochrome *c* oxidase (complex IV) is the **terminal oxidase,** transferring electrons directly to oxygen. This complex is therefore the critical link between aerobic respiration and the oxygen that makes it all possible. In fact, the reason cyanide and azide ions are highly toxic to nearly all aerobic cells is that they bind tightly to the Fe-Cu center of cytochrome *c* oxidase, thereby blocking electron transport.

Transport of two electrons from cytochrome *c* to O_2 plus the addition of two protons results in the production of a molecule of H_2O. However, studies have shown that complexes I and III also can transfer electrons directly to O_2, resulting in its incomplete reduction. This can generate toxic *superoxide anion* (O_2^-) or *hydrogen peroxide* (H_2O_2), compounds that can contribute to cellular aging under both normal and pathological conditions.

The Respiratory Complexes Move Freely Within the Inner Membrane

Static images of ETC components may mislead you into thinking that the protein complexes of the mitochondrial inner membrane are immobile. In fact, they are free to diffuse within the plane of the membrane. This diffusion can be demonstrated experimentally. In one study, for example, inner mitochondrial membranes were placed in an electric field and the distribution of protein complexes was assessed subsequently by freeze-fracture analysis. Particles corresponding to protein complexes were found to accumulate at one end of the membrane. When the electrical field was turned off, the particles returned to a random distribution within seconds, clearly demonstrating their freedom to diffuse in a fluid lipid bilayer.

The results of this and similar studies make it clear that the respiratory complexes are not lined up in the membrane in the orderly fashion often seen in textbook illustrations but exist in the membrane as mobile complexes. In fact, the inner mitochondrial membrane has a high ratio of unsaturated to saturated phospholipids and virtually no cholesterol, so its fluidity is very high, and the mobility of the respiratory complexes is correspondingly high.

Recent work suggests that these multiprotein respiratory complexes themselves are organized into supercomplexes, known as **respirasomes,** containing several individual respiratory complexes associated in defined ratios. The association of several of the citric acid cycle dehydrogenases with these respirasomes suggests that they function to minimize diffusion distances, facilitating electron flow between the respiratory complexes. They may have other regulatory functions as well. Respirasomes have been found in systems as diverse as bacteria, yeast, plants, and humans.

As Figure 10-18 indicates, NADH, coenzyme Q, and cytochrome *c* are key intermediates in the electron transfer process. NADH links the ETC to the dehydrogenation (oxidation) reactions of the citric acid cycle and to most other oxidation reactions in the matrix of the mitochondrion. Coenzyme Q and cytochrome *c* transfer electrons between the respiratory complexes. Coenzyme Q accepts electrons from both complexes I and II and is, in fact, the "funnel" that collects electrons from virtually every oxidation reaction in the cell. Coenzyme Q and cytochrome *c* are both relatively small molecules that can diffuse rapidly, either within the membrane (coenzyme Q) or on the membrane surface (cytochrome *c*). They are also quite numerous— there are about 10 molecules of cytochrome *c* and 50 molecules of coenzyme Q for every complex I. Because of their abundance and mobility, these carriers are able to transfer electrons between the major complexes frequently enough to account for the observed rates of electron transfer in actively respiring mitochondria.

CONCEPT CHECK 10.4

Why are the electron carriers in the ETC arranged in the respiratory complexes of the mitochondrial inner membrane in a very specific orientation?

10.5 The Electrochemical Proton Gradient: Key to Energy Coupling

So far, you have learned that coenzymes are reduced during the oxidative events of glycolysis, pyruvate oxidation, and the citric acid cycle (stages ❶, ❷, and ❸ of aerobic respiration; see Figure 10-1). You have also learned that reduced coenzymes are reoxidized by the exergonic transfer of electrons to oxygen via a system of reversibly oxidizable intermediates located within the inner mitochondrial membrane (stage ❹). Now we will consider how the free energy released during electron transport is used to generate an electrochemical proton gradient and how the energy of the gradient is then used to drive ATP synthesis in stage ❺.

Because this means of ATP synthesis involves phosphorylation events that are linked to oxygen-dependent electron transport, the process is called **oxidative phosphorylation**, thereby distinguishing it from *substrate-level phosphorylation*, which occurs as an integral part of a specific reaction in a metabolic pathway. (Reactions Gly-7 and Gly-10 of the glycolytic pathway and reaction CAC-5 of the citric acid cycle are examples of substrate-level phosphorylation; see Figure 9-7 and Figure 10-9.)

Electron Transport and ATP Synthesis Are Coupled Events

Mechanistically, oxidative phosphorylation is more complex than substrate-level phosphorylation. In contrast to direct transfer of phosphates to ADP to generate ATP, a key feature of oxidative phosphorylation is the crucial link between electron transport and ATP production mediated by an **electrochemical proton gradient** that is established by the directional pumping of protons across membranes in which electron transport is occurring.

Under normal cellular conditions, ATP synthesis is *coupled* to electron transport, meaning not only that ATP synthesis depends on electron flow but also that electron flow is possible only when ATP can be synthesized. However, treating isolated mitochondria with certain chemicals—known as *uncouplers*—abolishes the interdependence of these two processes. In other words, uncouplers allow continued electron transport (and O_2 consumption) even in the absence of ATP synthesis. In contrast, addition of respiratory complex inhibitors that stop electron flow also stops ATP synthesis. Therefore, ATP synthesis is strictly dependent on electron transport, but electron transport is not necessarily dependent on ATP synthesis. In brown adipose (fat) tissue in the bodies of newborn mammals, a similar uncoupling mechanism operates, using exergonic electron transport to generate heat rather than ATP (see Human Connections, pages 270–271).

Respiratory Control of Electron Transport. Because electron transport is coupled to ATP synthesis, the availability of ADP regulates the rate of oxidative phosphorylation and therefore of electron transport. This is called **respiratory control,** and its physiological significance is easy to appreciate: electron transport and ATP generation will be favored when the ADP concentration is high (that is, when the ATP concentration is low) and inhibited when the ADP concentration is low (when the ATP concentration is high). Oxidative phosphorylation is therefore regulated by cellular ATP needs, such that electron flow from organic fuel molecules to oxygen is adjusted to the energy needs of the cell. This regulatory mechanism becomes apparent during exercise, when the accumulation of ADP in muscle tissue causes an increase in electron transport rates, followed by a dramatic rise in the need for oxygen.

MAKE CONNECTIONS 10.1

Based on what you now know about electron transport and the molecules involved, why is electron transport crucial for allowing normal aerobic respiration instead of fermentation? (Ch. 9.5)

The Chemiosmotic Model: The "Missing Link" Is a Proton Gradient. How can ATP synthesis, a dehydration reaction, be tightly coupled to electron transport, which involves the sequential oxidation and reduction of various protein complexes in a lipid bilayer? Scientists puzzled over this question, but the answer eluded their best efforts for several decades. Most were convinced that some as-yet-undiscovered high-energy intermediate must be involved in oxidative phosphorylation just as in substrate-level phosphorylation. (Recall the ATP-generating steps of the glycolytic pathway and the citric acid cycle in reactions Gly-7 and Gly-10 in Figure 9-7 and reaction CAC-5 in Figure 10-9.)

But whereas others continued their feverish search for high-energy intermediates, the British biochemist Peter Mitchell made the revolutionary suggestion that such intermediates might not exist at all. In 1961, Mitchell proposed an alternative explanation, which he called the **chemiosmotic coupling model.** According to this model, the exergonic transfer of electrons through the respiratory complexes is accompanied by the unidirectional pumping of protons across the membrane in which the transport system is localized. The electrochemical proton gradient that is generated in this way represents potential energy that then provides the driving force for ATP synthesis. In other words, the "missing link" between electron transport and ATP synthesis was not a high-energy chemical intermediate but an electrochemical proton gradient. This mechanism is illustrated in Figure 10-19, which depicts both the outward pumping of protons that accompanies electron transport through complexes I, III, and IV and the proton-driven synthesis of ATP by the F_oF_1 complex.

As you might expect, Mitchell's theory met with considerable initial skepticism and resistance. Not only was it a radical departure from conventional wisdom on coupling, but Mitchell proposed it in the virtual absence of experimental data. Over the years, however, a large body of evidence has been amassed in its support, and in 1978 Mitchell was awarded a Nobel Prize for his pioneering work. Today the chemiosmotic coupling model is a well-verified concept that provides a unifying framework for understanding energy transformations not just in mitochondrial membranes but in chloroplast and bacterial membranes as well.

The essential feature of the chemiosmotic model is that *the link between electron transport and ATP formation is an electrochemical potential across a membrane.* It is, in fact, this feature that gives the model its name: the "chemi" part of the term refers to the chemical processes of oxidation and electron transfer, and the "osmotic" part comes from the Greek word *osmos,* meaning "to push"—to push protons across the membrane, in this case. The chemiosmotic model has turned out to be exceptionally useful not only because of the very plausible explanation it provides for coupled ATP generation but also because of its pervasive influence on the way we now think about energy conservation in biological systems.

Coenzyme Oxidation Pumps Enough Protons to Form Three ATP Molecules per NADH and Two ATP Molecules per $FADH_2$

Before considering the experimental evidence that has confirmed the chemiosmotic model, notice that Figure 10-19 provides us with estimates of the numbers of protons pumped outward by the three respiratory complexes involved in NADH oxidation, as well as the number of protons required to drive the synthesis of ATP by the F_oF_1 complex. The word *estimates* is important here because investigators still disagree on some of these numbers. This disagreement arises partly due to a structural feature that varies in different ATP synthase complexes. As we will see shortly, the number of protons required to produce three ATP molecules is directly related to the number of *c* subunits in the F_o portion of the F_oF_1 ATP synthase, and this number can differ among organisms.

As we have already seen, most of the dehydrogenases in the mitochondrial matrix transfer electrons from oxidizable substrates to NAD^+, generating NADH. NADH, in turn, transfers electrons to the FMN component of complex I, thereby initiating the electron transport chain. As Figure 10-19 shows, transfer of two electrons from NADH down the respiratory chain to oxygen is accompanied by the transmembrane pumping of four protons by complex I, four protons by CoQ plus complex III, and two protons by complex IV. This gives a total of ten protons per NADH.

The number of protons required to drive the synthesis of one molecule of ATP by the F_oF_1 ATP synthase is thought to be three or four, with three generally regarded as more likely. If we assume that ten protons are pumped per NADH oxidized and that three protons are required per ATP molecule, then we can conclude that oxidative phosphorylation yields about three molecules of ATP synthesized per molecule of NADH oxidized. This agrees with evidence dating back to the 1940s that a fixed relationship usually exists between the number of molecules of ATP generated and the number of oxygen atoms consumed in respiration as a pair of electrons passes along

HUMAN *Connections*

A Diet Worth Dying For?

Most people have been on a diet at some point in their life. Diet plans come and go, each one claiming to help you lose weight and keep it off. What about a diet that promised to help you lose weight and keep you warm at the same time? Throughout the mid-1930s, a compound called *dinitrophenol* (more precisely, *2,4-dinitrophenol*), or *DNP*, was prescribed to patients trying to lose weight. Indeed, these patients did lose pounds—but they also experienced rapid heartbeat, elevated body temperature, and sweating. In some cases, patients taking DNP died.

DNP is effective as a weight loss tool because it exploits the body's need for ATP. After reading this chapter, you are keenly aware that most of the ATP produced during cellular respiration is generated through oxidative phosphorylation in mitochondria. But what would happen if the efficiency of ATP production decreased?

Recall that the end-products of metabolic pathways can regulate these pathways (Figure 9-13). When the body requires more ATP, it can liberate glucose from glycogen stores in the liver, and if necessary, it can generate reducing power via β oxidation of fatty acids (Figure 10-12). Imagine a diet that could fool the body into thinking that although it was performing oxidative phosphorylation, it was nevertheless continually running short of ATP. The body would continuously seek to mobilize more glucose from glycogen and "burn" fats to supply the NADH and $FADH_2$ necessary for additional electron transport. DNP accomplishes this trick, but at a tremendous cost: it reduces the efficiency of ATP generation at the inner mitochondrial membrane. DNP is an *uncoupling agent*: it short-circuits the establishment of the proton gradient across the inner mitochondrial membrane. This in turn decreases the number of protons moving through ATP synthases in the inner mitochondrial membrane, thereby reducing the overall rate of ATP production within the cell.

How does DNP disrupt the mitochondrial proton gradient? DNP is a lipophilic, weak acid (**Figure 10B-1**), and it is able to cross membranes in both its protonated and unprotonated forms. By carrying a proton from an area of lower pH to one of higher pH, DNP acts to reduce differences in proton concentration between the matrix and intermembrane space. As a result, the proton motive force declines, and the production of ATP in mitochondria slows.

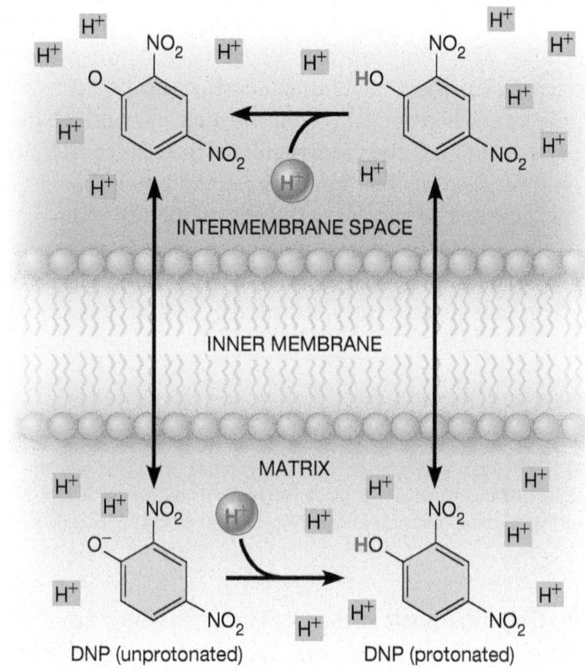

Figure 10B-1 Dinitrophenol (DNP) Can Shuttle Protons Across Membranes. DNP can acquire a proton, pass through a membrane, and then release its proton, effectively acting as a proton shuttle. In the case of the mitochondrial inner membrane, this proton shuttling circumvents the normal path of protons through ATP synthase.

Some of the energy normally devoted to driving ATP formation is invariably lost as heat in mitochondria exposed to DNP. Ingesting DNP distributes the uncoupler to every cell of the body, which explains how dangerous DNP can be as a diet drug. If body temperature increases too much, a person taking DNP can die of hyperthermia.

DNP is a chemical uncoupler, but naturally occurring molecules exist that act in the same way in certain mitochondria.

the sequence of carriers from reduced coenzyme to oxygen. Efraim Racker and his colleagues, for example, showed that synthetic phospholipid vesicles containing complex I, complex III, or complex IV could generate one molecule of ATP per pair of electrons that passed through the complex. These results are consistent with our current understanding that each of these three complexes serves as a proton pump, thereby contributing to the proton gradient that drives ATP synthesis.

For purposes of our discussion, we will assume the ATP yields to be three molecules for NADH and two molecules for $FADH_2$, recognizing that these values may be somewhat imprecise. These are reasonable numbers, considering that the $\Delta G^{\circ\prime}$ for the oxidation of NADH by molecular oxygen is -52.4 kcal/mol. This is enough of a free energy change to produce three ATP molecules (requiring about 10 kcal/mol each under cellular conditions), even assuming an efficiency of only 60%.

The Chemiosmotic Model Is Affirmed by an Impressive Array of Evidence

Since its initial formulation in 1961, the chemiosmotic model has come to be universally accepted as the link between electron transport and ATP synthesis. To understand why, we will first consider several of the most important lines of experimental evidence that support this model. In the process, we will also learn more about the mechanism of chemiosmotic coupling.

1. Electron Transport Causes Protons to Be Pumped out of the Mitochondrial Matrix. Shortly after proposing the chemiosmotic model, Mitchell and his colleague Jennifer Moyle demonstrated experimentally that the flow of electrons through the ETC is accompanied by the movement of protons across the inner mitochondrial membrane. They first suspended mitochondria in a medium in which

Thermogenin, also known as *UCP1* (for *uncoupling protein 1*) is a protein found in the mitochondria of brown fat tissue. Like DNP, it allows protons to flow back to the matrix without passing through an ATP synthase, dissipating the proton motive force, and generating heat (**Figure 10B-2**). Brown fat is typically found in infants on the upper half of the back up to the neck and is thought to help them generate heat because they lose heat more quickly than adults. Small remnant patches of brown fat also exist in adults, mainly at the back of the neck.

Despite the dangers of uncoupling agents, DNP and similar drugs continue to be used for rapid weight loss, for example, by body builders. Even though they cause several deaths each year, for some, the dramatic effects of DNP are worth the risks.

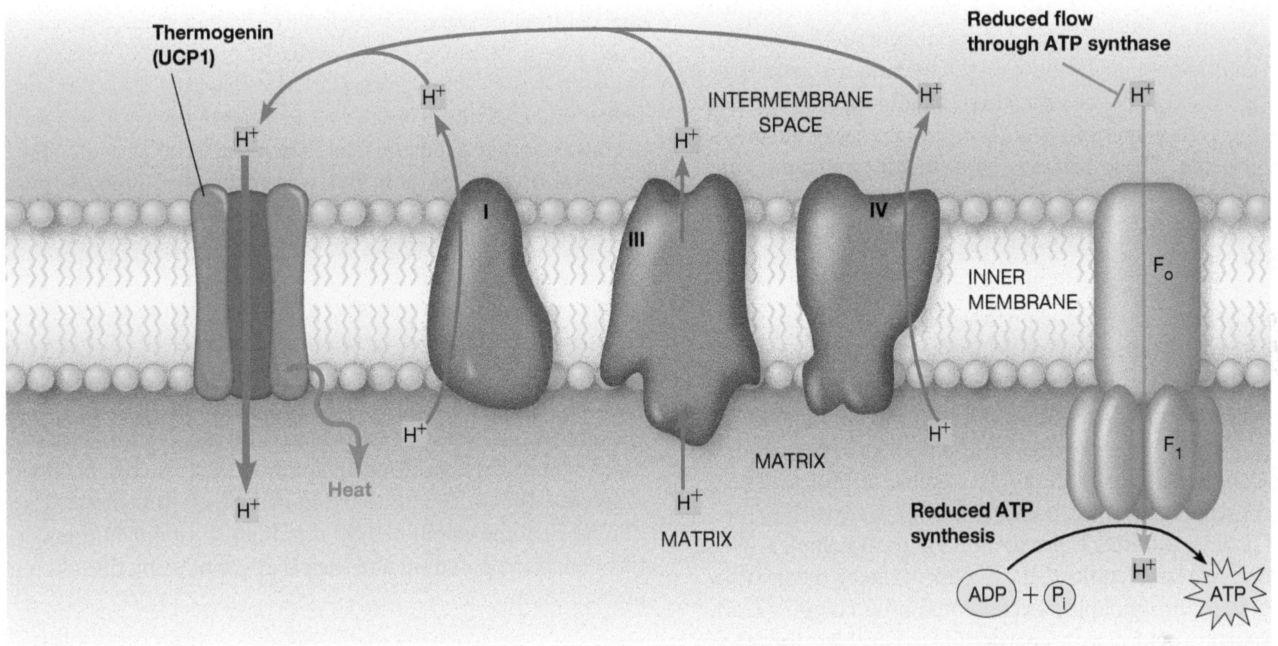

Figure 10B-2 Thermogenin (UCP1) Uncouples Oxidation from ATP Synthesis. When thermogenin is present, protons can pass directly through the inner mitochondrial membrane without passing through an ATP synthase. The result is reduced ATP synthesis, both because the proton gradient is reduced and because fewer protons pass through ATP synthase. Note: only the overall flow of protons, but not their precise numbers, are indicated.

electron transfer could not occur because oxygen was lacking. The proton concentration (that is, the pH) of the medium was then monitored as electron transfer was stimulated by the addition of oxygen. Under these conditions, the pH of the medium declined rapidly with either NADH or succinate as the electron source (**Figure 10-21**). Because a decline in pH reflects an increase in proton concentration, Mitchell and Moyle concluded that electron transfer within the mitochondrial inner membrane is accompanied by *unidirectional pumping of protons* from the mitochondrial matrix into the external medium.

Throughout our discussion, we have been using the term "pumping" to refer to the movement of protons across the inner mitochondrial membrane against a concentration gradient. However, it is important to recognize that the movement of protons is not as direct as what occurs during active transport of ions via transporters (as

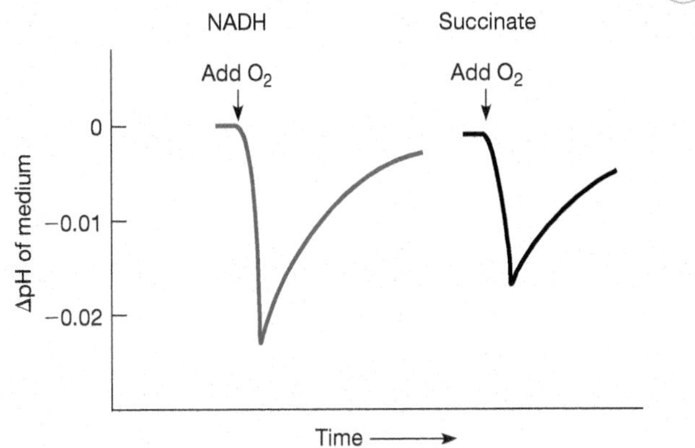

Figure 10-21 Experimental Evidence That Electron Transport Generates a Proton Gradient. Isolated mitochondria were incubated with either NADH or succinate as an electron source in the absence of oxygen. When electron transport was stimulated by the addition of oxygen, there was a rapid decline in pH in the external medium, indicating that protons were pumped out of the mitochondrial matrix when oxygen was available.

discussed in Chapter 8). In fact, the mechanism whereby the transfer of electrons from one carrier to another is coupled to the directional transport of protons is still not well understood, and there may be more than one such mechanism.

2. Components of the Electron Transport Chain Are Asymmetrically Oriented Within the Inner Mitochondrial Membrane. The unidirectional pumping demonstrated by Mitchell and Moyle requires that the electron carriers making up the ETC be asymmetrically oriented within the membrane. If not, protons would presumably be pumped randomly in both directions. Studies with a variety of antibodies, enzymes, and labeling agents designed to interact with various membrane components have shown clearly that some constituents of the respiratory complexes face the matrix side of the inner membrane, others are exposed on the opposite side, and some are transmembrane proteins. These findings confirm the prediction that ETC components are asymmetrically distributed across the inner mitochondrial membrane, as shown schematically in Figure 10-19.

3. Membrane Vesicles Containing Complexes I, III, or IV Establish Proton Gradients. Further support for the chemiosmotic model has come from experiments involving the reconstitution of membrane vesicles from mixtures of isolated components. The chemiosmotic model predicts that each of these complexes should be capable of pumping protons across the inner mitochondrial membrane, which in turn should lead to some ATP production via each complex. This prediction has been confirmed experimentally by reconstituting artificial phospholipid vesicles containing complex I, III, or IV. When provided with appropriate oxidizable substrates, each of these three complexes is able to pump protons across the membrane of the vesicle. As noted before, these vesicles are also capable of ATP synthesis.

4. Oxidative Phosphorylation Requires a Membrane-Enclosed Compartment. An obvious prediction of the chemiosmotic model is that oxidative phosphorylation occurs only within a compartment enclosed by an intact mitochondrial membrane. Otherwise, the proton gradient that drives ATP synthesis could not be maintained. This prediction has been verified by experimental demonstration that electron transfer carried out by isolated complexes cannot be coupled to ATP synthesis unless the complexes are incorporated into membranes that form enclosed, intact vesicles.

5. Uncoupling Agents Abolish Both the Proton Gradient and ATP Synthesis. Additional evidence for the role played by proton gradients in ATP formation has come from studies using agents such as *dinitrophenol (DNP)* that are known to uncouple ATP synthesis from electron transport. For example, Mitchell showed in 1963 that membranes become freely permeable to protons in the presence of DNP. Dinitrophenol, in other words, abolished *both* ATP synthesis and the capability of a membrane to maintain a proton gradient. This finding is clearly consistent with the concept of ATP synthesis driven by an electrochemical proton gradient.

Because DNP uncouples ATP synthesis from electron transport, it reduces the efficiency of ATP synthesis, causing cells to catabolize oxidizable fuels at a higher rate to maintain adequate ATP levels. For this reason, DNP has been used as a weight loss aid, but this can be dangerous for many reasons, including the excessive amount of heat released as a by-product of the accelerated catabolic reactions (see Human Connections, pages 270–271).

6. The Proton Gradient Has Enough Energy to Drive ATP Synthesis. For the chemiosmotic model to be viable, the proton gradient generated by electron transport must store enough energy to drive ATP synthesis. We can address this point with a few thermodynamic calculations. The electrochemical proton gradient across the inner membrane of a metabolically active mitochondrion involves both a membrane potential (the "electro" component of the gradient) and a concentration gradient (the "chemical" component). The equation used to quantify this electrochemical gradient has two terms, one for the membrane potential, V_m, and the other for the concentration gradient, which in the case of protons is a pH gradient.

A mitochondrion actively involved in aerobic respiration typically has a membrane potential of about +0.16 V (positive on the side that faces the intermembrane space) and a pH gradient of about 1.0 pH unit (higher on the matrix side). This electrochemical gradient exerts a **proton motive force (pmf)** that tends to drive protons back down their electrochemical gradient—that is, back into the matrix of the mitochondrion. The pmf can be calculated by summing the contributions of the membrane potential and the pH gradient using the following equation:

$$\text{pmf} = V_m + \frac{2.303RT\Delta\text{pH}}{F} \qquad \textbf{(10-16)}$$

where pmf is the proton motive force in volts, V_m is the membrane potential in volts, ΔpH is the difference in pH across the membrane ($\Delta\text{pH} = \text{pH}_{matrix} - \text{pH}_{cytosol}$), R is the gas constant (1.987 cal/mol-K), T is the temperature in kelvins, and F is the Faraday constant (23,062 cal/mol-V).

For a mitochondrion at 37°C with a V_m of 0.16 V and a pH gradient of 1.0 unit, the pmf can be calculated as follows:

$$\begin{aligned}\text{pmf} &= 0.16 + \frac{(2.303)(1.987)(37+273)(1.0)}{23{,}062} \qquad \textbf{(10-17)}\\ &= 0.16 + 0.06 = 0.22\text{V}\end{aligned}$$

Notice that the membrane potential accounts for more than 70% of the mitochondrial pmf.

Like the redox potential, pmf is an electrical force in volts and can be used to calculate $\Delta G^{\circ\prime}$, the standard free energy change for the movement of protons across the membrane, using the following equation:

$$\begin{aligned}\Delta G^{\circ\prime} &= -nF(\text{pmf}) = -(23.062)(0.22)\\ &= -5.1\ \text{kcal/mol} \qquad \textbf{(10-18)}\end{aligned}$$

As in Equation 10-11, n is the number of electrons transferred, which in this case is 1, and F is the Faraday constant. Thus, a proton motive force of 0.22 V across the inner mitochondrial membrane corresponds to a free energy change of about −5.1 kcal/mol of protons. This is the amount of energy that will be released as protons return to the matrix.

Is this enough to drive ATP synthesis? Not surprisingly, the answer is yes, though it depends on how many protons are required to drive the synthesis of one ATP molecule by the F_oF_1 complex. Mitchell's original model assumed two protons per ATP, which would provide about 10.2 kcal/mol, barely enough to drive ATP formation, assuming the $\Delta G'$ for phosphorylation of ADP to be about 10 to 14 kcal/mol under mitochondrial conditions. Differences of opinion still exist concerning the number of protons per ATP, but the real number is probably closer to three or four. That number would provide about 15 to 20 kcal of energy per ATP, enough to ensure that the reaction is driven strongly in the direction of ATP formation.

7. Artificial Proton Gradients Are Able to Drive ATP Synthesis in the Absence of Electron Transport. Direct evidence that a proton gradient can in fact drive ATP synthesis has been obtained by exposing mitochondria or inner membrane vesicles to artificial pH gradients. When mitochondria are suspended in a solution in which the external proton concentration is suddenly increased by the addition of acid, ATP is generated in response to the artificially created proton gradient. Because such artificial pH gradients induce ATP formation even in the absence of oxidizable substrates that would otherwise pass electrons to the ETC, it is evident that ATP synthesis can be induced, at least transiently, by a proton gradient even in the absence of electron transport.

CONCEPT CHECK 10.5

How is the chemical energy that is released as electrons travel through the ETC stored so that it can be used for ATP synthesis?

10.6 ATP Synthesis: Putting It All Together

We are now ready to take up the fifth, and final, stage of aerobic respiration: ATP synthesis. We have thus far seen that (1) some of the energy of glucose is transferred to reduced coenzymes during the oxidation reactions of glycolysis and the citric acid cycle and that (2) this energy is used to generate an electrochemical proton gradient across the inner membrane of the mitochondrion. Now we can ask how the pmf of that gradient is harnessed to drive ATP synthesis. For that, we return to the F_1 complexes that can be seen along the inner surfaces of cristae (see Figure 10-5a) and ask about the evidence that these particles are capable of synthesizing ATP.

F_1 Particles Have ATP Synthase Activity

Key evidence concerning the role of F_1 particles came from studies conducted by Efraim Racker and his colleagues to test the prediction of the chemiosmotic hypothesis: that a reversible, proton-translocating ATPase is present in membranes that are capable of coupled ATP synthesis. Beginning with intact mitochondria (**Figure 10-22a**), the investigators were able to disrupt the mitochondria in such a way that fragments of the inner membrane formed small vesicles, which they called *submitochondrial particles* (Figure 10-22b). These submitochondrial particles were capable of carrying out electron transport and ATP synthesis. By subjecting these particles to mechanical

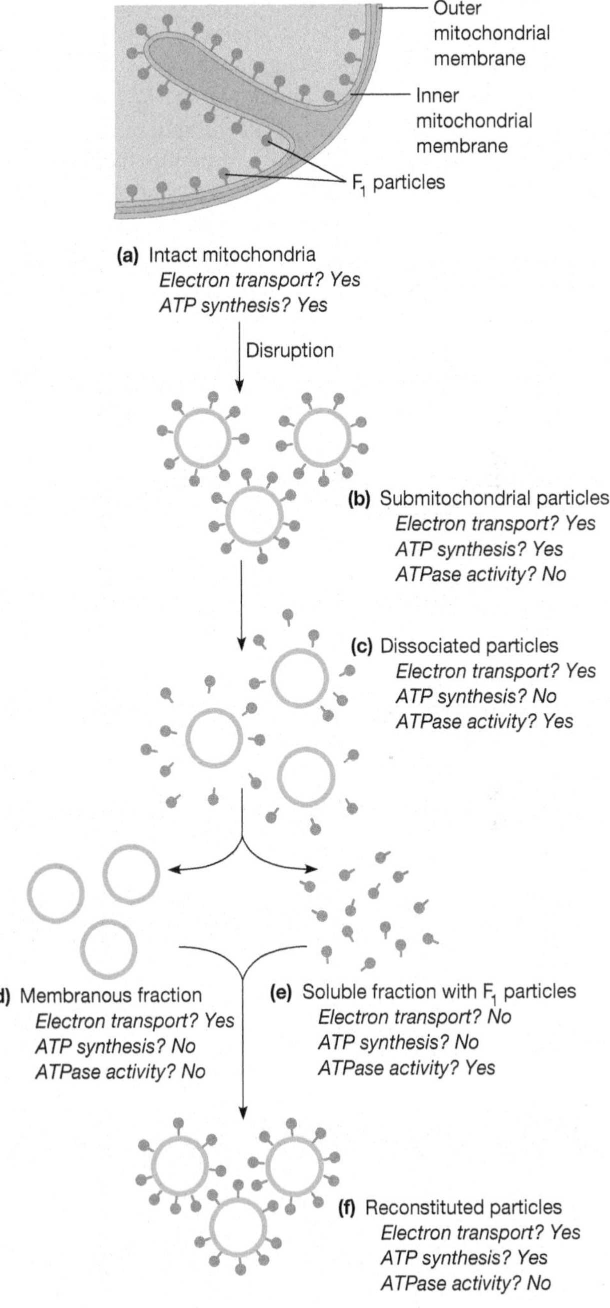

Figure 10-22 Dissociation and Reconstitution of the Mitochondrial ATP-Synthesizing System. (a) Intact mitochondria were disrupted so that fragments of the inner membrane formed **(b)** submitochondrial particles, capable of both electron transport and ATP synthesis. **(c)** When these particles were dissociated by mechanical agitation or enzyme treatment, the components could be separated into **(d)** a membranous fraction devoid of ATP-synthesizing capacity and **(e)** a soluble fraction with F_1 particles having ATPase activity. **(f)** Mixing the two fractions reconstituted the structure and restored ATP synthase activity.

agitation or protease treatment, they were able to dislodge the F_1 structures from the membranous vesicles (Figure 10-22c).

When the F_1 particles and the membranous vesicles were separated from each other by centrifugation, the membranous fraction could still carry out electron transport but could no longer synthesize ATP; the two functions had become uncoupled (Figure 10-22d). The isolated F_1 particles, on the other hand, were not capable of either electron transport or ATP synthesis but had ATPase activity (Figure 10-22e), a property consistent with the F-type ATPase that we now know the mitochondrial F_oF_1 complex to be (see Table 8-2, page 203). The ATP-generating capability of the membranous fraction was restored by adding the F_1 particles back to the membranes, suggesting that the spherical projections seen on the inner surface of the inner mitochondrial membrane are an important part of the ATP-generating complex of the membrane. These F_1 particles were therefore referred to as *coupling factors* (the "F" in F_1) and are now known to be the structures responsible for the ATP-synthesizing activity of the inner mitochondrial membrane or the bacterial plasma membrane.

Why in these experiments does an isolated F_1 fragment catalyze the hydrolysis of ATP (that is, act as an ATPase) when the intact complex synthesizes ATP? For a given enzyme, whether the reaction favors products or reactants at equilibrium depends on the concentration of products versus reactants and whether the reaction is coupled to other reactions occurring at the same time (see Chapter 6). An isolated F_1 subunit in which enzymatic activity is not coupled to movement of protons favors ATP hydrolysis when ATP is plentiful.

In fact, there are situations in which the entire, intact F_oF_1 complex can run in reverse as well. In anaerobic bacteria, which produce ATP by fermentation, the ATP synthase works in reverse: it uses ATP to generate a proton gradient, which the bacterium uses for ion transport and for moving its swimming appendage (flagellum). To understand how a proton gradient is normally coupled to ATP synthesis, we need to turn to a detailed look at the F_o and F_1 subunits.

MAKE CONNECTIONS 10.2

The enzyme that normally catalyzes formation of ATP during oxidative phosphorylation is called ATP synthase, but it can also catalyze the hydrolysis of ATP. Based on what you know about enzymes, is this surprising? Why or why not? (Ch. 6)

Proton Translocation Through F_o Drives ATP Synthesis by F_1

Although the F_1 portion of the F_oF_1 ATP synthase complex is not directly membrane bound, it is attached to the F_o complex that is embedded in the inner mitochondrial membrane (see Figure 10-5c). We now know that the F_o complex serves as the **proton translocator**, the channel through which protons flow across the mitochondrial inner membrane (or the bacterial plasma membrane, as in the case of the *Escherichia coli* complex discussed below). Thus, the F_oF_1 complex is the complete, functional **ATP synthase**. The F_o component provides a channel for the exergonic flow of protons through the membrane, and the F_1 component carries out the actual synthesis of ATP, driven by the energy of the proton gradient.

Table 10-4 presents the polypeptide composition of the bacterial F_oF_1 ATP synthase complex from *E. coli*. **Figure 10-23a** illustrates the four major structural components in this functional complex, and **Figure 10-23b** shows its three-dimensional structure. The F_o and F_1 complexes each contain both a static component that remains stationary and a mobile component that moves during proton translocation. Together, these components of the F_oF_1 complex form one of the wonders of the molecular world—a miniature motor in which proton flow turns a microscopic gear that drives ATP synthesis.

The F_o complex is embedded in the bacterial membrane (or the inner mitochondrial membrane in eukaryotic cells) and consists of one *a* subunit, two *b* subunits, and ten *c* subunits. The *a* and *b* subunits form the static component that is immobilized in the membrane. The ten *c* subunits form a ring that acts as a miniature gear that can rotate in the membrane relative to the *a* and *b* subunits. The *a* subunit functions as the

Table 10-4 Polypeptide Composition of the *E. coli* F_oF_1 ATP Synthase (ATPase)*

Structure	Polypeptide	Molecular Weight	Number Present	Function
F_o	*a*	30,000	1	Proton channel
	b_2	17,000	2	Peripheral stator stalk connecting F_o and F_1
	c	8000	10	Rotating ring that turns γ subunit of F_1
F_1	α	52,000	3	Promotes activity of β subunit
	β	55,000	3	Catalytic site for ATP synthesis
	δ	19,000	1	Anchors $\alpha_3\beta_3$ ring to stator stalk of F_o
	γ	31,000	1	Rotates to transmit energy from F_o to F_1
	ε	15,000	1	Anchors γ subunit to c_{10} ring of F_o

*Mitochondrial F_oF_1 complexes are similar to the bacterial complex but with somewhat different polypeptide compositions for F_o and F_1.

11 Phototrophic Energy Metabolism: Photosynthesis

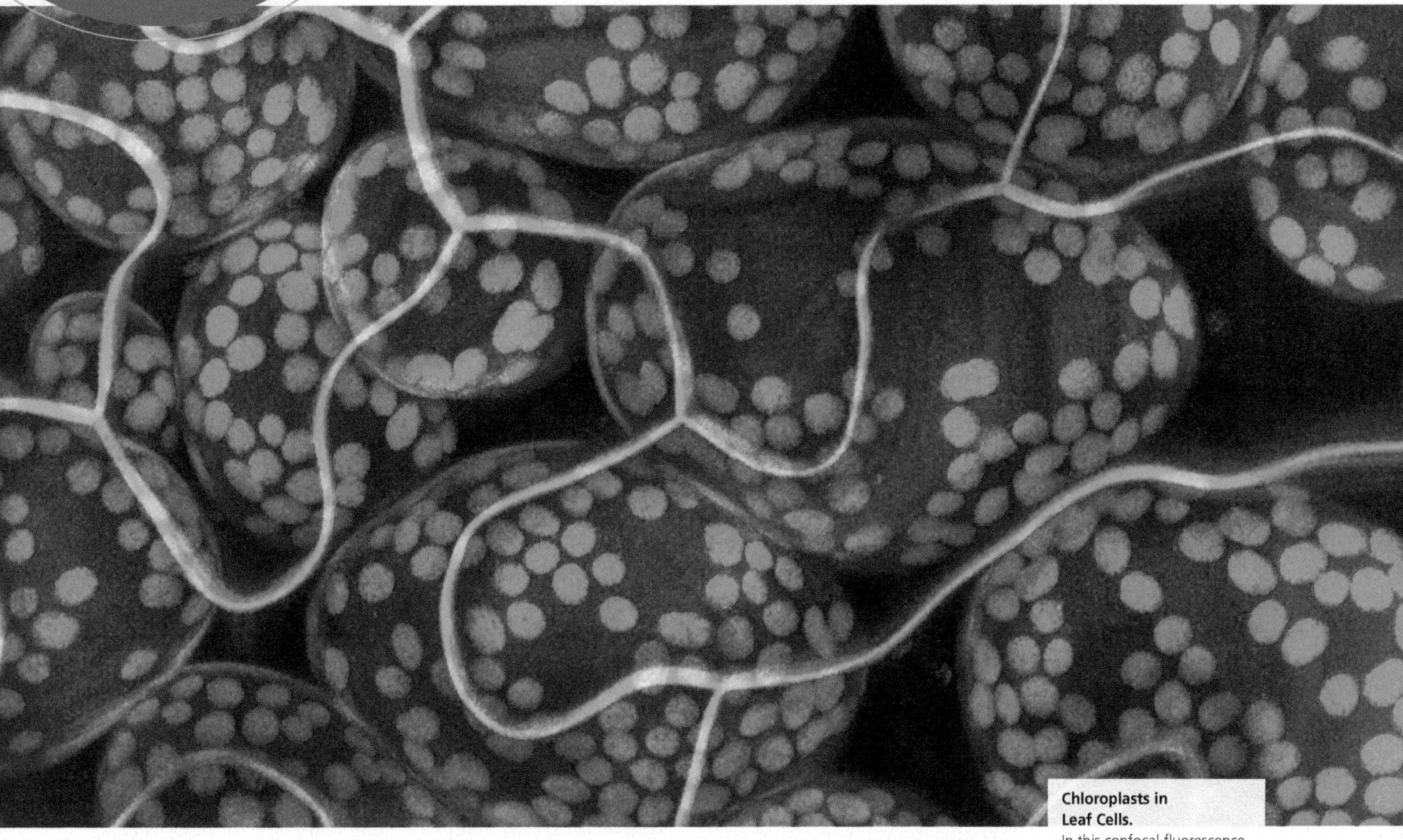

Chloroplasts in Leaf Cells.
In this confocal fluorescence micrograph, numerous red autofluorescing chloroplasts are visible inside oval mesophyll cells. The wavy structures above them are a layer of epidermal cells. All cell walls have been labeled with a blue fluorescent dye.

In the two preceding chapters, we studied fermentation and cellular respiration, which are two chemotrophic solutions to the universal challenge of meeting the energy and carbon needs of a living cell. Most chemotrophs (humans included) depend on an external source of organic substrates for survival. As you may recall, chemotrophs obtain energy by oxidizing high-energy, reduced compounds such as carbohydrates, fats, and proteins in catabolic pathways. They can also use intermediates of these oxidative pathways as sources of carbon skeletons for biosynthesis via anabolic pathways. However, left alone, chemotrophs would soon perish without an ongoing supply of these reduced organic compounds.

In this chapter, we will learn how photosynthetic organisms produce the chemical energy and organic carbon that are drained from the biosphere by chemotrophs. These photosynthetic organisms use solar energy to drive the reduction of CO_2 to produce carbohydrates, fats, and proteins—the reduced forms of carbon that all chemotrophs depend upon.

The use of solar energy to drive the anabolic pathways that produce these building blocks of life is aptly named **photosynthesis**—the conversion of light energy to chemical energy and its subsequent use in synthesizing organic molecules. Nearly all life on Earth is sustained by the energy that arrives at the planet as sunlight. **Phototrophs** are organisms that convert solar energy to chemical energy in the form of ATP and the reduced coenzyme NADPH. NADPH is the electron and hydrogen carrier for a large number of anabolic pathways and is closely related to NADH, which we have previously encountered in catabolic pathways (Chapter 9).

Some phototrophs (such as the halobacteria described in Chapter 7) are **photoheterotrophs,** organisms that acquire energy from sunlight but depend on organic sources of reduced carbon. Most other phototrophs—including plants, algae, and most photosynthetic bacteria—are known as **photoautotrophs,** organisms that use solar energy to synthesize energy-rich organic molecules from simple inorganic starting materials such as carbon dioxide and water. Many photoautotrophs release molecular oxygen as a by-product of photosynthesis. Thus, phototrophs not only replenish reduced carbon in the biosphere but also provide the oxygen in the atmosphere used by aerobic organisms to oxidize these reduced compounds for energy.

In this chapter, we will study two general aspects of photosynthesis: how photoautotrophs capture solar energy and convert it to chemical energy and how this energy is used to transform energy-poor carbon dioxide and water into energy-rich organic molecules such as carbohydrates, fats, and proteins.

11.1 An Overview of Photosynthesis

Photosynthesis involves two major biochemical processes: **energy transduction** and **carbon assimilation** (**Figure 11-1**). During the *energy transduction reactions,* light energy is captured by chlorophyll molecules and converted to chemical energy in the form of ATP and NADPH. The ATP and NADPH generated by the energy transduction reactions subsequently provide energy (ATP) and reducing power (NADPH) for the carbon assimilation reactions, which synthesize carbohydrates from CO_2. During the *carbon assimilation reactions,* commonly known as the *Calvin cycle,* carbon atoms from carbon dioxide are "fixed"—covalently joined to organic compounds, forming carbohydrates. Thus, these reactions are often referred to as *carbon fixation reactions.* Both the energy transduction reactions and the carbon assimilation reactions occur in chloroplasts of eukaryotic phototrophs, such as plants and algae, or in specialized proteinaceous membrane systems and cytosolic microcompartments of photosynthetic bacteria.

The Energy Transduction Reactions Convert Solar Energy to Chemical Energy

Light energy from the sun is captured by a variety of green pigment molecules called *chlorophylls,* which are present in the green leaves of plants and in the cells of algae and photosynthetic bacteria. Light absorption by a chlorophyll molecule excites one of its electrons, which is then ejected from the molecule and flows energetically downhill through an electron transport chain (ETC) much like the ETC we saw previously in mitochondria. As in mitochondria, this flow of electrons is coupled to unidirectional proton pumping, which first stores energy in an electrochemical proton gradient that subsequently drives an ATP synthase. As you may recall, in mitochondria this process is known as *oxidative phosphorylation*—ATP synthesis driven by energy derived from the oxidation of organic compounds. In photosynthetic organisms, ATP synthesis driven by energy derived from the sun is called *photophosphorylation.*

To incorporate fully oxidized carbon atoms from carbon dioxide into organic molecules, these carbon atoms must be reduced. Therefore, photoautotrophs need not only energy in the form of ATP but also reducing power, which is provided by NADPH, a coenzyme related to NADH (see Figure 9-5). In **oxygenic phototrophs**—plants, algae, and cyanobacteria—water is the electron donor, and light energy absorbed by chlorophyll powers the movement of two electrons from water to $NADP^+$, which is reduced to NADPH. Molecular oxygen is released as water is oxidized. In **anoxygenic phototrophs**—green and purple photosynthetic bacteria—compounds such as sulfide (S^{2-}), thiosulfate ($S_2O_3^{2-}$), or succinate serve as electron donors, and oxidized forms of these compounds are released. In both oxygenic and anoxygenic phototrophs, the light-dependent generation of NADPH is called *photoreduction.*

Usually, it is the oxygenic phototrophs such as trees, grasses, and other common plants that come to mind when we think of photosynthesis. However, it was work with the anoxygenic green and purple photosynthetic bacteria conducted by C. B. van Niel in the 1930s and 1940s that led to our understanding that photosynthesis is a light-driven oxidation-reduction process.

The Carbon Assimilation Reactions Fix Carbon by Reducing Carbon Dioxide

Most of the energy accumulated within photosynthetic cells by the light-dependent generation of ATP and NADPH is used for carbon dioxide fixation and reduction. A general reaction for the complete process of photosynthesis may be written as

$$\text{light} + CO_2 + 2H_2A \rightarrow [CH_2O] + 2A + H_2O \qquad \textbf{(11-1)}$$

where H_2A is a suitable *electron donor,* $[CH_2O]$ represents a carbohydrate, and A is the oxidized form of the electron donor. By expressing photosynthesis in this way, we avoid perpetuating the incorrect notion that all phototrophs use water as an electron donor.

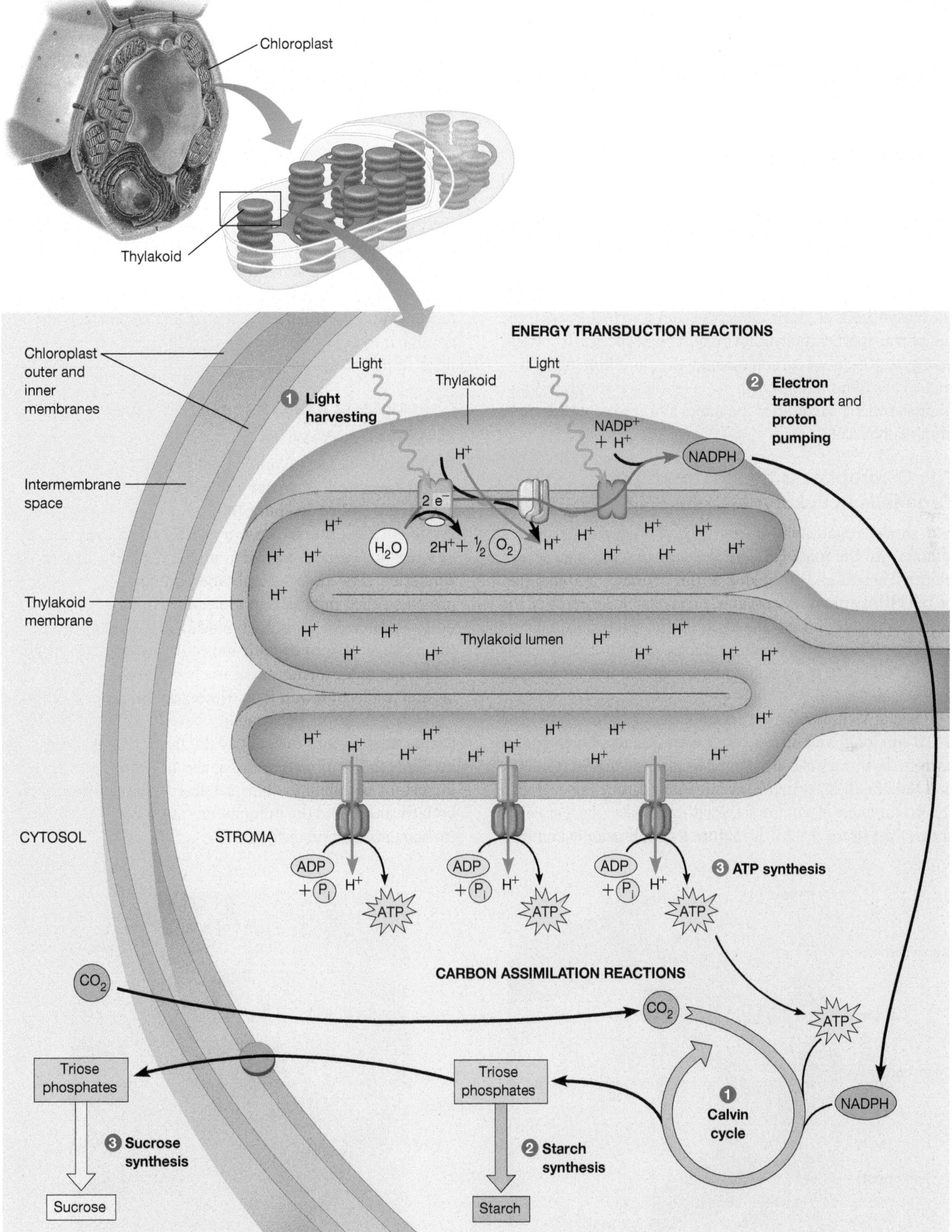

Figure 11-1 An Overview of Photosynthesis. This diagram of a chloroplast shows the intracellular location of major photosynthetic processes. Photosynthesis has two major stages: energy transduction and carbon assimilation. The energy transduction reactions occur in the chloroplast thylakoids and include ❶ light harvesting, ❷ electron transport to NADPH with simultaneous proton pumping, and ❸ ATP synthesis. The carbon assimilation reactions include ❶ the Calvin cycle and ❷ starch biosynthesis in the chloroplast stroma, plus ❸ sucrose synthesis in the cytosol.

When we focus on oxygenic phototrophs, which do use water as an electron donor, we can rewrite Reaction 11-1 by substituting H_2O for H_2A, multiplying all reactants and products by six, and canceling water molecules that appear on both sides of the reaction. This gives the following more specific and familiar form of the general reaction for photosynthesis:

$$\text{light} + 6CO_2 + 6H_2O \rightarrow C_6H_{12}O_6 + 6O_2 \qquad \textbf{(11-2)}$$

We will soon see, however, that the *immediate* product of photosynthetic carbon fixation is a three-carbon sugar (a *triose*), not a six-carbon sugar as in Reaction 11-2. These trioses are then used for a variety of biosynthetic pathways, including the biosynthesis of glucose, sucrose, and starch. Sucrose is the major transport carbohydrate in most plant species. It conveys energy and reduced carbon through the plant from photosynthetic cells to nonphotosynthetic cells. Starch (or glycogen in photosynthetic bacteria) is the major storage carbohydrate in phototrophic cells.

The Chloroplast Is the Photosynthetic Organelle in Eukaryotic Cells

Our model organisms for studying photosynthesis in this chapter are the most familiar oxygenic phototrophs—green plants. Therefore, we will look at the structure and function of the **chloroplast,** the organelle responsible for most of the events of photosynthesis in eukaryotic phototrophs. In plants and algae, the primary events of photosynthetic energy transduction and carbon assimilation are confined to these specialized organelles.

Because chloroplasts are usually large (1–5 μm wide and 1–10 μm long) and opaque, they were described and studied early in the history of cell biology by Antonie van Leeuwenhoek and Nehemiah Grew in the seventeenth century. For example, the prominence of chloroplasts within a *Coleus* sp. leaf cell is shown in **Figure 11-2a**. A mature leaf cell usually contains 20–100 chloroplasts, whereas an algal cell typically contains only one or a few chloroplasts. The shapes of these organelles vary from the simple flattened spheres common in plants to the more elaborate forms found in green algae. A cell of the filamentous green alga *Spirogyra*, for example, contains one or more ribbon-shaped chloroplasts, visible as the corkscrew-shaped green structures in Figure 11-2b.

Not all plant cells contain chloroplasts. Newly differentiated plant cells have smaller organelles called **proplastids,** which can develop into any of several kinds of **plastids** equipped to serve different functions. Chloroplasts are only one example of a plastid. Some proplastids differentiate into *amyloplasts*, which are sites for storing starch. Other proplastids acquire red, orange, or yellow pigments, forming the *chromoplasts* that give flowers and fruits their distinctive colors. Proplastids can also develop into organelles for storing protein (*proteinoplasts*) and lipids (*elaioplasts*).

Chloroplasts Are Composed of Three Membrane Systems

A chloroplast, like a mitochondrion, has both an **outer membrane** and an **inner membrane,** often separated by an **intermembrane space** (**Figure 11-3a**). The inner membrane encloses the **stroma,** a gel-like matrix teeming with enzymes for carbon, nitrogen, and sulfur reduction and assimilation. The outer membrane contains transmembrane proteins called **porins** that are similar to those found in the outer membrane of mitochondria. Because porins permit the passage of solutes with molecular weights up to about 5000, the outer membrane is freely permeable to most small organic molecules and ions. However, the inner membrane forms a significant permeability barrier. Transport proteins in the inner membrane control the flow of most metabolites between the intermembrane space and the stroma. Three important metabolites that are able to diffuse freely across both the outer and the inner membranes, however, are water, carbon dioxide, and oxygen.

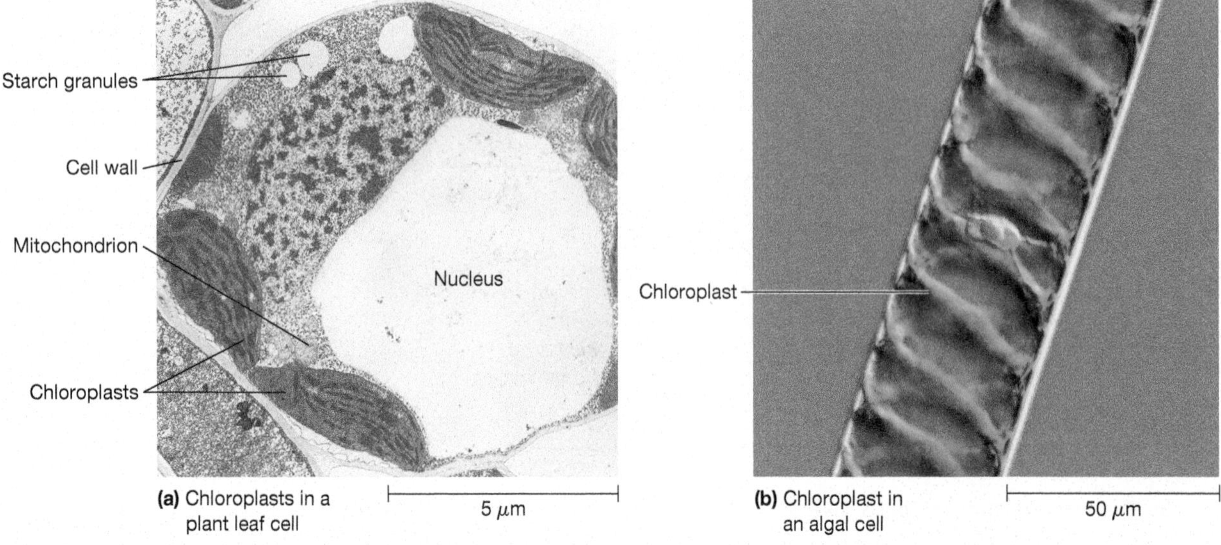

Figure 11-2 Chloroplasts. (a) This cross section of a *Coleus* sp. leaf cell shows two chloroplasts. The presence of large starch granules in the chloroplasts indicates that the cell was photosynthetically active (TEM). **(b)** This light micrograph reveals the unusual ribbon-shaped chloroplasts in a cell of the filamentous green alga *Spirogyra*.

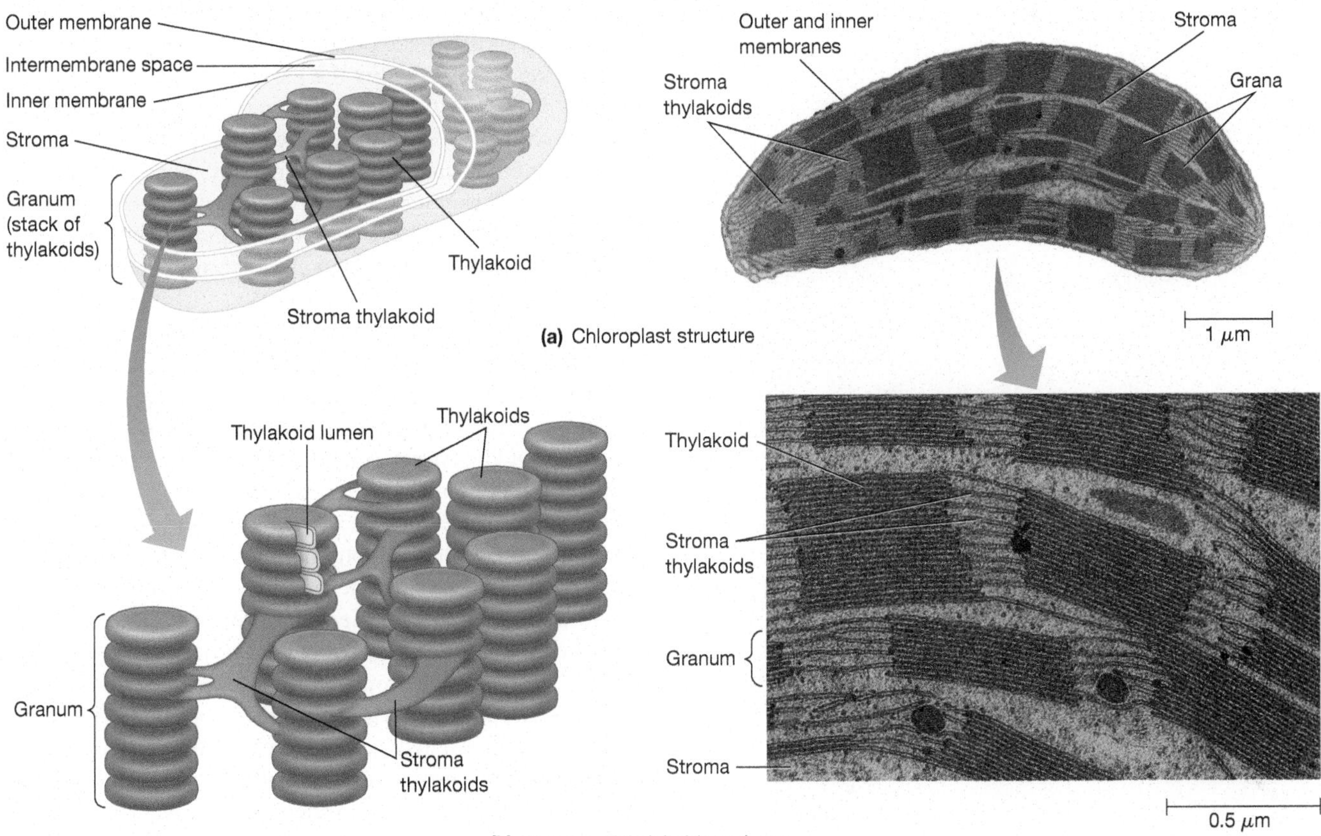

Figure 11-3 Structural Features of a Chloroplast. (a) Left: An illustration showing the three-dimensional structure of a typical chloroplast. Right: Micrograph of a chloroplast from a leaf of corn (*Zea* sp.) (colorized TEM). **(b)** Left: An illustration depicting the continuity of the thylakoid membranes that connect grana with the stroma thylakoids. Right: A more magnified view of the grana and stroma thylakoids in part a (colorized TEM).

In addition, chloroplasts have **thylakoids,** a third membrane system inside the chloroplast that creates an internal compartment. Thylakoids, illustrated in Figure 11-3b, are flat, saclike structures suspended in the stroma. They are usually arranged in stacks called **grana** (singular: **granum**). Grana are interconnected by a network of longer thylakoids called **stroma thylakoids.** A large number of the photosynthetic components, such as pigments, enzymes, and electron carriers, are localized on or in the thylakoid membranes.

Electron micrographs of serial sections show that the grana and stroma thylakoids enclose a single continuous compartment, the **thylakoid lumen.** The semipermeable barrier formed by the thylakoid membrane separates the lumen from the stroma and plays an important role in the generation of an electrochemical proton gradient and the synthesis of ATP. The high concentration of protons pumped across this membrane barrier into the thylakoid lumen during light-driven electron transport stores potential energy that later drives ATP synthesis when the protons return to the chloroplast stroma—in a manner very similar to the ATP synthesis that occurs in mitochondria during oxidative phosphorylation.

Photosynthetic bacteria do not have chloroplasts. In some of them, however, such as the cyanobacteria, the plasma membrane is folded inward and forms *photosynthetic membranes.* Such structures, shown in **Figure 11-4a**, are analogous to the thylakoids. Indeed, to some extent cyanobacteria appear to be free-living chloroplasts. Similarities among mitochondria, chloroplasts, and bacterial cells led to the *endosymbiont theory,* which suggests mitochondria and chloroplasts evolved from bacteria that were engulfed by primitive cells 1 to 2 billion years ago (see Chapter 4). Cyanobacteria also have remarkable polyhedral structures in which the carbon fixation reactions of photosynthesis are carried out called **carboxysomes** (Figure 11-4b). Carboxysomes contain the enzymes carbonic anhydrase and rubisco. Carbonic anhydrase catalyzes formation of carbon dioxide from bicarbonate ions, which enter the bacterium through transporter proteins. As we'll see later in the case of higher plants, the carbon from CO_2 is then fixed via rubisco as part of the Calvin cycle.

CONCEPT CHECK 11.1

Both mitochondria and chloroplasts are energy-generating organelles. Explain how the structure of their membrane systems is critical for their functions, emphasizing their similarities and differences.

11.2 Photosynthetic Energy Transduction I: Light Harvesting

The first stage of photosynthetic energy transduction is the capture of light energy from the sun. As with all electromagnetic radiation, light has both wave-like and particle-like

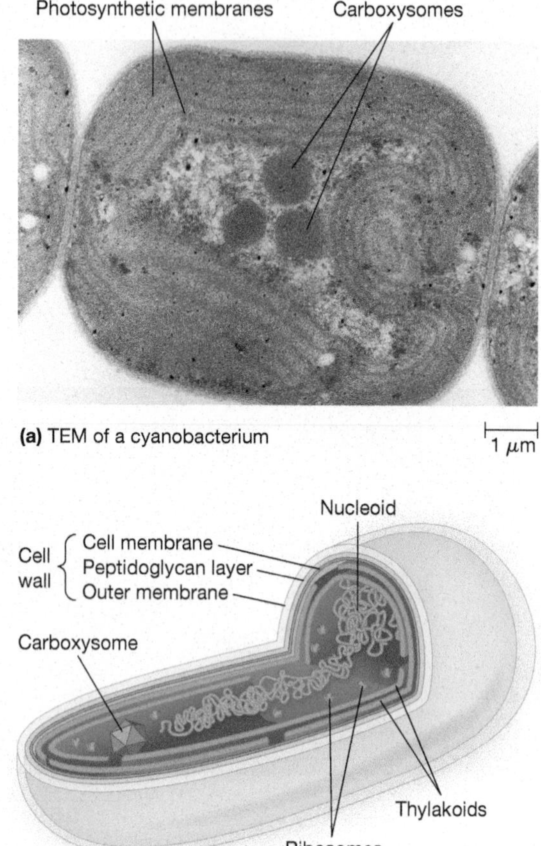

Figure 11-4 The Photosynthetic Structures of a Cyanobacterium. (a) This electron micrograph of a thin section of a *Pseudanabaena* cell shows its extensively folded photosynthetic membranes. Notice how these membranes in cyanobacterial cells are layered like the thylakoid membranes of chloroplasts in eukaryotic cells (TEM). **(b)** A diagram of a cyanobacterium. The light reactions of photosynthesis occur on thylakoid membranes. Carbon fixation occurs within polyhedral protein-containing microcompartments known as carboxysomes, which contain the enzymes carbonic anhydrase and rubisco.

properties. By this we mean that light consists of a stream of discrete particles called **photons**, each photon carrying a *quantum* (indivisible packet) of energy. The wavelength of a photon and the precise amount of energy it carries are inversely related (see Figure 2-3). A photon of ultraviolet or blue light, for example, has a shorter wavelength and carries a larger quantum of energy than a photon of red or infrared light does. The portion of the electromagnetic spectrum visible to us consists of light having wavelengths ranging from about 380 to 750 nm.

When a photon is absorbed by a **pigment** (light-absorbing molecule), such as chlorophyll, the energy of the photon is transferred to an electron, which is energized from its *ground state* in a low-energy orbital to an *excited state* in a high-energy orbital. This event, called **photoexcitation**, is the first step in photosynthesis.

Because each pigment has a different configuration of atoms and electrons, pigments display characteristic **absorption spectra** that describe the particular wavelengths of light each absorbs. The absorption spectra of several common pigments found in photosynthetic organisms, along with the spectrum of solar radiation reaching the Earth's surface, are shown in **Figure 11-5**. Note how the overlapping absorption spectra of various chlorophylls and accessory pigments effectively utilize almost the entire visible spectrum of sunlight that reaches the Earth. (**Key Technique, pages 290–291**, describes the use of a **spectrophotometer** for the determination of absorption and action spectra for biologically relevant compounds and processes.)

A photoexcited electron in a pigment molecule is unstable and must either return to its ground state in a low-energy orbital or undergo transfer to a relatively stable high-energy orbital, usually in a different molecule. When the electron returns to a low-energy orbital in the pigment molecule, the absorbed energy is lost either as heat released or as emitted light (fluorescence). Alternatively, the absorbed energy can be transferred from the photoexcited electron to an electron in an adjacent pigment molecule in a process known as **resonance energy transfer.** This process is very important for moving captured energy from light-absorbing molecules to molecules such as chlorophyll that are capable of passing an excited electron to an organic acceptor molecule. The transfer of the photoexcited electron itself to another molecule is called **photochemical reduction.** This transfer is essential for converting light energy to chemical energy.

Chlorophyll Is Life's Primary Link to Sunlight

Chlorophyll, which is found in nearly all photosynthetic organisms, is the primary energy transduction pigment that channels solar energy into the biosphere. The structure of two types of

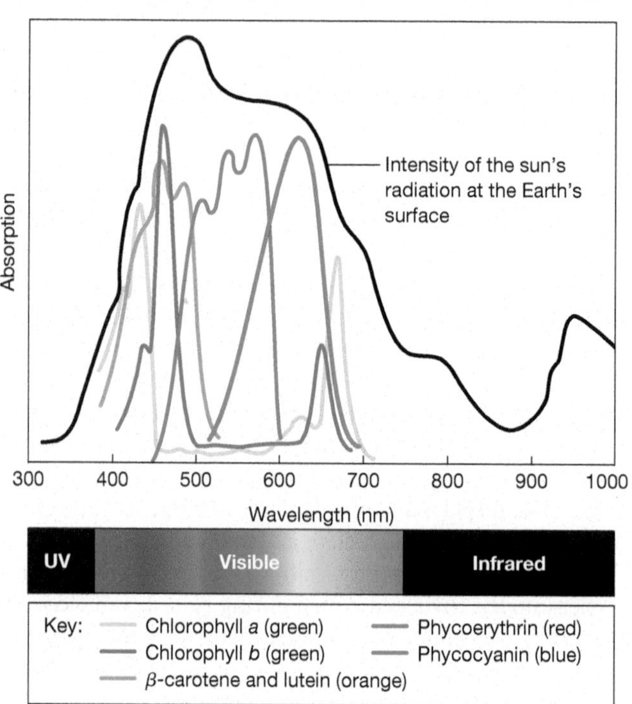

Figure 11-5 The Absorption Spectra of Common Photosynthetic Pigments. The graph shows absorption spectra of various chlorophylls and accessory pigments (colored lines) and compares them with the spectral distribution of solar energy reaching the Earth's surface (black line). Below the graph are the colors that each visible wavelength appears to the human eye.

chlorophyll—chlorophylls *a* and *b*—is shown in **Figure 11-6.** The skeleton of each molecule consists of a central *porphyrin ring* and a strongly hydrophobic *phytol* side chain. The alternating double bonds in the porphyrin ring are responsible for absorbing visible light, whereas the phytol side chain interacts with lipids of the thylakoid or cyanobacterial membranes, anchoring the light-absorbing molecules in these membranes.

The magnesium ion (Mg^{2+}) found in chlorophylls *a* and *b* affects the electron distribution in the porphyrin ring and ensures that a variety of high-energy orbitals are available. As a result, multiple specific wavelengths of light can be absorbed. Chlorophyll *a*, for example, has a broad absorption spectrum, with absorption *maxima* at about 420 and 660 nm (see Figure 11-5). Chlorophyll *b* is distinguished from chlorophyll *a* by the presence of a *formyl* (—CHO) group in place of one of the *methyl* (—CH_3) groups on the porphyrin ring. This minor structural alteration shifts the absorption maxima toward the center of the visible spectrum (see Figure 11-5).

All plants and green algae contain both chlorophyll *a* and *b*. The combined absorption spectra of the two forms of chlorophyll provide access to a broader range of wavelengths of sunlight and enable such organisms to collect more photons. Because chlorophylls absorb mainly blue and red light, they appear green. Other oxygenic photosynthetic organisms supplement chlorophyll *a* with either chlorophyll *c* (brown algae, diatoms, and dinoflagellates), chlorophyll *d* (red algae), or phycobilin (red algae and cyanobacteria).

Bacteriochlorophyll is a subfamily of chlorophyll molecules restricted to anoxygenic phototrophs (photosynthetic bacteria) and is characterized by a saturated site not found in other chlorophyll molecules (indicated by an arrow in Figure 11-6). In this case, structural alterations shift the absorption maxima of bacteriochlorophylls toward the near-ultraviolet and far-red regions of the spectrum.

Human color vision likewise depends on the absorption of specific wavelengths of light by photosensitive pigments. Our retinas contain color-sensing cells known as cone cells that contain the protein opsin, which is bound to retinal, a chromophore that absorbs light and allows us to see in color.

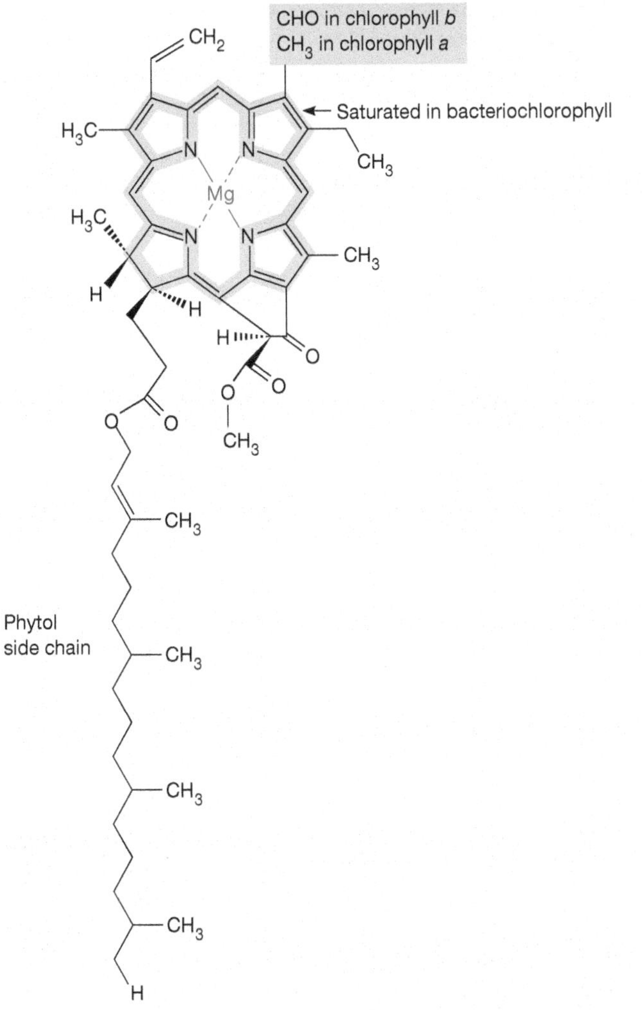

Figure 11-6 The Structure of Chlorophylls *a* and *b*. Each chlorophyll molecule has a Mg^{2+}-containing central porphyrin ring (outlined in green) and a hydrophobic phytol side chain. Chlorophylls *a* and *b* differ by only one ring substituent at the position indicated by the green box. Bacteriochlorophyll has a saturated carbon-carbon bond in the porphyrin ring at the location marked by an arrow.

Accessory Pigments Further Expand Access to Solar Energy

Most photosynthetic organisms also contain **accessory pigments,** which absorb photons that cannot be captured by chlorophyll. The pigments transfer the energy of these photons to a chlorophyll molecule by resonance energy transfer. This feature enables organisms to collect energy from a much larger portion of the sunlight reaching the Earth's surface—as shown in Figure 11-5, which plots the emission spectrum of solar energy in black and the absorption spectra of various accessory pigments in color.

Two types of accessory pigments are **carotenoids** and **phycobilins.** Two carotenoids that are abundant in the thylakoid membranes of most plants and green algae are *β-carotene* and *lutein*, which have very similar absorption spectra (orange line in Figure 11-5). When sufficiently abundant and not masked by chlorophyll, these pigments confer an orange or yellow tint to leaves (or to carrots, from which carotenoids got their name). With absorption maxima of about 420 and 480 nm, carotenoids absorb photons from a broad range of the blue region of the spectrum and thus appear yellow.

Phycobilins are found only in red algae and cyanobacteria. Two common examples are *phycoerythrin* and *phycocyanin* (red and blue lines in Figure 11-5). Phycoerythrin absorbs photons from the blue, green, and yellow regions of the spectrum (thus appearing red) and enables red algae to utilize the dim light that penetrates an ocean's surface water. Phycocyanin, on the other hand, absorbs photons from the orange region of the spectrum, thus appearing blue, and is characteristic of cyanobacteria living close to the surface of a lake or on land. Thus, variations in the amounts and properties of accessory pigments often reflect a phototroph's adaptation to a specific environment.

Light-Gathering Molecules Are Organized into Photosystems and Light-Harvesting Complexes

Chlorophyll molecules, accessory pigments, and associated proteins are organized into functional units, called **photosystems,** that are localized to thylakoid or photosynthetic

Key Technique: Determining Absorption and Action Spectra via Spectrophotometry

PROBLEM: Many biological processes, such as photosynthesis in plants and vision in animals, depend on perception of and responses to light of specific wavelengths. How can researchers determine the biologically relevant wavelengths that stimulate a light-regulated process and thus better understand the process and the key light-absorbing molecules involved?

A prism separates white light into its variously colored component wavelengths.

SOLUTION: Using *spectrophotometry*, researchers can determine which wavelengths of light are absorbed and used by biologically important light-absorbing molecules. In this technique, a *monochromatic* light—light of a very specific wavelength—is directed at a purified sample containing the light-absorbing molecule. After the light passes through the sample, a device measures the intensity of the light that reaches it, thus determining how much light was absorbed by the sample.

Key Tools: Spectrophotometer; a purified sample containing a light-absorbing molecule; a photodetector to record the results.

Details: Typically, specific wavelengths of light are absorbed by light-absorbing molecules in a cell to initiate a process in response to the light stimulus. A graph showing the amount of absorbance of each wavelength is known as an *absorption spectrum*. Often, researchers wish to measure the absorption spectrum associated with light-absorbing molecules because this provides information about the process that employs the light-absorbing molecules. A good example is the absorption spectra of the various pigments associated with photosynthesis.

Determining the absorption spectrum of a molecule such as a photosynthetic pigment requires a *spectrophotometer*, a device that measures the amount of light absorbed by a sample. A spectrophotometer consists of a light source, a prism, a monochromator (a device that transmits a narrow wavelength of light from a range of available wavelengths), a cuvette to hold the sample, and a photodetection system to measure the amount of light transmitted through the sample (**Figure 11A-1**). This allows measurement of the amount of light absorbed by the sample at a wide variety of wavelengths.

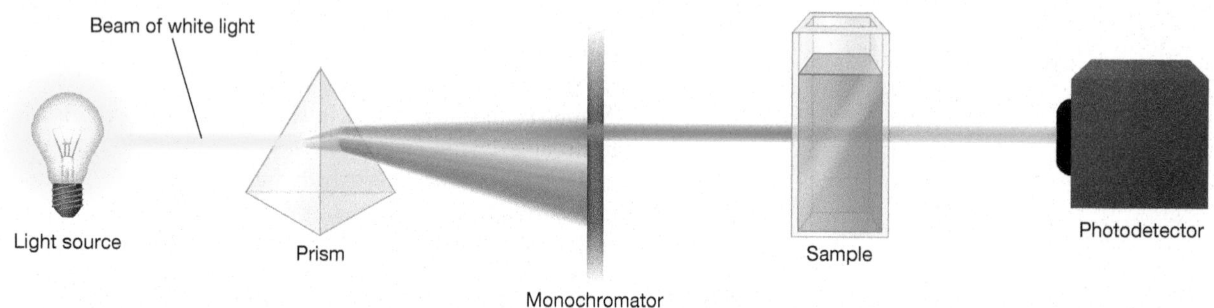

Figure 11A-1 A Spectrophotometer. A spectrophotometer consists of a light source, a prism, a monochromator to select specific wavelengths to illuminate the sample, and a photodetector to determine the amount of light absorbed by the sample.

bacterial membranes (**Figure 11-7** on page 292). The chlorophyll molecules are anchored to the membranes by the long hydrophobic phytol side chains. **Chlorophyll-binding proteins** stabilize the chlorophyll within a photosystem and modify the absorption spectra of specific chlorophyll molecules. Other proteins in the photosystems bind components of electron transport systems or catalyze oxidation-reduction reactions.

Most pigments of a photosystem serve as light-gathering **antenna pigments,** which collect light energy much like a radio antenna collects radio waves. These antenna pigments absorb photons and pass the energy to a neighboring chlorophyll molecule or accessory pigment by *resonance energy transfer*, which does not involve emission of photons (Figure 11-7c), eventually reaching the reaction center chlorophyll. However, work with photosynthetic algae suggests that this energy transfer involves quantum mechanical effects not previously seen in biology. Rather than a simple stepwise process from molecule to molecule, energy transfer can proceed in a wave-like manner, following several paths simultaneously!

The photochemical events that drive electron flow and proton pumping do not begin until the energy reaches the **reaction center** of a photosystem, where two distinct chlorophyll *a* molecules known as the *special pair* reside. Surrounded by other components of the reaction center, this pair of

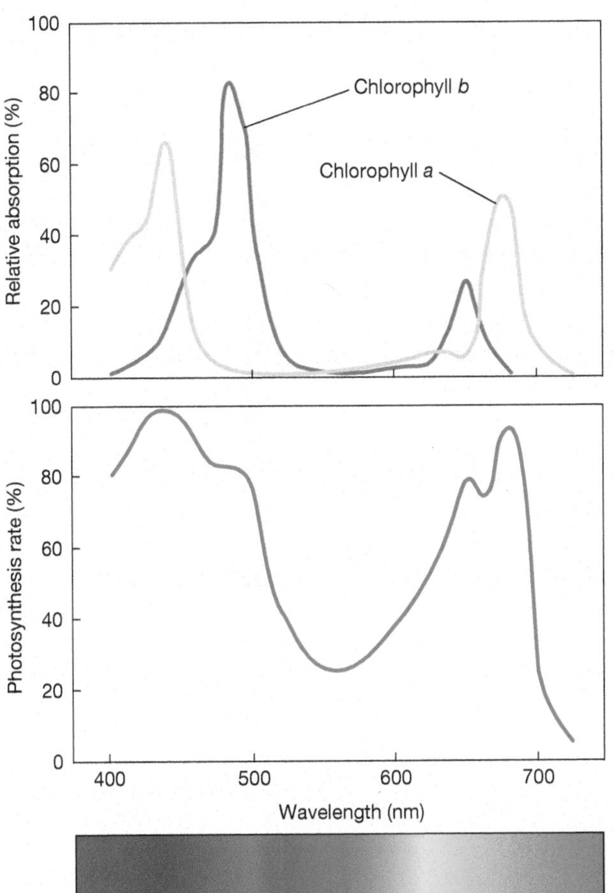

Figure 11A-2 Absorption and Action Spectra of Chlorophyll *a* and *b*. Note how the absorption spectra of chlorophylls *a* and *b* (top) are reflected in the action spectrum for photosynthesis (bottom).

The light source provides a mixture of wavelengths—typically in the ultraviolet (200–380 nm), visible (380–750 nm), and infrared (750–1000 nm) ranges for biologically relevant processes—and the monochromator selects a specific wavelength to direct through the sample. One type of monochromator uses a prism to separate white light into its spectral components. Another uses a *diffraction grating,* a surface etched with a series of very fine lines (1000 or more per millimeter) that produces light of a specific wavelength, depending on the spacing of the fine lines.

After passing through the sample, the light reaches a photodetector, which, as it absorbs a photon of light, produces an electrical signal. This signal is amplified and recorded for each wavelength used, and the results are displayed as relative absorption (*y*-axis) versus wavelength (*x*-axis). The absorption spectra for chlorophylls *a* and *b* are shown in **Figure 11A-2** (top). Interestingly, the absorption spectra for the various light-absorbing pigments involved in photosynthesis are remarkably similar to the intensity of various wavelengths of light at the Earth's surface.

In addition to determining absorption spectra, spectrophotometry can be used to determine the *action spectrum* for light-regulated biological processes. An action spectrum, as the name implies, is a measure of the biological response (or "action") of a biological system to a series of wavelengths of light. Determining an action spectrum requires irradiation of a sample, a cell, or an organism with various wavelengths of monochromatic light followed by measurement of the degree of response at that wavelength. A good example of an action spectrum is the photosynthetic rate of plants, measured as oxygen generation, at various wavelengths. As you can see in Figure 11A-2 (bottom), the photosynthetic action spectrum clearly shows that there is a strong correlation between wavelengths at which chlorophylls absorb light strongly and photosynthetic rate. Other pigments, such as carotenoids, also contribute to the photosynthetic action spectrum.

The action spectra for a number of plant physiological processes have been determined. For example, it was once thought that flowering of plants such as soybeans depended upon temperature rather than light perception. But determination of the action spectrum for flowering by irradiation of soybean plants with monochromatic light showed that only those exposed to red light flowered. This ultimately led to the identification of phytochrome, which absorbs photons of red light and initiates a signaling pathway resulting in the production of flowers.

QUESTION: Based on the absorption spectra in Figure 11A-2, which color of visible light would be least effective in driving photosynthesis?

chlorophyll molecules catalyzes the conversion of solar energy into chemical energy.

Each photosystem is generally associated with a **light-harvesting complex (LHC),** which, like a photosystem, collects light energy. The LHC does not, however, contain a reaction center. Instead, it passes the collected energy to a nearby photosystem by resonance energy transfer. These LHCs are mobile in the thylakoid membrane and can move to take advantage of changing light conditions. Plants and green algae have LHCs composed of about 80–250 chlorophyll *a* and *b* molecules, along with carotenoids and pigment-binding proteins. Red algae and cyanobacteria have a different type of LHC, called a *phycobilisome,* which contains phycobilins rather than chlorophyll and carotenoids. Together, a photosystem and the associated LHCs are referred to as a **photosystem complex.**

Oxygenic Phototrophs Have Two Types of Photosystems

In the 1940s, Robert Emerson and his colleagues at the University of Illinois discovered that two separate photoreactions are involved in oxygenic photosynthesis. Using the green alga *Chlorella,* they measured the rate of photosynthesis versus wavelength by monitoring the rate of oxygen release as they

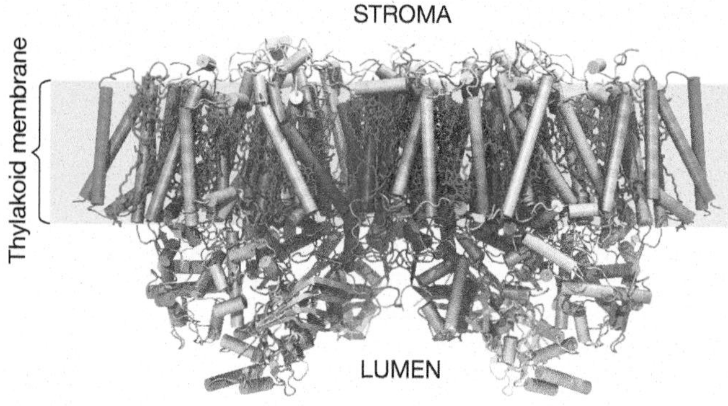

(a) Structure of photosystem II

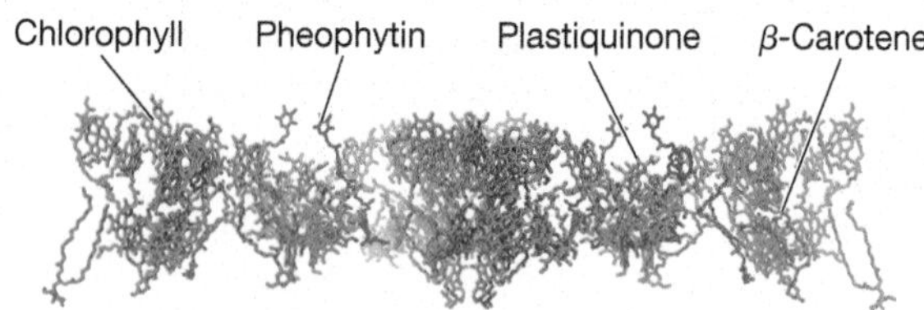

(b) Photosystem pigments

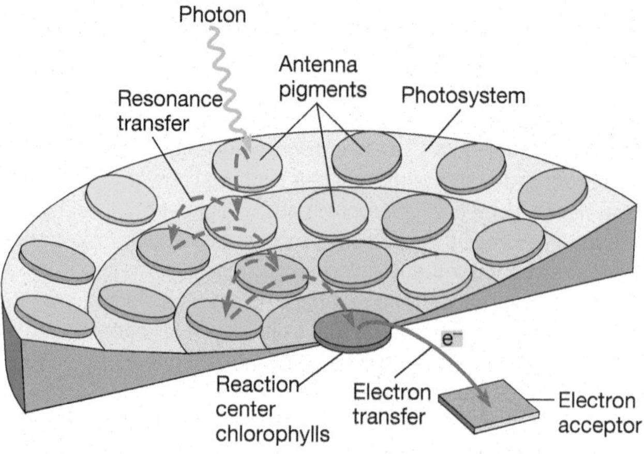

(c) Energy transfer within a photosystem

Figure 11-7 Structure and Function of a Photosystem. **(a)** Molecular model of a membrane-bound photosystem showing protein subunits (cylinders) and bound chlorophyll and other pigment molecules (sticks). **(b)** A view of photosystem pigment molecules with all proteins removed. **(c)** Light energy absorbed by antenna pigments is passed along by resonance energy transfer until it reaches the special pair of chlorophyll *a* molecules at the reaction center of a photosystem (not shown as separate subunits). The energy leaves as an excited electron that is ejected from the chlorophyll and transferred to an organic acceptor molecule.

changed the wavelength of light supplied to the cells. Initially, they observed a dramatic drop in photosynthesis rate above a wavelength of about 690 nm. This seemed odd to them because *Chlorella* contains chlorophyll molecules that strongly absorb light at wavelengths above 690 nm. When they supplemented the longer wavelengths of light with a shorter wavelength (about 650 nm), the decrease in photosynthesis was nearly eliminated. Indeed, they discovered that photosynthesis driven by a combination of long and short wavelengths of red light exceeded the sum of activities obtained with either wavelength alone. This synergistic phenomenon became known as the **Emerson enhancement effect.**

We now know that the Emerson enhancement effect is the result of two distinct photosystems working together in series. **Photosystem I (PSI)** has an absorption maximum of 700 nm, whereas **photosystem II (PSII)** has an absorption maximum of 680 nm. Each electron that passes from water to $NADP^+$ must be photoexcited twice, once by each photosystem. When illumination is restricted to wavelengths above 690 nm, PSII is not active, and photosynthesis is severely impaired.

As we will soon see, each electron is first excited by PSII and then by PSI. But don't be confused by the nomenclature. The photosystems were named based on the order of their discovery, so although PSI was discovered first, it was later shown that the initial photoexcitation event occurs in PSII.

In the reaction center of each photosystem is a distinctive **special pair** of chlorophyll *a* molecules that have the electrons that will be photoexcited following photon absorption. These pairs of chlorophyll molecules are designated **P680** for PSII and **P700** for PSI to reflect their specific absorption maxima. The granal and stromal thylakoid membranes have differing amounts of the two photosystems.

As we saw earlier with the associated light-harvesting complexes, photosystems are somewhat mobile in the membrane and thus able to move to allow the cell to adapt to changing conditions of light quantity (brightness) and quality (wavelength).

CONCEPT CHECK 11.2

How do the type and arrangement of the various photosynthetic pigments maximize the absorption and use of solar energy for photosynthesis?

11.3 Photosynthetic Energy Transduction II: NADPH Synthesis

The second stage of photosynthetic energy transduction uses a series of electron carriers to transport excited electrons from chlorophyll to the coenzyme **nicotinamide adenine dinucleotide phosphate ($NADP^+$),** forming NADPH, the reduced form of $NADP^+$. This process is known as **photoreduction** and uses a chloroplast **electron transport chain (ETC)** that resembles the mitochondrial ETC we previously studied (see Chapter 10). The complete photoreduction pathway (**Figure 11-8**) includes multiple components, and many of the molecules of the chloroplast ETC are similar to those found in the mitochondrial ETC—cytochromes, iron-sulfur proteins, and quinones.

Photosystem II Transfers Electrons from Water to a Plastoquinone

Our detailed tour of the photoreduction pathway begins at photosystem II, a complicated dimeric molecular complex that uses electrons from water to reduce a plastoquinone (Q_B) to a plastoquinol (Q_BH_2). Each PSII monomer, as shown in

(a) Photosystem II removes electrons from water and passes them sequentially to an oxygen-evolving complex (OEC), a tyrosine residue (Tyr) on protein D1, a special pair of chlorophyll *a* molecules (P680), pheophytin (Ph), and two plastoquinones (Q_A and Q_B). Oxygen is released as water is oxidized.

(b) The cytochrome b_6/f complex receives electrons from PSII and passes them to PSI via plastocyanin (PC). Simultaneously, protons are pumped from the chloroplast stroma into the thylakoid lumen.

(c) Photosystem I accepts electrons from PC and passes them to ferredoxin (Fd) via another special pair of chlorophyll *a* molecules (P700), a modified chlorophyll *a* molecule (A_0), phylloquinone (A_1), and three iron-sulfur centers (F_X, F_A, and F_B).

(d) Ferredoxin-NADP$^+$ reductase (FNR) transfers electrons from ferredoxin to NADP.

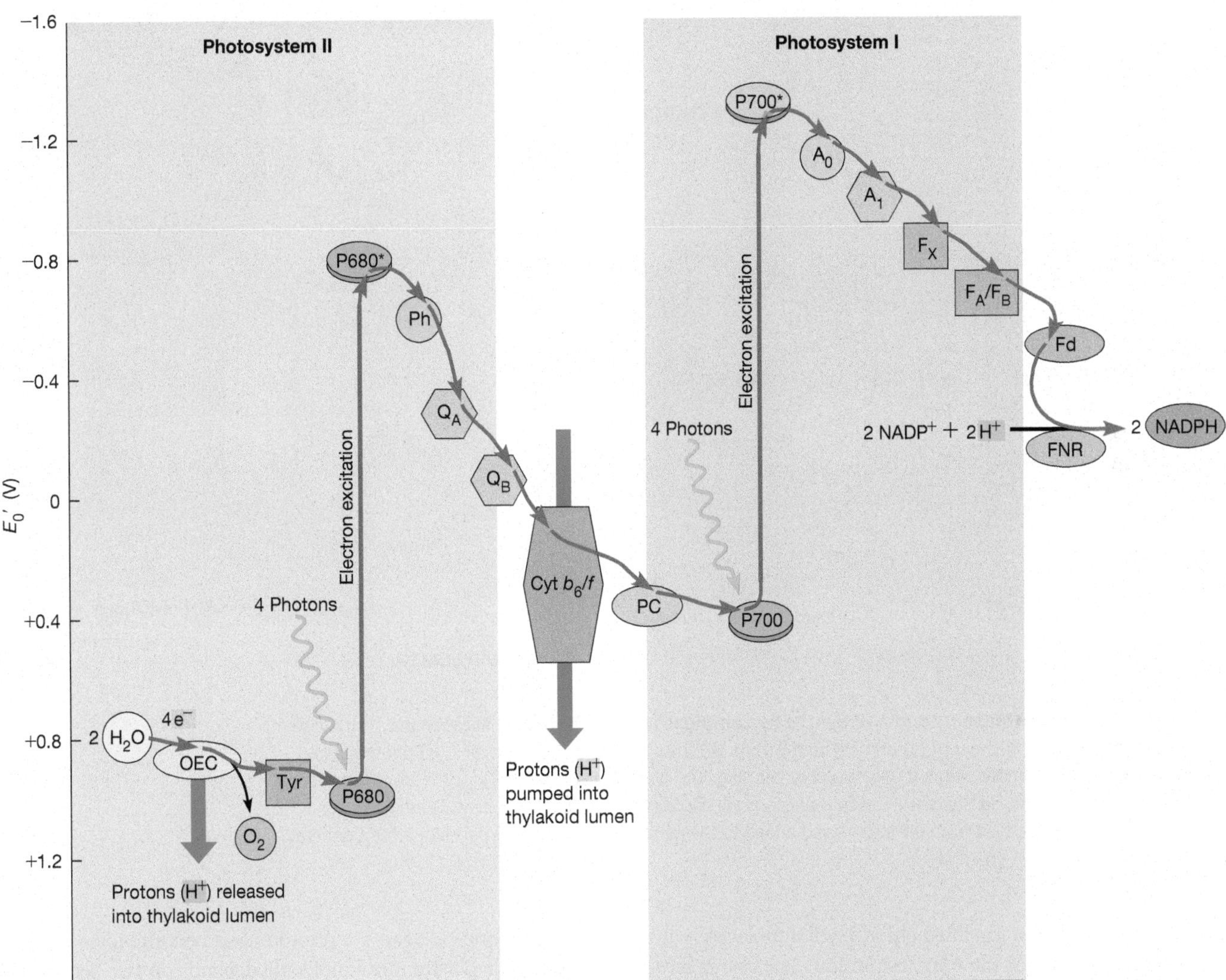

Figure 11-8 Noncyclic Electron Flow in Oxygenic Phototrophs. Red arrows show the flow of electrons from water to NADPH as they follow a linear pathway through **(a)** photosystem II (PSII), **(b)** the cytochrome b_6/f complex, **(c)** photosystem I (PSI), and **(d)** ferredoxin-NADP$^+$ reductase (FNR). Absorption of photons by the chlorophylls of P680 and P700 (asterisks indicate a photoexcited state) causes a large decrease in their reduction potentials (higher position on the vertical axis). This enables them to donate excited electrons to an acceptor with a highly negative reduction potential. Figure 11-9 shows the orientation of these components within a thylakoid membrane.

Figure 11-9 on page 294, contains approximately 20 unique polypeptides, including D1 and D2, which bind not only the P680 chlorophylls but also a water-splitting complex and the components of an ETC. Surrounding the reaction center are about 30 additional chlorophyll *a* and about 10 β-carotene molecules that enhance photon collection.

In plants and green algae, PSII is generally associated with multiple trimers of **light-harvesting complex II (LHCII)**, each of which contains numerous chlorophyll and carotenoid molecules. Energy captured by antenna pigments of PSII or LHCII is funneled to the PSII reaction center by resonance energy transfer. When the energy reaches the reaction center, it lowers the reduction potential (E'_0, a measure of affinity for electrons) of a P680 chlorophyll *a* molecule to about −0.80 V, making it a much better electron donor. In this state the chlorophyll is said to be "excited" and is designated by an asterisk (P680*; Figure 11-8a). The lowering of P680's reduction potential allows a photoexcited electron to be passed from

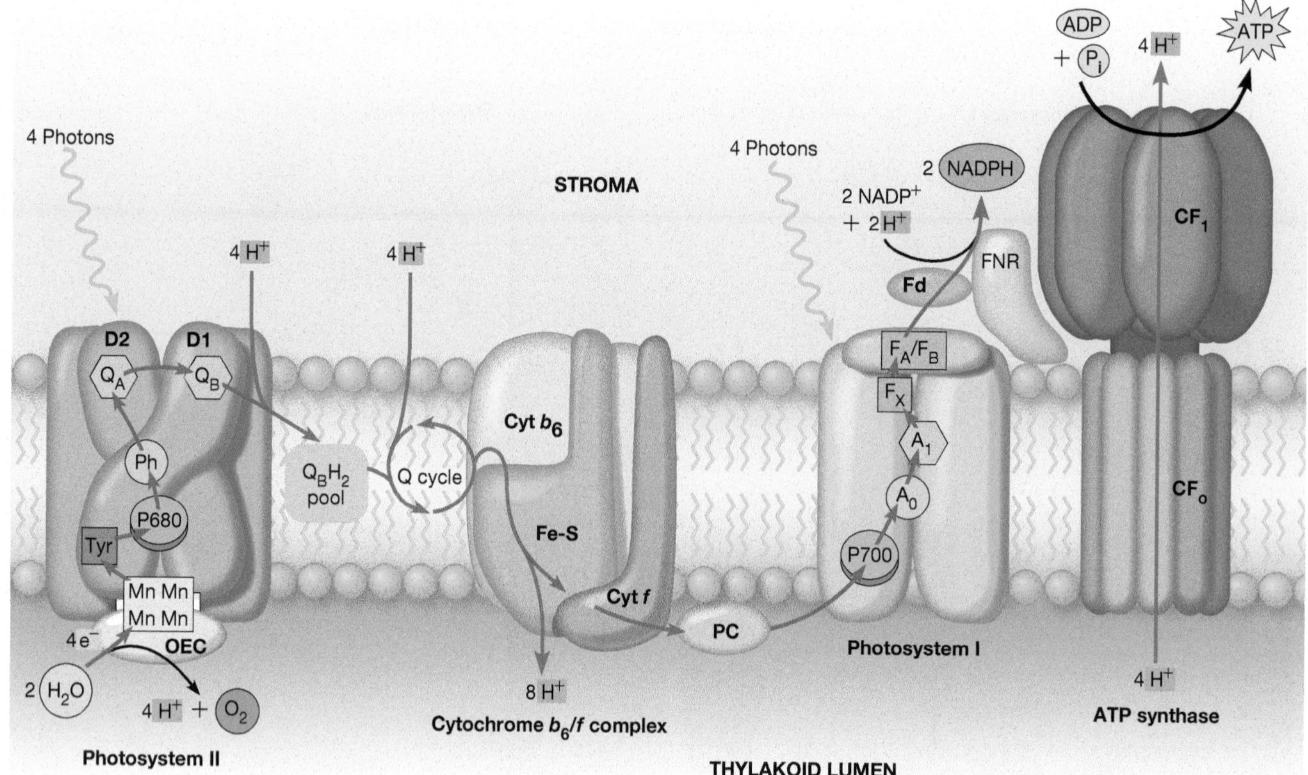

(a) PSII includes about 20 polypeptides, two of which (D1 and D2) bind P680 chlorophylls and components of the electron transport system. The four manganese atoms help to oxidize water.
(b) The cytochrome b6/f complex includes two cytochromes (b6 and f) and an iron-sulfur protein (Fe-S). It transfers electrons from PSII to PSI.
(c) PSI contains more than ten polypeptides and binds chlorophyll P700 and several additional electron carriers. Electrons from PSI are used to reduce NADP+ to NADPH.
(d) ATP synthase acts as a proton channel and uses the electrochemical proton gradient generated by electron transport to synthesize ATP.
4 Photons
STROMA
4 H+
4 H+
D2
D1
QA
QB
Ph
P680
Tyr
Mn Mn
Mn Mn
4e−
OEC
2 H2O
4 H+ + O2
Photosystem II
QBH2 pool
Q cycle
Cyt b6
Fe-S
Cyt f
8 H+
Cytochrome b6/f complex
PC
THYLAKOID LUMEN
4 Photons
2 NADP+
+ 2 H+
2 NADPH
FNR
Fd
FA/FB
FX
A1
A0
P700
Photosystem I
ADP
+ Pi
4 H+
ATP
CF1
CFo
4 H+
ATP synthase

Figure 11-9 Electron Transport and ATP Synthesis Complexes in the Thylakoid Membrane. This diagram shows the predicted arrangement of **(a)** photosystem II (PSII), **(b)** the cytochrome b_6/f complex, **(c)** photosystem I (PSI), and **(d)** the ATP synthase within the thylakoid membrane. The stoichiometry shown is for the transport of four electrons (red arrows) from H_2O to $NADP^+$ as four photons are absorbed by each photosystem. During electron transport, protons are removed from the stroma and added to the lumen. See Figure 11-8 part labels for a general key to symbols.

it to *pheophytin (Ph)*, a modified chlorophyll *a* molecule with two protons in place of the magnesium ion. The *charge separation* produced between the oxidized $P680^+$ and the reduced pheophytin (Ph^-) conserves the increased potential energy of the excited electron and prevents the electron from returning to its ground state in P680, where its energy would simply be lost as heat or fluorescence. *Thus, solar energy has been harvested and converted into electrochemical potential energy in the form of a charge separation.*

Next the electron is passed to Q_A, a **plastoquinone** that is tightly bound to PSII polypeptide D2 (Figure 11-9a). Plastoquinone is similar to *coenzyme Q*, a component of the mitochondrial electron transport system (see Figure 10-18). Another plastoquinone, Q_B, receives two electrons from Q_A and picks up two protons from the stroma, and thereby becomes reduced to **plastoquinol,** Q_BH_2. It then enters a mobile pool of Q_BH_2 in the interior lipid phase of the photosynthetic membrane. Q_BH_2 can then pass two electrons and two protons to the *cytochrome b_6/f complex* as it is oxidized back to Q_B (Figure 11-9b). Because a chlorophyll molecule transfers one electron per photon absorbed, the formation of one mobile plastoquinol molecule depends on two sequential photoreactions at the same reaction center:

$$2\ \text{photons} + Q_B + 2H^+_{\text{stroma}} + 2e^- \rightarrow Q_BH_2 \qquad \textbf{(11-3)}$$

To replace each electron lost to plastoquinone, oxidized $P680^+$ is reduced by an electron obtained from water. To do this, PSII includes an **oxygen-evolving complex (OEC),** an assembly of proteins and manganese ions that catalyzes the splitting and oxidation of water, producing molecular oxygen (O_2), electrons, and protons. Two water molecules donate four electrons, one at a time by way of a tyrosine residue on PSII protein D1, to four molecules of oxidized $P680^+$ (see Figure 11-8a and 11-9a). A cluster of four manganese ions (Mn_4) serves to accumulate the four electrons released as the two water molecules are oxidized. This prevents the formation of partly oxidized intermediates, such as hydrogen peroxide (H_2O_2) or superoxide anion O_2^-, that can be toxic to the cell if released. In the process, four protons and one oxygen molecule are released within the thylakoid lumen, as shown here:

$$2H_2O \rightarrow O_2 + 4e^- + 4H^+_{lumen} \qquad \textbf{(11-4)}$$

The protons accumulating in the lumen contribute to an electrochemical proton gradient across the thylakoid membrane, and the oxygen molecule diffuses out of the chloroplast. Because the complete oxidation of two water molecules to molecular oxygen depends on four photoreactions, the net reaction catalyzed by four photoexcitations at PSII can be summarized by doubling Reaction 11-3 and adding it to Reaction 11-4:

$$4\text{ photons} + 2H_2O + 2Q_B + 4H^+_{stroma} \rightarrow$$
$$O_2 + 2Q_BH_2 + 4H^+_{lumen} \qquad \textbf{(11-5)}$$

Note that the protons removed from the stroma are actually still in transit to the lumen as part of Q_BH_2, while the protons added to the lumen at this point are derived from oxidation of water. The light-dependent oxidation of water to protons and molecular oxygen, called *water photolysis,* is believed to have appeared in cyanobacteria between 2 and 3 billion years ago, thereby permitting exploitation of water as an abundant electron donor. The oxygen released by the process dramatically changed the Earth's early atmosphere, which did not originally contain free oxygen, and allowed the development of aerobic respiration.

Under very high-light conditions, plants often receive far more energy than they can use for photosynthesis, and this excess energy can be damaging. Many plants have a protective mechanism known as the *xanthophyll cycle,* which helps dissipate excess energy (see Human Connections, page 296–297).

The Cytochrome b_6/f Complex Transfers Electrons from a Plastoquinol to Plastocyanin

Electrons carried by Q_BH_2 flow through an ETC coupled to unidirectional proton pumping across the thylakoid membrane into the lumen. This happens by way of the **cytochrome b_6/f complex**, which is analogous to the mitochondrial respiratory Complex III (described in Chapter 10; see Figures 10-18 through 10-20). The cytochrome b_6/f complex is composed of seven distinct integral membrane proteins, including two cytochromes and an iron-sulfur protein. Cytochrome *f* donates electrons to a copper-containing protein called **plastocyanin (PC)**. Plastocyanin, like plastoquinol, is a mobile electron carrier, and it will carry electrons to PSI. Unlike plastoquinol, however, it is a peripheral membrane protein found on the lumenal side of the thylakoid membrane.

Like cytochrome *c* in the mitochondrial inner membrane, PC carries only one electron at a time. This means that, like coenzyme Q in the mitochondrion, once the mobile Q_BH_2 receives its two electrons from PSII it must donate those two electrons one at a time: one to *cytochrome b_6* and the other to an iron-sulfur protein. Ultimately, these electrons are passed to *cytochrome f* and then PC. A chloroplast *Q cycle,* similar to the one first suggested for mitochondria by Peter Mitchell in 1975 (Chapter 10) is involved in this process (Figure 11-9b).

Starting with the two Q_BH_2 molecules generated at PSII (see Reaction 11-5), the net reaction catalyzed by the cytochrome b_6/f complex for every electron directly transferred to PC can be summarized as

$$2Q_BH_2 + 4PC(Cu^{2+}) \rightarrow 2Q_B + 4H^+_{lumen} + 4PC(Cu^+) \qquad \textbf{(11-6)}$$

Thus, four photoreactions at PSII add eight net protons to the thylakoid lumen: four from the oxidation of water to oxygen and four by plastoquinol (Figure 11-9).

Photosystem I Transfers Electrons from Plastocyanin to Ferredoxin

The task of photosystem I is to transfer photoexcited electrons from reduced plastocyanin (Pc) to a protein known as *ferredoxin* (Fd), the immediate donor of electrons to $NADP^+$. As indicated in Figure 11-9, PSI includes a special pair of chlorophyll *a* molecules, designated P700, plus a third chlorophyll *a* called A_0 (instead of the pheophytin molecule in PSII). Other components of PSI include a *phylloquinone* (A_1), and three *iron-sulfur centers* (F_X, F_A, and F_B) that form an ETC linking A_0 to ferredoxin.

PSI in plants and green algae is associated with **light-harvesting complex I (LHCI)**, which contains fewer antenna molecules than LHCII does. As with the PSII complex, light energy captured by antenna pigments or LHCI is funneled to a reaction center containing a special pair of chlorophyll *a* molecules. Light energy lowers the reduction potential of the PSI special pair to about –1.30 V (see Figure 11-8c). A photoexcited electron is then rapidly passed from P700 to A_0, and charge separation between oxidized $P700^+$ and the reduced A_0 prevents the electron from returning to its original ground state. As in PSII, this charge separation preserves part of the light energy absorbed as electrochemical potential energy. The electron lost by P700 is replaced by an incoming electron from reduced plastocyanin.

From A_0, electrons flow exergonically through the ETC to ferredoxin, the final electron acceptor for PSI. **Ferredoxin (Fd)** is a mobile iron-sulfur protein found in the chloroplast stroma. Ferredoxin is an important reductant in several other metabolic pathways in the chloroplast, including those for nitrogen and sulfur assimilation. Starting with the four reduced plastocyanin molecules generated at the cytochrome b_6/f complex (Reaction 11-6), the net reaction catalyzed by PSI can be summarized as

$$4\text{ photons} + 4PC(Cu^+) + 4Fd(Fe^{3+}) \rightarrow$$
$$4PC(Cu^{2+}) + 4Fd(Fe^{2+}) \qquad \textbf{(11-7)}$$

Ferredoxin-$NADP^+$ Reductase Catalyzes the Reduction of $NADP^+$

The final step in the photoreduction pathway is the transfer of electrons from ferredoxin to $NADP^+$, thereby providing the NADPH essential for photosynthetic carbon reduction and assimilation. This transfer is catalyzed by the enzyme **ferredoxin-$NADP^+$ reductase (FNR)**, a peripheral membrane protein found on the stromal side of the thylakoid membrane. Starting with the four reduced ferredoxin

HUMAN *Connections*

How Do Plants Put On Sunscreen?

We need a little sun exposure on our skin each day to help with vitamin D production. But too much sunlight can be bad; sunburn results when UV light damage triggers inflammation and cell death. The skin may redden and become painful to the touch, and the upper layer of cells may die and peel away. If the pigments in the skin that absorb light cannot dissipate the extra energy, a reaction with oxygen may occur, producing harmful *reactive oxygen species (ROS)* that damage cell components and membranes. UV light can also cause thymine dimers to form in DNA (see Figure 17-25), which can cause errors in DNA replication. For most humans, of course, one way to avoid such damage is to move inside, out of the sun.

Plants are exposed to intense, potentially damaging sunlight every day, but they face a dilemma. On the one hand, plants need to absorb as much light as possible to maximize photosynthesis. On the other hand, prolonged exposure to intense sunlight can cause the plant version of a sunburn. UV light can cause DNA damage and the formation of ROS that harm enzymes and membranes. But most of the injury affects the photosynthetic machinery. If more light energy is absorbed than can be used by the light-dependent reactions, then the excess energy damages photosystem II protein D1 and inhibits photosynthesis. If D1 is damaged, the entire photosystem must be disassembled so that D1 can be removed and replaced (**Figure 11B-1**). Photosystem I can also suffer ROS damage when the electron carrier ferredoxin (Fd) reacts with O_2 under excess light.

During photosynthesis, light is absorbed by pigments in the antenna system, and the energy is channeled to the special pair of chlorophyll molecules at the heart of the two photosystems, PSII and PSI. When these chlorophyll molecules become excited, there are three main destinations for the excited electron. Ideally, the excited electron will be used for photosynthesis. But what happens to the excess energy if light absorption occurs faster than the excited electrons can be used by photosynthesis? A second pathway for dissipating energy involves the *carotenoids*. Carotenoids, for example, β-carotene, are orange pigments found in the light-harvesting complexes around the two photosystems. Normally, carotenoids absorb a photon of light and transfer the energy to a chlorophyll in the photosystem reaction center. But they can also receive energy from chlorophylls if the light-harvesting reactions

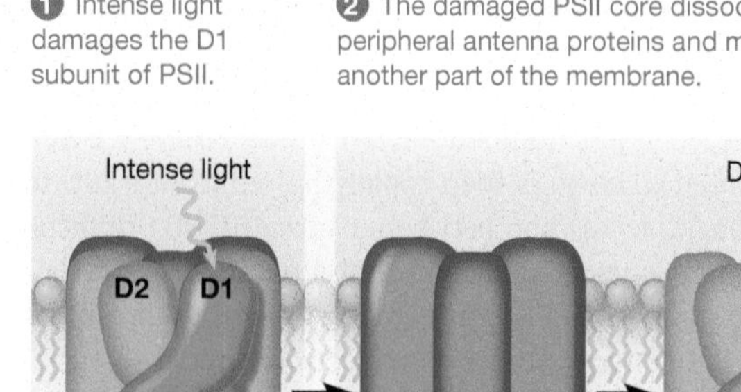
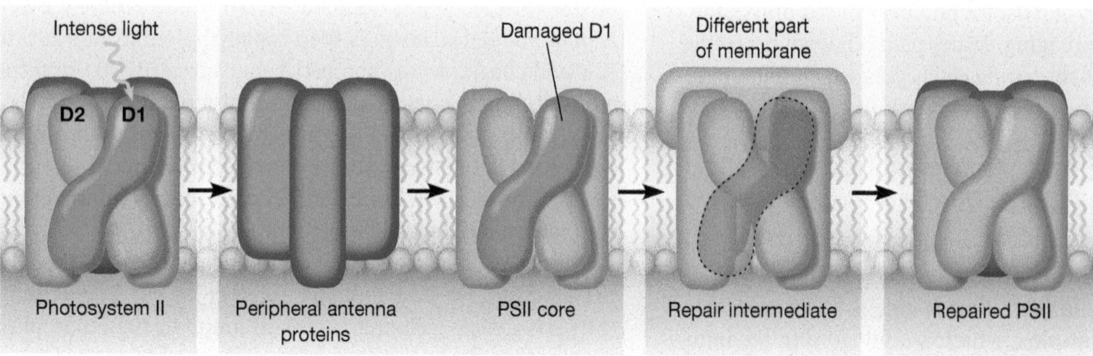

1 Intense light damages the D1 subunit of PSII.
2 The damaged PSII core dissociates from peripheral antenna proteins and moves to another part of the membrane.
3 D1 is degraded and replaced.
4 Repaired core moves back into place; reassociates with antenna proteins.
Intense light
D2
D1
Damaged D1
Different part of membrane
Photosystem II
Peripheral antenna proteins
PSII core
Repair intermediate
Repaired PSII

Figure 11B-1 Disassembly and Reassembly of PSII After UV Damage to D1. Damaged PSII moves away from the antenna complex to a region of the membrane where the damaged photosystem undergoes repair through removal and replacement of D1. It can then reassociate with the antenna complex and participate in photosynthesis.

molecules generated at PSI (see Reaction 11-7), we obtain the following:

$$4Fd(Fe^{2+}) + 2\,NADP^+ + 2H^+_{stroma} \rightarrow 4Fd(Fe^{3+}) + 2\,NADPH \qquad \textbf{(11-8)}$$

Notice that the reduction of one $NADP^+$ molecule consumes one H^+ from the stroma plus two electrons, each from a single reduced ferredoxin. Although not actually moving protons from the stroma to the lumen, this reaction lowers the proton concentration of the stroma and contributes to the electrochemical proton gradient across the thylakoid membrane.

We've now seen that, functioning together within the chloroplast, the various components of the ETC provide a continuous, unidirectional flow of electrons from water to $NADP^+$, as indicated in Figures 11-8 and 11-9. This is called **noncyclic electron flow**, primarily to distinguish it from the *cyclic electron flow* from ferredoxin back to cytochrome b_6/f, an optional process that we will encounter shortly.

The net result of the complete light-dependent oxidation of two water molecules to molecular oxygen can be obtained by summing Reactions 11-5 through 11-8, assuming ferredoxin does not donate electrons to other pathways:

$$8\text{ photons} + 2H_2O + 6H^+_{stroma} + 2\,NADP^+ \rightarrow 8H^+_{lumen} + O_2 + 2\,NADPH \qquad \textbf{(11-9)}$$

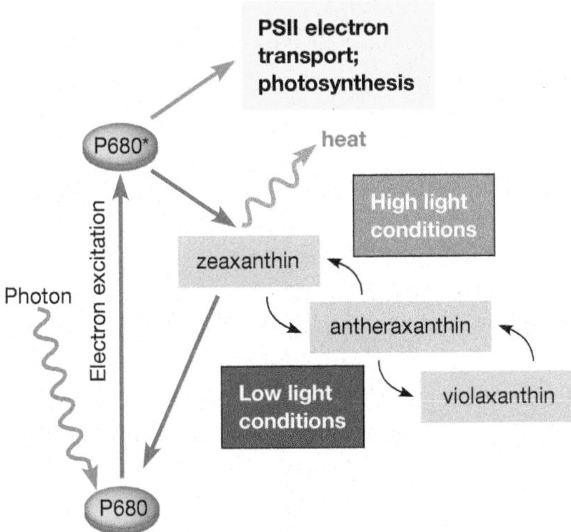

Figure 11B-2 The Xanthophyll Cycle. During periods of exposure to intense light, extra electrons from an excited chlorophyll *a* molecule at the photosynthetic reaction center (indicated by the asterisk) are transferred to zeaxanthin, allowing excess energy to be dissipated as heat and allowing electrons to be returned to chlorophyll. Under low light conditions zeaxanthin is converted back to violaxanthin.

cannot utilize electrons fast enough to allow chlorophylls to "relax" back to their "ground" state (their "unexcited" state). If an excited chlorophyll molecule does not give up its electron, the chlorophyll is much more likely to react with O_2 to make reactive oxygen species that can damage membranes. Carotenoids, in contrast, do not react with O_2 and do not produce ROS. Instead, the excess energy is released as heat. Carotenoids can also act as antioxidants that detoxify ROS produced by chlorophylls and convert them to nonreactive chemicals.

Plants have a third pathway for dissipating excess energy in their leaves. *Xanthophylls* are common red and yellow pigments in plants. Yellow xanthophylls are a type of carotenoid present in chloroplasts, and they act by dissipating heat and excess light energy. The *xanthophyll cycle* siphons off excited electrons from the reaction center when it is overburdened under high light conditions (**Figure 11B-2**). Under conditions of high light, violaxanthin is converted to antheraxanthin and then to zeaxanthin. Zeaxanthin is particularly good at accepting electrons from excited chlorophyll *a* molecules, reducing the risk that excited chlorophyll *a* will lead to ROS production. Once a plant returns to low light conditions, its cells can perform the reverse reactions, converting zeaxanthin back to antheraxanthin and ultimately to vioxanthin.

Plants grown under intense light accumulate higher amounts of the "plant sunscreens" xanthophyll and carotenoid to protect their photosynthetic machinery from light-induced damage (**Figure 11B-3**). Remarkably, zeaxanthin is also found in the human eye, where it protects the eye from damage by intense light, underscoring the fundamental conserved processes that protect all eukaryotic cells from damage.

Figure 11B-3 Xanthophylls on the Edges of Leaves. Some plants have increased concentrations of xanthophylls in regions of the leaves that receive the most light. At the hottest time of day when the sun is directly overhead, high concentrations of xanthophylls in leaf "edges" help protect the leaves and dissipate the excess energy as heat.

For every eight photons absorbed (four by PSII and four by PSI), two NADPH molecules are generated. Furthermore, the ETC for photoreduction is coupled to a mechanism for unidirectional proton pumping across the thylakoid membrane from the stroma to the lumen. Operation of the Q cycle would increase the number of protons accumulating in the lumen from eight to twelve. Thus, solar energy has been captured and stored as potential energy in two forms: the reductant NADPH and an electrochemical proton gradient.

How is the potential energy of this proton gradient used to synthesize ATP? The mechanism resembles mitochondrial ATP synthesis. Despite the differences between chloroplasts and mitochondria, both have an outer membrane, an inner membrane, and an internal space in which protons are accumulated—the intracristal space in mitochondria and the thylakoid lumen in chloroplasts. Both involve exergonic electron transport through an ordered array of membrane-bound carriers that are alternatively reduced and oxidized as they create the proton gradient used for ATP synthesis. However, in mitochondria, electrons travel from high-energy reduced coenzymes to oxygen, forming water, as chemical energy is converted from one form (NADH) to another (ATP). In chloroplasts, solar energy is used to remove electrons from water, forming oxygen, as solar energy is converted into chemical energy in the form of NADPH and ATP.

CONCEPT CHECK 11.3

An essential structural feature in both mitochondria and chloroplasts is the presence of an ordered series of electron carriers. Discuss similarities and differences in electron transport in these two organelles.

11.4 Photosynthetic Energy Transduction III: ATP Synthesis

Now we come to the final stage of photosynthetic energy transduction, in which the potential energy that has been stored in the proton gradient is used to synthesize ATP from ADP and P_i. Because the energy used to phosphorylate ADP originates from sunlight, this process is known as **photophosphorylation**.

Because the thylakoid membrane is virtually impermeable to protons, a substantial electrochemical proton gradient can develop across the membrane in illuminated chloroplasts. We can calculate the total proton motive force (pmf) across the thylakoid membrane by summing a proton concentration (pH) term and a membrane potential (V_m) term as we did for mitochondria.

The pmf can be calculated by summing the contributions of the membrane potential and the pH gradient using the following equation (introduced in Chapter 10):

$$\text{pmf} = V_m + \frac{2.303RT\Delta\text{pH}}{F} \qquad \textbf{(11-10)}$$

The pmf can be used to determine the standard free energy change ($\Delta G^{\circ\prime}$) for the movement of protons from the lumen to the stroma (as we did for mitochondria in Chapter 10):

$$\Delta G^{\circ\prime} = -nF(\text{pmf}) \qquad \textbf{(11-11)}$$

In mitochondria, the pH difference (ΔpH) across the inner membrane is only about 1 unit, contributing about 0.06 V (30%) of a total pmf of 0.22 V. The other 70% comes from the membrane potential difference of about 0.16 V. In chloroplasts, however, ΔpH is more important than V_m, contributing about 80% of the total pmf. Conservative estimates of light-induced proton pumping into the thylakoid lumen causes its pH to drop to about 5. Coupled with a rise in stromal pH to about 8 due to proton depletion, this creates a ΔpH of about 3 units, which is enough to generate a pmf of about 0.18 V at 25°C. Assuming current estimates of a membrane potential difference of approximately 0.03 V, this gives a total pmf of 0.21 V, representing a $\Delta G^{\circ\prime}$ of about 5 kcal/mol of protons moving across the thylakoid membrane. This result is quite reasonable in terms of thermodynamics. The $\Delta G'$ value for the synthesis of ATP within the chloroplast stroma is usually 10–14 kcal/mol, so moving 12 mol of protons is enough to drive formation of several moles of ATP.

A Chloroplast ATP Synthase Couples Transport of Protons Across the Thylakoid Membrane to ATP Synthesis

In chloroplasts, as in mitochondria and bacteria, the movement of protons across a membrane from high to low concentration drives the synthesis of ATP by an **ATP synthase**. The ATP synthase found in chloroplasts, designated the **CF_oCF_1 complex,** is remarkably similar to the F_oF_1-ATPases of mitochondria and bacteria (see Figure 10-23). The **CF_1** component is a hydrophilic assembly of polypeptides that protrudes from the stromal side of the thylakoid membrane and contains three catalytic sites for ATP synthesis. Like the bacterial F_1, CF_1 is composed of five distinct polypeptides with a stoichiometry of $\alpha_3\beta_3\gamma\delta\varepsilon$. CF_1 polypeptides have structures and functions similar to those of the analogous F_1 polypeptides.

The **CF_o** component is a hydrophobic assembly of polypeptides anchored to the thylakoid membrane, much like the F_o component of the F_oF_1-ATP synthase. CF_o subunits I and II form a peripheral stalk that connects CF_o with CF_1, thus serving the same function as the two *b* subunits of F_o (see Figure 10-23). CF_o subunit IV (similar to the *a* subunit of F_o) is the **proton translocator** through which protons flow from the lumen back to the stroma under the pressure of the pmf. Proton flow through subunit IV causes rotation of an adjacent ring composed of subunit III polypeptides (similar to the *c* subunits in the *c* ring of F_o). This rotation is coupled to ATP synthesis by the CF_1 component as for the F_oF_1-ATP synthase (see Section 10-6).

Biochemical experiments involving liposomes indicate that, unlike mammalian mitochondrial ATP synthase, through which about 3 protons pass for every ATP synthesized, four protons pass through the CF_oCF_1 for every ATP synthesized. Structural evidence indicates that there are 14 copies of subunit III in the rotating ring of spinach chloroplast CF_oCF_1. This suggests that translocation of 14 protons through this CF_oCF_1 causes one rotation. Differences in the number of subunit III polypeptides in different ATP synthases, combined with storage of elastic energy in the γ subunit may explain the different H^+/ATP ratios obtained by researchers using different experimental systems.

Assuming an H^+/ATP ratio of four, the reaction catalyzed by the ATP synthase complex can be summarized as

$$4H^+_{\text{lumen}} + \text{ADP} + P_i \rightarrow 4H^+_{\text{stroma}} + \text{ATP} + H_2O \qquad \textbf{(11-12)}$$

Thus, the flow of four electrons through the noncyclic pathway shown in Figures 11-8 and 11-9 not only generates two NADPH molecules but also leads to the synthesis of about three ATP molecules (4 electrons $\times$ 3 protons/electron $\times$ 1 ATP/4 protons = 3 ATP molecule) per oxygen molecule.

Cyclic Photophosphorylation Allows a Photosynthetic Cell to Balance NADPH and ATP Synthesis

Before we conclude our discussion of the photosynthetic energy transduction reactions, we must consider how phototrophs might balance NADPH and ATP synthesis to meet the precise energy needs of a living cell. Notice that noncyclic electron flow leads to the generation of two ATP for every two NADPH molecules. Because both ATP and NADPH are consumed by a variety of metabolic pathways, it is very unlikely that a photosynthetic cell will always require them in the precise ratio generated by noncyclic electron flow. Typically, cells will require more ATP than NADPH because many cellular activities, such as active transport across membranes, require ATP but not NADPH.

When additional ATP is needed, an optional process known as **cyclic electron flow** can divert the reducing power generated at PSI into ATP synthesis rather than $NADP^+$ reduction. Cyclic electron flow, illustrated in **Figure 11-10**, is distinct from the Q cycle discussed previously. In cyclic electron flow, the reduced ferredoxin generated by PSI transfers electrons back to a cytochrome b_6/f complex instead of donating them to $NADP^+$. The exergonic flow of electrons from ferredoxin through the cytochrome b_6/f complex to plastocyanin is coupled to proton pumping, thereby contributing to the pmf across the thylakoid membrane. Because cyclic electron flow is coupled to unidirectional proton pumping across the thylakoid membrane, excess reducing power can be channeled into ATP synthesis rather than production of NADPH.

From plastocyanin, electrons return to an oxidized $P700^+$ molecule in PSI, completing a closed circuit and allowing P700 to absorb another photon. The excess ATP synthesis resulting from this cyclic electron flow is called *cyclic photophosphorylation*. No water is oxidized and no oxygen is released because the flow of electrons from PSII is not involved.

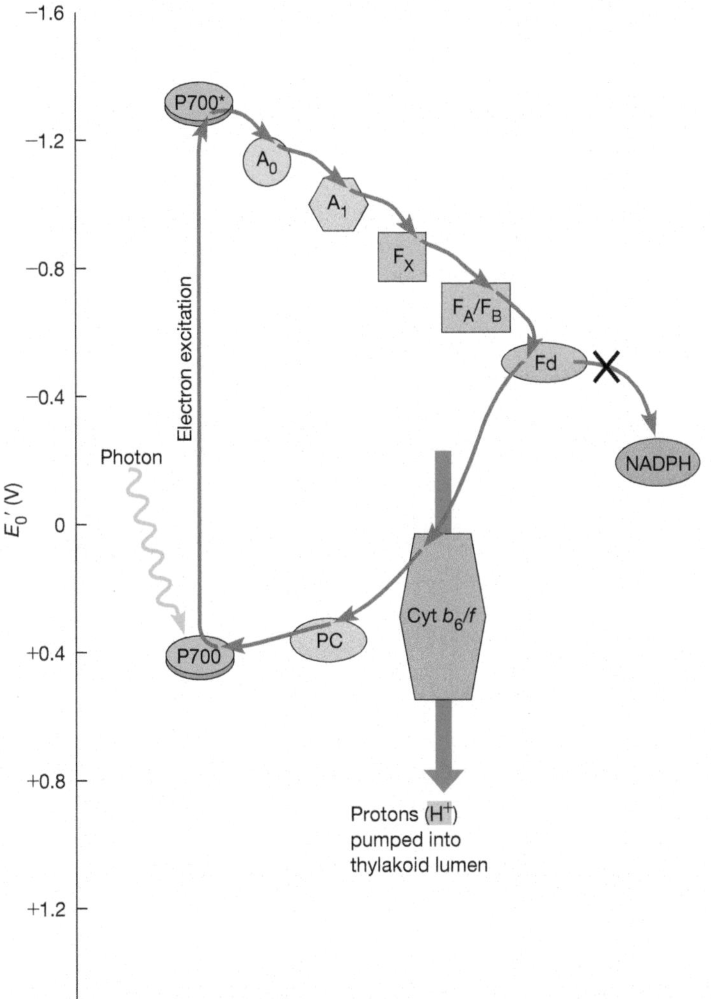

Figure 11-10 Cyclic Electron Flow. Cyclic electron flow through PSI enables oxygenic phototrophs to increase the ratio of ATP/NADPH production within photosynthetic cells. When extra ATP is needed, ferredoxin (Fd) donates electrons to the cytochrome b_6/f complex rather than to $NADP^+$. Electrons then return to P700 via plastocyanin (PC). Because this cyclic electron flow is coupled to unidirectional proton pumping into the thylakoid lumen, excess reducing power is channeled into ATP synthesis. See Figure 11-8 for a general key to symbols.

A Summary of the Complete Energy Transduction System

The model shown in Figure 11-9 represents the entire photosynthetic energy transduction system within a thylakoid membrane. The essential features of the complete system for the transfer of electrons from water to $NADP^+$ and the resulting ATP synthesis can be summarized in terms of the following component parts:

1. *Photosystem II complex:* An assembly of chlorophyll molecules, accessory pigments, and proteins that contains the P680 reaction center chlorophyll. Water is oxidized and split by the oxygen-evolving complex, and electrons flow from water to P680. Following photon absorption, photoexcited P680 molecules donate electrons to plastoquinone, reducing it to plastoquinol, a mobile electron carrier.
2. *Cytochrome b_6/f complex:* An electron transport system that transfers electrons from plastoquinol to plastocyanin, thereby linking PSII with PSI. Electron flow through this complex is coupled to unidirectional proton pumping across the thylakoid membrane, establishing an electrochemical proton gradient that drives ATP synthesis. Optional cyclic electron flow allows additional ATP synthesis.
3. *Photosystem I complex:* An assembly of chlorophyll molecules, accessory pigments, and proteins that contains the P700 reaction center chlorophyll, which accepts electrons from plastocyanin. Following photoexcitation, P700 donates electrons to ferredoxin, a stromal protein.
4. *Ferredoxin-$NADP^+$ reductase:* An enzyme on the stromal side of the thylakoid membrane that catalyzes the transfer of electrons from two reduced ferredoxin proteins along with a proton to a single $NADP^+$ molecule. The NADPH generated in the stroma is an essential reducing agent in many anabolic pathways.
5. *ATP synthase complex (CF_oCF_1):* A proton channel and ATP synthase that couples the exergonic flow of protons from the thylakoid lumen to the stroma with the synthesis of ATP. Like NADPH, the ATP accumulates in the stroma, where it provides energy for carbon assimilation.

Within a chloroplast, both noncyclic and cyclic pathways of electron flow operate (with or without the Q cycle), thereby providing flexibility in the relative amounts of ATP and NADPH generated. ATP can be produced on a close to equimolar basis with respect to NADPH if the noncyclic pathway is operating alone, or ATP can be generated in excess by using either the Q cycle or the cyclic pathway of PSI.

Bacteria Use a Photosynthetic Reaction Center and Electron Transport System Similar to Those in Plants

Much of our knowledge of light harvesting and photosynthetic reaction centers originally came from studies of reaction center complexes from photosynthetic bacteria. In the early 1980s, Hartmut Michel, Johann Deisenhofer, and Robert Huber were able to crystallize a reaction center complex from a photosynthetic purple bacterium, *Blastochloris viridis* (formerly *Rhodopseudomonas viridis*), and determined its molecular structure by X-ray crystallography. They not only provided the first detailed look at how pigment molecules are arranged to capture light energy but also were the first group to crystallize any membrane protein complex. For their exciting and groundbreaking contributions, Michel, Deisenhofer, and Huber shared a Nobel Prize in 1988.

As shown in **Figure 11-11**, the reaction center of *B. viridis* includes four protein subunits. The first subunit is a cytochrome *c* molecule extending from the outer surface of the bacterial membrane. The L and M subunits span the membrane and stabilize a total of four bacteriochlorophyll *b* molecules, two bacteriopheophytin molecules, and two quinones. Subunits L and M are similar to the PSII proteins D1 and D2 of plants. The H subunit extends out of the cytoplasmic surface of the membrane and is homologous to a PSII protein called CP43. Electron flow through this bacterial photosystem resembles the flow of electrons through PSII in oxygenic phototrophs, with one major difference: the bacterial photosystem does not release molecular oxygen and thus is an example of anoxygenic photosynthesis.

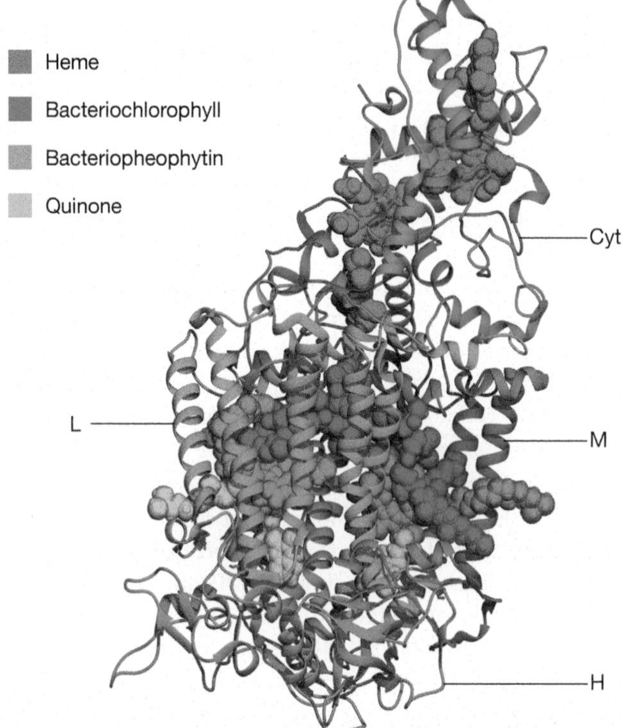

Figure 11-11 A Bacterial Photosynthetic Reaction Center. The structure for the reaction center from *Blastochlorus viridis* contains four protein subunits with bound cofactors: the cytochrome at the bacterial surface (Cyt) with four bound heme groups; the transmembrane L and M subunits containing bacteriochlorophyll, bacteriopheophytin, and two quinones; and the cytosolic H subunit.

A pair of bacteriochlorophyll *b* molecules, together designated *P960* (absorption maxima at 960 nm), initiate the light-dependent electron flow. Absorption of a photon by P960 lowers its reduction potential from about +0.5 to −0.7 V, exciting an electron. The photoexcited electron is immediately transferred to bacteriopheophytin, stabilizing the charge separation. From bacteriopheophytin, the electron flows exergonically through two quinones and a cytochrome b/c_1 complex to cytochrome *c*. This electron flow is coupled to unidirectional proton pumping across the bacterial membrane, creating an electrochemical proton gradient. Cytochrome *c* then returns the electron to oxidized P960.

Notice that the flow of electrons just described is cyclic, starting and ending at P960, so there is no net gain of reducing power. How, then, does the cell generate reductants such as NADH? In *B. viridis*, the cytochrome b/c_1 complex and cytochrome *c* can accept electrons from donors such as hydrogen sulfide, thiosulfate, or succinate. ATP generated by this process is then used to push some electrons energetically uphill from the cytochrome b/c_1 complex or from cytochrome *c* to NAD^+, generating the reductant NADH.

CONCEPT CHECK 11.4

Follow the path of energy interconversions from the moment of capture of solar energy to the (final) synthesis of ATP. Consider solar, chemical, electrochemical, mechanical, and other forms of energy.

11.5 Photosynthetic Carbon Assimilation I: The Calvin Cycle

With information about chloroplast structure and photosynthetic energy transduction in mind, we are now prepared to look closely at photosynthetic *carbon assimilation*. This biosynthetic process is also referred to as *carbon fixation* because inorganic carbon atoms from atmospheric carbon dioxide are *fixed*. This means that they are covalently joined to solid organic compounds, ultimately forming carbohydrates. More specifically, we will look at the primary events of carbon assimilation: the initial fixation and reduction of carbon dioxide to form simple three-carbon carbohydrates.

The fundamental pathway for the movement of inorganic carbon into the biosphere is the **Calvin cycle**, which is found in all oxygenic and most anoxygenic phototrophs. This pathway is named after Melvin Calvin, who received a Nobel Prize in 1961 for the work he and his colleagues Andrew Benson

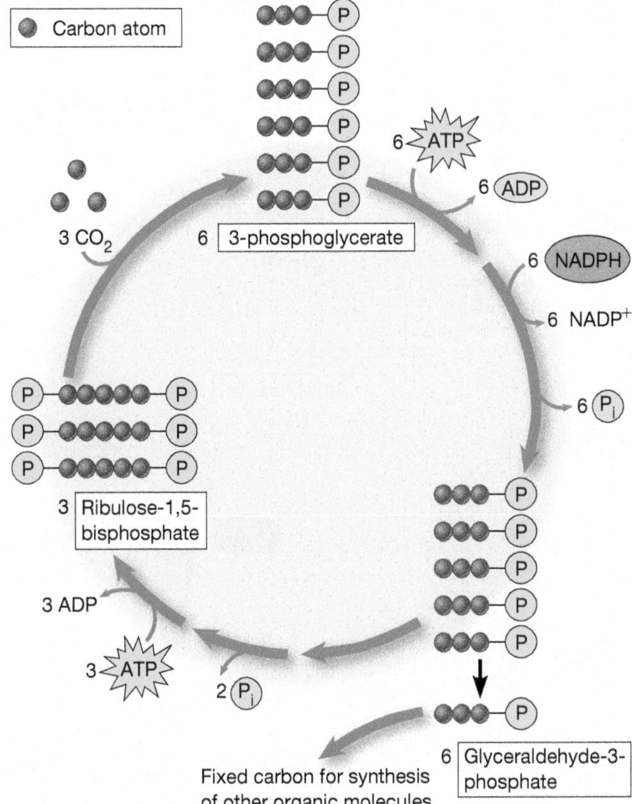

Figure 11-12 Overview of the Calvin Cycle. Three molecules of CO_2 are fixed onto three molecules of ribulose-1,5-bisphosphate (RuBP), forming six molecules of 3-phosphoglycerate, which are then reduced to glyceraldehyde-3-phosphate (G3P). Five of these G3P molecules are used to regenerate three molecules of RuBP, which accept three more CO_2. The sixth molecule of G3P represents the three carbons fixed per turn of the cycle.

and James Bassham did to elucidate the process. Taking advantage of the availability of radioactive isotopes following World War II, they were able to use $^{14}CO_2$ to show that the primary products of photosynthetic carbon fixation are *triose sugars*. These trioses enter a variety of metabolic pathways, the most important of these being sucrose and starch biosynthesis. An overview of the Calvin cycle is shown in **Figure 11-12**, with more biochemical and structural detail shown in **Figure 11-13** on page 302.

In plants and algae, the Calvin cycle is confined to the chloroplast stroma, where the ATP and NADPH generated by photosynthetic energy transduction reactions accumulate. In plants, carbon dioxide generally enters a leaf through special pores called **stomata** (singular: **stoma**). Once inside the leaf, carbon dioxide molecules diffuse into **mesophyll cells** and, in most plant species, travel unhindered into the chloroplast stroma, the site of carbon fixation.

For convenience, we divide the Calvin cycle into three stages:

1. The carboxylation (fixation) of the initial acceptor molecule *ribulose-1,5-bisphosphate* and immediate hydrolysis to generate two molecules of *3-phosphoglycerate*.
2. The reduction of 3-phosphoglycerate to form *glyceraldehyde-3-phosphate*.
3. The regeneration of initial acceptor ribulose-1,5-bisphosphate to allow continued carbon assimilation.

Carbon Dioxide Enters the Calvin Cycle by Carboxylation of Ribulose-1,5-Bisphosphate

The first stage of the Calvin cycle begins with the covalent attachment of carbon dioxide to the carbonyl carbon of ribulose-1,5-bisphosphate (reaction CC-1 in Figure 11-13). Carboxylation of this five-carbon acceptor molecule would seem to generate a six-carbon product, but no one has isolated such a molecule. As Calvin and his colleagues showed by exposing the alga *Chlorella* to radiolabeled $^{14}CO_2$, the first detectable product of carbon dioxide fixation by this pathway is a three-carbon molecule, 3-phosphoglycerate. Presumably, the six-carbon compound exists only as a transient enzyme-bound intermediate, which is immediately hydrolyzed to generate two molecules of 3-phosphoglycerate, as shown in Reaction 11-13.

$$\underset{\text{Ribulose-1,5-bisphosphate}}{\text{H-C(H)(O-P)-C(=O)-C(H)(OH)-C(H)(OH)-C(H)(H)-O-P}} \xrightarrow{CO_2\ \ H^+} \underset{\text{Enzyme-bound intermediate}}{\text{H-C(H)(O-P)-C(COO}^-\text{)(OH)-C(=O)-C(H)(OH)-C(H)(H)-O-P}} \xrightarrow{H_2O\ \ H^+} \underset{\text{Two molecules of 3-phosphoglycerate}}{2\ \ {}^-\text{O-C(=O)-C(H)(OH)-C(H)(H)-O-P}} \qquad \textbf{(11-13)}$$

The newly fixed carbon dioxide (shown in red) appears as a carboxyl group on one of the two 3-phosphoglycerate molecules.

The enzyme that catalyzes the capture of the carbon dioxide and the formation of 3-phosphoglycerate is called **rubisco (ribulose-1,5-bisphosphate carboxylase/oxygenase).** This is a relatively large enzyme (**Figure 11-14** on page 303). In plants and algae, it has 16 subunits and a molecular weight of about 560,000 kDa. Rubisco is unique to phototrophs and is found in all photosynthetic organisms except for a few photosynthetic bacteria. Considering its essential role in carbon dioxide fixation for virtually the entire biosphere, it is hardly surprising that rubisco is thought to be the most abundant protein on the planet. About 10–25% of soluble leaf protein is rubisco, and one estimate puts the total amount of rubisco on the Earth at 40 million tons, or almost 15 lb (about 7 kg) for each living person.

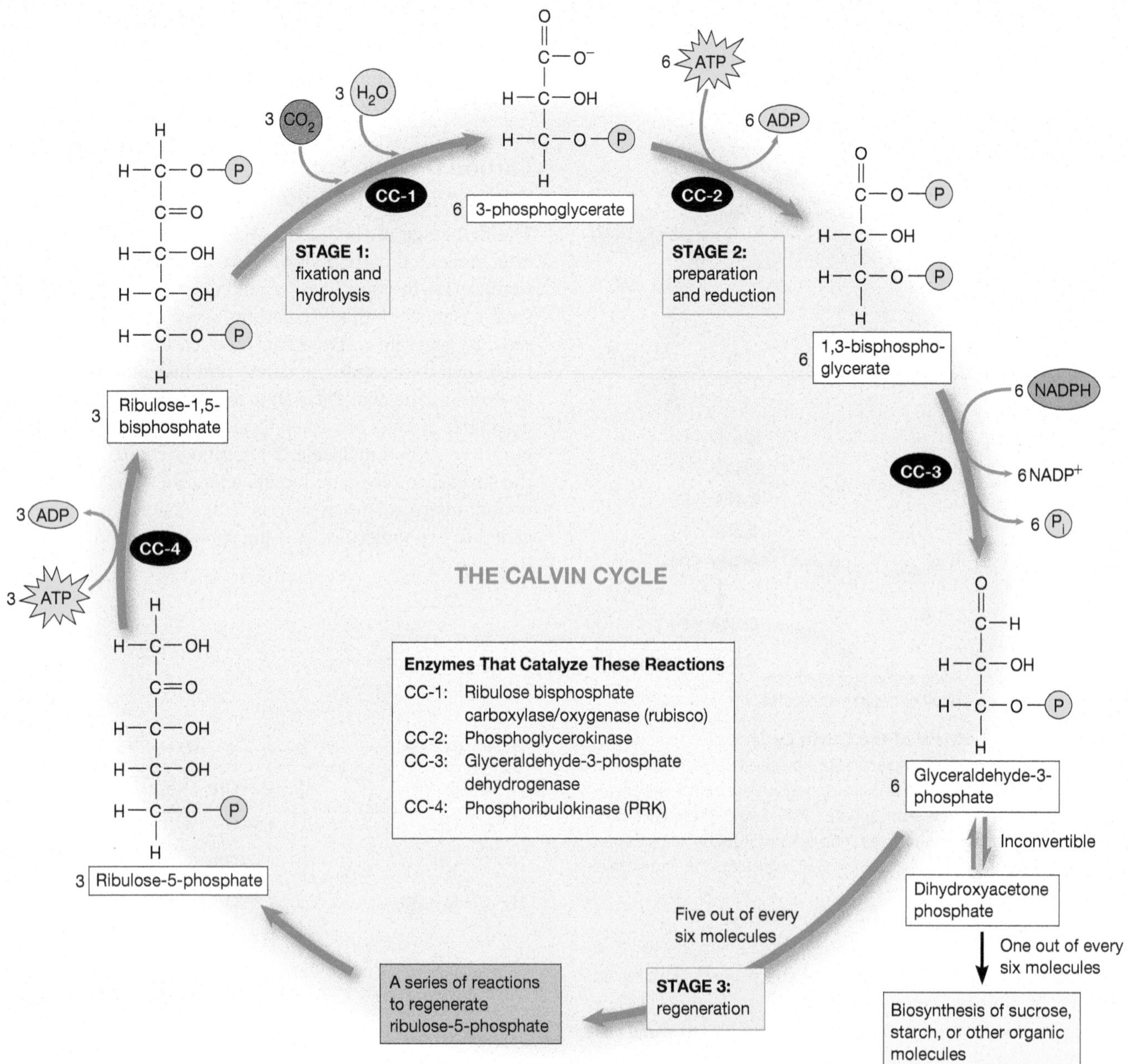

Figure 11-13 The Calvin Cycle for Photosynthetic Carbon Assimilation. During one turn of the Calvin cycle, three CO_2 molecules are fixed and one molecule of glyceraldehyde-3-phosphate (G3P) is produced. CO_2 is fixed in reaction CC-1, forming 3-phosphoglycerate (3PG). Reactions CC-2 and CC-3 use ATP and NADPH to reduce 3PG to G3P. One out of every six G3P molecules is used for the biosynthesis of sucrose, starch, or other organic molecules. The other five G3P molecules are used to regenerate three molecules of ribulose-5-phosphate. The ribulose-5-phosphate is then phosphorylated using ATP in reaction CC-4 to regenerate ribulose-1,5-bisphosphate (RuBP), the acceptor molecule for reaction CC-1, thus completing one turn of the cycle.

3-Phosphoglycerate Is Reduced to Form Glyceraldehyde-3-Phosphate

In the second stage of the Calvin cycle, the 3-phosphoglycerate molecules formed during carbon dioxide fixation are reduced to glyceraldehyde-3-phosphate. This involves a sequence of reactions that is essentially the reverse of the oxidative sequence of glycolysis (reactions Gly-6 and Gly-7 in Figure 9-7) except that the coenzyme involved is NADPH, not NADH. The steps for NADPH-mediated reduction are shown as reactions CC-2 and CC-3 in Figure 11-13. In the first reaction, *phosphoglycerokinase* catalyzes the transfer of a phosphate group from ATP to 3-phosphoglycerate. This reaction generates an activated intermediate, 1,3-bisphosphoglycerate. In the second reaction, *glyceraldehyde-3-phosphate dehydrogenase* catalyzes the transfer of two electrons and one proton from NADPH to 1,3-bisphosphoglycerate, reducing it to glyceraldehyde-3-phosphate (G3P).

Some accounting is in order at this point. For every carbon dioxide molecule that is fixed by rubisco (reaction CC-1), two 3-phosphoglycerate molecules are generated. The reduction of both of these molecules to glyceraldehyde-3-phosphate requires the hydrolysis of two ATP molecules and the

of carbon dioxide in leaf cells may decline. Moreover, water photolysis continues to generate oxygen, which accumulates because it cannot diffuse out of the leaf when the stomata are closed.

In some cases, the potential for energy and carbon drain through photorespiration is so overwhelming that plants must depend on adaptive strategies for solving the problem. One general approach is to confine rubisco to cells that contain a high concentration of carbon dioxide, thereby minimizing the enzyme's inherent oxygenase activity.

In many tropical grasses, including economically important plants such as maize, sorghum, and sugarcane, the isolation of rubisco is accomplished by a short carboxylation/decarboxylation pathway referred to as the **Hatch–Slack cycle,** after Marshall D. Hatch and C. Roger Slack, two plant physiologists who played key roles in the elucidation of the pathway. Plants containing this pathway are referred to as **C_4 plants** because the immediate product of carbon dioxide fixation by the Hatch–Slack cycle is the *four-carbon* organic acid oxaloacetate. This term distinguishes such plants from **C_3 plants,** in which the first detectable product of carbon dioxide fixation is the *three-carbon* compound 3-phosphoglycerate.

To appreciate the advantage of the Hatch–Slack cycle, we must first consider the arrangement of the Hatch–Slack and Calvin cycles within the leaf of a C_4 plant. As shown in **Figure 11-18**, C_4 plants, unlike C_3 plants, have two distinct types of photosynthetic cells in their leaves: In addition to mesophyll cells, the **bundle sheath cells** of C_4 plants have chloroplasts as well. The mesophyll and bundle sheath cells of C_4 plants differ in their enzyme composition and hence their metabolic activities. The carbon dioxide fixation step within a C_4 plant is accomplished by an enzyme other than rubisco in mesophyll cells, which are exposed to the carbon dioxide and oxygen that enter a leaf through its stomata. The carbon dioxide that is fixed in mesophyll cells is subsequently released in bundle sheath cells, which are relatively isolated from the atmosphere. The entire Calvin cycle, including rubisco, is confined to chloroplasts in the bundle sheath cells. Due to the activity of the Hatch–Slack cycle, the carbon dioxide concentration in C_4 bundle sheath cells may be as much as ten times the level in the atmosphere, strongly favoring rubisco's carboxylase activity and minimizing its oxygenase activity.

As detailed in **Figure 11-19** on page 310, the Hatch–Slack cycle begins with the carboxylation of *phosphoenolpyruvate (PEP)* to form oxaloacetate (reaction HS-1). Carboxylation is catalyzed by a specific cytosolic form of *PEP carboxylase,* which is particularly abundant in mesophyll cells of C_4 plants. Not only does this carboxylase lack rubisco's oxygenase activity, it is an excellent scavenger for carbon dioxide in the form of the bicarbonate that is produced when carbon dioxide dissolves in water.

In one version of the Hatch–Slack cycle, the oxaloacetate generated by PEP carboxylase is rapidly converted to malate by an *NADPH-dependent malate dehydrogenase* (reaction HS-2 in Figure 11-19). Malate is a stable four-carbon acid that carries carbon from mesophyll cells to chloroplasts of bundle sheath cells, where decarboxylation by *$NADP^+$ malic enzyme* releases CO_2 (reaction HS-3). The liberated carbon dioxide is then refixed and reduced by the Calvin cycle. Because decarboxylation of malate is accompanied by the generation of NADPH, the Hatch–Slack cycle also conveys reducing power from mesophyll to bundle sheath cells. This decreases the demand for noncyclic electron flow from water to $NADP^+$ in the bundle sheath cells, thereby minimizing the formation of oxygen by PSII complexes and further favoring rubisco's carboxylase activity.

The pyruvate generated by decarboxylation of malate diffuses into a mesophyll cell, where it is phosphorylated by

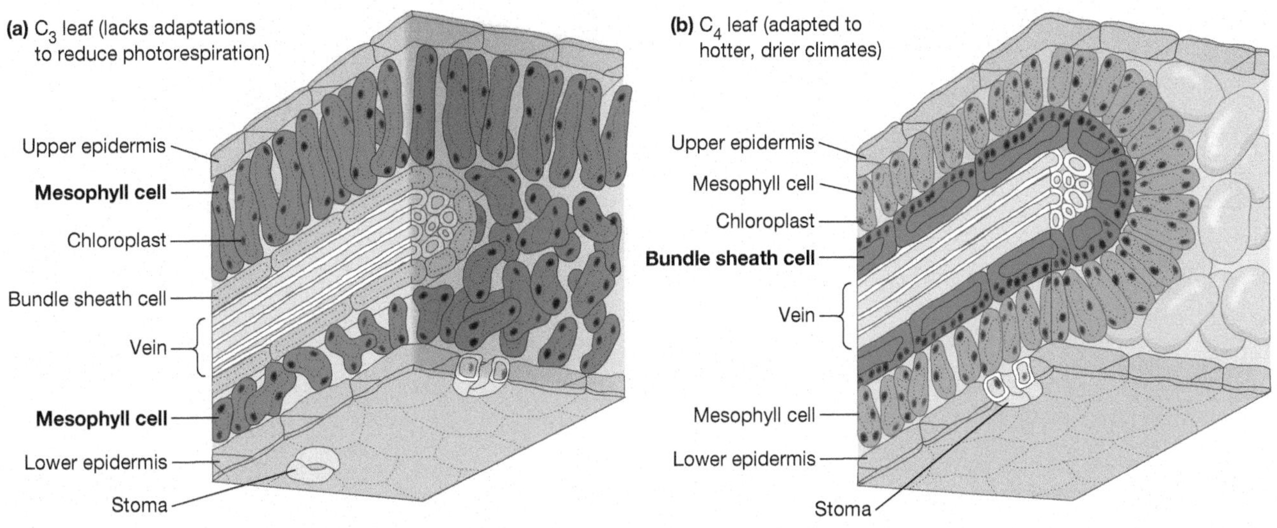

Figure 11-18 Structural Differences Between Leaves of C_3 and C_4 Plants. (a) In C_3 plants, the Calvin cycle occurs in mesophyll cells (dark green). **(b)** In C_4 plants, the Calvin cycle is confined to bundle sheath cells (dark green), which are relatively isolated from atmospheric carbon dioxide and oxygen. C_4 plants utilize the Hatch–Slack cycle in mesophyll cells (light green) to collect carbon dioxide, which is then concentrated in bundle sheath cells. The bundle sheath cells surround the vascular bundles (veins) of the leaf, which carry carbohydrates to other parts of the plant.

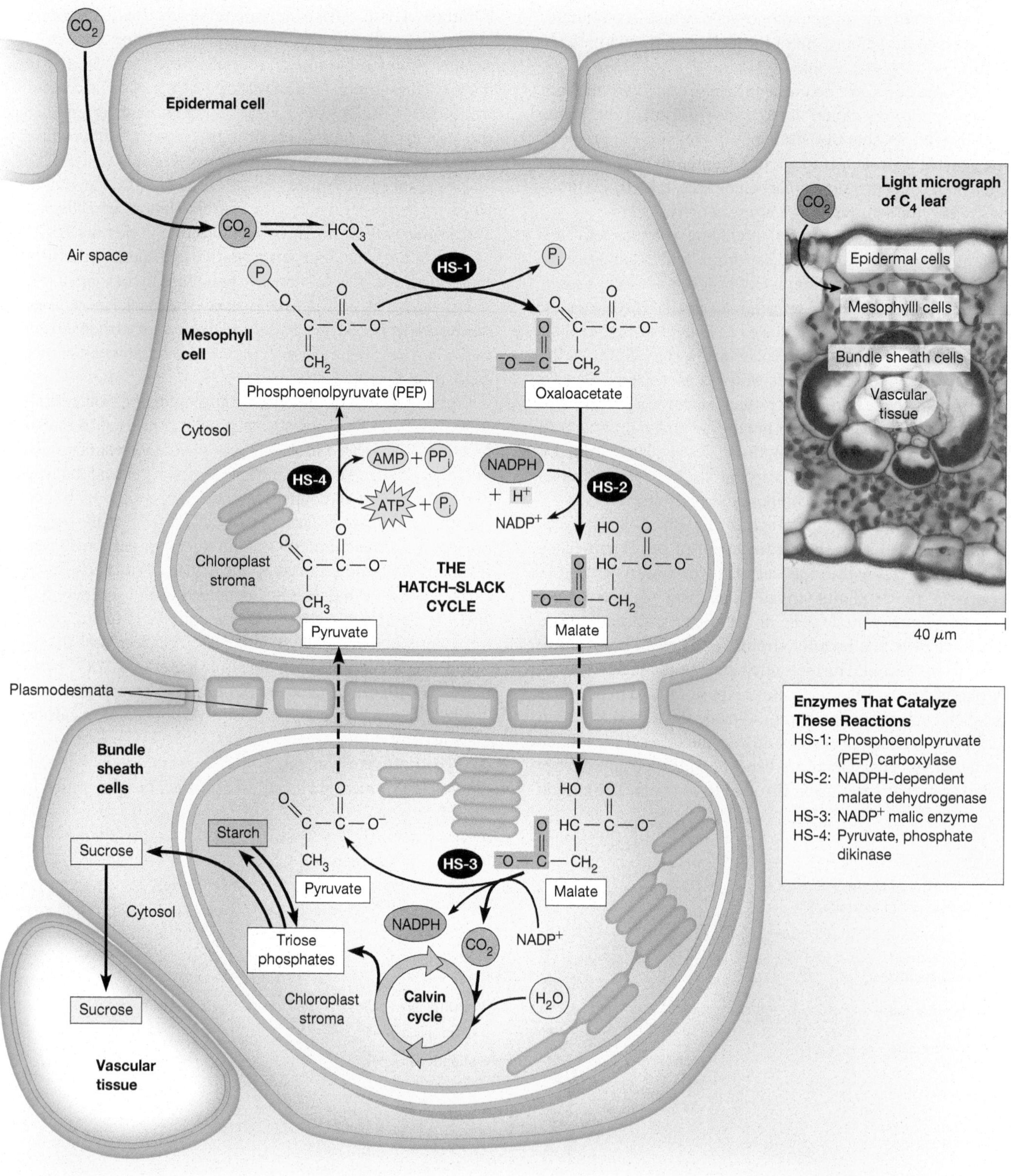

Figure 11-19 Localization of the Hatch–Slack Cycle Within Different Cells of a C_4 Leaf. Carbon dioxide fixation in C_4 plants occurs by the Hatch–Slack cycle within mesophyll cells, initially forming oxaloacetate, which is then reduced to malate. The malate moves to the bundle sheath cells, where it is decarboxylated, and the resulting carbon dioxide is refixed by the Calvin cycle.

the enzyme pyruvate *phosphate dikinase*, regenerating PEP, the original carbon dioxide acceptor of the Hatch–Slack cycle (reaction HS-4). (This dikinase catalyzes the phosphorylation of both pyruvate and phosphate and is unique to the Hatch–Slack cycle.) Thus, the overall process is cyclic, and the net result is a feeder system that captures carbon dioxide in mesophyll cells and passes it to the Calvin cycle in bundle sheath cells. The Hatch–Slack cycle is not a substitute for the Calvin cycle. It is simply a preliminary carboxylation/decarboxylation sequence that concentrates CO_2 in the bundle sheath cells where there is less oxygen to compete with rubisco's carboxylation activity.

Because two high-energy phosphoanhydride bonds are hydrolyzed in converting ATP to AMP in reaction HS-4, the energetic cost of moving carbon from mesophyll to bundle sheath cells is roughly equivalent to hydrolyzing two ATP molecules to ADP for each carbon dioxide molecule, in addition to the three ATP molecules needed during the Calvin cycle in the bundle sheath cell. In an environment that enhances rubisco's oxygenase activity, however, the energy required to prevent the formation of phosphoglycolate may be far less than the energy that would otherwise be lost through photorespiration.

When temperatures exceed about 30°C, the photosynthetic efficiency of a C_4 plant exposed to intense sunlight may be twice that of a C_3 plant. This is one reason that crabgrass, a C_4 plant, often outgrows other C_3 lawn grasses. Whereas the higher efficiency of a C_4 plant is largely due to reduced photorespiration, other factors are also important. For example, because of their ability to concentrate CO_2, C_4 plants are less affected by the lowered CO_2 concentrations that can exist in dense growth where CO_2 is being rapidly consumed.

Enrichment of carbon dioxide in the vicinity of rubisco by the Hatch–Slack cycle in bundle sheath cells confers an additional advantage on C_4 plants. Because PEP carboxylase (reaction HS-1) is an efficient scavenger of carbon dioxide, gas exchange through the stomata of C_4 plants can be substantially reduced to conserve water without adversely affecting photosynthetic efficiency. As a result, C_4 plants are able to assimilate over twice as much carbon as C_3 plants for each unit of water transpired. This adaptation makes C_4 plants suitable for regions of periodic drought, such as tropical savannas.

Although only a few percent of plant species investigated depend on the Hatch–Slack cycle, the pathway is of particular interest because several economically important species are in this group. Also, C_4 plants such as maize and sugarcane are characterized by net photosynthetic rates that are often two or three times those of C_3 plants such as cereal grains. Little wonder, then, that crop physiologists and plant breeders have devoted so much attention to C_4 species and to the question of whether it is possible to improve on the relatively inefficient carbon dioxide fixation pathway of the C_3 plants. Some genetic engineers even envision genetically converting C_3 plants into C_4 plants.

CAM Plants Minimize Photorespiration and Water Loss by Opening Their Stomata Only at Night

Finally, we will consider a third strategy used by some plants to cope with the wasteful oxygenase activity of rubisco. Certain plant species that live in deserts, salt marshes, and other environments where access to water is severely limited contain a preliminary carbon dioxide fixation pathway closely related to the Hatch–Slack cycle. The sequence of reactions is similar, but these plants segregate the carboxylation and decarboxylation reactions by *time* rather than by *space*. Because the pathway was first recognized in the family of succulent plants known as the Crassulaceae, it is called **crassulacean acid metabolism (CAM)**, and plants that take advantage of CAM photosynthesis are called **CAM plants**. CAM photosynthesis has been found in 5–10% of plant species investigated, including many succulents, cacti, orchids, and bromeliads such as pineapple.

CAM plants, unlike most C_3 and C_4 plants, generally open their stomata only at night, when the atmosphere is relatively cool and moist. As carbon dioxide diffuses into mesophyll cells, it is assimilated by the first two steps of a pathway similar to the Hatch–Slack cycle and accumulates within the cells as malate. Instead of being exported from mesophyll cells, however, the malate is stored in large vacuoles, which become very acidic. The process of moving malate into vacuoles consumes ATP but is necessary to protect cytosolic enzymes from a large drop in pH at night.

During the day, CAM plants close their stomata to conserve water. The malate then diffuses from vacuoles to the cytosol, where the Hatch–Slack-like cycle continues. Carbon dioxide released by decarboxylation of malate diffuses into the chloroplast stroma, where it is refixed and reduced by the Calvin cycle. The high carbon dioxide and low oxygen concentrations established when light is available for generating ATP and NADPH strongly favor rubisco's carboxylase activity and minimize the loss of carbon through photorespiration. Notice that the carboxylation of PEP and decarboxylation of malate occur in the same compartment. Because of this, the activity of PEP carboxylase in CAM plants must be strictly inhibited during the day to prevent a futile cycle from developing.

With their remarkable ability to conserve water, CAM plants may assimilate more than 25 times as much carbon as a C_3 plant does for each unit of water transpired. Moreover, some CAM plants display a process called *CAM idling*, whereby the plant keeps its stomata closed night *and* day. Carbon dioxide is simply recycled between photosynthesis and respiration, with virtually no loss of water. Such plants will not, of course, display a net gain of carbohydrate and will not show much if any growth. This ability, however, can enable them to survive droughts lasting up to several months and may contribute to their long life spans, which can exceed 100 years.

CONCEPT CHECK 11.8

The oxygenase activity of rubisco limits its photosynthetic efficiency by competing with its carboxylase activity. What three strategies do plants implement to minimize this limitation?

Summary of Key Points

Mastering™ Biology For activities, animations, and review quizzes, go to the study area at www.masteringbiology.com.

11.1 An Overview of Photosynthesis

- Photosynthesis is the single most vital metabolic process for virtually all forms of life on Earth because all of us, whatever our immediate sources of energy, ultimately depend on the energy radiating from the sun.
- The energy transduction reactions of photosynthesis convert solar energy into chemical energy in the form of NADPH and ATP. The carbon assimilation reactions use this chemical energy to fix and reduce carbon dioxide to carbohydrates.
- In eukaryotic phototrophs, photosynthesis occurs in the chloroplasts, which contain an internal membrane system known as the thylakoids that contains many of the required components.
- Cyanobacteria have internal membranes that function like thylakoids. Although they do not have organelles, they have polyhedral protein complexes called carboxysomes, in which carbon fixation occurs.

11.2 Photosynthetic Energy Transduction I: Light Harvesting

- Photons of light are absorbed by chlorophyll or accessory pigment molecules within the thylakoid or photosynthetic bacterial membranes. Their energy is rapidly passed to a special pair of chlorophyll molecules at the reaction center of a photosystem.
- At the reaction center, the energy is used to excite and eject an electron from chlorophyll and induce charge separation. In oxygenic phototrophs, this electron is replaced by an electron obtained from a water molecule, generating oxygen.

11.3 Photosynthetic Energy Transduction II: NADPH Synthesis

- Electron transfer from water to $NADP^+$ relies on two photosystems acting in series, with photosystem II responsible for the oxidation of water and photosystem I responsible for the reduction of $NADP^+$ to NADPH in the stroma.
- Electron flow between the two photosystems passes through a cytochrome $b_6/_f$ complex, which pumps protons into the thylakoid lumen. The resulting proton gradient represents the stored energy of sunlight.

11.4 Photosynthetic Energy Transduction III: ATP Synthesis

- The proton motive force across the thylakoid membrane is used to drive ATP synthesis by the CF_oCF_1 complex embedded in the membrane.
- As protons flow back from the lumen to the stroma through the CF_o proton channel in the membrane, ATP is synthesized by the CF_1 portion of the complex that extends into the stroma.
- Photosynthetic bacteria use a photosystem and electron transport system similar to those in plants and algae, using similar bound cofactors.

11.5 Photosynthetic Carbon Assimilation I: The Calvin Cycle

- In the stroma, ATP and NADPH are used for the fixation and reduction of carbon dioxide into organic form by enzymes of the Calvin cycle.
- The Calvin cycle involves three main stages: (1) fixation of carbon dioxide by rubisco to form 3-phosphoglycerate, (2) reduction of 3-phosphoglycerate to glyceraldehyde-3-phosphate, and (3) regeneration of ribulose-1,5-bisphosphate, the initial carbon dioxide acceptor.
- The net synthesis of one triose phosphate molecule requires the fixation of three CO_2 molecules and uses nine ATP and six NADPH molecules. For each three CO_2 molecules fixed, one molecule of glyceraldehyde-3-phosphate leaves the cycle to be used for further carbohydrate synthesis.

11.6 Regulation of the Calvin Cycle

- Key enzymes of the Calvin cycle are regulated to ensure maximum efficiency. Several of them are regulated at the level of synthesis, being made only in photosynthetic tissues that are exposed to the light. They are also activated by the high stromal pH and magnesium concentrations that occur in the light.
- Additional means of regulation involve thioredoxin, which senses the redox state of the cell, and rubisco activase, which removes inhibitors from the rubisco active site in the light.

11.7 Photosynthetic Carbon Assimilation II: Carbohydrate Synthesis

- The initial product of carbon dioxide fixation is glyceraldehyde-3-phosphate (G3P), which is interconvertible with a second triose phosphate called dihydroxyacetone phosphate (DHAP). Some G3P and DHAP are used for the biosynthesis of more complex carbohydrates, such as glucose, sucrose, starch, or glycogen, or as sources of energy or carbon skeletons for other metabolic pathways.
- In addition, ATP produced in the chloroplast is used for fatty acid and chlorophyll synthesis and for nitrogen and sulfur reduction and assimilation.

11.8 Rubisco's Oxygenase Activity Decreases Photosynthetic Efficiency

- Rubisco can use oxygen as well as CO_2, resulting in the production of phosphoglycolate, which cannot be used in the Calvin cycle. The glycolate pathway converts two molecules of phosphoglycolate into one molecule of 3-phosphoglycerate, which can enter the Calvin cycle.
- The glycolate pathway involves three organelles: the chloroplast, the leaf peroxisome, and the mitochondrion, each with a characteristic set of enzymes. Because CO_2 is released and oxygen is consumed in a light-dependent manner, this process is also called photorespiration.
- In C_4 and CAM plants, carbon dioxide is fixed by a preliminary carboxylation that does not involve rubisco. Then it is decarboxylated under conditions of low oxygen and concentrated CO_2—conditions that favor rubisco's carboxylase activity—either in a different cell type or at a different time of day.

Problem Set

Mastering™ Biology For activities, animations, and review quizzes, go to the study area at www.masteringbiology.com.

11-1 True or False Indicate whether each of the following statements is true (T) or false (F).

(a) Sucrose is synthesized in the chloroplast stroma and exported from photosynthetic cells to provide energy and reduced carbon for nonphotosynthetic plant cells.

(b) Ninety percent of the solar energy collected by a photosystem complex is absorbed when photons strike a special pair of chlorophyll molecules at the reaction center of the complex.

(c) The energy requirement expressed as ATP consumed per molecule of carbon dioxide fixed is higher for a C_3 plant than for a C_4 plant.

(d) The ultimate electron donor for the photosynthetic generation of NADPH is always water.

(e) The enzyme rubisco is unusual in that, depending on conditions, it exhibits two different enzymatic activities.

11-2 The Mint and the Mouse. Joseph Priestley, a British clergyman, was a prominent figure in the early history of research in photosynthesis. In 1771, Priestley wrote these words:

> One might have imagined that since common air is necessary to vegetable as well as to animal life, both plants and animals had affected it in the same manner; and I own that I had that expectation when I first put a sprig of mint into a glass jar standing inverted in a vessel of water; but when it had continued growing there for some months, I found that the air would neither extinguish a candle, nor was it at all inconvenient to a mouse which I put into it.

Explain the basis of Priestley's observations and indicate their relevance to the early understanding of the nature of photosynthesis.

11-3 DATA ANALYSIS Seeing the Light. In 1883 Theodor Engelmann illuminated filamentous algae with light that had been separated into its component colors using a prism. By adding aerobic bacteria that become more concentrated in regions with a higher concentration of oxygen he could assess which regions of the algae were releasing the most oxygen. Based on the data in Figure 11A-2, predict at which wavelengths the bacteria would be most numerous.

11-4 QUANTITATIVE In the Dark about Chloroplasts. (a) Suppose isolated thylakoids were made acidic by being placed in a solution at pH 4 until the thylakoid lumen reached pH 4. Then the thylakoids were placed into a solution at pH 8. The thylakoids were initially able to make ATP in the dark. Explain this result. **(b)** If the pH difference across a thylakoid membrane is normally 2 pH units, and the membrane potential is 0.03 V, what is the free energy available for movement of 1 mole of protons from the thylakoid lumen to the stroma at 25°C?

11-5 The *hcef* Mutant. Livingston et al. (*Plant Cell* 22[2010]:221) isolated an *Arabidopsis thaliana* mutant with unusually high cyclic electron flow (*hcef*). Can you predict the results they observed when comparing the *hcef* mutant plants with normal *Arabidopsis* plants?

(a) How was noncyclic electron flow affected in *hcef* mutant plants?

(b) What happened to the light-driven proton flux across the thylakoid membrane?

(c) How was the activity of PSII affected?

(d) The researchers observed a large increase in the level of fructose-1,6-bisphosphate in the stroma of the *hcef* mutant. What enzyme do you predict is defective?

(e) Predict the effect on starch synthesis in the stroma.

11-6 The Role of Sucrose. A plant uses solar energy to make ATP and NADPH, which then drive the synthesis of carbohydrates in the leaves. At least one carbohydrate, sucrose, is translocated to nonphotosynthetic parts of the plant (stems, roots, flowers, and fruits) for use as a source of energy. Thus, ATP is used to make sucrose, and the sucrose is then used to make ATP. It would seem simpler for the plant just to make ATP and translocate the ATP itself directly to other parts of the plant, thereby completely eliminating the need for a Calvin cycle, a glycolytic pathway, and a citric acid cycle. Suggest at least two major reasons why plants do not manage their energy economies in this way.

11-7 Effects on Photosynthesis. Assume that you have an illuminated suspension of *Chlorella* cells carrying out photosynthesis in the presence of 0.1% carbon dioxide and 20% oxygen. What will be the short-term effects of the following changes in conditions on the levels of 3-phosphoglycerate and ribulose-1,5-bisphosphate? Explain your answer in each case.

(a) Carbon dioxide concentration is suddenly reduced 1000-fold.

(b) Light is restricted to green wavelengths (510–550 nm).

(c) An inhibitor of photosystem II is added.

(d) Oxygen concentration is reduced from 20% to 1%.

11-8 Photosynthetic Efficiency. In this chapter (page 304), we estimated the maximum photosynthetic efficiency for the conversion of red light, carbon dioxide, and water to glyceraldehyde. Under laboratory conditions, a photosynthetic organism *might* convert 31% of the light energy striking it to chemical bond energy of organic molecules. In reality, however, photosynthetic efficiency is far lower, closer to 5% or less. Considering a plant growing in a natural environment, suggest four reasons for this discrepancy.

11-9 Chloroplast Structure. Where in a chloroplast are the following substances or processes localized? Be as specific as possible.

(a) Ferredoxin-$NADP^+$ reductase

(b) Cyclic electron flow

(c) Starch synthase

(d) Light-harvesting complex I

(e) Plastoquinol

(f) Proton pumping

(g) P700

(h) Reduction of 3-phosphoglycerate

(i) Carotenoid molecules

(j) Oxygen-evolving complex

11-10 Metabolite Transport Across Membranes. For each of the following metabolites, indicate whether you would expect it to be in steady-state flux across one or more membranes in a photosynthetically active chloroplast, and, if so, indicate which membrane(s) the metabolite must cross.

(a) CO_2

(b) P_i

(c) Electrons

(d) Starch

(e) Glyceraldehyde-3-phosphate

(f) NADPH

(g) ATP

(h) O_2

(i) Protons

(j) Pyruvate

12 The Endomembrane System and Protein Sorting

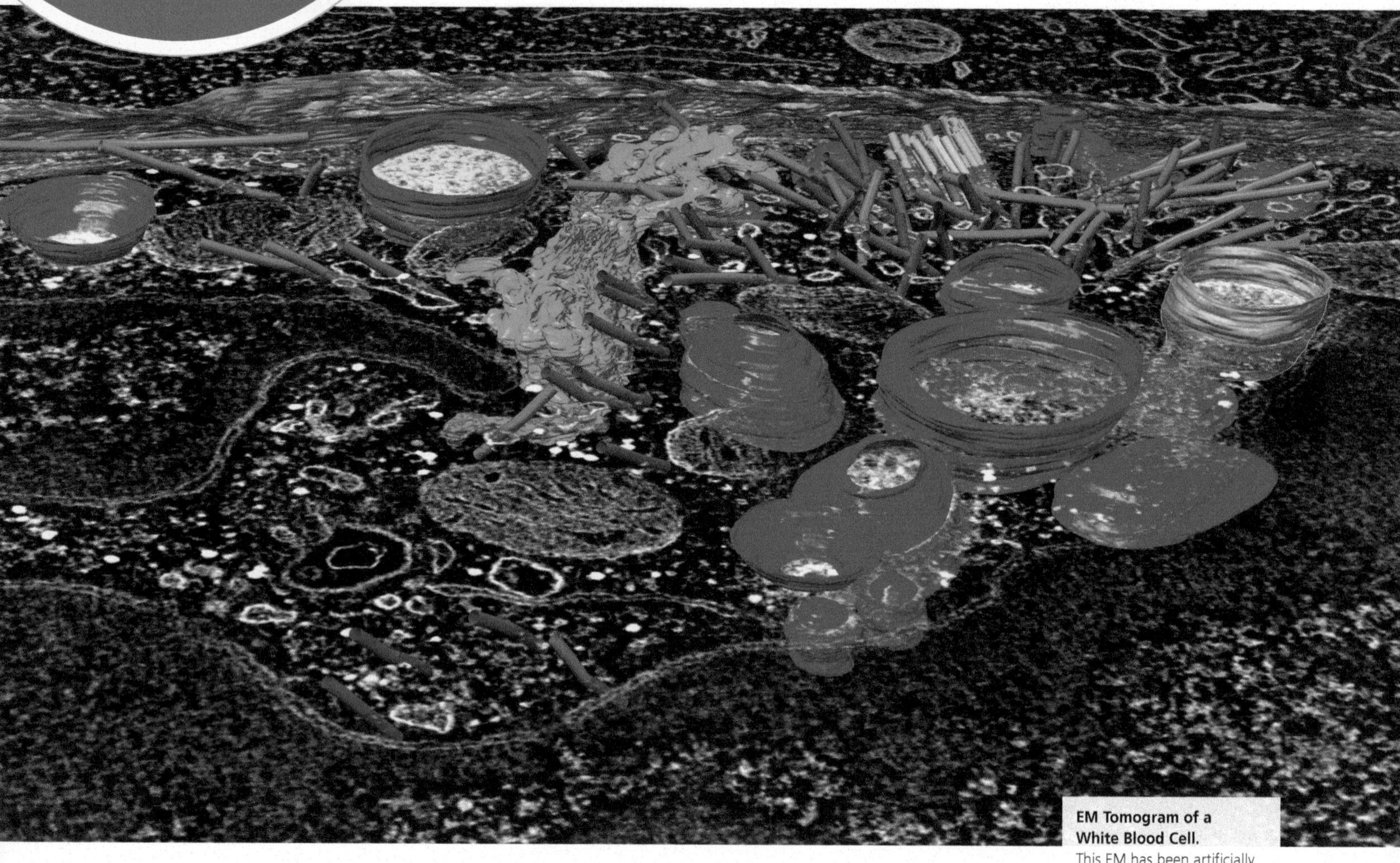

EM Tomogram of a White Blood Cell. This EM has been artificially colored to show lysosomes (blue), Golgi (green), microtubules (red), a centriole (yellow), and the plasma membrane (orange).

A full appreciation of eukaryotic cells depends on an understanding of the prominent role of intracellular membranes and the compartmentalization of function within *organelles*—intracellular membrane-bounded compartments that house various cellular activities. Whether we consider the storage and transcription of genetic information, the biosynthesis of secretory proteins, the breakdown of long-chain fatty acids, or any of the numerous other metabolic processes occurring within eukaryotic cells, many of the reactions of a particular pathway occur within a distinct type of organelle. Also, the movement of proteins and lipids between organelles, known as *trafficking,* must be tightly regulated to ensure that each organelle has the correct components for its proper structure and function.

We briefly encountered the major organelles found in eukaryotic cells in Chapter 4, and we then learned more about the mitochondrion and chloroplast in Chapters 10 and 11, respectively. We are now ready to consider several other individual organelles in more detail. We will begin with the *endoplasmic reticulum* and the *Golgi apparatus,* which are sites for protein synthesis, processing, and sorting; lipid synthesis; and drug detoxification. Next, we will look at *endosomes,* organelles that are important for carrying and sorting material brought

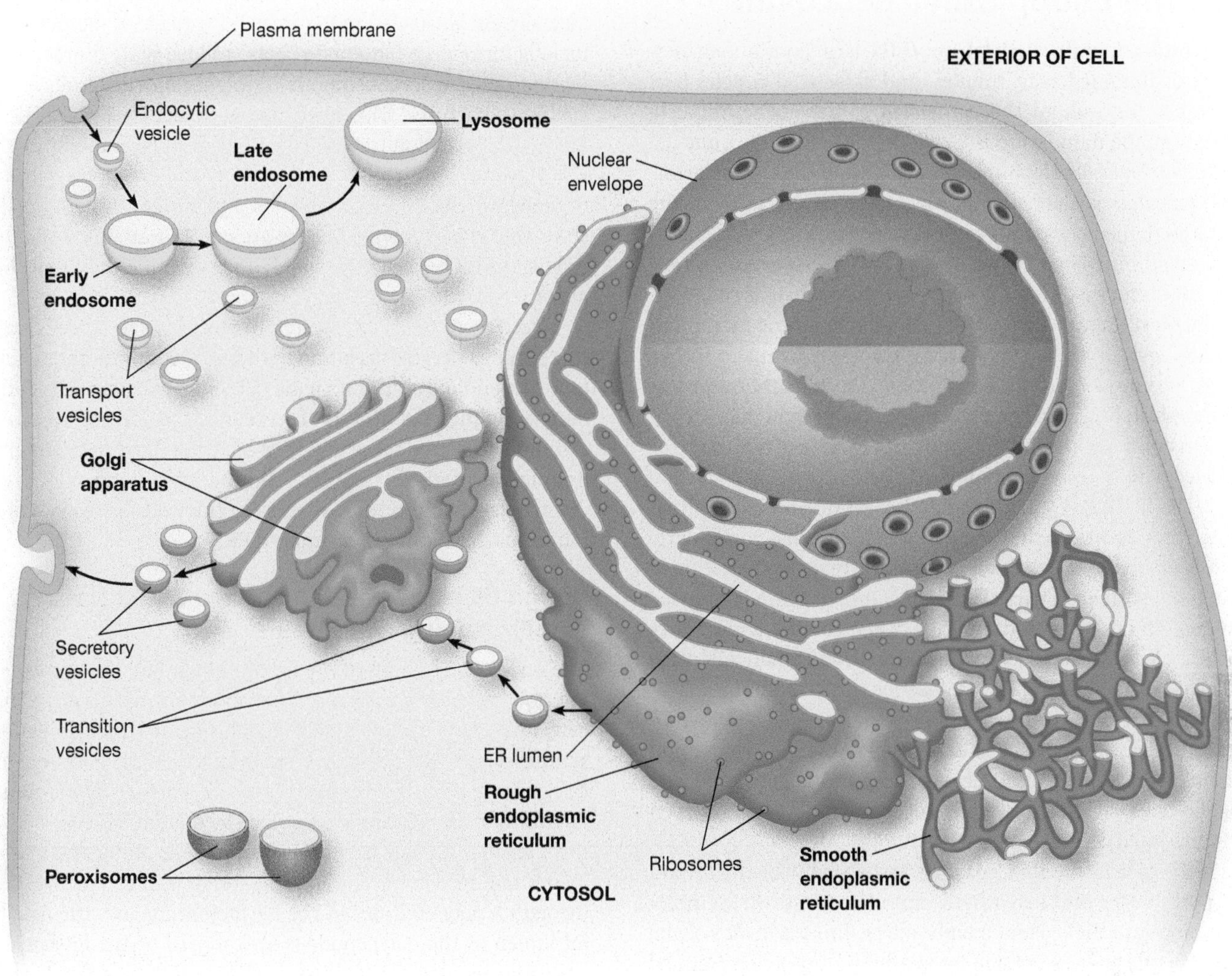

Figure 12-1 The Endomembrane System. The endomembrane system of the eukaryotic cell traditionally consists of the endoplasmic reticulum (ER), the Golgi apparatus, endosomes, and lysosomes. It is associated with both the nuclear envelope and the plasma membrane. The ER lumen is linked to the interiors of the Golgi apparatus, endosomes, and lysosomes by transport vesicles that shuttle material between organelles as well as to and from the plasma membrane. Peroxisomes may originate from the endoplasmic reticulum.

into the cell. Endosomes help form *lysosomes,* which are organelles responsible for digestion of both ingested material and unneeded intracellular components. We will conclude with a look at *peroxisomes,* which house hydrogen peroxide–generating reactions and perform diverse metabolic functions.

As you study the role of each organelle, keep in mind that the endoplasmic reticulum, the Golgi apparatus, endosomes, and lysosomes traditionally make up the **endomembrane system,** as shown in **Figure 12-1**. Because the nuclear envelope is continuous with the endoplasmic reticulum, it too is associated with the endomembrane system. Chapter 4 provided an overview of how the endomembrane system may have originated in early eukaryotic cells. While peroxisomes aren't generally considered part of the endomembrane system, recent work suggests that the endoplasmic reticulum may play a role in peroxisome biogenesis. Material flows from the endoplasmic reticulum to and from the Golgi apparatus, endosomes, and lysosomes by means of *transport vesicles* that shuttle between the various organelles. These transport vesicles carry membrane lipids and membrane-bound proteins to their proper destinations in the cell at the proper time, and they also carry soluble materials destined for secretion to the plasma membrane.

Thus, these organelles and the vesicles connecting them make up a single dynamic system of membranes and internal spaces. Currently, one of the most exciting questions in modern cell biology concerns *endomembrane trafficking:* how does each of the multitude of proteins and lipids in a cell manage to reach its proper destination at the proper time?

12.1 The Endoplasmic Reticulum

The **endoplasmic reticulum (ER)** is a continuous network of flattened sacs, tubules, and associated vesicles that stretches throughout the cytoplasm of the eukaryotic cell. Although the name sounds formidable, it is actually quite descriptive. *Endoplasmic* simply means "within the (cyto)plasm," and *reticulum* is a Latin word meaning "network." Thus, the ER is so named because it is a network within the cytoplasm. The membrane-bounded sacs are called **ER cisternae** (singular: **ER cisterna**), and the space enclosed by them is called the **ER lumen** (Figure 12-1). Of the total membrane in a mammalian cell, up to 50–90% surrounds the ER lumen. Unlike more prominent organelles, such as the mitochondrion or chloroplast, however, the ER is not visible by light microscopy unless one or more of its components are stained with a dye or labeled with a fluorescent molecule.

The ER was first observed in the late nineteenth century, when it was noted that some eukaryotic cells, particularly those involved in secretion, contained regions that stained intensely with basic dyes. The significance of these regions remained in doubt until the 1950s, when the resolving power of the electron microscope was improved dramatically. This allowed cell biologists to visualize for the first time the ER's elaborate network of intracellular membranes and to investigate the role of the ER in cellular processes. This is a common theme in scientific discovery—conceptual advances in one field often follow technological advances in a related (or even an unrelated) field.

We now know that enzymes associated with the ER are responsible for the biosynthesis of proteins destined for incorporation into the plasma membrane or into organelles of the endomembrane system and for synthesis of proteins destined for export from the cell. The ER also plays a central role in the biosynthesis of lipids, including triacylglycerols, cholesterol, and related compounds. The ER is the source of most of the lipids that are assembled to form intracellular membranes and the plasma membrane.

The Two Basic Kinds of Endoplasmic Reticulum Differ in Structure and Function

The two basic kinds of endoplasmic reticulum typically found in eukaryotic cells are distinguished from one another by the presence or absence of ribosomes attached to the ER membrane. **Rough endoplasmic reticulum (rough ER)** is characterized by ribosomes attached to the cytosolic side of the membrane (the side that faces away from the ER lumen; see Figure 4-16d). Translation by these ribosomes occurs in the cytosol, but the newly synthesized proteins will enter the ER lumen shortly. Because ribosomes contain RNA, it was this RNA that reacted strongly with the basic dyes originally used to identify the rough ER. A subdomain of rough ER, the *transitional elements (TEs)*, play an important role in the formation of **transition vesicles** that shuttle lipids and proteins from the ER to the Golgi apparatus. In contrast, **smooth endoplasmic reticulum (smooth ER)** appears smooth due to the absence of ribosomes attached to the membrane (see Figure 4.16c) and has other roles in the cell involving the processing and storage of nonprotein substances.

Rough and smooth ER are easily distinguished morphologically. As illustrated in Figure 12-1, rough ER membranes usually form large flattened sheets studded with characteristic ribosomes, whereas smooth ER membranes generally form tubular structures. The transitional elements of the rough ER are an exception to this rule; they often resemble smooth ER. However, the rough and smooth ER are not separate organelles—electron micrographs and studies in living cells show that their lumenal spaces are continuous. Thus, material can travel between the rough and smooth ER without the aid of vesicles.

Both types of ER are present in most eukaryotic cells, but there is considerable variation in the relative amounts of each type, depending on the activities of the particular cell. Cells involved in the biosynthesis of secretory proteins, such as liver cells and cells producing digestive enzymes, tend to have very prominent rough ER networks. On the other hand, cells producing steroid hormones, such as in the testis or ovary, contain extensive networks of smooth ER.

Rough ER Is Involved in the Biosynthesis and Processing of Proteins

The rough ER is a critical site for protein production, processing, folding, and trafficking. The ribosomes attached to the cytosolic side of the rough ER membrane are responsible for synthesizing both membrane-bound and soluble proteins for the endomembrane system. It is predicted that as many as one-third of all proteins synthesized by mammalian cells pass through the ER. Newly synthesized proteins enter the endomembrane system *cotranslationally*—that is, they are inserted through a pore complex in the ER membrane into the rough ER lumen as the polypeptide is synthesized by the ER-bound ribosome. In addition to its role in the biosynthesis of polypeptide chains, the rough ER is the site for several other processes, including the initial steps of addition and processing of carbohydrate groups to glycoproteins, the folding of polypeptides, the recognition and removal of misfolded polypeptides, and the assembly of multimeric proteins. Thus, ER-specific proteins include a host of enzymes that catalyze cotranslational and posttranslational modifications. The ER is also a site for quality control, where proteins improperly modified, folded, or assembled are exported from the ER for degradation by cytosolic *proteasomes* instead of moving on to the Golgi apparatus. Several human diseases, including familial hypercholesterolemia, are associated with defects in these processes. We will examine the role of the ER in protein sorting, folding, and trafficking in more detail later in this chapter.

Smooth ER Is Involved in Drug Detoxification, Carbohydrate Metabolism, Calcium Storage, and Steroid Biosynthesis

Whereas the rough ER is primarily involved in protein processing and the export of proteins from the cell, the smooth ER is primarily involved in processing or storing nonprotein molecules within cells. We will briefly consider a few of these functions here.

Drug Detoxification. Drug detoxification often involves enzyme-catalyzed **hydroxylation** because the addition of hydroxyl groups to hydrophobic drugs makes them more soluble and easier to excrete from the body. Hydroxylation of organic acceptor molecules is typically catalyzed by a member of the **cytochrome P-450** family of proteins. These proteins are especially prevalent in the smooth ER of hepatocytes (liver cells), in which many drugs are detoxified.

In hepatocytes, an electron transport system transfers electrons from NADPH or NADH to a heme group in a cytochrome P-450 protein, which then donates an electron to molecular oxygen. One atom from molecular oxygen is reduced as it gains two electrons and two H^+, forming H_2O. The other oxygen atom is added to the organic substrate molecule as part of a hydroxyl group. Because one of the two oxygen atoms of O_2 is incorporated into the reaction product, these cytochrome P-450 enzymes are often called *monooxygenases*. The net reaction is shown below, where R represents the drug or compound being hydroxylated to increase its solubility:

$$\mathrm{RH + NAD(P)H + H^+ + O_2 \rightarrow ROH + NAD(P)^+ + H_2O} \qquad \textbf{(12-1)}$$

The elimination of hydrophobic barbiturate drugs, for example, is enhanced by hydroxylation enzymes in the smooth ER. Injection of the sedative phenobarbital into a rat causes a rapid increase in the level of barbiturate-detoxifying enzymes in the liver, accompanied by a dramatic proliferation of smooth ER. However, this means that increasingly higher doses of the drug are necessary to achieve the same sedative effect, an effect known as *tolerance* that is seen in habitual users of phenobarbital. Furthermore, the enzyme induced by phenobarbital can hydroxylate and therefore solubilize a variety of other drugs, including such useful agents as antibiotics, anticoagulants, and steroids. As a result, the chronic use of barbiturates decreases the effectiveness of many other clinically useful drugs.

Another cytochrome P-450 protein found in the smooth ER is part of an enzyme complex called *aryl hydrocarbon hydroxylase*. This complex is involved in metabolizing *polycyclic hydrocarbons*, organic molecules composed of two or more linked benzene rings that are often toxic. Hydroxylation of such molecules is important for increasing their solubility in water, but the oxidized products are often more toxic than the original compounds. Aryl hydrocarbon hydroxylase converts some potential carcinogens into their chemically active forms. Mice synthesizing high levels of this hydroxylase have a higher incidence of spontaneous cancer than normal mice do, whereas mice treated with an inhibitor of aryl hydrocarbon hydrolase develop few tumors. Significantly, cigarette smoke is a potent inducer of aryl hydrocarbon hydroxylase.

Recent work shows that differences in activities and side effects of certain medications can result from differences in the presence or activity of particular cytochrome P-450 genes in different patients. This field of study is known as **pharmacogenetics** (also called *pharmacogenomics*), which investigates how inherited differences in genes like P-450 (and their resulting protein products) can lead to differential responses to drugs and medications.

Carbohydrate Metabolism. The smooth ER of hepatocytes (liver cells) is also involved in the enzymatic breakdown of stored glycogen, as evidenced by the presence of *glucose-6-phosphatase*, a membrane-bound enzyme that is unique to the ER. Thus, its presence is used as a marker to identify the ER during subcellular fractionation or to visualize the ER using fluorescent antibodies. Glucose-6-phosphatase hydrolyzes the phosphate group from glucose-6-phosphate to form free glucose and inorganic phosphate (P_i):

$$\text{glucose-6-phosphate} + H_2O \rightarrow \text{glucose} + P_i \qquad \textbf{(12-2)}$$

This enzyme is abundant in the liver because a major role of the liver is to keep the level of glucose in the blood relatively constant (**Figure 12-2** on page 318). The liver stores glucose as glycogen in granules associated with smooth ER (Figure 12-2a). When glucose is needed by the body, especially between meals and in response to increased muscular activity, liver glycogen is broken down by phosphorolysis (see Figure 9-11), producing glucose-1-phosphate, which is converted to glucose-6-phosphate by phosphoglucomutase (see Figure 9-10, Figure 12-2b). Because membranes are generally impermeable to phosphorylated sugars, the glucose-6-phosphate must be converted to free glucose by glucose-6-phosphatase to leave the cell and enter the bloodstream. Free glucose then leaves the liver cell via a glucose transporter (GLUT2) and moves into the blood for transport to other cells that need energy. Significantly, glucose-6-phosphatase activity is present in liver, kidney, and intestinal cells but not in muscle or brain cells. Muscle and brain cells retain glucose-6-phosphate and use it to meet their own substantial energy needs.

Calcium Storage. The *sarcoplasmic reticulum* found in muscle cells is an example of smooth ER that specializes in the storage of calcium. In these cells, the ER lumen contains high concentrations of calcium-binding proteins. Calcium ions are pumped into the ER by *ATP-dependent calcium ATPases* and are released in response to extracellular signals to aid in muscle contraction. Binding of neurotransmitter molecules to receptors on the surface of the muscle cell triggers a signaling cascade that leads to the release of calcium from the sarcoplasmic reticulum and causes the contraction of muscle fibers. (We will look at nerve impulse transmission in more detail in Chapter 22 and calcium regulation in Chapter 23, Figure 23-14.)

Steroid Biosynthesis. The smooth ER in certain cells is the site of biosynthesis of cholesterol and steroid hormones such as cortisol, testosterone, and estrogen. Large amounts of smooth ER are found in the cortisol-producing cells of the adrenal gland; the Leydig cells of the testes, which produce testosterone; the cholesterol-producing cells of the liver; and the follicular cells of the ovary, which produce estrogen. Smooth ER has also been found in close association with plastids in some plants, where it may be involved in plant hormone synthesis.

Cholesterol, cortisol, and the male and female steroid hormones just described share a common four-ring structure but differ in the number and arrangement of carbon side chains and hydroxyl groups (see Figure 3-27e and Figure 3-30). *Hydroxymethylglutaryl-CoA reductase (HMG-CoA reductase)*, the committed step in cholesterol biosynthesis, is present in

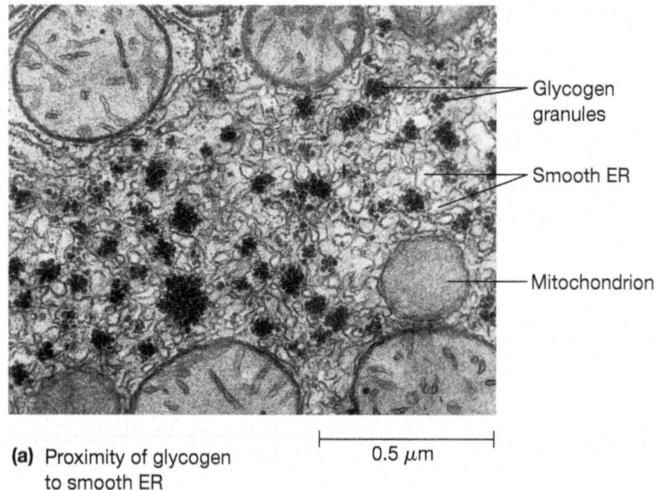

(a) Proximity of glycogen to smooth ER

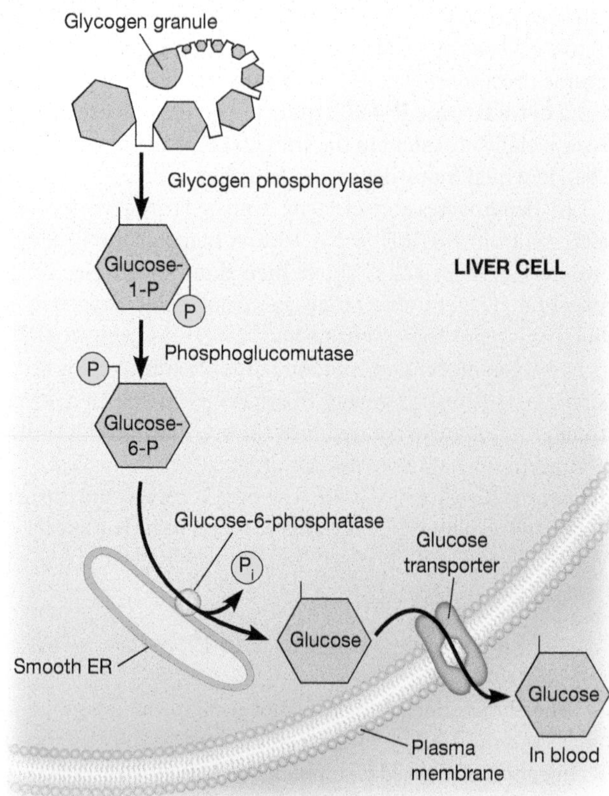

(b) Process of glycogen breakdown in liver

Figure 12-2 The Role of the Smooth ER in the Catabolism of Liver Glycogen. (a) This electron micrograph of a liver cell shows numerous granules of glycogen closely associated with smooth ER (TEM). **(b)** Glycogen degradation produces glucose-1-phosphate (glucose-1-P), which is converted to glucose-6-phosphate (glucose-6-P). Production of free glucose from glucose-6-P requires glucose-6-phosphatase, an enzyme associated with the smooth ER membrane. Free glucose is then transported out of the liver cell by a glucose transporter in the plasma membrane.

large amounts in the smooth ER of liver cells. This enzyme is targeted for inhibition by a class of cholesterol-lowering drugs known as *statins*. In addition, the smooth ER contains a number of P-450 monooxygenases that are important not only in the synthesis of cholesterol but also in its conversion into steroid hormones by hydroxylation.

The ER Plays a Central Role in the Biosynthesis of Membranes

In eukaryotic cells, the ER is the primary source of membrane lipids, including phospholipids and cholesterol. Indeed, most of the enzymes required for the biosynthesis of membrane phospholipids are found nowhere else in the cell. There are, however, important exceptions. Mitochondria synthesize phosphatidylethanolamine by decarboxylating imported phosphatidylserine. Peroxisomes have enzymes to synthesize cholesterol, and chloroplasts contain enzymes for the synthesis of chloroplast-specific lipids.

Biosynthesis of fatty acids for membrane phospholipid molecules occurs in the cytoplasm, and incorporation is restricted to the monolayer of the ER membrane facing the cytosol. Cellular membranes, of course, are phospholipid *bilayers*, with phospholipids distributed to both sides. Thus, there must be a mechanism for transferring phospholipids from one layer of the membrane to the other. Because it is thermodynamically unfavorable for phospholipids to flip spontaneously at a significant rate from one side of a bilayer to the other, transfer depends on **phospholipid translocators,** also known as **flippases,** which catalyze the translocation of phospholipids through ER membranes (see Figure 7-10).

Phospholipid translocators, like other enzymes, are quite specific and affect only the rate of a process. Therefore, the type of phospholipid molecules transferred across a membrane depends on the particular translocators present, contributing to the *membrane asymmetry* described in Chapter 7. For example, the ER membrane contains a translocator for phosphatidylcholine, and thus it is found in both monolayers of the ER membrane. In contrast, there is no translocator for phosphatidylethanolamine, phosphatidylinositol, or phosphatidylserine, which are therefore confined to the cytosolic monolayer. When vesicles from the ER membrane fuse with other organelles of the endomembrane system, the distinct compositions of the cytosolic and lumenal monolayers established in the ER are transferred to these other cellular membranes.

Movement of phospholipids from the ER to a mitochondrion, chloroplast, or peroxisome poses a unique problem. Unlike organelles of the endomembrane system, these organelles do not grow by fusion with ER-derived vesicles. Instead, cytosolic **phospholipid exchange proteins** (also called *phospholipid transfer proteins*) convey phospholipid molecules from the ER membrane to the outer mitochondrial and chloroplast membranes. Each exchange protein recognizes a specific phospholipid, removes it from one membrane, and carries it through the cytosol to another membrane. Such transfer proteins also contribute to the movement of phospholipids from the ER to other cellular membranes, including the plasma membrane.

Although the ER is the source of most membrane lipids, the compositions of other cellular membranes vary significantly from the composition of the ER membrane

Table 12-1 Composition of the ER and Plasma Membranes of Rat Liver Cells		
Membrane Components	**ER Membrane**	**Plasma Membrane**
Membrane components as % of membrane by weight		
Carbohydrate	10	10
Protein	62	54
Total lipid	27	36
Membrane lipids as % of total lipids by weight		
Phosphatidylcholine	40	24
Phosphatidylethanolamine	17	7
Phosphatidylserine	5	4
Cholesterol	6	17
Sphingomyelin	5	19
Glycolipids	trace	7
Other lipids	27	22

(**Table 12-1**). A striking feature of the plasma membrane of hepatocytes is the relatively low amount of phosphoglycerides and high amounts of cholesterol, sphingomyelin, and glycolipids. Researchers have observed an increasing gradient of cholesterol content from the ER through the compartments of the endomembrane system to the plasma membrane. This correlates with an increasing gradient of membrane thickness. ER membranes are about 5 nm thick, whereas plasma membranes are about 8 nm thick. The observed change in membrane thickness has implications for sorting and targeting integral membrane proteins, which we will discuss after we look at the Golgi apparatus and its role in protein processing.

CONCEPT CHECK 12.1

How do differences in the structure and arrangement of rough and smooth ER reflect their different functions?

12.2 The Golgi Apparatus

We now turn our attention to the Golgi apparatus, a component of the endomembrane system that is closely linked, both physically and functionally, to the ER. In the Golgi apparatus, glycoproteins from the ER undergo further processing and, along with membrane lipids, are sorted and packaged for transport to their proper destinations inside or outside the cell. Thus, the Golgi apparatus plays a central role in *membrane* and *protein trafficking* in eukaryotic cells.

The **Golgi apparatus** (or *Golgi complex*) derives its name from Camillo Golgi, the Italian biologist who first described it in 1898. He reported that nerve cells soaked in osmium tetroxide showed deposits of osmium in a threadlike network surrounding the nucleus. The same staining reaction was demonstrated with a variety of cell types and other heavy metals. However, no cellular structure could be identified that explained the staining. As a result, the nature—actually, the very existence—of the Golgi apparatus remained controversial until the 1950s, when its existence was finally confirmed by electron microscopy.

The Golgi Apparatus Consists of a Series of Membrane-Bounded Cisternae

The Golgi apparatus is a series of flattened membrane-bounded *cisternae*, disk-shaped sacs that are stacked together as illustrated in **Figure 12-3**. A series of such cisternae is called a *Golgi stack* and can be visualized by electron microscopy (see Figure 4-17c). Usually, there are three to eight cisternae per stack, though the number and size of Golgi stacks vary with the cell type and with the metabolic activity of the cell. Some cells have one large stack, whereas others—especially active secretory cells—have hundreds or even thousands of Golgi stacks.

The static view of the ER and the Golgi apparatus presented by electron micrographs (such as in Figure 4-16c, Figure 4-16d, and Figure 4-17c) can be misleading. These organelles are actually dynamic structures. Both the ER and Golgi apparatus are typically surrounded by numerous vesicles that carry lipids and proteins from the ER to the Golgi apparatus, between the cisternae of a Golgi stack, and from the Golgi apparatus to various destinations in the cell, including endosomes, lysosomes, and secretory granules. Thus, the Golgi apparatus lumen, or *intracisternal space*, is part of the endomembrane system's network of internal spaces (see Figure 12-1).

The Two Faces of the Golgi Stack. Each Golgi stack has two distinct sides, or *faces* (Figure 12-3). The *cis face* is oriented toward the ER. The Golgi compartment closest to the ER is a network of flattened, membrane-bounded tubules referred to as the ***cis*-Golgi network (CGN).** Vesicles containing newly synthesized lipids and proteins from the ER continuously arrive at the CGN, where they fuse with CGN membranes.

The opposite side of the Golgi apparatus is called the *trans face*. The compartment on this side of the Golgi apparatus

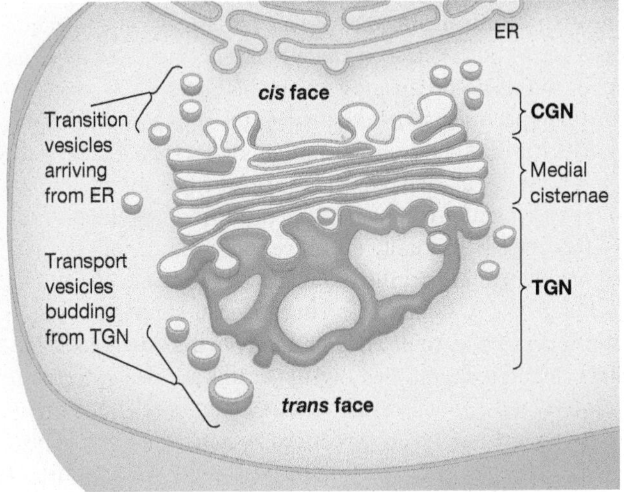

Figure 12-3 Golgi Structure. A Golgi stack consists of a small number of flattened cisternae. At the *cis* face, transition vesicles from the ER fuse with the *cis*-Golgi network (CGN). At the *trans* face, transport vesicles bud from the *trans*-Golgi network (TGN) and carry lipids and proteins to other components of the endomembrane system.

has similar morphology and is referred to as the ***trans*-Golgi network (TGN).** Here, proteins and lipids leave the Golgi in **transport vesicles** that continuously bud from the tips of TGN cisternae. These transport vesicles carry lipids and proteins from the Golgi apparatus to secretory granules, endosomes, lysosomes, and the plasma membrane. The central sacs between the CGN and TGN comprise the **medial cisternae** of the Golgi stack, in which much of the processing of proteins occurs.

The CGN, TGN, and medial cisternae of the Golgi apparatus are biochemically and functionally distinct. Each compartment contains specific receptor proteins and enzymes necessary for specific steps in protein and membrane processing, as shown by immunological and cytochemical staining techniques. For example, staining to detect *N-acetylglucosamine transferase I*, an enzyme that modifies carbohydrate side chains of glycoproteins, localizes the enzyme to medial cisternae of the Golgi apparatus.

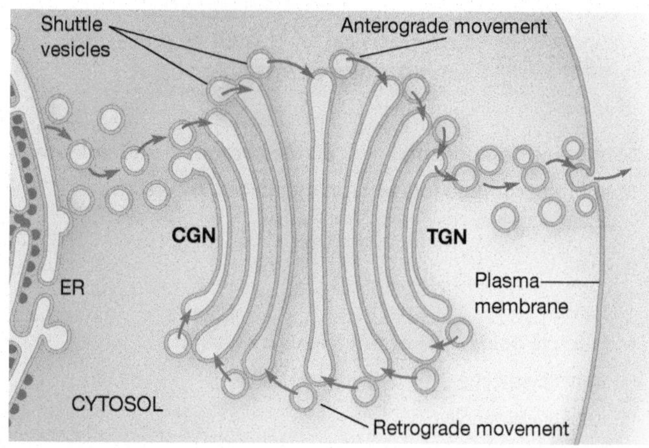

(a) Stationary cisternae model

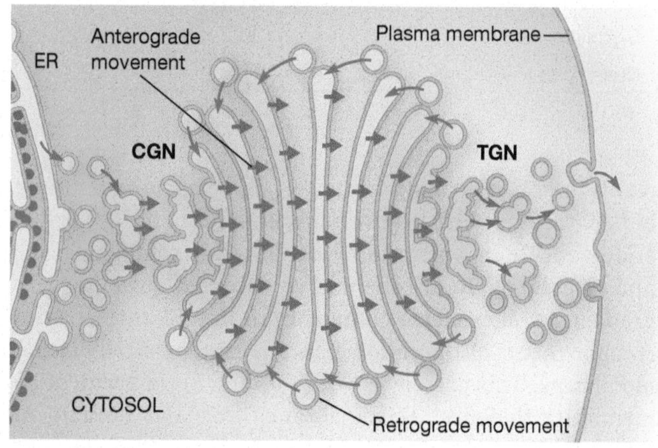

(b) Cisternal maturation model

Figure 12-4 Two Models for Movement Through the Golgi. **(a)** In the stationary cisternae model, shuttle vesicles carry material forward from the ER to successive Golgi compartments, which remain in place. **(b)** In the cisternal maturation model, the cisternae gradually change in composition as they themselves move forward. In both models, enzymes and lipids needed in earlier compartments move backward in retrograde fashion. Green arrows show anterograde movement from the CGN toward the TGN, and purple arrows show retrograde movement in the opposite direction.

Two Models Account for the Flow of Lipids and Proteins Through the Golgi Apparatus

Two models have been proposed to explain the movement of lipids and proteins from the CGN to the TGN via the medial cisternae of the Golgi apparatus (**Figure 12-4**). According to the **stationary cisternae model,** each compartment of the Golgi stack is a stable structure (Figure 12-4a). Trafficking between successive cisternae is mediated by *shuttle vesicles* that bud from one cisterna and fuse with the next cisterna in the *cis*-to-*trans* sequence. Proteins destined for the TGN are simply carried forward by shuttle vesicles, whereas molecules that belong in the ER and successive Golgi compartments are actively retained or retrieved.

According to the second model, known as the **cisternal maturation model,** the Golgi cisternae are transient compartments that gradually change from CGN cisternae through medial cisternae to TGN cisternae (Figure 12-4b). In this model, transition vesicles from the ER converge to form the CGN, which accumulates specific enzymes for the early steps of protein processing. Step by step, each *cis* cisterna is transformed first into an intermediate medial cisterna and then into a *trans* cisterna as it acquires additional enzymes. Enzymes that are no longer needed in late compartments return in vesicles to early compartments. In both models, the TGN forms transport vesicles or secretory granules containing sorted cargo targeted for various destinations beyond the Golgi apparatus.

Experimental results suggest that these two models are not necessarily mutually exclusive. Both may apply to some degree, depending on the organism and the role of the cell. Although the stationary cisternae model is supported by substantial evidence, some cellular components observed in medial compartments of the Golgi apparatus are clearly too large to travel by the small shuttle vesicles found in cells. For example, polysaccharide scales produced by some algae first appear in early Golgi compartments. Too large to fit inside transport vesicles, the scales nevertheless reach late Golgi compartments on their way to the plasma membrane for incorporation into the cell wall.

Recently, time-lapse fluorescence microscopy has been used in live yeast cells to study individual Golgi cisternae in real time. Three-dimensional analysis of images supports the cisternal maturation model and suggests that the cisternae mature at a constant rate. In addition, the rate of movement of labeled secretory proteins through the Golgi apparatus was measured and was shown to match the rate of cisternal maturation.

Because experimental evidence fails to completely support either of these models, it remains possible that another model explains traffic through the Golgi. An alternative model that has been proposed is the diffusion model, which suggests that transport through the Golgi occurs when cargo freely diffuses from *cis* to *trans* cisternae. This diffusion would require that the cisternae are continuous with each other. Although these connections have not been observed, some have postulated that small tubules could exist, making the Golgi stacks connected and continuous. Again, the answer may be a combination of models. A variation of the diffusion model, called the kiss-and-run model, suggests that vesicles could transiently form tubules with the Golgi cisternae. Cargo could then pass through these tubules before they are broken. Clearly, much work is left to do in order to elucidate the precise mechanism of transport within the Golgi.

Anterograde and Retrograde Transport. The movement of material from the ER through the Golgi apparatus toward the plasma membrane is called **anterograde transport** (*antero* is derived from a Latin word meaning "front," and *grade* is related to a word meaning "step"). Every time a secretory granule fuses with the plasma membrane and discharges its contents by exocytosis, a bit of membrane that originated in the ER becomes a part of the plasma membrane. To balance the flow of lipids toward the plasma membrane and to ensure a supply of components for forming new vesicles, the cell recycles lipids and proteins no longer needed during the late stages of anterograde transport. This is accomplished by **retrograde transport** (*retro* is a Latin word meaning "back"), the flow of vesicles from Golgi cisternae back toward the ER.

In the stationary cisternae model, retrograde flow facilitates both the recovery of ER-specific lipids and proteins that are passed from the ER to the CGN and the transport of compartment-specific proteins back to distinct medial cisternae of the Golgi stack (Figure 12-4a). Material destined for the TGN continues forward. Whether such retrograde traffic occurs directly from all medial cisternae of the Golgi stack back to the ER or by reverse flow through successive cisternae is not yet clear. In the cisternal maturation model, retrograde flow carries material back toward newly forming compartments after receptors and enzymes are no longer needed in the more mature compartments.

CONCEPT CHECK 12.2

Why is it necessary for material flowing through the Golgi to move in both the anterograde and the retrograde directions?

12.3 Roles of the ER and Golgi Apparatus in Protein Processing

The protein processing carried out within the ER and Golgi apparatus includes protein folding, quality control, and **glycosylation**—the addition of carbohydrate side chains to specific amino acid residues of proteins, forming **glycoproteins.** Two general kinds of glycosylation are observed in cells (see Figure 7-21). **N-linked glycosylation** (or **N-glycosylation**) involves the addition of a specific oligosaccharide unit to the *nitrogen* atom on the terminal amino group of certain asparagine residues. **O-linked glycosylation** involves addition of an oligosaccharide to the *oxygen* atom on the hydroxyl group of certain serine, threonine, and, in rare cases, tyrosine residues. Each step of glycosylation is strictly dependent on preceding modifications. An error at one step, perhaps due to a defective enzyme, can block further modification of a carbohydrate side chain and can lead to disease in the organism. Before looking at glycosylation in detail, let's start by examining the first protein processing steps that occur in the ER—folding and quality control.

Protein Folding and Quality Control Take Place Within the ER

After polypeptides are released into the ER lumen, they fold into their final shape and, in some cases, assemble with other polypeptides to form multisubunit proteins. *Molecular chaperones* facilitate these folding and assembly events (see Chapter 19). The most abundant chaperone in the ER lumen is a member of the Hsp70 family of chaperones known as **BiP** (an abbreviation for *bi*nding *p*rotein). Like other Hsp70 chaperones, BiP acts by binding to hydrophobic regions of polypeptide chains, especially to regions enriched in the amino acids tryptophan, phenylalanine, and leucine.

BiP prevents aggregation by transiently binding to the hydrophobic regions of unfolded polypeptides as they emerge into the ER lumen, stabilizing them and preventing them from interacting with other unfolded polypeptides. BiP then releases the polypeptide chain, accompanied by ATP hydrolysis, giving the polypeptide a brief opportunity to fold (perhaps aided by other chaperones). If the polypeptide folds correctly, its hydrophobic regions become buried in the molecule's interior and can no longer bind to BiP. But if the hydrophobic segments fail to fold properly, BiP binds again to the polypeptide and the cycle is repeated. In this way, BiP uses energy from ATP hydrolysis to promote proper protein folding.

Folding is often accompanied by the formation of disulfide bonds between cysteines located in different regions of a polypeptide chain. This reaction is facilitated by **protein disulfide isomerase,** an enzyme present in the ER lumen that catalyzes the formation and breakage of disulfide bonds between cysteine residues. Protein disulfide isomerase starts acting before the synthesis of a newly forming polypeptide has been completed, allowing various disulfide bond combinations to be tested until the most stable arrangement is found.

Proteins that repeatedly fail to fold properly can activate several types of quality control mechanisms. One such mechanism, called the **unfolded protein response (UPR),** uses sensor molecules in the ER membrane to detect misfolded proteins. These sensors activate signaling pathways that shut down the synthesis of most proteins while enhancing the production of those required for protein folding and degradation. Another type of quality control, known as **ER-associated degradation (ERAD),** recognizes misfolded or unassembled proteins and exports, or "retrotranslocates," them back across the ER membrane to the cytosol, where they are degraded by *proteasomes* (see Chapter 20). The precise mechanism utilized in ERAD for recognizing mislabeled or misfolded proteins is poorly understood.

Initial Glycosylation Occurs in the ER

We will focus here on N-glycosylation. **Figure 12-5** on page 322 describes the steps of glycosylation that may occur as a glycoprotein travels from the ER to the CGN and through the Golgi apparatus to the TGN. Note that specific enzymes that catalyze various steps of glycosylation and subsequent modifications are present in specific compartments of the ER and Golgi apparatus. The initial steps of N-glycosylation take

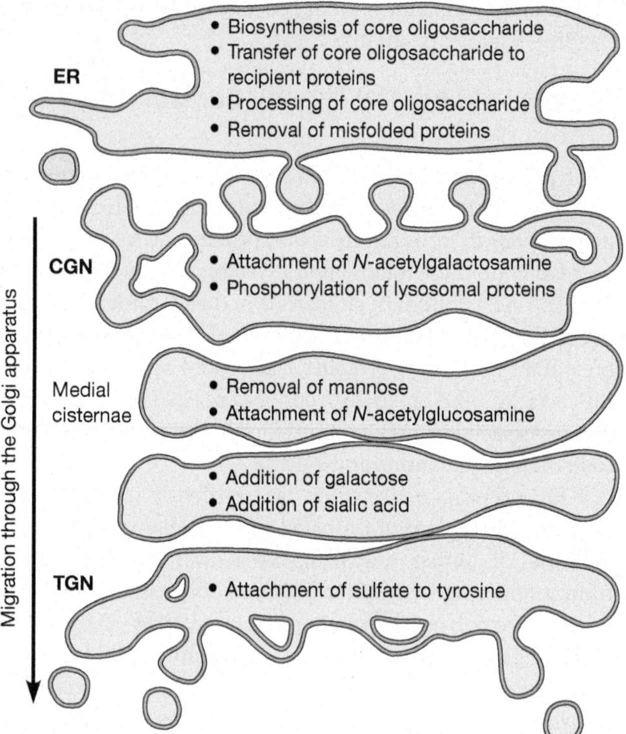

Figure 12-5 Compartmentalization of the Steps of Glycosylation and Subsequent Modification of Proteins. Enzymes that catalyze specific steps of glycosylation and further modification of proteins reside in different compartments of the ER and Golgi apparatus. Processing occurs sequentially as proteins travel from the ER to the CGN and on to the TGN. These potential modifications do not necessarily occur with all glycoproteins.

place on the cytosolic surface of the ER membrane, and later steps occur in the ER lumen (**Figure 12-6**). Despite the variety of oligosaccharides found in mature glycoproteins, all the carbohydrate side chains added to proteins in the ER initially have a common **core oligosaccharide** consisting of two units of N-acetylglucosamine (GlcNAc; see Figure 3-26a), nine mannose units, and three glucose units.

Glycosylation begins as *dolichol phosphate,* an oligosaccharide carrier, is ❶ inserted into the ER membrane. ❷ GlcNAc and mannose groups are then added to the phosphate group of dolichol phosphate. ❸ The growing core oligosaccharide is then translocated from the cytosol to the ER lumen by a *flippase.* ❹ Once inside the ER lumen, more mannose and glucose units are added. ❺ The completed core oligosaccharide is then transferred as a single unit from dolichol to an asparagine residue of the recipient protein. ❻ Finally, the core oligosaccharide attached to the protein is trimmed and modified.

Usually, the core oligosaccharide is added to the protein as the polypeptide is being synthesized by a ribosome bound to the ER membrane. This *cotranslational glycosylation* helps promote proper protein folding; experimental inhibition of glycosylation leads to the appearance of misfolded, aggregated proteins. Addition of a single glucose unit allows other ER proteins to interact with the newly synthesized glycoprotein to ensure its proper folding. One of two ER proteins, **calnexin** (membrane-bound) or **calreticulin** (soluble), can bind to the monoglucosylated glycoprotein and promote proper folding by forming a complex with the glycoprotein and a thiol oxidoreductase known as *ERp57,* which catalyzes disulfide bond formation (see Figures 3-5a and 3-7). The protein complex then dissociates, and the final glucose unit is removed by an enzyme named *glucosidase II.*

At this point, a specific glucosyl transferase in the ER known as *UGGT (UDP-glucose:glycoprotein glucotransferase)* acts as a sensor for proper folding of the newly synthesized glycoprotein. UGGT binds to improperly folded proteins and adds back a single glucose unit, making the protein a substrate for another round of calnexin/calreticulin binding and disulfide bond formation. Once the proper conformation is achieved, UGGT no longer binds the new glycoprotein, which is then free to exit the ER and move to the Golgi.

Further Glycosylation Occurs in the Golgi Apparatus

Further processing of N-glycosylated proteins happens in the Golgi apparatus as the glycoproteins move from the *cis* face through the medial cisternae to the *trans* face of the Golgi stack. These **terminal glycosylations** in the Golgi show remarkable variability among proteins and account for much of the great diversity in structure and function of protein oligosaccharide side chains.

Terminal glycosylation always includes the removal of a few of the carbohydrate units of the core oligosaccharide. In some cases, no further processing occurs in the Golgi apparatus. In other cases, more complex oligosaccharides are generated by the further addition of GlcNAc and other monosaccharides, including galactose and sialic acid. Some glycoproteins contain galactose units that are added by *galactosyl transferase,* a marker enzyme unique to the Golgi.

Given the role of the Golgi apparatus in glycosylation, it is not surprising that two of the most important categories of enzymes present in Golgi stacks are *glucan synthetases,* which produce oligosaccharides from monosaccharides, and *glycosyl transferases,* which attach carbohydrate groups to proteins. The ER and Golgi apparatus contain hundreds of different glycosyl transferases, which indicates the potential complexity of oligosaccharide side chains. Within the Golgi stack, each cisterna contains a distinctive set of processing enzymes.

Notice in the preceding discussion that the mature oligosaccharides in glycoproteins are found only on the lumenal side (interior) of the ER and Golgi apparatus membranes and thus contribute to membrane asymmetry. Because the lumenal side of the ER membrane is topologically equivalent to the exterior surface of the cell, it is easy to see why all plasma membrane glycoprotein oligosaccharides are found on the extracellular side of the membrane.

CONCEPT CHECK 12.3

What problems would you likely see in cells defective in ER glycosylation of proteins?

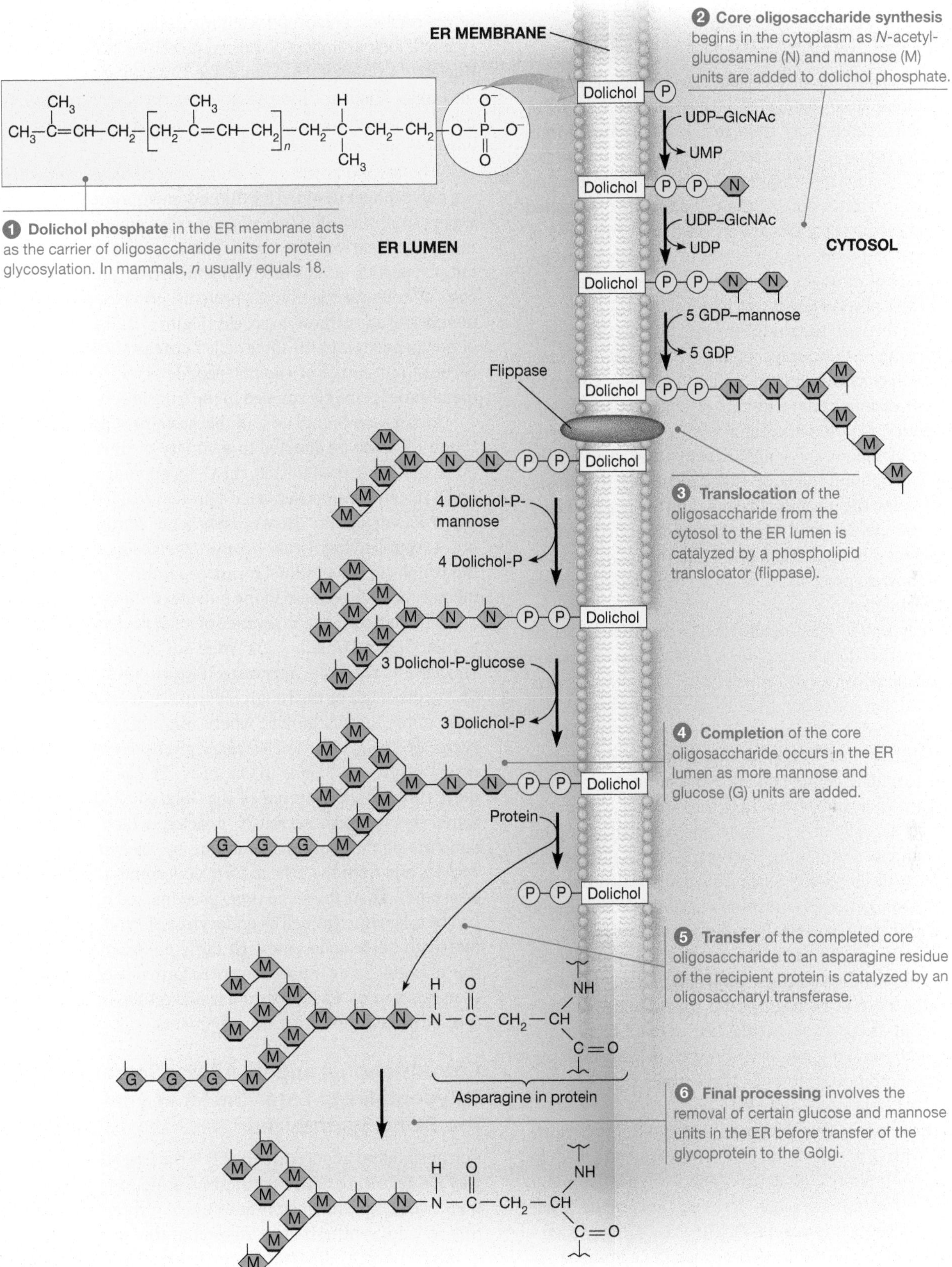
ER MEMBRANE
2 Core oligosaccharide synthesis begins in the cytoplasm as N-acetyl-glucosamine (N) and mannose (M) units are added to dolichol phosphate.
Dolichol
UDP–GlcNAc
UMP
UDP–GlcNAc
UDP
5 GDP–mannose
5 GDP
1 Dolichol phosphate in the ER membrane acts as the carrier of oligosaccharide units for protein glycosylation. In mammals, n usually equals 18.
ER LUMEN
CYTOSOL
Flippase
3 Translocation of the oligosaccharide from the cytosol to the ER lumen is catalyzed by a phospholipid translocator (flippase).
4 Dolichol-P-mannose
4 Dolichol-P
3 Dolichol-P-glucose
3 Dolichol-P
4 Completion of the core oligosaccharide occurs in the ER lumen as more mannose and glucose (G) units are added.
Protein
5 Transfer of the completed core oligosaccharide to an asparagine residue of the recipient protein is catalyzed by an oligosaccharyl transferase.
Asparagine in protein
6 Final processing involves the removal of certain glucose and mannose units in the ER before transfer of the glycoprotein to the Golgi.

Figure 12-6 N-Linked Glycosylation of Proteins in the ER. Synthesis of core oligosaccharides begins in the cytoplasm, using a dolichol phosphate molecule as a carrier. The partially synthesized oligosaccharide is translocated to the ER lumen, where additional monosaccharides are added. The completed oligosaccharide is then transferred to the target protein, and several monosaccharides are removed in final processing.

12.4 Roles of the ER and Golgi Apparatus in Protein Trafficking

Think for a moment about a typical eukaryotic cell with its diversity of organelles, each containing its own unique set of proteins. Such a cell is likely to have billions of protein molecules, representing at least 10,000 kinds of polypeptides. And each polypeptide must traffic to the appropriate location within the cell, or even outside the cell altogether. A limited number of these polypeptides are encoded by the genome of the mitochondrion (and, for plant cells, by the chloroplast genome as well), but most are encoded by nuclear genes and are synthesized by a process beginning in the cytosol.

Before we focus on the role of the ER and Golgi apparatus in the trafficking of these polypeptides, let's review how most nuclear proteins reach their final destination. We can begin by grouping the various compartments of eukaryotic cells into three categories: (1) the endomembrane system, the interrelated system of membrane compartments that includes the endoplasmic reticulum (ER), the Golgi apparatus, lysosomes, secretory vesicles, the nuclear envelope, and the plasma membrane, (2) the cytosol, and (3) mitochondria, chloroplasts, peroxisomes (and related organelles), and the interior of the nucleus. So how do proteins transit both between and within these compartments?

Each polypeptide must be directed to its proper destination and must therefore have some sort of molecular "postal code" ensuring its delivery to the correct place. Nuclear proteins destined for mitochondria use a *transit sequence* for proper targeting (see Chapter 10). Therefore, each protein contains a specific "tag" targeting that protein to its location within each cellular compartment. For example, some tags direct proteins to a transport vesicle that will carry material from one specific cellular location to another. Depending on the protein and its destination, the tag may be a short amino acid sequence, an oligosaccharide side chain, a hydrophobic domain, or some other structural feature. Tags may also be involved in excluding material from certain vesicles.

Tagged polypeptides encoded by nuclear genes are routed to cellular compartments using several different mechanisms. In each case, the process begins with transcription of DNA into RNAs that are processed in the nucleus and then transported through nuclear pores for translation in the cytoplasm, where most ribosomes are found. (These topics are covered in great detail in Chapters 18 and 19.)

Upon arriving in the cytoplasm, these mRNAs become associated with *free ribosomes* (ribosomes not attached to any membrane). Shortly after translation begins, two main pathways for sorting the newly forming polypeptide products begin to diverge (**Figure 12-7**). The first pathway is employed for polypeptides destined either for the cytosol or for mitochondria, chloroplasts, peroxisomes, and the nuclear interior (Figure 12-7a). Ribosomes synthesizing these types of polypeptides remain free in the cytosol, unattached to any membrane. After translation has been completed, the polypeptides are released from the ribosomes and either remain in the cytosol as their final destination or are taken up by the appropriate organelle. The uptake by organelles of such completed polypeptides requires the presence of special targeting signals and is called **posttranslational import.** In the case of the nucleus, polypeptides enter through the nuclear pores. (We will look at import of macromolecules in and out of the nucleus in Chapter 16, Figure 16-32.) Polypeptide entrance into mitochondria, chloroplasts, and peroxisomes involves a different kind of mechanism. (See the coverage of polypeptide entrance into mitochondria in Chapter 10.)

The second pathway is utilized by ribosomes synthesizing polypeptides destined for the endomembrane system or for export from the cell. Such ribosomes become attached to ER membranes early in the translational process, and the growing polypeptide chains are then transferred across (or, in the case of integral membrane proteins, inserted into) the ER membrane as synthesis proceeds (Figure 12-7b). This transfer of polypeptides into the ER is called **cotranslational import** because movement of the polypeptide across or into the ER membrane is directly coupled to the translation process.

Membrane-bound and soluble proteins synthesized in the rough ER must be directed to a variety of intracellular locations, including the ER itself, the Golgi apparatus, endosomes, and lysosomes. Moreover, once a protein reaches an organelle where it is to remain, there must be a mechanism for preventing it from leaving. Other groups of proteins synthesized in the rough ER are destined for incorporation into the plasma membrane or for release to the outside of the cell.

The subsequent conveyance of such proteins from the ER to their final destinations is carried out by various membrane vesicles and the Golgi apparatus (**Figure 12-8** on page 326). ❶ As ribosomes of the rough ER synthesize proteins, the proteins enter the ER lumen, where initial glycosylation steps occur. ❷ Transition vesicles carry glycosylated proteins and newly synthesized lipids to the CGN. ❸ Lipids and proteins move through the cisternae of the Golgi stack. 4a At the TGN, some vesicles form secretory vesicles, which release their contents at the plasma membrane by exocytosis. 4b Other vesicles bud from the TGN to form endosomes that help make lysosomes. ❶ At the same time, proteins and other materials can be taken into the cell by endocytosis, forming vesicles that fuse with early endosomes. ❷ Early endosomes containing this material for digestion mature to form late endosomes and then lysosomes. ❸ Retrograde traffic returns compartment-specific proteins to earlier compartments.

Cotranslational Import Allows Some Polypeptides to Enter the ER as They Are Being Synthesized

Cotranslational import into the ER is the first step in the pathway for delivering newly synthesized proteins to various locations within the endomembrane system. Proteins trafficked in this way are synthesized on ribosomes that become attached to the ER shortly after translation begins. The role of the ER in this process was first suggested by experiments in which Colvin Redman and David Sabatini studied protein synthesis in isolated vesicles of rough ER (ER vesicles with attached ribosomes). Such vesicles, known as *microsomes,* can be isolated using subcellular fractionation and centrifugation. After briefly incubating the rough ER vesicles in the presence of radioactive amino acids and other components needed for protein synthesis, they stopped the reaction by adding *puromycin,*

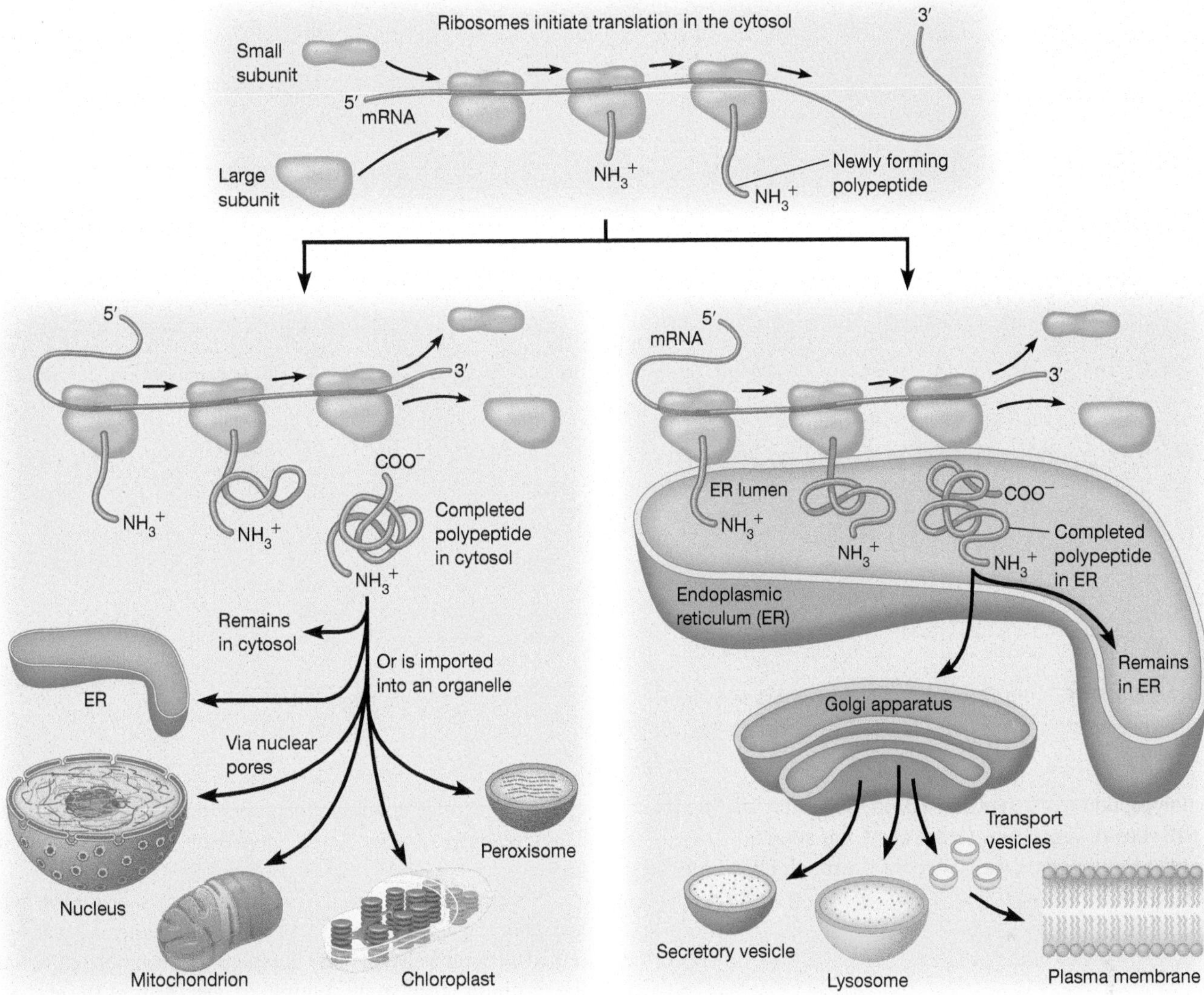

Figure 12-7 Intracellular Sorting of Proteins. Polypeptide synthesis begins in the cytosol but takes one of two alternative routes when the polypeptide is about 30 amino acids long. **(a)** In posttranslational import, ribosomes remain free in the cytosol if they are synthesizing polypeptides destined for the cytosol or for import into the nucleus, mitochondria, chloroplasts, or peroxisomes. When the polypeptide is complete, it is released from the ribosome and either remains in the cytosol or is transported into the appropriate organelle. Polypeptide uptake by the nucleus occurs via the nuclear pores, using a mechanism different from that involved in posttranslational uptake by other organelles. **(b)** In cotranslational import, ribosomes attach to ER membranes if they are synthesizing polypeptides destined for the endomembrane system or for export from the cell. As synthesis continues, the newly forming polypeptide is transferred across the ER membrane. The completed polypeptide either remains in the ER or is transported via various vesicles to another compartment of the endomembrane system. (Integral membrane proteins are inserted into the ER membrane as they are made, rather than being released into the ER lumen and delivered to the plasma membrane in transport vesicles.)

an antibiotic that causes partially completed polypeptide chains to be released from ribosomes. When the ribosomes and membrane vesicles were then separated and analyzed to see where the newly made, radioactive polypeptide chains were located, a substantial fraction of the radioactivity was found inside the ER lumen (**Figure 12-9** on page 326). Such results suggested that newly forming polypeptides pass into the lumen of the ER *as they are being synthesized,* allowing them to be routed through the ER to their correct destinations.

If some polypeptides move directly into the lumen of the ER as they are being synthesized, how does the cell determine which polypeptides are targeted this way? An answer was first suggested in 1971 by Günter Blobel and David Sabatini, whose model was called the *signal hypothesis* because it proposed that some sort of intrinsic molecular signal distinguishes such polypeptides from the many polypeptides destined to be released into the cytosol. This hypothesis has so profoundly influenced the field of cell biology that Blobel was awarded a Nobel Prize in 1999 for his work in demonstrating that proteins have intrinsic signals governing their transport and localization within the cell. The signal hypothesis stated that for polypeptides destined for the ER, the first segment of

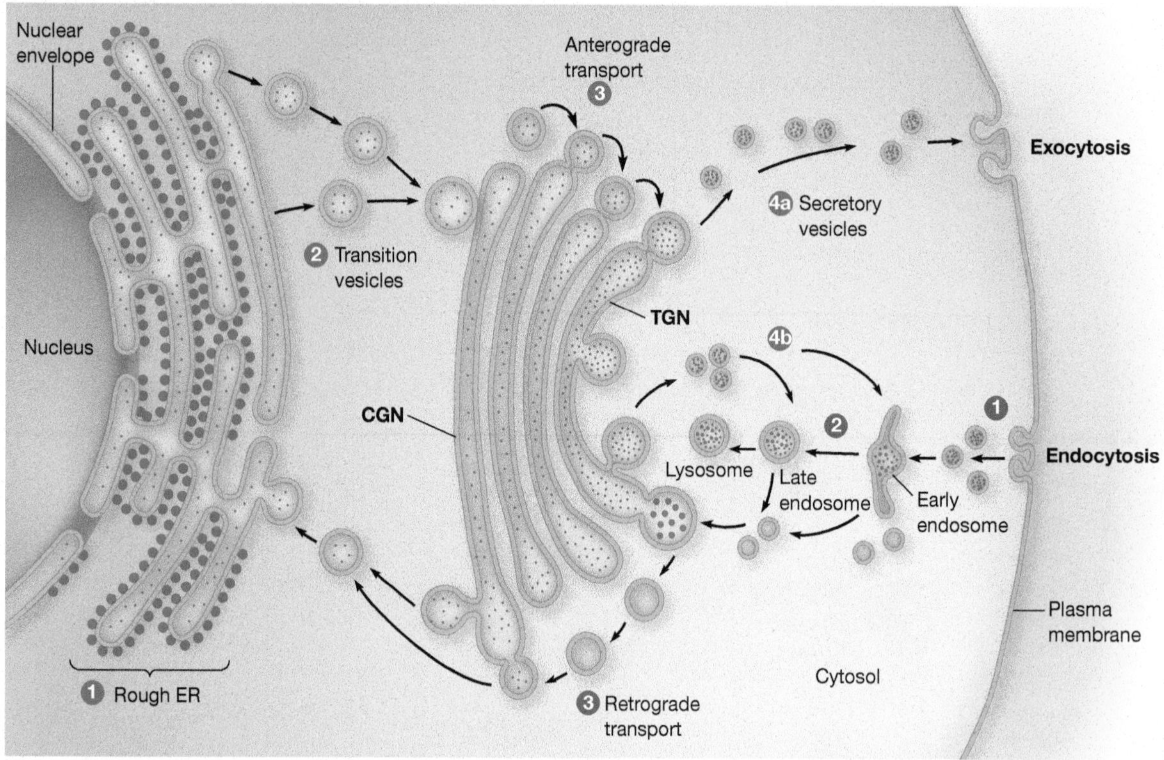

Figure 12-8 Trafficking Through the Endomembrane System. Vesicles carry lipids and proteins through the endomembrane system in both (1–4) anterograde and (1–3) retrograde directions.

the polypeptide to be synthesized, the N-terminus, contains an **ER signal sequence** that directs the ribosome-mRNA-polypeptide complex to the surface of the rough ER, where the complex anchors at a protein "dock" on the ER surface. Then,

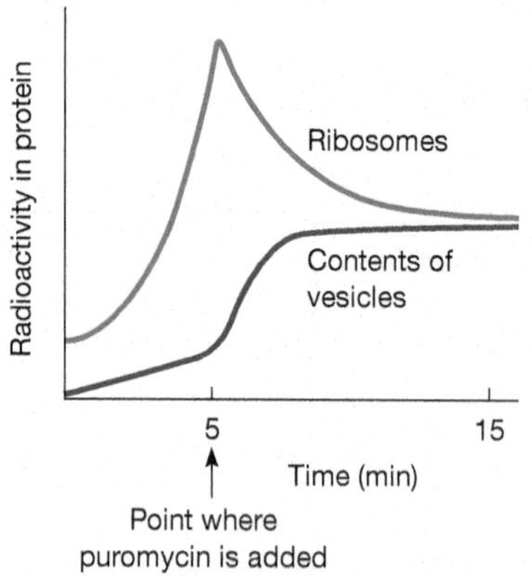

Figure 12-9 Evidence That Proteins Synthesized on Ribosomes Attached to ER Membranes Pass Directly into the ER Lumen. ER vesicles containing attached ribosomes were isolated and incubated with radioactive amino acids to label newly made polypeptide chains. Next, protein synthesis was halted by adding puromycin, which also causes the newly forming polypeptide chains to be released from the ribosomes. The ribosomes were then removed from the membrane vesicles, and the amount of radioactive protein associated with the ribosomes and in the membrane vesicles was measured. The graph shows that after the addition of puromycin, radioactivity is lost from the ribosomes and appears inside the vesicles. This observation suggests that the newly forming polypeptide chains are inserted through the ER membrane as they are being synthesized, and puromycin causes the chains to be prematurely released into the vesicle lumen.

as the polypeptide chain elongates during mRNA translation, it progressively crosses the ER membrane and enters the ER lumen.

Shortly after the signal hypothesis was first proposed, evidence for the actual existence of ER signal sequences was obtained by César Milstein and his associates, who were studying the synthesis of the small subunit, or *light chain*, of the protein *immunoglobulin G*. In cell-free systems containing purified ribosomes and the components required for protein synthesis, the mRNA encoding the immunoglobulin light chain directs the synthesis of a polypeptide product that is 20 amino acids longer at its N-terminal end than the authentic light chain itself. Adding ER membranes (microsomes) to this system leads to the production of an immunoglobulin light chain of the correct size (**Figure 12-10**). Such findings suggested that the extra 20-amino-acid segment is functioning as an ER signal sequence and that this signal sequence is removed when the polypeptide moves into the ER. Subsequent studies revealed that other polypeptides destined for the ER also possess an N-terminal sequence that is required for targeting the protein to the ER and that is removed as the polypeptide moves into the ER. Proteins containing such ER signal sequences at their N-terminus are often referred to as *preproteins* (for example, prelysozyme, preproinsulin, and pretrypsinogen).

Sequencing studies have revealed that the amino acid compositions of ER signal sequences are surprisingly variable, but several unifying features have been noted. ER signal sequences are typically 15–30 amino acids long and consist of three domains: a positively charged N-terminal region, a central hydrophobic region, and a polar region adjoining the site where cleavage from the mature protein will take place. The positively charged end may promote interaction with the

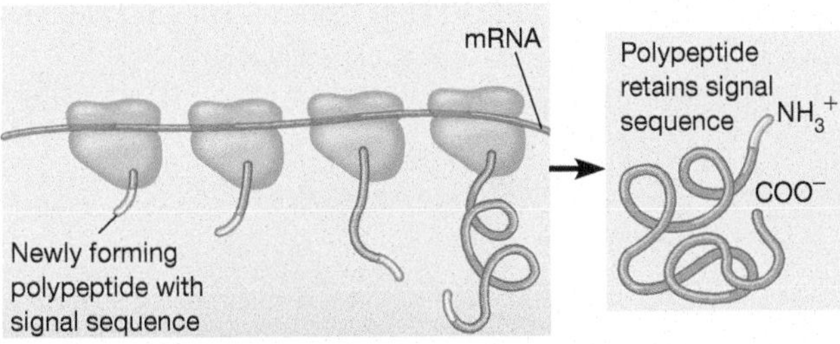

(a) Microsome-free ribosomes

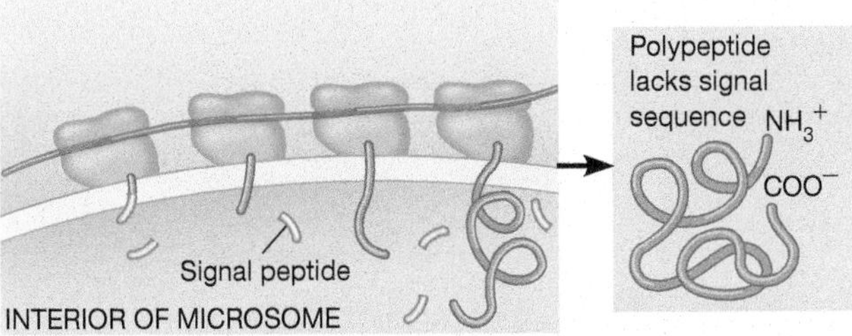

(b) Microsomes present

Figure 12-10 Evidence That Cotranslational Insertion into the ER Is Required for Normal Processing of Secreted Proteins. **(a)** Protein synthesis can be carried out in a cell-free system that includes ribosomes and other components but no membranes. When a messenger RNA encoding a protein that is normally secreted is added, the resulting protein is abnormally large because it retains its signal sequence. **(b)** When microsomes, which consist of ER membranes and attached ribosomes, are isolated and the same mRNA is added, the resulting protein is transported across the vesicle membrane and the signal sequence is cleaved.

hydrophilic exterior of the ER membrane, and the hydrophobic region may facilitate interaction of the signal sequence with the membrane's lipid interior. In any case, it is now established that only polypeptides with ER signal sequences can be inserted into or across the ER membrane as their synthesis proceeds. In fact, when recombinant DNA methods are used to add ER signal sequences to polypeptides that do not usually have them, the recombinant polypeptides are directed to the ER.

The Signal Recognition Particle (SRP) Attaches the Ribosome-mRNA-Polypeptide Complex to the ER Membrane

Once the existence of ER signal sequences was established, it quickly became clear that newly forming polypeptides must become attached to the ER membrane before very much of the polypeptide has emerged from the ribosome. If translation were to continue without attachment to the ER, the folding of the growing polypeptide chain might bury the signal sequence. To understand what prevents this from happening, we need to look at the signal mechanism in further detail.

Contrary to the original signal hypothesis, the ER signal sequence does not itself initiate contact with the ER. Instead, the contact is mediated by a **signal recognition particle (SRP)**, which recognizes and binds to the ER signal sequence of the newly forming polypeptide and then binds to the ER membrane. At first the SRP was thought to be purely protein (the *P* in its name originally stood for protein). Later, however, the SRP was shown to consist of six different polypeptides complexed with a 300-nucleotide (7S) molecule of RNA. The protein components have three main active sites: one that recognizes and binds to the ER signal sequence, one that interacts with the ribosome to block further translation, and one that binds to the ER membrane.

Figure 12-11 on page 328 illustrates the role played by the SRP in cotranslational import. The process begins when an mRNA encoding a polypeptide destined for the ER starts to be translated on a free ribosome. Polypeptide synthesis proceeds until the ER signal sequence has been formed and emerges from the surface of the ribosome. At this stage, ❶ SRP (shown in orange) binds to the signal sequence and blocks further translation. The SRP then binds the ribosome to a special structure in the ER membrane called a **translocon** because it carries out the translocation of polypeptides across the ER membrane. (Note that the term *translocation*, which literally means "a change of location," is used to describe both the movement of proteins through membranes and, earlier in the chapter, the movement of mRNA across the ribosome.)

The translocon is a protein complex composed of several components involved in cotranslational import, including an *SRP receptor* to which the SRP binds, a *ribosome receptor* that holds the ribosome in place, a *pore protein* that forms a channel through which the growing polypeptide can enter the ER lumen, and *signal peptidase*, an enzyme that removes the ER signal sequence. As ❷ shows, SRP (bringing an attached ribosome) first binds to the SRP receptor, allowing the ribosome to become attached to the ribosome receptor. Next, ❸ GTP binds to both SRP and the SRP receptor, unblocking translation and causing transfer of the signal sequence to the pore protein. The core of the pore is formed by three subunits that form the *Sec61 complex;* Sec61 has a central channel that opens as the signal sequence is inserted. ❹ GTP is then hydrolyzed, accompanied by release of the SRP. ❺ As the polypeptide elongates, it passes into the ER lumen, and signal peptidase cleaves the signal sequence, which is quickly degraded. After polypeptide synthesis is completed, ❻ the final polypeptide is released into the ER lumen, the translocon channel is closed, and the ribosome detaches from the ER membrane and dissociates into its subunits, releasing the mRNA.

One final note is in order about ribosomes and the rough ER. For simplicity, so far we have been discussing cotranslational import as if only single ribosomes are engaged with an mRNA during this process. In fact, as is true for free ribosomes, ER-associated ribosomes can form polyribosomes. The shapes of ER-bound polyribosomes are often very complex, presumably because it is important to maintain regular spacing between ribosomes attached to adjacent translocons.

MAKE CONNECTIONS 12.1

What are the similarities and differences between the cotranslational insertion of proteins into the ER and the posttranslational insertion of proteins into the mitochondria? (Fig. 10-4)

Proteins Released into the ER Lumen Are Routed to the Golgi Apparatus, Secretory Vesicles, Lysosomes, or Back to the ER

Most of the proteins synthesized on ribosomes attached to the ER are *glycoproteins*—that is, proteins with covalently bound carbohydrate groups. The initial glycosylation reactions that

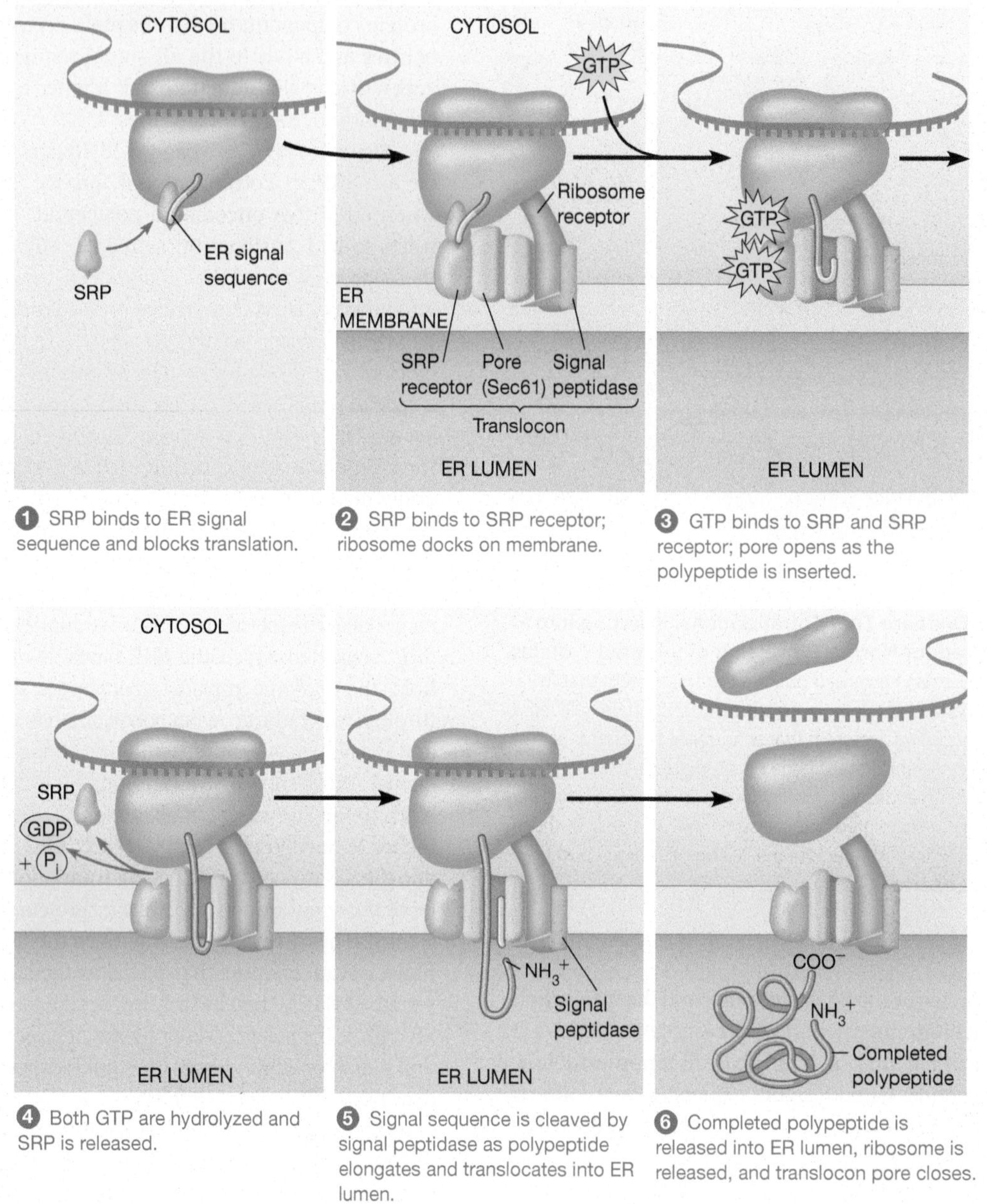
CYTOSOL
SRP
ER signal sequence
CYTOSOL
GTP
Ribosome receptor
ER MEMBRANE
SRP receptor
Pore (Sec61)
Signal peptidase
Translocon
ER LUMEN
GTP
GTP
ER LUMEN
1 SRP binds to ER signal sequence and blocks translation.
2 SRP binds to SRP receptor; ribosome docks on membrane.
3 GTP binds to SRP and SRP receptor; pore opens as the polypeptide is inserted.
CYTOSOL
SRP
GDP
+ Pi
ER LUMEN
NH3+
Signal peptidase
ER LUMEN
COO−
NH3+
Completed polypeptide
4 Both GTP are hydrolyzed and SRP is released.
5 Signal sequence is cleaved by signal peptidase as polypeptide elongates and translocates into ER lumen.
6 Completed polypeptide is released into ER lumen, ribosome is released, and translocon pore closes.

Figure 12-11 The Signal Mechanism of Cotranslational Import. The growing polypeptide translocates through a hydrophilic pore whose core is formed by the Sec61 complex. The complex of membrane proteins that carry out translocation is called the *translocon*.

add these carbohydrate side chains take place in the ER, often while the growing polypeptide is still being synthesized. After polypeptides have been released into the ER lumen, glycosylated, and folded, they are delivered by various types of transport vesicles to their destinations within the cell (see Figure 12-8). The first stop in this transport pathway is the Golgi apparatus, where further glycosylation and processing of carbohydrate side chains may occur. The Golgi apparatus then serves as a site for sorting and distributing proteins to other locations.

For soluble proteins, the default pathway takes them from the Golgi apparatus to secretory vesicles that move to the cell surface and fuse with the plasma membrane, leading to secretion of such proteins from the cell. Soluble proteins entering the Golgi apparatus that are not destined for secretion from the cell possess specific carbohydrate side chains and/or short amino acid signal sequences that target each protein to its appropriate location within the endomembrane system. A different signaling mechanism is used for proteins whose final destination is the ER. The C-terminus of these proteins usually contains a specific type of sequence tag. The Golgi apparatus employs a receptor protein that binds to this tag and delivers the targeted protein back to the ER. Protein disulfide isomerase and BiP—two ER-resident proteins whose role in protein folding was described in the preceding section—are examples of proteins possessing such a sequence that confines them to the ER. Let's examine this retention mechanism in more detail.

ER-Specific Proteins Contain Retention and Retrieval Tags. The protein composition required in the ER is maintained both by preventing some proteins from escaping when vesicles bud from the ER membrane and by retrieving other proteins that have left the ER and reached the CGN. It is not entirely clear how proteins that never leave the ER are retained, but one theory proposes that they form extensive complexes that are physically excluded from vesicles budding from the ER.

Several proteins localized to the ER contain the tripeptide sequence RXR (Arg-X-Arg, where X is any amino acid), which appears to promote retention in the ER. This *retention tag* is also found in some multisubunit proteins that are destined for the plasma membrane. For example, the N-methyl-D-aspartate (NMDA) receptor, which is important in neurotransmission in the mammalian brain, contains such a tag. But why, you may ask, does this plasma membrane protein have an ER retention signal? It is thought that the tripeptide causes individual subunits of NMDA to be retained in the ER until assembly of the multisubunit complex in the ER is complete. Proper assembly masks the RXR sequence, allowing the assembled complex to leave the ER. Other evidence suggests that phosphorylation of this protein near the RXR sequence promotes ER release.

Retrieval by retrograde flow of vesicles from the CGN to the ER is better understood than retention. Many soluble ER-specific proteins contain *retrieval tags* that bind to specific transmembrane receptors facing the Golgi apparatus lumen. The tags are short C-terminal amino acid sequences such as KDEL (Lys-Asp-Glu-Leu) or KKXX (where X can be any amino acid) in mammals and HDEL (His-Asp-Glu-Leu) in yeast. When a protein containing such a tag binds to its receptor, the receptor undergoes a conformational change, and the receptor-ligand complex is packaged into a transport vesicle for return to the ER.

Evidence for the importance of retrieval tags comes from experiments involving artificial *chimeric proteins*, also known as *fusion proteins*. They are made by joining the DNA sequences encoding two polypeptide segments from different proteins, allowing a hybrid polypeptide to be produced from the joined DNAs. When proteins normally secreted from the cell were altered in this manner to include amino acids representing an ER retrieval tag, these proteins were subsequently found in the ER rather than being secreted. Interestingly, some of these chimeric proteins found in the ER had undergone partial processing by enzymes found only in the Golgi apparatus. This finding indicates that the suspected tag did not simply prevent escape of the protein from the ER but actively promoted the retrieval of ER-specific proteins from the Golgi apparatus.

Golgi Apparatus Proteins May Be Sorted According to the Lengths of Their Membrane-Spanning Domains. Like resident proteins required in the ER, some resident proteins of the Golgi apparatus contain retention or retrieval tags. Moreover, in the Golgi apparatus as in the ER, the formation of large complexes that are excluded from transport vesicles may play a role in maintaining the protein composition of the Golgi apparatus. In addition, we will next see that a third, distinctly different mechanism is also likely at work in this organelle—a mechanism involving hydrophobic regions of Golgi proteins.

All known Golgi-specific proteins are integral membrane proteins that have one or more hydrophobic membrane-spanning domains anchoring them to Golgi membranes. The length of the hydrophobic domains may determine into which cisterna of the Golgi apparatus each membrane-bound protein is incorporated as it moves through the organelle. Recall that the thickness of cellular membranes increases progressively from the ER (about 5 nm) to the plasma membrane (about 8 nm). Among Golgi-specific proteins, there is a corresponding increase in the lengths of hydrophobic membrane-spanning domains going from the CGN to the TGN. Such proteins tend to move from compartment to compartment until the thickness of the membrane exceeds the length of their membrane-spanning domains, thereby blocking further migration.

Targeting of Soluble Lysosomal Proteins to Endosomes and Lysosomes Is a Model for Protein Sorting in the TGN. During their journey through the ER and early compartments of the Golgi apparatus, soluble lysosomal enzymes, like other glycoproteins, undergo N-glycosylation followed by removal of glucose and mannose units. Within the Golgi apparatus, however, mannose residues on the carbohydrate side chain of lysosomal enzymes are phosphorylated, forming an oligosaccharide containing mannose-6-phosphate. This oligosaccharide tag distinguishes soluble lysosomal proteins from other glycoproteins and ensures their delivery to lysosomes (**Figure 12-12** on page 330).

❶ In the ER, soluble lysosomal enzymes undergo N-glycosylation followed by removal of glucose and mannose units. ❷ Within the Golgi apparatus, mannose residues on the lysosomal enzymes are phosphorylated by two Golgi-specific enzymes. The first, located in an early compartment of the Golgi stack, is a phosphotransferase that adds GlcNAc-1-phosphate to carbon atom 6 of mannose. The second, located in a mid-Golgi compartment, removes GlcNAc, leaving behind the mannose-6-phosphate residue.

❸ In the TGN, these tagged lysosomal enzymes bind to mannose-6-phosphate receptors (MPRs) in the TGN membrane. The pH of the TGN is around 6.4, favoring binding of soluble lysosomal enzymes to these receptors. Next, the receptor-ligand complexes are packaged into transport vesicles and conveyed to an endosome. In animal cells, these lysosomal enzymes are needed for degradation of material brought into the cell by endocytosis and are transported from the TGN to organelles known as **late endosomes.** These develop from **early endosomes,** which are formed by the coalescence of vesicles from the TGN and plasma membrane.

In some cells, degradation and recycling of unneeded or damaged cell components are carried out by a specialized subset of late endosomes known as *multivesicular endosomes (MVEs)*, which are also sometimes called *multivesicular bodies (MVBs)*. MVEs serve as an intermediate between early endosomes and lysosomes and contain intraluminal vesicles that form by budding into the interior of the MVEs. This sequesters material destined for degradation or recycling. A MVE can later fuse with a lysosome to degrade its contents, or material in the MVE (such as membrane-bound receptors proteins) can be recycled.

❹ As an early endosome matures to form a late endosome, the pH of the lumen decreases to about 5.5, causing the bound lysosomal enzymes to dissociate from the MPRs. This prevents the retrograde movement of these enzymes back to

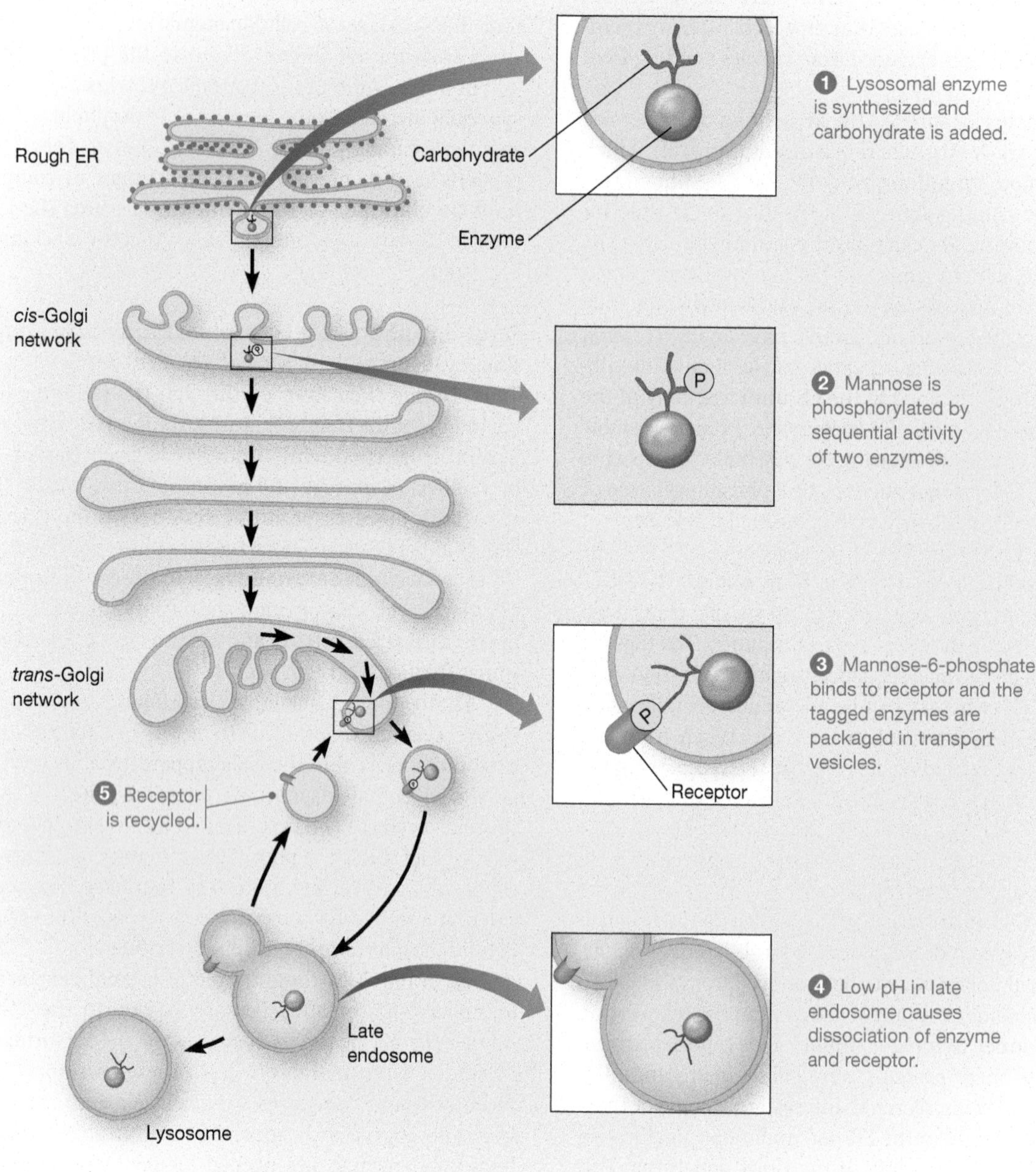

Rough ER
Carbohydrate
Enzyme
1 Lysosomal enzyme is synthesized and carbohydrate is added.
cis-Golgi network
P
2 Mannose is phosphorylated by sequential activity of two enzymes.
trans-Golgi network
P
Receptor
3 Mannose-6-phosphate binds to receptor and the tagged enzymes are packaged in transport vesicles.
5 Receptor is recycled.
Late endosome
Lysosome
4 Low pH in late endosome causes dissociation of enzyme and receptor.

Figure 12-12 Targeting of Soluble Lysosomal Enzymes to Endosomes and Lysosomes by a Mannose-6-Phosphate Tag. Lysosomal enzymes carry distinctive mannose-6-phosphate tags that direct them to lysosomes.

the Golgi as 5 the receptors are recycled in vesicles that return back to the TGN. Finally, the late endosome either matures to form a new lysosome or delivers its contents to an active lysosome.

Strong support for this model of lysosomal enzyme targeting come from studies of a human genetic disorder called *I-cell disease.* Cultured fibroblast cells from patients with I-cell disease synthesize all the expected lysosomal enzymes but then release most of the soluble proteins to the extracellular medium instead of incorporating them into lysosomes. I-cell disease results from a defective phosphotransferase that is needed to add the mannose-6-phosphate to oligosaccharide chains on lysosomal enzymes, showing that the mannose-6-phosphate tag is essential for targeting soluble lysosomal glycoproteins to the lysosome. Without the necessary signal, the hydrolytic enzymes are not transported into the lysosomes. Thus, the lysosomes become engorged with undegraded polysaccharides, lipids, and other material. This causes irreversible damage to the cells and tissues (**see Human Connections in Chapter 4, pages 94–95**). Although mannose-6-phosphate residues are important for directing traffic from the ER to the lysosome, other pathways appear to be involved as well. For example, in patients with I-cell disease, lysosomal acid hydrolases somehow still reach the lysosomes in liver cells.

Stop-Transfer Sequences Mediate the Insertion of Integral Membrane Proteins

So far, we have focused on the cotranslational import and sorting of *soluble proteins* that are destined either for secretion from the cell or for the lumen of endomembrane components, such as the ER, Golgi apparatus, lysosomes, and related vesicles. The other major group of polypeptides synthesized on ER-attached ribosomes consists of molecules destined to become *integral membrane proteins.* Polypeptides of this type are synthesized by a mechanism similar to the one illustrated in Figure 12-11 for soluble proteins, except that the completed polypeptide chain remains embedded in the ER membrane rather than being released into the ER lumen.

Recall that integral membrane proteins are typically anchored to the lipid bilayer by one or more α-helical *transmembrane segments* consisting of 20–30 hydrophobic amino acids (see Chapter 7). How are such proteins retained within the ER membrane after synthesis rather than being released into the ER lumen? There are two main mechanisms by which hydrophobic transmembrane segments anchor newly forming polypeptide chains to the lipid bilayer of the ER membrane (**Figure 12-13** on page 332).

Stop-Transfer Sequences. The first mechanism involves polypeptides with a typical ER signal sequence at their N-terminus, which allows an SRP to bind the ribosome-mRNA complex to the ER membrane. Elongation of the polypeptide chain then continues until the hydrophobic transmembrane segment of the polypeptide is synthesized. As shown in Figure 12-13a, this stretch of amino acids functions as a **stop-transfer sequence** that halts translocation of the polypeptide through the ER membrane. Translation continues, but the rest of the polypeptide chain remains on the cytosolic side of the ER membrane, resulting in a transmembrane protein with its N-terminus in the ER lumen and its C-terminus in the cytosol. Meanwhile, the hydrophobic stop-transfer signal moves laterally out through a side opening in the translocon and into the lipid bilayer, forming the permanent transmembrane segment that anchors the protein to the membrane. Such proteins are often called *type I transmembrane proteins.*

Internal Start-Transfer Sequences. The second mechanism involves membrane proteins that lack a typical signal sequence at their N-terminus and instead possess an *internal* **start-transfer sequence** that performs two functions: it first acts as an ER signal sequence that allows an SRP to bind the ribosome-mRNA complex to the ER membrane, and then its hydrophobic region functions as a membrane anchor that moves out through a side opening in the translocon and permanently embeds the polypeptide in the lipid bilayer (Figure 12-13b). The orientation of the start-transfer sequence at the time of insertion determines which terminus of the polypeptide ends up in the ER lumen versus the cytosol. Proteins whose C-terminus is in the ER lumen and N-terminus is in the cytosol are known as *type II transmembrane proteins.*

Many transmembrane proteins have a single membrane-spanning region, but many others have more than one membrane-spanning region. Many important types of receptor proteins, for example, have *seven* membrane-spanning domains! How are such complicated proteins threaded through the membrane? Transmembrane proteins with multiple membrane-spanning regions are formed in a manner similar to single-pass transmembrane proteins that use the start- and stop-transfer strategy, except that an alternating pattern of start-transfer and stop-transfer sequences creates a polypeptide containing multiple transmembrane segments that pass back and forth across the membrane. The basic mechanism is shown in Figure 12-13c. ❶ The first transmembrane segment is inserted via an internal start-transfer sequence; ❷ a stop-transfer sequence halts insertion and releases this first section of membrane-spanning amino acids laterally into the ER membrane. ❸ Now the additional start- and stop-transfer sequences come into play, as the next start-transfer sequence is threaded into the translocon. Threading continues until the next stop-transfer sequence is encountered; at this point the next transmembrane section is released laterally into the ER membrane. This process continues until all of the membrane-spanning sections are inserted through the membrane.

Once a newly formed polypeptide has been incorporated into the ER membrane by one of the preceding mechanisms, it can either remain in place to function as an ER membrane protein or be transported to other components of the endomembrane system, such as the Golgi apparatus, lysosomes, nuclear envelope, or plasma membrane. Transport is carried out by a series of membrane budding and fusing events in which membrane vesicles pinch off from one compartment of the endomembrane system and fuse with another compartment (see Figure 12-8). The net result of transport of transmembrane proteins to the surface is that the regions of the protein that were originally in the *lumen* of the ER eventually come to lie *outside* the cell. The regions of the protein that were in the cytosol remain there.

Posttranslational Import Is an Alternative Mechanism for Import into the ER Lumen

In contrast to the cotranslational import of proteins into the ER discussed in the previous sections, some proteins are synthesized in the cytosol and subsequently transported into the ER lumen. Such *posttranslational import* is accomplished by some of the same components as cotranslational import, but other components differ. The basic process is shown in **Figure 12-14** on page 333. ❶ As the polypeptide is synthesized, it associates with chaperones of the Hsp70 family, which keep it unfolded so that it can pass through the ER membrane. A complex of ER membrane proteins (called Sec62, 63, 71, and 72) associated with the Sec61 pore targets the protein for translocation. ❷ Then the protein moves through the pore, losing its associated chaperones as it does so. The Sec protein complex associated with the pore also recruits the chaperone BiP to the site where the protein emerges into the ER lumen. ❸ BiP then attaches to the polypeptide. BiP is thought to couple ATP hydrolysis to pulling the polypeptide into the ER lumen, using a ratchet-like, stepwise mechanism.

CONCEPT CHECK 12.4

What features of membrane lipids and proteins contribute to the proper trafficking of vesicles and targeting of proteins in cells?

Figure 12-13 Cotranslational Insertion of Transmembrane Proteins into the ER Membrane. For clarity, the SRP, ribosome, and many other parts of the translocational apparatus have been omitted. **(a)** A type I transmembrane protein with its C-terminus in the cytosol. **(b)** A type II transmembrane protein, whose N-terminus is in the cytosol. **(c)** A multipass transmembrane protein. Multiple start- and stop-transfer sequences allow threading of more than one portion of a polypeptide through the ER membrane. For proteins destined for the plasma membrane, the regions in the ER lumen ultimately come to lie outside the cell after the patch of membrane within which the protein resides fuses with the plasma membrane.

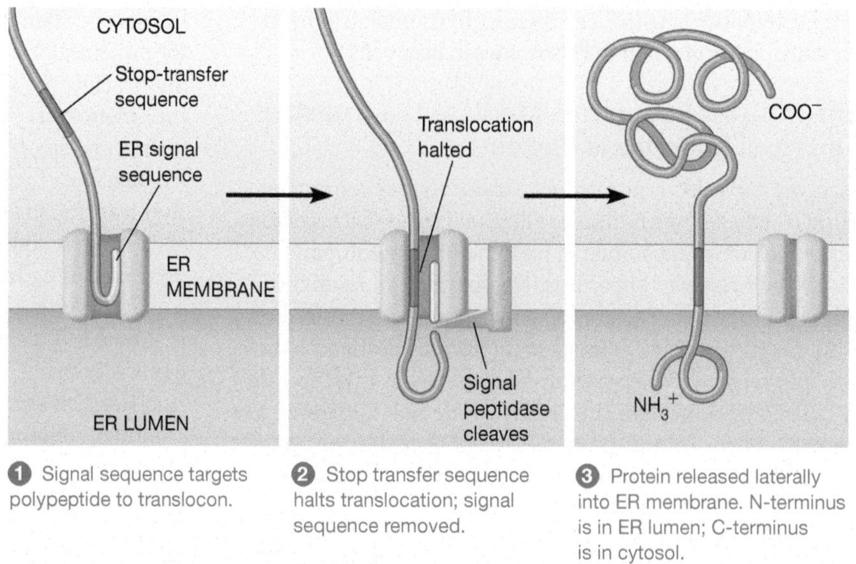
CYTOSOL
Stop-transfer sequence
ER signal sequence
ER MEMBRANE
ER LUMEN
Translocation halted
Signal peptidase cleaves
COO⁻
NH3+
1 Signal sequence targets polypeptide to translocon.
2 Stop transfer sequence halts translocation; signal sequence removed.
3 Protein released laterally into ER membrane. N-terminus is in ER lumen; C-terminus is in cytosol.

(a) Type I transmembrane protein: Internal stop-transfer sequence and terminal ER signal sequence

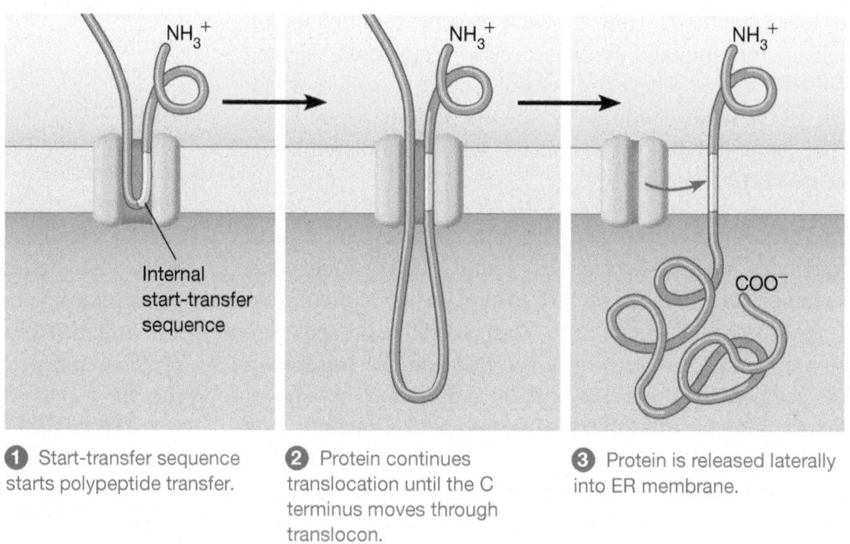
NH3+
Internal start-transfer sequence
COO⁻
1 Start-transfer sequence starts polypeptide transfer.
2 Protein continues translocation until the C terminus moves through translocon.
3 Protein is released laterally into ER membrane.

(b) Type II transmembrane protein: Polypeptide with single internal start-transfer sequence

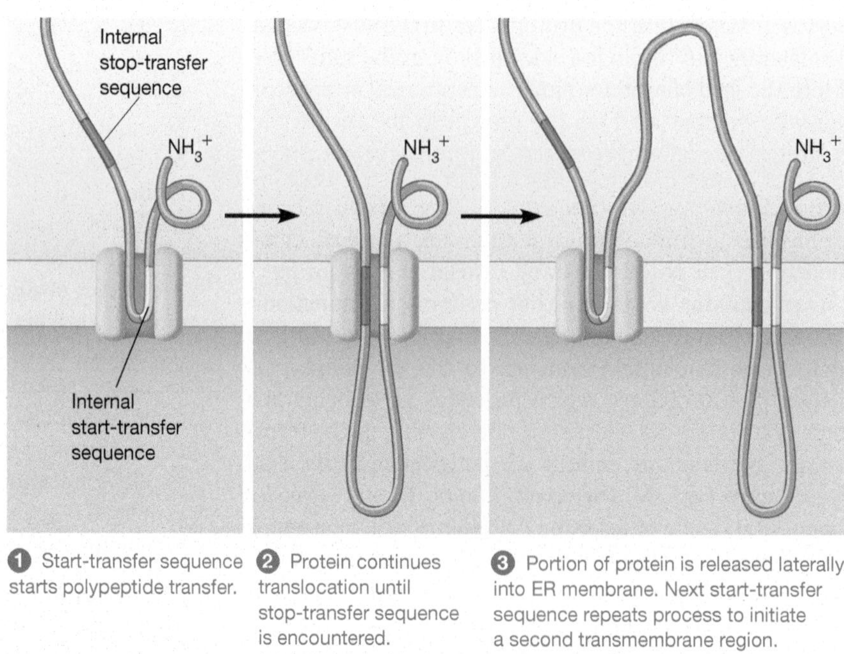
Internal stop-transfer sequence
NH3+
Internal start-transfer sequence
1 Start-transfer sequence starts polypeptide transfer.
2 Protein continues translocation until stop-transfer sequence is encountered.
3 Portion of protein is released laterally into ER membrane. Next start-transfer sequence repeats process to initiate a second transmembrane region.

(c) Multipass transmembrane protein: Multiple internal start- and stop-transfer sequences

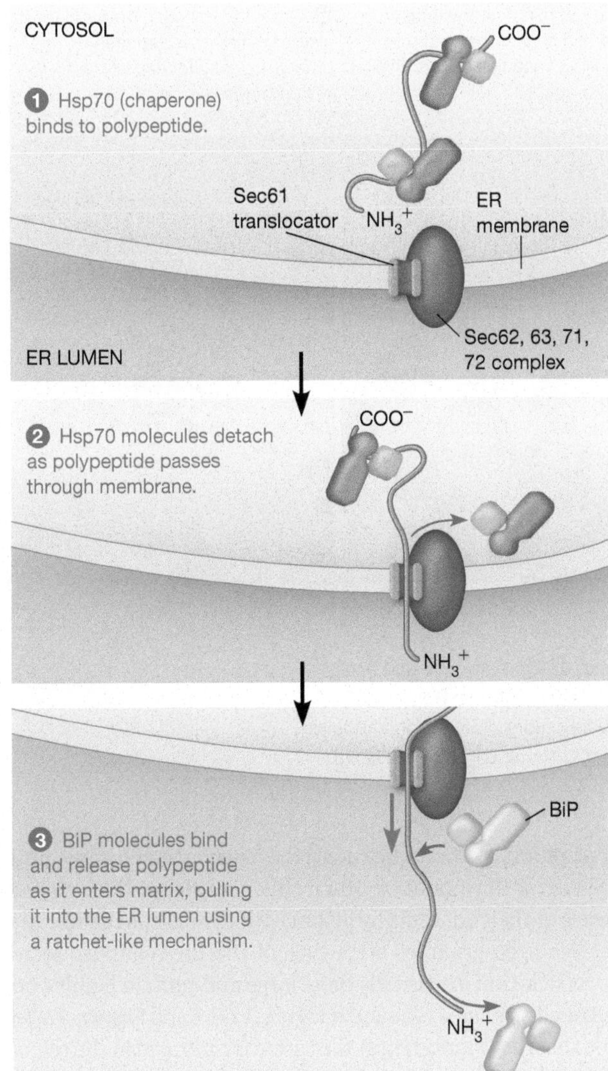
CYTOSOL
COO⁻
1 Hsp70 (chaperone) binds to polypeptide.
Sec61 translocator
NH3+
ER membrane
Sec62, 63, 71, 72 complex
ER LUMEN
COO⁻
2 Hsp70 molecules detach as polypeptide passes through membrane.
NH3+
BiP
3 BiP molecules bind and release polypeptide as it enters matrix, pulling it into the ER lumen using a ratchet-like mechanism.
NH3+

Figure 12-14 Posttranslational Import of Polypeptides into the ER Lumen. For simplicity, ATP hydrolysis and exchange are not shown. ❶ A recently translated polypeptide associates with Hsp70 chaperones, keeping it unfolded. ❷ The polypeptide docks with the translocator, whose core is the Sec61 complex. Hsp70 proteins are released as it begins to thread through the pore. ❸ The Sec62, 63, 71, 72 complex recruits an ER-resident chaperone (BiP) to the end of the polypeptide. BiP pulls the polypeptide into the lumen.

12.5 Exocytosis and Endocytosis: Transporting Material Across the Plasma Membrane

So far, we have discussed movement of membrane-bounded components within cells. Two methods of transporting materials across the plasma membrane are *exocytosis*, the process by which secretory granules release their contents to the exterior of the cell, and *endocytosis*, the process by which cells internalize external materials. Both processes are unique to eukaryotic cells and are also involved in the delivery, recycling, and turnover of membrane proteins. We will first consider exocytosis because it is the final step in a secretory pathway that began with the ER and the Golgi apparatus.

Secretory Pathways Transport Molecules to the Exterior of the Cell

A key feature of the vesicular traffic shown in Figure 12-8 is the **secretory pathways** by which proteins move from the ER through the Golgi apparatus to **secretory vesicles** and **secretory granules**, which then discharge their contents to the exterior of the cell. The concerted roles of the ER and the Golgi apparatus in secretion were demonstrated in 1967 by James Jamieson and George Palade in secretory cells from guinea pig pancreatic slices. They treated slices of guinea pig pancreatic tissue with a small amount of a radioactive amino acid for a brief period to "pulse label" newly synthesized proteins. After washing away unincorporated amino acids, they used microscopic autoradiography to determine the location of radioactively labeled protein in pancreatic secretory cells at several time points following treatment.

Results from this classic experiment are presented in **Figure 12-15** on page 334. Three minutes after brief exposure of tissue slices to a radioactive amino acid, newly synthesized protein labeled with the radioactive amino acid (irregular dark coloration) was found primarily in the rough ER ❶. A few minutes later, labeled protein began to appear in the Golgi apparatus ❷. By 37 minutes, labeled protein was detected in vesicles budding from the Golgi that Jamieson and Palade called *condensing vacuoles* ❸. After 117 minutes, labeled protein began to accumulate in dense *zymogen granules*, vesicles that discharge secretory proteins to the exterior of the cell ❹. Some of the radioactively labeled protein was observed in the exterior lumen adjacent to the cell, demonstrating that some of the secretory granules had released their contents from the cell.

Based on this classic experiment and numerous similar studies since then, the secretory pathways shown in Figure 12-8 are now understood in considerable detail. The molecules that allow movement through the secretory pathway were in part identified using yeast as a genetic model system. Randy Schekman and colleagues identified a number of *Sec* proteins required at various stages of the secretion process. These proteins were so named because mutations in the genes that encode them led to defective secretion of glycoproteins. For this pioneering work, Schekman shared a Nobel Prize with James Rothman, whose work with vesicle targeting we will discuss shortly.

We now distinguish several different modes of secretion by eukaryotic cells. *Constitutive secretion* involves the continuous discharge of vesicles at the plasma membrane surface, *regulated secretion* involves controlled, rapid releases that happen in response to an extracellular signal, and *polarized secretion* involves secretion at one specific end of the cell.

Constitutive Secretion. After budding from the TGN, some secretory vesicles move directly to the cell surface, where they immediately fuse with the plasma membrane and release their contents by exocytosis. This unregulated process, which is continuous and independent of specific extracellular signals, occurs in most eukaryotic cells and is called

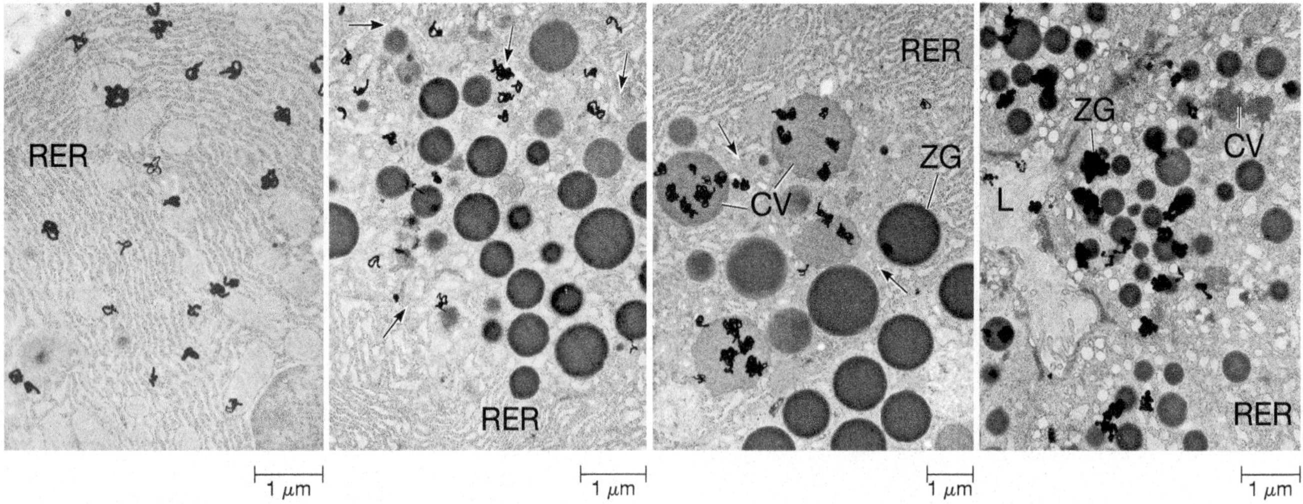
1 After 3 minutes, most of the labeled protein is found in the rough ER where it has just been synthesized.
2 After 7 minutes, most of the new labeled protein has moved into the adjacent Golgi apparatuses (arrows).
3 After 37 minutes, the labeled protein is being concentrated in condensing vacuoles next to the Golgi.
4 After 117 minutes, the labeled protein is found in zymogen granules, ready for export from the cell to the lumen.
RER
RER
RER
CV
ZG
ZG
CV
L
RER
1 μm
1 μm
1 μm
1 μm

Figure 12-15 Autoradiographic Evidence of a Secretory Pathway. In steps 1–4, Jamieson and Palade used electron microscopic autoradiography to trace the movement of radioactively labeled protein from the ER through the Golgi apparatus to its subsequent packaging in secretory vesicles. Abbreviations: RER, rough endoplasmic reticulum; CV, condensing vacuole; ZG, zymogen granule; L, lumen. Arrows point to the edges of the Golgi apparatus (TEMs).

constitutive secretion. One example is the continuous release of mucus by cells that line your intestine.

Constitutive secretion was once assumed to be a *default pathway* for proteins synthesized by the rough ER. According to this model, all proteins destined to remain in the endomembrane system must have a tag that diverts them from constitutive secretion. Otherwise, they will move through the endomembrane system and be released outside the cell by default. Support for this model came from studies in which removal of the KDEL retrieval tags on resident ER proteins led to secretion of the modified protein. However, more recent evidence suggests that a variety of short amino acid tags may identify specific proteins for constitutive secretion.

Regulated Secretion. Whereas vesicles containing constitutively secreted proteins move continuously and directly from the TGN to the plasma membrane, secretory vesicles involved in **regulated secretion** accumulate in the cell and then fuse with the plasma membrane only in response to specific extracellular signals. An important example is the release of neurotransmitters. (We will look in detail at how neurotransmitter release allows nerve signals to propagate in Chapter 22, Figure 22-15.) Two additional well-known examples of regulated secretion are the release of insulin from pancreatic β cells in response to glucose and the release of *zymogens*—inactive precursors of hydrolytic enzymes—from pancreatic acinar cells in response to calcium or endocrine hormones.

Regulated secretory vesicles form by budding from the TGN as immature secretory vesicles, which undergo a subsequent maturation process. Maturation of secretory proteins involves concentration of the proteins—referred to as *condensation*—and frequently also some proteolytic processing. The mature secretory vesicles then move close to the site of secretion and remain near the plasma membrane until receiving a hormonal or other chemical signal that triggers release of their contents by fusion with the plasma membrane.

Zymogen granules are a type of mature regulated secretory vesicle that are usually quite large and contain highly concentrated protein, as shown in Figure 12-15 and **Figure 12-16**. Note that the zymogen granules are concentrated in the region of the cell between the Golgi stacks from which they arise and

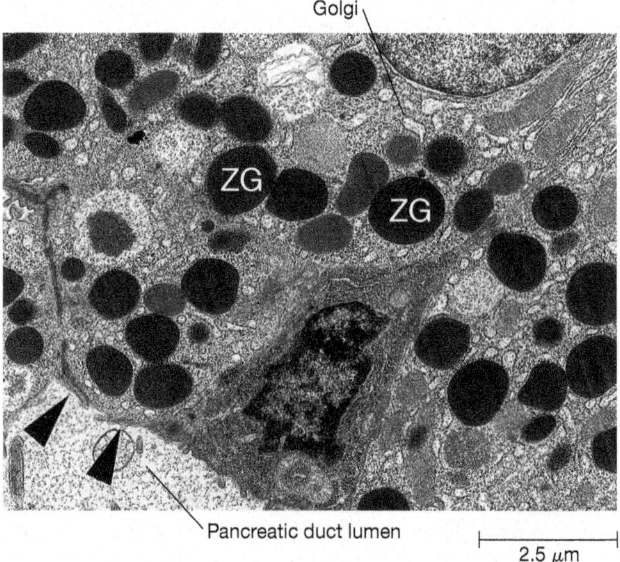
Golgi
ZG
ZG
Pancreatic duct lumen
2.5 μm

Figure 12-16 Zymogen Granules. This electron micrograph of an acinar (secretory) cell from the exocrine pancreas of illustrates the prominence of zymogen granules (ZG), which contain enzymes destined for secretion. Arrows indicate where the granules will ultimately be secreted into the pancreatic duct lumen (TEM).

the portion of the plasma membrane bordering the lumen into which the contents of the granules are eventually discharged.

The information needed to direct a protein to a regulated secretory vesicle is presumably inherent in the amino acid sequence of the protein, though the precise signals and mechanisms are not yet known. Current evidence suggests that high concentrations of secretory proteins in secretory granules promote the formation of large *protein aggregates* that exclude nonsecretory proteins. This could occur in the TGN, where only aggregates would be packaged in vesicles destined for secretory granules, or it could occur in the secretory granule itself. The pH of the TGN and the secretory granule lumens may serve as a trigger favoring aggregation as material leaves the TGN. The soluble proteins that do not become part of an aggregate in the TGN or a secretory granule would be carried by transport vesicles to other locations.

Polarized Secretion. In many cases, secretion of specific proteins is limited to a specific surface of the cell. For example, the secretory cells that line your intestine release digestive enzymes only on the side of the cell that faces the interior of the intestine. This phenomenon, called **polarized secretion,** is also seen in nerve cells, which secrete neurotransmitter molecules only at junctions with other nerve cells (see Figure 22-13). Proteins destined for polarized secretion, as well as lipid and protein components of the two different membrane layers, are sorted into vesicles that bind to localized recognition sites on subdomains of the plasma membrane.

Exocytosis Releases Intracellular Molecules Outside the Cell

Ultimately, material in vesicles destined for secretion must leave the cell. In **exocytosis,** proteins or other materials in a vesicle are released to the exterior of the cell as the membrane of the vesicle fuses with the plasma membrane. A variety of proteins are exported from both animal and plant cells by exocytosis. Animal cells secrete peptide and protein hormones, mucus, and several digestive enzymes in this manner. Plant and fungal cells secrete enzymes and structural proteins associated with the cell wall, and carnivorous plants secrete hydrolytic enzymes that are used to digest trapped insects.

The process of exocytosis is illustrated in **Figure 12-17.** In part a, ❶ vesicles containing cellular products destined for secretion move to the cell surface, where ❷ the membrane of the vesicle fuses with the plasma membrane. ❸ Fusion with the plasma membrane discharges the vesicle contents to the exterior of the cell. In the process, ❹ the membrane of the vesicle becomes integrated into the plasma membrane, with the *inner* (lumenal) surface of the vesicle becoming the *outer* (extracellular) surface of the plasma membrane. Thus, glycoproteins and glycolipids that were originally formed in the ER and Golgi lumens will face the extracellular space.

The mechanism underlying the movement of exocytic vesicles to the cell surface and their fusion with the plasma membrane is under intensive investigation. (See Key Technique, pages 336–337, for a discussion of total internal reflection microscopy, a useful technique for monitoring vesicle dynamics at the cell surface.) Current evidence points to the involvement of microtubules in vesicle movement toward the plasma membrane. For example, in some cells, vesicles appear to move from the Golgi apparatus to the plasma membrane along "tracks" of microtubules that are oriented parallel to the direction of vesicle movement. Moreover, vesicle movement stops when the cells are treated with *colchicine,* a plant alkaloid that prevents microtubule assembly. (We will discuss the intracellular movement of vesicles along microtubules in detail in Chapter 14.)

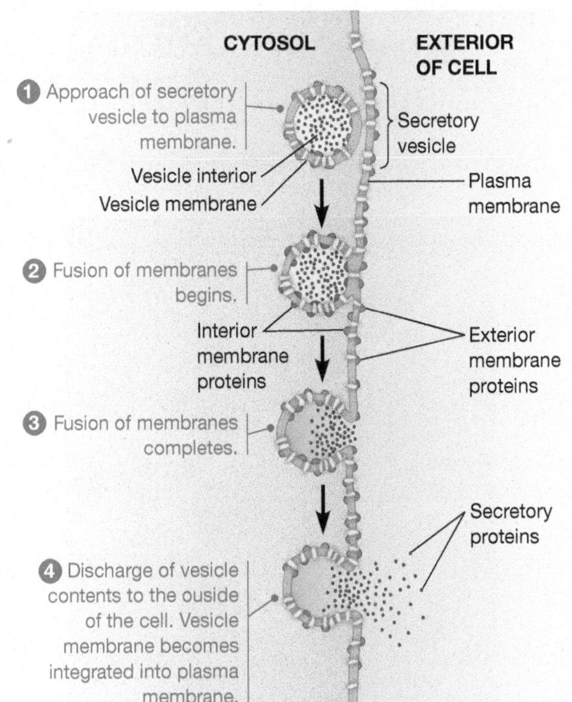

(a)

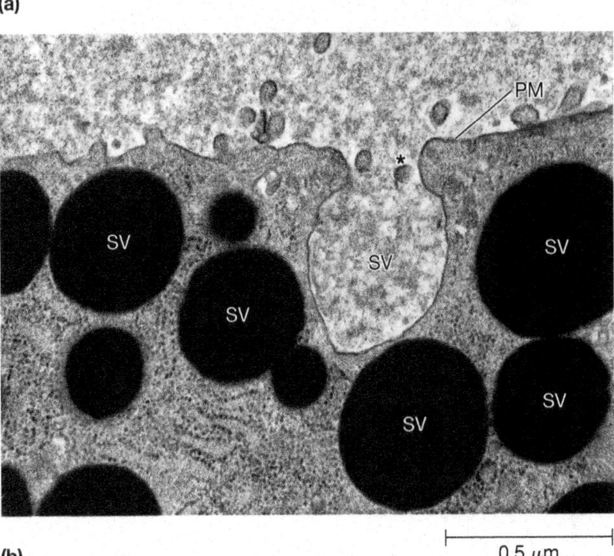

(b)

Figure 12-17 Exocytosis. (a) As a secretory vesicle approaches the plasma membrane, it fuses with it and releases its contents to the exterior of the cell by exocytosis. **(b)** In this cell, secretory vesicles (SV) approach the plasma membrane (PM), where one vesicle has fused with the membrane and secreted its contents at the asterisk (TEM).

The Role of Calcium in Triggering Exocytosis. Fusion of regulated secretory vesicles with the plasma membrane is

Key Technique | Visualizing Vesicles at the Cell Surface Using Total Internal Reflection (TIRF) Microscopy

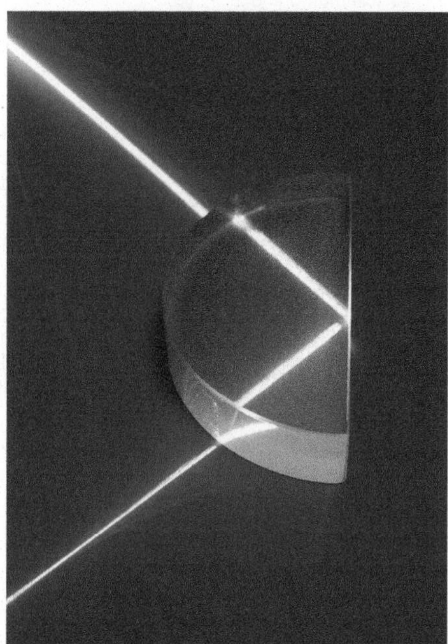

Total Internal Reflection.

PROBLEM: Although many techniques exist for introducing fluorescent proteins into cells using dyes or by genetically engineering cells to express a fluorescent protein, these molecules are often difficult to visualize against the backdrop of fluorescence that comes from molecules that are not in focus. This out-of-focus light poses a serious problem for visualizing events at the cell surface, such as fusion of secretory vesicles with the plasma membrane during exocytosis. Being able to assess where and when such fusion events occur is valuable to cell biologists and neuroscientists as they study the signals that trigger vesicle release, such as during transmission of signals between neurons (see Chapter 22) or release of histamine by cells in response to allergens.

SOLUTION: Using *total internal reflection fluorescence (TIRF) microscopy,* researchers can use a light microscope to follow dynamic changes in fluorescent molecules only a very short distance into the specimen without causing other out-of-focus fluorescent molecules to glow.

Key Tools: Fluorescence microscope with special TIRF lens; a laser to illuminate the specimen; a fluorescent cell or fluorescently labeled proteins; a camera to record images.

Details: TIRF relies on a useful property of light: when light moves from a medium with a high refractive index (such as glass) to a medium with a lower refractive index (such as water or a cell), if the angle of incidence of the light exceeds a certain angle (called the *critical angle*), the light is reflected. You may be familiar with fiber-optic cables, which rely on the same physical principle. The curvature of the fiber causes almost all of the light that passes into the cable to be internally reflected, allowing such fibers to serve as "light pipes." Microscopists can use special lenses to exploit total internal reflection in a similar way. When fluorescent light shines on a cell such that all of the light exceeds the critical angle, it will undergo total internal reflection.

If all of the light is reflected, why is TIRF useful? It turns out that a very small layer of light, called the *evanescent field,* extends into the water or cell (**Figure 12A-1**). This layer is very thin, about 100 nm, making TIRF extremely useful for studying the release of secretory vesicles or for watching other events in which fluorescence from out-of-focus light would cause a haze of "light pollution" that would impede viewing events very near the coverslip. Because only a small region is illuminated, the intensity of the light required is high; typically, lasers are used to provide the needed intense light.

TIRF is up to ten times better than a confocal microscope for resolving small objects very close to the surface of a coverslip (see the Appendix for more on confocal microscopy), so it is often the method of choice for viewing processes that occur at the cell surface. To see what a difference this technique makes, compare the same HeLa cells imaged using conventional fluorescence microscopy and using TIRF (**Figure 12A-2**). TIRF is also useful for high-speed imaging, which is crucial for observing individual vesicles as they fuse with the plasma membrane during exocytosis. Although technology is changing rapidly in this area, TIRF microscopy is currently dramatically faster than super-resolution techniques (see the Appendix) and so is the method of choice for monitoring vesicle fusion events at the plasma membrane.

generally triggered by a specific extracellular signal. In most cases, the signal is a hormone or a neurotransmitter that binds to specific receptors on the cell surface and triggers the synthesis or release of a *second messenger* within the cell (as we will see in Chapter 23). During regulated secretion, a transient elevation of the intracellular concentration of calcium ions often appears to be an essential step in the signaling cascade leading from the receptor on the cell surface to exocytosis. For example, microinjection of calcium into pancreatic cells induces mature secretory granules to discharge their contents to the extracellular medium. The specific role of calcium is not yet clear, but it appears that an elevation in the intracellular calcium concentration leads to the activation of protein kinases whose target proteins are components of either the vesicle membrane or the plasma membrane.

Endocytosis Imports Extracellular Molecules by Forming Vesicles from the Plasma Membrane

For many eukaryotic cells, vesicular traffic is a two-way affair. Some components are exported via exocytosis; others are imported via one or more forms of **endocytosis** (**Figure 12-18**

By combining information gathered from electron microscopy and X-ray crystallography, researchers have assembled the following model for the organization of triskelions into the characteristic hexagons and pentagons of clathrin-coated pits and vesicles (Figure 12-22c). One clathrin triskelion is located at each vertex of the polygonal lattice. Each polypeptide leg extends along two edges of the lattice, with the knee of the clathrin heavy chain located at an adjacent vertex. This arrangement of triskelions into overlapping networks ensures extensive longitudinal contact between clathrin polypeptides and may confer the mechanical strength needed when a coated vesicle forms.

The components of the **adaptor protein (AP) complexes** were originally identified simply by their ability to promote the assembly of clathrin coats around vesicles. We now know that eukaryotic cells contain at least four types of AP complexes, each composed of four polypeptides—two adaptin subunits, one medium chain, and one small chain. The four polypeptides, which are slightly different in each type of AP complex, bind to different transmembrane receptor proteins and confer specificity during vesicle budding and targeting. For example, the *AP-2 complex* is involved in endocytosis.

In addition to ensuring that appropriate macromolecules will be concentrated in coated pits, AP complexes mediate the attachment of clathrin to proteins embedded in the plasma membrane. Considering the central role of APs, it is not surprising that AP complexes are sites for regulation of clathrin assembly and disassembly. For example, the ability of AP complexes to bind to clathrin is affected by pH, phosphorylation, and dephosphorylation.

The Assembly of Clathrin Coats Drives the Formation of Vesicles from the Plasma Membrane and TGN

The binding of AP complexes to the plasma membrane and the concentration of receptors or receptor-ligand complexes in coated pits require ATP and GTP—though perhaps only for regulation of the process. The assembly of a clathrin coat on the cytosolic side of a membrane appears to provide part of the driving force for formation of a vesicle at the site (**Figure 12-23**). Initially, all clathrin units are hexagonal and form a planar, two-dimensional structure (Figure 12-23a, c). As more clathrin triskelions are incorporated into the growing lattice, a combination of hexagonal and pentagonal units allows the new clathrin coat to curve around the budding vesicle (Figure 12-23b, d).

As clathrin accumulates around the budding vesicle, at least one more protein—**dynamin**—participates in the process. Dynamin is a cytosolic GTPase required for coated pit constriction and closing of the budding vesicle. This essential protein was first identified in the fruit fly, *Drosophila melanogaster*. Flies expressing a temperature-sensitive form of dynamin were instantly paralyzed after a temperature shift disrupted dynamin function. Further investigation revealed an accumulation of coated pits in the membranes of neuromuscular junctions in the affected flies. Formation of the closed vesicle occurs as dynamin forms helical rings around the neck of the coated pit. As GTP is hydrolyzed, the dynamin rings tighten and separate the fully sealed endocytic vesicle from the plasma membrane.

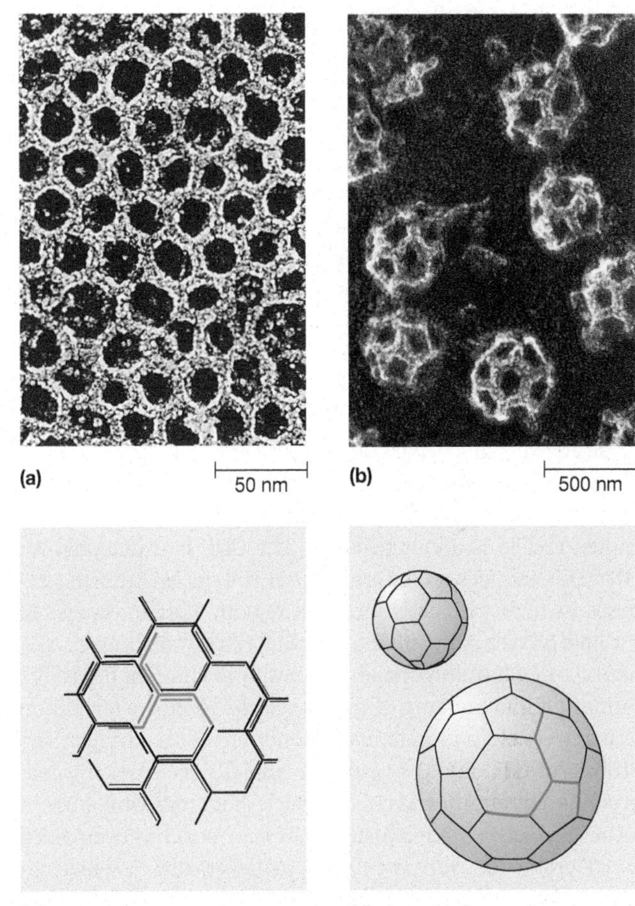

Figure 12-23 Clathrin Lattices. Each vesicle is surrounded by a cage of overlapping clathrin complexes. **(a)** Freeze-etch electron micrograph of a hexagonal clathrin lattice in a human carcinoma cell (TEM). **(b)** Electron micrograph of clathrin cages showing both pentagonal and hexagonal units (TEM). **(c, d)** Interpretive drawings of the lattice and clathrin cages shown in (a) and (b).

Some mechanism is also required to *uncoat* clathrin-coated membranes. Moreover, uncoating must be done in a regulated manner because, in most cases, the clathrin coat remains intact as long as the membrane is part of a coated pit or budding vesicle but dissociates rapidly once the vesicle is fully formed. Like assembly, dissociation of the clathrin coat is an energy-consuming process, accompanied by the hydrolysis of about three ATP molecules per triskelion. At least one protein, an *uncoating ATPase,* is essential for this process, though the uncoating ATPase releases only the clathrin triskelions from the APs.

Clathrin-coated vesicles readily dissociate into soluble clathrin complexes, adaptor protein complexes, and uncoated vesicles that can spontaneously reassemble under appropriate conditions. In a slightly acidic solution containing calcium ions, clathrin complexes will reassemble independently of adaptor protein and membrane-bounded vesicles, resulting in empty shells called *clathrin cages.* Assembly occurs remarkably fast—within seconds, under favorable conditions. Ease of assembly and disassembly is an important feature of the clathrin coat because fusion of the underlying membrane with the membrane of another structure appears to require partial or complete uncoating of the vesicle.

COPI- and COPII-Coated Vesicles Travel Between the ER and Golgi Apparatus Cisternae

COPI-coated vesicles have been found in all eukaryotic cells examined, including mammalian, insect, plant, and yeast cells. They are involved in retrograde transport from the Golgi apparatus back to the ER as well as bidirectional transport between Golgi apparatus cisternae. COPI-coated vesicles are surrounded by coats composed of **COPI** protein and **ADP ribosylation factor (ARF)**, a small GTP-binding protein. The major component of the coat, COPI, is a protein multimer composed of seven subunits.

Assembly of a COPI coat is mediated by ARF. In the cytosol, ARF occurs as part of an ARF-GDP complex. However, when ARF encounters a specific *guanine nucleotide exchange factor* associated with the membrane (from which a new coated vesicle is about to form), the GDP is exchanged for GTP. This induces a conformational change in ARF that exposes its hydrophobic N-terminal region, which attaches to the lipid bilayer of the membrane. Once firmly anchored, ARF binds to COPI multimers, and assembly of the coat drives the formation and budding of a new vesicle. After the formation of a free vesicle, a protein in the donor membrane triggers hydrolysis of GTP, and the resulting ARF-GDP releases the coat proteins for another cycle of vesicle budding. Our current knowledge of COP-mediated vesicle transport has been aided considerably by using the fungal toxin *brefeldin A*, which inhibits this process by interfering with the ability of the guanine nucleotide exchange factor to produce ARF-GTP from ARF-GDP.

COPII-coated vesicles were first discovered in yeast, where they have a role in transport from the ER to the Golgi apparatus. Mammalian and plant homologues of some of the components of **COPII** coats have been identified, and the COPII-mediated mechanism of ER export appears to be highly conserved between organisms as different as yeast and humans. The COPII coat found in yeast is assembled from two protein complexes—called *Sec13/31* and *Sec23/24*—and a small GTP-binding protein called *SarI*, which is similar to ARF. By a mechanism resembling formation of a COPI coat, a SarI molecule with GDP bound to it approaches the membrane from which a vesicle is about to form. A peripheral membrane protein then triggers exchange of GTP for GDP, enabling SarI to bind to Sec13/31 and Sec23/24. After the formation of a free vesicle, a component of the COPII coat triggers GTP hydrolysis, and SarI releases Sec13/31 and Sec23/24.

SNARE Proteins Mediate Fusion Between Vesicles and Target Membranes

Much of the intracellular traffic mediated by coated vesicles is highly specific. As we have seen, the final sorting of proteins synthesized in the ER occurs in the TGN when lipids and proteins are packaged into vesicles for transport to various destinations. Recall that when clathrin-coated vesicles form from the TGN, the adaptor complexes include two adaptin subunits. The two adaptin subunits are partly responsible for the specificity displayed when receptors are concentrated for inclusion in a budding vesicle.

Once a vesicle forms, however, additional proteins are needed to ensure delivery of the vesicle to the appropriate destination. Therefore, there must be a mechanism to keep the various vesicles in the cell from accidentally fusing with the wrong membrane. The **SNARE hypothesis** provides a working model for this important sorting and targeting step in intracellular transport (**Figure 12-24**). This hypothesis was proposed by James Rothman and colleagues, who took a biochemical approach to dissecting the secretion pathway in mammalian cells. They developed an in vitro system to study Golgi transport and identified proteins necessary for secretion, including **N-ethylmaleimide-sensitive factor (NSF)** and associated **soluble NSF attachment proteins (SNAPs)**. This work nicely complemented the genetic work in yeast by Schekman and colleagues, and it was soon discovered that some of the essential proteins were conserved in yeast and mammals, representing an ancient origin for the secretory pathway in cells. For this work, the two shared a Nobel Prize in 2013.

According to the SNARE hypothesis, the proper sorting and targeting of vesicles in eukaryotic cells involve two families of **SNARE (SNAP receptor) proteins**: the **v-SNAREs (vesicle-SNAP receptors)** found on transport vesicles and the **t-SNAREs (target-SNAP receptors)** found on target membranes. The v-SNAREs and t-SNAREs are complementary molecules that, along with additional tethering proteins, allow a vesicle to recognize and fuse with a target membrane. Both v-SNAREs and t-SNAREs were originally investigated because of their role in neuronal exocytosis. Since their discovery in brain tissue, both families of proteins have also been implicated in transport from the ER to the Golgi apparatus in yeast and other organisms.

When a vesicle reaches its destination, a third family of proteins, the **Rab GTPases**, comes into play. Rab GTPases are also specific: vesicles fated for different destinations have distinct members of the Rab family associated with them. The affinity of complementary v-SNAREs and t-SNAREs for one another enables them to form a stable complex by forming a bundle of parallel α helices, with some of the helices contributed by each protein (Figure 12-24). This ensures that, when these proteins collide, they will remain in contact long enough for a Rab protein associated with the vesicle to lock the complementary t-SNARE and v-SNARE together, facilitating membrane fusion.

Recently, scientists used optical tweezers to analyze how the v- and t-SNAREs associate. They attached individual molecules of v- and t-SNARE proteins to opposite ends of the tweezers, allowed the proteins to interact, and then measured the amount of force necessary to dissociate the proteins from each other. These studies support the hypothesis that, after initial contact, the rapid "zippering" together of the v- and t-SNARE proteins provides the force necessary to drive membrane fusion.

Following vesicle fusion, NSF and a group of SNAPs mediate release of the v- and t-SNAREs of the donor and target membranes. ATP hydrolysis may be involved at this step, but its precise role is unclear. NSF and SNAPs are involved in fusion between a variety of cellular membranes, indicating they are not responsible for specificity during targeting.

SNARE proteins are required for the fusion of neurotransmitter-containing vesicles with the plasma membrane

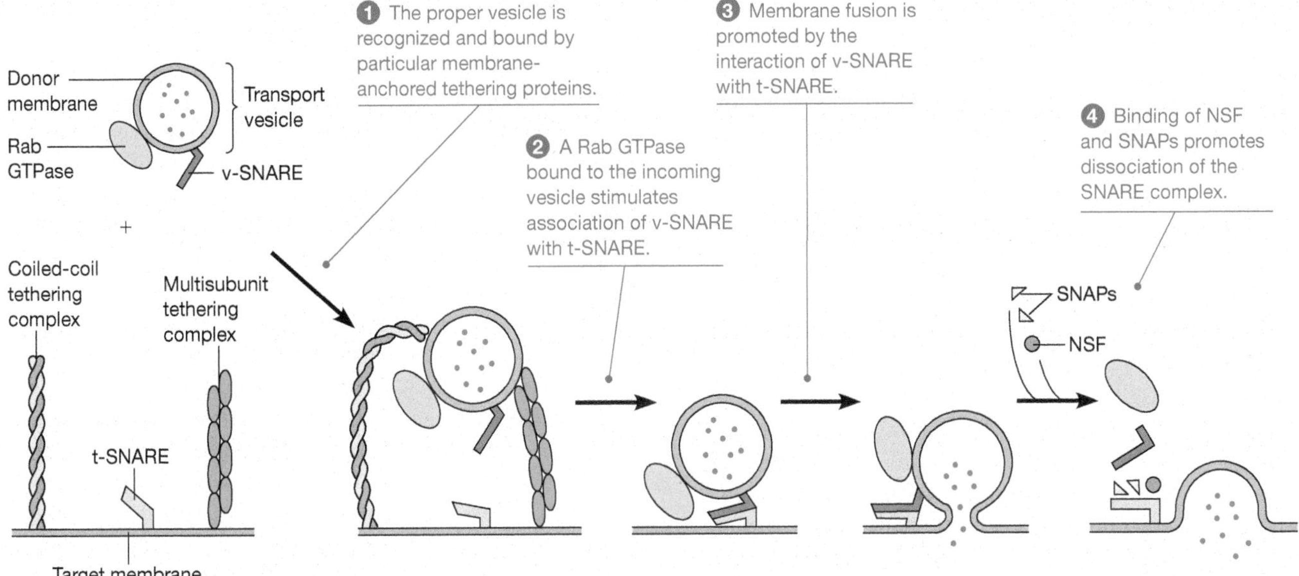

Figure 12-24 The SNARE Hypothesis for Transport Vesicle Targeting and Fusion. The basic molecular components that mediate sorting and targeting of vesicles in eukaryotic cells include tethering proteins, v-SNAREs on transport vesicles, t-SNAREs on target membranes, Rab GTPase, NSF, and several SNAPs. The exact timing of GTP or ATP hydrolysis is still unclear, but it most likely occurs after vesicle fusion.

of nerve cells. This fusion event releases the neurotransmitter by exocytosis (a process you will learn more about in Chapter 22). When the vesicle contacts a muscle cell, it can lead to an electrical impulse that initiates muscle contraction (see Chapter 14). Botulinum toxin (Botox), produced by the bacterium *Clostridium botulinum*, is a protease that cleaves a SNARE protein that is required for this fusion. Thus, the toxin interferes with muscle contraction and can cause paralysis. Although it is one of the most potent biological toxins known, in very small doses Botox can be used therapeutically to control muscle spasms or correct crossed eyes. It is also used cosmetically to remove wrinkles caused by muscle contractions in the skin. Recently, the use of Botox was approved as a treatment for migraine headaches.

Since the discovery of SNARE proteins, it has become clear that they alone cannot account for the specificity observed in vesicle targeting. A distinct class of proteins known as **tethering proteins** acts over longer distances and provides specificity by connecting vesicles to their target membranes prior to v-SNARE/t-SNARE interaction (Figure 12-24). Evidence for such long-range action in vivo comes from experiments showing that toxin-induced cleavage of SNARE proteins blocks SNARE complex formation without blocking vesicle association with the target membrane. In addition, in an in vitro reconstituted system, ER-derived vesicles can attach to Golgi membranes without addition of SNARE proteins.

Two main groups of tethering proteins are known at present. The first group contains coiled-coil proteins, including the **golgins.** Golgins are important in the initial recognition and binding of COPI- or COPII-coated vesicles to the Golgi. Golgins are also important in connecting Golgi cisternae to one other; antibodies directed against certain golgins block the action of the golgins and disrupt the structure of the Golgi medial cisternae. The second class of tethering proteins is multisubunit protein complexes containing four to eight or more individual polypeptides. For example, the *exocyst* complex of yeast and mammals is important for protein secretion, binding both to the plasma membrane and to vesicles from the TGN whose contents are destined for export. Other multisubunit tethering complexes are implicated in the initial recognition and specificity of vesicle–target membrane interactions. Identifying the functions of these complexes is one of the most intriguing frontiers of modern cell biology.

CONCEPT CHECK 12.6

What intracellular transport processes would you expect to be affected when cells are treated with a GTPase inhibitor?

12.7 Lysosomes and Cellular Digestion

The **lysosome** is an organelle of the endomembrane system that contains digestive enzymes capable of degrading all the major classes of biological macromolecules—lipids, carbohydrates, nucleic acids, and proteins. These hydrolytic enzymes degrade extracellular materials brought into the cell by endocytosis and digest intracellular structures and macromolecules that are damaged or no longer needed. We will first look at the organelle itself and then consider lysosomal digestive processes, as well as some of the diseases that result from lysosomal malfunction. Finally, we will examine the plant vacuole and compare it to the lysosome.

Lysosomes Isolate Digestive Enzymes from the Rest of the Cell

Lysosomes were discovered in the early 1950s by Christian de Duve and his colleagues (see Key Technique in Chapter 4, pages 100–101). Differential centrifugation led researchers to realize that an acid phosphatase initially thought to be located in the mitochondrion was in fact associated with a class of particles that had never been reported before. Along with the acid phosphatase, the new organelle contained several other hydrolytic enzymes, including β-glucuronidase, a deoxyribonuclease, a ribonuclease, and a protease. Because of its apparent role in cellular lysis, de Duve called this newly discovered organelle a *lysosome.*

Only after the existence of the lysosome had been predicted, its properties described, and its enzyme content specified was the organelle actually observed by electron microscopy and recognized as a normal constituent of most animal cells. Final confirmation came from cytochemical staining reactions capable of localizing the acid phosphatase and other lysosomal enzymes to specific structures that can be seen by electron microscopy (see Figure 4-19b).

Lysosomes vary considerably in size and shape but are generally about 0.5 μm in diameter. Like the ER and Golgi apparatus, the lysosome is bounded by a single membrane. This membrane protects the rest of the cell from the hydrolytic enzymes in the lysosomal lumen. The portion of lysosomal membrane protein that faces the lumen is highly glycosylated, forming a nearly continuous carbohydrate coating that appears to protect these proteins from lysosomal proteases. ATP-dependent proton pumps in the membrane maintain an acidic environment (pH 4.0–5.0) within the lysosome. This favors enzymatic digestion of macromolecules both by activating acid hydrolases and by partially denaturing the macromolecules targeted for degradation. The products of digestion are then transported across the membrane to the cytosol, where they enter various synthetic pathways or are exported from the cell.

The list of lysosomal enzymes has expanded considerably since de Duve's original work, but all have the common property of being *acid hydrolases*—hydrolytic enzymes with a pH optimum around 5.0. The list includes at least 5 phosphatases, 14 proteases and peptidases, 2 nucleases, 6 lipases, 13 glycosidases, and 7 sulfatases. Taken together, these lysosomal enzymes can digest all the major classes of biological molecules. No wonder, then, that they are sequestered from the rest of the cell, where they cannot quickly destroy the cell itself.

Lysosomes Develop from Endosomes

Lysosomal enzymes are synthesized by ribosomes attached to the rough ER and are translocated through a pore in the ER membrane into the ER lumen before transport to the Golgi apparatus. After modification and processing in the ER and Golgi apparatus compartments, the lysosomal enzymes are sorted from other proteins in the TGN. Earlier in the chapter, you learned about the addition of a unique mannose-6-phosphate tag to soluble lysosomal enzymes. Distinctive sorting signals are also present on membrane-bound lysosomal proteins. The lysosomal enzymes are packaged in clathrin-coated vesicles that bud from the TGN, lose their protein coats, and travel to one of the endosomal compartments (see Figure 12-12).

Lysosomal enzymes are delivered from the TGN to endosomes in transport vesicles, as shown in **Figure 12-25**. Recall that early endosomes are formed by the coalescence of vesicles from the TGN and vesicles from the plasma membrane. Over time, the early endosome matures to form a late endosome, an organelle having a full complement of acid hydrolases but not engaged in digestive activity. As the pH of the early endosomal lumen drops from about 6.0 to 5.5, the organelle loses its capacity to fuse with endocytic vesicles. The late endosome is essentially a collection of newly synthesized digestive enzymes as well as extracellular and intracellular material fated for digestion, packaged in a way that protects the cell from hydrolytic enzymes.

The final step in lysosome development is the activation of the acid hydrolases, which occurs as the enzymes and their substrates encounter a more acidic environment. There are two ways that eukaryotic cells accomplish this step. ATP-dependent proton pumps may lower the pH of the late endosomal lumen to 4.0–5.0, transforming the late endosome into a lysosome, thereby generating a new organelle. Alternatively, the late endosome may transfer material to the acidic lumen of an existing lysosome.

Lysosomal Enzymes Are Important for Several Different Digestive Processes

Lysosomes are important for cellular activities as diverse as nutrition, defense, recycling of cellular components, and differentiation. The digestive processes that depend on lysosomal enzymes can be distinguished by the site of their activity and by the origin of the material that is digested, as shown in Figure 12-25. Usually, the site of activity is intracellular. In some cases, though, lysosomes may release their enzymes to the outside of the cell by exocytosis. The materials to be digested are often of extracellular origin, although there are also important processes known to involve lysosomal digestion of internal cellular components.

To distinguish between mature lysosomes of different origins, those containing substances of extracellular origin are known as **heterophagic lysosomes,** whereas those with materials of intracellular origin are known as **autophagic lysosomes.** The specific processes in which lysosomal enzymes are involved are *phagocytosis, receptor-mediated endocytosis, autophagy,* and *extracellular digestion.* These are illustrated in Figure 12-25 as pathways Ⓐ, Ⓑ, Ⓒ, and Ⓓ, respectively.

Phagocytosis and Receptor-Mediated Endocytosis: Lysosomes in Defense and Nutrition. One of the most important functions of lysosomal enzymes is the degradation of foreign material brought into eukaryotic cells by *phagocytosis* and *receptor-mediated endocytosis* (see Figures 12-19 and 12-20). Phagocytic vacuoles are transformed into lysosomes by fusing with early endosomes (pathway Ⓐ). Depending on the material ingested, these lysosomes can vary considerably in size, appearance, content, and the stage of digestion. Vesicles containing material brought into a cell by receptor-mediated endocytosis also form early endosomes (pathway Ⓑ). As early endosomes fuse with vesicles from the TGN containing acid hydrolases, they mature to form late endosomes and lysosomes, in which the ingested material is digested.

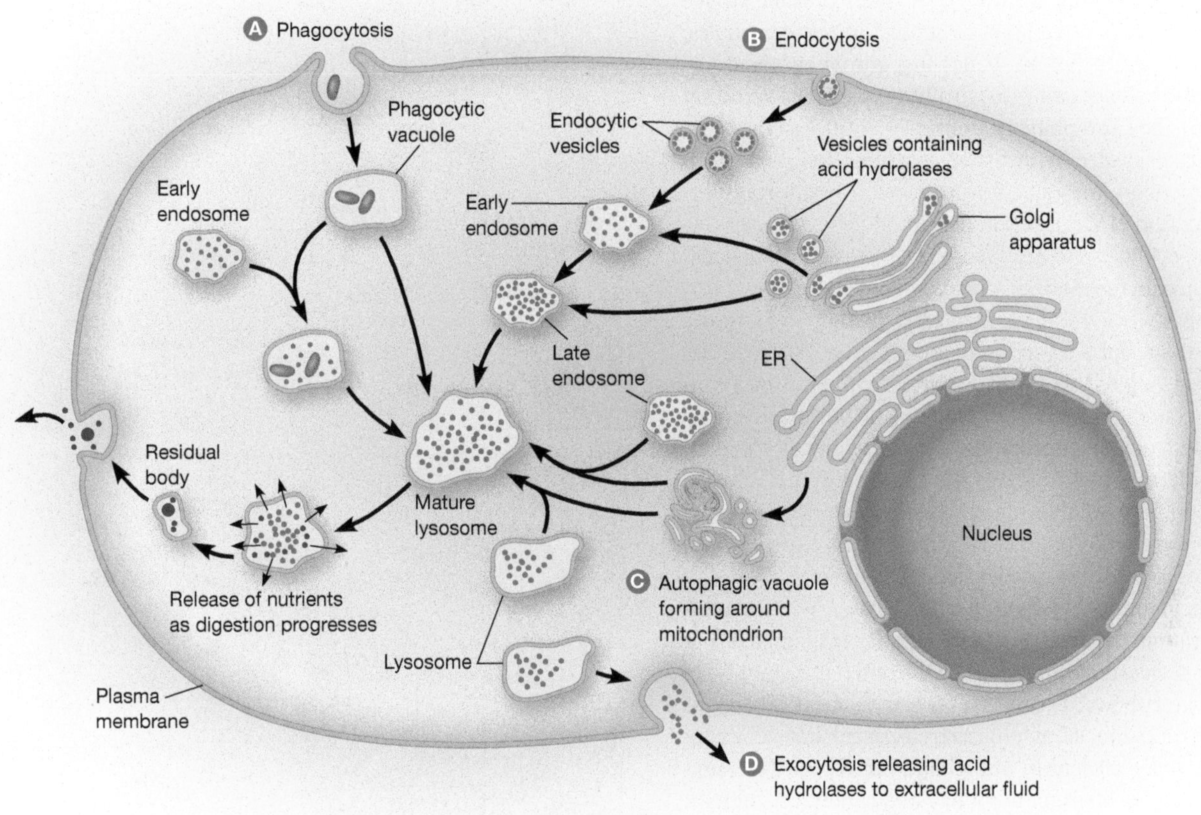

Figure 12-25 The Formation of Lysosomes and Their Roles in Cellular Digestive Processes. Illustrated in this composite cell are the major processes in which lysosomes are involved. The pathways depicted are Ⓐ phagocytosis, Ⓑ receptor-mediated endocytosis, Ⓒ autophagy, and Ⓓ extracellular digestion.

Soluble products of digestion, such as sugars, amino acids, and nucleotides, are then transported across the lysosomal membrane into the cytosol and are used as a source of nutrients by the cell. Some may cross by facilitated diffusion, whereas others undergo active transport. The acidity of the lysosomal lumen contributes to an electrochemical proton gradient across the lysosomal membrane, which can provide energy for driving transport to and from the cytosol.

Eventually, however, only indigestible material remains in the lysosome, which becomes a **residual body** as digestion ceases. In protozoa, residual bodies routinely fuse with the plasma membrane and expel their contents to the outside by exocytosis, as illustrated in Figure 12-25. In vertebrates, residual bodies may accumulate in the cytoplasm. This accumulation of debris is thought to contribute to cellular aging, particularly in long-lived cells such as those of the nervous system.

Certain white blood cells of the immune system, however, use this residual material. After neutrophils digest invading microorganisms via phagocytosis and lysosome action, they release the debris, which is picked up by scavenging macrophages. These macrophages transport the debris to the lymph nodes and present it to other immune system cells to "educate" them regarding what foreign material has been found in the body. This process causes the lymphocytes not only to initiate an immune response specific to the invading microorganism, but also to form memory cells that will quickly respond in case the same microorganism is encountered in the future.

Autophagy: A Biological Recycling System. A second important task for lysosomes is the breakdown of cellular structures and components that are damaged or no longer needed. Most cellular organelles are in a state of dynamic flux, with new organelles continuously being synthesized while old organelles are destroyed. Removal of damaged organelles is extremely important, as mitochondria can release dangerous compounds, like reactive oxygen species, when they are breaking down. Therefore, failure to remove these failing organelles will commonly cause mutation or even cell death. The digestion of old or unwanted organelles or other cell structures is called **autophagy,** which is Greek for "self-eating" (pathway Ⓒ in Figure 12-25).

There are two types of autophagy—*macrophagy* and *microphagy*. **Macrophagy** begins when an organelle or other structure becomes wrapped in a double membrane derived from the ER. The resulting vesicle is called an **autophagic vacuole** (or **autophagosome**). It is often possible to see identifiable remains of cellular structures in these vacuoles, as shown in **Figure 12-26**. Following autophagosome formation, these vesicles can fuse with lysosomes. The resulting vesicles are called autophagolysosomes or autolysosomes. The contents are degraded by digestive enzymes, and the broken down components are generally recycled. **Microphagy** involves formation of a much smaller autophagic vacuole, surrounded by a single phospholipid bilayer that encloses small bits of cytoplasm rather than whole organelles.

Autophagy occurs at varying rates in most cells under most conditions, but it is especially prominent in red blood cell development. As a red blood cell matures, virtually all of the intracellular content is destroyed, including all of the mitochondria. This destruction is accomplished by autophagic digestion. A marked increase in autophagy is also noted in cells stressed by starvation. Presumably, the process represents a desperate attempt by the cell to continue providing for its energy needs, even if it has to consume its own structures to do so.

Defects in autophagy have been implicated in human disease. For example, recent genetic studies have suggested that autophagy defects may contribute to the onset of certain types of inflammatory bowel disease (see Human Connections, pages 350–351). In addition, it has long been suspected that some cancer cells may be deficient in autophagy, but recent work suggests a direct link between autophagy and cancer. The human version of a gene required for autophagy in yeast cells has been shown to be frequently deleted in human breast and ovarian tumors. Knocking out the function of this gene in mice caused a decrease in autophagy and an increase in the number of breast and lung tumors. However, some oncogenes that promote cancer have the opposite effect: they suppress autophagy during tumor development. Clearly, more research is needed.

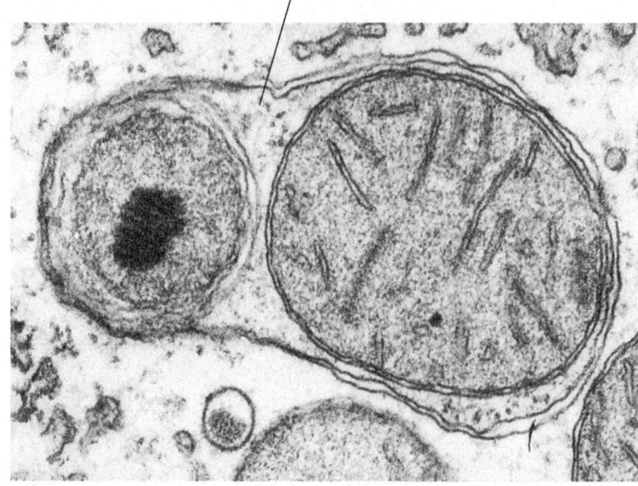

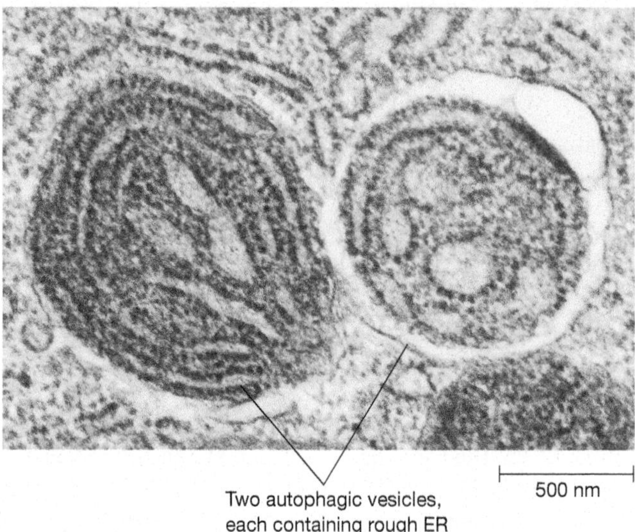

Figure 12-26 Autophagic Digestion. Early and late stages of autophagic digestion are shown here in a rat liver cell. At the top, a single autophagic vacuole contains both a peroxisome as well as a mitochondrion. The contents are sequestered by a double-membrane derived from the ER. At the bottom are two autophagic vacuoles, each containing remnants of rough ER (TEM).

Extracellular Digestion. Whereas most digestive processes involving lysosomal enzymes occur intracellularly, in some cases lysosomes discharge their enzymes to the outside of the cell by exocytosis, resulting in **extracellular digestion** (see Figure 12-25, pathway Ⓓ). One example of extracellular digestion occurs during fertilization of animal eggs. The head of the sperm releases lysosomal enzymes capable of degrading barriers that would otherwise prevent contact between the sperm and egg membranes. Certain inflammatory diseases, such as rheumatoid arthritis, may result from the inadvertent release of lysosomal enzymes by white blood cells in the joints, damaging the joint tissue. The steroid hormones cortisone and hydrocortisone are thought to be effective anti-inflammatory agents because of their role in stabilizing lysosomal membranes and thereby inhibiting enzyme release.

Lysosomal Storage Diseases Are Usually Characterized by the Accumulation of Indigestible Material

The essential role of lysosomes in the recycling of cellular components is clearly seen in disorders caused by deficiencies of specific lysosomal proteins. More than 40 such **lysosomal storage diseases** are known, each characterized by the harmful accumulation of specific substances, usually polysaccharides or lipids (see Human Connections in Chapter 4, pages 94–95). In most cases, the substances accumulate because specific hydrolytic enzymes are defective or missing or because a protein that transports degradation products out of the lysosome to the cytosol is defective. In other cases, the requisite enzymes are synthesized in normal amounts but are secreted into the extracellular medium rather than being targeted to the lysosomes. In all of these cases, the cells in which material accumulates are severely impaired or destroyed. Skeletal deformities, muscle weakness, and mental retardation commonly result, and the disease is often fatal. Unfortunately, most lysosomal storage diseases are not yet treatable.

The first storage disease to be understood was *type II glycogenosis*, in which young children accumulate excessive amounts of glycogen in the liver, heart, and skeletal muscles and die at an early age. The problem turned out to be a defective form of the lysosomal enzyme *α-1,4-glucosidase,* which catalyzes glycogen hydrolysis in normal cells. Although glycogen metabolism occurs predominantly in the cytosol, a small amount of glycogen can enter the lysosome through autophagy and will accumulate to a damaging level if not broken down to glucose.

Two of the best-known lysosomal storage diseases are *Hurler syndrome* and *Hunter syndrome.* Both arise from defects in the degradation of glycosaminoglycans, which are the major carbohydrate components of the extracellular matrix (see Chapter 15). The defective enzyme in a patient with Hurler syndrome is *α-L-iduronidase,* which is required for the degradation of glycosaminoglycans. Electron microscopic observation of sweat gland cells from a patient with Hurler syndrome reveals large numbers of atypical vacuoles that stain for both acid phosphatase and undigested glycosaminoglycans. These vacuoles are apparently abnormal late endosomes filled with indigestible material.

Mental retardation is a common feature of lysosomal storage diseases. One particularly well-known example is *Tay-Sachs disease,* which is quite rare in the general population but has a higher incidence among Ashkenazi Jews of Eastern European ancestry. At about 6 months of age, afflicted children exhibit mental and motor deterioration as well as skeletal, cardiac, and respiratory dysfunction. This is followed by dementia, paralysis, blindness, and death, usually within 3 years. The disease results from the accumulation in nervous tissue of a particular glycolipid called a *ganglioside* (see Figure 7-6b). The missing lysosomal enzyme in this case is *β-N-acetylhexosaminidase,* which is responsible for cleaving the terminal *N*-acetylgalactosamine from the carbohydrate portion of the ganglioside. Lysosomes from children afflicted with Tay-Sachs disease are filled with membrane fragments containing undigested gangliosides.

All of the known lysosomal storage diseases can be diagnosed prenatally, leading to the hope that treatments will be developed that can replace the missing enzyme using purified protein, a treatment known as *enzyme replacement therapy,* or by directly manipulating the DNA in the cells of patients (an approach known as *gene therapy*). One case in which some progress has been made is the lysosomal disorder called *Gaucher disease,* characterized by the absence or deficiency of a specific lysosomal hydrolase called *glucocerebrosidase.* Glucocerebroside accumulation typically leads to liver and spleen enlargement, anemia, and mental retardation. Enzyme replacement therapy has been used to successfully treat Gaucher disease. Recently, a synthetic form of glucocerebrosidase has been produced using recombinant DNA technology; macrophages that take up this enzyme are able to degrade glucocerebrosides as needed, thereby effectively treating what would otherwise be a fatal disease.

The Plant Vacuole: A Multifunctional Digestive Organelle

Plant cells contain acidic membrane-enclosed compartments called *vacuoles* that resemble the lysosomes found in most animal cells but generally serve additional roles (see Figure 4-22). The biogenesis of a vacuole parallels that of a lysosome. Most of the components are synthesized in the ER and transferred to the Golgi apparatus, where proteins undergo further processing. Coated vesicles then convey lipids and proteins destined for the vacuole to a **provacuole**, which is analogous to an endosome. The provacuole eventually matures to form a functional vacuole that can fill as much as 90% of the volume of a plant cell.

In addition to confining hydrolytic enzymes, plant vacuoles have various other essential functions. Most of these functions reflect the plant's lack of mobility and consequent susceptibility to changes in the surrounding environment. A major role of the vacuole is to accumulate solutes, causing water to enter the cell by osmosis (see Figure 8-4b). This helps the cell maintain *turgor pressure,* which prevents plant cells from collapsing and prevents a plant from wilting. In addition, turgor pressure drives the remodeling of plant cells. Softening of the cell wall—accompanied by higher turgor pressure—allows plant cells to expand.

Another role of the plant vacuole is the regulation of cytosolic pH. ATP-dependent proton pumps in the vacuolar membrane can compensate for a decline in cytosolic pH (perhaps due to a change in the extracellular environment) by transferring protons from the cytosol to the lumen of the vacuole.

A different kind of vacuole serves as a protein storage compartment in seeds. Seed storage proteins are generally synthesized by ribosomes attached to the rough ER and cotranslationally inserted into the ER lumen. Some of the storage proteins remain in the ER, and others are transferred to protein storage vacuoles, either by autophagy of vesicles budding from the ER or by way of the Golgi. When the seeds germinate, the storage proteins are available for hydrolysis by vacuolar proteases, thereby releasing amino acids for the biosynthesis of new proteins needed by the growing plant.

Other substances found in vacuoles include malate stored in CAM plants (see Chapter 11), the anthocyanins that impart color to flowers, toxic substances that deter predators, inorganic and organic nutrients, compounds that shield cells from ultraviolet light, and residual indigestible waste. Storage of soluble as well as insoluble waste is an important function of plant vacuoles. Unlike animals, most plants do not have a mechanism for excreting soluble waste from the organism. The large vacuoles found in plant cells enable them to accumulate solutes to a degree that would inhibit or restrict metabolic processes if the material were to remain in the cytosol.

MAKE CONNECTIONS 12.2

Explain how accumulating a high solute concentration within the vacuole prevents plants from wilting. (Fig. 8.5)

CONCEPT CHECK 12.7

What problems would a cell have if it could not produce lysosomes?

12.8 Peroxisomes

During the course of their early studies on lysosomes, Christian de Duve and his colleagues also discovered another organelle that sedimented at a slightly different density in a sucrose density gradient than lysosomes. Because of its apparent involvement in hydrogen peroxide metabolism, the new organelle became known as a **peroxisome**. Peroxisomes, like the Golgi apparatus, endosomes, and lysosomes, are bounded by single membranes. Because early

HUMAN *Connections*

A Bad Case of the Munchies? (Autophagy In Inflammatory Bowel Disease)

Genome-wide association studies (GWASs) are large-scale genetic investigations that attempt to link gene mutations to human disease susceptibility. Instead of sequencing entire genomes, these studies examine thousands of single nucleotide polymorphisms (SNPs). The idea is to compare a pool of healthy individuals to a pool of diseased individuals. Geneticists attempt to identify any SNPs that have a higher prevalence in the diseased group than in the healthy group. When statistical tests reveal an association, the investigators try to determine which gene located in the vicinity of the SNP might be responsible for causing the disease. In most cases the polymorphism is a marker that flags a general area within the genome that causes disease, not the actual mutation itself. The first major GWAS was published in 2007 and examined 3000 healthy individuals along with 14,000 cases divided among seven common human diseases thought to be linked to genetic causes. One possible outcome of these large GWAS reports is the identification of novel associations between cellular pathways and disease. A great example of this was when genes involved in the process of autophagy (see Section 12.7) were found to be associated with susceptibility to Crohn's disease.

Crohn's disease is a type of inflammatory bowel disease (IBD) that can affect any portion of the gastrointestinal tract (**Figure 12B-1**). The tissue damage and underlying inflammation can trigger the clinical symptoms associated with the disease, which include bloody diarrhea, abdominal cramping, fever, and weight loss. Most work examining the pathophysiology of Crohn's disease has focused on the interactions between the epithelium of the intestine and resident gut bacteria. Therefore, it was unexpected when autophagy proteins were first implicated in susceptibility to Crohn's disease. Prior to this, autophagy was not thought to play a significant role in the symbiosis between commensal bacteria and the cells that line the intestinal tract.

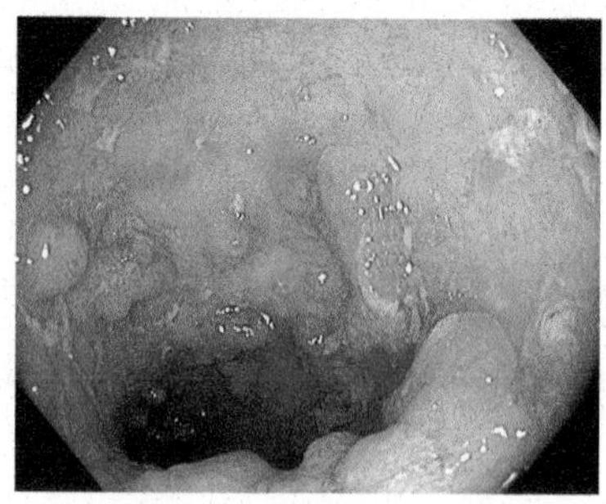

Figure 12B-1 Endoscopic View of IBD Pathology. Inflammatory polyps from the transverse colon (part of the large intestine) of a patient with Crohn's disease.

Along with several other genes, GWASs have identified at least two autophagy genes (*IRGM* and *ATG16L1*) in Crohn's disease progression. The implication is that mutated versions of these proteins are affecting some aspect of autophagy that ultimately gives these individuals a higher likelihood of developing the disease (**Figure 12B-2**). Normally, both of these proteins are involved in the recognition of intracellular bacteria and in the initial formation of the autophagosome. Once surrounded by the ER-derived double membrane structure, the bacteria-containing autophagosome will fuse with a lysosome. This fusion causes bacterial degradation (Figure 12B-2a).

It isn't entirely clear how mutations in autophagy proteins can lead to Crohn's disease. Studies have suggested that the autophagy process within intestinal epithelial cells normally regulates the inflammation that can be induced by commensal and/or pathogenic bacteria within the intestinal tract. By inducing autophagy,

work on peroxisome biogenesis suggested that they are not derived from the endoplasmic reticulum, they were not thought to be part of the endomembrane system that includes the other organelles discussed in this chapter. However, recent studies have demonstrated a distinct role of the ER in supplying membrane directly for peroxisome growth and division. Therefore, more and more scientists are including peroxisomes within the endomembrane system. Peroxisomes are found in all eukaryotic cells but are especially prominent in mammalian kidney and liver cells, in algae and photosynthetic cells of plants, and in germinating seedlings of plant species that store fat in their seeds. Peroxisomes are somewhat smaller than mitochondria, though there is considerable variation in size, depending on their function and the tissue where they are found.

Regardless of location or size, the defining characteristic of a peroxisome is the presence of *catalase*, an enzyme essential for the degradation of hydrogen peroxide (H_2O_2). Hydrogen peroxide is a potentially toxic compound that is formed by a variety of oxidative reactions catalyzed by *oxidases*. Both catalase and the oxidases are confined to peroxisomes. Thus, the generation and degradation of H_2O_2 occur within the same organelle, thereby protecting other parts of the cell from exposure to this harmful compound.

As seen in **Figure 12-27** on page 352, animal peroxisomes often contain a distinct crystalline core, which usually consists of a crystalline form of urate oxidase. Crystalline cores consisting of catalase may be present in the peroxisomes of plant leaves (see Figure 4-21b). When such crystals are present, it is easy to identify peroxisomes. In the absence of a

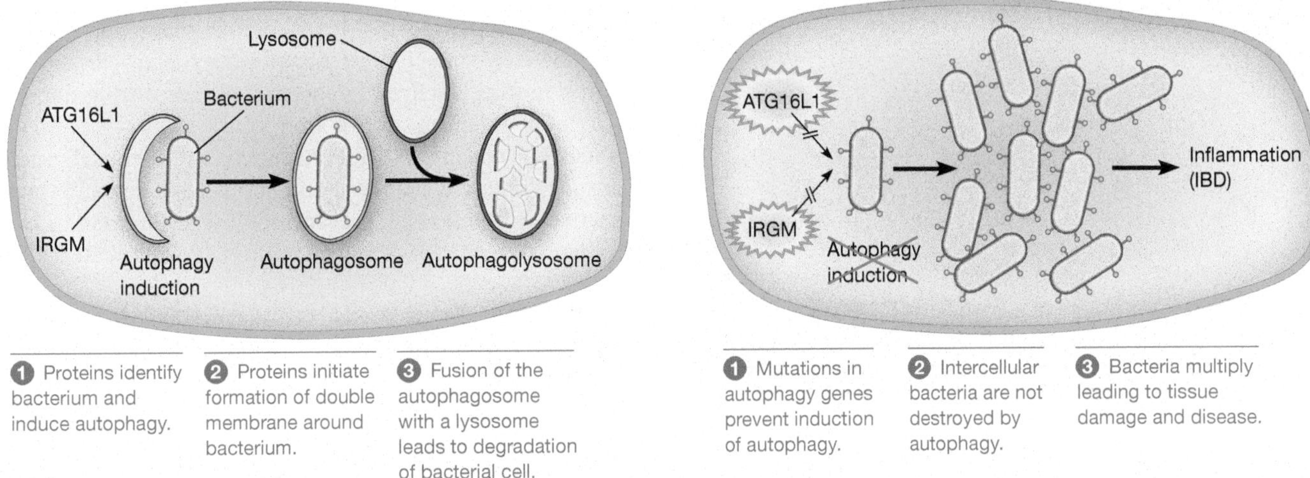
Lysosome
Bacterium
ATG16L1
IRGM
Autophagy induction
Autophagosome
Autophagolysosome
ATG16L1
IRGM
Autophagy induction
Inflammation (IBD)
1 Proteins identify bacterium and induce autophagy.
2 Proteins initiate formation of double membrane around bacterium.
3 Fusion of the autophagosome with a lysosome leads to degradation of bacterial cell.
1 Mutations in autophagy genes prevent induction of autophagy.
2 Intercellular bacteria are not destroyed by autophagy.
3 Bacteria multiply leading to tissue damage and disease.
(a) Autophagy induction and control of intracellular bacteria.
(b) IBD as a result of mutations in autophagy genes.

Figure 12B-2 How Defective Autophagy Could Lead to IBD. (a) When autophagy works normally, intracellular bacteria can be cleared so that no disease occurs. **(b)** Mutations in autophagy proteins may prevent destruction of intracellular bacteria. The resulting inflammation could explain the tissue damage associated with IBD.

invading bacteria can be destroyed as described above, and any resulting inflammation can be dampened. Mutations in autophagy proteins (like IRGM and ATG16L1) result in more inflammation, as intracellular bacteria are no longer processed within autophagosomes. This prolonged inflammation can lead to the tissue damage associated with Crohn's disease symptoms (Figure 12B-2b).

Understanding how autophagy defects cause different types of IBD will lead to the development of new therapeutics. Scientists are already exploring ways to enhance autophagy in an effort to treat cases of IBD. Using GWAS to uncover the novel cell pathways associated with disease susceptibility makes possible these and other such treatment options for human genetic disorders.

crystalline core, however, it is not always easy to spot peroxisomes ultrastructurally.

A useful technique in such cases is a cytochemical test for catalase called the *diaminobenzidine (DAB) reaction.* Catalase oxidizes DAB to a polymeric form that causes deposition of electron-dense osmium atoms when the tissue is treated with osmium tetroxide that can be readily seen in cells from stained tissue.

Most Peroxisomal Functions Are Linked to Hydrogen Peroxide Metabolism

The essential roles of peroxisomes in eukaryotic cells have become more apparent in recent years, stimulating new research into the metabolic pathways and the disorders arising from defective components of the pathways found in these organelles. There are five general categories of peroxisomal functions that we will discuss in the sections below: hydrogen peroxide metabolism, detoxification of harmful compounds, oxidation of fatty acids, metabolism of nitrogen-containing compounds, and catabolism of unusual substances.

Hydrogen Peroxide Metabolism. A significant role of peroxisomes in eukaryotic cells is the detoxification of H_2O_2 by catalase, which can comprise up to 15% of the total protein in the peroxisome. The oxidases that generate H_2O_2 in peroxisomes transfer electrons plus hydrogen ions (hydrogen atoms) from their substrates to molecular oxygen (O_2),

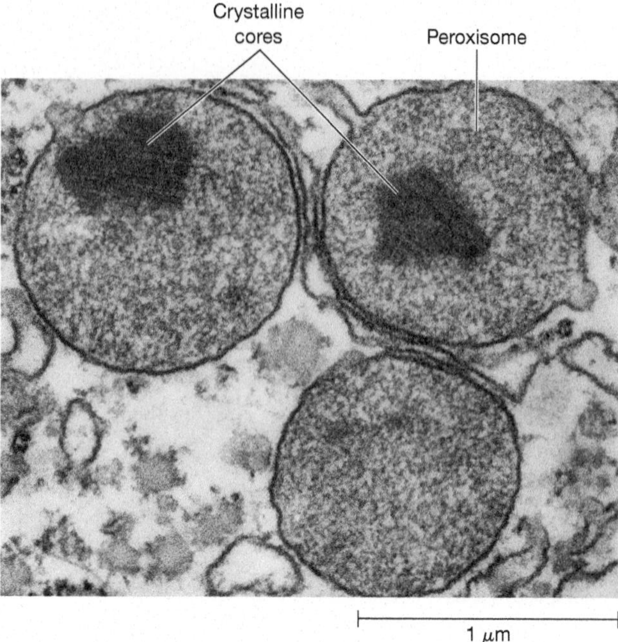

Figure 12-27 Peroxisomes in Animal Cells. This electron micrograph shows several peroxisomes in the cytoplasm of a rat liver cell. A crystalline core is readily visible in each peroxisome. In animal peroxisomes, the cores are almost always crystalline urate oxidase (TEM).

reducing it to H_2O_2. Using RH_2 to represent an oxidizable substrate, the general reaction catalyzed by oxidases can be written as

$$RH_2 + O_2 \rightarrow R + H_2O_2 \qquad \textbf{(12-3)}$$

The hydrogen peroxide formed in this manner is broken down by catalase in one of two ways. Usually, catalase functions by detoxifying two molecules of H_2O_2 simultaneously—one is oxidized to oxygen, and a second is reduced to water:

$$2H_2O_2 \rightarrow O_2 + 2H_2O \qquad \textbf{(12-4)}$$

Alternatively, catalase can function as a *peroxidase*, in which electrons derived from an organic donor are used to reduce hydrogen peroxide to water:

$$R'H_2 + H_2O_2 \rightarrow R' + 2H_2O \qquad \textbf{(12-5)}$$

(The prime on the R group simply indicates that this substrate is likely to be different from the substrate in Reaction 12-3.)

The result is the same in either case: hydrogen peroxide is degraded without ever leaving the peroxisome. Given the toxicity of hydrogen peroxide (which is the main active ingredient in a variety of disinfectants), it makes good sense for the enzymes responsible for peroxide generation to be compartmentalized together with the catalase that catalyzes its degradation.

Detoxification of Harmful Compounds. As a peroxidase (Reaction 12-5), catalase can use a variety of toxic substances as electron donors, including methanol, ethanol, formic acid, formaldehyde, nitrites, and phenols. Because all of these compounds are harmful to cells, their oxidative detoxification by catalase may be a vital peroxisomal function. The prominent peroxisomes of liver and kidney cells are thought to be important in such detoxification reactions.

In addition, peroxisomal enzymes are important in the detoxification of **reactive oxygen species** such as H_2O_2, superoxide anion (O_2^-), hydroxyl radical (OH•, where the dot signifies an unpaired, highly reactive electron), and organic peroxide conjugates. These reactive oxygen species can be formed in the presence of molecular oxygen during normal cellular metabolism, and if they accumulate the cell can suffer from *oxidative stress.* Peroxisomal enzymes such as *superoxide dismutase,* catalase, and peroxidases detoxify these reactive oxygen species, preventing their accumulation and subsequent oxidative damage to cellular components.

Oxidation of Fatty Acids. Peroxisomes found in animal, plant, and fungal cells contain enzymes necessary for oxidizing fatty acids via *β oxidation* to provide energy for the cell (see Figure 10-12). About 25–50% of fatty acid oxidation in animal tissues occurs in peroxisomes, with the remainder localized in mitochondria. In plant and yeast cells, on the other hand, all *β* oxidation occurs in peroxisomes.

In animal cells, peroxisomal *β* oxidation is especially important for degrading long-chain (16–22 carbons), very long chain (24–26 carbons), and branched fatty acids. The primary product of *β* oxidation, acetyl-CoA, is then exported to the cytosol, where it enters biosynthetic pathways or the citric acid cycle (see Chapter 10). In plants and yeast, on the other hand, peroxisomes are essential for the complete catabolism of all fatty acids to acetyl-CoA.

Metabolism of Nitrogen-Containing Compounds. Except for primates, most animals require *urate oxidase* (also called *uricase*) to oxidize urate, a purine that is formed during the catabolism of nucleic acids and some proteins. Like other oxidases, urate oxidase catalyzes the direct transfer of hydrogen atoms from the substrate to molecular oxygen, generating H_2O_2:

$$\text{urate} + O_2 \rightarrow \text{allantoin} + H_2O_2 \qquad \textbf{(12-6)}$$

As noted earlier, the H_2O_2 is immediately degraded in the peroxisome by catalase. The allantoin is further metabolized and excreted by the organism, either as allantoic acid or—in the case of crustaceans, fish, and amphibians—as urea.

Additional peroxisomal enzymes involved in nitrogen metabolism include *aminotransferases.* Members of this collection of enzymes catalyze the transfer of amino groups ($-NH_3^+$) from amino acids to *α*-keto acids:

$$\underset{\text{Amino acid}}{R-\overset{\overset{+}{H_3N}}{\underset{|}{C}}-\overset{O}{\overset{\|}{C}}-O^-} + \underset{\alpha\text{-Keto acid}}{R'-\overset{O}{\overset{\|}{C}}-\overset{O}{\overset{\|}{C}}-O^-} \rightleftharpoons \underset{\alpha\text{-Keto acid}}{R-\overset{O}{\overset{\|}{C}}-\overset{O}{\overset{\|}{C}}-O^-} + \underset{\text{Amino acid}}{R'-\overset{\overset{+}{H_3N}}{\underset{|}{C}}-\overset{O}{\overset{\|}{C}}-O^-} \qquad \textbf{(12-7)}$$

Such enzymes play important roles in the biosynthesis and degradation of amino acids by moving amino groups from one molecule to another (see Figure 10-13).

Catabolism of Unusual Substances. Some of the substrates for peroxisomal oxidases are rare compounds for which the cell has no other degradative pathways. Such compounds include D-amino acids, which are not recognized by enzymes capable of degrading the L-amino acids found in polypeptides. In some cells, the peroxisomes also contain enzymes that break down unusual substances called **xenobiotics**, chemical compounds foreign to biological organisms such as *alkanes*, short-chain hydrocarbon compounds found in oil and other petroleum products. Fungi containing enzymes capable of metabolizing such xenobiotics may turn out to be useful for cleaning up oil spills that would otherwise contaminate the environment.

Peroxisomal Disorders. Considering the variety of metabolic pathways found in peroxisomes, it is not surprising that a large number of disorders arise from defective peroxisomal proteins. The most common peroxisomal disorder is *X-linked adrenoleukodystrophy*. The defective protein causing this disorder is an integral membrane protein that may be responsible for transporting very long chain fatty acids into the peroxisome for β oxidation. Accumulation of these long-chain fatty acids in body fluids destroys the myelin sheath in nervous tissues.

Plant Cells Contain Types of Peroxisomes Not Found in Animal Cells

In plants and algae, peroxisomes are involved in several specific aspects of cellular energy metabolism (discussed in Figure 10-15 and Figure 11-17). Here, we will simply introduce several plant-specific peroxisomes and briefly describe their functions.

Leaf Peroxisomes. Cells of leaves and other photosynthetic plant tissues contain characteristic large, prominent **leaf peroxisomes**, which often appear in close contact with chloroplasts and mitochondria (see Figure 4-21). The spatial proximity of the three organelles probably reflects their mutual involvement in the *glycolate pathway* (see Figure 11-17), also called the *photorespiratory pathway* because it involves the light-dependent uptake of O_2 and release of CO_2. Several enzymes of this pathway, including a peroxide-generating oxidase and two aminotransferases, are confined to leaf peroxisomes.

Glyoxysomes. Another functionally distinct type of plant peroxisome occurs transiently in seedlings of plant species that store carbon and energy reserves in the seed as fat (primarily triacylglycerols). In such species, stored triacylglycerols are mobilized and converted to sucrose during early postgerminative development by a sequence of events that includes β oxidation of fatty acids as well as a pathway known as the *glyoxylate cycle*. All of the enzymes needed for these processes are localized to specialized peroxisomes called **glyoxysomes** (see Chapter 10, pages 259–260).

Glyoxysomes are found only in the tissues where the fat is stored (endosperm or cotyledons, depending on the species) and are present only for the relatively short period of time required for the seedling to deplete its supply of stored fat. Once they fulfill their role in the seedling, the glyoxysomes are converted to peroxisomes. Glyoxysomes have been reported to appear again in the senescing (aging) tissues of some plant species, presumably to degrade lipids derived from the membranes of the senescent cells.

Peroxisome Biogenesis Can Occur by Division of Preexisting Peroxisomes or by Vesicle Fusion

Like other organelles, peroxisomes increase in number as cells grow and divide. This proliferation of organelles is called *biogenesis*, and the peroxisomal proteins required for this process are known as *peroxins*. Biogenesis of endosomes and lysosomes occurs by fusion of vesicles budding from the Golgi apparatus, and peroxisomes were once thought to form solely from vesicles in a similar manner. Later, most investigators believed that peroxisome biogenesis occurred solely from the division of preexisting peroxisomes. Recent evidence suggests that new peroxisomes can be formed by either method, or perhaps by a combination of the two. Either way, their biogenesis raises two important questions.

First, where do lipids in new peroxisomes come from? Some are synthesized by peroxisomal enzymes, whereas others are synthesized in the ER and carried to the peroxisome by phospholipid exchange proteins. Second, where are the new peroxisomal enzymes and other proteins synthesized? Proteins destined for peroxisomes are synthesized on free cytosolic ribosomes and are incorporated into preexisting peroxisomes after translation, using posttranslational import. This passage of polypeptides across the peroxisomal membrane is an ATP-dependent process mediated by specific membrane peroxins.

Figure 12-28 illustrates both the incorporation of membrane components, matrix enzymes, and enzyme cofactors from the cytosol into a peroxisome and the formation of new peroxisomes by division of a preexisting organelle. ❶ Lipids and membrane proteins can be added to existing peroxisomes via

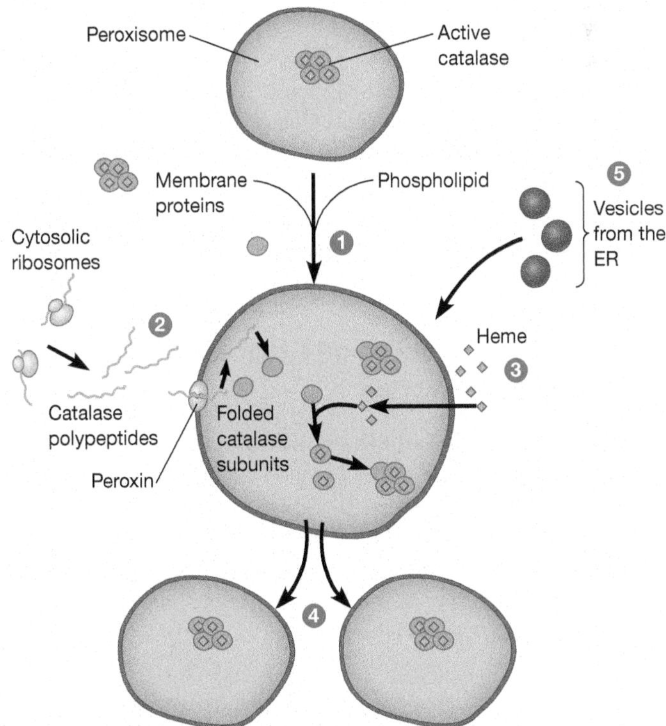

Figure 12-28 Biogenesis of Peroxisomes and Protein Import. New peroxisomes can arise by the division of existing peroxisomes, fusion of vesicles from the ER, or perhaps a combination of both.

cytosolic sources. ❷ Polypeptides for peroxisomal matrix enzymes are synthesized on cytosolic ribosomes and are threaded through the membrane via a transmembrane peroxin transport protein. The protein depicted in the figure is catalase, a tetrameric protein with a heme group bound to each subunit. ❸ Heme enters the peroxisomal lumen via a separate pathway, and the catalase polypeptides are folded and assembled along with heme to form the active tetrameric protein. Note that in addition to single, unfolded polypeptides, peroxisomes can import larger, folded polypeptides and even native oligomeric proteins.

After lipids and protein are added, ❹new peroxisomes are formed by the division of the existing peroxisomes. ❺ Recent evidence suggests that peroxisomes can obtain proteins or form de novo from vesicles derived from the ER. Evidence for this route includes experiments involving the toxin brefeldin A, which prevents formation of ER-derived vesicles and causes accumulation of the peroxin *Pex3p* in the ER. Tracking one peroxin (peroxin 3) in yeast in real time by tagging it with the yellow fluorescent protein also supports this route: the peroxin was first seen in the ER near the nucleus but was later associated with peroxisomal matrix proteins. Similar studies have shown that more than 20 different peroxisomal proteins have been shown to move from the ER to the peroxisome.

CONCEPT CHECK 12.8

Why is it important for the biochemical reactions occurring in peroxisomes to be isolated from the cytoplasm in a separate organelle?

Summary of Key Points

Mastering™ Biology For activities, animations, and review quizzes, go to the study area at www.masteringbiology.com.

12.1 The Endoplasmic Reticulum

- Especially prevalent within most eukaryotic cells is the endomembrane system, an elaborate array of membrane-bounded organelles derived from the endoplasmic reticulum (ER). The ER itself is a network of sacs, tubules, and vesicles surrounded by a single membrane that separates the ER lumen from the surrounding cytosol.
- The rough ER has ribosomes that synthesize proteins destined for the plasma membrane, for secretion, or for various organelles of the endomembrane system— the nuclear envelope, Golgi apparatus, endosomes, and lysosomes. Both the rough and smooth ER synthesize lipids for cellular membranes. The smooth ER is also the site of drug detoxification, carbohydrate metabolism, calcium storage, and steroid biosynthesis in some cells.

12.2 The Golgi Apparatus

- The Golgi apparatus plays an important role in the glycosylation of proteins and in the sorting of proteins for transport to other organelles, for transport to the plasma membrane, or for secretion.
- Transition vesicles that bud from the ER fuse with the *cis*-Golgi network (CGN), delivering lipids and proteins to the Golgi apparatus. Proteins then move through the Golgi cisternae toward the *trans*-Golgi network (TGN).

12.3 Roles of the ER and Golgi Apparatus in Protein Processing

- Before proteins leave the ER in transition vesicles, they undergo the first few steps of protein modification. ER-specific proteins catalyze core glycosylation and folding of polypeptides, elimination of misfolded proteins, and the assembly of multimeric proteins.
- During their journey through the Golgi apparatus, proteins are further modified as oligosaccharide side chains are trimmed or further glycosylated in the Golgi lumen.

12.4 Roles of the ER and Golgi Apparatus in Protein Trafficking

- Many polypeptides possess special amino acid sequences that target them to their appropriate location.
- Polypeptides destined for the endomembrane system, or for secretion from the cell, have an N-terminal ER signal sequence that causes them to enter translocon channels in the ER membrane while the polypeptide chain is still being synthesized.
- Some of these polypeptides pass completely through the translocon and are released into the ER lumen. Others possess one or more internal stop-transfer sequences that cause the polypeptide to remain anchored to the membrane. In either case, the resulting proteins may stay in the ER or be transported to other locations within the endomembrane system, such as the Golgi apparatus, lysosomes, plasma membrane, or secretory vesicles that expel the protein from the cell.
- Some ER-specific proteins have a retention signal that prevents them from leaving the ER as vesicles move from the ER to the CGN. Other ER proteins have a retrieval tag that allows them to return to the ER in vesicles that return from the CGN.
- Polypeptides destined for the nuclear interior, mitochondria, chloroplasts, or peroxisomes are synthesized on cytosolic ribosomes (as are polypeptides that remain in the cytosol) and are then imported posttranslationally into the targeted organelle. Polypeptides destined for peroxisomes or the nucleus contain targeting sequences that promote their uptake into these structures.
- Hydrolytic enzymes destined for the lysosome are phosphorylated on a mannose residue. Vesicles containing specific receptors for this mannose phosphate group carry these enzymes to the lysosome. During acidification in the lysosome, the enzymes are released from the receptors, which are then recycled back to the TGN.

12.5 Exocytosis and Endocytosis: Transporting Material Across the Plasma Membrane

- Exocytosis adds lipids and proteins to the plasma membrane when secretory granules release their contents to the extracellular medium by fusing with the plasma membrane. This addition of material to the plasma membrane is balanced by endocytosis, which removes lipids and proteins from the plasma membrane as extracellular material is internalized in vesicles.
- Phagocytosis is a type of endocytosis involving the ingestion of extracellular particles through invagination of the plasma membrane. Receptor-mediated endocytosis depends on highly specific binding of ligands to corresponding receptors on the cell surface. In both cases, after sorting of receptors and other necessary

proteins back to the plasma membrane, ingested material is sent to lysosomes for digestion or to other locations for reuse.

12.6 Coated Vesicles in Cellular Transport Processes

- Transport vesicles carry material throughout the endomembrane system. Coat proteins—which include clathrin, COPI, COPII, and caveolin—participate in the sorting of molecules fated for different destinations as well as in the formation of vesicles.
- The specific coat proteins covering a vesicle indicate its origin and help determine its destination within the cell. Clathrin-coated vesicles deliver material from the TGN or plasma membrane to endosomes. COPII-coated vesicles carry materials from the ER to the Golgi, while COPI-coated vesicles transport material from the Golgi back to the ER.
- Once a transport vesicle nears its destination, it is recognized and bound by tethering proteins attached to the target membrane. At this point the v-SNAREs in the transport vesicle membrane and the t-SNAREs in the target membrane interact physically, helping promote membrane fusion.

12.7 Lysosomes and Cellular Digestion

- Extracellular material obtained from phagocytosis or receptor-mediated endocytosis is sorted in early endosomes, which mature to form late endosomes and lysosomes as they fuse with vesicles containing inactive hydrolytic enzymes packaged in the TGN.
- The late endosomal membrane contains ATP-dependent proton pumps that lower the pH of the endosomal lumen and help transform the late endosome into a lysosome. Then latent acid hydrolases capable of degrading most biological molecules become active due to the low pH.
- Lysosomes also function in autophagy, the turnover and recycling of cellular structures that are damaged or no longer needed. In some cells, extracellular material is digested by enzymes that lysosomes discharge out of the cell by exocytosis.
- The plant vacuole is an acidic compartment resembling the animal lysosome. In addition to having hydrolytic enzymes for digestion of macromolecules, it helps the plant cell maintain positive turgor pressure and serves as a storage compartment for a variety of plant metabolites.

12.8 Peroxisomes

- Peroxisomes, which are not part of the endomembrane system, appear to increase in number either by the division of preexisting organelles or through the coalescence of vesicles from the ER or Golgi. Evidence supports both of these models of peroxisome synthesis.
- Some peroxisomal membrane lipids are synthesized by peroxisomal enzymes, whereas others are carried from the ER by phospholipid exchange proteins. Most peroxisomal proteins are synthesized by cytosolic ribosomes and are imported posttranslationally. Others are believed to travel from a subdomain of the ER.
- The defining enzyme of a peroxisome is catalase, an enzyme that degrades the toxic hydrogen peroxide that is generated by various oxidases in the peroxisome. Animal cell peroxisomes are important for detoxification of harmful substances, oxidation of fatty acids, and metabolism of nitrogen-containing compounds. In plants, peroxisomes play distinctive roles in the conversion of stored lipids into carbohydrate by glyoxysomes and in photorespiration by leaf peroxisomes.

Problem Set

Mastering™ Biology For activities, animations, and review quizzes, go to the study area at www.masteringbiology.com.

12-1 Compartmentalization of Function. Each of the following processes is associated with one or more specific eukaryotic organelles. In each case, identify the organelle or organelles, and suggest one advantage of confining the process to the organelle or organelles.

(a) β oxidation of long-chain fatty acids
(b) Biosynthesis of cholesterol
(c) Biosynthesis of insulin
(d) Biosynthesis of testosterone or estrogen
(e) Degradation of damaged organelles
(f) Glycosylation of membrane proteins
(g) Hydroxylation of phenobarbital
(h) Sorting of lysosomal proteins from secretory proteins

12-2 Endoplasmic Reticulum. For each of the following statements, indicate if the statement is true of the rough ER only (R), of the smooth ER only (S), or of both rough and smooth ER (RS).

(a) Contains less cholesterol than does the plasma membrane
(b) Has ribosomes attached to its outer (cytosolic) surface
(c) Is involved in the detoxification of drugs
(d) Is involved in the breakdown of glycogen
(e) Is the site for biosynthesis of secretory proteins
(f) Is the site for the folding of membrane-bound proteins
(g) Tends to form tubular structures
(h) Usually consists of flattened sacs
(i) Visible only by electron microscopy

12-3 Biosynthesis of Integral Membrane Proteins. In addition to their role in cellular secretion, the rough ER and the Golgi apparatus are also responsible for the biosynthesis of integral membrane proteins. More specifically, these organelles are the source of glycoproteins commonly found in the outer phospholipid monolayer of the plasma membrane.

(a) In a series of diagrams, depict the synthesis and glycosylation of glycoproteins of the plasma membrane.
(b) Explain why the carbohydrate groups of membrane glycoproteins are always found on the outer surface of the plasma membrane.
(c) What assumptions did you make about biological membranes in order to draw the diagrams in part a and answer the question in part b?

12-4 Protein Folding. The role of BiP in protein folding was briefly described in this chapter. Answer the following questions about observations and situations involving BiP.

(a) BiP is found in high concentration in the lumen of the ER but is not present in significant concentrations elsewhere in the cell. How do you think this condition is established and maintained?
(b) If the gene encoding BiP acquires a mutation that disrupts the protein's binding site for hydrophobic amino acids, what kind of impact might this have on the cell?

12-5 QUANTITATIVE Cotranslational Import. You perform a series of experiments on the synthesis of the pituitary hormone prolactin, which is a single polypeptide chain 199 amino acids long. The mRNA coding for prolactin is translated in a cell-free protein-synthesizing system containing ribosomes, amino acids, tRNAs, aminoacyl-tRNA synthetases, ATP, GTP, and the appropriate initiation, elongation, and termination factors. Under these conditions, a polypeptide chain 227 amino acids long is produced.

(a) How might you explain the discrepancy between the normal length of prolactin (199 amino acids) and the length of the polypeptide synthesized in your experiment (227 amino acids)?

(b) You perform a second experiment, in which you add SRP to your cell-free protein-synthesizing system, and find that translation stops after a polypeptide about 70 amino acids long has been produced. How can you explain this result? Can you think of any purpose this phenomenon might serve for the cell?

(c) You perform a third experiment, in which you add both SRP and ER membrane vesicles to your protein-synthesizing system, and find that translation of the prolactin mRNA now produces a polypeptide 199 amino acids long. How can you explain this result? Where would you expect to find this polypeptide?

12-6 Coated Vesicles in Intracellular Transport. For each of the following statements, indicate for which coated vesicle the statement is true: clathrin- (C), COPI- (I), or COPII-coated (II). Each statement may be true for one, several, or none (N) of the coated vesicles discussed in this chapter.

(a) Binding of the coat protein to an LDL receptor is mediated by an adaptor protein complex.

(b) Fusion of the vesicle (after dissociation of the coat) with the Golgi membrane is facilitated by specific t-SNARE and Rab proteins.

(c) Has a role in bidirectional transport between the ER and Golgi apparatus

(d) Has a role in sorting proteins for intracellular transport to specific destinations

(e) Has a role in transport of acid hydrolases to late endosomes

(f) Is essential for all endocytic processes

(g) Is important for retrograde traffic through the Golgi apparatus

(h) Is involved in the movement of membrane lipids from the TGN to the plasma membrane

(i) The basic structural component of the coat is called a triskelion.

(j) The protein coat dissociates shortly after formation of the vesicle.

(k) The protein coat always includes a specific small GTP-binding protein.

12-7 QUANTITATIVE Interpreting Data. Each of the following statements summarizes the results of an experiment related to exocytosis or endocytosis. In each case, explain the relevance of the experiment and its result to our understanding of these processes.

(a) Addition of the drug colchicine to cultured fibroblast cells inhibits movement of transport vesicles.

(b) Certain pituitary gland cells secrete laminin continuously but secrete adrenocorticotropic hormone only in response to specific signals.

(c) Certain adrenal gland cells can be induced to secrete epinephrine when their intracellular calcium concentration is experimentally increased.

(d) Cells expressing a temperature-sensitive form of dynamin do not display receptor-mediated endocytosis after a temperature shift, yet they continue to ingest extracellular fluid (at a reduced level initially, and then at a normal level within 30–60 minutes).

(e) Brefeldin A inhibits cholesterol efflux in adipocytes (fat cells) without affecting the rate of cellular uptake and resecretion of apolipoprotein A-I in adipocytes.

12-8 Cellular Digestion. For each of the following statements, indicate the specific digestion process or processes for which the statement is true: phagocytosis (P), receptor-mediated endocytosis (R), autophagy (A), or extracellular digestion (E). Each statement is true for one or more of these processes.

(a) Can involve exocytosis

(b) Can involve fusion of vesicles or vacuoles with a lysosome

(c) Digested material is of extracellular origin.

(d) Digested material is of intracellular origin.

(e) Essential for sperm penetration of the egg during fertilization

(f) Important for certain developmental processes

(g) Involves acid hydrolases

(h) Involves fusion of endocytic vesicles with an early endosome

(i) Involves fusion of lysosomes with the plasma membrane

(j) Occurs within lysosomes

(k) Serves as a source of nutrients within the cell

12-9 DATA ANALYSIS Virus Entry by Endocytosis. Australian bat lyssavirus (ABLV), a virus closely related to rabies virus, causes neurodegeneration in humans. It has a surface glycoprotein that mediates attachment to and internalization into host cells via receptor-mediated endocytosis. Certain inhibitors have been tested to see if they block viral entry into cells (**Figure 12-29**).

(a) Investigators have tested whether dynamin is required for ABLV infection of cells. Cultured human embryonic kidney cells were treated with *dynasore*, a specific inhibitor of dynamin, before being infected with a virus expressing the ABLV glycoprotein. Do the resulting data (Figure 12-29a, left) indicate dynamin involvement? Explain your answer.

(b) As a control, the researchers monitored cell viability at varying dynasore concentrations (Figure 12-29a, right). Why was this control necessary?

(c) They then used chlorpromazine, a chemical that blocks the assembly of clathrin-coated pits, in similar tests. Are the results (Figure 12-29b) consistent with those results in part a? Why or why not?

12-10 Lysosomal Storage Diseases. Despite a bewildering variety of symptoms, lysosomal storage diseases have several properties in common. For each of the following statements, indicate whether you would expect the property to be common to most lysosomal storage diseases (M), to be true of a specific lysosomal storage disease (S), or not to be true of any lysosomal storage diseases (N).

(a) Impaired metabolism of glycolipids that causes mental deterioration

(b) Leads to accumulation of degradation products in the lysosome

(c) Leads to accumulation of excessive amounts of glycogen in the lysosome

(d) Results from an inability to regulate the synthesis of glycosaminoglycans

(e) Results from an absence of functional acid hydrolases

(f) Results in accumulation of lysosomes in the cell

(g) Symptoms include muscle weakness and mental retardation.

(h) Triggers proliferation of organelles containing catalase

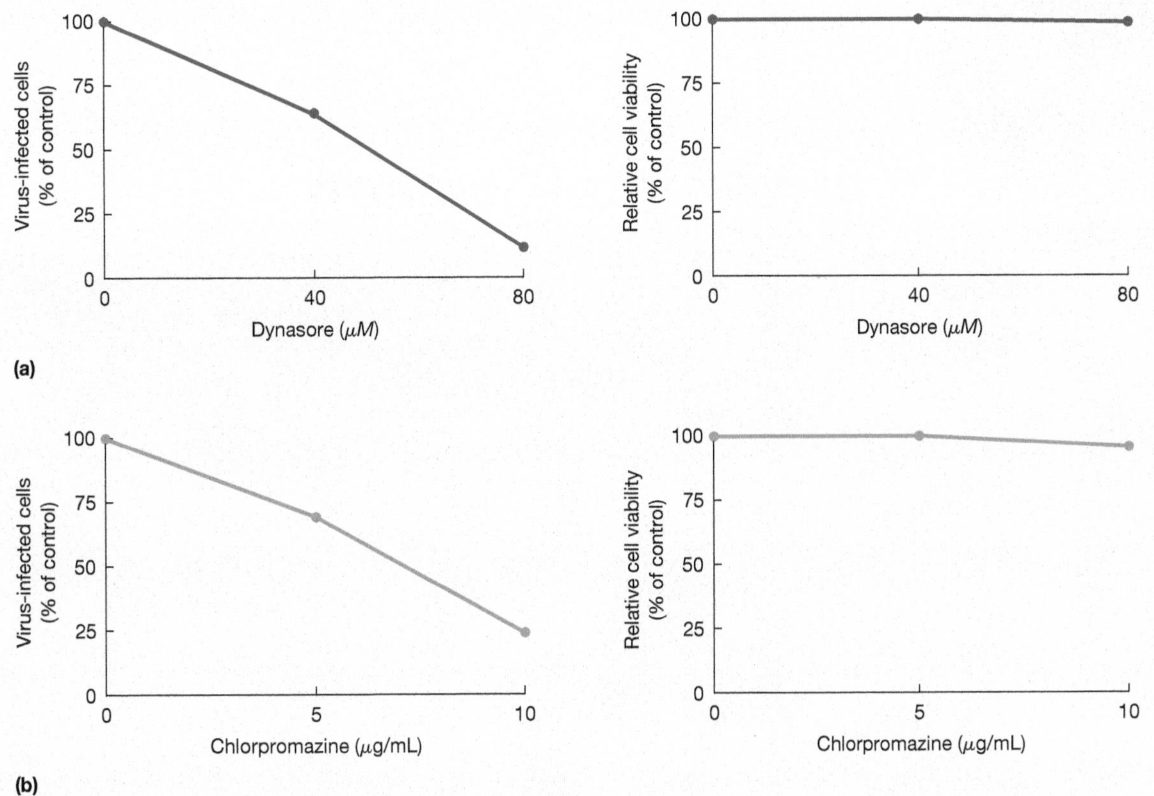

Figure 12-29 Effect of Inhibitors on Viral Entry into Cells. See Problem 12-9.

12-11 Sorting Proteins. Specific structural features tag proteins for transport to various intracellular and extracellular destinations. Several examples were described in this chapter, including (1) the short peptide Lys-Asp-Glu-Leu, (2) characteristic hydrophobic membrane-spanning domains, and (3) mannose-6-phosphate residues attached to oligosaccharide side chains. For each structural feature, answer the following questions.

(a) Where in the cell is the tag incorporated into the protein?

(b) How does the tag ensure that the protein reaches its destination?

(c) Where would the protein likely go if you were to remove the tag?

12-12 Silicosis and Asbestosis. *Silicosis* is a debilitating miner's disease that results from the ingestion of silica particles (such as sand or glass) by macrophages in the lungs. *Asbestosis* is a similarly serious disease caused by inhalation of asbestos fibers. In both cases, the particles or fibers are found in lysosomes. Fibroblasts, which secrete collagen, are stimulated to deposit nodules of collagen fibers in the lungs, leading to reduced lung capacity, impaired breathing, and eventually death.

(a) How do you think the fibers get into the lysosomes?

(b) What effect do you think fiber or particle accumulation has on the lysosomes?

(c) How might you explain the death of silica-containing or asbestos-containing cells?

(d) What do you think happens to the silica particles or asbestos fibers when such cells die? How can cell death continue almost indefinitely, even after prevention of further exposure to silica dust or asbestos fibers?

(e) Cultured fibroblast cells will secrete collagen and produce connective tissue fibers after the addition of material from a culture of lung macrophages that have been exposed to silica particles. What does this tell you about the deposition of collagen nodules in the lungs of silicosis patients?

12-13 What's Happening? Researchers have discovered a group of plant proteins that are related to the exocyst proteins in yeast (*Plant Cell* 20 (2008): 1330). Explain how the following observations made by these researchers suggest that these plant proteins form a tethering complex similar to the exocyst complex of yeast and mammals.

(a) Following size fractionation of plant protein extracts, antibodies recognizing each of several different plant exocyst proteins bind to the same high-molecular-weight protein fraction.

(b) Mutations in four of the proteins each cause defective pollen germination.

(c) Plants lacking more than one of these proteins have more serious defects in pollen germination than plants lacking only one.

(d) The exocyst proteins all colocalize at the growing tip of elongating pollen cells.

(e) Pollen cells of plants with mutations in exocyst genes are defective in tip growth or germinating pollen cells.

13 Cytoskeletal Systems

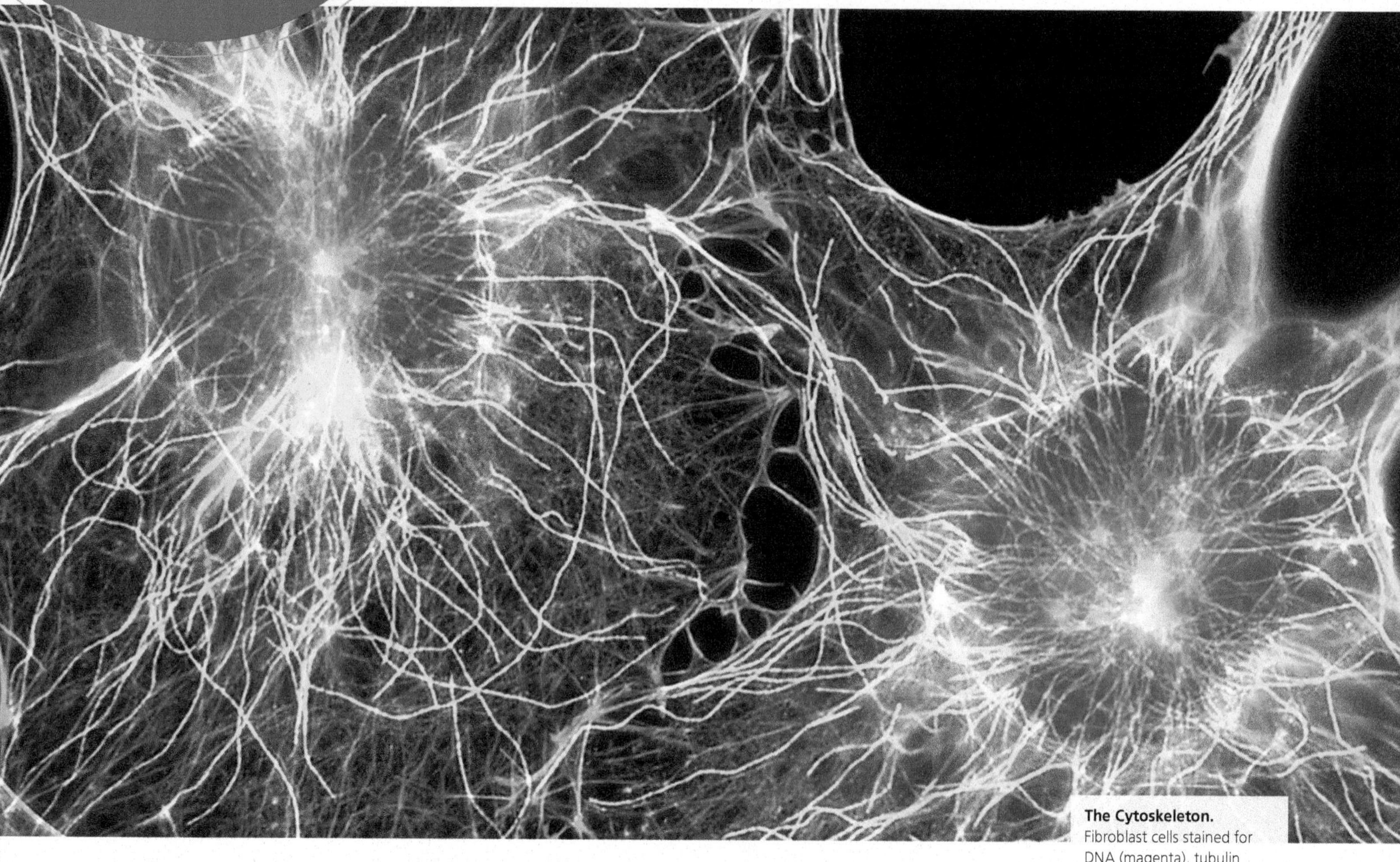

The Cytoskeleton. Fibroblast cells stained for DNA (magenta), tubulin (yellow), and F-actin (blue) (fluorescence microscopy).

The cytosol of the eukaryotic cell was originally regarded as the generally uninteresting, gel-like substance in which the nucleus and other organelles were suspended. Advances in microscopy and other investigative techniques have revealed that the cytosol is anything but uninteresting: The interior of a eukaryotic cell is highly structured and dynamic. Part of this structure is provided by the **cytoskeleton**, a complex network of interconnected filaments and tubules that extends throughout the cytosol, from the nucleus to the inner surface of the plasma membrane.

The term *cytoskeleton* accurately expresses the role of this polymer network in providing an architectural framework for cellular function. It confers a high level of internal organization on cells and enables them to assume and maintain complex shapes that would not otherwise be possible. The electron micrograph in **Figure 13-1** shows just how dense and fibrous this network can be. The name does not, however, convey the dynamic, changeable nature of the cytoskeleton and its critical involvement in a great variety of cellular processes.

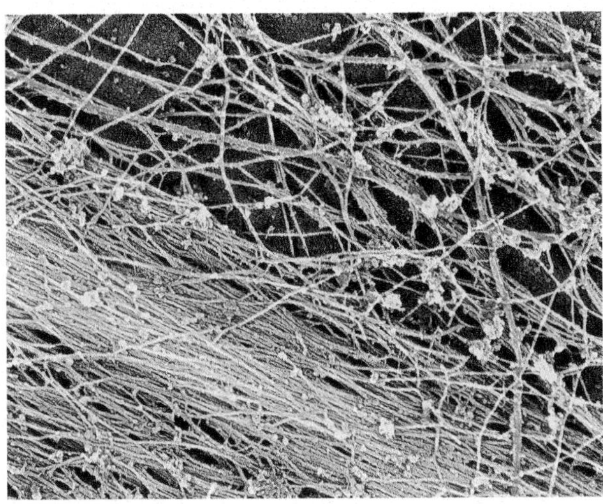

Figure 13-1 The Cytoskeleton. In this colorized electron micrograph, the three major elements of the eukaryotic cytoskeleton can be seen together. Yellow: actin microfilaments; blue: intermediate filaments; pink: microtubules (deep-etch TEM).

The cytoskeleton plays important roles in cell movement and cell division, and in eukaryotes it actively moves membrane-bounded organelles within the cytosol. It plays a similar role for messenger RNA and other cellular components. The cytoskeleton is also intimately related to other processes, such as cell signaling and cell-cell adhesion. The cytoskeleton is altered by events at the cell surface and, at the same time, appears to participate in and modulate these events.

13.1 Major Structural Elements of the Cytoskeleton

A fundamental feature of the cytoskeleton in both eukaryotes and prokaryotes is its *modularity*; that is, a small number of cytoskeletal elements are deployed in different locations and arranged in different ways to meet the needs of a particular cellular structure. In this way the cytoskeleton is very much like the skeleton of the human body, in which only a few elements (bone, cartilage, tendons, and ligaments) are used repeatedly to give the body its overall shape and structure. We now turn to the basic elements of the cytoskeleton in eukaryotes and prokaryotes.

Eukaryotes Have Three Basic Types of Cytoskeletal Elements

The three major structural elements of the cytoskeleton in eukaryotes are *microtubules*, *microfilaments*, and *intermediate filaments*. The existence of three distinct systems of filaments and tubules was first revealed by electron microscopy. Biochemical and cytochemical studies then identified the distinctive proteins of each system. The technique of *immunofluorescence* (see Figure 1-A3 and the Appendix) was especially important in localizing specific proteins to the cytoskeleton.

Each structural element of the cytoskeleton has a characteristic size, structure, and intracellular distribution, and each element is formed by the polymerization of a different kind of protein subunit (**Table 13-1** on page 360). Microtubules are composed of the protein *tubulin* and are about 25 nm in diameter. Microfilaments, with a diameter of about 7 nm, are polymers of the protein *actin*. Intermediate filaments have diameters in the range of 8–12 nm. Intermediate filament subunits differ depending on the cell type. Each type of cytoskeletal filament has a number of other proteins associated with it. These accessory proteins account for some of the remarkable structural and functional diversity of cytoskeletal networks.

In addition to the three fundamental polymer systems within eukaryotic cells, another polymer network is composed of proteins called *septins*. Originally identified in yeast, septins are sometimes called "the fourth cytoskeleton." They are closely associated with the contractile ring, a structure involved in the pinching off of daughter cells during cell division.

Bacteria Have Cytoskeletal Systems That Are Structurally Similar to Those in Eukaryotes

Cytoskeletal proteins were once thought to be unique to eukaryotes. However, it is now clear that bacteria and archaea have polymer systems that function similarly to microfilaments, microtubules, and intermediate filaments. **Figure 13-2** on page 361 illustrates three examples. Based on the effects of mutating prokaryotic cytoskeletal proteins, it is clear that they play roles similar to those of their eukaryotic counterparts. For example, the actin-like *MreB* protein forms filaments perpendicular to the long axis of rod-shaped bacteria and associates with proteins involved in synthesis of the peptidoglycan layer (bacterial cell wall). MreB is important in maintaining bacterial cell shape and during bacterial cell division. The tubulin-like *FtsZ* protein is involved in determining where bacterial cells will divide; and the intermediate filament-like *crescentin* protein is an important regulator of cell shape in some bacteria.

Although prokaryotic cytoskeletal proteins are not very similar to their eukaryotic counterparts at the level of amino acid sequence, X-ray crystallography has shown that when they are assembled into polymers, their overall structure is remarkably similar (Figure 13-2b). Moreover, the equivalent proteins bind the same phosphonucleotides (for example, GTP, ATP) as their eukaryotic equivalents, indicating striking similarities at the biochemical level.

MAKE CONNECTIONS 13.1

The *FtsZ* protein is produced by certain eukaryotic organelles, such as chloroplasts and mitochondria, and localizes to sites where these organelles divide. How is this finding consistent with current theories of the origin of eukaryotes? (Ch. 4.3)

The Cytoskeleton Is Dynamically Assembled and Disassembled

The cytoskeleton is perhaps best known for its roles in cell motility. For example, microfilaments are essential components of *muscle fibrils*, and microtubules are the structural elements of *cilia* and *flagella*, appendages that enable certain

Table 13-1 Properties of Microtubules, Microfilaments, and Intermediate Filaments

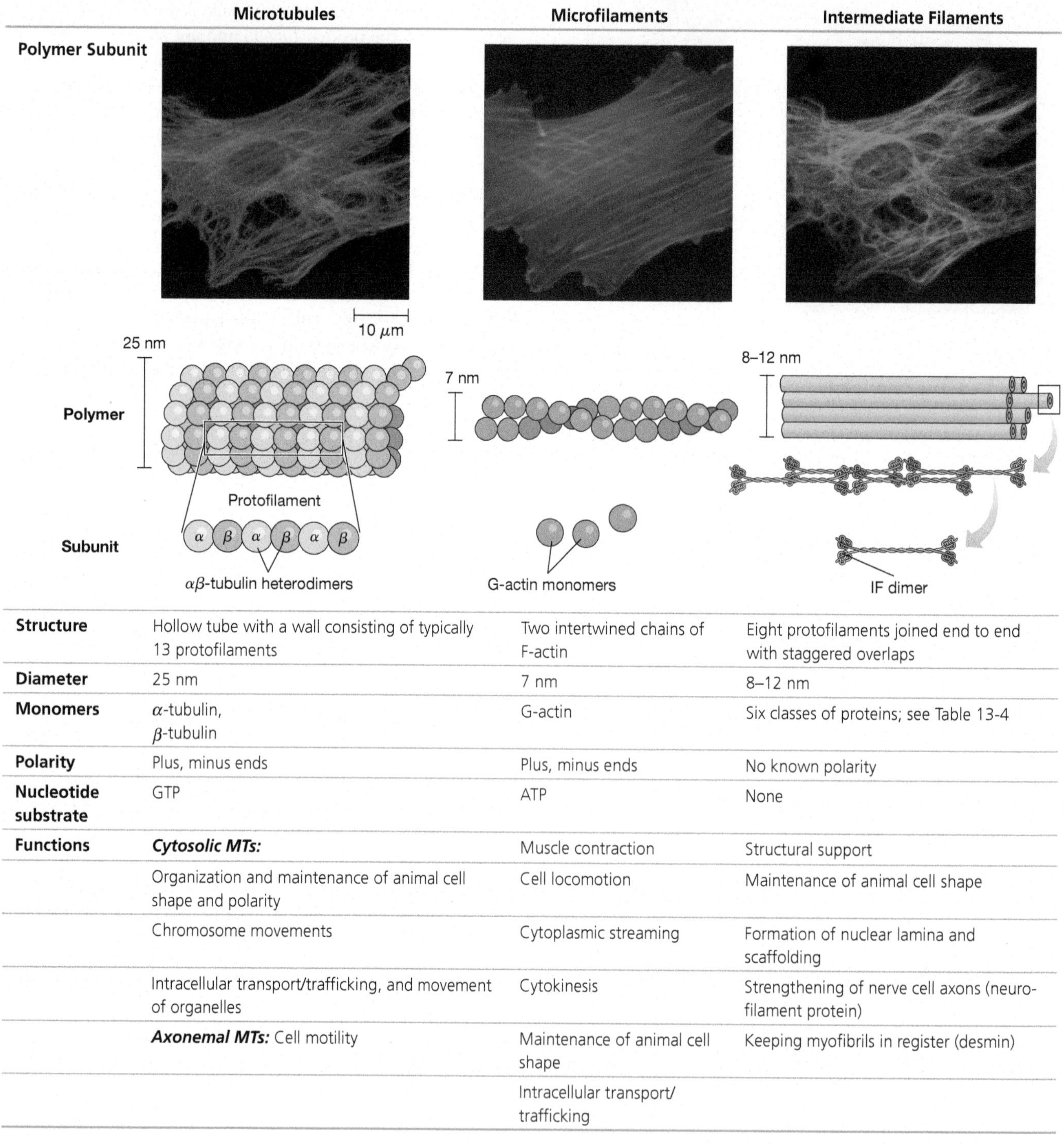

	Microtubules	Microfilaments	Intermediate Filaments
Structure	Hollow tube with a wall consisting of typically 13 protofilaments	Two intertwined chains of F-actin	Eight protofilaments joined end to end with staggered overlaps
Diameter	25 nm	7 nm	8–12 nm
Monomers	α-tubulin, β-tubulin	G-actin	Six classes of proteins; see Table 13-4
Polarity	Plus, minus ends	Plus, minus ends	No known polarity
Nucleotide substrate	GTP	ATP	None
Functions	***Cytosolic MTs:***	Muscle contraction	Structural support
	Organization and maintenance of animal cell shape and polarity	Cell locomotion	Maintenance of animal cell shape
	Chromosome movements	Cytoplasmic streaming	Formation of nuclear lamina and scaffolding
	Intracellular transport/trafficking, and movement of organelles	Cytokinesis	Strengthening of nerve cell axons (neurofilament protein)
	Axonemal MTs: Cell motility	Maintenance of animal cell shape	Keeping myofibrils in register (desmin)
		Intracellular transport/trafficking	

cells to either propel themselves through a fluid environment or move fluids past the cell. These structures are large enough to be seen by light microscopy and were therefore known and studied long before it became clear that the same structural elements are also integral parts of the cytoskeleton in most cells.

With the advent of sophisticated microscopy techniques, it eventually became clear that most cells dynamically regulate where and when specific cytoskeletal structures are assembled and disassembled. Recent progress in understanding cytoskeletal structure relies heavily on a combination of powerful techniques, including biochemistry and chemical disruption of the cytoskeleton (see we discuss below). Modern cell biologists also use *fluorescence microscopy, digital video microscopy,* and various types of *electron microscopy* to study the cytoskeleton (**Table 13-2** on page 362). (Each of these techniques is described in more detail in the Appendix.) These modern visualization techniques have revealed the incredibly

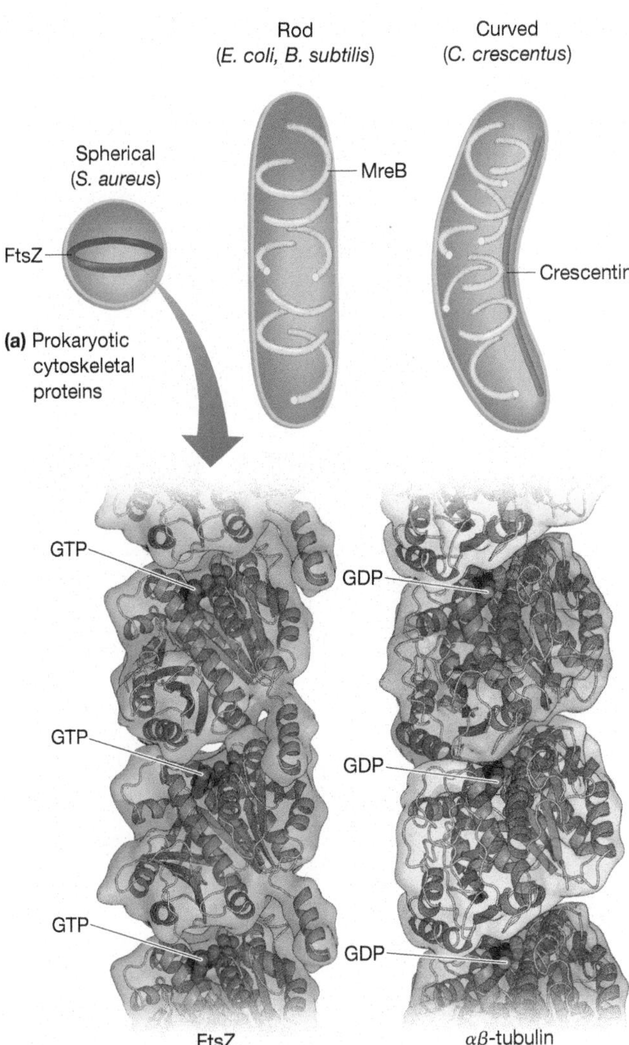

Figure 13-2 Cytoskeletal Proteins in Prokaryotes Are Similar to Those in Eukaryotes. (a) The distribution of several bacterial cytoskeletal proteins is shown. Blue: the microtubule-like FtsZ protein; yellow: the actin-like MreB protein; red: the intermediate-filament-like protein crescentin. **(b)** Comparison of the structure of FtsZ homodimer (left) and an $\alpha\beta$-tubulin heterodimer (based on X-ray crystallography). Note the similarity in monomer folding and filament structure.

dynamic nature of the cytoskeleton and the remarkably elaborate structures it comprises.

In this chapter, we will focus on the structure of the cytoskeleton and how its components are dynamically assembled and disassembled. In each case, we will consider the chemistry of the subunit(s), the structure of the polymer and how it is polymerized, the role of accessory proteins, and some of the structural and functional roles each component plays within the cell. In the last section of this chapter we will see that the components of the cytoskeleton are linked together both structurally and functionally.

CONCEPT CHECK 13.1

Cytoskeletal polymers are modular, and their assembly is reversible. What advantages do you think there are to using repeating structural elements to construct the cytoskeleton?

13.2 Microtubules

Microtubules (MTs) are the largest structural elements of the eukaryotic cytoskeleton (see Table 13-1). They function in a wide array of cellular processes whose common theme is movement—either moving components within cells or moving fluid outside of cells. In this chapter, we will consider MTs structurally (in Chapter 14, we will consider the mechanisms by which MT-dependent movement is achieved).

Two Types of Microtubules Are Responsible for Many Functions in the Cell

Microtubules can be classified into two general groups, which differ in both degree of organization and structural stability. The first group comprises an often loosely organized, dynamic network of **cytosolic microtubules**. The existence of cytosolic MTs was not recognized until the early 1960s, when better fixation techniques permitted direct visualization of the network of MTs now known to pervade the cytosol of most eukaryotic cells. Since then, fluorescence microscopy has revealed the diversity and complexity of MT networks in different cell types.

Cytosolic MTs are responsible for a variety of cellular functions. In animals, for example, crosslinked bundles of cytosolic MTs are required to maintain axons (specialized extension of nerve cells). In plant cells, cytosolic MTs govern the orientation of cellulose microfibrils deposited during the growth of cell walls. Significantly, cytosolic MTs form the mitotic and meiotic spindles that are essential for the movement of chromosomes during mitosis and meiosis. Cytosolic microtubules also contribute to the spatial disposition and directional movement of vesicles and organelles by providing an organized system of fibers to guide their movement.

The second group of microtubules, **axonemal microtubules**, includes the highly organized, stable microtubules found in subcellular structures specialized for cellular movement, including cilia, flagella, and the basal bodies to which these appendages are attached. The central shaft, or *axoneme*, of a cilium or flagellum consists of a highly ordered bundle of axonemal MTs and associated proteins. Given their order and stability, it is not surprising that the axonemal MTs were the first of the two groups to be recognized and studied.

Tubulin Heterodimers Are the Protein Building Blocks of Microtubules

MTs are straight, hollow cylinders with an outer diameter of about 25 nm and an inner diameter of about 15 nm (**Figure 13-3** on page 363). Microtubules vary greatly in length. Some are less than 200 nm long; others, such as axonemal MTs, can be many micrometers in length. The MT wall consists of longitudinal arrays of linear polymers called **protofilaments**. There are usually 13 protofilaments arranged side by side around the hollow center, or lumen, of the MT. Although some MTs in some animals contain more or fewer protofilaments, 13 is by far the most common number.

As shown in Figure 13-3, the basic subunit of a protofilament is a *heterodimer* (from *hetero*, "different" and *dimer*, referring to two subunits) of the protein **tubulin**. Protofilaments are composed of one molecule of **α-tubulin** and one

Table 13-2 Techniques for Visualizing the Cytoskeleton

Technique	Description	Example
Fluorescence microscopy on fixed specimens*	Fluorescent compounds directly bind to cytoskeletal proteins, or antibodies are used to indirectly label cytoskeletal proteins in chemically preserved cells, causing them to glow in the fluorescence microscope.	A fibroblast stained with fluorescent antibodies directed against actin shows bundles of actin filaments.
Live-cell fluorescence microscopy*	Fluorescent versions of cytoskeletal proteins are made and introduced into living cells. Fluorescence microscopy and video or digital cameras are used to view the proteins as they function in cells.	Fluorescent tubulin molecules were microinjected into living fibroblast cells. Inside the cell, the tubulin dimers become incorporated into microtubules, which can be seen easily with a fluorescence microscope. (0 sec → 145 sec)
Computer-enhanced digital video microscopy	High-resolution images from a video or digital camera attached to a microscope are computer processed to increase contrast and remove background features that obscure the image.	The two micrographs show several microtubules. The image on the right was processed to make the details of the microtubules more visible. (Unenhanced / Enhanced)
Electron microscopy	Electron microscopy can resolve individual filaments. Cells to be visualized may be prepared by thin section, quick-freeze deep-etch, or direct-mount techniques.	A fibroblast cell prepared by the quick-freeze deep-etch method. Bundles of actin microfilaments are visible.

*Confocal, deconvolution, multiphoton, and total internal reflection fluorescence (TIRF) microscopy are often used to improve detection of fluorescent signals. (See the Appendix for more details.)

molecule of **β-tubulin.** As soon as individual α- and β-tubulin molecules are synthesized, they bind noncovalently to each other to produce an **$\alpha\beta$-heterodimer** that does not dissociate under normal conditions.

Individual α- and β-tubulin molecules have diameters of about 4–5 nm and molecular weights of 55 kDa. Structural studies show that α- and β-tubulins have nearly identical three-dimensional structures, even though they share only 40% amino acid sequence identity. Each has (1) a binding domain at the N-terminus for the phosphonucleotide *guanosine triphosphate (GTP)*, (2) a domain in the middle to which colchicine, an MT poison, can bind, and (3) a third domain at the C-terminus that interacts with MT-associated proteins (MAPs). (You will learn about MAPs later in this chapter.)

This uniform orientation of tubulin dimers means that one end of the protofilament differs structurally from the other. Because the orientation of tubulin dimers is the same for all of the protofilaments in an MT, the MT itself is also a polarized structure, with a *plus (+) end* and a *minus (−) end*, whose properties we will examine shortly.

Most organisms have several closely related but nonidentical genes for each of the α- and β-tubulin subunits. These slightly different forms of tubulin are called *tubulin isoforms*. In the mammalian brain, for example, there are five α- and five β-tubulin isoforms. These isoforms differ mainly in the C-terminal domain, which suggests that various tubulin isoforms may interact with different proteins. In addition to different isoforms, tubulin can be chemically modified. For example, *acetylated* tubulin tends to form more stable MTs than nonacetylated tubulin does.

Microtubules Can Form as Singlets, Doublets, or Triplets

Cytosolic MTs are simple tubes, or *singlet* MTs, built from 13 protofilaments. Some axonemal MTs are more complex, however: they can contain *doublet* or *triplet* MTs. Doublets and triplets contain one complete 13-protofilament microtubule (the *A tubule*) and one or two additional incomplete tubules (called *B* and *C tubules*) consisting of 10 or 11 protofilaments (see Figure 13-3c). Doublets are found in cilia and flagella;

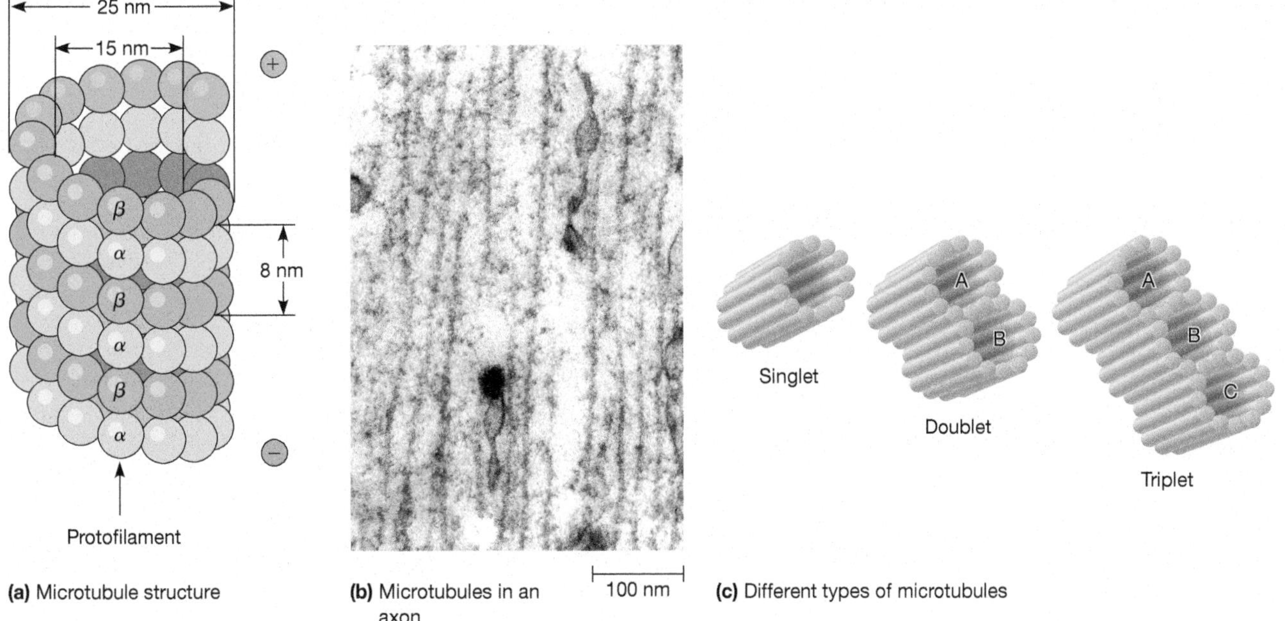

(a) Microtubule structure

(b) Microtubules in an axon

(c) Different types of microtubules

Figure 13-3 Microtubule Structure. (a) A schematic diagram showing a microtubule (MT) as a cylinder enclosing a hollow lumen. The outside diameter is about 25 nm, the inside about 15 nm. The wall of the cylinder consists of 13 protofilaments, one of them indicated by an arrow. A protofilament is a linear polymer of tubulin dimers, each consisting of two polypeptides: α-tubulin and β-tubulin. All heterodimers in the protofilaments have the same orientation; the plus (+) and minus (−) ends are indicated. **(b)** MTs in a longitudinal section of an axon (TEM). **(c)** MTs can form as singlets (13 protofilaments around a hollow lumen), doublets, and triplets. Doublets and triplets contain one complete 13-protofilament microtubule (the *A tubule*) and one or two additional incomplete tubules (called *B* and *C tubules*) consisting of 10 protofilaments.

triplets are found in basal bodies and centrioles. (You will learn about cilia and flagella in much more detail in Chapter 14.) Doublets and triplets are stabilized by proteins that line the interior of the fused microtubules, collectively known as *microtubule inner proteins (MIPs)*.

Microtubules Form by the Addition of Tubulin Dimers at Their Ends

Microtubules form by the reversible polymerization of tubulin dimers. The polymerization process has been studied extensively in vitro; a schematic representation of MT assembly in vitro is shown in **Figure 13-4** on page 364. When a solution containing a sufficient concentration of tubulin dimers, GTP, and Mg^{2+} is warmed from 0°C to 37°C, the polymerization reaction begins. (MT formation in the solution can be readily measured as an increase in light scattering with an instrument called a spectrophotometer.) A critical step in the formation of MTs is the aggregation of tubulin dimers into clusters called *oligomers.* These oligomers serve as "seeds" from which new microtubules can grow; hence this process is referred to as **nucleation**. Once an MT has been nucleated, it grows by addition of subunits at either end, via a process called **elongation**.

Microtubule formation is initially slow, a period referred to as the *lag phase* of MT assembly. This period reflects the relatively slow process of MT nucleation. The *elongation phase* of MT assembly—the addition of tubulin dimers—is relatively fast compared with nucleation. Eventually, the mass of MTs increases to a point where the concentration of free tubulin becomes limiting. This leads to the *plateau phase,* where MT assembly is balanced by disassembly.

Microtubule growth in vitro depends on the concentration of tubulin dimers. The tubulin heterodimer concentration at which MT assembly is exactly balanced with disassembly is called the overall **critical concentration**. MTs tend to grow when the tubulin concentration exceeds the critical concentration and depolymerize when the tubulin concentration falls below the critical concentration.

Recent work involving electron tomography (see **Key Technique in Chapter 10, pages 248–249**, for further details of this technique) indicates that MTs in vitro and in vivo elongate by addition of curved protofilaments to the growing tips of MTs, which then straighten as the protofilament becomes part of the wall of the MT.

> **MAKE CONNECTIONS 13.2**
>
> If you think about polymerization like a chemical reaction at equilibrium, how would the concentration of tubulin heterodimers influence the likelihood of microtubule growth versus shrinkage? (Ch. 5.3)

Addition of Tubulin Dimers Occurs More Quickly at the Plus Ends of Microtubules

The inherent structural polarity of microtubules also means that the two ends differ chemically. An important difference between the two ends of an MT is that one can grow or shrink much faster than the other. This difference

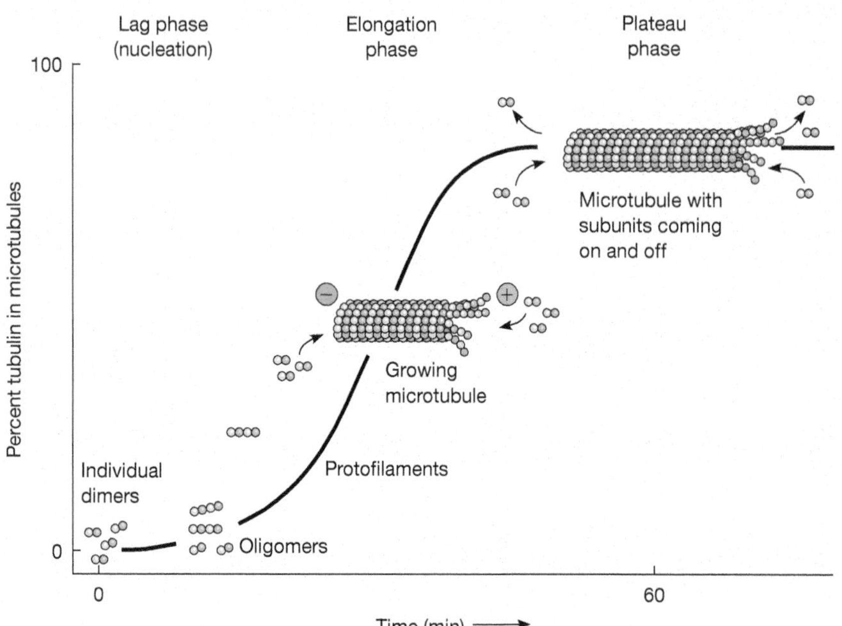

Figure 13-4 The Kinetics of Microtubule Assembly In Vitro. The kinetics of MT assembly can be monitored by observing the amount of light scattered by a solution containing GTP-tubulin after it is warmed from 0°C to 37°C. (Microtubule assembly is inhibited by cold and activated by warming.) Light-scattering measurements reflect changes in the MT population as a whole, not the assembly of individual microtubules. MT assembly exhibits three phases: lag, elongation, and plateau. The lag phase is the period of nucleation. During the elongation phase, MTs grow rapidly, causing the concentration of tubulin subunits in the solution to decline. When this concentration is low enough to limit further assembly, the plateau phase is reached, during which subunits are added and removed from MTs at equal rates.

in assembly rate can readily be visualized by mixing the MT-associated structures found at the base of cilia, known as *basal bodies*, with tubulin heterodimers. Microtubule nucleation and assembly occur at both ends of the basal body, but the MTs grow much faster from one end than from the other (**Figure 13-5**). The rapidly growing end of a microtubule is called the **plus end**, and the other end is the **minus end**. As will be discussed below, minus ends of MTs are often anchored at the centrosome; in this case, MT dynamics are confined to plus ends.

The different growth rates of the plus and minus ends of microtubules reflect the different critical concentrations required for assembly at the two ends of an MT. The critical concentration for the plus end is lower than that for the minus end. If the free tubulin concentration is higher than the critical concentration for the plus end but lower than the critical concentration for the minus end, assembly will occur at the plus end while disassembly takes place at the minus end. This simultaneous assembly and disassembly produce the phenomenon known as *treadmilling*. Treadmilling arises when a given tubulin molecule incorporated at the plus end is displaced progressively along the MT and eventually lost by depolymerization at the opposite end (**Figure 13-6**). By examining fluorescently labeled MTs, treadmilling has been observed in living cells, although it is uncertain how important it is to overall MT dynamics.

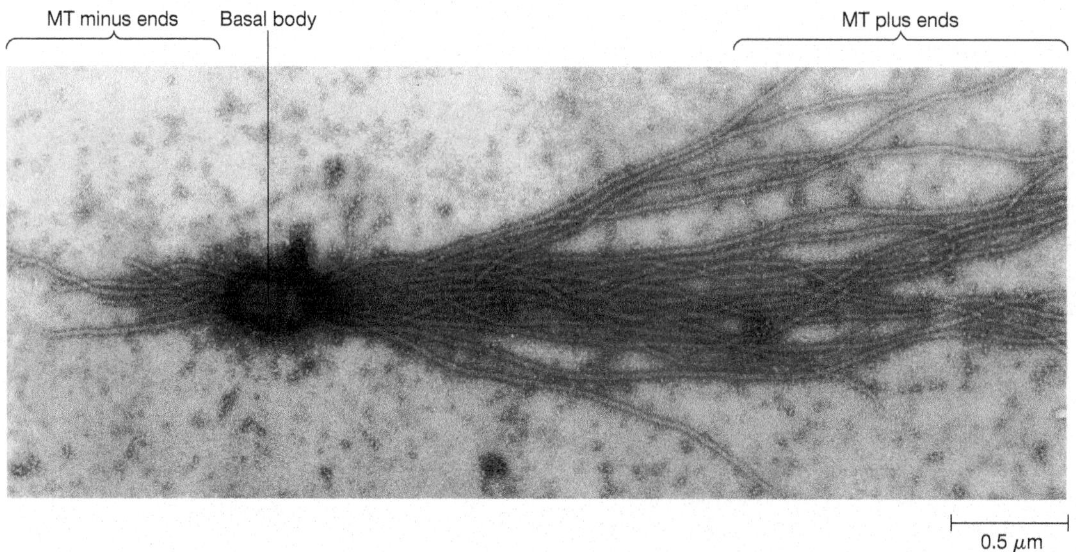

Figure 13-5 Polar Assembly of Microtubules In Vitro. The polarity of MT assembly can be demonstrated by adding basal bodies to a solution of tubulin dimers. The dimers add to the plus and minus ends of the MTs in the basal body. However, MTs that grow from the plus end are much longer than those growing from the minus end.

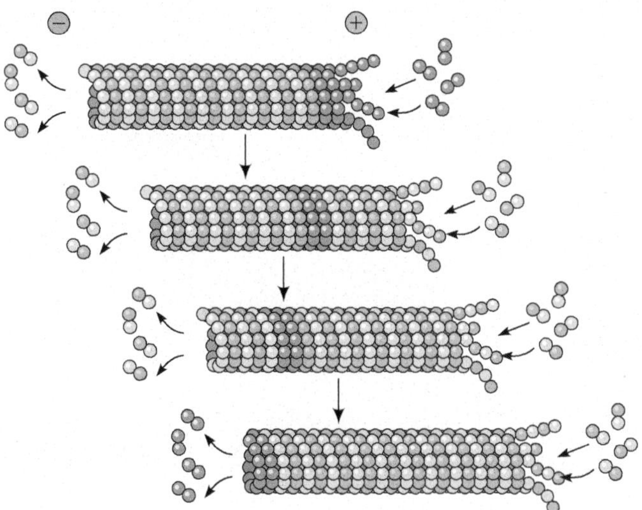

Figure 13-6 Treadmilling of Microtubules. Microtubule assembly occurs more readily at the plus end of an MT than at the minus end. When the tubulin concentration is higher than the critical concentration for the plus end but lower than the critical concentration for the minus end, the MT can add tubulin heterodimers to its plus end while losing them from its minus end.

Drugs Can Affect the Assembly and Stability of Microtubules

A number of drugs affect microtubule assembly (**Table 13-3**). One well-known drug of this sort is **colchicine**, a naturally occurring compound from the autumn crocus, *Colchicum autumnale.* Colchicine binds to the β-subunit of tubulin heterodimers. Colchicine binding strongly inhibits incorporation of tubulin heterodimers into microtubules. The resulting tubulin-colchicine complex can still add to the growing end of an MT, but it then prevents any further addition of tubulin molecules and destabilizes the structure, thereby promoting MT disassembly. *Vinblastine* and *vincristine* are related compounds from the periwinkle plant (*Vinca rosea*) that bind and sequester tubulin dimers inside the cell. **Nocodazole** (a synthetic agent) is another compound that inhibits MT assembly and is frequently used in experiments instead of colchicine because its effects are more readily reversible when the drug is removed.

These compounds are often called *antimitotic drugs* because they disrupt the mitotic spindle of dividing cells, blocking the further progress of mitosis. The sensitivity of the mitotic spindle to these drugs is understandable because spindle fibers are composed of microtubules. Indeed, vinblastine and vincristine find application in medical practice as anticancer drugs because cancer cells divide in an uncontrolled fashion and are therefore preferentially susceptible to drugs that interfere with the mitotic spindle.

In contrast, **paclitaxel**, or **Taxol** (originally isolated from the bark of the Pacific yew tree, *Taxus brevifolia*), binds tightly to microtubules and stabilizes them, causing much of the free tubulin in the cell to be locked up within microtubules. Within mitotic cells, paclitaxel blocks MT disassembly and arrests cell division. Thus, both paclitaxel and colchicine block cells in mitosis, but they do so by opposing effects on MTs and hence on the fibers of the mitotic spindle. Paclitaxel is also used in the treatment of some cancers, especially breast cancer.

GTP Hydrolysis Contributes to the Dynamic Instability of Microtubules

In the previous section, you saw that tubulin can assemble in vitro in the presence of Mg^{2+} and GTP. In fact, GTP is required for MT assembly. Each tubulin heterodimer binds two GTP molecules. The α-tubulin binds one GTP; β-tubulin binds the other. Although the GTP bound to the α subunit is not hydrolyzed, the GTP bound to the β subunit can be hydrolyzed to GDP sometime after the heterodimer is added to an MT. Hydrolysis of GTP is not necessary for assembly because MTs polymerize from tubulin heterodimers bound to a nonhydrolyzable analogue of GTP. However, MT assembly apparently requires that GTP be present in both the α and

Table 13-3 Chemical Agents Used to Perturb the Cytoskeleton

Agent	Source	Effect
Agents Affecting Microtubules		
Colchicine, colcemid	Autumn crocus, *Colchicum autumnale*	Binds β-tubulin, inhibiting assembly
Nocadazole	Synthetic compound	Binds β-tubulin, inhibiting polymerization
Vinblastine, vincristine	Periwinkle plant, *Vinca rosea*	Aggregates tubulin heterodimers
Paclitaxel (Taxol)	Symbiotic fungi in bark of the Pacific yew tree, *Taxus brevifolia*	Stabilizes microtubules
Agents Affecting Microfilaments		
Cytochalasin D	Fungal metabolite	Prevents addition of new monomers to plus ends
Latrunculin A	Red sea sponge, *Latrunculia magnifica*	Sequesters actin monomers
Phalloidin	Death cap fungus, *Amanita phalloides*	Binds and stabilizes assembled microfilaments
Agents Affecting Intermediate Filaments		
Acrylamide	Synthetic compound	Causes loss of intermediate filament networks

β subunits because the association of GDP-bound tubulin heterodimers with each other is thought to be too weak to support polymerization.

Studies of MT assembly in vitro using isolated centrosomes (a structure we will discuss in detail below) as nucleation sites show that some microtubules can grow by polymerization at the same time that others shrink by depolymerization. As a result, some MTs effectively enlarge at the expense of others. To explain how both polymerization and depolymerization might occur simultaneously, Tim Mitchison and Marc Kirschner proposed the **dynamic instability model**. This model presumes two populations of microtubules, one growing in length by continued polymerization at plus ends and the other shrinking in length by plus-end depolymerization. The distinction between the two populations is that growing MTs have GTP bound to the β-tubulin at their plus ends, whereas shrinking MTs have GDP instead.

Figure 13-7 shows how GTP-bound tubulin functions at the plus end. Heterodimeric tubulin complexed with two GTPs, referred to as *GTP-tubulin*, is thought to protect MTs by preventing tubulin from peeling away from their plus ends. This *GTP cap* provides a stable MT tip to which further dimers can be added (Figure 13-7a). Hydrolysis of GTP by β-tubulin eventually results in an unstable tip, at which point depolymerization may occur rapidly.

The concentration of GTP-tubulin is crucial to the dynamic instability model. When GTP-tubulin is readily available, it is added to the microtubule quickly, creating a large GTP-tubulin cap. If the concentration of GTP-tubulin falls, however, the rate of tubulin addition decreases. At a sufficiently low concentration of GTP-tubulin, the rate of hydrolysis of GTP on the β-tubulin subunits near the tip of the MT exceeds the rate of addition of new GTP-tubulin. This results in shrinkage of the GTP cap. When the GTP cap disappears, the MT becomes unstable, and loss of subunits from its tip is favored.

Direct evidence for dynamic instability comes from observation of individual microtubules in vitro via light microscopy, which shows that dynamic instability is a key feature of MTs in living cells. An individual MT can undergo alternating periods of growth and shrinkage (Figure 13-7b). When an MT switches from growth to shrinkage, an event called *microtubule catastrophe*, the MT can disappear completely, or it can abruptly switch back to a growth phase, a phenomenon known as *microtubule rescue*. The frequency of catastrophe is inversely related to the free tubulin concentration. High tubulin concentrations make catastrophe less likely, but it can still occur. When catastrophe does occur, higher tubulin concentrations make the rescue of a shrinking MT more likely. At any tubulin concentration, catastrophe is more likely at the plus end of an MT—that is, dynamic instability is more pronounced at the plus end of the MT.

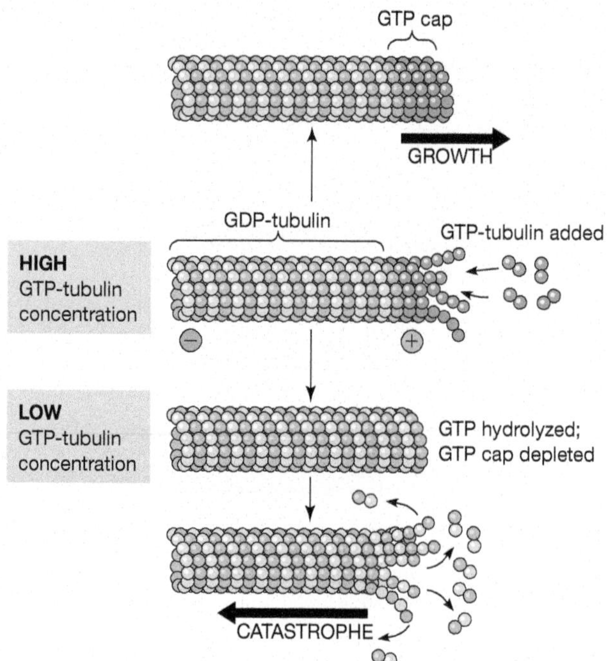

(a) Model for how the GTP cap functions

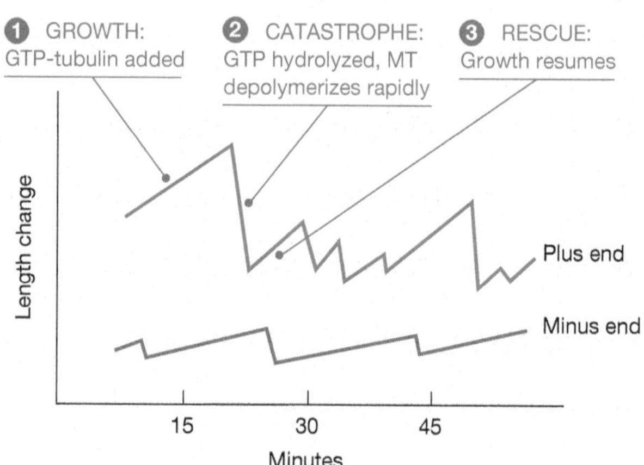

(b) Evidence for dynamic instability

Figure 13-7 The GTP Cap and Its Role in the Dynamic Instability of Microtubules. (a) When the tubulin concentration is high, GTP-tubulin is added to the microtubule tip faster than the incorporated GTP can be hydrolyzed. The resulting GTP cap stabilizes the MT tip and promotes further growth. At lower tubulin concentrations, the rate of growth decreases, thereby allowing GTP hydrolysis to catch up. This creates an unstable tip (no GTP cap) that favors MT depolymerization. **(b)** In an individual MT observed by light microscopy, ❶ growth and ❷ catastrophic shrinkage can occur. Plus and minus ends grow and shrink independently; changes in length are much more dynamic at the plus end. ❸ Rescue involves the switch from shrinkage to growth.

Microtubules Originate from Microtubule-Organizing Centers Within the Cell

In previous sections, we primarily focused on the properties of tubulin and microtubules in vitro, providing a foundation for understanding how MTs function in the cell. MT formation in vivo, however, is a more ordered and regulated process that produces sets of MTs in specific locations for specific cellular functions.

Microtubules commonly originate from a structure in the cell called a **microtubule-organizing center (MTOC)**. An MTOC serves as a site at which MT assembly is initiated and acts as an anchor for one end of these MTs. During interphase of the cell cycle, many cells have an MTOC called the **centrosome** that is positioned near the nucleus (**Figure 13-8**). During cell division, centrosomes are duplicated, creating new MTOCs for each of the daughter cells.

The centrosome in an animal cell is normally associated with two structures called **centrioles** surrounded by a diffuse granular matrix known as *pericentriolar material* (Figure 13-8a). In electron micrographs of the centrosome, MTs can be seen to originate from the pericentriolar material (Figure 13-8b). The symmetrical structure of centrioles is remarkable: as Figure 13-8c shows, their walls are formed by nine sets of triplet microtubules. In most cases, pairs of centrioles are oriented at right angles to one another; the significance of this arrangement is still unknown.

Centrioles are known to be involved in the formation of basal bodies, which are important for the formation of cilia and flagella (see Chapter 14). The role of centrioles in nonciliated cells is less clear. In animal cells, centrioles may serve to recruit pericentriolar material to the centrosome, which then nucleates growth of microtubules. When centrioles are removed from many animal cells, microtubule-nucleating material disperses, and the MTOCs disappear. Cells lacking centrioles can still divide, probably because chromosomes can recruit MTs to some extent on their own. However, the resulting spindles are poorly organized. In contrast to animal cells, the cells of higher plants and most fungi lack centrioles; their absence indicates that centrioles are not essential for the formation of MTOCs.

γ-Tubulin, a form of tubulin found only in centrosomes, has been implicated in MT nucleation. In conjunction with a number of other proteins called *GRiPs* (*g*amma tubulin *ri*ng *p*roteins), γ-tubulin forms large, ring-shaped structures called **γ-tubulin ring complexes (γ-TuRCs),** which can be seen at the base of MTs emerging from the centrosome (**Figure 13-9**).

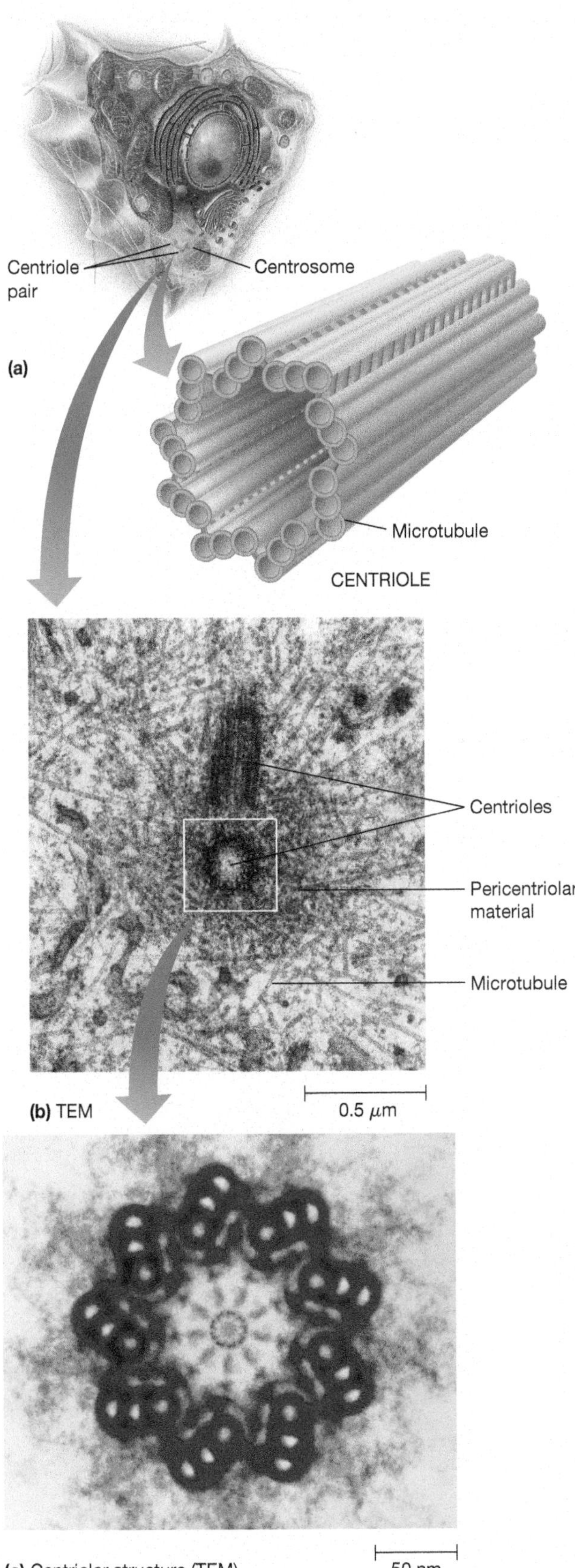

Figure 13-8 The Centrosome. (a) In animal cells, the centrosome contains two centrioles and associated pericentriolar material. The walls of centrioles are composed of nine sets of triplet microtubules. **(b)** A centrosome showing the centrioles and the pericentriolar material. Notice that microtubules originate from the pericentriolar material. **(c)** A closeup of the centriole. Centrioles form a characteristic "pinwheel" structure.

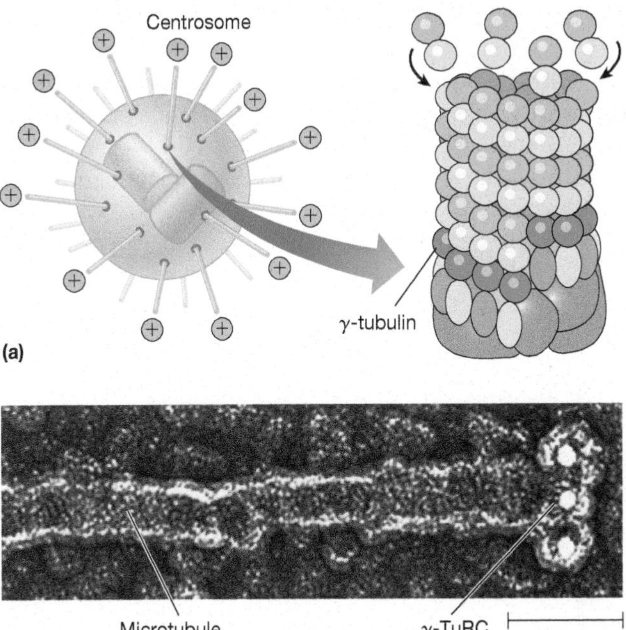

Figure 13-9 γ-Tubulin Ring Complexes (γ-TuRCs) Nucleate Microtubules. (a) γ-TuRCs are found at centrosomes, where the minus ends (−) of MTs are anchored. The enlargement shows a single γ-TuRC. The plus end of the MT is oriented away from the γ-TuRC. **(b)** An MT in vitro, visualized by metal shadowing. Here a GRiP component of the γ-TuRC (Xgrip109) was labeled with antibodies to which small particles of gold were attached. These particles appear as bright spheres (TEM).

The importance of γ-TuRCs has been demonstrated by depleting cells of γ-tubulin or other components of the γ-TuRC. In the absence of these proteins, centrosomes can no longer nucleate MTs. In addition to the centrosome, some types of cells have other MTOCs. For example, the basal body at the base of each cilium in ciliated cells also serves as an MTOC.

Other MTOC proteins, such as the protein *ninein*, are more important in anchoring, rather than promoting polymerization, at the minus ends of MTs. In addition, branched MTs have recently been discovered. New branches can form at 30-40° angles on the sides of existing MTs by recruiting new -TuRCs, anchored by the protein *augmin*.

MTOCs Organize and Polarize Microtubules Within Cells

The most important role MTOCs play is probably their ability to nucleate and anchor MTs. Because of this ability, in cells that rely on centrosomes as MTOCs, MTs extend outward toward the periphery of the cell. Furthermore, they grow out from an MTOC with a fixed polarity—their minus ends are usually anchored in the MTOC, so their plus ends point outward toward the cell membrane. This is why dynamic growth and shrinkage of MTs tend to occur at the periphery of cells.

Not all cells have a centrally located MTOC organized by a centrosome. Some cells have numerous noncentrosomal MTOCs. The MTOCs and the distribution and polarity of MTs within several types of nondividing cells are shown in **Figure 13-10**. One important noncentrosomal MTOC is the Golgi apparatus. MTs anchored at the Golgi associate with many, but not all, of the same proteins found at centrosomes. The Golgi is thought to be important for organizing MTs associated with vesicle transport (see Chapter 14).

MTOCs influence the number of microtubules in a cell. Each MTOC has a limited number of nucleation and anchorage sites at which MTs can form. However, the MT-nucleating activity of the MTOC can be modulated during certain processes such as mitosis, during which the organization of MTs changes dramatically (**Figure 13-11**). For example, centrosomes associated with spindle poles in mitotic cells have the highest MT-nucleating activity during prophase and metaphase, when the mitotic spindle is forming.

The picture we have developed thus far is that MT networks are highly organized. Equally important is that they are *dynamic*. Techniques such as electron microscopy, however, are limited in that they cannot be used to analyze the dynamic cytoskeleton. Fortunately, modern cell biologists have other tools to study the dynamic assembly and disassembly of arrays of MTs. These include the use of filament-perturbing chemical agents, high-resolution fluorescence microcopy, and genetically engineered cells expressing fluorescent (see Key Technique, pages 372–373).

Microtubule Stability Is Tightly Regulated in Cells by a Variety of Microtubule-Binding Proteins

You have seen that cellular microtubules exhibit dynamic instability—they can grow out from the centrosome and then disassemble. This process could account for randomly distributed and short-lived MTs, but not for organized and stable arrays of MTs within cells. Indeed, cells regulate the assembly and structure of MTs with great precision. To do so, they use a variety of MT-binding proteins. Some MT-binding proteins act as ATP-driven motors that transport vesicles and organelles or generate sliding forces between MTs. (These proteins will be discussed in detail in Chapter 14.) Here we will focus on proteins that regulate MT organization and structure.

Microtubule-Stabilizing/Bundling Proteins. **Microtubule-associated proteins (MAPs)** account for 10–15% of the mass of MTs isolated from cells. MAPs bind at regular intervals along the walls of MTs, allowing interaction with other filaments and cellular structures. Several different types of

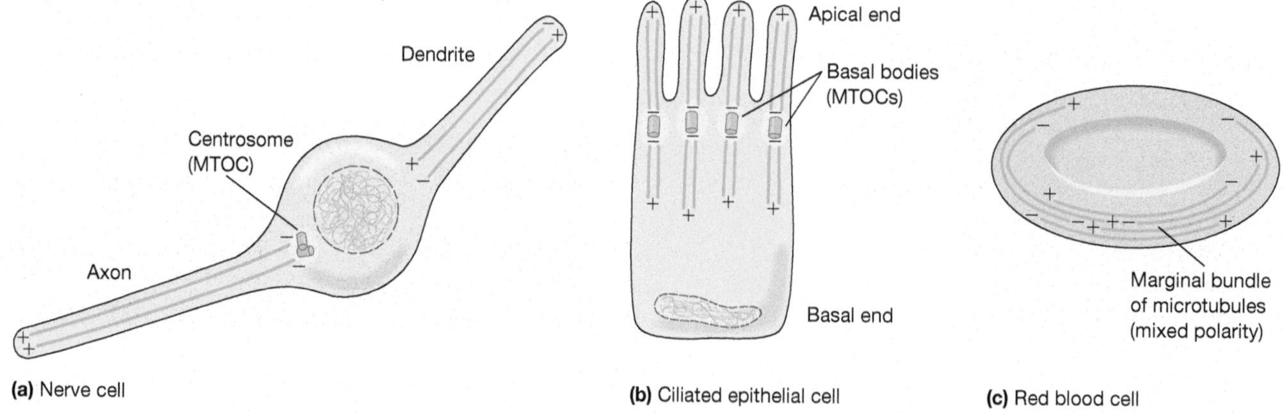

Figure 13-10 Microtubule Polarity in Nondividing Animal Cells. In the cell, the distribution of most microtubules is determined by microtubule-organizing centers (MTOCs). MT orientation in a cell (shown in orange) may vary with that cell's function. **(a)** Nerve cells contain two distinct sets of MTs, those of the axon and those of the dendrite. Axonal MTs are attached at their minus ends to the centrosome, with their plus ends at the tip of the axon. Dendritic MTs are not associated with the centrosome and are of mixed polarities. **(b)** Ciliated epithelial cells have many MTOCs called basal bodies, one at the base of each cilium. Ciliary MTs originate with their minus ends in the basal bodies and elongate with their plus ends toward the tips of the cilia. **(c)** Mature human red blood cells have no nucleus or MTOC. MTs of mixed polarities persist as a circular band at the periphery of the cell. This band helps maintain the cell's disklike shape.

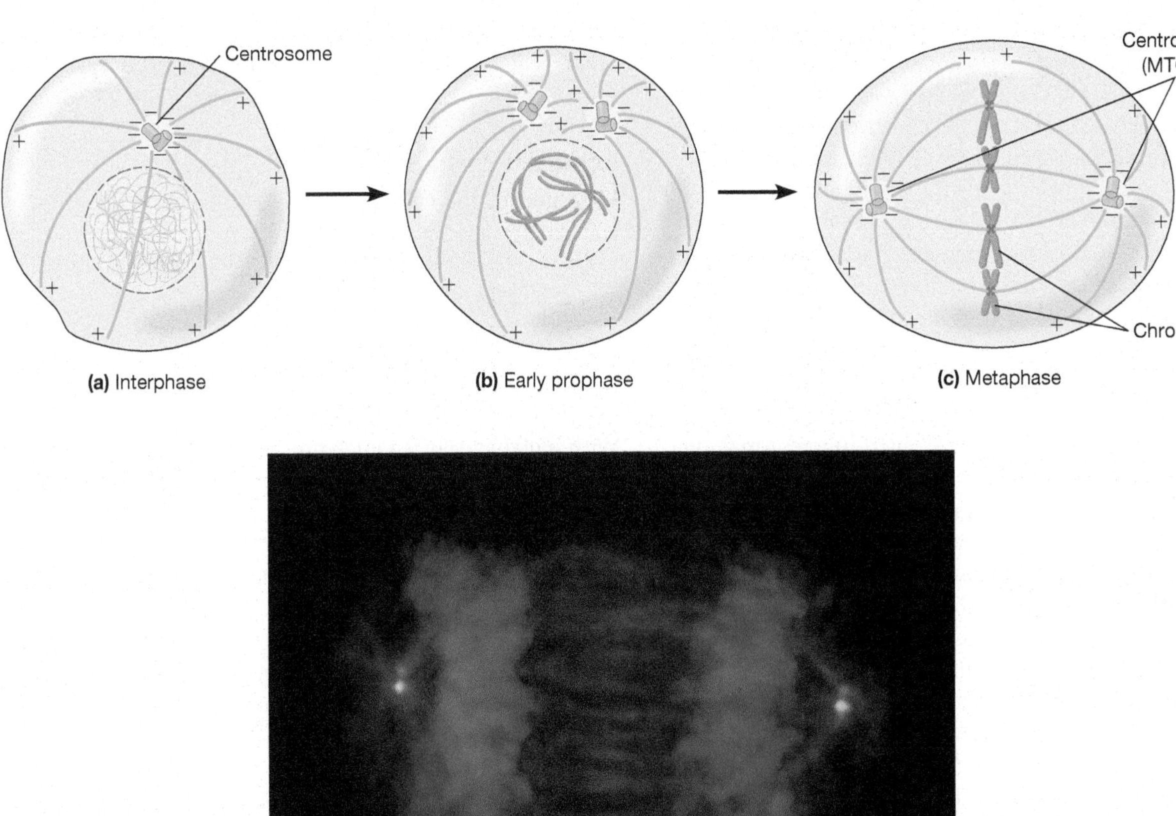

Figure 13-11 Changes in Microtubule Orientation During Mitosis. MTs in a dividing cell are oriented with their minus ends anchored in the centrosome and their plus ends pointing away from the centrosome. **(a)** Cell division is preceded by the division of the centrosome. **(b)** The daughter centrosomes then separate, each forming one pole of the mitotic spindle. **(c)** At metaphase, the two centrosomes are at opposite sides of the cell. Each spindle pole anchors half of the spindle MTs. **(d)** A cell stained for spindle components. γ-tubulin (yellow) marks centrosomes, α-tubulin (red) marks spindle MTs, and DNA is stained blue (fluorescence micrograph).

MAPs are shown in **Figure 13-12** on page 370. Most MAPs have been shown to promote MT stability by preventing the severing of MTs and by keeping heterodimers locked into the MT. They can also affect the density of bundles of MTs.

MAP function has been studied extensively in brain cells, as they are the most abundant source of these proteins. Neurons have axons, which carry electrical signals away from the cell body of the neuron, and dendrites, which receive signals from neighboring cells and carry them to the cell body. Within these projecting structures groups of microtubules are crosslinked into bundles. The bundles are characteristically denser in axons than they are in dendrites. A MAP called *Tau* causes MTs to form tight bundles in axons. Another MAP, *MAP2,* is present in dendrites and causes the formation of looser bundles of MTs. One portion of MAPs such as Tau and MAP2 binds along the length of an MT; another portion protrudes away from the microtubule, where it can interact with other proteins (Figure 13-12a). The length of these "arms" controls the spacing of MTs within bundles. MAP2 has a longer arm than does Tau, and so MAP2 forms MT bundles that are less densely packed than those formed with Tau.

The importance of MAPs can be demonstrated by forcing non-neuronal cells to make large amounts of Tau protein. Such cells extend single long processes that look remarkably similar to the axons of neurons (Figure 13-12b). Tau is also important in human disease. Dense tangles of neurites, known as *neurofibrillary tangles,* are a hallmark of several diseases that result in dementia, such as Alzheimer disease, Pick disease, and several types of palsy. In the case of Alzheimer disease, these tangles contain large amounts of hyperphosphorylated Tau protein, which forms paired helical filaments (see Human Connections, pages 48–49). Mutations that result in defective Tau protein lead to hereditary predisposition to form neurofibrillary tangles. Such diseases are therefore sometimes called *tauopathies.*

+-TIP Proteins. MTs are generally too unstable to remain intact for long periods of time and will depolymerize unless they are stabilized in some way. One way to stabilize MTs is to "capture" and protect their growing plus ends. To do so, **+-TIP proteins** (+- end *t*ubulin *i*nteracting *p*roteins) associate with MT plus ends. One important example of MT capture involves kinetochores during mitosis. Other +-TIPs associate with the *cell cortex*, an actin-based network underneath the plasma membrane, and can stabilize MTs that extend there. Some +-TIPs, including *end-binding protein 1 (EB1)*, directly bind to and stabilize plus ends, decreasing the likelihood that they will undergo catastrophic subunit loss (Figure 13-12c).

EB1 associates with GTP-tubulin at MT plus ends, yielding a characteristic "fireworks" pattern (Figure 13-12d). EB1 protein that is tagged with the green fluorescent protein (GFP) is an incredibly useful tool for following the plus ends of MTs dynamically (see Key Technique, pages 372–373).

Microtubule-Destabilizing/Severing Proteins. Whereas some MT-binding proteins stabilize microtubules, making them less likely to depolymerize, others *promote* depolymerization of MTs. For example, the protein *stathmin/Op18* binds to tubulin heterodimers, preventing them from polymerizing. Other proteins act at the ends of MTs once they have polymerized, promoting the peeling of protofilaments and eventual loss of subunits from their ends. Several proteins of the *kinesin* family, called *catastrophins* (Figure 13-12e), act in this way. By tightly regulating where catastrophins act, a cell can precisely control where and when MTs form and depolymerize. A prime example of such regulation is the mitotic spindle. Indeed, a catastrophin known as MCAK (mitotic centromere-associated kinesin) plays just such a role. Still other proteins sever MTs; one example of such proteins is *katanin*.

CONCEPT CHECK 13.2

The amount of dynamic instability exhibited by microtubules varies depending on the kind of cell and whether it is dividing. Under what circumstances might you expect dynamic instability to be greatest, and why?

13.3 Microfilaments

With a diameter of about 7 nm, **microfilaments (MFs)** are the smallest of the cytoskeletal filaments (see Table 13-1). Microfilaments are best known for their role in the contractile fibrils of muscle cells, where they interact with thicker filaments of myosin to cause the contractions characteristic of muscle (see Chapter 14). MFs are not confined to muscle cells, however. They occur in almost all eukaryotic cells and are involved in numerous other phenomena, including a variety of locomotory and structural functions.

Examples of cell movements in which microfilaments play a role include *cell migration* via lamellipodia and filopodia, *amoeboid movement*, and *cytoplasmic streaming*, a regular

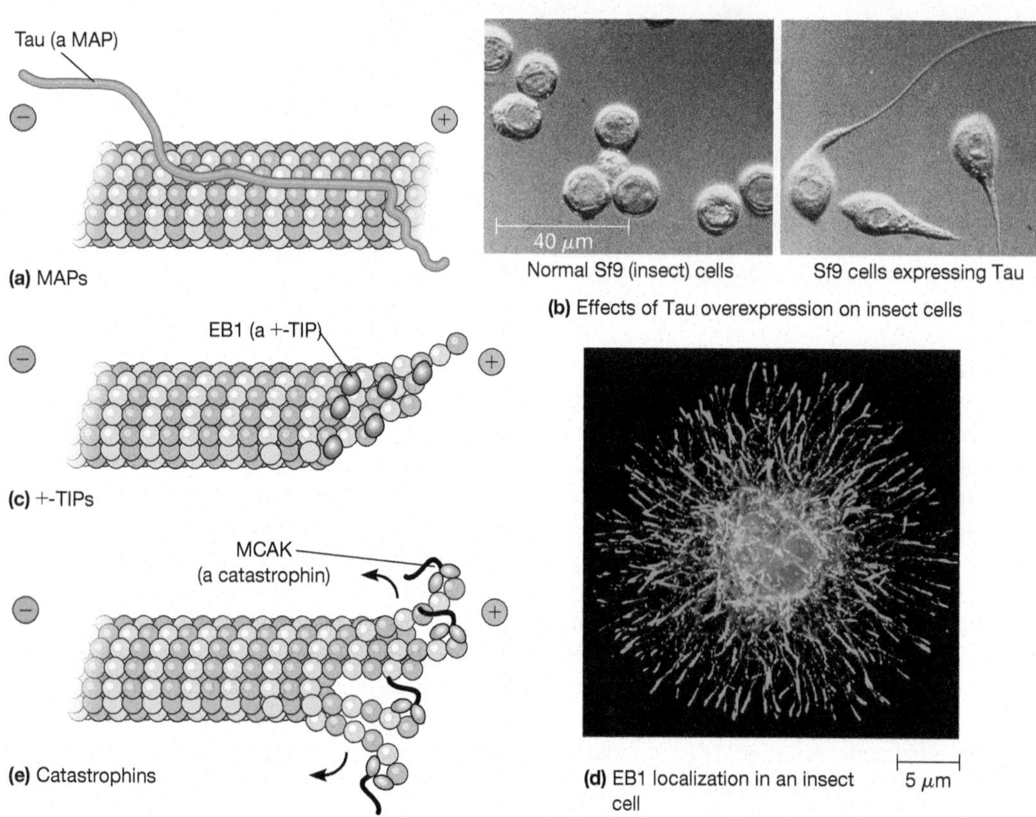

Figure 13-12 Microtubule-Binding Proteins Regulate Microtubule Function in Vivo. (a) Tau is a MAP that binds along the length of an MT; another portion of Tau extends away from the MT, regulating MT spacing in bundles. **(b)** The effects of Tau overexpression in a line of cultured non-neuronal cells called Sf9. **(c)** +-TIP proteins, such as EB1, bind at or near the plus ends of MTs, stabilizing them. **(d)** An S2 cell stained for EB1 (green), MTs (red), and DNA (blue). **(e)** Catastrophins, such as MCAK (mitotic centromere-associated kinesin), are kinesin family proteins that destabilize MTs.

pattern of flow in the cytosol of some plant and animal cells. (We will discuss all of these phenomena in detail in Chapter 14.) MFs also produce the cleavage furrows that divide the cytosol of animal cells during cytokinesis, and MFs are found at sites of attachment of cells to one another and to the extracellular matrix.

In addition to mediating a variety of cell movements, MFs are important in developing and maintaining cell shape. Most animal cells, for example, have a dense network of microfilaments within the cell cortex just beneath the plasma membrane. Cortical actin confers structural rigidity to the cell surface and facilitates shape changes and cell movement. Parallel bundles of MFs also make up the structural core of *microvilli*, the fingerlike extensions found on the surface of many animal cells (see Figure 4-5).

Actin Is the Protein Building Block of Microfilaments

Actin is an extremely abundant protein in virtually all eukaryotic cells, including those of plants, algae, and fungi. Actin is synthesized as a single polypeptide consisting of 375 amino acids, with a molecular weight of about 42 kDa. Once synthesized, it folds into a roughly U-shaped globular molecule, with a central cavity that binds ATP or ADP. Individual actin molecules are referred to as **G-actin** (globular actin). Under the right conditions, G-actin molecules polymerize to form microfilaments; in this form, actin is referred to as **F-actin** (filamentous actin; **Figure 13-13**). Actin in the G or F form is also bound by a wide variety of other proteins, collectively known as *actin-binding proteins*.

Different Types of Actin Are Found in Cells

Of the three types of cytoskeletal proteins, actin is the most highly conserved in amino acid sequence. Yet, despite a high degree of sequence similarity, actins do differ among different organisms and among tissues of the same organism. Based on sequence similarity, actins can be broadly divided into two major groups: the *muscle-specific actins (α-actins)* and the *nonmuscle actins (β- and γ-actins)*. β- and γ-actins localize to different regions of the cell and appear to have different interactions with actin-binding proteins. For example, in epithelial cells, one end of the cell, the apical end, contains microvilli, whereas the opposite side of the cell, known as the basal end, is attached to the extracellular matrix. β-actin is predominantly found at the apical end of epithelial cells, whereas γ -actin is concentrated at the basal end and sides of the cell.

G-Actin Monomers Polymerize into F-Actin Microfilaments

The kinetics of actin polymerization can be studied in solution using fluorescent G-actin. The fluorescence of the labeled F-actin can be measured to yield data similar to those for tubulin. Like tubulin dimers, G-actin monomers can polymerize reversibly into filaments with a lag phase corresponding to filament nucleation, which begins with formation of dimers or trimers, called *nucleation seeds*, followed by a more rapid polymer elongation phase. The F-actin filaments that form are composed of two linear strands of polymerized G-actin wound around each other in a helix, with roughly 13.5 actin monomers per half-turn of the helix (Figure 13-13a, b).

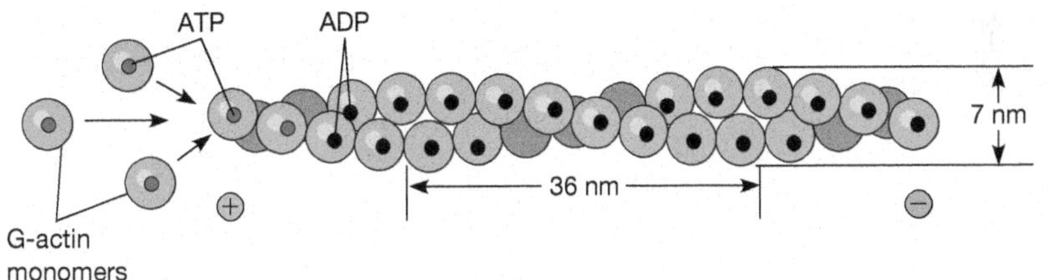

(a) Microfilament (MF) assembly

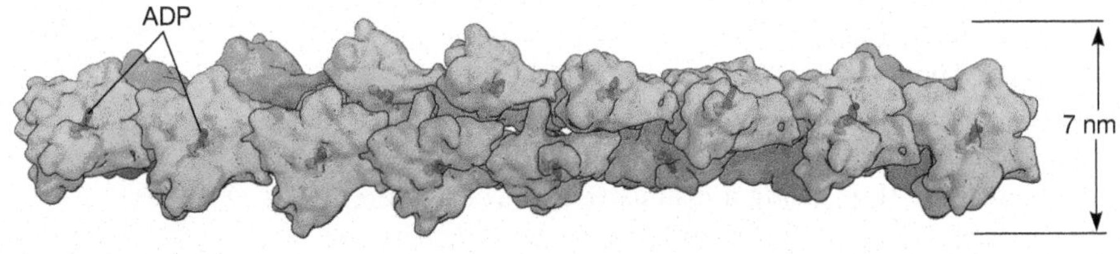

(b) Molecular model

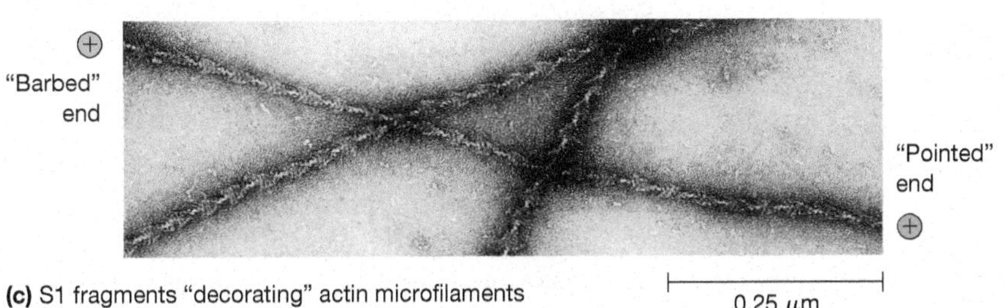

(c) S1 fragments "decorating" actin microfilaments

Figure 13-13 A Model for Microfilament Assembly in Vitro. **(a)** Monomers of G-actin polymerize into long helical filaments of F-actin with a diameter of about 7 nm. A half-turn of the helix occurs every 36–37 nm, requiring about 13.5 monomers (total number of monomers for both strands). Addition of each G-actin monomer is usually accompanied or followed by hydrolysis of bound ATP molecules, but this is not required to drive the polymerization reaction. **(b)** A theoretical model of F-actin, based on electron microscopy and X-ray crystal data. Two strands of G-actin monomers are shown. **(c)** When MFs are incubated with myosin S1 and examined with an electron microscope, the S1 fragments "decorate" the microfilaments like arrowheads. The S1 arrowheads point toward the minus ends of MFs.

Key Technique | Studying the Dynamic Cytoskeleton

PROBLEM: Cytoskeletal systems are dynamically involved in many cellular functions (summarized in Table 13-1). But to determine exactly how a particular cytoskeletal system is required for a particular process, researchers need to be able to follow where and when cytoskeletal polymers assemble in living cells.

SOLUTION: Chemical agents that disrupt specific cytoskeletal elements can quickly help identify whether or not a cytoskeletal system is required for a particular process.

Key Tools: (1) Chemicals that disrupt the cytoskeleton, (2) fluorescently labeled cytoskeletal monomers and/or cytoskeletal binding proteins, and (3) a fluorescence microscope and sensitive camera.

Details: Different chemical agents are known to affect the assembly or disassembly of specific cytoskeletal elements at the biochemical level (see Table 13-3). Some bind to monomers, taking them out of commission, or block addition of new monomers to a polymer. These agents are useful for studying whether new polymer assembly is necessary for a particular cellular event. Conversely, others bind to assembled polymers, preventing them from being disassembled. Such agents are useful for determining whether a preexisting cytoskeletal element must be dismantled for a cellular process to proceed.

Each kind of agent can be used in a "wash-out" experiment. When the chemical is applied, a particular cytoskeletal system is blocked from forming (or, conversely, is blocked from disassembling). When the chemical is washed out of the medium surrounding the cells, normal cytoskeletal dynamics will resume over time.

Consider how researchers learned that the centrosome is a microtubule-organizing center (MTOC) in the mid-1970s. When a microtubule-depolymerizing drug called colcemid was applied to cells, their cytosolic microtubules disappeared, as verified by immunofluorescence microscopy using MT-specific antibodies (**Figure 13A-1a**). After the colcemid was washed out, MTs began to reform, but always at the centrosome (Figure 13A-1b). Such experiments provided powerful evidence that the plus ends of MTs are oriented away from the centrosome in cells.

Antibody staining, however, has a drawback: it requires killing and chemically preserving cells before they can be analyzed. An even more powerful approach is to observe formation and disassembly of cytoskeletal elements in living cells. In this method, fluorescently labeled building blocks of a particular cytoskeletal element are introduced into cells, either by injecting labeled monomers directly or by genetically engineering cells to express monomers to which a fluorescent protein such as GFP has been attached. As **Figure 13A-2** shows, this approach was crucial for establishing that dynamic instability of MTs occurs in living cells.

Labeling all polymers in a system can be effective for studying cytoskeletal dynamics, but sometimes it is easier to make sense of the fluorescent signal if a cytoskeletal element is only partially labeled.

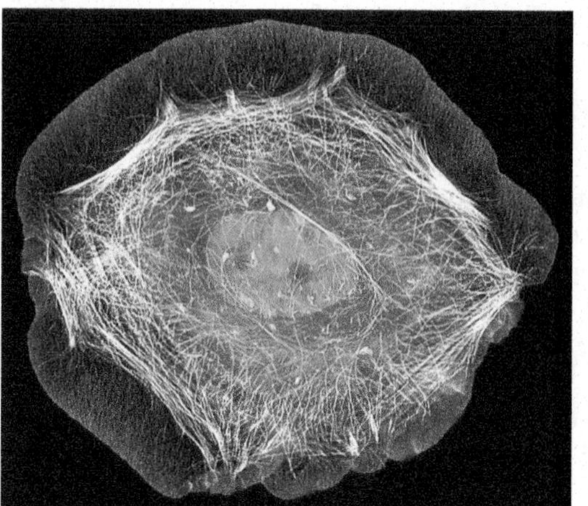

Human keratinocyte (green, DNA; yellow, microtubules; white, actin)

Within a microfilament, all the actin monomers are oriented in the same direction so that an MF, like a microtubule, has an inherent polarity, with one end differing chemically and structurally from the other end. This polarity can be demonstrated by incubating MFs with *myosin subfragment 1 (S1)*, a proteolytic fragment of myosin. S1 fragments bind to, or "decorate," actin MFs, with all the S1 molecules pointing in the same direction, thus giving a distinctive arrowhead pattern to MFs (Figure 13-13c). Based on this arrowhead pattern, the terms *"pointed" end* and *"barbed" end* are often used to identify the minus and plus ends of an MF, respectively. The polarity of the MF is important because it allows independent regulation of actin assembly or disassembly at each end of the filament.

The polarity of microfilaments is reflected in how G-actin monomers are added to them. If G-actin is polymerized onto short fragments of S1-decorated F-actin in vitro, polymerization can be seen to proceed much faster at the barbed end, indicating that the barbed end of the filament is also the plus end. Thus, even when conditions are favorable for adding monomers to both ends of the filament, the plus end will grow faster than the minus end. In vivo, addition of G-actin only occurs at the plus end. This is because another protein in the cytosol called profilin binds G-actin and prevents it from adding at minus ends (see below).

As G-actin monomers assemble onto a microfilament, the ATP bound to them is slowly hydrolyzed to ADP, much the same as the GTP bound to tubulin is hydrolyzed to GDP. Thus, the ends of a growing MF tend to have ATP-F-actin, whereas the bulk of the MF is composed of ADP-F-actin. However, ATP hydrolysis is not a strict requirement for MF elongation because MFs can also assemble from ADP-G-actin or from nonhydrolyzable analogues of ATP-G-actin.

Specific Drugs Affect Polymerization of Microfilaments

As we saw with microtubules, several drugs have been used to perturb the assembly of actin into microfilaments (see Table 13-3). Processes that are disrupted in cells treated with these drugs are likely to depend in some way on microfilaments. Several drugs result in depolymerization of microfilaments.

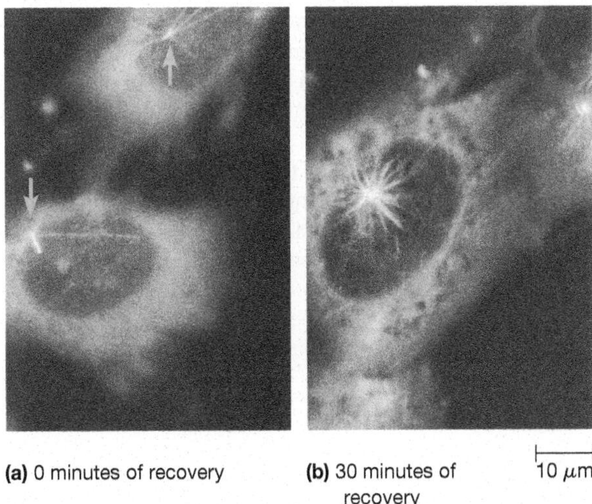

Figure 13A-1 A Wash-out Experiment Shows That Centrosomes Polarize Microtubules in Cells. (a) Cells treated with colcemid, then immunostained to visualize MTs. Only MTs associated with the centrosome (which are resistant to colcemid) persist (arrows). **(b)** Another cell fixed 30 minutes after washing out the colcemid. New MTs are growing away from the centrosome.

One such technique, *speckle microscopy*, involves introducing a limited amount of fluorescent monomer, which results in speckles of fluorescence along the length of a cytoskeletal element. This technique has been very useful for analyzing the flow of cytoskeletal elements. If, for example, treadmilling of MTs or MFs occurs, speckles of labeled monomer incorporated near the plus end of a cytoskeletal filament will be displaced rearward as subunits are lost at the minus end and additional monomers are added at the plus end.

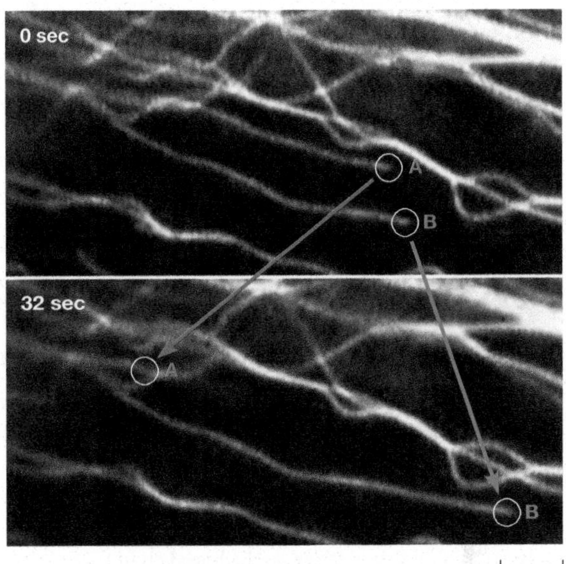

Figure 13A-2 Dynamic Instability of Microtubules in Vivo. MTs in a rat embryo cell expressing GFP-β-tubulin visualized by live-cell fluorescence microscopy exhibit dynamic instability in vivo. Here, two individual MTs (A, B) have been labeled. Over a 32-second time span, MT B grows whereas A shrinks.

QUESTION: How could you use paclitaxel, which makes MTs less likely than normal to depolymerize, to study whether the completion of mitosis requires disassembly of MTs?

The **cytochalasins**, such as *cytochalasin D*, are fungal metabolites that effectively cap the plus ends of existing polymerized MFs, preventing the addition of new monomers as well as loss of old ones. As subunits are gradually lost from the minus ends of MFs in cytochalasin-treated cells, the MFs eventually depolymerize. In contrast, **latrunculin A**, a marine toxin isolated from the Red Sea sponge *Latrunculia magnifica*, acts by sequestering actin monomers, thus preventing their addition to the plus ends of growing MFs. In either case, the net result is the loss of MFs within the treated cells. Conversely, the drug *phalloidin*, a cyclic peptide from the death cap fungus (*Amanita phalloides*), stabilizes MFs, preventing their depolymerization. Fluorescently labeled phalloidin is also useful for visualizing F-actin via fluorescence microscopy.

Cells Can Dynamically Assemble Actin into a Variety of Structures

As with microtubules, cells can dynamically regulate where and how G-actin is assembled into microfilaments; the variety of microfilament-based networks in a crawling cell is shown in **Figure 13-14** on page 374. Relatively thick and stable bundles of actin, called *stress fibers*, stretch from the tail, or trailing edge, of the cell to the front (Figure 13-14a). In vivo, structures that correspond to stress fibers help cells exert strong forces on their surroundings. Highly motile cells, in contrast, typically contain fewer, thinner, and more dynamic stress fibers. In such cells, an actin network immediately underneath the plasma membrane, known as the **cell cortex**, is crosslinked into a very loosely organized meshwork, or *gel*, of microfilaments (Figure 13-14b).

Cells that crawl have specialized structures called *lamellipodia* and *filopodia* at their leading edge that allow them to move along a surface (you will learn about these specialized structures in more detail in Chapter 14). The form of the protrusion appears to depend on the nature of the cell's movement and on the organization of the actin filaments within the cell.

Lamellipodia are characterized by a branched network of actin (Figure 13-14c). In filopodia, microfilaments form highly oriented, polarized bundles, with their barbed (plus) ends oriented toward the tip of the protrusion (Figure 13-14d). Electron

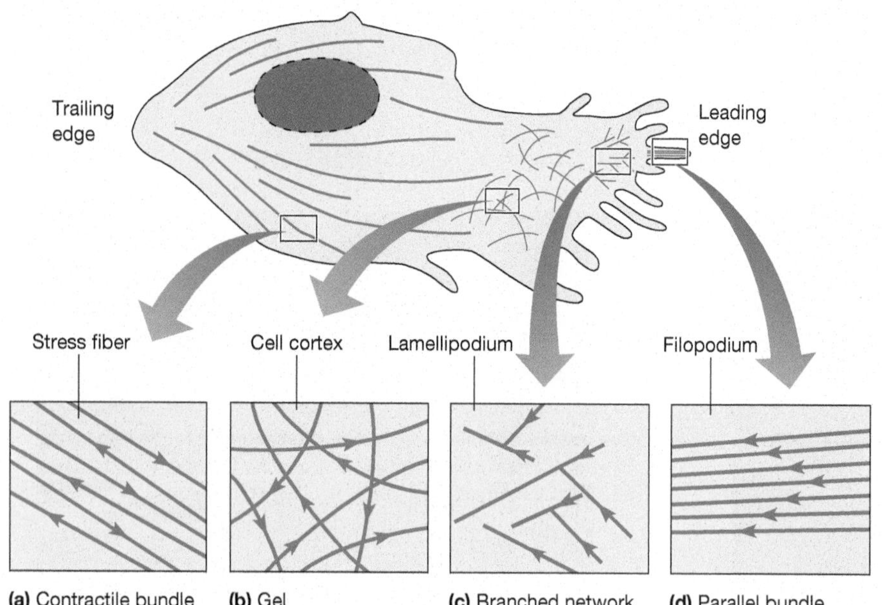

Figure 13-14 The Architecture of Actin in Crawling Cells. Actin is found in a variety of structures in crawling cells. The polarity of actin is shown schematically in the insets using arrowheads to show barbed (plus) and pointed (minus) ends of actin filaments in blue. **(a)** Running from the trailing edge of the cell to the leading edge are contractile bundles of actin (stress fibers). **(b)** At the periphery of the cell is the cortex, containing a three-dimensional meshwork of actin filaments crosslinked into a gel. **(c)** The leading edge of lamellipodia contains a branched actin network. **(d)** Filopodia contain parallel bundles of actin filaments with their plus ends directed outward.

micrographs show how these two types of actin networks actually look within a cell (**Figure 13-15**). Understanding how cells regulate such a wide variety of actin-based structures requires understanding both how cells can regulate the polymerization of MFs and how MFs, once polymerized, assemble into networks.

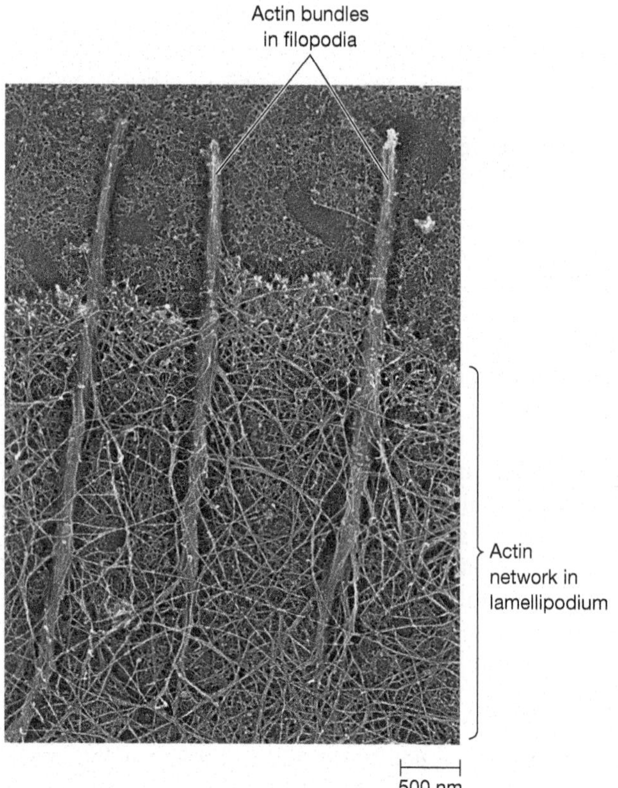

Figure 13-15 Actin Bundles in Filopodia. The periphery of a mouse melanoma cell shows two prominent actin networks. Actin filaments in the filopodia merge with a network of actin filaments lying just beneath the plasma membrane of the lamellipodium (deep-etch TEM).

Actin-Binding Proteins Regulate the Polymerization, Length, and Organization of Microfilaments

As we saw with microtubules, cells can precisely control where actin assembles and the structure of the resulting actin networks. To do so, cells use a variety of **actin-binding proteins** (**Figure 13-16**). Control of the process of MF polymerization occurs at several steps, including the nucleation of new MFs, the elongation and severing of preexisting MFs, and the association of MFs into networks.

Proteins That Regulate Monomers and Their Polymerization. In the absence of other factors, the nucleation and growth of microfilaments depend on the concentration of ATP-bound G-actin. If the concentration of ATP-G-actin is high, monomer nucleation and microfilament assembly will proceed until the G-actin level becomes limiting. In the cell, however, a large amount of free G-actin is not available for polymerization because it is bound by the actin-sequestering protein *thymosin* $\beta 4$.

A second protein called *profilin* appears to compete with thymosin $\beta 4$ for binding to G-actin monomers. Profilin does not sequester G-actin, but instead promotes the addition of G-actin to plus ends of actin filaments. As a result, when the profilin concentration is high, polymerization is favored—but only if free filament ends are available. Another protein, known as *ADF/cofilin*, is known to bind to ADP-G-actin and F-actin. ADF/cofilin is thought to increase the rate of turnover of ADP-actin at the minus ends of MFs. The ADP on these G-actin monomers can then be exchanged for a new ATP, and the ATP-G-actin can be recycled for addition to the growing plus ends of MFs. ADF/cofilin also severs filaments, creating new plus ends as it does so.

Once G-actin monomers form small aggregates (nucleation "seeds") and are ready to polymerize into filaments, other proteins can aid the polymerization process. Different types of actin networks depend on different proteins to aid polymerization, such as the Arp2/3 complex or formins (see below).

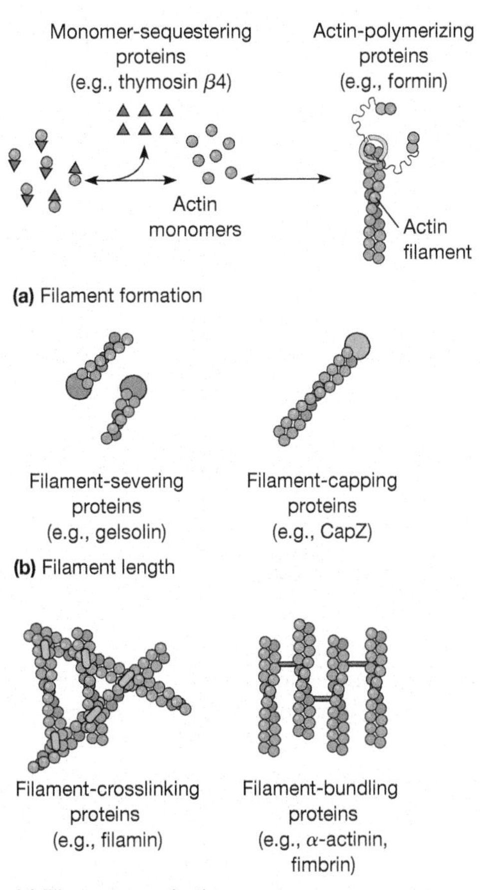

Figure 13-16 Actin-Binding Proteins Regulate the Organization of Actin. Actin-binding proteins are responsible for converting actin filaments from one form to another. **(a)** Some proteins affect monomer availability or monomer addition; **(b)** others affect severing or growth of existing filaments; **(c)** still others affect filament organization.

Proteins That Cap Actin Filaments. Whether microfilament ends are available for further growth depends on whether the filament end is *capped.* Capping occurs when a **capping protein** binds the end of a filament and prevents further addition or loss of subunits, thereby stabilizing it. One such protein, which functions as a cap for the plus ends of microfilaments, is appropriately named *CapZ.* When CapZ is bound to a filament, further addition of subunits at the plus end is prevented; when CapZ is removed, addition of subunits can resume. Another class of proteins called *tropomodulins* binds to and prevents loss of subunits from the minus ends of F-actin.

Proteins That Crosslink Actin Filaments. In many cases, actin forms loose meshworks of crisscrossing, crosslinked MFs. One of the crosslinking proteins that is important for such networks is *filamin,* a long molecule consisting of two identical polypeptides joined head to head, with an actin-binding site at each tail. Molecules of filamin "splice" crisscrossing MFs together where they intersect. In this way, actin MFs can be linked to form large three-dimensional networks.

Proteins That Sever Actin Filaments. Other proteins play the opposite role, breaking up MF networks and causing the cortical actin gel to soften and liquefy. They do this by severing and/or capping MFs. In some cases, such proteins can serve both functions. One of these severing and capping proteins is *gelsolin,* which functions by breaking actin MFs and capping their newly exposed plus ends, thereby preventing further polymerization.

Proteins That Bundle Actin Filaments. In contrast to the loose organization of actin in the cell cortex, other actin-containing structures can be highly ordered. In such cases, actin may be bundled into tightly organized arrays, and a number of actin-binding proteins mediate such bundling. One such protein is *α-actinin,* a protein that is prominent within structures known as *focal contacts* and *focal adhesions,* which are required for cells to make adhesive connections to the extracellular matrix as they migrate (see Chapter 15). Another bundling protein, *fascin,* is found in filopodia. Fascin keeps the actin within the core of a filopodium tightly bundled, contributing to the spike-like appearance of such protrusions.

Perhaps the best-studied example of ordered actin arrays is the actin bundles found in microvilli, which are shown in **Figure 13-17**. **Microvilli** (singular: **microvillus**) are especially prominent features of intestinal mucosal cells (Figure 13-17a). A single mucosal cell in your small intestine, for example, has several thousand microvilli, each about 1–2

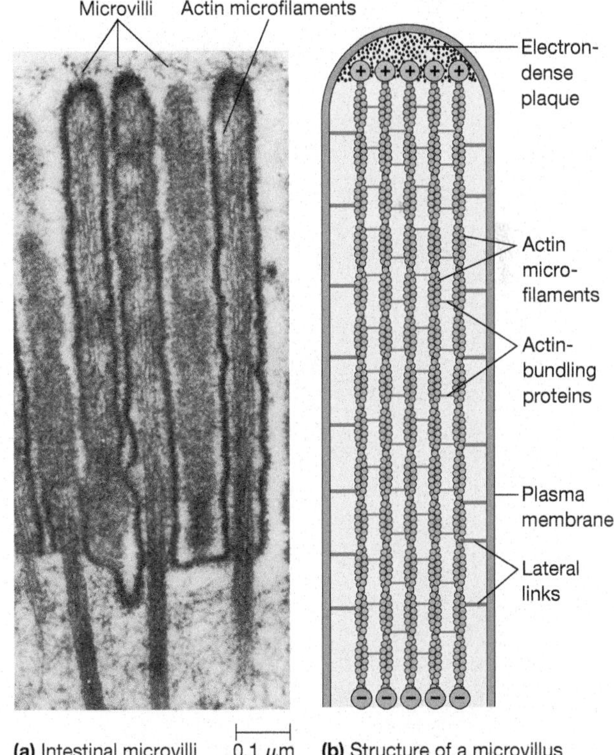

Figure 13-17 Microvillus Structure. (a) Microvilli from intestinal mucosal cells (TEM). **(b)** A schematic diagram of a single microvillus, showing a core consisting of several dozen MFs oriented with their plus ends facing outward toward the tip and their minus ends facing the cell. The plus ends are embedded in an amorphous, electron-dense plaque. MFs are held together tightly by actin-bundling (crosslinking) proteins and are also connected to the inner surface of the plasma membrane by lateral links.

μm long and about 0.1 μm in diameter, which increase the surface area of the cell about 20-fold. This large surface area is essential to intestinal function because the uptake of digested food depends on an extensive absorptive surface.

As illustrated in Figure 13-17b, the core of the intestinal microvillus consists of a tight bundle of microfilaments. The plus ends point toward the tip, where they are attached to the membrane through an amorphous, electron-dense plaque. Adjacent MFs in the bundle are crosslinked tightly together at regular intervals by the actin-bundling proteins *fimbrin* and *villin*. MFs on the outside of the bundle are also connected to the plasma membrane by lateral links consisting of the proteins *myosin I* and *calmodulin*.

At the base of the microvillus, the MF bundle extends into a network of filaments called the **terminal web** (**Figure 13-18**). The filaments of the terminal web are composed mainly of the proteins *myosin II* and *spectrin*, which connect the MF bundles to each other, to proteins within the plasma membrane, and perhaps also to the network of intermediate filaments beneath the terminal web. By anchoring MF bundles securely so that they project straight out from the cell surface, the terminal web apparently gives rigidity to microvilli.

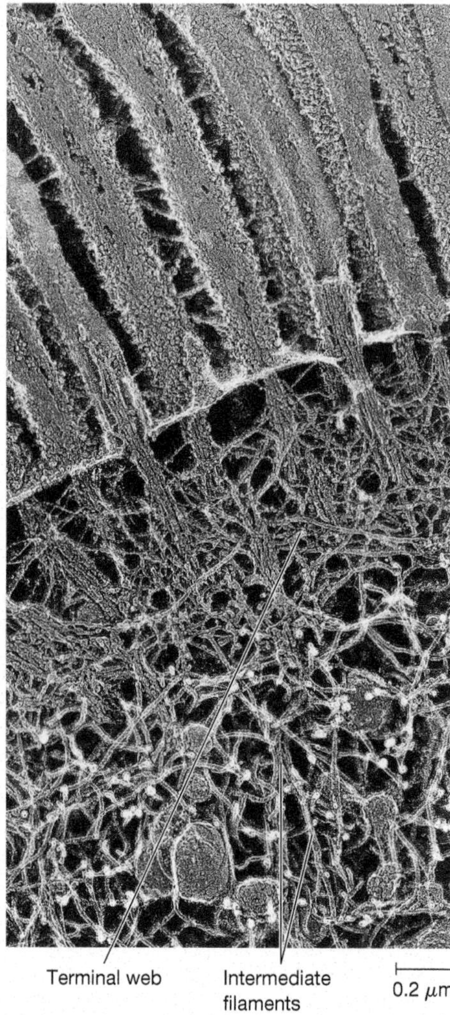

Figure 13-18 The Terminal Web of an Intestinal Epithelial Cell. Bundles of microfilaments that form the cores of microvilli extend into the terminal web beneath the plasma membrane of an intestinal epithelial cell (deep-etch TEM).

Proteins That Link Actin to Membranes

For MFs to exert force on the plasma membrane during events such as cell movement and cytokinesis, they must be connected to the plasma membrane. This connection is indirect and requires one or more linker proteins that anchor MFs to transmembrane proteins embedded within the plasma membrane. One highly conserved group of proteins that appears to function widely in linking MFs to membranes is the band 4.1 superfamily of proteins. Members of the family include *band 4.1* and the ERM proteins *ezrin*, *radixin*, and *moesin*. When members of this family are mutated, a wide variety of cellular processes are affected, including cytokinesis, secretion, and the formation of microvilli.

Another example of how actin can be linked to membranes mentioned in Chapter 7 involves the proteins *spectrin* and *ankyrin* (see Figure 7-28). As **Figure 13-19** shows, the plasma membrane of a red blood cell is supported by a polygonal network of spectrin filaments that gives the cell both strength and flexibility. The spectrin filaments are crosslinked by very short chains of actin and connected to specific transmembrane proteins in the plasma membrane by molecules of the proteins ankyrin and band 4.1. Based on mutations in the genes encoding spectrins and ankyrins in *Drosophila melanogaster* and in the nematode *Caenorhabditis elegans*, these proteins are important in a wide variety of cells for maintenance of cell shape.

Proteins That Promote Actin Branching and Filament Growth. In addition to loose networks and bundles, cells can assemble actin into branched networks that form a treelike, or *dendritic*, network, as shown in **Figure 13-20**. As shown in the

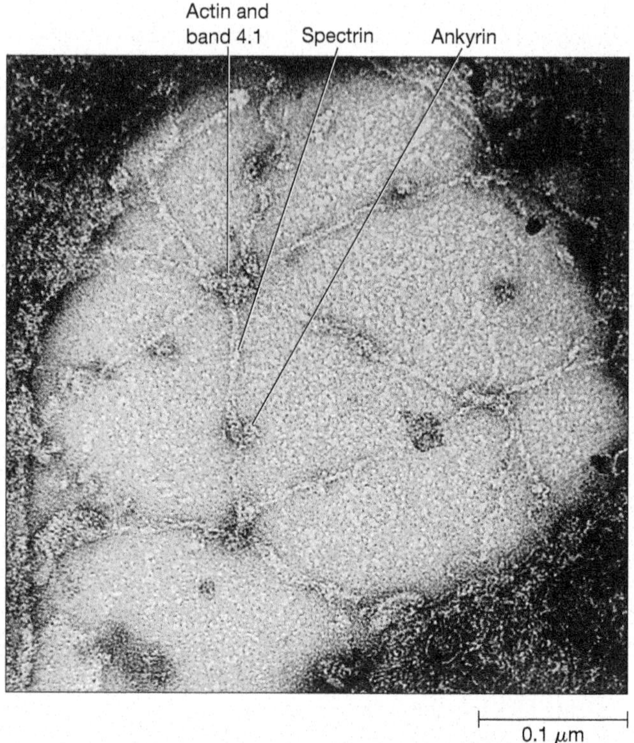

Figure 13-19 A Spectrin-Ankyrin-Actin Network Supports Red Blood Cell Membranes. An erythrocyte membrane, showing actual spectrin network components (TEM). For a schematic diagram, see Figure 7-28.

image of a frog keratocyte in Figure 13-20a, such dendritic networks are a prominent feature of lamellipodia. A complex of actin-related proteins, the **Arp2/3 complex**, helps branches form by nucleating new branches on the sides of existing filaments (Figure 13-20b). Proteins we have already discussed, such as profilin, cofilin, and capping proteins, regulate the length of filaments that polymerize from branch points. As new branches form and elongate from their plus ends near the plasma membrane, filaments are severed and dismantled at the base of the lamellipodium, allowing actin to be recycled. Based on fluorescence microscopy in living cells, actin in lamellipodia can undergo treadmilling similar to that discussed earlier in this chapter for microtubules.

Arp2/3 branching is activated by the WASP family of proteins, which includes the *Wiskott-Aldrich syndrome protein* (for which the family is named) and *WAVE/Scar.* Human patients who cannot produce functional WASP have defects in the ability of their blood platelets to undergo changes in shape and so have difficulties in forming blood clots. A very different kind of disease also involves the Arp2/3 complex: pathogenic bacteria can "hijack" the actin polymerization machinery of the cell. When they do so, they can spread by propelling themselves using actin "rockets" (see Human Connections, page 378).

Branched actin networks are only one type of actin-based structure that cells can produce. For some cellular events, long actin filaments are more useful. Actin polymerization in this case can be regulated independently of the Arp2/3 complex. For example, members of the *Ena/VASP* (*va*sodilator-*s*timulated *p*hosphoprotein) family of proteins promote filament growth at the tips of filopodia.

Another family of proteins known as **formins** is required to assemble structures such as actin bundles and the contractile rings observed during cell division. Formins appear to be able to act "processively," moving along the end of a growing filament as they stimulate filament growth (Figure 13-20c). Formins can form dimers, binding to the barbed (plus) ends of actin filaments. Some formins have extensions that can bind profilin and are therefore thought to act as "staging areas" for the addition of actin monomers to growing filaments.

Phospholipids and Rho Family GTPases Regulate Where and When Actin-Based Structures Assemble

We have seen that actin-binding proteins regulate the types of actin-based structures that cells assemble. Other molecules, in turn, regulate the activity of these proteins. Both plasma membrane lipids and regulatory proteins affect the formation, stability, and breakdown of MFs.

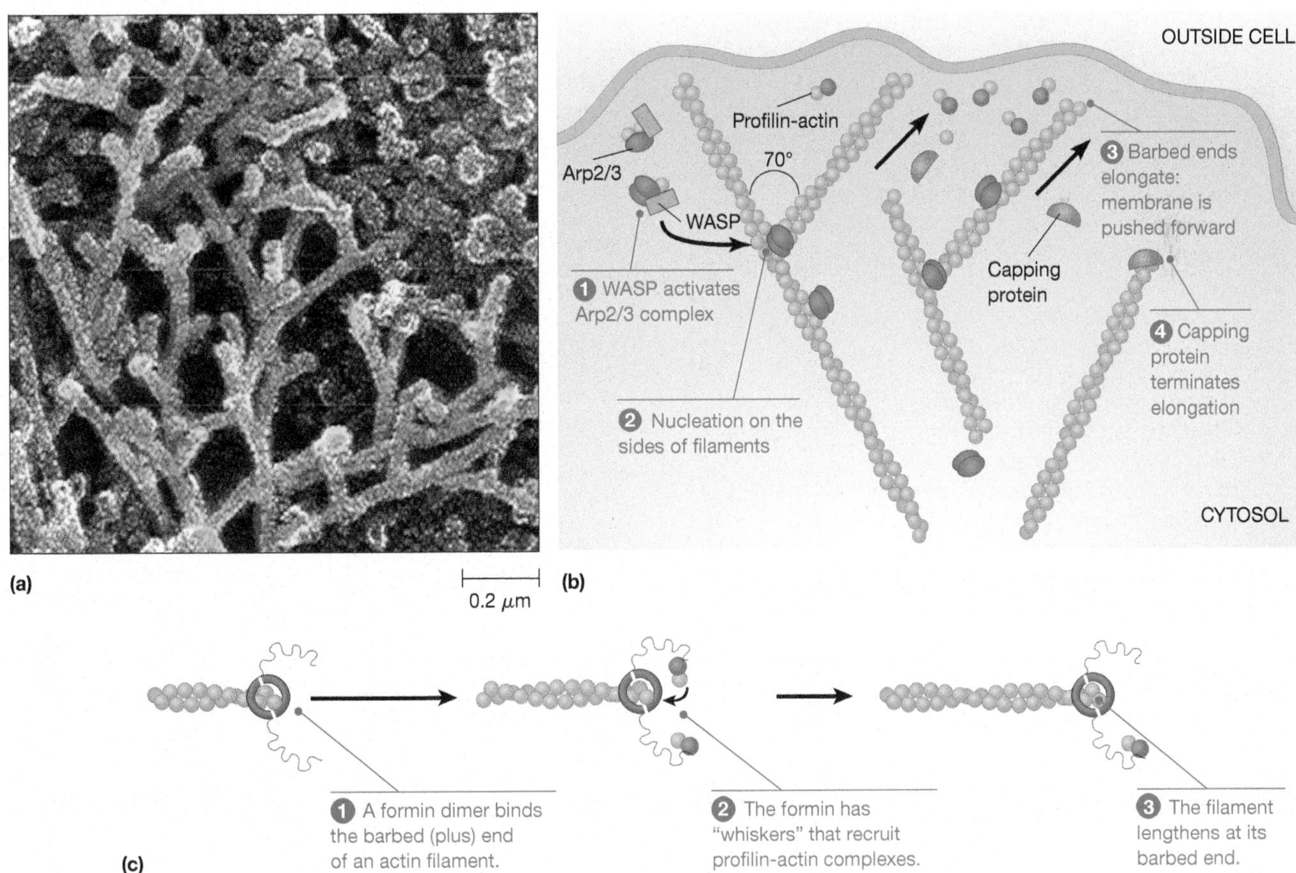
OUTSIDE CELL
Profilin-actin
70°
Arp2/3
WASP
❶ WASP activates Arp2/3 complex
❷ Nucleation on the sides of filaments
Capping protein
❸ Barbed ends elongate: membrane is pushed forward
❹ Capping protein terminates elongation
CYTOSOL
(a)
0.2 μm
(b)
❶ A formin dimer binds the barbed (plus) end of an actin filament.
❷ The formin has "whiskers" that recruit profilin-actin complexes.
❸ The filament lengthens at its barbed end.
(c)

Figure 13-20 Formation of Actin Networks by Directed Actin Polymerization. (a) Branched actin filaments in a frog keratocyte. Individual branched filaments are colored to make them easier to distinguish (deep-etch TEM). **(b)** Arp2/3-dependent branching is stimulated by WASP family proteins; capping protein helps regulate the length of new branches. **(c)** Formin dimers bind profilin-actin and move with the plus end of an MF, serving as a "staging area" for actin polymerization.

HUMAN *Connections*

When Actin Kills

Pregnancy can bring incredible feelings of joy, but also worry. Because the fetus is dependent on the mother for its nutrition and support during development, pregnant mothers have to be careful about what they do and eat. Some of the many recommended dietary restrictions are no deli sandwiches, sliced cold cuts, or unpasteurized cheeses. Why? These foods are sometimes colonized with the bacterium *Listeria monocytogenes*. *Listeria* is an *enteropathogen*, a microorganism that invades the body through the intestine. The resulting disease, known as *listeriosis*, is actually quite rare, with ~2000 cases per year. One reason for this rarity is that *Listeria* does not affect people with a normally functioning immune system. Pregnancy depresses the mother's immune system, however, so she is more susceptible to colonization by this pernicious bacterium. Once a pregnant woman is infected with *Listeria*, the bacteria can also harm the fetus, resulting in miscarriage or an infant born with skin or organ lesions, blood infection, or even meningitis.

Pregnancy is a sensitive time for mother and fetus.

Cell biology has provided critical understanding of the various mechanisms killer pathogens use to invade host cells. Foreign invaders often use the very same proteins that healthy cells require to gain entry into cells and spread throughout the body. *Listeria* is no different. *Listeria* expresses a protein called *internalin A* on its surface that can bind to E-cadherin, a cell adhesion protein on the surface of cells in the gut. (You will learn more about cadherins in Chapter 15.) Binding triggers phagocytosis of the bacterium by the host cell (**Figure 13B-1**). *Listeria*, however, also produces another cell surface protein, *listeriolysin O*, that allows it to break down the membrane of phagosomes and escape back into the cytosol, thereby evading destruction.

Once back in the cytosol, the bacteria continue to divide. *Listeria* spreads infection to other nearby cells by manipulating the host cell once again. *Listeria* normally swims using an appendage known as a *flagellum* in the external environment. Inside the host, however, it does not produce flagella, but instead moves through the host cell by coaxing it to do the work.

How can *Listeria* do this? It uses a form of "jet propulsion" to move within infected cells that depends on "fuel" supplied by the host. *Listeria* polymerizes an actin "rocket" at its rear through Arp2/3-mediated polymerization of actin filaments from the host cell. Polymerization is promoted by a protein on the surface of *Listeria* known as *ActA*. The microfilament growth promoted by ActA is strikingly similar to that found at the leading edge of migrating cells. As the new F-actin filaments form, they push the bacterium forward. This actin-based jet propulsion is very fast: *Listeria* can rocket at a rate of 22 μm/min.

As bacteria rocket about, they can push the surface of an infected cell outward, causing a neighboring cell to become indented. The neighboring cell then incorporates the bacterium. The result is a double-layered membrane with the bacterium inside. Because the bacterium is surrounded by a membrane that contains normal host cell proteins, the newly infected cell does not detect it as an intruder. The bacterium eventually sheds its double-layered membrane, and the infection process continues. Because the rocketing happens within the safe confines of infected cells and their neighbors, *Listeria* evades the infected patient's immune system. This is one reason *Listeria* is so insidious once it gains a foothold within the patient's tissues.

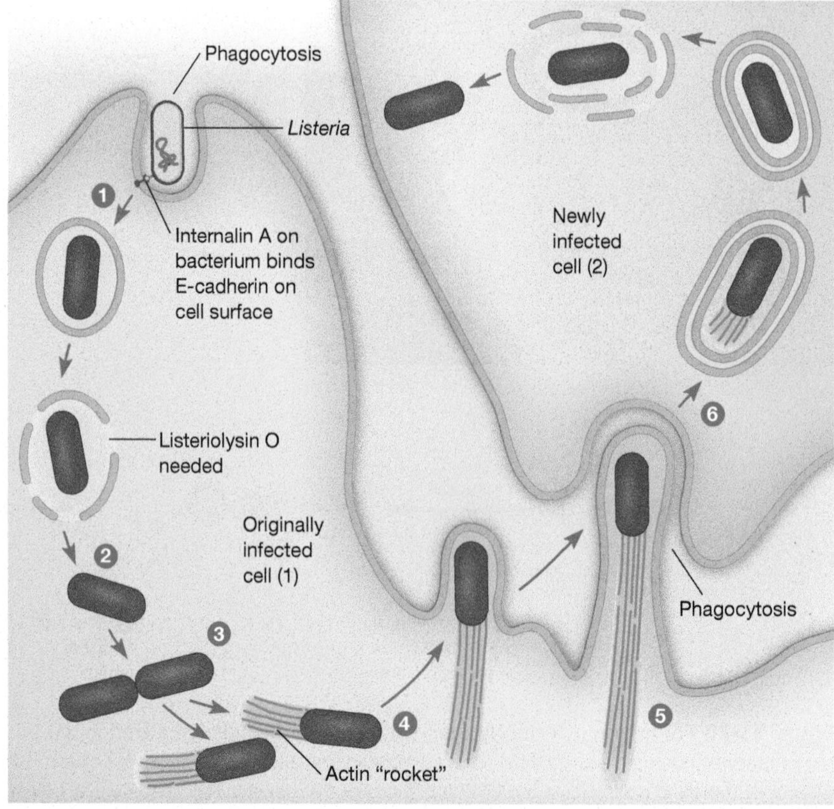

Figure 13B-1 Initial and Subsequent Infection of Host Cells by *Listeria*. ❶ *L. monocytogenes* enters the cell by using internalin A to bind to E-cadherin on the surface of the intestinal cell and enters the cell through the phagosome. ❷ Production of lysteriolysin O bursts the phagosome and releases the bacterium into the cell where ❸ it divides and reproduces. ❹ Movement within the cell is achieved through "rocketing" on actin filaments. ❺ Infection spreads when the bacterium is propelled into a neighboring cell, and ❻ the infection cycle begins again.

Inositol Phospholipids. Inositol phospholipids (also called *phosphoinositides*) are one type of membrane phospholipid that regulates actin assembly. One species called *phosphatidylinositol-4,5-bisphosphate* (PIP_2) can bind to profilin, CapZ, and proteins such as ezrin and recruit them to the plasma membrane. PIP_2 can also modulate the ability of these proteins to interact with actin. For example, CapZ binds tightly to PIP_2, resulting in its removal from the end of a microfilament, thereby permitting the filament to be disassembled and making its monomers available for assembly into new filaments.

Rho Family GTPases. One striking case of regulation of the actin cytoskeleton is the dramatic change seen in the cytoskeleton of cells exposed to certain extracellular signaling molecules called *growth factors.* (Growth factors and their range of biological roles are introduced in Chapter 23.) How do such extracellular signals result in such dramatic reorganization of the actin cytoskeleton? Many of these signals exert their effects through their action on phospholipids, but they also act through a family of intracellular proteins known as **Rho GTPases.** Rho GTPases are monomeric *G proteins.* Their activity is localized to the plasma membrane through a lipid modification that keeps them attached to the inner leaflet of the plasma membrane. G proteins act like molecular switches, whose "on" or "off" state depends on whether the G protein is bound to GTP (guanosine triphosphate) or GDP (guanosine diphosphate). They are active when they are bound to GTP. When they hydrolyze their bound GTP to GDP + P_i, they become inactive. Three key members of this family are **Rho**, **Rac**, and **Cdc42**. Originally identified in yeast, these proteins are important regulators of the actin cytoskeleton in all eukaryotes.

Rho GTPases perform a vast array of functions within cells, ranging from formation of protrusions to assembly and disassembly of the cytokinetic furrow to regulating endo- and exocytosis. Each Rho GTPase has profound and different effects on the actin cytoskeleton (**Figure 13-21**). For example, activation of the Rho pathway can result in the formation of stress fibers (Figure 13-21b). Rac activation often results in extension of lamellipodia (Figure 13-21c). Finally, activation of Cdc42 results in the formation of filopodia (Figure 13-21d).

Rho family GTPases are stimulated by *guanine-nucleotide exchange factors (GEFs),* which foster exchange of a bound GDP for GTP (Figure 13-21e). Corresponding *GTPase-activating proteins (GAPs)* stimulate Rho GTPases to hydrolyze bound GTP, thereby inactivating them. In addition, proteins known as *guanine-nucleotide dissociation inhibitors (GDIs)* can sequester inactive Rho GTPases in the cytosol. They do so by preventing the addition of lipid groups to Rho GTPases that target them to the inner leaflet of the plasma membrane, where they normally act. All of these events can be modulated by cell signals, allowing fine-tuning of where and when actin-based structures are assembled.

CONCEPT CHECK 13.3

A friend of yours who is interested in how cells crawl notices that when she treats cultured fibroblasts with phalloidin, they stop moving. She asks you what conclusions you would draw from this experiment. How would you answer?

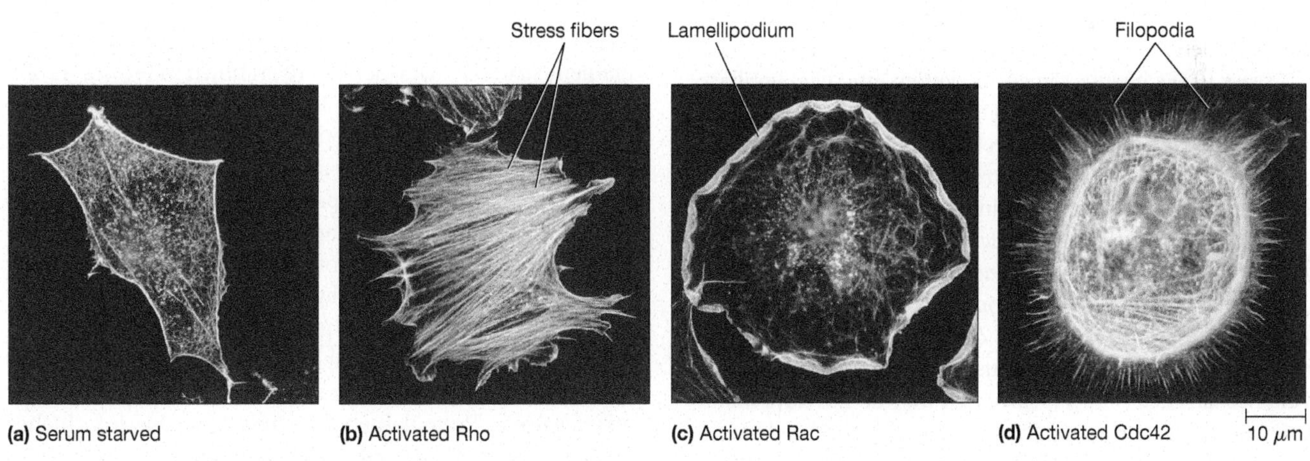

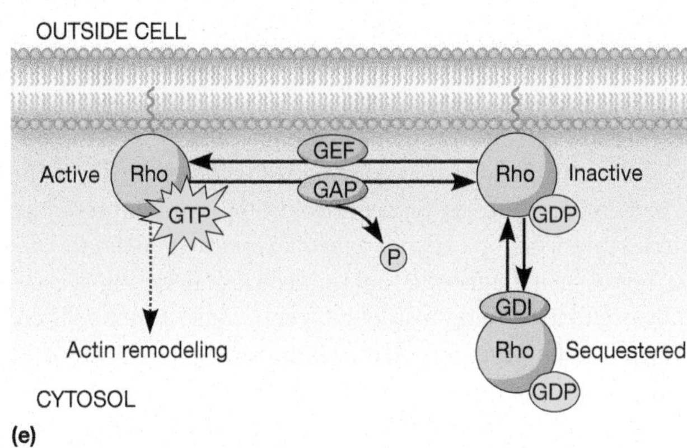

Figure 13-21 Regulation of Protrusions by Rho Family Proteins. **(a)** A cultured fibroblast in the absence of growth factors ("serum starved") has few actin bundles and shows little protrusive activity. **(b)** Under conditions that activate Rho protein (for example, addition of lysophosphatidic acid, LPA), stress fibers form. **(c)** When Rac is activated (in this case by injecting mutated Rac that is always active), lamellipodia form. **(d)** When Cdc42 is activated (for example, by injecting a guanine-nucleotide exchange factor that activates Cdc42), filopodia form. **(e)** Guanine-nucleotide exchange factors (GEFs) stimulate exchange of a bound GDP for GTP, activating Rho family proteins and allowing them to stimulate actin remodeling. GTPase-activating proteins (GAPs) stimulate Rho GTPases to hydrolyze their bound GTP, thereby inactivating them. Guanine-nucleotide dissociation inhibitors (GDIs) can sequester inactive Rho family G proteins.

13.4 Intermediate Filaments

Intermediate filaments (IFs) have a diameter of about 8–12 nm, which makes them intermediate in size between microtubules and microfilaments (see Table 13-1). Intermediate filaments are not found in the cytosol of plant cells, but they are abundant in many animal cells, where IFs occur singly or in bundles and appear to play a structural or tension-bearing role. One well-known and abundant intermediate filament protein is *keratin.* Keratin is an important component of structures that grow from skin in animals, including hair, claws and fingernails, horns and beaks, turtle shells, feathers, scales, and the outermost layer of the skin. **Figure 13-22** is an electron micrograph of purified keratin IFs.

Intermediate filaments are the most stable and the least soluble constituents of the cytoskeleton. Treatment of cells with detergents or with solutions of high or low ionic strength removes most of the microtubules, microfilaments, and other proteins of the cytosol but leaves networks of IFs that retain their original shape. Given their stability, IFs likely serve as a scaffold to support the entire cytoskeletal framework.

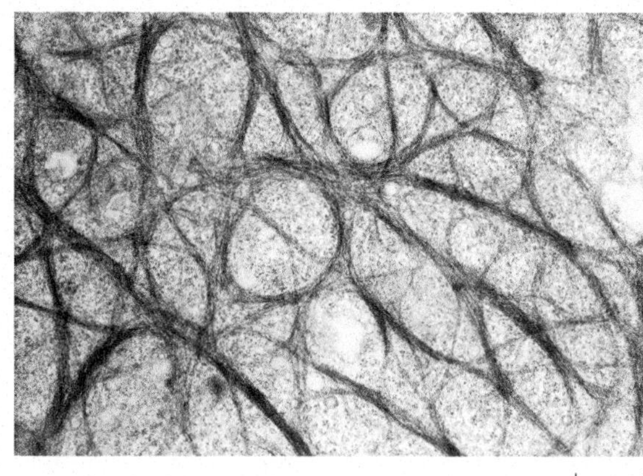

Figure 13-22 Intermediate Filaments. Negatively stained keratin filaments in vitro (TEM).

Intermediate Filament Proteins Are Tissue Specific

In contrast to microtubules and microfilaments, intermediate filaments differ markedly in amino acid composition from tissue to tissue. IFs and their proteins can be grouped into six classes (**Table 13-4**). Classes I and II comprise the *keratins,* proteins found in the epithelial cells covering the body surfaces and lining its cavities. (The IFs visible beneath the terminal web in the intestinal mucosal cell of Figure 13-18 consist of keratin.) Class I keratins are *acidic keratins,* whereas class II are *basic* or *neutral keratins;* each of these classes contains at least 15 different keratins.

Class III IFs include vimentin, desmin, and glial fibrillary acidic protein. *Vimentin* is present in connective tissue and other cells derived from nonepithelial cells. *Desmin* is found in muscle cells, and *glial fibrillary acidic protein (GFAP)* is characteristic of the glial cells that surround and insulate nerve cells.

Class IV IFs are the *neurofilament (NF) proteins* found in nerve cells. Class V IFs are *nuclear lamins* A, B, and C, which form a filamentous scaffold along the inner surface of the nuclear membrane of animal cells. Neurofilaments found in cells in the embryonic nervous system are made of *nestin,* which constitutes class VI. Although they do not have lamin genes, electron microscopy shows that plants and fungi have a nuclear lamina; it must consist of non-lamin proteins. Strong candidates for playing a role analogous to lamins in higher plants are *CRWN* (for CRoWded Nucleus) *proteins.* Mutations in CRWN proteins cause defects in plant cell nuclei similar to the effects of disrupting lamins in animal cells.

Different tissues can be distinguished on the basis of the IF protein present, as determined by immunofluorescence microscopy. This technique, called *intermediate filament typing,* serves as a diagnostic tool in medicine. IF typing is especially useful in the diagnosis of cancer because tumor cells are known to retain the IF proteins characteristic of the tissue of origin, regardless of where the tumor occurs in the body.

Intermediate Filaments Assemble from Fibrous Subunits

All IF proteins have some common features, although they differ significantly in size and chemical properties. The fundamental subunits of IFs are dimers. In contrast to actin and tubulin, IF dimers are fibrous, rather than globular, proteins. Each of the subunits of an IF dimer has a homologous central rodlike domain of 310–318 amino acids. This central domain consists of four α-helical segments interspersed with three short linker segments. Flanking the central domain are N- and C-terminal domains that differ greatly in size, sequence, and function among IF proteins, presumably accounting for their functional diversity.

In contrast to MTs and MFs, IFs do not appear to be polarized. Several models have been proposed for intermediate filament structure. One possibility is shown in **Figure 13-23**. The basic structural unit of IFs consists of a dimer of two polypeptides intertwined via their central domains. The two polypeptides are identical for all IFs except keratin filaments, which are heterodimers of type I and type II polypeptides. The two polypeptides of a dimer are aligned in parallel, with the N- and C-terminal regions protruding as globular domains at each end. Two dimers then align laterally in an offset fashion to form a tetramer. Tetramers interact with each other and appear to associate in an overlapping manner to build up a filamentous structure both laterally and longitudinally to form *protofilaments.* Protofilaments in turn assemble into higher order filaments. When fully assembled, an intermediate filament is thought to be eight protofilaments thick at any point, with protofilaments probably joined end to end in staggered overlaps.

Intermediate Filaments Confer Mechanical Strength on Tissues

Because they often occur in areas of the cell that are subject to mechanical stress, intermediate filaments are thought to play a tension-bearing role. For example, when keratin filaments are genetically modified in the keratinocytes (a type of epithelial cell in skin) of transgenic mice, the resulting skin cells

Table 13-4 Classes of Intermediate Filaments

Class	IF Protein	Molecular Mass (kDa)	Tissue	Function
I	Acidic keratins	40–56.5	Epithelial cells	Mechanical strength
II	Basic or neutral keratins	53–67	Epithelial cells	Mechanical strength
III	Vimentin	54	Fibroblasts; cells of mesenchymal origin; lens of eye	Maintenance of cell shape
III	Desmin	53–54	Muscle cells, especially smooth muscle	Structural support for contractile machinery
III	GFA protein	50	Glial cells and astrocytes	Maintenance of cell shape
IV	Neurofilament proteins		Central and peripheral nerves	Axon strength; determines axon size
	NF-L (major)	62		
	NF-M (minor)	102		
	NF-H (minor)	110		
V	Nuclear lamins		All cell types	Form a nuclear scaffold to give shape to nucleus
	Lamin A	70		
	Lamin B	67		
	Lamin C	60		
VI	Nestin	240	Neuronal stem cells	Unknown

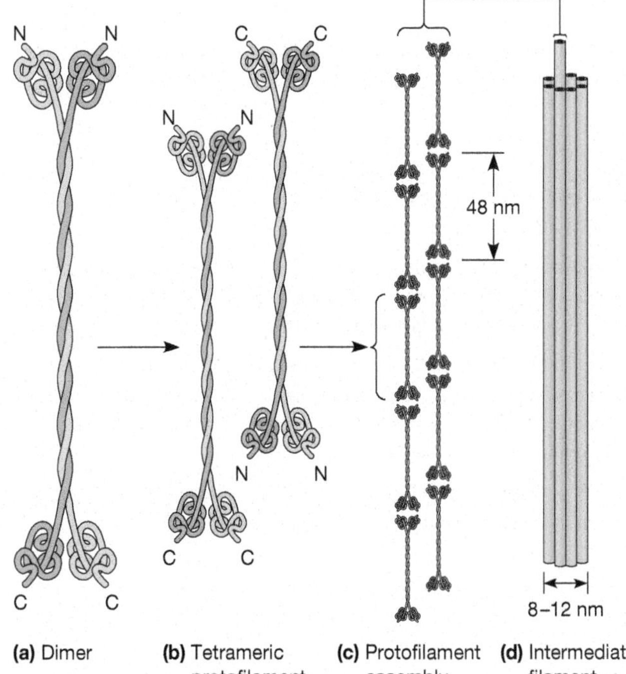

Figure 13-23 A Model for Intermediate Filament Assembly. **(a)** The starting point for assembly is a pair of intermediate filament (IF) polypeptides. The central domains of the two polypeptides twist around each other, with their N- and C-terminal ends aligned. **(b)** Two dimers align laterally to form a tetrameric protofilament. **(c)** Protofilaments assemble into larger filaments by end-to-end and side-to-side alignment. **(d)** A fully assembled IF is thought to be eight protofilaments thick at any point.

are fragile and rupture easily. In humans, naturally occurring mutations of keratins give rise to a blistering skin disease called *epidermolysis bullosa simplex*.

Although our discussion of IFs may give the impression that they are static structures, this is not the case. Imaging of fluorescent keratin IFs in crawling cells indicates that IFs can be dynamically remodeled. The nuclear lamina, on the inner surface of the nuclear envelope (see Chapter 16) is composed of three separate IF proteins (*nuclear lamins A, B,* and *C*). These lamins become phosphorylated and disassemble as a part of nuclear envelope breakdown at the onset of mitosis. After mitosis, lamin phosphatases remove the phosphate groups, allowing the nuclear envelope to reform.

IFs are less susceptible to chemical attack than are microtubules and microfilaments. One compound, however, *acrylamide*, has been shown to disrupt some classes of IFs. Acrylamide is used as a thickener in some familiar commercial products such as paints and paper, and it is also an important reagent for performing polyacrylamide gel electrophoresis. Acrylamide has received attention as a human health hazard because it has been discovered in processed foods. How acrylamide causes loss of intermediate networks is not well understood.

The Cytoskeleton Is a Mechanically Integrated Structure

In the preceding sections, we discussed the individual components of the cytoskeleton as separate entities. In fact, cellular architecture depends on the unique properties of the different cytoskeletal components working together. Microtubules are generally thought to resist bending when a cell is compressed, whereas microfilaments serve as contractile elements that generate tension. Intermediate filaments are elastic and can withstand tensile forces. The mechanical integration of IFs, MFs, and MTs is made possible by linker proteins known as *spectraplakins*. One versatile spectraplakin, called *plectin*, is found at sites where IFs are connected to MFs or MTs (**Figure 13-24**). Plectin and other spectraplakins contain binding sites for IFs, MFs, and MTs. By linking these polymers, spectraplakins help integrate them into a mechanically resilient cytoskeletal network.

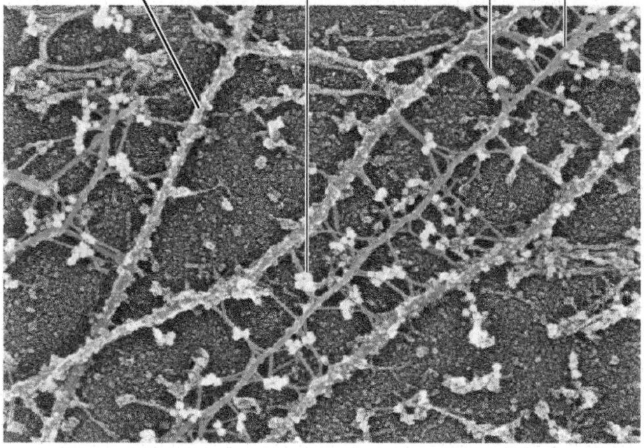

MT Gold particle labeling plectin Plectin IF

0.1 μm

Figure 13-24 Connections Between Intermediate Filaments and Other Components of the Cytoskeleton. A protein called plectin (red) links IFs (green) to MTs (blue). Plectins can also bind actin MFs (not shown). Gold particles (yellow) label plectin. Here, IFs serve as strong but elastic connectors between the different cytoskeletal filaments (deep-etch TEM).

CONCEPT CHECK 13.4

In which of the following cells would you expect to find extensive intermediate filament networks? For each cell type, explain your answer, based on the "lifestyle" of each cell type: human epidermis (skin), plant leaf epidermis, snake intestinal smooth muscle cell, sea urchin sperm cell.

Summary of Key Points

Mastering™ Biology For activities, animations, and review quizzes, go to the study area at www.masteringbiology.com.

13.1 Major Structural Elements of the Cytoskeleton

- Both prokaryotes and eukaryotes possess an interconnected network of proteins, called the cytoskeleton, that polymerize from smaller subunits. In eukaryotes, the cytoskeleton consists of an extensive three-dimensional network of microtubules (MTs), microfilaments (MFs), and intermediate filaments (IFs) that determines cell shape and allows a variety of cell movements.
- A variety of drugs can be used to perturb the assembly and disassembly of microtubules and microfilaments in eukaryotes. Such drugs are useful for determining which cellular processes require different cytoskeletal filaments.
- The cytoskeleton undergoes dynamic changes to meet the differing needs of various cell types.

13.2 Microtubules

- Microtubules (MTs) are hollow tubes with walls consisting of heterodimers of α- and β-tubulin polymerized linearly into protofilaments. Both α- and β-tubulin can bind GTP.
- MTs are polarized structures that elongate preferentially from one end, known as the plus end. MT growth occurs when the concentration of tubulin subunits rises above the critical concentration—that is, the concentration of $\alpha\beta$-heterodimers at which subunit addition is exactly balanced by subunit loss from an MT.
- MTs can undergo cycles of catastrophic shortening and elongation known as dynamic instability, which involves hydrolysis of GTP by β-tubulin near the plus end, followed by recovery of the GTP cap at the plus end.
- Within cells, MT dynamics and growth are organized by microtubule-organizing centers (MTOCs). The centrosome is a major MTOC in animal cells, which contains nucleation sites rich in γ-tubulin where MT growth is initiated and anchors the minus ends of MTs.
- Microtubule-associated proteins (MAPs) stabilize MTs along their length, +-TIP proteins stabilize and anchor their plus ends, and catastrophins hasten their catastrophic depolymerization.

13.3 Microfilaments

- Microfilaments (MFs), or F-actin, are double-stranded polymers of G-actin monomers, which bind ATP.
- Like microtubules, MFs are polarized structures; G-actin monomers preferentially add to the plus (barbed) ends of MFs; their minus (pointed) ends display far slower subunit addition and loss.
- Actin-binding proteins tightly regulate F-actin. These include monomer-binding proteins, proteins that stimulate actin polymerization, and proteins that cap, crosslink, sever, bundle, and anchor F-actin in various ways.
- MF assembly within cells is regulated by the activity of inositol phospholipids and by the monomeric G proteins Rho, Rac, and Cdc42.

13.4 Intermediate Filaments

- Intermediate filaments (IFs) are the most stable constituents of the cytoskeleton. They appear to play a structural or tension-bearing role within cells.
- IFs are tissue specific and can be used to identify cell type. Such typing is useful in the diagnosis of cancer.
- All IF proteins have a highly conserved central domain flanked by different terminal regions, presumably accounting for their functional diversity.
- IFs, MTs, and MFs are interconnected within cells to form cytoskeletal networks that provide mechanical strength and rigidity to cells.

Problem Set

Mastering™ Biology For activities, animations, and review quizzes, go to the study area at www.masteringbiology.com.

13-1 Filaments and Tubules. Indicate whether each of the following descriptions is true of microtubules (MT), microfilaments (MF), or intermediate filaments (IF). More than one response may be appropriate for some statements.

(a) Involved in muscle contraction

(b) Involved in the movement of cilia and flagella

(c) More important for chromosome movements than for cytokinesis

(d) More important for cytokinesis than for chromosome movements in animal cells

(e) Most likely to remain when cells are treated with solutions of nonionic detergents or solutions of high ionic strength

(f) Structurally similar proteins are found in bacterial cells

(g) Their subunits can bind and catalyze hydrolysis of phosphonucleotides

(h) Can be detected by immunofluorescence microscopy

(i) Play well-documented roles in cell movement

(j) The fundamental repeating subunit is a dimer

13-2 True or False. Identify each of the following statements as true (T) or false (F). Provide a brief justification for your answer.

(a) The minus end of microtubules and microfilaments is so named because subunits are lost and never added there.

(b) The energy required for tubulin and actin polymerization is provided by hydrolysis of a nucleoside triphosphate.

(c) Microtubules, microfilaments, and intermediate filaments all exist in a typical eukaryotic cell in dynamic equilibrium with a pool of subunit proteins.

(d) Latrunculin A treatment, which blocks actin polymerization, would block intracellular movements of *Listeria.*

(e) An algal cell contains neither tubulin nor actin.

(f) All of the protein subunits of intermediate filaments are encoded by genes in the same gene family.

(g) All microtubules within animal cells have their minus ends anchored at the centrosome.

(h) As long as the actin monomer concentration is above the overall critical concentration, actin monomers to which CapZ are attached will continue to elongate.

13-3 QUANTITATIVE Microtubule Mania. For each of the following studies involving microtubules, explain the results.

(a) You repeat the experiment shown in Figure 13-4 in your cell biology laboratory class, but you add a small amount of nocodazole to the preparation. Describe the graph you expect, comparing it to Figure 13-4.

(b) When an animal cell is treated with colchicine, its microtubules depolymerize and virtually disappear. If the colchicine is then washed away, the MTs appear again, beginning at the centrosome and elongating outward.

(c) Extracts from nondividing frog eggs contain structures that can induce the polymerization of tubulin into microtubules in vitro. When examined by immunostaining, the induced microtubules contain γ-tubulin.

13-4 Centrosomes and the Critical Concentration. You determine the overall critical concentration for a sample of purified tubulin. Then you add centrosomes, which nucleate microtubules with their minus ends bound to centrosomes and stabilized against disassembly. When you again determine the overall critical concentration, you find it is different. Would the new overall critical concentration be higher or lower? Explain your answer.

13-5 DATA ANALYSIS Actin' Up. Polymerization of actin in a test tube can be followed using pyrene-labeled G-actin and a device to measure the fluorescence of polymerized actin. The increase in fluorescence can then be plotted over time, similar to using light-scattering to measure tubulin polymerization. Examine **Figure 13-25**, and identify which curve corresponds to each of the following situations. In each case, state your reasoning.

(a) Pyrene-labeled actin is added alone in the presence of buffer.

(b) Pyrene-labeled actin is added along with purified Arp2/3 complex proteins and a purified protein fragment that corresponds to active N-WASP (a WASP family protein).

13-6 Spongy Actin. You are interested in the effects of cytochalasin D on microfilaments over time. Based on what you know about the molecular mechanism of action of cytochalasins, describe what happens to MFs in cells treated with the drug. In particular, explain why existing F-actin eventually deploymerizes.

13-7 Crawling to a Halt. Migrating fibroblasts can be treated with various chemical agents while they are migrating. Explain what you expect for each of the following conditions.

(a) The cells are injected with large amounts of purified gelsolin, and the live cells are observed periodically.

(b) The cells are treated with latrunculin A, the drug is washed out, and the live cells are observed periodically.

(c) The cells are treated with a chemical fixative and prepared for electron microscopy. Their actin filaments are decorated with myosin S1 subfragments, and you examine in which direction the arrowhead-shaped decorations are pointing in the filopodia of the fixed cells.

13-8 Stressed Out. It is now possible, via nanoengineering, to attach magnetic beads to the surface of large cells. Using the beads, you apply force to measure how mechanically stiff the cells are. What effect would you predict acrylamide treatment would have on the mechanical rigidity of a keratinocyte (skin cell)? Explain your answer.

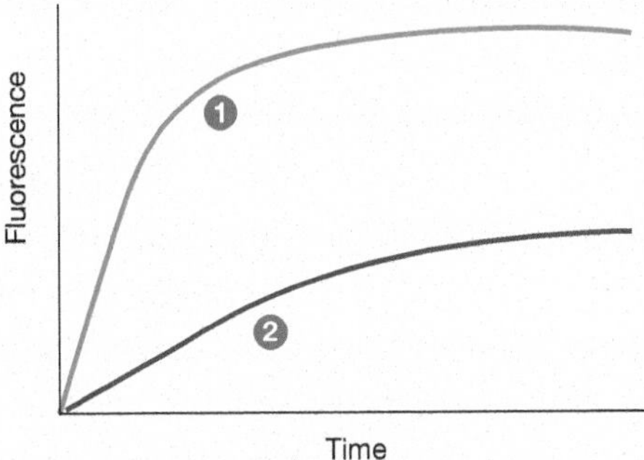

Figure 13-25 Microfilament Polymerization Kinetics Measured by Pyrene-Labeled Actin Incorporation Under Various Experimental Conditions. See Problem 13-5.

14 Cellular Movement: Motility and Contractility

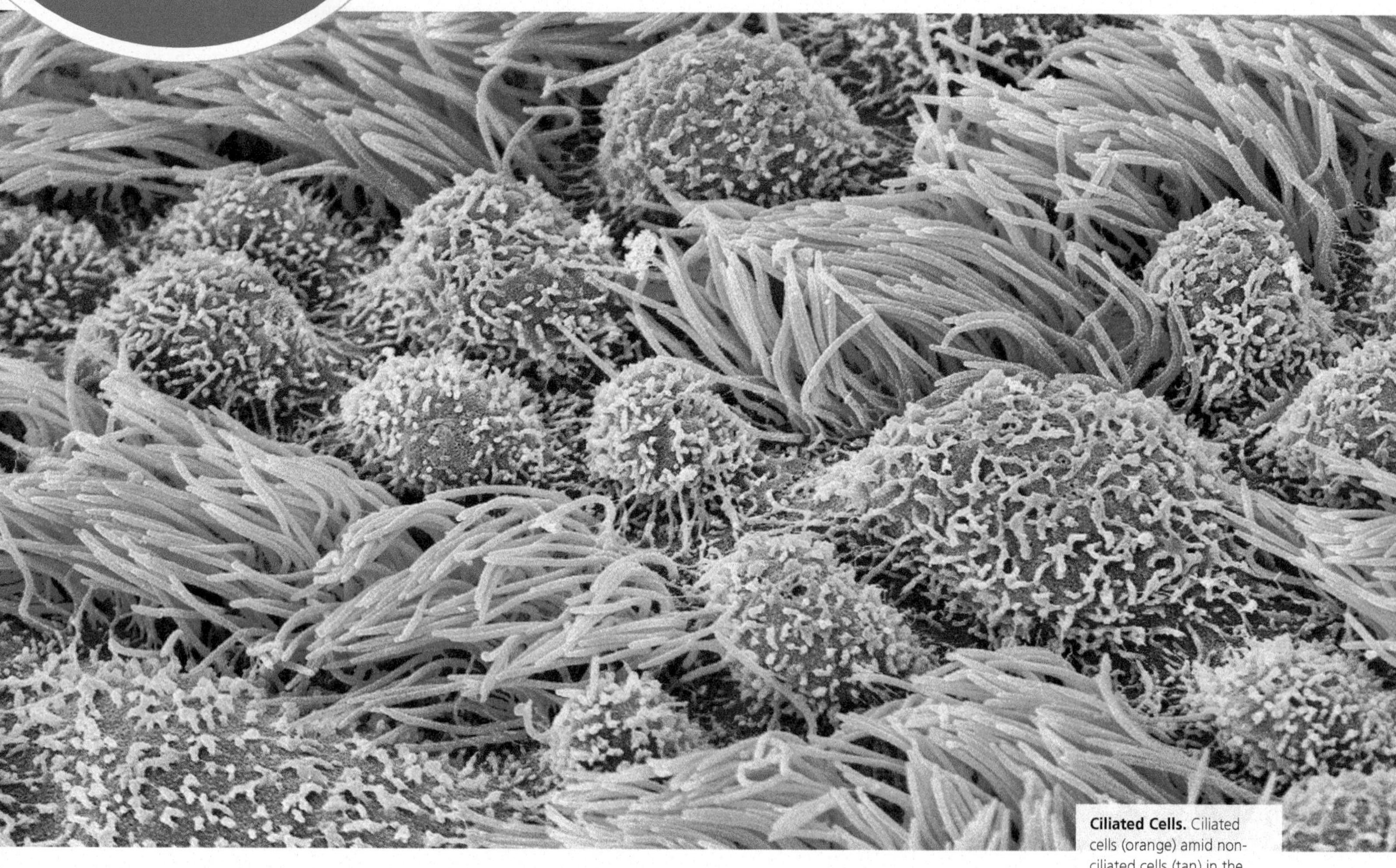

Ciliated Cells. Ciliated cells (orange) amid non-ciliated cells (tan) in the fallopian tube epithelium (colorized SEM).

The cytoskeleton of eukaryotic cells (introduced in Chapter 13) serves as an intracellular scaffold that organizes structures within the cell and shapes the cell itself. In this chapter, we will explore the role of these cytoskeletal elements in cellular **motility.**

Motility may involve the movement of a cell (or a whole organism) through its environment, the movement of the environment past or through a cell, the movement of components within a cell, or the shortening of a cell itself. (**Contractility**, a related term often used to describe the shortening of muscle cells, is a specialized form of motility.) Motility occurs at the tissue, cellular, and subcellular levels.

The most conspicuous examples of motility, particularly in the animal world, take place at the tissue level. The muscle tissues common to most animals consist of cells specifically adapted for contraction. The movements they produce are often obvious, whether manifested as the bending of a limb, the beating of a heart, or a uterine contraction during childbirth.

At the cellular level, motility occurs in single cells or in organisms consisting of one or only a few cells. It occurs among cell types as diverse as ciliated protozoa, motile sperm, and crawling cells.

Equally important is the movement of *intracellular* components. For example, highly ordered microtubules of the mitotic spindle play a key role in the separation of chromosomes during cell division (as you will see in Chapter 24). Nondividing cells continuously shuttle cell components—such as RNAs, multiprotein complexes, and membrane-bounded vesicles—from one location to another.

Cellular motility typically involves the conversion of chemical energy directly to mechanical energy via specialized **motor proteins**. The microfilaments and microtubules of the cytoskeleton provide a basic scaffold for these motor proteins, which interact with the cytoskeleton to produce motion at the molecular level. In cases such as muscle contraction, the combined effects of many motors moving simultaneously produce movement at the tissue level.

In eukaryotes, there are two major motility systems, each based on a specific type of cytoskeletal filament and associated motor proteins (**Table 14-1**). Microtubule-based motility is mediated by *kinesins* and *dyneins*. Microfilament-based motility involves the *myosin* family of motor molecules. **Figure 14-1** on page 386 shows representative examples of each of the three motor families and highlights some common structural elements we will return to throughout this chapter.

Molecular motors have several common features. First, all the motors you will learn about couple *ATP hydrolysis* to changes in shape and attachment of the motor to its associated cytoskeletal filament. You may initially think that ATP hydrolysis and changes in protein conformation are directly linked in a one-to-one fashion. However, the situation is actually a bit more complicated than that, as the products of ATP hydrolysis ($ADP + P_i$) are typically not released from the motor immediately after hydrolysis, or even at the same time.

Second, motors undergo *cycles* of ATP hydrolysis, release of ADP and P_i, and acquisition of a new ATP molecule. This cyclical process allows motors to attach, move, and detach over and over again as they move along a filament. Third, to transduce changes in shape to movement of a filament (or movement along a stationary filament), motors have common structural features. In addition to ATPase domains, which use the energy of ATP hydrolysis to power changes in protein shape, they all have regions that bind to microtubules or F-actin. They also often have a *mechanical transducer* that couples the ATPase domain through some sort of linker or lever arm to the region that binds the cytoskeleton.

Table 14-1 Selected Motor Proteins of Eukaryotic Cells

Motor Protein	Typical Function
Microtubule (MT)-Associated Motors	
Dyneins	
Cytoplasmic dyneins	Moves cargo toward minus ends of MTs
Axonemal dyneins	Activates MT sliding in flagella and cilia
Kinesins*	
Kinesin-1 (classic kinesin)	Dimer; moves cargo toward plus ends of MTs
Kinesin-3	Monomer; movement of synaptic vesicles in neurons
Kinesin-5	Bipolar, tetrameric; bidirectional sliding of MTs during anaphase of mitosis
Kinesin-6	Completion of cytokinesis
Kinesin-13 (catastrophins)	Dimer; destabilization of plus ends of MTs**
Kinesin-14	Spindle dynamics in meiosis and mitosis; moves toward minus end of MTs
Microfilament (MF)-Associated Motors	
Myosins*	
Myosin I	Motion of membranes along MFs; endocytosis
Myosin II	Slides MFs in muscle; other contractile events such as cytokinesis, cell migration
Myosin V	Vesicle positioning and trafficking
Myosin VI	Endocytosis; moves toward minus ends of MFs
Myosin VII	Base of stereocilia in inner ear
Myosin X	Tips of filopodia
Myosin XV	Tips of stereocilia in inner ear

*Kinesins and myosins comprise large families of proteins. There are many families of kinesins and myosins.

**Kinesin-13s do not have motor function. Instead, they foster depolymerization of microtubules.

Finally, because most motors have more than one subunit that binds a cytoskeletal filament at one time, they are *processive;* that is, they can move along a cytoskeletal filament for significant distances. To visualize such processive movement, it may help to think of a child moving along the "monkey bars" at a playground, detaching her rear hand and extending it forward to the next bar while remaining attached to the monkey bars with her other hand. Once the forward hand is attached, she swings what is now her rear hand forward, and so on.

With these general concepts in mind, it is time for a detailed examination of motility systems, beginning with the proteins essential for microtubule-based movement.

MAKE CONNECTIONS 14.1

What type of biological work do motor proteins exhibit? (Ch. 5)

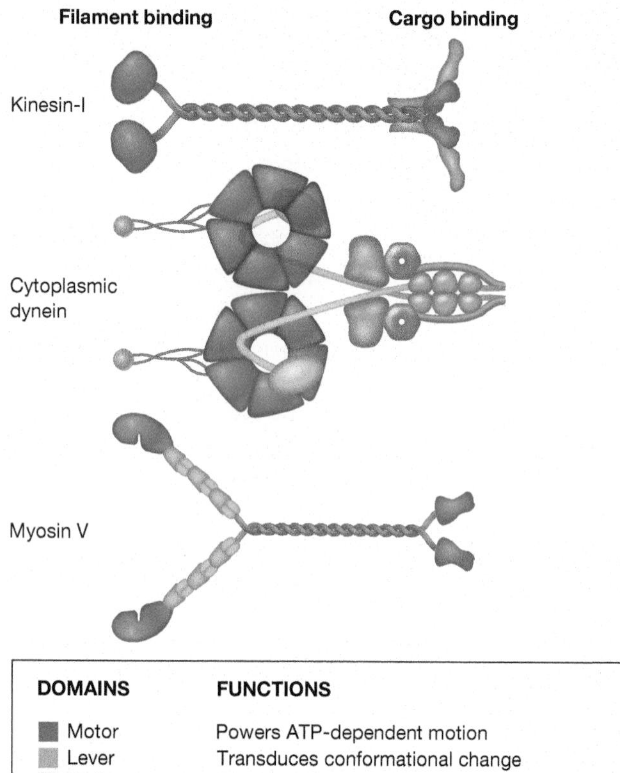

DOMAINS	FUNCTIONS
Motor	Powers ATP-dependent motion
Lever	Transduces conformational change
Stalk	Connects cargo- and filament-binding domains
Tail	Binds cargo
Light chain	

Figure 14-1 Representative Examples of Major Classes of Motor Proteins and Their Functional Domains. Models are based in part on X-ray crystallography data and shown to scale. Motor domains are shown in dark blue. Lever domains, which undergo changes in position and/or shape, are shown in light blue. Cargo-binding regions are shown in purple and green.

14.1 Microtubule-Based Movement Inside Cells: Kinesins and Dyneins

Microtubules (MTs) provide a rigid set of tracks for the transport of a variety of membrane-enclosed organelles and vesicles. The centrosome, in many cases located near the center of the cell (see Figure 13-11), organizes and orients MTs within the cell because the minus ends of most MTs are located at the centrosome. As a result, traffic toward the minus ends of MTs might be considered "inbound" traffic. Traffic directed toward the plus ends might likewise be considered "outbound," meaning that it is directed toward the periphery of the cell.

Although microtubules provide a set of tracks along which traffic can move, they do not directly generate the force necessary for movement. The mechanical work needed for movement depends on microtubule-associated motor proteins—**kinesins** and **dyneins**—that attach to vesicles or organelles and then move along the MT, using ATP to provide the needed energy. These motor proteins recognize the polarity of MTs, with each motor having a preferred direction of movement.

Motor Proteins Move Cargoes Along MTs During Axonal Transport

A historically important system for studying microtubule-dependent intracellular movement is the giant axons of squid motor neurons (see Figure 22-1 for details regarding the squid giant axon). The need for such transport arises because ribosomes are present only in the cell body, so no protein synthesis occurs in axons or synaptic knobs. Instead, proteins are synthesized in the cell body, packaged into membranous vesicles, and transported along the axons to the synaptic knobs over distances of up to a meter. Some form of energy-dependent transport is clearly required, and MT-based movement provides the mechanism in a process called **fast axonal transport**. (We will not discuss slow axonal transport, which involves somewhat different processes.)

The role of microtubules in axonal transport was initially suggested because the process is inhibited by drugs that depolymerize MTs but is insensitive to drugs that affect microfilaments. Since then, MTs have been visualized by light and electron microscopy and shown to be prominent features of the axonal cytoskeleton. Moreover, cargoes including small, membranous vesicles are observed to be associated with them (**Figure 14-2**).

Further insight into axonal transport was obtained when investigators found that, in the presence of ATP, organelles could move along fine filamentous structures present in exuded *axoplasm* (the cytosol of axons). The rate of organelle movement was measured to be about 2 μm/sec,

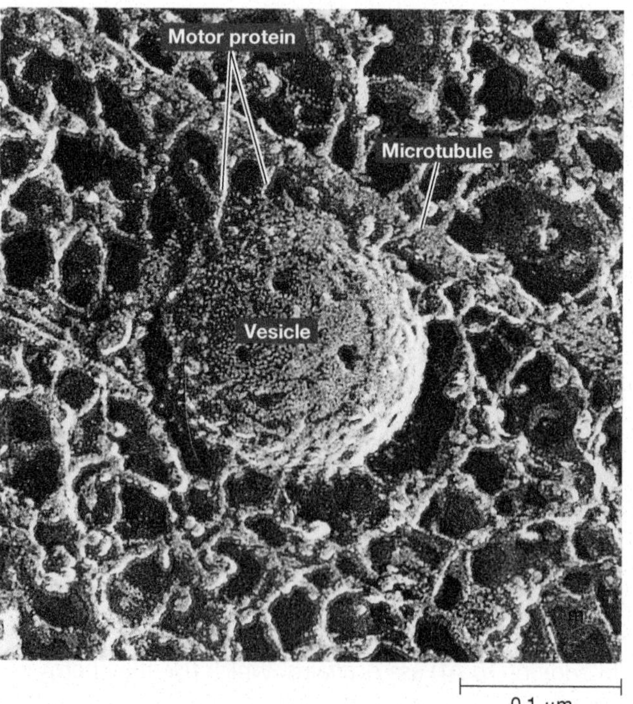

Figure 14-2 A Vesicle Attached to a Microtubule in a Crayfish Axon. Motor proteins connect the membrane vesicle (round structure at center) to a microtubule (deep-etch TEM).

comparable to the axonal transport rate in intact neurons. A combination of fluorescence and electron microscopy demonstrated that the fine filaments the organelles moved along were single microtubules. The fact that motility required both MTs and ATP suggested that one or more ATP-driven motors were responsible for movement. Since that time, two MT motor proteins responsible for fast axonal transport, kinesin-1 and cytoplasmic dynein, have been purified and characterized.

To determine the direction of transport by these motors, an in vitro system of microtubules polymerized from purified centrosomes was employed. When purified kinesin, ATP, and polystyrene beads (an artificial cargo) were added to MTs, the beads moved toward the plus ends (that is, away from the centrosome). This finding means that in a nerve cell, kinesin mediates transport from the cell body down the axon to the nerve ending (called *anterograde [forward] axonal transport*). When similar experiments were carried out with purified cytoplasmic dynein, particles were moved in the opposite direction, toward the minus ends of the MTs (called *retrograde [backward] axonal transport*). Thus, these two motors shuttle cargo along microtubules in opposite directions within the cytosol.

This discussion may suggest that cargo always moves along MTs in one direction or another. In the case of fast axonal transport, such unidirectional transport is the rule. However, some vesicles are capable of changing direction as they move along an MT. In these cases, cargo is attached to both kinesins and dyneins; the direction of motion seems to be determined by which type of motor predominates.

Classic Kinesins Move Toward the Plus Ends of Microtubules

The first kinesin (now called kinesin-1), originally identified in the cytosol of squid giant axons, consists of two dimerized heavy chains and two light chains (that is, the two basic kinds of subunits differ in size and molecular weight). Together, these form four functional domains. The heavy chains contain *globular* domains that attach to microtubules and are involved in hydrolysis of ATP, as well as a lever-like *neck* that connects the globular domains to a coiled-coil *stalk*, and a *tail* with associated light chains that is involved in attaching the kinesin to a variety of cargo, including other proteins and membranous organelles (**Figure 14-3**). In some cases, such as mitochondria, adapter proteins allow the light chains to attach to cargo indirectly.

The movements of single motor molecules along cytoskeletal filaments have been studied by tracking the movement of attached beads or by measuring the force exerted by single motor domains using calibrated glass fibers or a special device known as an "optical tweezer" (see Key Technique, pages 388–389, for more details). Classic kinesins have been measured to move along MTs in 8-nm steps. Recall the "monkey bars" analogy earlier in this chapter. This is precisely how kinesin-1 moves: one of the two globular domains moves forward to make an attachment to a new β-tubulin subunit, followed by dissociation of the trailing globular domain, which

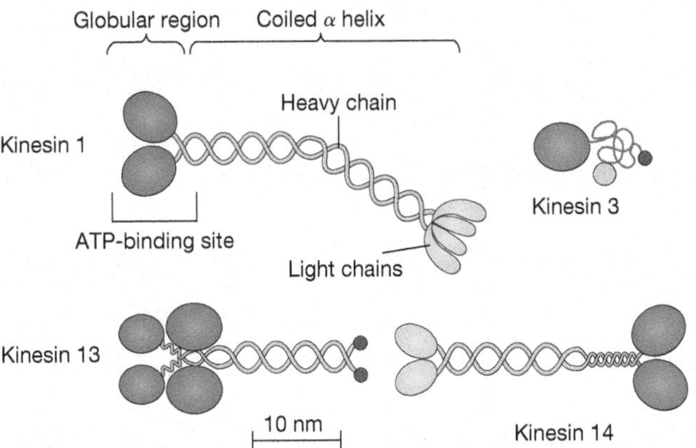

Figure 14-3 Several Different Kinesin Family Members. Kinesins contain globular regions (blue) that bind ATP and α-helical heavy chains (brown), which connect to light chains (pink) or other domains involved in binding to cargo.

can now bind to a new region of the MT. Movement is coupled to the hydrolysis of ATP bound at specific sites within the globular domains, which also have ATPase activity (**Figure 14-4** on page 390).

A single kinesin-1 molecule exhibits *processivity*. It can cover long distances (more than 100 steps) before detaching from an MT. A single kinesin-1 molecule can travel as far as 1 μm, which is a great distance relative to its size. As a molecular motor, kinesins appear to be quite efficient; estimates of their efficiency in converting the energy of ATP hydrolysis to useful work are on the order of 60–70%.

Kinesins Are a Large Family of Proteins

Since the discovery of the original ("classic") kinesin involved in anterograde transport in neurons, many other kinesins have been discovered. They can be grouped into families based on their amino acid sequence (see Table 14-1). Some kinesins form homodimers with another identical protein or heterodimers with another, different kinesin. One kinesin (kinesin-14) acts as a minus-end-directed motor; another (kinesin-5) is bidirectional.

Kinesins are involved in many different processes within cells. One important function is moving and localizing materials, including RNAs, multiprotein complexes, and membranous vesicles and organelles. One family of kinesins, the catastrophins (kinesin-13 family proteins), aid depolymerization of MTs (see Chapter 13). Several kinesins localize to the mitotic or meiotic spindle or to kinetochores, where they play roles in various phases of mitosis and meiosis (see Chapter 24).

Dyneins Are Found in Axonemes and the Cytosol

Dyneins can be divided into two basic categories: cytoplasmic dyneins and axonemal dyneins (see Table 14-1).

Two types of **cytoplasmic dynein** (**Figure 14-5** on page 390) have been identified. They consist of two heavy chains (which interact with microtubules), two intermediate chains,

Key Technique: Watching Motors Too Small to See

PROBLEM: Using visible light to watch proteins presents a seemingly insurmountable challenge: the wavelength of visible light is too long to resolve even large protein complexes.

SOLUTION: Advanced modes of microscopy allow the study of individual molecules in motion. These techniques are being used to watch dynamic molecules as diverse as RNA polymerase (the enzyme that synthesize RNA molecules from a DNA template), mitochondrial ATPases (see Chapter 10), as well as cytoskeletal motor proteins discussed in this chapter. Different experimental set-ups can be adopted to measure characteristics of motors ranging from step size to speed, motor force, and conformational changes.

Key Tools: (1) Highly purified motor proteins, alone, attached to beads, or labeled with a fluorescent dye. (2) Purified cytoskeletal filaments as appropriate (microtubules or F-actin). (3) ATP to activate motors. (4) Sophisticated microscopes to record how the activated motor moves.

Details: **Optical Trapping.** *Optical (laser) trapping* is a widely used approach for measuring motor motility. In a very common kind of optical trap experiment called a "three-bead" assay, a single cytoskeletal filament (such as actin) is held taut between two small beads and allowed to interact with a larger, immobilized bead that is sparsely coated with motor proteins (such as myosin II) (**Figure 14A-1**). The two beads at the ends of the filament are held weakly in place by pressure exerted by two laser beams—the traps (or "tweezers") that give this technique its name. Each power stroke by a myosin displaces the trapped filament. The apparatus is designed so that the beads on the ends of the filament snap back into the centers of the trapping laser beams. The lasers are calibrated so that the force required to reposition the filament and the distance it moved can be measured and recorded by a computer. Such a set-up allowed researchers to determine that a single myosin II motor domain exerts 3–5 piconewtons of force per power cycle and that its power stroke is approximately 8 nm in length.

Fluorescent Spot Tracking. Many motors are highly processive. That is, they can travel a significant distance before falling off of a filament, using the globular domains of their two heavy chains. How do the two heavy chains of processive motors work together? This question can be answered by labeling motors with a spot of fluorescent dye at a position near the filament-binding region of one of its two heavy chains. A computer and a special fluorescence microscope track the position of the fluorescent spot.

The results of one such experiment, involving myosin V, are shown in **Figure 14A-2**. Spots were found to advance in increments of approximately 72–74 nm (Figure 14A-2a). Optical trap experiments had previously shown that one myosin V motor on its own can move 36–37 nm—half the distance observed in the fluorescence experiment. In combination, the two sets of results suggest that myosin V moves by alternately swinging the trailing globular domain forward to take the lead, followed by the rear domain—in other words, in a "hand-over-hand" fashion (Figure 14-A2b).

Atomic Force Microscopy. Indirect methods, like the spot-tracking technique just described, have helped researchers test hypotheses about how motor proteins work. More recently, however, advances in a technique called *atomic force microscopy (AFM)* have made it possible to directly visualize entire myosin fragments moving along actin filaments in three dimensions (**Figure 14A-3**). AFM uses a strategy akin to an old-fashioned record player: an extremely tiny needle (about 2 nm wide at its tip; Figure 14A-3a) is moved across the surface of a prepared sample, reading out a topographic map of the molecular landscape. At very high speeds, these tiny probes can even trace and record the shape of proteins in motion. The results of a high-speed AFM experiment are shown in Figure 14A-3b. (You can learn more about AFM in the Appendix.)

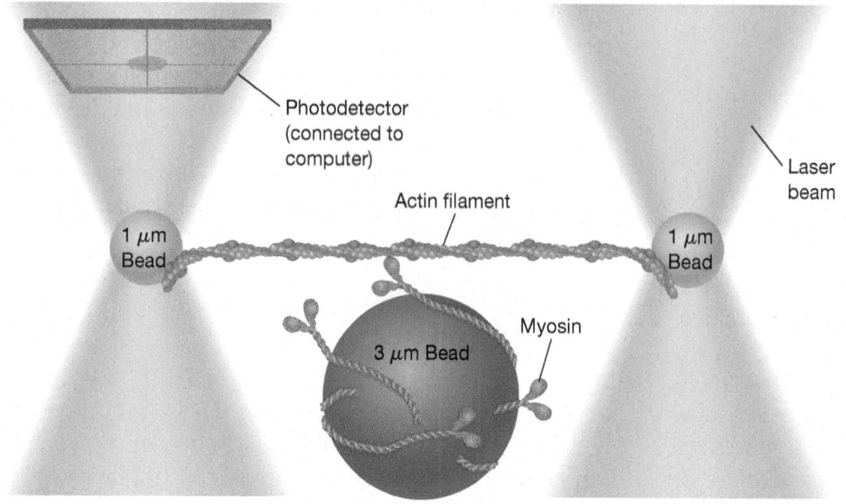

Figure 14A-1 A Three-Bead Optical Trap to Measure Motor Step Size and Force.

QUESTION: If myosin V moved more like an "inchworm" (meaning it dragged its rear filament-binding domain forward, but not in a hand-over-hand fashion), how would a graph of spot movement over time differ from the one shown in Figure 14A-2b?

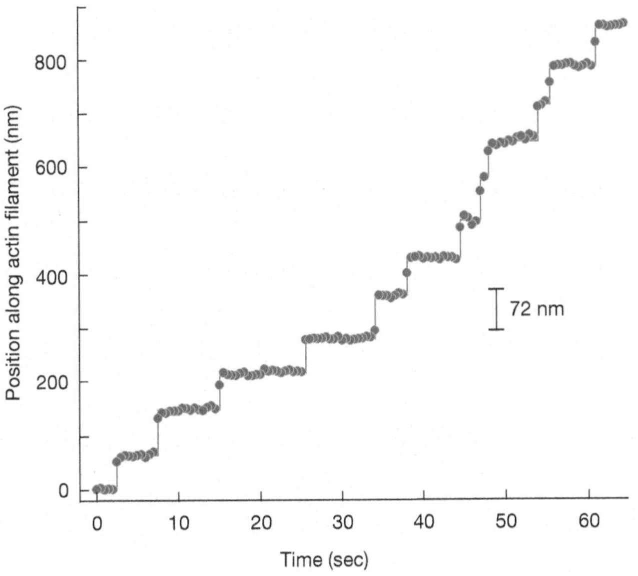

(a) Recorded traces of stepping myosin V molecules

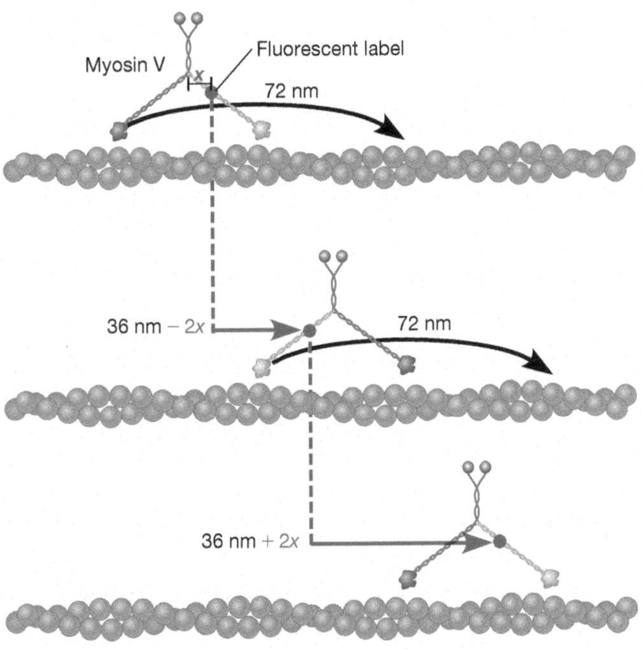

(b) Tracing hand-over-hand motility

Figure 14A-2 "Hand-over-Hand" Motility Measured by Fluorescent Spot Tracking.

(a) Color-enhanced scanning electron micrograph of probes used in atomic force microscopy (AFM)

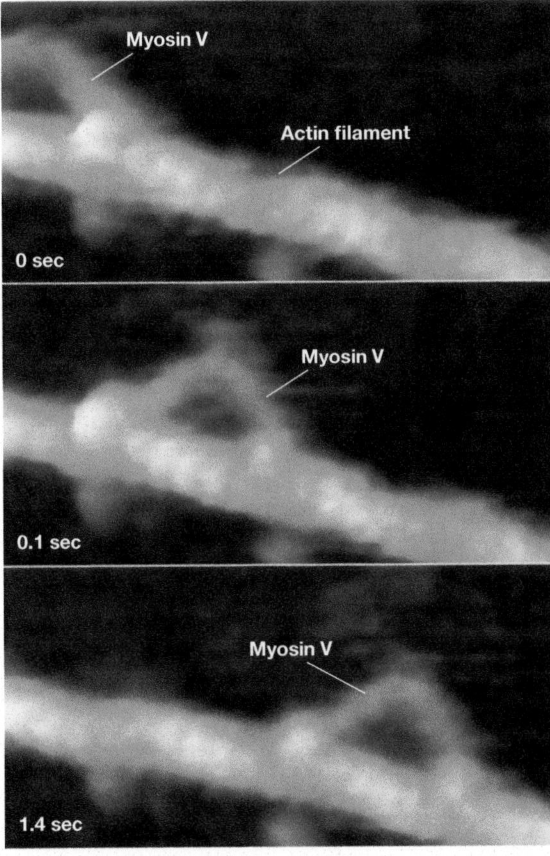

(b) Results of high-speed AFM experiment

Figure 14A-3 Motor Movement Visualized Using Atomic Force Microscopy (AFM). High-speed AFM was used to generate successive images of a myosin V molecule moving along an actin filament attached to a mica surface.

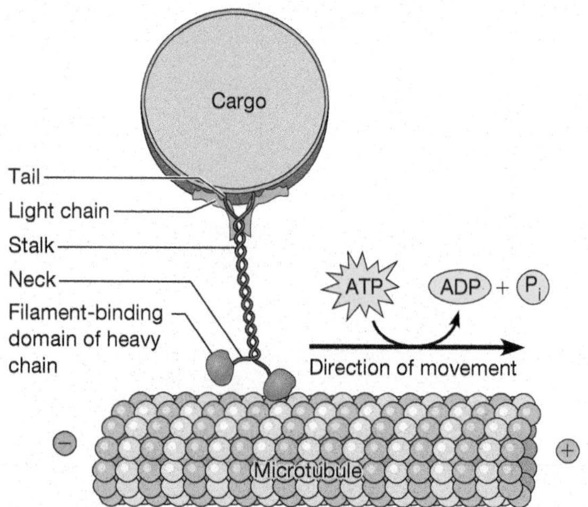

(a) Schematic representation of kinesin-1

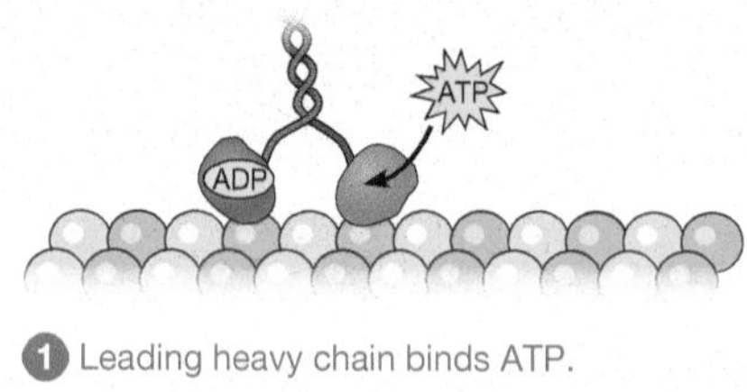

1 Leading heavy chain binds ATP.

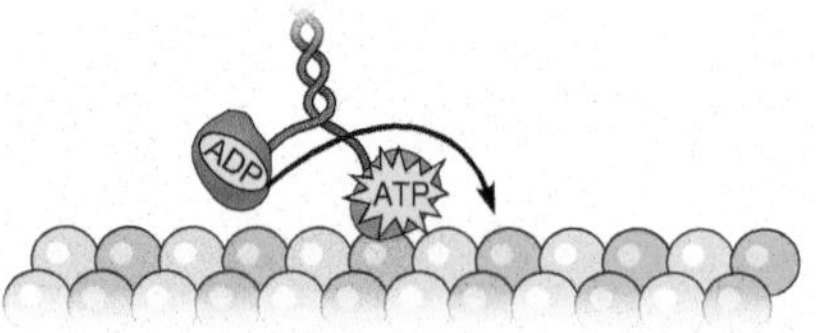

2 ATP binding causes conformational change; trailing heavy chain swings forward.

3 Trailing heavy chain finds new MT binding site.

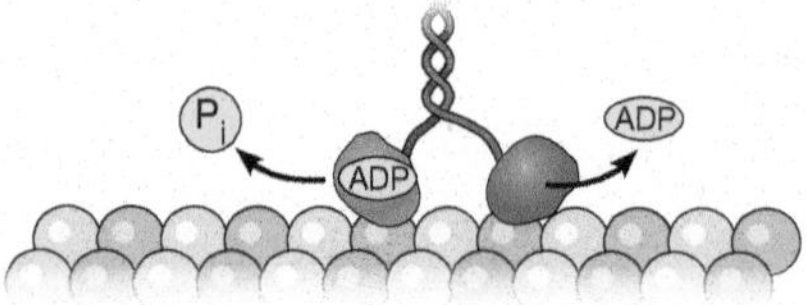

4 New leading heavy chain releases ADP; new trailing head hydrolyzes ATP to ADP + P_i. P_i released.

(b) Kinesins move along MTs in an ATP-dependent fashion.

Figure 14-4 How Kinesins Move Cargo Along Microtubules. **(a)** Kinesin-1, shown here, uses the globular filament-binding domains of its heavy chains to bind to microtubules (MTs). Its light chains are involved in cargo binding. It moves toward the plus ends of MTs. **(b)** Kinesins move using a hand-over-hand mechanism along microtubules.

two light intermediate chains, and various light chains (Figure 14-5a). The largest part of each heavy chain folds into six

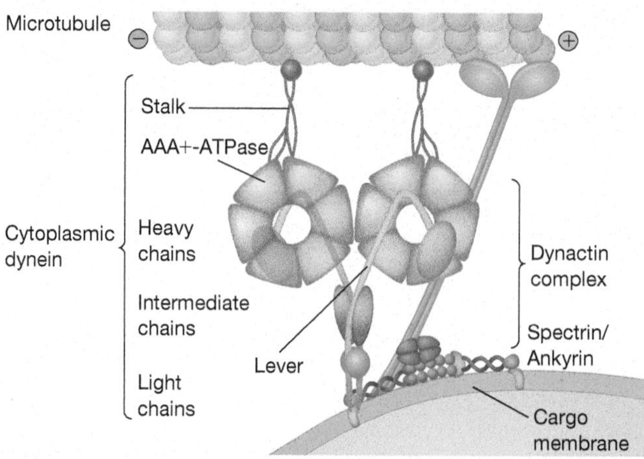

(a) Dynein/dynactin structure

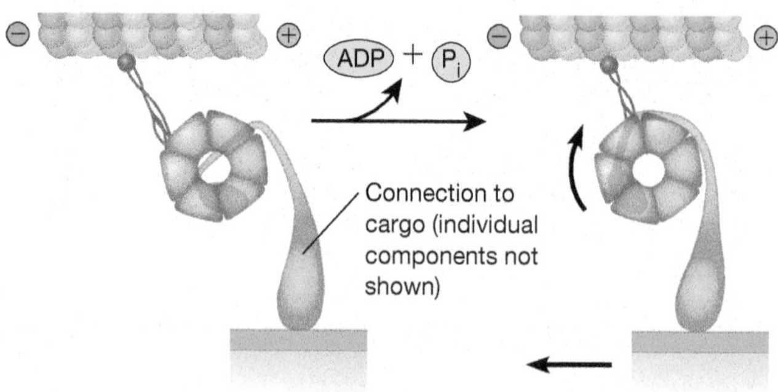

(b) Power stroke

Figure 14-5 Cytoplasmic Dynein and the Dynactin Complex. **(a)** Cytoplasmic dynein is linked to cargo membranes indirectly through the dynactin multiprotein complex. The dynactin complex binds to a complex containing spectrin and ankyrin associated with the membrane of the cargo vesicle. One component of the dynactin complex, p150Glued (aqua), can also bind MTs. **(b)** During the power stroke, dyneins move along MTs through changes in the conformation of the lever and AAA + domains of the heavy chain.

AAA+ domains ("AAA+" stands for "*A*TPases *a*ssociated with diverse cellular *a*ctivities") that form a hexagonally shaped ring. A small region near the C-terminal end of the heavy chain binds MTs. In contrast to most kinesins, cytoplasmic dyneins move toward the minus ends of MTs.

As Figure 14-5a shows, cytoplasmic dyneins associate with a protein complex known as **dynactin**. The dynactin complex helps link cytoplasmic dynein to its cargo (such as membranous vesicles) by binding to proteins such as spectrin attached to the membrane of its cargo.

Seven types of **axonemal dyneins** have been identified. Their function in cilia and flagella is discussed in some detail later in this chapter.

Unlike kinesins, whose MT-binding domains also couple ATP hydrolysis to changes in protein shape, the AAA+ domains in dyneins act at a distance. The ATPase is crucial for the *power stroke* that moves the dynein along an MT (Figure 14-5b). When the ATPase is bound to ADP, the AAA+ ring becomes "cocked." Release of the ADP causes a large shift in a lever attached to the ATPase. As a result, the dynein and its attached cargo move toward the minus end of the MT to which the dynein is attached.

Microtubule Motors Direct Vesicle Transport and Shape the Endomembrane System

The cell has an elaborate transportation system of vesicles and membranous organelles—the endomembrane system. Proteins are processed in the endoplasmic reticulum and eventually are packaged into membrane vesicles for distribution to the correct cellular destinations (see Chapter 12). Vesicular traffic is not one-way, however: some vesicles move in the opposite direction. Thus, there is a continuous flow of vesicles from the endoplasmic reticulum (ER) to the Golgi apparatus to the cell surface (*anterograde transport*) and back from the surface (*retrograde transport*).

Recall that in many cells the centrosome anchors a polarized array of microtubules, with the plus ends oriented toward the cell periphery. MT motors, moving cargoes along this MT array, are crucial to vesicular trafficking (**Figure 14-6**). Kinesins move toward MT plus ends, carrying vesicles away from the Golgi apparatus toward the cell periphery. Cytoplasmic dyneins perform the opposite function, moving vesicles that form via endocytosis away from the cell surface into the cell's interior toward MT minus ends.

MT motors appear to be important for dynamically shaping the endomembrane system as well. For example, live imaging of MTs and ER membrane in vitro and in living cells indicates that extensions of the ER can be moved along MTs. Other experiments show that MTs are required for assembly and maintenance of the Golgi apparatus. If MTs are depolymerized using the drug nocodazole (see Chapter 13), the Golgi apparatus disperses. Similarly, disruption of the function of the dynactin complex results in disruption of transport from the ER to the Golgi and collapse of the Golgi apparatus.

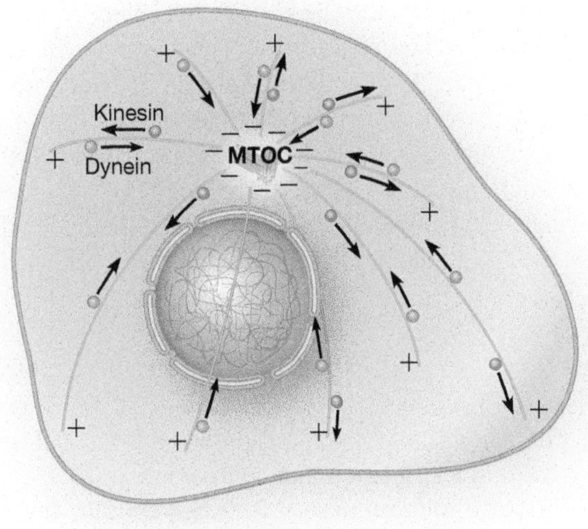

Figure 14-6 Microtubules and Vesicle Trafficking. Vesicles going toward and away from the cell surface are attached to microtubules and carried by MT motors. The microtubule-organizing center (MTOC) in the cell anchors minus ends of MTs at the centrosome; plus ends are oriented toward the cell periphery. Vesicles are carried by kinesins (purple) away from the MTOC. Dyneins (blue) carry cargoes from the cell surface toward the MTOC.

CONCEPT CHECK 14.1

Kinesins and dyneins are both microtubule-based motors. How are they similar? How are they different?

14.2 Microtubule-Based Cell Motility: Cilia and Flagella

Microtubules are crucial not only for movement within cells but also for the movements of *cilia* and *flagella*, the motile appendages of eukaryotic cells.

Cilia and Flagella Are Common Motile Appendages of Eukaryotic Cells

Cilia and flagella share a common structural basis and differ only in relative length, number per cell, and mode of beating. Both are surrounded by an extension of the plasma membrane and are therefore intracellular structures. **Cilia** (singular: **cilium**) have a diameter of about 0.25 μm, are about 2–10 μm long, and tend to occur in large numbers on the surface of ciliated cells (**Figure 14-7** on page 392). Cilia occur in both unicellular and multicellular eukaryotes. Unicellular organisms, such as *Paramecium*, a protozoan, use cilia for both locomotion and the collection of food particles. In multicellular organisms, cilia serve primarily to move the environment past the cell rather than to propel the cell through the environment. The cells that line the air passages of the human respiratory tract, for example, have several hundred cilia each, which means that every square centimeter of epithelial tissue lining the respiratory tract has about a billion cilia (Figure 14-7a). The coordinated, wavelike beating of these cilia carries mucus, dust, dead cells, and other foreign matter out of the lungs. One of the health hazards of cigarette smoking lies in the inhibitory effect that smoke has on normal ciliary beating. Certain respiratory ailments can also be traced to defective cilia.

Cilia exhibit three-dimensional beat strokes. They display an oarlike pattern of beating, with a power stroke perpendicular to the cilium that generates a force parallel to the cell surface. The recovery stroke is initiated at the base and is propagated toward the tip. The cycle of beating for an epithelial cilium is shown in Figure 14-7b. Each cycle requires about 0.1–0.2 sec.

Flagella (singular: **flagellum**) move cells through a fluid environment. Although they have the same diameter as cilia, eukaryotic flagella are often much longer—from 1 μm to several millimeters, though usually in the range of 10–200 μm—and may be limited to one or a few per cell (Figure 14-7c). Some flagella are remarkably prodigious. If human sperm were the size of a fully grown human, the distance they must swim would be equivalent to a human swimming from the east coast of the United States across the Atlantic Ocean to England!

Flagella differ from cilia in the nature of their beat. They move with a propagated bending motion that is usually symmetrical and undulatory (Figure 14-7c) and may

(a) Cilia on mammalian tracheal cells

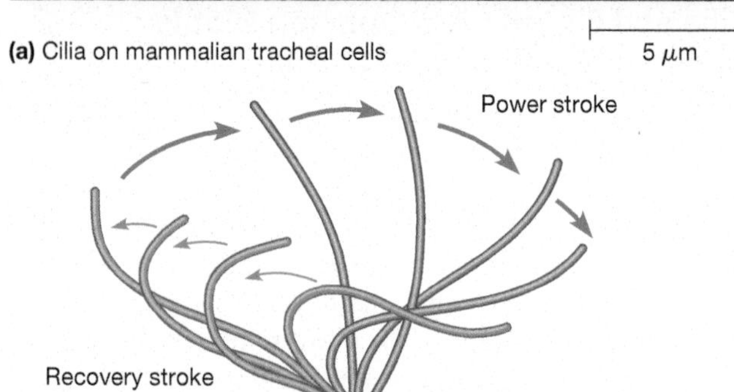

(b) Beating of a cilium

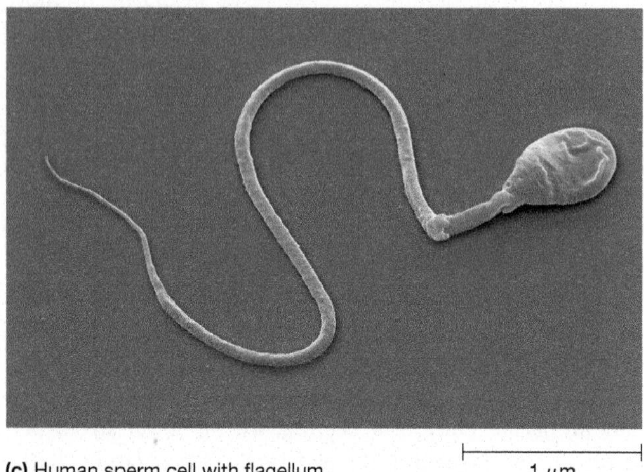

(c) Human sperm cell with flagellum

Figure 14-7 Cilia and Flagella. (a) A micrograph of cilia on mammalian lung cells (SEM). **(b)** The pattern of beating of a single cilium on the surface of an epithelial cell from the human respiratory tract. A beat begins with a power stroke (blue) that sweeps fluid over the cell surface. A recovery stroke (pink) follows, leaving the cilium poised for the next beat. Each cycle requires about 0.1–0.2 sec. **(c)** A micrograph of a human sperm cell (SEM).

even have a helical pattern. This type of beat generates a force parallel to the flagellum, such that the cell moves in approximately the same direction as the axis of the flagellum. The locomotory pattern of most flagellated cells, such as many sperm cells, involves propulsion of the cell by a trailing flagellum, but sometimes the flagellum has been found to precede the cell.

The flagella of bacteria are constructed from an entirely different set of proteins, and their mechanism of movement is completely different. These flagella are fascinating, but we will not consider them further here.

Cilia and Flagella Consist of an Axoneme Connected to a Basal Body

Cilia and flagella share a common structure, known as the **axoneme**, that is about 0.25 μm in diameter (**Figure 14-8**). The axoneme is connected to a **basal body** and surrounded by an extension of the cell membrane (Figure 14-8a). Between the axoneme and the basal body is a *transition zone* in which the arrangement of microtubules in the basal body takes on the pattern characteristic of the axoneme. Cross-sectional views of the axoneme and basal body are shown in Figure 14-8b and c.

The basal body is identical in appearance to a centriole (see Figure 13-8). It consists of nine sets of tubular structures arranged around its circumference. Each tubule set is called a *triplet* because it consists of three microtubules that share common walls—one complete (A) tubule and two incomplete tubules (B and C). The triplets are interconnected by a "cartwheel" structure, most of which forms through the polymerization of a protein known as SAS-6. As a cilium or flagellum forms, a centriole migrates to the cell surface and makes contact with the plasma membrane. That centriole then acts as a nucleation site for MT assembly, initiating polymerization of the nine outer doublets of the axoneme. After the process of tubule assembly has begun, the centriole is then referred to as a basal body.

The axonemes of propulsive cilia and flagella have a characteristic "9 + 2" pattern, with nine **outer doublets** of tubules and two additional microtubules in the center, often called the **central pair**. Other specialized cilia, called *primary cilia*, are used in sensory structures. These cilia often have a "9 + 0" structure; that is, they lack the central pair.

The drawing in Figure 14-8a illustrates the structural features of a typical "9 + 2" cilium in detail. The nine outer doublets of the axoneme are thought to be extensions of two of the three subfibers from each of the nine triplets of the basal body. Each outer doublet therefore consists of one complete MT, called the **A tubule**, and one incomplete MT, the **B tubule**. The A tubule has 13 protofilaments, whereas the B tubule has only 10 or 11. The tubules of the central pair are both complete, with 13 protofilaments each. All of these structures contain tubulin, together with a second protein called *tektin*. Tektin is related to intermediate filament proteins (see Chapter 13); the A and B tubules share a wall that appears to contain tektin as a major component.

In addition to microtubules, axonemes contain several other key components. The most important of these components is the sets of **sidearms** that project out from each of the A tubules of the nine outer doublets. When viewed from the tip (distal end) of the axoneme, each sidearm reaches out counterclockwise toward the B tubules of the adjacent doublet. These sidearms consist of axonemal dynein, which is responsible for sliding doublets past one another to bend the axoneme. The dynein arms occur in pairs, one inner arm and one outer arm, spaced along the A tubule at regular intervals. At less frequent intervals, a protein called *nexin* links adjacent

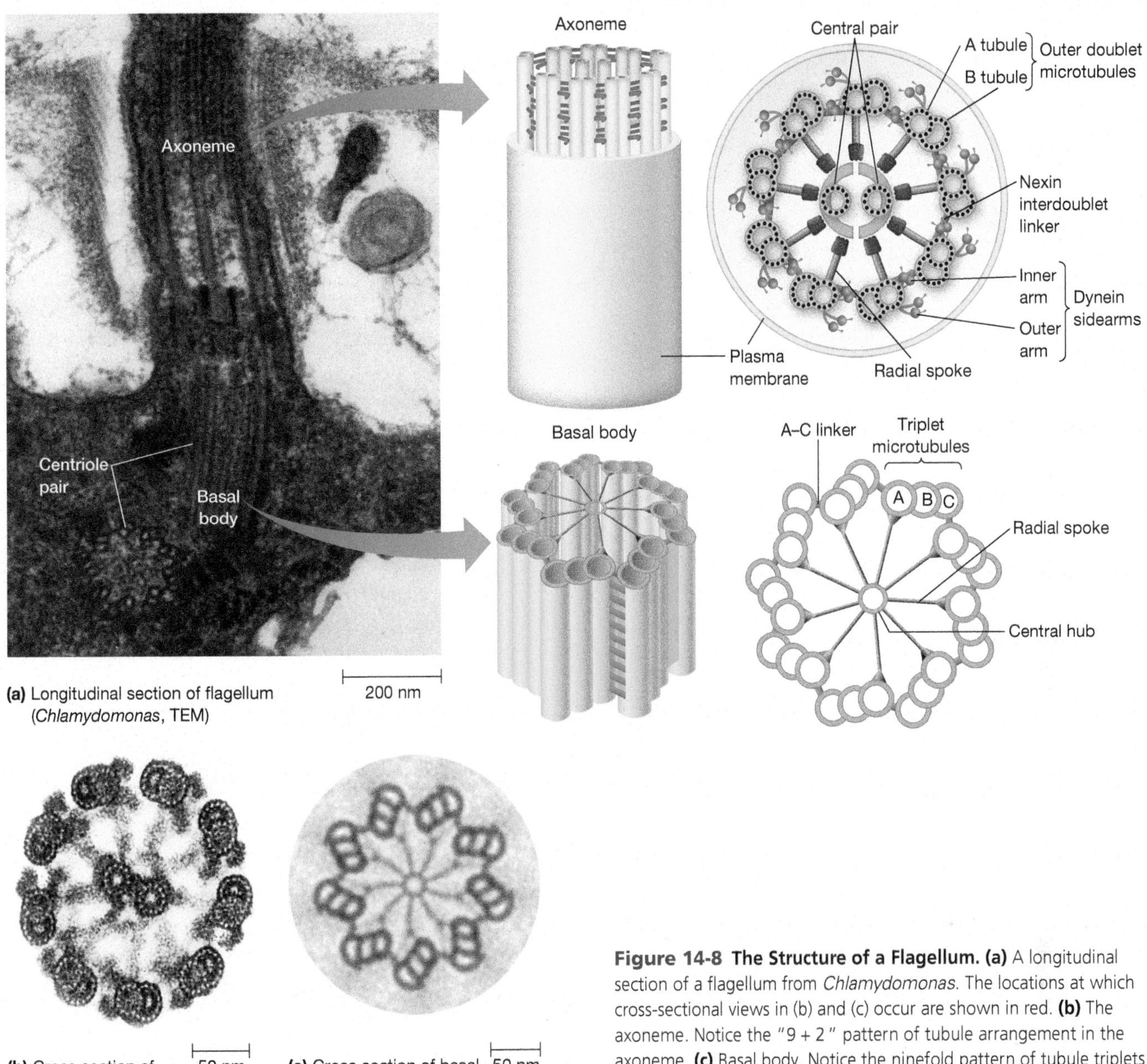

(a) Longitudinal section of flagellum (*Chlamydomonas*, TEM)

(b) Cross section of axoneme (TEM)

(c) Cross section of basal body (TEM)

Figure 14-8 The Structure of a Flagellum. (a) A longitudinal section of a flagellum from *Chlamydomonas*. The locations at which cross-sectional views in (b) and (c) occur are shown in red. **(b)** The axoneme. Notice the "9 + 2" pattern of tubule arrangement in the axoneme. **(c)** Basal body. Notice the ninefold pattern of tubule triplets and the characteristic "cartwheel" structure connecting them.

doublets to one another. Nexin **interdoublet links** are thought to limit the extent to which doublets can move with respect to each other as the axoneme bends.

At regular intervals, **radial spokes** project inward from each of the nine MT doublets, terminating near a set of projections that extend outward from the central pair of microtubules. These spokes are thought to be important in translating the sliding motion of adjacent doublets into the bending motion that characterizes the beating of cilia and flagella.

Doublet Sliding Within the Axoneme Causes Cilia and Flagella to Bend

How does the elaborate structure that constitutes an axoneme generate the characteristic bending of cilia and flagella? The overall length of MT doublets in cilia and flagella does not change during beating. Instead, adjacent outer doublets slide relative to one another. This sliding movement is converted to a localized bending because the doublets of the axoneme are connected radially to the central pair and circumferentially to one another and therefore cannot slide past each other freely.

Crosslinks and Spokes Are Responsible for Bending. Strong evidence that radial and circumferential connections restrain sliding and lead to bending comes from a classic experiment (**Figure 14-9** on page 394). When isolated axonemes are treated with a protease to cleave the nexin linkages between adjacent outer doublets and then treated with ATP, the doublets slide past each other (Figure 14-9a). When the linkages are left intact, however, the axoneme bends when treated with ATP. The combined, coordinated effect of many axonemal doublets sliding simultaneously on alternating sides of the axoneme leads to local bending of a cilium or flagellum (Figure 14-9b). Bends propagate along the length of

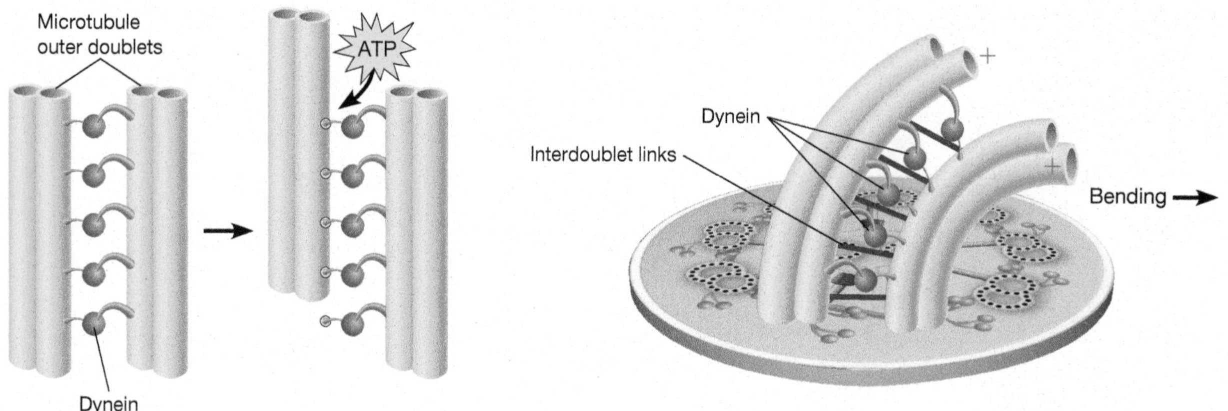

Figure 14-9 Evidence That Bending in Cilia and Flagella Depends on Crosslinks in the Axoneme. **(a)** Outer doublet microtubules and associated proteins can be isolated from cilia or flagella. When ATP is added to such preparations after treating with a protease to remove crosslinks, MTs slide apart. **(b)** In an intact axoneme, bending occurs via the coordinated action of many dyneins, which slide MTs in an organized pattern. A pair of doublets within an axoneme is shown.

cilia or flagella, leading to a wave-like undulation that begins at the base of the organelle and proceeds toward the tip.

Dynein Sidearms Are Responsible for Doublet Sliding. Axonemal dynein is a very large protein. It has multiple subunits, the three largest having ATPase activity and a molecular weight of about 450 kDa each. The dynein arms provide the driving force for axonemal bending. Two kinds of evidence point to this conclusion. First, when dynein is removed from isolated axonemes, the arms disappear from the outer doublets, and the axonemes lose both their ability to hydrolyze ATP and their capacity to beat. Furthermore, the effect is reversible: if purified dynein is added to isolated outer doublets, the sidearms reappear, and in the presence of ATP, sliding is restored. A second kind of evidence comes from studies of mutant flagella in such species as *Chlamydomonas*, a green alga. In some of these mutants, flagella are present but nonfunctional. Depending on the particular mutant, their nonmotile flagella lack dynein arms, radial spokes, or the central pair. These structures are therefore essential for flagellar bending.

During the sliding process, the stalks of the dynein arms attach to and detach from the B tubule in a cyclic manner. Each cycle requires the hydrolysis of ATP and shifts the dynein arms of one doublet relative to the adjacent doublet. In this way, the dynein arms of one doublet move the neighboring doublet, resulting in a relative displacement of the two.

In addition to their role in flagellar and ciliary beating, axonemal dyneins have been implicated in a surprising process: positioning the internal organs of the body (see Human Connections, page 396).

Intraflagellar Transport Adds Components to Growing Flagella and Cilia. The elaborate structure of flagella and cilia raises an interesting question: how are tubulin subunits and other components added to the growing ends of cilia or flagella? Both plus- and minus-end-directed microtubule motors are involved in shuttling components to and from the tips of cilia and flagella. This process, known as **intraflagellar transport (IFT)**, is somewhat analogous to the process of axonal transport in nerve cells: kinesins move material out to the tips, and a cytoplasmic dynein brings material back toward the base. Several human disease syndromes known as *ciliopathies* are now known to result from mutations in IFT components. One ciliopathy, *Bardet Biedl Syndrome (BBS)*, leads to highly variable symptoms, including retinal degeneration, obesity, kidney defects, and extra digits in the hands or feet.

CONCEPT CHECK 14.2

Human sperm swim using a flagellum. They also have mitochondria at the base of the flagellum (see Figure 4.13). What advantage do you think this structure provides, based on what you know about flagella?

14.3 Microfilament-Based Movement Inside Cells: Myosins

Movements of molecules and other intracellular components also occur along another major filament system within the cell—the actin cytoskeleton.

Myosins Are a Large Family of Actin-Based Motors with Diverse Roles in Cell Motility

As with microtubules, ATP-dependent motors interact with and can exert force on actin microfilaments. These motors are all members of a large superfamily of proteins known as **myosins**. Currently, there are 24 known classes of myosins (for a list of some major types of myosins, see Table 14-1).

All myosins have at least one heavy chain, which forms a globular domain at one end attached to a tail of varying length (**Figure 14-10**). The globular domain binds to actin and uses the energy of ATP hydrolysis to move along an actin filament. Many myosins move toward the plus (barbed) ends of actin filaments. (Myosin VI is a well-studied exception: it moves toward the minus, or pointed, ends of actin filaments.)

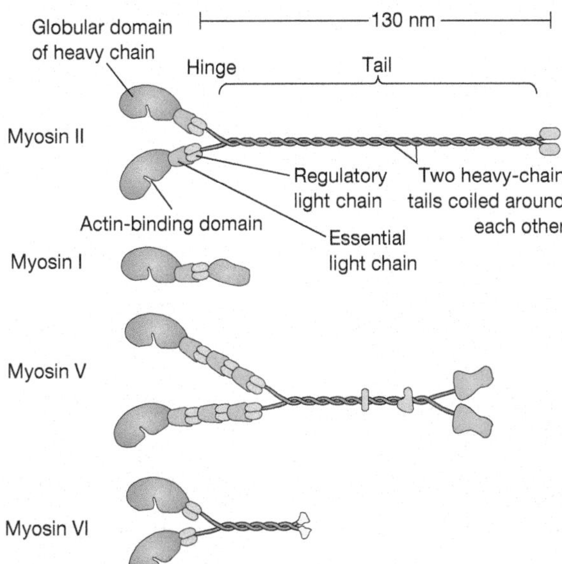

Figure 14-10 Myosin Family Members. All myosins have an actin- and ATP-binding heavy chain, some sort of hinge (lever), and typically two or more regulatory light chains, including essential and regulatory light chains. Some myosins, like myosin I, have one actin-binding domain. Other myosins, like myosins II, V, and VI, associate via their tails, which coil around one another.

The structure of the tail region varies among the different kinds of myosin, giving them the ability to bind to various cargoes and to associate to form dimers or larger arrays. Myosin light chains often play a role in regulating the activity of the myosin ATPase.

Myosins have functions in many events, including muscle contraction (muscle myosin II), cell migration (nonmuscle myosin II), endocytosis (myosin VI), and vesicle transport or other membrane-associated events (myosins I, V).

The best-understood myosins are the **type II myosins**. They are composed of two heavy chains, each featuring a globular domain; a hinge, or lever, region; a long, rodlike tail; and a total of four light chains. In the case of myosin II, one *essential light chain* and one *regulatory light chain* are associated with each heavy chain. These myosins are found in skeletal, cardiac (heart), and smooth muscle cells, as well as in nonmuscle cells. In addition, nonmuscle type II myosins are involved in formation of the contractile ring during cytokinesis (see Chapter 24). Type II myosins are distinctive in that they can assemble into long filaments such as the thick filaments of muscle cells. The basic function of myosin II in all cell types is to pull arrays of actin filaments together, resulting in the contraction of a cell or group of cells.

Many Myosins Move Along Actin Filaments in Short Steps

It is useful at this point to compare two of the best-studied cytoskeletal motor proteins, "classic" kinesin and myosin II. Both have two globular domains that they use to move along a protein filament, and both utilize ATP hydrolysis to change their shape. The force individual myosin motor domains exert on actin is similar to that measured for kinesins. Like kinesin, myosin II is an efficient motor. When myosins must pull against moderate loads, they are about 50% efficient. Despite these similarities, there are profound differences as well. Conventional kinesins operate alone or in small numbers over large distances; a single kinesin can move hundreds of nanometers along a single microtubule. In contrast, a single myosin II molecule slides an actin filament only about 12–15 nm per power stroke. (Not all myosins move such short distances. Myosin V, for example, which is involved in organelle and vesicle transport, is much more processive.)

If myosin II cannot move large distances along actin, how can it be involved in movement? Myosin II molecules often operate in large arrays. In the case of myosin II filaments in muscle, these arrays can contain billions of motors working together to mediate contraction of skeletal muscle, the process to which we now turn.

CONCEPT CHECK 14.3

A single myosin II motor domain can exert about 1–5 piconewtons (pN) of force. The gravitational force of 1 kg near Earth's surface is 9.8 newtons (N) ($1\,pN = 1.0 \times 10^{-12}$ N). How is it that myosin can move kilograms of tissue in large animals?

14.4 Microfilament-Based Motility: Muscle Cells In Action

Muscle contraction is the most familiar example of mechanical work mediated by intracellular filaments. Mammals have several kinds of muscles, including skeletal muscle, cardiac muscle, and smooth muscle. We will first consider skeletal muscle.

Skeletal Muscle Cells Contain Thin and Thick Filaments

Skeletal muscles are responsible for voluntary movement. The structural organization of skeletal muscle is shown in **Figure 14-11** on page 397. A muscle consists of bundles of parallel **muscle fibers** (muscle cells) joined by tendons to the bones that the muscle must move. Each fiber is actually a long, thin, multinucleate cell that is highly specialized for its contractile function. The multinucleate state arises from the fusion of embryonic cells called *myoblasts* during muscle differentiation to produce a *syncytium*. This cell fusion also accounts at least in part for the striking length of muscle cells, which may be many centimeters in length.

At the subcellular level, each muscle cell contains numerous **myofibrils**. Myofibrils are 1–2 μm in diameter and may extend the entire length of the cell. Each myofibril is subdivided along its length into repeating units called **sarcomeres**. The sarcomere is the fundamental contractile unit of the muscle cell. Each sarcomere of the myofibril contains bundles of **thick filaments** and **thin filaments**. Thick filaments consist of myosin, whereas thin filaments consist mainly of F-actin, tropomyosin, and troponin. The thin filaments are arranged around the thick filaments in a hexagonal pattern, as can be seen when the myofibril is viewed in cross section (**Figure 14-12** on page 398).

The filaments in skeletal muscle are aligned in lateral register, giving the myofibrils a pattern of alternating dark

HUMAN *Connections*

Dyneins Help Us Tell Left From Right

Imagine you are a medical student, and you are examining a young patient. You place your stethoscope over the left side of her chest, but you can't find a heartbeat! Because your patient is clearly alive, a little more searching shows that your patient's heart is beating vigorously on the *right* side of her chest. After a more thorough medical examination, you find that all of your patient's internal organs are reversed relative to those of a normal patient.

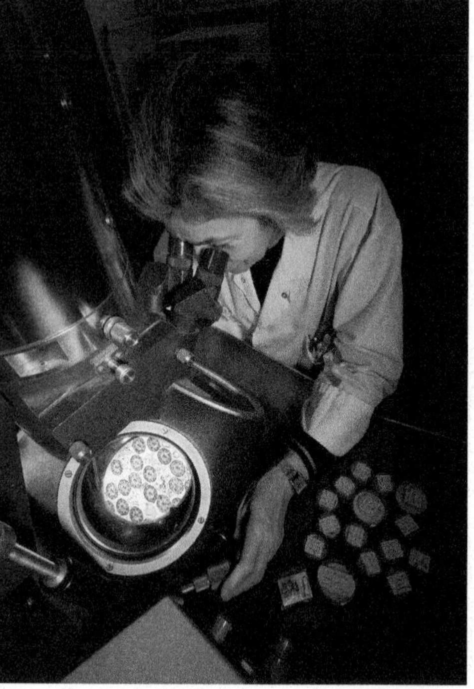

A PCD researcher uses transmission electron microscopy to examine defective cilia.

At least 1 in 10,000 live human births result in an individual whose organs are completely reversed, left to right. Remarkably, this condition, known as *situs inversus totalis* (**Figure 14B-1**), usually has no medical consequences and may not even be recognized until a patient undergoes medical tests for an unrelated condition. In contrast, when only some organs are reversed (*heterotaxia*), serious health complications, and often death, result.

How can this remarkable organ reversal occur? Modern cell biology and genetics have found one culprit: ciliary dyneins. Half of newborn mice that are homozygous recessive for a mutation in the gene for *left-right dynein,* an axonemal dynein, exhibit situs inversus totalis. Left-right dyneins are found in cilia, the beating of which may create the fluid flow necessary to shuttle signaling molecules important for establishing left from right during early embryonic development.

In human patients who inherit the autosomal recessive condition known as *Kartagener syndrome,* there is also a 50% probability of situs inversus totalis. In addition, such patients suffer from various respiratory problems and infertility. What is the connection among this triad of symptoms? Like the mice mentioned above, left-right asymmetry fails to be established reliably. Infertility in affected men is due to a structural defect in the outer dynein sidearms of the sperm tail axoneme, causing the sperm tail to be nonfunctional. Fertility problems also occur in some affected women, probably because faulty cilia in the fallopian tubes cannot sweep eggs from the ovaries to the uterus. The same defects also affect the cilia that move mucus and foreign matter out of the lungs and sinuses (**Figure 14B-2**), causing recurrent bronchitis and sinusitis.

Kartagener syndrome is a classic example of a *primary ciliary dyskinesia (PCD),* so named because cilia display defective movement. PCD can result from a defect in, or lack of, inner or outer dynein sidearms, radial spoke heads, or one or both microtubules of the central pair. This modern detective story is being solved through the powerful techniques of modern cell biology, which are beginning to uncover the underlying causes of situs inversus and how humans tell their left from their right.

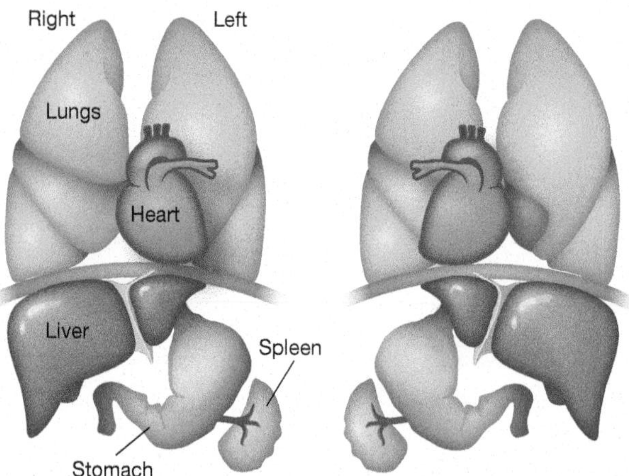

(a) Situs solitus (normal organ assymetry)

(b) Situs inversus totalis (mirror reversal)

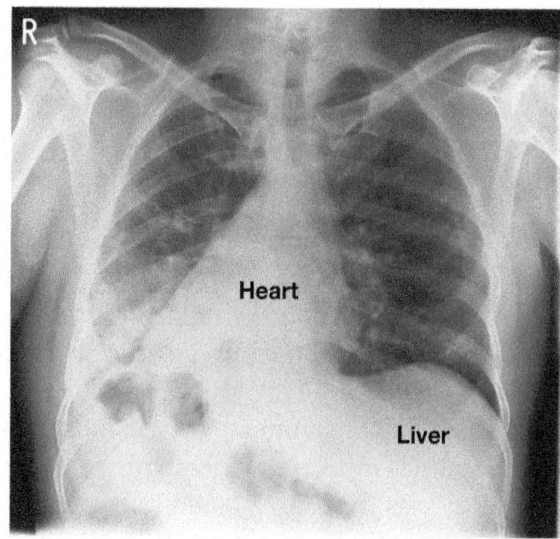

(c) Frontal X-ray of a male with situs inversus totalis

Figure 14B-1 Humans Can Exhibit Left-Right Organ Reversal. **(a)** The normal location of the internal organs is heart on left, liver on right. **(b)** In situs inversus totalis, all internal organs show mirror-image reversal: heart on right, liver on left. The X-ray in part **(c)** comes from such a patient.

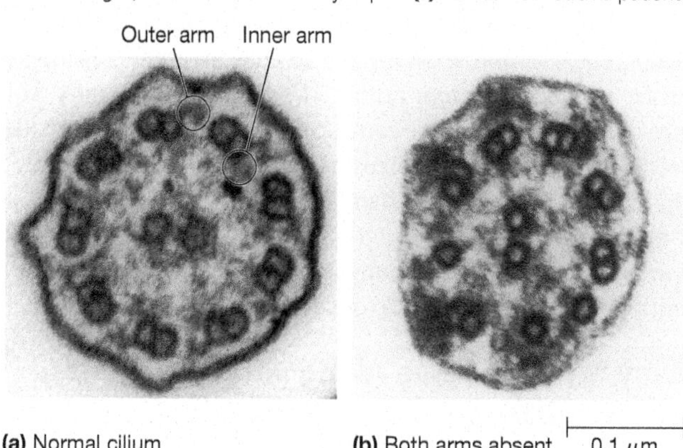

(a) Normal cilium

(b) Both arms absent

Figure 14B-2 Kartagener Syndrome Patients Have Defective Dynein Sidearms. **(a)** Normal respiratory cilia have inner and outer dynein sidearms. **(b)** Respiratory cilia from Kartagener syndrome patients lack both inner and outer sidearms (TEM).

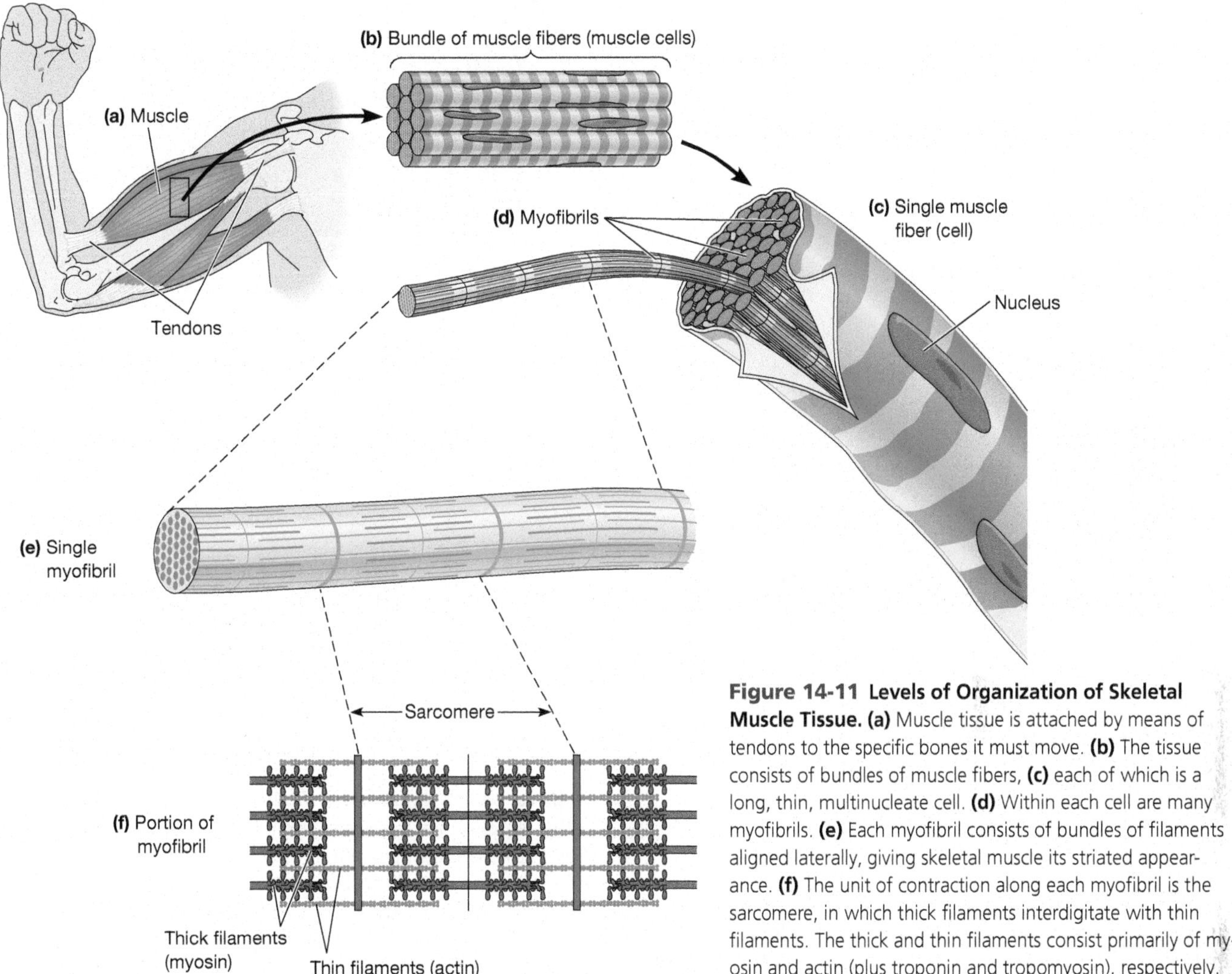

Figure 14-11 Levels of Organization of Skeletal Muscle Tissue. **(a)** Muscle tissue is attached by means of tendons to the specific bones it must move. **(b)** The tissue consists of bundles of muscle fibers, **(c)** each of which is a long, thin, multinucleate cell. **(d)** Within each cell are many myofibrils. **(e)** Each myofibril consists of bundles of filaments aligned laterally, giving skeletal muscle its striated appearance. **(f)** The unit of contraction along each myofibril is the sarcomere, in which thick filaments interdigitate with thin filaments. The thick and thin filaments consist primarily of myosin and actin (plus troponin and tropomyosin), respectively.

and light bands (**Figure 14-13** on page 398). This pattern of bands, or *striations*, is characteristic of skeletal and cardiac muscle, which are therefore referred to as **striated muscle** (Figure 14-13a). The dark bands are called **A bands**, and the light bands are called **I bands**. (This terminology was developed from observations originally made with the polarizing light microscope. I stands for *isotropic* and A for *anisotropic*, terms related to the appearance of these bands when illuminated with plane-polarized light.)

As illustrated in Figure 14-13b, the lighter region in the middle of each A band is called the **H zone** (from the German word *hell*, meaning "light"). Running down the center of the H zone is the **M line**, which contains *myomesin*, a protein that links myosin filaments together. In the middle of each I band appears a dense **Z line** (from the German word *zwischen*, meaning "between"). The distance from one Z line to the next defines a single sarcomere. A sarcomere is about 2.5–3.0 μm long in the relaxed state but shortens progressively as the muscle contracts.

Sarcomeres Contain Ordered Arrays of Actin, Myosin, and Accessory Proteins

The striated pattern of skeletal muscle and the observed shortening of the sarcomeres during contraction are due to the arrangement of thick and thin filaments in myofibrils, which we now examine in some detail.

Thick Filaments. The thick filaments of myofibrils are about 15 nm in diameter and about 1.6 μm long (**Figure 14-14** on page 398). They lie parallel to one another in the middle of the sarcomere (see Figure 14-13). Every thick filament consists of hundreds of myosin molecules organized in a staggered array such that the globular domains of successive molecules protrude from the thick filament in a repeating pattern, facing away from the center (Figure 14-14a). Globular domains are spaced 14.3 nm apart along the thick filament, with each pair displaced one-third of the way around the filament from the previous pair (Figure 14-14b), allowing them to form the cross-bridges between thick and thin filaments that are essential for muscle contraction.

Thin Filaments. The thin filaments of myofibrils interdigitate with the thick filaments. The thin filaments are about 7 nm in diameter and about 1 μm long. Each I band consists of *two* sets of thin filaments, one set on either side of the Z line, with each filament attached to the Z line and extending toward and into the A band in the center of the sarcomere. This accounts for the length of almost 2 μm for I bands in relaxed

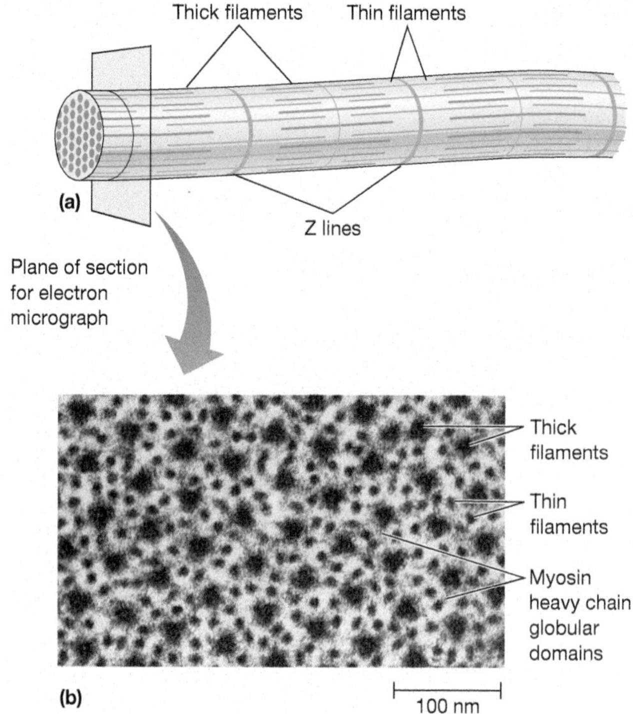

Figure 14-12 Arrangement of Thick and Thin Filaments in a Myofibril. (a) A myofibril consists of interdigitated thick and thin filaments. **(b)** The thin filaments are arranged around the thick filaments in a hexagonal pattern, as seen in this cross section of a flight muscle from the fruit fly *Drosophila melanogaster* viewed by high-voltage electron microscopy (HVEM).

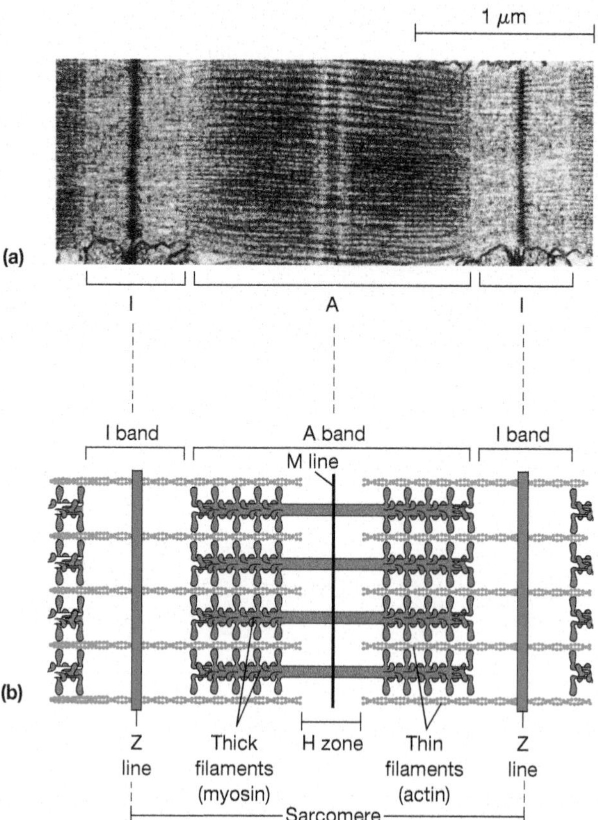

Figure 14-13 Appearance of and Nomenclature for Skeletal Muscle. (a) An electron micrograph of a single sarcomere (TEM). **(b)** A schematic diagram of a sarcomere. An A band corresponds to the length of the thick filaments, and an I band represents the portion of the thin filaments that does not overlap with thick filaments. The lighter area in the center of the A band is called the H zone; the line in the middle is known as the M line. The dense zone in the center of each I band is called the Z line. A sarcomere, the basic repeating unit along the myofibril, is the distance between two successive Z lines.

muscle. The structure of an individual thin filament is shown in **Figure 14-15**. It consists of at least three proteins. The most important component is an F-actin helix, intertwined with the proteins **tropomyosin** and **troponin**. One turn of the actin helix, as shown in Figure 14-15, is a bit longer than 70 nm.

Tropomyosin is a long, rodlike molecule that fits in grooves along the sides of the F-actin helix. Each tropomyosin molecule stretches for about 40 nm along the filament and associates along its length with seven G-actin monomers.

Troponin is actually a complex of three polypeptide chains, called *TnT, TnC,* and *TnI.* (*Tn* stands for troponin, *T* for tropomyosin, *C* for calcium, and *I* for inhibitory because TnI inhibits muscle contraction.) TnT binds to tropomyosin and is responsible for positioning the complex on the tropomyosin molecule. TnC binds calcium ions, and TnI binds to actin. One

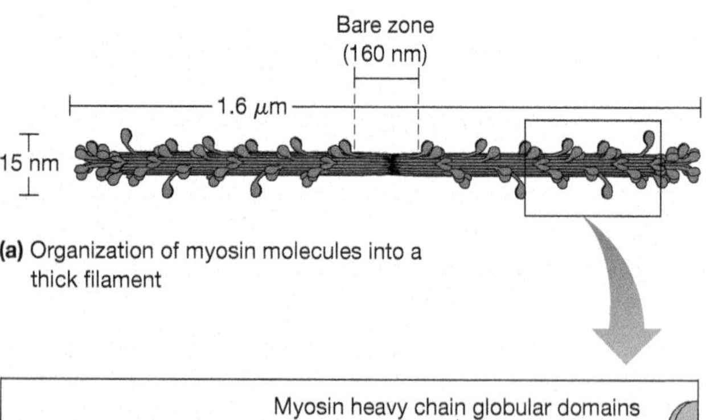

(a) Organization of myosin molecules into a thick filament

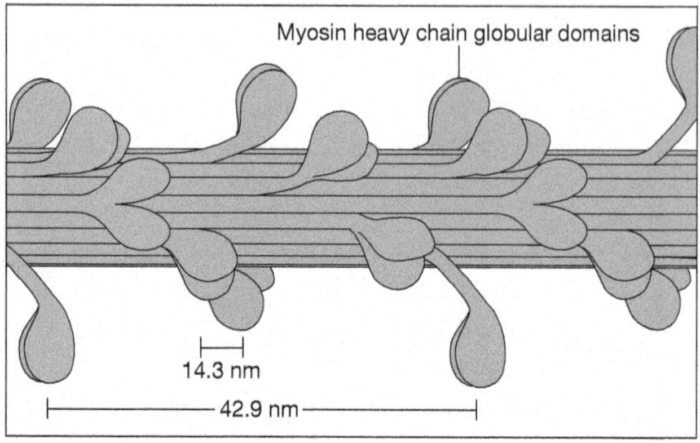

(b) Portion of a thick filament

Figure 14-14 The Thick Filament of Skeletal Muscle. (a) The thick filament of the myofibril consists of hundreds of myosin molecules organized in a repeating, staggered array. A typical thick filament is about 1.6 μm long and about 15 nm in diameter. Individual myosin molecules are integrated into the filament longitudinally, with their ATPase-containing heads oriented away from the center of the filament. The central region of the filament is a bare zone containing no heads. **(b)** An enlargement of a portion of the thick filament shows that pairs of myosin heads are spaced 14.3 nm apart.

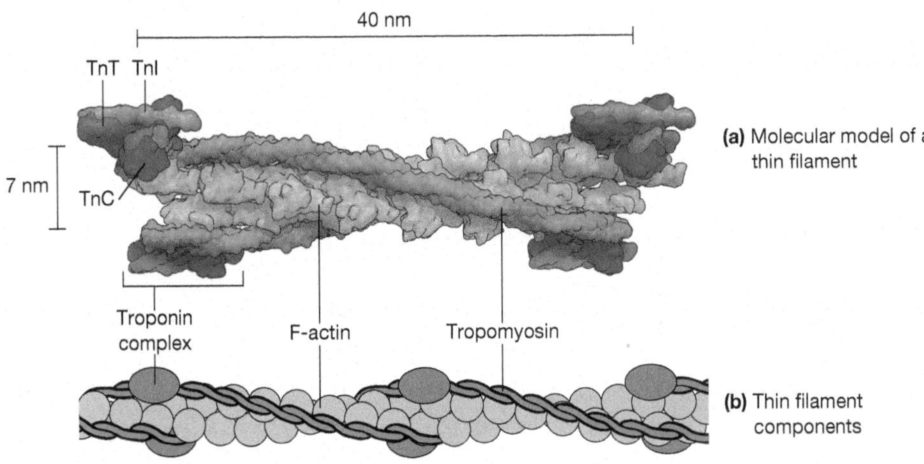

Figure 14-15 The Thin Filament of Skeletal Muscle. Each thin filament is a single strand of F-actin in which the G-actin monomers are staggered and give the appearance of a double-stranded helix. One result of this arrangement is that two grooves run along both sides of the filament. Long, ribbon-like molecules of tropomyosin lie in these grooves. Each tropomyosin molecule consists of two α helices wound about each other to form a coiled-coil "rope." Associated with each tropomyosin molecule is a troponin complex consisting of the three subunits TnT, TnC, and TnI.

troponin complex is associated with each tropomyosin molecule, so the spacing between successive troponin complexes along the thin filament is about 40 nm.

As discussed in a later section, troponin and tropomyosin constitute a calcium-sensitive switch that activates contraction in both skeletal and cardiac muscle.

Organization of Muscle Filament Proteins. How can the filamentous proteins of muscle fibers maintain such a precise organization when filaments in other cell types are relatively disorganized? First, the actin in the thin filaments is oriented such that all of the plus (barbed) ends are anchored at Z lines. Because myosin II moves toward the plus end of F-actin, this guarantees that the thick filaments will move toward the Z lines.

Second, structural proteins play a central role in maintaining the architectural relationships of muscle proteins (**Figure 14-16**). For instance, *α-actinin* keeps actin filaments bundled into parallel arrays at the Z line. The capping protein, *CapZ*, maintains the attachment of the barbed (plus) ends of actin filaments to the Z line and simultaneously caps the actin in the thin filaments. At the other end of the thin filaments is *tropomodulin*, which helps maintain the length and stability of thin filaments by binding their pointed (minus) ends. *Myomesin* is present at the M line of thick filament arrays and bundles the myosin molecules composing the arrays. A third structural protein, *titin*, attaches the thick filaments to the Z lines. Titin is highly flexible; during contraction-relaxation cycles, it can keep thick filaments in the correct position relative to thin filaments. This titin scaffolding also keeps the thick and thin filament arrays from being pulled apart when a muscle is stretched. Another protein, *nebulin*, stabilizes the organization of thin filaments.

The various protein components of vertebrate skeletal muscle are summarized in **Table 14-2** on page 400. The contractile process involves the complex interaction of all these proteins.

The Sliding-Filament Model Explains Muscle Contraction

With our understanding of muscle structure, we can now consider what happens during the contraction process. Based on electron microscopic studies, it is clear that the A bands of the myofibrils remain fixed in length during contraction, whereas the I bands shorten progressively and virtually disappear in the fully contracted state. To explain these observations, the **sliding-filament model** illustrated in **Figure 14-17** on page 400 was proposed in 1954 independently by Andrew Huxley and Rolf Niedergerke and by Hugh Huxley and Jean Hanson. According to this model, contraction is due to thin filaments sliding past thick filaments, with no change in the length of either type of filament. The sliding-filament model not only proved to be correct but also was instrumental in focusing attention on the molecular interactions between thick and thin filaments that underlie the sliding process.

As Figure 14-17a indicates, contraction involves the sliding of thin filaments such that they are drawn progressively toward the H zone, into the spaces between adjacent thick

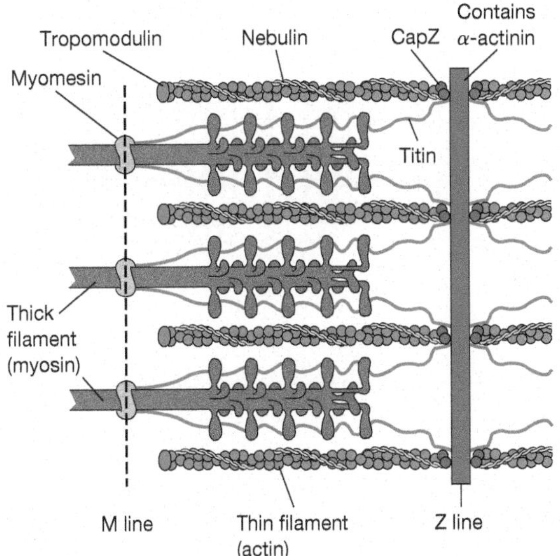

Figure 14-16 Structural Proteins of the Sarcomere. The thick and thin filaments require structural support to maintain their precise organization. The support is provided by α-actinin and myomesin, which bundle actin and myosin filaments, respectively. Titin attaches thick filaments to the Z line, thereby maintaining their position. Nebulin stabilizes the organization of thin filaments. For clarity, tropomyosin is not shown.

Table 14-2 Major Protein Components of Vertebrate Skeletal Muscle

Protein	Molecular Weight (kDa)	Function
G-actin	42	Major component of thin filaments
Myosin	510	Major component of thick filaments
Tropomyosin	64	Binds along the length of thin filaments
Troponin	78	Positioned at regular intervals along thin filaments; mediates calcium regulation of contraction
Titin	2500	Links thick filaments to Z line
Nebulin	700	Links thin filaments to Z line; stabilizes thin filaments
Myomesin	185	Myosin-binding protein present at the M line of thick filaments
α-actinin	190	Bundles actin filaments and attaches them to Z line
Ca^{2+} ATPase	115	Major protein of sarcoplasmic reticulum (SR); transports Ca^{2+} into SR to relax muscle
CapZ	68	Attaches actin filaments to Z line; caps actin
Tropomodulin	41	Maintains thin filament length and stability

filaments, overlapping more and more with the thick filaments and narrowing the I band in the process. The result is a shortening of individual sarcomeres (and the myofibrils to which they belong) and a contraction of the muscle cell and the whole tissue. This, in turn, causes the movement of the body parts attached to the muscle.

The sliding of the thin and thick filaments past each other as a means of generating force suggests that there should be a relationship between the force generated during contraction and the degree of shortening of the sarcomere. In fact, when the relationship between shortening and force is measured in isolated muscle fibers, it is exactly what the sliding-filament

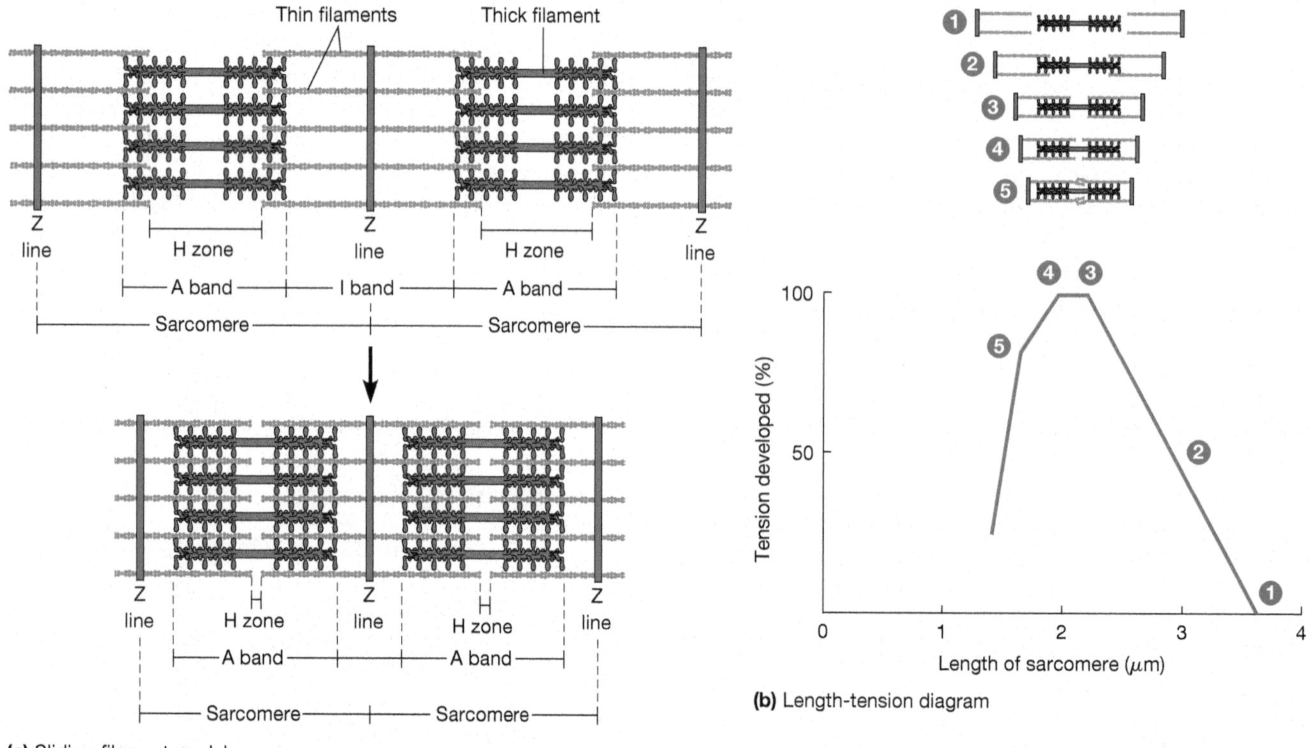

Figure 14-17 The Sliding-Filament Model of Muscle Contraction. (a) Sarcomeres of a myofibril during the contraction process. A myofibril contracts by the progressive sliding of thick and thin filaments past each other, leading to a progressive decrease in the width of the I band. **(b)** This graph shows that the amount of tension developed by the sarcomere is proportional to the amount of overlap between the thin filament and the region of the thick filament containing myosin heads. ❶ When a muscle fiber is stretched so much that there is no overlap between thin and thick filaments, no tension can be developed. ❷ When a normal, relaxed sarcomere begins to contract, the Z lines move closer together, increasing the amount of overlap between thin and thick filaments so, the muscle can develop more tension. ❸–❹ This proportional relationship continues until the ends of the thin filaments move into the H zone. Here thin filaments encounter no further myosin heads, so tension remains constant. ❺ Any further shortening of the sarcomere results in a dramatic decline in tension as the filaments crowd into one another.

model predicts: the amount of force the muscle can generate during a contraction depends on the number of actin-binding domains from the thick filament that can make contact with the thin filament (Figure 14-17b).

Cross-Bridges Hold Filaments Together, and ATP Powers Their Movement

The sliding of thin filaments past thick filaments depends on the elaborate structural features of the sarcomere, and it requires energy. These basic observations raise important questions. First, by what mechanism are the thin filaments pulled progressively into the spaces between thick filaments to cause contraction? Second, how is the energy of ATP used to drive this process? These questions are addressed next.

Cross-Bridge Formation. Regions of overlap between thick and thin filaments, whether extensive (in contracted muscle) or minimal (in relaxed muscle), are always characterized by the presence of transient **cross-bridges**. The cross-bridges are formed by interactions between the actin-binding domains of myosins in the thick filaments and the F-actin of the thin filaments (**Figure 14-18**). For contraction, the cross-bridges must form and dissociate repeatedly, so that each cycle of cross-bridge formation causes the thin filaments to interdigitate with the thick filaments more and more, thereby shortening individual sarcomeres and the muscle fiber.

A given myosin heavy chain on the thick filament undergoes a cycle of events in which it binds actin on the thin filament, undergoes an energy-requiring change in shape that pulls the thin filament toward the H zone, and then breaks its association with the thin filament and associates with another site farther toward the Z line along the thin filament. Muscle contraction is the net result of the repeated making and breaking of many such cross-bridges.

The driving force for cross-bridge formation is the hydrolysis of ATP, catalyzed by the myosin heavy chains. The requirement for ATP can be demonstrated in vitro because isolated muscle fibers contract in response to added ATP.

The Contraction Cycle. The mechanism of muscle contraction is depicted in **Figure 14-19** on page 402. ❶ Beginning at its high-energy, "cocked" configuration, a myosin globular domain contains an ADP and a P_i molecule (a hydrolyzed ATP, in effect). ❷ The transition of myosin to the more tightly bound state triggers a conformational change in the myosin heavy chain, which involves release of P_i. ❸ This conformational change is associated with the *power stroke*, a movement of the myosin heavy chain that causes the thick filament to pull against the thin filament, which then moves with respect to the thick filament. Next, ADP is released from the myosin heavy chain. ❹ At this point, the myosin heavy chain remains attached to the thin filament. ❺ The next step, cross-bridge dissociation, requires that a new ATP molecule bind to the myosin heavy chain. Binding of ATP causes the myosin heavy chain to change its conformation in a way that weakens its binding to actin, triggering detachment. (In the absence of adequate ATP, cross-bridge dissociation does not occur, and the muscle becomes locked in a stiff, rigid state called *rigor.* The *rigor mortis* associated with death results from the depletion of

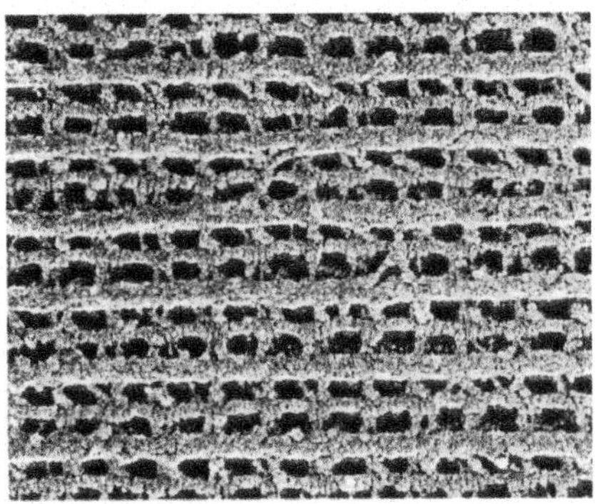

Figure 14-18 Cross-Bridges. The cross-bridges between thick and thin filaments formed by the projecting heads of myosin molecules can be readily seen in this high-resolution electron micrograph (TEM).

ATP and the progressive accumulation of cross-bridges in the configuration shown at the end of ❹.)

Note that, once detached, the thick and thin filaments would be free to slip back to their previous positions, but they are held together at all times by the many other cross-bridges along their length at any given moment—just as at least some legs of a millipede are always in contact with the surface it is walking on. In fact, each thick filament has about 350 myosin heavy chains, and each one attaches and detaches about five times per second during rapid contraction, so there are always many cross-bridges intact at any time.

❻ Finally, ATP hydrolysis returns the myosin heavy chain to the high-energy configuration necessary for the next round of cross-bridge formation and filament sliding.

> **MAKE CONNECTIONS 14.2**
>
> Skeletal muscle myosin heavy chain is a mechanoenzyme that requires ATP for its function. Other enzymes, such as phosphofructokinase-2, also rely on ATP. How are PFK-2 and myosin II different in their use of ATP? (Fig. 9-14)

The Regulation of Muscle Contraction Depends on Calcium

So far, our description of muscle contraction has implied that skeletal muscle ought to contract continuously as long as there is sufficient ATP. Yet experience tells us that most skeletal muscles spend more time in the relaxed state than in contraction. Contraction and relaxation must therefore be regulated to result in the coordinated movements associated with muscle activity.

The Role of Calcium in Contraction. Regulation of muscle contraction depends on free calcium ions (Ca^{2+}) and on the muscle cell's ability to rapidly raise and lower the calcium level in the cytosol (called the *sarcoplasm* in muscle cells) around the myofibrils. The regulatory proteins *tropomyosin* and *troponin* act together to regulate the availability of myosin-binding sites on actin filaments in a way that depends critically on the level of calcium in the sarcoplasm.

To understand how this process works, it is important to recognize that the myosin-binding sites on actin are

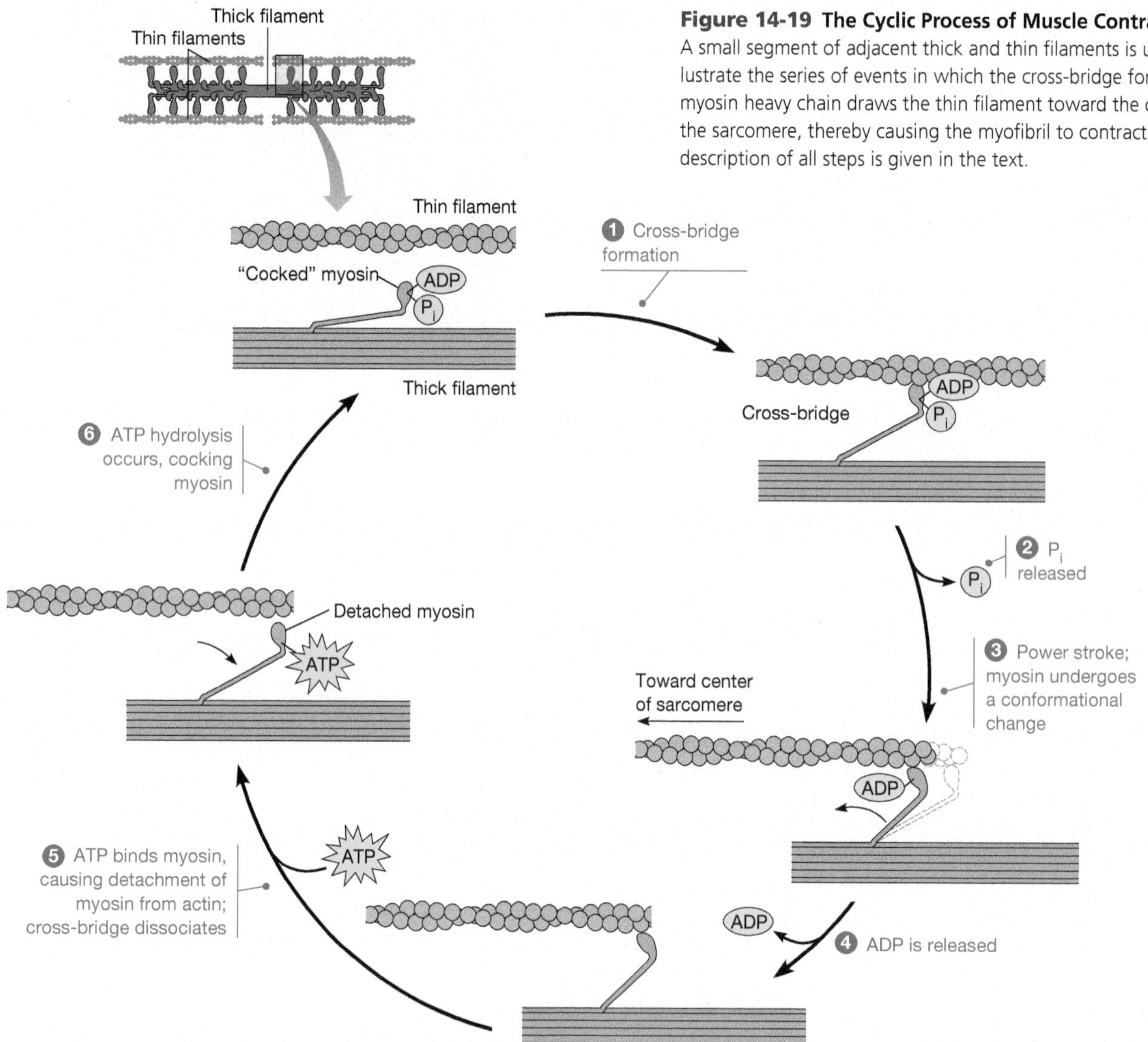

Figure 14-19 The Cyclic Process of Muscle Contraction. A small segment of adjacent thick and thin filaments is used to illustrate the series of events in which the cross-bridge formed by a myosin heavy chain draws the thin filament toward the center of the sarcomere, thereby causing the myofibril to contract. A detailed description of all steps is given in the text.

normally blocked by tropomyosin. For myosin to bind to actin and initiate the cross-bridge cycle, the tropomyosin molecule must be moved out of the way. The calcium dependence of muscle contraction is due to troponin C (TnC), which binds calcium ions. When a calcium ion binds to TnC, it undergoes a conformational change that is transmitted to the tropomyosin molecule, causing it to move toward the center of the helical grooves of the thin filament, out of the blocking position. The binding sites on actin are then accessible to the myosin heavy chains, allowing contraction to proceed.

Figure 14-20 illustrates how the troponin-tropomyosin complex regulates the interaction between actin and myosin. When the calcium concentration in the sarcoplasm is low (0.1 μM), tropomyosin blocks the binding sites on the F-actin filament, effectively preventing their interaction with myosin (Figure 14-20a). As a result, cross-bridge formation is inhibited, and the muscle becomes or remains relaxed. At higher calcium concentrations (1 μM), calcium binds to TnC, causing tropomyosin molecules to shift their position, which allows myosin heavy chains to make contact with the binding sites on the F-actin filament and thereby initiate contraction (Figure 14-20b).

When the calcium concentration falls again as it is pumped out of the sarcoplasm (discussed next), calcium dissociates from TnC and the tropomyosin moves back to the blocking position. Myosin binding is therefore inhibited, further cross-bridge formation is prevented, and the contraction cycle ends.

Regulation of Calcium Levels in Skeletal Muscle Cells. Muscle contraction is regulated by the concentration of calcium ions in the sarcoplasm. But how is the level of calcium controlled? Think for a moment about what must happen when you move any part of your body—when you flex an index finger, for instance. A nerve impulse is generated in the brain and transmitted down the spinal column to the nerve cells, or *motor neurons,* that control a small muscle in the forearm. The motor neurons activate the appropriate muscle cells, which contract and relax, all within about 100 msec. When nerve impulses to the muscle cell cease, the calcium level declines quickly and the muscle relaxes. Therefore, to

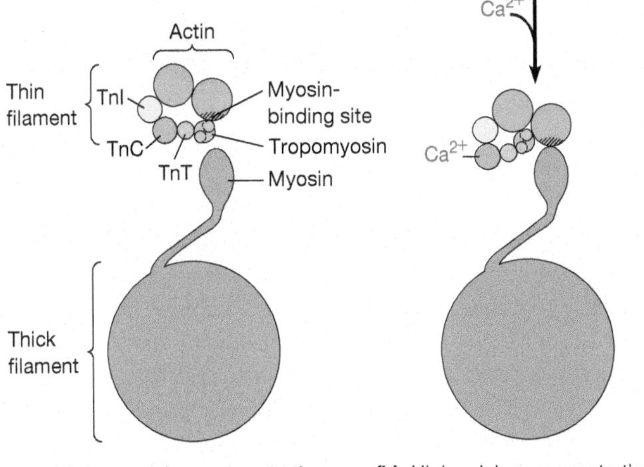

Figure 14-20 Regulation of Contraction in Striated Muscle. **(a)** At low concentrations (< 0.1μMCa^{2+}), calcium is not bound to the TnC subunit of troponin, and tropomyosin blocks the binding sites on actin, preventing access by myosin and thereby maintaining the muscle in the relaxed state. **(b)** At high concentrations (> 1μMCa^{2+}), calcium binds to the TnC subunit of troponin, inducing a conformational change that is transmitted to tropomyosin. The tropomyosin is displaced, allowing myosin to gain access to the binding sites on actin and thereby initiating contraction.

understand how muscle contraction is regulated, we need to know how nerve impulses cause calcium level in the sarcoplasm to rise and fall.

Events at the Neuromuscular Junction. The signal for a muscle cell to contract is conveyed by a nerve cell. The site where the nerve *innervates*, or nearly makes contact with, the muscle cell is called the **neuromuscular junction**. At the neuromuscular junction, the axon branches out and forms *axon terminals* that come in very close proximity to the muscle cell. These terminals contain the neurotransmitter *acetylcholine*, which is stored in membrane-enclosed vesicles and secreted by axon terminals in response to an action potential. The area of the muscle cell plasma membrane under the axon terminals is called the *motor end plate*. There, in the plasma membrane (called the *sarcolemma* in muscle cells), clusters of acetylcholine receptors are associated with each axon terminal. The receptor is a regulated sodium channel; when the receptor binds acetylcholine, it allows sodium ions to flow into the muscle cell. The sodium influx in turn causes a membrane depolarization to be transmitted away from the motor end plate.

Transmission of an Impulse to the Interior of the Muscle. Once a membrane depolarization occurs at the motor end plate, it spreads throughout the sarcolemma via the **transverse (T) tubule system** (**Figure 14-21**), a series of regular inpocketings of the muscle membrane that penetrate the interior of a muscle cell. The T tubules conduct membrane depolarization deep into the muscle cell, and they are part of the reason that muscle cells can respond so quickly to a nerve impulse.

Inside the muscle cell, the T tubule system comes into contact with the **sarcoplasmic reticulum (SR)**. As the name suggests, the SR is similar to the endoplasmic reticulum (ER) found in nonmuscle cells except that it is highly specialized. The SR runs along the myofibrils, where it is poised to release calcium ions directly into the myofibril and cause contraction and then to remove calcium from the myofibril and cause relaxation. This close proximity of the SR to the myofibrils facilitates the rapid response of muscle cells to nerve signals.

SR Function in Calcium Release and Uptake. The SR can be functionally divided into two components: the *medial element* and the *terminal cisternae* (singular: *terminal cisterna;* Figure 14-21). The terminal cisternae of the SR contain a high concentration of an ATP-dependent calcium pump that is a member of a family of a type of ATPase pump

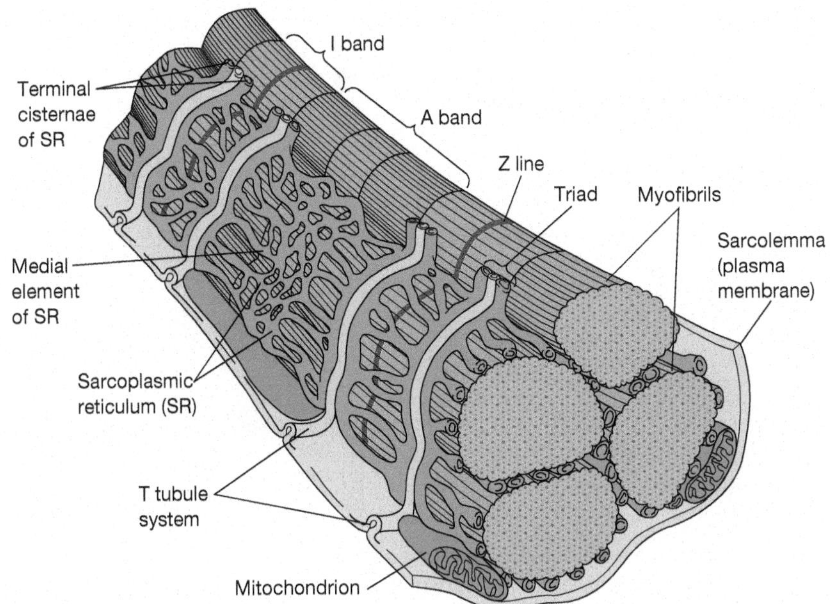

Figure 14-21 The Sarcoplasmic Reticulum and the Transverse Tubule System of Skeletal Muscle Cells. The sarcoplasmic reticulum (SR) is an extensive network of specialized ER that accumulates calcium ions and releases them in response to nerve signals. T tubules are invaginations of the sarcolemma (plasma membrane) that relay changes in membrane potential to the interior of the cell. Where the T tubule passes near the terminal cisternae of the SR, a triad structure is formed that appears in electron micrographs as the cross section of three adjacent tubes. The T tubule is in the middle, and on each side is one of the SR terminal cisternae.

called **SERCAs**, or **sarco/endoplasmic reticulum Ca^{2+}-ATPases**, that continually pump calcium into the lumen of the SR. The SERCA in skeletal muscle, called *SERCA1*, is very abundant; it accounts for as much as 90% of membrane proteins in the SR of skeletal muscle. Calcium pumping produces a high calcium concentration in the lumen of the SR (up to several millimolar). This calcium can then be released from the terminal cisternae of the SR when needed. Figure 14-21 shows how the terminal cisternae of the SR are positioned adjacent to the contractile apparatus of each myofibril (at the junctions between A and I bands).

Terminal cisternae are typically found right next to a T tubule, giving rise to a structure called a **triad**. In electron micrographs, a triad appears as three circles in a row. The central circle is the membrane of the T tubule, and the circles on each side are the membranes of the terminal cisternae.

The proximity of the T tubule, the terminal cisternae of the SR, and the contractile machinery of the myofibril provides the basis for how muscle cells can respond so rapidly to a nerve impulse. Membrane depolarization travels from the motor end plate, spreads out over the sarcolemma, and enters the T tubule (**Figure 14-22**). As the membrane depolarization travels down the T tubule, it activates voltage-gated calcium channels in the T tubule that are adjacent to other calcium channels, known as *ryanodine receptors*, in the terminal cisternae of the SR, causing them to open. When the ryanodine receptor channels open, calcium rushes into the sarcoplasm immediately adjacent to the myofibrils, causing contraction.

Letting calcium out of the SR causes a muscle cell to contract. For the muscle cell to relax, the calcium level must be brought back down to the resting level. This is accomplished by SERCA, which pumps calcium back into the SR.

The pumping of calcium from the sarcoplasm back into the SR cisternae quickly lowers the sarcoplasmic calcium level to a point at which troponin releases calcium, tropomyosin moves back to the blocking position on actin, and further cross-bridge formation is prevented. Cross-bridges therefore disappear rapidly as myosin dissociates from actin and becomes blocked by tropomyosin. This leaves the muscle relaxed and allows the thin filaments to slide out from between the thick filaments, due to passive stretching of the muscle by other tissues.

The Coordinated Contraction of Cardiac Muscle Cells Involves Electrical Coupling

Cardiac (heart) muscle is responsible for the beating of the heart and the pumping of blood through the body's circulatory system. Cardiac muscle functions continuously. Astoundingly, in one year, your heart beats about 40 million times. Cardiac muscle is very similar to skeletal muscle in the organization of actin and myosin filaments and has the same striated appearance (**Figure 14-23**).

An important difference between cardiac and skeletal muscle is that heart muscle cells are not multinucleate. Instead, during embryonic development, cells called *cardiomyocytes* are joined end to end through structures called **intercalated discs**. The discs have a high density of desmosomes and *gap junctions* (see Chapter 15); the gap junctions act as small conduits that electrically couple neighboring cells, allowing depolarization waves to spread throughout the heart during its contraction cycle. The heart contracts spontaneously once every second or so. Heart rate is controlled by a "pacemaker" region in an upper portion of the heart (right atrium). The depolarization wave initiated by the pacemaker then spreads to the rest of the heart to produce the heartbeat. Whereas in skeletal muscle voltage-gated calcium channels are directly involved in opening ryanodine receptor channels, in cardiac cells the voltage-gated calcium channels release a small amount of calcium that indirectly leads to a large release of calcium from the ryanodine receptors.

Smooth Muscle Is More Similar to Nonmuscle Cells than to Skeletal Muscle

Smooth muscle is responsible for involuntary contractions such as those of the stomach, intestines, uterus, and blood vessels. In general, such contractions are slow, taking up to

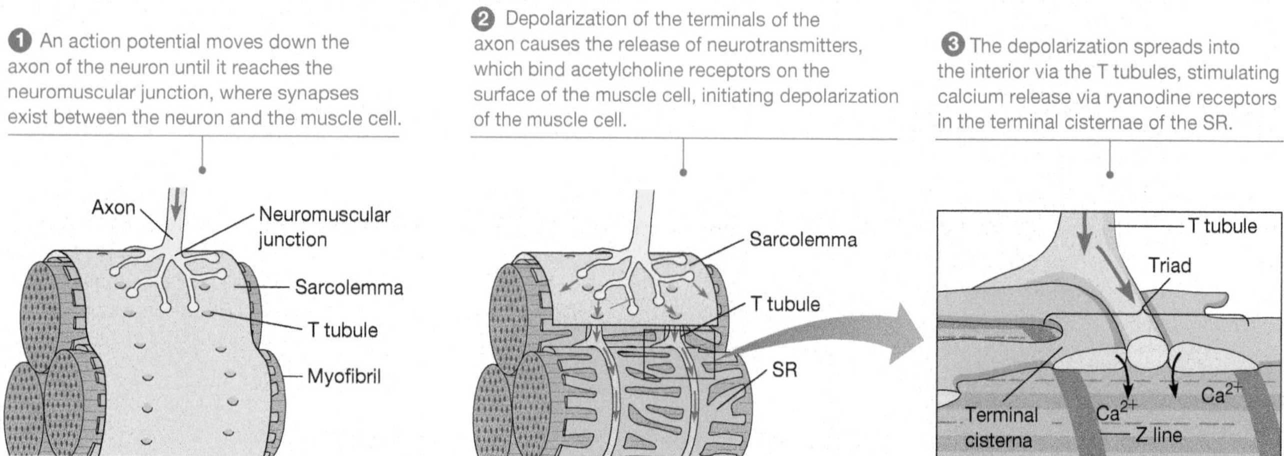

1 An action potential moves down the axon of the neuron until it reaches the neuromuscular junction, where synapses exist between the neuron and the muscle cell.
2 Depolarization of the terminals of the axon causes the release of neurotransmitters, which bind acetylcholine receptors on the surface of the muscle cell, initiating depolarization of the muscle cell.
3 The depolarization spreads into the interior via the T tubules, stimulating calcium release via ryanodine receptors in the terminal cisternae of the SR.
Axon
Neuromuscular junction
Sarcolemma
T tubule
Myofibril
Sarcolemma
T tubule
SR
T tubule
Triad
Ca2+
Ca2+
Terminal cisterna
Z line

Figure 14-22 Stimulation of a Muscle Cell by a Nerve Impulse. The nerve causes a depolarization of the muscle cell membrane, which spreads into the cell interior via the T tubule system, stimulating calcium release from the terminal cisternae of the sarcoplasmic reticulum (SR).

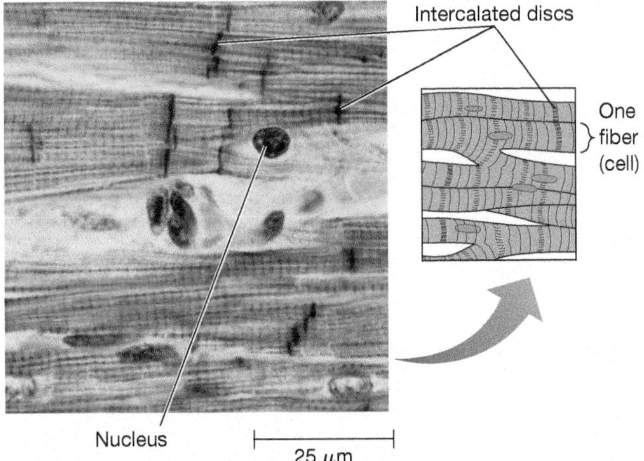

Figure 14-23 Cardiac Muscle Cells. Cardiac muscle cells have a contractile mechanism and sarcomeric structure similar to those of skeletal muscle cells. However, unlike skeletal muscle cells, cardiac muscle cells are joined together end to end at intercalated discs, which allow ions and electrical signals to pass from one cell to the next. This ionic permeability enables a contraction stimulus to spread evenly to all the cells of the heart (LM).

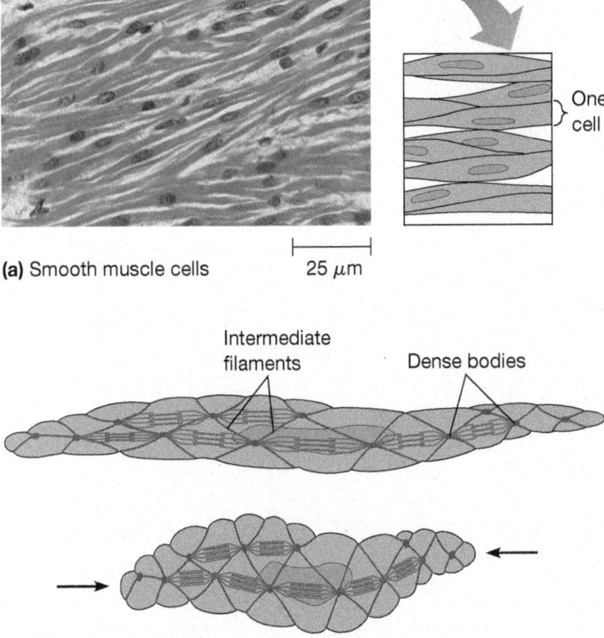

Figure 14-24 Smooth Muscle and Its Contraction. (a) Individual smooth muscle cells are long and spindle shaped, with no Z lines or sarcomeric structure (LM). **(b)** In a smooth muscle cell, contractile bundles of actin and myosin are anchored to plaque-like structures called dense bodies. The dense bodies are connected to each other by intermediate filaments, thereby orienting the actin and myosin bundles obliquely to the long axis of the cell. When the actin and myosin bundles contract, they pull on the dense bodies and intermediate filaments, producing the cellular contraction shown here.

5 sec to reach maximum tension. Smooth muscle contractions are also of greater duration than those of skeletal or cardiac muscle, as is required in these organs and tissues.

The Structure of Smooth Muscle. Smooth muscle cells are long and thin, with pointed ends (**Figure 14-24**). Unlike skeletal or heart muscle, smooth muscle has no striations (Figure 14-24a). Smooth muscle cells do not contain Z lines, which are responsible for the periodic organization of the sarcomeres found in skeletal and cardiac muscle cells. Instead, smooth muscle cells contain *dense bodies,* plaque-like structures in the sarcoplasm and on the cell membrane (Figure 14-24b) that are connected by a network of intermediate filaments. Contractile bundles of actin and myosin filaments are also anchored at both ends to the dense bodies. As a result, contractile bundles are aligned obliquely to the long axis of the cell. Myosin cross-bridges connect thick and thin filaments in smooth muscle but not in the regular, repeating pattern seen in skeletal muscle.

Regulation of Contraction in Smooth Muscle Cells. Smooth muscle contraction is regulated in a manner distinct from that of skeletal muscle cells. Although the focus is on smooth muscle here, nonmuscle cells that contract use a similar mechanism. When calcium concentrations increase in smooth muscle and nonmuscle cells, a cascade of events takes place that includes the activation of **myosin light-chain kinase (MLCK)**. (Recall from Chapter 6 that kinases are enzymes that add phosphate groups to substrates.) Activated MLCK phosphorylates one type of myosin light chain known as a **regulatory light chain** (see Figure 14-10).

Myosin light-chain phosphorylation affects myosin in two ways (**Figure 14-25** on page 406). First, some myosin molecules are curled up so that they cannot assemble into filaments. When the myosin light chain is phosphorylated, the myosin tail uncurls and becomes capable of assembly (Figure 14-25a).

Second, light-chain phosphorylation activates myosin, enabling it to interact with actin filaments. The cascade of events involved in myosin activation is shown in Figure 14-25b. In response to a nerve impulse or hormonal signal reaching a smooth muscle cell, an influx of extracellular calcium ions occurs, increasing the intracellular calcium concentration and causing contraction. Elevation of intracellular calcium can activate the calcium-sensitive protein *calmodulin* (discussed in greater detail in Chapter 23). The resulting *calcium-calmodulin complex* can bind to myosin light-chain kinase, activating the enzyme. As a result, myosin light chains become phosphorylated, and myosin can interact with actin to cause contractions.

When the calcium level within smooth muscle cells eventually drops, MLCK is inactivated, and a second enzyme, *myosin light-chain phosphatase,* removes the phosphate group from the myosin light chain. Because the dephosphorylated myosins can no longer bind to actin, the muscle cell relaxes.

Thus, both skeletal muscle and smooth muscle are activated to contract by calcium ions but from different sources

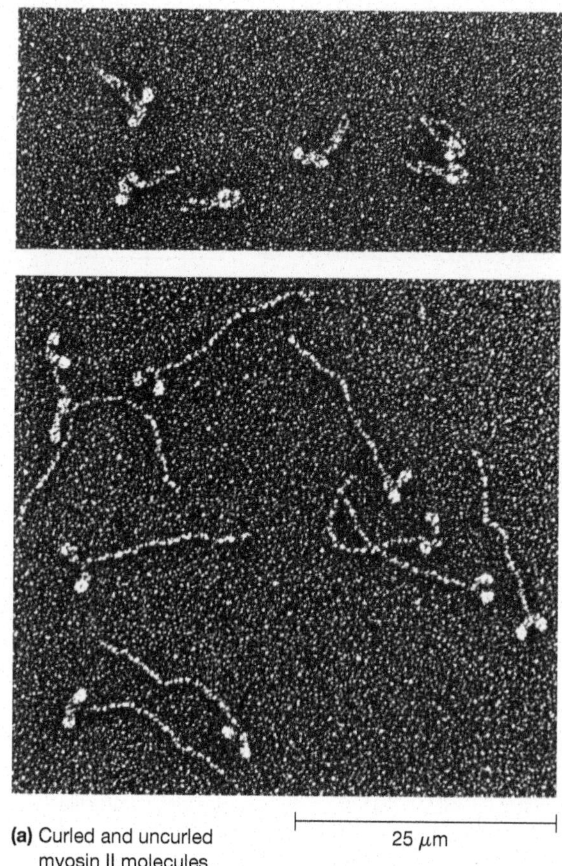

(a) Curled and uncurled myosin II molecules

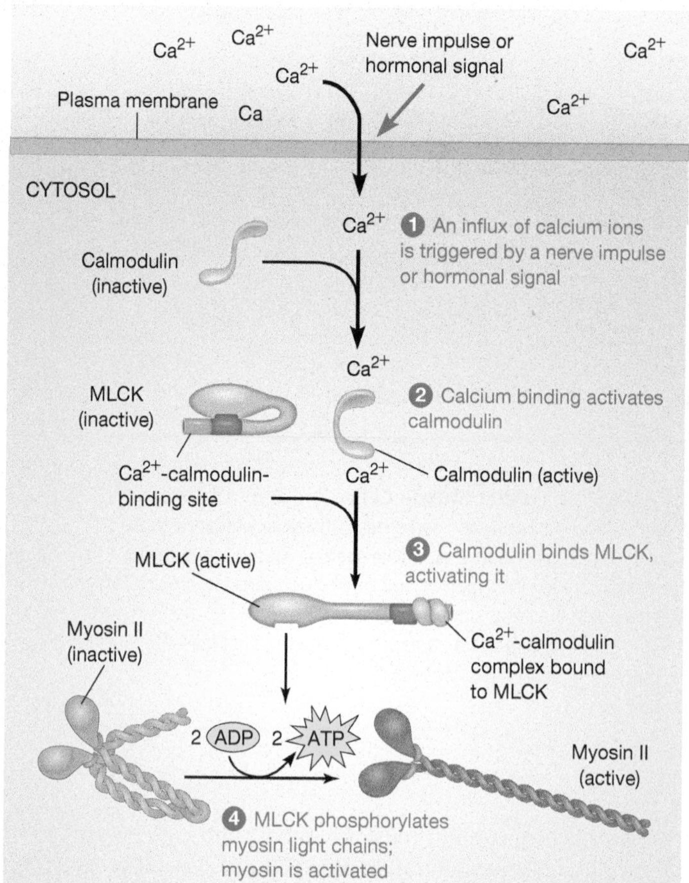

(b) Phosphorylation of myosin II by myosin light-chain kinase (MLCK)

Figure 14-25 Phosphorylation of Smooth Muscle and Nonmuscle Myosin. (a) Electron micrographs of curled and uncurled myosin II molecules (TEMs). **(b)** Both smooth muscle and nonmuscle myosin II are regulated by phosphorylation of the regulatory light chains. An influx of calcium ions into the cell allows the calcium-calmodulin complex to bind myosin light-chain kinase (MLCK), which in turn phosphorylates the myosin light chains. The activated (and uncurled) myosin can then bind to actin.

and by different mechanisms. In skeletal muscle, the calcium comes from the sarcoplasmic reticulum. Its effect on actin-myosin interaction is mediated by troponin and is very rapid because it depends on conformational changes only. In smooth muscle, the calcium comes from outside the cell, and its effect is mediated by calmodulin. The effect is much slower in this case because it involves a covalent modification (phosphorylation) of the myosin molecule.

CONCEPT CHECK 14.4

Based on what you know about the structure of sarcomeres, why do myosin heavy chains in thick filaments always move in such a way that they cause contraction of the sarcomere?

14.5 Microfilament-Based Motility In Nonmuscle Cells

Although they are the best known, in fact, muscle cells represent only one specialized case of cell movement driven by the interactions of actin and myosin. Actins and myosins have now been discovered in almost all eukaryotic cells and are known to play important roles in various types of nonmuscle motility. One example of actin-dependent, nonmuscle motility occurs during cytokinesis (see Chapter 24). In this section, we examine several other examples.

Cell Migration via Lamellipodia Involves Cycles of Protrusion, Attachment, Translocation, and Detachment

Many nonmuscle cells, such as the connective tissue cells known as fibroblasts, growing neurons, and many embryonic cells in animals, are capable of crawling over a substrate using lamellipodia and/or filopodia (specialized motile structures we explored in Chapter 13). **Figure 14-26** shows a fibroblast crawling in vitro.

Cell crawling involves several distinct events: (1) extension of a protrusion at the cell's leading edge, (2) attachment of the protrusion to the substrate, and (3) generation of tension, which pulls the cell forward as its "tail," or trailing edge, releases its attachments and retracts. These events are summarized in **Figure 14-27** and discussed more extensively in the sections that follow.

Cell Protrusion. To crawl, cells must produce specialized extensions, or *protrusions*, at their front or *leading edge*. One type of protrusion is a thin sheet of cytosol called a **lamellipodium** (plural: **lamellipodia**). Another type of protrusion consists

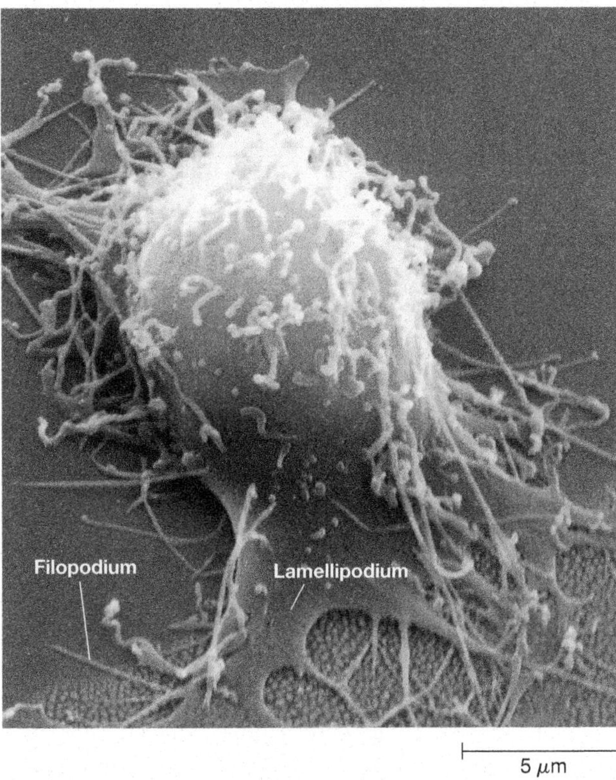
Filopodium
Lamellipodium
5 μm

Figure 14-26 A Crawling Cell. A mouse fibroblast forms a lamellipodium and numerous filopodia toward its direction of motion (SEM). In vivo, fibroblasts are the most common cells of connective tissue in animals; they play a critical role in wound healing.

of thin, fingerlike projections known as **filopodia** (singular: **filopodium**). Forward assembly, especially in lamellipodia, is driven by Arp2/3-dependent branching (Figure 13-20). Recall from Chapter 13 that the small GTPases *Rho, Rac*, and *Cdc42* regulate the types of polymerized actin that cells produce. Activated Rac promotes lamellipodia, whereas Cdc42 promotes filopodia (see Figure 13-21). Crawling cells often exhibit interconversion of these two types of protrusions as they migrate.

Fundamental to the dynamics of protrusions is the phenomenon of **retrograde flow**. During normal retrograde flow, there is bulk backward movement of F-actin toward the rear of the protrusion at the same time as it extends forward. Normal retrograde flow results from two simultaneous processes: *forward assembly* of filaments at the tip of the growing lamellipodium or filopodium and *rearward translocation* of filaments toward the base of the protrusion, powered by myosin. In a typical cell, forward assembly and rearward translocation balance one another; if one or the other predominates, a protrusion can be extended or retracted. Polymerized F-actin is drawn rearward by myosins situated at the base of the protrusion, where the actin is disassembled. Released actin monomers are then available for addition to the barbed (plus) ends of new or growing microfilaments as the cell continues to crawl forward.

Microtubules are also involved in the production of protrusions, although how MTs are involved is not entirely clear. In cultured cells MTs can be seen polymerizing in a polarized manner near the leading edge. Moreover, treating some cells with MT depolymerizing agents causes them to lose their polarized appearance, and they make protrusions simultaneously all around their perimeter.

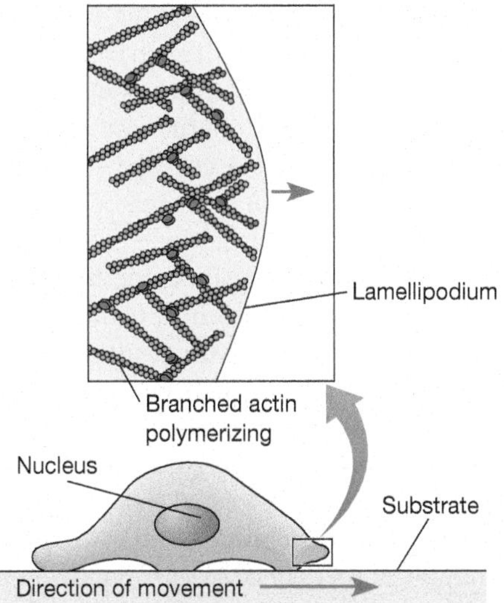
Lamellipodium
Branched actin polymerizing
Nucleus
Substrate
Direction of movement
❶ The leading edge extends via polymerization of actin at its tip.

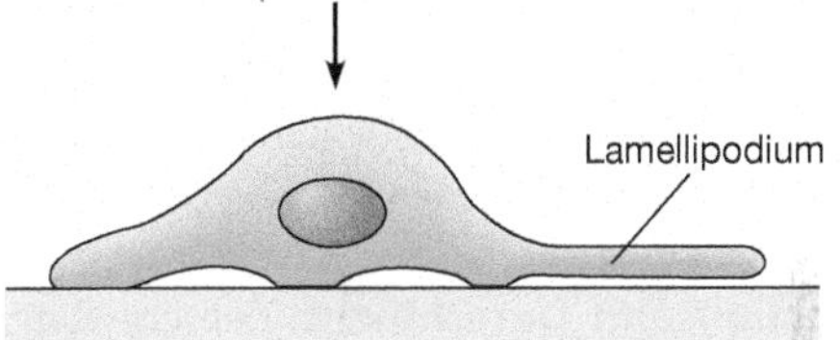
Lamellipodium
❷ New adhesions, anchored by actin, form on the undersurface of the lamellipodium.

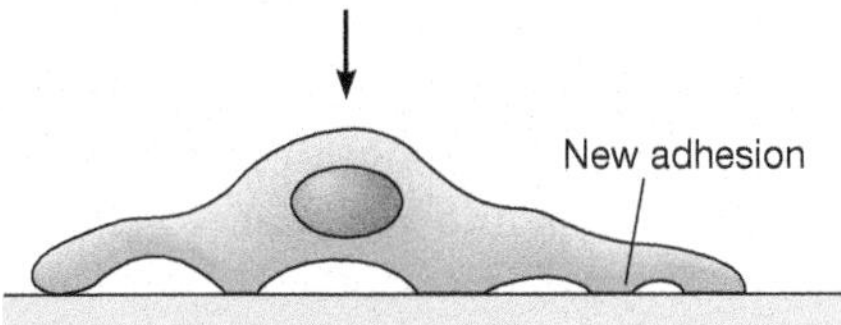
New adhesion
❸ The trailing edge (tail) of the cell detaches and is drawn forward by contraction of the cell body.

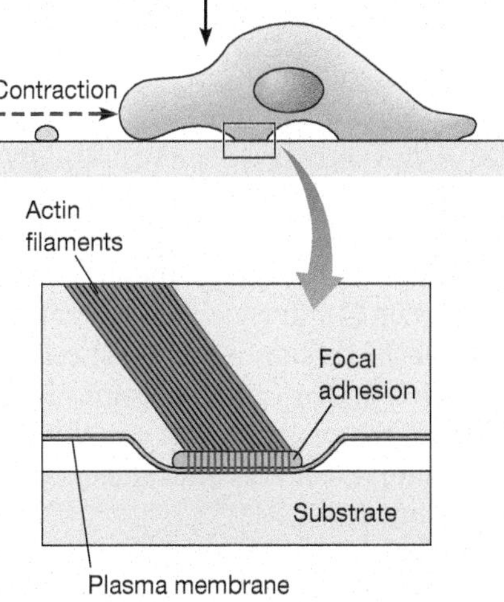
Contraction
Actin filaments
Focal adhesion
Substrate
Plasma membrane

Figure 14-27 The Steps of Cell Crawling. Several different processes are involved in cell crawling, including cell protrusion, attachment, and contractile activities. Protrusion is accompanied by Arp2/3-dependent actin polymerization at the leading edge. Attachments connect actin to the cell surface, typically via integrins, which cluster at focal adhesions.

Cell Attachment. Attachment of the cell to its substrate is also necessary for cell crawling. If cells only polymerized actin at the tips of protrusions, pulled it rearward via myosin, and then depolymerized it, they would not be able to move. Something must couple retrograde flow to forward movement of the cell as a whole. New sites of attachment must be formed at the front of a cell, and contacts at the rear must be broken.

Attachment sites between a cell and its substrate are complex structures consisting of transmembrane proteins attached to other proteins both outside and inside the cell. One family of such attachment proteins is the *integrins*. On the outside of the cell, integrins attach to extracellular matrix proteins. Inside the cell, integrins are connected to actin filaments through linker proteins. Such integrin-dependent attachments are known as *focal adhesions* and are crucial for cell migration. Signals inside the cell regulate such attachments; one such regulator is a protein known as *focal adhesion kinase* (which you will learn about in more detail in Chapter 15).

How firmly a cell is attached to the underlying substrate helps determine whether or not a cell will move forward. In this sense, actin polymerization at the leading edge is analogous to a car with its transmission in neutral. Once the car is shifted into a forward gear, it rapidly moves forward. In the same way, firm attachment of the leading edge shifts the balance in favor of forward movement.

Cell Contraction and Detachment. Cell crawling coordinates protrusion and attachment with forward movement of the entire cell body. Contraction of the rear of the cell squeezes the cell body forward and releases the cell from attachments at its rear.

Contraction at the rear of migrating cells is due to the action of myosin and actin under the control of the protein Rho (a GTPase introduced in Chapter 13). In mutant cells from the cellular slime mold *Dictyostelium* that lack myosin II, the ability of the trailing edge of the cell to retract is reduced. Similarly, when Rho activity is impaired in migrating monocytes (a type of white blood cell), they are unable to withdraw attachments at their rear and contract their trailing edge.

Contraction of the cell body also requires detachment of the trailing edge of the cell. Detachment requires breaking adhesive contacts. Interestingly, adhesions at the rear of some cells are occasionally too tight to be detached; in such cases the tail of the cell actually breaks off as the cell moves forward (see Figure 14-27, step ③). In general, how firmly a cell attaches to a substrate affects how quickly that cell can crawl. If cells adhere too tightly, they cannot dynamically make and break connections to the substrate, and motility is impeded. Thus, for movement to occur, new attachments must be balanced by loss of old ones.

Chemotaxis Is a Directional Movement in Response to a Graded Chemical Stimulus

A key feature of most migrating cells is *directional* migration. One way that directional migration occurs is through the formation of protrusions predominantly on one side of the cell. Diffusible molecules can act as important cues for such directional migration. When a migrating cell moves toward a greater or lesser concentration of a diffusible chemical, the response is known as **chemotaxis**. The molecule(s) that elicit this response are called *chemoattractants* (when a cell moves toward higher concentrations of the molecule) or *chemorepellants* (when a cell moves away from higher concentrations of the molecule). In eukaryotes, chemotaxis has been studied most intensively in white blood cells and in the *Dictyostelium* amoeba. In both cases, increasing the local concentration of a chemoattractant results in dramatic changes in the actin cytoskeleton, induced through local activation of chemoattractant receptors on the cell surface. These receptors are G protein–linked receptors (which we will discuss in Chapter 23). Activation of these receptors leads to polarized recruitment of the cell's cytoskeletal machinery to form a protrusion in the direction of migration.

Amoeboid Movement Involves Cycles of Gelation and Solation of Actin

Amoebas and white blood cells exhibit a different, specialized type of crawling movement known as **amoeboid movement** (**Figure 14-28**). This type of movement is accompanied by protrusions of the cytosol called **pseudopodia** (singular: **pseudopodium**, from the Greek for "false foot"). Cells that undergo amoeboid movement have an outer layer of thick, gelatinous, actin-rich cytosol ("gel") and an inner, more fluid layer of cytosol ("sol"), known as *endoplasm*.

In an amoeba, as a pseudopodium is extended, more fluid material streams forward in the direction of extension and congeals at the tip of the pseudopodium (this event is often called *gelation*). Meanwhile, at the rear of the moving cell, gelatinous cytosol changes into a more fluid state and streams toward the pseudopodium (*solation*). Specific proteins, such as gelsolin, that are present within these gels may be activated by calcium to convert the gel to a more fluid state.

Pressure exerted on the endoplasm, due to contraction of an actomyosin network in the trailing edge of the cell, may play a role in squeezing the endoplasm forward, aiding formation of a protrusion at the leading edge. Experiments have shown, however, that the forward streaming in the pseudopodium can also occur without squeezing from the rear of the cell: when a pseudopodium's cell membrane is removed using detergent, the remaining components can still stream forward if the appropriate mixture of ions and other chemicals is added.

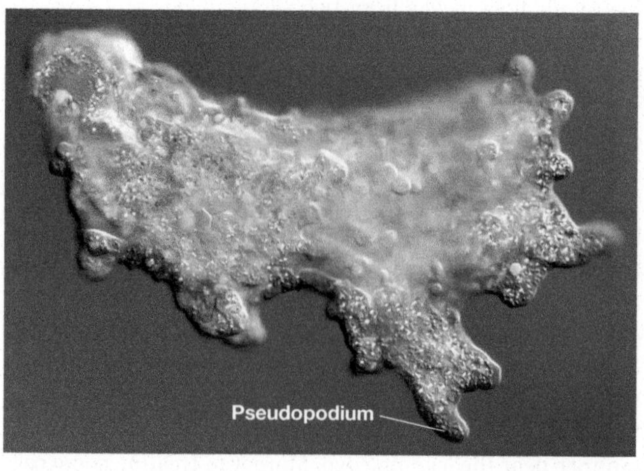

Figure 14-28 Amoeboid Movement. A micrograph of *Amoeba proteus*, a protozoan that moves by extending pseudopodia (LM).

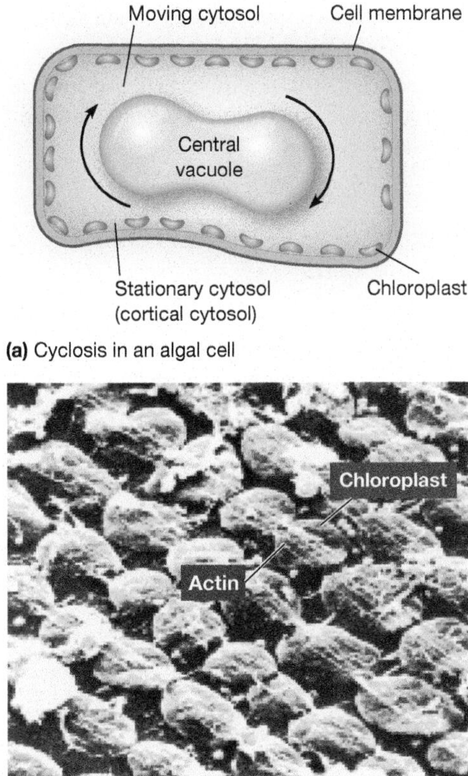

(b) Aligned microfilaments 10 μm

Figure 14-29 Cytoplasmic Streaming. (a) In cyclosis, the cytosol moves in a circular path around a central vacuole, driven by myosin motors that interact with actin anchored to chloroplasts near the cell wall. **(b)** Chloroplasts and tracks of aligned actin filaments in an algal cell (SEM).

Actin-Based Motors Move Components Within the Cytosol of Some Cells

Cytoplasmic streaming, an actomyosin-dependent movement of the cytosol within a cell, is seen in a variety of organisms. For example, plasmodial slime molds such as *Physarum polycephalum* use myosin motors to stream cytosol back and forth in the branched network of actin filaments that constitutes most of the cell mass.

Many plant cells display a circular flow of cell contents, including organelles, around a central vacuole, a process known as *cyclosis* (**Figure 14-29**). Cyclosis has been studied most extensively in the giant cells of the green alga *Nitella*. In this case, the movement seems to circulate and mix cell contents (Figure 14-29a). Dense sets of aligned microfilaments form tracks around the perimeter of the region where cyclosis occurs (Figure 14-29b). Myosins provide the force for movement of components within the streaming cytosol. When latex beads coated with various types of myosin are added to *Nitella* cells that have been broken open, the beads move along the actin filaments in an ATP-dependent manner in the same direction as normal organelle movement.

In animal cells, actomyosin systems may be involved in vesicular transport. Careful observation of vesicles from the cytosol of squid giant axons in vitro makes it clear that individual vesicles can jump from microtubules to microfilaments. Myosins V, VI, and X may be particularly important for such actin-based vesicle movement. Myosin V may physically interact with kinesins on the surface of vesicles, and it has been shown to interact with the plus ends of MTs, making it well suited to a "handoff" between MTs and MF-based vesicle trafficking. A human genetic disorder called *Griscelli's disease*, which involves partial albinism and neurological defects, has been shown to result from a mutation in this class of myosins.

CONCEPT CHECK 14.5

In addition to being found at the rear of migrating cells, non-muscle myosin II is also typically found in cellular protrusions, where it "reels in" actin at the base of lamellipodia. Based on what you know about the direction in which myosin II moves along actin and the polarity of actin filaments in lamellipodia, why does myosin have this effect on the actin in lamellipodia?

Summary of Key Points

Mastering™ Biology For activities, animations, and review quizzes, go to the study area at www.masteringbiology.com.

14.1 Microtubule-Based Movement Inside Cells: Kinesins and Dyneins

- Cell motility and intracellular movements of cellular components are driven by motor proteins, which couple ATP hydrolysis to movements along microtubules (MTs) or microfilaments (MFs).
- MT motors are important for the shaping and transport of the endomembrane system within cells and for intraflagellar transport.
- Kinesins generally move toward the plus ends of MTs; dyneins move toward the minus ends.
- There are many families of kinesins, which can act as highly processive motors, traveling great distances along MTs. One major function of kinesins is moving intracellular cargo. Kinesins connect to adapter proteins and/or their cargo mainly via their light chains.
- There are relatively few dyneins, which fall into two basic classes: cytoplasmic and axonemal. Both classes move along MTs by coupling changes in the shape of their AAA+ domain to ATP hydrolysis.
- Cytoplasmic dyneins move cargo toward the minus ends of MTs. The dynactin complex serves as an adapter between dyneins and their cargo.

14.2 Microtubule-Based Cell Motility: Cilia and Flagella

- Axonemal dyneins mediate the bending of cilia and eukaryotic flagella. Dynein sidearms project out from one MT doublet to the next and slide one set of microtubules past the next. The nine outer doublets of the axoneme are connected laterally to one another and radially to the central pair of single MTs. These connections allow the sliding movement powered by dynein to be converted into bending of a cilium or flagellum.

14.3 Microfilament-Based Cell Movement Inside Cells: Myosins

- There are many families of myosins, and many of them move toward the plus ends of MFs. The best studied is myosin II, one form of which is found in skeletal muscle. Other myosins are involved in events as diverse as cytokinesis, vesicular trafficking, and endocytosis.

14.4 Microfilament-Based Motility: Muscle Cells in Action

- Skeletal muscle contraction involves progressive sliding of actin thin filaments past myosin thick filaments, driven by the interaction between the ATPase of the myosin heavy chains and successive myosin-binding sites on thin filaments. Contraction is triggered by the release of calcium from the sarcoplasmic reticulum (SR), which is bound by troponin, causing a conformational change in tropomyosin that in turn opens myosin-binding sites on the thin filament. Contraction ceases again as calcium is actively pumped back into the SR.
- In smooth muscle, the effect of calcium is mediated by calmodulin, which activates myosin light-chain kinase, leading to the phosphorylation of myosin.

14.5 Microfilament-Based Motility in Nonmuscle Cells

- Actin and myosin are involved in various sorts of motility, including cell crawling. In crawling cells, polymerization of actin extends cellular protrusions; attachment of the protrusions to the substrate and contraction of the cell drive forward movement.
- Chemotaxis is the guided migration of cells toward a higher concentration of a diffusible substance.
- Amoeboid movement, cytoplasmic streaming, cytokinesis, and some vesicle movements are powered by actomyosin.

Problem Set

Mastering™ Biology For activities, animations, and review quizzes, go to the study area at www.masteringbiology.com.

14-1 Ciliobrevins. Ciliobrevins are specific small molecule inhibitors of the AAA+ domain of cytoplasmic dynein. For each of the following processes, would ciliobrevins disrupt the process or not? Explain your answers.

(a) Swimming of human sperm

(b) Contraction of smooth muscle cells

(c) Movement of materials along cilia and flagella

(d) Movement of vesicle-packaged cargo from the Golgi to the endoplasmic reticulum or cell periphery

(e) Movement of vesicles toward the minus ends of microtubules

14-2 A Moving Experience. For each of the following statements, indicate whether it is true of the motility system that you use to lift your arm (A), to cause your heart to beat (H), to move ingested food through your intestine (I), or to sweep mucus and debris out of your respiratory tract (R). More than one response may be appropriate in some cases.

(a) It depends on muscles that have a striated appearance when examined with an electron microscope.

(b) It would probably be affected by the same drugs that inhibit the motility of a flagellated protozoan.

(c) It requires ATP.

(d) It involves calmodulin-mediated calcium signaling.

(e) It involves interaction between actin and myosin filaments.

(f) It depends heavily on fatty acid oxidation for energy.

(g) It is under the control of the voluntary nervous system.

14-3 QUANTITATIVE **Muscle Structure.** Frog skeletal muscle consists of thick filaments that are about 1.6 μm long and thin filaments that are about 1 μm long.

(a) What is the length of the A band and the I band in a muscle with a sarcomere length of 3.2 μm? Describe what happens to the length of both bands as the sarcomere length decreases during contraction from 3.2 to 2.0 μm.

(b) The H zone is a specific portion of the A band. If the H zone of each A band decreases in length from 1.2 to 0 μm as the sarcomere length contracts from 3.2 to 2.0 μm, what can you deduce about the physical meaning of the H zone?

14-4 Rigor Mortis and the Contraction Cycle. After death, the muscles of the body become very stiff and inextensible, and the corpse is said to go into rigor.

(a) Explain the basis of rigor. Where in the contraction cycle is the muscle arrested? Why?

(b) What effect do you think the addition of ATP might have on muscles in rigor?

14-5 AMP-PNP and the Contraction Cycle. AMP-PNP is the abbreviation for a structural analogue of ATP in which the second and third phosphate groups are linked by an NH group instead of an oxygen atom. AMP-PNP binds to the ATP-binding site of virtually all ATPases, including myosin. It differs from ATP, however, in that its terminal phosphate cannot be removed by hydrolysis. When isolated myofibrils are placed in a flask containing a solution of calcium ions and AMP-PNP, contraction is quickly arrested.

(a) Where in the contraction cycle will contraction be arrested by AMP-PNP? Draw the arrangement of a thin filament, a thick filament, and a cross-bridge in the arrested configuration.

(b) Do you think contraction would resume if ATP were added to the flask containing the AMP-PNP-arrested myofibrils? Explain.

(c) What other processes in a muscle cell do you think are likely to be inhibited by AMP-PNP?

14-6 DATA ANALYSIS **Straight-Laced.** Examine Figure 14-17b again. Why is the line connecting points ❶, ❷, and ❸ a straight line (that is, the line has a constant slope)?

14-7 Under Stress. The following questions focus on stress fiber contractility. (Before answering (b) and (c), you may wish to refer to Chapter 13 for a refresher on the protein Rho.)

(a) Stress fibers of nonmuscle cells contain contractile bundles of actin and myosin II. For stress fibers to contract or develop tension, how would actin and myosin have to be oriented within the stress fibers?

(b) Fibroblasts can be placed on thin sheets of silicone rubber. Under normal circumstances, fibroblasts exert sufficient tension on the rubber so that it visibly wrinkles. Explain how the ability of fibroblasts to wrinkle rubber would change if the cells were treated with a permeable form of C3 transferase, a toxin (from the bacterium *Clostridium botulinum*) that covalently modifies an amino acid in Rho. The modification renders Rho unable to bind to proteins that it would normally bind to when active.

(c) The experiment in part b is repeated, but instead of C3 a form of Rho that is always active is introduced into the fibroblasts. Predict what would happen.

14-8 Nervous Twitching. Recent political and military conflicts have involved discussions of "weapons of mass destruction," including nerve gas. One such nerve gas contains the chemical sarin. Sarin inhibits the enzyme that leads to destruction of the neurotransmitter acetylcholine at neuromuscular junctions. What effect would you expect sarin to have on muscle function in individuals exposed to the nerve gas, and why? In your answer, discuss in detail how sarin would be expected to affect the neuromuscular junction, and how it would affect signaling and cytoskeletal events within affected muscle cells.

14-9 QUANTITATIVE **Tipped Off.** Polarized cytoskeletal structures and intraflagellar transport (IFT) are involved in formation of cilia and flagella.

(a) Observation of flagella in the biflagellate alga *Chlamydomonas reinhardtii* indicates that tubulin subunits and other flagellar materials move toward the tips of flagella at a rate of 2.5 μm/min, but materials moving back toward the base of flagella move at 4 μm/min. How do you explain this difference in rate of movement?

(b) Temperature-sensitive mutations in a kinesin II required for intraflagellar transport (IFT) have been identified in *Chlamydomonas.* Such mutations only lead to defects when the temperature is raised above a certain threshold, called the *restrictive temperature.* When algae with fully formed flagella are grown at the restrictive temperature, their flagella degenerate. What can you conclude about the necessity of IFT from this experiment?

(c) Based on your knowledge of the directionality of microtubule motors and the information in part b, where would you predict that the plus ends of flagellar microtubules are? State your reasoning.

14-10 AMP-PNP Again. AMP-PNP (see Problem 14-5) can be used to study microtubule (MT) motors as well as myosins.

(a) What effects would you predict in a sperm flagellum to which AMP-PNP is added? In your explanation, be specific about which molecule's function would be inhibited and what the effect on overall flagellar function would be.

(b) When researchers incubated purified vesicles, axoplasm from squid giant axons, and MTs in the presence of AMP-PNP, the vesicles bound tightly to the microtubules and did not move. This observation suggested that AMP-PNP causes the formation of rigor complexes consisting of vesicles, microtubules, and microtubule motor proteins. Then they collected MTs with bound proteins by centrifugation. The main protein purified in this way promotes movement of vesicles away from the cell body, where the nucleus resides. What was the motor protein?

15 Beyond the Cell: Cell Adhesions, Cell Junctions, and Extracellular Structures

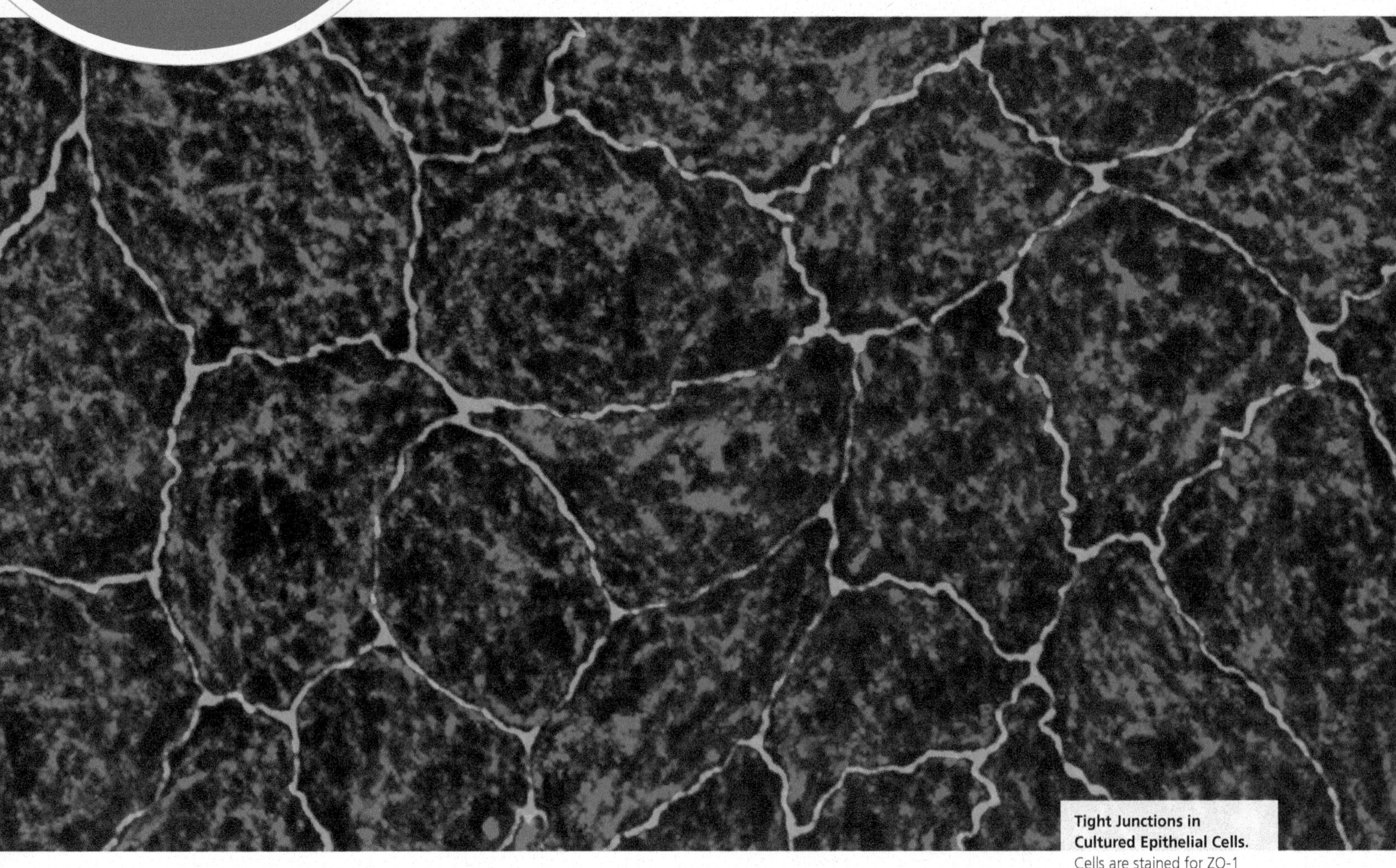

Tight Junctions in Cultured Epithelial Cells. Cells are stained for ZO-1 (green) and F-actin (purple) (confocal microscopy).

In the preceding chapters, you learned about cells as if they exist in isolation and as if they "end" at the plasma membrane. However, most organisms are *multicellular*: they consist of many—sometimes trillions—of cells. Many kinds of cells—including most of those in your own body—spend their entire lives linked to neighboring cells. It is the organization of cells into *tissues* that allows multicellular organisms to adopt complex structures. These tissues, in turn, are arranged in precise ways to generate the form of an organism. How do these tissues achieve their structures? Consider the two types of animal tissues shown in **Figure 15-1.** One is a sheet of cells known as an *epithelium,* such as the cells that line your small intestine. Such cells are clearly *polarized*—that is, they contain discrete specializations at their opposite ends. The end of the cell in contact with the external environment (for example, your lumen of the small intestine) is often called the *apical* side of the cell. The *basolateral* region of epithelial cells includes the surfaces in contact with the *basal lamina,* a component of a large extracellular structure known as the *basement membrane*. The other cell type in Figure 15-1 is from a more loosely organized *connective tissue,* such as might be found in the dermis of the skin. In each case, cells must be attached to one another, to a mechanically rigid scaffolding, or to

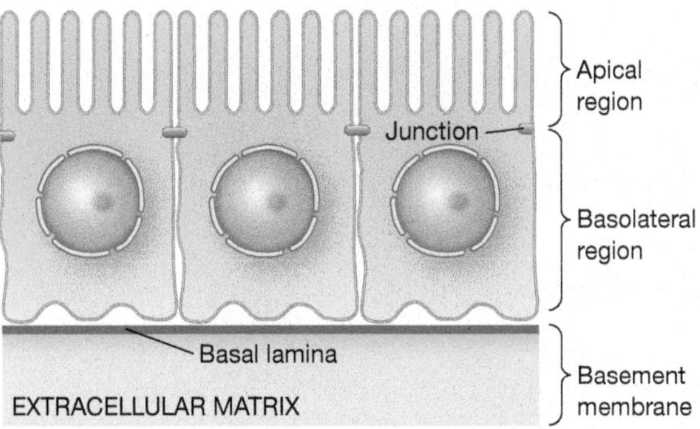

(a) Epithelial tissue

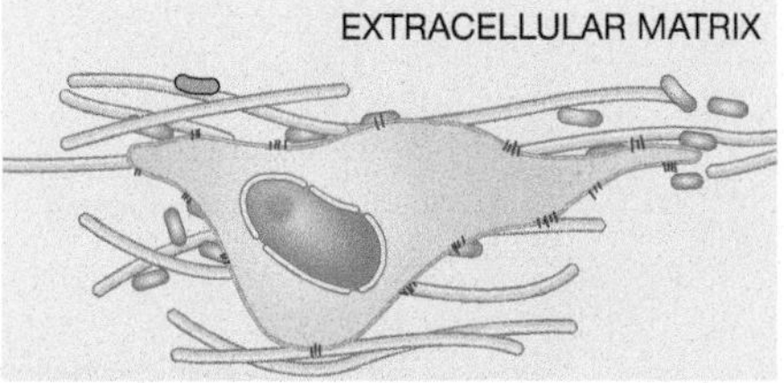

(b) Cell embedded in connective tissue

Figure 15-1 Different Types of Tissues in Animals. Tissues are multicellular structures that include cells and extracellular material. **(a)** In a polarized epithelium, such as the cells lining the small intestine, cells are connected together by cell-cell adhesions. The apical surfaces of these cells are very different from the basal surfaces, which lie on top of an extracellular matrix known as the basal lamina. **(b)** In a connective tissue, such as the dermis of the skin, loosely organized cells are embedded in extracellular matrix fibers.

both. Thus, to understand how multicellular organisms are constructed, we must consider both connections between cells, or *cell-cell adhesions,* and the *extracellular structures* to which cells attach. Cells use a variety of elaborate molecular complexes, or *junctions,* to attach to one another, and most of these involve transmembrane proteins that link the cell surface to the cytoskeleton (which you learned about in Chapters 13 and 14). Cells use other types of molecular complexes to attach to extracellular structures, and these also involve specific linkages between the cell surface and the cytoskeleton.

The extracellular structures themselves consist mainly of macromolecules that are secreted by the cell. Animal cells have an *extracellular matrix* that takes on a variety of forms and plays important roles in cellular processes as diverse as division, motility, differentiation, and adhesion. Epithelial cells, such as the ones shown in Figure 15-1, produce a specialized extracellular matrix called a *basal lamina.* Connective tissues, on the other hand, produce a more loosely organized matrix. In plants, fungi, algae, and prokaryotes, the extracellular structure is a *cell wall*—although its chemical composition differs considerably among these organisms. Cell walls confer rigidity on the cells they encase, serve as permeability barriers, and protect cells from physical damage and from attack by viruses and infectious organisms.

Our initial focus in this chapter is on the adhesions animal cells make with one another, the junctions that characterize these adhesions, and how animal cells interact with the extracellular matrix. Then we will turn to the plant cell wall and the specialized structures that allow direct cell-to-cell communication between plant cells, despite the presence of a cell wall.

15.1 Cell-Cell Junctions

By definition, unicellular organisms have no permanent associations (although they can form temporary associations, such as during bacterial swarming, the aggregation of slime mold amoebae, or during mating). Multicellular organisms, in contrast, have specific means of joining cells in long-term associations to form tissues and organs. Animal cells use specialized structures called **cell-cell junctions** for this purpose. These include *adhesive junctions* (such as adherens junctions and desmosomes), *tight junctions,* and *gap junctions.* **Figure 15-2** on page 414 illustrates each of these kinds of junctions, whose main features are summarized in **Table 15-1** on page 414. (The cell–extracellular matrix connections shown in Figure 15-2 are discussed later in this chapter.) As Figure 15-2 suggests, in epithelia, adherens junctions, desmosomes, and gap junction occur in spatially distinct areas of the cell.

In plants, the presence of a cell wall between the plasma membranes of adjacent cells precludes the kinds of cell junctions that link animal cells. However, the cell wall and special structures called *plasmodesmata* carry out similar functions, as you will see a bit later in this chapter.

Adhesive Junctions Link Adjoining Cells

Adhesive junctions link cells together into tissues, thereby enabling the cells to function as a unit. All junctions in this category anchor the cytoskeleton to the cell surface. The resulting interconnected cytoskeletal network helps maintain tissue integrity and withstands mechanical stress.

Adhesive junctions rely on specialized adhesion proteins. Many of these adhesion proteins are transmembrane proteins; the extracellular portion of these proteins can interact with the extracellular portion of similar proteins on the surface of a neighboring cell. In some cases, cells interact with identical molecules on the surface of the cell to which they adhere. Such interactions are said to be **homophilic interactions** (from the Greek *homo,* meaning "like," and *philia,* meaning "friendship"). In other cases, a cell adhesion receptor on one cell interacts with a different molecule on the surface of the cell to which it attaches. Such interactions are said to be **heterophilic interactions** (from the Greek *hetero,* meaning "different"). Many transmembrane adhesion proteins attach to the cytoskeleton via *linker proteins,* which differ depending on the class of molecule and its location within the cell.

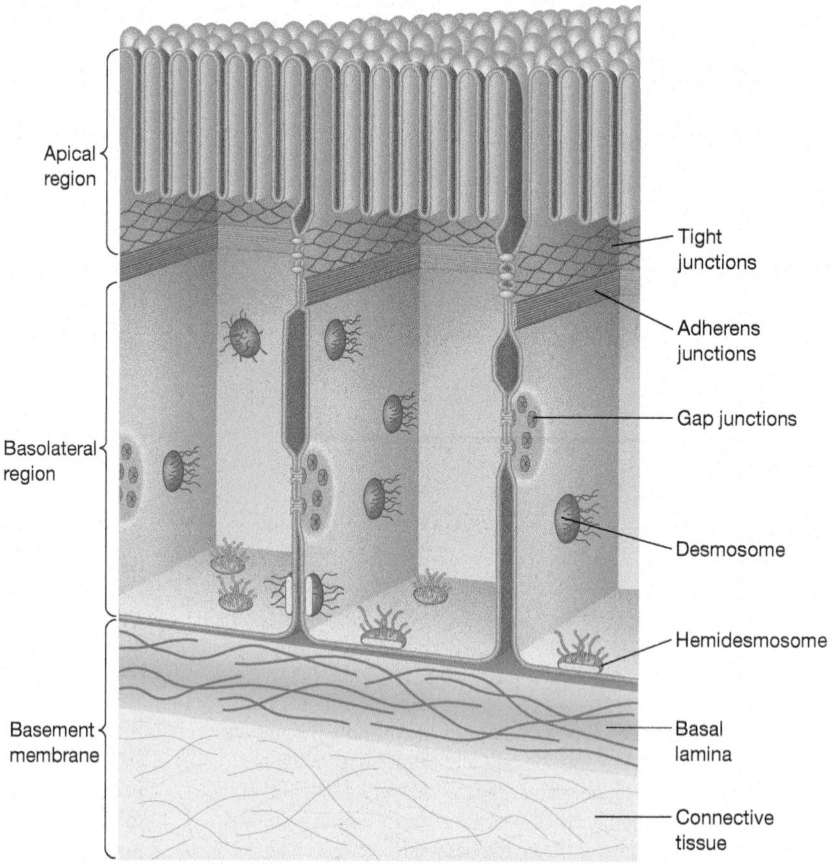

Figure 15-2 Major Types of Cell Attachments in Epithelial Cells. From top to bottom: Tight junctions create an impermeable seal between cells, thereby preventing fluids, molecules, or ions from crossing a cell layer via the intercellular space. Adherens junctions provide adhesive attachments between cells and connect to F-actin. Note that tight junctions can also recruit actin, but for clarity this is not shown. Gap junctions provide direct chemical and electrical communication between cells by allowing the passage of small molecules and ions from one cell to another. Desmosomes provide mechanically strong "spot welds" between cells and connect to intermediate filaments. Hemidesmosomes attach the basal surfaces of epithelial cells to the basal lamina; they are anchored to intermediate filaments.

Although pictures of adhesive structures may suggest that they are static once assembled, they are anything but! Cells can dynamically assemble and disassemble adhesions in response to a variety of events. Many adhesion proteins are continuously recycled: protein at the cell surface is internalized by endocytosis, and new protein is deposited at the surface via exocytosis. In addition, adhesion proteins serve as key sites for assembly of signaling complexes in cells and for dynamically assembling cytoskeletal structures at sites of cell adhesion. In this way, cell adhesion is coordinated with cell signaling, cell movement, cell proliferation, and cell survival.

The two main kinds of cell-cell adhesive junctions are *adherens junctions* and *desmosomes* (Figure 15-2). Despite structural and functional differences, both types of junction rely on *intracellular attachment proteins*, which link the junction to the appropriate cytoskeletal filaments on the inside of the plasma membrane, and *cadherins*, transmembrane proteins that protrude on the outer surface of the membrane and bind cells to each other.

Table 15-1 Cell-Cell and Cell-ECM Attachments in Animal Cells

Type of Attachment	Main Function	Intermembrane Features	Space	Associated Structures
Adhesive junctions				
Adherens junction	Cell-cell adhesion	Continuous zones of attachment	20–25 nm	Actin microfilaments
Desmosome	Cell-cell adhesion	Localized points of attachment	25–35 nm	Intermediate filaments (tonofilaments)
Tight junction	Sealing spaces between cells	Membranes joined along ridges	None	Transmembrane junctional proteins, actin
Gap junction	Exchange of ions and molecules between cells	Connexons (transmembrane protein complexes with 3-nm pores)	2–3 nm	Connexins in one membrane align with those in another to form channels between cells
Cell-ECM attachments				
Focal adhesion	Cell-ECM adhesion	Localized points of attachment	20–25 nm	Actin microfilaments
Hemidesmosome	Cell–basal lamina adhesion	Localized points of attachment	25–35 nm	Intermediate filaments (tonofilaments)

membranes. Lipid movement is blocked in the outer monolayer (leaflet) only; the movement of integral membrane proteins is also blocked. As a result, different kinds of integral membrane proteins can be maintained in the region of plasma membrane surrounded by a TJ belt.

Gap Junctions Allow Direct Electrical and Chemical Communication Between Cells

A **gap junction** is a region where the plasma membranes of two cells are aligned and brought into intimate contact, with a gap of only 2–3 nm in between, spanned by small molecular "pipelines." The gap junction thus provides a point of cytosolic contact between two adjacent cells through which ions and small molecules can pass. It allows adjacent cells to be in direct electrical and chemical communication with each other.

The structure of gap junctions is illustrated in **Figure 15-10**. Two plasma membranes from adjacent cells are joined by tightly packed, hollow cylinders called **connexons**. A single gap junction may consist of just a few or as many as thousands of clustered connexons. In vertebrates, each connexon is a circular assembly of six subunits of the protein *connexin*. Invertebrates do not have connexins. Instead, they produce proteins called *innexins* that appear to serve the same function.

Many different connexins (more than a dozen types) are found in different tissues, but each one functions similarly in forming connexons. The assembly spans the membrane and protrudes into the space between the two cells (Figure 15-10a). Each connexon has a diameter of about 7 nm and a hollow center that forms a very thin hydrophilic channel through the membrane. The channel is about 3 nm in diameter at its narrowest point—just large enough to allow the passage of ions and small molecules but too small to allow proteins, nucleic acids, and organelles through. By injecting fluorescent molecules into cells connected by gap junctions, researchers have shown that gap junctions allow the passage of solutes with molecular weights up to about 1200. Included in this range are single sugars, amino acids, and nucleotides—molecules involved in cellular metabolism.

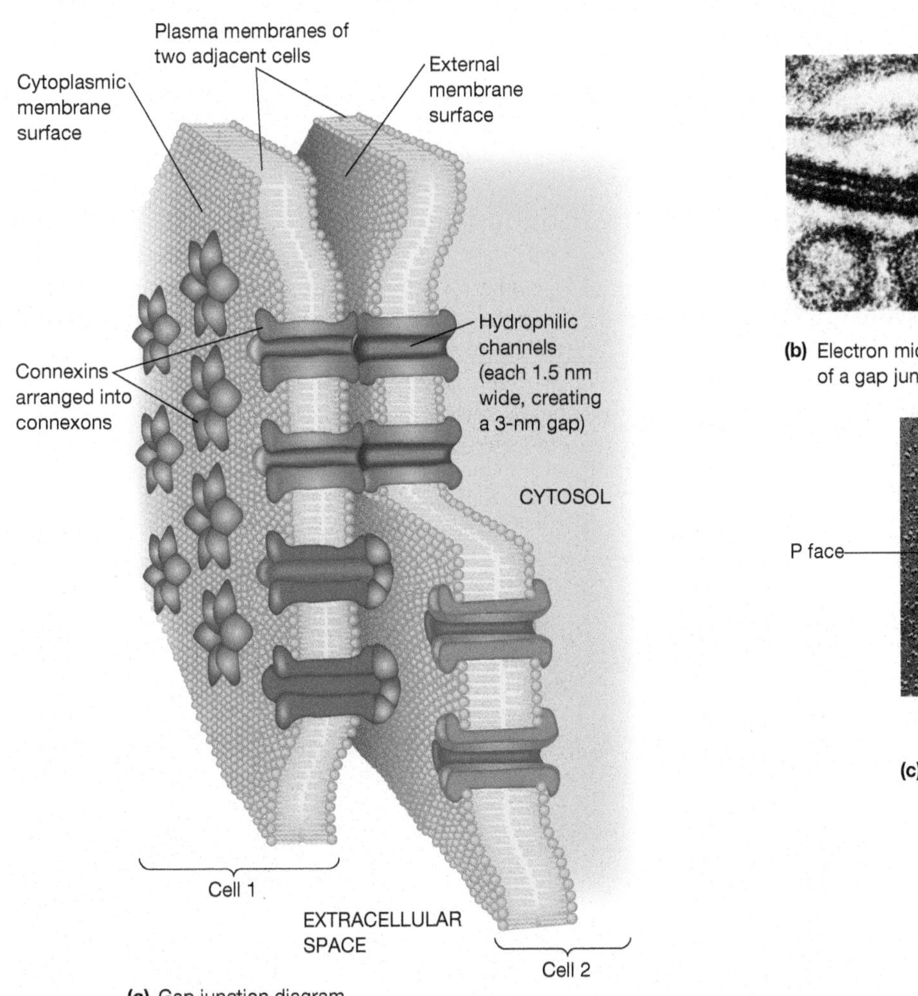

(a) Gap junction diagram

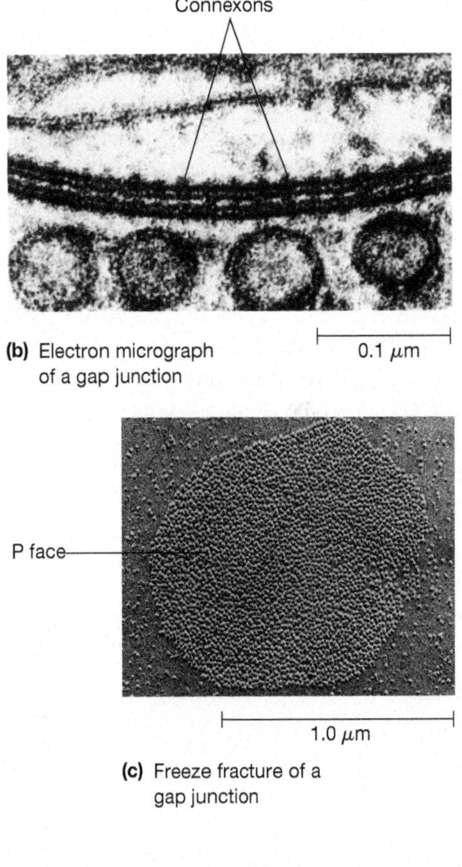

(b) Electron micrograph of a gap junction

(c) Freeze fracture of a gap junction

Figure 15-10 Gap Junction Structure. (a) A schematic representation of a gap junction. A gap junction consists of hydrophilic channels formed by the alignment of connexons in the plasma membranes of two adjoining cells. **(b)** A gap junction between two adjacent nerve cells. The connexons are visible as beadlike projections spaced about 17 nm apart on either side of the membrane-membrane junction (TEM). **(c)** A gap junction as revealed by the freeze-fracture technique. The junction appears as an aggregation of intramembranous particles on the protoplasmic (P) face (TEM).

Although they are formed in a closed state initially, once connexons from adjacent cells meet, the cylinders in the two membranes join end to end. They form direct channels between the two cells that can be seen with an electron microscope (Figure 15-10b and c). Once fully formed, electrical potential, concentration of second messengers, and other conditions can influence whether gap junctions are open or closed.

Gap junctions occur in many cell types, but they are especially abundant in tissues such as muscle and nerve, where extremely rapid communication between cells is required (for example, in electrical synapses; see Chapter 22). In heart tissue, gap junctions facilitate the flow of electrical current that causes the heart to beat. These roles are confirmed by analyzing mutants. For example, mice lacking one type of connexin have defects in conducting electrical impulses in the heart. Several human disorders have been directly linked to defects in gap junctions, including demyelinating neurodegenerative diseases, various skin disorders, formation of cataracts, and deafness.

CONCEPT CHECK 15-1

What are the two main types of cell-cell adhesive junctions in vertebrates, and how do their attachments to the cytoskeleton differ? What advantages might there be for these junctions to be attached to the cytoskeleton?

15.2 The Extracellular Matrix of Animal Cells

You might get the impression from the discussion so far that tissues are only composed of cells, but tissue have another key ingredient materials outside of cells that are crucial for tissue structure and function. In animal cells, this **extracellular matrix (ECM)** takes on a remarkable variety of forms in different tissues. **Figure 15-11** illustrates just three examples. *Bone* consists largely of a rigid extracellular matrix with a tiny number of interspersed cells. *Cartilage* is another tissue constructed almost entirely of matrix materials, although the matrix is much more flexible than in bone. As you already saw (Figure 15-1), epithelial cells produce an ECM known as a basal lamina. In contrast, the *connective tissue* surrounding glands and blood vessels has a relatively gelatinous extracellular matrix containing numerous interspersed fibroblast cells.

These examples illustrate the diverse roles that the ECM plays in determining the shape and mechanical properties of organs and tissues. Despite this diversity of function, the ECM of animal cells almost always consists of the same three classes of molecules: (1) structural proteins such as *collagens* and *elastins*, which give the ECM its strength and flexibility, (2) protein-polysaccharide complexes called *proteoglycans* that provide the matrix in which structural molecules are embedded, and (3) adhesive glycoproteins such as *fibronectins* and *laminins*, which allow cells to attach to the ECM (**Table 15-2**). The considerable variety in the properties of the ECM in different tissues results from differences in the types and amounts of these molecules that are present. We now consider each of these classes of ECM constituents in turn.

Collagens Are Responsible for the Strength of the Extracellular Matrix

The most abundant component of the ECM in animals is a large family of closely related proteins called **collagens**, which form fibers with high tensile strength and thus account for much of the strength of the ECM. Considered collectively, collagen is the most abundant protein in vertebrates, accounting for as much as 25–30% of total body protein. Collagen is secreted by several types of cells in connective tissues, including *fibroblasts*. Without collagen, cells in these and other

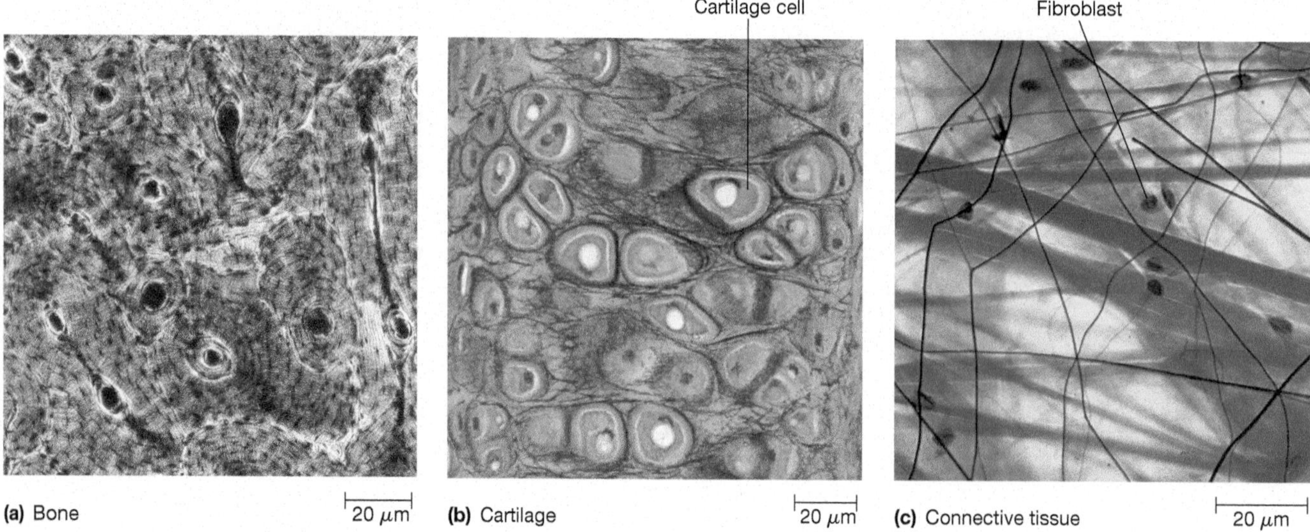

Figure 15-11 Different Kinds of Extracellular Matrix. The ECM takes on different forms in different tissues. **(a)** In bone, a hard, calcified ECM is laid down in concentric rings around central canals. The small elliptical depressions are regions where bone cells are found. **(b)** In cartilage, the cells are embedded in a flexible matrix that contains large amounts of proteoglycans. **(c)** In the connective tissue found under skin, fibroblasts are surrounded by an ECM that contains large numbers of collagen fibers.

Table 15-2 Extracellular Structures of Eukaryotic Cells

Kind of Organism	Extracellular Structure	Structural Fiber	Components of Hydrated Matrix	Adhesive Molecules
Animals	Extracellular matrix (ECM)	Collagens and elastins	Proteoglycans	Fibronectins and laminins
Plants	Cell wall	Cellulose	Hemicelluloses and extensins	Pectins

tissues would not have sufficient adhesive strength to maintain a given form. Indeed, several human diseases result from mutations in collagens. For example, Ehlers–Danos syndrome is a group of inherited disorders characterized by excessive looseness (laxity) of the joints, hyperelastic skin that is fragile and bruises easily, and/or easily damaged blood vessels. These disorders arise from mutations in collagens. Vitamin C is an essential cofactor for collagen synthesis. Vitamin C deficiency leads to *scurvy*, which was a historically important disease among sailors.

Two defining characteristics are shared by all collagens: their structure, which features a rigid *triple helix* of three intertwined polypeptide chains, and their unusual amino acid composition. Specifically, collagens are high in both the common amino acid glycine and the unusual amino acids hydroxylysine and hydroxyproline, which rarely occur in other proteins. (For the structures of hydroxylysine and hydroxyproline, see Figure 7-21c.) The high glycine content makes the triple helix possible because the spacing of the glycine residues places them in the axis of the helix, and glycine is the only amino acid small enough to fit in the interior of a triple helix.

In most animal tissues, **collagen fibers** form bundles throughout the extracellular matrix (**Figure 15-12**). They often exhibit a characteristic pattern of dark crossbands, or *striations*, when viewed by scanning electron microscopy (Figure 15-12a). One of the most striking features of collagen fibers is their enormous physical strength. For example, it takes a load of more than 20 pounds (about 9 kg) to tear a collagen fiber just 1 mm in diameter! As illustrated in Figure 15-12b, each collagen fiber is composed of numerous *fibrils*. A fibril, in turn, is made up of many collagen molecules, each consisting of three polypeptides called *α chains* that are twisted together into a rigid, right-handed triple helix (Figure 15-12c and d). Collagen molecules are about 270 nm in length and 1.5 nm in diameter and are aligned both laterally and end to end within the fibrils. A typical collagen fiber has about 270 collagen molecules in cross section.

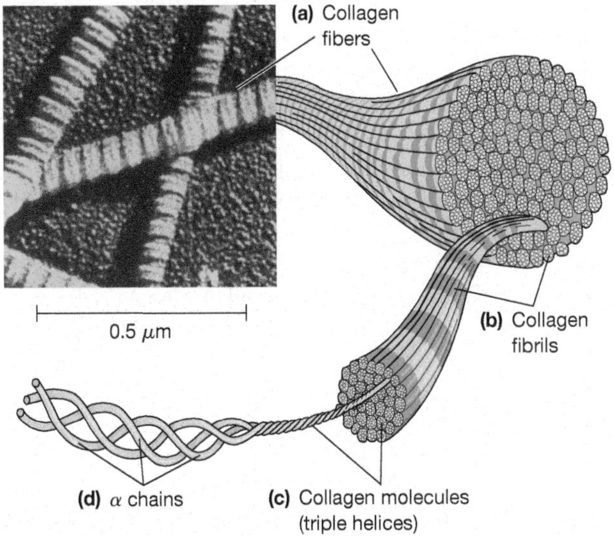

Figure 15-12 The Structure of Collagen. (a) Collagen fibers as seen by SEM. **(b)** A collagen fiber contains many fibrils, each of which is a bundle of collagen molecules, also called tropocollagen. **(c)** Each collagen molecule is a triple helix consisting of **(d)** three entwined α chains. The repeating bands visible on the fibers in the SEM partially reflect the regular but offset way that collagen molecules associate laterally to form fibrils; see Figure 15-13 for details of collagen assembly.

Figure 15-13 on page 424 shows how collagen fibers form. In the lumen of the endoplasmic reticulum (ER), three α chains assemble to form a triple helix called **procollagen**. At both ends of the triple-helical structure, short nonhelical regions prevent the formation of collagen fibrils, as long as the procollagen remains within the cell. Once procollagen is secreted into the intercellular space, it is converted to collagen by *procollagen peptidase*, an enzyme that removes the extra amino acids from both its N- and C-terminal ends. The resulting collagen molecules spontaneously associate to form mature collagen fibrils, which then assemble into fibers.

The stability of the collagen fibril is reinforced by hydrogen bonds that involve the hydroxyl groups of hydroxyproline and hydroxylysine residues in the α chains. These hydrogen bonds form crosslinks both within and between the individual collagen molecules in a fibril. In some specialized collagens this triple-helical structure is interrupted at intervals, allowing the molecules to bend and hence to serve as flexible bridges between adjacent collagen fibrils or between collagen fibrils and other matrix components.

Vertebrates have about 25 kinds of α chains, each encoded by its own gene and having its own unique amino acid sequence. These different α chains combine in various ways to form at least 15 types of collagen molecules. **Table 15-3** on page 424 lists the types and the tissues where they are found. Types I, II, and III are the most abundant forms. Type I alone makes up about 90% of the collagen in the human body.

Fibrils containing type I, II, III, or V collagen molecules exhibit characteristic striations that repeat at intervals of about 67 nm. These bands reflect the regular but offset way that triple helices associate laterally to form fibrils (see Figure 15-13). Type IV collagen forms very fine, unstriated fibrils. The structures of other collagen types are less well characterized.

Elastins Impart Elasticity and Flexibility to the Extracellular Matrix

Although collagen fibers give the ECM great tensile strength, their rigid, rodlike structure is not very well suited to the elasticity and flexibility required by some tissues, such as lungs,

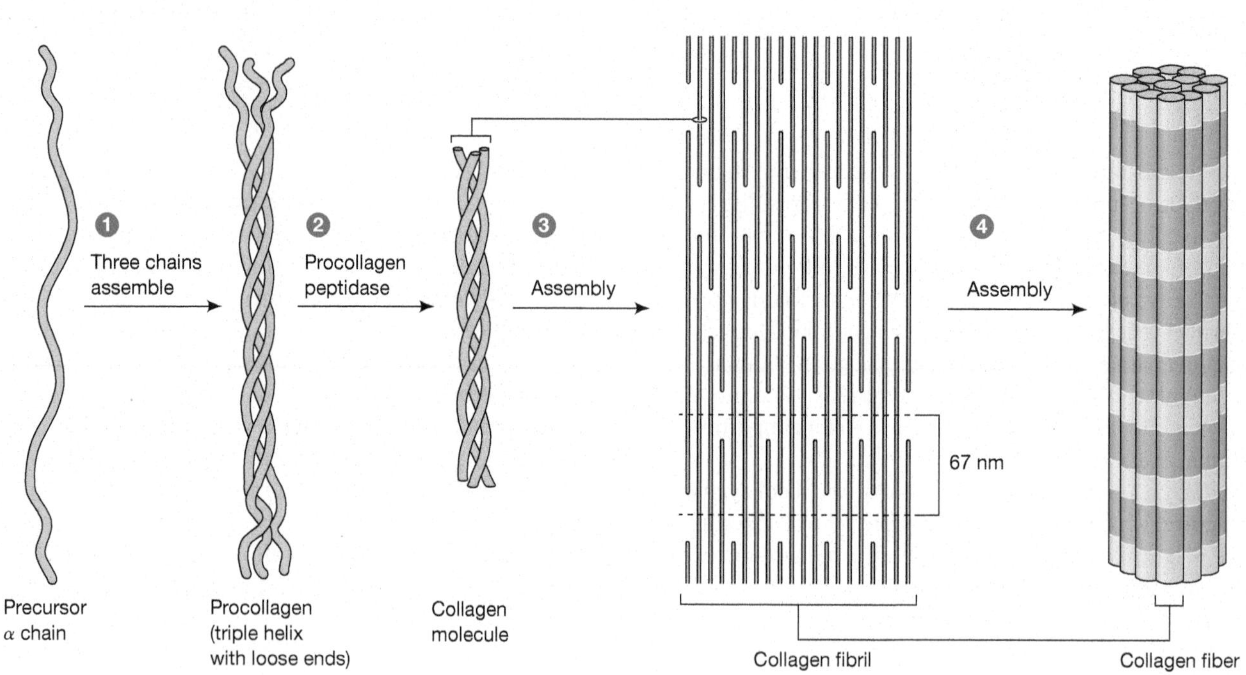

Figure 15-13 Collagen Assembly. ❶ Collagen precursor chains are assembled in the ER lumen to form triple-helical procollagen molecules. ❷ After secretion from the cell, procollagen is converted to collagen in a peptide-cleaving reaction catalyzed by the enzyme procollagen peptidase. ❸ The molecules of collagen then bind to each other and self-assemble into collagen fibrils. ❹ The fibrils assemble laterally into collagen fibers. In striated collagen, the 67-nm repeat distance is created by packing together rows of collagen molecules in which each row is displaced by one-fourth the length of a single molecule.

arteries, skin, and the intestine, which change shape continuously. Elasticity is provided by stretchable elastic fibers, whose principle constituent is a family of ECM proteins called **elastins**. Like collagens, elastins are rich in the amino acids glycine and proline. However, the proline residues are not hydroxylated, and no hydroxylysine is present. Elastin molecules are crosslinked to one another by covalent bonds between lysine residues (**Figure 15-14**). Tension on an elastin network causes the overall network to stretch (Figure 15-14a). When the tension is released, the individual molecules relax, returning to a, less extended conformation. The crosslinks between molecules then stabilize the network as it recoils to its original shape (Figure 15-14b).

The important roles of collagens and elastins are demonstrated as people age. Over time, collagens become increasingly crosslinked and inflexible, and elastins are lost from tissues like skin. As a result, older people find that their bones and joints are less flexible, and their skin becomes wrinkled.

Collagen and Elastin Fibers Are Embedded in a Matrix of Proteoglycans

The hydrated, gel-like network in which the collagen and elastin fibrils of the ECM are enmeshed consists primarily of proteoglycans, glycoproteins in which a large number of glycosaminoglycans are attached to a single protein molecule.

Glycosaminoglycans (GAGs) are large carbohydrates characterized by repeating disaccharide units, as illustrated in **Figure 15-15** for the three most common types: *chondroitin sulfate, keratan sulfate,* and *hyaluronate.* In each case, one of the two sugars in the disaccharide repeating

Table 15-3 Types of Collagens, Their Occurrence, and Their Structure

Type of Structure	Collagen Type(s)	Representative Tissues
Long fibrils	I, II, III, V, XI	Skin, bone, tendon, cartilage, muscle
Fibril-associated, with interrupted triple helices	IX, XII, XIV	Cartilage, embryonic skin, tendon
Fibril-associated, forms beaded filaments	VI	Interstitial tissues
Sheets	IV, VIII, X	Basal laminae, cartilage growth plates
Anchoring fibrils	VII	Epithelia
Transmembrane	XVII	Skin
Other	XIII, XV, XVI, XVIII, XIX	Basement membranes, assorted tissues

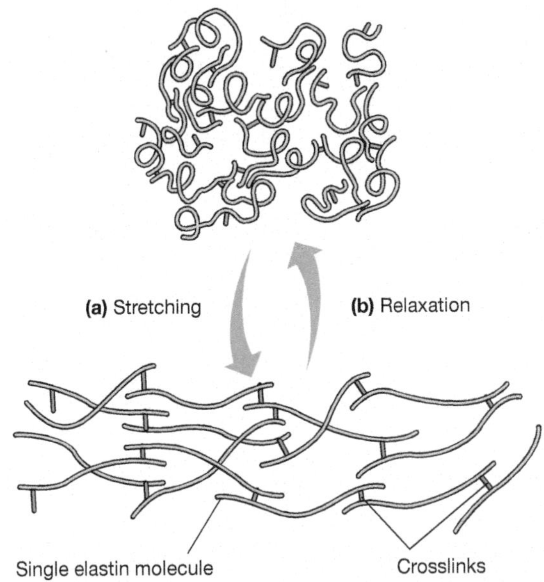

Figure 15-14 Stretching and Recoiling of Elastin Fibers. Each elastin molecule in the crosslinked network can assume either an extended (bottom) or compact (top) configuration. The fiber **(a)** stretches to its extended form when tension is exerted on it and **(b)** recoils to its compact form when the tension is released.

unit is an amino sugar, either *N-acetylglucosamine (GlcNAc)* or *N-acetylgalactosamine (GalNAc)*. The other sugar in the disaccharide repeating unit is usually a sugar or a sugar acid, commonly *galactose (Gal)* or *glucuronate (GlcUA)*. In most cases, the amino sugar has one or more sulfate groups attached. Of the GAG repeating units shown in Figure 15-15, only hyaluronate contains no sulfate groups. Because GAGs are hydrophilic molecules with many negatively charged sulfate and carboxyl groups, they attract both water and cations, thereby forming the hydrated, gelatinous matrix in which collagen and elastin fibrils become embedded.

Most glycosaminoglycans in the ECM are covalently bound to protein molecules to form **proteoglycans**. Each proteoglycan consists of numerous GAG chains attached along the length of a **core protein**, as shown in Figure 15-15. In some cases, proteoglycans are integral components of the plasma membrane, with their core polypeptides embedded within the membrane. In other cases, proteoglycans are linked covalently to membrane phospholipids.

Many kinds of proteoglycans can be formed by the combination of different core proteins and GAGs of varying types and lengths. Proteoglycans vary greatly in size, depending on the molecular weight of the core protein (which ranges from about 10,000 to more than 500,000) and the number and length of the carbohydrate chains (1–200 per molecule, with an average length of about 800 monosaccharide units). Most proteoglycans are huge, with molecular weights in the range of 0.25–3 million! Proteoglycans are linked directly to collagen fibers in ECM networks.

In many tissues, proteoglycans are present as individual molecules. In cartilage, however, numerous proteoglycans become attached to long molecules of hyaluronate, forming large complexes as shown in Figure 15-15. A single such complex can have a molecular weight of many millions and may exceed several micrometers in length. The remarkable resilience and pliability of cartilage are due mainly to the properties of these complexes. Smaller proteoglycans are also known

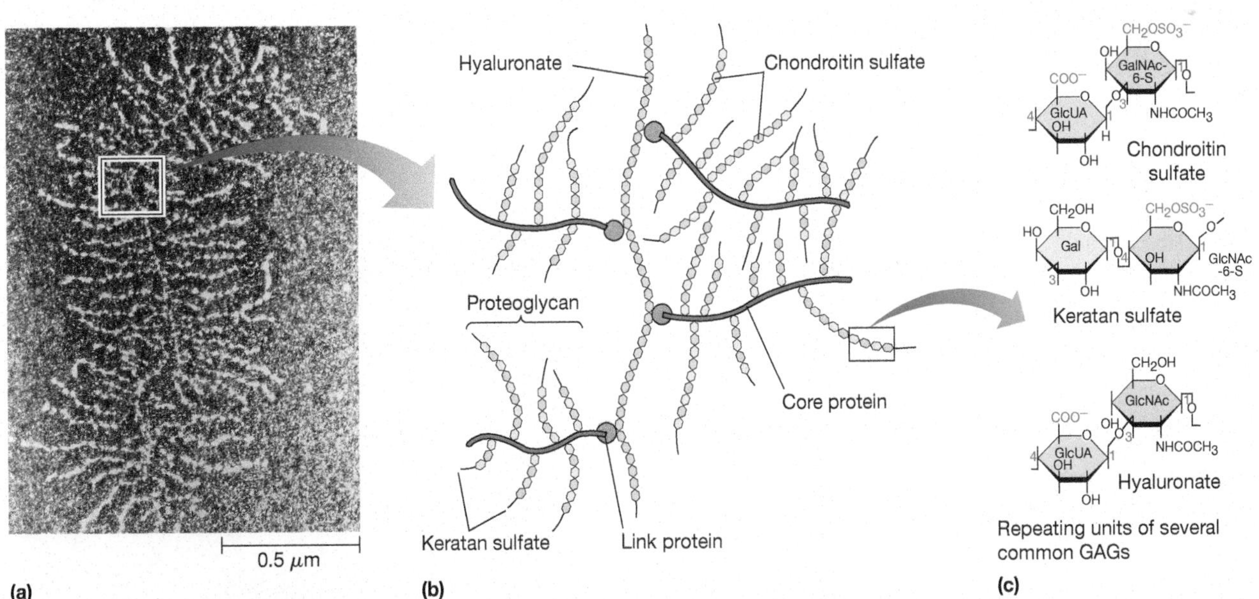

Figure 15-15 Proteoglycan Structure in Cartilage. In cartilage, many proteoglycans associate with a hyaluronate backbone to form a complex readily visible in the electron microscope. **(a)** A hyaluronate-proteoglycan complex isolated from bovine cartilage (TEM). **(b)** A small portion of the structure showing core proteins of proteoglycans attached via linker proteins to a long hyaluronate molecule. Short keratan sulfate and chondroitin sulfate chains are linked covalently along the length of the core proteins. Proteoglycans have a carbohydrate content of about 95%. **(c)** Structures of the disaccharide repeating units in three common extracellular glycosaminoglycans (GAGs) found in the ECM of animal cells. The repeating unit of chondroitin sulfate (top) consists of glucuronate, the ionized form of glucuronic acid (GlcUA), and *N*-acetylgalactosamine (GalNAc). The repeating unit of keratan sulfate (middle) is galactose (Gal) and *N*-acetylglucosamine (GlcNAc). Hyaluronate (bottom) has a repeating unit consisting of GlcUA and GlcNAc.

to be involved in facilitating growth factor signaling (which you will learn about in Chapter 23). For example, a proteoglycan called *syndecan* is a modulator of fibroblast growth factor signaling.

Free Hyaluronate Lubricates Joints and Facilitates Cell Migration

Although most of the GAGs found in the extracellular matrix exist only as components of proteoglycans and not as free glycosaminoglycans, **hyaluronate** is an exception. In addition to its role as the backbone of the proteoglycan complex in cartilage, free hyaluronate, consisting of hundreds or even thousands of repeating disaccharide units, has lubricating properties. It is most abundant where friction must be reduced, such as the joints between movable bones.

Adhesive Glycoproteins Anchor Cells to the Extracellular Matrix

Direct links between the ECM and the plasma membrane are reinforced by a family of adhesive glycoproteins that bind proteoglycans and collagen molecules and receptors on the membrane surface. The two most common kinds of adhesive glycoproteins are the *fibronectins* and *laminins.* Many of the membrane receptors to which these glycoproteins bind belong to a family of transmembrane proteins called *integrins.* The following sections discuss each of these families of proteins.

Fibronectins Bind Cells to the ECM and Foster Cellular Movement

Fibronectins are a family of closely related adhesive glycoproteins that occur in soluble form in blood and other body fluids, as insoluble fibrils in the extracellular matrix, and as an intermediate form loosely associated with cell surfaces. These different forms of the protein are generated because the RNA transcribed from the fibronectin gene is processed to generate many different mRNAs.

Fibronectin consists of two very large polypeptide subunits that are linked near their carboxyl ends by a pair of disulfide bonds (**Figure 15-16**). Each subunit has about 2500 amino acids and is folded into a series of rodlike domains connected by short, flexible segments (Figure 15-16a). Several of the domains bind to one or more specific kinds of macromolecules located in the ECM, including several types of collagen (I, II, and IV), heparin, and the blood-clotting protein fibrin. Other domains are recognized by cell surface receptors. The receptor-binding activity of these domains has been localized to a specific tripeptide sequence, RGD (arginine-glycine-aspartate). This *RGD sequence* is a common motif among extracellular adhesive proteins and is recognized by various integrins on the cell surface. Fibronectin *fibrils* can form a loose network to which cells attach. Figure 15-16b shows an example of such a meshwork associated with muscle cells in culture.

Effects of Fibronectin on Cell Movement. Fibronectin functions as a bridging molecule that attaches cells to the ECM. One important process that requires fibronectin is cell migration. In living embryos, for example, the paths taken by migrating cells are often rich in fibronectin. Direct evidence for the crucial importance of fibronectin in embryos comes from studying genetically engineered mice that cannot produce it. Such mice have severe defects in some cells that make the musculature and the vasculature. The involvement of fibronectin in uncontrolled migration of cancer cells is suggested by the observation that many kinds of cancer cells either synthesize too much or too little fibronectin compared to the normal tissues from which they arise. In many carcinomas, for example, a poor prognosis is correlated with overexpression of fibronectin.

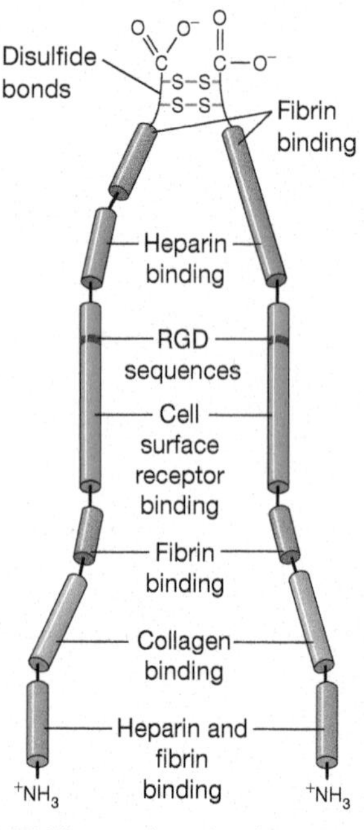

(a) Fibronectin molecule

(b) Fibronectin fibrils

Figure 15-16 Fibronectin Structure. (a) A fibronectin molecule consists of two nearly identical polypeptide chains joined by two disulfide bonds near their carboxyl ends. Each polypeptide chain is folded into a series of domains linked by short, flexible segments. These domains have binding sites for ECM components or for specific receptors on the cell surface, including the tripeptide sequence RGD (arginine-glycine-aspartate), which is recognized by integrins. Besides the binding activities noted, fibronectin has binding sites for heparan sulfate, hyaluronate, and gangliosides (glycosphingolipids that contain sialic acid groups). **(b)** Myoblast cells on an ECM containing fibronectin in vitro, immunostained for fibronectin (green) and for DNA in the nucleus (blue).

Effects of Fibronectin on Blood Clotting. The soluble form of fibronectin present in the blood, called *plasma fibronectin*, is involved in blood clotting. Fibronectin promotes blood clotting because it has several binding domains that recognize *fibrin*, the blood-clotting protein, and it can attach blood platelets to fibrin as the blood clot forms.

Laminins Bind Cells to the Basal Lamina

As shown in Figure 15-1, the ECM can be loosely organized, as is the case of fibroblasts and other connective tissues, or it can be highly organized, as is the case when the ECM is in contact with epithelial cells. In the latter case, a thin, highly specialized ECM known as the **basal lamina** attaches to the basal surfaces of the epithelial cells. This thin sheet of specialized ECM, typically about 50 nm thick, underlies epithelial cells, thereby separating them from connective tissues (**Figure 15-17**). Basal laminae also surround muscle cells, fat cells, and the Schwann cells that form myelin sheaths around nerve cells. The major adhesive glycoproteins present in the basal lamina are a family of proteins called **laminins**, which are conserved from simple invertebrates to humans.

Properties of the Basal Lamina. The basal lamina serves as a structural support that maintains tissue organization and as a permeability barrier that regulates the movement of molecules as well as cells. In the kidney, for example, an extremely thick basal lamina functions as a filter that allows small molecules but not blood proteins to move from the blood into the urine. The basal lamina beneath epithelial cells prevents the passage of underlying connective tissue cells into the epithelium but permits the migration of the white blood cells needed to fight infections. The effect of the basal lamina on cell migration is of special interest because some cancer cells show increased binding to the basal lamina, which may facilitate their movement through it as they become metastatic

Despite differences in function and specific molecular composition from tissue to tissue, all forms of basal lamina contain type IV collagen, proteoglycans, laminins, and another glycoprotein called *nidogen* or *entactin*. Fibronectins may be present, but laminins are the most abundant adhesive glycoproteins in the basal lamina. Laminins are localized mainly on the surface of the lamina that faces the overlying epithelial cells, where they help bind the cells to the lamina. Fibronectins, on the other hand, are located on the other side of the lamina, where they help anchor cells of the connective tissue.

Cells can alter the properties of the basal lamina by secreting enzymes that catalyze changes in the basal lamina. One important class of such enzymes is the *matrix metalloproteinases (MMPs)*. These enzymes, which require metal ions as cofactors, degrade the ECM locally, allowing cells to pass through the ECM. Such activity is important for cells such as leukocytes to invade injured tissues during inflammation. MMPs are also involved in abnormal invasive behavior. The MMP activity of invasive cancer cells, such as metastatic melanoma cells, is very high. (You will learn more about these enzymes in Chapter 26.)

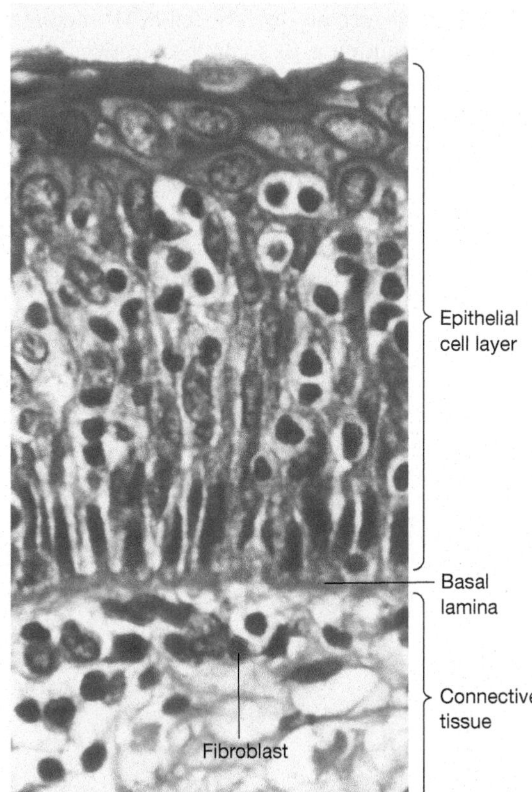

Figure 15-17 The Basal Lamina. The basal lamina is a thin sheet, typically about 50 nm, of matrix material that separates an epithelial cell layer from underlying connective tissue (light micrograph).

Properties of Laminin. Laminin, a very large protein with a molecular weight of about 850,000, consists of three long polypeptides, denoted α, β, and γ (**Figure 15-18** on page 428). Disulfide bonds hold the polypeptide chains together in the shape of a cross, with part of the long arm wound into a three-stranded coil (Figure 15-18a). Like fibronectin, laminin has several domains with binding sites for type IV collagen, heparin, heparan sulfate, and nidogen, as well as for laminin receptors on the surface of overlying cells. Nidogen molecules have binding sites for both laminin and type IV collagen and are therefore thought to reinforce the binding of type IV collagen and laminin networks in the basal lamina.

Together, the components of the basal lamina form a complex, crisscrossing meshwork (Figure 15-18b). In many epithelia, a more gelatinous ECM lies beneath the basal lamina; together, they form an extracellular structure known as the *basement membrane*. Because the basal lamina is so complex, it is difficult to perform experiments to determine how the precise makeup of the basal lamina affects cellular behaviors in animals. Instead, cell biologists often examine the effects of defined artificial basal laminae on cells in tissue culture (see Key Technique, page 429.)

Integrins Are Cell Surface Receptors That Bind ECM Components

Fibronectins and laminins can bind to animal cells because the plasma membranes of most cells have specific receptors on their surfaces that recognize and bind to specific regions of the fibronectin or laminin molecule. These receptors—and those for a variety of other ECM constituents—belong to a large family of transmembrane proteins that are called

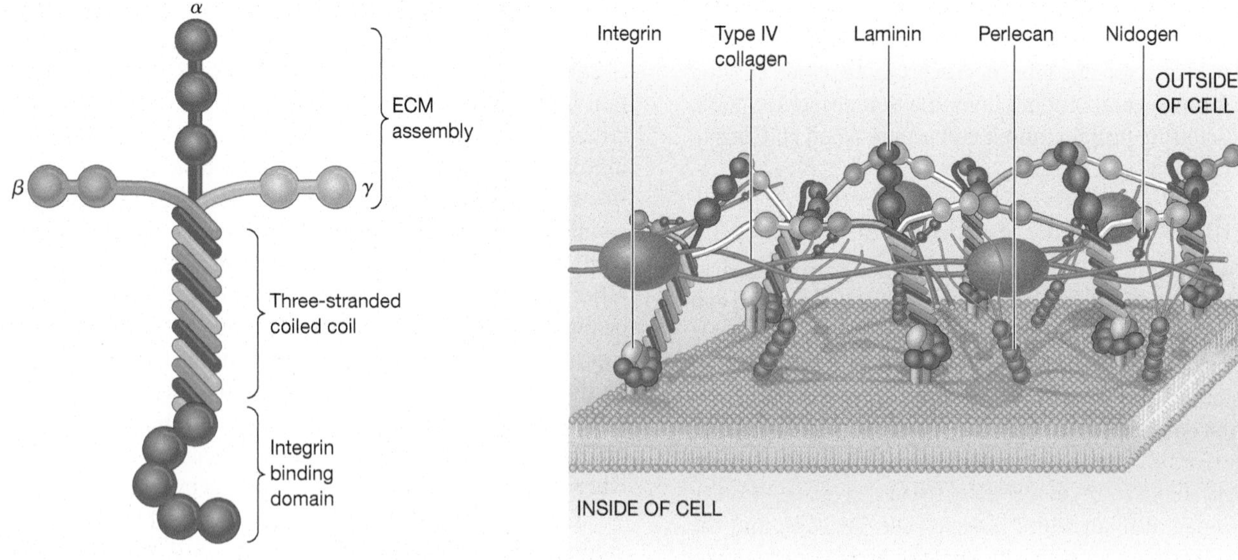

Figure 15-18 Laminin and the Basal Lamina. (a) A laminin molecule consists of three large polypeptides— α, β, and γ —joined by disulfide bonds into a crosslinked structure. A portion of the long arm consists of a three-stranded coil. Domains on the ends of the α chain are recognized by cell surface receptors; those at the ends of the two arms of the cross are specific for type IV collagen. The cross-arms also contain laminin-laminin binding sites, which enable laminin to form large aggregates. Laminin also contains binding sites for heparin, heparan sulfate, and entactin (not shown). (Adapted with permission from Macmillan Publishers Ltd: Fig. 1 from M. P. Marinkovich, "Laminin 332 in Squamous-Cell Carcinoma," *Nature Reviews Cancer* 7:370-380. Copyright 2007.) **(b)** Laminin associates with type IV collagen, perlecan, nidogen, and other components to form a mat of extracellular matrix. Cells attach to the basal lamina using integrins.

integrins because of their role in *integrating* the cytoskeleton with the extracellular matrix.

Structure of Integrins. An integrin consists of two large transmembrane polypeptides, the α and β subunits, that associate with each other noncovalently (**Figure 15-19**). Integrins differ from one another in their binding specificities and in the sizes of their subunits (molecular weight ranges are 110,000–140,000 for the α subunit and 85,000–91,000 for the β subunit). For a specific integrin, the extracellular portions of the α and β subunits interact to form the binding site for a particular ECM protein, with most of the binding specificity apparently depending on the α subunit. On the cytosolic side of the membrane, integrins have binding sites for specific linker proteins that mechanically link the cytoskeleton and the ECM across the plasma membrane.

The presence of multiple types of both α and β subunits results in a large number of different integrin heterodimers, which vary in their binding specificities. For example, integrins containing a β_1 subunit are found on the surfaces of most vertebrate cells and mediate mainly cell-ECM interactions, whereas those containing a β_2 subunit are restricted to the surface of white blood cells and are involved primarily in cell-cell interactions. The most common integrin that binds fibronectin is $\alpha_5\ \beta_1$, whereas $\alpha_6\ \beta_1$ binds laminin. Many integrins

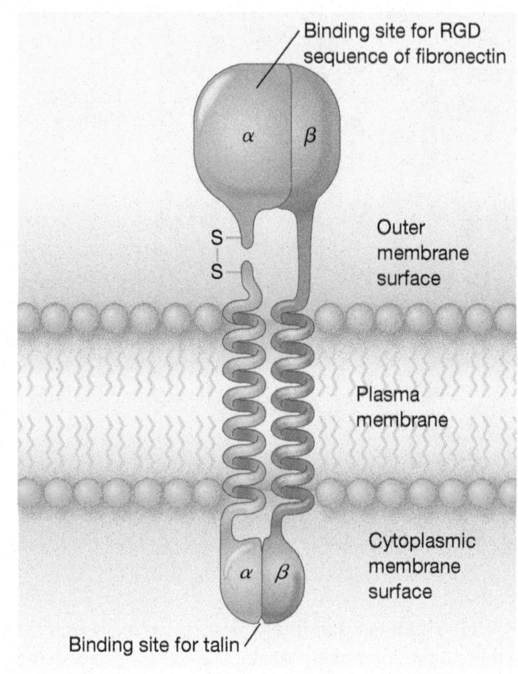

Figure 15-19 An Integrin: The Fibronectin Receptor. An integrin consists of α and β subunits, transmembrane polypeptides that associate with each other noncovalently to form a binding site for the ligand on the outer membrane surface and a binding site for linker proteins on the inner membrane surface. Shown here is the fibronectin receptor $\alpha_5\ \beta_1$), which has a binding site for fibronectin on the outer surface and a binding site for talin on the cytosolic side of the membrane. In this and several other integrins, the α subunit is split into two segments held together by a disulfide bond. Both the α and β subunits are glycosylated on the exterior side, although the sugar side chains are not shown here.

Key Technique

Building an ECM from Scratch

PROBLEM: The extracellular matrix is a crucial component of tissues throughout the body. Studying how cells interact with the ECM is therefore crucial for understanding how multicellular organisms function normally and for understanding certain types of human disease, such as cancer metastasis. However, the ECM is a complex network of proteins with an intricate geometry that is difficult to manipulate in vivo. How then is it studied?

SOLUTION: Artificial extracellular matrices can be produced. Adding cultured cells to such systems allows cellular behavior to be studied on or in a substrate that mimics characteristics of a natural ECM.

Key Tools: Cultured cells, Matrigel (an artificial ECM).

Details: Cell biologists have increasingly turned to tissue culture and artificial extracellular matrices to study how cells respond to the ECM. In the early days of tissue culture, researchers learned that although cells could sometimes bind to naked glass or plastic, they behaved differently when plated onto ECM of defined composition. These artificial substrates usually contained some type of collagen and an adhesive glycoprotein, such as laminin (for epithelial cells) or fibronectin (for non-epithelial cells). A great deal was learned about how cells interact with different ECM "recipes" by combining ECM proteins in varying amounts.

Matrigel: Results from early experiments with artificial matrices were found to be limited, however, because the substrates were missing trace amounts of key proteins, such as growth factors, that are normally associated with the ECM in vivo. (You will learn more about growth factors in Chapter 23.)

Today, one common way to circumvent this problem is to use a commercially available basement membrane extract known by the trade name *Matrigel*. Matrigel is a gelatinous mixture of ECM proteins secreted by Engelbreth-Holm-Swarm (EHS) mouse sarcoma cells and is considered to be equivalent to a reconstituted natural basement membrane. A small amount of Matrigel will assemble into a thin layer when added to a plastic tissue culture dish. Cells cultured on this type of substrate show behavior not seen when simpler ECM mixtures are used.

Three-Dimensional Cultures: Using Matrigel on a flat tissue culture dish still has drawbacks. The ECM is limited to a thin, two-dimensional shape, whereas cells in living organisms are rarely confined to two dimensions. To address this issue, cell biologists have developed ways to culture cells in thick, three-dimensional matrices. The difference in cell behavior in a three-dimensional situation can sometimes be remarkable. **Figure 15B-1** illustrates this for a commonly used human breast epithelial cell line known as MCF-10A. Instead of forming a sheet, as they do in a tissue culture dish (Figure 15B-1a), MCF-10A cells embedded in a thick slab of Matrigel organize into hollow-centered balls called *cysts*. Moreover, the cells become polarized, with apical and basal ends like epithelial cells in vivo (Figure 15B-1b). Because these are mammary cells, they can even produce milk!

Because MCF-10A cells are derived from breast epithelial cells, which often go on to become cancerous, the fact that this one change in their growth conditions elicits such a dramatic change toward a more normal appearance suggests that fundamental

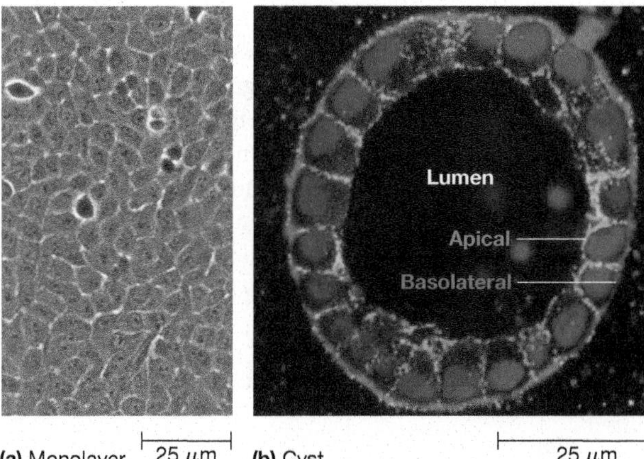

Figure 15B-1 Different ECM Conditions Dramatically Affect Cultured Human Breast Epithelial Cells. (a) MCF-10A cells cultured in a dish visualized using phase contrast microscopy. The cells adopt a characteristic "cobblestone" appearance after they have proliferated to completely cover the surface of the dish. **(b)** MCF-10A cells after 20 days in a preparation of three-dimensional Matrigel, visualized using confocal microscopy. The cells have generated a polarized epithelial cyst, as revealed by immunostaining with antibodies to mark the apical surfaces (green) and the basolateral surfaces (red). A DNA stain marks nuclei (blue).

insights into what controls the switch to the metastatic state will come from this work.

Tissue Engineering: Another exciting possibility raised by artificial extracellular matrices involves the burgeoning field of *tissue engineering*. By adding cells to combinations of artificial ECM and inert scaffolds (such as plastic or even pieces of coral), scientists working at the interface between cell biology and engineering are manipulating the behavior of cells to coax them to form a variety of structures and to become specialized in precise ways. Although in its infancy, tissue engineering has already been used to produce replacement parts for human patients, including heart valves, windpipes, skin for burn patients, and structures as exotic as the exterior part of the human ear (see photo inset).

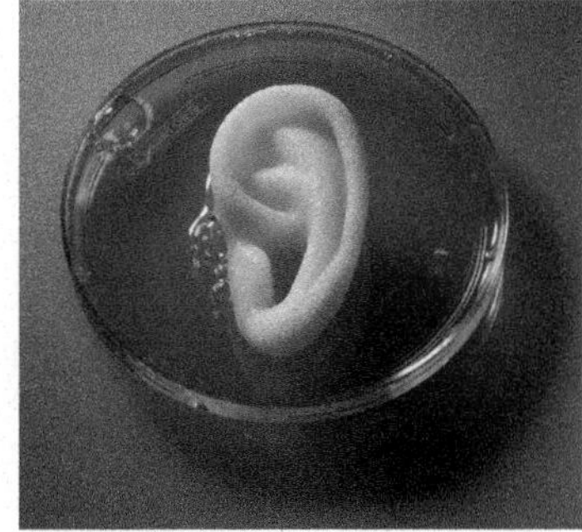

Tissue Engineering.

QUESTION: Cells behave differently from those shown in Figure 15B-1a not just when they are cultured in three-dimensional ECMs as in Figure 15B-1b, but also when they are cultured on flat ECMs that are not attached to the bottom of a culture dish. What do you think three-dimensional cultures and these detached ECM cultures have in common that is different from a plastic dish?

recognize the RGD sequence in the specific ECM glycoproteins that they bind. However, the binding site must recognize other parts of the glycoprotein molecule as well because integrins display a greater specificity of glycoprotein binding than can be accounted for by the RGD sequence alone.

Integrins and the Cytoskeleton. Although integrins link the extracellular matrix and the cytoskeleton, they do not do so directly. Instead, the tails of integrins interact with proteins in the cytosol that link integrins to cytoskeletal proteins. Integrins make two main types of connections to the cytoskeleton (**Figure 15-20**). Migratory and non-epithelial cells, such as fibroblasts, attach to extracellular matrix molecules via **focal adhesions**. Focal adhesions contain clustered integrins that interact with bundles of actin microfilaments via several linker proteins (Figure 15-20a, b). These include *talin*, which can bind to the actin-binding protein *vinculin*, and *α-actinin*, which can bind directly to actin microfilaments. Because they bind to components of the ECM on the outside of the cell and to actin microfilaments on the inside, integrins that connect to actin play important roles in regulating cell movement and cell attachment. For example, mice lacking a particular integrin subunit required for attachment to laminin develop a distinctive form of progressive muscular dystrophy. Humans that carry the same mutation develop progressive muscle degeneration.

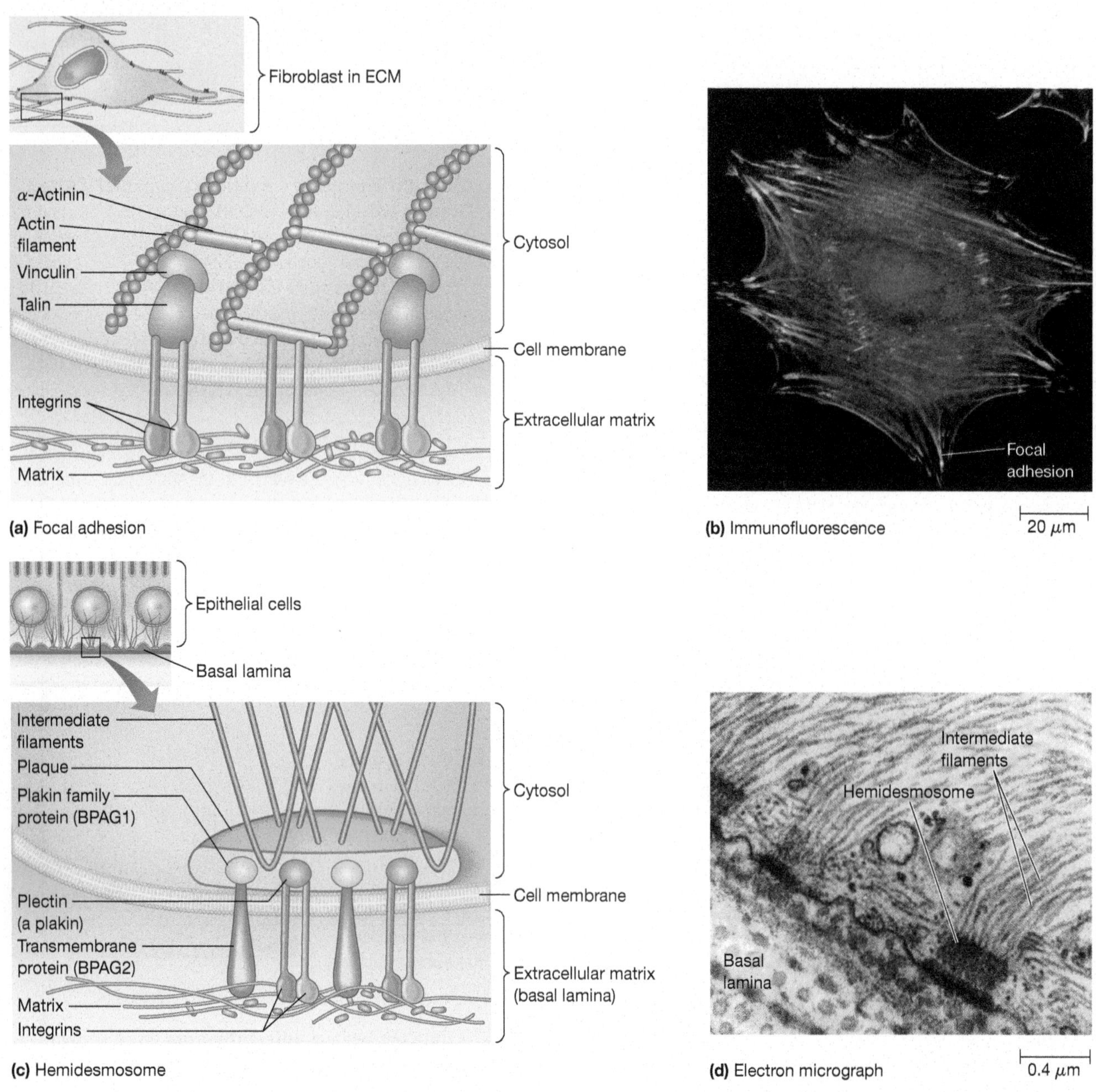

Figure 15-20 Integrins, Focal Adhesions, and Hemidesmosomes. (a) Migrating cells attach to the extracellular matrix via focal adhesions. Focal adhesions contain integrins, such as $\alpha_5\beta_1$ integrin, and integrin-associated linker proteins, such as α-actinin, vinculin, and talin, that attach integrins to the actin cytoskeleton. **(b)** An invasive human breast cancer cell stained for F-actin (red) and vinculin to mark integrin attachment sites (blue). Note how the actin microfilaments terminate at focal adhesions. **(c)** Epithelial cells attach to the extracellular matrix via hemidesmosomes, which contain $\alpha_6\beta_4$ integrins, and connect to intermediate filaments via linker proteins such as plectin. **(d)** The external surface of a hemidesmosome directly abuts the basal lamina (TEM).

The other major type of integrin-mediated attachment is found in epithelial cells. Epithelial cells attach to laminin in the basal lamina via **hemidesmosomes** (so called because they resemble "half desmosomes"; desmosomes were discussed earlier in this chapter). The integrin found in hemidesmosomes contains one α_6 and one β_4 subunit. In this case, the integrins are not attached to actin but to intermediate filaments, typically *keratin* (Figure 15-20c, d). The linker proteins in hemidesmosomes form a dense **plaque** that connects clustered integrins to the cytoskeleton. Among the linker proteins, members of the **plakin** family of proteins are prominent. A plakin known as *plectin* attaches keratin filaments to the integrins. Another transmembrane protein, called *BPAG2*, and its associated plakin, *BPAG1*, can serve as a bridge between keratin and laminin. Like desmosomes, hemidesmosomes are crucial for the mechanical strength of skin; devastating effects on human health can result when desmosome or hemidesmosome components are defective or missing (see Human Connections, pages 418–419).

Integrins and Signaling. Although a key role of integrins is their ability to link the cytoskeleton and the extracellular matrix, they also interact with intracellular signaling pathways. For example, cells exposed to growth factors (see Chapter 23) often exhibit integrin clustering. (Such effects are often called "inside-out" signaling because internal changes in the cell result in effects on integrins at the surface.) Integrins can also act as receptors that activate intracellular signaling themselves (sometimes called "outside-in" signaling). The original observations that suggested this role for integrins came from studies of cancer cells. If normal cells are prevented from attaching to an extracellular matrix layer, they will cease dividing and undergo a kind of programmed cell death known as apoptosis (you will learn about apoptosis in detail in Chapter 24). Such behavior is referred to as **anchorage-dependent growth**. In contrast, cancer cells continue to grow even when they are not firmly attached to an ECM layer, apparently because they no longer need to transduce signals that result from attachment to the ECM.

Several kinases are activated at focal adhesions following integrin clustering. They are recruited to focal adhesions by adapter proteins such as *paxillin*. These include *focal adhesion kinase (FAK)*, *kindlins*, and *integrin-linked kinase (ILK)*. ILK probably functions as a scaffolding protein during integrin-based signaling because "kinase dead" versions of the protein still function. FAK appears to be important in regulating anchorage-dependent growth. Cancer cells contain activated FAK even when they are not attached, and cells can be transformed into cancerlike cells by the expression of an activated form of FAK.

The Dystrophin/Dystroglycan Complex Stabilizes Attachments of Muscle Cells to the ECM

In addition to focal adhesions and hemidesmosomes, a third type of ECM attachment is important in human disease. The *costamere* is an attachment structure at the surface of striated muscle (**Figure 15-21**). Costameres align in register with Z discs around the surface of muscle cells (Figure 15-21a). They physically attach muscle fibers at the surface of muscle cells to their plasma membranes, which in turn are anchored to

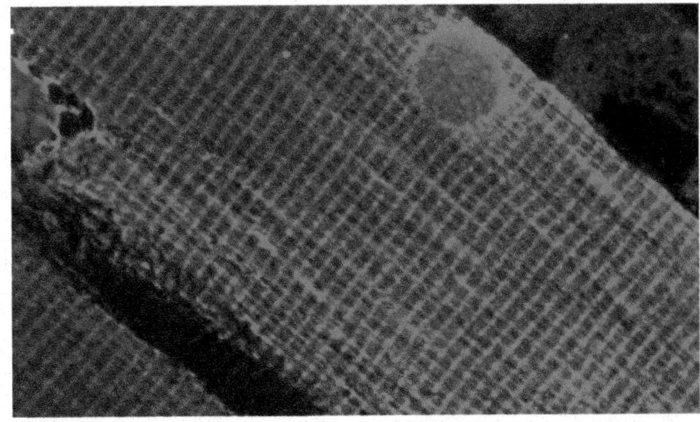

(a)

(b)

Figure 15-21 Costameres and the Dystrophin/Dystroglycan Complex of Muscle Cells. (a) Muscle cells immunostained for dystrophin (green). Blue-stained DNA marks the locations of cell nuclei. Costameres are visible as stripes of dystrophin. **(b)** Dystrophin, an enormous cytosolic protein, interacts with actin in the cytosol. Dystrophin is linked, via a series of proteins, to the dystroglycan complex at the cell surface. The dystroglycan complex in turn interacts with extracellular matrix proteins, such as laminin.

the ECM surrounding the muscle cells. Costameres contain many of the same proteins found at focal contacts, including β_1-integrin, vinculin, talin, and α-actinin. In addition, costameres contain specialized protein complexes that include the huge cytosolic protein **dystrophin** (Figure 15-21b). Mutations in the human dystrophin gene cause the two most common types of muscular dystrophy, *Duchenne muscular dystrophy (DMD)* and *Becker muscular dystrophy (BMD)*. Together, BMD and DMD affect 1 in 3500 patients worldwide each year. (Because the dystrophin gene is on the X chromosome, DMD and BMD virtually always affect only boys.) Although the symptoms are similar, because the mutation in DMD results in nearly complete loss of function of the dystrophin protein, DMD is more severe than BMD. DMD patients undergo progressive muscle degeneration, eventually leading to loss of the ability to walk and ultimately to death in the teenage or young adult years. Death is caused by injury to the heart muscle, known as *cardiomyopathy*, or because of difficulties in breathing due to damage to the diaphragm.

Why do defects in dystrophin result in muscular dystrophy? The specialized complex to which dystrophin belongs includes the integral membrane protein *dystroglycan*. Dystroglycan binds to laminin, linking costameres to the extracellular matrix (Figure 15-21b). Dystrophin probably acts like a stiff spring, preventing damage to muscle cells where they attach to the ECM and exert force on the ECM during contraction. When dystrophin is absent, muscle cells are much more susceptible to damage and eventual degeneration.

CONCEPT CHECK 15-2

Hemidesmosomes and focal contacts are not found on the same cells at the same time. Why?

15.3 The Plant Cell Surface

So far, the discussion has focused on animal cell surfaces. The surfaces of plant, algal, fungal, and bacterial cells exhibit some of the same properties, but they also have several unique features of their own. The rest of this chapter considers some distinctive features of the plant cell surface.

Cell Walls Provide a Structural Framework and Serve as a Permeability Barrier

One of the most impressive features of plants is that they have no bones or related skeletal structures and yet exhibit remarkable strength. This strength is provided by the rigid **cell walls** that surround all plant cells except sperm and some eggs. The rigidity of the cell wall makes cell movements virtually impossible. At the same time, sturdy cell walls enable plant cells to withstand the considerable *turgor pressure* that is exerted by the uptake of water. Turgor pressure is vital to plants because it accounts for much of the turgidity, or firmness, of plant tissues and provides the driving force behind cell expansion.

MAKE CONNECTIONS 15.2

How does a cell wall change how a plant cell responds to changes in osmolarity compared with an animal cell? (Ch. 8.2)

The wall that surrounds a plant cell is also a permeability barrier for large molecules. For water, gases, ions, and small water-soluble molecules such as sugars or amino acids, the cell wall is not a significant obstacle. These substances diffuse through the wall readily.

The Plant Cell Wall Is a Network of Cellulose Microfibrils, Polysaccharides, and Glycoproteins

Like the extracellular matrix of animal cells, plant cell walls consist predominantly of long fibers embedded in a network of branched molecules (**Figure 15-22**). Instead of collagen and proteoglycans, however, plant cell walls contain *cellulose microfibrils* enmeshed in a complex network of branched polysaccharides and glycoproteins called *extensins* (see Table 15-2). The two main types of polysaccharides are *hemicelluloses* and *pectins*. On a dry-weight basis, cellulose typically makes up about 40% of the cell wall, hemicelluloses account for another 20%, pectins represent about 30%, and glycoproteins make up about 10%. Figure 15-22a illustrates the relationships among these cell wall components. Cellulose, hemicelluloses, and glycoproteins are linked together to form a rigid interconnected network that is embedded in a pectin matrix. Not shown are lignins, which in woody tissues are localized between the cellulose fibrils and make the wall especially strong and rigid. The next several sections consider each major cell wall component in turn.

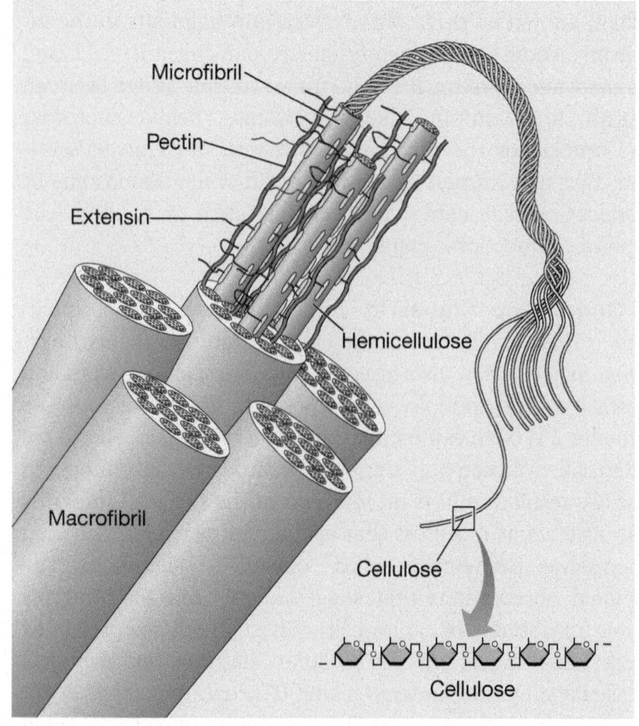

(a)

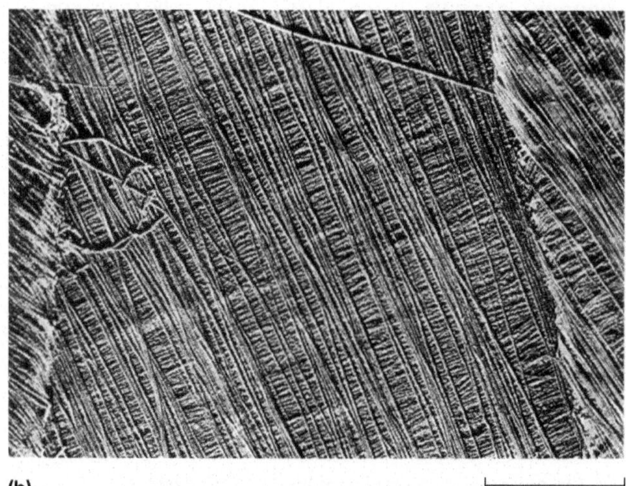

(b)

Figure 15-22 Structural Components of Plant Cell Walls. **(a)** Cellulose microfibrils are linked by hemicelluloses and glycoproteins called extensins to form a rigid interconnected network embedded in a matrix of pectins. Cellulose microfibrils are often twisted together to form larger structures called macrofibrils. (Not shown are the lignins that are localized between the cellulose microfibrils in woody tissues.) **(b)** This electron micrograph shows individual cellulose microfibrils in the cell wall of a green alga. Each microfibril consists of many cellulose molecules aligned laterally (TEM).

Cellulose and Hemicellulose. The predominant polysaccharide of the plant cell wall is **cellulose**, the single most abundant organic macromolecule on Earth. Cellulose is an unbranched polymer consisting of thousands of β-D-glucose units linked together by β $(1 \rightarrow 4)$ bonds (see Figure 3-25). Cellulose molecules are long, ribbonlike structures that are stabilized by intramolecular hydrogen bonds. Many such molecules (50–60, typically) associate laterally to form the **microfibrils** found in cell walls. Cellulose microfibrils are often twisted together in a ropelike fashion to generate even larger structures, called *macrofibrils* (Figure 15-22a). Cellulose macrofibrils are as strong as an equivalently sized piece of steel!

Despite their name, **hemicelluloses** are chemically and structurally distinct from cellulose. The hemicelluloses are a heterogeneous group of polysaccharides, each consisting of a long, linear chain of a single kind of sugar (glucose or xylose) with short side chains bonded into a rigid network (Figure 15-22a). The side chains usually contain several kinds of sugars, including the hexoses glucose, galactose, and mannose and the pentoses xylose and arabinose.

Other Cell Wall Components. In addition to cellulose and hemicellulose, cell walls have several other major components. **Pectins** are branched polysaccharides but with backbones called *rhamnogalacturonans* that consist mainly of negatively charged galacturonic acid and rhamnose. The side chains attached to the backbone contain some of the same monosaccharides found in hemicelluloses. Pectins form the matrix in which cellulose microfibrils are embedded (Figure 15-22a). They also bind adjacent cell walls together. Because of their highly branched structure and their negative charge, pectins trap and bind water molecules. As a result, pectins have a gel-like consistency, and this explains why they are added to fruit juice in the process of making fruit jams and jellies.

In addition to hemicelluloses and pectins, cell walls contain a group of related glycoproteins called **extensins**. Extensins are rigid, rodlike molecules that are tightly woven into the complex polysaccharide network of the cell wall (Figure 15-22a). Extensins are initially deposited in the cell wall in a soluble form. Once deposited, however, extensin molecules become covalently crosslinked to one another and to cellulose, generating a reinforced protein-polysaccharide complex. Extensins are most abundant in the cell walls of tissues that provide mechanical support to the plant.

Lignins are very insoluble polymers of aromatic alcohols that occur mainly in woody tissues. (The Latin word for "wood" is *lignum*.) Lignin molecules are localized mainly between the cellulose fibrils, where they function to resist compression forces. Lignin accounts for as much as 25% of the dry weight of woody plants, making it second only to cellulose as the most abundant organic compound on Earth.

Cell Walls Are Synthesized in Several Discrete Stages

The plant cell wall components are secreted from the cell stepwise, creating a series of layers in which the first layer to be synthesized ends up farthest away from the plasma membrane. The first structure to be laid down is called the **middle lamella**. It is shared by neighboring cell walls and holds adjacent cells together (**Figure 15-23**). The next structure to be formed is called the **primary cell wall**, which forms when the cells are still growing. Primary walls are about 100–200 nm thick, only several times the thickness of the basal lamina of animal cells. The primary cell wall consists of a loosely organized network of cellulose microfibrils associated with hemicelluloses, pectins, and glycoproteins (**Figure 15-24** on page 434). The cellulose microfibrils are generated by cellulose-synthesizing enzyme complexes called *rosettes* that are localized within the plasma membrane (Figure 15-24a). Because the microfibrils are anchored to other wall components, the rosettes must move in the plane of the membrane as they lengthen the growing cellulose microfibrils.

Several lines of evidence indicate that rosette deposition depends on microtubules. First, in chemically fixed specimens, microtubules are aligned with cellulose fibrils in the cell wall. Second, disrupting microtubules using chemical inhibitors leads to disrupted orientation of cellulose fibrils. Simultaneous, direct observation of rosette movement and microtubules in genetically engineered plant cells indicates that rosette and microtubule movement are often coupled.

When the shoots or roots of a plant are growing, cell walls must be remodeled. A family of proteins called **expansins** helps cell walls retain their pliability. One way in which expansins may act is by disrupting the normal hydrogen bonding of glycans within the microfibrils of the cell wall to allow for rearrangement of the microfibrils. The plant hormone *auxin* (indole-3-acetic acid) likely stimulates expansin activity. One way it may do so is by stimulating proton pumps that locally decrease the pH in a section of cell wall that is destined to elongate. The low pH activates the expansins, leading to cell wall loosening and cell elongation.

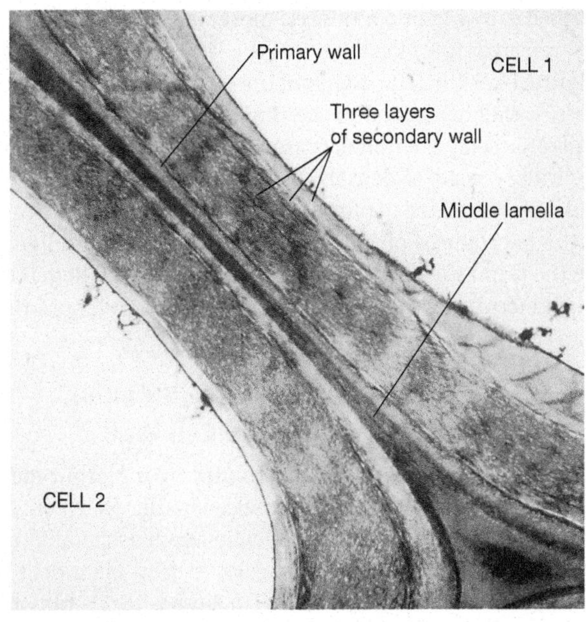

Figure 15-23 The Middle Lamella. The middle lamella is a layer of cell wall material that is shared by two adjacent plant cells. Consisting mainly of sticky pectins, the middle lamella binds the cells together tightly.

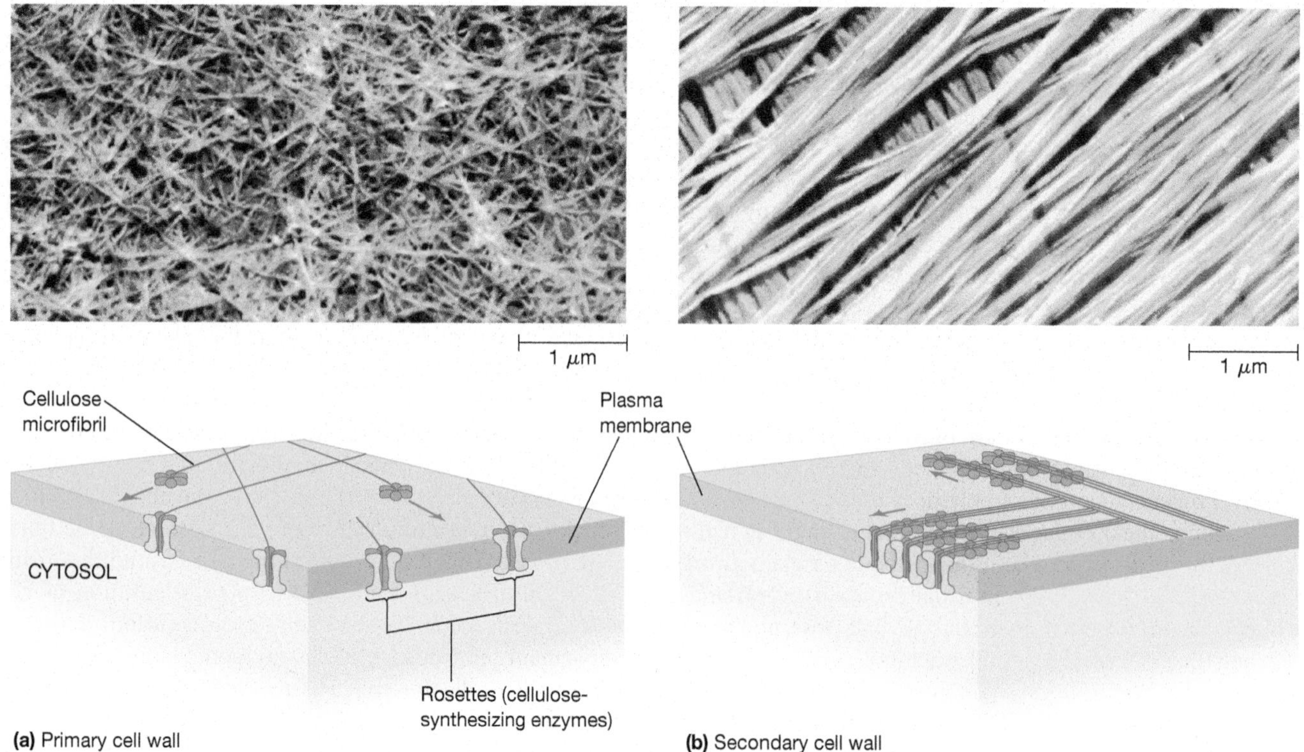

Figure 15-24 Cellulose Microfibrils of Primary and Secondary Plant Cell Walls. (a) The electron micrograph shows the loosely organized cellular microfibrils of a primary cell wall, and the drawing depicts how cellular microfibrils are synthesized by rosettes, each of which is a cluster of cellulose-synthesizing enzymes embedded in the plasma membrane. As a rosette synthesizes a bundle of cellulose molecules, it moves through the plasma membrane in the direction indicated by the arrows. **(b)** The electron micrograph shows densely packed cellulose macrofibrils of a secondary cell wall oriented in parallel; the drawing depicts how rosettes form dense aggregates that synthesize large numbers of microfibrils in parallel, generating cellular macrofibrils. (TEMs.)

The loosely textured organization of the primary cell wall creates a relatively thin, flexible structure. In some plant cells, development of the cell wall does not proceed beyond this point. However, many cells that have stopped growing add a thicker, more rigid set of layers that are referred to collectively as the **secondary cell wall** (Figure 15-24b). The components of the multilayered secondary wall are added to the inner surface of the primary wall after cell growth has ceased. Each layer of the secondary wall consists of densely packed bundles of cellulose microfibrils, arranged in parallel and oriented to lie at an angle to the microfibrils of adjacent layers. Cellulose and lignins are the primary constituents of the secondary wall, making this structure significantly stronger, harder, and more rigid than the primary wall.

Plasmodesmata Permit Direct Cell-Cell Communication Through the Cell Wall

Because every plant cell is surrounded by a plasma membrane and a cell wall, you may wonder whether plant cells are capable of intercellular communication such as that afforded by the gap junctions of animal cells. In fact, plant cells do possess such structures. As shown in **Figure 15-25**, **plasmodesmata** (singular: **plasmodesma**) are cytosolic channels through relatively large openings in the cell wall, allowing continuity of the plasma membranes from two adjacent cells. Each plasmodesma is therefore lined with plasma membrane common to the two connected cells (Figure 15-25a). A plasmodesma is cylindrical in shape, with the cylinder narrower in diameter at both ends. The channel diameter varies from about 20 to about 200 nm. A single tubular structure, the *desmotubule*, usually lies in the central channel of the plasmodesma. Endoplasmic reticulum (ER) cisternae are often seen near the plasmodesmata on either side of the cell wall. ER membranes from adjoining cells are continuous with the desmotubule and with the ER of the other cells, as depicted in Figure 15-25b.

The ring of cytosol between the desmotubule and the membrane that lines the plasmodesma is called the *annulus*. The annulus is thought to provide cytosolic continuity between adjacent cells, thereby allowing molecules to pass freely from one cell to the next. Even after cell division and deposition of new cell walls between the two daughter cells, cytosolic continuities are maintained between the daughter cells by plasmodesmata that pass through the newly formed walls. In fact, most plasmodesmata are formed at the time of cell division, when the new cell wall is being formed. Minor changes may occur later, but the number and location of plasmodesmata are largely fixed at the time of division.

In many respects, plasmodesmata appear to function somewhat like gap junctions. They reduce electrical resistance between adjacent cells by about 50-fold compared with cells that are completely separated by plasma membranes.

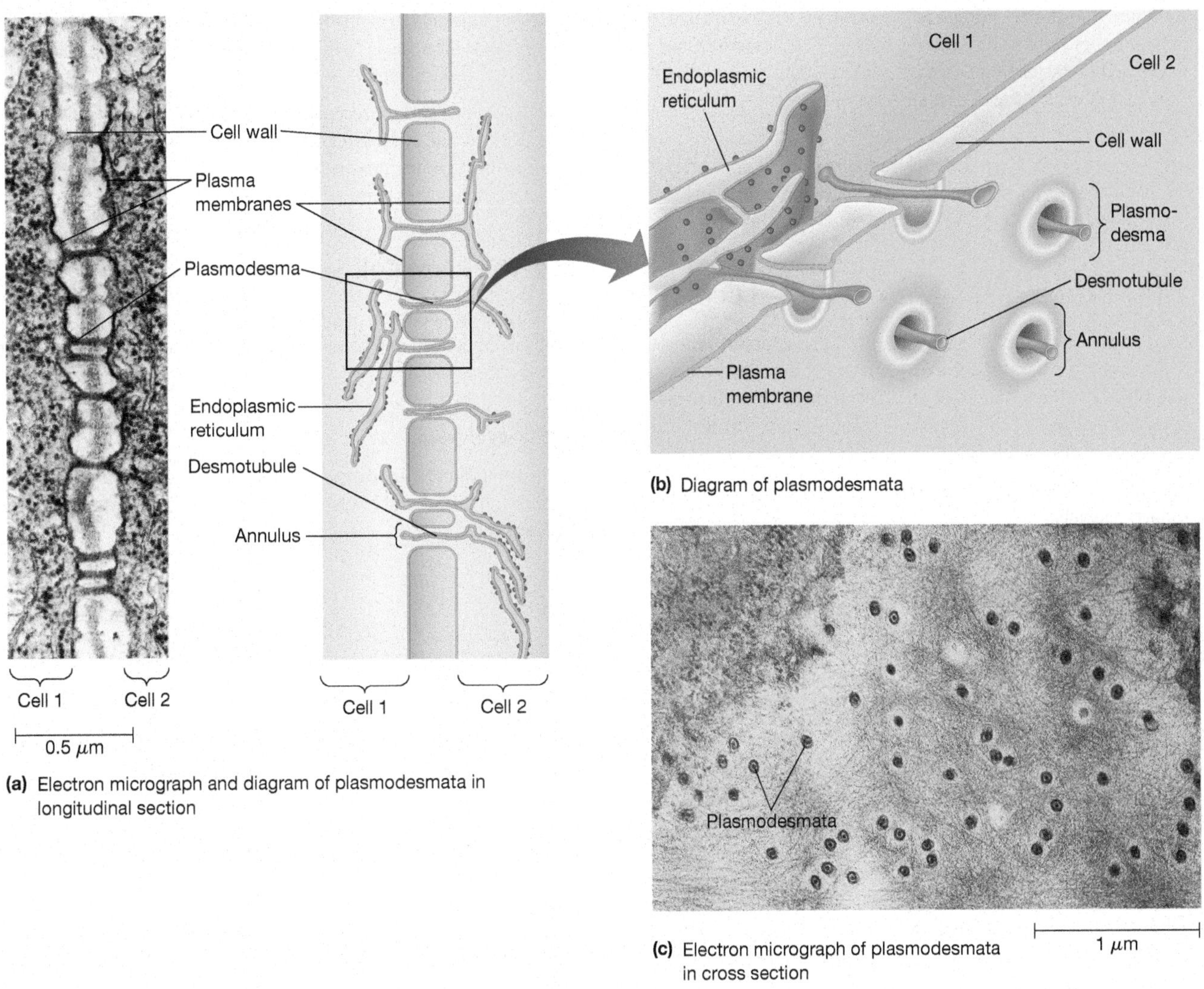

Figure 15-25 Plasmodesmata. A plasmodesma is a channel through the cell wall between two adjacent plant cells, allowing cytosolic exchange between the cells. The plasma membrane of one cell is continuous with that of the other cell at each plasmodesma. Most plasmodesmata have a narrow cylindrical desmotubule at the center that is derived from the ER and appears to be continuous with the ER of both cells. Between the desmotubule and the plasma membrane that lines the plasmodesma is a narrow ring of cytosol called the annulus. **(a)** This electron micrograph and drawing show the cell wall between two adjacent root cells of timothy grass, with numerous plasmodesmata (TEM). **(b)** A drawing of a cell wall with numerous plasmodesmata, illustrating the continuity of the ER and cytosol between adjacent cells. **(c)** This electron micrograph shows many plasmodesmata in cross section (TEM).

In fact, the movement of ions between adjacent cells (measured as current flow) is proportional to the number of plasmodesmata that connect the cells. Plasmodesmata also allow the passage of much larger molecules, including signaling molecules, RNAs, transcription factors, and even viruses.

CONCEPT CHECK 15-3

Cell walls can be dynamically remodeled to allow for plant growth through the action of proteins like expansins. Based on what you know about plant cells, how might they exert forces on parts of the cell wall to remodel them?

Summary of Key Points

Mastering™ Biology For activities, animations, and review quizzes, go to the study area at www.masteringbiology.com.

15.1 Cell-Cell Junctions

- Junctions link many animal cells together. There are three general types: adhesive junctions, tight junctions, and gap junctions.
- Adhesive junctions, such as adherens junctions and desmosomes, use cadherins to hold cells together.
- Adhesive junctions are anchored to the cytoskeleton by linker proteins that attach to actin microfilaments (adherens junctions) or intermediate filaments (desmosomes).
- Desmosomes are particularly prominent in tissues that must withstand considerable mechanical stress.
- Other cell-cell adhesion events utilize other plasma membrane glycoproteins, such as IgSF proteins and selectins.

- Tight junctions form a permeability barrier between epithelial cells, and they prevent the lateral movement of membrane proteins, thereby partitioning the membrane into discrete functional domains. They also mediate paracellular transport across epithelia.
- Gap junctions form open channels between cells, allowing direct chemical and electrical communication between cells. Gap junction permeability is limited to ions and small molecules.

15.2 The Extracellular Matrix of Animal Cells

- Plant as well as animal cells have extracellular structures consisting of long, rigid fibers embedded in an amorphous, hydrated matrix of branched molecules. In animals, the extracellular matrix (ECM) consists of collagen and/or elastin fibers embedded in a network of glycosaminoglycans and proteoglycans.
- Collagen is responsible for the strength of the ECM, and elastin imparts elasticity.
- The ECM is held in place by adhesive glycoproteins such as fibronectins, which link cells to the ECM, and laminins, which attach cells to the basal lamina.
- Cells use cell surface receptor glycoproteins called integrins to connect to the ECM. Integrins, like cadherins, attach to the cytoskeleton via linker proteins. Focal adhesions connect to actin, and hemidesmosomes connect to intermediate filaments.

15.3 The Plant Cell Surface

- The primary cell wall of a plant cell consists mainly of cellulose fibers embedded in a complex network of hemicelluloses, pectins, and extensins.
- The secondary cell wall is reinforced with lignins, a major component of wood.
- Plasmodesmata are membrane-lined cytosolic channels that allow chemical and electrical communication between adjacent plant cells.

Problem Set

Mastering™ Biology For instructor-assigned tutorials and problems, go to www.masteringbiology.com.

15-1 Beyond the Membrane: ECM and Cell Walls. Compare and contrast the extracellular matrix (ECM) of animal cells with the walls around plant cells.

(a) What basic organizational principle underlies both ECM and cell walls?

(b) What are the chemical constituents in each case?

(c) What functional roles do the ECM and cell wall have in common?

(d) What functional roles are unique to the ECM? To the cell wall?

15-2 Anchoring Cells to the ECM. Animal cells attach to several different kinds of proteins within the ECM.

(a) Briefly explain how the various domains of the fibronectin molecule (see Figure 15-16) or the laminin molecule (see Figure 15-18) are important for their function.

(b) Historically, an important strategy for disrupting the adhesion of integrins to their ligands is by using a synthetic peptide that mimics the binding site on the ECM molecule to which the integrin attaches. In the case of fibronectin, the amino acid sequence is arginine-glycine-aspartate (when written using the single letter designation for each amino acid, this sequence becomes RGD). Explain why addition of such synthetic peptides would disrupt binding of cells to their normal substratum.

15-3 Compare and Contrast. For each of the terms in list A, choose a related term in list B, and explain the relationship between the two terms by comparing or contrasting them structurally or functionally.

List A	List B
(a) Collagen	Basolateral surface
(b) Fibronectin	Focal adhesion
(c) Claudin	Connexin
(d) IgSF	Laminin
(e) Hemidesmosome	Cadherin
(f) Apical surface	Cellulose

15-4 Compaction. In mammalian embryos such as the mouse, the fertilized egg divides three times to form eight loosely packed cells, which become tightly adherent in a process known as *compaction.* In the late 1970s, several laboratories made antibodies against mouse cell surface proteins. The antibodies prevented compaction, as did removal of Ca^{2+} from the medium. What sort of protein do the antibodies probably recognize, and why?

15-5 Cellular Junctions and Plasmodesmata. Indicate whether each of the following statements is true of adhesive junctions (A), tight junctions (T), gap junctions (G), and/or plasmodesmata (P). A given statement may be true of any, all, or none (N) of these structures.

(a) Associated with filaments that confer either contractile or tensile properties

(b) Sites of membrane fusion that are limited to abutting ridges of adjacent membranes

(c) Require the alignment of connexons in the plasma membranes of two adjacent cells

(d) Seal membranes of two adjacent cells tightly together

(e) Allow the exchange of metabolites between the cytosol of two adjacent cells

15-6 Junction Proteins. Indicate whether each of the following is a component of adherens junctions (A), desmosomes (D), tight junctions (T), gap junctions (G), or plasmodesmata (P), and describe briefly the role each plays.

(a) Connexin

(b) E-cadherin

(c) Desmocollin

(d) Desmotubule

(e) Desmoplakin

(f) Annulus

(g) α-Catenin

(h) Claudin

Table 15-4 **Movement of Small Molecules Between Cells in *Elodea.* "F"= fluorescein conjugated. See Problem 15-10**

Dye Molecule	Size (Da)	Number of Trials in Which Movement Between Cells Occurred
6-Carboxyfluorescein	376	9/9
Lissamine rhodamine B	559	16/16
F-glutamic acid	536	9/9
F-glutamylglutamic acid	665	9/9
F-hexaglycine	749	7/9
F-leucyldiglutamylleucine	874	3/17
F-(prolyl-prolyl-glycine)$_5$	1678	0/12
F-microperoxidase	2268	0/10

Source: From Goodwin, P.B. (1983). Molecular size limit for movement in the symplast of the *Elodea* leaf. *Planta* 2, 124–130.

15-7 QUANTITATIVE **Mind the Gap.** Gap junctional communication can be examined using a technique pioneered by Werner Loewenstein and his colleagues known as *dye coupling*, the ability of small fluorescent molecules to pass from one cell to another through gap junctions. What can you conclude about gap junctions from each of the following experiments?

(a) The research team injected cells with fluorescent molecules of different molecular weights, and a fluorescence microscope was then used to observe the movement of the molecules into adjacent cells. When molecules with molecular weight of 1926 were injected, they did not pass from cell to cell, but molecules with a molecular weight of 1158 did.

(b) Microglia (support cells in the brain) are connected by gap junctions. When the fluorescent dye Lucifer yellow was injected into individual microglia that are part of a large group of cells, little dye passed to other cells. When microglia were treated with a drug that increases intracellular calcium concentration, there was a dramatic increase in dye passage between cells.

15-8 Claudin Selectivity. Claudin-4 normally prohibits movement of Na^+ ions across epithelial cells that express it. However, when positively charged amino acids near the tip of the large extracellular loop of claudin-4 are changed to negatively charged amino acids, Na^+ can pass through the epithelium. How do you account for this change in permeability to Na^+?

15-9 Scurvy and Collagen. Scurvy is a disease that until the nineteenth century was common among sailors and others whose diets were deficient in vitamin C (ascorbic acid). Individuals with scurvy suffer from various disorders, including extensive bruising, hemorrhages, and breakdown of supporting tissues. Ascorbic acid serves as a reducing agent responsible for maintaining the activity of prolyl hydroxylase, the enzyme that catalyzes hydroxylation of proline residues within the collagen triple helix, which is required for helix stability.

(a) Why does dietary vitamin C deficiency lead to symptoms such as bruising and breakdown of supporting tissues?

(b) Can you guess why sailors are no longer susceptible to scurvy? Why are British sailors called "limeys" to this day?

15-10 DATA ANALYSIS **Size Matters.** In a classic set of experiments, Paul Goodwin injected fluorescent dyes of various molecular weights into cells of the plant *Elodea canadensis* (**Table 15-4**). What do you conclude about the size selectivity of plasmodesmata in *Elodea*?

15-11 Microtubules and Cell Walls. Genetically engineered plant cells have been produced in which cellulose synthase is labeled with yellow fluorescent protein and α-tubulin is labeled with the cyan fluorescent protein (YFP and CFP are variants of the green fluorescent protein, GFP, that can be distinguished in the fluorescence microscope).

(a) Design an experiment that would show that the positions of cellulose synthase rosettes and microtubules are strongly correlated.

(b) Design an experiment to confirm that microtubules are *necessary* for correct positioning of cellulose synthase rosettes.

16 The Structural Basis of Cellular Information: DNA, Chromosomes, and the Nucleus

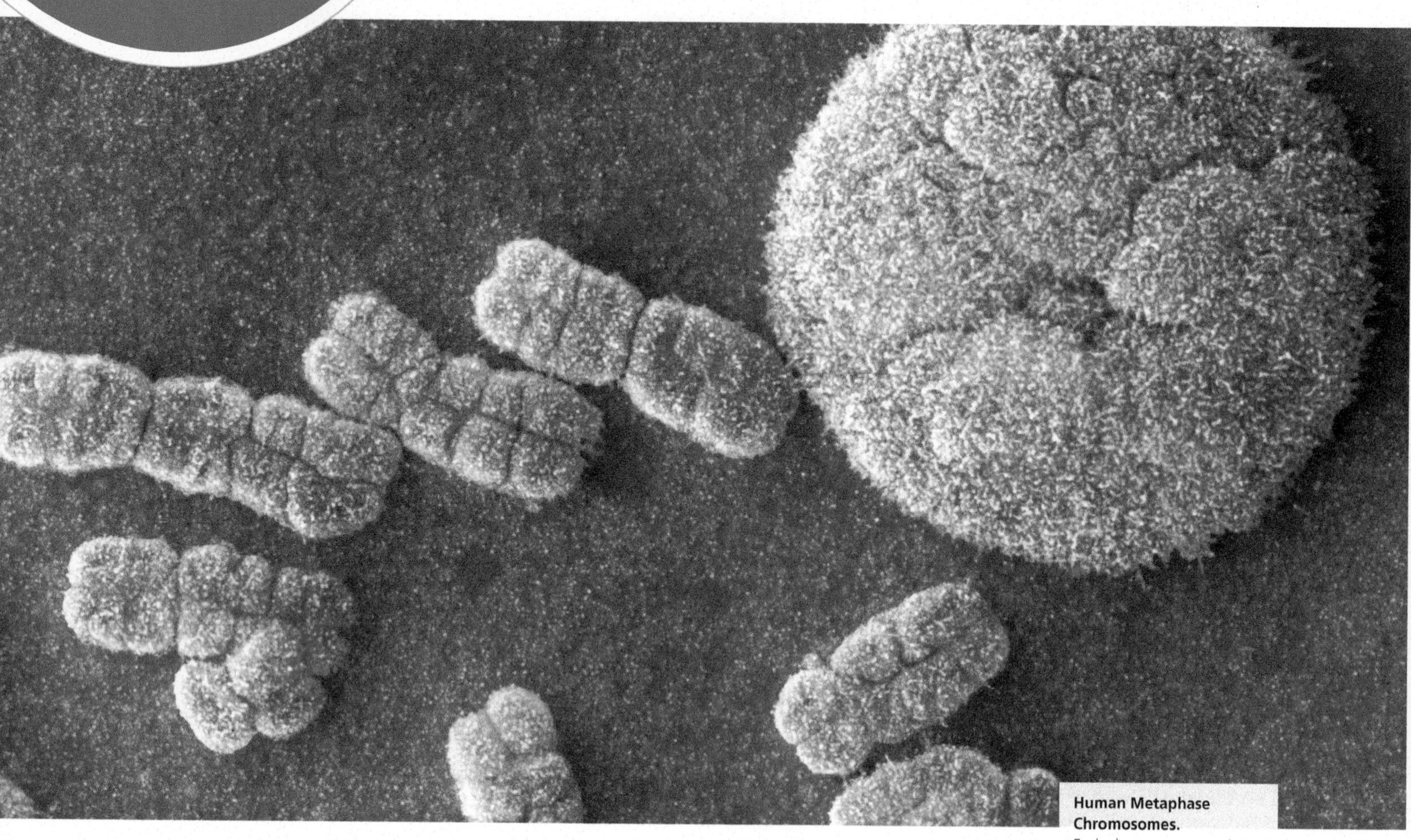

Human Metaphase Chromosomes. Each chromosome consists of two identical strands (chromatids), joined at the centromere with terminal regions called telomeres. Here they are pictured in their most compact form (center and lower left). A nucleus can be seen in the upper right (colorized SEM).

Implicit in our earlier discussions of cellular structure and function has been a sense of predictability, order, and control. We have come to expect that organelles and other cellular structures will have a predictable appearance and function, that metabolic pathways will proceed in an orderly fashion in specific intracellular locations, and that all of a cell's activities will be carried out in a carefully controlled, highly efficient, and heritable manner.

Such expectations express our confidence that cells possess a set of "instructions" that can be passed on faithfully to daughter cells. Just over 150 years ago, the Augustinian monk Gregor Mendel worked out rules accounting for the inheritance patterns he observed in pea plants, although he had little inkling of the cellular basis for these rules. These studies led Mendel to conclude that hereditary information is transmitted in the form of distinct units that we now call **genes.** We now know that genes consist of DNA sequences that encode functional products—usually proteins, but in some important cases RNAs.

Figure 16-1 presents a preview of how DNA carries out its instructional role in cells and provides a framework for describing how the four chapters on information flow are organized. The figure highlights the fact that the information carried by DNA flows both *between* generations of cells and *within* each individual cell. During the first of these two

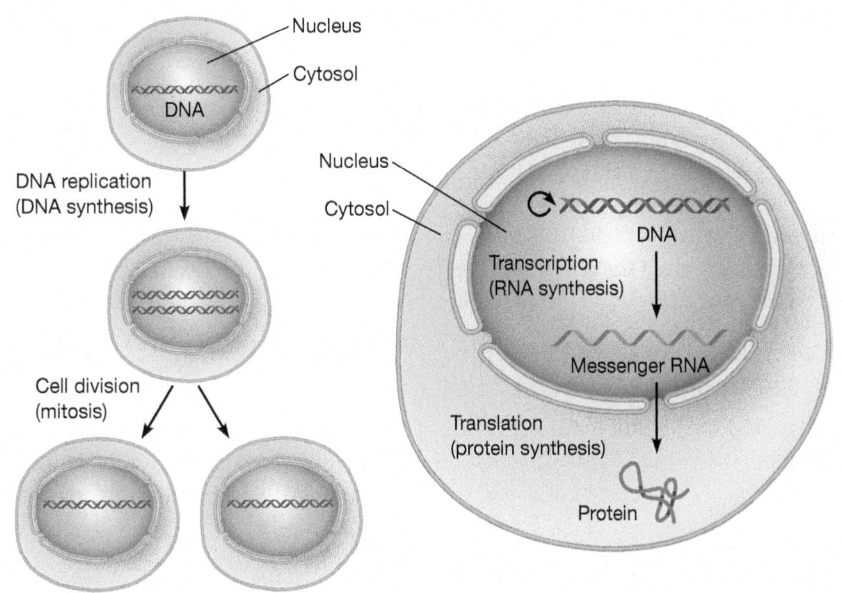

(a) The flow of genetic information between generations of cells

(b) The flow of genetic information within a cell: the expression of genetic information

Figure 16-1 The Flow of Information in Cells. The diagrams here feature eukaryotic cells, but DNA replication, cell division, transcription, and translation are processes that occur in prokaryotic cells as well. **(a)** Genetic information encoded in DNA molecules is passed on to successive generations of cells by DNA replication and cell division (in eukaryotic cells, by means of mitosis). The DNA is first duplicated and then divided equally between the two daughter cells. In this way, it is ensured that each daughter cell will have the same genetic information as the cell from which it arose. **(b)** Within each cell, genetic information encoded in the DNA is expressed through the processes of transcription (RNA synthesis) and translation (protein synthesis). Transcription involves the use of selected segments of DNA as templates for the synthesis of messenger RNA and other RNA molecules. Translation is the process whereby amino acids are joined together in a sequence dictated by the sequence of nucleotides in messenger RNA.

processes (Figure 16-1a), a cell's DNA molecules undergo *replication*, generating two DNA copies that are distributed to the daughter cells when the cell divides. This chapter and the next focus on the structures and events associated with this aspect of information flow. The present chapter covers the structural organization of DNA and the chromosomes in which it is packaged; it also discusses the nucleus, which is the organelle that houses the chromosomes of eukaryotic cells. Chapter 17 then discusses DNA replication, repair, and recombination.

Figure 16-1b summarizes how information residing in DNA is used *within* a cell via a two-stage process called *transcription* and *translation*. During transcription, RNA is synthesized in an enzymatic reaction that copies information from DNA. During translation, the base sequences of the resulting *messenger RNA* molecules are used to determine the amino acid sequences of proteins. *Thus, the information initially stored in DNA base sequences is ultimately used to synthesize specific proteins*. It is the particular proteins synthesized by a cell that ultimately determine most of a cell's structural features as well as the functions it performs. Transcription and translation, which together constitute the expression of genetic information, are the subjects of Chapters 18 and 19.

We open this chapter by describing the discovery of DNA, whose informational role lies at the heart of this group of chapters.

16.1 Chemical Nature of the Genetic Material

When Mendel first postulated the existence of genes, he did not know the identity of the molecule that allows them to store and transmit inherited information. But a few years later, this molecule was unwittingly discovered by Johann Friedrich Miescher, a Swiss physician. Miescher reported the discovery of the substance now known as DNA in 1869, just a few years before the cell biologist Walther Flemming first observed chromosomes as he studied dividing cells under the microscope.

The Discovery of DNA Led to Conflicting Proposals Concerning the Chemical Nature of Genes

Miescher isolated nuclei from white blood cells obtained from pus found on surgical bandages. Extracting these nuclei with alkali led to the discovery of a novel, slightly acidic substance that he called "nuclein," which we now know was largely DNA. In the early 1880s a botanist, Eduard Zacharias, reported that extracting DNA from cells causes the staining of the chromosomes to disappear. Zacharias and others therefore inferred that DNA is the genetic material. This view prevailed until the early 1900s, when incorrectly interpreted staining experiments led to the false conclusion that the amount of DNA changes dramatically within cells. Because cells would be expected to maintain a constant amount of the substance that stores their hereditary instructions, these mistaken observations led to a repudiation of the idea that DNA carries genetic information.

As a result, from around 1910 to the 1940s, most scientists believed that genes were made of protein rather than DNA. It was argued that proteins are constructed from 20 different amino acids that can be assembled in a vast number of combinations, thereby generating the sequence diversity and complexity expected of a molecule that stores and transmits genetic information. In contrast, DNA was thought to be a simple polymer consisting of the same sequence of four bases repeated over and over, thereby lacking the variability expected of a genetic molecule. Such a simple polymer was thought to serve merely as a structural support for the genetic material in the form of protein. This view prevailed until two lines of evidence resolved the matter in favor of DNA (and *not* protein) as the genetic material, described next.

Avery, MacLeod, and McCarty Showed That DNA Is the Genetic Material of Bacteria

The background to this story was provided in 1928 by a series of experiments by the British physician and microbiologist Frederick Griffith, who was studying a pathogenic strain of a bacterium, then called "pneumococcus," that causes a fatal pneumonia in animals (**Figure 16-2**). Griffith discovered that this bacterium (now called *Streptococcus pneumoniae*) exists in two forms, called the *S strain* and the *R strain*. When grown on a solid agar medium, the S strain produces colonies that are smooth and shiny because of the mucous polysaccharide coat each cell secretes, whereas the R strain lacks the ability to manufacture a mucous coat and therefore produces colonies exhibiting a rough boundary.

When injected into mice, S-strain (but not R-strain) bacteria trigger a fatal pneumonia. The S strain's ability to cause disease is directly related to the presence of its polysaccharide coat, which protects the bacterial cell from attack by the mouse's immune system. One of Griffith's most intriguing discoveries, however, was that this fatal pneumonia can also be induced by injecting animals with a mixture of live R-strain

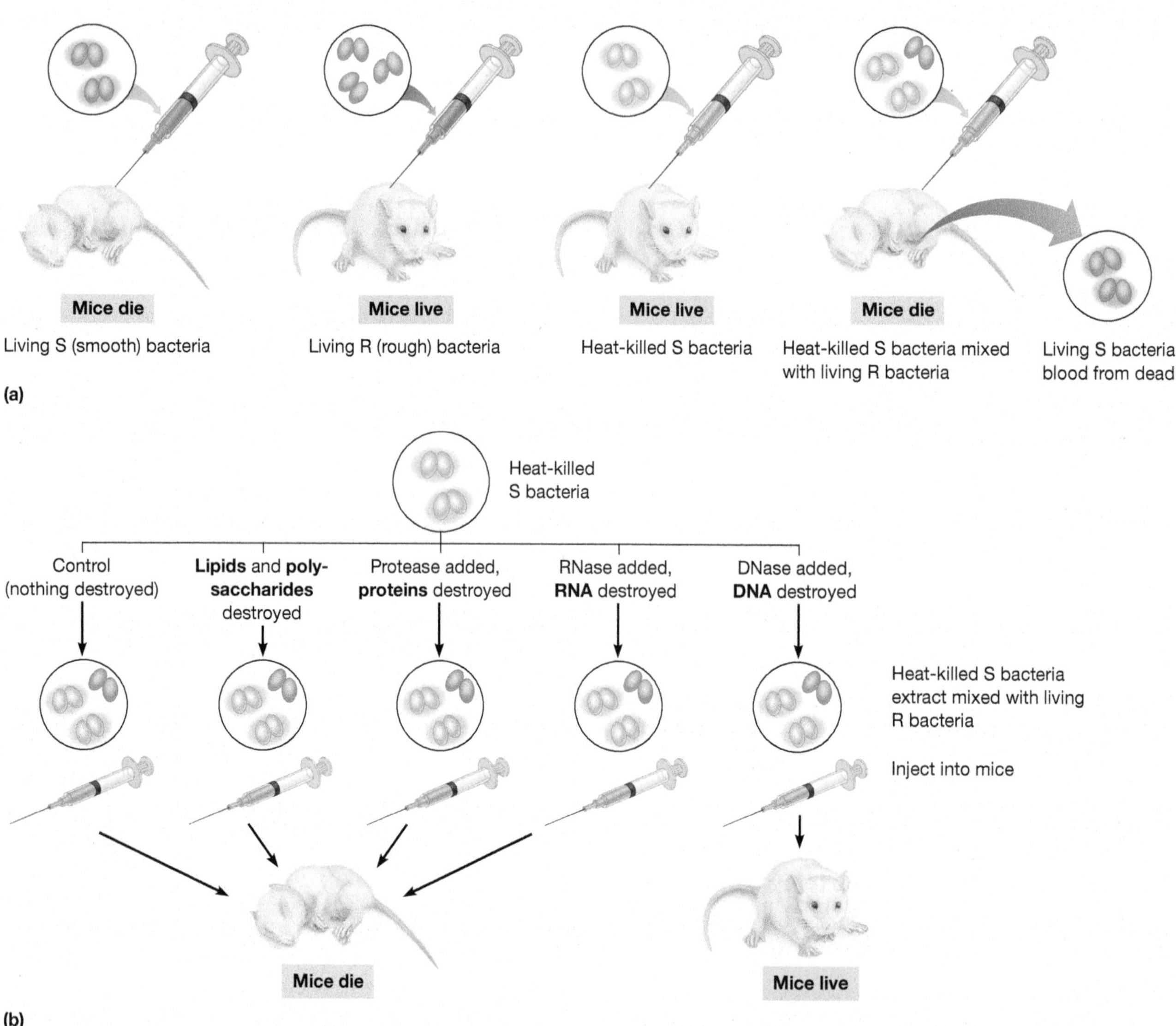

Figure 16-2 Griffith's Experiment on Genetic Transformation in Pneumococcus. S (smooth) cells of the pneumococcus bacterium *(Streptococcus pneumoniae)* are pathogenic in mice; R (rough) cells are not. **(a)** Injection of living S bacteria into mice causes pneumonia and death. Injection of living R bacteria leaves mice healthy. Heat-killed S bacteria have no effect when injected alone. When a mixture of living R bacteria and heat-killed S bacteria is injected, the result is pneumonia and death. The discovery of living S-strain bacteria in the blood of the mice suggested to Griffith that a substance in the heat-killed S cells caused a heritable change (transformation) of nonpathogenic R bacteria into pathogenic S bacteria. **(b)** The chemical substance was later identified as DNA by treating the heat-killed S bacteria to remove various types of macromolecules. Only when DNA was destroyed was the extract no longer virulent.

bacteria and dead S-strain bacteria (Figure 16-2a). This finding was surprising because neither live R-strain nor dead S-strain organisms cause pneumonia if injected alone. When Griffith autopsied the animals that had been injected with the mixture of live R-strain and dead S-strain bacteria, he found them teeming with live S-strain bacteria. Because the animals had not been injected with any live S-strain cells, he concluded that the nonpathogenic R bacteria were somehow converted or transformed into pathogenic S bacteria by a substance present in the heat-killed S bacteria that had been co-injected. He called this substance a *transforming principle*.

Griffith's discoveries set the stage for work by Oswald Avery and his colleagues at the Rockefeller Institute in New York. These researchers pursued the investigation of bacterial transformation to its logical conclusion by asking which component of the heat-killed S bacteria was the transforming principle and thus the source of heredity. They fractionated cell-free extracts of S-strain bacteria and found that only the nucleic acid fraction was capable of causing transformation. Moreover, the activity was specifically eliminated by treatment with deoxyribonuclease (DNase), an enzyme that degrades DNA (Figure 16-2b). This and other evidence convinced them that the transforming substance of pneumococcus was DNA—a conclusion published by Avery, Colin MacLeod, and Maclyn McCarty in 1944.

Despite the rigor of these experiments, the assignment of a genetic role to DNA was not immediately accepted. However, most remaining doubts were alleviated eight years later when DNA was also shown to be the genetic material of a virus, the bacteriophage T2.

Hershey and Chase Showed That DNA Is the Genetic Material of Viruses

Bacteriophages—or **phages**, for short—are viruses that infect bacteria. Much of our early understanding of molecular genetics came from experiments involving these viruses. Some of the most thoroughly studied phages are the T2, T4, and T6 (the so-called T-even) bacteriophages, which infect the bacterium *Escherichia coli*. The three T-even phages have similar structures and life cycles; T4 is shown in **Figure 16-3** on page 442. The *head* of the phage T4 is a protein capsule shaped like a hollow icosahedron (a 20-sided object) and filled with DNA (Figure 16-3a). The head is attached to a protein *tail*, which is used to attach to a bacterium and inject DNA. When phages are mixed with bacterial cells growing in liquid medium and then spread onto agar growth medium in a Petri dish, they produce clear spots called *plaques*. The clear spots appear in the "lawn" of bacteria on the plate because the bacterial cells there have been lysed by the multiplying phage population. Figure 16-3b shows plaques formed by T4 bacteriophage on a lawn of *E. coli* cells.

Figure 16-3c depicts the main events in the replication cycle of the T4 phage that leads to the death of bacteria. The process begins with the adsorption of a phage particle to the wall of a bacterial cell, which then punctures the bacterial cell wall and injects its DNA into the interior (Figure 16-3c). Once this DNA has gained entry to the bacterial cell, the genetic information of the phage is transcribed and translated. The phage DNA and capsid proteins then self-assemble into hundreds of new phage particles. Within about half an hour, the infected cell lyses (breaks open), releasing the new phage particles into the medium.

The course of events shown in Figure 16-3c is called *lytic growth* and is characteristic of a *virulent phage*. Lytic growth results in lysis of the host cell and the production of many progeny phage particles. In contrast, a *temperate phage* can either produce lytic growth, as a virulent phage does, or integrate its DNA into the bacterial chromosome without causing any immediate harm to the host cell. An especially well-studied example of a temperate phage is bacteriophage *λ (lambda)*, which, like the T-even phages, infects *E. coli* cells. In the integrated or *lysogenic state*, the DNA of the temperate phage is called a *prophage*. The prophage is replicated along with the bacterial DNA, often through many generations of host cells (Figure 16-3d). During this time, the phage genes, though potentially lethal to the host, are inactive, or *repressed*. Under certain conditions the prophage DNA is excised from the bacterial chromosome in a process called induction. The phage again enters the lytic cycle, producing progeny phage particles and lysing the host cell.

One of the most thoroughly studied of the phages that infect *E. coli* is bacteriophage T2. In 1952, Alfred Hershey and Martha Chase designed an experiment to address what kind of molecule carries the genetic information to make new phage particles. There are only two possibilities because the T2 virus is constructed from only two kinds of molecules: DNA and protein. To distinguish between these two alternatives, Hershey and Chase took advantage of the fact that the proteins of the T2 virus, like most proteins, contain the element sulfur (in the amino acids methionine and cysteine) but not phosphorus, whereas the viral DNA contains phosphorus (in its sugar-phosphate backbone) but not sulfur. Hershey and Chase therefore prepared two batches of T2 phage particles (as intact phages are called) with different kinds of radioactive labeling. In one batch, they labeled the phage proteins with the radioactive isotope ^{35}S; in the other batch, they labeled the phage DNA with the radioactive isotope ^{32}P.

By using radioactive isotopes in this way, Hershey and Chase were able to trace the fates of both protein and DNA during the infection process (**Figure 16-4** on page 443). First, they mixed radioactive phages with intact bacterial cells and allowed the phage particles to attach to the bacterial cell surface and inject their genetic material into the cells. At this point, Hershey and Chase found that the empty protein coats (or phage "ghosts") could be effectively removed from the surface of the bacterial cells by agitating the suspension in an ordinary kitchen blender and recovering the bacterial cells by centrifugation. They then measured the radioactivity in the supernatant liquid and in the pellet of bacteria at the bottom of the tube.

The data revealed that most (65%) of the ^{32}P remained with the bacterial cells, whereas the bulk (80%) of the ^{35}S was released into the surrounding medium. Because the ^{32}P labeled the viral DNA and the ^{35}S labeled the viral protein, Hershey and Chase concluded that DNA, not protein, had been injected into the bacterial cells. Furthermore, the newly produced phage offspring contained some radioactivity, but only when ^{32}P was used; hence, DNA must function as the genetic material of phage T2.

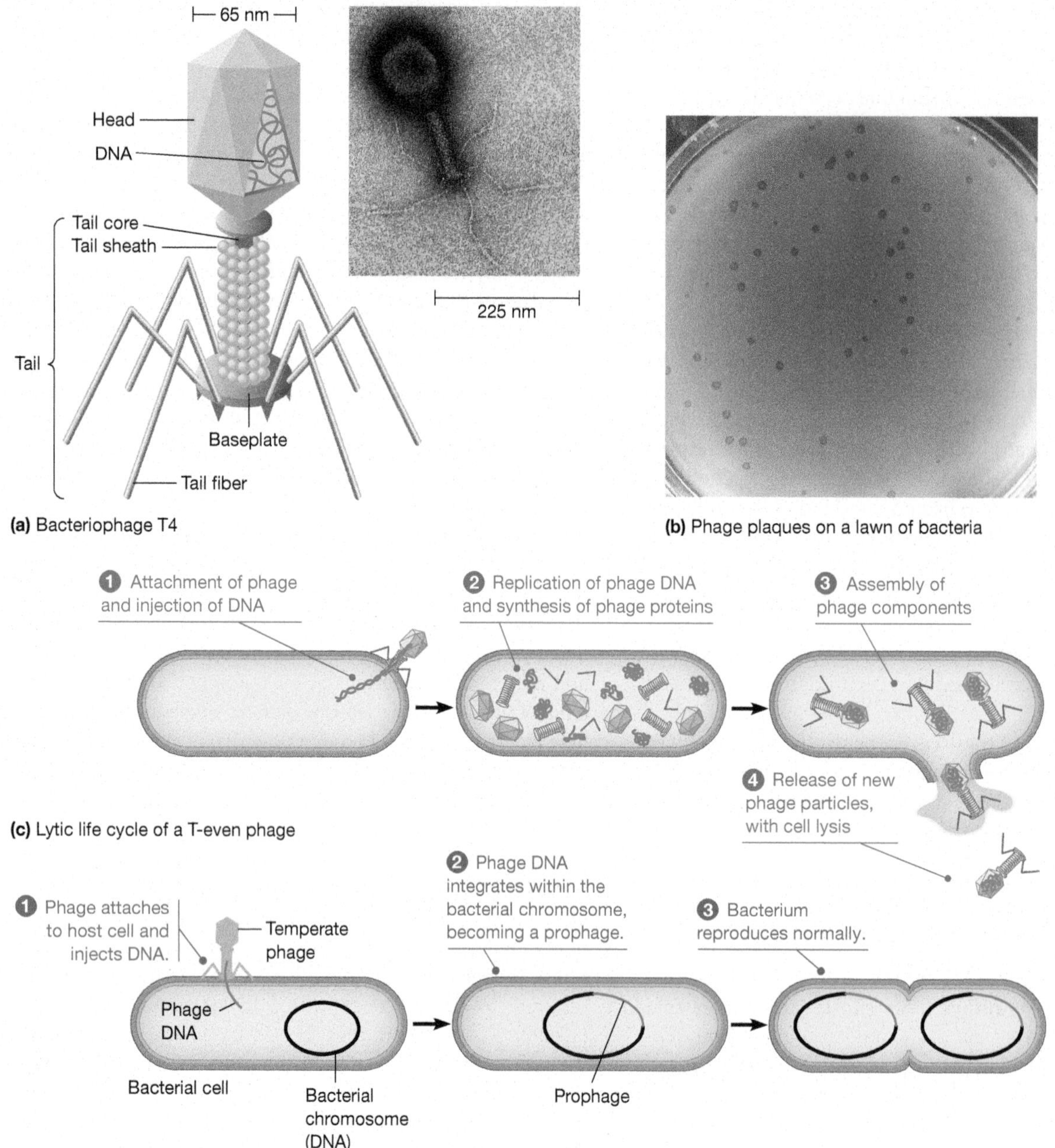
65 nm
Head
DNA
Tail core
Tail sheath
Tail
Baseplate
Tail fiber
225 nm
(a) Bacteriophage T4
(b) Phage plaques on a lawn of bacteria
1 Attachment of phage and injection of DNA
2 Replication of phage DNA and synthesis of phage proteins
3 Assembly of phage components
4 Release of new phage particles, with cell lysis
(c) Lytic life cycle of a T-even phage
1 Phage attaches to host cell and injects DNA.
Temperate phage
2 Phage DNA integrates within the bacterial chromosome, becoming a prophage.
3 Bacterium reproduces normally.
Phage DNA
Bacterial cell
Bacterial chromosome (DNA)
Prophage
(d) Lysogenic state of a prophage within a bacterial chromosome

Figure 16-3 The Structure and Life Cycle of Bacteriophage T4. (a) The drawing identifies the main structural components of this phage; not all of them are visible in the micrograph (TEM). **(b)** Phage plaques have formed on a lawn of *E. coli* infected with phage T4. Each plaque arises from the reproduction of a single phage particle in the original mixture. **(c)** The replication cycle of a T-even phage begins when a phage particle ❶ becomes adsorbed to the surface of a bacterial cell and injects its DNA into the cell. ❷ The phage DNA replicates in the host cell; it encodes phage proteins, which are synthesized. ❸ These components assemble into new phage particles. ❹ Eventually, the host cell lyses, releasing offspring phage particles that can infect additional bacteria. **(d)** The DNA injected by a temperate phage ❶ can become integrated into the DNA of the bacterial chromosome ❷. The integrated phage DNA, called a prophage, is replicated along with the bacterial DNA each time the bacterium reproduces ❸.

As a result of these experiments, by the early 1950s most biologists came to accept the view that genes are made of DNA, not protein.

RNA Is the Genetic Material in Some Viruses

Although double-stranded DNA is the predominant carrier of genetic information in most cell types, some viruses violate this general rule. Some types of bacteriophages carry their genetic material as single-stranded DNA, and a much larger group of viruses carry their genetic material as RNA. A good example is the *tobacco mosaic virus (TMV)*, which infects tobacco plants. In the 1950s, it was found that when RNA purified from TMV is spread on tobacco leaves, they become infected and produce lesions, suggesting that RNA is the genetic material in TMV. Somewhat later, more elegant experiments were performed to confirm this suggestion. Purified RNA and coat proteins from

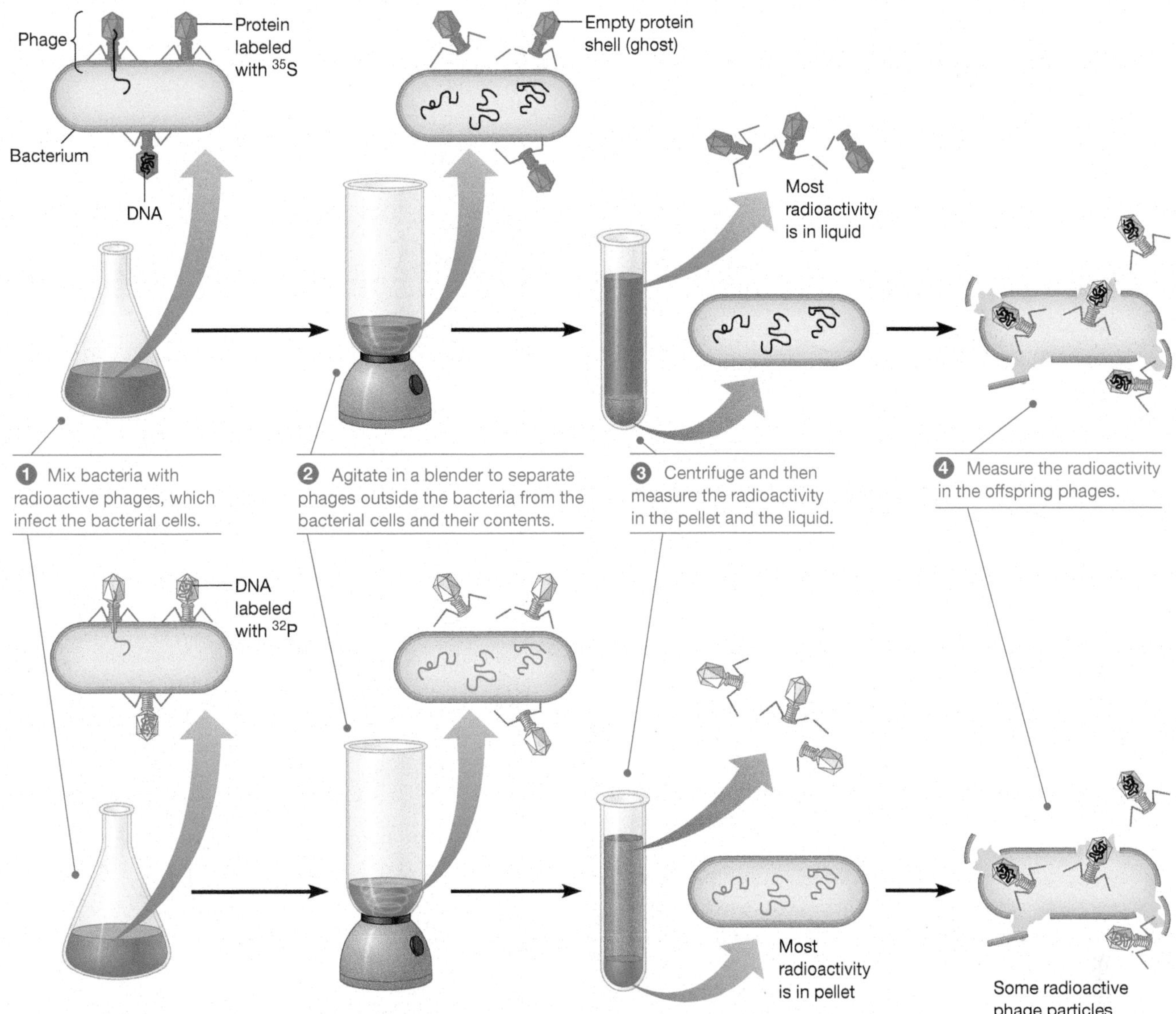

Figure 16-4 The Hershey–Chase Experiment: DNA as the Genetic Material of Phage T2. ❶ T2 labeled with either ^{35}S (to label protein) or ^{32}P (to label DNA) is used to infect bacteria. The phages adsorb to the cell surface and inject their DNA. ❷ Agitation of the infected cells in a blender dislodges most of the ^{35}S from the cells, whereas most of the ^{32}P remains. ❸ Centrifugation causes the cells to form a pellet; any free phage particles, including ghosts, remain in the supernatant liquid. ❹ When the cells in each pellet are incubated further, the phage DNA within them dictates the synthesis and eventual release of new phage particles. Some of these phages contain ^{32}P in their DNA (because the old, labeled phage DNA is packaged into some of the new particles), but none contain ^{35}S in their coat proteins.

TMV can self-assemble (Chapter 3). This means that the exterior protein coat and the internal RNA of TMV and similar viruses can be isolated and then mixed. When RNA and coat proteins are separated and isolated from TMV and a second type of virus, Holmes ribgrass (HR), and then mixed, the reconstituted virus can be used to infect tobacco leaves. The type of lesion that develops consistently corresponds to the type of RNA in the reconstituted virus (**Figure 16-5** on page 444). This provides strong evidence that RNA is the genetic material in TMV and similar viruses.

Another group of RNA-containing viruses is important in human health: the **retroviruses.** In retroviruses, such as the human immunodeficiency virus (HIV), RNA serves as a template for making complementary DNA inside the infected cells, using a special enzyme called reverse transcriptase. **Figure 16-6** on page 444 depicts the reproductive cycle of a typical retrovirus. In the virus particle, two copies of the RNA genome are enclosed within a protein capsid that is surrounded by a membranous envelope. Each RNA copy has a molecule of reverse transcriptase attached to it. The virus first ❶ binds to the surface of the host cell, and its envelope fuses with the plasma membrane, releasing the capsid and its contents into the cytosol. Once inside the cell, the viral reverse transcriptase ❷ catalyzes the synthesis of a DNA strand that is complementary to the viral RNA and then ❸ catalyzes the formation of a second DNA strand complementary to the first.

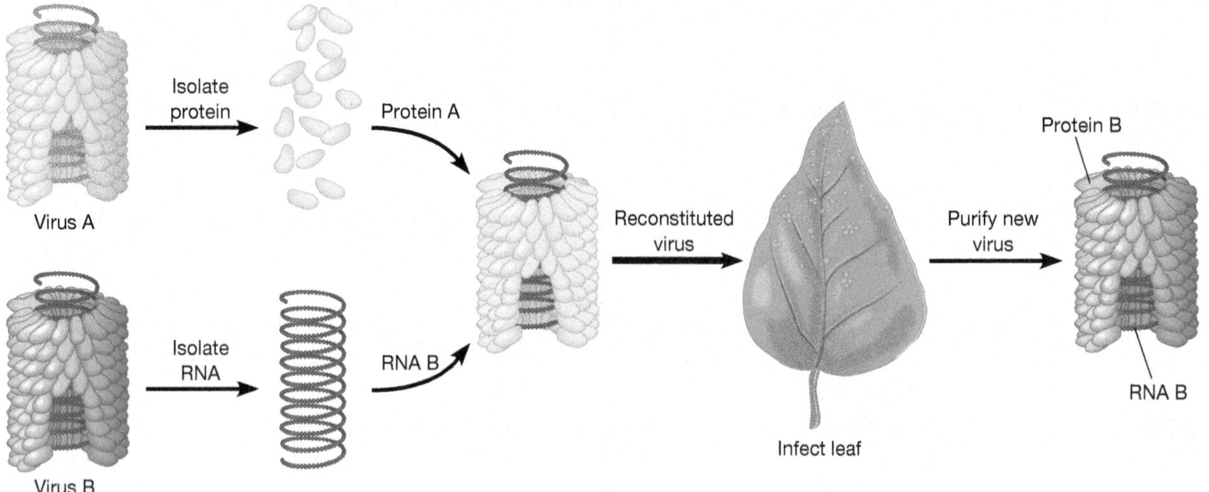

Figure 16-5 RNA Is the Genetic Material of Tobacco Mosaic Virus. Reconstitution of hybrid tobacco mosaic viruses. In the recombined virus particles, the protein subunits are derived from one strain of virus (virus A), and the RNA is derived from another (virus B). Following infection, viruses are produced with protein subunits characteristic of virus B and not A. This result indicates that RNA is the genetic material for making new viruses.

The result is a double-stranded DNA version of the viral genome. ④ This double-stranded DNA then enters the nucleus and integrates into the host cell's chromosomal DNA, much as the DNA genome of a lysogenic phage integrates into the DNA of the bacterial chromosome (see Figure 16-3d). The integrated viral genome, called a *provirus*, is replicated every time the cell replicates its own DNA. ⑤ Transcription of the proviral DNA (by cellular enzymes in a process described in detail in Chapter 18) produces RNA transcripts that function in two ways. First, they serve as ⑥ mRNA molecules that direct the synthesis of viral proteins (capsid protein, envelope protein, and reverse transcriptase). Second, ⑦ some of these same RNA transcripts are packaged with the viral proteins into new virus particles. ⑧ The new viruses then "bud" from the plasma membrane without necessarily killing the infected cell.

The ability of a retroviral genome to integrate into host cell DNA helps explain how some retroviruses can cause cancer. These viruses, called *RNA tumor viruses*, are of two types. Viruses of the first type carry a cancer-causing oncogene in their genomes, along with the genes encoding viral proteins. An oncogene is a mutated version of a normal cellular gene (a proto-oncogene) that encodes a protein used to regulate cell growth and division (as we will see in Chapter 26). RNA tumor viruses of the second type do not themselves carry oncogenes, but integration of their genomes into a host chromosome alters the cellular DNA such that a normal proto-oncogene is converted into an oncogene.

CONCEPT CHECK 16.1

Based on what you know about protein and DNA, why did Hershey and Chase pick ^{32}P and ^{35}S to label their phage?

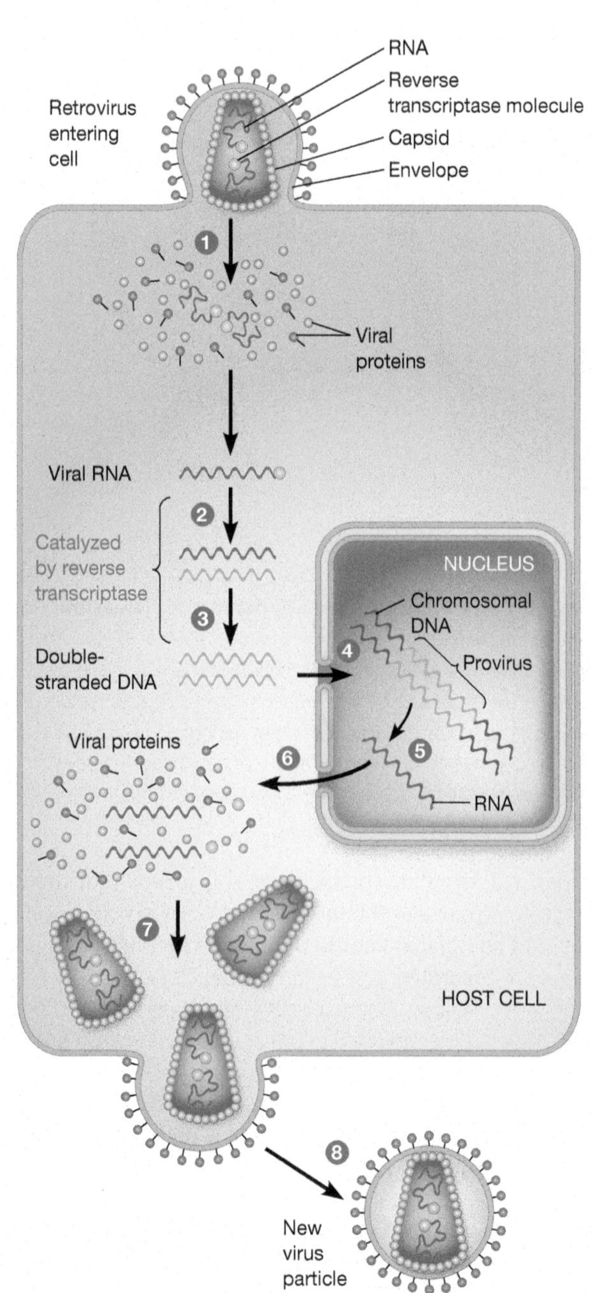

Figure 16-6 The Reproductive Cycle of a Retrovirus. The virus carries its genetic material as RNA.

16.2 DNA Structure

As the scientific community gradually came to accept the conclusion that DNA stores genetic information, a new set of questions began to emerge concerning how DNA performs this function. One of the first questions to be addressed was how cells accurately replicate their DNA so that duplicate copies of the genetic information can be passed on from cell to cell during cell division and from parent to offspring during reproduction. Answering this question required an understanding of the three-dimensional structure of DNA, which was provided in 1953 when Watson and Crick formulated their double-helical model of DNA. (We described the structure of the double helix in Chapter 3, but we return to it now for review and some further details.)

Chargaff's Rules Reveal That A = T and G = C

Despite the initial lukewarm reaction to Avery's work, it had an important influence on several other scientists. Among them was Erwin Chargaff, who was interested in the base composition of DNA. Between 1944 and 1952, Chargaff used chromatographic methods to separate and quantify the relative amounts of the four bases—adenine (A), guanine (G), cytosine (C), and thymine (T)—found in DNA. Several important discoveries came from his analyses. First, he showed that DNA isolated from different cells of a given species has the same percentage of each of the four bases (Table 16-1, rows 1–4) and that this percentage does not vary with individual, tissue, age, nutritional state, or environment. This is exactly what would be expected of a chemical substance that stores genetic information because the cells of a given species would be expected to have similar genetic information. However, Chargaff did find that DNA base composition varies from species to species. This can be seen by examining the last column of **Table 16-1**, which shows the relative amounts of the bases A and T versus G and C in the DNAs of various organisms. The data also reveal that DNA preparations from closely related species have similar base compositions, whereas those from very different species tend to exhibit quite different base compositions. Again, this is what would be expected of a molecule that stores genetic information.

But Chargaff's most striking observation was his discovery that, for all DNA samples examined, the number of adenines is equal to the number of thymines (A = T), and the number of guanines is equal to the number of cytosines (G = C). This meant that the number of purines is equal to the number of pyrimidines (A + G = C + T). The significance of these equivalencies, known as **Chargaff's rules**, became clear when Watson and Crick proposed the double-helical model of DNA in 1953.

Watson and Crick Discovered That DNA Is a Double Helix

In 1952, James Watson and Francis Crick, working at Cambridge University in England, were among a handful of scientists who were convinced that DNA is the genetic material and that knowing its three-dimensional structure would provide valuable clues to how it functions. Therefore, they worked to model the structure of DNA (**Figure 16-7** on page 446). DNA had been known for years to be a long polymer having a backbone of repeating sugar (deoxyribose) and phosphate units, with a nitrogenous base attached to each sugar. Watson and Crick were aided by knowing Chargaff's rules. In addition, they knew that the particular forms in which the bases A, G, C, and T exist at physiological pH permit specific hydrogen bonds to form between pairs of them. The crucial experimental evidence, however, came from an X-ray diffraction

Table 16-1 **DNA Base Composition Data That Led to Chargaff's Rules**

	Number of Each Type of Nucleotide*				Nucleotide Ratios**			
Source of DNA	**A**	**T**	**G**	**C**	**A/T**	**G/C**	**(A + G) / (C + T) Purines/Pyrimidines**	**(A + T) / (G + C)**
Bovine thymus	28.4	28.4	21.1	22.1	1.00	0.95	0.98	1.31
Bovine liver	28.1	28.4	22.5	21.0	0.99	1.07	1.02	1.30
Bovine kidney	28.3	28.2	22.6	20.9	1.00	1.08	1.04	1.30
Bovine brain	28.0	28.1	22.3	21.6	1.00	1.03	1.01	1.28
Human liver	30.3	30.3	19.5	19.9	1.00	0.98	0.99	1.53
Locust	29.3	29.3	20.5	20.7	1.00	1.00	1.00	1.41
Sea urchin	32.8	32.1	17.7	17.3	1.02	1.02	1.02	1.85
Wheat germ	27.3	27.1	22.7	22.8	1.01	1.00	1.00	1.19
Marine crab	47.3	47.3	2.7	2.7	1.00	1.00	1.00	17.50
Aspergillus (mold)	25.0	24.9	25.1	25.0	1.00	1.00	1.00	1.00
Saccharomyces cerevisiae (yeast)	31.3	32.9	18.7	17.1	0.95	1.09	1.00	1.79
Clostridium (bacterium)	36.9	36.3	14.0	12.8	1.02	1.09	1.04	2.73

*The values in these four columns are the average number of each type of nucleotide found per 100 nucleotides in DNA.

**The A/T, G/C, and Purine (A + G) / Pyrimidine (T + C) ratios are not all exactly 1.00 because of experimental error.

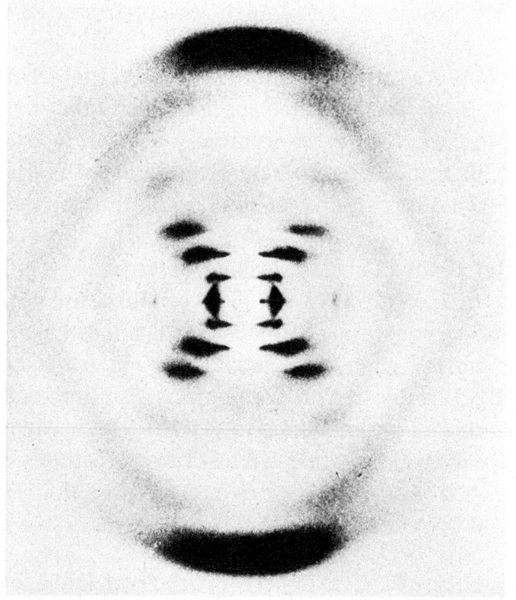

(a)

(b)

Figure 16-7 Modeling DNA Structure. (a) The original X-ray diffraction data generated by Rosalind Franklin and used by Watson and Crick to construct their model. **(b)** One of Watson and Crick's original models, made with brass plates and wire. The base has a hand-drawn scale indicating 10 nm.

picture of DNA produced by Rosalind Franklin (Figure 16-7a), who was working at King's College in London. Franklin's painstaking analysis of the diffraction pattern revealed that DNA was a long, thin, helical molecule with one type of structural feature repeated every 0.34 nm and another repeated every 3.4 nm. Watson and Crick approached the puzzle by building wire models of possible structures. Based on the information provided by Franklin's picture, Watson and Crick eventually produced a DNA model consisting of two intertwined strands—a **double helix** (Figure 16-7b).

The Watson–Crick Model. In the Watson–Crick double helix, illustrated in **Figure 16-8**, the sugar-phosphate backbones of the two strands are on the outside of the helix, and the bases face inward toward the center of the helix, forming the "steps" of the "circular staircase" that the structure resembles. The helix is right-handed, meaning that it curves "upward" to the right (notice that this is true even if you turn the drawing upside down). It contains ten nucleotide pairs per turn and advances 0.34 nm per nucleotide pair. Consequently, each complete turn of the helix adds 3.4 nm to the length of the molecule. The diameter of the helix is 2 nm. This distance turns out to be too small for two purines and too great for two pyrimidines, but it accommodates a purine and a pyrimidine well, consistent with Chargaff's rules. The two strands are held together by hydrogen bonding between the bases in opposite strands. Moreover, the hydrogen bonds holding together the two strands of the double helix fit *only when they form between the base adenine (A) in one chain and thymine (T) in the other or between the base guanine (G) in one chain and cytosine (C) in the other.* This means that the base sequence of one chain determines the base sequence of the opposing chain; the two chains of the DNA double helix are therefore said to be **complementary** to each other. You may have heard these standard pairings referred to as *Watson–Crick base pairings.* Such a model explains why Chargaff had observed that DNA molecules contain equal amounts of the bases A and T and equal amounts of the bases G and C.

The most profound implication of the Watson–Crick model was that it suggested a mechanism by which cells can faithfully replicate their genetic information: the two strands of the DNA double helix could simply separate from each other before cell division so that each strand could function as a *template,* dictating the synthesis of a new complementary DNA strand using Watson–Crick base-pairing rules. In other words, the base A in the template strand would specify insertion of the base T in the newly forming strand, the base G would specify insertion of the base C, the base T would specify insertion of the base A, and the base C would specify insertion of the base G. (In the next chapter, we will discuss DNA replication in detail.)

Key Features of DNA Structure. Several other important features of the DNA double helix are illustrated in Figure 16-8. For example, notice that the two strands are twisted around each other so that there is a *major groove* and a *minor groove.* Regulatory proteins often bind to the major groove and recognize specific base sequences without unfolding the DNA double helix.

Another important feature is the *antiparallel* orientation of the two DNA strands, illustrated in Figure 16-8b. This drawing shows that the phosphodiester bonds, which join the 5′ carbon of one nucleotide to the 3′ carbon of the adjacent nucleotide, are oriented in *opposite* directions in the two DNA strands. Starting at the top of the diagram, the first (top) nucleotide of the strand on the left has a free 5′ end, and its final nucleotide has a free 3′ end. Conversely, the strand on the right has the reverse orientation: its first nucleotide at its 5′ end is at the bottom, whereas its free 3′ end is at the top. The opposite orientation of the two strands has important

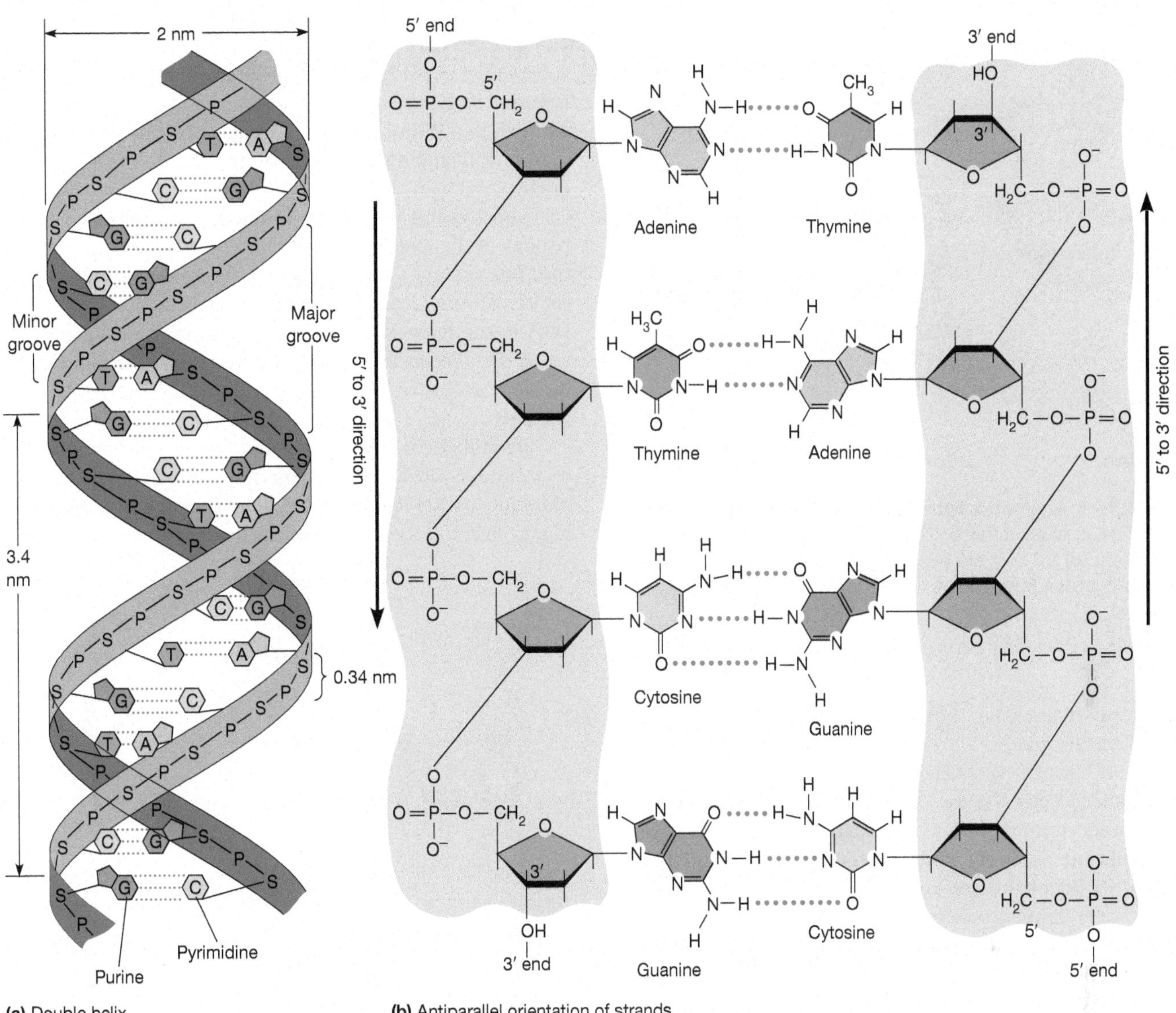

Figure 16-8 The DNA Double Helix. (a) This schematic diagram shows the sugar-phosphate chains of the DNA backbone, the complementary base pairs, the major and minor grooves, and several important dimensions. A = adenine, G = guanine, C = cytosine, T = thymine, P = phosphate, and S = sugar (deoxyribose). **(b)** One strand of a DNA molecule has its 5′ and 3′ ends oriented in one direction, whereas the 5′ and 3′ ends of its complement are in the opposite orientation. This illustration also shows the hydrogen bonds that connect the bases in AT and GC pairs.

implications for both DNA replication and transcription (as we will see in Chapters 17 and 18).

The sequence of nucleotides found along the length of a piece of DNA is a defining feature of DNA. In addition, DNA (and RNA) can be characterized by size. Because each nucleotide contains a nitrogenous base, and nucleotides in DNA pair across the double helix, the length of DNA is measured in **base pairs (bp).** Larger stretches of DNA are measured in multiples of a single base pair. For example, a **kilobase (kb)** is 1000 bp, and a **megabase (Mb)** is 1,000,000 bp. RNA is measured using similar terminology, although because it is often single-stranded we use bases instead of base pairs.

Structural Variants of DNA. DNA is theoretically able to form different versions of the double helix (**Figure 16-9** on page 448). The right-handed helix described by Watson and Crick is an idealized version of what is called *B-DNA* (Figure 16-9a). Naturally occurring B-DNA double helices are flexible molecules whose exact shapes and dimensions depend on the local nucleotide sequence. Although B-DNA is the main form of DNA in cells (and in test tube solutions of DNA), other forms may also exist, perhaps in short segments interspersed in molecules that are mostly B-DNA. The most important of these alternative forms are Z-DNA and A-DNA. As shown in Figure 16-9b, *Z-DNA* is a *left-handed* double helix. Its name derives from the zigzag pattern of its sugar-phosphate backbone, and it is longer and thinner than B-DNA. The Z form arises most readily in DNA regions that either contain alternating purines and pyrimidines or have modified cytosines with extra methyl groups. Although the biological significance of Z-DNA is not well understood, some evidence suggests that short stretches of DNA transiently flip into the Z

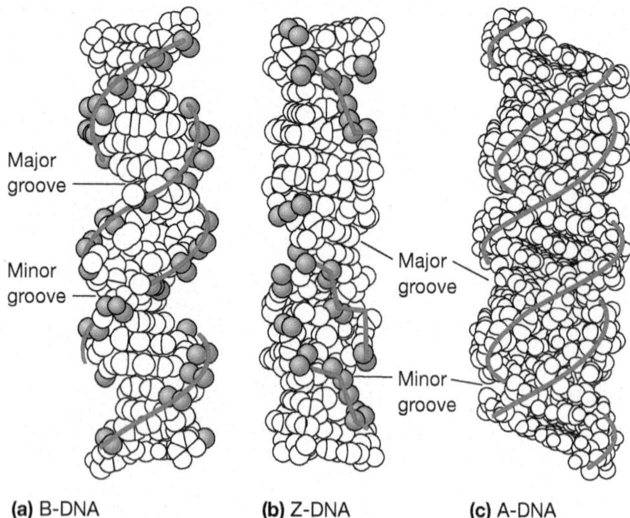

Figure 16-9 Alternative Forms of DNA. (a) In the normal B form of DNA, the sugar-phosphate backbone forms a smooth right-handed double helix. **(b)** In Z-DNA, the backbone forms a zigzag left-handed helix. **(c)** In A-DNA, the backbone is a right-handed double helix similar to the B-DNA, but shorter and more compact. Color is used to highlight the backbones.

configuration as part of the process that activates the expression of certain genes.

A-DNA, which is shorter and thicker than B-DNA, is a right-handed helix (Figure 16-9c) and can be created artificially by dehydrating B-DNA. Although A-DNA does not exist in significant amounts under normal cellular conditions, most RNA double helices are of the A type. A-type helices have a wider minor groove and a narrower major groove than B-type helices, so A-RNA is not well suited for base recognition by RNA-binding proteins from the major groove. To recognize specific base sequences in A-RNA, regulatory proteins generally need to unwind the duplex.

DNA Can Be Interconverted Between Relaxed and Supercoiled Forms

In many situations, the DNA double helix can be twisted upon itself to a more compact form called **supercoiled DNA.** Supercoiling was first identified in the DNA of certain small viruses containing circular DNA molecules that exist as closed loops. Circular DNA molecules are also found in bacteria, mitochondria, and chloroplasts. Although supercoiling is not restricted to circular DNA, it is easiest to study in such molecules.

A DNA molecule can go back and forth between the supercoiled state and the nonsupercoiled, or *relaxed,* state. To understand the basic idea, you might perform the following exercise. Start with a length of rope consisting of two strands twisted together into a right-handed coil; this is the equivalent of a relaxed, linear DNA molecule. Just joining the ends of the rope together changes nothing; the rope is now circular but still in a relaxed state. But before sealing the ends, if you first give the rope an extra twist in the direction in which the strands are already entwined around each other, the rope is thrown into a *positive supercoil.* Conversely, if before sealing, you give the rope an extra twist in the opposite direction, the rope is thrown into a *negative supercoil.* Like the rope in this example, a relaxed DNA molecule can be converted to a positive supercoil by twisting in the same direction as the double helix is wound and into a negative supercoil by twisting in the opposite direction (**Figure 16-10**). Circular DNA molecules found in nature, including those of bacteria, viruses, and eukaryotic organelles, are invariably negatively supercoiled.

Supercoiling also occurs in linear DNA molecules when regions of the molecule are anchored to some cellular structure (for example, the nuclear matrix) and so cannot freely rotate. At any given time, significant portions of the linear DNA in the nucleus of eukaryotic cells may be supercoiled; when DNA is packaged into chromosomes at the time of cell division, extensive supercoiling helps make the DNA more compact.

By influencing both the spatial organization and the energy state of DNA, supercoiling affects the ability of a DNA molecule to interact with other molecules. Positive supercoiling involves tighter winding of the double helix and therefore

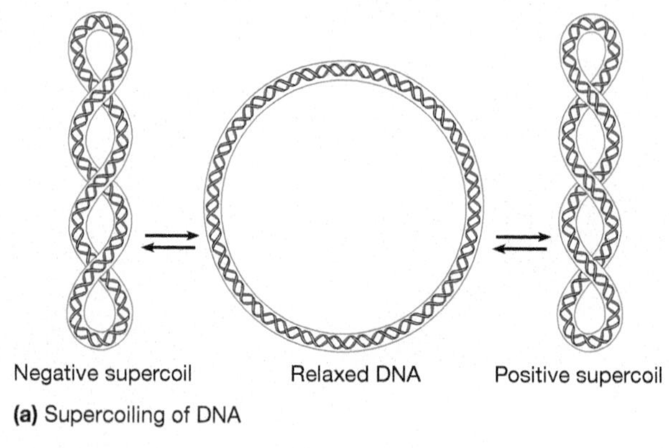

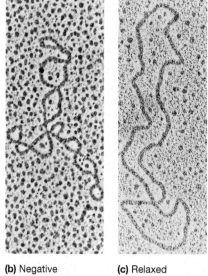

(b) Negative supercoiling

(c) Relaxed

Figure 16-10 Interconversion of Relaxed and Supercoiled DNA. (a) Conversion of a relaxed circular DNA molecule into a negatively supercoiled form (by twisting it in the opposite direction as the double helix is wound) and into a positively supercoiled form (by twisting it in the same direction as the double helix is wound). **(b, c)** Electron micrographs of circular DNA molecules from bacteriophage PM2, showing a molecule with negative supercoils (b) and a relaxed molecule (c; TEMs).

reduces opportunities for interaction. In contrast, negative supercoiling is associated with unwinding of the double helix, which gives its strands increased access to proteins involved in DNA replication or transcription. This explains why negative supercoiling is favored in a cell.

The interconversion between relaxed and supercoiled forms of DNA is catalyzed by enzymes known as **topoisomerases,** which are classified as either *type I* or *type II*. Both types catalyze the relaxation of supercoiled DNA; type I enzymes do so by introducing transient single-strand breaks in DNA, whereas type II enzymes introduce transient double-strand breaks. **Figure 16-11** shows how these temporary breaks affect DNA supercoiling. Type I topoisomerases induce DNA relaxation by cutting one strand of the double helix, thereby allowing the DNA to rotate and the uncut strand to be passed through the break before the broken strand is resealed. In contrast, type II topoisomerases induce relaxation by cutting both DNA strands and then passing a segment of uncut double helix through the break before resealing.

Type I and type II topoisomerases are able to remove both positive and negative supercoils from DNA. In addition, bacteria have a type II topoisomerase called **DNA gyrase,** which can induce as well as relax supercoiling. DNA gyrase is one of several enzymes involved in DNA replication (as you will learn in Chapter 17). It can relax the positive supercoiling that results from partial unwinding of a double helix, or it can actively introduce negative supercoils that promote strand separation, thereby facilitating access of other proteins involved in DNA replication. Like other type II topoisomerases, DNA gyrase requires ATP to generate supercoiling but not to relax an already supercoiled molecule.

The Two Strands of a DNA Double Helix Can Be Denatured and Renatured

In addition to altering the topology of intact double-stranded DNA, cells can also regulate the separation of each single strand of a DNA double helix. As we will see in coming chapters, strand separation is an integral part of both DNA replication and RNA synthesis. In the laboratory, the two strands of a DNA molecule can also be readily separated from each other by raising the temperature or pH because the two strands are bound together by relatively weak, noncovalent bonds. When DNA strands are separated experimentally, the phenomenon is known as **DNA denaturation (melting);** the reverse process, which reestablishes a double helix from separated DNA strands, is called **DNA renaturation (reannealing).**

When denaturation is induced by slowly raising the temperature, the DNA initially retains its double-stranded, or native, state. As the temperature is raised further, however, more and more of the DNA separates into single strands, until eventually the entire piece of DNA "melts" into its component strands. The melting process is easy to monitor because double-stranded and single-stranded DNA differ in their light-absorbing properties. All DNA absorbs ultraviolet light, with an absorption maximum around 260 nm. When the temperature of a DNA solution is slowly raised, the absorbance at 260 nm remains constant until the double helix begins to melt into its component strands. As the strands separate, the absorbance at 260 nm increases rapidly due to the increasing amount of single-stranded DNA (**Figure 16-12** on page 452).

The temperature at which one-half of the absorbance change has been achieved is typically called the **DNA melting**

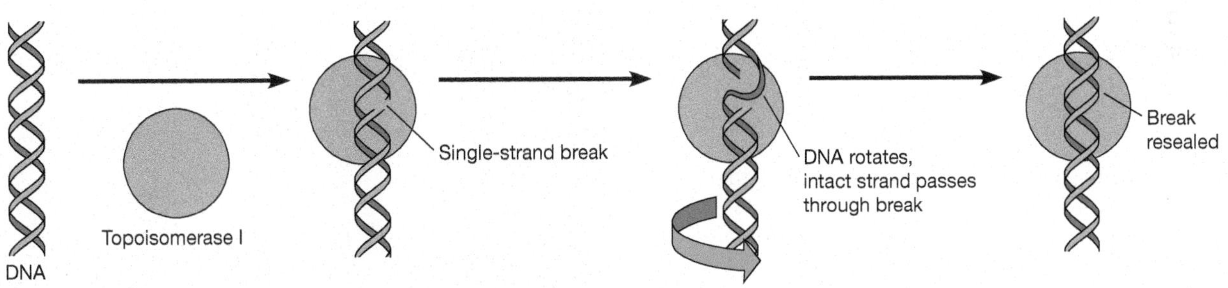

(a) Topoisomerase I. Supercoils are removed by transiently cleaving one strand of the DNA double helix and passing the unbroken strand through the break.

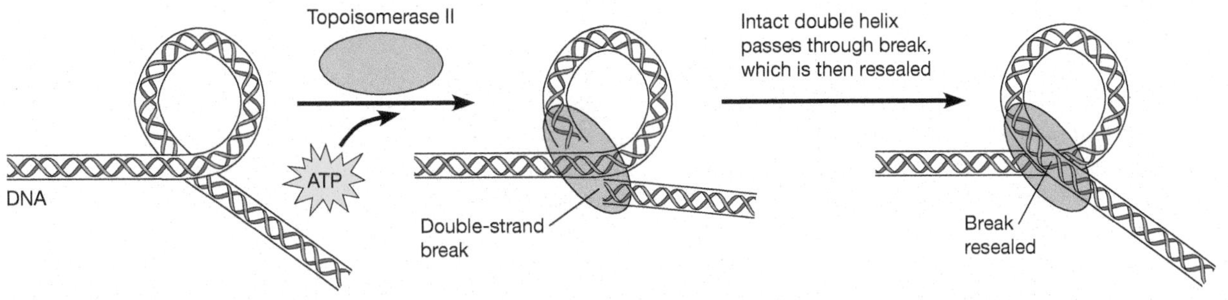

(b) Topoisomerase II. Supercoils are removed by transiently cleaving both strands of the DNA double helix and passing an unbroken region of the DNA double helix through the break.

Figure 16-11 Reactions Catalyzed by Topoisomerases I and II. (a) Type I and **(b)** type II topoisomerases are used for removing both positive and negative supercoils from DNA.

Key Technique | FISHing for Specific Sequences

PROBLEM: Stains can identify some differences in chromosome structure or various types of nucleic acids but cannot identify specific DNA sequences. How can specific DNA sequences be identified along chromosomes?

SOLUTION: *Fluorescence in situ hybridization (FISH)* allows detection of specific nucleic acid sequences in an intact cellular sample.

Key Tools: Fluorescently labeled nucleic acid probe; preserved tissue sample; incubator or oven for heating samples; fluorescence microscope.

Details: Nucleic acids have several key properties that allow them to be used as *probes* for detecting specific sequences in DNA or RNA. First, they can undergo *hybridization*: hydrogen bonding with complementary bases in a sequence-specific fashion. Second, hybridization is *reversible*: it depends on pH and temperature. As shown in **Figure 16A-1**, FISH capitalizes on these properties ❶ by creating fluorescent DNA using molecular biology techniques. ❷ The labeled probe is then denatured into single strands by heating to near boiling. ❸ After the tissue sample itself has been treated and heated to allow access to its nucleic acids, ❹ the probe is applied to the tissue sample, which is cooled somewhat to allow for optimal hybridization. ❺ After washing away the unbound probe, the bound fluorescent probe can then be detected with a sensitive camera attached to a fluorescence microscope.

The example in Figure 16A-1 involves direct labeling of the probe. In some FISH applications, the probe DNA is labeled with a tag that allows a second fluorescent molecule to bind to the probe. This adds extra steps but can amplify weak signals in a manner very similar to indirect immunostaining (**see Key Technique in Chapter 1, pages 8–9**).

In some cases, FISH is used to detect very specific DNA sequences within chromosomes. Two examples you will encounter in this chapter are centromeres and telomeres (see Figure 16-22). In the cytogenetics laboratory, FISH is often performed on metaphase-arrested cells to detect whole chromosomes. In cells treated with microtubule inhibitors, chromosomes remain condensed, making them

Probe
Target
❶ Label probe with fluorescent tag
❷ Denature probe
❸ Denature nucleic acid in sample
❹ Cool slightly; allow probe to hybridize with sample
❺ Wash unbound probe; view in microscope

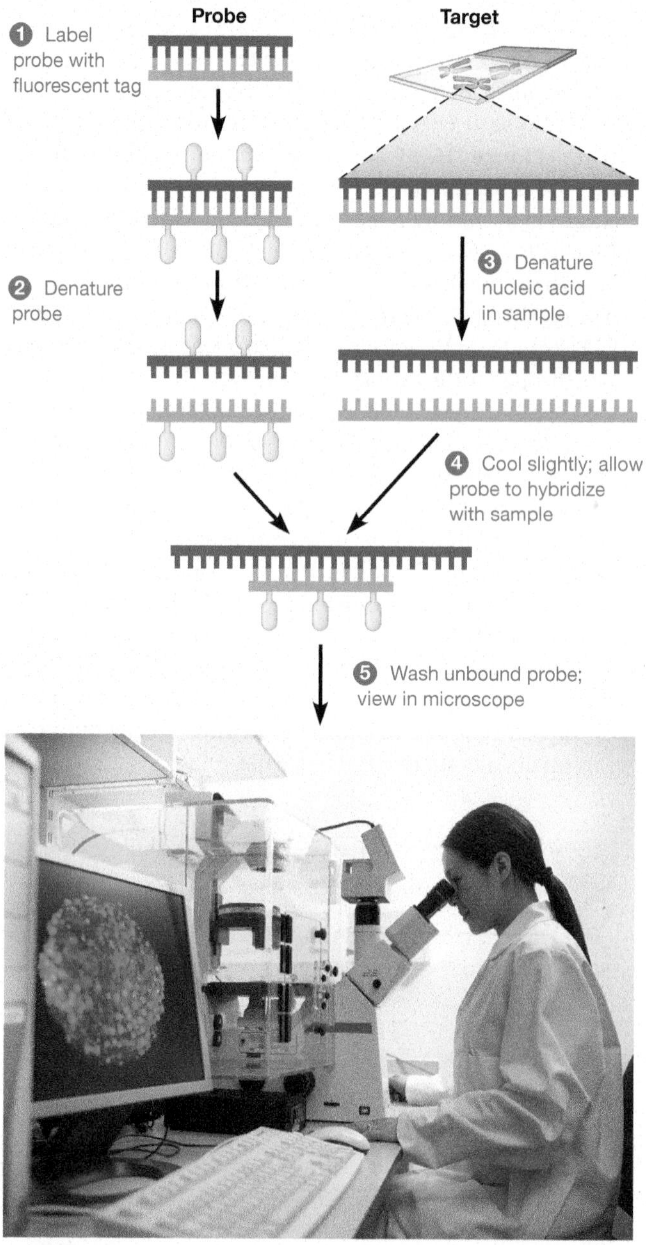

Figure 16A-1 Fluorescent in Situ Hybridization (FISH). See text for description of steps.

temperature (T_m), or *thermal midpoint of denaturation*. The value of T_m reflects how tightly the DNA double helix is held together. One influence on T_m is the percentage of the DNA that contains G's and C's. GC base pairs, held together by three hydrogen bonds, are more resistant to separation than are AT base pairs, which have only two (see Figure 16-8b). The melting temperature therefore increases in direct proportion to the relative number of GC base pairs in the DNA (**Figure 16-13** on page 452).

Hydrogen bonding across the two strands of a piece of DNA stabilizes the DNA and accounts for the specificity of base pairing across the double helix. Another major interaction occurs *within* each of the single strands that also stabilizes the DNA double helix. Known as **base stacking**, it involves interactions between adjacent aromatic rings in organic compounds via hydrophobic and van der Waals interactions. In the case of DNA, the most energetically favorable conformation is achieved by tilting the DNA backbone by an angle of 30°, which allows the rings of adjacent base pairs within a strand to stack one above the other to maximize base stacking interactions. GC dinucleotides have the greatest base-stacking energies; when they are found next to one another within a single strand of DNA, they tend to stabilize its shape.

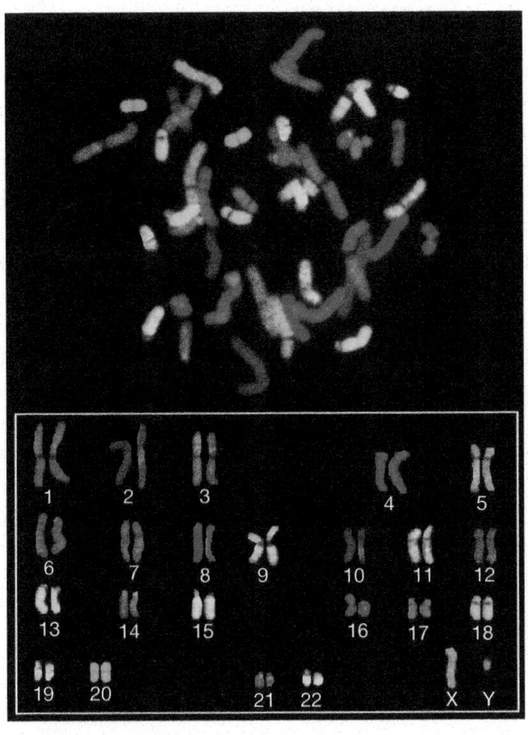

(a) Metaphase chromosomes

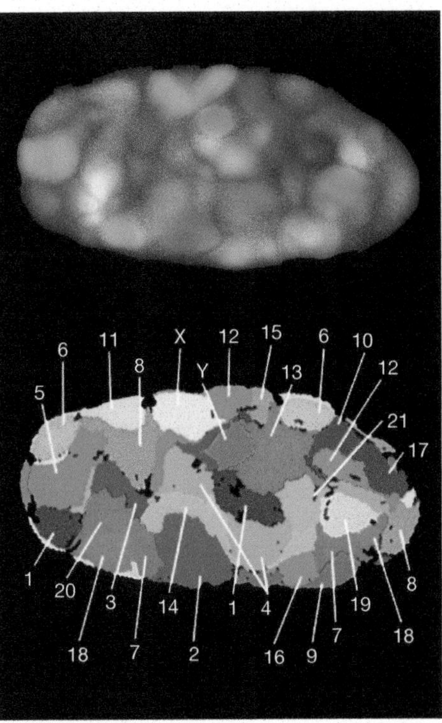

(b) Interphase chromosomes

Figure 16A-2 Painting Chromosomes Using FISH. **(a)** The 46 chromosomes of a metaphase-arrested cell from a human male have been "painted" using multicolor FISH and a combination of specific probes. After processing using a computer, chromosome pairs (1–22) and sex chromosomes (X and Y) can be arranged for clarity, as in the bottom image. **(b)** The chromosomal territories of the chromosomes of a human male interphase cell can be visualized using similar techniques. All different types of decondensed chromosomes can be identified in the bottom image.

easy to visualize by FISH. When the probes are selected so that they can hybridize over large portions of a specific chromosome, the result is a karyotype in which each chromosome can be "painted" using a different color (**Figure 16A-2**). Sophisticated analysis of the resulting fluorescence images can be performed to distinguish many different probes applied to single tissue sample. Each chromosome can then be identified, and the images of the painted chromosomes can be arranged using a computer (Figure 16A-2a). Such data provide even more specific information about the karyotype of a patient than through the use of stains (see Figure 16-23) and can be useful in diagnosing chromosomal rearrangements in cancer cells or in human genetic disorders.

FISH has also proven useful for identifying the positions of interphase chromosomes, that is, the positions of chromosomes in the nucleus when cells are not dividing (Figure 16A-2b). FISH has revealed that interphase chromosomes occupy distinct *chromosomal territories*. Although the positions vary from cell to cell, FISH has been essential in showing that the nucleus is a much more orderly place than might have been suspected based on general staining techniques. This realization is spurring current research on how this spatial organization affects the function of specific genes.

QUESTION: Suppose you are trying FISH for the first time, and you forget to heat your probe before applying it to your tissue sample. What do you expect to happen, and why?

A final factor that affects T_m is the extent to which DNA molecules in the two strands of the double helix are properly base-paired at each position. Proper pairing leads to a higher melting temperature than for DNA in which the two strands are not perfectly complementary. This property leads to a very useful feature of DNA. Denatured DNA can be renatured (reannealed) by lowering the temperature to permit hydrogen bonds between the two strands to be reestablished (**Figure 16-14** on page 452). The ability to renature nucleic acids has a variety of important scientific applications. Most importantly, it forms the basis for **nucleic acid hybridization,** a family of procedures for identifying nucleic acids based on the ability of single-stranded chains with complementary base sequences to bind, or *hybridize,* to each other. Nucleic acid hybridization can lead to the formation of DNA-DNA, DNA-RNA, and even RNA-RNA hybrids. In DNA-DNA hybridization, for example, the DNA being examined is denatured and then incubated with a purified, single-stranded radioactive DNA fragment, called a **probe,** whose sequence is complementary to the base sequence one is trying to detect. A key use of this approach is **fluorescence in situ hybridization (FISH)** (see Key Technique, pages 450–451).

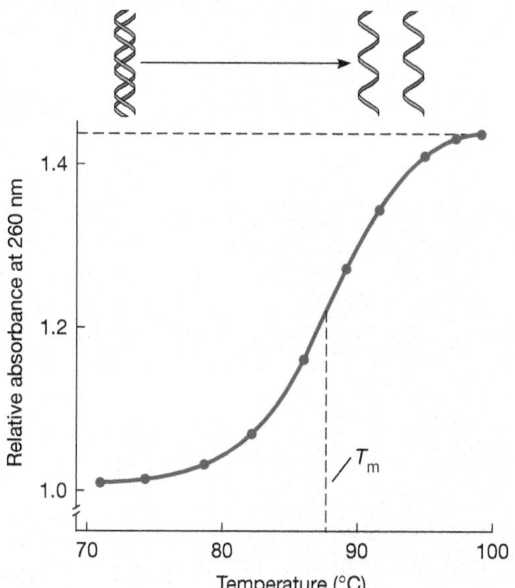

Figure 16-12 A Thermal Denaturation Profile for DNA. When the temperature of a solution of double-stranded (native) DNA is raised, the heat causes the DNA to denature. The conversion to single strands is accompanied by an increase in the absorbance of light at 260 nm. The temperature at which the midpoint of this increase occurs is called the melting temperature, or thermal denaturation midpoint, T_m. For the sample shown, the T_m is about 87°C.

Nucleic acid sequences do not need to be perfectly complementary to be able to hybridize. Changing the temperature, salt concentration, and pH used during hybridization can permit pairing to take place between partially complementary sequences exhibiting numerous mismatched bases. Under such conditions, hybridization will occur between DNAs that are related to one another but not identical. This approach is useful for identifying families of related genes, both within a given type of organism and among different kinds of organisms.

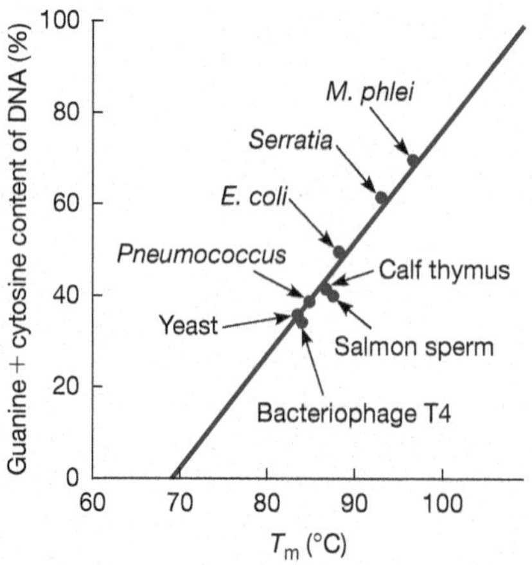

Figure 16-13 Dependence of DNA Thermal Denaturation on Base Composition. The melting temperature (thermal denaturation midpoint) of DNA increases with its G+C content, as shown by the relationship between T_m and G+C content for DNA samples from a variety of sources.

CONCEPT CHECK 16.2

The GC content of the DNA from a newly discovered bacterium is 48%. What are the percentages of each of the four bases (A, T, G, C) in this organism?

16.3 DNA Packaging

Cells must accommodate an awesome amount of DNA, even in species with modestly sized genomes. For example, the typical *E. coli* cell measures about 1 μm in diameter and 2 μm in length, yet it must accommodate a (circular) DNA molecule with a length of about 1600 μm —enough DNA to encircle the cell more than 400 times. Eukaryotic cells face an even greater challenge. A human cell of average size contains enough DNA to wrap around the cell more than 15,000 times. In more accessible terms, this feat is equivalent to stuffing 10,000 miles of spaghetti into a space the size of a basketball! Somehow, all of this DNA must be efficiently packaged into cells and still be accessible to the cellular machinery for both DNA replication and the transcription of specific genes. Clearly, DNA packaging is a challenging problem for all forms of life. We will look first at how bacteria accomplish this task of organizing their DNA and then consider how eukaryotes address the same problem.

Bacteria Package DNA in Bacterial Chromosomes and Plasmids

The genome of a bacterium such as *E. coli* was once thought to be a "naked" DNA molecule lacking any elaborate organization and having only trivial amounts of protein associated with it. We now know that the organization of the bacterial genome is more like the chromosomes of eukaryotes than previously realized. Bacterial geneticists therefore refer to

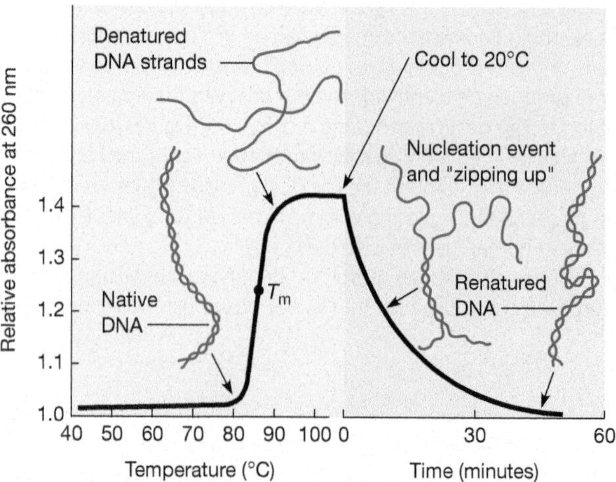

Figure 16-14 DNA Denaturation and Renaturation. If a solution of native (double-stranded) DNA is heated slowly under carefully controlled conditions, the DNA "melts" over a narrow temperature range, with an increase in absorbance at 260 nm. When the solution is allowed to cool, the separated DNA strands reassociate by random collisions, followed by a rapid "zipping up" of complementary base pairs in the two strands. The reassociation requires varying amounts of time, depending on both the DNA concentration in the solution and the number of unique sequences within the DNA strands.

the structure that contains the main bacterial genome as the **bacterial chromosome.**

Bacterial Chromosomes. Bacteria can have single or multiple, circular or linear chromosomes; the most common arrangement, however, is a single, circular, double-stranded DNA molecule that is bound to small amounts of protein and localized to a special region of the bacterial cell called the **nucleoid** (**Figure 16-15**). The bacterial DNA residing in this region forms a threadlike mass of fibers packed together in a way that keeps the nucleoid distinct from the rest of the cell. The DNA of the bacterial chromosome is negatively supercoiled and folded into an extensive series of loops averaging about 20,000 bp in length. Because the two ends of each loop are anchored to structural components that lie within the nucleoid, the supercoiling of any individual loop can be altered without influencing the supercoiling of adjacent loops.

The loops are thought to be held in place by RNAs and proteins. Evidence for a structural role for RNA in the bacterial chromosome has come from studies showing that treatment with *ribonuclease*, an enzyme that degrades RNA, releases some of the loops, although it does not relax the supercoiling. Nicking the DNA with a topoisomerase, on the other hand, relaxes the supercoiling but does not disrupt the loops. The supercoiled DNA that forms each loop is organized into beadlike packets containing small, basic proteins, analogous to the histones of eukaryotic cells (discussed below). Current evidence suggests that the DNA molecule is wrapped around particles of these basic proteins. Thus, from what we know so far, the bacterial chromosome consists of supercoiled DNA that is bound to small, basic proteins and then folded into looped domains.

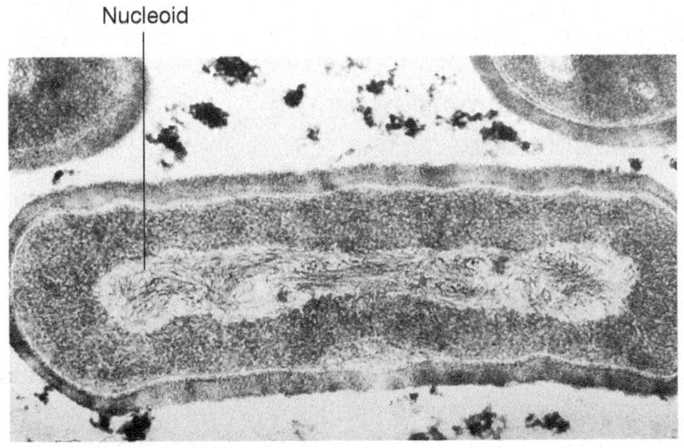

(a) Bacterial nucleoid 0.25 μm

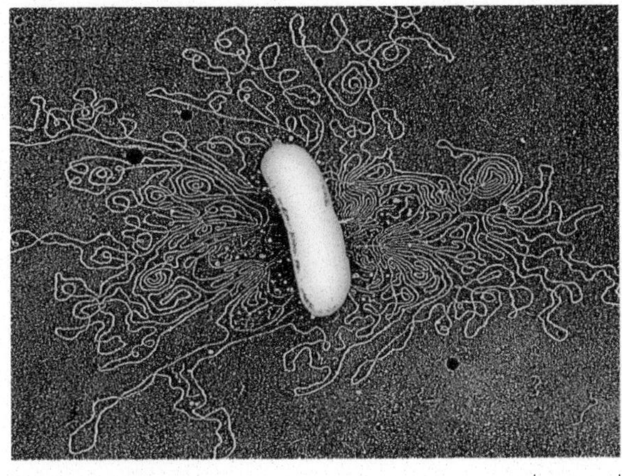

(b) Released DNA 1 μm

Figure 16-15 The Bacterial Nucleoid. The electron micrograph at the top shows a bacterial cell with a distinct nucleoid, the area in which the bacterial chromosome resides. When bacterial cells are ruptured, their chromosomal DNA is released from the cell. The bottom micrograph shows that the released DNA forms a series of loops that remain attached to a structural framework within the nucleoid (TEMs).

Bacterial Plasmids. In addition to its chromosome, a bacterial cell may contain one or more plasmids. **Plasmids** are relatively small, usually circular molecules of double-stranded DNA that carry genes both for their own replication and, often, for one or more cellular functions (usually nonessential ones). Most plasmids are supercoiled, giving them a condensed, compact form. Although plasmids replicate autonomously, the replication is usually sufficiently synchronized with the replication of the bacterial chromosome to ensure a roughly comparable number of plasmids from one cell generation to the next. In *E. coli* cells, several classes of plasmids are recognized: *F (fertility) factors* are involved in the process of conjugation (a sexual process we will discuss in Chapter 25); *R (resistance) factors* carry genes that impart drug resistance to the bacterial cell; *col (colicinogenic) factors* allow the bacterium to secrete *colicins*, compounds that kill other bacteria lacking the col factor; *virulence factors* enhance the ability to cause disease by producing toxic proteins that cause tissue damage or enzymes that allow the bacteria to enter host cells; and *metabolic plasmids* produce enzymes required for certain metabolic reactions. Some strains of *E. coli* also possess *cryptic plasmids*, which have no known function and possess no genes other than those needed for the plasmid to replicate and spread to other cells.

Eukaryotes Package DNA in Chromatin and Chromosomes

When we turn from bacteria to eukaryotes, DNA packaging becomes more complicated. First, substantially larger amounts of DNA are involved. Each eukaryotic chromosome contains a single, linear DNA molecule of enormous size. In human cells, for example, just *one* of these DNA molecules may be 10 cm or more in length—roughly 100 times the size of the DNA molecule found in a typical bacterial chromosome. Second, greater structural complexity is introduced by the association of eukaryotic DNA with greater amounts and numbers of proteins. When bound to these proteins, the DNA is known as **chromatin,** which forms fibers 10–30 nm in diameter that are normally dispersed throughout the nucleus. At the time of cell division (and in a few other situations), these fibers condense and fold into much larger, compact structures that become recognizable as individual **chromosomes.**

The proteins with the most important role in chromatin structure are the **histones,** a group of relatively small proteins whose high content of the amino acids lysine and arginine gives them a strong positive charge. The binding of histones to DNA, which is negatively charged, is therefore stabilized by ionic bonds. In most cells, the mass of histones in chromatin is approximately equal to the mass of DNA. Histones are divided into five main types, designated H1, H2A, H2B, H3, and H4. Chromatin contains roughly equal numbers of H2A, H2B, H3, and H4 molecules, and about half that number of H1 molecules. These proportions are remarkably constant among different kinds of eukaryotic cells, regardless of the type of cell or its physiological state. In addition to histones, chromatin contains a diverse group of *nonhistone proteins* that play a variety of enzymatic, structural, and regulatory roles.

Nucleosomes Are the Basic Unit of Chromatin Structure

The DNA contained within a typical nucleus would measure a meter or more in length if it were completely extended, whereas the nucleus itself is usually no more than 5–10 μm in diameter. The folding of such an enormous length of DNA into a nucleus that is almost a million times smaller presents a significant topological problem. One of the first insights into the folding process emerged in the late 1960s, when X-ray diffraction studies carried out by Maurice Wilkins revealed that purified chromatin fibers have a repeating structural subunit that is seen in neither DNA nor histones alone. Wilkins therefore concluded that histones impose a repeating structural organization upon DNA. A clue to the nature of this structure was provided in 1974, when Ada Olins and Donald Olins published electron micrographs of chromatin fibers isolated from cells in a way that avoided the harsh solvents used in earlier procedures for preparing chromatin for microscopic examination. Chromatin fibers viewed in this way appear as a series of tiny particles attached to one another by thin filaments. This "beads-on-a-string" appearance led to the suggestion that the beads consist of protein (presumably histones) and the thin filaments connecting the beads correspond to DNA. We now refer to each bead, along with its associated short stretch of DNA, as a **nucleosome** (**Figure 16-16**).

Based on electron microscopy alone, it would have been difficult to determine whether nucleosomes are a normal component of chromatin or an artifact generated during sample preparation. Fortunately, independent evidence for the existence of a repeating structure in chromatin was reported at about the same time by Dean Hewish and Leigh Burgoyne, who discovered that rat liver nuclei contain a nuclease that is capable of cleaving the DNA in chromatin fibers. In one crucial set of experiments, these investigators exposed chromatin to this nuclease and then purified the partially digested DNA to remove chromatin proteins. Upon examining the purified DNA by gel electrophoresis, they found a distinctive pattern of fragments in which the smallest piece of DNA measured about 200 bp in length, and the remaining fragments were exact multiples of 200 bp (**Figure 16-17**). Because nuclease digestion of protein-free DNA does not generate this fragment pattern, they concluded that (1) chromatin proteins are clustered along the DNA molecule in a regular pattern that repeats at intervals of roughly 200 bp, and (2) the DNA located between these protein clusters is susceptible to nuclease digestion, yielding fragments that are multiples of 200 bp in length.

Subsequent experiments using a combination of nuclease digestion and electron microscopy led to the conclusion that the spherical particles observed in electron micrographs are each associated with 200 bp of DNA. This basic repeat unit, containing an average of 200 bp of DNA associated with a protein particle, is the nucleosome.

A Histone Octamer Forms the Nucleosome Core

The first insights into the molecular architecture of the nucleosome emerged from the work of Roger Kornberg, who was awarded a Nobel Prize in 2006 for a series of fundamental discoveries concerning DNA packaging and transcription in eukaryotes. In their early studies, Kornberg and his colleagues showed that chromatin fibers composed of nucleosomes can be generated by combining purified DNA with a mixture of all five histones. However, they discovered that nucleosomes could be assembled only when histones were isolated using

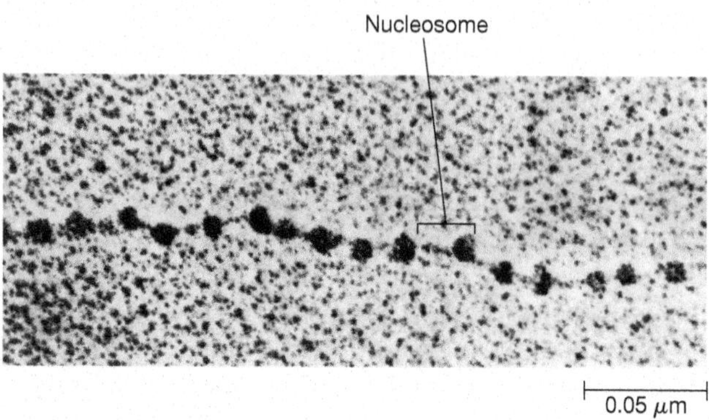

Figure 16-16 Nucleosomes. This electron micrograph shows the characteristic "beads-on-a-string" formation of nucleosomes in chromatin fibers first observed by Ada and Donald Olins in 1974 (TEM).

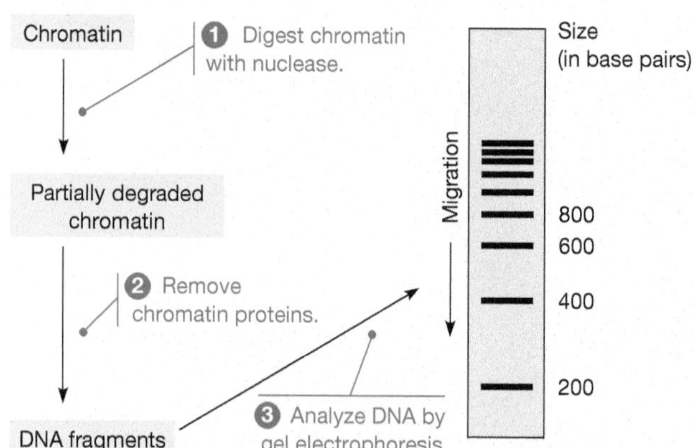

Figure 16-17 Evidence That Proteins Are Clustered at 200-Base-Pair Intervals Along the DNA Molecule in Chromatin Fibers. In these experiments, DNA fragments generated by partial nuclease digestion of rat liver chromatin were analyzed by gel electrophoresis. The discovery that the DNA fragments are multiples of 200 base pairs suggests that histones are clustered at 200-base-pair intervals along the DNA, thereby conferring a regular pattern of protection against nuclease digestion.

gentle techniques that left histone H2A bound to histone H2B, and histone H3 bound to histone H4. Kornberg therefore concluded that histone H3-H4 and H2A-H2B complexes are an integral part of the nucleosome.

To investigate the nature of these histone interactions in more depth, Kornberg and his colleague Jean Thomas treated isolated chromatin with a chemical reagent that forms covalent crosslinks between protein molecules that are located next to each other. After being treated with this reagent, the chemically crosslinked proteins were isolated and analyzed by polyacrylamide gel electrophoresis. Protein complexes the size of eight histone molecules were prominent in such gels, suggesting that the nucleosomal particle contains an *octamer* of eight histones. Given the knowledge that histones H3 and H4 and histones H2A and H2B each form tight complexes and that these four histones are present in roughly equivalent amounts in chromatin, Kornberg and Thomas proposed that histone octamers are created by joining together two H2A-H2B dimers and two H3-H4 dimers and that the DNA double helix is then wrapped around the resulting octamer (**Figure 16-18**). Each core histone has a "tail" that protrudes from the octameric core. These tails contain several key lysine and arginine amino acids. The positive charges on these tails interact with the negative charges in the DNA backbone, facilitating its packaging. The amino acids in the tails are also subject to modifications that alter how tightly chromatin is packed, as you'll see shortly.

One issue not addressed by the preceding model concerns the significance of histone H1, which is not part of the octamer. If individual nucleosomes are isolated by briefly digesting chromatin with nuclease, histone H1 is still present (along with the four other histones and 200 bp of DNA). When digestion is carried out for longer periods, the DNA fragment is further degraded until it reaches a length of about 146 bp; during the final stages of the digestion process, histone H1 is released. The remaining particle, consisting of a histone octamer associated with 146 bp of DNA, is referred to as a *core particle*. The DNA that is degraded during digestion from 200 to 146 bp in length is referred to as *linker DNA* because it joins one nucleosome to the next (Figure 16-18). Because histone H1 is released upon degradation of the linker DNA, histone H1 molecules are thought to be associated with the linker region. A recent study using cryo-EM (a technique described in the Appendix) suggests that H1 interacts with the linker DNA where it enters and leaves the histone core, reducing the flexibility of the linker DNA. The length of the linker DNA varies somewhat among organisms, but the DNA associated with the core particle always measures close to 146 bp, which is enough to wrap around the core particle roughly 1.7 times.

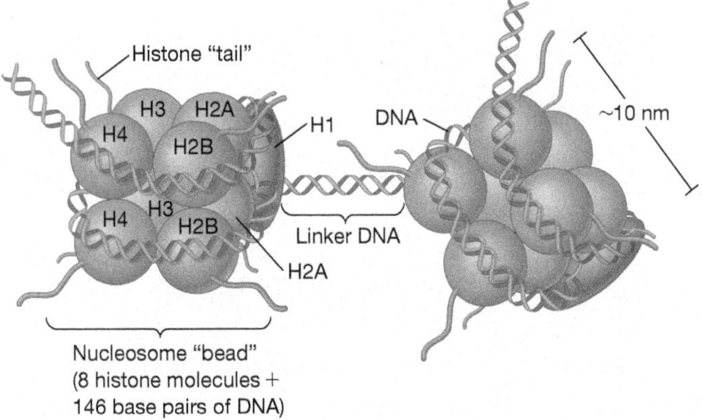

Figure 16-18 A Closer Look at Nucleosome Structure. Each nucleosome consists of eight histone molecules (two each of histones H2A, H2B, H3, and H4) associated with 146 base pairs of DNA and a stretch of linker DNA about 50 base pairs in length. The diameter of the nucleosome "bead," or core particle, is about 10 nm. The histones in the core particle each have "tails," which are subject to chemical modification that can affect nucleosome packing. Histone H1 is thought to bind to the linker DNA and facilitate the packing of nucleosomes into 30-nm fibers. Although not depicted in the figure, experimental evidence suggests that histone H1 is bound to the linker DNA both entering and exiting the core particle.

Nucleosomes Are Packed Together to Form Chromatin Fibers and Chromosomes

The formation of nucleosomes is only the first step in the packaging of nuclear DNA (**Figure 16-19** on page 456). Isolated chromatin fibers exhibiting the beads-on-a-string appearance measure about 10 nm in diameter, but the chromatin of intact cells often forms a slightly thicker fiber, about 30 nm in diameter, called the **30-nm chromatin fiber.** In preparations of isolated chromatin, the 10-nm and 30-nm forms of the chromatin fiber can be interconverted by changing the salt concentration of the solution. However, the 30-nm fiber does not form in chromatin preparations whose histone H1 molecules have been removed, suggesting that histone H1 facilitates the packing of nucleosomes into the 30-nm fiber. Current models suggest that the nucleosomes of the 30-nm fiber seem to be packed together to form an irregular, three-dimensional zigzag structure that can interdigitate with its neighboring fibers.

The next level of chromatin packaging is the folding of the 30-nm fibers into **DNA loops** averaging 50,000–100,000 bp in length. The bases of the loops are stabilized via proteins that in mammals include *cohesin*, which also plays an important role in keeping chromosomes attached to one another prior to anaphase in dividing cells (see Chapter 24), and a protein known as *CTCF*. The spatial arrangement of the loops is maintained by the periodic attachment of DNA to an insoluble network of nonhistone proteins that form a chromosomal *scaffold* to which the long loops of DNA are attached. The loops can be most clearly seen in electron micrographs of chromosomes isolated from dividing cells and treated to remove all the histones and most of the nonhistone proteins (**Figure 16-20** on page 457). Loops can also be seen in specialized types of chromosomes that are not associated with the process of cell division (called polytene chromosomes; Chapter 20). In these cases, the chromatin loops turn out to contain "active" regions of DNA—that is, DNA that is being transcribed. It makes sense that active DNA would be less tightly packed than inactive DNA because it would allow easier access by proteins involved in gene transcription.

The extent to which a DNA molecule has been folded in chromatin and chromosomes can be quantified using the *DNA packing ratio*, which is calculated by determining the

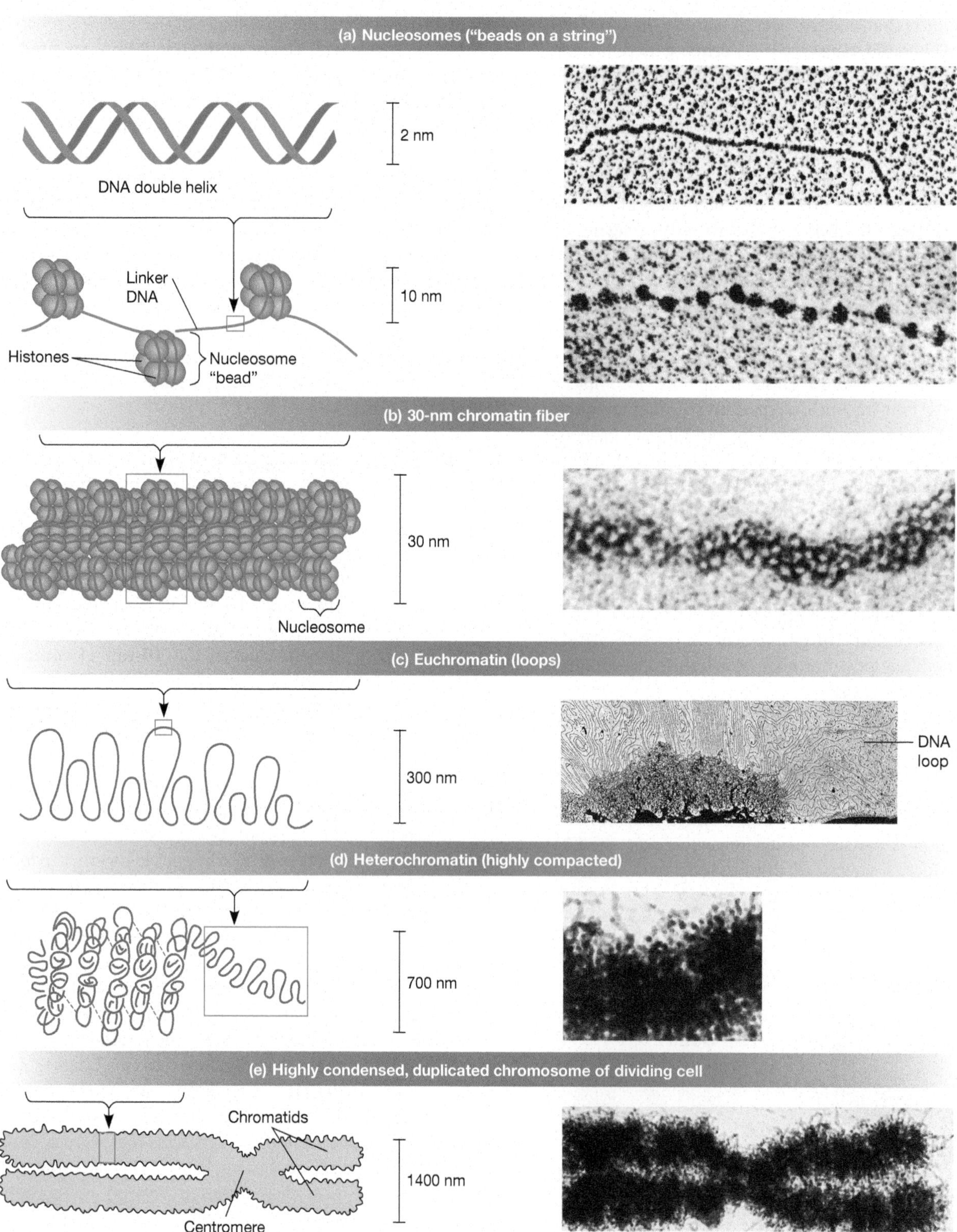

Figure 16-19 Levels of Chromatin Packing. These illustrations and TEMs show a current model for progressive stages of DNA coiling and folding, culminating in the highly compacted chromosome of a dividing cell. **(a)** "Beads on a string," an extended configuration of nucleosomes formed by the association of DNA with four types of histones. **(b)** The 30-nm chromatin fiber, shown here as a tightly packed collection of nucleosomes. **(c)** Loops of 30-nm fibers, visible in the TEM here because a mitotic chromosome has been experimentally unraveled. **(d)** Highly folded chromatin. **(e)** A replicated chromosome (two attached chromatids) from a dividing cell, with all the DNA of the chromosome in the form of very highly compacted chromatin. A constriction marks the centromere.

total extended length of a DNA molecule and dividing it by the length of the chromatin fiber or chromosome into which it has been packaged. The initial coiling of the DNA around the histone cores of the nucleosomes reduces the length by a factor of about seven, and formation of the 30-nm fiber results in a further sixfold condensation. Further folding and

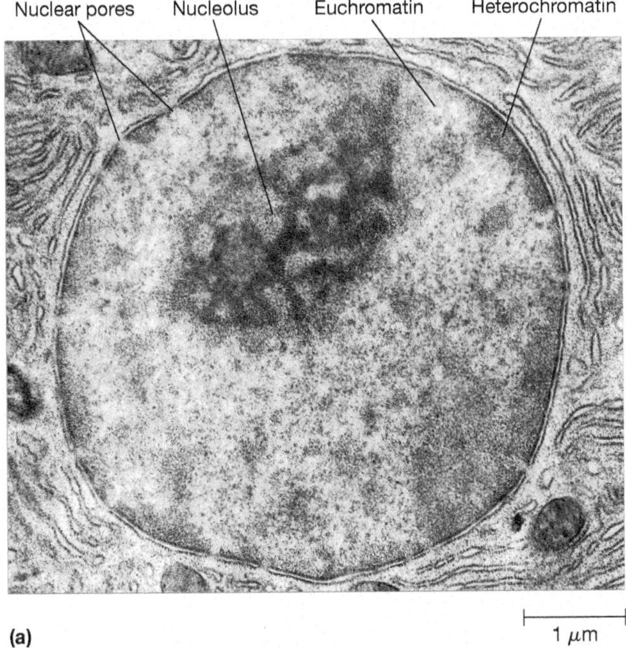

Figure 16-29 Structural Organization of the Nucleus and Nuclear Envelope. (a) An electron micrograph of the nucleus from a pancreatic acinar cell, with prominent structural features labeled (TEM). The nuclear envelope is a double membrane perforated by nuclear pores. Internal structures include the nucleolus, euchromatin, and heterochromatin. **(b)** A drawing of a typical nucleus. Structural features included here but not visible in the micrograph include the nuclear lamina, ribosomes on the outer nuclear membrane, and the continuity between the outer nuclear membrane and the rough ER.

the lumen of the ER. Like membranes of the rough ER, the outer membrane is often studded on its outer surface with ribosomes engaged in protein synthesis. Several proteins in the outer membrane bind to the cell's cytoskeleton, thereby anchoring the nucleus within the cell and providing a mechanism for nuclear movements. Tubular invaginations of the nuclear envelope may project into the internal nuclear space and increase the area of the nucleus that makes direct contact with the inner nuclear membrane.

One of the most distinctive features of the nuclear envelope is the presence of specialized channels called **nuclear pores,** which are especially easy to see when the nuclear envelope is examined by freeze-fracture microscopy (**Figure 16-30**). Each pore is a small cylindrical channel extending through both membranes of the nuclear envelope, thereby providing direct continuity between the cytosol and the **nucleoplasm** (the name for the interior space of the nucleus other than the region occupied by the nucleolus). The number of pores varies greatly with cell type and activity. A typical mammalian nucleus has about 3000–4000 pores, or about 10–20 pores per square micrometer of membrane surface area. These pores function as the primary nuclear gatekeeper, controlling the flow of substances into and out of the nucleus.

At each pore, the inner and outer membranes of the nuclear envelope are fused together, forming a channel

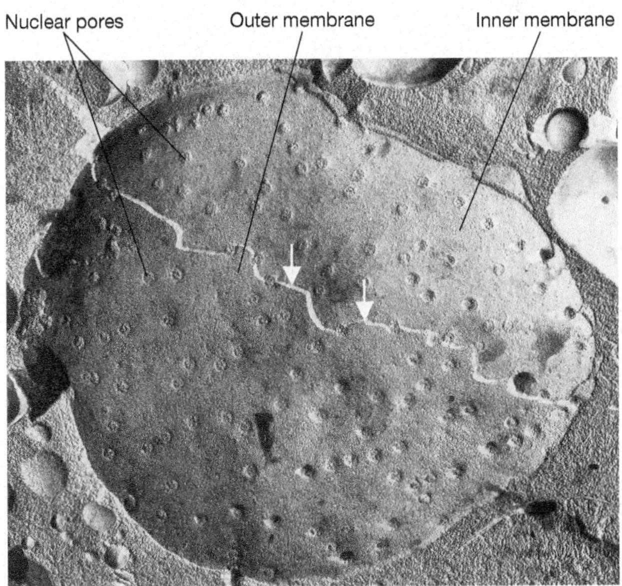

Figure 16-30 Nuclear Pores. Numerous nuclear pores are visible in this freeze-fracture micrograph of the nuclear envelope of an epithelial cell from a pig kidney. The fracture plane reveals faces of both the inner membrane and the outer membrane. The arrows point to ridges that represent the perinuclear space delimited by the two membranes (TEM).

that is lined with an intricate protein structure called the **nuclear pore complex (NPC).** The NPC has an outer diameter of ~120 nm and is built from about 30 different proteins called *nucleoporins.* In electron micrographs, the most striking feature of the pore complex is its octagonal symmetry. Micrographs such as the one in **Figure 16-31** show rings of eight subunits arranged in an octagonal pattern. Notice that central granules can be seen in some of the nuclear pore complexes in Figure 16-31a. Although these granules were once thought to consist solely of particles in transit through the pores, they are now thought to also contain components that are an integral part of the pore complex.

Figure 16-31b illustrates the main components of the nuclear pore complex, which is exquisitely complex. The pore complex as a whole is shaped somewhat like a wheel lying on its side within the nuclear envelope. Two parallel rings, outlining the rim of the wheel, each consist of the eight subunits seen in electron micrographs. Eight spokes (shown in green) extend from the rings to the wheel's hub (purple), which is the "central granule" seen in many electron micrographs. This granule is sometimes called the *transporter* because it is thought to contain components that are involved in moving macromolecules across the nuclear envelope. The nucleoporin subunits associated with the transporter are called *FG nucleoporins* because they contain repeating elements containing the amino acids phenylalanine (F) and glycine (G). FG nucleoporins appear to be intrinsically unstructured, flexible proteins that associate with molecules moving in or out of the nucleus through nuclear pores. Proteins extending from the rim into the perinuclear space may help anchor the pore complex to the envelope. Fibers also extend from the rings into the cytosol and nucleoplasm, with those on the nucleoplasm side forming a basket (sometimes called a "cage" or "fishtrap").

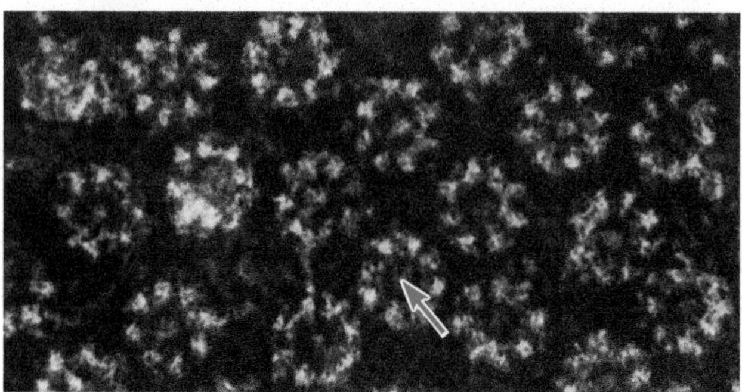

(a) Nuclear pores in the envelope

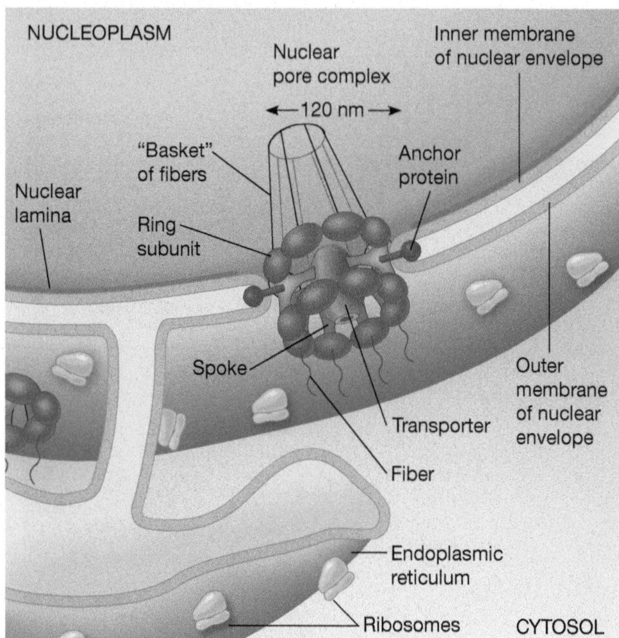

(b) Location of nuclear pores in nuclear membrane

Figure 16-31 Structure of the Nuclear Pore. (a) Negative staining of an oocyte nuclear envelope reveals the octagonal pattern of the nuclear pore complexes. The arrow shows a central granule (TEM). **(b)** A nuclear pore is formed by fusion of the inner and outer nuclear membranes and is lined by an intricate protein structure called the nuclear pore complex. See the text for a description.

Molecules Enter and Exit the Nucleus Through Nuclear Pores

The nuclear envelope solves one problem but creates another. You will learn much more about how cells produce new DNAs, RNAs, and protein in subsequent chapters. For now, however, consider the general consequences of segregating DNA in a nucleus. As a means of localizing chromosomes and their activities to one region of the cell, it is an example of the general eukaryotic strategy of compartmentalization. Presumably, it is advantageous for a nucleus to possess a barrier that keeps in the chromosomes and keeps out organelles such as ribosomes, mitochondria, and lysosomes. For example, the nuclear envelope protects newly synthesized RNA from being acted upon by cytosolic organelles or enzymes before it has been fully processed.

But in the act of separating immature RNA molecules and chromosomes from the cytosol, the nuclear envelope creates several formidable transport problems for eukaryotic cells that are unknown in prokaryotes. All the enzymes and other proteins required for chromosome replication and transcription of DNA in the nucleus must be imported from the cytosol, and all the RNA molecules and partially assembled ribosomes needed for protein synthesis in the cytosol must be obtained from the nucleus (**Figure 16-32**).

To get some idea of how much traffic must travel through the nuclear pores, consider the flow of ribosomal subunits from the nucleus to the cytosol. Ribosomes are partially assembled in the nucleus as two classes of subunits, each of which is a complex of RNA and proteins. These subunits move to the cytosol and, when needed for protein synthesis, are combined into functional ribosomes containing one of each type of subunit. An actively growing mammalian cell can easily be synthesizing 20,000 ribosomal subunits per minute. We already know that such a cell has about 3000–4000 nuclear pores, so ribosomal subunits must be transported to the cytosol at a rate of about five to six subunits per minute per pore. Traffic in the opposite direction is, if anything, even heavier. When chromosomes are being replicated, histones are needed at the rate of about 300,000 molecules per minute. The rate of inward movement must therefore be about 100 histone molecules per minute per pore in an actively dividing cell!. In addition to all this macromolecular traffic, the pores mediate the transport of smaller particles, molecules, and ions.

skeletal structure that is well understood. This structure, called the **nuclear lamina,** is a thin, dense meshwork of fibers that lines the inner surface of the inner nuclear membrane and confers mechanical strength to the nucleus. In animals, the nuclear lamina is about 10–40 nm thick and is constructed from intermediate filaments made of proteins called **lamins** (intermediate filaments are described in more detail in Chapter 13). Inherited abnormalities in these proteins have been linked to more than a dozen human diseases, including several involving severe muscle wasting or premature aging. One striking example is a single base mutation in one lamin gene, which results in *Hutchinson–Gilford progeria syndrome (HGPS),* a disease in which symptoms of old age—such as hair loss, cardiovascular disease, and degeneration of skin, muscle, and bone—appear in young children and usually cause death by the early teenage years (see Human Connections, page 468). Lamins may be the ancestral type of intermediate filament, as these are the only type of intermediate filament in some animals, such as insects. Interestingly, plants and fungi lack lamins entirely; they presumably use other proteins to structurally reinforce their nuclei. A leading candidate for this role in plants is a family of proteins called nuclear matrix constituent proteins (NMCPs).

Chromatin Is Located Within the Nucleus in a Nonrandom Fashion

Other than during cell division, a cell's chromatin tends to be highly extended and dispersed throughout the nucleus. You might therefore guess that the chromatin threads corresponding to each individual chromosome are randomly distributed and highly intertwined within the nucleus. Perhaps surprisingly, this seems not to be the case. Instead, the chromatin of each chromosome has its own discrete location. This idea was first proposed in 1885, but evidence that it is true in a variety of cells awaited the techniques of modern molecular biology. Experiments using *in situ hybridization* have shown that the chromatin fibers corresponding to individual chromosomes occupy discrete compartments within the nucleus, referred to as **chromosomal territories** (see Key Technique, pages 450–451). The positions of these territories do not seem to be fixed, however. They vary from cell to cell of the same organism and change during a cell's life cycle, likely reflecting changes in gene activity of the different chromosomes.

The nuclear envelope helps organize chromatin by binding parts of it to the inner nuclear envelope at sites that are closely associated with the nuclear pores. These chromatin regions are highly compacted—that is, they are heterochromatin. In electron micrographs, this material appears as a dark irregular layer around the nuclear periphery, as we saw in Figure 16-29a. Most of it seems to be constitutive heterochromatin, probably centromeric or telomeric regions of chromosomes; for example, in yeast there is good evidence that centromeres are located near the nuclear envelope at times other than during cell division.

The Nucleolus Is Involved in Ribosome Formation

A prominent structural component of the eukaryotic nucleus is the **nucleolus** (plural **nucleoli**), the ribosome factory of the cell. Typical eukaryotic cells contain one or two nucleoli,

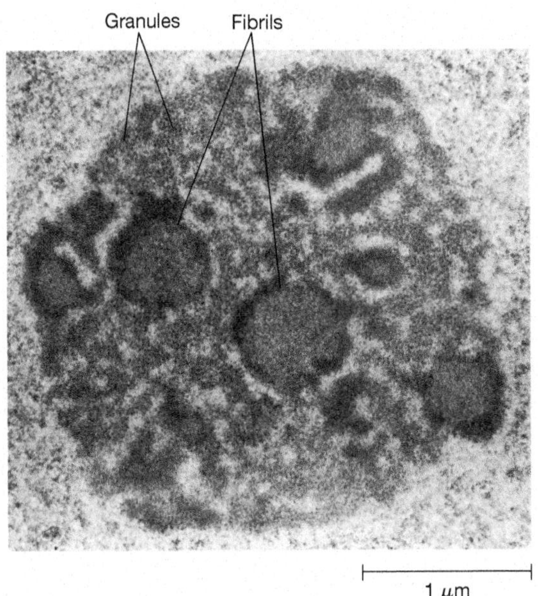

Figure 16-36 The Nucleolus. The nucleolus is a prominent intranuclear structure composed of a mass of fibrils and granules. The fibrils are DNA and rRNA; the granules are newly forming ribosomal subunits. Shown here is a nucleolus of a spermatogonium cell (TEM).

but the occurrence of hundreds or even thousands is not uncommon. The nucleolus is usually spherical, measuring several micrometers in diameter, but wide variations in size and shape are observed. Because of their relatively large size, nucleoli are easily seen with the light microscope and were first observed nearly 240 years ago.

It was not until the advent of electron microscopy in the 1950s, however, that the structural components of the nucleolus were clearly identified. Each nucleolus appears as a membrane-free organelle consisting of fibrils and granules (**Figure 16-36**). The fibrils contain DNA that is being transcribed into *ribosomal RNA (rRNA),* the RNA component of ribosomes (see Chapter 19). The granules are rRNA molecules being packaged with proteins (imported from the cytosol) to form ribosomal subunits. As we saw earlier, the ribosomal subunits are subsequently exported through the nuclear pores to the cytosol.

The size of the nucleolus is correlated with its level of activity. In cells having a high rate of protein synthesis and hence a need for many ribosomes, nucleoli tend to be large and can account for 20–25% of the total volume of the nucleus. In less-active cells, nucleoli are much smaller.

In addition to the nucleolus, microscopists have also identified several kinds of small, nonmembrane-enclosed nuclear structures. These nuclear bodies are thought to play a variety of roles related to the processing and handling of RNA molecules produced in the nucleus (see Chapter 19).

CONCEPT CHECK 16.4

You are studying a cytosolic protein, and your friend is studying nuclear localization sequences in her favorite protein. How could you work together to show that her NLS is sufficient to cause any protein, including yours, to move into the nucleus?

Summary of Key Points

Mastering™ Biology For activities, animations, and review quizzes, go to the study area at www.masteringbiology.com.

16.1 Chemical Nature of the Genetic Material

- DNA was discovered in 1869, but it was not until the mid-twentieth century that studies with pneumococcal bacteria and bacteriophage T2 revealed that genes are made of DNA.

16.2 DNA Structure

- DNA is a double helix in which the base A is paired with T, and the base G is paired with C. Supercoiling of the double helix affects the ability of DNA to interact with other molecules.
- Under experimental conditions, the two strands of the double helix can be separated from each other (denaturation) and rejoined (renaturation).

16.3 DNA Packaging

- The DNA (or RNA in some viruses) that contains one complete set of an organism's genetic information is called its genome. Viral or prokaryotic genomes consist of one or a small number of DNA molecules. Eukaryotes have a nuclear genome divided among multiple chromosomes, each containing one long DNA molecule plus a mitochondrial genome; plants and algae possess a chloroplast genome as well.
- Due to their enormous size, DNA molecules need to be efficiently packaged by proteins that bind to the DNA. In eukaryotic chromosomes, short stretches of DNA are wrapped around protein particles composed of eight histone molecules to form a basic structural unit called the nucleosome. Chains of nucleosomes ("beads on a string") are packed together to form a 30-nm chromatin fiber, which can then loop and fold further.
- In eukaryotic cells that are actively transcribing DNA, much of the chromatin is in an extended, uncoiled form called euchromatin. Other portions are in a highly condensed, transcriptionally inactive state called heterochromatin. During cell division, all the chromatin becomes highly compacted, forming discrete chromosomes with internal structure that can be seen with a light microscope. Some heterochromatin, including centromeres and telomeres, serves a structural role.
- In the nuclear genomes of multicellular eukaryotes, much of the DNA consists of repeated sequences that do not code for RNA or proteins. Some of this noncoding DNA performs structural or regulatory roles, but most of it has no obvious function.

16.4 The Nucleus

- Eukaryotic chromosomes are contained within a nucleus that is bounded by a double-membrane nuclear envelope.
- The nuclear envelope is perforated with nuclear pores that mediate two-way transport of materials between the nucleoplasm and the cytosol. Ions and small molecules diffuse passively through aqueous channels in the pore complex; larger molecules and particles are actively transported through it.
- The nucleus appears to contain a fibrous skeleton. The nuclear membranes are connected to the cellular cytoskeleton. One well-defined skeletal structure is the nuclear lamina—a thin, dense meshwork of fibers lining the inner surface of the inner nuclear membrane that confers mechanical strength to the nucleus.
- The nucleolus is a specialized nuclear structure involved in the synthesis of ribosomal RNA and the assembly of ribosomal subunits. Other nuclear bodies perform functions related to processing of RNA.

Problem Set

Mastering™ Biology For activities, animations, and review quizzes, go to the study area at www.masteringbiology.com.

16-1 Prior Knowledge. Virtually every experiment performed by biologists builds on knowledge provided by earlier experiments.

(a) Of what significance to Avery and his colleagues was the finding (made in 1932 by J. L. Alloway) that the same kind of transformation of R cells into S cells that Griffith observed to occur in mice could also be demonstrated in culture with isolated pneumococcal cells?

(b) Of what significance to Hershey and Chase was the following suggestion (made in 1951 by R. M. Herriott)? "A virus may act like a little hypodermic needle full of transforming principles; the virus as such never enters the cell; only the tail contacts the host and perhaps enzymatically cuts a small hole through the outer membrane and then the nucleic acid of the virus head flows into the cell."

(c) Of what significance to Watson and Crick were the data of their colleagues at Cambridge suggesting that the particular forms in which A, G, C, and T exist at physiological pH permit the formation of specific hydrogen bonds?

(d) How did the findings of Hershey and Chase help explain an earlier report (by T. F. Anderson and R. M. Herriott) that bacteriophage T2 loses its ability to reproduce when it is burst open osmotically before adding them to a bacterial culture?

16-2 DNA Base Composition. Based on your understanding of the rules of complementary base pairing, answer the following questions (dsDNA = double-stranded DNA).

(a) You analyze a DNA sample and find that its base composition is 30% A, 20% T, 30% G, and 20% C. What can you conclude about the structure of this DNA?

(b) For a dsDNA molecule in which 40% of the bases are either G or C, what can you conclude about its content of the base A?

(c) For a dsDNA molecule in which 40% of the bases are either G or T, what can you conclude about its content of the base A?

(d) For a dsDNA molecule in which 15% of the bases are A, what can you conclude about its content of the base C? Would you expect the T_m of this DNA to be above or below 90°C? (See Figure 16-13.)

16-3 DNA Structure. Carefully inspect the double-stranded DNA molecule shown here, and notice that it has twofold rotational symmetry:

3′ A—G—C—G—C—T—A—T—A—G—C—G—C—T 5′
5′ T—C—G—C—G—A—T—A—T—C—G—C—G—A 3′

Label each of the following statements as T if true or F if false.

(a) There is no way to distinguish the right end of the double helix from the left end.

(b) If a solution of these molecules were heated to denature them, every single-stranded molecule in the solution would be capable of hybridizing with every other molecule.

(c) If the molecule were cut at its midpoint into two halves, it would be possible to distinguish the left half from the right half.

(d) If the two single strands were separated from each other, it would not be possible to distinguish one strand from the other.

(e) In a single strand from this molecule, it would be impossible to determine which is the 3′ end and which is the 5′ end.

16-4 DATA ANALYSIS DNA Melting. Figure 16-37 shows the melting curves for two DNA samples that were thermally denatured under the same conditions.

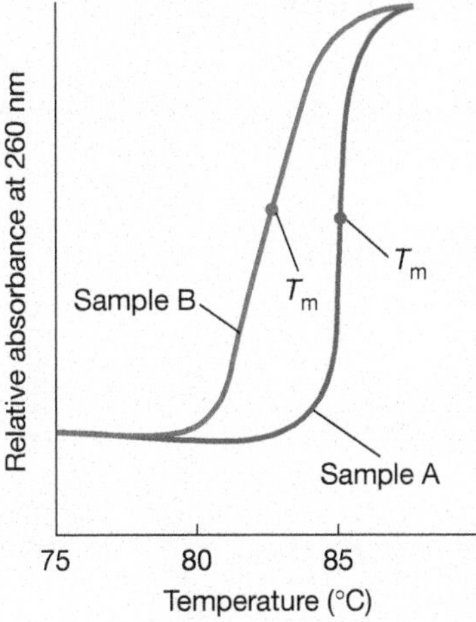

Figure 16-37 Thermal Denaturation of Two DNA Samples. See Problem 16-4.

(a) What conclusion can you draw concerning the base compositions of the two samples? Explain.

(b) How might you explain the steeper slope of the melting curve for sample A?

(c) Formamide and urea are agents known to form hydrogen bonds with pyrimidines and purines. What effect, if any, would the inclusion of a small amount of formamide or urea in the incubation mixture have on the melting curves?

16-5 DNA Renaturation. You are given two samples of DNA, each of which melts at 92°C during thermal denaturation. After denaturing the DNA, you mix the two samples together and then cool the mixture to allow the DNA strands to reassociate. When the newly reassociated DNA is denatured a second time, the sample now melts at 85°C.

(a) How might you explain the lowering of the melting temperature from 92°C to 85°C?

(b) If the newly reassociated DNA had melted at 92°C instead of 85°C, what conclusions might you have drawn concerning the base sequences of the two initial DNA samples?

16-6 Nucleosomes. You perform an experiment in which chromatin is isolated from sea urchin sperm cells and briefly digested with micrococcal nuclease. When the chromatin proteins are removed and the resulting purified DNA is analyzed by gel electrophoresis, you observe a series of DNA fragments that are multiples of 260 base pairs in length (that is, 260 bp, 520 bp, 780 bp, and so forth).

(a) Although these results differ somewhat from the typical results discussed in the chapter, explain why they still point to the likely existence of nucleosomes in this cell type.

(b) What can you conclude about the amount of DNA that is associated with each nucleosome?

(c) Suppose you perform an experiment in which the chromatin is digested for a much longer period of time with micrococcal nuclease before removal of chromatin proteins. When the resulting DNA preparation is analyzed by electrophoresis, all of the DNA appears as fragments 146 bp in length. What does this suggest to you about the length of the linker DNA in this cell type?

16-7 Nuclear Structure and Function. Indicate the implications for nuclear structure or function of each of the following experimental observations.

(a) Sucrose crosses the nuclear envelope so rapidly that its rate of movement cannot be accurately measured.

(b) Colloidal gold particles with a diameter of 5.5 nm equilibrate rapidly between the nucleus and cytosol when injected into an amoeba, but gold particles with a diameter of 15 nm do not.

(c) Nuclear pore complexes sometimes stain heavily for RNA and protein.

(d) If gold particles up to 26 nm in diameter are coated with a polypeptide containing a nuclear localization signal (NLS) and are then injected into the cytosol of a living cell, they are transported into the nucleus. If they are injected into the nucleus, however, they remain there.

(e) Many of the proteins of the nuclear envelope appear from electrophoretic analysis to be the same as those found in the endoplasmic reticulum.

(f) Ribosomal proteins are synthesized in the cytosol but are packaged with rRNA into ribosomal subunits in the nucleus.

(g) Treatment of nuclei with the nonionic detergent Triton X-100 dissolves the nuclear envelope but leaves an otherwise intact nucleus.

16-8 Nuclear Transport. Budding yeast with temperature-sensitive mutations in NTF2 have been identified. In such mutants, NTF2 functions at 25°C. Raising the temperature to 37°C, however, causes the NTF2 protein to stop functioning. What effects would you predict raising these mutant yeast to 37°C would have on nuclear transport? Explain your answer.

16-9 Going Nuclear. Experiments have been performed with the nuclear localization sequence (NLS) of nucleoplasmin. In each case, a string of amino acids was added to pyruvate kinase, a glycolytic enzyme normally found in the cytosol. The sequence of each added string of amino acids is shown using single-letter amino acid designations. Consult the single-letter amino acid code table (Table 3-2, page 45) if you feel it is necessary. Many alanines in a row are denoted by $(A)_n$, where n is the number of alanines. The following results were obtained:

Name of added sequence	Sequence added to pyruvate kinase	Localization
Normal sequence	KRPAATKKAGQAKKKKLD	Nuclear
Artificial sequence #1	$AAP(A)_{18}KKKKLD$	Cytosolic
Artificial sequence #2	$KRP(A)_{20}KKKKLD$	Nuclear

(a) What sort of NLS does nucleoplasmin appear to have, based on these results? Explain your answer.

(b) How does your conclusion about the NLS of nucleoplasmin compare to the NLS for SV40 large T antigen?

17 DNA Replication, Repair, and Recombination

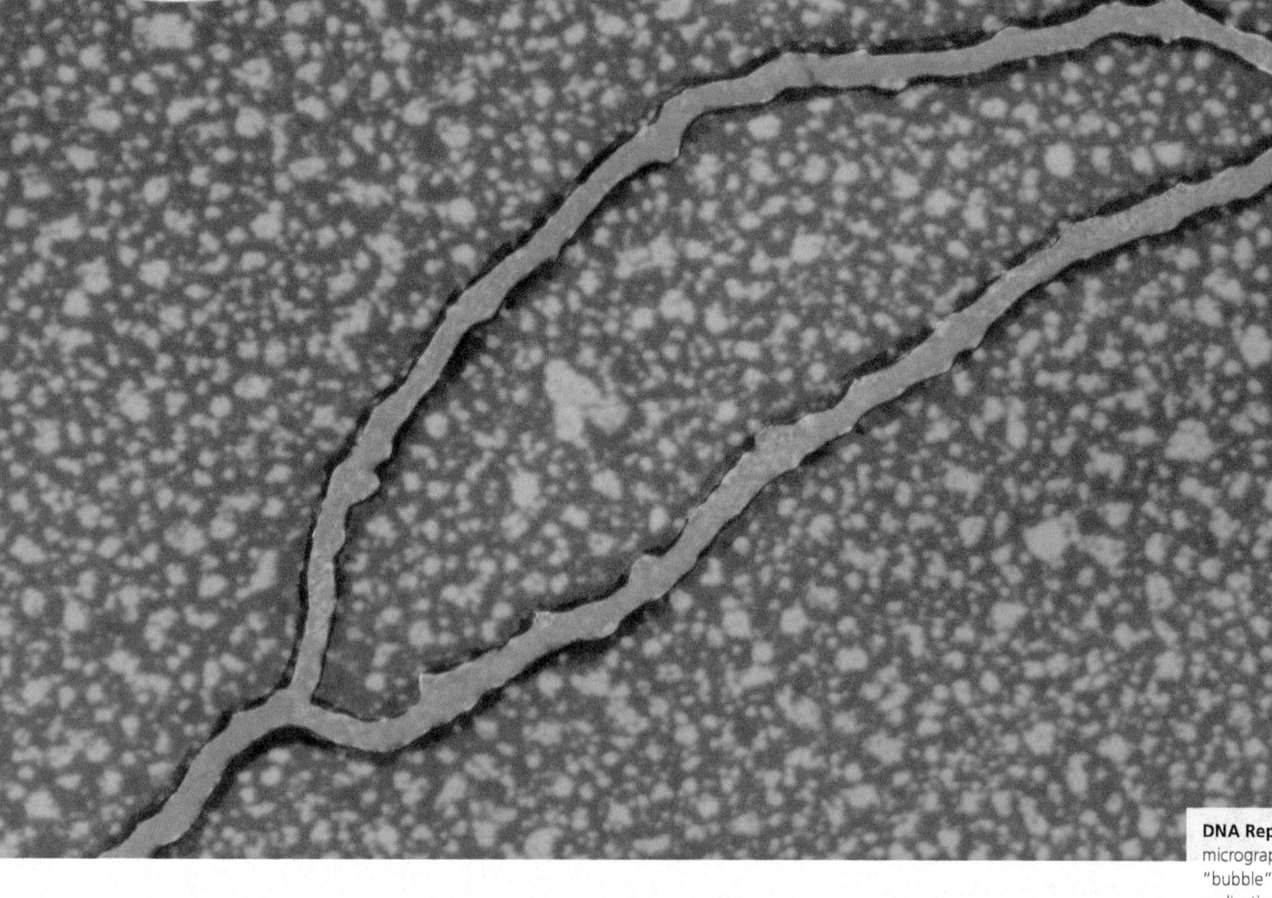

DNA Replication. Colored micrograph of a replication "bubble" showing two replication forks in DNA (orange) from a HeLa cell (TEM).

The discovery that DNA carries the genetic information in cells placed DNA at the center of two fundamental problems that cells must solve. First, cells must be able to accurately reproduce, or *replicate,* their genetic material at each cell division. Without the ability to faithfully duplicate genetic information in this way, the genetic information contained in even the simplest cells would quickly become corrupted with each successive cell division. Consider the human body. The average human has 100 trillion (10^{14}) or so cells, but these are all derived from a single cell, the fertilized egg. With so many cell divisions, if even a small number of errors occurred at each cell division, the results would be catastrophic. The same problem exists for rapidly dividing single cells, such as the bacterium *Escherichia coli*. Without accurate replication, the genetic material of the resulting cells would be riddled with errors, making cell division a risky proposition. The second problem cells must face is how to repair *damage* to their genetic material. Environmental challenges, such as chemical exposure, ultraviolet light, ionizing radiation, or the occasional errors that occur during replication of DNA, must be identified and repaired. Cells capitalize on the complementary base pairing of the two strands of DNA to faithfully replicate and repair their DNA. This chapter discusses these two key processes, as well as the related processes by which genetic material is exchanged (*homologous recombination*). It concludes by examining special mobile genetic elements within the genome known as *transposable elements.*

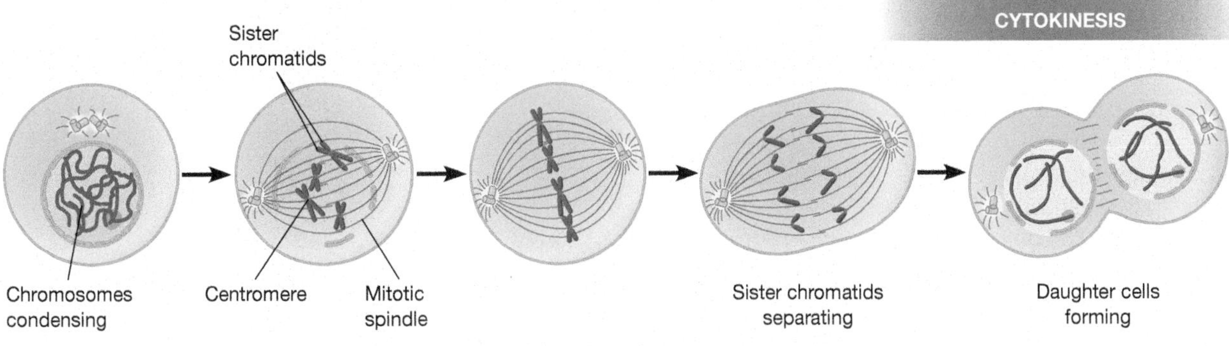

(a) M (mitotic) phase

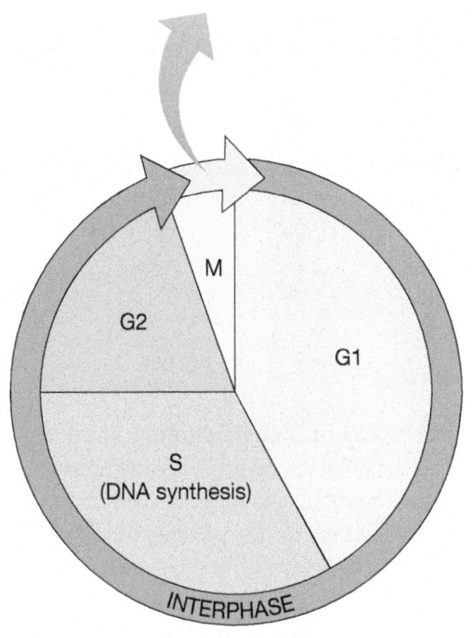

(b) The cell cycle

Figure 17-1 The Eukaryotic Cell Cycle. (a) M (mitotic) phase consists of two overlapping processes, mitosis and cytokinesis. In mitosis, the mitotic spindle segregates the duplicated, condensed chromosomes into two daughter nuclei; in cytokinesis, the cytoplasm divides to yield two genetically identical daughter cells. **(b)** Between divisions, the cell is in interphase, which is made up of S phase (the period of nuclear DNA replication) and two "gap" phases, called G1 and G2. The cell continues to grow throughout interphase, a time of high metabolic activity.

17.1 DNA Replication

When cells grow and divide, the newly formed daughter cells are usually genetic duplicates of the parent cell, containing the same (or virtually the same) DNA sequences. Therefore, all the genetic information in the nucleus of the parent cell must be duplicated and equally divided between the daughter cells during the division process (**Figure 17-1**). The division process involves two overlapping events in which the nucleus divides first and the cytoplasm second. Nuclear division is called **mitosis**, and the division of the cytoplasm to produce two daughter cells is termed **cytokinesis.**

The stars of the mitotic drama are the chromosomes. As you can see in Figure 17-1a, the beginning of mitosis is marked by condensation (coiling and folding) of the cell's chromatin, which generates chromosomes that are thick enough to be individually discernible under the microscope. Because DNA replication has already taken place, each chromosome actually consists of two copies that remain attached to each other until the cell divides. As long as they remain attached, the two new chromosomes are referred to as **sister chromatids.** As the chromatids become visible, the nuclear envelope breaks into fragments. Then, in a process guided by the microtubules of the *mitotic spindle*, the sister chromatids separate and—each now a full-fledged chromosome—move to opposite ends of the cell. By this time, cytokinesis has usually begun, and new nuclear membranes envelop the two sets of daughter chromosomes as cell division is completed.

DNA Synthesis Occurs During S Phase

Although visually striking, the events of mitosis are only one phase of the *cell cycle*; for a typical mammalian cell, this period, known as *M phase* (M is for "mitosis"), usually lasts less than an hour. Cells spend most of their time in between divisions, a period known as **interphase** (Figure 17-1b). During interphase, the amount of nuclear DNA doubles, and experiments using radioactive DNA precursors have shown that the new nuclear DNA is synthesized during a specific portion of interphase named **S phase** (S is for "synthesis"). A time gap called *G1 phase* separates S phase from the preceding M phase; a second gap, *G2 phase*, separates the end of S phase from the onset of the next M phase.

This chapter focuses on the mechanisms of DNA synthesis. How DNA synthesis is coordinated with other events during the cell cycle and how chromosomes are physically segregated into daughter cells during mitosis are covered in Chapter 24.

DNA Replication Is Semiconservative

The underlying mechanisms of DNA replication crucially depend on the double-helical structure of DNA. In fact, a month after Watson and Crick published their now-classic paper postulating a double helix for DNA structure, they followed up with an equally important paper suggesting how such a base-paired structure might duplicate itself. Here, in their own words, is the basis of that suggestion:

> *Now our model for deoxyribonucleic acid is, in effect, a pair of templates, each of which is complementary to the other. We imagine that prior to duplication the hydrogen bonds are broken, and the two chains unwind and separate. Each chain then acts as a template for the formation onto itself of a new companion chain, so that eventually we shall have two pairs of chains, where we only had one before. Moreover, the sequence of the pairs of bases will have been duplicated exactly.* (Watson and Crick, 1953, p. 966.)

The model Watson and Crick proposed for DNA replication is shown in **Figure 17-2**. Their key suggestion was that one of the two strands of every newly formed DNA molecule is derived from the double-stranded DNA (dsDNA) parent molecule, whereas the other strand is newly synthesized. This is called **semiconservative replication** because half of the parent molecule is retained (or conserved) in each daughter molecule. Although the Watson–Crick structure suggested this semiconservative model, at least two other models are conceivable. These are shown in **Figure 17-3**. In the *conservative model*, the parental DNA double helix remains intact, and a second, all-new double-stranded copy is made. In the *dispersive model*, each strand of both daughter molecules contains a mixture of old and newly synthesized segments.

Within five years of its publication, the Watson–Crick model of semiconservative DNA replication was tested and proved correct by Matthew Meselson and Franklin Stahl, in collaboration with Jerome Vinograd. Their ingenious studies utilized two isotopic forms of nitrogen, ^{14}N and ^{15}N, to distinguish newly synthesized strands of DNA from old strands. Bacterial cells were first grown for many generations in a medium containing ^{15}N-labeled ammonium chloride to incorporate this *heavy* (but nonradioactive) isotope of nitrogen into their newly synthesized DNA molecules as the cells divided. Cells containing ^{15}N-labeled DNA were then transferred to a growth medium containing the normal *light* isotope of nitrogen, ^{14}N. Any new strands of DNA synthesized after this transfer would therefore incorporate ^{14}N rather than ^{15}N.

Because ^{15}N-labeled DNA is significantly denser than ^{14}N-labeled DNA, the old and new DNA strands can be distinguished from each other by *equilibrium density centrifugation*, a technique we encountered when discussing the purification of organelles in Chapter 4 (see Key Technique, pages 100–101). For DNA analysis, equilibrium density centrifugation often uses cesium chloride (CsCl), a heavy-metal salt that forms solutions of very high density. The DNA to be analyzed is simply mixed with a solution of cesium chloride and then centrifuged at high speed (generating a centrifugal force of several hundred thousand times gravity for 8 hours, for example). The difference in density between heavy (^{15}N-containing) DNA and light (^{14}N-containing) DNA causes each to come to rest at a different position in the gradient.

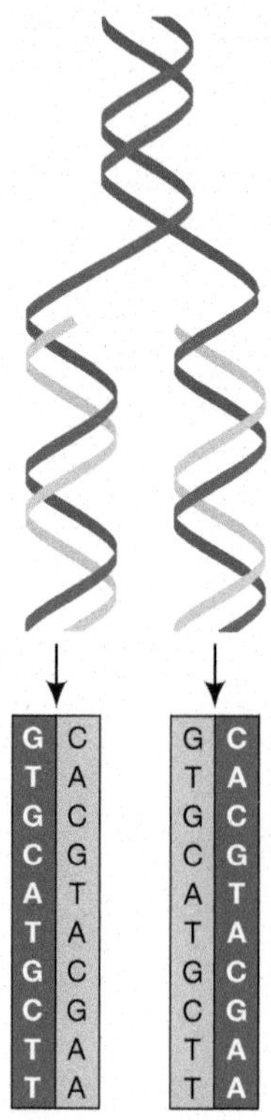

Figure 17-2 Watson–Crick Model of DNA Replication. The double-stranded helix unwinds, and each parent strand serves as a template for the synthesis of a complementary daughter strand, assembled according to base-pairing rules. A, T, C, and G stand for the adenine, thymine, cytosine, and guanine nucleotides. A pairs with T, and G with C.

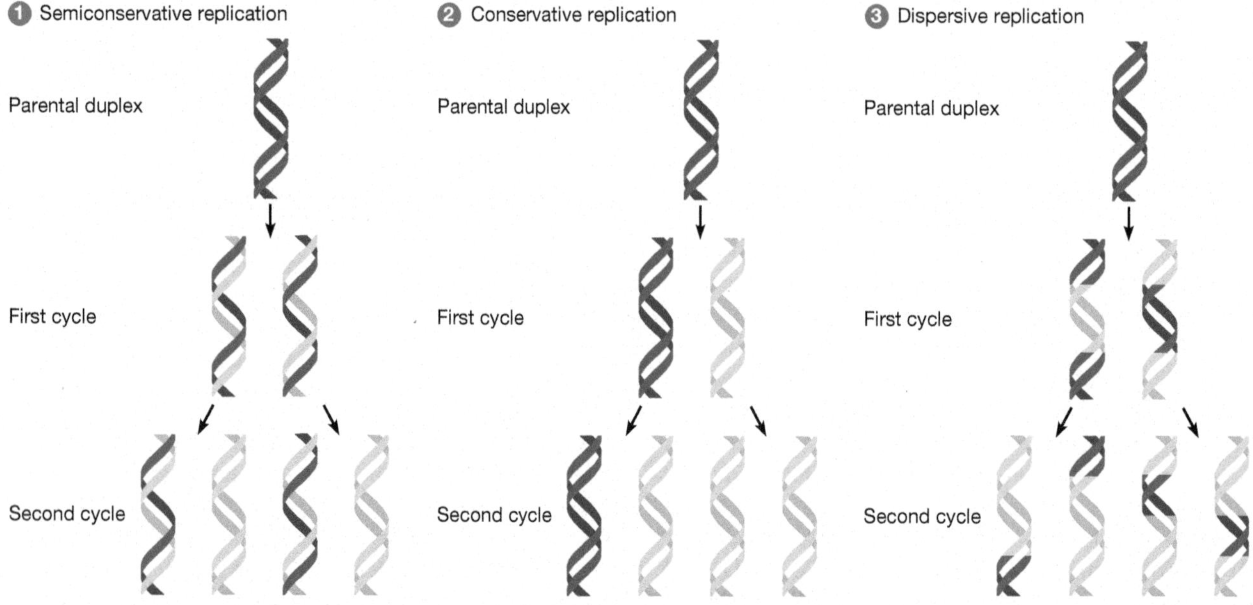

Figure 17-3 Alternative Models for DNA Replication. The results expected for two cycles of DNA replication are shown for each model, which were tested in the experiments of Meselson and Stahl.

Using this approach, Meselson and Stahl analyzed DNA obtained from bacterial cells that were first grown for many generations in ^{15}N and then transferred to ^{14}N for additional cycles of replication (**Figure 17-4**). What does the semiconservative model predict? After one replication cycle in ^{14}N, each DNA molecule should consist of one ^{15}N strand (the old strand) and one ^{14}N strand (the new strand), so the overall density would be intermediate between heavy DNA and light DNA. The experiments clearly supported this model. After one replication cycle in the ^{14}N medium, centrifugation in cesium chloride revealed a single band of DNA whose density was *exactly halfway* between that of ^{15}N-DNA and ^{14}N-DNA (Figure 17-4b). Because they saw no band at the density expected for heavy DNA, Meselson and Stahl concluded that the original double-stranded parental DNA was not preserved intact in the replication process. Similarly, the absence of a band at the density expected for light DNA indicated that no daughter DNA molecules consisted exclusively of new DNA. Instead, it appeared that a part of every daughter DNA molecule was newly synthesized and another part was derived from the parent molecule. In fact, the density halfway between that of ^{14}N-DNA and ^{15}N-DNA meant that the $^{14}N/^{15}N$ hybrid DNA molecules were one-half parental and one-half newly synthesized. Although this finding ruled out the conservative replication model, it did not rule out semiconservative or dispersive replication.

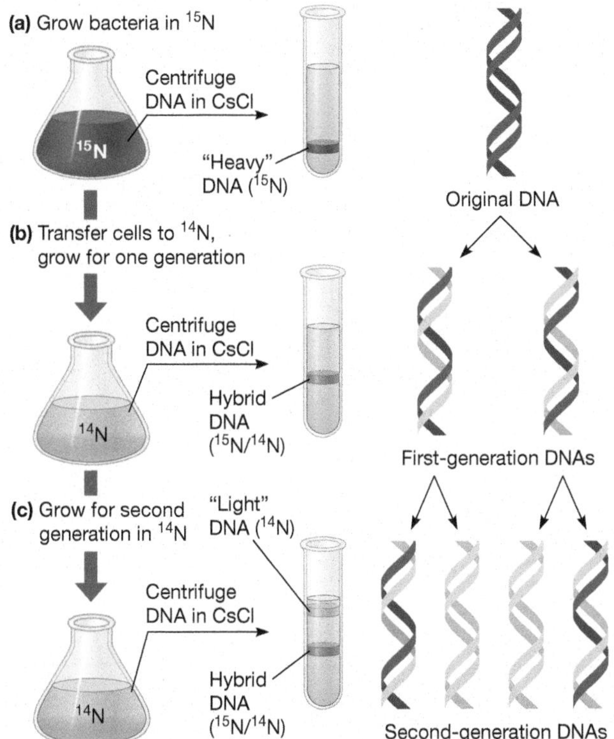

Figure 17-4 Semiconservative Replication of Density-Labeled DNA. Meselson and Stahl **(a)** grew bacteria for many generations on a ^{15}N-containing medium and then transferred the cells to a ^{14}N-containing medium for **(b)** one or **(c)** two further cycles of replication. In each case, DNA was extracted from the cells and centrifuged to equilibrium in cesium chloride (CsCl). As shown in the model on the right (where dark blue strands contain ^{15}N and light blue strands contain ^{14}N), the data are compatible with a semiconservative model of DNA replication.

When the $^{14}N/^{15}N$ hybrid DNA was heated to separate its two strands, one strand exhibited the density of a ^{15}N-containing strand and the other exhibited the density of a ^{14}N-containing strand, just as predicted by the semiconservative model of replication. Data obtained from cells grown for additional generations in the presence of ^{14}N provided further confirmation. After the second cycle of DNA replication, Meselson and Stahl saw two equal bands, one at the hybrid density of the previous cycle and one at the density of purely ^{14}N-DNA (Figure 17-4c). As the figure illustrates, this is only consistent with the semiconservative model of replication.

A similar conclusion emerged from experiments in eukaryotes performed in 1957 (a year before the work of Meselson and Stahl was published) by J. Herbert Taylor, Philip Woods, and Walter Hughes. In these studies, dividing cells from the root tips of the broad bean, *Vicia faba*, were briefly exposed to ^{3}H-thymidine and then returned to a nonradioactive medium containing the microtubule inhibitor colchicine. Colchicine blocks mitosis, so that at each round of DNA replication, duplicated chromosomes (chromatids) fail to separate into daughter cells and remain attached to one another. This allowed Taylor and colleagues to keep track of how many rounds of replication had occurred. To determine which pieces of DNA in this experiment contained radioactive nucleotides, they used a technique known as *autoradiography*. Autoradiography relies on the use of a photographic emulsion that is layered on top of a cellular sample. A chemical reaction in the emulsion involving silver nitrate causes the appearance of black grains, which can be seen in a light microscope, in areas of the sample where radioactivity is present. The result of these experiments is shown in **Figure 17-5** on page 476. After the first replication, each chromatid of each chromosome contained radioactivity. After the next replication, however, only one of each pair of chromatids was radioactive, with the exception of occasional sister chromatid exchanges due to recombination (see Chapter 25). This is precisely the result one would expect if replication occurs in a semiconservative fashion.

DNA Replication Is Usually Bidirectional

The experiments discussed in the previous section provided strong support for the idea that during DNA replication, each strand of the DNA double helix serves as a template for the synthesis of a new complementary strand. As biologists proceeded to unravel the molecular details of this process, it gradually became clear that DNA replication is a complex event involving numerous enzymes and other proteins—and even the participation of RNA. We will first examine the general features of this replication mechanism and then focus on some molecular details. We will frequently refer to the bacterium *E. coli*, for which DNA replication is especially well understood. DNA replication seems to be a drama whose plot and molecular actors are basically similar in bacterial and eukaryotic cells. This is perhaps not surprising for such a fundamental process—one that must have arisen early in the evolution of life. Along the way, we will also discuss a few important differences between bacteria and eukaryotes.

Early work attempted to study the mechanism of circular DNA replication in bacterial cells (**Figure 17-6** on page 476).

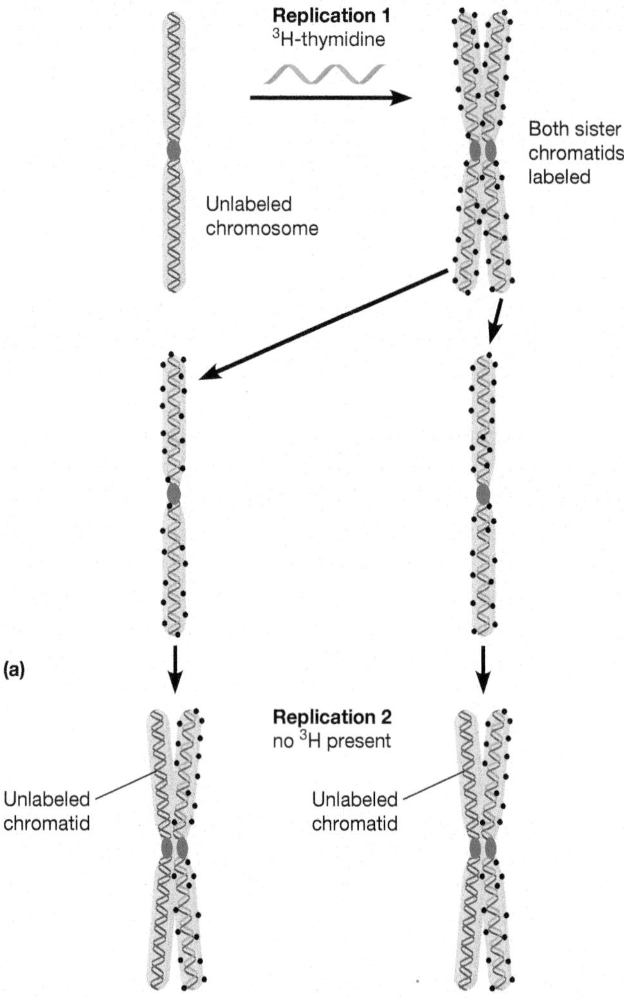

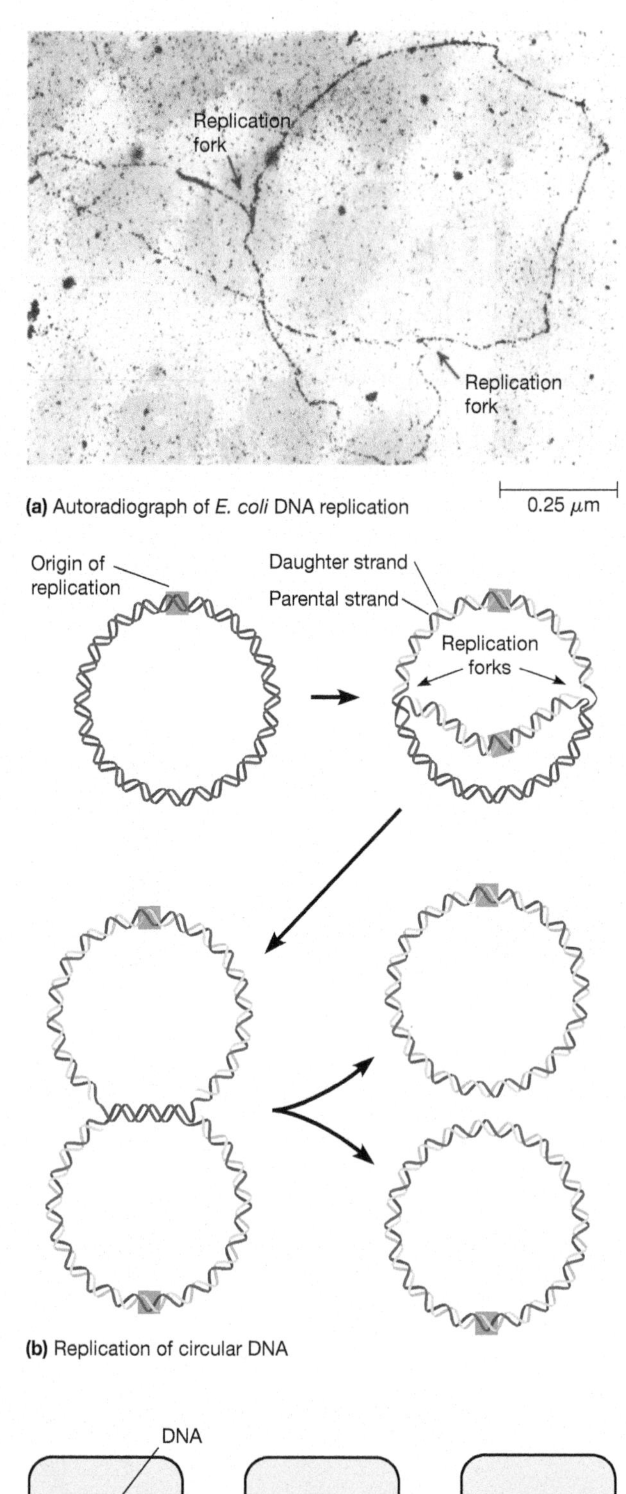

Figure 17-5 Evidence for Semiconservative Replication in Eukaryotes. J. Herbert Taylor and colleagues performed autoradiography to track radioactive DNA (indicated by the black dots) in root tip cells of the broad bean, *Vicia faba*. **(a)** An unlabeled chromosome proceeds through replication in the presence of ^{3}H-thymidine. As chromosomes enters mitosis, both sister chromatids are labeled. **(b)** In cells from part a allowed to proceed to the next mitosis, the labeled sister chromatids have separated and a second round of replication has occurred. Only one chromatid is labeled, consistent with semiconservative replication.

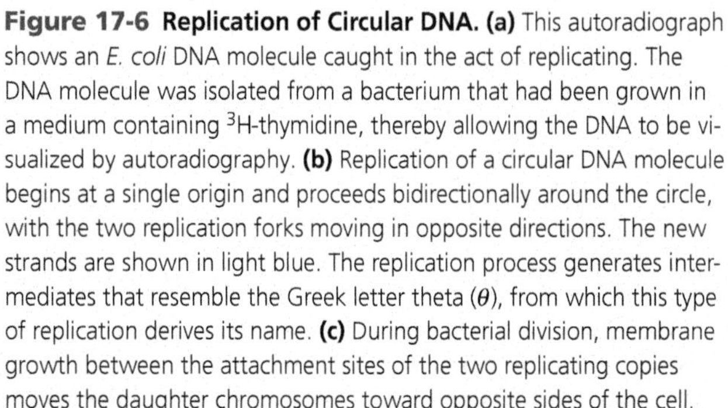

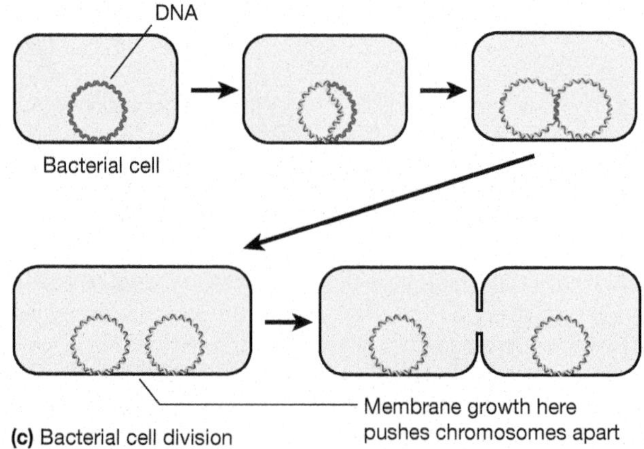

Figure 17-6 Replication of Circular DNA. (a) This autoradiograph shows an *E. coli* DNA molecule caught in the act of replicating. The DNA molecule was isolated from a bacterium that had been grown in a medium containing ^{3}H-thymidine, thereby allowing the DNA to be visualized by autoradiography. **(b)** Replication of a circular DNA molecule begins at a single origin and proceeds bidirectionally around the circle, with the two replication forks moving in opposite directions. The new strands are shown in light blue. The replication process generates intermediates that resemble the Greek letter theta (θ), from which this type of replication derives its name. **(c)** During bacterial division, membrane growth between the attachment sites of the two replicating copies moves the daughter chromosomes toward opposite sides of the cell.

The first experiments to directly visualize this process were carried out by John Cairns, who grew *E. coli* cells in a medium containing the DNA precursor ^{3}H-thymidine and then used autoradiography to examine the cell's single circular DNA molecule caught in the act of replication. One such molecule is shown in Figure 17-6a. The two Y-shaped structures indicated by the arrows represent the sites where the DNA duplex is being replicated. These **replication forks** are created by a DNA replication mechanism that begins at a single point within the DNA and proceeds in a *bidirectional* fashion away from this origin. In other words, two replication forks are created that move in opposite directions away from the point of origin, unwinding the helix and copying both strands as they proceed. For circular DNA, this process is sometimes called *theta replication* because it generates intermediates that look like the Greek letter theta (θ), as you can see in Figure 17-6b. Theta replication occurs not only in bacterial genomes such as that of *E. coli* but also in the circular DNAs of mitochondria, chloroplasts, bacterial plasmids, and some viruses. At the end of a round of theta replication, the two replication forks meet opposite the replication origin. This results in two interlinked dsDNA circles, and the action of a topoisomerase (see Figure 16-11) is required to disconnect the two circles from each other.

Figure 17-6c shows how these events relate to the division of bacteria that contain a single circular chromosome. In such bacteria, the two copies of the replicating chromosome bind to the plasma membrane at their replication origins. As the cell grows in preparation for cell division, new plasma membrane and cell wall components are added to the region between these chromosome attachment sites, thereby pushing the chromosomes toward opposite ends of the cell. When DNA replication is complete and the cell has doubled in size, *binary fission* partitions the cell down the middle and segregates the two chromosomes into two daughter cells.

In contrast to circular bacterial chromosomes—where DNA replication is initiated at a single origin—replication of the linear DNA molecules of eukaryotic chromosomes is initiated at multiple sites, creating multiple replication units called **replicons** (**Figure 17-7**). The DNA of a typical large eukaryotic chromosome may contain several thousand replicons, each about 50,000–300,000 base pairs in length.

After DNA synthesis has been initiated at an origin of replication, two replication forks begin to synthesize DNA in opposite directions away from the origin, creating a *replication bubble* that grows in size as replication proceeds in both directions (Figure 17-7a, b). When the growing replication bubble of one replicon encounters the replication bubble of an adjacent replicon, the DNA synthesized by the two replicons is joined together (Figure 17-7c). In this way, DNA synthesized at numerous replication sites is ultimately linked together to form two double-stranded daughter molecules, each composed of one parental strand and one new strand.

Why does DNA replication involve multiple replication sites in eukaryotes but not in bacteria? Because eukaryotic chromosomes contain significantly more DNA than bacterial chromosomes do, it would take eukaryotes much longer to replicate their chromosomes if DNA synthesis were initiated from only a single replication origin. Moreover, the rate

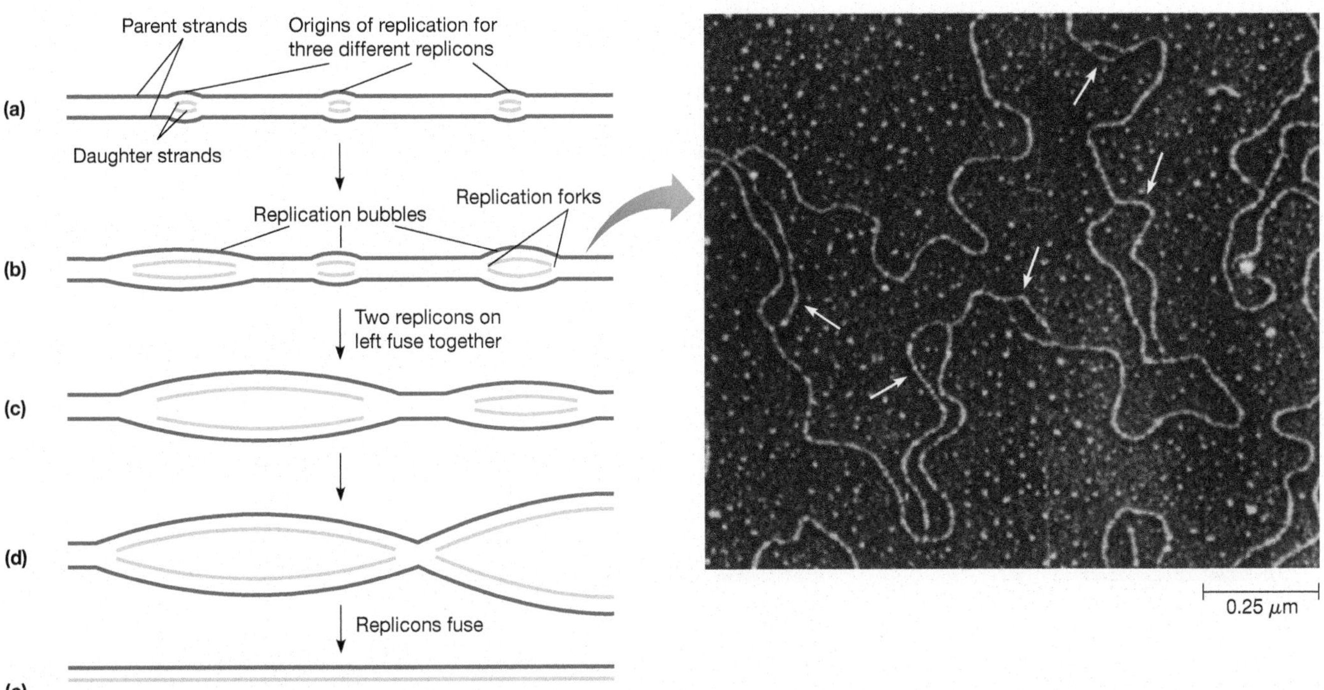

Figure 17-7 Multiple Replicons in Eukaryotic DNA. Replication of linear eukaryotic DNA molecules is initiated at multiple origins along the DNA; the timing of initiation is specific for each cluster of origins. **(a)** Replication bubbles form at origins. **(b)** The bubbles grow as the replication forks move along the DNA in both directions from each origin. The micrograph shows replication bubbles in DNA (arrows) from *Drosophila melanogaster* cells (TEM). **(c)** Eventually, individual bubbles meet and fuse. **(d)** A Y-shaped structure forms as a replication fork reaches the end of a DNA molecule. **(e)** When all bubbles have fused, replication is complete, and the two daughter molecules separate.

at which each replication fork synthesizes DNA is slower in eukaryotes than in bacteria, presumably because the presence of nucleosomes slows down the replication process (see below). Measurements of the length of radioactive DNA synthesized by cells exposed to ^{3}H-thymidine for varying periods of time have revealed that eukaryotic replication forks synthesize DNA at a rate of about 2000 base pairs/minute, compared with 50,000 base pairs/minute in bacteria. The average human chromosome contains about 10^8 base pairs of DNA, so it would take more than a month to duplicate a chromosome if there were only a single replication origin!

Replication Initiates at Specialized DNA Elements

A site where DNA replication initiates in either eukaryotes or bacteria is known as an **origin of replication** (**Figure 17-8**). The best-characterized origin of replication is from *E. coli* and is known as *oriC* (Figure 17-8a). This sequence contains approximately 245 bp of DNA and is rich in AT base pairs. *oriC* also has a characteristic set of repeated sequences lying end to end (tandem repeats): three 13-bp elements ("13-mers") and four 9-bp elements ("9-mers"). These sequences vary in other bacteria but are recognizable and have similar sequences. Such conserved sequences are known as **consensus sequences** because these nucleotides are most often found in DNA that carries out a specific function.

The DNA sequences that act as replication origins exhibit a great deal of variability in eukaryotes. The best understood sequences were first identified in budding yeast cells (*S. cerevisiae*) by isolating chromosomal DNA fragments and inserting them into DNA molecules that lack the ability to replicate. If the inserted DNA fragment gave the DNA molecule the ability to replicate within the yeast cell, it was called an *autonomously replicating sequence*, or *ARS* (Figure 17-8b). The number of ARS elements detected in normal yeast chromosomes is similar to the total number of replicons, suggesting that ARS sequences function as replication origins. The ARS elements of *S. cerevisiae* are 100–150 base pairs in length and contain a common 11-nucleotide core sequence, consisting largely of AT base pairs, flanked by auxiliary sequences containing additional AT-rich regions. The replication origins of multicellular eukaryotes are generally larger and more variable in sequence than the ARS elements of *S. cerevisiae*, but they also tend to contain regions that are AT-rich.

Replication Initiation in Bacteria. Replication initiates when the many proteins that facilitate the unwinding of DNA and its subsequent replication are recruited to origins of replication (**Figure 17-9**). In *E. coli*, three enzymes, known as DnaA, DnaB, and DnaC, bind *oriC* and initiate

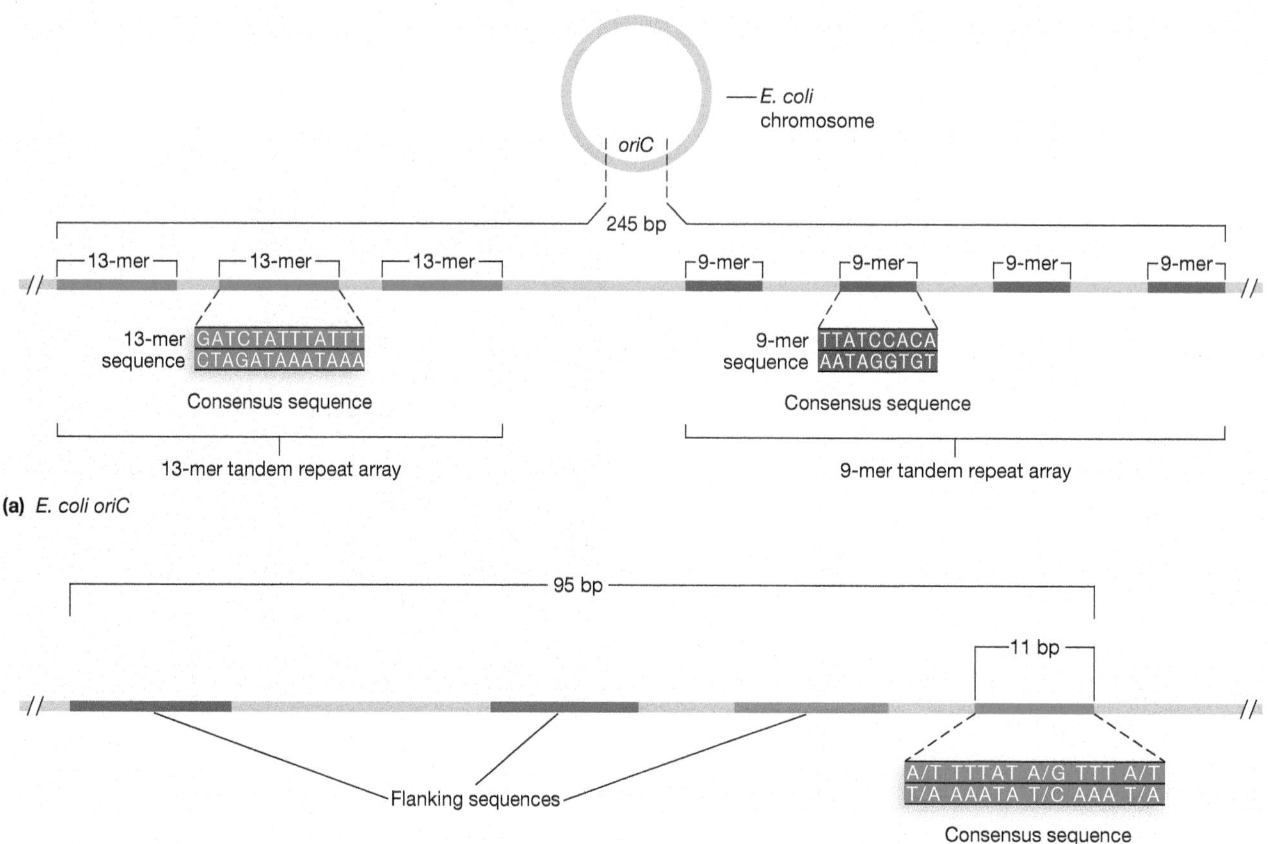

Figure 17-8 Origins of Replication in a Bacterium and a Eukaryote. (a) *oriC* in *E. coli* contains three 13-base-pair (bp) repeats (13-mers) and four 9-bp repeats (9-mers), each in regions that are highly similar in other bacteria (consensus sequences). **(b)** The yeast ARS1 origin of replication spans 95 bp. It contains an 11-bp consensus sequence and additional flanking regions (dark blue). The slash notation (for example, A/T) indicates that each of the two bases is equally likely at that position in the consensus sequences in other yeast origins of replication.

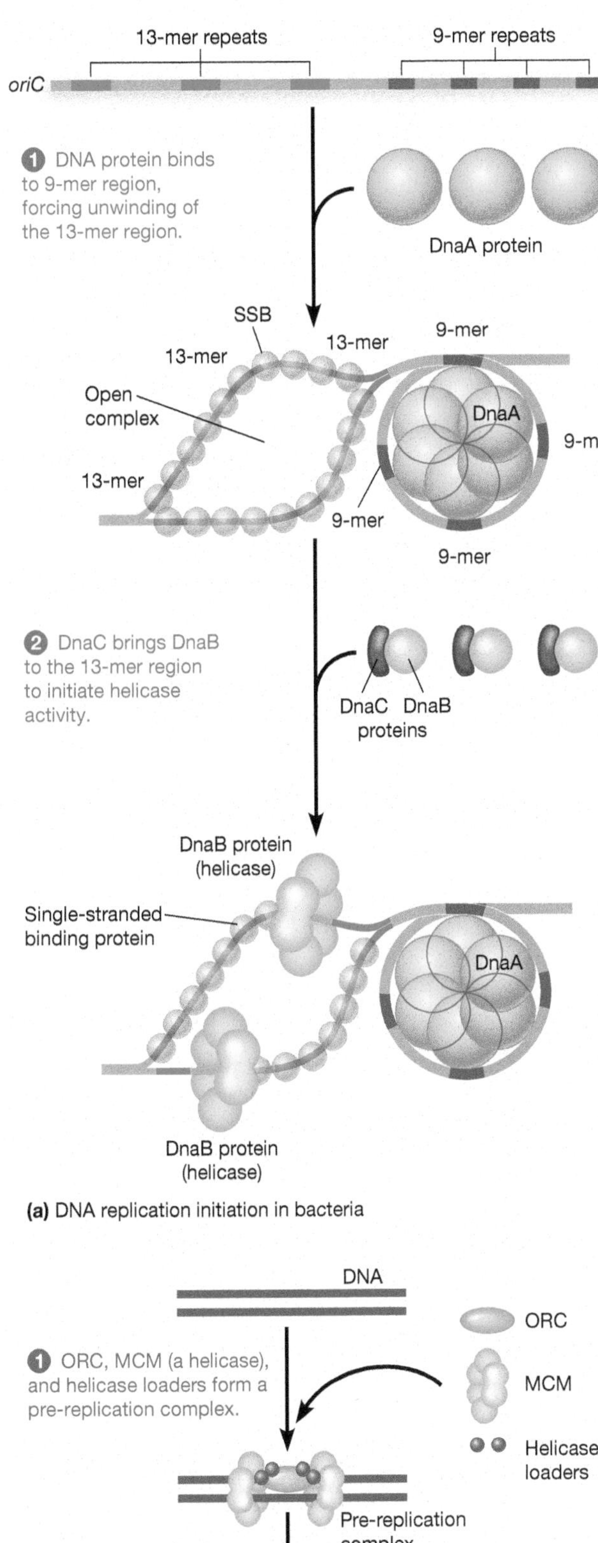

Figure 17-9 Initiation of Replication in Bacteria and Eukaryotes. (a) Replication initiation in bacteria requires DnaA, DnaB, and DnaC. DnaB proteins assemble into helicases. **(b)** In eukaryotes, DNA is made available for replication during the G1 phase of the cell cycle by the binding of MCM proteins (helicases) to replication origins, a licensing event that requires both ORC and helicase loaders.

replication (Figure 17-9a). ❶ DnaA binds to the 9-mer of *oriC*, resulting in the unwinding of DNA at the 13-mer sites in *oriC*. To stabilize the unwound single strands of DNA, **single-stranded DNA-binding proteins (SSBs)** bind to the unwound regions. The opening of the 13-mer regions allows ❷ DnaB, carried by DnaC, to attach and assemble. DnaB is a **DNA helicase.** DNA helicases couple the energy of ATP hydrolysis to the unwinding of the strands of DNA as replication proceeds.

Replication Initiation in Eukaryotes. In eukaryotes, origins of replication also recruit proteins that initiate unwinding and replication of DNA (Figure 17-9b). ❶ A multisubunit protein complex known as the *origin recognition complex (ORC)* binds to a replication origin. The next components to bind are the *minichromosome maintenance (MCM) proteins*, which include several DNA helicases that facilitate DNA replication by unwinding the double helix. The recruitment of MCM proteins to the replication origin requires the participation of a third set of proteins, known as *helicase loaders*, which mediate binding of the MCM proteins to the ORC. At this point, the complete group of DNA-bound proteins is called a **pre-replication complex.** However, replication does not begin until ❷ several more proteins, including the enzymes that catalyze DNA synthesis, are added. The formation of this complex is also called **licensing,** as only "licensed" origins are able to replicate. This ensures that DNA is replicated only once per each cell cycle (control of replication and the rest of the cell cycle are discussed in much more detail in Chapter 24).

During the S phase of a typical eukaryotic cell cycle, replicons are not all activated at the same time. Instead, certain clusters of replicons tend to replicate early during S phase, whereas others replicate later. Information concerning the order in which replicons are activated has been obtained by incubating cells at various points during S phase with *5-bromodeoxyuridine (BrdU)*, a substance that is incorporated into DNA in place of thymidine. Because DNA that contains BrdU is denser than normal DNA, it can be separated from the remainder of the DNA by equilibrium density centrifugation. Studying such DNA has revealed that genes that are actively expressed in a given tissue are replicated early during S phase, whereas inactive genes are replicated later during S phase. If the same gene is analyzed in two different cell types, one in which the gene is active and one in which it is inactive, then early replication is observed only in the cell type where the gene is being transcribed.

DNA Polymerases Catalyze the Elongation of DNA Chains

When the semiconservative model of DNA replication was first proposed in the early 1950s, biologists thought that DNA replication was so complex that it could only be carried out by intact cells. But just a few years later, Arthur Kornberg showed that an enzyme he had isolated from bacteria can copy DNA molecules in a test tube, work for which he received a Nobel Prize in 1959. This enzyme, which he named **DNA polymerase,** requires that a small amount of DNA be initially present to act as a template. Guided by this template, DNA polymerase catalyzes the elongation of DNA chains using as substrates the triphosphate deoxynucleoside derivatives of the four bases found in DNA (dATP, dTTP, dGTP, and dCTP). As each of these substrates is incorporated into a newly forming DNA chain, its two terminal phosphate groups are released. Because deoxynucleoside triphosphates are high-energy compounds (comparable to ATP), the energy released as these phosphate bonds are broken drives what would otherwise be a thermodynamically unfavorable polymerization reaction.

In the DNA polymerase reaction, incoming nucleotides are covalently bonded to the 3′-hydroxyl end of the growing DNA chain. Each successive nucleotide is linked to the growing chain by a phosphoester bond between the phosphate group on its 5′ carbon and the hydroxyl group on the 3′ carbon of the nucleotide added in the previous step (**Figure 17-10**). In other words, chain elongation occurs at the 3′ end of a DNA strand, and the new strand is therefore said to grow in the 5′ → 3′ direction.

Soon after Kornberg's initial discovery, several other forms of DNA polymerase were detected. (**Table 17-1** on page 482 lists the main DNA polymerases used in DNA replication, along with other key proteins involved in the process.) In *E. coli*, the enzyme discovered by Kornberg turned out not to be the primary DNA polymerase. This fact first became apparent when it was discovered that mutant strains of bacteria lacking the Kornberg enzyme can still replicate their DNA and reproduce normally. Eventually, several other bacterial enzymes that synthesize DNA were identified. These additional enzymes are named using Roman numerals (for example, DNA polymerases II, III, IV, and V) to distinguish them from the original Kornberg enzyme, now called *DNA polymerase I*. When the rates at which the various DNA polymerases synthesize DNA in a test tube were first compared, only *DNA polymerase III* was found to work fast enough to account for the rate of DNA replication in intact cells, which averages about 50,000 base pairs/minute in bacteria.

Such observations suggested that DNA polymerase III is the main enzyme responsible for DNA replication in bacterial cells, but the evidence would be more convincing if it could be shown that cells lacking DNA polymerase III are unable to replicate their DNA. One powerful approach to do so involves the use of *temperature-sensitive mutants*, which are bacteria that produce proteins that function properly at normal temperatures but become seriously impaired when the temperature is altered slightly. For example, mutant bacteria have been isolated in which DNA polymerase III behaves normally at 37°C but loses its function when the temperature is raised to 42°C. Such bacteria grow normally at 37°C but lose the ability to replicate their DNA when the temperature is elevated to 42°C, indicating that DNA polymerase III plays an essential role in the process of normal DNA replication.

Although the preceding observations indicate that DNA polymerase III is central to bacterial DNA replication, it is not the only enzyme involved. As we will see shortly, a variety of other proteins are required for DNA replication, including DNA polymerase I. The other main types of bacterial DNA polymerase (II, IV, and V) play more specialized roles in events associated with *DNA repair*, a process described in detail later in the chapter.

Like bacteria, eukaryotic cells contain several types of DNA polymerases. More than a dozen different enzymes have been identified thus far, each named with a different Greek letter. Of this group, DNA polymerases α (alpha), δ (delta), and ε (epsilon) are involved in nuclear DNA replication. DNA polymerase γ (gamma) is present only in mitochondria and is the main polymerase used in mitochondrial DNA replication. Most of the remaining eukaryotic DNA polymerases are involved either in DNA repair or in replication across regions of DNA damage.

In addition to their biological functions inside cells, DNA polymerases have found important practical applications in the field of biotechnology. In a technique called the polymerase chain reaction (PCR), a special type of DNA polymerase is the prime tool (see Key Technique in Chapter 21, pages 624–625). Used experimentally for the rapid amplification of tiny samples of DNA, PCR is a powerful adjunct to the process of DNA fingerprinting (see Human Connections in Chapter 21, page 642).

MAKE CONNECTIONS 17.1

The formation of a phosphodiester bond in DNA replication is an endergonic reaction, with a positive ΔG_0 of approximately +5.3 kcal/mol. What drives this reaction forward? (Ch. 5.2)

DNA Is Synthesized as Discontinuous Segments That Are Joined Together by DNA Ligase

The discovery of DNA polymerase was just the first step in unraveling the mechanism of DNA replication. An early conceptual problem arose from the finding that DNA polymerases catalyze the addition of nucleotides *only to the 3′ end of an existing DNA chain*; in other words, DNA polymerases synthesize DNA exclusively in the 5′ → 3′ direction. Yet the two strands of the DNA double helix run in opposite directions. How does an enzyme that functions solely in the 5′ → 3′ direction manage to synthesize both DNA strands at a moving replication fork when one chain runs in the 5′ → 3′ direction and the other chain runs in the 3′ → 5′ direction?

An answer to this question was first proposed in 1968 by Reiji Okazaki, whose studies suggested that DNA is synthesized as small fragments that are later joined together. Okazaki isolated DNA from bacterial cells that had been briefly exposed to a radioactive substrate that is incorporated into newly made DNA. Analysis of this DNA revealed that much of the radioactivity was located in small DNA fragments measuring about 1000 nucleotides in length. After longer labeling periods, the radioactivity became associated with larger DNA molecules. These findings suggested to Okazaki that the smaller DNA pieces, now known as **Okazaki fragments**, are precursors of newly forming larger DNA molecules. Later research revealed that the conversion of Okazaki fragments into larger DNA molecules fails to take place in mutant bacteria that lack the enzyme **DNA ligase**, which joins DNA fragments together by catalyzing the ATP-dependent formation of a phosphodiester bond between the 3′ end of one nucleotide chain and the 5′ end of another.

The preceding observations suggest a model of DNA replication that is consistent with the fact that DNA polymerase synthesizes DNA only in the 5′ → 3′ direction. According to this model, DNA synthesis at each replication fork is *continuous* in

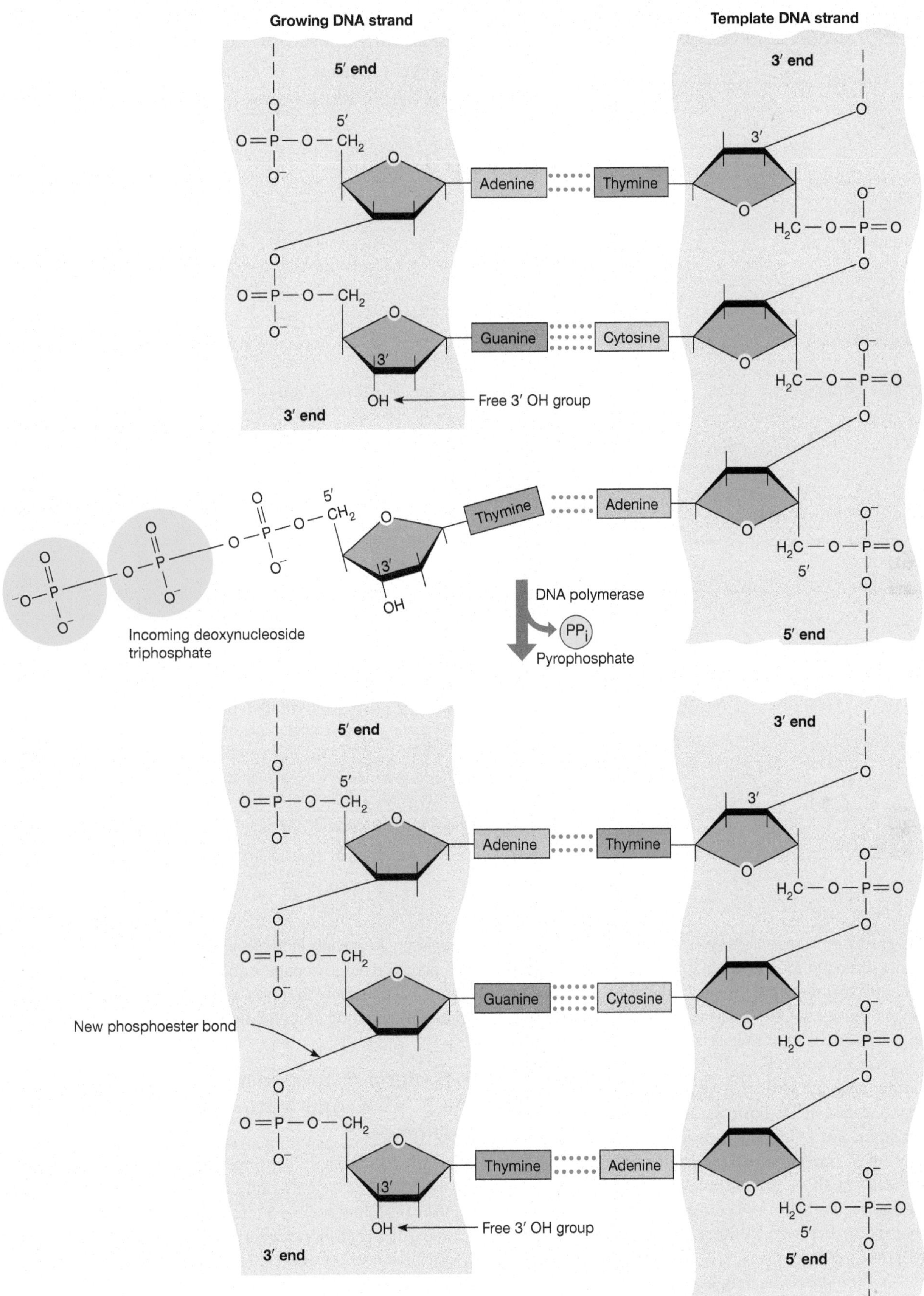

Figure 17-10 The Directionality of DNA Synthesis. Addition of the next nucleotide to a growing DNA strand is catalyzed by DNA polymerase and always occurs at the 3′ end of the strand. A phosphoester bond is formed between the 3′-hydroxyl group of the terminal nucleotide and the 5′ phosphate of the incoming deoxynucleoside triphosphate, extending the growing chain by one nucleotide, liberating pyrophosphate (PP$_i$) and leaving the 3′ end of the strand with a free hydroxyl group to accept the next nucleotide.

Table 17-1 Some Important DNA Replication Proteins in Bacteria and Eukaryotes

Protein	Cell Type	Main Activities and/or Functions
Initiator proteins	Both	Bind to origin of replication and initiate unwinding of DNA double helix
DNA polymerase I	Bacteria	DNA synthesis; $3' \rightarrow 5'$ exonuclease (for proofreading); $5' \rightarrow 3'$ exonuclease; removes and replaces RNA primers used in DNA replication (also functions in excision repair of damaged DNA)
DNA polymerase III	Bacteria	DNA synthesis; $3' \rightarrow 5'$ exonuclease (for proofreading); used in synthesis of both DNA strands
DNA polymerase α (alpha)	Eukaryotes	Nuclear DNA synthesis; forms complex with primase and begins DNA synthesis at the 3′ end of RNA primers for both leading and lagging strands (also functions in DNA repair)
DNA polymerase γ (gamma)	Eukaryotes	Mitochondrial DNA synthesis
DNA polymerase δ (delta)	Eukaryotes	Nuclear DNA synthesis; $3' \rightarrow 5'$ exonuclease (for proofreading); involved in lagging and leading strand synthesis (also functions in DNA repair)
DNA polymerase ε (epsilon)	Eukaryotes	Nuclear DNA synthesis; $3' \rightarrow 5'$ exonuclease (for proofreading); thought to be involved in leading and lagging strand synthesis (also functions in DNA repair)
Primase	Both	RNA synthesis; makes RNA oligonucleotides that are used as primers for DNA synthesis
DNA helicase	Both	Unwinds double-stranded DNA
Sliding clamp (PCNA in eukaryotes)	Both	Binds core polymerase subunit and keeps it on DNA
Single-stranded DNA-binding protein (SSBs in bacteria)	Both	Binds to single-stranded DNA; stabilizes strands of unwound DNA in an extended configuration that facilitates access by other proteins
DNA topoisomerase (type I and type II)	Both	Makes single-strand cuts (type I) or double-strand cuts (type II) in DNA; induces and/or relaxes DNA supercoiling; can serve as a swivel to prevent overwinding ahead of the DNA replication fork; can separate linked DNA circles at the end of DNA replication
DNA gyrase	Bacteria	Type II DNA topoisomerase that serves as a swivel to relax supercoiling ahead of the DNA replication fork in *E. coli*
DNA ligase	Both	Makes covalent bonds to join together adjacent DNA strands, including the Okazaki fragments in lagging strand DNA synthesis and the new and old DNA segments in excision repair of DNA
Telomerase	Eukaryotes	Using an integral RNA molecule as template, synthesizes DNA for extension of telomeres (sequences at ends of chromosomal DNA)
RNA endonuclease (RNase H), RNA exonuclease (FEN1)	Eukaryotes	RNase H recognizes RNA/DNA strands and nicks them; FEN1 then digests the RNA

the direction of fork movement for one strand but *discontinuous* in the opposite direction for the other strand (**Figure 17-11**). The two daughter strands can therefore be distinguished based on their mode of growth. One of the two new strands, called the **leading strand,** is synthesized as a continuous chain because it is growing in the $5' \rightarrow 3'$ direction. Due to the opposite orientation of the two DNA strands, the other newly forming strand, called the **lagging strand,** must grow in the $3' \rightarrow 5'$ direction. But DNA polymerase cannot add nucleotides in the $3' \rightarrow 5'$ direction, so the lagging strand is instead formed as a series of short, discontinuous Okazaki fragments that are synthesized in the $3' \rightarrow 5'$ direction. These fragments are then joined together by DNA ligase to make a continuous new $3' \rightarrow 5'$ DNA strand. Okazaki fragments are generally about 1000–2000 nucleotides long in viral and bacterial systems but only about one-tenth this length in eukaryotic cells. In *E. coli*, the same DNA polymerase, polymerase III, is used for synthesizing both the Okazaki fragments of the lagging strand and the continuous DNA chain of the leading strand. In eukaryotes, DNA polymerases δ and ε are both thought to be involved in synthesizing leading and lagging strands. Recall that replication is typically bidirectional in bacteria and eukaryotes, as replication forks propagate away from origins of replication. Figure 17-11b shows a more "zoomed-out" view of such bidirectional spreading of replication forks.

In Bacteria, Proofreading Is Performed by the 3′→5′ Exonuclease Activity of DNA Polymerase

Given the complexity of the preceding model, you might wonder why cells have not simply evolved an enzyme that synthesizes DNA in the $3' \rightarrow 5'$ direction. One possible answer is related to the need for error correction during DNA replication. About 1 out of every 100,000 nucleotides incorporated during DNA replication is incorrectly base-paired with the template DNA strand, an error rate that would yield more than 120,000 errors every time a human cell replicates its DNA. Fortunately, such mistakes are usually fixed by a **proofreading** mechanism that uses the same DNA polymerase molecules that catalyze DNA synthesis. Proofreading is made possible by the fact that almost all DNA polymerases possess $3' \rightarrow 5'$ exonuclease

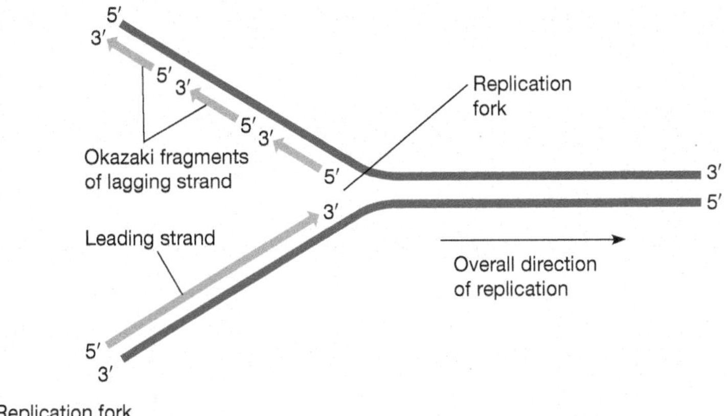

(a) Replication fork

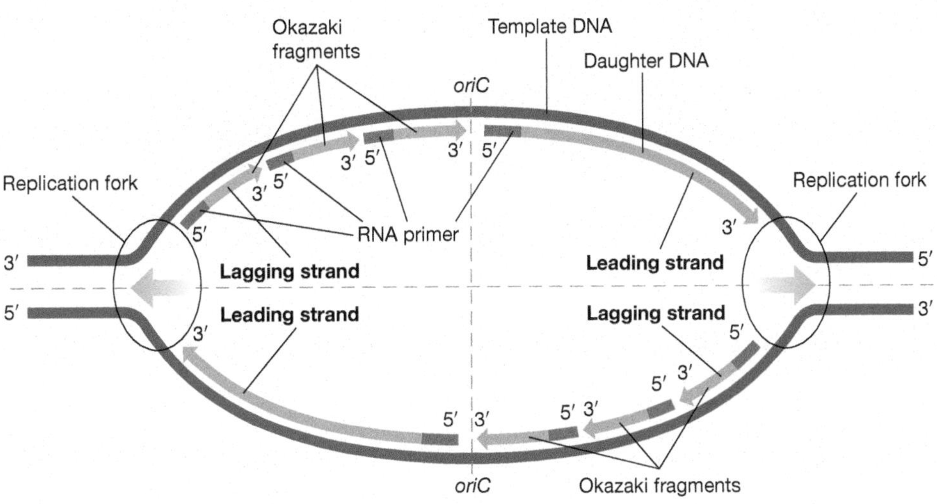

(b) Replication bubble

Figure 17-11 Directions of DNA Synthesis at a Replication Fork. (a) Replication fork structure. Parental DNA is shown in dark blue and newly synthesized DNA in lighter blue. Here and in subsequent figures, arrows indicate the direction in which the nucleic acid chain is being elongated. Because DNA polymerases synthesize DNA chains only in the 5′ → 3′ direction, synthesis at each replication fork is continuous in the direction of fork movement for the leading strand but discontinuous in the opposite direction for the lagging strand. Discontinuous synthesis involves short intermediates called Okazaki fragments, which are later joined together by the enzyme DNA ligase. **(b)** Relation of replication forks to replication bubble progression. Bidirectional expansion is driven by DNA synthesis at each replication fork.

activity (in addition to their ability to catalyze DNA synthesis). **Exonucleases** are enzymes that degrade nucleic acids (usually DNA) from one end, rather than making internal cuts, as **endonucleases** do. A 3′ → 5′ exonuclease is one that clips off nucleotides from the 3′ end of a nucleotide chain (**Figure 17-12** on page 484). Hence, the 3′ → 5′ exonuclease activity of DNA polymerase allows it to remove improperly base-paired nucleotides from the 3′ end of a growing DNA chain (Figure 17-12a). This ability to remove incorrect nucleotides improves the fidelity of DNA replication to an average of only a few errors for every billion base pairs replicated.

DNA polymerases like DNA polymerases I and III are shaped somewhat like a hand. When a replication error occurs, the resulting mismatched bases are unable to establish correct hydrogen bonding. This results in the displacement of the new daughter strand from the active site where polymerization is catalyzed to a second site that carries out exonuclease activity (Figure 17-12b).

If cells did happen to possess an enzyme capable of synthesizing DNA in the 3′ → 5′ direction, proofreading could not work because a DNA chain growing in the 3′ → 5′ direction would have a nucleotide triphosphate at its growing 5′ end. What if this 5′ nucleotide contained an incorrect base that needed to be removed during proofreading? Removing this 5′ nucleotide would eliminate the triphosphate group that provides the free energy that allows DNA polymerase to add nucleotides to a growing DNA chain, so the chain could not elongate further.

RNA Primers Initiate DNA Replication

Because DNA polymerase can only add nucleotides to an existing nucleotide chain, how is replication of a DNA double helix initiated? Shortly after Okazaki fragments were first discovered, researchers implicated RNA in the initiation process through the following observations: (1) Okazaki fragments often have short stretches of RNA, usually three to ten nucleotides in length, at their 5′ ends; (2) DNA polymerase can catalyze the addition of nucleotides to the 3′ end of RNA chains as well as to DNA chains; (3) cells contain an enzyme called **primase** that synthesizes RNA fragments about ten bases long using DNA as a template; and (4) unlike DNA polymerase, which adds nucleotides only to the ends of existing chains, primase can initiate RNA synthesis from scratch by joining two nucleotides together.

These observations led to the conclusion that DNA synthesis is initiated by the formation of short **RNA primers** (**Figure 17-13** on page 484). ❶ RNA primers are synthesized by primase, which uses a single DNA strand as a template to guide the synthesis of a complementary stretch of RNA.

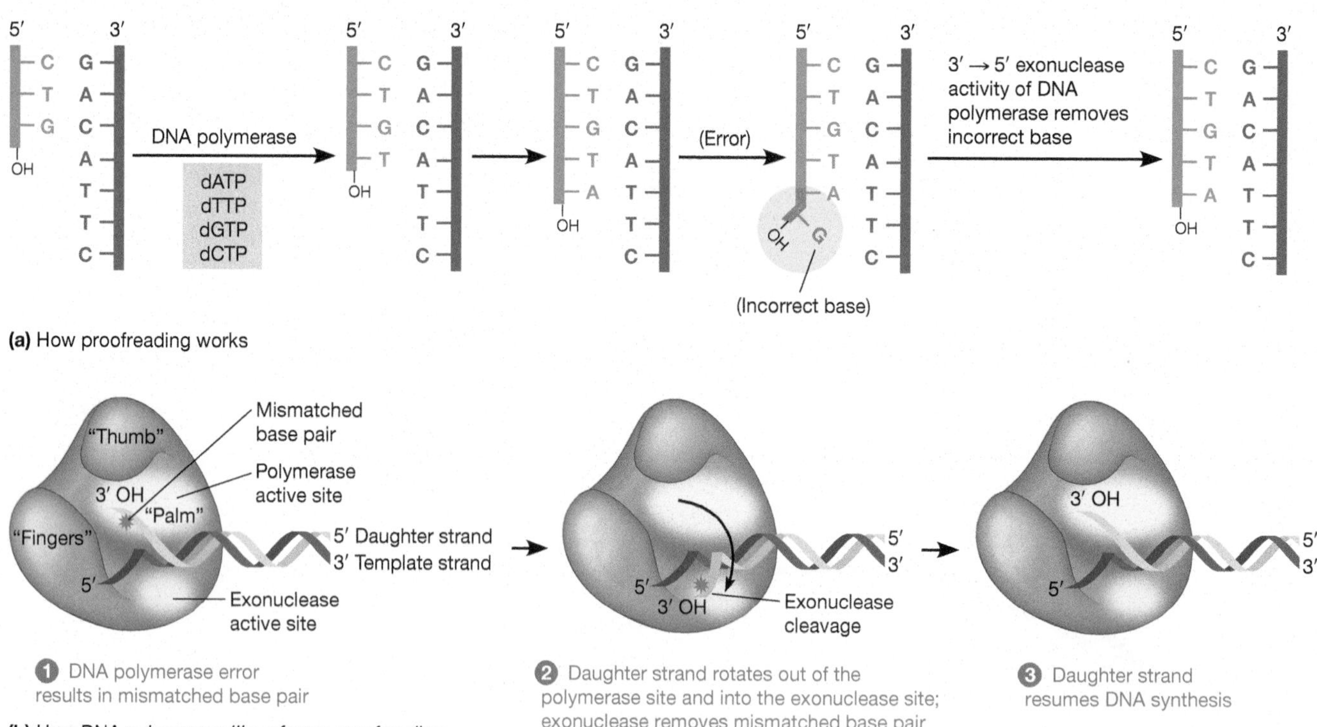
5′
3′
C
T
G
OH
G
A
C
A
T
T
C
DNA polymerase
dATP
dTTP
dGTP
dCTP
(Error)
(Incorrect base)
3′ → 5′ exonuclease activity of DNA polymerase removes incorrect base
(a) How proofreading works
"Thumb"
Mismatched base pair
Polymerase active site
3′ OH
"Palm"
"Fingers"
5′ Daughter strand
3′ Template strand
Exonuclease active site
Exonuclease cleavage
1 DNA polymerase error results in mismatched base pair
2 Daughter strand rotates out of the polymerase site and into the exonuclease site; exonuclease removes mismatched base pair
3 Daughter strand resumes DNA synthesis
(b) How DNA polymerase III performs proofreading

Figure 17-12 Proofreading by 3′ → 5′ Exonuclease. (a) If an incorrect base is inserted during DNA replication, the 3′ → 5′ exonuclease activity that is part of the DNA polymerase molecule catalyzes its removal so that the correct base can be inserted. **(b)** When DNA polymerase makes an error during replication, ① the mismatch is detected, ② causing rotation of the daughter strand out of the polymerase active site and into the exonuclease site. After removal of the incorrect base, ③ the daughter strand continues to be synthesized.

Primase is a specific kind of RNA polymerase used only in DNA replication. Like other RNA polymerases, but unlike DNA polymerases, primases can *initiate* the synthesis of a new polynucleotide strand complementary to a template strand; they do not themselves require a primer.

In *E. coli*, primase is relatively inactive unless it is accompanied by six other proteins, forming a complex called a **primosome.** The other primosome proteins function in unwinding the parental DNA and recognizing target DNA sequences where replication is to be initiated. The situation in eukaryotic cells is slightly different, so the term *primosome* is not used. The eukaryotic primase is not as closely associated with unwinding proteins, but it is very tightly bound to DNA polymerase α, the main DNA polymerase involved in initiating DNA replication.

Once an RNA primer has been created, ② DNA synthesis can proceed, with DNA polymerase III (or DNA polymerase α followed by polymerase δ or ε in eukaryotes) adding successive deoxynucleotides beginning at the hydroxyl group on the 3′ end of the primer. For the leading strand, initiation using an RNA primer needs to occur only once, when a replication fork first forms; DNA polymerase can then add nucleotides to the chain continuously in the 5′ → 3′ direction. In contrast,

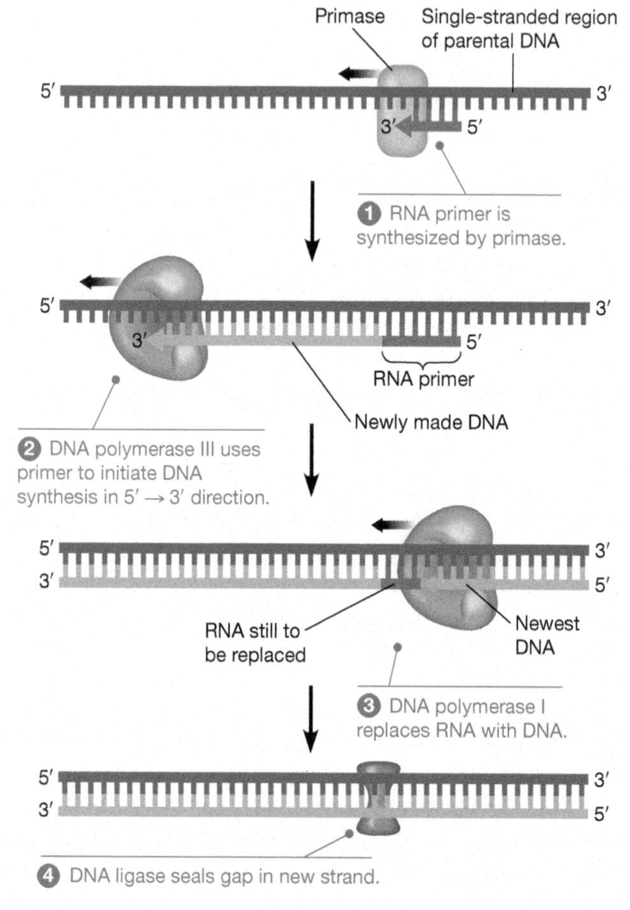
Primase
Single-stranded region of parental DNA
5′
3′
1 RNA primer is synthesized by primase.
RNA primer
Newly made DNA
2 DNA polymerase III uses primer to initiate DNA synthesis in 5′ → 3′ direction.
RNA still to be replaced
Newest DNA
3 DNA polymerase I replaces RNA with DNA.
4 DNA ligase seals gap in new strand.

Figure 17-13 The Role of RNA Primers in DNA Replication. DNA synthesis is initiated with a short RNA primer in both bacteria and eukaryotes. This figure shows the process in *E. coli*.

the lagging strand is synthesized as a series of discontinuous Okazaki fragments, and each of them must be initiated with a separate RNA primer. For each primer, DNA nucleotides are added by DNA polymerase III until the growing fragment reaches the adjacent Okazaki fragment. No longer needed at that point, ③ the RNA segment is removed, and DNA nucleotides are polymerized to fill its place. In *E. coli*, the RNA primers are removed by a 5′ → 3′ exonuclease activity inherent to the DNA polymerase I (distinct from the 3′ → 5′ exonuclease activity involved in proofreading). At the same time, the DNA polymerase I synthesizes DNA in the normal 5′ → 3′ direction to fill in the resulting gaps. ④ Adjacent fragments are subsequently joined together by DNA ligase.

Why do cells employ RNA primers that must later be removed rather than simply using a DNA primer in the first place? Again, the answer may be related to the need for error correction. We have already seen that DNA polymerase possesses a 3′ → 5′ exonuclease activity that allows it to remove incorrect nucleotides from the 3′ end of a DNA chain. In fact, DNA polymerase will elongate an existing DNA chain only if the nucleotide present at the 3′ end is properly base-paired. But an enzyme that *initiates* the synthesis of a new chain cannot perform such a proofreading function because it is not adding a nucleotide to an existing base-paired end. As a result, enzymes that initiate nucleic acid synthesis are not very good at correcting errors. By using RNA rather than DNA to initiate DNA synthesis, cells ensure that any incorrect bases inserted during initiation are restricted to RNA sequences destined to be removed by DNA polymerase I.

The DNA Double Helix Must Be Locally Unwound During Replication

During DNA replication, the two strands of the double helix must unwind at each replication fork to expose the single strands to the enzymes responsible for copying them. Three classes of proteins with distinct functions facilitate this unwinding process: *DNA helicases, topoisomerases,* and *single-stranded DNA-binding proteins* (**Figure 17-14**).

The proteins responsible for unwinding DNA are the **DNA helicases.** Using energy derived from ATP hydrolysis, these proteins unwind the DNA double helix in advance of the replication fork, breaking the hydrogen bonds as they go. In *E. coli*, at least two different DNA helicases are involved in DNA replication; one attaches to the lagging strand template and moves in a 5′ → 3′ direction, and the other attaches to the leading strand template and moves 3′ → 5′. Both are part of the primosome, but the 5′ → 3′ helicase is more important for unwinding DNA at the replication fork.

Once strand separation has begun, SSBs quickly attach to the exposed single strands to keep the DNA unwound and therefore accessible to the DNA replication machinery. After a particular segment of DNA has been replicated, the SSB molecules fall off and are recycled, attaching to the next single-stranded segment.

The unwinding associated with DNA replication would create an intolerable amount of supercoiling and possibly tangling in the rest of the DNA were it not for the actions of **topoisomerases** (see Chapter 16). These enzymes create swivel points in the DNA molecule by making and then quickly resealing breaks in the double helix (Figure 17-14b).

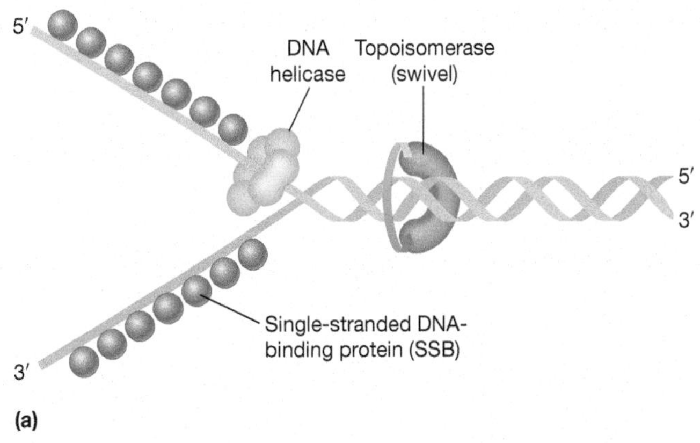

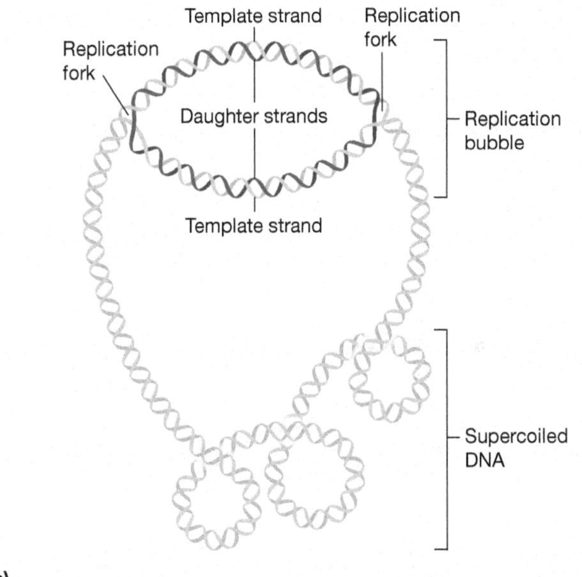

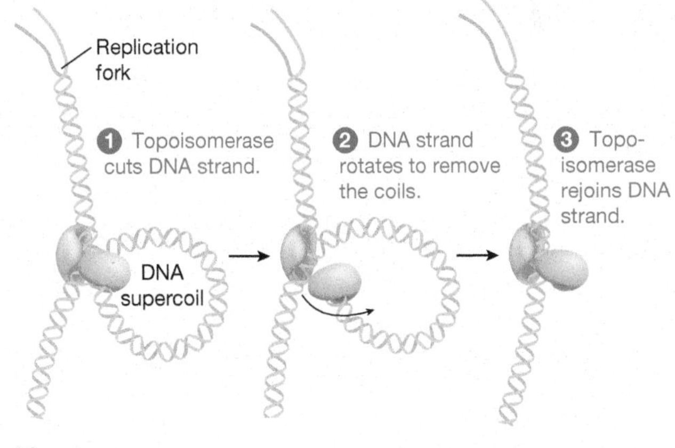

Figure 17-14 Proteins Involved in Unwinding DNA at the Replication Fork. (a) Three types of proteins are involved in DNA unwinding. The actual unwinding proteins are the DNA helicases; the principal one in *E. coli*, which is part of the primosome, operates 5′→3′ along the template for the lagging strand, as shown here. Single-stranded DNA-binding proteins (SSBs) stabilize the unwound DNA in an extended position. A topoisomerase forms a swivel ahead of the replication fork; in *E. coli*, this topoisomerase is DNA gyrase. **(b)** The importance of topoisomerases during replication. As replication forks spread, supercoiled DNA piles up ahead of each fork. **(c)** Topoisomerases relieve supercoiling during replication.

Topoisomerases are classified based on whether the generated break is single- (type I) or double-strand (type II). Of the ten or so topoisomerases found in *E. coli*, the key enzyme for DNA replication is *DNA gyrase*, a type II topoisomerase. Using energy derived from ATP, DNA gyrase introduces negative supercoils and thereby relaxes positive ones. DNA gyrase serves as the main swivel that prevents overwinding (positive supercoiling) of the DNA ahead of the replication fork (Figure 17-14c). In addition, this enzyme has a role in both initiating and completing DNA replication in *E. coli*—in opening up the double helix at the origin of replication and in separating the linked circles of daughter DNA at the end. Similar topoisomerases of both types have been isolated in eukaryotes.

DNA Unwinding and DNA Synthesis Are Coordinated on Both Strands via the Replisome

Figure 17-15 reviews the highlights of what we currently understand about the mechanics of DNA replication in *E. coli*. Starting at the origin of replication, the machinery at the replication fork sequentially adds the different proteins required for synthesizing DNA—that is, DNA helicase, DNA gyrase, SSB, primase, DNA polymerase, and DNA ligase. Figure 17-15 shows many of the key players in replication, but it may give the impression that DNA polymerases on the leading and lagging strands of replication forks chug along strands of DNA rather like railroad locomotives. This is an incomplete picture, however. The various proteins involved in DNA replication are all closely associated in one large complex, called a **replisome**, that is about the size of a ribosome. The activity and movement of the replisome are powered by the hydrolysis of nucleoside triphosphates. These include both nucleoside triphosphates (used by DNA polymerases and primase as building blocks for DNA and RNA synthesis) and the ATP hydrolyzed by several other DNA replication proteins (including DNA helicase, DNA gyrase, and DNA ligase). As the replisome moves along the DNA in the direction of the replication fork, it must accommodate the fact that DNA is being synthesized in opposite directions along the template on the two strands.

Figure 17-16 on page 488 provides a schematic snapshot illustrating how this might be accomplished in the case of bacterial DNA polymerase III. A key element of the replisome is the folding of the lagging strand template into a loop and the connection of the catalytic subunits of the DNA polymerase working along each strand. This model for how the replisome works is often called the *trombone model* because the pushing out of the lagging strand loop looks rather like the slide of a trombone. Although the details of how this complicated, looping structure is maintained are still being worked out, the function of this structure is clear: it allows leading and lagging strand synthesis to occur in a coordinated fashion, even though the two template strands are oriented with opposite polarity.

One look at Figure 17-16 is probably enough to convince you that the movement of the replisome along DNA is a complicated process! There are several features of this process that are worth pointing out. First, the replisome relies on the assembly of a complete multisubunit protein complex known as a *holoenzyme* (*holo-* means "whole" or "complete"). In the case of bacterial DNA polymerase III, the polymerase holoenzyme appears to include three catalytic subunits that carry out DNA polymerization, linked together by *τ (tau) protein*. Another key component is a *clamp loader* protein, which contains five subunits. The clamp loader, as the name implies, in turn feeds a ring-shaped **sliding clamp** protein onto the DNA. The sliding clamp forms a "doughnut," whose hole is approximately 35 Å in diameter, fitting snugly around the DNA. The sliding clamp, which attaches to a DNA polymerase catalytic subunit, allows the catalytic subunit to remain on DNA for prolonged periods of time. As a result, the clamp is often called a *processivity factor*; it allows the polymerase to "process" along the DNA longer without falling off.

A second feature of the replisome is that the leading and lagging strands differ regarding how long the sliding clamp and its associated catalytic subunit stay attached. Once loaded onto the leading strand DNA at the start of replication, the sliding clamp and its associated polymerase can remain associated with the leading strand throughout replication. This is not the case for the lagging strand, however. There, as an Okazaki fragment is completed, the polymerase detaches, and its sliding clamp is picked up by the clamp loader protein. As a result, a sliding clamp protein is held in reserve by the clamp loader, rather like a spare tire on the back of a sport utility vehicle, and one of the polymerase subunits detaches from the DNA. When a new Okazaki fragment is initiated via RNA priming, a sliding clamp and polymerase can be loaded onto the new fragment by the clamp loader. The "snapshot" of the replisome shown in Figure 17-16 has caught the replisome in the middle of this process: one catalytic subunit is near the end of its travel as an Okazaki fragment is nearly completed, and the clamp loader is about to add a sliding clamp and polymerase to the region that has just had an RNA primer added.

Eukaryotes Disassemble and Reassemble Nucleosomes as Replication Proceeds

Eukaryotes possess much of the same replication machinery found in prokaryotes. For example, like prokaryotes, a DNA sliding clamp protein acts along with DNA polymerase during DNA synthesis. One such eukaryotic clamp protein, *proliferating nuclear cell antigen (PCNA)*, was originally identified as an antigen that is expressed in the nuclei of dividing cells during S phase. PCNA is a clamp protein for DNA polymerase δ. Unlike bacteria, however, eukaryotes do not rely on the ribonuclease activity of DNA polymerase I to remove RNA primers. Instead, an RNA endonuclease, *RNAse H*, nicks the backbone of the RNA-DNA hybrids, and an RNA exonuclease, *FEN1*, removes the RNA "flap" thus created.

In eukaryotic cells, the enormous length and elaborate folding of the chromosomal DNA molecules pose additional challenges for DNA replication. For example, how are the many replication origins coordinated, and how is their activation linked to other key events in the cell cycle? Answering such questions requires a better understanding of the spatial organization of DNA replication within the nucleus. When cells are briefly incubated with DNA precursors that make the most recently formed DNA fluorescent, microscopic examination reveals that the new, fluorescent DNA is located in a series of discrete spots scattered throughout the nucleus. Such observations

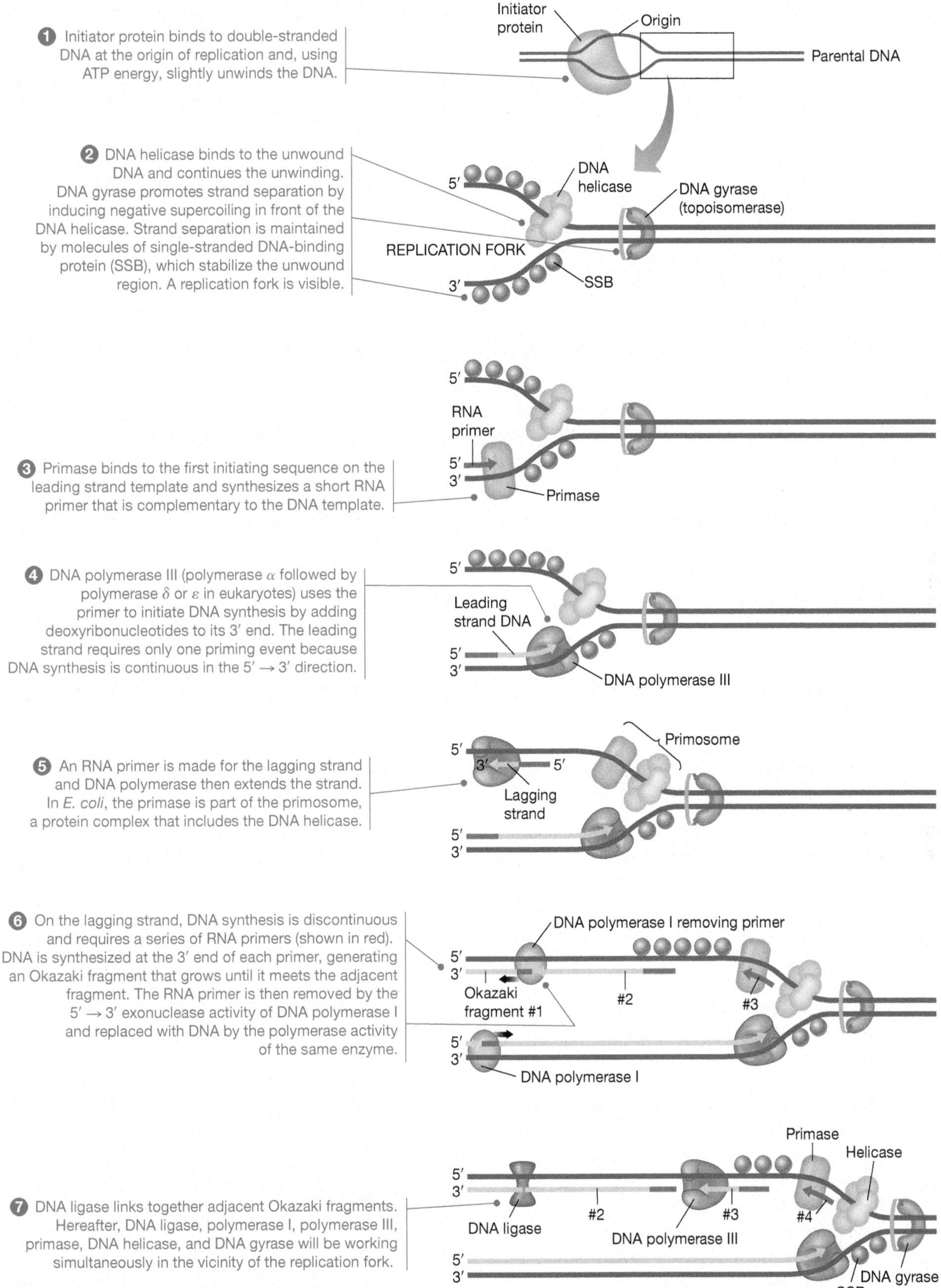

1 Initiator protein binds to double-stranded DNA at the origin of replication and, using ATP energy, slightly unwinds the DNA.
Initiator protein
Origin
Parental DNA
2 DNA helicase binds to the unwound DNA and continues the unwinding. DNA gyrase promotes strand separation by inducing negative supercoiling in front of the DNA helicase. Strand separation is maintained by molecules of single-stranded DNA-binding protein (SSB), which stabilize the unwound region. A replication fork is visible.
5′
DNA helicase
DNA gyrase (topoisomerase)
REPLICATION FORK
3′
SSB
3 Primase binds to the first initiating sequence on the leading strand template and synthesizes a short RNA primer that is complementary to the DNA template.
5′
RNA primer
5′
3′
Primase
4 DNA polymerase III (polymerase α followed by polymerase δ or ε in eukaryotes) uses the primer to initiate DNA synthesis by adding deoxyribonucleotides to its 3′ end. The leading strand requires only one priming event because DNA synthesis is continuous in the 5′ → 3′ direction.
5′
Leading strand DNA
5′
3′
DNA polymerase III
5 An RNA primer is made for the lagging strand and DNA polymerase then extends the strand. In E. coli, the primase is part of the primosome, a protein complex that includes the DNA helicase.
5′
3′
5′
Primosome
Lagging strand
5′
3′
6 On the lagging strand, DNA synthesis is discontinuous and requires a series of RNA primers (shown in red). DNA is synthesized at the 3′ end of each primer, generating an Okazaki fragment that grows until it meets the adjacent fragment. The RNA primer is then removed by the 5′ → 3′ exonuclease activity of DNA polymerase I and replaced with DNA by the polymerase activity of the same enzyme.
DNA polymerase I removing primer
5′
3′
Okazaki fragment #1
#2
#3
5′
3′
DNA polymerase I
7 DNA ligase links together adjacent Okazaki fragments. Hereafter, DNA ligase, polymerase I, polymerase III, primase, DNA helicase, and DNA gyrase will be working simultaneously in the vicinity of the replication fork.
Primase
Helicase
5′
3′
DNA ligase
#2
#3
#4
DNA polymerase III
5′
3′
DNA gyrase
SSB

Figure 17-15 A Summary of DNA Replication in Bacteria. Starting with the initiation event at the replication origin in step 1, this figure depicts DNA replication in *E. coli* in seven steps. Two replication forks move in opposite directions from the origin, but only one fork is illustrated for steps 2–7. The various proteins shown here as separate entities are actually closely associated (along with others) in a single large complex called a replisome. The primase and DNA helicase are particularly closely bound and, together with other proteins, form a primosome. Parental DNA is shown in dark blue, newly synthesized DNA in light blue, and RNA in red.

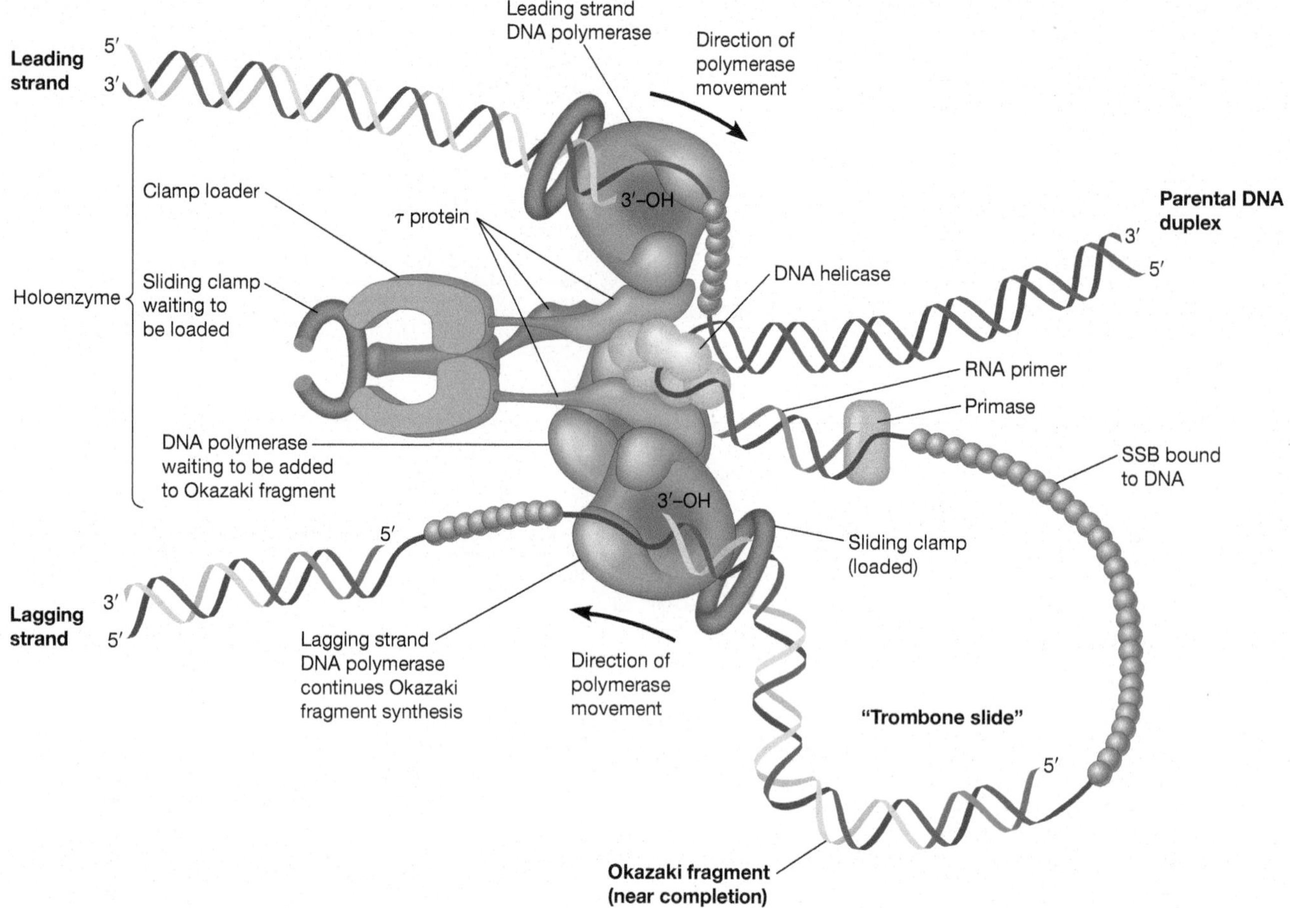

Figure 17-16 The Replisome. This model shows how some of the key replication proteins in bacteria illustrated in Figure 17-15 may actually be organized at the replication fork. The DNA polymerase holoenzyme contains three core polymerase subunits interconnected by τ subunits. The lagging strand DNA is folded into a loop resembling the slide of a trombone, thereby allowing the DNA polymerase subunits on the leading and lagging strands to remain connected even though the two template strands are oriented with opposite polarity in the parental DNA molecule. The sliding clamp and one of the catalytic subunits remain on the leading strand as replication continues. On the lagging strand, the sliding clamp and catalytic subunit detach as each Okazaki fragment is completed. The clamp loader loads a sliding clamp and catalytic subunit onto each new Okazaki fragment as it forms after RNA priming. A similar structure is found in eukaryotes.

suggest the existence of apparently immobile structures, known as *replication factories*, where chromatin fibers are fed through stationary replisomes that carry out DNA replication. These sites are closely associated with the inner surface of the nuclear envelope, although it is not clear whether they are anchored in the nuclear membrane or to some other nuclear support.

When a chromatin fiber is fed through a stationary replication factory, chromatin fibers must be unwound ahead of the replication machinery. This unwinding is facilitated by *chromatin remodeling* proteins (see Chapter 16), which loosen nucleosome packing to give the replication machinery access to the DNA template. After a stretch of DNA is replicated, nucleosomes are reassembled on the newly formed strands. Specific chromatin assembly proteins assist in this process. These include *nucleosome assembly protein-1 (Nap-1)* and *chromatin assembly factor-1 (CAF-1)*. Nap-1 transports histones back into the nucleus after they are synthesized in the cytosol. CAF-1 then targets the histones to replication forks by binding to PCNA. The dynamic nucleosome disassembly/reassembly process allows nucleosomes to remain associated with eukaryotic DNA throughout the replication process, as shown in the electron micrograph in **Figure 17-17**.

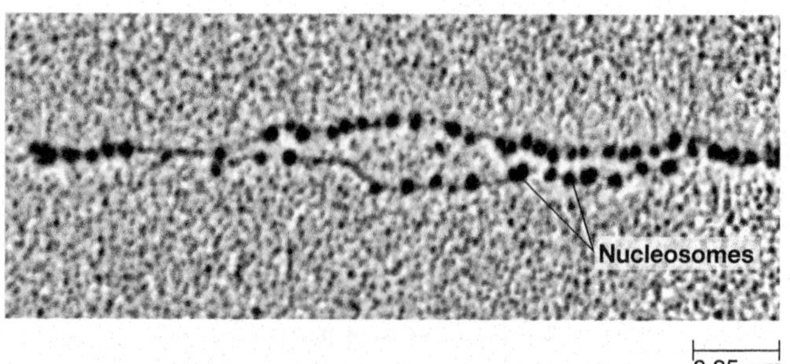

Figure 17-17 Nucleosomes Are Disassembled and Reassembled During Replication in Eukaryotes. An electron micrograph showing nucleosomes on both sides of the two replication forks of a replication bubble in DNA from *Drosophila* (TEM).

Telomeres Solve the DNA End-Replication Problem

If we continue the process summarized in Figure 17-15 for the circular genome of *E. coli* (or any other circular DNA), we will eventually complete the circle. The leading strand can simply continue to grow 5′ → 3′ until its 3′ end is joined to the 5′ end of the lagging strand coming around in the other direction. And for the lagging strand, the very last bit of DNA to be synthesized—the replacement for the RNA primer of the last Okazaki fragment—can be added to the free 3′-OH end of the leading strand coming around in the opposite direction.

For linear DNA molecules, however, the fact that DNA polymerases can add nucleotides only to the 3′-OH end of a *preexisting* DNA chain creates a serious problem, which is depicted in **Figure 17-18.** When a growing lagging strand (lightest blue) reaches the end of the DNA molecule and the last RNA primer is removed by a 5′ → 3′ exonuclease, the final gap cannot be filled because there is no 3′-OH end that deoxynucleotides can be added to. As a result, linear DNA molecules are in danger of yielding shorter and shorter daughter DNA molecules each time they replicate. Clearly, if this trend continued indefinitely, we would not be here today!

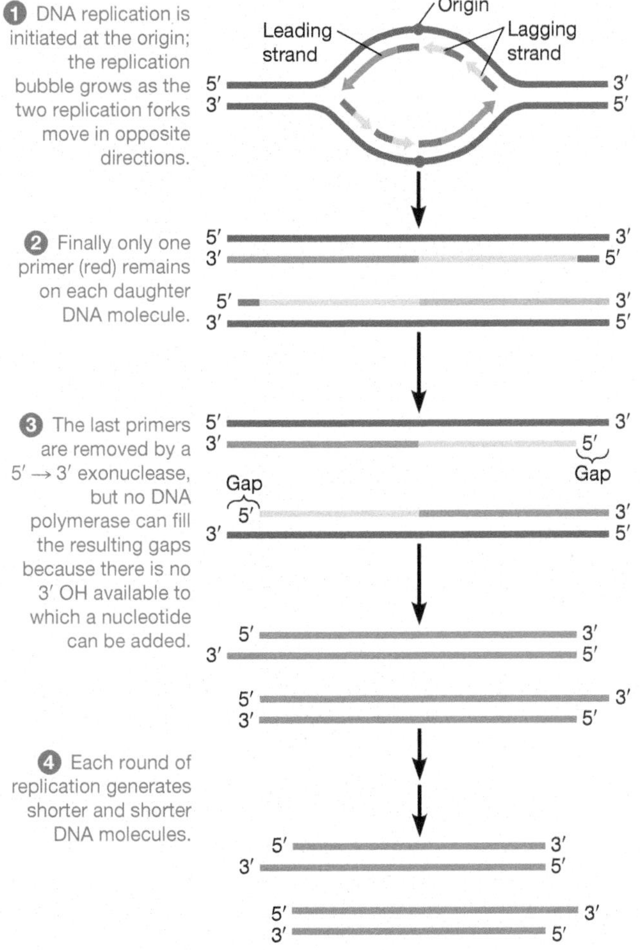

1 DNA replication is initiated at the origin; the replication bubble grows as the two replication forks move in opposite directions.
Origin
Leading strand
Lagging strand
2 Finally only one primer (red) remains on each daughter DNA molecule.
3 The last primers are removed by a 5′ → 3′ exonuclease, but no DNA polymerase can fill the resulting gaps because there is no 3′ OH available to which a nucleotide can be added.
Gap
Gap
4 Each round of replication generates shorter and shorter DNA molecules.
5′
3′

Figure 17-18 The End-Replication Problem. For a linear DNA molecule, such as that of a eukaryotic chromosome, the usual DNA replication machinery is unable to replicate the ends. As a result, with each round of replication, the DNA molecules will get shorter, with potentially disastrous consequences for the cell. In this illustration, the initial parental DNA strands are dark blue, daughter DNA strands are lighter blue, and RNA primers are red. For simplicity, only one origin of replication is shown, and, in the last two steps, only the two shortest of the single-stranded progeny molecules being used as the template are shown.

Viruses with linear DNA genomes solve this problem in various ways. In some cases, such as bacteriophage λ, the linear DNA forms a closed circle before it replicates. In other cases, the viruses use more exotic reproduction strategies, although the DNA polymerases always progress 5′ to 3′, adding nucleotides to a 3′ end.

Eukaryotes have solved the end-replication problem by locating highly repeated DNA sequences at the terminal ends, or **telomeres,** of each linear chromosome. These special telomeric elements consist of short, repeating sequences enriched in the base G in the 5′ → 3′ strand. The sequence TTAGGG, located at the ends of human chromosomes, is an example of such a telomeric, or *TEL, sequence.* Human telomeres typically contain between 100 and 1500 copies of the TTAGGG sequence repeated in tandem. Such noncoding sequences at the ends of each chromosome ensure that the cell will not lose any important genetic information if a DNA molecule is shortened slightly during the process of DNA replication. Moreover, a special DNA polymerase called **telomerase** can catalyze the formation of additional copies of the telomeric repeat sequence, thereby compensating for the gradual shortening that occurs at both ends of the chromosome during DNA replication. Elizabeth Blackburn, Carol Greider, and Jack Szostak received a Nobel Prize for their fundamental contributions to our understanding of telomeres and telomerase.

Telomerase is an unusual enzyme in that it is composed of RNA as well as protein. In the protozoan *Tetrahymena,* whose telomerase was the first to be isolated, the RNA component contains the sequence 3′-AACCCC-5′, which is complementary to the 5′-TTGGGG-3′ repeat sequence that makes up *Tetrahymena* telomeres. As shown in **Figure 17-19** on page 490, this enzyme-bound RNA acts as a template for creating the DNA repeat sequence that is added to the telomere ends. In many eukaryotes, the 3′ end of the DNA can loop back and base-pair with the opposite DNA strand, generating a closed loop that protects the end of the telomere (Figure 17-19, step 5).

In multicellular organisms, telomerase resides mainly in the *germ cells* that give rise to sperm and eggs and in a few other types of actively proliferating cells. The presence of telomerase allows these cells to divide indefinitely without telomere shortening. In most cells, however, telomerase is inactive. This places an upper bound on the number of times such cells can divide because at each successive division, telomeres become shorter and shorter. For primary cultured cells, this limits the number of times cells can be *passaged,* that is, over how many generations the cells can be propagated. This limit, which is 50–60 generations for many cells, is known as the *Hayflick limit,* for Leonard Hayflick, who investigated this growth limitation in the early 1960s (**Figure 17-20** on page 490). If a cell divides enough times, the telomeres in the descendants of that cell are in danger of disappearing entirely, and the cell would then be at risk of eroding its coding DNA. This potential danger is averted by an orchestrated cell destruction pathway set in motion by the shortened telomeric DNA, known as *apoptosis* (described

Telomere
5′ 3′
3′ 5′
Eukaryotic DNA
Repeat unit
Protein
-TTGGGGTTGGGGTTGGGGTTG 3′
-AACCCCAAC
AACCCCAAC
3′ 5′
Telomerase
1 Telomerase binds to DNA.
RNA
2 Nucleotides added to 3′ end of DNA.
-TTGGGGTTGGGGTTGGGGTTGGGGT
-AACCCCAAC
AACCCCAAC
-TTGGGGTTGGGGTTGGGGTTGGGGTTG
-AACCCCAAC CAAC
AACCCCAAC
3 Telomerase advances along the DNA strand as more nucleotides are added.
4 Standard replication machinery extends the complementary DNA strand.
3′
5′
30 nm
5′
3′
5 In many eukaryotes, telomere DNA loops back on itself.

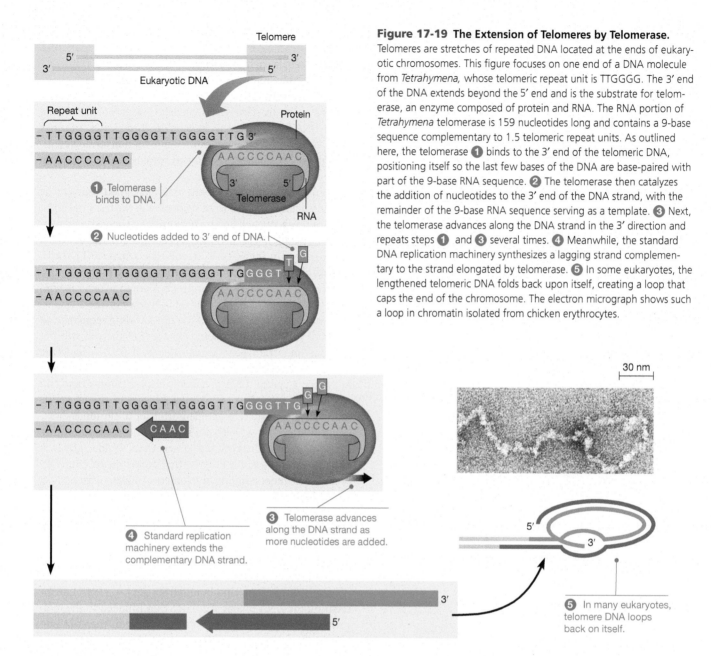

Figure 17-19 The Extension of Telomeres by Telomerase. Telomeres are stretches of repeated DNA located at the ends of eukaryotic chromosomes. This figure focuses on one end of a DNA molecule from *Tetrahymena,* whose telomeric repeat unit is TTGGGG. The 3′ end of the DNA extends beyond the 5′ end and is the substrate for telomerase, an enzyme composed of protein and RNA. The RNA portion of *Tetrahymena* telomerase is 159 nucleotides long and contains a 9-base sequence complementary to 1.5 telomeric repeat units. As outlined here, the telomerase 1 binds to the 3′ end of the telomeric DNA, positioning itself so the last few bases of the DNA are base-paired with part of the 9-base RNA sequence. 2 The telomerase then catalyzes the addition of nucleotides to the 3′ end of the DNA strand, with the remainder of the 9-base RNA sequence serving as a template. 3 Next, the telomerase advances along the DNA strand in the 3′ direction and repeats steps 1 and 3 several times. 4 Meanwhile, the standard DNA replication machinery synthesizes a lagging strand complementary to the strand elongated by telomerase. 5 In some eukaryotes, the lengthened telomeric DNA folds back upon itself, creating a loop that caps the end of the chromosome. The electron micrograph shows such a loop in chromatin isolated from chicken erythrocytes.

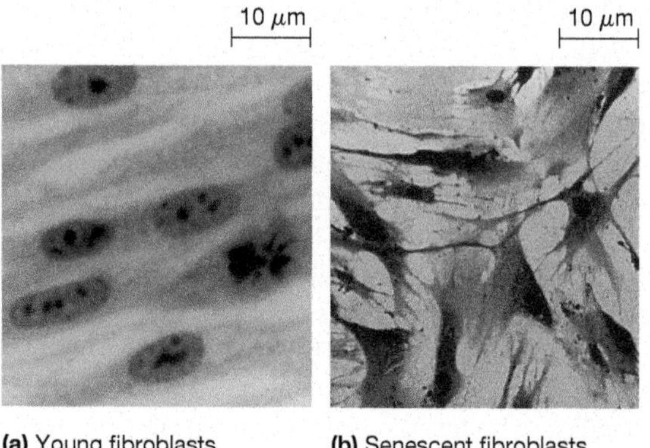

(a) Young fibroblasts **(b)** Senescent fibroblasts

Figure 17-20 The Importance of Telomeres During Cellular Aging. (a) Young fibroblasts that have divided only a few times in culture exhibit a thin, elongated shape. **(b)** After dividing about 50 times, the cells stop dividing and undergo degenerative changes. Note the striking difference in appearance between the young (dividing) and older (nondividing) cells.

in Chapter 24). Immortalized cell lines, such as the HeLa cells isolated in the 1950s (see Human Connections in Chapter 1, page 16), circumvent the Hayflick limit because they produce telomerase. This allows such cells to be passaged indefinitely.

Cell death triggered by a lifetime of telomere shortening is thought to contribute to some of the degenerative diseases associated with human aging—for example, increased susceptibility to infections due to the death of immune cells, inefficient wound healing caused by depletion of connective tissue cells, and ulcers triggered by loss of cells in the digestive tract. People who inherit mutations in telomerase (or in other proteins affecting telomere length) experience similar degenerative changes, show symptoms of premature aging, and usually die when they are relatively young. Based on these relationships, scientists have speculated that telomerase-based therapies may one day be used to combat the symptoms of human aging and thereby extend life span.

Besides being found in germ cells and a few other types of proliferating normal cells, telomerase has also been detected in human cancers. Because cancer cells divide an abnormally large number of times, their telomeres should become unusually short. This progressive telomere shortening would lead to self-destruction of the cancer unless telomerase were produced to stabilize telomere length. This is exactly what seems to happen. A survey of a large number of human cell samples—including more than 100 tumors—found telomerase activity in almost all of the cancer cells but in none of the samples from normal tissues. It isn't surprising that telomerase inhibitors have been investigated as potential chemotherapeutic agents.

Proteins that bind the tandem repeat DNA in telomeres recruit a host of proteins that protect the single-stranded DNA at their very ends from structural damage. These *telomere-capping proteins* bind to the exposed 3′ end of the DNA. Evidence that these proteins are important comes from hereditary diseases such as *Werner syndrome*. Werner syndrome patients lack a telomeric cap protein, WRN. They exhibit an adult-onset form of progeria, in which they exhibit premature signs of aging.

CONCEPT CHECK 17.1

Bacterial DNA replication and DNA replication in typical eukaryotes have many features in common but also several differences. What are at least three similarities of DNA replication in each type of organism? At least three differences?

17.2 DNA Damage and Repair

The faithful transmission of genetic information from one generation of cells to the next requires not just that DNA be replicated accurately. Provision must also be made for repairing DNA alterations that arise both spontaneously and from exposure to DNA-damaging environmental agents. Of course, DNA alterations are occasionally beneficial because DNA base-sequence changes, or **mutations**, provide the genetic variability that is the raw material of evolution. Still, the net rate at which organisms accumulate mutations is quite low; by some estimates, an average gene retains only one mutation every 200,000 years. The underlying mutation rate is far greater than this number suggests, but most DNA damage is repaired shortly after it occurs so does not affect future generations. Moreover, because most mutations occur in cells other than sperm and eggs, they are never passed on to offspring.

Mutations Can Occur Spontaneously During Replication

Mutations can arise spontaneously, without being induced by exposure of cells to physical, chemical, or biological agents. As we learned in our discussion of DNA replication, the DNA polymerase complex will occasionally incorporate the incorrect base in the newly synthesized strand. Such replication errors will usually be repaired through the intrinsic proofreading activity of the enzyme. In addition, other spontaneous mutations may arise in one of three main ways: (1) through spontaneous mispairing of bases via transient formation of tautomers, (2) through slippage during replication, or (3) through spontaneous damage to individual bases.

DNA Tautomers. Mispairing of DNA nucleotides due to the presence of **tautomers** is the most common form of spontaneous replication error. Tautomers are rare, alternate resonance structures of nitrogenous bases. When a base is in its rare tautomeric form, it can pair in a nonstandard way with other bases in a process known as a *tautomeric shift* (**Figure 17-21**). The result is a

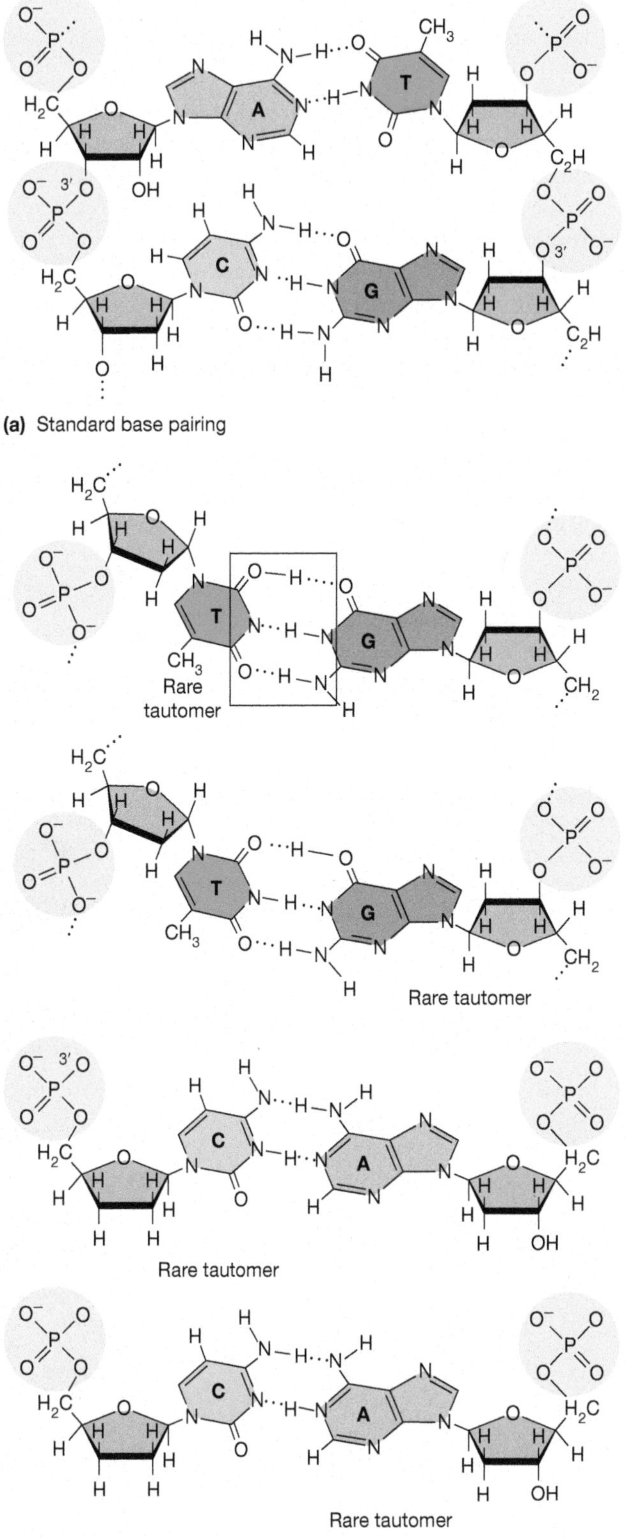

(a) Standard base pairing

(b) Base pairing involving rare tautomers

Figure 17-21 Tautomers and DNA Mismatch. (a) Standard base pairing of the common tautomers of nucleotides. **(b)** Base-pair mismatch can result when a common tautomer and a rare tautomer undergo hydrogen bonding.

new daughter strand that carries an incorrect base at that position. When that strand is replicated during the next cell division, it will transmit an incorrect base to its complementary strand.

MAKE CONNECTIONS 17.2

Explain how rare tautomers (like those seen in Figure 17-21) would affect the ratios of Chargaff's rules if they were instead present in substantial numbers in a given genome. (Table 16-1)

Trinucleotide Repeats. A second type of spontaneous error that can arise during replication involves regions of chromosomes that contain highly repetitive DNA. One well-characterized example of this sort of error involves **trinucleotide repeats.** Trinucleotide repeats seem to be especially susceptible to *strand slippage,* a process whereby DNA polymerase replicates a short stretch of DNA twice. If DNA polymerase detaches from a newly replicated region of repetitive DNA, then the repeats may form a transient hairpin, exposing a stretch of the template that had already been replicated. The result is that when the DNA polymerase reattaches and resumes replication, the same region is re-replicated (**Figure 17-22**), and additional copies of the trinucleotide repeat accumulate. Several well-characterized human diseases, now collectively referred to as *trinucleotide repeat disorders,* involve accumulation of various trinucleotide repeats (**Table 17-2**). Although the specifics for each disorder may vary, the accumulation of extra trinucleotide repeats in specific genes can result in abnormal proteins that lead to cellular defects or to the reduction of levels of expression of the associated genes.

Chemical Modifications. Another type of reaction that can occur spontaneously involves chemical modification of bases. The most common are depurination and deamination reactions, which are spontaneous hydrolysis reactions caused by random interactions between DNA and the water molecules around it (**Figure 17-23**). *Depurination* refers to the loss of a purine base (either adenine or guanine) by spontaneous hydrolysis of the glycosidic bond that links it to deoxyribose (Figure 17-23a). This glycosidic bond is intrinsically unstable and is in fact so susceptible to hydrolysis that the DNA in a human cell may lose thousands of purine bases every day. When an apurinic site is encountered by DNA polymerase, the enzyme typically adds an adenine to the daughter strand, which can lead to a change in the base at that position in subsequent copies of the DNA.

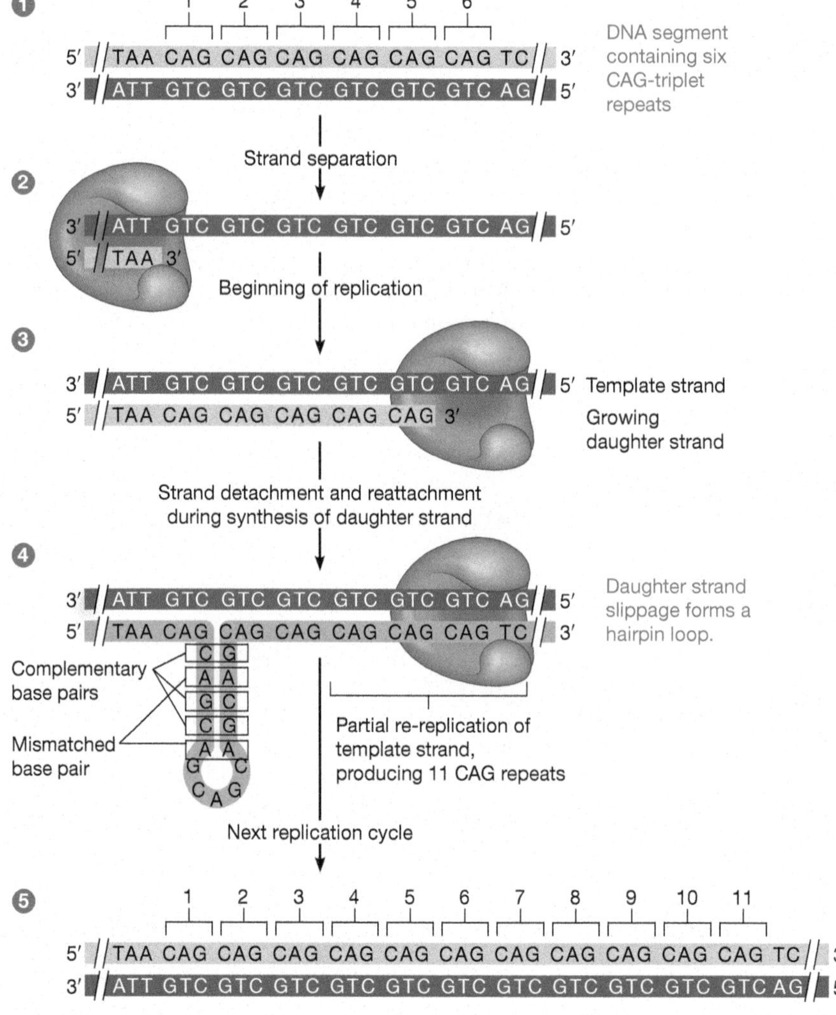

Figure 17-22 Trinucleotide Repeat Expansion. In this example, a region of DNA containing six CAG trinucleotide repeats is shown (step 1). 2 Replication of the two strands then begins (in this example, replication of the original bottom strand is shown). 3 The daughter strand undergoes slippage during replication, creating a hairpin loop. Partial re-replication of the CAG repeats then occurs, adding additional CAG repeats. 4 At the next replication cycle, these extra CAG repeats are replicated, creating double-stranded DNA with extra CAG repeats.

Table 17-2	Human Trinucleotide Repeat Disorders*			
		Repeat Number		
Disease	**Repeat Sequence**	**Normal**	**Disease**	**Primary Symptom**
Fragile X syndrome	CGG	6–50	200–2000	Mental retardation
Friedrich ataxia	GAA	6–29	200–900	Loss of motor coordination
Huntington disease	CAG	10–34	40–200	Uncontrolled movement, dementia
Myotonic dystrophy (type I)	CTG	5–37	80–1000	Muscle weakness
Spinal and bulbar muscular atrophy	CAG	14–32	40–55	Muscle wasting
Spinocerebellar ataxia	CAG	4–44	45–140	Loss of motor coordination

*Source: From Sanders and Bowman, *Genetic Analysis: An Integrated Approach*, 2d ed., Table 12.4.

Deamination is the removal of a base's amino group (—NH_2). Like depurination, deamination is a hydrolytic reaction. This type of alteration, which can involve cytosine, adenine, or guanine, changes the base-pairing properties of the affected base. Of the three bases, cytosine is most susceptible to deamination, giving rise to uracil (Figure 17-23b).

Spontaneous damage can also be caused by reactive oxygen species (for example, hydrogen peroxide, superoxide anions, or hydroxyl radicals), generated from within the cell. These can be released from damaged mitochondria (due to defects within the electron transport chain), generated from free radicals, or formed by ionizing radiation. These chemicals cause oxidation damage to the bases of DNA. This is seen most commonly in the oxidation of guanine to oxoguanine.

If a DNA strand with oxidized bases, missing purines, or deaminated bases is not repaired, an erroneous base sequence may be propagated when the strand serves as a template in the next round of DNA replication. For example, where a cytosine has been converted to a uracil by deamination, the uracil behaves like thymine in its base-pairing properties; that is, it directs the insertion of an adenine in the opposite strand, rather than guanine, the correct base. Oxidized guanine can hydrogen-bond with adenine, which would pair with thymine in the subsequent replication cycle. The ultimate effect of such changes in base sequence may be a change in the amino acid sequence and function of a protein encoded by the affected gene. Fortunately, cells use several repair processes to dramatically reduce the frequency with which permanent defects in DNA are passed along during replication (see below).

Mutagens Can Induce Mutations

In addition to spontaneous mutations, DNA damage can also be caused by mutation-inducing agents, or **mutagens.** Environmental mutagens fall into two major categories: chemicals and radiation. Mutation can also be induced by mobile genetic elements, such as those found in viruses, or in transposable elements (transposons). DNA transposons are discussed later in this chapter.

Mutagenic chemicals alter DNA structure by a variety of mechanisms. *Base analogues* resemble nitrogenous bases in structure and are incorporated into DNA; *base-modifying*

Figure 17-23 Two Common Types of DNA Damage. Two common kinds of chemical changes that can damage DNA are **(a)** depurination and **(b)** deamination. Depurination and deamination are spontaneous hydrolytic reactions that lead to loss or alteration of bases in DNA.

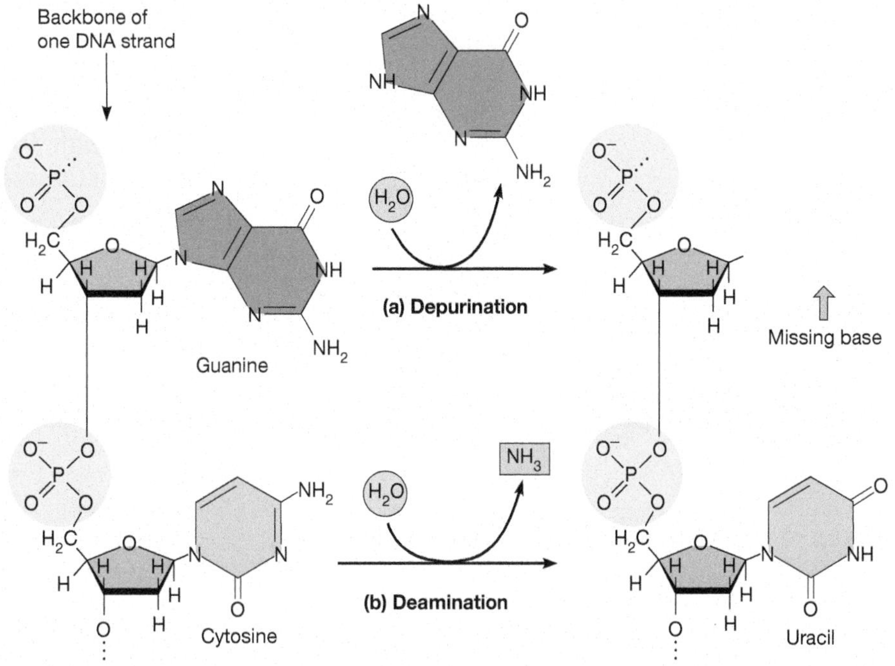

agents react chemically with DNA bases to alter their structure, including those that form *DNA adducts*; and *intercalating agents* insert themselves between adjacent bases of the double helix, thereby distorting DNA structure and increasing the chance that a base will be deleted or inserted during DNA replication. Not surprisingly, mutagens can cause cancer (see Chapter 26). Analysis of DNA from heavy cigarette smokers, for example, shows a direct correlation between how much a patient smokes and the frequency of DNA adducts in such patients.

Base Analogues. Base analogues are chemical compounds that have a structure similar to one of the four DNA nucleotides. DNA polymerase cannot distinguish such analogues from normal nucleotides; when they are incorporated into a DNA molecule during replication, they lead to errors in replication because they pair with a nucleotide during the next replication that would not normally be found at that position. One example is *5-bromodeoxyuridine* (BrdU). A derivative of uracil, its base-pairing properties are like thymine. However, the bromine group allows a potential tautomeric shift, resulting in the pairing of this base analogue with guanine. In the next round of replication, the guanine will pair with cytosine, replacing the original AT pairing.

Base-Modifying Agents. Several mutagens act by chemically modifying a base so that it will form a specific mispairing at the next replication (**Figure 17-24**). Examples of such agents are the alkylating agents *ethyl methansulfonate (EMS)*, which adds ethyl groups to bases, and nitrosoguanidine (which adds methyl groups). Other chemical agents, such as *nitrous acid* (HNO_2), dramatically increase the likelihood of deamination. A third type of chemical modifying agent causes the addition of large, bulky *DNA adducts* to bases. One such mutagen, *aflatoxin* B_1, attaches to the N-7 position of guanine (Figure 17-24a). This leads to the weakening of the bond between the guanine and its sugar, leading to depurination. Aflatoxin B_1, an extremely powerful carcinogen, is produced by the fungus *Aspergillus*, which can grow on peanuts and maize (corn).

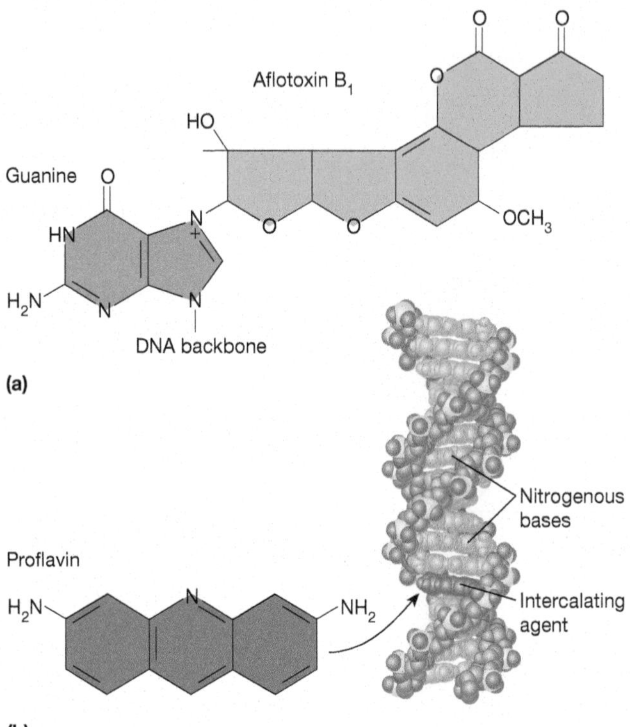

Figure 17-24 Two Types of Chemical Mutagens. (a) Bulky DNA adducts, such as aflatoxin B_1, can bind to bases, causing them to become subject to breakage. **(b)** Intercalating agents, such as proflavin, can insert into the DNA double helix and distort its shape, leading to mutations during DNA repair or replication.

Intercalating Agents. Intercalating agents, which include proflavin, acridine orange, and benzo(*a*)pyrene, act as mutagens not by chemically modifying individual bases, but by inserting into the DNA double helix (Figure 17-24b). Intercalating agents cause mutations by altering the shape of the DNA double helix, leading to small nicks in the DNA. When repaired, the DNA may have additions or deletions of bases. Intercalating dyes that are fluorescent are very useful for gel electrophoresis of DNA. *Ethidium bromide* is a very common fluorescent DNA dye that acts in this way (see Chapter 21).

Radiation. DNA mutations can also be caused by several types of radiation. Sunlight is a strong source of ultraviolet radiation, which alters DNA by triggering *pyrimidine dimer formation*—that is, the formation of covalent bonds between adjacent pyrimidine bases, often two thymines (**Figure 17-25**). Both replication and transcription are blocked by such dimers, since these template bases have difficulty hydrogen bonding with incoming free nucleotides in the active site of the polymerase. Mutations can also be caused by X-rays and related forms of radiation emitted by radioactive substances. This type of radiation is called *ionizing radiation* because it removes electrons from biological molecules, thereby generating highly reactive intermediates that cause various types of DNA damage, including double-strand breaks in DNA. As a result, ionizing radiation is considered a *clastogenic* agent, from the Greek word for "broken." This dangerous form of DNA damage can lead to cancer (see Chapter 26).

DNA Repair Systems Correct Many Kinds of DNA Damage

As is perhaps not surprising for a molecule so important to an organism's health and survival, a variety of mechanisms have evolved for repairing damaged DNA. In fact, in 1974, Francis Crick wrote, "We totally missed the possible role of enzymes in DNA repair. . . . I later came to realize the DNA is so precious that probably many distinct repair mechanisms would exist. Nowadays one could hardly discuss mutation without considering repair at the same time." Both prokaryotic and eukaryotic cells adopt a variety of strategies to repair damaged DNA, depending on how severe that damage is and whether the cell is actively undergoing cell division (and hence DNA replication).

Light-Dependent Repair. Pyrimidine dimers can be directly repaired in a light-dependent process known as **photoactive**

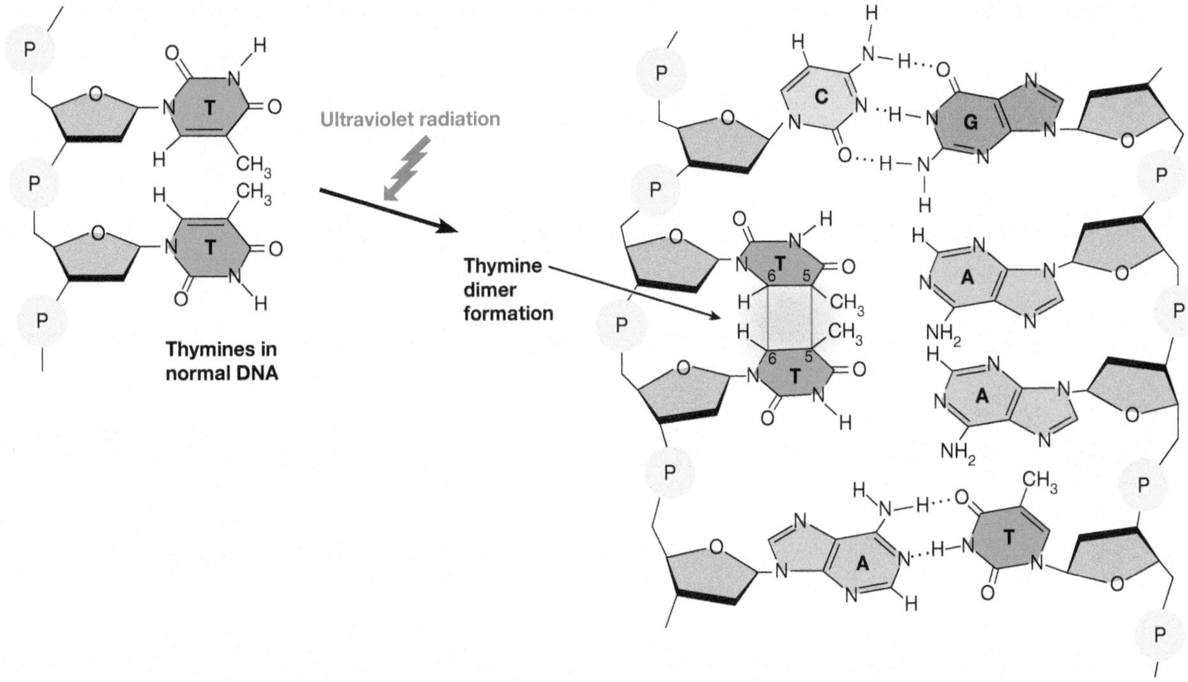

Figure 17-25 Pyrimidine Dimer Formation. UV irradiation can form photoproducts through bond formation between adjacent pyrimidines (for example, thymine), which causes a distortion in the double helix and can block replication or transcription.

repair. Photoactive repair is best understood in bacteria, but appears to occur in some eukaryotes (but not humans). It depends on the enzyme *photolyase*, which catalyzes the breakage of bonds between thymine dimers. The energy needed for photolyase to repair DNA is provided by visible light, which is why the enzyme has the name it does.

Base Excision Repair. Excision repair pathways are classified into two main types, *base excision repair* and *nucleotide excision repair.* The first of these pathways, **base excision repair**, corrects single damaged bases in DNA (**Figure 17-26** on page 496). Deaminated or depurinated bases are detected by specific *DNA glycosylases,* which ❶ recognize a specific deaminated base and remove it from the DNA molecule by cleaving the bond between the base and the sugar it is attached to. The sugar with the missing base is then recognized by a repair endonuclease that detects depurination called *AP endonuclease* (AP stands for "apurinic," that is, DNA in which purine bases have been lost). This repair endonuclease ❷ breaks the phosphodiester backbone on one side of the sugar lacking a base, and a second enzyme then completes the removal of the sugar-phosphate unit. ❸ DNA polymerase then uses the intact complementary strand to synthesize a new base using the 3′-OH created in the previous step. Finally, ❹ DNA ligase seals the nick in the DNA to complete the repair.

For many years, it was not clear why DNA contains thymine instead of the uracil found in RNA. Both bases pair with adenine, but thymine has a methyl group that is not present on uracil. Because the methylation step used to create thymine is energetically expensive, it might seem more efficient for DNA to contain uracil. Now that deamination damage repair mechanisms are better understood, the advantages of thymine in DNA are more apparent. When a deamination event that converts cytosine to uracil (see Figure 17-23b) is detected, uracil is removed by the DNA repair enzyme *uracil-DNA glycosylase.* But if uracil were present as a normal component of DNA, DNA repair would not work because these normal uracils would not be distinguishable from the uracils generated by the accidental deamination of cytosine. By using thymine in DNA in place of uracil, cells ensure that DNA damage caused by the deamination of cytosine can be recognized and repaired without causing other changes in the DNA molecule. This is critical, as estimates suggest that spontaneous deamination of cytosine to uracil occurs anywhere from 100 to 500 times per cell each day!

Nucleotide Excision Repair. For removing pyrimidine dimers and other bulky lesions in DNA, cells employ a second type of excision repair, known as **nucleotide excision repair (NER).** This repair system utilizes proteins that recognize major distortions in the DNA double helix and recruit an enzyme, called an *NER endonuclease* (or *excinuclease*), that makes two cuts in the DNA backbone, one on either side of the distortion. Then a DNA helicase binds to the stretch of DNA between the nicks (12 nucleotides long in *E. coli,* 29 in humans) and unwinds it, freeing it from the rest of the DNA. Finally, the resulting gap is filled in by DNA polymerase and sealed by DNA ligase. In *E. coli*, NER requires several proteins called *Uvr proteins* (for *UV* repair). The process is shown in **Figure 17-27** on page 496.

The NER system is the most versatile of a cell's DNA repair systems. In some cases, the NER system is specifically recruited to DNA regions where transcription has been halted because the transcription machinery encountered an area of DNA damage. This mechanism, known as *transcription-coupled repair,* permits active genes to be

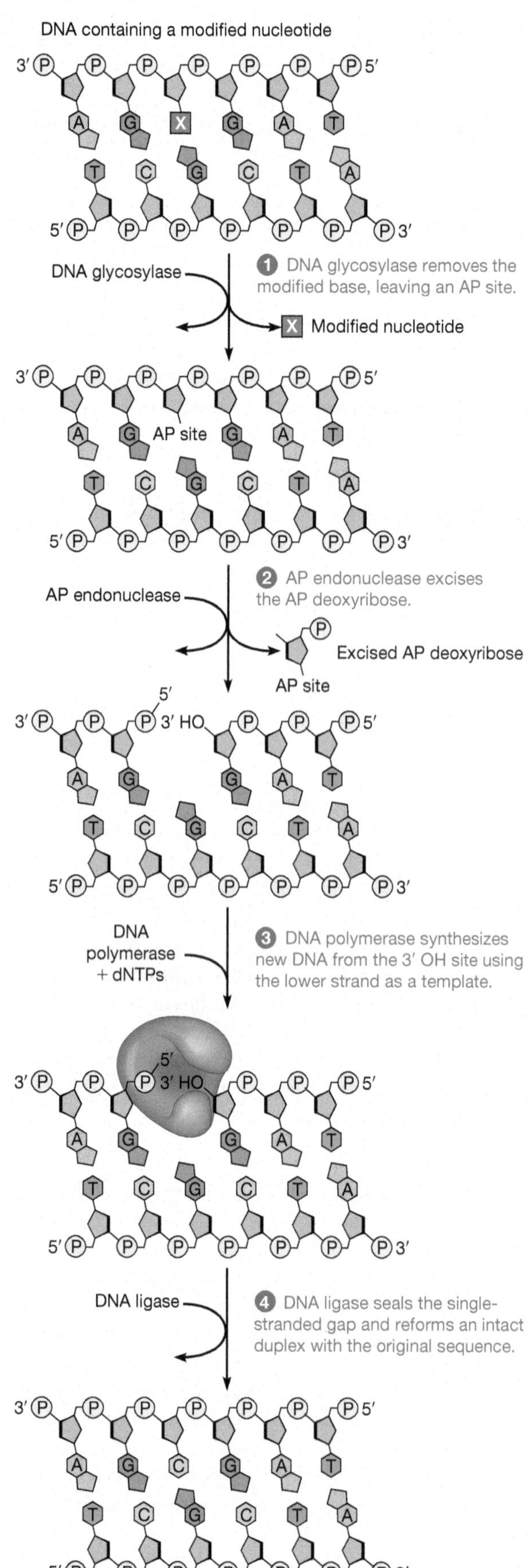

DNA containing a modified nucleotide
DNA glycosylase
1 DNA glycosylase removes the modified base, leaving an AP site.
Modified nucleotide
AP site
AP endonuclease
2 AP endonuclease excises the AP deoxyribose.
Excised AP deoxyribose
AP site
DNA polymerase + dNTPs
3 DNA polymerase synthesizes new DNA from the 3′ OH site using the lower strand as a template.
DNA ligase
4 DNA ligase seals the single-stranded gap and reforms an intact duplex with the original sequence.

Figure 17-26 Base Excision Repair. DNA glycosylase removes a modified base to create an apurinic (AP) nucleotide. AP endonuclease then removes the remainder of the nucleotide, and DNA polymerase and ligase repair the gap.

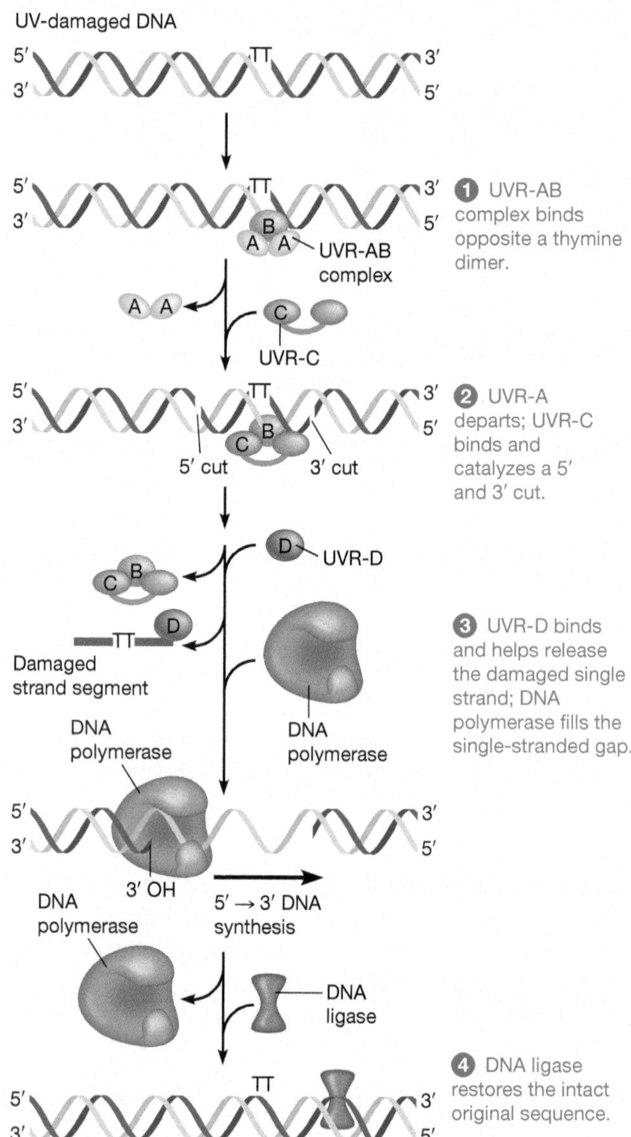

UV-damaged DNA
1 UVR-AB complex binds opposite a thymine dimer.
UVR-AB complex
UVR-C
2 UVR-A departs; UVR-C binds and catalyzes a 5′ and 3′ cut.
5′ cut
3′ cut
UVR-D
Damaged strand segment
DNA polymerase
3 UVR-D binds and helps release the damaged single strand; DNA polymerase fills the single-stranded gap.
3′ OH
5′ → 3′ DNA synthesis
DNA ligase
4 DNA ligase restores the intact original sequence.

Figure 17-27 Nucleotide Excision Repair in Bacteria. UVR proteins remove UV-induced damage in DNA.

repaired faster than DNA sequences located elsewhere in the genome. The importance of NER is underscored by the plight of people who have mutations affecting the NER pathway. Individuals with the disease *xeroderma pigmentosum*, for example, usually carry a mutation in any of seven genes encoding components of the NER system. As a result, they cannot repair the DNA damage caused by the ultraviolet radiation in sunlight and so have a high risk of developing skin cancer (see Human Connections, page 497).

Mismatch Repair. Excision repair is a powerful mechanism for correcting damage involving the presence of abnormal nucleotides during replication. But how can cells repair incorrect pairing of standard bases *after* replication? A postreplication repair pathway, called **mismatch repair**, targets improperly base-paired nucleotides that sometimes escape the normal proofreading mechanisms. Because mismatched base pairs do not hydrogen-bond properly, their presence can be detected and corrected by the mismatch repair system.

HUMAN *Connections*

Children of the Moon

For most children, playing outside at recess or after school is a part of growing up. For some, however, exposure to the sun can be extremely dangerous, or even deadly. At a minimum, sunlight causes DNA damage in these children that puts them at extremely high risk for developing cancer.

What is the cause of such extreme sensitivity to the sun? These children suffer from a rare inherited disorder called *xeroderma pigmentosum (XP)* in which the process of DNA repair is defective. The bulky DNA lesions that result from exposure to UV radiation are normally removed through nucleotide excision repair (NER). In the 1960s it was discovered that cells in most individuals with XP lack the ability to perform NER. Further studies revealed that mutations in seven genes, named *XPA* through *XPG,* disrupt NER (for a similar process in bacteria, see Figure 17-27). Each of these seven genes encodes an enzyme that acts at a different step in the NER pathway. The mutations in *XPA* through *XPG* are inherited, and the failure of this repair pathway means that the DNA mutations in the genome accumulate over time, eventually leading to cancer.

Inherited mutations in an eighth gene, *XPV,* produce an alternate form of XP in which the excision repair enzymes remain intact, but affected individuals are still susceptible to sunlight-induced cancers. This particular mutation affects the gene for *DNA polymerase* η (eta), a special form of DNA polymerase that catalyzes translesion synthesis of a new stretch of DNA across regions in which the template strand is damaged. Unlike other translesion polymerases that add bases in an error-prone, random fashion, DNA polymerase η is capable of accurately replicating DNA in regions where thymine dimers are present and correctly inserts the proper adenine bases in the newly forming DNA strand. Thus, inherited defects in DNA polymerase η, like inherited defects in excision repair, hinder the ability to correct pyrimidine dimers created by UV radiation.

The defects in DNA repair inherited by children with XP lead to skin cancer rates that are 2000-fold higher than normal. Affected individuals develop skin cancer at an average age of 8 years old, compared to age 60 for the general population. Although XP is most likely to increase risk of skin cancer, the DNA repair defects seen in XP also cause a 20-fold increase in risk for leukemia and cancers of the brain, lung, stomach, breast, uterus, and testes. Obviously, lack of repair of damaged DNA has grave consequences.

Because individuals who inherit XP are highly susceptible to UV damage from sunlight, managing the disease primarily involves avoiding the sun. In most cases, changing the schedule of an individual with XP from being active during the day to being active during nighttime can help immensely with decreasing the levels of pyrimidine dimers formed in the skin. Because of their nocturnal habits, individuals with XP are often referred to as "children of the moon." In upstate New York, a special camp, called Camp Sundown, allows affected children to play outside, but on a different schedule. Outdoor recreational activities begin after sundown and take place throughout the night, allowing the children to be safely back indoors and behind drawn curtains by sunrise. Scientists from NASA have even designed a special space suit that provides enough sunlight protection to allow children with XP to occasionally play outdoors (**Figure 17A-1**).

Although not a cure, a topical treatment that reduces incidence of skin cancers in XP individuals is available. Treatment involves the delivery, via specially engineered liposomes, of T4 endonuclease V into the skin. This endonuclease is able to repair a specific type of cyclobutane pyrimidine dimer in damaged DNA. In early trials, this treatment resulted in a reduction of 60% for some types of skin cancers. Gene therapy is in its infancy, but it is hoped that as research into new and novel methods to restore DNA repair mechanisms to XP individuals continues, more permanent treatments will be found.

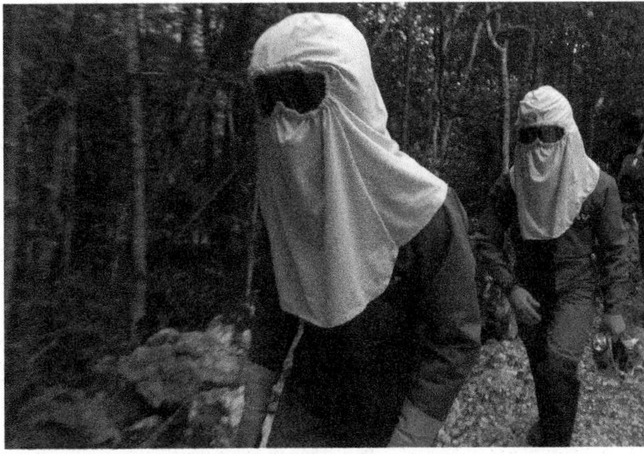

Figure 17A-1 Children with Xeroderma Pigmentosum. When outdoors, these twins with XP must wear special NASA-designed suits to protect their skin from sun exposure. Since they are able to safely remove their suits in the dark of caves, the teens are on their way to an underground camping adventure.

But to operate properly, this repair system must solve a problem that puzzled biologists for many years: how is the *incorrect* member of an abnormal base pair distinguished from the *correct* member? Unlike the situation in excision repair, neither of the bases in a mismatched pair exhibits any structural alteration that would allow it to be recognized as an abnormal base.

To solve this problem, the mismatch repair system must be able to recognize which of the two DNA strands was newly synthesized during the previous round of DNA replication (the new strand would be the one that contains the incorrectly inserted base). The bacterium *E. coli* employs a detection system that is based on the fact that a methyl group is normally added to the base adenine (A) wherever it appears in the sequence GATC in DNA. This process of *DNA methylation* does not take place until a short time after a new DNA strand has been synthesized; hence, the bacterial mismatch repair system can detect the new DNA strand

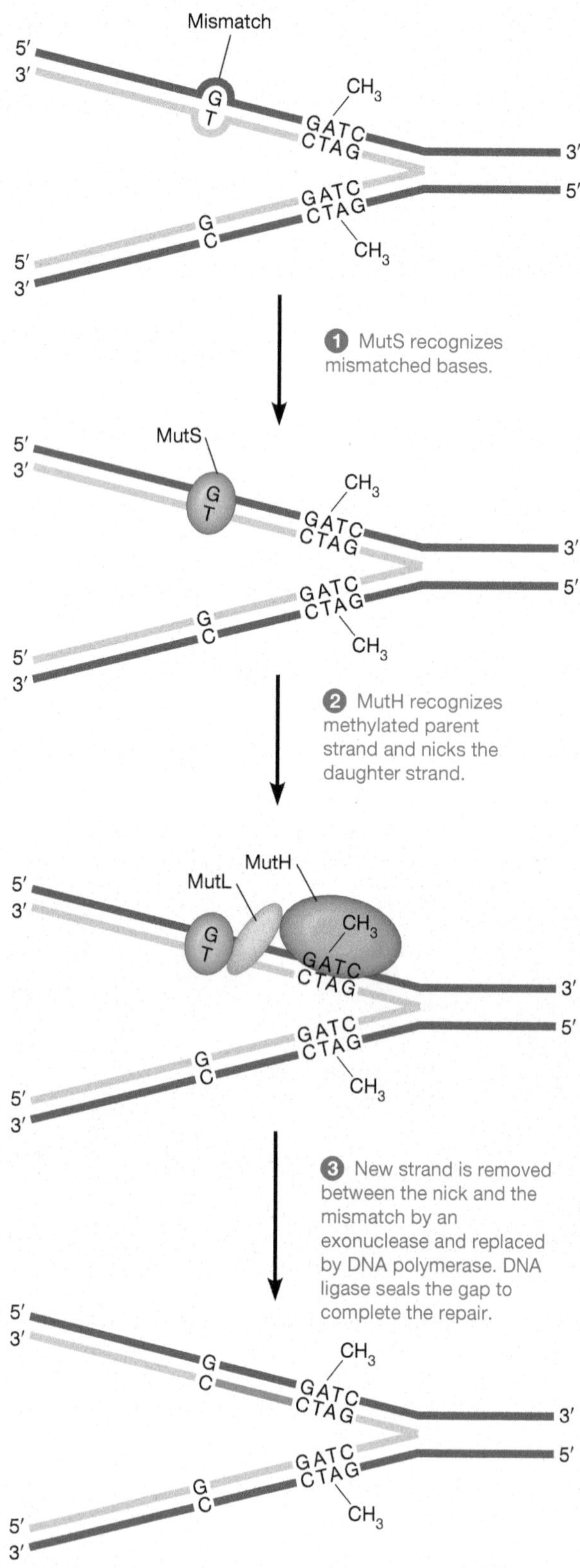
Mismatch
5′
3′
G
T
CH3
GATC
CTAG
3′
5′
GATC
CTAG
G
C
CH3
5′
3′
1 MutS recognizes mismatched bases.
MutS
2 MutH recognizes methylated parent strand and nicks the daughter strand.
MutH
MutL
3 New strand is removed between the nick and the mismatch by an exonuclease and replaced by DNA polymerase. DNA ligase seals the gap to complete the repair.

Figure 17-28 Mismatch Repair in Bacteria. In the example shown, the A in the sequence GATC is methylated, marking the parent strands. When a mismatch occurs (a T is put into the top daughter strand instead of the correct C), ❶ MutS recognizes the mismatch. ❷ MutL is then recruited by MutS and activates the nicking activity of MutH. MutH uses methylation to identify the parent strand, leading to ❸ removal of bases near the mismatch and replacement with the correct sequence.

by its unmethylated state (**Figure 17-28**). A protein known as *MutS* detects the mismatch; a repair endonuclease called *MutH* then introduces a single nick in the unmethylated strand, and an exonuclease removes the incorrect nucleotides from the nicked strand, followed by replacement with the correct sequence.

Although the initial evidence for the existence of mismatch repair came largely from studies of bacterial cells, comparable repair systems have been detected in eukaryotes as well. Eukaryotic mismatch repair differs from the bacterial process in that eukaryotes use mechanisms other than DNA methylation for distinguishing the newly synthesized DNA strand. Although it's somewhat unclear how this distinction takes place, current evidence suggests that the nicks on the lagging strand or sliding clamp proteins like PCNA on the leading strand may be involved. The importance of mismatch repair for eukaryotes is highlighted by the discovery that one of the most common hereditary cancers, *hereditary nonpolyposis colon cancer (HNPCC)*, results from mutations in genes encoding proteins involved in mismatch repair.

Error-Prone Repair: Translesion Synthesis. The DNA repair mechanisms described thus far are error-free, in the sense that they either directly reverse damage to DNA or use the complementary sequence from the undamaged to faithfully repair the damaged strand. In some cases, however, damage to one strand of DNA is so severe that the mechanisms described thus far cannot repair the damage, and a repair process must be used that itself introduces errors into the repaired strand. Both bacteria and eukaryotes have evolved such mechanisms, apparently as a last-ditch effort to repair DNA before more serious outcomes, such as the death of the cell, occur (for more detail on the connection between DNA damage and cell death in eukaryotes, see Chapter 24). In *E. coli*, this process involves activation of the **SOS system,** so named because this pathway is only activated in an emergency. The SOS system is activated when a DNA polymerase III in the replication fork stalls due to damage of the template strand. When the DNA polymerase stalls, the DNA in front of the polymerase continues to be unwound and is bound by single-stranded binding protein. When the polymerase behind this single-stranded DNA stalls, another protein known as *RecA* joins SSB on the single strand to form a protein/DNA filament. The presence of RecA leads to the expression of special DNA polymerases, known as **bypass polymerases,** in the bacterium. These special DNA polymerases are more tolerant of DNA damage and can continue DNA synthesis at the damage site when DNA polymerase III cannot, albeit with the trade-off that the daughter strand that is produced contains errors. Once the bypass polymerase moves through the damaged region, DNA polymerase III can resume synthesis of the remainder of the daughter strand in an error-free fashion.

DNA polymerases that can bypass damage have now been found in many organisms. In eukaryotes, specialized bypass polymerases carry out a process similar to the SOS response known as **translesion synthesis.** Although this type of DNA synthesis is sometimes prone to error, it is also capable of synthesizing new DNA strands in which the damage has been eliminated. For example, eukaryotic DNA polymerase η (eta) can catalyze DNA synthesis across a region containing a thymine dimer, correctly inserting two new adenines in the newly forming DNA strand. Because the mutation is eliminated from the new strand but not the template strand, translesion synthesis is a damage-tolerance mechanism that can prevent an initial mutation from being passed on to newly forming DNA strands.

Double-Strand Break Repair. The repair mechanisms described thus far are effective in correcting DNA damage involving chemically altered or incorrect bases. In such cases, a "cut-and-patch" pathway removes the damaged or incorrect nucleotides from one DNA strand, and the resulting gap is filled using the intact strand as a template. For more severe damage in which both strands of the DNA are broken, however, repair is more difficult. With other types of DNA damage, one strand of the double helix remains undamaged and can serve as a template for aligning and repairing the defective strand. In contrast, double-strand breaks completely cleave the DNA double helix into two separate fragments; the repair machinery is therefore confronted with the problem of identifying the two fragments and rejoining their broken ends without losing any nucleotides. Two main pathways are employed in such cases: nonhomologous end-joining and mechanisms that rely on homologous sequences to faithfully repair the DNA. The generation and subsequent repair of double-strand breaks is the foundation of the latest genome modification techniques like CRISPR (see **Key Technique, pages 500–501**).

In **nonhomologous end-joining (NHEJ)**, a set of proteins binds to the ends of the two broken DNA fragments and joins them together (**Figure 17-29**). NHEJ begins after

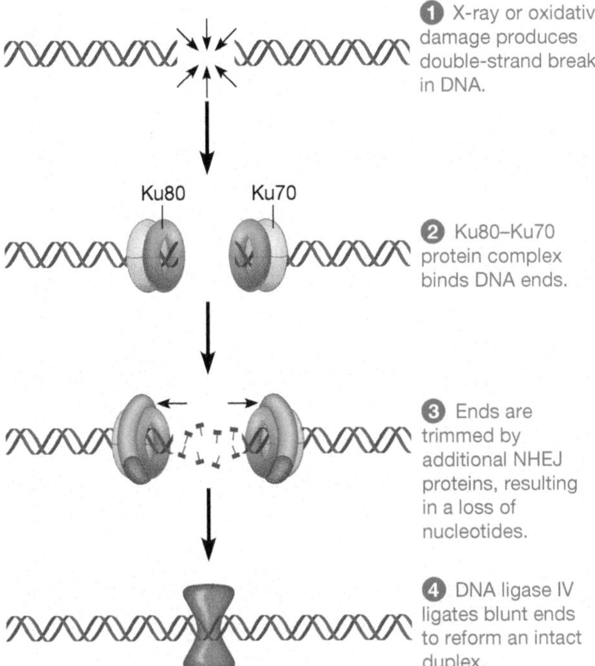

1 X-ray or oxidative damage produces double-strand break in DNA.
Ku80
Ku70
2 Ku80–Ku70 protein complex binds DNA ends.
3 Ends are trimmed by additional NHEJ proteins, resulting in a loss of nucleotides.
4 DNA ligase IV ligates blunt ends to reform an intact duplex.

Figure 17-29 Nonhomologous End-Joining (NHEJ). NHEJ relies on Ku70/80 proteins, which bind damaged ends, leading to error-prone repair.

(1) X-rays or oxidative damage produces a double-strand break in the DNA. (2) The break is then recognized by two proteins, Ku70 and Ku80, which bind one another and fit like a basket over the ends of the broken DNA. (3) Next, the ends of the break are trimmed by nucleases. (4) Finally, the break is sealed via DNA ligase IV. Unfortunately, this mechanism is error-prone because it cannot prevent the loss of nucleotides from the broken ends and has no way of ensuring that the correct two DNA fragments are being joined to each other. NHEJ proteins are also involved in the synthesis of the lymphocyte receptors used by adaptive immune cells.

NHEJ is dangerous for a cell, given its very error-prone nature. A cell uses it only when it cannot use existing DNA sequences as a template to repair double-strand breaks with high fidelity. A more precise method for fixing double-strand breaks relies on the availability of the corresponding sequences to the damaged region from sister chromatids during cell division and requires temporary or permanent exchange between regions of chromosomes. Let's consider two such mechanisms (**Figure 17-30** on page 502). The first is called **synthesis-dependent strand annealing (SDSA)**, and it capitalizes on the fact that once DNA synthesis is complete, each chromosome is composed of two sister chromatids. Thus, if the DNA molecule in one chromosome incurs a double-strand break, another intact copy of the chromosomal DNA is still available to serve as a template for guiding the repair of the broken chromosome. The process of SDSA is shown in Figure 17-30a. SDSA repairs a double-strand break in one sister chromatid; the other sister, which is intact, is used to repair the damage. SDSA begins with (1) the trimming of the break by nucleases and the recruitment of the protein *Rad51* to strands at the break site. Rad51 is a member of the same protein family as the bacterial protein RecA (see the discussion of SOS repair above). Rad51 forms DNA/protein filaments that stabilize single strands and allow strands to search for homologous sequences on a sister chromatid during the repair process, a process known as *strand invasion*. During strand invasion, (2) the single strand causes displacement of the DNA within the sister chromatid that corresponds to the invading strand, creating a *D loop* (D stands for "displacement"). (3) The D loop structure now allows replication of the single strand from the broken DNA to occur. The bits of repaired DNA then (4) dissociate from the D loop and are ligated into the broken chromatid, (5) leading to its repair.

The second mechanism of faithful repair of double-strand breaks is known as **homologous recombination.** Homologous recombination involves the process of *crossing over*, in which genetic information is exchanged between DNA molecules exhibiting extensive sequence similarity (Figure 17-30b). You may be familiar with this process during normal meiotic recombination; in fact, much of the same molecular machinery is used. We will consider the details of that process in more detail in the next section.

Comparing homologous recombination to SDSA reveals that the main difference is the involvement of *both* strands of the broken DNA in the repair process, whereas in

Key Technique | CRISPR/Cas9 Genome Editing

PROBLEM: The ability to change bases within genomes with precision has been the holy grail for geneticists looking to correct deleterious mutations and as a powerful tool for individualized medicine. Early gene therapy attempts utilized retroviruses and relied on highly inefficient recombination. This approach has achieved very limited success to date. More recently, the introduction of double-strand breaks in DNA using special enzymes called zinc-finger nucleases or transcription activator-like effector nucleases (*TALENS*) has proven successful in some cases in editing the genome. However, these approaches required synthesizing novel proteins for each new manipulation, require a great deal of time to carry out, and are somewhat limited in terms of targeting precision and flexibility. As more is understood about the genetic contributions to human disease, the ability to rapidly and accurately manipulate genomes will be a powerful tool for individualized medicine in the future.

SOLUTION: A solution to the need for genome editing has been found in an unlikely place. The *CRISPR* system was first described as an adaptive immune system in bacteria as a way to protect the cell from infection by bacteriophage (see Chapter 20). Before its function was known, CRISPR was named by Francisco Mojica based on his discovery of *c*lustered *r*egularly *i*nterspaced *s*hort *p*alindromic *r*epeats in the genomes of both archaeal and eubacterial species. The application of CRISPR in mammalian genome editing has been refined and improved over the past several years from techniques pioneered in the laboratories of Emmanuelle Charpentier and Jennifer Doudna. It has quickly emerged as one of the most exciting scientific advances in recent memory.

Key Tools: Cas9 nuclease, synthetic guide RNA (made from tracrRNA and a CRISPR gRNA array), repair dsDNA template.

Details: The key to the success of CRISPR/Cas9 genome editing is the introduction of a double-stranded DNA (dsDNA) break in the area to be modified (**Figure 17B-1**). This activates one of the cell's dsDNA repair pathways and leads to one of two potential outcomes. If the goal is to knock out the function of the gene of interest, the *n*on-*h*omologous *e*nd *j*oining (NHEJ) repair pathway is initiated. This pathway is error-prone and often joins the two strands of DNA back together with one or more missing bases, causing catastrophic frameshift mutations. When this occurs in the

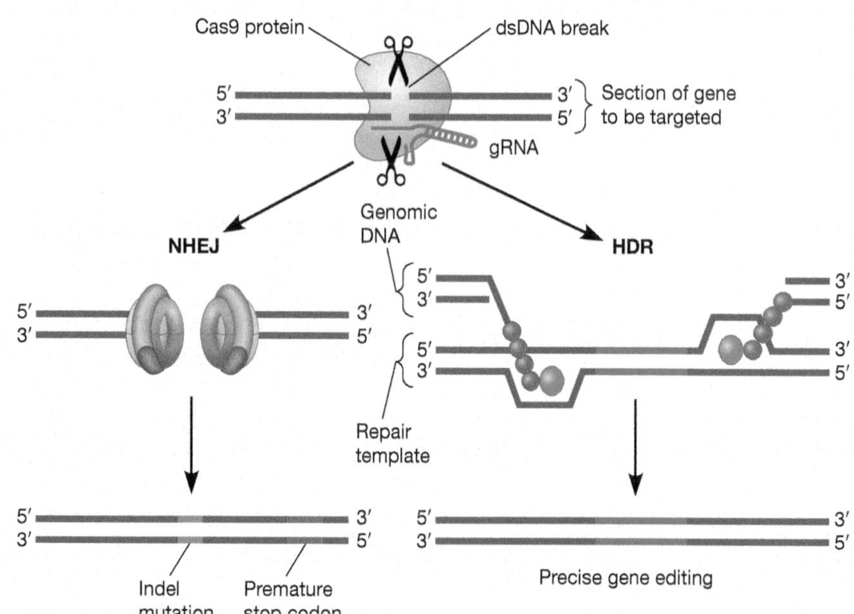

Figure 17B-1 Generation and Repair of dsDNA Breaks in CRISPR Editing. The Cas9 nuclease generates a dsDNA break in the genome locus to be modified. The activation of the non-homologous end joining (NHEJ) repair pathway can cause insertion or deletion of bases (an indel mutation). The resulting frameshift often produces a premature stop codon and loss of protein function. If a dsDNA repair template is added, homology directed repair (HDR) will repair the break using the repair template as a guide. This will lead to precise editing of the gene, based on the sequence of bases within the repair template.

SDSA it is only one strand that is initially repaired using a template. As with SDSA, ❶ the double-strand break is recognized and enzymes trim the ends of the break. As with SDSA, ❷ strand invasion occurs, but this time it involves both strands. ❸ DNA synthesis then fills in DNA on both strands using the intact template pieces of DNA. ❹ The result is a striking feature of homologous recombination: a crossed structure, called a **Holliday junction**, named after Robin Holliday, who proposed a detailed model regarding how DNA strands are involved in recombination. ❺ There are two possible ways in which this crossed structure is "resolved" to generate two separated strands of repaired DNA. In one case, there is a permanent exchange (crossing over) of DNA between the broken DNA from one chromosome and the corresponding piece of intact DNA from the other chromosome. The other possibility leads to repair of the broken coding region of a gene, the activity of the protein is often reduced or lost entirely (usually through the presence of a new premature stop codon). If the goal is to edit part of the genome, then a DNA template containing the edited sequence needs to be added. This guides the homologous recombination machinery to repair the dsDNA break using this sequence as a guide to copy from.

So how is the dsDNA break created in order to start the process? The elegance of the CRISPR system is that the nuclease responsible for the break is targeted to a precise location using Watson–Crick base pairing (**Figure 17B-2**). This is accomplished through the design of a synthetic guide RNA (gRNA) for the desired target sequence. The gRNA includes 20 nucleotides complementary to the target sequence in the genome known as the CRISPR RNA (crRNA) array (also called a protospacer). The only limitation in the design of this sequence is that it must immediately precede a *p*roto*s*pacer *a*djacent *m*otif (PAM). Although this may appear to limit the usefulness of the technique, it is not the limitation that it appears to be. PAM sequences are species specific. For example, the PAM sequence in *Streptococcus pyogenes* (one of the more commonly used CRISPR systems) is 5′-NGG-3′ (where N is any base). Such a short PAM sequence should be found once every 8–12 bases in a given genome, giving great flexibility in the design of this part of the gRNA. The other half of the gRNA is the trans-activating CRISPR RNA (tracrRNA), which interacts with the Cas9 nuclease and appears to activate the nuclease activity of the enzyme.

The last component required for CRISPR editing is the nuclease. Most of the recent work uses the Cas9 nuclease (short for CRISPR-associated protein 9). Because there is minor variation between bacterial species, care should be taken to select the appropriate Cas9 protein. For example, if the PAM sequence used to design the gRNA is from *S. pyogenes*, then *S. pyogenes* Cas9 is required. The standard Cas9 protein will cut both strands of the target sequence between the 17th and 18th bases of the target sequence. Mutated forms of Cas9 are being used to develop alternative genome editing strategies.

The coding sequence for Cas9 and the gRNA can be delivered to the target cell in the form of a dsDNA plasmid. Once in the cell, the Cas9 sequence is transcribed and translated into protein, while the gRNA sequence is transcribed from DNA into the functional RNA. Therefore, it becomes a relatively simple task to quickly design and engineer a CRISPR editing plasmid for precise genome manipulation. It is certainly an exciting time for this emerging technology, as several clinical trials are currently underway using CRISPR to treat patients with conditions like sickle cell anemia, HIV, and cancer. Charpentier and Doudna were awarded the 2020 Nobel Prize in Chemistry for their role in developing this novel method of genome editing.

QUESTION: What would be the effect of a "broken" Cas9 protein that targets a gene promoter and binds, rather than cuts, the DNA?

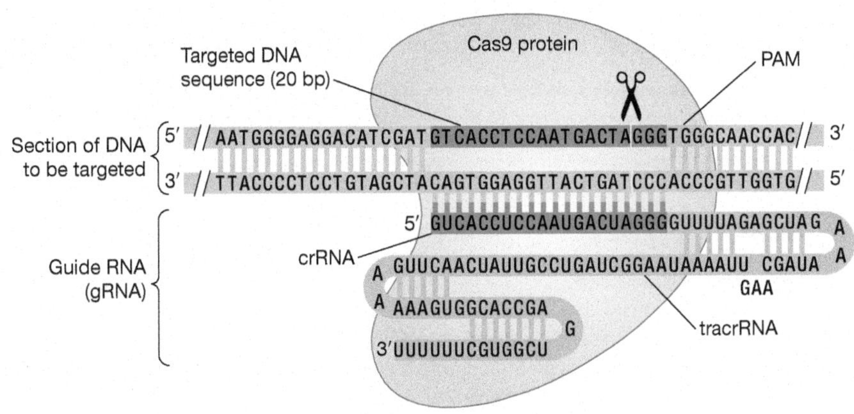

Figure 17B-2 Targeting of Cas9 Nuclease to a Specific Genome Locus. Successful targeting of Cas9 requires the design of a synthetic guide RNA (gRNA). This consists of RNA complementary to the targeted region (blue), which is often called the CRISPR RNA (crRNA). This sequence must immediately precede the protospacer adjacent motif (PAM) sequence for the species from which the Cas9 enzyme was isolated. The gRNA also contains the trans-activating CRISPR RNA (tracrRNA) sequence (red), which helps recruit and activate the nuclease activity of the enzyme. Once bound, Cas9 will cut both strands of the DNA at the indicated sites.

DNA without such exchange. When the DNA is slightly different on the two chromosomes (for example, the two homologous chromosomes carry a different allele of a particular gene), the result is that the repaired chromosome now has the same sequence as the undamaged chromosome. This process is known as **gene conversion** because the sequence of the damaged chromosome has been changed in the repair process.

CONCEPT CHECK 17.2

Nonhomologous end-joining and synthesis-dependent strand annealing are both strategies eukaryotes use to repair double-strand breaks in DNA. Because SDSA is error-free and NHEJ is error-prone, why don't eukaryotic cells always use SDSA?

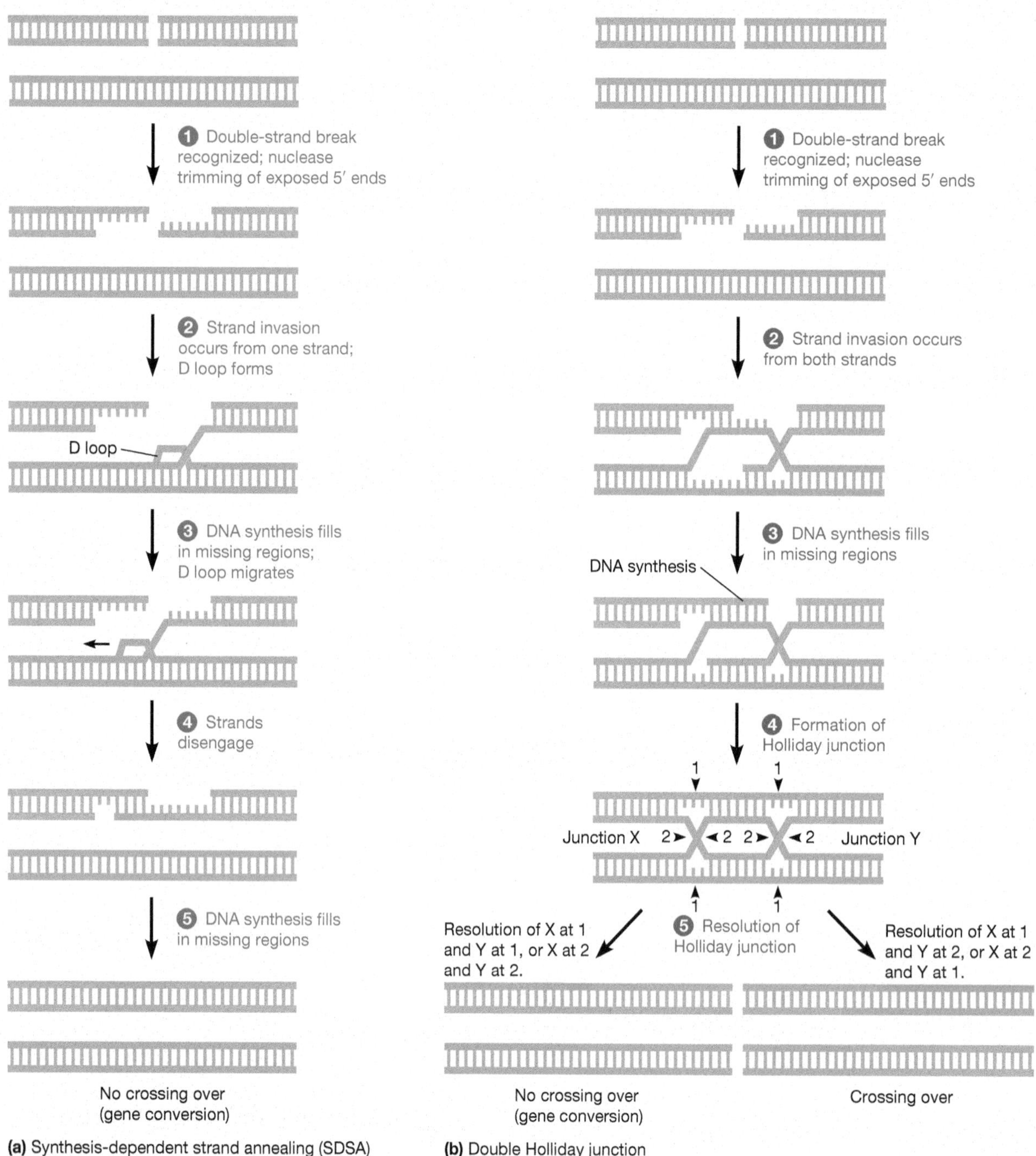

Figure 17-30 DNA Repair Using Homologous Sequences. **(a)** Synthesis-dependent strand annealing (SDSA). SDSA repairs a double-strand break using strand invasion and D loop formation by a single strand from the DNA that needs repair. DNA synthesis causes the D loop to migrate, followed by strand disengagement and DNA synthesis and ligation to fill in gaps. **(b)** Double Holliday junction formation. A double-strand break leads to strand invasion from both strands, DNA synthesis, Holliday junction formation, and resolution of the Holliday junction in one of two ways.

17.3 Homologous Recombination and Mobile Genetic Elements

The recognition that sister chromatids can be used in DNA repair processes naturally raises a question regarding the mechanisms of a widespread exchange of chromatin that occurs during another cell division process, *meiosis*. (For a detailed discussion of the cellular events of meiosis, see Chapter 25.) Here the focus is on the molecular events that allow chromatin exchange during meiosis in eukaryotes. Another related type of recombination occurs in prokaryotes (also see Chapter 25). The principles involved appear to be quite similar in both cases.

Homologous Recombination Is Initiated by Double-Strand Breaks in DNA

Homologous recombination is more complicated than can be explained by a simple breakage-and-exchange model in which crossing over is accomplished by cleaving two double-stranded

DNA molecules and then exchanging and rejoining the cut ends. Although Robin Holliday was the first to propose a molecular mechanism of recombination, his original ideas have now been superseded by a model involving double-strand breaks in the DNA.

Double-Strand Breaks and Formation of Crossovers. A current model of genetic recombination, proposed in the 1980s by Jack Szostak, Terry Orr-Weaver, Rodney Rothstein, and Franklin Stahl, is illustrated in **Figure 17-31**. According to this model, the initial step in recombination in eukaryotes

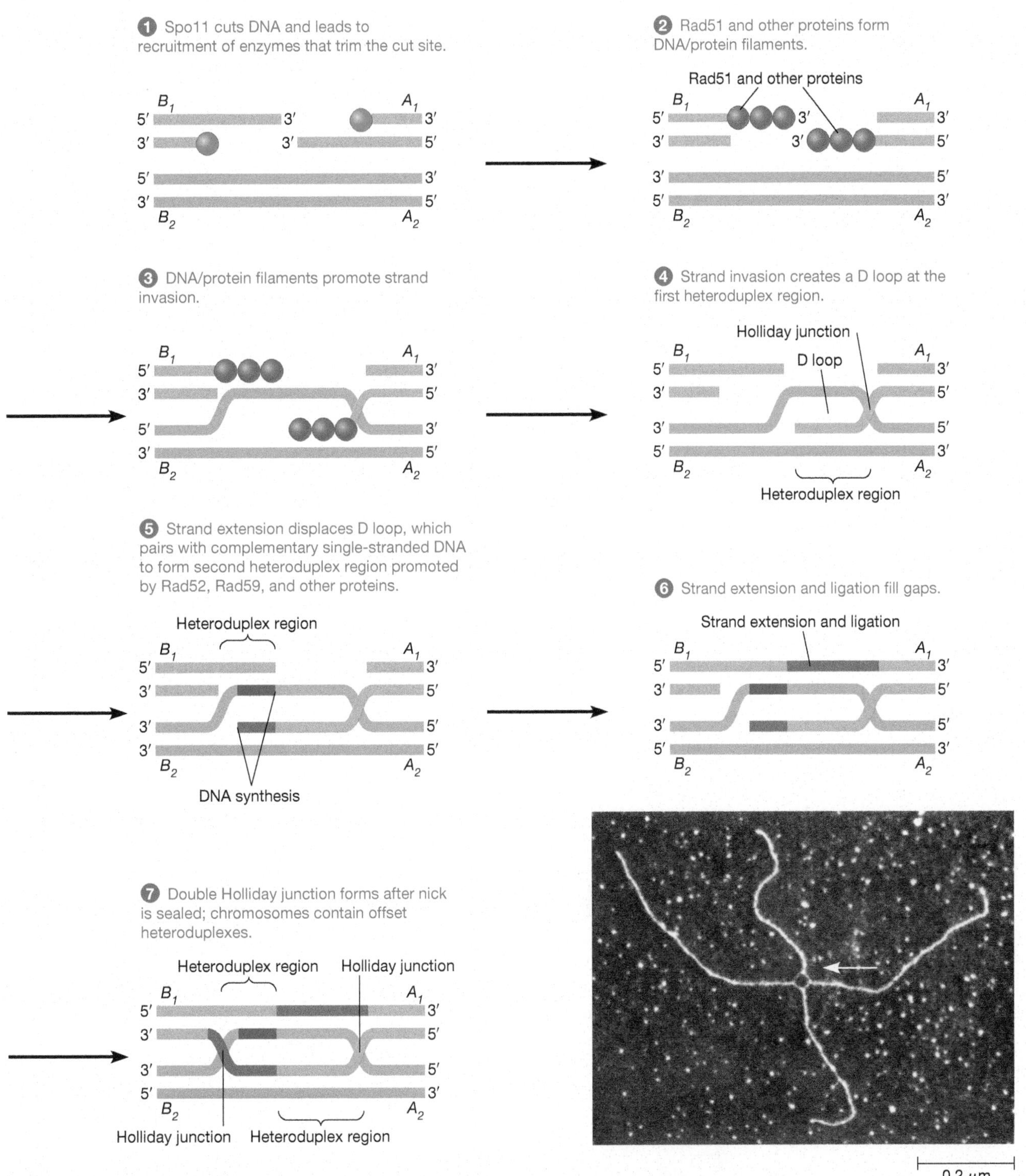

Figure 17-31 A Model for Homologous Recombination via Double-Strand Breaks. Meiotic recombination occurs between nonsister chromatids of homologous chromosomes and is mediated by a double-strand break initiated by Spo11. Single strands stabilized by Rad51 and Dmc1 undergo invasion to create a D loop, leading to Holliday junction formation. Strand extension, which involves proteins such as Rad52 and Rad59, forms a second Holliday junction. In the transmission electron micrograph, the arrow points to a region where two plasmid DNA molecules are in the process of undergoing homologous recombination, forming a Holliday junction.

is ❶ the highly asymmetric cleavage of one chromatid by the dimeric protein *Spo11* (pronounced "Spo-eleven"). Spo11 recruits two other proteins that trim the resulting single strands. Recall that the protein Rad51, which is related to the bacterial protein RecA, is involved in stabilizing single-stranded DNA during DNA repair. Rad51, along with other proteins, performs a similar function during recombination and ❷ is the next to join the trimmed region. Rad51 and its associates aid formation of DNA/protein filaments that can engage in strand exchange. Although not shown here, RecA acts in a similar way during bacterial recombination (see Chapter 25). The single stabilized DNA strands can then ❸ invade a complementary region on another homologous piece of DNA, displacing one of the two strands to form a D loop; ❹ the invading strand then pairs with the complementary DNA in the D loop. The Holliday junction is an interim structure in which a single strand from each DNA double helix has crossed over and joined the opposite double helix. Electron microscopy has provided direct support for the existence of Holliday junctions, revealing the presence of DNA double helices joined by single-strand crossovers at sites of genetic recombination (Figure 17-31, EM).

Once a Holliday junction has been formed, ❺ DNA synthesis leads to *strand extension*, or branch migration, displacing the D loop and allowing it to pair with more single-stranded DNA, which involves other recombination proteins, such as Rad52 and Rad59. This phenomenon can rapidly increase the length of single-stranded DNA that is exchanged between two DNA molecules. After strand extension has occurred, ❻ the gaps in the single-stranded DNA paired with the D loop are filled in via DNA polymerase and ligase. Finally, ❼ the 3′ end of the invading strand connects with the 5′ end of one of the original single-stranded segments. The result is a *double Holliday junction*.

Holliday Junction Resolution. Holliday junctions are transitory structures; homologous chromosomes must be disengaged before cell division. A Holliday junction can be cleaved and rejoined, or *resolved*, in either of two ways (**Figure 17-32**). If the Holliday junction is resolved via *same-sense resolution*, crossing over does not occur, but the DNA molecules exhibit a noncomplementary region near the site where the Holliday junction had formed (Figure 17-32a). The other type of resolution, *opposite-sense resolution*, results in crossing over; that is, the chromosomal DNA beyond the point where recombination occurred will have been completely exchanged between the two chromosomes (Figure 17-32b). In either case, there will be DNA between the Holliday junctions where the two paired strands are derived from different chromosomes. These are known as **heteroduplex regions** and often contain single-base mismatches due to sequence differences between the homologous pairs. These mismatches are repaired using a mismatch repair system, but the strand picked for repair is random (there is no "newly synthesized strand" like we see in standard mismatch repair). Therefore, the process of fixing these heteroduplex regions in meiosis often results in gene conversion.

Transposons Are Mobile Genetic Elements

Recombination is one important way that organisms maintain genetic diversity and involves exchange of DNA from one chromosome to another. Another important type of DNA exchange involves mobile genetic elements known as **transposable elements**, or **transposons.** The first evidence for such movable genetic elements came from pioneering work by the maize (corn) geneticist Barbara McClintock in the 1940s and 1950s. McClintock studied changes in the pigmentation of kernels that could only be explained if the genetic elements responsible were

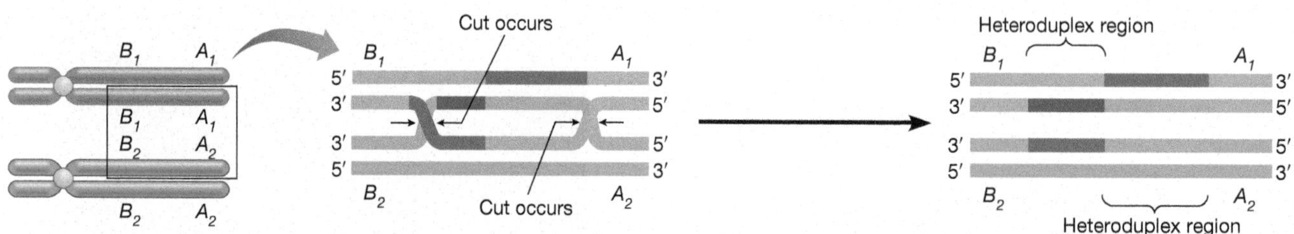

(a) Same-sense resolution: offset heteroduplex region, no recombination of flanking genes (less common)

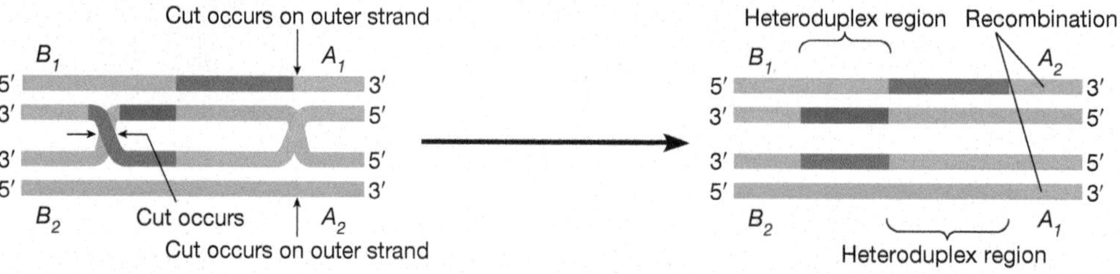

(b) Opposite-sense resolution: offset heteroduplex region, recombination of flanking genes (much more common)

Figure 17-32 Resolution of Holliday Junctions During Recombination. Holliday junctions must be resolved before cells divide and can undergo either **(a)** same-sense or **(b)** opposite-sense resolution. The latter is much more common.

capable of moving from one chromosomal location to another. For her incisive studies McClintock received a Nobel Prize in 1983. Since that time, transposable elements have become recognized to be very widespread in prokaryotes and eukaryotes.

Unlike homologous recombination, the movement of transposons does not require sequence similarity with the sites into which they embed themselves. Instead, transposons encode all or many of the components they need to move within the genome. There are two major kinds of transposons. Some act only through DNA and the proteins encoded within their DNA sequences and are known as *DNA-only transposons.* DNA-only transposons are common in bacteria but are also found in eukaryotes. Other transposons act through an RNA intermediate step and are known as **retrotransposons.** These are common in eukaryotes and make up a significant proportion of the human genome (see Figure 16-25).

Transposons Differ Based on Their Autonomy and Mechanism of Movement

In order to move, transposons require the activity of the enzyme *transposase.* Transposons that encode their own transposases are known as **autonomous transposable elements.** Conversely, **nonautonomous transposable elements** lack a functional transposase gene and may lack other DNA elements required for transposition. Such elements need the help of other autonomous elements to move, a process geneticists sometimes refer to as "mobilizing" the element. One of the original transposons in maize studied by Barbara McClintock (called *Activator*) was autonomous, whereas the other (*Dissociation*) was not.

DNA-only transposons can also be classified based on how they move (**Figure 17-33**). In **replicative transposition**, the transposon is copied from its current site and the copy is inserted at a new site. To use an analogy from word processing, replicative transposition is a "copy-and-paste" mechanism. In contrast, **conservative transposition** is a "cut-and-paste" mechanism: the transposon is excised from its original site and moves to a new site in the genome.

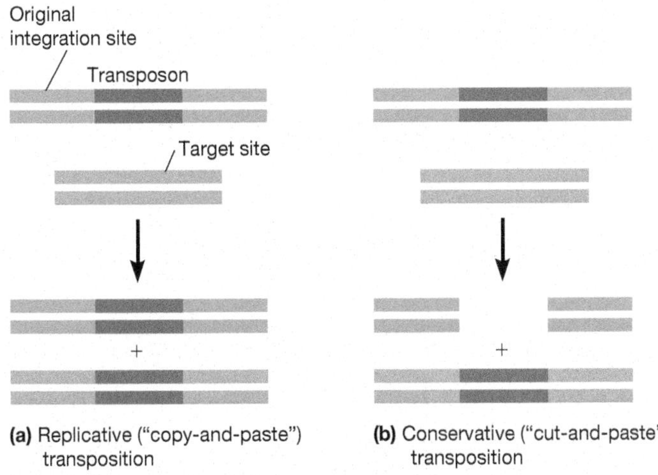

Figure 17-33 Two Types of Transposition. (a) Replicative transposition. Using an analogy from word processing, replicative transposition is a "copy-and-paste" mechanism. **(b)** Conservative transposition (sometimes called "cut-and-paste" transposition).

Bacterial DNA-Only Transposons Can Be Composite or Noncomposite

Types of Bacterial Transposons. There are several types of bacterial transposons (**Figure 17-34**). **Composite transposons** contain *IS (insertion-sequence) elements.* Although different IS elements vary in length and structure, they all encode transposases. In addition, all IS elements begin and end with short repeated sequences that are in reverse order at either end (such repeated sequences are called *inverted repeats*; Figure 17-34a). In this sense, IS elements are themselves very simple transposons. In composite transposons, a number of genes reside between two identical IS elements oriented in opposite directions on either end (Figure 17-34b). Often, one of the genes in the center confers resistance to an antibiotic. Other bacterial transposons, known as **noncomposite transposons**, do not rely on IS elements. Instead, they have short, inverted repeat sequences at their ends and a transposase gene in the middle (Figure 17-34c).

Mechanism of Insertion at a New Location. A transposase gene is a key element in bacterial transposons; when this gene is used to produce the transposase enzyme, the transposon can move to a new site. **Figure 17-35** on page 506 shows the steps in this process. ❶ First, the transposase makes a staggered cut in the DNA at the new insertion site. ❷ Then the transposon inserts between the staggered ends. Finally, ❸ the DNA repair machinery of the bacterium seals the remaining single-stranded gaps. The result is two duplicate sequences,

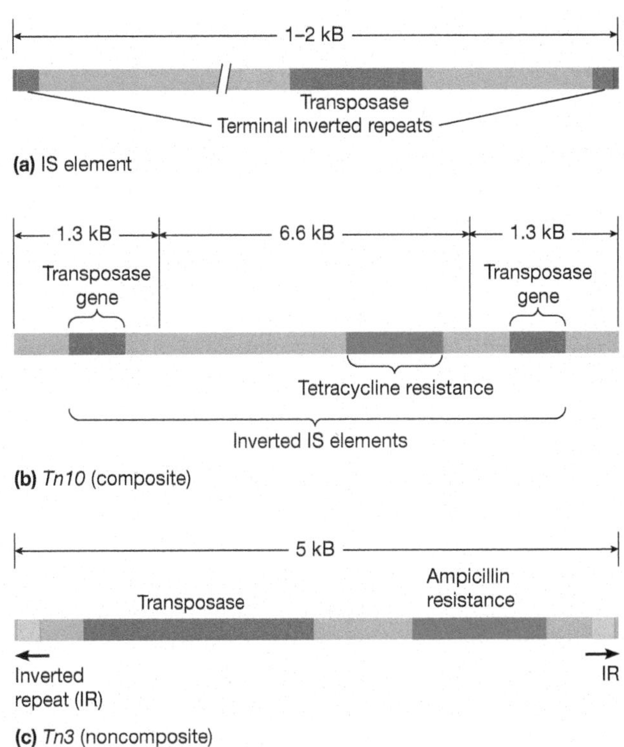

Figure 17-34 Types of Transposable Elements in Bacteria. (a) The basic structure of an insertion sequence (IS) element in *E. coli.* **(b)** *Tn10,* an example of a composite transposon. IS elements are found on either end as inverted repeats. **(c)** *Tn3,* a simple noncomposite transposon. Transposase is encoded within the transposon; the ends are short inverted repeats.

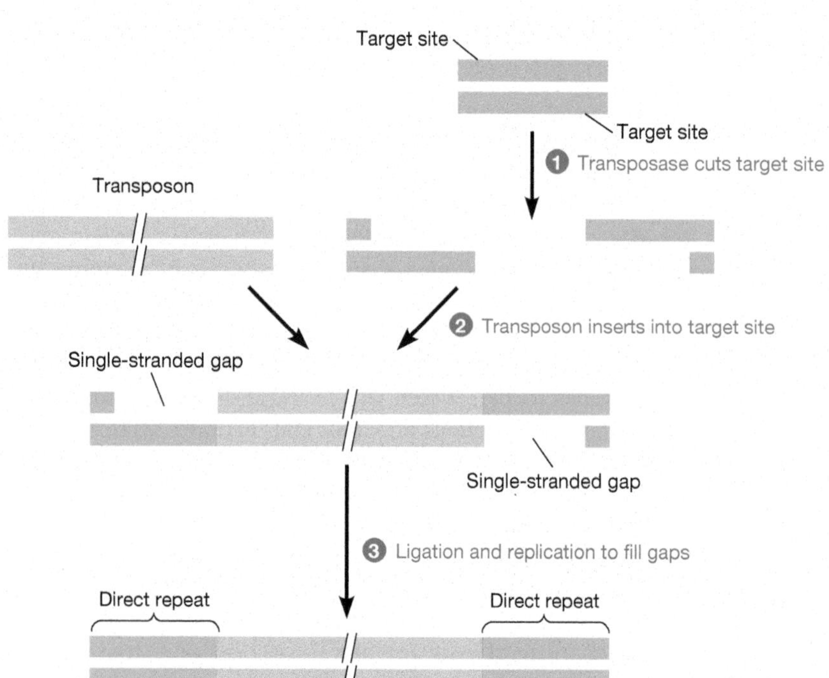

Figure 17-35 Target Site Duplication During Transposon Insertion. Transposase creates a staggered cut site into which the transposon inserts. The gaps are then filled in, creating duplicate direct repeat sequences on either end of the insertion.

one at either end of the inserted transposon. This process is known as *target-site duplication* and is a common feature of transposon insertion.

Eukaryotes Also Have DNA-Only Transposons

The pioneering experiments of Barbara McClintock identified transposons in maize. Other eukaryotes also possess DNA-only transposons. One of the best studied of these is a transposon in the fruit fly, *Drosophila*, known as a *P element*, originally identified by Margaret Kidwell because it inserted into an eye color gene known as *white*, causing unstable eye color phenotypes similar to those seen by McClintock in maize. P elements are very useful tools in the hands of geneticists. When P elements insert into genes at random, they disrupt the function of these genes and allow rapid cloning of the associated gene in a process known as *transposon tagging*. Similar transposons have been also identified in the nematode *Caenorhabditis elegans*.

Retrotransposons

Retrotransposons are a special type of transposable element that use reverse transcription to carry out this movement. As outlined in **Figure 17-36**, the transposition mechanism begins with ❶ transcription of the retrotransposon DNA followed by ❷ translation of the resulting RNA, which produces a protein exhibiting both reverse transcriptase and endonuclease activities (the basic mechanisms of transcription and translation are covered in detail in Chapters 18 and 19, respectively). ❸ Next, the retrotransposon RNA and protein bind to chromosomal DNA at some other location, and ❹ the endonuclease cuts one of the DNA strands. ❺ The reverse transcriptase then uses the retrotransposon RNA as a template to make a DNA copy that is ❻ integrated into the target DNA site.

Although retrotransposons do not transpose themselves very often, they can attain very high copy numbers within a

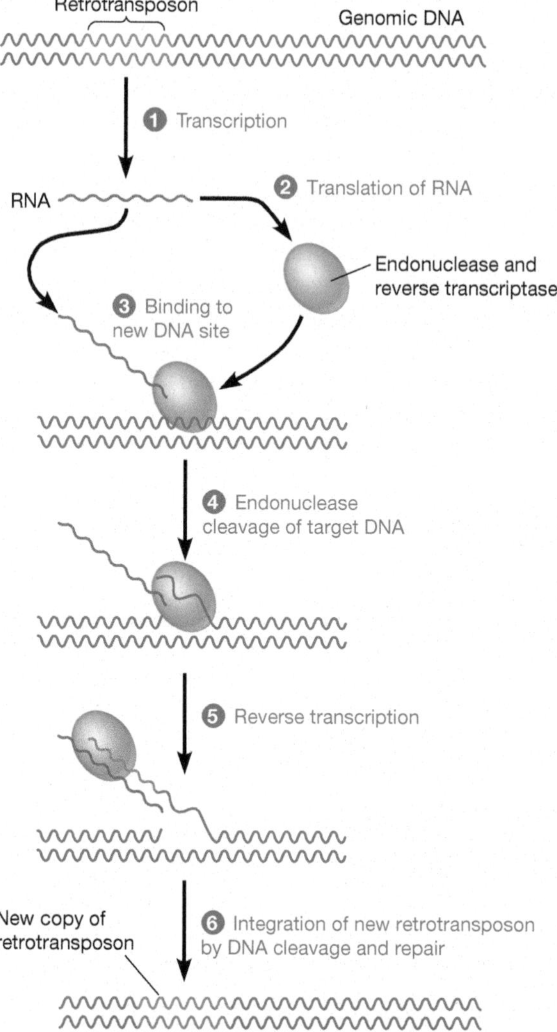

Figure 17-36 Movement of a Retrotransposon. Retrotransposons are duplicated via an RNA intermediate.

genome. One example is the *Alu* family of sequences (Chapter 16). *Alu* sequences are only 300 base pairs long, and they do not encode a reverse transcriptase. But by using a reverse transcriptase encoded elsewhere in the genome, they have sent copies of themselves throughout the genomes of humans and other primates. The human genome contains about a million *Alu* sequences that together represent about 11% of the total DNA. Another type of retrotransposon, called an *L1 element*, is even more prevalent, accounting for roughly 17% of human DNA. The L1 retrotransposon is larger than *Alu* and encodes its own reverse transcriptase and endonuclease, as illustrated in Figure 17-36. The reason that genomes retain so many copies of retrotransposon sequences such as L1 and *Alu* is not well understood, but they are thought to contribute to evolutionary flexibility and variability.

CONCEPT CHECK 17-3

You work in a laboratory that studies *E. coli* and have isolated a strain that lacks the function of the *RecA* gene. Recalling that RecA is similar to the eukaryotic protein Rad51, what molecular defects do you expect this strain to exhibit?

Summary of Key Points

Mastering™ Biology For activities, animations, and review quizzes, go to the study area at www.masteringbiology.com.

17.1 DNA Replication

- DNA is replicated by a semiconservative mechanism in which the two strands of the double helix unwind and each strand serves as a template for the synthesis of a complementary strand.
- Bacterial chromosome replication is typically initiated at a single point and moves in both directions around a circular DNA molecule. In contrast, eukaryotes initiate DNA replication at multiple replicons, and replication proceeds bidirectionally at each replicon. A licensing mechanism involving the binding of MCM proteins to replication origins allows eukaryotes to ensure that DNA is replicated only once prior to each mitosis.
- During replication, DNA polymerases synthesize DNA in the $5' \rightarrow 3'$ direction. Synthesis is continuous along the leading strand but discontinuous along the lagging strand, generating small Okazaki fragments that are subsequently joined by DNA ligase.
- DNA replication is initiated by primase, which synthesizes short RNA primers that are later removed and replaced with DNA. During replication, the double helix is unwound through the combined action of DNA helicases, topoisomerases, and single-stranded DNA-binding proteins. As replication proceeds, a proofreading mechanism based on the $3' \rightarrow 5'$ exonuclease activity of DNA polymerase allows incorrectly base-paired nucleotides to be removed and replaced.
- The ends of linear chromosomal DNA molecules are synthesized by telomerase, an enzyme that uses an RNA template for creating short, repeated DNA sequences at the ends of each chromosome.

17.2 DNA Damage and Repair

- DNA damage arises both spontaneously and through the action of mutagenic chemicals and radiation.
- Excision repair is used to correct DNA damage involving abnormal bases, whereas mismatch repair fixes improperly base-paired nucleotides.
- In emergencies, bypass DNA polymerases carry out translesion synthesis of new DNA across regions where the template DNA is damaged.
- Nonhomologous end-joining can achieve error-prone repair of double-strand breaks in DNA; synthesis-dependent strand annealing and homologous recombination are used to faithfully repair double-strand DNA breaks.

17.3 Homologous Recombination and Mobile Genetic Elements

- Homologous recombination involves breakage and exchange of DNA strands that display extensive sequence similarity.
- Transposable elements are mobile elements that can move to new places in the genome.

Problem Set

Mastering™ Biology For activities, animations, and review quizzes, go to the study area at www.masteringbiology.com.

17-1 Meselson and Stahl Revisited. For each of the alternative models of DNA replication shown in Figure 17-3, use a sketch to indicate the distribution of DNA bands that Meselson and Stahl would have found in their cesium chloride gradients after one and two rounds of replication.

17-2 DNA Replication. Sketch a replication fork of bacterial DNA in which one strand is being replicated discontinuously and the other is being replicated continuously. List six different enzyme activities associated with the replication process, identify the function of each activity, and show where each would be located on the replication fork. In addition, identify the following features on your sketch: DNA template, RNA primer, Okazaki fragments, and single-stranded DNA-binding protein.

17-3 More DNA Replication. The following are observations from five experiments carried out to determine the mechanism of DNA replication in the hypothetical organism *Fungus mungus*. For each experiment, indicate whether the results support (S), refute (R), or have no bearing (NB) on the hypothesis that this fungus replicates its DNA by the same mechanism as that known for *E. coli*. Explain your reasoning in each case.

(a) Neither of the two DNA polymerases of *F. mungus* appears to have an exonuclease activity.

(b) Replicating DNA from *F. mungus* shows discontinuous synthesis on both strands of the replication fork.

(c) Some of the DNA sequences from *F. mungus* are present in multiple copies per genome, whereas other sequences are unique.

(d) Short fragments of *F. mungus* DNA isolated during replication contain both ribose and deoxyribose.

(e) *F. mungus* cells are grown in the presence of the heavy isotopes ^{15}N and ^{13}C for several generations and then grown for one generation in normal (^{14}N, ^{12}C) medium; then DNA is isolated from these cells and denatured. The single strands yield a single band in a cesium chloride density gradient.

17-4 QUANTITATIVE Still More DNA Replication. Suppose you are given a new temperature-sensitive bacterial mutant that grows normally at 37°C but cannot replicate its chromosomes properly at 42°C. To investigate the nature of the underlying defect, you incubate the cells at 42°C with radioactive substrates required for DNA synthesis. After 1 hour you find that the cell population has doubled its DNA content, suggesting that DNA replication can still occur at 42°C. Moreover, centrifugation reveals that all of this DNA has the same large molecular weight as the original DNA present in the cells. When the DNA is denatured, however, you discover that 75% of the resulting single-stranded DNA has a molecular weight that is half that of the original double-stranded DNA, and the remaining 25% of the DNA has a much lower molecular weight. Based on the preceding results, what gene do you think is defective in these cells? Explain how such a defect would account for the experimental results that you observed.

17-5 The Minimal Chromosome. To enable it to be transmitted intact from one cell generation to the next, the linear DNA molecule of a eukaryotic chromosome must have appropriate nucleotide sequences making up three special kinds of regions: origins of replication (at least one), a centromere, and two telomeres. What would happen if such a chromosomal DNA molecule somehow lost

(a) all of its origins of replication?

(b) all of the DNA constituting its centromere?

(c) one of its telomeres?

17-6 DNA Damage and Repair. Indicate whether each of the following statements is true of depurination (DP), deamination (DA), or pyrimidine dimer formation (DF). A given statement may be true of any, all, or none of these processes.

(a) This process is caused by spontaneous hydrolysis of a glycosidic bond.

(b) This process is induced by ultraviolet light.

(c) This can happen to guanine but not to cytosine.

(d) This can happen to thymine but not to adenine.

(e) This can happen to thymine but not to cytosine.

(f) Repair involves a DNA glycosylase.

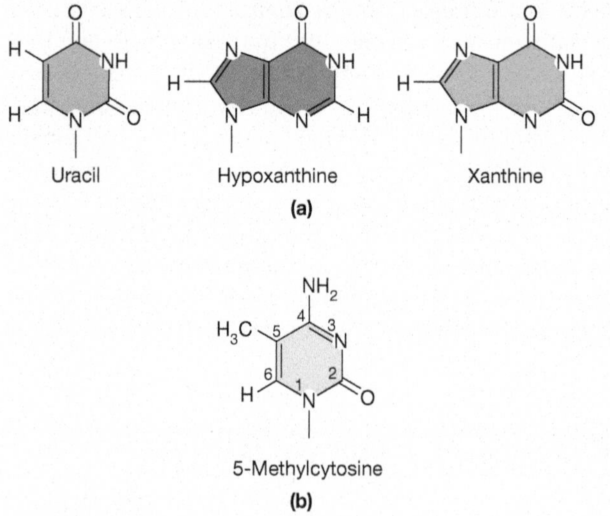

Uracil Hypoxanthine Xanthine

(a)

5-Methylcytosine

(b)

Figure 17-37 Structures of Several Nonstandard Purines and Pyrimidines. See Problem 17-7.

(g) Repair involves an endonuclease.

(h) Repair involves DNA ligase.

(i) Repair depends on the existence of separate copies of the genetic information in the two strands of the double helix.

(j) Repair depends on cleavage of both strands of the double helix.

17-7 Nonstandard Purines and Pyrimidines. Shown in **Figure 17-37** are four nonstandard nitrogenous bases that are formed by the deamination or methylation of naturally occurring bases in DNA.

(a) Indicate which base in DNA must be deaminated to form each of these bases.

(b) Why are there only three bases shown, when DNA contains four bases?

(c) Why is it important that none of the bases shown in Figure 17-37a occurs naturally in DNA?

(d) Figure 17-37b shows 5-methylcytosine, a pyrimidine that arises naturally in DNA when cellular enzymes methylate cytosine. Why is the presence of this base likely to increase the probability of a mutation?

17-8 Mismatch Repair. Given the following mismatch (underlined) following replication in *E. coli*, what will be the sequence of the correctly repaired DNA (CH_3 = methyl group)?

Mutant sequence:

```
   CH3
    |
5'-GATCTCAGGC-3'
3'-CTAGAGGCCG-5'
```

(The underlined mismatched base is the G at position 7 of the lower strand, 3'-CTAGAG**G**CCG-5'.)

17-9 Homologous Recombination. Bacterial cells use at least three different pathways for carrying out genetic recombination. One of these three pathways utilizes an enzyme complex called RecBCD, which binds to double-strand breaks in DNA and exhibits both helicase and single-strand nuclease activities. Briefly describe a model showing how the RecBCD enzyme complex might set the stage for genetic recombination.

17-10 Mutatis Mutandis. You work in a lab that uses the nematode *C. elegans* as a model organism. You find that one of the strains you are using has an inexplicably high incidence of mutations. It turns out that this is because the strain is carrying an active transposon that was accidentally introduced into the strain. Such strains are called "mutator" strains by labs that work on worms. Why would a transposon cause such a high incidence of mutation?

17-11 DATA ANALYSIS **Healing Light.** The data shown in **Figure 17-38** are adapted from work that studied whether human cells possess the DNA photolyase enzyme. Extracts of protein from *E. coli* (EC) or human HeLa cells (HC) were mixed with a short oligonucleotide (5′-GCTCGAGCTA**T<>T**AACGTCAG-3′) containing a thymine dimer (**T<>T**). The extracts were treated with the indicated number of light pulses designed to activate the photolyase enzyme if it is present. These samples were digested with the restriction enzyme MseI, which cuts the sequence T/TAA in the oligonucleotide (only if no thymine dimer is present), and the resulting DNA fragments were separated using polyacrylamide gel electrophoresis.

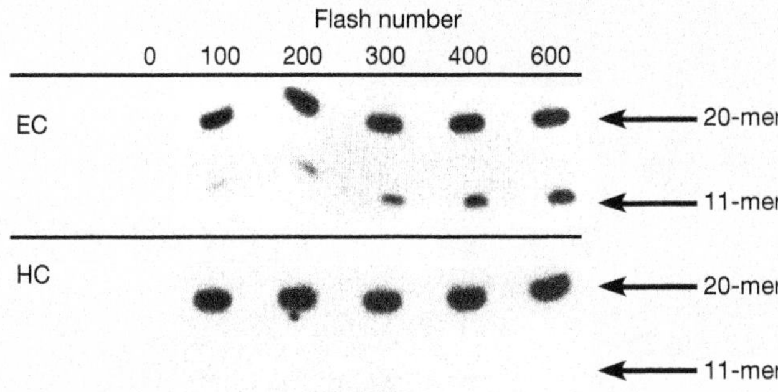

Figure 17-38 Repair of Thymine Dimers using *E. coli* (EC) and Human Cell Extracts (HC). See Problem 17-11.

(a) If the thymine dimer is successfully repaired, what size band(s) would you expect to see when the oligonucleotide is digested with MseI?

(b) Which species shows evidence of containing DNA photolyase, and why?

(c) Does the number of light pulses affect the activity of the enzyme? Explain using the data.

18 Gene Expression: I. Transcription

Transcription of RNA.
mRNA in the process of being transcribed (colorized TEM).

So far, we have described DNA as the genetic material of cells and organisms. We have come to understand its structure, chemistry, and replication, as well as the way it is packaged into chromosomes and replicated during cell divisions. Now we are ready to explore how DNA is *expressed*—that is, how the coded information it contains is used to guide the production of RNA and protein molecules. Our discussion of this important subject is divided among three chapters. The present chapter deals with how information stored in DNA guides the synthesis of RNA molecules in the process we call *transcription*. Chapter 19 describes the nature of the genetic code and how RNA molecules are then used to guide the synthesis of specific proteins in the process known as *translation*. Finally, Chapter 20 elaborates on the various mechanisms used by cells to control transcription and translation, thereby leading to the *regulation* of gene expression. To put these topics in context, we start here with an overview of the roles played by DNA, RNA, and proteins in gene expression.

Mastering™ Biology www.masteringbiology.com

18.1 The Directional Flow of Genetic Information

The flow of genetic information in cells generally proceeds from DNA to RNA to protein (see Figure 16-1). DNA (more precisely, a segment of one DNA strand) first serves as a template for the synthesis of an RNA molecule, which in most cases then directs the synthesis of a particular protein. (In a few cases, the RNA is the final product of gene expression and functions as such within the cell.) The principle of directional information flow from DNA to RNA to protein is known as the **central dogma of molecular biology**, a term coined in 1958 by Francis Crick soon after the double-helical model of DNA was first proposed. This principle is summarized as follows:

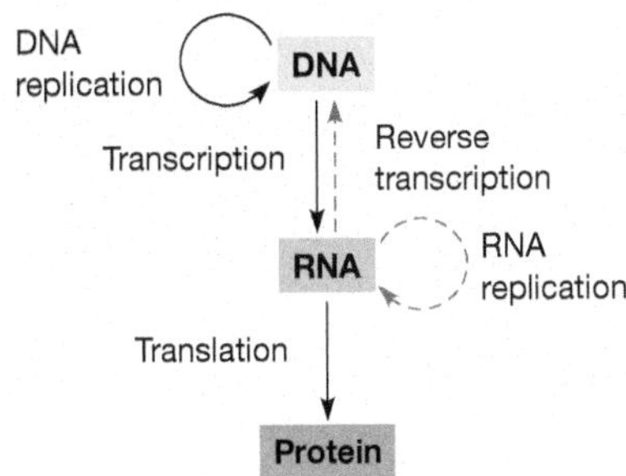

The main flow of genetic information involves replication of DNA, transcription of information carried by DNA in the form of RNA, and translation of this information from RNA into protein. The term **transcription** is used when referring to RNA synthesis using DNA as a template to emphasize that this phase of gene expression is simply a transfer of information from one nucleic acid to another, so the basic "language" remains the same. In contrast, protein synthesis is called **translation** because it involves a language change—from the nucleotide sequence of an RNA molecule to the amino acid sequence of a polypeptide chain.

In the years since it was first formulated by Crick, the central dogma has been refined in various ways. For example, many viruses with RNA genomes have been found to synthesize RNA molecules using RNA as a template. In the central dogma diagram pictured above, the dashed arrow from RNA back to itself has been added to represent RNA replication. Other RNA viruses, such as HIV, carry out *reverse transcription*, whereby the viral RNA is used as a template for DNA synthesis—a "backward" flow of genetic information. Likewise, the dashed arrow from RNA back to DNA in the central dogma diagram has been added to represent reverse transcription. But despite these variations on the original model, the principle that information flows from DNA to RNA to protein remains the main operating principle by which all cells use their genetic information.

Transcription and Translation Involve Many of the Same Components in Prokaryotes and Eukaryotes

RNA that is translated into protein is called **messenger RNA (mRNA)** because it carries a genetic message from DNA to the ribosomes, where protein synthesis actually takes place. In addition to mRNA, two other types of RNA are directly involved in protein synthesis: **ribosomal RNA (rRNA)** (an integral component of the ribosome) and **transfer RNA (tRNA)**

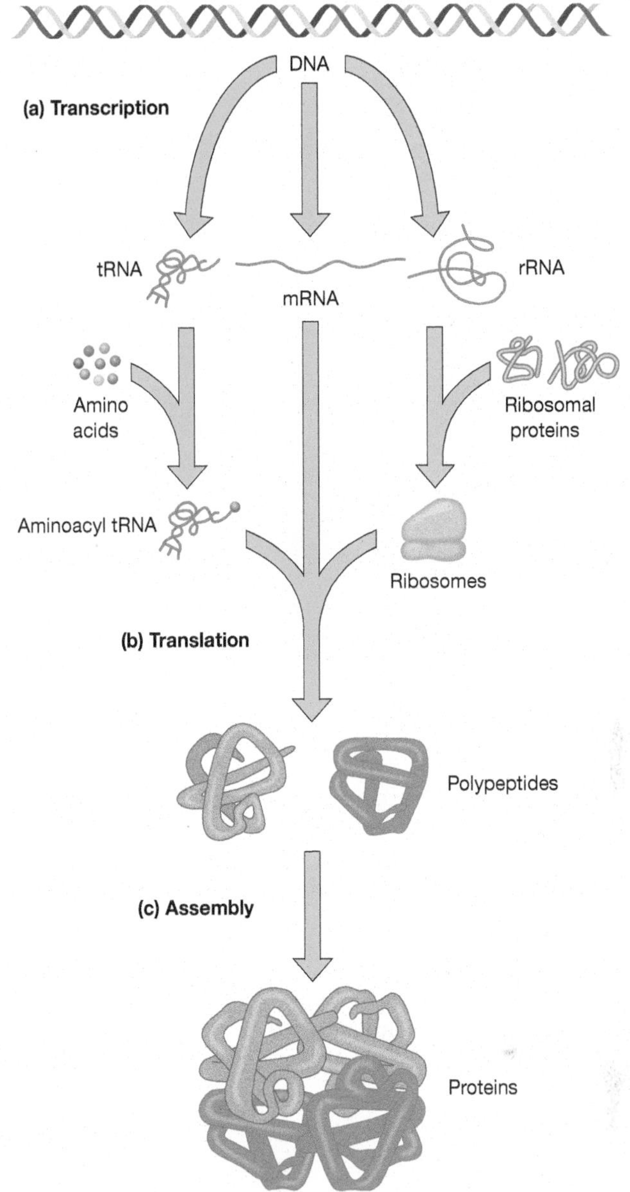

Figure 18-1 RNAs as Intermediates in the Flow of Genetic Information. All three major classes of RNA—tRNA, mRNA, and rRNA—are **(a)** synthesized by transcription of the appropriate DNA sequences (genes) and **(b)** involved in the subsequent process of translation (polypeptide synthesis). The appropriate amino acids are brought to the mRNA and ribosome by tRNA. A tRNA molecule carrying an amino acid is called an aminoacyl tRNA. Polypeptides then fold and **(c)** assemble into functional proteins. The specific polypeptides shown here are the globin chains of the protein hemoglobin.

(an intermediary that translates the coded base sequence of messenger RNA and brings the appropriate amino acids to the ribosome). Note that ribosomal and transfer RNAs do not themselves encode proteins; genes encoding these two types of RNA are examples of genes whose final products are RNA molecules rather than protein chains. The involvement of all three major classes of RNA in the overall flow of genetic information from DNA to protein is outlined in **Figure 18-1**.

Where Transcription and Translation Occur Differs in Prokaryotes and Eukaryotes

The basic steps—and the fundamental molecular details—of the flow of genetic information are similar in prokaryotes and eukaryotes, but there are important differences in *where*

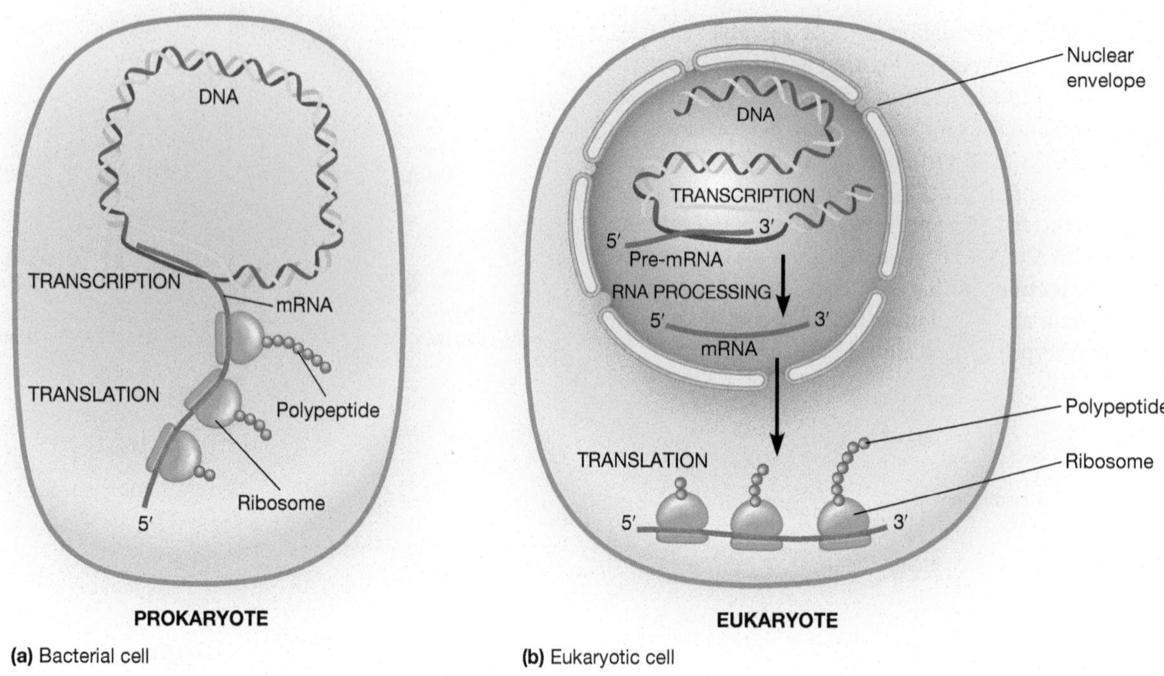

Figure 18-2 Prokaryotic Versus Eukaryotic Transcription. (a) In a prokaryotic cell, which lacks a nuclear envelope, ribosomes can begin translating mRNAs into protein while they are being transcribed. **(b)** In eukaryotic cells, compartmentalization separates transcription and RNA processing, which occur in the nucleus, from translation, which occurs in the cytosol.

transcription and translation occur (**Figure 18-2**). Because prokaryotes do not have a nuclear envelope, their DNA is not physically segregated from ribosomes and the translational machinery (Figure 18-2a). In contrast, the compartmentalization of eukaryotic cells leads to spatial separation of transcription and translation (Figure 18-2b). The lack of compartmentalization in prokaryotic cells allows translation of an mRNA to begin *before* its transcription is complete (**Figure 18-3**). An electron micrograph that shows this coupling of transcription with translation is seen in Figure 18-3a. Notice

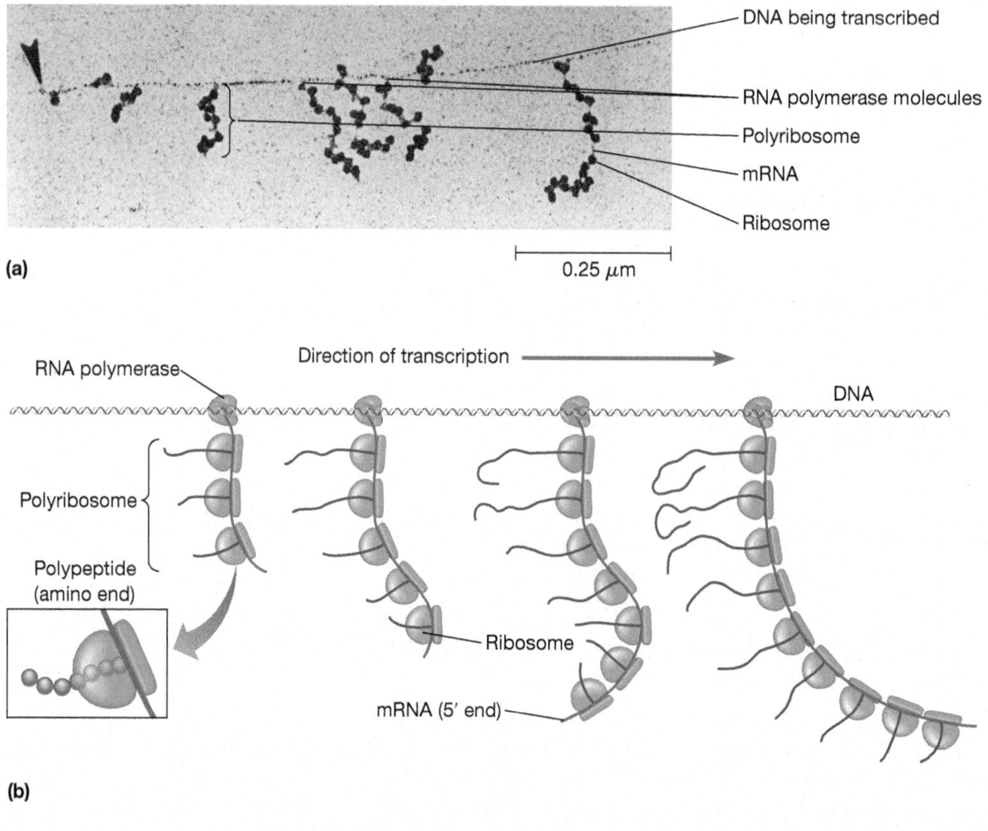

Figure 18-3 Coupling of Transcription and Translation in Bacterial Cells. (a) This electron micrograph shows *Escherichia coli* DNA being transcribed by RNA polymerase molecules moving from left to right. Attached to each polymerase molecule is a strand of mRNA still being transcribed. The large dark particles attached to each growing mRNA strand are ribosomes that are actively translating the partially complete mRNA. (The polypeptides being synthesized are not visible.) The arrowhead marks the transcriptional start site. A cluster of ribosomes attached to a single mRNA strand is called a polyribosome (TEM). **(b)** An illustration showing the relationship of polyribosomes to DNA.

that multiple ribosomes can be bound to each mRNA, forming what is called a *polyribosome* (illustrated in Figure 18-3b). Because transcription takes place in the nucleus, the resulting RNA can be modified into mature mRNA prior to nuclear export. Once in the cytosol, this mRNA has access to ribosomes, and translation can occur.

In Some Cases RNA Is Reverse Transcribed into DNA

Transcription generally proceeds in the direction described in the central dogma, with DNA serving as a template for RNA synthesis. In certain cases, however, the process can be reversed and RNA serves as a template for DNA synthesis. This process of *reverse transcription* is catalyzed by the enzyme **reverse transcriptase**, first discovered by Howard Temin and David Baltimore in certain viruses with RNA genomes. Viruses that carry out reverse transcription are called *retroviruses* (see Chapter 16). Examples of retroviruses include some important pathogens, such as the *human immunodeficiency virus (HIV)*, which causes *acquired immune deficiency syndrome (AIDS)*, and a number of viruses that cause cancers in animals. Reverse transcription also occurs in normal eukaryotic cells in the absence of viral infection. Much of it involves the mobile DNA elements called *retrotransposons* (see Chapter 17).

CONCEPT CHECK 18-1

Compare and contrast bacterial and eukaryotic transcription, focusing on the general components used and the location where it occurs relative to translation. What are the advantages gained by the differences that you've highlighted?

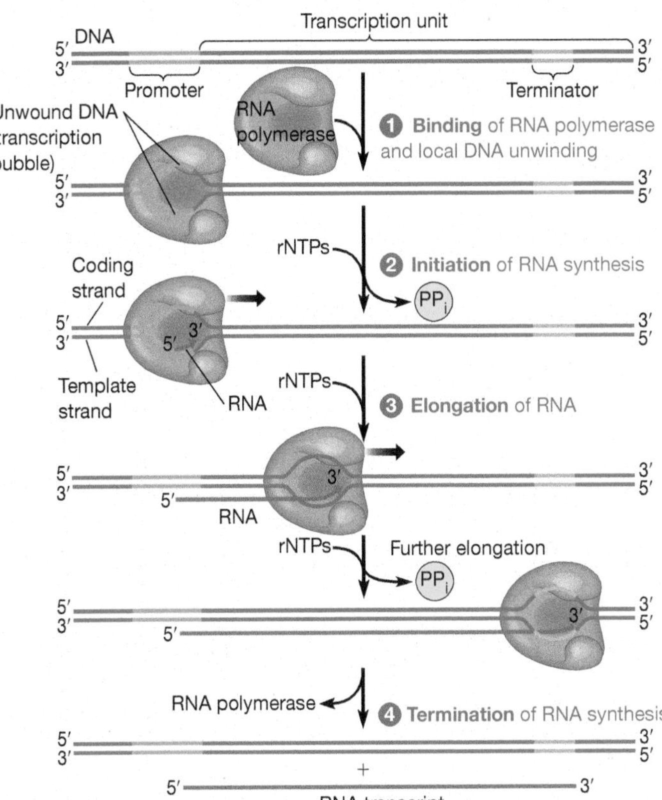

Figure 18-4 An Overview of Transcription. Transcription of DNA occurs in four main stages: ❶ binding of RNA polymerase to DNA at a promoter, ❷ initiation of transcription on the template DNA strand, ❸ subsequent elongation of the RNA chain, and ❹ eventual termination of transcription, accompanied by the release of RNA polymerase and the completed RNA product from the DNA template. RNA polymerase moves along the template strand of the DNA in the 3′ → 5′ direction, and the RNA molecule grows in the 5′ → 3′ direction. The general scheme illustrated here holds for transcription in all organisms. rNTPs (ribonucleoside triphosphate molecules) = rATP, rGTP, rCTP, and rUTP.

18.2 Mechanisms of Transcription

Now that you have been introduced to the directional flow of genetic information, we can discuss the specific steps involved in the pathway from DNA to protein. The first stage in this process is the transcription of a nucleotide sequence in DNA into a sequence of nucleotides in RNA. RNA is chemically similar to DNA, but it contains ribose instead of deoxyribose as its sugar, has the base uracil (U) in place of thymine (T), and is usually single stranded. As in other areas of molecular genetics, the fundamental principles of RNA synthesis were first elucidated in bacteria. For that reason, we will start with transcription in bacteria and then move on to consider eukaryotes.

Transcription Involves Four Stages: RNA Polymerase Binding, Initiation, Elongation, and Termination

Transcription is the synthesis of an RNA molecule whose base sequence is complementary to the base sequence of a template DNA strand. **Figure 18-4** provides an overview of RNA synthesis from a single **transcription unit**, a segment of DNA whose transcription gives rise to a single, continuous RNA molecule. Transcription of DNA is carried out by the enzyme **RNA polymerase**, which catalyzes the synthesis of RNA using DNA as a template. Although transcription is a complicated process, it can be thought of in four distinct stages: binding, initiation, elongation, and termination. The process of transcription begins with the binding of RNA polymerase to a DNA **promoter** sequence, which triggers local unwinding of the DNA double helix. Using one of the two DNA strands as a template, RNA polymerase then initiates the synthesis of an RNA chain. After initiation has taken place, the RNA polymerase molecule moves along the DNA template, unwinding the double helix and elongating the RNA chain as it goes. The unwound region of DNA is known as a *transcription bubble*. During this process, the enzyme catalyzes the polymerization of nucleotides in an order determined by their base pairing with the DNA template strand. Eventually, RNA polymerase dissociates from the DNA template, which leads to *termination* of RNA synthesis and release of the completed RNA molecule.

Bacterial Transcription Involves σ Factor Binding, Initiation, Elongation, and Termination

We will now look at each stage in detail as it occurs in *Escherichia coli*. You can refer to Figure 18-4 throughout this discussion to see how each step fits into the overall process.

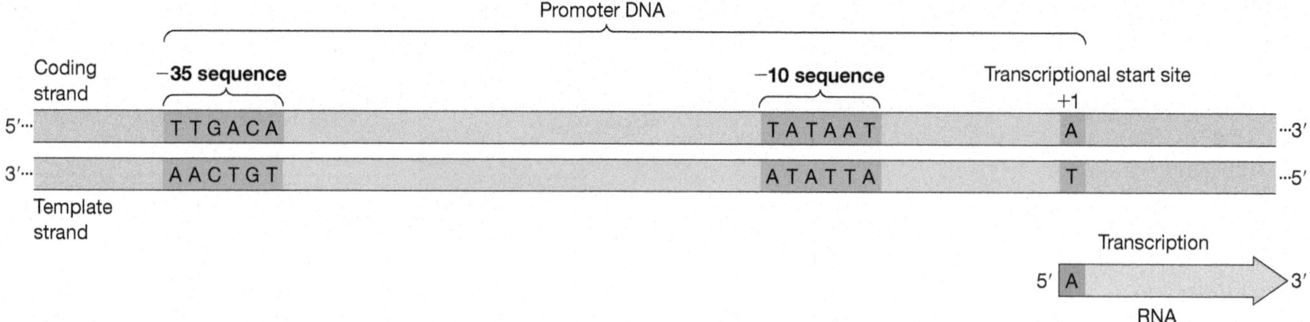

Figure 18-5 Organization of a Bacterial Promoter Sequence. The promoter region in bacteria is a stretch of about 40 bp adjacent to and including the transcription start site. By convention, the critical DNA sequences are given as they appear on the coding strand. Essential features of the promoter are the start site (designated +1 and usually an A), the six-nucleotide −10 sequence, and the six-nucleotide −35 sequence. As their names imply, the two key sequences are located approximately 10 nucleotides and 35 nucleotides upstream from the start site. The sequences shown here are consensus sequences, which means they have the most commonly found base at each position. The numbers of nucleotides separating the consensus sequences from each other and from the start site are important for promoter function, but the identity of these nucleotides is not.

Binding of RNA Polymerase to a Promoter Sequence. The first step in RNA synthesis is the *binding* of RNA polymerase to a specific DNA sequence of several dozen base pairs at the promoter site that determines where RNA synthesis starts and which DNA strand is to serve as the template strand. Each transcription unit has a promoter site located near the beginning of the DNA sequence to be transcribed. By convention, promoter sequences are described in the $5' \rightarrow 3'$ direction on the coding strand, which is the strand that lies opposite the template strand. The terms **upstream** and **downstream** are used to refer to DNA sequences located toward the 5′ end and the 3′ end of the coding strand, respectively. Therefore, the region where the promoter is located is said to be upstream of the transcribed sequence. Binding of RNA polymerase to the promoter is mediated by the sigma subunit (discussed below) and leads to unwinding of a short stretch of DNA in the area where transcription will begin, exposing the two separate strands of the double helix.

Promoter sequences were initially identified by *DNA footprinting* and *electrophoretic mobility shift assays (EMSAs)*, techniques for locating the DNA region to which a DNA-binding protein has become bound. More recently, *chromatin immunoprecipitation (ChIP)* has been used to assess binding of proteins to specific genomic DNA sequences in eukaryotes (see Key Technique, pages 516–517). Sequences essential to the promoter region have also been identified by deleting or adding specific base sequences to cloned genes and then testing the ability of the altered DNA to be transcribed by RNA polymerase. Such techniques have revealed that DNA promoter sites differ significantly among bacterial transcription units. How, then, does a single kind of RNA polymerase recognize them all? Enzyme recognition and binding, it turns out, depend only on several very short sequences located at specific positions within each promoter site. The identities of the nucleotides making up the rest of the promoter are not important for this purpose.

Figure 18-5 highlights the essential sequences in a typical bacterial promoter. The transcriptional *start site* is almost always a purine and often an adenine. Approximately 10 bases upstream of the start site is the six-nucleotide sequence TATAAT, called the *−10 sequence*, or the *Pribnow box* after its discoverer. By convention, the nucleotides are numbered from the transcription start site (+1), with positive numbers to the right (downstream) and negative ones to the left (upstream). The −1 nucleotide is immediately upstream of the start site (there is no "0"). At or near the −35 position is the six-nucleotide sequence TTGACA, called the *−35 sequence*.

The −10 and the −35 sequences (and their positions relative to the start site) have been conserved during evolution, but they are not identical in all bacterial promoters or even in all the promoters in a single genome. For example, the −10 sequence in the promoter for one of the *E. coli* tRNA genes is TATGAT, whereas the −10 sequence for a group of genes needed for lactose breakdown in the same organism is TATGTT, and the sequence given in Figure 18-5 is TATAAT. As with origins of replication (described in Chapter 17), the particular promoter sequences shown in the figure are *consensus sequences*, which consist of the most common nucleotides at each position within a given sequence. Mutations that cause significant deviations from the consensus sequences tend to interfere with promoter function and may even eliminate promoter activity entirely. In addition to the −35 and −10 sequences, many bacterial promoters also contain *upstream (UP) elements*. UP elements are associated with particularly strong promoters, such as those directing expression of ribosomal RNA (rRNA) genes.

Bacterial cells have a single kind of RNA polymerase that synthesizes all three major classes of RNA—mRNA, tRNA, and rRNA. The enzymes from different bacteria are quite similar, and the RNA polymerase from *E. coli* has been especially well characterized. It is a large protein consisting of two α subunits, two β subunits that differ enough to be identified as β and β', and a dissociable subunit called the **sigma (σ) factor** (**Figure 18-6**). Although the *core enzyme* lacking the sigma subunit is competent to carry out RNA synthesis, the *holoenzyme* (complete enzyme containing all of its subunits) is required to ensure initiation at the proper sites within a DNA molecule. As shown in step 1 of the figure, parts of the

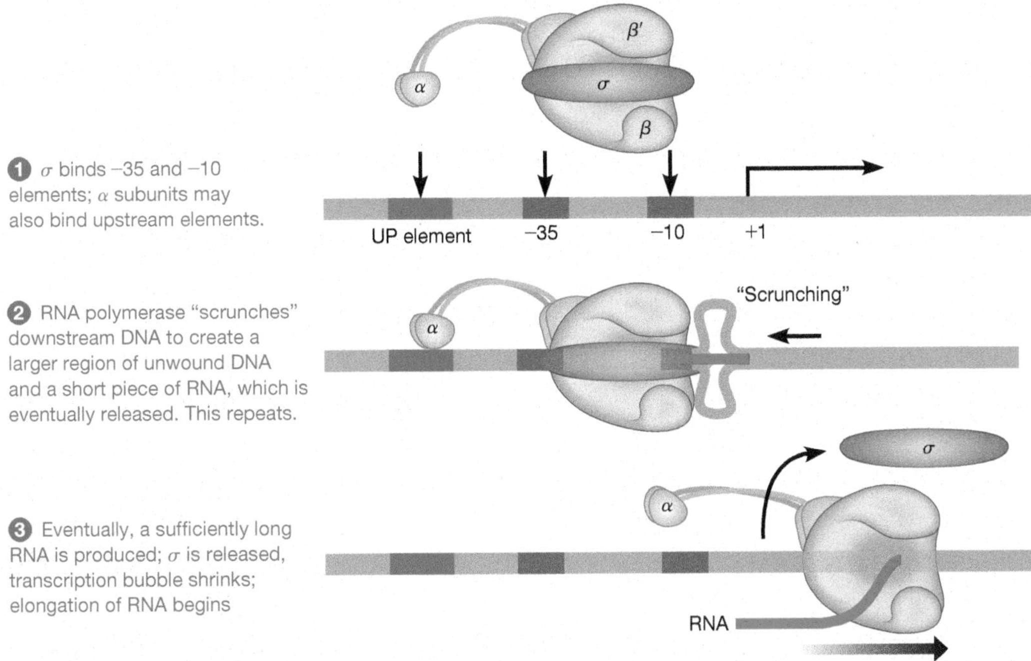
β′
α
σ
β
1 σ binds –35 and –10 elements; α subunits may also bind upstream elements.
UP element
–35
–10
+1
2 RNA polymerase "scrunches" downstream DNA to create a larger region of unwound DNA and a short piece of RNA, which is eventually released. This repeats.
"Scrunching"
3 Eventually, a sufficiently long RNA is produced; σ is released, transcription bubble shrinks; elongation of RNA begins
RNA

Figure 18-6 Transcriptional Initiation in Bacteria. RNA polymerase is a large, multisubunit protein that binds to –35 and –10 sequences in the promoter via the σ subunit. An additional upstream (UP) element may be bound by the α subunits. While tightly bound to the promoter, RNA polymerase pulls additional DNA toward itself ("scrunching"). Once a sufficiently long RNA is produced, the polymerase escapes the promoter, releasing the σ subunit.

sigma subunit play a critical role in this process by promoting the binding of RNA polymerase to the –35 and –10 elements. Bacteria contain a variety of different sigma factors that selectively initiate the transcription of specific categories of genes (as we will discuss in Chapter 20). In addition, the α subunits can interact with UP elements to enhance promoter binding. Another element, known as the *discriminator element*, lies just downstream of the –10 site; it also stabilizes RNA polymerase at the promoter site.

Initiation of RNA Synthesis. Once an RNA polymerase molecule has bound to a promoter site and locally unwound the DNA double helix, *initiation* of RNA synthesis can take place (see step 2 in Figure 18-4). One of the two exposed segments of single-stranded DNA serves as the template for the synthesis of RNA, using incoming ribonucleoside triphosphate molecules (rNTPs) as substrates. The DNA strand that carries the promoter sequence determines which way the RNA polymerase faces, and the enzyme's orientation in turn determines which DNA strand it transcribes (see steps 2–3 in Figure 18-4). As soon as the first two incoming rNTPs are hydrogen-bonded to the complementary bases of the DNA template strand at the start site, RNA polymerase catalyzes the formation of a phosphodiester bond between the 3′-hydroxyl group of the first rNTP and the 5′-phosphate of the second, accompanied by the release of pyrophosphate (PP_i), which can be broken down further. It is the energy released by the breaking of these phosphate bonds that fuels the catalysis of RNA by the enzyme. The polymerase then advances along the template strand as additional nucleotides are added one by one, the 5′-phosphate of each new nucleotide joining to the 3′-hydroxyl group of the growing RNA chain. Thus, unlike DNA polymerase (see Chapter 17), RNA polymerase does not need primers to initiate polynucleotide synthesis. Therefore, we say that RNA polymerases are capable of *de novo* (from the beginning) RNA synthesis.

RNA polymerase initiates synthesis until the chain is nine nucleotides long or shorter. Based on a number of studies, however, the polymerase repeats this seemingly futile initial transcription repeatedly, producing and releasing short pieces of RNA in a process known as *abortive synthesis*. During this process, the polymerase remains stationary, firmly attached to the promoter. It pulls a bit of downstream DNA into its interior, unwinding it to produce a bulge of 22–24 nucleotides. This movement is known as *scrunching* of the DNA (step 2 in Figure 18-6).

Eventually, a piece of RNA ten nucleotides or longer is produced, and this weakens the interactions of RNA polymerase with the promoter, allowing it to escape the promoter. Once it escapes, the RNA polymerase releases sigma factor, allowing the rest of the polymerase molecule to move along the DNA to transcribe an elongating RNA (step 3 in Figure 18-6).

Elongation of the RNA Chain. Chain *elongation* now continues as RNA polymerase moves along the DNA molecule, untwisting the helix bit by bit and adding one complementary nucleotide at a time to the growing RNA chain (**Figure 18-7** on page 518). The β subunits contain regions that act like "pincers" that bind DNA and move it into the active site of the enzyme, rather like the classic video game Pac-Man. As RNA polymerase moves along DNA, different channels in the enzyme allow components to enter (incoming DNA, ribonucleotides) and leave (DNA that has already been transcribed, newly formed

Key Technique | Hunting for DNA-Protein Interactions

PROBLEM: The interaction of proteins with DNA is central to the control of many cellular activities including DNA replication, DNA transcription, and DNA recombination and repair. One step toward piecing together the details of these processes is to identify the specific DNA sequences with which particular DNA-binding proteins interact.

SOLUTION: A variety of techniques are available for assessing where a protein and a particular DNA sequence can physically bind one another. These techniques vary in what they show, but they all use approaches that infer that a physical interaction occurs.

Key Tools: Purified DNA-binding proteins; purified DNA sequences; antibodies that recognize the protein of interest; DNA microarrays or other genome sequence information.

Details: Electrophoretic Mobility Shift Assay (EMSA): Proteins and nucleic acids can be separated by size by measuring how quickly they migrate through a gel to which an electric field has been applied—a technique known as electrophoresis (see Chapter 21). This approach can also be used to assess whether a particular piece of DNA and a protein interact. If a piece of DNA has a protein bound to it, it will migrate more slowly through a gel than the corresponding free DNA. This delayed migration is evidenced as a *gel shift,* a change in the position of the bound sequence relative to its protein-free form. For this reason, this technique is referred to as an *electrophoretic mobility shift assay (EMSA).*

In a typical EMSA, a short double-stranded DNA fragment is labeled (either radioactively or using a fluorescent molecule) so that it can be detected in small quantities. The resulting DNA is then mixed with the protein of interest, and the mixture is separated on a special nondenaturing gel. If the protein binds the DNA, a shift in migration will be detected (**Figure 18A-1**). A similar approach can be used for RNA-binding proteins.

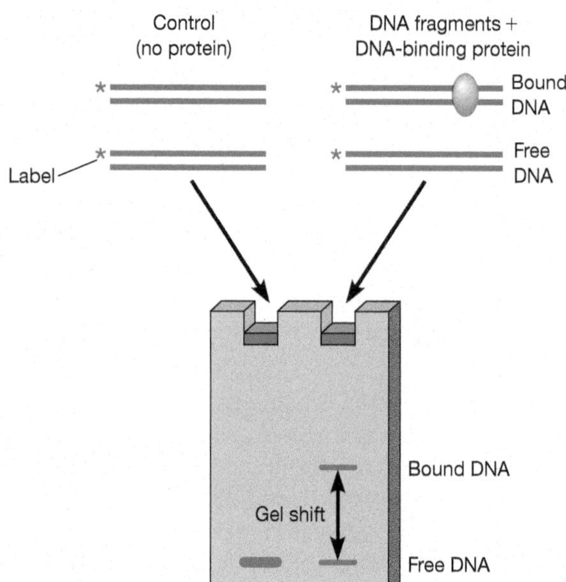

Figure 18A-1 Electrophoretic Mobility Shift Assay (EMSA). Binding of protein to DNA slows the rate of migration of the bound DNA in a gel.

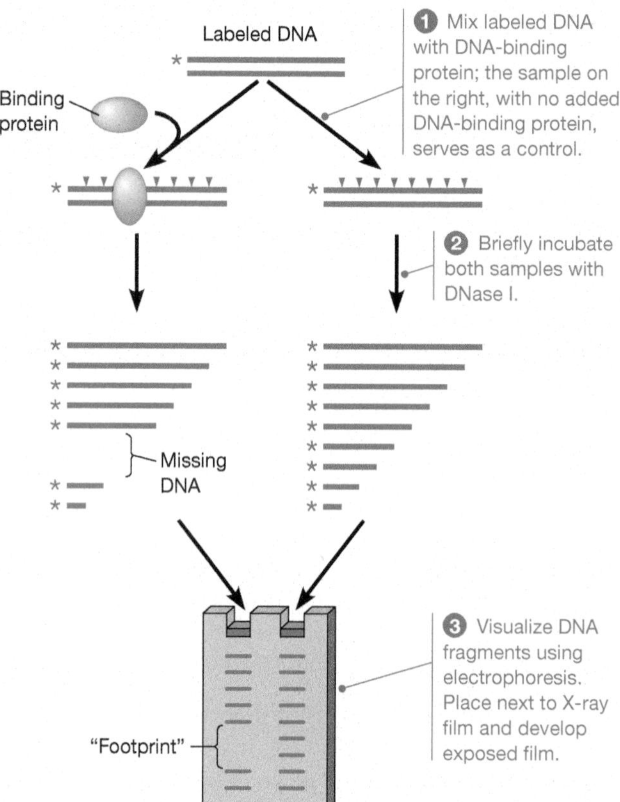

Figure 18A-2 DNA Footprinting. Binding of protein to DNA protects it from digestion by nucleases.

DNA "Footprinting": DNA footprinting is another technique that has been used to locate the DNA sites where specific proteins attach. The underlying principle is that the binding of a protein to a particular DNA sequence should protect that sequence from degradation by enzymes or chemicals. A common version of footprinting, outlined in **Figure 18A-2**, employs a DNA-degrading enzyme called *DNase I,* which attacks the phosphodiester linkages between nucleotides more or less at random. In this example, the starting material is a DNA fragment that has been isolated from genomic DNA and labeled at one end with a radioactive phosphate (indicated by a red asterisk). The fragment is known to contain the binding site of interest.

To begin, ❶ the labeled sample is mixed with the DNA-binding protein under study. Another sample, without the added protein, serves as the control. ❷ Both samples are briefly incubated with a low concentration of DNase I—conditions ensuring that most of the DNA molecules in each sample will be cleaved only once. The arrowheads indicate possible cleavage sites in the DNA.

The two incubation mixtures are submitted to electrophoresis and ❸ visualized by placing the gel next to X-ray film, which, when developed, shows the positions of the bands in the gel. The control lane (right) has nine bands because every possible cleavage site has been cut. However, the other lane (left) is missing some bands because protein bound to the DNA protected some cleavage sites during DNase treatment. The blank region is the "footprint" that identifies the location and length of the DNA sequence in contact with protein.

Chromatin Immunoprecipitation (ChIP): Although in vitro assays for protein-DNA binding such as EMSA and footprinting can be informative, cell biologists usually want to know whether a protein binds a particular DNA regulatory element in living cells. Although a typical DNA-binding protein can bind a large number of different DNA sites, in actuality DNA-binding proteins will only be attached to a fraction of these sites in any particular cell. This is because other proteins present in some cells, but not in others, may prevent its binding or because the protein needs the assistance of other proteins to bind to DNA tightly, and only some cells contain these other proteins. *Chromatin immunoprecipitation (ChIP)* is a powerful in vivo answer to this investigative challenge (**Figure 18A-3**). The key to ChIP is the availability of an antibody specific for the DNA-binding protein of interest.

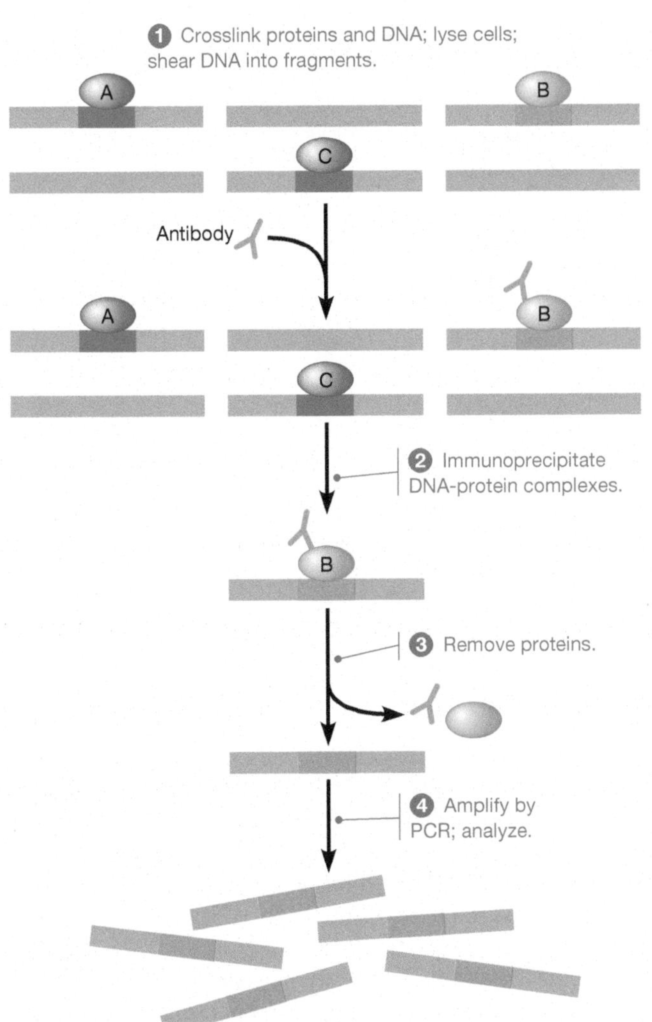
1 Crosslink proteins and DNA; lyse cells; shear DNA into fragments.
A
B
C
Antibody
A
B
C
2 Immunoprecipitate DNA-protein complexes.
B
3 Remove proteins.
4 Amplify by PCR; analyze.

Figure 18A-3 Chromatin Immunoprecipitation (ChIP). ChIP detects where a protein binds within cellular DNA. PCR can be used to amplify a specific sequence after it has been isolated.

Step 1: Because many protein-DNA interactions are transient in vivo, ChIP assays begin with a stabilization step. Cells are treated with formaldehyde to induce formation of covalent crosslinks between DNA and any proteins bound to it. The treated cells are then lysed, and their DNA is sheared into small fragments (200–300 bp each) using a special vibrating probe that operates at ultrasonic frequencies—a procedure known as *sonication*.
Step 2: An antibody specific for the protein of interest is then added to the mixture, allowing the protein and its interacting DNA to be separated out—a procedure known as *immunoprecipitation* (abbreviated IP).
Step 3: After IP, the crosslinks between the protein and DNA are reversed by heating, releasing the precipitated DNA sequence for analysis.
Step 4: If the DNA sequence of the target element is known, the polymerase chain reaction (PCR) can be performed **(see Key Technique in Chapter 21, pages 624–625)**. If the sequence was attached to the protein of interest, then it will be present in the IP sample. Primers specific for the sequence can be generated, and PCR will amplify it.

If the sequence of the element is unknown, special whole-genome DNA microarrays, called *tiling arrays,* can be used to locate its position in the genome. (You will learn more about microarray technology in Chapter 21.) Briefly, the total DNA in the cellular sample is labeled with one fluorescent tag, for example, red. The DNA isolated by IP is labeled with a different fluorescent color (for example, green). The two populations of DNA are mixed and hybridized to the tiling array. Regions on the chip with a high red/green ratio identify DNA fragments containing binding sites for the protein. This approach is powerful because no prior knowledge of which DNA sequences interact with the protein is needed. Because such microarrays have been dubbed "gene chips," this approach is often called "ChIP-chip."

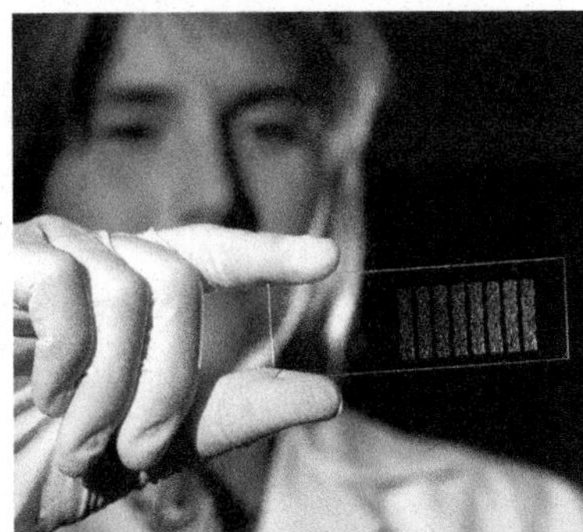

DNA microarray on a glass slide

Technology is moving quickly in this area, and several variations on the basic ChIP technique are now used. One example is "ChIP-seq," an approach in which the immunoprecipitated DNA is directly sequenced. ChIP-seq is sometimes easier than ChIP-chip because it avoids using a DNA array. It does, however, require that the genome sequence of the organism or cell being studied is known.

QUESTION: Of these three techniques, which one provides the most specific information about where a protein binds to DNA? Which one(s) require(s) specific knowledge about the DNA sequence bound by the protein before performing the procedure?

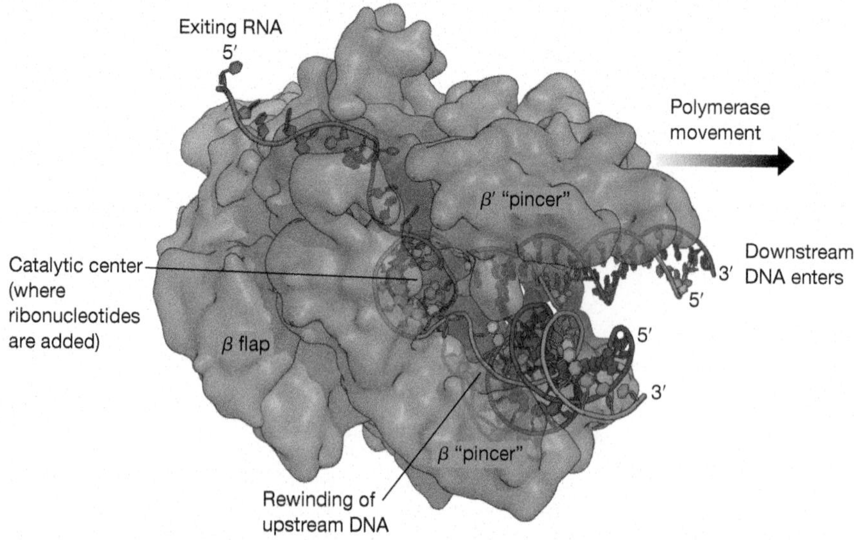

(a) Structure of bacterial RNA polymerase

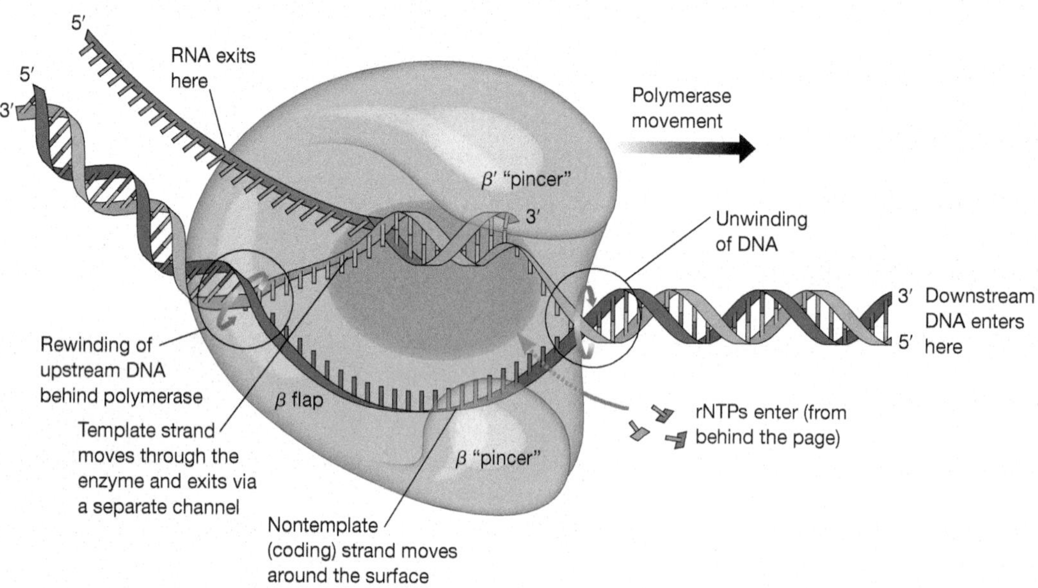

(b) Schematic of RNA polymerase, oriented in the same direction as in part a.

Figure 18-7 Closeup of a Bacterial Elongation Complex. (a) The β subunits form a clamp, which holds the incoming (downstream) DNA. Different channels allow entry of DNA (at the right), movement of the nontemplate (coding) strand around the outside of the polymerase (not labeled), movement of the template strand and ribonucleotides into the catalytic core, and exit of the template strand DNA (not labeled) and RNA at the back. **(b)** During elongation, RNA polymerase binds to about 30 bp of DNA. (Recall that each complete turn of the DNA double helix is about 10 bp.) At any given moment, about 18 bp of DNA are unwound, and the most recently synthesized RNA is still hydrogen-bonded to the DNA, forming a short RNA-DNA hybrid about 8–9 bp long. The total length of growing RNA bound to the enzyme and/or DNA is about 25 nucleotides.

RNA) the polymerase as it moves along (Figure 18-7a). The enzyme moves along the template DNA strand from the 3′ end toward the 5′ end. Because complementary base pairing between the DNA template strand and the newly forming RNA chain is antiparallel, *the RNA strand is elongated in the 5′ → 3′ direction* as each successive nucleotide is added to the 3′ end of the growing chain. (This is the same direction in which DNA strands are synthesized during DNA replication.) As the RNA chain grows, the most recently added nucleotides remain base-paired with the DNA template strand, forming a short RNA-DNA hybrid about 8–9 bp long. As the polymerase moves forward, the DNA ahead of the enzyme is unwound to permit the RNA-DNA hybrid to form. At the same time, the DNA behind the moving enzyme is rewound into a double helix (Figure 18-7b). The supercoiling that would otherwise be generated by this unwinding and rewinding is released through the action of topoisomerases, just as in DNA replication (see Figure 16-10).

RNA Proofreading. Like DNA polymerase, RNA polymerases appear to possess some weak proofreading ability. They accomplish proofreading in two ways. First, the reactions catalyzed by the active site of RNA polymerase can proceed in either direction, as is true for any enzyme. When the polymerase "runs in reverse," it catalyzes the removal of a nucleotide by reincorporation of a pyrophosphate. This reverse reaction might remove a correct nucleotide, so why does it tend to correct mistakes? When an incorrect ribonucleotide is put into the RNA, RNA polymerase waits at that position, making it more likely that the reverse reaction will occur. The second proofreading mechanism involves another site within the polymerase. When a noncomplementary nucleotide is incorporated into a growing RNA chain by mistake, the RNA polymerase backs up slightly, transcription stalls, and the noncomplementary nucleotide participates in catalyzing its own removal by RNA polymerase along with removal of the previously incorporated nucleotide, a process called **RNA backtracking**. Prokaryotic and eukaryotic RNA polymerases use proteins positioned in similar places to perform backtracking, although the proteins themselves are not similar in sequence. RNA proofreading, though less stringent than DNA proofreading, appears to be sufficient for correcting some of the mistakes that arise during transcription. This is largely because occasional errors in RNA molecules are not as critical as errors in DNA replication. Because numerous RNA copies are transcribed from each gene, a few inaccurate copies can be tolerated. In contrast, only one copy of each DNA molecule is made when DNA is replicated prior to cell division. Because each newly forming cell receives only one set of DNA molecules, it is crucial that the copying mechanism used in DNA replication be extremely accurate.

Termination of RNA Synthesis. Elongation of the growing RNA chain proceeds until RNA polymerase copies a special sequence, called a **termination signal**, that triggers the end of transcription. In bacteria, two classes of termination signals can be distinguished based on whether they require the participation of a protein called **rho (ρ) factor.** RNA molecules terminated without the aid of the rho factor contain a short GC-rich sequence followed by several U residues near their 3′ end (**Figure 18-8**). This configuration promotes termination in the following way. First, the GC region contains sequences that are complementary to each other, causing the RNA to spontaneously fold into a **hairpin loop** that tends to pull the RNA molecule away from the DNA and causes the polymerase to pause. The small number of hydrogen bonds between the sequence of U residues of the transcript and the A residues of the DNA template are then broken, releasing the newly formed RNA molecule.

In contrast, RNA molecules that do not form a GC-rich hairpin loop require participation of the rho factor for termination. Genes encoding such RNAs were first discovered in experiments in which purified DNA obtained from bacteriophage λ was transcribed with purified RNA polymerase. Some genes were found to be transcribed into RNA molecules that are longer than the RNAs produced in living cells, suggesting that transcription was not terminating properly. This problem could be corrected by adding rho factor, which binds to specific termination sequences 50–90 bases long located near the 3′ end of newly forming RNA molecules. The rho factor acts as an ATP-dependent unwinding enzyme, moving along the newly forming RNA molecule toward its 3′ end and unwinding it from the DNA template as it proceeds.

Whether termination depends on rho or not, it results in the release of the newly transcribed RNA molecule and the core RNA polymerase. The core polymerase can then bind sigma factor again and reinitiate RNA synthesis at the same or another promoter.

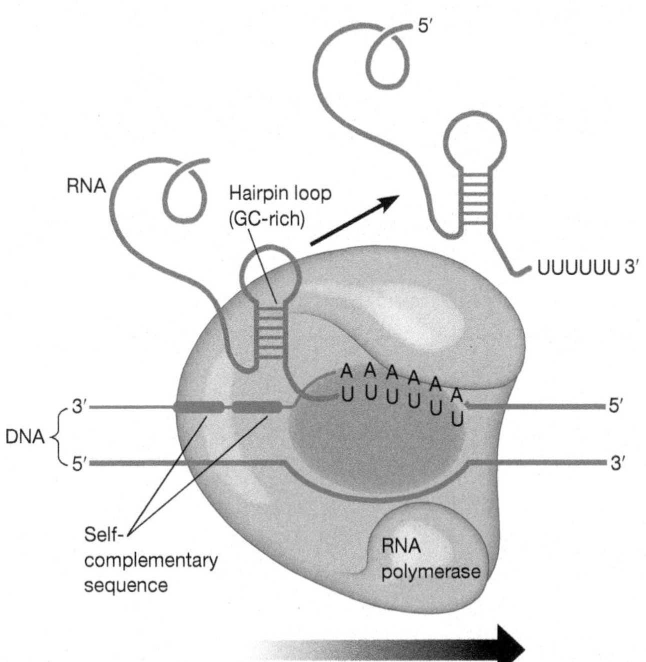

Figure 18-8 Termination of Transcription in Bacterial Genes That Do Not Require the Rho Termination Factor. A short self-complementary sequence near the end of the gene allows the newly formed RNA molecule to form a hairpin loop structure that helps dissociate the RNA from the DNA template.

Transcription in Eukaryotic Cells Has Additional Complexity Compared with Prokaryotes

Transcription in eukaryotic cells involves the same four stages described in Figure 18-4, but the process in eukaryotes is more complicated than that in bacteria. The main differences are as follows:

- *Three different RNA polymerases* transcribe the nuclear DNA of eukaryotes. Each synthesizes one or more classes of RNA.
- *Eukaryotic promoters* are more varied than bacterial promoters. Not only are different types of promoters employed for the three polymerases, but there is great variation within each type—especially among the ones for protein-coding genes. Furthermore, some eukaryotic promoters are actually located *downstream* from the transcription start site.
- Binding of eukaryotic RNA polymerases to DNA requires the participation of additional proteins, called *transcription factors.* Unlike the bacterial sigma factor, eukaryotic transcription factors are not part of the RNA polymerase molecule itself. Rather, some of them must bind to DNA *before* RNA polymerase can bind to the promoter and initiate transcription. Thus, transcription factors, rather than RNA polymerase itself, determine the specificity of transcription in eukaryotes. In this chapter, we limit our discussion to the class of factors that is essential for the transcription of all genes transcribed by an RNA polymerase. (We defer discussion of the regulatory class of transcription factors, which selectively act on specific genes, until Chapter 20.)
- *Protein-protein interactions* play a prominent role in the first stage of eukaryotic transcription. Although some

transcription factors bind directly to DNA, many attach to other proteins—either to other transcription factors or to RNA polymerase itself.

- *RNA cleavage* is more important than the site where transcription is terminated in determining the location of the 3′ end of the RNA product.
- Newly forming eukaryotic RNA molecules typically undergo extensive *RNA processing* (chemical modification) both during and, to a larger extent, after transcription.

We will now examine these various aspects of eukaryotic transcription, starting with the existence of multiple forms of RNA polymerase.

RNA Polymerases I, II, and III Carry Out Transcription in the Eukaryotic Nucleus

Table 18-1 summarizes some properties of the three RNA polymerases that function in the nucleus of the eukaryotic cell, along with two other polymerases found in mitochondria and chloroplasts. The nuclear enzymes are designated RNA polymerases I, II, and III. As the table indicates, these enzymes differ in their location within the nucleus and in the kinds of RNA they synthesize. The nuclear RNA polymerases also differ in their sensitivity to various inhibitors, such as *α-amanitin,* a deadly toxin produced by the mushroom *Amanita phalloides* (the "death cap" fungus; see Human Connections, page 522–523). (The F-actin binding drug phalloidin, introduced in Chapter 13, also comes from this organism.)

RNA polymerase I resides in the nucleolus and is responsible for synthesizing an RNA molecule that serves as a precursor for three of the four types of rRNA found in eukaryotic ribosomes (28S rRNA, 18S rRNA, and 5.8S rRNA). This enzyme is not sensitive to α-amanitin. Its association with the nucleolus is understandable because the nucleolus is the site of ribosomal RNA synthesis and ribosomal subunit assembly.

RNA polymerase II is found in the nucleoplasm and synthesizes precursors to mRNA, the class of RNA molecules that encode proteins. Rather than being diffusely distributed throughout the nucleus, active molecules of polymerase II are located in discrete clusters, called *transcription factories,* that may represent sites where active genes come together to be transcribed. In addition to producing mRNA precursors, RNA polymerase II synthesizes most of the *snRNAs,* small nuclear RNAs involved in posttranscriptional RNA processing, and the *microRNAs,* which regulate the translation and stability of specific mRNAs and, to a lesser extent, control the transcription of certain genes. Polymerase II is responsible for producing the greatest variety of RNA molecules and is extremely sensitive to α-amanitin, which explains the toxicity of this compound to humans and other animals.

RNA polymerase II differs from polymerases I and III at its C-terminus, where it has extra amino acids. The C-terminus of RNA polymerase II can be phosphorylated at a variety of locations to produce what is sometimes called a phosphorylation "code." This "code" dramatically affects the functions of polymerase II and correlates with where the enzyme is located along the DNA as it continues transcription. As a result, this most versatile of the RNA polymerases is also the most tightly regulated.

RNA polymerase III is also a nucleoplasmic enzyme, but it synthesizes a variety of small RNAs, including tRNA precursors and the smallest type of ribosomal RNA, 5S rRNA. Mammalian RNA polymerase III is sensitive to α-amanitin but only at higher levels of the toxin than are required to inhibit RNA polymerase II. (The comparable enzymes of some other eukaryotes, such as insects and yeasts, are insensitive to α-amanitin.)

Structurally, RNA polymerases I, II, and III are somewhat similar to each other as well as to bacterial core RNA polymerase. The three enzymes are all quite large, with multiple polypeptide subunits and molecular weights around 500,000. RNA polymerase II, for example, has more than ten subunits of at least eight different types. The three biggest subunits are evolutionarily related to the bacterial RNA polymerase subunits α, β, and β'. Three of the smaller subunits lack that relationship but are also found in RNA polymerases II and III. The RNA polymerases of mitochondria and chloroplasts resemble their bacterial counterparts closely, as you might expect from the probable origins of these organelles as endosymbiotic bacteria (see Figure 4-15). Like bacterial RNA polymerase, the mitochondrial and chloroplast enzymes are resistant to α-amanitin.

Three Classes of Promoters Are Found in Eukaryotic Nuclear Genes, One for Each Type of RNA Polymerase

The promoters that eukaryotic RNA polymerases bind to are even more varied than bacterial promoters, but they can be grouped into three main categories, one for each type of polymerase. **Figure 18-9** shows examples of the three types of promoters.

The promoter used by RNA polymerase I—that is, the promoter of the transcription unit that produces the precursor for

Table 18-1 Properties of Eukaryotic RNA Polymerases

RNA Polymerase	Location	Main Products	α-Amanitin Sensitivity
I	Nucleolus	Precursor for 28S rRNA, 18S rRNA, and 5.8S rRNA	Resistant
II	Nucleoplasm	Pre-mRNA, most snRNA, and microRNA	Very sensitive
III	Nucleoplasm	Pre-tRNA, 5S rRNA, and other small RNAs	Moderately sensitive*
Mitochondrial	Mitochondrion	Mitochondrial RNA	Resistant
Chloroplast	Chloroplast	Chloroplast RNA	Resistant

*In mammals.

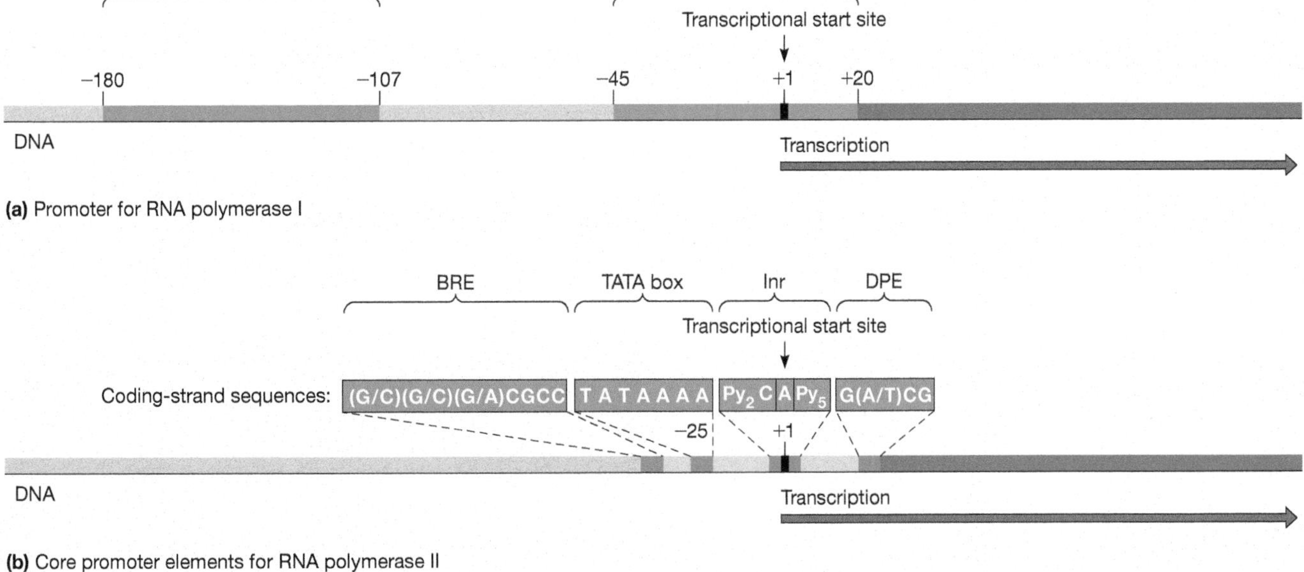

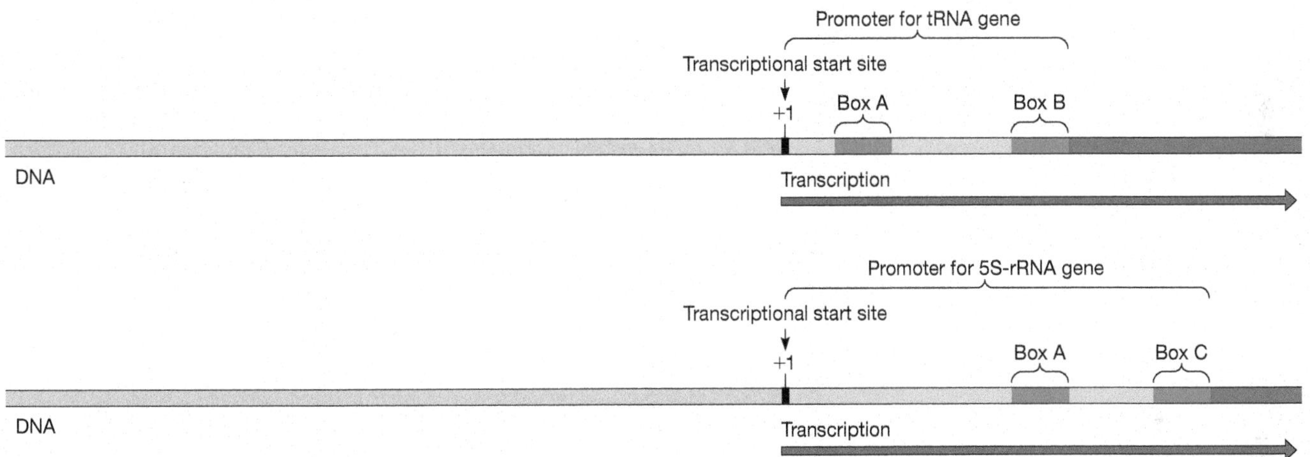

Figure 18-9 Examples of Eukaryotic Promoters for RNA Polymerases I, II, and III. (a) The promoter for RNA polymerase I has two parts, a core promoter surrounding the start site and an upstream control element. After the binding of appropriate transcription factors to both parts, the RNA polymerase binds to the core promoter. **(b)** The typical promoter for RNA polymerase II has a short initiator (Inr) sequence, consisting mostly of pyrimidines (Py), combined with either a TATA box or a downstream promoter element (DPE). Promoters containing a TATA box may also include a TFIIB recognition element (BRE) as part of the core promoter. **(c)** The promoters for RNA polymerase III vary in structure, but the ones for tRNA genes and 5S-rRNA genes are located entirely downstream of the start site, within the transcribed sequence. Boxes A, B, and C are DNA consensus sequences, each about 10 bp long. In tRNA genes, about 30–60 bp of DNA separate boxes A and B. In 5S-rRNA genes, about 10–30 bp separate boxes A and C.

in both cases the consensus sequences fall into two blocks of about 10 bp each (Figure 18-9c). The tRNA promoter has consensus sequences called *box A* and *box B*. The promoters for 5S-rRNA genes have box A (positioned farther from the start site than in tRNA-gene promoters) and another critical sequence, called *box C*. (Not shown in the figure is a third type of RNA polymerase III promoter, an upstream promoter that is used for the synthesis of other kinds of small RNA molecules.)

The promoters used by all the eukaryotic RNA polymerases must be recognized and bound by transcription factors before the RNA polymerase molecule can bind to DNA. We turn now to these transcription factors.

General Transcription Factors Are Involved in the Transcription of All Nuclear Genes

A **general transcription factor** is a protein that is always required for an RNA polymerase molecule to bind to its promoter and initiate RNA synthesis, regardless of the identity of the gene involved. Eukaryotes have many such transcription factors; their names usually include "TF" (for transcription factor), a roman numeral identifying the polymerase they aid, and a capital letter that identifies each individual factor (for example, TFIIA, TFIIB).

Using RNA polymerase II as an example, **Figure 18-10** on page 524 illustrates the involvement of general transcription

HUMAN *Connections*

Death by Fungus (*Amanita Phalloides* Poisoning)

The popularity of wild mushrooms in the human diet has been on the rise. Of the thousands of mushroom species identified, about 100 are known to be poisonous. Therefore, collectors of wild mushrooms need to be vigilant (and not too bold) in distinguishing edible from poisonous species. The wide degree of variability that exists in the appearance of mushrooms based on the local environment and weather makes this a challenging task. Although accurate disease statistics are difficult to gather, in 2012 there were approximately 6600 cases of *mycetism,* or mushroom toxicity, reported in the United States. Of these, the most serious and deadly arise from ingestion of *Amanita phalloides*—appropriately known as the death cap mushroom.

Death cap mushrooms are rather innocuous in appearance (**Figure 18B-1**). In fact, they resemble many varieties of edible white mushrooms. *Amanita* species are found throughout the world, including the northeastern United States, where they grow in the spring, summer, and fall months. As described by survivors, when first ingested, the mushrooms taste pleasant. However, a few mouthfuls from a single mushroom can be enough to cause death. Symptoms begin within hours after consumption, starting with gastrointestinal distress, abdominal pain, and diarrhea. While these symptoms can pass, the true danger occurs as the poison enters and affects the liver. In ~10–30% of patients, damage to the liver will result in organ failure and eventually death. Treatment options are limited and involve immediate pumping of the stomach and supportive care. In extreme cases, a liver transplant remains the best chance for survival. Trials are underway for different drug treatments, including a compound isolated from milk thistle that blocks entry of the poison into the liver. Despite tales to the contrary, there is no way to safely prepare death cap mushrooms for human consumption. So how do these mushrooms make us so sick?

Three different classes of toxins have been identified in *Amanita* species. These include phallotoxins and virotoxins, which bind to actin and disrupt cytoskeletal function. One type of phallotoxin called phalloidin is commonly used by scientists to stain actin filaments when performing fluorescence microscopy (see Figure A-13 in the Appendix). However, these compounds do not cross the gastrointestinal tract in high enough concentrations to cause systemic symptoms after eating. Instead, mycetism and death are primarily due to the third class of toxins: amatoxins. *Amanita phalloides* produces α-amanitin, a unique circular peptide containing eight amino acids (**Figure 18B-2**). This peptide is encoded in the genome of the fungus and then transcribed and translated on ribosomes just like any other protein. How does this small compound wreak so much havoc on human cells? The toxin is able to bind tightly and noncovalently within the RNA polymerase II enzyme. The toxin's affinity for RNA polymerase II is the strongest, but it can bind to the other families of RNA polymerases as well (see Table 18-1). Once bound, the toxin interferes with the ability of the polymerase to translocate along the DNA and elongate. In fact, elongation rates drop from thousands of bases transcribed per minute to only a handful. That is why amatoxins are considered transcription inhibitors. As a result, protein production is dramatically diminished, leading to eventual cell

Figure 18B-1 The Death Cap Mushroom, *Amanita phalloides.* You can see the various developmental stages of the mushroom as it matures.

---

the three largest rRNAs—has two parts (Figure 18-9a). The part called the **core promoter**, defined as the smallest set of DNA sequences able to direct the accurate initiation of transcription by RNA polymerase, actually extends into the nucleotide sequence to be transcribed. The core promoter is sufficient for proper initiation of transcription, but transcription is made more efficient by the presence of an *upstream control element,* which for RNA polymerase I is a fairly long sequence similar (though not identical) to the core promoter. Attachment of transcription factors to both parts of the promoter facilitates the binding of RNA polymerase I to the core promoter and enables it to initiate transcription at the start site.

In the case of RNA polymerase II, at least four types of DNA sequences are involved in core promoter function (Figure 18-9b). These four elements are (1) a short **initiator (Inr)** sequence surrounding the transcriptional start site (which is often an A, as in bacteria); (2) the **TATA box**, which consists of a consensus sequence of TATA followed by two or three more A's, usually located about 25 nucleotides upstream from the start site; (3) the **TFIIB recognition element (BRE)** located slightly upstream of the TATA box; and (4) the **downstream promoter element (DPE)** located about 30 nucleotides downstream from the start site. These four elements are organized into two general types of core promoters: *TATA-driven promoters,* which contain an Inr sequence and a TATA box with or without an associated BRE, and *DPE-driven promoters,* which contain DPE and Inr sequences but no TATA box or BRE. Besides being found in eukaryotes, TATA-driven promoters are also present in archaea, a key piece of evidence supporting the idea that in some ways, archaea resemble eukaryotes more closely than they resemble bacteria (see Table 4-1, page 81).

By itself, a core promoter (TATA-driven or DPE-driven) is capable of supporting only a *basal* (low) level of transcription. However, most protein-coding genes have additional short

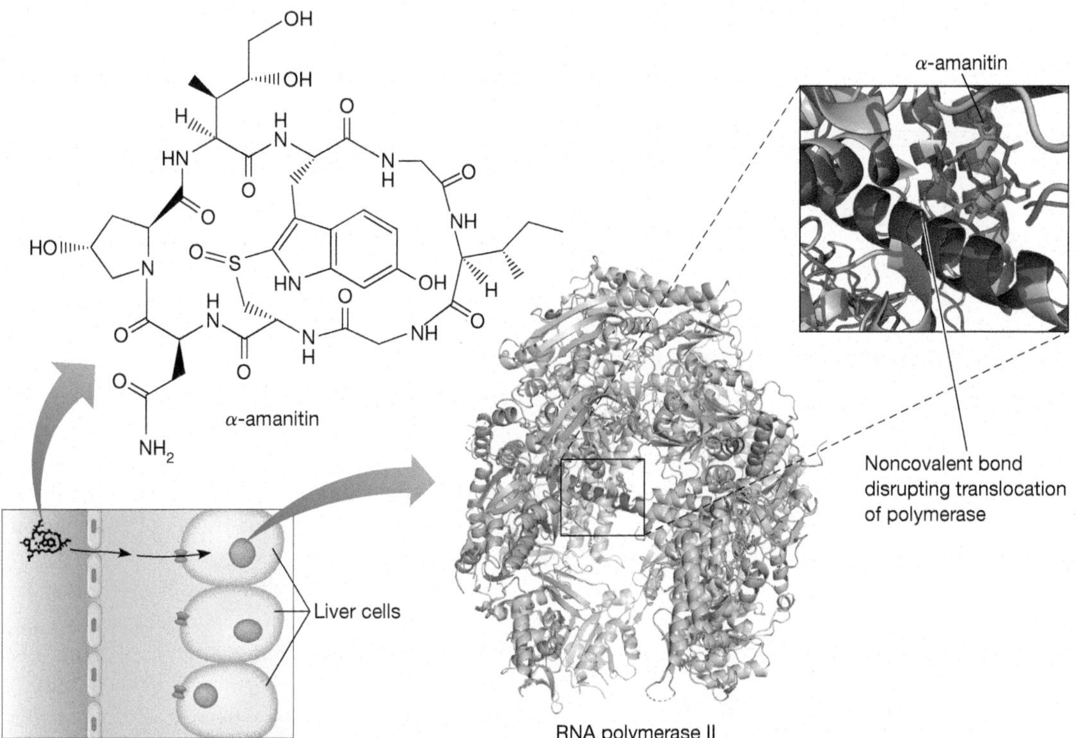

Figure 18B-2 Molecular Structure and Mechanism of α-Amanitin toxicity. The structure of the amatoxin derived from *Amanita phalloides* is shown at top left. Upon ingestion, the toxin can cross the gastrointestinal tract, enter the bloodstream, and be taken up by liver cells via a protein transporter. Once in hepatocytes, α-amanitin can bind tightly within the RNA polymerase II enzyme. This bond limits the ability of the polymerase to translocate down the DNA, thus drastically slowing down transcription. The resulting decrease in protein production leads to hepatic cell death and eventually liver failure.

death. The loss of hepatocytes can cause liver failure, coma, and death. Therefore, if you plan to gather and eat wild mushrooms, it's best to remember an old saying: there are old mushroom hunters and there are bold mushroom hunters; however, there are no old, bold mushroom hunters!

sequences further upstream—*upstream control elements*—that improve the promoter's efficiency. Some of these upstream elements are common to many different genes; examples include the *CAAT box* (consensus sequence GCCCAATCT in animals and yeasts) and the *GC box* (consensus sequence GGGCGG). The locations of these elements relative to a gene's transcriptional start site vary from gene to gene. The elements within 100–200 nucleotides of the start site are often called *proximal control elements* to distinguish them from *enhancer* elements, which tend to be farther away and can even be located downstream of the gene. When activating proteins bind to enhancers, they locally change the conformation of DNA near promoters by allowing chromatin remodeling proteins and histone-modifying enzymes to alter chromatin structure and by promoting assembly of the basal transcriptional machinery. Keep in mind that for genes transcribed using RNA polymerase II, these long-range elements are often crucial for determining whether a gene is "switched on" (that is, it is transcribed efficiently) or not (you will learn more about proximal control elements and enhancers in Chapter 20).

The sequences important in promoter activity are often identified by deleting specific sequences from a cloned DNA molecule, which is then tested for its ability to serve as a template for gene transcription, either in a test tube or after introduction of the DNA into cultured cells. For example, when transcription of the gene for β-globin (the β chain of hemoglobin) is investigated in this way, deletion of either the TATA box or an upstream CAAT box reduces the rate of transcription at least tenfold.

In contrast to RNA polymerases I and II, the RNA polymerase III molecule uses promoters that are entirely *downstream* of the transcription unit's transcriptional start site when transcribing genes for tRNAs and 5S-rRNA. The promoters used by tRNA and 5S-rRNA genes are different, but

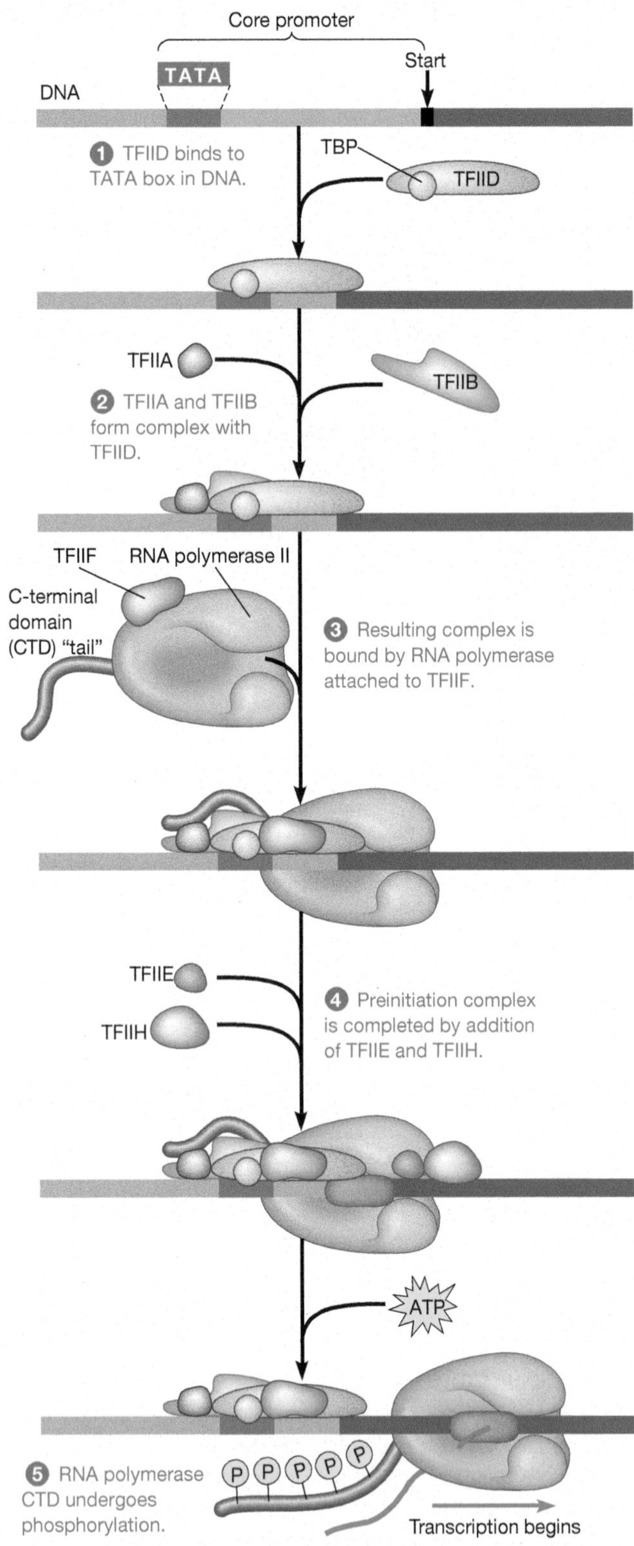

Figure 18-10 Role of General Transcription Factors in Binding RNA Polymerase II to DNA. This figure outlines the sequential binding of six general transcription factors (called TFII_, where _ is a letter identifying the particular factor) and RNA polymerase. After the final activation step involving ATP-dependent phosphorylation of the RNA polymerase, the polymerase can initiate transcription. In intact chromatin, the efficient binding of general transcription factors and RNA polymerase to DNA requires the participation of additional regulatory proteins that open up chromatin structure and facilitate assembly of the preinitiation complex at specific genes.

factors in the binding of RNA polymerase to a TATA-containing promoter site in DNA. General transcription factors bind to promoters in a defined order, starting with TFIID. Notice that although TFIID binds directly to a DNA sequence (the TATA box in this example or the DPE sequence in the case of DPE-driven promoters), the other transcription factors interact primarily with each other. Hence, protein-protein interactions play a crucial role in the binding stage of eukaryotic transcription. RNA polymerase II does not bind to the DNA until several steps into the process. Eventually, a large complex of proteins, including RNA polymerase, becomes bound to the promoter region to form a **preinitiation complex**.

Before RNA polymerase II can actually initiate RNA synthesis, it must be released from the preinitiation complex. A key role in this process is played by the general transcription factor TFIIH, which possesses both a helicase activity that unwinds DNA and a protein kinase activity that catalyzes the phosphorylation of RNA polymerase II. Phosphorylation allows RNA polymerase to detach from the transcription factors so that it can initiate RNA synthesis at the transcriptional start site. At the same time, the helicase activity of TFIIH is thought to unwind the DNA so that the RNA polymerase molecule can begin to move.

The region of RNA polymerase II that is phosphorylated by TFIIH is very important for regulating RNA polymerase II function. The large subunit of RNA polymerase II has a *C-terminal domain (CTD)* that functions as a "tail." The CTD is much longer than the core portion of the polymerase; it is the region of the polymerase that is phosphorylated to allow elongation of the RNA. The CTD contains a series of seven repeated amino acids; in yeast there are 27 of these repeats, whereas in humans there are 52. Each repeat contains sites that can be phosphorylated by specific kinases, including the kinase that is part of TFIIH.

TFIID, the initial transcription factor to bind to the promoter, is worthy of special note. Its ability to recognize and bind to DNA promoter sequences is conferred by one of its subunits, the **TATA-binding protein (TBP)**, which combines with a variable number of additional protein subunits to form TFIID. Despite its name, the ability of TBP to bind to DNA, illustrated in **Figure 18-11**, is not restricted to TATA-containing promoters. TBP can also bind to promoters lacking a TATA box, including promoters used by RNA polymerases I and III. Depending on the type of promoter, TBP associates with different proteins, and for promoters lacking a TATA box, much of TBP's specificity is probably derived from its interaction with these associated proteins. TBP binds the minor groove of DNA, much like a saddle on a horse. This causes a severe kink in the DNA that promotes the attachment of other components of the preinitiation complex.

In addition to general transcription factors and RNA polymerase II, several other kinds of proteins are required for the efficient transcription and regulated activation of specific genes. Some of these proteins are involved in opening up chromatin structure to facilitate the binding of RNA polymerase to DNA. Others are *regulatory transcription factors*, which activate specific genes by binding to upstream control elements and recruiting *coactivator proteins* that in turn facilitate assembly of the RNA polymerase preinitiation complex. (The identities

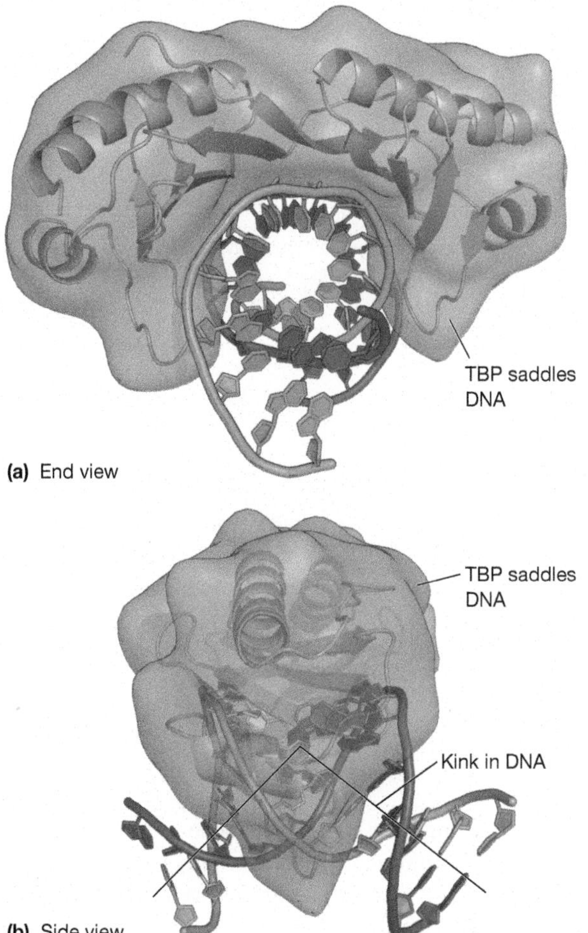

Figure 18-11 TATA-Binding Protein (TBP) Bound to DNA. In this computer graphic model, human TBP (purple) is shown bound to DNA (blue). TBP differs from most DNA-binding proteins in that it interacts with the minor groove of DNA, rather than the major groove, and imparts a sharp bend to the DNA. When TBP is bound to DNA, other transcription factors can interact with the convex surface of the TBP "saddle." TBP is involved in transcription initiation for all types of eukaryotic promoters.

and roles of these additional proteins will be described in Chapter 20, which covers the regulation of gene expression; see Figure 20-24.)

> **MAKE CONNECTIONS 18.1**
>
> The successful binding of RNA polymerases to eukaryotic promoters relies on successful recruitment of chromatin and histone-modifying enzymes. Explain how this is different from prokaryotic transcription, and based on your knowledge of genome packaging, why this recruitment is required. (Ch. 16.3)

Elongation, Termination, and RNA Cleavage Are Involved in Completing Eukaryotic RNA Synthesis

After initiating transcription, RNA polymerases move along the DNA and synthesize a complementary RNA copy of the DNA template strand. Special proteins facilitate the disassembly of nucleosomes in front of the moving polymerase and their immediate reassembly after the enzyme passes. If an area of DNA damage is encountered, RNA polymerase may become stalled temporarily while the damage is corrected by proteins that carry out nucleotide excision repair (see Figure 17-27).

Termination of transcription is governed by an assortment of signals that differ for each type of RNA polymerase. For example, transcription by RNA polymerase I is terminated by a protein factor that recognizes an 18-nucleotide termination signal in the growing RNA chain. Termination signals for RNA polymerase III are also known; they always include a short run of U's (as in bacterial termination signals), and no ancillary protein factors are needed for their recognition. Hairpin structures do not appear to be involved in termination by either polymerase I or polymerase III.

For RNA polymerase II, transcripts destined to become mRNA are cleaved at a specific site before transcription is actually terminated. The cleavage site is 10–35 nucleotides downstream from a special AAUAAA sequence in the growing RNA chain. The polymerase may continue transcription for hundreds or even thousands of nucleotides beyond the cleavage site, but this additional RNA is quickly degraded, as it lacks the cap modification described in the following section. The cleavage site is also the site for the addition of a *poly(A) tail*, a string of adenine nucleotides found at the 3′ end of almost all eukaryotic mRNAs. Addition of the poly(A) tail is part of RNA processing, our next topic.

> **CONCEPT CHECK 18-2**
>
> Compare and contrast bacterial and eukaryotic transcription, focusing on initiation and termination. What features are similar? What features are different?

18.3 RNA Processing and Turnover

An RNA molecule newly produced by transcription, called a **primary transcript**, frequently must undergo chemical changes before it can function in the cell. We use the term **RNA processing** to mean all the chemical modifications necessary to generate a final, "mature" RNA product from the primary transcript that serves as its precursor. Processing typically involves removal of portions of the primary transcript, and it may also include the addition or chemical modification of specific nucleotides. For example, methylation of bases or ribose groups is a common modification of individual nucleotides. In addition to chemical modifications, other posttranscriptional events, such as association with specific proteins or (in eukaryotes) passage from the nucleus to the cytoplasm, are often necessary before the RNA can function.

Thus far, you have learned about transcription primarily in the context of production of mRNAs, RNAs that usually encode proteins. However, mRNAs are only one of many RNAs found in cells (**Table 18-2** on page 526). As you can see, RNAs perform many different functions within cells. You will learn more about these other types of RNA in this and the next two chapters. In this section, we examine the most important processing steps involved in the production of three types of RNA (rRNA, tRNA, and mRNA) from their respective primary transcripts. Although the term *RNA processing* is most often associated with eukaryotic systems, bacteria process some of their RNA as well. We therefore include examples involving both eukaryotic and bacterial RNAs.

Table 18-2 Classes of Naturally Occurring RNA

Type of RNA	Function(s)
I. Messenger RNA	Provides information for production of protein
II. RNAs involved in translation	
Transfer RNAs (tRNAs)	Bring the correct amino acid to mRNA during translation
Ribosomal RNAs (rRNAs)	Major components of ribosomes; guide and catalyze assembly of polypeptides from mRNA during translation
III. RNAs involved in RNA processing	
Small nuclear RNAs (snRNAs)	Form components of the spliceosome; involved in splicing of eukaryotic RNAs
Small nucleolar RNAs (snoRNAs)	Process rRNAs in nucleolus
IV. RNAs involved in regulation of gene expression and other events	
MicroRNAs (miRNAs)	Regulate stability and translation of mRNAs
Small interfering RNAs (siRNAs)*	Inhibit production of viruses; suppress spread of transposable elements in plants
Piwi-interacting RNAs (piRNAs)	Suppress spread of transposable elements in germ cells in animals
Long noncoding RNAs (lncRNAs)	Some regulate chromatin; others may regulate transcription or RNA processing

*Not to be confused with small inhibitory RNAs, which are synthetic RNAs used in gene knockdown experiments (see Chapter 21).

The Nucleolus Is Involved in Ribosome Formation

A prominent structural component of the eukaryotic nucleus is the **nucleolus** (plural: **nucleoli**), the ribosome factory of the cell (**Figure 18-12**). Typical eukaryotic cells contain one or two nucleoli, but the occurrence of several more is not uncommon; in certain situations, hundreds or even thousands may be present. The nucleolus is usually spherical, measuring several micrometers in diameter, but wide variations in size and shape are observed. Because of their relatively large size, nucleoli are easily seen with the light microscope and were first observed more than 200 years ago. However, it was not until the advent of electron microscopy in the 1950s that the structural components of the nucleolus were clearly identified. In thin-section electron micrographs, each nucleolus appears as a membrane-free organelle consisting of fibrils and granules (Figure 18-12a). The fibrils contain DNA that is being transcribed into *ribosomal RNA (rRNA)*, the RNA component of ribosomes. The granules are rRNA molecules being packaged with proteins (imported from the cytoplasm) to form ribosomal subunits. The ribosomal subunits are subsequently exported through the nuclear pores to the cytoplasm (see Chapter 16).

If rRNA is synthesized in the nucleolus, then the DNA sequences coding for this RNA must reside in the nucleolus as well. This prediction has been verified by showing that isolated nucleoli contain the **nucleolus organizer region (NOR)**—a stretch of DNA carrying multiple copies of rRNA genes. These multiple rRNA genes occur in all genomes and are thus an important example of repeated DNA that carries genetic information. The ability to biochemically purify DNA from nucleoli was historically important because the purified DNA and associated rRNA could be visualized with the electron microscope (Figure 18-12b). Such images show the repeating structure of the rRNA genes, and they show something else: along each gene, many RNA polymerases are simultaneously transcribing rRNAs. They do so directionally; the polymerases that have moved the farthest along the DNA from the transcriptional start site produce progressively longer RNA, leading to a characteristic repeating "bottle brush" appearance in electron micrographs.

The number of copies of the rRNA genes varies greatly from species to species, but animal cells generally contain hundreds of copies and plant cells often contain thousands. The multiple copies are grouped into one or more NORs, which may reside on more than one chromosome; in each NOR, the multiple gene copies are tandemly arranged. A single nucleolus may contain rRNA genes derived from more than one NOR. For example, the human genome has five NORs per haploid chromosome set—or ten per diploid nucleus—each located near the tip of a different chromosome. But instead of ten separate nucleoli, the typical human nucleus has a single large nucleolus containing loops of chromatin derived from ten separate chromosomes.

The size of the nucleolus is correlated with its level of activity. In cells having a high rate of protein synthesis and hence a need for many ribosomes, nucleoli tend to be large and can account for 20–25% of the total volume of the nucleus. In less-active cells, nucleoli are much smaller.

The nucleolus disappears during mitosis, at least in the cells of higher plants and animals. As the cell approaches division, chromatin condenses into compact chromosomes accompanied by the shrinkage and then disappearance of the nucleoli. With our current knowledge of nucleolar composition and function, this makes perfect sense: the extended chromatin loops of the nucleolus cease being transcribed as they are coiled and folded, and any remaining rRNA and ribosomal protein molecules disperse or are degraded. Then, as mitosis is ending, the chromatin uncoils, the NORs loop

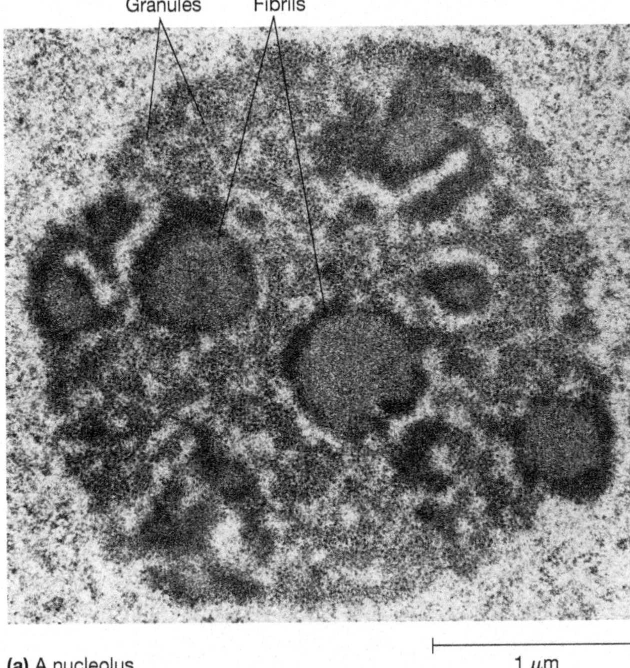

(a) A nucleolus

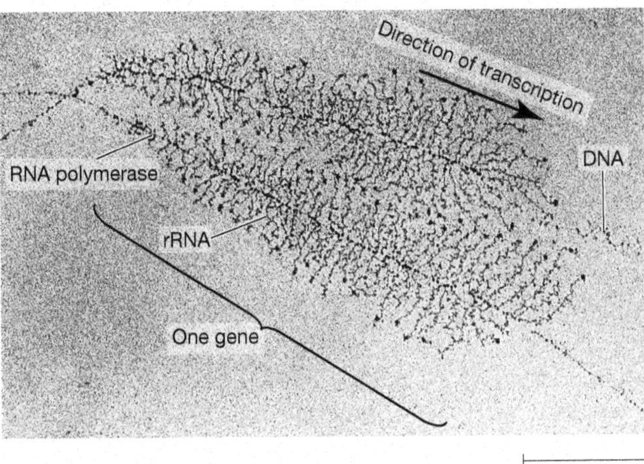

(b) rRNA being transcribed from an rRNA gene

Figure 18-12 The Nucleolus and rRNA Synthesis. (a) The nucleolus is a prominent intranuclear structure composed of a mass of fibrils and granules. The fibrils are DNA and rRNA; the granules are newly forming ribosomal subunits. Shown here is a nucleolus of a spermatogonium, a cell that gives rise to sperm cells (TEM). **(b)** rRNA genes and associated transcriptional complexes can be visualized. Two rRNA transcription units are shown from active nucleolar chromatin isolated from primary nuclei of the green alga *Acetabularia mediterranea*. Multiple transcripts are being produced from each gene, as more than one RNA polymerase is transcribing RNA simultaneously. RNA can be seen elongating away from the DNA; the shorter RNAs represent RNA earlier in the transcription process (TEM).

out again, and rRNA synthesis resumes. In human cells, this is the only time the ten NORs of the diploid nucleus are apparent; as rRNA synthesis begins again, ten tiny nucleoli become visible, one near the tip of each of ten chromosomes. As these nucleoli enlarge, they quickly fuse into the single large nucleolus found in human cells that are not in the process of dividing.

MAKE CONNECTIONS 18.2

Describe how the nuclear pore complexes are involved in the formation of the nucleolus, and how the structure of the nucleolus might change if nuclear import through these pores were temporarily blocked. (Fig. 16-32)

Ribosomal RNA Processing Involves Cleavage of Multiple rRNAs from a Common Precursor

Ribosomal RNA (rRNA) is by far the most abundant and most stable form of RNA found in cells. Typically, rRNA represents about 70–80% of the total cellular RNA, tRNA represents about 10–20%, and mRNA accounts for less than 10%. In eukaryotes, cytoplasmic ribosomes contain four types of rRNA, usually identified by their differing sedimentation rates during centrifugation. **Table 18-3** lists the sedimentation coefficients (S values) of these different types of rRNA. The smaller of the two ribosomal subunits has a single 18S-rRNA molecule. The larger subunit contains three rRNA molecules, one of about 28S (as low as 25S in some species) and the other two of about 5.8S and 5S. In bacterial ribosomes, only three species of rRNA are present: a 16S molecule associated with the small subunit and molecules of 23S and 5S associated with the large subunit.

Of the four kinds of rRNA in eukaryotic ribosomes, the three larger ones (28S, 18S, and 5.8S) are encoded by a single transcription unit that is transcribed by RNA polymerase I in the nucleolus to produce a single primary transcript called **pre-rRNA** (**Figure 18-13** on page 528). The DNA sequences that encode these three rRNAs are separated within the transcription unit by segments of DNA called *transcribed spacers*. The presence of three different rRNA genes within a single transcription unit ensures that the cell makes these three rRNAs in equal quantities. Most eukaryotic genomes have multiple copies of the pre-rRNA transcription unit, arranged in one or more tandem arrays (Figure 18-13a, b). These multiple copies facilitate production of the large amounts of ribosomal RNA typically needed by cells. The human haploid genome, for example, has 150–200 copies of the pre-rRNA transcription unit, distributed among five chromosomes. *Nontranscribed spacers* separate the transcription units within each cluster.

After RNA polymerase I has transcribed the pre-rRNA transcription unit, the resulting pre-rRNA molecule is processed

Table 18-3 RNA Components of Cytoplasmic Ribosomes

Source	Ribosomal Subunit	rRNA Sedimentation Coefficient	rRNA Nucleotides
Bacterial cells	Large (50S)	23S	2900
		5S	120
	Small (30S)	16S	1540
Eukaryotic cells	Large (60S)	25–28S	≤ 4700
		5.8S	160
		5S	120
	Small (40S)	18S	1900

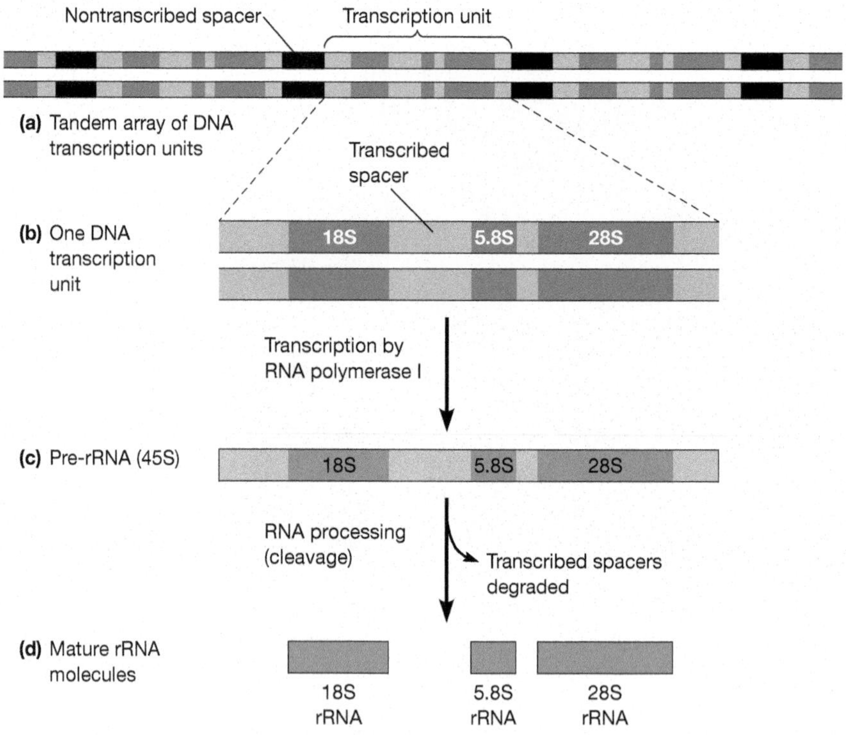

Figure 18-13 Eukaryotic rRNA Genes: Processing of Primary Transcripts. (a) The eukaryotic transcription unit that includes the genes for the three largest rRNAs occurs in multiple copies, arranged in tandem arrays. Nontranscribed spacers (black) separate the units. **(b)** Each transcription unit includes the genes for the three rRNAs (darker blue) and four transcribed spacers (lighter blue). **(c)** The transcription unit is transcribed by RNA polymerase I into a single long transcript (pre-rRNA) with a sedimentation coefficient of about 45S. **(d)** RNA processing yields mature 18S-, 5.8S-, and 28S-rRNA molecules. RNA cleavage actually occurs in a series of steps. The order of steps varies with the species and cell type, but the final products are always the same three types of rRNA molecules.

by a series of cleavage reactions that remove the transcribed spacers and release the mature rRNAs (Figure 18-13c, d). The transcribed spacer sequences are then degraded. The pre-rRNA is also processed by the addition of methyl groups. The main site of methylation is the 2′-hydroxyl group of the sugar ribose, although a few bases are methylated as well. The methylation process, as well as pre-rRNA cleavage, is guided by a special group of RNA molecules, called **snoRNAs** (small nucleolar RNAs), which bind to complementary regions of the pre-rRNA molecule and target specific sites for methylation or cleavage.

Pre-rRNA methylation has been studied by incubating cells with radioactive *S-adenosyl methionine*, which is the methyl group donor for cellular methylation reactions. When human cells are incubated with radioactive S-adenosyl methionine, all of the radioactive methyl groups initially incorporated into the pre-rRNA molecule are eventually found in the finished 28S-, 18S-, and 5.8S-rRNA products, indicating that the methylated segments are selectively conserved during rRNA processing. Methylation may help guide RNA processing by protecting specific regions of the pre-rRNA molecule from cleavage. In support of this hypothesis, it has been shown that depriving cells of one of the essential components required for the addition of methyl groups leads to disruption of pre-rRNA cleavage patterns.

In mammalian cells, the pre-rRNA molecule has about 13,000 nucleotides and a sedimentation coefficient of 45S. The three mature rRNA molecules generated by cleavage of this precursor contain only about 52% of the original RNA. The remaining 48% (about 6200 nucleotides) consists of transcribed spacer sequences that are removed and degraded during the cleavage steps. The rRNA precursors of some other eukaryotes contain smaller amounts of spacer sequences, but in all cases the pre-rRNA is larger than the aggregate size of the three rRNA molecules made from it. Thus, some processing is always required.

Processing of pre-rRNA in the nucleolus is accompanied by assembly of the RNA with proteins to form ribosomal subunits. In addition to the 28S, 18S, and 5.8S rRNAs generated by pre-rRNA processing, the ribosome assembly process also requires 5S rRNA. The gene for 5S rRNA constitutes a separate transcription unit that is transcribed by RNA polymerase III rather than RNA polymerase I. It, too, occurs in multiple copies arranged in long, tandem arrays. However, 5S-rRNA genes are not usually located near the genes for the larger rRNAs and so do not tend to be associated with the nucleolus. Unlike pre-rRNA, the RNA molecules generated during transcription of 5S-rRNA genes require little or no processing.

As in eukaryotes, ribosome formation in prokaryotic cells involves processing of multiple rRNAs from a larger precursor. *E. coli*, for example, has seven rRNA transcription units scattered about its genome. Each contains genes for all three bacterial rRNAs—23S rRNA, 16S rRNA, and 5S rRNA—plus several tRNA genes. Processing of the primary transcripts produced from these transcription units involves two sets of enzymes, one for the rRNAs and one for the tRNAs. The three mature rRNAs combine with approximately 55 proteins to form an *E. coli* ribosome that is smaller than, but structurally similar to, the eukaryotic ribosome.

Transfer RNA Processing Involves Removal, Addition, and Chemical Modification of Nucleotides

Cells synthesize several dozen kinds of tRNA molecules, each designed to bring a particular amino acid to one or more codons in mRNA. However, all tRNA molecules share a common general structure, as illustrated in **Figure 18-14**. A mature tRNA molecule contains only 70–90 nucleotides, some of

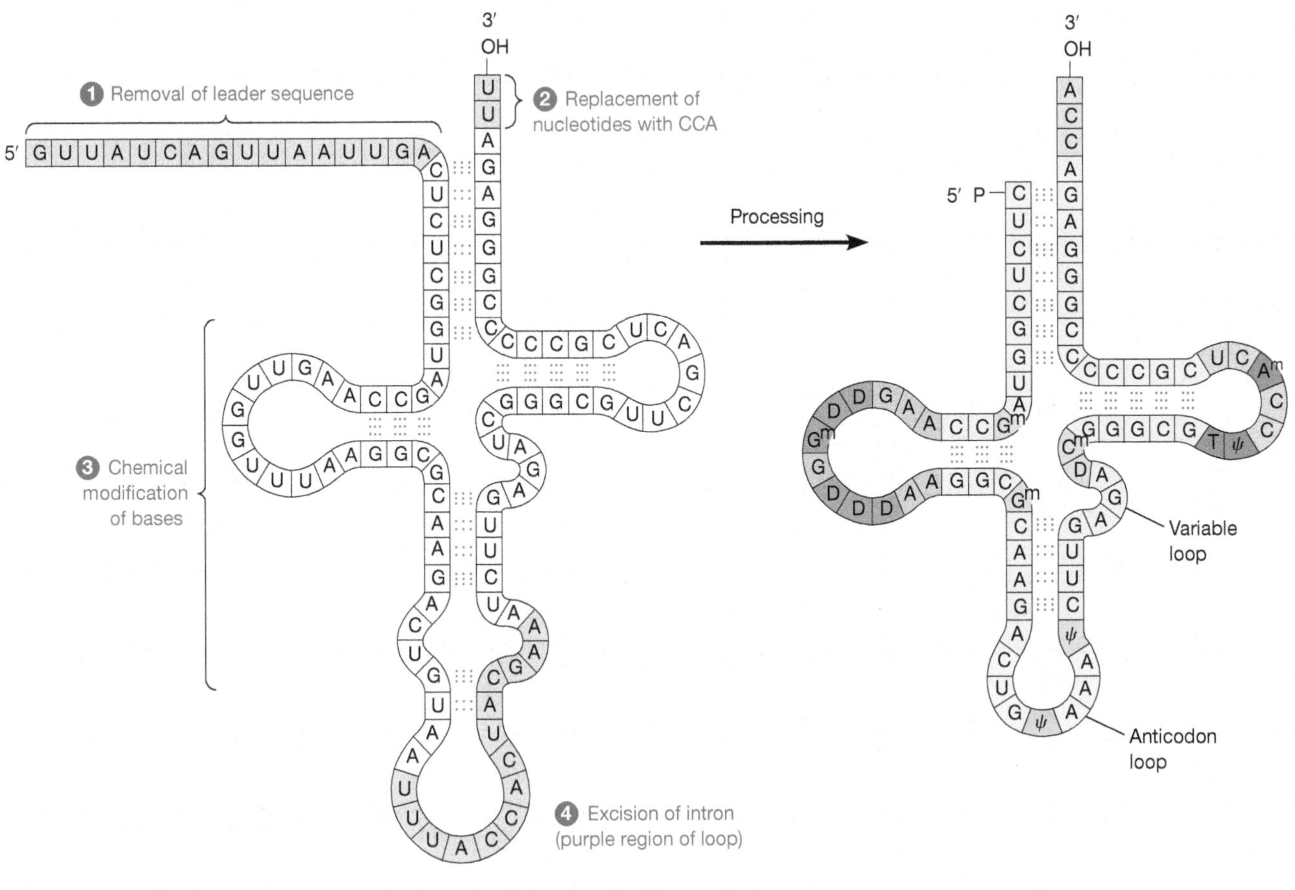

Figure 18-14 Processing and Secondary Structure of Transfer RNA. Every tRNA gene is transcribed as a precursor that must be processed into a mature tRNA molecule. In the primary transcript for yeast tyrosine tRNA, all regions highlighted in purple are removed during processing. Processing for this tRNA involves removal of the leader sequence at the 5′ end, replacement of two nucleotides at the 3′ end by the sequence CCA (which serves as an attachment site for amino acids in all mature tRNAs), chemical modification of certain bases, and excision of an intron. The mature tRNA is depicted in a flattened cloverleaf representation, which clearly shows the base pairing between self-complementary stretches in the molecule. Modified bases (darker colors) are abbreviated as A^m for methyladenine, G^m for methylguanine, C^m for methylcytosine, D for dihydrouracil, T for ribothymidine, and ψ for pseudouridine.

which are chemically modified. Base pairing between complementary sequences located in different regions causes each tRNA molecule to fold into a secondary structure containing several *hairpin loops*, illustrated in the figure. Most tRNAs have four base-paired regions, indicated by the light blue dots in Figure 18-14. In some tRNAs, a fifth such region is present at the *variable loop*. Each of these base-paired regions is a short stretch of RNA double helix. Molecular biologists call the tRNA secondary structure a *cloverleaf* structure because it resembles a cloverleaf when drawn in two dimensions. However, in its normal three-dimensional tertiary structure, the molecule is folded so that the overall shape actually resembles the letter "L." (The structure of tRNA is covered in detail in Chapter 19; see Figure 19-8b.)

Like ribosomal RNA, transfer RNA is synthesized in a precursor form in both eukaryotic and prokaryotic cells. Processing of these **pre-tRNA** molecules involves several different events, as shown in Figure 18-14 for yeast tyrosine tRNA: 1 At the 5′ end, a short *leader sequence* of 16 nucleotides is removed from the pre-tRNA. 2 At the 3′ end, the two terminal nucleotides of the pre-tRNA are removed and replaced with the trinucleotide CCA, which is a common structural feature of all tRNA molecules. (Some tRNAs already have CCA in their primary transcripts and therefore do not require modification at the 3′ end.) 3 In a typical tRNA molecule, about 10–15% of the nucleotides are chemically modified during pre-tRNA processing. The principal modifications include methylation of bases and sugars and creation of unusual bases such as dihydrouracil, ribothymidine, pseudouridine, and inosine.

The processing of yeast tyrosine tRNA is also characterized by 4 the removal of an internal 14-nucleotide sequence, although the transcripts for most tRNAs do not require this kind of excision. An internal segment of an RNA transcript that must be removed to create a mature RNA product is called an RNA *intron*. We will consider introns in more detail during our discussion of mRNA processing because they are a nearly universal feature of mRNA precursors in eukaryotic cells. For the present, we simply note that some eukaryotic tRNA precursors contain introns that must be eliminated by a precise mechanism that cuts and splices the precursor molecule at exactly the same location every time. The cutting-splicing mechanism involves two separate enzymes, an RNA endonuclease and an RNA ligase, which are similar from species to species, even among organisms that are evolutionarily distant from one another. In an experiment that demonstrates this point vividly, cloned genes for the yeast tyrosine tRNA shown in Figure 18-14 were microinjected into eggs of *Xenopus laevis*, the African clawed frog. Despite the long evolutionary divergence between fungi and amphibians, the yeast genes placed in the frog eggs were transcribed and processed properly, including removal of the 14-nucleotide intron.

Messenger RNA Processing in Eukaryotes Involves Capping, Addition of Poly(A), and Removal of Introns

Some bacterial RNAs, such as tRNAs, are processed. Most bacterial mRNAs, however, are exceptions to the generalization that RNA requires processing before it can be used by the cell. They are synthesized in a form ready for translation, even before the entire RNA molecule has been completed (see Figure 18-3). In contrast, transcription and translation in eukaryotic cells are separated in both time and space: transcription takes place in the nucleus, whereas translation occurs in the cytoplasm. Substantial processing is required in the nucleus to convert primary transcripts into mature mRNA molecules that are ready to be transported out of the nucleus to the cytoplasm and translated. Primary transcripts are often very long, typically ranging from 2000 to 20,000 nucleotides. This size heterogeneity is reflected in the term *heterogeneous nuclear RNA (hnRNA)*, which refers to the nonribosomal, nontransfer RNA found in eukaryotic nuclei. hnRNA consists of a mixture of mRNA molecules and their precursors, **pre-mRNA**. Conversion of pre-mRNA molecules into functional mRNAs usually requires the removal of intronic nucleotide sequences and the addition of 5′ caps and 3′ tails, as will be described in the following sections. The C-terminal domain of one of the subunits of RNA polymerase II plays a key role in coupling these RNA-processing events to transcription, presumably by acting as a platform for the assembly of the enzymatic protein complexes involved in pre-mRNA processing.

5′ Caps and 3′ Poly(A) Tails. Most eukaryotic mRNA molecules bear distinctive modifications at both ends. At the 5′ end, they all possess a modified nucleotide called a **5′ cap**, and at the 3′ end they usually have a long stretch of adenine ribonucleotides known as a **poly(A) tail**.

A 5′ cap is simply a guanosine nucleotide that has been methylated at position 7 of the purine ring and is attached "backward"—that is, the bond joining it to the 5′ end of the RNA molecule is a 5′ → 5′ linkage rather than the usual 3′ → 5′ bond (**Figure 18-15**). This distinctive feature of eukaryotic mRNA is added to the primary transcript shortly after initiation of RNA synthesis. As part of the capping process, in a separate enzymatic step the ribose rings of the first, and often the second, nucleotides of the RNA chain can also become methylated, as shown in Figure 18-15. The 5′ cap contributes to mRNA stability by protecting the molecule from degradation by nucleases that attack RNA at the 5′ end.

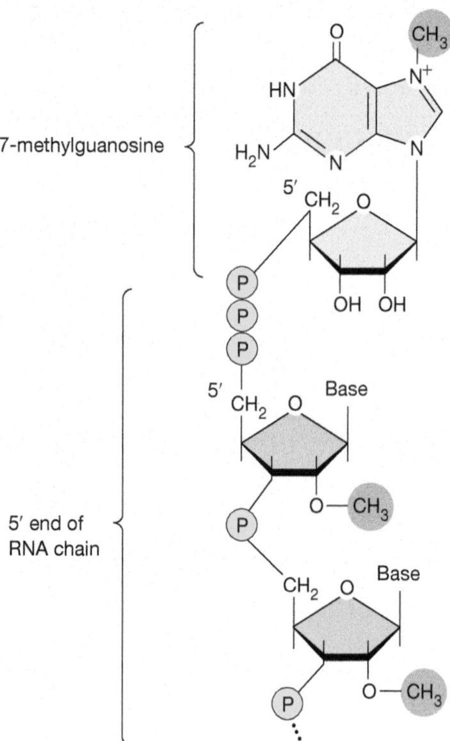

Figure 18-15 Cap Structure Located at the 5′ End of Eukaryotic Pre-mRNA and mRNA Molecules. The methyl groups attached to the first two riboses at the 5′ end of the RNA chain are not always present. Notice that the bond joining the RNA to 7-methylguanosine is a 5′-to-5′ linkage rather than the usual 3′-to-5′ linkage.

The cap also plays an important role in positioning mRNA on the ribosome for the initiation of translation.

In addition to the 5′ cap, a poly(A) tail ranging from 50 to 250 nucleotides in length is present at the 3′ end of most eukaryotic mRNA molecules. (In animal cells, mRNAs encoding the major histones are among the few mRNAs known to lack such a poly(A) tail.) It is clear that poly(A) must be added after transcription because genes do not contain long stretches of thymine (T) nucleotides that could serve as a template for the addition of poly(A). Direct support for this conclusion has come from the isolation of the enzyme *poly(A) polymerase*, which catalyzes the addition of poly(A) sequences to mRNA without requiring a DNA template.

The addition of poly(A) is part of the termination process that creates the 3′ end of most eukaryotic mRNA molecules. Unlike bacteria, where specific termination sequences halt transcription at the 3′ end of newly forming mRNAs, the transcription of eukaryotic pre-mRNAs often proceeds hundreds or even thousands of nucleotides beyond the site destined to become the 3′ end of the final mRNA molecule. A special signal—consisting of an AAUAAA sequence located slightly upstream from this site and a GU-rich and/or U-rich element downstream from the site—determines where the poly(A) tail should be added. As shown in **Figure 18-16**, this signaling element triggers cleavage of the primary transcript 10–35 nucleotides downstream from the AAUAAA sequence, and poly(A) polymerase catalyzes formation of the poly(A) tail. In addition to creating the poly(A) tail, the processing events

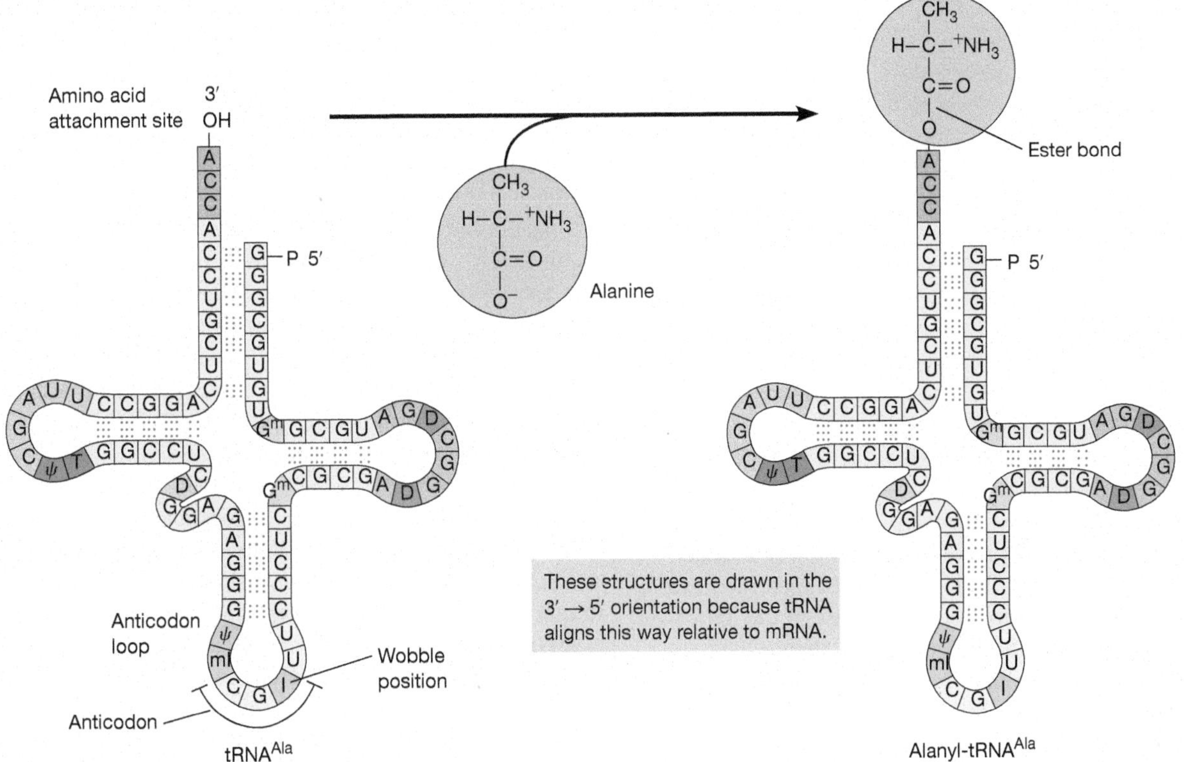
Amino acid attachment site
3′ OH
P 5′
Alanine
Ester bond
These structures are drawn in the 3′ → 5′ orientation because tRNA aligns this way relative to mRNA.
Anticodon loop
Wobble position
Anticodon
$tRNA^{Ala}$
Alanyl-$tRNA^{Ala}$

(a) Secondary structure of tRNA, before and after amino acid attachment

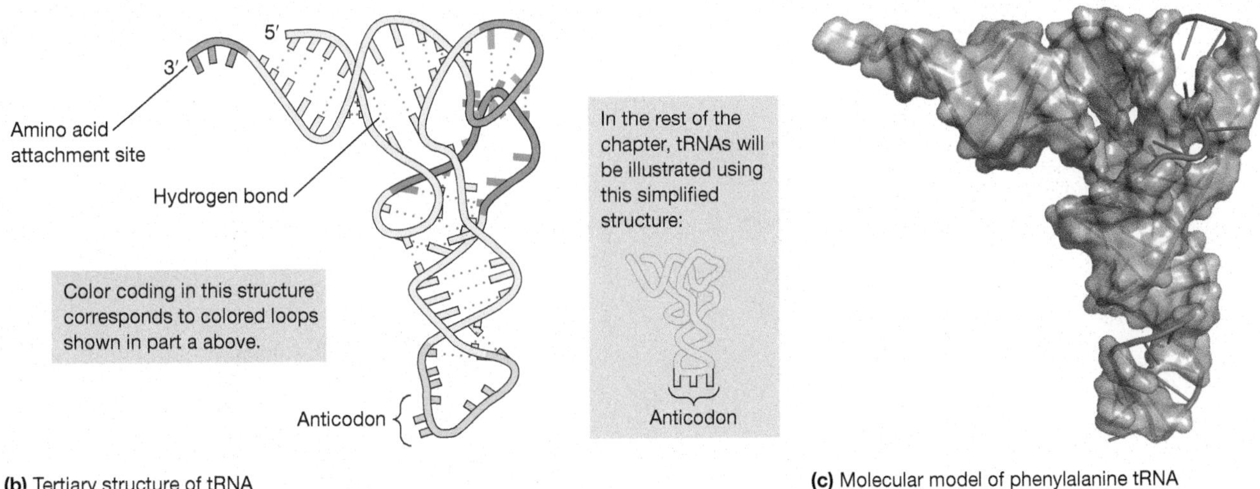
5′
3′
Amino acid attachment site
Hydrogen bond
Color coding in this structure corresponds to colored loops shown in part a above.
Anticodon
In the rest of the chapter, tRNAs will be illustrated using this simplified structure:
Anticodon

(b) Tertiary structure of tRNA

(c) Molecular model of phenylalanine tRNA

Figure 19-8 Structure and Aminoacylation of a tRNA. (a) Yeast alanine tRNA, like all tRNA molecules, contains three major loops, four base-paired regions, an anticodon triplet, and a 3′ terminal sequence of CCA, to which the appropriate amino acid can be attached by an ester bond. Modified bases are darker colored, and their names (as nucleosides) are abbreviated I for inosine, mI for methylinosine, D for dihydrouridine, T for ribothymidine, ψ for pseudouridine, and G^m for methylguanosine. (For the significance of the wobble position in the anticodon, see Figure 19-9.) **(b)** In the L-shaped tertiary structure of a tRNA, the amino acid attachment site is at one end and the anticodon at the other. **(c)** This image shows a three-dimensional model of the yeast phenylalanine tRNA, with the molecular surface in transparent gray and the RNA backbone in blue.

convention, the name of the amino acid that attaches to a given tRNA is indicated by a superscript. For example, tRNA molecules specific for alanine are identified as $tRNA^{Ala}$. Once the amino acid is attached, the tRNA is called an **aminoacyl tRNA** (for example, alanyl-$tRNA^{Ala}$). The tRNA is then said to be in its *charged* form, and the amino acid is said to be *activated*.

Transfer RNA molecules can recognize codons in mRNA because each tRNA possesses an **anticodon**, a special trinucleotide sequence located within one of the loops of the tRNA molecule (Figure 19-8a). The anticodon of each tRNA is complementary to one or more mRNA codons that specify the amino acid being carried by that tRNA. Therefore, *anticodons permit tRNA molecules to recognize codons in mRNA by complementary base pairing.* Take careful note of the convention used in representing codons and anticodons: codons in mRNA are written in the 5′ → 3′ direction, whereas anticodons in tRNA are usually represented in the 3′ → 5′ orientation. Thus, one of the codons for alanine is 5′-GCC-3′, and the corresponding anticodon in tRNA is 3′-CGG-5′.

Because the genetic code employs 61 codons to specify amino acids, you might expect to find 61 different tRNA molecules involved in protein synthesis, each recognizing a different codon. However, the number of different tRNAs is significantly less than 61 because many tRNA molecules recognize more than one codon. You can see why this is possible by examining the table of the genetic code (see Figure 19-5). Codons differing in the third base often encode the same amino acid. For example, UUU and UUC both encode phenylalanine; UCU, UCC, UCA, and UCG all encode serine; and so forth. In such cases, the same tRNA can bind to more than one codon without introducing mistakes. For example, a single tRNA can recognize the codons UUU and UUC because both encode the same amino acid, phenylalanine.

Such considerations led Francis Crick to propose that mRNA and tRNA line up on the ribosome in a way that permits flexibility, or "wobble," in the pairing between the third base of the codon and the corresponding base in the anticodon. According to this **wobble hypothesis**, the flexibility in codon-anticodon binding allows some unexpected base pairs to form. A closer examination of the possible hydrogen bonds that can form between various bases helps explain how wobble base pairing occurs (**Figure 19-9**). For example, although cytosine pairing with guanine is a "normal" pairing, guanine can also pair with uracil (Figure 19-9a). Remember that adenosines are converted to inosines at specific locations in tRNAs through the process of RNA editing (which you learned about in Chapter 18). Inosine (I) is extremely rare in other RNA molecules but occurs often in the wobble position of tRNA anticodons. Inosine is the "wobbliest" of all third-position bases because it can pair with U, C, or A (Figure 19-9b). For example, a tRNA with the anticodon 3′-UAI-5′ can recognize the codons AUU, AUC, and AUA, all of which encode the amino acid isoleucine.

It is because of wobble that fewer tRNA molecules are required for some amino acids than the number of codons that specify those amino acids, as shown in the example in **Figure 19-10**. Figure 19-10a shows two different arginine codons in the mRNA recognized by a single tRNA anticodon. In the case of isoleucine, a cell can translate all three codons with a single tRNA molecule containing 3′-UAI-5′ as its anticodon. Similarly, the six codons for the amino acid leucine (UUA, UUG, CUU, CUC, CUA, and CUG) require only three tRNAs because of wobble. All of the possible wobble combinations are listed in Figure 19-10b for reference. Although the existence of wobble means that a single tRNA molecule can recognize more than one codon, the different codons recognized by a given tRNA always encode the same amino acid, so wobble does not cause insertion of incorrect amino acids.

Aminoacyl-tRNA Synthetases Link Amino Acids to the Correct Transfer RNAs

Before a tRNA molecule can bring its amino acid to the ribosome, that amino acid must be attached covalently to the tRNA. The enzymes responsible for linking amino acids to their corresponding tRNAs are called **aminoacyl-tRNA synthetases**.

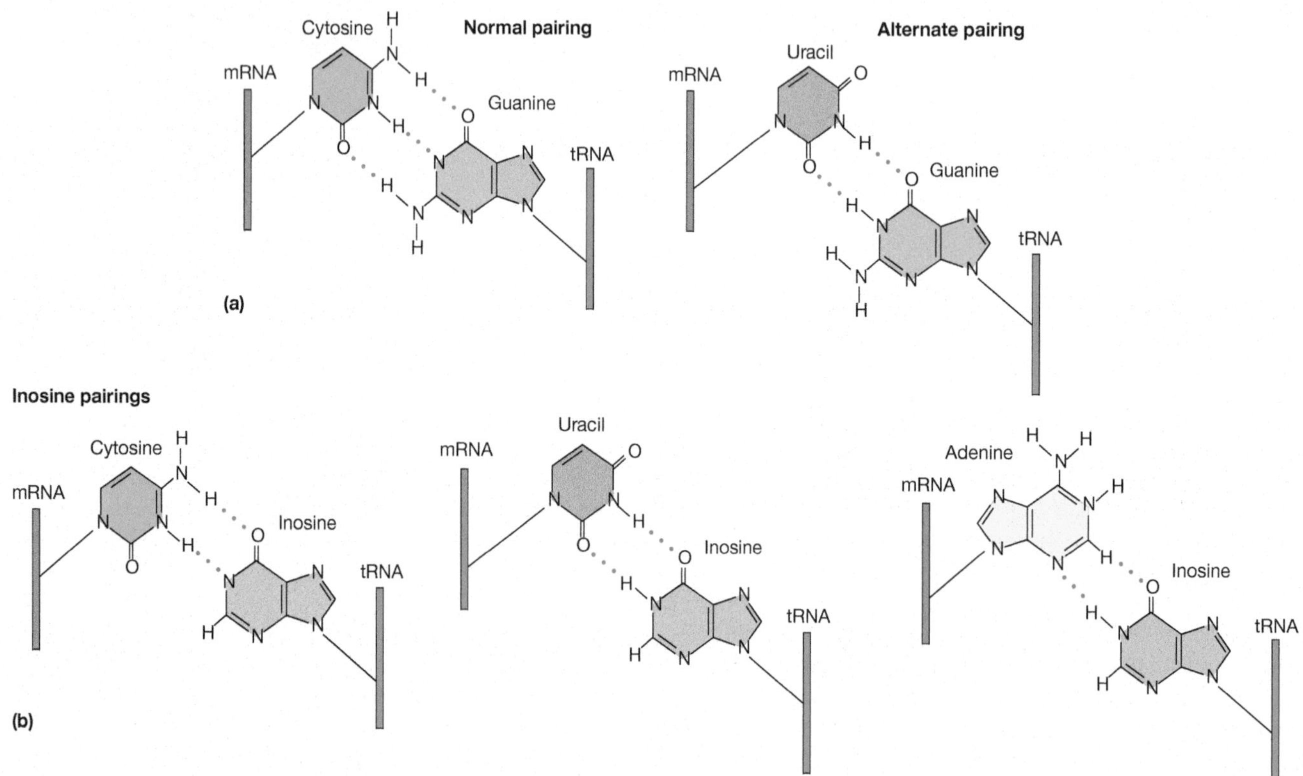

Figure 19-9 "Wobble" as a Result of Alternate Base Pairing Between Codons and Anticodons. (a) The same base (in this case, guanine) in a tRNA anticodon can pair with uracil (right) or its typical complementary base, cytosine (left). **(b)** Inosine can pair with several different bases, including cytosine, uracil, and adenine.

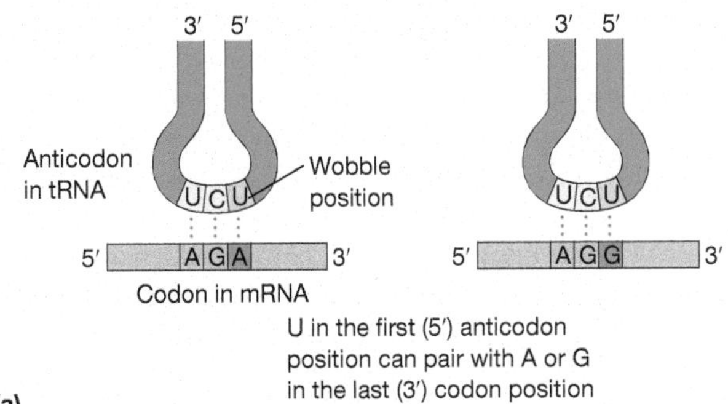
3′ 5′
Anticodon in tRNA
Wobble position
U C U
5′ A G A 3′
Codon in mRNA
3′ 5′
U C U
5′ A G G 3′
U in the first (5′) anticodon position can pair with A or G in the last (3′) codon position
(a)

Bases Recognized in Codon (third position only)	Base in Anticodon (first position only)
U	A
G	C
A or G	U
C or U	G
U, C, or A	I (Inosine)

(b)

Figure 19-10 Wobble Base Pairing at the Codon-Anticodon Interface. (a) A codon and anticodon, showing the wobble position and two permitted codons with which the same tRNA can interact. **(b)** The table summarizes the base pairs permitted at the third position of a codon by the wobble hypothesis.

Cells typically have 20 different aminoacyl-tRNA synthetases, one for each of the 20 amino acids commonly used in protein synthesis. Cells that utilize the unusual 21st and 22nd amino acids, *selenocysteine* and *pyrrolysine*, contain special tRNAs and aminoacyl-tRNA synthetases for these amino acids as well.

Aminoacyl-tRNA synthetases catalyze the attachment of amino acids to their corresponding tRNAs via an ester bond, accompanied by the hydrolysis of ATP to adenosine monophosphate (AMP) and the two terminal phosphates as a molecule of pyrophosphate.

$$\underset{\text{Amino acid}}{NH_3^+\text{—}\overset{\overset{\displaystyle R}{|}}{C}H\text{—}\overset{\overset{\displaystyle O}{\|}}{C}\text{—}O^+} + HO\text{—}\boxed{tRNA} \xrightarrow[\text{Aminoacyl-tRNA synthetase}]{ATP \quad AMP + PP_i} NH_3^+\text{—}\overset{\overset{\displaystyle R}{|}}{C}H\text{—}\overset{\overset{\displaystyle O}{\|}}{C}\text{—}O\text{—}\boxed{tRNA} \quad \textbf{(19-1)}$$

Figure 19-11 outlines the steps by which this reaction occurs. The driving force for the reaction is provided by the hydrolysis of pyrophosphate to 2 P_i shown in step ❷.

In the resulting charged product—an aminoacyl tRNA—the ester bond linking the amino acid to the tRNA is said to be a "high-energy" bond. This simply means that hydrolysis of the bond releases sufficient energy to drive formation of the peptide bond that will eventually attach the amino acid to a growing polypeptide chain. The process of aminoacylation of a tRNA molecule is therefore also called *amino acid activation* because it links an amino acid to its proper tRNA as well as activating it for subsequent peptide bond formation.

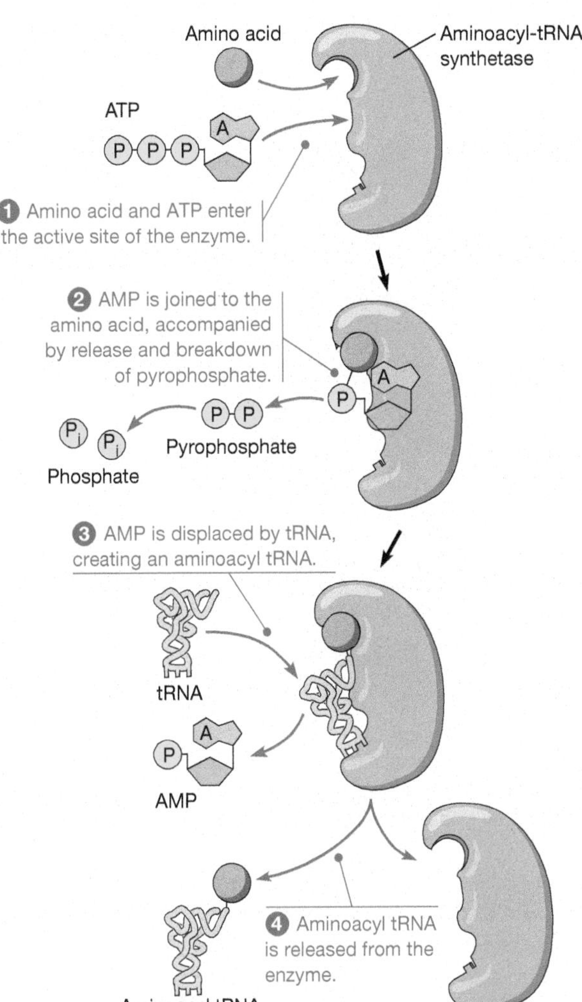
Amino acid
Aminoacyl-tRNA synthetase
ATP
P P P
A
❶ Amino acid and ATP enter the active site of the enzyme.
❷ AMP is joined to the amino acid, accompanied by release and breakdown of pyrophosphate.
P P
Pyrophosphate
Phosphate
❸ AMP is displaced by tRNA, creating an aminoacyl tRNA.
tRNA
AMP
❹ Aminoacyl tRNA is released from the enzyme.
Aminoacyl tRNA

Figure 19-11 Amino Acid Activation by Aminoacyl-tRNA Synthetase. This enzyme catalyzes the formation of an ester bond between the carboxyl group of an amino acid and the 3′ OH of the appropriate tRNA, generating an aminoacyl tRNA.

How do aminoacyl-tRNA synthetases identify the correct tRNA for each amino acid? Differences in the base sequences of the various tRNA molecules allow them to be distinguished, and, surprisingly, the anticodon is not the only feature to be recognized. Changes in the base sequence of either the anticodon triplet or the 3′ end of a tRNA molecule can alter the amino acid to which a tRNA attaches. Thus, aminoacyl-tRNA synthetases recognize nucleotides located in at least two different regions of tRNA molecules when they pick out the tRNA that is to become linked to a particular amino acid. When more than one tRNA exists for a given amino acid, the aminoacyl-tRNA synthetase specific for that particular amino acid recognizes each of the tRNAs. Some cells possess fewer than 20 aminoacyl-tRNA synthetases. In such cases, the same aminoacyl-tRNA synthetase may catalyze the attachment of two different amino acids to their corresponding tRNAs.

After linking an amino acid to a tRNA molecule, aminoacyl-tRNA synthetases proofread the final product to make sure that the correct amino acid has been used. This proofreading function is performed by a site on the aminoacyl-tRNA synthetase molecule that recognizes incorrect amino

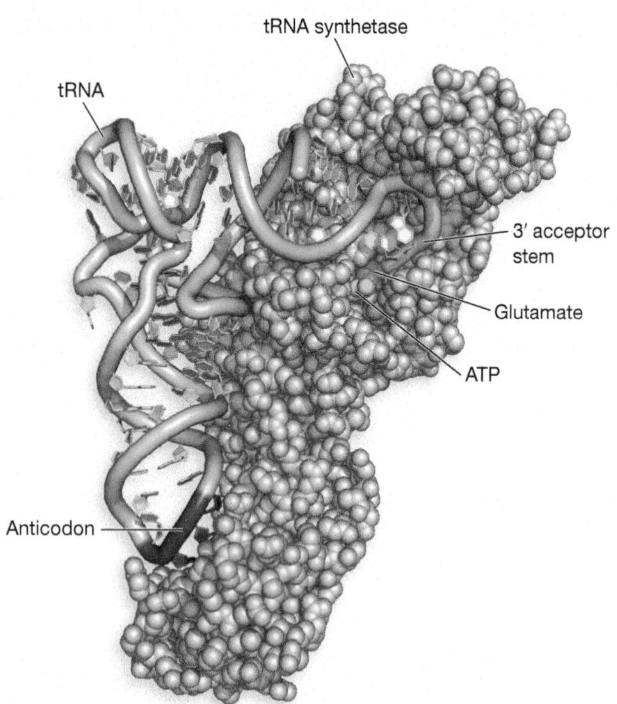

Figure 19-12 Molecular Model of an Aminoacyl-tRNA Synthetase in Complex with tRNA. In the example shown here, the tRNA is being charged with the amino acid glutamate.

acids and releases them by hydrolyzing the bond that links the amino acid to the tRNA. Proofreading is critical because the ribosome has no way to distinguish tRNAs carrying the incorrect amino acid from those carrying the correct one.

Once the correct amino acid has been joined to its tRNA, it is the tRNA itself (and not the amino acid) that recognizes the appropriate codon in mRNA. The first evidence for this was provided by François Chapeville and Fritz Lipmann, who designed an elegant experiment involving the tRNA that carries the amino acid cysteine. They took the tRNA after its cysteine had been attached and treated it with a nickel catalyst, which converts the attached cysteine into the amino acid alanine. The result was alanine covalently linked to a tRNA molecule that normally carries cysteine. When the researchers added this abnormal aminoacyl tRNA to a cell-free protein-synthesizing system, alanine was inserted into polypeptide chains in locations normally occupied by cysteine. Such results proved that codons in mRNA recognize tRNA molecules rather than their bound amino acids. Hence, the specificity of the aminoacyl-tRNA synthetase reaction is crucial to the accuracy of gene expression because it ensures that the proper amino acid is linked to each tRNA. **Figure 19-12** shows the structure of the complex between a tRNA and its aminoacyl-tRNA synthetase in detail. The 3′ acceptor stem of the tRNA tucks into a binding pocket in the synthetase enzyme, where the amino acid is added to the end of the tRNA.

Messenger RNA Brings Polypeptide Coding Information to the Ribosome

We saw earlier how the sequence of codons in mRNA directs the order in which amino acids are linked together during protein synthesis. Hence, the mRNA that happens to bind to a given ribosome will determine which polypeptide that ribosome will manufacture. In eukaryotes, where transcription takes place in the nucleus and protein synthesis is mainly a cytoplasmic event, the mRNA must first be exported from the nucleus. This step is not required in prokaryotes, which by definition have no nucleus. As a result, transcription and translation are often coupled in prokaryotic cells (see Figure 18-2).

At the heart of each messenger RNA molecule is, of course, its message—the sequence of nucleotides that encodes a polypeptide. However, mRNAs also possess sequences at either end that are not translated (**Figure 19-13**). The untranslated sequence at the 5′ end of an mRNA precedes the **start codon**, which is the first codon to be translated. AUG is the most common start codon, although a few other triplets are occasionally used for this purpose. The untranslated sequence at the 3′ end follows the **stop codon**, which signals the end of translation and can be UAG, UAA, or UGA. These *5′ and 3′ untranslated regions* range from a few dozen to hundreds of nucleotides in length. Although these sequences are

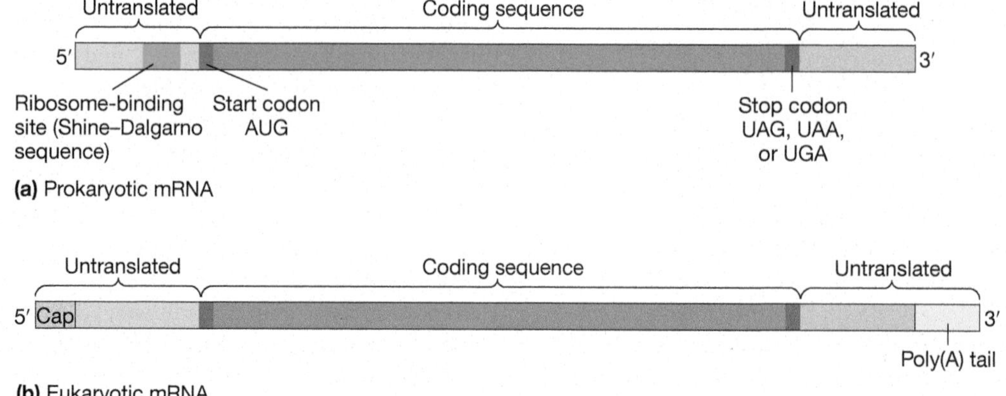

Figure 19-13 Comparison of Prokaryotic and Eukaryotic mRNA. (a) A prokaryotic mRNA molecule encoding a single polypeptide has the features shown here. (A polycistronic prokaryotic mRNA would generally have a set of these features for each gene.) Prokaryotic mRNAs include a ribosome-binding site (a nucleotide sequence also called a Shine–Dalgarno sequence, after its discoverers). **(b)** A eukaryotic mRNA molecule has, in addition, a 5′ methylguanine cap and a 3′ poly(A) tail. It lacks a Shine–Dalgarno sequence (but still contains a start codon).

not translated, their presence is essential for proper mRNA function. Included in the untranslated regions of eukaryotic mRNAs are a 5′ cap and a 3′ poly(A) tail (both of which were described in Chapter 18). As we will see, the 5′ cap is important in initiating translation in eukaryotes.

In eukaryotes, most mRNA molecules are *monocistronic* (that is, they encode a single polypeptide). In bacteria and archaea, however, some mRNAs are *polycistronic*—meaning they encode several polypeptides, usually with related functions in the cell. The clusters of genes that give rise to polycistronic mRNAs are single transcription units called *operons* (which we will discuss in Chapter 20 in the context of gene regulation). Although most often thought of as a feature of prokaryotes, eukaryotes also produce polycistronic RNAs. In some cases, such as the nematode *Caenorhabditis elegans*, it is only the pre-mRNA that is polycistronic; subsequent RNA processing results in monocistronic mRNAs that are translated separately. In other eukaryotes, dicistronic RNAs (that is, RNAs that encode two proteins) remain joined and are translated together.

Protein Factors Are Required for Translational Initiation, Elongation, and Termination

In addition to aminoacyl-tRNA synthetases and the protein components of the ribosome, translation requires the participation of several other kinds of protein molecules. Some of these *protein factors* are required for initiating the translation process, others for elongating the growing polypeptide chain, and still others for terminating polypeptide synthesis. The exact roles played by these factors will become apparent as we proceed to a discussion of the mechanism of translation.

CONCEPT CHECK 19-2

Suppose a tRNA has the anticodon 3′-CGU-5′. What two different codons could it bind to? How many different amino acids could this tRNA insert into a polypeptide?

19.3 The Mechanism of Translation

The translation of mRNAs into polypeptides begins the synthesis of a polypeptide chain at its amino-terminal end, or *N-terminus*, and sequentially adds amino acids to the growing chain until the carboxyl-terminal end, or *C-terminus*, is reached. The first experimental evidence for such a mechanism was provided in 1961 by Howard Dintzis, who investigated hemoglobin synthesis in developing red blood cells that had been incubated briefly with radioactive amino acids. Dintzis reasoned that if the time of incubation is kept relatively brief, then the radioactivity present in completed hemoglobin chains should be concentrated at the most recently synthesized end of the molecule. He found that the highest concentration of radioactivity in completed hemoglobin chains was at the C-terminal end, indicating that the C-terminus is the last part of the polypeptide chain to be synthesized. This allowed him to conclude that during mRNA translation, amino acids are added to the growing polypeptide chain beginning at the N-terminus and proceeding toward the C-terminus.

In theory, mRNA could be read in either the 5′ → 3′ direction or the 3′ → 5′ direction during this process. The first attempts to determine the direction in which mRNA is actually read involved the use of artificial RNA molecules. A typical example is the synthetic RNA that can be made by adding the base C to the 3′ end of poly(A), yielding the molecule 5′-AAAAAAAAAAAA…AAC-3′. When added to a cell-free protein-synthesizing system, this RNA stimulates the synthesis of a polypeptide consisting of a stretch of lysine residues with an asparagine at the C-terminus. Because AAA codes for lysine and AAC codes for asparagine, this means that mRNA is translated in the 5′ → 3′ direction. Confirming evidence has come from many studies in which the base sequences of naturally occurring mRNAs have been compared with the amino acid sequences of the polypeptide chains they encode. In all cases, the amino acid sequence of the polypeptide chain corresponds to the order of mRNA codons read in the 5′ → 3′ direction.

To understand how translation of mRNA in the 5′ → 3′ direction leads to the synthesis of polypeptides in the N-terminal to C-terminal direction, it is helpful to subdivide the translation process into three stages, as shown in **Figure 19-14** on page 554: ❶ an *initiation* stage, in which mRNA is bound to the ribosome and positioned for proper translation; ❷ an *elongation* stage, in which amino acids are sequentially joined together via peptide bonds in an order specified by the arrangement of codons in mRNA; and ❸ a *termination* stage, in which the mRNA and the newly formed polypeptide chain are released from the ribosome.

In the following sections we examine each of these stages in detail. Although our discussion focuses mainly on translation in bacterial cells, where the mechanisms are especially well understood, the comparable events in eukaryotic and archaeal cells are rather similar. The aspects of translation that differ between bacteria and eukaryotes are confined mostly to the initiation stage, as we describe in the next section.

Translational Initiation Requires Initiation Factors, Ribosomal Subunits, mRNA, and Initiator tRNA

Bacterial Initiation. The initiation of translation in bacteria is illustrated in **Figure 19-15** on page 555, which shows that initiation can be subdivided into three distinct steps. In step ❶, three **initiation factors**—proteins called IF1, IF2, and IF3—bind to the small (30S) ribosomal subunit, with GTP attaching to IF2. The presence of IF3 at this early stage prevents the 30S subunit from prematurely associating with the 50S subunit.

In step ❷, mRNA and the tRNA carrying the first amino acid bind to the 30S ribosomal subunit. The mRNA is bound to the 30S subunit in its proper orientation by means of a special nucleotide sequence called the **Shine–Dalgarno sequence**, named after its discoverers. This sequence consists of a stretch of three to nine purine nucleotides (often AGGA) located slightly upstream of the start codon (see Figure 19-15, inset). These purines in the mRNA form complementary base pairs with a pyrimidine-rich sequence at the 3′ end of 16S rRNA, which forms the ribosome's mRNA-binding site. The importance of the mRNA-binding site has been shown by studies involving colicins, which are proteins produced by certain

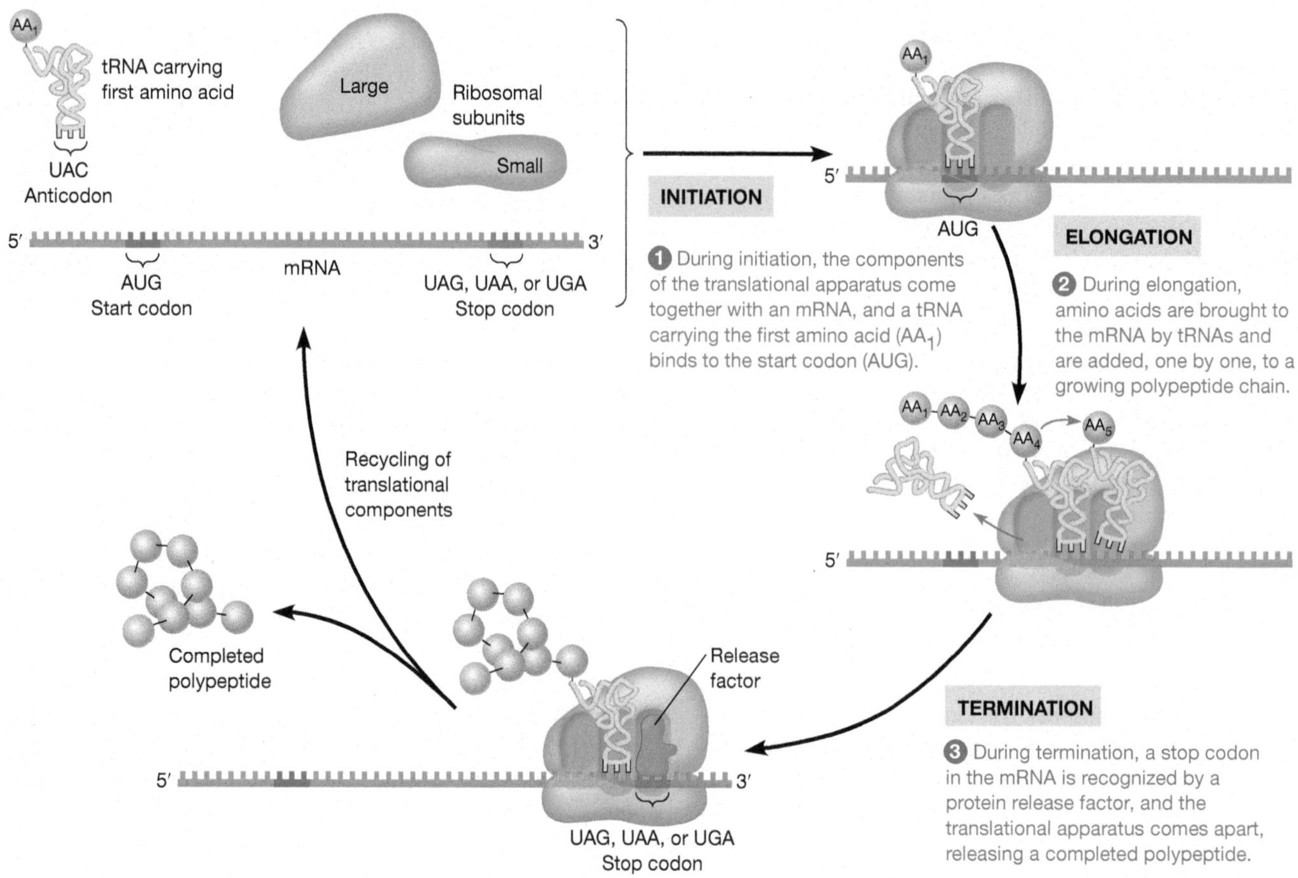
AA1
tRNA carrying first amino acid
UAC
Anticodon
Large
Ribosomal subunits
Small
5′
3′
AUG
Start codon
mRNA
UAG, UAA, or UGA
Stop codon
INITIATION
1 During initiation, the components of the translational apparatus come together with an mRNA, and a tRNA carrying the first amino acid (AA1) binds to the start codon (AUG).
AUG
ELONGATION
2 During elongation, amino acids are brought to the mRNA by tRNAs and are added, one by one, to a growing polypeptide chain.
AA1 AA2 AA3 AA4 AA5
Recycling of translational components
Completed polypeptide
Release factor
TERMINATION
3 During termination, a stop codon in the mRNA is recognized by a protein release factor, and the translational apparatus comes apart, releasing a completed polypeptide.
UAG, UAA, or UGA
Stop codon

Figure 19-14 An Overview of Translation. Translation occurs in three stages: initiation, elongation, and termination.

strains of *Escherichia coli* (*E. coli*) that can kill other types of bacteria. When one colicin, E3, enters the cytoplasm of susceptible bacteria, it catalyzes the removal of a 49-nucleotide fragment from the 3′ end of 16S rRNA. This action destroys the mRNA-binding site, thereby creating ribosomes that can no longer initiate polypeptide synthesis.

The binding of mRNA to the mRNA-binding site of the small ribosomal subunit places the mRNA's AUG start codon at the ribosome's P site, where it can bind to the anticodon of the appropriate tRNA. The first clue that a special kind of tRNA is involved in this step emerged when it was discovered that roughly half the proteins in *E. coli* contain methionine at their N-terminal ends. This was surprising because methionine is a relatively uncommon amino acid, accounting for no more than a few percent of the amino acids in bacterial proteins. The explanation for such a pattern became apparent when it was discovered that bacterial cells contain two different methionine-specific tRNAs. One, designated $tRNA^{Met}$, carries a normal methionine destined for insertion into the internal regions of polypeptide chains. The other, called $tRNA^{fMet}$, carries a methionine that is converted to the derivative *N*-formylmethionine (fMet) after linkage to the tRNA. In *N*-formylmethionine, the amino group of methionine is blocked by the addition of a formyl group and so cannot form a peptide bond with another amino acid; only the carboxyl group is available for bonding to another amino acid. Hence *N*-formylmethionine can be situated only at the N-terminal end of a polypeptide chain—suggesting that $tRNA^{fMet}$ functions as an **initiator tRNA** that starts the process of translation. This idea was soon confirmed by the discovery that bacterial polypeptide chains in the early stages of synthesis always contain *N*-formylmethionine at their N-terminus. Following completion of the polypeptide chain (and in some cases while it is still being synthesized), the formyl group, and often the methionine itself, is enzymatically removed.

During initiation, the initiator tRNA with its attached *N*-formylmethionine is bound to the P site of the 30S ribosomal subunit by the action of initiation factor IF2 (plus GTP), which can distinguish initiator $tRNA^{fMet}$ from other kinds of tRNA. This attribute of IF2 helps explain why AUG start codons bind to the initiator $tRNA^{fMet}$, whereas AUG codons located elsewhere in mRNA bind to the noninitiating $tRNA^{Met}$. Once $tRNA^{fMet}$ enters the P site, its anticodon becomes base-paired with the AUG start codon in the mRNA, and IF3 is released. At this point the 30S subunit with its associated IF1, IF2-GTP, mRNA, and *N*-formylmethionyl $tRNA^{fMet}$ is referred to as the **30S initiation complex** (Figure 19-15, step 2).

Once IF3 has been released, the 30S initiation complex can bind to a free 50S ribosomal subunit, generating the **70S initiation complex** (Figure 19-15, step 3). The 50S subunit

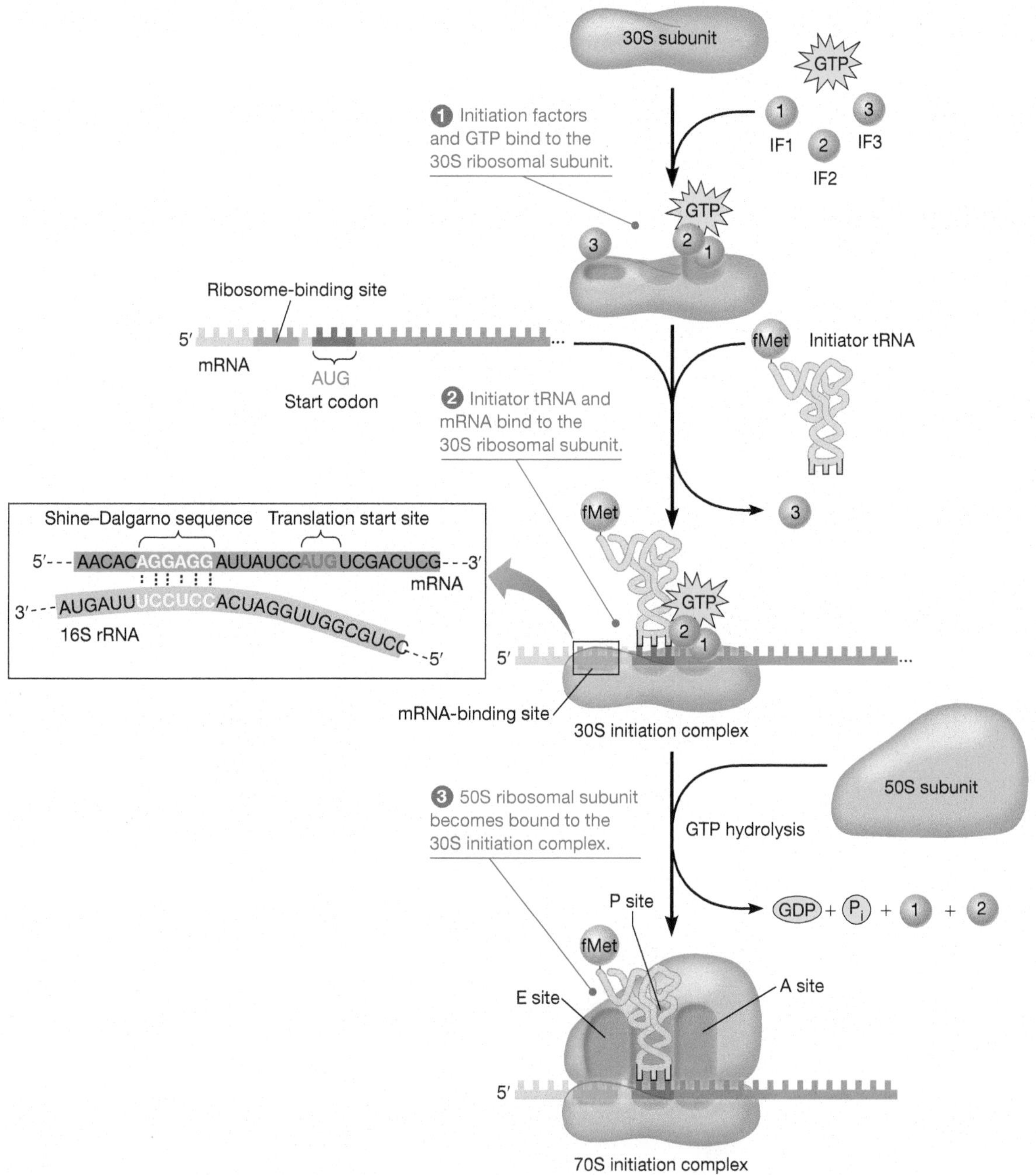

Figure 19-15 Initiation of Translation in Bacteria. In bacteria, assembly of the 70S initiation complex occurs in three steps. First, the initiation factors IF1, IF2, and IF3 plus GTP bind to the small ribosomal subunit; next, the initiator aminoacyl tRNA and mRNA are attached; and finally, the large ribosomal subunit joins the complex. The resulting 70S initiation complex has fMet-tRNA^{fMet} residing in the ribosome's P site. The inset in step 2 shows that the mRNA-binding site is composed, at least in part, of a portion of the 16S rRNA of the small ribosomal subunit, which is complementary to a Shine–Dalgarno sequence in the mRNA.

then promotes hydrolysis of the IF2-bound GTP, leading to the release of IF2 and IF1. At this stage, all three initiation factors have been released.

Eukaryotic Initiation. Unlike the situation in bacteria, the AUG start codon in eukaryotes (and archaea) specifies the amino acid methionine rather than *N*-formylmethionine. Eukaryotes also use a somewhat different pathway for assembling the initiation complex. First, they use a different set of roughly a dozen initiation factors known as eIFs (*e*ukaryotic *i*nitiation *f*actors), which are named with "eIF" followed by a number (eIF1, eIF2, and so forth). Second, eukaryotes use different genes to encode an initiator tRNA^{Met} and a tRNA^{Met} used during peptide elongation. The initiator tRNA^{Met} favors binding to eIF2 and prevents its binding to the elongation machinery.

Translational initiation in eukaryotes is summarized in **Figure 19-16** on page 556. At the beginning of eukaryotic initiation, 1 the initiation factor eIF2 (with GTP attached) binds to the initiator methionyl tRNA^{Met} *before* the tRNA. The tRNA binds to the small ribosomal subunit along with other initiation factors (eIF1, -3, and -5) and eIF1A (the eukaryotic counterpart of bacterial IF1). The resulting complex is known as the 43S preinitiation complex. 2 Meanwhile, the mRNA is readied for binding the 43S preinitiation complex. The 5′

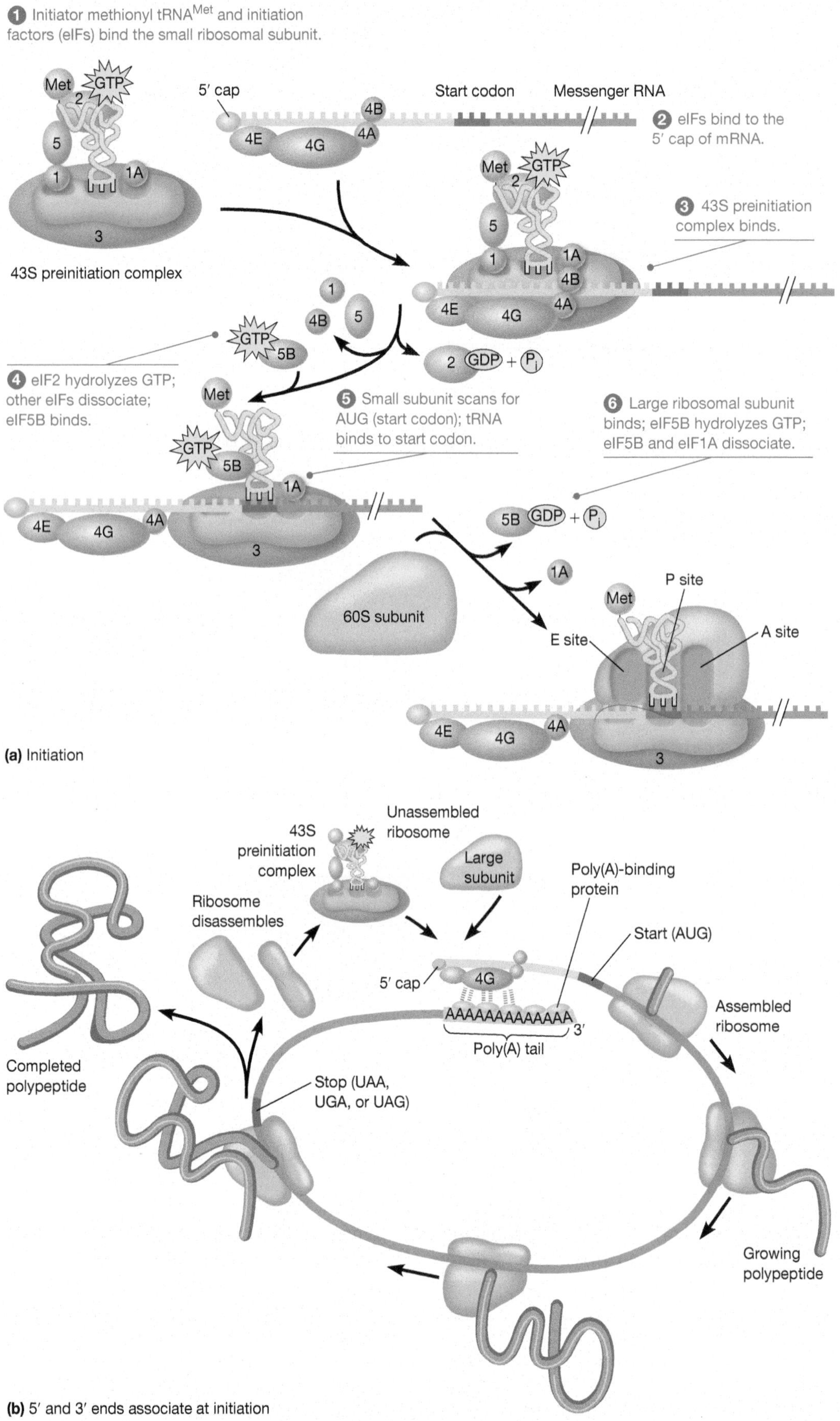

① Initiator methionyl tRNA^{Met} and initiation factors (eIFs) bind the small ribosomal subunit.

Met
GTP
2
5
1
1A
3
43S preinitiation complex

5′ cap
Start codon
Messenger RNA
4B
4E
4G
4A

② eIFs bind to the 5′ cap of mRNA.

③ 43S preinitiation complex binds.

④ eIF2 hydrolyzes GTP; other eIFs dissociate; eIF5B binds.

2 GDP + P_i

⑤ Small subunit scans for AUG (start codon); tRNA binds to start codon.

⑥ Large ribosomal subunit binds; eIF5B hydrolyzes GTP; eIF5B and eIF1A dissociate.

5B GDP + P_i

60S subunit
P site
E site
A site

(a) Initiation

43S preinitiation complex
Unassembled ribosome
Large subunit
Ribosome disassembles
Poly(A)-binding protein
Start (AUG)
5′ cap
4G
AAAAAAAAAAAAA 3′
Poly(A) tail
Assembled ribosome
Completed polypeptide
Stop (UAA, UGA, or UAG)
Growing polypeptide

(b) 5′ and 3′ ends associate at initiation

Figure 19-16 Initiation of Translation in Eukaryotes. (a) Assembly of initiation factors on the mRNA and assembly of the 43S preinitiation complex lead to assembly of a complex containing the mRNA and the small ribosomal subunit. **(b)** Poly(A)-binding protein attached to the poly(A) tail at the 3′ end of the mRNA promotes attachment of eIF4G and other proteins to the 5′ cap on the RNA, forming a circular RNA. Ribosomes can then attach and move along the mRNA to synthesize new polypeptides.

cap of the mRNA is recognized by the cap-binding initiation factor, eIF4E, which recruits eIF4G to bind the mRNA as well. This binding is an important step in regulating the efficiency of translation. eIF4A and eIF4B then join the complex. eIF4A has helicase activity that ensures that the mRNA has no secondary structure, which would prevent translation. Finally, ❸ the two complexes join together via interactions between eIF4G and eIF3 and between the small ribosomal subunit and the mRNA. ❹ eIF2 then hydrolyzes its associated GTP, allowing several eIFs to leave the complex, followed by binding of eIF5B with its attached GTP (note that this protein is different from eIF5, which is part of the 43S preinitiation complex).

After binding to mRNA, ❺ the small ribosomal subunit, with the initiator tRNA in tow, scans along the mRNA and usually begins translation at the first AUG triplet it encounters. The nucleotides on either side of the eukaryotic start codon appear to be involved in its recognition. A common start sequence is ACCAUGG (also called a **Kozak sequence**, after Marilyn Kozak, who discovered that many eukaryotic mRNAs have this sequence), where the underlined triplet is the actual start codon. After the initiator $tRNA^{Met}$ becomes base-paired with the start codon, ❻ the large ribosomal subunit joins the complex in a reaction facilitated by eIF5B. When eIF5B hydrolyzes its associated GTP, it helps other eIFs dissociate from the fully assembled ribosome.

So far, our discussion of translational initiation has focused on the 5′ end of mRNA. Based on biochemical interactions and other data, however, the 3′ end of eukaryotic mRNAs also promotes translational initiation. To see how this might be, recall that eukaryotic mRNAs are *polyadenylated* (see Figure 18-16). A protein known as **poly(A)-binding protein (PABP)** binds the stretch of A residues in the poly(A) tail. PABP in turn can bind the initiation factor eIF4G; in addition, eIF4G can bind to the 3′ end of eukaryotic mRNAs directly. eIF4G is also involved in translational initiation, which helps explain why the both the 5′ and 3′ ends of eukaryotic mRNAs are important for efficient translation: the 3′ stabilizes the 5′ end. Proteins can interfere with the ability of PABP and eIF4G to bind, thereby preventing translation (as you will learn in Chapter 20). Circularization of eukaryotic mRNA makes it easier for those ribosomes that have recently disassembled after the stop codon to reinitiate translation back at the end. A model for how these interactions contribute to translation is shown in Figure 19-16b.

Internal Ribosome Entry Sequences. Not all eukaryotic mRNAs must be capped to recruit a ribosome and then be translated. In some cases, the ribosome complex may instead bind to an **internal ribosome entry sequence (IRES)**, which lies directly upstream of the start codon of certain types of mRNA, especially viral mRNAs. Such sequences are very useful in molecular biology applications for efficient expression of proteins from cloned DNA sequences.

Chain Elongation Involves Cycles of Aminoacyl tRNA Binding, Peptide Bond Formation, and Translocation

In both prokaryotes and eukaryotes, once the initiation complex has been completed, a polypeptide chain is synthesized by the successive addition of amino acids in a sequence specified by codons in mRNA. As summarized in the case of bacteria in **Figure 19-17** on page 558, the elongation stage of polypeptide synthesis involves a repetitive three-step cycle in which ❶ binding of an aminoacyl tRNA to the ribosome brings a new amino acid into position to be joined to the polypeptide chain, ❷ peptide bond formation links this amino acid to the growing polypeptide, and ❸the mRNA is advanced a distance of three nucleotides by the process of translocation to bring the next codon into position for translation. Each of these steps is described in more detail in the following paragraphs.

Binding of Aminoacyl tRNA. At the onset of the elongation stage, the AUG start codon in the mRNA is located at the ribosomal P site, and the second codon (the codon immediately downstream from the start codon) is located at the A site. Elongation begins when an aminoacyl tRNA whose anticodon is complementary to the second codon binds to the ribosomal A site (Figure 19-17, ❶). The binding of this new aminoacyl tRNA to the codon in the A site requires two protein **elongation factors**, *EF-Tu* and *EF-Ts*, and is driven by the hydrolysis of GTP. From now on, every incoming aminoacyl tRNA will bind first to the A (aminoacyl) site—hence the site's name.

The function of EF-Tu, along with its bound GTP, is to convey the aminoacyl tRNA to the A site of the ribosome. The EF-Tu–GTP complex promotes the binding of all aminoacyl tRNAs *except the initiator tRNA* to the ribosome, thus ensuring that AUG codons located downstream from the start codon do not mistakenly recruit an initiator tRNA to the ribosome. As the aminoacyl tRNA is transferred to the ribosome, the GTP is hydrolyzed and the EF-Tu–GDP complex is released. The role of EF-Ts is to regenerate EF-Tu–GTP from EF-Tu–GDP for the next round of the elongation cycle.

Elongation factors do not recognize individual anticodons, which means that aminoacyl tRNAs of all types (other than initiator tRNAs) are randomly brought to the A site of the ribosome. Some mechanism must therefore ensure that only the correct aminoacyl tRNA is retained by the ribosome for subsequent use during peptide bond formation. A precise match of the tRNA with the codon positions the GTP of EF-Tu into a GTPase center of the ribosome for hydrolysis. If the anticodon of an incoming aminoacyl tRNA is not complementary to the mRNA codon exposed at the A site, the aminoacyl tRNA does not bind to the ribosome long enough for GTP hydrolysis to take place. When the match is close but not exact, transient binding may occur, and GTP is hydrolyzed. However, the mismatch between the anticodon of the aminoacyl tRNA and the codon of the mRNA results in the rejection of the bound aminoacyl tRNA as it attempts to rotate into position for peptide bond synthesis during a process known as accommodation. These mechanisms for selecting against incorrect aminoacyl tRNAs, combined with the proofreading capacity of aminoacyl-tRNA synthetases described earlier, ensure that the final error rate in translation is usually no more than 1 incorrect amino acid per 10,000 incorporated.

Peptide Bond Formation. After the appropriate aminoacyl tRNA has become bound to the ribosomal A site, the next step is formation of a peptide bond between the amino group of the amino acid bound at the A site and the carboxyl group that links the initiating amino acid (or growing polypeptide chain) to the tRNA at the P site. The formation of this peptide bond

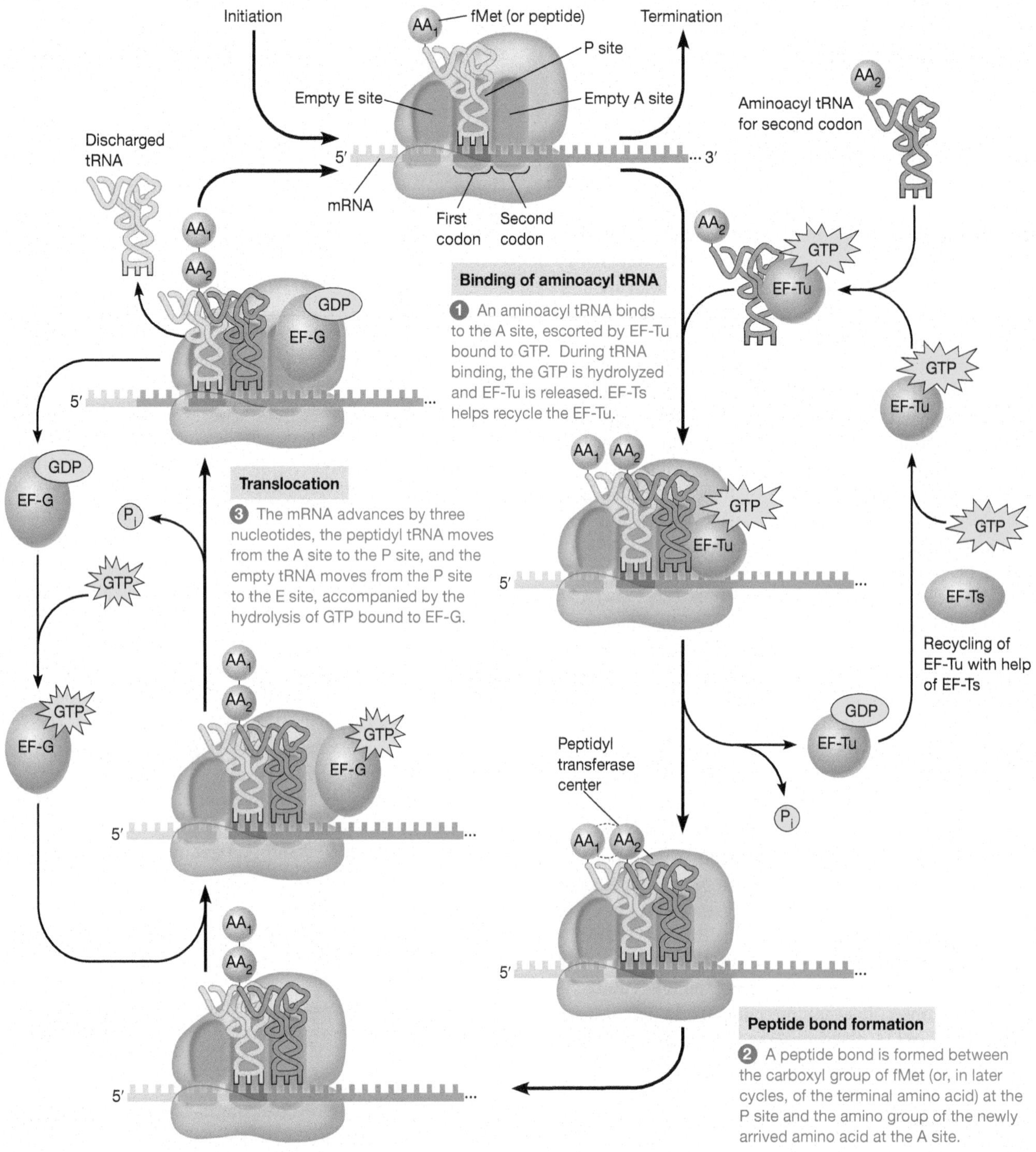

Figure 19-17 Polypeptide Chain Elongation in Bacteria. Chain elongation requires the presence of a peptidyl tRNA or, in the first elongation cycle shown here, an fMet-tRNA^{fMet} at the ribosomal P site. Binding of aminoacyl tRNA is followed by peptide bond formation, which is catalyzed by the peptidyl transferase activity of the 23S rRNA of the large ribosomal subunit. During translocation, the peptidyl tRNA moves from the A site to the P site, taking the mRNA along with it, and the empty tRNA moves from the P site to the E site and leaves the ribosome. The next mRNA codon is now located in the A site, where the same cycle of events can be repeated for the next amino acid.

causes the growing polypeptide chain to be transferred from the tRNA located at the P site to the tRNA located at the A site (Figure 19-17, ❷). Peptide bond formation is the only step in protein synthesis that requires neither nonribosomal protein factors nor an outside source of energy such as GTP or ATP. The necessary energy is provided by cleavage of the high-energy bond that joins the amino acid or peptide chain to the tRNA located at the P site.

For many years, peptide bond formation was thought to be catalyzed by a hypothetical ribosomal protein that was given the name "peptidyl transferase." However, in 1992 Harry Noller and his colleagues showed that the large subunit of bacterial ribosomes retains peptidyl transferase activity after all ribosomal proteins have been removed. In contrast, peptidyl transferase activity is quickly destroyed when rRNA is degraded by exposing ribosomes to ribonuclease. Such observations suggested that rRNA, rather than a ribosomal protein, is responsible for catalyzing peptide bond formation. In bacterial ribosomes, peptidyl transferase activity has been localized to the 23S rRNA of the large ribosomal subunit, and high-resolution X-ray data have pinpointed the catalytic site to a specific region of the RNA chain. Hence 23S rRNA is another example of a ribozyme, an enzyme made entirely of RNA.

Translocation. After a peptide bond has been formed, the P site contains an empty tRNA, and the A site contains a peptidyl tRNA (the tRNA to which the growing polypeptide chain is attached). The mRNA now advances a distance of three nucleotides relative to the small subunit, bringing the next codon into proper position for translation. During this process of **translocation**—which requires that an elongation factor called EF-G plus GTP become transiently associated with the ribosome—the peptidyl tRNA moves from the A site to the P site and the empty tRNA moves from the P site to the E (exit) site. Although these movements are shown in Figure 19-17 as occurring in a single step ❸, an intermediate "hybrid" state exists in which the anticodon of the peptidyl tRNA still resides at the A site while its aminoacyl end has rotated into the P site, and the anticodon of the empty tRNA still resides at the P site while its other end has rotated into the E site. Hydrolysis of the GTP bound to EF-G triggers a conformational change in the ribosome that completes the movement of the peptidyl tRNA from the A site to the P site, and the empty tRNA from the P site to the E site.

During translocation, the peptidyl tRNA remains hydrogen-bonded to the mRNA as the mRNA advances by three nucleotides. The central role played by peptidyl tRNA in the translocation process has been demonstrated using mutant tRNA molecules that have *four*-nucleotide anticodons. These tRNAs hydrogen-bond to four nucleotides in mRNA; when translocation occurs, the mRNA advances by four nucleotides rather than the usual three. Although this observation indicates that the size of the anticodon loop of the peptidyl tRNA bound to the A site determines how far the mRNA advances during translocation, the physical basis for the mechanism that actually translocates the mRNA over the surface of the ribosome is not well understood.

The net effect of translocation is to bring the next mRNA codon into the A site, so the ribosome is now set to receive the next aminoacyl tRNA and repeat the elongation cycle. The only difference between succeeding elongation cycles and the first cycle is that an initiator tRNA occupies the P site at the beginning of the first elongation cycle, and a peptidyl tRNA occupies the P site at the beginning of all subsequent cycles. As each successive amino acid is added, the mRNA is progressively read in the $5' \rightarrow 3'$ direction. The amino-terminal end of the growing polypeptide passes out of the ribosome through an *exit tunnel* in the 50S subunit, where it is met by molecular chaperones that help fold the polypeptide into its proper three-dimensional shape. Polypeptide synthesis is very rapid; in a growing *E. coli* cell, a polypeptide of 400 amino acids can be made in 10 seconds!

Most mRNAs Are Read by Many Ribosomes Simultaneously

Not only is translation a rapid process, but the cell makes efficient use of each mRNA molecule. Most messages are translated by many ribosomes at the same time, each ribosome following closely behind the next on the same mRNA molecule. A cluster of such ribosomes attached to a single mRNA molecule is called a **polyribosome**, or **polysome** (**Figure 19-18**). By allowing many polypeptides to be synthesized at the same time from a single mRNA molecule, polyribosomes maximize the efficiency of mRNA utilization.

Termination of Polypeptide Synthesis Is Triggered by Release Factors That Recognize Stop Codons

The elongation process depicted in Figure 19-17 continues in cyclic fashion, reading one codon after another and adding successive amino acids to the polypeptide chain, until one of the three possible stop codons (UAG, UAA, or UGA) in the mRNA arrives at the ribosome's A site (**Figure 19-19** on page 560). Unlike other codons, stop codons are not recognized by tRNA molecules. Instead, the stop codons are recognized by proteins called **release factors**, which possess special regions ("peptide anticodons") that bind to mRNA stop codons present at the ribosomal A site. The shape of these protein release factors closely matches the shape of tRNA—a phenomenon known as **molecular mimicry**. After binding to the A site along with GTP, the release factors terminate translation by triggering release of the completed polypeptide from the peptidyl tRNA. In essence, the reaction is a hydrolytic cleavage: the polypeptide transfers to a water molecule instead of to an activated amino acid, producing a free carboxyl group at the

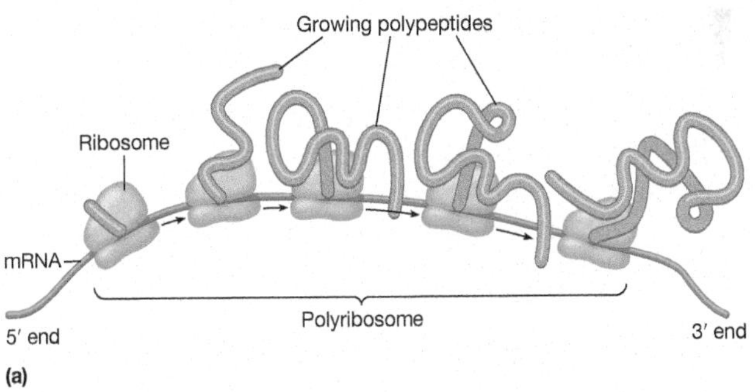

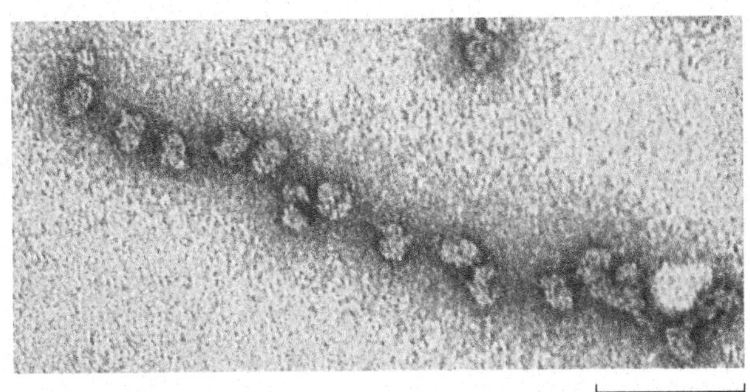

Figure 19-18 A Polyribosome. (a) Schematic of a polyribosome. Ribosomes bind near the 5′ end of the mRNA and move toward the 3′ end, synthesizing polypeptides as they do so. Many ribosomes can be attached to the same mRNA simultaneously. **(b)** An actual polyribosome (TEM).

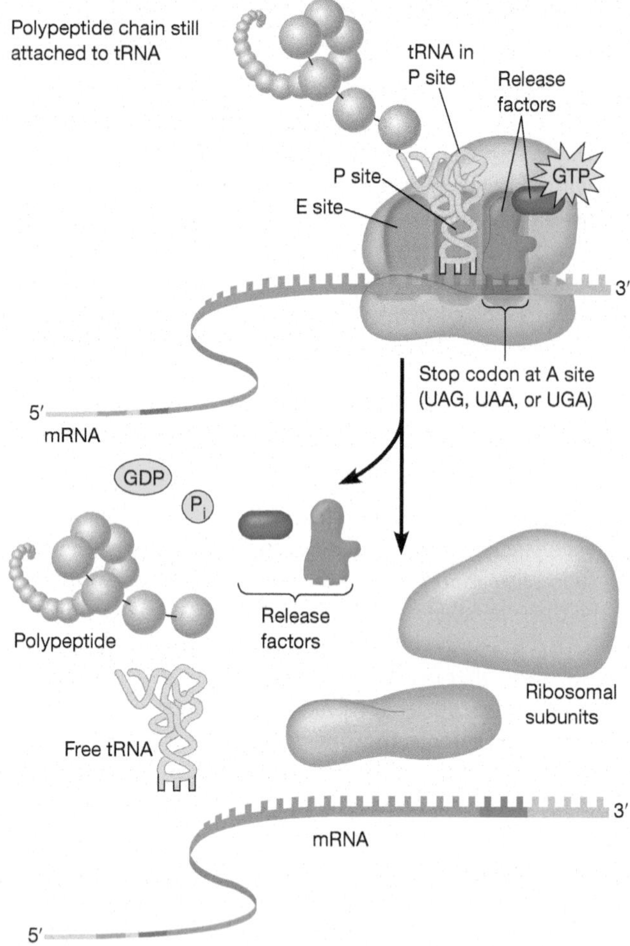

Figure 19-19 Termination of Translation. When a stop codon—UAG, UAA, or UGA—arrives at the A site, it is recognized and bound by protein release factors associated with GTP. Hydrolysis of the GTP is accompanied by release of the completed polypeptide, followed by dissociation of the tRNA, mRNA, ribosomal subunits, and release factors.

end of the polypeptide—its C-terminus. After the polypeptide is released accompanied by GTP hydrolysis, the ribosome dissociates into its subunits and the tRNAs and mRNAs are released. All of these components are now available for reuse in a new cycle of protein synthesis.

> **MAKE CONNECTIONS 19-1**
>
> Which type of eukaryotic mRNA modification is most likely to affect the size of a translated protein? (Ch. 18.3)

Polypeptide Folding Is Facilitated by Molecular Chaperones

Before newly synthesized polypeptides can function properly, they must fold into the correct three-dimensional shape. The primary sequence of a protein can be sufficient to specify its three-dimensional structure, and some polypeptides spontaneously fold into the proper shape in a test tube (see Chapters 2 and 3). However, protein folding inside cells is usually facilitated by proteins called **molecular chaperones**. In fact, proper folding often requires several chaperones to act in sequence, beginning when the growing polypeptide chain first emerges from the ribosome's exit tunnel.

A key function of molecular chaperones is to bind to polypeptide chains during the early stages of folding, thereby preventing them from interacting with other polypeptides before the newly folding chains have acquired the proper conformation. If the folding process goes awry, chaperones can sometimes rescue improperly folded proteins and help them fold properly, or the improperly folded proteins may be destroyed. However, some kinds of incorrectly folded polypeptides tend to bind to each other and form insoluble aggregates that become deposited both within and between cells. Such protein deposits disrupt cell function and may even lead to tissue degeneration and cell death.

Chaperones are found throughout the living world, from bacteria and archaea to the various compartments of eukaryotic cells. Recall that in a mature, fully folded protein, hydrophobic regions are buried in the interior of the protein molecule. In an unfolded polypeptide, these same hydrophobic regions are exposed to the surrounding aqueous environment, creating an unstable situation in which polypeptides tend to aggregate with one another. Chaperones detect these hydrophobic residues and repair misfolded proteins by interacting with them in an ATP-dependent manner (**Figure 19-20**).

Two of the most widely occurring chaperone families are Hsp70 and Hsp60. The "Hsp" comes from the original designation of these proteins as "heat shock proteins" because cells produce them in response to stressful conditions, such as exposure to high temperatures; under these conditions, chaperones facilitate the refolding of heat-damaged proteins. Hsp70 proteins (and their associated proteins) bound to ATP can attach to a newly forming, partially folded protein or to a misfolded protein (Figure 19-20a). ATP hydrolysis is then coupled to changes in the folding of the protein. Release of bound ADP then leads to release of the correctly folded protein. Hsp70 proteins can associate with nascent polypeptides, that is, polypeptides that are still in the process of being created during translation.

Hsp60 proteins function somewhat differently. Although the model shown in Figure 19-20b is from bacteria, in which the Hsp60 complex is known as the *GroEL/GroES complex*, many of the events are similar in archaea and eukaryotes. In Figure 19-20c, ❶ a partially folded or misfolded polypeptide enters one end of the GroEL/GroES complex. ❷ A GroES subunit then attaches to the open end; coupled with ATP hydrolysis, this causes a change in shape of the GroEL subunit, which creates a hydrophilic environment favorable for correct folding of the protein. Finally, ❸ the correctly folded protein is released, and a GroES subunit attaches to the opposite end, completing the cycle.

Molecular chaperones are also involved in activities other than protein folding. For example, they help assemble folded polypeptides into multisubunit proteins and (as we saw in Chapter 12) they facilitate protein transport into mitochondria and chloroplasts by maintaining polypeptides in an unfolded state prior to their transport into these organelles.

Protein Synthesis Typically Utilizes a Substantial Fraction of a Cell's Energy Budget

Polypeptide elongation involves the hydrolysis of at least four "high-energy" phosphoanhydride bonds per amino acid added. Two of these bonds are broken in the aminoacyl-tRNA

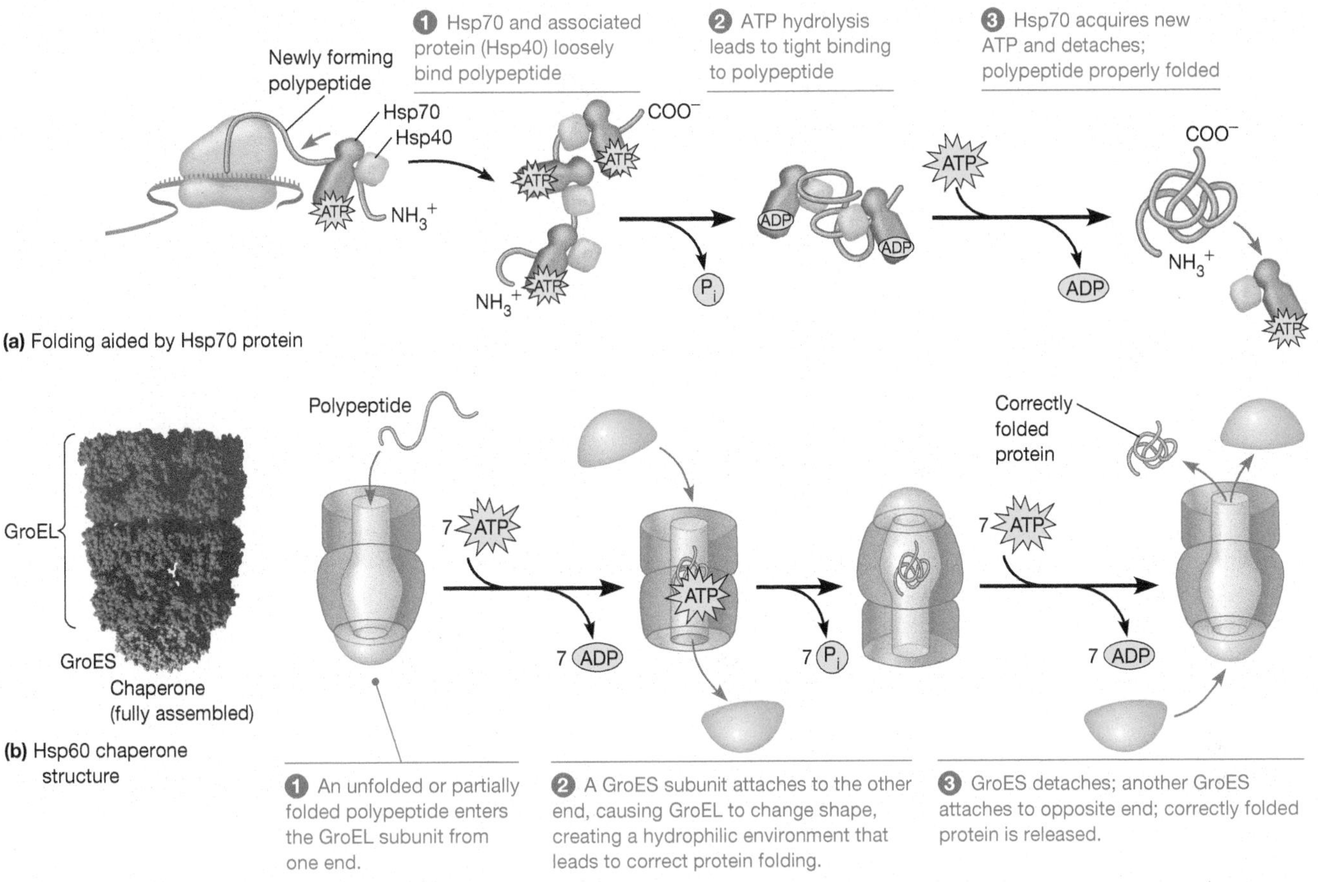

Figure 19-20 Chaperones and Protein Folding. (a) Hsp70 chaperones aid folding of nascent (newly forming) polypeptides during translation. Their association with polypeptides and their ability to aid proper folding of misfolded proteins depends on ATP hydrolysis. **(b)** Hsp60 chaperones form a two-chambered cage into which misfolded protein can enter for refolding. The bacterial Hsp60 chaperone contains two main subunits, GroEL (the "cage"), which can bind a GroES subunit at either end (the "lids"). **(c)** In chaperone-assisted protein folding aided by Hsp60, misfolded or partially folded polypeptides enter the GroEL subunit, followed by the detachment and attachment of GroES subunits at the ends of the chaperone. Conformational changes in GroEL promote the correct folding of the polypeptide. Hydrolysis of seven ATPs is required for these events.

synthetase reaction, where ATP is hydrolyzed to AMP accompanied by the release of two free phosphate groups (see Figure 19-11). The rest are supplied by two molecules of GTP: one used in binding the incoming aminoacyl tRNA at the A site and the other in the translocation step. Assuming each phosphoanhydride bond has a $\Delta G^{\circ\prime}$ (standard free energy) of 7.3 kcal/mol, the four bonds represent a standard free energy input of 29.2 kcal/mol of amino acid inserted. Thus, the elongation steps required to synthesize a polypeptide 100 amino acids long have a $\Delta G^{\circ\prime}$ value of about 2920 kcal/mol. Moreover, additional GTPs are utilized during formation of the initiation complex, during the transient binding of incorrect aminoacyl tRNAs to the ribosome, and during the termination step of polypeptide synthesis. Clearly, protein synthesis is an expensive process energetically. In fact, it accounts for a substantial fraction of the total energy budget of most cells. When we also consider the energy required to synthesize messenger RNA and the components of the translational apparatus, as well as the use of ATP by chaperone proteins, the cost of protein synthesis becomes even greater.

A Summary of Translation

Translation converts information stored in strings of mRNA codons into a chain of amino acids linked by peptide bonds. For a visual summary of the process, refer back to Figure 19-14. As the ribosome reads the mRNA codon by codon in the $5' \rightarrow 3'$ direction, successive amino acids are brought into place by complementary base pairing between the codons in the mRNA and the anticodons of aminoacyl-tRNA molecules. When a stop codon is encountered, the completed polypeptide is released, and the mRNA and ribosomal subunits become available for further cycles of translation. While there is a high degree of evolutionary conservation between translation in eukaryotic and prokaryotic cells, antibiotics have been developed that target the unique properties of bacterial translation. (**Human Connections, pages 562–563,** describes these fascinating drugs in more detail.)

RNA molecules play especially important roles in translation. The mRNA plays a central role, of course, as the carrier of the genetic message. The tRNA molecules serve as the adaptors that bring the amino acids to the appropriate codons. Last

HUMAN *Connections*

To Catch a Killer: The Problem of Antibiotic Resistance In Bacteria

Proteins are the workhorses of the cell. Take away the ability of a bacterium to produce proteins, and death of that bacterium will soon occur. Or will it?

The first antibiotics killed off bacteria by blocking cell wall synthesis (*penicillin* is a well-known example of such an antibiotic) or by stopping replication (a class of antibiotics known as quinolones). As more and more bacteria became resistant to these early antibiotics, however, researchers had to find new ways to stop the growth of infectious bacteria. Another category of potent antibiotics inhibits translation by specifically targeting prokaryotic ribosomes, leaving the eukaryotic ribosomes in the human patient unaffected.

Different classes of antibiotics disrupt various steps in the pathway to protein synthesis; **Figure 19A-1** shows a few of these mechanisms. *Chloramphenicol* binds to the large (50S) ribosomal subunit and inhibits peptidyl transferase activity of the rRNA. *Erythromycin* inhibits translation by blocking translocation of the ribosome to the next codon following peptide bond formation. *Tetracycline* binds to the small (30S) ribosomal subunit and inhibits aminoacyl tRNAs from associating with the mRNA; *streptomycin* is important for inhibiting the initiation of translation by binding to 16S rRNA. Once bound, it interferes with the appropriate association of fMet-tRNA^{fMet} and the mRNA. Streptomycin also interferes with how subsequent codons are read and ultimately stops protein synthesis.

Many people have had the experience of taking a course of antibiotics to fight a bacterial infection but not feeling any better after their course of treatment. A return trip to the doctor usually results in a prescription for a different antibiotic. Why is a second course of antibiotics necessary? Bacteria reproduce very quickly; some divide every 20 minutes! With each round of division, mutations can be introduced into the bacterial genome, or the bacteria may acquire a plasmid (a circular piece of DNA that is replicated separately from the bacterial chromosome; see Chapter 25), and either event can confer resistance.

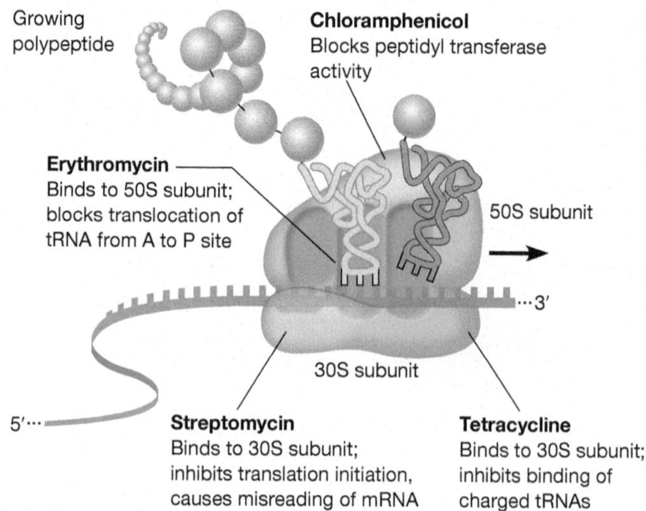

Figure 19A-1 Various Mechanisms of Antibiotic Interference with Protein Translation.

Although a second, different antibiotic will usually curb a persistent infection, more and more often such treatment proves ineffective. Infection by so-called "superbugs," bacteria that have developed resistance to multiple types of antibiotics, has become more prevalent. These resistant strains of bacteria often develop in antibiotic-rich environments such as hospitals—ironically, this means that hospital patients are at higher risk for infection by a superbug.

but not least, the rRNA molecules have multiple functions. Not only do they serve as structural components of the ribosomes, but one (the 16S rRNA of the small subunit) provides the binding site for incoming mRNA, and another (the 23S rRNA of the large subunit) catalyzes the formation of the peptide bond.

CONCEPT CHECK 19-3

Kanamycin is an antibiotic that binds to bacterial ribosomes and allows diverse tRNA with any anticodons to bind at the A site. What effect do you think kanamycin would have on bacterial translation, and why?

19.4 Mutations and Translation

Having described the *normal* process of translation, let's now consider what happens when mRNAs containing *mutant* codons are translated. In its broadest sense, the term *mutation* refers to any change in the nucleotide sequence of a genome. Now that we have examined the processes of transcription and translation, we can understand the effects of several kinds of mutations. Limiting our discussion to protein-coding genes, let's consider some of the main types of mutations and their impact on the polypeptide encoded by the mutant gene.

We have already encountered several types of mutations in which the DNA change involves only one or a few base pairs (**Figure 19-21** on page 564). A common example that we learned about at the start of this chapter is the genetic allele that, when homozygous, causes sickle-cell anemia. This allele originated from a type of mutation called a *base-pair substitution.* In this case, an AT base pair was substituted for a TA base pair in DNA. As a result, a GUA codon replaces a GAA in the mRNA transcribed from the mutant allele, and in the polypeptide (β-globin) a valine replaces a glutamic acid. This single amino acid change, caused by a single base-pair change, is enough to change the conformation of β-globin and, in turn, the hemoglobin tetramer, altering the way hemoglobin molecules pack into red cells and producing abnormally shaped cells that become trapped and damaged when they pass through small blood vessels (see Figure 19-1). Such a base-pair substitution is called a *missense* or *nonsynonymous mutation* because the

How do superbugs evade antibiotics? There are many possible mechanisms. Chloramphenicol resistance often arises due to the presence of a gene known as *cat* (chloramphenicol acetyltransferase), which encodes an enzyme that attaches two acetyl groups to chloramphenicol. When it is acetylated, chloramphenicol cannot bind to the ribosome and block translation. *cat* usually resides on a plasmid, which can be readily passed among bacteria. Resistance to erythromycin is afforded by genes encoding specific *methylases,* which are often contained within transposons (you learned about transposons in Chapter 17) that facilitate their spread among bacteria. These methylases modify 23S rRNA, part of the large ribosomal subunit (see Figure 19A-1), so that erythromycin cannot bind to it. In other cases, direct mutations in the nucleotides encoding the 23S rRNA disrupt an antibiotic's ability to inhibit peptidyl transferase activity. Still other bacteria have genes that produce *ribosomal protection proteins (RPPs).* RPPs are a common form of resistance against tetracycline because they are able to dislodge the antibiotic from its binding site. In fact, many of the well-characterized RPPs show some sequence similarity to the bacterial elongation factors EF-Tu and EF-G. The RPPs perform the same functions during translation as EF-Tu and EF-G but are able to displace tetracycline at the A site.

The rise of superbugs seems to call for new antibiotics; however, the development of new antibiotics is declining (**Figure 19A-2**). Although it may seem that there should be multiple ways to block the spread of bacteria, many have already been exploited, and resistant bacteria already exist. Older ideas, like exploiting bacteriophages that normally infect and lyse bacteria, are being revisited. Other research focuses instead on prescribing strategic combinations of antibiotics delivered simultaneously to rid the body of infection. It is clear that significant scientific resources must be allocated to develop novel antibiotics to address this emerging health safety concern.

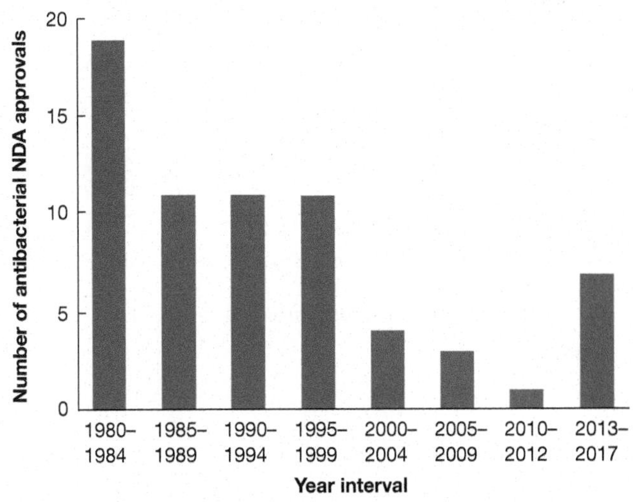

Figure 19A-2 The Decline in New Antibiotic Discovery. The graph illustrates that the number of new antibiotics being developed each year is in steady decline, although a new initiative seeks to reverse this trend. NDA, new drug application.

From George H Talbot, et al., The Infectious Diseases Society of America's 10 × '20 Initiative (10 New Systemic Antibacterial Agents US Food and Drug Administration Approved by 2020): Is 20 × '20 a Possibility?, *Clinical Infectious Diseases*, July 2019. https://doi.org/10.1093/cid/ciz089.

mutated codon continues to encode an amino acid—but the "wrong" one.

Alternatively, a base-pair substitution can create a *nonstop mutation* by converting a normal stop codon into an amino acid codon, or, conversely, it can create a *nonsense mutation* by converting an amino acid codon into a stop codon. In the latter case, the translation machinery will terminate the polypeptide prematurely. Unless the nonsense mutation is close to the end of the message or a suppressor tRNA (see below) is present, the polypeptide is not likely to be functional. Nonsense, nonstop, and missense codons can also arise from *base-pair insertions, deletions,* or a combination thereof, collectively known as indels, that cause *frameshift mutations.*

A single amino acid change (or even a change in several amino acids) does not always affect a protein's function in a major way. As long as the protein's three-dimensional conformation remains relatively unchanged, biological activity may be unaffected. Substitution of one amino acid for another of the same type—for example, valine for isoleucine—is especially unlikely to affect protein function. The nature of the genetic code actually reduces the effects of single base-pair alterations because many turn out to be *silent* or *synonymous mutations* that change the nucleotide sequence without changing the genetic message. For example, changing the third base of a codon often produces a new codon that still codes for the same amino acid. Here, the "mutant" polypeptide is exactly the same as the wild-type.

In addition to mutations affecting one or a few base pairs, some alterations involve longer stretches of DNA (Figure 19-21b). A few affect genome segments so large that the DNA changes can be detected by light microscopic examination of chromosomes. Some of these large-scale mutations are created by *indels* involving long DNA segments, but several other mechanisms also exist. In a *duplication,* a section of DNA is tandemly repeated. In an *inversion,* a chromosome segment is cut out and reinserted in its original position but in the reverse direction. A *translocation* involves the movement of a DNA segment from its normal location in the genome to another place, either in the same chromosome or in a different one. Because these large-scale mutations may or may not

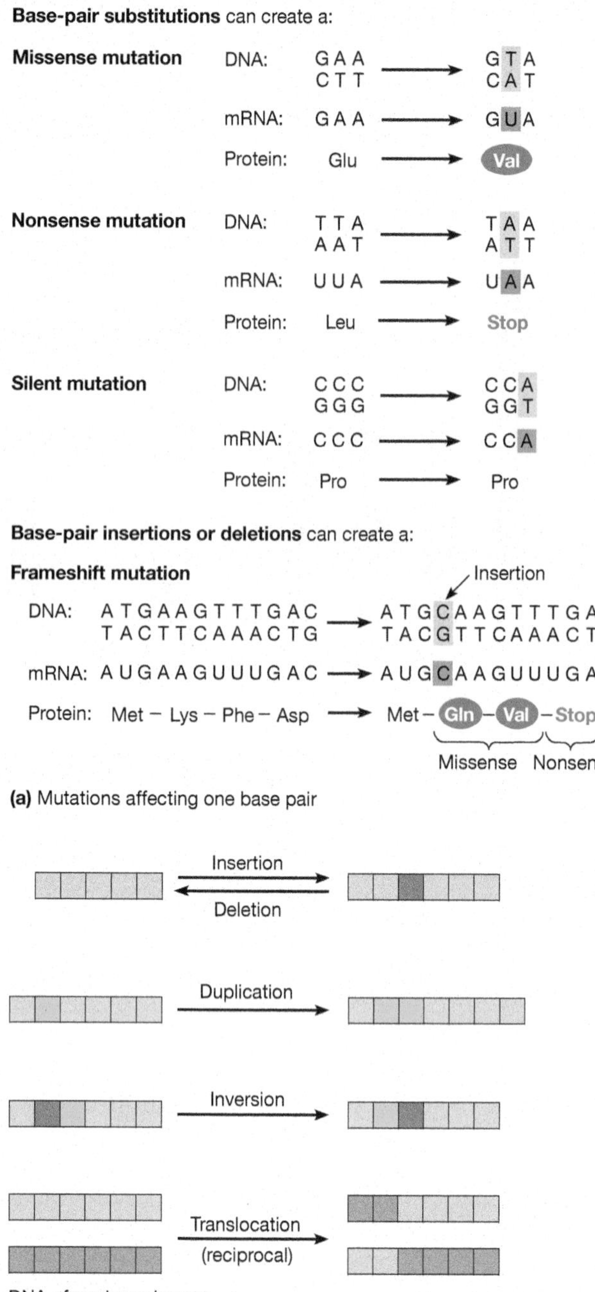

Figure 19-21 Types of Mutations and Their Effects. Mutations can affect one base pair or long DNA segments. **(a)** Single base-pair substitutions within the coding region of a gene can have several effects on the amino acid sequence of the resulting protein. Missense mutations result in the substitution of one amino acid for another. Nonsense mutations result in the substitution of a stop codon for an amino acid, causing production of a truncated protein. Silent mutations have no effect on amino acid sequence. Insertion or deletion of a single base (an "indel" mutation) results in a change in reading frame that causes alterations in amino acid sequence and typically a premature stop codon. **(b)** Larger alterations in DNA include larger indels, duplications, inversions, or translocations. In some cases, the latter are reciprocal; that is, pieces of chromosomes exchange large segments.

affect the expression of many genes, they have a wide range of phenotypic effects, from no effect at all to lethality.

As you think about the potential effects of mutations, it is useful to remember that genes have important noncoding components and that these, too, can be mutated in ways that seriously affect gene products. A mutation in a promoter, for example, can result in more frequent or less frequent transcription of the gene. Even a mutation in an intron can affect the gene product in a major way if it touches a critical part of a splice-site sequence. Finally, mutations in genes that encode regulatory proteins—that is, proteins that control the expression of other genes—can have far-reaching effects on many other proteins. (We will discuss this topic in Chapter 20.)

Now that you understand in detail how changes in the DNA sequence of a gene can affect the protein encoded by that gene, you are in a good position to understand how such knowledge can be used to great advantage to engineer proteins by combining DNA sequences. One powerful application of this technology is to fuse a protein to the **green fluorescent protein (GFP)** (see Key Technique, pages 566–567).

Suppressor tRNAs Overcome the Effects of Some Mutations

Mutations that convert amino acid-coding codons into stop codons are referred to as **nonsense mutations. Figure 19-22** shows a case in which mutation of a single base pair in DNA converts an AAG lysine codon in mRNA to a UAG stop signal. Nonsense mutations like this one typically lead to production of incomplete, nonfunctional polypeptides that have been prematurely terminated at the mutant stop codon.

Nonsense mutations in essential genes are often lethal, but sometimes their detrimental effects can be overcome by an independent mutation affecting a tRNA gene. Such mutant tRNA genes produce mutant tRNAs that recognize what would otherwise be a stop codon and insert an amino acid at that point. In the example shown in Figure 19-22c, a mutant tRNA has an altered anticodon that allows it to read the stop codon UAG as a codon for tyrosine. The inserted amino acid is almost always different from the amino acid that would be present at that position in the wild-type protein, but the important point is that chain termination is averted and a full-length polypeptide can be made.

A tRNA molecule that somehow negates the effect of a mutation is called a **suppressor tRNA**. As you might expect, suppressor tRNAs exist that negate the effects of various types of mutations in addition to nonsense mutations (see Problem 19-13 at the end of the chapter). For a cell to survive, suppressor tRNAs must be rather inefficient; otherwise, the protein-synthesizing apparatus would produce too many abnormal proteins. An overly efficient nonsense suppressor, for example, would cause normal stop codons to be read as if they coded for an amino acid, thereby preventing normal termination. In fact, the synthesis of most polypeptides is terminated properly in cells containing nonsense suppressor tRNAs, indicating that a stop codon located in its proper place at the end of an mRNA coding sequence still triggers termination, whereas the same codon in an internal location does not. The

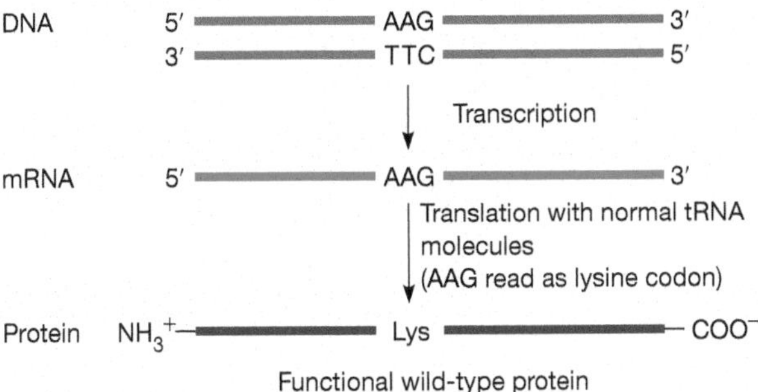

(a) Normal gene, normal tRNA molecules

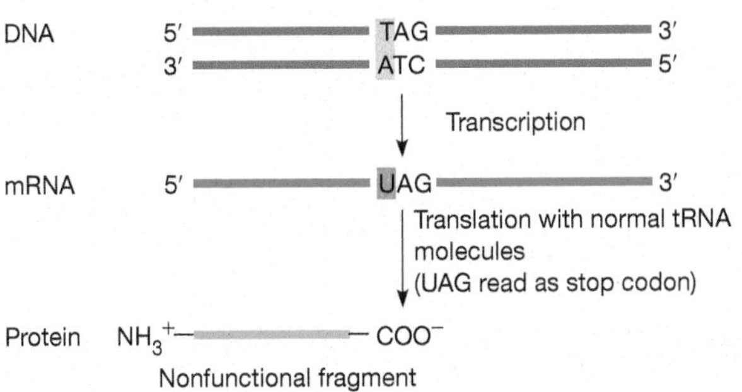

(b) Mutant gene, normal tRNA molecules

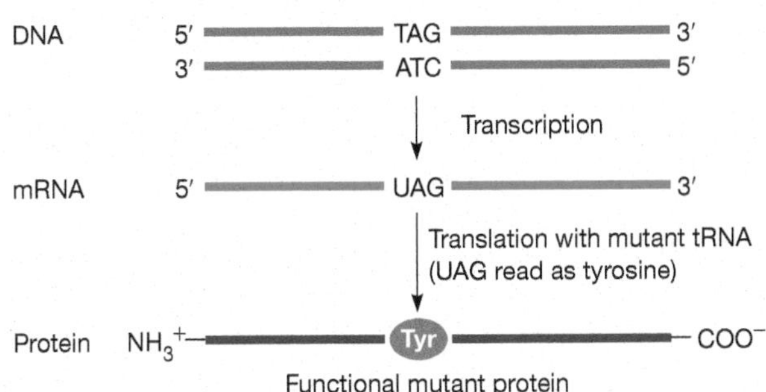

(c) Mutant gene, mutant (suppressor) tRNA molecule

Figure 19-22 Nonsense Mutations and Suppressor tRNAs. **(a)** A wild-type (normal) gene is transcribed into an mRNA molecule that contains the codon AAG at one point. Upon translation, this codon specifies the amino acid lysine (Lys) at this point in the functional, wild-type protein. **(b)** If a DNA mutation occurs that changes the AAG codon in the mRNA to UAG, the UAG codon will be read as a stop signal, and the translation product will be a short, nonfunctional polypeptide. Mutations of this sort are called nonsense mutations. **(c)** In the presence of a mutant tRNA that reads UAG as an amino acid codon instead of a stop signal, an amino acid will be inserted and polypeptide synthesis will continue. In the example shown, UAG is read as a codon for tyrosine because the mutant tyrosine tRNA has as its anticodon 3′-AUC-5′ instead of the usual 3′-AUG-5′ (which recognizes the tyrosine codon 5′-UAC-3′). The resulting protein will be mutant because a lysine has been replaced by a tyrosine at one point along the chain. However, the protein may still be functional if its biological activity is not significantly affected by the amino acid substitution.

most likely explanation is based on the behavior of the release factors that trigger normal termination. When a stop codon occurs in its proper location near the end of an RNA, release factors trigger termination because they are more efficient than suppressor tRNAs in binding to stop codons in this location—perhaps because release factor action is stimulated by a special sequence or three-dimensional configuration near the end of the mRNA.

Nonsense-Mediated Decay and Nonstop Decay Promote the Destruction of Defective mRNAs

In the absence of an appropriate suppressor tRNA, a nonsense stop codon will cause mRNA translation to stop prematurely, thereby generating an incomplete polypeptide chain that cannot function properly and may even harm the cell. To avoid wasting energy on the production of such useless products, eukaryotic cells invoke a quality control mechanism called **nonsense-mediated decay** to destroy mRNAs containing premature stop codons. In mammals, the method for identifying premature stop codons involves the *exon junction complex (EJC)*, a multiprotein complex deposited during mRNA splicing at each point where an intron is removed from pre-mRNA (see Figure 18-20). Thus, every newly spliced mRNA molecule will have one or more EJCs bound to it, one at each exon-exon junction. During translation, the distinction between normal and premature stop codons is made on the basis of their relationship to EJCs. If a stop codon is encountered in an mRNA prior to the last EJC—in other words, before the last exon—it must be a premature stop codon. The presence of such a stop codon will cause translation to be terminated while the mRNA still has one or more EJCs bound to it, and the presence of these remaining EJCs marks the mRNA for degradation.

How do cells handle the opposite situation, namely, an mRNA with no stop codons? In eukaryotic cells, translation becomes stalled when a ribosome reaches the end of an mRNA without encountering a stop codon. An RNA-degrading enzyme complex then binds to the empty A site of the ribosome and degrades the defective mRNA in a process called **nonstop decay**. The same problem is handled a bit differently in bacteria. When translation halts at the end of a bacterial mRNA lacking a stop codon, an unusual type of RNA called *tmRNA* ("transfer-messenger RNA") binds to the A site of the ribosome and directs the addition of about a dozen more amino acids to the growing polypeptide chain. This amino acid sequence creates a signal that targets the protein for destruction. At the same time, the mRNA is degraded by a ribonuclease associated with the tmRNA.

CONCEPT CHECK 19-4

You are working in a lab to identify new alleles of a gene in the nematode *Caenorhabditis elegans* that encodes the protein β-catenin. Based on the phenotype of the homozygous mutants, you believe that you have identified a nonsense mutation. What would the effect be on the total length of the β-catenin mRNA? On the length of the protein? Explain your answers.

19.5 Posttranslational Processing

After polypeptide chains have been synthesized, they often must be chemically modified before they can perform their normal functions. Such modifications are known collectively

Key Technique | Protein Localization Using Fluorescent Fusion Proteins

PROBLEM: Techniques such as immunostaining can localize proteins of interest in cells or whole organisms. But how can the location of a specific protein be monitored in a living cell?

SOLUTION: DNA encoding a protein of interest can be engineered so that the resulting protein is "tagged" with a fluorescent protein. The resulting DNA can be introduced into cells, which produce the tagged protein. Using fluorescence microscopy, the location of the fluorescent protein of interest can then be followed in living cells.

Key Tools: Fluorescence microscopy (see Appendix, Figure A-12); a known DNA sequence that encodes a protein of interest; standard techniques for cloning and expressing DNA (**see Key Technique in Chapter 21, pages 624–625**).

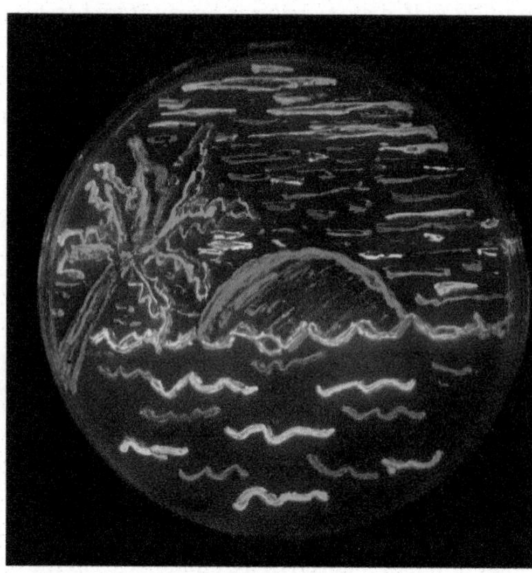

Bacteria expressing green fluorescent protein variants.

Details: In this and previous chapters you learned that the DNA sequence of the coding region of a gene indirectly provides the information cells need to produce a protein. When a DNA sequence with the appropriate promoter is transcribed, if the resulting RNA has the appropriate features, it can be translated into a protein. Scientists can capitalize on this relationship between DNA and protein sequences in the laboratory to produce *fusion proteins*. By fusing DNA sequences encoding a protein of interest with DNA encoding a portion of another protein that fluoresces, scientists can produce DNA that encodes a fluorescently tagged protein. When introduced into cells or an organism, the protein produced from this DNA can be visualized in living cells. Other types of fluorescent constructs are useful for studying transcriptional regulation (see Chapter 21). Fluorescent proteins produced in this way are said to be *genetically encoded,* which simply means that cells can be induced to make an engineered protein by adding in the appropriate DNA encoding it.

A common approach to making fluorescent fusion proteins involves the *green fluorescent protein (GFP)*. GFP is a fluorescent protein produced by the jellyfish *Aequorea victoria,* originally characterized by Osamu Shimomura and colleagues. In 1992, the gene for GFP was cloned. Shortly thereafter, GFP was expressed in living organisms by Martin Chalfie and colleagues. GFP fusions have since become a powerful tool in the arsenal of cell biologists.

The basic approach to making a GFP fusion protein is shown in **Figure 19B-1**. DNA encoding GFP is fused to DNA corresponding to the exons of a gene of interest (Figure 19B-1a). The DNA is then introduced into the cell or organism (methods for doing so in specific cases are discussed in Chapter 21), which transcribes the DNA into mRNA; the mRNA in turn is translated into a fluorescent protein.

Depending on the protein of interest and its particular structure, the GFP tag can be added to either the N-terminus or the C-terminus of the protein. If a functional assay is available, the fusion protein is typically tested to ensure that the GFP tag doesn't adversely affect the protein's function. The best test of a fusion protein's function is to see if it can replace the normal, nonfluorescent protein in living cells.

If the GFP fusion protein can be expressed in the desired cells, the location of the fluorescence will reveal the location of the protein. Figure 19B-1b shows a *C. elegans* embryo that is expressing a GFP-tagged protein that localizes to its cell-cell junctions. The GFP fusion reveals the detailed location of the tagged protein in a living embryo.

By making minor changes in the DNA of the GFP coding region and thus the resulting protein's sequence, a number of different color variants of GFP have been engineered—an approach pioneered by Roger Y. Tsien and colleagues. These different proteins absorb and emit light at a variety of wavelengths, thus producing a wide range of colors (**Figure 19B-2**). Shimomura, Chalfie, and Tsien shared a Nobel Prize for their work on GFP in 2008. In addition to GFP, other

as **posttranslational modifications**. In bacteria, for example, the N-formyl group located at the N-terminus of polypeptide chains is always removed. The methionine it was attached to is often removed also, as is the methionine that starts eukaryotic polypeptides. As a result, relatively few mature polypeptides have methionine at their N-terminus, even though they all started out that way. Sometimes whole blocks of amino acids are removed from the polypeptide. Certain enzymes, for example, are synthesized as inactive precursors that must be activated by the removal of a specific sequence at one end or the other. The transport of proteins across membranes also may involve the removal of a terminal signal sequence (as we saw in Chapter 12), and some polypeptides have internal stretches of amino acids that must be removed to produce an active protein. A well-studied example of posttranslational processing is the protein insulin. The mature

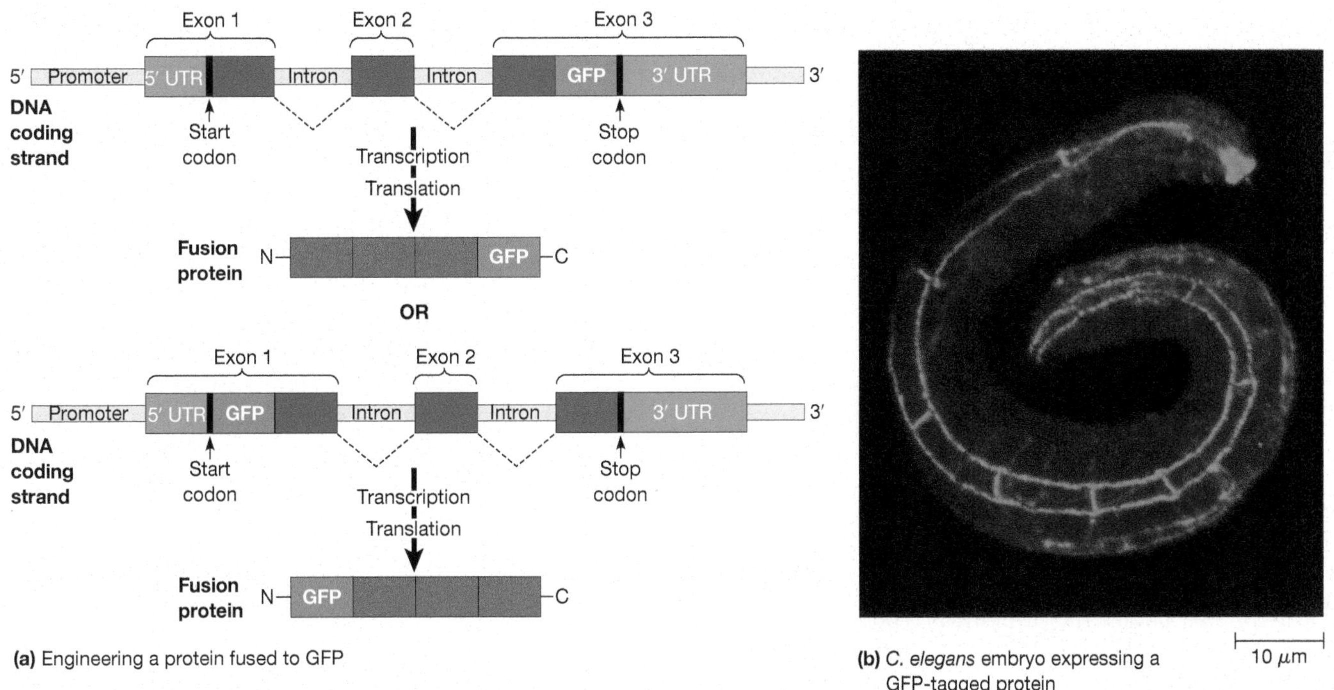

Figure 19B-1 Creating Fusion Proteins. (a) Fusion proteins are created by combining DNA encoding a protein of interest with DNA encoding another protein that can be visualized, such as GFP. The GFP-encoding DNA is often placed in a position so that the resulting fusion protein contains GFP after (that is, C-terminal to) or in front of (that is, N-terminal to) the rest of the protein. **(b)** A *C. elegans* embryo expressing a GFP-tagged protein that localizes to cell-cell junctions (confocal microscopy).

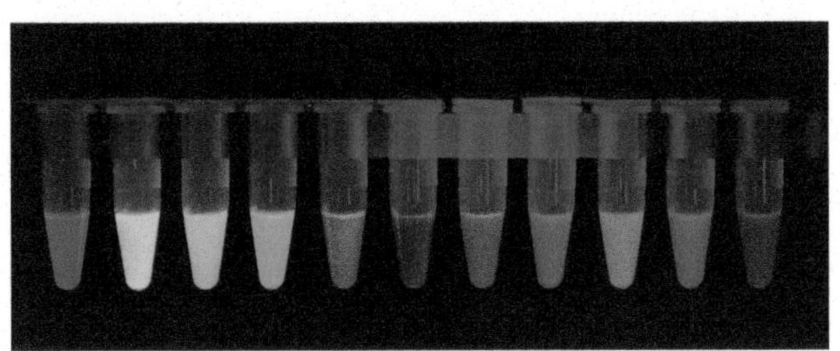

Figure 19B-2 Color Variants of GFP.

fluorescent fusion proteins have been produced from other organisms (such as coral), allowing cells to be engineered to express a wide variety of fusion proteins simultaneously. By using different wavelengths of fluorescent light, the locations of multiple fluorescent proteins can be determined at the same time in living cells.

QUESTION: Suppose you are using GFP fusion proteins to study a cell-cell junction protein. How could you show that a particular amino acid sequence in the protein is needed for targeting it to cell-cell junctions in cultured cells?

insulin protein is composed of two subunits held together by disulfide linkages. However, these two subunits come from a single longer protein known as preproinsulin (**Figure 19-23** on page 568). Preproinsulin undergoes several processing steps. ❶ The N-terminal ("pre-") amino acids are removed to produce proinsulin. ❷ Disulfide linkages then form between what will be the A and B subunits of the mature protein. Finally, ❸ intervening ("pro-") amino acids are removed to form the mature insulin protein.

Other common processing events include chemical modifications of individual amino acid groups—by methylation, phosphorylation, ubiquitination, or acetylation reactions, for example. In addition, a polypeptide may undergo glycosylation (the addition of carbohydrate side chains; see Chapter 12) or may be associated with a prosthetic group. Hemoglobin is a good example of the latter: each of its four globin subunits contains a heme group (see Figure 3-4). Finally, in the case of proteins composed of multiple

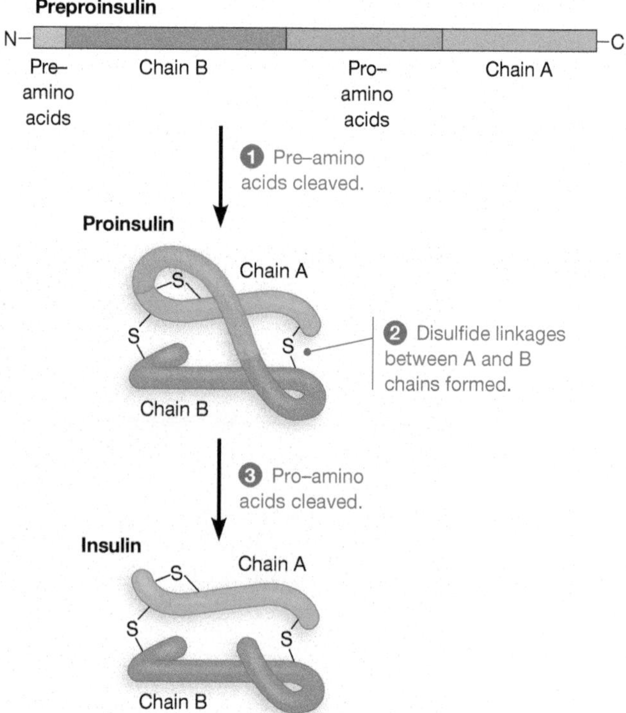

Figure 19-23 Posttranslational Processing of Insulin. Insulin is translated from mRNA as preproinsulin, which is cleaved to form proinsulin. Disulfide linkages then form between the A and B chains, followed by cleavage of intervening sequence to generate the mature insulin protein.

subunits, individual polypeptide chains must bind to one another to form the appropriate multisubunit proteins or multiprotein complexes.

In addition to the preceding posttranslational events, some proteins undergo a relatively unusual type of processing called *protein splicing*, which is analogous to the phenomenon of *RNA splicing* (discussed in Chapter 18). As you learned there, intron sequences are removed from RNA molecules during RNA splicing, and the remaining exon sequences are simultaneously spliced together. Likewise, during protein splicing, specific amino acid sequences called *inteins* are removed from a polypeptide chain, and the remaining segments, called *exteins*, are spliced together to form the mature protein. Protein splicing is usually intramolecular, involving the excision of an intein from a single polypeptide chain by a self-catalytic mechanism. However, splicing can also take place between two polypeptide chains arising from two different mRNAs. For example, in some photosynthetic bacteria a subunit of DNA polymerase III is produced from two separate genes, each encoding an intein-containing polypeptide that includes part of the DNA polymerase subunit. Once considered to be an oddity of nature, protein splicing has now been detected in dozens of different organisms, prokaryotes as well as eukaryotes.

MAKE CONNECTIONS 19-2

What posttranslational modification plays a key role during transcription of mRNA in eukaryotic cells? (Fig. 18-23)

CONCEPT CHECK 19-5

Enzymes can undergo allosteric regulation or regulation by covalent modification (see Chapter 6). Which category do posttranslational modifications fall into?

Summary of Key Points

Mastering™ Biology For activities, animations, and review quizzes, go to the study area at www.masteringbiology.com.

19.1 The Genetic Code

- The base sequence of each mRNA molecule dictates the sequence of amino acids in a polypeptide chain.
- When guiding the synthesis of a polypeptide chain, mRNA is read in units of three bases called codons.
- The table of the genetic code indicates which amino acid (or stop signal) is specified by each codon. The code is unambiguous, nonoverlapping, degenerate, and nearly universal.

19.2 Translation: The Cast of Characters

- Translation refers to the synthesis of polypeptide chains on ribosomes using a process that employs mRNA to determine the amino acid sequence.
- Ribosomes have a large and small subunit, each of which is assembled from a large number of proteins and rRNAs.
- The rRNA component of the ribosome helps position the mRNA and catalyzes peptide bond formation; aminoacyl-tRNA synthetases link amino acids to the tRNA molecules that bring amino acids to the ribosome; and various protein factors trigger specific events associated with the translation cycle.
- Wobble allows flexibility in pairing of a tRNA anticodon with more than one codon in the mRNA.

19.3 The Mechanism of Translation

- Translation involves initiation, elongation, and termination stages.
- During the initiation stage, initiation factors trigger the assembly of mRNA, ribosomal subunits, and initiator aminoacyl tRNA into an initiation complex. Ribosome binding sequences on the mRNA promote binding of the small ribosomal subunit. In eukaryotes, both the 5′ and 3′ ends of mRNAs promote translational initiation.
- Chain elongation involves sequential cycles of aminoacyl tRNA binding, peptide bond formation, and translocation, with each cycle driven by the action of elongation factors. The net result is that aminoacyl tRNAs add their amino acids to the growing polypeptide chain in an order specified by the codon sequence in mRNA.

- More than one ribosome can move along a single mRNA at the same time.
- Chain termination occurs when a stop codon in mRNA is recognized by release factors, which cause the mRNA and newly formed polypeptide to be released from the ribosome.
- GTP binding and hydrolysis are required for the action of several initiation, elongation, and release factors.
- Proper folding of newly produced polypeptide chains is assisted by molecular chaperones.

19.4 Mutations and Translation

- Nonsense mutations, which change an amino acid codon to a stop codon, can be overcome by suppressor mutations that allow a tRNA anticodon to read the stop codon as an amino acid.
- Defective mRNAs containing premature stop codons are destroyed by nonsense-mediated decay; defective mRNAs containing no stop codon are destroyed by nonstop decay.

19.5 Posttranslational Processing

- Newly made polypeptide chains often require chemical modification before they can function properly. Such modifications include cleavage of peptide bonds, phosphorylation, acetylation, ubiquitination, methylation, glycosylation, and protein splicing.

Problem Set

Mastering™ Biology For activities, animations, and review quizzes, go to the study area at www.masteringbiology.com.

19-1 QUANTITATIVE Triplets or Sextuplets? In his Nobel Prize lecture in 1962, Francis Crick pointed out that although the pioneering experiments he performed with Barnett, Brenner, and Watts-Tobin suggested that the DNA "code" is a triplet, their experiments did not rule out the possibility that the code could require six or nine bases.

(a) Assuming the code is a triplet, what effect would adding or removing six or nine bases have on the reading frame of a piece of DNA?

(b) If the code actually were a sextuplet, how would addition of three, six, or nine nucleotides affect the reading frame of a piece of DNA?

19-2 The Genetic Code in a T-Even Phage. A portion of a polypeptide produced by bacteriophage T4 was found to have the following sequence of amino acids:

...Lys-Ser-Pro-Ser-Leu-Asn-Ala...

Deletion of a single nucleotide from one location in the T4 DNA template strand with subsequent insertion of a different nucleotide nearby changed the sequence to:

...Lys-Val-His-His-Leu-Met-Ala...

(a) What was the nucleotide sequence of the mRNA segment that encoded this portion of the original polypeptide?

(b) What was the nucleotide sequence of the mRNA encoding this portion of the mutant polypeptide?

(c) Can you determine which nucleotide was deleted and which was inserted? Explain your answer.

19-3 Frameshift Mutations. Each of the mutants listed below this paragraph has a different mutant form of the gene encoding protein X. Each mutant gene contains one or more nucleotide insertions (+) or deletions (−) of the type caused by acridine dyes. Assume that all the mutations are located very near the beginning of the gene for protein X. In each case, indicate with an "OK" if you would expect the mutant protein to be nearly normal or with a "Not OK" if you would expect it to be obviously abnormal.

(a) −

(b) −/+

(c) −/−

(d) +/−/+

(e) +/−/+/−

(f) +/+/+

(g) +/+/−/+

(h) −/−/+/−/−

(i) −/−/−/−/−/−

19-4 Amino Acid Substitutions in Mutant Proteins. Although the codon assignments summarized in Figure 19-5 were originally deduced from experiments involving synthetic RNA polymers and triplets, their validity was subsequently confirmed by examining the amino acid sequences of normally occurring mutant proteins. The table below lists some examples of amino acid substitutions seen in mutant forms of hemoglobin, tryptophan synthase, and the tobacco mosaic virus coat protein. For each amino acid alteration, list the corresponding single base changes in mRNA that could have caused that particular amino acid substitution.

Protein	Amino Acid Substitution
Hemoglobin	Glu → Val His → Tyr Asn → Lys
Coat protein of tobacco mosaic virus	Glu → Gly Ile → Val
Tryptophan synthase	Tyr → Cys Gly → Arg Lys → Stop

19-5 QUANTITATIVE **Copolymer Analysis.** In their initial attempts to determine codon assignments, Nirenberg and Matthaei first used RNA homopolymers and then used RNA copolymers synthesized by the enzyme polynucleotide phosphorylase. This enzyme adds nucleotides randomly to the growing chain but in proportion to their presence in the incubation mixture. By varying the ratio of precursor molecules in the synthesis of copolymers, Nirenberg and Matthaei were able to deduce base compositions (but usually not actual sequences) of the codons that encode various amino acids. Suppose you carry out two polynucleotide phosphorylase incubations, with UTP and CTP present in both but in different ratios. In incubation A, the precursors are present in equal concentrations. In incubation B, there is three times as much UTP as CTP. The copolymers generated in both incubation mixtures are then used in a cell-free protein-synthesizing system, and the resulting polypeptides are analyzed for amino acid composition.

(a) What are the eight possible codons represented by the nucleotide sequences of the resulting copolymers in both incubation mixtures? What amino acids do these codons encode?

(b) For every 64 codons in the copolymer formed in incubation A, how many of each of the 8 possible codons would you expect on average? How many for incubation B?

(c) What can you say about the expected frequency of occurrence of the possible amino acids in the polypeptides obtained upon translation of the copolymers from incubation A? What about the polypeptides that result from translation of the incubation B copolymers?

(d) Explain what sort of information can be obtained by this technique.

(e) Would it be possible to use this technique to determine that codons with 2 U's and 1 C encode phenylalanine, leucine, and serine? Why or why not?

(f) Would it be possible to use this technique to decide which of the three codons with 2 U's and 1 C (UUC, UCU, CUU) correspond to each of the three amino acids mentioned in part e? Why or why not?

(g) Suggest a way to assign the three codons of part f to the appropriate amino acids of part e.

19-6 DATA ANALYSIS **The Genetic Code and Two Human Hormones.** The following is the actual sequence of a small stretch of human DNA:

3′-AATTATACACGATGAAGCTTGTGACAGGGTTTCCAATCATTAA-5′
5′-TTAATATGTGCTACTTCGAACACTGTCCCAAAGGTTAGTAATT-3′

(a) What are the two possible RNA molecules that could be transcribed from this DNA?

(b) Only one of these two RNA molecules can actually be translated. Explain why.

(c) The RNA molecule that can be translated is the mRNA for the hormone vasopressin. What is the apparent amino acid sequence for vasopressin? (The genetic code is given in Figure 19-5.)

(d) In its active form, vasopressin is a nonapeptide (that is, it has nine amino acids) with cysteine at the N-terminus. How can you explain this in light of your answer to part c?

(e) A related hormone, oxytocin, has the following amino acid sequence:

Cys-Tyr-Ile-Glu-Asp-Cys-Pro-Leu-Gly

Where and how would you change the DNA that encodes vasopressin so that it would encode oxytocin instead? Does your answer suggest a possible evolutionary relationship between the genes for vasopressin and oxytocin?

19-7 Tracking a Series of Mutations. The following diagram shows the amino acids that result from mutations in the codon for a particular amino acid in a bacterial polypeptide:

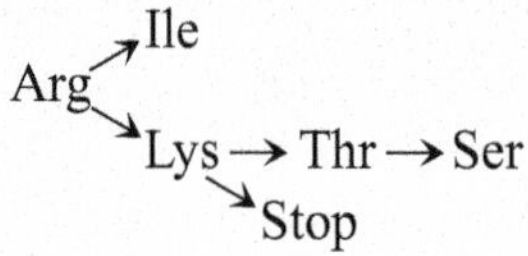

Assume that each arrow denotes a single base-pair substitution in the bacterial DNA.

(a) Referring to the genetic code table in Figure 19-5, determine the most likely codons for each of the amino acids and the stop signal in the diagram.

(b) Starting with a population of mutant cells carrying the nonsense mutation, another mutant is isolated in which the premature stop signal is suppressed. Assuming wobble does not occur and assuming a single-base change in the tRNA anticodon, what are all the possible amino acids that might be found in this mutant at the amino acid position in question?

19-8 Sleuthing Using Mutants. An mRNA contains the following sequence:

5′...CAACUGCCUGACCCACACUUAUCACUAAG
UAGCCUAGCAGUCUGA...3′

The wild-type amino acid sequence encoded by this piece of the mRNA is as follows:

N...Asn-Cys-Leu-Thr-His-Thr-Tyr-His-C

The C-terminal ends of three mutant proteins resulting from mutations in the gene encoding this mRNA are as follows:

Mutant 1: N...Asn-Cys-Leu-Thr-His-Thr-C

Mutant 2: N...Asn-Cys-Leu-Thr-His-Thr-Tyr-His-Lys-C

Mutant 3:
N...Asn-Cys-Leu-Thr-His-Thr-Tyr-His-Tyr-Ser-Ser-Leu-Ala-Val-C

Identify the mutations in each case that result in these mutant proteins.

19-9 More Sleuthing Using Mutants. You identify three independent missense mutations that all affect the same gene. In fact, all three mutations affect the same codon but do not cause the same amino acid substitution. The first mutation results in substitution of arginine (Arg), the second results in substitution of tyrosine (Tyr), and the third results in substitution of glutamine (Gln). Each mutation affects just a single base within this codon *but not necessarily the same base*. What was the original amino acid?

19-10 Bacterial and Eukaryotic Protein Synthesis Compared. For each of the following statements, indicate whether it applies to protein synthesis in bacteria (Ba), in eukaryotes (E), in both (Bo), or in neither (N).

(a) The mRNA has a ribosome-binding site within its leader region.

(b) AUG is a start codon.

(c) The enzyme that catalyzes peptide bond formation is an RNA molecule.

(d) The mRNA is translated in the 3′ → 5′ direction.

(e) The C-terminus of the polypeptide is synthesized last.

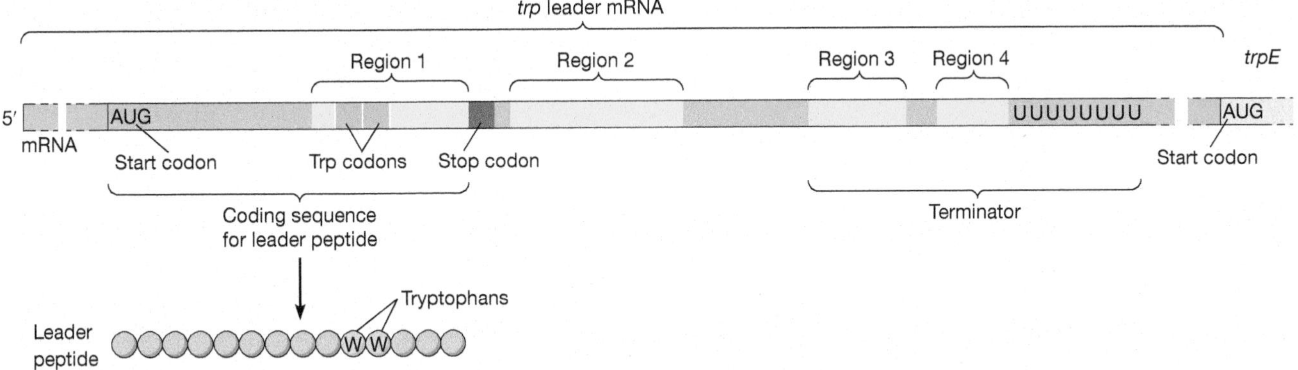

Figure 20-7 The *trp* mRNA Leader Sequence. The transcript of the *trp* operon includes a leader sequence of 162 nucleotides situated directly upstream of the start codon for *trpE*, the first operon gene. This leader sequence includes a section encoding a leader peptide of 14 amino acids. Two adjacent tryptophan (Trp/W) codons within the leader mRNA sequence play an important role in the operon's regulation by attenuation. The leader mRNA also contains four regions capable of base pairing in various combinations to form hairpin structures, as Figure 20-8 shows. A part of the leader mRNA containing regions 3 and 4 and a string of eight U's is called the terminator.

colleagues found that the *trp* operon of *E. coli* has a novel type of regulatory site located between the promoter/operator and the operon's first gene, *trpE*. This stretch of DNA, called the *leader sequence* (or *L*), is transcribed to produce a leader mRNA segment, 162 nucleotides long, located at the 5′ end of the polycistronic *trp* mRNA (see Figure 20-6b).

Analysis of *trp* operon transcripts made under various conditions revealed that, as expected, the full-length, polycistronic *trp* mRNA is produced when tryptophan is scarce. This allows the enzymes of the tryptophan biosynthetic pathway to be synthesized, and hence the pathway produces more tryptophan. On the other hand, when tryptophan is present, transcription of the genes encoding enzymes of the tryptophan pathway is inhibited—also as expected. An unexpected result, however, was that the DNA corresponding to most of the leader sequence may still be transcribed under such conditions, even though the full-length, polycistronic *trp* mRNA is not produced. Based on these findings, Yanofsky suggested that the leader sequence contains a control element that is more sensitive than the *trp* repressor mechanism in determining whether enough tryptophan is present to support protein synthesis. This control sequence somehow determines not whether *trp* operon transcription can begin, but whether it will continue to completion. Such control was called **attenuation** because of its role in attenuating, or prematurely stopping, the synthesis of mRNA. Although attenuation was once considered to be an unusual type of regulation, it turns out to be relatively common, especially in operons that produce enzymes involved in amino acid biosynthesis.

To see how attenuation works, let's take a closer look at the leader segment of the *trp* operon mRNA (**Figure 20-7**). This leader has two unusual features that enable it to play a regulatory role. First, in contrast to the untranslated leader sequences typically encountered at the 5′ end of mRNA molecules (see Figure 19-13), a portion of the *trp* leader sequence *is* translated, forming a *leader peptide* 14 amino acids long. Within the mRNA sequence encoding this peptide are two adjacent codons for the amino acid tryptophan; these will prove important. Second, the *trp* leader mRNA also contains four segments (labeled regions 1, 2, 3, and 4) whose nucleotides can base-pair with each other to form several distinctive hairpin loop structures (**Figure 20-8** on page 582). The region comprising regions 3 and 4 plus an adjacent string of eight U nucleotides is called the *terminator.* When base pairing between regions 3 and 4 creates a hairpin loop, it acts as a transcription termination signal (Figure 20-8a). Recall that the typical bacterial rho independent termination signal (shown in Figure 18-8) is also a hairpin followed by a string of U's. The formation of such a structure causes RNA polymerase and the growing RNA chain to detach from the DNA.

As Yanofsky's experiments suggested, translation of the leader RNA plays a crucial role in the attenuation mechanism. A ribosome attaches to its first binding site on the *trp* mRNA as soon as the site appears, and from there it follows close behind the RNA polymerase. When tryptophan levels are low (Figure 20-8b), the concentration of tryptophanyl-tRNA (tRNA molecules carrying tryptophan) is also low. Thus, when the ribosome arrives at the tryptophan codons of the leader RNA, it stalls briefly, awaiting the arrival of tryptophanyl-tRNA. The stalled ribosome blocks region 1, allowing an alternative hairpin structure to form by pairing regions 2 and 3, called an *antiterminator hairpin.* When region 3 is tied up in this way, it cannot pair with region 4 to create a termination structure, and so *transcription by RNA polymerase continues,* eventually producing a complete mRNA transcript of the *trp* operon. Ribosomes use this mRNA to synthesize the tryptophan pathway enzymes, and production of tryptophan therefore increases.

If, however, tryptophan is plentiful and tryptophanyl-tRNA levels are high, the ribosome does not stall at the tryptophan codons (Figure 20-8c). Instead, the ribosome continues to the stop codon at the end of the coding sequence for the leader peptide and pauses there, blocking region 2. This pause permits the formation of the 3–4 hairpin, which is the transcription termination signal. Transcription by RNA polymerase is therefore terminated (or attenuated) near the end of the leader sequence (after 141 nucleotides), and mRNAs encoding enzymes involved in tryptophan synthesis

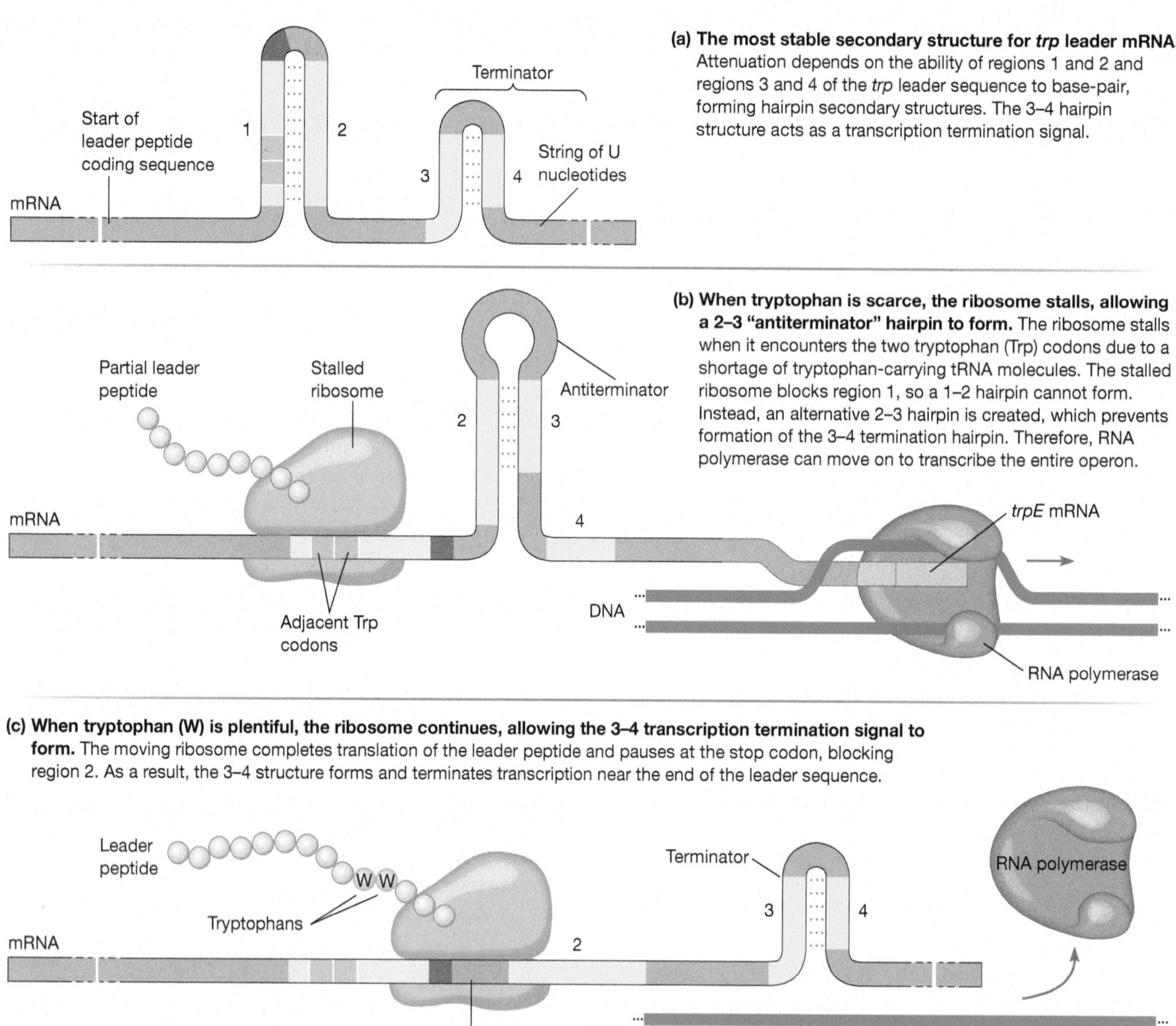
Terminator
Start of
leader peptide
coding sequence
1
2
3
4
String of U
nucleotides
mRNA
(a) The most stable secondary structure for trp leader mRNA. Attenuation depends on the ability of regions 1 and 2 and regions 3 and 4 of the trp leader sequence to base-pair, forming hairpin secondary structures. The 3–4 hairpin structure acts as a transcription termination signal.
Partial leader
peptide
Stalled
ribosome
Antiterminator
mRNA
Adjacent Trp
codons
DNA
trpE mRNA
RNA polymerase
(b) When tryptophan is scarce, the ribosome stalls, allowing a 2–3 "antiterminator" hairpin to form. The ribosome stalls when it encounters the two tryptophan (Trp) codons due to a shortage of tryptophan-carrying tRNA molecules. The stalled ribosome blocks region 1, so a 1–2 hairpin cannot form. Instead, an alternative 2–3 hairpin is created, which prevents formation of the 3–4 termination hairpin. Therefore, RNA polymerase can move on to transcribe the entire operon.
(c) When tryptophan (W) is plentiful, the ribosome continues, allowing the 3–4 transcription termination signal to form. The moving ribosome completes translation of the leader peptide and pauses at the stop codon, blocking region 2. As a result, the 3–4 structure forms and terminates transcription near the end of the leader sequence.
Leader
peptide
W W
Tryptophans
mRNA
Terminator
RNA polymerase
Stop codon
DNA

Figure 20-8 Attenuation in the *trp* Operon. Attenuation depends on the ability of regions 1 and 2 and regions 3 and 4 of the *trp* mRNA leader sequence to base-pair, forming hairpin secondary structures. The 3–4 hairpin acts as a transcription termination signal; as soon as it forms, RNA polymerase is released from the DNA.

are no longer produced. The probability of terminating versus continuing transcription is highly sensitive to small changes in the concentration of tryptophanyl-tRNA, providing a more responsive control system than can be achieved by the interaction of free tryptophan with the *trp* repressor as the only means of regulation.

The preceding mechanism, discovered in *E. coli*, requires that ribosomes begin translating the mRNA before transcription is completed. However, attenuation does not always require this coupling of transcription and translation. In the bacterium *Bacillus subtilis*, the leader RNA transcribed from the *trp* operon is not translated by ribosomes and yet still functions as an attenuator. In this case, formation of the terminator loop is controlled by a tryptophan-binding protein. When tryptophan is present, the protein-tryptophan complex binds to the leader RNA and exposes the terminator loop, thereby terminating transcription and shutting down the *trp* operon.

Riboswitches Allow Transcription and Translation to Be Controlled by Small-Molecule Interactions with RNA

The ability of small molecules to induce shape changes in allosteric proteins plays a central role in regulating gene expression. We have already seen, for example, how gene transcription is controlled by the binding of allolactose to the *lac* repressor protein or tryptophan to the *trp* repressor protein, or of cAMP to a cAMP receptor protein. In each case, mRNA production is altered by the binding of a small molecule to a regulatory protein.

Small molecules can also regulate gene expression by binding to special sites in mRNA called **riboswitches**. Binding of an appropriate small molecule to its corresponding riboswitch triggers changes in mRNA shape that can affect either transcription or translation. Riboswitches are typically found in the untranslated leader region of mRNAs transcribed from bacterial operons, although they have been detected in mRNAs from archaea and eukaryotes as well. One of the first riboswitches to be discovered is in the RNA transcribed from the *riboflavin (rib)* operon of the bacterium *B. subtilis*. The *rib* operon contains five genes encoding enzymes involved in synthesizing the essential coenzymes FMN and FAD. RNA transcribed from the *rib* operon possesses a leader sequence that, like the *trp* leader discussed earlier, can fold into a hairpin loop that terminates transcription. As shown in **Figure 20-9**, binding of FMN to the leader sequence promotes formation of this hairpin loop. Transcription of the *rib* operon is therefore terminated when FMN is present, halting the production of enzymes that are not necessary because they are involved in synthesizing FMN itself.

Besides regulating transcription, the binding of small molecules to riboswitches can also control mRNA translation. One example occurs in mRNAs encoding enzymes involved in the pathway for synthesizing FMN and FAD in *E. coli*. Unlike the situation in *B. subtilis*, the *E. coli* genes encoding these enzymes are not clustered in a single operon, but some of them are still controlled by riboswitches. In this case, binding of FMN to its mRNA riboswitch promotes the formation of a hairpin loop that includes the sequences required for binding the mRNA to ribosomes. By sequestering away these sequences, the bound FMN prevents the mRNA from interacting with ribosomes and initiating translation (see Figure 20-9b).

In addition to small molecules, riboswitches can also respond to environmental signals like changes in temperature. A conserved sequence called a ROSE element (repression of heat shock gene expression) forms hairpins in the mRNA, blocking important translation signals like the ribosome-binding site (see Chapter 19). When the hairpins are present (at about 30°C), the initiation of translation is inhibited. Once the temperature rises to ~37°C, the hydrogen bonds holding together the hairpins begin breaking. At 42°C the hairpins are completely disrupted, and the Shine–Dalgarno sequence becomes available for ribosomal binding. Translation begins, and genes associated with heat shock responses are produced. You will learn more about these heat shock responses later in this chapter. These unique riboswitches are aptly named RNA thermometers.

MAKE CONNECTIONS 20-1

Describe in detail how the formation of a hairpin that includes the Shine–Dalgarno sequence blocks the initiation of translation in bacterial cells. (Fig. 19-15)

The CRISPR/Cas System Protects Bacteria Against Viral Infection

The last example of prokaryotic control we will examine is not, strictly speaking, an example of self-regulated gene expression. It is instead a way that certain cells can limit the gene expression of invading viruses. During the examination of the DNA of bacteria and archaea, highly organized stretches of repeated DNA were identified; they were given the rather inelegant name **clustered regularly interspaced short palindromic repeats (CRISPRs)**. CRISPRs have been found in half of all bacterial genomes and in essentially all archaea.

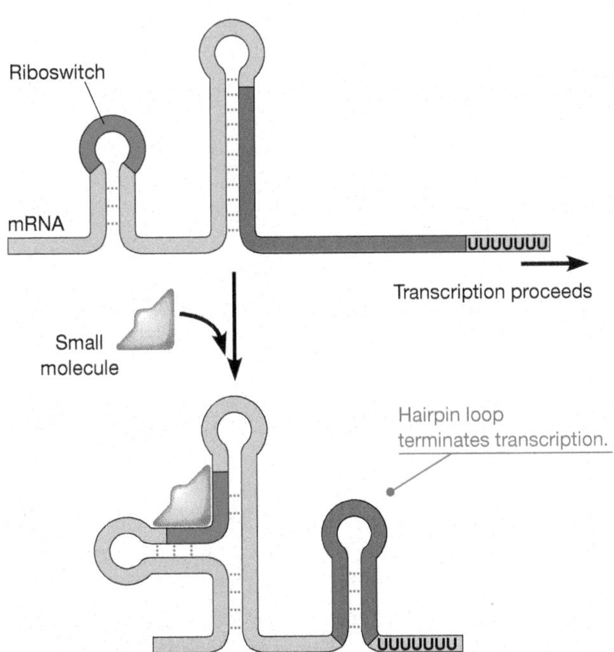

(a) Transcription termination. Binding of a small molecule to a riboswitch in the leader sequence of some mRNAs triggers the formation of a hairpin loop that terminates transcription. FMN binding to the leader sequence of the mRNA transcribed from the *rib* operon of *B. subtilis* works in this way.

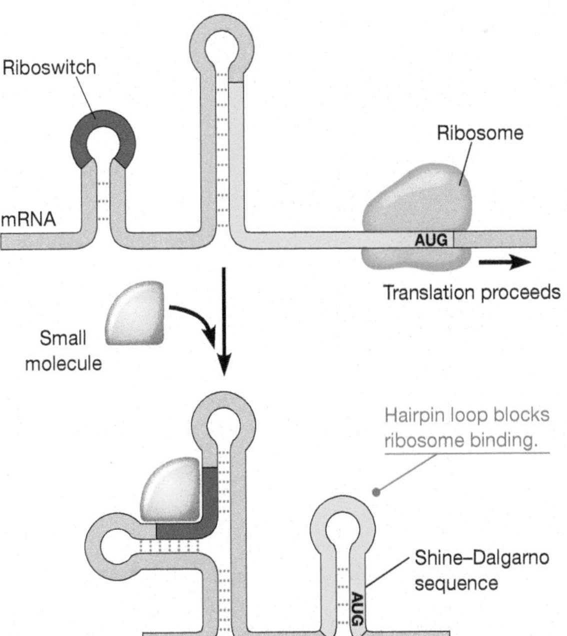

(b) Translation initiation. In other mRNAs, a small molecule binding to a riboswitch triggers formation of a hairpin loop containing the site where ribosomes normally bind, thereby interfering with translation initiation. In *E. coli*, this type of control is used by FMN to inhibit translation of mRNAs encoding enzymes involved in FMN synthesis.

Figure 20-9 Riboswitch Control of Gene Expression. These models show two ways in which the binding of small molecules to riboswitches in mRNA can exert control over **(a)** transcription and **(b)** translation.

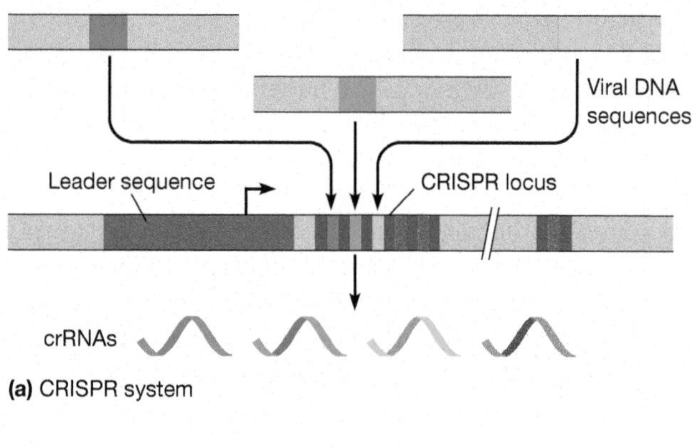

(a) CRISPR system

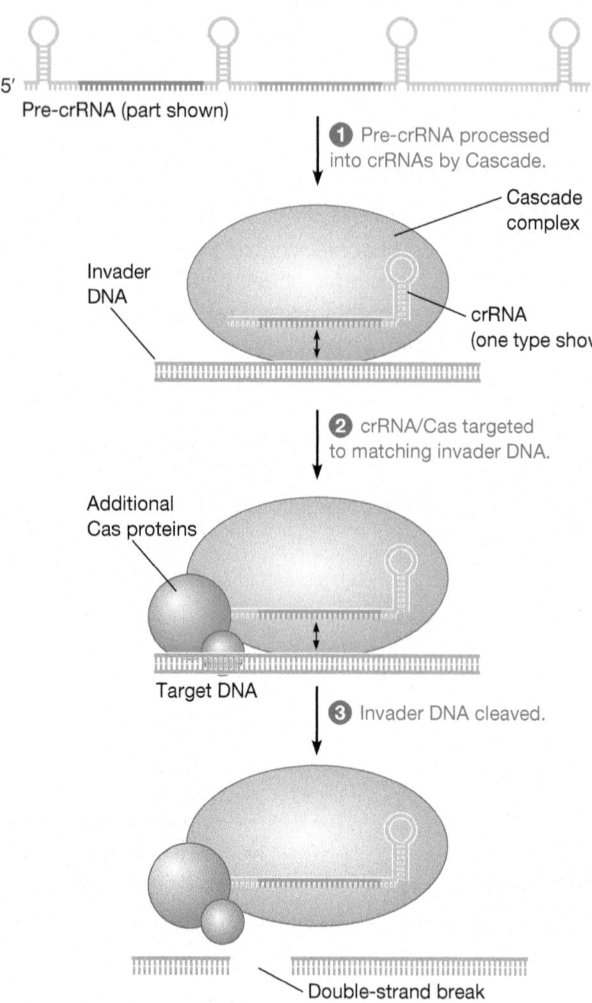

(b) crRNA processing and targeting

Figure 20-10 The CRISPR/Cas System in *E. coli*. **(a)** Viral DNA sequences are captured into the CRISPR locus, which contains spacer sequences that are highly similar to viral DNA sequences (colors), with intervening repeat sequences (gray). The CRISPR locus is transcribed as a long RNA, which is processed to produced short crRNAs. **(b)** During crRNA production, the repeat sequences in the single, long pre-crRNA lead to formation of hairpins (pink), with spacer sequences between them (colors). Cas (CRISPR-associated) proteins in the Cascade complex ❶ process the pre-crRNA into crRNAs and ❷ target them to viral DNAs, followed by recruitment of additional Cas proteins and ❸ cleavage of the foreign DNA.

What are CRISPRs? The CRISPR system (see Chapter 1) is a means to edit genomes in the laboratory (the details of which can be found in **Key Technique, Chapter 17, pages 500–501**). In bacteria and archaea, CRISPRs function as a form of antiviral immunity. They contain conserved, repeated sequences, each about 30 bp long, alternating with *spacer sequences*, each also about 30 bp in length but highly variable in sequence (**Figure 20-10**). At one end is a *leader sequence*, about 500 bp in length (Figure 20-10a). The sequences of the CRISPR spacers indicated that they come from the DNA of bacterial viruses known as bacteriophages (see Chapter 16). The more CRISPR repeats a bacterial strain contains, the more resistant it is to phage infection, which suggested that CRISPRs are somehow involved in protecting bacteria from viral infection. Tightly associated with the CRISPR region are genes that encode a set of *CRISPR-associated (Cas) proteins*. Some of the Cas proteins are involved in helping to incorporate the viral sequences into the CRISPR spacers; others help process and target RNAs made from the CRISPR locus.

CRISPR sequences protect bacteria from viral infection through the production of a long RNA called a *pre-crRNA* (Figure 20-10b). The next steps in the process are best understood in *E. coli*. In *E. coli*, the pre-crRNA is processed by a group of Cas proteins, collectively known as *Cascade*, into short, individual *crRNAs*, each of which contains one spacer and adjacent repeat sequences. These small RNAs, which stay associated with Cascade, direct the complex to the DNA that comes from foreign invaders. The foreign DNA that matches one of the CRISPR spacers is then cleaved. In some archaea, crRNAs, produced in a somewhat different way from a pre-crRNA, are used to target viral RNA sequences. This mechanism is strikingly similar to how eukaryotes use small RNAs to target mRNAs for destruction, which you will learn about later in this chapter.

CONCEPT CHECK 20-1

Suppose you are analyzing a haploid *E. coli* strain and discover that the genes of the *lac* operon are never transcribed and that the bacteria cannot grow in lactose-containing medium. Provide two possible genotypes for your bacteria that would explain this phenotype.

20.2 Eukaryotic Gene Regulation: Genomic Control

In the early days of molecular biology, there was a popular saying: "What is true of *E. coli* is also true of elephants." This maxim expressed the initial conviction that almost everything learned about bacterial cell function at the molecular level would also apply to eukaryotes. Such predictions have turned out to be only partly true, however. In terms of central metabolic pathways, mechanisms for transporting solutes across membranes, and such fundamental features as DNA structure, protein synthesis, and enzyme function, there are numerous similarities between the bacterial and eukaryotic worlds, and findings from studies with bacteria can often be extrapolated to eukaryotes.

When it comes to the regulation of gene expression, however, the comparison needs to be scrutinized more carefully. The DNA of eukaryotic cells is packaged into chromatin fibers and located in a nucleus that is separated from the protein synthetic machinery by a nuclear envelope. Eukaryotic genomes are usually much larger than those of bacteria, and multicellular eukaryotes must create a variety of different cell types from the same genome. Such differences require a diversity of genetic control mechanisms that in some cases differ significantly from and are more complex than those routinely observed in bacteria.

Multicellular Eukaryotes Are Composed of Numerous Specialized Cell Types

To set the stage for our discussion of eukaryotic gene control, let's briefly consider the enormous regulatory challenges faced by multicellular eukaryotes, which are often composed of hundreds of different cell types. A single organism consists of a complex mixture of specialized or *differentiated* cell types—for example, nerve, muscle, bone, blood, cartilage, and fat—brought together in various combinations to form tissues and organs. Differentiated cells are distinguished from each other based on differences in their microscopic appearances and in the products they manufacture. For example, red blood cells synthesize hemoglobin, nerve cells produce neurotransmitters, and lymphocytes make antibodies. Because these cells all share a common genome, such differences indicate that selectively controlling gene expression must play a central role in the mechanism responsible for creating differentiated cells. Differentiated cells are produced from populations of immature, nonspecialized cells by a process known as **cell differentiation**. The classic example occurs in embryos, in which cells of the early embryo produce all the cell types that make up the organism.

Eukaryotic Gene Expression Is Regulated at Five Main Levels

The fact that a multicellular plant or animal may need to produce hundreds of different cell types using a single genome underscores the difficulty of explaining eukaryotic gene regulation entirely in terms of known bacterial mechanisms—elephants are not just large *E. coli* after all! If we are to thoroughly understand gene regulation in eukaryotic cells, we must approach the topic from a eukaryotic perspective.

The pattern of genes being expressed in any given eukaryotic cell is ultimately reflected in the spectrum of functional gene products—usually proteins, but in some cases RNAs—produced by that cell. The overall pattern is the culmination of controls exerted at several different levels. These potential control points are highlighted in **Figure 20-11**, which traces the flow of genetic information from genomic DNA in the nucleus to functional proteins in the cytoplasm (many, but not all, of these control points also apply to prokaryotes). As you can see, control is exerted at five main levels: ❶ the genome, ❷ transcription, ❸ RNA processing and export from nucleus to cytoplasm, ❹ translation, and ❺ posttranslational events. Regulatory mechanisms in the last three categories are all examples of *posttranscriptional control*, a term that encompasses a wide variety of events. In the rest of the chapter, we will examine some mechanisms eukaryotes use to exert control over gene expression at each of these five levels.

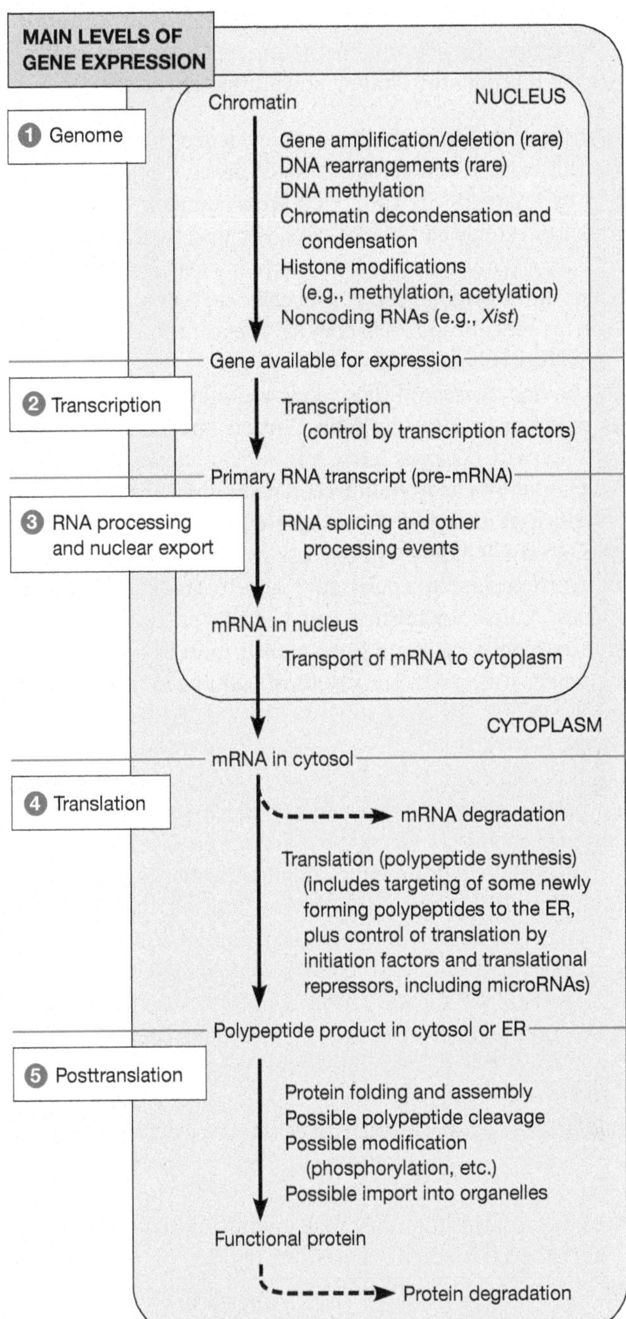

Figure 20-11 The Multiple Levels of Eukaryotic Gene Expression and Regulation. Gene expression in eukaryotic cells is regulated through a diverse array of mechanisms operating at five distinct levels. Controls operating at the last three levels are all considered to be *posttranscriptional control* mechanisms.

The Cells of a Multicellular Organism Usually Contain the Same Set of Genes

The first level of control is exerted at the level of the overall genome. In multicellular plants and animals, each specialized cell type uses only a small fraction of the total number of genes contained within its genome. Nevertheless, almost all cells (other than the haploid sperm and eggs) retain the same complete set of genes, a phenomenon known as **genomic equivalence**. Two clear demonstrations of genomic equivalence are reproductive cloning and pluripotent stem cells.

Reproductive Cloning. Dramatic evidence that specialized cells still carry a full complement of genes was provided in 1958 by Frederick Steward, who grew complete new carrot plants from single carrot root cells. Because the cells of each new carrot plant contained the same nuclear DNA that was present in the original carrot root cells, each new plant was a **clone** (genetically identical copy) of the plant from which the original root cells were obtained.

The first successful reproductive cloning of animal cells was reported in 1964 by John Gurdon and his colleagues. In studies with *Xenopus laevis*, the African clawed frog, they transplanted nuclei from differentiated tadpole cells into unfertilized eggs that had been deprived of their own nuclei. Although the frequency of success was low, some eggs containing transplanted nuclei gave rise to viable, swimming tadpoles. A new organism created by this process of *nuclear transplantation* is a clone of the organism from which the original nucleus was taken. The results of both the carrot and frog studies indicated that the nucleus of a differentiated cell can direct the development of an entire new organism. Such a nucleus is therefore said to be *totipotent:* it contains the complete set of genes needed to create a new organism of the same type as the organism from which the nucleus was taken.

An especially dramatic example of animal cloning was reported in 1997, when Ian Wilmut and his colleagues in Scotland made newspaper headlines by announcing the birth of a cloned lamb, Dolly—the first mammal ever cloned from a cell derived from an adult. Dolly was born from a sheep egg whose original nucleus had been replaced by a nucleus taken from a single cell of an adult sheep. Although this **nuclear transfer** technique had been used before to clone mammals, it had never been successful with cells taken from an adult.

Wilmut suspected that these previous failures were caused by the state of the chromatin in the donor cells. The secret to his team's success was that they took mammary gland cells from the udder of a 6-year-old female sheep and starved them in culture to force them into the dormant, G_0 phase of the cell cycle. When such a cell was fused with an egg cell lacking a nucleus, it delivered the diploid amount of DNA in a condition that allowed the egg cytoplasm to reprogram the DNA to support normal embryonic development. Implanting such an egg into the uterus of another female sheep led to the birth of a lamb with no father—that is, a lamb whose cells had the same nuclear DNA as the cells of the 6-year-old female sheep that provided the donor nucleus (**Figure 20-12**). Within a few years of these pioneering experiments, similar techniques had been used to clone other mammals, including livestock, mice, cats, dogs, and monkeys.

Why would researchers want to pursue such technology? One reason is to clone valuable genetically engineered animals that produce milk containing medically important proteins, such as blood-clotting factors. Such animals are difficult to produce in sufficient quantities by other means. Commercial applications include cloning the beefiest cattle or the fastest racehorses. An even more dramatic application of cloning involves attempts to clone animals that are on the verge of extinction or that have recently become extinct.

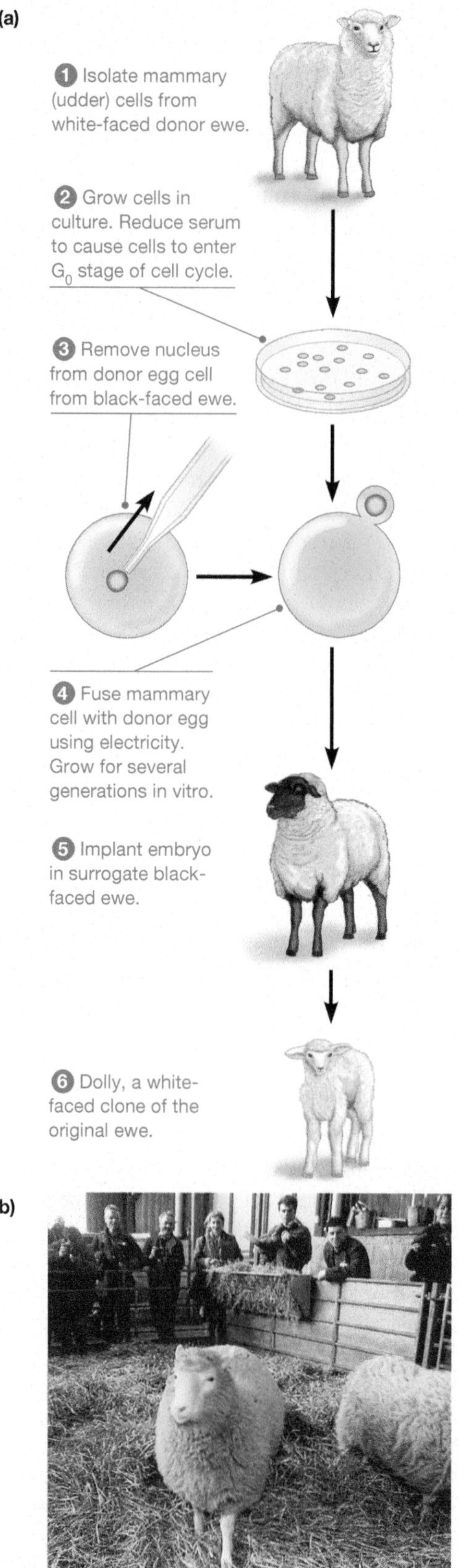

(a)
1 Isolate mammary (udder) cells from white-faced donor ewe.
2 Grow cells in culture. Reduce serum to cause cells to enter G_0 stage of cell cycle.
3 Remove nucleus from donor egg cell from black-faced ewe.
4 Fuse mammary cell with donor egg using electricity. Grow for several generations in vitro.
5 Implant embryo in surrogate black-faced ewe.
6 Dolly, a white-faced clone of the original ewe.
(b)

Figure 20-12 Dolly, the First Mammal Cloned from an Adult Cell. (a) Dolly was cloned from a serum-starved mammary (udder) cell that was fused with an egg cell from which the nucleus had been removed. **(b)** Dolly as an adult.

But the most controversial possibility concerns whether it would ever be ethical to attempt cloning with human cells. For the moment, it appears that producing healthy human clones would be exceedingly difficult. It took 277 attempts to produce Dolly, and many cloned animals that seem normal at birth later develop health problems and die prematurely. These common defects result from a failure of the DNA in the donated nucleus to undergo proper *epigenetic reprogramming*, including methylation of important genes related to growth. In the face of such problems and nearly universal unease about the prospect of human reproductive cloning, many individuals have called for laws banning the use of such technology for duplicating human beings. Others have suggested that society might eventually find cloning acceptable under certain restricted circumstances.

Pluripotent Stem Cells. The most frequently proposed application of the nuclear transfer technique in humans is for producing stem cells genetically matched to a patient. Some have hoped that such *therapeutic cloning* could be used to replace damaged cells in human patients. Successful cloning by somatic cell nuclear transfer, followed by stem cell production, was announced in humans in 2013, indicating that this approach will be technically feasible. To see why this approach has been proposed, we need to consider why stem cells have such useful properties and why they are another remarkable example of nuclear equivalence.

Stem cells are defined by their capacity to replenish themselves through continued division, as well as by their ability, in the presence of appropriate signals, to produce daughter cells that differentiate into a variety of specialized cell types, often accompanied by the cessation of cell division. Stem cells can be isolated from more mature organisms, such as umbilical cord blood of newborns, as well as from adults. These cells, especially those from blood and bone marrow, have been used successfully for years to treat human patients. Such stem cells are considered to be *multipotent:* they can form several cell types, but their repertoire is fairly limited. **Embryonic stem (ES) cells**, on the other hand, do not suffer this limitation. They are **pluripotent**; that is, they can form all types of cells that give rise to a human organism except those that form structures such as the placenta and other support membranes needed during gestation.

There are two major ways that pluripotent stem cells can be obtained (**Figure 20-13**). These cells are most effectively derived from a special group of cells called *inner cell mass cells* at one end of the early embryo, which is known as a *blastocyst* (Figure 20-13a). Inner cell mass cells, once isolated, behave as true stem cells. When treated with various soluble factors or in other ways, they can be coaxed to differentiate and become various types of cells.

The ability to isolate and grow stem cells in the laboratory raises the possibility that scientists will be able to produce healthy new cells to replace damaged tissues in patients with various diseases. For example, stem cells that can differentiate into nerve cells might eventually be used to repair the brain damage that occurs in patients suffering from stroke, Parkinson's disease, or Alzheimer's disease. Stem cells might be utilized to replace the defective pancreatic cells of patients with diabetes or the defective skeletal muscle cells present in individuals with muscular dystrophy. If the stem cells were to be produced using therapeutic cloning, they would have the further benefit of being a genetic match to the patient's own cells, avoiding issues of immune rejection.

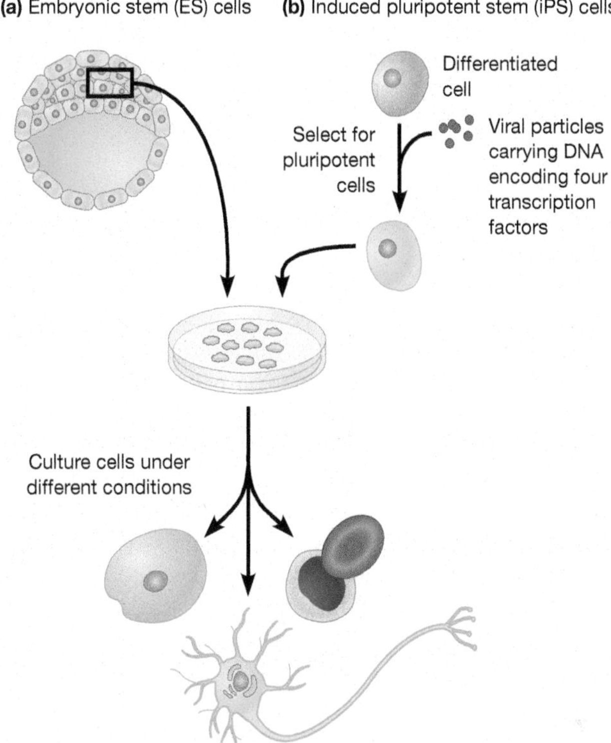

Figure 20-13 Two Different Ways to Produce Pluripotent Stem Cells. Pluripotent cells are selected using drugs that only allow pluripotent cells to survive. Once ES or iPS cells are isolated, they are cultured under a variety of conditions, causing them to differentiate into various cell types. **(a)** Embryonic stem (ES) cells are produced from inner cell mass cells of a blastocyst. The embryo is destroyed to harvest the ES cells. **(b)** Induced pluripotent stem (iPS) cells result when cells are forced to express four transcription factor proteins. Originally, this was accomplished by introducing viral particles carrying DNA encoding four transcription factors into differentiated cells.

Despite their developmental potential and their promise in therapies, there is a major ethical objection regarding harvesting of ES cells. Isolating ES cells necessarily involves the destruction of a human organism—the embryo. Many people view each human organism as an inviolable end in itself, worthy of profound moral respect, making procurement of ES cells ethically impermissible. Others' intuition is that the blastocyst has a different moral status from more advanced human organisms; such people view ES cell production as an important potential medical therapy. Clearly, it would be highly desirable to find cells that seem to have developmental potential similar to ES cells but without the associated ethical issues.

In 2006, Shinya Yamanaka announced just such a method: he and his research team had succeeded in coaxing differentiated cells from mice into reverting to a pluripotent state. Such **induced pluripotent stem (iPS) cells** seem to have many of the same properties as ES cells. To produce iPS cells, Yamanaka's team forced cells to express four proteins that are expressed by pluripotent cells: Oct4, Sox2, Klf4, and c-Myc (Figure 20-13b). For their work on genomic equivalence, Yamanaka and John Gurdon received a Nobel Prize in 2012. Further work by Yamanaka's group and by James Thomson and colleagues extended these studies to human cells. Technical progress has been rapid ever since.

iPS cells seem quite similar to ES cells, and so they seem to be an ideal source of genetically matched, pluripotent cells for medical treatments. There are some issues to be overcome before they can be used in therapies, however. First, in the original experiments the genes were introduced using a retrovirus, which incorporated its DNA into the host cell's genome. Similar effects can now be achieved without integration of viral DNA. Second, c-Myc is a *proto-oncogene*, which can promote tumor formation (see Chapter 26). However, some cell types can be induced to pluripotency without c-Myc. Finally, like histones (see Chapter 16), DNA can be methylated, which affects the likelihood that a gene will be expressed. Unlike ES cells, many iPS cells do not seem to have the methylation status of their DNA fully reset. For this reason, many stem cell biologists who are not opposed to ES cell harvesting favor continuing ES and iPS cell research in parallel because the two types of cells may exhibit somewhat different properties. Given the widely varying viewpoints about ES cells, cloning, and similar procedures, it is crucial that debate continues about the ethics and use of these technologies.

Gene Amplification and Deletion Can Alter the Genome

Although the preceding evidence indicates that the genome tends to be identical in all cells of an adult eukaryotic organism, a few types of gene regulation create exceptions to this rule. One example is the use of **gene amplification** to create multiple copies of the same gene. Amplification is accomplished by replicating the DNA in a specific chromosomal region many times in succession, thereby creating dozens, hundreds, or even thousands of copies of the same stretch of DNA. Gene amplification can be regarded as an example of **genomic control**—that is, a regulatory change in the makeup or structural organization of the genome.

One of the best-studied examples of gene amplification involves the ribosomal RNA genes in *Xenopus laevis*. The haploid genome of *Xenopus* normally contains about 500 copies of the genes that encode 5.8S, 18S, and 28S rRNA. However, the DNA of these genes is selectively replicated about 4000-fold during oogenesis (development of the egg prior to fertilization). As a result, the mature oocyte contains about 2 million copies of the genes for rRNA. This level of amplification is needed to sustain the enormous production of ribosomes that occurs during oogenesis, which in turn is required to sustain the high rate of protein synthesis needed for early embryonic development.

This particular example of gene amplification involves genes whose products are RNA rather than protein molecules. Enhanced production of certain proteins can occasionally be traced to gene amplification as well. One example is the synthesis of chorion proteins in insect eggs. Abnormal amplification of genes encoding proteins involved in cell cycle signaling is a frequently observed abnormality in cancer cells (see Chapter 26).

In addition to amplifying gene sequences whose products are in great demand, some cells delete genes whose products are not required. An extreme example of *gene deletion* (also called *DNA diminution*) occurs in mammalian red blood cells, which discard their nuclei entirely after adequate amounts of hemoglobin mRNA have been made. A less-extreme example occurs in some nematode worms and in a group of tiny crustaceans known as copepods. During the embryonic development of copepods, the heterochromatic (transcriptionally inactive) regions of their chromosomes are excised and discarded from all cells except those destined to become gametes. In this way, up to half of the organism's total DNA content is removed from its body cells.

DNA Rearrangements Can Alter the Genome

Recall that transposons are mobile genetic elements that can move from place to place within the genome (see Chapter 17). Other specific cases are known in which gene regulation is based on the movement of DNA segments from one location to another within the genome, a process known as *DNA rearrangement*. One particularly interesting example involves the mechanism used by vertebrates to produce millions of different antibodies.

Lymphocyte Receptor Gene Rearrangements. Antibodies are proteins composed of two kinds of polypeptide subunits, called *heavy chains* and *light chains*. Vertebrates make millions of different kinds of antibodies, each produced by a different B lymphocyte (and its descendants) and each capable of specifically recognizing and binding to a different foreign molecule. But this enormous diversity of antibody molecules creates a potential problem: if every antibody molecule were to be encoded by a different gene, virtually all of a person's DNA would be occupied by the millions of required antibody genes.

Lymphocytes get around this problem in two ways. First, by starting with a relatively small number of different DNA segments and rearranging them in various combinations they can produce millions of unique antigen-binding receptor genes, each one formed in a different, developing lymphocyte. The rearrangement process involves four kinds of DNA sequences, called *V, D, J,* and *C segments*. In B cells, the C segment encodes a heavy- or light-chain *constant region* whose amino acid sequence is the same among different antibodies. The V, J, and D segments together encode *variable regions* that differ among antibodies and give each one the ability to recognize and bind to a specific type of foreign molecule.

To see how this works, let's consider human antibody heavy chains, which are constructed from roughly 200 kinds of V segments, more than 20 kinds of D segments, and at least 6 kinds of J segments. As shown in **Figure 20-14**, the DNA regions containing the various V, D, and J segments are rearranged during lymphocyte development to randomly bring together one V, one D, and one J segment in each lymphocyte. This random rearrangement allows the immune system to create at least $200 \times 20 \times 6 = 24{,}000$ different kinds of heavy-chain variable regions. In a similar fashion, thousands of different kinds of light-chain variable regions can also be created (light chains have their own types of V, J, and C segments; they do not use D segments). Finally, any one of the thousands of different kinds of heavy chains can be assembled with any one of the thousands of different kinds of light chains, creating the possibility of millions of different types of antibodies. The net result is that millions of different antibodies are produced from the human genome by rearranging a few hundred different kinds of V, D, J, and C segments.

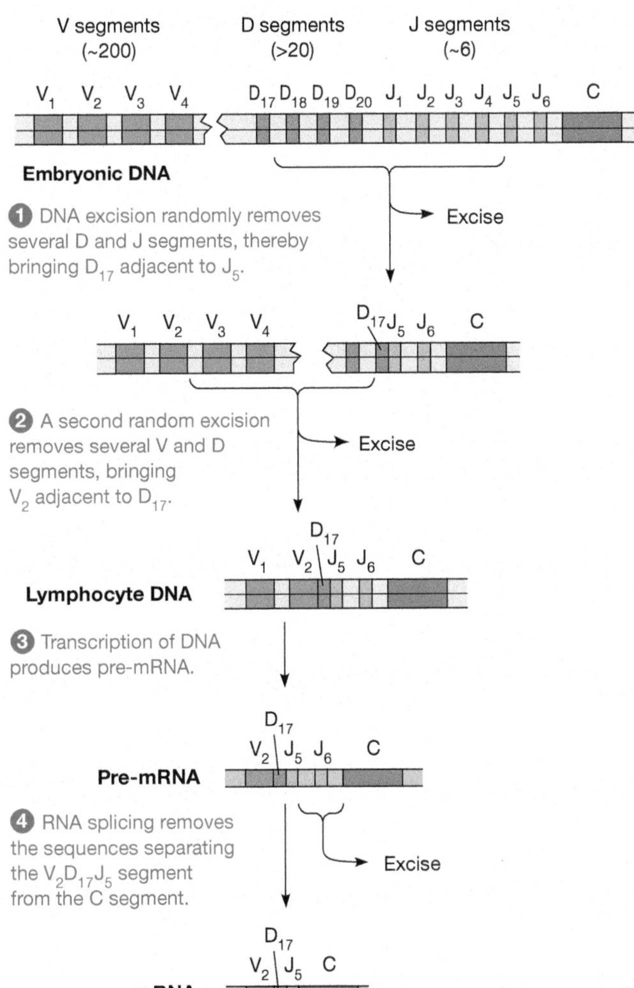

Figure 20-14 DNA Rearrangement During the Formation of Antibody Heavy Chains. Genes encoding human antibody heavy chains are created by DNA rearrangements that involve multiple types of V, D, and J segments. In this example, random DNA excisions bring together a unique combination of V, D, and J segments. After transcription, RNA splicing removes the RNA sequences that separate this VDJ segment from the C segment.

The DNA rearrangement process that creates antibody genes also activates transcription of these genes via a mechanism involving special DNA sequences called *enhancers*. As we will discuss shortly, enhancers increase the rate of transcription initiation. Prior to DNA rearrangement, the promoter and enhancer sequences of an antibody gene are so far apart that transcription does not occur. Only after rearrangement are they close enough for transcription to be activated.

A second way that antibody diversity is increased even further involves a mechanism known as *somatic hypermutation*. As B lymphocytes proliferate, they acquire mutations in their variable regions at a high rate—as much as a million times more frequent than the typical rate of mutation across the genome. This high mutation rate is due to an enzyme known as *activation-induced cytidine deaminase (AID)* that acts specifically in activated B lymphocytes and deaminates cytosine to uracil. Recall that uracil is not normally found in DNA; the uracil is removed by *base excision repair* (see Figure 17-26). Base excision repair is normally a high-fidelity process. In B cells, however, error-prone DNA polymerases repair the gap left when the uracil is removed, leading to higher mutation rates. Paradoxically, this high mutation rate is beneficial because it increases antibody diversity. Accordingly, people deficient in AID activity have poor antibody responses.

Why do these processes occur in DNA and not in RNA? Making changes like this in the DNA ensures that subsequent daughter cells produced through cell division will make the same antigen-binding receptor/antibody that the original cell made. It prevents those descendent cells from having to repeat this almost random process of combinatorial joining in precisely the same way (an extremely unlikely occurrence!). Thus, B cells derived from the original parental cell will produce antibodies with the same specificity. These clones are critical for successful clearance of invading microbes.

Chromatin Decondensation Is Involved in Genomic Control

DNA rearrangements during antibody production show that the position of DNA elements within a stretch of genomic DNA is important. In addition, however, the physical state of the chromatin is a key regulator of eukaryotic gene expression. Recall that, to initiate transcription, a eukaryotic RNA polymerase must interact with both DNA and a number of specific proteins (general transcription factors) in the promoter region of a gene (see Chapter 18). Except when a gene is being transcribed, its promoter region is embedded within a highly folded and ordered chromatin superstructure. Thus, some degree of chromatin decondensation (unfolding) appears to be necessary for the expression of eukaryotic genes.

Chromosome Puffs. The earliest evidence that chromatin decondensation is required for gene transcription came from microscopic visualization of certain types of insect chromosomes caught in the act of transcription. In the fruit fly *Drosophila melanogaster* and related insects, some metabolically active tissues (such as the salivary glands and intestines) grow by an enormous increase in the size of, rather than the number of, their constituent cells. The development of giant cells is accompanied by successive rounds of DNA replication, but because this replication occurs in cells that are not dividing, the newly synthesized chromatids accumulate in each nucleus and line up in parallel to form multistranded structures called **polytene chromosomes**. The four giant polytene chromosomes found in the salivary glands of *Drosophila* larvae, for example, are generated by ten rounds of chromosome replication.

As **Figure 20-15** on page 590 shows, polytene chromosomes are enormous structures—roughly 10 times longer and 100 times wider than the metaphase chromosomes of typical eukaryotic cells. Visible in each polytene chromosome is a characteristic pattern of dark bands. Each band represents a chromatin domain that is highly condensed compared with the chromatin in the "interband" regions between the bands. Activation of the genes of a given chromosome band causes the compacted chromatin strands to uncoil and expand outward, resulting in a **chromosome puff**. Though puffs are not the only sites of gene transcription along the polytene chromosome, the extent of chromosome decondensation at puffs

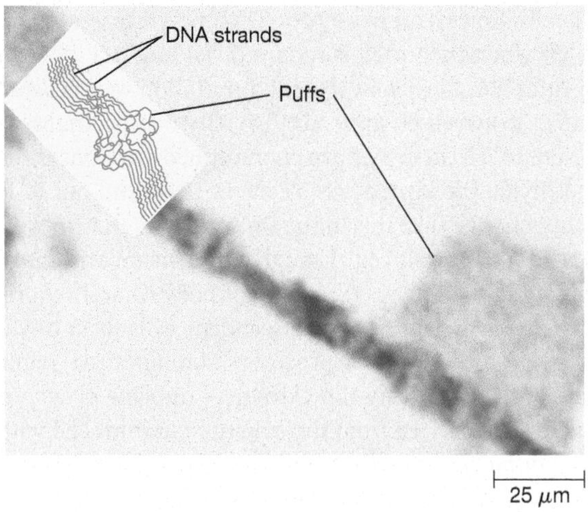

Figure 20-15 Puffs in Polytene Chromosomes. Puffs are regions in which transcriptionally active chromatin has become less condensed, as indicated diagrammatically. The light micrograph shows part of a polytene chromosome.

correlates well with the enhancement of transcriptional activity at these sites. That puffs are indeed sites of active transcription is confirmed by the fact that puffs are sites where RNA polymerase II, the key enzyme that carries out polymerization of protein-coding RNAs, accumulates. The characteristic puffing patterns seen during the development of *Drosophila* larvae are direct visual manifestations of the selective decondensation and transcription of specific DNA segments according to a precise developmental program.

DNase I Sensitivity. The absence of polytene chromosomes in most eukaryotic cells makes it difficult to visualize chromatin decondensation in regions of active genes. Nonetheless, other kinds of evidence support the idea that chromatin decondensation is generally associated with gene transcription. One particularly useful research tool is *DNase I*, an endonuclease isolated from the pancreas. In test tube experiments, low concentrations of DNase I preferentially degrade transcriptionally active DNA in chromatin. The increased sensitivity of these DNA regions to degradation by DNase I provides evidence that the DNA is uncoiled.

Figure 20-16 illustrates a classic DNase I sensitivity experiment focusing on the chicken gene for a globin, one of the polypeptide subunits of hemoglobin. The globin gene is actively expressed in the nuclei of chicken erythrocytes (red blood cells). In contrast to the mammalian erythrocytes with which you are familiar, avian erythrocytes retain their nuclei at maturity. If these nuclei are isolated and the chromatin is digested with DNase I, the globin gene is completely digested. As you might predict, a gene that is not active in

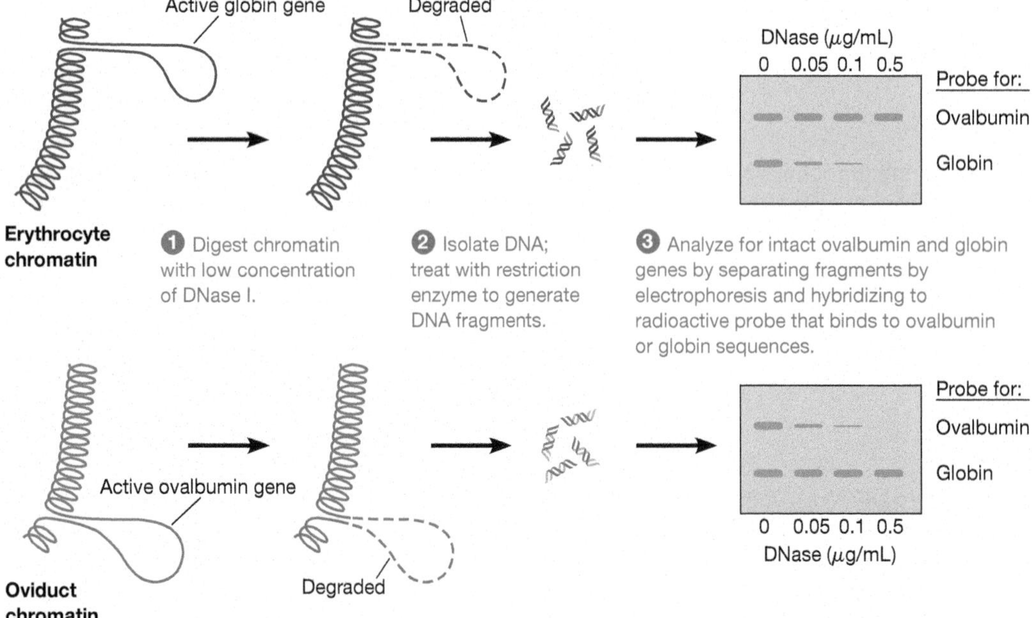

Figure 20-16 Sensitivity of Active Genes in Chromatin to Digestion with DNase I. The chromatin configuration of active genes can be studied by exposing cell nuclei to DNase I, which digests DNA by cutting internal phosphodiester bonds. DNA in condensed chromatin is protected from DNase I attack. The experiment shown here uses chromatin from two different chicken cell types, erythrocytes and oviduct cells, and focuses on genes for globin and ovalbumin, which are expressed in erythrocytes and oviduct cells, respectively. ❶ Digesting chromatin with a low concentration of DNase I preferentially digests the DNA in uncoiled regions. ❷ The DNA is then purified and digested with an enzyme (a restriction enzyme) that releases DNA fragments containing an intact globin or ovalbumin gene (if it has not been nicked by DNase I). ❸ The presence of fragments is detected by separating the DNA using electrophoresis, transferring the separated fragments to filter paper (a Southern blot), and hybridizing with radioactive DNA probes for the globin and ovalbumin genes. Note that DNA isolated from erythrocyte chromatin treated with increasing amounts of DNase I contains progressively smaller amounts of intact globin gene; comparable results are obtained for the ovalbumin gene in oviduct chromatin. In contrast, even high concentrations of DNase I have no effect on the globin gene in oviduct chromatin or on the ovalbumin gene in erythrocyte chromatin. In other words, genes are more susceptible to DNase I attack in tissues where they are actively transcribed.

erythrocytes (for example, the gene for ovalbumin, an egg white protein) is not digested by DNase I. The opposite result is obtained when the same procedure is carried out using cells from other tissues, such as the oviduct, where the ovalbumin gene is expressed and the globin gene is inactive. In this case, the ovalbumin gene is digested, while the globin gene is not. Such data demonstrate that transcription of eukaryotic DNA is correlated with an increased sensitivity to DNase I digestion.

DNase I has also been employed in other kinds of experiments. When nuclei are treated with very low concentrations of DNase I, it is possible to detect specific locations in the chromatin that are exceedingly susceptible to digestion. These **DNase I hypersensitive sites** tend to occur up to a few hundred bases upstream from the transcriptional start sites of active genes. These regions appear to correspond to regions in which the DNA is not part of a nucleosome.

Histone Modifications and Chromatin Remodeling Factors. What accounts for the changes in DNA identified through these classic experiments? Recall that chemical modifications of histones and the activity of chromatin remodeling proteins lead to changes in chromatin structure (see Chapter 16). *Histone deacetylation* and *methylation* favor compacted, inaccessible chromatin, whereas acetylated, nonmethylated histones favor accessibility of chromatin to the transcriptional machinery. One important group of proteins involved in specific histone modifications is the *Polycomb-group* proteins, which are involved in several well-studied examples of tissue-specific gene regulation. Polycomb proteins form two complexes, known as *Polycomb repressive complex (PRC) 1* and *2*. PRC2 contains histone methyltransferase activity and methylates specific histones, which allows PRC1 to attach and condense chromatin.

Histone modifications are not the only way in which chromatin packing can be changed. *Chromatin remodeling factors* can regulate the accessibility of DNA within nucleosomes to the transcriptional machinery, including the *Swi/Snf* family of proteins (mentioned in Chapter 16).

In addition to histones and chromatin remodeling factors, DNA itself can be regulated by methylation, which we examine next.

DNA Methylation Is Associated with Inactive Regions of the Genome

Another way of regulating the availability of specific regions of the genome is through **DNA methylation**, the addition of methyl groups to selected cytosine bases in DNA (**Figure 20-17** on page 592). Eukaryotes, especially mammals, rely on DNA methylation broadly to regulate accessibility of DNA to the transcriptional machinery. The net effect is *transcriptional repression*, that is, prevention of transcription of regions associated with methylated DNA.

CpG Islands. The DNA of most vertebrates contains small amounts of methylated cytosine (Figure 20-17a), which tend to cluster near the 5′ ends of genes where promoter sequences are located. The cytosines are typically next to guanines, so such regions are called **CpG islands**. CpG islands are associated with many eukaryotic promoters. When these islands are methylated, the net effect is either a localized or a regional silencing of gene expression. Changes in methylation can be detected using techniques that identify methylated sequences (see Chapter 21) and then correlated with transcriptional activation of the associated gene. Such studies confirm that, in general, CpG islands are methylated in tissues where a gene is inactive, but they are unmethylated in tissues where the gene is active or potentially active.

How does methylation of CpG sequences in promoter regions act to block transcription? There are several mechanisms, depending on the gene (Figure 20-17b). One way is by directly blocking the binding of transcription factors to the methylated bases. A second way involves changes in chromatin packing. DNA methylation is closely associated with histone modifications. Specific proteins bind to the methylated DNA; these proteins in turn recruit histone-modifying enzymes, such as histone deacetylases, histone methyltransferases, and inactivating chromatin remodeling proteins, to the methylation site. The result is tighter chromatin packing, which makes the DNA inaccessible to transcription factors. One such protein, *MeCP2*, is mutated in human patients with the autism spectrum disorder known as *Rett syndrome* (see Human Connections, page 593). The problems such human patients experience is presumably due to incorrect gene expression at methylated sites in the genome because these regions do not undergo transcriptional repression.

X Chromosome Inactivation. We receive much of our genetic endowment via independent assortment of alleles of genes on non-sex chromosomes, or *autosomes*, from our mothers and fathers. We also receive specialized sex chromosomes from our parents (see Chapter 25). A striking example of wholesale chromatin alteration and subsequent transcriptional repression involves the X chromosomes of female placental mammals, which inherit an X chromosome from each of their two parents. Because males have only one X chromosome, a potential imbalance exists in the expression of X-linked genes between males and females. Female mammals solve this problem by randomly inactivating most regions of one of the two X chromosomes in most cells during early embryonic development, a process known as **X-inactivation** (**Figure 20-18** on page 594). During X-inactivation, the DNA of one X chromosome becomes extensively methylated, the chromatin fibers condense into a tight mass of heterochromatin whose DNA is less accessible to transcription factors and RNA polymerase, and gene transcription largely ceases. When interphase cells are examined under a microscope, the inactivated X chromosome is visible as a dark spot called a *Barr body*, first discovered by Canadian researcher Murray Barr (Figure 20-18a). Once a given X chromosome has been inactivated in a particular cell, the same X chromosome remains inactivated in nearly all of the descendant cells produced by succeeding cell divisions. As a result, females, like males, contain only one active X chromosome per adult cell, but females are actually a *mosaic* of groups of cells in which one or the other X chromosome was initially inactivated. This can be seen in animals in which hair color is controlled by X-linked genes, such as tortoiseshell cats. In fact, *MeCP2*, mentioned above, is a gene that is found on the X chromosome; like other genes on

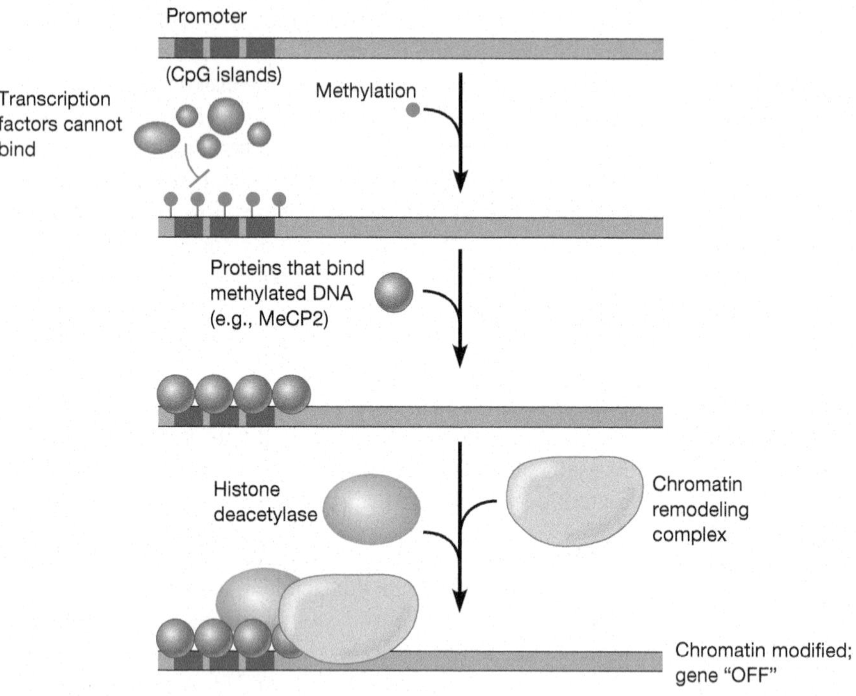

Figure 20-17 Roles of DNA Methylation in Silencing of Gene Expression. (a) The main target of DNA methylation is cytosine, often found in clusters with G residues (CpG islands). **(b)** Methylation of CpG islands can prevent regulatory transcription factors from binding regulatory elements in DNA and can also lead to binding of proteins such as MeCP2 that lead to chromatin condensation via histone deacetylases and chromatin remodeling proteins.

the X chromosome, one of the two copies that girls receive is inactivated at random during early embryonic development.

The mechanism of X-inactivation involves an RNA produced by the chromosome destined for inactivation, called *Xist*. How one of the two X chromosomes is chosen for inactivation is under intensive investigation. Once chosen, however, the pathway to inactivation is clear. *Xist* RNA is an example of a **long noncoding RNA (lncRNA)**; that is, once made, the transcript is not translated into a protein. *Xist* RNA is transcribed from X chromosomes that become inactive; once an X chromosome begins to transcribe *Xist* RNA, it eventually becomes largely (although not completely) transcriptionally inactive. Strikingly, the inactive X chromosome is the only chromosome that expresses *Xist*, as can be seen using in situ hybridization to detect *Xist* RNA (Figure 20-18b). How does *Xist* RNA lead to the shutting down of transcription of nearly an entire chromosome? Although some of the details are still being worked out, *Xist* RNA eventually spreads to "coat" the inactive chromosome, which in turn leads to recruitment of chromatin-modifying proteins that cause most of the chromosome to become transcriptionally repressed (Figure 20-18c). *Xist* is an example of a lncRNA that acts in *cis;* that is, it acts on the same chromosome from which it is expressed. As you will learn below, other lcnRNAs act in *trans*, meaning they are produced from one chromosome but can act on another.

Epigenetic Inheritance. Just as specific alleles are inherited from mother and father, it is now clear that chemical modifications to DNA that differ between the oocyte and sperm are also inherited and, in some cases, play important roles in normal embryonic development and subsequent gene expression in humans. The pervasive importance of DNA methylation is made clear by the discovery that a growing number of human diseases, including cancer, are associated with abnormalities in inherited patterns of DNA methylation.

How are differences in DNA methylation stabilized and transmitted? The enzyme responsible for DNA methylation acts preferentially on cytosines situated in 5′–CG–3′ sequences that are base-paired to complementary 3′–GC–5′ sequences that are themselves already methylated. Such pieces of DNA are said to be *hemimethylated* ("hemi" means "half") because only one strand carries methyl modifications. If the old DNA strand in a newly replicated DNA double helix has a methylated 5′–CG–3′ sequence, then the complementary 3′–GC–5′ sequence in the new strand will become a target for methylation. This phenomenon allows DNA methylation patterns to be inherited during successive rounds of DNA replication. The net result is that DNA methylation provides a means for creating **epigenetic changes**—stable alterations in gene expression that are transmitted from one cellular generation

HUMAN *Connections*

The Epigenome: Methylation and Disease

Imagine a young child, no more than 18 months old, trying to walk and speak as best she can. But then her newly gained abilities begin to slip away. Over time, she begins to develop symptoms of cerebral palsy, autism, epilepsy, and scoliosis (curvature of the spine). Does she have several separate disorders, or do these symptoms indicate a single disease?

In fact, the little girl has *Rett syndrome,* a neurodegenerative disorder of gray matter in the brain that overwhelmingly affects girls. Because the defective gene associated with Rett syndrome is X-linked, girls, with their two X chromosomes, are more likely to survive. Males can develop Rett syndrome but are more likely to die at a young age because they have only a single X chromosome.

Rett syndrome is most often caused by mutations in the gene that encodes *MeCP2* (methyl-CpG-binding protein 2). MeCP2 plays a crucial role in the transcriptional regulation of expression of neuron-specific genes during postnatal development. Normally MeCP2 binds to methylated CpG islands near promoters, recruiting other proteins, whose combined effects lead to chromatin condensation and inactivation (**Figure 20A-1**).

Many different mutations in *MeCP2* have been characterized. The typical result is the expression of genes that would normally be transcriptionally silenced, which causes the unique set of symptoms found in Rett syndrome patients.

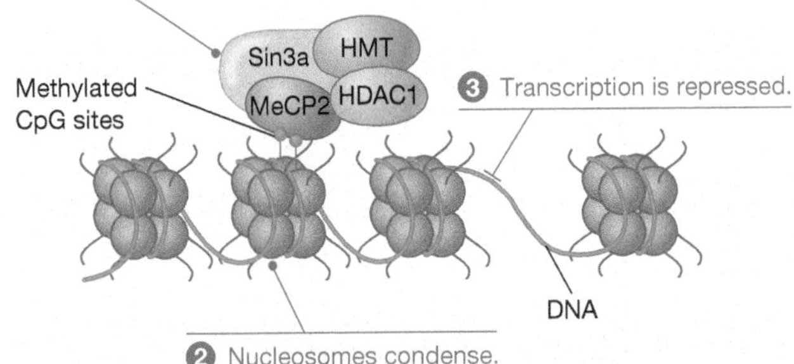

Figure 20A-1 Regulation of Transcription by MeCP2. 1 Bound MeCP2 recruits histone deacetylases, such as HDAC1, histone methyltransferases, and chromatin remodeling factors, such as Sin3a, that lead to 2 chromatin compaction and 3 transcriptional silencing.

As technology for examining the human genome advances, more and more diseases have been attributed to abnormal DNA methylation patterns. An example is *DNMT1,* a maintenance methylase that is active just after replication, ensuring that daughter strands are methylated in the same pattern as parent strands. Mutations in the *DNMT1* gene that render the protein nonfunctional can result in expression of genes that would otherwise be silenced. Various cancers and some instances of congenital heart defects are also associated with defects in DNA methylation.

Maintenance of DNA methylation patterns also plays an important role in *genomic imprinting,* the differential expression of genes in offspring based on the parental origin of the gene. In mammals, imprinting results when the copy of a gene from one parent is expressed and the copy from the other parent is silenced through extensive methylation. The process of imprinting in some genes is thought to minimize *dosage effects,* in which two active copies of the gene would be detrimental to the offspring.

One well-studied example of imprinting in humans involves chromosome 15 (**Figure 20A-2**). Only one of the two copies of several genes on chromosome 15 in any cell is active. The copy that is active is descended from either the maternal or paternal chromosome of the fertilized egg. *Prader–Willi syndrome* is caused by deletion of a region of chromosome 15 on the copy of the chromosome inherited from the father, resulting in minor mental retardation, compulsive eating, obesity, and gonadal atrophy. Conversely, *Angelman syndrome* is caused when a nearby deletion occurs in the maternally provided copy of chromosome 15, resulting in frequent laughter, smiling, and characteristic hand-wringing movements. One of the culprits in Angelman syndrome is the gene *UBE3A,* which encodes an *E3 ubiquitin ligase,* involved in protein degradation (see Figure 20-38). In people with Angelman syndrome, proteins accumulate in excess, including a protein found at synapses between neurons in the brain.

The symptoms of diseases that involve these kinds of *epigenetic* regulation of gene expression pose impressive obstacles to therapy. However, some strides have been made. Researchers have introduced a functional copy of the *MeCP2* gene into neurons in a mouse model of Rett syndrome, leading to improved neuronal structure. In a mouse model of Angelman syndrome, drugs that inhibit specific *topoisomerases,* enzymes that relieve DNA supercoiling (see Chapter 16), allow expression of the alternate copy of the *UBE3A* gene.

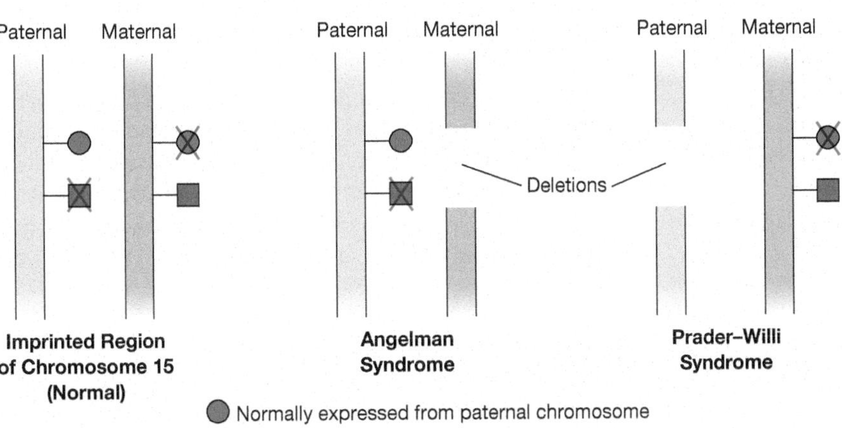

Figure 20A-2 Imprinting on Chromosome 15 and Human Disease. Two regions of chromosome 15 are critical to Angelman and Prader–Willi syndromes. Squares represent genes associated with Angelman syndrome, circles represent genes associated with Prader–Willi syndrome, and red Xs represent inactive genes due to imprinting. Angelman syndrome genes are inactive on the paternal copy of chromosome 15; conversely, Prader–Willi genes are inactive on the maternal copy. The syndromes result when the offspring inherits a chromosome with a deletion of the region of chromosome 15 normally expressed from the paternally derived (Prader–Willi) or maternally derived (Angelman) chromosomes.

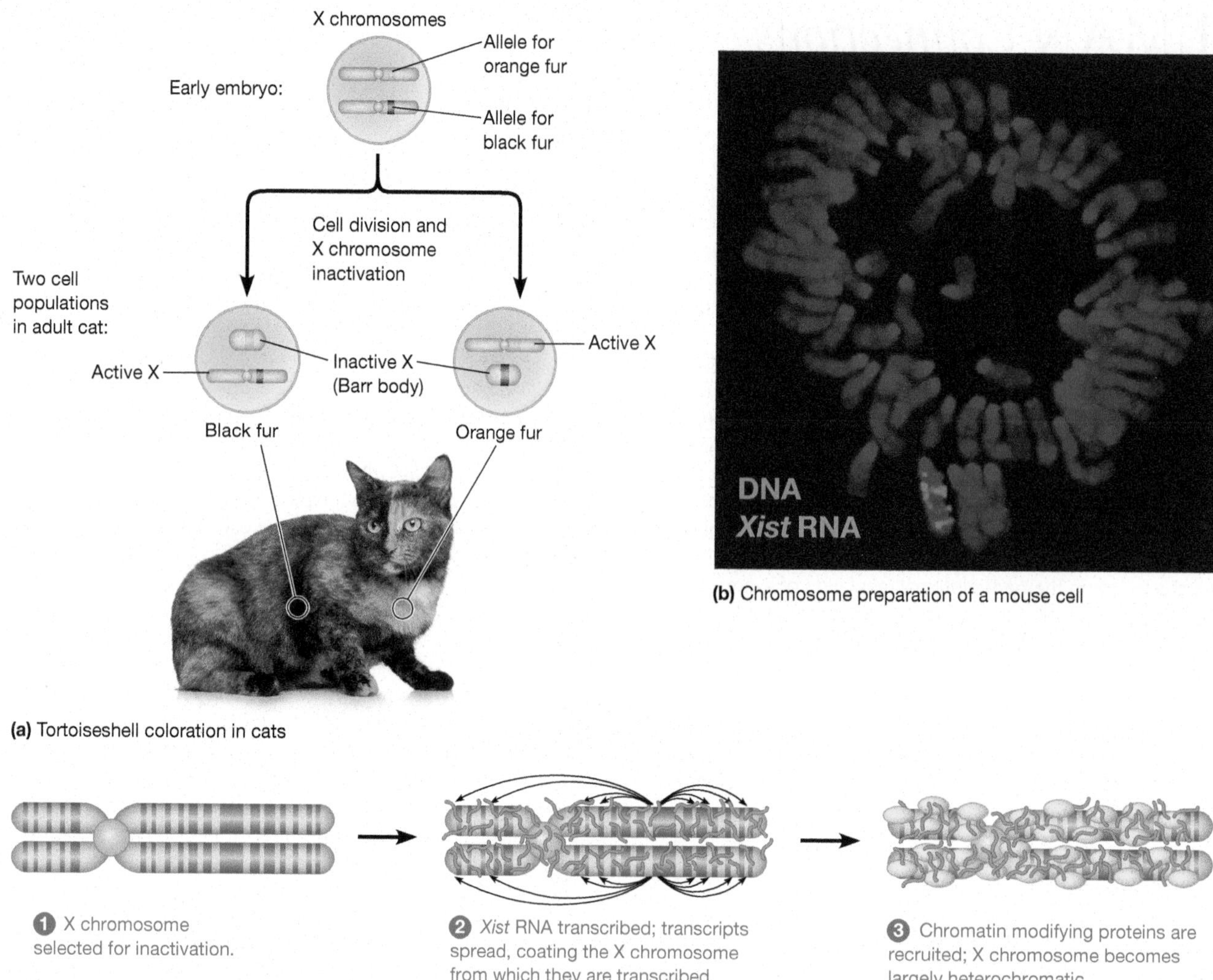

(a) Tortoiseshell coloration in cats

(b) Chromosome preparation of a mouse cell

(c) Mechanism of X-inactivation

Figure 20-18 X Chromosome Inactivation in Mammals. (a) The gene responsible for tortoiseshell coloration in cats is on the X chromosome. To display the tortoiseshell phenotype, a female cat must have one X chromosome with the orange fur allele and one with the black fur allele. Orange patches are formed from cells in the embryo derived from cells in which the X chromosome containing the orange allele is active, black patches from cells in which the black allele is active. The inactive X chromosome forms a Barr body. **(b)** A chromosome preparation from mouse cells (stained blue with a DNA dye) in which in situ hybridization has been performed to detect *Xist* transcripts (pink). One X chromosome expresses *Xist*. **(c)** In the mechanism of X-inactivation, ❶ one of the two X chromosomes is selected for inactivation, leading to ❷ *Xist* transcription and ❸ chromatin condensation.

to the next without requiring changes in a gene's underlying base sequence. Such epigenetic changes also underlie genomic **imprinting**, a process that causes certain genes to be expressed differently depending on whether they are inherited from a person's mother or father. (Human Connections, page 593, discusses the consequences of these sorts of epigenetic changes for human disease in more detail.)

Furthermore, modifications (such as methylation and acetylation) made to histone proteins (see Chapter 16) are also heritable. When cells copy their DNA and divide, the histones must disassemble ahead of the replication fork and reassemble behind it. Often, the posttranslational modifications made to the histones in the parental cell are "rewritten" onto the newly assembled nucleosomes. In this way, changes in gene expression that result from histone modifications can be inherited by daughter cells postreplication.

Inheritance is not the only factor that influences DNA methylation patterns: studies involving identical twins have revealed that environment plays a role as well. Identical twins exhibit DNA methylation patterns that are virtually indistinguishable when the twins are young, but significant differences emerge as twins grow older. The greatest differences in DNA methylation, and in the pattern of genes being expressed, are seen in twins who spend large amounts of time apart, suggesting that environmental factors influence DNA methylation and, in turn, gene expression.

MAKE CONNECTIONS 20-2

After cell division, a histone deacetylase is recruited to remove acetyl groups from histones associated with a specific gene. What can you reasonably conclude about how this epigenetic change affects the expression of this gene? (Fig. 16-21)

CONCEPT CHECK 20-2

How do epigenetic changes, such as DNA methylation, differ from genetic alterations to DNA, including various kinds of mutations (see Chapters 17 and 18)?

20.3 Eukaryotic Gene Regulation: Transcriptional Control

We have seen that the structure of the genome influences gene expression. However, the existence of structural changes associated with active regions of the genome does not address the underlying question of how the DNA sequences contained in these regions are actually selected for activation. The answer is to be found in the phenomenon of **transcriptional control**, the second main level for controlling eukaryotic gene expression (see Figure 20-11).

Different Sets of Genes Are Transcribed in Different Cell Types

Transcriptional control is the most common—and most crucial—level of control over gene expression in eukaryotes. The first direct evidence for the importance of transcriptional regulation in eukaryotes came from experiments comparing newly synthesized RNA in the nuclei of different mammalian tissues. Liver and brain cells, for example, produce different sets of proteins, although there is considerable overlap between the two. How can we determine the source of this difference? If the different proteins produced by the two cell types are a reflection of *differential gene transcription*—that is, transcription of different genes to produce different sets of RNAs—we should see corresponding differences between the populations of nuclear RNAs derived from brain and liver cells. On the other hand, if all genes are equally transcribed in liver and brain, we would find few, if any, differences between the populations of nuclear RNAs from the two tissues, and we would conclude that the tissue-specific differences in protein synthesis were due to posttranscriptional mechanisms that control the ability of various RNAs to be translated.

One way of distinguishing between these alternatives is to use the technique of *nuclear run-on transcription*, which provides a snapshot of the transcriptional activity occurring in a nucleus at a given moment in time (**Figure 20-19**). Transcriptionally active nuclei are gently isolated from cells and allowed to complete synthesis of RNA molecules in the presence of radioactively labeled nucleoside triphosphates. When such an experiment is performed using liver and brain cells, the newly transcribed (radioactively labeled) RNAs in the liver nuclei contain sequences from liver-specific genes, but these liver-specific sequences are not detected in the labeled RNAs synthesized by isolated brain nuclei. Likewise, the

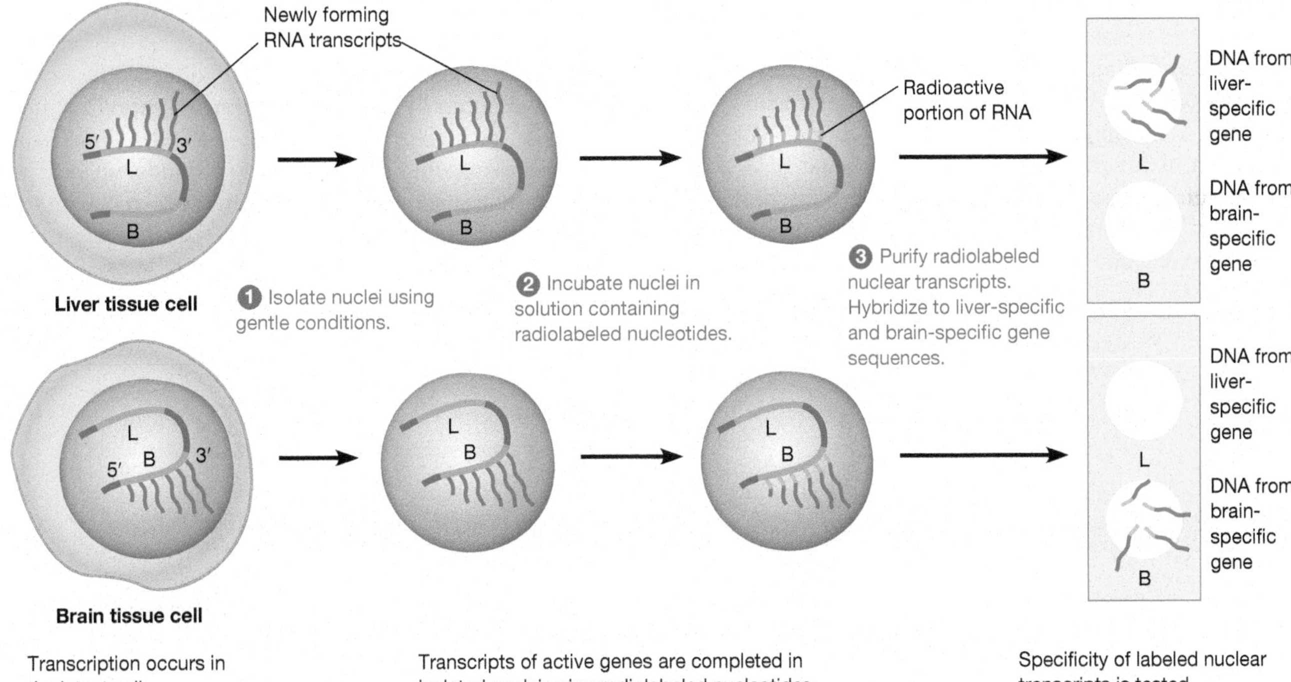

Figure 20-19 Demonstration of Differential Transcription by Nuclear Run-on Transcription Assays. In these studies, B represents a hypothetical gene that is expressed only in brain tissue, and L is a gene expressed only in liver tissue. ❶ Isolated nuclei are ❷ incubated in a solution containing radioactively labeled ribonucleotides, which become incorporated into the mRNA being synthesized by active genes. If different genes are active in liver and brain tissue, some labeled sequences in the liver nuclear transcripts will not be present in brain transcripts and vice versa. ❸ The composition of the labeled RNA population is assayed by allowing the labeled RNA to hybridize with DNA sequences representing different genes that have been attached to a filter-paper support. Labeled liver transcripts hybridize with a different set of genes than do labeled brain transcripts, indicating that the identities of the active genes in the two tissues differ.

newly transcribed (radioactively labeled) RNAs in the brain nuclei contain sequences from brain-specific genes, but these brain-specific sequences are not detected in the labeled RNAs synthesized by isolated liver nuclei.

These experiments indicate that gene expression is being regulated at the transcriptional level. To state this as an important general principle, *different cell types transcribe different sets of genes*, thereby allowing each cell type to produce those proteins needed for carrying out that cell's specialized functions. These classic experiments set the stage for what we now understand about transcriptional control. Modern molecular techniques can now analyze transcripts from many different genes rapidly and confirm this general conclusion (you will learn about these technologies in Chapter 21).

Proximal Control Elements Lie Close to the Promoter

How is transcription regulated such that only certain cell types transcribe particular genes? In our discussion we will focus our attention on protein-coding genes, which are transcribed by RNA polymerase II. The specificity of transcription—that is, where on the DNA it is initiated—is determined not by RNA polymerase II itself, but by proteins called *transcription factors* (see Chapter 18). (Unlike bacterial sigma factors, which also determine initiation specificity, none of the eukaryotic transcription factors are considered to be an integral part of an RNA polymerase molecule.) *General transcription factors* are essential for the transcription of *all* the genes transcribed by a given type of RNA polymerase.

For genes transcribed by RNA polymerase II, the general transcription factors assemble with RNA polymerase at the *core promoter*, a DNA region located in the immediate vicinity of the transcriptional start site (see Figure 18-9b). The interaction of general transcription factors and RNA polymerase with the core promoter often initiates transcription at only a low, "basal" rate so that few transcripts are produced. However, in addition to a core promoter, most protein-coding genes have short DNA sequences farther upstream (and, in some cases, downstream) to which other specific transcription factors bind, thereby improving the efficiency of the core promoter. If these additional DNA elements are deleted experimentally, the frequency and accuracy of transcription are reduced.

In discussing such regulatory DNA sequences, we will use the term **proximal control elements** to refer to sequences located upstream of the core promoter but within about 100–200 base pairs of it. The number, exact location, and identities of these proximal control elements vary with each gene, but three types are especially common: the *CAAT box*, the *GC box*, and the *octamer* (the first two were discussed in Chapter 18 and are illustrated in **Figure 20-20**). Transcription factors that selectively bind to one of these, or to other control sequences located outside the core promoter, are called **regulatory transcription factors**. They increase (or decrease) transcriptional initiation by interacting with components of the transcription apparatus.

Enhancers and Silencers Are DNA Elements Located at Variable Distances from the Promoter

Proximal control elements, like most bacterial control elements, lie close to the core promoter on its upstream side. A second class of DNA control sequences are located either upstream or downstream from the genes they regulate and often lie far away from the promoter (**Figure 20-21**). This second type of control region is called an **enhancer** if it stimulates gene transcription or a **silencer** if it inhibits transcription. Subsequently, regions that restrict the influence of enhancers and silencers, known as **insulators**, were identified. Originally, such control sequences were called *distal control elements* because they can function at distances up to several hundred thousand base pairs upstream or downstream from the promoter they regulate (the word *distal* means "away from"). Enhancers and silencers need not be located at great

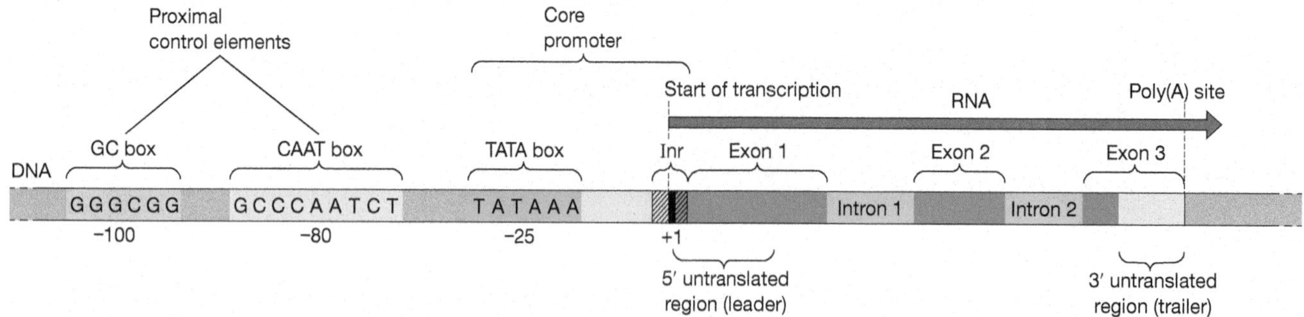

Figure 20-20 Anatomy of a Typical Eukaryotic Gene, with Its Core Promoter and Proximal Control Region. This diagram (not to scale) features a typical protein-coding eukaryotic gene, which is transcribed by RNA polymerase II. The core promoter is characterized by an initiator (Inr) sequence surrounding the transcriptional start site and a TATA box located about 25 bp upstream (to the 5′ side) of the start site. The core promoter is where the general transcription factors and RNA polymerase assemble for the initiation of transcription. Within about 100 nucleotides upstream from the core promoter lie several proximal control elements, which stimulate transcription of the gene by interacting with regulatory transcription factors. The number, identity, and exact location of the proximal elements vary from gene to gene. This proximal control region contains one copy of each of two common elements, the GC box and the CAAT box. The transcription unit includes a 5′ untranslated region (leader) and a 3′ untranslated region (trailer), which are transcribed and included in the mRNA but do not contribute sequence information for the protein product. At the end of the last exon is a site where, in the primary transcript, the RNA will be cleaved and given a poly(A) tail.

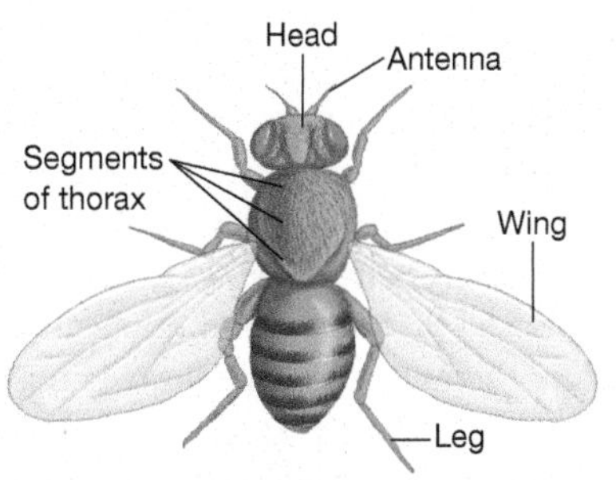

(a) **Wild-type *Drosophila*.** Two wings and six legs extend from its three thorax segments.

(b) ***Bithorax* mutant.** An additional set of wings is formed.

(c) ***Antennapedia* mutant.** Legs develop where the antennae should be.

Figure 20-30 Homeotic Mutants of *Drosophila*. **(a)** A non-mutated (wild-type) fly is pictured. **(b)** Mutations in the *bithorax* gene complex convert the third thoracic segment to a second thoracic segment (the wing-producing segment), so an additional set of wings is formed. **(c)** Mutations in the *antennapedia* gene complex cause legs to develop where the antennae should be.

is mediated by the binding of a protein called the *heat shock transcription factor* to the heat shock response element. The heat shock transcription factor is present in an inactive form in nonheated cells, but elevated temperatures cause a change in the structure of the protein that allows it to bind to the heat shock response element in DNA. The protein is then further modified by phosphorylation, which makes it capable of activating gene transcription.

The heat shock system again illustrates a basic principle of eukaryotic coordinate regulation: genes located at different chromosomal sites can be activated by the same signal if the same response element is located near each of them.

Homeotic Genes Encode Transcription Factors That Regulate Embryonic Development

Some especially striking examples of coordinated gene regulation in eukaryotes involve an unusual class of genes in *Drosophila melanogaster* known as **homeotic genes**. When mutations occur in one of these genes, a strange thing happens during embryonic development: one part of the body is replaced by a structure that normally occurs somewhere else (**Figure 20-30**). The discovery of homeotic genes can be traced back to the 1940s, when Edward B. Lewis discovered a cluster of genes, the *bithorax gene complex*, in which certain mutations cause drastic developmental abnormalities such as the growth of an extra pair of wings (Figure 20-30a, b). Later, Thomas C. Kaufman and his colleagues discovered a second group of genes that, when mutated, lead to different but equally bizarre developmental changes—for example, causing legs to grow from the fly's head in place of antennae (Figure 20-30a, c). This group of genes is the *antennapedia gene complex*. The *bithorax* and *antennapedia* genes are called homeotic genes because *homeo* means "alike" in Greek, and mutations in these genes change one body segment of *Drosophila* to resemble another. A clue to how homeotic genes work first emerged from the discovery that the *bithorax* and *antennapedia* genes each contain a similar, 180-bp sequence near their 3′ end that resembles a comparable sequence found in other homeotic genes. Termed the **homeobox**, this DNA sequence encodes a stretch of 60 amino acids called a **homeodomain** (**Figure 20-31**).

Although such phenomena might at first appear to be no more than oddities of nature, the growth of an extra pair of wings or legs in the wrong place suggests that homeotic genes play a key role in determining the body plan of developing fly embryos. The proteins encoded by the homeotic genes are a family of regulatory transcription factors that activate (or inhibit) the transcription of developmentally important genes by binding to specific DNA sequences. By binding to all copies of the appropriate DNA element in the genome, each homeotic transcription factor can influence the expression of dozens or even hundreds of genes in the growing embryo. The result of this coordinated gene regulation is the establishment of fundamental body characteristics such as appendage shape and location.

A remarkable result of modern genomics and molecular developmental biology is the high degree of conservation of

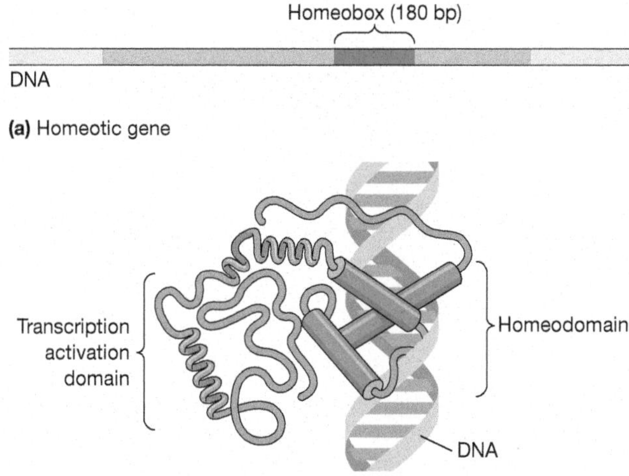

Figure 20-31 Homeotic Genes and Proteins. Homeotic mutants like the ones in Figure 20-30 have mutations in homeotic genes. **(a)** Homeotic genes all contain a 180-bp sequence, called a homeobox, that encodes a homeodomain in the homeotic protein. **(b)** The homeodomain (red) is a helix-turn-helix DNA-binding domain that, in combination with a transcription activation domain, enables the protein to function as a regulatory transcription factor. Notice that the homeodomain has three α helices (shown as cylinders).

homeotic genes in other animals. Sequences similar to those in the homeotic genes of flies have been detected in the genes of organisms as diverse as nematode worms, insects, sea squirts, frogs, mice, and humans—an evolutionary distance of more than 500 million years. Remarkably, with a few minor exceptions, the physical organization of these *Hox* genes has been retained across these great distances. In both flies and vertebrates, the 3′-5′ location of these genes within a chromosome corresponds to the anterior to posterior region along the body axis these proteins regulate (**Figure 20-32**). In vertebrates, there appears to have been a duplication of the original single set of these genes possessed by fruit flies and other simple animals: whereas fruit flies have one set of homeotic genes, mammals have four sets of *Hox* genes. Based on analysis of

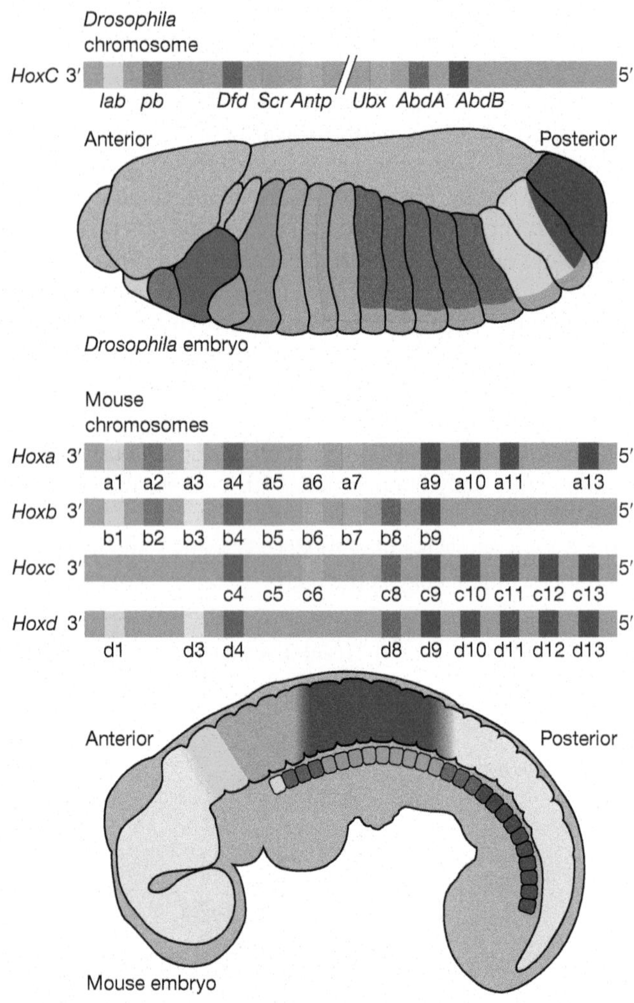

Figure 20-32 Organization and Expression of *Hox* Genes in *Drosophila* and the Mouse. In *Drosophila* the homeotic gene cluster is organized along the chromosome such that genes required for the identity of anterior structures are at the 3′ end of the cluster and genes required for the identity of posterior structure are at the 5′ end. The physical order of homeotic genes on the chromosome corresponds to the domain whose identity each gene controls. Note that the mouse has four *Hox* complexes, labeled a–d. Corresponding genes in each of the four *Hox* clusters in the mouse are numbered, with the lowest numbers being the most anteriorly expressed. Note that some genes are not found in some clusters. Genes with the same color represent similar genes. Note that the single *AbdB* gene in flies (dark purple) appears to have been duplicated multiple times in the mouse (genes 9–13).

the expression pattern of mouse *Hox* genes and the effects of making "knockout" mutants for individual and multiple *Hox* genes, it is clear that, like their fly counterparts, they control the identity of structures along the anterior-posterior axis of vertebrates. Mutations in human *Hox* genes have also been identified. For example, mutations in the *HoxD13* locus result in *synpolydactyly*, characterized by fusions and duplications of fingers and toes. A major theme of modern efforts to unite evolutionary and developmental biology is to understand how subtle changes in *Hox* genes lead to changes in the body plans of higher animals.

CONCEPT CHECK 20-3

Compare the basic "state" of a gene in bacteria versus that in eukaryotes with respect to activation of transcription. How are the states similar? How are they different?

20.4 Eukaryotic Gene Regulation: Posttranscriptional Control

We have now examined the first two levels of regulation for eukaryotic gene expression, namely, genomic controls and transcriptional controls. After transcription has taken place, the flow of genetic information involves a complex series of *posttranscriptional events*, any or all of which can turn out to be regulatory points (see Figure 20-11, ❸, ❹, and ❺). Posttranscriptional regulation may be especially useful in providing ways to rapidly fine-tune the pattern of gene expression, allowing cells to respond to transient changes in the intracellular or extracellular environment without changing their overall transcription patterns.

Control of RNA Processing and Nuclear Export Follows Transcription

Posttranscriptional control begins with the processing of primary RNA transcripts, which provides many opportunities for the regulation of gene expression. Virtually all RNA transcripts in eukaryotic nuclei undergo substantial processing, including addition of a 5′ cap and a 3′ poly(A) tail, chemical modifications such as methylation, splicing together of exons accompanied by the removal of introns, and RNA editing (see Chapter 18). Among these processing events, RNA splicing is an especially important control point because its regulation allows cells to create a variety of different mRNAs from the same pre-mRNA, thereby permitting a gene to generate more than one protein product.

This phenomenon, called **alternative splicing**, is based on a cell's ability to permit some splice sites to be skipped and others to become activated (see Figure 18-21). Roughly half of all human genes produce pre-mRNAs that are alternatively spliced, reinforcing the view that splicing control is an important component of gene regulation. Splicing patterns are determined by various factors, including the presence of regulatory proteins and small nucleolar RNA molecules (snoRNAs) that bind to *splicing enhancer* or *splicing silencer* sequences in pre-mRNA. By increasing the ways in which a given pre-mRNA can be spliced, these regulatory molecules make it possible for a single gene to yield dozens or even hundreds of alternate mRNAs. A remarkable example occurs in the inner ear of birds, where 576 alternatively spliced forms of mRNA are produced from a single potassium channel gene. Expression of varying subsets of these mRNAs in different cells of the inner ear facilitates perception of different sound frequencies.

Figure 20-33 illustrates another way in which the same gene can produce different transcripts that in turn lead to different proteins. This example involves mRNAs encoding a type of antibody called *immunoglobulin M (IgM)*. The IgM protein exists in two forms, a secreted version and a version that becomes incorporated into the plasma membrane of the B lymphocyte that makes it. Like all antibodies, IgM consists of four polypeptide subunits—two heavy chains and two light chains. It is the sequence of the heavy chain that determines whether IgM is secreted or membrane bound. The gene encoding the heavy chain has two alternative sites where RNA transcripts can terminate, yielding two kinds of pre-mRNA molecules that differ at their 3′ ends. The exons within these pre-mRNAs are then spliced together in two different ways, producing mRNAs that encode either the secreted version of the heavy chain or the version that is bound to the plasma membrane. Only the splicing pattern for the membrane-bound version includes the exons encoding the hydrophobic amino acid domain that anchors the heavy chain to the plasma membrane.

After RNA splicing, the next posttranscriptional step subject to control is the export of mRNA through nuclear pores to the cytoplasm. We know that this export step can be controlled because RNAs exhibiting defects in capping or splicing are not readily exported from the nucleus. Even with normal mRNAs, certain molecules are retained in the nucleus until their export is triggered by a stimulus signaling that a particular mRNA is needed in the cytoplasm. Export control has also been observed with viral RNAs produced in infected cells. For example, HIV—the virus responsible for AIDS—produces an RNA molecule that remains in the nucleus until a viral protein called Rev is synthesized. The amino acid sequence of the Rev protein includes a *nuclear export signal* that allows Rev to guide the viral RNA out through the nuclear pores and into the cytoplasm.

Translation Rates Can Be Controlled by Initiation Factors and Translational Repressors

Once mRNA molecules have been exported from the nucleus to the cytosol, several **translational control** mechanisms are available to regulate the rate at which each mRNA is translated into its polypeptide product. Some translational control mechanisms work by altering ribosomes or protein synthesis factors; others regulate the activity and/or stability of mRNA itself. Translational control takes place in prokaryotes as well as eukaryotes, but we will restrict our discussion to several eukaryotic examples.

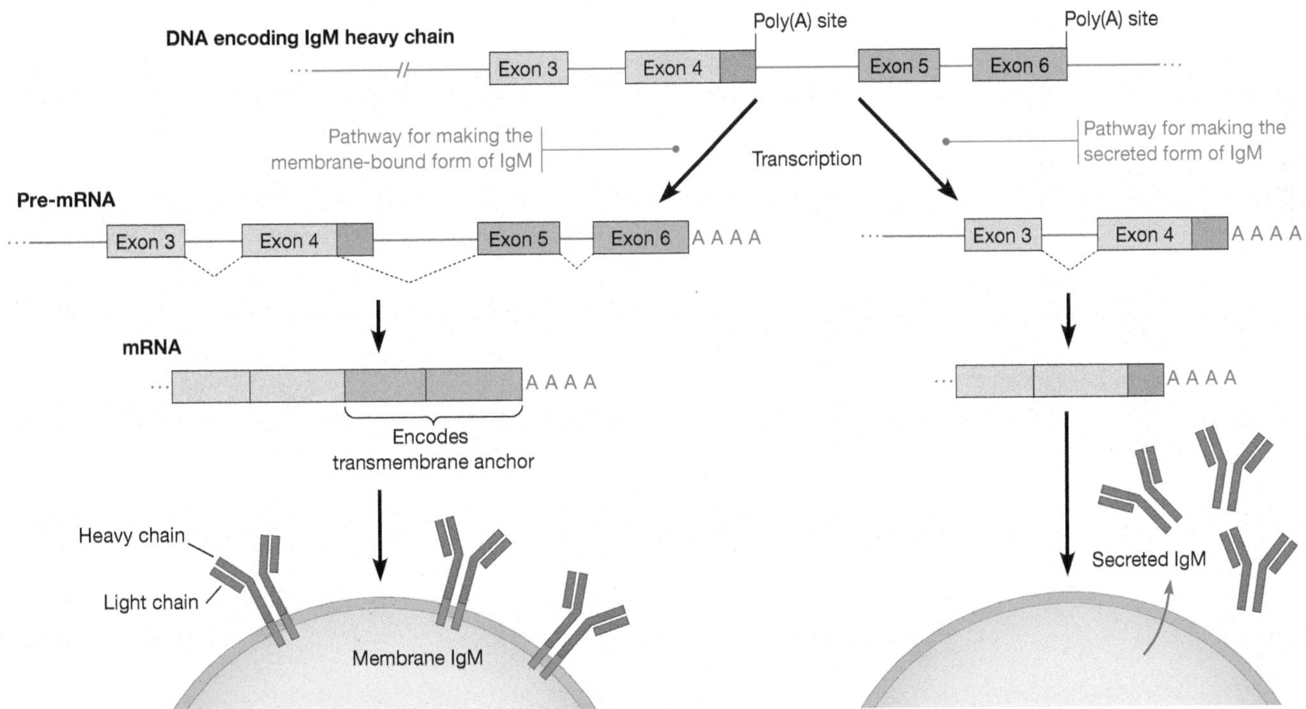

Figure 20-33 Using the Same Gene to Produce Variant Proteins. The antibody protein immunoglobulin M (IgM) exists in two forms, secreted IgM and membrane-bound IgM, which differ in the carboxyl ends of their heavy chains. A single gene carries the genetic information for both types of IgM heavy chain. The end of the gene corresponding to the polypeptide's C-terminus is shown at the top of the figure. This DNA has two possible poly(A) addition sites, where RNA transcripts can terminate, and four exons that can be spliced in two alternative configurations. The splices made in the primary transcripts are indicated by V-shaped, thin dashed lines beneath each pre-mRNA. A splicing pattern that uses a splice junction within exon 4 and retains exons 5 and 6 results in the synthesis of membrane IgM heavy chains that are held in the plasma membrane by a transmembrane anchor encoded by exons 5 and 6. The alternative product is secreted because a splice within exon 4 is not made and the transcript is terminated after exon 4.

One well-studied example of translational control occurs in developing erythrocytes, where *globin* polypeptide chains are the main product of translation. The synthesis of globin depends on the availability of *heme*—the iron-containing prosthetic group that attaches to globin chains to form the final protein product, hemoglobin. Developing erythrocytes normally synthesize globin at a high rate. However, globin synthesis would be wasteful if not enough heme were available to complete the formation of hemoglobin molecules, so erythrocytes have a mechanism for adjusting polypeptide synthesis to match heme availability.

This mechanism involves a protein kinase, called *heme-controlled inhibitor (HCI)*, whose activity is regulated by heme. HCI is inactive in the presence of heme. But when heme is absent, HCI phosphorylates and inhibits eIF2, one of several proteins required for initiating translation in eukaryotes (see Figure 19-16a). The reduced activity of eIF2 depresses all translational activity, but because globin chains account for more than 90% of the polypeptides made in developing erythrocytes, the main impact of HCI in these cells is on globin synthesis. Other kinases that phosphorylate eIF2 have been detected in a diverse array of cell types, suggesting that phosphorylation of eIF2 is a widely used mechanism for regulating translation.

The use of protein phosphorylation to control translation is not restricted to eIF2. Eukaryotic initiation factor eIF4F (see Figure 19-16a), a multiprotein complex (composed of eIF4E, eIF4G, and eIF4A) that binds to the 5′ mRNA cap, is also regulated by phosphorylation, although in this case phosphorylation activates rather than inhibits the initiation factor. An example of this type of control is observed in cells infected by *adenovirus*, which inhibits protein synthesis in infected cells by blocking phosphorylation of eIF4E, the cap-binding subunit of eIF4F. Other viruses inhibit the eIF4F complex by producing proteases that cleave eIF4F into two fragments, one containing the site that binds to the 5′ cap of mRNA molecules and the other containing the site that binds to ribosomes. Surprisingly, the ribosome-binding fragment can still initiate the translation of viral (but not cellular) mRNA by binding to an *internal ribosome entry sequence* that lies directly upstream of the start codon of some viral mRNAs. This allows a virus to hijack the translation machinery for translating its own mRNA while simultaneously shutting down the translation of normal cellular mRNAs.

The types of translational control we have discussed so far are relatively nonspecific in that they affect the translation rates of all mRNAs within a cell. An example of a more specific type of translational control is seen with *ferritin*, an iron-storage protein whose synthesis is selectively stimulated in the presence of iron. The key to this selective stimulation lies in the 5′ untranslated sequence of ferritin mRNA, which contains a 28-nucleotide segment—the **iron-response element (IRE)**—that forms a hairpin loop required for the stimulation of ferritin synthesis by iron. If the IRE sequence is experimentally inserted into a gene whose expression is not normally regulated by iron, translation of the resulting mRNA becomes iron sensitive.

Figure 20-34 shows how the IRE sequence works. When the iron concentration is low, a regulatory protein called the *IRE-binding protein* binds to the IRE sequence in ferritin mRNA, preventing the mRNA from forming an initiation complex with ribosomal subunits. But the IRE-binding protein is an allosteric protein whose activity can be controlled by the binding of iron. When more iron is available, the protein binds an iron atom and undergoes a conformational change that prevents it from binding to the IRE, thereby allowing the ferritin mRNA to be translated. The IRE-binding protein is therefore an example of a **translational repressor**, a protein that selectively controls the translation of a particular mRNA. This type of translational control allows cells to respond to specific changes in the environment faster than would be possible through the use of transcriptional control.

Other proteins act as translational repressors but bind the *3′ untranslated regions (3′-UTRs)* of specific mRNAs. One example of this type of control is the *PUF* proteins, which regulate the translation of proteins important during development. In addition to translational repressor proteins, we will see below that an important class of small RNAs—*microRNAs*—also interacts with the 3′-UTR to regulate translation.

Translation Can Also Be Controlled by Regulation of mRNA Degradation

Translation rates are also subject to control by alterations in mRNA stability—in other words, if an mRNA molecule is degraded more rapidly, then less time is available for it to be translated. Degradation rates can be measured by *pulse-chase*

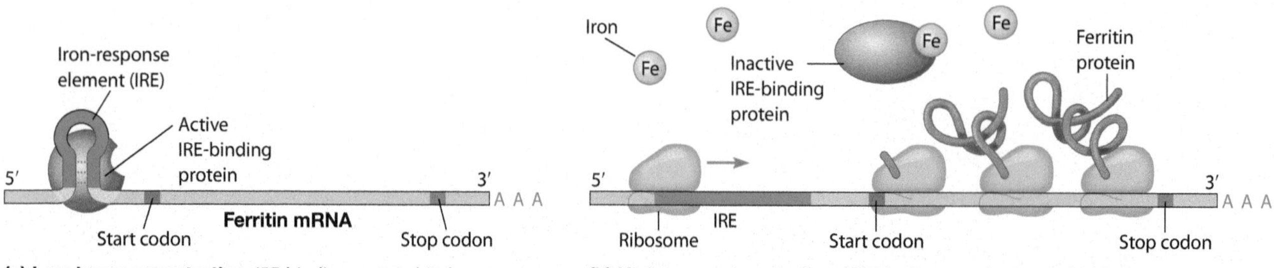

Iron-response element (IRE)
Active IRE-binding protein
5′
3′
A A A
Ferritin mRNA
Start codon
Stop codon
(a) Low iron concentration. IRE-binding protein binds to IRE, so translation of ferritin mRNA is inhibited.
Iron
Fe
Fe
Fe
Fe
Inactive IRE-binding protein
Ferritin protein
5′
3′
A A A
Ribosome
IRE
Start codon
Stop codon
(b) High iron concentration. IRE-binding protein cannot bind to IRE, so translation of ferritin mRNA proceeds.

Figure 20-34 Translational Control in Response to Iron. (a) An iron-response element (IRE) located in the 5′ untranslated region of ferritin mRNA can form a hairpin structure that is bound by the IRE-binding protein, which inhibits the initiation of translation. **(b)** When the iron concentration is high, iron binds to the IRE-binding protein and changes it to a conformation that does not recognize the IRE. Ribosomes can therefore assemble on the ferritin mRNA and translate it.

experiments, in which cells are first incubated for a brief period of time (the "pulse") in the presence of a radioactive compound that becomes incorporated into mRNA. The cells are then placed in a nonradioactive medium, and incubation is continued (the "chase"). No additional radioactivity is taken up by the cell during the chase period, so the fate of the radioactivity previously incorporated into mRNA can be measured over time. This approach allows researchers to measure an mRNA's *half-life,* which is the time required for 50% of the initial (radioactive) RNA to be degraded.

The half-life of eukaryotic mRNAs varies widely, ranging from 30 minutes or less for some growth factor mRNAs to over 10 hours for the mRNA encoding β-globin. The length of the poly(A) tail is one factor that plays a role in controlling mRNA stability. Messenger RNAs with short poly(A) tails tend to be less stable than mRNAs with longer poly(A) tails. In some cases, mRNA stability is also influenced by specific features of the 3′-UTR. For example, short-lived mRNAs for several growth factors have particular *AU-rich elements* in this region. The AU-rich sequences trigger removal of the poly(A) tail by degradative enzymes. When an AU-rich element is transferred to the 3′ end of a normally stable globin message using recombinant DNA techniques, the hybrid mRNA acquires the short half-life typical of a growth factor mRNA.

Another way of regulating mRNA stability is illustrated by the action of iron, which, in addition to participating in translational control of ferritin synthesis as just described, plays a role in controlling mRNA degradation as well. The uptake of iron into mammalian cells is mediated by a plasma membrane receptor protein called the *transferrin receptor.* When iron is scarce, synthesis of the transferrin receptor is stimulated by a mechanism that protects transferrin receptor mRNA from degradation, thereby making more mRNA molecules available for translation. As shown in **Figure 20-35,** this control mechanism involves an IRE (similar to the one in ferritin mRNA) located in the 3′-UTR of transferrin receptor mRNA. When the intracellular iron level is low and increased uptake of iron is necessary, the IRE-binding protein binds to this IRE and protects the mRNA from degradation. When iron levels in the cell are high and additional uptake is not necessary, the IRE-binding protein binds an iron atom and dissociates from the mRNA, allowing the mRNA to be degraded.

Two general mechanisms exist for degrading mRNAs that have been targeted for destruction. The main difference is whether the mRNA is degraded in the $3' \rightarrow 5'$ or the $5' \rightarrow 3'$ direction. In the *$3' \rightarrow 5'$ pathway,* shortening of the poly(A) tail and removal of poly(A)-binding proteins from the 3′ end of the mRNA is followed by degradation of the RNA chain by the cytoplasmic **exosome**, a complex of several $3' \rightarrow 5'$ exonucleases. When the exosome eventually arrives at the 5′ end of the mRNA, the 5′ cap is destroyed by a cap-degrading enzyme. In contrast, the *$5' \rightarrow 3'$ pathway* involves shortening of the poly(A) tail at the 3′ end followed by immediate removal of the cap from the 5′ end. The mRNA is then degraded by $5' \rightarrow 3'$ exonucleases. The enzymes involved in this $5' \rightarrow 3'$ pathway are concentrated in cytoplasmic structures called *mRNA processing bodies,* or simply **P bodies**. In addition to their role in mRNA degradation, P bodies can serve as temporary storage sites for nontranslated mRNAs and for recruitment of proteins involved in RNA interference, to which we now turn.

RNA Interference Utilizes Small RNAs to Silence Gene Expression

Regulatory proteins like IRE-binding and PUF proteins, which bind to specific mRNAs, are not the only molecules that control mRNA activity. Individual mRNAs are also controlled by a class of small RNA molecules that inhibit the expression of mRNAs containing sequences related to sequences found in the small RNAs. Such RNA-mediated inhibition, known as **RNA interference (RNAi)**, is based on the ability of small RNAs to trigger mRNA degradation, inhibit mRNA translation, or inhibit transcription of the gene encoding a particular mRNA.

The first type of RNA interference to be discovered occurs in response to the presence of double-stranded RNA.

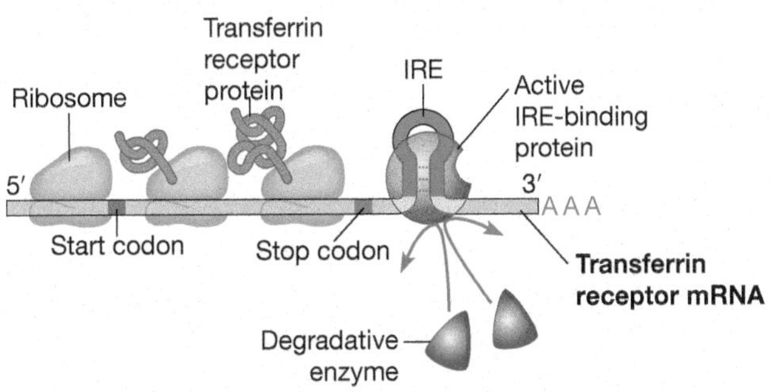

(a) Low iron concentration. IRE-binding protein binds to the IRE of transferrin receptor mRNA, thereby protecting the mRNA from degradation. Synthesis of transferrin receptor therefore proceeds.

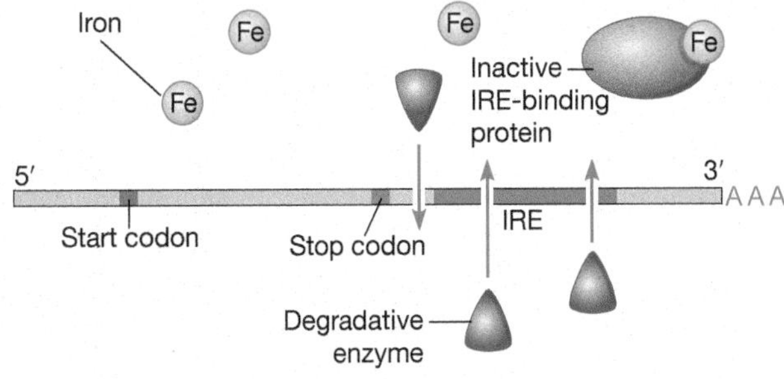

(b) High iron concentration. IRE-binding protein cannot bind to IRE, so mRNA is degraded and synthesis of transferrin receptor is thereby inhibited.

Figure 20-35 Control of mRNA Degradation in Response to Iron. Degradation of the mRNA encoding the transferrin receptor, a protein required for iron uptake, is regulated by the same IRE-binding protein shown in Figure 20-34. **(a)** When the intracellular iron concentration is low, the IRE-binding protein binds to an IRE in the 3′ untranslated region of the transferrin receptor mRNA, protecting the mRNA from degradation and allowing the transferrin receptor protein to be synthesized. **(b)** When the iron level is high, iron atoms bind to the IRE-binding protein, causing it to leave the IRE. The mRNA is then vulnerable to attack by degradative enzymes.

For example, if plants are infected with viruses that produce double-stranded RNA as part of their life cycle, the infected cells limit the infection by reducing the expression of viral genes. Moreover, the effect is not limited to viral genes. If a normal plant gene is inserted into a virus using genetic engineering techniques, cells infected with the virus reduce expression of *their own* copy of the same gene. The causal role played by double-stranded RNA in this type of gene silencing was demonstrated in 1998 by Andrew Fire and Craig Mello, who reported that injecting double-stranded RNA into *C. elegans* greatly reduces the levels of RNA transcripts that contain sequences complementary to those present in the double-stranded RNA. Because their studies laid the groundwork for a burgeoning new field of biology related to the roles played by small RNAs in the control of gene expression, Fire and Mello were awarded a Nobel Prize in 2006.

Figure 20-36 illustrates how double-stranded RNAs knock down the expression of specific genes. First, a cytoplasmic ribonuclease known as **Dicer** cleaves the double-stranded RNA into short fragments about 21–22 base pairs in length. The resulting double-stranded fragments, called **siRNAs** (small interfering, or silencing, RNAs), then combine with a group of proteins to form an inhibitor of gene expression called **RISC** (RNA-induced silencing complex). This particular type of RISC is known as an **siRISC** because its siRNA component determines which gene will be targeted for silencing.

After being incorporated into an siRISC, one of the two strands of the siRNA is degraded. The remaining single-stranded RNA then binds the siRISC to a target mRNA molecule through complementary base pairing. If pairing between the siRNA and the mRNA is a perfect match (or very close), mRNA degradation is initiated by an **Argonaute protein,** a ribonuclease component of the siRISC that cleaves the mRNA in the middle of the complementary site. If the match between the siRNA and mRNA is imperfect, translation of the mRNA may be inhibited without the mRNA being degraded. Additionally, in some cases, the siRISC may enter the nucleus and be guided by its siRNA to complementary nuclear DNA sequences. After associating with these gene sequences, the siRISC silences their expression by stimulating DNA methylation and/or recruiting an enzyme that adds methyl groups to histones, thereby triggering the formation of a transcriptionally inactive, condensed form of chromatin (heterochromatin).

RNAi is sequence-specific by virtue of the complementary base pairing between an siRNA and an mRNA. In addition, a surprisingly small amount of introduced RNA can lead to a very potent RNAi response. One reason for this is that in many cases cells possess an *RNA-dependent RNA polymerase (RdRP)* that amplifies the response. When an siRNA is produced, RdRP is recruited and more siRNA is produced using the original siRNA as a template.

RNA interference originally may have evolved to protect cells from viruses that utilize double-stranded RNA. However, it also turns out to be a powerful laboratory tool that allows scientists to selectively knock down expression of any gene they wish to study. Because complete genome sequences are now available for a variety of organisms, the function of each individual gene can be systematically explored by using RNA interference to turn it off. **Key Technique, pages 612–613,** describes RNAi as an experimental tool in more detail.

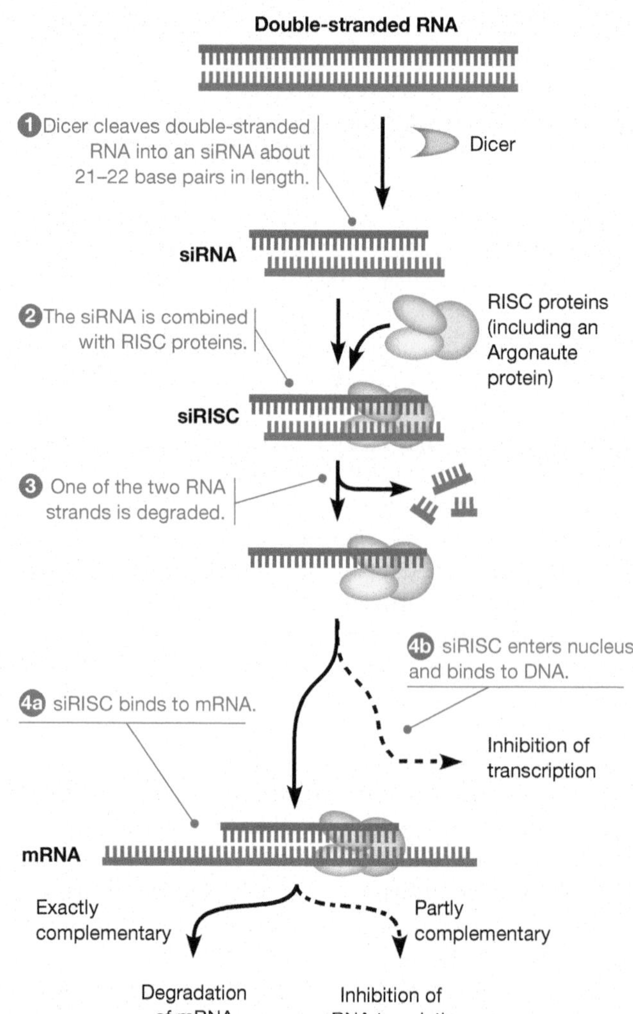

Figure 20-36 RNA Interference by Double-Stranded RNA. When a cell encounters double-stranded RNA, ① Dicer cleaves it into an siRNA, about 21–22 base pairs in length. ② It is subsequently combined with RISC proteins to form an siRISC. ③ After degradation of one of the two siRNA strands, the remaining strand binds the siRISC via complementary base pairing either ④a to a target mRNA molecule in the cytoplasm or ④b to a target DΩNA sequence in the nucleus, thereby silencing gene expression at either the translational or transcriptional level. The most common situation (indicated by the solid arrows) is an exact complementary match between the siRNA and a corresponding mRNA, which triggers mRNA cleavage by an Argonaute protein, an enzyme component of the siRISC.

Another potential application of siRNAs is to use them for silencing genes involved in various diseases. For example, studies in mice have shown that injecting animals with siRNAs directed against a gene involved in cholesterol metabolism can reduce cholesterol levels in the blood. Although questions concerning safety, side effects, and effectiveness remain to be answered, the therapeutic use of siRNAs as drugs for treating or preventing human diseases is currently under intensive investigation.

MicroRNAs Produced by Normal Cellular Genes Silence the Translation of mRNAs

The finding that gene expression can be silenced by introducing double-stranded RNAs into cells raises the question of whether any normal genes produce RNAs that function in a comparable fashion. The search for such molecules has led to the discovery of **microRNAs (miRNAs)**, a class of single-stranded RNAs about 21–22 nucleotides in length that are produced by genes found in almost all eukaryotes. These microRNAs bind to and regulate the expression of messenger RNAs produced by genes that are separate from the genes that produce the microRNAs.

As shown in **Figure 20-37**, genes that produce microRNAs are initially transcribed into longer RNA molecules called *primary microRNAs (pri-miRNAs)*, which fold into hairpin loops. These looped pri-miRNAs are converted into mature miRNAs by sequential processing. First, a nuclear enzyme called **Drosha** cleaves the pri-miRNAs into smaller hairpin RNAs, called *precursor microRNAs (pre-miRNAs)*, roughly 70 nucleotides long. Next, the pre-miRNAs are exported to the cytoplasm, where Dicer cleaves the hairpin loop to produce two complementary short RNA molecules 21–22 nucleotides long.

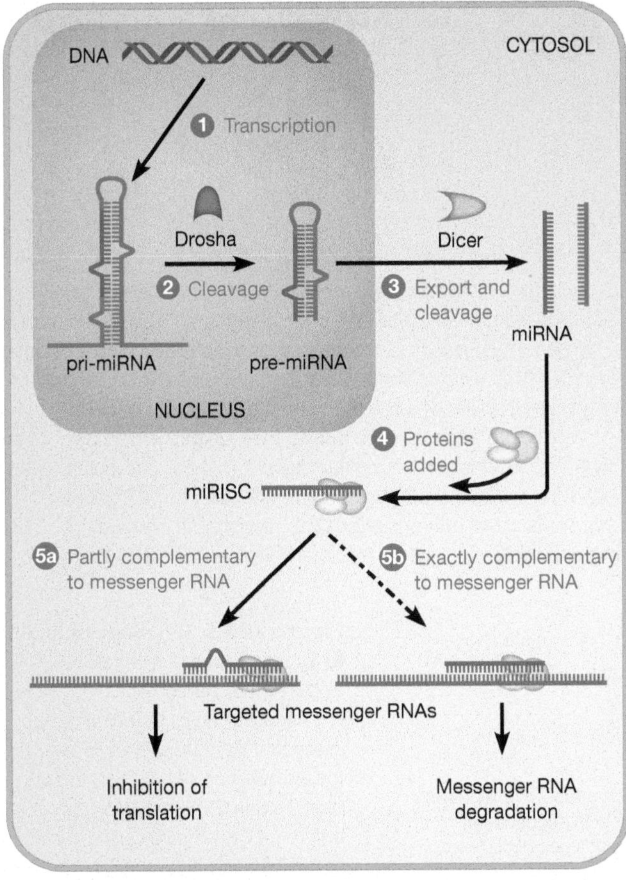

Figure 20-37 Translational Silencing by microRNAs. MicroRNAs (miRNAs) are produced by a multistep process in which ❶ a miRNA gene is first transcribed into a primary transcript (pri-miRNA) that folds into a hairpin loop. ❷ The nuclear enzyme Drosha cleaves pri-miRNA into a smaller (roughly 70-nucleotide) hairpin RNA (pre-miRNA), which is ❸ exported from the nucleus and cleaved by Dicer to release a miRNA molecule that is 21–22 nucleotides long. ❹ The miRNA then joins with RISC proteins to form an miRISC, which is directed by its miRNA to messenger RNAs containing sequences complementary to those in the miRNA. 5a Most commonly, multiple miRISCs bind to the same messenger RNA through partly complementary sequences and together inhibit translation. 5b Occasionally, the miRNA is exactly complementary to a site within a particular messenger RNA, resulting in messenger RNA degradation by a mechanism similar to that observed with siRNAs.

One of the two short RNAs is the mature microRNA, which joins with a group of proteins to form an **miRISC**. Each miRISC inhibits the expression of messenger RNAs containing a base sequence complementary to that of the microRNA residing within the miRISC. The interactions between miRISCs and their targeted messenger RNAs take place mainly in cytoplasmic P bodies, mentioned earlier in this chapter. Sometimes the microRNA in a particular miRISC is exactly complementary to a sequence present in a targeted messenger RNA. In such cases, the targeted message is destroyed by a mechanism similar to that observed with siRNAs. It is more common, however, for microRNAs to exhibit partial complementarity to sites in messenger RNA. Rather than triggering messenger RNA degradation, the binding of miRISCs to partially complementary sites inhibits translation of the targeted message. This inhibitory effect usually requires the binding of multiple miRISCs to different, partially complementary sites within a given messenger RNA. Such inhibition is not necessarily permanent, however. Changing cellular conditions can trigger the release of a messenger RNA from inhibiting miRISCs, accompanied by movement of the messenger RNA from P bodies to actively translating ribosomes.

Current estimates suggest that genes encoding microRNAs account for up to 5% of the total number of genes in eukaryotic genomes. In mammals, this corresponds to the presence of about 1000 microRNA genes. The microRNA produced by each of these genes is capable of binding to target sites in about 200 different messenger RNAs, and many messenger RNAs are targeted by several microRNAs acting in combination. MicroRNAs appear to play important roles during embryonic development. One way of uncovering the developmental roles played by microRNAs has been to use the gene knockout technique (see Figure 21-31) to eliminate specific microRNA genes from mouse embryos. In one such experiment, deleting a microRNA gene called *miR-1-2* led to severe heart defects that caused more than half the mice to die during embryonic development or within the first few months of life. At least a dozen of the mRNAs targeted for regulation by the *miR-1-2* microRNA have been tentatively identified. Included in this group are mRNAs encoding transcription factors that regulate muscle cell development and an mRNA encoding a subunit of a potassium channel that helps maintain the resting membrane potential.

Piwi-Interacting RNAs Are Small Regulatory RNAs That Protect the Germline of Eukaryotes

miRNAs arise from processing of hairpins of double-stranded RNA, which are then processed by Dicer into single-stranded RNA. In contrast, **Piwi-interacting RNAs (piRNAs)** come from long, single-stranded RNA transcribed from clustered regions of DNA. In this sense, they are strikingly similar to the

Key
Technique

Gene Knockdown via RNAi

PROBLEM: Isolating mutants using classical genetics or by targeting mutations in specific genes is laborious and time-consuming. Developing rapid techniques for gene knockdown that exploit the vast DNA sequence information available for humans and model organisms is highly desirable.

SOLUTION: Introduce double-stranded RNA into single cells or into the cells of entire organisms and then capitalize on defense mechanisms already present in the cells to target endogenous RNA for degradation.

Key Tools: Molecular biology reagents to produce double-stranded RNA (dsRNA) corresponding to a gene of interest; reagents or an apparatus to introduce the dsRNA into the cells or organism of choice.

Details: You learned in this chapter that eukaryotes have developed defenses against viruses or other foreign DNA elements. The normal, or endogenous, function of these defenses is to act like a genomic "immune system" to suppress the expression of invader DNA. However, surveillance of dsRNA also turns out to be a powerful tool in the laboratory because the same cellular machinery can be used to selectively reduce/eliminate mRNA derived from any gene.

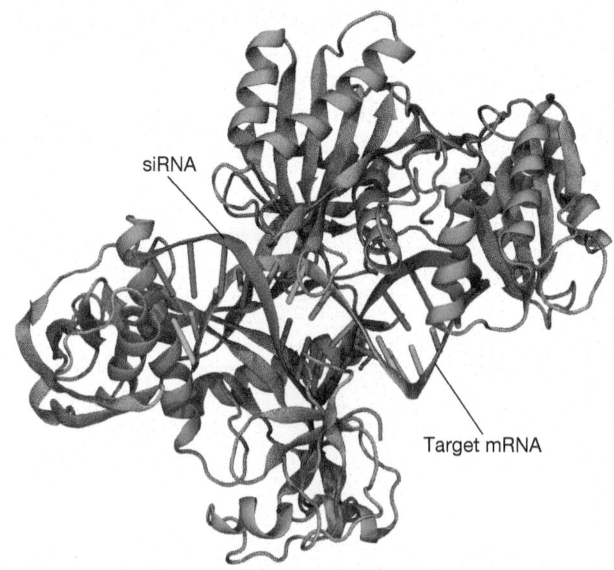

dsRNA-Argonaute complex.

Overview of RNAi. RNA-mediated interference (RNAi) exploits the dsRNA degradation pathway to target mRNAs that are complementary to introduced dsRNA. The dsRNA is cleaved via the activity of Dicer, the resulting small RNAs are used to target RISC to endogenous RNAs, and then an Argonaute protein cleaves the endogenous RNA. One technical challenge is that the dsRNA has to be introduced into the cells of interest. How this is done varies depending on the cell type or organism (**Figure 20B-1**). dsRNA can be introduced directly into cells or organisms, such as the nematode *C. elegans,* by injection or in other cases by infection with a double-stranded RNA virus. In some cases, it is even possible to bathe cells or animals in RNA, which is then taken up and induces RNAi.

Worms are a special case due to their unique physiology. When RNAi was attempted in a similar manner in mammalian cells, it was soon realized that long dsRNA molecules trigger a response that shuts down all translation. This response is aimed at blocking viral replication, given that many viruses have RNA genomes.

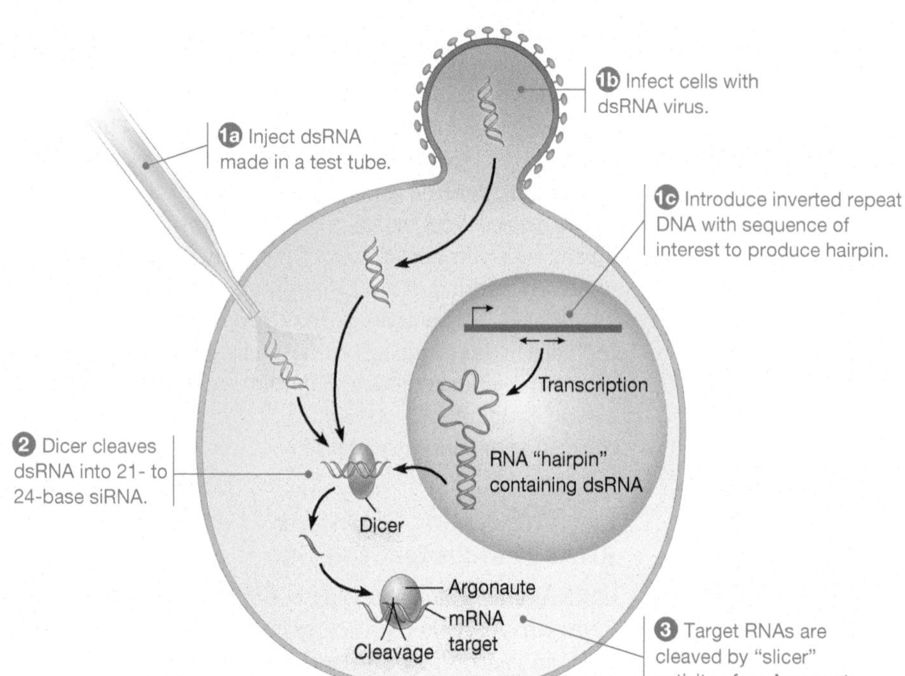

Figure 20B-1 Overview of RNAi. ❶ Double-stranded RNA (dsRNA) is introduced into cells or whole organisms in a variety of ways, including direct injection, infection via RNA viruses, or the introduction of DNA that encodes short hairpin RNAs (shRNAs). ❷ dsRNA is then processed by Dicer, which is part of the RISC complex, and ❸ the target mRNA is degraded via an Argonaute protein's activity.

In addition, direct addition of dsRNA proved very inefficient. Fortunately, however, DNA can be introduced into the cell(s) of interest. One type of DNA is designed so that when it is transcribed, it produces a piece of RNA that can pair with itself to form a *short "hairpin" RNA (shRNA)*. One major advantage of this last approach is that if the DNA becomes integrated into the genome, it can be passed on from one cell to its progeny; that is, it is *heritable*. In all cases, the dsRNA can be cleaved by Dicer into *small interfering RNAs (siRNAs)*. These siRNAs pair with endogenous mRNA targets that are subsequently degraded by Argonaute.

In general, the longer RNAs that can be used in organisms like *C. elegans* lead to highly specific RNAi because multiple siRNAs derived from the long introduced RNA all target the same mRNA. shRNAs or other directly introduced siRNAs, however, can sometimes target not only the desired mRNA, but also other mRNAs that contain stretches of similar sequences. Such *off-target effects* are a serious practical issue, and several controls are usually performed to ensure that the effects observed using a particular introduced siRNA are specific.

Some organisms do not make the Argonaute proteins required for efficient RNAi responses. Remarkably, in some cases, simply supplying the cells of such organisms with RNA encoding Argonaute proteins is enough to allow them to be susceptible to RNAi.

Genome-wide RNAi. RNA is very useful for targeting specific sequences. But what if you don't know which genes (and the proteins they encode) are required for a particular cellular process? In that case, you might wish to perform a *genome-wide* RNAi screen. Here again the details vary, but two examples illustrate the general principles: the nematode *C. elegans* and tissue culture cells.

One of the methods for introducing dsRNA into *C. elegans* is surprisingly simple (**Figure 20B-2**). Any desired DNA sequence can be expressed in *E. coli*, often by PCR amplification of DNA and insertion of the DNA into an engineered bacterial plasmid. The plasmids can then be introduced into *E. coli*, which can be induced to express dsRNA corresponding to the inserted DNA sequence. Large "libraries" of *E. coli* carrying DNAs corresponding to most genes in the *C. elegans* genome exist. This allows researchers to march through the genome, performing RNAi to look for interesting phenotypes following knockdown of specific genes.

In the laboratory, *C. elegans* normally eats *E. coli* as food. Remarkably, when *C. elegans* is fed *E. coli* producing dsRNA, the dsRNA in the *E. coli* is transferred to the worm's tissues via specialized transporters, including the gonad. There is also an amplification step, which is part of the worm's defense system: an RNA-dependent RNA polymerase makes more of the original RNA. The overall effect is that RNAi spreads throughout most tissues of the worm's body. This results in knockdown of the gene of interest in that worm's offspring as well.

Worms can be fed dsRNAs that are fairly long; their own cellular machinery creates small dsRNAs that are very specific and that recruit RISC and mediate knockdown. This approach is extraordinarily powerful: RNAi has been used to study the roles of 98% of the roughly 20,000 protein-coding genes in *C. elegans* by turning them off one by one.

For human tissue culture cells, researchers use a similar approach. In this case, short shRNAs that are complementary to sequences present in the genes the researchers wish to silence are synthesized (or purchased from a biotechnology company). The shRNAs are then introduced into cells after they have been made transiently permeable to the shRNAs. Then the effects of expression of the shRNAs, which induce an RNAi response, can be assessed.

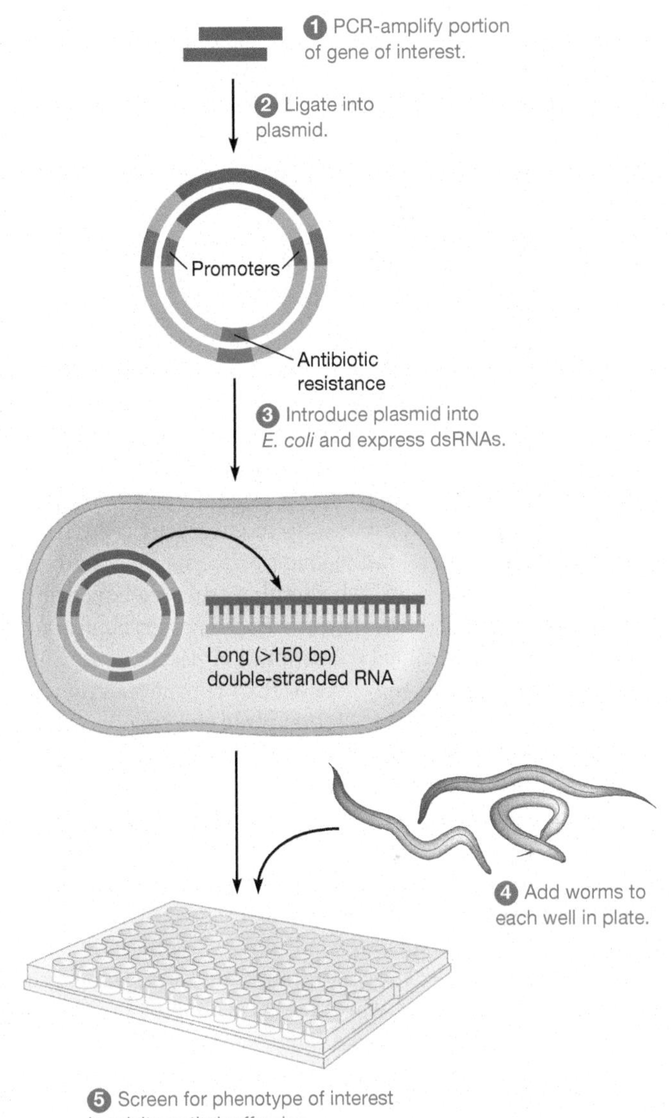

1 PCR-amplify portion of gene of interest.
2 Ligate into plasmid.
Promoters
Antibiotic resistance
3 Introduce plasmid into E. coli and express dsRNAs.
Long (>150 bp) double-stranded RNA
4 Add worms to each well in plate.
5 Screen for phenotype of interest in adults or their offspring.

Figure 20B-2 Feeding RNAi in *C. elegans*. RNA interference can be induced in worms by feeding them bacteria expressing dsRNAs. 1–2 The dsRNAs are introduced using a plasmid, which contains a DNA fragment produced by PCR (see Chapter 21 for more details on how plasmids work). 3 The plasmid contains two promoters, which are recognized by an RNA polymerase from a phage called T7, allowing RNA to be transcribed from each of the two strands of DNA in opposite orientations. 4 Bacteria expressing dsRNA and worms are added to agar that coats the bottom of wells in a plate. 5 The progeny of the worms that ingest the bacteria are then screened for phenotypes.

QUESTION: You want to use shRNAs to knock down mRNA corresponding to a gene of interest. You examine the genome and find that the exonic DNA of another gene is 90% identical to your gene. Based on this information, explain your strategy for designing your shRNA to yield an interpretable result.

CRISPR RNAs of bacteria. The main threat that piRNAs fight are not viruses, however, but transposons. Recall that transposons are very common in eukaryotes and that some kinds make up large portions of the human genome (see Chapter 17). piRNA clusters contain bits and pieces of transposon DNA, which are thought to have hopped into these regions of the genome. piRNAs get their name because they bind to members of the *Piwi family* of Argonaute proteins, thereby forming RISC complexes that mainly target RNAs transcribed from transposons. piRNAs are predominantly expressed in the germline (the cells that give rise to gametes). These cells are especially in need of protection from transposons, as the cells will pass along their DNA to the next generation.

Long Noncoding RNAs Play a Variety of Roles in Eukaryotic Gene Regulation

miRNAs and piRNAs are two examples of short RNAs that are not translated into proteins that regulate other RNAs. Earlier in this chapter you learned that a *long noncoding RNA (lncRNA), Xist,* leads to inactivation of an entire chromosome. Modern methods that allow rapid sequencing of many RNAs (described in Chapter 21) have revealed the presence of many noncoding RNAs longer than 200 nucleotides in eukaryotic cells. There is substantial debate at the moment regarding how many of these RNAs actually work, whether they truly are noncoding, and exactly how prevalent they are. The transcription of some lncRNAs may act to inhibit transcription of nearby genes transcribed in the opposite direction, as RNA polymerases moving in opposite directions physically bump into one another. Others appear to bind to DNA and result in chromatin alterations. One lncRNA relates to a human *Hox* cluster, the *HoxC* cluster (you learned about *Hox* genes earlier in this chapter). A 2200-bp lncRNA is produced from the *HoxC* cluster known as HOTAIR (for *Hox* antisense intergenic RNA). HOTAIR regulates expression of genes in another *Hox* cluster, the *HoxD* cluster, which resides on another chromosome. Therefore, unlike *Xist*, it acts in *trans*. HOTAIR RNA recruits Polycomb group proteins and other proteins to the *HoxD* region, which results in changes in chromatin packing that inactivate the *HoxD* complex. HOTAIR is not found in mice, suggesting that at least some lncRNAs may have evolved rapidly in mammals.

Posttranslational Control Involves Modifications of Protein Structure, Function, and Degradation

After translation of an mRNA molecule has produced a polypeptide chain, there are still many ways of regulating the activity of the polypeptide product. This brings us to the final level controlling gene expression, namely, the various **posttranslational control** mechanisms that are available for modifying protein structure and function (see Figure 20-11, step ⑤). Included in this category are reversible structural alterations that influence protein function, such as protein phosphorylation and dephosphorylation, as well as permanent alterations such as proteolytic cleavage. Other posttranslational events subject to regulation include the guiding of protein folding by chaperone proteins, the targeting of proteins to intracellular or extracellular locations, and the interaction of proteins with regulatory molecules or ions, such as cAMP or Ca^{2+}.

So far, we have focused on control mechanisms that influence gene expression by affecting the rate of protein production from each gene. But the amount of any given protein present in a cell is influenced by its rate of degradation as well as its rate of production. Thus, regulating the rate of protein degradation is another important means for influencing gene expression. The relative contributions made by the rates of protein synthesis and degradation in determining the final concentration of any protein in a cell are summarized by the equation

$$P = \frac{k_{syn}}{k_{deg}} \qquad \textbf{(20-1)}$$

where P is the concentration of a given protein present, k_{syn} is the rate constant for its synthesis, and k_{deg} is the rate constant for its degradation. As we saw with mRNAs (see page 536), rate constants for degradation are often expressed in the form of a half-life. The half-lives of cellular proteins range from as short as a few minutes to as long as several weeks. Such differences in half-life can dramatically affect the ability of different proteins to respond to changing conditions.

A striking example is provided by the behavior of liver cells exposed to the steroid hormone cortisone, which stimulates the synthesis of various metabolic enzymes. When researchers examined the impact of cortisone on two liver enzymes involved in amino acid metabolism, *tryptophan pyrrolase* and *arginase*, they noticed a pronounced difference: the concentration of tryptophan pyrrolase quickly increased tenfold, whereas the concentration of arginase increased only slightly. At first glance, such data would seem to indicate that cortisone selectively stimulates the synthesis of tryptophan pyrrolase. In fact, cortisone stimulates the synthesis of both tryptophan pyrrolase and arginase to about the same extent. The explanation for the apparent paradox is that tryptophan pyrrolase has a much shorter half-life (larger k_{deg}) than does arginase. Using the mathematical relationship summarized in Equation 20-1, calculations show that the concentration of a protein exhibiting a short half-life (large k_{deg}) changes more dramatically in response to alterations in synthetic rate than does the concentration of a protein with a longer half-life. For this reason, enzymes that are important in metabolic regulation tend to have short half-lives, allowing their intracellular concentrations to be rapidly increased or decreased in response to changing conditions.

Although the factors determining the rates at which various proteins are degraded are not completely understood, it is clear that the process is subject to regulation. In rats deprived of food, for example, the amount of the enzyme arginase in the liver doubles within a few days without any corresponding change in the rate of arginase synthesis. The explanation for the observed increase is that the arginase degradation rate slows dramatically in starving rats. This effect is highly selective; most proteins are degraded more rapidly, not more slowly, in starving animals.

Ubiquitin Targets Proteins for Degradation by Proteasomes

The preceding example raises the question of how the degradation of individual proteins is selectively regulated. The most common method for targeting specific proteins for destruction is to link them to **ubiquitin**, a small protein containing 76 amino acids. Ubiquitin is joined to target proteins by a process that involves three components: a *ubiquitin-activating enzyme* (E1), a *ubiquitin-conjugating enzyme* (E2), and a *substrate recognition protein*, or *ubiquitin ligase* (E3). As shown in **Figure 20-38**, ❶ ubiquitin is first activated by attaching it to E1 in an ATP-dependent reaction. ❷ The activated ubiquitin is then transferred to E2 and ❸ subsequently linked, in a reaction facilitated by E3, to a lysine residue in a target protein. ❹ Additional molecules of ubiquitin are then added in sequence (to internal lysine residues on ubiquitin itself), forming short polyubiquitin chains.

One form of these ubiquitin chains serves as a targeting signal that is recognized by large, protein-degrading structures called **proteasomes.** ❺ Proteasomes serve as a "cellular recycling center" for proteins, allowing amino acids to be liberated as the proteins are degraded. They are the predominant proteases (protein-degrading enzymes) of the cytosol and are often present in high concentration, accounting for up to 1% of all cellular protein. Each proteasome has a molecular weight of roughly 2 million and consists of half a dozen proteases associated with several ATPases and a binding site for ubiquitin chains. Although the proteasomes from prokaryotes and eukaryotes differ somewhat in structure, they all have a remarkably similar "trash can" structure, with a central region shaped like a short cylinder with caps on either end. The 26S proteasome from yeast, which has a central 20S central region and 19S caps on either end, is shown in **Figure 20-39** on page 616. The cap at one end of the proteasome binds to ubiquitylated proteins and removes their ubiquitin chains. The proteins are then fed into the central channel of the proteasome, and their peptide bonds are hydrolyzed in an ATP-dependent process, generating small peptide fragments that are released from the other end of the cylinder. Some proteasomes can degrade short peptides in an ATP-independent fashion; these proteasomes have smaller 11S caps.

The key to regulating protein degradation lies in the ability to select particular proteins for ubiquitylation. This selectivity stems in part from the existence of multiple E3 ligases, each directing the attachment of ubiquitin to different target proteins. One feature recognized by the various forms of E3 is the amino acid present at the N-terminus of a potential target protein. Some N-terminal amino acids cause proteins to be rapidly ubiquitylated and degraded; other amino acids make proteins less susceptible. In other cases, internal amino acid sequences called **degrons** also allow particular proteins to be selected for destruction. Finally, **deubiquitinating enzymes (DUBs)** can regulate protein degradation by removing ubiquitin chains from target proteins.

Besides participating in such selective mechanisms for degrading specific proteins at appropriate times, proteasomes play a role in general, ongoing mechanisms for eliminating defective proteins from cells. It has been reported that up to 30% of newly synthesized proteins are defective and so are immediately tagged with ubiquitin, triggering their destruction by proteasomes.

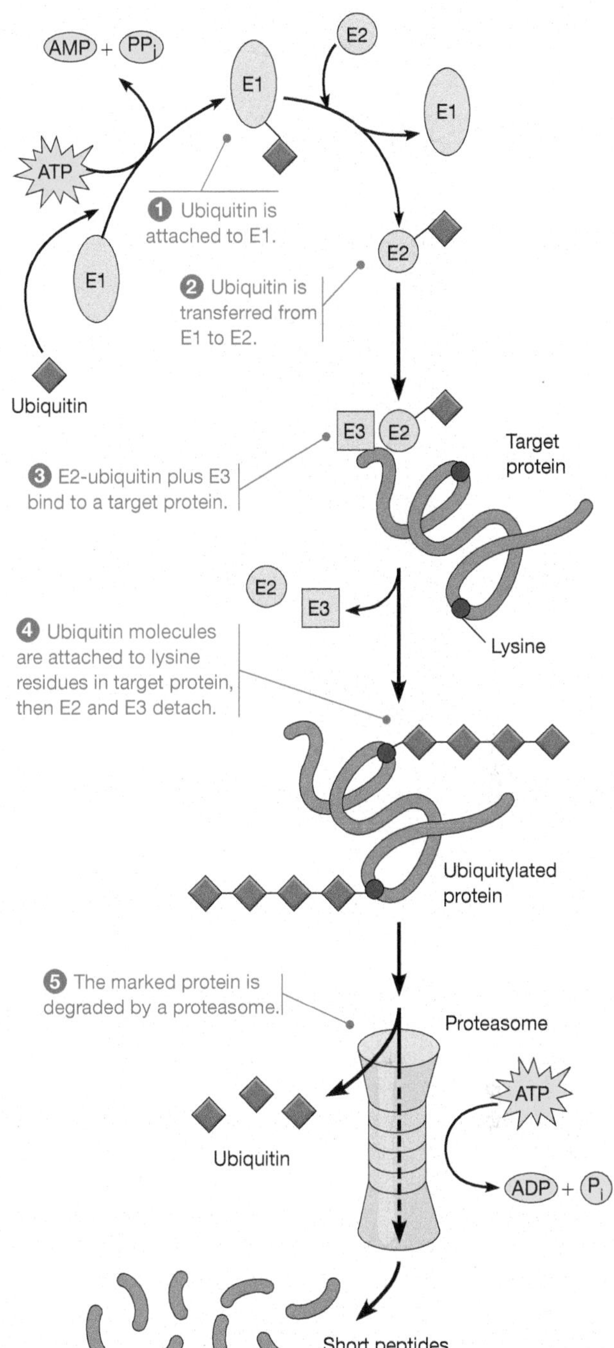

Figure 20-38 Ubiquitin-Proteasome System for Regulated Protein Degradation. Ubiquitin is attached to lysine in targeted proteins through the sequential action of a ubiquitin-activating enzyme (E1), a ubiquitin-conjugating enzyme (E2), and a ubiquitin ligase (E3). A proteasome then degrades the targeted protein into short peptides.

In addition to ubiquitylation, proteins can undergo a related posttranslational modification, through the addition of *small ubiquitin-related modifiers (SUMOs)*. These peptides are ubiquitin-like polypeptides that are conjugated to cellular

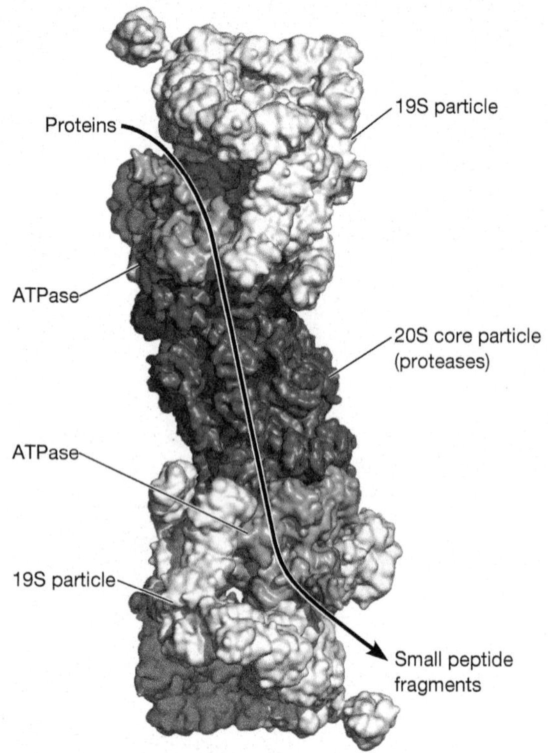

Figure 20-39 The Yeast 26S ATP-Dependent Proteasome. This molecular model from X-ray crystallographic and cryoEM data shows that the 26S proteasome from the budding yeast, *S. cerevisiae*, has a central 20S region in which proteins are degraded and two 19S regulatory caps on either end.

proteins in a manner similar to ubiquitylation, using similar enzymes. SUMOylation appears to have numerous effects, including altering protein stability, movement into and out of the nucleus, and regulation of transcription factor function.

Although the ubiquitin-proteasome system is the primary mechanism used by cells for degrading proteins, it is not the only means available. Lysosomes contain digestive enzymes that degrade all major classes of macromolecules, including proteins (see Chapter 12). Lysosomes can take up and degrade cytosolic proteins by an infolding of the lysosomal membrane, creating small vesicles that are internalized within the lysosome and broken down by the organelle's hydrolytic enzymes. This process of **microautophagy** tends to be rather nonselective in the proteins it degrades. The result is a slow, continual recycling of the amino acids found in most of a cell's protein molecules. Under certain conditions, however, such nonselective degradation of proteins would be detrimental to the cell. For example, during prolonged fasting, nonselective degradation of cellular proteins could lead to depletion of critical enzymes or regulatory proteins. Under these conditions, lysosomes preferentially degrade proteins containing a targeting sequence that consists of glutamine flanked on either side by a tetrapeptide composed of very basic, very acidic, or very hydrophobic amino acids. Proteins exhibiting this sequence are targeted for selective degradation, presumably because they are dispensable to the cell.

A Summary of Eukaryotic Gene Regulation

You have now seen that eukaryotic gene regulation encompasses a broad array of control mechanisms operating at five distinct levels (refer again to Figure 20-11 for a visual summary). The first level of regulation, *genomic control*, includes DNA alterations (amplification, deletion, rearrangements, and methylation), chromatin decondensation and condensation, and histone modifications such as methylation and acetylation. The second level, *transcriptional control*, involves interactions between regulatory transcription factors and various types of DNA control elements, permitting specific genes to be turned on and off in different tissues and in response to changing conditions. The third level involves control of *RNA processing and nuclear export* by mechanisms such as alternative splicing and regulated export of mRNAs through nuclear pores. The fourth level, *translational control*, includes mechanisms for modifying the activity of protein synthesis factors and for controlling the translation and degradation of specific mRNAs through the use of translational repressors and microRNAs. Finally, the fifth level involves various *posttranslational control* mechanisms for reversibly or permanently altering protein structure and function, as well as mechanisms for controlling protein degradation rates. Thus, when you read that a particular alteration in cellular function or behavior is based on a "change in gene expression," you should understand that the underlying explanation might involve mechanisms operating at any one or more of these levels of control.

Because eukaryotes have larger genomes than prokaryotes do, and because they package their DNA into chromatin fibers that are separated from ribosomes by a nuclear envelope, several of the control mechanisms summarized in the preceding paragraph are unique to eukaryotes. Most of the control mechanisms discovered in eukaryotes, however, have counterparts in at least some prokaryotes as well.

CONCEPT CHECK 20-4

You are studying a protein known as GLP-1 in *C. elegans* that is normally found in only two of the four cells of the early embryo. *glp-1* mRNA is found in all four cells. You identify a mutation in a region of the *glp-1* gene that encodes an exon, but the mutation does not affect the sequence of the GLP-1 protein. Now GLP-1 is expressed in all four cells. What part of the mRNA must be affected, and why?

Summary of Key Points

Mastering™ Biology For activities, animations, and review quizzes, go to the study area at www.masteringbiology.com.

20.1 Bacterial Gene Regulation

- Rather than being active at all times, most genes are expressed as needed. In bacteria, genes with related functions are often clustered into operons that can be turned on and off in response to changing cellular needs.
- Operons that encode catabolic enzymes are generally inducible—that is, their transcription is turned on by the presence of substrate. Operons that encode anabolic enzymes are typically repressible; transcription is turned off in the presence of end-product. Both types of regulation involve allosteric repressor proteins that bind to the operator and prevent transcription. Some operons also have control sites where the binding of activator proteins, such as catabolite activator protein (CAP), can activate transcription.
- Certain operons, such as those for amino acid biosynthesis, produce mRNAs with a leader sequence that attenuates transcription when the corresponding amino acid is available. Leader sequences in mRNAs may also contain riboswitches that bind to small molecules, thereby altering hairpin loop configurations that affect the termination of transcription or the initiation of translation.
- Prokaryotes can develop "molecular immunity" against future viral attacks through the Cas protein by incorporation of viral DNA into CRISPR DNA elements.

20.2 Eukaryotic Gene Regulation: Genomic Control

- Most of the cells in a multicellular organism contain the same genes, although amplification, deletion, rearrangement, and epigenetic regulation via methylation can cause localized changes in DNA sequence and expression.
- Gene transcription is associated with uncoiling of chromatin fibers, which is regulated by histone modifications and chromatin remodeling proteins and maintained by DNA methylation.

20.3 Eukaryotic Gene Regulation: Transcriptional Control

- The initiation of eukaryotic transcription is controlled by regulatory transcription factors that bind to DNA control sequences, including proximal control elements that lie close to the core promoter and enhancers and silencers that act at distances up to several hundred thousand base pairs away from the promoter. Insulators protect other nearby genes from regulation by enhancer and silencers.
- Activator proteins bind to enhancer elements in DNA, locally changing DNA conformation. Activator proteins in turn interact with coactivator proteins, including Mediator, that loosen chromatin packing and serve as a bridge to the RNA polymerase complex at the promoter.
- DNA response elements allow nonadjacent genes to be regulated in a coordinated fashion. Examples include hormone response elements and heat shock response elements.

20.4 Eukaryotic Gene Regulation: Posttranscriptional Control

- Eukaryotes utilize a diverse array of posttranscriptional controls, including alternative splicing to generate multiple mRNAs from the same gene; general, as well as mRNA-specific, mechanisms to regulate translation and degradation of mRNA; and post-translational modifications of protein structure, activity, and degradation.
- Two classes of small RNA molecules (siRNAs and microRNAs) regulate the expression of specific genes through a process known as RNA interference. By binding to complementary sequences in specific messenger RNAs (or in some cases, DNA), these small RNAs silence the translation and/or transcription of individual genes.

Problem Set

Mastering™ Biology For activities, animations, and review quizzes, go to the study area at www.masteringbiology.com.

20-1 Laboring with *lac*. Most of what we know about the *lac* operon of *E. coli* has come from genetic analysis of various mutants. In the following list are the genotypes of seven strains of *E. coli*. For each strain, indicate whether the Z gene product will be expressed in the presence of lactose and whether it will be expressed in the absence of lactose. Explain your reasoning in each case.

(a) $I^+P_{lac}{}^+O^+Z^+$

(b) $I^SP_{lac}{}^+O^+Z^+$

(c) $I^+P_{lac}{}^+O^cZ^+$

(d) $I^-P_{lac}{}^+O^+Z^+$

(e) $I^SP_{lac}{}^+O^cZ^+$

(f) $I^+P_{lac}{}^-O^+Z^+$

(g) Same as part a, but with glucose present

20-2 The Pickled Prokaryote. *Pickelensia hypothetica* is an imaginary prokaryote that converts a wide variety of carbon sources to ethanol when cultured anaerobically in the absence of ethanol. When ethanol is added to the culture medium, however, the organism obligingly shuts off its own production of ethanol and makes lactate instead. Several mutant strains of *Pickelensia* have been isolated that differ in their ability to synthesize ethanol. Class I mutants cannot synthesize ethanol at all. Mutations of this type map at two loci, *A* and *B*. Class II mutants, on the other hand, are constitutive for ethanol synthesis: they continue to produce ethanol whether it is present in the medium or not. Mutations of this type map at loci *C* and D. Strains of *Pickelensia* constructed to be diploid for the ethanol operon have the following genotypes and phenotypes:

$A^+B^-C^+D^+/A^-B^+C^+D^+$	inducible
$A^+B^+C^+D^-/A^+B^+C^+D^+$	inducible
$A^+B^+C^+D^+/A^+B^+C^-D^+$	constitutive

(a) Identify each of the four genetic loci of the ethanol operon.

(b) Indicate the expected phenotype for each of the following partially diploid strains.

i. $A^-B^+C^+D^+/A^-B^+C^+D^-$

ii. $A^+B^-C^+D^+/A^-B^+C^+D^-$

iii. $A^+B^+C^+D^-/A^-B^-C^-D^+$

iv. $A^-B^+C^-D^+/A^+B^-C^+D^+$

20-3 Regulation of Bellicose Catabolism. The enzymes bellicose kinase and bellicose phosphate dehydrogenase are coordinately regulated in the bacterium *Hokus focus*. The genes encoding these proteins, *belA* and *belB*, are contiguous segments on the genetic map of the organism. In her pioneering work on this system, Professor Jean X. Pression established that the bacterium can grow with the monosaccharide bellicose as its only carbon and energy source, that the two enzymes involved in bellicose catabolism are synthesized by the bacterium only when bellicose is present in the medium, and that enzyme production is turned off in the presence of glucose. She identified a number of mutations that reduce or eliminate enzyme production and showed that these could be grouped into two classes. Class I mutants are in *cis*-acting elements, whereas those in class II all map at a distance from the genes encoding the enzymes and are *trans*-acting. Thus far, the only constitutive mutations Pression has found are deletions that connect *belA* and *belB* to new DNA at the upstream side of *belA*. The following is a list of conclusions she would like to draw from her observations. Indicate in each case whether the conclusion is consistent with the data (C), inconsistent with the data (I), or irrelevant to the data (X).

(a) Bellicose can be metabolized by *H. focus* cells to yield ATP.

(b) Enzyme production by genes *belA* and *belB* is under positive control.

(c) Phosphorylation of bellicose makes the sugar less permeable to transport across the plasma membrane.

(d) The operator for the bellicose operon is located upstream from the promoter.

(e) Some of the mutations in class I may be in the promoter.

(f) Some of the mutations in class II may be in the operator.

(g) Class II mutations may include mutations in the gene that encodes CAP.

(h) The constitutive deletion mutations connect genes *belA* and *belB* to a new promoter.

(i) Expression of the bellicose operon is subject to regulation by attenuation.

20-4 Attenuation in 25 or Fewer Words. Complete each of the following statements about attenuation of the *trp* operon of *E. coli* in 25 or fewer words.

(a) The leader sequence can be considered to be a *conditional terminator* because ...

(b) When the gene product of the operon is not needed, the 5′ end of the mRNA forms a secondary structure that ...

(c) Implicit in our understanding of the mechanism of attenuation is the assumption that ribosomes follow RNA polymerase very closely. This "tailgating" is essential to the model because ...

(d) Measurements of rates of synthesis indicate that translation normally proceeds faster than transcription. This supports the proposed mechanism for attenuation because ...

(e) If the operon encoding the enzymes in the biosynthetic pathway for amino acid X is subject to this type of attenuation, the leader sequence probably encodes a polypeptide that ...

(f) Attenuation makes sense as a mechanism for regulating operons that encode enzymes involved in amino acid biosynthesis because ...

(g) Attenuation of this type is not a likely mechanism for regulating gene expression in eukaryotes because ...

20-5 Positive and Negative Control. Assume you have a culture of *E. coli* cells growing on medium B, which contains both lactose and glucose. At time *t*, 33% of the cells are transferred to medium L, which contains lactose but not glucose; 33% are transferred to medium G, which contains glucose but not lactose; and the remaining cells are left in medium B. For each of the following statements, indicate with an L if it is true of the cells transferred to medium L, with a G if it is true of the cells transferred to medium G, with a B if it is true of the cells left in medium B, and with an N if it is true of none of the cells.

(a) The rate of glucose consumption per cell is approximately the same after time *t* as before.

(b) The rate of lactose consumption is higher after time *t* than before.

(c) The intracellular cAMP level is lower after time *t* than before.

(d) Most of the *lac* operator sites have *lac* repressor proteins bound to them.

(e) Most of the catabolite activator proteins exist as CAP-cAMP complexes.

(f) Most of the *lac* repressor proteins exist as repressor-glucose complexes.

(g) The rate of transcription of the *lac* operon is greater after time *t* than before.

(h) The *lac* operon has both CAP and the repressor protein bound to it.

20-6 Enhancers. An enhancer may increase the frequency of transcription initiation for its associated gene when ... (Indicate true or false for each statement, and explain your answer.)

(a) ... it is located 1000 nucleotides upstream of the gene's core promoter.

(b) ... it is in the gene's coding region.

(c) ... no promoter is present.

(d) ... it causes looping out of the intervening DNA.

(e) ... it causes alternative splicing of the DNA.

20-7 QUANTITATIVE Gene Amplification. The best-studied example of gene amplification involves the genes encoding ribosomal RNA during oogenesis in *Xenopus laevis*. The unamplified number of ribosomal RNA genes is about 500 per haploid genome. After amplification, the oocyte contains about 500,000 ribosomal RNA genes per haploid genome. This level of amplification is apparently necessary to allow the egg cell to synthesize the 10^{12} ribosomes that accumulate during the 2 months of oogenesis in this species. Each ribosomal RNA gene consists of about 13,000 base pairs, and the genome size of *Xenopus* is about 2.7×10^9 base pairs per haploid genome.

(a) What fraction of the total haploid genome do the 500 copies of the ribosomal RNA genes represent?

(b) What is the total size (in base pairs) of the genome after amplification of the ribosomal RNA genes? What proportion of this do the amplified ribosomal RNA genes represent?

(c) Assume that all the ribosomal RNA genes in the amplified oocyte are transcribed continuously to generate the number of ribosomes needed during the 2 months of oogenesis. How long would oogenesis have to extend if the genes had not been amplified?

(d) Why do you think genes have to be amplified when the gene product needed by the cell is an RNA but not usually when the desired gene product is a protein?

20-8 Steroid Hormones. Steroid hormones are known to increase the expression of specific genes in selected target cell types. For example, testosterone increases the production of a protein called α2-microglobulin in the liver, and hydrocortisone (a type of glucocorticoid) increases the production of the enzyme tyrosine aminotransferase in the liver.

(a) Based on your general knowledge of steroid hormone action, explain how these two steroid hormones are able to selectively influence the production of two different proteins in the same tissue.

(b) Prior to the administration of testosterone or hydrocortisone, liver cells are exposed to either puromycin (an inhibitor of protein synthesis) or α-amanitin (an inhibitor of RNA polymerase II). How would you expect such treatments to affect the actions of testosterone and hydrocortisone?

(c) Suppose you carry out a domain-swap experiment in which you use recombinant DNA techniques to exchange the zinc finger domains of the testosterone receptor and glucocorticoid receptors with each other. What effects would you now expect testosterone and hydrocortisone to have in cells containing these altered receptors?

(d) If recombinant DNA techniques are used to substitute a testosterone response element for the glucocorticoid response element that is normally located adjacent to the tyrosine aminotransferase gene in liver cells, what effects would you expect testosterone and hydrocortisone to have in such genetically altered cells?

20-9 Homeotic Genes in *Drosophila*. Homeotic genes are considered crucial to early development in *Drosophila* because . . . (Indicate true or false for each statement, and explain your answer.)

(a) . . . they encode proteins containing zinc finger domains.

(b) . . . mutations in homeotic genes are always lethal.

(c) . . . they control the expression of many other genes required for development.

(d) . . . homeodomain proteins act by influencing mRNA degradation.

20-10 *Hox* Genes in Mammals. *Hox* genes are known to influence the identity of body parts in mammals, similar to their role in flies. Mario Capecchi's laboratory has made many knockout mouse strains in which individual *Hox* genes are deleted. The phenotype of many of these knockouts is extremely mild, especially when compared with the corresponding mutants in flies, which often die. How can you explain this difference?

20-11 Protein Degradation. A protein known as a mitotic cyclin is an example of a protein that is selectively degraded at a particular point in the cell cycle, namely, the onset of the portion of the cell cycle known as anaphase. Suppose that you use recombinant DNA techniques to create a mutant form of mitotic cyclin that is not degraded at the onset of anaphase.

(a) Sequence analysis of the mutant mitotic cyclin shows that it is missing a stretch of nine amino acids near one end of the molecule. Based on this information, provide at least three possible explanations of why this form of mitotic cyclin is not degraded at the onset of anaphase.

(b) What kinds of experiments could you carry out to distinguish among these various possible explanations?

(c) Suppose that you discover cells containing a mutation in a second protein and learn that this mutation also prevents mitotic cyclin from being degraded at the onset of mitosis, even when mitotic cyclin is normal. The degradation of other proteins appears to proceed normally in such cells. A mutation in what kind of protein might explain such results?

20-12 RNA Noninterference. You are helping a new student in a lab that studies *C. elegans*. The student has generated engineered DNA for use in RNAi from genomic DNA for a gene of interest to the lab. She is puzzled that the clone does not have an effect, even though she knows, based on available mutants, that loss of the function of this gene causes worms to die as larvae. You look at the sequence of the DNA she used to make the feeding clone and discover that it corresponds to an intron in the gene. What would you say to the student to explain why her experiment did not succeed?

20-13 Levels of Control. If sea urchin zygotes are exposed to *actinomycin D*, an inhibitor of RNA synthesis, during the early stages of embryonic development, protein synthesis and development continue until the embryos form a hollow ball of cells, or blastula, with hundreds of cells. However, if the same inhibitor is added later during embryonic development, protein synthesis is severely depressed, and normal embryonic development halts. How might you explain these observations?

20-14 DATA ANALYSIS Mitochondrial Interference. Transcription in mitochondria occurs bidirectionally as two large, complementary, polycistronic transcripts. Therefore, these transcripts often form complementary double-stranded RNA (dsRNA). When dsRNA is detected in the cytoplasm, your cells respond with an inflammatory response. Why? Many viruses have dsRNA genomes, so your cell knows to respond because they normally have very little cytosolic dsRNA. SUV3 is an RNA helicase that can convert double-stranded RNA to single-stranded RNA and is part of the mitochondrial degradosome complex, which is responsible for RNA surveillance and degradation.

In **Figure 20-40**, the nuclei of HeLa cells are stained blue, dsRNA is stained green, and mitochondria red. (The last column merges all three images.) The cells in the upper row (control) are untreated; cells in the bottom row have been treated with SUV3 siRNA.

(a) What effect is the siRNA treatment having, and why?

(b) Based on the function of SUV3 and the data from this figure, come up with an explanation for why individuals with mutations in SUV3 may have autoimmune inflammation.

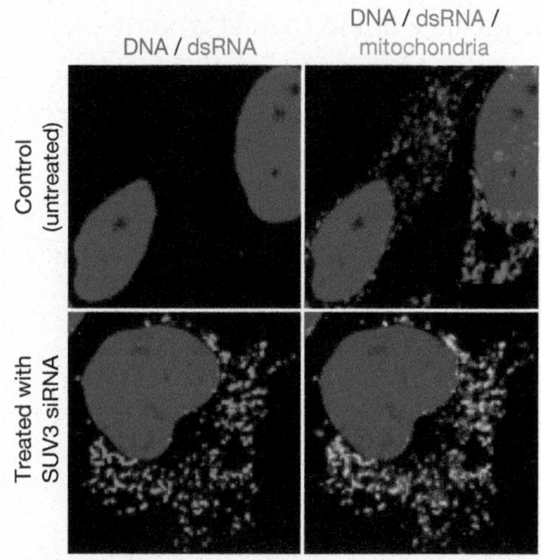

Figure 20-40 Treatment of HeLa cells with siRNA Against SUV3. See Problem 20-14.

21 Molecular Biology Techniques for Cell Biology

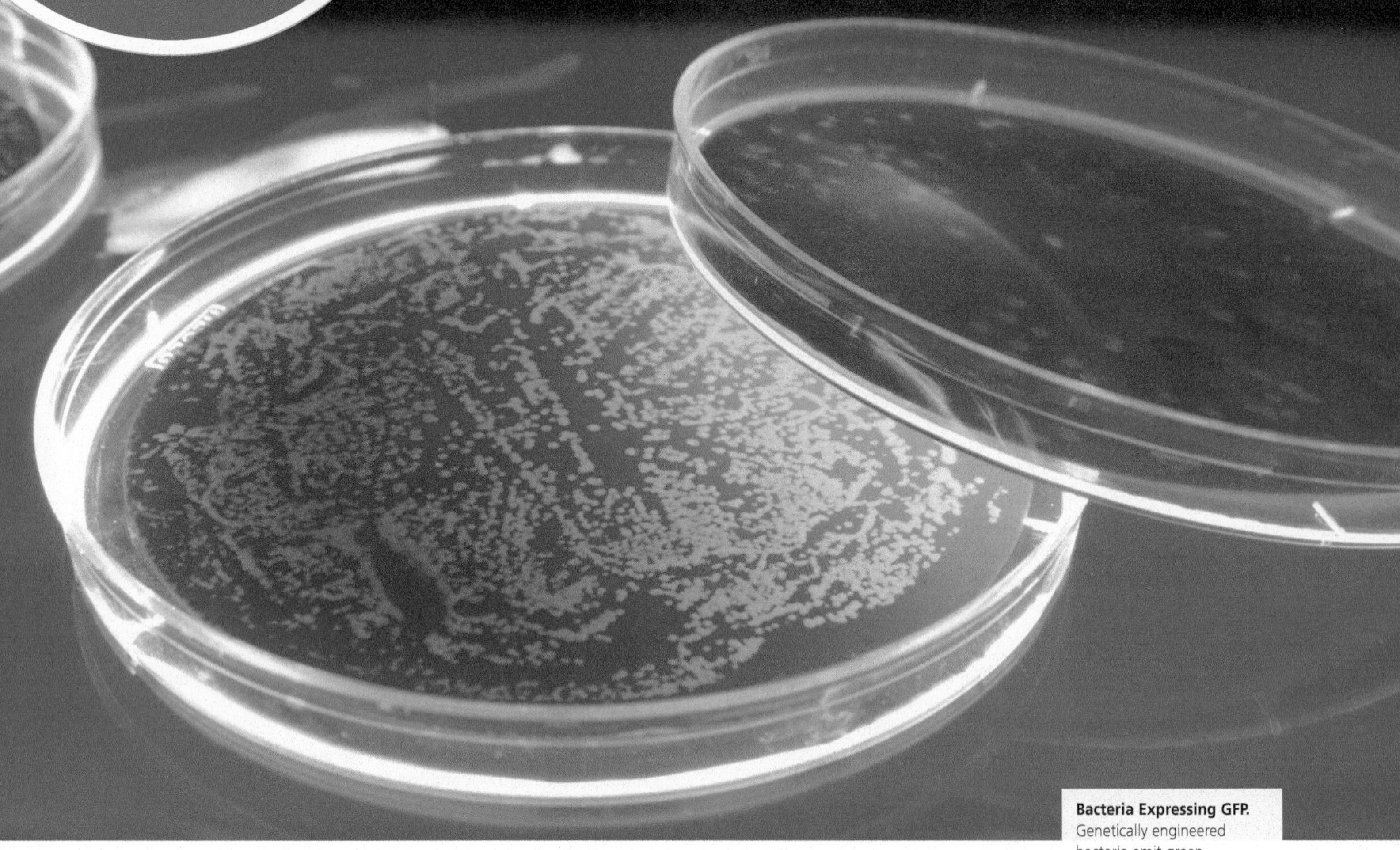

Bacteria Expressing GFP. Genetically engineered bacteria emit green fluorescence when exposed to blue light.

As you learned at the very beginning of this book, cell biology has been revolutionized by the tools of molecular biology and biochemistry (see Chapter 1). This chapter focuses on these tools—the key technologies cell biologists use to study DNA, RNA, and proteins. The goal of this chapter is not to provide a detailed protocol for each of these key technologies, but instead to provide an overview of when these technologies are used and why they are useful in specific contexts. In addition, some of these techniques are presented here because of their historical significance and not necessarily because they are actively used today. Technical advancements in the field are happening at an amazing pace! At the heart of many of these technologies is the ability to study and manipulate DNA, with which we begin our discussion.

21.1 Analyzing, Manipulating, and Cloning DNA

In nature, genetic recombination usually takes place between two DNA molecules derived from organisms of the same species (see Chapter 25). In animals and plants, for example, an individual's two parents are the original sources of the DNA that recombines during meiosis. A naturally arising recombinant DNA molecule usually differs from the parental DNA molecules only in the combination of alleles it contains; the fundamental identities of its genes remain the same. In the laboratory, such limitations do not exist. Since the development of **recombinant DNA technology** in the 1970s, scientists have been able to splice segments of DNA from one source together with any other piece of DNA.

A central feature of recombinant DNA technology is the ability to produce specific pieces of DNA in large enough quantities for research and other uses. This process of generating many copies of specific DNA fragments is called **DNA cloning**. (In biology, a *clone* is a population of organisms that is derived from a single ancestor and hence is genetically homogeneous, and a *cell clone* is a population of cells derived from the division of a single cell. By analogy, a *DNA clone* is a population of DNA molecules that are derived from the replication of a single molecule and hence are identical to one another.) The advent of recombinant DNA technology made it possible to isolate individual eukaryotic genes in quantities large enough to permit them to be thoroughly studied, ushering in a new era in molecular biology.

PCR Is Widely Used to Clone Genes

The simplest and quickest means of amplifying a DNA sequence for cloning is to use the *polymerase chain reaction (PCR)*. The PCR method (see Key Technique, pages 624–625) simply requires that you know part of the base sequence of the gene you wish to amplify. The next step is to synthesize short, single-stranded DNA primers that are complementary to sequences located at opposite ends of the gene; these primers are then used to target the intervening DNA for amplification. PCR can be used to rapidly generate sufficient quantities of DNA for cloning. Depending on the application, the resulting PCR products can be cloned in a variety of ways. Commonly, the fact that the polymerases typically used in PCR generate overhanging single-stranded regions (containing T's) at the ends of amplified DNA provides ready-made ends that can be exploited for rapid cloning into vectors with overhanging ends containing A's. In other cases, the PCR primers themselves can be designed so that they contain specific restriction sites (described next). After performing PCR, the amplified DNA that is now flanked by these sites is digested with the appropriate enzyme and then can be cloned into a vector of choice.

Restriction Endonucleases Cleave DNA Molecules at Specific Sites

Despite the convenience and power of working with DNA, many DNA molecules are far too large to be studied intact. In fact, until the early 1970s, DNA was the most difficult biological molecule to analyze biochemically, yet in less than a decade, DNA became one of the easiest biological molecules to work with. This breakthrough was made possible by the discovery of **restriction endonucleases** (also called *restriction enzymes*), proteins isolated from bacteria that cut foreign DNA molecules at specific internal sites.

Restriction enzymes help bacteria protect themselves against invasion by foreign DNA molecules, particularly the DNA of bacteriophages. In fact, the name "restriction" endonuclease came from the discovery that these enzymes *restrict* the ability of foreign DNA to take over the transcription and translation machinery of the bacterial cell. To protect its own DNA from being degraded, the bacterial cell has enzymes that add methyl groups (—CH_3) to specific nucleotides that its own restriction enzymes would otherwise recognize. Once they have been methylated, the nucleotides are no longer recognized by the restriction enzymes, so the bacterial DNA is not attacked by the cell's own restriction enzymes. Restriction enzymes are therefore said to be part of the cell's *restriction/methylation system:* foreign DNAs (like bacteriophage genomes that are unmethylated) are cleaved by the restriction enzymes, while bacterial genomes are protected by prior methylation.

The cutting action of a restriction enzyme generates a specific set of DNA pieces called *restriction fragments.* Each restriction enzyme cleaves double-stranded DNA only in places where it encounters a specific recognition sequence, called a **restriction site**, that is usually four, six, or eight (or more) nucleotides long. For example, here is the restriction site recognized by the widely used *Escherichia coli* restriction enzyme called *Eco*RI:

5′ G ↓ A — A — T — T — C 3′
3′ C — T — T — A — A ↑ G 5′

The arrows indicate where *Eco*RI cuts the DNA. This restriction enzyme, like many others, makes a *staggered* cut in the double-stranded DNA molecule.

Restriction sites occur frequently enough in DNA to permit typical restriction enzymes to cleave DNA into fragments ranging from a few hundred to a few thousand base pairs in length. Fragments of these sizes are far more amenable to further manipulation than are the enormously long DNA molecules from which they are generated.

Restriction enzymes are named after the bacteria from which they are obtained. Each enzyme name is derived by combining the first letter of the bacterial genus with the first two letters of the species. The strain of the bacterium may also be indicated, and if two or more enzymes have been isolated from the same species, the enzymes are numbered (using Roman numerals) in order of discovery. *Eco*RI, for example, was the first restriction enzyme isolated from *E. coli* strain R, whereas the third enzyme isolated from *Haemophilus aegyptius* is called *Hae*III.

Restriction enzymes are specific for double-stranded DNA and cleave both strands. Each restriction enzyme recognizes a specific DNA sequence (**Figure 21-1** on page 622). For example, *Hae*III recognizes the tetranucleotide sequence GGCC and cleaves both strands of the DNA double helix at the same point, generating restriction fragments with *blunt ends* (Figure 21-1a).

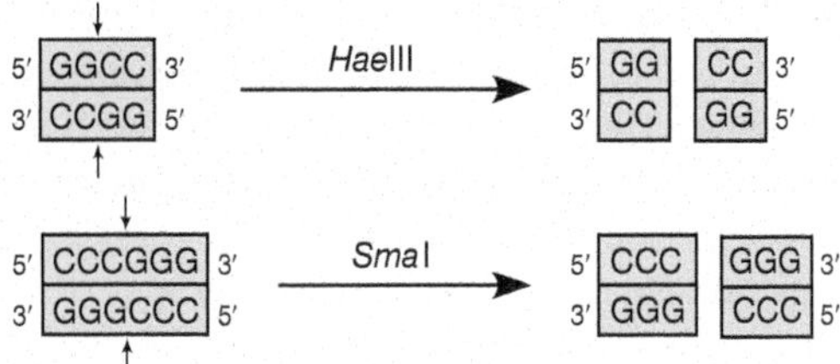

(a) Cleavage by enzymes producing blunt ends

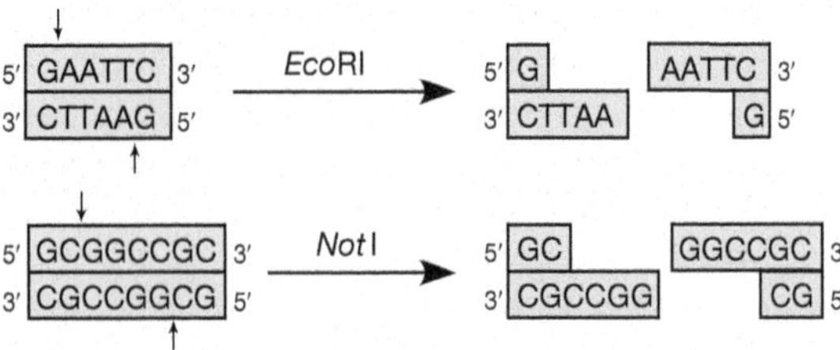

(b) Cleavage by enzymes producing sticky ends

Figure 21-1 Cleavage of DNA by Restriction Enzymes. **(a)** *Hae*III and *Sma*I are examples of restriction enzymes that cut both DNA strands in the same location, generating fragments with blunt ends. **(b)** *Eco*RI and *Not*I are examples of enzymes that cut DNA in a staggered fashion, generating fragments with sticky ends. Such sticky ends are useful for joining DNA fragments from different sources.

Table 21-1 Some Common Restriction Enzymes and Their Recognition Sequences

Enzyme	Source Organism	Recognition Sequence*
*Bam*HI	*Bacillus amyloliquefaciens*	5′ G↓G—A—T—C—C 3′ 3′ C—C—T—A—G↑G 5′
*Eco*RI	*Escherichia coli*	5′ G↓A—A—T—T—C 3′ 3′ C—T—T—A—A↑G 5′
*Hae*III	*Haemophilus aegyptius*	5′ G—G↓C—C 3′ 3′ C—C↑G—G 5′
*Hin*dIII	*Haemophilus influenzae*	5′ A↓A—G—C—T—C 3′ 3′ T—T—C—G—A↑G 5′
*Pst*I	*Providencia stuartii* 164	5′ C↓T—G—C—A—G 3′ 3′ G—A—C—G—T↑C 5′
*Pvu*I	*Proteus vulgaris*	5′ C↓G—A—T—C—G 3′ 3′ G—C—T—A—G↑C 5′
*Pvu*II	*Proteus vulgaris*	5′ C↓A—G—C—T—G 3′ 3′ G—T—C—G—A↑C 5′
*Sal*I	*Streptomyces albus* G	5′ C↓T—C—G—A—C 3′ 3′ C—A—G—C—T↑G 5′

*The arrows within the recognition sequence indicate the points at which each restriction enzyme cuts the two strands of the DNA molecule.

Many other restriction enzymes cleave the two strands in a staggered manner, generating short, single-stranded tails or overhangs on both fragments. As we have seen, *Eco*RI is an example of such an enzyme; it recognizes the sequence GAATTC and cuts the DNA molecule in an offset manner, leaving an AATT tail on both fragments (Figure 21-1b). The restriction fragments generated by enzymes with this staggered cleavage pattern always have **sticky ends** (also called *cohesive ends*). This term derives from the fact that the single-stranded tail at the end of each such fragment can base-pair with the tail at either end of any other fragment generated by the same enzyme, causing the fragments to stick to one another by hydrogen bonding. Enzymes that generate such fragments are particularly useful because they can be employed experimentally to create recombinant DNA molecules, as we will see later in this chapter. The restriction sites for several other restriction enzymes are summarized in **Table 21-1**.

The restriction sites for most restriction enzymes are *palindromes*, which means that the sequence reads the same in either direction. (The English word *radar* is a palindrome, for example.) The palindromic nature of a restriction site is due to its twofold rotational symmetry, which means that rotating the double-stranded sequence 180° in the plane of the paper yields a sequence that reads the same as it did before rotation. Palindromic restriction sites have the same base sequence on both strands when each strand is read in the $5' \rightarrow 3'$ direction.

The frequency with which any particular restriction site is likely to occur within a DNA molecule can be predicted statistically. For example, in a DNA molecule containing equal amounts of the four bases (A, T, C, and G), we can predict that, on average, a recognition site with four nucleotide pairs will occur once every 256 (that is, 4^4) nucleotide pairs, whereas the likely frequency of a six-nucleotide sequence is once every 4096 (that is, 4^6) nucleotide pairs. Restriction enzymes therefore tend to cleave DNA into fragments that typically vary in length from several hundred to a few thousand nucleotide pairs—gene-sized pieces, essentially. Because each restriction enzyme cleaves only a single, specific nucleotide sequence, it will always cut a given DNA molecule in the same predictable manner, generating a reproducible set of restriction fragments. This property makes restriction enzymes powerful tools for generating manageable-sized pieces of DNA for further study.

Gel Electrophoresis Allows DNA to Be Separated by Size

DNA samples, such as those derived from treatment of DNA with restriction enzymes, often include DNA fragments of varying size. To determine the number and lengths of such fragments and to isolate individual fragments for further study, a researcher must be able to separate the fragments from one another. The technique of choice for this purpose is **gel electrophoresis**, a method used to separate nucleic acids, proteins, and polypeptides by their relative size. In general, electrophoresis is a group of related techniques that utilize an electrical field to separate charged molecules. How quickly any given molecule moves during electrophoresis depends on its charge as well as its size. Electrophoresis can be carried out using various support media, such as paper, cellulose acetate, starch, *polyacrylamide*, or *agarose* (a polysaccharide obtained from seaweed). Of these media, gels made of polyacrylamide or agarose provide the best separation and are most commonly employed for the electrophoresis of nucleic acids and proteins. As you will see later in the chapter, proteins are first treated with a negatively charged detergent to make them move toward the positively charged anode. The procedure for DNA is even simpler than that for proteins because DNA molecules have an inherent negative charge (due to the phosphate groups present in the backbone) and therefore do not need to be pre-treated with this detergent to make them migrate through the gel. Small DNA fragments (1–1000 bp) are usually separated in polyacrylamide gels, whereas larger DNA fragments (200–20,000 bp) are separated in more porous gels made of agarose.

Figure 21-2 illustrates the separation of DNA fragments of different sizes by gel electrophoresis; in the figure,

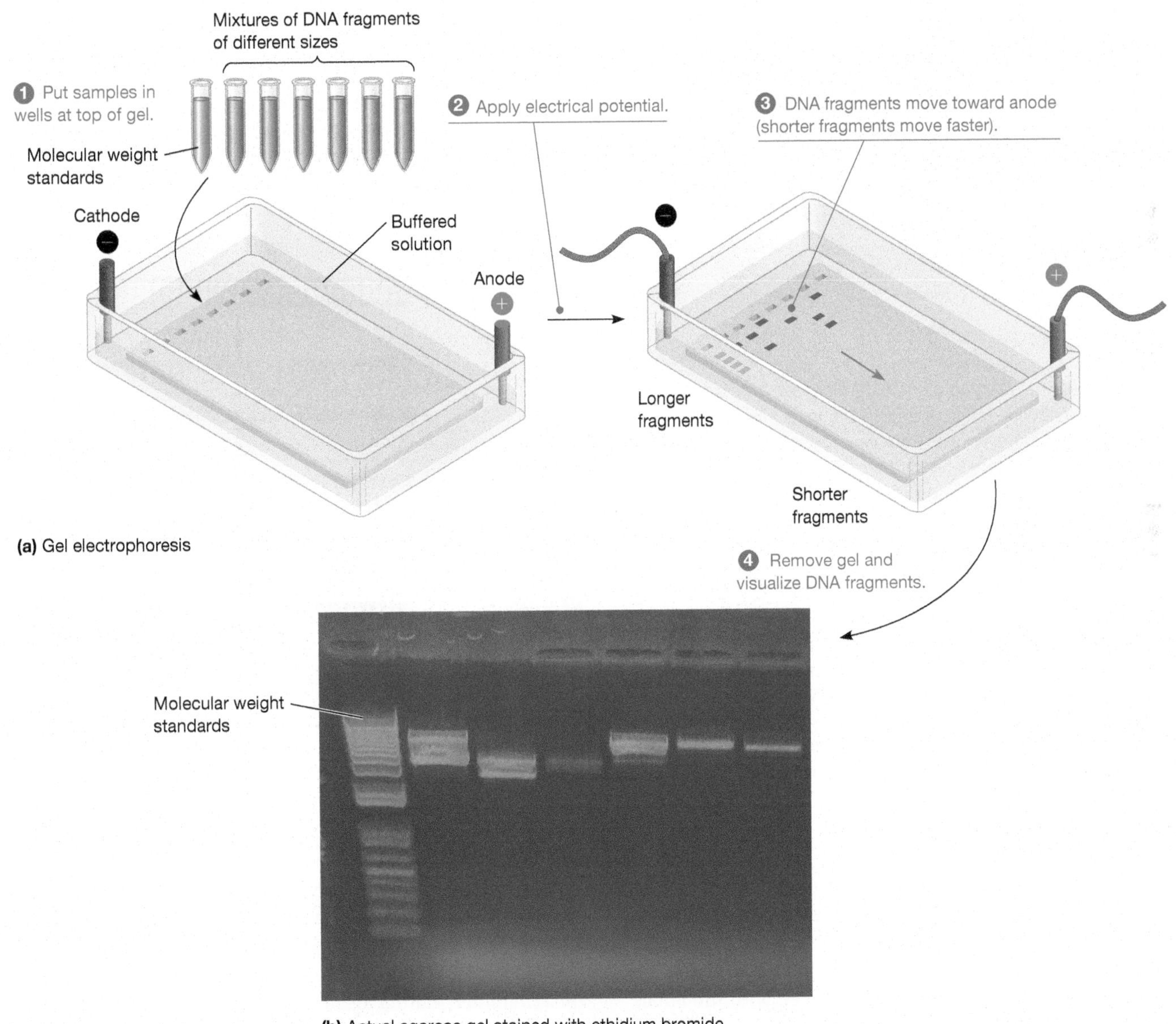

(a) Gel electrophoresis

(b) Actual agarose gel stained with ethidium bromide

Figure 21-2 Gel Electrophoresis of DNA. (a) Six test tubes contain mixtures of DNA fragments produced by incubating DNA with different restriction enzymes (blue). A seventh (teal) contains molecular weight standards. A small sample of each is applied to the wells of a horizontal gel, and an electrical potential of several hundred volts is applied. This causes the DNA fragments to migrate toward the anode, with shorter fragments migrating faster than larger ones. After allowing time for the fragments to separate, the gel is removed and stained with a dye such as ethidium bromide, which binds to the DNA fragments and causes them to fluoresce under ultraviolet light. **(b)** An ethidium bromide–stained agarose gel exposed to ultraviolet light reveals the DNA.

Key Technique: The Polymerase Chain Reaction (PCR)

PCR thermal cycler.

PROBLEM: The ability to work with minuscule amounts of DNA is invaluable in a wide range of endeavors, from paleontology to criminology. But in many cases—for example, at a crime scene—the source of DNA is very limited (a drop of blood, a few body hairs, skin under a victim's fingernails, the tiny number of cells contained in a human fingerprint, etc.). To analyze DNA, a large amount of sample DNA is often needed. It therefore becomes necessary to take the small amount of DNA collected and increase, or amplify, its numbers of copies.

SOLUTION: In such cases, a method called the polymerase chain reaction (PCR) can come to the rescue. With PCR, it is possible to rapidly replicate selected DNA segments that are initially present in extremely small amounts. In only a few hours, PCR can make millions or even billions of copies of a particular DNA sequence, thereby producing enough material for DNA fingerprinting, DNA sequencing, and numerous other applications. PCR is widely held to be one of the most important inventions of the twentieth century in molecular biology. Biochemist Kary Mullis developed the technique in the 1980s and received a Nobel Prize for doing so in 1993.

Key Tools: Template DNA (the DNA to be copied); heat-stable DNA polymerase; DNA primers; free deoxynucleotides

Details: To perform PCR, you first need to know part of the base sequence of the DNA segment you wish to amplify. Based on this information, short, single-stranded DNA primers are chemically synthesized; these primers generally consist of DNA segments 15–20 nucleotides long that are complementary to sequences located at opposite ends of the target segment. If the sequences that naturally flank the segment of interest are unknown, artificial sequences can be attached before running the procedure. PCR

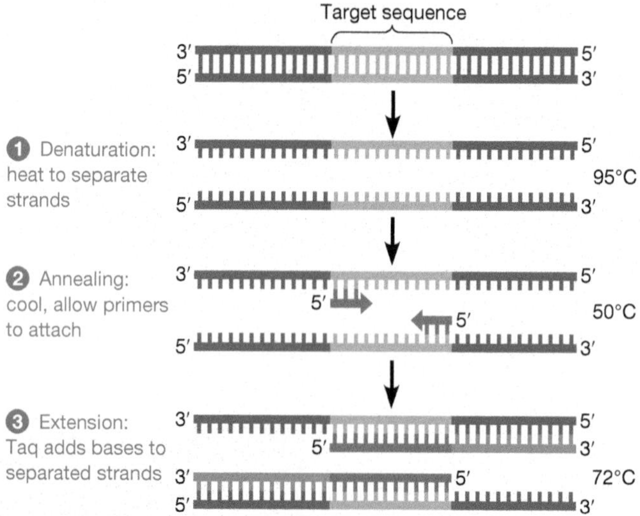

Figure 21A-1 PCR: First Cycle of Amplification. See the description in the box text. After this first cycle, additional cycles of denaturation, annealing, and extension will exponentially amplify the target sequence.

six different samples are shown. The samples are mixed with a dye to allow easy visualization of the progress of the experiment. ❶ They are then placed ("loaded") into separate compartments ("wells") at one end of the gel. ❷ Next, an electrical potential is applied across the gel, with the anode situated at the opposite end of the gel from the samples. Because DNA has a negative charge, ❸ the fragments migrate toward the anode. (The anode is usually the red lead, so a useful pneumonic is "run to red.") Smaller fragments (that is, those with lower molecular weight) move through the gel with relative ease and therefore migrate rapidly, whereas larger fragments move more slowly. The current is left on until the fragments are well spaced out on the gel. The final result is a series of DNA fragments that have been separated based on differences in size.

❹ DNA fragments in the gel are usually visualized by staining the DNA with a fluorescent dye such as *ethidium bromide*, an intercalating agent that binds to DNA and fluoresces orange when exposed to ultraviolet light. Due to the potential mutagenicity of ethidium bromide, safer intercalating dyes have been developed that emit visible light when excited with various wavelengths of light. After individual DNA fragments are located in this way, they can be removed from the gel for further study.

Standard gel electrophoresis cannot be used to resolve differences in the size of very large DNA molecules (greater works best when the DNA segment to be amplified—the region flanked by the two primers—is ~50–2000 nucleotides long.

DNA polymerase is then added to catalyze the synthesis of complementary DNA strands using the two primers as starting points. The DNA polymerase routinely used for this purpose was first isolated from the bacterium *Thermus aquaticus,* an inhabitant of thermal hot springs where the waters are normally 70–80°C. The optimal temperature for this enzyme, called *Taq polymerase,* is 72°C, and it is relatively stable at even higher, near boiling temperatures. Free dATP, dTTP, dCTP, and dGTP deoxynucleotides are also added to the reaction mixture.

Figure 21A-1 summarizes how the PCR procedure works. ❶ Each reaction cycle begins with a short period of heating to near boiling (95°C) to denature the DNA double helix into its two strands. ❷ The DNA solution is then cooled to 50°C to allow the primers to base-pair to complementary regions on the DNA strands being copied. ❸ The temperature is then raised to 72°C, and the Taq polymerase goes to work, adding nucleotides beginning at the 3′ end of the primer. The specificity of the primers ensures the selective copying of the stretches of downstream template DNA. It takes no more than a few minutes for the Taq polymerase to completely copy the targeted DNA sequence. The reaction mixture is then heated again to melt the new double helices, more primer binds to the DNA, and steps ❶–❸ are repeated. **Figure 21A-2** follows the reaction for three successive cycles to show how the PCR can rapidly produce copies of the sequence between the primers.

QUESTION: The theoretical amplification accomplished by *n* cycles of PCR is 2*n*. Starting from a single DNA segment, how much amplification would you expect to see after 20 cycles of PCR? After 30 cycles?

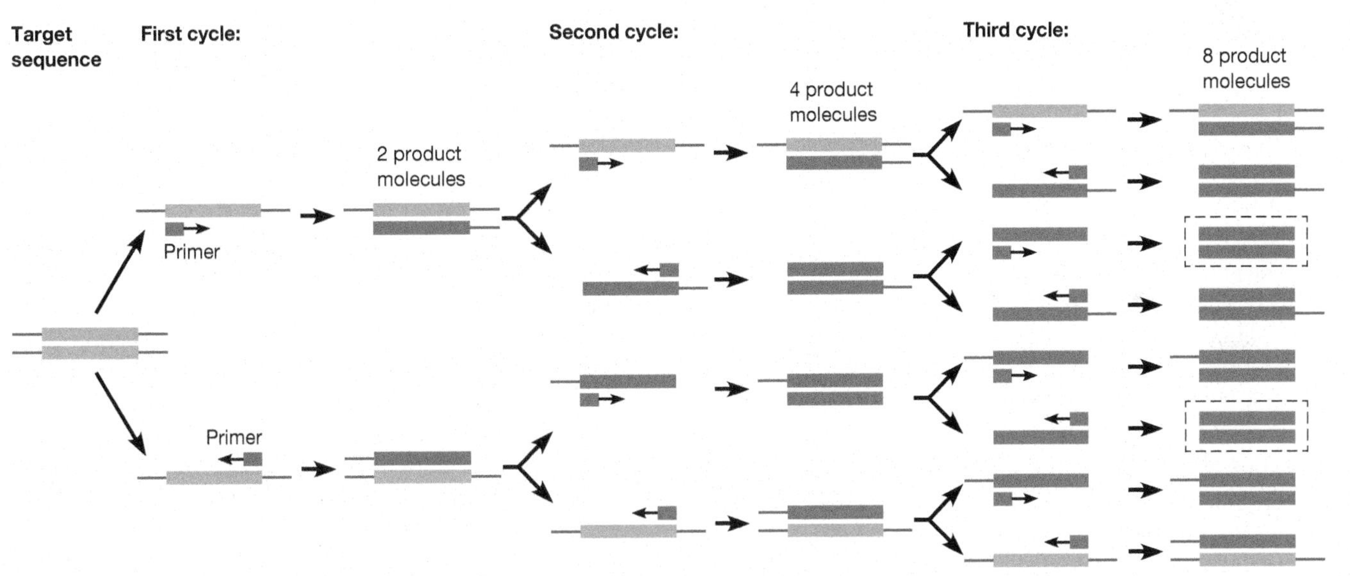

Figure 21A-2 PCR: Three Cycles of Amplification. Products that consist of only the target sequence are outlined.

than about 30 kb) because the migration of very long DNAs is limited by their ability to move through the pores in an agarose gel. For some applications involving very large pieces of DNA, a specialized type of electrophoresis known as **pulsed-field electrophoresis** is used. As the name suggests, this technique involves applying pulsed electric fields, one along the long axis of the gel and two additional fields that are directed at about ±60° relative to the long axis. The longer the piece of DNA, the longer it takes the DNA to reorient to the new direction of the electric field as the pulses are applied, allowing efficient size separation of large pieces of DNA. DNA as large as small whole chromosomes from bacteria and fungi (up to several million base pairs in length) can be separated in this way.

Restriction Mapping Can Characterize DNA

How does a researcher determine the order in which a set of restriction fragments is arranged in a DNA molecule? One approach involves treating the DNA with two or more restriction enzymes, alone and in combination, followed by gel electrophoresis to determine the size of the resulting DNA fragments. The sizes of the resulting fragments can be used to construct a **restriction map** of the molecule, which depicts the location of all the restriction sites in the original DNA. **Figure 21-3** on page 626 shows how this process would work for a simple DNA molecule cleaved with the restriction enzymes *Eco*RI and *Hae*III. In this example, each individual restriction enzyme cleaves the DNA into two fragments, indicating that the DNA contains one restriction site for each enzyme. Based

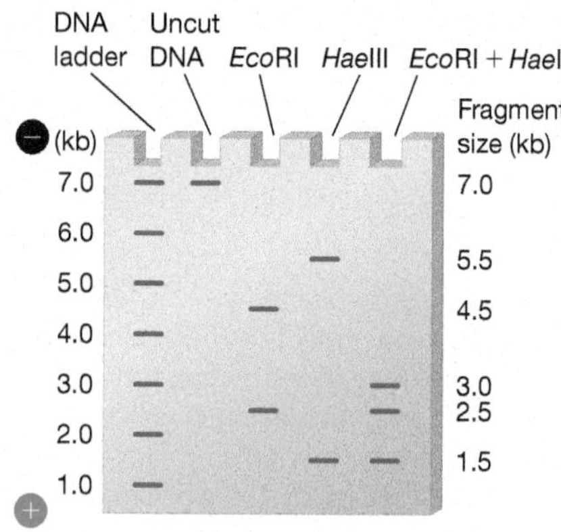

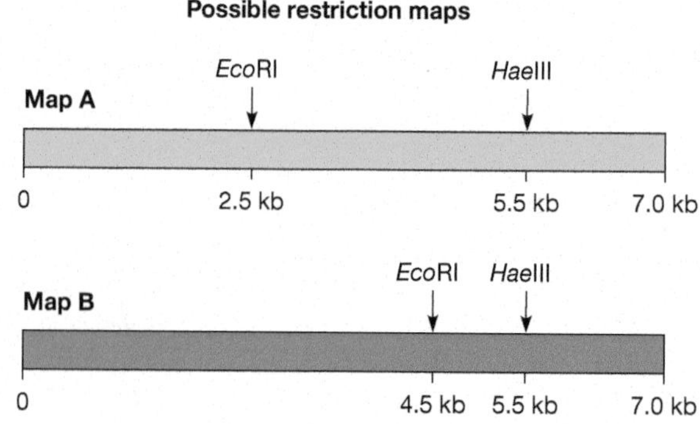

Figure 21-3 Restriction Mapping. In this hypothetical example, the location of restriction sites for *Eco*RI and *Hae*III is determined in a DNA fragment 7.0 kb long. The gel on the left shows that *Eco*RI cleaves the DNA into two fragments measuring 2.5 kb and 4.5 kb, indicating that DNA has been cleaved at a single point located 2.5 kb from one end. These fragment sizes were estimated by comparing their position in the gel to a ladder sample containing DNA fragments of known sizes. Treatment with *Hae*III cleaves the DNA into two fragments measuring 1.5 kb and 5.5 kb, indicating that DNA has been cleaved at a single point located 1.5 kb from one end. Based on this information alone, two possible restriction maps can be proposed. If map A were correct, simultaneous digestion of the DNA with *Eco*RI and *Hae*III should yield three fragments measuring 3.0 kb, 2.5 kb, and 1.5 kb. If map B were correct, simultaneous digestion of the DNA with *Eco*RI and *Hae*III should yield three fragments measuring 4.5 kb, 1.5 kb, and 1.0 kb. The experimental data reveal that map A must be correct.

on such information alone, two possible restriction maps can be proposed (see maps A and B in Figure 21-3). To determine which of the two maps is correct, an experiment must be done in which the starting DNA molecule is cleaved *simultaneously* with *Eco*RI and *Hae*III. The size of the fragments produced by simultaneous digestion with both enzymes reveals that map A must be the correct one.

In practice, restriction mapping usually involves data that are considerably more complex than in our simple example. In addition, inferring the location of restriction sites from the sizes of restriction fragments depends on whether the starting DNA is linear (as in the example in Figure 21-3) or circular, as is often the case when the DNA being characterized is carried in a plasmid vector (which you will learn about in more detail later in this section). Restriction digests can also be used to analyze human DNA as part of identifying specific features of an individual patient's (or suspect's) DNA called restriction fragment length polymorphisms, a type of analysis you will learn about later in this chapter.

Southern Blotting Identifies Specific DNAs from a Mixture

In addition to analyzing fragments of DNA based on size, it is often desirable to identify fragments of DNA of a certain size that contain a specific DNA sequence of interest. The particular technique used to identify a piece of DNA in this way depends on what the researcher knows about the DNA and its base sequence. *PCR* and *DNA sequencing* are very commonly used to determine whether a piece of DNA contains a sequence of interest. If a sequence can be amplified using sequence-specific primers, then its ends contain sequences corresponding to the primers, and the nature of the intervening sequencing can be determined by sequencing.

Another historically important way to characterize DNA is to use a nucleic acid *probe*, a single-stranded molecule of DNA that can identify a desired DNA sequence by base-pairing with it. Nucleic acid probes are labeled either with radioactivity or with some other chemical group (like a fluorescent tag) that allows the probe to be easily visualized by producing light that can expose X-ray film or be detected using an imaging scanner. The use of nucleic acid probes to study nucleic acids was originally developed for studying DNA by Edwin M. Southern, so the technique as applied to DNA is known as **Southern blotting** in his honor. Modern Southern blotting involves several distinct steps (**Figure 21-4**). First, DNA is digested with a restriction enzyme. The resulting fragments are then separated from each other by gel electrophoresis. Because the DNA often contains many, many types of DNA fragments of different sizes, hundreds or thousands of bands might appear at this stage if all were made visible. Here is where Southern's procedure comes in. A special kind of "blotter" paper (nitrocellulose or nylon) is pressed against the completed gel, allowing the separated DNA fragments to be transferred to the paper. Then a single-stranded, fluorescent nucleic acid probe is added to the blot. The bound probe is made visible by scanning the blot with a fluorescence imager. Southern blotting has in many cases been replaced by PCR-based techniques for studying DNA, but related techniques are still very important for analyzing RNAs.

Restriction Enzymes Allow Production of Recombinant DNA

Much of what we call recombinant DNA technology was made possible by the discovery of restriction enzymes, which you learned about earlier in this chapter. Restriction enzymes that make staggered cuts in DNA are especially useful because they generate single-stranded sticky ends that provide a simple means for joining DNA fragments obtained from different sources. In essence, any two DNA fragments generated by the

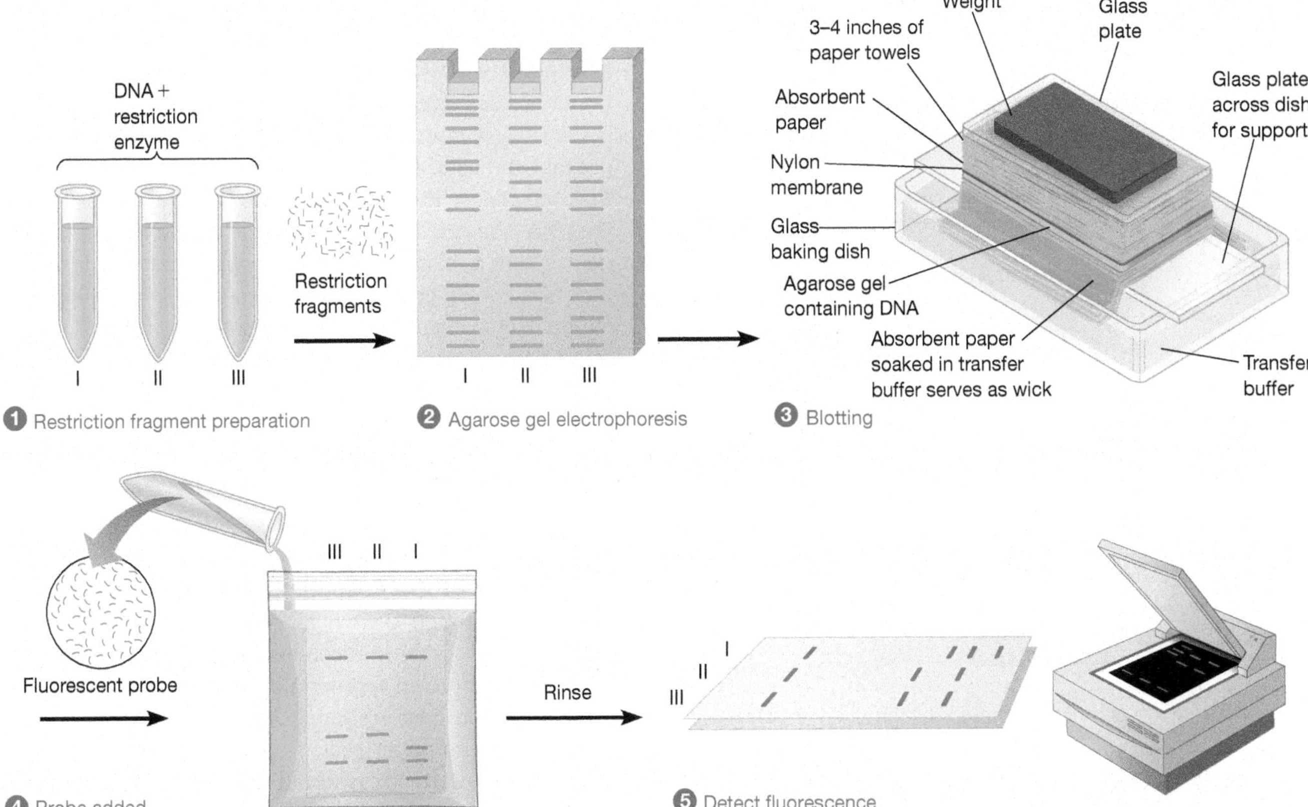

Figure 21-4 Southern Blotting. ❶ DNA is extracted from different types of cell (I, II, and III). A restriction enzyme is added to the three samples of DNA to produce restriction fragments, ❷ which are separated by electrophoresis. ❸ After the DNA on the gel is denatured by raising the pH, the single strands are transferred onto special paper by blotting. ❹ A solution of single-stranded, fluorescent DNA probe is added to the blot that is complementary to the DNA of interest. After excess probe is rinsed off, ❺ fluorescence is detected by scanning the blot on an imaging system. An image can then be captured corresponding to specific DNA bands that base-pair with the probe.

same restriction enzyme can be joined together by complementary base pairing between their single-stranded, sticky ends.

Figure 21-5 on page 628 illustrates how this general approach works. ❶ DNA molecules from two sources are first treated with a restriction enzyme known to generate fragments with sticky ends. ❷ The fragments are then mixed together under conditions that favor base pairing between these sticky ends. Once joined in this way, ❸ the DNA fragments are covalently sealed together by DNA ligase, an enzyme normally involved in DNA replication and repair (see Chapter 17). The final product is a *recombinant DNA molecule* containing DNA sequences derived from two different sources.

The combined use of restriction enzymes and DNA ligase allows any two (or more) pieces of DNA to be spliced together, regardless of their origins. A piece of human DNA, for example, can be joined to bacterial or phage DNA just as easily as it can be linked to another piece of human DNA. In other words, it is possible to form recombinant DNA molecules that never existed in nature, without regard for the natural barriers that otherwise limit recombination to genomes of the same or closely related species. Therein lies the power of (and, for some, the concern about) recombinant DNA technology. In the laboratory, DNA cloning proceeds using specially engineered DNA specifically suited to the organism in which the cloned DNA will be propagated.

DNA Cloning Can Use Bacterial Cloning Vectors

The use of bacterial cloning vectors is the stock and trade of traditional DNA cloning. DNA cloning is accomplished by splicing the DNA of interest to the DNA of a genetic element, called a **cloning vector**, which can replicate autonomously when introduced into a cell grown in culture—in many cases, a bacterium such as *E. coli*. The cloning vector can be a circular piece of DNA known as a *plasmid* or the DNA of a virus, usually a bacteriophage; in either case, the vector's DNA "passenger" is copied every time it replicates.

Although the specific details vary, the following five steps are typically involved in the process of DNA cloning using bacterial cloning vectors: (1) insertion of DNA into a cloning vector, (2) introduction of the recombinant vector into cells, usually bacteria, (3) amplification of the recombinant vector in the bacteria, (4) selection of cells containing recombinant DNA, and (5) identification of clones containing the DNA of interest.

Insertion of DNA into a Cloning Vector. To examine these events in more detail, consider cloning an insert into a historically important plasmid vector called *pUC19* ("puck-19"), developed in the mid-1980s (**Figure 21-6** on page 629). This plasmid carries the amp^R gene, which confers resistance to the antibiotic ampicillin (Figure 21-6a). Bacteria containing such plasmids

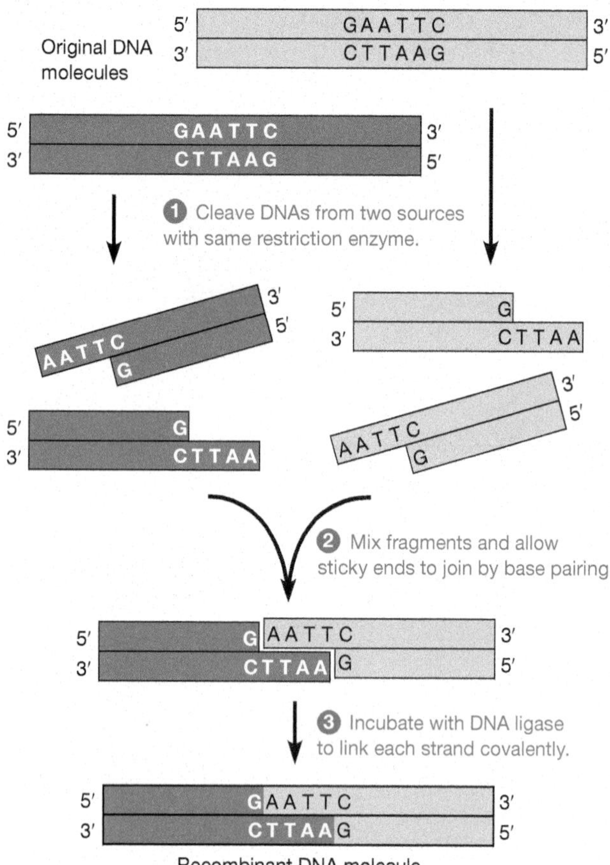

Figure 21-5 Creating Recombinant DNA Molecules. A restriction enzyme that generates sticky ends (in this case *Eco*RI) is used to cleave DNA molecules from two different sources. The complementary ends of the resulting fragments join by base pairing to create recombinant molecules containing segments from both of the original sources.

can therefore be identified by their ability to grow in the presence of ampicillin. The pUC19 plasmid also has 11 different restriction sites clustered in a region of the plasmid containing the *lacZ* gene, which encodes the enzyme β-galactosidase. Integration of foreign DNA at any of these restriction sites will disrupt the *lacZ* gene, thereby blocking the production of β-galactosidase. As we will see shortly, this disruption in β-galactosidase production can be used later in the cloning process to detect the presence of plasmids containing foreign DNA.

Figure 21-6b illustrates how a specific gene of interest residing in a foreign DNA source is inserted into a plasmid cloning vector, using pUC19 as the vector and a restriction enzyme that cleaves pUC19 at a single site within the *lacZ* gene. ❶ Incubation with the restriction enzyme cuts the plasmid at that site, making the DNA linear (opening the circle). ❷ The same restriction enzyme is used to cleave the DNA molecule containing the gene to be cloned. The sticky-ended fragments of foreign DNA are then ❸ incubated with the linearized vector molecules under conditions that favor base pairing, ❹ followed by treatment with DNA ligase to link the molecules covalently. Often an alkaline phosphatase (from shrimp or from calf intestinal cells) is used to remove phosphates on the ends of the linearized vector. Dephosphorylation dramatically reduces the likelihood of re-ligation of the cut ends of the linearized plasmid.

A final issue related to cloning has to do with the *orientation* of the inserted DNA. This might be quite important, for example, if you wish to use your clone to express protein in bacteria. The correct coding sequence will be produced only if the DNA is inserted in one way, so that starting with the plasmid and continuing into the insert, an RNA is produced that can encode the correct protein. One way to ensure that the DNA is inserted in the proper orientation is via *directional cloning*, that is, cloning in which the ligation of the insert occurs in a known orientation. One commonly used approach for directional cloning is to cut the two ends of the vector and insert with different restriction enzymes, producing different sticky ends. The different identity of the 5' and 3' sticky ends ensures that there is only one only possible orientation for the insert. It is equally common, however, to sequence DNA from many positive clones to determine clones that have the correct orientation.

Introduction of the Recombinant Vector into Bacterial Cells. Once foreign DNA has been inserted into a cloning vector, ❺ the resulting recombinant vector is replicated by *transforming* it into an appropriate host cell, usually the bacterium *E. coli*. Plasmids are simply introduced into the medium surrounding the target cells. Prokaryotic cells will take up plasmid DNA from the external medium, although special treatments are usually necessary to enhance the efficiency of the process. Adding calcium ions, for example, markedly increases the rate at which cells take up DNA from the external environment.

Amplification of the Recombinant Vector in Bacteria. After they have taken up the recombinant plasmids, the host bacteria are plated out on a nutrient medium so that the recombinant DNA can be replicated, or *amplified*. Bacteria proliferate and form colonies, each derived from a single cell. Under favorable conditions, *E. coli* will divide every 22 minutes, giving rise to a billion cells in less than 11 hours. As the bacteria multiply, the recombinant plasmids also replicate, producing an enormous number of vector molecules containing foreign DNA fragments. Under such conditions, a single recombinant plasmid introduced into one cell will be amplified several hundred billion-fold in less than half a day.

Selection of Cells Containing Recombinant DNA. During amplification of the cloning vector, procedures are introduced that preferentially select for the growth of those cells that have successfully incorporated the vector. For plasmid vectors such as pUC19, the selection method is based on the plasmid's antibiotic-resistance genes. For example, all bacteria carrying the recombinant plasmids generated in Figure 21-6b will be resistant to the antibiotic ampicillin because all plasmids have an intact ampicillin-resistance gene. The *amp*^{R} gene is a **selectable marker**, which allows only the cells carrying plasmids to grow on culture medium containing ampicillin (the medium "selects for" the growth of the ampicillin-resistant cells).

Identification of Clones Containing the DNA of Interest. To keep the diagram simple, Figure 21-6b shows only the recombinant plasmid containing the desired

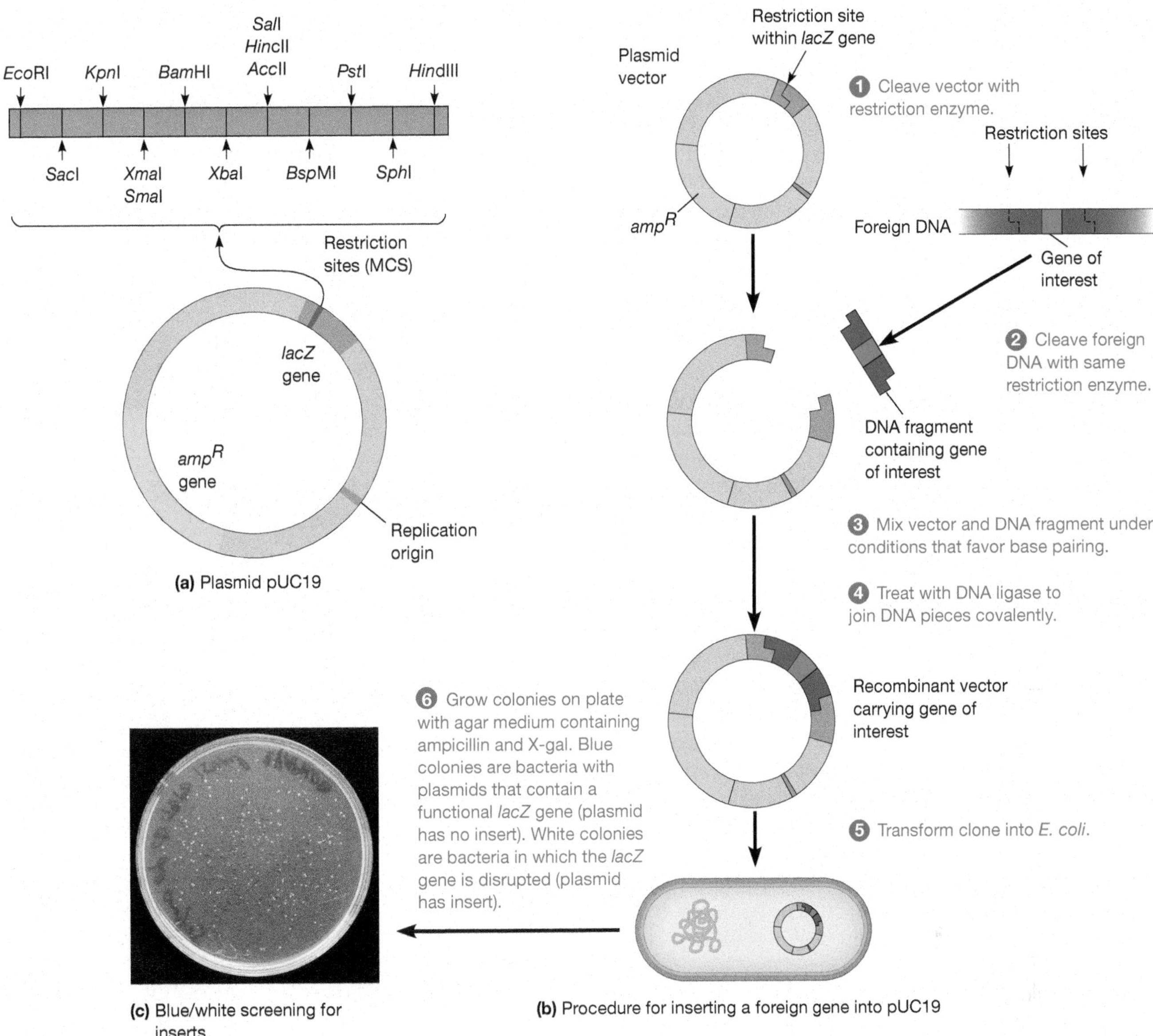
SalI
HincII
AccII
EcoRI
KpnI
BamHI
PstI
HindIII
SacI
XmaI
SmaI
XbaI
BspMI
SphI
Restriction sites (MCS)
lacZ gene
ampR gene
Replication origin
(a) Plasmid pUC19
Plasmid vector
Restriction site within lacZ gene
ampR
1 Cleave vector with restriction enzyme.
Restriction sites
Foreign DNA
Gene of interest
2 Cleave foreign DNA with same restriction enzyme.
DNA fragment containing gene of interest
3 Mix vector and DNA fragment under conditions that favor base pairing.
4 Treat with DNA ligase to join DNA pieces covalently.
Recombinant vector carrying gene of interest
5 Transform clone into E. coli.
6 Grow colonies on plate with agar medium containing ampicillin and X-gal. Blue colonies are bacteria with plasmids that contain a functional lacZ gene (plasmid has no insert). White colonies are bacteria in which the lacZ gene is disrupted (plasmid has insert).
(c) Blue/white screening for inserts
(b) Procedure for inserting a foreign gene into pUC19

Figure 21-6 Cloning Using a Plasmid Vector. (a) A historically important engineered plasmid, pUC19. Plasmid cloning vectors carry antibiotic resistance (for example, against ampicillin), have an engineered polylinker with several restriction sites (or multiple cloning site, MCS), and may have features that allow visual screening for inserts (for example, *lacZ* for blue/white screening). **(b)** To insert foreign DNA into the plasmid, the foreign and plasmid DNAs are cleaved with a restriction enzyme that recognizes the same site, in this case a site within the *lacZ* gene. The fragments of foreign DNA are incubated with the linearized plasmid DNA under conditions that favor base pairing and ligation between sticky ends. Among the expected products will be plasmid molecules recircularized by base pairing with a single fragment of foreign DNA, some containing the gene of interest. Bacterial cells carrying such plasmids will be resistant to ampicillin and are then screened for the presence of the correct insert via blue/white screening or via colony PCR. Further verification typically occurs via restriction mapping and/or DNA sequencing. **(c)** A blue/white screen for cloned inserts. DNA inserted into the MCS of the plasmid disrupts the *lacZ* gene so that bacteria containing recombinant DNA are unable to metabolize X-gal, resulting in white colonies. Blue colonies do not contain inserts.

fragment of foreign DNA. In practice, however, a variety of DNA products will be present, including nonrecombinant plasmids and recombinant plasmids containing other fragments generated by the action of the restriction enzyme. However, not all the ampicillin-resistant bacteria will carry *recombinant* plasmids—that is, plasmids containing spliced-in DNA. But those bacteria that do contain recombinant plasmids can be readily identified because the *lacZ* gene has been disrupted by the foreign DNA, and as a result, β-galactosidase will no longer be produced. Step 6 shows how the lack of β-galactosidase can be detected by a simple color test in which bacteria are exposed to a substrate that is normally cleaved by β-galactosidase into a blue-staining compound. Bacterial colonies containing the normal pUC19 plasmid will therefore stain blue, whereas colonies containing recombinant plasmids with inserted DNA fragments will appear white (Figure 21-6c). Such a *blue/white screen* has historically been a routine procedure in the laboratory.

The final stage in any recombinant DNA cloning procedure is screening the bacterial colonies to identify those containing the specific DNA fragment of interest. For standard bacterial clones, this is usually done by isolating DNA from the bacteria and then using restriction mapping to confirm that the cloned DNA has the expected pattern, or by directly using PCR to confirm the presence of the correct insert (a technique called *colony PCR*). The cloned DNA is then often sequenced to provide precise verification of the nature of the inserted DNA (covered later in this chapter).

Genomic and cDNA Libraries Are Both Useful for DNA Cloning

Genomic Versus cDNA Libraries. As we've just seen, cloning foreign DNA in bacterial cells is now a routine procedure. In practice, obtaining a good source of DNA to serve as starting material is often one of the most difficult steps. Two different approaches are commonly used for producing the DNA starting material. In one approach, an organism's entire *genome* (or some substantial portion of it) is cleaved into a large number of restriction fragments, which are then inserted into cloning vectors for introduction into bacterial cells (or phage particles). The resulting group of clones is called a **genomic library** because it contains cloned fragments representing most, if not all, of the genome. Genomic libraries of eukaryotic DNA are valuable resources from which specific genes can be isolated, provided that a sufficiently sensitive identification technique is available. Once a rare bacterial colony containing the desired DNA fragment has been identified, it can be grown on a nutrient medium to generate as many copies of the fragment as may be needed. Of course, the DNA cuts made by a restriction enzyme do not respect gene boundaries, and some genes may be divided among two or more restriction fragments. This problem can be circumvented by carrying out a *partial DNA digestion* in which the DNA is briefly exposed to a small quantity of restriction enzyme. Under such conditions, some restriction sites remain uncut, increasing the probability that at least one intact copy of each gene will be present in the genomic library. Genomic libraries contain DNA that corresponds to all of the DNA in the genome of an organism, including DNA that is not transcribed, such as enhancers and promoters (see Chapter 20), and DNA that corresponds to the introns of primary RNA transcripts. As such, genomic libraries are crucial for studying which DNA elements regulate gene expression.

An alternative DNA source for cloning experiments is DNA that has been generated by copying messenger RNA (mRNA) with the enzyme *reverse transcriptase* (see Figure 16-6). This reaction generates a population of **complementary DNA (cDNA)** molecules that are complementary in sequence to the mRNA employed as template (**Figure 21-7**). If the entire mRNA population of a cell is isolated and copied into cDNA for cloning, the resulting group of clones is called a **cDNA library**. The advantage of a cDNA library is that it contains only those DNA sequences that are transcribed into mRNA—presumably, the active genes in the cells or tissue from which the mRNA was prepared.

Besides being limited to transcribed genes, a cDNA library has another important advantage as a starting point for the cloning of eukaryotic genes. Using mRNA to make cDNA guarantees that the cloned genes will contain only gene-coding sequences, without introns. Introns can be so extensive that the overall length of a eukaryotic gene can become too unwieldy for recombinant DNA manipulation. Using cDNA eliminates this problem. In addition, because bacteria lack mRNA splicing machinery, they cannot synthesize the correct protein product of an intron-containing eukaryotic gene unless the introns have been removed—as they are in cDNA.

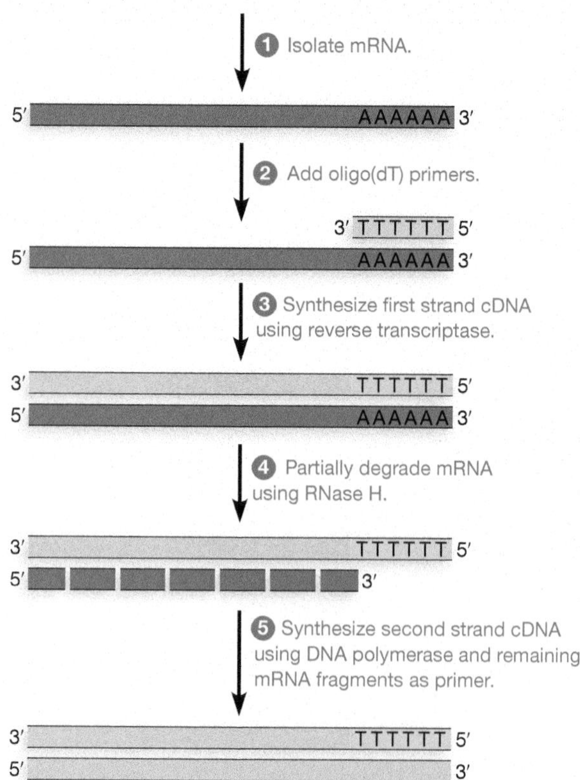

Figure 21-7 Preparation of Complementary DNA (cDNA) for Cloning. 1 mRNA is isolated. 2 Oligo(dT), a short chain of thymine deoxynucleotides, can be used as a primer because eukaryotic mRNA always has a stretch of adenine nucleotides at its 3′ end. 3 mRNA is incubated with reverse transcriptase (isolated from a retrovirus) to create a complementary DNA (cDNA) strand. 4 The resulting mRNA-cDNA hybrid is treated with alkali or an enzyme (RNase) to hydrolyze the RNA, leaving the single-stranded cDNA. 5 DNA polymerase can now synthesize the complementary DNA strand, using the partially degraded RNA as a primer.

MAKE CONNECTIONS 21.1

Reverse transcriptase requires a primer for generating the first cDNA strand. When making eukaryotic cDNA, a primer consisting of multiple consecutive T bases called an oligo(dT) primer is often used. Will this primer also be useful when generating prokaryotic cDNA? Why or why not? (Ch. 18.3)

Vectors for Cloning Genomic DNA. Bacterial plasmid vectors are useful for cloning cDNA, but they have an important limitation: the foreign DNA fragments cloned in these vectors cannot exceed about 15,000–20,000 base pairs (bp) in length. Eukaryotic genes (containing both introns and exons) are often larger than this and hence cannot be cloned in an intact form using such vectors. For genome-mapping projects, the availability of clones containing even longer stretches of DNA is desirable because the more DNA per clone, the fewer the number of clones needed to cover the entire genome.

Fortunately, there are many types of vectors available for cloning regions of genomic DNA (**Figure 21-8**). All of these vectors contain a replication origin so they can be copied, and most carry a gene to allow for drug selection. Small regions (20–25 kb in size) can be packaged in bacteriophage vectors like phage λ. A second such vector, known as a **cosmid**, uses elements derived from bacteriophages and carries DNA inserts from 40 to 45 kb in size. Cosmids contain *cos* sites of phage λ inserted into a plasmid. The *cos* sites allow the recombinant DNA molecules to be inserted, followed by packaging into phage protein coats; after infection of bacterial cells, the recombinant molecules are replicated and can be maintained as a plasmid.

Another type of vector used for cloning large DNA fragments is the **bacterial artificial chromosome (BAC)**, a derivative of the *F factor* plasmid that some bacteria employ for transferring DNA between cells during bacterial conjugation (see Chapter 25). BAC vectors are modified forms of the F factor plasmid that can hold up to 350,000 bp of foreign DNA and have all the components required for a bacterial cloning vector, such as replication origins, antibiotic-resistance genes, and insertion sites for foreign DNA.

Even larger pieces of DNA can be cloned using a **yeast artificial chromosome (YAC)**. A YAC is a "minimalist" eukaryotic chromosome—it contains all the DNA sequences needed for normal chromosome replication and segregation to daughter cells—and very little else. As you might guess from your knowledge of chromosome structure and replication (see Chapters 16 and 17), a eukaryotic chromosome requires three kinds of DNA sequences: (1) a eukaryotic origin of DNA replication, (2) two telomeres to allow periodic extension of the shrinking ends by telomerase, and (3) a centromere to ensure proper attachment, via a kinetochore, to spindle microtubules during cell division. If yeast versions of these three kinds of DNA sequences are combined with a segment of foreign DNA, the resulting YAC will replicate in yeast and segregate into daughter cells with each round of cell division, just like a

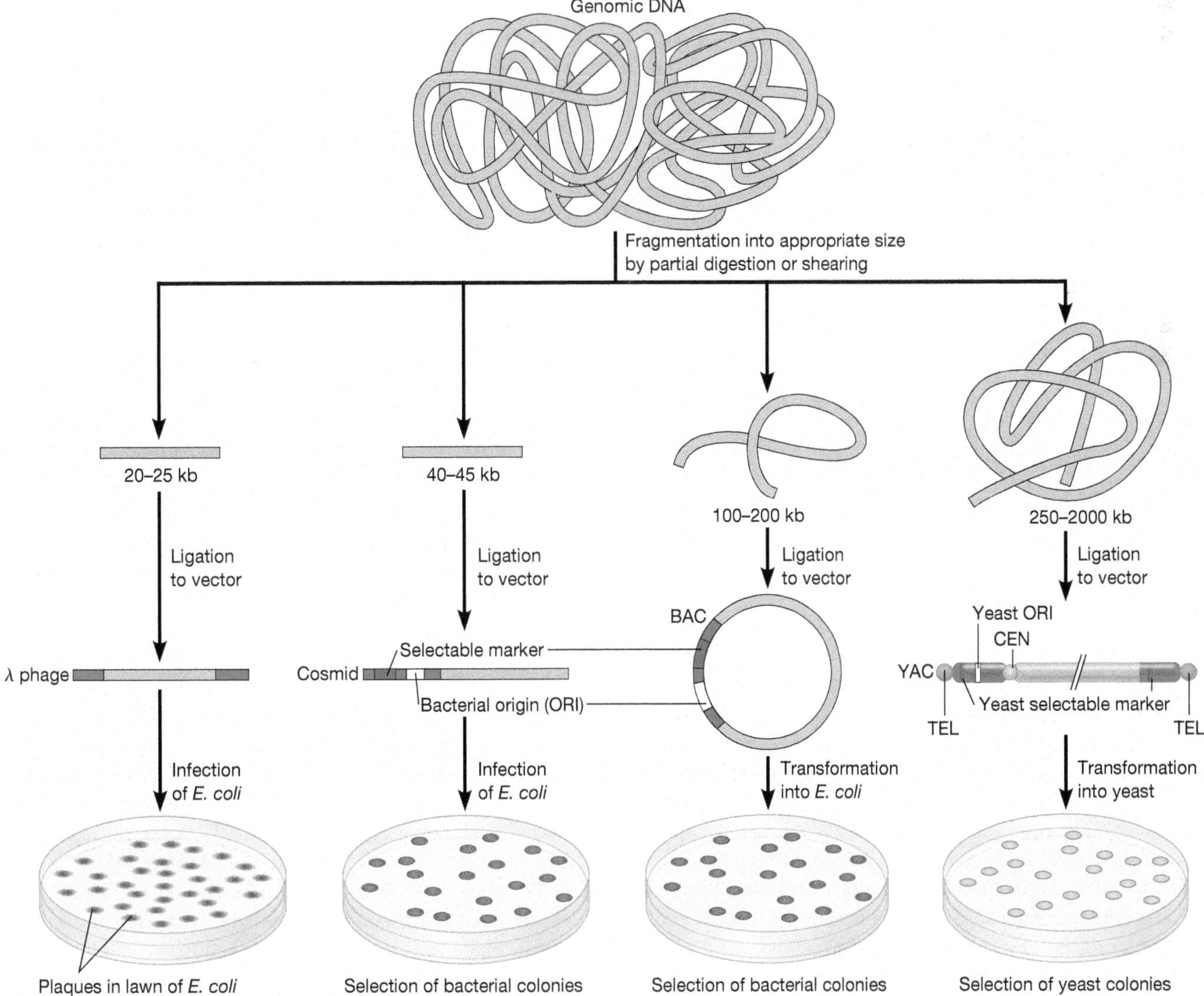

Figure 21-8 Constructing Genomic Libraries. Different vectors are used to construct genomic libraries whose inserts vary in size. Genomic libraries have individual pieces of DNA from the genome represented roughly equally.

natural chromosome. And under appropriate conditions, its foreign genes may be expressed.

In addition to a replication origin (ORI), centromere sequence (CEN), and two telomeres (TEL), YAC vectors carry two genes that function as selectable markers as well as convenient restriction sites. As with cloning in plasmids, the YAC vector and foreign DNAs are cleaved with the appropriate restriction enzymes, mixed together, and joined by DNA ligase. The resulting products, which include a variety of YACs carrying different fragments of foreign DNA, are introduced into yeast cells whose cell walls have been removed. The presence of two selectable markers makes it easy to select for yeast cells containing YACs with both chromosomal "arms." YACs can carry very large inserts; the YAC vector alone is only about 10,000 bp, but the inserted foreign DNA usually ranges from 300,000 to 1.5 million bp in length. In fact, YACs must carry at least 50,000 bp to be reliably replicated and segregated.

Screening Libraries for Clones of Interest. Once a library is constructed, it can be used for several purposes. Genomic libraries can serve as the starting point for whole-genome sequencing (see below). In other cases, however, a clone containing specific genomic or cDNA sequences is desired. To identify such clones of interest the library must be "screened." A historically important approach involved colony hybridization, a technique similar in approach to Southern blotting. By layering a nitrocellulose or nylon filter over a plate containing individual bacterial colonies or phage plaques and then probing the filter with a DNA probe, colonies containing the clone of interest could be identified. Today, PCR is much more commonly performed to extract DNA of interest from a sample of a library.

CONCEPT CHECK 21.1

By weight, spider silk is stronger than steel, so it is of great interest for potential uses in biotechnology. Suppose you are trying to clone DNA with the ultimate goal of producing spider silk proteins in bacteria. Would you start with spider genomic DNA or cDNA? Explain your answer.

21.2 Sequencing and Analyzing Genomes

The **genome** of an organism or virus consists of the DNA (or for some viruses, RNA) that contains one complete copy of all the genetic information of that organism or virus. For many viruses and prokaryotes, the genome resides in a single linear or circular DNA molecule or in a small number of them. Eukaryotic cells have a nuclear genome, a mitochondrial genome, and, in the case of plants and algae, a chloroplast genome as well. Mitochondrial and chloroplast genomes are single, usually circular DNA molecules resembling those of prokaryotes. The nuclear genome generally consists of multiple DNA molecules dispersed among a haploid set of chromosomes. (As we will explore in more detail in Chapter 25, a *haploid* set of chromosomes consists of one representative of each type of chromosome, whereas a *diploid* set consists of two copies of each type of chromosome, one copy from the mother and one from the father. Sperm and egg cells each have a haploid set of chromosomes, whereas most other types of eukaryotic cells are diploid.) DNA sequencing technologies have made it possible to analyze an organism's entire genome.

Rapid Procedures Exist for DNA Sequencing

At about the same time that techniques for preparing restriction fragments were developed, two methods were devised for rapid **DNA sequencing**—that is, determining the linear order of bases in DNA. One method was devised by Allan Maxam and Walter Gilbert, the other by Frederick Sanger and his colleagues. The Maxam–Gilbert method, called the *chemical method*, was based on the use of (nonprotein) chemicals that cleave DNA preferentially at specific bases. The Sanger procedure, called the *chain termination method*, utilizes dideoxynucleotides to interfere with the normal enzymatic synthesis of DNA. Sanger's method became the method of choice, and a version of this technique is still used today for routine sequencing.

The Sanger (Dideoxy) Method. In this procedure, a single-stranded DNA fragment is employed as a template to guide the synthesis of new complementary DNA strands. DNA synthesis is carried out in the presence of the *deoxynucleotides* dATP, dCTP, dTTP, and dGTP, which are the normal substrates that provide the bases A, C, T, and G to growing DNA chains. Also included, at lower concentrations, are four *dideoxynucleotides* (ddATP, ddCTP, ddTTP, and ddGTP), which lack the hydroxyl group attached to the 3′ carbon of normal deoxynucleotides. When a dideoxynucleotide is incorporated into a growing DNA chain in place of the normal deoxynucleotide, *DNA synthesis is prematurely halted* because the absence of the 3′ hydroxyl group makes it impossible to form a bond with the next nucleotide. Hence, a series of incomplete DNA fragments are produced whose sizes provide information concerning the linear sequence of bases in the DNA. By separating the fragments using techniques that can distinguish the size of each fragment with single-base precision, it is possible to infer which base resides at each position along a piece of DNA.

One historically important version of the Sanger method relied on adding radioactive bases to the dideoxy reactions, separating them by size on a polyacrylamide gel, and analyzing the results using X-ray film. **Figure 21-9** illustrates how Sanger sequencing is typically performed today. The unwieldy polyacrylamide gels have been replaced with single reaction tubes, and sequencing products are resolved using ultra-thin tubes of gel called *capillary gels*. Instead of spiking the sequencing reactions with radioactive nucleotides, fluorescent chain-terminating nucleotides are used. In Figure 21-9a step ❶, a reaction mixture is assembled that includes the dideoxynucleotides ddATP, ddCTP, ddTTP, and ddGTP, each labeled with a fluorescent dye of a different color (for example, ddATP = red, ddCTP = blue, ddTTP = orange, and ddGTP = green). These colored dideoxynucleotides are mixed with the normal deoxynucleotide substrates for DNA synthesis, along with a single-stranded DNA molecule to be sequenced and a short, single-stranded DNA *primer* that is complementary to the 3′ end of the DNA strand being sequenced. When DNA polymerase is added, it catalyzes the attachment of nucleotides, one by one,

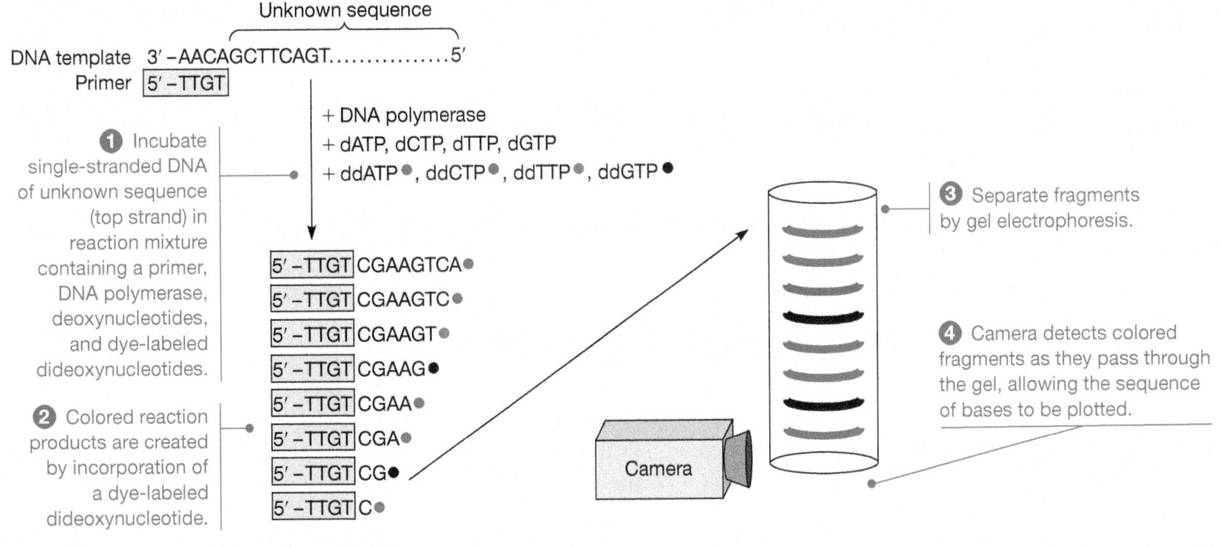

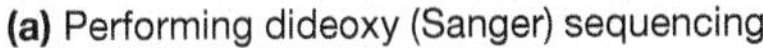

(a) Performing dideoxy (Sanger) sequencing

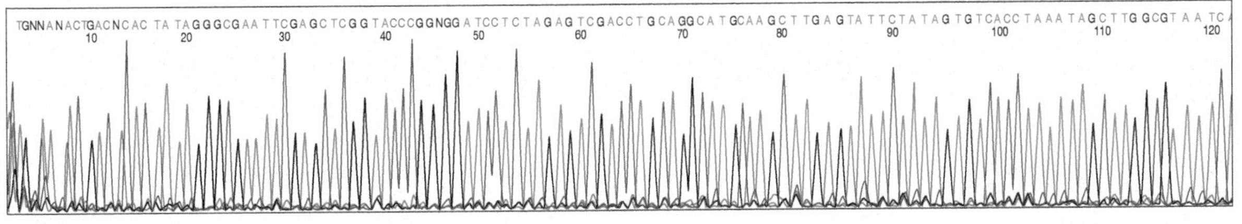

(b) Automated DNA sequencing readout

Figure 21-9 DNA Sequencing Using Dideoxy Chain Termination (Sanger Method). (a) The chain termination technique illustrated here, which employs dye-labeled dideoxynucleotides, has been adapted for use in high-speed, automated sequencing machines. Although this example summarizes the results obtained for only the first eight bases of a DNA sequence, experiments of this type typically determine the sequence of DNA fragments that are 500–800 bases long. The four main steps involved in the procedure are described in more detail in the text. **(b)** In a readout from an automated DNA sequencing run, peaks indicate which fluorescent nucleotide passed the detector at a given time.

to the 3′ end of the primer, producing a growing DNA strand that is complementary to the template DNA strand. Most of the nucleotides inserted are the normal deoxynucleotides because they are the preferred substrates for DNA polymerase. But every so often, at random, a colored dideoxynucleotide is inserted instead of its normal equivalent. Each time a dideoxynucleotide is incorporated, it halts further DNA synthesis for that particular strand. Consequently, ❷ a mixture of strands of varying lengths is generated, each containing a colored base at the end where DNA synthesis was prematurely terminated by incorporation of a dideoxynucleotide.

Next, in step ❸, the sample is subjected to electrophoresis in a capillary gel, which allows the newly synthesized DNA fragments to be separated from one another because the shorter fragments migrate through the gel more quickly than the longer fragments. As the fragments move through the gel, a special camera detects the color of each fragment as it passes by. ❹ This information allows the DNA base sequence to be determined. In this particular example, the shortest DNA fragment is blue, and the next shortest fragment is green. Because blue and green are the colors of ddCTP and ddGTP, respectively, the first two bases added to the primer must have been C followed by G. In automatic sequencing machines, such information is collected for hundreds of bases in a row and fed into a computer, allowing the complete sequence of the initial DNA fragment to be quickly determined. An actual trace from a commonly used automated sequencing device is shown in Figure 21-9b. The peaks of fluorescence correspond to the base in the resulting DNA polymer at which polymerization was halted.

Next-Generation and Third-Generation Sequencing Technologies. New sequencing technologies have dramatically increased the speed and decreased the cost of sequencing a given DNA sequence. To put into perspective how dramatic the technological advances have been in recent years, consider the sequencing of the human genome. This monumental achievement took more than a decade to complete and cost nearly \$3 billion. Today, continued advancements in automated sequencing technologies have made it possible to sequence a comparable-sized genome for less than \$1000 in under a day, and rapid progress continues to be made.

These newer technologies, known collectively as **next-generation DNA sequencing** technologies, have supplanted Sanger sequencing for *high-throughput sequencing* of complete genomes, as well as multi-organism communities (like microbiomes). These technologies can be uniformly characterized by the number of bases sequenced, either millions of shorter sequences or thousands of larger sequences called *reads*. They can also be broadly categorized as either *sequencing by synthesis* or *sequencing by ligation*.

Like the Sanger method, most of these emerging techniques rely on sequencing by synthesis. In other words, the sequence identity of the template DNA is determined as a result of monitoring new DNA synthesis. Most of these methods begin by fragmentation of the target genome, followed by amplification of the fragments (via PCR), and then parallel sequencing in millions of small fluid channels (each containing a unique piece of template DNA).

The first next-generation technique called *pyro-sequencing* was developed by the company 454 Life Sciences. In this technique, solutions containing each of the four labeled nucleotides and other chemicals, including the firefly enzyme, *luciferase*, are introduced. When the solution containing the nucleotide complementary to the next template base is added, incorporation of this nucleotide results in the release of pyrophosphate, which ultimately produces a flash of light from the luciferase that can be detected by a sensor. This technique is similar to ion semiconductor sequencing, which detects the H^+ ions that are released whenever a base is incorporated into a growing DNA strand. Pyro-sequencing, however, has been supplanted by higher-throughput technologies.

Another sequencing by synthesis technique, developed by the company Illumina, utilizes special dNTPs that are tagged with a unique fluorescent signal, using a different color for each base. In addition, a reversible terminator blocks the addition of subsequent bases. Therefore, once a base is added it can be identified in real time by its fluorescence emission. The terminator and fluorescent tag are then removed so that the next base can be added and recorded. This method can produce short read lengths ranging from 50 to 300 nucleotides. Illumina sequencers vary in the read lengths and number of reads produced. Furthermore, these reads can be either single-end or paired-end (sequenced from both ends of the DNA template fragment). Within a matter of hours, around 10 million reads of sequence data can be produced; the Illumina HiSeq platform is now capable of producing 8 billion reads of sequence data in just 6 days!

An example of sequencing by ligation is called SOLiD (supported oligonucleotide ligation and detection). Instead of relying on polymerization of new DNA, two-base probes associate with single-stranded template DNA and are detected with fluorescence. Then the next probe can be ligated and detected. Read lengths are small, averaging 50–75 bases in size.

Third-generation DNA sequencing technologies are focused on producing longer reads lengths than 454 Life Sciences, Illumina, and SOLiD sequencers. A technique called single-molecule real-time sequencing was developed by Pacific Biosciences (PacBio). Like Illumina, it also measures incorporation of single, fluorescent nucleotides. However, the template is not amplified prior to the run. This technique produces longer reads that are 30,000 bases in size on average, with some reads over 100 kb in length. A single PacBio sequencing run can result in 20–50 Gb of data. Another third-generation technology called *nanopore sequencing* produces long reads but does not commonly involve a synthesis step. Instead, special protein pores are embedded within an electronically resistant, artificial membrane. As single-stranded DNA passes through a pore, an applied electrical field is disrupted. Each base that passes through the pore changes the current in a distinct way so that each base can be identified. Nanopore sequencing reads are routinely 10 to 100 kb in length on average, with some produced reads up to 1 Mb in length. The nanopore MinION sequencer is the only portable sequencer on the market, fitting in the palm of the hand. This sequencer can produce 10–30 Gb of data in hours.

A huge advantage of these modern technologies over the traditional Sanger method is that they can perform millions of reactions in parallel. Another advantage of next-generation sequencing technologies over dideoxy sequencing is that it is not necessary to clone the DNA before sequencing because the template DNA is typically directly amplified by PCR. In addition to increasing speed, these technologies allow the sequencing of tiny amounts of DNA. Next-generation techniques have been used, for example, to sequence DNA obtained from extinct animals such as woolly mammoths preserved in the Siberian permafrost and extinct *Homo* species, such as Neanderthals. The portable, third-generation nanopore minION sequencer has been used in the field to do real-time monitoring of Ebola and Zika outbreaks in Africa.

These new sequencing technologies are also creating societal challenges. As the price of sequencing a human genome continues to drop, the cost is becoming so affordable that doctors can now order a copy of a patient's genome sequence to facilitate treatment decisions best suited to the genetic makeup of that patient. But with this power comes challenges. One relates to data storage, as sequencing machines continue to churn out billions and billions of base pairs of sequence. Another is the confidentiality of DNA sequence data corresponding to individual patients, which, in the wrong hands, could be used to deny health coverage or to discriminate in other ways. As has been true in the past, society is struggling to keep pace with the ethical challenges presented by these new technologies.

Bisulfite Sequencing for Detecting DNA Methylation. DNA methylation is important in gene regulation (see Chapter 20), so identifying methylated DNA in a sequence can be valuable. One way to identify methylated bases involves treating DNA with *bisulfite*. This converts cytosine residues to uracils but leaves 5-methylcytosine residues unaffected. By comparing the DNA sequences obtained from untreated and bisulfite-treated DNA, it is possible to identify individual cytosines that have been methylated.

MAKE CONNECTIONS 21.2

You perform bisulfite sequencing within a CpG island from a gene promoter and find no differences when comparing untreated DNA with the bisulfite-treated DNA. What can you conclude about the expression of this gene? (Fig. 20-17)

Whole Genomes Can Be Sequenced

It is easy today to take for granted the vast amount of genetic information available for performing molecular comparisons between organisms and for analyzing the genomes of individual organisms, including human patients. It is also easy to take for granted the speed with which computers can wade

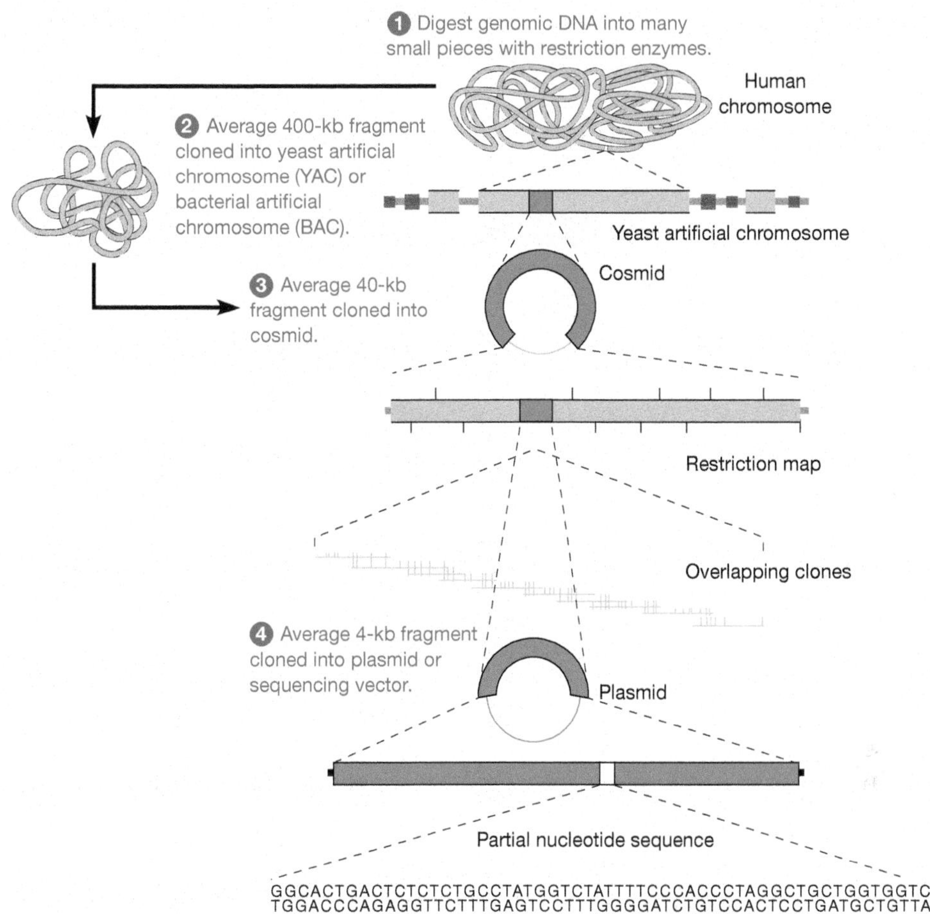

1 Digest genomic DNA into many small pieces with restriction enzymes.
Human chromosome
2 Average 400-kb fragment cloned into yeast artificial chromosome (YAC) or bacterial artificial chromosome (BAC).
Yeast artificial chromosome
3 Average 40-kb fragment cloned into cosmid.
Cosmid
Restriction map
Overlapping clones
4 Average 4-kb fragment cloned into plasmid or sequencing vector.
Plasmid
Partial nucleotide sequence
GGCACTGACTCTCTCTGCCTATGGTCTATTTTCCCACCCTAGGCTGCTGGTGGTCTACCC
TGGACCCAGAGGTTCTTTGAGTCCTTTGGGGATCTGTCCACTCCTGATGCTGTTATGG

Figure 21-10 Map-Based (Clone-by-Clone) Sequencing of Genomic DNA. Genomic DNA is cloned prior to sequencing. Restriction maps are used to help align and assemble clones into contigs. Clones are then "subcloned" into smaller pieces and sequenced.

through vast quantities of DNA sequence data to find matching sequences. Such commonplaces are, however, relatively recent in the history of cell biology; indeed, the sequencing of the human genome began in earnest roughly 30 years ago. Historically, efforts to sequence whole genomes have started with using one or more of the automated sequence techniques mentioned previously. Then the resulting data (often millions of short sequences) are assembled into the complete build of the genome.

Sequencing Using Map-Based Cloning. Before the development of so-called shotgun sequencing (see below), genomes were assembled using a clone-by-clone, or **map-based cloning**, approach (**Figure 21-10**). For example, the first sequenced genome of a multicellular eukaryote, the nematode *Caenorhabditis elegans*, was sequenced in this way. The initial stage of the *Human Genome Project*, begun in 1990 and coordinated by Francis Collins at the National Institutes of Health in the United States, was a cooperative international effort involving hundreds of scientists who shared their data in an attempt to determine the entire sequence of the human genome. It relied on this map-based approach.

In map-based cloning, individual DNA fragments, generated from restriction digests of chromosomes, are aligned using known chromosomal markers to create a physical "map" of a chromosome. Recall that bacterial artificial chromosomes (BACs) and yeast artificial chromosomes (YACs) can carry very large inserts of genomic DNA. BAC and YAC libraries from individual chromosomes could then be made. Before high-throughput sequencing became available, larger clones were typically "subcloned" into smaller fragments in cosmid or plasmid vectors, and their inserts were then completely sequenced. The sequenced clones could be painstaking aligned and ultimately sequenced to create overlapping, or *contiguous*, sequences, called **contigs** for short. Bioinformatics approaches (described later in this section) could then be used to analyze the assembled sequences.

Whole-Genome (Shotgun) Sequencing. Map-based cloning is time-consuming but meticulous. Another more widely used strategy for rapidly sequencing and assembling an entire genome involves variations of a method called **whole-genome sequencing**, or **shotgun sequencing**. **Figure 21-11** on page 636 provides an overview of whole-genome sequencing. ❶ Entire chromosomes are cut into thousands to millions of short, overlapping fragments, either mechanically (traditionally done by sonication), using a transposon-dependent system, or using restriction enzymes. In theory, these fragments can be assembled into a giant contig, but how? J. Craig Venter and colleagues at The Institute for Genomic Research (TIGR) pioneered the basic approach, which relies on ❷ sequencing the random fragments and then ❸ using computer algorithms to assemble the resulting sequences into contigs. In 1995, Venter and colleagues used this approach to sequence the genome of the bacterium *Haemophilus influenzae*, which has a genome of 1.8 million bp. This was the first completed genome sequence from a free-living organism and served as a proof of concept for the whole-genome approach.

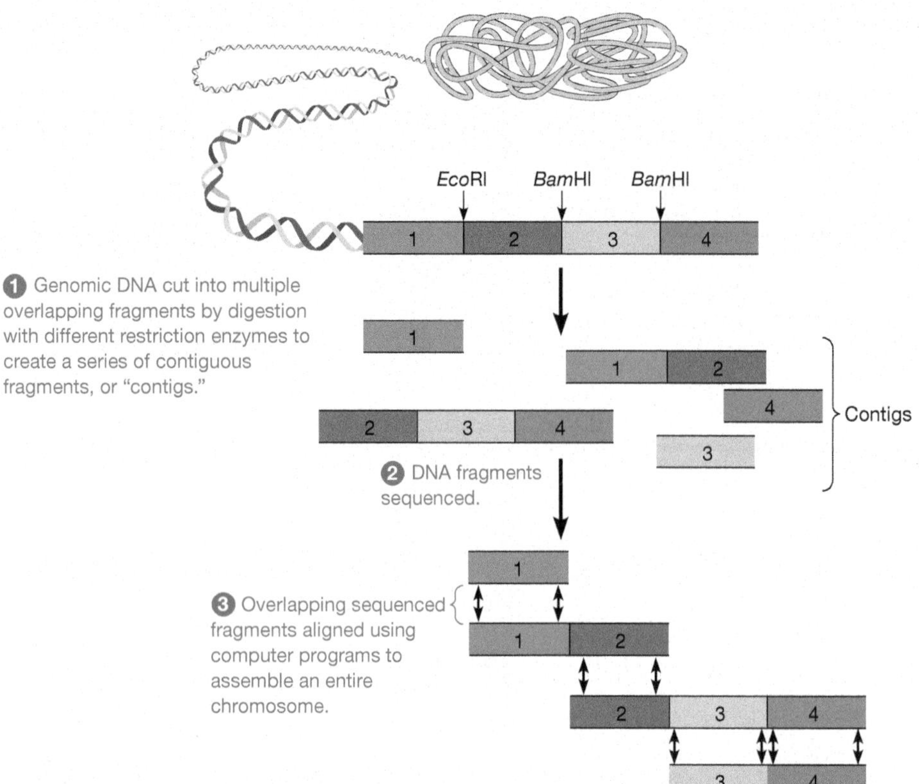

EcoRI
BamHI
BamHI
1
2
3
4
1 Genomic DNA cut into multiple overlapping fragments by digestion with different restriction enzymes to create a series of contiguous fragments, or "contigs."
Contigs
2 DNA fragments sequenced.
3 Overlapping sequenced fragments aligned using computer programs to assemble an entire chromosome.

Figure 21-11 "Shotgun" Sequencing of Genomic DNA. In this example, two different restriction enzymes (*Bam*HI and *Eco*RI) are used to fragment genomic DNA, which is then sequenced and aligned using bioinformatics to identify overlapping fragments (contigs).

As high-throughput sequencing became more and more affordable and computing power steadily increased, whole-genome sequencing replaced map-based sequencing as the method of choice. Variations of the shotgun approach were subsequently used for sequencing many genomes, including those of flies, dogs, and many bacteria. By comparison, the clone-by-clone approach is time consuming because it requires cloning of DNA fragments into vectors, transforming bacteria or yeast using the cloned DNA, and so on. Map-based sequencing still has its uses in the final, "cleanup" phase of assembling whole genomes. It is especially useful in regions where DNA sequences are highly repetitive, and hence random sequences are hard to align correctly, or for filling in small gaps left after whole-genome approaches have been applied.

Exactly how one approaches whole-genome sequencing today depends on whether the goal is to sequence a new genome built from scratch (de novo assembly) or through the use of an existing reference genome. For de novo assembly, a hybrid approach is applied. Because long-read techniques have a higher error rate than short-read techniques, long reads (using a method like single-molecule real-time sequencing) are usually used to provide the orientation and general spacing among contigs containing unique sequences. Think of this as building the "skeleton" for the genome. Short reads (using an approach like Illumina sequencing) can then be used to capture areas of high variability and repeated DNA sequences, filling in the gaps. If a reference genome exists, the short reads generated with Illumina sequencing are usually sufficient to piece together the final build (using the reference sequence as a guide). This is referred to as *reference-based assembly*, or simply *mapping*.

To us as human beings, the ultimate goal of DNA sequencing was to sequence the human genome. How great a challenge was that? To answer this question, we need to realize that the human nuclear genome contains about 3.2 billion bases, which is roughly 1000-fold more DNA than is present in an *E. coli* cell. One way to comprehend the magnitude of such a challenge is to note that in the early 1990s, when genome sequencing efforts began in earnest, it required almost 6 years for the laboratory of Frederick Blattner to determine the complete base sequence of the *E. coli* genome. At that rate, it would have taken a single lab almost 6000 years to sequence the entire human genome! In the late 1990s Craig Venter's approach was applied by the company he founded, Celera Genomics, to the human genome, complementing work done by the teams coordinated by Francis Collins and others. Through these efforts, the complete sequence of the human genome was determined by 2003, roughly 2 years ahead of schedule.

Comparative Genomics Allows Comparison of Genomes and Genes Within Them

The significance of automated DNA sequencing and genome-assembly technology for genomics can scarcely be overestimated. These technologies have enabled comparisons of the sequences of individual genes and the analysis of entire genomes of hundreds of individual species, as well as within microbial communities (a field called *metagenomics*). A lot of metagenomic work has focused on the study of microbial communities (or the microbiome) within the human body. Two examples of the power of this approach are analysis of genome size and phylogenetic analysis using specific DNA sequences.

Genome Size. Many of the initial successes in genome sequencing involved bacteria and bacterial viruses because they have relatively small genomes. The first complete genome produced by Fredrick Sanger in 1977 was that of the bacteriophage phiX174, which is just 5386 nucleotides long. Genome size is usually expressed as the total number of base-paired nucleotides, or **base pairs (bp).** For example, the circular DNA molecule that constitutes the genome of an *E. coli* cell has 4,639,221 bp. Because such numbers tend to be rather large, the abbreviations **kb** (kilobases), **Mb** (megabases), and **Gb** (gigabases) are used to refer to a thousand, or million, or billion base pairs, respectively. Thus, the size of the *E. coli* genome can be expressed simply as 4.6 Mb. Complete DNA sequences are now available for more than 200,000 different bacteria, including those that cause a variety of human diseases. DNA sequencing has also been successfully applied to much larger genomes, including those from several dozen organisms that are most important in biological research (**Table 21-2**).

A closer look at the range of genome sizes in a variety of organisms (including nonmodel organisms, whose genomes are not selected by virtue of being compact) raises a puzzle, however. As shown in **Figure 21-12**, these data reveal a spread of almost eight orders of magnitude in genome size, from a few thousand base pairs for the simplest viruses to more than 100 billion base pairs for certain plants, amphibians, and protists. Broadly speaking, genome size increases with the complexity of the organism. Most viruses contain enough nucleic acid to encode only a few or a few dozen proteins; bacteria can specify a few thousand proteins; and eukaryotic cells have enough DNA (at least in theory) to encode hundreds of thousands of proteins. But closer examination of the data reveals that the genome sizes of eukaryotes exhibit great variations that do not clearly correlate with any known differences in organismal complexity. Some amphibians and plants, for example, have gigantic genomes that are tens or even hundreds of times larger than those of other amphibians or plants or of mammalian species. *Trillium*, for example, is a member of the lily family that has no obvious need for exceptional amounts of genetic information. Yet its genome size is more than 20 times that of pea plants and 30 times that of humans. One kind of amoeba has a genome that is 200 times the size of the human genome!

Table 21-2 Examples of Sequenced Genomes

Organism	Genome Size*	Estimated Gene Number**
Bacteria		
Mycoplasma genitalium	0.6 Mb	470
Haemophilus influenza	1.8 Mb	1740
Streptococcus pneumoniae	2.2 Mb	2240
Escherichia coli	4.6 Mb	4400
Eukaryotes		
Yeast *(S. cerevisiae)*	12.2 Mb	6700
Roundworm *(C. elegans)*	100 Mb	20,500
Mustard plant *(A. thaliana)*	120 Mb	27,000
Fruit fly *(D. melanogaster)*	170 Mb	13,900
Rice *(O. sativa)*	374 Mb	36,000
Mouse *(Mus musculus)*	2700 Mb	23,000
Human *(H. sapiens)*	3100 Mb	21,000

*Genome sizes and gene numbers are rounded for simplicity.
**Gene numbers are subject to change and based on Ensembl predictions of coding genes (minus pseudogenes and short noncoding predictions when available).

Historically, these variations in genome size became known as the *C-value paradox:* genome size does not appear to obviously reflect the total number of genes in eukaryotes. The reasons for these wide variations have become clearer due to whole-genome sequencing. In some species, large-scale duplications of parts of the genome have occurred. In addition, the total amount of repetitive DNA (see Chapter 16) and other noncoding DNA in eukaryotic genomes varies widely (see Chapter 20); the functions of these genomic regions are still being worked out.

The Tree of Life. Large-scale sequencing of the genomes of organisms has dramatically altered how biologists think about the "tree of life" (see Figure 4-3). Prior to the availability of DNA sequence information, the morphological and other visible features of organisms were the primary tools for establishing evolutionary relationships. DNA sequence information has been able to clarify the relationships between organisms that appearance alone was unable to resolve. This has been especially true for bacteria and archaea.

One of the most useful approaches for establishing where branches lie in the tree of life is to consider how sequences vary within the same gene in many different organisms. To construct a basic tree (see Figure 4-3), *homologous nucleotides* were compared and aligned among many DNA sequences. Homologous nucleotides are those that are inferred to be descended from the same nucleotide in a common ancestor of the two species

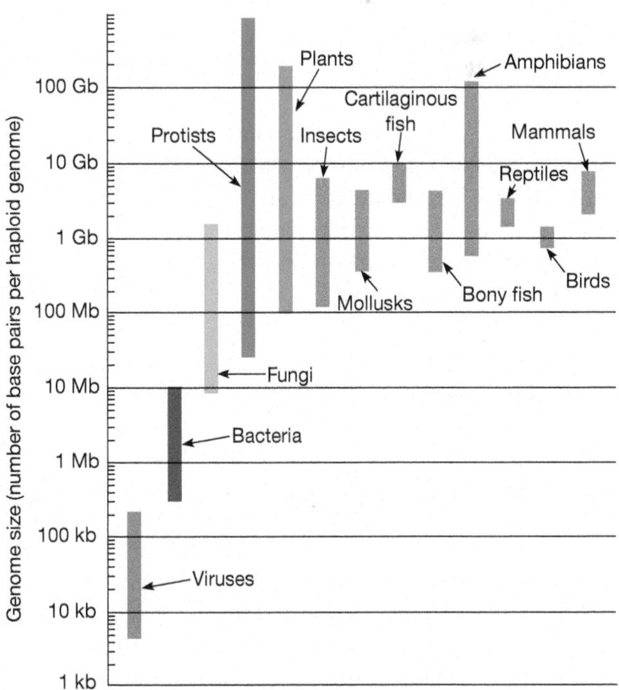

Figure 21-12 Relationship Between Genome Size and Type of Organism. For each group of organisms shown, the bar represents the approximate range in genome size measured as the number of base pairs per haploid genome. The same color (purple) is used for groups that involve members of the animal kingdom.

being compared. The tree is constructed by examining the most "parsimonious" set of changes in DNA sequences that could generate the set of sequences being examined, that is, the shortest path to the observed diversity of sequences. More detailed comparisons are then performed to generate finer and finer branches on the tree. Conserved genes that are specific to particular groups and comparison of regulatory elements and noncoding regions, which are known to change rapidly over time, are especially useful in this phase of the process.

Some surprising results have emerged from this sort of analysis that are relevant for cell biologists. For example, fungi (including budding and fission yeast, two commonly used model organisms in cell biology) are relatively more closely related to metazoans (multicellular animals) than previously realized; conversely, fungi are less closely related to plants than previous suspected. Other key insights include genetic evidence that bolsters the endosymbiont theory of the origins of mitochondria and chloroplasts (see Chapter 4) and widespread evidence for lateral gene transfer between prokaryotes. This has led some to say that the "tree" of life is, more properly, a "bush" of life!

The Field of Bioinformatics Helps Decipher Genomes

Because of its sheer scale, as well as its potential impact on our understanding of human evolution, physiology, and disease, sequencing the human genome is one of the crowning achievements of modern biology. And yet, unraveling the sequence of bases was the "easy" part. Next comes the hard part: figuring out the meaning of this sequence of 3 billion A's, G's, C's, and T's. Many questions remain: Which stretches of DNA correspond to genes? When and in what tissues are these genes expressed? What kinds of proteins do they encode? How do all these proteins function, and how do they interact?

Bioinformatics. The prospect of analyzing such a vast amount of data has led to the emergence of a new discipline, called **bioinformatics**, which merges computer science, statistics, and biology in an attempt to make sense of it all. Such analyses suggest the presence of about 20,000 protein-coding genes in the human genome, roughly half of which were not known to exist prior to genome sequencing. The fascinating thing about this estimate is that it means humans have only about twice the number of genes as a fruit fly, barely more genes than a worm, and 12,000 fewer genes than a rice plant! Computer analysis has also revealed that less than 2% of the human genome actually encodes proteins. The remaining 98% contains many important regulatory elements, some genes that encode noncoding RNA products instead of proteins (see Chapter 20), transposons and transposon-related sequences (see Chapter 17), and other nonfunctional sequences. Because proteins are encoded by portions of the genome that are retained in exons (that is, those parts of the RNA transcripts that are retained after splicing), the exonic portion of the genome is often called the **exome**. Sequencing of the exome in human patients has shown that a much higher percentage of mutations associated with human disease are found in the exome than in other parts of the genome.

Determining the DNA sequence of an organism's genome can provide only a partial understanding of the functions a genome performs. Scientists must look beyond the genome to examine the molecules it produces. Because the first step in gene expression involves transcription of genome sequences into RNA, techniques have been developed for identifying **transcriptomes**—that is, the entire set of RNA molecules produced by a genome. You will learn more about the technology that lies behind analyzing transcriptomes later in this chapter.

Most RNAs, in turn, are used to guide the production of proteins, so scientists are also studying **proteomes**—the structure and properties of every protein produced by a genome. An organism's proteome is considerably more complex than its genome. For example, the roughly 20,000 genes in human cells produce hundreds of thousands of different proteins. You have learned that one way that many more proteins can be produced than there are genes is via *alternative splicing* (see Chapter 20). Moreover, the resulting proteins are subject to subsequent posttranslational biochemical modifications that produce either new proteins or multiple versions of the same protein. Later in this chapter you will see how interactions between proteins in the cell (the cell's *interactome*) are investigated.

Bioinformatics Tools. The enormous amount of DNA and protein sequence data being collected present a daunting challenge to scientists who wish to locate information about a particular gene or protein. To cope with this problem, the most recent DNA and protein sequences from thousands of organisms are stored in several online databases, and software has been developed to help researchers find the information they need. Among the more widely used tools is **BLAST (Basic Local Alignment Search Tool)**, a computer program originally developed in 1990 by Stephen Altschul, David Lipman, and colleagues to search for homologous sequences, typically using the GenBank database. Most end-users access the BLAST search engine through a web-based portal. One of the most highly accessed of these portals is hosted by the U.S. National Center for Biotechnology Information (NCBI; http://blast.ncbi.nlm.nih.gov/Blast.cgi). Several types of BLAST searches can be performed. These include (1) *nucleotide blast (blastn)*, in which a nucleotide sequence is compared to nucleotide sequences in the database; (2) *protein blast (blastp)*, in which an amino acid sequence is compared to protein sequences in the database; (3) *tblastn*, in which a protein sequence is compared with nucleotide sequences in the database by producing hypothetical translations of the nucleotides into all six possible reading frames; (4) *tblastx*, in which a nucleotide query sequence is translated into all six possible reading frames and compared against the nucleotide sequences in the database that have also been translated into all six possible reading frames; and (5) *blastx*, in which a nucleotide sequence is translated into all six possible reading frames and compared to protein sequences in the database. Using these tools, for example, if you identify a new gene in an organism that has not been previously studied and determine its base sequence, a BLAST search could quickly determine whether humans (or any other organism in the database) possess a similar gene.

information is obtained about the abundance of the transcript from which that cDNA was derived. In addition, variation in the form of single nucleotide polymorphisms, rare alternative splice forms, and RNA editing can be detected. Like other shotgun methods we encountered in discussing DNA genome sequencing, RNAseq does not require any cloning, and so it, too, is rapid.

Improvements in the quality and sensitivity of RNAseq have led to the development of techniques to allow analysis of single cells. A key to such techniques is successful isolation of the single cell, which can be done using approaches that involve limiting dilution plating, laser-capture microscopy, or flow cytometry. Current and future applications of single-cell transcriptome analysis are truly remarkable. The individual cells from a tumor could be analyzed for their heterogeneity and response to chemotherapeutic drugs. Single-cell RNAseq could also be used to diagnose certain tumor types or other disorders. Finally, precisely how certain cell lineages develop could be teased out using this technique. The Human Cell Atlas is an international conglomerate whose goal is to sequence millions of single-cell transcriptomes to eventually characterize and classify all 35 trillion cells in the human body.

Proteins Can Be Studied Using Electrophoresis

Proteins can be separated by size using *SDS–polyacrylamide gel electrophoresis* (*SDS-PAGE*), a powerful technique crucial for characterizing a mixture of proteins (**Figure 21-19**). ❶ Proteins are first solubilized with the anionic detergent SDS along with heat, which disrupts most protein-protein associations. The proteins denature, unfolding into stiff polypeptide rods that cannot refold because their surfaces are

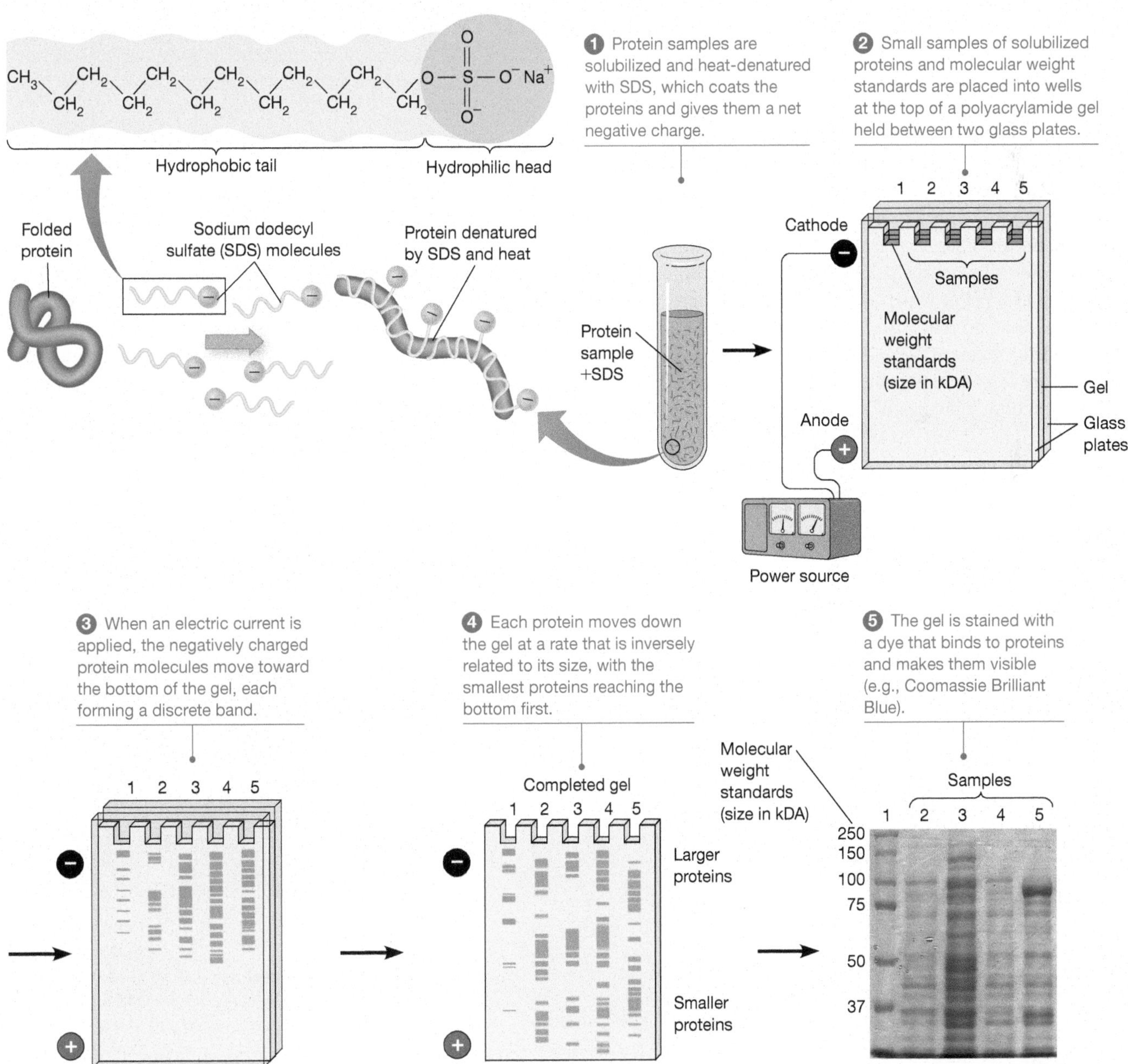

Figure 21-19 SDS–Polyacrylamide Gel Electrophoresis (SDS-PAGE). Steps ❶ through ❺ show the general procedure used to isolate and separate proteins using SDS-PAGE.

coated with negatively charged detergent molecules. ❷ The solubilized, SDS-coated polypeptides are then loaded into wells on the top of a polyacrylamide gel. Typically, a set of purified proteins of known molecular weights is run in one lane of the gel alongside the other samples to determine the molecular weights of the other polypeptides. ❸ An electrical potential is applied across the gel, such that the bottom of the gel is the positively charged anode. Because the polypeptides are coated with negatively charged SDS molecules, they migrate down the gel toward the anode. The polyacrylamide gel can be thought of as a fine meshwork that impedes the movement of large molecules more than that of small molecules. As a result, polypeptides move down the gel at a rate that is inversely related to their size. ❹ When the smallest polypeptides approach the bottom of the gel, the process is terminated. ❺ The gel is then stained with a dye (usually using a blue dye called Coomassie Brilliant Blue) that binds to polypeptides and makes them visible.

Variations on the SDS-PAGE technique provide additional tools that are part of the cell biologist's toolkit. SDS-PAGE relies on SDS to minimize differences in charge among the separated proteins. Sometimes, however, it is advantageous to use charge differences to separate proteins prior to the SDS-PAGE step. This technique, known as **two-dimensional (2D) gel electrophoresis,** is shown in **Figure 21-20**. In the first "dimension," ❶ proteins are separated in a polyacrylamide tube gel using *isoelectric focusing*. Isoelectric focusing relies on separating proteins by charge using a pH gradient; proteins stop migrating at a pH at which their net charge is zero, a point known as that protein's *isoelectric point*. ❷ The tube gel is then loaded into an SDS-PAGE gel, and the charge-separated proteins are forced to migrate by size through the SDS-PAGE gel (the second "dimension"). ❸ The proteins can then be detected by Coomassie staining the gel. Two-dimensional gels can resolve subtle differences in various isoforms of a protein, which often have differing isoelectric points, or detect post-translational modifications of a protein. As is the case for other types of electrophoresis, standards that have well-characterized charges and sizes are often used to help align protein spots from gel to gel.

Antibodies Can Be Used to Study Specific Proteins

Gel electrophoresis is powerful, but SDS-PAGE and even 2D gel electrophoresis have limitations similar to those you encountered with nucleic acid gels: how can a specific protein within a complex mixture be identified? In the case of nucleic acids, probes could be used to identify the DNA or RNA of interest amid a very complicated mixture of other DNAs or RNAs. Nucleic acid probes rely on hybridization of complementary sequences via hydrogen bonding, which is highly sequence-specific. Because proteins lack this convenient feature, another kind of probe is needed. Fortunately, such probes exist: *antibodies*.

Antibodies are soluble proteins that bind to substances, referred to as *antigens*, that provoke an immune response. Antibody molecules recognize and bind to specific antigens with extraordinary precision, making them ideally suited for targeting specific antigens, including proteins, and so they are incredibly valuable for cell biologists. In mammals, antibodies are produced by *B lymphocytes*, or *B cells*. Each lymphocyte produces one unique type of antibody molecule. During an immune response to a particular antigen, B lymphocytes that produce an antibody specific for that antigen rapidly divide to produce a group of cells derived from the original cell. Such

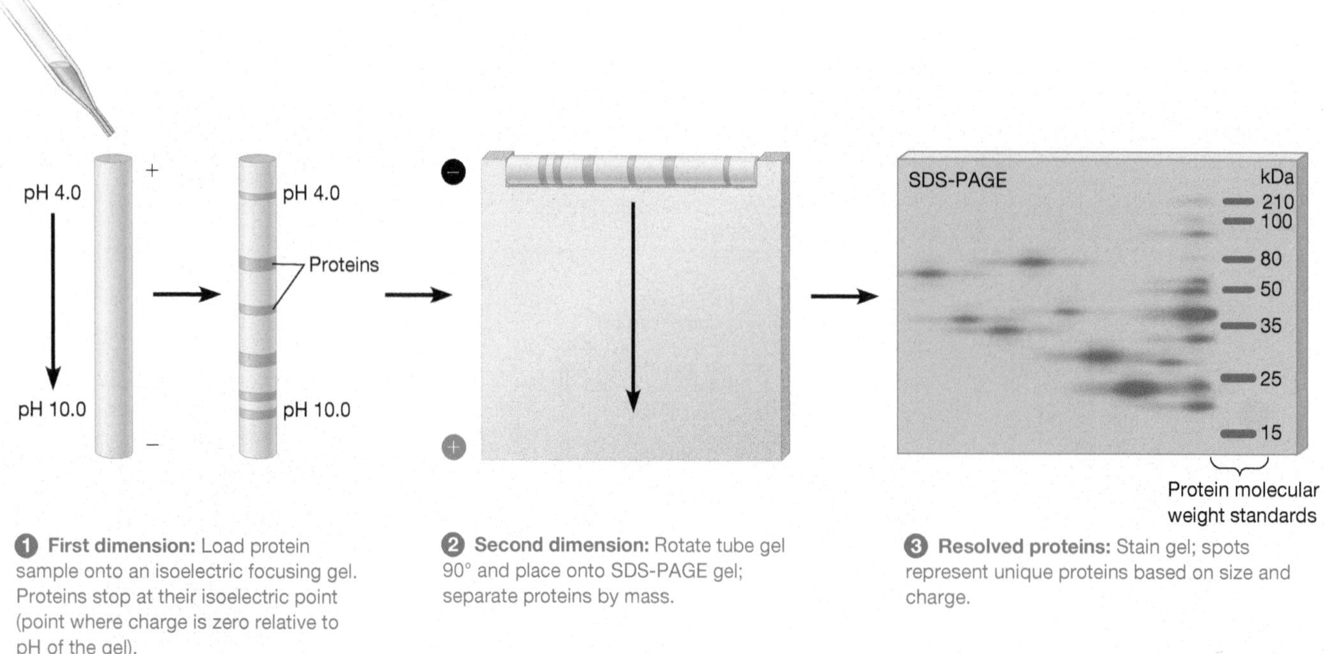
pH 4.0
+
pH 10.0
–
pH 4.0
Proteins
pH 10.0
SDS-PAGE
kDa
210
100
80
50
35
25
15
Protein molecular weight standards
❶ First dimension: Load protein sample onto an isoelectric focusing gel. Proteins stop at their isoelectric point (point where charge is zero relative to pH of the gel).
❷ Second dimension: Rotate tube gel 90° and place onto SDS-PAGE gel; separate proteins by mass.
❸ Resolved proteins: Stain gel; spots represent unique proteins based on size and charge.

Figure 21-20 Two-Dimensional (2D) Gel Electrophoresis of Proteins. Two-dimensional gel electrophoresis is useful for separating a complex mixture of proteins isolated from cells or tissues.

cells are said to be *clonally related.* B lymphocytes are the starting point for two different types of antibodies used in cell biology.

Types of Antibodies. The most straightforward way to produce antibodies for use in cell biology applications is to inject an antigen into a host animal. The animal's immune system produces antibodies against the antigen, which can then be purified from the plasma in the animal's bloodstream. Such antibodies are said to be **polyclonal antibodies;** that is, they represent a mixture of antibodies, produced by different groups of clonally related B cells. Such antibodies are useful because they are a mixture of different antibody molecules that recognize different antigenic sites along a protein and are not directed against a single region of the protein. Polyclonal antibodies are limited, however, because they are nonrenewable; that is, once the antibodies produced in the original immune response from the original animal are expended, the procedure must be repeated by immunizing new animals.

Another type of antibody has the advantage of being renewable. These antibodies are grown from cultured cells, which can be maintained in the laboratory. Because they are grown from single cells (rather than different sets of clonally related cells), they are called **monoclonal antibodies.** The procedure for growing cells that produce monoclonal antibodies, originally developed in 1975 by Georges Köhler and César Milstein, is shown in **Figure 21-21**. In this technique, ❶ mice or rats are first injected with an antigen of interest, and then ❷ antibody-producing B lymphocytes are isolated from the animals a few weeks later. Within such a lymphocyte population, each lymphocyte produces a single type of antibody directed against one particular antigen. ❸ To facilitate selection and growth of individual lymphocytes, they are fused with cells that divide rapidly and have an unlimited life span when grown in culture, which were originally derived from cancer cells of the immune system called *myeloma cells.* ❹ Individual hybrid cells are then selected and grown to form a series of clones called *hybridomas.* The antibodies produced by hybridomas are referred to as "monoclonal" because each one is a pure antibody produced by a group of cells originally derived from one lymphocyte. Because they are derived from cells that can be grown indefinitely in the laboratory, monoclonal antibodies are a renewable resource. They are limited, however, in that they recognize a single region of the antigen. Sometimes such high specificity is useful, but in other cases it means that monoclonal antibodies may not react against a portion of a protein that is of interest.

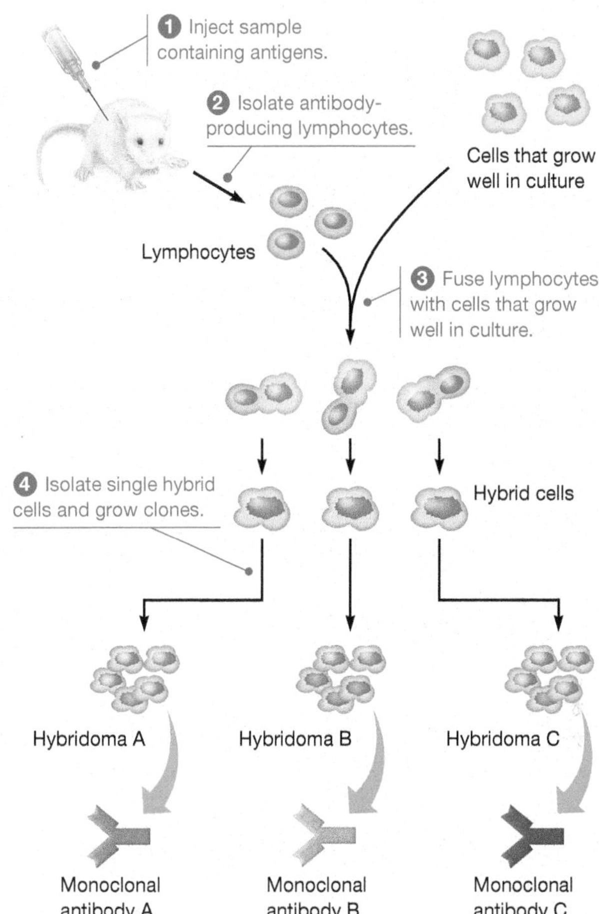

Figure 21-21 Monoclonal Antibody Technique for Producing Pure Antibodies. A sample containing an antigen of interest is injected into mice to stimulate antibody formation. Antibody-producing lymphocytes isolated from the animal are then fused with cells that grow well in culture, and the resulting hybrid cells are used to create a series of cloned cell populations (hybridomas) that each make a single type of antibody. Extensive screening may be required to find a hybridoma that makes an antibody directed against a particular antigen of interest.

Using Antibodies. Antibodies can be used in several ways in cell biology applications. A key use of antibodies is in *immunostaining* (see Key Technique in Chapter 1, pages 8–9). Another technique in which antibodies are useful is the protein equivalent of a Southern or Northern blot and so has creatively enough been called **Western blotting** (or immunoblotting). Western blots are performed by solubilizing a mixture of proteins (**Figure 21-22** on page 648) and then ❶ separating the proteins by size using the SDS-PAGE process shown in Figure 21-19. ❷ The proteins are then blotted onto a nitrocellulose or nylon membrane. The blotting process is a bit different than for nucleic acids, where capillary action was sufficient. In the case of Western blots, the protein must be moved from the gel onto the membrane using an electric current. ❸ The next step is similar in logic to immunostaining: after the transfer is complete, the blot is washed, incubated with a primary antibody, washed again, and then exposed to a secondary antibody. ❹ The secondary antibody is conjugated to an enzyme whose presence can be detected either using a colored precipitation reaction product or using *chemiluminescence,* the generation of light through a chemical reaction.

Proteins Can Be Isolated by Size, Charge, or Affinity

Proteins from whole cells or complex mixtures can be characterized by size and charge using electrophoresis, and antibodies can be used to recognize specific protein on gels or in fixed cells. But how can large amounts of specific proteins be purified for biochemical experiments? One common method used in protein purification is *column chromatography.* The basic idea in this approach is to pass a solution containing a mixture of proteins through a glass or plastic column. Inside

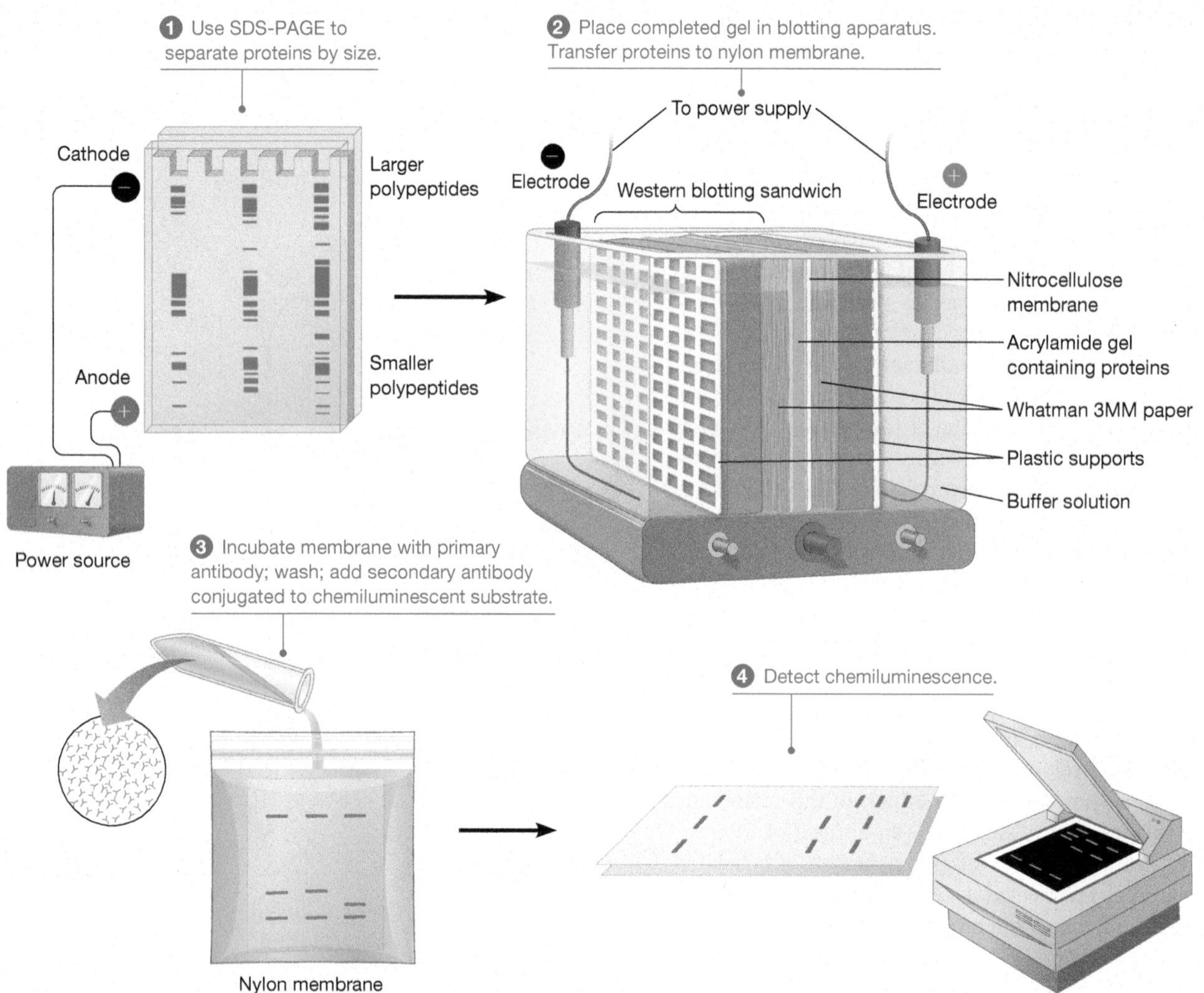

Figure 21-22 Western (Immuno) Blotting. SDS-PAGE is performed and the resulting proteins are blotted onto a nitrocellulose or nylon membrane using an electric field. The resulting blot is probed using primary antibodies, and the locations where the primary antibodies have bound are detected using secondary antibodies and a reaction to visualize the bound secondary antibodies.

the column are small beads (usually made of acrylamide or agarose) that have different properties depending on the type of column being used. **Figure 21-23** shows three common approaches.

Ion-Exchange Chromatography. The first is **ion-exchange chromatography**. In this approach, the beads are chemically modified so that they carry a weak positive or negative charge. Charged proteins interact with the beads to varying degrees (in the example in Figure 21-23a, the beads are negatively charged and the proteins that bind would therefore be positively charged); the interactions can be systematically manipulated using solutions of varying pH and ionic strength. Because different proteins have different charges on their surfaces, they will each be eluted (that is, they will fall off of the beads and pass through the column) at a characteristic pH and salt concentration.

Gel Filtration Chromatography. A second approach, **gel filtration chromatography**, relies on differences in the size and shape of proteins. The beads used in this case have a variety of different-sized pores that allow proteins to percolate through them (Figure 21-23b). Smaller proteins can enter all of the pores and therefore take longer to move through the column. Larger proteins pass through more rapidly because they are unable to transit through each bead.

Affinity Chromatography. If specific information is available about what molecules a protein binds to, then **affinity chromatography** can be used. The word "affinity" refers to the fact that the beads in this case have a specific molecule attached to their surfaces that preferentially binds to the protein being purified (Figure 21-23c). One very common form of affinity chromatography is *immunoaffinity chromatography*. In this approach, an antibody that is specific for a protein of interest is attached to beads. The bound protein can then be eluted from the column using salt, a change in pH, or, in some cases, even a mild detergent. If the beads are in a test tube, rather than a column, they can be used to precipitate a specific protein (and any proteins tightly associated with it) from a crude cellular mixture. After the proteins bind to the beads, they can be easily centrifuged, collected, and washed to purify the bound proteins. This process is therefore called **immunoprecipitation**. The chromatin immunoprecipitation technique for identifying DNA-binding proteins (discussed in Chapter 18) relies on this approach.

In other cases, proteins can be modified by adding "tags" that facilitate their purification. Sometimes cells can be exposed to agents such as *biotin* that chemically alter surface proteins. *Streptavidin* can then be added to the beads to purify the biotinylated proteins. Alternatively, molecular biology techniques can be used to engineer proteins so that they have amino acid tags at their N- or C-terminus. For example, adding six histidine residues in a row to the beginning or end of a protein makes the protein bind tightly to a column with Ni^{2+} ions attached to beads.

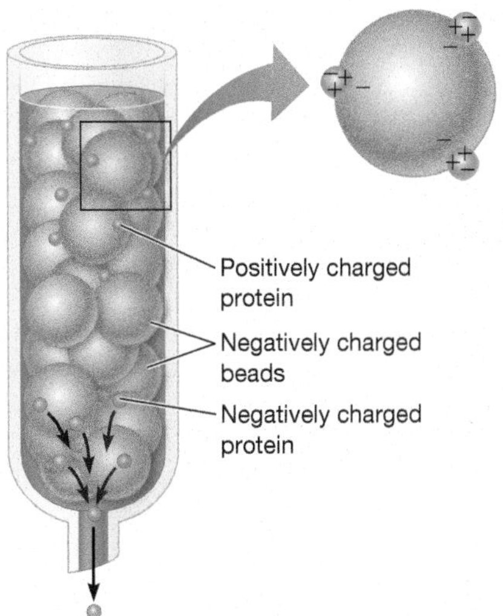

(a) Ion-exchange chromatography

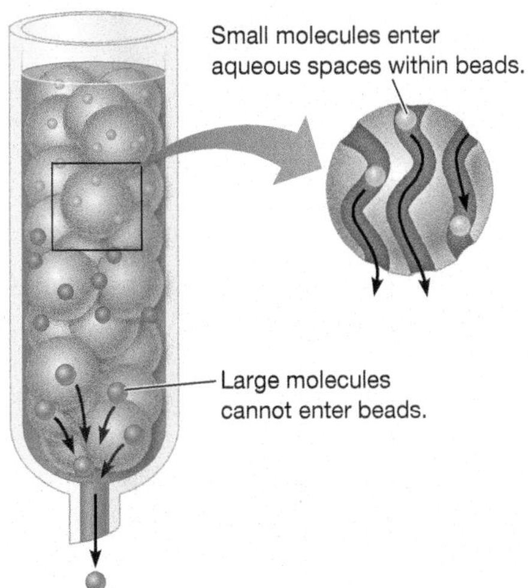

(b) Gel filtration chromatography

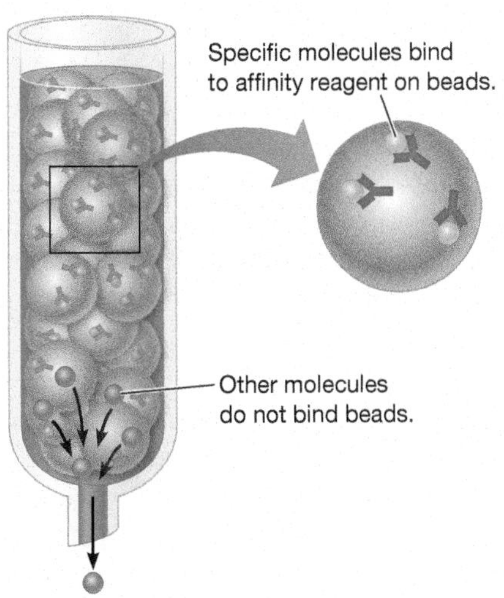

Specific molecules bind to affinity reagent on beads.
Other molecules do not bind beads.

(c) Affinity chromatography

Figure 21-23 Using Chromatography to Purify Proteins. (a) In ion-exchange chromatography, the beads are negatively charged. Positively charged proteins bind and are retained on the column; negatively charged proteins pass through. Increasing the concentration of salt in the surrounding buffer can elute bound proteins by competing for the negative charges on the column. **(b)** In gel filtration chromatography, small proteins can pass through the beads, slowing down their progress through the column. Larger proteins cannot enter the beads, so they pass more quickly through the column. **(c)** In affinity chromatography, beads are coated with antibodies that recognize a specific protein. In other cases, proteins might be tagged with a label such as biotin; in this case a molecule that recognizes the tag (for example, streptavidin) is attached to the beads.

Proteins Can Be Identified from Complex Mixtures Using Mass Spectrometry

Most of the protein purification techniques mentioned thus far are used for isolating single types of proteins. They would be too laborious for identifying thousands of proteins simultaneously. Identifying the vast number of proteins produced by a tissue sample has been facilitated by *mass spectrometry* (see Key Technique in Chapter 2, pages 26–27). Recall that mass spectrometry uses magnetic and electric fields to separate molecules based on charge. It turns out that the movement of particles in the vacuum chamber of a mass spectrometer reflects differences in not only charge, but also mass. The *mass/charge (m/z) ratio* determines the properties of a charged particle in the spectrometer. This makes mass spectrometry very useful for analyzing proteins.

Although mass spectrometry has been used for decades, it was not until the late 1980s that procedures were developed for using mass spectrometry with proteins because the charged particles used in mass spectrometry must be in the gaseous phase, which until then had caused destruction of proteins. At that time, two techniques were developed that circumvent this problem. One, known as **matrix-assisted laser desorption/ionization mass spectrometry (MALDI MS)**, places proteins in a light-absorbing matrix. A laser beam is used to vaporize the protein in the matrix, which can then enter the vacuum system of the mass spectrometer. In the other method, known as **electrospray ionization mass spectrometry (ESI MS)**, a solution of protein in solvent is forced into the gaseous phase by passage through a small-diameter, highly charged needle, which causes the protein solution to disperse into tiny droplets. The solvent then evaporates, leaving protein in the gaseous phase. Often, a prior purification step, such as chromatography, can be directly coupled to mass spectrometry. One example is shown in **Figure 21-24** on page 650, which couples liquid chromatography to the electrospray ionization method. These two common methods can also be combined with *time-of-flight*

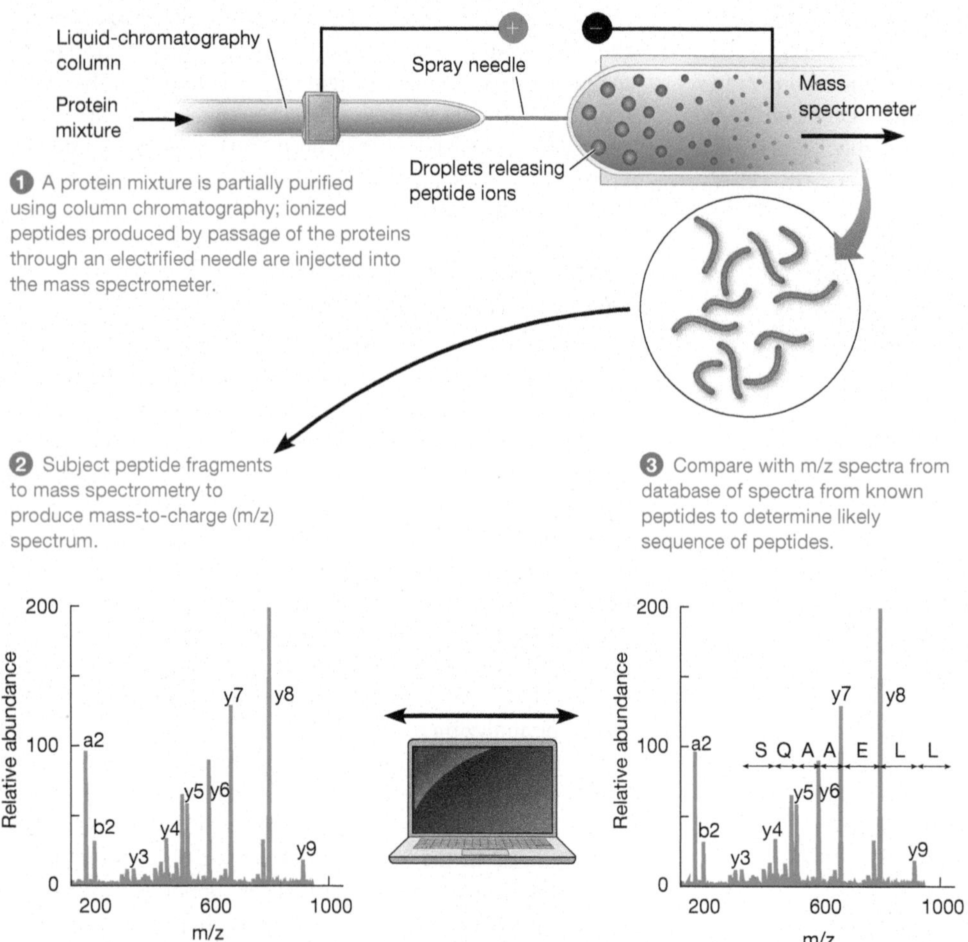
Liquid-chromatography column
Spray needle
Protein mixture
Mass spectrometer
Droplets releasing peptide ions
1 A protein mixture is partially purified using column chromatography; ionized peptides produced by passage of the proteins through an electrified needle are injected into the mass spectrometer.
2 Subject peptide fragments to mass spectrometry to produce mass-to-charge (m/z) spectrum.
3 Compare with m/z spectra from database of spectra from known peptides to determine likely sequence of peptides.
Relative abundance
200
100
0
a2
b2
y3
y4
y5
y6
y7
y8
y9
200
600
1000
m/z
S Q A A E L L

Figure 21-24 Mass Spectrometry for Peptide Analysis. In this example, a mixture of proteins is subjected to liquid chromatography, followed by tandem mass spectrometry. In other cases, spots are cut directly out of a two-dimensional gel and subjected to mass spectrometry. ❶ Proteins that exit the column at certain times and under certain chemical conditions are fragmented, ionized, and enter the mass spectrometer where they are separated by their mass/charge ratio (m/z). ❷ Selected sets of related peptides (the differences between these closely related peaks are due to the presence of different atomic isotopes in the peptide) are fragmented, and the resulting peptide fragments are analyzed in a second round of mass spectroscopy. ❸ Computer analysis is then performed to compare the spectrum obtained for a given sample with the possible theoretical spectra obtained from the amino acid sequences of proteins encoded by the organism from which the proteins were isolated.

(TOF) analysis. TOF relies on the fact that each charged peptide has a speed that varies with its mass. Differences in the kinetic energy of charged particles moving in the vacuum can be used to develop a mass-dependent velocity profile for each charged particle.

What makes mass spectrometry especially useful for proteomic analysis is that a variation on the basic technique can be used as a part of a "shotgun" approach to analyze the sequences of complex mixtures of peptides. In this technique, called **tandem mass spectrometry (MS/MS)**, two mass spectrometers are placed in line. Peptide fragments are injected into the first, which is used to separate specific peptides with particular properties, which are then allowed to enter a *collision cell* placed between the first and second spectrometers. The peptides are further fragmented in this chamber and then injected into the second spectrometer for further analysis. The resulting spectrum is a "fingerprint" of the peptides in the original protein sample that can be analyzed by computer for matches in the database.

Protein Function Can Be Studied Using Molecular Biology Techniques

In the last section you learned about many techniques that cell biologists use for studying proteins directly. These are powerful and very useful. But remember that the central dogma of molecular biology and the flow of information it entails means that working with DNA is a key part of the toolkit cell biologists use to study proteins. Once a DNA clone that encodes a protein or protein fragment is in hand, there are many uses to which the clone can be put.

Dissecting Protein Function via Mutations. One of the huge benefits of working with DNA is that it allows relatively easy removal or rearrangement of pieces of a protein or the production of new proteins by taking pieces of one protein and adding them to another. For example, pieces of the DNA encoding a protein can be removed (deleted), leaving a piece of DNA that encodes only part of the protein. Examining the function of the protein fragment can help determine which parts of a protein are important for its biochemical functions. In other cases, a single base (or several bases) can be specifically mutated so that the resulting DNA encodes a mutated protein. This approach is known as **site-directed mutagenesis**. Site-directed mutagenesis can often be performed using PCR by deliberately altering the primers so that the resulting DNA carries the desired mutation. There are also commercial kits available for rapid introduction of point mutations into existing pieces of DNA.

Expressing Proteins Using Expression Vectors. Once a clone corresponding to a desired normal or mutated protein is in hand, there are specialized *expression vectors* for use in bacteria and eukaryotic cells that allow for high levels of expression of protein from the cloned DNA. One commonly used vector in bacteria relies on the *lac* operon (see Chapter 20). In this case, however, the *lac* operon has been engineered into a plasmid vector.

The cloned DNA can be inserted into the vector and introduced into bacteria lacking a functional *lacZ* gene. In the presence of the artificial substrate *isopropyl-D-1-thiogalactopyranoside,* the operon in the plasmid is activated, leading to transcription of the cloned gene and translation of the resulting mRNA. Expressed proteins can then be purified using the approaches mentioned earlier, and the purified proteins can be used to produce antibodies or can be used in biochemical studies. Using the tagging methods mentioned earlier in this chapter is a key technique that aids purification of expressed proteins.

Expressing GFP-Tagged Proteins in Cells. A final technique to consider in this section is the engineering of proteins to express them in cells. One common use for such tagged proteins is to visualize them. This is possible through the use of the *green fluorescent protein (GFP)* and its variants (see Key Technique in Chapter 19, pages 566–567) as well as other genetically encoded fluorescent tags. ("Genetically encoded" means that the DNA for the fluorescent tag can be attached to the gene for a protein; expression of this construct will produce a tagged protein.) It is now routine to clone DNA that encodes a GFP-tagged version of a protein of interest. Such proteins are often called *translational fusions* because they fuse GFP to another protein. Other types of GFP constructs are useful for studying transcriptional regulation, as you will learn below.

Protein-Protein Interactions Can Be Studied in a Variety of Ways

So far, you have learned how to study individual types of proteins in cells. You have learned throughout this book, however, that proteins frequently bind other proteins to form complicated multiprotein complexes. Cell biologists often want to know whether two proteins bind one another and which parts of each protein allow them to bind. One way to assess the likelihood that two proteins bind is by immunostaining; if two proteins appear to be in the same spot at the same time, then this increases the likelihood that they actually attach to one another. "Guilt by association" is not enough for cell biologists, however. Just because two proteins seem to be close to one another does not necessarily mean that they actually bind one another directly or indirectly. In fact, the light microscope cannot provide sufficient resolution to determine that two proteins actually bind. To determine whether proteins physically interact requires other techniques.

"Pull-Downs" and Co-immunoprecipitation. You already learned above that affinity purification of proteins can be performed using affinity tags or antibodies and that proteins can be expressed in bacterial expression systems to produce large quantities of protein for performing biochemical analyses. These sorts of techniques can also be used to assess whether two proteins bind or whether they are part of a multiprotein complex. Suppose you want to determine whether a protein (let's call it protein A) and another protein (let's call it protein B) bind one another. If protein A carries an affinity tag, it can be attached to agarose beads and placed in buffer in a test tube. Protein B is then added to the tube, and the beads are removed by centrifugation and boiled to remove attached proteins. The protein sample is then subjected to SDS-PAGE to visualize the proteins present. If both protein A and protein B are present in the gel, it indicates that protein A "pulled" protein B down with the beads. A **pull-down assay** is a common approach for testing protein-protein binding.

Sometimes two proteins do not bind directly, but do so indirectly. Another technique, a variation on the immunoprecipitation approach you learned about earlier, can be used. Let's go back to our hypothetical proteins, protein A and protein B. Suppose they do not bind directly, but are part of a large complex. Now suppose that you have an antibody that recognizes protein A. If you isolate proteins from the cells you are studying and perform an immunoprecipitation, protein A will be precipitated, but so will the protein complex of which protein A is a part. If protein B is part of the complex, it will also be present. In other words, protein B was precipitated along with protein A. For this reason, this approach is called **co-immunoprecipitation** (or coIP for short). If the precipitated proteins are run on an SDS-PAGE gel, the presence of protein B can be determined by staining the gel and looking for a protein of the right size, or by performing a Western blot of the precipitated proteins to detect protein B.

The Yeast Two-Hybrid System. Another approach for studying protein-protein interactions involves genetically modified yeast. Recall that transcriptional activators have two physically separate, essential domains that lead to their ability to regulate transcription: a DNA-binding domain and an activation domain (see Chapter 20). The ability to separate these two parts of a yeast transcriptional activator known as *Gal4* was exploited by Stanley Fields and colleagues to study protein-protein interactions in a technique called the **yeast two-hybrid system** (**Figure 21-25** on page 652), also sometimes called an *interaction trap assay*. Gal4 activates transcription of a gene *(Gal1)* that encodes a galactose metabolic enzyme. To do so, Gal4 binds to a sequence, the *upstream activating sequence (UAS)*, of *Gal1*. The key to the two-hybrid technique is that it is possible to use genetic engineering to separate the DNA that encodes the DNA-binding domain of Gal4 from its activation domain and to place each piece of DNA on a different plasmid. The DNA encoding the DNA-binding domain is placed next to DNA encoding a protein of interest. Such a construct is called a *bait construct* because it is used to "catch" interactions with other proteins. A separate plasmid contains a sequence encoding the activation domain of Gal4, plus DNA encoding a protein we wish to test for interaction with the bait protein. This construct is called a *prey construct* because it is "caught" by the assay.

The two plasmids are introduced into yeast carrying a *reporter gene,* often the *lacZ* gene, fused to the UAS of Gal1. When this gene is transcribed and the mRNA is translated in the yeast, its presence can be detected through a color reaction. Yeast carrying both plasmids is identified by virtue of selectable markers carried by the two plasmids. If the bait protein produced in the yeast binds the prey protein, then the DNA-binding and activation domains of Gal4 attached to each protein fragment are brought sufficiently close together to reconstitute the transcriptional activating activity of Gal4. When this happens, the reporter gene is expressed, indicating that the bait and the prey proteins bind one another.

The yeast two-hybrid assay is a powerful way to rapidly assess whether proteins interact. If a library of prey plasmids is used to transform yeast, this method can be used to screen

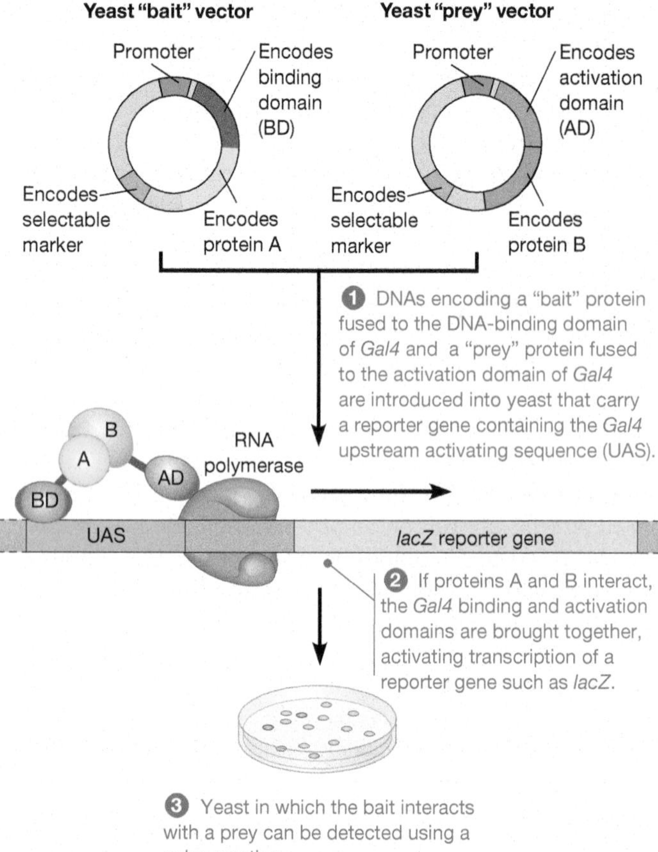
Yeast "bait" vector
Yeast "prey" vector
Promoter
Encodes binding domain (BD)
Encodes selectable marker
Encodes protein A
Promoter
Encodes activation domain (AD)
Encodes selectable marker
Encodes protein B
1 DNAs encoding a "bait" protein fused to the DNA-binding domain of Gal4 and a "prey" protein fused to the activation domain of Gal4 are introduced into yeast that carry a reporter gene containing the Gal4 upstream activating sequence (UAS).
B
A
AD
BD
RNA polymerase
UAS
lacZ reporter gene
2 If proteins A and B interact, the Gal4 binding and activation domains are brought together, activating transcription of a reporter gene such as lacZ.
3 Yeast in which the bait interacts with a prey can be detected using a color reaction.

Figure 21-25 The Yeast Two-Hybrid System. The yeast two-hybrid system detects protein-protein interactions by using two different DNA constructs: a "bait" vector that fuses DNA encoding the DNA-binding domain of *Gal4* to DNA encoding a protein of interest (protein A) and a "prey" vector that fuses DNA encoding the activation domain of *Gal4* to DNA encoding a second protein (protein B). If the proteins interact, a reporter gene (in this case, *lacZ*) will be expressed.

for unknown proteins that bind to a known bait. In addition, pairwise tests using thousands of proteins can also be performed, and the network of interactions between most of the proteins in an organism's proteome can be determined and visually mapped to determine an *interactome*. The results of one such study in budding yeast are shown in **Figure 21-26**. The dots in this figure represent individual proteins, which have been grouped together into functional modules based on their ability to interact with one another. Similar genome-wide two-hybrid screens have been carried out in flies, worms, and humans. The information from these large screens can suggest that two proteins interact, which can be followed up using more traditional biochemical techniques.

Other techniques make it feasible to study the interactions and functional properties of the vast number of proteins found in a proteome. For example, it is possible to manufacture *protein microarrays* (or protein "chips"), which are analogous to the DNA microarrays we discussed previously. These protein arrays can be used to study a variety of protein properties, such as the ability of each individual spot to bind to other molecules added to the surrounding solution.

Visualizing Protein-Protein Binding. The techniques we have discussed so far for assessing whether two proteins bind have been biochemical (or, in the case of the yeast two-hybrid system, genetic). With the advent of fluorescence microscopy, however, it is now possible to harness the power of microcopy to see where and when proteins interact in a cell. Techniques are becoming available for analyzing protein-protein binding. These include *Förster resonance energy transfer* (*FRET*; see Figure A-21) and *bimolecular fluorescence complementation* (*BiFC*). Each of these techniques relies on the production of fluorescent signals between two engineered proteins only when they physically bind one another. (These methods

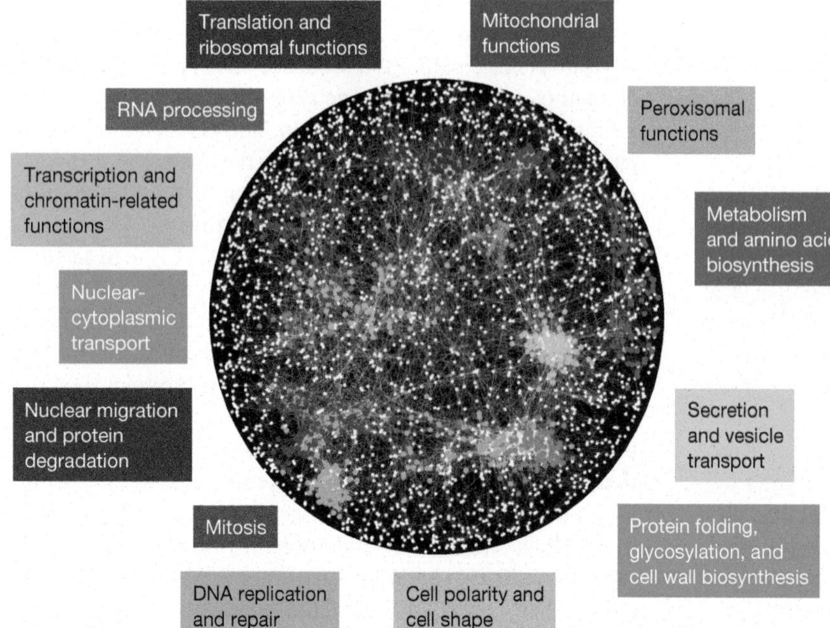
Translation and ribosomal functions
Mitochondrial functions
RNA processing
Peroxisomal functions
Transcription and chromatin-related functions
Metabolism and amino acid biosynthesis
Nuclear-cytoplasmic transport
Nuclear migration and protein degradation
Secretion and vesicle transport
Mitosis
Protein folding, glycosylation, and cell wall biosynthesis
DNA replication and repair
Cell polarity and cell shape

Figure 21-26 A Sample Interactome. A global study of protein-protein interactions in budding yeast, *Saccharomyces cerevisiae,* shown as a visual map. The dots represent about 4500 proteins in yeast. Lines represent interactions between proteins. Dots of the same color represent proteins involved in one of 13 selected cellular processes as shown.

are discussed in the Appendix, which covers many different microscopy techniques.) These technologies are part of an exciting new array of techniques for studying dynamic protein interactions in cells.

CONCEPT CHECK 21.3

Compare and contrast each of the following in terms of the probe(s) used and the sort of information you can obtain from the technique: (1) a Northern blot versus a Western blot; (2) a yeast two-hybrid assay versus co-immunoprecipitation.

21.4 Analyzing and Manipulating Gene Function

Recombinant DNA technology has had an enormous impact on the field of cell biology, leading to many new insights into the organization, behavior, and regulation of genes and their protein products. Many of these discoveries would have been virtually inconceivable without such powerful techniques for isolating gene sequences. The rapid advances in our ability to manipulate genes have also opened up the field of **genetic engineering**, which involves the application of recombinant DNA technology to practical problems, primarily in medicine and agriculture. In concluding the chapter, we will briefly examine some of the areas where these practical benefits of recombinant DNA technology are beginning to be seen.

Transgenic Organisms Carry Foreign Genes That Are Passed on to Subsequent Generations

Tissue culture is a powerful technique, and you have learned about many uses of cultured cells throughout this book. However, most cells in nature do not act in isolation; if they are part of an organism, for example, they interact with other cells, and their functions are carried out in the context of tissues. As a result, it is important to study engineered proteins in these multicellular, whole-organism contexts. Genetically engineering proteins that can be studied in living organisms requires engineering the organisms themselves. When a foreign piece of DNA is added to an organism in such a way that it can be passed along through the germ line (eggs and sperm) and hence to subsequent generations, the resulting organism is said to be a **transgenic** organism, and the process used to create such organisms is called *transgenesis*. Transgenic organisms have been extremely useful in the laboratory, where they have allowed the study of gene expression and the function of specific genes in living animals. Transgenic organisms are also being used in practical applications in agriculture and biomedicine, as you will learn below.

Transgenesis by Direct Injection or Bombardment. The techniques used to produce transgenic organisms vary significantly depending on the organism of interest. In some cases, DNA can be directly injected. For example, in *C. elegans*, DNA can be injected into the gonad; in fruit flies, eggs can be injected. In mammals, the basic approach was originally developed in mice by Richard Palmiter and Ralph Brinster, who transferred an engineered version of the rat gene encoding *growth hormone* into fertilized mouse eggs, thereby creating a transgenic mouse that carries a gene from another organism in its cells (**Figure 21-27** on page 654). In these experiments, DNA was injected into the male *pronucleus*, the haploid sperm nucleus that has not yet fused with the haploid egg nucleus (Figure 21-27a). The injected, fertilized eggs were then implanted back into the reproductive tracts of surrogate mothers. In at least one case, a transgenic mouse transmitted the engineered gene faithfully to about half of its offspring, suggesting that the gene had become stably integrated into one of its chromosomes. A resulting "supermouse" is shown in Figure 21-27b. It weighed almost twice as much as its littermates. The accomplishment was heralded as a significant breakthrough because it proved the feasibility of applying genetic engineering to mammals.

There are other ways of physically introducing DNA into organisms besides injection. In yeast, for example, chemicals can be used to facilitate passage of DNA through the cell wall into the interior. Alternatively, the cell wall can be degraded by enzymes, and foreign DNA can be introduced into the resulting, somewhat fragile *spheroplasts*. DNA can be introduced into many cells (plant and animal) by driving it with an electric current, a technique known as *electroporation*. In many cases, from worms to plants, small metal particles coated with DNA are used to introduce foreign DNA. The particles are propelled at high velocity using a specialized gun in a technique known as *biolistic transformation* or, more colloquially, a "gene gun."

Transgenesis by Infection. Another very common approach for making transgenic organisms is via genetically engineered viruses or bacteria. This approach is often used with mammalian cells, for example, by using genetically engineered *retroviruses* (you learned about retroviruses in Chapter 16). Cloned genes are often transferred into plants by inserting them first into the **Ti plasmid**, a naturally occurring DNA molecule carried by the bacterium *Agrobacterium tumefaciens*. The general approach for transferring genes into plants using these plasmids is summarized in **Figure 21-28** on page 654. In nature, infection of plant cells by this bacterium leads to insertion of a small part of the plasmid DNA, called the *T DNA region*, into the plant cell chromosomal DNA; expression of the inserted DNA then triggers the formation of an uncontrolled growth of tissue called a *crown gall tumor*. In the laboratory, the DNA sequences that trigger tumor formation can be removed from the Ti plasmid without stopping the transfer of DNA from the plasmid to the host cell chromosome. Inserting genes of interest into such modified plasmids produces vectors that can transfer foreign genes into plant cells. When the recombinant plasmid enters the plant cell, its T DNA, along with the inserted DNA, becomes stably integrated into the plant genome and is passed on to both daughter cells at every cell division.

Transcriptional Reporters Are Useful for Studying Regulation of Gene Expression

One of the main goals of modern molecular genetics is to understand how the expression of particular genes is regulated. This can occur in many ways, and transcriptional regulation

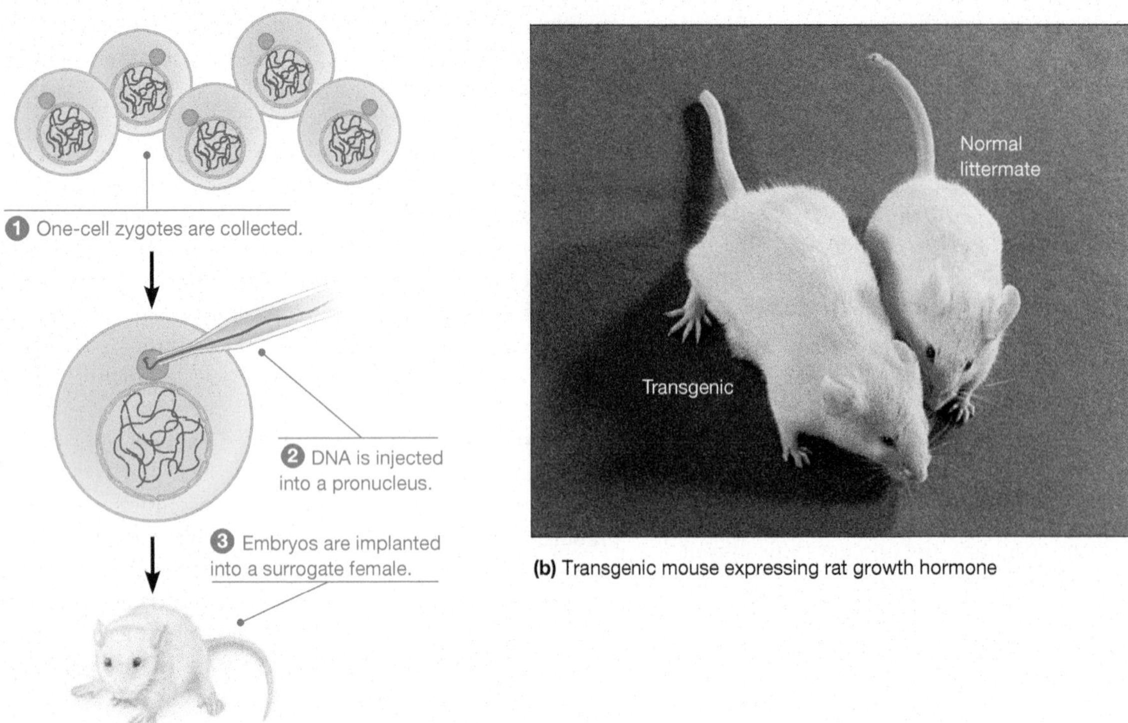

1 One-cell zygotes are collected.
2 DNA is injected into a pronucleus.
3 Embryos are implanted into a surrogate female.
(a) Producing transgenic mice
Normal littermate
Transgenic
(b) Transgenic mouse expressing rat growth hormone

Figure 21-27 Genetic Engineering in Mice. (a) Fertilized mouse eggs (one-cell zygotes) are injected with DNA, typically into a pronucleus (the nucleus derived from either the sperm or egg). The injected zygotes are transferred into a surrogate mother during subsequent gestation. **(b)** A "supermouse" (on the left) is significantly larger than its littermate because it was engineered to carry, and express at high levels, the gene for rat growth hormone.

plays a crucial role in this process (see Chapter 20). *Enhancers* are central to the regulation of tissue-specific transcription in higher eukaryotes. How is information obtained about which enhancer elements are important for a gene's expression? One way transgenic organisms are useful in this regard is by the production of *transcriptional reporters*. As you saw earlier, it is possible to make translational fusions between a protein such as GFP and a protein of interest to follow where the protein goes and when it is expressed (see page 651). But suppose that you want to study how an essential gene, that is, one that is absolutely necessary for early embryonic development or for the ongoing life and health of an organism, is regulated. If you were to perturb the expression of such a gene, it would likely lead to the death of the organism, and your experiment would be over! In these cases, transcriptional reporters are extremely useful. The idea behind such reporters is shown in

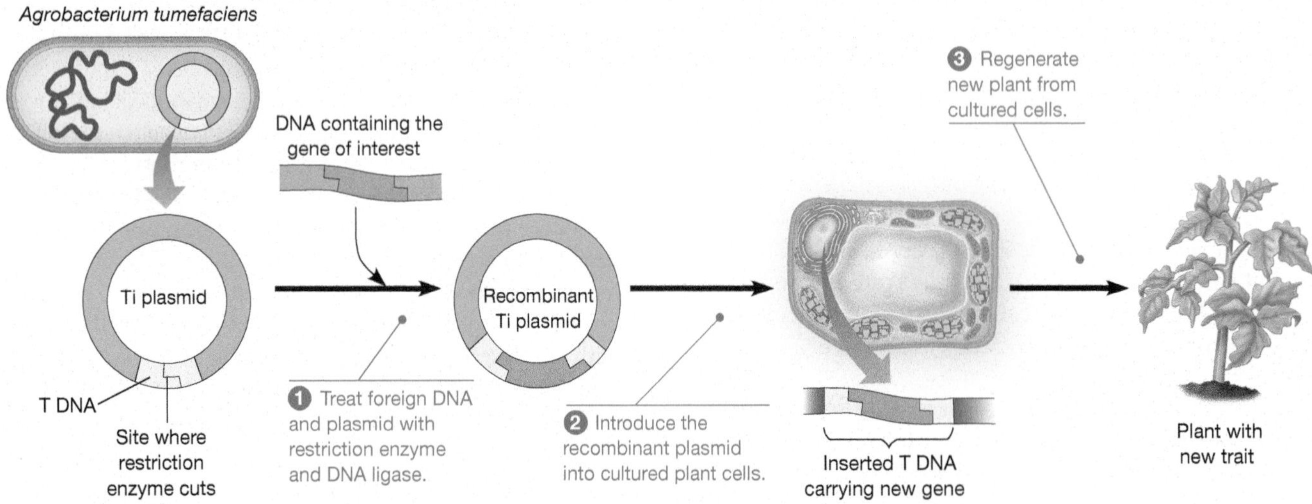

Agrobacterium tumefaciens
DNA containing the gene of interest
Ti plasmid
T DNA
Site where restriction enzyme cuts
Recombinant Ti plasmid
1 Treat foreign DNA and plasmid with restriction enzyme and DNA ligase.
2 Introduce the recombinant plasmid into cultured plant cells.
Inserted T DNA carrying new gene
3 Regenerate new plant from cultured cells.
Plant with new trait

Figure 21-28 Using the Ti Plasmid to Transfer Genes into Plants. Most genetic engineering in plants uses the Ti plasmid as a vector. A DNA fragment containing a gene of interest is inserted into a restriction site located in the T DNA region of the plasmid. The recombinant plasmid is then introduced into plant cells, which regenerate a new plant containing the recombinant T DNA stably incorporated into the genome of every cell.

Figure 21-29. To make a transcriptional reporter, the coding region of a gene (that is, that part that encodes the protein normally produced from its gene) is removed, and the coding region of a "reporter" is put in its place (Figure 21-29a). The reporter is so called because it encodes a protein that is harmless to the organism but that can be seen easily, either by the production of a colored reaction product or by a fluorescent protein such as GFP. Some common reporters are the *lacZ* gene from bacteria, which can act on a substrate to produce color (Figure 21-29b), the *β-glucuronidase* gene (abbreviated GUS; often used in plants, Figure 21-29c), or GFP (Figure 21-29d). The engineered gene is then put into an otherwise normal organism to make a transgenic organism.

Why is this necessary? Altering the promoter or other regulatory elements of an essential gene would lead to an organism's death, but altering the expression of the transcriptional reporter in an organism that retains its normal copies of the gene only leads to altered expression of an innocuous protein. By studying this "substitute" version of the gene, which has the normal gene's regulatory elements, it is then possible to dissect the regulatory components of the gene. A celebrated example of this approach is the *even-skipped* gene in the fruit fly, *Drosophila* (**Figure 21-30** on page 656). This gene, which encodes a regulatory transcription factor, is essential; without *even-skipped* expression, the embryo dies. However, by examining the expression of a reporter when various *even-skipped* enhancers are removed, it is possible to identify specific enhancer elements required for the seven stripes of *even-skipped* expression that normally occur along the body axis of the embryo.

The Role of Specific Genes Can Be Assessed By Identifying Mutations and by Knockdown

Although adding an engineered piece of DNA to an animal using transgenic technology can be very useful, it is often even more useful to *remove* a gene of interest. Removing gene function can be done at several levels. Parts of the gene itself can be removed or replaced using several methods. Other approaches

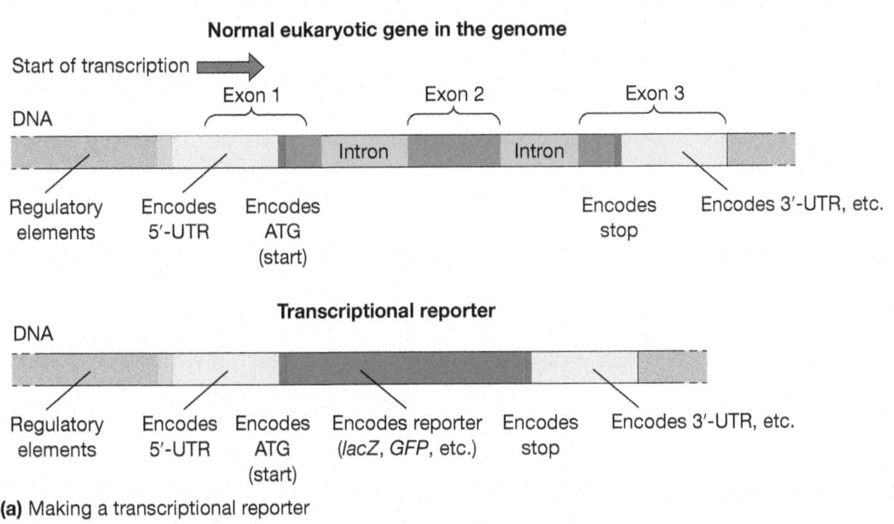

(a) Making a transcriptional reporter

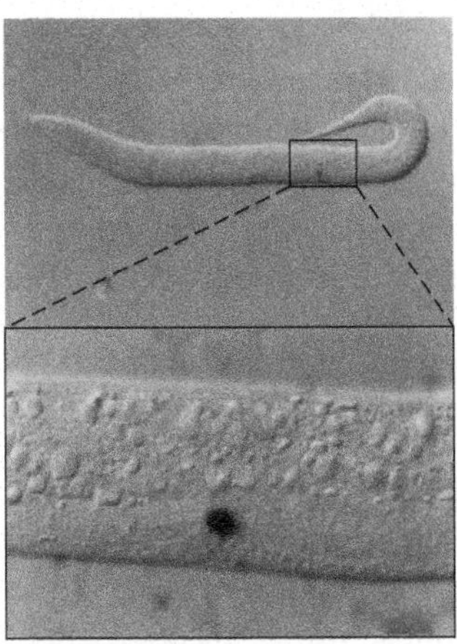

(b) *Lin-3* regulatory sequences driving *lacZ* reporter gene in *C. elegans*

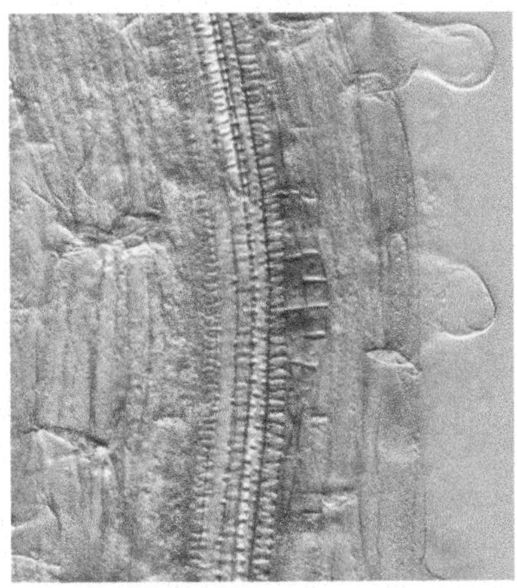

(c) *PHABULOSA* regulatory sequences driving *β-glucuronidase* reporter gene in *Arabidopsis*

(d) *RHODOPSIN* regulatory sequences driving *GFP* reporter gene in *Mus musculus*

Figure 21-29 Transcriptional Reporters. (a) Making a transcriptional reporter. The coding region of a eukaryotic gene is replaced with a reporter that allows transcription to be visualized (the reporter). The regulatory elements associated with the gene are retained. The DNA is then used to make a transgenic organism or cell. **(b)** A *C. elegans* worm expressing a *β*-galactosidase (*lacZ*) reporter. **(c)** An *Arabidopsis* plant expressing a *β-glucuronidase* (GUS) reporter. **(d)** A mouse expressing a GFP reporter.

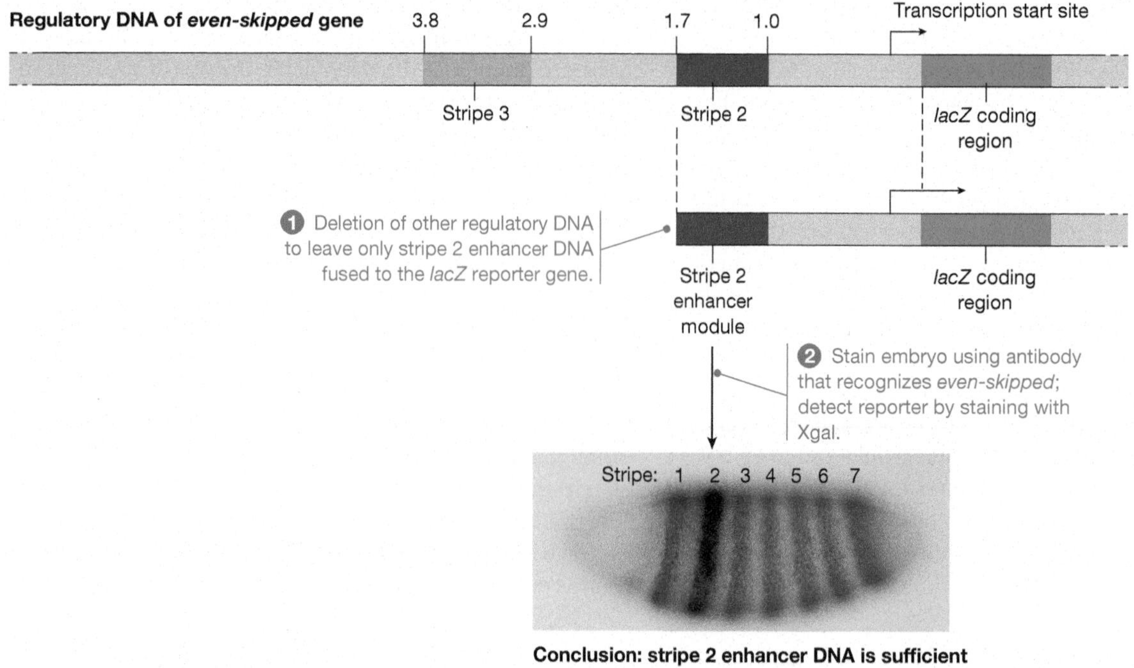

Figure 21-30 Using Transcriptional Reporters to Dissect Regulatory DNA Function. The *Drosophila even-skipped* gene is expressed in a pattern of seven stripes from head to tail in embryos. Normal even-skipped protein can be detected using immunostaining (brown). A transcriptional reporter containing only the stripe 2 enhancer region of the *even-skipped* regulatory DNA is expressed only in stripe 2. This result indicates that the stripe 2 enhancer is sufficient for normal expression of *even-skipped* in stripe 2.

target the corresponding RNA or protein. Some of these approaches are discussed in the next few paragraphs.

Finding Mutants by TILLING. One approach to identifying mutations in a gene of interest is via *targeted induced local lesions in genomes (TILLING)*. TILLING, originally developed in the plant *Arabidopsis thaliana*, is a PCR-based strategy that looks for specific mutations in a gene starting with a population. PCR-amplified DNA from a group of cells or organisms exposed to a mutagenic chemical is analyzed to look for DNA mismatches that result when a mutation occurs in the gene of interest. The mismatches can be detected using a sophisticated form of electrophoresis to detect differences in migration speed of mismatched DNA or through use of an endonuclease that specifically cuts mismatched DNA. Once a mutation is detected, siblings of the organisms from which the DNA was extracted that have the mutation can be recovered.

Making "Knockouts" by Homologous Recombination. This method uses *homologous recombination* to transfer base sequences between related DNA molecules to "knock out" individually targeted genes by inducing recombination with related pieces of nonfunctional DNA. This procedure works well in yeast, in the slime mold *Dictyostelium*, and in embryonic stem cells. Because this procedure works well in ES cells, it has been used to create hundreds of unique strains of **knockout mice**, each defective in a single gene, and efforts are under way to create knockout strains for every one of the mouse's roughly 25,000 genes. In 2007, Mario Capecchi, Oliver Smithies, and Martin Evans were awarded a Nobel Prize in recognition of their pioneering efforts in making the creation of knockout mice possible.

The traditional method for producing a knockout mouse is shown in **Figure 21-31.** ❶ The first step is to synthesize an artificial DNA that is similar in base sequence to the target gene and its flanking sequences, but with two important changes. First, an antibiotic-resistance gene (for example, one that confers resistance to neomycin) is inserted into the middle of the target gene sequence. This simultaneously renders the engineered copy of the gene nonfunctional and also allows cells that carry the DNA to survive in the presence of antibiotic. Second, DNA encoding a viral enzyme, *thymidine kinase*, is attached to the end of the DNA. If this DNA is present in a cell, it will die in the presence of an antiviral drug (for example, ganciclovir). ❷ This DNA is then introduced into mouse *embryonic stem cells (ES cells)*, which are cells that can differentiate into all the cell types of an adult mouse. In very rare cases, the DNA enters the nucleus and, using homologous recombination-based DNA repair mechanisms (see Chapter 17), the artificial DNA aligns with complementary sequences flanking the targeted gene. Homologous recombination then replaces the targeted gene with the new, nonfunctional copy. Only if homologous recombination occurs is the thymidine kinase gene removed from the engineered DNA and degraded by nucleases in the ES cell. In this case, ❸ the cell will survive in the presence of an antibiotic, but ❹ it will *not* be sensitive to the antiviral drug. ❺ Cells identified using this double drug selection are then introduced into mouse embryos, which develop into adult mice containing tissues in which the gene of interest has been inactivated. ❻ Crossbreeding such animals eventually yields strains of pure knockout mice in which both copies of the target gene (one on each chromosome) are knocked out in all tissues. By allowing scientists to study what happens when specific genes are disrupted, knockout mice

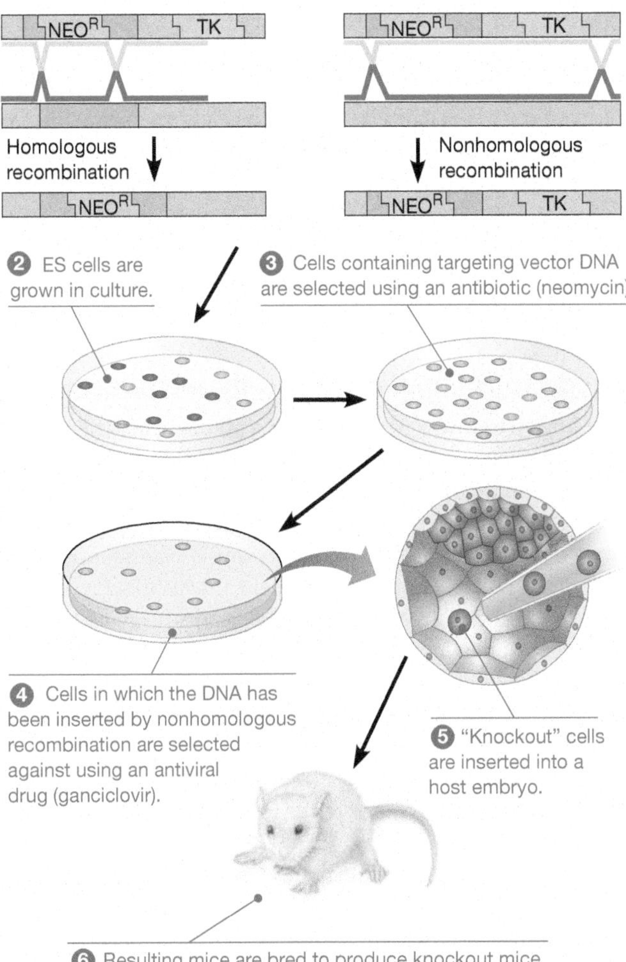
1 DNA is introduced into embryonic stem (ES) cells. The DNA contains a nonfunctional copy of the gene of interest, an antibiotic-resistance gene (NEOR), and a gene encoding a viral enzyme (TK).
NEOR
TK
Homologous recombination
Nonhomologous recombination
2 ES cells are grown in culture.
3 Cells containing targeting vector DNA are selected using an antibiotic (neomycin).
4 Cells in which the DNA has been inserted by nonhomologous recombination are selected against using an antiviral drug (ganciclovir).
5 "Knockout" cells are inserted into a host embryo.
6 Resulting mice are bred to produce knockout mice.

Figure 21-31 Making "Knockout" Mice by Homologous Recombination. The production of "knockout" mice involves the disruption of a target gene. Through the process of homologous recombination in embryonic stem cells (ES cells), part (or all) of this target gene is replaced by a drug resistance cassette. The detailed steps of this process are shown.

have shed light on the roles played by individual genes in numerous human conditions, including cancer, obesity, heart disease, diabetes, arthritis, and aging.

In some cases, homozygous knockout mice are unable to survive through development to birth. This occurs when the targeted gene is necessary for a key developmental checkpoint. This occurrence of embryonic lethality led to the engineering of conditional knockout mice. In these mice, special recombination sites are inserted on either side of the gene to be targeted. Then a specific recombinase is introduced, but only in one cell or tissue type. This leads to deletion of the target gene, but only in the cell or tissue type that expresses the recombinase. Therefore, the targeted mice will be born (assuming that the gene targeted is not required in the cell or tissue type studied). Today, both traditional knockout mice and conditional knockout mice have largely been replaced with mice that have been edited using CRISPR/Cas systems (described in the following section). These new genome-editing techniques are faster, cheaper, and more reliable than previous methods.

Knockout technology has now been extended to other mammals besides mice. One use of this technology relates to *xenotransplantation*, the transplantation of tissues from one species to another. Pigs have been a major focus of such research, which aims to provide a source of temporary organs for human patients awaiting transplants. To avoid immune rejection, pigs have been produced that lack a key enzyme called *α-1,3-galactosyltransferase*, which normally catalyzes the addition of sugar residues to cell surfaces in pigs that contributes to the immune response.

Genome Editing. Knockouts via homologous recombination are a powerful technology for gene deletion and replacement, but for largely technical reasons this approach is efficient only in selected organisms. New technologies stand to transform the ability of cell biologists to alter the genomes of cells and organisms. Collectively, these technologies involve **genome editing**. In all of these cases, specific sequences are used to target molecular complexes to specific genomic sites, where the genomic DNA is directly altered. This occurs by intentionally inducing double-stranded breaks (DSBs), which are then repaired either by nonhomologous end-joining (see Chapter 17) or, if homologous sequences are provided, by homology-directed repair. The result is removal of most of the genomic DNA or replacement of the normal DNA with another sequence. If these alterations occur in the germ line, they can be transmitted to subsequent generations.

Two genome-editing technologies are especially popular. The first involves a special class of "designer" zinc finger nucleases called *transcriptional activator-like effector nucleases (TALENs)*, which are based on bacterial nucleases with known DNA-binding specificity. DNA encoding TALENs can be designed so that the resulting protein encodes an enzyme that will bind to a sequence contained in a gene of interest. When the TALEN DNA (or sometimes RNA made in a test tube) is introduced into an organism, the designer protein is made and targets the gene of interest for double-strand breakage. The DSB is then repaired, producing alterations in the gene.

A second genome editing technology applies the bacterial *CRISPR/Cas system* (see Chapter 20 and Key Technique in Chapter 17, pages 500–501). Recall that the CRISPR (clustered regularly interspaced short palindromic repeat) pathway essentially acts like an immune system in bacteria. Short DNA spacer sequences at the CRISPR locus, acquired due to previous viral infection, are used to make short *crRNAs*, which then find matching sequences in viral DNAs. In the CRISPR system used in molecular biology applications, a crRNA acts along with a *trans-activating crRNA (tracrRNA)* and a *CRISPR-associated (Cas)* protein, *Cas9*, to introduce double-stranded breaks in the foreign DNA. In molecular biology applications, an artificial CRISPR sequence that encodes a *guide RNA* is introduced into cells along with DNA encoding Cas9, often all in one DNA clone. This leads to cutting of the genomic DNA corresponding to the sequence contained in the CRISPR clone (see Figure 1-9).

Gene Knockdown at the Level of RNA or Protein. Producing knockouts by homologous recombination and introducing changes in a gene by genome editing are examples of *targeted gene disruption:* they act at the level of the DNA in chromosomes, replacing or removing the DNA associated with a gene of interest. Although it may be confusing, cell biologists typically refer to approaches that intentionally reduce levels of the mRNA or protein corresponding to a gene as gene "knockdown" because these approaches reduce, or *knock down*, the level of mRNA and/or protein associated with a gene of interest. One key knockdown approach is the introduction of double-stranded RNA into cells, which results in *RNA-mediated interference* (*RNAi*; **see Key Technique in Chapter 20, pages 612–613**).

Although RNAi is extremely powerful, it does not seem to work well in some organisms, such as frogs, zebrafish, and sea urchins, because they do not produce the correct types or sufficient amounts of the correct Argonautes to allow for a robust RNAi response (see Chapter 20). An alternative approach in these organisms is to use chemically modified nucleotides to make nucleic acids composed of *morpholino antisense oligonucleotides*, or *morpholinos*. These molecules can bind to complementary RNA sequences, but they block normal molecular functions associated with the RNA. For example, some morpholinos can be targeted to splice sites in the primary RNA transcript. The result is that the splicing machinery cannot access the splice site, leading to a processed RNA that does not encode a proper protein. Other morpholinos bind just upstream of the translation start site in the mRNA and block translation.

Genetic Engineering Can Produce Valuable Proteins That Are Otherwise Difficult to Obtain

For basic cell biologists, many of the techniques discussed so far are very useful for perturbing cellular functions associated with particular gene products. However, molecular techniques can also be harnessed to intentionally engineer cells or organisms to benefit society. One practical benefit to emerge from recombinant DNA technology is the ability to clone genes that encode medically useful proteins that are difficult to obtain by conventional means. Among the first proteins to be produced by genetic engineering was human *insulin*, which is used by roughly 7 million people with diabetes in the United States to treat their disease. Supplies of insulin purified from human blood or pancreatic tissue are extremely scarce; thus, for many years diabetic patients were treated with insulin obtained from pigs and cattle, which can cause immune reactions in some individuals. Now there are several ways of producing human insulin from genetically engineered bacteria containing the human insulin gene. As a result, people with diabetes can be treated with insulin molecules that are identical to the insulin produced by the human pancreas.

Like insulin, a variety of other medically important proteins that were once difficult to obtain in adequate amounts are now produced using recombinant DNA technology. Included in this category are the *blood-clotting factors* needed for the treatment of hemophilia, *growth hormone* utilized for treating pituitary dwarfism, *tissue plasminogen activator (TPA)* used for dissolving blood clots in heart attack patients, *erythropoietin* employed for stimulating the production of red blood cells in patients with anemia, and *tumor necrosis factor, interferon*, and *interleukin*, which are used in treating certain kinds of cancer. Traditional methods for isolating and purifying such proteins from natural sources are quite cumbersome and tend to yield only tiny amounts of protein. Now that the genes for these proteins have been cloned in bacteria and yeast, large quantities of protein can be produced in the laboratory at a reasonable cost.

One goal of genetic engineering in animals is to produce farm animals that can synthesize medically important human proteins (for example, in the milk of female mammals), which can then be easily purified. Another is the production of engineered livestock as a food source. Many of the same issues we consider in the next section regarding crops have also been raised concerning genetically modified animals grown for food.

Food Crops Can Be Genetically Modified

The ability to insert new genes into plants using the Ti plasmid has allowed scientists to create *genetically modified (GM)* crops exhibiting a variety of new traits. For example, plants can be made more resistant to insect damage by introducing a gene cloned from the soil bacterium *Bacillus thuringiensis (Bt)*. This *Bt* gene encodes a protein that is toxic to certain insects—especially caterpillars and beetles that cause crop damage by chewing on plant leaves. Putting the *Bt* gene into plants such as cotton and corn has permitted farmers to limit their use of more hazardous pesticides for controlling insects, leading to improved crop yields and a significant return of wildlife to crop fields. Crops engineered to resist weed-killing herbicides likewise require fewer toxic chemicals and exhibit higher yields.

Recently, another strategy has been explored to fight crop pests. Scientists have engineered plant chloroplasts to express long, double-stranded RNAs (dsRNAs). When larvae of the Colorado potato beetle eat the leaves of potato plants expressing dsRNAs that target mRNA encoding the protein β-actin, the dsRNAs are taken up by midgut cells of the larvae, and they die.

Another goal of genetic modification is to improve the nutritional value of food. Consider rice, for example, which is the most common food source in the world. More than 3 billion people currently eat rice daily, so fortifying rice with essential nutrients and vitamins could benefit millions. As one example, the genes required for the synthesis of β-carotene, a precursor of vitamin A, were genetically engineered into rice in 2001, and the β-carotene content of such rice was increased more than 20-fold a few years later. The resulting product, called "Golden Rice" because of the color imparted by β-carotene, has the potential to help alleviate a global vitamin A deficiency that now causes blindness and disease in millions of children (**Figure 21-32**).

As the prevalence of GM crops has begun to increase, some people have expressed concerns about the possible risks associated with this technology, especially because it permits genes to cross species barriers that cannot be crossed by traditional breeding techniques. For consumers, the main focus has been on safety because it is well known that toxic and allergic reactions to things we eat can be serious and even life-threatening. Thus far, GM plants have not posed a greater allergy risk than those generated through conventional plant breeding, which has already created massive changes in the genetic makeup of crop plants.

Figure 21-32 A Field of "Golden Rice." These rice plants have been genetically engineered to produce significantly higher amounts of β-carotene compared to normal rice, leading to their golden appearance. Since β-carotene is a precursor of vitamin A, this fortified rice can help prevent blindness and disease associated with diets low in vitamin A.

The possibility that GM crops may pose environmental risks has also raised concerns. In a 1999 laboratory experiment, scientists reported that monarch butterfly larvae died after being fed leaves dusted with pollen from GM corn containing the *Bt* gene. This observation led to fears that the *Bt* toxin produced by GM plants can harm friendly insects. However, the lab bench is not the farm field, and the initial studies were carried out under artificial laboratory conditions in which butterfly larvae consumed far higher doses of *Bt* toxin than they would in the real world. Subsequent data collected from farm fields containing GM crops suggest that the amount of *Bt* corn pollen encountered under real-life conditions does not pose a significant hazard to monarch butterflies.

Despite legitimate concerns about potential hazards, the GM experience has thus far revealed little evidence of significant risks to either human health or the environment. Of course, no new technology is entirely without risk, so the safety and environmental impact of GM crops must be continually assessed. At the same time, GM crops have allowed reductions in pesticide use, and they could make a unique contribution to the fight against hunger and disease in developing countries.

Gene Therapies Are Being Developed for the Treatment of Human Diseases

Humans suffer from many diseases that might conceivably be cured by transplanting normal, functional copies of genes into people who possess defective, disease-causing genes. The success of transgenic and knockout mice, along with similar experiments, raises the question of whether gene transplantation techniques might eventually be applied to the problem of repairing defective genes in humans. Obvious candidates for such an approach, called *gene therapy*, include the inherited genetic diseases cystic fibrosis, hemophilia, hypercholesterolemia, hemoglobin disorders, muscular dystrophy, lysosomal storage diseases, and an immune disorder called *severe combined immunodeficiency (SCID)*.

The first person to be treated using gene therapy was a 4-year-old girl with a type of SCID caused by a defect in the gene encoding *adenosine deaminase (ADA)*. Loss of ADA activity leads to an inability to produce sufficient numbers of immune cells called *T lymphocytes*. As a result, the girl suffered from frequent and potentially life-threatening infections. In 1990, she underwent a series of treatments in which a normal copy of the cloned ADA gene was inserted into a virus, the virus was used to infect T lymphocytes obtained from the girl's blood, and the lymphocytes were then injected back into her bloodstream. The result was a significant improvement in her immune function, although the effect diminished over time, and the treatment did not seem to help most SCID patients.

In the years since these pioneering studies, considerable progress has been made in developing better techniques for delivering cloned genes into target cells and getting the genes to function properly. In the year 2000, French scientists finally reported what seemed to be a successful treatment for children with SCID (in this case, an especially severe form of SCID caused by a defective receptor gene rather than a defective ADA gene). By using a virus that was more efficient at transferring cloned genes and by devising better conditions for culturing cells during the gene transfer process, these scientists were able to restore normal levels of immune function to the children they treated. In fact, the outcome was so dramatic that, for the first time, the treated children were able to leave the protective isolation "bubble" that had been used in the hospital to shield them from infections.

It was therefore a great disappointment when three of the ten children treated in the initial study developed leukemia a few years later. Examination of the leukemia cells revealed that the virus used to deliver the corrective gene sometimes inserts itself next to a normal gene that, when expressed abnormally, can cause cancer to arise. (In Chapter 26 you will learn how such an event, called *insertional mutagenesis*, can initiate cancer development.) The associated cancer risks must be better understood before such treatments will become practical.

One tactic for addressing the problem of cancer risk is to change the type of virus being used to ferry genes into target cells. The SCID studies employed retroviruses, which randomly insert themselves into chromosomal DNA and possess sequences that inadvertently activate adjacent host genes. Another type of virus being investigated as a vehicle for gene therapy, called *adeno-associated virus (AAV)*, is less likely to insert directly into chromosomal DNA and less likely to inadvertently activate host genes when it does become inserted. The first gene therapy utilizing an AAV vector for gene delivery (Luxturna) was approved by the FDA in 2017 for the treatment of a rare, inherited form of blindness.

Given the challenges of using DNA in gene therapy, others are turning to RNA-based molecular therapies as an alternative. These therapies exploit RNA-mediated interference (RNAi). DNA encoding short hairpin RNAs (shRNAs) can be delivered in similar ways to other DNAs, but faces similar technical challenges. Instead, some therapies are attempting to directly deliver short interfering RNAs (*siRNAs*; **see Key Technique in Chapter 20, pages 612–613**). Because introduction of double-stranded RNA triggers an innate immune response (an alternative immune response that does not involve antibody production) in human patients, siRNAs must be packaged to avoid detection by a patient's immune system. Promising methods for packaging include cholesterol or lipid-based carriers, or "nanoparticles" that are endocytosed by cells.

Most recently, the advent of genome-editing technologies, such as the CRISPR/Cas9 system mentioned earlier in this chapter, coupled with the use of *induced pluripotent stem cells* (*iPS cells*; see Chapter 20) has led to great excitement regarding the ability to directly repair defective genes in iPS cells. Once repaired, iPS cells could be differentiated and reintroduced into a patient. Technical improvements in these methods will likely be rapid in the years to come.

In the years since the enormous potential of gene therapy was first publicized in the early 1980s, the field has been criticized for promising too much and delivering too little. But most new technologies take time to be perfected and encounter disappointments along the way, and gene therapy is no exception. Despite the setbacks, it appears likely that using normal genes to treat genetic diseases is a reachable goal that may one day become common practice, at least for a few genetic diseases that involve single gene defects. Of course, the ability to alter people's genes raises important ethical, safety, and legal concerns.

Ethical Issues. Detailed knowledge of the human genome and the availability of methods for altering the germ line of mammals increases the potential for using recombinant DNA techniques to *alter* people's genes, not only to correct diseases in malfunctioning body tissues but also to change genes in sperm and eggs, thereby altering the genetic makeup of future generations. In 2018, a Chinese scientist made the shocking announcement that he had used CRISPR to alter the genome of twin girls. This news was met with universal disapproval from scientific and medical communities throughout the world and resulted in a 3-year prison sentence for "illegal medical practices." It also led the World Health Organization to establish an expert panel and to the formation of a separate international commission for oversight of human genome-editing techniques. Being able to identify potentially harmful genes also raises ethical concerns because all of us are likely to carry a few dozen genes that place us at risk for something. Such information could be misused—for example, in genetic discrimination against individuals or groups of people by insurance companies, employers, or even government agencies. The ultimate question of how society will control our growing power to change the human genome is an issue that will need to be thoroughly discussed not just by scientists and physicians, but by society as a whole.

CONCEPT CHECK 21.4

Many human proteins have now been produced in *E. coli* for use in the biotechnology and biomedical fields. However, simply inserting a human gene into *E. coli* will not work. In your new job in a biotechnology firm, you are hoping to produce a human protein in *E. coli*. What steps would you need to take to do so? Based on what you learned in earlier chapters about how proteins are made, what limitations might you expect regarding the produced protein?

Summary of Key Points

Mastering™ Biology For activities, animations, and review quizzes, go to the study area at www.masteringbiology.com.

21.1 Analyzing, Manipulating, and Cloning DNA

- DNA can be separated by size using electrophoresis, and specific DNA fragments can be identified using restriction digests or by probing a mixture of DNA using Southern blotting.
- Recombinant DNA technology makes it possible to combine DNA from any two (or more) sources into a single DNA molecule, including DNA produced by the polymerase chain reaction (PCR).
- Combining DNA of interest with a plasmid or phage cloning vector allows the gene to be cloned (amplified) in bacterial cells. Larger pieces of DNA can be inserted into other vectors, including artificial chromosomes.

21.2 Sequencing and Analyzing Genomes

- Techniques exist for rapid sequencing of DNA, including variations of the Sanger (dideoxy) chain termination technique.
- Rapid next-generation and third-generation DNA sequencing methods and advances in computer technology have allowed sequencing, assembly, and annotation of entire genomes of thousands of organisms, including humans. Online databases and software tools help researchers sort through the enormous amount of DNA and protein sequence data being collected. Bioinformatics and comparative genomics have allowed comparison of protein coding and other DNA sequences among many organisms, aiding phylogenetic analysis.
- DNA polymorphisms (including SNPs and RFLPs) and variations in tandem repeat sequences (in VNTRs) are useful for identifying a human individual's genotype.

21.3 Analyzing RNA and Proteins

- The expression of RNA can be assessed using Northern blotting, in situ hybridization, and reverse transcriptase (RT)-PCR. The expression of many genes can be simultaneously assessed using DNA microarrays and RNAseq.
- Proteins can be separated using SDS-PAGE and 2D gel electrophoresis. Antibodies can be used to identify specific proteins following electrophoresis using Western blotting. Monoclonal antibodies are highly specific for a particular part of a protein and are renewable; polyclonal antibodies typically recognize more sites along a protein but are not renewable.
- Proteins can be isolated using various types of chromatography, by affinity purification using antibodies, and based on their charge and mass using mass spectrometry.
- Proteins can be studied using the tools of molecular biology by engineering proteins in various ways, including tagging them to aid their purification and visualization.
- Protein-protein interactions can be studied using yeast two-hybrid assays, pull-downs, and co-immunoprecipitation.

are called **axons**. The cytosol within an axon is commonly referred to as **axoplasm**. Many vertebrate axons are surrounded by a discontinuous **myelin sheath**, which insulates the segments of axon between **nodes of Ranvier**. Axons can be very long—up to several thousand times longer than the diameter of the cell body. For example, a motor neuron that innervates your foot has its cell body in your spinal cord, and its axon extends approximately a meter down your leg! A **nerve** is simply a bundle of nerve cell processes outside the central nervous system.

As Figure 22-1 illustrates, the axon of a typical neuron is much longer than its dendrites and forms multiple branches. The branches terminate in structures called **synaptic boutons** (also called *terminal bulbs* or *synaptic knobs;* the French word *bouton* means "button"). The boutons are responsible for transmitting the signal to the next cell, which may be another neuron or a muscle or gland cell. In each case, the junction is called a **synapse**. For neuron-to-neuron junctions, synapses often occur between an axon and a dendrite, but they can occur between two dendrites. Typically, neurons make synapses with many other neurons. They can do so not only at the ends of their axons but also at other points along their length.

Neurons Undergo Changes in Membrane Potential

Membrane potential (denoted V_m) is a fundamental property of all cells (see Chapter 8). It results from an excess of negative charge on one side of the plasma membrane and an excess of positive charge on the other side. Cells at rest normally have an excess of negative charge inside and an excess of positive charge outside the cell; in the case of neurons the steady-state electrical potential is usually called the **resting membrane potential**.

To understand how the resting membrane potential forms, recall several principles related to ion transport (see Chapter 8). First, solutes tend to diffuse from an area where they are more highly concentrated to an area where they are less concentrated. For example, cells normally have a high concentration of potassium ions inside and a low concentration of potassium ions outside. This uneven distribution of potassium ions is known as a *potassium ion concentration gradient.* Given the large potassium concentration gradient, potassium ions will tend to diffuse out of the cell.

The second basic principle is that of *electroneutrality:* in solution, positively and negatively charged ions balance one another. For any given ion, let's call it A, there must be an oppositely charged ion B in the solution; B is referred to as the *counterion* for A. As we will see later, in the cytosol potassium ions (K^+) serve as counterions for trapped anions. Outside the cell, sodium (Na^+) is the main cation and chloride (Cl^-) is its counterion.

Third, although a solution must have overall electroneutrality, charges can be *locally* separated so that one region has more positive charges whereas another region has more negative charges. Such charge separation, which requires work, is called an *electrical potential,* or *voltage.*

Finally, in neurons we also need to consider the *movement* of negative or positive ions, which creates **current**; current is measured in amperes (A). Given these principles, we can understand how membrane potential changes as nerve impulses are transmitted.

Neurons Display Electrical Excitability

A great technical advance in understanding how membrane potential changes as nerve signals are transmitted came with the discovery in the 1930s of very large axons in some nerve fibers of squid. These nerves stimulate the explosive expulsion of water from the mantle cavity of the squid, enabling it to propel itself quickly to escape predators (**Figure 22-2**). The **squid giant axon** has a diameter of about 0.5–1.0 mm, allowing the easy insertion of tiny electrodes, or *microelectrodes,* to measure and control electrical potentials and ionic currents across the axonal membrane. **Figure 22-3** on page 666 shows several ways to measure membrane potential. The resting membrane potential can be measured by placing one microelectrode inside the cell and another outside the cell (Figure 22-3a); its value is generally given in millivolts (mV). (Do not confuse the potential V_m and the unit mV!) The electrodes compare the ratio of negative to positive charge inside the cell and outside the cell. Because the area just inside the plasma membrane typically has excess negative charge, cells have a negative resting membrane potential. For example, the resting membrane potential is approximately –60 mV for the squid giant axon.

Nerve, muscle, and certain other cell types, such as the islet cells of the pancreas of vertebrates, exhibit a special property called **electrical excitability**. In electrically excitable cells, certain types of stimuli trigger a rapid sequence of changes in membrane potential known as an *action potential.* During an action potential, the membrane potential changes from negative to positive values and then back to negative values again, all in as little as a few milliseconds. Microelectrodes can be used to measure these dynamic changes in membrane potential. An electrode called the *stimulating electrode* is connected to a power source and inserted into the axon some distance from the recording electrode (see Figure 22-3b). A brief impulse from this stimulating electrode depolarizes

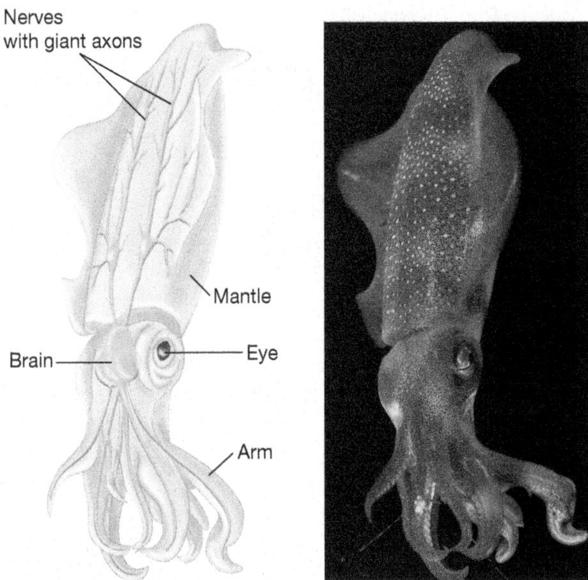

Figure 22-2 Squid Giant Axons. The squid nervous system includes motor nerves that control swimming movements. The nerves contain giant axons (fibers) with diameters ranging up to 1 mm, providing a convenient system for studying resting and action potentials in a biological membrane.

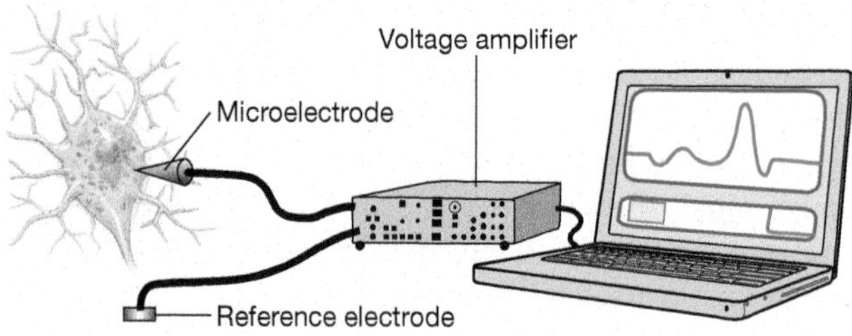

(a) Measuring the resting membrane potential

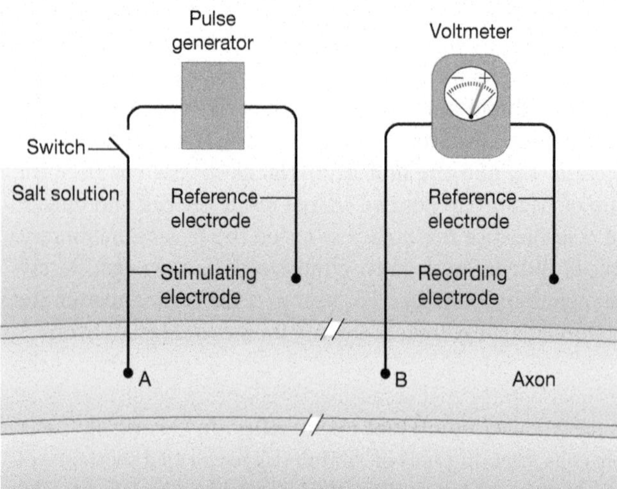

(b) Measuring an action potential in a squid axon

Figure 22-3 Measuring Membrane Potentials. (a) Measurement of the resting membrane potential requires two electrodes, one inserted inside the cell (the recording electrode) and one placed in the fluid surrounding the cell (the reference electrode). Differences in potential between the recording and reference electrodes are amplified by a voltage amplifier and displayed on a voltmeter, an oscilloscope, or a computer monitor. **(b)** Measurement of an action potential requires four electrodes, one in the axon for stimulation, another in the axon for recording, and two in the fluid surrounding the cell for reference. The stimulating electrode is connected to a pulse generator, which delivers a pulse of current to the axon when the switch is momentarily closed. The nerve impulse this generates is propagated down the axon and can be detected a few milliseconds later by the recording electrode.

the membrane by about 20 mV (that is, from −60 to about −40 mV), bringing the neuron to a state that triggers an action potential that propagates away from the stimulating electrode. As the action potential passes the site of the recording electrode, a characteristic pattern of membrane potential changes is recorded. Alan Hodgkin and Andrew Huxley used microelectrodes and the squid giant axon to learn how action potentials are generated; for their work, they received a share of a Nobel Prize in 1963.

To understand how nerve cells use action potentials to transmit signals, let's first consider how cells generate a resting membrane potential, which is the topic of the rest of this section. How membrane potential changes during an action potential is the topic of the following section.

Resting Membrane Potential Depends on Ion Concentrations and Selective Membrane Permeability

The resting membrane potential develops because the cytosol of the cell and the extracellular fluid contain different compositions of cations and anions (**Figure 22-4**). Extracellular fluid contains dissolved salts, including sodium chloride and lesser amounts of potassium chloride. The anions in the cytosol consist largely of trapped macromolecules such as proteins, RNA, and a variety of other molecules that are not present, or much less abundant, outside the cell. These negatively charged macromolecules cannot pass through the plasma membrane and therefore remain inside the cell. The concentrations of important ions for two types of well-studied neurons, the squid axon and one type of mammalian axon, are shown in **Table 22-1**.

Note from Table 22-1 that there is a large difference in concentration of potassium and sodium ions inside versus outside of cells. What contributes to this difference? Recall that ion channels allow ions to pass in and out of the plasma membrane (see Chapter 8). The properties of two specific types of ion channels are especially important for maintaining resting membrane potential.

Figure 22-4 Steady-State Ion Concentrations. Steady-state ion concentrations are maintained by leak channels, which are not voltage regulated, and by the Na^+/K^+ pump, which maintains low resting Na^+ levels. Impermeable negatively charged macromolecules (M^-) inside the cell contribute to the net negative resting potential of most cells.

Table 22-1 Ionic Concentrations Inside and Outside Axons and Neurons

	Squid Axon		Mammalian Neuron (cat motor neuron)	
Ion	Outside (m*M*)	Inside (m*M*)	Outside (m*M*)	Inside (m*M*)
Na^+	440	50	145	10
K^+	20	400	5	140
Cl^-	560	50	125	10

Leak Channels. *Ion channels* are integral membrane proteins that form ion-conducting pores through the lipid bilayer (see Chapter 8). The neuronal plasma membrane is normally somewhat permeable to potassium and sodium ions due to **leak channels** specific for these ions. Leak channels are not *gated*, that is, they are always open, which permits ions to diffuse in or out of the cell depending on the membrane voltage and local ion concentration. The density of potassium leak channels is relatively high; as a result, potassium ions are the single greatest contributing component to the resting membrane potential. One additional feature of Figure 22-4 is important: There are no channels for large, negatively charged macromolecules. As potassium leaves the cytosol, increasing numbers of these trapped anions are left behind without counterions. Excess negative charge therefore accumulates in the cytosol near the plasma membrane, whereas excess positive charge accumulates on the outside of the cell, resulting in a negative resting membrane potential.

The Na^+/K^+ Pump. Although the plasma membrane is relatively impermeable to sodium ions, there is always a small amount of movement due to leak channels. To compensate for this leakage, the *Na^+/K^+ pump* continually uses energy from ATP hydrolysis to move sodium out of the cell while carrying potassium inward (see Figure 8-13). On average, the pump transports three sodium ions out of the cell and two potassium ions into the cell for every molecule of ATP that is hydrolyzed, further contributing to the negative resting potential of the cell. The Na^+/K^+ pump maintains the large potassium ion gradient across the membrane that provides the basis for the resting membrane potential.

The Nernst Equation Describes the Relationship Between Membrane Potential and Ion Concentration

At steady state, an equilibrium is reached in which the force of attraction due to the membrane potential balances the tendency of ions to diffuse down their concentration gradients. This type of equilibrium, in which a chemical gradient is balanced with an electrical potential, is referred to as an **electrochemical equilibrium**. The membrane potential at the point of equilibrium is known as an **equilibrium potential**, also known as a **reversal potential**.

The **Nernst equation** provides a mathematical description of the membrane potential at equilibrium and makes it possible to estimate the value of equilibrium potential. The equation is named for the German physical chemist and Nobel laureate Walther Nernst, who first formulated it in the late 1880s in the context of his work on electrochemical cells (forerunners of modern batteries). The Nernst equation describes the mathematical relationship between an ion gradient and the equilibrium potential that will form when the membrane is permeable only to that ion:

$$E_X = \frac{RT}{zF} \ln \frac{[X]_{\text{outside}}}{[X]_{\text{inside}}} \qquad \textbf{(22-1)}$$

Here, E_X is the equilibrium potential for ion X (in volts), R is the gas constant (1.987 cal/mol-degree), T is the absolute temperature in kelvins (recall that 0°C = 273 K), z is the valence of the ion, F is the Faraday constant (23,062 cal/mol-V), $[X]_{\text{outside}}$ is the concentration of X outside the cell (in *M*), and $[X]_{\text{inside}}$ is the concentration of X inside the cell (in *M*). This equation can be simplified by assuming that the temperature is 293 K (20°C)—a value appropriate for the squid giant axon—and that X is a monovalent cation and therefore has a valence of +1. Substituting these values for R, T, F, and z into the Nernst equation and converting from natural logs to $\log_{10}$ ($\log_{10} = \ln/2.303$), Equation 22-1 reduces to

$$E_X = 0.0581 \log_{10} \frac{[X]_{\text{outside}}}{[X]_{\text{inside}}} \qquad \textbf{(22-2)}$$

In this simplified form, it is easy to see that for every tenfold increase in the cation gradient, the membrane potential changes by −0.058 V, or −58 mV.

MAKE CONNECTIONS 22.1

Plugging in numbers, Equation 22-2 calculates the "proton motive force" (really a potential) across a membrane (see the right side of Equation 10-16). How much change in voltage occurs with a pH difference of 1 across a membrane if the pH inside is higher? (Ch. 10.5)

Steady-State Ion Concentrations Affect Resting Membrane Potential

Although Equation 22-2 explains much about membrane potential, it is incomplete, mainly because it does not account for the effects of anions. The main anion in the extracellular fluid is chloride. As we have seen, sodium, potassium, and chloride ions are the major ionic components present in both the cytosol and the extracellular fluid. Because of their unequal distributions across the cell membrane, each ion has a different impact on the membrane potential. The qualitative concentration of each ion for a typical neuron is illustrated in Figure 22-4. Each ion will tend to diffuse down its electrochemical gradient and thereby produce a change in the membrane potential.

Potassium ions tend to diffuse out of the cell, which makes the membrane potential more negative. Sodium ions tend to flow into the cell, driving the membrane potential in the positive direction and thereby causing a **depolarization** of the membrane (that is, causing the membrane potential to be less negative). Chloride ions tend to diffuse into the cell, which should in principle make the membrane potential more negative. However, chloride ions are also repelled by the negative membrane potential, so that chloride ions usually enter the cell at the same time as positively charged sodium ions. This simultaneous movement nullifies the depolarizing effect of sodium entry. Increasing the permeability of cells to chloride can have two effects, both of which decrease neuronal excitability. First, the net entry of chloride ions (chloride entry without a matching cation) causes *hyperpolarization* of the membrane (that is, the membrane potential becomes more highly negative than usual). Second, when the membrane becomes permeable to sodium ions, some chloride will tend to enter the cell along with sodium. As you will learn later in this chapter, this effect of chloride entry is a prominent feature of the way inhibitory neurotransmitters act.

The Goldman Equation Describes the Combined Effects of Ions on Membrane Potential

The relative contributions of the major ions are important to the resting membrane potential because, even in its resting state, the cell has some permeability to sodium and chloride ions as well as to potassium ions. To account for the leakage of sodium and chloride ions into the cell, the Nernst equation cannot be used because it deals with only one type of ion at a time, and it assumes that this ion is in electrochemical equilibrium. A more accurate accounting for the effects of ions on dynamic changes in membrane potential requires that we move from the more static concept of equilibrium potential to a consideration of *steady-state ion movements* across the membrane.

The concept of steady-state ion movements can be illustrated by returning once more to the model of a cell in electrochemical equilibrium (see Figure 22-4). Remember that a cell permeable only to potassium would have a membrane potential equal to the equilibrium potential for potassium ions. Under these conditions, there would be no net movement of potassium out of the cell. Assuming the membrane were also slightly permeable to sodium ions, what would happen? The cell would have both a large sodium gradient across the membrane and a negative membrane potential corresponding to the potassium equilibrium potential. These forces would tend to drive sodium ions into the cell. As sodium ions leak inward, the membrane would be partially depolarized. At the same time, as the membrane potential became neutralized, there would be less restraining force preventing potassium from leaving the cell, so potassium ions would diffuse outward, balancing the inward movement of sodium. The inward movement of sodium ions would shift the membrane potential in the positive direction, whereas the outward movement of potassium ions would shift the membrane potential back in the negative direction.

Movements of sodium and potassium ions across the membrane, therefore, have essentially opposite effects on membrane potential. For one type of squid axon, the sodium ion gradient tends toward a cell membrane potential of about +55 mV, whereas the potassium ion gradient tends toward a membrane potential of about −75 mV. At what value will the membrane potential come to rest when we consider multiple ions at the same time? The pioneering neurobiologists David E. Goldman, Alan Lloyd Hodgkin, and Bernard Katz were the first to describe how gradients of several different ions each contribute to the membrane potential as a function of relative ionic permeabilities. The Goldman-Hodgkin-Katz equation, more commonly known as the **Goldman equation**, is as follows:

$$V_m = \frac{RT}{F} \ln \frac{(P_K)[K^+]_{out} + (P_{Na})[Na^+]_{out} + (P_{Cl})[Cl^-]_{in}}{(P_K)[K^+]_{in} + (P_{Na})[Na^+]_{in} + (P_{Cl})[Cl^-]_{out}} \qquad \textbf{(22-3)}$$

Because chloride ions have a negative valence, $[Cl^-]_{inside}$ appears in the numerator and $[Cl^-]_{outside}$ in the denominator.

A key difference between the Nernst equation and the Goldman equation is the incorporation of terms for permeability. Here, P_K, P_{Na}, and P_{Cl} are the *relative permeabilities* of the membrane for the respective ions. The use of relative permeabilities circumvents the complicated task of determining the absolute permeability of each ion. Although the equation shown here takes into account only the contributions of potassium, sodium, and chloride ions, other ions could be added as well. Except under special circumstances, however, the permeability of the plasma membrane to other ions is usually so low that their contributions are negligible.

A squid axon can be used to illustrate how useful the Goldman equation is. We assign K^+ a permeability value of 1.0, and the permeability values of all other ions are determined relative to that of K^+. For squid axons, the permeability of sodium ions is only about 4% of that for potassium ions; for chloride ions, the value is 45%. Relative values of P_K, P_{Na}, and P_{Cl} are therefore 1.0, 0.04, and 0.45, respectively. Using these values, a temperature of 20°C, and the intracellular and extracellular concentrations of Na^+,K^+, and Cl^- for the squid axon from Table 22-1, you should be able to estimate the resting membrane potential of a squid axon as −60.3 mV. Typical measured values for the resting membrane potential of a squid axon are about −60 mV, which is remarkably close to the calculated potential.

When the relative permeability for one ion is very high, the Goldman equation reduces to the Nernst equation for that particular ion. For example, if we ignore the contribution due to other ions when P_{Na} is very high, Equation 22-3 reduces to

$$V_m = \frac{RT}{F} \ln \frac{[Na^+]_{out}}{[Na^+]_{in}}, \qquad \textbf{(22-4)}$$

which is just the Nernst equation for sodium. Because the resting potential for sodium is about +55 mV, making a cell highly permeable to sodium will cause it to depolarize. Similarly, if P_K is much higher than the permeabilities for other ions, then the Goldman equation reduces to the Nernst equation for potassium. Because the resting potential for potassium is about −75 mV, high permeability to potassium will tend to return a cell to a polarized state. Neurons use precisely this strategy to transiently depolarize and then repolarize their plasma membranes, as you will see in the next section.

CONCEPT CHECK 22.1

Ouabain is an African plant derivative that has been used historically to make poison-tipped hunting arrows. It disables the main Na^+/K^+ ATPase in neurons. How would the resting potential of neurons in an organism exposed to ouabain change relative to the normal situation? Explain your answer.

22.2 Electrical Excitability and the Action Potential

The establishment of a resting membrane potential and its dependence on ion gradients and ion permeability are properties of almost all cells. The unique feature of electrically excitable cells is their response to membrane depolarization. Whereas a nonexcitable cell that has been temporarily and slightly depolarized will simply return to its original resting membrane potential, an electrically excitable cell that is depolarized to the same degree will respond with an *action potential*.

Electrically excitable cells produce an action potential because, in addition to the leak channels and Na^+/K^+ pumps we have discussed thus far, they contain *voltage-gated channels* in their plasma membranes. To see how nerve cells communicate signals electrically, we need to understand the characteristics of the ion channels in the nerve cell membrane, which can be studied using several important techniques.

Patch Clamping and Molecular Biological Techniques Allow Study of Single Ion Channels

Several modern techniques are used to examine ion channels in neuronal cells. Used together, these techniques provide a powerful set of tools for studying ion channels in unprecedented detail.

Patch Clamping. The clearest picture of how channels operate comes from using a technique that permits the recording of ion currents passing through individual channels. This technique, known as *single-channel recording*, or more commonly as **patch clamping**, was developed by Erwin Neher and Bert Sakmann, who earned a Nobel Prize in 1991 for their discovery. Patch clamping is an extremely important tool in modern neurobiology (see Key Technique, pages 670–671).

Frog Oocytes. Much of the present research on ion channels combines patch clamping with molecular biology. It is now possible to synthesize large amounts of channel proteins in the laboratory and to study their functions in lipid bilayers or in frog eggs. Specific molecular modifications or mutations of the channel can be used to determine how various regions of the channel protein are involved in channel function. This approach has been used to study the domains of the sodium and potassium channels responsible for voltage gating.

Optogenetics. The ability to measure the electrical properties of single ion channels is powerful, but it would be even more powerful to be able to locally manipulate ion channels in specific regions of single cells. This is now becoming possible using genetically engineered channel proteins that are sensitive to light. This approach, known as *optogenetics*, uses molecular biology techniques to introduce light-sensitive ion channels into neurons. One such protein is bacteriorhodopsin (see Chapter 8), which is used to inhibit some neurons. Another family of bacterial light-sensitive channels known as *channelrhodopsins* is used to activate some neurons. By expressing one of these light-sensitive channels in a neuronal cell in culture or in a transgenic animal (see Chapter 21) and then shining light in a defined region of the cell, the ion concentration can be changed locally and the effects assessed under the microscope. Light-sensitive molecules are now being engineered to study many events in neurons. Optogenetics clearly has a "bright" future.

Specific Domains of Voltage-Gated Channels Act as Sensors and Inactivators

Recall that *leak channels* contribute to the steady-state ionic permeabilities of membranes, allowing resting cells to be somewhat permeable to cations, especially potassium ions. In contrast, **voltage-gated ion channels**, as the name suggests, respond to changes in the voltage across a membrane. Voltage-gated sodium and potassium channels are responsible for the action potential. **Ligand-gated ion channels**, by contrast, open when a particular molecule binds to the channel. (*Ligand* comes from the Latin root *ligare*, meaning "to bind.")

The basic structure of a voltage-gated ion channel is shown in **Figure 22-5** on page 672. Such channels fall into two different structural categories. *Voltage-gated potassium channels* are *multimeric* proteins—that is, they consist of four separate protein subunits that come together in the membrane, forming a central pore that ions can pass through. *Voltage-gated sodium channels*, by contrast, are large *monomeric* proteins; they consist of a single polypeptide with four separate domains. Each domain is similar to one of the subunits of the voltage-gated potassium channel. In both kinds of channels, each subunit or domain contains six transmembrane α helices (called subunits S1–S6; Figure 22-5a).

The size of the central pore and, more importantly, the way it interacts with an ion, give a channel its ion selectivity. Figure 22-5b shows why this is so for the bacterial potassium channel KcsA. Vertebrate voltage-gated potassium channels have a similar structure. Oxygen atoms in amino acids lining the center of the channel are precisely positioned to interact with ions as they move through the *selectivity filter*, allowing them to give up their waters of hydration. The fit between potassium ions and oxygens lining the channel is remarkably precise. Na^+, which is smaller than K^+, can only interact with oxygen atoms on one side of the channel. This makes it energetically unfavorable for Na^+ to give up its waters of hydration and enter the channel. Roderick MacKinnon received a Nobel Prize in 2003 in part for his work on the pore structure of potassium channels.

Voltage-gated sodium channels have the ability to open rapidly in response to some stimulus and then to close again, a phenomenon known as **channel gating**. This open or closed state is an all-or-none phenomenon—that is, gates do not

Key Technique | Patch Clamping

PROBLEM: Standard electrical measurement of neurons measures changes in overall cellular voltage, which reflects the opening and closing of many channels at any given time. How can the electrical properties of a single channel be studied?

SOLUTION: Patch clamping allows a tiny patch of membrane to be isolated and its electrical properties studied at the level of the opening and closing of a single channel protein.

Key Tools: A polished glass microelectrode (micropipette); sensitive voltage recording equipment; access to the neuronal membrane being studied.

Details: To record single-channel currents, a glass micropipette with a tip diameter of approximately 1 μm is carefully pressed against the surface of a cell such as a neuron. Gentle suction is then applied so that a tight seal forms between the pipette and the plasma membrane (**Figure 22A-1**, ❶). There is now a patch of membrane under the mouth of the micropipette that is sealed off from the surrounding medium (❷). This patch is small enough that it usually contains only one or perhaps a few ion channels. Current can enter and leave the pipette only through these channels, thereby enabling an experimenter to study various properties of the individual channels. The channels can be studied in the intact cell, or the patch can be pulled away from the cell so that the researcher has access to the cytosolic side of the membrane.

During the experimental process, an amplifier maintains voltage across the membrane with the addition of a sophisticated electronic feedback circuit (a voltage clamp, hence the term *patch clamp*). The voltage clamp keeps the cell at a fixed membrane potential, regardless of changes in the electrical properties of the plasma membrane, by injecting current as needed to hold the voltage constant. The voltage clamp then measures tiny changes in current flow—actual ionic currents through individual channels—from the patch pipette. The patch-clamp method has been used to show that each time a voltage-gated sodium channel opens, it conducts the same amount of *current*—that is, the same number of ions per unit of time. In other words, voltage-gated sodium channels tend to be either open or closed.

Based on these properties of ion channels, a particular channel can be characterized in terms of its conductance. *Conductance* is an indirect measure of the permeability of a channel when a specified voltage is applied across the membrane. In electrical terms, conductance is the inverse of resistance. For voltage-gated sodium channels, when a voltage of 50 mV is applied across the membrane, a current of approximately 1 picoampere (pA) is generated. This current corresponds to about 6 million sodium ions flowing through the channel per second. This can be seen in the traces shown in Figure 22A-1, ❸.

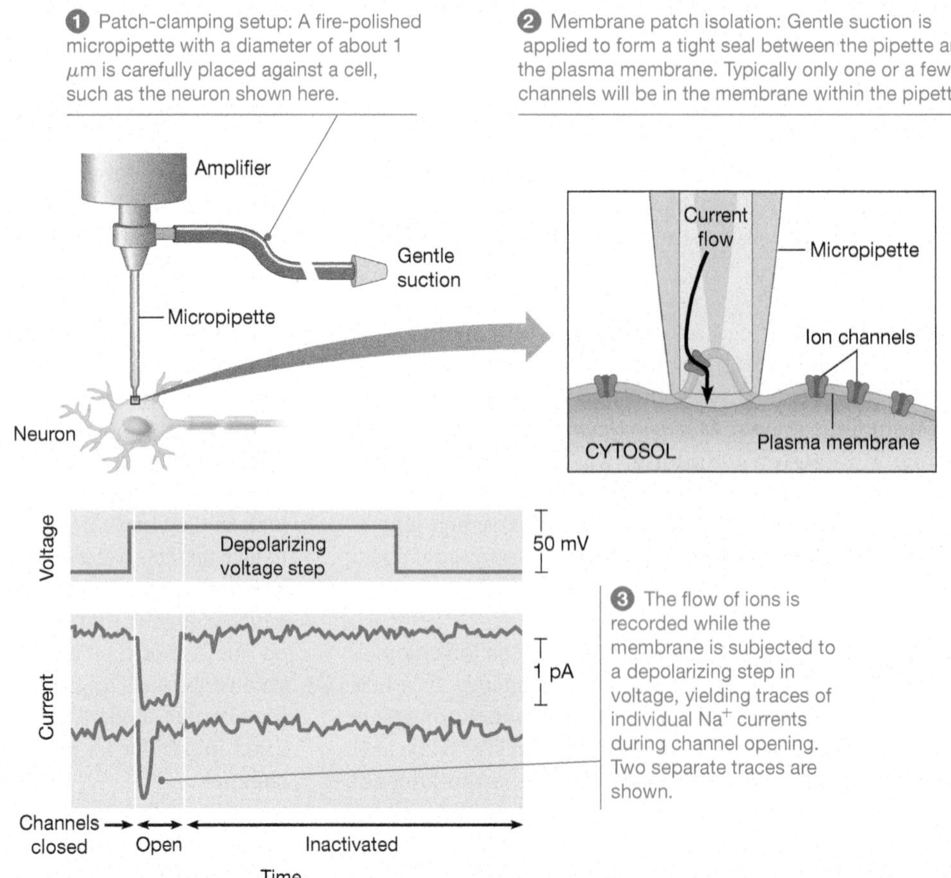

❶ Patch-clamping setup: A fire-polished micropipette with a diameter of about 1 μm is carefully placed against a cell, such as the neuron shown here.
❷ Membrane patch isolation: Gentle suction is applied to form a tight seal between the pipette and the plasma membrane. Typically only one or a few channels will be in the membrane within the pipette.
Amplifier
Gentle suction
Micropipette
Neuron
Current flow
Micropipette
Ion channels
CYTOSOL
Plasma membrane
Voltage
Depolarizing voltage step
50 mV
Current
1 pA
❸ The flow of ions is recorded while the membrane is subjected to a depolarizing step in voltage, yielding traces of individual Na+ currents during channel opening. Two separate traces are shown.
Channels closed
Open
Inactivated
Time

Figure 22A-1 The Basic Patch-Clamping Method. Patch clamping makes it possible to study the behavior of individual ion channels in a small patch of membrane. When each channel opens, the amount of current that flows through it is always the same (1 pA for the sodium channels shown here). Following the burst of channel opening, a quiescent period occurs due to channel inactivation.

Patch clamping is a very versatile technique. By applying differing amounts of suction to the pipette, patch clamping can be used to isolate and study patches of membrane in various orientations. Starting with a tight membrane seal (**Figure 22A-2a**), the *whole-cell* mode (Figure 22A-2b), a commonly used patch-clamp mode, involves ripping a patch of membrane away with strong suction so that the inside of the pipette is continuous with the cytosol. This allows measurement of current from the entire cell. To record currents from a small patch of membrane, which may include only a single channel, the pipette can be retracted to break loose a small patch of membrane. The patch can be orientated in one of two ways inside the patch pipette. In the *inside-out* configuration the pipette is attached to the cell and then retracted to break off a patch of membrane (Figure 22A-2c). In this case the cytosolic side of membrane is exposed, and chemicals that might normally act in the cytosol can be added to test their effects on the channel.

If the goal is to study neurotransmitters or other molecules that bind to the extracellular part of a channel, the *outside-out* configuration (Figure 22A-2d) is used. In this case the pipette is retracted as in the whole-cell mode, but a small patch of membrane is captured, which reseals to produce a patch with an orientation opposite to that of the inside-out mode. The extracellular side of the patch can then be bathed with chemicals.

Pipette

Cell body of neuron

10 μm

A Patch Clamped Neuron.

QUESTION: You have isolated a toxin from a previously undiscovered poisonous snake. You believe that it binds to sodium channels in nerve cell membranes that are normally tightly regulated, keeping them permanently open. Explain how you could use patch clamping to test your idea, using one of the specific variants of the patch-clamp technique in your answer.

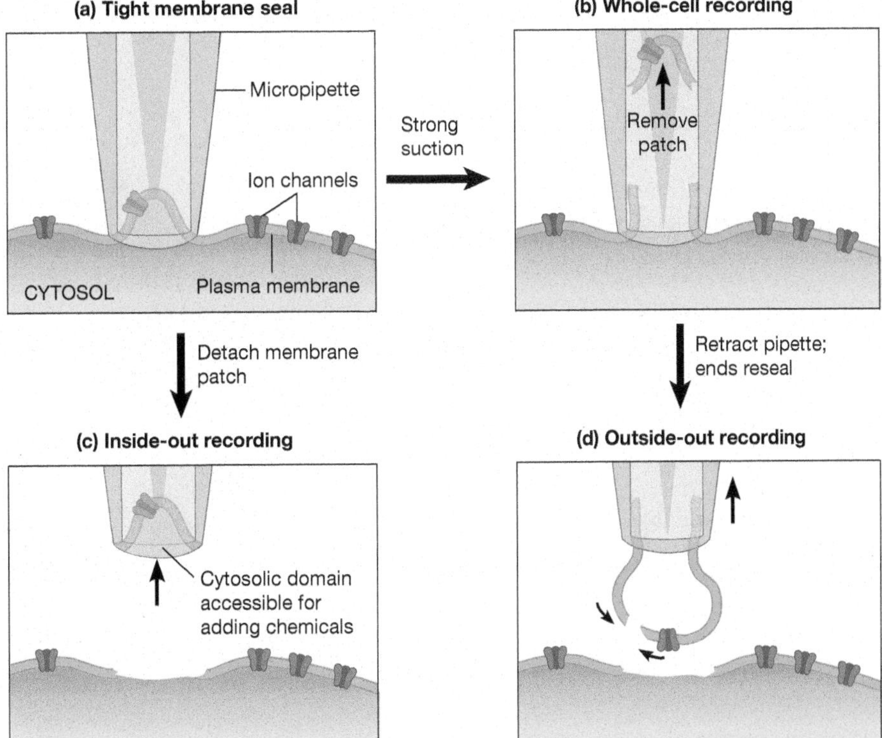

Figure 22A-2 Different Uses of Patch Clamping. Variations on the basic patch-clamp method, shown in part a, include **(b)** the whole-cell method, which allows measurement of current from an entire cell; **(c)** the inside-out method, which allows current across a small patch whose cytosolic surface is facing outward; and **(d)** the outside-out method, which measures current across a patch whose extracellular side is facing outward.

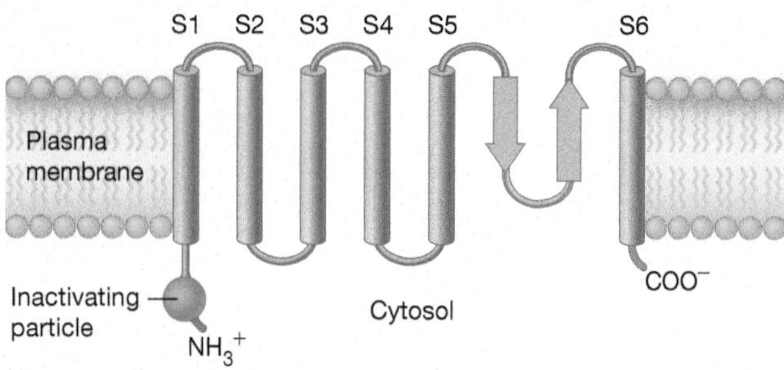

(a) Domain structure of an individual subunit

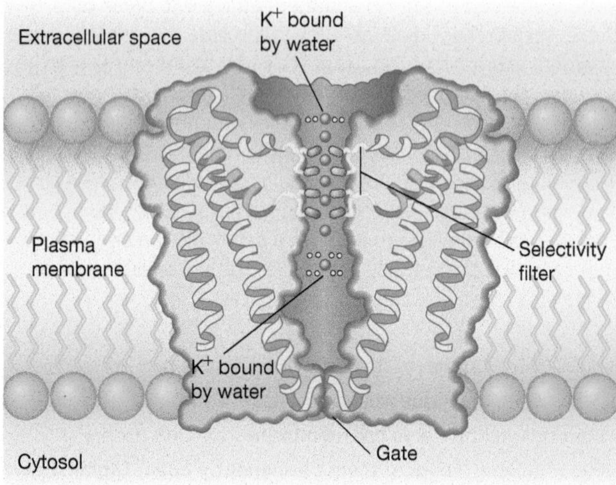

(b) Pore structure

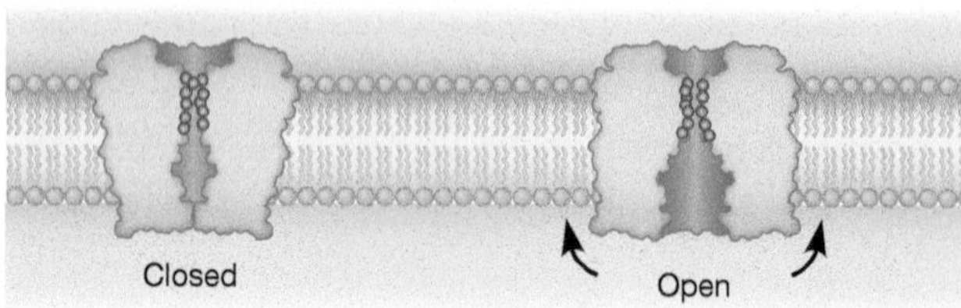

(c) Channel gating

Figure 22-5 The General Structure of Voltage-Gated Ion Channels. (a) Domain structure. Voltage-gated channels for sodium, potassium, and calcium ions all share the same basic structural themes. The channel is essentially a rectangular tube whose four walls are formed from either four subunits (for example, potassium channels) or four domains of a single polypeptide (for example, sodium channels). Each subunit or domain contains six transmembrane helices, labeled S1–S6. The fourth transmembrane helix, S4, is a voltage sensor and part of the gating mechanism. For voltage-gated sodium channels and some types of potassium channels, a region near the N-terminus protrudes into the cytosol and forms an inactivating particle. **(b)** Pore structure. The transmembrane region of a potassium channel in the closed position. This diagram is based on the bacterial KcsA channel, but vertebrate potassium channels are similar. Two of the four subunits of a voltage-gated potassium channel are shown here. Only the transmembrane part of the channel is shown. Hydrated potassium ions (teal) enter the channel, where they give up their water and bind oxygen atoms, precisely positioned in the amino acids lining the selectivity filter (brown). **(c)** Channel gating. The channel is regulated by a gate, which can open or close, depending on the state of voltage sensor domains in the channel. The channel gate opens and closes depending on the conformational state of channel subunits.

appear to remain partially open (Figure 22-5c). One of the transmembrane α helices of vertebrate voltage-gated channels, S4, acts as a **voltage sensor** during channel gating. When positively charged amino acids in S4 are replaced with neutral amino acids, the channel does not open, indicating that S4 makes these channels responsive to changes in potential.

Most voltage-gated channels can also adopt a second type of closed state, referred to as **channel inactivation**, which is an important feature of voltage-gated sodium and potassium channels (**Figure 22-6**). When a channel is inactivated, it cannot reopen immediately, even if stimulated to do so. Channel inactivation functions like placing a padlock on a closed gate; only when the padlock is unlocked can the

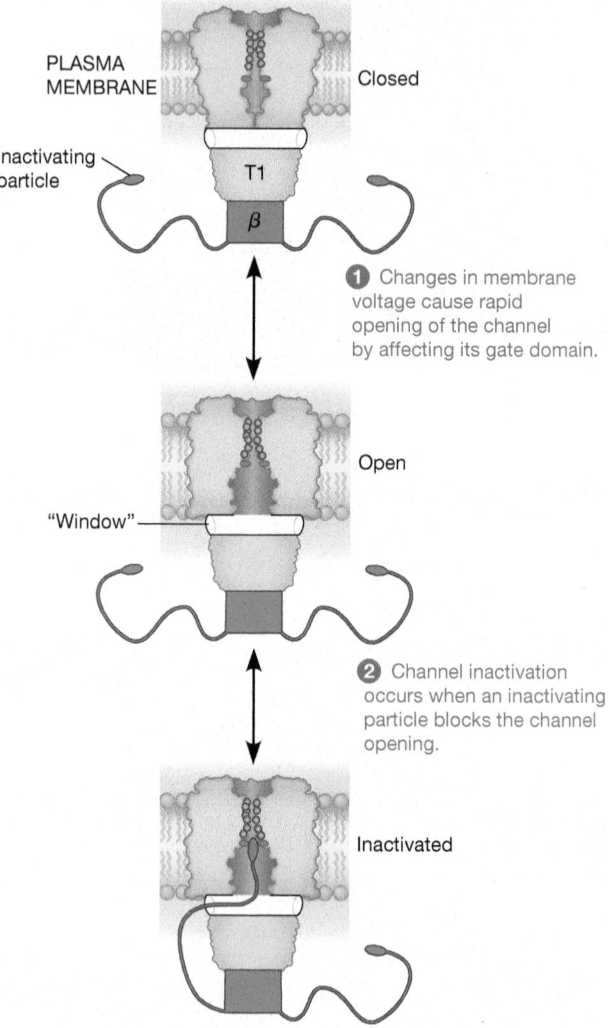

Figure 22-6 The Function of a Voltage-Gated Ion Channel. 1 Mammalian potassium channels have three major domains: the transmembrane (S) and cytosolic (T1) domains of the α subunit and the β subunit. The N-terminal region of the β subunit contains an inactivating particle. Channel gating occurs because a portion of the transmembrane domain changes conformation when the membrane potential changes. 2 Channel inactivation. Here, two of the four inactivating particles are shown for a voltage-gated potassium channel. Channels are inactivated when an inactivating particle moves through a "window" region of the channel, blocking the opening of the pore.

gate be opened again. Inactivation is caused by a portion of the channel called the *inactivating particle.* Common channels have four such particles. During inactivation, a particle inserts into the opening of the channel. For a channel to reactivate and open in response to a stimulus, the inactivating particle must move away from the pore. The importance of the inactivating particle can be demonstrated by treating channels with a protease or with antibodies prepared against the fragment of the channel responsible for inactivation. When the inactivating particle no longer functions, channels can no longer be inactivated.

The regulation of ion channels is crucial for the proper functioning of neurons. Defects in several voltage-gated ion channels have been linked to human neurological diseases (such defects have been termed *channelopathies*). For example, humans carrying mutations in certain K^+ channels suffer from ataxia (a defect in muscle coordination), and one form of epilepsy is caused by a mutation in one type of voltage-gated Na^+ channel.

Action Potentials Propagate Electrical Signals Along an Axon

Now we are ready to explore how the coordinated opening and closing of ion channels can lead to an action potential. Let's begin by examining how membrane potential changes during an action potential. Because of its historical importance, we will use the squid giant axon as the model for our discussion.

A resting neuron is a system poised for electrical action. As we have seen, the membrane potential of the cell is set by a delicate balance of ion gradients and ion permeability. Depolarization of the membrane upsets this balance. If the level of depolarization is small—less than about +20 mV—the membrane potential will normally drop back to resting levels without further consequences. Further depolarization brings the membrane to the **threshold potential**. Above the threshold potential, the nerve cell membrane undergoes rapid and dramatic alterations in its electrical properties and permeability to ions, and an action potential is initiated.

An **action potential** is a brief but large electrical depolarization and repolarization of the neuronal plasma membrane caused by the inward movement of sodium ions and the subsequent outward movement of potassium ions. These ion movements are controlled by the opening and closing of voltage-gated sodium and potassium channels. Once an action potential is initiated in one region of the membrane, it will travel along the membrane away from the site of origin by a process called **propagation**.

Action Potentials Involve Rapid Changes in the Membrane Potential of the Axon

The apparatus we encountered in Figure 22-3b can be used to measure the ion currents that flow through the membrane at different phases of an action potential in squid giant axons. To do so, an additional electrode known as the *holding electrode* is inserted into the cell and connected to a *voltage clamp,* thereby enabling the investigator to set and hold the membrane at a particular potential. Using the voltage-clamp apparatus, a researcher can measure the current flowing through the membrane at any given membrane potential. Such experiments have contributed fundamentally to our present understanding of the mechanism that causes an action potential.

Figure 22-7 on page 674 shows the sequence of membrane potential changes associated with an action potential. In less than a millisecond, the membrane potential rises dramatically from the resting membrane potential to about +40 mV—the interior of the membrane actually becomes positive for a brief period. The potential then falls somewhat more slowly, dropping to about −75 mV (called *undershoot* or *hyperpolarization*) before stabilizing again at the resting potential of about −60 mV. As Figure 22-7 indicates, the complete sequence of events during an action potential takes place within a few milliseconds.

Action Potentials Result from the Rapid Movement of Ions Through Axonal Membrane Channels

In a resting neuron, the voltage-dependent sodium and potassium channels are usually closed. Because of leak channels, at steady state the membrane is roughly 100 times more permeable to potassium than to sodium ions. When a region of the nerve cell is slightly depolarized, a fraction of the sodium channels respond and open. As they do, the increased sodium current acts to further depolarize the membrane. Increasing depolarization causes an even larger sodium current to flow, which depolarizes the membrane even more. This relationship between depolarization, the opening of voltage-gated sodium channels, and an increased sodium current constitutes a positive feedback loop known as the *Hodgkin cycle.*

Subthreshold Depolarization. Under resting conditions, the outward movement of potassium ions through leak channels restores the resting membrane potential. When the membrane is depolarized by a small amount, the membrane potential recovers and no action potential is generated. Levels of depolarization that are too small to produce an action potential are referred to as *subthreshold depolarizations.*

The Depolarizing Phase. If all the voltage-dependent sodium channels in the membrane were to open at once, the cell would suddenly become ten times more permeable to sodium than to potassium. As a result, the membrane potential would be largely a function of the sodium ion gradient. This is effectively what happens when the membrane is depolarized past the threshold potential (Figure 22-7, ❶ and ❷). Once the threshold potential is reached, a significant number of gated sodium channels begin activating. At this point, the membrane potential shoots rapidly upward. When the rate of sodium entry slightly exceeds the maximum rate of movement of potassium through leak channels, an action potential is triggered. When the membrane potential peaks, at approximately +40 mV, the action potential approaches—although it does not actually reach—the equilibrium potential for sodium ions (about +55 mV). It never actually reaches this value because the membrane remains permeable to other ions during this time.

The Repolarizing Phase. Once the membrane potential has risen to its peak, the membrane quickly repolarizes (Figure 22-7, ❸). This is due to a combination of the inactivation of sodium

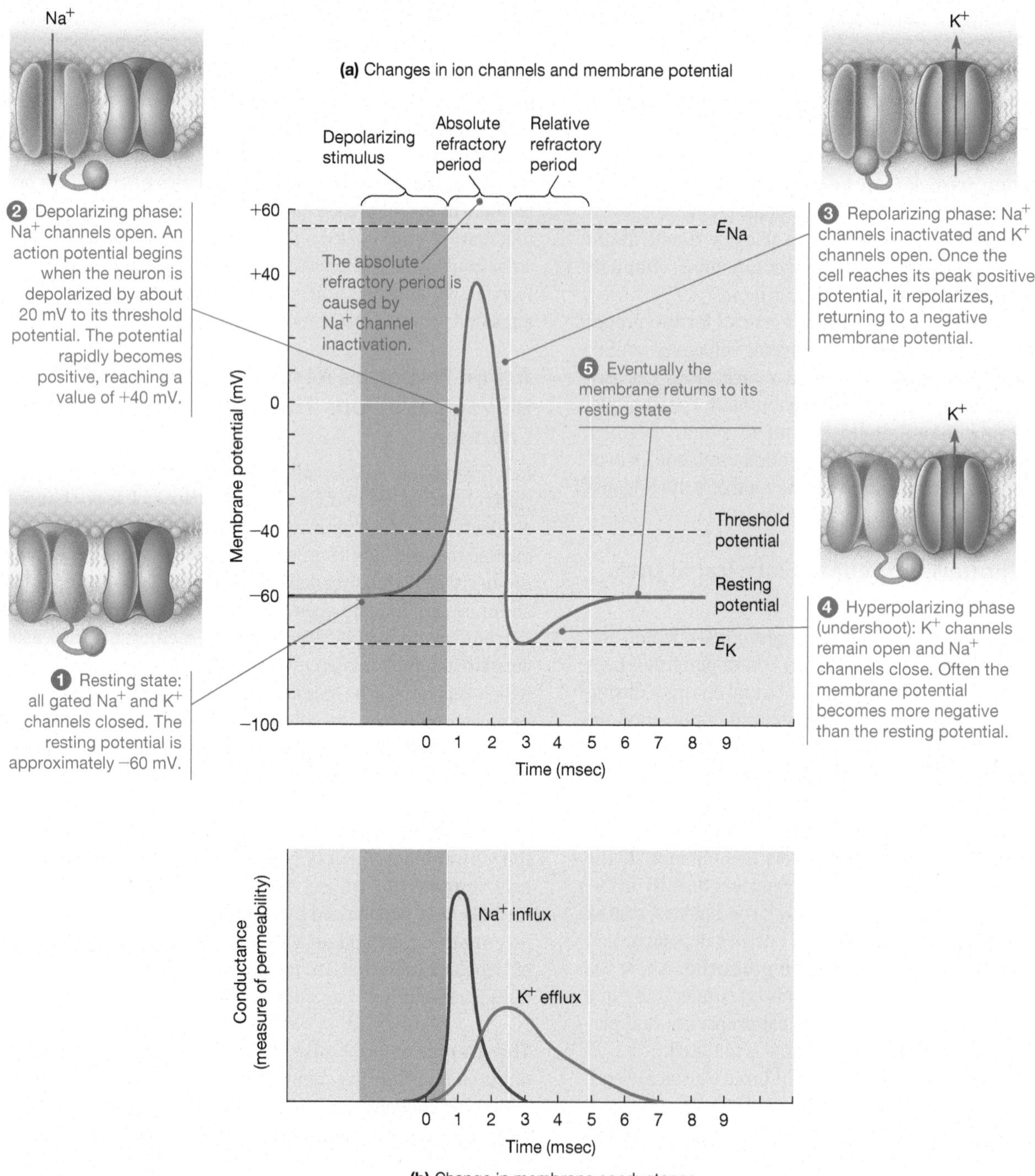

(a) Changes in ion channels and membrane potential
Na+
K+
Depolarizing stimulus
Absolute refractory period
Relative refractory period
2 Depolarizing phase: Na+ channels open. An action potential begins when the neuron is depolarized by about 20 mV to its threshold potential. The potential rapidly becomes positive, reaching a value of +40 mV.
The absolute refractory period is caused by Na+ channel inactivation.
3 Repolarizing phase: Na+ channels inactivated and K+ channels open. Once the cell reaches its peak positive potential, it repolarizes, returning to a negative membrane potential.
5 Eventually the membrane returns to its resting state
E_{Na}
Threshold potential
Resting potential
E_K
Membrane potential (mV)
+60
+40
0
−40
−60
−100
4 Hyperpolarizing phase (undershoot): K+ channels remain open and Na+ channels close. Often the membrane potential becomes more negative than the resting potential.
1 Resting state: all gated Na+ and K+ channels closed. The resting potential is approximately −60 mV.
0 1 2 3 4 5 6 7 8 9
Time (msec)
Conductance (measure of permeability)
Na+ influx
K+ efflux
0 1 2 3 4 5 6 7 8 9
Time (msec)
(b) Change in membrane conductance

Figure 22-7 Changes in Ion Channels and Currents in the Membrane of a Squid Axon During an Action Potential. (a) The change in membrane potential caused by movement of Na^+ and K^+ through their voltage-gated channels, which are shown at each step of the action potential. The absolute refractory period is caused by sodium channel inactivation. Notice that at the peak of the action potential, the membrane potential approaches the E_{Na} (sodium equilibrium potential) value of about +55 mV; similarly, the potential undershoots nearly to the E_K (potassium equilibrium potential) value of about −75 mV. **(b)** The change in membrane conductance (permeability of the membrane to specific ions). The depolarized membrane initially becomes very permeable to sodium ions, facilitating inward movement of sodium. Thereafter, as permeability to sodium declines, the permeability of the membrane to potassium increases transiently, causing the membrane to hyperpolarize.

channels and the opening of voltage-gated potassium channels. When sodium channels are inactivated, they close and remain closed until the membrane potential becomes negative again. Channel inactivation thus stops the inward flow of sodium ions. The cell now repolarizes as potassium ions leave the cell.

The difference in response speed between voltage-gated potassium channels and voltage-gated sodium channels plays an important role in generating the action potential. An action potential begins with an increase in the membrane's permeability to sodium, which depolarizes the membrane, followed by an increased permeability to potassium, which repolarizes the membrane.

The Hyperpolarizing Phase (Undershoot). At the end of an action potential, most neurons show a transient **hyperpolarization**, or **undershoot**, in which the membrane potential briefly becomes even more negative than it normally is at rest (Figure 22-7, ❹). The undershoot occurs because of the increased potassium permeability that exists while voltage-gated potassium channels remain open. Note that the potential of the undershoot closely approximates the equilibrium potential for potassium ions (about −75 mV for the squid axon). As the voltage-gated potassium channels close, the membrane potential returns to its original resting state (Figure 22-7, ❺). Notice that *the rapid restoration of the resting potential following an action potential does not use the* Na^+/K^+ *pump but instead involves the passive movements of ions.* In cells that have been treated with a metabolic inhibitor so that they cannot produce ATP (and hence their Na^+/K^+ pumps cannot use it), action potentials can still be generated. The pump helps maintain a negative potential once an action potential has passed and the membrane has returned to its resting state.

The Refractory Periods. For a few milliseconds after an action potential, it is impossible to trigger a new action potential. During this interval, known as the **absolute refractory period** (see Figure 22-7, ❸), sodium channels are inactivated and cannot be opened by depolarization. During the period of undershoot, when the sodium channels have reactivated and are capable of opening again, it is possible but difficult to trigger an action potential. This is because both potassium leak channels and voltage-gated potassium channels are open during this time. This tends to drive the membrane potential to a very negative value, far from the threshold for triggering another round of sodium channel opening. This interval is known as the **relative refractory period** (see Figure 22-7, ❹).

Changes in Ion Concentrations Due to an Action Potential. Our discussion of ion movements might give the impression that an action potential involves large changes in the cytosolic concentrations of sodium and potassium ions, but this is not the case. In fact, *during a single action potential, the cellular concentrations of sodium and potassium ions hardly change at all.* Remember that the membrane potential is due to a slight excess of negative charge on one side and a slight excess of positive charge on the other side of the membrane. The number of excess charges is a tiny fraction of the total ions in the cell, and the number of ions that must cross the membrane to neutralize or alter the balance of charge is likewise small.

Nevertheless, intense neuronal activity can lead to significant changes in overall ion concentrations. For example, as a neuron continues to generate large numbers of action potentials, the concentration of potassium outside the cell will begin to rise perceptibly. This can affect the membrane potential of both the neuron itself and surrounding cells. Astrocytes, the glial cells that form the blood-brain barrier, are thought to control this problem by taking up excess potassium ions.

Action Potentials Are Propagated Along the Axon Without Losing Strength

For neurons to transmit signals to one another, the transient depolarization and repolarization that occur during an action potential must travel along the neuronal membrane. The depolarization at one point on the membrane spreads to adjacent regions through a process called the **passive spread of depolarization.** As a wave of depolarization spreads passively away from the site of origin, it also decreases in magnitude. This fading of the depolarization with distance from the source makes it difficult for signals to travel very far by passive means only. For signals to travel longer distances, an action potential must be *propagated,* or actively generated, from point to point along the membrane.

To understand the difference between the passive spread of depolarization and the propagation of an action potential, consider how a signal travels along the neuron from the site of origin at the dendrites to the end of the axon (**Figure 22-8**). Incoming signals are transmitted to a neuron at synapses that form points of contact between the synaptic boutons of the transmitting neuron and the dendrites of the receiving neuron. When these incoming signals depolarize the dendrites of the receiving neuron, the depolarization spreads passively over the membrane from the dendrites to the base of the axon—the

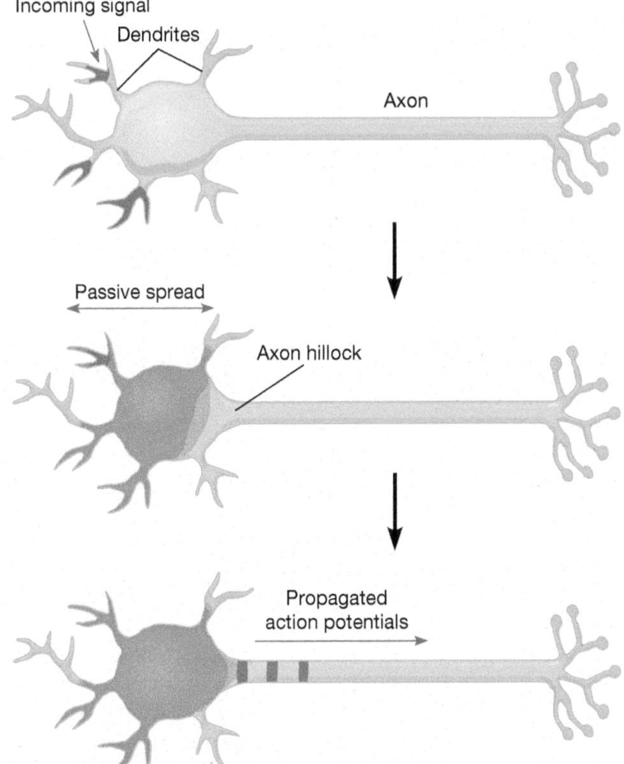

Figure 22-8 The Passive Spread of Depolarization and Propagated Action Potentials in a Neuron. The transmission of a nerve impulse along a neuron depends on both the passive spread of depolarization and the propagation of action potentials. A neuron is stimulated when its dendrites receive a depolarizing stimulus from other neurons. A depolarization starting at a dendrite spreads passively over the cell body to the axon hillock, where an action potential forms. The action potential is then propagated down the axon.

axon hillock. The axon hillock is the region where action potentials are initiated most easily. This is because sodium channels are distributed sparsely over the dendrites and cell body but are concentrated at the axon hillock and nodes of Ranvier; a given amount of depolarization will produce the greatest amount of sodium entry at sites where sodium channels are abundant. The action potentials initiated at the axon hillock are then propagated along the axon.

The mechanism for propagating an action potential in nonmyelinated nerve cells is illustrated in **Figure 22-9**. ❶ Stimulation of a resting membrane results in a depolarization of the membrane and movement of sodium ions into the axon at that location. ❷ Membrane polarity is temporarily reversed at that point, and this depolarization then spreads to an adjacent point. ❸ The depolarization at this adjacent point is sufficient to bring it above the threshold potential, triggering the inward movement of sodium ions. By this time, the original region of membrane has become highly permeable to potassium ions. ❹ As potassium ions move out of the cell, negative polarity is restored and that portion of the membrane returns to its resting state.

Meanwhile, ❺ the depolarization has spread to a new region, initiating the same sequence of events there. In this way, the signal moves along the membrane as a ripple of depolarization-repolarization events; the membrane polarity is reversed in the immediate vicinity of the signal but returned to normal again as the signal travels down the axon. The propagation of this cycle of events along the nerve fiber is called a *propagated action potential*, or **nerve impulse**. The nerve impulse can move only *away* from the initial site of depolarization because the sodium channels that have just been depolarized are in the inactivated state and cannot respond immediately to further stimulation.

Because an action potential is actively propagated, it does not fade as it travels; it is generated anew, as an all-or-none event, at each successive point along the membrane. Thus, a nerve impulse can be transmitted over essentially any distance with no decrease in strength.

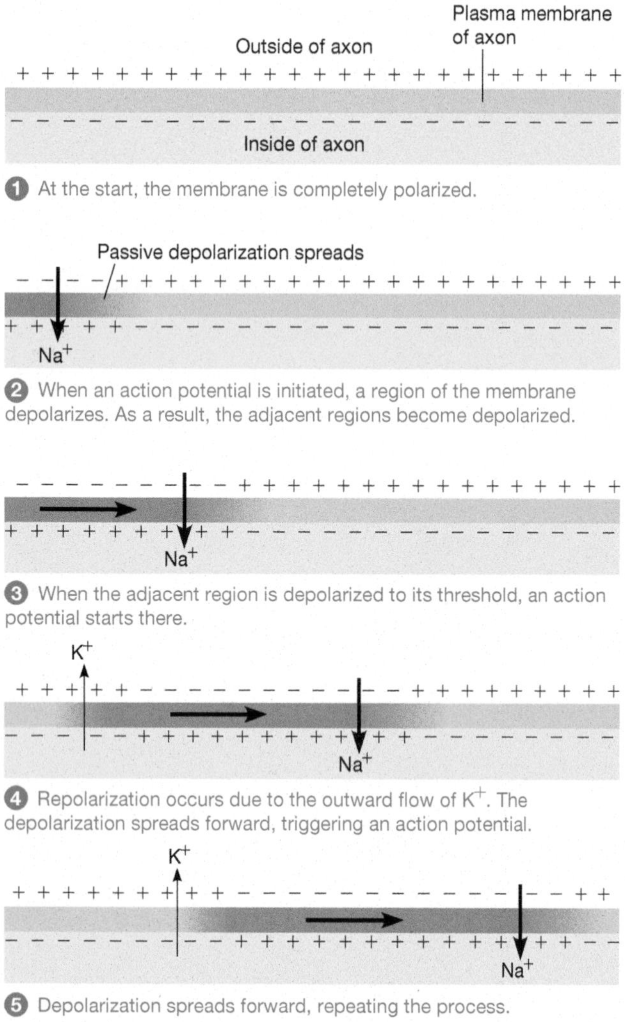

Outside of axon

Plasma membrane of axon

Inside of axon

❶ At the start, the membrane is completely polarized.

Passive depolarization spreads

Na^+

❷ When an action potential is initiated, a region of the membrane depolarizes. As a result, the adjacent regions become depolarized.

Na^+

❸ When the adjacent region is depolarized to its threshold, an action potential starts there.

K^+

Na^+

❹ Repolarization occurs due to the outward flow of K^+. The depolarization spreads forward, triggering an action potential.

K^+

Na^+

❺ Depolarization spreads forward, repeating the process.

Figure 22-9 The Transmission of an Action Potential Along a Nonmyelinated Axon. A nonmyelinated axon can be viewed as a string of points, each capable of undergoing an action potential. Notice that no backward propagation occurs near sites where action potentials form because sodium channels are in an inactivated state and the membrane is hyperpolarized.

The Myelin Sheath Acts Like an Electrical Insulator Surrounding the Axon

Most axons in vertebrates have an additional specialization: they are surrounded by a discontinuous *myelin sheath* consisting of many concentric layers of membrane. The myelin sheath acts as electrical insulation for the segments of the axon that it envelops. The myelin sheath of neurons in the CNS is formed by **oligodendrocytes**; in the peripheral nervous system (PNS), the myelin sheath is formed by **Schwann cells** (see Figure 22-1). Members of both classes of glial cells wrap layer after layer of their own plasma membrane around the axon in a tight spiral (**Figure 22-10**). Because each Schwann cell surrounds a short segment (about 1 mm) of a single axon, many Schwann cells are required to encase a PNS axon with a discontinuous sheath of myelin. Myelination decreases the ability of the neuronal membrane to retain electric charge (that is, myelination decreases its *capacitance*), permitting a depolarization event to spread farther and faster than it would along a nonmyelinated axon.

Myelination does not eliminate the need for propagation, however. For depolarization to spread from one site to the rest of the neuron, the action potential must still be renewed periodically down the axon. This happens at the *nodes of Ranvier,* interruptions in the myelin layer that are spaced just close enough together (1–2 mm) to ensure that the depolarization spreading out from an action potential at one node is still strong enough to bring an adjacent node above its threshold potential (**Figure 22-11**). The nodes of Ranvier are the only places on a myelinated axon where an action potential can be generated because current flow through the membrane is restricted elsewhere and because voltage-gated sodium channels are concentrated there. Thus, action potentials jump from node to node along myelinated axons rather than moving as a steady ripple along the membrane. This so-called *saltatory propagation* is much more rapid than the continuous propagation that occurs in nonmyelinated axons (*saltatory* is derived from the Latin word for "dancing"; see Figure 22-11). Myelination is a crucial feature of mammalian axons. Loss of myelination results in a dramatic decrease in the electrical resistance of

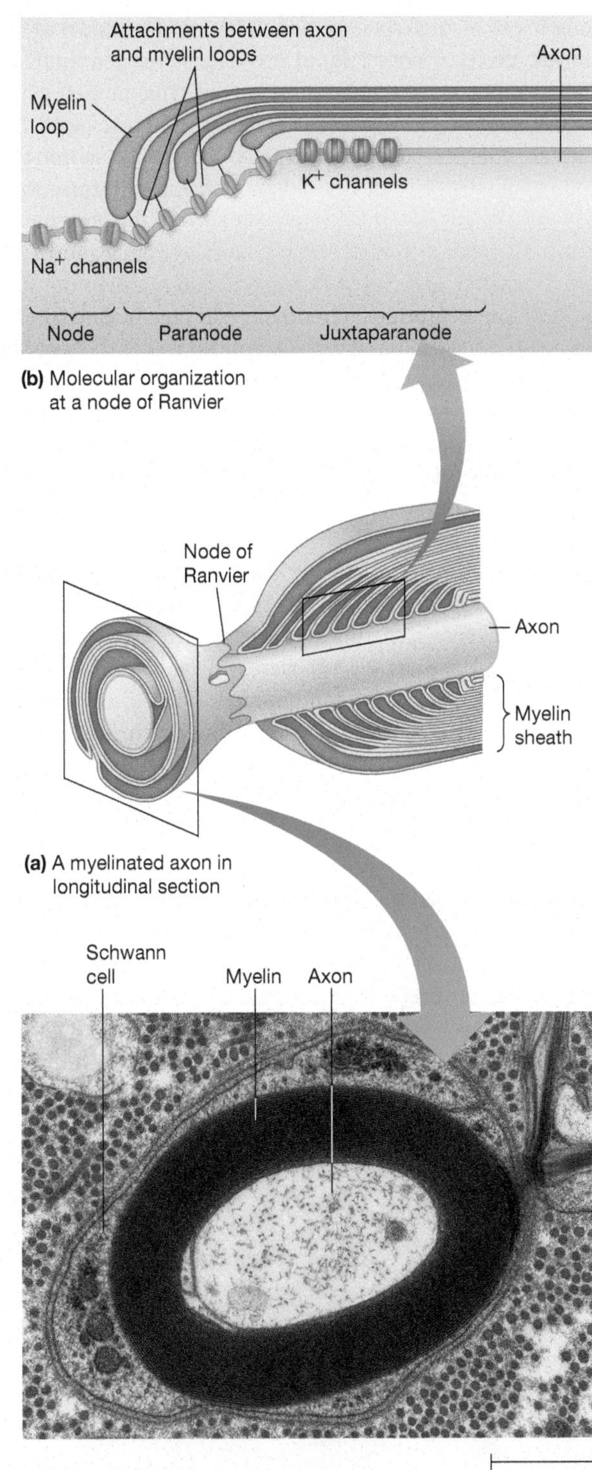

(b) Molecular organization at a node of Ranvier

(a) A myelinated axon in longitudinal section

(c) A myelinated axon in cross section

Figure 22-10 Myelination of Axons. (a) An axon of the peripheral nervous system that has been myelinated by a Schwann cell. Each Schwann cell gives rise to one segment of myelin sheath by wrapping its own plasma membrane concentrically around the axon. **(b)** Organization of a typical node of Ranvier in the peripheral nervous system. Sodium channels (purple) are concentrated in the node. Myelin loops attach to the regions next to the node ("paranodal" regions) via proteins on the axonal membrane and on the myelin loops (green). Potassium channels (teal) cluster next to the paranodal regions. **(c)** This cross-sectional view of a myelinated axon from the nervous system of a cat shows the concentric layers of membrane that have been wrapped around the axon by the Schwann cell that envelops it (TEM).

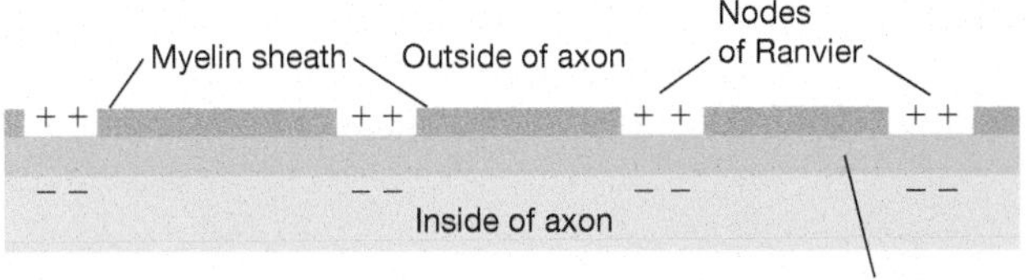

1 In myelinated neurons, an action potential is usually triggered at the axon hillock, just before the start of the myelin sheath. The depolarization then spreads along the axon.

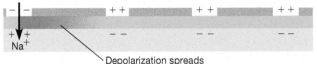

2 Because of myelination, the depolarization spreads passively to the next node.

3 The next node reaches its threshold, and a new action potential is generated.

4 This cycle is repeated, triggering an action potential at the next node.

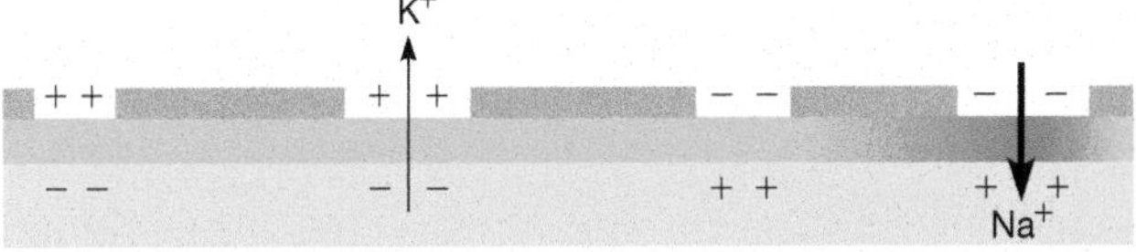

5 The process continues.

Figure 22-11 The Transmission of an Action Potential Along a Myelinated Axon. In a myelinated axon, action potentials can be generated only at nodes of Ranvier. Myelination reduces membrane capacitance, thereby allowing a given amount of sodium current, entering at one point of the membrane, to spread much farther along the membrane than it would in the absence of myelin. The result is a wave of depolarization-repolarization events that are propagated along the axon from node to node.

the axonal membrane. Much like the flow of water through a leaky garden hose, this loss of resistance drastically reduces conduction velocity along an axon. The debilitating human disease *multiple sclerosis* results when a patient's immune system attacks his or her own myelinated nerve fibers, causing demyelination. If the affected nerves innervate muscles, the patient's capacity for movement can be severely compromised.

Nodes of Ranvier are highly organized structures that involve close contact between the loops of glial or Schwann cell membrane and the plasma membrane of the axon(s) they myelinate. Three distinct regions are associated with these specialized sites of contact. In the node of Ranvier itself, voltage-sensitive sodium channels are highly concentrated. In the adjacent regions, called *paranodal regions* (*para–* means "alongside"), the axonal and glial cell membranes contain specialized adhesive proteins. Finally, in the region next to the paranodal areas, called *juxtaparanodal regions* (*juxta*– means "next to"), potassium channels are highly concentrated (see Figure 22-10b). The organization of nodes of Ranvier prevents free movement of the sodium and potassium channels within the axon's plasma membrane in regions around the nodes.

CONCEPT CHECK 22.2

The poison produced by some species of Central and South American poison dart frogs, called batrachotoxin, binds to voltage-gated sodium channels and maintains them in an open state. Predict what would happen if a neuron were exposed to batrachotoxin. Explain your answer.

22.3 Synaptic Transmission and Signal Integration

How do nerve cells communicate with one another and with glands and muscles at synapses? They do so at structurally distinct close contacts known as *synapses*. In an **electrical synapse**, one neuron, called the **presynaptic neuron**, is connected to a second neuron, the **postsynaptic neuron**, by gap junctions (**Figure 22-12**; see Chapter 15 for more on gap junctions). As ions move back and forth between the two cells, the depolarization in one cell spreads passively to the connected cell. Electrical synapses provide for transmission with virtually no delay and occur in places in the nervous system where speed of transmission is critical. Similar electrical connections can be found between nonneuronal cells, such as the cardiac muscle cells in the heart (see Chapter 14).

In a **chemical synapse** (**Figure 22-13**), the presynaptic and postsynaptic neurons are not connected by gap junctions, although they are connected by cell adhesion proteins. Instead, the presynaptic plasma membrane is separated from the postsynaptic membrane by a small space of about 20–50 nm, known as the **synaptic cleft**. A nerve signal arriving at the terminals of the presynaptic neuron cannot bridge the synaptic cleft as an electrical impulse. For synaptic transmission to take place, the electrical signal must be converted at the presynaptic neuron to a chemical signal carried by a *neurotransmitter*. Neurotransmitter molecules are stored in the synaptic boutons of the presynaptic neuron. An action potential arriving at the terminal causes the neurotransmitter to be secreted into and diffuse across the synaptic cleft. The neurotransmitter molecules then bind to specific proteins embedded within the plasma membrane of the postsynaptic neuron (*receptors*) and are converted back into electrical signals, setting in motion a sequence of events that either stimulates or inhibits the production of an action potential in the postsynaptic neuron, depending on the kind of synapse.

Neurotransmitter receptors fall into two broad groups: *ligand-gated ion channels* (sometimes called *ionotropic* receptors), in which activation directly affects the cell, and receptors that exert their effects indirectly through a system of intracellular messengers (sometimes called *metabotropic* receptors; see Chapter 23; **Figure 22-14**). Here, we focus on *ligand-gated channels*. These membrane ion channels open in response to the binding of a neurotransmitter, and they can mediate either excitatory or inhibitory responses in the postsynaptic cell.

Neurotransmitters Relay Signals Across Nerve Synapses

A **neurotransmitter** is essentially any signaling molecule released by a neuron. Many kinds of molecules act as neurotransmitters. Most are detected by the postsynaptic cell via a specific type of receptor; most neurotransmitters have more than one type of receptor. When a neurotransmitter molecule binds to its receptor, the properties of the receptor are altered, and the postsynaptic neuron responds accordingly.

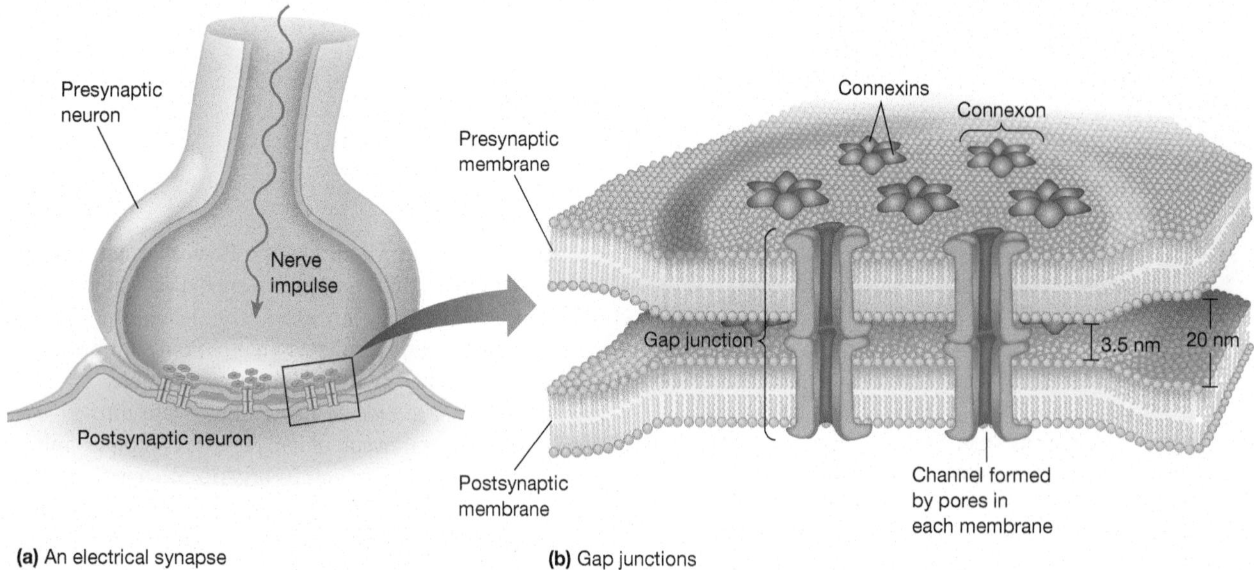

Figure 22-12 An Electrical Synapse. (a) In electrical synapses, the presynaptic and postsynaptic neurons are coupled by gap junctions, which allow small molecules and ions to pass freely from the cytosol of one cell to the next. When an action potential arrives at the presynaptic side of an electrical synapse, the depolarization spreads passively due to the flow of positively charged ions across the gap junction. **(b)** The gap junction is composed of sets of channels. A channel is made up of six protein subunits, each called a connexin. The entire set of six subunits together is called a connexon. Two connexons, one in the presynaptic membrane and one in the postsynaptic membrane, make up a gap junction.

Synaptic vesicles
Presynaptic neuron
Postsynaptic neuron
0.1 μm
(c)

Presynaptic axon
Direction of presynaptic nerve impulse
Synaptic bouton
Presynaptic membrane
Postsynaptic membrane
Synaptic vesicles containing neurotransmitter molecules
Mitochondrion
Synaptic cleft
Postsynaptic dendrite
(a)

Synaptic vesicles
Presynaptic membrane
Neurotransmitter molecules
Synaptic cleft
Postsynaptic membrane receptors
(b)

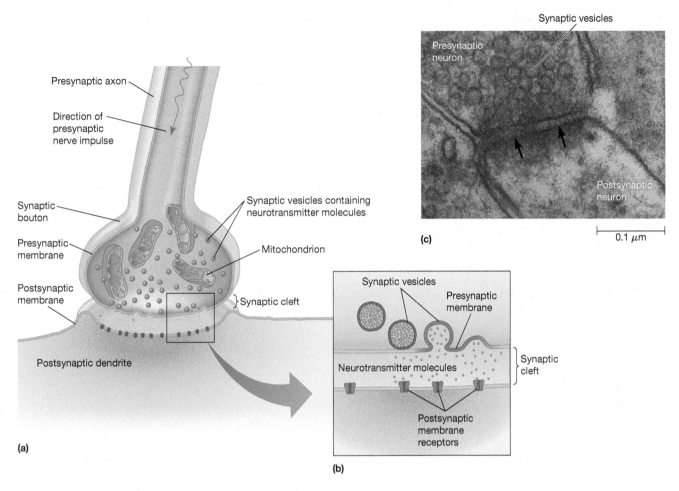

Figure 22-13 A Chemical Synapse. (a) When a nerve impulse from the presynaptic axon arrives at the synapse (red arrow), it causes synaptic vesicles containing neurotransmitter in the synaptic bouton to fuse with the presynaptic membrane, releasing their contents into the synaptic cleft. **(b)** Neurotransmitter molecules diffuse across the cleft from the presynaptic (axonal) membrane to the postsynaptic (dendritic) membrane, where they bind to specific membrane receptors and change the polarization of the membrane, either exciting or inhibiting the postsynaptic cell. **(c)** Electron micrograph of a chemical synapse (TEM). Arrows indicate a postsynaptic density, where membrane receptors and other proteins cluster.

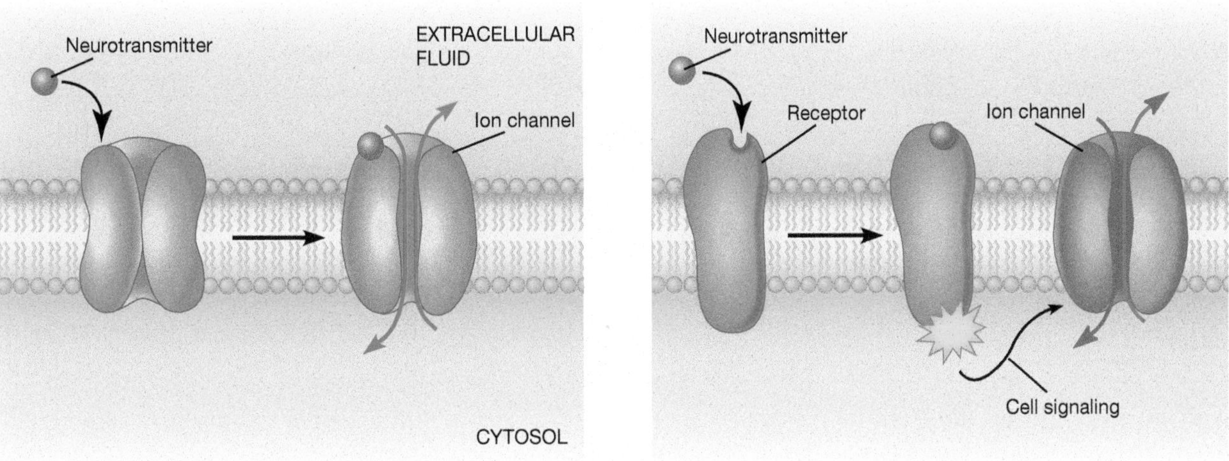

(a) Direct neurotransmitter action (ionotropic receptor)

(b) Indirect neurotransmitter action (metabotropic receptor)

Figure 22-14 Different Kinds of Receptors That Act at Chemical Synapses. (a) Direct neurotransmitter action. Ionotropic receptors act directly as ion channels. When they bind a neurotransmitter, they undergo a conformational change, and ions can pass through them. **(b)** Indirect neurotransmitter action. When metabotropic receptors bind neurotransmitters, they set in motion a series of cell signaling events that indirectly lead to the opening of an ion channel. Because they act indirectly, metabotropic receptors act more slowly than ionotropic receptors.

An *excitatory receptor* causes depolarization of the postsynaptic neuron, whereas an *inhibitory receptor* typically causes the postsynaptic cell to hyperpolarize.

Although definitions vary, to qualify as a neurotransmitter, a compound must satisfy three criteria: (1) it must elicit the appropriate response when introduced into the synaptic cleft, (2) it must occur naturally in the presynaptic neuron, and (3) it must be released at the right time when the presynaptic neuron is stimulated. Many molecules meet these criteria; **Table 22-2** lists several common neurotransmitters, some of which we consider here.

Acetylcholine. In vertebrates, **acetylcholine** is the most common neurotransmitter for synapses between neurons outside the central nervous system (CNS), as well as for neuromuscular junctions (see Chapter 14). Acetylcholine is an excitatory neurotransmitter. Bernard Katz and his collaborators were the first to make the important observation that acetylcholine increases the permeability of the postsynaptic membrane to sodium within 0.1 msec of binding to its receptor. Synapses that use acetylcholine as their neurotransmitter are called **cholinergic synapses**.

Catecholamines. The **catecholamines** include *dopamine* and the hormones *norepinephrine* and *epinephrine* (adrenaline), all derivatives of the amino acid tyrosine. Because the catecholamines are also synthesized in the adrenal gland, synapses that use them as neurotransmitters are termed **adrenergic synapses**. Adrenergic synapses are found at the junctions between nerves and smooth muscles in internal organs such as the intestines, as

Table 22-2 Different Kinds of Neurotransmitters

Neurotransmitter	Structure	Functional Class	Secretion Site(s)
Acetylcholine	$H_3C-C(=O)-O-CH_2-CH_2-N^+-[CH_3]_3$	Excitatory to vertebrate skeletal muscles; excitatory or inhibitory at other sites	CNS; PNS; vertebrate neuromuscular junction
Catecholamines			
Norepinephrine	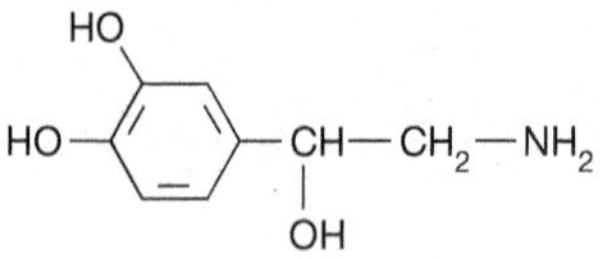 (HO)$_2$–benzene ring–$CH(OH)-CH_2-NH_2$	Excitatory or inhibitory	CNS; PNS
Dopamine	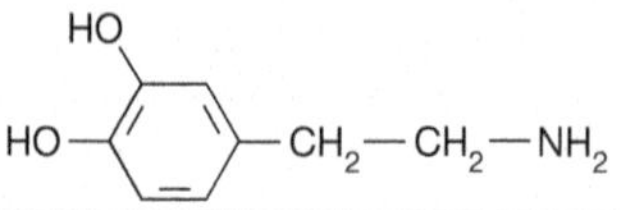 (HO)$_2$–benzene ring–$CH_2-CH_2-NH_2$	Generally excitatory; may be inhibitory at some sites	CNS; PNS
Amino Acids and Derivatives			
GABA (γ-aminobutyric acid)	$H_2N-CH_2-CH_2-CH_2-COOH$	Inhibitory	CNS; invertebrate neuromuscular junction
Glutamate	$H_2N-CH(COOH)-CH_2-CH_2-COOH$	Excitatory	CNS; invertebrate neuromuscular junction
Glycine	H_2N-CH_2-COOH	Inhibitory	CNS
Serotonin	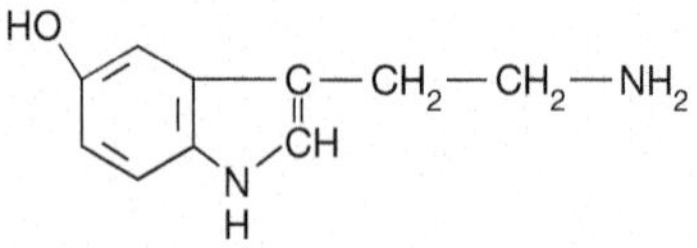 HO–indole ring (N–H)–$C(=CH)-CH_2-CH_2-NH_2$	Generally inhibitory	CNS
Neuropeptides			
Substance P	Arg—Pro—Lys—Pro—Gln—Gln—Phe—Phe—Gly—Leu—Met	Excitatory	CNS; PNS
Met-enkephalin (an endorphin)	Tyr—Gly—Gly—Phe—Met	Generally inhibitory	CNS
Endocannabinoids			
Anandamide	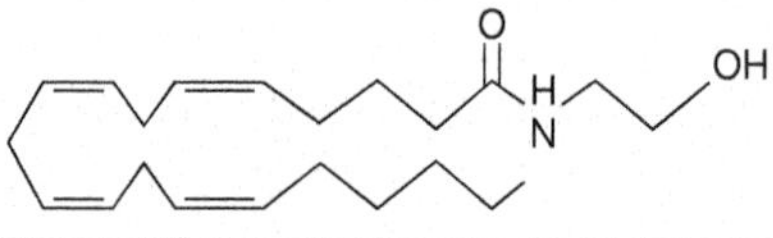 (arachidonoyl chain)–C(=O)–NH–CH_2CH_2–OH	Inhibitory	CNS
Gases			
Nitric oxide	N=O	Excitatory or inhibitory	PNS

well as at nerve-nerve junctions in the brain. (The mode of action of adrenergic hormones will be considered in Chapter 23.)

Amino Acids and Derivatives. Other neurotransmitters that consist of amino acids and derivatives include *histamine, serotonin,* and *γ-aminobutyric acid (GABA),* as well as *glycine* and *glutamate.* GABA and glycine are inhibitory neurotransmitters, whereas glutamate has an excitatory effect. Glutamate is a key excitatory neurotransmitter in vertebrates; neurons that release glutamate as a neurotransmitter are known as *glutamatergic* neurons. Serotonin functions in the CNS. It is considered an excitatory neurotransmitter because it indirectly causes potassium channels to close, which has an effect similar to opening sodium channels in that the postsynaptic cell is depolarized. However, its effect is exerted much more slowly than that of sodium channels.

Neuropeptides. Short chains of amino acids called **neuropeptides** are formed by proteolytic cleavage of precursor proteins. Hundreds of different neuropeptides have been identified. Some neuropeptides exhibit characteristics similar to neurotransmitters in that they excite, inhibit, or modify the activity of other neurons in the brain. However, they differ from most neurotransmitters in that they act on groups of neurons and have long-lasting effects.

Examples of neuropeptides include the *enkephalins,* which are naturally produced in the mammalian brain and inhibit the activity of neurons involved in the perception of pain. The modification of neural activity by these neuropeptides appears to be responsible for the insensitivity to pain experienced by individuals under conditions of great stress or shock. The analgesic (that is, pain-killing) effectiveness of drugs such as morphine, codeine, Demerol, and heroin derives from their ability to bind to the same sites within the brain that are normally targeted by enkephalins.

Endocannabinoids. Other substances released at synapses include the lipid derivatives known as *endocannabinoids,* which inhibit the activity of presynaptic neurons. The main endocannabinoid receptor found in the brain is also stimulated by tetrahydrocannabinol, or THC, a substance found in plants of the genus *Cannabis.* Marijuana is derived from the leaves of the species *Cannabis sativa* and owes its effects to THC.

Elevated Calcium Levels Stimulate Secretion of Neurotransmitters from Presynaptic Neurons

Secretion of neurotransmitters by the presynaptic cell is directly controlled by the concentration of calcium ions in the synaptic bouton (**Figure 22-15**). Each time an action potential arrives, the depolarization causes the calcium concentration in the synaptic bouton to increase temporarily due to the opening of **voltage-gated calcium channels** in the synaptic boutons. Normally, the cell is relatively impermeable to calcium ions, so that the cytosolic calcium concentration remains low (about 0.1 μM). However, there is a very large concentration gradient of calcium across the membrane because the calcium concentration outside the cell is about 10,000 times higher than that of the cytosol. As a result, calcium ions will move into the cell when the calcium channels open.

Before they are released, neurotransmitter molecules are stored in small, membrane-bounded **neurosecretory vesicles** in the synaptic boutons (see Figure 22-15). The

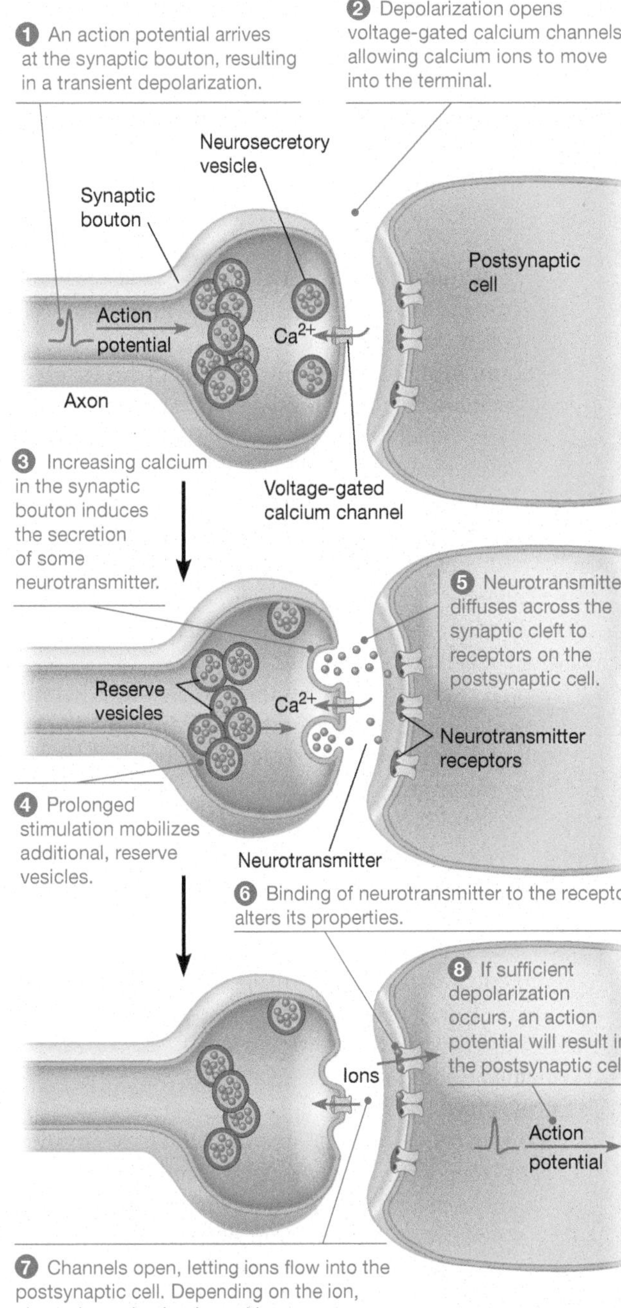

1 An action potential arrives at the synaptic bouton, resulting in a transient depolarization.
2 Depolarization opens voltage-gated calcium channels, allowing calcium ions to move into the terminal.
Neurosecretory vesicle
Synaptic bouton
Postsynaptic cell
Action potential
Ca^{2+}
Axon
3 Increasing calcium in the synaptic bouton induces the secretion of some neurotransmitter.
Voltage-gated calcium channel
5 Neurotransmitter diffuses across the synaptic cleft to receptors on the postsynaptic cell.
Reserve vesicles
Ca^{2+}
Neurotransmitter receptors
4 Prolonged stimulation mobilizes additional, reserve vesicles.
Neurotransmitter
6 Binding of neurotransmitter to the receptor alters its properties.
8 If sufficient depolarization occurs, an action potential will result in the postsynaptic cell.
Ions
Action potential
7 Channels open, letting ions flow into the postsynaptic cell. Depending on the ion, channel opening leads to either depolarization or hyperpolarization.

Figure 22-15 The Transmission of a Signal Across a Synapse. When an action potential arrives at the presynaptic bouton, a transient depolarization occurs, which leads to opening of voltage-gated calcium channels. Calcium elevation results in secretion of a neurotransmitter, which moves across the synapse and binds to receptors on the postsynaptic cell. The resulting depolarization can trigger an action potential.

release of calcium within the synaptic bouton has two main effects on neurosecretory vesicles. First, vesicles held in storage are mobilized for rapid release. Second, vesicles that are ready for release rapidly dock and fuse with the plasma membrane in the synaptic bouton region. During this process, the membrane of a vesicle moves into close contact with the plasma membrane of the axon terminal and then fuses with it to release the contents of the vesicle. We will now examine this process in more detail.

Secretion of Neurotransmitters Involves the Docking and Fusion of Vesicles with the Plasma Membrane

For the neurotransmitter to act on a postsynaptic cell, it must be secreted by the process of *exocytosis* (see Chapter 12 for details), in which the neurosecretory vesicle fuses with the plasma membrane, discharging the vesicle's contents into the synaptic cleft.

When an action potential arrives at an axon terminal and triggers the opening of voltage-gated calcium channels, calcium enters the synaptic bouton. Neurosecretory vesicles near the plasma membrane are now capable of fusing with the plasma membrane of the presynaptic neuron. Because it can occur very rapidly, vesicle fusion is thought to involve vesicles that are already "docked" at the plasma membrane (**Figure 22-16**). Docking and fusion of neurosecretory vesicles with the plasma membrane of the active zone are mediated by t- and v-SNARE proteins (just as we saw with other exocytosis events in Chapter 12). The calcium "sensor" involved in coordinated docking of vesicles appears to be the protein *synaptotagmin*, which can bind calcium. When it does, it undergoes a conformational change that allows the t- and v-SNARE complexes to interact efficiently. For his key contributions to our understanding of calcium-induced regulation of vesicle docking and neurotransmitter secretion, Thomas Südhof received a portion of a Nobel Prize in 2013.

Docking takes place at a specialized site, called the **active zone**, within the membrane of the presynaptic neuron (see Figure 22-16). At the active zone synaptic vesicles and the calcium channels that elicit their release are found in close proximity to one another, which helps explain the extremely rapid fusion of docked vesicles with the presynaptic neuron's plasma membrane when that neuron is stimulated.

Two familiar and potentially deadly illnesses result from interference with vesicle docking and fusion events. Both tetanus and botulism result from interference with vesicle docking and release by compounds known as neurotoxins. *Tetanus toxin* prevents the release of neurotransmitter from inhibitory neurons in the spinal cord, resulting in uncontrolled muscle contraction (which is why tetanus has been colloquially referred to as "lockjaw"). *Botulinum toxin* prevents release of neurotransmitter from motor neurons, resulting in muscle weakness and paralysis (see Human Connections, page 684).

Recall that exocytosis of vesicles involves fusion of the vesicle's membrane with the plasma membrane (see Chapter 12). When neurons release many neurosecretory vesicles in rapid succession, this has the potential to lead to accumulation of excess membrane in the presynaptic nerve terminal. Neurons solve this problem by *compensatory endocytosis*. Compensatory endocytosis relies on the formation of clathrin-dependent vesicles (see Chapter 12), which allows the recycling of membranes and thereby maintains the size of the nerve terminal.

In cases in which neurons need to fire very rapidly, they may use a more transient method for release of neurotransmitter, called *kiss-and-run exocytosis*. In this case, a vesicle may temporarily fuse with the plasma membrane via a tiny opening, causing release of some neurotransmitter from the vesicle. The vesicle then rapidly reseals without the added step of complete fusion with the plasma membrane.

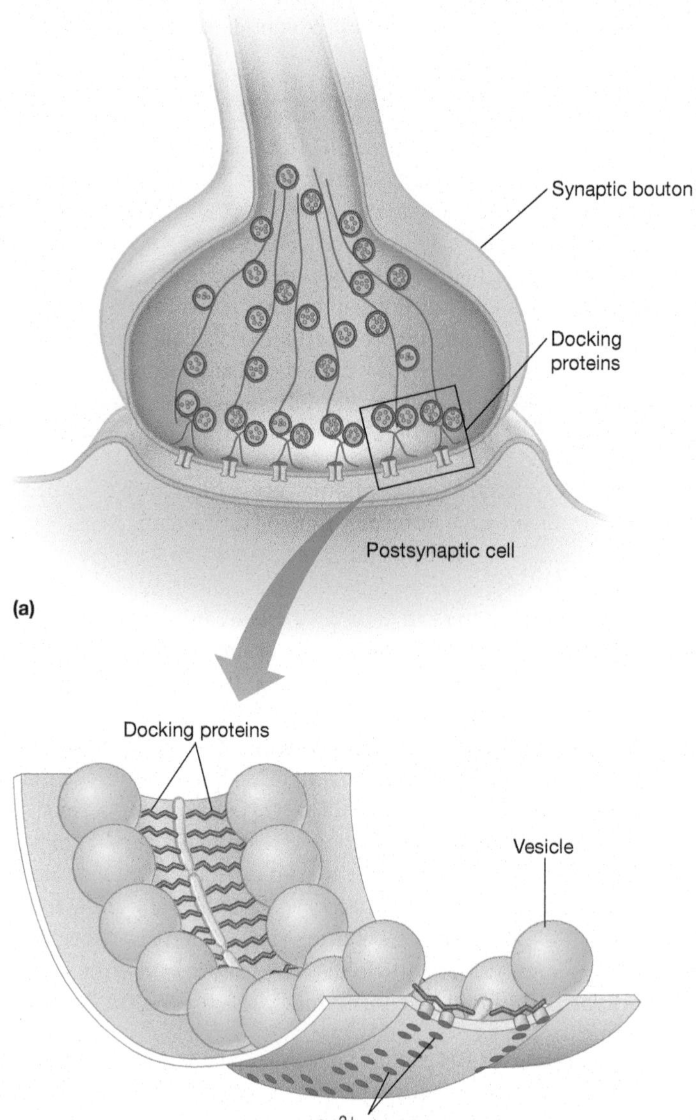

Figure 22-16 Docking of Synaptic Vesicles with the Plasma Membrane of the Presynaptic Neuron. (a) In response to local elevation of calcium in the presynaptic neuron, some synaptic vesicles become tightly associated (or docked) with the plasma membrane. When nearby calcium channels open, such docked vesicles fuse with the plasma membrane, releasing their contents. **(b)** A drawing based on an actual TEM reconstruction of the active zone of a motor neuron from a frog. Docked vesicles are arranged in rows and connected by a complex of proteins associated with docking. Calcium channels lie immediately beneath the vesicles.

MAKE CONNECTIONS 22.2

Shibire mutants in *Drosophila* have a temperature-sensitive mutation in the gene that encodes dynamin. At elevated temperature, dynamin stops functioning in *shibire* mutants, and they become paralyzed. Why? (Fig. 12-15, step 3)

Neurotransmitters Are Detected by Specific Receptors on Postsynaptic Neurons

When neurotransmitters are secreted across a synapse, their presence must be detected by the postsynaptic cell. This response requires a *receptor*, which binds the neurotransmitter and mediates the response of the postsynaptic neuron. Different neurotransmitters are bound by specific receptors. Here we discuss several well-studied receptors that function specifically during synaptic transmission (we will discuss receptors more generally in Chapter 23).

The Nicotinic Acetylcholine Receptor. One type of receptor to which acetylcholine binds is a ligand-gated sodium channel known as the *nicotinic acetylcholine receptor (nAchR;* **Figure 22-17**). It is called "nicotinic" because the actions of acetylcholine on this receptor can be mimicked by nicotine; muscarinic AchRs, which are also activated by acetylcholine, are activated by the mushroom toxin muscarine and not by nicotine. When two molecules of acetylcholine bind, the channel opens and lets sodium ions move into the postsynaptic neuron, causing depolarization.

Our understanding of the nAchR has been greatly aided by study of the electric organs of the electric ray (*Torpedo californica*). The electric organ consists of *electroplaxes*—stacks of cells that are innervated on one side but not on the other. The innervated side of the stack can undergo a potential change from about −90 mV to about +60 mV upon excitation, whereas the noninnervated side stays at −90 mV. Therefore, at the peak of an action potential, a potential difference of about 150 mV can be built up across a single electroplax. Because the electric organ contains thousands of electroplaxes arranged in series, their voltages are additive, allowing the ray to deliver a jolt of several hundred volts.

When examined under the electron microscope, electroplax membranes are found to be rich in rosette-like particles about 8 nm in diameter (Figure 22-17a); these particles are the nicotinic acetylcholine receptors.

Biochemical purification of the acetylcholine receptor was greatly aided by the availability of several neurotoxins from snake venom, including *α-bungarotoxin* and *cobratoxin.* These neurotoxins serve as a highly specific means of locating and quantifying nAchRs because they can be easily radiolabeled and bind tightly and specifically to the receptor protein.

The purified nAchR has a molecular weight of about 300,000 and consists of four kinds of subunits—α, β, γ, and δ—each containing about 500 amino acids (Figure 22-17b, c). These receptors play a key role in transmitting nerve impulses to muscle. In some cases, human patients develop an autoimmune response to their own acetylcholine receptors (in other words, their immune system attacks their own receptors). When this happens, the patients can develop a condition known as *myasthenia gravis,* in which degenerative muscle weakness occurs.

Nerve transmission at cholinergic synapses can be blocked in several other ways. One method is through substances that compete with acetylcholine for binding to its receptor on the postsynaptic membrane. A notorious example of such a poison is *curare,* a plant extract once used by native South Americans to poison arrows. Among the active factors in curare is *d-tubocurarine.* Snake venoms act in the same way. Both α-bungarotoxin (from kraits, snakes of the genus *Bungarus*) and *cobratoxin* (from cobra snakes) are small, basic proteins that bind covalently to the acetylcholine receptor, thereby blocking depolarization of the postsynaptic membrane.

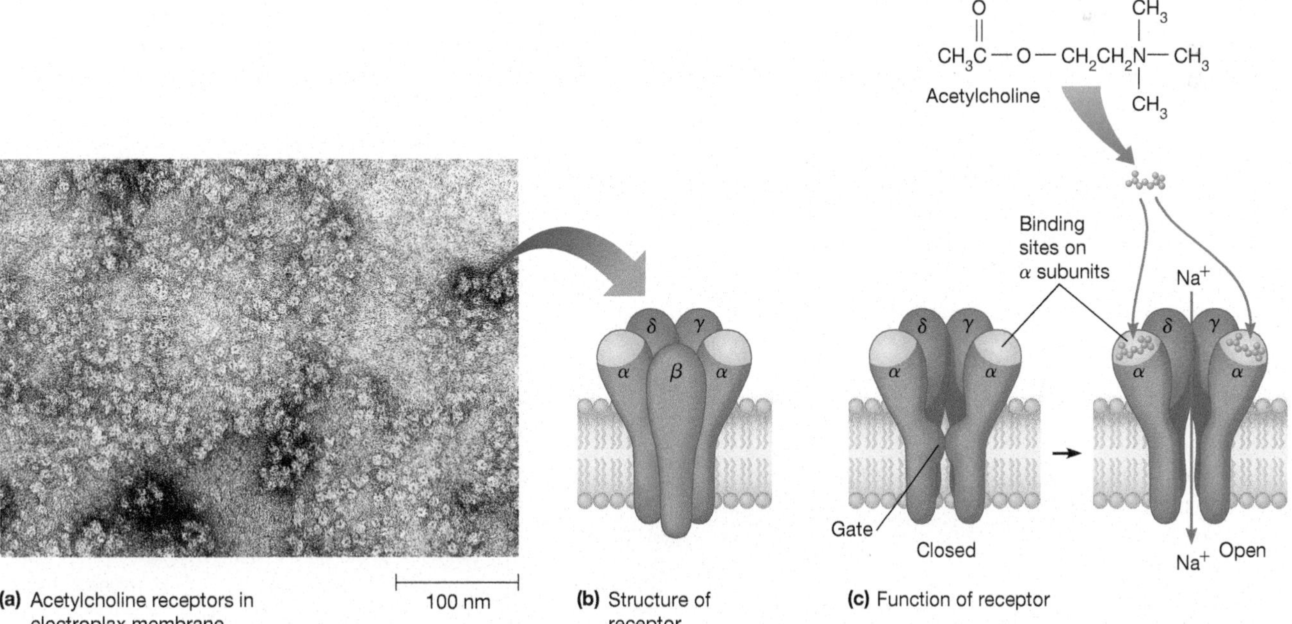

(a) Acetylcholine receptors in electroplax membrane

(b) Structure of receptor

(c) Function of receptor

Figure 22-17 The Nicotinic Acetylcholine Receptor. The nicotinic acetylcholine receptor (nAchR) is an important excitatory receptor of the central nervous system. **(a)** This micrograph of an electroplax postsynaptic membrane shows rosette-like nAchR particles (TEM). **(b)** The nAchR contains five subunits, including two α subunits with binding sites for acetylcholine and one each of β, γ, and δ. The subunits aggregate in the lipid bilayer in such a way that the transmembrane portions form a channel. **(c)** The channel (shown here with the β subunit removed) is normally closed; however, when acetylcholine binds to the two sites on the α subunits, the subunits are altered so that the channel opens to allow sodium ions to cross.

HUMAN *Connections*

The Toxic Price of the Fountain of Youth

Some people will go to great lengths to hold on to a youthful appearance for as long as possible. For those who want to avoid plastic surgery to reduce the signs of aging, there are other options. One popular alternative is injection with a paralytic toxin produced by the bacterium *Clostridium botulinum*. The commercial form of the toxin (toxin A) is known as *Botox*. Botox is big business; worldwide sales of Botox top $3 billion per year. But what is *C. botulinum,* and how does botulinum toxin act?

C. botulinum can be found just about anywhere, from soil to fruits and vegetables to the air. When food isn't preserved properly,

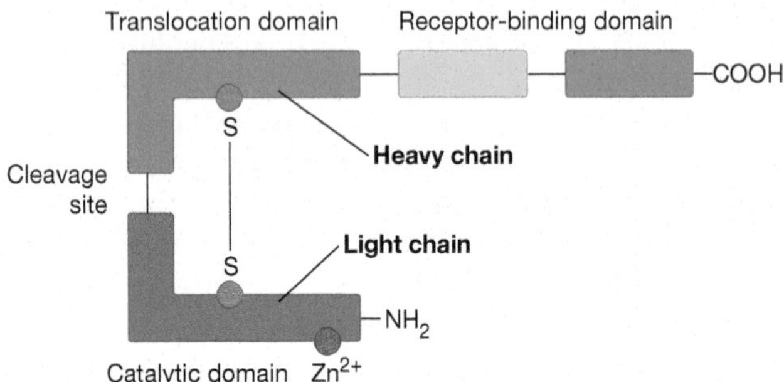

Figure 22B-1 Botulinum Toxin A Structure. Protein structure of botulinum toxin A. The toxin is present initially in an inactive form. The translocation domain facilitates uptake into the cell through endocytosis. The reducing environment of the endosome cleaves the disulfide bond, releasing the catalytic light chain into the cell.

C. botulinum can grow in the anaerobic environment provided by food storage and produce large amounts of botulinum toxin. Botulism, the disease that results from ingestion of the toxin, often occurs when people have eaten foods that were improperly canned or preserved.

Different *C. botulinum* strains produce different toxins. There are seven main types of toxins, with different cellular targets. When a toxin is ingested, it is in an inactive state, with the heavy and light chains joined by a disulfide bond (**Figure 22B-1**). It is what happens when the toxin enters a neuron that makes it so deadly (**Figure 22B-2**). Toxins are internalized through endocytosis (see Chapter 12 for details on endocytosis). The heavy chain of each toxin has a high affinity for receptors found on the presynaptic membrane of cholinergic neurons, and the toxin-receptor complex is readily taken up into the cell. Once internalized, the reducing environment inside the endosome cleaves the disulfide bond between the two chains, allowing the light chain to enter the cytoplasm. The light chain, now in an activated form, can interact with proteins found in the presynaptic nerve terminal. The catalytic light chain of each of the various toxins preferentially cleaves the v-SNARE protein *synaptobrevin,* the t-SNARE protein *SNAP-25,* and/or *syntaxin*. The cleavage of one or more of these proteins means that the machinery necessary for vesicle fusion is no longer present. This results in inhibition of acetylcholine release into the synaptic cleft. If the neuron innervates a muscle, prevention of acetylcholine release in turn blocks muscle contraction.

How can such a deadly toxin be used to give a more "youthful" appearance? Using small, local injections of Botox, plastic surgeons can induce local paralysis of specific regions of the face. Wrinkles under the skin are diminished because the muscles that cause the wrinkles no longer contract. Although the effect is ephemeral, lasting for a few weeks to a few months, the skin does appear smoother, with fewer wrinkles. So for those who emulate the endeavors of Ponce de Leon, a toxin may indeed be their "fountain of youth."

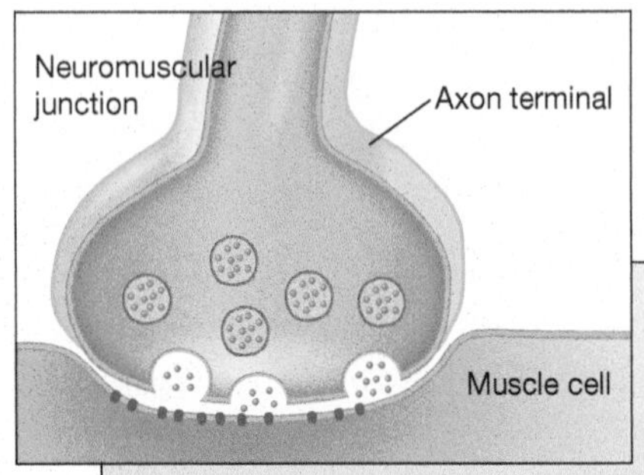

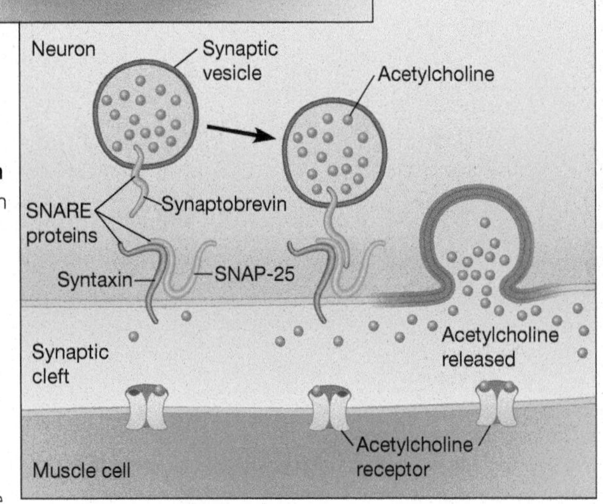

(a) Normal transmitter release

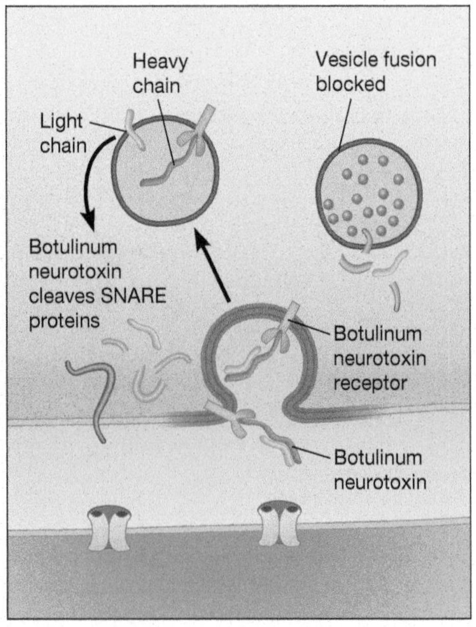

(b) Action of botulinum neurotoxins

Figure 22B-2 Botulinum Toxin A Activation in the Cell. Botulinum toxin A cleaves SNAP-25, inhibiting vesicular fusion with the neuronal membrane and release of acetylcholine into the synaptic cleft. Other botulinum toxins cleave other vesicle docking proteins, such as syntaxin, but the ultimate result is the same.

Substances that function in this way are called *antagonists* of cholinergic systems. Other compounds, called *agonists*, have just the opposite effect. Agonists also bind to the acetylcholine receptor, but in doing so they mimic acetylcholine and cause depolarization of the postsynaptic membrane. Unlike acetylcholine, however, agonists cannot be rapidly inactivated, so the membrane does not regain its polarized state.

The GABA Receptor. The *γ-aminobutyric acid (GABA) receptor* is also a ligand-gated channel, but when open, it conducts chloride ions rather than sodium ions. Because chloride ions are normally found at a higher concentration in the medium surrounding a neuron (see Table 22-1), opening GABA receptor channels produces an influx of chloride into the postsynaptic neuron, causing it to hyperpolarize. Hyperpolarization of the postsynaptic nerve terminal decreases the chance that an action potential will be initiated in the postsynaptic neuron. *Benzodiazepine drugs*, such as Valium and Librium, can enhance the effects of GABA on its receptor. Presumably, this produces the tranquilizing effects of these drugs.

The NMDA Receptor. The *N-methyl-D-aspartate (NMDA) receptor* is one of several ionotropic receptors for the amino acid glutamate. When it binds glutamate, the NMDA receptor becomes permeable to cations, including Na^+ and Ca^{2+}. NMDA receptors are important in memory and plasticity of neuronal connections. NMDA antagonists are often used as anesthetics.

Neurotransmitters Must Be Inactivated Shortly After Their Release

For neurons to transmit signals effectively, it is just as important to turn a stimulus off as it is to turn it on. Once a neurotransmitter has been secreted, it must be rapidly removed from the synaptic cleft. If it were not, stimulation or inhibition of a postsynaptic neuron would be abnormally prolonged, even without further signals from presynaptic neurons. Neurotransmitters are removed from the synaptic cleft by two specific mechanisms: degradation into inactive molecules or reuptake. Acetylcholine is a good example of the first mechanism.

Acetylcholinesterase. Excess acetylcholine must be hydrolyzed to restore the postsynaptic membrane to its polarized state; otherwise, major problems arise. The enzyme *acetylcholinesterase* hydrolyzes acetylcholine into acetic acid (or acetate ion) and choline, neither of which stimulates the acetylcholine receptor. Purine nucleotide neurotransmitters are also degraded by specific enzymes in a similar manner.

If excess acetylcholine is not rapidly hydrolyzed by acetylcholinesterase, the membrane cannot be restored to its polarized state, and further neuronal transmission is not possible. Substances that inhibit the activity of acetylcholinesterase are therefore usually very toxic. One family of acetylcholinesterase inhibitors is the *carbamoyl esters*, which inhibit acetylcholinesterase by covalently blocking the active site of the enzyme. An example of such an inhibitor is *physostigmine* (also called *eserine*), a naturally occurring alkaloid produced by the Calabar bean. Many synthetic organic phosphates are even more potent inhibitors. These include the widely used insecticides *parathion* and *malathion*, as well as nerve gases such as *tabun* and *sarin*. The primary effect of these compounds is muscle paralysis, caused by an inability of the postsynaptic membrane to regain its polarized state.

Neurotransmitter Reuptake. The second, very common method of terminating synaptic transmission is **neurotransmitter reuptake**. Reuptake involves pumping neurotransmitters back into the presynaptic axon terminals or nearby support cells. The rate of neurotransmitter reuptake can be rapid; for some neurons, the synapse may be cleared of stray neurotransmitter within as little as a millisecond. Some antidepressant drugs act by blocking the reuptake of specific neurotransmitters. For example, *Prozac* blocks the reuptake of serotonin at serotonergic synapses, leading to a local increase in the level of serotonin available to postsynaptic neurons.

Postsynaptic Potentials Integrate Signals from Multiple Neurons

Sending a signal across a synapse does not automatically generate an action potential in the postsynaptic cell. There is not necessarily a one-to-one relationship between an action potential arriving at the presynaptic neuron and one initiated in the postsynaptic neuron. A single action potential can cause the secretion of enough neurotransmitter to produce a detectable depolarization in the postsynaptic neuron but usually is not enough to cause the firing of an action potential in the postsynaptic cell. These incremental changes in potential due to the binding of neurotransmitter are referred to as *postsynaptic potentials (PSPs)*. If a neurotransmitter is excitatory, it will cause a small amount of depolarization known as an **excitatory postsynaptic potential (EPSP)**. Likewise, if the neurotransmitter is inhibitory, it will hyperpolarize the postsynaptic neuron by a small amount; this is called an **inhibitory postsynaptic potential (IPSP)**.

For a presynaptic neuron to stimulate the formation of an action potential in the postsynaptic neuron, the EPSP must rise to a point at which the postsynaptic membrane reaches its threshold potential. EPSPs can do this in two different ways. First, if two action potentials fire in rapid succession at the presynaptic neuron, the postsynaptic neuron will not have time to recover from the first EPSP before experiencing a second EPSP. The result is that the postsynaptic neuron will become more depolarized. A rapid sequence of action potentials effectively sums EPSPs over time and brings the postsynaptic neuron to its threshold. This process is called *temporal summation*.

Second, when signals coming from many different neurons with which a neuron has synaptic connections cause the release of neurotransmitter simultaneously, their effects can combine. Sometimes this results in a large depolarization of the postsynaptic cell. This process is known as *spatial summation* because the postsynaptic neuron integrates the

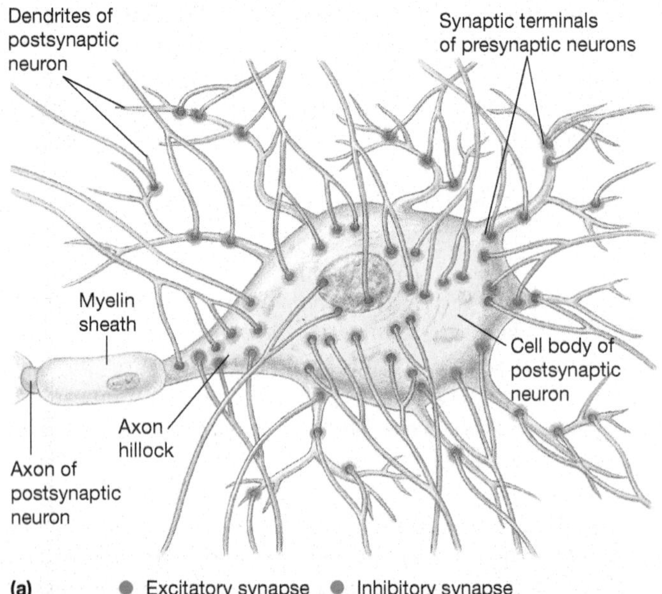

(b) 10 μm

Figure 22-18 Integration of Synaptic Inputs. (a) Neurons, particularly those in the CNS, receive inputs from thousands of synapses, some of them excitatory (green) and others inhibitory (red). An action potential may be generated in such a neuron at the axon hillock if the combined effects of membrane potentials induced by these synapses result in depolarization above the threshold potential. Both temporal and spatial summation contribute to the likelihood that an action potential will be initiated. **(b)** Synaptic terminals of many presynaptic neurons (stained for synapsin I, red) can make contact with a single postsynaptic neuron (stained for PSD-95, green; immunofluorescence micrograph).

numerous small depolarizations that occur over its surface into one large depolarization.

Besides receiving stimuli of varying strength, postsynaptic neurons can receive inputs from both excitatory and inhibitory neurons (**Figure 22-18**). Neurons can receive literally thousands of synaptic inputs from other neurons. When these different neurons fire at the same time, they exert combined effects on the membrane potential of the postsynaptic neuron. Thus, by physically summing EPSPs and IPSPs, an individual neuron effectively integrates incoming signals (excitatory or inhibitory).

CONCEPT CHECK 22.3

Drugs that affect the reuptake of neurotransmitters are in widespread use for the treatment of attention-deficit/hyperactivity disorder and clinical depression. In molecular terms, explain what effect(s) such reuptake inhibitors have on postsynaptic neurons, assuming that the neurotransmitter in question is excitatory for that postsynaptic neuron.

Mastering™ Biology For activities, animations, and review quizzes, go to the study area at www.masteringbiology.com.

Summary of Key Points

22.1 Neurons and Membrane Potential

- Cells in the nervous system are highly specialized to transmit electric impulses, using slender processes (dendrites and axons) that either receive transmitted impulses (dendrites) or conduct them to the next cell (axons).
- The membrane of an axon may or may not be encased in a myelin sheath, which provides electrical insulation that allows faster propagation of nerve impulses.
- Cells develop a membrane potential due to the separation of positive and negative charges across the plasma membrane. This potential develops as each ion to which the membrane is permeable moves down its electrochemical gradient.
- The Goldman equation is used to calculate the resting membrane potential of a cell, which depends on the permeability of the membrane to particular ions. The resting potential for the plasma membrane of most animal cells is from –60 to –75 mV, quite near the equilibrium potential for potassium ion.

22.2 Electrical Excitability and the Action Potential

- An action potential is a transient depolarization and repolarization of the neuronal membrane due to the sequential opening and closing of voltage-gated sodium and potassium channels. In voltage-gated ion channels, the probability of opening, and consequently their conductance, depends on membrane potential.
- The properties of ion channels have been studied using molecular techniques combined with patch clamping to measure the conductance of single channels.
- An action potential is initiated when the membrane is depolarized to its threshold, at which point the opening of voltage-gated sodium channels allows sodium ions to enter the cells, driving the membrane potential to approximately +40 mV. Eventually, voltage-gated sodium channels inactivate.
- Repolarization of the membrane involves the opening of slower, voltage-gated potassium channels, which leads to repolarization

is a "primary messenger." The binding of ligand to receptor often results in the production of additional molecules or ions within the cell receiving the signal, which is how a cell is able to "sense" that the appropriate ligand has successfully bound to the receptor. Such **second messengers** relay the signals from one location in the cell, such as the plasma membrane, to the interior of the cell, initiating a cascade of changes within the receiving cell. Often these events affect the expression of specific genes within the receiving cell. The ultimate result is a change in the identity or function of the cell. The ability of a cell to translate a receptor-ligand interaction to changes in its behavior or gene expression is known as **signal transduction**.

Preprogrammed Responses. When a ligand binds to its receptor, the receptor can be altered in two main ways. The binding of a ligand can induce a change in receptor conformation, can cause receptors to cluster together, or can do both. Once these changes take place, the receptor initiates a preprogrammed sequence of signal transduction events inside the cell. By *preprogrammed*, we mean that cells have a greater repertoire of functions than are in use at any particular time. Some of these cellular processes remain unused until particular signals are received that trigger them. The specific preprogrammed responses of a cell depend on that cell's past history. For example, a cell may have changed its gene expression due to signals it has received previously. This in turn may lead that cell to express receptors that make it sensitive to a new set of signals it receives in the future.

Receptor Binding Involves Quantitative Interactions Between Ligands and Their Receptors

With an overview of basic principles of signaling in hand, we can now consider in more detail how a ligand interacts with its receptor. In most cases, the binding reaction between a ligand and its receptor is similar to the binding of an enzyme to its substrate. When a receptor binds its ligand, the receptor is said to be *occupied*. For soluble ligands, the amount of receptor that is occupied by its ligand is proportional to the concentration of free ligand in solution. As the ligand concentration increases, more and more of its receptors become occupied, until most receptors are occupied—a condition known as *saturation*. Further increases in ligand concentration will, in principle, have no further effect on the target cell.

Receptor Affinity. The relationship between the concentration of ligand in solution and the number of receptors occupied can be described qualitatively in terms of **receptor affinity**. When almost all of the receptors are occupied at low concentrations of free ligand, the receptor has a high affinity for its ligand. Conversely, when it takes a relatively high concentration of ligand for most receptors to be occupied, the receptor has a low affinity for its ligand. Receptor affinity can be described quantitatively in terms of the **dissociation constant (K_d)**, the concentration of free ligand needed to produce a state in which half the receptors are occupied. The methods for analyzing the simplest cases of receptor-ligand binding turn out to be similar to the Michaelis–Menten equation (see Equation 6-7 on page 141). Although some receptor-ligand interactions deviate from this simple behavior, such analysis allows us to sketch the basic way in which receptor-ligand affinity is analyzed.

Imagine a hormone, H, and its receptor, R, interacting at equilibrium in solution. They associate in the following way:

$$[\mathrm{H}]+[\mathrm{R}] \rightleftharpoons [\mathrm{HR}] \quad \textbf{(23-1)}$$

where H represents free hormone and HR represents hormone bound to the receptor. The equilibrium constant (also known as the *association constant* or *affinity constant*, K_a) for the binding of the hormone to the receptor is described by the following equation:

$$K_a = [\mathrm{HR}]/[\mathrm{H}][\mathrm{R}] \quad \textbf{(23-2)}$$

The dissociation constant, K_d, is just the inverse of K_a:

$$K_d = [\mathrm{H}][\mathrm{R}]/[\mathrm{HR}] \quad \textbf{(23-3)}$$

The fraction of receptor bound by ligand, or the *fractional occupancy*, for the simplest situation is the same as the ratio of ligand bound/total concentration of receptor. In the form of an equation, this is

$$\text{fractional occupancy} = [\mathrm{HR}]/([\mathrm{R}]+[\mathrm{HR}]) \quad \textbf{(23-4)}$$

We can rewrite this, after a little algebra, using K_d, as

$$\text{fractional occupancy} = [\mathrm{H}]/(K_d+[\mathrm{H}]) \quad \textbf{(23-5)}$$

When the ligand concentration (in our case, [H]) is zero, the fractional occupancy is 0. When [H] is many times K_d, fractional occupancy approaches 1. When the receptor is half saturated with hormone, that hormone concentration (let's call it $[\mathrm{H}]_{1/2}$) is equal to K_d:

$$1/2 = [\mathrm{H}]_{1/2}/(K_d+[\mathrm{H}]_{1/2})$$

$$(K_d+[\mathrm{H}]_{1/2}) = 2[\mathrm{H}]_{1/2}$$

$$K_d = [\mathrm{H}]_{1/2} \quad \textbf{(23-6)}$$

Thus K_d provides a measure of affinity of a receptor for its ligand, just as K_m in enzyme kinetics provides a measure of the affinity of an enzyme for its substrate. A sample binding curve that illustrates these concepts is shown in **Figure 23-3** on page 692.

As you learned with K_m and enzyme kinetics, reciprocal plots can be used to estimate K_d. By dividing the fractional occupancy above by the free hormone concentration and after some more algebra, we can write out a linear equation:

$$\left(\frac{\text{fractional ocupancy}}{[H]}\right) = \frac{1}{K_d}(1-\text{fractional occupancy}) \quad \textbf{(23-7)}$$

This equation, which is called the *Scatchard equation*, results in a graph with the same basic form as the Eadie–Hofstee plot used in studying enzymes (which you learned about in Chapter 6). In this case the slope of the resulting line is $-1/K_d$ and the y-intercept is $1/K_d$.

In practice, measuring real-world K_d values can be technically challenging and is often done by measuring indirect biological responses elicited by a known concentration of a ligand or by measuring the quantity of radioactive ligand that binds to cells and correcting for nonspecific binding that does not involve real receptor-ligand interactions. The basic concept behind K_d is what is important for our purposes.

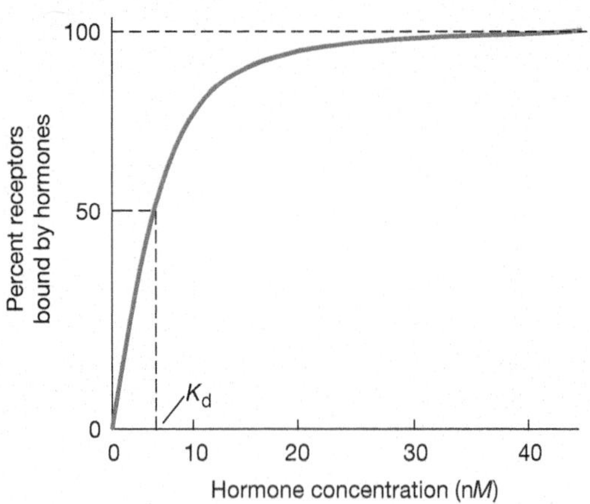

Figure 23-3 Concentration Dependence of Ligand Binding for a Simple Receptor-Ligand System. This graph depicts the relationship between hormone concentration and the amount bound to its receptor. The percentage of receptors bound by hormone increases as the concentration of hormone increases, eventually approaching 100%. The dissociation constant, K_d, is the hormone concentration at which 50% of the receptors are bound.

Values for K_d range from roughly 10^{-7} to 10^{-10} *M*. As with the Michaelis constant in enzyme kinetics (K_m; see Chapter 6), the importance of the value is that it tells us at what concentration a particular ligand will be effective in producing a cellular response. Thus, receptors with a high affinity for their ligands have a very low K_d, and, conversely, low-affinity receptors have a high K_d. Typically, the ligand concentration must be in the range of the K_d value of the receptor for the ligand to affect the target tissue.

Understanding the nature of receptor-ligand binding has provided great opportunities for researchers and pharmaceutical companies. Although receptors have binding sites that fit the messenger molecule quite closely, it is possible to make similar synthetic ligands that bind even more tightly or selectively. Such drugs that activate the receptor to which they bind are known as **agonists**. In contrast, both synthetic and natural compounds have been discovered that can bind to receptors without triggering the changes that occur when the normal ligand binds. These **antagonists** inhibit the receptor by preventing the naturally occurring second messenger from binding and activating the receptor.

MAKE CONNECTIONS 23.1

In what ways is the ligand-receptor interaction similar to the enzyme-substrate interaction, and in what ways do they differ? (Ch. 6.2–6.3)

Turning Receptors "Off." It is usually quite harmful for cells to remain continuously stimulated by a signal. Cells can shut down signaling in several ways. Two common ways are (1) by reducing the amount of free ligand, which reduces the total amount of signal, and (2) by reducing the sensitivity of the receptor or the amount of receptor the cell possesses. The first approach is one that is used in neurons (as you learned in Chapter 22): neurons can reduce the amount of neurotransmitter at synapses via neurotransmitter reuptake.

To appreciate the second approach, it is important to realize that although receptors have a characteristic affinity for their ligands, cells are geared to sense *changes* in ligand concentration rather than fixed ligand concentrations. When a ligand is present and receptors are occupied for prolonged periods of time, the cell adapts so that it no longer responds to the ligand. To further stimulate the cell, the ligand concentration must be increased. Such changes are known as *receptor desensitization*. There are two main ways in which such adaptation occurs. First, cells can change the density of receptors in response to a signal. When the receptor is on the cell surface, the removal of receptors takes place through the process of *receptor-mediated endocytosis*, in which small portions of the plasma membrane containing receptors invaginate and are internalized. (Receptor-mediated endocytosis was discussed in detail in Chapter 12.) The reduced number of receptors on the cell surface results in a diminished cellular response to ligand.

Cells can also adapt to signals through biochemical changes to receptors that alter the receptor's affinity for ligand or render it unable to initiate changes in cellular function. For example, the addition or removal of phosphate groups to specific amino acids within the cytosolic portion of a receptor can dramatically affect its affinity for ligand or its ability to transduce signals. Eventually, once a signal is no longer present, cells can reset their sensitivity by adding more receptors or by resetting the biochemical alterations to their existing receptors.

Coreceptors. In addition to the intrinsic properties of the receptor, receptor-ligand interactions can also be affected by *coreceptors* on the cell surface. Coreceptors help facilitate the interaction of the receptor with its ligand through their physical interaction with the receptor. One well-studied class of such molecules is *heparan sulfate proteoglycans*, including *glypicans* and *syndecans*. Coreceptors provide another layer of regulation of receptor-ligand interactions.

Cells Can Amplify Signals Once They Are Received

Signal transduction pathways allow another important aspect of a cell's response to an external signal: *signal amplification*. As a result, exceedingly small quantities of a ligand are often sufficient to elicit a response from a target cell. Often the strong response of the target cell results from a signaling cascade with the responding cell. At each step in the cascade, a signaling intermediate persists long enough to stimulate the production of many molecules required for the next step in the cascade, thereby multiplying the effects of a single receptor-ligand interaction on the cell surface. One example involves the breakdown of glycogen in liver cells in response to the hormone epinephrine (**Figure 23-4**). As a result of the cascade, a single epinephrine ligand can stimulate the release of hundreds of millions of glucose molecules from glycogen, a glucose polymer that stockpiles glucose for storage within cells.

Cell-Cell Signals Act Through a Limited Number of Receptors and Signal Transduction Pathways

With these basic principles of cell-cell signaling in mind, we are now in a position to consider specific cell-cell signaling pathways. Although the details may vary, cells use a limited number of basic signaling pathways. Throughout much of the rest of this chapter, you will learn about the details of these pathways.

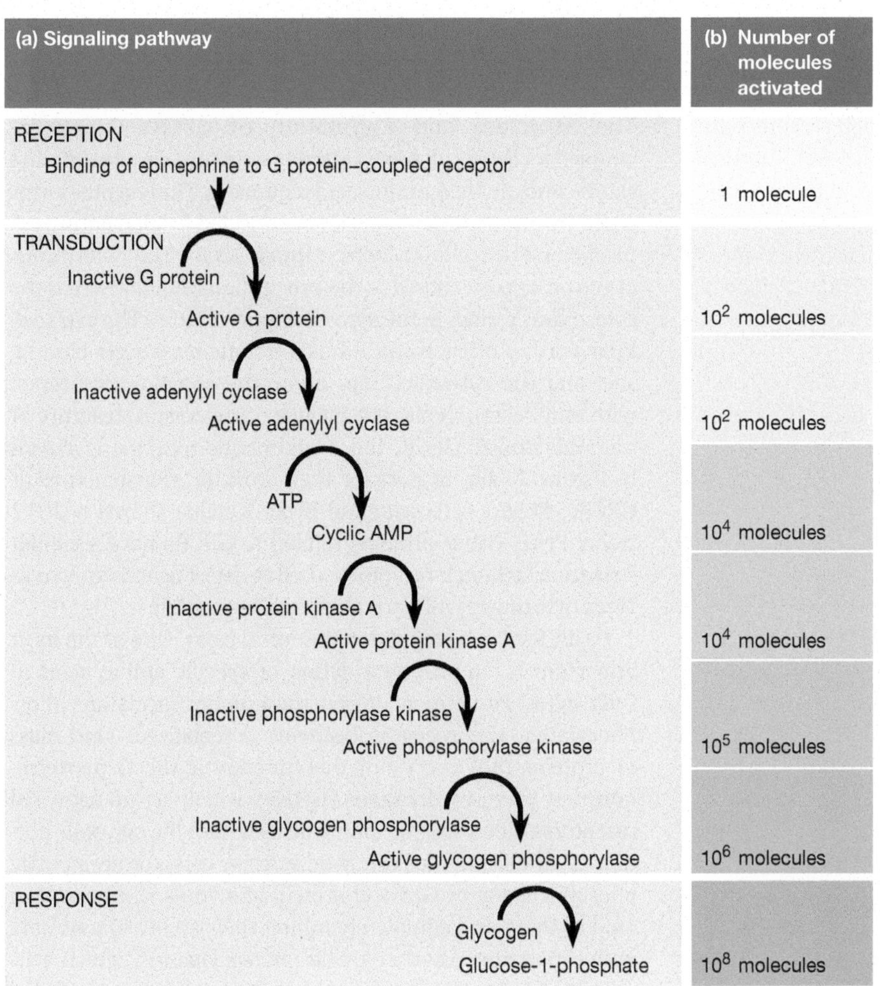

Figure 23-4 Signal Transduction Pathways Can Amplify the Cellular Response to an External Signal. Liver cells respond to the hormone epinephrine by breaking down glycogen to liberate glucose-1-phosphate. **(a)** The epinephrine receptor is a G protein–coupled receptor, which activates an enzyme known as adenylyl cyclase. Adenylyl cyclase catalyzes formation of a second messenger, cAMP, which activates a protein kinase (protein kinase A), which in turn activates another kinase (phosphorylase kinase). Ultimately, the enzyme glycogen phosphorylase is activated, which catalyzes the breakdown of glycogen. **(b)** The approximate number of molecules produced at each step is shown on the right. One epinephrine molecule is capable of triggering the production of hundreds of millions of glucose-1-phosphate molecules.

Figure 23-5 provides a basic overview of some common signal transduction pathways. Many ligands are hydrophilic compounds whose function lies entirely in their ability to bind to one or more specific receptors on a target cell. Hydrophilic ligands include proteins or small peptides, amino acids and their derivatives, and nucleotides or nucleosides. Hydrophilic ligands bind to transmembrane receptor proteins. One example of this type of signaling is *ligand-gated ion channels* (see Chapter 22), such as the acetylcholine receptor (Figure 23-5a). There are several other major cell surface

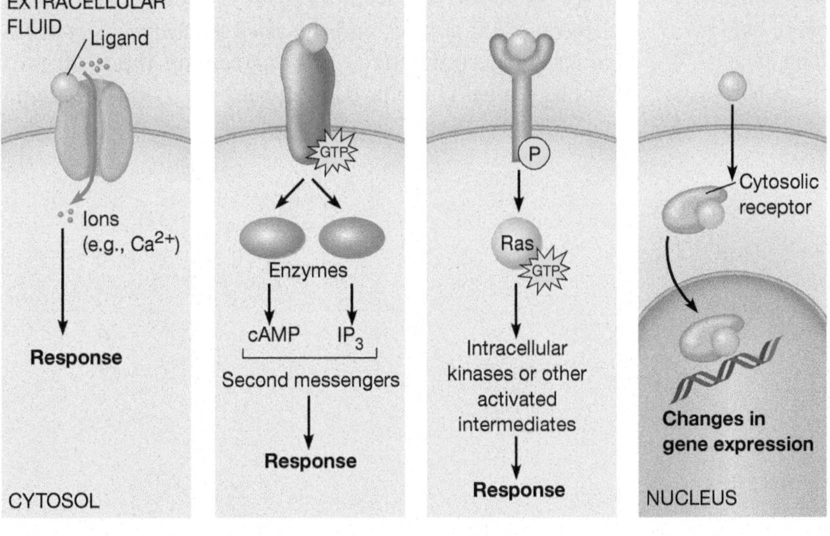

Figure 23-5 Some Basic Types of Signaling Pathways. There are many different types of signaling pathways, each beginning with the binding of a ligand by its corresponding receptor. These include **(a)** ligand-gated ion channels (see Chapter 22), **(b)** G protein–coupled receptors (GPCRs), **(c)** enzyme-coupled receptors, such as receptor tyrosine kinases, and **(d)** nuclear receptors, such as steroid hormone receptors.

signaling systems that animal cells use. We will focus on two main examples in much of the rest of this chapter. One involves cell surface receptors that pass through the membrane seven times and rely on hydrolysis of GTP; such receptors are known as *G protein–coupled receptors* (Figure 23-5b). Another involves receptors that are coupled with and activate cytosolic enzymes, often *protein kinases* (Figure 23-5c).

Hydrophobic ligands, on the other hand, act on receptors in the cytosol (Figure 23-5d). When the receptor binds ligand, it can move into the nucleus and regulate the transcription of specific genes. Among the hydrophobic messengers that bind to intracellular receptors are *steroid hormones*, which are derived from the compound cholesterol, and *retinoids*, derived from vitamin A. In the next several sections, you will learn about the details of some of these pathways, beginning with G protein–coupled receptors.

CONCEPT CHECK 23.1

One effect of the hormone insulin is to cause liver cells to make glycogen. Assume the dissociation constant (K_d) for the interaction of insulin with its receptor is 10^{-7} *M*. A hypoglycemic patient's insulin receptors have a K_d of 10^{-8}. Why would this patient have low blood sugar?

23.2 G Protein–Coupled Receptors

G Protein–Coupled Receptors Act via Hydrolysis of GTP

The **G protein–coupled receptors (GPCRs)** are so named because ligand binding causes a change in receptor conformation that activates a particular **G protein** (an abbreviation for *guanine-nucleotide binding protein*). A portion of the activated G protein in turn binds to a target protein, such as an enzyme or a channel protein, thereby altering the target's activity. Examples of GPCRs include olfactory receptors (responsible for our sense of smell), β-adrenergic receptors, and hormone receptors such as those for thyroid-stimulating hormone and follicle-stimulating hormone. A class of GPCRs of great clinical importance is the *opioid receptors*. Narcotic drugs such as morphine bind to these receptors, which accounts for their pain-killing benefits. Unfortunately, morphine and related drugs, such as heroin, also induce long-term changes in synaptic function in the brain that are responsible for their addictive effects.

The Structure and Regulation of GPCRs. G protein–coupled receptors all have a similar structure despite differing significantly in their amino acid sequences. The receptor forms seven transmembrane α helices connected by alternating cytosolic or extracellular loops (**Figure 23-6**). The N-terminus of the protein is exposed to the extracellular fluid, whereas the C-terminus resides in the cytosol (Figure 23-6a). The extracellular portion of each GPCR has a unique messenger-binding site, and the cytosolic loops allow the receptor to interact with only certain types of G proteins. The crystal structure of one well-studied GPCR, the β-adrenergic receptor, is shown in Figure 23-6b. In part for their work on the structure of GPCRs, Robert Lefkowitz and Brian Kobilka shared a 2012 Nobel Prize. Other proteins related to GPCRs have a similar structure. One such receptor is used to detect ligands known as *Wnts* (which you will learn about in Chapter 26).

GPCRs can be regulated in several ways. One of the most important is via phosphorylation of specific amino acids in their cytosolic domain. When these amino acids are phosphorylated, the receptor becomes desensitized. One class of proteins that carry out this function is the **G protein–coupled receptor kinases (GRKs)**, which act on activated receptors. When specific amino acids within the cytosolic portion of GPCRs such as the β-adrenergic receptor are heavily phosphorylated by GRKs, a protein known as *β-arrestin* can bind to them and completely inhibit their ability to associate with G proteins. Another kinase, *protein kinase A*, which is itself activated by G protein–mediated signaling (as discussed later in the chapter), can phosphorylate other amino acids on the receptor. Such inhibitory action is a good example of negative feedback during cell signaling.

The Structure, Activation, and Inactivation of G Proteins. G proteins act very much like molecular switches, whose "on" or "off" state depends on whether the G protein is bound to GTP (guanosine triphosphate) or GDP (guanosine diphosphate). There are two distinct classes of G proteins: the *large heterotrimeric G proteins* and the *small monomeric G proteins*. The large heterotrimeric G proteins contain three different subunits, called G alpha (G_α), G beta (G_β), and G gamma

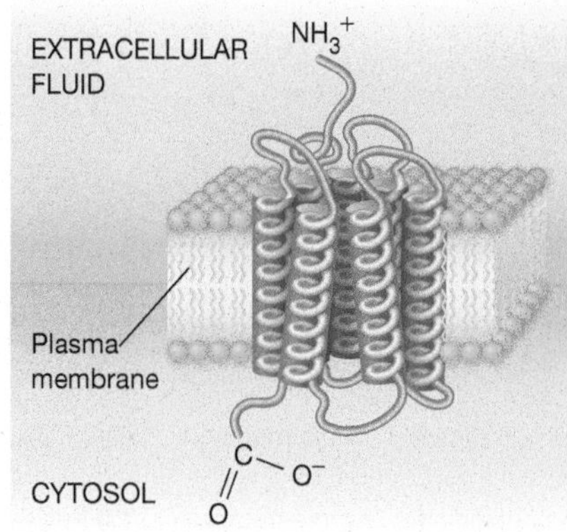

(a) GPCR schematic

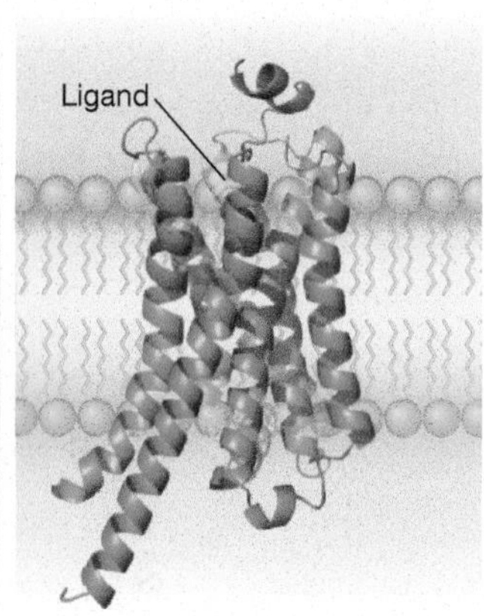

(b) Structure of β-adrenergic receptor (a GPCR)

Figure 23-6 The Structure of G Protein–Coupled Receptors (GPCRs). (a) Each GPCR has seven transmembrane α helices. A ligand binds to the extracellular portion of the receptor, causing an intracellular portion of the receptor to bind and activate a G protein. Besides the regions shown, the second cytosolic loop is also involved in G protein interactions in some cases. Specific amino acids in the cytosolic region are also targets for phosphorylation by G protein–coupled receptor kinases (GRKs) and protein kinase A. **(b)** A molecular model of the β-adrenergic receptor.

1 In the absence of TGFβ, the type I and type II receptors for TGFβ are not clustered or phosphorylated. R-Smads and Smad4 are in the cytosol.

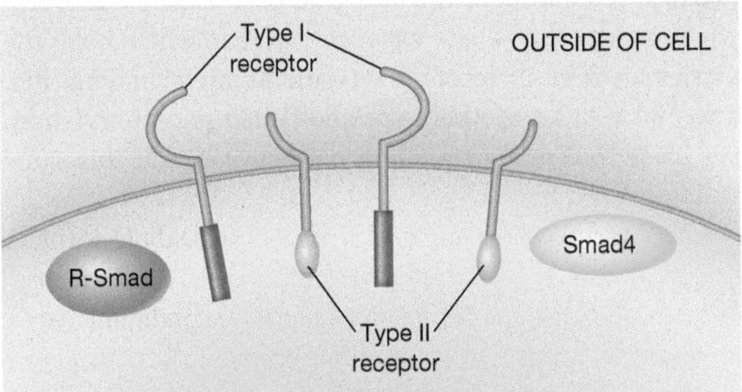

2 Binding of TGFβ results in clustering of type I and type II receptors, followed by phosphorylation of type I receptors by type II receptors.

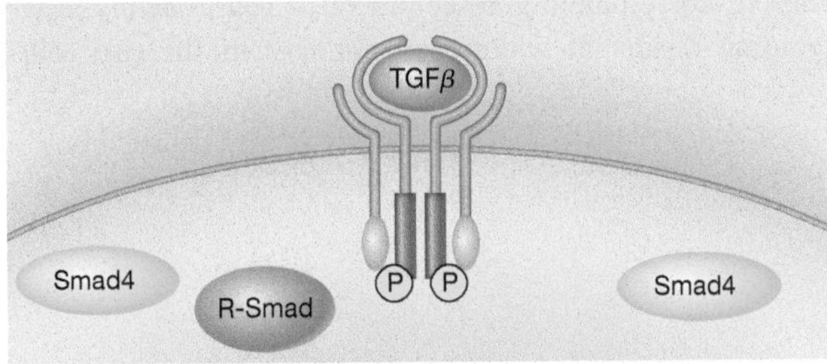

3 The activated type I receptors bind a complex of an anchoring protein and an R-Smad, resulting in R-Smad phosphorylation.

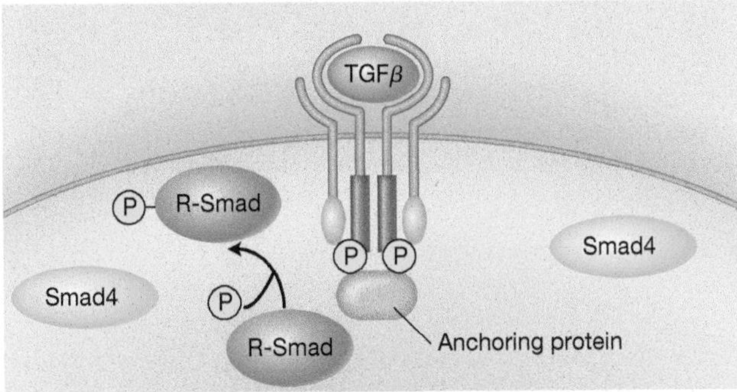

4 The phosphorylated R-Smad binds Smad4, and the complex enters the nucleus. Along with other proteins, they activate or repress gene expression. Eventually, the R-Smad is degraded or leaves the nucleus, and Smad4 returns to the cytosol, terminating the signal.

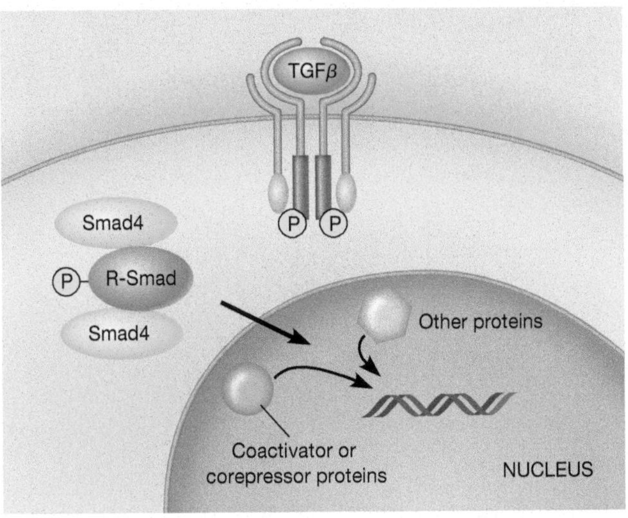

Figure 23-22 Signal Transduction by TGF β Receptor Family Proteins. TGF β receptors transduce their signals to the nucleus via proteins called Smads. After ligand binding activates the receptor, receptor-regulated Smads (R-Smads) become phosphorylated, bind Smad4, and enter the nucleus, leading to changes in gene expression. A variety of inhibitory proteins can block receptor activation, interactions of R-Smads with receptors, or interaction of R-Smads with Smad4.

bound, these receptors are usually unclustered and not phosphorylated. However, some of the TGF β family members form dimers with one another before binding to the appropriate receptors. 2 When ligand is bound, the type II receptor phosphorylates the type I receptor. The type I receptor then initiates a signal transduction cascade within the cell, phosphorylating a class of proteins known as **Smads** (a name coined from two of the founding members of this class of proteins). There are three types of Smads. 3 Those that are phosphorylated by a complex of anchoring proteins and activated receptors are known as receptor-regulated, or *R-Smads.* Another Smad, *Smad4,* forms a multiprotein complex with phosphorylated R-Smads. 4 When Smad4 molecules bind R-Smads, the entire complex can move into the nucleus, where it can associate with other cofactors and DNA-binding proteins to regulate gene expression. Still other Smads act at various points in TGF β signaling to inhibit the pathway; one type can bind to and inhibit receptors, whereas another can bind and inhibit Smad4. Smad signals are terminated when the R-Smad is degraded or the R-Smad moves back into the cytosol, where it can be reused when the cell receives another signal.

Other Enzyme-Coupled Receptor Families

In addition to kinases, there are two other types of enzyme-coupled receptors. *Tyrosine phosphatase* receptor family members are broadly responsible for removing phosphate groups from tyrosine residues. Therefore, it is not surprising that they have been implicated in the regulation of receptor tyrosine kinases. Thus, they play a role in diverse processes including neuronal development and thymocyte differentiation. What is puzzling to scientists is that many of these receptors don't have clear ligands yet. Therefore, there is still much left to learn about how these receptors are regulated.

The final group of enzyme-coupled receptors include the **guanylyl cyclase** family. The function of this enzyme is similar to adenylyl cyclase, which you learned about earlier in the chapter. However, instead of producing cAMP, guanylyl cyclase generates the closely related molecule **cyclic GMP (cGMP)**. As you'll learn later, this group of receptors has been shown to play a role in the photoreceptor cells of the retina, fluid regulation in the gut, and vasodilation (see Human Connections, page 716).

CONCEPT CHECK 23.3

You are studying cells that normally respond to epidermal growth factor (EGF) by increasing their rate of cell division. You measure the cell division rate in normal cells and in cells in which a GTPase-activating protein (GAP) for Ras has been knocked down using siRNA. When the two types of cells are exposed to the same concentration of EGF, what differences in response to EGF do you expect? Explain your answer.

23.4 Putting It All Together: Signal Integration

Cells in the human body can be exposed to a multitude of signals at any given moment. How do cells respond to such complexity? For most of this chapter, we have considered each signal transduction pathway as if it occurs in isolation within a cell. But in reality, cells must *coordinate* responses to many simultaneous signals. Cells have many different ways in which they do so. These include localizing signaling complexes via *scaffolds* and *integration* of different signaling pathways using signaling crosstalk.

Scaffolding Complexes Can Facilitate Cell Signaling

In many cases, cells need to regulate precisely where signaling occurs so that they can mount a response in a very specific spot. To do so, signaling components are sometimes assembled into large multiprotein complexes, which make such cascading reactions more efficient and confine the signals to a small area in the cell. One good example is a protein known as *Ksr (kinase suppressor of Ras)*, which holds the kinases involved in receptor tyrosine kinase signaling, Raf, Mek, and MAPK, together to promote their phosphorylation. It is likely that many signaling pathways exploit this same strategy.

A well-studied example of a signaling scaffold involves the mating pathway of the budding yeast, *Saccharomyces cerevisiae* (**Figure 23-23**). Under favorable conditions, yeast are haploid. In times of stress, however, cells of mating type *a* secrete a chemical signal called *a factor*, which can bind to specific G protein–coupled receptors on nearby α cells (Figure 23-23a). At the same time, α cells secrete α *factor*, which binds to corresponding receptors on *a* cells. *Mating factor* signaling results in widespread changes in the two cells,

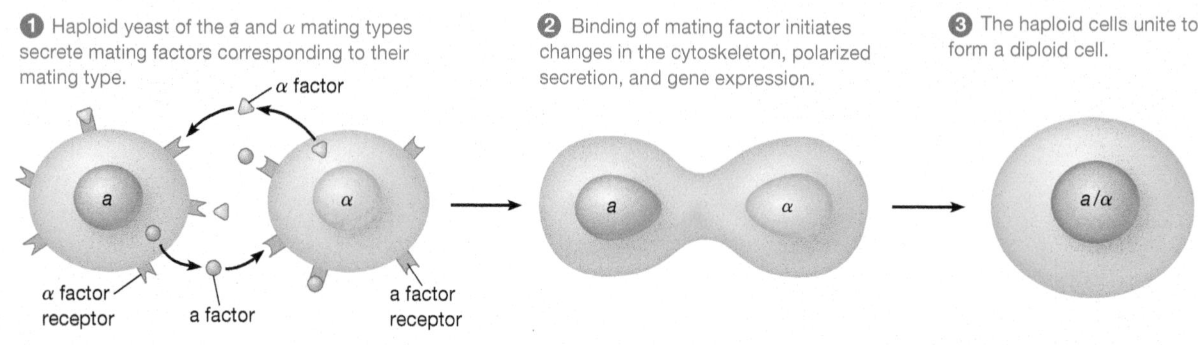

1 Haploid yeast of the a and α mating types secrete mating factors corresponding to their mating type.
α factor
a
α
α factor receptor
a factor
a factor receptor
2 Binding of mating factor initiates changes in the cytoskeleton, polarized secretion, and gene expression.
a
α
3 The haploid cells unite to form a diploid cell.
a/α

(a) Mating in yeast

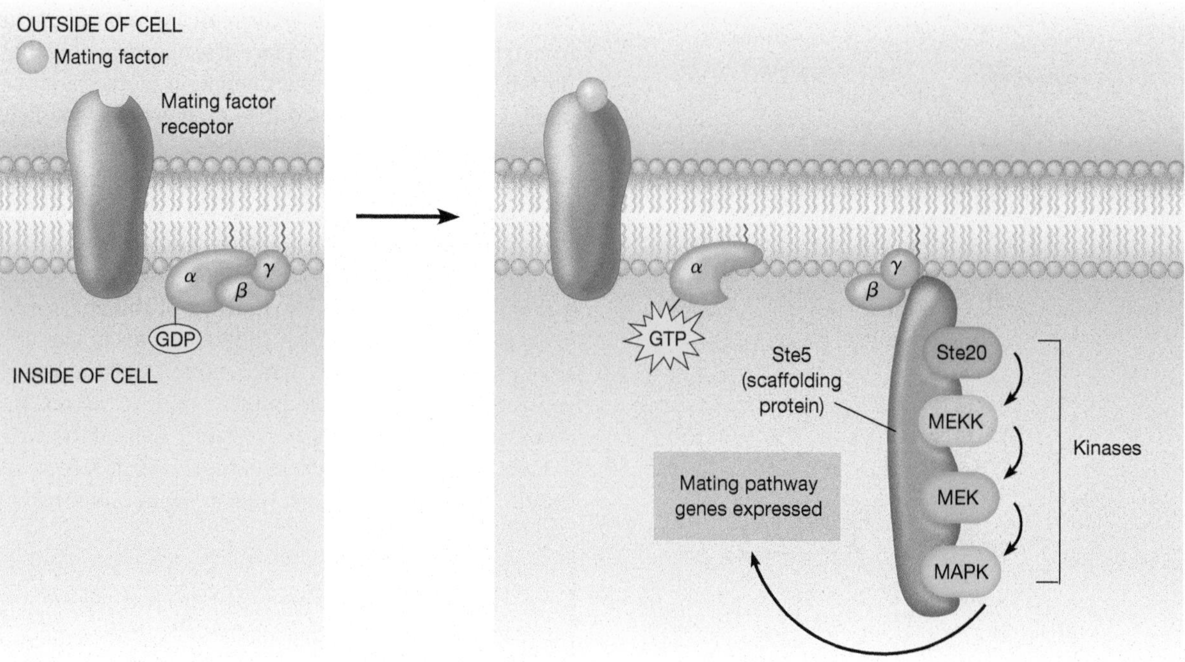

(b) Mating factor signals and scaffolding complexes

Figure 23-23 Scaffolding Complexes Can Facilitate Cell Signaling. (a) Mating in yeast involves the exchange of mating factors by haploid cells of different mating types. Mating factor receptors are G protein–coupled receptors. **(b)** Mating factor signaling involves the recruitment of a scaffolding complex in response to the $G_{\beta\gamma}$ subunit of the G protein associated with mating factor receptors.

including polarized secretion, alterations in the cytoskeleton, and changes in gene expression. Ultimately, the two cells fuse to create a diploid a/α cell.

Mating factor signaling is mediated through the activated $G_{\beta\gamma}$ subunit of a G protein. $G_{\beta\gamma}$ recruits a large scaffolding protein, known as *Ste5*, to the plasma membrane. Ste5 increases the efficiency of signal transduction through the mating pathway by assembling the kinases involved into a large complex. Interestingly, the activated $\beta\gamma$ subunits in this case initiate a series of phosphorylation events that lead to activation of a type of kinase you encountered as part of receptor tyrosine kinase-based signaling: *MAP kinase* (Figure 23-23b). MAPK in turn phosphorylates other proteins, leading to changes in gene expression that are important for mating. Other similar scaffolding complexes have been identified in yeast that regulate responses to changes in osmolarity, nutrients, and other environmental signals.

The mating pathway illustrates not only the importance of scaffolding, but also another important principle: cells can "mix and match" various parts of signaling pathways. The mating pathway is a variation on the theme of GPCRs at the surface, mixed with a MAPK signaling cascade in the cytosol.

Different Signaling Pathways Are Integrated Through Crosstalk

In addition to localizing and organizing signaling complexes via scaffolds, a cell must integrate its response to the multiple signals to which it is exposed. How can it do so? Sometimes a single receptor can activate multiple pathways. In other cases, different ligands bind their corresponding receptors at the cell surface, activating specific signaling pathways within the cell. Activated components (for example, kinases) from one pathway can then affect components in another pathway. In other cases, different pathways converge onto the same molecules, such as second messengers. Such interactions are collectively known as *signaling crosstalk*.

To illustrate the complexities of signaling crosstalk, consider **Figure 23-24** (this figure actually also contains a pathway involving nitric oxide, which you will learn about later in this chapter). There are several things to notice about Figure 23-24. First, several pathways lead to production of *second messengers*, such as IP_3 and calcium ions. Second, many pathways ultimately lead to *phosphorylation* of target proteins. These targets then carry out many different cellular functions, such as altering the cytoskeleton or changes

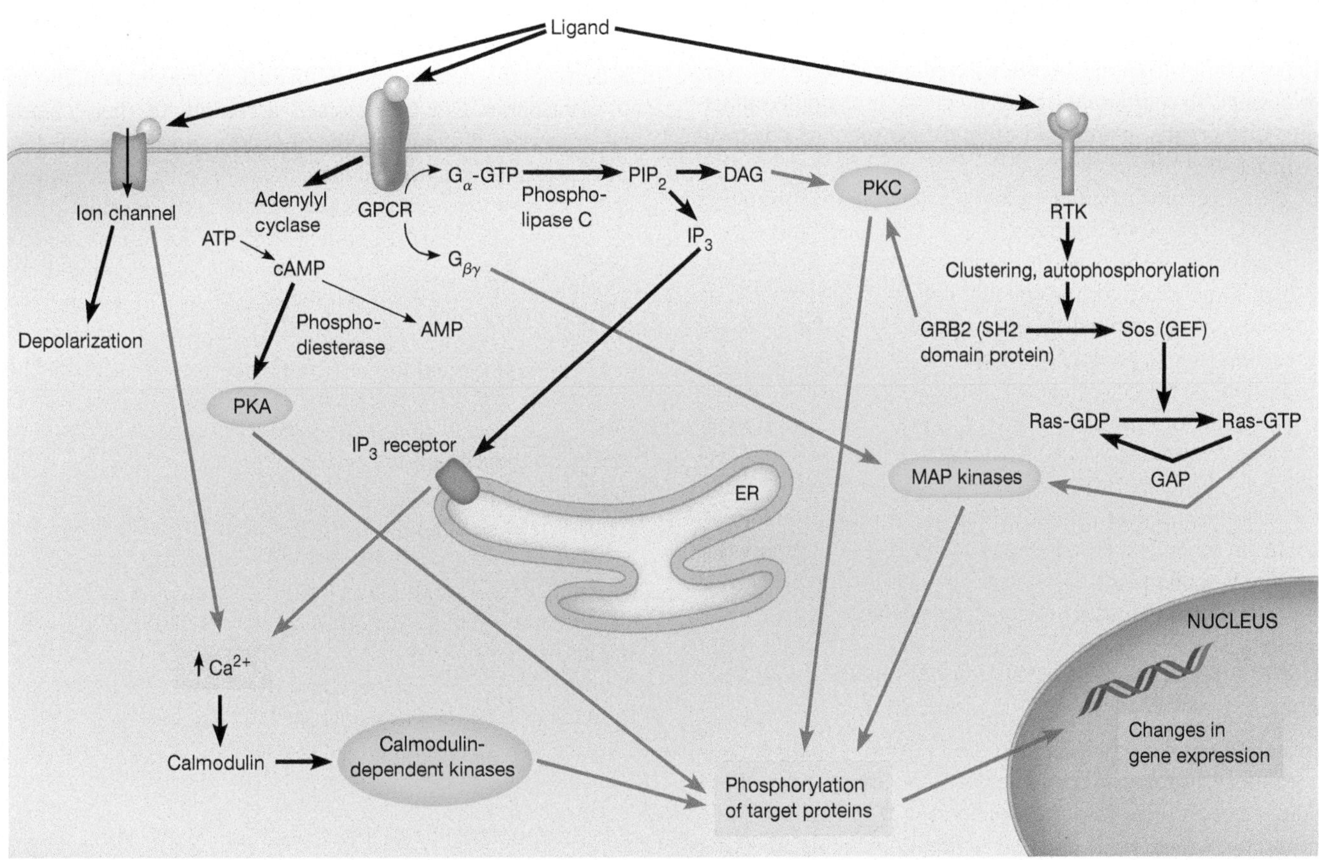

Figure 23-24 Signaling Crosstalk. This diagram of a hypothetical cell shows several possible signaling pathways (discussed in Chapter 22 and this chapter) and several possible places where crosstalk can occur (red arrows). Crosstalk can occur at the level of second messengers (for example, Ca^{2+}, pink box), protein kinases (yellow ovals), the proteins phosphorylated by these kinases (green box), and the level of gene expression (purple box).

in gene expression. Because different pathways can feed into these events at different levels, the opportunities for crosstalk are numerous! It is probably best to think of cell signaling as a complex network of biochemical pathways leading to changes in a cell's properties at any given moment in response to the signals it receives.

CONCEPT CHECK 23.4

When some cells are treated with growth factors that act through receptor tyrosine kinases, they divide rapidly. These cells are also known to be able to respond to the hormone epinephrine, which leads to activation of adenylyl cyclase. In addition, it is known that protein kinase A (PKA) inhibits the protein Raf in these cells. Based on this information, how do you predict that cells treated with growth factor alone will compare with cells treated *simultaneously* with both epinephrine and growth factor? Explain your reasoning.

23.5 Hormones and Other Long-Range Signals

We've just seen how growth factors and other ligands operate at short range to regulate cellular function. However, large organisms must also be able to coordinate the function of various cells and tissues over long distances. Let's return to the example of the pop quiz at the beginning of this chapter. In that case, organs that may be a meter apart (digestive system, brain, heart) must all simultaneously respond to a new environmental situation (the unexpected quiz initiated by your professor). To coordinate signals involving many different tissues and organs, plants and animals use secreted chemical signals called **hormones**.

Endocrine hormones are an important class of hormones in animals. Endocrine hormones are synthesized by the *endocrine tissues* of the body and are secreted directly into the bloodstream. Once secreted into the circulatory system, endocrine hormones have a limited life span, ranging from a few seconds for epinephrine (a product of the adrenal gland) to many hours for insulin. As they circulate in the bloodstream, hormone molecules come into contact with receptors in tissues throughout the body. A tissue that is specifically affected by a particular hormone is called a *target tissue* for that hormone (**Figure 23-25**). For example, the heart and the liver are target tissues for epinephrine, whereas the liver and skeletal muscles are targets for insulin. Vascular plants also produce hormones that are carried throughout the plant and function in a similar way.

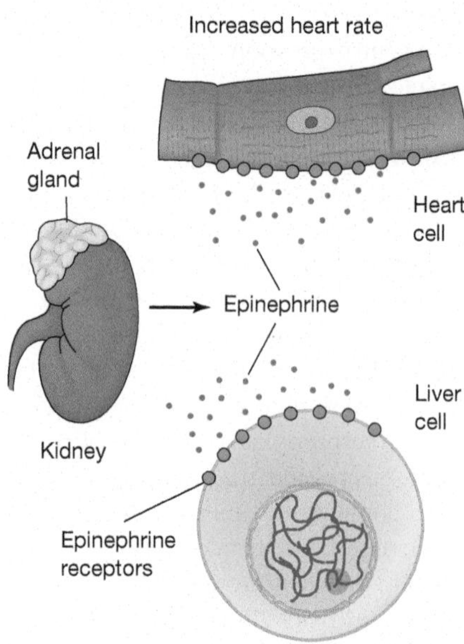

Figure 23-25 Target Tissues for Endocrine Hormones. Cells in a target tissue have hormone-specific receptors embedded in their plasma membranes (or, in the case of the steroid hormones, present in the nucleus or cytosol). Heart and liver cells can respond to epinephrine synthesized by the adrenal glands because these cells have epinephrine-specific receptors on their outer surfaces. A specific hormone may elicit different responses in different target cells. Epinephrine causes an increase in heart rate but stimulates glycogen breakdown in the liver.

Hormones Can Be Classified by Their Chemical Properties

Although we can consider them as a group based on their regulatory functions, hormones differ in many ways. Both plants and animals produce a wide array of hormones. Some hormones are steroids or other hydrophobic molecules that are targeted to intracellular receptors. Other hormones, such as the adrenergic hormones discussed later in this section, are targeted to a wide variety of G protein–coupled receptors. Still others, such as insulin, are ligands for receptor tyrosine kinases. Plants produce steroid hormones called *brassinosteroids* that regulate leaf growth. Similarly, *abscisic acid* causes the closing of stomata during drought conditions.

Hormones can be classified according to their chemical properties. Chemically, the endocrine hormones fall into four categories: amino acid derivatives, peptides, proteins, and lipid-like hormones such as the steroids. An example of an amino acid derivative is epinephrine, derived from tyrosine. *Antidiuretic hormone* (also called *vasopressin*) is an example of a peptide hormone, whereas insulin is a protein. *Testosterone* is an example of a *steroid hormone*. Steroid hormones are derivatives of cholesterol (see Figure 3-30) that are synthesized either in the gonads (*sex hormones*) or in the adrenal cortex (*corticosteroids*). Hundreds of different hormones regulate a wide range of physiological functions, including growth and development, rates of body processes, concentrations of substances, and responses to stress and injury. A few are listed in **Table 23-4**.

Because animal hormones are so much better understood, we will focus on them for the remainder of our discussion. First, we will focus on hydrophilic hormones that act in the endocrine system. In this case, the very same pathways you have learned about are used; the main difference is the distance over which hormones act. We will conclude our discussion of cell-cell signaling by considering steroid hormones, which act via nuclear receptors.

The Endocrine System Controls Multiple Signaling Pathways to Regulate Glucose Levels

Because hormones modulate the function of particular target tissues, an important aspect of studying hormones is understanding the specific functions of those target tissues. To

Table 23-4 Chemical Classification and Function of Hormones

Chemical Classification	Example	Regulated Function
Endocrine Hormones		
Amino acid derivatives	Epinephrine (adrenaline) and norepinephrine (both derived from tyrosine)	Stress responses: regulation of heart rate and blood pressure; release of glucose and fatty acids from storage sites
	Thyroxine (derived from tyrosine)	Regulation of metabolic rate
Peptides	Antidiuretic hormone (vasopressin)	Regulation of body water and blood pressure
	Hypothalamic hormones (releasing factors)	Regulation of tropic hormone release from pituitary gland
Proteins	Anterior pituitary hormones	Regulation of other endocrine systems
Steroids	Sex hormones (androgens and estrogens)	Development and control of reproductive capacity and secondary sexual characteristics
	Corticosteroids	Stress responses; control of blood electrolytes
Paracrine Hormones		
Amino acid derivative	Histamine	Local responses to allergens, stress and injury
Arachidonic acid derivatives	Prostaglandins	Local responses to stress and injury

illustrate how endocrine hormones act, let's look more closely at the **adrenergic hormones** *epinephrine* and *norepinephrine.* (Epinephrine is also called *adrenaline;* the two words are of Greek and Latin derivation, respectively, and mean "above, or near, the kidney," referring to the location in the body of the *adrenal glands,* which synthesize this hormone.) The overall strategy of adrenergic hormone action is to put many of the normal bodily functions on hold and to deliver vital resources to the heart and skeletal muscles instead, as well as to produce a heightened state of alertness. When secreted into the bloodstream, epinephrine and norepinephrine stimulate changes in many different tissues or organs, all aimed at preparing the body for dangerous or stressful situations (the so-called "fight-or-flight" response). Overall, adrenergic hormones trigger increased cardiac output, shunting blood from the visceral organs to the muscles and the heart, and cause dilation of arterioles to facilitate oxygenation of the blood. In addition, these hormones stimulate the breakdown of glycogen to supply glucose to the muscles.

Adrenergic hormones bind to a family of G protein–coupled receptors known as **adrenergic receptors.** They can be broadly classified into *α- and β-adrenergic receptors.* The α-adrenergic receptors bind both epinephrine and norepinephrine. These receptors are located on the smooth muscles that regulate blood flow to visceral organs. The β-adrenergic receptors bind epinephrine much better than norepinephrine. These receptors are found on smooth muscles associated with arterioles that feed the heart, smooth muscles of the bronchioles in the lungs, and skeletal muscles.

The α- and β-adrenergic receptors stimulate different signal transduction pathways because they are linked to different G proteins. For example, α-adrenergic receptors act through G_q proteins, whereas the β-adrenergic receptors activate G_s. As we discussed earlier, activation of G_s stimulates the cAMP signal transduction pathway, leading to relaxation of certain smooth muscles. Activation of G_q stimulates phospholipase C, leading to the production of IP_3 and DAG, which in turn elevates intracellular calcium levels. As a specific example of cAMP-mediated regulation, we will consider the control of glycogen degradation by the hormone epinephrine in liver or muscle cells.

Adrenergic Hormones and Control of Glycogen Degradation. One action of adrenergic hormones is to stimulate the breakdown of glycogen to provide muscle cells with an adequate supply of glucose. The breakdown of glycogen is facilitated by the enzyme *glycogen phosphorylase,* which catalyzes a reaction in which the glycosidic linkage between two glucose subunits undergoes attack by inorganic phosphate. This results in release of a molecule of glucose-1-phosphate. The glycogen phosphorylase system was the first cAMP-mediated regulatory sequence to be elucidated. The original work was published in 1956 by Earl Sutherland, who received a Nobel Prize in 1971 for this discovery.

The sequence of events that leads from hormonal stimulation to enhanced glycogen degradation is shown in **Figure 23-26** on page 714. It begins when ❶ an epinephrine molecule binds to a β-adrenergic receptor on the plasma membrane of a liver or muscle cell. The receptor activates a neighboring G_s protein, and the G_s protein in turn stimulates adenylyl cyclase, which generates cAMP from ATP. The resulting transient increase in the concentration of cAMP in the cytosol activates protein kinase A. ❷ PKA then activates another cascade of events that begins with the phosphorylation of the enzyme *phosphorylase kinase.* This leads to ❸ the conversion of *glycogen phosphorylase* from phosphorylase *b,* the less active form, to phosphorylase *a,* the more active form, and thus to ❹ an increased rate of glycogen breakdown.

cAMP also stimulates the inactivation of the enzyme system responsible for glycogen synthesis because PKA also phosphorylates the enzyme *glycogen synthase.* Rather than activating this enzyme, however, phosphorylation inactivates it. Thus, the overall effect of cAMP involves both an increase in glycogen breakdown and a decrease in its synthesis.

α_1-Adrenergic Receptors and the IP_3 Pathway. Another important adrenergic pathway is represented by the α_1-adrenergic receptors, which stimulate the formation of IP_3 and DAG. The α_1-adrenergic receptors are found mainly on smooth muscles in blood vessels, including those controlling blood flow to the intestines. When α-adrenergic receptors are stimulated, the formation of IP_3 causes an increase in the

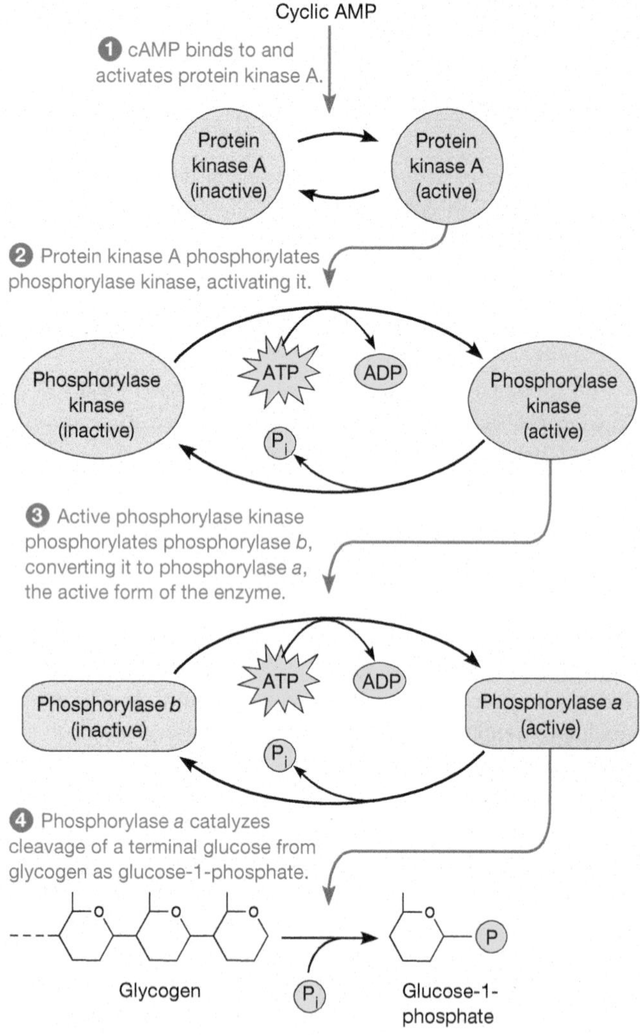

Cyclic AMP
1 cAMP binds to and activates protein kinase A.
Protein kinase A (inactive)
Protein kinase A (active)
2 Protein kinase A phosphorylates phosphorylase kinase, activating it.
Phosphorylase kinase (inactive)
ATP
ADP
Pi
Phosphorylase kinase (active)
3 Active phosphorylase kinase phosphorylates phosphorylase b, converting it to phosphorylase a, the active form of the enzyme.
Phosphorylase b (inactive)
ATP
ADP
Pi
Phosphorylase a (active)
4 Phosphorylase a catalyzes cleavage of a terminal glucose from glycogen as glucose-1-phosphate.
Glycogen
Pi
Glucose-1-phosphate
P

Figure 23-26 Stimulation of Glycogen Breakdown by Epinephrine. Muscle and liver cells respond to an increased concentration of epinephrine in the blood by increasing their rate of glycogen breakdown. The stimulatory effect of extracellular epinephrine on intracellular glycogen catabolism is mediated by a G protein–cyclic AMP regulatory cascade.

intracellular calcium concentration. The elevated level of calcium causes smooth muscle contraction, resulting in constriction of the blood vessels and diminished blood flow. Thus, the activation of α_1-adrenergic receptors affects smooth muscle cells in a manner opposite to that of β-adrenergic activation, which causes smooth muscles to relax.

Insulin Signaling and PI 3-Kinase. We have seen how epinephrine, acting via G protein–mediated signals, regulates metabolic processes—including the production and use of glucose in liver and muscle cells—in stressful situations. During periods of normal activity, however, two peptide hormones, produced by specialized cells in the pancreas known as *islets of Langerhans*, regulate blood glucose levels. One of these, the peptide hormone *glucagon*, acts via the same G_s protein activated by epinephrine, so it acts to increase blood glucose by the breakdown of glycogen when the level of glucose in the blood is too low. The other hormone, **insulin**, acts in a manner opposite to that of epinephrine and glucagon, *reducing* blood glucose levels. It does so by stimulating uptake of glucose into muscle and adipose (fat) cells and by stimulating glycogen synthesis. A disease with major worldwide consequences is *type 1 diabetes*, which results in loss of insulin-producing beta cells in the islets of Langerhans. The discovery that type 1 diabetes can be successfully treated with insulin led to a Nobel Prize for Frederick Banting and John Macleod in 1923. Unfortunately, insulin treatment is less effective in treating *type 2 diabetes*, which has reached epidemic proportions in the United States and afflicts hundreds of millions of adults worldwide. Type 2 diabetes appears to result from resistance to insulin, rather than an inability to produce it.

Insulin has both very rapid and longer-lasting effects. In muscle and adipose cells, insulin causes uptake of glucose within a few minutes and does not require the synthesis of new proteins. Long-term effects of insulin, such as production of enzymes involved in glycogen synthesis, require higher levels of insulin sustained over many hours. To exert its effects, insulin binds to receptor tyrosine kinases. Unlike the epidermal growth factor receptor, however, the insulin receptor has two α subunits and two β subunits (**Figure 23-27**). When insulin binds the receptor, β subunits of the receptor phosphorylate a protein called *insulin receptor substrate 1 (IRS-1)*, which can stimulate two different pathways. ❶ IRS-1 can recruit GRB2, activating the Ras pathway, or ❷ IRS-1 can bind an enzyme known as phosphatidylinositol 3-kinase (abbreviated as **PI 3-kinase**, or **PI3K**). PI 3-kinase then catalyzes the addition of a phosphate group to the plasma membrane lipid *PIP_2* (phosphatidylinositol-4,5-bisphosphate), which converts PIP_2 into *PIP_3* (phosphatidylinositol-3,4,5-trisphosphate). ❸ PIP_3 in turn binds to a protein kinase called **Akt** (also called *protein kinase B*), which can also be activated by phosphorylation from other kinases. How much PIP_3 is present is also regulated by enzymes that act to oppose PI 3-kinase. One such protein is **PTEN**, a phosphatase that removes a phosphate group from PIP_3 and thereby prevents the activation of Akt.

❹ Activation of Akt has two important consequences. First, it leads to movement of a glucose transporter protein known as *GLUT4* from vesicles in the cytosol to the plasma membrane, allowing glucose uptake. Second, Akt can phosphorylate a protein known as *glycogen synthase kinase 3 (GSK3)*, reducing its activity. This leads to an increase in the amount of the unphosphorylated, more active form of glycogen synthase, which we have seen enhances the production of glycogen. By stimulating absorption of glucose and its polymerization into glycogen, insulin signaling is thus a key component of glucose homeostasis.

Steroid Hormones Bind Hormones in the Cytosol and Carry Them into the Nucleus

Not all hormone receptors act at the cell surface. **Steroid hormones**, an important class of hormones, bind to receptors that act primarily in the nucleus rather than at the cell surface. These **steroid receptor** proteins mediate the actions of steroid hormones such as progesterone, estrogen, testosterone, and glucocorticoids. (Related receptors for thyroid hormone, vitamin D, and retinoic acid work in a similar way.) Steroid hormones are hydrophobic molecules synthesized by cells in endocrine tissues. They are released into the bloodstream, where they bind to blood plasma proteins and travel

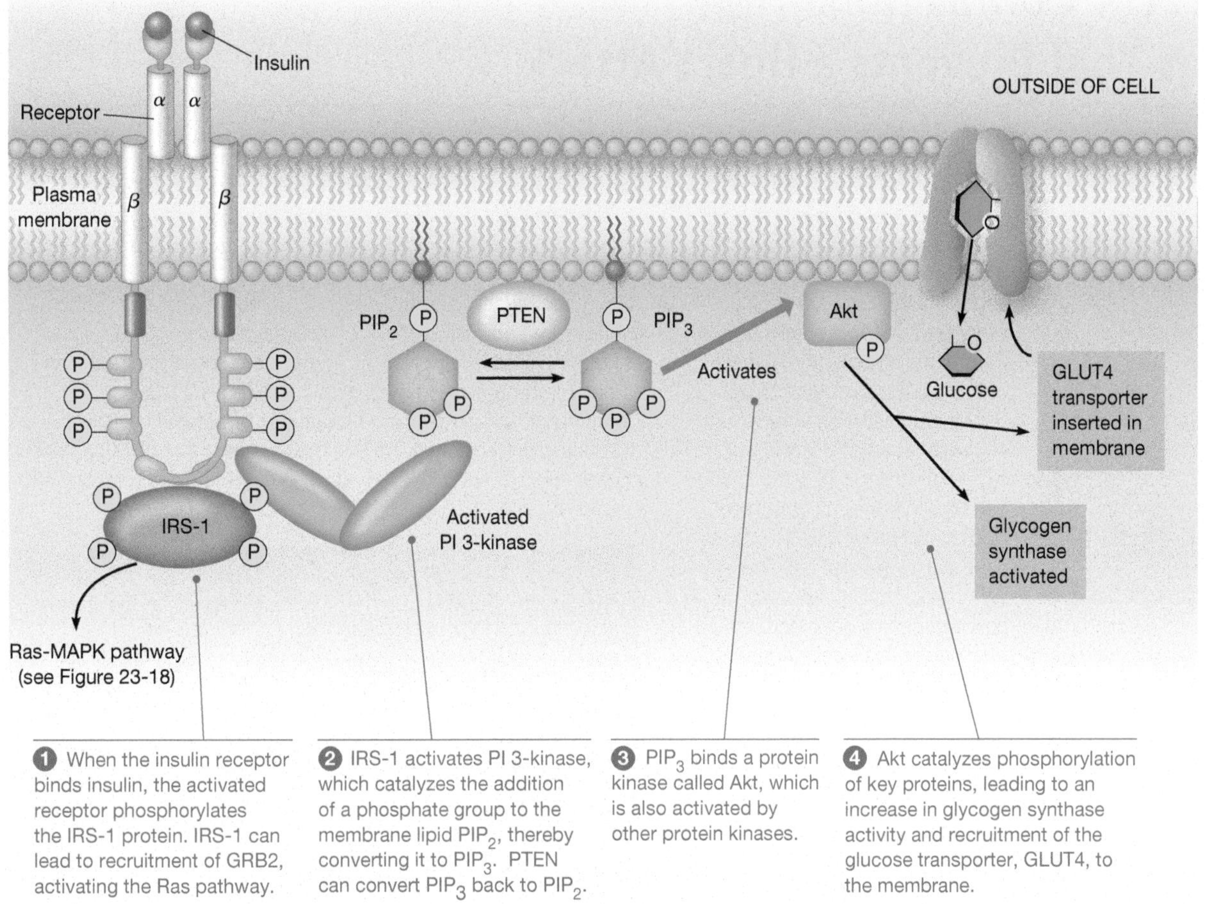
Insulin
Receptor
α
α
OUTSIDE OF CELL
Plasma membrane
β
β
PIP2
PTEN
PIP3
Akt
Activates
Glucose
GLUT4 transporter inserted in membrane
IRS-1
Activated PI 3-kinase
Glycogen synthase activated
Ras-MAPK pathway (see Figure 23-18)
1 When the insulin receptor binds insulin, the activated receptor phosphorylates the IRS-1 protein. IRS-1 can lead to recruitment of GRB2, activating the Ras pathway.
2 IRS-1 activates PI 3-kinase, which catalyzes the addition of a phosphate group to the membrane lipid PIP2, thereby converting it to PIP3. PTEN can convert PIP3 back to PIP2.
3 PIP3 binds a protein kinase called Akt, which is also activated by other protein kinases.
4 Akt catalyzes phosphorylation of key proteins, leading to an increase in glycogen synthase activity and recruitment of the glucose transporter, GLUT4, to the membrane.

Figure 23-27 The Insulin Signaling Pathway. The insulin signaling pathway influences glucose homeostasis by regulating multiple signaling pathways. The insulin receptor is a multisubunit receptor tyrosine kinase. When it binds insulin, recruitment and activation of the IRS-1 protein initiate signal transduction, leading to glucose import, stimulation of glycogen synthesis, and regulation of gene expression.

throughout the body. Because they are lipophilic, they can pass relatively easily through the plasma membranes of target cells. After entering a target cell, a steroid hormone binds to its corresponding receptor, triggering a series of events that ultimately activates, or in a few cases inhibits, the transcription of a specific set of genes (**Figure 23-28**).

Gases Can Act as Cell Signals

The final example of cell signaling in this chapter is another case of long-range signaling. In addition to proteins and soluble molecules, dissolved gases can sometimes serve as signals. In animals, dissolved gases such as O_2 and CO_2 are key long-range signals in respiration that allow for homeostatic regulation of the concentration of dissolved gases in the bloodstream. In plants, the ripening of fruits is mediated

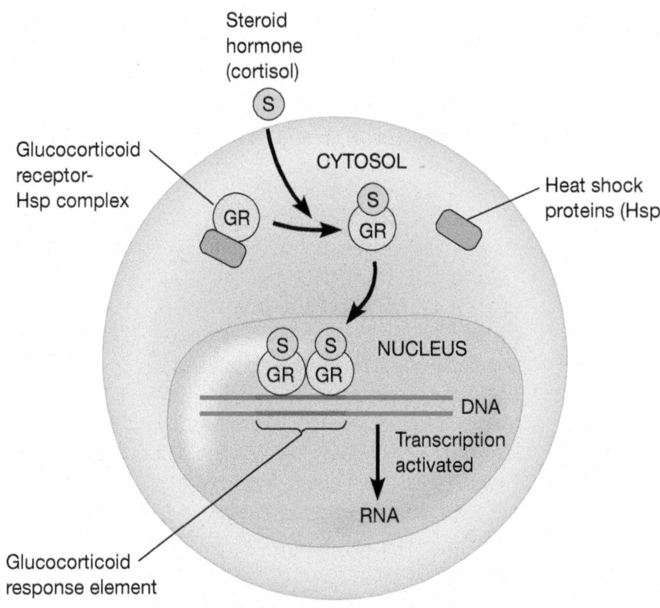
Steroid hormone (cortisol)
S
Glucocorticoid receptor-Hsp complex
CYTOSOL
Heat shock proteins (Hsp)
GR
S
GR
S
S
GR
GR
NUCLEUS
DNA
Transcription activated
RNA
Glucocorticoid response element

Figure 23-28 Activation of Gene Transcription by Steroid Hormone Receptors. The example shown here is the steroid hormone cortisol (S). Cortisol diffuses through the plasma membrane and binds to the glucocorticoid receptor (GR), causing the release of heat shock proteins (Hsp) and activating the GR molecule's DNA-binding site. The GR molecule then enters the nucleus and binds to a glucocorticoid response element in DNA, which in turn causes a second GR molecule to bind to the same response element. The resulting GR dimer activates transcription of the adjacent gene.

HUMAN *Connections*

The Gas That Prevents a Heart Attack

Heart disease is an enormous health challenge in the developed world. Diets rich in fats, lack of sufficient exercise, and other lifestyle issues make heart disease the leading cause of death for both men and women in the United States: it is responsible for the deaths of approximately 600,000 people every year, or 1 in 4 deaths. Central to the devastating effects of heart disease is loss of blood flow through *coronary arteries,* which supply the heart with oxygenated blood so that it can perform its crucial muscular contractions (**Figure 23B-1**). You have already learned about one factor that can impede blood flow through blood vessels: the buildup of arterial plaques due to "bad" cholesterol, or low-density lipoproteins (see Chapter 12). The characteristic narrowing of a coronary artery in a human patient is shown in Figure 23B-1a.

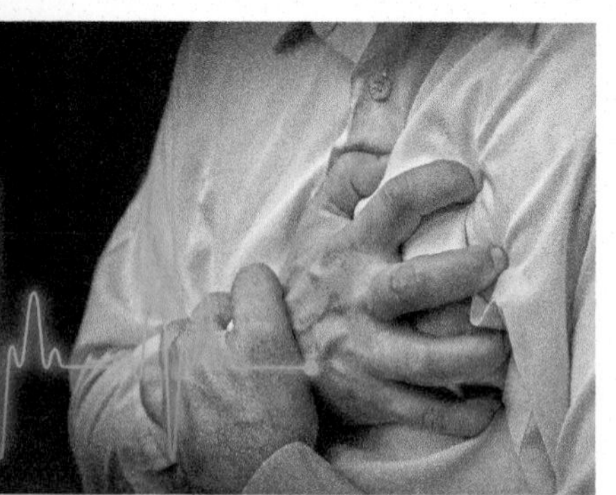

For patients with heart disease, keeping coronary arteries as open as possible is important. Cell signaling is crucial in the regulation of blood flow; as you learned in this chapter, adrenergic receptors, which are activated by hormones such as epinephrine, are good examples of regulation of blood flow via signaling. Another signaling pathway is also important for regulating cardiac blood vessels that feed the heart. In this case, instead of a protein ligand, the signaling ligand is the gas *nitric oxide (NO).*

NO is a toxic, short-lived molecule produced by the enzyme *NO synthase,* which converts the amino acid arginine to NO and citrulline. How does it affect blood vessels? It was known for many years that acetylcholine dilates blood vessels by causing their smooth muscles to relax. In 1980, Robert Furchgott demonstrated that acetylcholine dilated blood vessels only if the *endothelium* (the inner lining of the blood vessel) was intact. He concluded that blood vessels are dilated because the endothelial cells produce a signal (a *vasodilator*) that makes vascular smooth muscle cells relax. In 1986, work by Furchgott and parallel work by Louis Ignarro identified NO as the signal released by endothelial cells that causes relaxation of the vascular smooth muscle.

Figure 23B-1b illustrates how the binding of acetylcholine to the surface of vascular endothelial cells results in release of NO. There are six steps to this process. ❶ Acetylcholine binds to G protein–coupled receptors, causing IP_3 to be produced by the endothelial cells. ❷ IP_3 causes the release of calcium from the endoplasmic reticulum. ❸ The calcium ions then bind calmodulin, forming a complex that stimulates NO synthase to produce nitric oxide. Nitric oxide readily diffuses from the endothelial cells into the adjacent smooth muscle cells. ❹ Once inside smooth muscle cells, NO activates *guanylyl cyclase,* leading to ❺ production of *cyclic GMP (cGMP),* a molecule similar to cAMP. ❻ The increase in cGMP concentration activates *protein kinase G,* which induces muscle relaxation by catalyzing the phosphorylation of the appropriate muscle proteins.

The mechanism in Figure 23B-1b also explains the mechanism of action of the chemical *nitroglycerin.* Nitroglycerin is often taken by patients with *angina* (chest pain due to inadequate blood flow to the heart) to relieve constriction of coronary arteries. In 1977, Ferid Murad found that nitroglycerin and similar vasodilators elicit release of nitric oxide. In 1998, Furchgott, Ignarro, and Murad received a Nobel Prize for their elucidation of NO's effects on the cardiovascular system.

Nitric oxide is also used by neurons to signal nearby cells. For example, NO released by neurons in the penis results in the blood vessel dilation responsible for penile erection. The drug *sildenafil,* sold under the trade name Viagra, is an inhibitor of a phosphodiesterase that normally catalyzes the breakdown of cGMP. By maintaining elevated levels of cGMP in erectile tissue, this pathway is stimulated for a longer time period following NO release.

Although the genuine health benefits of sildenafil are heavily debated, there is no denying the importance of the cGMP pathway in patients with heart disease. NO truly is the gas that can prevent a heart attack.

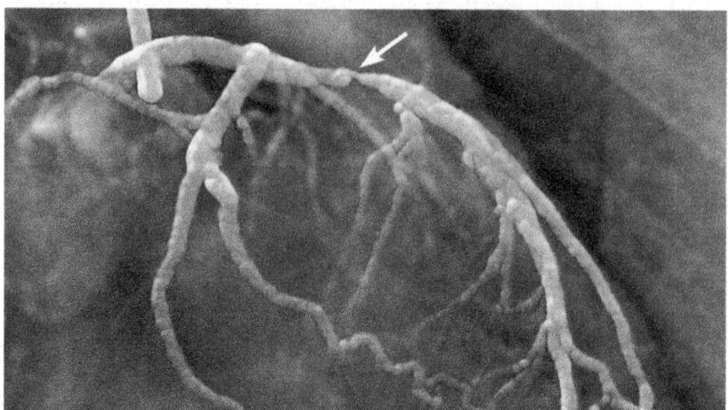

(a) Narrowing (stenosis) of a coronary artery

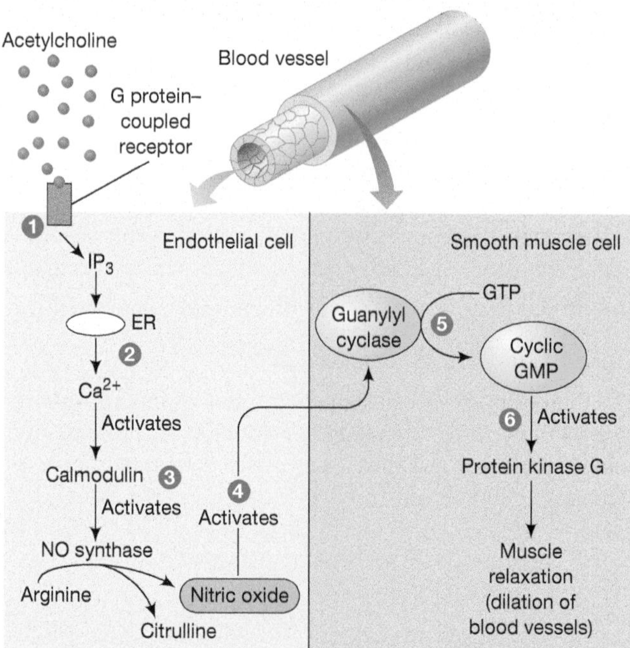

(b) Nitric oxide signaling in blood vessels

Figure 23B-1 The Action of Nitric Oxide on Blood Vessels. **(a)** Narrowing (stenosis) of the left coronary artery in a patient who is a heavy smoker. The arrow shows marked narrowing. **(b)** The binding of acetylcholine to endothelial cells triggers the production of nitric oxide, which diffuses into the adjacent smooth muscle cells and stimulates guanylyl cyclase, thereby leading to muscle relaxation.

by the gas *ethylene*. The spread of ethylene from ripening fruit causes nearby unripe fruit to ripen (for example, placing unripe bananas next to ripe apples in your kitchen will cause the former to rapidly ripen).

One important example of gaseous signaling is *nitric oxide (NO)*. NO is an especially important signaling molecule in the nervous system. **Human Connections, page 716**, considers NO signaling in one particularly interesting context in endothelial cells in some detail. This example involves G protein–mediated signaling that involves not cAMP, but the related small molecule we were introduced to earlier: cGMP.

CONCEPT CHECK 23.5

What might be some advantages for a multicellular organism when it uses local mediators versus hormones to mediate a particular physiological response? What are some disadvantages?

Summary of Key Points

Mastering™ Biology For activities, animations, and review quizzes, go to the study area at www.masteringbiology.com.

23.1 Chemical Signals and Cellular Receptors

- Cells use a variety of specific receptors to respond to hormones, growth factors, and other substances (ligands) present in the extracellular fluid. Many receptors are transmembrane proteins.
- Ligand binding is followed by transmission of the signal to the interior of the cell, thereby regulating specific intracellular events. Signal transmission is often carried out by second messengers, which can dramatically amplify signaling responses.
- One ligand can trigger multiple signaling pathways, and cells often integrate multiple signals at any given time. Ultimately, signals must be terminated. Different pathways accomplish this action in different ways.
- The affinity of a receptor for its ligand can be described using the dissociation constant, K_d.
- Drugs or other chemicals that bind a receptor can be used to stimulate the receptors artificially (agonists) or inhibit them (antagonists).

23.2 G Protein–Coupled Receptors

- A heterotrimeric G protein is activated when a ligand binds to its associated receptor, resulting in exchange of GDP for GTP on the G_α subunit and dissociation of G_α and $G_{\beta\gamma}$. The G_α or $G_{\beta\gamma}$ subunits then activate various signaling pathways. Once G_α hydrolyzes its bound GTP, G_α and $G_{\beta\gamma}$ can reassociate. RGS proteins regulate G_α activity by stimulating more rapid hydrolysis of GTP.
- The second messenger cyclic AMP is synthesized when adenylyl cyclase is activated by a G_α protein. cAMP can bind and activate protein kinase A (PKA), which can catalyze phosphorylation of many different proteins. Phosphodiesterase can catalyze cleavage and inactivation of cAMP. Activation of other G protein–coupled receptors results in production of cyclic GMP instead.
- Inositol trisphosphate (IP_3) and diacylglycerol (DAG) are produced from phosphatidylinositol bisphosphate when other G proteins activate the enzyme phospholipase C_β. $G_{\beta\gamma}$ subunits can activate signaling pathways as well, including those regulating potassium channels in neurons.
- Release of Ca^{2+} from internal stores within the endoplasmic reticulum is triggered by IP_3.
- Ca^{2+} effects are often mediated by calcium-binding proteins, such as calmodulin.
- Calcium-induced calcium release allows rapid propagation of calcium signals, such as during fertilization of animal eggs.

23.3 Enzyme-Coupled Receptors

- Growth factors regulate cell growth and behavior. Many growth factors bind receptor tyrosine kinases; others bind receptor serine-threonine kinases.
- Receptor tyrosine kinases become phosphorylated on specific tyrosines via autophosphorylation after binding ligand. Phosphorylated receptors recruit SH2 domain-containing proteins, which activate major signal transduction pathways (including the Ras and phospholipase C_γ pathways).
- Ras is a monomeric G protein, regulated by GTPase-activating proteins (GAPs) and by the guanine-nucleotide exchange factor (GEF) Sos. Ras activates a cascade of phosphorylation events, which results in activation of transcription factor proteins that regulate gene expression.
- Receptor serine-threonine kinases act via receptor-regulated Smads and their binding partner, Smad4, which enter the nucleus as a complex following receptor activation.
- Other enzyme-coupled receptors include phosphatase receptors and guanylyl receptors.

23.4 Putting It All Together: Signal Integration

- In some cases, such as the yeast mating pathway, components of signaling pathways are held in close proximity by scaffolding proteins. Signaling crosstalk allows signal integration.

23.5 Hormones and Other Long-Range Signals

- Endocrine hormones regulate the activities of body tissues distant from the tissues that secrete them.
- Adrenergic hormones, secreted by the adrenal medulla, bind G protein–coupled β-adrenergic receptors, which stimulate the formation of cAMP, and α-adrenergic receptors, which stimulate phospholipase C_β, resulting in elevation of intracellular Ca^{2+}.
- Insulin regulates glucose homeostasis by stimulating multiple signaling pathways, including the Ras and PI 3-kinase pathways.
- Steroid hormones act by binding cytosolic receptor proteins. The hormone-receptor complex then acts in the nucleus to regulate gene expression.
- Gases can also act as long-range signals; NO acts via a cGMP-mediated pathway in vascular endothelial cells.

Problem Set

Mastering™ Biology For activities, animations, and review quizzes, go to the study area at www.masteringbiology.com.

23-1 Chemical Signals and Second Messengers. Fill in the blanks with the appropriate terms.

(a) _____ is an intracellular protein that binds calcium and activates enzymes.

(b) Glucagon is an example of an _____ hormone.

(c) A substance that fits into a specific binding site on the surface of a protein molecule is called a _____.

(d) Two products of phospholipase C activity that serve as second messengers are _____ and _____.

(e) Cyclic AMP is produced by the enzyme _____ and degraded by the enzyme _____.

(f) Calcium ions are released within cells mainly from the _____.

23-2 QUANTITATIVE **Pure Agony.** Agonists are drugs that bind to receptors and activate them, like their normal ligands. Suppose you work for a pharmaceutical corporation, and you are testing the receptor affinity of a new preparation of *phenylephrine*, which is primarily used as a selective agonist of α_1-adrenergic receptors in over-the-counter decongestants. Physiological ligands for α_1-adrenergic receptors are epinephrine and norepinephrine. Assume that phenylephrine can saturate the receptors in this assay to the same extent as its normal ligands if enough phenylephrine is added (that is, phenylephrine can bind nearly 100% of the receptors if enough is added). You measure the K_d for epinephrine, norepinephrine, and phenylephrine and find it is 63, 196, and 239 n*M*, respectively. Draw a graph that represents the change in percent of α_1-adrenergic receptors bound versus concentration for each ligand. Be sure to label your axes and the curve for each ligand. Indicate the K_d for each on your graph.

23-3 Heterotrimeric and Monomeric G Proteins. G protein–coupled receptors interact with heterotrimeric G proteins to activate them.

(a) Upon binding to the receptor, the $G_{\beta\gamma}$ subunit catalyzes GDP/GTP exchange by the G_α subunit. How is this similar to the activation of Ras by a receptor tyrosine kinase?

(b) What similarities are there at the molecular level between how cells regulate the rate of GTP hydrolysis by the catalytically active portion of a heterotrimeric G protein and by Ras?

23-4 Calcium Chelators and Ionophores. In addition to calcium ionophores, another tool that has aided studies of the role of Ca^{2+} in triggering different cellular events is *calcium chelators*. Chelators are compounds such as EGTA and EDTA that bind very tightly to calcium ions, thereby reducing nearly to zero the free (or available) Ca^{2+} concentration outside of the cell. Using these tools, describe how you could demonstrate that a hormone exerts its effect by (1) causing Ca^{2+} to enter the cell through channels or (2) releasing Ca^{2+} from intracellular stores such as the ER.

23-5 The Eyes Have It. For this question, please consult Figure 23-21. Recall that each ommatidium in the compound eye contains a photoreceptor cell, R7, that must receive a signal from a neighboring cell, R8, to differentiate. This pathway depends on the Ras signaling pathway. Also, recall that mutants provide insight into this pathway. For each of the following situations, indicate whether the R7 cell would be expected to differentiate. In each case, clearly state your reasoning, based on your understanding of the molecular pathway involved.

(a) R7 cells are mutated so that their Sevenless receptors lack the SH2-binding domain.

(b) R7 cells are mutated so that they lack functional Sos but contain a constitutively active, dominant mutation in Ras.

(c) R7 cells are mutated so that they have no functional MAP kinase.

(d) R8 cells are mutated so that they produce too much of the Bride of sevenless ligand, and R7 cells lack Sevenless.

23-6 Membrane Receptors and Medicine. *Hypertension*, or high blood pressure, is often seen in elderly people. A typical prescription to reduce a patient's blood pressure includes compounds called *beta-blockers*, which block β-adrenergic receptors throughout the body. Why do you think beta-blockers are effective in reducing blood pressure?

23-7 DATA ANALYSIS/QUANTITATIVE **Following Ras Activation.** It is possible to use engineered molecules to follow the activation of individual Ras molecules in stimulated cells using a technique called fluorescence resonance energy transfer (FRET; see the Appendix for further details). **Figure 23-29** shows the time course of Ras activation following stimulation by epidermal growth factor (EGF).

(a) Based on the graph, how long does it take for maximal Ras activation to be achieved?

(b) Why does Ras activity decline after a few minutes, even when EGF is still present?

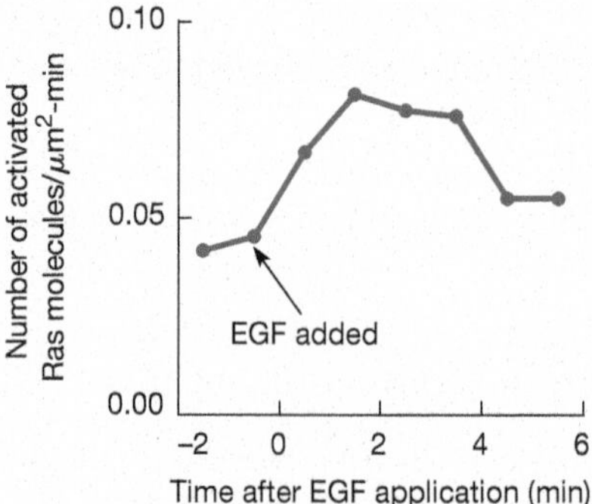

Figure 23-29 Activation of Ras Following Epidermal Growth Factor (EGF) Stimulation. See Problem 23-7.

23-8 Chemoattractant Receptors on Neutrophils. *Neutrophils* are blood cells normally responsible for killing bacteria at sites of infection. Neutrophils are able to find their way toward sites of infection by the process of *chemotaxis*. In this process, neutrophils sense the presence of bacterial proteins and then follow the trail of these proteins toward the site of infection. Suppose you find that chemotaxis is inhibited by pertussis toxin. What kind of receptor is likely to be involved in responding to bacterial proteins?

23-9 Scrambled Eggs. Unfertilized starfish eggs can be induced to produce foreign proteins either by injecting them with mRNA encoding the protein of choice or by directly injecting them with purified molecules. What do you predict would happen in the following cases, and why?

(a) "Caged calcium," a combination of calcium ions and a chelator, is injected into an egg. A flash of light is then used to induce the chelator to release its Ca^{2+} in a small region of the egg.

(b) mRNA for IP_3 receptors that do not allow Ca^{2+} release is injected, and then the egg is fertilized.

(c) Fluo3, a molecule that fluoresces in response to elevated Ca^{2+} levels, is injected into normal eggs, which are then fertilized.

24 The Cell Cycle and Mitosis

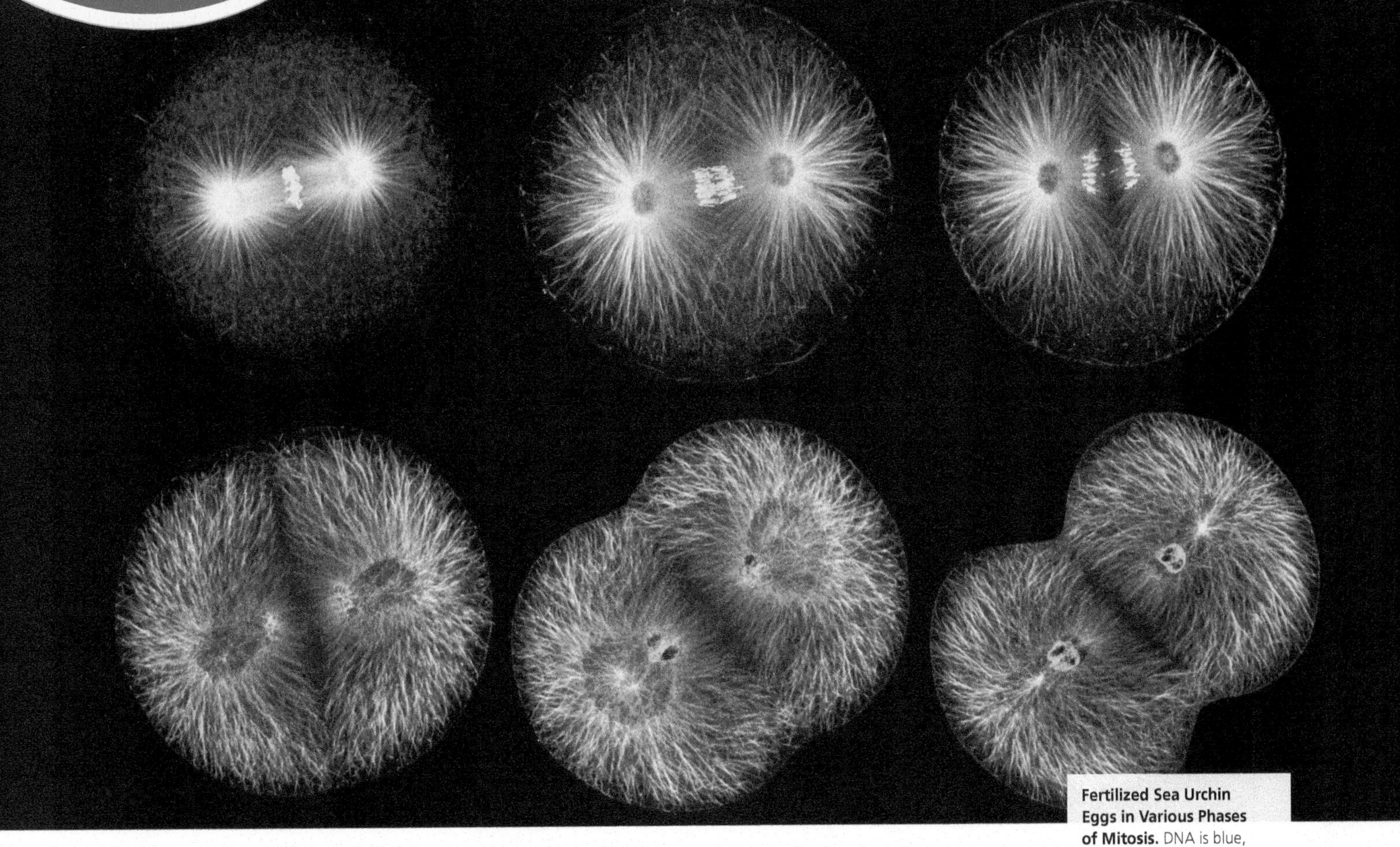

Fertilized Sea Urchin Eggs in Various Phases of Mitosis. DNA is blue, microtubules gold (confocal micrograph).

The ability to grow and reproduce is a fundamental property of living organisms. However, growth of single cells is fundamentally limited. As new proteins, nucleic acids, carbohydrates, and lipids are synthesized, their accumulation causes the volume of a cell to increase, forcing the plasma membrane to expand to prevent the cell from bursting. But cells cannot continue to enlarge indefinitely; as a cell grows larger, there is an accompanying decrease in its surface area/volume ratio and hence in its capacity for effective exchange with the environment. Therefore, cell growth is generally accompanied by **cell division,** whereby one cell gives rise to two new daughter cells. (The term *daughter* is used by convention and does not indicate that cells have gender.) For single-celled organisms, cell division increases the total number of individuals in a population. In multicellular organisms, cell division usually either increases the number of cells, leading to growth of the organism, or replaces cells that have died. In an adult human, for example, about 2 million stem cells in bone marrow divide every second to maintain a constant number of red blood cells in the body. A notable exception is the fertilized animal egg, which undergoes many divisions without growth, dividing the volume of the egg into smaller and smaller parcels.

When cells divide, the newly formed daughter cells are usually genetic duplicates of the parent cell, containing the same (or virtually the same) DNA sequences. Therefore, all the genetic information in the nucleus of the parent cell must be duplicated. The molecular mechanisms of this process are known as *DNA replication* (see Chapter 17). Once genetic material has been duplicated, it must be carefully distributed to the daughter cells during the division process. In accomplishing this task, a cell passes through a series of discrete stages, collectively known as the *cell cycle*. In this chapter, we will examine the events associated with the cell cycle, focusing first on the mechanisms that ensure that each new cell receives a complete set of genetic instructions during *mitosis* and *cytokinesis*, and then examining how the cell cycle is regulated.

24.1 Overview of the Cell Cycle

The **cell cycle** begins when two new cells are formed by the division of a single parental cell and ends when one of these cells divides again into two cells (**Figure 24-1**). To early cell biologists studying eukaryotic cells under the microscope, the most dramatic events in the life of a cell were those associated with the point in the cycle when a cell actually divides. This stage, called **M phase,** involves two overlapping events: the nucleus divides first and the cytoplasm second. Nuclear division is called **mitosis,** and the division of the cytoplasm to produce two daughter cells is termed **cytokinesis.**

The stars of the mitotic drama are chromosomes. As you can see in Figure 24-1a, the beginning of mitosis is marked by condensation (coiling and folding) of the cell's chromatin, which generates chromosomes that are thick enough to be individually discernible under the microscope. Because DNA replication has already taken place, each chromosome actually consists of two chromosome copies that remain attached to each other until the cell divides. As long as they remain attached, the two new chromosomes are referred to as **sister chromatids.** As the chromatids become visible, the nuclear envelope breaks into fragments. Then, in a remarkable movement guided by the microtubules of the *mitotic spindle*, the sister chromatids separate and—each now a full-fledged chromosome—move to opposite ends of the cell. By this time, cytokinesis has usually begun, and new nuclear membranes envelop the two groups of daughter chromosomes as cell division is completed.

Although visually striking, the events of M phase account for a relatively small portion of the total cell cycle; for a typical

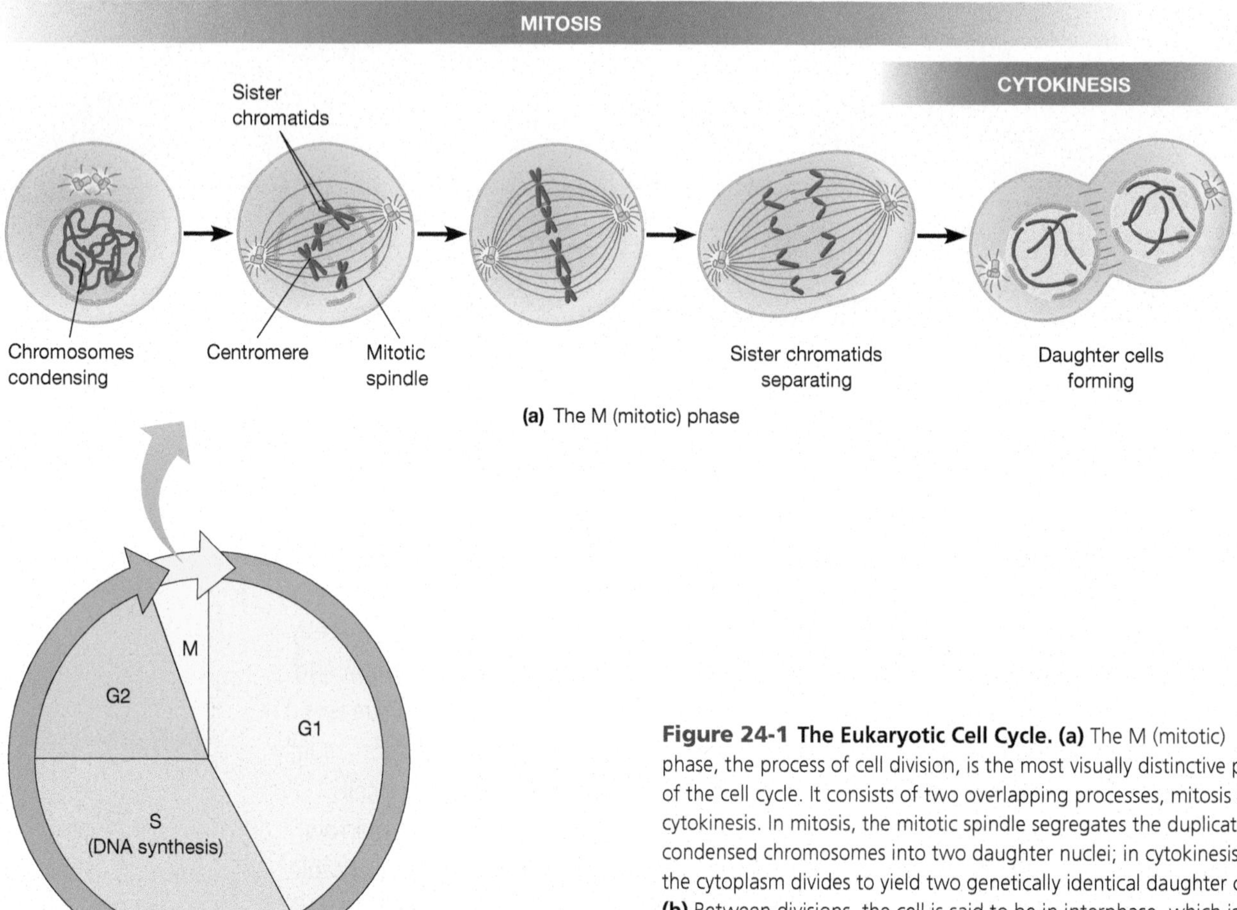

Figure 24-1 The Eukaryotic Cell Cycle. (a) The M (mitotic) phase, the process of cell division, is the most visually distinctive part of the cell cycle. It consists of two overlapping processes, mitosis and cytokinesis. In mitosis, the mitotic spindle segregates the duplicated, condensed chromosomes into two daughter nuclei; in cytokinesis, the cytoplasm divides to yield two genetically identical daughter cells. **(b)** Between divisions, the cell is said to be in interphase, which is made up of S phase (the period of nuclear DNA replication) and two "gap" phases, called G1 and G2. The cell continues to grow throughout interphase, a time of high metabolic activity.

mammalian cell, M phase usually lasts less than an hour. Cells spend most of their time in the phase between divisions, called **interphase** (Figure 24-1b). Most cellular contents are synthesized continuously during interphase, so cell mass gradually increases as the cell approaches division. The amount of nuclear DNA doubles during a specific portion of interphase known as **S phase** (S for synthesis). A time gap called **G1 phase** separates S phase from the preceding M phase; a second gap, the **G2 phase,** separates the end of S phase from the onset of the next M phase.

It is easy to determine the overall length of the cell cycle—the *generation time*—for cultured cells by counting the cells under a microscope and determining how long it takes for the cell population to double. In cultured mammalian cells, for example, the cell cycle usually takes about 18–24 hours. Once the total length of the cycle is determined, it is possible to determine the length of specific phases. Classically, this was done by exposing cells to a radioactively labeled DNA precursor (usually ^{3}H-thymidine) for a short period of time and then examining the cells by *autoradiography* (see Figure 17-5). The fraction of cells with silver grains over their nuclei represents the fraction of cells that were somewhere in S phase when the radioactive compound was available. When this fraction is multiplied by the total length of the cell cycle, the result is an estimate of the average length of the S phase. For mammalian cells in culture, this fraction is often around 0.33; based on a 24-hour cell cycle, this indicates that S phase is about 6–8 hours in length. Similarly, the length of M phase can be estimated by multiplying the generation time by the percentage of the cells that are actually in mitosis at any given time. This percentage is called the **mitotic index.** For cultured mammalian cells this is often about 3–5%, which means that M phase lasts less than an hour (usually 30–45 minutes).

In contrast to the S and the M phases, whose lengths tend to be similar for different mammalian cells, the length of G1 is quite variable, depending on the cell type. Although a typical G1 phase lasts 8–10 hours, some cells spend only a few minutes or hours in G1, whereas others are delayed for long periods of time. During G1, a major "decision" is made as to whether and when the cell should divide again. Cells arrested in G1, awaiting a signal that will trigger reentry into the cell cycle and a commitment to divide, are said to be in **G0 (G zero)**. Other cells exit from the cell cycle entirely and undergo *terminal differentiation,* which means they are destined never to divide again; most of the nerve cells in your body are in this state. In some cells, transient arrest of the cell cycle can also occur in G2. In general, however, G2 is shorter than G1 and more uniform in duration, usually lasting 4–6 hours.

Modern cell biologists no longer use radioactive thymidine. Instead, they use *thymidine analogues,* such as *5-bromodeoxyuridine (BrdU)* or *5-ethynyl-2′-deoxyuridine (EdU),* substances that are incorporated into DNA in place of thymidine. Which cells contain BrdU- or EdU-labeled DNA can be assessed via fluorescence microscopy, using either immunostaining (in the case of BrdU) or a chemical reaction (in the case of EdU).

Cell cycle studies have also been greatly facilitated by the use of **flow cytometry,** a standard technology that permits automated analysis of the chemical makeup of millions of individual cells. Flow cytometry has become a standard technology in basic cell biology research and in molecular medicine. (This important technique and a related technique, known as **fluorescence-activated cell sorting (FACS),** which allows physical sorting of cells, are described in more detail in **Key Technique, pages 722–723.**)

CONCEPT CHECK 24.1

Which DNA sequences are more alike: a pair of sister chromatids or a pair of homologous chromosomes?

24.2 Nuclear and Cell Division

With this introduction, we can now turn to the question of how the two copies of each chromosomal DNA molecule created during the S phase of the cell cycle are subsequently separated from each other and partitioned into daughter cells. These events occur during M phase, which encompasses both nuclear division (mitosis) and cytoplasmic division (cytokinesis).

Mitosis Is Subdivided into Prophase, Prometaphase, Metaphase, Anaphase, and Telophase

Mitosis has been studied since the nineteenth century, but only in the past few decades has significant progress been made toward understanding the mitotic process at the molecular level. We will begin by surveying the morphological changes that occur in a cell as it undergoes mitosis; we will then examine the underlying molecular mechanisms.

Mitosis is subdivided into five stages based on the changing appearance and behavior of the chromosomes. These five phases are *prophase, prometaphase, metaphase, anaphase,* and *telophase.* (An alternative term for prometaphase is *late prophase.*) The micrographs and schematic diagrams of **Figure 24-2** on page 724–725 illustrate the phases in a typical animal cell; **Figure 24-3** on page 726 depicts the comparable stages in plant cells.

Prophase. After completing DNA replication, cells exit S phase and enter G2 phase (see Figure 24-1b), where final preparations are made for the onset of mitosis. Toward the end of G2, the chromosomes start to condense from the extended, highly diffuse interphase chromatin fibers into the compact, extensively folded structures that are typical of mitosis. Chromosome condensation is important because interphase chromatin fibers are so long and intertwined that in an uncompacted form, they would become impossibly tangled during distribution of the chromosomal DNA at the time of cell division. Although the transition from G2 to prophase is not sharply defined, a cell is considered to be in **prophase** when individual chromosomes have condensed to the point of being visible as discrete objects in the light microscope. Because the chromosomal DNA molecules have replicated during S phase, each prophase chromosome is composed of two sister chromatids that are tightly attached to each other. In animal cells, the nucleoli usually disperse as the chromosomes condense; plant cell nucleoli may remain as discrete entities, undergo partial disruption, or disappear entirely.

Key Technique | Measuring Cells Millions at a Time

PROBLEM: Cell biologists and physicians often want to measure many cells, such as cells in the immune system, very rapidly to determine how many cells have a given set of properties. Such measurements are difficult and tedious when done by hand.

SOLUTION: Flow cytometry allows the rapid measurements of thousands or millions of cells based on specific properties. When combined with fluorescence-activated cell sorting, purified populations of cells can be obtained and studied further.

Key Tools: Cells labeled with fluorescent dyes or antibodies; a flow cytometer; and/or a fluorescence-activated cell sorter.

Details: *Flow cytometry* is a powerful technique for automated measurements of cells (cytometry comes from two Greek roots: *cyto*–, having to do with cells, and *metry*, "measurement"). Flow cytometry starts with a population of cells that have been isolated as a slurry in culture medium. The cellular slurry is then placed into a machine that can pass cells as a thin stream through a device that uses light to measure their properties, one cell at a time, very quickly. *Fluorescence-activated cell sorting (FACS)* is an additional step during which cells can be separated based on their unique properties.

Flow Cytometry. For most flow cytometry applications, cells are first stained with one or more fluorescent dyes—for example, a purple dye that stains DNA might be combined with a green dye that binds specifically to a particular cellular protein on the surface. By analyzing fluctuations in the intensity and color of the fluorescent light emitted by each cell as it passes through the laser beam, researchers can assess the concentration of DNA and specific proteins in individual cells. In addition, the laser light is often scattered as it interacts with the cells in ways that allow scientists to infer the size and shape of the cells at the same time (**Figure 24A-1**).

Fluorescence-Activated Cell Sorting (FACS). By adding a special device in the flow path of a flow cytometer, it is possible not only to measure how many cells have particular properties, but also to physically sort them. In a FACS machine, cells are released in tiny droplets, and the droplets can be deflected by applying a small electric charge to each drop after the properties of the cell inside the droplet are measured. Droplets with a particular charge can be sorted by deflecting them using an electromagnet, so that they are placed into separate "bins," allowing them to be collected (**Figure 24A-2**). The isolated cells can then be analyzed to see which genes they express or to examine their migratory or cell adhesive properties.

Example: Measuring Cells in the Immune System. For some medical applications, it is important to measure the proportion of particular white blood cells (cells that do not contain hemoglobin, such as leukocytes, macrophages, and various lymphocytes) in a patient's blood. This provides key information about what illness a patient might have and whether a patient is responding to treatment. FACS can also assess the amount of DNA in cells using a

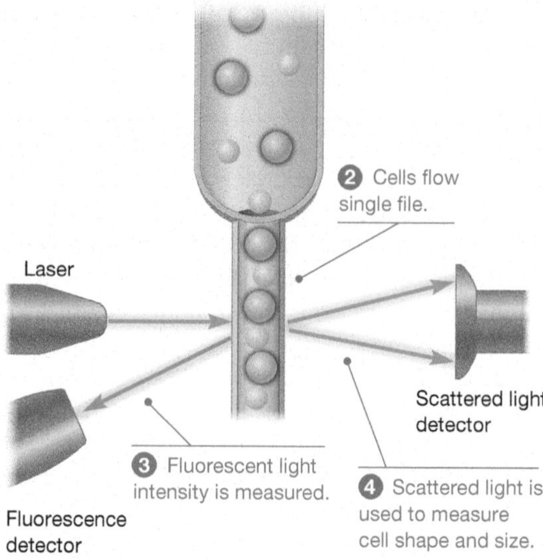

1 A mixture of cells is treated with fluorescent antibodies or a DNA dye.
2 Cells flow single file.
Laser
Scattered light detector
3 Fluorescent light intensity is measured.
4 Scattered light is used to measure cell shape and size.
Fluorescence detector

Figure 24A-1 Flow Cytometry. 1 A mixture of cells is labeled with antibodies or dyes. 2 A thin stream of cells passes single file through a detector, which 3 relies on a laser beam to assess the fluorescence of the cells (and hence how much of a specific molecule they possess) and 4 how they scatter light (which provides information about the shape and size of the cells). The results are then sent to a computer for further analysis.

The Laser Beam Within a FACS Machine.

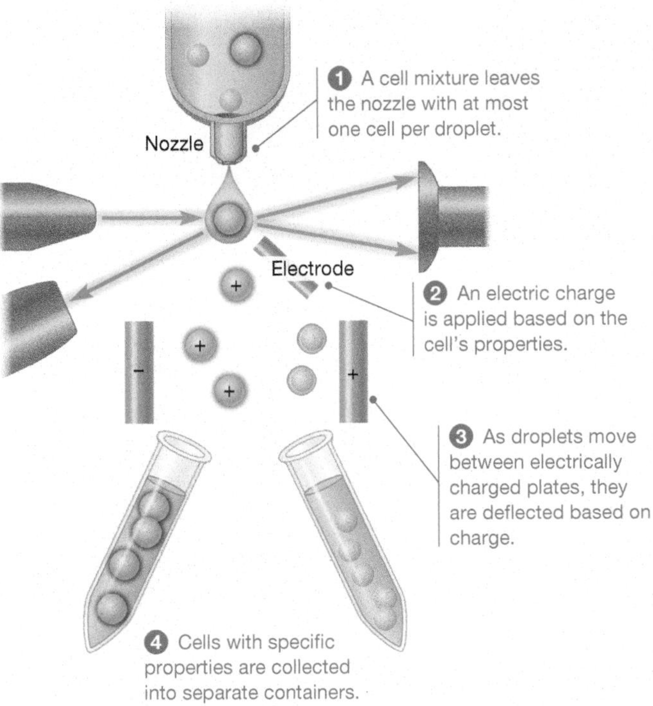

Figure 24A-2 Fluorescence-Activated Cell Sorting (FACS). 1 An apparatus similar to a flow cytometer is used, but cells are released into tiny droplets. 2 An electric charge is applied to droplets based on the properties of the cell contained in the droplet. 3 Charged plates then separate the droplets (and the cells they contain) based on electric charge. 4 The separated cells are then collected.

fluorescent DNA stain. Cells that have replicated their DNA (that is, they have completed S phase) have twice as much fluorescent DNA as cells that have not entered S phase (**Figure 24A-3a**). When DNA staining is combined with size and shape information, or other dyes or antibodies that distinguishes specific cell types, a complex profile of which cells are present and which phase of the cell cycle they are in can be obtained. Such measurements are often plotted on a log plot, with the *x*-axis representing the fluorescence of one dye or antibody, and the *y*-axis representing the fluorescence due to another. Figure 24A-3b shows such a plot that clearly distinguishes cells in the immune system based on their DNA content and fluorescent staining for histones, which shows which cells are in M phase.

QUESTION: *Lymphoma* is a cancer of white blood cells known as lymphocytes. *B lymphocytes* are antibody-producing cells that express the cell surface protein CD19. You predict that a human patient has a B cell lymphoma. How could you use flow cytometry to provide support for your prediction?

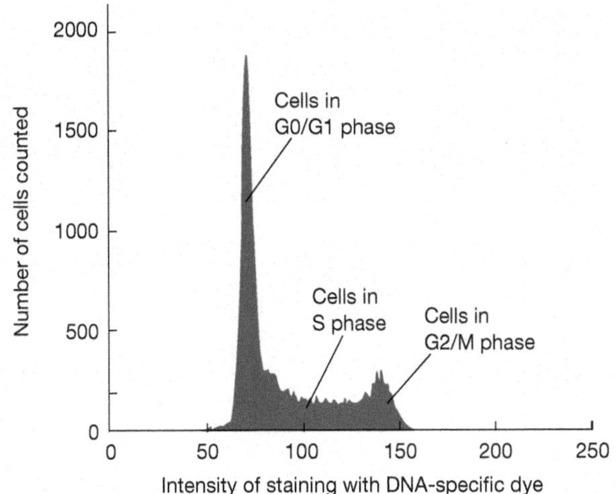

(a) Measuring DNA content of cells

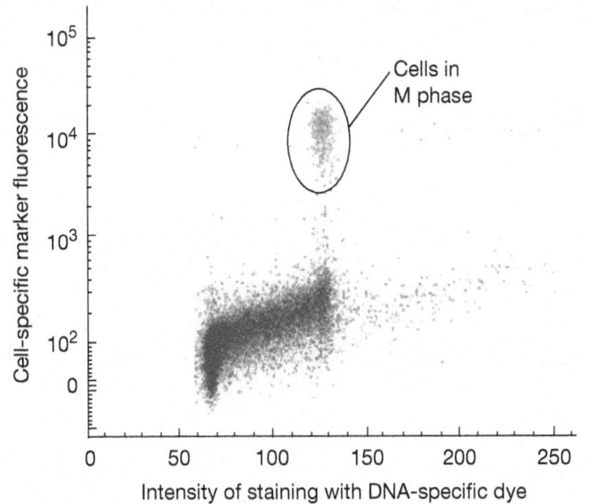

(b) Identifying cells in M phase

Figure 24A-3 Tracking the Cell Cycle. (a) Tracking the cell cycle using a DNA-specific dye. Isolated Jurkat cells (a T lymphocyte cell line) were treated with a fluorescent violet dye that binds DNA. Cells that have completed S phase and are in G2/M phase have twice as much signal per cell as cells in G0/G1 phase; cells beginning to synthesize DNA have an intermediate level of signal. **(b)** Dual-color flow cytometry. Red blood cell precursor cells were fixed and stained with anti-histone H3 antibodies tagged with a fluorescent green color (*y*-axis) and with a fluorescent violet dye that binds DNA (*x*-axis). Cells that show a high intensity of both green and violet fluorescence are in M phase (data points colored red on the graph).

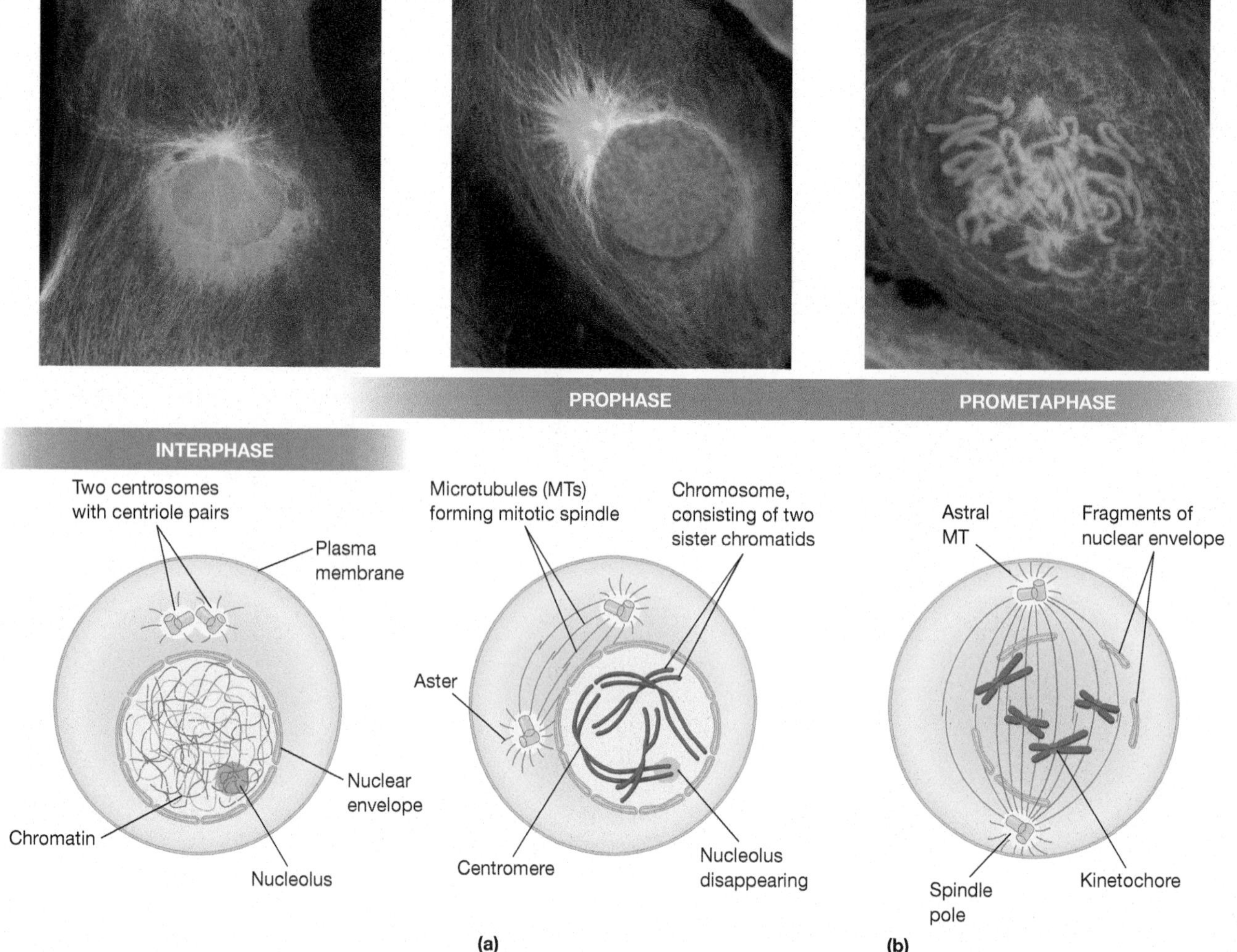

Figure 24-2 The Phases of Mitosis in an Animal Cell. The micrographs show mitosis in newt lung cells viewed by fluorescence microscopy. Chromosomes are stained blue; microtubules green; intermediate filaments red. At this low magnification (about 600-fold), we see spindle "fibers" rather than individual microtubules; each fiber consists of a number of microtubules. The drawings are schematic and include details not visible in the micrographs; for simplicity, only four chromosomes are drawn (MT = microtubule).

Meanwhile, another important multiprotein complex has sprung into action, the **centrosome**. Centrosomes function as *microtubule-organizing centers (MTOCs)* where *microtubules* (MTs) are assembled and anchored (see Chapter 13). The positioning of centrosomes in turn influences the position of the mitotic spindle during cell division. During each cell cycle, the centrosome is duplicated prior to mitosis, usually during S phase. At the beginning of prophase the two centrosomes then separate from each other and move toward opposite sides of the nucleus.

Embedded within the centrosome of animal cells is a pair of small, cylindrical, microtubule-containing *centrioles* (see Figure 13-8), typically oriented at right angles to each other. Because centrioles are absent in certain cell types, including most plant cells, they do not appear to be essential to mitosis. They do, however, play an essential role in the formation of cilia and flagella (see Figure 14-8). At the same time that centrosomes are duplicated, the centrioles embedded within the centrosomal material also replicate. "Daughter" centrioles begin to grow near one end of each "mother" centriole, growing at right angles to its mother centriole.

As they move apart, each centrosome acts as a nucleation site for MT assembly and the region between the two centrosomes begins to fill with MTs destined to form the **mitotic spindle,** the structure that distributes the chromosomes to the daughter cells later in mitosis. During this process, cytoskeletal MTs disassemble and their tubulin subunits are added to the growing mitotic spindle. At the same time, a dense starburst of MTs called an *aster* forms in the immediate vicinity of each centrosome.

Prometaphase. The onset of **prometaphase** is marked by fragmentation of the membranes of the nuclear envelope. As centrosomes complete their movement toward opposite sides of the nucleus (see Figure 24-2b), the breakdown of the nuclear envelope allows spindle microtubules to make contact with the chromosomes, which still consist of paired chromatids at this stage. The spindle MTs are destined to attach to the chromatids in the region of the *centromere,* a constricted area where the two members of each chromatid pair are held together. The DNA of each centromere consists of simple-sequence, tandemly repeated *CEN sequences,* whose

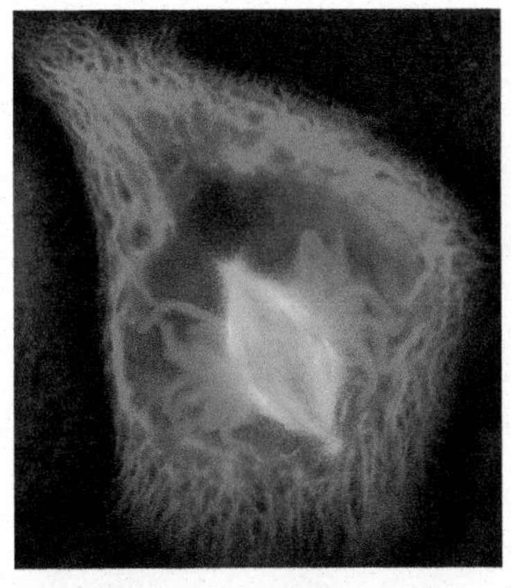

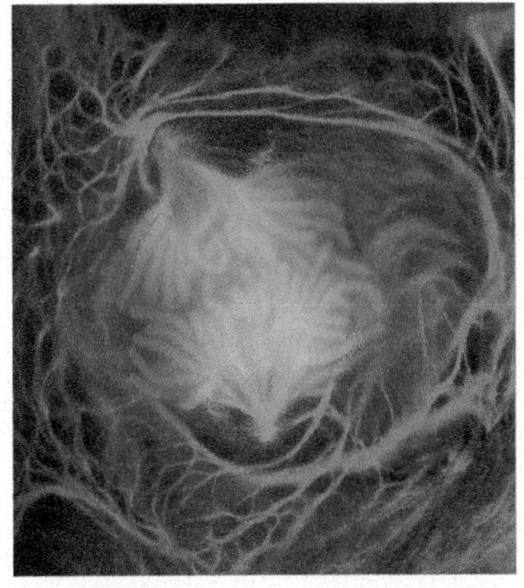

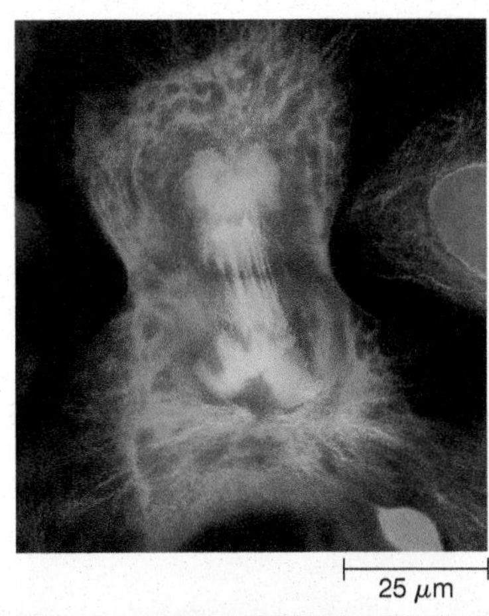

makeup varies considerably among species (see Chapter 16). An additional common feature of centromeres is the presence of specialized nucleosomes in which histone H3 is replaced by a related protein, which in humans is called *CENP-A (centromere protein A)*.

CENP-A plays a key role in recruiting additional proteins to the centromere to form the **kinetochore,** the structure that attaches the paired chromatids to spindle microtubules (**Figure 24-4** on page 727). Kinetochore proteins begin to associate with the centromere shortly after DNA is replicated during S phase. As shown in Figure 24-4a, each chromosome eventually acquires two kinetochores facing in opposite directions, one associated with each of the two chromatids. During prometaphase some spindle MTs bind to these kinetochores, thereby attaching the chromosomes to the spindle. Forces exerted by these **kinetochore microtubules,** or **K-fibers,** then gradually move chromosomes toward the center of the cell. When a kinetochore is not attached, a specialized structure is visible in the electron microscope called the *fibrous corona*. The corona enlarges by polymerization of the *RZZ complex* (for *ROD–Zwilch–ZW10*), which helps coordinate binding of the microtubule motor protein cytoplasmic dynein and recruitment of the spindle assembly checkpoint (SAC) complex proteins Mad1 and Mad2 to unattached kinetochores, a role we will return to below.

Figure 24-4b is an electron micrograph of a metaphase chromosome with two sets of attached microtubules embedded in the two kinetochores. Each kinetochore is a platelike, layered structure. The *inner kinetochore* contains proteins that bind to centromeric DNA. The *outer kinetochore* contains proteins that attach to the plus ends of MTs and allow the kinetochore to sense tension transmitted from MTs to the kinetochore. Kinetochores of different species vary in size. In yeast, for example, they are small and bind only one spindle MT each, whereas the kinetochores of mammalian cells are much larger, each binding 15-30 MTs.

The kinetochore is a complex structure, with more than a hundred proteins in vertebrates. Proteins are added sequentially after S phase until mature kinetochores are assembled. This is an impressive self-assembly process; by combining the roughly 30 core kinetochore proteins and DNA in a test tube, researchers have been able to reconstitute kinetochores. Mature kinetochores have two main parts that correspond to the regions seem under the electron microscope. In addition to DNA and CENP-A, the centromere-associated, inner part of the kinetochore is formed by the **constitutive centromere-associated network (CCAN),** which contains CENP and other proteins (Figure 24-4d). The outer kinetochore is formed by the **KMN network** (for *Knl1, Mis12 complex*, and *Ndc80 complex*). Because the Ndc80 complex of the KMN

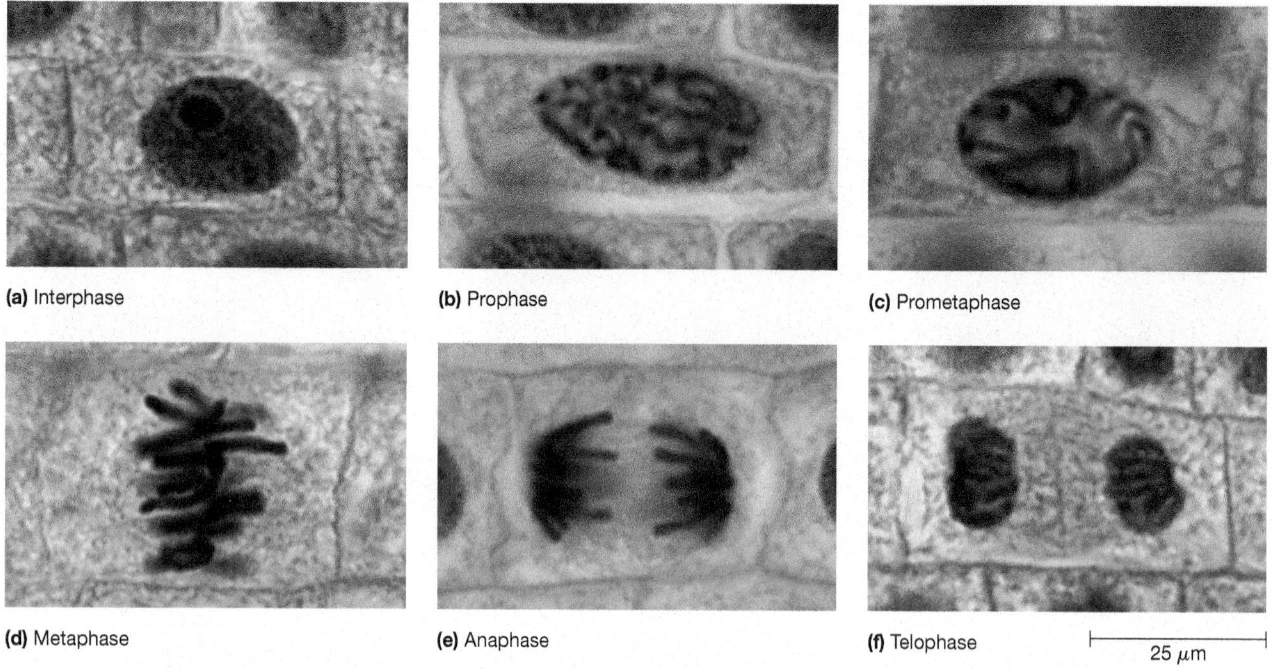

Figure 24-3 The Phases of Mitosis in a Plant Cell. These micrographs show mitosis in cells of an onion root viewed by light microscopy.

network directly interacts with microtubules and the CCAN binds centromeric chromatin, the two complexes form a molecular sandwich that bridges chromosomes and microtubules during mitosis.

In addition to kinetochore microtubules, there are two other kinds of MTs in the spindle. Those that interact with MTs from the opposite spindle pole are called **polar microtubules;** the shorter ones that form the asters (from the Greek word for "star") at each pole are called **astral microtubules.** Some of the astral MTs interact with proteins at the plasma membrane.

Metaphase. A cell is said to be in **metaphase** when the fully condensed chromosomes all become aligned at the *metaphase plate,* the plane equidistant between the two poles of the mitotic spindle (see Figure 24-2c). Agents that interfere with microtubules, such as the drug *colchicine,* can be used to arrest cells at metaphase. At metaphase the chromosomes appear to be relatively stationary, but this appearance is misleading. Actually, the two sister chromatids of each chromosome are already being actively tugged toward opposite poles. They appear stationary because the forces acting on them are equal in magnitude and opposite in direction; the chromatids are the prizes in a tug-of-war between two equally strong opponents. (We will discuss the source of these opposing forces shortly.)

Anaphase. The shortest phase of mitosis, **anaphase,** typically lasts only a few minutes. At the beginning of anaphase, the two sister chromatids of each chromosome abruptly separate and begin moving toward opposite spindle poles at a rate of about 1 μm/min (see Figure 24-2d).

Anaphase is characterized by two kinds of movements, called anaphase A and anaphase B (**Figure 24-5** on page 728). In **anaphase A,** the chromosomes are pulled, centromere first, toward the spindle poles as the kinetochore microtubules get shorter and shorter. In **anaphase B,** the poles themselves move away from each other as the polar microtubules lengthen. Depending on the cell type involved, anaphase A and B may take place at the same time, or anaphase B may follow anaphase A.

Telophase. At the beginning of **telophase,** the daughter chromosomes have arrived at the poles of the spindle (see Figure 24-2e). Next, the chromosomes uncoil into the extended fibers typical of interphase chromatin, nucleoli develop at the nucleolar organizing sites on the DNA, the spindle disassembles, and nuclear envelopes form around the two groups of daughter chromosomes. During this period most cells undergo cytokinesis, which divides the cell into two daughter cells.

The Mitotic Spindle Is Responsible for Chromosome Movements During Mitosis

The central purpose of mitosis is to separate the two sets of daughter chromosomes and partition them into the two newly forming daughter cells. To understand the mechanisms that allow this to be accomplished, we need to take a closer look at the mitotic spindle.

Spindle Assembly and Chromosome Attachment. Recall that microtubules have inherent *polarity* (see Chapter 13); that is, the two ends of each MT are chemically different (**Figure 24-6** on page 728). The end where MT assembly is initiated—located at the centrosome for spindle MTs—is the minus (–) end. The more dynamic end, located away from the centrosome, is the plus (+) end; increases in MT length come mainly from addition of subunits to the plus end.

During late prophase, microtubules become much more dynamic, and initiation of new MTs at the centrosomes increases. Once the nuclear envelope disintegrates at the

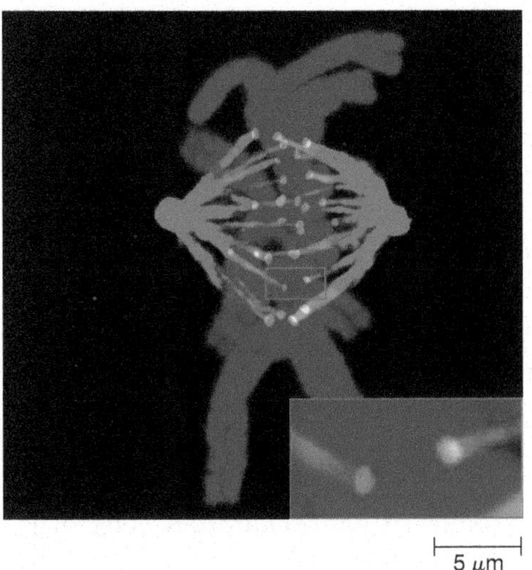

(a) Metaphase spindle with chromosomes (fluorescence)

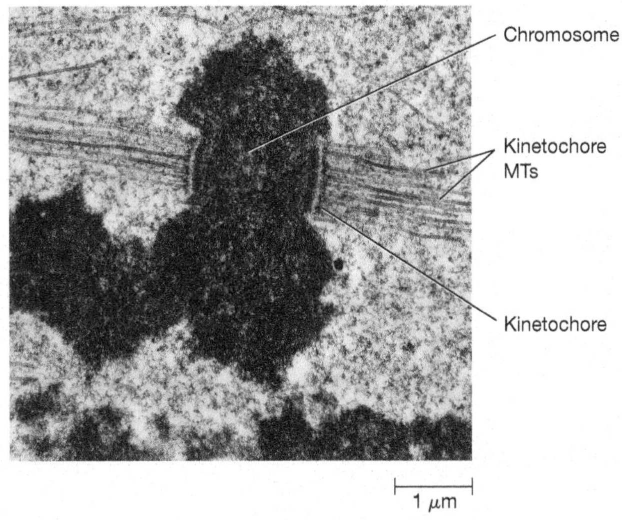

(b) Single chromosome attached to the spindle (TEM)

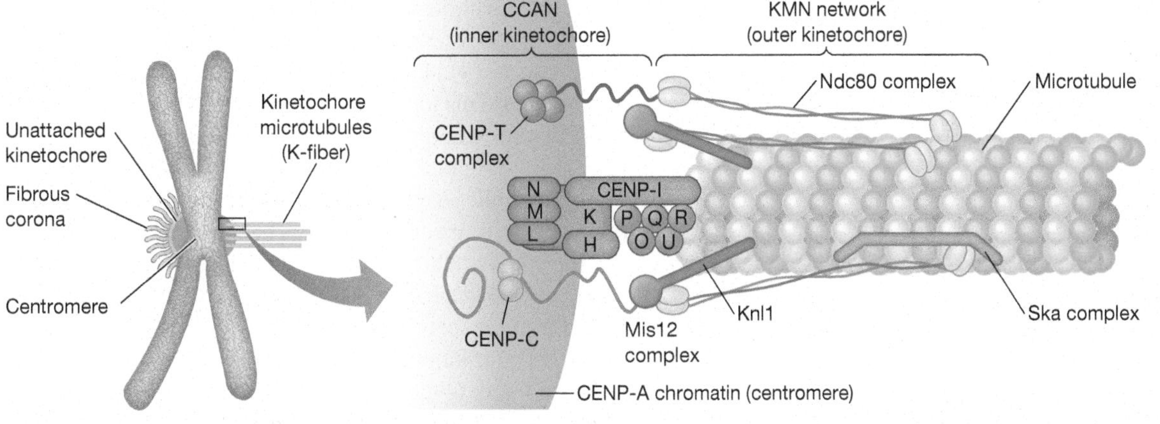

(c) Single chromosome attached via one kinetochore

(d) Schematic of a single microtubule attached to a kinetochore

Figure 24-4 Attachment of Chromosomes to the Mitotic Spindle. (a) This micrograph shows the spindle of a female rat kangaroo kidney epithelial (PtK1) cell stained for chromosomes (blue), kinetochores (pink) and microtubules (green; fluorescence microscopy). The inset is a magnification of the boxed region. Microtubules can be seen to insert at kinetochores. **(b)** The striped structures on either side of this metaphase chromosome are its kinetochores, each associated with one of the two sister chromatids. Numerous kinetochore MTs are attached to each kinetochore. The two sets of microtubules come from opposite poles of the cell (TEM). **(c)** In this schematic diagram of the centromere, kinetochores, and kinetochore microtubules, note that the unattached kinetochores possess a fibrous corona. **(d)** The inner kinetochore attaches to centromeric DNA via the constitutive centromere-associated network (CCAN), which contains numerous CENP proteins, including the CENP-T complex and CENP-C. The outer kinetochore attaches plus ends of microtubules to the CCAN through the KMN (Knl1, Mis12 complex, Ndc80 complex) network. The Ska complex stabilizes MT attachment to the KMN. For simplicity only two Ndc80 complexes are shown, showing ways these complexes can attach to the inner kinetochore. In a mammalian kinetochore there might be as many as 14 Ndc80 complexes.

beginning of prometaphase, contact between MTs and chromosomal kinetochores becomes possible. When contact is made between a kinetochore and the plus end of a MT, they bind to each other. This binding slows depolymerization at the plus end of the MT, although polymerization and depolymerization can still occur there.

Because the two kinetochores are located on opposite sides of a chromosome, they usually attach to microtubules emerging from centrosomes located at opposite poles of the cell. (The orientation of each chromosome appears random; either kinetochore can end up facing either pole.) Meanwhile, the other main group of MTs—the polar microtubules—make direct contact with polar MTs coming from the opposite centrosome. When the plus ends of two polar MTs of opposite polarity start to overlap, crosslinking proteins bind them to each other (Figure 24-6) and stabilize them. Thus, we can picture a barrage of microtubules rapidly shooting out from each centrosome during late prophase and prometaphase. The ones that successfully hit a kinetochore or a microtubule of opposite polarity are stabilized; the others retreat by disassembling.

One shortcoming of the preceding mechanism is that it does not explain how spindles are assembled in cells lacking centrosomes, which includes most of the cells of higher plants and the oocytes (immature eggs) of many animals. Moreover,

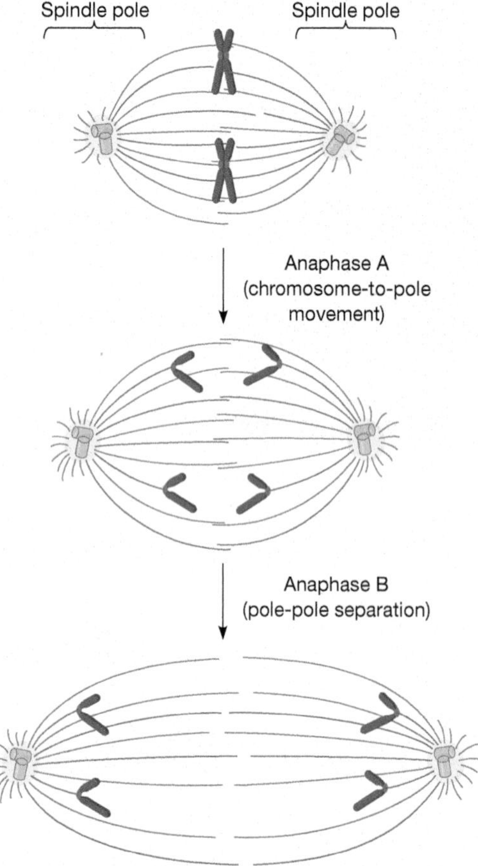

Figure 24-5 The Two Types of Movement Involved in Chromosome Separation During Anaphase. Anaphase A involves the movement of chromosomes toward the spindle poles. Anaphase B is the movement of the two spindle poles away from each other. Anaphase A and anaphase B may occur simultaneously.

experiments using a laser microbeam to destroy centrosomes have shown that animal cells that normally contain centrosomes can nevertheless assemble spindles. In cells lacking centrosomes, chromosomes rather than centrosomes promote microtubule assembly and spindle formation. Chromosome-induced MT assembly requires the involvement of *Ran*, a GTP-binding protein with a role in nuclear transport (see Figure 16-34). Mitotic chromosomes possess a protein that promotes the binding of GTP to Ran. The Ran-GTP complex then interacts with the protein *importin*, just as it does during nuclear transport, leading to release of importin-bound proteins that promote MT assembly. Even when cells use centrosomes to generate spindle microtubules, Ran-GTP is thought to help organize the newly forming spindle and guide attachment of MTs to chromosomal kinetochores.

MAKE CONNECTIONS 24.1

Because the spindle is largely composed of MTs, why would increased dynamic instability during prometaphase be advantageous? (Fig. 13-7)

Chromosome Alignment and Separation. After spindle microtubules attach to chromosomal kinetochores during early prometaphase, chromosomes move toward the central region of the spindle through a series of agitated, back-and-forth motions known as **congression**. Several mechanisms underlie congression and rely on microtubule motor proteins (see Chapter 14). First, kinetochore microtubules exert a pulling force. This force can be demonstrated experimentally by using glass microneedles to tear individual chromosomes away from the spindle. A chromosome that has been removed from the spindle remains motionless until new MTs attach to its kinetochore, at which time the chromosome is drawn back into the spindle. This force is mediated by the minus-end-directed motor *cytoplasmic dynein*, which moves chromosomes toward the pole that the microtubules are attached to.

A second kinetochore-dependent force balances this poleward force, tending to push chromosomes away if they approach either spindle pole. This pushing force was first identified by careful observation of monopolar spindles, which could be observed to move away from the spindle pole to which they were attached and back toward the metaphase plate. This force is mediated by the kinesin *CENP-E*.

A final force exerted on chromosomes does not involve kinetochores. This second pushing force involves polar microtubules, and so it is sometimes called the *polar ejection force*. It has been demonstrated by using a laser microbeam to break off one end of a chromosome. Once the broken chromosome fragment has been cut free from its associated centromere and kinetochore, the fragment tends to move away from the

Figure 24-6 Microtubule Polarity in the Mitotic Spindle. This diagram shows only a few representatives of the many microtubules making up a spindle. The orientation of the tubulin subunits constituting a microtubule (MT) make the two ends of the MT different. The minus end is at the initiating centrosome; the plus end points away from the centrosome. MTs lengthen by adding tubulin subunits and shrink by losing subunits. In general, lengthening is due to addition at the plus ends and shortening is due to loss at the minus ends, but subunits can also be removed from the plus ends. The red structures between the plus ends of the polar MTs shown here represent proteins that crosslink them.

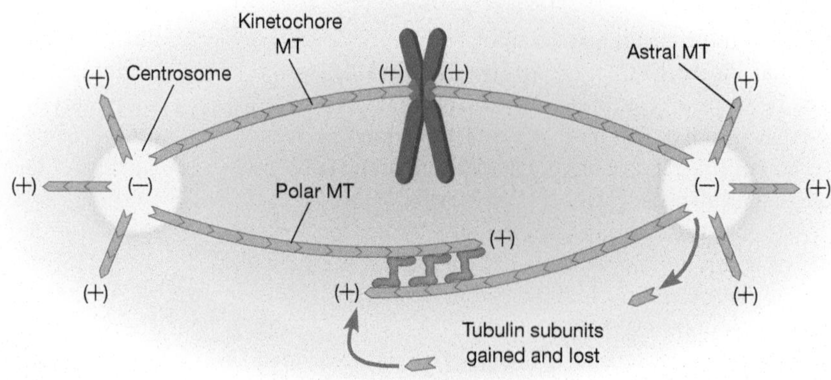

Most of these variations in generation time are based on differences in the length of G1, although S and G2 can also vary. Cells that divide slowly may spend days, months, or even years in the offshoot of G1 called G0, whereas cells that divide very rapidly have a short G1 phase or even eliminate G1 entirely. The embryonic cells of insects, amphibians, and several other nonmammalian animals are dramatic examples of cells that have very short cell cycles, with no G1 phase and a very short S phase. For example, during early embryonic development of the frog *Xenopus laevis* the cell cycle takes less than 30 minutes. The rapid rate of DNA synthesis needed to sustain such a quick cell cycle is achieved in part by increasing the total number of replicons (see Figure 17-7), thereby decreasing the amount of DNA that each replicon must synthesize. In addition, all replicons are activated at the same time, in contrast to the sequential activation observed in adult tissues. By increasing the number of replicons and activating them all simultaneously, S phase is completed in less than 3 minutes, at least 100 times faster than in adult tissues of the same organism.

Although we know from such examples that cell growth is not essential to the cell cycle, the two are generally linked so that cells can divide without getting progressively smaller. A protein kinase called *TOR (target of rapamycin)* plays a central role in the signaling network that controls cell size and coordinates it with cell cycle progression. This signaling network activates TOR in the presence of nutrients and growth factors, and the activated TOR then stimulates molecules that control the rate of protein synthesis. The resulting increase in protein production leads to an increase in cell mass. Some of the molecules activated by TOR also facilitate entry into S phase, making TOR an important regulator of both cell growth and cell cycle progression.

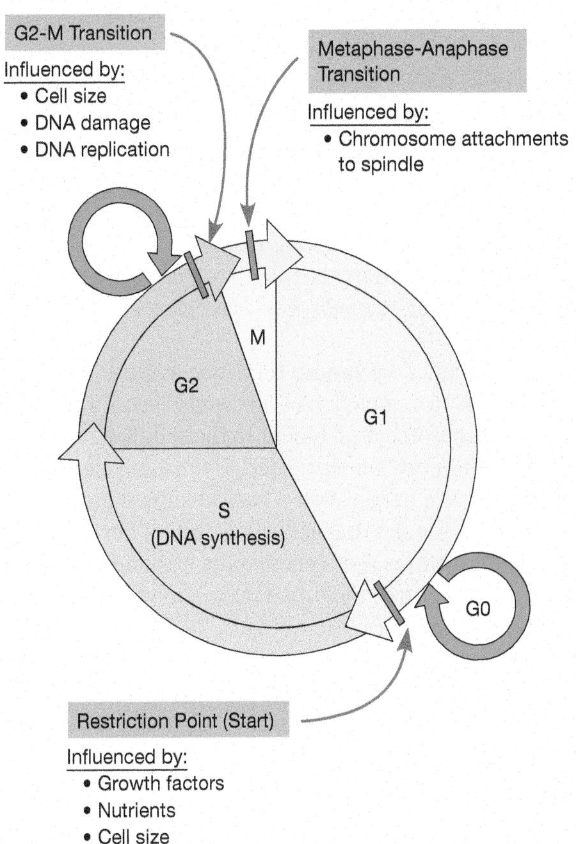

Figure 24-13 Key Transition Points in the Cell Cycle. The red bars mark three important transition points in the eukaryotic cell cycle where control mechanisms determine whether the cell will continue to proceed through the cycle. That determination is based on chemical signals reflecting both the cell's internal state and its external environment. The two circular, dark green arrows indicate locations in late G1 and late G2 where the cell can exit from the cycle and enter a nondividing state.

Cell Cycle Progression Is Controlled at Several Key Transition Points

The control system that regulates progression through the cell cycle must accomplish several tasks. First, it must ensure that the events associated with each phase of the cell cycle are carried out at the appropriate time. Second, it must make sure that each phase of the cycle has been properly completed before the next phase is initiated. Finally, it must be able to respond to external conditions that trigger cell proliferation (for example, the quantity of nutrients available or the presence of growth-signaling molecules).

These objectives are accomplished by a group of molecules that act at key transition points in the cell cycle (**Figure 24-13**). At each of these points, conditions within the cell determine whether the cell will proceed to the next stage of the cycle.

Restriction Point (G1-S Transition). The first such control point occurs during late G1. We have already seen that G1 is the phase that varies most among cell types, and mammalian cells that have stopped dividing are almost always arrested during G1. This suggests that progression from G1 into S is a critical control point in the cell cycle. In yeast, this control point is called **Start**; yeast cells must have sufficient nutrients and must reach a certain size before they can pass through Start. In animal cells, the comparable control point is called the **restriction point**. The ability to pass through the restriction point is influenced by the presence of extracellular *growth factors*, proteins that multicellular organisms use to stimulate or inhibit cell proliferation (see Chapter 23). Cells that have successfully passed through the restriction point are committed to S phase, whereas those that do not pass the restriction point enter G0 and reside there for variable periods of time, awaiting a signal that will allow them to reenter G1 and pass through the restriction point.

G2-M Transition. A second important transition point occurs at the G2-M boundary, when the commitment is made to enter mitosis. In certain cell types, the cell cycle can be indefinitely arrested at the end of G2 if cell division is not necessary; under such conditions, the cells enter a nondividing state analogous to G0. In general, arresting the cell cycle in late G1 (at the restriction point) is the more prevalent type of control in multicellular organisms, but in a few cases G2 arrest is more important.

Metaphase-Anaphase Transition. A third key transition point occurs during M phase at the junction between metaphase and anaphase, when the commitment is made to move the two sets of chromosomes into the newly forming daughter cells. Before cells pass through this transition point and begin anaphase, it is important that all chromosomes are properly attached to the spindle. If the two chromatids that make up each chromosome are not properly attached to opposite spindle poles, cell cycle progression is temporarily halted to allow spindle attachment to occur. Without such a mechanism, there would be no guarantee that each of the newly forming daughter cells would receive a complete set of chromosomes.

Cell behavior at the various transition points is influenced both by successful completion of preceding events in the cycle (such as chromosome attachment to the spindle) and by factors in the cell's environment (such as nutrients and growth factors). Cell cycle progression is mediated by a group of related control molecules that activate or inhibit one another in chains of interactions that can be quite elaborate. Let's now see how these control molecules were identified and what functions they perform.

Cell Fusion Experiments and Cell Cycle Mutants Identified Molecules That Control the Cell Cycle

Heterokaryons. The first hints concerning the identity of the molecules that drive progression through the cell cycle came from cell fusion experiments performed in the early 1970s. In these studies, two cultured mammalian cells in different phases of the cell cycle were fused to form a single cell with two nuclei—a *heterokaryon*. As **Figure 24-14a** indicates, if one of the original cells is in S phase and the other is in G1, the G1 nucleus in the heterokaryon quickly initiates DNA synthesis, even if it would not normally have reached S phase until many hours later. Such observations indicate that S phase cells contain molecules that trigger progression from G1 into S. The controlling molecules are not simply the enzymes involved in DNA replication because these enzymes can be present in high concentration in cells that do not enter S phase.

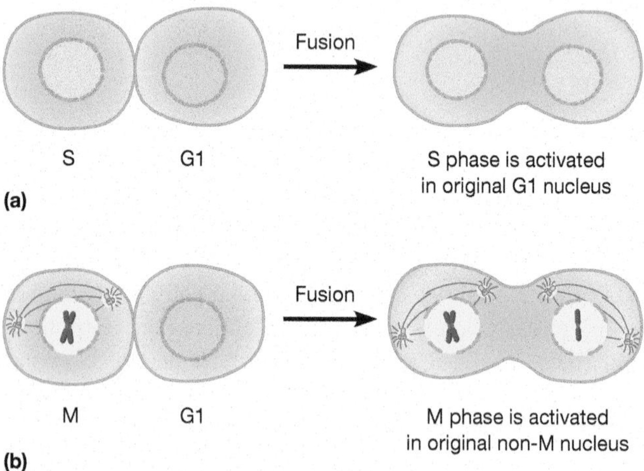

Figure 24-14 Evidence for the Role of Chemical Signals in Cell Cycle Regulation. Evidence was obtained from studies in which cells at two different points in the cell cycle were fused, forming a single cell with two nuclei. Cell fusion can be induced by adding certain viruses or polyethylene glycol or by applying a brief electrical pulse, which causes plasma membranes to destabilize momentarily. **(a)** When cells in S phase and G1 phase are fused, DNA synthesis begins in the original G1 nucleus, suggesting that a substance that activates S phase is present in the S phase cell. **(b)** When a cell in M phase is fused with one in any other phase, the latter cell immediately enters mitosis. If the cell was in G1, the condensed chromosomes that appear have not replicated and therefore are analogous to single chromatids.

Cell fusion experiments were also performed in which cells undergoing mitosis are fused with interphase cells in G1, S, or G2. After fusion, the nucleus of such interphase cells is immediately driven into the early stages of mitosis, including chromatin condensation, spindle formation, and fragmentation of the nuclear envelope. If the interphase cell was in G1, the condensed chromosomes remained unduplicated (Figure 24-14b). Taken together, these experiments suggested that molecules present in the cytoplasm are responsible for driving cells from G1 into S phase and from G2 into M phase.

Cell Cycle Mutants in Yeast. Progress in identifying cell cycle control molecules was greatly facilitated by genetic studies in yeasts. Because they are single-celled organisms that can be readily grown and studied under defined laboratory conditions, yeasts are particularly convenient for investigating the genes involved in cell cycle control.

Working with the budding yeast *Saccharomyces cerevisiae*, geneticist Leland Hartwell identified yeast mutants that are "stuck" at some point in the cell cycle. It might be expected that most such mutants would be difficult or impossible to study because their blocked cell cycle would prevent them from reproducing. But Hartwell overcame this potential obstacle with a powerful strategy—the use of *temperature-sensitive mutants*. Yeast cells carrying a temperature-sensitive mutation can be successfully grown at a lower ("permissive") temperature, even though their cell cycles would be blocked at higher temperatures. The structure and function of the protein encoded by the mutated gene in such cases are similar to those of the normal gene product at the permissive temperature, whereas the increased thermal energy at higher temperatures disrupts its active conformation (the molecular shape needed for function) more readily than that of the normal protein.

Using this approach, Hartwell and colleagues identified many genes involved in the cell cycle of *S. cerevisiae* and established when during the cycle their products operate. Predictably, some of these genes produce DNA replication proteins, but others seemed to function in cell cycle regulation. A breakthrough discovery was made by Paul Nurse, who carried out similar research with the fission yeast *Schizosaccharomyces pombe*. He identified a gene called *cdc2*, whose activity is needed for initiating mitosis—that is, for moving cells through the G2-M transition. (The acronym *cdc* stands for cell division cycle.) The *cdc2* gene was soon found to have counterparts in all eukaryotic cells. When the properties of the protein produced by the *cdc2* gene were examined, it was discovered to be a *protein kinase*—that is, an enzyme that catalyzes the transfer of a phosphate group from ATP to other target proteins. This discovery opened the door to unraveling the mysteries of the cell cycle.

The Cell Cycle Is Controlled by Cyclin-Dependent Kinases (Cdks)

The phosphorylation of target proteins by protein kinases, and their dephosphorylation by enzymes called *protein phosphatases*, is a common mechanism for regulating protein activity that turns out to be widely used in controlling the cell cycle. Progression through the cell cycle is driven by a series of protein kinases—including the protein kinase produced by the *cdc2* gene—that exhibit enzymatic activity only when they are bound to a special type of activator protein called a **cyclin**. Such protein kinases are therefore referred to as **cyclin-dependent kinases**, or simply **Cdks**. The eukaryotic cell cycle is controlled by several different Cdks that bind to different cyclins, thereby creating a variety of Cdk-cyclin complexes.

As originally shown by Tim Hunt using sea urchin embryos, cyclins are so named because their concentration in the cell oscillates up and down with the different phases of the cell cycle. Cyclins required for the G2-M transition and the early events of mitosis are called *mitotic cyclins*, and the Cdks to which they bind are known as *mitotic Cdks*. The activated Cdk-cyclin complex in this case is called **MPF (mitosis-promoting factor)**. MPF controls meiotic division as well (see Chapter 25).

Other Cdk-cyclin complexes are required for passage through other phases of the cell cycle. For example, the cyclins required for progression through the G1 restriction point (or Start) are called *G1 cyclins*, and the Cdks to which they bind are *G1 Cdks*. Yet another group of cyclins, called *S cyclins*, are required for events associated with DNA replication during S phase, and so on.

The similarities between the Cdks and cyclins in diverse eukaryotes are remarkable. In fact, in yeast cells with a defective or missing *cdc2* gene, the human gene encoding mitotic Cdk can substitute perfectly well, even though the last ancestor common to yeasts and humans probably lived about a billion years ago! The pioneering work of Hartwell, Nurse, and Hunt that led to our current understanding of the pivotal role of Cdks and cyclins was honored by a Nobel Prize in 2001.

Although our focus here is mainly on Cdks, other kinases act alongside Cdks to control the cell cycle. These include proteins known as *polo-like kinases* and a protein known as *Aurora B kinase*. Aurora B kinase is part of a complex of proteins that transiently associates with chromosomes and then with the mitotic spindle at the end of cell division. Because these proteins transiently "ride" along with chromosomes, they are called the *chromosomal passenger complex*. Their regulatory roles vary with different phases of the cell cycle. Together with Cdks, polo-like kinases, and other regulatory proteins, these proteins exercise tight control over cell division.

Cdk-Cyclin Complexes Are Tightly Regulated

If progression through critical points in the cell cycle is controlled by an assortment of different Cdks and cyclins interacting in various combinations, how is the activity of these protein complexes regulated? One level of control is exerted by the availability of cyclin molecules, which are required for activating the protein kinase activity of Cdks. A second type of regulation involves phosphorylation of Cdks.

Cyclin Availability. After it was established that MPF is a mitotic Cdk-cyclin that triggers the onset of mitosis in many cell types, the question arose as to how mitotic Cdk-cyclin is controlled so that it functions only at the proper time—that is, at the end of G2 (**Figure 24-15**). The answer does involve the availability of mitotic Cdk itself because its concentration remains relatively constant throughout the cell cycle. However, mitotic Cdk is active as a protein kinase only when it is bound to mitotic cyclin, and mitotic cyclin is not always present in adequate amounts. Instead, the concentration of mitotic cyclin gradually increases during G1, S, and G2; eventually, it reaches a critical threshold at the end of G2 that permits it to activate mitotic Cdk and thereby trigger the onset of mitosis (Figure 24-15a). Halfway through mitosis, mitotic cyclin protein is abruptly destroyed. The resulting decline in mitotic Cdk activity prevents another mitosis from occurring until the mitotic cyclin concentration builds up again during the next cell cycle.

As other cyclins were subsequently discovered, each was assigned a letter. The original mitotic cyclin is now known as *cyclin B*. Although the profile of their accumulation and

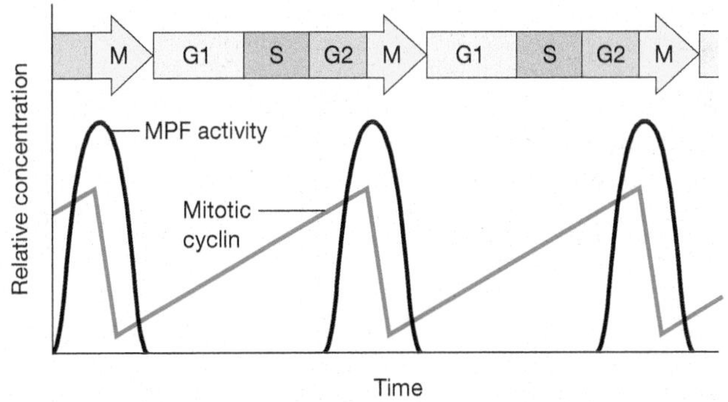

(a) Level of M phase cyclin (cyclin B) during the cell cycle

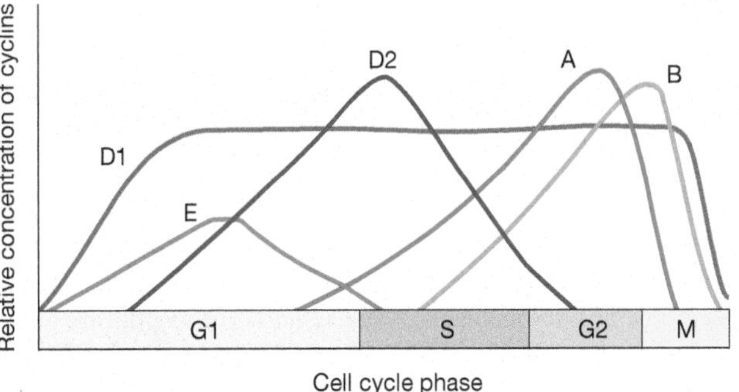

(b) Level of different cyclins during the cell cycle

Figure 24-15 Fluctuating Levels of Cyclins and MPF Activity During the Cell Cycle. (a) The cellular level of mitotic cyclin (cyclin B) rises during interphase (G1, S, and G2), then falls abruptly during M phase. The peaks of MPF activity (assayed by testing for the ability to stimulate mitosis) and cyclin concentration correspond, although the rise in MPF activity is not significant until a threshold concentration of cyclin is reached. Active MPF has been found to consist of a combination of mitotic cyclin and mitotic Cdk. The mitotic Cdk itself is present at a constant concentration (not shown on the graph) because the amount of mitotic Cdk increases at a rate corresponding to the overall growth of the cell. **(b)** The levels of various cyclins rise and fall during specific phases of the cell cycle. Various cyclins form complexes with various Cdks to regulate key steps in the cell cycle.

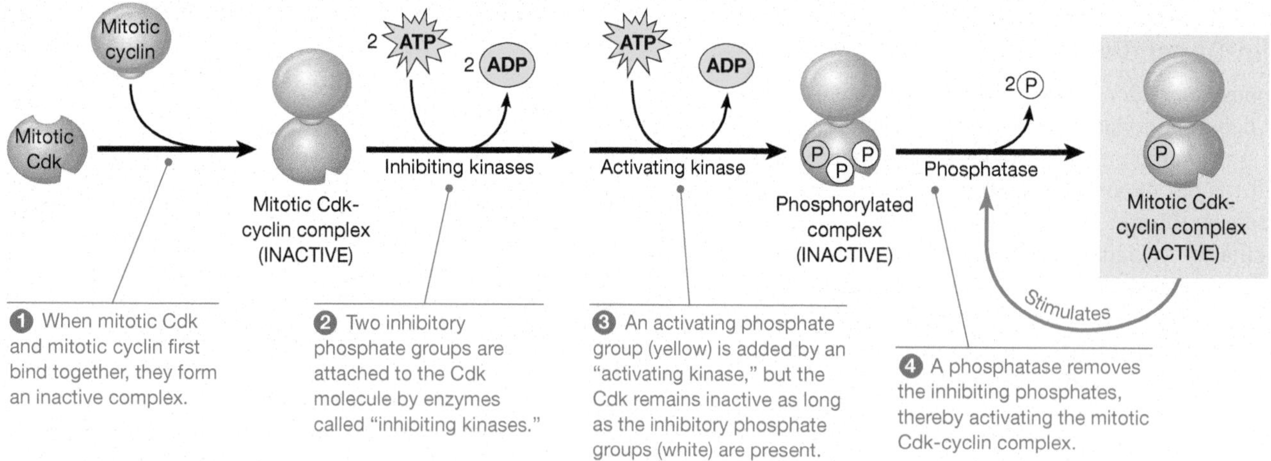
Mitotic cyclin
Mitotic Cdk
2 ATP
2 ADP
Inhibiting kinases
ATP
ADP
Activating kinase
2 P
Phosphatase
Mitotic Cdk-cyclin complex (INACTIVE)
Phosphorylated complex (INACTIVE)
Mitotic Cdk-cyclin complex (ACTIVE)
Stimulates
1 When mitotic Cdk and mitotic cyclin first bind together, they form an inactive complex.
2 Two inhibitory phosphate groups are attached to the Cdk molecule by enzymes called "inhibiting kinases."
3 An activating phosphate group (yellow) is added by an "activating kinase," but the Cdk remains inactive as long as the inhibitory phosphate groups (white) are present.
4 A phosphatase removes the inhibiting phosphates, thereby activating the mitotic Cdk-cyclin complex.

Figure 24-16 Regulation of Mitotic Cdk-Cyclin by Phosphorylation and Dephosphorylation. Activation of mitotic Cdk-cyclin involves the addition of inhibiting and activating phosphate groups, followed by removal of the inhibiting phosphate groups by a phosphatase. Once removal of the inhibiting phosphate groups has begun, a positive feedback loop is set up: the activated Cdk-cyclin complex generated by this reaction stimulates the phosphatase, thereby causing the activation process to proceed more rapidly.

degradation varies somewhat from mitotic cyclin, the other cyclins undergo cyclical changes in their levels as well (Figure 24-15b). The availability of these various cyclins to form complexes with various Cdks thereby regulates other key phases of the cell cycle.

Cdk Phosphorylation. In addition to requiring mitotic cyclin, the activation of mitotic Cdk involves phosphorylation and dephosphorylation of the Cdk molecule itself. As shown in **Figure 24-16**, ❶ the binding of mitotic cyclin to mitotic Cdk yields a Cdk-cyclin complex that is initially inactive. To trigger mitosis, the complex requires the addition of an activating phosphate group to a particular amino acid in the Cdk protein. Before this phosphate is added, however, ❷ *inhibiting* kinases phosphorylate the Cdk molecule at two other locations, causing the active site to be blocked. The activating phosphate group, highlighted with yellow after ❸, is then added by a specific *activating* kinase. The last step in the activation sequence is ❹ the removal of the inhibiting phosphates by a specific *phosphatase* enzyme. Once the phosphatase begins removing the inhibiting phosphates, a positive feedback loop is set up: the activated mitotic Cdk generated by this reaction stimulates the phosphatase, thereby causing the activation process to proceed more rapidly.

After mitotic Cdk-cyclin has been activated, its protein kinase activity triggers the onset of mitosis (**Figure 24-17**). We have already seen that the early events of mitosis include chromosome condensation, assembly of the mitotic spindle, and nuclear envelope breakdown. How are these changes triggered by mitotic Cdk-cyclin? In the case of nuclear envelope breakdown, the mitotic Cdk-cyclin phosphorylates (and stimulates other kinases to phosphorylate) the *lamin* proteins of the *nuclear lamina*, to which the inner nuclear membrane is attached (see Figure 16-29). Phosphorylation causes the lamins to depolymerize, resulting in breakdown of the nuclear

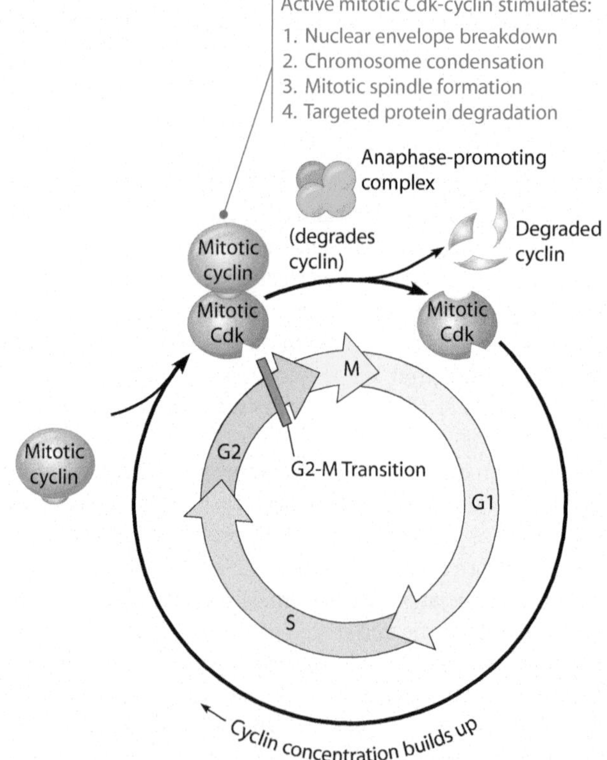
Active mitotic Cdk-cyclin stimulates:
1. Nuclear envelope breakdown
2. Chromosome condensation
3. Mitotic spindle formation
4. Targeted protein degradation
Anaphase-promoting complex
(degrades cyclin)
Degraded cyclin
Mitotic cyclin
Mitotic Cdk
Mitotic Cdk
M
G2
G2-M Transition
G1
S
Mitotic cyclin
Cyclin concentration builds up

Figure 24-17 The Mitotic Cdk Cycle. This diagram illustrates the activation and inactivation of the mitotic Cdk protein during the cell cycle. In G1, S, and G2, mitotic Cdk is made at a steady rate as the cell grows, while the mitotic cyclin concentration gradually increases. Mitotic Cdk and cyclin form an active complex whose protein kinase activity drives the cell cycle through the G2-M transition and into mitosis by stimulating the mitotic events listed. By activating a protein-degradation pathway that degrades cyclin, the mitotic Cdk-cyclin complex also brings about its own demise, allowing the completion of mitosis and entry into G1 of the next cell cycle.

lamina and destabilization of the nuclear envelope. The integrity of the nuclear envelope is further disrupted by phosphorylation of envelope-associated proteins, and the membranes of the envelope are soon torn apart.

Phosphorylation of other proteins by mitotic Cdk-cyclin is important for other mitotic events. Phosphorylation of the *condensin* complex is involved in condensing chromatin fibers into compact chromosomes. In addition, phosphorylation of microtubule-associated proteins by mitotic Cdk-cyclin is thought to facilitate assembly of the mitotic spindle. Finally, recall that cells have an elaborate set of DNA repair pathways (see Chapter 17). Key proteins that initiate DNA strand repair are phosphorylated by mitotic Cdk, preventing the assembly of DNA break repair complexes during mitosis. Why is this necessary? Apparently the ends of telomeres are not protected during mitosis; DNA repair suppression seems to be important for preventing telomeres at the ends of chromosomes from being joined end to end.

The Anaphase-Promoting Complex Allows Exit from Mitosis

Besides triggering the onset of mitosis, mitotic Cdk-cyclin also plays an important role later in mitosis when the decision is made to separate the sister chromatids during anaphase. Mitotic Cdk-cyclin exerts its influence on this event by phosphorylating and thereby contributing to the activation of the **anaphase-promoting complex**, or **APC/C** (for *anaphase-promoting complex/cyclosome*), a multiprotein complex that coordinates mitotic events by promoting the destruction of several key proteins at specific points during mitosis. The anaphase-promoting complex functions as a *ubiquitin ligase*, a type of enzyme that targets specific proteins for degradation by joining them to the small protein *ubiquitin* (see Chapter 20).

One crucial protein targeted for destruction by the anaphase-promoting complex is **securin**, an inhibitor of sister chromatid separation. As shown in **Figure 24-18**, sister chromatids are held together prior to anaphase by proteins called **cohesins**, which become bound to newly replicated chromosomal DNA in S phase following the movement of the replication forks. Securin maintains this sister chromatid attachment by inhibiting a protease called **separase**, which would otherwise degrade the cohesins. At the beginning of anaphase, however, the APC/C attaches ubiquitin to securin and thereby triggers its destruction, releasing separase from inhibition. The activated separase then cleaves cohesin, which frees sister chromatids to separate and begin their movements toward the spindle poles.

Besides initiating anaphase by causing cohesins to be destroyed, the APC/C induces events associated with the end of mitosis by targeting mitotic cyclin for destruction, thereby inactivating the mitotic Cdk/cyclin. Evidence suggests that many changes associated with the exit from mitosis—such as cytokinesis, chromosome decondensation, and reassembly of the nuclear envelope—depend on cyclin degradation and the associated reduction in Cdk activity. For example, it has been shown that introducing a nondegradable form of mitotic cyclin into cells inhibits cytokinesis, blocks nuclear envelope reassembly, and stops chromosomes from decondensing, thereby preventing completion of mitosis.

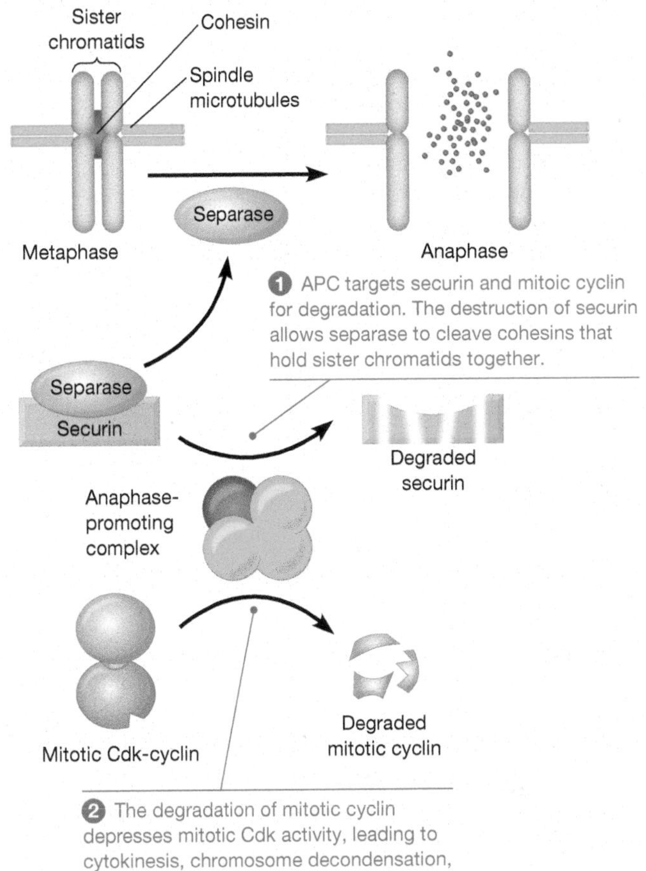

Figure 24-18 The Anaphase-Promoting Complex. The anaphase-promoting complex controls the final stages of mitosis by targeting selected proteins, including securin and mitotic cyclin, for destruction.

Checkpoint Pathways Monitor Key Steps in the Cell Cycle

It would obviously create problems if cells proceeded from one phase of the cell cycle to the next before the preceding phase had been properly completed. For example, if chromosomes were to start moving toward the spindle poles before they had all been properly attached to the spindle, the newly forming daughter cells might receive extra copies of some chromosomes and no copies of others, a situation known as **aneuploidy** (an = "not," eu = "good," and ploidy refers to chromosome number). Similarly, it would be potentially hazardous for a cell to begin mitosis before all of its chromosomal DNA had been replicated. To minimize the possibility of such errors, cells utilize a series of **checkpoint** mechanisms that monitor conditions within the cell and transiently halt the cell cycle if conditions are not suitable for continuing.

G1-S: G1 Cdk-Cyclin and the Rb Protein. As one example, let's briefly see how another type of Cdk-cyclin regulates entry into S phase. As mentioned earlier, the restriction point (Start in yeast) controls whether a cell will enter S phase, proceed through the rest of the cell cycle, and divide. Because passing through the restriction point is the main step that commits a cell to the cell division cycle, it is subject to tight control by a variety of factors such as cell size, the availability of nutrients, and the presence of growth factors that signal the need for cell proliferation.

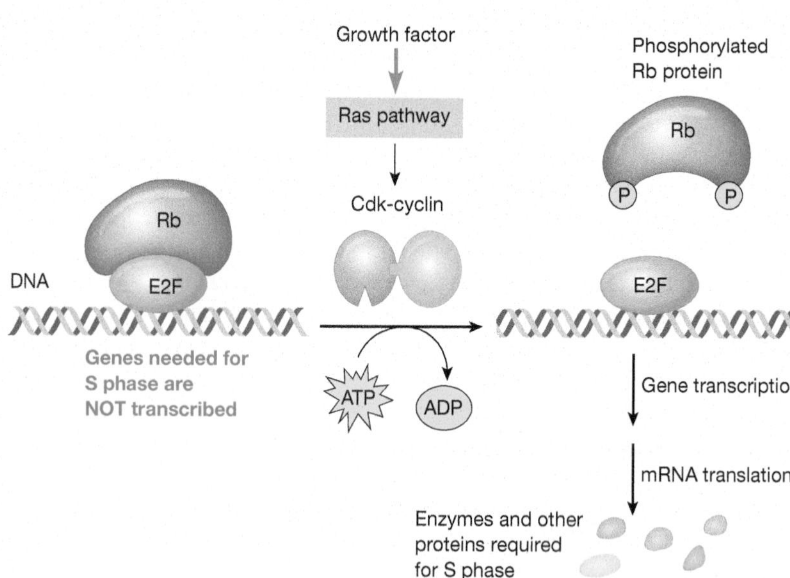

Figure 24-19 Role of the Rb Protein in Cell Cycle Control. In its dephosphorylated state, the Rb protein binds to the E2F transcription factor. This binding prevents E2F from activating the transcription of genes coding for proteins required for DNA replication, which are needed before the cell can pass through the restriction point into S phase. In cells stimulated by growth factors, the Ras pathway is activated (see Figure 24-25), which leads to the production and activation of a G1 Cdk-cyclin complex that phosphorylates the Rb protein. Phosphorylated Rb can no longer bind to E2F, thereby allowing E2F to activate gene transcription and trigger the onset of S phase. During the subsequent M phase (not shown), the Rb protein is dephosphorylated so that it can once again inhibit E2F.

Such signals exert their effects by activating G1 Cdk-cyclin, whose protein kinase activity triggers progression through the restriction point by phosphorylating several target proteins. A key target is the **Rb protein**. The molecular mechanism by which Rb exerts this control is summarized in **Figure 24-19**. Prior to being phosphorylated by G1 Cdk-cyclin, Rb binds to and inhibits the **E2F transcription factor**, a protein that would otherwise activate the transcription of genes encoding proteins required for initiating DNA replication. As long as Rb remains bound to E2F, E2F cannot activate transcription of these genes, thereby preventing the cell from entering into S phase. When cells are stimulated to divide by the addition of growth factors, however, eventually G1 Cdk-cyclins are activated. They in turn catalyze Rb phosphorylation. Phosphorylation of Rb abolishes its ability to bind to E2F, thereby freeing E2F to activate the transcription of genes whose products are needed for entry into S phase.

Because the Rb protein regulates such a key cell cycle event, it is not surprising to discover that defects in Rb can have disastrous consequences. For example, such defects can lead to both hereditary and environmentally induced forms of cancer (see Chapter 26).

S Phase: Replication Licensing. The G1-to-S transition is a key gatekeeper during the eukaryotic cell cycle. Once the cell enters S phase, however, another key regulatory step is needed. During the eukaryotic cell cycle it is crucial that nuclear DNA molecules undergo replication once, and only once, prior to cell division. To enforce this restriction, a process called **licensing** ensures that after DNA is replicated at any given replication origin during S phase, the DNA at that site does not become competent ("licensed") for a further round of DNA replication until the cell has first passed through mitosis. The license is provided by the binding of *MCM proteins* to origins of replication (**Figure 24-20**), an event that requires both ORC and helicase loaders (see Chapter 17). Once replication begins, the MCM proteins are displaced from the origins by the traveling replication fork. Thus the license to replicate is never associated with replicated DNA.

A critical player in this mechanism is a Cdk that is produced at the beginning of S phase, which functions both in activating DNA synthesis at licensed origins and in ensuring that these same origins cannot become licensed again. This Cdk blocks relicensing by catalyzing the phosphorylation, and thereby inhibiting the function, of proteins required for licensing, such as ORC and the helicase loaders. Multicellular eukaryotes contain another inhibitor of relicensing called *geminin*, a protein made during S phase that blocks the binding of MCM proteins to DNA. After the cell completes mitosis, geminin is degraded and Cdk activity falls, so the proteins required for DNA licensing can function again for the next cell cycle.

G2-M: DNA Replication Checkpoint. Another checkpoint mechanism, called the **DNA replication checkpoint**, monitors the state of DNA replication to help ensure that

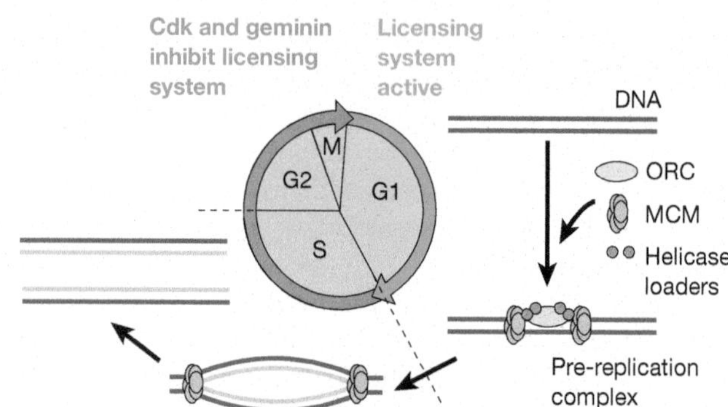

Figure 24-20 Licensing of DNA Replication During the Eukaryotic Cell Cycle. DNA is licensed for replication during G1 by the binding of MCM proteins to replication origins, an event that requires both ORC and helicase loaders. The licensing system is turned off at the end of G1 by the production of Cdk and/or geminin, whose activities block the functions of the proteins required for licensing (ORC, helicase loaders, and MCM). After the cell completes mitosis, geminin is degraded and Cdk activity falls, so the licensing system becomes active again for the next cell cycle.

DNA synthesis is completed before the cell exits from G2 and begins mitosis. The existence of this checkpoint has been demonstrated by treating cells with inhibitors that prevent DNA replication from completing. Under such conditions, the phosphatase that catalyzes the final dephosphorylation step involved in the activation of mitotic Cdk-cyclin (see Figure 24-16) is inhibited through events triggered by proteins associated with replicating DNA. The resulting lack of mitotic Cdk-cyclin activity halts the cell cycle at the end of G2 until all DNA replication is completed.

Spindle Assembly Checkpoint: Mad and Bub Proteins. The checkpoint pathway that prevents anaphase chromosome movements from beginning before the chromosomes are all attached to the spindle is called the **spindle assembly checkpoint** (abbreviated *SAC*). The SAC acts through kinetochores that remain *unattached* to spindle microtubules, which produce a "wait" signal that inhibits the anaphase-promoting complex (**Figure 24-21**). As long as the anaphase-promoting complex is inhibited, it cannot trigger destruction of the cohesins that hold sister chromatids together. The molecular basis of the wait signal involves members of the *Mad* and *Bub protein families* (Figure 24-21a). Mad and Bub proteins accumulate at unattached chromosomal kinetochores (Figure 24-21b), where they become part of a multiprotein complex called the **mitotic checkpoint complex (MCC).** The MCC inhibits the anaphase-promoting complex by binding to and blocking the action of one of its essential activators, the *Cdc20* protein, so that the APC/C cannot degrade its protein targets. After all the chromosomes have become attached to the spindle, Mad and Bub proteins no longer persist at kinetochores. A dynein motor complex is thought to strip several proteins, including Mad protein and CENP-F, from the kinetochore. Eventually, the inhibitory complex of which the Mad protein is a part turns over, thereby freeing the APC/C to initiate anaphase.

DNA Damage: p53. A final important checkpoint involves preventing cells with damaged DNA from proceeding through the cell cycle unless the DNA damage is first repaired. In this case, a multiple series of **DNA damage checkpoints** exist that monitor for DNA damage and halt the cell cycle at various points—including late G1, S, and late G2—by inhibiting different Cdk-cyclin complexes. A protein called **p53**, sometimes referred to as the "guardian of the genome," plays a central role in these checkpoint pathways. As shown in **Figure 24-22** on page 742, when cells encounter agents that cause extensive double-stranded breaks in DNA, the altered DNA triggers the activation of an enzyme called *ATM protein kinase* (for *ataxia telangiectasia mutated;* mutations in the ATM gene can cause defects in the cerebellum that lead to uncoordinated movement and prominent blood vessels that form in the whites of the eyes). ATM catalyzes the phosphorylation of kinases known as *checkpoint kinases,* which in turn phosphorylate p53 (and several other target proteins). Phosphorylation of p53 prevents it from interacting with *Mdm2,* a protein that would otherwise mark p53 for destruction by linking it to ubiquitin. ATM-catalyzed phosphorylation of p53 therefore protects it from degradation and leads to a buildup of p53 in the presence of damaged DNA. A protein related to ATM, called *ATR* (ATM-related), acts similarly but instead causes cell cycle arrest as a result of extensive single-, as opposed to double-stranded, breaks in DNA.

The accumulating p53 in turn activates two types of events: *cell cycle arrest* and *cell death.* Both responses are based on the ability of p53 to bind to DNA and act as a transcription

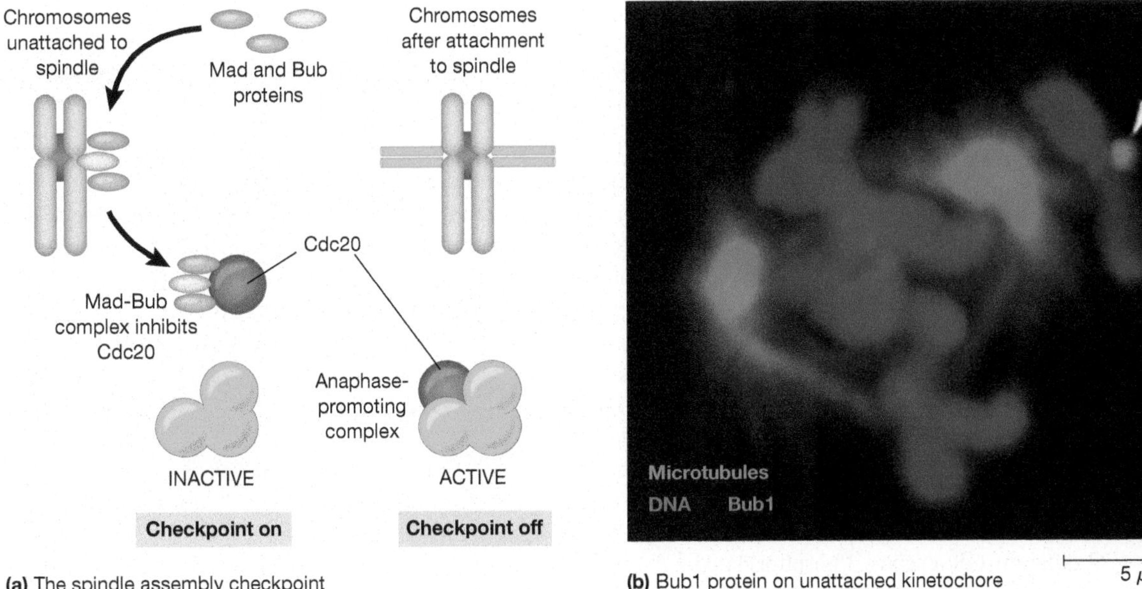

Figure 24-21 The Spindle Assembly Checkpoint. (a) A model for the spindle assembly checkpoint shows how chromosomes that are not attached to the spindle may organize Mad and Bub proteins into a complex that inhibits the anaphase-promoting complex, thereby delaying the onset of anaphase until all chromosomes are attached to the spindle. **(b)** When a cultured *Drosophila* cell is treated with a low dose of paclitaxel, a chromosome is slow to arrive at the metaphase plate (arrow); Bub1 is found at its kinetochore (red).

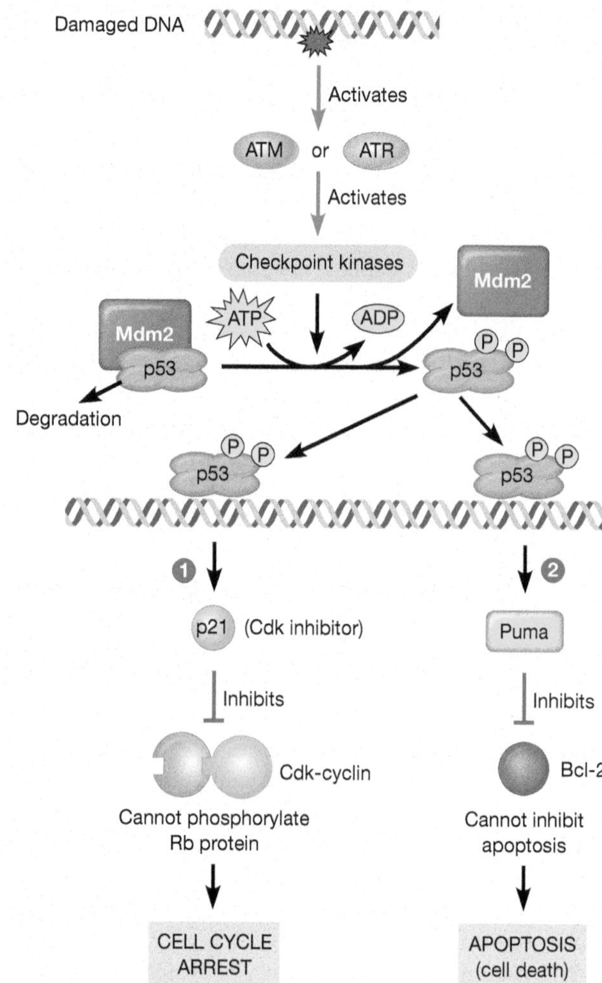

Figure 24-22 Role of the p53 Protein in Responding to DNA Damage. Damaged DNA activates the ATM or ATR protein kinase, leading to activation of checkpoint kinases, which leads to phosphorylation of the p53 protein. Phosphorylation stabilizes p53 by blocking its interaction with Mdm2, a protein that would otherwise mark p53 for degradation. (The degradation mechanism is not shown, but it involves Mdm2-catalyzed attachment of p53 to ubiquitin, which targets molecules to the cell's main protein destruction machine, the proteasome.) When the interaction between p53 and Mdm2 is blocked by p53 phosphorylation, the phosphorylated p53 protein accumulates and triggers two events. ❶ The p53 protein binds to DNA and activates transcription of the gene coding for the p21 protein, a Cdk inhibitor. The resulting inhibition of Cdk-cyclin prevents phosphorylation of the Rb protein, leading to cell cycle arrest at the restriction point. ❷ When the DNA damage cannot be repaired, p53 then activates genes encoding a group of proteins that trigger cell death by apoptosis. A key protein is Puma, which promotes apoptosis by binding to, and blocking the action of, the apoptosis inhibitor Bcl-2.

factor that stimulates the transcription of specific genes. One such gene encodes **p21**, a *Cdk inhibitor* protein that halts progression through the cell cycle at multiple points by inhibiting the activity of several different Cdk-cyclins. Phosphorylated p53 also stimulates the production of enzymes involved in DNA repair. But if the damage cannot be successfully repaired, p53 then activates a group of genes encoding proteins involved in triggering cell death by apoptosis (see later in this chapter). A key protein in this pathway, called **Puma (p53 upregulated modulator of apoptosis),** promotes apoptosis by binding to and inactivating a normally occurring inhibitor of apoptosis known as *Bcl-2*.

The ability of p53 to trigger cell cycle arrest and cell death allows it to function as a "molecular stoplight" that protects cells with damaged DNA from proliferating and passing the damage to daughter cells. (The importance of this role in cancer will be highlighted in Chapter 26.)

CONCEPT CHECK 24.3

Cyclin D is part of the G1 Cdk-cyclin complex. Cyclin D is often overexpressed in cancers. Why would overexpression of cyclin D tend to cause cancer?

24.4 Growth Factors and Cell Proliferation

Simple unicellular organisms, such as bacteria and yeast, often live under conditions in which the presence of sufficient nutrients in the external environment is the primary factor determining whether cells grow and divide. In multicellular organisms, the situation is usually reversed; cells are typically surrounded by nutrient-rich extracellular fluids, but the organism as a whole would be quickly destroyed if every cell were to continually grow and divide just because it had access to adequate nutrients. Cancer is a potentially lethal reminder of what happens when cell proliferation continues unabated without being coordinated with the needs of the organism as a whole.

To overcome this potential problem, multicellular organisms utilize extracellular signaling proteins called *growth factors* to control the rate of cell proliferation (see Table 23-3, page 704). Many growth factors are *mitogens*, which means that they stimulate cells to pass through the restriction point and subsequently divide by mitosis.

Stimulatory Growth Factors Activate the Ras Pathway

If mammalian cells are placed in a culture medium containing nutrients and vitamins but lacking growth factors, they normally become arrested in G1 despite the presence of adequate nutrients. Growth and division can be triggered by adding small amounts of blood serum, which contains several stimulatory *growth factors*. Among them is *platelet-derived growth factor (PDGF)*, a protein produced by blood platelets that stimulates the proliferation of connective tissue cells and smooth muscle cells. Another important growth factor, *epidermal growth factor (EGF)*, is widely distributed in many tissues and body fluids. EGF was initially isolated from the salivary glands of mice by Stanley Cohen, who received a Nobel Prize in 1987 for his pioneering work.

Growth factors such as PDGF and EGF act by binding to plasma membrane receptors located on the surface of target cells that exhibit tyrosine kinase activity (see Chapter 23). The binding of a growth factor to its receptor activates this tyrosine kinase activity, which in turn triggers a complex cascade of events that culminates in the cell passing through the restriction point and entering S phase. The *Ras pathway* plays a central role in these events (see Chapter 23), as shown by studies involving cells that have stopped dividing because growth factor is not present. When mutant, hyperactive forms of the Ras protein are injected into such cells, the cells enter S phase and begin dividing, even in the absence of growth factor. Conversely, injecting cells with antibodies that inactivate the Ras protein prevents cells from entering S phase and dividing in response to growth factor stimulation.

How does the Ras pathway affect the cell cycle? As shown in **Figure 24-23**, the process involves several steps. ❶ First, binding of a growth factor to its receptor at the plasma membrane leads to activation of Ras. ❷ Next, activated Ras leads to phosphorylation and activation of a protein kinase called *Raf*, which sets in motion a cascade of phosphorylation events. Activated Raf phosphorylates serine and threonine residues in a protein kinase called *MEK*, which in turn phosphorylates threonine and tyrosine residues in a group of protein kinases called *MAP kinases (mitogen-activated protein kinases; MAPKs)*. ❸ The activated MAPKs enter the nucleus and phosphorylate several regulatory proteins that activate the transcription of specific genes. Among these proteins are *Jun* (a component of the AP-1 transcription factor) and members of the *Ets family* of transcription factors. These activated transcription factors turn on the transcription of "early genes" that encode other transcription factors, including Jun, Fos, and Myc, which then activate the transcription of a family of "delayed genes." One of these latter genes encodes the E2F transcription factor, whose role in controlling entry into S phase was described earlier in the chapter. ❹ Also included in the delayed genes are several genes encoding Cdks or cyclins, whose production leads to the formation of Cdk-cyclin complexes that phosphorylate Rb and hence trigger passage from G1 into S phase.

Thus, in summary, the Ras pathway is a multistep signaling cascade in which the binding of a growth factor to a receptor on the cell surface ultimately causes the cell to pass through the restriction point and into S phase, thereby starting the cell on the road to cell division. The importance of this pathway for the control of cell proliferation has been highlighted by the discovery that mutations affecting the Ras pathway appear frequently in cancer cells. For example, mutant Ras proteins that provide an ongoing stimulus for the cell to proliferate, independent of growth factor stimulation, are commonly encountered in pancreatic, colon, lung, and bladder cancers, and they occur in about 25–30% of all human cancers overall. (In Chapter 26, the role played by such Ras pathway mutations in the development of cancer will be described in detail.)

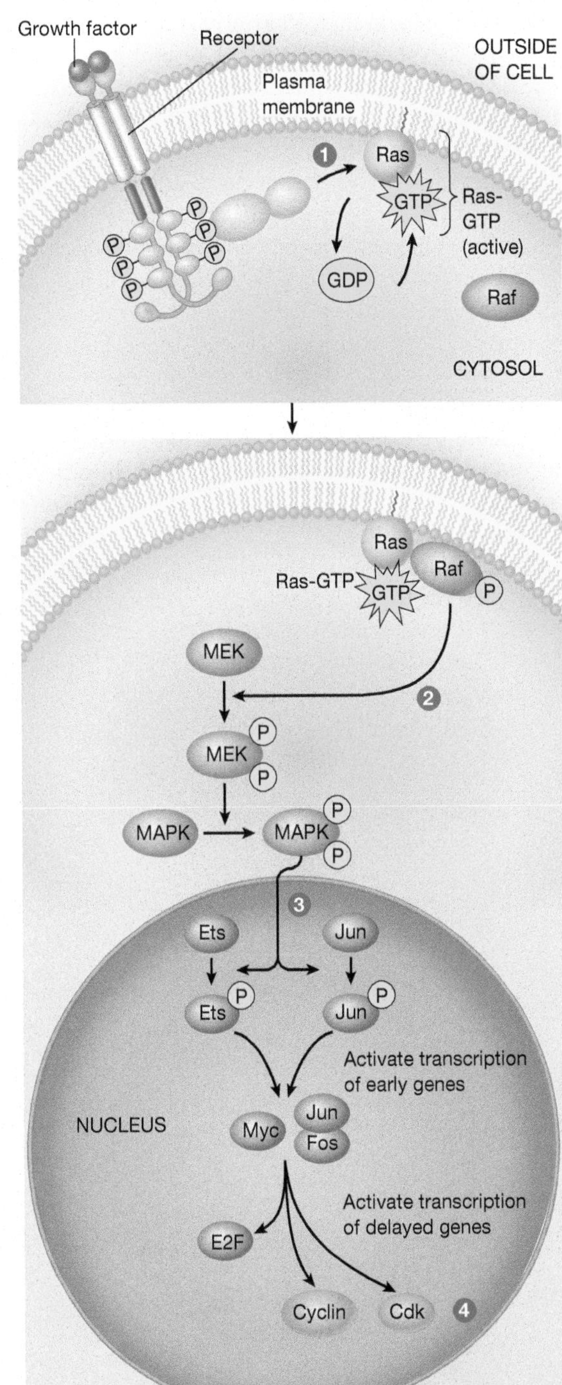

Figure 24-23 Regulation of the Cell Cycle via the Ras Pathway. Regulation of the cell cycle by Ras consists of four steps: binding of a growth factor to its receptor, leading to activation of Ras protein; activation of a cascade of cytoplasmic protein kinases (Raf, MEK, and MAP kinase, or MAPK); activation or production of nuclear transcription factors (Ets, Jun, Fos, Myc, E2F); and synthesis of cyclin and Cdk molecules. The resulting Cdk-cyclin complexes catalyze the phosphorylation of Rb and hence trigger passage from G1 into S phase.

MAKE CONNECTIONS 24.2

Many cancers have a mutation in Ras GAP, a Ras GTP-activating protein. Based on what you know, why might you expect this? (Fig. 23-18)

Stimulatory Growth Factors Can Also Activate the PI 3-Kinase–Akt Pathway

When a growth factor binds to a receptor that triggers the Ras pathway, the activated receptor may simultaneously trigger other pathways as well. One example (examined in Chapter 23 in the context of insulin signaling; Figure 23-27) is the *PI 3-kinase–Akt pathway*. This pathway begins with receptor-induced activation of phosphatidylinositol 3-kinase (abbreviated as *PI 3-kinase* or *PI3K*), which catalyzes formation of PIP_3 (phosphatidylinositol-3,4,5-trisphosphate), ultimately leading to phosphorylation and activation of *Akt*. Through its ability to catalyze the phosphorylation of several key target proteins, Akt suppresses apoptosis (see the next section of this chapter) and inhibits cell cycle arrest (**Figure 24-24**). One way in which the latter happens is through activation of a monomeric G protein called Rheb. Activation of Rheb leads to activation of TOR, a key regulator of cell growth mentioned earlier in this chapter. The net effect of the PI 3-kinase–Akt signaling pathway is therefore to promote cell survival and proliferation.

As in the case of the Ras pathway, mutations that disrupt the normal behavior of the PI 3-kinase–Akt pathway are associated with many cancers. In some cases, such mutations cause excessive activity of the Akt protein that leads to increased cell proliferation and survival. In other cases, hyperactivity of the PI 3-kinase–Akt pathway is caused by mutations in inhibitor proteins. One such protein is *PTEN*, a phosphatase that removes a phosphate group from PIP_3 and thereby prevents the activation of Akt. When mutations disrupt PTEN, the cell cannot degrade PIP_3 efficiently, and its concentration rises. The accumulating PIP_3 activates Akt in an uncontrolled fashion, even in the absence of growth factors. Mutations that reduce PTEN activity are found in up to 50% of prostate cancers, 35% of uterine cancers, and varying extents in ovarian, breast, liver, lung, kidney, thyroid, and lymphoid cancers.

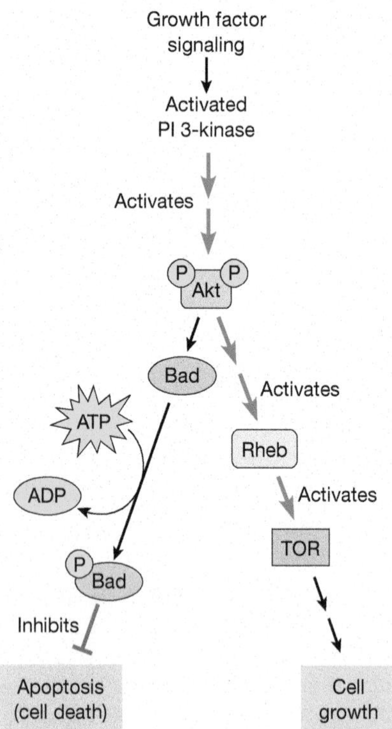

Figure 24-24 The PI 3-Kinase–Akt Signaling Pathway. Growth factors that bind to receptor tyrosine kinases activate several pathways in addition to the Ras pathway illustrated in Figure 24-23. The pathway shown here leads to activation of the protein kinase Akt. Akt suppresses apoptosis, in part by phosphorylating and inactivating a protein called Bad that normally promotes apoptosis. Akt also inhibits cell cycle arrest by leading to activation of TOR. Thus, the net effect of PI 3-kinase–Akt signaling is to promote cell survival and proliferation. The PI 3-kinase–Akt pathway is inhibited by PTEN (see Figure 23-27).

Inhibitory Growth Factors Act Through Cdk Inhibitors

Although we usually think of growth factors as being growth-stimulating molecules, the function of some growth factors is actually to *inhibit* cell proliferation. One example is **transforming growth factor β (TGFβ)**, a protein that can exhibit either growth-stimulating or growth-inhibiting properties depending on the target cell type. When acting as a growth inhibitor, the binding of TGFβ; to its cell surface receptor triggers a series of events in which the receptor catalyzes the phosphorylation of *Smad* proteins that move into the nucleus and regulate gene expression (see Figure 23-22). Once inside the nucleus, Smads activate the expression of genes encoding proteins that inhibit cell proliferation. Two key genes encode proteins called *p15* and *p21*, which are **Cdk inhibitors** that suppress the activity of Cdk-cyclin complexes and thereby block progression through the cell cycle. (The p21 protein was mentioned earlier in the chapter when we discussed the mechanism of p53-mediated cell cycle arrest.)

Growth-inhibiting signals that act on cells by triggering the production of Cdk inhibitors help protect normal tissues from excessive cell divisions that might otherwise produce more cells than are needed. (In Chapter 26, we will see how defects in such pathways can lead to uncontrolled cell proliferation and the development of cancer.)

Putting It All Together: The Cell Cycle Regulation Machine

Figure 24-25 is a simplified summary of the main features of the molecular "machine" that regulates the eukaryotic cell cycle. The operation of this machine can be described in terms of two fundamental, interacting mechanisms. One mechanism is an autonomous clock that cycles over and over again. The molecular basis of this clock is the synthesis and degradation of cyclins, alternating in a rhythmic fashion. These cyclins in turn bind to Cdk molecules, creating various Cdk-cyclin complexes that trigger the passage of cells through the main cell cycle transition points. The second mechanism adjusts the clock as needed by providing feedback from the cell's internal and external environments to influence the activity of Cdks and cyclins. Many of these additional proteins are protein kinases or phosphatases. It is this part of the cell cycle machine that relays information about the state of the cell's metabolism—including DNA damage and replication—and about conditions outside the cell, such as growth factors, thereby influencing whether or not the cell should commit to the process of cell division.

CONCEPT CHECK 24.4

Both EGF (a mitogen) and TGF β (when acting as an inhibitor) affect Cdk-cyclin complex function. Compare and contrast the ways in which EGF and TGF β operate. What would happen if a cell were exposed to both at the same time?

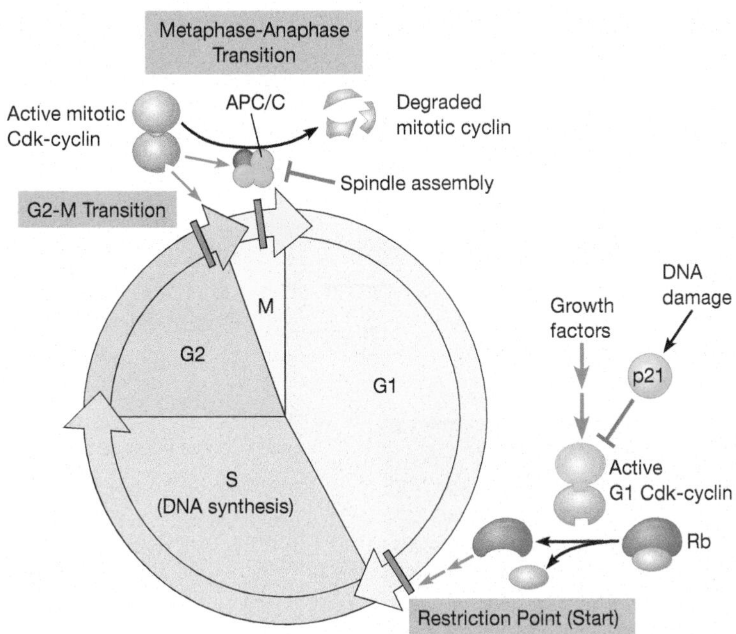

Figure 24-25 Summary of Cell Cycle Regulation by Checkpoints. Passage through the three main transition points in the cell cycle is triggered by protein complexes made of cyclin and Cdk, whose phosphorylation of other proteins induces progression through the cycle. For simplicity, other Cdk-cyclin complexes are not shown. G1 Cdk-cyclin acts at the restriction point by catalyzing phosphorylation of the Rb protein. Mitotic Cdk-cyclin acts at the G2-M boundary by catalyzing the phosphorylation of proteins involved in chromosome condensation, nuclear envelope breakdown, and spindle assembly. The same mitotic Cdk-cyclin also influences the metaphase-anaphase transition by catalyzing the phosphorylation of the anaphase-promoting complex (APC/C), which in turn triggers chromosome separation and the breakdown of mitotic cyclin. Checkpoint pathways that monitor the cell for DNA damage, DNA replication, and chromosome attachment to the spindle can send signals that halt the cell cycle at one or more of these key transition points.

24.5 Apoptosis

As you saw in previous sections of this chapter, organisms tightly regulate when their cells enter mitosis. When a cell divides is often regulated by growth factors, which can stimulate or inhibit division, depending on the circumstances. At other times, however, organisms need to regulate whether cells stay alive, such as damaged or diseased cells. This presents a challenge, however. Dismantling cells that are destined for death must occur in such a way that the internal contents of the dead cell—which include digestive enzymes in organelles such as lysosomes—do not wreak havoc on other cells around them. Multicellular organisms accomplish this feat through an important kind of programmed cell death called **apoptosis**. Apoptosis is a key event in many biological processes. In embryos, apoptosis removes the webbing between the digits (fingers and toes) and mediates the "pruning" of neurons that occurs in human infants during the first few months of life as connections mature within the developing brain. In adult humans, apoptosis occurs continually; for example, when cells become infected by pathogens or when white blood cells reach the end of their life span, they are eliminated through apoptosis. When cells that should die via apoptosis do not, the consequences can be dire. Mutations in some of the proteins that participate in apoptosis can lead to cancer. For example, melanoma frequently results from a mutation in Apaf-1, a protein we discuss later in this section.

Apoptosis is very different from another type of cell death, known as *necrosis*, which sometimes follows massive tissue injury. Whereas necrosis involves the swelling and rupture of the injured cells, apoptosis involves an orderly dismantling of the internal contents of the cell (**Figure 24-26** on page 746). ❶ During the early phases of apoptosis, the cell's DNA segregates near the periphery of the nucleus, and the volume of the cytoplasm decreases. ❷ Next, the cell begins to produce small, bubble-like, cytoplasmic extensions ("blebs"), and the nucleus and organelles begin to fragment. The cell's DNA is cleaved by an apoptosis-specific DNA endonuclease, or *DNase* (an enzyme that digests DNA), at regular intervals along the DNA. As a result, the DNA fragments, which are multiples of 200 base pairs in length, form a diagnostic "ladder" of fragments (see Problem 24-8). ❸ Eventually, the cell is dismantled into small pieces called *apoptotic bodies.* During apoptosis, inactivation of a phospholipid translocator, or *flippase* (see Chapter 7), results in accumulation of phosphatidylserine in the outer leaflet of the plasma membrane. The phosphatidylserine serves as an "eat me" signal for the remnants of the affected cell to be engulfed by other nearby cells (typically macrophages) via phagocytosis (see Chapter 12). The macrophages act as scavengers to remove the resulting cellular debris.

That cells have a "death program" was first conclusively demonstrated in the nematode *Caenorhabditis elegans,* in which key genes that control apoptosis were first identified. *C. elegans* is uniquely suited to studying cell death. Its life cycle is very short, it is optically transparent, and its embryos are so remarkably consistent in their development that the sequence of cell divisions that results in each of the 1090 cells produced during development can be traced back to the single-celled fertilized egg! This feat was achieved largely through the work of John Sulston at the Medical Research Council in Cambridge, England. In his analysis, Sulston showed that 131 cells undergo precisely timed apoptosis during normal embryonic development in *C. elegans.* Largely through the work of Robert Horvitz and colleagues at the Massachusetts Institute of Technology, mutants defective in various aspects of cell death, called *ced* mutants (for cell death abnormal), were identified. For example, Horvitz and colleagues identified several mutations that block phagocytosis of dead cells so that their corpses persist, making them easy to see in the light microscope (**Figure 24-27** on page 746). One *ced* gene, *ced-10,* encodes a member of the Rac family of proteins (see Chapter 13), which is required for phagocytosis of dead cells. *ced-3* encodes a member of the caspase family of proteins, and *ced-9* encodes the *C. elegans* version of Bcl-2, which plays a key role in regulating

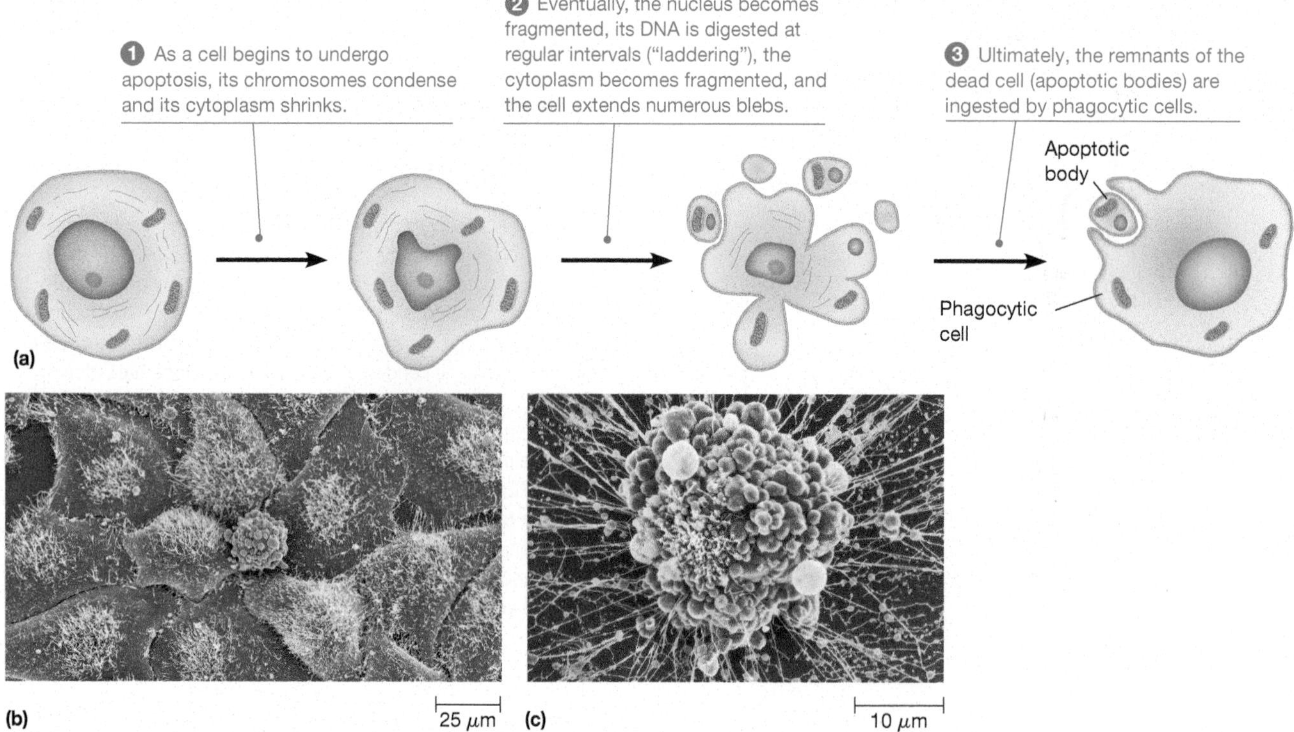
1 As a cell begins to undergo apoptosis, its chromosomes condense and its cytoplasm shrinks.
2 Eventually, the nucleus becomes fragmented, its DNA is digested at regular intervals ("laddering"), the cytoplasm becomes fragmented, and the cell extends numerous blebs.
3 Ultimately, the remnants of the dead cell (apoptotic bodies) are ingested by phagocytic cells.
Apoptotic body
Phagocytic cell
(a)
(b)
25 μm
(c)
10 μm

Figure 24-26 Major Steps in Apoptosis. (a) Cells in the process of apoptosis undergo a series of characteristic changes. Ultimately, the remnants of the dead cell (apoptotic bodies) are ingested by phagocytic cells. **(b–c)** SEMs of epithelial cells undergoing apoptosis. **(b)** Epithelial cells in contact with one another in culture form flat sheets. The cell in the center is undergoing apoptosis and has rounded up. **(c)** A closeup of a single dying cell with many apoptotic bodies.

the leakage of molecules from mitochondria that can trigger apoptosis (see the next section). For this work—and his work on cell signaling in *C. elegans*—Horvitz, along with Sulston and geneticist Sydney Brenner, shared a Nobel Prize in 2002.

Subsequent research showed that many other organisms, including mammals, use similar proteins during apoptosis. A key event in apoptosis is the activation of a series of enzymes called **caspases.** (Caspases get their name because they contain a *c*ysteine at their active site, and they cleave proteins at sites that contain an *asp*artic acid residue followed by four amino acids that are specific to each caspase.) Caspases are produced as inactive precursors known as **procaspases**, which are subsequently cleaved to create active enzymes, often by other caspases, in a proteolytic cascade. Once they are activated, caspases cleave other proteins. The apoptosis-specific DNase is a good example; it is bound to an inhibitory protein that is cleaved by a caspase.

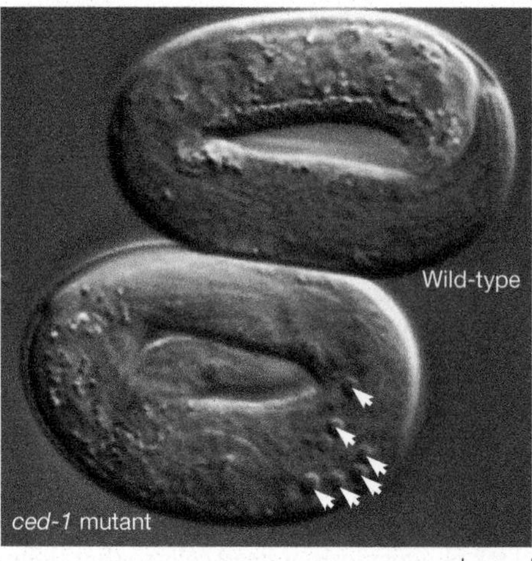
Wild-type
ced-1 mutant
10 μm

Figure 24-27 Cell Death in *C. elegans*. Wild-type (top) and *ced-1* mutant (bottom) embryos (DIC microscopy). Although cell deaths can be seen in both embryos as small "buttons," these cell corpses accumulate only in the *ced-1* embryo because phagocytosis of dead cells fails (arrows).

Apoptosis Is Triggered by Death Signals or Withdrawal of Survival Factors

There are two main routes by which cells can activate caspases and enter the apoptotic pathway. In some cases, activation of caspases occurs directly. Such activation is triggered when cells receive *cell death signals.* Two well-known death signals are *tumor necrosis factor* and *CD95/Fas.* Here, we will focus on CD95, a protein on the surface of infected cells in the human body. When cells are infected by certain viruses, a population of *cytotoxic T lymphocytes* is activated and induces the infected cells to initiate apoptosis. Lymphocytes have a protein on their surfaces that binds to CD95, causing the CD95 within the infected cell to aggregate (**Figure 24-28,** 1). CD95 aggregation results in the attachment of adaptor proteins to the clustered CD95, which in turn recruits a procaspase (*procaspase-8*) to sites of receptor clustering. When the procaspase is activated, 2 it acts as an initiator of the caspase cascade. A key action of such *initiator caspases* is 3 the activation of an *executioner*

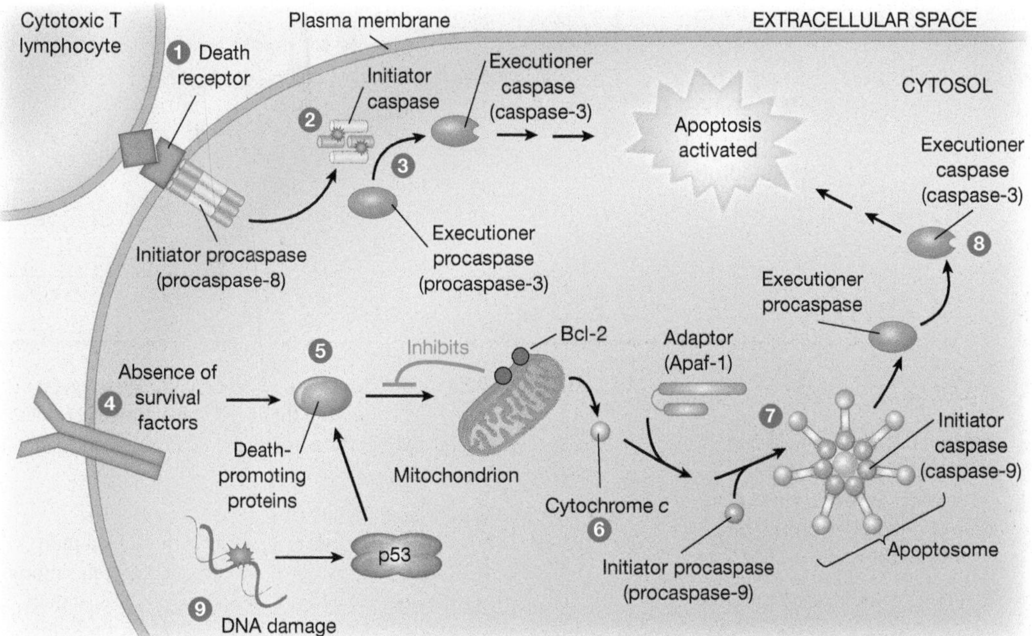

Figure 24-28 Induction of Apoptosis by Cell Death Signals or by Withdrawal of Survival Factors. Cell death signals, such as ligands on the surface of a cytotoxic T lymphocyte, can lead to apoptosis. ❶ Ligand binds to a "death receptor" on the surface of a target cell. Binding causes clustering of receptors and recruitment of adaptor proteins in the target cell, resulting in clustering of initiator procaspase (procaspase-8) protein. ❷ Initiator caspases then become activated. ❸ The initiator caspases in turn activate the executioner caspase, caspase-3, a key initiator of apoptosis. ❹ When survival factors are no longer present, ❺ death-promoting (pro-apoptotic) proteins accumulate, counterbalancing anti-apoptotic proteins (such as Bcl-2) at the mitochondrial outer membrane, ❻ causing release of cytochrome c. ❼ Cytochrome *c* forms a complex with other proteins, resulting in activation of an initiator caspase (caspase-9). ❽ The initiator caspase in turn activates the executioner caspase, caspase-3, triggering apoptosis. ❾ DNA damage can also lead to apoptosis through the activity of the p53 protein.

caspase, known as *caspase-3*. Active caspase-3 is important for activating many steps in apoptosis.

In other cases, apoptosis is triggered indirectly. One of the best-studied cases of this second type of apoptosis involves **survival factors**. ❹ When such factors are withdrawn, a cell may enter apoptosis. Surprisingly, a key site of action of this second pathway is the mitochondrion. The connection between mitochondria and cell death may be surprising, but it is clear that, in addition to their role in energy production, mitochondria are important in apoptosis. If withdrawal of survival factors is the sentence of execution, then the executioners are mitochondria.

How do mitochondria hasten cell death? In a healthy cell that is not committed to apoptosis, there are several *anti-apoptotic* proteins in the outer mitochondrial membrane that prevent apoptosis, but only as long as a cell continues to be exposed to survival factors. These proteins are structurally related to a protein known as **Bcl-2**, the best understood of these anti-apoptotic proteins. Bcl-2 and other anti-apoptotic proteins exert their effects by counteracting other proteins that are themselves structurally similar to Bcl-2. These proteins, however, *promote* apoptosis, so they are collectively referred to as *pro-apoptotic* proteins. Thus, ❺ pro- and anti-apoptotic proteins, influenced by cell signals, wage an ongoing battle; when the balance shifts toward pro-apoptotic proteins, a cell is more likely to undergo apoptosis. For example, stimulation of the Akt pathway, which we discussed earlier in this chapter, can lead to phosphorylation and inactivation of a pro-apoptotic protein called *Bad* (for Bcl-2-associated death promoter; see Figure 24-24).

Surprisingly, mitochondria trigger apoptosis by ❻ releasing **cytochrome *c*** into the cytosol. Eventually, the accumulation of pro-apoptotic protein at the surface of the mitochondrion leads to the formation of channels in the outer mitochondrial membrane, allowing cytochrome *c* to escape into the cytosol. Although cytochrome *c* is normally involved in electron transport (see Chapter 10), it is important in triggering apoptosis in at least two ways. First, cytochrome *c* stimulates calcium release from adjacent mitochondria and from the endoplasmic reticulum, where it binds inositol-1-4-5-trisphosphate (IP_3) receptors. Second, it can activate an initiator procaspase associated with mitochondria, known as *procaspase-9*. It does this by ❼ recruiting a cytosolic adaptor protein (known as *Apaf-1*) that assembles procaspase-9 into a complex sometimes called an *apoptosome;* the apoptosome promotes the production of active caspase-9. Like other initiator caspases, ❽ caspase-9 activates the executioner caspase, caspase-3. Thus, in the end, both cell death mechanisms lead to the activation of a common caspase that sets apoptosis in motion.

There is another situation that can trigger the mitochondrial pathway to apoptosis. When a cell suffers so much damage that it is unable to repair itself, it may trigger its own demise. In particular, ⑨ when a cell's DNA is damaged (for example, by radiation or ultraviolet light), it can enter apoptosis via the activity of *p53*. As we saw earlier in this chapter, p53 acts through the protein Puma, which binds to and inhibits Bcl-2 (see Figure 24-22, ②). In the end, just like withdrawal of survival factors, the p53 pathway activates pro-apoptotic proteins to trigger apoptosis.

CONCEPT CHECK 24.5

A "knockout" mouse has been produced using genetic techniques (described in Chapter 21) that is unable to make caspase-9. What sorts of defects do you predict might be present in the brains of knockout embryos?

Summary of Key Points

Mastering™ Biology For activities, animations, and review quizzes, go to the study area at www.masteringbiology.com.

24.1 Overview of the Cell Cycle

- The eukaryotic cell cycle is divided into G1, S, G2, and M phases. Chromosomal DNA replication takes place during S phase, whereas cell division (mitosis and cytokinesis) occurs in M phase. Interphase (G1, S, and G2) is a time of cell growth and metabolism that typically occupies about 95% of the cycle.
- The length of the cell cycle varies greatly, ranging from cells that divide rapidly and continuously to cells that do not divide at all.

24.2 Nuclear and Cell Division

- Mitosis is subdivided into prophase, prometaphase, metaphase, anaphase, and telophase. During prophase, replicated chromosomes condense into paired sister chromatids, and centrosomes initiate assembly of the mitotic spindle. In prometaphase, the nuclear envelope breaks down, and chromosomes then attach to spindle microtubules and move to the spindle equator, where they line up at metaphase. At anaphase, sister chromatids separate, and the resulting daughter chromosomes move toward opposite spindle poles. During telophase, the chromosomes decondense, and a nuclear envelope reassembles around each daughter nucleus.
- Chromosome movements are driven by three groups of motor proteins: (1) Motor proteins located at the kinetochores and spindle poles move chromosomes toward the spindle poles, accompanied by disassembly of the microtubules at their plus and minus ends. (2) Motor proteins that crosslink the polar microtubules move overlapping microtubules in opposite directions, thereby pushing the spindle poles apart. (3) Motor proteins move astral microtubules toward the plasma membrane, thereby pulling the spindle poles apart.
- Cytokinesis usually begins before mitosis is complete. In animal cells, an actomyosin filament network forms a cleavage furrow that constricts the cell at the midline and separates the cytoplasm into two daughter cells. In plant cells, a cell wall forms through the middle of the dividing cell.
- Cell division in bacteria and eukaryotic organelles such as chloroplasts involves the microtubule homology FtsZ and a complex of protein called the divisome.

24.3 Regulation of the Cell Cycle

- Progression through the eukaryotic cell cycle is regulated by various Cdk-cyclin complexes.
- At the G1-S boundary (restriction point, or Start in yeast), a Cdk-cyclin complex catalyzes the phosphorylation of the Rb protein to trigger passage into S phase.
- At the G2-M boundary, another Cdk-cyclin complex triggers entry into mitosis by catalyzing the phosphorylation of proteins that promote nuclear envelope breakdown, chromosome condensation, and spindle formation.
- At the metaphase-anaphase boundary, activation of the anaphase-promoting complex triggers a protein degradation pathway that initiates chromatid separation and targets mitotic cyclin for breakdown. The resulting loss of mitotic Cdk activity leads to events associated with the exit from mitosis, including cytokinesis, chromatin decondensation, and reassembly of the nuclear envelope.
- Checkpoint pathways monitor intracellular conditions and temporarily halt the cell cycle if conditions are not suitable for proceeding. The DNA replication checkpoint verifies that DNA synthesis has been completed before allowing the cell to exit from G2 and begin mitosis. DNA licensing ensures that replication only happens once per cell cycle. DNA damage checkpoints involving the p53 protein halt the cell cycle at various points if DNA damage is detected. The spindle assembly checkpoint prevents chromosomes from moving to the spindle poles before all chromosomes are attached to the spindle.

24.4 Growth Factors and Cell Proliferation

- Most cells in multicellular organisms do not proliferate unless they are stimulated by an appropriate growth factor.
- Many growth factors bind to receptors that activate the Ras pathway, allowing passage through the restriction point and into S phase. Growth factor receptors can also activate the PI 3-kinase–Akt pathway, which promotes cell survival and proliferation by phosphorylating and thereby activating target proteins that suppress apoptosis and inhibit cell cycle arrest.
- Some growth factors inhibit (rather than stimulate) cell proliferation by triggering the production of Cdk inhibitors.

24.5 Apoptosis

- Apoptosis is a form of cellular death triggered by activation of death receptors, by withdrawal of survival factors, or as a result of DNA damage.
- Apoptosis involves the orderly dismantling of a dying cell's contents.
- Proteases called caspases are key mediators of apoptosis. Initiator caspases activate executioner caspases, which in turn activate other apoptosis proteins.
- Initiator caspases can be activated in several ways, including through the release of cytochrome *c* from mitochondria. Pro- and anti-apoptotic proteins at the mitochondrion regulate release of cytochrome *c*.

Problem Set

Mastering™ Biology For activities, animations, and review quizzes, go to the study area at www.masteringbiology.com.

24-1 Cell Cycle Phases. Indicate whether each of the following statements is true of the G1 phase, S phase, G2 phase, or M phase of the cell cycle. A given statement may be true of any, all, or none of the phases.

(a) The amount of nuclear DNA in the cell doubles.

(b) The nuclear envelope breaks into fragments.

(c) Sister chromatids separate from each other.

(d) Cells that will never divide again are likely to be arrested in this phase.

(e) The primary cell wall of a plant cell forms.

(f) Chromosomes are present as diffuse, extended chromatin.

(g) This phase is part of interphase.

(h) Mitotic cyclin is at its lowest level.

(i) Cdk proteins do not play a role in this phase.

(j) A cell cycle checkpoint has been identified in this phase.

24-2 QUANTITATIVE The Mitotic Index and the Cell Cycle. An alternative to flow cytometry is to directly examine cells under the microscope to determine the mitotic index, the percentage of cells in mitosis at any one time. Assume that upon examining a sample of 1000 cells, you find 30 cells in prophase, 20 in prometaphase, 20 in metaphase, 10 in anaphase, 20 in telophase, and 900 in interphase. Of those in interphase, 400 are found (by staining the cells with a DNA-specific stain) to have X amount of DNA, 200 to have $2X$, and 300 cells to be somewhere in between. Analysis using timed applications of BrdU indicates that the G2 phase lasted 4 hours.

(a) What is the mitotic index for this population of cells?

(b) Specify the proportion of the cell cycle spent in each of the following phases: prophase, prometaphase, metaphase, anaphase, telophase, G1, S, and G2.

(c) What is the total length of the cell cycle?

(d) What is the actual amount of time (in hours) spent in each of the phases of part b?

(e) To measure the G2 phase, you add BrdU to the culture at some time *t*, and you analyze samples of the culture for the presence of labeled nuclei at regular intervals thereafter. In which sorts of cells should you look for the first appearance of BrdU to assess the length of the G2 phase?

(f) What proportion of the interphase cells would you expect to exhibit labeled nuclei shortly after exposure to BrdU? (Assume a labeling period just long enough to allow the BrdU to get into the cells and begin to be incorporated into DNA.)

24-3 DATA ANALYSIS Chromosome Movement in Mitosis. It is possible to mark the microtubules of a spindle by photobleaching with a laser microbeam (**Figure 24-29**). When this is done, chromosomes move *toward* the bleached area during anaphase. Are the following statements consistent or inconsistent with this experimental result? In each case, explain your answer.

(a) Microtubules move chromosomes solely by disassembling at the spindle poles.

(b) Chromosomes move by disassembling microtubules at their kinetochore ends.

(c) Chromosomes are moved along microtubules by a kinetochore motor protein that moves along the surface of the microtubule and "pulls" the chromosome with it.

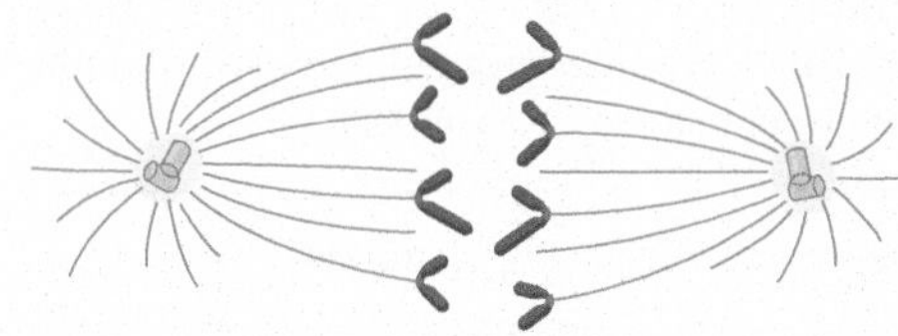

Microtubules are labeled with a fluorescent dye during anaphase.

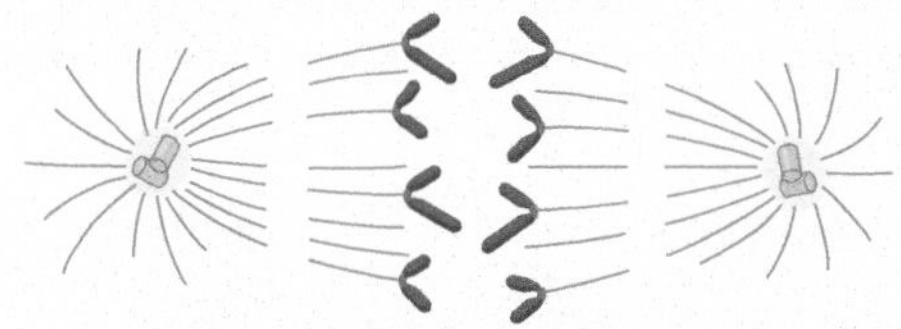

A laser microbeam is used to mark two areas by bleaching the fluorescent dye.

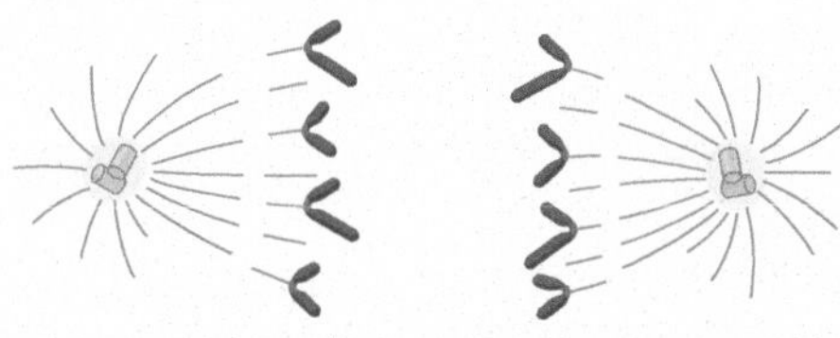

The chromosomes are observed to move toward the bleached areas.

Figure 24-29 Use of Laser Photobleaching to Study Chromosome Movement During Mitosis. See Problem 24-3.

24-4 Cytokinesis. Predict what will happen in each of the following situations, based on your knowledge of cytokinesis. In each case, explain your answer.

(a) A fertilized sea urchin egg is injected with C3 transferase, a bacterial toxin that ADP ribosylates and inhibits Rho, 30 minutes prior to first cleavage.

(b) A one-celled *C. elegans* zygote lacks *anillin*, a protein that is required to assemble myosin efficiently at the cell surface.

24-5 More on Cell Cycle Phases. For each of the following pairs of phases from the cell cycle, indicate how you could tell in which of the two phases a specific cell is located.

(a) G1 and G2

(b) G1 and S

(c) G2 and M

(d) G1 and M

24-6 Cell Cycle Regulation. Recall that one approach to the study of cell cycle regulation has been to fuse cultured cells that are at different stages of the cell cycle and observe the effect of the fusion on the nuclei of the fused cells (heterokaryons). When cells in G1 are fused with cells in S, the nuclei from the G1 cells begin DNA replication earlier than they would have if they had not been fused. In fusions of cells in G2 and S, however, nuclei continue their previous activities, apparently uninfluenced by the fusion. Fusions between mitotic cells and interphase cells always lead to chromatin condensation in the nonmitotic nuclei. Based on these results, identify each of the following statements about cell cycle regulation as probably true (T), probably false (F), or not possible to conclude from the data (NP).

(a) The activation of DNA synthesis may result from the stimulatory activity of one or more cytoplasmic factors.

(b) The transition from S to G2 may result from the presence of a cytoplasmic factor that inhibits DNA synthesis.

(c) The transition from G2 to mitosis may result from the presence in the G2 cytoplasm of one or more factors that induce chromatin condensation.

(d) G1 is not an obligatory phase of all cell cycles.

(e) The transition from mitosis to G1 appears to result from the disappearance or inactivation of a cytoplasmic factor present during M phase.

24-7 Role of Cyclin-Dependent Protein Kinases. Based on your understanding of the regulation of the eukaryotic cell cycle, how could you explain each of the following experimental observations?

(a) When mitotic Cdk-cyclin is injected into cells that have just emerged from S phase, chromosome condensation and nuclear envelope breakdown occur immediately, rather than after the normal G2 delay of several hours.

(b) When an abnormal, indestructible form of mitotic cyclin is introduced into cells, they enter into mitosis but cannot emerge from it and reenter G1 phase.

(c) Mutations that inactivate the main protein phosphatase used to catalyze protein dephosphorylations cause a long delay in the reconstruction of the nuclear envelope that normally takes place at the end of mitosis.

24-8 Apoptosis and DNA. When cells undergo apoptosis, their DNA becomes fragmented. When the DNA is run out on a gel, the resulting fragments differ in size by a standard amount, about 200 bp, so that the resulting separated DNA has a characteristic "ladder" appearance (**Figure 24-30**). Explain why this pattern of DNA fragments is obtained, based on what you know about eukaryotic chromatin.

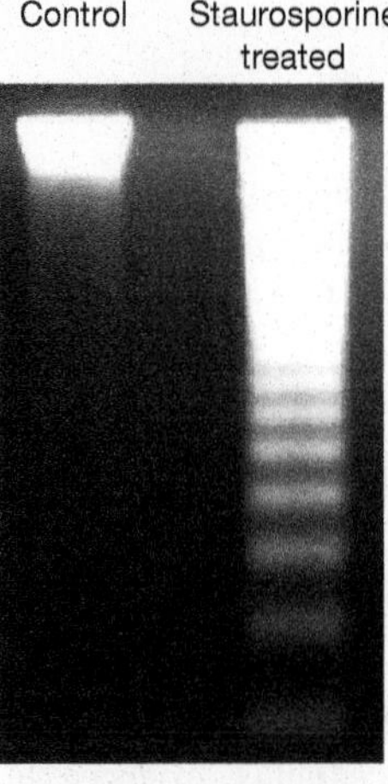

Figure 24-30 DNA "Laddering" During Apoptosis. Chromosomal DNA was isolated from control (untreated) and staurosporine-treated cultured Burkitt lymphoma cells and analyzed by gell electrophoresis. Staurosporin induces apoptosis. See Problem 24-8.

24-9 Apoptosis and Medicine. A current focus of molecular medicine is to trigger or prevent apoptosis in specific cells. Several components of the apoptotic pathway are being targeted using this approach. For each of the following, state specifically how the treatment would be expected to stimulate or inhibit apoptosis.

(a) Cells are treated with a small molecule called pifithrin-α, which was originally isolated for its ability to reversibly block p53-dependent transcriptional activation.

(b) Cells are exposed to recombinant TRAIL protein, a ligand for the tumor necrosis factor family of receptors.

(c) Cells are treated with organic compounds that enter the cell and bind with high affinity to the active site of caspase-3.

25 Sexual Reproduction, Meiosis, and Genetic Recombination

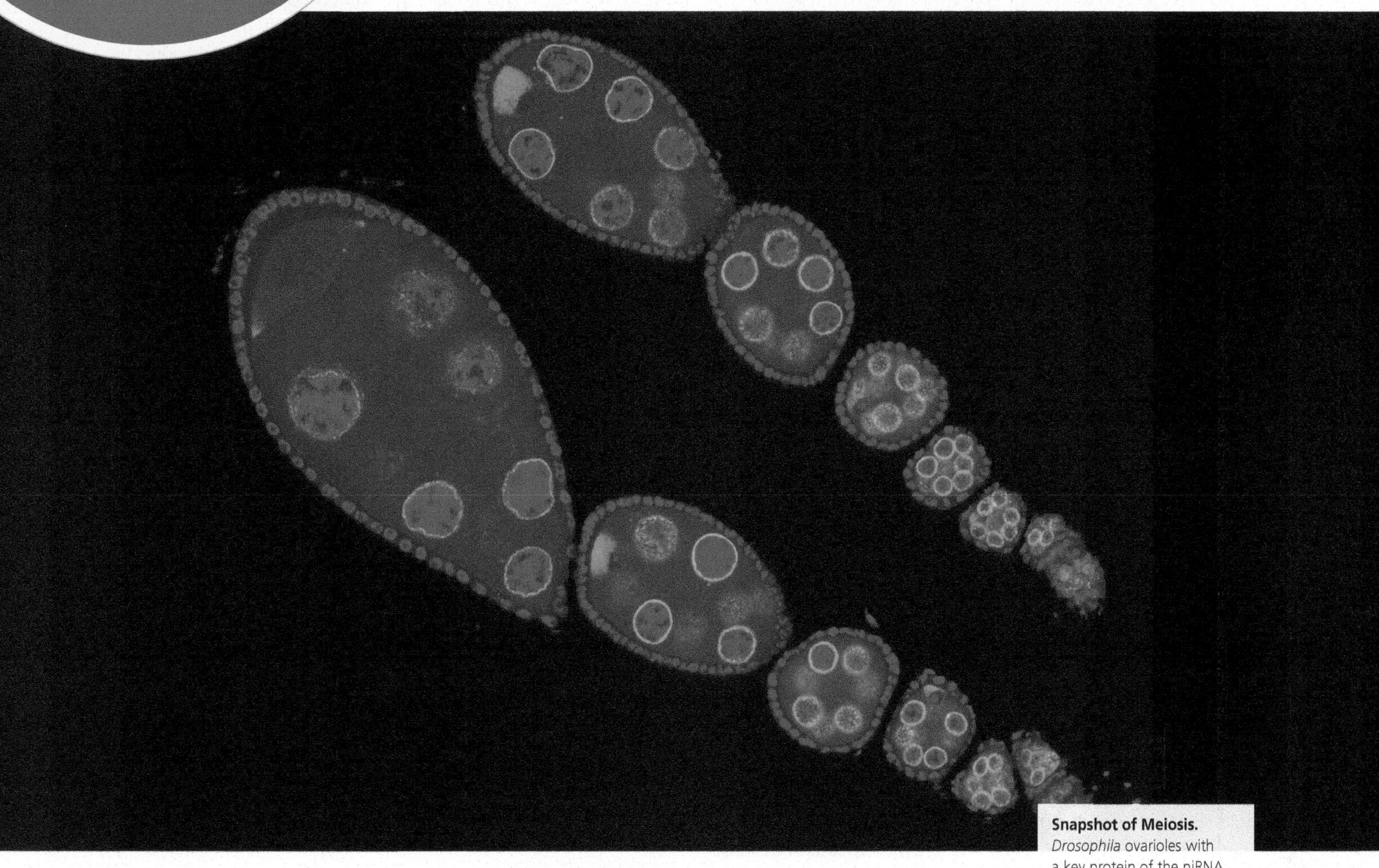

Snapshot of Meiosis. *Drosophila* ovarioles with a key protein of the piRNA pathway stained green and DNA stained blue (confocal microscopy).

Mitotic cell division (which we discussed in the preceding chapter) is used for the proliferation of most eukaryotic cells. Because a mitotic cell division cycle involves one round of DNA replication followed by the segregation of identical chromatids into two daughter cells, mitotic division produces cells that are genetically identical, or very nearly so. This ability to perpetuate genetic traits faithfully allows mitotic division to form the basis of **asexual reproduction** in eukaryotes. During asexual reproduction, new individuals are generated by mitotic division of cells in a single parent organism, either unicellular or multicellular. Although the details vary among organisms, asexual reproduction is widespread in nature. Examples include *mitotic division* of unicellular organisms, *budding* of offspring from a multicellular parent's body, and *regeneration* of whole organisms from pieces of a parent organism. In plants, entire organisms can be regenerated from single cells taken from an adult plant.

Asexual reproduction can be an efficient and evolutionarily successful mode of perpetuating a species. But if the environment changes, a population that reproduces asexually may not be able to adapt to the new conditions. Under such conditions, organisms that reproduce sexually rather than asexually can have an advantage, as we now discuss.

Mastering™ Biology www.masteringbiology.com

25.1 Sexual Reproduction

In contrast to asexual reproduction, **sexual reproduction** allows genetic information from two parents to be mixed together, thereby producing offspring that are genetically dissimilar, both from each other and from the parents.

Sexual Reproduction Produces Genetic Variety

Sexual reproduction allows genetic traits found in different individuals to be combined in various ways in newly developing offspring, thereby generating enormous variety among the individuals that make up a population. Because sexual reproduction combines genetic information from two different parents into a single offspring, at some point in its life cycle every sexually reproducing organism has cells that contain two copies of each type of chromosome, one inherited from each parent. The two members of each chromosome pair are called **homologous chromosomes**. Two homologous chromosomes generally carry their genetic information in the same order, although the two versions may differ slightly in base sequence. Not surprisingly, homologous chromosomes usually look alike when viewed with a microscope (see Figure 16-23). An exception to this rule is the **sex chromosomes**, which can determine whether an individual is male or female. The two kinds of sex chromosomes differ significantly in genetic makeup and appearance. In mammals, for example, females have two X chromosomes of the same size, whereas males have one X chromosome and a Y chromosome that is much smaller. Nonetheless, parts of the X and Y chromosomes are actually homologous, and, during sexual reproduction, the X and Y chromosomes behave as homologues.

A cell or organism with two sets of chromosomes is said to be **diploid** (from the Greek word *diplous*, meaning "double") and contains two copies of its genome. A cell or organism with a single set of chromosomes, and therefore a single copy of its genome, is **haploid** (from the Greek word *haplous*, meaning "single"). By convention, the haploid chromosome number for a species is designated n (or $1n$) and the diploid number $2n$. For example, in humans $n = 23$, which means that most human cells contain two sets of 23 chromosomes, yielding a diploid total of 46. The diploid state is an essential feature of the life cycle of sexually reproducing species. If, for example, a mutation occurs in one copy of a gene, this usually will not threaten the survival of the organism, even if the mutation disrupts the original function of that particular gene copy. The diploid state also provides some protection against chromosome damage; if one chromosome is accidentally broken, it can sometimes be repaired using the DNA sequence of the homologous chromosome as a template (see Figure 17-30).

Gametes Are Haploid Cells Specialized for Sexual Reproduction

The hallmark of sexual reproduction is that genetic information contributed by two parents is brought together in a single individual. Because the offspring of sexual reproduction are diploid, the contribution from each parent must be haploid. The haploid cells produced by each parent that fuse together to form the diploid offspring are called **gametes**, and the process that produces them is **gametogenesis**. Biologists distinguish between male and female individuals on the basis of the gametes they produce. Gametes produced by males, called **sperm** (or *spermatozoa*), are usually quite small and may be inherently motile. Female gametes, called **eggs**, or **ova** (singular: **ovum**), are specialized for the storage of nutrients and tend to be quite large and nonmotile. For example, in sea urchins the volume of an egg cell is more than 10,000 times greater than that of a sperm cell; in birds and amphibians, which have massive yolky eggs, the size difference is even greater. But despite their differing sizes, sperm and egg bring equal amounts of chromosomal DNA to the offspring.

The union of sperm and egg during sexual reproduction is called **fertilization**. The resulting fertilized egg, or **zygote**, is diploid, having received one chromosome set from the sperm and a homologous set from the egg. In the life cycles of multicellular organisms, fertilization is followed by a series of mitotic divisions and progressive specialization of various groups of cells to form a multicellular embryo and eventually an adult.

In a few unusual situations, eggs can develop into offspring without the need for sperm—a phenomenon known as *parthenogenesis*. A striking case in point is the Komodo dragon, the world's largest lizard, which lives on the central islands of Indonesia. When male Komodos are unavailable for sexual reproduction, female Komodos give birth to offspring through a self-fertilization process involving the fusion of a haploid egg cell with a haploid *polar body*—a "mini-egg" produced at the same time eggs are formed (see Figure 25-8). Consequently, the genes of each offspring derive solely from the mother, though the offspring are not exact duplicates of the mother because different combinations of alleles are present in the haploid cells that fuse together to create the embryo.

In some organisms, gametes cannot be categorized as being male or female. Certain fungi and unicellular eukaryotes, for example, produce gametes that are identical in appearance but differ slightly at the molecular level. Such gametes are said to differ in *mating type*. The union of two of these gametes requires that they be of different mating types, but the number of possible mating types in a species may be greater than two—in some cases, more than ten!

CONCEPT CHECK 25.1

Cloning can be done by somatic cell nuclear transfer (SCNT) (see Figure 20-12). Some conservation biologists have proposed using SCNT technology to preserve highly endangered animal species. What might be some of the genetic disadvantages of this approach?

25.2 Meiosis

Because gametes are haploid, they cannot be produced from diploid cells by mitosis. If they were, both sperm and egg would have a diploid chromosome number, just like the parent diploid cells. The hypothetical zygote created by the fusion of such diploid gametes would be *tetraploid* (that is, possess *four* homologous sets of chromosomes). Moreover, chromosome number would continue to double in each succeeding generation—an impossible scenario. Thus, for chromosome number to remain constant from generation to generation, a different type of cell division must occur during the formation of gametes. That special type of division, called **meiosis**, reduces the chromosome number from diploid ($2n$) to haploid ($1n$).

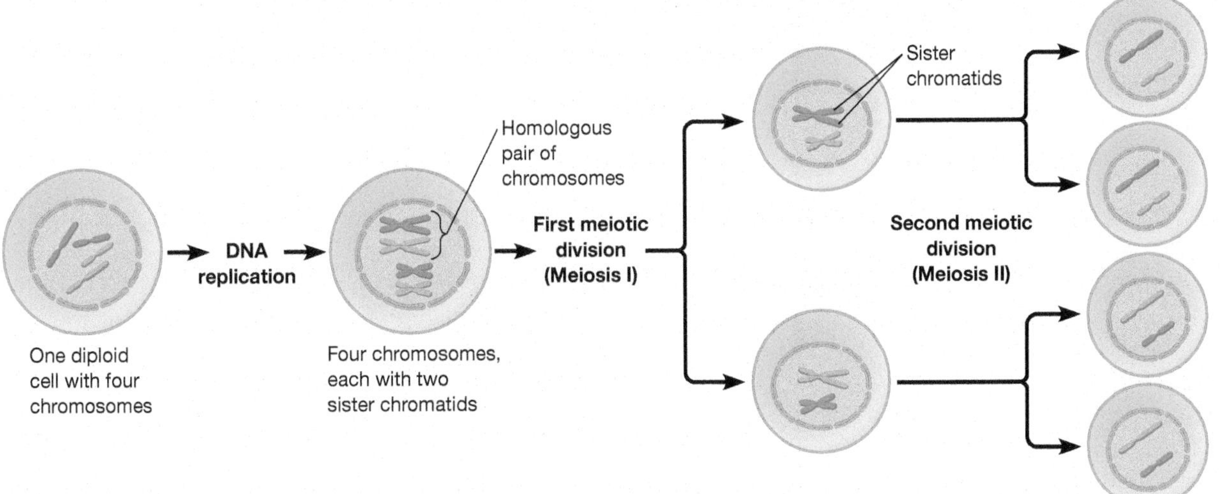

Figure 25-1 The Principle of Meiosis. Meiosis involves a single round of DNA replication (chromosome duplication) in a diploid cell followed by two successive cell division events. In this example, the diploid cell has only four chromosomes, which can be grouped into two homologous pairs. After DNA replication, each chromosome consists of two sister chromatids. In the first meiotic division (meiosis I), homologous chromosomes separate, but sister chromatids remain attached. In the second meiotic division (meiosis II), sister chromatids separate, resulting in four haploid daughter cells with two chromosomes each. Notice that each haploid cell has one chromosome from each homologous pair that was present in the diploid cell. For simplicity, the effects of genetic recombination are not shown in this diagram.

Meiosis involves one round of chromosomal DNA replication followed by *two* successive nuclear divisions. This results in the formation of four daughter nuclei (usually in separate daughter cells) containing one haploid set of chromosomes per nucleus. **Figure 25-1** outlines the principle of meiosis starting with a diploid cell containing four chromosomes ($2n = 4$). A single round of DNA replication is followed by two cell divisions, meiosis I and meiosis II, leading to the formation of four haploid cells.

The Life Cycles of Sexual Organisms Have Diploid and Haploid Phases

Meiosis and fertilization are indispensable components of the life cycle of every sexually reproducing organism because the doubling of chromosome number that takes place at fertilization is balanced by the halving that occurs during meiosis. As a result, the life cycle of sexually reproducing organisms is divided into two phases: a diploid ($2n$) phase and a haploid ($1n$) phase. The diploid phase begins at fertilization and extends until meiosis, whereas the haploid phase is initiated at meiosis and ends with fertilization.

As shown for some representative groups in **Figure 25-2**, organisms vary greatly in the relative prominence of the haploid and diploid phases of their life cycles. Bacteria are always haploid (Figure 25-2a). Some fungi are examples of sexually reproducing organisms whose life cycles are primarily haploid but include a brief diploid phase that begins with gamete

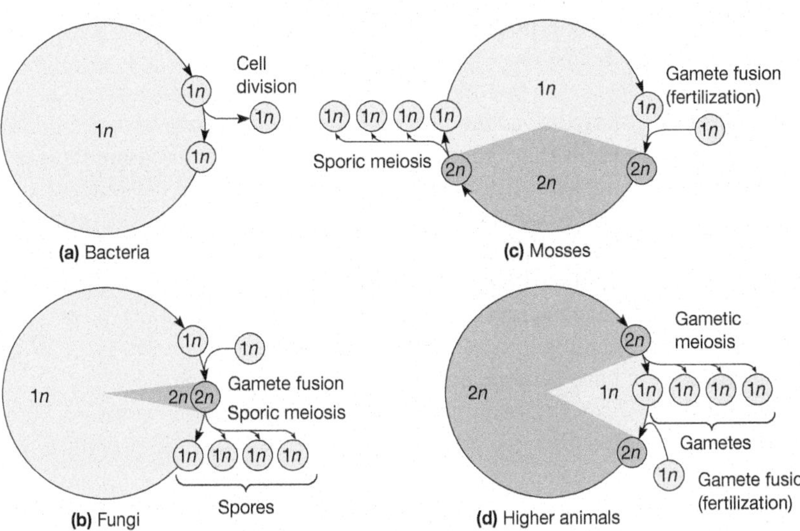

Figure 25-2 Types of Life Cycles. The relative prominence of the haploid ($1n$) and diploid ($2n$) phases of the life cycle differs greatly, depending on the organism. **(a)** Bacteria exist exclusively in the haploid state. **(b)** Many fungi exemplify a life form that is predominantly haploid but has a brief diploid phase. Because the products of meiosis in fungi are haploid spores, this type of meiosis is called sporic meiosis. **(c)** Mosses (and ferns as well) alternate between haploid and diploid forms, both of which are significant components in the life cycles of these organisms. Sporic meiosis produces haploid spores, which in this case grow into haploid plants. Eventually, some of the haploid plant's cells differentiate into gametes. **(d)** Higher animals are the best examples of organisms that are predominantly diploid, with only the gametes representing the haploid phase of the life cycle. Animals are said to have a gametic meiosis because the immediate products of meiosis are haploid gametes.

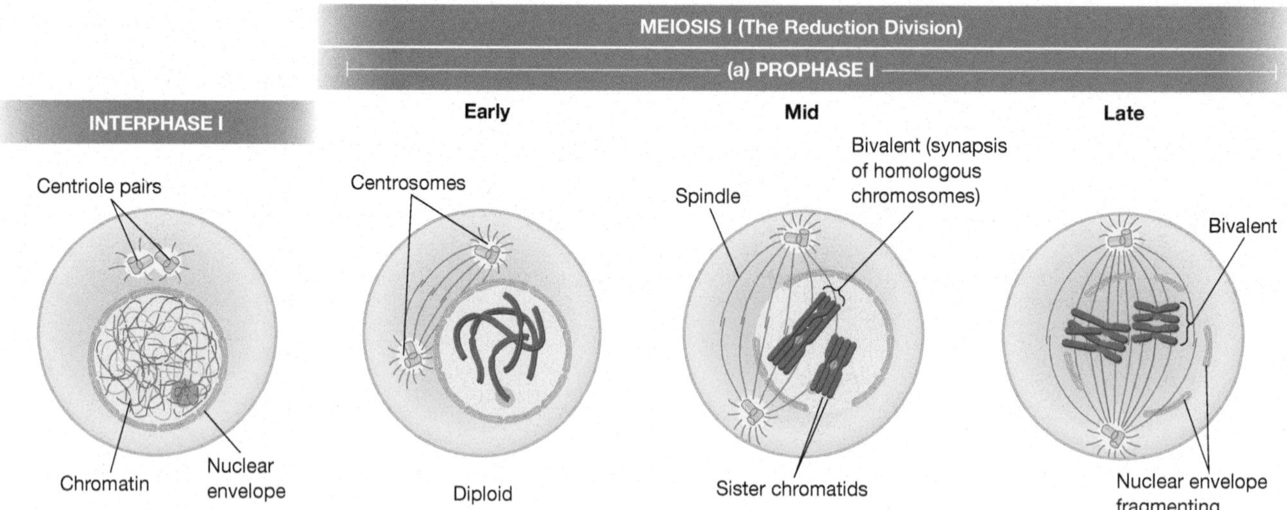

Figure 25-3 Meiosis in an Animal Cell. Meiosis consists of two successive divisions, called meiosis I and II, with no intervening DNA synthesis or chromosome duplication. **(a)** During prophase I, the chromosomes (duplicated during the previous S phase) condense and the two centrosomes migrate to opposite poles of the cell. Each chromosome (four in this example) consists of two sister chromatids. Homologous chromosomes pair to form bivalents. **(b)** Bivalents become aligned at the spindle equator (metaphase I). **(c)** Homologous chromosomes separate during anaphase I, but sister chromatids remain attached at the centromere. **(d)** Telophase and cytokinesis follow. Although not illustrated here, there may then be a short interphase (interphase II). In meiosis II, **(e)** chromosomes recondense (prophase II), **(f)** chromosomes align at the spindle equator (metaphase II), and **(g)** sister chromatids at last separate (anaphase II). **(h)** After telophase II and cytokinesis, the result is four haploid daughter cells, each containing one chromosome of each homologous pair.

fusion (the fungal equivalent of fertilization) and ends with meiosis (Figure 25-2b). Meiosis usually takes place almost immediately after gamete fusion, so the diploid phase is very short. Accordingly, only a very small fraction of nuclei in such fungi are diploid at any one time. Fungal gametes develop, without meiosis, from cells that are already haploid.

Mosses and ferns are probably the best examples of organisms in which both the haploid and diploid phases are prominent features of the life cycle (Figure 25-2c). For mosses, the haploid form of the organism is larger and more prominent, and the diploid form is smaller and more short-lived. For ferns, it is the other way around. In both cases, gametes develop from preexisting haploid cells.

Organisms that alternate between haploid and diploid multicellular forms in this way are said to display an **alternation of generations** in their life cycles. In addition to mosses and ferns, eukaryotic algae and other plants exhibit an alternation of diploid and haploid generations. In all such organisms, the products of meiosis are **haploid spores**, which, after germination, give rise by mitotic cell division to the haploid form of the plant or alga. The haploid form in turn produces the gametes by specialization of cells that are already haploid. The gametes, upon fertilization, give rise to the diploid form. Because the diploid form produces spores, it is called a **sporophyte** ("spore-producing plant"). The haploid form produces gametes and is therefore called a **gametophyte**. Although all plants exhibit an alternation of generations, in most cases the sporophyte generation predominates. In flowering plants, for example, the gametophyte generation is an almost vestigial structure located in the flower (female gametophyte in the *carpel*, male gametophytes in the flower's *anthers*).

The best examples of life cycles dominated by the diploid phase are found in animals (Figure 25-2d). In such organisms, including humans, meiosis gives rise not to spores but to gametes directly, so the haploid phase of the life cycle is represented only by the gametes.

Meiosis Converts One Diploid Cell into Four Haploid Cells

Wherever it occurs in an organism's life cycle, meiosis is always preceded by chromosome duplication in a diploid cell and involves two successive divisions that convert the diploid nucleus into four haploid nuclei. **Figure 25-3** illustrates the various phases of meiosis; refer to it as you read the following discussion.

The first meiotic division, or **meiosis I**, is sometimes referred to as the *reduction division* of meiosis because it is the event that reduces the chromosome number from diploid to haploid. Early during meiosis I, the two chromosomes of each homologous pair come together during prophase to exchange some of their genetic information (using a mechanism to be discussed shortly). This pairing of homologous chromosomes, called **synapsis**, is unique to meiosis; at all other times, including mitosis, the chromosomes of a homologous pair behave independently.

The two chromosomes of each homologous pair bind together so tightly during the first meiotic prophase that

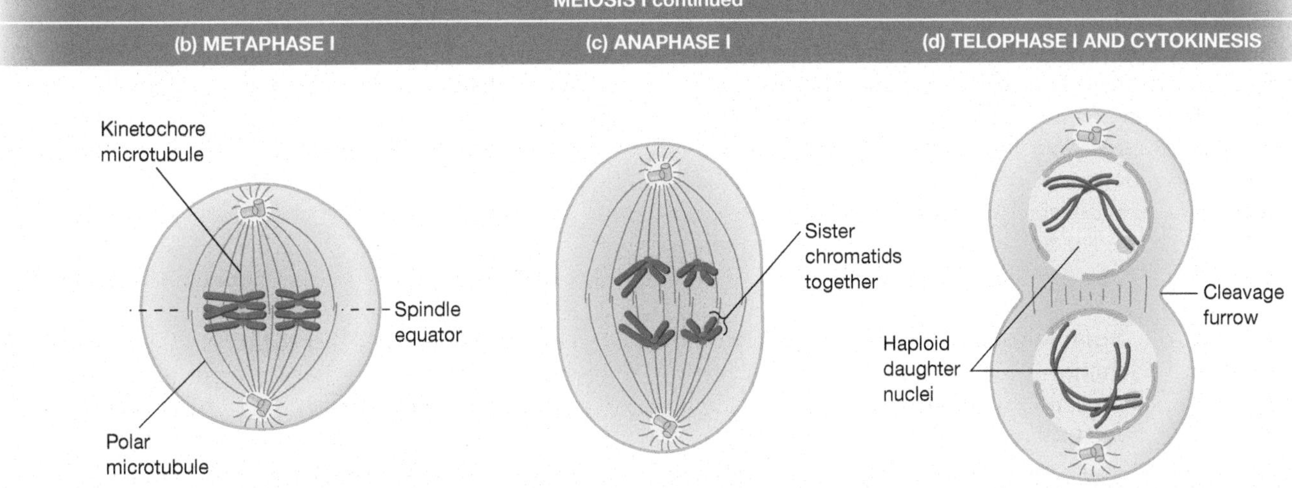

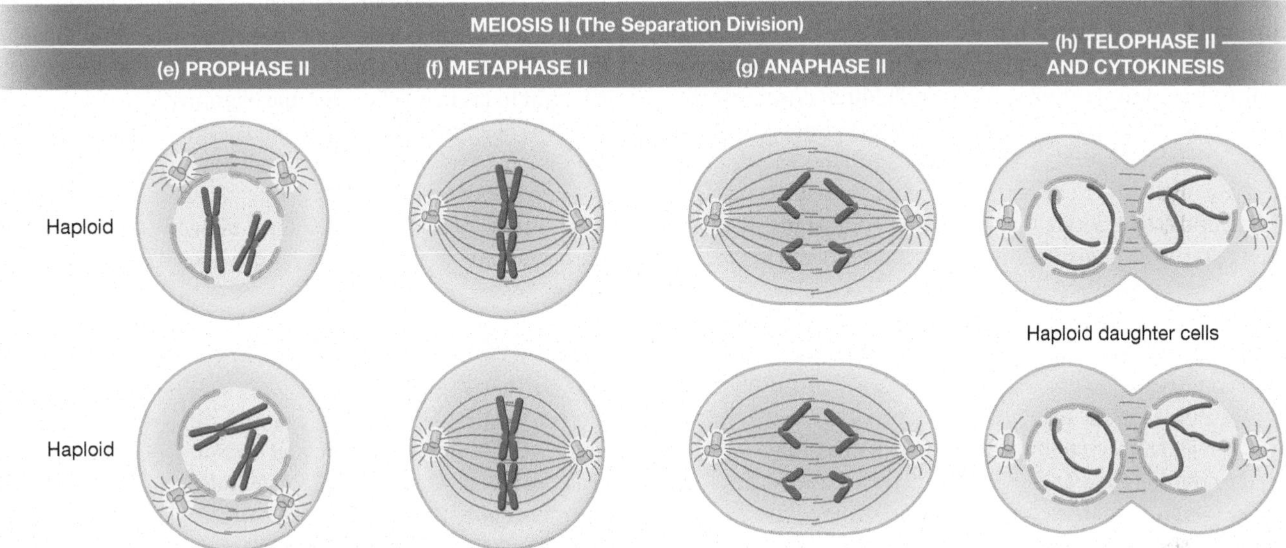

they behave as a single unit called a **bivalent** (or *tetrad*, which emphasizes that each of the two homologous chromosomes consists of two sister chromatids, yielding a total of four chromatids). After aligning at the spindle equator, each bivalent splits apart in such a way that its two homologous chromosomes move to opposite spindle poles. Because each pole receives only one member of each homologous pair, the daughter nuclei produced by meiosis I are considered to be haploid (even though the individual chromosomes in these nuclei are composed of two sister chromatids). During the second meiotic division (meiosis II), which closely resembles a mitotic division, the two sister chromatids of each chromosome separate into two daughter cells. Hence, the events unique to meiosis happen during the first meiotic division: the synapsis of homologous chromosomes and their subsequent segregation into different daughter nuclei.

Meiotic cell divisions involve the same basic stages as mitosis, although cell biologists do not usually distinguish prometaphase as a separate phase. Thus, the meiotic phases are *prophase, metaphase, anaphase,* and *telophase.* Prophase I is much longer and more complicated than mitotic prophase, whereas prophase II tends to be quite short. Another important difference from mitosis is that a normal interphase does not intervene between the two meiotic divisions. If an interphase does take place, it is usually very short and—most importantly—does *not* include DNA replication because each chromosome already consists of a pair of replicated sister chromatids that had been generated prior to the first meiotic division. The purpose of the second meiotic division, like that of a typical mitotic division, is to parcel these sister chromatids into two daughter nuclei.

Meiosis I Produces Two Haploid Cells That Have Chromosomes Composed of Sister Chromatids

The first meiotic division segregates homologous chromosomes into different daughter cells. This feature of meiosis has special genetic significance because it represents the point in an organism's life cycle when the two alleles for each gene part company. And it is this separation of alleles that makes possible the eventual remixing of different pairs of alleles at fertilization. Also of great significance during the first meiotic division are events involving the physical exchange of parts of DNA molecules. When such an exchange of DNA segments occurs between two homologous chromosomes it is called **homologous recombination**. This type of DNA exchange takes place when the chromosomes are synapsed during prophase I. The steps of prophase I are summarized in **Figure 25-4**.

Prophase I: Homologous Chromosomes Become Paired and Exchange DNA. Prophase I is a particularly long and complex phase. Based on light microscopic observations, early cell biologists divided prophase I into five stages called *leptotene, zygotene, pachytene, diplotene,* and *diakinesis* (Figure 25-4).

The **leptotene** stage begins with the condensation of chromatin fibers into long, threadlike structures, similar to what occurs at the beginning of mitosis. At **zygotene**, continued condensation makes individual chromosomes distinguishable, and homologous chromosomes become closely paired with each other via the process of synapsis, forming bivalents. Keep in mind that each bivalent has four chromatids, two derived from each chromosome. Bivalent formation is of considerable genetic significance because the close proximity between homologous chromosomes allows DNA segments to be exchanged by a process called **crossing over**. It is this physical exchange of genetic information between corresponding regions of homologous chromosomes that accounts for genetic recombination. You will learn more about the nature of crossing over later in this chapter. Crossing over occurs during the **pachytene** stage, which is marked by a dramatic compacting process that reduces each chromosome to less than a quarter of its previous length.

At the **diplotene** stage, the homologous chromosomes of each bivalent begin to separate from each other, particularly near the centromere. However, the two chromosomes of each homologous pair remain attached by connections known as **chiasmata** (singular: **chiasma**). Such connections are situated in regions where homologous chromosomes have exchanged DNA segments and hence provide visual evidence that crossing over has occurred between two chromatids, one derived from each chromosome.

In some organisms—female mammals, for instance—the chromosomes decondense during diplotene, transcription resumes, and the cells "take a break" from meiosis for a prolonged period of growth, sometimes lasting for years. (We will consider this situation at the end of our discussion of meiosis.) With the onset of **diakinesis**, the final stage of prophase I, the chromosomes recondense to their maximally compacted state. Now the centromeres of the homologous chromosomes separate farther, and the chiasmata eventually become the only remaining attachments between the homologues. At this stage, the nucleoli disappear, the spindle forms, and the nuclear envelope breaks down, marking the end of prophase I.

With the advent of modern tools, especially the electron microscope, cell biologists have been able to refine our picture of what happens during prophase I (**Figure 25-5** on page 758). They have found that what holds homologous chromosomes in tight apposition during synapsis is the **synaptonemal complex**, an elaborate protein structure resembling a zipper (Figure 25-5a). The *lateral elements* of the synaptonemal complex start to attach to individual chromosomes during leptotene, but the *central element*, which actually joins homologous chromosomes together, does not form until zygotene (Figure 25-5b). How do the members of each pair of homologous chromosomes find each other so they can be joined by a synaptonemal complex? During early zygotene, the ends (telomeres) of each chromosome become clustered on one side of the nucleus and attach to the nuclear envelope, with the body of each chromosome looping out into the nucleus. To picture this, imagine holding all the ends of four ropes (two long and two short) together. If you give the ropes a strong shake, they will settle into four loops, arranged according to length. This type of chromosome configuration, called a *bouquet*, is thought to promote chromosome alignment.

The alignment of similar-sized chromosomes facilitates formation of synaptonemal complexes, which become fully developed during pachytene. The synaptonemal complexes then disassemble during diplotene, allowing the homologous chromosomes to separate (except where they are joined by chiasmata).

Metaphase I: Bivalents Align at the Spindle Equator. During metaphase I, the bivalents attach via their kinetochores to spindle microtubules and migrate to the spindle equator. The presence of *paired* homologous chromosomes (that is, bivalents) at the spindle equator during metaphase I is a crucial difference between meiosis I and a typical mitotic division, where such pairing is not observed (**Figure 25-6** on page 759). Because each bivalent contains four chromatids (two sister chromatids from each chromosome), four kinetochores are also present. The kinetochores of sister chromatids lie side by side—in many species appearing as a single mass—and face the same pole of the cell. Such an arrangement allows the kinetochores derived from the sister chromatids of one homologous chromosome to attach to microtubules emanating from one spindle pole and the kinetochores derived from the sister chromatids of the other homologous chromosome to attach to microtubules emanating from the opposite spindle pole. This orientation sets the stage for separation of the homologous chromosomes during anaphase. The bivalents are randomly oriented at this point, in the sense that, for each bivalent, either the maternal or paternal homologue may face a given pole of the cell. As a result, each spindle pole (and hence each daughter cell) will receive a random mixture of maternal and paternal chromosomes when the two members of each chromosome pair move toward opposite spindle poles during anaphase.

At this stage, homologous chromosomes are held together solely by chiasmata. If for some reason prophase I had occurred without crossing over, and hence without chiasma

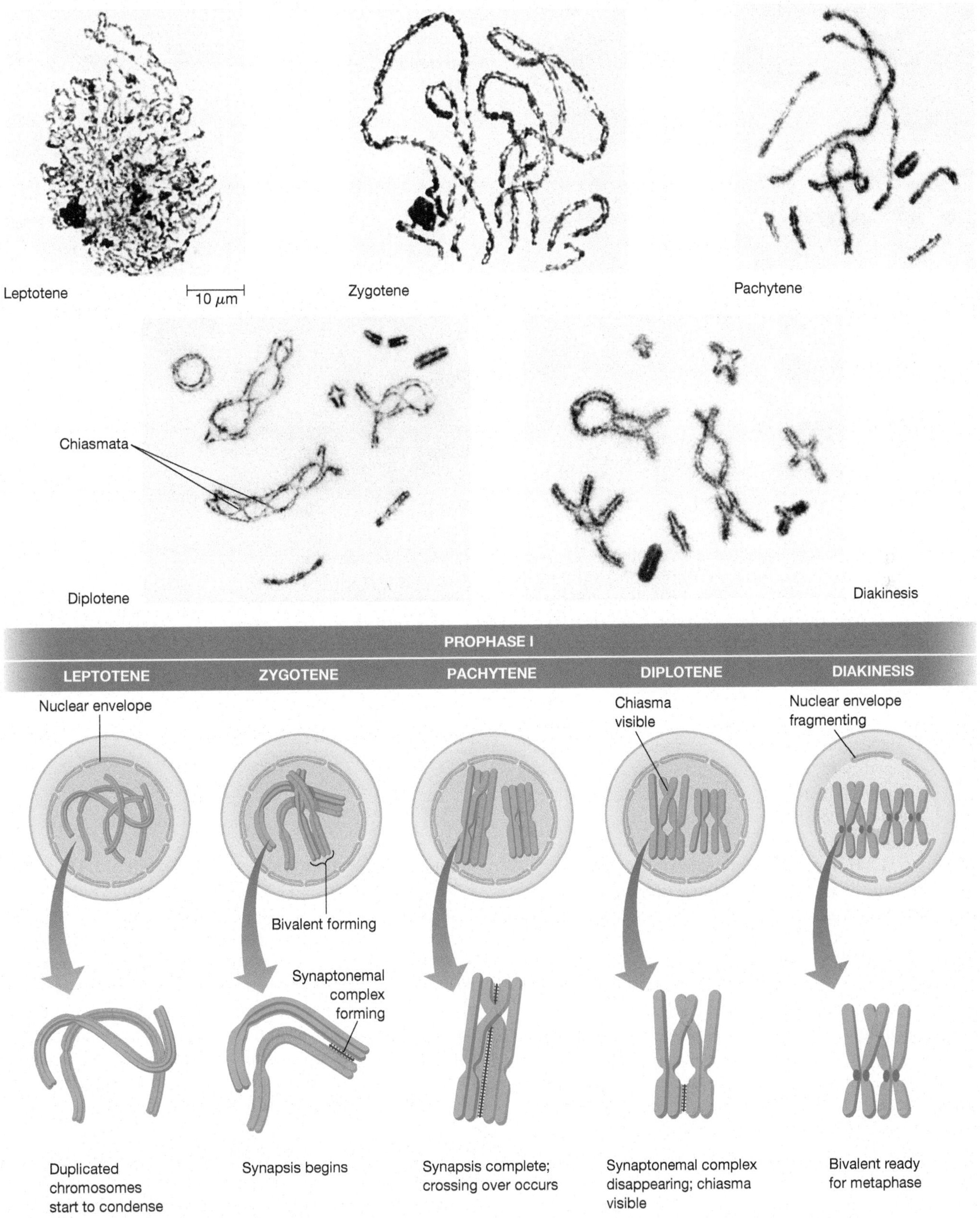

Leptotene 10 μm Zygotene Pachytene

Chiasmata

Diplotene Diakinesis

PROPHASE I

LEPTOTENE ZYGOTENE PACHYTENE DIPLOTENE DIAKINESIS

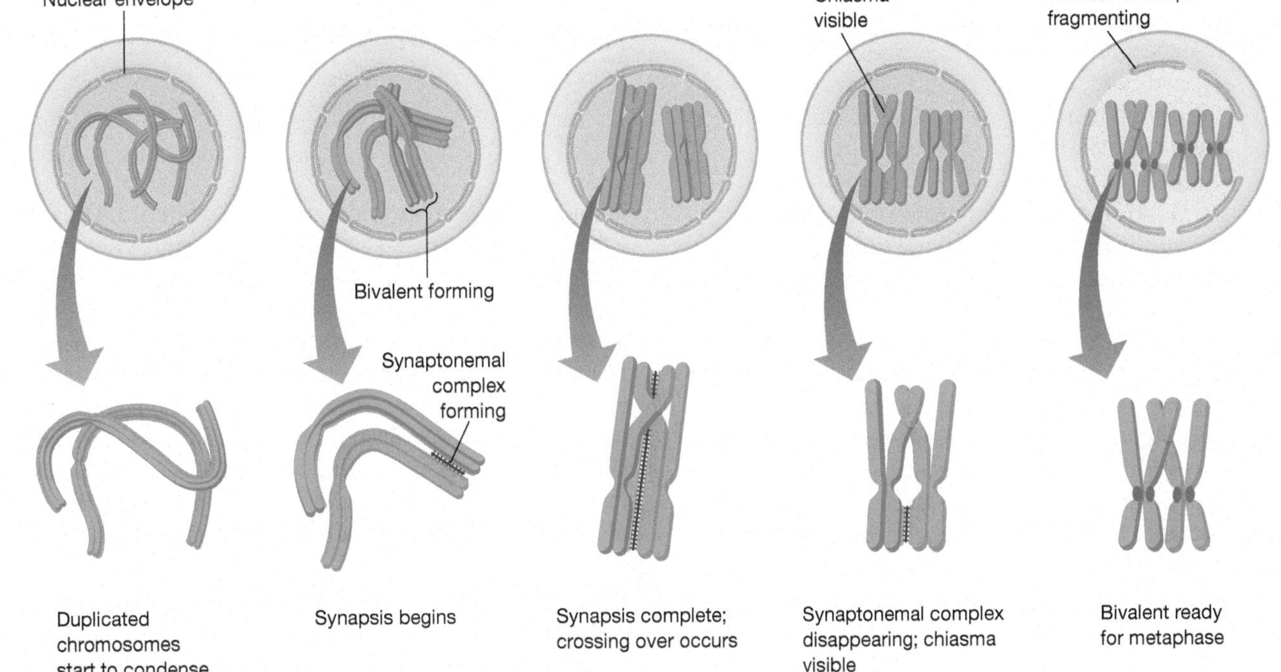

Figure 25-4 Meiotic Prophase I. Based on changes in chromosome behavior and appearance, prophase I is subdivided into the five stages shown in these electron micrographs and schematic diagram. The diagram depicts the cell nucleus at each stage for a diploid cell containing four chromosomes (two homologous pairs). The lower part of the diagram focuses on a single homologous pair in greater detail, revealing the formation and subsequent disappearance of the synaptonemal complex, a protein structure that holds homologous chromosomes in close lateral apposition during pachytene. Red and blue distinguish the paternal and maternal chromosomes of each homologous pair; the synaptonemal complex is shown in purple.

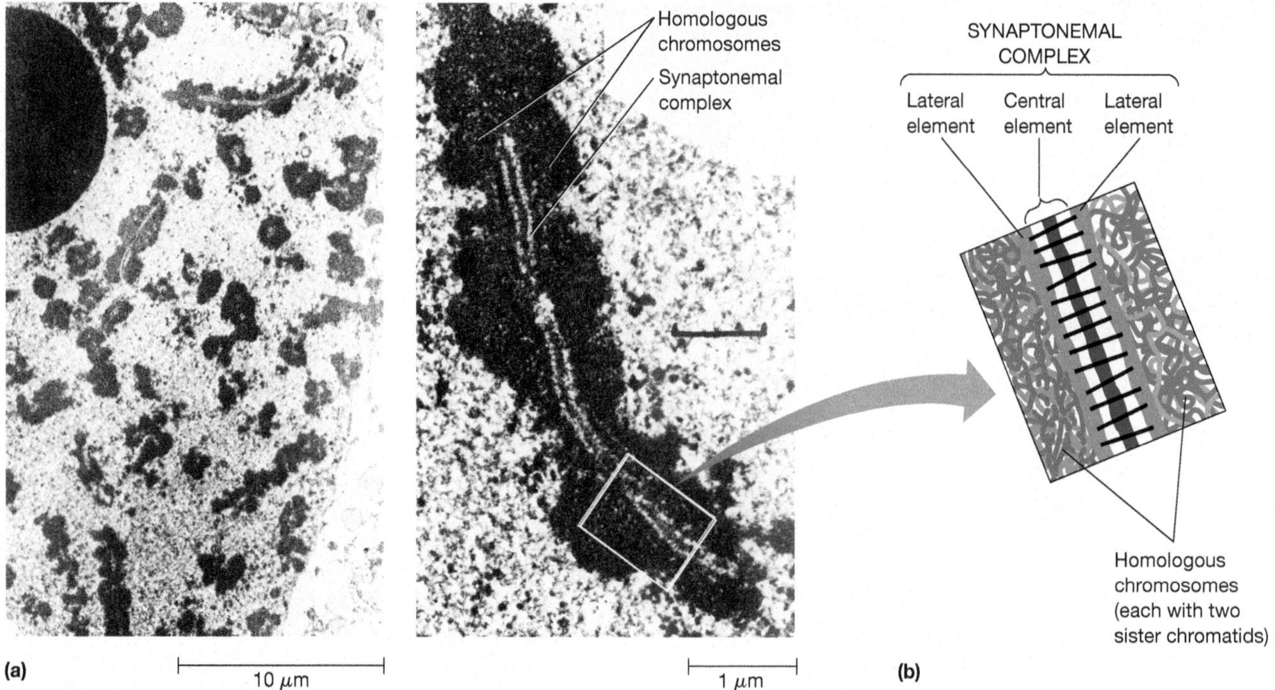

Figure 25-5 The Synaptonemal Complex. (a) These electron micrographs show, at two magnifications, synaptonemal complexes in the nuclei of cells from a lily. The cells are at the pachytene stage of prophase I. **(b)** The diagram identifies the complex's *lateral elements* (light purple), which seem to form on the chromosomes during leptotene, and its *central* (or *axial*) *element* (dark purple), which starts to appear during zygotene and "zips" the homologous chromosomes together. At pachytene, the homologues are held tightly together all along their lengths.

formation, the chromosomes might not pair properly at the spindle equator, and homologous chromosomes might not separate properly during anaphase I. This is exactly what happens during meiosis in mutant yeast cells that exhibit deficiencies in genetic recombination.

Anaphase I: Homologous Chromosomes Move to Opposite Spindle Poles. At the beginning of anaphase I, the members of each pair of homologous chromosomes separate from each other and start migrating toward opposite spindle poles, pulled by their respective kinetochore microtubules. Again, note the fundamental difference between meiosis and mitosis (Figure 25-6). During mitotic anaphase, *sister chromatids separate* and move to opposite spindle poles, whereas in anaphase I of meiosis, *homologous chromosomes separate* and move to opposite spindle poles while sister chromatids remain together. Because the two members of each pair of homologous chromosomes move to opposite spindle poles during anaphase I, each pole receives a haploid set of chromosomes.

How can we explain the fact that sister chromatids separate from each other during mitosis but not during meiosis I? Mitotic anaphase is associated with activation of the enzyme *separase* by the anaphase-promoting complex (see Figure 24-18); the activated separase in turn cleaves the cohesins that hold sister chromatids together, thereby allowing sister chromatids to separate. To prevent chromatid separation from taking place during anaphase I of meiosis, a protein called *shugoshin* (Japanese for "guardian spirit") protects the cohesins located at the chromosomal centromere from being degraded by separase.

MAKE CONNECTIONS 25.1

Recall the role of separase in anaphase of mitosis. How would mitotic cells be affected if shugoshin was overexpressed in them? (Fig. 24-18)

Telophase I and Cytokinesis: Two Haploid Cells Are Produced. The onset of telophase I is marked by the arrival of a haploid set of chromosomes at each spindle pole. Which member of a homologous pair ends up at which pole is determined entirely by how the chromosomes happened to be oriented at the spindle equator during metaphase I. After the chromosomes have arrived at the spindle poles, nuclear envelopes sometimes form around the chromosomes before cytokinesis ensues, generating two haploid cells whose chromosomes consist of sister chromatids. In most cases, the chromosomes do not decondense before meiosis II begins.

Meiosis II Resembles a Mitotic Division

After meiosis I has been completed, a brief interphase may intervene before **meiosis II** begins. However, this interphase is not accompanied by DNA replication because each chromosome already consists of a pair of replicated sister chromatids that were generated by DNA synthesis during the interphase preceding meiosis I. *So DNA is replicated only once in conjunction with meiosis, and that is prior to the first meiotic division.* The purpose of meiosis II, like that of a typical mitotic division, is to parcel the sister chromatids created by this initial round of DNA replication into two newly forming cells. As a result, meiosis II is sometimes referred to as the *separation division* of meiosis.

MEIOSIS I

MITOSIS

$2n$
4C

Bivalent

Prophase

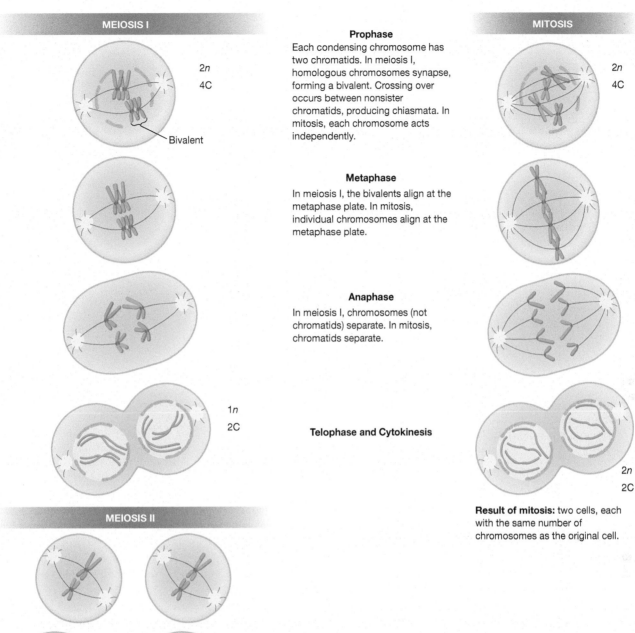

Each condensing chromosome has two chromatids. In meiosis I, homologous chromosomes synapse, forming a bivalent. Crossing over occurs between nonsister chromatids, producing chiasmata. In mitosis, each chromosome acts independently.

$2n$
4C

Metaphase

In meiosis I, the bivalents align at the metaphase plate. In mitosis, individual chromosomes align at the metaphase plate.

Anaphase

In meiosis I, chromosomes (not chromatids) separate. In mitosis, chromatids separate.

$1n$
2C

Telophase and Cytokinesis

$2n$
2C

Result of mitosis: two cells, each with the same number of chromosomes as the original cell.

MEIOSIS II

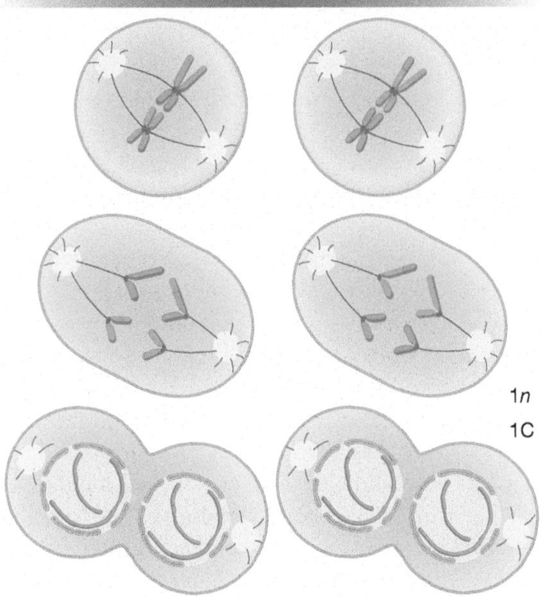

In meiosis II, sister chromatids separate.

Result of meiosis: four haploid cells, each with half as many chromosomes as the original cell. Each haploid cell contains a random mixture of maternal and paternal chromosomes.

Figure 25-6 Meiosis and Mitosis Compared. Meiosis and mitosis are both preceded by DNA replication, resulting in two sister chromatids per chromosome at prophase. Meiosis, which occurs only in sex cells, includes two nuclear (and cell) divisions, halving the chromosome number to the haploid level. Moreover, during the elaborate prophase of the first meiotic division, homologous chromosomes synapse, and crossing over occurs between nonsister chromatids. Mitosis involves only a single division, producing two diploid cells, each with the same number of chromosomes as the original cell. In mitosis, the homologous chromosomes behave independently; at no point do they come together in pairs. The meaning of the C values in this figure (4C, 2C, 1C) is described on page 760.

Prophase II is very brief. If detectable at all, it is much like a mitotic prophase. Metaphase II also resembles the equivalent stage in mitosis, except that only half as many chromosomes are present at the spindle equator. The kinetochores of sister chromatids now face in opposite directions, allowing the sister chromatids to separate and move (as new daughter chromosomes) to opposite spindle poles during anaphase II. The remaining phases of the second meiotic division resemble the comparable stages of mitosis. The final result is the formation of four daughter cells, each containing a haploid set of chromosomes. Because the two members of each homologous chromosome pair were randomly distributed to the two cells produced by meiosis I, each of the haploid daughter cells produced by meiosis II contains a random mixture of maternal and paternal chromosomes. Moreover, each of these chromosomes is composed of a mixture of maternal and paternal DNA sequences created by crossing over during prophase I.

At this point, you may want to study Figure 25-6 in its entirety to review the similarities and differences between meiosis and mitosis. In this diagram, the amount of DNA present at various stages is indicated using the **C value**, which corresponds to the amount of DNA present in a single (haploid) set of chromosomes. In a diploid cell prior to S phase, the chromosome number is $2n$ and the DNA content is 2C because two sets of chromosomes are present. When DNA undergoes replication during S phase, the DNA content is doubled to 4C because each chromosome now consists of two chromatids (but is still considered $2n$). In meiosis I, segregation of homologous chromosomes into different daughter cells reduces the chromosome number from $2n$ to $1n$ and the DNA content from 4C to 2C. Sister chromatid separation during meiosis II then reduces the DNA content from 2C to 1C, while the chromosome number remains at $1n$. In contrast, a normal mitosis reduces the DNA content from 4C to 2C (by sister chromatid separation), while the chromosome number remains at $2n$.

Defects in Meiosis Lead to Nondisjunction

The two successive divisions of meiosis start with a single diploid meiotic precursor cell. Two divisions later, the result is four haploid cells. In human males, all four of these cells give rise to gametes (sperm), whereas the highly asymmetric meiotic divisions in females lead to one large gamete (the oocyte). Although these divisions are remarkably faithful, what would happen if mistakes in chromosome segregation occur during meiosis? When a zygote results from a sperm-oocyte union in which one of the two gametes has an incorrect number of chromosomes, the zygote in turn will have an abnormal number of chromosomes, a situation known as *aneuploidy*. These occasional mishaps in chromosomal partitioning are referred to as **nondisjunction**. Depending on the chromosome(s) involved, nondisjunction can be lethal or lead to lifelong consequences.

Nondisjunction can occur during either meiosis I or meiosis II (**Figure 25-7**). In either case, one daughter cell receives two of the same type of chromosome and the other receives none. As we have seen, the normal complement of chromosomes in a diploid cell is written as $2n$ by geneticists. Fertilization involving a gamete with an extra copy of one particular chromosome will result in a zygote with *three* copies of the chromosome, a condition known as *trisomy*, usually indicated as $2n + 1$. The converse case, in which a gamete is lacking one chromosome entirely, is indicated with $2n - 1$. In this case the zygote is said to be *monosomic*. As the cells of the zygote divide during early embryonic development, all successive cells will inherit the chromosomal abnormality. As one example, consider an abnormal human sperm containing two

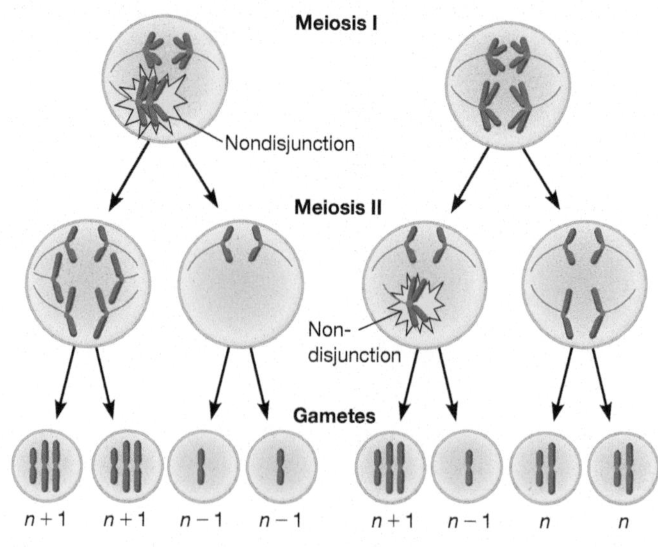

Figure 25-7 Meiotic Nondisjunction. Gametes with an abnormal chromosome number can arise by nondisjunction in either meiosis I or meiosis II.

copies of chromosome 21. If this sperm fertilizes a normal egg containing one copy of chromosome 21, the resulting embryo, possessing three copies of chromosome 21, can develop fully and lead to the birth of a child, but this child will exhibit a series of developmental abnormalities that together constitute *Down syndrome*. (Trisomy 21 and other nondisjunction events in humans are covered in detail in **Human Connections, pages 762–763.**)

Monosomy and trisomy are not the only possibilities. In some cases, organisms contain complete, multiple sets of *all* chromosomes. The general term for this situation is *polyploidy* (note that "ploidy" refers to whole sets of chromosomes, whereas "–somy" refers to single chromosomes). *Triploidy (3n)* and *tetraploidy (4n)* refer to situations in which, instead of the standard two sets of chromosomes, an organism's cells contain three or four sets of chromosomes, respectively. In many well-studied cases, especially in plants, polyploids have fewer abnormalities than aneuploids. Apparently having only one extra (or missing) chromosome is more harmful to overall gene dosage, the number of copies found in the genome for a given gene, than having an entire extra set of chromosomes.

How might polyploidy arise? A triploid cell might arise when one gamete has two sets of chromosomes due to wholesale nondisjunction of all its chromosomes. Although there are other possibilities, one way in which tetraploidy can arise is from failure of a $2n$ zygote to divide after replicating its chromosomes at the first mitotic division. Subsequent normal mitotic divisions would then produce a $4n$ embryo. Polyploidy is fairly common in the plant kingdom, where it is thought to have been an important driving force in plant evolution. Although clear cases of polyploidy are uncommon among animals, genome sequencing of bony fish and some amphibians indicates that some of them have undergone complete genome duplication at some point in the past.

Sperm and Egg Cells Are Generated by Meiosis Accompanied by Cell Differentiation

Meiosis lies at the heart of gametogenesis, which, as we saw earlier, is the process of forming haploid gametes from diploid precursor cells. But male and female gametes differ significantly in structure, which means that gametogenesis must consist of more than just meiosis. **Figure 25-8** on page 764 is a schematic depiction of gametogenesis in animals. In males, meiosis converts a diploid *spermatocyte* into four haploid *spermatids* of similar size (Figure 25-8a). After meiosis has been completed, the spermatids then differentiate into sperm cells by discarding most of their cytoplasm and developing flagella and other specialized structures.

In females, meiosis converts a diploid *oocyte* into four haploid cells, but only one of the four survives and gives rise to a functional egg cell (Figure 25-8b). This outcome is generated by two meiotic divisions that divide the cytoplasm of the oocyte unequally, with one of the four daughter cells receiving the bulk of the cytoplasm of the original diploid oocyte. The other three, smaller cells, called **polar bodies**, usually degenerate (although recall that the Komodo dragon can use a polar body to "fertilize" an egg). The advantage of having only one of the four haploid products of meiosis develop into a functional egg cell is that the cytoplasm that would otherwise have been distributed among four cells is instead concentrated into one egg cell, maximizing the content of stored nutrients in each egg.

An important difference between sperm and egg formation concerns the stage at which the cells acquire the specialized characteristics that make them functionally mature gametes. During sperm cell development, meiosis creates haploid spermatids that must then discard most of their cytoplasm and develop flagella before they are functionally mature. In contrast, developing egg cells acquire their specialized features *during* the process of meiosis. Many of the specialized features of the egg cell are acquired during prophase I, when meiosis is temporarily halted to allow time for extensive cell growth. During this *growth phase*, the cell also develops various types of external coatings designed to protect the egg from chemical and physical injury. The amount of growth that takes place during this phase can be quite extensive. A human egg cell, for example, has a diameter of about 100 μm, giving it a volume more than 100 times as large as that of the diploid oocyte from which it arose. And consider the gigantic size of a bird egg!

By the time meiosis is completed, the egg cell is fully mature and may even have been fertilized. The mature egg is a highly differentiated cell that is specialized for the task of producing a new organism in much the same sense that a muscle cell is specialized to contract or a red blood cell is specialized to transport oxygen. This inherent specialization of the egg is vividly demonstrated by the observation that even in the absence of fertilization by a sperm cell, many kinds of animal eggs can be stimulated to begin dividing and progress through significant phases of early embryonic development by artificial treatments as simple as a pinprick. Hence, virtually everything needed for programming the early stages of development must already be present in the egg.

Normally, of course, this developmental sequence is activated by the interaction between sperm and egg, which triggers important biochemical changes in the egg, including a burst of metabolic activity in preparation for embryonic development. Fertilization in plants involves similarly complex cellular and biochemical events.

Meiotic Maturation of Oocytes Is Tightly Regulated

In reproductively mature vertebrates, sperm are generated continuously through the male's lifetime, as populations of cells in the testis continue to divide via mitosis. These cells are *stem cells*, able to produce more of themselves or more differentiated cells that enter meiosis. In contrast, oocytes use a very different strategy during their maturation. Amphibian oogenesis is probably the best understood and is the focus here, but many of the events are similar in mammals.

After their growth phase has been completed, developing frog oocytes remain arrested in prophase I until resumption of meiosis is initiated by an appropriate stimulus, which in frogs turns out to be the hormone progesterone. The arrest of oocytes while they await this hormonal signal implies that something inside the oocytes stimulates resumption of meiosis once progesterone is present. Crucial experiments by Yoshio Masui and colleagues identified this stimulatory molecular complex (**Figure 25-9** on page 765). They demonstrated that if cytoplasm taken from a mature egg cell is injected into the cytoplasm of an immature oocyte that is awaiting hormonal stimulation, the oocyte immediately proceeds through meiosis. Masui therefore hypothesized that a cytoplasmic chemical, which he named *MPF (maturation-promoting factor)*, induces oocyte maturation (that is, meiotic division).

Subsequent experiments demonstrated that this acronym was very apt: biochemical studies of MPF revealed that it is just like mitosis-promoting factor (also called MPF; see Chapter 24). It consists of two subunits: a *cyclin-dependent kinase (Cdk)* and a *cyclin*. Progesterone exerts its control over MPF by stimulating the production of *Mos*, a protein kinase that activates a series of other protein kinases, which in turn leads to the activation of MPF (**Figure 25-10** on page 765).

❶ In response to the activation of MPF, the first meiotic division is completed. In most vertebrate eggs, including frogs, the second meiotic division is arrested at metaphase II until fertilization takes place. In frogs, metaphase II arrest is triggered by *cytostatic factor (CSF)*, a biochemical activity present in the cytoplasm of mature eggs. One important protein that seems to be involved in the activities of CSF is the protein *Emi2/Erp1* (Emi stands for early mitotic inhibitor). ❷ Emi2 works by inhibiting the *anaphase-promoting complex (APC)*, whose activity is normally required for the transition from metaphase to anaphase (see Figure 24-18). ❸ After an egg is fertilized, kinases phosphorylate Emi2, which leads to its destruction. ❹ In the absence of Emi2, the APC is free to trigger the transition from metaphase to anaphase, thereby allowing meiosis to be completed.

CONCEPT CHECK 25.2

If the DNA content of a diploid cell in the G1 phase of the cell cycle is *X*, what would the DNA content be at metaphase of meiosis I if that cell entered meiosis? Assuming that the cell divided and its daughters entered meiosis II, what would their DNA content be at metaphase of meiosis II?

HUMAN *Connections*

When Meiosis Goes Awry

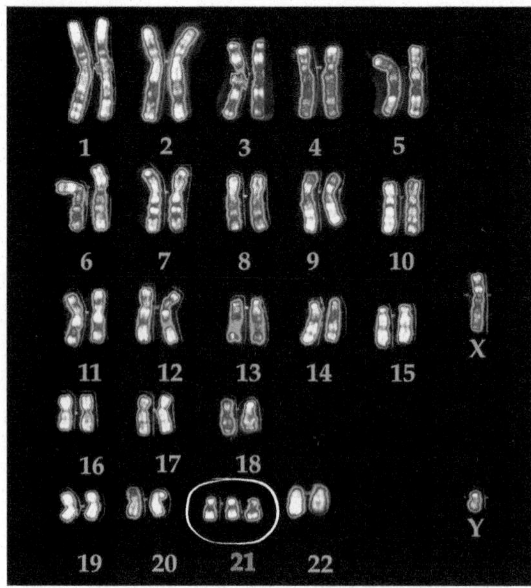

(a) Karyotype of an individual with trisomy 21

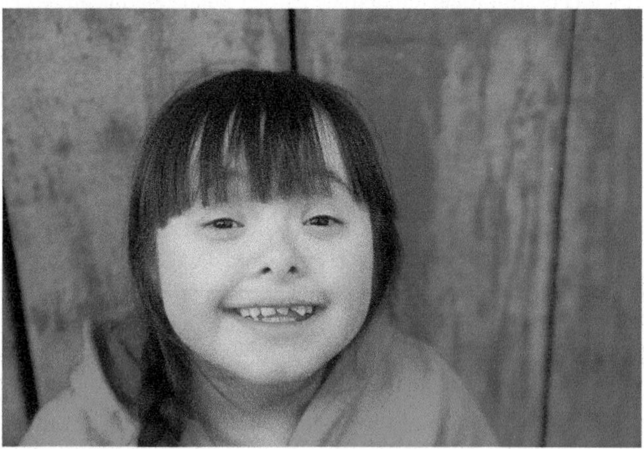

(b) An individual with Down syndrome

Figure 25A-1 Down Syndrome. (a) This karyotype shows trisomy 21, the most common cause of Down syndrome. **(b)** This child exhibits facial features characteristic of this disorder.

Perhaps you know a couple expecting their first child. Perhaps, as seems all too common today, your friends had been trying to conceive for some time, and the expectant mother is now in her late thirties, or even her early forties. For most couples like this, the first ultrasound is an amazing experience, as the expectant parents get their first glimpse of their child. But sometimes that exhilaration quickly turns to concern, as the ultrasound reveals developmental problems, such as heart or intestinal defects that require neonatal surgery after the baby is born. Often *amniocentesis* is next performed, which involves withdrawal and culturing of cells from the amniotic fluid surrounding the developing fetus to assess whether there are any genetic anomalies. In some cases, these tests reveal a special type of aneuploidy that causes a condition known as *Down syndrome*.

Down Syndrome. Down syndrome affects roughly 1 of every 700 children born in the United States. It usually results from an extra chromosome 21; as a result, Down syndrome is often called *trisomy 21*. Trisomy 21 is by far the most common type of trisomy observed among live births. This is likely a direct result of chromosome 21 being one of the smallest chromosomes with the fewest active genes. Thus, effects of increased gene dosage will be the most limited. Individuals with Down syndrome have characteristic features, affecting the face, head, neck, and tongue, and short stature (**Figure 25A-1**). Roughly half of those with Down syndrome have heart defects, and a smaller percentage have intestinal defects, which are now usually correctable by surgery. In addition, individuals with Down syndrome have varying degrees of cognitive deficit. Almost all affected males and roughly half of females are sexually underdeveloped and sterile. People with Down syndrome, on average, have a life span shorter than $2n$ individuals, but most live to middle age and beyond and are loved and valuable members of society.

What causes Down syndrome? The answer to this question remains unclear, but several observations provide clues. First, in general, trisomy in humans is much more common when it involves smaller chromosomes (for example, chromosomes 13–22) than for larger chromosomes (chromosomes 1–12). Second, errors during female meiosis seem to account for most (~90%) of Down syndrome cases. Finally, the frequency of meiotic errors increases dramatically with the age of the mother. Although the frequency of Down syndrome is very low among children born to women under age 30, the risk climbs sharply by age 40, and continues to climb thereafter (**Figure 25A-2**). The reasons for this correlation between maternal age and likelihood of Down syndrome remain unclear, but clearly something happens in the ovaries of older women that leads to a higher frequency of nondisjunction. Most cases of

25.3 Genetic Variability: Segregation and Assortment of Alleles

As we have seen, one of the main functions of meiosis is to preserve the chromosome number in sexually reproducing organisms. Equally important, however, is the role meiosis plays in generating genetic diversity in sexually reproducing populations. Meiosis is the point in the flow of genetic information where novel combinations of chromosomes (and the alleles they carry) are assembled in gametes for potential passage to offspring.

Meiosis Generates Genetic Diversity

Although every gamete is haploid, and hence possesses one member of each pair of homologous chromosomes, the particular combination of paternal and maternal chromosomes in any given gamete is random. This randomness is generated

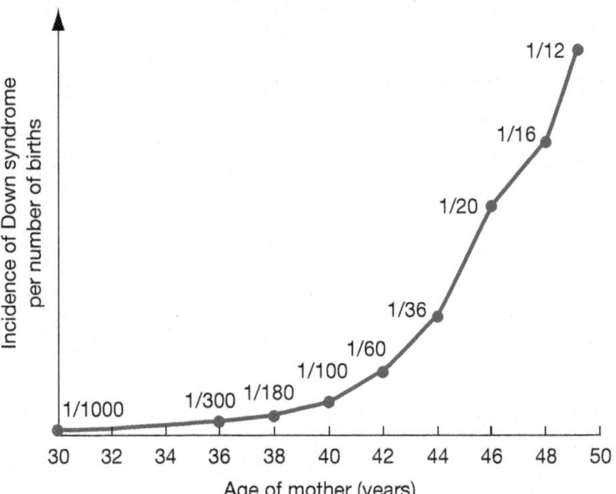

Figure 25A-2 Frequency of Down Syndrome Increases Dramatically with Maternal Age.

trisomy 21 appear to result from nondisjunction during meiosis I. In a minority of cases, (about 5%), Down syndrome is not caused by a nondisjunction event. Instead, a third copy of chromosome 21 becomes fused to chromosome 14. If a gamete carries this abnormal chromosome 14, it carries along with it an extra copy of the genes from chromosome 21, resulting in Down syndrome.

It is the presence of extra copies of genes on chromosome 21 that seems to cause the cluster of symptoms in individuals with Down syndrome. But which genes are important? Several candidates have been identified in a region of chromosome 21 known as the *Down syndrome critical region (DSCR)*. These include a protein kinase known as DYRK and a cell surface protein known as Dscam. Identifying how incorrect gene dosage at these or other loci in the DSCR lead to the syndrome's characteristic symptoms is a current challenge in understanding the disease and proposing possible treatments.

Aneuploidy of Sex Chromosomes. Down syndrome involves an *autosome,* that is, a non-sex chromosome. Except for Down syndrome, individuals who survive to birth with autosomal aneuploidies are rare. Aneuploid conditions involving *sex chromosomes* are much more common among live births. One reason for this higher prevalence may be that relatively few genes reside on the human Y chromosome; in addition, one of the two X chromosomes is inactivated in somatic (non-sex) cells in humans as part of the process of dosage compensation.

An extra X chromosome in a male, producing an individual who is XXY, results in *Klinefelter syndrome*. Even though the extra X chromosome is inactivated, such individuals have male sex organs, but they are sterile and their testes are small. The resulting low testosterone during development can sometimes cause a noticeable change in normal digit ratios on the right hand (**Figure 25A-3**). In addition, some breast enlargement and other female body characteristics, such as reduced body and facial hair, are common. About 1 of every 1000 males is born with an extra Y chromosome (XYY), and 1 of every 1000 females has an extra X chromosome. Remarkably, in both cases there are no clearly defined abnormalities, but such individuals tend to be taller than average. Monosomy X, known as *Turner syndrome,* occurs about once in every 5000 female births. Although phenotypically female, such individuals are sterile because their sex organs do not mature. When provided with estrogen replacement therapy, however, individuals with Turner syndrome can develop secondary sex characteristics.

Although the precise reasons for the specific observed patterns of aneuploidies in the human population remain under investigation, successful meiosis is clearly key for human health and well-being.

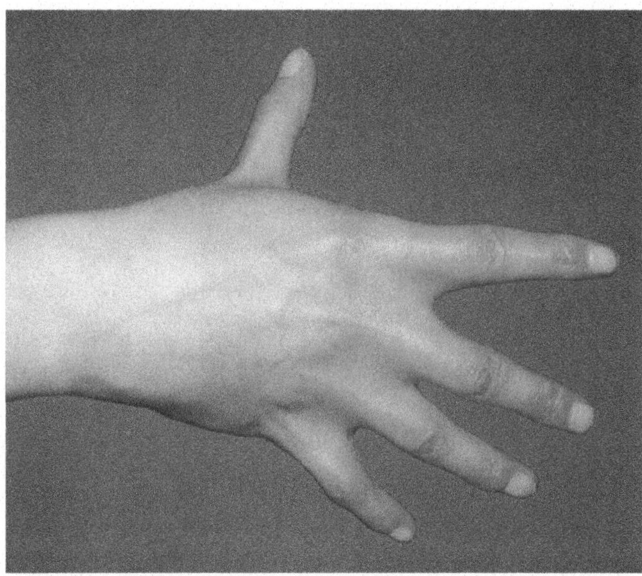

Figure 25A-3 Right Hand of an Individual with Klinefelter Syndrome. Notice the altered digit ratio between the second and fourth fingers. Low testosterone in utero will sometimes cause the index finger to be significantly longer than the ring finger.

by the random orientation of bivalents at metaphase I, where the paternal and maternal homologues of each pair can face either pole, independent of the orientations of the other bivalents. In human gametes, which contain 23 chromosomes, the end result is more than 8 million different combinations of maternal and paternal chromosomes!

Moreover, the crossing over that takes place between homologous chromosomes during prophase I generates additional genetic diversity among the gametes. By allowing the exchange of DNA segments between homologous chromosomes, crossing over generates more combinations of alleles than the random assortment of maternal and paternal chromosomes would create by itself. First, we turn to the historic experiments that revealed the genetic consequences of chromosome segregation and random assortment during meiosis.

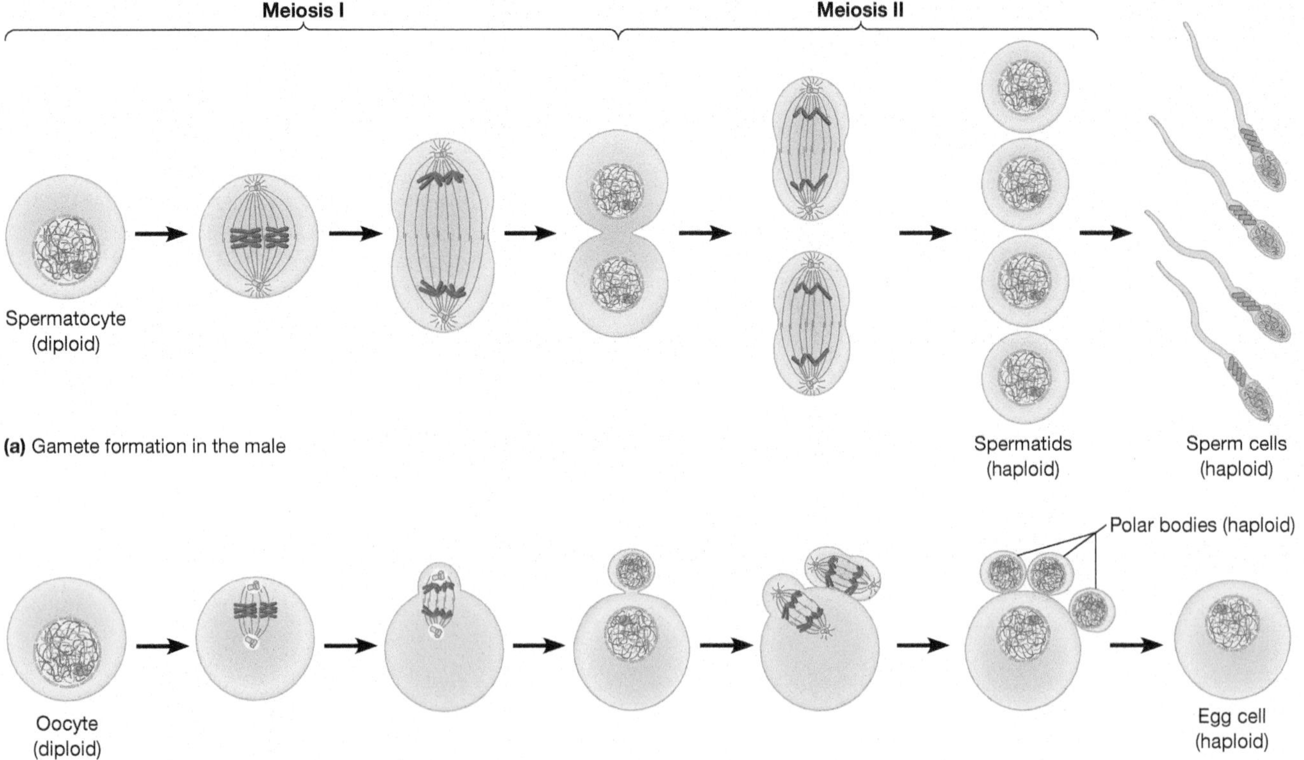

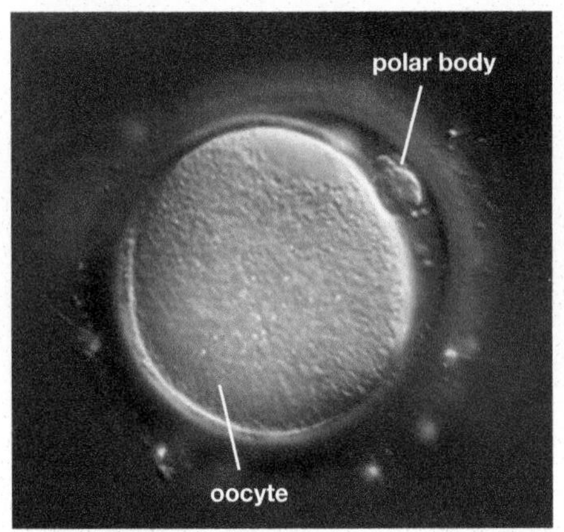

(c) Polar body formation in a human secondary oocyte

Figure 25-8 Gamete Formation. **(a)** In the male, all four haploid products of meiosis are retained and differentiate into sperm. **(b)** In the female, both meiotic divisions are asymmetric, forming one large egg cell and three (in some cases, only two) small cells called polar bodies that do not give rise to functional gametes. Although not indicated here, the mature egg cell usually grows much larger than the oocyte it arose from. **(c)** A human oocyte with polar body; the oocyte/polar bodies lie underneath a protein shell known as the zona pellucida (differential interference contrast micrograph).

Information Specifying Recessive Traits Can Be Present Without Being Displayed

Most students of biology have heard of Gregor Mendel and the classic genetics experiments he conducted in a monastery garden. Mendel's findings, first published in 1865, laid the foundation for what we now know as *Mendelian genetics.* Working with the common garden pea, Mendel chose seven readily identifiable characters of pea plants and selected in each case two varieties of plants that displayed different forms of the character. For example, flower color was one character Mendel chose because he had one strain of peas with purple flowers and another with white flowers. He first established that each of the plant strains was *true-breeding* upon self-fertilization, which means that plants with purple flowers, when self-pollinated, produced seeds that only gave rise to purple-flowered plants, and plants with white flowers produced only white-flowered plants. Once this had been established, Mendel was ready to investigate the principles that govern the inheritance of such traits.

In his first set of experiments, Mendel *cross-fertilized* the true-breeding parental plants (the **P generation**) to produce **hybrid** strains (**Figure 25-11** on page 766). The outcome of this experiment must have seemed mystifying at first: in every case, the resulting offspring—called the **F_1 generation**—exhibited one or the other of the parental traits, but never both. In other words, one parental trait was always *dominant* and the other was always *recessive*. In the case of flower color, for example, all the F_1 plants had purple flowers, indicating that purple flower color is dominant.

During the next summer, Mendel allowed all the F_1 hybrids to self-fertilize. For each of the seven characters under study, he made the same surprising observation:

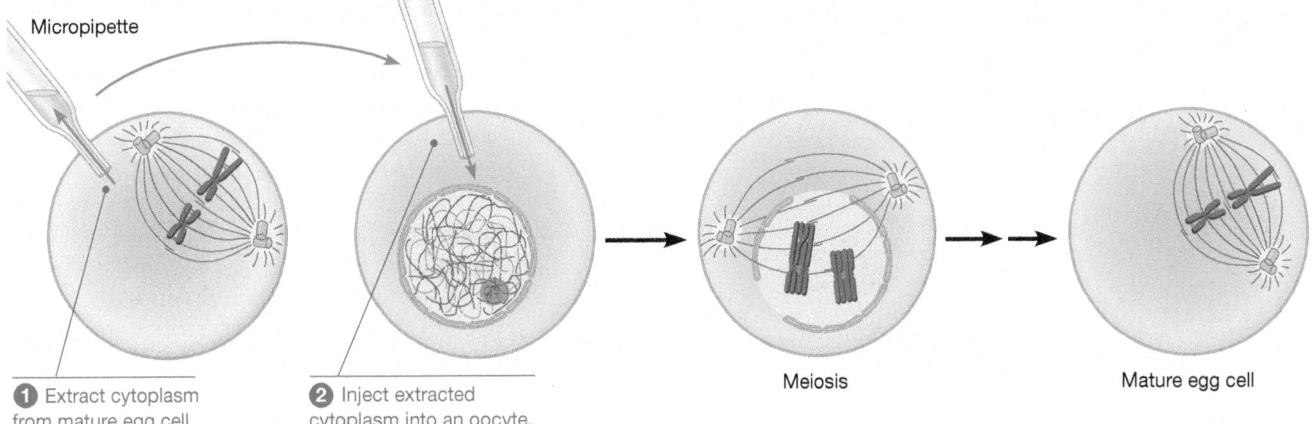
Micropipette
1 Extract cytoplasm from mature egg cell.
2 Inject extracted cytoplasm into an oocyte.
Meiosis
Mature egg cell

Figure 25-9 Evidence for the Existence of MPF. Hormones act on frog oocytes to trigger meiosis and development into mature frog eggs, which are arrested (until fertilization) in metaphase of the second meiotic division. In 1971 Y. Masui and C. L. Markert established the existence of a substance they called maturation-promoting factor (MPF) involved in this process. They used a micropipette to remove cytoplasm from a mature egg cell (arrested in metaphase of the second meiotic division) and injected it into an immature oocyte. The oocyte then proceeded through meiosis and became a mature egg cell, under the influence of MPF.

the recessive trait that had seemingly disappeared in the F_1 generation reappeared among the progeny in the next generation, the **F_2 generation**. Moreover, for each of the seven characters, the ratio of dominant to recessive phenotypes in the progeny was always about 3:1. In the case of flower color, for example, plants grown from the seeds produced by the F_1 plants produced about 75% purple-flowered plants and 25% white-flowered plants (bottom of Figure 25-11).

This outcome was quite different from the behavior of the true-breeding purple flowers of the parent strain, which had produced only purple flowers when self-fertilized. Clearly, there was an important difference between the purple-flowered plants of the P stock and those of the F_1 generation. They looked alike, but the former bred true, whereas the latter did not.

Next, Mendel investigated the F_2 plants through self-fertilization. The F_2 plants exhibiting the recessive trait (white, in the case of flower color) always bred true, suggesting they were genetically identical to the white-flowered parental strain that Mendel had begun with. F_2 plants with the dominant trait yielded a more complex pattern. One-third bred true for the dominant trait and therefore seemed identical to the parental plants with the dominant (purple-flowered) trait. The other two-thirds of the purple-flowered F_2 plants produced progeny with both dominant and recessive phenotypes,

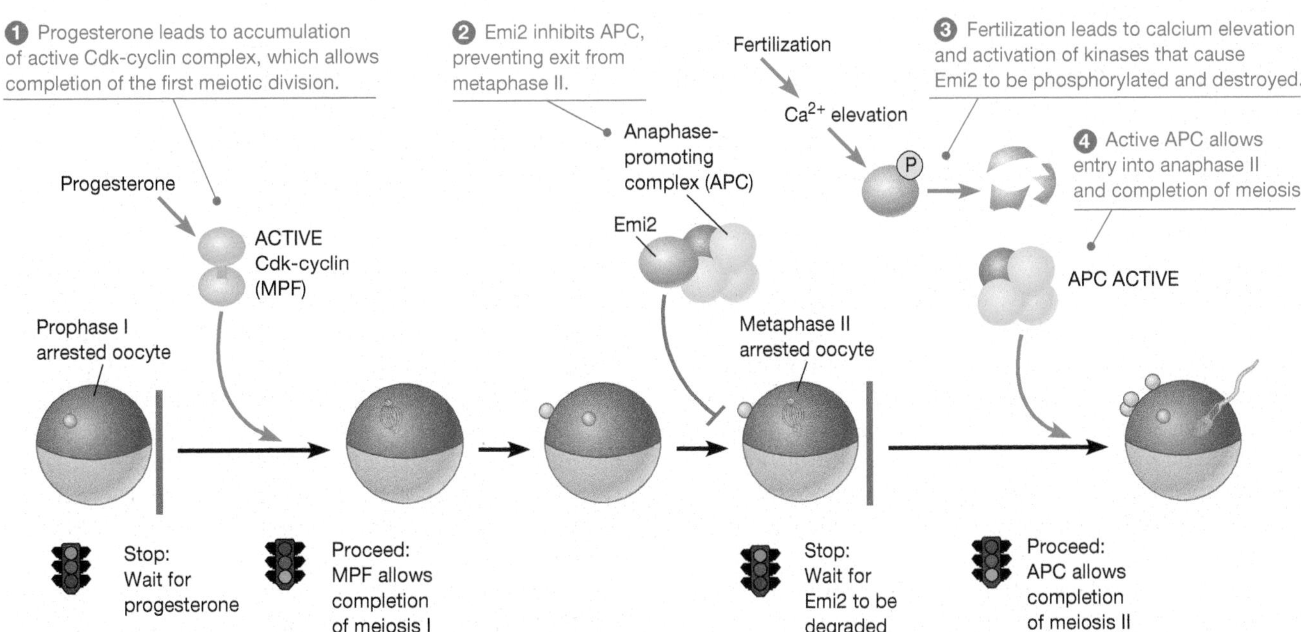
1 Progesterone leads to accumulation of active Cdk-cyclin complex, which allows completion of the first meiotic division.
2 Emi2 inhibits APC, preventing exit from metaphase II.
Fertilization
3 Fertilization leads to calcium elevation and activation of kinases that cause Emi2 to be phosphorylated and destroyed.
Ca2+ elevation
Anaphase-promoting complex (APC)
4 Active APC allows entry into anaphase II and completion of meiosis.
P
Progesterone
ACTIVE Cdk-cyclin (MPF)
Emi2
APC ACTIVE
Prophase I arrested oocyte
Metaphase II arrested oocyte
Stop: Wait for progesterone
Proceed: MPF allows completion of meiosis I
Stop: Wait for Emi2 to be degraded
Proceed: APC allows completion of meiosis II

Figure 25-10 Regulation of Meiotic Division in Amphibian Oocytes. Entry into and exit from meiosis are regulated by cell signaling. These various checkpoints are regulated as shown by the hormone progesterone, the accumulation of MPF, the degradation of Emi2, and the activation of APC.

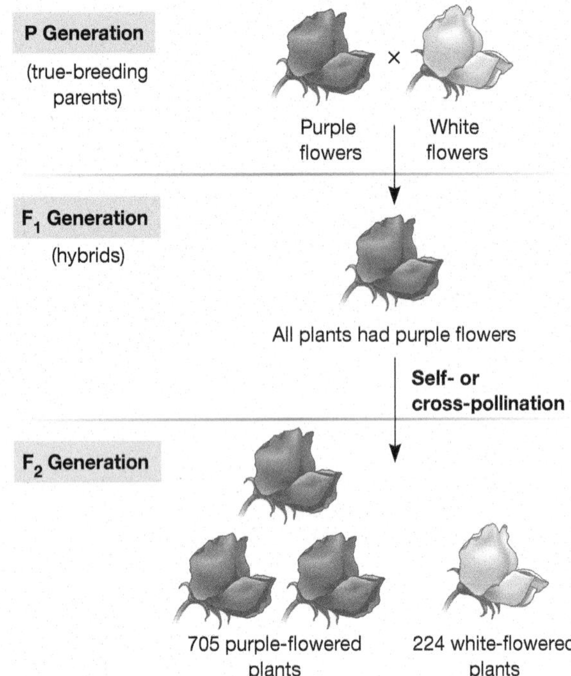

Figure 25-11 Genetic Analysis of Flower Color in Pea Plants. Mendel performed genetic crosses starting with two true-breeding parental (P) strains of pea plants, one having purple flowers and one having white flowers. When crossed, the parent stocks yield F_1 plants (hybrids) that all have purple flowers. Upon self-fertilization, the F_1 plants produce purple- and white-flowered F_2 plants in a ratio of roughly 3:1.

in a 3:1 ratio—the same ratio that had arisen from the F_1 self-fertilization.

Alleles of Each Gene Segregate from Each Other During Gamete Formation

After a decade of careful work documenting the preceding patterns of inheritance and those obtained from other related breeding experiments, Mendel formulated several principles—now known as **Mendel's laws of inheritance**—that explained the results he had observed. The first of these principles was that *phenotypic traits are determined by inheritance of alternative versions of discrete factors*. We now call these factors **genes**, and the different versions of a gene are called **alleles**. The place on a chromosome that contains the DNA sequence for a particular gene is termed an individual genetic **locus** (plural: **loci**). These genetic terms and others are illustrated in **Figure 25-12**, which depicts how the flowers from Figure 25-11 produce their purple coloring. For simplicity, we will assume that the gene controls a single, clear-cut characteristic—or *character*, as geneticists usually say—in the organism. Let's also assume that only one copy of this gene is present per haploid genome, so that a diploid organism will have two copies of the gene, which may be either identical or slightly different.

The second principle is that for each character, *an organism inherits two alleles of a gene, one from each parent*. An organism with two identical alleles for a given gene is said to be **homozygous** for that gene or for the character it determines. Thus, a pea plant that inherited the same allele for purple flower color from each of its parents is said to be homozygous for flower color. An organism with two different alleles for a gene is said to be **heterozygous** for that gene or character. A pea plant with one allele that specifies purple flower color and a second allele that specifies white flower color is therefore heterozygous for flower color. Mendel's experiments further indicated that in a heterozygous individual, one of the two alleles is often **dominant** and the other **recessive**. These terms convey the idea that the dominant allele determines how the trait will appear in a heterozygous individual. For flower color in peas, the purple trait is dominant over white; this means that pea plants heterozygous for flower color have purple flowers. This also means that having one allele for purple flower color is sufficient for a pea plant to have purple flowers. A dominant allele is usually designated by an uppercase letter that stands for the trait, whereas a recessive allele is represented by the same letter in lowercase. In both cases, italics are used. Thus, alleles for flower color in peas are represented by *P* for purple and *p* for white because the purple trait is dominant.

It is important to distinguish between the **genotype**, or genetic makeup of an organism, and its **phenotype**, or the physical expression of its genotype. The phenotype of an organism can usually be determined by inspection (for example, by looking to see whether flowers are purple or white). Genotype, on the other hand, can be *directly* determined only by studying an organism's DNA. In the case shown in Figure 25-12, one allele has a particular DNA sequence at the locus specifying flower color, whereas the other allele differs in DNA sequence. DNA sequencing techniques can in fact be used to confirm this difference, but the genotype also can be deduced from indirect evidence, such as the organism's phenotype and information about the phenotypes of its parents and/or offspring. Organisms exhibiting the same phenotype do not necessarily have identical genotypes. A good example is shown in Figure 25-11; pea plants in the F_1 generation, which all have purple flowers (phenotype) can be either *PP* or *Pp* (genotype).

Mendel's conclusion seems almost self-evident to us, but it was an important assertion in his day. At that time, most scientists favored a *blending theory of inheritance*, a theory that viewed traits such as purple and white flower color rather like cans of paint that are poured together to yield intermediate results. Other investigators had described the nonblending nature of inheritance before Mendel but without the accompanying quantitative analysis that was Mendel's special contribution. Mendel's breakthrough is especially impressive when we recall that he formulated his theory before anyone had seen chromosomes.

Of special importance to the development of genetics was Mendel's conclusion regarding the way genes are parceled out during gamete formation. According to his **law of segregation**, *the two alleles of a gene are distinct entities that segregate, or separate from each other, during the formation of gametes*. In other words, the two alleles retain their identities even when both are present in a hybrid organism, and they are then parceled out into separate gametes so they can emerge unchanged in later generations.

Mendel's principle of segregation accounted nicely for the results he obtained in his crosses with purple- and

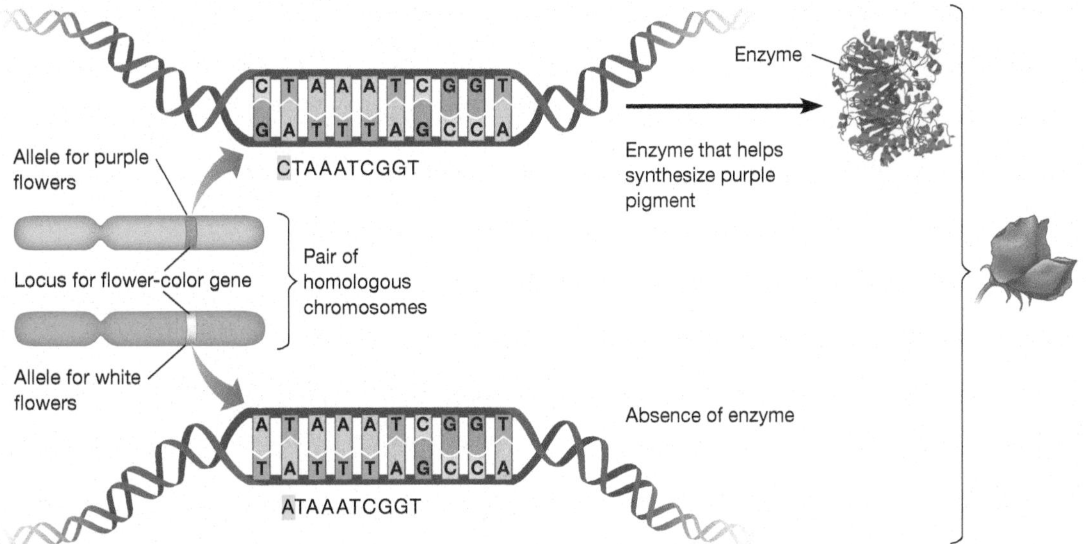

Figure 25-12 Some Genetic Terminology. This diagram shows a homologous pair of chromosomes from a diploid cell of an F_1 pea plant from Figure 25-11. The chromosomes are the same size and shape and carry genes for the same characters (characteristics) in the same order. Red and blue are used here and in many later figures to distinguish the different parental origins of the two chromosomes. The site of the flower-color gene (locus) is shown on each chromosome. The paternally inherited chromosome (blue) has one version of the flower color gene (allele) for purple flowers, which encodes a protein that indirectly controls synthesis of purple pigment. The maternally inherited chromosome (red) has an allele for white flowers, which results in no functional protein being made. The DNA sequence from the flower-color allele is shown for each. In the case of flower color, plants that possess one allele for purple flowers make sufficient quantities of the relevant enzyme to produce purple flowers.

white-flowered plants (**Figure 25-13** on page 768). If he assumed that the true-breeding purple-flowered parental plants were homozygous for a dominant allele at the flower-color locus (*P*) and the white-flowered parents were homozygous for a recessive allele (*p*), then all of the F_1 plants would have the genotype *Pp*. If he then assumed that one or the other allele found its way into each gamete with equal likelihood, then when these plants self-pollinated, there were four equally likely possible combinations of gametes in the F_2 plants. These possibilities can be represented using a *Punnett square*, a diagram representing all possible outcomes of a genetic cross. The genotypes of all possible gametes from the male and female parents are arranged along two adjacent sides of a square, and each box in the matrix is then used to represent the genotype resulting from the union of the two gametes at the heads of the intersecting rows. What color will the flowers of these F_2 plants be? One-fourth will inherit two purple-flower alleles (*PP*), one-half will inherit one purple-flower and one white-flower allele (*Pp*), and the remainder will inherit two white-flower alleles (*pp*). If purple is dominant, all plants with at least one purple allele will be purple, accounting for the 3:1 ratio Mendel observed.

Alleles of Each Gene Segregate Independently of the Alleles of Other Genes

In addition to the crosses already described, each focusing on a single pair of alleles, Mendel studied *multifactor* crosses between plants that differed in *several* characters. Besides differing in flower color, for example, the plants he crossed might differ in seed shape and color or flower position. As in his single-factor (monohybrid) crosses, Mendel used parent plants that were true-breeding (homozygous) for the characters he was testing and generated F_1 hybrids heterozygous for each character. He then self-fertilized these dihybrids and determined how frequently the dominant and recessive forms of the various characters appeared among the progeny.

Mendel found that all possible combinations of traits appeared in the F_2 progeny, and he concluded that all possible combinations of the different alleles must therefore have been present among the F_1 gametes. Furthermore, based on the proportions of the various phenotypes detected in the F_1 generation, Mendel deduced that all possible combinations of alleles occurred in the gametes with equal frequency. In other words, *the two alleles of each gene segregate independently of the alleles of other genes.* This is the **law of independent assortment**, another cornerstone of genetics. Later, it would be shown that this law applies only to genes on different chromosomes, or ones far enough apart on the same chromosome for crossing over to always occur between them.

Chromosome Behavior Explains the Laws of Segregation and Independent Assortment

By 1875 microscopists had identified chromosomes with the help of stains produced by the developing aniline dye industry. At about the same time, fertilization was shown to involve the fusion of sperm and egg nuclei, suggesting that the nucleus carries genetic information. The first proposal that chromosomes might be the bearers of this genetic information was made in 1883. Within 10 years, chromosomes had been studied in dividing cells and had been seen to split longitudinally into two apparently identical daughter chromosomes. With

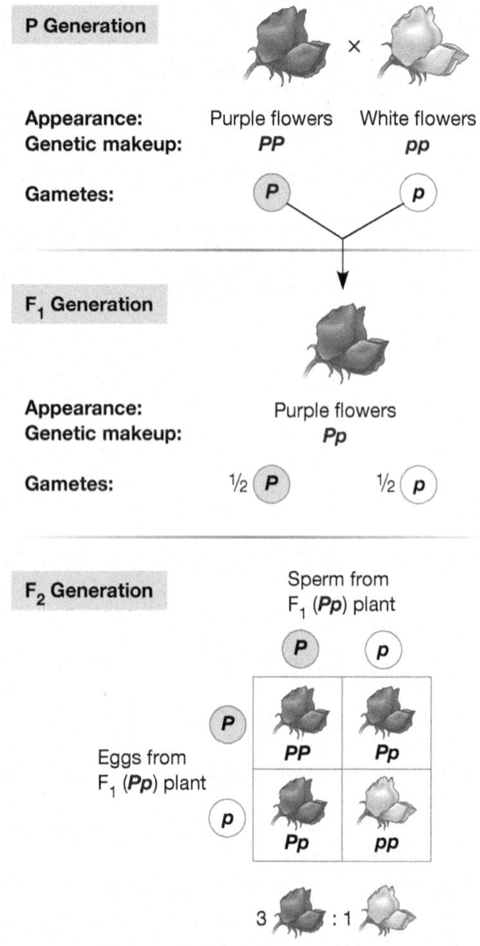

Figure 25-13 Mendel's Law of Segregation. This diagram shows the genetic makeup of the generations in Figure 25-11. Each plant has two alleles for the gene controlling flower color, one allele inherited from each of the plant's parents. The parental plants are homozygous for the purple (*P*) or white flower (*p*) trait. The F_1 plants are all heterozygous (*Pp*). To construct a Punnett square that predicts the F_2 generation offspring, we list all the possible gametes from one parent (here, the F_1 female) along the left side of the square and all the possible gametes from the other parent (here, the F_1 male) along the top. The boxes represent the offspring resulting from all the possible unions of male and female gametes. Random combination of the gametes results in the 3:1 ratio of purple- and white-flowered plants Mendel observed.

the invention of better optical systems, more detailed analysis of chromosomes became possible. Mitosis was shown to involve the movement of identical daughter chromosomes to opposite poles, thereby ensuring that daughter cells would have exactly the same complement of chromosomes as their parent cell.

Mendel's findings lay dormant in the scientific literature until 1900, when his paper was rediscovered almost simultaneously by three other European biologists: Edward Montgomery, Theodor Boveri, and Walter Sutton. Montgomery recognized the existence of pairs of homologous chromosomes, with one member of the pair of maternal origin and the other of paternal origin, which come together during synapsis in meiosis I. Boveri added the crucial observation that each chromosome plays a unique genetic role by studying sea urchin eggs fertilized by more than one sperm. Sutton, meanwhile, was studying meiosis in grasshoppers. In 1902, he made the important observation that the orientation of each pair of homologous chromosomes (bivalent) at the spindle equator during metaphase I is purely a matter of chance. Many different combinations of maternal and paternal chromosomes are therefore possible in the gametes produced by any given individual.

During 1902–1903, Sutton put the preceding observations together into a coherent *chromosomal theory of inheritance*, which can be summarized as five main points:

1. Nuclei of all cells except those of the *germ line* (sperm and eggs) contain two sets of homologous chromosomes, one set of maternal origin and the other of paternal origin.
2. Chromosomes retain their individuality and are genetically continuous throughout an organism's life cycle.
3. The two sets of homologous chromosomes in a diploid cell are functionally equivalent, each carrying a similar set of genes.
4. Maternal and paternal homologues synapse during meiosis and then move to opposite poles of the division spindle, thereby becoming segregated into different cells.
5. The maternal and paternal members of different homologous pairs segregate independently during meiosis.

The chromosomal theory of inheritance provided a physical basis for understanding how Mendel's genetic factors could be carried, transmitted, and segregated. **Figure 25-14** illustrates the chromosomal basis of Mendel's laws, using an example from his pea plants, the genes for seed color (alleles *Y* and *y*, where *Y* stands for yellow seeds, a dominant trait, and *y* for green seeds, a recessive trait) and seed shape (alleles *R* and *r*, where *R* stands for round seeds, which is dominant, and *r* for wrinkled seeds, which is recessive). Peas have seven pairs of chromosomes, but only the pairs bearing the alleles for seed color and shape are shown. The explanation for the segregation of alleles and independent assortment of genes is that there are two possible, equally likely arrangements of the chromosome pairs at metaphase I. Of the *YyRr* cells that undergo meiosis, half will produce gametes like the four at the bottom left of the figure, and half will produce gametes like the four at the bottom right. Therefore, the *YyRr* plant will produce equal numbers of gametes of the eight types.

The rules that Mendel devised are remarkably powerful and universal. Even in an era in which direct sequencing of DNA is becoming commonplace, Mendel's rules for segmentation of traits are still important. The same laws of segregation and independent assortment that Mendel discovered in his peas apply to traits in humans. (**Key Technique, pages 770–771**, shows how these elegant rules can provide insight into inherited human diseases.)

As powerful as Mendel's rules are, as we will discuss shortly, the law of independent assortment holds only for genes on *different* chromosomes (or for genes located far apart on the same chromosome). It is remarkable that Mendel happened to choose seven independently assorting characters in an organism that has only seven pairs of chromosomes. Mendel was also fortunate that each of the seven characters he chose to study turned out to be controlled by a single gene

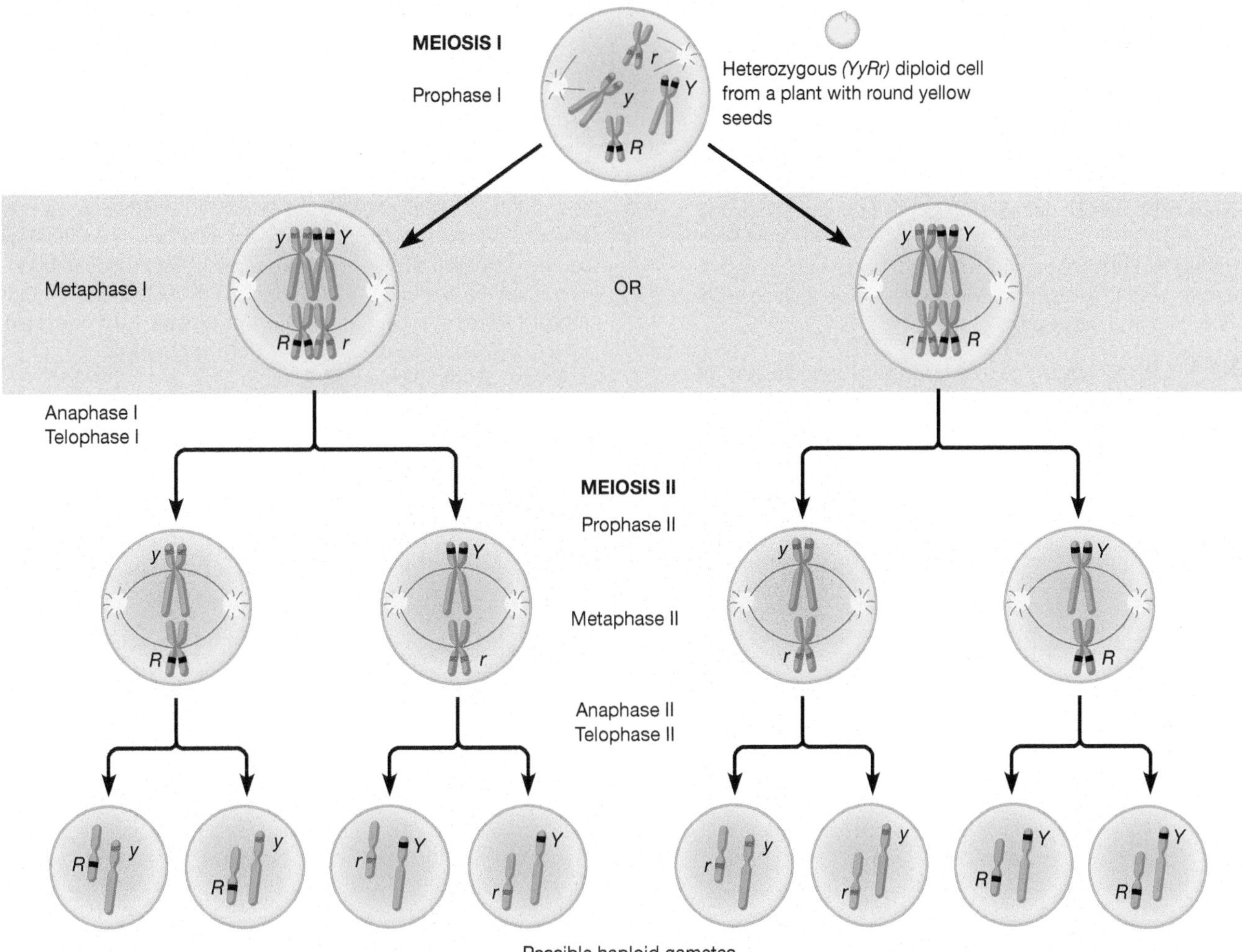

Figure 25-14 The Meiotic Basis for Mendel's Laws. Independent assortment of the alleles of two genes on different chromosomes is illustrated by meiosis in a pea plant heterozygous for seed color *(Yy)* and seed shape *(Rr)*. Pea plants have seven pairs of chromosomes, but only the two pairs bearing the seed-color alleles and the seed-shape alleles are shown. During meiosis, segregation of the seed-color alleles occurs independently of segregation of the seed-shape alleles. The basis of this independent assortment is found at metaphase I, when the homologous pairs (bivalents) align at the spindle equator. Because each bivalent can face in either direction, there are two alternative situations: either the paternal homologues (blue) can both face the same pole of the cell, with the maternal homologues (red) facing the other pole, or they can face different poles. The arrangement at metaphase I determines which homologues subsequently go to which daughter cells. Because the alternative arrangements occur with equal probability, all the possible combinations of the alleles therefore occur with equal probability in the gametes. For simplicity, chiasmata and crossing over are not shown in this diagram.

(pair of alleles). Perhaps he concentrated on pea strains that, in preliminary experiments, gave him the most consistent and comprehensible results.

The DNA Molecules of Homologous Chromosomes Have Similar Base Sequences

One of the five main points in the chromosomal theory of inheritance is the idea that homologous chromosomes are functionally equivalent, each carrying a similar set of genes. What does this mean in terms of our current understanding of the molecular organization of chromosomes? Homologous chromosomes typically carry the same genes in exactly the same order. However, minor differences in base sequence along the chromosomal DNA molecule can create different alleles of the same gene. Such differences arise by mutation, and the different alleles for a gene that we find in a population—whether of pea plants or people—arise from mutations that have gradually occurred in an ancestral gene. Alleles are usually "expressed" by transcription into RNA and translation into proteins, and it is the behavior of these proteins that ultimately creates an organism's phenotype. A change as small as a single DNA base pair can create an allele that encodes an altered protein that is different enough to cause an observable change in an organism's phenotype (see Chapter 19)—in other cases, such a DNA alteration can be lethal.

The underlying similarity in the base sequences of their DNA molecules—*DNA homology*—is thought to explain the ability of homologous chromosomes to undergo synapsis (close pairing) during meiosis and is essential for normal crossing over. One popular model for synapsis holds that the correct alignment of homologous chromosomes is brought

Key Technique — Using Mendel's Rules to Predict Human Disease

PROBLEM: Human diseases often arise from mutations in a single locus within the human genome. Many of these mutations are recessive. Potential parents who are heterozygous at such a locus may wish to assess the likelihood that their children will be homozygous and hence develop the disease.

SOLUTION: Mendelian genetics provides a way to assess the probability that children will have a particular genotype. Once they know the risk, parents can decide whether to seek to have biological children, to seek further genetic screening and/or counseling, or to pursue other parenting options.

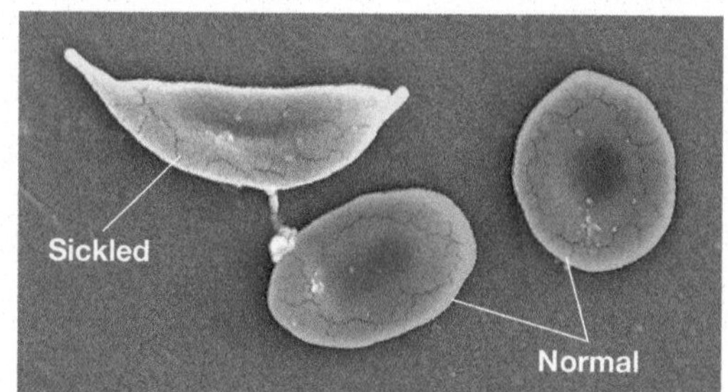

Sickled vs. Normal Red Blood Cells (SEM).

Key Tools: The law of independent assortment, along with knowledge of the genotypes of the parents and/or their relatives.

Details: In an era in which direct sequencing of DNA of human individuals is commonplace and becoming ever cheaper, it may seem that the indirect inference of genotypes made possible by Gregor Mendel's laws would have an ever-diminishing role to play in human medical genetics. This would perhaps be true if we wanted to establish the genotype of a *living* person. However, it is not currently possible to assess the genotype of individual oocytes or sperm in a way that does not injure the sperm or oocyte. How then can we predict, *before* a zygote is created at fertilization, how likely it is that the zygote (or resulting child) will be homozygous for a disease-causing mutation, starting from two parents of a known genotype?

The Online Mendelian Inheritance in Man. The Online Mendelian Inheritance in Man (OMIM; http://omim.org) is a public database that is continually updated with information on tens of thousands of human traits that are genetically inherited. As the name suggests, the traits in the OMIM are inherited in just the same way that traits of Mendel's peas are inherited. Using the basic tools of Mendelian analysis is therefore crucial for making predictions about the likelihood that the children of two parents will carry particular alleles at a locus associated with disease. Each trait listed in the OMIM catalog has a unique identifying number.

A Case Study: Sickle-Cell Anemia. One trait in the OMIM is number 603903, sickle-cell anemia. Recall that sickle-cell anemia results when an individual is homozygous for a particular mutation in one of the subunits of hemoglobin, β-globin (the β^S variant), which causes deleterious aggregation of hemoglobin inside of red blood cells, leading to their characteristic sickled shape (see image at left). The normal allele is called β^A. The mutant allele can be detected using molecular markers in human patients (see Figure 21-14). Here we focus on how to predict the likelihood that two heterozygous parents will have a child that is homozygous for the mutant β^S allele (**Figure 25B-1**).

Using Mendel's laws, the inheritance of the β^S allele and sickle-cell anemia can be traced by identifying the phenotypes of family members (that is, whether they have the disease) and displaying them in a *pedigree,* or family tree. The pedigree shown in Figure 25B-1a identifies females with circles and males with squares and shows typical results for a family in which sickle-cell anemia is prevalent. Blue circles and squares indicate family members who appear healthy; a pink circle or square indicates a person has the disease. Mating is indicated by a horizontal line connecting a male symbol and a female symbol. The children of a union are connected to the horizontal line by vertical branches. The individuals in a pedigree are assigned numbers that indicate their generation (Roman numerals) and their birth order in a generation (Arabic numerals). In Figure 25B-1a, the father and mother are identified as

about before completion of the synaptonemal complex by base-pairing interactions between matching regions of DNA in the two chromosomes. Support for this idea has come from studies showing that the chromosomes of some organisms possess special DNA sequences, called *pairing sites,* that promote synapsis between chromosomes when the same pairing site is present in two chromosomes. Protein components of the synaptonemal complex also play a role in facilitating the pairing between matching regions in the DNA of homologous chromosomes.

CONCEPT CHECK 25.3

A recessive allele of a particular gene causes tigers to lack pigmentation (that is, they are "white") and to have crossed eyes. If two heterozygous tigers, which appear normal, mate, what percentage will be expected to be white? What percentage of the white tigers will be cross-eyed?

25.4 Genetic Variability: Recombination and Crossing Over

Segregation and independent assortment of homologous chromosomes during the first meiotic division lead to random assortment of the alleles carried by different chromosomes, as was shown in Figure 25-14 for seed shape and color in pea plants. **Figure 25-15** on page 772 summarizes the outcome of meiosis for a generalized version of the same situation. Here we have a diploid organism heterozygous for two genes on nonhomologous chromosomes, with the allele pairs *Aa* and *Bb* (Figure 25-15a). Meiosis in such an organism will produce gametes in which allele *A* is just as likely to occur with allele *B* as it is with allele *b,* and *B* is just as likely to occur with *A* as it is with *a.* But what happens if two genes, *D* and *E,* reside on the same chromosome? In that case, alleles *D* and *E* will routinely be linked together in the same gamete, as will alleles *d* and *e* (Figure 25-15b).

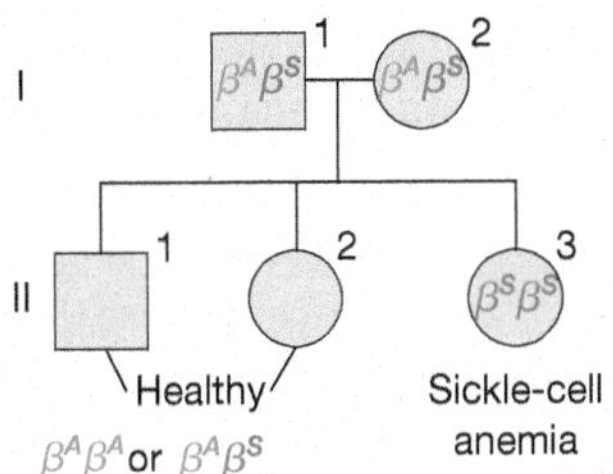

(a) A typical pedigree of two heterozygous carrier parents

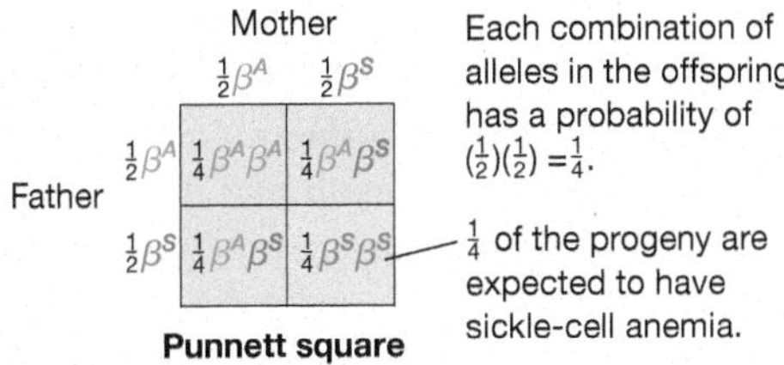

The children are expected to have three genotypes with proportions $\frac{1}{4}\beta^A\beta^A : \frac{1}{2}\beta^A\beta^S : \frac{1}{4}\beta^S\beta^S$.

(b) Determining the probability that a child will have sickle-cell anemia

Figure 25B-1 Hereditary Transmission of Sickle-Cell Anemia. **(a)** In this pedigree, squares are male and circles female. Individuals with sickle-cell anemia appear in pink. Each of the heterozygous parents passes one allele to each child. **(b)** Three genotypes are expected to occur among the children in the proportions shown. Only individuals homozygous for the β^S allele have the disease.

I-1 and I-2; each is heterozygous at the β-globin locus. One daughter (II-3) has sickle-cell anemia. Her siblings, individuals II-1 and II-2, are healthy. Each healthy child could carry either two copies of the β^A allele or one β^A allele and one β^S allele. To know more, we would have to know more about *their* mates and whether *their* children do or do not have sickle-cell anemia. Often this is enough to infer the genotypes of the children.

Remember, however, that our goal was to predict the *likelihood* that children born to the two parents in the pedigree would be homozygous and hence develop sickle-cell anemia. Figure 25B-1b illustrates how we can use Mendel's laws to determine the probability of transmission of alleles from the parents to their children. To do so we can use a Punnett square, which conveniently lays out the possible combinations of alleles, donated through the parents' gametes (Problem 25-8 gives you more practice using this convenient tool). Because each of the two alleles carried by a heterozygous parent has an equal chance of being transmitted to his or her children, there are four different possible combinations of alleles that can be transmitted from the parents to their children. Three of the possibilities result in children with one non-mutant copy of the β-globin gene; these children would be healthy. The other possibility, which occurs on average ¼ of the time, results in a child with sickle-cell anemia. This 3:1 ratio of phenotypes is the classic result when one allele is dominant and the other recessive. In contrast, the genotypes are expected to occur, on average, in a distinctive ratio of 1:2:1 (that is, 1 $\beta^A\beta^A$: 2 $\beta^A\beta^S$: and 1 $\beta^S\beta^S$). Actual genotypes of real patients can be determined using molecular genetic techniques (see Chapter 21).

QUESTION: Some inherited human disorders result from *dominant,* rather than recessive, mutations; some of these dominant mutations are lethal when two copies are present. One such mutation leads to polydactyly (extra fingers or toes). Suppose a mother with polydactyly and a father without it want to have children. What is the likelihood that they will have a child with polydactyly? Explain your answer.

The pattern of segregation shown in Figure 25-15b is commonly observed, especially for genes that are close together on the same chromosome. But studies with the fruit fly *Drosophila melanogaster,* conducted by Thomas Hunt Morgan and his colleagues beginning around 1910, uncovered an additional feature of meiosis: *genetic recombination*. We therefore now turn to Morgan's work, beginning with his discovery of linkage groups.

Chromosomes Contain Groups of Linked Genes That Are Usually Inherited Together

The fruit flies used by Morgan and his colleagues had certain advantages over Mendel's pea plants as objects of genetic study, not the least of which was the fly's relatively brief generation time (about 2 weeks versus several months for pea plants). Unlike Mendel's peas, however, the fruit flies did not come with a ready-made variety of phenotypes and genotypes. Whereas Mendel was able to purchase seed stocks of different true-breeding varieties, the only type of fruit fly initially available to Morgan was what has come to be known as the **wild type**, or "normal" organism. Morgan and his colleagues therefore had to generate variants—that is, mutants—for their genetic experiments.

Morgan and his coworkers began by breeding large numbers of flies and then selecting mutant individuals having phenotypic modifications that were heritable. Later, X-irradiation was used to enhance the mutation rate, but in their early work Morgan's group depended entirely on spontaneous mutations. Within 5 years, they were able to identify about 85 different mutants, each carrying a mutation in a different gene. Each of these mutants could be propagated as a laboratory stock and used for matings as needed.

One of the first discoveries made by Morgan and his colleagues as they began analyzing their mutants was that, unlike the genes Mendel had studied in peas, the mutant fruit fly genes

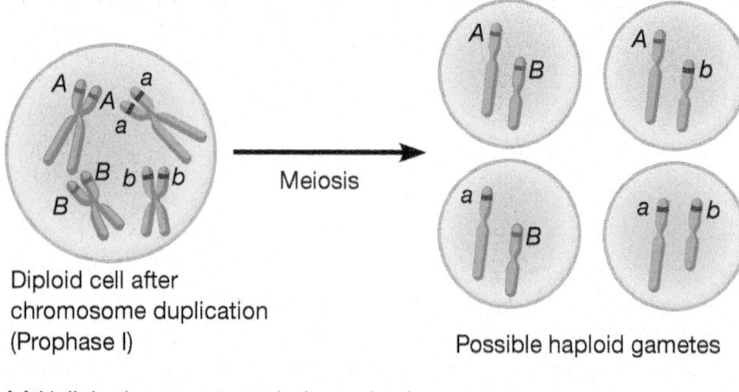

(a) Unlinked genes assort independently

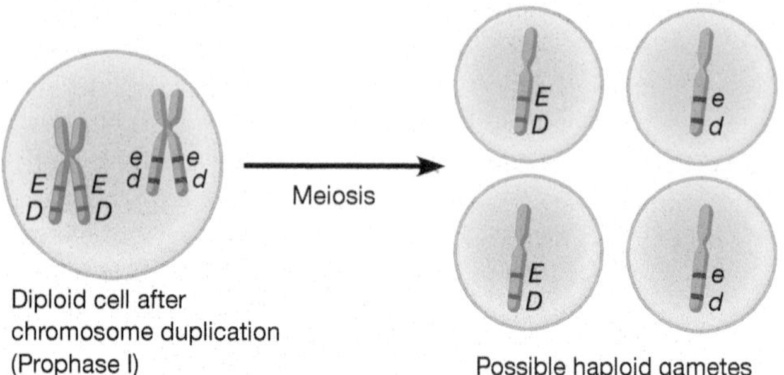

(b) Linked genes end up together in the absence of crossing over

Figure 25-15 Segregation and Assortment of Linked and Unlinked Genes. In this figure, *A* and *a* are alleles of one gene, *B* and *b* are alleles of another gene, and so forth. **(a)** Alleles of genes on different chromosomes segregate and assort independently during meiosis; allele *A* is as likely to occur in a gamete with allele *B* as it is with allele *b*. **(b)** Alleles of genes on the same chromosome remain linked during meiosis in the absence of crossing over; in this case, allele *D* will occur in the same gamete with allele *E* but not with allele *e*.

did not all assort independently. Instead, some genes behaved as if they were linked together, and for such genes, the new combinations of alleles predicted by Mendel were infrequent or even nonexistent. In fact, it was soon recognized that fruit fly genes can be classified into four **linkage groups**, each group consisting of a collection of **linked genes** that are usually inherited together. Morgan quickly realized that the number of linkage groups was the same as the number of different chromosomes in the organism (the haploid chromosome number for *Drosophila* is four). The conclusion he drew was profound: *each chromosome is the physical basis for a specific linkage group.* Mendel had not observed linkage because the genes he studied either resided on different chromosomes or were far apart on the same chromosome and therefore behaved as if they were not linked to each other.

Homologous Chromosomes Exchange Segments During Crossing Over

Although the genes they discovered could all be organized into linkage groups, Morgan and his colleagues found that linkage within such groups was incomplete. Most of the time, genes known to be linked (and therefore on the same chromosome) assorted together, as would be expected. Sometimes, however, two or more such traits would appear in the offspring in nonparental combinations. This phenomenon of less-than-complete linkage was called *recombination* because the different alleles appeared in new and unexpected associations in the offspring.

To explain such recombinant offspring, Morgan proposed that homologous chromosomes can exchange segments by a breakage-and-fusion event, as illustrated in **Figure 25-16**.

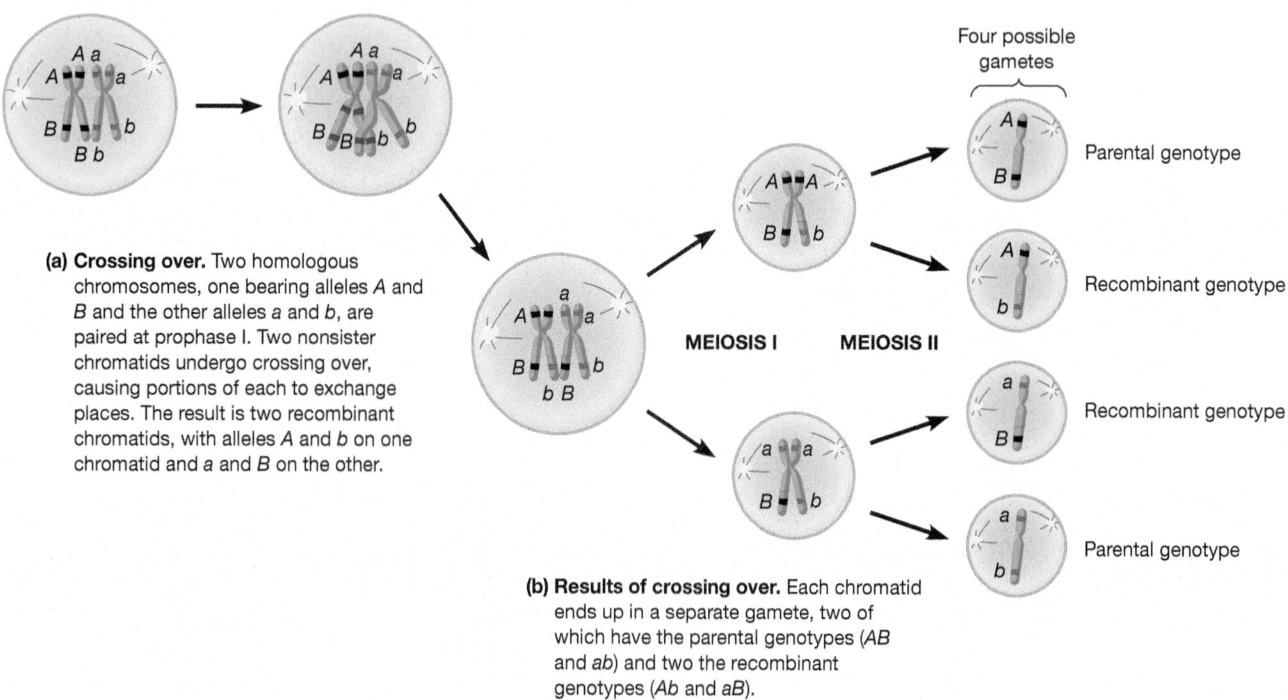

Four possible gametes
Parental genotype
Recombinant genotype
Recombinant genotype
Parental genotype
MEIOSIS I
MEIOSIS II
(a) Crossing over. Two homologous chromosomes, one bearing alleles *A* and *B* and the other alleles *a* and *b*, are paired at prophase I. Two nonsister chromatids undergo crossing over, causing portions of each to exchange places. The result is two recombinant chromatids, with alleles *A* and *b* on one chromatid and *a* and *B* on the other.
(b) Results of crossing over. Each chromatid ends up in a separate gamete, two of which have the parental genotypes (*AB* and *ab*) and two the recombinant genotypes (*Ab* and *aB*).

Figure 25-16 Crossing Over and Genetic Recombination. After a single crossover event, the results of meiosis include two parental gametes and two recombinant gametes.

By this process of genetic recombination, which Morgan termed *crossing over*, a particular allele or group of alleles initially present on one member of a homologous pair of chromosomes could be transferred to the other chromosome in a reciprocal manner. The exchange of genetic material between homologous chromosomes occurs during prophase I of meiosis when the homologues are synapsed, creating chromosomes exhibiting new combinations of alleles.

In the example of Figure 25-16a, two homologous chromosomes, one with alleles *A* and *B* and the other with alleles *a* and *b*, lie side by side at synapsis. Portions of an *AB* chromatid and a nonsister *ab* chromatid then exchange DNA segments, thereby producing two recombinant chromatids—one with alleles *A* and *b* and the other with alleles *a* and *B*. Each of the four chromatids ends up in a different gamete at the end of the second meiotic division, so the products of meiosis will include two *parental* gametes and two *recombinant* gametes, assuming a single crossover event (Figure 25-16b).

We now know that crossing over takes place during the pachytene stage of meiotic prophase I (see Figure 25-4), when sister chromatids are packed tightly together and it is difficult to observe what is happening. As the chromatids begin to separate at diplotene, each of the four chromatids in a bivalent can be identified as belonging to one or the other of the two homologues. Wherever crossing over has taken place between nonsister chromatids, the two homologues remain attached to each other, forming a chiasma.

At the first meiotic metaphase, homologous chromosomes are almost always held together by at least one chiasma; if not, they may not segregate properly. Many bivalents contain multiple chiasmata. Human bivalents, for example, typically have two or three chiasmata because multiple crossover events routinely occur between paired homologues. To be genetically significant, crossing over must involve nonsister chromatids. In some species, exchanges between sister chromatids are also observed, but such exchanges have no genetic consequences because sister chromatids are genetically identical.

Gene Locations Can Be Mapped by Measuring Recombination Frequencies

Eventually, it became clear to Morgan and others that the frequency of recombinant progeny differed for different pairs of genes within each linkage group. This observation suggested that the frequency of recombination between two genes might be a measure of how far the genes are located from each other because genes located near each other on a given chromosome would be less likely to become separated by a crossover event than would genes that are far apart. This insight led Alfred Sturtevant, an undergraduate student in Morgan's lab, to suggest in 1911 that recombination data could be used to determine where genes are located along the chromosomes of *Drosophila*. In other words, a chromosome had come to be viewed as a linear string of genes whose positions can be determined on the basis of recombination frequencies.

Determining the sequential order and spacing of genes on a chromosome based on recombinant frequencies is one form of **genetic mapping**. In the construction of such maps, the recombinant frequency is the map distance expressed in *map units (centimorgans)*. If, for example, the alleles in Figure 25-16b appear among the progeny in their parental combination (*AB* and *ab*) 85% of the time and in the recombinant combinations (*Ab* and *aB*) 15% of the time, we conclude that the two genes are linked (on the same chromosome) and are 15 map units apart. This approach has been used to map the chromosomes of many species of plants and animals, as well as bacteria and viruses. Problem 25-9 provides you with some practice in applying these ideas. Because bacteria and viruses do not reproduce sexually, the methods used to generate recombinants are somewhat different, but the same basic principles apply (see the next section).

With the advent of affordable genome sequencing (see Chapter 21), scientists now often dispense with recombination-based genetic mapping of an organism's genome. However, in those cases in which genetic and physical mapping (that is, sequencing of DNA directly) have been performed, the results are in remarkable agreement regarding the ordering of genes along chromosomes. The distances between genes obtained by direct sequencing sometimes varies from estimates based on genetic mapping, however, because there are a number of factors that affect the frequency of recombination along a chromosome in addition to the physical distance between loci.

CONCEPT CHECK 25.4

Suppose you are repeating Sturtevant's experiments in fruit flies and are studying genes (*A*, *B*, and *C*) known to be on the same chromosome in fruit flies. Suppose that the recombination frequency between genes *A* and *B* is 25% and that between *A* and *C* is 10%. Can you use this information to determine the order of these genes on the chromosome? Explain your answer.

25.5 Genetic Recombination in Bacteria and Viruses

Because it requires crossing over between homologous chromosomes, you might expect genetic recombination to be restricted to sexually reproducing organisms. Sexual reproduction provides an opportunity for crossing over once every generation when homologous chromosomes become closely juxtaposed during the meiotic divisions that produce gametes. In contrast, bacteria and viruses have haploid genomes and reproduce asexually, with no obvious mechanism for regularly bringing together genomes from two different parents.

Nonetheless, viruses and bacteria are still capable of genetic recombination. In fact, recombination data permitted the extensive mapping of viral and bacterial genomes well before the advent of modern DNA technology. To understand how recombination takes place despite a haploid genome, we need to examine the mechanisms that allow two haploid genomes, or portions of genomes, to be brought together within the same cell.

Co-infection of Bacterial Cells with Related Bacteriophages Can Lead to Genetic Recombination

Much of our early understanding of genes and recombination at the molecular level, as well as the vocabulary we use to express that information, came from experiments involving bacteriophages, particularly T-even phages and phage λ (see Figure 16-3). Although phages are haploid and do not reproduce sexually, genetic recombination between related phages can take place when individual bacterial cells are simultaneously infected by different versions of the same phage. **Figure 25-17** illustrates an experiment in which bacterial cells are co-infected by two related types of T4 phage. As the phages replicate in the bacterial cell, their DNA molecules occasionally become juxtaposed in ways that allow DNA segments to be exchanged between homologous regions. The resulting recombinant phage arises at a frequency that depends on the distance between the genes under study, just as in diploid organisms: the farther apart the genes, the greater the likelihood of recombination between them.

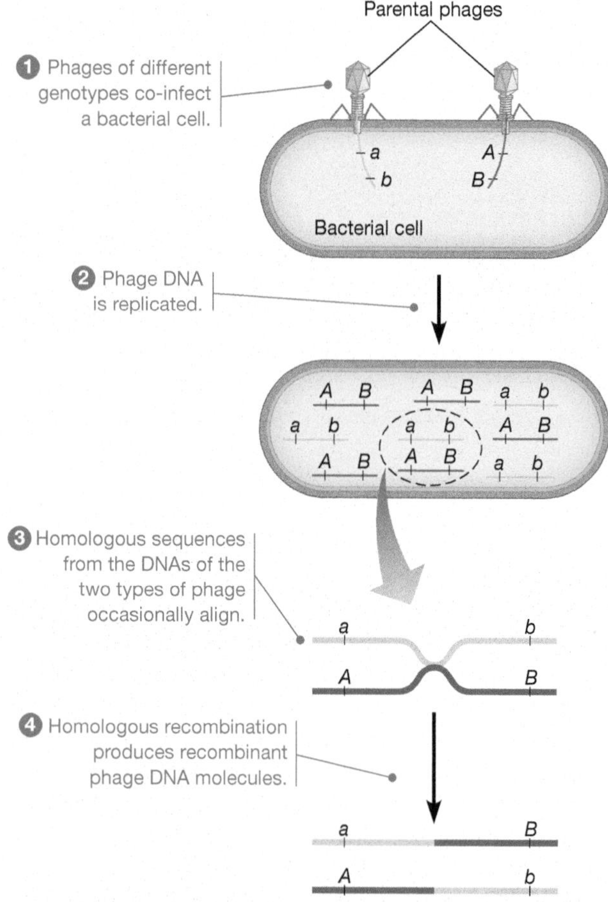

Figure 25-17 Genetic Recombination of Bacteriophages. Two phage populations with different genotypes co-infect a bacterial cell, thereby ensuring the simultaneous presence of both phage genomes in the same cell. One phage carries the mutant alleles *a* and *b*; the other has the wild-type alleles *A* and *B*. As the phage DNAs replicate, they occasionally become aligned, and crossing over takes place, generating recombinant phage DNA molecules exhibiting *aB* and *Ab* genotypes. The frequency of occurrence of these recombinant genotypes provides a measure of the distance between genes—the frequency of crossover is low for two genes close together.

Recombination in Bacteria Can Occur Via Transformation or Transduction

In bacteria, several mechanisms exist for recombining genetic information (**Figure 25-18**). Experiments with rough and smooth strains of pneumococcal bacteria led Oswald Avery to conclude that bacterial cells can be transformed from one genetic type to another by exposing them to purified DNA (see Chapter 16). This ability of a bacterial cell to take up DNA molecules and to incorporate some of that DNA into its own genome is called **transformation** (Figure 25-18a). Although initially described as a laboratory technique for artificially introducing DNA into bacterial cells, transformation is now recognized as a natural mechanism by which some (though by no means all) kinds of bacteria acquire genetic information when they have access to DNA from other cells. In transformation, exogenous DNA is taken up by competent bacteria and can recombine into the bacterial genome if the sequences are similar enough.

A second mechanism for genetic recombination in bacteria, called **transduction**, involves DNA that has been brought into a bacterial cell by a bacteriophage. Most phages contain only their own DNA, but occasionally a phage will incorporate some bacterial host cell DNA sequences into its progeny particles. Such a phage particle can then infect another bacterium, acting like a syringe carrying DNA from one bacterial cell to the next (Figure 25-18b). Phages capable of carrying host cell DNA from one cell to another are called *transducing phages*. Similar to transformation, once inside the cell this DNA can integrate into similar sequences in the bacterial genome. The transducing phage known as P1, which infects *Escherichia coli*, has been especially useful for a type of gene mapping known as *cotransductional mapping*, in which the proximity of one marker to another is determined by measuring how frequently the markers accompany each other in a transducing phage particle (markers that are closer together cotransduce at a higher frequency than those farther apart).

Conjugation Is a Modified Sexual Activity That Facilitates Genetic Recombination in Bacteria

In addition to transformation and transduction, some bacteria also transfer DNA from one cell to another by **conjugation**. As the name suggests, conjugation resembles a mating in that one bacterium is clearly identifiable as the donor (often called a "male") and another as the recipient ("female"). Although conjugation resembles a sexual process, this mode of DNA transfer is not an inherent part of the bacterial life cycle, and it usually involves only a portion of the genome; therefore, conjugation does not qualify as true sexual reproduction. The existence of conjugation was postulated in 1946 by Joshua Lederberg and Edward L. Tatum, who were the first to show that genetic recombination occurs in bacteria. They also established that physical contact between two cells is necessary for conjugation to take place. We now understand that conjugation involves the directional transfer of DNA from the donor bacterium to the recipient bacterium. Let's look at some details of how the process works.

The F Factor. The presence of a DNA sequence called the **F factor** (F for fertility) enables an *E. coli* cell to act as a donor during conjugation. The F factor can take the form of either

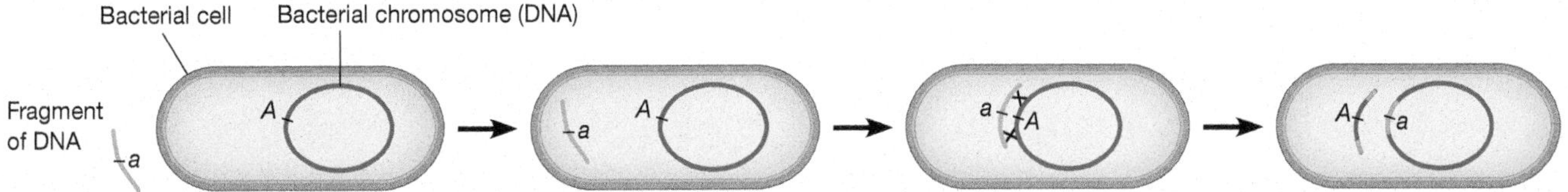

(a) Transformation. Transformation involves uptake by the bacterial cell of exogenous DNA, which occasionally becomes integrated into the bacterial genome by two crossover events (indicated by X's). The exogenous DNA will be detectable in progeny cells only if integrated into the bacterial chromosome because the fragment of DNA initially taken up does not normally have the capacity to replicate itself autonomously in the cell. (The main exception is an intact plasmid.)

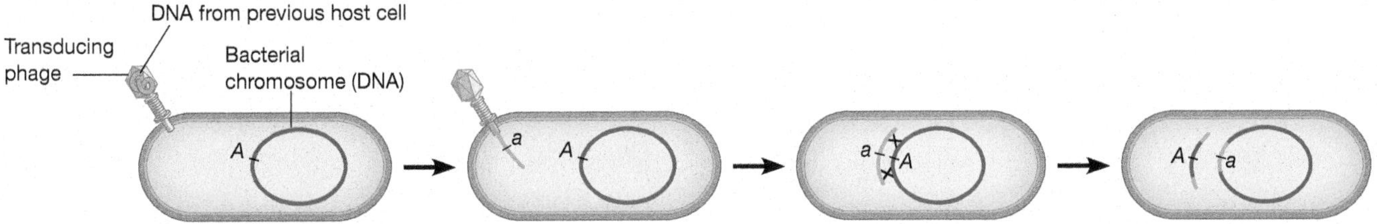

(b) Transduction. Transduction involves the introduction of exogenous DNA into a bacterial cell by a phage. Once injected into the host cell, the DNA can become integrated into the bacterial genome in the same manner as in transformation. In both cases, linear fragments of DNA that end up outside the bacterial chromosome are eventually degraded by nucleases.

Figure 25-18 Transformation and Transduction in Bacterial Cells. Transformation and transduction are mechanisms by which bacteria incorporate external genetic information derived either from free DNA or from DNA brought into bacterial cells by bacteriophages. In this figure, the letters *A* and *a* represent alleles of the same gene.

an independent, circular piece of DNA known as a *plasmid* or a segment of DNA within the bacterial chromosome. Donor bacteria containing the F factor in its plasmid form are designated F^+, whereas recipient cells, which usually lack the F factor completely, are designated F^-. The F^+ cells produce the apparatus required for conjugation (**Figure 25-19**). Donor cells develop long, hairlike projections called **sex pili** (singular: **pilus**) that emerge from the cell surface (Figure 25-19a). The end of each sex pilus contains molecules that selectively bind to the surface of recipient cells, thereby leading to the formation of a transient cytoplasmic **mating bridge** through which DNA is transferred from donor cell to recipient cell (Figure 25-19b).

The conjugation process is outlined in **Figure 25-20** on page 776. When a donor cell contains an F factor in its plasmid form, a copy of the plasmid is quickly transferred to the recipient cell during conjugation, converting the recipient cell from F^- to F^+ (Figure 25-20a). Transfer always begins at a point on the plasmid called its **origin of transfer**, represented in the figure by an arrowhead. During transfer of an

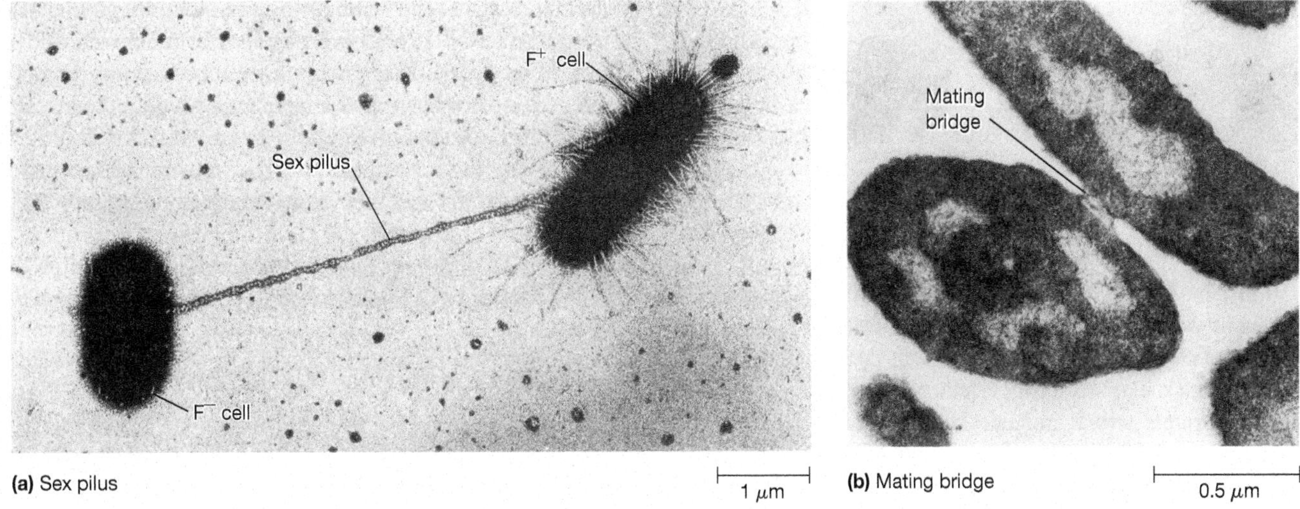

Figure 25-19 The Cellular Apparatus for Bacterial Conjugation. (a) The donor bacterial cell on the right, an F^+ cell, has many slender appendages, called pili, on its surface. Some of these pili are sex pili, including the very long pilus leading to the other cell, an F^- cell. Made of protein encoded by a gene on the F factor, sex pili enable a donor cell to attach to a recipient cell. **(b)** Subsequently, a cytoplasmic mating bridge forms, through which DNA is passed from the donor cell to the recipient cell (TEMs).

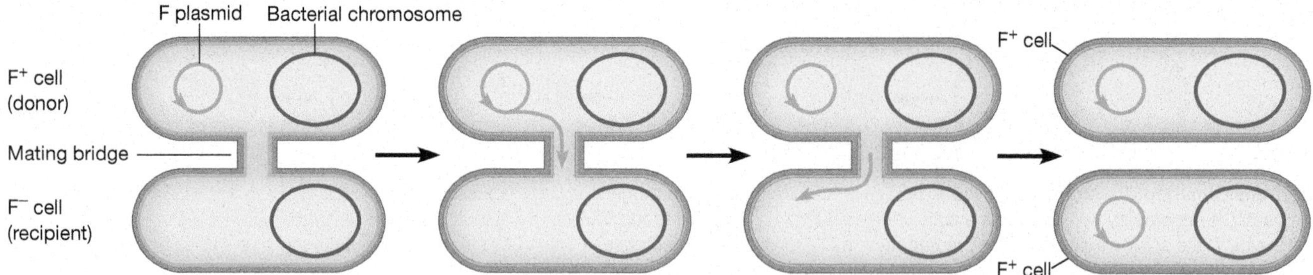

(a) Conjugation between F⁺ and F⁻ cells. The transfer of a copy of the F factor plasmid from an F⁺ donor bacterium to an F⁻ recipient converts the F⁻ cell into an F⁺ cell. Plasmid transfer occurs through a mating bridge and begins at the F factor's origin of transfer, indicated by the arrowhead.

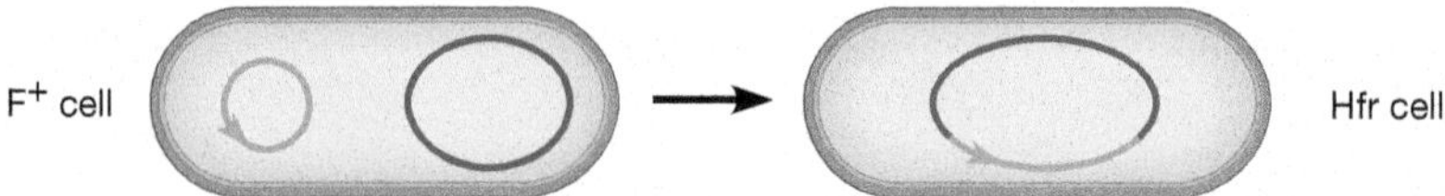

(b) Conversion of an F⁺ cell into an Hfr cell. Integration of the F factor into the bacterial chromosome converts an F⁺ cell into an Hfr cell.

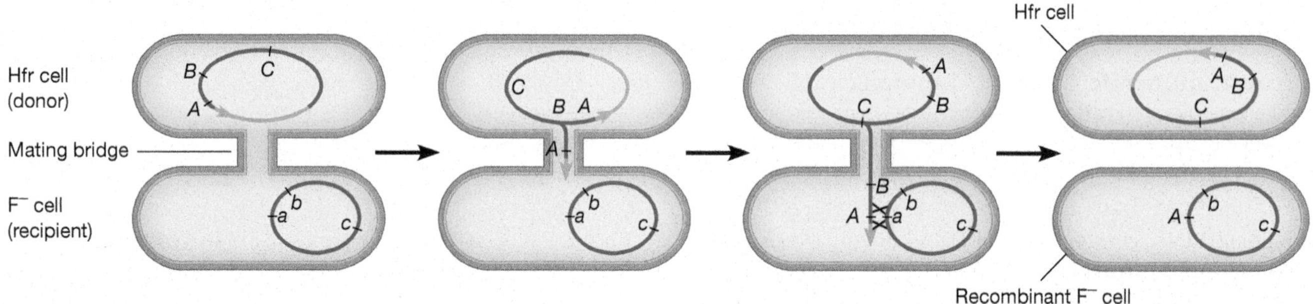

(c) Conjugation between an Hfr cell and an F⁻ cell. Transferring a copy of the Hfr genome into an F⁻ cell begins with the origin of transfer on the integrated F factor. Cells rarely remain in contact long enough for the entire bacterial chromosome to be transferred. Once inside the F⁻ cell, parts of the Hfr DNA recombine with the DNA of the F⁻ cell. Uppercase letters represent alleles carried by the Hfr; lowercase letters represent corresponding alleles in the F⁻ cell. In the last step, allele *A* from the Hfr is recombined into the F⁻ cell's DNA in place of its *a* allele.

Figure 25-20 DNA Transfer by Bacterial Conjugation. (a) The presence of the F factor plasmid enables an *E. coli* cell to act as a plasmid donor during conjugation. **(b)** Chromosomal integration of the F factor creates an Hfr cell that can **(c)** transfer genomic DNA during conjugation.

F factor to an F⁻ cell, the donor cell does not lose its F⁺ status because the F factor is replicated in close association with the transfer process, allowing a copy of the F plasmid to remain behind in the donor cell. As a result, mixing an F⁺ population of bacteria with F⁻ cells will eventually lead to a population of cells that is entirely F⁺. "Maleness" is in a sense infectious, and the F factor is responsible for this behavior.

Some bacteria carry special plasmids called *R factors* that typically have the same components as F plasmids and also contain genes that confer antibiotic resistance. Engineered versions of some R plasmids are used routinely in molecular biology for cloning because bacteria that contain them survive antibiotics. In many cases, R factors carry multiple transposable elements, which increases their genetic variability.

Hfr Cells and Bacterial Chromosome Transfer. Thus far, we have seen that donor and recipient cells are defined by the presence or absence of the F factor, which is in turn transmitted by conjugation. But how do recombinant bacteria arise by this means? The answer is that the F factor—although usually present as a plasmid—can sometimes become integrated into the bacterial chromosome, as shown in Figure 25-20b. (Integration results from crossing over between short DNA sequences in the chromosome and similar sequences in the F factor.) Chromosomal integration of the F factor converts an F⁺ donor cell into an **Hfr cell**, which is capable of producing a *h*igh *f*requency of *r*ecombination in further matings because it can now transfer genomic DNA during conjugation.

When an Hfr bacterium is mated to an F⁻ recipient, DNA is transferred into the recipient cell (Figure 25-20c). But instead of transferring just the F factor itself, the Hfr cell transfers at least part (and occasionally all) of its chromosomal DNA, retaining a copy, as in F⁺ DNA transfer. Transfer begins at the origin of transfer within the integrated F factor and proceeds in a direction dictated by the orientation of the F factor within the chromosome. Note that the chromosomal DNA is transferred in a linear form, with a small part of the F factor at the leading end and the remainder at the trailing end. Because the F factor is split in this way, only recipient cells that receive a complete bacterial chromosome from the Hfr donor actually become Hfr cells themselves. Transfer of the whole chromosome is extremely rare, however, because it takes about 90 minutes. Usually, mating contact is spontaneously disrupted with the mating bridge breaking before transfer is complete, leaving the recipient cell with only a portion of the Hfr chromosome (Figure 25-20c). As a result, genes located downstream of the origin of transfer on the Hfr chromosome are most likely to be transmitted to the recipient cell.

Once a portion of the Hfr chromosome has been introduced into a recipient cell by conjugation, it can recombine with regions of the recipient cell's chromosomal DNA that are homologous (similar) in sequence. The recombinant bacterial chromosomes generated by this process contain some genetic information derived from the donor cell and some from the recipient. Only donor DNA sequences that are successfully integrated by this recombination mechanism survive in the recipient cell and its progeny. Donor DNA that is not integrated during recombination, as well as DNA removed from the recipient chromosome during the recombination process, is eventually degraded by nucleases.

The correlation between the position of a gene within the bacterial chromosome and its likelihood of transfer can be used to map genes with respect to the origin of transfer and therefore with respect to one another. For example, if gene *A* of an Hfr cell is transferred in conjugation 95% of the time, gene *B* 70% of the time, and gene *C* 55% of the time, then the sequence of genes is *A-B-C*, with gene *A* closest to the origin of transfer.

CONCEPT CHECK 25.5

What do you think would happen to a pathogenic (disease-causing) bacterial population in which some of the bacteria contain an R factor plasmid in a region in which antibiotics are heavily used to treat routine human infections?

25.6 Mechanisms of Homologous Recombination

We have now described five different situations in which genetic information can be exchanged between homologous DNA molecules: (1) prophase I of meiosis associated with gametogenesis in eukaryotes, (2) co-infection of bacteria with related bacteriophages, (3) transformation of bacteria with DNA, (4) transduction of bacteria by transducing phages, and (5) bacterial conjugation. Despite their obvious differences, all five situations share a fundamental feature: each involves *homologous recombination* in which genetic information is exchanged between DNA molecules exhibiting extensive sequence similarity. Recall that homologous recombination can occur in the context of repair of double-stranded breaks in DNA (see Figure 17-31); here we focus on homologous recombination at a more cellular level.

DNA Breakage and Exchange Underlie Homologous Recombination Between Chromosomes

Homologous recombination between chromosomes involves breaks in the DNA of two adjoining chromosomes, followed by exchange and rejoining of the broken segments. The first experimental evidence providing support for the breakage-and-exchange model was obtained in 1961 by Matthew Meselson and Jean Weigle, who employed phages of the same genetic type labeled with either the heavy (^{15}N) or light (^{14}N) isotope of nitrogen. Simultaneous infection of bacterial cells with these two labeled strains of the same phage resulted in the production of recombinant phage particles containing genes derived from both phages. When the DNA from these recombinant phages was examined, it was found to contain a mixture of ^{15}N and ^{14}N (**Figure 25-21**). Because these experiments were performed under conditions that prevented any new DNA from being synthesized, the recombinant DNA molecules must have been produced by breaking and rejoining DNA molecules derived from the two original phages. Subsequent experiments revealed that a similar exchange mechanism operates during genetic recombination between bacterial chromosomes as well.

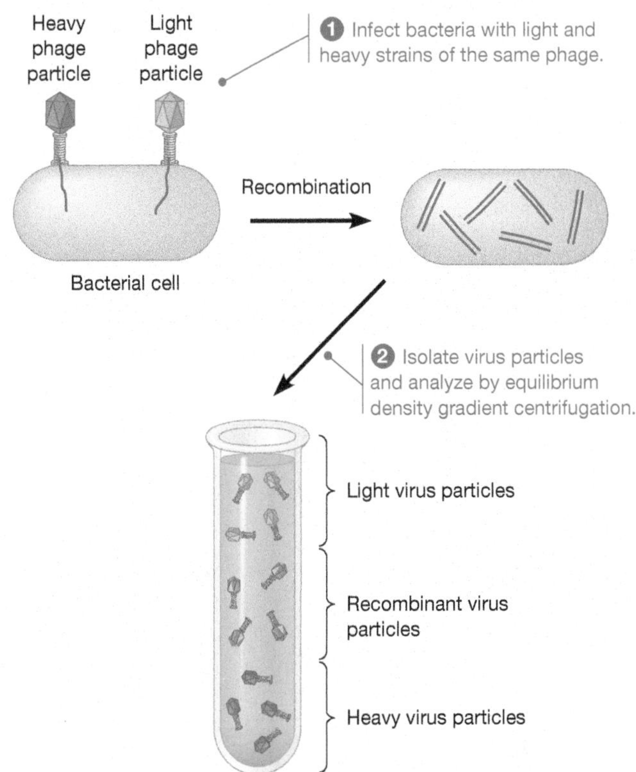

Figure 25-21 Evidence for DNA Breakage and Exchange During Bacteriophage Recombination. Bacterial cells were infected with two strains of the same phage, one labeled with ^{15}N and the other with ^{14}N. After recombination, the DNA from the recombinant phages was found to contain both ^{15}N and ^{14}N, supporting the idea that recombination involves the breaking and rejoining of DNA molecules.

A similar conclusion emerged from experiments performed shortly thereafter on eukaryotic cells by J. Herbert Taylor. In these studies, cells were briefly exposed to ^{3}H-thymidine during the S phase preceding the last mitosis prior to meiosis, producing chromatids containing one radioactive DNA strand per double helix. During the following S phase, DNA replication in the absence of ^{3}H-thymidine generated chromosomes containing one labeled chromatid and one unlabeled chromatid (**Figure 25-22** on page 778). But during the subsequent meiosis, individual chromatids exhibited

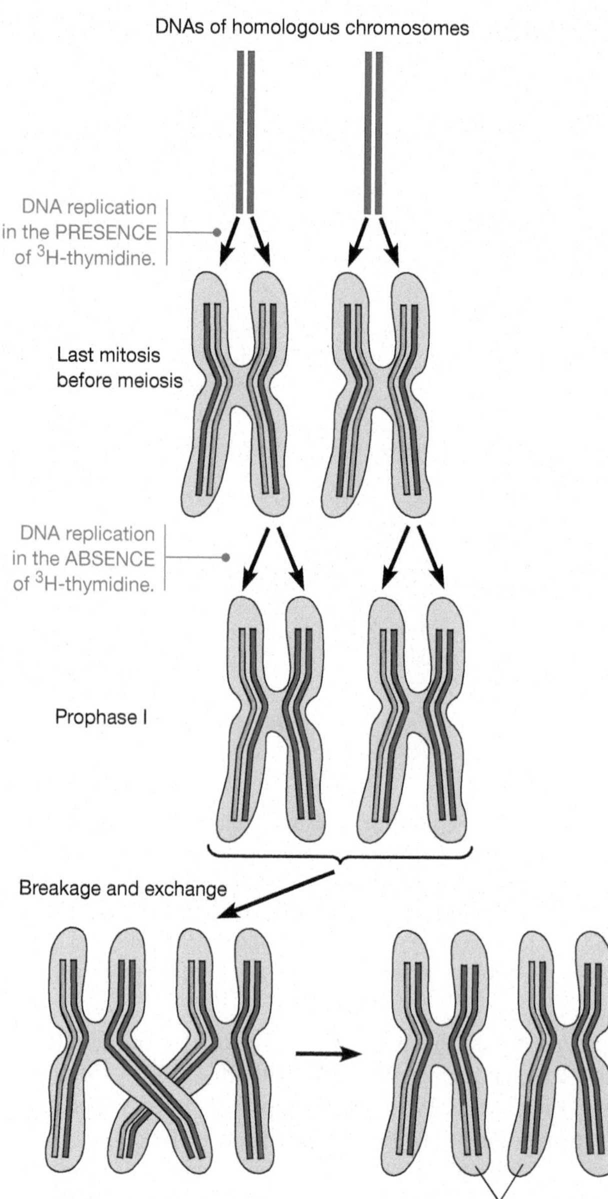

Figure 25-22 Experimental Demonstration of Breakage and Exchange During Eukaryotic Recombination. In this experiment, DNA was radioactively labeled by briefly exposing eukaryotic cells to ^{3}H-thymidine before the last mitosis prior to meiosis. When autoradiography was employed to examine the chromatids during meiosis, some were found to contain a mixture of labeled (orange) and unlabeled (dark blue) segments as predicted by the breakage-and-exchange model.

a mixture of radioactive and nonradioactive segments, as would be predicted by the breakage-and-exchange model. Moreover, the frequency of such exchanges was directly proportional to the frequency with which the genes located in these regions underwent genetic recombination. Such observations provided strong support for the notion that genetic recombination in eukaryotic cells, as in prokaryotes, involves DNA breakage and exchange. These experiments also showed that most DNA exchanges arise between homologous chromosomes rather than between the two sister chromatids of a given chromosome. This selectivity is important because it ensures that genes are exchanged between paternal and maternal chromosomes.

The Synaptonemal Complex Facilitates Homologous Recombination During Meiosis

Earlier in the chapter, you learned that during prophase I of meiosis, homologous chromosomes are joined together by a zipperlike, protein-containing structure called the synaptonemal complex. Several observations suggest that this structure plays an important role in genetic recombination. First, the synaptonemal complex appears at the time when recombination takes place. Second, its location between the opposed homologous chromosomes corresponds to the region where crossing over occurs. And finally, synaptonemal complexes are absent from organisms—such as male fruit flies—that fail to carry out meiotic recombination.

Presumably, the synaptonemal complex facilitates recombination by maintaining a close pairing between adjacent homologous chromosomes along their entire length. One current model for how this process occurs is shown in **Figure 25-23.** Recall that pairing of chromosomes during mitosis is mediated by proteins known as *cohesins* (see Chapter 24). ❶ Cohesins play a similar role during meiosis, allowing sister chromatids within the same chromosome to adhere to one another. During early prophase I, homologous chromosomes align along their length such that each locus on each of the two homologues is precisely aligned. During recombination, the DNA of adjacent nonsister chromatids is broken at precisely corresponding points. ❷ The formation of the synaptonemal complex, which involves a complex of many proteins, stabilizes the arrangement of these broken chromosomes during synapsis, holding the two homologues snugly against one other. ❸ During synapsis, the DNA breaks are repaired such that breaks are rejoined, but often by joining to the corresponding segment of the *nonsister* chromatid. ❹ Later in prophase I, after the synaptonemal complexes have disassembled, the homologues have moved apart, and the chromosomes have undergone further condensation, the crossover sites become visible as chiasmata. At least one crossover per chromosome must occur for the homologous pair to stay together as it moves to the metaphase I plate. In reality, many crossovers typically occur between homologues.

If the synaptonemal complex facilitates recombination, how is the precise alignment of chromosomes along their length achieved such that crossovers form only between homologous chromosomes? Evidence suggests the existence of a process called *homology searching*, in which a single-strand break in one DNA molecule produces a free strand that checks for the presence of complementary sequences. If extensive homology is not found, the free DNA strand invades another DNA molecule and checks for complementarity, repeating the process until a homologous DNA molecule is detected. Only then does a synaptonemal complex develop, bringing the homologous chromosomes together throughout their length to facilitate the recombination process.

A second model of meiotic recombination begins with the introduction of a double-strand break in one of the homologous chromosomes. In yeast, this is carried out by the Spo11 protein. These intentional double-strand breaks are then repaired using the double-strand break model of homologous recombination introduced in Chapter 17. When the two Holliday junctions are resolved, crossover recombinants can result (see Figure 17-32b).

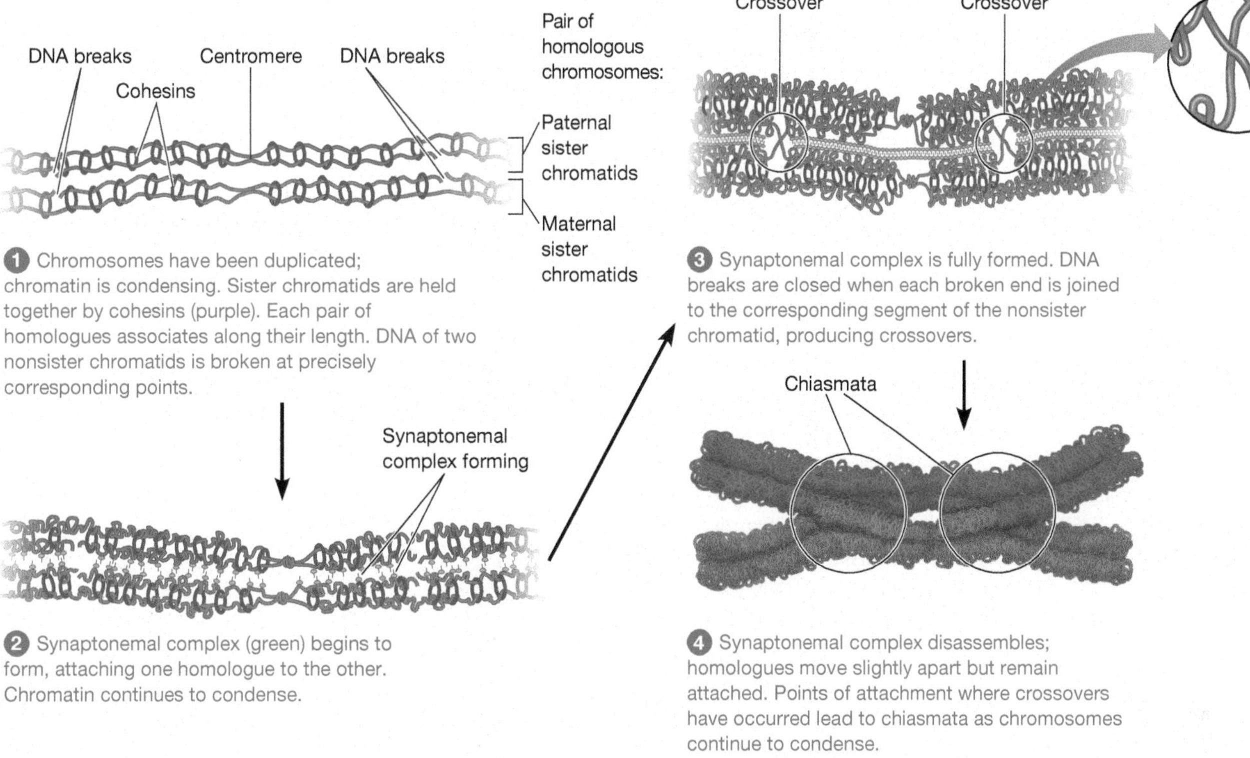

DNA breaks
Centromere
DNA breaks
Cohesins
Pair of homologous chromosomes:
Paternal sister chromatids
Maternal sister chromatids
1 Chromosomes have been duplicated; chromatin is condensing. Sister chromatids are held together by cohesins (purple). Each pair of homologues associates along their length. DNA of two nonsister chromatids is broken at precisely corresponding points.
Synaptonemal complex forming
2 Synaptonemal complex (green) begins to form, attaching one homologue to the other. Chromatin continues to condense.
Crossover
Crossover
3 Synaptonemal complex is fully formed. DNA breaks are closed when each broken end is joined to the corresponding segment of the nonsister chromatid, producing crossovers.
Chiasmata
4 Synaptonemal complex disassembles; homologues move slightly apart but remain attached. Points of attachment where crossovers have occurred lead to chiasmata as chromosomes continue to condense.

Figure 25-23 One Model for Homologous Recombination During Prophase I of Meiosis. Sister chromatids, held together by cohesins, pair by forming synaptonemal complexes, allowing orderly DNA strand breakage and resealing, leading to formation of chiasmata.

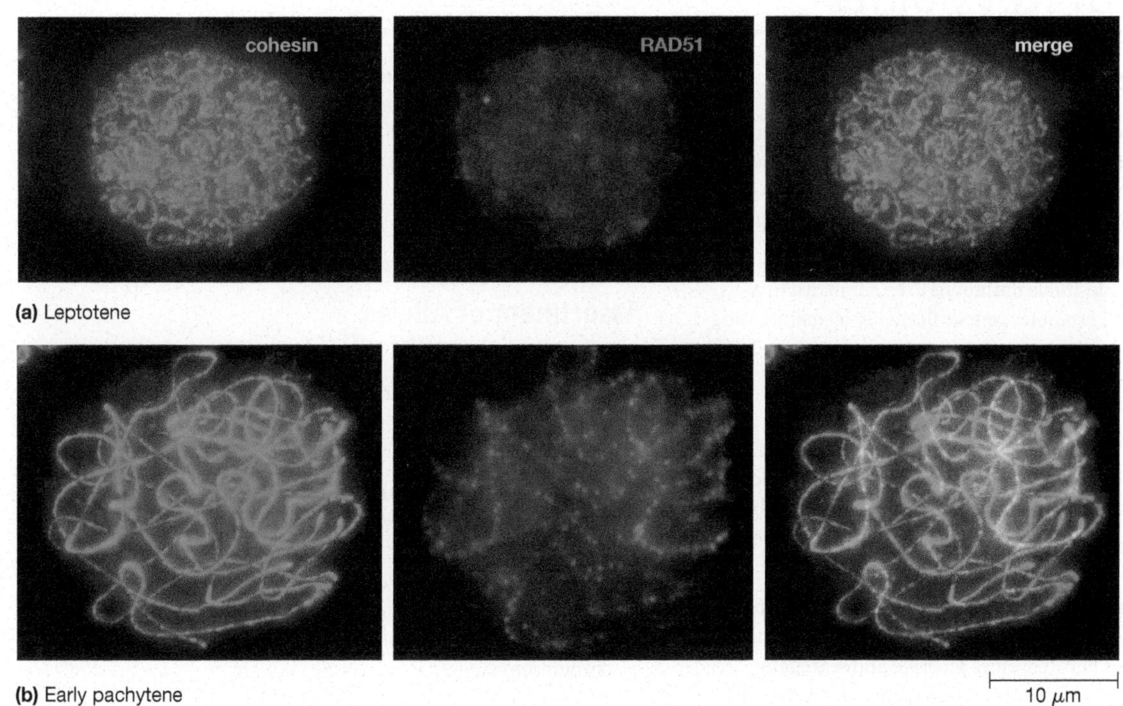

Figure 25-24 DNA Repair Proteins Operate During Prophase I of Meiosis. Cohesins (green) mark condensed meiotic chromosomes; RAD51 (red) marks sites of DNA strand breakage and repair during recombination. In the merged images on the right, yellow indicates where both proteins are found together. **(a)** At the leptotene stage, few RAD51 foci are evident. **(b)** At the early pachytene stage, numerous sites of strand breaks and repair are evident (fluorescence microscopy).

Homologous Recombination Between Chromosomes Relies on High-Fidelity DNA Repair

DNA double-stranded break (DSB) repair can occur using a homologous stretch of DNA as a repair template (see Chapter 17). This kind of DNA repair faithfully repairs the damaged strand and resolves the repaired strands using a specialized junction known as a *Holliday junction*. In some ways, the normal process of recombination that occurs in eukaryotes during meiosis can be considered a very large-scale DNA "repair" process that relies on homologous recombination. In this case, however, strand breakage and resealing occur on a massive scale, as a tightly regulated part of normal meiosis.

A key enzyme involved in homologous recombination is called *RecA* in prokaryotes; the comparable protein in eukaryotes is called *RAD51* (see Chapter 17). During homologous recombination, RecA or RAD51 first coats the single-stranded DNA region; the coated, single-stranded DNA then interacts with a DNA double helix, moving along the target DNA until it reaches a complementary sequence it can pair with. If meiotic recombination uses the very same DNA repair machinery used to repair breaks in DNA due to DNA damage, then we might expect RAD51 to appear at the time recombination occurs. Staining meiotic cells to look for RAD51 in association with synapsed chromosomes confirms this prediction (**Figure 25-24** on page 779), once again emphasizing the underlying conserved mechanisms involved in DNA repair and meiotic recombination. In addition to the general requirements for DNA repair during meiotic recombination, geneticists have uncovered several additional key features of meiotic recombination. First, the frequency of recombination is not uniform across the genome. Some chromosomal regions show very high recombination frequency; these regions are called *recombination hotspots*. Current research is aimed at understanding why such regions are so active. Second, too much recombination can lead to chromosomal defects. One example involves a DNA helicase, associated in humans with a disease known as *Fanconi's anemia*. This enzyme is associated with DSB repair during replication. Plants that lack the activity of this enzyme show abnormally high rates of recombination and genome instability. Future research should continue to unravel the details of how double-stranded break repair is regulated during meiosis.

MAKE CONNECTIONS 25.2

Recall that the success of genome editing with CRISPR relies on the introduction of a repair template. Describe how the homologous recombination pathway uses this template to cause genome editing. (Fig. 1-9)

CONCEPT CHECK 25.6

What effects on meiosis do you think there might be if a meiotic cell were unable to make proteins of the synaptonemal complex? Explain your answer.

Summary of Key Points

Mastering™ Biology For activities, animations, and review quizzes, go to the study area at www.masteringbiology.com.

25.1 Sexual Reproduction

- Asexual reproduction is based on mitotic cell division and produces offspring that are genetically identical (or nearly so) to the single parent. In contrast, sexual reproduction involves two parents and leads to a mixture of parental traits in the offspring.
- Sexual reproduction allows populations to adapt to environmental changes, enables desirable mutations to be combined in a single individual, and promotes genetic flexibility by maintaining a diploid genome.

25.2 Meiosis

- The life cycle of sexually reproducing eukaryotes includes both haploid and diploid phases. Haploid gametes are generated by meiosis and fuse at fertilization to restore the diploid chromosome number.
- Meiosis consists of two successive cell divisions without an intervening duplication of chromosomes. During the first meiotic division, homologous chromosomes separate and segregate into the two daughter cells. During the second meiotic division, sister chromatids separate and four haploid daughter cells are produced.
- In addition to reducing the chromosome number from diploid to haploid, meiosis differs from mitosis in that homologous chromosomes synapse during prophase of the first meiotic division, thereby allowing crossing over and genetic recombination between nonsister chromatids.
- Meiotic divisions of spermatocytes in animals are symmetric, yielding four equally sized haploid sperm, whereas the meiotic divisions of oocytes are highly asymmetric, such that a single cell receives almost all of the cytosol from the starting meiotic precursor cell.

25.3 Genetic Variability: Segregation and Assortment of Alleles

- Mendel's laws of inheritance describe the genetic consequences of chromosome behavior during meiosis, even though chromosomes had not yet been discovered at the time of Mendel's experiments.
- Mendel's law of segregation states that the two (maternal and paternal) alleles of a gene are distinct entities that separate into different gametes during meiosis. The law of independent assortment states that the alleles of each gene separate independently of the alleles of other genes (now known to apply only to genes located on different chromosomes or very far apart on the same chromosome).

25.4 Genetic Variability: Recombination and Crossing Over

- The genetic variability among an organism's gametes arises partly from the independent assortment of chromosomes during anaphase I and partly from genetic recombination during prophase I.

- The frequency of recombination between genes on the same chromosome is a measure of the distance between the two genes and can therefore be used to map their chromosomal locations.

25.5 Genetic Recombination in Bacteria and Viruses

- Besides occurring during meiosis, homologous recombination also takes place in viruses (during co-infection) and when DNA is transferred into bacterial cells by transformation, transduction, or conjugation.

25.6 Mechanisms of Homologous Recombination

- Synaptonemal complexes facilitate homologous recombination between chromosomes during meiosis; crossover sites eventually become visible as chiasmata as chromosomes become highly condensed.
- Recombination during meiosis involves the same molecular mechanisms of DNA strand breakage and exchange between homologous DNA molecules that occurs during high-fidelity DNA repair, but on a massive scale.

Problem Set

Mastering™ Biology For activities, animations, and review quizzes, go to the study area at www.masteringbiology.com.

25-1 The Truth About Sex. For each of the following statements, indicate with an S if it is true of sexual reproduction, with an A if it is true of asexual reproduction, or with a B if it is true of both.

(a) Traits from two different parents can be combined in a single offspring.

(b) Each generation of offspring is virtually identical to the previous generation.

(c) Mutations are propagated to the next generation.

(d) Some offspring in every generation will be less suited for survival than the parents, but others may be better suited.

(e) Mitosis is involved in the life cycle.

25-2 Ordering the Phases of Meiosis. Drawings of several phases of meiosis in an organism, labeled A through F, are shown in **Figure 25-25**.

(a) What is the diploid chromosome number in this species?

(b) Place the six phases in chronological order, and name each one.

(c) Between which two phases do homologous centromeres separate?

(d) Between which two phases does recombination occur?

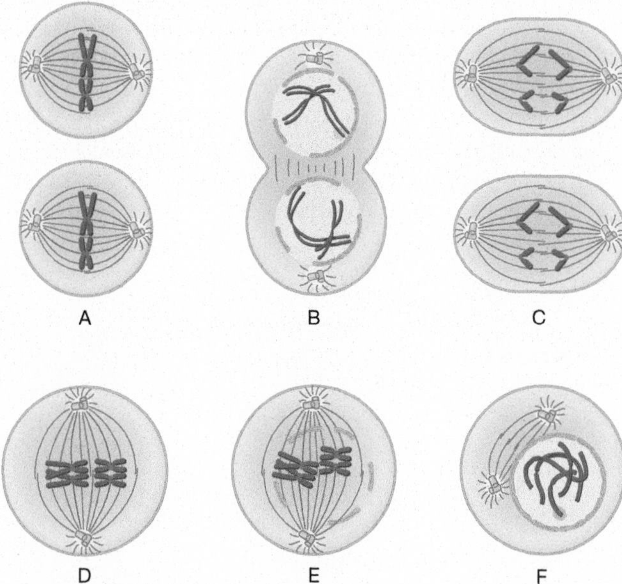

Figure 25-25 Six Phases of Meiosis to Be Ordered and Identified. See Problem 25-2.

25-3 Telling Them Apart. Briefly describe how you might distinguish between each of the following pairs of phases in the same organism.

(a) Metaphase of mitosis and metaphase I of meiosis

(b) Metaphase of mitosis and metaphase II of meiosis

(c) Metaphase I and metaphase II of meiosis

(d) Telophase of mitosis and telophase II of meiosis

(e) Pachytene and diplotene stages of meiotic prophase I

25-4 Your Centromere Is Showing. Suppose you have a diploid organism in which all the chromosomes contributed by the sperm have cytological markers on their centromeres that allow you to distinguish them visually from the chromosomes contributed by the egg.

(a) Would you expect all the somatic cells (cells other than gametes) to have equal numbers of maternal and paternal centromeres in this organism? Explain.

(b) Would you expect equal numbers of maternal and paternal centromeres in each gamete produced by that individual? Explain.

25-5 More About DNA. Let X be the amount of DNA present in the gamete of an organism that has a diploid chromosome number of 4. Assuming all chromosomes to be of approximately the same size, how much DNA (X, $2X$, $1/2X$, and so on) would you expect in each of the following?

(a) A zygote immediately after fertilization

(b) A single sister chromatid

(c) A daughter cell following mitosis

(d) A single chromosome following mitosis

(e) A nucleus in mitotic prophase

(f) The cell during metaphase II of meiosis

(g) One bivalent

25-6 Meiotic Mistakes. Infants born with Patau syndrome have an extra copy of chromosome 13, which leads to developmental abnormalities such as cleft lip and palate, small eyes, and extra fingers and toes. Another type of genetic disorder, Turner syndrome, results from the presence of only one sex chromosome—an X chromosome. Individuals born with one X chromosome are females exhibiting few noticeable defects until puberty, when they fail to develop normal breasts and internal sexual organs. Describe the meiotic events that could lead to the birth of an individual with either Patau syndrome or Turner syndrome.

25-7 Down Syndrome Revisited. About 5% of individuals with Down syndrome (see Human Connections, pages 762–763) have a chromosomal translocation in which a third copy of chromosome 21 is attached to chromosome 14. Such individuals receive this chromosome from one of their parents due to a malformation in a gonadal cell. What is the ploidy of such individuals with Down syndrome? Why does receiving the modified chromosome from one of their parents lead to Down syndrome in such children?

25-8 QUANTITATIVE Punnett Squares as Genetic Tools. The genetic characters of seed color (where *Y* is the allele for yellow seeds and *y* for green seeds) and seed shape (where *R* is the allele for round seeds and *r* for wrinkled seeds) were used by Mendel in some of his crosses.

(a) Mendel performed a one-factor cross between parent plants that were both heterozygous for seed color, $Yy \times Yy$. Using a Punnett square, explain the 3:1 phenotypic ratio Mendel observed for the offspring of such a cross.

(b) Mendel went on to perform a two-factor cross between plants heterozygous for both seed color (*Yy*) and seed shape (*Rr*). How does the Punnett square of **Figure 25-26** reflect Mendel's law of independent assortment?

(c) Complete the Punnett square of Figure 25-26 by writing in each of the possible progeny genotypes. How many different genotypes will be found in the progeny? In what ratios?

(d) For the case of Figure 25-26, how many different phenotypes will be found in the progeny? In what ratios?

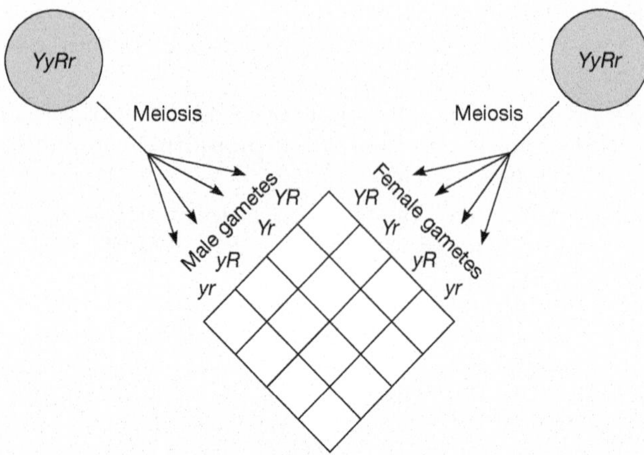

Figure 25-26 Punnett Square of a Two-Factor Cross Between Plants Heterozygous for Both Seed Color (*Yy*) and Seed Shape (*Rr*). See Problem 25-8.

25-9 DATA ANALYSIS Genetic Mapping. The following table provides data concerning the frequency with which four genes (*w*, *x*, *y*, and *z*) located on the same chromosome recombine with each other.

Genes	Recombination Frequency
w and *x*	25%
w and *y*	29%
w and *z*	17%
x and *y*	50%
x and *z*	9%
y and *z*	44%

(a) Construct a genetic map indicating the order in which these four genes occur and the number of map units that separate the genes from each other.

(b) In constructing this map, you may have noticed that the map distances are not exactly additive. Can you provide an explanation for this apparent discrepancy?

25-10 Homologous Recombination. Bacterial cells use at least three different pathways for carrying out genetic recombination. All three pathways require the RecA protein, but in each case, a different set of steps precedes the action of RecA in catalyzing strand invasion. One of these three pathways utilizes an enzyme complex called RecBCD, which binds to double-strand breaks in DNA and exhibits both helicase and single-strand nuclease activities.

(a) Briefly describe a model showing how the RecBCD enzyme complex might set the stage for genetic recombination.

(b) When bacterial cells are co-infected with two different strains of bacteriophage λ, genes located near certain regions of the phage DNA, called *CHI sites*, recombine at much higher frequencies than other genes do. However, in mutant bacteria lacking the RecBCD protein, genes located near CHI sites do not recombine any more frequently than do other genes. How can you modify your model to accommodate this additional information?

25-11 Gene Cloning and Recombination. Not only are the plasmid vectors used in molecular biology engineered, but so are the strains of *E. coli* used in cloning.

(a) Nearly all strains of *E. coli* used in DNA cloning carry mutations in the *recA* gene that result in loss of RecA activity. Why would a *recA* mutation make an *E. coli* cell a better host for propagating recombinant plasmid DNA?

(b) Recall that restriction endonucleases are normally made by bacteria such as *E. coli* (page 621). *E. coli* used in molecular biology also carry mutations in restriction endonucleases. Why do you think these mutations would be useful?

26 Cancer Cells

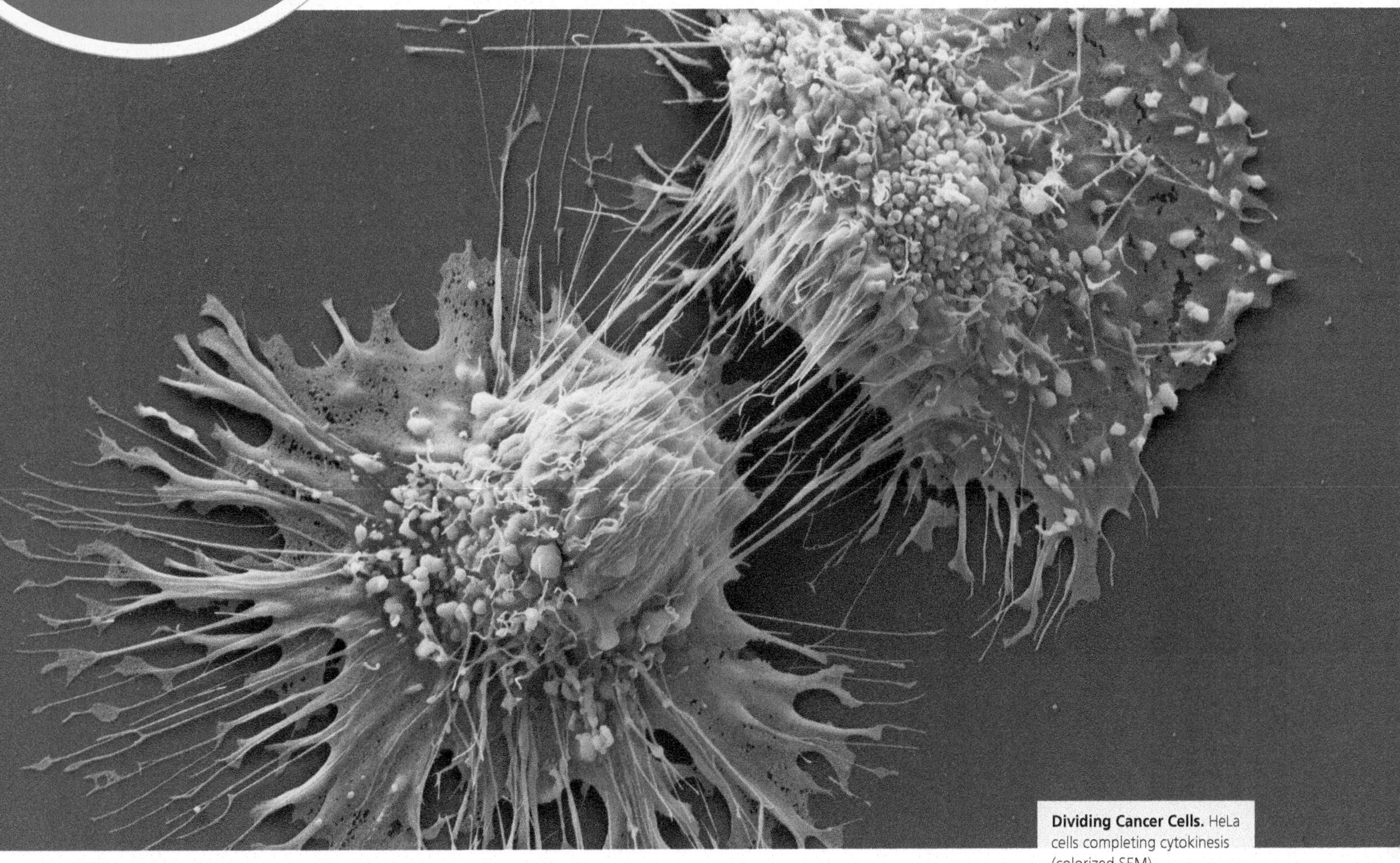

Dividing Cancer Cells. HeLa cells completing cytokinesis (colorized SEM).

Anyone familiar with the events occurring inside living cells must feel a sense of awe at the complexities involved. Given the vast number of activities that need to be coordinated in every cell, it is not surprising that malfunctions occasionally arise. Cancer—the second-leading cause of death after cardiovascular disease—is a prominent example of a disease that arises from such abnormalities in cell function. If current trends continue, almost half the population of the United States will eventually develop some form of the disease, and cancer will become the most common cause of death.

We now know that gene mutations and changes in gene expression play a central role. In this chapter we will see how these genetic changes contribute to the aberrant properties of cancer cells. As we do so, it will become apparent that explaining the behavior of cancer cells requires an intimate knowledge of how normal cells behave and, conversely, that investigating the biology of cancer cells has deepened our understanding of normal cells.

26.1 How Cancers Arise

The term *cancer*, which means "crab" in Latin, was coined by Hippocrates in the fifth century B.C. to describe diseases in which tissues grow and spread unrestrained throughout the body, eventually choking off life. Cancers can originate in almost any organ. Depending on the cell type involved, they are grouped into several different categories. **Carcinomas**, which account for about 90% of all cancers, arise from the *epithelial cells* that cover external and internal body surfaces. Lung, breast, and colon cancer are the most frequent cancers of this type. **Sarcomas** develop from supporting tissues such as bone, cartilage, fat, and muscle. Finally, **lymphomas** and **leukemias** arise from cells of blood and lymphatic origin, with the term *lymphoma* being used for tumors that grow as solid masses of tissue and the term *leukemia* being reserved for cases in which the cancer cells proliferate mainly in the bloodstream.

Based on such differences in their sites of origin and the specific cell type involved, more than 100 different types of cancer can be distinguished from one another. But no matter where cancer arises and what cell type is involved, the danger posed by the disease comes from a combination of two properties: the ability of cells to proliferate in an uncontrolled way and their ability to spread through the body. We begin this chapter by describing these two deadly properties.

Tumors Arise When the Balance Between Cell Division and Cell Differentiation or Death Is Disrupted

A cancer occurs when some cells in a tissue divide and accumulate in an uncontrolled, relatively autonomous way. The resulting mass of growing tissue is called a **tumor** (or *neoplasm*). Although tumors escape from the normal controls on cell proliferation, tumor cells do not always divide more rapidly than normal cells. The crucial issue is not the rate of cell division but rather the balance between cell division and cell differentiation or cell death.

Cell differentiation is the process by which cells acquire specialized traits; as cells do so, they often cease dividing. To illustrate, consider cell division and differentiation in the skin, where new cells continually replace aging cells that are being shed from the outer body surfaces (**Figure 26-1**). The new replacement cells are generated by cell divisions occurring in the *basal layer* of the skin (Figure 26-1a). On average, each time a basal cell divides, it gives rise to two cells with different fates. One cell stays in the basal layer and retains the capacity to divide, whereas the other loses the capacity to divide and differentiates as it leaves the basal layer and moves toward the outer skin surface. During the differentiation process, the migrating cell gradually flattens and begins to make *keratin*, the structural protein that imparts mechanical strength to the outer layers of the skin. Eventually, the cell dies and is shed from the outer skin surface.

Basal cells are a good example of a *stem cell:* a cell that can divide to form more cells like itself or cells that are destined to become more specialized. Embryonic and induced pluripotent stem cells are other examples of stem cells (see Chapter 20). In some cases, such as hair follicles or the mammalian testis, stem cells can produce daughter cells that are more specialized, but that continue to divide. Such cells are sometimes called *transit-amplifying cells* and represent a variation on the stem cell theme.

In normal skin, one of the two cells produced by each cell division retains the ability to divide, whereas the other cell leaves the basal layer, loses the capacity to divide, and finally dies. This arrangement ensures that there is no increase in the number of dividing cells. A similar phenomenon occurs in bone marrow, where cell division creates new blood cells to replace aging blood cells that are being destroyed. It also occurs in the gastrointestinal tract lining, where cell division creates new cells to replace the ones being shed from the inner

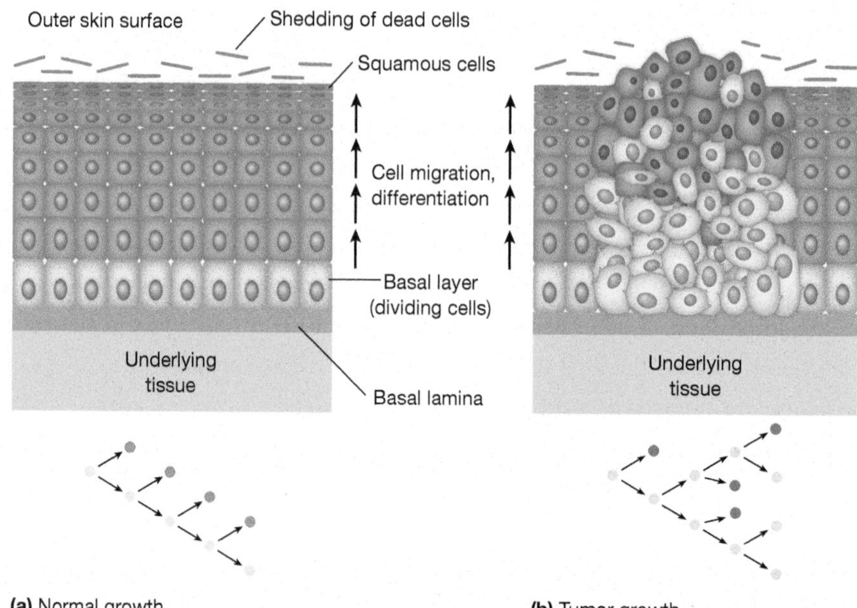

Figure 26-1 Comparison of Normal and Tumor Growth in the Epithelium of the Skin. (Top left) In normal skin, cell division in the basal layer gives rise to new cells that migrate toward the outer surface of the skin, changing shape and losing the capacity to divide. **(a)** In tumor growth, this orderly process is disrupted, and some of the cells that migrate toward the outer surface retain the capacity to divide. In both diagrams, a lighter color is used to identify cells that retain the capacity to divide. **(b)** Schematic diagrams illustrating the fate of dividing cells. In normal skin, each cell division in the basal layer gives rise on average to one cell that retains the capacity to divide (lighter color) and one cell that differentiates, thereby losing the capacity to divide (darker color). As a result, no net accumulation of dividing cells occurs. In tumor growth, cell division is not appropriately balanced with cell death or differentiation, thereby leading to a progressive increase in the number of dividing cells.

surface of the stomach and intestines. In each of these situations, cell division is carefully balanced with cell differentiation and death so that no net accumulation of dividing cells takes place.

In tumors, this finely balanced arrangement is disrupted, and cell division is uncoupled from cell differentiation and death. As a result, some cell divisions give rise to two cells that both continue to divide, thereby feeding a progressive increase in the number of dividing cells (see Figure 26-1b). If the cells are dividing rapidly, the tumor will grow quickly; if the cells are dividing more slowly, tumor growth will be slower. But regardless of how fast or slow the cells divide, the tumor will continue to grow because new cells are being produced in greater numbers than needed. As the dividing cells accumulate, the normal organization and function of the tissue gradually become disrupted.

Based on differences in their growth patterns, tumors are classified as either benign or malignant. **Benign tumors** grow in a confined local area and are rarely dangerous. In contrast, **malignant tumors** are capable of invading surrounding tissues, entering the bloodstream, and spreading to distant parts of the body. The term **cancer** refers to any malignancy, but especially malignant cells that can spread from their original location to other sites. Because the ability to grow in an uncontrolled way and spread to distant locations makes cancer a potentially life-threatening disease, it is important to understand the mechanisms that give rise to these traits.

Cancer Cell Proliferation Is Anchorage Independent and Insensitive to Population Density

Cancer cell proliferation exhibits several traits that distinguish it from proliferation of normal cells. One trait, of course, is the ability to form tumors. This trait can be demonstrated experimentally by injecting cells into *nude mice*, which lack a functional immune system that would normally attack and destroy foreign cells. Normal human cells injected into such animals will not grow, whereas cancer cells will proliferate and form tumors.

Cancer cells also exhibit other growth properties that distinguish them from normal cells. For example, normal cells do not grow well in culture if they are suspended in a liquid medium or in a semisolid material such as soft agar. But when provided with a solid surface they can attach to, normal cells become anchored to the surface, spread out, and begin to proliferate. In contrast, cancer cells grow well not only when they are anchored to a solid surface but also when they are freely suspended in a liquid or semisolid medium. Cancer cells are therefore said to exhibit **anchorage-independent growth**.

When growing in the body, most normal cells meet the anchorage requirement by binding to the extracellular matrix through cell surface proteins called *integrins* (see Figure 15-19). If attachment to the matrix is prevented using chemicals that block the binding of integrins to the matrix, normal cells usually lose the ability to divide and often self-destruct by *apoptosis* (for the process of apoptosis, see Figure 24-26). Triggering apoptosis in the absence of proper anchorage is one of the safeguards that prevents normal cells from floating away and setting up housekeeping in another tissue. Cancer cells, in contrast, circumvent this safeguard.

Cancer cells also differ from normal cells in their response to crowded conditions in culture. When normal cells are grown in culture, they divide until the surface of the culture vessel is covered by a single layer of cells. Once this *monolayer stage* is reached, cell division stops—a phenomenon referred to as **density-dependent inhibition of growth**. Cancer cells exhibit reduced sensitivity to density-dependent inhibition of growth; instead of ceasing division, cancer cells continue to divide and gradually begin piling up on one another.

Cancer Cells Are Immortalized by Mechanisms That Maintain Telomere Length

When most normal cells are grown in culture, they divide a limited number of times. For example, freshly isolated human fibroblasts divide about 50–60 times in culture and then stop dividing, undergo various degenerative changes, and may even die. Under similar conditions, cancer cells exhibit no such limit and continue dividing indefinitely, behaving as if they were immortal. A striking example is provided by *HeLa cells* (see Human Connections in Chapter 1, page 16), which were obtained in 1951 from a uterine cervical cancer from Henrietta Lacks. Although she died of cancer that same year, some of her tumor cells had been placed in culture after surgery. These cancer cells began to proliferate and have continued to do so for 70 years, dividing more than 20,000 times with no signs of stopping.

Why can cancer cells divide indefinitely in culture when most normal human cells divide only 50–60 times? The answer is related to the telomeric DNA sequences that are lost from the ends of each chromosome every time DNA replicates. If a normal cell divides too many times, its telomeres become too short to protect the ends of the chromosomes, triggering apoptosis. In cancer cells, the telomere-imposed limit is overcome because most cancer cells produce *telomerase*, the enzyme that adds telomeric DNA sequences to the ends of DNA molecules (see Figure 17-19). An alternative mechanism for maintaining telomere length employs enzymes that exchange DNA sequence information between chromosomes. By one mechanism or the other, cancer cells maintain telomere length above a critical threshold and thereby retain the capacity to divide indefinitely.

Defects in Signaling Pathways, Cell Cycle Controls, and Apoptosis Contribute to Cancer

Mechanisms that maintain telomere length play a permissive role in allowing cancer cells to continue dividing, but they do not actually cause cells to divide. The defects driving the uncontrolled proliferation of cancer cells can be traced to a variety of signaling pathways and control mechanisms that normally maintain the proper balance between cell division and death. For example, recall that cell proliferation is regulated by protein *growth factors* that bind to cell surface receptors and activate signaling pathways within the targeted cells (see Chapters 23 and 24). Cells do not usually divide unless they are stimulated by the proper growth factor, but cancer cells circumvent this restraining mechanism through alterations in signaling pathways that create a constant signal to divide.

Disruptions in cell cycle control also contribute to the unrestrained proliferation of cancer cells. The commitment to proceed through the cell cycle is made at the *restriction point*, which controls progression from G1 into S phase (see Chapter 24). If normal cells are grown under suboptimal conditions (for example, insufficient growth factors, high cell density, lack of anchorage, or inadequate nutrients), the cells become arrested at the restriction point and stop dividing. In comparable situations, cancer cells continue to proliferate. This abnormal behavior occurs because cell cycle controls do not function properly in cancer cells. Besides failing to respond appropriately to external signals, cancer cells are also unresponsive to internal conditions, such as DNA damage, that would normally trigger checkpoint mechanisms to halt the cell cycle.

For certain kinds of cancer, uncontrolled growth arises primarily from a failure to undergo apoptosis rather than increased cell division. Apoptosis is typically triggered to get rid of unnecessary or defective cells. Why aren't cancer cells killed by apoptosis? The answer is that cancer cells block apoptosis in various ways, allowing them to survive and proliferate under conditions that would normally cause cell death.

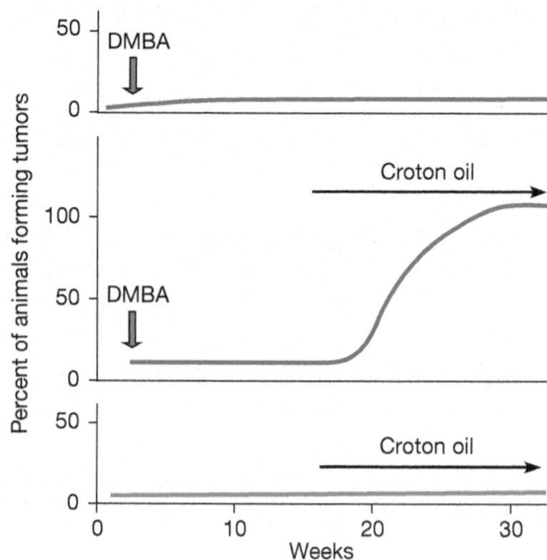

Figure 26-2 Evidence for Initiation and Promotion Stages During Cancer Development. (Top) Mice given a single dose of *DMBA (dimethylbenz[a]anthracene)* do not form tumors. (Center) Painting the mice's skin with croton oil twice a week after DMBA treatment leads to the appearance of skin tumors. If the croton oil application is stopped a few weeks into the treatment (data not shown), the tumors regress. (Bottom) Croton oil alone does not produce skin tumors. These data are consistent with the conclusion that DMBA is an initiator and croton oil is a promoting agent.

Cancer Arises Through a Multistep Process Involving Initiation, Promotion, and Tumor Progression

Many pieces of evidence suggest that the development of cancer requires multiple steps. First, normal cells are converted to a precancerous state, which sensitizes them to further changes that lead to cancer in a process known as **initiation**. Second, there is a long, gradual process of **promotion**, involving repeated exposure of sensitized cells to cancer-promoting agents. Finally, once a tumor has formed, the tumor grows and differentiates in the process of **tumor progression**.

Initiation. Early evidence for the idea of initiation and promotion phases came from studying the ability of coal tar components, such as *dimethylbenz[a]anthracene (DMBA)*, to cause cancer in laboratory animals. These studies showed that feeding mice a single dose of DMBA rarely causes tumors to develop. However, if a mouse that has been fed a single dose of DMBA is later treated with a substance that causes skin irritation, cancer develops in the treated area (**Figure 26-2**). The irritant most commonly used for triggering tumor formation is a plant-derived substance called *croton oil*, which is enriched in compounds called *phorbol esters*. Croton oil does not cause cancer by itself, nor does cancer arise if DMBA is administered *after* the croton oil.

Promotion. A year or more can transpire after feeding animals a single dose of DMBA and yet tumors will still develop if an animal's skin is then irritated with croton oil. This means that a single DMBA treatment can create a permanently altered, initiated state in cells located throughout the body. The ability of chemicals to act as initiators correlates with their ability to cause DNA damage, suggesting that the permanently altered, initiated state is based on the ability of initiators to cause *DNA mutation*. The subsequent administration of croton oil then acts on these altered cells to promote tumor development.

In contrast to initiation, promotion is a gradual process requiring prolonged or repeated exposure to a promoting agent. Investigation of a wide variety of promoting agents has revealed that their main shared property is the ability to stimulate *cell proliferation*. Note that not all tumor promoters are foreign substances. Hormones and growth factors that stimulate normal cell proliferation may inadvertently behave as tumor promoters if they act on a cell that has already sustained an initiating mutation.

When a cell with an initiating mutation is exposed to a promoting agent (or natural growth regulator) that causes the initiated cell to proliferate, the number of mutant cells increases. The time required for promotion contributes to the lengthy delay that often transpires between exposure to an initiating carcinogen and the development of cancer.

Tumor Progression. The basic pattern of cancer progression is shown in **Figure 26-3**. That cancers grow from a group of cells that have lost the normal controls on cell proliferation is clear. What is less clear is exactly how, from its tiny beginnings, a tumor grows to contain millions of cells, and how a tumor comes to have the proportion of different types of cells that it has. Several ideas are being intensively investigated.

One idea is that tumor cells are much like organisms in a population under Darwinian selection. As proliferation continues, natural selection tends to favor cells exhibiting enhanced growth rate and invasive properties, eventually leading to the formation of a malignant tumor. Just as evolution selects for organisms with enhanced fitness and reproductive success within a particular environment, cells in a growing tumor may thus gain a growth advantage over their neighbors. The result is a large population of

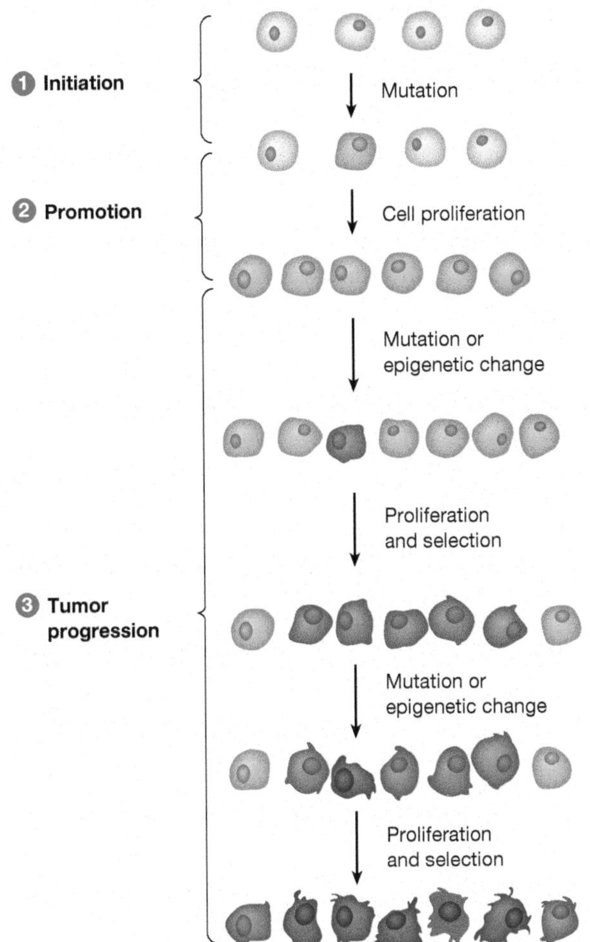

Figure 26-3 Main Stages in Cancer Development. Cancer arises by a multistep process involving an initiation event based on DNA mutation, a promotion stage in which the initiated cell is stimulated to proliferate, and tumor progression, in which mutations and changes in gene expression create variant cells exhibiting enhanced growth rates or other aggressive properties that give them a selective advantage. Such cells tend to outgrow their companions and become the predominant cell population in the tumor. During tumor progression, repeated cycles of this selection process create a population of cells whose properties gradually change over time.

cells descended from these more successful cells, a process known as *clonal expansion*.

Although clonal expansion is likely a key ingredient in tumor growth, it is probably not the whole story. Experiments involving several different types of cancer indicate that not all cells in a tumor are equally capable of rapid proliferation. When tumor cells are separated using fluorescence-activated cell sorting (a technique described in **Key Technique in Chapter 24, pages 722–723**), only a tiny minority—as few as 1–2%—of the cells can rapidly form tumors when introduced into nude mice. Moreover, the new tumors formed by these cells are similar to the original tumor: only a small minority of cells within the new tumors have potent tumor-forming properties. This suggests that tumors may, like the skin cells we discussed earlier in this chapter, contain small groups of cells that function like *cancer stem cells* to generate large numbers of descendants.

One final property of tumors is worth mentioning. Thus far, you may have developed the impression that tumors are homogeneous, that is, they are composed of one type of cell, rather like a bag of marbles. Most tumors are not at all like this, however; they are complicated, multicellular tissues. If clonal expansion accounts for how tumors grow, different types of cells probably lead to successful expansions over time in a single growing tumor. The result is a mixture of genetically distinct regions in a single tumor, a property known as *intratumor heterogeneity*. These genetically related parts of a tumor can be very large. In one type of pancreatic cancer, for example, the genetically distinct regions each contain more than 100 million cells. A type of tumor resulting from cancerous support cells in the brain, known as a *glioblastoma*, is well known to contain many different cell types in a complicated mixture.

CONCEPT CHECK 26.1

As you will learn later in this chapter, cancer cells from patients with Burkitt lymphoma carry a reciprocal chromosomal translocation between chromosome 8 and another chromosome, often 22. Each patient exhibits a unique pattern of chromosome breakage ("breakpoints") and reattachment, but each cell from that patient carries the same breakpoints. From this result, what can you conclude regarding how the lymphoma arose?

26.2 How Cancers Spread

Although uncontrolled proliferation is a defining feature of cancer cells, it is not what makes cancer so dangerous. After all, the cells of benign tumors also proliferate in an uncontrolled way, but such tumors remain in their original location and are usually easy to remove (unless they arise in surgically inaccessible locations). The hazards posed by cancer cells come from *uncontrolled* proliferation combined with the ability to spread throughout the body, which makes complete surgical removal impractical. Roughly 90% of all cancer deaths are caused by the spread of cancer rather than by the primary tumor itself. We will now examine the mechanisms that make this spreading of cancer cells possible.

Angiogenesis Is Required for Tumors to Grow Beyond a Few Millimeters in Diameter

For more than 100 years, scientists have known that tumors possess a dense network of blood vessels. Initially, these were thought to be preexisting vessels that had expanded in response to the tumor's presence or were thought to be part of an inflammatory response designed to defend the host against the tumor. But in 1971, Judah Folkman proposed a new idea regarding the role of these blood vessels. He suggested that tumors release signaling molecules that trigger **angiogenesis**—that is, growth of blood vessels—in the surrounding host tissues and that these new vessels are required for tumors to grow beyond a tiny, localized clump of cells.

This idea initially emerged from studies involving cancer cells grown in isolated organs under artificial laboratory conditions. In one experiment, a normal thyroid gland was removed from a rabbit and placed in a glass chamber, a small number of cancer cells were injected into the gland, and a nutrient solution was pumped into the organ to keep it alive (**Figure 26-4a**). Under these conditions, the cancer cells divided for a few days but suddenly stopped growing when the tumor reached a diameter of 1–2 mm. Virtually every tumor stopped growing at exactly the same size, suggesting that some kind of limitation allowed them to grow only so large.

When tumor cells were removed from the thyroid gland and injected back into animals, however, cell division resumed and massive tumors developed. Why did the tumors stop growing at a small size in the isolated thyroid gland and yet grow in an unrestrained way in live animals? On closer examination, a possible explanation became apparent. The tiny tumors, alive but dormant in the isolated thyroid gland, had failed to link up with the organ's blood vessels. As a result, the tumors stopped growing when they reached a diameter of 1–2 mm. When injected into live animals, these same tumors became infiltrated with blood vessels and grew to an enormous size.

To test the idea that blood vessels are needed to sustain tumor growth, Folkman implanted cancer cells in the anterior chamber of a rabbit's eye, where there is no blood supply. As shown in Figure 26-4b, these cancer cells survived and formed tiny tumors. But blood vessels from the nearby iris could not reach the cells, and the tumors soon stopped growing. When the same cells were implanted directly on the iris tissue, blood vessels from the iris quickly infiltrated the tumor cells, and each tumor grew to thousands of times its original size. Once again, it appeared that tumors need a blood supply to grow beyond a very small mass.

Blood Vessel Growth Is Controlled by a Balance Between Angiogenesis Activators and Inhibitors

If tumors require blood vessels to sustain their growth, how do they ensure that this need is met? The first hint came from studies in which cancer cells were placed inside a chamber surrounded by a filter possessing tiny pores that cells cannot pass through. When such chambers are implanted into animals, new capillaries proliferate in the surrounding host tissue. In contrast, normal cells placed in the same type of chamber do not stimulate blood vessel growth. Such results suggest that cancer cells produce molecules that diffuse through the tiny pores in the filter and activate angiogenesis in the surrounding host tissue.

Subsequent investigations have revealed that the main angiogenesis-activating molecules are two growth factors: *vascular endothelial growth factor (VEGF)* and *fibroblast growth factor (FGF)*. VEGF and FGF are important during normal embryonic development, when they stimulate formation of blood vessels in the developing embryo, but they are also produced by many kinds of cancer cells. When cancer cells release these proteins into the surrounding tissue, they bind to receptor proteins on the surface of the *endothelial cells* that form the lining of blood vessels. This binding activates a signaling pathway that causes the endothelial cells to divide and to secrete protein-degrading enzymes called *matrix metalloproteinases (MMPs)*. The MMPs break down the extracellular matrix, aiding the migration of the endothelial cells into the surrounding tissues. As they migrate, the proliferating endothelial cells become organized into hollow tubes that develop into new blood vessels.

Although many tumors produce VEGF and/or FGF, these signaling molecules are not the sole explanation for the activation of angiogenesis. For angiogenesis to proceed, these molecules must overcome the effects of angiogenesis *inhibitors* that normally restrain the growth of blood vessels. More than a dozen naturally occurring inhibitors of angiogenesis have been identified, including the proteins *angiostatin, endostatin,* and *thrombospondin.* A finely tuned balance between the concentration of angiogenesis inhibitors and of angiogenesis activators determines whether a tumor will induce the growth of new blood vessels. When tumors trigger angiogenesis, it is

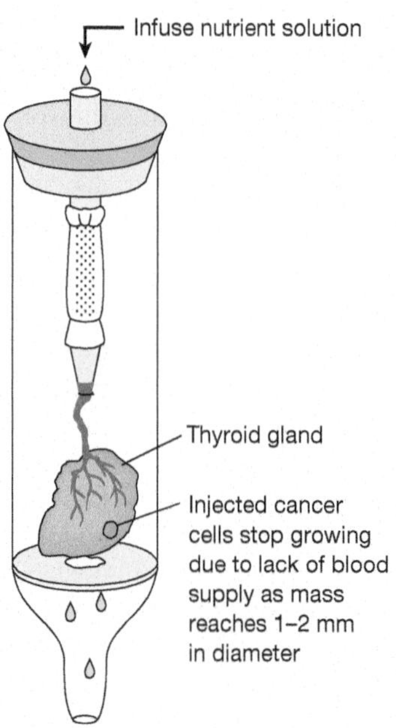

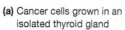

(a) Cancer cells grown in an isolated thyroid gland

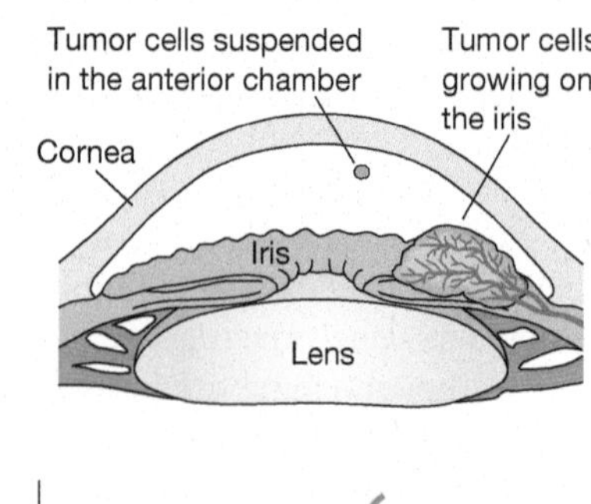

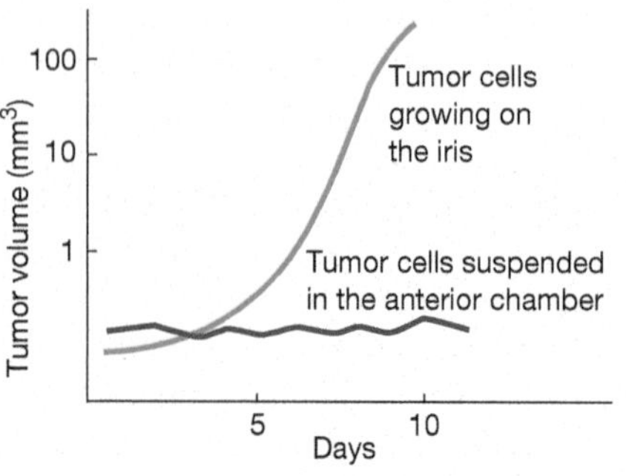

(b) Cancer cells grown on the iris or in the anterior chamber of the eye

Figure 26-4 Two Experiments Showing the Requirement for Angiogenesis. (a) Cancer cells were injected into an isolated rabbit thyroid gland that was kept alive by pumping a nutrient solution into its main blood vessel. The tumor cells failed to link up to the organ's blood vessels, and the tumor stopped growing when it reached a diameter of roughly 1–2 millimeters. **(b)** Cancer cells were either injected into the liquid-filled anterior chamber of a rabbit's eye, where there are no blood vessels, or placed directly on the iris. Tumor cells in the anterior chamber, nourished solely by diffusion, remained alive but stopped growing before the tumor reached a millimeter in diameter. In contrast, blood vessels quickly infiltrated the cancer cells implanted on the iris, allowing the tumors to grow to thousands of times their original mass.

Adapted from "The Vascularization of Tumors" by Judah Folkman, Scientific American (May 1976). Graph: Reproduced with permission. Drawings: Adapted by permission of the illustrator, Carol Donner.

usually accomplished by an increase in the production of angiogenesis activators and a simultaneous decrease in the production of angiogenesis inhibitors.

Cancer Cells Spread by Invasion and Metastasis

Once angiogenesis has been triggered at an initial tumor site, the stage is set for the cancer cells to spread throughout the body. This ability to spread is based on two distinct mechanisms: invasion and metastasis. **Invasion** refers to the direct migration and penetration of cancer cells into neighboring tissues, whereas **metastasis** involves the ability of cancer cells to enter the bloodstream (or other body fluids) and travel to distant sites, where they form tumors—called *metastases*—that are not physically connected to the primary tumor.

The ability of a tumor to metastasize depends on a complex cascade of events, beginning with angiogenesis. The events following angiogenesis can be grouped into three main steps (**Figure 26-5**). First, cancer cells invade surrounding tissues and penetrate through the walls of lymphatic and blood vessels, thereby gaining access to the bloodstream. Second, the cancer cells are transported by the circulatory system throughout the body. Third, cancer cells leave the bloodstream and enter various organs, where they establish new metastatic tumors. If cells from the initial tumor fail to complete any of these steps, or if any of the steps can be prevented, metastasis will not occur. It is therefore crucial to understand how the properties of cancer cells make these three steps possible.

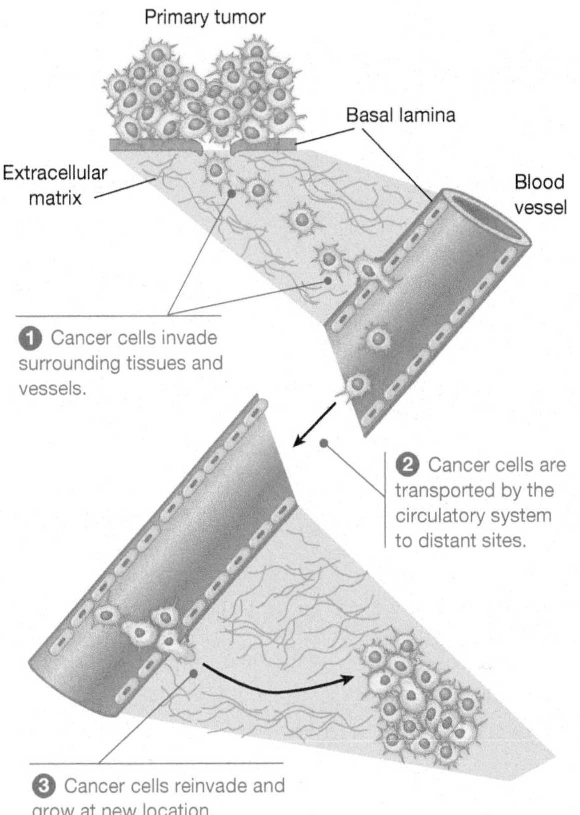

Figure 26-5 Stages in the Process of Metastasis. Only a small fraction of the cells in a typical cancer successfully carry out all three steps involved in metastasis: invasion into surrounding tissues and vessels, transportation via the circulatory system, and reinvasion and growth at a distant site.

Changes in Cell Adhesion, Motility, and Protease Production Promote Metastasis

The first step in metastasis involves the invasion of surrounding tissues and vessels by cancer cells (Figure 26-5, 1). Unlike the cells of benign tumors or most normal cells, which remain together at the site where they are formed, cancer cells can leave their initial location and invade surrounding tissues, eventually entering the circulatory system.

Several mechanisms make this invasive behavior possible. The first involves changes in cell-cell adhesion proteins. Such proteins are often missing or defective in cancer cells, which allows cells to separate from the main tumor more readily. One crucial molecule is *E-cadherin* (see Figure 15-3), a cell-cell adhesion protein whose loss underlies the reduced adhesiveness of many epithelial cancers. Highly invasive cancers usually have less E-cadherin than do normal cells. Restoring E-cadherin to isolated cancer cells lacking this molecule has been shown to inhibit their ability to form invasive tumors when the cells are injected back into animals.

A second factor underlying invasive behavior is the increased *motility* of cancer cells, which is stimulated by signaling molecules produced by surrounding host tissues or by the cancer cells themselves. Some of these signaling molecules can act as chemoattractants that attract migrating cancer cells. Activation of the Rho family GTPases (see Figure 13-21) also plays a central role in the stimulation of cell motility.

Another trait contributing to invasion is the ability of cancer cells to produce *proteases* that degrade protein-containing structures that would otherwise act as barriers to cancer cell movement. A critical barrier encountered by most cancers is the *basal lamina*, a dense layer of protein-containing material that separates epithelial layers from underlying tissues (see Figure 15-2). Before epithelial cancers, which account for 90% of all human malignancies, can invade adjacent tissues, the basal lamina must first be breached. Cancer cells break through this barrier by secreting proteases, which degrade the proteins that form the backbone of the basal lamina.

One such protease is *plasminogen activator*, an enzyme that converts the inactive precursor *plasminogen* into the active protease *plasmin*. Because high concentrations of plasminogen are present in most tissues, small amounts of plasminogen activator released by cancer cells can quickly catalyze the formation of large quantities of plasmin. The plasmin performs two tasks: (1) it degrades components of the basal lamina and the extracellular matrix, thereby facilitating tumor invasion, and (2) it cleaves inactive precursors of matrix metalloproteinases, produced mainly by surrounding host cells, into active enzymes that also degrade the basal lamina and extracellular matrix.

After proteases allow cancer cells to penetrate the basal lamina, they facilitate cancer cell migration by degrading the extracellular matrix of the underlying tissues. The cancer cells migrate until they reach tiny blood or lymphatic vessels, which are also surrounded by a basal lamina. The cancer cells then squeeze through the layer of endothelial cells that form the vessel's inner lining to finally gain entry into the circulatory system.

Relatively Few Cancer Cells Survive the Voyage Through the Bloodstream

Once in the bloodstream, cancer cells can be transported to distant parts of the body (Figure 26-5, ❷). If cancer cells instead penetrate the walls of lymphatic vessels, the cells are first carried to regional lymph nodes, where they may become lodged and grow. For this reason, regional lymph nodes are a common site for the initial spread of cancer. However, lymph nodes have numerous interconnections with blood vessels, so cancer cells that initially enter into the lymphatic system eventually find their way into the bloodstream.

Regardless of their initial entry route, the net result is often large numbers of cancer cells in the bloodstream. Even a tiny malignant tumor weighing only a few grams can release several million cancer cells into the circulation each day. However, the bloodstream is a relatively inhospitable place for most cancer cells, and fewer than one in a thousand cells survive the trip to a potential site of metastasis. We have already discussed the idea that natural selection favors certain cells in tumors. Are the few successful cells that metastasize similar, or are they simply a random representation of the original tumor cell population? **Figure 26-6** illustrates an experiment designed to address these questions. In this experiment, mouse melanoma cells were injected into the bloodstream of healthy mice to study their ability to metastasize. A few weeks later, metastases appeared in a variety of locations but mainly in the lungs. Cells from the lung metastases were removed and injected into another mouse, leading to the production of more lung metastases. By repeating this procedure many times in succession, researchers eventually obtained a population of cancer cells that formed many more metastases than did the original tumor cell population.

The most straightforward interpretation is that the initial melanoma consisted of a heterogeneous population of cells with differing properties and that repeated isolation and reinjection of cells derived from successful metastases gradually selected for those cells that were best suited for metastasizing. To test this hypothesis, further experiments have been performed in which single cells isolated from a primary melanoma were allowed to proliferate in culture into a separate population of cells. When injected into animals, some of the cells from single colonies produced few metastases, some produced numerous metastases, and some fell in between. Because each colony was derived from a different cell in the original tumor, the results confirmed that the cells of the primary tumor differed in their ability to metastasize and that a relatively small population of cells in a tumor are mainly responsible for the spread of cancer.

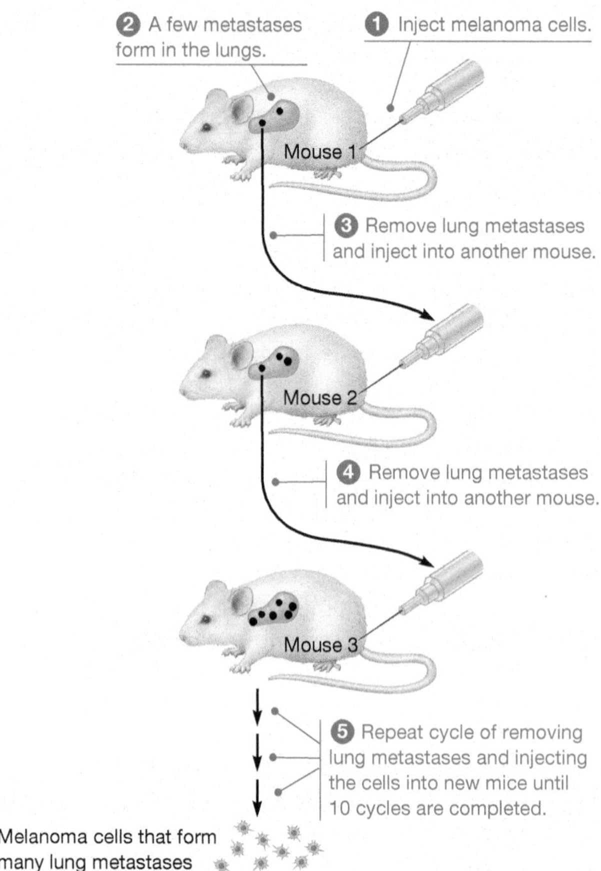

Figure 26-6 Selection of Melanoma Cells Exhibiting an Enhanced Ability to Metastasize. Mouse melanoma cells were injected into the tail vein of a mouse. After a small number of metastases arose in the lungs, cells from these lung tumors were removed and injected into another mouse. When this cycle was repeated ten times, the final population of melanoma cells produced many more lung metastases than were produced by the original cells.

Blood Flow and Organ-Specific Factors Determine Sites of Metastasis

Some cancer cells circulating in the bloodstream eventually exit through the wall of a tiny vessel and invade another organ, where they form metastases that may be located far from the initial tumor (see Figure 26-5, ❸). Although the bloodstream carries cancer cells throughout the entire body, metastases develop preferentially at certain sites. One factor responsible for this specificity is related to blood-flow patterns. Based solely on size considerations, circulating cancer cells are most likely to become lodged in *capillaries* (tiny vessels with a diameter no larger than a single blood cell). After becoming stuck in capillaries, cancer cells penetrate the walls of these tiny vessels, enter the surrounding tissues, and seed the development of new tumors. After cancer cells enter the bloodstream, the first capillary bed they typically encounter is in the lungs. As a result, the lungs are a frequent site of metastasis for many kinds of cancer. However, blood-flow patterns do not always favor the lungs. For cancers of the stomach and colon, cancer cells entering the bloodstream are first carried to the liver, where the vessels break up into a bed of capillaries. As a result, the liver is a common site of metastasis for these cancers.

Although blood-flow patterns are important, they do not always explain the observed distribution of metastases. As early as 1889, Stephen Paget proposed that circulating cancer cells have a special affinity for the environment provided by particular organs. Paget's idea, referred to as the "seed and soil" hypothesis, is based on the analogy that when a plant produces seeds, they are carried by the wind in all directions but grow only if they fall on congenial soil. According to this view, cancer cells are carried to a variety of organs by the bloodstream, but only a few sites provide an optimal growth environment for each type of cancer. Supporting evidence has come from a systematic analysis of the sites where metastases tend to arise. For roughly two-thirds of the human cancers examined, the rates of metastasis to various organs can be explained solely on the basis of blood-flow patterns.

Why do cancer cells grow best at particular sites? The answer is thought to involve interactions between cancer cells and the microenvironment in the organs they are delivered to. For example, prostate cancer commonly metastasizes to bone (a pattern that would not be predicted by blood-flow patterns). The reason for this preference was uncovered by experiments in which prostate cancer cells were mixed with cells from various organs—including bone, lung, and kidney—and the cell mixtures were then injected into animals. The ability of prostate cancer cells to develop into tumors is stimulated by the presence of cells derived from bone but not by those from lung or kidney. Subsequent studies uncovered the explanation: bone cells produce specific growth factors that stimulate the proliferation of prostate cancer cells.

The Immune System Influences the Growth and Spread of Cancer Cells

Does the body have any mechanisms for defending against the growth and spread of cancer cells? One possibility is the immune system, which can attack and destroy foreign cells. Of course, cancer cells are not literally "foreign," but they often exhibit molecular changes that might allow the immune system to recognize the cells as being abnormal. The *immune surveillance theory* postulates that immune destruction of cancer cells is a common event and that cancer simply reflects the occasional failure of an adequate immune response to be mounted against aberrant cells.

One piece of evidence that supports the idea of immune surveillance comes from organ transplant patients, who take immunosuppressive drugs to depress immune function and thereby decrease the risk of immune rejection of the transplanted organ. Such individuals develop many cancers at higher rates than normal. The immune surveillance concept has been tested more directly in animals that are genetically altered to introduce specific defects in the immune system. One study used mutant mice containing disruptions in *Rag2*, a gene expressed only in lymphocytes. The mutant mice, which have no functional lymphocytes (and hence no immune response), exhibit an increased cancer rate both for spontaneous cancers and for cancers induced by injecting animals with chemicals that cause cancer.

Although such results indicate that the immune system helps protect mice from developing cancer, how relevant are these findings to humans? If the immune system plays a major role in protecting us from cancer, we would expect a dramatic increase in cancer rates in humans infected with the human immunodeficiency virus (HIV), which leads to acquired immune deficiency syndrome (AIDS), a severe depression of immune function. Although AIDS does cause higher rates for a few types of cancer, especially Kaposi's sarcoma and lymphomas, increases are not seen for the more common forms of cancer. Most of the cancers that occur at increased rates in people with AIDS are caused by viruses (whose role in cancer is covered later in the chapter). Such observations suggest that the immune system may help protect humans from virus-induced cancers, but it is less successful in defending against the more common types of cancer.

The reason for this failure is that cancers find ways of evading destruction by the immune system. Those cells containing cell surface molecules that elicit a strong immune response are much more likely to be attacked and destroyed, whereas cells that either lack or produce smaller quantities of such molecules are more likely to survive and proliferate. So as tumors grow, there is a continual selection for cells that elicit a weaker immune response.

Cancer cells have also devised ways of actively confronting and overcoming the immune system. For example, some cancer cells produce molecules that kill or inhibit the function of T lymphocytes (immune cells involved in destroying foreign or defective cells). Tumors may also surround themselves with dense supporting tissue that shields them from immune attack. And some cancer cells simply divide so quickly that the immune system cannot kill them fast enough to keep tumor growth in check.

Scientists are now using several approaches to sensitize the immune system so that it is more effective in attacking cancer cells. We will discuss these technologies later in this chapter.

The Tumor Microenvironment Influences Tumor Growth, Invasion, and Metastasis

We have now encountered several examples in which tumor behavior is influenced by interactions between tumor cells and the surrounding *tumor microenvironment*, which includes various kinds of normal cells, extracellular molecules, and components of the extracellular matrix. From angiogenesis to release of proteases to stimulation of motility and metastasis, the microenvironment plays pivotal roles.

The tumor microenvironment can also hinder invasion and metastasis. For example, some cells in the tumor microenvironment produce TGFβ, a signaling protein that acts as a potent inhibitor of proliferation for many cell types. Cancer cells may in turn acquire mutations that allow them to continue growing in the presence of TGFβ. Sometimes cancer cells even start secreting TGFβ themselves, which inhibits the growth of surrounding normal cells and allows the cancer cells to reproduce and invade surrounding tissues more rapidly because of the decreased competition from neighboring cells. These examples illustrate the complexities regarding how the tumor microenvironment influences the ability of cancer cells to grow, invade neighboring tissues, and metastasize to distant sites.

CONCEPT CHECK 26.2

The protein *p120catenin* stabilizes the cadherin complex by binding to the cytosolic tail of E-cadherin. You work in a laboratory that has produced mouse cells that lack p120catenin function. You inject these cells into a mouse and find that they promote tumors. Explain this result.

26.3 What Causes Cancer?

The uncontrolled proliferation of cancer cells, combined with their ability to metastasize to distant sites, makes cancer a potentially life-threatening disease. What causes the emergence of cells with such destructive properties? People often view cancer as a mysterious disease that strikes randomly and without known cause, but this misconception fails to consider the results of thousands of scientific investigations, some dating back more than 200 years. The inescapable conclusion emerging from these studies is that *cancers are commonly caused by random errors in DNA replication and repair, environmental agents, lifestyle factors, and heredity.* At its heart, cancer is a *genetic* disease, in which cells that have acquired mutations move down the path toward uncontrolled proliferation and metastasis.

Epidemiological Data Have Allowed Many Causes of Cancer to Be Identified

The first indication that a particular agent may cause cancer usually comes from *epidemiology*—the branch of science that investigates the frequency and distribution of diseases in human populations. Epidemiological studies have revealed several key findings. First, cancers arise with differing frequencies in different parts of the world. For example, stomach cancer is frequent in Japan, breast cancer is prominent in the United States, and liver cancer is common in Africa and Southeast Asia. To determine whether differences in heredity or environment are responsible for such differences, scientists have examined cancer rates in people who move from one country to another. For example, in Japan the incidence of stomach cancer is greater and the incidence of colon cancer is lower than in the United States; however, when Japanese families move to the United States, their cancer rates come to resemble those in the United States, indicating that cancer rates are strongly influenced by environmental and lifestyle factors.

Epidemiological data have played an important role in identifying these environmental and lifestyle factors. A striking example involves lung cancer, which saw a more than tenfold increase in frequency in men in the United States during the twentieth century (**Figure 26-7**). Investigations into the possible causes for this lung cancer epidemic revealed that most people who develop lung cancer have one trait in common: a history of smoking cigarettes. As might be expected if cigarettes were responsible, heavy smokers were found to develop lung cancer more frequently than light smokers and long-term smokers more frequently than short-term smokers, and lung cancer rates were found to fall after cigarette smokers quit smoking (Figure 26-7a). Smoking is also linked to most cancers of the mouth, pharynx, and larynx, as well as to some cancers of the esophagus, stomach, pancreas, uterine cervix, kidney, bladder, and colon. About half of all people who smoke will be killed by the cancers and cardiovascular diseases caused by smoking, losing an average of 10–15 years of life. This premature loss of life includes about 135,000 deaths per year from lung cancer in the United States alone.

Although the connection between cigarette smoking and cancer was first suggested by epidemiological studies, additional direct experimental evidence that smoking promotes cancer has come from laboratory studies showing that cigarette smoke contains several dozen chemicals that cause cancer when administered to animals. Armed with this information, the U.S. government began to publicize the dangers of tobacco smoke in the 1960s; smoking rates among men in the United States subsequently stopped growing and began to decline (Figure 26-7b). This decline is one of the main reasons why cancer death rates have slowly begun to fall during the past 20 years.

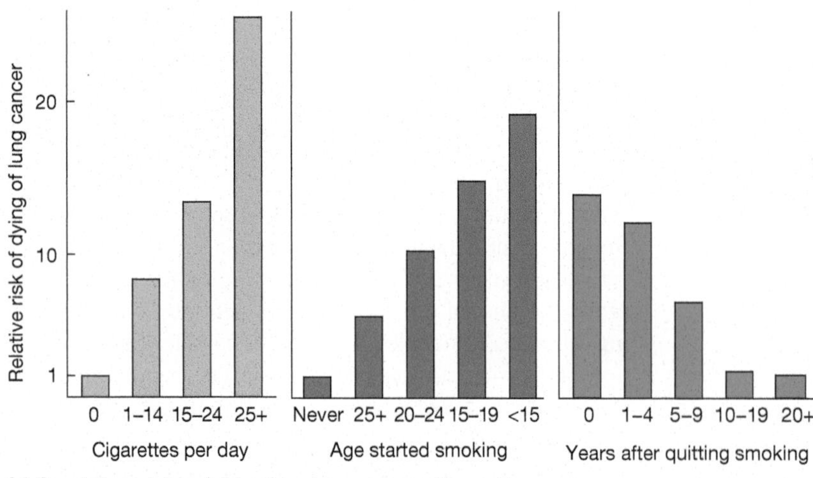

(a) Correlation of risk of dying from lung cancer with smoking

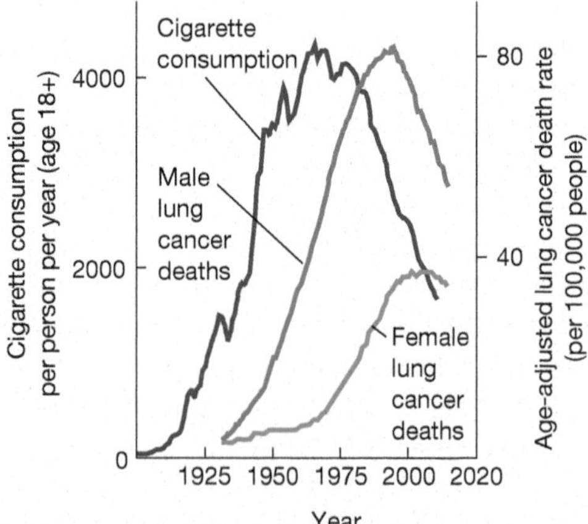

(b) Cigarette consumption and lung cancer death rates for men and women over the last century

Figure 26-7 Cigarette Smoking and Lung Cancer. (a) These bar graphs show that lung cancer rates increase as a function of the number of cigarettes smoked per day, that long-term smokers develop lung cancer more frequently than do short-term smokers, and that lung cancer rates fall after cigarette smokers quit smoking. **(b)** An increase in cigarette smoking in men in the United States during the first half of the twentieth century was followed by an explosive growth in lung cancer deaths. A time lag of about 25 years transpired between the increase in smoking rates and the subsequent increase in cancer deaths, which is typical of the time required for most human cancers to develop after exposure to a cancer-causing agent.

Errors in DNA Replication or Repair Explain Many Cancers

Environmental factors explain important aspects of cancer prevalence, and many public campaigns to combat cancer rightly focus on these factors, which we can control. But just how many cases of cancer can be ascribed to environmental or lifestyle factors? Clearly not all. Recent epidemiological studies suggest that another major explanation for cancer lies beyond the environment: in failures in DNA replication or repair. The prevalence of certain tumors shows no correlation between which patients have these tumors and the patients' environments or lifestyles. Recall that there is a low, but measurable, failure rate of DNA replication and DNA repair (see Chapter 17). The prediction from this fact is that when cells undergo many rounds of cell division, the likelihood of such errors, while still low, increases. Consistent with this prediction, Bert Vogelstein and Cristian Tomasetti at Johns Hopkins University found a correlation between the number of stem cell divisions that cells in many tissues undergo and the lifetime incidence of cancer in those tissues.

What is the interplay between increased errors of replication in stem cells and environmental cues? The effects are likely additive. In addition to the increased accumulation of innate mutations, environmental agents that cause mutations (called *mutagens*) or conditions (such as cellular stress) that negatively impact DNA repair are also more likely to act on rapidly dividing cells.

Inborn Errors Explain Some Cancers

Mutations do not just act through replication errors in mature tissues. Humans inherit between 50 and 100 mutations from their parents at birth. Many of these mutations do not occur in protein-coding sequences; of those that do, only a few occur in genes that promote cancer. However, some hereditary mutations lead to a dramatically higher incidence of certain cancers; not surprisingly, the mutated genes encode proteins that act to control cell proliferation or DNA repair. Later in this chapter we will discuss several of these genes. Any new mutations acquired at any stage of life act on the background of these inherited mutations.

Many Chemicals Can Cause Cancer, Often After Metabolic Activation in the Liver

The idea that certain chemicals, such as those found in tobacco smoke, can cause cancer was first proposed 250 years ago. In 1761 a London doctor, John Hill, reported that people who routinely used snuff (a powdered form of tobacco that is inhaled) experience an abnormally high incidence of nasal cancer. A few years later another British physician, Percivall Pott, observed an elevated incidence of scrotum cancer among men who had served as chimney sweeps in their youth. Pott speculated that the chimney soot became dissolved in the natural oils of the scrotum, irritated the skin, and eventually triggered the development of cancer. This conjecture led to the discovery that wearing protective clothing and bathing regularly could prevent scrotum cancer in chimney sweeps.

In the years since these pioneering observations, the list of known and suspected **carcinogens** (cancer-causing agents) has grown to include hundreds of different chemicals. Chemicals are usually labeled as carcinogens because humans or animals develop cancer when exposed to them. This does not mean, however, that each chemical causes cancer through its own direct action. For example, consider the behavior of *2-naphthylamine,* a potent carcinogen that causes bladder cancer in industrial workers and is present in tobacco smoke. As might be expected, feeding 2-naphthylamine to laboratory animals induces a high rate of bladder cancer. But if 2-naphthylamine is implanted directly into an animal's bladder, cancer rarely develops. The explanation for the apparent discrepancy is that when 2-naphthylamine is ingested (by animals) or inhaled (by humans), it passes through the liver and is metabolized into other chemicals that are the actual causes of cancer. Placing 2-naphthylamine directly in an animal's bladder bypasses this metabolic activation, and so cancer does not arise.

Many carcinogens share this need for metabolic activation before they can cause cancer. Substances exhibiting such behavior are more accurately called **precarcinogens**, a term applied to any chemical that is capable of causing cancer only after it has been metabolically activated. Most precarcinogens are activated by liver proteins that are members of the *cytochrome P-450* enzyme family. Members of this enzyme family catalyze the oxidation of ingested foreign chemicals, such as drugs and pollutants, to make the molecules less toxic and easier to excrete from the body. However, in some cases these oxidation reactions inadvertently convert foreign chemicals into carcinogens—a phenomenon known as *carcinogen activation.*

DNA Mutations Triggered by Chemical Carcinogens Lead to Cancer

Once it was known that chemicals can cause cancer, the question arose as to how they work. The idea that carcinogenic chemicals act by triggering DNA mutations was first proposed around 1950, but there was little supporting evidence because no one had systematically compared the mutagenic potency of different chemicals with their ability to cause cancer. The need for such information inspired Bruce Ames to develop a simple laboratory test for measuring a chemical's mutagenic activity. This procedure, called the **Ames test**, uses bacteria as a test organism because they can be quickly grown in enormous numbers in culture (**Figure 26-8** on page 794). The bacteria are a special strain that cannot synthesize the amino acid *histidine.* As shown in Figure 26-8a, the bacteria are placed in a culture dish containing a growth medium lacking histidine, along with the chemical being tested for mutagenic activity. Normally, the bacteria would not grow in the absence of histidine. However, if the chemical being tested is mutagenic, it will trigger random mutations, some of which might restore the ability to synthesize histidine. Each bacterium acquiring such a mutation will grow into a visible colony, so the total number of colonies is a measure of the mutagenic potency of the substance being tested (Figure 26-8b).

Because many chemicals that cause cancer do not become carcinogenic until after they have been modified by liver enzymes, the Ames test includes a step in which the chemical being tested is first incubated with an extract of liver cells to mimic the reactions that normally occur in the liver.

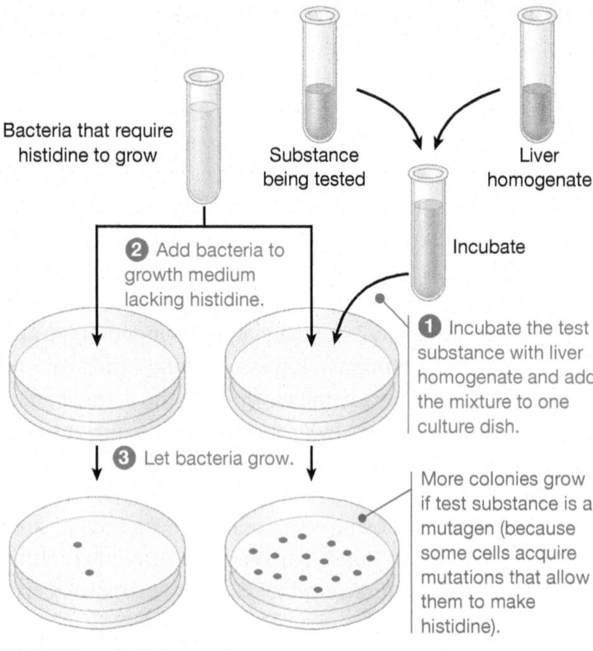

(a) The Ames test

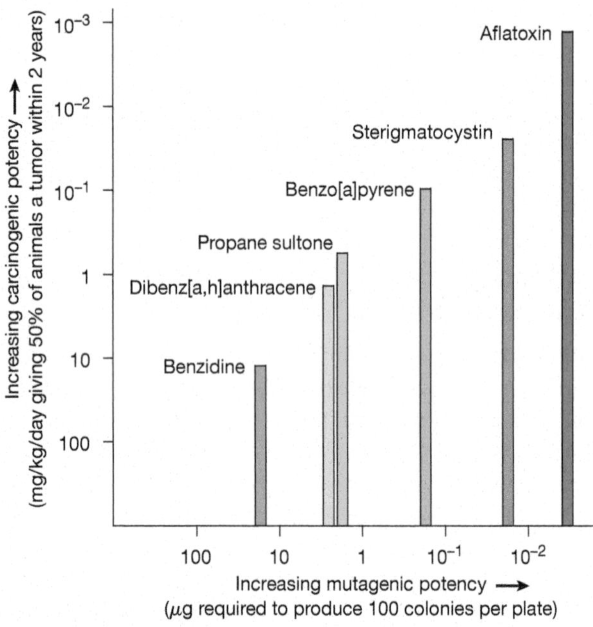

(b) Correlation between mutagenic and carcinogenic activity for selected carcinogenic chemicals

Figure 26-8 Ames Test for Identifying Potential Carcinogens. The Ames test is based on the rationale that most carcinogens are mutagens. **(a)** The ability of chemicals to induce mutations is measured in bacteria that cannot synthesize the amino acid histidine. When the bacteria are placed in a growth medium lacking histidine, the only bacteria that can grow are those that have acquired a mutation allowing them to make histidine. The number of bacterial colonies that grow is therefore related to the mutagenic potency of the substance being tested. Chemicals studied with the Ames test are first incubated with a liver homogenate because many chemicals become carcinogenic only after they have undergone biochemical modification in the liver. **(b)** The bar graph shows that substances exhibiting strong mutagenic activity in the Ames test also tend to be strong carcinogens. Among this particular group of substances, aflatoxin is the most potent mutagen and the most potent carcinogen. Note that the data are plotted on logarithmic scales to permit aflatoxin to be shown on the same graph as benzidine, which is about 10,000 times weaker than aflatoxin as a mutagen and carcinogen.

Carcinogenic chemicals inflict DNA damage in several ways, including binding to DNA and disrupting normal base pairing; generating crosslinks between the two strands of the double helix; creating chemical linkages between adjacent bases; hydroxylating or removing individual DNA bases; and causing breaks in one or both DNA strands. In some cases, the mutational role of specific chemicals in causing cancer has been linked to effects on specific genes. For example, the *polycyclic aromatic hydrocarbons* found in tobacco smoke preferentially bind to specific regions of the *p53* gene and trigger unique mutations in which the base T is substituted for the base G. As we will see later in the chapter, mutations in the *p53* gene play a key role in the development of many kinds of cancer.

Ionizing and Ultraviolet Radiation Also Cause DNA Mutations That Lead to Cancer

Chemicals are not the only DNA-damaging agents that can cause cancer. Shortly after the discovery of X-rays by Wilhelm Roentgen in 1895, it was noticed that people working with this type of radiation developed cancer at abnormally high rates. Animal studies subsequently confirmed that X-rays create DNA mutations and cause cancer in direct proportion to the dose administered.

A similar type of radiation is emitted by many radioactive elements. An early example of the cancer hazards posed by radioactivity occurred in the 1920s in a New Jersey factory that produced glow-in-the-dark watch dials. A luminescent paint containing the radioactive element *radium* was used for painting the dials, and this paint was applied with a fine-tipped brush that the workers frequently wetted with their tongues. As a result, tiny quantities of radium were inadvertently ingested and became concentrated in their bones, triggering bone cancer.

Elevated cancer rates caused by exposure to radioactivity have also been reported in people exposed to radioactive fallout from nuclear explosions, such as the Japanese cities of Hiroshima and Nagasaki after atomic bombs were dropped there in 1945, or leakage from nuclear power plants, such as the Chernobyl nuclear power plant in what is now the Ukraine in 1986.

X-rays and related forms of radiation emitted by radioactive elements are called **ionizing radiation** because they remove electrons from molecules, thereby generating highly reactive ions that create various types of DNA damage, including single- and double-strand breaks. **Ultraviolet radiation (UV)** is another type of radiation that causes cancer by damaging DNA. The ability of the UV radiation in sunlight to cause cancer was first deduced from the observation that skin cancer is most prevalent in people who spend long hours in the sun, especially in tropical regions where the sunlight is intense. UV radiation is absorbed mainly by the skin, where it imparts enough energy to trigger *pyrimidine dimer* formation—the formation of covalent bonds between adjacent pyrimidine bases in DNA (see Figure 17-25). If the damage is not repaired, improper base pairing occurs during DNA replication, which in turn produces distinctive mutation patterns. For example, a CC → TT mutation (conversion of two adjacent cytosines to thymines) is a unique product of UV exposure and can be used as a distinctive "signature" to identify mutations caused by sunlight.

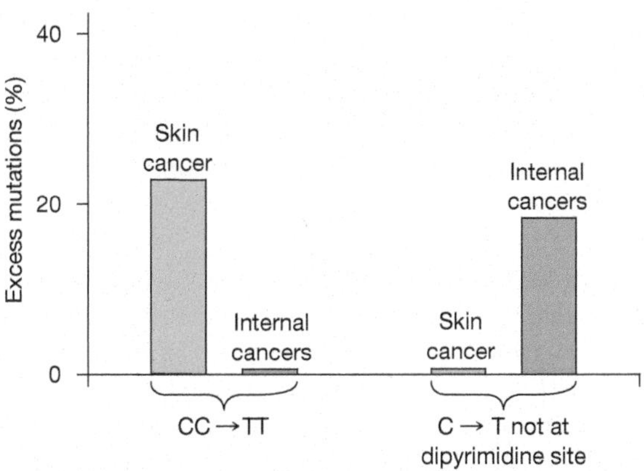

Figure 26-9 Incidence of Two Types of *p53* Mutations in Skin Cancer and Internal Cancers. The two bars on the left represent the frequency of CC → TT mutations, which are triggered by UV radiation. The two bars on the right represent the frequency of C → T mutations not located at dipyrimidine sites, which are not caused by UV radiation. Note that the UV-triggered type of mutation occurs in the *p53* gene of skin cancers (squamous cell carcinomas) but not in cancers of internal organs. Mutation frequencies are plotted relative to what would be expected to occur randomly.

The existence of such signature mutations strengthens the connection between UV-induced mutations and skin cancer. One of the first genes to be studied was the *p53* gene, which is mutated in many human cancers. When the *p53* gene of skin cancer cells is examined using DNA sequencing techniques, mutations exhibiting a distinctive UV signature (such as CC → TT) are frequently observed. In contrast, when *p53* mutations are detected in other types of cancer, they do not exhibit the UV signature (**Figure 26-9**).

Viruses and Other Infectious Agents Trigger the Development of Some Cancers

Although the carcinogenic properties of chemicals and radiation were recognized by the early 1900s, the possibility that infectious agents might also cause cancer was not widely appreciated because cancer does not usually behave like a contagious disease. However, in 1911 Peyton Rous performed experiments on sick chickens brought to him by local farmers that showed for the first time that cancer can be caused by a virus. These chickens had cancers of connective tissue origin, or *sarcomas*. To investigate the origin of the tumors, Rous ground up the tumor tissue and passed it through a filter with pores so small that not even bacterial cells could pass through. When he injected the cell-free extract into healthy chickens, they developed sarcomas. Because no cancer cells had been injected into the healthy chickens, Rous concluded that sarcomas can be transmitted by an agent that is smaller than a bacterial cell. This was the first time anyone had detected an **oncogenic virus**—a virus that causes cancer. Although Rous's findings were initially greeted with skepticism, an 87-year-old Rous finally received a Nobel Prize in 1966—more than 50 years after his discovery of the first cancer virus.

It is now clear that dozens of viruses can cause cancer in animals, and a few have been linked to human cancers. The first human example was uncovered by Denis Burkitt, a British surgeon working in Africa in the late 1950s. At certain times of the year, Burkitt noted large outbreaks of lymphocytic cancers of the neck and jaw. Because this cancer, now known as **Burkitt lymphoma**, occurred in periodic epidemics in specific geographical regions, Burkitt proposed that it was transmitted by an infectious agent. Burkitt's ideas attracted the attention of two virologists, named Epstein and Barr, whose electron microscopic studies of Burkitt lymphoma cells revealed a virus now called the **Epstein-Barr virus (EBV)**. The following evidence supports the idea that EBV can play a role in Burkitt lymphoma: (1) EBV DNA and proteins are often found in tumor cells obtained from patients with Burkitt lymphoma but not in normal cells from the same individuals, (2) adding EBV to normal human lymphocytes in culture causes the cells to acquire some of the properties of cancer cells, and (3) injecting EBV into monkeys causes lymphomas to arise.

Several other viruses have been linked to human cancers. Among these are the *hepatitis B* and *hepatitis C* viruses, which trigger some liver cancers; *human T-cell lymphotropic virus-I (HTLV-I)*, which causes adult T-cell leukemia and lymphoma; and the sexually transmitted *human papillomavirus (HPV)*, which is associated with uterine cervical cancer (see later in this chapter). Viruses are not the only infectious agents that can cause cancer. Chronic infection with the bacterium *Helicobacter pylori*—a common cause of stomach ulcers—can trigger stomach cancer, and flatworm infections have been linked to a small number of bladder and bile duct cancers.

Knowing the identity of such infectious agents opens the door for new cancer prevention strategies. For example, antibiotics that kill *H. pylori* are helpful in preventing stomach cancer, and vaccines against hepatitis B and HPV can reduce the incidence of liver and cervical cancers, respectively. The HPV vaccine, which first became available in 2006, illustrates the difficulties that often hinder the development of such vaccines. Scientists have known for more than 100 years that cervical cancer is caused by a sexually transmitted agent, but HPV was not identified as the responsible agent until the 1980s. The difficulty in making this identification arose because HPV is not a single virus but rather a family of related viruses containing more than 100 different subtypes. After tests were developed for distinguishing the various subtypes, it became clear that cervical cancer is linked to infection with certain high-risk forms of the virus—mainly HPV 16, HPV 18, and a few others (**Figure 26-10**). The first HPV vaccine was

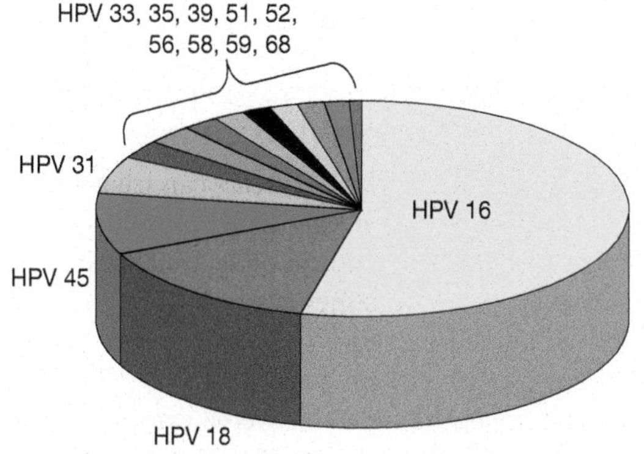

Figure 26-10 Prevalence of Various Subtypes of HPV in Cervical Cancers. About 90% of all cancers of the uterine cervix involve infection with at least one of the subtypes of HPV shown in this chart.

therefore targeted against HPV subtypes 16 and 18, which gives the vaccine the potential to prevent about 70% of all cervical cancers. This vaccine also protects against two other HPV subtypes that cause genital warts. Interestingly, recall that *HeLa* cells were produced from cervical cancer cells from Henrietta Lacks (see Human Connections in Chapter 1, page 16). Sequencing of the DNA in HeLa cells revealed the insertion of HPV 18 DNA at a position on chromosome 8 500 kb upstream of the *Myc* proto-oncogene.

Although the infectious agents that cause cancer include a diverse collection of viruses, bacteria, and parasites, their mechanisms of action can be grouped into two main categories. One involves those agents—such as the hepatitis B and C viruses, *H. pylori*, and parasitic flatworms—that cause tissue destruction and chronic inflammation. Under these conditions, cells of the immune system infiltrate the tissue and attempt to kill the infectious agent. Unfortunately, the mechanisms used by immune cells to fight infections often produce mutagenic chemicals, such as oxygen *free radicals* (highly reactive forms of oxygen containing an unpaired electron). The net result is an increased likelihood that cancer-causing mutations will arise. The other mechanism of action is based on the ability of certain viruses to stimulate proliferation of infected cells. The types of genes involved play roles not just in viral cancers but in cancers caused by chemicals and radiation as well. We therefore now proceed to a discussion of cancer-associated genes and the ways in which they act.

CONCEPT CHECK 26.3

Although tobacco smoking is responsible for a large number of human cancers, not all smokers develop cancer. Suggest an explanation for this finding.

26.4 Oncogenes and Tumor Suppressor Genes

A large body of evidence points to the central role played by DNA mutations in cancer. Spontaneous mutations, DNA replication errors, inherited mutations, and mutations triggered by chemicals, radiation, or infectious agents can all trigger cancer. Despite differences in origin, the result is always the mutation of genes involved in controlling cell proliferation and survival. We now turn to a discussion of the two main classes of affected genes: *oncogenes* and *tumor suppressor genes.*

Oncogenes Are Genes Whose Products Can Trigger the Development of Cancer

An **oncogene** is a gene whose *presence* can trigger the development of cancer. Some oncogenes are introduced into cells by cancer-causing viruses; others arise from the mutation of normal cellular genes. In either case, oncogenes encode proteins that stimulate excessive cell proliferation and/or promote cell survival by inhibiting apoptosis.

The first oncogene to be discovered was in the *Rous sarcoma virus*, mentioned previously. Mutational studies revealed that mutant viruses with defects in one of these genes, called *src*, are still able to infect cells and reproduce normally but can no longer cause cancer. In other words, a functional copy of the *src* gene must be present for cancer to arise. Similar approaches subsequently led to the identification of oncogenes in dozens of other viruses.

Evidence for the existence of oncogenes in cancers not caused by viruses first came from studies in which DNA isolated from human bladder cancer cells was introduced into a strain of cultured mouse cells. When these cells were injected back into mice, the animals developed cancer. Scientists therefore suspected that a human gene taken up by the mouse cells had caused the cancer. To confirm the suspicion, DNA was isolated and cloned from the mouse cancer cells, which led to the identification of the first human oncogene: a mutant *RAS* gene encoding an abnormal form of Ras, a protein that has a role in growth signaling (see Chapters 23 and 24).

The *RAS* gene was just the first of more than 200 human oncogenes to be discovered. Although these oncogenes are defined as genes that can cause cancer, a single oncogene is usually not sufficient. In the transfection experiments described in the preceding paragraph, introducing the *RAS* oncogene caused cancer only because the mouse cells used in these studies already possess a mutation in another cell cycle control gene. If freshly isolated normal mouse cells are used instead, introducing the *RAS* oncogene by itself will not cause cancer. This observation illustrates an important principle: *multiple mutations are usually required to convert a normal cell into a cancer cell.*

Proto-oncogenes Are Converted into Oncogenes by Several Distinct Mechanisms

How do human cancers, most of which are not caused by viruses, come to acquire oncogenes? The answer is that oncogenes arise by mutation from normal cellular genes called **proto-oncogenes**. Despite their harmful-sounding name, proto-oncogenes are not "bad" genes simply waiting for an opportunity to foster the development of cancer. Rather, they are normal cellular genes that make essential contributions to the regulation of cell growth and survival. The term *proto-oncogene* simply implies that if and when the structure or activity of a proto-oncogene is disrupted by certain kinds of mutations, the mutant form of the gene can cause cancer. The mutations that convert proto-oncogenes into oncogenes are created through several distinct mechanisms, which are summarized in **Figure 26-11** and briefly described next.

MAKE CONNECTIONS 26.1

Cancers commonly carry mutations in the Ras protein that make Ras hyperactive. Why would hyperactive Ras promote cancer? (Figs. 23-18, 24-13, 24-25)

Point Mutation. The simplest mechanism for converting a proto-oncogene into an oncogene is a *point mutation*—a single nucleotide substitution in DNA that causes a single amino acid substitution in the protein encoded by the proto-oncogene. The most frequently encountered oncogenes of this type are *RAS* oncogenes, which encode hyperactive forms of the Ras protein (see below).

Gene Amplification. A second mechanism for creating oncogenes utilizes *gene amplification* to increase the number of

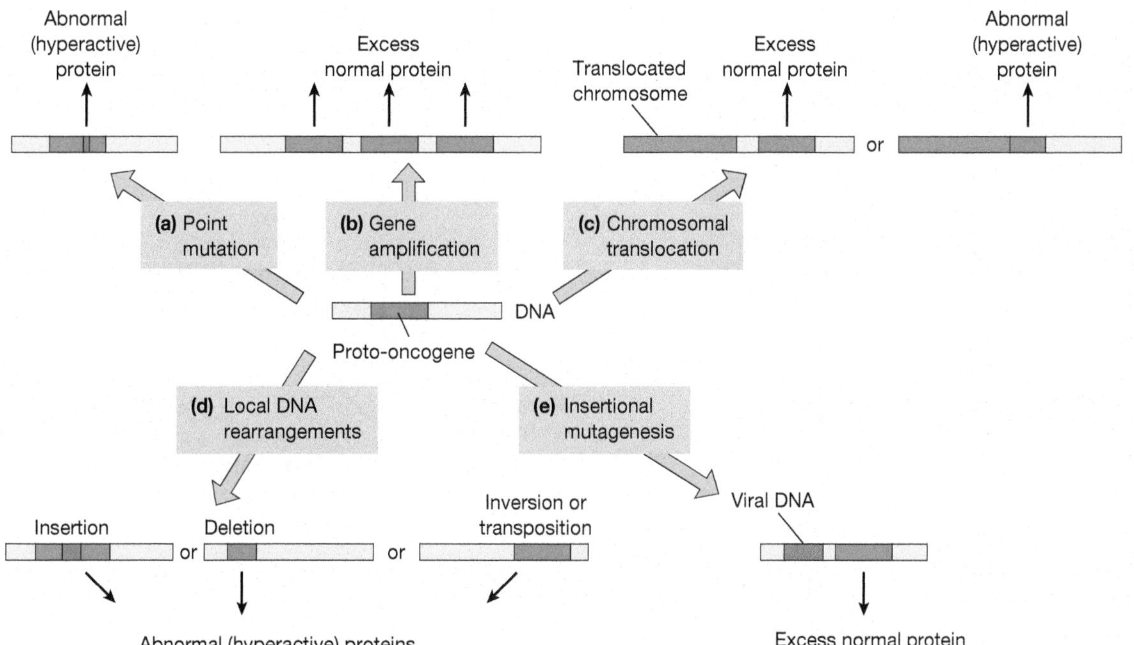

Figure 26-11 Five Mechanisms for Converting Proto-oncogenes into Oncogenes. Some oncogenes produced by these mechanisms encode abnormal proteins, whereas others produce normal proteins in excessive amounts. **(a)** Point mutation involves a single nucleotide substitution that creates an oncogene encoding an abnormal protein differing in a single amino acid from the normal protein produced by the proto-oncogene. **(b)** Gene amplification creates multiple gene copies, thereby leading to excessive production of a normal protein. **(c)** Chromosomal translocations move chromosome segments from one chromosome to another. This may fuse two genes together to form an oncogene encoding an abnormal protein, or it may place a proto-oncogene next to a highly active gene, thereby causing the translocated proto-oncogene to become more active. **(d)** Local DNA rearrangements such as insertions, deletions, inversions, and transpositions can disrupt the structure of proto-oncogenes and cause them to produce abnormal proteins. **(e)** Insertional mutagenesis occurs when viral DNA is integrated into a host chromosome near a proto-oncogene. The inserted DNA may stimulate the expression of the proto-oncogene and cause it to produce too much protein.

copies of a proto-oncogene. When the number of gene copies is increased, it causes the protein encoded by the proto-oncogene to be produced in excessive amounts, although the protein itself is normal. For example, about 25% of human breast and ovarian cancers have amplified copies of the *ERBB2* gene, which encodes a growth factor receptor. The existence of multiple copies of the gene leads to the production of too much receptor protein, which in turn causes excessive cell proliferation.

Chromosomal Translocation. During *chromosomal translocation,* a portion of one chromosome is physically removed and joined to another chromosome. A classic example occurs in Burkitt lymphoma (see above). In the most frequent translocation, a proto-oncogene called *MYC* is moved from chromosome 8 to 14, where it becomes situated next to an intensely active region of chromosome 14 containing genes encoding antibody molecules (**Figure 26-12** on page 798). Moving the *MYC* gene so close to the highly active antibody genes causes the *MYC* gene to be highly transcribed, thereby leading to overproduction of the Myc protein—a transcription factor that stimulates cell proliferation. Although the translocated *MYC* gene sometimes retains its normal structure and encodes a normal Myc protein, it is still an oncogene because its new location on chromosome 14 causes the gene to be overexpressed (Figure 26-12a).

Translocations can also disrupt gene structure and cause abnormal proteins to be produced. One example involves the *Philadelphia chromosome,* an abnormal version of chromosome 22 commonly associated with chronic myelogenous leukemia. The Philadelphia chromosome is created by DNA breakage near the ends of chromosomes 9 and 22, followed by reciprocal exchange of DNA between the two chromosomes (Figure 26-12b). This translocation creates an oncogene called *BCR-ABL,* which contains DNA sequences derived from two different genes (*BCR* and *ABL*). As a result, the oncogene produces a *fusion protein* that functions abnormally because it contains amino acid sequences derived from two different proteins.

Local DNA Rearrangements. Another mechanism for creating oncogenes involves local rearrangements in which the base sequences of proto-oncogenes are altered by *deletions, insertions, inversions* (removal of a sequence followed by reinsertion in the opposite direction), or *transpositions* (movement of a sequence from one location to another). An example encountered in thyroid and colon cancers illustrates how a simple rearrangement can create an oncogene from two normal genes. This example involves two genes, named *NTRK1* and *TPM3,* which reside on the same chromosome. *NTRK1* encodes a receptor tyrosine kinase, and *TPM3* encodes a

Figure 26-12 Chromosome Translocations and Cancer. **(a)** In the lymphocytes that give rise to Burkitt lymphoma, a segment of chromosome 8 containing the normal *MYC* gene is frequently exchanged with a segment of chromosome 14. This reciprocal translocation places the normal *MYC* gene adjacent to a very active region of chromosome 14, which contains genes encoding antibody molecules. Moving the *MYC* gene so close to the highly active antibody genes results in activation of the *MYC* gene, leading to an overproduction of Myc protein that stimulates cell proliferation. **(b)** In nearly all chronic myelogenous leukemias, an abnormally short chromosome 22, called the Philadelphia chromosome, and an abnormally long chromosome 9 are found. These chromosomes result from a reciprocal translocation, which presumably occurred in a single white blood cell precursor undergoing mitosis and was then passed along to all descendant cells. The result is the creation of a fusion gene that encodes a mutated tyrosine kinase known as BCR-ABL.

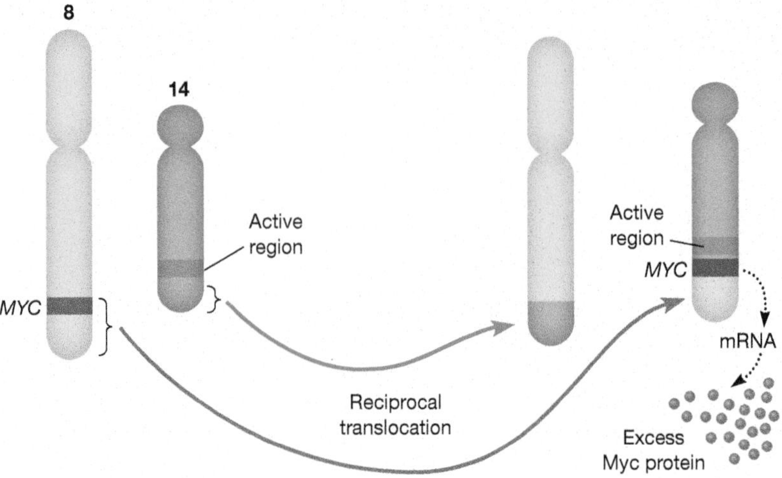

(a) Overexpression of Myc in Burkitt lymphoma

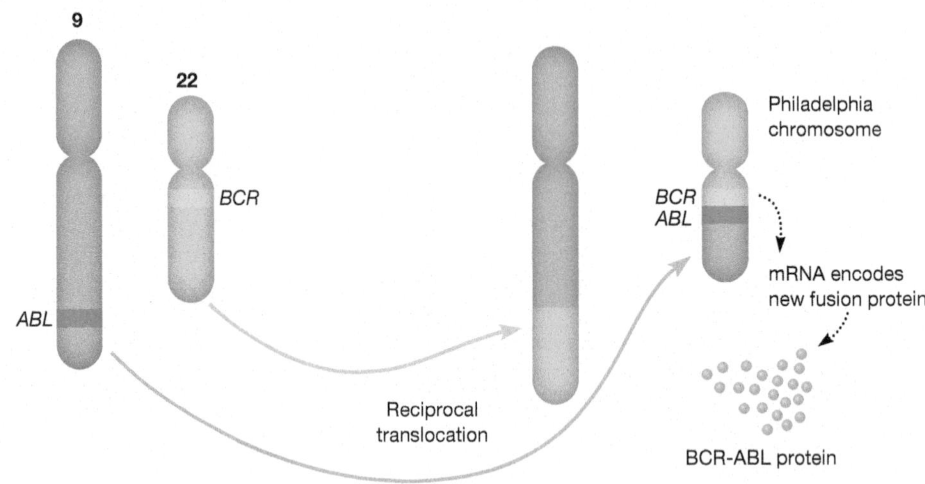

(b) The Philadelphia chromosome and the BCR-ABL kinase

completely unrelated protein, nonmuscle tropomyosin. In some cancers, a DNA inversion occurs that causes one end of the *TPM3* gene to fuse with the opposite end of the *NTRK1* gene (**Figure 26-13**). The resulting gene, called the *TRK oncogene*, produces a fusion protein containing the tyrosine kinase site of the receptor joined to a region of the tropomyosin molecule that forms a *coiled coil*—a structure that causes two polypeptide chains to join together as a dimer. As a result, the fusion protein forms a permanent dimer and its tyrosine kinase domain is permanently activated. (Recall from Chapter 23 that receptor tyrosine kinases are normally activated by bringing together two receptor molecules to form a dimer.)

Insertional Mutagenesis. Retroviruses (see Figure 16-6) can sometimes cause cancer even if they have no oncogenes of their own. Retroviruses accomplish this task by integrating their genes into a host chromosome in a region where a proto-oncogene is located. Integration of the viral DNA then converts the host cell proto-oncogene into an oncogene by causing the gene to be overexpressed. This phenomenon, called **insertional mutagenesis**, is frequently encountered in animal cancers but is rare in humans. However, some human cancers may have been inadvertently created this way in early gene therapy trials that used retroviruses as vectors for repairing defective genes.

Most Oncogenes Encode Components of Growth-Signaling Pathways

We have just seen that alterations in proto-oncogenes can convert them into oncogenes, which in turn encode proteins that either are structurally abnormal or are produced in excessive amounts. How do these oncogene-encoded proteins cause cancer? Although more than 200 oncogenes have been identified to date, many of the proteins they produce fit into one of a few categories: *growth factors, receptors, plasma membrane GTP-binding proteins, nonreceptor protein kinases, transcription factors,* and *cell cycle or apoptosis regulators* (**Table 26-1** on page 800). These categories are all related to steps in growth-signaling pathways (for example, see the steps in the Ras pathway shown in Figure 24-23). The following sections provide examples of how oncogene-produced proteins in each of these groups contribute to the development of cancer.

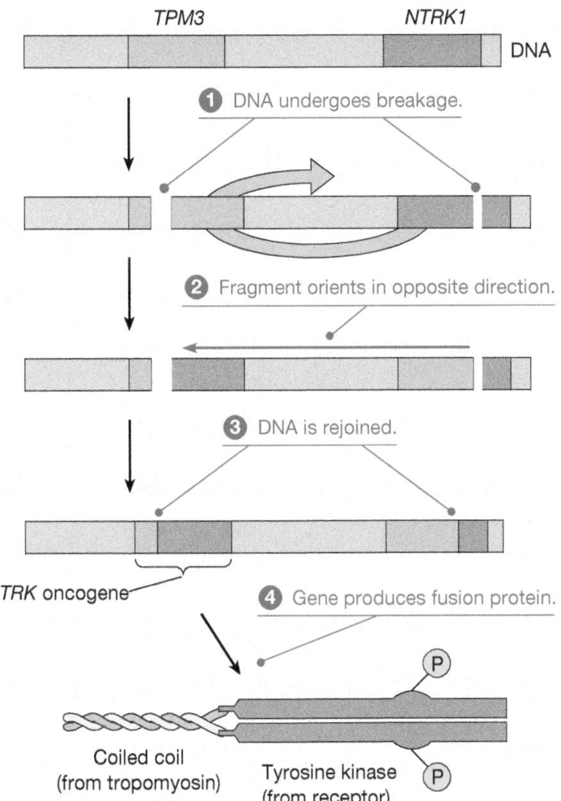

Figure 26-13 Origin of the *TRK* Oncogene. The *TRK* oncogene is created by a chromosomal inversion that brings together segments of two genes residing on the same chromosome, one gene encoding a growth factor receptor with tyrosine kinase activity *(NTRK1)* and the other gene encoding nonmuscle tropomyosin *(TPM3)*. The inversion causes one end of the *TPM3* gene to become fused to the opposite end of the *NTRK1* gene. The resulting *TRK* oncogene produces a fusion protein in which the tropomyosin segment causes the receptor region to form a dimer, thereby permanently activating its tyrosine kinase site.

Growth Factors. Normally, cells will not divide unless they have been stimulated by an appropriate growth factor. But if a cell possesses an oncogene that produces such a growth factor, the cell may stimulate its own proliferation. One oncogene that functions in this way is the *v-sis* gene ("*v*" means viral) found in the *simian sarcoma virus*, which causes cancer in monkeys. The *v-sis* oncogene encodes a mutant form of *platelet-derived growth factor (PDGF)*. When the virus infects a monkey cell whose growth is normally controlled by PDGF, the PDGF produced by the *v-sis* oncogene continually stimulates the cell's own proliferation (in contrast to the normal situation, in which cells are exposed to PDGF only when it is released from surrounding blood platelets). A PDGF-related oncogene has also been detected in some human sarcomas. These tumors possess a chromosomal translocation that creates a gene in which part of the *PDGF* gene is joined to part of an unrelated gene (a gene encoding a collagen).

Receptors. Several dozen oncogenes encode receptors involved in growth-signaling pathways. Many receptors exhibit intrinsic tyrosine kinase activity that is activated only when a growth factor binds to the receptor (see Chapter 23). Oncogenes sometimes encode mutant versions of such receptors whose tyrosine kinase activity is permanently activated, regardless of the presence or absence of a growth factor. The *TRK* oncogene, which was described in the section on DNA rearrangements, is one example (see Figure 26-13). Another example is the *v-erb-b* oncogene, which is found in a virus that causes a red blood cell cancer in chickens (**Figure 26-14** on page 801). The *v-erb-b* oncogene produces an altered version of the *epidermal growth factor (EGF) receptor*. Unlike the normal form of the receptor, which exhibits tyrosine kinase activity only when bound to EGF (Figure 26-14a), the mutant protein retains tyrosine kinase activity but lacks the EGF binding site. Consequently, the receptor is *constitutively active*—that is, it stays active as a tyrosine kinase whether EGF is present or not (Figure 24-16b).

Other oncogenes produce normal receptors but in excessive quantities, which can also lead to hyperactive growth signaling (see Figure 26-14c). An example is the human *ERBB2* gene. As discussed earlier in the chapter, amplification of the *ERBB2* gene in certain breast and ovarian cancers causes it to overproduce a growth factor receptor. The presence of too many receptor molecules causes a magnified response to growth factor and hence excessive cell proliferation.

Some growth-signaling pathways, such as the Jak-STAT pathway (described in Chapter 20), utilize receptors that do not possess protein kinase activity. With such receptors, binding of growth factor causes the activated receptor to stimulate the activity of an independent tyrosine kinase molecule. An example of an oncogene that encodes such a receptor occurs in the *myeloproliferative leukemia virus*, which causes leukemia in mice. The oncogene, called *v-mpl*, encodes a mutant version of the receptor for *thrombopoietin*, a growth factor that uses the Jak-STAT pathway to stimulate the production of blood platelets.

Plasma Membrane GTP-Binding Proteins. In many growth-signaling pathways, the binding of a growth factor to its receptor leads to activation of Ras. Oncogenes encoding mutant Ras proteins are one of the most common types of genetic abnormality detected in human cancers. The point mutations that create *RAS* oncogenes usually cause a single incorrect amino acid to be inserted at one of three possible locations within the Ras protein. The net result is a hyperactive Ras protein that retains bound GTP instead of hydrolyzing it to GDP, thereby maintaining the protein in a permanently activated state. In this hyperactive state, the Ras protein continually sends a growth-stimulating signal to the rest of the Ras pathway, regardless of whether growth factor is bound to the cell's growth factor receptors.

4. Nonreceptor Protein Kinases. A common feature shared by many growth-signaling pathways is the presence of protein phosphorylation cascades that transmit signals within the cell. The enzymes that catalyze these intracellular phosphorylation reactions are referred to as **nonreceptor protein kinases** to distinguish them from the protein kinases that are intrinsic to cell surface receptors. For example, in the case of the Ras pathway, the activated Ras protein triggers a cascade of intracellular protein phosphorylation reactions, beginning with phosphorylation of the Raf protein kinase and eventually leading to the phosphorylation of MAP kinases (see Figure 23-18). Several oncogenes encode protein kinases involved in this

Table 26-1 A Few Examples of Oncogenes Grouped by Protein Function

Oncogene Name	Protein Produced	Oncogene Origin	Common Cancer Type*
Growth factors			
v-sis	PDGF	Viral	Sarcomas (monkeys)
COL1A1-PDGFB	PDGF	Translocation	Fibrosarcoma
Receptors			
v-erb-b	Epidermal growth factor receptor	Viral	Leukemia (chickens)
TRK	Nerve growth factor receptor	DNA rearrangement	Thyroid
ERBB2	Epidermal growth factor receptor 2	Amplification	Breast
v-mpl	Thrombopoietin receptor	Viral	Leukemia (mice)
Plasma membrane GTP-binding proteins			
KRAS	Ras	Point mutation	Pancreas, colon, lung, others
HRAS	Ras	Point mutation	Bladder
NRAS	Ras	Point mutation	Leukemias
Nonreceptor protein kinases			
BRAF	Raf kinase	Point mutation	Melanoma
v-src	Src kinase	Viral	Sarcomas (chickens)
SRC	Src kinase	DNA rearrangement	Colon
TEL-JAK2	Jak kinase	Translocation	Leukemias
BCR-ABL	Abl kinase	Translocation	Chronic myelogenous leukemia
Transcription factors			
MYC	Myc	Translocation	Burkitt lymphoma
MYCL	Myc	Amplification	Small cell lung cancer
c-myc	Myc	Insertional mutagenesis	Leukemia (chickens)
v-jun	Jun	Viral	Sarcomas (chickens)
v-fos	Fos	Viral	Bone (mice)
Cell cycle or apoptosis regulators			
CYCD1	Cyclin	Amplification, translocation	Breast, lymphoma
CDK4	Cdk	Amplification	Sarcomas, glioblastoma
BCL2	Bcl-2	Translocation	Non-Hodgkins lymphoma
MDM2	Mdm2	Amplification	Sarcomas, lung, breast, others

*Cancers are in humans unless otherwise specified. Only the most frequent cancer types are listed.

cascade. Oncogenes encoding nonreceptor protein kinases involved in other signaling pathways have been identified as well. Included in this group are oncogenes that produce abnormal versions of the Src, Jak, and Abl protein kinases.

5. Transcription Factors. Some growth-signaling pathways subsequently trigger changes in transcription factors, thereby altering gene expression. Oncogenes that produce mutant forms or excessive quantities of various transcription factors have been detected in a broad range of cancers. Among the most common are oncogenes encoding Myc transcription factors, which control the expression of numerous genes involved in cell proliferation and survival. We have already seen how the chromosomal translocation associated with Burkitt lymphoma creates a *MYC* oncogene that produces excessive amounts of Myc protein (see Figure 26-12). Burkitt lymphoma is only one of several human cancers in which the Myc protein is overproduced. In other cancers, gene amplification rather than chromosomal translocation is responsible. For example, *MYC* gene amplification is frequently observed in small-cell lung cancers and to a lesser extent in a wide range of other carcinomas, including 20–30% of breast and ovarian cancers.

6. Cell Cycle and Apoptosis Regulators. In the final step of growth-signaling pathways, transcription factors activate genes encoding proteins that control cell proliferation and survival. The activated genes include those encoding *cyclins* and *cyclin-dependent kinases (Cdks)*, which have roles in triggering passage through key points in the cell cycle (see Chapter 24). Several human oncogenes produce proteins of this type. For example, a cyclin-dependent kinase gene, *CDK4*, is amplified in some sarcomas, and the cyclin gene *CYCD1* is commonly amplified in breast cancers and is altered by chromosomal translocation in some lymphomas. Such oncogenes produce excessive amounts or hyperactive versions of Cdk-cyclin

(a) Normal cell (without Wnt proteins). β-catenin is targeted for degradation by the APC-axin-GSK3 destruction complex, which catalyzes phosphorylation of β-catenin. Phosphorylated β-catenin is linked to ubiquitin and degraded by proteasomes. The resulting absence of β-catenin maintains pathway in the OFF position.

(b) Normal cell (with Wnt proteins). The Wnt pathway is turned ON when Wnt proteins activate cell surface Wnt receptors, which bind axin. This prevents assembly of the destruction complex. β-catenin enters the nucleus and binds to TCF, forming a complex that activates genes that control cell proliferation, including *MYC* and *CYCD1* (a cyclin gene).

(c) Cancer cell (with or without Wnt proteins). Some cancer cells have loss-of-function mutations in the *APC* gene. In the absence of functional APC protein, the destruction complex cannot form, β-catenin accumulates, enters the nucleus, and locks the Wnt pathway in the ON position.

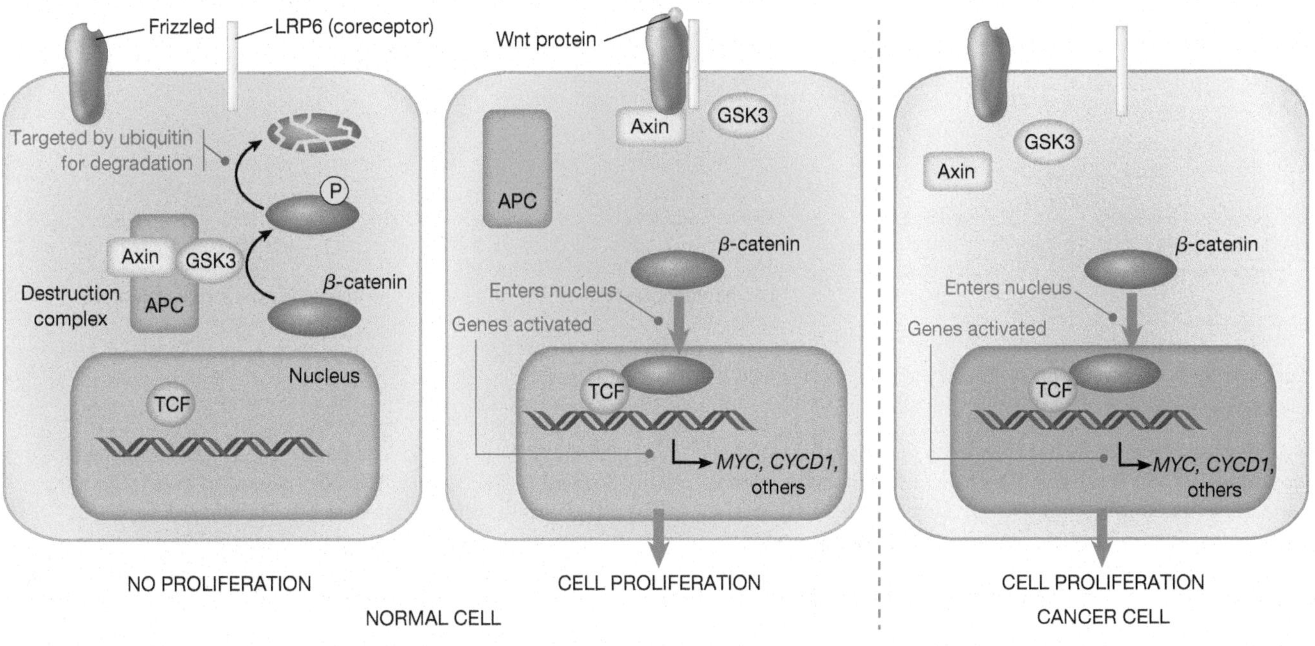

Figure 26-18 The Wnt Signaling Pathway. (a, b) In normal cells, the Wnt pathway is active only in the presence of Wnt proteins. **(c)** In cancer cells, the Wnt pathway is active regardless of the presence or absence of Wnt proteins.

mutations are necessary to trigger a cancer. How do cancer cells manage to acquire so many mutations if mutation is such a rare event? The apparent explanation is that *mutation rates in cancer cells are hundreds or even thousands of times higher than normal.*

This state, called **genetic instability**, can arise in several different ways. One group of mechanisms involves *disruptions in DNA repair.* For example, inherited defects in genes required for *mismatch repair* (see Chapter 17) are responsible for *hereditary nonpolyposis colon cancer (HNPCC)*, an inherited syndrome that elevates a person's cancer risk by allowing mutations to accumulate rather than being corrected by mismatch repair. Another hereditary disease, *xeroderma pigmentosum*, is caused by inherited defects in genes needed for *excision repair* (**see Human Connections in Chapter 17, page 497**). Children who inherit this condition develop an extremely high skin cancer risk because they are unable to repair DNA damage triggered by exposure to sunlight.

Faulty DNA repair is also involved in hereditary forms of breast cancer, which account for about 10% of all breast cancer cases. Most hereditary forms of breast (and ovarian) cancer arise in women who inherit a mutant copy of either the *BRCA1* or *BRCA2* gene, both of which encode proteins involved in repairing double-strand DNA breaks. Breast and ovarian cells defective in either of the BRCA proteins exhibit many chromosomal abnormalities, including translocations, deletions, and broken or fused chromosomes. As a result, women inheriting *BRCA* mutations exhibit a 40–80% lifetime risk for breast cancer and a 15–65% risk for ovarian cancer. Because the risks are so high, women with a family history of breast or ovarian cancer may wish to undergo genetic testing to determine whether they carry a mutant *BRCA1* or *BRCA2* gene. (**Human Connections, page 806**, discusses this and other genetic diagnostic techniques in more detail.)

Genetic instability is not restricted to hereditary cancers. Most cancers are nonhereditary, but they still exhibit genetic instability. In some cases, the instability can be traced to mutations in DNA repair genes that either arise spontaneously or are caused by environmental mutagens. Another explanation is that the p53 pathway is defective in most cancer cells, which removes a protective mechanism that would arrest cells at cell cycle checkpoints to allow DNA repair to occur or would trigger apoptosis to destroy cells containing damaged DNA. The result of such instability can be dramatic, wholesale rearrangements of pieces of chromosomes, as shown in **Figure 26-19** on page 807.

Genetic instability can also arise from defects in mitosis that cause *disruptions in chromosome sorting* during cell division, resulting in broken chromosomes and *aneuploidy* (an abnormal number of chromosomes). One reason for improper chromosome sorting is the presence of extra *centrosomes*, the structures that help guide microtubule assembly at spindle

HUMAN *Connections*

Molecular Sleuthing in Cancer Diagnosis

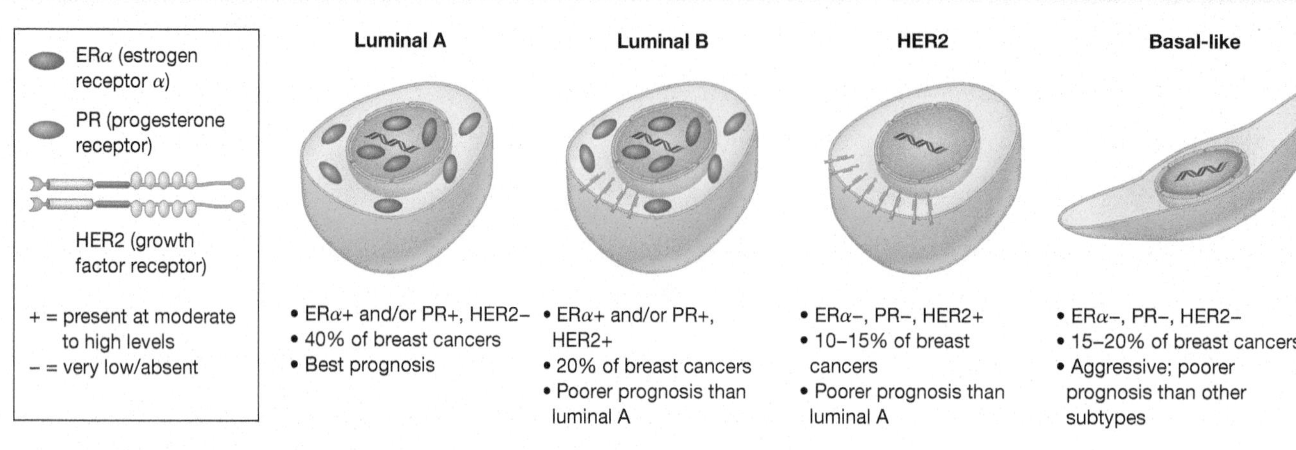

Figure 26A-1 Classification of Breast Cancers. Breast cancer can be classified based on appearance and differences in expression of key receptor proteins. The incidence and prognosis are shown for each.

You or someone you know has likely been through a familiar set of procedures: routine screening identifies a small growth, which may or may not be cancerous. The first step in diagnosis is usually biopsy of the tissue. A pathologist then carefully examines the tissue sample. If the news is good—the growth is benign—then the tissue is simply removed. If, however, the tissue sample turns out to be cancerous, then determining the exact nature of the cancer becomes crucial for successful treatment. Breast cancer is a good example of the utility of molecular diagnostics in this process.

Typing Breast Cancers. Breast cancers are often categorized into four main subtypes (**Figure 26A-1**), which show different expression patterns of three key proteins: the steroid hormone receptors estrogen receptor α and *progesterone receptor,* and *HER2,* a member of the epidermal growth factor (EGF) family of receptor tyrosine kinases.

The luminal A and B subtypes can be successfully treated with drugs that block steroid hormone receptors, such as the drug *tamoxifen*. In contrast, HER2 subtype tumors are more effectively treated using antibodies that recognize the HER2 receptor (for more details, **see Key Technique, pages 812–813**). The fourth subtype, basal-like tumors, is more difficult to treat; aggressive chemotherapy is often used.

Breast cancer diagnosis has benefited greatly from modern molecular analysis. Using a variety of techniques (discussed in Chapter 21), a list of mRNAs misexpressed in breast cancers has been determined. Commercial diagnostic kits have been developed to assay for these mRNAs. In one commonly used kit, the *Oncotype DX* kit, RNA is extracted from chemically preserved tissue samples. Reverse transcription is then performed, followed by quantitative PCR. The expression of 21 genes (16 cancer-related genes and five reference genes) is measured in triplicate. The results of the Oncotype DX test can be converted into a single number known as a *recurrence score.* Women whose breast tumors have a high recurrence score are more likely to have their cancers recur after surgery. This information can help doctors determine which patients are most likely to benefit from chemotherapy.

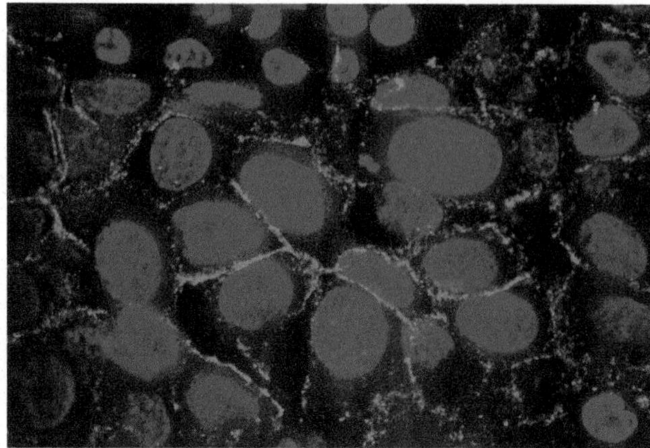

HER2-positive breast cancer cells (DNA stained blue, HER2 stained red).

BRCA1 and BRCA2. The basal-like subtype, also known as the *triple negative subtype,* of breast cancer is difficult to treat. What if there were a way to test whether a patient has a higher likelihood of developing such a cancer *before* it ever develops? Genetic tests offer just such a possibility. In a few cases, single-gene mutations dramatically increase hereditary susceptibility to cancer. These include mutations in the genes encoding Rb, p53, and the example we will discuss, BRCA1.

The genetic test for *BRCA1* and a very similar gene, *BRCA2,* was originally developed by Myriad Genetics as a screening tool in human patients. The test was based on the work of Mary-Claire King and colleagues in the early 1990s, who identified a gene on human chromosome 17 tightly associated with increased familial risk of breast and ovarian cancer, which they named *Breast Cancer 1 (BRCA1).* Subsequent studies showed that *BRCA1* and *BRCA2* are involved in repair of double-stranded DNA breaks.

The Myriad Genetics test involves sequencing of the *BRCA1* and *BRCA2* genes. The advantage of this approach is that specific mutations in each gene greatly increase the risk of cancer, whereas other mutations do not.

These tests, while powerful, create dilemmas for patients. A key decision is whether to undergo such testing at all; an appreciable percentage of patients at risk choose not to do so. A positive result leads women to consider more drastic intervention, including surgical removal of breasts and ovaries as a preventative measure or treatment with drugs such as tamoxifen. Molecular sleuthing also raises key questions for health-care professionals, the government, and insurance companies regarding confidentiality and the uses of such test results.

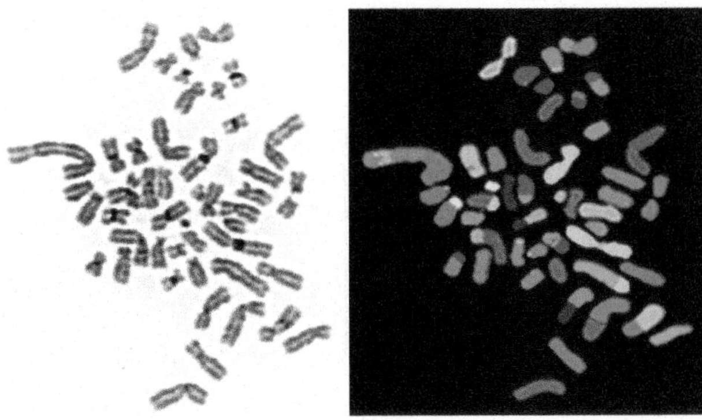

Figure 26-19 Chromosomal Instability in a Cancer Cell. A chromosomal preparation (left) processed via in situ hybridization using probes that normally recognize single chromosomes (right). The in situ shows that widespread chromosomal translocations have occurred.

poles during mitosis (Figure 13-11). **Figure 26-20** shows a cancer cell with three centrosomes that have assembled a mitotic spindle with three poles. Such spindles are rare in normal tissues but common in cancer cells; they contribute to aneuploidy because such spindles cannot sort the two sets of chromosomes accurately. Cells that are missing certain chromosomes are deficient in any tumor suppressor genes that normally reside on the missing chromosomes.

Cancer cells may also exhibit defects in proteins involved in attaching chromosomes to the spindle or in the *spindle assembly checkpoint*, whose normal function is to prevent anaphase chromosome separation from beginning before the chromosomes are all attached to the spindle (Figure 24-21). For example, some cancers have mutations in genes encoding Mad or Bub proteins, which are central components of the mitotic spindle checkpoint. If Mad or Bub proteins are defective or in short supply, chromosome movements toward the spindle poles may begin before all the chromosomes are properly attached to the mitotic spindle. The result is inaccurate chromosome sorting and the production of aneuploid cells.

The genes encoding proteins involved in DNA repair and chromosome sorting discussed in this section fit the definition of tumor suppressor genes because their loss or inactivation contributes to cancer development. But such genes are not in the same category as the *RB*, *p53*, and *APC* genes, which produce proteins that restrain cell proliferation and whose loss can directly lead to cancer. To distinguish the two classes of tumor suppressors, genes like *RB*, *p53*, and *APC* are called **gatekeeper genes** because their loss directly opens the "gates" to excessive cell proliferation and tumor formation. In contrast, genes involved in DNA repair and chromosome sorting are referred to as **caretaker genes** because they maintain genetic stability but are not directly involved in controlling cell proliferation and survival. Defects in caretaker genes lead to genetic instability, which in turn allows the accumulation of mutations in other genes (including gatekeepers) that then trigger excessive cell proliferation and cause cancer. **Table 26-2** lists some common tumor suppressor genes, organized on the basis of whether they are caretakers or gatekeepers and grouped according to the pathways they affect.

Cancers Develop by the Stepwise Accumulation of Mutations Involving Oncogenes and Tumor Suppressor Genes

Genome sequencing studies have revealed that a given type of cancer (for example, breast, lung, or colon cancer) typically involves mutations in about 50–75 different protein-coding genes. A small handful of these genes are mutated frequently in cancer samples of the same type taken from different people, whereas the rest of the genes are infrequently mutated. The commonly mutated genes affect about a dozen different pathways, most of which are discussed at one point or another in this chapter.

Figure 26-20 Abnormal Mitosis in a Cancer Cell. A cancer cell stained with a dye to visualize DNA (blue) and antibodies that recognize tubulin (red) and imaged by confocal microscopy shows a cell with an abnormal mitotic spindle containing three spindle poles.

Table 26-2 Examples of Tumor Suppressor Genes

Gene	Pathway Affected
Gatekeeper Genes	
APC	Wnt signaling
CDKN2A	Rb and p53 signaling
PTEN	PI 3-kinase–Akt signaling
RB	Restriction point control
SMAD4	TGF β-Smad signaling
TGF β receptor	TGF β-Smad signaling
p53	DNA damage response
Caretaker Genes	
BRCA1, BRCA2	DNA double-strand break repair
MSH2, MSH3, MSH4, MSH5, MSH6, PMS1, PMS2, MLH1	DNA mismatch repair
XPA, XPB, XPC, XPD, XPE, XPF, XPG	DNA excision repair
XPV (POLη)	DNA translesion synthesis

Common mutations involve the inactivation of tumor suppressor genes as well as the conversion of proto-oncogenes into oncogenes. In other words, creating a cancer cell usually requires that the "brakes" on cell growth (tumor suppressor genes) be released and the "accelerators" for cell growth (oncogenes) be activated. This principle is well illustrated by the stepwise progression toward malignancy observed in colon cancer. The most common pattern is the presence of a *KRAS* oncogene (a member of the *RAS* gene family) accompanied by loss-of-function mutations in the tumor suppressor genes *APC*, *SMAD4*, and *p53*. Rapidly growing colon cancers tend to exhibit all four genetic alterations, whereas benign tumors have only one or two.

As shown in **Figure 26-21**, the earliest mutation to be routinely detected is loss of function of the *APC* gene—a mutation that frequently occurs in small polyps before cancer has even arisen. Mutations in *KRAS* and *SMAD4* tend to be seen when the polyps get larger, and mutations in *p53* usually appear when cancer finally develops. However, these mutations do not always occur in the same sequence or with the same exact set of genes. For example, *APC* mutations are found in about two-thirds of all colon cancers, but this means that the *APC* gene is normal in one out of every three cases. Analysis of tumors containing normal *APC* genes has revealed that many of them possess oncogenes that produce an abnormal, hyperactive form of *β-catenin*, a protein that—like the APC protein—is involved in Wnt signaling (see Figure 26-18). Because APC inhibits the Wnt pathway and β-catenin stimulates it, mutations leading to the loss of APC and mutations that create hyperactive forms of β-catenin have the same effect: both enhance cell proliferation by increasing the activity of the Wnt pathway.

Another pathway frequently disrupted in colon cancer is the *TGF β-Smad pathway* (see Figure 23-22), which *inhibits* rather than stimulates epithelial cell proliferation. Loss-of-function mutations in genes encoding components of this pathway, such as the TGF β receptor or Smad4, are commonly detected in colon cancers. Such mutations disrupt the growth-inhibiting activity of the TGF β-Smad pathway and thereby contribute to enhanced cell proliferation.

With the advent of rapid DNA sequencing technologies (see Chapter 21), a much more comprehensive analysis of mutated genomes in common types of cancers is emerging. One of the coordinated efforts using this approach is the *Cancer Genome Atlas*, whose web portal is managed by the U.S. National Institutes of Health (http://cancergenome.nih.gov). A number of papers that are a part of this project, as well as a large number of other studies, are examples of **genome-wide association studies (GWAS)**. By sequencing the entire genomes of various cancers, GWAS analysis seeks to correlate specific mutations with particular diseases, including cancer. These comprehensive GWAS analyses have refined the basic picture that has already emerged regarding how cancer arises at the molecular level.

Epigenetic Changes in Gene Expression Influence the Properties of Cancer Cells

Mutations that create oncogenes or inactivate tumor suppressor genes play central roles in cancer development, but they do not explain all the properties of cancer cells. Many traits exhibited by cancer cells arise not from gene mutations but from *epigenetic changes*—that is, changes in gene expression that do not involve changes in a gene's underlying base sequence.

One mechanism for creating epigenetic changes involves *DNA methylation* at –CG– sites located near gene promoters (see Chapter 20). Most of these sites are unmethylated in normal cells, but extensive methylation is common in cancer cells, where it leads to *epigenetic silencing* of numerous genes—including tumor suppressor genes. In fact, the tumor suppressor genes of cancer cells are inactivated by DNA methylation at least as often as they are inactivated by DNA mutation. Epigenetic silencing of a tumor suppressor gene can be a critical, initiating event on the road to cancer. For example, individuals who inherit an extensively methylated form of a DNA mismatch repair gene known as *MLH1* are highly susceptible to developing multiple cancers. Inheriting a methylated tumor suppressor gene may predispose a person to developing cancer just like inheriting a mutated form of the gene does.

Another mechanism for altering gene expression involves *microRNAs*, which bind to and silence the translation

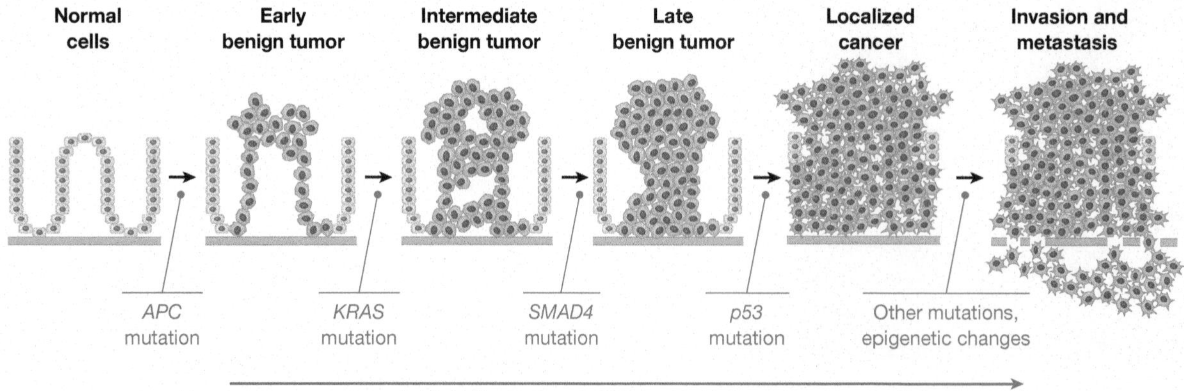

Figure 26-21 Stepwise Model for the Development of Colon Cancer. Colon cancer often arises through a stepwise series of mutations involving the *APC*, *KRAS*, *SMAD4*, and *p53* genes. Each successive mutation is associated with increasingly abnormal cell behavior. The early stages in this process produce benign tumors called *polyps*, which protrude from the inner surface of the colon (see Figure 26-17 on page 804).

of thousands of individual mRNAs (see Figure 20-37). Cancer cells can produce either too many or too few microRNAs. One example of the former is the microRNA produced by *miR-17-92*, which inhibits translation of the messenger RNA encoding PTEN, a phosphatase that inhibits PI 3-kinase–Akt signaling pathways (see Figure 23-27). The *miR-17-92* gene is frequently amplified in certain types of cancer, which leads to excessive production of its microRNA and a resulting inhibition of the synthesis of PTEN. Depletion of PTEN allows the PI 3-kinase–Akt pathway to be continually activated, thereby leading to enhanced cell proliferation and survival. Unproduced microRNAs can contribute to cancer development by acting as tumor suppressors. For example, the *miR-15a/miR-16-1* gene cluster is frequently deleted in certain forms of leukemia. One of the normal functions of *miR-15a/miR-16-1* is to inhibit translation of the messenger RNA encoding Bcl-2. Deletion of *miR-15a/miR-16-1* can therefore lead to excessive production of Bcl-2, which in turn interferes with the ability to carry out apoptosis. Some microRNAs indirectly influence *histone modification* reactions, which have a role in regulating gene expression (see Chapter 20), including that of tumor suppressor genes.

Summing Up: Carcinogenesis and the Hallmarks of Cancer

You might wonder at this point whether it is possible to provide a unifying overview of the process of *carcinogenesis*—that is, the multistep series of events that convert normal cells into cancer cells. One way of presenting such an overview is provided by **Figure 26-22**, which begins with the main causes of cancer through DNA mutations we have already discussed: replication and DNA repair errors, radiation, chemicals, infectious agents, and heredity. For cancer to develop, these mutations must involve a stepwise series of changes involving the inactivation of tumor suppressor genes and the conversion of proto-oncogenes into oncogenes. Eventually, cancer cells develop a group of traits referred to by Douglas Hanahan and Robert Weinberg as the *hallmarks of cancer.* These six traits are common to all forms of cancer, but each trait can be acquired through a variety of genetic and epigenetic mechanisms.

1. Self-Sufficiency in Growth Signals. Cells do not normally proliferate unless they are stimulated by an appropriate growth factor. Cancer cells escape this requirement through the action of oncogenes that produce excessive quantities or mutant versions of proteins involved in growth-stimulating pathways (see Table 26-1). As we have seen, one such pathway commonly activated in cancer cells is the Ras pathway. About 25–30% of all human cancers have mutant Ras proteins that provide an ongoing stimulus for the cell to proliferate independently of growth factors. Mutations affecting other components of the Ras pathway are common as well.

2. Insensitivity to Antigrowth Signals. Normal tissues are protected from excessive cell proliferation by a variety of growth-inhibiting mechanisms. Cancer cells must evade such antigrowth signals if they are to continue proliferating. Most antigrowth signals act during late G1 and exert their effects through the Rb protein, whose phosphorylation regulates passage through the restriction point and into S phase. For example, TGF β normally inhibits proliferation by triggering the TGF β-Smad pathway, which produces Cdk inhibitors that block Rb phosphorylation and thereby prevent passage from G1 into S phase. In cancer cells, the TGF β-Smad pathway is disrupted by a variety of mechanisms. Mutations in the *RB* gene also make cells insensitive to the antigrowth effects of TGF β or any other growth inhibitor that exerts its effects through the Rb protein.

3. Evasion of Apoptosis. The survival of cancer cells depends on their ability to evade the normal fate of genetically damaged cells—destruction by apoptosis. The ability to evade

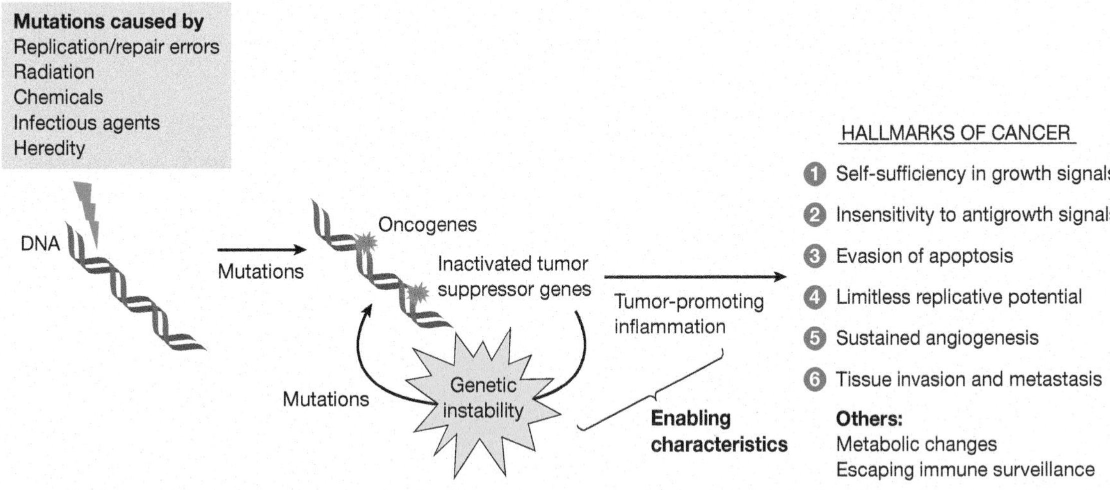

Figure 26-22 Overview of Carcinogenesis. The main causes of cancer (replication and DNA repair errors, radiation, chemicals, infectious agents, and heredity) trigger initial DNA alterations that can create oncogenes and disrupt tumor suppressor genes. An early result of these changes is genetic instability, which in turn facilitates the acquisition of additional mutations that (along with accompanying epigenetic changes and inflammation) lead to the six hallmark traits: self-sufficiency in growth signals, insensitivity to antigrowth signals, evasion of apoptosis, limitless replicative potential, sustained angiogenesis, and tissue invasion and metastasis.

apoptosis is often imparted by mutations in the *p53* tumor suppressor gene, which disrupt the ability of the p53 pathway to trigger apoptosis in response to DNA damage. The p53 pathway can also be disrupted by certain oncogenes. An example is the *MDM2* gene, which produces the Mdm2 protein that targets p53 for destruction (see Figure 24-22). Mutant forms of the *BCL2* gene also act as oncogenes, leading to production of excessive quantities of Bcl-2 protein. This suppresses apoptosis (see Figure 24-28), allowing abnormal cells to continue proliferating.

4. Limitless Replicative Potential. Uncontrolled growth would not ensure unlimited proliferation in the absence of a mechanism for replenishing the telomere sequences that are lost from the ends of each chromosome during DNA replication. Telomere maintenance is usually accomplished by activating the gene encoding telomerase, but a few cancer cells activate an alternative mechanism for maintaining telomeres that involves the exchange of sequence information between chromosomes. In either case, cancer cells maintain telomere length above a critical threshold and thereby retain the ability to divide indefinitely.

5. Sustained Angiogenesis. Without a blood supply, tumors will not grow beyond a few millimeters in size. Thus, at some point during early tumor development, cancer cells must trigger angiogenesis. A common strategy involves the activation of genes encoding angiogenesis stimulators combined with the inhibition of genes encoding angiogenesis inhibitors. In some cases the activities of known tumor suppressor genes or oncogenes are involved. For example, the p53 protein activates the gene encoding the angiogenesis inhibitor thrombospondin; hence the loss of p53 function, common in human cancers, can cause thrombospondin levels to fall. Conversely, *RAS* oncogenes trigger increased expression of the gene encoding the angiogenesis activator VEGF.

6. Tissue Invasion and Metastasis. The ability to invade surrounding tissues and metastasize to distant sites is the defining trait that distinguishes a cancer from a benign tumor. Three properties exhibited by cancer cells play a crucial role in these events: decreased cell-cell adhesion, increased motility, and the production of proteases that degrade the extracellular matrix and basal lamina.

Other Hallmarks. Other features of some cancers represent additions to Hanrahan and Weinberg's original list. The first involves an observation made by Otto Warburg in the 1920s and 1930s that cancer cells often bypass aerobic metabolism, relying instead on high rates of glycolysis even when oxygen is plentiful, a process called *aerobic glycolysis* (see Chapter 9). Why this promotes cancer remains unclear, but one popular idea is that this metabolic strategy allows cancer cells to divert their cellular resources, including macromolecules, to processes required for rapid cell division. Another hallmark has to do with the immune system. As mentioned previously, tumors appear to evade attack by the body's immune system. That evasion of the immune system is widespread is suggested by the effectiveness of treatments designed to stimulate the immune system to become more sensitive to cancer cells (as we will discuss later in Section 26.5).

Enabling Characteristics: Genetic Instability and Tumor-Promoting Inflammation. In addition to the previously described hallmarks, the road to aggressive cancers is paved by two *enabling characteristics*. The first is *genetic instability*. It is placed in a category separate from the six hallmark traits—which are directly involved in cancer cell proliferation and spread—because genetic instability is a crucial underlying trait that enables cancer cells to accumulate the mutations that permit the six hallmark traits to arise. Genetic instability arises most commonly from mutations that disrupt the ability of the p53 pathway to trigger the destruction of genetically damaged cells. However, mutations in genes encoding proteins involved in DNA repair and chromosome sorting also play a role.

The second enabling characteristic is *tumor-promoting inflammation*. Most tumors are infiltrated with cells from the immune system. This may be due to the immune system's attempt to attack the tumor, but it now seems clear that the inflammatory response in many cases has the paradoxical effect of enhancing tumor growth, allowing tumors to acquire the hallmarks of cancer. Inflammation probably promotes cancer by supplying tumor-promoting molecules, including growth factors, survival factors that reduce cell death, and factors that stimulate angiogenesis.

CONCEPT CHECK 26.4

Compare and contrast the types of mutations that would have to occur in tumor suppressor genes and proto-oncogenes to lead to cancer in terms of the effects these mutations would have on the products of each kind of gene.

26.5 Diagnosis, Screening, and Treatment

Much progress has been made in recent years in elucidating the genetic and biochemical abnormalities underlying cancer development. One of the hopes for such research is that our growing understanding of the molecular alterations exhibited by cancer cells will eventually lead to improved strategies for cancer diagnosis and treatment.

Cancer Is Diagnosed by Microscopic and Molecular Examination of Tissue Specimens

Because cancer can arise in almost any tissue, few generalizations are possible regarding disease symptoms. A definitive diagnosis typically requires a *biopsy*, which involves surgical removal of a tiny tissue sample for microscopic examination. Although no single trait is sufficient for visually identifying cancer cells, they usually display several features that together indicate the presence of cancer (**Table 26-3**). For example, cancer cells often exhibit large, irregularly shaped nuclei, prominent nucleoli, a high ratio of nuclear-to-cytoplasmic volume, significant variations in cell size and shape, and a loss of normal tissue organization. To varying extents, cancer cells lose the specialized structural and biochemical properties of the cells normally residing in the tissue of origin. Cancers also have more dividing cells than normal. And finally, cancers typically have a poorly defined outer boundary, with signs of tumor cells penetrating into surrounding tissues.

Table 26-3 Some Differences in the Microscopic Traits of Benign and Malignant Tumors

Trait	Benign	Malignant
Nuclear size	Small	Large
N/C ratio (ratio of nuclear to cytoplasmic volume)	Low	High
Nuclear shape	Regular	Pleomorphic (irregular shape)
Mitotic index	Low	High
Tissue organization	Normal	Disorganized
Differentiation	Well differentiated	Poorly differentiated
Tumor boundary	Well defined	Poorly defined
	↓	↓

If a sufficient number of these abnormal traits are observed, it can be concluded that cancer is present—*even if invasion and metastasis have not yet occurred.* However, the severity of the abnormal traits varies significantly among cancers. This variability forms the basis for **tumor grading**, the assignment of numbered grades to tumors based on differences in their microscopic appearance. Lower numbers (for example, grade 1) are assigned to tumors whose cells exhibit largely normal differentiated features, divide slowly, and display only modest abnormalities in the traits listed in Table 26-3. Higher numbers (for example, grade 4) are assigned to tumors containing rapidly dividing, poorly differentiated cells that bear less and less resemblance to normal cells and exhibit severe abnormalities in the traits listed in Table 26-3. High-grade cancers tend to grow and spread more aggressively and respond less to therapy than lower-grade cancers do.

A major enhancement in the ongoing effort to more precisely diagnose tumors has come from molecular biology. The use of molecular diagnostics is described in more detail in Human Connections, page 806.

Screening Techniques for Early Detection Can Prevent Cancer Deaths

When cancer is detected before it has spread, cure rates tend to be relatively high, even for cancers that would otherwise have a poor prognosis. A great need therefore exists for screening techniques that can detect cancers at an early stage. One of the most successful screening procedures is the **Pap smear**, a technique for early detection of cervical cancer developed in the early 1930s by George Papanicolaou (for whom it is named). A Pap smear is performed by taking a tiny sample of a woman's vaginal secretions and examining it with a microscope. If the cells in the fluid exhibit unusual features—such as large, irregular nuclei or prominent variations in cell size and shape (**Figure 26-23**)—it is a sign that cancer may be present and further tests need to be done.

The success of the Pap smear has led to the development of screening techniques for other cancers. For example, *mammography* utilizes a special X-ray technique to look for early signs of breast cancer, and *colonoscopy* uses a slender fiber-optic instrument to examine the colon for early signs of colon cancer. The ideal screening test would allow doctors to detect cancers anywhere in the body with one simple procedure, such as a blood test. Prostate cancer is an example of a cancer that can sometimes be detected this way. Men over the age of 50 are often advised to get a **PSA test**, which measures how much *prostate-specific antigen (PSA)* is present in the bloodstream. PSA, which is a protein produced by cells of the prostate gland, normally appears in only tiny concentrations in the blood. If a PSA test reveals a high concentration of PSA, it indicates a possible prostate problem. Further tests are then performed to determine whether cancer is actually present. Other cancers also release specific proteins into the bloodstream; the development of methods for reliably detecting these proteins is under continuing development.

Surgery, Radiation, Chemotherapy Are Standard Treatments for Cancer

Strategies for treating cancer depend both on the type of cancer involved and how far it has spread. The most common approach involves surgery to remove the primary tumor, followed (if necessary) by radiation therapy and/or chemotherapy to destroy any remaining cancer cells.

Radiation therapy uses high-energy X-rays or other forms of ionizing radiation to kill cancer cells. Earlier in the chapter we saw that the DNA damage created by ionizing radiation can cause cancer. Ironically, the same type of radiation is used in higher doses to destroy cancer cells in people who already have the disease. Ionizing radiation kills cells in two different ways. First, DNA damage caused by radiation activates the p53 signaling pathway, which then triggers apoptosis. However, many cancers have mutations that disable the p53 pathway, so p53-induced apoptosis plays only a modest role

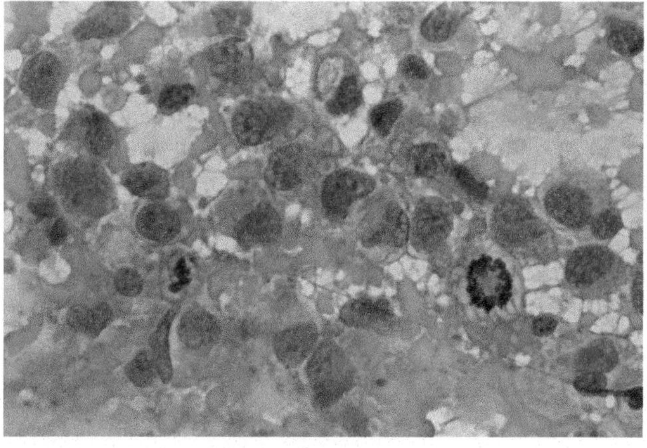

Figure 26-23 A Pap Smear Showing Endometrial Cancer. Cells from a cervical smear test showing endometrial sarcoma, a cancer of the lining of the uterus, the endometrium. The cancerous cells have enlarged nuclei (dark purple). The abnormalities in these isolated cells suggest that further examination of the uterus is required (light microscopy).

Key Technique | Targeting Molecules in the Fight Against Cancer

PROBLEM: Although chemotherapies that target rapidly dividing cells are a major tool in the fight against cancer, they are not specific to cancer cells themselves, but to all rapidly dividing cells.

SOLUTION: Targeting proteins unique to, or highly expressed in, cancer cells provides a more specific approach to attacking cancer.

Key Tools: "Humanized" monoclonal antibodies that recognize cell surface proteins; "rational" drugs that inactivate mutated cytosolic or nuclear proteins.

Details: A key weapon in the war on cancer is the use of chemotherapies that target rapidly dividing cells. Such chemotherapies can be remarkably effective, but at a heavy cost: they target all rapidly dividing cells, including hair follicles, blood cell precursors, and germ cell precursors in the ovaries and testes. Moreover, such treatments are not very specific to particular types of cancer. The modern molecular battle against cancer seeks to capitalize on the unique molecular "signature" of specific cancers by finding drugs or other agents that uniquely target only cancerous cells. In most cases, these specific approaches are combined with chemotherapy to attack cancers on multiple fronts simultaneously; such combined treatments have been shown to be much more successful at holding cancers at bay than either type of treatment alone.

Here we discuss two such specific approaches: the use of monoclonal antibodies to target cell surface proteins and "rational" drug design.

Monoclonal Antibodies. Recall that antibodies are proteins that recognize individual *antigens,* the molecular determinants to which they bind (see Chapter 21). In particular, *monoclonal antibodies* bind very specifically to one type of antigen. Such antibodies can be useful in cancer treatment. If a monoclonal antibody binds to a cell surface antigen, it is often the case that, once bound to its antigen, the monoclonal antibody interferes with the function of the antigen.

As potentially useful as mouse monoclonal antibodies might be, they suffer from a major technical problem: when they are injected into a human patient, the patient's immune system generates antibodies that attack the mouse proteins, neutralizing them and short-circuiting the intended therapeutic benefits. Fortunately, modern molecular biology can circumvent this problem. When useful types of monoclonal antibodies are identified, it is now possible to clone the antibody gene expressed in that particular monoclonal antibody-producing cell. Molecular engineering now makes it possible to combine the bit of DNA that encodes the antigen-binding part of the mouse monoclonal antibody with the human DNA that encodes the rest of the antibody (**Figure 26B-1**). Such *humanized antibodies* have been shown to prevent an immune response in patients.

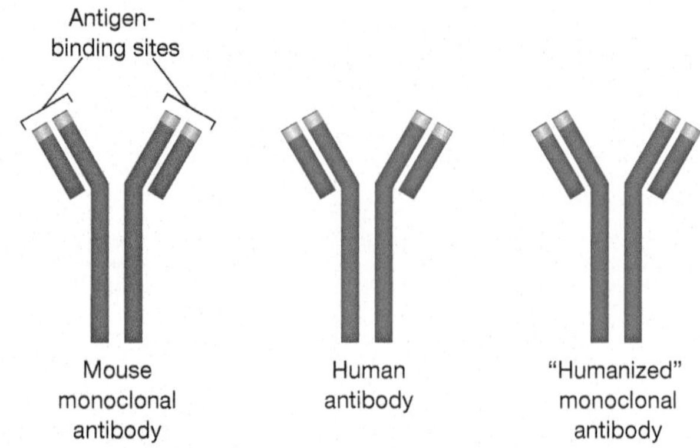

Figure 26B-1 "Humanized Antibodies." Antibodies like Herceptin were originally generated as monoclonal antibodies in mice. The genes encoding the antibody protein subunits were cloned, and the portion that encodes the antigen-binding region—that is, the region that binds the HER2 protein, in the case of Herceptin—was combined with human DNA sequences that encode the rest of the human antibody protein, which, when expressed, creates a humanized monoclonal antibody.

How can humanized antibodies fight cancer? First, when injected into individuals with cancer, they can bind to cancer cells and their presence can trigger an immune attack against the cells containing the bound antibody. Although this approach does not yet work for most cancers, one promising success has been achieved with *non-Hodgkin lymphoma.* Antibodies that target the *CD20 antigen* found on the surface of these lymphoma cells are now among the standard treatments for this particular type of cancer.

A second use for humanized antibodies is as a delivery vehicle; by linking them to radioactive molecules or to other toxic substances that are too lethal to administer alone, the radioactivity or toxins can be selectively concentrated at tumor sites without accumulating to toxic levels elsewhere in the body.

Finally, many antibodies have been developed that bind to and inactivate the proteins to which they bind. The first humanized antibody approved for use in cancer patients, called trastuzumab (*Herceptin*), acts in this way. It binds to and inactivates the growth factor receptor produced by the *ERBB2* gene (also known as *HER2,* for human epidermal growth factor receptor 2), which is amplified in about 25% of all breast and ovarian cancers. When individuals with such cancers are treated with Herceptin, the Herceptin antibody binds to the receptor and inhibits its ability to stimulate cell proliferation, thereby slowing or stopping tumor growth (**Figure 26B-2**). Herceptin likely acts through multiple mechanisms. Herceptin probably directly blocks HER2 receptors at the surface, preventing dimerization and signaling. HER2 bound to antibody is also more

in the radiation response. Radiation also kills cells by causing such severe chromosomal damage that mitosis cannot proceed, and the cells therefore die while trying to divide.

Most forms of *chemotherapy* use drugs that, like radiation, kill dividing cells. Such drugs can be subdivided into four major categories. (1) *Antimetabolites* inhibit metabolic pathways required for DNA synthesis by acting as competitive inhibitors that bind to enzyme active sites in place of normal substrate molecules. Examples include *fluorouracil, methotrexate, fludarabine, pemetrexed,* and *gemcitabine.* (2) *Alkylating*

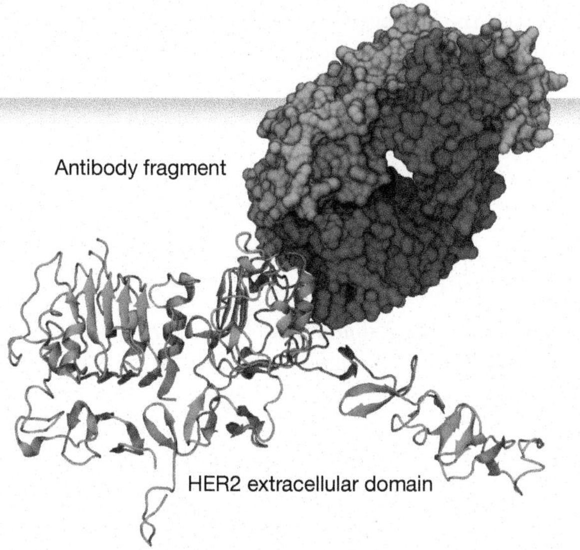

Figure 26B-2 Herceptin-HER2 Interaction.

likely to undergo receptor-mediated endocytosis, which leads to receptor down-regulation. When signaling is blocked, cancer cells are much more likely to die via apoptosis.

Many other monoclonal antibodies are now used in cancer therapy, including cetuximab (*Erbitux;* directed against the epidermal growth factor receptor, also known as *HER1*) and bevacizumab (*Avastin;* directed against the angiogenesis stimulating growth factor VEGF).

Rational Drug Design. An alternative way of targeting molecules for inactivation, called *rational drug design,* involves the laboratory synthesis of *small molecule inhibitors* that are designed to inactivate specific target proteins. This approach is called "rational" because rather than seeing how such chemicals affect cells, animals with tumors, or human patients, the chemicals are targeted against one particular protein, whose activity is assayed biochemically. This approach usually begins with a *chemical library* of potential small molecular inhibitors, which can be tested, one by one, to assess their biochemical effects on a protein.

One of the first anticancer drugs developed in this way was *Gleevec* (**Figure 26B-3**). Gleevec binds to and inhibits an abnormal tyrosine kinase produced by the *BCR-ABL* oncogene, whose association with chronic myelogenous leukemia was described earlier in the chapter (Figure 26-12b). The *BCR-ABL* oncogene arises from the fusion of two unrelated genes and produces a structurally abnormal tyrosine kinase that represents an ideal drug target because it is present only in cancer cells. The effectiveness of Gleevec as a treatment for early stage myelogenous leukemia is quite striking. More than half of the patients treated with Gleevec have no signs of the cancer 6 months after treatment, a response rate that is ten times better than that observed with earlier treatments. Other small-molecule inhibitors of protein kinases that have subsequently been found to be useful in cancer therapy include *Iressa, Tarceva, Sutent,* and *Nexavar.*

Another promising set of chemical inhibitors targets the G1 cyclin-dependent kinases Cdk4 and Cdk6. Cdk4/6 inhibitors include *palbociclib, ribociclib*, and *abemaciclib*. They have been shown to be especially effective at treating patients with hormone receptor-positive, human epidermal growth factor receptor 2-negative (HR+/HER2−) advanced breast cancer.

Rational drugs are a powerful new weapon in the arsenal in the fight against cancer. Many cancers eventually develop secondary mutations that circumvent the effectiveness of rational drugs. In the case of Gleevec, for example, many such mutations are actually secondary mutations in the BCR-ABL protein that render it resistant to Gleevec. Fortunately, additional small molecules have been identified that target these resistant versions of BCR-ABL. Using such drugs in combination, along with conventional chemotherapy, increases the likelihood that a treatment will be successful.

QUESTION: Monoclonal antibodies directed against the CD20 antigen are used to treat certain forms of lymphoma. However, the CD20 antigen is present on both mature normal and malignant lymphocytes, and treatment with the antibody therefore kills normal cells as well as cancer cells. Why do you think this antibody still turns out to be an effective treatment for lymphoma?

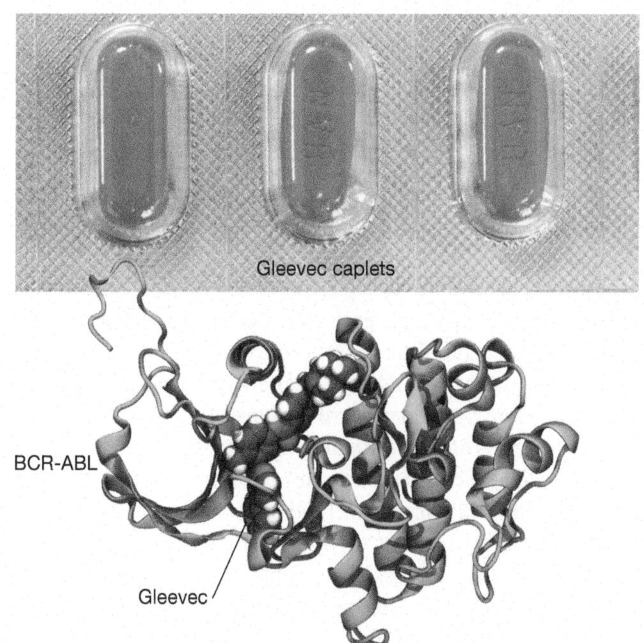

Figure 26B-3 Small Molecule Inhibitors. (Top) Caplets of the anti-cancer drug imatinib, marketed as Gleevec. (Bottom) Gleevec bound to the BCR-ABL kinase.

agents inhibit DNA function by chemically crosslinking the DNA double helix. Examples include *cyclophosphamide, chlorambucil,* and *cisplatin.* (3) *Antibiotics* are substances made by microorganisms that inhibit DNA function by either binding to DNA or inhibiting topoisomerases required for DNA replication. Examples include *doxorubicin* and *epirubicin.* (4) *Plant-derived* drugs either inhibit topoisomerases or disrupt the microtubules of the mitotic spindle. Examples include the topoisomerase-inhibitor *etoposide* and the microtubule-disrupting drug *paclitaxel* (*Taxol*).

One problem with such drug (and radiation) treatments is that they are toxic to normal dividing cells as well as to cancer cells. Less toxic approaches are possible for those cancers whose growth requires a specific hormone. For example, many breast cancers require estrogen for their growth. Steroid hormones such as estrogen exert their effects by binding to steroid receptor proteins that activate the expression of specific genes (see Chapter 23). The drug *tamoxifen*, which binds to estrogen receptors in place of estrogen and prevents the receptors from being activated, is useful both in treating breast cancer and in reducing the occurrence of breast cancer in women at high risk for the disease.

A frequent complication with drug therapies is that tumors tend to acquire mutations that make them resistant to the drugs being used. Even when only a single drug is employed, tumors often become resistant to the administered drug and to the effects of several unrelated drugs at the same time. This state, known as multiple drug resistance, arises because cells start producing *multidrug resistance transport proteins (ABC transporters)* that actively pump a wide range of chemically dissimilar, hydrophobic drugs out of the cell.

Drug resistance may also arise because, as we have already seen, tumors consist of a heterogeneous mixture of cells. In some types of cancer only a small population of cells—cancer stem cells—proliferate indefinitely and produce the rest of the cells found in tumors. If treatments that destroy most of the tumor cells leave behind a small number of cancer stem cells, these cells may be enough to regenerate the tumor after treatment is stopped. Researchers are therefore trying to identify drugs that selectively destroy cancer stem cells because those are the cells that fuel tumor growth.

Molecular Targeting Can Attack Cancer Cells More Specifically Than Chemotherapy

Until the early 1980s, the development of new cancer drugs focused largely on agents that disrupt DNA and interfere with cell division. Although such drugs are useful in treating cancer, their effectiveness is often limited by toxic effects on normal dividing cells. In the past three decades, the identification of individual genes that are mutated or abnormally expressed in cancer cells has created a new possibility—*molecular targeting*—in which drugs are designed to specifically target those proteins that are critical to the cancer cell.

One approach for molecular targeting involves the use of monoclonal antibodies that bind to proteins involved in the signaling pathways that drive cancer cell proliferation. One well studied and successful example of this approach is the monoclonal antibody *Herceptin*. A second approach, known as *rational drug design*, involves identifying chemical agents, such as *Gleevec*, that disrupt the functions of specific molecules mutated in cancer cells. (Herceptin and Gleevec are the focus of **Key Technique, pages 812–813.**)

Molecular targeting is not always aimed at cancer cells directly, however. Earlier we saw that sustained tumor growth depends on *angiogenesis* (growth of new blood vessels), so it is logical to expect that inhibitors that target blood vessel growth might be useful for treating cancer. Initial support for this concept of *anti-angiogenic therapy* came from the studies of Judah Folkman, who found that angiogenesis inhibitors make tumors shrink in mice.

The first anti-angiogenic drug to be approved for use in humans was *Avastin*, a monoclonal antibody that binds to and inactivates the angiogenesis-stimulating growth factor VEGF. Avastin has been found to improve short-term survival rates for several types of cancer, but the benefits are usually temporary and tumor growth resumes. Many other drugs that target angiogenesis are currently being evaluated, but the value of such anti-angiogenic agents remains unclear.

Using the Immune System to Target Cancer Cells

The use of surgery, radiation, or chemotherapy—either alone or in various combinations—can cure or prolong survival times for many types of cancer, especially when cancer is diagnosed early. However, some of the more aggressive types of cancer (including aggressive leukemias, melanoma, and those involving the lung, pancreas, or liver) are difficult to control in these ways. In an effort to develop better approaches for treating such cancers, scientists have been trying to find ways to selectively seek out and destroy cancer cells without damaging normal cells in the process.

One strategy for introducing such selectivity into cancer treatment is to exploit the ability of the immune system to recognize cancer cells. This approach, called *immunotherapy*, was first proposed in the 1800s after doctors noticed that tumors occasionally regress in people who develop bacterial infections. Because infections trigger an immune response, subsequent attempts were made to build on this observation by utilizing live or dead bacteria to provoke the immune system of cancer patients. This approach has not worked as well as initially hoped, although some success was seen with *bacillus Calmette-Guérin (BCG)*—a bacterial strain that does not cause disease but elicits a strong immune response at the site where it is introduced into the body. BCG has been used in treating early stage bladder cancers.

Although it demonstrates the potential usefulness of immune stimulation, BCG must be administered directly into the bladder to provoke an immune response at the primary tumor site. To treat cancers that have already metastasized to unknown locations, an immune response must be stimulated wherever cancer cells might have traveled. Normal proteins that the body produces to stimulate the immune system are sometimes useful for this purpose. *Interferon alpha* and *interleukin-2 (IL-2)* are two such proteins that have been successfully used in cancer treatment.

Attempts are also under way to develop therapeutic *vaccines* that introduce cancer cell antigens into patients to stimulate the immune system to selectively attack cancer cells. One such vaccine, called *Provenge*, uses an antigen commonly found in prostate cancer cells combined with antigen-processing cells obtained from the patient's blood; it was approved by the FDA in 2010 for use as a standard treatment for prostate cancer.

In general, however, because tumors are derived from the host's own cells, they are not highly antigenic, and so they do not provoke a strong immune response. Current strategies for boosting immune surveillance in such cases involve several approaches. In one approach, monoclonal antibodies (see Figure 21-21) are used to target cell surface proteins that partially inhibit the immune response of subsets of *T cells*, immune system cells involved in triggering death of foreign cells. These antibodies bind to proteins on the surface of T cells or to the proteins on tumor cells that the T cell proteins bind to. The normal function of the T-cell surface proteins is to moderate the immune response to prevent autoimmunity. By making T cells more "aggressive," immune surveillance can be enhanced. This strategy is known as **immune checkpoint therapy**.

Immune checkpoint therapy stems from work in the 1990s by James Allison and Tasuku Honjo. Working independently, they and their coworkers discovered proteins on the surface of T cells that suppress their ability to react to cancer cells. Allison and colleagues discovered that a protein called CTLA-4 on the surface of T cells binds to a protein known as B7 found on the surface of some cancer cells. Honjo and coworkers found that the protein PD-1 found on the surface of some T cells binds a protein called PD-L1 on some cancer cells.

Monoclonal antibodies that target these proteins are now widely used to treat cancer. Ipilimumab (*Yervoy*), which is used to treat melanoma and some other cancers, binds CTLA-4. Pembrolizumab (*Keytruda*) and nivolumab (*Opdivo*) bind PD-1, while atezolizumab (Tecentriq), avelumab (*Bavencio*), and durvalumab (*Imfinzi*) target PD-L1. These drugs are now widely used to treat melanoma, non-small cell lung cancer, kidney cancer, bladder cancer, head and neck cancers, and Hodgkin lymphoma. For their work Allison and Honjo received the Nobel Prize for Physiology or Medicine in 2018.

Another promising approach is called *adoptive cell transfer (ACT)*. ACT involves collecting patients' own immune cells, genetically engineering them, and reintroducing them into the patient. There are several types of ACT, but the most popular involves engineered T cells that express a genetically engineered protein, known as **chimeric antigen receptor (CAR) T cells**. ("Chimeric" comes from the chimera of Greek mythology, a monster with different body parts from different animals.) One of the CARs is shown in **Figure 26-24**. The extracellular portion of the CAR contains the portions of an antibody protein that are known to bind to a particular cell surface molecule (antigen) on the cells of the cancer being targeted. The antigen-binding portion of the chimeric protein is fused to a linker and a transmembrane region. The cytosolic portion of the CAR contains regions that promote activation of T cells when the extracellular portion of the receptor binds its target. In some CAR T cells an additional feature is added that causes expression of molecules, such as cytokines, that stimulate the recruitment of other immune cells to a tumor once the CAR T cells bind it.

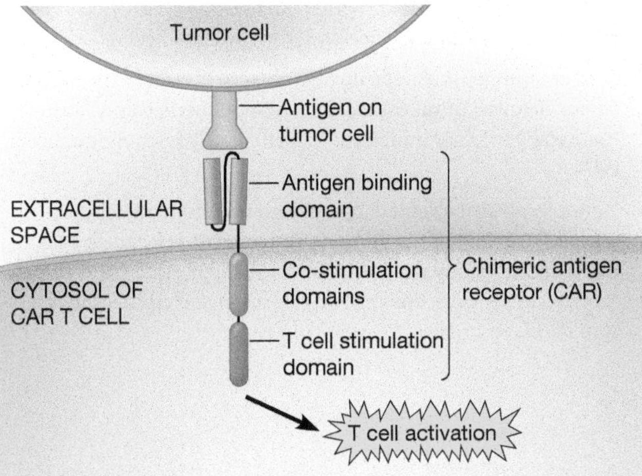

Figure 26-24 CAR T Cell Therapy. CAR T cells are genetically engineered to express a chimeric antigen receptor (CAR). CARs contain an extracellular antigen-recognition domain and a cytosolic signaling domain that activate the T cell when the CAR binds it target. Often, more than one co-stimulatory signaling domain is added.

Because CAR T cells can attack healthy cells and induce overly aggressive immune system responses, they can cause potentially serious, even life-threatening, side effects. Nevertheless, early clinical trials have led to some remarkable results. Receiving injections of their own genetically engineered T cells helps up to 90% of patients with B-cell acute lymphoblastic leukemia (B-ALL) recover. CAR T cell therapy also works for more than half of patients with non-Hodgkin lymphoma. The use of CRISPR/Cas9 and other genome-editing approaches promises to allow rapid development of this technology.

Cancer Treatments Can Be Tailored to Individual Patients

CAR T cell therapy is an example of a general strategy known as *personalized medicine*—an approach in which treatment choices are based on the individual characteristics of each patient. It has been known for many years that people with cancers that appear to be identical by traditional criteria often exhibit different outcomes after receiving the same treatment. *Transcriptome analysis*—the use of DNA microarrays or RNAseq to determine which genes are being transcribed into RNA—has provided an explanation: cancers involving the same cell type often exhibit differing patterns of gene expression that cause tumors to behave differently. These differing patterns of gene expression allow predictions about tumor behavior to be made. (One common example of this approach involving diagnosis of breast cancer is described in Human Connections, page 806.)

Analyzing gene expression profiles and testing for the presence of specific mutations may also help guide the choice of which drug to use. A striking example involves *Iressa*, a drug that acts by inhibiting the tyrosine kinase activity of the epidermal growth factor (EGF) receptor. Iressa causes tumors to shrink in only 10% of lung cancer patients, but when the drug does work, it works extremely well. The reason has been traced to the presence of a mutant EGF receptor gene in the cancers of the patients who receive the most benefit from the drug. This discovery opens the door to a personalized type of cancer therapy in which genetic testing of cancer cells is used to identify patients who are most likely to benefit from a particular type of treatment.

CONCEPT CHECK 26.5

Patients who have received organ transplants typically receive immunosuppressant drugs. Such patients also have an increased risk of developing solid tumors and cancers of the blood. Explain this result.

Summary of Key Points

Mastering™ Biology For activities, animations, and review quizzes, go to the study area at www.masteringbiology.com.

26.1 How Cancers Arise

- Cancer cells proliferate in an uncontrolled way and are capable of spreading by invasion and metastasis.
- The balance between cell division and cell differentiation or death is disrupted in cancers, leading to a progressive increase in the number of dividing cells.
- Cancer cells are anchorage-independent, exhibit a decreased susceptibility to density-dependent inhibition of growth, and are able to replenish their telomeres.
- Tumors arise through a multistep process of initiation, promotion, and tumor progression. Initiation is based on DNA mutation, whereas promotion involves a prolonged period of cell proliferation accompanied by selection of cells exhibiting enhanced growth properties. During tumor progression, additional mutations as well as epigenetic changes in gene expression produce cells with increasingly aberrant traits.
- Tumors can be heterogeneous; there is evidence that a small population of cancer stem cells is highly proliferative, whereas other cells are less so.

26.2 How Cancers Spread

- Sustained tumor growth requires a network of blood vessels whose growth is triggered by an increased production of angiogenesis activators and a decreased production of angiogenesis inhibitors.
- Cancer cells invade surrounding tissues, enter the circulatory system, and metastasize to distant sites. Invasion is facilitated by decreased cell-cell adhesion, increased motility, and secretion of proteases that degrade the extracellular matrix and basal lamina. Only a tiny fraction of the cancer cells that enter the bloodstream establish successful metastases.
- Sites of metastasis are determined by the location of the first capillary bed as well as by organ-specific conditions that influence cancer cell growth.
- Cancers have various ways of evading destruction by the immune system. Interactions between tumor cells and components of the surrounding microenvironment influence tumor growth, invasion, and metastasis.

26.3 What Causes Cancer?

- Some cancers are caused by certain kinds of chemicals, including those found in tobacco smoke.
- Cancer can also be caused by ionizing or UV radiation, both of which trigger DNA mutations, as well as by certain viruses, bacteria, and parasites.
- Some cancer-causing viruses trigger cell proliferation directly, either through the action of viral genes or by altering the behavior of cellular genes. Other viruses and infectious agents create tissue destruction that indirectly stimulates cell proliferation under conditions in which DNA damage is likely.

26.4 Oncogenes and Tumor Suppressor Genes

- Oncogenes are genes whose presence can cause cancer. Although oncogenes are sometimes brought into cells by viruses, more often they arise from normal cellular genes (proto-oncogenes) by point mutation, gene amplification, chromosomal translocation, local DNA rearrangements, or insertional mutagenesis.
- Many of the proteins produced by oncogenes are signaling pathway components, such as growth factors, receptors, plasma membrane GTP-binding proteins, nonreceptor protein kinases, transcription factors, and cell cycle and apoptosis regulators. Oncogenes encode abnormal forms or excessive quantities of such proteins, thereby leading to excessive cell proliferation and survival.
- Tumor suppressor genes are genes whose loss or inactivation can lead to cancer. Susceptibility to cancer is increased in people who inherit defective tumor suppressor genes.
- Common tumor suppressor genes include the *RB* gene, which produces a protein that restrains passage from G1 into S phase; the *p53* gene, which produces a protein that prevents proliferation and triggers apoptosis in cells with damaged DNA; and the *APC* gene, which produces a protein that inhibits the Wnt pathway.
- The genetic instability of cancer cells facilitates the acquisition of multiple mutations. Genetic instability arises from defects in DNA repair mechanisms, disruptions in pathways that trigger apoptosis, and failures in chromosome-sorting mechanisms.
- Cancers arise through a stepwise accumulation of mutations and epigenetic changes involving oncogenes and tumor suppressor genes. Most of the affected genes code for proteins, but some produce microRNAs.

26.5 Diagnosis, Screening, and Treatment

- Screening techniques, such as the Pap smear, can prevent cancer deaths by detecting cancer before it has spread.
- Molecular technologies can greatly refine diagnosis of cancer and can lead to more specific and effective treatments.
- Cancer treatments usually involve surgery to remove the primary tumor, followed (if necessary) by radiation therapy and/or chemotherapy to kill or inhibit the growth of any remaining cancer cells.
- Newer treatment approaches include immunotherapies that exploit the ability of the immune system to attack cancer cells, molecular targeting drugs that bind to proteins that play critical roles in cancer cells, and anti-angiogenic agents that attack a tumor's blood supply.

Problem Set

Mastering™ Biology For activities, animations, and review quizzes, go to the study area at www.masteringbiology.com.

26-1 Normal Cells and Cancer Cells. You are given two test tubes, one containing cells from a human cancer and the other containing normal cells. Before you begin your studies, the labels fall off the test tubes. Describe at least four experiments you could carry out to determine which sample contains the cancer cells.

26-2 Phorbol Esters and Cancer. The active ingredient in croton oil is a phorbol ester, which mimics the effects of the molecule diacylglycerol (DAG). Based on what you know about DAG (see Chapter 23), explain in detail why it might be an effective promoter, but not initiator, of tumor formation.

26-3 DATA ANALYSIS **Smoking and Cancer Revisited.** Examine Figure 26-7 again.

(a) There is a strong correlation between deaths due to lung cancer in men and overall cigarette consumption in the United States. Approximately how much of a time lag is there between these two sets of data, and why is this sobering regarding when symptoms appear in heavy smokers?

(b) How do you explain that the lung cancer death rate for women in the 20th century is so dramatically different than for men based on these data?

26-4 Rats, Guinea Pigs, and Humans. The chemical substance 2-acetylaminofluorene (AAF) causes bladder cancer when injected into rats but not guinea pigs. If normal bladder cells obtained from rats and guinea pigs are grown in culture and exposed to AAF, neither are converted into cancer cells. How can you explain these findings? Does your explanation suggest how to predict whether AAF is carcinogenic in humans without actually exposing humans to AAF?

26-5 Oncogenes and Tumor Suppressor Genes. Indicate whether each of the following descriptions applies to an oncogene, a proto-oncogene, or a tumor suppressor gene. Some descriptions may apply to more than one of these gene types. Explain your answers.

(a) A type of gene found in normal cells

(b) A gene that could encode a normal growth factor

(c) A type of gene found in cancer cells

(d) A type of gene found only in cancer cells

(e) A gene whose presence can cause cancer

(f) A gene whose absence can cause cancer

(g) A type of gene found in normal cells and cancer cells

26-6 Viruses and Cancer. Based on what you know about viruses and their means of reproduction, explain why many oncogenic viruses interact with tumor suppressor proteins.

26-7 Children of the Moon, Revisited. Children with *xeroderma pigmentosum* usually cannot carry out excision repair (see Human Connections in Chapter 17, page 497). Why does this make them so susceptible to developing cancer? Why was the word "usually" included in the first sentence?

26-8 QUANTITATIVE **Cancer Screening.** The annual incidence of colon cancer in the United States is about 55 cases per 100,000. Because colon cancer often causes bleeding, doctors sometimes use a screening procedure called the fecal occult blood test (FOBT) to look for tiny amounts of blood in the feces. One form of this test has a specificity of about 98%, which means that when it indicates the presence of blood in the feces, the result is an error (that is, cancer is not present) only 2% of the time. Although this may seem like excellent specificity, a 2% "false positive" rate makes this test almost useless as a tool for colon cancer screening. Why?

APPENDIX Visualizing Cells and Molecules

Cell biologists often need to examine the structure of cells and their components in detail. The microscope is an indispensable tool for this purpose because most cellular structures are too small to be seen by the unaided eye. In fact, the beginnings of cell biology can be traced to the invention of the **light microscope**, which made it possible for scientists to see enlarged images of cells for the first time. The first generally useful light microscope was developed in 1590 by Z. Janssen and his nephew, H. Janssen. Many important microscopic observations were reported during the next century, notably those of Robert Hooke, who observed the first cells (see Figure 1-1), and Antonie van Leeuwenhoek, whose improved microscopes provided our first glimpses of internal cell structure. Since then, the light microscope has undergone many improvements and modifications, right up to the present time.

Light microscopy has continued to develop through advances that involve the merging of technologies from physics, engineering, chemistry, and molecular biology. These advances have greatly expanded our ability to study cells using the light microscope. To overcome the limitations of a brightfield microscope, a variety of special optical techniques have been developed for observing living cells directly. These techniques include phase-contrast microscopy, differential interference contrast microscopy, fluorescence microscopy, and confocal microscopy. In **Table A-1**, you can see images acquired using each of these techniques and compare them with the images seen with brightfield microscopy for both unstained and stained specimens. You will learn about these techniques in more detail in subsequent sections of this appendix.

Just as the invention of the light microscope heralded a wave of scientific achievement by allowing us to see cells for the first time, the development of the **electron microscope** in the 1930s revolutionized our ability to explore cell structure and function. Because it is at least 100 times better at visualizing objects than the light microscope, the electron microscope ushered in a new era in cell biology, opening our eyes to an exquisite subcellular architecture never before seen and changing forever the way we think about cells.

In this appendix, we explore the fundamental principles of light and electron microscopy, emphasizing the various specialized techniques used to adapt these two types of microscopy for a variety of specialized purposes. We also examine other related technologies that extend the range of view beyond the microscope to visualization of single molecules at high resolution.

Optical Principles of Microscopy

Although light and electron microscopes differ in many ways, they make use of similar optical principles to form images. Therefore, we begin our discussion of microscopy by examining these underlying common principles, placing special emphasis on the factors that determine how small an object can be seen with current technologies.

The Illuminating Wavelength Sets a Limit on How Small an Object Can Be Seen

Regardless of the type of microscope being used, three elements are always needed to form an image: a *source of illumination*, a *specimen* to be examined, and a system of *lenses* that focuses the illumination on the specimen and forms the image. **Figure A-1** illustrates these features for light and electron microscopes. In a light microscope, the source of illumination is *visible light* (wavelength approximately 400–700 nm), and the lens system consists of a series of glass lenses. The image can be either viewed directly through an eyepiece or focused on a detector, such as an electronic camera. In an electron microscope, the illumination source is a *beam of electrons* emitted by a heated tungsten filament, and the lens system consists of a series of electromagnets. The electron beam is then digitally imaged using a detector.

Despite these differences in illumination source and instrument design, both types of microscopes depend on the same principles of optics and form images in a similar manner. When a specimen is placed in the path of a light or electron beam, physical characteristics of the beam are changed in a way that creates an image that can be interpreted by the human eye or recorded on a photographic detector. To understand this interaction between the illumination source and the specimen, we need to understand the concept of wavelength, which is illustrated in **Figure A-2** on page A-3 using the following simple analogy.

If two people hold onto opposite ends of a slack rope and wave the rope with a rhythmic up-and-down motion, they will generate a long, regular pattern of movement in the rope called a *waveform* (Figure A-2a). The distance from the crest of one wave to the crest of the next is called the **wavelength**. If someone standing to one side of the rope tosses a large object such as a beach ball toward the rope, the ball may interfere with, or perturb, the waveform of the rope's motion (Figure A-2b). If, however, a small object, such as a softball, is tossed toward the rope, the movement of the rope will probably not be affected at all (Figure A-2c). If the rope holders move the rope faster, the motion of the rope will still have a waveform, but the wavelength will be shorter (Figure A-2d). In this case, a softball tossed toward the rope is quite likely to perturb the rope's movement (Figure A-2e).

This simple analogy illustrates an important principle: *the ability of an object to perturb a wave's motion depends crucially on the size of the object in relation to the wavelength of the motion.* This principle is of great importance in microscopy because it means that the wavelength of the illumination source sets a limit on how small an object can be seen. In fact, both light and electrons have wave-like properties. Returning to our analogy, the moving rope of Figure A-2 is analogous to the beam of light (photons) or the electrons used as an illumination source in a light or electron microscope, respectively. When a beam of light (or electrons) encounters a specimen, the specimen alters the physical characteristics of the illuminating beam, just as the beach ball or softball alters the motion of the rope.

Table A-1 Different Types of Light Microscopy: A Comparison

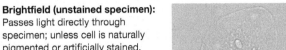

Type of Microscopy	Light Micrographs of Human Cheek Epithelial Cells		Type of Microscopy
Brightfield (unstained specimen): Passes light directly through specimen; unless cell is naturally pigmented or artificially stained, image has little contrast.	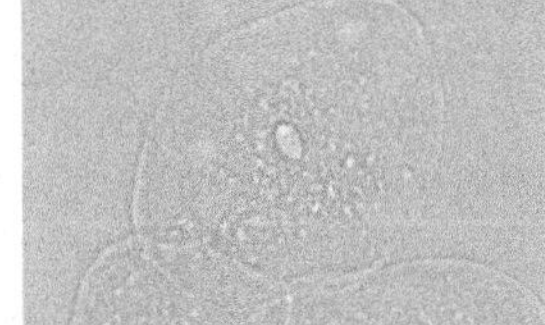	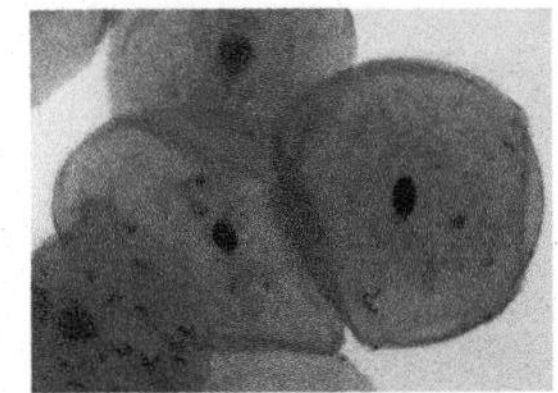	**Brightfield (stained specimen):** Staining with various dyes enhances contrast, but most staining procedures require that cells be fixed (preserved).
Phase contrast: Enhances contrast in unstained cells by amplifying variations in refractive index within specimen; especially useful for examining living, unpigmented cells.	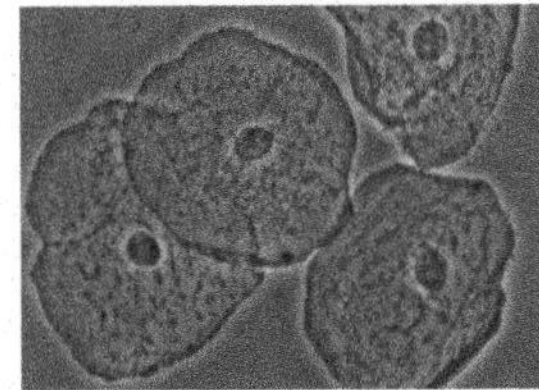	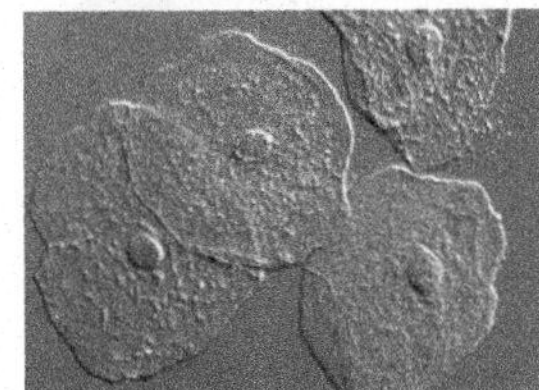	**Differential interference contrast:** Also uses optical modifications to exaggerate differences in refractive index.
Fluorescence: Shows the locations of specific molecules in the cell. Fluorescent substances absorb light and emit light of longer wavelength. The fluorescing molecules may occur naturally in the specimen but more often are made by tagging the molecules of interest with fluorescent dyes or antibodies.	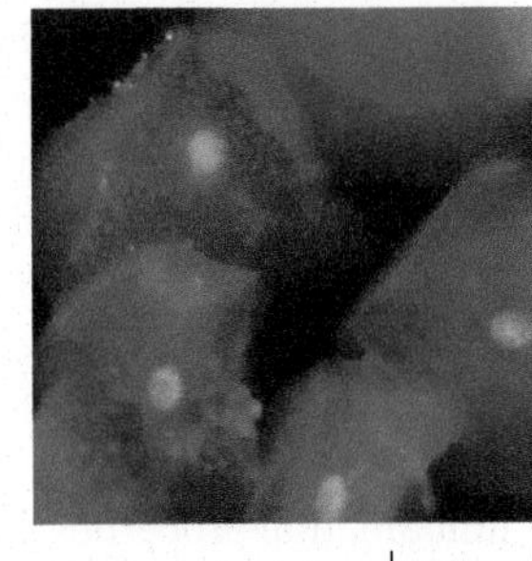 20 μm	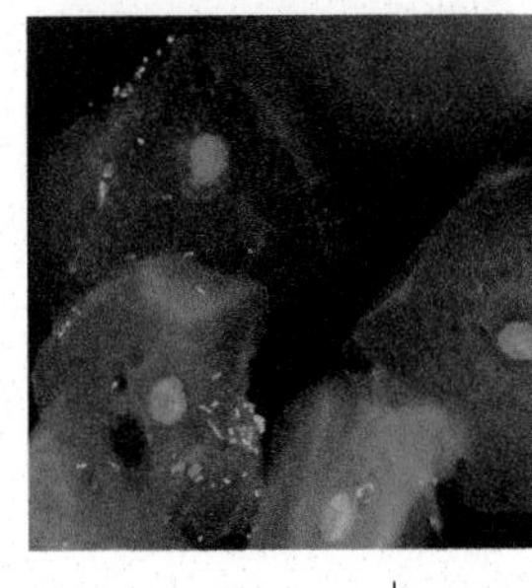 20 μm	**Confocal:** Uses lasers and special optics to visualize a single plane within the specimen. Only those regions within a narrow depth of focus are imaged. Regions above and below the selected plane of view appear black rather than blurry.

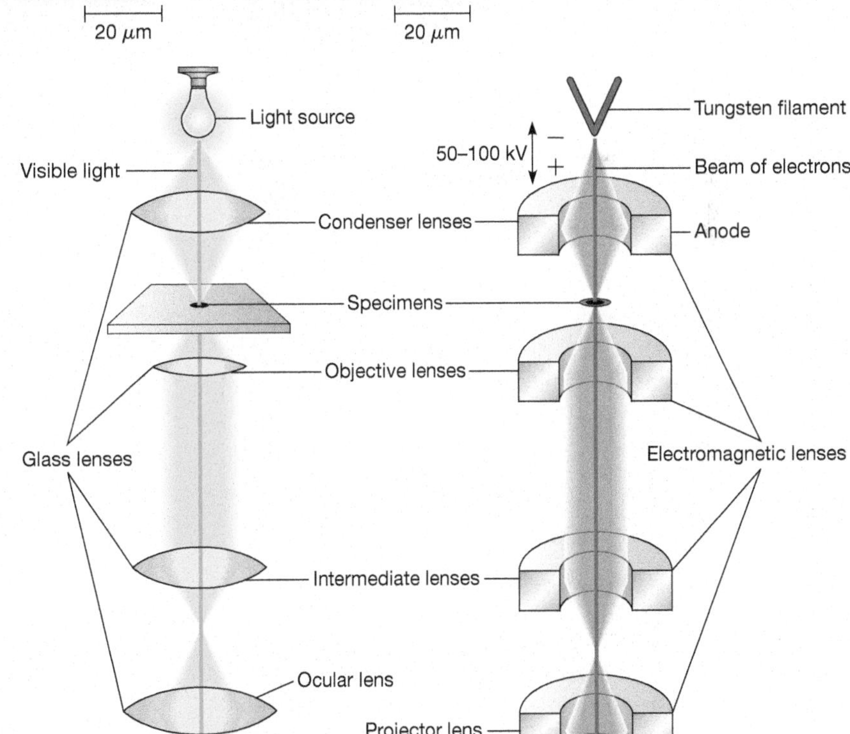

Figure A-1 Optical Systems of the Light Microscope and the Electron Microscope. (a) The light microscope uses visible light (wavelength approximately 400–700 nm) and glass lenses to form an image of the specimen that can be seen by the eye, focused on photographic film, or received by an electronic detector such as a digital camera. **(b)** The electron microscope uses a beam of electrons emitted by a tungsten filament and focused by electromagnetic lenses to form an image of the specimen on a fluorescent screen, a digital detector, or photographic film. (These diagrams have been drawn to emphasize the similarities in overall design between the two types of microscope. In reality, a light microscope is designed with the light source at the bottom and the ocular lens at the top, as shown in Figure A-5b.)

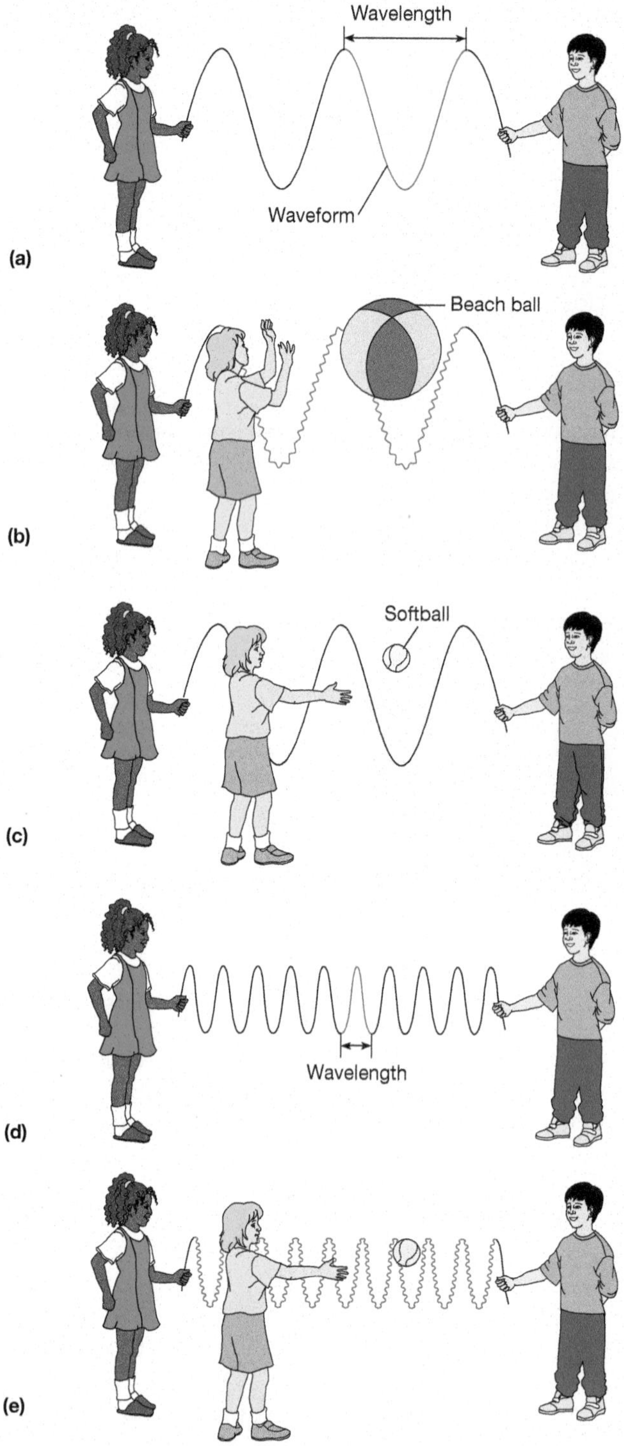

Figure A-2 Wave Motion, Wavelength, and Perturbations. The wave motion of a rope held between two people is analogous to the waveform of both photons and electrons, and it can be used to illustrate the effect of the size of an object on its ability to perturb wave motion. **(a)** Moving a slack rope up and down rhythmically will generate a waveform with a characteristic wavelength. **(b)** When thrown against a rope, an object with a diameter comparable to the wavelength of the rope (e.g., a beach ball) will perturb the motion of the rope. **(c)** An object with a diameter significantly less than the wavelength of the rope (e.g., a softball) will probably cause little or no perturbation of the rope because, with its smaller diameter, it is not likely to strike the rope when tossed toward it. **(d)** If the rope is moved more rapidly, the wavelength will be reduced substantially. **(e)** A softball can now perturb the motion of the rope because its diameter is comparable to the wavelength of the rope.

And because an object can be detected only by its effect on the wave, the wavelength must be comparable in size to the object that is to be detected.

Once we understand this relationship between wavelength and object size, we can readily appreciate why very small objects can be seen only by electron microscopy: the wavelengths of electrons are very much shorter than those of photons. Thus, objects such as viruses and ribosomes are too small to perturb the waveform of photons, but they can readily interact with electrons.

Resolution Refers to the Ability to Distinguish Adjacent Objects as Separate from One Another

When waves of light or electrons pass through a lens and are focused, the image that is formed results from a property of waves called **interference**—the process by which two or more waves combine to reinforce or cancel one another, producing a wave equal to the sum of the two combining waves. Thus, the image you see when you look at a specimen through a series of lenses is really just a pattern of either additive or canceling interference of the waves that went through the lenses, a phenomenon known as **diffraction.**

In a light microscope, glass lenses are used to direct photons, whereas an electron microscope uses electromagnets as lenses to direct electrons. Yet both kinds of lenses have two fundamental properties in common: focal length and angular aperture. The **focal length** is the distance between the midline of the lens and the point at which rays passing through the lens converge to a focus (**Figure A-3**). The focal length is determined by the index of refraction of the lens itself, the medium in which it is immersed, and the geometry of the lens. The lens magnifying strength, measured in diopters, is the inverse of the focal length, measured in meters. The **angular aperture** is the half-angle α of the cone of light entering the objective lens of the microscope from the specimen (**Figure A-4**). Angular aperture is therefore a measure of how much of the illumination that leaves the specimen actually passes through the lens. This in turn determines the sharpness of the interference pattern and therefore the ability of the lens to convey information about the specimen. In the best light microscopes, the angular aperture is about 70°.

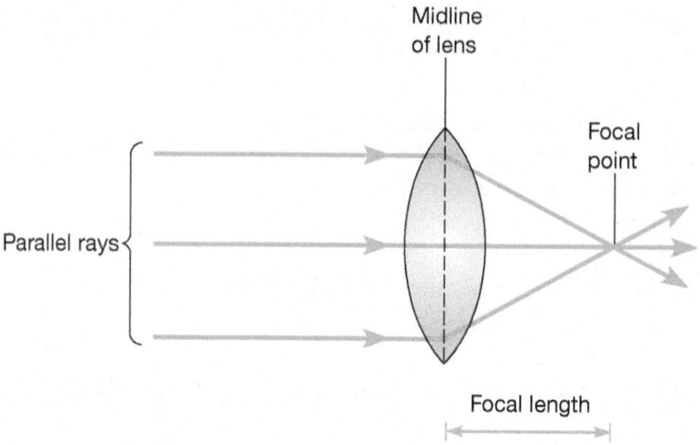

Figure A-3 The Focal Length of a Lens. Focal length is the distance from the midline of a lens to the point where parallel rays passing through the lens converge to a focus.

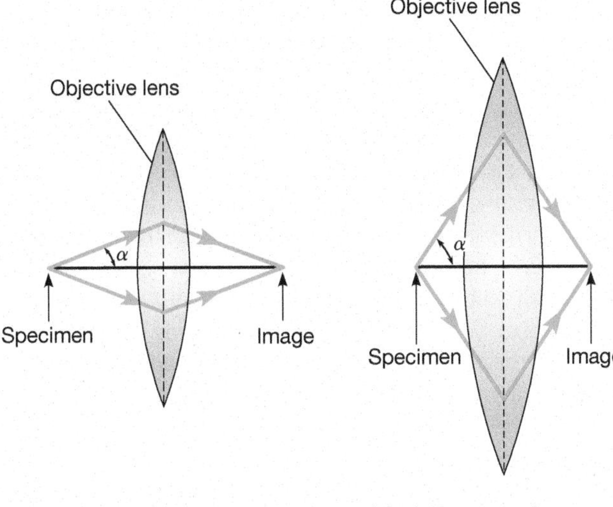

Figure A-4 The Angular Aperture of a Lens. The angular aperture is the half-angle α of the cone of light entering the objective lens of the microscope from the specimen. **(a)** A low-aperture lens (α is small). **(b)** A high-aperture lens (α is large). The larger the angular aperture, the more information the lens can transmit. The best glass lenses have an angular aperture of about 70°.

The angular aperture of a lens is one of the factors influencing a microscope's **resolution**, which is defined as the minimum distance two points can be apart and yet still remain identifiable as separate points when viewed through the microscope: the higher the resolution, the smaller the objects that can be distinguished using a particular lens.

Resolution is governed by three factors: the wavelength of the light used to illuminate the specimen, the angular aperture, and the refractive index of the medium surrounding the specimen. (**Refractive index** is a measure of the change in the velocity of light as it passes from one medium to another.) The effect of these three variables on resolution is described quantitatively by an equation known as the *Abbé equation:*

$$r = \frac{0.61\lambda}{n \sin \alpha} \quad \textbf{(A-1)}$$

where r is the resolution, λ is the wavelength of the light used for illumination, n is the refractive index of the medium between the specimen and the objective lens of the microscope, and α is the angular aperture as previously defined. The constant 0.61 represents the degree to which image points can overlap and still be recognized as separate points by an observer.

In the preceding equation, the quantity $n \sin \alpha$ is called the **numerical aperture** of the objective lens, abbreviated **NA**. An alternative expression for resolution is therefore

$$r = \frac{0.61\lambda}{\text{NA}} \quad \textbf{(A-2)}$$

The Practical Limit of Resolution Is Roughly 200 nm for Standard Light Microscopes and 2 nm for Electron Microscopes

Maximizing resolution is an important goal in both light and electron microscopy. Because r is a measure of how close two points can be and still be distinguished from each other, resolution improves as r becomes smaller. Thus, for the best resolution, the numerator of Equation A-2 should be as small as possible and the denominator should be as large as possible.

Consider a glass lens that uses visible light as an illumination source. First, we need to make the numerator as small as possible. The wavelength for visible light falls in the range of 400–700 nm, so the minimum value for λ is set by the shortest wavelength in this range that is practical to use for illumination, which turns out to be blue light of approximately 450 nm. To maximize the denominator of Equation A-2 on this page, recall that the numerical aperture is the product of the refractive index and the sine of the angular aperture. Both of these values must therefore be maximized to achieve optimal resolution. Because the angular aperture for the best objective lenses is approximately 70°, the maximum value for sin α is about 0.94. The refractive index of air is approximately 1.0, so for a lens designed for use in air, the maximum numerical aperture is approximately 0.94. Putting all of these numbers together, the resolution in air for a sample illuminated with blue light of 450 nm can be calculated as follows:

$$\begin{aligned} r &= \frac{0.61\lambda}{\text{NA}} \\ &= \frac{(0.61)(450)}{0.94} = 292\,\text{nm} \end{aligned} \quad \textbf{(A-3)}$$

As a rule of thumb, then, the limit of resolution for a glass lens in air is roughly 300 nm.

As a means of increasing the numerical aperture, some microscope lenses are designed to be used with a layer of *immersion oil* between the lens and the specimen. Immersion oil has a higher refractive index than air and therefore allows the lens to receive more of the light transmitted through the specimen. Because the refractive index of immersion oil is about 1.5, the maximum numerical aperture for an oil immersion lens is about $1.5 \times 0.94 = 1.4$. If we perform the same calculations as before, we find that the resolution of an oil immersion lens is approximately 200 nm. Thus, the **limit of resolution** (best possible resolution) for a microscope that uses visible light is roughly 300 nm in air and 200 nm with an oil immersion lens. In actual practice, such limits can rarely be reached because of aberrations (technical flaws) in the lenses. By using ultraviolet light as an illumination source, the resolution can be enhanced to approximately 100 nm because of the shorter wavelength (200–300 nm) of this type of light. However, special cameras must be used because ultraviolet light is invisible to the human eye. Moreover, ordinary glass is opaque to ultraviolet light, so expensive quartz lenses must be used.

Because the limit of resolution measures the ability of a lens to distinguish between two objects that are close together, it sets an upper boundary on the **useful magnification** that is possible with any given lens. In practice, the greatest useful magnification that can be achieved with a light microscope is approximately 1000 times the numerical aperture of the lens being used. Because numerical aperture ranges from approximately 1.0 to 1.4, this means that the useful magnification of a light microscope is limited to roughly 1000 × in air and 1400 × with immersion oil. Magnification greater than these limits is referred to as "empty magnification" because it provides no additional information about the object being studied.

The most effective way to achieve better magnification is to switch from visible light to electrons as the illumination source. Because the wavelength of an electron is approximately 100,000 times shorter than that of a photon of visible light, the theoretical limit of resolution of the electron microscope (0.002 nm) is orders of magnitude better than that of the light microscope (200 nm). However, practical problems in the design of the electromagnetic lenses used to focus the electron beam prevent the electron microscope from achieving this theoretical potential. The main problem is that electromagnets produce considerable distortion when the angular aperture is more than a few tenths of a degree. This tiny angle is several orders of magnitude less than that of a good glass lens (about 70°), giving the electron microscope a numerical aperture considerably smaller than that of the light microscope. The limit of resolution for good, standard electron microscopes is therefore only approximately 0.2 nm, far from the theoretical limit of 0.002 nm. Moreover, when viewing biological samples, problems with specimen preparation and contrast are such that the practical limit of resolution is often closer to 2 nm. Practically speaking, therefore, resolution in a routine electron microscope is generally about 100 times better than that of the light microscope. As a result, the useful magnification of an electron microscope is approximately 100 times that of a light microscope, or approximately 100,000 ×.

The Light Microscope

It was the light microscope that first opened our eyes to the existence of cells. A pioneering name in the history of light microscopy is that of Antonie van Leeuwenhoek, the Dutch shopkeeper who is generally regarded as the father of light microscopy. Leeuwenhoek's lenses, which he manufactured himself during the late 1600s, were of surprisingly high quality for his time. They were capable of 300-fold magnification—a tenfold improvement over previous instruments. This improved magnification made the interior of cells visible for the first time, and Leeuwenhoek's observations over a period of more than 25 years led to the discovery of cells in various types of biological specimens and set the stage for the formulation of the cell theory.

Compound Microscopes Use Several Lenses in Combination

In the centuries since Leeuwenhoek's pioneering work, considerable advances in the construction and application of light microscopes have been made. Today, the instrument of choice for light microscopy uses several lenses in combination and is therefore called a **compound microscope** (**Figure A-5**). The optical path through a compound microscope, illustrated in Figure A-5b, begins with a source of illumination, often a light source located in the base of the instrument. The light from the source first passes through **condenser lenses**, which direct the light toward a specimen mounted on a glass slide and positioned on the **stage** of the microscope. The **objective lens**, located immediately above the specimen, is responsible for forming the *primary image*. Most compound microscopes have several objective lenses of differing magnifications mounted on a rotatable turret.

The primary image is further enlarged by the **ocular lens**, or *eyepiece*. In some microscopes, an **intermediate lens** is positioned between the objective and ocular lenses to accomplish still further enlargement. Overall magnification of the image can be calculated by multiplying the enlarging powers of the objective lens, the ocular lens, and the intermediate lens (if present). Thus, a microscope with a 10 × objective lens, a 2.5 × intermediate lens, and a 10 × ocular lens will magnify a specimen 250-fold.

The elements of the microscope described so far enable a basic form of light microscopy called **brightfield microscopy**. Compared with other microscopes, the brightfield microscope is inexpensive and simple to align and use. However, the only specimens that can be seen directly by brightfield microscopy are those possessing color or some other property that affects the amount of light that passes through. Many biological specimens lack these characteristics and must therefore be stained with dyes or examined with specialized types of light microscopes. These special microscopes have various advantages that make them especially well-suited for visualizing specific types of specimens. These include phase-contrast, differential interference contrast, fluorescence, and confocal microscopes.

Phase-Contrast Microscopy Detects Differences in Refractive Index and Thickness

As we will describe in more detail later, cells are often killed, sliced into thin sections, and stained before being examined by brightfield microscopy. Although such procedures are useful for visualizing the details of a cell's internal architecture, little can be learned about the dynamic aspects of cell behavior by examining cells processed in this way. Therefore, various techniques have been developed to observe cells that are intact and, in many cases, still living. One such technique, **phase-contrast microscopy**, improves contrast without sectioning and staining by exploiting differences in the thickness and refractive index of various regions of the cells being examined.

To understand the basis of phase-contrast microscopy, we must first recognize that a beam of light is made up of many individual rays of light. As the rays pass from the light source through the specimen, their velocity may be affected by the physical properties of the specimen. Usually, the velocity of the rays is slowed down to varying extents by different regions of the specimen, resulting in a change in phase relative to light waves that have not passed through the specimen. (Light waves are said to be traveling *in phase* when the crests and troughs of the waves match each other.)

Although the human eye cannot detect such phase changes directly, the phase-contrast microscope overcomes this problem by converting phase differences into alterations in brightness. This conversion is accomplished using a *phase plate* (**Figure A-6** on page A-7), which is an optical material inserted into the light path above the objective lens. On aver-

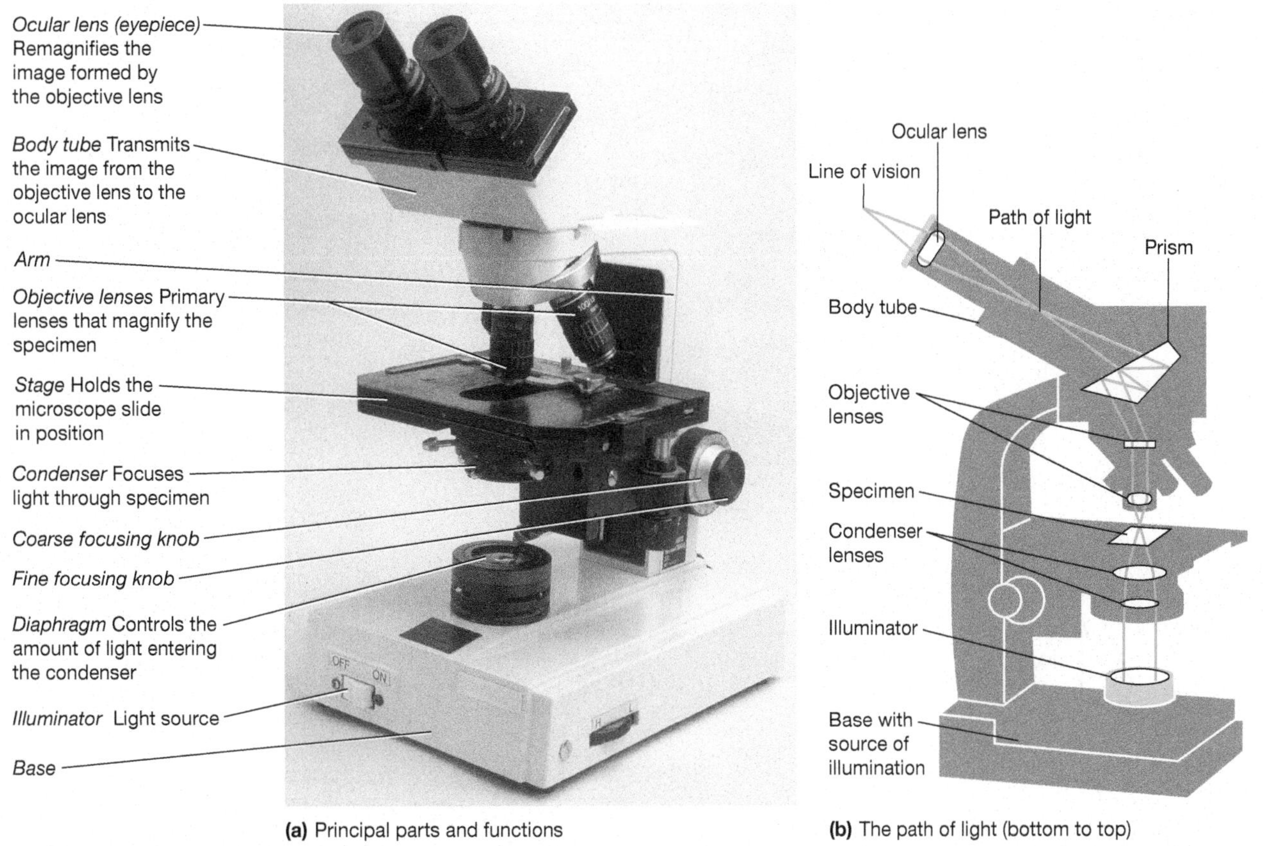

Figure A-5 The Compound Light Microscope. (a) A compound light microscope. **(b)** The path of light through the compound microscope.

age, light passing through transparent specimens is retarded by approximately $\frac{1}{4}$ wavelength. The direct, undiffracted light passes through a portion of the phase plate that speeds it up by approximately $\frac{1}{4}$ wavelength. Now the two types of light interfere with one another, producing an image with highly contrasting bright and dark areas against an evenly illuminated background (**Figure A-7** on page A-7). As a result, internal structures of cells are often better visualized by phase-contrast microscopy than with brightfield optics. Phase-contrast microscopy is widely used in microbiology and tissue culture research to detect bacteria, cellular organelles, and other small entities in living specimens.

Differential Interference Contrast (DIC) Microscopy Utilizes a Split Light Beam to Detect Phase Differences

Differential interference contrast (DIC) microscopy, or *Nomarski* microscopy (named for its inventor), resembles phase-contrast microscopy in principle but is more sensitive because it employs a special prism to split the illuminating light beam into two separate rays (**Figure A-8** on page A-7). When the two beams are recombined, any changes that occurred in the phase of one beam as it passed through the specimen cause it to interfere with the second beam.

The optical components required for DIC microscopy consist of a *polarizer*, an *analyzer*, and a pair of *Wollaston prisms*. The polarizer and the first Wollaston prism split a beam of light, creating two beams that are separated by a small distance along one direction. After traveling through the specimen, the beams are recombined by the second Wollaston prism. If no specimen is present, the beams recombine to form one beam that is identical to the one that initially entered the polarizer and first Wollaston prism. In the presence of a specimen, the two beams do not recombine in the same way (that is, they interfere with each other), and the resulting beam's polarization becomes rotated slightly compared with the original.

The net effect makes this technique especially useful for studying living, unstained specimens. Because the largest phase changes usually occur at cell edges (the refractive index is more constant within the cell), the outline of the cell typically gives a strong signal. The image appears three-dimensional due to a shadow-casting illusion that arises because differences in phase are positive on one side of the cell but negative on the opposite side (**Figure A-9** on page A-7).

Other contrast enhancement methods are also used by cell biologists. *Hoffman modulation contrast* increases contrast by detecting optical gradients across a transparent specimen using special filters and a rotating polarizer. Hoffman modulation contrast results in a shadow-casting effect similar to that in DIC microscopy.

Digital Microscopy Can Enhance Captured Images

The advent of solid-state light detectors made it possible to replace photographic film with an electronic equivalent—the technique of **digital microscopy**. Microscopic images are

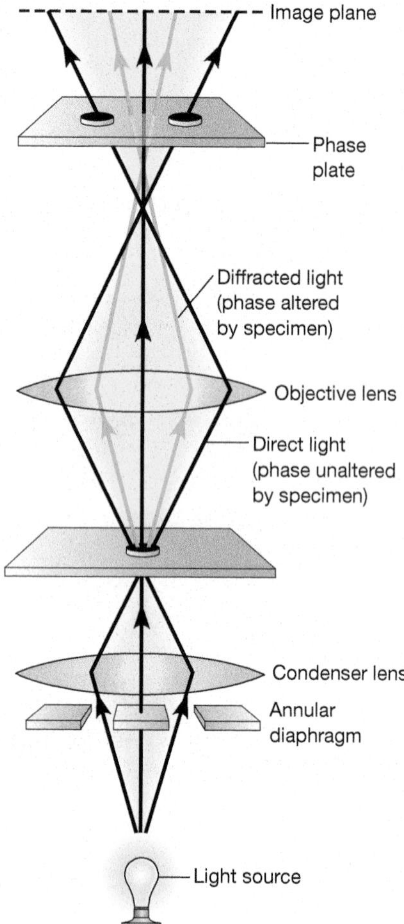

Figure A-6 Optics of the Phase-Contrast Microscope. Configuration of the optical elements and the paths of light rays through the phase-contrast microscope. Orange lines represent light diffracted by the specimen, and black lines represent direct light.

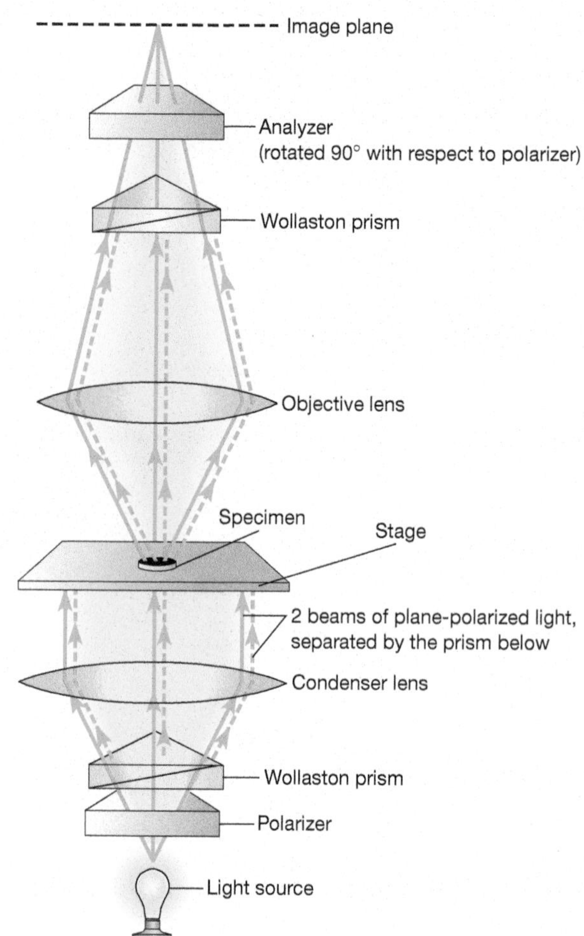

Figure A-8 Optics of the Differential Interference Contrast (DIC) Microscope. Configuration of the optical elements and the paths of light rays through the DIC microscope.

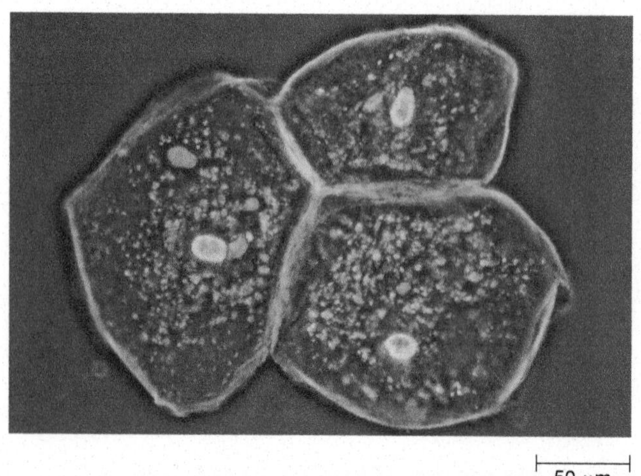

Figure A-7 Phase-Contrast Microscopy. A phase-contrast micrograph of epithelial cells. The cells were observed in an unprocessed and unstained state, which is a major advantage of phase-contrast microscopy.

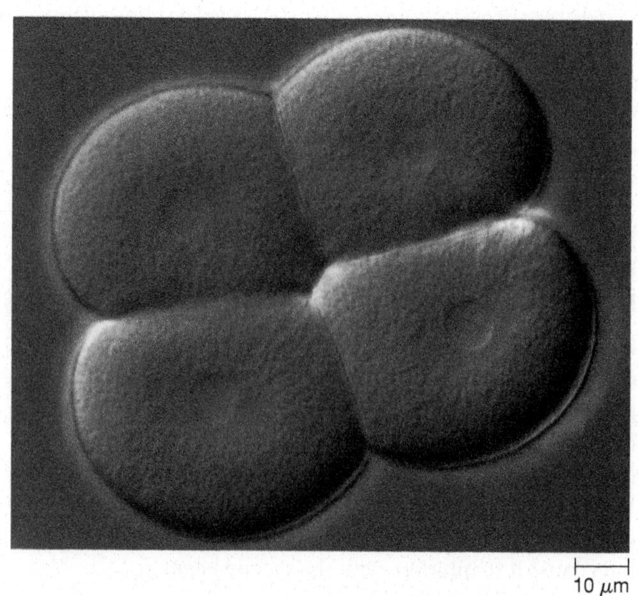

Figure A-9 DIC Microscopy. A DIC micrograph of a four-cell sea urchin embryo. Notice the shadow-casting effect that makes these cells appear dark at the bottom and light at the top.

recorded and stored electronically by placing a video or digital camera in the image plane produced by the ocular lens. The resulting digital images can then be *enhanced* by a computer to increase contrast and remove background features that obscure the image of interest. The resulting enhancement allows the visualization of structures that are an order of magnitude smaller than those that can be seen with a conventional light microscope (**Figure A-10**). Digital techniques can be applied to conventional brightfield light microscopy, as well as to DIC and fluorescence images, thereby creating a powerful set of approaches for improving the effectiveness of light microscopy.

Along with advances in computing technology, special, sensitive cameras have been developed that can detect extremely dim images, thereby facilitating the ability to record a rapid series of time-lapse pictures of cellular events as they proceed.

Digital microscopy is not only useful for examining events in one focal plane. In a variation of this technique, a computer is used to control a focus motor attached to a microscope. Images are then collected throughout the thickness of a specimen. When a series of images is collected at specific time intervals, such microscopy is called *four-dimensional microscopy* (this phrase is borrowed from physics; the four dimensions are the three dimensions of space plus the additional dimension of time). Analyzing four-dimensional data requires special computer software that can navigate between focal planes over time to display specific images.

Fluorescence Microscopy Can Detect the Presence of Specific Molecules or Ions Within Cells

Although the microscopic techniques described so far are quite effective for visualizing cell structures, they provide relatively little information concerning the location of specific molecules. One way of obtaining such information is through the use of **fluorescence microscopy**. To understand how fluorescence microscopy works, it is first necessary to understand the phenomenon of fluorescence.

The Nature of Fluorescence. The term **fluorescence** refers to a process that begins with the absorption of light by a molecule and ends with emission of light with a longer wavelength. This phenomenon is best approached by considering the quantum behavior of light, as opposed to its wave-like behavior **Figure A-11.** Figure A-11a on page A-9 is a diagram of the various energy levels of a simple atom. When an atom absorbs a photon (or *quantum*) of light of the proper energy, one of its electrons jumps from its ground state to a higher-energy, or *excited*, state. Eventually, this electron loses some of its energy and drops back down to the original ground state, emitting another photon as it does so. The emitted photon is always of lower energy (longer wavelength) than the original photon that was absorbed. Thus, for example, shining blue light on the atom may result in green light being emitted. (The energy of a photon is inversely proportional to its wavelength; therefore, green light, being longer in wavelength than blue light, is lower in energy.)

Real fluorescent molecules have energy diagrams that are more complicated than that depicted in Figure A-11a. The number of possible energy levels in real molecules is much greater, so the different energies that can be absorbed and emitted are correspondingly greater. Every fluorescent molecule has its own characteristic absorption and emission spectra. The absorption and emission spectra of a typical fluorescent molecule are shown in Figure A-11b.

The Fluorescence Microscope. Fluorescence microscopy is a specialized type of light microscopy that employs light to excite fluorescence in the specimen. A standard fluorescence microscope has an *excitation filter* between the light source and the rest of the light path that transmits only light of a particular wavelength (**Figure A-12** on page A-9). A *dichroic mirror*, which reflects light below a certain wavelength and transmits light above a certain wavelength, deflects the incoming light toward the objective lens, which focuses the light onto the specimen (because the light is routed through the objective lens onto the specimen in this way, this technique is often called *epifluorescence microscopy*, from the Greek *epi-*, "upon"). The incoming light causes fluorescent compounds in the specimen to emit

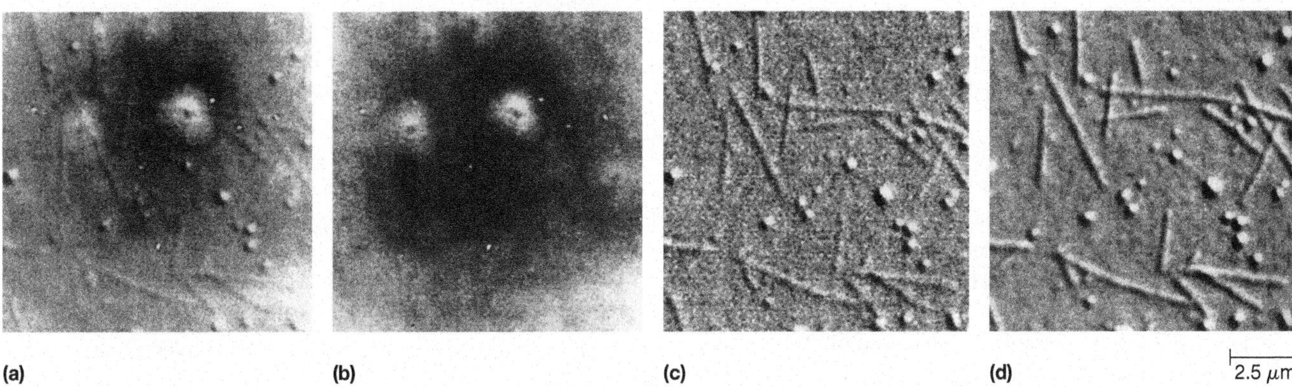

Figure A-10 Computer-Enhanced Digital Microscopy. This classic series of micrographs shows how computers can be used to enhance images obtained with light microscopy. In this example, an image of several microtubules—too small to be seen with unenhanced light microscopy—are processed to make them visible in detail. **(a)** The image resulting from electronic contrast enhancement of the original image (which appeared to be empty). **(b)** The background of the enhanced image in part a, which is then **(c)** subtracted from image a, leaving only the microtubules. **(d)** The final, detailed image resulting from electronic averaging of the separate images processed as shown in parts a–c.

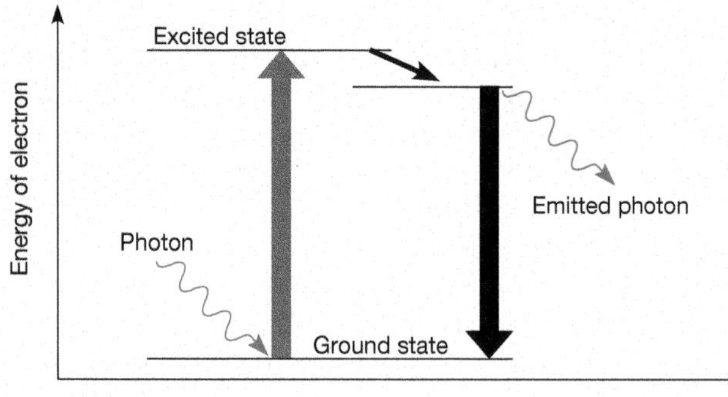

(a) Energy diagram

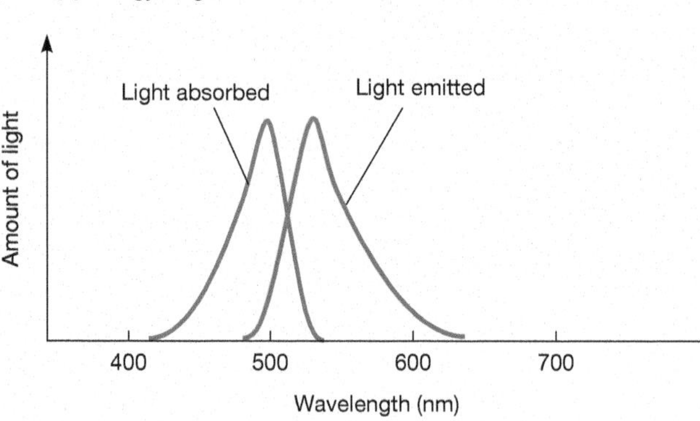

(b) Absorption and emission spectra

Figure A-11 Principles of Fluorescence. (a) An energy diagram of fluorescence from a simple atom. Light of a certain energy is absorbed (in this case blue light). The electron jumps from its ground state to an excited state. It returns to the ground state by emitting a photon of lower energy and hence longer wavelength (in this case green light). **(b)** The absorption and emission spectra of a typical fluorescent molecule. The blue curve represents the amount of light absorbed as a function of wavelength, and the green curve shows the amount of emitted light as a function of wavelength.

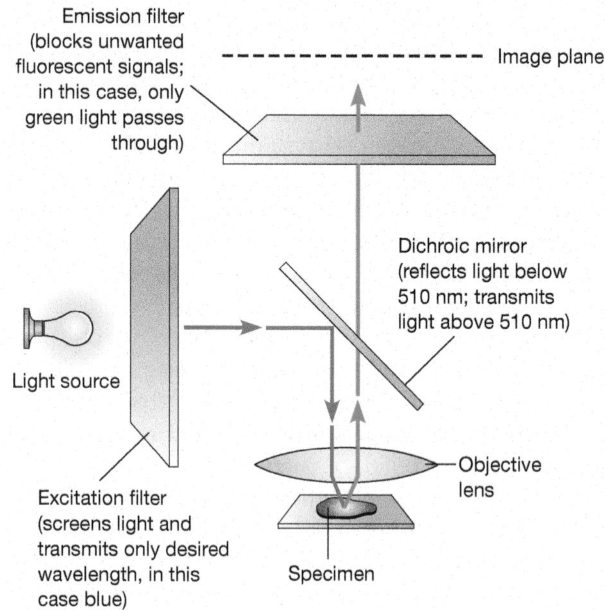

Figure A-12 Optics of the Fluorescence Microscope. Configuration of the optical elements and the paths of light rays through the fluorescence microscope. Light from the source passes through an excitation filter that transmits only excitation light (solid blue lines). Illumination of the specimen with this light induces fluorescent molecules in the specimen to emit longer-wavelength light (green lines). The emission filter subsequently removes the excitation light, while allowing passage of the emitted light. The image is therefore formed exclusively by light emitted by fluorescent molecules in the specimen.

light of longer wavelength, which passes back through the objective lens. The longer-wavelength light, instead of being reflected, passes through the dichroic mirror and encounters an *emission filter* that specifically prevents light that does not match the emission wavelength of the fluorescent molecules of interest from exiting the microscope. This leaves only the emission wavelength to form the final fluorescent image, which therefore appears bright against a dark background.

Fluorescent Antibodies. To use fluorescence microscopy for locating specific molecules or ions within cells, researchers must employ special indicator molecules called *fluorescent probes.* A fluorescent probe is a molecule capable of emitting fluorescent light that can be used to indicate the presence of a specific molecule or ion.

One of the most common applications of fluorescent probes is in **immunostaining**, a technique based on the ability of antibodies to recognize and bind to specific molecules (see Key Technique in Chapter 1, pages 8–9). Antibodies can be generated in the laboratory by injecting a foreign protein or other macromolecule into an animal such as a rabbit or mouse. In this way, it is possible to produce antibodies that will bind selectively to virtually any protein that a scientist wishes to study. Antibodies are not directly visible using light microscopy, however, so they are linked to a fluorescent dye such as *fluorescein,* which emits green light, or to *rhodamine,* which emits red light. More recently, antibodies have been linked to *quantum dots*—tiny, light-emitting crystals that are chemically more stable than traditional dyes and can be tuned to very specific wavelengths. To identify the subcellular location of a specific protein, cells are simply stained with a fluorescent antibody directed against that protein. The location of the fluorescence is then detected by viewing the cells with light of the appropriate wavelength.

Immunofluorescence microscopy is more commonly performed using **indirect immunofluorescence** (see Key Technique in Chapter 1, pages 8–9). In indirect immunofluorescence, a tissue or cell is treated with an antibody that is not labeled with dye. This antibody, called the *primary antibody,* attaches to specific antigenic sites within the tissue or cell. A second type of antibody, called the *secondary antibody,* is then added. The secondary antibody is labeled with a fluorescent dye, and it attaches to the primary antibody. Because more than one primary antibody molecule can attach to an antigen and more than one secondary antibody molecule can attach to each primary antibody, more fluorescent molecules are concentrated near each molecule being detected. As a result, indirect immunofluorescence is much more sensitive than the use of a primary antibody alone. The method is "indirect" because it does not examine where antibodies are bound to antigens; technically, the fluorescence reflects where the secondary antibody is located. This, of course, provides an indirect measure of where the original molecule of interest is located.

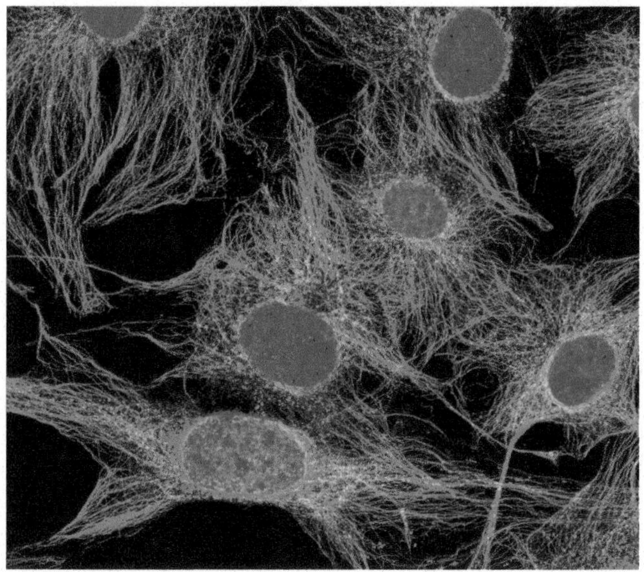

Figure A-13 Immunostaining Using Fluorescent Antibodies. Fibroblasts immunostained to detect microtubules (red), a dye that stains mitochondria (green), and DAPI (blue) for labeling DNA in nuclei.

By using different combinations of antibodies or other fluorescent probes, more than one molecule in a cell can be labeled at the same time. Different probes can be imaged using different combinations of fluorescent filters, and the different images can be combined to generate striking pictures of cellular structures. **Figure A-13** shows one example of this approach, in which fibroblast cells are labeled with antibodies that recognize microtubules (red) and stained for mitochondria (green) and DNA (blue).

Other Fluorescent Probes. Naturally occurring proteins that selectively bind to specific cell components are also used in fluorescence microscopy. For example, when fluorescently tagged, the mushroom toxin, *phalloidin,* binds to and specifically detects actin microfilaments. Another powerful fluorescence technique utilizes the *green fluorescent protein (GFP)* (see Key Technique in Chapter 19, pages 566–567). Using recombinant DNA techniques, scientists can fuse DNA encoding GFP to a gene coding for a particular cellular protein. The resulting recombinant DNA can then be introduced into cells, where it is expressed to produce a fluorescently tagged version of the normal cellular protein that can be observed in living cells (**Figure A-14**). Molecular biologists have produced mutated forms of GFP that absorb and emit light at a variety of wavelengths. Other naturally fluorescent proteins have also been identified from other organisms. These tools have expanded the repertoire of fluorescent molecules at the disposal of cell biologists.

Besides detecting macromolecules such as proteins, fluorescence microscopy can be used to monitor the subcellular distribution of various ions (see Key Technique in Chapter 23, page 702) or membrane voltage. To accomplish this task, chemists have synthesized molecules whose fluorescent properties are sensitive to the concentrations of ions such as Ca^{2+}, H^+, Na^+, Zn^{2+}, and Mg^{2+}, or to the electrical potential across the plasma membrane or the membranes of organelles. For example, a fluorescent probe called *fura-2* is commonly used to track the Ca^{2+} concentration inside living cells because fura-2 emits a yellow fluorescence in the presence of low concentrations of Ca^{2+} and a green and then blue fluorescence in the presence of progressively higher concentrations of this ion. Other probes for calcium involve engineered proteins. One family of such proteins is called GCaMPs; they involve a fusion of GFP, calmodulin, and M13, a peptide sequence from myosin light-chain kinase.

Confocal Microscopy Minimizes Blurring by Excluding Out-of-Focus Light from an Image

As powerful as it is, using conventional fluorescence microscopy to view intact tissues has its limitations. To overcome this problem, cell biologists often turn to the **confocal microscope**—a specialized type of light microscope that employs a laser beam to produce an image of a single plane of the specimen at a time (**Figure A-15**). Because fluorescence is emitted throughout the entire depth of the specimen, light emitted from regions of the specimen above and below the focal plane causes a blurring of the image (Figure A-15a on page A-11).

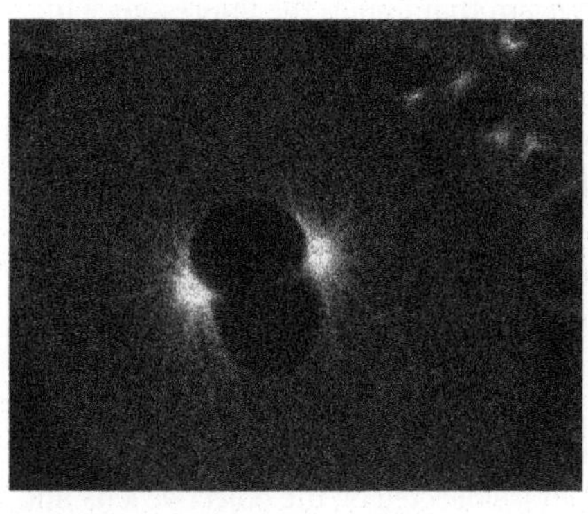

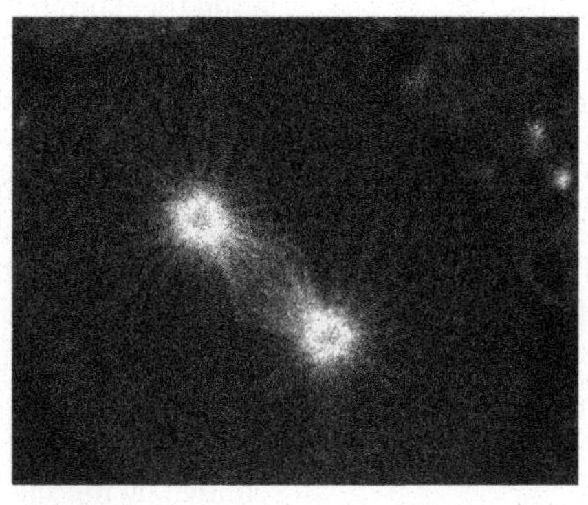

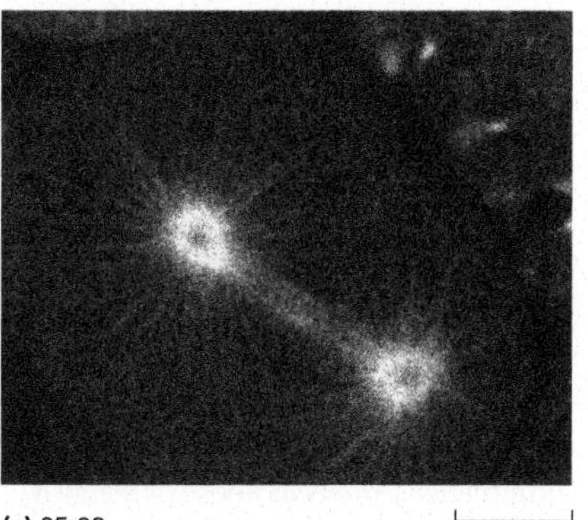

Figure A-14 Using Green Fluorescent Protein to Visualize Proteins. An image series of a living, one-cell nematode worm embryo undergoing mitosis. The embryo is expressing β-tubulin that is tagged with green fluorescent protein (GFP). Elapsed time from the first frame is shown in minutes:seconds.

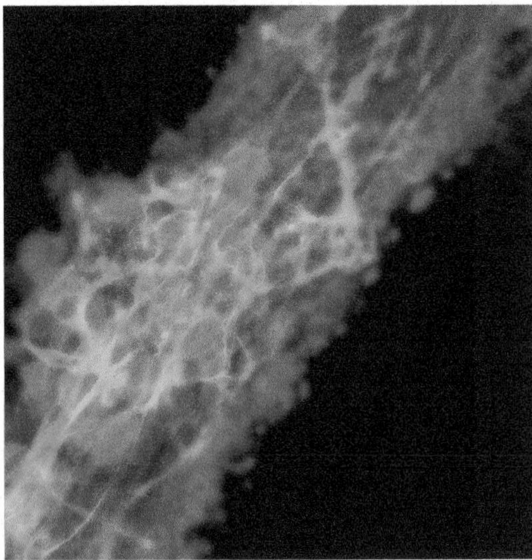

(a) Traditional fluorescence microscopy

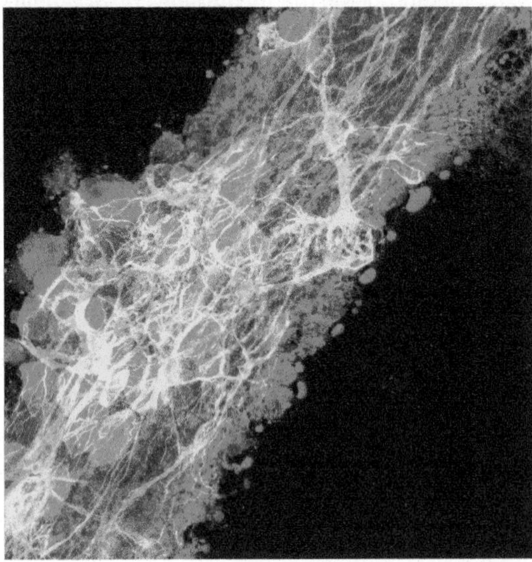

(b) Confocal fluorescence microscopy

Figure A-15 Comparison of Confocal Fluorescence Microscopy with Traditional Fluorescence Microscopy. These fluorescence micrographs show fluorescently labeled glial cells (red) and nerve cells (green) stained with two different fluorescent markers. **(a)** In traditional fluorescence microscopy, signal from the entire specimen is detected, so fluorescent material above and below the plane of focus tends to blur the image. **(b)** In confocal fluorescence microscopy, light emitted from a single plane is detected, and out-of-focus fluorescence from the specimen is excluded. The resulting image is therefore much sharper.

When the same specimen is imaged using a confocal microscope, the resolution along the optical axis of the microscope is greatly improved (Figure A-15b)—that is, structures in the middle of the tissue may be distinguished from those on the top or bottom. Likewise, a cell in the middle of a dense tissue can be distinguished from cells above or below it.

To understand this type of microscopy, it is first necessary to consider the paths of light taken through a simple lens. **Figure A-16** illustrates how a simple lens forms an image of a point source of light. To understand what your eye would see, imagine placing a digital camera in the plane of focus (image plane). Now ask how the images of other points of light placed farther away or closer to the lens contribute to the original image (Figure A-16b). As you might guess, there is a precise relationship between the distance of the object from the lens (o), the distance from the lens to the image of that object brought into focus (i), and the focal length of the lens (f). This relationship is given by the equation

$$\frac{1}{f} = \frac{1}{o} + \frac{1}{i} \tag{A-4}$$

As Figure A-16b shows, light arising from the points that are not in focus covers a greater surface area because the rays are still either converging or diverging. Thus, the image captured by the camera now has the original point source that is in focus, with a superimposed halo of light from the out-of-focus objects.

If we were interested only in seeing the original point source, we could mask out the extraneous light by placing an aperture, or *pinhole,* in the same plane as the camera. Exactly this principle is used in a confocal microscope to discriminate against out-of-focus rays. In a real specimen, of course, we have more than one extraneous source of light on each side of the object we wish to see; in fact, we have a continuum of points. To understand how this affects our image, imagine that instead of three points of light, our specimen consists of a long, thin tube of light, as in Figure A-16c. Now consider obtaining an image of some arbitrary small section dx. If the tube sends out the same amount of light per unit length, then even with a pinhole, the image of interest will be obscured by the halos arising from other parts of the tube. This occurs because there is a small contribution from each out-of-focus section, and the sheer number of small sections will create a large background over the section of interest.

This situation is very close to the one we face when dealing with real biological samples that have been stained with a fluorescent probe. In general, the distribution of the probe is three-dimensional, and the image is often marred by the halo of background light that arises mostly from probes above and below the plane of interest. To circumvent this, we can preferentially illuminate the plane of interest, thereby biasing the contributions in the image plane so that they arise mostly from a single plane (Figure A-16d). Thus, the essence of confocal microscopy is to bring the illumination beam that excites the fluorescence into sharp focus and to use a pinhole to ensure that the light we collect in the image plane arises mainly from that plane of focus.

Figure A-17 on page A-13 illustrates how these principles are put to work in a *laser scanning confocal microscope,* which illuminates specimens using a laser beam focused by an objective lens down to a diffraction-limited spot. The position of the spot is controlled by scanning mirrors, which allow the beam to be swept over the specimen in a precise pattern. As the beam is scanned over the specimen, an image of the specimen is formed in the following way. First, the fluorescent light emitted by the specimen is collected by the objective lens and returned along the same path as the original incoming light. The path of the fluorescent light is then separated from the laser light using a dichroic mirror, which reflects one color but transmits another. Because the fluorescent light has a longer wavelength than the excitation beam, the fluorescence color is shifted (for example, from blue to green). The fluorescent light passes through a pinhole placed at an image plane in front of a camera or photomultiplier tube, which acts as a detector. The signal from the detector is then digitized and displayed by a computer. To see the enhanced resolution that results from confocal microscopy, look back to Figure A-15, which shows images of the same cells visualized by conventional fluorescence microscopy and by laser scanning confocal microscopy.

As an alternative to laser scanning confocal microscopy, a *spinning disc confocal microscope* uses rapidly spinning, aligned discs containing a series of small lenses and a corresponding series of pinholes. Although it cannot produce optical sections as thin as those produced by laser scanning microscopes, it can generate confocal images that can be acquired rapidly using sensitive digital cameras. Such speed is useful for visualizing very rapid events within living cells.

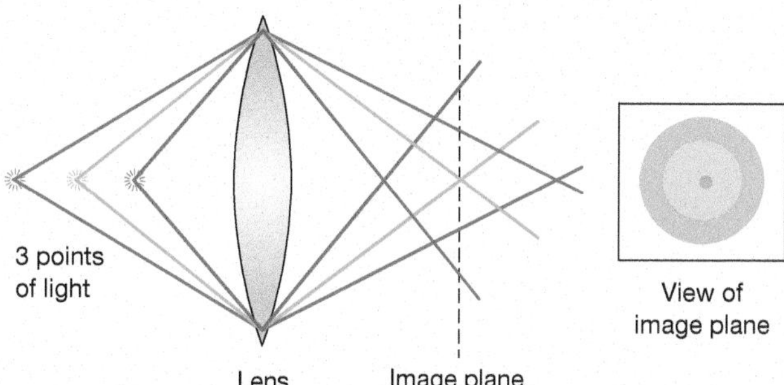

(a) Formation of an image of a single point of light by a lens

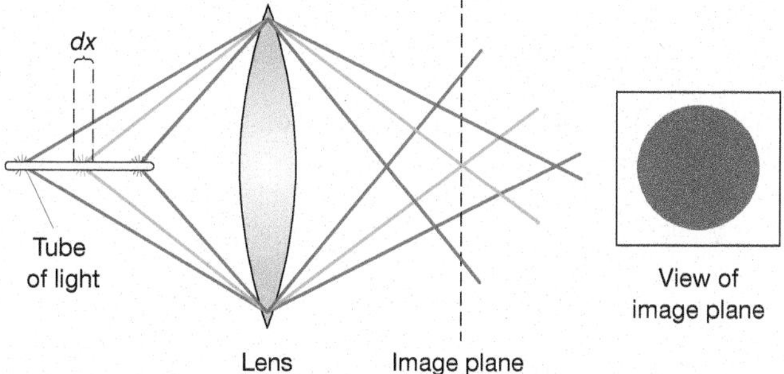

(b) Formation of an image of a point of light in the presence of two other points

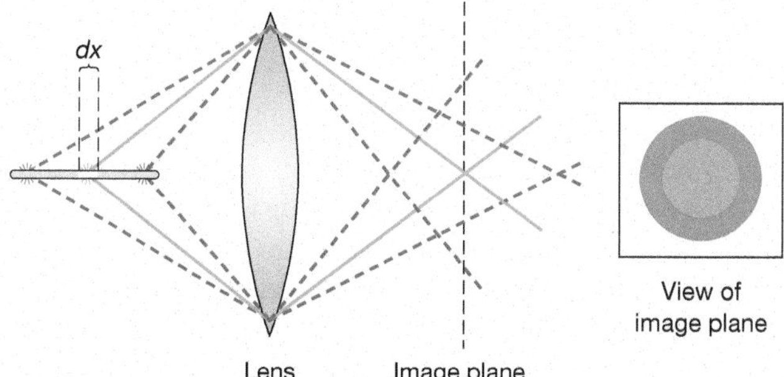

(c) Formation of an image of a section of an equally bright tube of light

(d) Formation of an image of a brightened section of a tube of light

Other Techniques Minimize Blurring by Exciting a Thin Strip of Fluorescent Molecules

In confocal microscopy, a pinhole is used to exclude out-of-focus light. The result is a sharp image, but molecules above and below the focal plane of the objective lens are still excited by the incoming light. This can result in rapid bleaching of the fluorescent molecules. In some cases, especially when viewing living cells that contain fluorescent molecules, such bleaching releases toxic radicals that can cause the cells to die. To reduce such "photodamage," it would be desirable if only the fluorescent molecules very close to the focal plane being examined were excited. There are several ways this can be achieved.

Light Sheet Microscopy. One method that is useful for observing living embryos for long periods of time is called **light sheet microscopy**. Unlike in standard epifluorescence microscopy, in which the illuminating light is routed through the same lens that will detect the emitted fluorescent light, in a light sheet microscope the illumination is brought in from the side, using a low-magnification objective lens (**Figure A-18a** on page A-13). The light sheet allows only a thin strip of the specimen to be excited, which reduces photodamage.

Multiphoton Excitation Microscopy. A second way to localize where fluorescent molecules are excited depends on the physics of light. This technique, known as **multiphoton excitation microscopy**, uses a laser that rapidly emits pulses of high-amplitude light to irradiate the specimen. When two (or in some cases, three or more) photons arrive at the specimen almost simultaneously, a consequence of the

Figure A-16 Paths of Light Through a Single Lens. (a) The image of a single point of light formed by a lens. **(b)** The paths of light from three points of light at different distances from the lens. In the image plane, the in-focus image of the central point is superimposed on the out-of-focus rays of the other points. A pinhole or aperture around the central point can be used to discriminate against out-of-focus rays and maximize the contributions from the central point. **(c)** The paths of light originating from a continuum of points, represented as a tube of light. This is similar to a uniformly illuminated sample. In the image plane, the contributions from an arbitrarily small in-focus section, *dx*, are completely obscured by the other out-of-focus rays; here a pinhole does not help. **(d)** By illuminating only a single section of the tube strongly and the rest weakly, we can recover information in the image plane about the section *dx*. Now a pinhole placed around the spot will reject out-of-focus rays. Because the rays in the middle are almost all from *dx*, we have a means of discriminating against the dimmer, out-of-focus points.

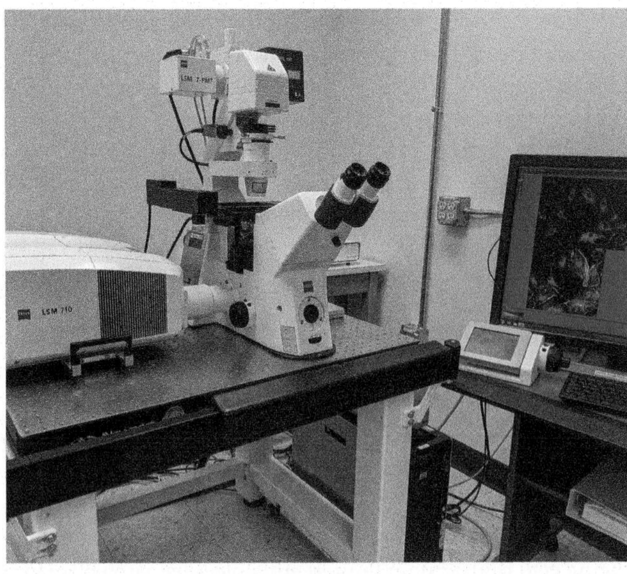

(a)

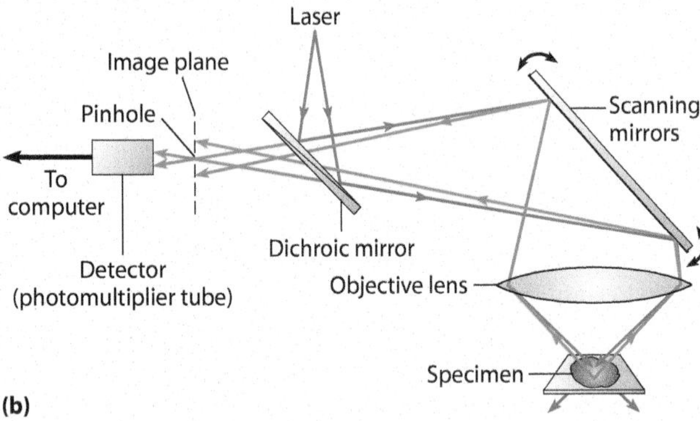

(b)

Figure A-17 A Laser Scanning Confocal Microscope. (a) A photograph and **(b)** a schematic of a laser scanning confocal microscope (LSCM). A laser is used to illuminate one spot at a time in the specimen (blue lines). Scanning mirrors move the spot in a given plane of focus through a precise pattern. The fluorescent light being emitted from the specimen (green lines) bounces off the same scanning mirrors and returns along the original path of the illumination beam. The emitted light does not return to the laser, but instead is transmitted through the dichroic mirror (which in this example reflects blue light but transmits green light). A pinhole in the image plane blocks the extraneous rays that are out of focus. The light is detected by a photomultiplier tube, whose signal is digitized and stored by a computer.

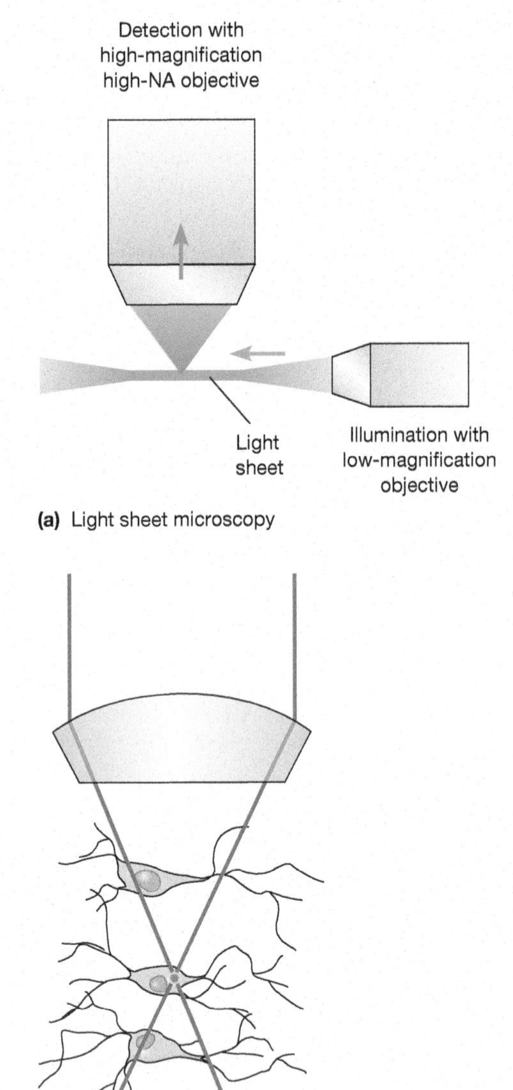

Figure A-18 Illuminating a Small Region of a Fluorescent Sample. (a) Light sheet microscopy. The illuminating light is brought in from the side using a low-magnification objective lens. The fluorescent light that is emitted from the specimen is detected using a high-magnification, high numerical aperture lens positioned at a right angle to the light sheet. **(b)** Multiphoton excitation microscopy. Fluorescence is limited to a spot at the focus of the pulsed laser beam. Another advantage is that the intense laser light penetrates more deeply into the specimen than in confocal microscopy.

physics of light is that these photons can combine to effectively approximate a photon of shorter wavelength. When this approximate wavelength is near the absorption wavelength for a fluorescent molecule, the fluorescent molecule absorbs the light and fluoresces (**Figure A-18b**). The likelihood of this happening is very low except near the focal plane of the objective lens. As a result, only the fluorescent molecules that are in focus fluoresce. The result is very similar in sharpness to confocal microscopy, but no pinhole is needed because there is no out-of-focus light that needs to be excluded. Photodamage is also dramatically reduced. As an example, the classic image series in Figure A-14 was captured using multiphoton excitation microscopy.

Total Internal Reflection Fluorescence (TIRF) Microscopy. An even more spatially precise method for observing fluorescent molecules is **total internal reflection fluorescence (TIRF) microscopy** (see Key Technique in Chapter 12, pages 336–337). TIRF relies on the production of a very thin layer of light, the "evanescent field," that extends about 100 nm into a specimen, making TIRF up to ten times better than a confocal microscope for resolving small fluorescent objects very close to the surface of a coverslip, such as secretory vesicles or polymerized cytoskeletal proteins.

Digital Deconvolution Microscopy Can Be Used to Generate Sharp Three-Dimensional Images After Acquisition

The various techniques we have discussed so far for generating crisp images of three-dimensional fluorescent objects use different physical principles to either exclude out-of-focus light (confocal microscopy) or illuminate a thin region of a specimen (light sheet, multiphoton, and TIRF microscopy). **Digital deconvolution microscopy** relies on a completely different principle. In this case, normal fluorescence microscopy is used to acquire a series of images throughout the thickness of a specimen. Then a computer is used to digitally process, or *deconvolve*, each focal plane to mathematically remove the contribution due to the out-of-focus light. In many cases, digital deconvolution can produce images comparable to those obtained by confocal microscopy (**Figure A-19**). One advantage of deconvolution is that a conventional fluorescence microscope can be used, which can be less expensive and more versatile in terms of wavelengths.

Optical Methods Can Be Used to Measure and Manipulate Macromolecules

Optical microscopy can be used not only to help visualize where molecules reside within cells but also to study the dynamic movements and properties of biological molecules. We briefly consider some modern techniques in this section.

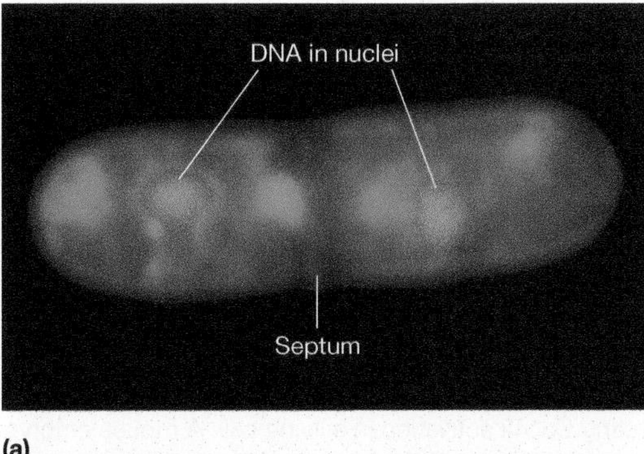

(a)

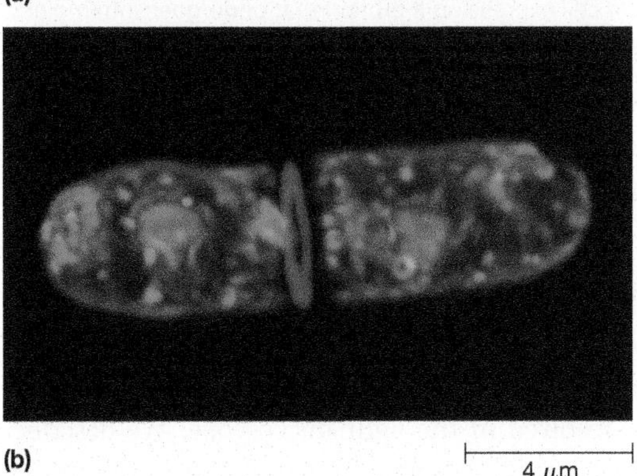

(b)

Figure A-19 Digital Deconvolution Microscopy. A fission yeast cell stained with a dye specific for DNA, an antibody that stains the dividing wall (septum) between cells (both red), and a membrane-specific dye (green). The image in **(a)** is an unprocessed optical section through the center of the cell, whereas the image in **(b)** is a projection of all the sections following three-dimensional image processing. The ring of the developing medial septum (red) is forming between the two nuclei (red) that arose by nuclear division during the previous mitosis.

Photobleaching, Photoactivation, and Photoconversion. When fluorescent molecules are irradiated with light at the appropriate excitation wavelength for long periods of time, they undergo **photobleaching** (that is, the irradiation induces the molecules to cease fluorescing). If a cell is exposed to intense light in a small region, such bleaching results in a characteristic decrease in fluorescence (**Figure A-20a**). As unbleached molecules move into the bleach zone, the fluorescence gradually returns to normal levels. Such *fluorescence recovery after photobleaching (FRAP)* is therefore one useful measure of how fast molecules diffuse or undergo directed transport (see Figure 7-A1 for a classic use of FRAP).

In other cases, fluorescent molecules are chemically modified so that they do not fluoresce until they are irradiated with a specific wavelength of light, usually ultraviolet light or short-wavelength light in the visible range. The light induces the release of the chemical modifier, allowing the molecule of interest to fluoresce. Such compounds are often called *caged compounds* because they are "freed" only by this light-induced cleavage. As an alternative, some forms of GFP only fluoresce after light of a particular wavelength is shined on them, so that they undergo **photoactivation**. Uncaging or photoactivation produces the converse situation to photobleaching: local uncaging of a fluorescent molecule produces a bright spot of fluorescence that can be followed as the molecules diffuse throughout the cell (**Figure A-20b**). Uncaging can also be used to convert other kinds of inert molecules to their active state. For example, "caged calcium" is actually a calcium chelator that is bound to calcium ions. When the caged compound is irradiated, it gives up its calcium, causing a local elevation of calcium within the cell. A third technique causes a fluorescent molecule to permanently change its fluorescence properties after it is exposed to ultraviolet or other short wavelengths of light. Such *photoconversion* can, for example, cause a green fluorescent protein to become a red fluorescent

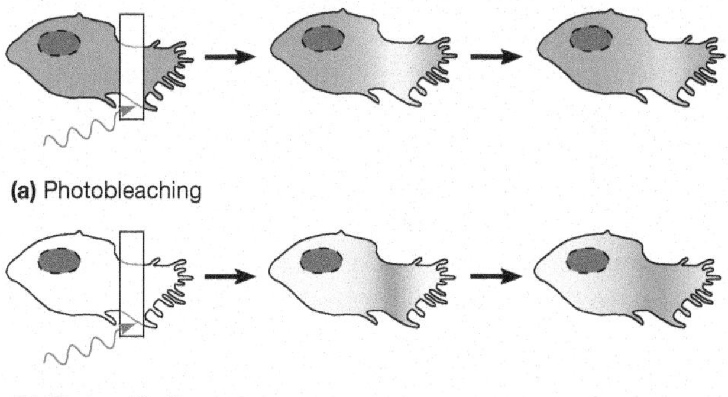

Figure A-20 Techniques for Studying Dynamic Movements of Molecules Within Cells. (a) Photobleaching of fluorescent molecules. A well-defined area of a cell is irradiated, causing bleaching of fluorescent molecules. Gradually, new, unbleached molecules invade the bleached zone. **(b)** Photoactivation of fluorescent proteins. Photoactivation of a selected region (indicated by the box) results in an increase in fluorescence. Subsequent movements of the fluorescent molecules can then be monitored.

protein and has uses that are similar to photoactivation. Other fluorescent molecules can be temporarily switched "off" so that they do not fluoresce and then back "on" again, a process called *photoswitching*.

Förster Resonance Energy Transfer (FRET). Fluorescence microscopy is useful not only for observing the movements of proteins but also for measuring their physical interactions with one another. When two fluorescent molecules whose fluorescence properties are matched are brought very close to one another, it is possible for them to experience **Förster resonance energy transfer (FRET; Figure A-21)**. In FRET, illuminating a cellular structure at the excitation wavelength of the first, or *donor*, fluorophore results in energy transfer to the second, or *acceptor*, fluorophore. This energy transfer does not itself involve a photon of light; however, once it occurs, it causes the acceptor to emit light at its characteristic wavelength. A commonly used pair of fluorescent molecules used in FRET is derived from GFP: one glows with bluish (cyan) fluorescence (so it is called *cyan fluorescent protein*, or *CFP*); the other is excited by bluish light and glows with a yellowish color (so it is called *yellow fluorescent protein*, or *YFP*). FRET can be measured between two separate proteins (*inter*molecular FRET; Figure A-21a) or between fluorescent side chains on the same protein (*intra*molecular FRET; Figure A-21b). FRET acts only at a very short range (100 Å or less); as a result, intermolecular FRET provides a readout regarding where within a cell the two fluorescent proteins are essentially touching one another. Intramolecular FRET is used in several kinds of molecular biosensors, including those that detect calcium elevation or local activation of small G proteins, such as Cdc42 (Figure A-21c).

"Optical Tweezers." Another application of light microscopy relies on well-established ways in which light interacts with small objects. When photons strike small objects, they exert "light pressure" on them. When a small object is highly curved, such as a small plastic bead, the differential forces exerted by a tightly focused laser beam channeled through a microscope tend to cause the bead to remain in the center of the beam, trapping it via light pressure. Such **optical tweezers** can be used to move objects, or they can be used to exert exceedingly small forces on beads that are attached to proteins or other molecules. By measuring the forces exerted on beads coupled to myosin, for example, scientists have been able to measure the forces produced by the power stroke of a single myosin protein. In one application of this technology, it is used to measure the forces exerted by single myosin motor proteins (see Key Technique in Chapter 14, pages 388–389).

Optogenetics. It is now becoming possible to use genetically engineered ion channel proteins that are sensitive to light to control the opening and closing of channels using a beam of light (see Chapter 22). This approach, known as *optogenetics*, uses molecular biology techniques to engineer light-sensitive ion channels. Once they are introduced into cells, exposing the cells to intense light of the appropriate wavelength induces a conformational change in the engineered protein, allowing a channel to be opened or closed on command. Other light-sensitive molecules are now being engineered to study

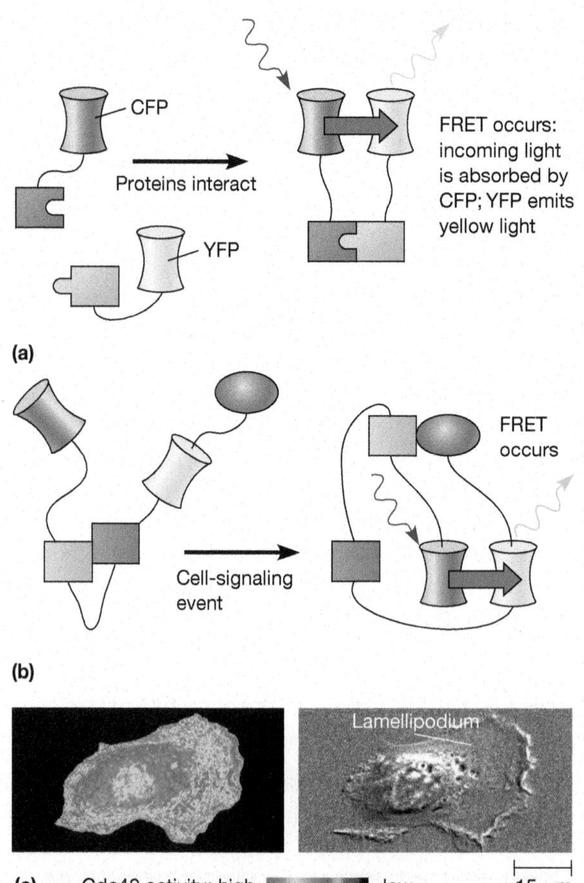

Figure A-21 Fluorescence Resonance Energy Transfer (FRET). **(a)** Intermolecular FRET. When two proteins attached to two different fluorescent side chains (such as cyan fluorescent protein, or CFP, and yellow fluorescent protein, or YFP) are in close proximity, the CFP can transfer energy directly to the YFP when the CFP is excited. The YFP then fluoresces. **(b)** Intramolecular FRET. In this case, CFP and YFP are attached to the same molecule, but they can interact only when the protein undergoes a conformational change. Such engineered proteins can be used as cellular "biosensors." **(c)** Detecting Cdc42 activation in a living cell. A mouse embryonic fibroblast cell expressing a protein that undergoes intramolecular FRET shows local regions of higher Cdc42 activity.

many cellular events. This technology is progressing rapidly, adding to the cell biologist's technical arsenal.

Superresolution Microscopy Has Broken the Diffraction Limit

The wavelength of visible light seems to place a limit on what can be resolved in the light microscope. We described this limit, due to diffraction, using the Abbé equation. Recent ingenious techniques, collectively called **superresolution microscopy** (**Figure A-22**), have broken this "diffraction barrier." One of these techniques is called *stimulated emission depletion (STED) microscopy*. STED uses very short pulses of laser light to cause molecules in a specimen to fluoresce. The first pulse is immediately followed by a ring-shaped "depletion" pulse, which causes stimulated emission, moving electrons from the excited state (which causes fluorescence to occur) to a lower energy state before they can fluoresce. This

Ultrathin Sectioning and Staining. After fixation the tissue is next passed through a series of alcohol solutions to dehydrate it, and then it is placed in a solvent such as acetone or propylene oxide to prepare it for embedding in liquefied plastic epoxy resin. After the plastic has infiltrated the specimen, it is put into a mold and heated in an oven to harden the plastic. The embedded specimen is then sliced into ultrathin sections by an instrument called an **ultramicrotome** (**Figure A-27**). The specimen is mounted firmly on the arm of the ultramicrotome, which advances the specimen in small increments toward a glass or diamond knife, which cuts sections from the block face. The sections float from the blade onto a water surface, where they can be picked up on a circular copper specimen grid. The grid consists of a meshwork of very thin copper strips, which support the specimen while still allowing openings between adjacent strips through which the specimen can be observed.

Once in place on the grid, the sections are usually stained again, this time with solutions containing lead and uranium. This step enhances the contrast of the specimen because the lead and uranium give still greater electron density to specific parts of the cell. After poststaining, the specimen is ready for viewing or photography with the TEM.

A recent, powerful technique called *serial block-face scanning electron microscopy* generates high-resolution three-dimensional images by mounting an ultramicrotome inside the vacuum chamber of a scanning electron microscope. Samples are prepared by methods similar to those used in TEM. The surface of the block is then imaged in the SEM. After the surface is imaged, the ultramicrotome cuts a thin section from the surface of the block. After the section is cut, the block is raised back into focus and imaged again. Images of serial sections can then be aligned and reconstructed using a computer.

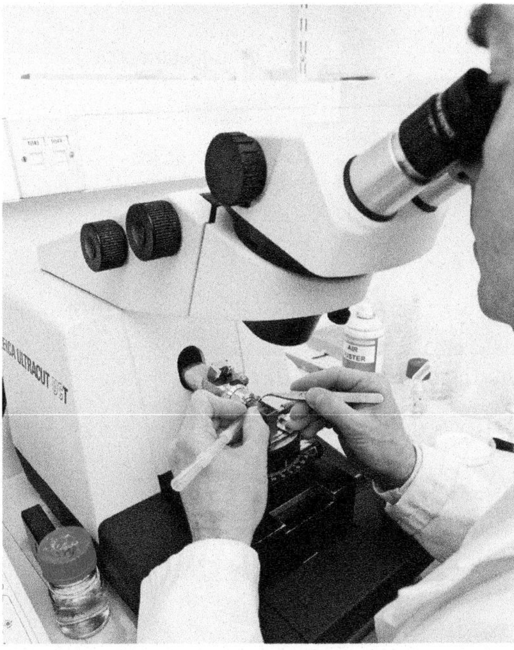

(a) Ultramicrotome

(b) Microtome arm of ultramicrotome

Figure A-27 An Ultramicrotome. (a) A photograph of an ultramicrotome. **(b)** A close-up view of the ultramicrotome arm, showing the specimen in a plastic block mounted on the end of the arm. As the ultramicrotome arm moves up and down, the block is advanced in small increments, and ultrathin sections are cut from the block face by the diamond knife.

Radioisotopes and Antibodies Can Localize Molecules in Electron Micrographs

In our discussion of light microscopy, we described how fluorescently labeled antibodies can be used in conjunction with light microscopy to locate specific cellular components. Antibodies are likewise used in the electron microscopic technique called **immunoelectron microscopy (immunoEM)**; fluorescence cannot be seen in the electron microscope, so antibodies are instead visualized by linking them to substances that are electron dense. One of the most common approaches is to couple antibody molecules to colloidal gold particles. When ultrathin tissue sections are stained with gold-labeled antibodies directed against various proteins, electron microscopy can reveal the subcellular location of these proteins with great precision (**Figure A-28**).

A powerful approach that unites light microscopy and immunoEM is **correlative microscopy**. In correlative microscopy, dynamic images of a cell are acquired using the light microscope. The very same cell is then processed and viewed

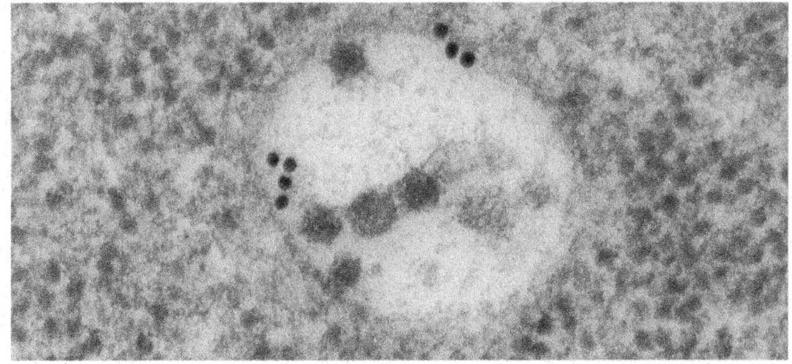

100 nm

Figure A-28 The Use of Gold-Labeled Antibodies in Electron Microscopy. Root cells of the plant *Arabidopsis thaliana* were stained with gold-labeled antibodies directed against the protein Rab5. The dark granules distributed at the periphery of a multivesicular endosome are gold particles conjugated to antibodies that recognize Rab5.

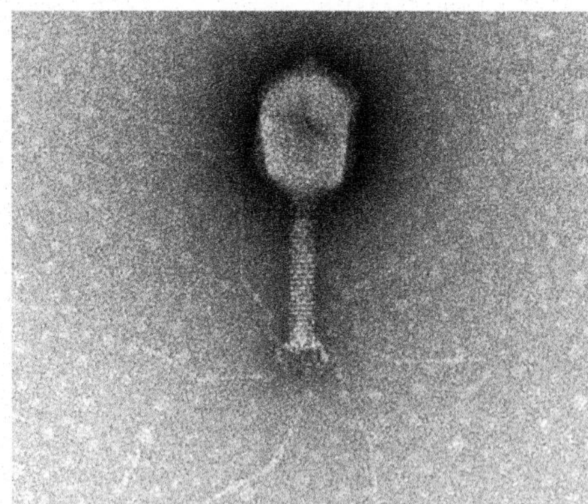

Figure A-29 Negative Staining. An electron micrograph of a bacteriophage as seen in a negatively stained preparation. This specimen was simply suspended in an electron-dense stain, allowing it to be visualized in relief against a darkly stained background (TEM).

using electron microscopy (EM). Commonly, immunoEM is used to determine where a protein is found at very high resolution. Correlative microscopy thus bridges the gap between dynamic imaging using the light microscope and the detailed images that can be acquired only via EM.

Negative Staining and Shadowing Can Highlight Small Objects in Relief

Although cutting tissues into ultrathin sections is the most common way of preparing specimens for transmission electron microscopy, other techniques are suitable for particular purposes. For example, the shape and surface appearance of very small objects, such as viruses or isolated organelles, can be examined without cutting the specimen into sections. In **negative staining**, a copper specimen grid must first be overlaid with an ultrathin plastic film. The specimen is then suspended in a small drop of liquid, applied to the overlay, and allowed to dry in air. After the specimen has dried on the grid, a drop of stain such as uranyl acetate or phosphotungstic acid is applied to the film surface. When viewed in the TEM, the specimen is seen in *negative contrast* because the background is dark and heavily stained, whereas the specimen itself is lightly stained (**Figure A-29**).

Isolated particles or macromolecules can also be visualized by the technique of **shadowing** (**Figure A-30**), which involves spraying a thin layer of an electron-dense metal such as gold or platinum at an angle across the surface of a biological specimen. As **Figure A-31** shows, shadowing can yield striking images.

1 The specimen is spread on a mica surface and dried.

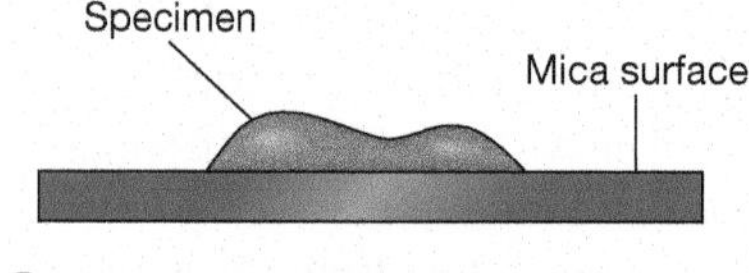

2 The specimen is shadowed by coating it with atoms of a heavy metal that are evaporated from a heated filament and applied at an angle of 10–45° relative to the specimen.

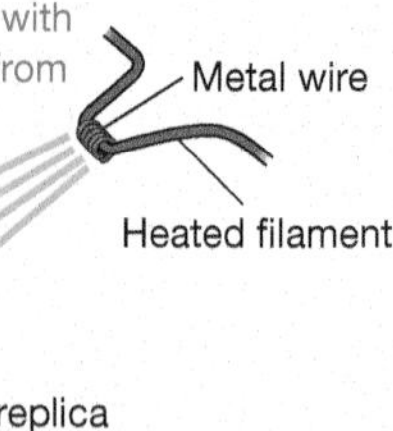

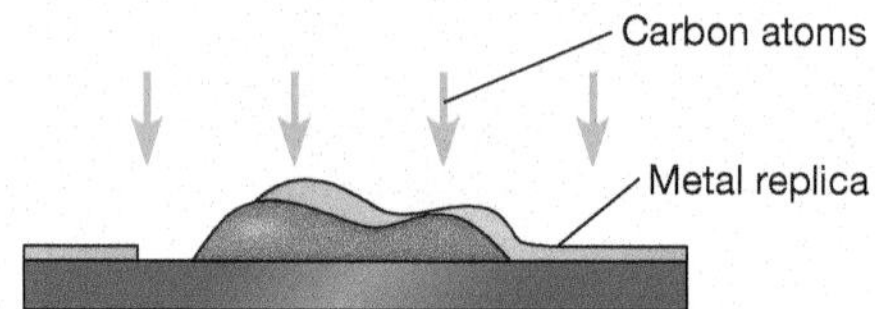

3 The specimen is coated with carbon atoms evaporated from an overhead electrode.

Carbon atoms

Metal replica

4 The replica is floated onto the surface of an acid bath to dissolve away the specimen, leaving a clean metal replica.

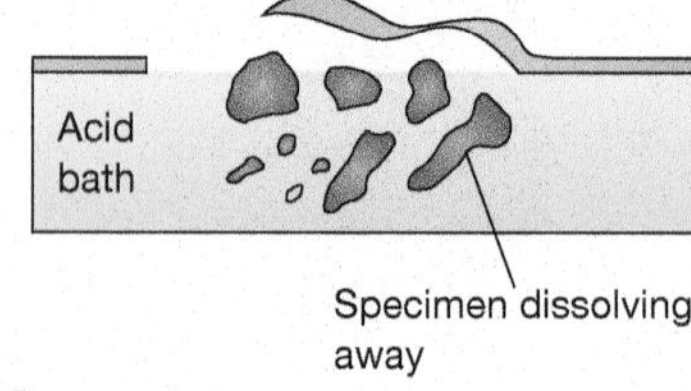

5 The replica is washed and picked up on a copper grid for examination in the TEM.

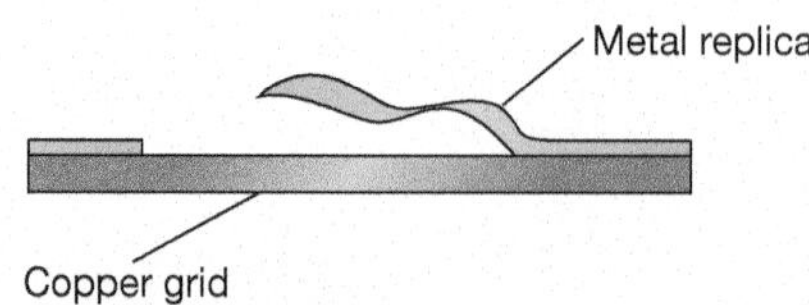

(a) Shadowing technique

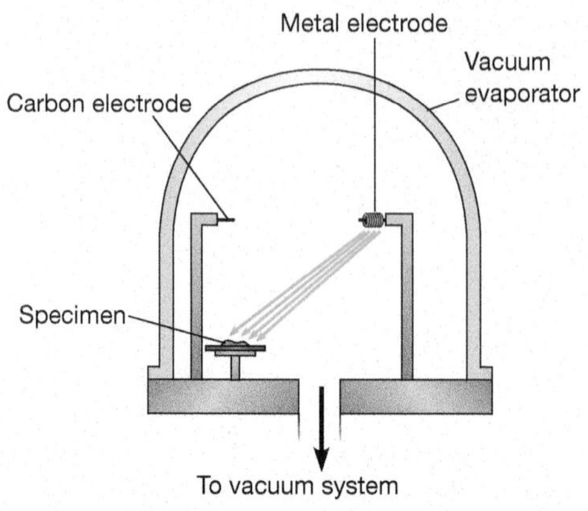

(b) Vacuum evaporator

Figure A-30 The Technique of Shadowing. (a) The specimen is spread on a mica surface and shadowed by coating it with atoms of a heavy metal (platinum or gold, shown in orange). This generates a metal replica (orange) whose thickness reflects the surface contours of the specimen. The replica is coated with carbon atoms to strengthen it, and it is then floated away and washed before viewing in the TEM. **(b)** The vacuum evaporator in which shadowing is done. The carbon electrode is located directly over the specimen; the heavy metal electrode is off to the side.

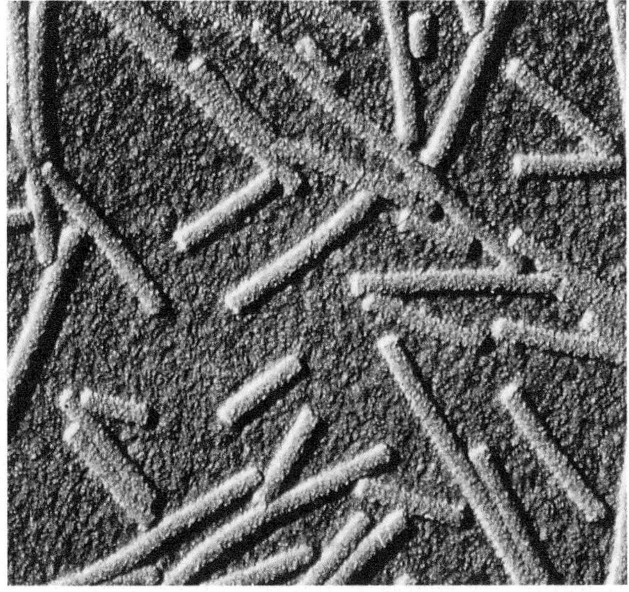
0.1 μm

Figure A-31 TMV Visualized Using Shadowing. An electron micrograph of tobacco mosaic virus particles visualized by shadowing. In this technique, heavy metal vapor was sprayed at an angle across the specimen, causing an accumulation of metal on one side of each virus particle and a shadow region lacking metal on the other side (TEM).

Freeze Fracturing and Freeze Etching Are Useful for Examining Membranes

Freeze fracturing is an approach to sample preparation that is fundamentally different from the methods described so far. Instead of cutting uniform slices through a tissue sample (or staining unsectioned material), specimens are rapidly frozen at the temperature of liquid nitrogen or liquid helium, placed in a vacuum, and struck with a sharp knife edge. Samples frozen at such low temperatures are too hard to be cut. Instead, they fracture along lines of natural weakness—the hydrophobic interior of membranes, in most cases. Platinum/carbon shadowing is then used to create a replica of the fractured surface.

Freeze fracturing is illustrated in **Figure A-32**. Specimens are generally fixed prior to freeze fracturing, although some living tissues can be frozen fast enough to keep them in almost lifelike condition. Because cells contain large amounts of water, fixed specimens are usually treated with an antifreeze such as glycerol to provide *cryoprotection*—that is, to reduce the formation of ice crystals during freezing. Freeze-fractured membranes appear as smooth surfaces studded with **intramembranous particles (IMPs)**. These particles are integral membrane proteins that have remained with one lipid monolayer or the other as the fracture plane passes through the interior of the membrane.

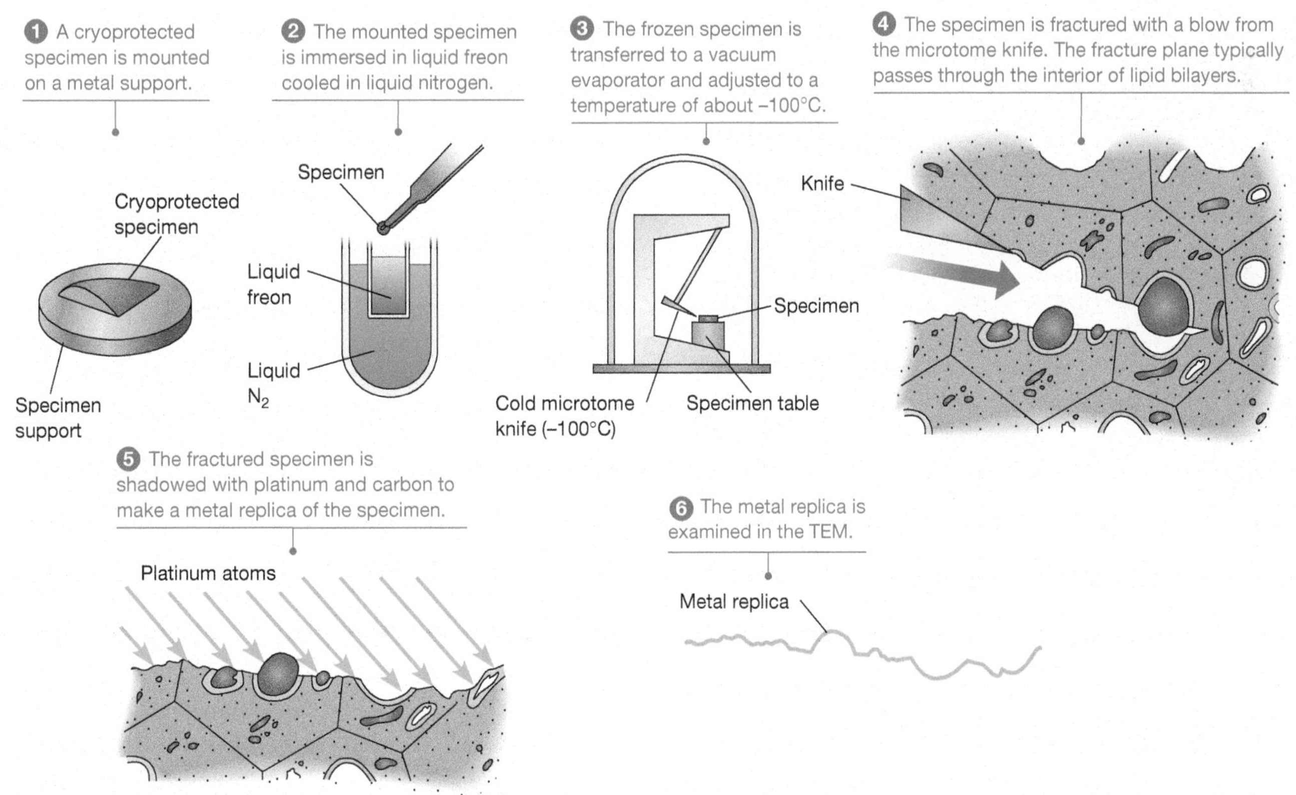
1 A cryoprotected specimen is mounted on a metal support.
Cryoprotected specimen
Specimen support
2 The mounted specimen is immersed in liquid freon cooled in liquid nitrogen.
Specimen
Liquid freon
Liquid N_2
3 The frozen specimen is transferred to a vacuum evaporator and adjusted to a temperature of about –100°C.
Knife
Specimen
Cold microtome knife (–100°C)
Specimen table
4 The specimen is fractured with a blow from the microtome knife. The fracture plane typically passes through the interior of lipid bilayers.
5 The fractured specimen is shadowed with platinum and carbon to make a metal replica of the specimen.
Platinum atoms
6 The metal replica is examined in the TEM.
Metal replica

Figure A-32 The Technique of Freeze Fracturing. The result is a replica of a specimen, fractured through its lipid bilayer, that can be viewed in the TEM. The shadowing is performed as in Figure A-31.

The electron micrograph in **Figure A-33** shows the two faces of a plasma membrane revealed by freeze fracturing. The **P face** is the interior face of the inner monolayer; it is called the P face because this monolayer is on the *protoplasmic* side of the membrane. The **E face** is the interior face of the outer monolayer; it is called the E face because this monolayer is on the *exterior* side of the membrane. Notice that the P face has far more intramembranous particles than does the E face. In general, most of the particles in the membrane stay with the inner monolayer when the fracture plane passes down the middle of a membrane.

To have a P face and an E face appear side by side as in Figure A-33, the fracture plane must pass through two neighboring cells, such that one cell has its cytoplasm and the inner monolayer of its plasma membrane removed to reveal the E face, whereas the other cell has the outer monolayer of its plasma membrane and the associated intercellular space removed to reveal the P face. Accordingly, E faces are always separated from P faces of adjacent cells by a "step" (marked by the arrows in Figure A-33) that represents the thickness of the intercellular space.

In a closely related technique called **freeze etching**, a further step is added to the conventional freeze-fracture procedure. Following the fracture of the specimen but prior to shadowing, the microtome arm is placed directly over the specimen for a short period of time (a few seconds to several minutes). This maneuver causes a small amount of water to evaporate (sublime) from the surface of the specimen to the cold knife surface, and this sublimation produces an *etching* effect—that is, an accentuation of surface detail. By using ultrarapid freezing techniques to minimize the formation of ice crystals during freezing, and by including a volatile cryoprotectant such as aqueous methanol, which sublimes very readily to a cold surface, the etching period can be extended and a deeper layer of ice can be removed, thereby exposing structures that are located deep within the cell interior. This modification, called **deep etching**, provides a fascinating look at cellular structure. Deep etching has been especially useful in exploring the cytoskeleton and examining its connections with other structures of the cell (see Figure 13-15).

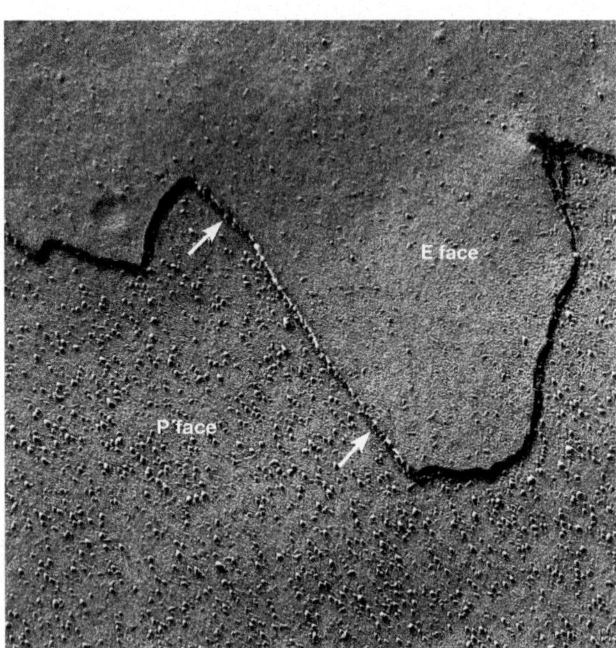

Figure A-33 Freeze Fracturing of the Plasma Membrane. This electron micrograph shows the exposed faces of the plasma membranes of two adjacent endocrine cells from a rat pancreas as revealed by freeze fracturing. The P face is the inner surface of the lipid monolayer on the protoplasmic side of the plasma membrane. The E face is the inner surface of the lipid monolayer on the exterior side of the plasma membrane. The P face is much more richly studded with intramembranous particles than the E face. The arrows indicate the "step" along which the fracture plane passed from the interior of the plasma membrane of one cell to the interior of the plasma membrane of a neighboring cell. The step therefore represents the thickness of the intercellular space (TEM).

Stereo Electron Microscopy and 3-D Electron Tomography Allow Specimens to Be Viewed in Three Dimensions

Electron microscopists frequently want to visualize specimens in three dimensions. Shadowing, freeze fracturing, and scanning electron microscopy are useful for this purpose, as is another specialized technique called **stereo electron microscopy**. In stereo electron microscopy, three-dimensional information is obtained by imaging the same specimen at two slightly different angles using a special specimen stage that can be tilted relative to the electron beam. The two micrographs are then mounted side by side as a *stereo pair.* When you view a stereo pair through a stereoscopic viewer (or by crossing your eyes), your brain uses the two independent images to construct a three-dimensional view that gives a striking sense of depth. **Figure A-34a** is a stereo pair of a *Drosophila* polytene chromosome imaged by high-voltage electron microscopy.

More recently, electron microscopists have used computer-based methods to make three-dimensional reconstructions of structures imaged in the TEM. In a process known as **three-dimensional (3-D) electron tomography**, serial thin sections containing a specimen are rotated and imaged at several different orientations; the resulting rotated views are then used to construct very thin, computer-generated "slices" of the specimen. Structures within these highly resolved slices can then be traced and reconstructed to provide three-dimensional information about the specimen (for more on this process, see Key Technique in Chapter 10, pages 248–249). Three-dimensional electron tomography has revolutionized our view of cellular structures, such as the organelles shown in Figure A-34b, c.

Specimen Preparation for Scanning Electron Microscopy Involves Fixation but Not Sectioning

When preparing a specimen for scanning electron microscopy, the biologist's goal is to preserve the structural features of the cell surface and to treat the tissue in a way that minimizes damage by the electron beam. The procedure is actually similar to preparing ultrathin sections for transmission electron microscopy but without the sectioning step. The tissue is fixed

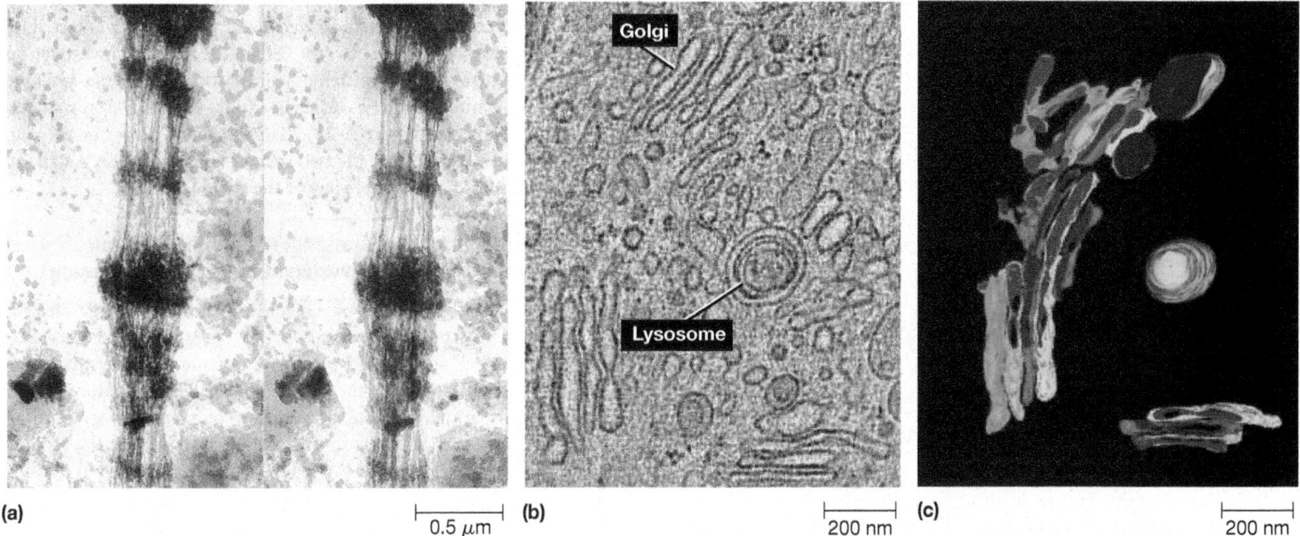

Figure A-34 Stereo Electron Microscopy and 3-D Electron Tomography. (a) The polytene chromosome in this micrograph is shown as a stereo pair. The two photographs were taken by tilting the specimen stage first 5° to the right and then 5° to the left of the electron beam. For a three-dimensional view, the stereo pair can be examined with a stereoscopic viewer. Alternatively, simply let your eyes cross slightly, fusing the two micrographs into a single image (HVEM). **(b)** A thin (3.2 nm) digital slice computed from a series of tilted images of a dendritic cell (TEM). A region of the Golgi apparatus and a lysosome are clearly visible. **(c)** A contour model of some of the structures visible in the tomogram shown in part b.

in aldehyde, postfixed in osmium tetroxide, and dehydrated by processing through a series of alcohol solutions. The tissue is then placed in a fluid such as liquid carbon dioxide in a heavy metal canister called a **critical point dryer**, which is used to dry the specimen under conditions of controlled temperature and pressure. This helps keep structures on the surfaces of the tissue in almost the same condition as they were before dehydration.

The dried specimen is then attached to a metal specimen mount with a metallic paste. The mounted specimen is coated with a layer of metal using a modified form of vacuum evaporation called **sputter coating**. Once the specimen has been mounted and coated, it is ready to be examined in the SEM.

Other Imaging Methods

Light and electron microscopy are direct imaging techniques in that they use photons or electrons to produce actual images of a specimen. However, some imaging techniques are indirect. To understand what we mean by indirect imaging, suppose you are given an object to handle with your eyes closed. You might feel 6 flat surfaces, 12 edges, and 8 corners. If you then draw what you have felt, it would turn out to be a box. This is an example of an indirect imaging procedure.

The two indirect imaging methods we describe here are *scanning probe microscopy* and *X-ray crystallography*. Both approaches have the potential for showing molecular structures at near-atomic resolution, ten times better than the best electron microscope. They do have some shortcomings that limit their usefulness with biological specimens. But when these techniques can be applied successfully, the resulting images provide unique information about molecular structure that cannot be obtained using conventional microscopic techniques.

Scanning Probe Microscopy Reveals the Surface Features of Individual Molecules

Although "scanning" is involved in both scanning electron microscopy and scanning probe microscopy, the two methods are in fact quite different. The first example of a **scanning probe microscope**, called the *scanning tunneling microscope (STM)*, was developed in the early 1980s. The STM utilizes a tiny probe that does not emit an electron beam, but instead possesses a tip made of a conducting material such as platinum-iridium. The tip of the probe is extremely sharp; ideally, its point is composed of a single atom. It is under the precise control of an electronic circuit that can move it in three dimensions over a surface. The x and y dimensions scan the surface, and the z dimension governs the distance of the tip above the surface (**Figure A-35** on page A-25).

As the tip of the STM is moved across the surface of a specimen, voltages from a few millivolts to several volts are applied. If the tip is close enough to the surface and the surface is electrically conductive, electrons will begin to leak or "tunnel" across the gap between the probe and the sample. The tunneling is highly dependent on the distance, so that even small irregularities in the surface will affect the rate of electron tunneling. As the probe scans the sample, the tip of the probe is automatically moved up and down to maintain a constant rate of electron tunneling across the gap. A computer measures this movement and uses the information to generate a map of the sample's surface, which is viewed on a computer screen.

Despite the enormous power of the STM, it suffers from two limitations: the specimen must be an electrical conductor, and the technique provides information only about electrons associated with the specimen's surface. Researchers have therefore developed other kinds of scanning probe

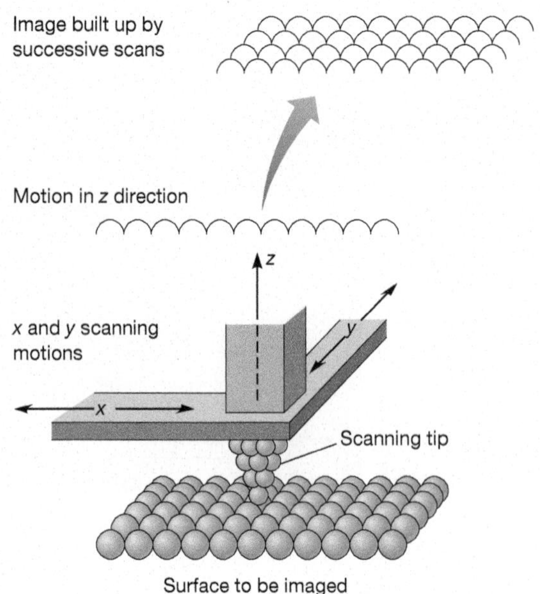

Figure A-35 Scanning Tunneling Electron Microscopy. The scanning tunneling microscope (STM) uses electronic methods to move a metallic tip across the surface of a specimen. The tip is not drawn to scale in this illustration; the point of the tip is ideally composed of only one or a few atoms, shown here as balls. An electrical voltage is produced between the tip and the specimen surface. As the tip scans the specimen in the *x* and *y* directions, electron tunneling occurs at a rate dependent on the distance between the tip and the first layer of atoms in the surface. The instrument is designed to move the tip in the *z* direction to maintain a constant current flow. The movement is therefore a function of the tunneling current and is presented on a computer screen. Successive scans then build up an image of the surface at atomic resolution.

microscopes. For example, in the *atomic force microscope (AFM)*, the scanning tip is pushed right up against the surface of the sample. When it scans, it moves up and down as it runs into the microscopic hills and valleys formed by the atoms present at the sample's surface. A variety of scanning probe microscopes have been designed to detect other properties, such as friction, magnetic force, electrostatic force, van der Waals forces, heat, and sound.

One of the most important potential applications of scanning probe microscopy is the measurement of dynamic changes in the conformation of functioning biomolecules, such as visualizing the movements of single myosin proteins as they change their shape during their power stroke (see Key Technique in Chapter 14, pages 388–389). It is even possible to use a modified form of atomic force microscopy to directly stretch large biomolecules to measure their mechanical properties.

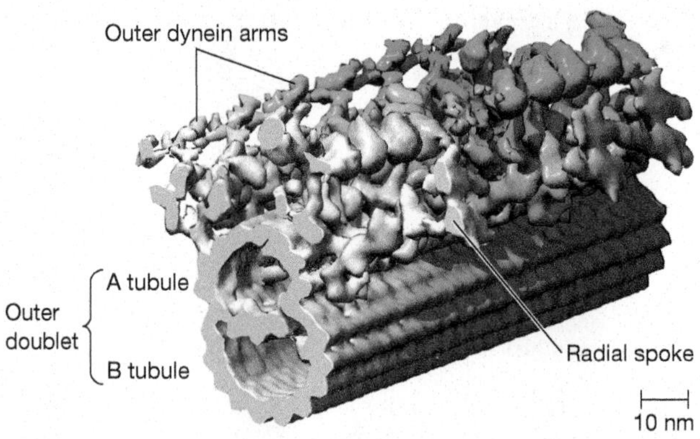

Figure A-36 CryoEM Bridges the Gap Between the Atomic and Molecular Levels. A three-dimensional reconstruction of the axoneme of a flagellum from *Chlamydomonas* shows the internal connections between an outer doublet microtubule and outer dynein arms (cryoEM).

CryoEM Bridges the Gap Between X-Ray Crystallography and Electron Microscopy

X-ray crystallography has provided unprecedented views of biological molecules down to the atomic level (for more on this important technique, see Key Technique in Chapter 3, pages 58–59). However, crystallography requires large amounts of purified molecules. In addition, X-ray crystallography is not currently well suited for analyzing very large macromolecular assemblies. For very large structures, another technique is needed. A technique that helps bridge this gap is *cryo-electron microscopy*, or **cryoEM** for short. In cryoEM, purified molecules or macromolecular assemblies are rapidly frozen via cryofixation. The rapid freezing prevents ice crystals from forming. Instead, the molecules are embedded in *vitreous ice*, noncrystalline frozen water, which better preserves the structures of the embedded molecules. The sample is then directly imaged at a very low temperature (–170°C) in an electron microscope. Often the specimens are imaged using three-dimensional electron tomography. To see how cryoEM can work together with X-ray crystallography, consider **Figure A-36**, which shows a three-dimensional reconstruction of the axoneme of a flagellum.

ANSWER KEY TO CONCEPT CHECK AND KEY TECHNIQUE QUESTIONS

Chapter 1 **Concept Check 1.1,** Using the light microscope, Robert Hooke and others showed that tissues are composed of many small compartments now known as cells. Examination of both plant and animal tissues showed that cells are the fundamental building blocks of all organisms examined. Studies of cell nuclei and cell division convinced scientists that all cells come from the division of preexisting cells. Technological advancements in microscope design and lens production allowed scientists to observe cells clearly enough to make these important observations. **Concept Check 1.2,** Our earliest understanding of cancer cell biology relied heavily on cytological observations that showed how cancer cells differ from normal cells. The light microscope made it possible to see individual cells, and the electron microscope made it possible to see subcellular components. Biochemical contributions include identification of the pathways by which cells use glucose for fuel, an understanding of ATP as the energy molecule in cells, and a number of techniques to study cells, including subcellular fractionation by centrifugation, chromatography, and electrophoresis. Genetic contributions showed us that DNA is the genetic material and how its defective replication can lead to uncontrolled, cancerous cell division. Genetics also provided us with a number of techniques to analyze DNA structure and function, including recombinant DNA technology and bioinformatics. **Concept Check 1.3,** The experimental process should begin by testing the hypothesis that the heartburn is due to the pizza. You can test the hypothesis by deliberately consuming pizza on certain nights and refraining from pizza on others. If you find a good correlation between pizza consumption and heartburn and are able to confirm this link repeatedly—note the requirement for repeatable observations—it seems reasonable to conclude that your heartburn is, in fact, caused by eating pizza. Having tentatively confirmed that hypothesis, you are now ready to test the further hypothesis that the heartburn is caused specifically by one or more of the pizza toppings. To do so, you will need a "control," consisting of plain, no-topping pizza, and pizzas with only one of the three toppings. Again, repeatability of the observations is an important criterion before you can conclude that one, two, or all three toppings cause heartburn. **Key Technique,** A high degree of antibody-antigen specificity ensures that the only molecules labeled by the covalently linked fluorophore will be the specific target antigen molecule. Knowing the target antigen of a specific antibody allows us to confidently determine the location of that specific antigen in the cell. Antibody-antigen specificity also ensures that a secondary antibody will only detect molecules in the cell that have been bound by the primary antibody.

Chapter 2 **Concept Check 2.1,** The large amount of energy required to break covalent bonds containing carbon makes carbon-containing molecules highly stable. In addition, carbon has four outer electrons available for bonding, so it can form a total of four covalent bonds. These bonds can be to other carbon atoms or to different types of atoms. Carbon can form single, double, and triple bonds and bonds readily to hydrogen, oxygen, nitrogen, and other important atoms in biology. Because it can form both linear chains and closed rings, it can be used to form an incredible variety of different compounds. **Concept Check 2.2,** If water molecules were linear, they would not be polar and could not form hydrogen bonds with each other. This would give water less cohesiveness and would result in a lower boiling point, a lower specific heat, and a lower heat of vaporization. This would give water a lower temperature-stabilizing ability, making it less suitable as the basis of living cells. Also, water molecules could not form hydrogen bonds with other molecules, resulting in very low (if any) solubility of polar and charged substances, which make up the majority of substances needed by cells. **Concept Check 2.3,** The amphipathic nature of phospholipids allows the hydrophobic fatty acid chains to associate with each other as the polar head groups interact with water, forming the bilayer structure that is the basis for all biological membranes. This bilayer can thus interact with the aqueous environment on either side of the membrane while providing a nonpolar barrier in the interior of the bilayer that is not permeable to polar molecules and ions. Proteins embedded in the hydrophobic interior of the membrane can act as transport channels to allow the passage of polar molecules and ions. **Concept Check 2.4,** *Similarities:* Proteins, nucleic acids, and carbohydrates are all long polymers composed of monomers. All are produced by polymerization of activated monomers in a condensation reaction that requires energy. All three also have directionality, meaning that the ends are different and monomer units are added to only one end of the growing polymer during synthesis. *Differences:* The monomer units are different for each of the three types of macromolecules, and each has a different number of monomers. Each has a different carrier, or in the case of nucleic acids, no carrier. Each also has a different function in the cell. **Concept Check 2.5,** Infectious virus particles were formed by self-assembly of viral RNA and coat protein. Presumably, the RNA and protein of the different strains are similar enough to interact. All the information needed for self-assembly is present in the molecules themselves. The hybrid viruses containing RNA from strain A would infect the normal host cells of strain A because RNA is an informational molecule and contains the information needed to take over the host cell. The importance of the RNA is suggested by its being internal and protected by the protein coat. **Key Technique,** The fragment whose path is shown in red has the higher m/z ratio. It tends to be deflected less because it has higher mass (so is harder to deflect than a lighter fragment with the same charge) or less charge (so it is less susceptible to the electromagnet compared to a more highly charged fragment with the same mass), or a combination of the two.

Chapter 3 **Concept Check 3.1,** By definition, the primary structure is altered if three amino acid residues are missing. It would disrupt secondary structure if the missing residues were part of an α helix or β sheet. It could be that the hemoglobin cannot fold into its final, functional conformation because the missing region contributed critical interactions that are necessary for proper folding. Deletion of charged residues would alter ionic bonding, and deletion of residues that are hydrogen-bond donors or acceptors would disrupt hydrogen bonding. Also, if nonpolar amino acids were deleted, this could alter hydrophobic interactions and proper folding. These changes could affect tertiary structure and interfere with proper folding of one of the polypeptides. Deletion of a cysteine residue might be even more detrimental, as it could eliminate a disulfide bond that keeps the protein in its proper conformation. If the missing amino acids were on the surface, they may have been necessary for the four polypeptides to associate and form the correct quaternary structure. Alternatively, a change in the amino acid residues near the heme-binding site could decrease the hemoglobin's affinity for this oxygen-binding group. **Concept Check 3.2,** Like proteins, nucleic acids are polymers composed of a linear series of monomer units that are similar for each macromolecule. Both are synthesized by dehydration, and both are degraded by hydrolysis. Energy is required for synthesis of both proteins and nucleic acids. They differ in that nucleic acids have only five different nucleotide monomers, whereas proteins have 20 different amino acid monomers. Nucleic acids function primarily to store genetic information (DNA) and to allow that information to be expressed in the production of proteins (RNA). Proteins, by contrast, have a wide variety of enzymatic, structural, transport, and other functions. Also, the overall structure of a double-stranded helical DNA fiber does not depend on the sequence of the monomers, whereas the structure of a protein depends critically on the sequence of its monomers. **Concept Check 3.3,** Polysaccharides are also polymers of similar monomer units but do not carry specific information as do proteins and nucleic acids. They are usually formed from only one or two different monomers, rather than 20 (proteins) or 5 (nucleotides). Like some proteins (but different from nucleic acids), polysaccharides such as chitin and cellulose serve structural roles. Unlike either proteins or nucleic acids, some polysaccharides function in energy storage, producing glucose when needed by the cell. (Note that some proteins can be used for energy, but that is typically only when stores of sugars, polysaccharides, and fats have been exhausted.) **Concept Check 3.4,** Lipids differ from proteins, nucleic acids, and polysaccharides in two main respects. First, they tend to lack oxygen and nitrogen atoms and charged regions and thus are predominantly nonpolar and insoluble in water. Second, although they are large molecules, they are not long polymers of monomer units. Similar to the other classes of macromolecules, however, lipids such as phospholipids and triglycerides are made by dehydration synthesis (from glycerol and fatty acids) and are important in cellular structures. Like proteins, they have a variety of roles in the cell, and like some polysaccharides, they function in energy storage for the cell. **Key Technique,** Crystals are characterized by a repeating pattern of individual units. Each protein molecule adopts the same unique conformation based on its primary sequence, consisting of repeating units linked together by peptide bonds. Therefore, each repeating unit in a protein can form the three-dimensional matrix in a crystal. Because the intermolecular forces holding different protein molecules together are also consistent, proteins will associate in a repetitive, consistent manner, either singly or in pairs. This repetitive association during crystallization is critical for the production of usable X-ray diffraction data, which depends on the reinforcement of numerous identical diffraction events within the crystal.

Chapter 4 **Concept Check 4.1,** Although in today's cells RNA is synthesized from a DNA template, this process requires that proteins act as enzymes. RNA, on the other hand, can in some cases be synthesized and replicated without the need for DNA or proteins. Moreover, the monomer units of RNA are derived from some of the simple organic compounds that may have been formed abiotically in the early Earth and that are used to make the monomers needed for DNA synthesis. **Concept Check 4.2,** A microscope can be used to examine the cytology of the new organism for the presence of a nucleus and organelles, which would indicate a eukaryote. Useful biochemical techniques that could be used include centrifugation, electrophoresis, and chromatography to isolate specific components of the cell. For example, you could isolate membranes, which differ in phospholipid composition for bacteria, archaea, and eukaryotes. Perhaps the most definitive answer would be obtained from a genetic analysis of ribosomal RNA, as each of these three groups is defined by similarities in rRNA sequence within each group. **Concept Check 4.3,** To build a eukaryotic cell from a bacterial cell, you might first surround its genetic material with a nuclear membrane. You could also add mitochondria to provide energy by oxidizing sugars. Chloroplasts would not be absolutely necessary, but would make the cell independent of an external source of sugar. You might also want lysosomes to digest ingested food molecules and peroxisomes to break down fatty acids and detoxify toxic compounds. Replacing the bacterial ribosomes with eukaryotic ribosomes might be the biggest challenge. **Concept Check 4.4,** Although viruses cannot reproduce on their own, once they infect a host cell, they can use the host cell synthetic machinery, such as enzymes and ribosomes, to make copies of their own nucleic acid and coat proteins. After reproducing, they typically destroy the host cell, and this cellular damage can lead to serious disease in the host organism. **Key Technique,** First, homogenize the tissue of interest. Centrifuge the homogenate at 1000 *g* for 10 minutes to concentrate nuclei and unbroken cells in the pellet, which is discarded. Collect and centrifuge the supernatant at 20,000 *g* for 20 minutes to concentrate large particles and organelles in the pellet, which again is discarded. Collect and centrifuge the supernatant at 80,000 *g* for 1 hour to concentrate membrane fragments in the pellet. Discard the supernatant and retain this last pellet, which contains a mixture of membranes fragments from the cells and organelles in the tissue (step ❹ in Figure 4B-2).

Chapter 5 **Concept Check 5.1,** Energy flows unidirectionally through the open system we call the biosphere, originating in the sun and leaving mainly as heat. Chemotrophs depend on phototrophs for both energy and matter. Phototrophs harvest energy from the environment during photosynthesis, and they use this energy to reduce low free energy compounds like CO_2, H_2O, and N_2 to high free energy compounds like carbohydrates, lipids, and amino acids. Chemotrophs then use these high-energy compounds as a source of both chemical energy and molecular building blocks. Many chemotrophs also use the oxygen produced by phototrophs as a by-product of photosynthesis. Because matter flows in a cyclic manner, phototrophs depend on chemotrophs to oxidize the high-energy organic compounds back to the low-energy inorganic compounds that they require to continue photosynthesis. **Concept Check 5.2,** All spontaneous processes lead to an increase in the entropy of the universe. With all other variables such as T, P, V, and H constant, a process will be spontaneous if $\Delta S >$ 0 (the process results an increase in randomness). By contrast, with all other variables constant, a process will be spontaneous if $\Delta H < 0$ (a decrease in heat content). Mathematically, we can see that the effects of changes in entropy and enthalpy are opposite because ΔG, which always decreases in a spontaneous process, equals $\Delta H - T\Delta S$. Thus, ΔG varies directly with ΔH and inversely with ΔS, so that ΔG decreases (a spontaneous process) when either ΔH decreases or ΔS increases. **Concept Check 5.3,** ΔG represents the change in free energy during a chemical reaction or biological process as it occurs in actual cellular conditions under prevailing concentrations of reactants and products. $\Delta G°$ and $\Delta G°'$ are used to describe free energy changes (and thus determine spontaneity) of chemical reactions under specified standardized conditions. $\Delta G°$ represents the change in free energy when all reactants and products (except water) are at a standard concentration of 1 *M* (rarely used directly in biochemistry), and $\Delta G°'$ is the free energy change at standard concentrations but at pH 7, where $[H^+] = 10^{-7}$ *M* rather than 1 *M*. **Key Technique,** During the first few rounds of injection of ligand, there is a large excess of the target molecule. At first, all of the newly injected ligand binds to the target, so there is a maximal release of heat. As more and more ligand is injected, however, there is less and less free (unbound) target, and eventually not all of the newly injected ligand can find an unoccupied target to bind to. The result is less heat released and hence a less dramatic spike.

Chapter 6 **Concept Check 6.1,** Gasoline is a high-energy compound but normally exists in a metastable state. Although thermodynamically feasible, the exothermic reaction of gasoline + O_2 to form $CO_2 + H_2O$ + heat will only occur if the reacting molecules have enough energy to reach the more energetic transition state. A match provides the energy to allow some molecules to reach this higher-energy transition state and undergo combustion. The resulting exothermic combustion reaction then provides the energy to continue combustion by raising other reactant molecules to this energy level, and this "chain reaction" continues until either gasoline or oxygen is depleted. By contrast, a catalyst allows molecules to reach the transition state by lowering the activation energy. **Concept Check 6.2,** Each enzyme has a uniquely shaped active site into which will fit only one specific substrate or group of closely related substrates. This specificity is provided both by the size and shape of the active site and by noncovalent interactions between atoms of the substrate and amino acid residues in the active site. Low temperatures will have little effect on substrate binding, but extremely high temperatures cause the enzyme to denature, destroying the structure of the active site. By contrast, both extremes of pH will likely have a marked effect because very high or very low pH will change the ionization state of key residues in the active site and alter any ionic bonding that is necessary for substrate binding. **Concept Check 6.3,** You would want to increase V_{max} if possible to ensure the highest possible reaction rate at saturating substrate concentrations. Similarly, you would want k_{cat} to be as high as possible to produce the maximum amount of product molecules per minute. By contrast, you would want a lower K_m, so that a lower [S] would achieve half maximal velocity. **Concept Check 6.4,** Enzymes must be properly regulated to ensure that they are active only when the cell is in need of the product of the reaction. For example, you would not want an enzyme that synthesizes glucose and one that degrades it to be active at the same time. Regulation is also necessary to ensure the appropriate rate of each reaction, as unregulated enzymes may not operate at a rate that is optimal given their biological function. In substrate-level regulation, the rate of the reaction is determined by the concentration of substrate, whereas in feedback regulation, the rate is regulated by the concentration of a product molecule that binds to an allosteric site. Additional regulation can be achieved by covalent modification of the enzyme, such as by phosphorylation/dephosphorylation or by methylation/demethylation. **Key Technique,** When the equally spaced points in Figure 6B-1 are used to produce the double-reciprocal plot in Figure 6B-2, points representing the higher concentrations of substrate become bunched together toward the left side of the graph. To improve the usefulness of these assays, more assays should be done at the lower substrate concentrations. This will generate more points to the right in the second graph and increase the accuracy of the line drawn through the points.

Chapter 7 **Concept Check 7.1,** Without proteins, a pure phospholipid bilayer composed solely of phospholipids would be unable to transport ions or large hydrophilic molecules in or out of the cell. Movement of these materials requires a membrane-based transport protein because alone they cannot pass through the hydrophobic interior of the bilayer. The cell would be well defined and separated from the environment, but import and export of many essential substances (such as ions and nutrients), response to environmental signals, and cell-to-cell communication and adhesion would be impossible, as these functions are carried out by membrane proteins. **Concept Check 7.2,** Most of the relevant insights leading to our current understanding of membrane structure have been based on biochemical analyses, for example, discovery of the ability of lipids to enter cells easily, discovery of the amphipathic nature of the molecular structure of membrane phospholipids, and the discovery that proteins act like "pores" to allow polar molecules and ions through membranes. Also important have been insights from cytological investigations, for example, recognition of the occurrence of membranes around every cell and organelle and the railroad track appearance of all membranes when viewed in the electron microscope. A landmark discovery was the use of electron microscopy to determine the three-dimensional structure of the protein bacteriorhodopsin, showing how this integral membrane protein is anchored in the membrane by α-helical transmembrane segments. There were not any substantial genetic studies that led to our current understanding of membrane structures. **Concept Check 7.3,** Although phospholipids only rarely move from one monolayer of the bilayer, they can rapidly move laterally within a monolayer. This ability is influenced mainly by the types of fatty acid chains attached to the phospholipid. Unsaturated fatty acids have a bent shape and do not pack as tightly as saturated fatty acids, increasing membrane fluidity. Longer fatty acids chains have the opposite effect—they increase packing and decrease fluidity. Sterols have a dual effect on fluidity depending on temperature. At high temperatures, the rigidity of the sterol ring decreases fluidity, but at lower temperatures it interferes with packing of fatty acid chains and increases fluidity. This minimizes changes in fluidity as the temperature fluctuates. **Concept Check 7.4,** Integral membrane proteins are embedded in the hydrophobic membrane interior and have hydrophilic regions exposed on both sides of the membrane facing aqueous environments. They can serve as transporters moving materials through the membrane or as signaling molecules sensing an extracellular cue and causing an intracellular response. Peripheral membrane proteins typically are involved in processes on only one side of the membrane. The glycosylated proteins on the outside of the membrane, for example, are important in cell recognition. Proteins attached to cytoskeletal elements on the inner monolayer may be important in sequestering signaling proteins into discrete membrane microdomains. **Key Technique,** Increased temperature will cause molecules to diffuse more rapidly due to increased thermal motion. More diffusion will usually mean faster recovery, which means that the half-life will decrease. Thus, we would expect the cells at 37°C to show the shorter half-life.

Chapter 8 Concept Check 8.1, The magnitude of the concentration gradient is important for all three types of transport. It determines which direction (and at what rate) simple diffusion or facilitated diffusion will proceed. It also determines how much energy will be required for active transport to proceed. The electrochemical potential, which is the sum of the concentration and charge gradients, is relevant only for facilitated diffusion and active transport, which can move ions. Because ions cannot cross a membrane by simple diffusion due to the hydrophobic membrane interior, charge differences on opposite sides of the membrane do not affect simple diffusion. **Concept Check 8.2,** Osmosis is different from diffusion of O_2 because osmosis refers specifically to the diffusion of water across a semipermeable membrane. Also, the processes differ in that water is a very small polar molecule and molecular oxygen is nonpolar. Osmosis and the diffusion of gases are similar in the following ways: in both, material moves from an area of higher concentration to an area of lower concentration; both processes move very small molecules; neither requires a transport protein; and neither process requires an input of energy. **Concept Check 8.3,** Carriers and channels are both integral membrane proteins capable of moving solutes down a concentration gradient. Neither requires an external source of energy other than the concentration gradient. They are both specific for a particular solute (molecule or ion) and discriminate against others. They differ primarily in their mechanism of action. Carrier proteins typically undergo a conformational change upon binding the solute, and this results in the release of the solute on the other side of the membrane. A specific number of solute molecules or ions is transported each time. Channel proteins do not undergo a conformational change, but instead provide a narrow opening through which a continuous flow of solute moves. **Concept Check 8.4,** A distinguishing feature of active transport (as compared to facilitated diffusion) is the requirement for energy because active transport moves solutes up a concentration gradient in an endergonic, nonspontaneous process. The transporter could be expressed in frog oocytes (see Key Technique, pages 196–197) or could be purified and incorporated into artificial liposomes. These can then be placed in solutions containing varying concentrations of solute and/or ATP. By measuring the final concentrations in the oocytes or liposomes versus the surrounding solution, we could determine whether solutes move with or against a concentration gradient. Movement against a concentration gradient that requires ATP would indicate active transport. Equalization of solute concentrations across the membrane in an ATP-dependent manner would indicate facilitated diffusion. **Concept Check 8.5,** The Na^+/K^+ pump uses the energy derived from the exergonic hydrolysis of ATP to drive the endergonic movement of Na^+ and K^+ up their respective concentration gradients. The Na^+/glucose symporter uses the energy derived from the exergonic movement of sodium ions down their concentration gradient to drive the simultaneous, endergonic active transport of glucose up its concentration gradient. **Concept Check 8.6,** If the side chain is in the neutral state, the molecule will have no net charge due to the balancing of the charges on the amino (positive) and acid (negative) ends of the amino acid. You would only need to consider the concentration gradient for aspartic acid. However, if the side chain is ionized, the molecule will have a net negative charge, and you would need to consider both the concentration gradient for aspartic acid and the electrochemical gradient involving all ions. **Key Technique,** The lack of transport may be due to the inability of the oocytes to synthesize the arsenic transport protein. Or perhaps the protein is produced but is not properly inserted in the membrane. It may be that the oocyte lacks some stimulatory molecule that regulates the protein's activity. Alternatively, the oocyte may contain some inhibitory chemical that blocks transport function.

Chapter 9 Concept Check 9.1, Both types of pathways consist of a series of reactions occurring in an ordered sequence, with each reaction being carried out by a specific enzyme. However, anabolic pathways are typically involved in biosynthesis and are endergonic. They require an input of energy to synthesize large molecules from smaller precursors. Catabolic pathways, on the other hand, are typically involved in degradation, often by oxidizing molecules to provide energy for the cell. They are exergonic and release energy when large molecules are degraded to smaller molecules. **Concept Check 9.2,** Because the standard free energy change for ATP hydrolysis ($\Delta G^{\circ\prime}$ = −7.3 kcal/mol) is intermediate among phosphorylated compounds, it not only can phosphorylate other cellular molecules, but also can be phosphorylated by a number of cellular molecules (see Table 9-1). If its $\Delta G^{\circ\prime}$ were too low (more negative), ATP would be able to phosphorylate many other molecules, but its synthesis by phosphorylation of ADP would be difficult. Conversely, if its $\Delta G^{\circ\prime}$ were too high, ATP would not be able to easily transfer its phosphate group to other cellular compounds. **Concept Check 9.3,** Oxidation is the removal of electrons from an atom or molecule, and reduction is the gain of electrons. Because electrons do not exist alone, when any compound loses an electron during oxidation, some other compound gains this electron and becomes reduced. When the carbon atoms in glucose are oxidized to CO_2, they lose electrons; these electrons are used to reduce molecular oxygen to H_2O (see Reaction 9-12). **Concept Check 9.4,** Because oxidation is the removal of electrons, a compound can be oxidized by any suitable electron acceptor, which in the process is reduced. Despite the name "oxidation," an electron acceptor during oxidation does not have to be molecular oxygen. In glycolysis, the electron acceptor is NAD^+, which is reduced to NADH. **Concept Check 9.5,** In the absence of oxygen, NADH accumulates as glucose is oxidized and transfers electrons to NAD^+ (reaction Gly-6). For glycolysis to continue, NAD^+ must be available, so the NADH produced during glycolysis must be oxidized back to NAD^+. In the absence of oxygen (which accepts these electrons from NADH in aerobic conditions), the electrons from NADH are transferred to pyruvate. This reduces pyruvate to lactate or to ethanol plus CO_2 and regenerates NAD^+ for continued glycolysis. **Concept Check 9.6,** In the absence of glucose, other cellular carbohydrates can be degraded for energy via conversion into molecules that are substrates in glycolysis. Carbohydrates such as lactose, sucrose, and glycogen and glycerol from fatty acid degradation can be converted either to glucose directly or to intermediates in the glycolytic pathway, as shown in Figure 9-10. **Concept Check 9.7,** If the reaction to synthesize glucose in gluconeogenesis were simply the reverse of the glycolytic reactions, it would require synthesis of ATP at steps Gly-1 and Gly-3 and a large input of energy. This energetic requirement is avoided by simple hydrolysis of glyceraldehyde-3-phosphate and fructose-6-phosphate. These reactions release inorganic phosphate instead of using the phosphate group to synthesize ATP from ADP. **Concept Check 9.8,** The key regulatory enzymes in glycolysis and gluconeogenesis are unique to each pathway so that these pathways can be regulated independently. Reciprocal regulation ensures that when energy is plentiful (high [ATP] and low [AMP]), glycolysis will slow down and gluconeogenesis will be activated. When energy is needed (low [ATP] and high [AMP]), glycolysis will be activated to provide energy for the cell, and gluconeogenesis will be inhibited. **Key Technique,** Note that the dihydroxyacetone phosphate produced during glycolysis can be converted to glyceraldehyde-3-phosphate, which produces phosphoenolpyruvate. If three carbons in phosphoenolpyruvate come from carbons 4, 5, and 6 of glucose via glyceraldehyde-3-phosphate, the phosphorylated carbon will be ^{12}C. However, if these three carbons come from carbons 1, 2, and 3 of glucose via dihydroxyacetone phosphate (that was subsequently converted to glyceraldehyde-3-phosphate), then the phosphorylated carbon of phosphoenolpyruvate will be ^{14}C.

Chapter 10 Concept Check 10.1, During respiration, electrons received from the oxidation of glucose are used to reduce an external electron acceptor outside the pathway, such as molecular oxygen. By contrast, in fermentation, an internal electron acceptor that is an intermediate in the pathway, such as pyruvate, accepts the electrons removed from glucose. But then pyruvate is reduced to lactate and cannot be completely oxidized to CO_2 in the citric acid cycle. Much more energy is produced from glucose during respiration because use of an external electron acceptor such as O_2 allows the complete oxidation of pyruvate to CO_2. All of the available chemical energy in pyruvate can be used to synthesize ATP. **Concept Check 10.2,** With the exception of components of the succinate dehydrogenase complex (Complex II), the citric acid cycle enzymes are soluble enzymes found in the mitochondrial matrix in a large multienzyme complex. Association of the various enzymes in a pathway allows the product of one reaction to immediately encounter the next enzyme in the pathway, where it will be the substrate. ATP synthase, on the other hand, is an integral membrane protein. The position of the F_o portion in the membrane allows it to take advantage of a transmembrane proton gradient to provide the energy by which the F_1 portion facing the matrix synthesizes ATP. **Concept Check 10.3,** As pyruvate is oxidized in the citric acid cycle, most of its chemical energy is conserved in the high-energy reduced coenzymes NADH and $FADH_2$, which accept electrons from pyruvate and its breakdown products. Most of the ATP generated by aerobic respiration is produced after NADH and $FADH_2$ transfer these electrons to molecular oxygen in the electron transport system. **Concept Check 10.4,** To function effectively, all the electron carriers in the ETS must be alternately reduced (as they accept electrons from the previous carrier) and oxidized (as they donate the electrons to the next carrier). They are spatially arranged in order of increasing reduction potential, the ability to accept electrons, so that electrons flow smoothly step by step from carriers with low reduction potentials (good electron donors such as NADH) to carriers with high reduction potentials (good electron acceptors such as oxygen). **Concept Check 10.5,** The transfer of electrons between several of the carriers is coupled to the unidirectional pumping of protons out of the mitochondrial membrane across the inner membrane and into the intermembrane space. This produces an electrochemical proton gradient across the membrane that is both necessary and sufficient to provide the energy for ATP synthesis as protons flow exergonically back across the membrane through the F_oF_1 ATP synthase. **Concept Check 10.6,** Protons cannot flow directly from the intermembrane space (high H^+ concentration) back to the matrix (low H^+ concentration). Their only path through the membrane is through the F_o proton channel. As they pass through, the integral membrane *c* ring rotates, turning the γ subunit within the $\alpha_3\beta_3$ catalytic head of the F_1 portion of the ATPase. This rotation uses the energy of the proton gradient to change the conformation of the $\alpha_3\beta_3$ catalytic head and allows the endergonic synthesis of ATP from ADP and P_i. **Concept Check 10.7,** Two ATP are produced during glycolysis; two more are produced directly in the citric acid cycle. The remainder are produced from reduced compounds. As glucose is oxidized completely to 6 CO_2 in glycolysis and the citric acid cycle, 10 NADH, 2 $FADH_2$, and 4 ATP are produced. In theory, the 10 NADH contain enough chemical energy to produce 30 ATP (3 per NADH), and the 2 $FADH_2$ allow production of 4 more ATP (2 per $FADH_2$), for a total of 4 + 30 + 4 = 38. In most eukaryotic cells, one of the NADH molecules is produced in the cytoplasm, and the energy required to transport it into the mitochondrion "costs" 2 ATP, reducing the total theoretical energy yield to 36 ATP molecules per glucose. The actual experimental estimate is lower (30 or 32 ATP per glucose). **Key Technique,** For a cylinder resting on its side, images would progress as follows: from a line the length of the cylinder, to rectangles of the same length but of increasing widths, up to a rectangle the width of the cylinder, back through rectangles the length of the cylinder but of progressively decreasing widths, and back to a line. If the cylinder were resting on its end, all images would be circular, with the same diameter as the cylinder.

Chapter 11 **Concept Check 11.1,** Both mitochondria and chloroplasts generate chemical energy in the form of ATP. Mitochondria derive energy from pyruvate oxidation, whereas chloroplasts get their energy from sunlight. In addition, the chloroplast generates reducing power in the form of NADPH, allowing anabolic fixation of CO_2 to sugars. In both organelles, energy released during electron transport is used to produce a transmembrane proton gradient by pumping protons into an enclosed space—the intermembrane space in mitochondria or the thylakoid lumen in chloroplasts. **Concept Check 11.2,** Photosynthetic pigments are arranged into photosystems that have a special pair of chlorophyll molecules at the center. The special pair absorbs light of one particular wavelength, but a variety of antenna pigments such as carotenoids surround the special pair and absorb light of higher energy (shorter wavelength) and transmit this energy to the special pair via resonance energy transfer. This allows antenna complexes to harvest light from multiple wavelengths, increasing efficiency. **Concept Check 11.3,** In both organelles, electron transport not only produces a transmembrane proton gradient but also reduces the terminal electron acceptor. In mitochondria, the electron source is typically NADH produced by aerobic respiration. In the chloroplast, the electron source is water, which is oxidized to O_2. The terminal electron acceptor in mitochondria is O_2, which gets reduced to H_2O. In chloroplasts, the final electron acceptor is $NADP^+$, producing the NADPH necessary for the Calvin cycle. **Concept Check 11.4,** Solar energy is absorbed by a chlorophyll molecule and produces a charge separation that represents potential energy. This initiates the flow of electrons through the photosystems and results in the energy being stored as a transmembrane electrochemical proton gradient. Proton flow back through the ATP synthase enzyme is transformed into mechanical energy as it drives rotation of the ring of ATP synthase subunit III polypeptides. This rotation, in turn, drives conformational changes in the $\alpha_3\beta_3$ catalytic head, allowing production of ATP from ADP and inorganic phosphate and conserving the solar energy as chemical energy in the form of ATP. **Concept Check 11.5,** Each CO_2 fixed combines with a five-carbon ribulose-1,5-bisphosphate molecule to form two molecules of 3-phosphoglycerate, each of which is reduced to glyceraldehyde-3-phosphate. After this happens three times, five of the six resulting glyceraldehyde-3-phosphate molecules (15 carbons) undergo a series of molecular rearrangements to form three molecules of the five-carbon ribulose-1,5-bisphosphate; the extra glyceraldehyde-3-phosphate is available for conversion to mono- and polysaccharides. **Concept Check 11.6,** The major regulatory points in the Calvin cycle are the three enzymes unique to the pathway: rubisco, sedoheptulose-1,7-bisphosphatase, and phosphoribulokinase (PRK). This is similar to the regulation of glycolysis and gluconeogenesis because key regulatory points in these two pathways also involve enzymes unique to each pathway. Regulation also occurs via metabolites such as ATP and NADPH, which activate Calvin cycle enzymes. The Calvin cycle is also regulated by the level of reduced ferredoxin, which in turn reduces and activates several Calvin cycle enzymes, and by the stimulatory effect of rubisco activase. **Concept Check 11.7,** Glucose is not actually the endpoint of sugar synthesis, but an intermediate. In the cytosol, glucose is used to produce sucrose (a disaccharide) from glucose and fructose, which is transported throughout the plant to tissues needing a source of energy. In the chloroplast stroma, glucose is used for starch synthesis, which provides a source of stored glucose polymer that can be used when photosynthesis is not active, such as at night. **Concept Check 11.8,** The use of molecular oxygen rather than CO_2 by rubisco results in the production of the two-carbon compound phosphoglycolate. One strategy to minimize loss of carbon is the glycolate pathway, which allows 3/4 of this carbon to be recovered when two molecules of phosphoglycolate are converted to one molecule of the three-carbon compound serine and one CO_2 molecule. C_4 plants use a second strategy, which is to confine rubisco and the Calvin cycle to bundle sheath cells, which have little O_2. A third strategy is found in CAM plants, which segregate the carboxylation and decarboxylation reactions in time rather than in different cell types. **Key Technique,** Green. Chlorophylls appear green because they *reflect* green light, not absorb it. Chlorophylls absorb light in the blue and red areas of the visible spectrum.

Chapter 12 **Concept Check 12.1,** Rough ER gets its name from the presence of numerous ribosomes on its cytosolic surface, reflecting its importance in the synthesis of membrane proteins and secreted proteins. Newly synthesized proteins enter the ER lumen, which is functionally equivalent to the interiors of vesicles, the Golgi, lysosomes, endosomes, and the cell exterior. Smooth ER lacks ribosomes and has several roles not involving protein synthesis that depend on the particular cell type. It is involved with drug detoxification and regulation of glycogen degradation in the liver, in storage of calcium in muscle cells, and in steroid hormone and plant hormone biosynthesis. **Concept Check 12.2,** Anterograde transport is needed to move proteins and lipids forward from the ER through the Golgi to their final destinations inside or outside the cell. Retrograde transport is required to return the membrane lipids of the vesicles back to the Golgi and ER for reuse in subsequent vesicles. In addition, retrograde transport returns ER- and Golgi-specific proteins back to the appropriate compartment once they are no longer needed in further stages of vesicle trafficking. **Concept Check 12.3,** Glycosylation of some proteins in the ER allows them to interact with the calnexin and calreticulin proteins that assist proper disulfide bond formation and protein folding. In cells defective in ER glycosylation of proteins, this interaction would not occur, and we would expect to see the accumulation of misfolded proteins in the ER. Also, UGGT, which normally binds to misfolded proteins, would not be able to glycosylate them and make them substrates for proper folding by calnexin and calreticulin. **Concept Check 12.4,** Membrane lipids are often tagged by the addition of one or more phosphate groups to a membrane phosphatidylinositol to assist with proper targeting of vesicles. In addition, the length and degree of saturation of some membrane lipids are also important. For proteins, several features ensure proper trafficking. For example, they may contain RXR sequences for ER retention, KDEL sequences to return them to the ER from the Golgi, or mannose phosphate tags to target them to the lysosome. **Concept Check 12.5,** Exocytosis would essentially be abolished because secretory granules from the Golgi could not fuse with the plasma membrane and discharge their contents outside the cell. Cells would accumulate these vesicles and eventually would have problems if the membranes and associated molecules were not returned to the ER and Golgi. Endocytosis could proceed to the point of formation of coated pits and nearly complete vesicles, but the final "pinching off" step to form a free vesicle could not occur. **Concept Check 12.6,** Most intracellular trafficking processes would be inhibited because most of them rely on GTP hydrolysis at some point. For example, closing of clathrin-coated pits requires dynamin, a GTPase. Production of COPI- and COPII-coated vesicles requires a GDP/GTP exchange factor and a GTPase. Also, association of v-SNARE and t-SNARE proteins prior to vesicle fusion requires Rab GTPases. **Concept Check 12.7,** Cells that depend upon phagocytosis for nutrient ingestion would be unable to feed and would likely die without supplied nutrition. Also, autophagy would not be possible, and old, defective organelles could not be recycled. Because so many hydrolytic enzymes would be absent, the cell would accumulate a large amount of macromolecular "garbage" such as misfolded proteins and undigested lipids and polysaccharides. **Concept Check 12.8,** In different cell types, peroxisomes carry out a variety of biochemical reactions such as fatty acid oxidation, urate degradation, detoxification reactions, and the glycolate pathway. Many of the reactions produce hydrogen peroxide, which is toxic and must be contained for degradation. A unifying feature of peroxisomes is the presence of catalase, which degrades the hydrogen peroxide to water and molecular oxygen. **Key Technique,** No, TIRF would not be useful for tracing the entire pathway. Translation of secretory proteins occurs in the rough ER, followed by processing in the Golgi and packaging into vesicles, which ultimately reach the plasma membrane. The initial steps of this pathway occur far too deep within the cell to be visualized using the evanescent field of a TIRF microscope. Only the last steps—especially vesicle fusion with the plasma membrane—would be visible and seen quite sharply. For visualizing the earlier steps, you would need a technique such as confocal microscopy or superresolution microscopy (see the Appendix).

Chapter 13 **Concept Check 13.1,** The modular assembly of the cytoskeleton allows a cell to assemble polymers of varying lengths to suit its needs. The reversibility of polymerization allows cells to assemble and later dismantle cytoskeletal elements as needs dictate. **Concept Check 13.2,** Greater dynamic instability leads to more rapid shrinkage and regrowth of MTs. This might be expected in situations where the cell is undergoing rapid, dynamic changes. One example is mitosis, when a dividing cell must construct the spindle. **Concept Check 13.3,** You might tell your friend that this experiment indicates that depolymerization of F-actin is required for continued movement of a crawling cell. In fact, crawling cells are continuously dismantling actin at the base of lamellipodia, thereby allowing it to be recycled into monomers, which can be added to the advancing leading edge. **Concept Check 13.4,** Extensive intermediate filament networks are found in cells that experience high degrees of mechanical stress. Skin cells are exposed to constant shear stress and have high levels of keratin (in fact, they are called keratinocytes). Smooth muscle cells in a snake intestine, which has to stretch dramatically to ingest prey, experience high levels of tension as well and express the IF proteins vimentin and desmin. Plants have cell walls that resist stress and don't have non-nuclear IFs. Sperm are likewise not subjected to mechanical stress and do not express IFs, especially given what little cytosol they have. **Key Technique,** Paclitaxel stabilizes microtubules. Adding paclitaxel too early in the cell cycle might perturb many processes, so a good experiment would be to add paclitaxel after the mitotic spindle has formed, but before completion of mitosis. If a cell is unable to disassemble its mitotic spindle, which contains numerous MTs, we might predict that it would be unable to progress past mitosis. In fact, paclitaxel is used as an anticancer drug for precisely this reason: it blocks rapidly dividing cells, such as cancer cells, from exiting mitosis and thereby prevents them from proliferating.

Chapter 14 **Concept Check 14.1,** Kinesins and dyneins are similar in that (a) they bind microtubules (MTs), (b) they couple ATP hydrolysis to changes in their shape, and (c) they can transport intracellular cargoes. They are also different: (a) most kinesins move toward the plus end of MTs, whereas dyneins move toward the minus end, and (b) the ATPase/motor of kinesins is in the globular domain of the heavy chain, whereas for dynein, the AAA+ part of the heavy chains transmits forces to the microtubule-binding domains via a linker ("stalk") region. **Concept Check 14.2,** The prominent mitochondrion in animal sperm provides large quantities of ATP that are used by the flagellar dynein that powers microtubule sliding (and hence flagellar bending) in the tail of the sperm. **Concept Check 14.3,** Even though a single myosin II exerts such a small amount of force, when millions of myosins work simultaneously, they are capable of moving large masses. Skeletal muscle fibers contain millions of myosin II motor domains, working together in this way. **Concept Check 14.4,** Actin thin filaments are uniformly oriented, with plus ends always anchored at Z lines. Myosin II always walks toward the plus end of the thin filament, thereby establishing the direction of contraction. In addition, the myosin thick filaments are assembled with their rodlike tail domains in the center of the filament and the globular heads located at the ends of the filament, pointing away from the center of the sarcomere. Therefore, the globular myosin heads at each end of the thick filament have opposite polarities. This results in the movement of the actin thin filaments in opposite directions such that thin filaments from each Z line are drawn to the center of the sarcomere. **Concept Check 14.5,** Myosin II moves toward the plus ends of actin filaments. Lamellipodia contain branched actin with the plus end of filaments oriented toward the periphery (i.e., the leading edge). Myosin II would therefore tend to try to pull the actin largely rearward, away from the leading edge. This is one contributing force in the phenomenon of retrograde flow. **Key Technique,** If myosin V moved like an "inchworm," that is, it jutted its front heavy chain forward first and then dragged its rear heavy chain forward, bringing them together and then repeating this cycle, the graph would look similar, but the step size would be half as big as it is in Figure 14A-2b.

Chapter 15 **Concept Check 15.1,** (1) Adherens junctions and (2) desmosomes. Adherens junctions and desmosomes both use cadherins as their transmembrane adhesive protein, but their intracellular connections differ. Adherens junctions couple cadherins to catenins and then actin, whereas desmosomes couple cadherins via plakoglobin and desmoplakins to intermediate filaments. Coupling junctions to the cytoskeleton has several potential advantages. First, at least in the case of actin, such coupling allows cells to exert forces on their surface by linking actin and myosin networks to adhesive structures. Second, by coupling cell-cell attachments to a rigid internal skeleton, cell-cell adhesion can be strengthened considerably, making them more resilient mechanically. **Concept Check 15.2,** Hemidesmosomes are associated with cells that make stable, relatively nondynamic attachments that experience high mechanical stress (for example, in skin), whereas focal adhesion are associated with cells that adopt a different cellular "lifestyle": motile cells. **Concept Check 15.3,** Plant cells can push on parts of the cell wall through turgor pressure. When their central vacuole is full, the plant cell plasma membrane pushes against the primary cell wall. If there is a local weak spot, cell wall remodeling will preferentially occur in the weakened regions. **Key Technique,** Unlike stiff, unmovable plastic, detached ECMs can be mechanically deformed by cells. There is apparently something about this ability that allows the cells to behave very differently than when cultured on plastic.

Chapter 16 **Concept Check 16.1,** DNA does not contain sulfur, but proteins have significant amounts of sulfur because they have the amino acids methionine and cysteine. Conversely, DNA bases have a phosphate backbone, so they are rich in phosphorus. Some proteins are phosphorylated, but the amount of phosphorus is low compared with DNA. **Concept Check 16.2,** Because %G + %C = 48% and Chargaff's rules require that %G = %C, %G = %C = 24%. This means that %A + %T = 100 − 48 = 52%, so %A = %T = 26%. **Concept Check 16.3,** Highly compacted chromatin presents a "steric hindrance" problem, i.e., its DNA is simply not spatially accessible to the enzymes that catalyze reactions such as transcription. Compacted chromatin undergoes other changes besides this steric issue, but this is an important contributing factor to its relative inactivity in terms of gene expression. **Concept Check 16.4,** To show that your friend's NLS is sufficient, you could use molecular biology techniques to produce a hybrid protein that contains your friend's NLS and the amino acids of your protein. If her NLS is sufficient, then this normally cytosolic protein should now move into the nucleus. The original experiments of Alan Smith and colleagues used pyruvate kinase, a glycolytic enzyme, for just this purpose. **Key Technique,** If you failed to heat your probe to denature it, the hydrogen bonds holding complementary strands together would not be broken. As a result, it would not be possible for the probe to bind to complementary sequences in the target tissue, and your attempt at FISH would not yield any detectable signal.

Chapter 17 **Concept Check 17.1,** There are many possibilities, including the following. Similarities: (1) both involve origins of replication; (2) both require helicases to unwind DNA to start replication; (3) both are typically bidirectional; and (4) both involve several key enzymes (DNA polymerases, primase, ligase, topoisomerases). Differences: (1) in bacteria, DNA polymerase I has RNAase activity, removing primers; special RNAses do this in eukaryotes; (2) nucleosomes must be reassembled in eukaryotes but not bacteria; and (3) eukaryotes have multiple origins of replication. **Concept Check 17.2,** SDSA only works when sister chromatids are available to direct the repair process. This is only true immediately after DNA synthesis and prior to division. If cells are not dividing, they need another way to repair double-strand breaks. NHEJ is less preferable, but an apparently acceptable alternative when a cell is faced with death. **Concept Check 17.3,** RecA, like Rad51, is involved in stabilizing single strands via formation of DNA/protein filaments required for both DNA repair and recombination. In its absence, bacteria will have difficulty performing certain types of DNA repair (for example, SOS system repair, double-strand break repair), and they will be defective in recombination. **Key Technique,** If the Cas9 protein was targeted to a gene promoter but did not generate a double-strand break in the DNA, no subsequent genome editing would occur. However, remaining bound to the promoter would prevent RNA polymerase from binding and initiating transcription. Therefore, the gene would be effectively switched "off." Scientists are coupling this with Cas9 joined with a transcription-activating protein targeted to the same promoter. This provides a rapid way to specifically turn genes on and off.

Chapter 18 **Concept Check 18.1,** Eukaryotic and prokaryotic transcription are similar in that they both utilize the same main components. These include mRNA, tRNA, and rRNA. However, the location of transcription relative to translation is different in these two cell types. In prokaryotic cells, both transcription and translation occur in the cytosol, allowing the two processes to be coupled. In other words, translation can begin prior to the completion of transcription. This is an advantage because these cells can begin producing protein almost immediately after transcription begins, resulting in rapid responses to certain stimuli. In eukaryotic cells, transcription occurs in the nucleus, while translation occurs in the cytosol. This compartmentalization is advantageous to eukaryotic cells, as it allows for mRNA to be modified prior to its use in translation. Splicing of introns, the addition of a 5′ cap and 3′ tail, and RNA editing affect the stability of the mRNA and can increase the diversity of protein produced from a single gene. In addition, regulating mRNA export from the nucleus allows for finer control over gene expression. **Concept Check 18.2,** The following are features similar in bacterial and eukaryotic transcription: (1) In initiation, RNA polymerase does not need a primer in both bacteria and eukaryotes. (2) In elongation, RNA polymerase moves along DNA, catalyzing RNA production in much the same way in bacteria and eukaryotes. (3) During proofreading, weak RNA proofreading occurs in both bacteria and eukaryotes. In addition to the distinct locations in which bacterial and eukaryotic transcription takes place (see Figure 18-2), there are several additional differences between them, including the following: (1) In initiation, bacteria use sigma factor (part of the RNA polymerase holoenzyme) to dock with a promoter site, whereas eukaryotes use general transcription factors that are not part of the RNA polymerase holoenzyme. (2) In initiation, eukaryotic promoters are more variable than those in bacteria. (3) In termination, bacteria use hairpins or rho factor to terminate transcripts, whereas eukaryotes transcribe long sequences that are cleaved off. (4) During processing, eukaryotic RNAs undergo extensive processing that is not the case in bacteria. **Concept Check 18.3,** snRNPs are essential for the splicing of primary transcripts in eukaryotes. If autoimmune antibodies interfere with snRNPs and mRNAs occasionally fail to undergo splicing, the results might be catastrophic. A defective mRNA that contains an intron, for example, might result in "garbage" amino acids that are not normally part of the protein or, more commonly, early termination of translation due to stop codons, or the defective mRNA might be degraded by RNA surveillance systems (a topic of Chapter 20). **Key Technique,** DNA footprinting is more specific than EMSA because it provides sequence information on exactly which bases are protected from digestion (the "footprint"). Both require knowledge of the specific DNA that interacts with a protein of interest, either for performing a gel shift (EMSA) or for looking for nuclease protection of a specific piece of DNA (footprinting). ChIP has the same requirement if PCR is then used to amplify a specific sequence. ChIP-chip and ChIP-seq do not require such prior knowledge and are therefore useful for identifying a wide range of DNA sequences to which a protein binds. Individual "hits" can then be followed up using footprinting or EMSA.

Chapter 19 **Concept Check 19.1,** The template strand is used to synthesize RNA, starting with the 3′ end of DNA. To get the RNA sequence, we need to generate the reverse complement to the DNA, substituting U's for T's. The codon RNA corresponding to ATA would therefore be UAU, which encodes tyrosine. The mutations work out as follows (nucleic acids are listed 5′→3′ in each case):

DNA	RNA	Amino acid
ATA	UAU	Tyrosine
GTA	UAC	Tyrosine
TTA	UAA	Stop codon
GCA	UGC	Cysteine

The first mutation is a "silent" mutation and would have no effect on the resulting protein. The second might have a drastic effect because it would lead to a truncated protein with a premature stop codon. The third might alter protein function because tyrosine and cysteine have very different R groups that might affect protein function. **Concept Check 19.2,** Due to wobble, this anticodon can bind to two different codons: 5′-GCA-3′ and 5′-GCG-3′. These each encode the amino acid alanine; this tRNA therefore brings Ala into the growing polypeptide. **Concept Check 19.3,** Kanamycin would be predicted to allow any tRNA at random to bind the A site and then transfer its attached amino acid to a growing polypeptide. It would therefore cause incorporation of incorrect amino acids into a polypeptide. **Concept Check 19.4,** Nonsense mutations do not add or remove bases, but change a base so that the resulting codon is a stop codon. Therefore, there would be no change in length of the mRNA. The protein, however, would be expected to be shorter than normal because the stop codon would cause premature termination of translation. **Concept Check 19.5,** Posttranslational modifications all involve covalent modification. **Key Technique,** To show that specific amino acids are necessary to target the protein to junctions, you could use molecular biology techniques to remove or mutate the DNA encoding the relevant amino acids. After introducing this DNA into cells and allowing them to form cell-cell junctions in vitro, you could then use fluorescence microscopy to see if the mutated, GFP-tagged protein still localizes to junctions. If it does not, the amino acids are necessary. Conversely, if you wished to show that these amino acids are sufficient, you could make a DNA construct that encodes only these amino acids fused to GFP. If the protein localizes to junctions, these amino acids are sufficient to target the protein to junctions.

Chapter 20 **Concept Check 20.1,** One possibility is that the promoter has been mutated so that RNA polymerase cannot bind; thus under no circumstances could the genes of the operon be transcribed. Another possibility is that there has been a mutation in the gene encoding the *lac* repressor protein so that it can no longer bind lactose or undergo a conformational change once lactose binds. This would cause the repressor to always remain bound to the operator (i.e., it becomes a "superrepressor"). **Concept Check 20.2,** Mutations involve changes to the identity of a base at a given position along the DNA (these include missense, nonsense, or indel mutations), whereas as methylation retains the same base (e.g., a C), but that base is chemically altered by covalent modification to add a methyl group. **Concept Check 20.3,** They are similar in that the transcriptional state of both bacterial and eukaryotic genes can be regulated by DNA-binding proteins (i.e., regulatory transcription factors). They are different in that bacterial genes are usually in the "on" position unless actively shut off. A good example is the *lac* operon, which would normally be "on," except that a repressor protein binds to the operator when lactose is not present. In contrast, eukaryotic proteins are in the "off" position unless activated by a transcription activator protein. **Concept Check 20.4,** The 3′ untranslated region (3′ -UTR) is likely affected. This is the region that is bound by translational regulatory proteins and/or microRNAs. If this region is mutated, it is possible that normal translational control will be perturbed, leading to translation in all four cells of the early embryo. **Key Technique,** RNAi is not gene-specific, but sequence-specific. You should use an exonic region of the gene of interest that is unique, rather than something in the region that is identical. Otherwise, your shRNA would knock down mRNA corresponding to both genes.

Chapter 21 **Concept Check 21.1,** You should start with cDNA. Bacteria cannot recognize and remove intronic sequences from mRNA. If you were to include genomic DNA in your clone, the bacteria would try to transcribe the intronic sequences, likely leading to early stop codons or nonrelevant amino acids. **Concept Check 21.2,** SNPs are single nucleotide polymorphisms, that is, they lead to changes in a single base. Sometimes, but not always, these can lead to creation or removal of restriction sites that could generate restriction fragment length polymorphisms (RFLPs). Thus, some SNPs can lead to RFLPs, but these genetic markers are partially nonoverlapping in terms of what they reveal. SNPs must be detected by sequencing; RFLPs can be detected using Southern blotting or PCR (in either case different sized fragments would result). **Concept Check 21.3,** (1) Similarities: Both separate molecules by size using electrophoresis. Both can identify the size of a specific molecules amid a complex mixture of molecules. Differences: Northern blots detect mRNA; Western blots detect proteins. Northern blots use nucleic acid probes; Western blots use antibodies. The types of gels used also differ (agarose versus SDS-PAGE). (2) Similarities: Both assess protein-protein interactions. Differences: Yeast two-hybrid assays assess whether two proteins *directly* bind; coIPs assess whether two proteins are in the same *macromolecular complex* (but they need not bind directly to be part of the same complex identified via coIP). coIPs isolate native protein complexes from cells; yeast two-hybrid assays are done in a heterologous (different) system (yeast). coIPs rely on antibodies for affinity purification of a complex; yeast two-hybrid assays do not require antibodies to be available. **Concept Check 21.4,** cDNA clones should be used for this purpose. See Concept Check 21.1 for more on the issue of introns. In addition, the cDNA would have to be cloned into the appropriate expression vector plasmid. The plasmid would then have to be used to transform *E. coli*. Once in *E. coli*, the bacteria would have to be induced to express the protein. One limitation of bacterial expression systems is that if the protein is normally glycosylated or otherwise posttranslationally modified (often in the endoplasmic reticulum or Golgi apparatus), this will not occur in bacteria. **Key Technique,** The theoretical amplification accomplished by n cycles is 2^n, so 20 cycles yields an amplification of over a millionfold (2^{20} = 1,048,576), and 30 cycles yields over a billionfold (2^{30} = 1,073,741,824).

Chapter 22 **Concept Check 22.1,** The Na^+/K^+ ATPase is responsible for maintenance of the resting potential of neurons. If this pump did not maintain the normal concentration gradients of Na^+ and K^+ across the membrane, the resting potential would be much more positive than the standard −70 mV. **Concept Check 22.2,** If voltage-gated Na^+ channels are permanently open, the membrane will tend to move toward (but not completely reach, due to Na^+/K^+ ATPase activity, channel permeability, etc.) the equilibrium potential for Na^+. This will cause the membrane potential to become positive, that is, the cell will become chronically depolarized. This will ultimately lead to the death of the neuron. **Concept Check 22.3,** If an excitatory neurotransmitter remains in the synaptic cleft for prolonged periods of time, the postsynaptic neuron will experience an excitatory postsynaptic potential (EPSP) for a longer period of time than normal. This will result in a prolonged period during which a rapid succession of action potentials will be generated in the postsynaptic neuron. **Key Technique,** ACh receptors are ionotropic receptors that act as ligand-gated Na^+ channels. You could use the whole-cell (for measuring effects on the entire cell) or outside-out patch-clamp methods. In the presence of standard buffer, you would measure current before and after exposure to ACh. In the whole-cell case, there would be a transient depolarization of the cell (that is, its voltage would become temporarily positive but would not remain so). In the outside-out case, which seeks to measure current through a single channel, there would be a step change in current similar to that in Figure 22A-1, ❸, which would eventually drop back to steady state because the channel does not stay open indefinitely. In contrast, the toxin would cause a permanent depolarization (whole-cell) or a persistent increase in current through a single channel (outside-out).

Chapter 23 **Concept Check 23.1,** This patient's insulin receptors are too sensitive (lower K_d means a higher affinity for insulin), so a lower concentration of insulin stimulates the same production of glycogen. If glucose is removed by adding it to glycogen more readily than normal, the glucose concentration will decrease, leading to hypoglycemia. **Concept Check 23.2,** Cells respond to *changes* in second messenger concentration. If the concentration of a second messenger is low, a smaller change in the total amount of calcium ions is required to cause a large relative change in calcium concentration and hence a biological effect. **Concept Check 23.3,** GAPs cause the more rapid *inactivation* of Ras because they stimulate Ras to hydrolyze its associated GTP to GDP. When the GAP is knocked down, Ras will remain active for a longer period of time, leading to an increase in cell division compared with untreated control cells. **Concept Check 23.4,** PKA is normally stimulated by cAMP in response to epinephrine. Epinephrine should therefore lead to more activated PKA, which should *inhibit* Raf, which is part of receptor tyrosine kinase signaling pathways that are activated by growth factors. You would therefore expect cells to divide less rapidly when treated with both epinephrine and growth factor. **Concept Check 23.5,** Local mediators allow spatially refined control of the effects of a signal once it is produced. They cannot diffuse very far, so the effects of the mediator are tightly controlled. A disadvantage is that local mediators cannot cause changes throughout an organism as rapidly as hormones can because the latter are carried through the circulatory system. In addition, local mediators tend to be short-lived, so their effects are limited. **Key Technique,** Assuming that a calcium indicator can be injected into the egg, you could inject the eggs with indicator and then expose them to ionophore while examining them using a fluorescence microscope. You would then expect that ionophore would cause local pockets of calcium-induced calcium release that spread throughout the ionophore-treated eggs, even when they have not been fertilized by a sperm cell.

Chapter 24 **Concept Check 24.1,** Sister chromatids. They are duplicates generated via DNA replication (except for occasional errors in replication), whereas different alleles of different genes can reside on each of the two homologous chromosomes. **Concept Check 24.2,** Myosin is involved in contractile ring closure during cytokinesis but is not involved in spindle assembly and chromosome movements during mitosis. As a result, you might expect that cytokinesis would fail but that DNA replication and mitosis would occur. This is in fact what does happen in most cases. The result is a multinucleate cell with two nuclei in a common cytosol. **Concept Check 24.3,** The G1 Cdk-cyclin complex regulates the restriction point/Start. One way in which it does so is via phosphorylating the Rb protein. There are at least two ways in which excess cyclin D could feed into this pathway: (1) With excess cyclin D, it is more likely that a cell will move past the restriction point into S phase inappropriately in the absence of a normal growth factor signal. (2) The DNA damage checkpoint might be less likely to prevent DNA synthesis because one way it acts is via p21 to inhibit the G1 Cdk-cyclin complex. It will be more difficult for p21 to inactivate the complex if there is more of it present. Either situation might lead to excess proliferation and other problems associated with cancer. **Concept Check 24.4,** EGF acts via Ras to activate the AP1 complex, leading to transcription of mRNAs encoding cyclins and Cdks; that is, it positively regulates G1 Cdk-cyclin action at the transcriptional level. In contrast, TGFβ acts via Smads, leading to transcription of mRNAs encoding Cdk inhibitors, such as p15 and p21; that is, it negatively regulates G1 Cdk-cyclin action, also at the transcriptional level. When a cell is exposed to both, the opposing actions tend to counteract one another. In this sense, the level of multiple growth factors to which a cell is exposed fine-tunes its proliferative response. **Concept Check 24.5,** Normally, there is significant "pruning" of neurons via apoptosis in the developing mammalian brain. Knockout mice lacking the initiator caspase, caspase-9, would be expected to undergo less apoptosis, leading to excessive cells in the brain. This is in fact what happens, leading to fetal death. **Key Technique,** You could perform immunostaining using anti-CD19 antibodies and a fluorescent secondary antibody (see Figure 1A-3) and stain with a fluorescent DNA dye that can be detected using a different wavelength. If you set up the flow cytometer so that it detects these two fluorescent probes, you can determine how many cells are both CD19-positive and in S/M phase, indicating they are rapidly proliferating. A patient with lymphoma should have proportionately more of these cells than a patient with normal white blood cells.

Chapter 25 **Concept Check 25.1,** Propagating a species via SCNT would lead to decreased genetic diversity. This in turn might lead to greater susceptibility to disease or a higher incidence of homozygosity, which might be deleterious. **Concept Check 25.2,** By metaphase of meiosis I, the DNA would have been replicated, that is, 2*X*. After meiosis I, each daughter cell would have half of this amount of DNA (that is, *X*). Because there is no DNA replication during meiosis II, but the second division has not occurred, the cells would still have *X* DNA. **Concept Check 25.3,** The two phenotypes result from a single allele. Half of the gametes from each parent will have the allele, half will not. Thus, one-quarter of the offspring are expected to be homozygous for the allele, leading to white coloration. Because only a single gene is involved and leads to both phenotypes, all white tigers will have crossed eyes. **Concept Check 25.4,** No; you would need more information. There are two possible arrangements for genes *A*, *B*, and *C* (map distances are shown as numbers):

*A*____10____*C*____15____*B* or *B*____25____*A*____10____*C*

You would need to know the recombination frequency between genes *B* and *C* to determine which possibility is correct. **Concept Check 25.5,** Artificial selection due to antibiotics would favor bacteria with the R factor plasmid. This would lead, over time, to an increase in the percentage of bacteria that carry the R factor plasmid and therefore to an increase in antibiotic-resistant bacteria. This is in fact a current medical problem in countries with widespread use of antibiotics. **Concept Check 25.6,** Loss of synaptonemal complexes might be expected to lead to loss of stabilization of chiasmata and a higher frequency of incorrectly repaired double-strand breaks. This is precisely what is observed in mutant flies, worms, and mice in which synaptonemal proteins are absent. **Key Technique,** The mother has one dominant allele, which led to her polydactyly; the father has normal alleles. Half of the mother's oocytes, on average, would therefore carry the dominant allele, so half of her offspring would be expected to receive one dominant allele. All offspring would receive a normal allele from their father. Thus we expect half of the children to have polydactyly, and for any given child, the chance is 50%.

Chapter 26 **Concept Check 26.1,** These data indicate that the lymphoma in any particular patient arose from a single cell in which the particular pattern of chromosomal breakage and reattachment occurred. This supports the clonal origin of Burkitt lymphoma. **Concept Check 26.2,** This experiment has actually been performed. The p120catenin knockout cells possess cadherin complexes that are less stable, which might lead to various defects. One possibility is that the cells will escape normal growth suppression due to decreased adhesion. Another possibility is that because they have weakened adhesion, they are more likely to dissociate and metastasize to form secondary tumors. **Concept Check 26.3,** Although smoking exposes air passages to carcinogens that might serve as initiators of cancer, additional mutations or changes in gene expression must occur to promote cancer and foster tumor growth. Not all smokers develop these secondary conditions. In addition, there is significant patient-to-patient genetic variation in many genes (and hence proteins) that impinge on cancer susceptibility. These genetic variations also affect the likelihood of growth defects, etc. that promote cancer. **Concept Check 26.4,** A cancer-causing mutation related to a proto-oncogene would be expected to *enhance* its activity. Consider one example, the epidermal growth factor receptor Her2. Overexpression of the normal Her2 protein would make cells too sensitive to growth factor stimulation. Alternatively, a mutation that makes the Her2 protein act as if it is always "on," that is, constitutively activated, would promote proliferation without the presence of a signal. Tumor suppressors, on the other hand, must be *inactivated* to promote cancer. Consider the Rb protein, which normally suppresses transition from the G1 phase to S phase in the absence of a growth-promoting signal. If a cell lacks Rb function, then it will pass into S phase whether or not it receives growth factor signals. **Concept Check 26.5,** A patient's own immune system appears to be involved in attacking cancer cells, which would normally be recognized by virtue of the different cell surface proteins expressed by the cancer cells. Suppression of the immune response via immunosuppressant drugs reduces the effectiveness of this response and hence is like the effects of immune suppression through AIDS or similar diseases. In either case, the result is an increased cancer risk. **Key Technique,** Although antibodies directed against CD20 are toxic to normal lymphocytes, CD20 is not present on the precursor cells whose proliferation gives rise to these lymphocytes. As a result, proliferation of these precursor cells can replenish the normal lymphocytes inadvertently destroyed by antibodies directed against CD20.

GLOSSARY

Note: The letter "A" preceding a page number refers to a page in the Appendix: *Visualizing Cells and Molecules.*

A

A: see *adenine.*

A band: region of a striated muscle myofibril that appears as a dark band when viewed by microscopy; contains thick myosin filaments and those regions of the thin actin filaments that overlap the thick filaments. (p. 397)

A site (aminoacyl site): site on the ribosome that binds each newly arriving tRNA with its attached amino acid. (p. 547)

A tubule: a complete microtubule that is fused to an incomplete microtubule (the B tubule) to make up an outer doublet in the axoneme of a eukaryotic cilium or flagellum. (p. 392)

ABC transporter: see *ABC-type ATPase.* (p.204)

ABC-type ATPase: type of transport ATPase characterized by an "ATP-binding cassette" (hence the "ABC"), with the term cassette used to describe catalytic domains of the protein that bind ATP as an integral part of the transport process; also called ABC transporters. (p. 204) Also see *multidrug resistance transport protein.*

abscission: the final step of cytokinesis in animal cells, which involves the severing of the narrow connection, or midbody, between the two daughter cells. (p. 670)

absolute refractory period: brief time during which the sodium channels of a nerve cell are inactivated and cannot be opened by depolarization. (p. 731)

absorption spectrum: relative extent to which light of different wavelengths is absorbed by a pigment. (p. 288)

accelerating voltage: difference in voltage between the cathode and anode of an electron microscope, responsible for accelerating electrons prior to their emission from the electron gun. (p. A-17)

accessory pigments: molecules such as a carotenoid or phycobilin that confer enhanced light-gathering properties on photosynthetic tissue by absorbing light of wavelengths not absorbed by chlorophyll; accessory pigments give distinctive colors to plant tissue, depending on their specific absorption properties. (p. 289)

acetyl CoA: high-energy, two-carbon compound generated by glycolysis and fatty acid oxidation; employed for transferring carbon atoms to the tricarboxylic acid cycle. (p. 251)

acetylcholine: the most common excitatory neurotransmitter used at synapses between neurons outside the central nervous system. (p. 675)

actin: principal protein of the microfilaments found in the cytoskeleton of nonmuscle cells and in the thin filaments of skeletal muscle; synthesized as a globular monomer (G-actin) that polymerizes into long, linear filaments (F-actin). (p. 365)

actin-binding proteins: proteins that bind to actin microfilaments, thereby regulating the length or assembly of microfilaments or mediating their association with each other or with other cellular structures, such as the plasma membrane. (p. 374)

action potential: brief change in membrane potential involving an initial depolarization followed by a rapid return to the normal resting potential; caused by the inward movement of Na^+ followed by the subsequent outward movement of K^+; serves as the means of transmission of a nerve impulse. (p. 668)

activated monomer: a monomer whose free energy has been increased by being linked via a high-energy bond to a carrier molecule. (p. 35)

activation domain: region of a transcription factor, distinct from the DNA-binding domain, that is responsible for activating transcription. (p. 597)

activation energy (E_A): energy required to initiate a chemical reaction. (p. 131)

active site: region of an enzyme molecule at which the substrate binds and the catalytic event occurs; also called the catalytic site. (p. 130)

active transport: membrane protein-mediated movement of a substance across a membrane against a concentration or electrochemical gradient; an energy-requiring process. (p. 202)

active zone: region of the presynaptic membrane of an axon where neurosecretory vesicles dock. (p. 677)

adaptor protein (AP) complexes: protein complexes found along with clathrin in the coats of clathrin-coated vesicles. (p. 343)

adenine (A): nitrogen-containing aromatic base, chemically designated as a purine, that serves as an informational monomeric unit when present in nucleic acids with other bases in a specific sequence; forms a complementary base pair with thymine (T) or uracil (U) by hydrogen bonding. (p. 60)

adenosine diphosphate (ADP): adenosine with two phosphates linked to each other by a phosphoanhydride bond and to the 5′ carbon of the ribose by a phosphoester bond. (p. 61)

adenosine monophosphate (AMP): adenosine with a phosphate linked to the 5′ carbon of ribose by a phosphoester bond. (p. 61)

adenosine triphosphate (ATP): adenosine with three phosphates linked to each other by phosphoanhydride bonds and to the 5′ carbon of the ribose by a phosphoester bond; principal energy storage compound of most cells, with energy stored in the high-energy phosphoanhydride bonds. (pp. 60, 216)

adenylyl cyclase: enzyme that catalyzes the formation of cyclic AMP from ATP; located on the inner surface of the plasma membrane of many eukaryotic cells and activated by specific ligand-receptor interactions on the outer surface of the membrane. (p. 691)

adherens junction: junction for cell-cell adhesion that is connected to the cytoskeleton by actin microfilaments. (p. 415)

ADP: see *adenosine diphosphate.*

ADP ribosylation factor (ARF): a protein associated with COPI in the "fuzzy" coats of COPI-coated vesicles. (p. 344)

adrenergic hormone: epinephrine or norepinephrine. (p. 707)

adrenergic receptor: any of a family of G protein–coupled receptors that bind to one or both of the adrenergic hormones, epinephrine and norepinephrine. (p. 708)

adrenergic synapse: a synapse that uses norepinephrine or epinephrine as the neurotransmitter. (p. 676)

aerobic respiration: exergonic process by which cells oxidize glucose to carbon dioxide and water using oxygen as the ultimate electron acceptor, with a significant portion of the released energy conserved as ATP. (p. 222)

affinity chromatography: a type of chromatography in which ions (e.g., Ni^{2+}) or other molecules (e.g., maltose, glutathione, or antibodies) are attached to the surface of beads, which can then be used to purify tagged or endogenous proteins. (p.648)

agonist: substance that binds to a receptor and activates it. (p. 687)

Akt: protein kinase involved in the PI 3-kinase–Akt pathway; catalyzes the phosphorylation of several target proteins that suppress apoptosis and inhibit cell cycle arrest. (p. 709)

alcoholic fermentation: anaerobic catabolism of carbohydrates with ethanol and carbon dioxide as the end-products. (p. 228)

allele: one of two or more alternative forms of a gene. (p. 761)

allosteric activator: a small molecule whose binding to an enzyme's allosteric site shifts the equilibrium to favor the high-affinity state of the enzyme. (p. 147)

allosteric effector: small molecule that causes a change in the state of an allosteric protein by binding to a site other than the active site. (p. 147)

allosteric enzyme: an enzyme exhibiting two alternative forms, each with a different biological property; interconversion of the two states is mediated by the reversible binding of a specific small molecule (allosteric effector) to a regulatory site called the allosteric site. (p. 147)

allosteric inhibitor: a small molecule whose binding to an enzyme's allosteric site shifts the equilibrium to favor the low-affinity state of the enzyme. (p. 148)

allosteric regulation: control of a reaction pathway by the effector-mediated reversible interconversion of the two forms of an allosteric enzyme. (pp. 140, 235, 255)

allosteric (or regulatory) site: region of a protein molecule that is distinct from the active site at which the catalytic event occurs and that binds selectively to a small molecule, thereby regulating the protein's activity. (p. 145)

alpha beta heterodimer ($\alpha\beta$-heterodimer): protein dimer composed of one α-tubulin molecule and one β-tubulin molecule that forms the basic building block of microtubules. (p. 362)

alpha helix (α helix): spiral-shaped secondary structure of protein molecules, consisting of a backbone of peptide bonds with R groups of amino acids jutting out. (p. 52)

alpha tubulin (α-tubulin): protein that joins with β-tubulin to form a heterodimer that is the basic building block of microtubules. (p. 361)

alternating conformation model: membrane transport model in which a carrier protein alternates between two conformational states, such that the solute-binding site of the protein is open or accessible first to one side of the membrane and then to the other. (p. 193)

alternation of generations: occurrence of alternating haploid and diploid multicellular forms within the life cycle of an organism. (p. 754)

alternative splicing: utilization of different combinations of intron/exon splice junctions in pre-mRNA to produce messenger RNAs that differ in exon composition, thereby allowing production of more than one type of polypeptide from the same gene. (pp. 529, 606)

Ames test: screening test for potential carcinogens that assesses whether a substance causes mutations in bacteria. (p. 793)
amino acid: monomeric unit of proteins, consisting of a carboxylic acid with an amino group and one of a variety of R groups attached to the α carbon; 20 different kinds of amino acids are normally found in proteins. (p. 43)
amino terminus: see *N-terminus.*
aminoacyl site: see *A site.*
aminoacyl tRNA: a tRNA molecule containing an amino acid attached to its 3′ end. (p. 549)
aminoacyl-tRNA synthetase: enzyme that joins an amino acid to its appropriate tRNA molecule using energy provided by the hydrolysis of ATP. (p. 550)
amoeboid movement: mode of cell locomotion that depends on pseudopodia and involves cycles of gelation and solation of the actin cytoskeleton. (p. 408)
AMP: see *adenosine monophosphate.*
amphibolic pathway: series of reactions that can function both in a catabolic mode and as a source of precursors for anabolic pathways. (p. 259)
amphipathic molecule: molecule having spatially separated hydrophilic and hydrophobic regions. (p. 31)
amylopectin: branched-chain form of starch consisting of repeating glucose subunits linked together by α (1 → 4) glycosidic bonds, with occasional α (1 → 6) linkages creating branches every 12 to 25 units that commonly consist of 20 to 25 glucose units. (p. 64)
amylose: straight-chain form of starch consisting of repeating glucose subunits linked together by α (1 → 4) glycosidic bonds. (p. 64)
anabolic pathway: series of reactions that results in the synthesis of cellular components. (p. 216)
anaerobic respiration: cellular respiration in which the ultimate electron acceptor is a molecule other than oxygen. (p. 222)
anaphase: stage during mitosis (or meiosis) when the sister chromatids (or homologous chromosomes) separate and move to opposite spindle poles. (p. 726)
anaphase A: movement of sister chromatids toward opposite spindle poles during anaphase. (p. 726)
anaphase B: movement of the spindle poles away from each other during anaphase. (p. 726)
anaphase-promoting complex (APC/C): large multiprotein complex that targets selected proteins (e.g., securin and mitotic cyclin) for degradation, thereby initiating anaphase and the subsequent completion of mitosis. (p. 739)
anchorage-dependent growth: requirement that cells be attached to a solid surface such as the extracellular matrix before they can grow and divide. (p. 431)
anchorage-independent growth: a trait exhibited by cancer cells, which grow well not just when they are attached to a solid surface, but also when they are freely suspended in a liquid or semisolid medium. (p. 785)
anchoring junction: see *adhesive junction.*
aneuploidy: abnormal state in which a cell possesses an incorrect number of chromosomes. (p. 734)
angiogenesis: growth of new blood vessels. (p. 783)
Aneuploidy of Sex Chromosomes 763
angular aperture: half-angle of the cone of light entering the objective lens of a microscope from the specimen. (p. A-2)
anion exchange protein: antiport carrier protein that facilitates the reciprocal exchange of chloride and bicarbonate ions across the plasma membrane. (p. 195)
anoxygenic phototroph: a photosynthetic organism that uses an oxidizable substrate other than water as the electron donor in photosynthetic electron transduction. (p. 284)
antagonist: substance that binds to a receptor and prevents it from being activated. (p. 687)
antenna pigment: light-absorbing molecule of a photosystem that absorbs photons and passes the energy to a neighboring chlorophyll molecule or accessory pigment by resonance energy transfer. (p. 290)
anterograde transport: movement of material from the ER through the Golgi apparatus toward the plasma membrane. (p. 321)
antibody: class of proteins produced by lymphocytes that bind with extraordinary specificity to substances, referred to as antigens, that provoke an immune response. (p. 6)
anticodon: triplet of nucleotides located in one of the loops of a tRNA molecule that recognizes the appropriate codon in mRNA by complementary base pairing. (p. 539)
antigen: a foreign or abnormal substance that can trigger an immune response. (p. 646)
antiport: coupled transport of two solutes across a membrane in opposite directions. (p. 194)
AP: see *adaptor protein.*
***APC* gene:** a tumor suppressor gene, frequently mutated in colon cancers, that codes for a protein involved in the Wnt pathway. (p. 799)
APC/C: see *anaphase promoting complex.*
apoptosis: cell suicide mediated by a group of protein-degrading enzymes called caspases; involves a programmed series of events that leads to the dismantling of the internal contents of the cell. (p. 745)
AQP: see *aquaporin.*
aquaporin (AQP): any of a family of membrane channel proteins that facilitate the rapid movement of water molecules into or out of cells in tissues that require this capability, such as the proximal tubules of the kidneys. (p. 199)
Archaea: one of the two domains of prokaryotes, the other being Bacteria; many archaea thrive under harsh conditions, such as salty, acidic, or hot environments, that would be fatal to most other organisms. (p. 80) Also see *Bacteria.*
ARF: see *ADP ribosylation factor.*
Argonaute protein: ribonuclease present in the RISC that cleaves mRNA at the site where the RISC has become bound. (p. 610)
Arp2/3 complex: complex of actin-related proteins that allows actin monomers to polymerize as new "branches" on the sides of existing microfilaments. (p. 377)
asexual reproduction: form of reproduction in which a single parent is the only contributor of genetic information to the new organism. (p. 751)
astral microtubule: type of microtubule that forms asters, which are dense starbursts of microtubules that radiate in all directions from each spindle pole. (p. 726)
asymmetric carbon atom: carbon atom that has four different substituents. Two different stereoisomers are possible for each asymmetric carbon atom in an organic molecule. (p. 25)
ATP: see *adenosine triphosphate.*
ATP synthase: alternative name for an F-type ATPase when it catalyzes the reverse process in which the exergonic flow of protons down their electrochemical gradient is used to drive ATP synthesis; examples include the CF_oCF_1 complex found in chloroplast thylakoid membranes, and the F_oF_1 complex found in mitochondrial inner membranes and bacterial plasma membranes. (pp. 204, 249, 274, 298)
attenuation: mechanism for regulating bacterial gene expression based on the premature termination of transcription. (p. 581)
autonomous transposable element: a transposable element that encodes transposase and can initiate its own transposition. (p. 505)
autophagic lysosome: mature lysosome containing hydrolytic enzymes involved in the digestion of materials of intracellular origin. (p. 346) Also see *heterophagic lysosome.*
autophagic vacuole (autophagosome): vacuole formed when an old or unwanted organelle or other cellular structure is wrapped in membranes derived from the endoplasmic reticulum prior to digestion by lysosomal enzymes. (p. 348)
autophagosome: see *autophagic vacuole.*
autophagy: intracellular digestion of old or unwanted organelles or other cell structures as it occurs within autophagic lysosomes; "self-eating." (p. 347)
autophosphorylation: phosphorylation of a receptor molecule by a receptor molecule of the same type. (p. 700)
axon: extension of a nerve cell that conducts impulses away from the cell body. (p. 665)
axon hillock: region at the base of an axon where action potentials are initiated most easily. (p. 671)
axonal transport: see *fast axonal transport.*
axonemal dynein: motor protein in the axonemes of cilia and flagella that generates axonemal motility by moving along the surface of microtubules driven by energy derived from ATP hydrolysis. (p. 390)
axonemal microtubules: microtubules present in highly ordered bundles in the axonemes of eukaryotic cilia and flagella. (p. 361)
axoneme: group of interconnected microtubules that form the backbone of a eukaryotic cilium or flagellum, usually arranged as nine outer doublet microtubules surrounding a pair of central microtubules. (p. 392)
axoplasm: cytoplasm within the axon of a nerve cell. (p. 665)

B

B tubule: an incomplete microtubule that is fused to a complete microtubule (the A tubule) to make up an outer doublet in the axoneme of a eukaryotic cilium or flagellum. (p. 392)
BAC: see *bacterial artificial chromosome.*
Bacteria: one of the two domains of prokaryotes, the other being Archaea; include most of the commonly encountered single-celled organisms with no nucleus that have traditionally been called bacteria (singular, bacterium). (p. 80) Also see *Archaea.*
bacterial artificial chromosome (BAC): bacterial cloning vector derived from the F-factor plasmid that is useful in cloning large DNA fragments. (p. 631)
bacterial chromosome: circular (or linear) DNA molecule with bound proteins that contains the main genome of a bacterial cell. (p. 453)
bacteriochlorophyll: type of chlorophyll found in bacteria that is able to extract electrons from donors other than water. (p. 289)
bacteriophage (phage): virus that infects bacterial cells. (p. 441)
bacteriorhodopsin: transmembrane protein complexed with rhodopsin, capable of transporting protons across the bacterial cell membrane to create a light-dependent electrochemical proton gradient. (p. 208)
basal body: microtubule-containing structure located at the base of a eukaryotic flagellum or cilium that consists of nine sets of triplet microtubules; identical in appearance to a centriole. (p. 392)

basal lamina: thin sheet of specialized extracellular matrix material that separates epithelial cells from underlying connective tissues. (p. 427)
base excision repair: DNA repair mechanism that removes and replaces single damaged bases in DNA. (p. 495)
base pair (bp): pair of nucleotides joined together by complementary hydrogen bonding. (p. 637)
base pairing: complementary relationship between purines and pyrimidines based on hydrogen bonding that provides a mechanism for nucleic acids to recognize and bind to each other; involves the pairing of A with T or U, and the pairing of G with C. (p. 62)
base stacking: phenomenon in which hydrophobic and van der Waals interactions between aromatic ring structures exert forces on adjacent chemical structures. In DNA, base stacking between adjacent nitrogenous bases on the same strand leads to a 30° tilt in the DNA backbone that stabilizes the structure of DNA. (p. 450)
Bcl-2: protein located in the outer mitochondrial membrane that blocks cell death by apoptosis. (p. 742)
benign tumor: tumor that grows only locally, unable to invade neighboring tissues or spread to other parts of the body. (p. 785)
beta oxidation (β oxidation): pathway involving successive cycles of fatty acid oxidation in which the fatty acid chain is shortened each time by two carbon atoms released as acetyl CoA. (p. 257)
beta sheet (β sheet): extended sheetlike secondary structure of proteins in which adjacent polypeptides are linked by hydrogen bonds between amino and carbonyl groups. (p. 52)
beta tubulin (β-tubulin): protein that joins with α-tubulin to form a heterodimer that is the basic building block of microtubules. (p. 362)
bimetallic iron-copper (Fe-Cu) center: a complex formed between a single copper atom and the iron atom bound to the heme group of an oxygen-binding cytochrome such as cytochrome a_3; important in keeping an O_2 molecule bound to the cytochrome until it picks up four electrons and four protons, resulting in the release of two molecules of water. (p. 262)
binding change model: a mechanism involving the physical rotation of the γ subunit of an F_oF_1 ATP synthase postulated to explain how the exergonic flow of protons through the F_o component of the complex drives the otherwise endergonic phosphorylation of ADP to ATP by the F_1 component. (p. 276)
biochemistry: study of the chemistry of living systems; same as biological chemistry. (p. 4)
bioenergetics: area of science that deals with the application of thermodynamic principles to reactions and processes in the biological world. (p. 109)
bioinformatics: using computers to analyze the vast amounts of data generated by sequencing and expression studies on genomes and proteomes. (pp. 13, 638)
biological chemistry: study of the chemistry of living systems; called biochemistry for short. (p. 22)
bioluminescence: production of light by an organism as a result of the reaction of ATP with specific luminescent compounds. (p. 109)
biosynthesis: generation of new molecules through a series of chemical reactions within the cell. (p. 108)
BiP: member of the Hsp70 family of chaperones; present in the ER lumen, where it facilitates protein folding by reversibly binding to the hydrophobic regions of polypeptide chains. (p. 321, 561)
bivalent: pair of homologous chromosomes that have synapsed during the first meiotic division; contains four chromatids, two from each chromosome. (p. 55)
BLAST (Basic Local Alignment Search Tool): software program that can search databases and locate DNA or protein sequences that resemble any known sequence of interest. (p. 638)
bond energy: amount of energy required to break one mole of a particular chemical bond. (p. 23)
bp: see *base pair.*
***BRCA1* and *BRCA2* genes:** tumor suppressor genes in which inheritance of a single mutant copy creates a high risk for breast and ovarian cancer; code for proteins involved in repairing double-strand DNA breaks. (p. 800)
BRE: see *TFIIB recognition element.*
brightfield microscopy: light microscopy of specimen that possesses color, has been stained, or has some other property that affects the amount of light that passes through, thereby allowing an image to be formed. (p. A-5)
bundle sheath cell: internal cell of the leaf of a C_4 plant located in close proximity to the vascular bundle (vein) of the leaf; site of the Calvin cycle in such plants. (p. 309)
Burkitt lymphoma: a lymphocyte cancer associated with infection by Epstein-Barr virus along with a chromosome translocation in which the *MYC* gene is activated by moving it from chromosome 8 to 14. (p. 795)
bypass polymerase: special DNA polymerase that is more tolerant of defects in DNA template strands than standard polymerases. Used during SOS repair in bacteria and translesion synthesis in eukaryotes. (p. 498)

C

C: see *cytosine.*
C value: amount of DNA in a single (haploid) set of chromosomes. (p. 760)
C_3 plant: plant that depends solely on the Calvin cycle for carbon dioxide fixation, creating the three-carbon compound 3-phosphoglycerate as the initial product. (p. 309)
C_4 plant: plant that uses the Hatch–Slack pathway in mesophyll cells to carry out the initial fixation of carbon dioxide, creating the four-carbon compound oxaloacetate; the assimilated carbon is subsequently released again in bundle sheath cells and recaptured by the Calvin cycle. (p. 309)
cadherins: any of a family of plasma membrane glycoproteins that mediate Ca^{2+}-dependent adhesion between cells. (p. 415)
cal: see *calorie.*
calcium indicator: a dye or genetically engineered protein whose fluorescence changes in response to the local cytosolic calcium concentration. (p. 695)
calcium ionophore: a molecule that increases the permeability of membranes to calcium ions. (p. 695)
calmodulin: calcium-binding protein involved in mediating many of the intracellular effects of calcium ions in eukaryotic cells. (p. 696)
calnexin: a membrane-bound ER protein that forms a protein complex with a newly synthesized glycoprotein and assists in its proper folding. (p. 322)
calorie (cal): unit of energy; amount of energy needed to raise the temperature of 1 gram of water 1°C. (pp. 23, 113)
calreticulin: a soluble ER protein that forms a protein complex with a newly synthesized glycoprotein and assists in its proper folding. (p. 322)
Calvin cycle: cyclic series of reactions used by photosynthetic organisms for the fixation of carbon dioxide and its reduction to form carbohydrates. (p. 300)
CAM: see *crassulacean acid metabolism.*
CAM plant: plant that carries out crassulacean acid metabolism. (p. 311)
cAMP: see *cyclic AMP.*
cancer: uncontrolled, growing mass of cells that is capable of invading neighboring tissues and spreading via body fluids, especially the bloodstream, to other parts of the body; also called a malignant tumor. (p. 785)
CAP: see *catabolite activator protein.*
capping protein: protein that binds to the end of an actin filament, thereby preventing the further addition or loss of subunits. (p. 375)
cap-binding complex (CBC): protein complex bound to 5′ cap on mRNA transcripts. (p.536)
carbohydrate: general name given to molecules that contain carbon, hydrogen, and oxygen in a ratio C_n $(H_2O)_n$; examples include starch, glycogen, and cellulose. (p. 65)
carbon assimilation: process in photosynthesis by which fully oxidized carbon atoms from carbon dioxide are fixed (covalently attached) to organic acceptor molecules and then reduced and rearranged to form carbohydrates and other organic compounds required for building a living cell. Also called carbon fixation. (p. 284)
carbon atom: the most important atom in biological molecules, capable of forming up to four covalent bonds. (p. 22)
carbon fixation: see *carbon assimilation.*
carboxyl terminus: see *C-terminus.*
carboxysome: polyhedral proteinaceous structure that contains enzymes important for carbon fixation in cyanobacteria, including carbonic anhydrase and rubisco (p. 287).
carcinogen: any cancer-causing agent. (p. 793)
carcinoma: a malignant tumor (cancer) arising from the epithelial cells that cover external and internal body surfaces. (p. 784)
cardiac (heart) muscle: striated muscle of the heart, highly dependent on aerobic respiration. (p. 404)
caretaker gene: a tumor suppressor gene involved in DNA repair or chromosome sorting; loss-of-function mutations in such genes contribute to genetic instability. (p. 807)
carotenoid: any of several accessory pigments found in most plant species that absorb in the blue region of the visible spectrum (420–480 nm) and are therefore yellow or orange in color. (p. 289)
carrier molecule: a molecule that joins to a monomer, thereby activating the monomer for a subsequent reaction. (p. 35)
carrier protein: membrane protein that transports solutes across the membrane by binding to the solute on one side of the membrane and then undergoing a conformational change that transfers the solute to the other side of the membrane. (p. 193)
caspase: any of a family of proteases that degrade other cellular proteins as part of the process of apoptosis. (p. 746)
catabolic pathway: series of reactions that results in the breakdown of cellular components. (p. 216)
catabolite activator protein (CAP): bacterial protein that binds cyclic AMP and then activates the transcription of catabolite-repressible genes. (p. 578)
catalyst: agent that enhances the rate of a reaction by lowering the activation energy without itself being consumed; catalysts change the rate at which a reaction approaches equilibrium, but not the position of equilibrium. (p. 129)
catalytic subunit: a subunit of a multisubunit enzyme that contains the enzyme's catalytic site. (p. 148)

catecholamine: any of several compounds derived from the amino acid tyrosine that functions as a hormone and/or neurotransmitter. (p. 676)
caveolae: small invaginations of the plasma membrane that are coated with the protein caveolin and enriched in cholesterol that may be involved in cholesterol uptake or signal transduction. (p. 342)
CBP: transcriptional coactivator that exhibits histone acetyltransferase activity and associates with RNA polymerase to facilitate assembly of the transcription machinery at gene promoters. (p. 600)
Cdc42: member of a family of monomeric G proteins, which also includes Rac and Rho, that stimulate formation of various actin-containing structures within cells. (p. 379)
Cdk: see *cyclin-dependent kinase.*
Cdk inhibitor: any of several proteins that restrain cell growth and division by inhibiting Cdk-cyclin complexes. (p. 744)
cDNA: see *complementary DNA.*
cDNA library: collection of recombinant DNA clones produced by copying the entire mRNA population of a particular cell type with reverse transcriptase and then cloning the resulting cDNAs. (p. 623)
cell: the basic structural and functional unit of living organisms; the smallest structure capable of performing the essential functions characteristic of life. (p. 1)
cell body: portion of a nerve cell that contains the nucleus and other organelles and has extensions called axons and dendrites projecting from it. (p. 659)
cell-cell junction: specialized connection between the plasma membranes of adjoining cells for the purpose of adhesion, sealing, or communication. (p. 413)
cell cortex: network of microfilaments immediately beneath the plasma membrane of eukaryotic cells that stabilizes cell shape and structure (p. 413)
cell cycle: stages involved in preparing for and carrying out cell division; begins when two new cells are formed by the division of a single parental cell and is completed when one of these cells divides again into two cells. (p. 715)
cell differentiation: process by which cells acquire the specialized properties that distinguish different types of cells from each other. (p. 585)
cell division: process by which one cell gives rise to two. (p. 719)
cell membrane: see *plasma membrane.*
cell plate: flattened sac representing a stage in plant cell wall formation, leading to separation of the two daughter nuclei during plant cell division. (p. 733)
cell theory: theory of cellular organization stating that all organisms consist of one or more cells, that the cell is the basic unit of structure for all organisms, and that all cells arise only from preexisting cells. (p. 3)
cell wall: rigid, nonliving structure exterior to the plasma membrane of bacterial, algal, fungal, and plant cells; plant cell walls consist of cellulose microfibrils embedded in a noncellulosic matrix. (pp. 99, 432)
cellular respiration (respiration): oxidation-driven flow of electrons from reduced coenzymes to an electron acceptor, usually accompanied by the generation of ATP. (p. 243)
cellulose: structural polysaccharide present in plant cell walls, consisting of repeating glucose units linked by β (1 $\rightarrow$ 4) bonds. (pp. 64, 433)
central dogma of molecular biology: term originally coined by Francis Crick; the idea that the basic, typical flow of genetic information moves from DNA to RNA via transcription and then to protein via translation. (p. 511)
central pair: two parallel microtubules located in the center of the axoneme of a eukaryotic cilium or flagellum. (p. 392)
central vacuole: large membrane-bounded organelle present in many plant cells; helps maintain turgor pressure of the plant cell, plays a limited storage role, and is also capable of a lysosome-like function in intracellular digestion. (p. 97)
centrifugation: process of rapidly spinning a tube containing a fluid to subject its contents to a centrifugal force. (p. 10)
centrifuge: machine for rapidly spinning a tube containing a fluid to subject its contents to a centrifugal force. (p. 98)
centriole: structure consisting of nine sets of triplet microtubules embedded within the centrosome of animal cells, where two centrioles lie at right angles to each other; identical in structure to the basal body of eukaryotic cilia and flagella. (p. 367)
centromere: chromosome region where sister chromatids are held together prior to anaphase and where kinetochores are attached; contains simple-sequence, tandemly repeated DNA. (pp. 458, 720)
centrosome: small zone of granular material surrounding two centrioles located adjacent to the nucleus of animal cells; functions as a cell's main microtubule-organizing center. (pp. 359, 720)
cerebroside: an uncharged glycolipid containing the amino alcohol sphingosine. (p. 162)
CF: see *cystic fibrosis.*
CF_1: component of the chloroplast ATP synthase complex that protrudes from the stromal side of thylakoid membranes and contains the catalytic site for ATP synthesis. (p. 298)
CF_0: component of the chloroplast ATP synthase complex that is embedded in thylakoid membranes and serves as the proton translocator. (p. 298)
CF_0CF_1 complex: ATP synthase complex found in chloroplast thylakoid membranes; catalyzes the process by which the exergonic flow of protons down their electrochemical gradient is used to drive ATP synthesis. (p. 298)
CFTR: see *cystic fibrosis transmembrane conductance regulator.*
CGN: see *cis-Golgi network.*
channel gating: closing of a membrane ion channel in such a way that it can reopen immediately in response to an appropriate stimulus. (p. 665)
channel inactivation: closing of a membrane ion channel in such a way that it cannot reopen immediately. (p. 665)
channel protein: membrane protein that forms a hydrophilic channel through which solutes can pass across the membrane without any change in the conformation of the channel protein. (p. 193)
chaperone: see *molecular chaperone.*
Chargaff's rules: observation, first made by Erwin Chargaff, that in DNA the number of adenines is equal to the number of thymines (A = T) and the number of guanines is equal to the number of cytosines (G = C). (p. 445)
charge repulsion: force driving apart two ions, molecules, or regions of molecules of the same electric charge. (p. 217)
checkpoint: pathway that monitors conditions within the cell and transiently halts the cell cycle if conditions are not suitable for continuing. (p. 734) Also see *DNA damage checkpoint, DNA replication checkpoint,* and *spindle assembly checkpoint.*
chemical synapse: junction between two nerve cells where a nerve impulse is transmitted between the cells by neurotransmitters that diffuse across the synaptic cleft from the presynaptic cell to the postsynaptic cell. (p. 673)
chemiosmotic coupling model: model postulating that electron transport pathways establish proton gradients across membranes and that the energy stored in such gradients can then be used to drive ATP synthesis. (p. 269)
chemotaxis: cell movement toward a chemical attractant or away from a chemical repellent. (p. 408)
chemotroph: organism that is dependent on the bond energies of organic molecules such as carbohydrates, fats, and proteins to satisfy energy requirements. (p. 110)
chemotrophic energy metabolism: reactions and pathways by which cells catabolize nutrients such as carbohydrates, fats, and proteins, conserving as ATP some of the free energy that is released in the process. (p. 220)
chiasma (plural, chiasmata): connection between homologous chromosomes produced by crossing over during prophase I of meiosis. (p. 754)
chimeric antigen receptor (CAR) T cells: T cells of a cancer patient genetically modified in the laboratory to express a special hybrid receptor (a chimeric antigen receptor, or CAR) that binds to a certain molecule on the surface of the patient's cancer cells and stimulates the T cells to attack the cancer cells. (p.815)
chitin: structural polysaccharide found in insect exoskeletons and crustacean cells; consists of *N*-acetylglucosamine units linked by β (1 $\rightarrow$ 4) bonds. (p. 66)
chlorophyll: light-absorbing molecule that donates photoenergized electrons to organic molecules, initiating photochemical events that lead to the generation of the NADPH and ATP required for the Calvin cycle; because of its absorption properties, chlorophyll gives plants their characteristic green color. (p. 288)
chlorophyll-binding protein: any of several proteins that bind to and stabilize the arrangement of chlorophyll molecules within a photosystem. (p. 290)
chloroplast: double membrane-enclosed cytoplasmic organelle of plants and algae that contains chlorophyll and the enzymes needed to carry out photosynthesis. (pp. 87, 286)
cholesterol: lipid constituent of animal cell plasma membrane; serves as a precursor to the steroid hormones. (pp. 70, 162)
cholinergic synapse: a synapse that uses acetylcholine as the neurotransmitter. (p. 676)
chromatid: see *sister chromatid.*
chromatin: DNA-protein fibers that make up chromosomes; constructed from nucleosomes spaced regularly along a DNA chain. (pp. 87, 453)
chromatin fiber (30 nm): fiber formed by packing together the nucleosomes of a 10-nm chromatin fiber. (p. 447)
chromatin remodeling protein: protein that induces alterations in nucleosome structure, packing, and/or position designed to give transcription factors access to DNA target sites in the promoter region of a gene. (p. 457)
chromatography: a group of related techniques that utilize the flow of a fluid phase over a nonmobile absorbing phase to separate molecules based on their relative affinities for the two phases, which in turn reflect differences in size, charge, hydrophobicity, or affinity for a particular chemical group. (p. 10)
chromosomal territory: region occupied by specific decondensed chromosomes during interphase in the eukaryotic nucleus. (p. 469)
chromosome: in eukaryotes a single DNA molecule, complexed with histones and other proteins, that becomes condensed into a compact structure at the time of mitosis or meiosis. (pp. 11, 83) Also see *bacterial chromosome.*

chromosome puff: uncoiled region of a polytene chromosome that is undergoing transcription. (p. 586)

chromosome theory of heredity: theory stating that hereditary factors are located in the chromosomes within the nucleus. (p. 11)

cilium (plural, cilia): membrane-bounded appendage on the surface of a eukaryotic cell, composed of a specific arrangement of microtubules and responsible for motility of the cell or the fluids around cells; shorter and more numerous than closely related organelles called flagella. (p. 391) Also see *flagellum*.

***cis*-acting element:** a DNA sequence to which a regulatory protein can bind. (p. 578)

***cis*-Golgi network (CGN):** region of the Golgi apparatus consisting of a network of membrane-bounded tubules that are located closest to the transitional elements. (p. 319)

cisterna (plural, cisternae): membrane-bounded flattened sac, such as in the endoplasmic reticulum or Golgi apparatus. (p. 92)

cisternal maturation model: model postulating that Golgi cisternae are transient compartments that gradually change from *cis*-Golgi network cisternae into medial cisternae and then into *trans*-Golgi network cisternae. (p. 320)

***cis-trans* test:** analysis used to determine whether a mutation in a bacterial operon affects a regulatory protein or the DNA sequence to which it binds. (p. 577)

citric acid cycle: cyclic metabolic pathway that oxidizes acetyl CoA to carbon dioxide in the presence of oxygen, generating ATP and the reduced coenzymes NADH and $FADH_2$; a component of aerobic respiration; also called the tricarboxylic acid (TCA) cycle or the Krebs cycle. (p. 251)

clathrin: large protein that forms a "cage" around the coated vesicles and coated pits involved in endocytosis and other intracellular transport processes. (p. 342)

claudin: transmembrane protein that forms the main structural component of a tight junction. (p. 420)

cleavage: process of cytoplasmic division in animal cells, in which a band of actin microfilaments lying beneath the plasma membrane constricts the cell at the midline and eventually divides it in two. (p. 730)

cleavage furrow: groove formed during the division of an animal cell that encircles the cell and deepens progressively, leading to cytoplasmic division. (p. 725)

clone: organism (or cell or molecule) that is genetically identical to another organism (or cell or molecule) from which it is derived. (pp. 582, 621)

cloning vector: DNA molecule to which a selected DNA fragment can be joined prior to rapid replication of the vector in a host cell (usually a bacterium); phage and plasmid DNA are the most common cloning vectors. (p. 621)

closed system: a system that is sealed from its environment and can neither take in nor release energy in any form. (p. 112)

clustered regularly interspaced short palindromic repeat (CRISPR): DNA element in bacteria and archaea that contains spacer sequences containing homology to viral DNA interspersed with repeat sequences; CRISPRs are transcribed and the resulting RNA processed into crRNAs, which can degrade corresponding foreign (e.g., viral) RNA. The CRISPR system has been utilized by scientists for genome editing. (p. 583)

CoA: see *coenzyme A*.

coactivator: class of proteins that mediate interactions between activators and the genes they regulate; include histone-modifying enzymes such as HAT, chromatin-remodeling proteins such as SWI/SNF, and Mediator. (p. 595)

coated vesicle: any of several types of membrane vesicles involved in vesicular traffic within the endomembrane system; surrounded by a coat protein such as clathrin, COPI, COPII, or caveolin. (p. 336)

coding strand: the nontemplate strand of a DNA double helix, which is base-paired to the template strand; identical in sequence to the single-stranded RNA molecules transcribed from the template strand, except that RNA has uracil (U) where the coding strand has thymine (T). (p. 544)

codon: triplet of nucleotides in an mRNA molecule that serves as a coding unit for an amino acid (or a start or stop signal) during protein synthesis. (p. 544)

codon usage bias: the preference for the use of one (or more) synonymous codons in a given species' mRNA. (p.546)

coenzyme: small organic molecule that functions along with an enzyme by serving as a carrier of electrons or functional groups. (p. 221)

coenzyme A (CoA): organic molecule that serves as a carrier of acyl groups by forming a high-energy thioester bond with an organic acid. (p. 251)

coenzyme Q (CoQ): nonprotein (quinone) component of the mitochondrial electron transport chain that serves as the collection point for electrons from both FMN- and FAD-linked dehydrogenases; also called ubiquinone. (p. 262)

coenzyme Q–cytochrome *c* oxidoreductase: see *complex III*.

cohesin: protein that holds sister chromatids together prior to anaphase. (p. 734)

co-immunoprecipitation (coIP): a variant of immunoprecipitation designed to purify a protein or molecule of interest, along with other molecules bound to it that may reside together in a multimolecular complex. (p. 651) Also see *immunoprecipitation*.

colchicine: plant-derived drug that binds to tubulin and prevents its polymerization into microtubules. (p. 365)

collagens: a family of closely related proteins that form high-strength fibers found in high concentration in the extracellular matrix of animals. (p. 422)

collagen fiber: extremely strong fibers measuring several micrometers in diameter found in the extracellular matrix; constructed from collagen fibrils that are, in turn, composed of collagen molecules lined up in a staggered array. (p. 423)

combinatorial model for gene regulation: model proposing that complex patterns of tissue-specific gene expression can be achieved by a relatively small number of DNA control elements and their respective transcription factors acting in different combinations. (p. 596)

competitive inhibitor: a compound that reduces enzyme activity by binding to the active site and competing directly with substrate binding. (p. 146)

complementary: in nucleic acids, the ability of guanine (G) to form a hydrogen-bonded base pair with cytosine (C) and adenine (A) to form a hydrogen-bonded base pair with thymine (T) or uracil (U). (p. 436)

complementary DNA (cDNA): DNA molecule copied from an mRNA template by the enzyme reverse transcriptase. (p. 623)

complex I (NADH–coenzyme Q oxidoreductase): multiprotein complex of the electron transport chain that catalyzes the transfer of electrons from NADH to coenzyme Q. (p. 265)

complex II (succinate–coenzyme Q oxidoreductase): multiprotein complex of the electron transport chain that catalyzes the transfer of electrons from succinate to coenzyme Q. (p. 265)

complex III (coenzyme Q–cytochrome *c* oxidoreductase): multiprotein complex of the electron transport chain that catalyzes the transfer of electrons from coenzyme Q to cytochrome *c*. (p. 265)

complex IV (cytochrome *c* oxidase): multiprotein complex of the electron transport chain that catalyzes the transfer of electrons from cytochrome *c* to oxygen. (p. 267)

composite transposon: in bacteria, a transposable element containing multiple genes located between two insertion sequence (IS) elements.

compound microscope: light microscope that uses several lenses in combination; usually has a condenser lens, an objective lens, and an ocular lens. (p. A-4)

concentration gradient: transmembrane gradient in the concentration of a molecule or ion, expressed as a ratio of the concentration of the substance on one side of the membrane to its concentration on the other side of the membrane; the sole driving force for transport of molecules across a membrane, but only one of two components of the electrochemical potential that serves as the driving force for transport of ions across a membrane. (p. 186)

concentration work: use of energy to transport ions or molecules across a membrane against an electrochemical or concentration gradient. (p. 109)

condensation reaction: chemical reaction that results in the joining of two molecules by the removal of a water molecule. (p. 35)

condenser lens: lens of a light microscope (or electron microscope) that is the first lens to direct the light rays (or electron beam) from the source toward the specimen. (pp. A-5, A-17)

confocal microscope: specialized type of light microscope that employs a laser beam to illuminate a single plane of the specimen at a time. (p. A-10)

conformation: three-dimensional shape of a polypeptide or other biological macromolecule. (p. 32, 37)

congression: process during cell division by which chromosomes align at the spindle equator, occupying a position roughly equidistant from both spindle poles.

conjugation: cellular mating process by which DNA is transferred from one bacterial cell to another. (p. 774)

connexon: assembly of six protein subunits with a hollow center that forms a channel through the plasma membrane at a gap junction. (p. 421)

consensus sequence: the most common version of a DNA base sequence when that sequence occurs in slightly different forms at different sites. (p. 478)

conservative transposition: transposition of a transposable element by removal from its original location followed by insertion into a new location. (p. 505)

constitutive centromere-associated network (CCAN): a protein complex that contains CENP and other proteins that form a complex with chromosomal DNA and CENP-A on the chromosome side of a kinetochore.

constitutive heterochromatin: chromosomal regions that are condensed in all cells of an organism at virtually all times and are therefore genetically inactive. (p. 458) Also see *facultative heterochromatin*.

constitutive secretion: continuous fusion of secretory vesicles with the plasma membrane and expulsion of their contents to the cell exterior, independent of specific extracellular signals. (p. 334)

constitutively active mutation: mutation that causes a receptor to be active even when no activating ligand is present. (p. 702)

contig: abbreviated from "contiguous," a contig is a set of overlapping DNA fragments that together form a single large stretch of DNA with a single sequence of bases. Contigs are assembled during sequencing projects involving very large pieces of DNA, including genomic DNA. (p. 635)

contractile ring: beltlike bundle of actin microfilaments that forms beneath the plasma membrane and acts to constrict the cleavage furrow during the division of an animal cell. (p. 731)

contractility: shortening of muscle or other cells. (p. 384)

cooperativity: property of enzymes possessing multiple catalytic sites in which the binding of a substrate molecule to one catalytic site causes conformational changes that influence the affinity of the remaining sites for substrate. (p. 148)

COPI: main protein component of the "fuzzy" coats of COPI-coated vesicles, which are involved in retrograde transport from the Golgi apparatus back to the ER and between Golgi apparatus cisternae. (p. 344)

COPII: main protein component of COPII-coated vesicles, which are involved in transport from the ER to the Golgi apparatus. (p. 344)

CoQ: see *coenzyme Q.*

core oligosaccharide: initial oligosaccharide segment joined to an asparagine residue during N-glycosylation of a polypeptide chain; consists of two N-acetylglucosamine units, nine mannose units, and three glucose units. (p. 322)

core promoter: minimal set of DNA sequences sufficient to direct the accurate initiation of transcription by RNA polymerase. (p. 522)

core protein: a protein molecule to which numerous glycosaminoglycan chains are attached to form a proteoglycan. (p. 425)

correlative microscopy: combination of light and electron-based microscopy that allows the fine structure associated with a fluorescent signal to be examined at high resolution, often using immunoelectron microscopy. (p. A-20)

cosmid: an engineered bacterial plasmid that contains *cos* sequences from λ phage; cosmids can harbor inserts that are much larger than standard plasmids and so are useful for producing genomic DNA libraries. (p. 623)

cotranslational import: transfer of a growing polypeptide chain across (or, in the case of integral membrane proteins, into) the ER membrane as polypeptide synthesis proceeds. (p. 324, 556)

coupled transport: coordinated transport of two solutes across a membrane in such a way that transport of either stops if the other is stopped or interrupted; the two solutes may move in the same direction (symport) or in opposite directions (antiport). (p. 194)

covalent bond: strong chemical bond in which two atoms share two or more electrons. (p. 23)

covalent modification: type of regulation in which the activity of an enzyme (or other protein) is altered by the addition or removal of specific chemical groups. (p. 148)

CpG islands: regions of DNA containing cytosines next to guanines; are often associated with promoters, especially in mammals, and are targets of DNA methylation. (p. 591)

crassulacean acid metabolism (CAM): pathway in which plants use PEP carboxylase to fix CO_2 at night, generating the four-carbon acid malate. The malate is then decarboxylated during the day to release CO_2, which is fixed by the Calvin cycle. (p. 311)

CREB: transcription factor that activates the transcription of cyclic AMP-inducible genes by binding to cAMP response elements in DNA. (p. 600)

CRISPR: see *clustered regularly interspaced short palindromic repeat.*

crista (plural, cristae): infolding of the inner mitochondrial membrane into the matrix of the mitochondrion, thereby increasing the total surface area of the inner membrane; contains the enzymes of electron transport and oxidative phosphorylation. (pp. 89, 246)

critical concentration: concentration at which the rate of assembly of cytoskeletal protein subunits into a polymer is exactly balanced with the rate of disassembly. (p. 363)

critical point dryer: heavy metal canister used to dry a specimen under conditions of controlled temperature and pressure. (p. A.24)

cross-bridge: structure formed by contact between the myosin heads of thick filaments and the thin filaments in muscle myofibrils. (p. 401)

crossing over: exchange of DNA segments between homologous chromosomes. (p. 756)

cryoEM: technique in which cryofixed biological samples are directly imaged in a transmission electron microscope at low temperature; often used to examine suspensions of isolated macromolecules. (p. A-25)

cryofixation: rapid freezing of small samples so that cellular structures can be immobilized in milliseconds; often followed by freeze substitution, in which an organic solvent replaces the frozen water in the sample. (p. A-19)

C-terminus (carboxyl terminus): the end of a polypeptide chain that contains the last amino acid to be incorporated during mRNA translation; usually retains a free carboxyl group. (p. 45)

current: movement of positive or negative ions. (p. 665)

cyclic AMP (cAMP): adenosine monophosphate with the phosphate group linked to both the 3′ and 5′ carbons by phosphodiester bonds; functions in both prokaryotic and eukaryotic gene regulation; in eukaryotes, acts as a second messenger that mediates the effects of various signaling molecules by activating protein kinase A. (p. 691)

cyclic electron flow: light-driven transfer of electrons from photosystem I through a sequence of electron carriers that returns them to a chlorophyll molecule of the same photosystem, with the released energy used to drive ATP synthesis. (p. 298)

cyclic GMP (cGMP): guanosine monophosphate with the phosphate group linked to both the 3′ and 5′ carbons by phosphodiester bonds; in eukaryotes, acts as a second messenger that plays a role in several processes, including vasodilation.

cyclin: any of a group of proteins that activate the cyclin-dependent kinases (Cdks) involved in regulating progression through the eukaryotic cell cycle. (p. 731)

cyclin-dependent kinase (Cdk): any of several protein kinases that are activated by different cyclins and that control progression through the eukaryotic cell cycle by phosphorylating various target proteins. (p. 37)

cyclosis: see *cytoplasmic streaming.*

cystic fibrosis (CF): a disease in which symptoms result from an inability to secrete chloride ions that is, in turn, caused by a genetic defect in a membrane protein that functions as a chloride ion channel. (p. 198)

cystic fibrosis transmembrane conductance regulator (CFTR): a membrane protein that functions as a chloride ion channel, a mutant form of which can lead to cystic fibrosis. (p. 199)

cytochalasins: family of drugs produced by certain fungi that inhibit a variety of cell movements by preventing actin polymerization. (p. 373)

cytochrome: family of heme-containing proteins of the electron transport chain; involved in the transfer of electrons from coenzyme Q to oxygen by the oxidation and reduction of the central iron atom of the heme group. (p. 262)

cytochrome b_6/f complex: multiprotein complex within the thylakoid membrane that transfers electrons from a plastoquinol to plastocyanin as part of the energy transduction reactions of photosynthesis. (p. 295)

cytochrome *c*: heme-containing protein of the electron transport chain involved in the transfer of electrons from coenzyme Q to oxygen by the oxidation and reduction of the central iron atom of the heme group. Also involved in triggering apoptosis. (p.747)

cytochrome *c* oxidase: see *complex IV.*

cytochrome P-450: family of heme-containing proteins, located mainly in the liver, that catalyze hydroxylation reactions involved in drug detoxification and steroid biosynthesis. (p. 317)

cytokinesis: division of the cytoplasm of a parent cell into two daughter cells; usually follows mitosis. (p. 473)

cytology: study of cellular structure, based primarily on microscopic techniques. (p. 4)

cytoplasm: that portion of the interior of a eukaryotic cell that is not occupied by the nucleus; includes organelles, such as mitochondria and components of the endomembrane system, as well as the cytosol. (p. 82)

cytoplasmic dynein: cytoplasmic motor protein that moves along the surface of microtubules in the plus to minus direction, driven by energy derived from ATP hydrolysis; associated with dynactin, which links cytoplasmic dynein to cargo vesicles. (p. 387)

cytoplasmic microtubules: microtubules arranged in loosely organized, dynamic networks in the cytoplasm of eukaryotic cells. (p. 354)

cytoplasmic streaming: movement of the cytoplasm driven by interactions between actin filaments and specific types of myosin; also called cyclosis in plant cells. (p. 409)

cytosine (C): nitrogen-containing aromatic base, chemically designated as a pyrimidine, that serves as an informational monomeric unit when present in nucleic acids with other bases in a specific sequence; forms a complementary base pair with guanine (G) by hydrogen bonding. (p. 59)

cytoskeleton: three-dimensional, interconnected network of microtubules, microfilaments, and intermediate filaments that provides structure to the cytoplasm of a eukaryotic cell and plays an important role in cell movement. (pp. 98, 358)

cytosol: the semifluid substance in which the organelles of the cytoplasm are suspended. (p. 82)

cytosolic microtubule (Ch 13, p. 361)

D

DAG: see *diacylglycerol.*

deep etching: modification of freeze-etching technique in which ultrarapid freezing and a volatile cryoprotectant are used to extend the etching period, thereby removing a deeper layer of ice and allowing the interior of a cell to be examined in depth. (p. A-22)

degenerate code: ability of the genetic code to use more than one triplet code to specify the same amino acid. (p. 543)

degron: amino acid sequence within a protein that is used in targeting the protein for destruction. (p. 615)

dehydrogenation: removal of electrons plus hydrogen ions (protons) from an organic molecule; oxidation. (p. 220)

denaturation: loss of the natural three-dimensional structure of a macromolecule, usually resulting in a loss of its biological activity; caused by agents such as heat, extremes of pH, urea, salt, and other chemicals. (p. 37) Also see *DNA denaturation.*

dendrite: extension of a nerve cell that receives impulses and transmits them inward toward the cell body. (p. 660)

density-dependent inhibition of growth: tendency of cell division to stop when cells growing in culture reach a high population density. (p. 785)

deoxyribonucleic acid: see *DNA.*

deoxyribose: five-carbon sugar present in DNA. (p. 59)

dephosphorylation: removal of a phosphate group. (p. 148)

depolarization: change in membrane potential to a less-negative value. (p. 668)

desmosome: junction for cell-cell adhesion that is connected to the cytoskeleton by intermediate filaments; creates buttonlike points of strong adhesion between adjacent animal cells that give tissue structural integrity and allow the cells to function as a unit and resist stress. (p. 416)

deubiquitinating enzyme (DUB): enzyme that plays a role in the reversible posttranslational process of ubiquitination. It is capable of removing ubiquitin or polyubiquitin chains from specific target proteins. (p. 615)

diacylglycerol (DAG): glycerol esterified to two fatty acids; formed, along with inositol trisphosphate (IP_3), upon hydrolysis of phosphatidylinositol-4, 5-bisphosphate by phospholipase C; remains membrane-bound after hydrolysis and functions as a second messenger by activating protein kinase C, which then phosphorylates specific serine and threonine groups in a variety of target proteins. (p. 693)

diakinesis: final stage of prophase I of meiosis; associated with chromosome condensation, disappearance of nucleoli, breakdown of the nuclear envelope, and initiation of spindle formation. (p. 755)

DIC microscopy: see *differential interference contrast microscopy.*

Dicer: enzyme that cleaves double-stranded RNAs into short fragments about 21–22 base pairs in length. (p. 606)

differential interference contrast (DIC) microscopy: technique that resembles phase-contrast microscopy in principle, but is more sensitive because it employs a special prism to split the illuminating light beam into two separate rays. (p. A-6)

differential scanning calorimetry: technique for determining a membrane's transition temperature by monitoring the uptake of heat during the transition of the membrane from the gel to fluid state. (p. 165)

differentiation: see *cell differentiation.*

diffraction: pattern of either additive or canceling interference exhibited by light waves. (p. A-2)

diffusion: free, unassisted movement of a solute, with direction and rate dictated by the difference in solute concentration between two different regions. (p. 82)

digital deconvolution microscopy: technique in which fluorescence microscopy is used to acquire a series of images through the thickness of a specimen, followed by computer analysis to remove the contribution of out-of-focus light to the image in each focal plane. (p. A-13)

digital microscopy: technique in which microscopic images are recorded and stored electronically by placing a video camera in the image plane produced by the ocular lens. (p. A-7)

diploid: containing two sets of chromosomes and therefore two copies of each gene; can describe a cell, nucleus, or organism composed of such cells. (p. 752)

diplotene: stage during prophase I of meiosis when the two homologous chromosomes of each bivalent begin to separate from each other, revealing the chiasmata that connect them. (p. 756)

direct active transport: membrane transport in which the movement of solute molecules or ions across a membrane is coupled directly to an exergonic chemical reaction, most commonly the hydrolysis of ATP. (p. 201)

directionality: having two ends that are chemically different from each other; used to describe a polymer chain such as a protein, nucleic acid, or carbohydrate; also used to describe membrane transport systems that selectively transport solutes across a membrane in one direction. (pp. 35, 201)

disaccharide: carbohydrate consisting of two covalently linked monosaccharide units. (p. 65)

dissociation constant (K_d): the concentration of free ligand needed to produce a state in which half the receptors are bound to ligand. (p. 686)

disulfide bond: covalent bond formed between two sulfur atoms by oxidation of sulfhydryl groups. The disulfide bond formed between two cysteines is important in stabilizing the tertiary structure of proteins. (p. 48)

DNA (deoxyribonucleic acid): macromolecule that serves as the repository of genetic information in all cells; constructed from nucleotides consisting of deoxyribose phosphate linked to adenine, thymine, cytosine, or guanine; forms a double helix held together by complementary base pairing between adenine and thymine, and between cytosine and guanine. (p. 58)

DNA-binding domain: region of a transcription factor that recognizes and binds to a specific DNA base sequence. (p. 600)

DNA cloning: generating multiple copies of a specific DNA sequence, either by replication of a recombinant plasmid or bacteriophage within bacterial cells or by use of the polymerase chain reaction. (p. 621)

DNA damage checkpoint: mechanism that monitors for DNA damage and halts the cell cycle at various points, including late G1, S, and late G2, if damage is detected. (p. 741)

DNA denaturation (melting): separation of the two strands of the DNA double helix caused by disruption of complementary base pairing. (p. 449)

DNA fingerprinting: technique for identifying individuals based on small differences in DNA fragment patterns detected by electrophoresis. (p. 638)

DNA gyrase: a type II topoisomerase that can relax positive supercoiling and induce negative supercoiling of DNA; involved in unwinding the DNA double helix during DNA replication. (p. 449)

DNA helicase: any of several enzymes that unwind the DNA double helix, driven by energy derived from ATP hydrolysis. (p. 479, 485)

DNA ligase: enzyme that joins two DNA fragments together by catalyzing the formation of a phosphoester bond between the 3′ end of one fragment and the 5′ end of the other fragment. (p. 480)

DNA loop: a loop 50,000–100,000 bp in length formed from the folding of a 30-nm chromatin fiber by periodic attachment of the DNA to an insoluble network of nonhistone proteins. (p. 455)

DNA melting temperature (T_m): temperature at which the transition from double-stranded to single-stranded DNA is halfway complete when DNA is denatured by increasing the temperature. (p. 449)

DNA methylation: addition of methyl groups to nucleotides in DNA; associated with suppression of gene transcription when selected cytosine groups are methylated in eukaryotic DNA. (p. 591)

DNA microarray: tiny chip that has been spotted at fixed locations with thousands of different DNA fragments for use in gene expression studies. (p. 640)

DNA polymerase: any of a group of enzymes involved in DNA replication and repair that catalyze the addition of successive nucleotides to the 3′ end of a growing DNA strand, using an existing DNA strand as template. (p. 479)

DNA renaturation (annealing): binding together of the two separated strands of a DNA double helix by complementary base pairing, thereby regenerating the double helix. (p. 449)

DNA replication checkpoint: mechanism that monitors the state of DNA replication to help ensure that DNA synthesis is completed prior to permitting the cell to exit from G2 and begin mitosis. (p. 740)

DNase I hypersensitive site: location near an active gene that shows extreme sensitivity to digestion by the nuclease DNase I; thought to correlate with binding sites for transcriptional factors or other regulatory proteins. (p. 591)

DNA sequencing: technology used to determine the linear order of bases in DNA molecules or fragments. (pp. 12, 632)

domain: a discrete, locally folded unit of protein tertiary structure, often containing regions of α helices and β sheets packed together compactly. (p. 56)

dominant (allele): allele that determines how the trait will appear in an organism, whether present in the heterozygous or homozygous form. (p. 766)

dominant negative mutation: a loss-of-function mutation involving proteins consisting of more than one copy of the same polypeptide chain, in which a single mutant polypeptide chain disrupts the function of the protein even though the other polypeptide chains are normal. (p. 702)

double bond: chemical bond formed between two atoms as a result of the sharing of two pairs of electrons. (p. 23)

double helix (model): two intertwined helical chains of a DNA molecule, held together by complementary base pairing between adenine (A) and thymine (T) and between cytosine (C) and guanine (G). (pp. 63, 446)

double-reciprocal plot: graphic method for analyzing enzyme kinetic data by plotting $1/v$ versus $1/[S]$. (p. 141)

downstream: located toward the 3′ end of the DNA coding strand. (p. 514)

downstream promoter element: see *DPE.*

DPE (downstream promoter element): component of core promoters for RNA polymerase II, located about 30 nucleotides downstream from the transcriptional start site. (p. 522)

Drosha: nuclear enzyme that cleaves the primary transcript of microRNA genes (pri-microRNAs) to form the smaller precursor microRNAs (pre-microRNAs). (p. 607)

DUB: see *deubiquitinating enzyme.*

dynactin: protein complex that helps link cytoplasmic dynein to the cargo (e.g., a vesicle) that it transports along microtubules. (p. 390)

dynamic instability model: model for microtubule behavior that presumes two populations of microtubules, one growing in length by continued polymerization at their plus ends and the other shrinking in length by depolymerization. (p. 366)

dynamin: cytosolic GTPase required for coated pit constriction and the closing of a budding, clathrin-coated vesicle. (p. 343)

dynein: motor protein that moves along the surface of microtubules in the plus to minus direction, driven by energy derived from ATP hydrolysis; present both in the cytoplasm and in the arms that reach between adjacent microtubule doublets in the axoneme of a flagellum or cilium. (p. 386)

dystrophin: a large protein found at muscle costameres that is part of a complex of proteins that attaches the muscle cell plasma membrane to the extracellular matrix. (p. 431)

E

E: see *internal energy.*

E face: interior face of the outer monolayer of a membrane as revealed by the technique of freeze fracturing; called the E face because this monolayer is on the exterior side of the membrane. (p. A-22)

E site (exit site): site on the ribosome to which the empty tRNA is moved during translocation prior to its release from the ribosome. (p. 547)

E'_0: see *standard reduction potential.*

E2F transcription factor: protein that, when not bound to the Rb protein, activates the transcription of genes coding for proteins required for DNA replication and entrance into the S phase of the cell cycle. (p. 740)

E_A: see *activation energy.*

early endosome: vesicles budding off the *trans-*Golgi network that are sites for the sorting and recycling of extracellular material brought into the cell by endocytosis. (p. 329)

EBV: see *Epstein-Barr virus.*

ECM: see *extracellular matrix.*

effector: see *allosteric effector.*

egg (ovum): haploid female gamete, usually a relatively large cell with many stored nutrients. (p. 752)

EJC: see *exon junction complex.*

elastin: protein subunit of the elastic fibers that impart elasticity and flexibility to the extracellular matrix. (p. 424)

electrical excitability: ability to respond to certain types of stimuli with a rapid series of changes in membrane potential known as an action potential. (p. 665)

electrical synapse: junction between two nerve cells where nerve impulses are transmitted by direct movement of ions through gap junctions without the involvement of chemical neurotransmitters. (p. 673)

electrical work: use of energy to transport ions across a membrane against a potential gradient. (p. 109)

electrochemical equilibrium: condition in which a transmembrane concentration gradient of a specific ion is balanced with an electrical potential across the same membrane, such that there is no net movement of the ion across the membrane. (p. 667)

electrochemical gradient: see *electrochemical potential.*

electrochemical potential: transmembrane gradient of an ion, with both an electrical component due to the charge separation quantified by the membrane potential and a concentration component; also called electrochemical gradient. (p. 186)

electrochemical proton gradient: transmembrane gradient of protons, with both an electrical component due to charge separation and a chemical component due to a difference in proton concentration (pH) across the membrane. (p. 268)

electron gun: assembly of several components that generates the electron beam in an electron microscope. (p. A-17)

electron micrograph: photographic image of a specimen produced by the exposure of a photographic plate to the image-forming electron beam of an electron microscope. (p. A-17)

electron microscope: instrument that uses a beam of electrons to visualize cellular structures and thereby examine cellular architecture; the resolution is much greater than that of the light microscope, allowing detailed ultrastructural examination. (pp. 7, A-17)

electron shuttle system: any of several mechanisms whereby electrons from a reduced coenzyme such as NADH are moved across a membrane; consists of one or more electron carriers that can be reversibly reduced, with transport proteins present in the membrane for both the oxidized and reduced forms of the carrier. (p. 278)

electron tomography: see *three-dimensional electron tomography.*

electron transport: process of coenzyme reoxidation under aerobic conditions, involving stepwise transfer of electrons to oxygen by means of a series of electron carriers. (p. 261)

electron transport chain (ETC): group of membrane-bound electron carriers that transfer electrons from the coenzymes NADH and $FADH_2$ to oxygen. (pp. 261, 292)

electronegative: property of an atom that tends to draw electrons toward it. (p. 28)

electrophoresis: a group of related techniques that utilize an electrical field to separate electrically charged molecules. (pp. 10, 171)

electrospray ionization mass spectrometry (ESI MS): a type of mass spectrometry useful for analyzing macromolecules, including peptides, based on their mass/charge ratio. ESI MS relies on producing a fine, ionized aerosol containing the molecule(s) of interest, which can then be introduced into a spectrometer. (p. 649)

elongation (of microtubules): growth of microtubules by addition of tubulin heterodimers to either end. (p. 363)

elongation factors: group of proteins that catalyze steps involved in the elongation phase of protein synthesis; examples include EF-Tu and EF-Ts. (p. 557)

embryonic stem (ES) cell: cultured cell derived from the inner cell mass of a mammalian blastocyst that can divide indefinitely and become many different types of cells. (p. 587)

Emerson enhancement effect: achievement of greater photosynthetic activity with red light of two slightly different wavelengths than is possible by summing the activities obtained with the individual wavelengths separately. (p. 292)

endergonic: an energy-requiring reaction characterized by a positive free energy change ($\Delta G > 0$). (p. 117)

endocytic vesicle: membrane vesicle formed by pinching off of a small segment of plasma membrane during the process of endocytosis. (p. 337)

endocytosis: uptake of extracellular materials by infolding of the plasma membrane, followed by pinching off of a membrane-bound vesicle containing extracellular fluid and materials. (p. 336)

endomembrane system: interconnected system of cytoplasmic membranes in eukaryotic cells composed of the endoplasmic reticulum, Golgi apparatus, endosomes, lysosomes, and nuclear envelope. (p. 315)

endonuclease: an enzyme that degrades a nucleic acid (usually DNA) by cutting the molecule internally. (p. 483) Also see *restriction endonuclease.*

endoplasmic reticulum (ER): network of interconnected membranes distributed throughout the cytoplasm and involved in the synthesis, processing, and transport of proteins in eukaryotic cells. (pp. 92, 316)

endosome: see *early endosome* or *late endosome.*

endosymbiont theory: theory postulating that mitochondria and chloroplasts arose from ancient bacteria that were ingested by ancestral eukaryotic cells about a billion years ago. (pp. 90, 287)

endothermic: a reaction or process that absorbs heat. (p. 112)

end-product repression: regulation of an anabolic pathway based on the ability of an end-product to repress further synthesis of enzymes involved in production of that end-product. (p. 573)

energy: capacity to do work; ability to cause specific changes. (p. 105)

energy transduction: process in photosynthesis by which light energy is converted to chemical energy in the form of ATP and the coenzyme NADPH, which subsequently provides energy and reducing power for the carbon assimilation reactions. (p. 284)

enhancer: DNA sequence containing a binding site for transcription factors that stimulates transcription and whose position and orientation relative to the promoter can vary significantly without interfering with the ability to regulate transcription. (p. 596)

enthalpy (*H*): the heat content of a substance, quantified as the sum of its internal energy, *E*, plus the product of pressure and volume: $H = E + PV$. (p. 112)

entropy (*S*): measure of the randomness or disorder in a system. (p. 116)

enzyme: biological catalyst; protein (or in certain cases, RNA) molecule that acts on one or more specific substrates, converting them to products with different molecular structures. (pp. 9, 130) Also see *ribozyme.*

enzyme catalysis: involvement of an organic molecule, usually a protein but in some cases RNA, in speeding up the rate of a specific chemical reaction or class of reactions. (p. 130) Also see *catalyst.*

enzyme kinetics: quantitative analysis of enzyme reaction rates and the manner in which they are influenced by a variety of factors. (p. 140)

epigenetic change: alteration in the expression of a gene rather than a change in the structure of the gene itself. (p. 592)

EPSP: see *excitatory postsynaptic potential.*

Epstein-Barr virus (EBV): virus associated with Burkitt lymphoma (as well as the noncancerous condition infectious mononucleosis). (p. 795)

equilibrium potential

equilibrium constant (K_{eq}): ratio of product concentrations to reactant concentrations for a given chemical reaction when the reaction has reached equilibrium. (p. 119)

equilibrium density centrifugation: technique used to separate cellular components by subjecting them to centrifugation in a solution that increases in density from the top to the bottom of the centrifuge tube; during centrifugation, an organelle or molecule sediments to the density layer equal to its own density, at which point movement stops because no further net force acts on the material. (pp. 96, 466)

equilibrium (or reversal) potential: membrane potential that exactly offsets the effect of the concentration gradient for a given ion. (p. 662)

ER: see *endoplasmic reticulum.*

ER-associated degradation (ERAD): quality control mechanism in the ER that recognizes misfolded or unassembled proteins and exports or "retrotranslocates" them back across the ER membrane to the cytosol, where they are degraded by proteasomes. (p. 321, 561)

ER cisterna (plural, cisternae): flattened sac of the endoplasmic reticulum. (p. 316)

ergosterol: a lipid resembling cholesterol that is found in fungal cell membranes and that is the target of some antifungal medications. (p. 161)

ER lumen: the internal space enclosed by membranes of the endoplasmic reticulum. (p. 316)

ER signal sequence: amino acid sequence in a newly forming polypeptide chain that directs the ribosome-mRNA-polypeptide complex to the surface of the rough ER, where the complex becomes anchored. (p. 326, 558)

ERAD: see *ER-associated degradation.*

ETC: see *electron transport chain.*

euchromatin: loosely packed, uncondensed form of chromatin present during interphase; contains DNA that is being actively transcribed. (p. 458) Also see *heterochromatin.*

Eukarya: one of the three domains of organisms, the other two being Bacteria and Archaea; the domain consisting of one-celled and multicellular organisms called eukaryotes whose cells are characterized by a membrane-bounded nucleus and other membrane-bounded organelles. (p. 80)

eukaryote: category of organisms whose cells are characterized by the presence of a membrane-bounded nucleus and other membrane-bounded organelles; includes plants, animals, fungi, algae, and protozoa. (p. 80)

excision repair: DNA repair mechanism that removes and replaces abnormal nucleotides. (p. 488)

excitatory postsynaptic potential (EPSP): small depolarization of the postsynaptic membrane triggered by binding of an excitatory neurotransmitter to its receptor; if the EPSP exceeds a threshold level, it can trigger an action potential. (p. 680) Also see *inhibitory postsynaptic potential.*

exergonic: an energy-releasing reaction characterized by a negative free energy change ($\Delta G < 0$). (p. 117)

exit site: see *E site.*

exocytosis: fusion of vesicle membranes with the plasma membrane so that contents of the vesicle can be expelled or secreted to the extracellular environment. (p. 335)

exome: the part of a eukaryotic genome formed by exons, i.e., that part of the genome that, when transcribed and following processing of RNA transcripts, is present in mRNA. (p. 638)

exon: nucleotide sequence in a primary RNA transcript that is preserved in the mature, functional RNA molecule. (p. 532) Also see *intron.*

exon junction complex (EJC): protein complex deposited at exon-exon junctions created by pre-mRNA splicing. (p. 533)

exonuclease: an enzyme that degrades a nucleic acid (usually DNA) from one end, rather than cutting the molecule internally. (p. 483)

exosome: macromolecular complex involved in mRNA degradation in eukaryotic cells; contains multiple $3' \rightarrow 5'$ exonucleases. (p. 609)

exothermic: a reaction or process that releases heat. (p. 115)

expansin: a protein that allows local remodeling of plant cell walls. Expansins may act by weakening hydrogen bonds within the cell wall and are activated at low pH. (p. 426)

exportin: nuclear receptor protein that binds to the nuclear export signal of proteins in the nucleus and then transports the bound protein out through the nuclear pore complex and into the cytosol. (p. 467)

extensin: group of related glycoproteins that form rigid, rodlike molecules tightly woven into the cell walls of plants and fungi. (p. 433)

extracellular digestion: degradation of components outside a cell, usually by lysosomal enzymes that are released from the cell by exocytosis. (p. 348)

extracellular matrix (ECM): material secreted by animal cells that fills the spaces between neighboring cells; consists of a mixture of structural proteins (e.g., collagen, elastin) and adhesive glycoproteins (e.g., fibronectin, laminin) embedded in a matrix composed of protein-polysaccharide complexes called proteoglycans. (pp. 99, 422)

F

F factor: DNA sequence that enables an *E. coli* cell to act as a DNA donor during bacterial conjugation. (p. 445)

F_1 complex: knoblike sphere protruding into the matrix from mitochondrial inner membranes (or into the cytosol from the bacterial plasma membrane) that contains the ATP-synthesizing site of aerobic respiration. (p. 248)

F_1 generation: offspring of the P_1 generation in a genetic breeding experiment. (p. 760)

F_2 generation: offspring of the F_1 generation in a genetic breeding experiment. (p. 765)

facilitated diffusion: membrane protein–mediated movement of a substance across a membrane that does not require energy because the ion or molecule being transported is moving down an electrochemical gradient. (p. 193)

F-actin: component of microfilaments consisting of G-actin monomers that have been polymerized into long, linear strands. (p. 371)

facultative heterochromatin: chromosomal regions that have become specifically condensed and inactivated in a particular cell type at a specific time. (p. 458) Also see *constitutive heterochromatin.*

facultative organism: organism that can function in either an anaerobic or aerobic mode. (p. 222)

FAD: see *flavin adenine dinucleotide.*

familial hypercholesterolemia (FH): genetic predisposition to high blood cholesterol levels and heart disease caused by an inherited defect in the gene coding for the LDL receptor. (p. 332)

fast axonal transport: microtubule-mediated movement of vesicles and organelles back and forth along a nerve cell axon. (p. 386)

fatty acid: long, unbranched hydrocarbon chain that has a carboxyl group at one end and is therefore amphipathic; usually contains an even number of carbon atoms and may be of varying degrees of unsaturation. (p. 163)

fatty acid–anchored membrane protein: protein located on a membrane surface that is covalently bound to a fatty acid embedded with the lipid bilayer. (p. 174)

Fd: see *ferredoxin.*

feedback (end-product) inhibition: ability of the end-product of a biosynthetic pathway to inhibit the activity of the first enzyme in the pathway, thereby ensuring that the functioning of the pathway is sensitive to the intracellular concentration of its product. (p. 147)

fermentation: partial oxidation of carbohydrates by oxygen-independent (anaerobic) pathways, resulting often (but not always) in the production of either ethanol and carbon dioxide or lactate. (p. 222)

ferredoxin (Fd): iron-sulfur protein in the chloroplast stroma involved in the transfer of electrons from photosystem I to $NADP^+$ during the energy transduction reactions of photosynthesis. (p. 295)

ferredoxin-$NADP^+$ reductase (FNR): enzyme located on the stroma side of the thylakoid membrane that catalyzes the transfer of electrons from ferredoxin to $NADP^+$. (p. 295)

fertilization: union of two haploid gametes to form a diploid cell, the zygote, that develops into a new organism. (p. 752)

FH: see *familial hypercholesterolemia.*

fibronectin: adhesive glycoprotein found in the extracellular matrix and loosely associated with the cell surface; binds cells to the extracellular matrix and is important in determining cell shape and guiding cell migration. (p. 426)

fibrous protein: protein with extensive α helix or β sheet structure that confers a highly ordered, repetitive structure. (p. 54)

filopodium (plural, filopodia): thin, pointed cytoplasmic protrusion that transiently emerges from the surface of eukaryotic cells during cell movements. (p. 407)

first law of thermodynamics: law of conservation of energy; principle that energy can be converted from one form to another but can never be created or destroyed. (p. 113)

Fischer projection: model depicting the chemical structure of a molecule as a chain drawn vertically with the most oxidized atom on top and horizontal projections that are understood to be coming out of the plane of the paper. (p. 65)

5′ cap: methylated structure at the 5′ end of eukaryotic mRNAs created by adding 7-methylguanosine and methylating the ribose rings of the first, and often the second, nucleotides of the RNA chain. (p. 530)

fixative: chemical that kills cells while preserving their structural appearance for microscopic examination. (p. A-16)

flagellum (plural, flagella): membrane-bounded appendage on the surface of a eukaryotic cell composed of a specific arrangement of microtubules and responsible for motility of the cell; longer and less numerous (usually limited to one or a few per cell) than closely related organelles called cilia. (p. 391) Also see *cilium.*

flavin adenine dinucleotide (FAD): coenzyme that accepts two electrons and two protons from an oxidizable organic molecule to generate the reduced form, $FADH_2$; important electron carrier in energy metabolism. (p. 254)

flavoprotein: protein that has a tightly bound flavin coenzyme (FAD or FMN) and that serves as a biological electron donor or acceptor. Several examples occur in the mitochondrial electron transport chain. (p. 261)

flippase: see *phospholipid translocator.*

flow cytometry: method for automated measurement of large numbers of cells in liquid suspension using a laser-based computerized detector that can measure multiple cellular properties simultaneously, including presence of fluorescence signal of various wavelengths and cell size and shape based on light scattering. (p. 719)

fluid mosaic model: model for membrane structure consisting of a lipid bilayer with proteins associated as discrete globular entities that penetrate the bilayer to various extents and are free to move laterally in the membrane. (p. 159)

fluid-phase endocytosis: nonspecific uptake of extracellular fluid by infolding of the plasma membrane, followed by budding off of a membrane vesicle. (p. 340)

fluorescence: property of molecules that absorb light and then reemit the energy as light of a longer wavelength. (p. 109)

fluorescence-activated cell sorting (FACS): method, combined with flow cytometry, for automated physical separation of cells based on fluorescence properties and cell shape and size; single cells in tiny drops containing are measured, and an electrostatic charge is applied to the drops based on the measurement; the drops can then be differentially deflected using a charged plate and collected for further study. (p. 719) Also see *flow cytometry.*

fluorescence in situ hybridization (FISH): technique for identifying specific DNA or RNA sequences in an intact, preserved tissue or cellular preparation. FISH uses a fluorescent nucleic acid probe (usually DNA) to hybridize to a sample. The bound probe is then detected using fluorescence microscopy. (p. 451)

fluorescence microscopy: light microscopic technique that focuses light of a specific wavelength on the specimen, thereby causing fluorescent compounds in the specimen to emit light of a specific, longer wavelength. (p. A-8)

fluorescence recovery after photobleaching (FRAP): technique for measuring the lateral diffusion rate of membrane lipids or proteins in which such molecules are linked to a fluorescent dye and a tiny membrane region is then bleached with a laser. (p. 165)

fluorescent antibody: antibody containing covalently linked fluorescent dye molecules that allow the antibody to be used to locate antigen molecules microscopically. (p. 179)

FNR: see *ferredoxin-NADP$^+$ reductase.*

F_o complex: group of hydrophobic membrane proteins that anchor the F_1 complex to either the inner mitochondrial membrane or the bacterial plasma membrane; serves as the proton translocator channel through which protons flow when the electrochemical gradient across the membrane is used to drive ATP synthesis. (p. 248)

F_oF_1 complex: protein complex in the mitochondrial inner membrane and the bacterial plasma membrane that consists of the F_1 complex bound to the F_o complex; the flow of protons through the F_o component leads to the synthesis of ATP by the F_1 component. (p. 249)

focal adhesion: localized points of attachment between cell surface integrin molecules and the extracellular matrix; contain clustered integrin molecules that interact with bundles of cytoskeletal actin microfilaments via several linker proteins. (p. 430)

focal length: distance between the midline of a lens and the point at which rays passing through the lens converge to a focus. (p. A-2)

formin: a protein involved in actin polymerization that associates with the fast-growing end (barbed end) of actin filaments and can remain at the tip of the growing filament as it elongates; most formins are Rho-GTPase effector proteins. (p. 377)

Förster resonance energy transfer (FRET): an extremely short-distance interaction between two fluorescent molecules in which excitation is transferred from a donor molecule to an acceptor molecule directly; can be used to determine whether two molecules are in contact, or to produce "biosensors" that detect where proteins are activated or identify changes in local ion concentration. (p.217)

frameshift mutation: insertion or deletion of one or more base pairs in a DNA molecule, causing a change in the reading frame of the mRNA molecule that usually garbles the message. (p. 542)

free energy (G): thermodynamic function that measures the extractable energy content of a molecule; under conditions of constant temperature and pressure, the change in free energy is a measure of the system's ability to do work. (p. 117)

free energy change (ΔG): thermodynamic parameter used to quantify the net free energy liberated or required by a reaction or process; measure of thermodynamic spontaneity. (p. 117)

freeze etching: cleaving a quick-frozen specimen, as in freeze fracturing, followed by the subsequent sublimation of ice from the specimen surface to expose small areas of the true cell surface. (p. A-22)

freeze fracturing: sample preparation technique for electron microscopy in which a frozen specimen is cleaved by a sharp blow followed by examination of the fractured surface, often the interior of a membrane. (pp. 171, A-21)

FRET: see *fluorescence resonance energy transfer.*

F-type ATPase: type of transport ATPase found in bacteria, mitochondria, and chloroplasts that can use the energy of ATP hydrolysis to pump protons against their electrochemical gradient; can also catalyze the reverse process, in which the exergonic flow of protons down their electrochemical gradient is used to drive ATP synthesis. (p. 204) Also see *ATP synthase.*

functional group: group of chemical elements covalently bonded to each other that confers characteristic chemical properties upon any molecule to which it is covalently linked. (p. 24)

G

G: see *guanine* or *free energy.*

ΔG: see *free energy change.*

$\Delta G^{\circ\prime}$: see *standard free energy change.*

G_0 (G zero): designation applied to eukaryotic cells that have become arrested in the G1 phase of the cell cycle and thus are no longer proliferating. (p. 721)

G_1 phase: stage of the eukaryotic cell cycle between the end of the previous division and the onset of chromosomal DNA synthesis. (p. 721)

G_2 phase: stage of the eukaryotic cell cycle between the completion of chromosomal DNA replication and the onset of cell division. (p. 718)

G-actin: globular monomeric form of actin that polymerizes to form F-actin. (p. 371)

GAG: see *glycosaminoglycan.*

gamete: haploid cells produced by each parent that fuse together to form the diploid offspring; for example, sperm or egg. (p. 752)

gametogenesis: the process that produces gametes. (p. 752)

gametophyte: haploid generation in the life cycle of an organism that alternates between haploid and diploid forms; form that produces gametes. (p. 754)

gamma tubulin (γ-tubulin): form of tubulin located in the centrosome, where it functions in the nucleation of microtubules. (p. 367)

gamma tubulin ring complex (γ-TuRC): ring of γ-tubulin that emerges from centrosomes and nucleate the assembly of new microtubules. (p. 359)

gamma-TuRC (γ-TuRC): see *gamma tubulin ring complex.*

ganglioside: a charged glycolipid containing the amino alcohol sphingosine and negatively charged sialic acid residues. (p. 160)

GAP: see *GTPase activating protein.*

gap junction: type of cell junction that provides a point of intimate contact between two adjacent cells through which ions and small molecules can pass. (p. 414)

gatekeeper gene: a tumor suppressor gene directly involved in restraining cell proliferation; loss-of-function mutations in such genes can lead to excessive cell proliferation and tumor formation. (p. 807)

Gb: gigabases; a billion base pairs. (p. 637)

GEF: see *guanine-nucleotide exchange factor.*

gel electrophoresis: technique in which proteins or nucleic acids are separated in gels made of polyacrylamide or agarose by placing the gel in an electric field. (p. 617)

gel filtration chromatography: a type of chromatography that separates molecules based on size by virtue of their ability to percolate at different rates through a resin containing beads perforated with pores of differing sizes. (p. 648)

gene: "hereditary factor" that specifies an inherited trait; consists of a DNA base sequence that codes for the amino acid sequence of one or more polypeptide chains or, alternatively, for one of several types of RNA that perform functions other than coding for polypeptide chains (e.g., rRNA, tRNA, snRNA, or microRNA). (pp. 11, 541)

gene amplification: mechanism for creating extra copies of individual genes by selectively replicating specific DNA sequences. (p. 588)

gene conversion: phenomenon in which genes undergo nonreciprocal recombination during meiosis, such that one of the recombining genes ends up on both chromosomes of a homologous pair rather than being exchanged from one chromosome to the other. (p. 501)

general transcription factor: a protein that is always required for RNA polymerase to bind to its promoter and initiate RNA synthesis, regardless of the identity of the gene involved. (p. 520)

genetic code: set of rules specifying the relationship between the sequence of bases in a DNA or mRNA molecule and the order of amino acids in the polypeptide chain encoded by that DNA or mRNA. (p. 540)

genetic engineering: application of recombinant DNA technology to practical problems, primarily in medicine and agriculture. (p. 653)

genetic instability: trait of cancer cells in which abnormally high mutation rates are caused by defects in DNA repair and/or chromosome sorting mechanisms. (p. 800)

genetic mapping: determining the sequential order and spacing of genes on a chromosome. This was first done by calculating recombination frequencies. (p. 773)

genetics: study of the behavior of genes, which are the chemical units involved in the storage and transmission of hereditary information. (p. 4)

genome: the DNA (or for some viruses, RNA) that contains one complete copy of all the genetic information of an organism or virus. (p. 4)

genome editing: type of genetic engineering in which genomic DNA is removed, replaced, or altered using engineered nucleases that recognize specific sequences in a gene of interest. (p. 653)

genome-wide association study (GWAS): study of genetic variants in individuals (or tissue samples from individuals) in a population to determine if specific genetic elements are statistically associated with a trait or disease; often performed by whole-genome sequencing of DNA. (p. 640, 808)

genomic control: a regulatory change in the makeup or structural organization of the genome. (p. 588)

genomic equivalence: the notion that all somatic cells of a multicellular organism contain the same DNA content, but that different specialized cell types express subsets of genes. (p. 582)

genomic library: collection of recombinant DNA clones produced by cleaving an organism's entire genome into fragments, usually using restriction endonucleases, and then cloning all the fragments in an appropriate cloning vector. (p. 630)

genotype: genetic makeup of an organism. (p. 766)

globular protein: protein whose polypeptide chains are folded into compact structures rather than extended filaments. (p. 55)

gluconeogenesis: synthesis of glucose from precursors such as amino acids, glycerol, or lactate; occurs in the liver via a pathway that is essentially the reverse of glycolysis. (p. 231)

glucose: six-carbon sugar that is widely used as the starting molecule in cellular energy metabolism. (p. 222)

glucose transporter (GLUT): membrane carrier protein responsible for the facilitated diffusion of glucose. (p. 194)

GLUT: see *glucose transporter.*

glycerol: three-carbon alcohol with a hydroxyl group on each carbon; serves as the backbone for triacylglycerols. (p. 69)

glycerol phosphate shuttle: mechanism for carrying electrons from cytosolic NADH into the mitochondrion, where the electrons are delivered to FAD in the respiratory complexes. (p. 278)

glycocalyx: carbohydrate-rich zone located at the outer boundary of many animal cells. (p. 179)

glycogen: highly branched storage polysaccharide in animal cells; consists of glucose repeating subunits linked by $\alpha(1 \rightarrow 4)$ bonds and $\alpha(1 \rightarrow 6)$ bonds. (p. 64)

glycolate pathway: light-dependent pathway that decreases the efficiency of photosynthesis by oxidizing reduced carbon compounds without capturing the released energy; occurs when oxygen substitutes for carbon dioxide in the reaction catalyzed by rubisco, thereby generating phosphoglycolate that is then converted to 3-phosphoglycerate in the peroxisome and mitochondrion; also called photorespiration. (p. 307)

glycolipid: a lipid molecule with a carbohydrate attached by a glycosidic bond. (pp. 70, 162)

glycolysis (glycolytic pathway): series of reactions by which glucose or some other monosaccharide is catabolized to pyruvate without the involvement of oxygen, generating two molecules of ATP per molecule of monosaccharide metabolized. (p. 222)

glycoprotein: protein with one or more carbohydrate groups linked covalently to amino acid side chains. (pp. 176, 321)

glycosaminoglycan (GAG): polysaccharide constructed from a repeating disaccharide unit containing one sugar with an amino group and one sugar that usually has a negatively charged sulfate or carboxyl group; component of the extracellular matrix. (p. 424)

glycosidic bond: bond linking a sugar to another molecule, which may be another sugar molecule. (p. 63)

glycosylation: addition of carbohydrate side chains to specific amino acid residues of proteins, usually beginning in the lumen of the endoplasmic reticulum and completed in the Golgi apparatus. (pp. 178, 321)

glyoxylate cycle: modified version of the citric acid cycle occurring in plant glyoxysomes; an anabolic pathway that converts two molecules of acetyl CoA to one molecule of succinate, thereby permitting the synthesis of carbohydrates from lipids. (p. 259)

glyoxysome: specialized type of plant peroxisome that contains some of the enzymes responsible for the conversion of stored fat to carbohydrate in germinating seeds. (pp. 96, 259, 353)

Goldman equation: modification of the Nernst equation that calculates the resting membrane potential by adding together the effects of all relevant ions, each weighted for its relative permeability. (p. 663)

Golgi apparatus: stacks of flattened, disk-shaped membrane cisternae in eukaryotic cells that are important in the processing and packaging of secretory proteins and in the synthesis of complex polysaccharides. (pp. 93, 319)

golgin: one of a class of tethering proteins that connect vesicles to the Golgi and Golgi cisternae to each other. (p. 345)

GPI-anchored membrane protein: protein bound to the outer surface of the plasma membrane by linkage to glycosylphosphatidylinositol (GPI), a glycolipid found in the external monolayer of the plasma membrane. (p. 174)

G protein: any of numerous GTP-binding regulatory proteins located in the plasma membrane that mediate signal transduction pathways, usually by activating a specific target protein such as an enzyme or channel protein. (p. 689)

G protein–coupled receptor family: group of plasma membrane receptors that activate a specific G protein upon binding of the appropriate ligand. (p. 689)

G protein–coupled receptor kinase (GRK): any of several protein kinases that catalyze the phosphorylation of activated G protein–coupled receptors, thereby leading to receptor desensitization. (p. 689)

grading: see *tumor grading.*

granum (plural, grana): stack of thylakoid membranes in a chloroplast. (pp. 89, 287)

green fluorescent protein (GFP): a protein that exhibits bright green fluorescence when exposed to blue light; using genetic engineering techniques, GFP can be fused to nonfluorescent proteins, allowing them to be followed using fluorescence microscopy. (p. 564)

GRK: see *G protein–coupled receptor kinase.*

group specificity: ability of an enzyme to act on any of a whole group of substrates as long as they possess some common structural feature. (p. 135)

growth factor: any of a number of extracellular signaling proteins that stimulate cell division in specific types of target cells; examples include platelet-derived growth factor (PDGF) and epidermal growth factor (EGF). (p. 699)

GTPase-activating protein (GAP): a protein that speeds up the inactivation of Ras by facilitating the hydrolysis of bound GTP. (p. 701)

guanine (G): nitrogen-containing aromatic base, chemically designated as a purine, which serves as an informational monomeric unit when present in nucleic acids with other bases in a specific sequence; forms a complementary base pair with cytosine (C) by hydrogen bonding. (p. 60)

guanine-nucleotide exchange factor (GEF): a protein that triggers the release of GDP from the Ras protein, thereby permitting Ras to acquire a molecule of GTP. (p. 701)

guanylyl cyclase: enzyme that catalyzes the formation of cyclic GMP from GTP. It makes up one family of enzyme-coupled receptors involved in several processes including vasodilation.

guide RNA (gRNA): relatively short RNA sequences that are utilized in RNA editing and DNA editing (e.g. using the CRISPR/Cas9 system). gRNAs are partially complementary to their target sequence. (p. 13)

gyrase: see *DNA gyrase.*

H

H: see *enthalpy.*

H zone: light region located in the middle of the A band of striated muscle myofibrils. (p. 397)

hairpin loop: looped structure formed when two adjacent segments of a nucleic acid chain are folded back on one another and held in that conformation by base pairing between complementary base sequences. (p. 515)

haploid: containing a single set of chromosomes and therefore a single copy of the genome; can describe a cell, nucleus, or organism composed of such cells. (p. 752)

haploid spore: haploid product of meiosis in organisms that display an alternation of generations; gives rise upon spore germination to the haploid form of the organism (the gametophyte, in the case of higher plants). (p. 748)

haplotype: group of SNPs located near one another on the same chromosome that tend to be inherited as a unit. (p. 641)

HAT: see *histone acetyltransferase.*

Hatch–Slack cycle: series of reactions in C_4 plants in which carbon dioxide is fixed in the mesophyll cells and transported as a four-carbon compound to the bundle sheath cells, where subsequent decarboxylation results in a higher concentration of carbon dioxide and therefore a higher rate of carbon fixation by rubisco. (p. 309)

Haworth projection: model that depicts the chemical structure of a molecule in a way that suggests the spatial relationship of different parts of the molecule. (p. 65)

HDL: see *high-density lipoprotein.*

heart muscle: see *cardiac muscle.*

heat: transfer of energy as a result of a temperature difference. (p. 106)

heat shock gene: gene whose transcription is activated when cells are exposed to elevated temperatures or other stressful conditions; some heat shock genes code for molecular chaperones that, in addition to guiding normal protein folding, can facilitate the refolding of heat-damaged proteins. (p. 601)

heat shock response element: DNA base sequence located adjacent to heat shock genes that functions as a binding site for the heat shock transcription factor. (p. 604)

helicase: see *DNA helicase.*

helix: see *alpha helix or double helix.*

helix-loop-helix: DNA-binding motif found in transcription factors that is composed of a short α helix connected by a loop to a longer α helix. (p. 601)

helix-turn-helix: DNA-binding motif found in many regulatory transcription factors; consists of two regions of α helix separated by a bend in the polypeptide chain. (p. 597)

hemicellulose: heterogeneous group of polysaccharides deposited along with cellulose in the cell walls of plants and fungi to provide added strength; each consists of a long, linear chain of a single kind of sugar with short side chains. (p. 426)

hemidesmosome: point of attachment between cell surface integrin molecules of epithelial cells and the basal lamina; contains integrin molecules that are anchored via linker proteins to intermediate filaments of the cytoskeleton. (p. 431)

heterochromatin: highly compacted form of chromatin present during interphase; contains DNA that is not being transcribed. (p. 458) Also see *euchromatin.*

heterodimer ($\alpha\beta$-heterodimer): see *alpha beta heterodimer.*

heteroduplex region: a stretch of DNA in which each strand is derived from a separate homologous chromosome, often resulting in mismatched bases. Heteroduplex DNA forms in the process of homologous recombination during double-stranded DNA break repair or meiotic recombination (p. 504).

heterophagic lysosome: mature lysosome containing hydrolytic enzymes involved in the digestion of materials of extracellular origin. (p. 346) Also see *autophagic lysosome.*

heterophilic interaction: binding of two different molecules to each other. (p. 413) Also see *homophilic interaction.*

heterozygous: having two different alleles for a given gene. (p. 766) Also see *homozygous.*

Hfr cell: bacterial cell in which the F factor has become integrated into the bacterial chromosome, allowing the cell to transfer genomic DNA during conjugation. (p. 776)

hierarchical assembly: synthesis of biological structures from simple starting molecules to progressively more complex structures, usually by self-assembly. (p. 40)

high-density lipoprotein (HDL): cholesterol-containing protein-lipid complex that transports cholesterol through the bloodstream and is taken up by cells; exhibits high density because of its low cholesterol content. (p. 333)

high-voltage electron microscope (HVEM): an electron microscope that uses accelerating voltages up to a thousand or more kilovolts, thereby allowing the examination of thicker samples than is possible with a conventional electron microscope. (p. A-18)

histone: class of basic proteins found in eukaryotic chromosomes; an octamer of histones forms the core of nucleosomes. (p. 453)

histone acetyltransferase (HAT): enzyme that catalyzes the addition of acetyl groups to histones. (p. 457)

histone code: the hypothesis that posttranslational modifications to histone proteins, primarily involving methylation and/or acetylation within their N-terminal tails, can modulate gene expression; specific histone modifications lead to recruitment of other proteins that locally destabilize or stabilize chromatin packing, thereby promoting or suppressing gene expression. (p. 457)

histone deacetylase (HDAC): enzyme that catalyzes removal of acetyl groups from histones. (p. 457)

histone methyltransferase: enzyme that catalyzes the addition of methyl groups to histones. (p. 457)

Holliday junction: X-shaped structure produced when two DNA molecules are joined together by single-strand crossovers during genetic recombination. (p. 500)

homeobox: highly conserved DNA sequence found in homeotic genes; codes for a DNA-binding protein domain present in transcription factors that are important regulators of gene expression during development. (p. 605)

homeodomain: amino acid sequence, about 60 amino acids long, found in transcription factors encoded by homeotic genes; contains a helix-turn-helix DNA-binding motif. (p. 605)

homeotic gene: family of genes that control formation of the body plan during embryonic development; they code for transcription factors possessing homeodomains. (p. 605)

homeoviscous adaptation: alterations in membrane lipid composition that keep the viscosity of membranes approximately the same despite changes in environmental temperature. (p. 169)

homologous chromosomes: two copies of a specific chromosome, one derived from each parent, that pair with each other and exchange genetic information during meiosis. (p. 752)

homologous recombination: exchange of genetic information between two DNA molecules exhibiting extensive sequence similarity; used in repairing double-strand DNA breaks and during meiosis. (pp. 499, 756)

homophilic interaction: binding of two identical molecules to each other. (p. 413) Also see *heterophilic interaction.*

homozygous: having two identical alleles for a given gene. (p. 761) Also see *heterozygous.*

hopanoid: lipid resembling cholesterol that is found in some bacterial cell membranes and is thought to add stability to the membrane. (p. 163)

hormone: chemical that is synthesized in one organ, secreted into the vasculature and able to cause a physiological change in cells or tissues of another organ. (p. 707)

hormone response element: DNA base sequence that selectively binds to a hormone-receptor complex, resulting in the activation (or inhibition) of transcription of nearby genes. (p. 603)

HPV: see *human papillomavirus.*

human papillomavirus (HPV): virus responsible for cervical cancer; possesses oncogenes that block the actions of the proteins produced by the RB and *p53* tumor suppressor genes. (p. 802)

HVEM: see *high-voltage electron microscope.*

hyaluronate: a glycosaminoglycan found in high concentration in the extracellular matrix where cells are actively proliferating or migrating, and in the joints between movable bones. (p. 426)

hybrid: product of the cross of two genetically different parents. (p. 760)

hybridization: see *nucleic acid hybridization.*

hydrocarbon: an organic molecule consisting only of carbon and hydrogen atoms; not generally compatible with living cells. (p. 24)

hydrogen bond: weak attractive interaction between an electronegative atom and a hydrogen atom that is covalently linked to a second electronegative atom. (p. 28, 48)

hydrogenation: addition of electrons plus hydrogen ions (protons) to an organic molecule; reduction. (p. 220)

hydrolysis: reaction in which a chemical bond is broken by the addition of a water molecule. (p. 36)

hydropathy index: a value representing the average of the hydrophobicity values for a short stretch of contiguous amino acids within a protein. (p. 175)

hydropathy plot: graph showing the location of hydrophobic amino acid clusters within the primary sequence of a protein molecule; used to determine the likely locations of transmembrane segments within integral membrane proteins; also called hydrophobicity plot. (p. 175)

hydrophilic: describing molecules or regions of molecules that readily associate with or dissolve in water because of a preponderance of polar groups; "water-loving." (p. 29)

hydrophobic: describing molecules or regions of molecules that are poorly soluble in water because of a preponderance of nonpolar groups; "water-hating." (p. 29)

hydrophobic interaction: tendency of hydrophobic groups to be excluded from interactions with water molecules. (p. 50)

hydrophobicity plot: see *hydropathy plot.*

hydroxylation: chemical reaction in which a hydroxyl group is added to an organic molecule. (p. 316)

hyperpolarization (undershoot): transient hyperpolarization that occurs at the end of an action potential during which the membrane potential briefly becomes even more negative than it normally is at rest. The undershoot occurs because of the transient increase in potassium permeability.

hypertonic solution: solution that has a higher concentration of solutes than is found inside a cell in that solution. In a hypertonic solution, water will move out of a cell, causing the cell to dehydrate. (p. 190)

hypothesis: a statement or explanation that is consistent with most of the observational and experimental evidence to date. (p. 14)

hypotonic solution: solution that has a lower concentration of solutes than is found inside a cell in that solution. In a hypotonic solution, water will move into a cell, creating osmotic pressure that can burst the cell in the absence of a cell wall. (p. 190)

I

I band: region of a striated muscle myofibril that appears as a light band when viewed by microscopy; contains those regions of the thin actin filaments that do not overlap the thick myosin filaments. (p. 397)

IF: see *intermediate filament.*

IgSF: see *immunoglobulin superfamily.*

immune checkpoint therapy: use of a drug that blocks proteins on the surface of T cells or cancer cells that moderate the immune response, thereby enhancing the ability of T cells to kill cancer cells. T cells can kill cancer cells better. (p. 815)

immunoelectron microscopy (immunoEM): electron microscopic technique for visualizing antibodies within cells by linking them to substances that are electron dense and therefore visible as opaque dots. (p. A-20)

immunoEM: see *immunoelectron microscopy.*

immunoglobulin superfamily (IgSF): family of cell surface proteins involved in cell-cell adhesion that are structurally related to the immunoglobulin subunits of antibody molecules. (p. 416)

immunoprecipitation: process whereby a protein can be precipitated out of a solution using antibodies, typically attached to agarose or similar beads; often used to concentrate or purify proteins from a solution. (p. 648)

immunostaining: technique in which antibodies are labeled with a fluorescent dye to enable them to be identified and localized microscopically based on their fluorescence. (p. A-8)

IMP: see *intramembranous particle.*

importin: receptor protein that binds to the nuclear localization signal of proteins in the cytosol and then transports the bound protein through the nuclear pore complex and into the nucleus. (p. 458)

imprinting: a process that causes some genes to be expressed differently depending on whether they are inherited from a person's mother or father; such effects can be caused by differences in DNA methylation. (p. 594)

indel: an insertion or deletion of nucleotides in DNA that leads to a shift in reading frame. (p. 541)

indirect active transport: Type of transport that requires energy and depends on the simultaneous transport of two solutes, with the favorable movement of one solute down its gradient driving the unfavorable movement of the other solute up its gradient (p. 202)

indirect immunofluorescence: type of fluorescence microscopy in which a specimen is first exposed to an unlabeled antibody that attaches to specific antigenic sites within the specimen and is then stained with a secondary, fluorescent antibody that binds to the first antibody. (p. A-9)

induced-fit model: model postulating that the active site of an enzyme is relatively specific for its substrate before it binds, but even more so thereafter because of a conformational change in the enzyme induced by the substrate. (p. 137)

induced pluripotent stem (IPS) cells: differentiated cells forced to express transcription factor proteins found in embryonic stem cells, allowing them to become many different types of cells. (p. 587)

inducer: any effector molecule that activates the transcription of an inducible operon. (p. 576)

inducible enzyme: enzyme whose synthesis is regulated by the presence or absence of its substrate. (p. 573)

inducible operon: group of adjoining genes whose transcription is activated in the presence of an inducer. (p. 576)

inhibition (of enzyme activity): decreasing the catalytic activity of an enzyme, either by a change in conformation or by chemical modification of one of its functional groups. (p. 141)

inhibitory postsynaptic potential (IPSP): small hyperpolarization of the postsynaptic membrane triggered by binding of an inhibitory neurotransmitter to its receptor, thereby reducing the amplitude of subsequent excitatory postsynaptic potentials and possibly preventing the firing of an action potential. (p. 680) Also see *excitatory postsynaptic potential.*

initial reaction velocity (*v*): reaction rate measured over a period of time during which the substrate concentration has not decreased enough to affect the rate and the accumulation of product is still too small to cause any measurable back reaction. (p. 141)

initiation 786

initiation complex (30S): complex formed by the association of mRNA, the 30S ribosomal subunit, an initiator aminoacyl tRNA molecule, and initiation factor IF2. (p. 543)

initiation complex (70S): complex formed by the association of a 30S initiation complex with a 50S ribosomal subunit; contains an initiator aminoacyl tRNA at the P site and is ready to commence mRNA translation. (p. 554)

initiation factors: group of proteins that promote the binding of ribosomal subunits to mRNA and initiator tRNA, thereby initiating the process of protein synthesis. (p. 542)

initiation (stage of carcinogenesis): irreversible conversion of a cell to a precancerous state by agents that cause DNA mutation. (p. 781)

initiator (Inr): short DNA sequence surrounding the transcription start site that forms part of the promoter for RNA polymerase II. (p. 522)

initiator tRNA: type of transfer RNA molecule that starts the process of translation; recognizes the AUG start codon and carries formylmethionine in prokaryotes or methionine in eukaryotes. (p. 554)

inner membrane: the inner of the two membranes that surround a mitochondrion, chloroplast, or nucleus. (pp. 246, 286)

inositol-1,4,5-trisphosphate (IP_3): triply phosphorylated inositol molecule formed as a product of the cleavage of phosphatidylinositol-4,5-bisphosphate catalyzed by phospholipase C; functions as a second messenger by triggering the release of calcium ions from storage sites within the endoplasmic reticulum. (p. 693)

Inr: see *initiator.*

insertional mutagenesis: change in gene structure or activity resulting from the integration of DNA derived from another source, usually a virus. (p. 798)

insulator: DNA element that prevents an adjacent gene from being regulated by a nearby enhancer or silencer. (p. 596)

insulin: peptide hormone, produced by the islets of Langerhans in the pancreas, that acts to decrease blood glucose by stimulating glucose uptake into muscle and adipose cells and by stimulating glycogen synthesis. (p. 709)

integral membrane protein: hydrophobic protein localized within the interior of a membrane but possessing hydrophilic regions that protrude from one or both membrane surfaces. (p. 168)

integral monotopic protein: an integral membrane protein embedded in only one side of the lipid bilayer. (p. 173)

integrin: any of several plasma membrane receptors that bind to extracellular matrix components at the outer membrane surface and interact with cytoskeletal components at the inner membrane surface; includes receptors for fibronectin, laminin, and collagen. (p. 421)

intercalated disc: membrane partition enriched in gap junctions that divides cardiac muscle into separate cells containing single nuclei. (p. 404)

interdoublet link: link between adjacent doublets in the axoneme of a eukaryotic cilium or flagellum; believed to limit the extent of doublet movement with respect to each other as the axoneme bends. (p. 393)

interference: process by which two or more waves of light combine to reinforce or cancel one another, producing a wave equal to the sum of the combining waves. (p. A-2)

intermediate filament (IF): group of protein filaments that are the most stable components of the cytoskeleton of eukaryotic cells; exhibits a diameter of 8–12 nm, which is intermediate between the diameters of actin microfilaments and microtubules. (pp. 98, 372)

intermediate lens (electron microscope): electromagnetic lens positioned between the objective and projector lenses in a transmission electron microscope. (p. A-17)

intermediate lens (light microscope): lens positioned between the ocular and objective lenses in a light microscope. (p. A-15)

intermembrane space: region of a mitochondrion or chloroplast between the inner and outer membranes. (pp. 245, 286)

internal energy (*E*): total energy stored within a system; cannot be measured directly, but the change in internal energy, ΔE, is measurable. (p. 115)

internal ribosome entry sequence (IRES): sequence in the interior of an mRNA that allows ribosome binding and initiation of translation without the presence of a 5′ cap. (p. 557)

interphase: growth phase of the eukaryotic cell cycle situated between successive division phases (M phases); composed of G1, S, and G2 phases. (p. 721)

interspersed repeated DNA: repeated DNA sequences whose multiple copies are scattered around the genome. (p. 460)

intracellular membrane: any cellular membrane internal to the plasma membrane; such membranes serve to compartmentalize functions within eukaryotic cells. (p. 153)

intraflagellar transport (IFT): movement of components to and from the tips of flagella driven by both plus- and minus-end-directed microtubule motor proteins. (p. 394)

intramembranous particle (IMP): integral membrane protein that is visible as a particle when the interior of a membrane is visualized by freeze-fracture microscopy. (p. A-22)

intron: nucleotide sequence in an RNA molecule that is part of the primary transcript but not the mature, functional RNA molecule. (p. 526) Also see *exon.*

invasion: direct spread of cancer cells into neighboring tissues. (p. 789)

inverted repeat: DNA segment containing two copies of the same base sequence oriented in opposite directions. (p. 603)

ion: an atom or molecule that has gained or lost a proton or an electron and is thus charged. (p. 24)

ion channel: membrane protein that allows the passage of specific ions through the membrane; generally regulated by either changes in membrane potential (voltage-gated channels) or binding of a specific ligand (ligand-gated channels). (pp. 198, 664)

ion-exchange chromatography: type of chromatography in which molecules passing through a column are separated based on their charge. (p. 648)

ionic bond: attractive force between a positively charged chemical group and a negatively charged chemical group. (p. 49)

ionizing radiation: high-energy forms of radiation that remove electrons from molecules, thereby generating highly reactive ions that cause DNA damage; includes X-rays and radiation emitted by radioactive elements. (p. 789)

IP_3: see *inositol-1,4,5-trisphosphate.*

IP_3 receptor: ligand-gated calcium channel in the ER membrane that opens when bound to IP_3, allowing calcium ions to flow from the ER lumen into the cytosol. (p. 694)

IPSP: see *inhibitory postsynaptic potential.*

IRE: see *iron-response element.*

iron-response element (IRE): short base sequence found in mRNAs whose translation or stability is controlled by iron; binding site for an IRE-binding protein. (p. 608)

iron-sulfur protein: protein that contains iron and sulfur atoms complexed with four cysteine groups and that serves as an electron carrier in the electron transport chain. Also called a nonheme iron protein. (p. 261)

irreversible inhibitor: molecule that binds to an enzyme covalently, causing an irrevocable loss of catalytic activity. (p. 146)

isoenzyme: any of several physically distinct proteins that catalyze the same reaction; also called an isozyme. (p. 305)

isoprenylated membrane protein: protein located on a membrane surface that is covalently bound to a prenyl group embedded within the lipid bilayer. (p. 174)

isotonic solution: solution that has the same concentration of solutes as is found inside a cell in that solution; in an isotonic solution, there is no net movement of water into or out of a cell. (p. 190)

isotopes: atoms of a particular chemical element that have different numbers of neutrons in the nucleus and thus have slightly different atomic weights. (p. 229)

isotopic labeling: technique that uses chemicals containing isotopes of normal elements in order to monitor their presence, distribution, or amount. (p. 229)

isozyme: see *isoenzyme.* (p. 305)

J

J: see *joule.*

Jak: see *Janus kinase.*

Janus kinase (Jak): cytoplasmic protein kinase that, after activation by cell surface receptors, catalyzes the phosphorylation and activation of STAT transcription factors. (p. 64)

joule (J): a unit of energy corresponding to 0.239 calories. (p. 113)

K

karyotype: picture of the complete set of chromosomes for a particular cell type, organized as homologous pairs arranged on the basis of differences in size and shape. (p. 459)

k_{cat}: see *turnover number.*

K_d: see *dissociation constant.*

K_{eq}: see *equilibrium constant.*

kb: kilobase; a thousand base pairs. (p. 637)

K-fiber: see *kinetochore microtubule.*

kilocalorie (kcal): unit of energy; 1kcal = 1000 calories. (p. 113)

kinesin: family of motor proteins that generate movement along microtubules using energy derived from ATP hydrolysis. (p. 386)

kinetochore: multiprotein complex located at the centromere region of a chromosome that provides the attachment site for spindle microtubules during mitosis or meiosis. (p. 720)

kinetochore microtubule: spindle microtubule that attaches to a kinetochore. Also called a K-fiber.

K_m: see *Michaelis constant.*

KMN network: large protein complex of the outer kinetochore that includes KNL1, the Mis12 complex, and the Ndc80 complex. Attaches to the inner kinetochore and, via the Ncd80 complex, to the plus ends of microtubules. Also associates with spindle assembly checkpoint (SAC) complex proteins.

knockout mice: mice that have been genetically engineered to contain a deletion of DNA sequences in a particular gene using homologous recombination in embryonic stem cells. (p. 651)

Kozak sequence: consensus sequence in eukaryotic mRNAs that contains a translation start site (AUG) and other nearby bases that is recognized as the translational start site after ribosome binding to the mRNA. (p. 557)

Krebs cycle: see *citric acid cycle.*

L

***lac* operon:** group of adjoining bacterial genes that code for enzymes involved in lactose metabolism and whose transcription is inhibited by the *lac* repressor. (p. 571)

lactate fermentation: anaerobic catabolism of carbohydrates with lactate as the end-product. (p. 227)

lagging strand: strand of DNA that grows in the 3′ → 5′ direction during DNA replication by discontinuous synthesis of short fragments in the 5′ → 3′ direction, followed by ligation of adjacent fragments. (p. 482) Also see *leading strand.*

lamellipodium (plural, lamellipodia): thin sheet of flattened cytoplasm that transiently protrudes from the surface of eukaryotic cells during cell crawling; supported by actin filaments. (p. 406)

lamin: member of a family of intermediate filament proteins conserved in most higher eukaryotes that forms part of the nuclear lamina. (p. 469)

laminin: adhesive glycoprotein of the extracellular matrix, localized predominantly in the basal lamina of epithelial cells. (p. 427)

large ribosomal subunit: component of a ribosome with a sedimentation coefficient of 60S in eukaryotes and 50S in prokaryotes; associates with a small ribosomal subunit to form a functional ribosome. (p. 98)

late endosome: vesicle containing newly synthesized acid hydrolases plus material fated for digestion; activated either by lowering the pH of the late endosome or by transferring its material to an existing lysosome. (p. 329)

lateral diffusion: diffusion of a membrane lipid or protein in the plane of the membrane. (p. 163)

latrunculin A: chemical compound derived from marine sponges that causes depolymerization of actin microfilaments. (p. 373)

law of independent assortment: principle stating that the alleles of each gene separate independently of the alleles of other genes during gamete formation. (p. 767)

law of segregation: principle stating that the alleles of each gene separate from each other during gamete formation. (p. 766)

LDL: see *low-density lipoprotein.*

LDL receptor: plasma membrane protein that serves as a receptor for binding extracellular LDL, which is then taken into the cell by receptor-mediated endocytosis. (p. 333)

leading strand: strand of DNA that grows as a continuous chain in the 5′ → 3′ direction during DNA replication. (p. 482) Also see *lagging strand.*

leaf peroxisome: special type of peroxisome found in the leaves of photosynthetic plant cells that contains some of the enzymes involved in photorespiration. (pp. 96, 307, 353)

leak channel: non-gated ion channel that helps to maintain steady-state ion concentrations of cells at electrochemical equilibrium, including potassium and sodium ions. (p. 667)

lectin: any of numerous carbohydrate-binding proteins that can be isolated from plant or animal cells and that promote cell-cell adhesion. (p. 418)

leptotene: first stage of prophase I of meiosis, characterized by condensation of chromatin fibers into visible chromosomes. (p. 756)

leucine zipper: DNA-binding motif found in many transcription factors; formed by an interaction between α helices in two polypeptide chains that are "zippered" together by hydrophobic interactions between leucine residues. (p. 601)

leukemia: cancer of blood or lymphatic origin in which the cancer cells proliferate and reside mainly in the bloodstream rather than growing as solid masses of tissue. (p. 779)

LHC: see *light-harvesting complex.*

LHCI: see *light-harvesting complex I.*

LHCII: see *light-harvesting complex II.*

licensing: process of making DNA competent for replication; normally occurs only once per cell cycle. (p. 479)

ligand: substance that binds to a specific receptor, thereby initiating the particular event or series of events for which that receptor is responsible. (p. 685)

ligand-gated ion channel: an integral membrane protein that forms an ion-conducting pore that opens when a specific molecule (ligand) binds to the channel. (p. 664)

light microscope: instrument consisting of a source of visible light and a system of glass lenses that allows an enlarged image of a specimen to be viewed. (pp. 9)

light-harvesting complex (LHC): collection of light-absorbing pigments, usually chlorophylls and carotenoids, linked together by proteins; unlike a photosystem, does not contain a reaction center, but absorbs photons of light and funnels the energy to a nearby photosystem. (p. 291)

light-harvesting complex I (LHCI): the light-harvesting complex associated with photosystem I. (p. 295)

light-harvesting complex II (LHCII): the light-harvesting complex associated with photosystem II. (p. 293)

light sheet microscopy: a type of fluorescence microscopy in which the illumination is brought in from the side, using a low-magnification objective lens; the light sheet allows only a thin strip of the specimen to be excited, which reduces photodamage. (p. A-12)

lignin: insoluble polymers of aromatic alcohols that occur mainly in woody plant tissues, where they contribute to the hardening of the cell wall and the structural strength we associate with wood. (p. 433)

limit of resolution: measurement of how far apart adjacent objects must be in order to be distinguished as separate entities. (pp. 7, A-4)

LINEs (long interspersed nuclear elements): class of interspersed repeated DNA sequences 6000–8000 base pairs in length that function as transposable elements and account for roughly 20% of the human genome; contain genes coding for enzymes needed for copying LINE sequences (and other mobile elements) and inserting the copies elsewhere in the genome. (p. 460)

Lineweaver–Burk equation: linear equation obtained by inverting the Michaelis–Menten equation, useful in determining parameters V_{max} and K_m and in the analysis of enzyme inhibition. (p. 143)

linkage group: group of genes that are transmitted, inherited, and assorted together. (p. 772)

linked genes: genes that are usually inherited together because they are located relatively close to each other on the same chromosome. (p. 772)

lipid: any of a large and chemically diverse class of organic compounds that are poorly soluble or insoluble in water but soluble in organic solvents. (p. 66)

lipid bilayer: unit of membrane structure, consisting of two layers of lipid molecules (mainly phospholipid) arranged so that their hydrophobic tails face toward each other and the polar region of each faces the aqueous environment on one side or the other of the bilayer. (pp. 32, 86, 158)

lipid raft: localized region of membrane lipids, often characterized by elevated levels of cholesterol and glycosphingolipids, that sequester proteins involved in cell signaling; also called lipid microdomain. (p. 167)

lipid-anchored membrane protein: protein located on a membrane surface that is covalently bound to one or more lipid molecules residing within the lipid bilayer. (p. 174) Also see *fatty acid–anchored membrane protein, isoprenylated membrane protein,* and *GPI-anchored membrane protein.*

liposome: hollow membrane-bound vesicle of varying size that forms spontaneously when lipids are mixed with water. (p. 79)

locus (plural, loci): location within a chromosome that contains the DNA sequence for a particular gene. (p. 761)

long noncoding RNA (lncRNA): RNA > 200 bp in length that is not translated into protein; known to regulate chromatin and other aspects of gene expression. (p. 592)

loss of heterozygosity (LOH): deletion of a chromosomal region that contains a tumor suppressor gene; when the other copy of the gene contains a loss-of-function mutation, this predisposes the cells that have suffered the deletion to a higher likelihood of becoming cancerous. (p. 802)

low-density lipoprotein (LDL): cholesterol-containing protein-lipid complex that transports cholesterol through the bloodstream and is taken up by cells; exhibits low density because of its high cholesterol content. (p. 333)

lumen: internal space enclosed by a membrane, usually the endoplasmic reticulum or related membrane systems. (p. 92)

lymphoma: cancer of lymphatic origin in which the cancer cells grow as solid masses of tissue. (p. 784)

lysosomal storage disease: disease resulting from a deficiency of one or more lysosomal enzymes and characterized by the undesirable accumulation of excessive amounts of specific substances that would normally be degraded by the deficient enzymes. (p. 348)

lysosome: membrane-bounded organelle containing digestive enzymes capable of degrading all the major classes of biological macromolecules. (pp. 94, 345)

M

M line: dark line running down the middle of the H zone of a striated muscle myofibril. (p. 397)

M phase: stage of the eukaryotic cell cycle when the nucleus and the rest of the cell divide. (p. 720)

macromolecule: polymer built from small repeating monomer units, with molecular weights ranging from a few thousand to hundreds of millions. (p. 33)

macrophagy: process by which an organelle becomes wrapped in a double membrane derived from the endoplasmic reticulum, creating an autophagic vacuole that acquires lysosomal enzymes which degrade the organelle. (p. 348)

malate-aspartate shuttle: mechanism for carrying electrons from cytosolic NADH into the mitochondrion, where the electrons are delivered to NAD^+ in the respiratory complexes. (p.278)

malignant tumor: tumor that can invade neighboring tissues and spread through the body via fluids, especially the bloodstream, to other parts of the body; also called a cancer. (p. 785)

MAP: see *microtubule-associated protein.*

map-based cloning: method used to assemble overlapping cloned fragments of DNA that are correlated with a genetic map. (p. 635)

MAP kinase: see *mitogen-activated protein kinase.*

MAPK: see *mitogen-activated protein kinase.*

mass spectrometry: high-speed, extremely sensitive technique that uses magnetic and electric fields to separate proteins or protein fragments based on differences in mass and charge. (pp. 11, 645)

***MAT* locus:** site in the yeast genome where the active allele for mating type resides. (p. 598)

mating bridge: transient cytoplasmic connection through which DNA is transferred from a male bacterial cell to a female cell during conjugation. (p. 775)

matrix: unstructured semifluid substance that fills the interior of a mitochondrion. (pp. 87, 246)

matrix-assisted laser desorption/ionization mass spectrometry (MALDI MS): type of mass spectrometry useful for analyzing macromolecules based on their mass/charge ratio; relies on embedding molecules in a light-absorbing matrix, which can then be ionized by laser irradiation, allowing them to enter the spectrometer. (p. 649)

maturation-promoting factor (MPF): see *mitosis-promoting factor (MPF)* (p.737).

maximum velocity (V_{max}): upper limiting reaction rate approached by an enzyme-catalyzed reaction as the substrate concentration approaches infinity. (p. 139)

Mb: megabase; a million base pairs. (p. 637)

MCC: See *mitotic checkpoint complex*.

MDR: see *multidrug resistance transport protein*.

mechanical work: use of energy to bring about a physical change in the position or orientation of a cell or some part of it. (p. 108)

medial cisterna (plural, cisternae): flattened membrane sac of the Golgi apparatus located between the membrane tubules of the *cis*-Golgi network and the *trans*-Golgi network. (p. 320)

Mediator: large multiprotein complex that functions as a transcriptional coactivator by binding both to activator proteins associated with an enhancer and to RNA polymerase; acts as a central coordinating unit of gene regulation, receiving both positive and negative inputs and transmitting the information to the transcription machinery. (p. 598)

meiosis: series of two cell divisions, preceded by a single round of DNA replication, that converts a single diploid cell into four haploid cells (or haploid nuclei). (p. 752)

meiosis I: the first meiotic division, which produces two haploid cells with chromosomes composed of sister chromatids. (p. 754)

meiosis II: the second meiotic division, which separates the sister chromatids of the haploid cell generated by the first meiotic division. (p. 754)

membrane: permeability barrier surrounding and delineating cells and organelles; consists of a lipid bilayer with associated proteins. (pp. 31, 154)

membrane asymmetry: a membrane property based on differences between the molecular compositions of the two lipid monolayers and the proteins associated with each. (p. 162)

membrane potential (V_m): voltage (charge difference) across a membrane created by ion gradients; usually the inside of a cell is negatively charged with respect to the outside. (pp. 186, 660) Also see *resting membrane potential*.

Mendel's laws of inheritance: principles derived by Mendel from his work on the inheritance of traits in pea plants. (p. 766). Also see *law of segregation* and *law of independent assortment*.

mesophyll cell: outer cell in the leaf of plants that serves as the site of carbon fixation. (p. 301)

messenger RNA (mRNA): RNA molecule containing the information that specifies the amino acid sequence of one or more polypeptides. (p. 511)

metabolic pathway: series of cellular enzymatic reactions that convert one molecule to another via a series of intermediates. (p. 216)

metabolism: all chemical reactions occurring within a cell. (p. 216)

metaphase: stage during mitosis or meiosis when the chromosomes become aligned at the spindle equator. (p. 726)

metastable state: condition where potential reactants are thermodynamically unstable but have insufficient energy to exceed the activation energy barrier for the reaction. (p. 131)

metastasis: spread of tumor cells to distant organs via the bloodstream or other body fluids. (p. 784)

MF: see *microfilament*.

Michaelis constant (K_m): substrate concentration at which an enzyme-catalyzed reaction is proceeding at one-half of its maximum velocity. (p. 141)

Michaelis–Menten equation: widely used equation describing the relationship between velocity and substrate concentration for an enzyme-catalyzed reaction $V = V_{max}[S]/(K_m + [S])$. (p. 141)

microautophagy: process by which lysosomes take up and degrade cytosolic proteins. (p. 616)

microfibril: aggregate of several dozen cellulose molecules laterally crosslinked by hydrogen bonds; serves as a structural component of plant and fungal cell walls. (p. 433)

microfilament (MF): polymer of actin, with a diameter of about 7 nm, that is an integral part of the cytoskeleton, contributing to the support, shape, and mobility of eukaryotic cells. (pp. 98, 370)

micrometer (μm): unit of measure: 1 micrometer = 10^{-6} meters. (p. 4)

microphagy: process by which small bits of cytoplasm are surrounded by ER membrane to form an autophagic vesicle whose contents are then digested either by accumulation of lysosomal enzymes or by fusion of the vesicle with a late endosome. (p. 348)

microRNA (miRNA): class of single-stranded RNAs, about 21–22 nucleotides long, produced by cellular genes for the purpose of inhibiting the translation of mRNAs produced by other genes. (p. 611)

microsatellite DNA: see *short tandem repeat (STR)*.

microtome: instrument used to slice an embedded biological specimen into thin sections for light microscopy. (p. A-16)

microtubule (MT): polymer of the protein tubulin, with a diameter of about 25 nm, that is an integral part of the cytoskeleton and that contributes to the support, shape, and motility of eukaryotic cells; also found in eukaryotic cilia and flagella. (pp. 98, 361)

microtubule-associated protein (MAP): any of various accessory proteins that bind to microtubules and modulate their assembly, structure, and/or function. (p. 368)

microtubule-organizing center (MTOC): structure that initiates the assembly of microtubules, the primary example being the centrosome. (p. 366)

microvillus (plural, microvilli): fingerlike projection from the cell surface that increases membrane surface area; important in cells that have an absorption function, such as those that line the intestine. (p. 375)

midbody: transient structure found in dividing animal cells near the end of cytokinesis prior to the complete separation of the dividing cells; contains bundled microtubules derived from the mitotic spindle's midzone and electron-dense material that may aid in final separation of daughter cells. (p. 731)

middle lamella: first layer of the plant cell wall to be synthesized; ends up farthest away from the plasma membrane, where it functions to hold adjacent cells together. (p. 426)

minisatellite DNA: repeat region in chromosomal DNA, 102–105 bp in total length, composed of a tandem repeat unit of roughly 10–100 bp. (p. 460)

minus end (of microtubule): slower-growing (or nongrowing or shrinking) end of a microtubule. (p. 364)

miRISC: complex between microRNA and several proteins that together silence the expression of messenger RNAs containing sequences complementary to those of the microRNA. (p. 611). Also see *siRISC* and *RISC*.

miRNA: see *microRNA*.

mismatch repair: DNA repair mechanism that detects and corrects base pairs that are improperly hydrogen-bonded. (p. 496)

mitochondrion (plural, mitochondria): double membrane-enclosed cytoplasmic organelle of eukaryotic cells that is the site of aerobic respiration and hence of ATP generation. (pp. 87, 243)

mitogen-activated protein kinase (MAP kinase, or MAPK): a family of protein kinases that are activated when cells receive a signal to grow and divide. (p. 701)

mitosis: process by which two genetically identical daughter nuclei are produced from one nucleus as the duplicated chromosomes of the parent cell segregate into separate nuclei; usually followed by cell division. (p. 473)

mitosis-promoting factor (MPF): a mitotic Cdk-cyclin complex that drives progression from G2 into mitosis by phosphorylating proteins involved in key stages of mitosis. (p. 732)

mitotic checkpoint complex (MCC): a protein complex composed of Mad and Bub proteins that binds to and inhibits the Cdc20 protein, so that it cannot activate the anaphase promoting complex (APC/C). (p.741)

mitotic index: percentage of cells in a population that are in any stage of mitosis at a certain point in time; used to estimate the relative length of the M phase of the cell cycle. (p. 721)

mitotic spindle: microtubular structure responsible for separating chromosomes during mitosis. (p. 720)

mixed inhibitor

MLCK: see *myosin light-chain kinase*.

model organism: an organism that is widely studied, well characterized, easy to manipulate, and has particular advantages for various experimental studies; examples include *E. coli*, yeast, *Drosophila*, *C. elegans*, *Arabidopsis*, and mice. (p. 15)

molecular chaperone: protein that facilitates the folding of other proteins but is not a component of the final folded structure. (pp. 38,50 561)

molecular mimicry: the presence of a high degree of structural similarity between two distinct macromolecules. (p.559)

molecular targeting: development of drugs designed to specifically target molecules that are critical to cancer cells. (p. 810)

monoclonal antibody: a highly purified antibody, directed against a single antigen, that is produced by a cloned population of antibody-producing cells. (p. 641)

monomer: small organic molecule that serves as a subunit in the assembly of a macromolecule. (p. 33)

monomeric protein: protein that consists of a single polypeptide chain. (p. 47)

monosaccharide: simple sugar; the repeating unit of polysaccharides. (p. 64)

motif: region of protein secondary structure consisting of small segments of α helix and/or β sheet connected by looped regions of varying length. (p. 53)

motility (cellular): movement or shortening of a cell, movement of components within a cell, or movement of environmental components past or through a cell. (p. 384)

motor protein: protein that uses energy derived from ATP to change shape in a way that exerts force and causes attached structures to move; includes three families of proteins (myosin, dynein, and kinesin) that interact with cytoskeletal elements (microtubules and microfilaments) to produce movements. (pp. 385, 724)

MPF: see *mitosis promoting factor*.

mRNA: see *messenger RNA*.

mRNA-binding site: place on the ribosome where mRNA binds during protein synthesis. (p. 547)

MS/MS: see *tandem mass spectrometry.*
MT: see *microtubule.*
MTOC: see *microtubule-organizing center.*
multidrug resistance (MDR) transport protein: an ABC-type ATPase that uses the energy of ATP hydrolysis to pump hydrophobic drugs out of cells. (p. 204)
multimeric protein: protein that consists of two or more polypeptide chains. (p. 47)
multiphoton excitation microscopy: specialized type of fluorescence light microscope employing a laser beam that emits rapid pulses of light; images are similar in sharpness to confocal microscopy, but photodamage is minimized because there is little out-of-focus light. (p. A-12)
multiprotein complex: two or more proteins (usually enzymes) bound together in a way that allows each protein to play a sequential role in the same multistep process. (p. 57)
muscle contraction: generation of tension in muscle cells by the sliding of thin (actin) filaments past thick (myosin) filaments. (p. 395)
muscle fiber: long, thin, multinucleate cell specialized for contraction. (p. 395)
mutagen: chemical or other environmental agent (e.g., ultraviolet light) that induces mutations in DNA. (p. 541)
mutation: change in the base sequence of a DNA molecule. (p. 491)
myelin sheath: concentric layers of membrane that surround an axon and serve as electrical insulation that allows rapid transmission of nerve impulses. (p. 665)
myofibril: cylindrical structure composed of an organized array of thin actin filaments and thick myosin filaments; found in the cytoplasm of skeletal muscle cells. (p. 395)
myosin: family of motor proteins that create movements by exerting force on actin microfilaments using energy derived from ATP hydrolysis; makes up the thick filaments that move the actin thin filaments during muscle contraction. (p. 394)
myosin light-chain kinase (MLCK): enzyme that phosphorylates myosin light chains, thereby triggering smooth muscle contraction. (p. 405)

N

NA: see *numerical aperture.*
Na^+/glucose symporter: a membrane transport protein that simultaneously transports glucose and sodium ions into cells, with the movement of sodium ions down their electrochemical gradient driving the transport of glucose against its concentration gradient. (p. 206)
Na^+/K^+ ATPase: see *Na^+/K^+ pump.*
Na^+/K^+ pump: membrane carrier protein that couples ATP hydrolysis to the inward transport of potassium ions and the outward transport of sodium ions to maintain the Na^+ and K^+ gradients that exist across the plasma membrane of most animal cells. (pp. 205, 662)
NAD^+: see *nicotinamide adenine dinucleotide.*
NADH-coenzyme Q oxidoreductase: see *complex I.*
$NADP^+$: see *nicotinamide adenine dinucleotide phosphate.*
nanometer (nm): unit of measure: 1 nanometer = 10^{-9} meters. (p. 4)
native conformation: three-dimensional folding of a polypeptide chain into a shape that represents the most stable state for that particular sequence of amino acids. (p. 54)
negative staining: technique in which an unstained specimen is visualized in a transmission electron microscope against a darkly stained background. (p. A-20)
NER: see *nucleotide excision repair.*
Nernst equation: equation for calculating the equilibrium membrane potential for a given ion: $E_x = (RT/zF)\ln[X]_{outside}/[X]_{inside}$. (p. 662)
nerve: a tissue composed of bundles of axons. (p. 660)
nerve impulse: signal transmitted along nerve cells by a wave of depolarization-repolarization events propagated along the axonal membrane. (p. 671)
NES: see *nuclear export signal.*
N-ethylmaleimide-sensitive factor (NSF): soluble cytoplasmic protein that acts in conjunction with several soluble NSF attachment proteins (SNAPs) to mediate the fusion of membranes brought together by interactions between v-SNAREs and t-SNAREs. (p. 344)
neuromuscular junction: site where a nerve cell axon makes contact with a skeletal muscle cell for the purpose of transmitting electrical impulses. (p. 403)
neuron: specialized cell directly involved in the conduction and transmission of nerve impulses; nerve cell. (p. 659)
neuropeptide: a molecule consisting of a short chain of amino acids that is involved in transmitting signals from neurons to other cells (neurons as well as other cell types). (p. 676)
neurosecretory vesicle: a small vesicle containing neurotransmitter molecules; located in the terminal bulb of an axon. (p. 677)
neurotoxin: toxic substance that disrupts the transmission of nerve impulses. (p. 677)
neurotransmitter: chemical released by a neuron that transmits nerve impulses across a synapse. (p. 673)
neurotransmitter reuptake: mechanism for removing neurotransmitters from the synaptic cleft by pumping them back into the presynaptic axon terminals or nearby support cells. (p. 680)
nexin: protein that connects and maintains the spatial relationship of adjacent outer doublets in the axoneme of eukaryotic cilia and flagella. (p. 385)
next-generation DNA sequencing: high-throughput methods of DNA sequencing using any of a number of technologies that replace Sanger (dideoxy) sequencing technologies and focus on the production of short-read sequences.
N-glycosylation: see *N-linked glycosylation.*
NHEJ: see *nonhomologous end-joining.* (p.633)
nicotinamide adenine dinucleotide (NAD^+): coenzyme that accepts two electrons and one proton to generate the reduced form, NADH; important electron carrier in energy metabolism. (p. 221)
nicotinamide adenine dinucleotide phosphate ($NADP^+$): coenzyme that accepts two electrons and one proton to generate the reduced form, NADPH; important electron carrier in the Calvin cycle and other biosynthetic pathways. (p. 221)
nitric oxide (NO): a gas molecule that transmits signals to neighboring cells by stimulating guanylyl cyclase. (p. 712)
N-linked glycosylation (N-glycosylation): addition of oligosaccharide units to the terminal amino group of asparagine residues in protein molecules. (p. 321)
NLS: see *nuclear localization signal.*
NO: see *nitric oxide.*
nocodazole: synthetic drug that inhibits microtubule assembly; frequently used instead of colchicine because its effects are more readily reversible when the drug is removed. (p. 365)
node of Ranvier: small segment of bare axon between successive segments of myelin sheath. (p. 665)
nonautonomous transposable element: transposable element that does not encode transposase and cannot initiate its own transposition. (p. 505)
noncompetitive inhibitor: compound that reduces enzyme activity by binding to a site on the enzyme other than the active site, resulting in diminished substrate binding or catalytic activity. Noncompetitive inhibitors can bind free enzyme or enzyme bound to substrate. (p.144)
noncomposite transposon: in bacteria, a transposable element that does not contain insertion sequence (IS) elements and contains multiple genes located between two simple inverted repeat sequences. (p. 505)
noncovalent bonds and interactions: binding forces that do not involve the sharing of electrons; examples include ionic bonds, hydrogen bonds, van der Waals interactions, and hydrophobic interactions. (pp. 37, 47)
noncyclic electron flow: continuous, unidirectional flow of electrons from water to $NADP^+$ during the energy transduction reactions of photosynthesis, with light providing the energy that drives the transfer. (p. 296)
nondisjunction: failure of the two members of a homologous chromosome pair to separate during anaphase I of meiosis, resulting in a joined chromosome pair that moves into one of the two daughter cells. (p. 760)
nonhomologous end-joining (NHEJ): mechanism for repairing double-strand DNA breaks that uses proteins that bind to the ends of the two broken DNA fragments and join the ends together. (p. 499)
nonreceptor protein kinase: any protein kinase that is not an intrinsic part of a cell surface receptor. (p. 799)
nonrepeated DNA: DNA sequences present in single copies within an organism's genome. (p. 459)
nonsense-mediated decay: mechanism for destroying mRNAs containing premature stop codons. (p. 565)
nonsense mutation: change in base sequence converting a codon that previously coded for an amino acid into a stop codon. (p. 564)
nonstop decay: mechanism for destroying mRNAs containing no stop codons. (p. 565)
NOR: see *nucleolus organizer region.*
Northern blotting: technique in which RNA molecules are size-fractionated by gel electrophoresis vand then transferred to a special type of "blotter" paper (nitrocellulose or nylon), which is then hybridized with a radioactive DNA probe. (p. 643)
NPC: see *nuclear pore complex.*
NSF: see *N-ethylmaleimide-sensitive factor.*
N-terminus (amino terminus): the end of a polypeptide chain that contains the first amino acid to be incorporated during mRNA translation; usually retains a free amino group. (p. 45)
nuclear envelope: double membrane around the nucleus that is interrupted by numerous small pores. (pp. 86, 462)
nuclear export signal (NES): amino acid sequence that targets a protein for export from the nucleus. (p. 467)
nuclear lamina: thin, dense meshwork of fibers that lines the inner surface of the inner nuclear membrane and helps support the nuclear envelope. (p. 469)
nuclear localization signal (NLS): amino acid sequence that targets a protein for transport into the nucleus. (p. 465)
nuclear matrix (nucleoskeleton): insoluble fibrous network that provides a supporting framework for the nucleus. (p. 467)
nuclear pore: small opening in the nuclear envelope through which molecules enter and exit the nucleus; lined by an intricate protein structure called the nuclear pore complex (NPC). (p. 463)

nuclear pore complex (NPC): intricate protein structure, composed of 30 or more different polypeptide subunits, that lines the nuclear pores through which molecules enter and exit the nucleus, both by simple diffusion of smaller molecules and by active transport of larger molecules. (p. 464)

nuclear RNA export factor 1 (NXF1): protein that aids nuclear export of mature mRNA through nuclear pore complexes. (p.536)

nuclear transfer: experimental technique in which the nucleus from one cell is transferred into another cell (usually an egg cell) whose own nucleus has been removed. (p. 582)

nucleation: act of providing a small aggregate of molecules from which a polymer can grow. (p. 363)

nucleic acid: a linear polymer of nucleotides joined together in a genetically determined order. Each nucleotide is composed of ribose or deoxyribose, a phosphate group, and the nitrogenous base guanine, cytosine, adenosine, or thymine (for DNA) or uracil (for RNA). (p. 58) Also see *DNA* and *RNA*.

nucleic acid hybridization: family of techniques in which single-stranded nucleic acids are allowed to bind to each other by complementary base pairing; used for assessing whether two nucleic acids contain similar base sequences. (p. 451)

nucleic acid probe: see *probe*.

nucleoid: region of cytoplasm in which the genetic material of a prokaryotic cell is located. (p. 453)

nucleolus (plural, nucleoli): large, spherical structure present in the nucleus of a eukaryotic cell; the site of ribosomal RNA synthesis and processing and of the assembly of ribosomal subunits. (pp. 87, 526)

nucleolus organizer region (NOR): stretch of DNA in certain chromosomes where multiple copies of the genes for ribosomal RNA are located and where nucleoli form. (p. 526)

nucleoplasm: the interior space of the nucleus, other than that occupied by the nucleolus. (p. 463)

nucleoside: molecule consisting of a nitrogen-containing base (purine or pyrimidine) linked to a five-carbon sugar (ribose or deoxyribose); a nucleotide with the phosphate removed. (p. 60)

nucleoside monophosphate: see *nucleotide*.

nucleoskeleton (nuclear matrix): insoluble fibrous network that provides a supporting framework for the nucleus. (p. 459)

nucleosome: basic structural unit of eukaryotic chromosomes, consisting of about 200 base pairs of DNA associated with an octamer of histones. (p. 454)

nucleotide: molecule consisting of a nitrogen-containing base (purine or pyrimidine) linked to a five-carbon sugar (ribose or deoxyribose) attached to a phosphate group; also called a nucleoside monophosphate. (p. 60)

nucleotide excision repair (NER): DNA repair mechanism that recognizes and repairs damage involving major distortions of the DNA double helix, such as that caused by pyrimidine dimers. (p. 495)

nucleus: large, double membrane-enclosed organelle that contains the chromosomal DNA of a eukaryotic cell. (pp. 86, 462)

null hypothesis: a statement that proposes the opposite of the hypothesis being tested and that, if unable to be proven, provides good evidence that the hypothesis is correct. (p. 14)

numerical aperture (NA): property of a microscope corresponding to the quantity $n \sin \alpha$, where n is the refractive index of the medium between the specimen and the objective lens and α is the aperture angle. (p. A-3)

NXF1: see *nuclear RNA export factor 1*.

O

objective lens (electron microscope): electromagnetic lens within which the specimen is placed in a transmission electron microscope. (p. A-17)

objective lens (light microscope): lens located immediately above the specimen in a light microscope. (p. A-5)

obligate aerobe: organism that has an absolute requirement for oxygen as an electron acceptor and therefore cannot live under anaerobic conditions. (p. 222)

obligate anaerobe: organism that cannot use oxygen as an electron acceptor and therefore has an absolute requirement for an electron acceptor other than oxygen. (p. 222)

ocular lens: lens through which the observer looks in a light microscope; also called the eyepiece. (p. A-5)

OEC: see *oxygen-evolving complex*.

Okazaki fragment: short fragment of newly synthesized, lagging-strand DNA that is joined together by DNA ligase during DNA replication. (p. 480)

oligodendrocyte: cell type in the central nervous system that forms the myelin sheath around nerve axons. (p. 671)

O-linked glycosylation: addition of oligosaccharide units to hydroxyl groups of serine or threonine residues in protein molecules. (p. 321)

oncogene: any gene whose presence can cause cancer; arises by mutation from normal cellular genes called proto-oncogenes. (p. 791)

oncogenic virus: virus that can cause cancer. (p. 795)

open system: a system that is not sealed from its environment and can exchange energy with the environment. (p. 113)

operator (*O*): base sequence in an operon to which a repressor protein can bind. (p. 571)

operon: cluster of genes with related functions that is under the control of a single operator and promoter, thereby allowing transcription of these genes to be turned on and off together. (p. 574)

optical tweezers: technique in which laser light focused through the objective lens of a microscope traps a small plastic bead, which is then used to manipulate the molecules it is attached to. (p. 380)

optogenetics: use of light to alter the conformation and/or activity of proteins; often used to control ion channels in the neurons of genetically engineered organisms. (p. 664)

organelle: any membrane-bounded, intracellular structure that is specialized for carrying out a particular function. Eukaryotic cells contain several kinds of membrane-enclosed organelles, including the nucleus, mitochondria, Golgi apparatus, endoplasmic reticulum, lysosomes, peroxisomes, secretory vesicles, and, in the case of plants, chloroplasts. (p. 82)

organic chemistry: the study of carbon-containing compounds. (p. 22)

origin of replication: specific base sequence within a DNA molecule where replication is initiated. (p. 478)

origin of transfer: point on an F-factor plasmid where the transfer of the plasmid from an F^+ donor bacterial cell to an F^- recipient cell begins during conjugation. (p. 775)

osmosis: movement of water through a semipermeable membrane driven by a difference in solute concentration on the two sides of the membrane. (p. 28)

outer doublet: pair of fused microtubules, arranged in groups of nine around the periphery of the axoneme of a eukaryotic cilium or flagellum. (p. 392)

outer membrane: the outer of the two membranes that surround a mitochondrion, chloroplast, or nucleus. (pp. 245, 286)

ovum (plural, ova): see *egg*.

oxidation: chemical reaction involving the removal of electrons; oxidation of organic molecules frequently involves the removal of both electrons and hydrogen ions (protons) and is therefore also called a dehydrogenation reaction. (p. 25) Also see *beta oxidation*.

oxidative phosphorylation: formation of ATP from ADP and inorganic phosphate by coupling the exergonic oxidation of reduced coenzyme molecules by oxygen to the phosphorylation of ADP, with an electrochemical proton gradient as the intermediate. (p. 268)

oxygen-evolving complex (OEC): assembly of manganese ions and proteins included within photosystem II that catalyzes the oxidation of water to oxygen. (p. 294)

oxygenic phototroph: organism that utilizes water as the electron donor in photosynthesis, with release of oxygen. (p. 284)

P

P face: interior face of the inner, or cytoplasmic, monolayer of a membrane as revealed by the technique of freeze fracturing; called the P face because this monolayer is on the protoplasmic side of the membrane. (p. A-22)

P generation: the first parental generation of a genetic breeding experiment. (p. 764)

P site (peptidyl site): site on the ribosome that contains the growing polypeptide chain at the beginning of each elongation cycle. (p. 547)

p21 protein: Cdk inhibitor that halts progression through the cell cycle by inhibiting several different Cdk-cyclins. (p. 742)

p53 gene: tumor suppressor gene that codes for the p53 protein, a transcription factor involved in preventing genetically damaged cells from proliferating; most frequently mutated gene in human cancers. (p. 802)

p53 protein: transcription factor that accumulates in the presence of damaged DNA and activates genes whose products halt the cell cycle and trigger apoptosis. (p. 741)

P680: the pair of chloroplast molecules that make up the reaction center of photosystem II. (p. 292)

P700: the pair of chloroplast molecules that make up the reaction center of photosystem I. (p. 292)

PABP: see *poly(A)-binding protein*.

pachytene: stage during prophase I of meiosis when crossing over between homologous chromosomes takes place. (p. 756)

paclitaxel: drug originally derived from the bark of the Pacific yew tree (*Taxus brevifolia*) that binds tightly to microtubules and stabilizes them, causing much of the free tubulin in the cell to assemble into microtubules. Its original trade name was taxol. (p. 365)

Pap smear: screening technique for early detection of cervical cancer in which cells obtained from a sample of vaginal secretions are examined with a microscope. (p. 811)

paracellular transport: a type of cellular transport in which ions move between, rather than through, cells. (p. 420)

passive spread of depolarization: process in which cations (mostly K^+) move away from the site of membrane depolarization to regions of membrane where the potential is more negative. (p. 670)

patch clamping: technique in which a tiny micropipette placed on the surface of a cell is used to measure the movement of ions through individual ion channels. (p. 664)

P body: microscopic structure present in the cytoplasm of eukaryotic cells that is involved in the storage and degradation of mRNAs. (p. 609)

PC: see *plastocyanin.*

PCR: see *polymerase chain reaction.*

pectin: branched polysaccharides, rich in galacturonic acid and rhamnose; found in plant cell walls, where they form a matrix in which cellulose microfibrils are embedded. (p. 433)

peptide bond: a covalent bond between the amino group of one amino acid and the carboxyl group of a second amino acid. (p. 45)

peptidyl site: see *P site.*

perinuclear space: space between the inner and outer nuclear membranes that is continuous with the lumen of the endoplasmic reticulum. (p. 462)

peripheral membrane protein: hydrophilic protein bound through weak ionic interactions and hydrogen bonds to a membrane surface. (p. 174)

peroxisome: single membrane-bounded organelle that contains catalase and one or more hydrogen peroxide–generating oxidases and is therefore involved in the metabolism of hydrogen peroxide. (pp. 95, 349) Also see *leaf peroxisome.*

phage: see *bacteriophage.*

phagocyte: specialized white blood cell that carries out phagocytosis as a defense mechanism. (p. 338)

phagocytic vacuole: membrane-bounded structure containing ingested particulate matter that fuses with a late endosome or matures directly into a lysosome, forming a large vesicle in which the ingested material is digested. (p. 338)

phagocytosis: type of endocytosis in which particulate matter or even an entire cell is taken up from the environment and incorporated into vesicles for digestion. (p. 9, 338)

pharmacogenetics: study of how inherited differences in genes cause people to respond differently to drugs and medications. (p. 317)

phase transition: change in the state of a membrane between a fluid state and a gel state. (p. 165)

phase-contrast microscopy: light microscopic technique that improves contrast without sectioning and staining by exploiting differences in thickness and refractive index; produces an image using an optical material that is capable of bringing undiffracted rays into phase with those that have been diffracted by the specimen. (p. A-5)

phenotype: observable physical characteristics of an organism attributable to the expression of its genotype. (p. 766)

phosphatidic acid: basic component of phosphoglycerides; consists of two fatty acids and a phosphate group linked by ester bonds to glycerol; key intermediate in the synthesis of other phosphoglycerides. (p. 69)

phosphoanhydride bond: high-energy bond between phosphate groups. (p. 216)

phosphodiester bridge (3′, 5′ phosphodiester bridge): covalent linkage in which two parts of a molecule are joined through oxygen atoms to the same phosphate group. (p. 59)

phosphodiesterase: enzyme that catalyzes the hydrolysis of cyclic AMP to AMP. (p. 692)

phosphoester bond: covalent linkage in which a molecule is joined through an oxygen atom to a phosphate group. (p. 216)

phosphoglyceride: predominant phospholipid component of cell membranes, consisting of a glycerol molecule esterified to two fatty acids and a phosphate group. (pp. 69, 161)

phosphoglycolate: two-carbon compound produced by the oxygenase activity of rubisco. Because it cannot be metabolized during the next step of the Calvin cycle, the production of phosphoglycolate decreases photosynthetic efficiency. (p. 307)

phospholipase C: enzyme that catalyzes the hydrolysis of phosphatidylinositol-4,5-bisphosphate into inositol-1,4,5-trisphosphate (IP_3) and diacylglycerol (DAG). (p. 693)

phospholipid: lipid possessing a covalently attached phosphate group and therefore exhibiting both hydrophilic and hydrophobic properties; main component of the lipid bilayer that forms the structural backbone of all cell membranes. (pp. 69, 160)

phospholipid exchange protein: any of a group of proteins located in the cytosol that transfer specific phospholipid molecules from the ER membrane to the outer mitochondrial, chloroplast, or plasma membranes. (p. 318)

phospholipid translocator (flippase): membrane protein that catalyzes the flip-flop of membrane phospholipids from one monolayer to the other. (pp. 164, 318)

phosphorolysis: reaction in which a chemical bond is broken by the addition of inorganic phosphate. (p. 230)

phosphorylation: addition of a phosphate group. (p. 148)

phosphosphingolipid: amphipathic, phosphate-containing sphingolipid that is a major phospholipid in some cell membranes. (p. 161)

photoactivation: light-induced activation of an inert molecule to an active state; often associated with the ultraviolet light-induced release of a caging group that had blocked the fluorescence of a molecule that it had been attached to. (p. A-14)

photoactive repair: light-activated repair of UV-induced DNA damage via the enzyme photolyase. (p. 494)

photoautotroph: organism capable of obtaining energy from the sun and using this energy to drive the synthesis of energy-rich organic molecules, using carbon dioxide as a source of carbon. (p. 284)

photobleaching: technique in which an intense beam of light within a well-defined area is used to render fluorescent molecules non-fluorescent; the rate at which unbleached fluorescent molecules repopulate the bleached area provides information about the dynamic movements of the molecule of interest. (p. A-14)

photochemical reduction: transfer of photoexcited electrons from one molecule to another. (p. 288)

photoexcitation: excitation of an electron to a higher energy level by the absorption of a photon of light. (p. 288)

photoheterotroph: organism capable of obtaining energy from the sun but dependent on organic compounds, rather than carbon dioxide, for carbon. (p. 284)

photon: fundamental particle of light with an energy content that is inversely proportional to its wavelength. (p. 288)

photophosphorylation: light-dependent generation of ATP driven by an electrochemical proton gradient established and maintained as excited electrons of chlorophyll return to their ground state via an electron transport chain. (p. 298)

photoreduction: light-dependent generation of NADPH by the transfer of energized electrons from photoexcited chlorophyll molecules to $NADP^+$ via a series of electron carriers. (p. 292)

photorespiration: light-dependent pathway that decreases the efficiency of photosynthesis by oxidizing reduced carbon compounds without capturing the released energy; occurs when oxygen substitutes for carbon dioxide in the reaction catalyzed by rubisco, thereby generating phosphoglycolate that is then converted to 3-phosphoglycerate in the peroxisome and mitochondrion; also called the glycolate pathway. (p. 307)

photosynthesis: process by which plants and certain bacteria convert light energy to chemical energy that is then used in synthesizing organic molecules. (p. 284)

photosystem: assembly of chlorophyll molecules, accessory pigments, and associated proteins embedded in thylakoid membranes or bacterial photosynthetic membranes; functions in the light-requiring reactions of photosynthesis. (p. 289)

photosystem I (PSI): photosystem containing a pair of chlorophyll molecules (P700) that absorbs 700-nm red light maximally; light of this wavelength can excite electrons derived from plastocyanin to an energy level that allows them to reduce ferredoxin, from which the electrons are then used to reduce $NADP^+$ to NADPH. (p. 292)

photosystem II (PSII): photosystem containing a pair of chlorophyll molecules (P680) that absorb 680-nm red light maximally; light of this wavelength can excite electrons donated by water to an energy level that allows them to reduce plastoquinone. (p. 292)

photosystem complex: a photosystem plus its associated light-harvesting complexes. (p. 291)

phototroph: organism that is capable of using the radiant energy of the sun to satisfy its energy requirements. (pp. 110, 284)

phragmoplast: parallel array of microtubules that guides vesicles containing polysaccharides and glycoproteins toward the spindle equator during cell wall formation in dividing plant cells. (p. 733)

phycobilin: accessory pigment found in red algae and cyanobacteria that absorbs visible light in the green-to-orange range of the spectrum, giving these cells their characteristic colors. (p. 289)

phytosterol: any of several sterols that are found uniquely, or primarily, in the membranes of plant cells; examples include campesterol, sitosterol, and stigmasterol. (p. 162)

PI 3-kinase (PI3K): enzyme that adds a phosphate group to PIP_2 (phosphatidylinositol-4,5-bisphosphate), thereby converting it to PIP_3 (phosphatidylinositol-3,4,5-trisphosphate); key component of the PI 3-kinase–Akt pathway, which is activated in response to the binding of certain growth factors to their receptors. (p. 709)

PI3K: see *PI 3-kinase.*

pigment: light-absorbing molecule responsible for the color of a substance. (p. 288)

pilus: see *sex pilus.* (p.775)

piRNA: see *Piwi-interacting RNA.*

Piwi-interacting RNA (piRNA): small single-stranded RNA that binds Piwi family Argonaute proteins and leads to suppression of transcripts from transposable elements, especially in germ cells. (p. 611)

PKA: see *protein kinase A.*

PKC: see *protein kinase C.*

plakin: family of proteins involved in linking the integrin molecules of a hemidesmosome to intermediate filaments of the cytoskeleton. (p. 431)

plaque: dense layer of fibrous material located on the cytoplasmic side of adhesive junctions such as desmosomes, hemidesmosomes, and adherens junctions; composed of intracellular attachment proteins that link the junction to the appropriate type of cytoskeletal filament. (p. 431) The same term can also refer to the clear zone produced when bacterial cells in a small region of a culture dish are destroyed by infection with a bacteriophage. (p. 625)

plasma membrane: bilayer of lipids and proteins that defines the boundary of the cell and regulates the flow of materials into and out of the cell; also called the cell membrane. (pp. 86, 153)

plasmid: small circular DNA molecule in bacteria that can replicate independent of chromosomal DNA; useful as cloning vectors. (p. 453)
plasmodesma (plural, plasmodesmata): cytoplasmic channel through pores in the cell walls of two adjacent plant cells, allowing fusion of the plasma membranes and chemical communication between the cells. (pp. 99, 434)
plasmolysis: outward movement of water that causes the plasma membrane to pull away from the cell wall in cells that have been exposed to a hypertonic solution. (p. 190)
plastid: any of several types of plant cytoplasmic organelles derived from proplastids, including chloroplasts, amyloplasts, chromoplasts, proteinoplasts, and elaioplasts. (pp. 90, 286)
plastocyanin (PC): copper-containing protein that donates electrons to chlorophyll P700 of photosystem I in the light-requiring reactions of photosynthesis. (p. 295)
plastoquinol: fully reduced form of plastoquinone, involved in the light-requiring reactions of photosynthesis; present in the lipid phase of the photosynthetic membrane, where it transfers electrons to the cytochrome b_6/f complex. (p. 294)
plastoquinone: nonprotein (quinone) molecule associated with photosystem II, where it receives electrons from a modified type of chlorophyll called pheophytin during the light-requiring reactions of photosynthesis. (p. 294)
pluripotent: capable of forming a wide variety of cell types; embryonic stem cells are pluripotent. (p. 587)
plus end (of microtubule): rapidly growing end of a microtubule. (p. 364)
+-TIP protein: see *plus-end tubulin interacting protein.* (p. 369)
plus-end tubulin interacting protein (+-TIP protein): protein that stabilizes the plus ends of microtubules, decreasing the likelihood that microtubules will undergo catastrophic subunit loss. (p. 364)
pmf: see *proton motive force.*
polar body: tiny haploid cell produced during the meiotic divisions that create egg cells. Polar bodies receive a disproportionately small amount of cytoplasm and usually degenerate. (p. 761)
polar microtubule: spindle microtubule that interacts with spindle microtubules from the opposite spindle pole. (p. 726)
polarity: property of a molecule that results from part of the molecule having a partial positive charge and another part having a partial negative charge, usually because one region of the molecule possesses one or more electronegative atoms that draw electrons toward that region. (p. 28)
polarized secretion: fusion of secretory vesicles with the plasma membrane and expulsion of their contents to the cell exterior specifically localized at one end of a cell. (p. 335)
poly(A)-binding protein (PABP): protein that binds to the poly(A) tail of eukaryotic mRNAs; can form a complex with proteins that are involved in translational initiation. (p. 557)
poly(A) tail: stretch of about 50–250 adenine nucleotides added to the 3′ end of most eukaryotic mRNAs after transcription is completed. (p. 530)
polycistronic mRNA: an mRNA molecule that codes for more than one polypeptide. (p. 572)
polyclonal antibody: antibodies produced by different groups (lineages) of B lymphocytes that recognize a specific molecule. (p. 647)
polymer: long compound that is built from one or more types of monomeric subunits. (p. 33)
polymerase chain reaction (PCR): reaction in which a specific segment of DNA is amplified by repeated cycles of (1) heat treatment to separate the two strands of the DNA double helix, (2) incubation with primers that are complementary to sequences located at the two ends of the DNA segment being amplified, and (3) incubation with DNA polymerase to synthesize DNA using the primers as starting points. (p. 476)
polynucleotide: linear chain of nucleotides linked by phosphodiester bonds. (p. 61)
polypeptide: linear chain of amino acids linked by peptide bonds. (pp. 37, 46)
polyribosome (polysome): cluster of two or more ribosomes simultaneously translating a single mRNA molecule. (p. 549)
polysaccharide: polymer consisting of sugars and sugar derivatives linked together by glycosidic bonds. (p. 64)
polytene chromosome: giant chromosome containing multiple copies of the same DNA molecule generated by successive rounds of DNA replication in the absence of cell division. (p. 589)
porin: transmembrane protein that forms pores for the facilitated diffusion of small hydrophilic molecules; found in the outer membranes of mitochondria, chloroplasts, and many bacteria. (pp. 198, 245, 286)
postsynaptic neuron: a neuron that receives a signal from another neuron through a synapse. (p. 673)
posttranslational control: mechanisms of gene regulation involving selective alterations in polypeptides that have already been synthesized; includes covalent modifications, proteolytic cleavage, protein folding and assembly, import into organelles, and protein degradation. (p. 610)
posttranslational import: uptake by organelles of completed polypeptide chains after they have been synthesized, mediated by specific targeting signals within the polypeptide. (p. 324, 556)
posttranslational modifications: alterations in polypeptides that have already been synthesized; includes covalent modifications such as glycosylation, phosphorylation, or ubiquitylation, as well as proteolytic cleavage and folding. (p. 566)
precarcinogen: substance capable of causing cancer only after it has been metabolically activated by enzymes in the liver. (p. 793)
preinitiation complex: complex of general transcription factors and RNA polymerase that binds a promoter prior to the onset of transcription in eukaryotes. (p. 524)
pre-mRNA: primary transcript whose processing yields a mature mRNA. (p. 530)
pre-replication complex: group of proteins that bind to eukaryotic DNA and license it for replication; includes the origin recognition complex, MCM complex, and helicase loaders. (p. 479)
pre-rRNA: primary transcript whose processing yields mature rRNAs. (p. 527)
presynaptic neuron: a neuron that transmits a signal to another neuron through a synapse. (p. 673)
pre-tRNA: primary transcript whose processing yields a mature tRNA. (p. 529)
primary cell wall: flexible portion of the plant cell wall that develops beneath the middle lamella while cell growth is still occurring; contains a loosely organized network of cellulose microfibrils. (p. 426)
primary structure: sequence of amino acids in a polypeptide chain. (p. 51)
primary transcript: any RNA molecule newly produced by transcription, before any processing has occurred. (p. 525)
primase: enzyme that uses a single DNA strand as a template to guide the synthesis of the RNA primers that are required for initiation of replication of both the lagging and leading strands of a DNA double helix. (p. 483)
primosome: complex of proteins in bacterial cells that includes primase plus six other proteins involved in unwinding DNA and recognizing base sequences where replication is to be initiated. (p. 484)
prion: infectious, protein-containing particle responsible for neurological diseases such as scrapie in sheep and goats, kuru in humans, and mad cow disease in cattle and its human form, variant Creutzfeldt-Jakob disease (vCJD). (pp. 38)
probe (nucleic acid): single-stranded nucleic acid that is used in hybridization experiments to identify nucleic acids containing sequences that are complementary to the probe. (pp. 451, 620)
procaspase: an inactive precursor form of a caspase. (p. 746)
procollagen: a precursor molecule that is converted to collagen by proteolytic cleavage of sequences at both the N- and C-terminal ends. (p. 423)
progression: see *tumor progression.*
projector lens: electromagnetic lens located between the intermediate lens and the viewing screen in a transmission electron microscope. (p. A-17)
prokaryote: category of organisms characterized by the absence of a true nucleus and other membrane-bounded organelles; includes bacteria and archaea. (p. 80)
prometaphase: stage of mitosis characterized by nuclear envelope breakdown and attachment of chromosomes to spindle microtubules; also called late prophase. (p. 720)
promoter: base sequence in DNA to which RNA polymerase binds when initiating transcription. (pp. 513, 574)
promotion (stage of carcinogenesis): gradual process by which cells previously exposed to an initiating carcinogen are subsequently converted into cancer cells by agents that stimulate cell proliferation. (p. 786)
proofreading: removal of mismatched base pairs during DNA replication by the exonuclease activity of DNA polymerase. (p. 482)
propagation: movement of an action potential along a membrane away from the site of origin. (p. 668)
prophase: initial phase of mitosis, characterized by chromosome condensation and the beginning of spindle assembly. Prophase I of meiosis is more complex, consisting of stages called leptotene, zygotene, pachytene, diplotene, and diakinesis. (p. 720)
proplastid: small, double-membrane-enclosed, plant cytoplasmic organelle that can develop into several kinds of plastids, including chloroplasts. (p. 286)
prosthetic group: small organic molecule or metal ion component of an enzyme that plays an indispensable role in the catalytic activity of the enzyme. (p. 134)
proteasome: multiprotein complex that catalyzes the ATP-dependent degradation of proteins linked to ubiquitin. (p. 615)
protein: macromolecule that consists of one or more polypeptides folded into a conformation specified by the linear sequence of amino acids. Proteins play important roles as enzymes, structural proteins, motility proteins, and regulatory proteins. (p. 44)
protein disulfide isomerase: enzyme in the ER lumen that catalyzes the formation and breakage of disulfide bonds between cysteine residues in polypeptide chains. (p. 321, 561)
protein kinase: any of numerous enzymes that catalyze the phosphorylation of protein molecules. (p. 148)

protein kinase A (PKA): a protein kinase, activated by the second messenger cyclic AMP, that catalyzes the phosphorylation of serine or threonine residues in target proteins. (p. 692)
protein kinase C (PKC): enzyme that phosphorylates serine and threonine groups in a variety of target proteins when activated by diacylglycerol. (p. 694)
protein phosphatase: any of numerous enzymes that catalyze the dephosphorylation, or removal by hydrolysis, of phosphate groups from a variety of target proteins. (p. 149)
proteoglycan: complex between proteins and glycosaminoglycans found in the extracellular matrix. (p. 425)
proteolysis: degradation of proteins by hydrolysis of the peptide bonds between amino acids. (p. 258)
proteolytic cleavage: removal of a portion of a polypeptide chain, or cutting a polypeptide chain into two fragments, by an enzyme that cleaves peptide bonds. (p. 149)
proteome: the structure and properties of all the proteins produced by a genome. (p. 638)
protoeukaryote: the ancestor of eukaryotic cells, presumed to have entered into a symbiotic relationship with the bacterial forerunners of mitochondria. (p. 90)
protofilament: linear polymer of tubulin subunits; usually arranged in groups of 13 to form the wall of a microtubule. (p. 361)
proton motive force (pmf): force across a membrane exerted by an electrochemical proton gradient that tends to drive protons back down their concentration gradient. (p. 272)
proton translocator: channel through which protons flow across a membrane driven by an electrochemical gradient; examples include CF_o in thylakoid membranes and F_o in mitochondrial inner membranes. (pp. 274, 298)
proto-oncogene: normal cellular gene that can be converted into an oncogene by point mutation, gene amplification, chromosomal translocation, local DNA rearrangement, or insertional mutagenesis. (p. 796)
provacuole: vesicle in plant cells comparable to an endosome in animal cells; arises either from the Golgi apparatus or by autophagy. (p. 349)
proximal control element: DNA regulatory sequence located upstream of the core promoter but within about 100–200 base pairs of it. (p. 596)
PSA test: screening technique for early detection of prostate cancer that measures how much prostate-specific antigen (PSA) is present in the blood. (p. 811)
pseudopodium (plural, pseudopodia): large, blunt-ended cytoplasmic protrusion involved in cell crawling by amoebas, slime molds, and leukocytes. (p. 408)
PSI: see *photosystem I.*
PSII: see *photosystem II.*
PTEN: phosphatase that removes a phosphate group from PIP_3 (phosphatidylinositol-3,4,5-trisphosphate), thereby converting it to PIP_2 (phosphatidylinositol-4,5-bisphosphate); component of the PI 3-kinase–Akt pathway. (p. 709)
P-type ATPase: type of transport ATPase that is reversibly phosphorylated by ATP as part of the transport mechanism. (p. 203)
pull-down assay: assay for assessing whether two purified proteins can bind one another; two proteins are mixed, and then the tag on one of the proteins is used to affinity purify the tagged protein. If the other protein binds to the protein with the tag, then when the mixture is separated using SDS-PAGE, both proteins will appear on the gel. (p. 651)
pulsed-field electrophoresis: type of gel electrophoresis in which the direction of the electric field is periodically altered; can separate macromolecules, including large fragments of genomic or chromosomal DNA, that cannot be separated by standard electrophoresis in agarose gels. (p. 625)
Puma (p53 upregulated modulator of apoptosis): protein that triggers apoptosis by binding to and inactivating Bcl-2, an inhibitor of apoptosis. (p. 742)
purine: two-ringed nitrogen-containing molecule; parent compound of the bases adenine and guanine. (p. 60)
pyrimidine: single-ringed nitrogen-containing molecule; parent compound of the bases cytosine, thymine, and uracil. (p. 60)

Q

Q cycle: proposed pathway for recycling electrons during mitochondrial or chloroplast electron transport to allow additional proton pumping across the membrane containing the electron carriers. (pp. 266, 295)
quaternary structure: level of protein structure involving interactions between two or more polypeptide chains to form a single multimeric protein. (p. 57)
q-PCR (quantitative polymerase chain reaction): a technique for detecting and quantifying mRNA expression. It is more quantitative than RT-PCR or Northern blotting. cDNA is first made from the mRNA in a sample, followed by amplification of specific cDNA sequences using PCR. A unique third primer is used that emits fluorescence when each new product is made. The amount of fluorescence is measured each cycle in real time. 643

R

Rab GTPase: GTP-hydrolyzing protein involved in locking v-SNAREs and t-SNAREs together during the binding of a transport vesicle to an appropriate target membrane. (p. 344)
Rac: member of a family of monomeric G proteins, which also includes Rho and Cdc42, that stimulate formation of various actin-containing structures within cells. (p. 379)
radial spokes: inward projections from each of the nine outer doublets to the center pair of microtubules in the axoneme of a eukaryotic cilium or flagellum, believed to be important in converting the sliding of the doublets into a bending of the axoneme. (p. 393)
Ras (protein): a small, monomeric G protein bound to the inner surface of the plasma membrane; a key intermediate in transmitting signals from receptor tyrosine kinases to the cell interior. (p. 700)
rate constants
***RB* gene:** tumor suppressor gene coding for the Rb protein. (p. 797)
Rb protein: protein whose phosphorylation controls passage through the restriction point of the cell cycle. (p. 740)
reaction center: portion of a photosystem containing the two chlorophyll molecules that initiate electron transfer, utilizing the energy gathered by other chlorophyll molecules and accessory pigments. Also see *P680* and *P700.* (p. 290)
reactive oxygen species: highly reactive oxygen-containing compounds such as H_2O_2, superoxide anion, and hydroxyl radical that are formed in the presence of molecular oxygen and can damage cells by oxidizing cellular components. (p. 352)
receptor: a protein that contains a binding site for a specific signaling molecule. (p. 685)
receptor affinity: a measure of the chemical attraction between a receptor and its ligand. (p. 686)
receptor tyrosine kinase (RTK): a receptor whose activation causes it to catalyze the phosphorylation of tyrosine residues in proteins, thereby triggering a chain of signal transduction events inside cells that can lead to cell growth, proliferation, and differentiation. (p. 699)
receptor-mediated (clathrin-dependent) endocytosis: type of endocytosis initiated at coated pits and resulting in coated vesicles; believed to be a major mechanism for selective uptake of macromolecules and peptide hormones. (p. 338)
recessive (allele): allele that is present in the genome but is phenotypically expressed only in the homozygous form; masked by a dominant allele when heterozygous. (p. 766)
recombinant DNA technology: group of laboratory techniques for joining DNA fragments from multiple sources using the techniques of molecular biology to produce a DNA molecule not found in an organism. (p. 12)
redox pair: two molecules or ions that are interconvertible by the loss or gain of electrons; also called a reduction-oxidation pair. (p. 263)
reduction: chemical reaction involving the addition of electrons; reduction of organic molecules frequently involves the addition of both electrons and hydrogen ions (protons) and is therefore also called a hydrogenation reaction. (p. 25)
reduction-oxidation pair: see *redox pair.*
refractive index: measure of the change in the velocity of light as it passes from one medium to another. (p. A-2)
regulated secretion: fusion of secretory vesicles with the plasma membrane and expulsion of their contents to the cell exterior in response to specific extracellular signals. (p. 334)
regulators of G protein signaling (RGS) proteins: group of proteins that stimulate GTP hydrolysis by the G_α subunit of G proteins. (p. 689)
regulatory light chain: type of myosin light chain that is phosphorylated by myosin light-chain kinase in smooth muscle cells, thereby enabling myosin to interact with actin filaments and triggering muscle contraction. (p. 405)
regulatory site: see *allosteric (regulatory) site.*
regulatory subunit: a subunit of a multisubunit enzyme that contains an allosteric site. (p. 148)
regulatory transcription factor: protein that controls the rate at which one or more specific genes are transcribed by binding to DNA control elements located outside the core promoter. (p. 596)
relative refractory period: time during the hyperpolarization phase of an action potential when the sodium channels of a nerve cell are capable of opening again, but it is difficult to trigger an action potential because Na^+ currents are opposed by larger K^+ currents. (p. 670)
release factors: group of proteins that terminate translation by triggering the release of a completed polypeptide chain from peptidyl tRNA bound to a ribosome's P site. (p. 559)
renaturation: return of a protein from a denatured state to the native conformation determined by its amino acid sequence, usually accompanied by restoration of physiological function. (p. 37) Also see *DNA renaturation.*
repeated DNA: DNA sequences present in multiple copies within an organism's genome. (p. 459)
replication fork: Y-shaped structure that represents the site at which replication of a DNA double helix is occurring. (p. 477)
replicative transposition: transposition of a transposable element by replication of a copy of the element followed by insertion of the copy into a new location. (p. 505)

replicon: total length of DNA replicated from a single origin of replication. (p. 469)
replisome: large complex of proteins that work together to carry out DNA replication at the replication fork; about the size of a ribosome. (p. 486)
repressible operon: group of adjoining genes that are normally transcribed, but whose transcription is inhibited in the presence of a corepressor. (p. 579)
repressor protein (bacterial): protein that binds to the operator site of an operon and prevents transcription of adjacent structural genes. (p. 574)
repressor protein (eukaryotic): regulatory transcription factor whose binding to DNA control elements leads to a reduction in the transcription rate of nearby genes. (p. 594)
residual body: mature lysosome in which digestion has ceased and only indigestible material remains. (p. 347)
resolution: minimum distance that can separate two points that still remain identifiable as separate points when viewed through a microscope. (p. A-2)
resolving power: ability of a microscope to distinguish adjacent objects as separate entities. (p. 7)
resonance energy transfer: mechanism whereby the excitation energy of a photoexcited molecule is transferred to an electron in an adjacent molecule, exciting that electron to a high-energy orbital; important means of passing energy from one pigment molecule to another in photosynthetic energy transduction. (p. 288)
resonance stabilization: achievement of the most stable configuration of a molecule by maximal delocalization of an unshared electron pair over all possible bonds. (p. 217)
respirasome: group of respiratory complexes associated together in defined ratios. (p. 268)
respiration: see *cellular respiration*. (p.243)
respiratory complex: subset of carriers of the electron transport chain consisting of a distinctive assembly of polypeptides and prosthetic groups, organized together to play a specific role in the electron transport process. (p. 264)
respiratory control: regulation of oxidative phosphorylation and electron transport by the availability of ADP. (p. 264)
response element: DNA base sequence located adjacent to physically separate genes whose expression can then be coordinated by binding a regulatory transcription factor to the response element wherever it occurs. (p. 602)
resting membrane potential (V_m): electrical potential (voltage) across the plasma membrane of an unstimulated nerve cell. (p. 665)
restriction endonuclease: any of a large family of enzymes isolated from bacteria that cut foreign DNA molecules at or near a palindromic recognition sequence that is usually four or six (but may be eight or more) base pairs long; used in recombinant DNA technology to cleave DNA molecules at specific sites. (p. 621)
restriction fragment length polymorphism (RFLP): difference in restriction maps between individuals caused by small differences in the base sequences of their DNA. (p. 640)
restriction map: map of a DNA molecule indicating the location of cleavage sites for various restriction endonucleases. (p. 619)
restriction point: control point near the end of G1 phase of the cell cycle where the cycle can be halted until conditions are suitable for progression into S phase; regulated to a large extent by the presence or absence of extracellular growth factors; called Start in yeast. (p. 735)
restriction site: DNA base sequence, usually four or six (but may be eight or more) base pairs long, that is cleaved by a specific restriction endonuclease. (p. 621)
retrograde flow (of F-actin): bulk movement of actin microfilaments toward the rear of a cell protrusion (e.g., lamellipodium) as the protrusion extends. (p. 407)
retrograde transport: movement of vesicles from Golgi cisternae back toward the endoplasmic reticulum. (p. 321)
retrotransposon: type of transposable element that moves from one chromosomal site to another by a process in which the retrotransposon DNA is first transcribed into RNA, and reverse transcriptase then uses the RNA as a template to make a DNA copy that is integrated into the chromosomal DNA at another site. (p. 505)
retrovirus: any RNA virus that uses reverse transcriptase to make a DNA copy of its RNA. (p. 443)
reverse transcriptase: enzyme that uses an RNA template to synthesize a complementary molecule of double-stranded DNA. (p. 13)
reverse transcription polymerase chain reaction: see *RT-PCR*.
reversible inhibitor: molecule that causes a reversible loss of catalytic activity when bound to an enzyme; upon dissociation of the inhibitor, the enzyme regains biological function. (p. 146)
RFLP: see *restriction fragment length polymorphism*.
R group: the side chain attached to the alpha carbon of an amino acid. It can be simply a hydrogen atom, as in glycine, or a more complex arrangement of atoms, as in tryptophan. (p. 44)
RGS proteins: see *regulators of G protein signaling (RGS) proteins*.
Rho GTPases: a family of monomeric G proteins, which includes Rho, Rac, and Cdc42, that stimulate formation of various actin-containing structures within cells. (p. 379)
Rho: member of a family of monomeric G proteins, which also includes Rac and Cdc42, that stimulate formation of various actin-containing structures within cells. (p. 379)
rho (ρ) factor: bacterial protein that binds to the 3′ end of newly forming RNA molecules, triggering the termination of transcription. (p. 519)
ribonucleic acid: see *RNA*.
ribose: five-carbon sugar present in RNA and in important nucleoside triphosphates such as ATP and GTP. (p. 59)
ribosomal RNA (rRNA): any of several types of RNA molecules used in the construction of ribosomes. (p. 511)
ribosome: small particle composed of rRNA and protein that functions as the site of protein synthesis in the cytoplasm of prokaryotes and in the cytoplasm, mitochondria, and chloroplasts of eukaryotes; composed of large and small subunits. (pp. 97, 536)
riboswitch: site in mRNA to which a small molecule can bind, triggering changes in mRNA conformation that impact either transcription or translation. (p. 579)
ribozyme: an RNA molecule with catalytic activity. (p. 76)
ribulose-1,5-bisphosphate carboxylase/oxygenase: see *rubisco*.
RISC: complex between either siRNA or micro-RNA and several proteins that together silence the expression of messenger RNAs or genes containing sequences complementary to those of these RNAs; abbreviation for RNA-induced silencing complex. (p. 610) Also see *miRISC* and *siRISC*.
RNA (ribonucleic acid): nucleic acid that plays several different roles in the expression of genetic information; constructed from nucleotides consisting of ribose phosphate linked to adenine, uracil, cytosine, or guanine. (p. 58) Also see *messenger RNA, ribosomal RNA, transfer RNA, microRNA, snRNA,* and *snoRNA*.
RNA backtracking: slight backward movement of RNA polymerase associated with proofreading of RNA during transcription. (p. 519)
RNA editing: altering the base sequence of an mRNA molecule by the insertion, removal, or modification of nucleotides. (p. 535)
RNA interference (RNAi): ability of short RNA molecules (siRNAs or microRNAs) to inhibit gene expression by triggering the degradation or inhibiting the translation of specific mRNAs, or inhibiting transcription of the gene coding for a particular mRNA. (p. 609)
RNA polymerase: any of a group of enzymes that catalyze the synthesis of RNA using DNA as a template; function by adding successive nucleotides to the 3′ end of the growing RNA strand. (p. 513)
RNA polymerase I: type of eukaryotic RNA polymerase present in the nucleolus that synthesizes an RNA precursor for three of the four types of rRNA. (p. 520)
RNA polymerase II: type of eukaryotic RNA polymerase present in the nucleoplasm that synthesizes pre-mRNA, microRNA, and most of the snRNAs. (p. 520)
RNA polymerase III: type of eukaryotic RNA polymerase present in the nucleoplasm that synthesizes a variety of small RNAs, including pre-tRNAs and 5S rRNA. (p. 520)
RNA primer: short RNA fragment, synthesized by DNA primase, that serves as an initiation site for DNA synthesis. (p. 483)
RNA processing: conversion of an initial RNA transcript into a final RNA product by the removal, addition, and/or chemical modification of nucleotide sequences. (p. 521)
RNAseq: technology that uses next-generation sequencing technologies to sequence large numbers of cDNAs produced from a complex mixture of mRNAs representing the entire set of transcripts produced by the cells in the sample. (p. 641)
RNA splicing: excision of introns from a primary RNA transcript to generate the mature, functional form of the RNA molecule. (p. 532)
RNAi: see *RNA interference*.
rotation (of lipid molecules): turning of a molecule about its long axis; occurs freely and rapidly in membrane phospholipids. (p. 164)
rough endoplasmic reticulum (rough ER): endoplasmic reticulum that is studded with ribosomes on its cytosolic side because of its involvement in protein synthesis. (pp. 92, 316)
rough ER: see *rough endoplasmic reticulum*.
rRNA: see *ribosomal RNA*.
RTK: see *receptor tyrosine kinase*.
RT-PCR (reverse transcription polymerase chain reaction): a technique for detecting and quantifying mRNA expression. cDNA is first made from the mRNA in a sample, followed by amplification of specific cDNA sequences using PCR. (p. 639)
rubisco (ribulose-1,5-bisphosphate carboxylase/oxygenase): enzyme that catalyzes the CO_2-capturing step of the Calvin cycle; joins CO_2 to ribulose-1,5-bisphosphate, forming two molecules of 3-phosphoglycerate. (p. 301)
rubisco activase: protein that stimulates photosynthetic carbon fixation by rubisco by removing inhibitory sugar phosphates from the rubisco active site. (p. 305)

S

S: see *entropy*.
S: see *substrate concentration* and *Svedberg unit*.
S phase: stage of the eukaryotic cell cycle in which DNA is synthesized. (p. 721)
sarcoma: any cancer arising from a supporting tissue, such as bone, cartilage, fat, connective tissue, and muscle. (p. 784)
sarcomere: fundamental contractile unit of striated muscle myofibrils that extends from one

Z line to the next and that consists of two sets of thin (actin) and one set of thick (myosin) filaments. (p. 395)

sarcoplasmic reticulum (SR): endoplasmic reticulum of a muscle cell, specialized for accumulating, storing, and releasing calcium ions. (p. 403)

sarco/endoplasmic reticulum Ca^{2+}-ATPase (SERCA): membrane protein that transports calcium ions across a membrane using energy derived from ATP hydrolysis; prominent in the sarcoplasmic reticulum (SR), where it pumps calcium ions into the SR lumen. (p. 404)

satellite DNA: highly repetitive tandemly repeated DNA, originally so named because it appears as a "satellite" band during isolation of DNA by centrifugation. (p. 460)

saturated fatty acid: fatty acid without double or triple bonds such that every carbon atom in the chain has the maximum number of hydrogen atoms bonded to it. (p. 163)

saturation: inability of higher substrate concentrations to increase the velocity of an enzyme-catalyzed reaction beyond a fixed upper limit determined by the finite number of enzyme molecules available. (p. 141)

scanning electron microscope (SEM): microscope in which an electron beam scans across the surface of a specimen and forms an image from electrons that are deflected from the outer surface of the specimen. (pp. 8, A-24)

scanning probe microscope: instrument that visualizes the surface features of individual molecules by using a tiny probe that moves over the surface of a specimen. (p. A-24)

Schwann cell: cell type in the peripheral nervous system that forms the myelin sheath around nerve axons. (p. 659)

SDSA: see *synthesis-dependent strand annealing.*

second law of thermodynamics: the law of thermodynamic spontaneity; principle stating that all physical and chemical changes proceed in a manner such that the entropy of the universe increases. (p. 116)

second messenger: any of several substances, including cyclic AMP, calcium ion, inositol trisphosphate, and diacylglycerol, that transmit signals from extracellular signaling ligands to the cell interior. (p. 686)

secondary cell wall: rigid portion of the plant cell wall that develops beneath the primary cell wall after cell growth has ceased; contains densely packed, highly organized bundles of cellulose microfibrils. (p. 434)

secondary structure: level of protein structure involving hydrogen bonding between atoms in the peptide bonds along the polypeptide backbone, creating two main patterns called the α helix and β sheet conformations. (p. 52)

secretory granule: a large, dense secretory vesicle. (p. 333)

secretory pathway: pathway by which newly synthesized proteins move from the ER through the Golgi apparatus to secretory vesicles and secretory granules, which then discharge their contents to the exterior of the cell. (p. 333)

secretory vesicle: membrane-bounded compartment of a eukaryotic cell that carries secretory proteins from the Golgi apparatus to the plasma membrane for exocytosis and that may serve as a storage compartment for such proteins before they are released; large, dense vesicles are sometimes referred to as secretory granules. (pp. 93, 333)

securin: protein that prevents sister chromatid separation by inhibiting separase, the enzyme that would otherwise degrade the cohesins that hold sister chromatids together. (p. 734)

sedimentation coefficient: a measure of the rate at which a particle or macromolecule moves in a centrifugal force field; expressed in Svedberg units. (p. 98)

selectable marker: gene whose expression allows cells to grow under specific conditions that prevent the growth of cells lacking this gene. (p. 628)

selectin: plasma membrane glycoprotein that mediates cell-cell adhesion by binding to specific carbohydrate groups located on the surface of target cells. (p. 417)

self-assembly: principle that the information required to specify the folding of macromolecules and their interactions to form more complicated structures with specific biological functions is inherent in the polymers themselves. (p. 37)

SEM: see *scanning electron microscope.*

semiautonomous organelle: organelle that can divide on its own and contains not only its own DNA but also its own mRNA, tRNAs, and ribosomes. Examples include mitochondria and chloroplasts. (p. 90)

semiconservative replication: mode of DNA replication in which each newly formed DNA molecule consists of one old strand and one newly synthesized strand. (p. 474)

separase: protease that initiates anaphase by degrading the cohesins that hold sister chromatids together. (p. 739)

SERCA: see *sarco/endoplasmic reticulum Ca^{2+}-ATPase.*

serine-threonine kinase receptor: a receptor that, upon activation, catalyzes the phosphorylation of serine and threonine residues in target protein molecules. (p. 704)

70S initiation complex: complex formed by the association of a 30S initiation complex with a 50S ribosomal subunit; contains an initiator aminoacyl tRNA at the P site and is ready to commence mRNA translation. (p. 544)

sex chromosome: chromosome involved in determining whether an individual is male or female. (p. 752)

sex pilus (plural, pili): projection emerging from the surface of a bacterial donor cell that binds to the surface of a recipient cell, leading to the formation of a transient cytoplasmic mating bridge through which DNA is transferred from donor cell to recipient cell during bacterial conjugation. (p. 775)

sexual reproduction: form of reproduction in which two parent organisms each contribute genetic information to the new organism; reproduction by the fusion of gametes. (p. 752)

SH2 domain: a region of a protein molecule that recognizes and binds to phosphorylated tyrosines in another protein. (p. 700)

shadowing: deposition of a thin layer of an electron-dense metal on a biological specimen from a heated electrode, such that surfaces facing toward the electrode are coated while surfaces facing away are not. (p. A-20)

sheet (β sheet): see *beta sheet.*

Shine–Dalgarno sequence: sequence found in bacterial mRNAs that is complementary to sequences in the small ribosomal subunit and serves as a ribosomal binding site on the mRNA. (p. 553)

short interfering RNA: see *siRNA.*

short tandem repeat (STR): short repeated DNA sequence whose variation in length between individuals forms the basis for DNA fingerprinting; also called microsatellite DNA. (p. 460)

sidearm: structure composed of axonemal dynein that projects out from each of the A tubules of the nine outer doublets in the axoneme of a eukaryotic cilium or flagellum. (p. 392)

sigma (σ) factor: subunit of bacterial RNA polymerase that ensures the initiation of RNA synthesis at the correct site on the DNA strand. (p. 514)

signal recognition particle (SRP): cytoplasmic RNA-protein complex that binds to the ER signal sequence located at the N-terminus of a newly forming polypeptide chain and directs the ribosome-mRNA-polypeptide complex to the surface of the ER membrane. (p. 327, 560)

signal transduction: mechanisms by which signals detected at the cell surface are transmitted into the cell's interior, resulting in changes in cell behavior and/or gene expression. (p. 686)

silencer: DNA sequence containing a binding site for transcription factors that inhibit transcription and whose position and orientation relative to the promoter can vary significantly without interfering with the ability to regulate transcription. (p. 596)

simple diffusion: unassisted net movement of a solute from a region where its concentration is higher to a region where its concentration is lower. (p. 188)

SINEs (short interspersed nuclear elements): class of interspersed, repeated DNA sequences fewer than 500 base pairs in length that function as transposable elements, relying on enzymes made by other mobile elements for their movement; include Alu sequences, the most prevalent SINE in humans. (p. 460)

single bond: chemical bond formed between two atoms as a result of sharing a pair of electrons. (p. 23)

single nucleotide polymorphism (SNP): variation in DNA base sequence involving single base changes that occurs among individuals of the same species. (p. 641)

single-stranded DNA-binding protein (SSB): protein that binds to single strands of DNA at the replication fork to keep the DNA unwound and therefore accessible to the DNA replication machinery. (p. 479)

siRISC: complex between siRNA and several proteins that together silence the expression of messenger RNAs or genes containing sequences complementary to those of the siRNA. (p. 606) Also see *miRISC* and *RISC.* (p.610)

siRNA: class of double-stranded RNAs 21–22 nucleotides in length that silence gene expression; act by either promoting the degradation of mRNAs with precisely complementary sequences or inhibiting the transcription of genes containing precisely complementary sequences. (p. 610)

sister chromatid: one of the two replicated copies of each chromosome that remain attached to one another prior to anaphase of mitosis. (p. 473)

site-directed mutagenesis: molecular biology procedure that introduces specific mutations into a DNA sequence; often used to introduce point mutations into a DNA sequence that encodes a protein. (p. 645)

skeletal muscle: type of muscle, striated in microscopic appearance, that is responsible for voluntary movements. (p. 395)

sliding-filament model: model stating that muscle contraction is caused by thin actin filaments sliding past thick myosin filaments, with no change in the length of either type of filament. (p. 399)

sliding-microtubule model: model of motility in eukaryotic cilia and flagella that proposes that microtubule length remains unchanged but adjacent outer doublets slide past each other, thereby causing a localized bending because lateral connections between adjacent doublets and radial links to the center pair prevent free sliding of the microtubules past each other. (p. 392)

sliding clamp: a protein that helps DNA polymerase remain attached to the DNA template being copied. It serves to increase the processivity of the polymerase. Also called a DNA clamp. (p. 486)

Smad (protein): class of proteins involved in the signaling pathway triggered by transforming growth factor β; upon activation, Smads enter the nucleus and regulate gene expression. (p. 704)

small ribosomal subunit: component of a ribosome with a sedimentation coefficient of 40S in eukaryotes and 30S in prokaryotes; associates with a large ribosomal subunit to form a functional ribosome. (p. 98)

smooth endoplasmic reticulum (smooth ER): endoplasmic reticulum that has no attached ribosomes and plays no direct role in protein synthesis; involved in packaging of secretory proteins and synthesis of lipids. (pp. 93, 316)

smooth ER: see *smooth endoplasmic reticulum.*

smooth muscle: muscle lacking striations that is responsible for involuntary contractions such as those of the stomach, intestines, uterus, and blood vessels. (p. 404)

SNAP: see *soluble NSF attachment protein.*

SNAP receptor protein: see *SNARE (SNAP receptor) protein.*

SNARE (SNAP receptor) protein: two families of proteins involved in targeting and sorting membrane vesicles; include the v-SNARES found on transport vesicles and the t-SNARES found on target membranes. (p. 344)

SNARE hypothesis: model explaining how membrane vesicles fuse with the proper target membrane; based on specific interactions between v-SNAREs (vesicle-SNAP receptors) and t-SNAREs (target-SNAP receptors). (p. 344)

snoRNA: group of small nucleolar RNAs that bind to complementary regions of pre-rRNA and target specific sites for methylation or cleavage. (p. 528)

SNP: see *single nucleotide polymorphism.*

snRNA: a small nuclear RNA molecule that binds to specific proteins to form a snRNP, which in turn assembles with other snRNPs to form a spliceosome. (p. 532)

snRNP: RNA-protein complex that assembles with other snRNPs to form a spliceosome; pronounced "snurp." (p. 532)

sodium/potassium pump: see *Na^+/K^+ pump.*

soluble NSF attachment protein (SNAP): soluble cytoplasmic protein that acts in conjunction with NSF (N-ethylmaleimide-sensitive factor) to mediate the fusion of membranes brought together by interactions between v-SNAREs and t-SNAREs. (p. 344)

solute: substance that is dissolved in a solvent, forming a solution. (p. 28)

solvent: substance, usually liquid, in which other substances are dissolved, forming a solution. (p. 28)

Sos (protein): a guanine-nucleotide exchange factor that activates Ras by triggering the release of GDP, thereby permitting Ras to acquire a molecule of GTP. The Sos protein is activated by interacting with a GRB2 protein molecule that has become bound to phosphorylated tyrosines in an activated tyrosine kinase receptor. (p. 701)

SOS system: in bacteria, a global response to DNA damage that initiates multiple DNA repair pathways. The SOS response involves the single-strand binding protein RecA. (p. 498)

Southern blotting: technique in which DNA fragments separated by gel electrophoresis are transferred to a special type of "blotter" paper (nitrocellulose or nylon), which is then hybridized with a radioactive DNA probe. (p. 626)

special pair: two chlorophyll *a* molecules, located in the reaction center of a photosystem, that catalyze the conversion of solar energy into chemical energy. (p. 292)

spectrophotometer: an instrument that measures the intensity of light in different parts of the electromagnetic spectrum. (p. 288)

sperm: haploid male gamete, usually flagellated. (p. 752)

sphingolipid: class of lipids containing the amine alcohol sphingosine as a backbone. (p. 70)

sphingosine: amine alcohol that serves as the backbone for sphingolipids; contains an amino group that can form an amide bond with a long-chain fatty acid; also contains a hydroxyl group that can attach to a phosphate group. (p. 70)

spindle assembly checkpoint: mechanism that halts mitosis at the junction between metaphase and anaphase if chromosomes are not properly attached to the spindle. (p. 741)

spliceosome: protein-RNA complex that catalyzes the removal of introns from pre-mRNA. (p. 532)

spore: see *haploid spore.*

sporophyte: diploid generation in the life cycle of an organism that alternates between haploid and diploid forms; form that produces spores by meiosis. (p. 754)

sputter coating: vacuum evaporation process used to coat the surface of a specimen with a layer of gold or a mixture of gold and palladium prior to examining the specimen by scanning electron microscopy. (p. A-24)

squid giant axon: an exceptionally large axon emerging from certain squid nerve cells; its wide diameter (0.5–1.0 mm) makes it relatively easy to insert microelectrodes that can measure and control electrical potentials and ionic currents. (p. 665)

SR: see *sarcoplasmic reticulum.*

SRP: see *signal recognition particle.*

SSB: see *single-stranded DNA-binding protein.*

stage: platform on which the specimen is placed in a microscope. (p. A-5)

staining: incubation of tissue specimens in a solution of dye, heavy metal, or other substance that binds specifically to selected cellular constituents, thereby giving those constituents a distinctive color or electron density. (p. A-16)

standard free energy change ($\Delta G^{\circ\prime}$): free energy change accompanying the conversion of 1 mole of reactants to 1 mole of products, with the temperature, pressure, pH, and concentration of all relevant species maintained at standard values. (p. 121)

standard reduction potential (E'_0): convention vused to quantify the electron transport potential of oxidation-reduction couples relative to the H^+/H_2 redox pair, which is assigned an E'_0 value of 0.0 V at pH 7.0. (p. 263)

standard state: set of arbitrary conditions defined for convenience in reporting free energy changes in chemical reactions. For systems consisting of dilute aqueous solutions, these conditions are usually a temperature of 25°C (298 K), a pressure of 1 atmosphere, and reactants other than water present at a concentration of 1 *M*. (p. 121)

starch: storage polysaccharide in plants consisting of repeating glucose subunits linked together by $\alpha(1 \rightarrow 4)$ bonds and, in some cases, $\alpha(1 \rightarrow 6)$ bonds. The two main forms of starch are the unbranched polysaccharide amylose and the branched polysaccharide amylopectin. (p. 64)

Start: control point near the end of G1 phase of the yeast cell cycle where the cycle can be halted until conditions are suitable for progression into S phase; known as the restriction point in other eukaryotes. (p. 735)

start codon: the codon AUG in mRNA when it functions as the starting point for protein synthesis. (pp. 508, 545)

start-transfer sequence: amino acid sequence in a newly forming polypeptide that acts as both an ER signal sequence that directs the ribosome-mRNA-polypeptide complex to the ER membrane and as a membrane anchor that permanently attaches the polypeptide to the lipid bilayer. (p. 331, 562)

STAT: type of transcription factor activated by phosphorylation in the cytoplasm catalyzed by Janus-activated kinase, followed by migration of the activated STAT molecules to the nucleus. (p. 600)

state: condition of a system defined by various properties, such as temperature, pressure, and volume. (p. 113)

stationary cisternae model: model postulating that each compartment of the Golgi stack is a stable structure and that traffic between successive cisternae is mediated by shuttle vesicles that bud from one cisterna and fuse with another. (p. 320)

steady state: nonequilibrium condition of an open system through which matter is flowing, such that all components of the system are present at constant, nonequilibrium concentrations. (p. 126)

stem cell: cell capable of unlimited division that can differentiate into a variety of other cell types. (p. 587)

stereo electron microscopy: microscopic technique for obtaining a three-dimensional view of a specimen by photographing it at two slightly different angles. (p. A-23)

stereoisomers: two molecules that have the same structural formula but are not superimposable; stereoisomers are mirror images of each other. (p. 25)

steroid: any of numerous lipid molecules that are derived from a four-membered ring compound called phenanthrene. (p. 70)

steroid hormone: lipophilic hormone synthesized from a cholesterol backbone that is a ligand for nuclear hormone receptors; examples include sex hormones, such as testosterone and estrogen, glucocorticoids, and mineralocorticoids. (p. 709)

steroid receptor: protein that functions as a transcription factor after binding to a specific steroid hormone. (p. 709)

sterol: any of numerous compounds consisting of a 17-carbon four-ring system with at least one hydroxyl group and a variety of other possible side groups; includes cholesterol and a variety of other biologically important compounds, such as the male and female sex hormones, that are related to cholesterol. (p. 162)

sticky end: single-stranded end of a DNA fragment generated by cleavage with a restriction endonuclease that tends to reassociate with another fragment generated by the same restriction endonuclease because of base complementarity. (p. 622)

stomata (singular, stoma): pores on the surface of a plant leaf that can be opened or closed to control gas and water exchange between the atmosphere and the interior of the leaf. (p. 301)

stop codon: sequence of three bases in mRNA that instructs the ribosome to terminate protein synthesis. UAG, UAA, and UGA generally function as stop codons. (pp. 505, 552)

stop-transfer sequence: hydrophobic amino acid sequence in a newly forming polypeptide that halts translocation of the chain through the ER membrane, thereby anchoring the polypeptide within the membrane. (p. 331, 562)

storage macromolecule: polymer that consists of one or a few kinds of subunits in no specific order and that serves as a storage form of monosaccharides; examples include starch and glycogen. (p. 35)

storage polysaccharides: highly branched sugars and sugar derivatives that are stored in the cell for future use in energy production. Starch and glycogen are the most familiar types of storage polysaccharides.

STR: see *short tandem repeat.*

striated muscle: muscle whose myofibrils exhibit a pattern of alternating dark and light bands when viewed microscopically; includes both skeletal and cardiac muscle. (p. 397)
stroma: unstructured semifluid matrix that fills the interior of the chloroplast. (pp. 90, 286)
stroma thylakoid: membrane that interconnects stacks of grana thylakoids with each other. (pp. 89, 287)
structural macromolecule: polymer that consists of one or a few kinds of subunits in no specific order and that provides structure and mechanical strength to the cell; examples include cellulose and pectin. (p. 35)
structural polysaccharides: highly branched sugars and sugar derivatives that play a role in maintaining cellular shape and structure. Cellulose and chitin are the most familiar types of structural polysaccharides.
subcellular fractionation: technique for isolating organelles from cell homogenates using various types of centrifugation. (pp. 10, 96)
substrate
substrate activation: role of an enzyme's active site in making a substrate molecule maximally reactive by subjecting it to the appropriate chemical environment for catalysis. (p. 134)
substrate analogue: compound that resembles the usual substrate of an enzyme-catalyzed reaction closely enough to bind to the active site but cannot be converted to a functional product. (p. 141)
substrate concentration ([S]): amount of substrate present per unit volume at the beginning of a chemical reaction. (p. 141)
substrate induction: regulatory mechanism for catabolic pathways in which the synthesis of enzymes involved in the pathway is stimulated in the presence of the substrate and inhibited in the absence of the substrate. (p. 573)
substrate specificity: ability of an enzyme to discriminate between very similar molecules. (p. 131)
substrate-level phosphorylation: formation of ATP by direct transfer to ADP of a high-energy phosphate group derived from a phosphorylated substrate. (p. 134)
substrate-level regulation: enzyme regulation that depends directly on the interactions of substrates and products with the enzyme. (p. 147)
succinate-coenzyme Q oxidoreductase: see *complex II.*
supercoiled DNA: twisting of a DNA double helix upon itself, either in a circular DNA molecule or in a DNA loop anchored at both ends. (p. 448)
superresolution microscopy: a set of related techniques that allows objects to be visualized in the light microscope at a resolution greater than the theoretical limit predicted due to diffraction by the Abbé equation. (p. A-15)
suppressor tRNA: mutant tRNA molecule that inserts an amino acid where a stop codon generated by another mutation would otherwise have caused premature termination of protein synthesis. (p. 564)
surface area/volume ratio: mathematical ratio of the surface area of a cell to its volume; decreases with increasing linear dimension of the cell (length or radius), thereby increasing the difficulty of maintaining adequate surface area for import of nutrients and export of waste products as cell size increases. (p. 81)
surroundings: the remainder of the universe when one is studying the distribution of energy within a given system. (p. 112)
survival factor: a secreted molecule whose presence prevents a cell from undergoing apoptosis. (p. 742)
Svedberg unit (S): unit for expressing the sedimentation coefficient of biological macromolecules: One Svedberg unit (S) = 10^{-13} second. In general, the greater the mass of a particle, the greater the sedimentation rate, though the relationship is not linear. (p. 98)
symbiotic relationship: a mutually beneficial association between two organisms. (p. 90)
symport: coupled transport of two solutes across a membrane in the same direction. (p. 194)
synapse: tiny gap between a neuron and another cell (neuron, muscle fiber, or gland cell), across which the nerve impulse is transferred by direct electrical connection or by chemicals called neurotransmitters. (p. 660)
synapsis: close pairing between homologous chromosomes during the zygotene phase of prophase I of meiosis. (p. 754)
synaptic bouton: region near the end of an axon where neurotransmitter molecules are stored for use in transmitting signals across the synapse. (p. 665)
synaptic cleft: gap between the presynaptic and postsynaptic membranes at the junction between two nerve cells. (p. 673)
synaptonemal complex: zipperlike, protein-containing structure that joins homologous chromosomes together during prophase I of meiosis. (p. 756)
synthesis-dependent strand annealing (SDSA): an error-free mechanism for repairing double-strand breaks in DNA after completion of DNA replication that uses strand invasion via stabilized single DNA strands. (p. 499)
system: the restricted portion of the universe that one decides to study at any given time when investigating the principles that govern the distribution of energy. (p. 112)

T

T: see *thymine.*
T tubule system: see *transverse (T) tubule system.*
tandemly repeated DNA: repeated DNA sequences whose multiple copies are adjacent to one another. (p. 460)
tandem mass spectrometry (MS/MS): type of mass spectrometry in which molecules are sorted and then separated based on their mass/charge properties with a fragmentation step between the first and second rounds of mass spectrometry; often useful for separating a complex mixture of peptide fragments. (p. 650)
target-SNAP receptor: see *t-SNARE.*
TATA box: part of the core promoter for many eukaryotic genes transcribed by RNA polymerase II; consists of a consensus sequence of TATA followed by two or three more A's, located about 25 nucleotides upstream from the transcriptional start site. (p. 522)
TATA-binding protein (TBP): component of transcription factor TFIID that confers the ability to recognize and bind the TATA box sequence in DNA; also involved in regulating transcription initiation at promoters lacking a TATA box. (p. 524)
tautomer: rare alternative resonance structure of a DNA base that can lead to nonstandard base pairing during replication. (p. 491)
Taxol: see *paclitaxel* (p. 365)
TBP: see *TATA-binding protein.*
telomerase: special type of DNA polymerase that catalyzes the formation of additional copies of a telomeric repeat sequence. (p. 489)
telomere: DNA sequence located at either end of a linear chromosome; contains simple-sequence, tandemly repeated DNA. (pp. 458, 489)
telophase: final stage of mitosis or meiosis, when daughter chromosomes arrive at the poles of the spindle accompanied by reappearance of the nuclear envelope. (p. 726)
TEM: see *transmission electron microscope.*
template: a nucleic acid whose base sequence serves as a pattern for the synthesis of another (complementary) nucleic acid. (p. 61)
template strand: the strand of a DNA double helix that serves as the template for RNA synthesis via complementary base pairing. (p. 544)
terminal bulb: see *synaptic bouton.*
terminal glycosylation: modification of glycoproteins in the Golgi apparatus involving removal and/or addition of sugars to the carbohydrate side chains formed by prior core glycosylation in the endoplasmic reticulum. (p. 322)
terminal oxidase: electron transfer complex that is capable of transferring electrons directly to oxygen. Complex IV (cytochrome *c* oxidase) of the mitochondrial electron transport chain is an example. (p. 266)
terminal web: dense meshwork of spectrin and myosin molecules located at the base of a microvillus; the bundle of actin microfilaments that make up the core of the microvillus is anchored to the terminal web. (p. 376)
termination signal: DNA sequence located near the end of a gene that triggers the termination of transcription. (p. 515)
terpene: a lipid constructed from the five-carbon compound isoprene and its derivatives, joined together in various combinations. (p. 71)
tertiary structure: level of protein structure involving interactions between amino acid side chains of a polypeptide, regardless of where along the primary sequence they happen to be located; results in three-dimensional folding of a polypeptide chain. (p. 53)
tethering protein: a coiled-coil protein or a multisubunit protein complex that recognizes and binds vesicles to their target membranes. (p. 345)
tetrahedral (carbon atom): an atom of carbon from which four single bonds extend to other atoms, each bond equidistant from all other bonds, causing the atom to resemble a tetrahedron with its four equal faces. (p. 25)
TFIIB recognition element (BRE): component of core promoters for RNA polymerase II, located immediately upstream from the TATA box. (p. 522)
TGFβ: see *transforming growth factor β.*
TGN: see *trans-Golgi network.* (p.744)
thermodynamic spontaneity: measure of whether a reaction can occur, but it says nothing about whether the reaction actually will occur. Reactions with a negative free energy change are thermodynamically spontaneous. (p. 115)
thermodynamics: area of science that deals with the laws governing the energy transactions that accompany all physical processes and chemical reactions. (p. 112)
thick filament: myosin-containing filament, found in the myofibrils of striated muscle cells, in which individual myosin molecules are arranged in a staggered array with the heads of the myosin molecules projecting out in a repeating pattern. (p. 395)
thin filament: actin-containing filament, found in the myofibrils of striated muscle cells, in which two F-actin molecules are arranged in a helix associated with tropomyosin and troponin. (p. 395)
thin-layer chromatography (TLC): procedure for separating compounds by chromatography in a medium, such as silicic acid, that is bound as a thin layer to a glass or metal surface. (p. 164)
third-generation DNA sequencing: high-throughput method of DNA sequencing using any of a number of technologies that replace Sanger (dideoxy) sequencing technologies and focus on the production of long-read sequences.
30-nm chromatin fiber: fiber formed by packing together the nucleosomes of a 10-nm chromatin fiber. (p. 455)

30S initiation complex: complex formed by the association of mRNA, the 30S ribosomal subunit, an initiator aminoacyl tRNA molecule, and initiation factor IF2. (p. 554)

three-dimensional (3-D) electron tomography: computer-based method for making three-dimensional reconstructions of structures visualized in serial thin sections by transmission electron microscopy. (p. A-23)

threshold potential: value of the membrane potential that must be reached before an action potential is triggered. (p. 668)

thylakoid: flattened membrane sac suspended in the chloroplast stroma, usually arranged in stacks called grana; contains the pigments, enzymes, and electron carriers involved in the light-requiring reactions of photosynthesis. (pp. 89, 287)

thylakoid lumen: compartment enclosed by an interconnected network of grana and stroma thylakoids. (p. 287)

thymine (T): nitrogen-containing aromatic base, chemically designated as a pyrimidine, which serves as an informational monomeric unit when present in DNA with other bases in a specific sequence; forms a complementary base pair with adenine (A) by hydrogen bonding. (p. 59)

Ti plasmid: DNA molecule that causes crown gall tumors when transferred into plants by bacteria; used as a cloning vector for introducing foreign genes into plant cells. (p. 653)

TIC: see *translocase of the inner chloroplast membrane.*

tight junction: type of cell junction in which the adjacent plasma membranes of neighboring animal cells are tightly sealed, thereby preventing molecules from diffusing from one side of an epithelial cell layer to the other by passing through the spaces between adjoining cells. (p. 417)

TIM: see *translocase of the inner mitochondrial membrane.* (p.247)

+-TIP protein: see *plus-end tubulin interacting protein.*

TIRF: see *total internal reflection fluorescence microscopy.*

TLC: see *thin-layer chromatography.*

T_m: see *transition temperature* or *DNA melting temperature.*

TOC: see *translocase of the outer chloroplast membrane.*

TOM: see *translocase of the outer mitochondrial membrane.* (p.247)

topoisomerase: enzyme that catalyzes the interconversion of the relaxed and supercoiled forms of DNA by making transient breaks in one or both DNA strands. (pp. 449, 485)

total internal reflection fluorescence (TIRF) microscopy: technique in which a light beam strikes the interface of two media of different indices of refraction at an angle beyond the critical angle; an "evanescent field" that develops at the interface allows selective excitation of fluorescent molecules located within ~100 nm of the interface. (p. A-13)

***trans*-acting factor:** regulatory protein that exerts its function by binding to specific DNA sequences. (p. 578)

transcription: process by which RNA polymerase utilizes one DNA strand as a template for guiding the synthesis of a complementary RNA molecule. (p. 511)

transcriptional activator: regulatory protein whose binding to DNA leads to an increase in the transcription rate of specific nearby genes. (p. 597)

transcriptional control: group of regulatory mechanisms involved in controlling the rates at which specific genes are transcribed. (p. 591)

transcriptional repressor: regulatory protein whose binding to DNA leads to a decrease in the transcription rate of specific nearby genes. (p. 598)

transcription factor: protein required for the binding of RNA polymerase to a promoter and for the optimal initiation of transcription. (p. 519) Also see *general transcription factor* and *regulatory transcription factor.*

transcription regulation domain: region of a transcription factor, distinct from the DNA-binding domain, that is responsible for regulating transcription. (p. 596)

transcription unit: segment of DNA whose transcription gives rise to a single, continuous RNA molecule. (p. 510)

transcriptome: the entire set of RNA molecules produced by a genome. (p. 638)

transcytosis: endocytosis of material into vesicles that move to the opposite side of the cell and fuse with the plasma membrane, releasing the material into the extracellular space. (p. 340)

transduction: transfer of bacterial DNA sequences from one bacterium to another by a bacteriophage. (p. 774)

transfer RNA (tRNA): family of small RNA molecules, each binding a specific amino acid and possessing an anticodon that recognizes a specific codon in mRNA. (pp. 511, 548)

transformation: change in the hereditary properties of a cell brought about by the uptake of foreign DNA. (pp. 433, 769)

transforming growth factor β (TGF β): family of growth factors that can exhibit either growth-stimulating or growth-inhibiting properties, depending on the target cell type; regulate a wide range of activities in both embryos and adult animals, including effects on cell growth, division, differentiation, and death. (pp. 704, 740)

transgenic: any organism whose genome contains a gene that has been experimentally introduced from another organism using the techniques of genetic engineering. (p. 653)

***trans*-Golgi network (TGN):** region of the Golgi apparatus consisting of a network of membrane-bounded tubules that are located on the opposite side of the Golgi apparatus from the *cis*-Golgi network. (p. 320)

transit sequence: amino acid sequence that targets a completed polypeptide chain to either mitochondria or chloroplasts. (p. 564)

transition state: intermediate stage in a chemical reaction, of higher free energy than the initial state, through which reactants must pass before giving rise to products. (p. 131)

transition state analogue: compound that resembles the intermediate transition state of an enzyme-catalyzed reaction but cannot be converted to a functional product. (p. 141)

transition temperature (T_m): temperature at which a membrane undergoes a sharp decrease in fluidity ("freezes") as the temperature is decreased and becomes more fluid again ("melts") when it is then warmed; determined by the kinds of fatty acid side chains present in the membrane. (p. 165)

transition vesicle: membrane vesicle that shuttles lipids and proteins from the endoplasmic reticulum to the Golgi apparatus. (p. 316)

translation: process by which the base sequence of an mRNA molecule guides the sequence of amino acids incorporated into a polypeptide chain; occurs on ribosomes. (p. 500)

translational control: mechanisms that regulate the rate at which mRNA molecules are translated into their polypeptide products; includes control of translation rates by initiation factors, selective inhibition of specific mRNAs by translational repressor proteins or microRNAs, and control of mRNA degradation rates. (p. 607)

translational repressor: regulatory protein that selectively inhibits the translation of a particular mRNA. (p. 608)

translesion synthesis: DNA replication across regions where the DNA template is damaged. (p. 498)

translocase of the inner chloroplast membrane (TIC): a transport complex involved in the uptake of specific polypeptides into the chloroplast. (p. 564)

translocase of the inner mitochondrial membrane (TIM): a transport complex involved in the uptake of specific polypeptides into the mitochondrion. (p. 564)

translocase of the outer chloroplast membrane (TOC): a transport complex involved in the uptake of specific polypeptides into the chloroplast. (p. 564)

translocase of the outer mitochondrial membrane (TOM): a transport complex involved in the uptake of specific polypeptides into the mitochondrion. (p. 564)

translocation: movement of mRNA across a ribosome by a distance of three nucleotides, bringing the next codon into position for translation. (p. 559) (Note: The same term, translocation, which literally means "a change of location," can also refer to the movement of a protein molecule through a membrane channel or to the transfer of a segment of one chromosome to another nonhomologous chromosome.)

translocon: structure in the ER membrane that carries out the translocation of newly forming polypeptides across (or into) the ER membrane. (p. 327, 561)

transmembrane protein: an integral membrane protein possessing one or more hydrophobic regions that span the membrane plus hydrophilic regions that protrude from the membrane on both sides. (p. 173)

transmembrane segment: hydrophobic segment about 20–30 amino acids long that crosses the lipid bilayer in a transmembrane protein. (p. 173)

transmission electron microscope (TEM): type of electron microscope in which an image is formed by electrons that are transmitted through a specimen. (pp. 8, A-16)

transport: selective movement of substances across membranes, both into and out of cells and into and out of organelles. (p. 186)

transport protein: membrane protein that recognizes substances with great specificity and assists their movement across a membrane; includes both carrier proteins and channel proteins. (p. 193)

transport vesicle: vesicle that buds off from a membrane in one region of the cell and fuses with other membranes; includes vesicles that convey lipids and proteins from the ER to the Golgi apparatus, between the Golgi stack cisternae, and from the Golgi apparatus to various destinations in the cell, including secretory vesicles, endosomes, and lysosomes. (p. 320)

transposable element (transposon): mobile DNA elements that can move from one chromosomal location to another, either via excision and insertion into a new site or by duplication and insertion to create a copy of the element. (p. 504)

transposon: see *transposable element.*

transverse diffusion: movement of a lipid molecule from one monolayer of a membrane to the other, a thermodynamically unfavorable and therefore infrequent event; also called "flip-flop." (p. 164)

transverse (T) tubule system: invaginations of the plasma membrane that penetrate into a muscle cell and conduct electrical impulses into the cell interior, where T tubules make close contact with the sarcoplasmic reticulum and trigger the release of calcium ions. (p. 403)

triacylglycerol: a glycerol molecule with three fatty acids linked to it; also called a triglyceride. (pp. 69, 257)

triad: region where a T tubule passes between the terminal cisternae of the sarcoplasmic reticulum in skeletal muscle. (p. 404)

triglyceride: see *triacylglycerol.*

trinucleotide repeat: repetitive three-base sequences in DNA that can lead to strand slippage during replication and the accumulation of additional repeats in some genes, leading to trinucleotide repeat disorders. (p. 492)

triple bond: chemical bond formed between two atoms as a result of sharing three pairs of electrons. (p. 23)

triplet code: a coding system in which three units of information are read as a unit; a reference to the genetic code, which is read from mRNA in units of three bases called codons. (p. 541)

triskelion: structure formed by clathrin molecules consisting of three polypeptides radiating from a central vertex; the basic unit of assembly for clathrin coats. (p. 342)

tRNA: see *transfer RNA.*

tropomyosin: long, rodlike protein associated with the thin actin filaments of muscle cells, functioning as a component of the calcium-sensitive switch that activates muscle contraction; blocks the interaction between actin and myosin in the absence of calcium ions. (p. 398)

troponin: complex of three polypeptides (TnT, TnC, and TnI) that functions as a component of the calcium-sensitive switch that activates muscle contraction; displaces tropomyosin in the presence of calcium ions, thereby activating contraction. (p. 398)

***trp* operon:** group of adjoining bacterial genes that code for enzymes involved in tryptophan biosynthesis and whose transcription is selectively inhibited in the presence of tryptophan. (p. 576)

true-breeding (plant strain): organism that, upon self-fertilization, produces only offspring of the same kind for a given genetic trait. (p. 760)

t-SNARE (target-SNAP receptor): protein associated with the outer surface of a target membrane that binds to a v-SNARE protein associated with the outer surface of an appropriate transport vesicle. (p. 344)

tubulin: family of related proteins that form the main building block of microtubules. (p. 361) Also see *alpha tubulin, beta tubulin, gamma tubulin,* and *gamma tubulin ring complexes.*

tumor: growing mass of cells caused by uncontrolled cell proliferation. (p. 784) Also see *benign tumor* and *malignant tumor.*

tumor grading: assignment of numerical grades to tumors based on differences in their microscopic appearance; higher-grade cancers tend to grow and spread more aggressively, and to be less responsive to therapy, than lower-grade cancers. (p. 811)

tumor progression: gradual changes in tumor properties observed over time as cancer cells acquire more aberrant traits and become increasingly aggressive. (p. 786)

tumor suppressor gene: gene whose loss or inactivation by deletion or mutation can lead to cancer. (p. 796) Also see *gatekeeper gene* and *caretaker gene.*

γ-TuRCs: see *gamma tubulin ring complexes.* (p. 367)

turgor pressure: pressure that builds up in a cell due to the inward movement of water that occurs because of a higher solute concentration inside the cell than outside; accounts for the firmness, or turgidity, of fully hydrated cells or tissues of plants and other organisms. (p. 190)

turnover number (k_{cat}): rate at which substrate molecules are converted to product by a single enzyme molecule when the enzyme is operating at its maximum velocity. (p. 142)

two-dimensional (2-D) gel electrophoresis: technique for separating proteins based on pH (isoelectric focusing), followed by separation in the second dimension by size via SDS-polyacrylamide gel electrophoresis (SDS-PAGE). (p.646)

two-hit hypothesis: hypothesis that loss-of-function mutations in each of the two copies of a tumor suppressor gene are required to predispose cells to become cancerous. (p. 802)

type II myosin: form of myosin composed of four light chains and two heavy chains, each having a globular myosin head, a hinge region, and a long rodlike tail; found in skeletal, cardiac, and smooth muscle cells, as well as in nonmuscle cells. (p. 395)

U

U: see *uracil.*

ubiquitin: small protein that is linked to other proteins as a way of marking the targeted protein for degradation by proteasomes. (p. 611)

ultracentrifuge: instrument capable of generating centrifugal forces that are large enough to separate subcellular structures and macromolecules on the basis of size, shape, and density. (pp. 10, 98)

ultramicrotome: instrument used to slice an embedded biological specimen into ultrathin sections for electron microscopy. (p. A-19)

ultraviolet radiation (UV): mutagenic type of radiation present in sunlight that triggers the formation of pyrimidine dimers in DNA. (p. 789)

uncompetitive inhibitor: compound that reduces enzyme activity by binding to a site on the enzyme other than the active site, resulting in diminished substrate binding or catalytic activity. Uncompetitive inhibitors can only bind enzyme when the enzyme is bound to substrate. (p.144)

undershoot: see *hyperpolarization.*

unfolded protein response (UPR): quality control mechanism in which sensor molecules in the ER membrane detect misfolded proteins and trigger a response that inhibits the synthesis of most proteins while enhancing the production of those required for protein folding and degradation. (p. 321, 561)

uniport: membrane protein that transports a single solute from one side of a membrane to the other. (p. 194)

unsaturated fatty acid: fatty acid molecule containing one or more double bonds. (p. 163)

UPR: see *unfolded protein response.*

upstream: located toward the 5′ end of the DNA coding strand. (p. 514)

upstream activating sequence (UAS): a DNA element distinct from the core promoter associated with several genes in the yeast *Saccharomyces cerevisiae* that activates transcription of a nearby gene. It is bound by the regulatory transcription factor Gal4. (p. 600)

uracil (U): nitrogen-containing aromatic base, chemically designated as a pyrimidine, that serves as an informational monomeric unit when present in RNA with other bases in a specific sequence; forms a complementary base pair with adenine (A) by hydrogen bonding. (p. 60)

useful magnification: measurement of how much an image can be enlarged before additional enlargement provides no additional information. (p. A-4)

UV: see *ultraviolet radiation.*

V

v: see *initial reaction velocity.*

vacuole: membrane-bounded organelle in the cytoplasm of a cell, used for temporary storage or transport; acidic membrane-enclosed compartment in plant cells. (pp. 96, 342)

vacuum evaporator: bell jar containing a metal electrode and a carbon electrode in which a vacuum can be created; used in preparing metal replicas of the surfaces of biological specimens. (p. A-20)

van der Waals interaction: weak attractive interaction between two atoms caused by transient asymmetries in the distribution of charge in each atom. (p. 50)

valence: in chemistry, the maximum number of chemical bonds that an atom can form with other atoms in order to fill its outer orbital shell. (p. 22)

variable number tandem repeat (VNTR): repeat region in chromosomal DNA containing tandem repeats; two categories are microsatellite (STR) and minisatellite DNA. (p. 641)

vesicle-SNAP receptor: see *v-SNARE.*

viroid: small, circular RNA molecule that can infect and replicate in host cells even though it does not code for any protein. (p. 102)

virus: subcellular parasite composed of a protein coat and DNA or RNA, incapable of independent existence; invades and infects cells and redirects the host cell's synthetic machinery toward the production of more virus particles. (p. 38, 102)

V_m: see *resting membrane potential.*

VNTR: see *variable number tandem repeat.*

voltage: see *electrical potential.*

voltage sensor: amino acid segment of a voltage-gated ion channel that makes the channel responsive to changes in membrane potential. (p. 665)

voltage-gated calcium channel: an integral membrane protein in the terminal bulb of presynaptic neurons that forms a calcium ion-conducting pore whose permeability is regulated by the membrane potential; action potentials cause the calcium channel to open and calcium ions rush into the cell, stimulating the release of neurotransmitters. (p. 677)

voltage-gated ion channel: an integral membrane protein that forms an ion-conducting pore whose permeability is regulated by changes in the membrane potential. (p. 664)

v-SNARE (vesicle-SNAP receptor): protein associated with the outer surface of a transport vesicle that binds to a t-SNARE protein associated with the outer surface of the appropriate target membrane. (p. 344)

V-type ATPase: type of transport ATPase that pumps protons into such organelles as vesicles, vacuoles, lysosomes, endosomes, and the Golgi apparatus. (p. 204)

W

wavelength: distance between the crests of two successive waves. (p. A-1)

Western blotting: technique in which polypeptides separated by gel electrophoresis are transferred to a special type of "blotter" paper (nitrocellulose or nylon), which is then reacted with labeled antibodies that are known to bind to specific polypeptides. (p. 647)

whole-genome (shotgun) sequencing: sequencing of small, random fragments of DNA from a complex mixture, whose sequences are then assembled using computers to produce a longer sequence derived from ordering the sequences of the small fragments; often used to assemble genomic DNA sequences. (p. 635)

wild type: normal, nonmutant form of an organism, usually the form found in nature. (p. 771)

Wnt pathway: signaling pathway that plays a prominent role in controlling cell proliferation and differentiation during embryonic development; abnormalities in this pathway occur in some cancers. (p. 804)

wobble hypothesis: flexibility in base pairing between the third base of a codon and the corresponding base in its anticodon. (p. 550)

work: transfer of energy from one place or form to another place or form by any process other than heat flow. (p. 113)

X

X-inactivation: process whereby one of the two X chromosomes in the cells of female mammals is transcriptionally inactivated via widespread DNA methylation and chromatin condensation. (p. 591)

xenobiotic: chemical compound that is foreign to biological organisms. (p. 353)

xeroderma pigmentosum: inherited susceptibility to cancer (mainly skin cancer) caused by defects in DNA excision repair or translesion synthesis of DNA. (p. 489)

X-ray crystallography: technique for determining the three-dimensional structure of macromolecules based on the pattern produced when a beam of X-rays is passed through a sample, usually a crystal or fiber. (pp. 175, A-24)

Y

YAC: see *yeast artificial chromosome.*

yeast artificial chromosome (YAC): yeast cloning vector consisting of a "minimalist" chromosome that contains all the DNA sequences needed for normal chromosome replication and segregation to daughter cells, and very little else. (p. 631)

yeast two-hybrid system: technique for determining whether two proteins interact by introducing DNA encoding one protein fused to DNA encoding the DNA-binding domain of a transcription factor (bait), and another DNA encoding the second protein plus sequence encoding the activation domain of the transcription factor (prey), followed by assessing expression of a reporter. (p. 651)

Z

Z line: dark line in the middle of the I band of a striated muscle myofibril; defines the boundary of a sarcomere. (p. 397)

zinc finger: DNA-binding motif found in some transcription factors; consists of an α helix and a two-segment β sheet held in place by the interaction of precisely positioned cysteine or histidine residues with a zinc atom. (p. 601)

zygote: diploid cell formed by the union of two haploid gametes. (p. 752)

zygotene: stage during prophase I of meiosis when homologous chromosomes become closely paired by the process of synapsis. (p. 756)

PHOTO, ILLUSTRATION, AND TEXT CREDITS

Photo Credits

Chapter 1 Opener: Dr. Jan Schmoranzer/Science Source; **1-1a:** Science & Society Picture Library/Getty Images; **1-1b:** World History Archive/Alamy Stock Photo; **1-2a:** Susumu Nishinaga/Science Source; **1-2b:** Science History Images/Alamy Stock Photo; **1-2c:** Science Source; **1-2d:** M. I. Walker/Science Source; **1-2e:** De Agostini Picture Library/Science Source; **1-2f:** David M. Phillips/Science Source; **1-2g:** Aaron J. Bell/Science Source; **1-2h:** Steve Gschmeissner/Science Source; **1-2i:** David Becker/Science Source; **1A-4:** Dr. Gopal Murti/Science Source; **1-6a:** Biophoto Associates/Science Source; **1-6b:** Keith R. Porter/Science Source; **1-6c, d:** Eye of Science/Science Source; **1-7a:** Richard Megna/Fundamental Photographs; **1-7b:** Pascal Goetgheluck/Science Source; **1-10a:** Daniela Beckmann/Science Source; **1-10b:** SCIMAT/Science Source; **1-10c:** Roblan/Shutterstock; **1-10d:** Sinclair Stammers/Science Source; **1-10e:** Jasmin Merdan/123RF; **01-10f:** Nigel Cattlin/Alamy Stock Photo; **1B-1a:** Published by permission of the Lacks family; **1B-1b:** Steve Gschmeissner/Science Source; **1-11:** Shao, X. et al. (2017). Cell–cell adhesion in metazoans relies on evolutionarily conserved features of the-catenin-catenin–binding interface. *The Journal of Biological Chemistry*. 292, 16477-16490.

Chapter 2 2A-1: Sebastiano Volponi/MARKA/Alamy Stock Photo; **2-9:** Nigel Cattlin/Alamy Stock Photo; **2B-2:** Du Cane Medical Imaging Ltd/Science Source; **2-14:** (left) Biophoto Associates/Science Source (right) Dr. G. F. Bahr/Armed Forces Institute of Pathology.

Chapter 3 Opener: Dr. Alex McPherson, University of California, Irvine/NASA; **3A-1:** Dr. Cecil H. Fox/Science Source; **3A-2:** Thomas Deerinck, NCMIR/Science Source; **3A-3:** Susan Landau/Lawrence Berkeley National Lab; **3B-1:** Science History Images/Alamy Stock Photo; **3B-2:** Volker Steger/Science Source; **3B-3:** James King-Holmes/Science Source; **3-24a:** Dr. Jeremy Burgess/Science Source; **3-24b:** Don W. Fawcett/Science Source; **3-25:** Biophoto Associates/Science Source.

Chapter 4 Opener: Talley Lambert/Science Source; **4-1:** Scripps Institution of Oceanography Archives, UC San Diego Library; **4-2a:** Menger, F. M. & Gabrielson, K. (1994). Chemically-Induced Birthing and Foraging in Vesicle Systems. *J. Am. Chem. Soc. 116*, 4, 1567–1568; **4-2b:** Jack W. Szostak; **4-5:** Biophoto Associates/Science Source; **4-6b:** Biophoto Associates/Science Source; **4-9:** Bruce J. Schnapp; **4-10:** Power and Syred/Science Source; **4-12b:** Science History Images/Alamy Stock Photo; **4-12c:** Biophoto Associates/Science Source; **4-13c:** Keith R Porter/Science Source; **4-14b:** Science History Images/Alamy Stock Photo; **4-16c:** Don W. Fawcett/Science Source; **4-16d:** Biophoto Associates/Science Source; **4-17c:** Biophoto Associates/Science Source; **4A-1:** Dr. Dag Malm; **4-19b:** Biophoto Associates/Science Source; **4-20:** Don W. Fawcett/Science Source; **4-21b:** Sue Ellen Frederick and Eldon H. Newcomb, *Journal of Cell Biology*, 1969, 43:343-353. doi: 10.1083/jcb.43.2.343; Figure 6. The Rockefeller University Press; **4-22b:** Dr Jeremy Burgess/Science Source; **4-24:** Guillaume T. Charras, Mike A. Horton. (2002). Single Cell Mechanotransduction and Its Modulation Analyzed by Atomic Force Microscope Indentation. *Biophysical Society*, 82:6 P2970-2981; **4-25b:** Biophoto Associates/Science Source; **4B-1:** wavebreakmedia/Shutterstock; **4-26:** (top, left to right) Biophoto Associates/Science Source, Science History Images/Alamy Stock Photo, CDC/Maureen Metcalfe, Tom Hodge; (bottom, left to right) Health Protection Agency Centre for Infections/Science Source, Science History Images/Alamy Stock Photo, M. Wurtz/Biozentrum, University of Basel/Science Source.

Chapter 5 Opener: Ben Nottidge/Alamy Stock Photo; **5-2:** Aaron J. Bell/Science Source; **5-3a:** Lisa Werner/Alamy Stock Photo; **5-3b:** Martin Shields/Alamy Stock Photo; **5-5:** tacojim/Getty Images; **5A-1:** PHOTO RF/Science Source; **p. 122:** Andrew Brookes, National Physical Laboratory/Science Source.

Chapter 6 6A-1: MAURICIO LIMA/Getty Images; **p. 144:** Crown Copyright Food and Environment Research Agency/Science Source.

Chapter 7 7-1a: Keith R. Porter/Science Source; **7-1b:** Eldon H. Newcomb; **7-3e:** Elizabeth Robertson; **7-4:** Don W. Fawcett/Science Source; **7-A3a:** Jeff Hardin, Univ. of Wisconsin-Madison; **7-16a:** David W. Deamer; **7-16b:** Dan Branton/Omikron/Science Source; **7-22:** Don W. Fawcett/Science Source; **7-24a:** Cheryl Power/Science Source.

Chapter 8 Opener: Cheryl Power/Science Source; **p. 196:** Pablo Bou Mira/Alamy Stock Photo; **8-10a:** Borgnia, M. et al. (1999). Cellular and Molecular Biology of the Aquaporin Water Channels. *Annual Review of Biochemistry*. Vol. 68, p. 429; **8-B2a:** BSIP SA/Alamy Stock Photo; **8-B2b:** Glen Stubbe/Minneapolis Star Tribune/ZUMAPRESS.com/Newscom; **8-16a:** Yann Arthus-Bertrand/Getty Images.

Chapter 9 Opener: Dr. Tim Evans/Science Source; **9-9:** Courtesy of Annick D. Van den Abbeele, MD, FACR, FICIS, Dana-Farber Cancer Institute, Boston, MA, USA; **p. 232:** Sinclair Stammers/Science Source; **p. 234:** Dan Kosmayer/Shutterstock.

Chapter 10 Opener: Biophoto Associates/Science Source; **10-2a:** Keith R. Porter/Science Source; **10-3a:** Talley Lambert/Science Source; **10-3b:** T.G. Frey and M. Ghochani, San Diego State University; **10A-1:** Courtesy of Pacific Northwest National Laboratory; **10A-2b, c:** T.G. Frey and M. Ghochani, San Diego State University; **10-5a:** Alexander Tzagoloff; **10-14:** Trelease, R.N. et al. (1971). Microbodies (Glyoxysomes and Peroxisomes) in Cucunber Cotyledons. *Plant Physiology*, 48, 461-475.

Chapter 11 Opener: Fernan Federici, Jim Haseloff; **11-2a:** Biophoto Associates/Science Source; **11-2b:** M.I. Walker/Science Source; **11-3a:** Omikron/Science Source; **11-4a:** Dr. Kari Lounatmaa/Science Source; **11A-1:** GIPhotoStock/Science Source; **11B-3:** Jose Gil/Shutterstock; **11-19b:** Mike Clayton.

Chapter 12 Opener: Achilleas Frangakis EMBL; **12-2a:** Barry F. King/Biological Photo Service; **12-15:** James D. Jamieson, George E. Palade; INTRACELLULAR TRANSPORT OF SECRETORY PROTEINS IN THE PANCREATIC EXOCRINE CELL: IV. Metabolic Requirements. *J Cell Biol* 1 December 1968; 39 (3): 589–603. doi: https://doi.org/10.1083/jcb.39.3.589; **12-16:** Nature's Faces/Science Source; **12-17b:** JOSE CALVO/Science Source; **p. 336:** GIPhotoStock/Science Source; **12A-2:** Joshua Z. Rappoport; **12-19b:** Biophoto Associates/Science Source; **12-21:** M. M. Perry and A. B. Gilbert. (1979). Yolk transport in the ovarian follicle of the hen (Gallus domesticus): lipoprotein-like particles at the periphery of the oocyte in the rapid growth phase. *The Journal of Cell Science*, 39, 257-272; **12-22a:** Heuser Lab, Washington University, St. Louis, MO; **12-23a, b:** Heuser Lab, Washington University, St. Louis, MO; **12-26:** Don W. Fawcett/Science Source; **12B-1:** Gastrolab/Science Source; **12-27:** Don W. Fawcett/Science Source.

Chapter 13 Opener: Dr. Torsten Wittmann/Science Source; **13-1:** Tatyana Svitkina; Table 13-1: Charras, G. T. & Horton, M. A. (2002). Single Cell Mechanotransduction and Its Modulation Analyzed by Atomic Force Microscope Indentation. *Biophysical Journal*, Volume 82, Issue 6, pp. 2970–2981; Table 13-2: (top to bottom) © 2015 Life Technologies Corporation/Thermo Fisher Scientific Inc., Yasushi Okada (RIKEN), E.D. Salmon. (1995). VE-DIC light microscopy and the discovery of kinesin. *Trends in Cell Biology* 5:154-58., Don W. Fawcett/Science Source; **13-3b:** Biology Pics/Science Source; **13-5:** LI Binder, JL Rosenbaum. (1978). The in vitro assembly of flagellar outer doublet tubulin. *The Journal of Cell Biology*, 79, 500-515; **13-8b:** Kent L. McDonald; **13-8c:** Don W. Fawcett/Science Source; **13-9b:** Dr. Thomas J. Keating, Ph.D; **13-11d:** Maric, I. Et al. (2011). Centrosomal and mitotic abnormalities in cell lines derived from papillary thyroid cancer harboring specific gene alterations. *Molecular Cytogenetics*, 4:26; **13-12b:** Kenneth S. Kosik; **13-12d:** Rogers, S.L. (2002). Drosophila EB1 is important for proper assembly, dynamics, and positioning of the mitotic spindle. *The Journal of Cell Biology*, 158(5), 873–884; **13-13c:** L. E. Roth, Y. Shigenaka and D. J. Pihlaja; **p. 372:** Dr. Torsten Wittmann/Science Source; **13A-1:** Osborn, M., Weber, K. (1975). Cytoplasmic microtubules in tissue culture cells appear to grow from an organizing structure towards the plasma membrane. *Cell Biology*, 73 (3), pp. 867–871; **13A-2:** Prof. Andrew Matus; **13-15:** Tatyana M. Svitkina; **13-17a:** Mooseker, M. S. & Tilney, S. G. (1975). Organization of an actin filament-membrane complex. Filament polarity and membrane attachment in the microvilli of intestinal epithelial cells. *J Cell Biol.* 67 (3): 725–743; **13-18:** John E. Heuser M.D.; **13-19:** Daniel Branton/Harvard; **13-20a:** Gary G. Borisy and Tatyana M. Svitkina; **p. 378:** Andrii Malkov/Shutterstock; **13-21:** Jeff Hardin, Univ. of Wisconsin-Madison; **13-22:** David M. Phillips/Science Source; **13-24:** Elaine Fuchs, Don W. Cleveland. (1998). *A Structural Scaffolding of Intermediate Filaments in Health and Disease*. 279 (5350), pp. 514–519.

Chapter 14 Opener: Steve Gschmeissner/Science Source; **14-2:** Nobutaka Hirokawa; **14A-3b:** Dr. Toshio Ando; **14-7a:** Charles Daghlian/Science Source; **14-7c:** Steve Gschmeissner/Science Source; **14-8a:** Dartmouth Electron Microscope Facility, Dartmouth College; **14-8b:** Don W. Fawcett/Science Source; **14-8c:** Dr. Paul Guichard; **p. 396:** Pascal Goetgheluck/Science Source; **14B-1c:** Zephyr/Science Source; **14B-2a:** Hirst, R. A. et al. (2010). Ciliated Air-Liquid Cultures as an Aid to Diagnostic Testing of Primary Ciliary Dyskinesia. *Chest*, 138(6), pp. 1441–1447; **14B-2b:** Biophoto Associates/Science Source; **14-12b:** Biophoto Associates/Science Source; **14-13a:** Clara Franzini-Armstrong; **14-18:** John E. Heuser M.D.; **14-23:** Manfred Kage/Science Source; **14-24a:** Biophoto Associates/Science Source; **14-25a:** K.M. Trybus, S. Lowey (1984). Conformational States of Smooth Muscle Myosin. *The Journal of Biological Chemistry*, 259(13) pp. 8564–8571; **14-26:** Guenter Albrecht-Buehler, Ph.D; **14-28:** Panther Media GmbH/Alamy Stock Photo; **14-29b:** Y.M. Kersey, N.K. Wessells; Localization of actin filaments in internodal cells of characean algae. A scanning and transmission electron microscope study. *J Cell Biol* 1 February 1976; 68 (2): 264–275. doi: https://doi.org/10.1083/jcb.68.2.264.

Chapter 15 Opener: Jeff D. Hildebrand; **15-3b:** Dr. Kenneth Dunn; **15-4:** Masatoshi Takeichi/Takeichi lab/RIKEN BioResource Center; **15-5:** Heasman, J. et al. (1994). A functional test for maternally inherited cadherin in Xenopus shows its importance in cell adhesion at the blastula stage. *Development*. Vol. 120: 49-57; **15-6a:** Don W. Fawcett/Science Source; **15A-1:** Biophoto Associates/Science Source; **15A-3:** Dr Harout Tanielian/Science Source; **15-8b:** Friend, D. S. (1972). VARIATIONS IN TIGHT AND GAP JUNCTIONS IN MAMMALIAN TISSUES. *Journal of Cell Biology*. Vol.53, No. 3, pp. 758–776; **15-9b:** Don W. Fawcett/Science Source; **15-10b:** C. Peracchia and A. F. Dulhunty. (1976). Low resistance junctions in crayfish. Structural changes with functional uncoupling. *Journal Cell Biology*, 70, pp. 419–439; **15-10c:** Don W. Fawcett/Science Source; **15-11c:** Ed Reschke/Getty Images; **15-12a:** J. Gross/Biozentrum, University of Basel/Science Source; **15-15a:** Rosenberg, L., Hellman, W. & Kleinschmidt. (1975). Electron microscopic studies of proteoglycan aggregates from bovine articular cartilage. *The Journal of Biological Chemistry*. 250, 1877-1883; **15-16b:** DANIEL SCHROEN, CELL APPLICATIONS INC/Science Source; **15-17:** Biophoto Associates/Science Source; **15B-1:** Debnath, J. (2003). Morphogenesis and oncogenesis of MCF-10A mammary epithelial acini grown in three-dimensional basement membrane cultures. *Methods*, 30:3, pp. 256–268; **p. 429:** National Geographic Image Collection/Alamy Stock Photo;

15-20b: Lindsy Boateng and Anna Huttenlocher; **15-20d:** Kelly, D. (1966). FINE STRUCTURE OF DESMOSOMES, HEMIDESMOSOMES, AND AN ADEPIDERMAL GLOBULAR LAYER IN DEVELOPING NEWT EPIDERMIS. *J Cell Biol.* Vol. 28 (1): 51–72; **15-21a:** Bunnell, T.M. et al. (2008). Destabilization of the Dystrophin-Glycoprotein Complex without Functional Deficits in α-Dystrobrevin Null Muscle. *PLoS ONE* 3(7): e2604; **15-22b:** Biophoto Associates/Science Source; **15-23:** Biophoto Associates/Science Source; **15-24:** F. C. STEWARD and K. MÜHLETHALER (1953). The Structure and Development of the Cell-Wall in the Valoniaceae as Revealed by the Electron Microscope. *Annals of Botany*. New Series, Vol. 17, No. 66, pp. 295–316; **15-25a:** W. P. Wergin/Eldon H. Newcomb; **15-25c:** Biophoto Associates/Science Source.

Chapter 16 Opener: Power and Syred/Science Source; **16-3a:** Lee D. Simon/Science Source; **16-3b:** Michael Feiss; **16-7a:** Omikron/Science Source; **16-7b:** Science & Society Picture Library/Getty Images; **16-10b, c:** James C. Wang; **16A-1:** Monty Rakusen/Getty Images; **16A-2a:** Schröck, E. et al. (1996). Multicolor Spectral Karyotyping of Human Chromosomes. *Science*. Vol. 273, No. 5274, pp. 494–497; **16A-2b:** Bolzer A. et al. (2005). Three-Dimensional Maps of All Chromosomes in Human Male Fibroblast Nuclei and Prometaphase Rosettes. *PLoS Biol* 3(5): e157. https://doi.org/10.1371/journal.pbio.0030157; **16-15a:** H. Kobayashi, K. Kobayashi, and Y. Kobayashi/American Society for Microbiology Archives; **16-15b:** G. Murti/Science Source; **16-16:** Victoria Foe; **16-19:** (top to bottom) Victoria Foe, Barbara Hamkalo, U. Laemmli/Science Source, G. F. Bahr/Armed Forces Institute of Pathology; **16-20:** Ulrich K. Laemmli; **16-22a:** Johannes Wienberg; **16-22b:** Peter M. Lansdorp/University Medical Center Groningen; **16-23b:** Darryl Leja/National Human Genome Research Institute; **16-26:** Kasamatsu, H., Robberson, D. & Vinograd, J. (1971). A Novel Closed-Circular Mitochondrial DNA with Properties of a Replicating Intermediate. *PNAS*. Vol. 68 (9) 2252-2257; **16-28a:** Don W. Fawcett/Science Source; 16-28b: Human Protein Atlas, www.proteinatlas.org/Uhlen et al (2015). "Tissue-based map of the human proteome."/*Science*/.DOI:10.1126/science.1260419; **16-29a:** Don W. Fawcett/Science Source; **16-30:** Biophoto Associates/Science Source; **16-31a:** Don W. Fawcett/Science Source; **16-33:** Kalderon, D., Roberts, B. L., Richardson, W. D., & Smith, A. E. (1984). A short amino acid sequence able to specify nuclear location. *Cell*, 39(3), 499-509; **16-35a:** Nickerson, J.A. et al. (1997). The nuclear matrix revealed by eluting chromatin from a cross-linked nucleus. *Cell Biology*, 94(9), 4446-4450; **16-35b:** Ueli Aebi Ph.D.; **p. 468:** Andrew H. Walker/Getty Images; **16B-2:** National Institute of Health and Consensus Development Conference; **16-36:** David M. Phillips/Science Source.

Chapter 17 Opener: Dr Gopal Murti/Science Source; **17-6a:** Cold Spring Harbor Laboratory Archives; **17-7b:** Cold Spring Harbor Laboratory Archives; **17-17:** Victoria Foe; **17-19:** Nikitina, T. & Woodcock, C. (2004). Closed chromatin loops at the ends of chromosomes. *J Cell Biol*. Vol. 166, No. 2, pp. 161–165; **17-20a:** Don W. Fawcett/Science Source; **17-20b:** L. Hayflick/Jeff Hardin; **17B-1:** DAMOURETTE VINCENT/SIPA/Newscom; **17-31:** COLD SPRING HARBOR LABORATORY ARCHIVES; **17-31:** Potter, H. & Dressler, D. (1977). On the mechanism of genetic recombination: the maturation of recombination intermediates. *PNAS*. Vol. 74, No. 10, pp. 4168–4172; **17-38:** Li, Y. F., Kim, S. T. & Sancar, A (1993). Evidence for lack of DNA photoreactivating enzyme in humans. *PNAS*. Vol. 90 (10) 4389-4393.

Chapter 18 Opener: DR ELENA KISELEVA/Science Source; **18-3a:** O. L. Miller Jr., Barbara A. Hamkalo, C. A. Thomas Jr. (1970). Visualization of Bacterial Genes in Action. *Science*, 169(3943), pp. 392–395; **p. 517:** Patrick Dumas/Science Source; **18B-1:** Vladyslav Siaber/Alamy Stock Photo; **18-12a:** David M. Phillips/Science Source; **18-12b:** Professor Oscar Miller/Science Source; **18-17a:** Bert W. O'Malley, M.D.; **18-19a:** Ann Beyer; **18-19b:** Jack Griffith, University of North Carolina.

Chapter 19 19-1: Janice Carr/CDC; **19-6a:** OMIKRON/Science Source; **19-18b:** Barbara Hamkalo; **p. 566:** J. A. Lake, *Scientific American* (1981) 245:86. © J. A. Lake; **19B-1b:** Jeff Hardin, Univ. of Wisconsin-Madison; **19B-2:** Paul Steinbach/Tsien Lab.

Chapter 20 Opener: Steve Paddock, Jim Langeland; **20-12b:** PA/AP Images; **20-15:** LPLT/Wikimedia; **20-18a:** fotojagodka/123RF; **20-18b:** Karen Ng, Dieter Pullirsch, Martin Leeb, Anton Wutz/European Molecular Biology Organization (EMBO); **20-30b:** Caltech Archives; **20-30c:** Matthew P. Scott; **20-40:** Silva, S., Camino, L. & Aguilera, A. (2018). Human mitochondrial degradosome prevents harmful mitochondrial R loops and mitochondrial genome instability. *PNAS*.115 (43) 11024–11029.

Chapter 21 Opener: Deco Images II/Alamy Stock Photo; **21-2b:** LOUISE MURRAY/Science Source; **p. 624:** Vit Kovalcik/Alamy Stock Photo; **p. 642:** Image Source/Alamy Stock Photo; **21-6c:** John Bowman; **21-17:** Jeff Hardin, Univ. of Wisconsin-Madison; **21-27b:** Ralph Brinster; **21-29b:** Dr. Paul Sternberg; **21-29c:** Nathaniel P. Hawker and John L. Bowman; **21-29d:** John H. Wilson; **21-30:** Dr. Stephen Small; **21-32:** Arcansel/Alamy Stock Photo.

Chapter 22 Opener: Thomas Deerinck, NCMIR/Science Source; **22-1b:** STEVE GSCHMEISSNER/Science Source; **22-2:** David Fleetham/Alamy Stock Photo; **22A-2a:** Dr. Jürg Streit; **22-10:** STEVE GSCHMEISSNER/Science Source; **22-13c:** Joseph F. Gennaro Jr./Science Source; **22-17a:** J. Cartaud, E. L. Bendetti, A. Sobel, and J. P. Changeux, *Journal of Cell Science* 29 (1978): 313. © 1978 The Company of Biologists, Ltd.; **22-18b:** Mary B. Kennedy, Davis Professor of Biology, Caltech.

Chapter 23 Opener: Michael Whittaker. (2006). Calcium at Fertilization and in Early Development. *Physiological Reviews*, 86, pp. 25–88; **23A-1:** Hamamatsu Corporation, Camera name ORCA-Flash4.0, Type number C11440-22CU; **23-15a:** Y. Hiramato/Jeff Hardin; **23-20a, c:** Thomas, Barbara J., Wassarman, David A.; A fly's eye view of biology. *Trends Genet*. 1999 May; 15(5):184–90. doi: 10.1016/s0168-9525(99)01720-5; **23-20b, d:** Rosemary Reinke, S. Lawrence Zipursky. (1988). Cell-cell interaction in the drosophila retina: The bride of sevenless gene is required in photoreceptor cell R8 for R7 cell development. *Cell*, 55:2, pp. 321–330; **p. 716:** Hriana/Shutterstock; **23B-1a:** Zephyr/Science Source.

Chapter 24 Opener: George Von Dassow; **24A-1:** Sigrid Gombert/Science Source; **24-2:** Conly L. Rieder, Department of Biology, Rensselaer Polytechnic Institute, Troy, New York; **24-4a:** Aussie Suzuki; **24-4b:** M. J. Schibler, J. D. Pickett-Heaps. (1987). The Kinetochore Fiber Structure in the Acentric Spindles of Oedogonium. *Protoplasma*, 137, pp. 29–44; **24-7b, c:** Courtesy of Jeremy Pickett-Heaps, University of Melbourne; **24-8:** Michael V. Danilchik; **24-9a:** Victoria Foe; **24-9b:** William Bement; **24-9c:** Ahna Skop, PhD, DSc; **24B-1:** inga spence/Alamy Stock Photo; **24B-2:** M. A. Jordan. Mechanism of Action of Antitumor Drugs that Interact with Microtubules and Tubulin. *Current Medicinal Chemistry - Anti-Cancer Agents*, 1 January 2002, pp.1–17(17); **24-10:** B. A. Palevitz/Eldon H. Newcomb; **24-11a:** Michael V. Danilchik; **24-11b:** George Von Dassow; **24-12b:** Tochiyuki Mori; **24-21b:** Logarinho, E. et al. (2003). Different spindle checkpoint proteins monitor microtubule attachment and tension at kinetochores in Drosophila cells. *Journal of Cell Science* 117, 1757-1771; **24-26:** Thomas Deerinck, NCMIR/Science Source; **24-27:** Bob Goldstein/University of North Carolina; **24-30:** Mizuta R, Araki S, Furukawa M, Furukawa Y, Ebara S, Shiokawa D, et al. (2013) DNase γ Is the Effector Endonuclease for Internucleosomal DNA Fragmentation in Necrosis. *PLoS ONE* 8(12): e80223. https://doi.org/10.1371/journal.pone.0080223.

Chapter 25 Opener: Dominik Handler; **25-4:** Bernard John (2005), *Meiosis*. Cambridge University Press. Vol. 3, No. 2; **25-5a:** P. B. Moens (1968). The Structure and Function of the Synaptinemal Complex in Lilium longiflorum Sporocytes. *Chromosoma*, 23, pp. 418–451; **25A-1a:** Phanie/Science Source; **25A-1b:** Denis Kuvaev/Shutterstock; **25A-3:** Clinical Photography, Central Manchester University Hospitals NHS Foundation Trust, UK/Science Source; **25-8c:** M. I. Walker/Science Source; **p. 770:** Janice Carr/CDC; **25-19a:** Charles C. Brinton, Jr. and Judith Carnahan; **25-19b:** Omikron/Science Source; **25-24:** Calvente, A. et al. (2005). DNA double-strand breaks and homology search: inferences from a species with incomplete pairing and synapsis. *Journal of Cell Science* 118, 2957–2963.

Chapter 26 Opener: Eye of Science/Science Source; **26-17:** Jeff Hardin, Univ. of Wisconsin-Madison; **26A-1:** National Institute of Standards and Technology; **26-19:** Dr. Paul A. W. Edwards; **26-20:** Dr. Beth A. Weaver, Ph.D.; **26-23:** CNRI/Science Source; **26B-3:** Dr P. Marazzi/Science Source.

Appendix A-5: Peter Skinner/Science Source; **A-7:** Biophoto Associates/Science Source; **A-9:** Dr. Timothy Ryan; **A-10:** Salmon, E.D. (1995). VE-DIC light microscopy and the discovery of kinesin. *Trends in Cell Biology*. 5:154-58, Fig. 3, Elsevier Science; **A-13:** Dr. Gopal Murti/Science Source; **A-14:** Susan Strome et al. (2001). Spindle Dynamics and the Role of g-Tubulin in Early Caenorhabditis elegans Embryos. *Molecular Biology of the Cell*, 12, pp. 1751–1764; **A-15:** Karl Garsha; **A-17a:** Sarah Swanson, Newcombe Imaging Center, Department of Botany, Univ. of Wisconsin-Madison; **A-19:** Dr. Shelley Sazer; **A-21c:** Louis Hodgson, PhD; **A-22a:** Hein et al. Stimulated emission depletion (STED) nanoscopy of a fluorescent protein-labeled organelle inside a living cell. *Proc. Natl. Acad. Sci.* 105 (2008): 14271–14276. © 2008 National Academy of Sciences, U.S.A.; **A-22b:** Bates M, Huang B, Dempsey GT, Zhuang X. (2007). Multicolor super-resolution imaging with photo-switchable fluoresecent probes. *Science*, 317,5845:1749–1753; **A-24a:** Hemis/Alamy Stock Photo; A-25a: Don W. Fawcett, M.D.; **A-25b:** Steve Gschmeissner/Science Source; **A-26:** Sarah Swanson, Newcomb Imaging Center, Department of Botany, Univ. of Wisconsin-Madison; **A-27a:** Gustoimages/Science Source; **A-27b:** Jeff Hardin, Univ. of Wisconsin-Madison; **A-28:** Dr. Marisa Otegui, Department of Botany, Univ. of Wisconsin-Madison; **A-29:** Biophoto Associates/Science Source; **A-31:** Omikron/Science Source; **A-33:** Don W. Fawcett/Science Source; **A-34a:** Estate of Hans Ris; **A-34b, c:** Koster AJ, Klumperman, J. "Electron Microscopy in cell biology: integrating structure and function". *Nat Rev Mol Cell Biol*. 2003 Sep: Suppl: SS6–10, Fig. 1; **A-36:** Jeff Hardin, Univ. of Wisconsin-Madison.

Illustration and Text Credits

Chapter 1

Page 2 Matthias Schlieden, Theodore Schwann and Rudolf Virchow in 1839 in cell theory.

Page 3 Quoted by Rudolf Virchow in 1855.

Page 11 Aristotle quoted in *On the Parts of Animals* Written 350 B.C.E Translated by William Ogle (London.: K. Paul, French & Co, 1882.)

Page 12 Source: Crick FHC. The Central Dogma of Molecular Biology. (1958) *Nature*, 227:561–563

Fig. 1-8 Source: http://ghr.nlm.nih.gov/handbook/howgeneswork/makingprotein.

Fig. 1-9 Adapted from Charpentier, E. & Doudna, J. (2013). "Biotechnology: Rewriting a Genome," *Nature*, Vol. 495, pp. 50–51.

Chapter 2

Fig. 2A-1 Based on: http://www.chemguide.co.uk/analysis/masspec/howitworks.html.

Fig. 2A-2 Data from http://webbook.nist. gov/cgi/cbook.cgi?Spec=C56406&Index=0&Type=Mass&Large=on Glycine formula taken from WOC 8e, Fig. 3-2 Labels added.

Page 37 Ellis, R. J. & S. M. Van der Vies. (1991). "Molecular chaperones," *Annual Review of Biochemistry*, Vol. 60, pp. 321–347.

Table 2-1 Adapted from Fraenkel-Conrat, H. & Williams, R. (1955). "Reconstitution of active tobacco mosaic virus from its inactive protein and nucleic acid components," *Proceedings of the National Academy of Sciences of the United States of America*, Vol. 41, no. 10, p. 690.

Chapter 3

Table 3-1 Adapted from Wald, G. (1964) "The origins of life," *Proceedings of the National Academy of Sciences of the United States of America*, Vol. 52, no. 2, p.595.

Fig. 3-7 Based on Kleinsmith, Lewis J.; Kish, Valerie M., *Principles of Cell and Molecular Biology*. 2nd Ed., © 1995 Pearson Education, Inc.

Fig. 3-9 Based on Kleinsmith, Lewis J.; Kish, Valerie M., *Principles of Cell and Molecular Biology*. 2nd Ed., © 1995 Pearson Education, Inc.

Fig. 3-14 Based on Kleinsmith, Lewis J.; Kish, Valerie M., *Principles of Cell and Molecular Biology*. 2nd Ed., © 1995 Pearson Education, Inc.

Fig. 3-27 Based on Kleinsmith, Lewis J.; Kish, Valerie M., *Principles of Cell and Molecular Biology*. 2nd Ed., © 1995 Pearson Education, Inc.

Fig. 3-31 Autumn, K. et al. (2002). "Evidence for van der Waals adhesion in gecko setae," *Proceedings of the National Academy of Sciences*, Vol. 99, no. 19, pp. 12252–12256.

Chapter 4

Fig. 4-15 Adapted from Baum, D. A. & Baum, B. (2014). "An inside-out origin for the eukaryotic cell," *BMC Biology*, Vol. 12, no. 76.

Fig. 4A-2 Based on Jeyakumar et al (2005). Storage Solutions: Treating Lysosomal Disorders Of The Brain. *Nature Reviews Neuroscience* 6, 1–12 (Box 1).

Fig. 4-23 Based on P. J. Russell, *GENETICS*, 5th ed., Fig. 13.18. © 1998 Pearson Education, Inc.

Fig. 4-27 Based on Hanczyc, M. M., Fujikawa, S. M & Szostak, J. W. (2003). Experimental Models of Primitive Cellular Compartments: Encapsulation, Growth, and Division. *Science*. 302(5645): 618–622.

Chapter 5

Fig. 5A-3 Source: http://philschatz.com/anatomy-book/resources/2511_A_Triglyceride_Molecule_(a)_Is_Broken_Down_Into_Monoglycerides_(b).jpg.

Fig. 5B-1, 5B-2 Based on Freyer, M. W. & Lewis, E. A. (2008). Isothermal Titration Calorimetry: Experimental Design, Data Analysis, and Probing Macromolecule/Ligand Binding and Kinetic Interactions. *Methods in Cell Biology*. Vol. 84, pp/79-113.

Fig. 5-12 Adapted from OpenStax "Chemistry," Bccampus.

Chapter 6

Fig. 6-7 Appling, D. R., Anthony-Cahill, S. J. & Matthews C. K. (2018). *Biochemistry: Concepts and Connections*, Biochemical Genetics, 2nd ed.

Chapter 7

Fig. 7-6 Based on Kleinsmith, Lewis J.; Kish, Valerie M., *Principles of Cell and Molecular Biology*. 2nd ed. p. 163. © 1995 Pearson Education, Inc.

Fig. 7-7 Based on Kleinsmith, Lewis J.; Kish, Valerie M., *Principles of Cell and Molecular Biology*. 2nd ed. Fig. 5.6, p. 161. © 1995 Pearson Education, Inc.

Fig. 7-A1 Based on Kleinsmith, Lewis J.; Kish, Valerie M., *Principles of Cell and Molecular Biology*. 2nd Ed., Fig. 5.22, p. 176. © 1995 Pearson Education, Inc.

Fig. 7-17 Based on Kleinsmith, Lewis J.; Kish, Valerie M., *Principles of Cell and Molecular Biology*. 2nd Ed., Fig. 5.17, p. 169. © 1995 Pearson Education, Inc.

Fig. 7-19 Based on Kleinsmith, Lewis J.; Kish, Valerie M., *Principles of Cell and Molecular Biology*, 2nd ed., Fig. 5.18, p. 169. © 1995 Pearson Education, Inc.

Fig. 7-20 Based on Kleinsmith, Lewis J.; Kish, Valerie M., *Principles of Cell and Molecular Biology*. 2nd Ed., Fig. 5.16, p. 168. © 1995 Pearson Education, Inc.

Fig. 7-25 Jeff Hardin, Univ. of Wisconsin-Madison

Chapter 8

Fig. 8-4 Based on Kleinsmith, Lewis J.; Kish, Valerie M., *Principles of Cell and Molecular Biology*. 2nd Ed., © 1995 Pearson Education, Inc.

Fig. 8-6 Based on Kleinsmith, Lewis J.; Kish, Valerie M., *Principles of Cell and Molecular Biology*. 2nd Ed., © 1995 Pearson Education, Inc.

Fig. 8-9 Based on Protein Data Bank accession 2F1C. Original reference: Subbarao, G. V. and van den Berg, B. (2006). Crystal structure of the monomeric porin OmpG. J. MOL. BIOL. 360: 750–759 (from Fig. 1, p. 752).

Fig. 8-10a Adapted from Protein Data Bank accession 3D9S. Original reference: Horsefield et al. (2008). High-resolution x-ray structure of human aquaporin 5. *Proc. Natl. Acad. Sci.* USA 105: 13327–13332.

Fig. 8-10b From Andrews, S., S. L. Reichow, and T. Gonen. Electron crystallography of aquaporins. *IUBMB Life* 60 (2008): 430.

Fig. 8-12 Based on Kleinsmith, Lewis J.; Kish, Valerie M., *Principles of Cell and Molecular Biology*, 2nd ed., Fig. 5.43, p. 190. © 1995 Pearson Education, Inc.

Fig. 8-17 Data from Psakis, G. et al. (2009). "The sodium-dependent D-glucose transport protein of Helicobacter pylori," *Molecular Microbiology*, Vol. 71, no. 2, pp. 391–403.

Chapter 9

Fig. 9A-1 Based on Fig 1 from Winder, C. L., Dunn, W. B. & Goodacre, R. (2011). "TARDIS-based microbial metabolomics: time and relative differences in systems," *Trends in Microbiology*, Vol 19, no. 7, pp. 315–322.

Chapter 10

Fig. 10-19 Based on Kleinsmith, Lewis J.; Kish, Valerie M., *Principles of Cell and Molecular Biology*, 2nd Ed., © 1995 Pearson Education, Inc.

Fig. 10-20 Source: Figures are from *Horton Principles of Biochemistry* 4e - (a) from Figure 14-8, page 426 PDB ID# 1NEK. (b) from Figure 14-10, page 428 PDB ID# 1PP9. (c) from Figure 14-12a, page 430 PDB ID# 1OCC.

Fig. 10-21 Source: L. J. Kleinsmith, V. M. Kish, Experimental Evidence That Electron Transport Generates a Protein Gradient, *Principles of Cell and Molecular Biology*, 2nd ed., Pearson Education, 1995, 336.

Fig. 10-24 Source: Based on *Horton Principles of Biochemistry* 4e, Figure 14.15, p. 434.

Chapter 11

Fig. 11-4b Based on Kelvinsong/CC BY-SA.

Fig. 11A-2 Based on "Concepts in Photobiology: Photosynthesis and Photomorphogenesis", Edited by GS Singhal, G Renger, SK Sopory, K-D Irrgang and Govindjee, Narosa Publishers/New Delhi; and Kluwer Academic/Dordrecht, pp. 11–51.

Fig. 11-11 Based on Figure from https://crystallography365.wordpress.com/2014/05/15/pass-the-electron-the-first-membrane-protein-structure-shows-how-purple-bacteria-get-energy-from-light/. Structure is from Protein Data Bank entry 1PRC At http://www.rcsb.org/pdb/explore/explore.do?structureId=1prc.

Fig. 11-14 Based on Rubisco, November 2000 Molecule of the Month, by David Goodsell. RCSB Protein Data Bank.

Page 313 Joseph Priestley, *The Theological and Miscellaneous Works*. Ed. with Notes by John Towill Rutt, Volume 1 (George Shallfield, 1831).

Chapter 12

Fig. 12-4 Based on Alisa Zapp Machalek, *An Owner's Guide to the Cell*, Chapter 1, NIH National Institute of General Medical Sciences, 2005 http://publications.nigms.nih.gov/insidethecell/chapter1.html#4

Fig. 12-12 Based on Kleinsmith, Lewis J.; Kish, Valerie M., *Principles of Cell and Molecular Biology*. 2nd Ed., © 1995 Pearson Education, Inc.

Fig. 12-4a, b Based on Alisa Zapp Machalek, An Owner's Guide to the Cell, Chapter 1, NIH National Institute of General Medical Sciences, 2005 http://publications.nigms.nih.gov/insidethecell/chapter1.html#4.

Fig. 12-19a Based on Kleinsmith, Lewis J.; Kish, Valerie M., *Principles of Cell and Molecular Biology*. 2nd Ed., © 1995 Pearson Education, Inc.

Fig. 12-22b, c Based on Kleinsmith, Lewis J.; Kish, Valerie M., *Principles of Cell and Molecular Biology*. 2nd Ed., © 1995 Pearson Education, Inc.

Fig. 12-29 Based on Wier, D.L., Laing, E.D., Smith, I.L., Wang, L.F. and Broder, C.C., "Host cell virus entry mediated by Australian bat lyssavirus G envelope glycoprotein occurs through a clathrin-mediated endocytic pathway that requires actin and Rab5." *Virology Journal*. Vol. 11, No. 40. February 7, 2014, Figures 1 and 2.

Chapter 13

Fig. 13-2a Adapted from M. T. Cabeen and C. Jacobs-Wagner, "Bacterial Cell Shape," *Nature Reviews Microbiology* (August 2005), 3 (8): 601–10 (Fig. 4), and W. Margolin, "FtsZ and the Division of Prokaryotic Cells and Organelles," *Nature Reviews Molecular Cell Biology* (November 2005), 6 (11): 862–71 (Fig. 6).

Fig. 13-4 Based on Bruce Alberts et al., *Molecular Biology of the Cell*, 3rd ed., Fig. 16.33, p. 810. © 1994 by Garland Science-Books.

Fig. 13-7 Based on Kleinsmith, Lewis J.; Kish, Valerie M., *Principles of Cell and Molecular Biology*. 2nd Ed., © 1995 Pearson Education, Inc.

Fig. 13-9 Based on Bruce Alberts et al., *Molecular Biology of the Cell* 5e. Garland Science, 2007.

Fig. 13-14 Based on Bruce Alberts et al., *Molecular Biology of the Cell*, 3rd ed., Fig. 16.65, p. 835. © 1994 by Garland Science-Books.

Fig. 13-25 Based on Rohatgi et al, "The Interaction between N-WASP and the Arp2/3 Complex Links Cdc42-Dependent Signals to Actin Assembly" *Cell* 97(2) April 16, 1999, pp 221–231.

Chapter 14

Fig. 14-1 Based on Carter, A. P. (2013). Crystal clear insights into how the dynein motor moves. *Journal of Cell Science*. Vol. 126, pp. 1–9.

Fig. 14-3 Based on N. Hirokawa and R. Takemura, "Molecular Motors and Mechanisms of Directional Transport in Neurons," *Nature Reviews Neuroscience*, 6: 201–214, © 2005.

Fig. 14A-1 Based on The Debold Lab. Single Molecule Laser Trap Assay.

Fig. 14-5 Sources: Revised dynein in (a) adapted from Vallee et al (2012). Multiple modes of cytoplasmic dynein regulation. *Nat Cell Biol*. 14(3):224–30. (b) adapted from Fig. 6 from Roberts et al, 2012. ATP-driven remodeling of the linker domain in the dynein motor. *Structure* 20, 1670–1680.

Fig. 14-10 Based on T. Hodge and M. J. Cope, "The Myosin Family Tree," *Journal of Cell Science* (2000), 113 (19): 3353–3354, Fig. 1. © 2000. Reproduced with permission of The Company of Biologists, Ltd.

Fig. 14B-1a, b Based on Fliegauf, M., Benzing, T. & Omra, H. (2007). When cilia go bad: cilia defects and ciliopathies. *Nature Reviews Molecular Cell Biology*. Vol. 8, pp. 880–893.

Fig. 14-22 Based on *Cell Movements* by Dennis Bray, p. 166. © 1992 by Taylor & Francis Group LLC- Books. Reproduced with permission of Taylor & Francis Group LLC-Books.

Fig. 14-27 Based on Kleinsmith, Lewis J.; Kish, Valerie M., *Principles of Cell and Molecular Biology*. 2nd Ed., © 1995 Pearson Education, Inc.

Chapter 15

Fig. 15-2 Source: Campbell *Biology* 9e, Fig. 6-32. © Pearson Education, Inc.

Fig. 15-7 Based on D. Vestweber and J. E. Blanks, "Mechanisms That Regulate the Function of the Selectins and Their Ligands," *Physiological Reviews* 79 (1), January 1999:181–213, Fig. 1. Am Physiol Soc.

Fig. 15A-2 Based on Elaine Fuchs & Srikala Raghavan, "Getting under the skin of epidermal morphogeneisis." *Nature Reviews Genetics* 3, 199–209 (March 2002) doi:10.1038/nrg758.

Fig. 15-12 Source: Campbell, Neil A.; Reece, Jane B.; Mitchell, Lawrence G., *Biology*, 5th Ed., ©1999. Reprinted and Electronically reproduced by permission of Pearson Education, Inc.

Fig. 15-18a Based Fig. 1 from Marinkovich, M. P. (2007). "Laminin 332 in Squamous-Cell Carcinoma." *Nature Reviews Cancer*. Vol. 7, pp. 370–380.

Table 15-4 Data from Goodwin, P. B. (1983). Molecular size limit for movement in the symplast of the Elodea leaf. *Planta* 157, 124–130.

Chapter 16

Fig. 16A-2b Bolzer A, Kreth G, Solvei I, Koehler D, Saracoglu K, et el. (2005) "Three-dimensional maps of all chromosomes in human male fibroblast nuclei and prometaphase rosettes." *PLoS Biol* 3(5): e157. Figure 1.

Fig. 16-17 Based on Kleinsmith, Lewis J.; Kish, Valerie M., *Principles of Cell and Molecular Biology*. 2nd Ed., © 1995 Pearson Education, Inc.

Fig. 16-23 Sources: (a) is adapted from Fig. 12-13 and (b) is adapted from Fig. 12-22, Klug 10e, *Genetic Analysis*, © Pearson Education, Inc.

Fig. 16-24 Based on Kleinsmith, Lewis J.; Kish, Valerie M., *Principles of Cell and Molecular Biology*. 2nd Ed., © 1995 Pearson Education, Inc.

Fig. 16B-1 Source: Based on Eran Meshorer and Yosef Gruenbaum, "One with the Wnt/Notch: Stem cells in laminopathies, progeria, and aging" *The Journal of Biology*, Vol. 181, No. 1, April 7, 2008, 9–13.

Page 470 Herriott, R. M. (1951). "NUCLEIC-ACID-FREE T2 VIRUS "GHOSTS" WITH SPECIFIC BIOLOGICAL ACTION," *Journal of Bacteriology*, Vol. 61, no. 6, pp. 752–754.

Page 473 Source: Watson J.D. and Crick F.H.C. (1953, p. 966) Genetical implications of the structure of deoxyribonucleic acid. *Nature*, 171: 962–967.

Chapter 17

Fig. 17-2 Based on Kleinsmith, Lewis J.; Kish, Valerie M., *Principles of Cell and Molecular Biology*. 2nd Ed., © 1995 Pearson Education, Inc.

Fig. 17-9 Based on Sanders, M.F. and J.L. Bowman. *Genetic Analysis: An Integrated Approach*, 2e, Fig. 7.17, p. 239. © Pearson Education, Inc.

Fig. 17-11 Sanders, *Genetic Analysis*, Fig 7.19 page 241. © Pearson Education, Inc.

Fig. 17-12a Based on Kleinsmith, Lewis J.; Kish, Valerie M., *Principles of Cell and Molecular Biology*. 2nd Ed., © 1995 Pearson Education, Inc.

Fig. 17-22 Sanders and Bowman, *Genetic Analysis: An Integrated Approach*, 2d ed., Fig 12.7.

Table 17-2 Sanders and Bowman, *Genetic Analysis: An Integrated Approach*, 2d ed., Table 12.4. © Pearson Education, Inc.

Fig. 17-24a Based on Griffiths, A.J.F. et al. *Introduction to Genetic Analysis*, 10e. New York: Freeman (2012), Fig. 16-16, p. 567.

Fig. 17-24b Based on Griffiths, A.J.F. et al. *Introduction to Genetic Analysis*, 10e. New York: Freeman (2012), Fig. 16-14, p. 565.

Fig. 17-26 Based on Sanders, *Genetic Analysis*, Fig. 17.26 Base Excision Repair. Pearson Education, Inc.

Fig. 17-27 Based on Sanders, *Genetic Analysis*, Fig. 12.19, p. 404 Pearson Education, Inc.

Fig. 17-28 Based on Griffiths 10e, *Introduction to Genetic Analysis* 10e, Fig. 16-23, p. 574. Macmillan Publishers, Ltd.

Fig. 17A-1, 17A-2 Based on F Ann Ran, et al. (2013). Genome engineering using the CRISPR-Cas9 system. *Nature Protocols*, Vol. 8, pp. 2281–2308.

Fig. 17-30 Based on Neta Agmon, N., S. Pur, B. Liefshitz and Martin Kupiec (2009), Analysis of repair mechanism choice during homologous recombination, *Nucleic Acids Research* 37, 5081–5092, by permission of Oxford University Press.

Fig. 17-32 Sanders, Mark F., *Genetics: An Integrated Approach*, 1e. Fig. 12-25. © 2012 Pearson Education, Inc.

Fig. 17-33 Based on Alberts et al, *Essential Cell Biology*, 3e, Fig 6-33, p. 223 (Garland Science, 2009)

Fig. 17-35 Adapted & simplified from Sanders, *Genetic Analysis*, Fig. 13.24, p. 446. © Pearson Education, Inc.

Page 473 Watson J.D. and Crick F.H.C. (1953, p. 966) Genetical implications of the structure of deoxyribonucleic acid. *Nature*, 171: 962–967.

Page 494 U.S. Government Printing Office. (1974). *Genetic Engineering: Evolution of a Technological Issue: Supplemental Report I*. U.S. Government Printing Office.

Chapter 18

Fig. 18-4 Based on Watson, James, *Molecular Biology of the Gene*, 7e. Fig 13.11. © 2014 Pearson Education, Inc.

Fig. 18-6 Based on Watson, James, *Molecular Biology of the Gene*, 7e. Fig 13.7. © 2014 Pearson Education, Inc.

Fig. 18A-3 Based on Watson, James, *Molecular Biology of the Gene*, 7e. Fig 7.35 © 2014 Pearson Education, Inc.

Fig. 18-8 Based on Watson, James, *Molecular Biology of the Gene*, 7e. Fig 13.11. © 2014 Pearson Education, Inc.

Fig. 18-9 Based on Kleinsmith, Lewis J.; Kish, Valerie M., *Principles of Cell and Molecular Biology*. 2nd Ed., © 1995 Pearson Education, Inc.

Fig. 18-10 Based on Watson, James, *Molecular Biology of the Gene*, 7e. Fig13.16 © 2014 Pearson Education, Inc.

Fig. 18B-2 Jeff Hardin, Univ. of Wisconsin-Madison.

Fig. 18-10 Watson, James, *Molecular Biology of the Gene*, 7e. Fig 13.11. © 2014 Pearson Education, Inc.

Fig. 18-16 Based on Kleinsmith, Lewis J.; Kish, Valerie M., *Principles of Cell and Molecular Biology*. 2nd Ed., © 1995 Pearson Education, Inc.

Fig. 18-17b Based on Kleinsmith, Lewis J.; Kish, Valerie M., *Principles of Cell and Molecular Biology*. 2nd Ed., © 1995 Pearson Education, Inc.

Fig. 18-19 Based on Kleinsmith, Lewis J.; Kish, Valerie M., *Principles of Cell and Molecular Biology*. 2nd Ed., © 1995 Pearson Education, Inc.

Fig. 18-21 Based on Kleinsmith, Lewis J.; Kish, Valerie M., *Principles of Cell and Molecular Biology*. 2nd Ed., © 1995 Pearson Education, Inc.

Chapter 19

Opener Based on Sanders Mark F. and Bowman, John, *Genetic Analysis: An Integrated Approach*, 1e. Fig. 9.4a. Pearson Education, Inc.

Fig. 19-2 Based on Kleinsmith, Lewis J.; Kish, Valerie M., *Principles of Cell and Molecular Biology*. 2nd Ed., © 1995 Pearson Education, Inc.

Fig. 19-7 Based on Sanders, Mark F., *Genetics: An Integrated Approach*, 1e. Fig. 9-4. © 2012 Pearson Education, Inc.

Fig. 19-8 Based on Watson, James, *Molecular Biology of the Gene*, 7e. Fig 13.11. © 2014 Pearson Education, Inc.

Fig. 19-8c Based on NDB ID: TR0001 Shi, H. and Moore, P.B., (2000) The crystal structure of yeast phenylalanine tRNA at 1.93 A resolution: A classic structure revisited. RNA 6: 1091–1105.

Fig. 19-9 Based on Kleinsmith, Lewis J.; Kish, Valerie M., *Principles of Cell and Molecular Biology*. 2nd Ed., © 1995 Pearson Education, Inc.

Fig. 19-12 Based on Sanders, Mark F., *Genetics: An Integrated Approach*, 1e. Fig. 9-16. © 2012 Pearson Education, Inc.

Fig. 19-16b Based on Watson, James, *Molecular Biology of the Gene*, 7e. Fig 15-27. © 2014 Pearson Education, Inc.

Fig. 19A-2 From George H. Talbot, et al., The Infectious Diseases Society of America's 10 × '20 Initiative (10 New Systemic Antibacterial Agents US Food and Drug Administration Approved by 2020): Is 20 × '20 a Possibility? *Clinical Infectious Diseases*, July 2019. https://doi.org/10.1093/cid/ciz089.

Fig. 19-23 Based on Sanders, Mark F., *Genetics: An Integrated Approach*, 1e. Fig. 19-20c. © 2012 Pearson Education, Inc.

Chapter 20

Fig. 20-5 Republished with permission of American Association for the Advancement of Science, from M. Lewis, et al., "Crystal Structure of the Lactose Operon Repressor and Its Complexes with DNA and Inducer," *Science* 271 (1 March 1996): 1247–1254.

Fig. 20-15 Based on Kleinsmith, Lewis J.; Kish, Valerie M., *Principles of Cell and Molecular Biology*. 2nd Ed., © 1995 Pearson Education, Inc.

Fig. 20-16 Based on Kleinsmith, Lewis J.; Kish, Valerie M., *Principles of Cell and Molecular Biology*. 2nd Ed., © 1995 Pearson Education, Inc.

Fig. 20-29 Based on Kleinsmith, Lewis J.; Kish, Valerie M., *Principles of Cell and Molecular Biology*. 2nd Ed., © 1995 Pearson Education, Inc.

Fig. 20-31 Based on Kleinsmith, Lewis J.; Kish, Valerie M., *Principles of Cell and Molecular Biology*. 2nd Ed., © 1995 Pearson Education, Inc.

Fig. 20-32 Republished with permission of American Association for the Advancement of Science, from Kosman et al., *Science*, v. 305, no. 5685, p. 846, 6 August 2004.

Fig. 20-39 PDB 4cr2 from Unverdorben, P., et al., "Deep classification of a large cryo-EM dataset defines the conformational landscape of the 26S proteasome." *Proceedings of the National Academy of Sciences* 111:5544, 2014.

Chapter 21

Fig. 21-15 Source: Fig ST1.02, Butler, J. M., from STR DNA Internet Database at http://www.cstl.nist.gov/div831/strbase/fbicore.htm. Figure courtesy of Dr. John M. Butler, National Institute of Standards and Technology.

Fig. 21-26 Michael Costanzo, et al. (2010). The Genetic Landscape of a Cell. *Science*, 327(5964), pp. 425–431.

Chapter 22

Fig. 22A-2 Adapted from 8e Fig 13-7; http://www.leica-microsystems.com/science-lab/the-patch-clamp-technique/.

Fig. 22-5 Adapted from Advanced Information on the Nobel Prize in Chemistry, 8 October 2003, Fig. 2. The Royal Swedish Academy of Sciences.

Fig. 22-6 Adapted from F. Bezanilla, "RNA Editing of a Human Potassium Channel Modifies Its Inactivation." *Nature Structural & Molecular Biology* 11 (2004): 915–916, Fig. 1.

Fig. 22-10a Based on Kleinsmith, Lewis J.; Kish, Valerie M., *Principles of Cell and Molecular Biology*. 2nd Ed., © 1995 Pearson Education, Inc.

Fig. 22-10b Based on Mary E.T. Boyle et al., "Contactin Orchestrates Assembly of the Septate-like Junctions at the Paranode in Myelinated Peripheral Nerve," *Neuron* (May 2001), 30 (2), pp. 385–397, Fig. 6.

Fig. 22-16a Based on C.C. Garner et al., "Molecular Determinants of Presynaptic Active Zones," *Current Opinion in Neurobiology*, 10 (3), pp. 321–27, Fig. 1. Elsevier.

Fig. 22-16b Based on Harlow, M. L. et al. (2001). The architecture of active zone material at the frog's neuromuscular junction. *Nature* 409, 479–484. https://doi.org/10.1038/35054000

Chapter 23

Fig. 23-1 Based on Kleinsmith, Lewis J.; Kish, Valerie M., *Principles of Cell and Molecular Biology*. 2nd Ed., © 1995 Pearson Education, Inc.

Fig. 23-3 Based on http://www.studyblue. com/notes/note/n/ph1-2-receptor-and-dose-response-theory/deck/1424003.

Fig. 23-6b Based on Chung et al (2011). "Conformational changes in the G protein Gs induced by the b2 adrenergic reactor." *Nature* 477, 611

Fig. 23-15b Based on Kleinsmith, Lewis J.; Kish, Valerie M., *Principles of Cell and Molecular Biology*. 2nd Ed., © 1995 Pearson Education, Inc.

Fig. 23-17c "Epidermal Growth Factor," June 2010 Molecule of the Month by David Goodsell. © 2010 David Goodsell & RCSB Protein Data Bank.

Fig. 23-19 Based on S. F. Gilbert, *Developmental Biology*, 5th ed., p. 110, Fig. 3.34. © 1997 by Sinauer Associates, Inc. Reprinted by permission.

Chapter 24

Fig. 24A-3a Vybrant(R) Dye Cycle(TM) Stains for Live Cell Cycle Analysis. Figure 1 © 2015 ThermoFisher Scientific Inc.

Fig. 24A-3b FxCycle(TM) Stains for Fixed Cell Cycle Analysis. Figure 1. © 2015 ThermoFisher Scientific Inc.

Fig. 24-4d Aussie Suzuki

Fig. 24-5 Based on Kleinsmith, Lewis J.; Kish, Valerie M., *Principles of Cell and Molecular Biology*. 2nd Ed., © 1995 Pearson Education, Inc.

Fig. 24-8 Based on Campbell, Neil A.; Reece, Jane B., *Biology*, 6th Ed., ©2002, p. 235 Pearson Education, Inc.

Fig. 24-10 Based on Campbell, Neil A.; Reece, Jane B., *Biology*, 6th Ed., ©2002, p. 235 Pearson Education, Inc.

Chapter 25

Fig. 25-7 Based on Campbell, *Biology* 10e, Fig. 15-13 © Pearson Education, Inc.

Fig. 25A-2 Based on Freeman, *Biological Science*, 5e, Fig. 13.13. © Pearson Education, Inc.

Fig. 25-11 Based on Campbell, *Biology*, 10e, Fig. 14-3. © Pearson Education, Inc.

Fig. 25-12 Based on Campbell, *Biology* 10e, Fig. 14-4. © Pearson Education, Inc.

Fig. 25-13 Based on Campbell, *Biology* 10e, Fig. 14-5. © Pearson Education, Inc.

Fig. 25B-1 Based on Sanders and Bowman, *Genetic Analysis*: An Integrated Approach, Fig 2.21. Pearson Education.

Fig. 25-21 Based on From Christopher K. Mathews and K. E. van Holde, *Biochemistry*. ©Benjamin Cummings

Fig. 25-23 Based on *Campbell Biology* 10e, Figure 13.9 © Pearson Education, Inc.

Fig. 25-25 Campbell, Neil A.; Reece, Jane B.; Mitchell, Lawrence G., *Biology*, 5th Ed., © 1999, p. 214. Pearson Education, Inc.

Chapter 26

Fig. 26-2 Adapted from R. K. Boutwell, "Some Biological Aspects of Skin Carcinogenesis." *Prog. Exp. Tumor Res.* (1963) (4): 207.

Fig. 26-4 Data from "The Vascularization of Tumors" by Judah Folkman, *Scientific American* May 1976 © Scientific American, a division of Nature America, Inc.

Fig. 26-6 Based on Kleinsmith, Lewis J.; Kish, Valerie M., *Principles of Cell and Molecular Biology*. 2nd Ed., © 1995 Pearson Education, Inc.

Fig. 26-7a Data from Jha, P. (2020). The hazards of smoking and the benefits of cessation: a critical summation of the epidemiological evidence in high-income countries. *eLife*.

Fig. 26-7b Based on Kleinsmith, Lewis J.; Kish, Valerie M., *Principles of Cell and Molecular Biology*. 2nd Ed., © 1995 Pearson Education, Inc.

Fig. 26-8 Data from S. Meselson and L. Russell in *Origins of Human Cancers*, H. H. Hiatt et al., eds. (Cold Spring Harbor, NY: Cold Spring Harbor Laboratory, 1977), pp. 1473–82.

Fig. 26-9 D. E. Brash et al, "A Role for Sunlight in Skin Cancer: UV-Induced p53 Mutations in Squamous Cell Carcinoma," *PNAS* (1991) 88 (22): 10124–10128.

Fig. 26-24 Rafiq, S., Hackett, C. S. & Brentjens, R. J. (2019). Engineering strategies to overcome the current roadblocks in CAR T cell therapy. *Nature Reviews Clinical Oncology*, 17, pages 147–167.

Appendix

Fig A-18 Adapted from Pantazis and Supatto (2014). Advances in whole-embryo imaging: a quantitative transition is underway. *Nature Rev. Mol. Cell. Biol.* 15:327-339, Box 2c.

INDEX

Note: A *b* following a page number indicates a box, an *f* indicates a figure, and a *t* indicates a table. Page numbers in **bold** indicate pages on which key terms are defined.

F

G

M

Q

R

T